Conversion Factors

LENGTH

1 m = 100 cm = 1000 mm = 10^8 Å = 0.001 km
1 ft = 12 in.   5280 ft = 1 mi
1 in = 2.540 cm   1 cm = 0.3937 in.
1 ft = 30.48 cm   1 m = 39.37 in. = 3.281 ft
1 mi = 1.609 km   1 km = 0.6214 mi
1 nautical mile = 6076 ft = 1.151 statute mile = $\frac{50}{27}$ km

MASS

1 kg = 1000 g = 0.0685 slug
1 slug = 14.59 kg

WEIGHT

1 lb = 16 oz   1 ton = 2000 lb
A mass of 453.6 g = 0.4536 kg *weighs* 1 lb
A mass of 1 g weighs 0.002205 lb
A mass of 1 kg weighs 2.205 lb

TIME

1 min = 60 sec
1 hr = 60 min = 3600 sec
1 day = 24 hr = 1440 min = 86400 sec
1 yr = 365.2 days = 8766 hr = 525900 min = 3.156×10^7 sec

FORCE

1 newton = 100,000 dynes = 0.2248 lb
1 lb = 4.445 N = 444,500 dynes
A mass of 1 kilogram weighs 9.80 newtons
A mass of 1 gram weighs 980 dynes
A mass of 1 slug weighs 32.16 lb

PHYSICS FOR SCIENCE AND ENGINEERING

Isaac Newton

PHYSICS FOR SCIENCE AND ENGINEERING

JOHN P. McKELVEY

Clemson University

HOWARD GROTCH

Pennsylvania State University

HARPER & ROW, PUBLISHERS
New York   Hagerstown   San Francisco   London

PHOTO CREDITS

Frontispiece	Bettmann Archive
Figure 9.13a	Bashford-Thompson, UPI
9.13b	Wide World
Figure 12.1a	UPI
12.1b	Granger
12.1c	Wide World
Figure 12.2a	Culver
12.2b	UPI
Figure 12.3	Granger
Figure 15.2	Granger
Figure 19.6	Education Development Center
Figure 25.37	Hale Observatory
Figure 26.11	Education Development Center
Figure 26.21	Education Development Center
Figure 27.1	UPI
Figure 27.23	UPI
Figure 28.4	Culver
Figure 28.7	H. E. White, *Atomic and Nuclear Physics*. New York: Van Nostrand, 1964
Figure 28.8	Culver
Figure 28.14	Granger

Project Editor: Brigitte Pelner
Designer: Howard Leiderman
Production Supervisor: Stefania J. Taflinska
Photo Researcher: Myra Schachne
Compositor: Syntax International Pte. Ltd.
Printer: The Murray Printing Company
Binder: Halliday Lithograph Corporation
Art Studio: Danmark & Michaels, Inc.

PHYSICS FOR SCIENCE AND ENGINEERING

Library of Congress Cataloging in Publication Data

McKelvey, John Philip.
  Physics for science and engineering.

  1. Physics.   I. Grotch, Howard, joint author.
II. Title.
QC21.2.M33   530   77-16827
ISBN 0-06-044376-6

CONTENTS

WAVE MOTION   305

FLUID MECHANICS   342

HEAT, TEMPERATURE, AND THE FIRST LAW OF THERMODYNAMICS   370

THERMAL PROPERTIES OF IDEAL SYSTEMS AND THE ABSOLUTE TEMPERATURE SCALE   403

14 DISORDER, REVERSIBILITY, ENTROPY, AND THE SECOND LAW OF THERMODYNAMICS 439

15 ELECTROSTATICS AND COULOMB'S LAW 469

16 ELECTRIC FIELDS AND ELECTROSTATIC POTENTIAL 484

17 CAPACITANCE, DIELECTRICS, AND POLARIZATION 535

PHYSICAL OPTICS 850

RELATIVITY AND NUCLEAR PHYSICS 899

QUANTUM PHYSICS 933

PREFACE

This book is designed to be used as an introductory text for students of science and engineering. It presupposes a knowledge of elementary algebra and trigonometry, and it is intended for students who have had a semester of calculus, or are at least undertaking calculus concurrently with their physics sequence. It is conventional in format and order of presentation, and takes a predominantly classical approach, though the two final chapters are devoted to quantum physics and relativity. It is dedicated, in a sense, to the good student rather than the exceptional one. It will therefore probably not be used by the physics majors at Berkeley or Caltech, who will, in any event, find other texts that suit their needs. On the other hand, it will be found to provide a solid and rigorous foundation in classical physics for students of science and engineering who are intelligent and hardworking, but whose background and educational requirements are normal rather than exceptional.

Quite so. But then, why another conventionally organized and presented physics text in a market in which there are already dozens? A fair and obvious question; the answer is that the present text differs in some important ways, but in ways that are not immediately apparent, from texts that are ordinarily used for calculus-level introductory physics courses. The most important of these unconventional aspects is in the deliberate *avoidance* of a terse and compact mode of presentation. In a sense, this is a deliberate repudiation of motherhood, since the fashion of the past five or ten years has been toward "a short and simple book that concentrates on fundamentals." Unfortunately, these tenets lead all too easily to a superficial and casual approach to a subject in which substantive progress is traditionally achieved only with detailed study and continual practice.

This is a book that is quite a bit longer than the average introductory physics text, and it is filled with equations. It is constructed along the lines of Schubert's Ninth Symphony, rather than Mozart's Symphony No. 40, so to speak. It is recognized that (like Schubert's lengthy symphony) this may not create a very favorable first impression on teacher or student. Still, there is no more subject matter here than in other books, and the level at which the subject matter is treated is not exceptionally high. There is a conscious attempt, instead, to provide the words of explanation and discussion necessary to a complete and detailed understanding of the subject; a dedicated effort is made to anticipate sticky questions that are easy to ask but hard to answer, and to put to rest the more common misconceptions and apparent paradoxes that crop up. *The real understanding of physics is achieved in these areas of activity, not in the mere statement and illustration of straightforward fundamental principles and examples!* This book is therefore longer than many others, not because it covers any more material, but because it undertakes to cover what it does in the clearest, most thorough, and most unambiguous way possible.

Likewise, though there are thousands of equations in this book, and though no attempt has been made to lower the mathematical level below the usual standard required for a text of this kind, neither is it at all any *higher* than what is ordinarily used. Instead, a conscious and thorough attempt is made to present *all* the mathematical details, *all* the intermediate equations, needed to make it a simple matter for the student to get from one equation to the next—and thus from the beginning of the presentation to the end. It is hoped, then, that the student will be in a position to concentrate on conquering the physics, rather than on floundering through the mathematics, which is what frequently happens when only the bare minimum of detail is presented. It was mentioned previously that the book is aimed at the "good" student. True enough, but it is also designed to be easy for the average student to follow with little or no assistance.

Along the same line, the authors have allowed themselves the luxury of describing important topics more than once, and from different points of view. It is often observed (but less frequently admitted) that students usually do not learn physics very well at the "first pass." We have therefore intentionally returned to important topics, sometimes several times, to allow students more than one go at them. For example, we purposely avoid treating statics as a special case of dynamics, to allow for some additional practice at the methods of analysis, common to both, that are of such wide applicability to all sorts of problems. The approach is not consciously "spiral," but is purposely less economical than it could have been. The book, therefore, is as long as it needs to be, and ought to be, to accomplish its purposes, rather than a preconceived package of 600 or 800 or 900 pages.

This book gives a calculus-based presentation of elementary physics. It is therefore envisioned as being used concurrently with courses in differential and integral calculus. It is advantageous, though not absolutely necessary, for the student to have completed a semester of calculus before beginning his or her study of physics. In any event, in the first four chapters, the use of calculus is intentionally restricted, and what calculus is used is very simple and straightforward. Thereafter, calculus is used more freely, though with due regard to the amount of mathematics the average student can reasonably be expected to know. A brief summary of the basic principles of algebra, trigonometry, analytic geometry, and calculus is given in a series of appendices at the end of the book.

There is enough subject matter in this book for a three-semester sequence of physics courses, in which mechanics is assigned to the first semester, heat, electricity, and magnetism to the second, and electromagnetic waves, optics, and modern physics to the third. By making a reasonable selection of material, it can also be adapted to a two-semester format. It is possible, for example, to omit most of the material on dielectric and magnetic materials, fluid mechanics, ac circuits, geometrical optics, relativity and quantum physics, without doing too much violence to a course in basic classical physics. But an extended range of subject matter is there for those who can utilize it. There are a number of instances, geometrical optics being perhaps the most conspicuous, where a point of view very different from what has been traditionally used has been adopted. The number of worked examples included in the book is unusually large, and forms an integral part of the text. Except for the more difficult ones marked by an asterisk, they *cannot* be omitted without loss of textual continuity. Also, a large number of questions and problems are given at the end of each chapter. The latter includes exercises that are very simple, of moderate difficulty, and ones (marked by an asterisk) that are definitely hard. Answers to odd-numbered problems are presented in a separate section in the back of the book.

Not everyone will agree with the views expressed above that led the authors to write a text having the characteristics that this one has. For those who do not, there are many concise texts now available that will be quite satisfactory. For those who do, however, it is believed that this one may have some unique advantages not possessed by those that are now in common use, and therefore will offer a useful and interesting alternative.

John P. McKelvey
Howard Grotch

ACKNOWLEDGMENTS

So many persons have contributed in one way or another to the creation and production of this book, that it would be futile, if not impossible, to list everyone individually. The authors are grateful for the assistance of all of them, however, whether mentioned by name or not.

The helpful discussions and suggestions of our colleagues at Clemson and Penn State have helped a great deal to make this text what it is. In particular, we would like to thank Professors John R. Ray, Max D. Sherrill, W. E. Gettys, and F. J. Keller of the Department of Physics and Astronomy at Clemson, and Professors Wayne and Helen Webb, Gilbert Ward, Stanley J. Shepherd, John J. Gibbons, Roland Good, Emil Kazes, and Santiago Polo of the Penn State Physics Department. We are particularly grateful for the assistance of Prof. Thomas Wiggins of Penn State in supplying many of the illustrations for Chapter 26. We are also grateful to the many persons who have reviewed and criticized the manuscript during its long gestation period. One of us (J.P.M.) is grateful for the tolerance and understanding that Dean H. E. Vogel, and other administrators at Clemson, have extended during a period when the completion of the manuscript interfered seriously with the efficient performance of his duties as department head. In addition, both of the authors owe a great deal to students they have taught during their academic careers, and to the understanding and insights they have obtained in innumerable individual discussions with them. Without this interaction, this text could never have been written. The students enrolled in Physics 132 at Clemson during the Spring semester of 1977 have been very helpful in discovering errors in problems and answers, and in reassuring the authors that students can in fact learn from their text. We are especially appreciative of their assistance.

We appreciate also the encouragement and assistance of the editorial staff of Harper & Row, and would like to acknowledge the contributions of Walter Sears, Blake Vance, John Woods, Wayne Schotanus, Karen Judd, and Brigitte Pelner in steering the project to completion. We are no less grateful for the encouragement and tolerance of our wives and families during a long period when our work usurped much of the time we should have devoted to them.

Finally, but far from the least in importance, there are the valiant and dedicated efforts of the people who have so skillfully undertaken the huge task of typing, reproducing, editing, and assembling this complex manuscript. Foremost among these persons are Sylvia Moore and Sally Latham at Clemson and Frances Fogle at Penn State, without whose help this manuscript could not have been completed. Also, we are grateful for the help that Mary Helen Waites, Onaway Mulligan, Donna Neuhaus, Richard Reid, Alice Stone, Maxine Deesley, Sandi Wallace, and many others have so kindly contributed to the compilation of the manuscript.

J. P. M.
H. G.

PHYSICS FOR SCIENCE AND ENGINEERING

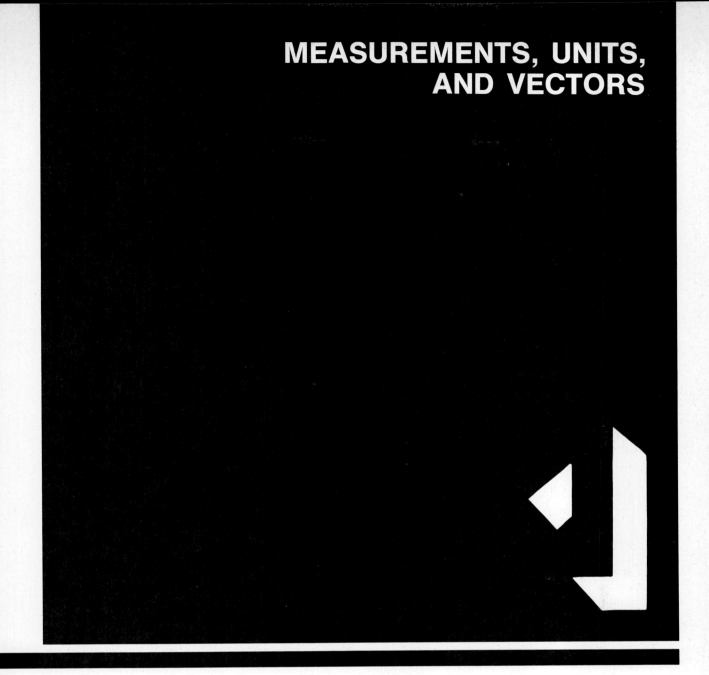

MEASUREMENTS, UNITS, AND VECTORS

1.1 Introduction and Advice to the Student

It is hard to give a simple and accurate definition of physics, and perhaps it is unnecessary to do so. But since we would like to show you where we are going before we start out, we shall try. Broadly speaking, physics is a science that seeks to understand *how things happen in our natural environment and why they happen as they do.* This statement has to be qualified in several ways. First of all, physics concerns itself with the events of the material world that we can see and experience and about which we can all agree on the basis of objective measurement and observation. Physicists use instruments of great power and exqui-

site sensitivity, such as particle accelerators and astronomical telescopes, to extend the range of our senses to the utmost. They have also invented devices such as galvanometers and oscilloscopes to explore certain natural phenomena not directly accessible to our senses, like currents of electricity and electric and magnetic fields. The operation of these instruments, however, is completely understood in terms of indicators perceptible to the senses. Physics is, therefore, an experimental science which relies heavily upon the *objective observation and measurement of natural phenomena.*

These characteristics apply equally well to other sciences, such as chemistry, biology, and geology. The distinguishing feature of physics is the breadth of its scope and its pervasive concern with general and fundamental principles. Physicists are continually trying to brush aside masses of irrelevant detail and to explain every aspect of the behavior of the natural world in terms of a very few laws of very wide applicability. In a sense, therefore, more specialized areas of science can be viewed as branches of physics, even though their study bears little resemblance to the study of physics and their practitioners do not usually think and talk like physicists. Thus, the study of basic molecular processes in biological systems is known as *biophysics*, the description of atomic interactions responsible for chemical reactions is embodied in *physical chemistry*, and our description of the properties of metals and alloys in terms of atomic structure and crystalline geometry is referred to as *physical metallurgy*.

Physics is a science of vast scope, ranging from the investigation of subatomic particles to the study of distant galaxies in the outer reaches of the known universe. Such a broad subject is not easy to classify in any way that is ideal for all purposes. In this book, we shall study mostly the basic principles of *classical physics*, that is, the physics that was known before the year 1900. This body of physics can be divided rather neatly into four well-defined areas as follows: (1) mechanics, (2) heat and thermodynamics, (3) electricity and magnetism, and (4) light and optics. The arrangement of subject matter in this book and the order in which it is presented is based on this scheme of classification. In *mechanics*, we shall study the effect of forces on material bodies and the motion that objects acquire as a result of forces acting on them. In *heat and thermodynamics*, the laws of mechanics are extended to molecular systems to explain thermal phenomena, including heat energy, temperature, and the thermal properties of matter. Under the heading of *electricity and magnetism*, we shall investigate effects arising from the presence of electric charge and the flow of electric current. Finally, the subject of *optics* deals with the propagation, reflection, refraction, and interference of light waves.

During the first decade of the twentieth century, the *theory of relativity* and the *quantum theory* were first proposed. These ideas soon gave rise to far-reaching changes in some of the basic concepts of physics and sharply altered existing patterns of scientific thought. Quantum physics and relativity did not in any sense destroy the framework of classical physics, but showed instead that there are limits to its validity and that beyond these limits new and very different concepts of physical reality are needed. It was found, for example, that the classical mechanics of 1900 was inadequate to describe the motion of systems in which velocities are comparable to the speed of light or to portray the mechanics of atoms, molecules, and atomic nuclei.

The year 1900, therefore, signifies a time of profound change in physics—change not only in substance but also in attitudes, concepts, and ideas. What has happened since then is referred to as *modern physics*. Modern physics has been successful in utilizing the ideas of quantum physics and relativity to describe the structure of atoms and molecules, to explain how nuclear reactions and elementary particle interactions take place, and to solve many other problems related to the fundamental structure of matter. Though modern physics is very different from the classical physics developed before 1900, its concepts are deeply rooted in the ideas of classical physics. For this reason, modern physics cannot be understood very well by anyone who is not reasonably well acquainted with the classical aspects of the subject.

An introductory physics text cannot discuss both classical and modern physics in great detail. In this book, we have deliberately chosen to give a thorough treatment of classical physics and a rather brief general survey of modern physics. The first 26 chapters, therefore, are devoted to classical physics, while the important ideas that have arisen since 1900 are discussed in Chapters 27 and 28. The authors have also tried to indicate the limitations of classical theory in every case where it is important to do so.

The object of physics is to arrive at a general description of how material systems will behave under any set of circumstances and then to predict their behavior in untried situations. The final judgment as to the correctness of a physical theory always rests upon the ability of that theory to explain phenomena which have already been observed experimentally and to successfully predict the results of initially unperformed experiments. The final criterion of success, in either case, is agreement with experimentally determined results, and therefore the role of laboratory work by experimental physicists is of basic importance to the science of physics.

The laws of physics are generally *quantitative* as well as qualitative; that is, they are capable of furnishing precise answers not only to questions such as "how" or "why" but also to questions such as "how much," "how far," or "how long." As Lord Kelvin[1] put it,

> I often say that when you can measure what you are speaking about, and express it in numbers, you know something about it; but when you cannot express it in

[1] William Thomson, Lord Kelvin (1824–1907), British scientist noted for his contributions to electricity and magnetism, mechanics, and thermodynamics.

numbers, your knowledge is of a meagre and unsatisfactory kind; it may be the beginning of knowledge, but you have scarcely, in your thoughts, advanced to the stage of Science, whatever the matter may be.

Under these circumstances, it is scarcely surprising that mathematics is an essential tool for the physicist, for he must be able to express all his laws in mathematical form if they are to yield quantitative answers. He must also be able to perform experiments in which numerical measurements of physical quantities are obtained in order that these may be compared with the numbers predicted by theory. The physicist, then, must acquire a working knowledge of the mathematics required to express thoughts in numbers. It is well to remember, however, that for the physicist mathematics is merely a tool that enables him to express the results of physical laws which have their own existence in nature, quite independent of any mathematics used to express them; while for the mathematician, mathematics and its logical structure and internal consistencies are ends unto themselves.

The physicist tends to regard mathematics as a practical method for expressing physical laws and working out the answers to problems involving those laws, rather than a logical system whose absolute rigor and internal consistency are all-important. In this respect his attitude toward mathematics somewhat resembles that of the engineer, whose paramount interest is in obtaining answers to problems of practical importance. The physicist may in many instances be satisfied with a mathematical result obtained by the use of intuitive processes which, while reasonable physically, would not meet the rigorous standards required by mathematicians for the proof of mathematical theorems. We shall, in fact, often appeal to intuition to guide us in formulating precise and reasonable ideas of certain physical concepts, notably that of force.

All this is not meant to disparage the methods and standards of other sciences but to point out the essential differences between them. The use of "physical intuition" is all-important to the physicist, for one cannot usually understand the essentials of any physical situation in mathematical terms alone. It is only when one understands the physical principles as well, independent of mathematics, that one may be confident in his understanding of the situation and in applying the relevant principles to other systems. It is quite possible to follow all the mathematical steps, for example, in the analysis of the simple pendulum and yet obtain very little awareness of the important physical facts about the system. These physical principles would include the factors that determine the period of oscillation, the constancy of the total energy, and the fact that the period of oscillation is independent of the amplitude. The ultimate truth, or absurdity, of any mathematical analysis of a physical system rests, in fact, upon how well it mirrors truths of nature external to the mathematics. To the pure mathematician, on the other hand, being interested only in certain axioms regarded as being true by definition, there are no external truths that matter. For him, the task is to establish the logical consequences of his axioms with absolute rigor, and in doing so, "physical intuition" quite rightly has no place. The student of physics, however, should *at all costs* avoid the pitfall of understanding a mathematical development without understanding the external physical truths upon which it is based nor the physical consequences it entails.

Physics is generally regarded as a particularly difficult and demanding subject, even on the elementary level. There are a number of reasons why this is so. First of all, the laws of physics must frequently be stated in terms of concepts—such as force, energy, temperature, and electric charge—that are abstract rather than tangible. We all have a vague general idea of what words like "force" and "energy" mean in a colloquial sense, but we may not understand the precise physical meaning of these concepts nor how to express them in numerical terms. The job of acquiring this understanding requires us to engage in precise, logical, and abstract thought. This is an activity that we sometimes find difficult and unpleasant and one that we may try to avoid whenever possible. But the fact of facing up to this sort of intellectual discipline is precisely what distinguishes the creative scientist, engineer, or mathematician from the routine mechanic or technician. Success in the study of physics—or any other worthwhile scientific discipline—is, in fact, made of such stuff.

In the best of all possible worlds, what we need to know about elementary physics would be contained in a number of equations or formulas, which, when memorized, would enable us to solve any practical problem simply by substituting given numerical data and solving for what we wish to find. This would, of course, relieve us of the necessity of facing that lonely struggle for understanding referred to in the preceding paragraph. Unfortunately, many science and engineering students have actually undertaken to do just that, only to suffer abrupt disillusionment at the first examination. Some even persist in this approach beyond that point. It is possible sometimes to get through physics courses with passing grades in this way. But in the long run, the student who takes this path cannot achieve the understanding of physics necessary for a successful career in pure or applied science, and thus only deprives himself of a vital section of his education.

Indeed, the most important objective of any course in elementary physics must be to acquire the habit of approaching problems through independent, rational use of a few simple—though abstract—physical laws, rather than by searching through one's notebook for the correct formula and "plugging in the right numbers." This is so important that we shall reiterate it here in bold letters:

The most important objective of the study of elementary physics is the acquisition of the ability to think in rational, though abstract, terms and to use this ability in applying a few fundamental physical principles to scientific problems. The practice of "memorizing all the formulas and plugging the numbers in" is to be avoided at all cost.

There are thousands of equations and formulas in this book, so many that nobody could hope to memorize all of them. Most of them are derived either as intermediate equations necessary to obtain some more important final result, or as examples to show you how to use the laws of physics in just the rational ways mentioned above. We have tried to distinguish fundamental relationships from those of subsidiary interest by enclosing them in "boxes." Likewise, the fundamental laws of physics upon which everything else follows are overlaid with gray shading. But though this will be useful in distinguishing what is of great importance from what is of lesser import, it is not sufficient simply to memorize everything that is shaded or boxed in. You must not only *know* this information but also *understand* it, which is quite a different thing. The understanding you need can be acquired, partially, by reading the text and following the examples and, beyond that, by *working the problems that are assigned at the end of each chapter.* The importance of working assigned problems cannot be overstated. If you encounter serious trouble in working them out, beware; there is then obviously some shortcoming in your understanding of the basic physics. Try to remedy this by rereading the text and examples.

The slide rule has been employed for over a century as a standard tool for working out numerical answers to physics problems such as those assigned as exercises in this book. Although it is still perfectly satisfactory in this application, it has been almost completely replaced by pocket-size electronic calculators that can execute the same calculational tasks much more quickly, accurately, and conveniently. Students will therefore probably find a small electronic calculator to be a valuable aid in doing the numerical work required in working out answers to problems. The units having built-in trigonometric, logarithmic, and exponential functions cost a bit more than the simplest machines but are generally to be preferred in view of their vastly extended calculational powers.

Throughout the history of science, physicists have been concerned with putting their knowledge to work in the development of devices that utilize physical principles. The pendulum clock, the chronometer, the gyroscope, the steam engine, the electric generator and motor, and radio and television broadcasting are examples of practical devices that were perfected that way. More recently, research in physics has been directly responsible for the development of nuclear power generation, solid-state electronic devices, and integrated circuits. Usually, physicists will be interested in technological development only in the early stages and, as soon as the application of the principles involved is well understood, will turn over the further perfection and manufacture of the devices to engineers. The engineers will then seek to incorporate further refinements and to solve the problems that arise in manufacturing and marketing. The various branches of engineering and technology all grew up this way; and, therefore, we may regard engineering primarily as applied physics.

Physics is frequently said to be a cold and objective discipline—quantitative, mathematical, dispassionate, abstract, far removed from the realm of human experience, and isolated from the relevant concerns of mankind. Like accounting or banking, it is held to be totally lacking in humanistic content. It is, of course, true enough that physics is abstract, and that its essential substance can most usefully be expressed and understood in the language of mathematics. But, after all, the ability to create abstract concepts and to employ them to the advantage of humanity is the most human of all activities—indeed, it is this unique capability that sets mankind apart from all other living species. In this sense, physics is perhaps the most human of all disciplines. Man's continuing struggle to utilize the natural environment for the satisfaction of his material and intellectual needs is a characteristically human endeavor and, moreover, forms a most important part of the history of his existence on this planet.

The study of physics has provided us with a successful scientific understanding of the natural universe. This understanding did not come easily but was the outcome of a long and intense effort over the whole span of recorded history. This effort has achieved notable success only in the last 300 years, since the time of Sir Isaac Newton, and represents in itself one of the finest intellectual accomplishments of mankind. It is not, and probably never will be, complete in the sense that a work of art is complete, but it is continually growing and developing. The mature growth of the science of physics took place largely in western Europe and North America. It comprises one of the important intellectual components of what we refer to as "western civilization."

It is in fact arguable that the development of physics made western civilization what it is and represents its most characteristic feature.

Mankind's struggle for survival in a natural environment that is not always benign but in many ways cruel, hostile, and unforgiving is a continuing one. In the present era, we are faced with materials and energy shortages, environmental degradation, and other serious threats to our security. In the long run, existing science and technology are probably inadequate to cope with these problems, and these threats will be countered only as a result of further growth of scientific understanding. Physics will certainly play an important if not dominant role in bringing this about. Can we then say that physics is something apart from human experience, devoid of human content?

Physics is also an exciting, challenging, and interesting subject in its own right. It requires some hard work and intellectual discipline from the student and the mastery of certain basic mathematical techniques that in themselves are not particularly interesting. But once these have been mastered, you will find that physics is interesting, important, very useful—and, yes, even fun.

1.2 Measurements, Units, and Dimensions

Underlying all of physics is the idea of measurement—that it is possible to arrive at consistent experimentally determined, quantitative values for certain properties of physical systems. This is done, ordinarily, by comparing the quantity to be measured with some arbitrarily defined standard which is regarded as the *unit* of that particular quantity. For example, we may measure the length, width, and height of a rectangular block with a meter stick. We are then comparing these dimensions of the object with the scales marked on the meter stick which, in turn, has been previously (if indirectly) calibrated against an international standard meter bar kept at the National Bureau of Standards. We may express the linear dimensions of the block as multiples (but not necessarily integral multiples) of the standard meter.

It is clear that numbers that express the linear dimensions of an object are not pure numbers but are multiples of some arbitrary standard of length, such as the standard meter. The numbers expressing such linear dimensions are said to be *dimensional quantities* having the dimensions of *length* in meters, feet, inches, or standard unit of length chosen. It is important to recognize that most of the quantities we shall have occasion to discuss are dimensional quantities of just this sort. It is also of great importance

to recognize that an understanding of the dimensions in which a physical quantity is expressed often leads to considerable physical insight into the nature of that quantity.

The units in which any given physical quantity is expressed can always be stated in terms of four fundamental units, such as *mass*,[2] *length*, *time*, and *electric charge*. For units of nonelectrical quantities only the first three of these are necessary. Thus, velocity may be expressed in units of length divided by units of time, as meters per second (m/sec, or m sec^{-1}) or miles per hour (mi/hr). In the same way, density may be expressed in units of mass divided by units of volume, the latter in turn being equivalent to units of length cubed, the unit of density being finally written, for example, as grams per cubic centimeter (g/cm^3, or g cm^{-3}). The choice of mass, length, time, and charge as a set of fundamental quantities by which to express all other units is not absolutely necessary; for example, one could equally well have chosen *force*, *length*, *time*, and *charge;* in fact, the British engineering system of units is based on such a choice. What is important is that the choice includes *four independent physical quantities.*

In this book, we shall use three different systems of units, the cgs-metric, the mks-rationalized metric,[3] and the British engineering system.[4] These are sometimes referred to simply as the cgs, (centimeter-gram-second) the mks, (meter-kilogram-second) and the English system, respectively. The cgs and the mks systems are both variants of the metric system, while the English system is based upon practical units used by engineers in Great Britain and the United States. In the cgs and mks systems, the four fundamental units chosen to represent all other units are *mass*,

---

[2] The precise meaning of the term *mass* will be established in Chapter 2. Roughly speaking, we may understand it as a measure of the quantity of matter contained in a given body. It is related to, but is not the same as, the body's weight.

[3] A comprehensive international system of units, based upon the mks metric system, has been widely adopted as a standard system of measurements by many countries in recent years. This scheme is referred to as the *Système Internationale*, or SI system, of units. We could, and perhaps should, have chosen to use these units exclusively in this book, for it is likely that the system will be universally adopted in the not-too-distant future, even in the United States. We chose, instead, to use SI units for the most part but to retain the freedom of choice to use other units in certain instances where it is particularly convenient or instructive to do so. There is no single system of units that is ideally suited to every requirement in all possible situations.

[4] Even though the British system of units is obsolescent, and will almost certainly be replaced by the SI system within the next decade or so, there are certain interesting and useful lessons to be learned in understanding how to work mechanics problems in both British and metric units. We, therefore, have included a discussion of this historically important system of measurement and have included a number of problems and examples that are stated and worked in British units.

TABLE 1.1. Standard Units of Fundamental Quantities in Various Dimensional Systems

System	Length	Mass	Time	Force	Charge
Metric cgs	centimeter	gram	second		esu of charge*
Metric mks	meter	kilogram	second		coulomb
British engineering	foot		second	pound	coulomb

* *Electrostatic unit* of charge, sometimes referred to as the *statcoulomb*. The units of electric charge will be discussed in detail in a later chapter.

length, *time*, and *electric charge*, while in the English system the corresponding choice is *force, length, time,* and *electric charge*. The standard unit quantities of each of these in each system are given in Table 1.1.

The *meter* (m) was originally defined as one ten-millionth of the arc of a great circle joining the equator and the north pole and passing through Paris, France. However, at the time the meter was first introduced as the legal standard of length in Europe, at which time standard meter bars were first constructed, the available surveys of the earth's surface were not completely accurate, and later measurements indicate that the earth's quadrant is actually very slightly more than 10,000,000 meters. In recent years, a new definition of the meter as 1,650,763.73 times the wavelength of a characteristic orange light emitted by atoms of the isotope krypton 86 has been adopted. Since the wavelength of this light can be determined spectroscopically with vast precision, it serves as a convenient and exact standard for units of length. The meter is conveniently divided into decimal units such as the *centimeter* (cm), equal to one hundredth of a meter, the *millimeter* (mm), equal to one thousandth of a meter, and the *micron* (μ), equal to one millionth of a meter, or a thousandth of a millimeter. Likewise, a distance of one thousand meters is referred to as a *kilometer* (km). The unit of distance in the English system is the foot (ft), and subsidiary units of inches (in.) ($\frac{1}{12}$ foot), yards (yd) (3 feet) and miles (mi) (5280 feet) are often used. Conversion between the metric and English systems is readily made by noting that

1 in. = 2.540 cm = 0.02540 m

1 ft = 30.48 cm = 0.3048 m

1 m = 39.37 in. = 3.281 ft

1 mi = 1609 m = 1.609 km

The unit of mass in the metric system is the *kilogram* (kg), corresponding to the mass of a standard cylinder of platinum-iridium which is stored at the Bureau of Standards. A kilogram is essentially the mass of 1000 cubic centimeters (cm^3) of pure water at a temperature of 4°C. One thousandth of this mass is referred to as a *gram* (g). There is no *fundamental* unit of mass in the British engineering system, the funda-

mental unit in that system having been chosen as the *pound of force*. A pound force is the force of gravity under specified conditions upon a standard pound (lb) weight. Such a standard pound weight contains 0.45359 kilogram of mass.

The standard unit of time in all three systems is the *second*, which is defined as 1/86,400 of the average length of the solar day throughout the year. Due to the motion of the earth in its orbit around the sun, and due to the fact that that orbit is elliptical rather than circular, and to the fact that the velocity of the earth along this orbital path is not constant but varies from day to day throughout the year, the length of the day, as measured by the time interval between successive transits of the sun across the north–south meridian in the sky, varies slightly during the course of a year. The unit of time must therefore be defined, as above, in terms of the average length of the solar day rather than the length of any particular day.

1.3 Mathematical Operations Involving Dimensional Quantities

In using dimensional quantities in algebraic equations, there are certain requirements which must be met that do not arise when only pure numbers are involved. These requirements may be summarized in the following way:

1. The dimensions of the quantities on both sides of an equation must be the same, unless the equation is one expressing a conversion factor between two different systems of units (for example, 1 in. = 2.54 cm).
2. Only quantities having the same dimensions may be added or subtracted. A corollary to this is that all the additive terms in any equation must have the same dimensions.
3. Any two dimensional quantities may be multiplied together or divided, one by the other, but the dimensions of the resulting product or quotient are then the *product or quotient of the dimensions of the individual factors.*

In connection with rule (1) above, it never makes

sense to say, for example, that 3 meters equals 3 kilograms, even though it is true that 3 equals 3 is a correct mathematical statement when only pure numbers are involved. Likewise, in equations expressing conversion factors between different systems of units, it certainly does make sense physically to say that 1 in. equals 2.54 cm, even though $1 = 2.54$ is an *incorrect* mathematical statement pertaining to pure numbers.

Requirement (2) may be illustrated by remarking that one cannot, for example, add three horses and five goats; the only answer that can be given is "three horses and five goats." One may, of course, add three horses and five horses to obtain eight horses, or three goats and five goats to obtain eight goats.[5] An example more closely related to physics is furnished by the equation giving the distance x covered by a particle starting at the origin ($x = 0$) at time $t = 0$ after time t, provided that the particle moves with constant acceleration a and starts with initial velocity v_0. This equation may be written

$$x = \tfrac{1}{2}at^2 + v_0 t \qquad (1.3.1)$$

Clearly, the dimensions of the two additive terms on the right-hand side of this equation must be the same. Using mks metric units for a (meter per sec per sec, or m/sec^2), v_0(m/sec), and t(sec), we see that the dimensions of the first term must be [according to requirement (3)]

$$\frac{\text{m}}{\text{sec}^2} \cdot \text{sec}^2 \qquad \text{or m}$$

while those of the second term are

$$\frac{\text{m}}{\text{sec}} \cdot \text{sec} \qquad \text{or m}$$

The two terms are dimensionally similar and may thus be added; Eq. (1.3.1) is then *consistent* in this respect. Also, since the dimensions of both terms on the right-hand side are meters and since the dimensions of the distance x on the left is meters, the dimensions of the quantities on both sides of the equation are the same, and requirement (1) is satisfied also. The form of the equation is clearly consistent with all three requirements, and the equation is said to be *dimensionally consistent*. If the second term on the right-hand side had the form $v_0 t^2$ rather than $v_0 t$, the dimensions of the two terms on this side of the equation would not have the same dimensions, and the equation would then be *dimensionally inconsistent*, and hence it could not possibly be correct. The fact that this second term does not have the dimensions of distance while the quantity on the left-hand side is incontrovertibly

[5] It can be contended that three horses and five goats can be added if the sum is said to be eight *animals*. The distinction is to a degree a semantic one and does not ordinarily arise in physical problems.

a distance indicates that it must be the term $v_0 t^2$ which is incorrect.

The violation of dimensional consistency is a sure sign that a mistake has been made somewhere, and it is always wise when deriving an expression for any physical quantity to be sure that the final result is dimensionally correct and consistent. The fact that it is is not a sure sign that the result is correct, but if it is not, the answer is surely wrong. Needless to say, when substituting dimensional quantities into an equation such as (1.3.1) it is necessary always to stick to the same system of units throughout. In other words, if one were to choose the mks system to begin with, one could not use units of m/sec^2 for a in the first term on the right and units of ft/sec for v_0 in the second. Instead, the mks unit of velocity, m/sec, would have to be substituted in the second term.

1.4 Scalars and Vectors: Graphic Methods for Vector Addition

Certain quantities of importance in physics can be completely described by stating their magnitude. Quantities of this sort—for example, time, mass, density, and volume—are called *scalars*, or *scalar quantities*. Other important quantities, of a more complex nature, have a *direction* associated with them as well as a magnitude. Such quantities, which include displacement, velocity, acceleration, and force, are called *vectors*. The velocity of an object, for example, is not completely specified by saying that its magnitude is 15 m/sec; one must say that it is 15 m/sec directed to the north to completely describe the situation. Since the laws of physics for the most part deal with vector quantities, we must familiarize ourselves with the various graphic and mathematical ways of representing vectors and with the rules of vector algebra.

Graphically, it is possible to represent a vector as a directed line segment pointed along the direction associated with the vector, whose magnitude or length is proportional to the magnitude of the vector quantity. Algebraically, a vector quantity is represented by a boldface symbol, thus, **A**, while the magnitude of the vector alone is represented by the corresponding lightface symbol A, or in some instances by $|\mathbf{A}|$. The *vector sum*, or *resultant*, of two vectors, **A** + **B**, is obtained graphically by placing the tail of vector **B** upon the head of vector **A** and then drawing a third vector whose tail rests upon the tail of **A** and whose head rests upon the head of **B**, as shown in Fig. 1.1a. This third vector represents the *vector sum* of **A** and **B**. From Figs. 1.1a and 1.1b it is obvious that it makes no difference in which order **A** and **B** are added; therefore,

$$\mathbf{A} + \mathbf{B} = \mathbf{B} + \mathbf{A} \qquad (1.4.1)$$

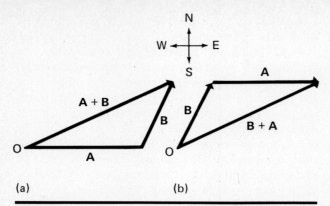

FIGURE 1.1. Vector addition as defined by graphical summation of two displacements. It is apparent that the resultant displacement is the same in magnitude and direction irrespective of whether **B** is added to **A**, as in (a), or **A** is added to **B**, as in (b).

It is evident from Fig. 1.1 that this law of vector addition is not simply an arbitrary mathematical rule but stems rather from the way in which the magnitude and direction of the vector describing an object's *total displacement* from an initial starting point O is related to the vectors **A** and **B** that give the length and direction of two separate legs of the journey. For example, if you start at O and walk east a distance corresponding to the length of **A** in Fig. 1.1a and subsequently walk along a northeasterly path a distance that corresponds to the length of **B**, you finally arrive at a point whose position with respect to the starting point is described by the vector **A** + **B**. This vector represents your total displacement from the point O.

The vector −**A** is defined as a vector having the same magnitude as **A** but pointing in the *opposite direction*. From this definition, and from the procedure for the addition of two vectors described previously, it is clear that the vector sum of any vector and its corresponding negative vector is zero, just as is the algebraic sum of any number and its algebraic negative. Two vectors may be subtracted by adding the negative of one of them to the other. Thus,

$$\mathbf{A} - \mathbf{B} = \mathbf{A} + (-\mathbf{B}) \tag{1.4.2}$$

The graphic procedure of arriving at the vector **A** − **B** is shown in Fig. 1.2a. From Fig. 1.2b it is easily seen that the vectors **A** + **B** and **A** − **B** can be represented as the two diagonals of the parallelogram determined by **A** and **B**. From elementary trigonometry (law of cosines), it follows that the magnitude of **A** + **B** is found from

$$|\mathbf{A} + \mathbf{B}|^2 = A^2 + B^2 + 2AB \cos \theta \tag{1.4.3}$$

where θ is the angle between the direction of **A** and the direction of **B**. The angle ϕ which **A** + **B** makes with **A** is given by

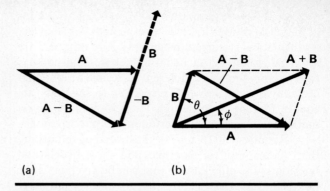

FIGURE 1.2. (a) Graphical procedure for vector subtraction. (b) Vector sum and difference represented as the two diagonals of the parallelogram defined by two vectors.

$$\tan \phi = \frac{B \sin \theta}{A + B \cos \theta} \tag{1.4.4}$$

The details of proving these results trigonometrically are assigned as an exercise. It is not necessary to remember Eqs. (1.4.3) and (1.4.4) in detail, since a simpler and more convenient way of arriving at the same answers will be discussed later.

The vector sum of three vectors, **A**, **B**, and **C**, may now be obtained by first summing **A** and **B** by the methods described previously and then finding the vector sum of the resulting vector with **C**. Thus,

$$\mathbf{A} + \mathbf{B} + \mathbf{C} = (\mathbf{A} + \mathbf{B}) + \mathbf{C} \tag{1.4.5}$$

This procedure is illustrated in Fig. 1.3a. It is clear from that figure that it makes no difference in what order the individual vectors are summed. It is also evident that three (or more) vectors will not generally lie in the same plane. Finally, from Fig. 1.3b, one may easily see that the resultant **A** + **B** + **C** is the vector forming the diagonal of a parallelopiped the edges of which are determined by **A**, **B**, and **C**. In arriving at the latter conclusion, it is necessary to transport the vector **C** (keeping it pointed at all times in the same direction) from location 1 to location 2. This is a permissible operation so far as any aspect of vector *algebra*, such as determining a vector sum, is concerned. But, as we shall see later, the line of action or point of application of vectors representing *physical* quantities is often important, and it will not in general be possible to do this with such vectors without altering the physical structure of the system. The graphic procedure for obtaining the vector sum of four or more vectors can be envisioned as a logical extension of the procedure outlined above for finding the sum of three vectors. It is illustrated in Fig. 1.4 for five vectors all lying in the same plane, although the methods are applicable for any number of vectors, irrespective of whether or not they are coplanar. The result can be summarized as the *polygon rule* for the

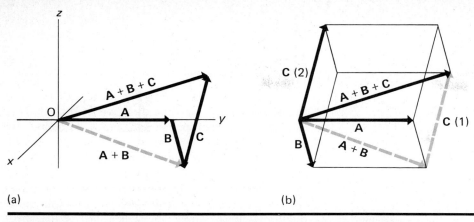

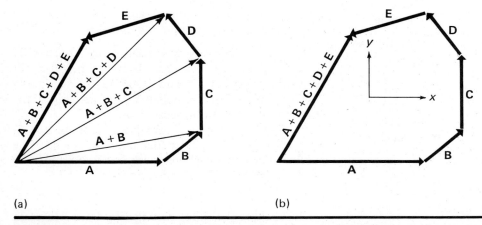

FIGURE 1.3. (a) Graphical summation of three vectors. (b) Sum of three vectors represented as the diagonal of the parallelopiped defined by the vectors.

FIGURE 1.4. Graphical representation of addition of several vectors, illustrating the polygon rule.

graphic addition of several vectors in the following way:

Place the tail of the second vector on the head of the first, the tail of the third on the head of the second, the tail of the fourth on the head of the third, etc. The vector sum of all the vectors so arranged is then a vector whose tail is on the tail of the first vector and whose head is on the head of the last vector of the series whose sum is sought.

The resultant vector is disposed in such a way as to close the open polygon formed when all the individual vectors whose sum is desired are arranged as described above. If the vector polygon formed by those vectors is a closed one to begin with, then the sum of all the individual vectors is zero, as shown by Fig. 1.5.

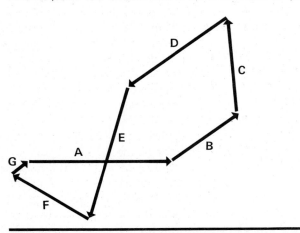

FIGURE 1.5. Vectors that form a closed polygon sum to zero.

1.5 Vector Components: Addition and Subtraction of Vectors by Components

Let us now consider the case of a single vector in the xy-plane, as shown in Fig. 1.6. The vector **A** shown there can be represented as the sum of the vector $\mathbf{A}_x$, which points along the positive x-axis, and $\mathbf{A}_y$, which points along the positive y-axis. The intercepts of the

9

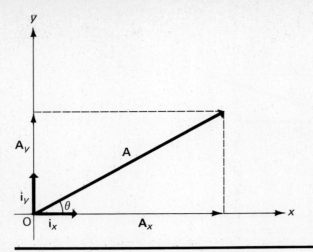

FIGURE 1.6. Vector **A** represented as sum of orthogonal vectors $\mathbf{A}_x$ and $\mathbf{A}_y$, which, in turn, can be represented as the components A_x and A_y times the unit vectors $\mathbf{i}_x$ and $\mathbf{i}_y$ in the x- and y-directions.

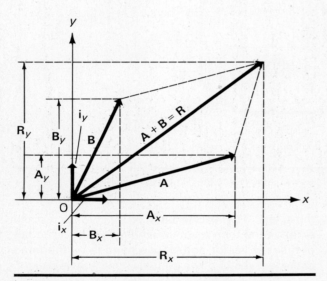

FIGURE 1.7. Vector addition by addition of components. The x- and y-components of the resultant **R** are obtained by adding algebraically the x- and y-components of vectors **A** and **B**.

projection of **A** along the x- and y-axes are

$$A_x = A \cos \theta \tag{1.5.1}$$
$$A_y = A \sin \theta \tag{1.5.2}$$

and these quantities are referred to as the x- and y-components of the vector **A**. The magnitude A and the angle θ can be obtained in terms of the components A_x and A_y very simply by squaring both sides of Eqs. (1.5.1) and (1.5.2) and adding the two resulting equations, recalling that $\sin^2 \theta + \cos^2 \theta = 1$, to obtain

$$A^2 = A_x{}^2 + A_y{}^2 \tag{1.5.3}$$

By dividing Eq. (1.5.2) by (1.5.1), it is then readily established that

$$\tan \theta = \frac{A_y}{A_x} \tag{1.5.4}$$

It is important to note that when the angle θ between the positive x-axis and the direction of the vector **A** exceeds $90°$, the components A_x and A_y as defined by Eqs. (1.5.1) and (1.5.2) can be negative. In general, whenever the projection of a vector like the one in Fig. 1.6 on the x- or y-direction lies along the negative x- or y-axis, the corresponding component A_x or A_y will be negative. A vector directed into the third quadrant, for example, for which $180° < \theta < 270°$, will have x- and y-components both of which are negative.

Another very useful way of viewing the situation is to envision the vectors $\mathbf{i}_x$ and $\mathbf{i}_y$ which have *unit magnitude* and which point along the positive x- and y-directions, respectively, and to note that the vector $\mathbf{A}_x$ is simply the component A_x times the unit vector $\mathbf{i}_x$ and that $\mathbf{A}_y$ is merely the component A_y times the unit vector[6] $\mathbf{i}_y$. Then, we may write

$$\mathbf{A} = \mathbf{A}_x + \mathbf{A}_y = A_x\mathbf{i}_x + A_y\mathbf{i}_y$$
$$= (A \cos \theta)\mathbf{i}_x + (A \sin \theta)\mathbf{i}_y \tag{1.5.5}$$

which expresses essentially the same sentiment as Eqs. (1.5.1) and (1.5.2).

Now suppose there are two vectors, **A** and **B**, lying in the xy-plane, as shown in Fig. 1.7, and we represent the resultant **A** + **B** as a third vector **R**, as illustrated. From the geometry of the parallelogram, it is obvious that

$$R_x = A_x + B_x$$
$$R_y = A_y + B_y \tag{1.5.6}$$

or, alternatively, expressing **A** and **B** in terms of the unit vectors $\mathbf{i}_x$ and $\mathbf{i}_y$ as in (1.5.5),

$$\mathbf{R} = \mathbf{A} + \mathbf{B} = A_x\mathbf{i}_x + A_y\mathbf{i}_y + B_x\mathbf{i}_x + B_y\mathbf{i}_y$$
$$= (A_x + B_x)\mathbf{i}_x + (A_y + B_y)\mathbf{i}_y \tag{1.5.7}$$

This equation is precisely equivalent to (1.5.6), because the coefficient of $\mathbf{i}_x$ has to be the x-component of **R** and that of $\mathbf{i}_y$, the y-component of **R**. In either way, we find that the x-component of the sum or resultant vector **R** is simply the sum of the x-components of **A** and **B**, and the y-component of the resultant is the sum of the y-components of the individual vectors.

---

[6] In making this statement, we are assuming that what we mean by multiplying a vector quantity by a scalar magnitude is a vector in the direction of the original vector whose magnitude is the original magnitude times the scalar multiplier. For example, the vector 6**A** is a vector in the direction of **A** six times as great in magnitude. We shall follow this practice consistently and indeed regard it as defining what is meant by a scalar times a vector.

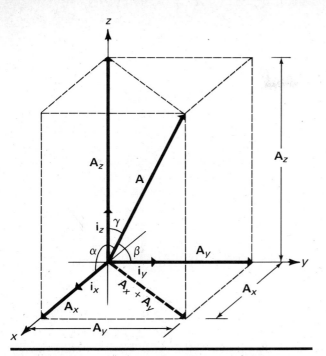

FIGURE 1.8. Components of a vector that does not lie in any of the planes defined by the coordinate axes. These components can always be expressed as multiples of the unit vectors $\mathbf{i}_x$, $\mathbf{i}_y$, and $\mathbf{i}_z$ along the axes.

That the result can be extended to sums involving more than two vectors is clearly evident. For example, in Fig. 1.4b, it is obvious that the x-component of the resultant vector $\mathbf{A} + \mathbf{B} + \mathbf{C} + \mathbf{D} + \mathbf{E}$ is simply the algebraic sum of the x-components of those individual vectors, noting, of course, that the x-components of $\mathbf{D}$ and $\mathbf{E}$ are negative. In the same way, the y-components of the individual vectors sum to the y-component of the resultant, provided that the y-component of $\mathbf{E}$ is regarded as negative.

If a vector is not constrained to lie within the xy-plane, as shown in Fig. 1.8, it may have a z-component as well as x- and y-components. In that figure, it is clear that the vector $\mathbf{A}$ may be represented as the vector sum of component vectors $\mathbf{A}_x$, $\mathbf{A}_y$, and $\mathbf{A}_z$ which point along the x-, y-, and z-directions, respectively. Also, from the geometry shown there, it is easily seen that

$$A_x = A \cos \alpha$$
$$A_y = A \cos \beta \qquad (1.5.8)$$
$$A_z = A \cos \gamma$$

where α, β, and γ are the direction angles between $\mathbf{A}$ and the x-, y-, and z-axes, respectively. If we define $\mathbf{i}_x$, $\mathbf{i}_y$, and $\mathbf{i}_z$ to be vectors of unit magnitude along the x-, y-, and z-axes, then the vector $\mathbf{A}_x$ is just $\mathbf{A}_x$ times the unit vector $\mathbf{i}_x$, $\mathbf{A}_y = A_y\mathbf{i}_y$, and $\mathbf{A}_z = A_z\mathbf{i}_z$. Then, since $\mathbf{A}$ is the vector sum of $\mathbf{A}_x$, $\mathbf{A}_y$, and $\mathbf{A}_z$,

we may write

$$\mathbf{A} = A_x\mathbf{i}_x + A_y\mathbf{i}_y + A_z\mathbf{i}_z$$
$$= (A \cos \alpha)\mathbf{i}_x + (A \cos \beta)\mathbf{i}_y + (A \cos \gamma)\mathbf{i}_z \qquad (1.5.9)$$

Also, from Fig. 1.8 it is evident that the magnitude A may always be expressed in terms of the components A_x, A_y, and A_z by

$$A^2 = A_x{}^2 + A_y{}^2 + A_z{}^2 \qquad (1.5.10)$$

Again, if we wish to find the resultant $\mathbf{R}$ which represents the sum of several vectors $\mathbf{A}_1, \mathbf{A}_2, \ldots, \mathbf{A}_n$, we may proceed by noting initially that each vector can be resolved into x-, y-, and z-components, whereby we may write

$$\mathbf{A}_1 = A_{1x}\mathbf{i}_x + A_{1y}\mathbf{i}_y + A_{1z}\mathbf{i}_z$$
$$\mathbf{A}_2 = A_{2x}\mathbf{i}_x + A_{2y}\mathbf{i}_y + A_{2z}\mathbf{i}_z$$
$$\vdots \qquad (1.5.11)$$
$$\mathbf{A}_n = A_{nx}\mathbf{i}_x + A_{ny}\mathbf{i}_y + A_{nz}\mathbf{i}_z$$

Adding all these vectors leads directly to the result

$$\mathbf{R} = R_x\mathbf{i}_x + R_y\mathbf{i}_y + R_z\mathbf{i}_z$$
$$= (A_{1x} + A_{2x} + A_{3x} + \cdots + A_{nx})\mathbf{i}_x$$
$$+ (A_{1y} + A_{2y} + A_{3y} + \cdots + A_{ny})\mathbf{i}_y$$
$$+ (A_{1z} + A_{2z} + A_{3z} + \cdots + A_{nz})\mathbf{i}_z \qquad (1.5.12)$$

In order that two vectors be equal, their x-, y-, and z-components must all be equal. Therefore:

The x-component of the resultant vector is the algebraic sum of all the x-components of the individual vectors, the y-component of the resultant is the algebraic sum of the y-components of the individual vectors, and the z-component of the resultant is the algebraic sum of the z-components of the individual vectors.

This is the rule for vector addition by components. It can be represented by Eq. (1.5.12) or by the equations

$$R_x = A_{1x} + A_{2x} + A_{3x} + \cdots + A_{nx} = \sum_{i=1}^{n} A_{ix}$$

$$R_y = A_{1y} + A_{2y} + A_{3y} + \cdots + A_{ny} = \sum_{i=1}^{n} A_{iy}$$

$$R_z = A_{1z} + A_{2z} + A_{3z} + \cdots + A_{nz} = \sum_{i=1}^{n} A_{iz}$$

$$(1.5.13)$$

The $\sum$ notation used on the right is the usual mathematical shorthand way of representing a *summation* of quantities of the form A_{ix}, A_{iy}, or A_{iz}, where i is a running index which goes through integer values from 1 to n, inclusive. It will be used extensively throughout this book.

In adding vectors by the component method

outlined above or in resolving vectors into components, it should be emphasized that the orientation of the x-, y-, and z-axes is completely arbitrary and can be chosen in any desired way. In most situations, the mathematical work to be done will be simplified if the orientation of these coordinate axes is chosen in such a way that as many vector components as possible are zero. For example, if all the vectors involved lie in a plane, it is obviously advantageous to call that plane the xy-plane. The z-axis will then be normal to the plane in which all the vectors lie, and none of the vectors will have a z-component, nor will the resultant. Also, the orientation of the x- and y-axes within the xy-plane can ordinarily be chosen in such a way that the x- or y-components of one or more of the vectors involved is zero. This technique is illustrated in Example 1.5.4.

Of particular importance in many instances is a *position vector* **r** from the origin to an arbitrary point P whose coordinates are x, y, and z. This vector describes the location of the point. If the point is envisioned as the location of a moving particle which started at the origin, such a vector represents the *displacement* of the particle from its initial position. The x-, y-, and z-components of the vector **r** are clearly just the coordinates x, y, and z of the point P, and therefore **r** can be written in terms of the unit vectors $\mathbf{i}_x$, $\mathbf{i}_y$, and $\mathbf{i}_z$ as

$$\mathbf{r} = x\mathbf{i}_x + y\mathbf{i}_y + z\mathbf{i}_z \qquad (1.5.14)$$

The concept of displacement of a body from an initial starting point is one of the most important ideas in the study of motion, as we shall see subsequently in Chapter 3.

EXAMPLE 1.5.1

An automobile is driven 30 miles east and then turns north to go 10 more miles. Describes its displacement from its starting point as a vector.

The total displacement **r** is the vector sum of the two individual displacement vectors representing the two legs of the journey, as shown in Fig. 1.9. Clearly, the vector r has the magnitude

$$r = \sqrt{30^2 + 10^2} = 31.6 \text{ mi}$$

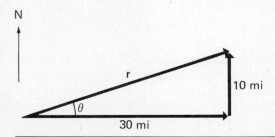

FIGURE 1.9

while the angle θ can be obtained by noting that

$$\tan \theta = \frac{10}{30} = 0.333 \qquad \theta = 18.4°$$

The magnitude and the direction as given by the angle θ, taken together, completely specify the displacement vector **r**. Note that despite the fact that the car has driven 40 miles, the magnitude of the displacement vector is only 31.6 miles. Indeed, were it to be driven back to its starting point, retracing the route whence it came, it would have been driven 80 miles, while the magnitude of its final total displacement vector would then be zero. Since the x-component of displacement is 30 miles and the y-component is 10 miles; the displacement vector **r** could equally well be specified by $\mathbf{r} = 30\mathbf{i}_x + 10\mathbf{i}_y$, where the numerical magnitudes are understood to be in units of miles.

EXAMPLE 1.5.2

Vector **A**, lying in the xy-plane, has magnitude 50 units, and its direction makes an angle of 60° with the x-axis, as shown in Fig. 1.10. Find its components along the x- and y-axes:

Trigonometrically, from Fig. 1.10 it is obvious that

$$A_x = A \cos 60° = (50)(0.500) = 25 \text{ units}$$
$$A_y = A \sin 60° = (50)(0.866) = 43.3 \text{ units}$$

These same results follow from Eqs. (1.5.1) and (1.5.2). This vector could be written in terms of the unit vectors $\mathbf{i}_x$ and $\mathbf{i}_y$ as $\mathbf{A} = 25\mathbf{i}_x + 43.3\mathbf{i}_y$.

EXAMPLE 1.5.3

An automobile is driven 30 miles east and then 20 miles northeast, as shown in Fig. 1.11a. Describe its displacement from the starting point as a vector.

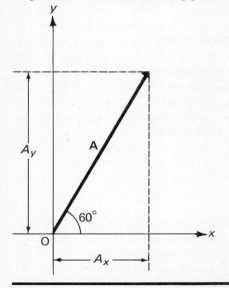

FIGURE 1.10

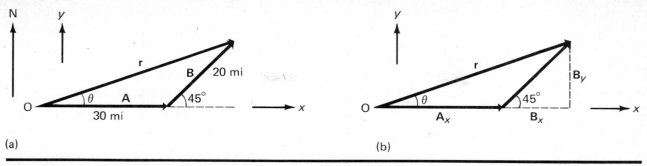

FIGURE 1.11. (a) Disposition of vectors discussed in Example 1.5.3. (b) Resolution of vectors shown in (a) into x- and y-components.

The total displacement **r** is the vector sum of two individual displacements **A** and **B**. It is easiest to calculate the resultant displacement by the method of components. For this purpose, it is advisable to choose the x-axis pointing east and the y-axis pointing north, as shown in Fig. 1.11a. Under these circumstances, the vector **A** has no y-component, while its x-component simply equals its total magnitude. Then, using the rule that the x- and y-components of the resultant equal the sum of the respective components of the individual vectors, we obtain

$$r_x = A_x + B_x = A + B \cos 45°$$
$$= 30 + (20)(0.707)$$
$$= 44.1 \text{ mi}$$

$$r_y = A_y + B_y = 0 + B \sin 45°$$
$$= (20)(0.707)$$
$$= 14.1 \text{ mi}$$

$$r = \sqrt{r_x^2 + r_y^2} = \sqrt{(44.1)^2 + (14.1)^2} = 46.4 \text{ mi}$$

$$\tan \theta = \frac{r_y}{r_x} = \frac{14.1}{44.1} = 0.320 \qquad \theta = 17.7°$$

Again, it is important to note that both the magnitude r and the angle θ are required to completely specify the vector **r**. Also, since r_x and r_y are simply distances measured from the origin along the x- and y-axes, and since they are components of a vector representing total displacement from the origin, they can (and should) simply be written as x and y; hereafter, in any such situation they will be so represented. The student should be able to rework this problem using the unit vectors $\mathbf{i}_x$ and $\mathbf{i}_y$ to express the individual vectors to be added and the components of the resultant vector.

EXAMPLE 1.5.4

Find a general expression for the magnitude of the vector **R** of Fig. 1.7 and for its direction relative to **A** and **B**, in terms of the magnitudes of **A** and **B** and the angle θ between **A** and **B**.

In this case, it is desirable to orient the xy-

coordinate axes in such a way that **A** lies along the x-axis, as shown in Fig. 1.12. Then, $A_y = 0$ and $A_x = A$ while, from Fig. 1.12, $B_x = B \cos \theta$ and $B_y = B \sin \theta$, and

$$R_x = A_x + B_x = A + B \cos \theta.$$
$$R_y = A_y + B_y = 0 + B \sin \theta$$

But

$$R = \sqrt{R_x^2 + R_y^2}$$
$$= \sqrt{A^2 + 2AB \cos \theta + B^2 \cos^2 \theta + B^2 \sin^2 \theta}$$
$$= \sqrt{A^2 + 2AB \cos \theta + B^2}$$

while the tangent of the angle φ between the direction of R and the direction of A is

$$\tan \phi = \frac{B_y}{A_x + B_x} = \frac{B \sin \theta}{A + B \cos \theta}$$

These results agree with those stated in Eqs. (1.4.3) and (1.4.4).

EXAMPLE 1.5.5

Find the resultant of the three vectors $\mathbf{A} = \mathbf{i}_x + 3\mathbf{i}_y + \mathbf{i}_z$, $\mathbf{B} = 3\mathbf{i}_x + 3\mathbf{i}_y + 3\mathbf{i}_z$, and $\mathbf{C} = \mathbf{i}_x + 3\mathbf{i}_y + 3\mathbf{i}_z$. What is

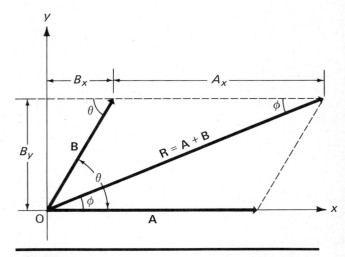

FIGURE 1.12

13

its magnitude and what are the direction angles it makes with the x-, y-, and z-coordinate axes?

In this case, the three vectors to be added do not lie in the same plane. There is no simple way in which to reorient the coordinate axes to make the vector components simpler, so we shall proceed straightaway. Using the rule for addition of components,

$$R_x = 1 + 3 + 1 = 5 \text{ units}$$
$$R_y = 3 + 3 + 3 = 9 \text{ units}$$
$$R_z = 1 + 3 + 3 = 7 \text{ units}$$

whence we may write

$$\mathbf{R} = R_x\mathbf{i}_x + R_y\mathbf{i}_y + R_z\mathbf{i}_z = 5\mathbf{i}_x + 9\mathbf{i}_y + 7\mathbf{i}_z$$

Alternatively, we could have directly summed the vectors as follows:

$$\mathbf{R} = \mathbf{A} + \mathbf{B} + \mathbf{C} = (\mathbf{i}_x + 3\mathbf{i}_y + \mathbf{i}_z) + (3\mathbf{i}_x + 3\mathbf{i}_y + 3\mathbf{i}_z)$$
$$+ (\mathbf{i}_x + 3\mathbf{i}_y + 3\mathbf{i}_z)$$
$$= 5\mathbf{i}_x + 9\mathbf{i}_y + 7\mathbf{i}_z$$

The same result is obtained in either case. The magnitude of **R** is given by

$$R = \sqrt{R_x{}^2 + R_y{}^2 + R_z{}^2} = \sqrt{155} = 12.44 \text{ units}$$

according to Eq. (1.5.10), while from (1.5.8),

$$\cos \alpha = \frac{R_x}{R} = \frac{5}{\sqrt{155}} = 0.402 \qquad \alpha = 66.3°$$

$$\cos \beta = \frac{R_y}{R} = \frac{9}{\sqrt{155}} = 0.723 \qquad \beta = 43.7°$$

$$\cos \gamma = \frac{R_z}{R} = \frac{7}{\sqrt{155}} = 0.562 \qquad \gamma = 55.8°.$$

1.6 The Multiplication of Vectors: Scalar and Vector Products

In addition to the algebraic rules we have devised for adding and subtracting vectors, it is possible, and very useful in some respects, to develop methods of *vector multiplication*. There are two ways of performing vector multiplication: one leading to a scalar quantity for the product, the other leading to a vector. The former process is referred to as *scalar* multiplication, the latter as *vector* multiplication.

The *scalar product* of two vectors **A** and **B** is written **A · B** and is often referred to as the *dot product*. The scalar product is defined as a scalar quantity given by the magnitude of **A** times the magnitude of **B** times the cosine of the angle between the direction of **A** and the direction of **B**. In the form of an equation, this can be stated as

$$\boxed{\mathbf{A} \cdot \mathbf{B} = AB \cos \theta} \tag{1.6.1}$$

where θ refers to the angle between the directions of **A** and **B**. From this definition, it is clear that *if two vectors are mutually perpendicular, their dot product is zero*, since then θ is 90° and $\cos \theta = 0$. Also, it is obvious that the dot product of a vector with itself is simply the square of its magnitude, that is,

$$\mathbf{A} \cdot \mathbf{A} = A^2 \tag{1.6.2}$$

since then $\cos \theta = 1$.

These two observations allow us to evaluate the scalar products of any pair of the unit vectors $\mathbf{i}_x, \mathbf{i}_y, \mathbf{i}_z$. According to Eq. (1.6.2),

$$\boxed{\mathbf{i}_x \cdot \mathbf{i}_x = \mathbf{i}_y \cdot \mathbf{i}_y = \mathbf{i}_z \cdot \mathbf{i}_z = 1} \tag{1.6.3}$$

since the magnitude of all these vectors is unity. Also, since any two *different* vectors of the set $\mathbf{i}_x, \mathbf{i}_y, \mathbf{i}_z$ are mutually perpendicular, we must have

$$\boxed{\begin{aligned} \mathbf{i}_x \cdot \mathbf{i}_y = \mathbf{i}_y \cdot \mathbf{i}_x = \mathbf{i}_z \cdot \mathbf{i}_x = \mathbf{i}_x \cdot \mathbf{i}_z \\ = \mathbf{i}_y \cdot \mathbf{i}_z = \mathbf{i}_z \cdot \mathbf{i}_y = 0 \end{aligned}} \tag{1.6.4}$$

If we express the vectors **A** and **B** in terms of their components and the unit vectors $\mathbf{i}_x, \mathbf{i}_y,$ and $\mathbf{i}_z$ as in (1.5.9) and take their dot product in this form, we find

$$\begin{aligned} \mathbf{A} \cdot \mathbf{B} &= (A_x\mathbf{i}_x + A_y\mathbf{i}_y + A_z\mathbf{i}_z) \cdot (B_x\mathbf{i}_x + B_y\mathbf{i}_y + B_z\mathbf{i}_z) \\ &= A_xB_x(\mathbf{i}_x \cdot \mathbf{i}_x) + A_xB_y(\mathbf{i}_x \cdot \mathbf{i}_y) + A_xB_z(\mathbf{i}_x \cdot \mathbf{i}_z) \\ &+ A_yB_x(\mathbf{i}_y \cdot \mathbf{i}_x) + A_yB_y(\mathbf{i}_y \cdot \mathbf{i}_y) + A_yB_z(\mathbf{i}_y \cdot \mathbf{i}_z) \\ &+ A_zB_x(\mathbf{i}_z \cdot \mathbf{i}_x) + A_zB_y(\mathbf{i}_z \cdot \mathbf{i}_y) + A_zB_z(\mathbf{i}_z \cdot \mathbf{i}_z) \end{aligned}$$

But many of these terms, according to Eq. (1.6.4), are zero; the others can be evaluated from (1.6.3). The final result is

$$\boxed{\mathbf{A} \cdot \mathbf{B} = A_xB_x + A_yB_y + A_zB_z} \tag{1.6.5}$$

Equation (1.6.5) expresses **A · B** in terms of the components of the two vectors **A** and **B**. We shall find this expression useful for many purposes in later sections. It is also important to note from the definition (1.6.1) that the quantities **B · A** and **A · B** are the same. In other words,

$$\mathbf{A} \cdot \mathbf{B} = \mathbf{B} \cdot \mathbf{A} \tag{1.6.6}$$

The definition of the scalar product, as given by (1.6.1), seems to be quite arbitrary. It is adopted because, as we shall see in due time, it allows us to express important physical quantities, such as work and electric and magnetic flux, in a particularly simple way and allows us to write many of the fundamental laws of physics in a concise and meaningful form.

The *vector product* of two vectors **A** and **B** is usually written as **A × B** and is often referred to as the *cross product*. The product is, by definition, a *vector* whose magnitude is the magnitude of **A** times the magnitude of **B** times the sine of the angle between the directions of **A** and **B**. This can be written as

$$\boxed{|\mathbf{A} \times \mathbf{B}| = AB \sin \theta} \qquad (1.6.7)$$

The direction of the vector $\mathbf{A} \times \mathbf{B}$ is perpendicular to both $\mathbf{A}$ and $\mathbf{B}$ (thus normal to the plane in which $\mathbf{A}$ and $\mathbf{B}$ lie), and the vector points along the direction in which a *right-handed screw* would advance when the vector $\mathbf{A}$ is rotated into vector $\mathbf{B}$ through the smallest possible angle, as shown in Fig. 1.13. Since the screw will move in the opposite direction when $\mathbf{B}$ is rotated into $\mathbf{A}$, the vector $\mathbf{B} \times \mathbf{A}$ will point in the direction opposite to that of $\mathbf{A} \times \mathbf{B}$. It follows from this that $\mathbf{B} \times \mathbf{A}$ and $\mathbf{A} \times \mathbf{B}$ are not equal. As a matter of fact, using the statement in the previous sentence along with (1.6.7), it is easy to see that

$$\mathbf{B} \times \mathbf{A} = -(\mathbf{A} \times \mathbf{B}) \qquad (1.6.8)$$

Also, from the definition (1.6.7) it is evident that the cross product of two vectors which are *parallel* is *zero*, since then $\sin \theta = 0$.

Using these facts in connection with the unit vectors $\mathbf{i}_x$, $\mathbf{i}_y$, and $\mathbf{i}_z$, it is easy to see that

$$\boxed{\mathbf{i}_x \times \mathbf{i}_x = \mathbf{i}_y \times \mathbf{i}_y = \mathbf{i}_z \times \mathbf{i}_z = 0} \qquad (1.6.9)$$

and from the definition (1.6.7) and the right-hand rule for the direction of the cross product vector, it is easily established that

$$\boxed{\begin{aligned} \mathbf{i}_x \times \mathbf{i}_y &= -(\mathbf{i}_y \times \mathbf{i}_x) = \mathbf{i}_z \\ \mathbf{i}_y \times \mathbf{i}_z &= -(\mathbf{i}_z \times \mathbf{i}_y) = \mathbf{i}_x \\ \mathbf{i}_z \times \mathbf{i}_x &= -(\mathbf{i}_x \times \mathbf{i}_z) = \mathbf{i}_y \end{aligned}} \qquad (1.6.10)$$

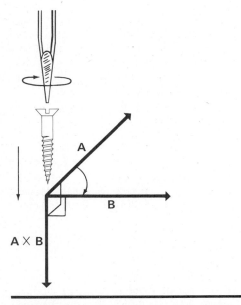

FIGURE 1.13. "Right-hand screw rule" for determining direction of cross product vector. The direction of the vector $\mathbf{A} \times \mathbf{B}$ is in the direction in which a right-handed screw advances as $\mathbf{A}$ is turned into $\mathbf{B}$ through the smallest possible angle.

Using these relations, it is possible to show, using the same general line of attack used in connection with (1.6.5), that

$$\boxed{\begin{aligned} \mathbf{A} \times \mathbf{B} &= (A_y B_z - A_z B_y)\mathbf{i}_x + (A_z B_x - A_x B_z)\mathbf{i}_y \\ &\quad + (A_x B_y - A_y B_x)\mathbf{i}_z \end{aligned}} \qquad (1.6.11)$$

This equation expresses $\mathbf{A} \times \mathbf{B}$ in terms of the components of $\mathbf{A}$ and $\mathbf{B}$ along the coordinate axes. We shall find it useful for future work.

In using the relations (1.6.10), it is always necessary to use a *right-handed system* of coordinate axes, that is, a system in which a right-handed screw advances along the positive z-direction when the x-axis is rotated into the y-axis through the smallest possible angle. These relations otherwise are not correct. In this book we shall invariably choose our coordinate systems to conform with this so-called *right-hand rule*.

Again, the definition chosen for the vector product seems arbitrary but it is one we shall find useful, since it enables us to express certain important physical quantities, such as torque and angular momentum as vector products, and since some important physical laws can be simply and clearly expressed and understood by its use.

1.7 Differentiation and Integration of Vectors

We shall discuss only a few of the essential facts about derivatives and integrals of vector quantities here, postponing detailed treatment of certain other points until the need arises. If we have a vector $\mathbf{A}$ of the form

$$\mathbf{A} = A_x \mathbf{i}_x + A_y \mathbf{i}_y + A_z \mathbf{i}_z \qquad (1.7.1)$$

we may differentiate $\mathbf{A}$ with respect to some scalar variable t (which may, but need not necessarily, refer to time) to obtain a *vector* $d\mathbf{A}/dt$ of the form

$$\frac{d\mathbf{A}}{dt} = \frac{dA_x}{dt}\mathbf{i}_x + \frac{dA_y}{dt}\mathbf{i}_y + \frac{dA_z}{dt}\mathbf{i}_z \qquad (1.7.2)$$

This vector has components dA_x/dt, dA_y/dt, and dA_z/dt and represents the *derivative* of the vector $\mathbf{A}$ with respect to the scalar variable t. In differentiating Eq. (1.7.1), the vectors $\mathbf{i}_x$, $\mathbf{i}_y$, and $\mathbf{i}_z$ are constant in magnitude and direction and can, therefore, be treated in the same way as algebraic constants in the differentiation.

The integral of $\mathbf{A}$ with respect to a scalar variable t may be calculated in essentially the same way:

$$\int \mathbf{A} \, dt = \mathbf{i}_x \int A_x \, dt + \mathbf{i}_y \int A_y \, dt + \mathbf{i}_z \int A_z \, dt \qquad (1.7.3)$$

The integral of a vector with respect to a scalar parameter is, therefore, a vector having components $\int A_x \, dt$, $\int A_y \, dt$, and $\int A_z \, dt$.

15

SUMMARY

Physics is a quantitative science in which the measurement of the properties of physical systems is of great importance. These properties are usually dimensional quantities, and measurements are made by comparing the measured quantities with standard unit quantities such as the standard meter or standard kilogram. In doing this, it is necessary to define four fundamental standard physical quantities—mass, length, time, and electric charge—though for the study of mechanics only the first three (or their equivalents) are required. In the mks metric system, these quantities are the meter, the kilogram, and the second; while in the cgs metric system, the centimeter, the gram, and the second are chosen. In the British engineering system, the fundamentally defined units are units of force, length, and time, expressed by the pound, the foot, and the second.

In performing mathematical operations with dimensional quantities, we may add or subtract only quantities having the same units. Quantities having different units may be multiplied or divided, the units of the resulting quantity then being the product or quotient of those of the individual factors. All the equations used to relate physical quantities must be dimensionally consistent, which means that all their terms must have the same dimensions. An awareness of this requirement frequently allows us to identify mistakes in our work that might otherwise be hard to find.

Physics deals not only with scalar quantities, which are completely specified by magnitude alone, but also with vectors, which are quantities that have direction as well as magnitude. A vector may be visualized as a directed line segment and may be specified by giving its magnitude and direction or, alternatively, by stating its projections along the coordinate axes, which are referred to as its components. Vectors may be added or subtracted graphically as in Fig. 1.4, or by adding or subtracting their components along the respective coordinate axes to obtain the components of the resultant. It is important to know how to express the magnitude and direction of a vector in terms of its components and to find the components when the magnitude and direction are known. Of particular importance is the vector that specifies the *displacement* of a point from the origin of coordinates; the components of this vector are the coordinates x, y, and z of the point. Also important are the unit vectors i_x, i_y, and i_z, which allow us to perform vector algebra very simply.

There are two ways of multiplying vectors. The first of these gives a "dot product" $A \cdot B$ that is a scalar quantity, while the second results in a "cross product" $A \times B$, which is a vector. It is important for future work to know how these products are defined. It is also useful to know how to evaluate scalar and vector products using the properties of the unit vectors i_x, i_y, and i_z.

Vector quantities may be differentiated and integrated by differentiating or integrating their individual components.

QUESTIONS

1. Can you suggest a way to measure the dimensions of an object roughly the size and shape of a brick to an accuracy of ± 0.5 cm? To an accuracy of ± 0.005 cm? To an accuracy of $\pm 5 \times 10^{-5}$ cm?

2. Can you suggest ways of measuring to an accuracy of 1% the length of an object roughly 1.0 m long? An object 1000 m long? An object 10^{-3} m long? An object 10^{-6} m long?

3. How might we measure, again to 1% accuracy, a time interval of length roughly 100 sec? Of length 1.0 sec? Of length 10^{-5} sec?

4. The speedometer of a police patrol car is certified by the Pennsylvania Department of Transportation as "100% accurate." Is this way of stating the possible error associated with its indications likely to be "100% accurate"?

5. It is obvious that if every component of a vector is zero, the vector quantity itself must vanish. But is the converse of this statement true? In other words, if a vector quantity is zero, does every component of the vector have to be zero?

6. A rabbit runs with constant speed around a circular track. Is its displacement vector, referred to an origin at the center of the track, a constant vector or does it change with time?

7. Can a vector quantity ever have a negative magnitude? Can the scalar product of two vectors ever be negative?

8. We are usually willing to accept the fact that quantities such as force, velocity, electric field strength, torque, and momentum are vectors. But how would you go about proving that they are, if you were asked to do so?

9. A set of three quantities have the values 5 units, 5 units, and 3 units referred to one coordinate system, and the values 4 units, 7 units, and 2 units when measured with respect to another coordinate system. Is it possible that these sets of numbers might represent the components of the same vector?

10. Does the derivative of a vector with respect to a scalar variable have to have the same direction as the original vector?

11. An equation expresses an equality between a vector quantity and a scalar quantity. How would you regard a relationship of this sort?

12. It is possible, according to section 1.5, to multiply a vector quantity by a scalar. But can you add or subtract a scalar and a vector?

13. A man starts out from a given point, walks south 1 mile, and shoots a bear. He then walks east 1 mile and then north 1 mile. He is now back at his original starting point. What color was the bear?

PROBLEMS

1. Show, from the basic definitions of both quantities in terms of the earth's dimensions that 1 nautical mile is equal, very nearly, to $\frac{50}{27}$ kilometers. *Hint:* How is the nautical mile defined in terms of the earth's size?; look it up!

2. Examine all aspects of the dimensional consistency of the equation

$$v^2 = v_0^2 + 2ax$$

for the velocity of a particle which starts from the origin at $t = 0$ with initial velocity v_0 and moves thereafter with constant acceleration a as a function of its displacement x from the starting point. Is the equation dimensionally consistent?

3. You are told that the equation giving the displacement of a particle starting with initial displacement x_0 and initial velocity v_0 at time $t = 0$ and moving thereafter with constant acceleration, as a function of time, is

$$x = \tfrac{1}{2}at^2 + v_0t + x_0{}^2$$

Is this equation correct or incorrect? Why?

4. A group of physicists and their wives (the latter scientifically untrained) were watching a science fiction program on television. A remark was made by one of the actors that "these space craft used by the creatures from alpha centauri are capable of moving with a speed of c^2" (where c refers to the speed of light). All the physicists were convulsed with laughter, while their wives stared at them blankly. What was the cause of the scientists' mirth?

5. Fifty cents squared is twenty-five hundred cents, or $25.00. But, also, a half-dollar squared is a quarter. What is the explanation of this apparent paradox?

6. A vector lying in the xy-plane has magnitude 25 units and makes an angle of $37°$ with the x-axis. What are its x- and y-components?

7. A vector lying in the xy-plane has an x-component of 12 units and a y-component of 16 units. What is its magnitude, and what angle does its direction make with the x-axis?

8. A vector **A** has magnitude 3 in. and points along the x-axis. Vector **B** lies in the xy-plane, has magnitude 2 in., and makes an angle of $45°$ with the x-axis. Vector **C** lies in the xy-plane, has magnitude 5 in., and makes an angle of $75°$ with the x-axis. Find the resultant of **A**, **B**, and **C** graphically, and measure its magnitude and the angle it makes with the x-axis.

9. Perform the vector sum in Problem 8 algebraically, using the method of components. Does the answer as obtained agree with the graphic result?

10. Verify, using trigonometric methods, Eqs. (1.4.3) and (1.4.4).

11. A car is driven 24 miles due east, then turns and travels 60 miles northwest. Find **(a)** the horizontal and vertical components of its total displacement, **(b)** the magnitude of the total displacement vector, **(c)** the angle between the total displacement vector and the positive x-axis (which points east). **(d)** The car drives back to the starting point. What is its final displacement?

12. A ship steers a course of $90°$ for 3 hours, and then alters course to $150°$ for 4 hours. Its speed is 12 knots (12 nautical miles per hour). What is its final displacement (magnitude and direction)? Note that a course of $0°$ is due north, $90°$ is due east, $180°$ is due south, and $270°$ is due west.

13. A car is driven east 40 miles, then in a direction $30°$ east of due north an additional 30 miles. Find the magnitude and direction of the vector representing the final displacement.

14. A ship proceeds due north for 60 miles, then changes course and steams in a general southeasterly direction until it reaches a position which is 50 miles away from the starting point on a bearing of $20.6°$ from that point. What was the length and course of the second leg of the trip?

15. Two forces, **A** and **B**, lying in the xy-plane, act on a small object at the origin. Force **A** is of magnitude 50 pounds and acts in a direction making an angle of $+30°$ with the positive x-axis. Force **B** is of magnitude 80 pounds and acts in a direction making an angle of $+135°$ with the positive x-axis. What should be the magnitude and direction of the force **C** that would have to be applied to make the resultant of all three forces zero?

16. Three horizontal forces act on an object. Force **A** is of magnitude 25 pounds and acts in a direction $30°$N of E. Force **B** is of magnitude 70 pounds and acts due north. Force **C** is of magnitude 55 pounds and acts southwest. What is the magnitude and direction of their resultant **R**? What are the x- and y-components of **R**, assuming that the x-axis points east and the y-axis north?

17. Consider the vector $\mathbf{i}_x + 3\mathbf{i}_y$. **(a)** What is the magnitude of the x-component? **(b)** The y-component? **(c)** What is the magnitude of the vector? **(d)** What is the angle the vector makes with the x-axis?

18. Consider the vector $\mathbf{i}_x + 3\mathbf{i}_y + 2\mathbf{i}_z$. Find **(a)** the magnitude of the x-component, **(b)** the magnitude of the y-component, **(c)** the magnitude of the z-component, **(d)** the magnitude of the vector, **(e)** the angle between the vector and the x-axis, **(f)** the angle between the vector and the y-axis, **(g)** the angle between the vector and the z-axis.

19. What would the answers to Problem 12 be if, in addition, there was an ocean current flowing **(a)** due south at a speed of 4 knots and **(b)** due north at a speed of 4 knots? In both instances, assume that the ship's speed with respect to the water is 12 knots.

20. What is the resultant of the vectors $\mathbf{A} = 3\mathbf{i}_x + 2\mathbf{i}_y$, $\mathbf{B} = \mathbf{i}_x - 3\mathbf{i}_y$, and $\mathbf{C} = -\mathbf{i}_x + 4\mathbf{i}_y$? What is the magnitude of the resultant and what angle does it make with the x-axis?

21. Two vectors, **A** and **B**, may assume any arbitrary orientation with respect to one another. Show that the magnitude of the resultant **R** must satisfy $|A - B| < R < |A + B|$.

22. What is the resultant of the vectors $\mathbf{A} = \mathbf{i}_x + 3\mathbf{i}_y + 5\mathbf{i}_z$, $\mathbf{B} = -3\mathbf{i}_x + \mathbf{i}_y - \mathbf{i}_z$, and $\mathbf{C} = 5\mathbf{i}_x - 2\mathbf{i}_y - 6\mathbf{i}_z$? Find the magnitude of the resultant and the direction angles α, β, and γ with respect to the three coordinate axes.

23. Find **(a)** the scalar product and **(b)** the vector product of the vectors $\mathbf{A} = \mathbf{i}_x - 2\mathbf{i}_y + 3\mathbf{i}_z$ and $\mathbf{B} = 3\mathbf{i}_x - \mathbf{i}_y - \mathbf{i}_z$.

24. Find the angle between the vectors $\mathbf{A} = 3\mathbf{i}_x + \mathbf{i}_y$ and $\mathbf{B} = \mathbf{i}_x + 4\mathbf{i}_y$.

25. Find a vector lying in the xy-plane perpendicular to the vector $A = 2\mathbf{i}_x + 3\mathbf{i}_y$.

17

26. Consider the vectors $\mathbf{A} = \mathbf{i}_x + 3\mathbf{i}_y$ and $\mathbf{B} = 3\mathbf{i}_x - 4\mathbf{i}_y$. Find **(a)** the magnitude and direction of $\mathbf{A} + \mathbf{B}$ and $\mathbf{A} - \mathbf{B}$, **(b)** the angle between vectors $\mathbf{A}$ and $\mathbf{B}$, **(c)** the angle between vectors $\mathbf{A} + \mathbf{B}$ and $\mathbf{A} - \mathbf{B}$, **(d)** the x- and y-components of a vector $\mathbf{C}$ of magnitude 6 units, which is perpendicular to vector $\mathbf{B}$.

27. Consider the vectors $\mathbf{A} = \mathbf{i}_x + 3\mathbf{i}_y$ and $\mathbf{B} = 3\mathbf{i}_x - 4\mathbf{i}_y$. Find **(a)** the magnitude and direction of the vector $-3\mathbf{B}$, **(b)** the magnitude and direction of the vector $3\mathbf{A} + 2\mathbf{B}$, **(c)** the x- and y-components of $3\mathbf{A} + 2\mathbf{B}$.

28. Consider the vector $\mathbf{C} = 12\mathbf{i}_x + 10\mathbf{i}_y$. Can $\mathbf{C}$ be expressed in terms of the vectors $\mathbf{A}$ and $\mathbf{B}$ of Problems 26 and 27 as some scalar multiple of $\mathbf{A}$ plus a scalar multiple of $\mathbf{B}$ of the form $\mathbf{C} = m\mathbf{A} + n\mathbf{B}$, where m and n are scalars? If so, what values of m and n must be chosen for this purpose?

29. A point $P(x, y)$ lies in the xy-plane. Its displacement relative to the origin is described by the radially directed displacement vector $\mathbf{r} = \mathbf{i}_x x + \mathbf{i}_y y$. Find a vector $\mathbf{i}_r$ of unit magnitude in the direction of the displacement vector $\mathbf{r}$ to the point P as illustrated in the diagram.

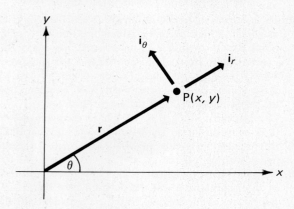

30. In the situation described in the preceding problem, find a tangential vector $\mathbf{i}_\theta$ of unit magnitude lying in the xy-plane normal to the vector $\mathbf{i}_r$ at point P.

31. Find the scalar and vector product of the vectors $\mathbf{A}$ and $\mathbf{B}$ of Problem 23.

32. Show that the vectors $\mathbf{A} = \mathbf{i}_x + 3\mathbf{i}_y - 6\mathbf{i}_z$ and $\mathbf{B} = 3\mathbf{i}_x + \mathbf{i}_y + \mathbf{i}_z$ are mutually perpendicular.

33. Find the angle between vectors $\mathbf{A} = \mathbf{i}_x + \mathbf{i}_y - \mathbf{i}_z$ and $\mathbf{B} = 2\mathbf{i}_x + 2\mathbf{i}_y + 5\mathbf{i}_z$.

34. Show that the vectors $\mathbf{A} = \mathbf{i}_x - 3\mathbf{i}_y + 2\mathbf{i}_z$ and $\mathbf{B} = -4\mathbf{i}_x + 12\mathbf{i}_y - 8\mathbf{i}_z$ are parallel.

35. Find a vector $\mathbf{B}$ lying in the xy-plane that is perpendicular to the vector $\mathbf{A} = 2\mathbf{i}_x + 3\mathbf{i}_y$ and is of magnitude 4 units.

36. Using methods similar to those used in the derivation of Eq. (1.6.5), verify the result (1.6.11) starting with Eqs. (1.6.10).

37. Show that the vectors $\mathbf{A}$ and $\mathbf{B}$ inscribed in a semicircle, as shown in the diagram, must always be perpendicular. *Hint:* Show that the scalar product is zero.

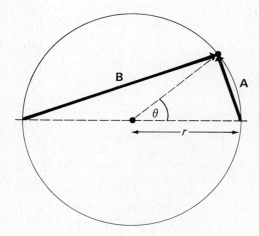

38. Referring once again to the vectors $\mathbf{A} = \mathbf{i}_x + 3\mathbf{i}_y$, $\mathbf{B} = 3\mathbf{i}_x - 4\mathbf{i}_y$, and $\mathbf{C} = 12\mathbf{i}_x + 10\mathbf{i}_y$ of Problems 26, 27, and 28, find **(a)** the vector $\mathbf{A} \times (\mathbf{B} \times \mathbf{C})$, **(b)** the vector $(\mathbf{A} \times \mathbf{B}) \times \mathbf{C}$. **(c)** Show that for these vectors $\mathbf{A} \times (\mathbf{B} \times \mathbf{C}) = 6[\mathbf{A} \times (\mathbf{B} \times \mathbf{A})]$ and $(\mathbf{A} \times \mathbf{B}) \times \mathbf{C} = 6[\mathbf{A} \times (\mathbf{B} \times \mathbf{A})] + 2[(\mathbf{A} \times \mathbf{B}) \times \mathbf{B}]$.

39. Consider the vectors $\mathbf{A} = \mathbf{i}_x + 4\mathbf{i}_y$, $\mathbf{B} = 2\mathbf{i}_x - 3\mathbf{i}_y$, $\mathbf{C} = \mathbf{i}_x + \mathbf{i}_z$. Find **(a)** the scalar quantity $\mathbf{A} \cdot (\mathbf{B} \times \mathbf{C})$ and **(b)** the quantity $(\mathbf{A} \times \mathbf{B}) \cdot \mathbf{C}$.

40. If $\mathbf{A} \cdot \mathbf{C} = \mathbf{B} \cdot \mathbf{C}$, are we justified in concluding that $\mathbf{A} = \mathbf{B}$? If not, why not, and what additional requirements must be satisfied?

41. Show that $(\mathbf{A} \times \mathbf{B}) \cdot (\mathbf{A} \times \mathbf{B}) + (\mathbf{A} \cdot \mathbf{B})^2 = A^2 B^2$.

NEWTON'S LAWS: FORCES AND TORQUES IN EQUILIBRIUM

2.1 Introduction—Definition of Force

In the preceding chapter, the algebra of vector quantities was discussed. We are now ready to see how the methods developed there may be applied in studying the effect of forces in systems which are at rest or, at any rate, unaccelerated. Before we can make progress in this direction, however, it is necessary to understand in physical terms what is meant by the term *force*. Intuitively speaking, a force is what we experience by our sense of touch when we push against or pull some material object. But this concept is purely qualitative and is hardly capable of expressing the magnitude of forces in precise numerical terms. In Lord Kelvin's language, we have perhaps "the beginning of knowl-edge," but we have hardly "advanced to the state of Science" in the matter of understanding and measuring forces.[1]

But there is a very common and uniform force that can be used as a *standard* against which all other forces may be compared quantitatively. This force is the downward attraction the earth exerts against all material objects near its surface—in other words, the force of *gravity*. Although the force of gravity varies slightly with elevation above sea level and is very slightly influenced by certain other terrestrial effects, a standard of force can be taken as the downward force of gravity upon some standard

[1] See p. 2.

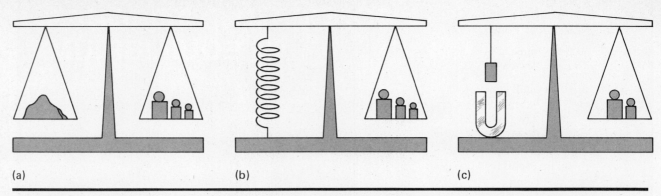

(a) **(b)** **(c)**

FIGURE 2.1. Comparison of forces of various physical origins with the gravitational force acting on a set of standard masses.

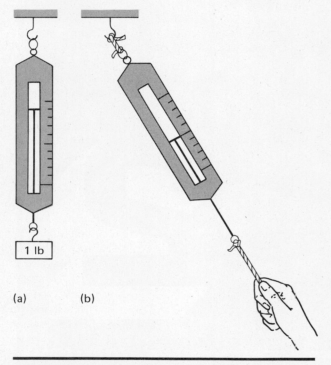

1 lb

(a) **(b)**

FIGURE 2.2. Use of a calibrated spring balance in measuring forces.

object at sea level and at a given location.[2] The force associated with the gravitational attraction of the earth for an object is called the *weight* of the object. By the use of an equal-arm balance, weights, and hence forces, can be compared, sets of standard weights can be constructed, and forces other than gravitational forces can be measured quantitatively in terms of weight forces, as illustrated in Fig. 2.1.

For convenience, forces are often measured by devices such as *spring balances* (Fig. 2.2). In the spring balance, a spring is stretched by an applied force, and

the magnitude of the force is then determined by observing the distance through which the spring is extended, the spring having been previously *calibrated* against known gravitational forces by observing how much it stretches when standard weights are attached. The use of the spring balance or of any other measuring device that has been calibrated against known gravitational forces is equivalent to a direct comparison like those illustrated in Fig. 2.1.

It is obvious that any force has a direction as well as a magnitude associated with it. It is not as obvious that the physical effect of two separate forces acting at a given point is the same as the effect of a *single force having the magnitude and direction of the vector sum* of the two separate forces, but this is nevertheless true and can easily be verified experimentally. Force is, therefore, a *vector* quantity, and the net resultant force applied to any system is the *vector sum* of all the individual forces that may act upon it.

2.2 Newton's Laws of Motion

The science of classical mechanics rests upon three fundamental laws which were first stated by Sir Isaac Newton in 1686. These three principles express the entire *physical* content of mechanics, and what else there is to learn is added for the sake of understanding and applying the physical principles contained in Newton's laws. We may regard Newton, therefore, as the founder of mechanics, and indeed as the originator of the science of physics as it is known today. Before Newton's time, mechanics was a tangle of contradictory hypotheses and beliefs and half-understood rules of thumb. After his work had been completed, it was a quantitative science, based on a firm conceptual foundation,[3] whose predictions were in exact agree-

---

[2] The slight local variations in the strength of the earth's gravitational field are well understood and easily corrected for. The factors that influence the strength of the gravitational force and the ways in which local variations can be accounted for will be described in subsequent chapters.

[3] It was a firm foundation, but it was found to be not perfect in every aspect, as we shall see later. Certain logical inconsistencies in the structure of classical mechanics were resolved only in 1905 by Einstein's special theory of relativity, which is discussed in detail later.

ment with experimental findings in practically every respect. In Alexander Pope's words,[4]

> Nature and Nature's laws lay hid in night:
> God said, Let Newton be! and all was light.

Although we shall generally avoid the axiomatic approach toward science, for simplicity and conciseness Newton's laws will be stated at this point. We shall find it necessary frequently to refer back to them to illustrate particular examples, and to convince outselves of their essential validity in various applications. They are:

First Law *When the resultant or vector sum of all the forces acting on a body is zero, the body's acceleration is zero, and it remains either at rest, or in uniform motion with a velocity that is constant in magnitude and direction.*
Second Law *When the resultant or vector sum of all the forces acting on a body is not zero, the body acquires an acceleration in the direction of the resultant force. The magnitude of this acceleration is directly proportional to that of the resultant force, and inversely proportional to the body's mass.*
Third Law *For every action there is an equal and opposite reaction; thus, the force exerted by one body on another is equal in magnitude and opposite in direction to the force exerted by the second body on the first.*

These laws immediately suggest a large number of questions, and it is not at all difficult to think of instances in which, to an uncritical observer, experience would seem flatly to contradict their predictions. For example, we certainly observe that objects that roll or slide upon level surfaces eventually come to rest, and thus do not seem to persist in a state of uniform motion. Also, what about a wheel rotating upon its axis? Even if it were to continue in a uniform rotation when once set into motion (which, ordinarily, it will not), does not the direction of motion of every particle of matter in it change continually, and is this not a contradiction of the first law? Without attempting to explain these specific objections in detail at this time, we may state that when we study these situations critically, examining precisely what is going on in each case,[5] we shall find that they are not at all in conflict with Newton's laws, but instead are actually in complete accord with what would be expected from them. Much of the work we shall undertake in the succeeding sections and chapters is, in fact, directed toward assuring ourselves that this is the case. The greatness of Newton's contribution lay in his ability to brush aside all superficial objections relating to frictional effects and similar extraneous matters, and extract what was really at the heart of the subject.

Newton's laws can in no way be derived mathematically. They are generalizations of experimental observations of the actual motion of material objects and how applied forces affect such motions. They are, therefore, natural laws that describe the behavior of the external world rather than mathematical axioms.

It should be noted that Newton's *second* law contains the crucial assertion about how objects move when subjected to forces. In a sense, therefore, the second law occupies a position of special importance, while the first and third laws serve to a degree to amplify the second law. For example, if the resultant of all the forces acting on a body is *zero*, then, according to the second law, its acceleration must also be zero, which means that it is either at rest or, at most, moving with a velocity that is unchanging both in magnitude and direction. But this is just what the first law tells us, so we may regard the first law as a special case of the second. Nevertheless, since it is difficult to describe how objects move under the influence of net external forces without understanding how they behave in their absence, and since situations in which the resultant of all forces on a body vanishes are very common, the first law is usually stated separately even though it follows directly from the second. The third law also supports the second, in that it tells us *how bodies exert forces on one another*, and enables us to distinguish forces exerted *upon* an object by external means from forces exerted *by* the object upon other bodies.

In this chapter, we wish to describe the action of forces upon bodies that are at rest or in a state of uniform, unaccelerated motion. For this purpose, strictly speaking, only the first and third laws are necessary. Nevertheless, in defining units of force and explaining the relationship between weight and mass, we need the second law as well, and therefore, to a certain extent, our discussion of unaccelerated systems will involve all of Newton's laws.

[4] Alexander Pope (1688–1744), *Epitaph Intended for Sir Isaac Newton.*
[5] In the case of the rotating wheel, we are required, in addition, to formulate precise and consistent notions of what is meant by "uniform in speed and direction" for a rotating body. In the case of the rolling and sliding objects, the bodies are really acted upon by frictional forces, and it is these external forces which serve to change their motion and ultimately to stop it altogether.

2.3 Newton's Second Law and the Definition of Mass: Units of Force and Mass

Newton's second law, as stated in section 2.2, tells us that if a net force acts on a body, the body undergoes an *acceleration* which is proportional to the force and

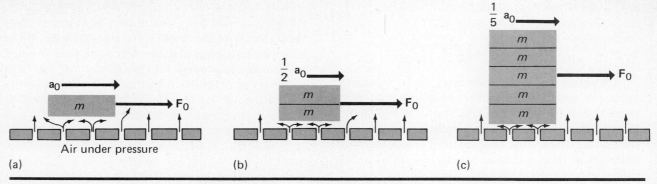

(a)   (b)   (c)

FIGURE 2.3. A set of identical metallic discs on the "frictionless" surface of an air table.

inversely proportional to its mass. Stated as an equation, this law can be written as

$$\mathbf{a} = k\,\frac{\mathbf{F}}{m} \quad \text{or} \quad \mathbf{F} = \frac{1}{k}\,m\mathbf{a} \tag{2.3.1}$$

where $\mathbf{a}$ is the acceleration, or the rate at which the body's velocity changes with time. In this expression, $\mathbf{F}$ is the resultant of all the forces acting on the body, m is its mass, and k is a constant of proportionality. In the metric systems, the fundamentally defined units are *mass*, *length*, and *time*. There is no leeway, therefore, in stating the units of mass or acceleration,[6] which must be distance/(time)^2. The unit of force, however, may be taken as large or small as desired; the proportionality constant k in Eq. (2.3.1) will then depend upon the size of the unit of force which is chosen. The most natural way of selecting a force unit is to require that it be of such a size that the proportionality constant k in Eq. (2.3.1) is *unity*, in which case (2.3.1) becomes

$$\mathbf{a} = \frac{\mathbf{F}}{m} \quad \text{or} \quad \mathbf{F} = m\mathbf{a} \tag{2.3.2}$$

In following this procedure, the unit of force that is chosen is (in the mks-metric system) just that force which imparts an acceleration of 1 m/sec^2 to a mass of 1 kilogram; such a force is called a *newton* (abbreviated N), and according to (2.3.1) has the dimensions of kg-m/sec^2. In the cgs-metric system, the force unit is that force which gives an acceleration of 1 cm/sec^2 to a mass of 1 gram; such a force is called a *dyne* (abbreviated dyn), and from (2.3.2) must have the dimensions of g-cm/sec^2. It is not difficult to show that a

newton is equal to 10^5 dynes; the proof is assigned as an exercise.

In the British system, the fundamentally defined units are *force*, *length*, and *time*. We have no freedom of choice for the units of force and acceleration that must be used, but now we are free to define the unit of *mass* as large or small as we may wish. The value of the proportionality constant k will now depend upon the size of the mass unit which is chosen, but again it makes sense to select a unit of mass of such a size that k has the value of unity. We shall always make our choice of units in this way, thus permitting the use of Newton's second law in the simple form of Eq. (2.3.2). In the British system, the unit of force—and of weight—is the *pound*, which is defined as the weight force exerted by a standard pound weight under standard gravitational conditions. Such a standard pound weight has a mass of 0.4536... kg, or 453.6... g. A mass of 0.4356 kg, therefore, weighs exactly 1 pound under standard gravitational conditions. The unit of mass in the British system is defined, naturally, so as to give the value of unity to the constant k in Eq. (2.4.1). Accordingly, in the British system, we define the unit of mass as *that mass which when subjected to a force of 1 pound acquires an acceleration of 1 ft/sec^2*. Such a mass is called a *slug*, and according to (2.4.2) must have units of lb-sec^2/ft.

The question of exactly what is meant by *mass* has been avoided in our previous discussions, but we must now arrive at some concrete understanding of this concept. Imagine a set of objects of precisely the same chemical composition, size, and shape, for example, a set of flat copper discs resting upon a frictionless surface such as an air table.[7] Now suppose

---

[6] Since velocity is the rate at which a distance changes with time, the units of velocity are distance/time. In the same way, since acceleration represents the change of velocity with time, the units of acceleration must be velocity/time or distance/time^2. In the mks system, then, acceleration has units of meters per second per second, or m/sec^2, while in the cgs and British system, the corresponding acceleration units are cm/sec^2 and ft/sec^2. The definitions of velocity and acceleration will be established more precisely and in greater detail in the next chapter.

[7] An air table is a flat surface pierced by many tiny holes through which air is blown under pressure. Flat objects resting on such a table are supported by a cushion of air and do not actually touch the surface of the table. Under these circumstances, friction is practically absent, the objects being set into motion at the slightest touch and "coasting" at practically constant velocity if allowed to move unhindered. The air table serves as a convenient apparatus to illustrate the laws of motion.

that a given constant force $\mathbf{F}_0$ is applied to such a body. A certain acceleration $\mathbf{a}_0$ is then observed, as in Fig. 2.3a. A second disc is then set on top of the first, as in Fig. 2.3b. When the same force $\mathbf{F}_0$ is applied, the resulting acceleration is now found to be only half as large. If five such objects are piled up, as in Fig. 2.3c, and the same force $\mathbf{F}_0$ is brought to bear, the resulting acceleration is now only one fifth the original value. Clearly, every object has an inherent property which determines the amount of acceleration produced by a given applied force, so that, roughly speaking, heavy objects are accelerated less readily than light ones.

But it is not the *weight* of the object which determines the force required to produce a given acceleration. After all, the weight of a body is the force it experiences when in a gravitational field, and hence the weight will vary according to the strength of the gravitational field in which the body may be located. In interstellar space, for example, far from any massive body having a strong gravitational field, an object would be essentially weightless. Suppose, for example, that the air table and the discs illustrated in Fig. 2.3 were transported from the earth's surface to such a region of interstellar space. The discs would now be practically weightless, but nevertheless we should find that the same accelerations observed previously would result from the application of the force $\mathbf{F}_0$ in these new surroundings.[8] The relation between applied force and acceleration is thus related not to the weight of the object but to some other inherent property thereof whose magnitude is independent of the particular gravitational environment of the body.

The inherent property of every material object that determines the relationship between its acceleration and the force required to produce the acceleration is called *inertia*, and the quantitative measure of the inertia of a body is referred to as its *mass*. For example, the inertia of the objects shown at (b) in Fig. 2.3 is twice as great and that of the objects at (c) is five times as great as that of the object at (a); likewise, the mass of the objects at (b) is twice as large and the mass of the objects at (c) is five times as large as that of the object at (a). This is all, of course, in accord with Newton's second law, as stated in Eq. (2.3.2). In fact, if that equation is written as a relation between the scalar magnitudes of the vectors $\mathbf{F}$ and $\mathbf{a}$ and solved for m to obtain

$$m = \frac{F}{a} \tag{2.3.3}$$

Newton's second law may be regarded as *defining* the

mass of a body as the ratio of the force applied and the resulting acceleration, and stating, furthermore, that this ratio is an *inherent* property of the body, that is, that it is a constant independent of the applied force or the body's environment. On the most fundamental level, the mass of a body is essentially the sum of the masses of all the atoms or elementary subatomic particles it may contain. It is an experimentally verified fact that every atom of the same isotopic species has the same mass as does every elementary subatomic particle of the same species.[9]

One of the simplest systems to which Newton's second law can be applied is an object falling freely subject to a gravitational force. In this case, the only force on the body is its own weight W which acts vertically downward. Substituting this into Newton's second law, as given by (2.3.3), we obtain

$$\mathbf{W} = m\mathbf{a} = m\mathbf{g} \tag{2.3.4}$$

where $\mathbf{g}$ is the acceleration acquired by a body in free fall due to gravitational attraction alone. This quantity is frequently referred to as the *acceleration of gravity*. Writing this as a relation between the scalar magnitudes of W and g, we obtain a useful equation:

$$\boxed{W = mg} \tag{2.3.5}$$

This expression gives the relation between the mass of a body and its weight in terms of an easily measurable quantity g which expresses the acceleration of a freely falling body.

It has been observed (first by Galileo's famous experiment at Pisa) that, so long as extraneous effects such as air resistance are negligible, all bodies in free fall near the earth's surface attain the same acceleration g irrespective of their size, shape, weight, or composition. The quantity g is, therefore, *the same for all objects*, and Eq. (2.3.5) expressing the relation between weight and mass is true for all bodies.

The fact that the acceleration of gravity is the same for light and heavy objects can easily be demonstrated by a simple experiment utilizing a feather and a lead weight in a glass tube, as illustrated in Fig. 2.4. If the tube contains air at normal atmospheric pressure, the feather takes much longer to fall through the length of the tube than does the lead weight when the tube is suddenly inverted. When the air is evacuated from the tube, however, there is no appreciable difference in the time required for the feather and the lead to fall to the bottom of the tube. The conclusion is, of course, that when air is present, forces arising from air resistance are comparable in magnitude to the weight of the feather and thus affect its motion

---

[8] If this were not true, it would be impossible for us to determine on the basis of Newton's laws the trajectories of man-made earth satellites and planetary probes. Actually, however, Newton's laws have succeeded in predicting the motion of such space probes with great accuracy as well as the motions of the planets themselves.

[9] These statements, while serving as excellent practical approximations under ordinary circumstances, are subject to certain relativistic modifications, which will be discussed in a later chapter.

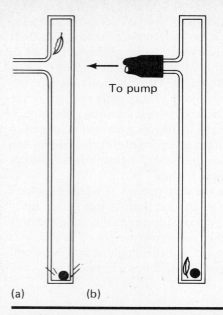

To pump

(a)   (b)

FIGURE 2.4. (a) A lead weight falls much more rapidly than a feather in a tube containing air. (b) When the air is pumped out, however, the feather and the lead weight attain the same acceleration and arrive at the bottom of the tube at the same time.

profoundly, while forces of the same order of magnitude acting on the lead are negligible in comparison to its much greater weight, and hence affect its motion hardly at all. When the air is evacuated, no air resistance is experienced by either object, and they both fall at the same rate, acquiring the constant acceleration g.

The experimentally measured value of g is about 980 cm/sec^2, 9.80 m/sec^2, or 32.16 ft/sec^2 at sea level and is essentially independent of height above the earth's surface for altitudes that are small compared with the earth's radius[10] (3900 miles, or 6400 kilometers). The fact that the acceleration of gravity is the same for all objects is in accord with Newton's law of universal gravitation, which will be discussed in a later chapter. The intrusion of effects such as air resistance and friction is largely responsible for Aristotle's incorrect assumption that heavier bodies fall faster than lighter ones and that force is required to maintain a state of constant motion. It is a tribute to the genius of Galileo and Newton that they were able to cut through the mass of extraneous circumstances and extract the essential truths that were hidden underneath.

[10] As we shall see in a later chapter, however, the acceleration of gravity can no longer be regarded as constant over vertical distances that are significant in comparison with the radius of the earth! Clearly, if we move upward into space a distance large compared to the earth's radius, the acceleration of gravity will be much less than on the earth's surface.

Equation (2.3.5) enables us to calculate the weight force, in dynes or newtons, associated with a body whose mass is given in grams or kilograms. For example, since $g = 980$ cm/sec$^2 = 9.8$ m/sec^2, setting m equal to unity in (2.3.5) tells us that near the earth's surface,

weight force associated with 1 gram mass
$$= 980 \text{ dyn} \qquad \text{(cgs-metric)}$$
or

weight force associated with 1 kilogram mass
$$= 9.80 \text{ N} \qquad \text{(mks-metric)}$$

In the British system, again setting m equal to unity in (2.3.5), but now using $g = 32.16$ ft/sec^2,

weight force associated with 1 slug mass
$$= 32.16 \text{ lb} \qquad \text{(British)}$$

It should be noted that *grams and kilograms are units of mass* and *dynes and newtons are units of force* in the metric system, while in the British system, *slugs are units of mass* and *pounds are units of force*. A pound weight at rest exerts a force of 1 pound on its supports; a newton weight at rest exerts a force of 1 newton on its supports, while a mass of 1 kilogram at rest exerts a force of 9.80 newtons on its supports.

2.4 The Application of Newton's First Law

From Newton's first law, as stated in section 2.2, it is clear that any body will remain at rest (or at constant velocity when in motion) *only so long as the resultant of all external forces upon it is zero.* Otherwise, according to the second law, it would undergo acceleration and its motion would change as a function of time. Furthermore, it is evident from the first law that *there is no essential difference in the conditions required either to keep a body at rest or to maintain it in a state of constant motion.* In many instances, the forces acting upon a system will be dependent upon its velocity, and under these circumstances a different set of forces will be required to maintain constant motion from those required to keep the system at rest. For example, it is obvious that a different set of forces must be brought to bear upon an automobile to keep it moving along the highway at a constant speed of 60 miles per hour from that required to keep it at rest. But this is only because certain frictional or resistive forces, which are absent when the car is at rest, appear when it is in motion. In either case, the vector sum, or resultant, of the forces that do exist must be *zero*.

When a system is either at rest or in a state of

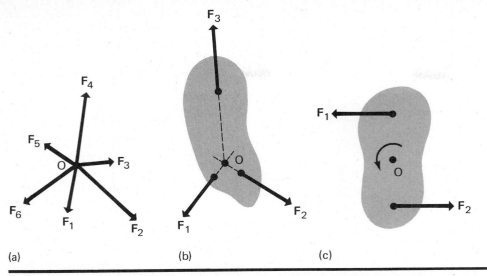

FIGURE 2.5. (a) Concurrent forces acting on a "point object" at O. (b) Rigid body in equilibrium under the action of three nonparallel coplanar forces. (c) Equal and opposite forces having different lines of action do not suffice to maintain the equilibrium of a rigid body.

constant motion, it is said to be *in equilibrium* with respect to the forces which act upon it. From Newton's first law, and from what has been said above, if we have a system acted upon by several external forces $\mathbf{F}_1, \mathbf{F}_2, \mathbf{F}_3, \ldots \mathbf{F}_i, \ldots \mathbf{F}_n$, then in order that the system be in equilibrium under the action of these forces,

$$\mathbf{F}_1 + \mathbf{F}_2 + \mathbf{F}_3 + \cdots + \mathbf{F}_i + \cdots + \mathbf{F}_n = 0$$

or, using the sum notation of section 1.5,

$$\sum_{i=1}^{n} \mathbf{F}_i = 0 \tag{2.4.1}$$

Since all three components of a vector must vanish in order that the vector be equal to zero, this equation can be written as three separate equations involving the x-, y-, and z-components of the forces, which have the form

$$\sum_{i=1}^{n} F_{ix} = 0$$

$$\sum_{i=1}^{n} F_{iy} = 0 \tag{2.4.2}$$

$$\sum_{i=1}^{n} F_{iz} = 0$$

The advantage of using the three separate equations (2.4.2) rather than the single equation (2.4.1) is, of course, that the summations in (2.4.2) are simply *algebraic* sums, while that referred to in (2.4.1) is a *vector* sum. In applying Newton's first law to practical examples, we usually know that the system is at equilibrium, and we may know, in addition, some but

not all of the forces that act upon it. Equations (2.4.2) may then be used to allow us to determine the unknown forces.

These equations are referred to as the equations of *translational equilibrium*. No system can be at equilibrium unless they are satisfied. For many, but not all, systems, they are all that is required to specify the equilibrium state. In other instances, additional equations must be satisfied before the equilibrium state of the system is completely determined. In any system where the lines of action of all the external forces intersect in a single point (that is, where all the forces are *concurrent*), the equations of translational equilibrium suffice by themselves to determine the state of the system at equilibrium. The most common examples of this state of affairs arise when the dimensions of the object to which the system of external forces is applied is so small that it can be considered as a mathematical point, or when three nonparallel *coplanar* forces are applied to a rigid body,[11] as shown in Figs. 2.5a and 2.5b.

Whenever the lines of action of all the applied forces do *not* intersect in a single point (are not concurrent), the equations of translational equilibrium are insufficient, by themselves, to specify the equilibrium state, and more information is needed. The simplest example of this situation is shown in Fig. 2.5c, where

[11] In general, objects are deformed by the application of external forces, but for many solid bodies this deformation is so small as to be completely negligible for all practical purposes. Objects of this latter type are what we mean when we use the term *rigid bodies*. The assertion that three nonparallel coplanar forces that maintain a rigid body in a condition of equilibrium must necessarily be concurrent is assigned as an exercise.

25

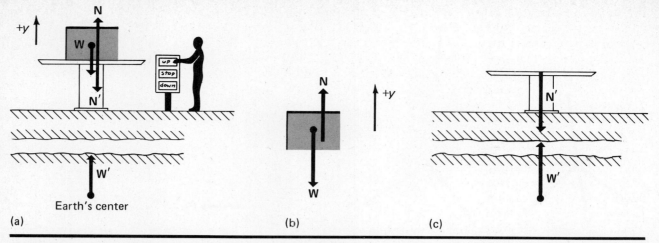

FIGURE 2.6. (a) System of forces on an object supported by a hydraulic lift. (b) Forces acting on the object. (c) Forces acting on the earth.

two forces, equal in magnitude and opposite in direction, having *different lines of action* are applied to a rigid body; a force system of this type is called a *couple*. Despite the fact that the vector sum of these two applied forces is zero and all the equations of translational equilibrium are satisfied, it is intuitively evident that this body is not in a state of equilibrium at all, and if left to itself under the action of these forces will acquire a *rotational* motion about some axis perpendicular to the plane of the page intersecting the body, for example, at O. A *rotational acceleration* about this axis will be produced by the action of the couple. The body, although satisfying all the conditions of translational equilibrium, is not in *rotational equilibrium*.[12] We shall postpone a complete discussion of rotational equilibrium until a later section, but we shall find that three additional equations referring to rotational motions about three independent directions must be satisfied in order that the system be in a state of rotational equilibrium. The three equations for translational equilibrium (2.4.2) and the three additional rotational equilibrium equations will then be found to be sufficient to determine the general equilibrium condition for any system.

2.5 Action and Reaction Forces: Newton's Third Law

Newton's third law states that for every force there must exist an equal and opposite reaction force. This is true because forces do not arise from nowhere but always appear as a result of *mutual interaction* between different bodies or systems. While Newton's third law appears to be a simple enough statement, it is frequently misunderstood and is often the source of certain misconceptions among students. The most common of these is exhibited by asserting that if every force is accompanied by another that is equal and opposite, the resultant of every such force pair is zero, in which case the total resultant force must also be zero. But then, according to Newton's first law, every system would be in equilibrium, and no set of forces could ever give rise to an acceleration! This is clearly at odds with what we know to be true from everyday experience.

To resolve this apparent paradox, we must be very careful to identify in every instance the body each force acts upon. You will note that the statement of the paradox is deliberately vague about this point. As a specific example, let us consider the case of a weight W supported by a hydraulic lift, as illustrated in Fig. 2.6. In Fig. 2.6a, all the forces acting on the weight and on the lift (which is assumed to be firmly embedded in the earth) are shown. The force W representing the weight of the object, that is, the force due to the earth's gravitational attraction, acts downward through its center of gravity.[13] According to Newton's third law, there must be an equal and oppositely directed reaction force W'. *The earth attracts the body* with a force W that acts upon it at its center of gravity; but likewise *the body attracts the earth* with an equal gravitational force that acts at

---

[12] The fact that the equations of translational equilibrium are satisfied does, however, guarantee that the center of mass of the body will remain in equilibrium and that the only possible accelerated motion is one of rotation about the center of mass.

[13] The center of gravity of an object, roughly speaking, is a point at which we may, for practical purposes, consider its weight force to be concentrated. The weight force on any body may be considered, therefore, to act downward through the body's center of gravity. We shall see in a later section how the center of gravity is defined in more precise terms and where it is located in any given object.

the center of the earth. It is this force, acting upward at the earth's center, 3900 miles below the earth's surface, that represents the reaction force $\mathbf{W}'$. It is important to note that though the forces $\mathbf{W}$ and $\mathbf{W}'$ are of equal magnitude, *they act on different objects*, the force $\mathbf{W}$ acting on the weight, and the force $\mathbf{W}'$ on the earth.

There is also an upwardly directed *contact force* or *normal force* $\mathbf{N}$ exerted by the surface of the lift upon the weight, while at the same time a downward force $\mathbf{N}'$ of the same character is exerted by the bottom of the weight on the surface of the lift. These are also forces of action and reaction, oppositely directed, and, according to Newton's third law, of equal magnitude. Once more, however, we observe that the action–reaction forces $\mathbf{N}$ and $\mathbf{N}'$ act on *different bodies*, the former acting on the weight, the latter on the lift, or, in effect, on the surface of the earth. Action–reaction force pairs always act on different bodies, the action force being a force that body A exerts on body B, the reaction force an equal but oppositely directed force exerted by body B on body A. Newton's third law tells us, then, that

$$\mathbf{W}' = -\mathbf{W} \tag{2.5.1}$$

and

$$\mathbf{N}' = -\mathbf{N} \tag{2.5.2}$$

While it is true in view of this that the vector sum of all the forces shown in Fig. 2.6a is zero, this tells us nothing about the accelerations that the bodies attain according to Newton's *second* law, because the forces that are shown in the diagram do not all act on the same body. Newton's second law informs us that any object that itself experiences a resultant force acquires an acceleration proportional to that resultant force, and, inversely, to its own mass. In order to use Newton's second law, therefore, we must regard each object as an isolated, free body and determine the resultant of all the forces exerted by the surroundings *on that object alone*. These same remarks apply also to the use of Newton's first law, since it can be regarded as a special case of the second.

In this example, there are two bodies that are acted upon by forces; the weight and the lift (or, actually, the earth). In using Newton's second law, we must focus our attention on each of these as an *isolated entity* and consider separately all the forces that are exerted upon each body by its surroundings. *This technique, which we use here for the first time, is of general usefulness in applying Newton's first and second laws, and we shall find ourselves employing it repeatedly.* In the case of the weight, shown as an isolated body in Fig. 2.6b, there are two forces, $\mathbf{N}$ and $\mathbf{W}$, that act on the object. Both these forces have components only along the vertical y-direction; their

resultant is, therefore, simply the algebraic sum of these y-components, or $N - W$.

Now $\mathbf{N}$ and $\mathbf{N}'$, and also $\mathbf{W}$ and $\mathbf{W}'$, are, according to Newton's *third* law, equal and opposite, since they are action–reaction force pairs acting on different bodies. But $\mathbf{N}$ and $\mathbf{W}$ are *different* forces that act on *the same body*. They may be equal and opposite, but there is no law of mechanics that says that they must be. If the weight is at rest (or is moving up or down with constant speed), then the system is in equilibrium, and the resultant force, according to Newton's first law, is zero. Under these circumstances, $\mathbf{N}$ and $\mathbf{W}$ are equal and opposite—but this state of affairs arises from the first law, not the third, since we are dealing with forces acting on a single body.

But by pressing the "up" button, the lift operator can actuate machinery that increases the upward force $\mathbf{N}$ acting on the bottom of the weight. When this happens, the upward force $\mathbf{N}$ exceeds the downward force $\mathbf{W}$, and the net resultant force $N - W$ in the y-direction is then greater than zero. This, according to Newton's *second* law, causes the weight to acquire an acceleration along the y-direction whose magnitude can be found by equating the magnitude of the net resultant force to the mass of the body times its acceleration. In this way, we can easily see from Eq. (2.3.2) that

$$\sum F_{iy} = N - W = ma \tag{2.5.3}$$

where m is the mass of the object and a is the magnitude of its upward acceleration. According to (2.3.5), m is related to W by

$$W = mg \tag{2.5.4}$$

where g is the acceleration of a freely falling body, equal to 9.80 m/sec². This acceleration will continue and the upward velocity of the weight will steadily increase so long as N remains larger than W. If, after a time, the value of N should be reduced until it is again equal to W, the acceleration, according to (2.5.3), would once more drop to zero; in this condition, the object's velocity would cease to increase and would instead remain constant. Under these circumstances the weight, now unaccelerated, is again in equilibrium, this time, however, moving with constant velocity rather than at rest.

At the same time, there are the forces $\mathbf{N}'$ and $\mathbf{W}'$ that act *on the earth*. These forces both act along the y-direction, as illustrated in Fig. 2.6c, and their resultant is a vector in that direction whose magnitude is $W' - N'$. Once more, according to Newton's second law, the resultant force acting on the earth must be equal to the earth's mass M times its acceleration $\mathbf{A}$. Therefore, we may write for the earth's motion the equation

$$\sum F_{iy} = W' - N' = MA \tag{2.5.5}$$

From Newton's third law, however, as we have already seen, $\mathbf{W'} = -\mathbf{W}$ and $\mathbf{N'} = -\mathbf{N}$, so this can be written

$$N - W = -MA \tag{2.5.6}$$

From this, we see that whenever N exceeds W, that is, whenever a net resultant force acts to accelerate the weight upward, the reaction forces $\mathbf{N'}$ and $\mathbf{W'}$ act to produce a downward acceleration of the earth. From (2.5.3) and (2.5.6) it follows that the earth's acceleration A is related to the weight's acceleration a by

$$A = -\frac{m}{M}a \tag{2.5.7}$$

Assuming that the object has a mass of 2000 kg, comparable to the mass of an automobile on a service station lift, we can compute the earth's acceleration by (2.5.7) using the known value 6.0×10^{24} kg for the earth's mass M. In this way, we find

$$\frac{A}{a} = -\frac{2.0 \times 10^3}{6.0 \times 10^{24}} = -3.3 \times 10^{-22}$$

The earth's downward acceleration is therefore 3.3×10^{-22} times smaller than the upward acceleration of the 2000-kg mass. This small acceleration is completely beyond our capacity to detect, and therefore we are never aware of the earth's motion in such a situation!

It is evident from this that action and reaction forces act in all cases on different bodies. Although it is true that the resultant force on *all* bodies of a system is inevitably zero as a consequence of Newton's third law, this is not what determines the motion of the separate bodies that are involved. The reason for this is that Newton's *second* law addresses itself to the resultant force acting on each body rather than to the resultant of all the forces that act on all the bodies in the system. To ascertain how each individual body moves, *we must regard that body as an isolated entity* and determine the resultant of all the forces that act *upon that body*. This, according to Newton's second law, may be equated to the body's mass times its acceleration. In doing this, the fact that the body is exerting forces on *other* objects that are equal and opposite to those it experiences itself is *irrelevant. It is only the forces exerted on a body, and not those exerted by it, that affect its motion!*

Another important illustration of the relationships between action and reaction forces is shown in Fig. 2.7, in which we see a weight suspended from an overhead support by a flexible rope. The weight force $\mathbf{W}$ and the attractive force $\mathbf{W'}$ exerted on the earth at its center, are much the same as in the previous example, as are the forces $\mathbf{W_r}$ and $\mathbf{W_r'}$ that represent the weight of the rope and its reaction force. As before,

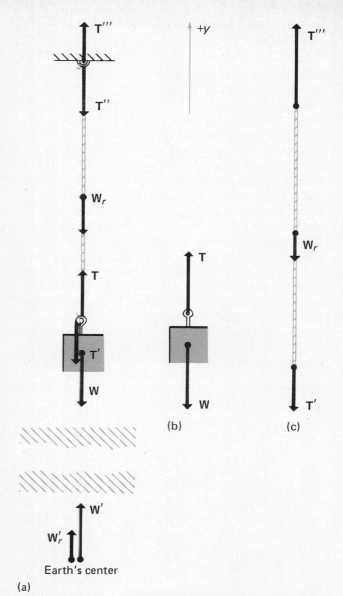

FIGURE 2.7. (a) System of forces acting on an object supported by a flexible rope. (b) Forces acting on the object. (c) Forces acting on the rope.

these forces are action–reaction pairs, and from Newton's third law,

$$\mathbf{W'} = -\mathbf{W} \quad \text{and} \quad \mathbf{W_r'} = -\mathbf{W_r} \tag{2.5.8}$$

The rope exerts an upward force $\mathbf{T}$ upon the weight, and it itself experiences a downward reaction force $\mathbf{T'}$ exerted upon it by the weight. These forces are another action–reaction pair, each acting on the other body, and again the third law tells us that

$$\mathbf{T'} = -\mathbf{T} \tag{2.5.9}$$

Finally, the rope exerts a downward force $\mathbf{T''}$ upon the overhead support and, in turn, experiences an upward force $\mathbf{T'''}$ exerted upon it by the support. This, also, is an action–reaction pair and consequently,

$$\mathbf{T}'' = -\mathbf{T}''' \qquad (2.5.10)$$

As before, the sum of all these individual forces acting on different bodies is zero as a consequence of Newton's third law.

Now let us consider the weight and the rope as isolated bodies and see what Newton's first and second laws tell us about their individual behavior. In the case of the weight, shown as an isolated object in Fig. 2.7b, we see that the only forces acting are $\mathbf{T}$ and $\mathbf{W}$, whose resultant is a vector having a magnitude $T - W$. According to Newton's second law, then,

$$\sum F_{iy} = T - W = ma \qquad (2.5.11)$$

where a is the acceleration of the weight and m is its mass. Once more, we know from (2.3.5) that m and W are related by $W = mg$. If the system is in equilibrium, the acceleration in Eq. (2.5.11) will be zero, and we are really dealing with a situation in which Newton's *first* law is applicable. Under these circumstances, T and W are equal. Their equality, however, has nothing to do with the third law; these forces are not an action–reaction pair, since they act on the same object. They are equal because of what Newton's first law tells us about a system that happens to be in equilibrium.

Suppose, now, that the system is not in equilibrium. This might happen if the support shown at the top in Fig. 2.7a were set in the roof of an elevator car rather than in the ceiling of a room. Under these circumstances, if the elevator were given an upward acceleration a, then the right-hand side of Eq. (2.5.11) would have the positive value ma rather than zero, and then, of course, since W does not change, we should find the force T no longer equal to W, but greater than W by the amount ma. The lesson to be learned here is that *the simple fact that a weight hangs on the end of a rope does not justify our assuming that the tension in the rope is equal to the weight!* This is true only if the system is at equilibrium. If the weight has an acceleration, the tension has to be different from W, or there would be no net force to cause the weight to accelerate. In the most extreme case, we might untie the knot that fastens the rope to the top support and allow the weight to fall freely,[14] acquiring a downward (negative) acceleration $-g$. Our intuition tells us—correctly—that now, despite the fact that the weight is still fastened to the lower end of the rope, the tension T in the rope has to be zero. This conclusion is verified by substituting $W = mg$ and $a = -g$

into Eq. (2.5.11), which is derived from Newton's second law.

Now let us turn to the rope and view it in the same way as an isolated body subject to certain forces. The forces that act on the rope are illustrated in Fig. 2.7c. The resultant force is a vector in the y-direction whose magnitude will be $T'' - T' - W_r$, and therefore Newton's second law tells us that

$$\sum F_{iy} = T''' - T' - W_r = m_r a \qquad (2.5.12)$$

where $m_r = W_r/g$ is the rope's mass and a is its acceleration. In using Newton's second law rather than the first law, we again allow for the possibility that the overhead support is (for example) in an elevator rather than on a fixed ceiling. This equation can be rearranged to read

$$T''' - T' = m_r a + W_r = W_r \left(\frac{a}{g} + 1 \right) \qquad (2.5.13)$$

since $m_r = W_r/g$. From this expression it is clear in general that forces acting on opposite ends of a rope, such as $\mathbf{T}'''$ and $\mathbf{T}'$, are not equal but differ by an amount that depends on the rope's mass and acceleration. If the overhead support is a fixed one, then $a = 0$ and the forces $\mathbf{T}'''$ and $\mathbf{T}'$ differ only by the weight of the rope, as we might expect intuitively.

It is very frequently the case, however, that the rope will be so light that its weight is completely negligible compared to the other forces that are present in the system. Under these circumstances, the right side of Eq. (2.5.13) can be set equal to zero, in which case we find

$$T''' = T' \qquad (2.5.14)$$

But from Eqs. (2.5.9) and (2.5.10), the magnitudes of $\mathbf{T}$ and $\mathbf{T}'$ are the same, as are the magnitudes of $\mathbf{T}''$ and $\mathbf{T}'''$, and therefore (2.5.14) may in this situation be written as

$$T''' = T'' = T' = T \qquad (2.5.15)$$

In the future, we shall always assume that the weight of ropes or cords that connect various objects is negligible unless some statement to the contrary is made, simply to avoid introducing needless computational difficulties.

The forces that act on opposite ends of a rope, such as $\mathbf{T}'''$ and $\mathbf{T}'$ in this example, are sometimes stated to be forces of action and reaction. This is clearly untrue, because, first of all, they act upon the same body rather than different bodies and because, according to Eq. (2.5.13), they are not, generally speaking, always exactly equal. The reaction to $\mathbf{T}'$ is the upward force $\mathbf{T}$ exerted by the rope on the weight, while the reaction to $\mathbf{T}'''$ is the downward force $\mathbf{T}''$ exerted by the rope on the upper support fixture.

---

[14] Alternatively, to accomplish the same result, we could cut the elevator's cable and allow the car to fall freely down the elevator shaft.

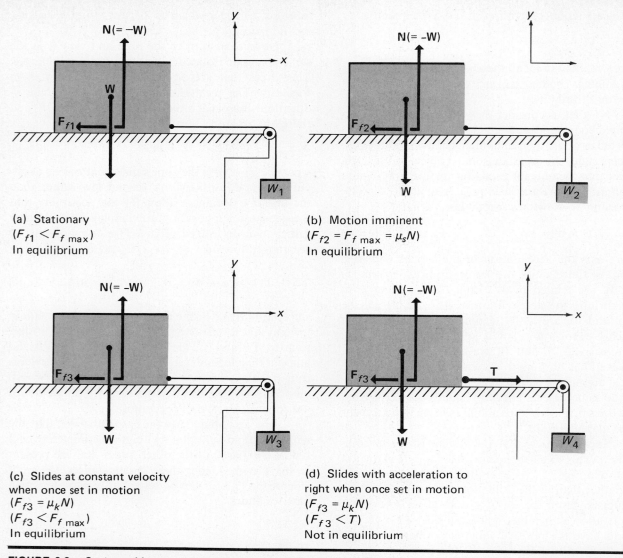

FIGURE 2.8. System of forces acting on an object resting on an interface at which frictional forces are present. Four different possibilities can arise, corresponding to cases (a), (b), (c), and (d) illustrated above.

2.6 Frictional Forces and Coefficients of Friction

In the study of mechanics, we often encounter forces that arise because of frictional resistance to motion at the interface between two bodies that are in contact. It is of some importance to understand the characteristics of such forces and to develop practical methods of incorporating them into mechanics problems. In order to see how these forces act, let us consider the situation illustrated in Fig. 2.8, where a rectangular object of weight W rests on a flat surface and is subjected to both vertical and horizontal forces. The easiest way to analyze what is happening is, as usual, to regard each part of the system as an *isolated free body* and write down all the forces acting on each such

isolated body. The way in which this can be done in this case is shown in Fig. 2.9, where all the forces upon both the supported and suspended weight are shown. In these diagrams, the tension in the cord is represented as vectors $\mathbf{T}$ acting in the horizontal x-direction on the supported object and $\mathbf{T}'$ acting in the vertical y-direction on the suspended object. If the weight of the cord is neglected (as it will be), the magnitude of $\mathbf{T}'$ is the same as the magnitude of $\mathbf{T}$. The pulley serves only to change the *direction* in which the tension acts and does not in any way alter its magnitude.[15] Also,

[15] This statement is true only insofar as frictional effects and inertial effects associated with the pulley are absent, and we shall always assume that this is the case, unless a statement to the contrary is made. In a later chapter, the inertial effects that arise when the rotating part of the pulley is very heavy will be analyzed in detail.

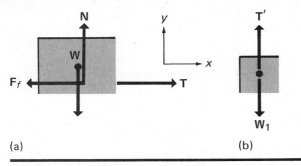

FIGURE 2.9. The set of forces acting (a) on the body illustrated in Fig. 2.8, viewed as an isolated free body, and (b) on the suspended weight, regarded in the same way.

in this example, it is assumed that the interface between the supported weight and the supporting surface is not perfectly slippery and thus that *frictional forces* arising from the contact of the supported weight and the surface on which it rests are present.

The frictional forces between the supported object and the surface arise from interatomic or intermolecular forces between the two surfaces. An exact description of friction in terms of these forces is very complex and cannot be attempted in detail here. Furthermore, even though the surfaces in contact may appear to be very smooth and flat, on an atomic scale such a high order of smoothness can rarely be achieved, and on this scale the surfaces are rough and irregular with mountainous high spots and valleys. As a result, the actual area of contact between the two objects occurs only at a relatively few spots, where high points on both surfaces are opposite one another; the contact area thus bears no direct relation to the total surface area of the bottom of the supported object but is actually much smaller. The pressure at the actual points of contact is therefore very large and sufficient in many cases to weld the two surfaces together. The maximum frictional force that can be supported by the interface is the force necessary to break these microscopic welds apart. If the contact is a sliding one, welds are formed and broken continuously, and material may be transferred from one surface to the other in the process. The same effects are found to play an important role in frictional forces associated with *rolling* contact between two bodies. In this case, the actual area of contact is still smaller, the result being that rolling friction is ordinarily smaller than sliding friction between the same materials. In the case of rolling friction, however, the deformation of the rolling object under the forces acting on it may also be important in determining the magnitude of the frictional forces.

Clearly, the physical mechanisms relevant to frictional effects are quite involved, and an analytical

description of these effects in fundamental terms is likely to be very complicated. It is quite simple, however, to describe *how frictional forces act*, without the necessity of referring to (or even knowing about) the physical mechanisms responsible for their action. This can be accomplished merely by observing that it has been ascertained experimentally that a *frictional force* $\mathbf{F}_f$ exists between an object and the surface upon which it rests. The way in which this force acts depends upon whether the body is at rest (*static friction*) or is sliding over the surface underneath it (*kinetic friction*). In all cases, however, its direction lies in the plane of the interface between the body and the surface on which it rests, as shown in Fig. 2.8.

In the case of static friction, since the body is at equilibrium, the vector sum of all forces upon it must be zero. This means that the force of friction has to be equal in magnitude and opposite in direction to the resultant of all the other forces acting on the object. But the force of static friction can only get so large before the body "breaks away" and starts to slide. The magnitude of the maximum possible friction force is observed to be directly proportional to the magnitude of the component of force exerted by the supporting plane on the body that is normal to the friction interface, which we usually refer to as the *normal force*. $\mathbf{N}$. These forces are illustrated in the simplest possible case in Fig. 2.9. Accordingly, the magnitude of the maximum possible force of static friction can be written

$$F_{f(\text{max})} = \mu_s N \qquad (2.6.1)$$

where μ_s is a constant of proportionality, called the *coefficient of static friction*. Its value obviously depends upon the materials that are in contact at the interface and upon whether they are rough or smooth. It takes less force to overcome the frictional forces between a cake of ice and a wood surface than those that exist when a block of wood having the same weight is put in its place. Once the coefficient of friction associated with a given frictional interface is known, the maximum static force it will support before "breaking away" and starting to slide can be ascertained from Eq. (2.6.1). In any situation where static frictional forces act, the condition of the system is one of equilibrium, in which the forces that act, *including* the force of static friction, are determined by the usual application of Newton's first law. There is really nothing new involved in doing this, except to remember that in all cases the calculated value of the friction force has to be *less than or equal to* the value $F_{f(\text{max})}$ given by (2.6.1). The force of static friction does not, therefore, have to be equal to $\mu_s N$. It can very well be less than that amount, but it cannot be *greater*. Equation (2.6.1), therefore, allows us to determine the limits within which forces of static friction can act to maintain a system in the state of static equilibrium.

In the case of kinetic friction, in which the object is not at rest but is sliding along the supporting surface, the force of friction acts upon the sliding object in the plane of the friction interface in a direction opposite to that of its motion. Its magnitude is once again proportional to that of the normal force **N**, but the coefficient of proportionality between the force of sliding friction *differs* from the coefficient of static friction that defines the maximum force the same interface can support in static equilibrium. It turns out, in fact, that the force of kinetic friction that acts when a body slides along a supporting surface is almost invariably *less* than the maximum static frictional force the same interface can support. We may, therefore, express the force of kinetic friction as

$$F_f = \mu_k N \qquad (2.6.2)$$

where μ_k is a constant of proportionality referred to as the *coefficient of kinetic friction* associated with the specific type of frictional interface involved. Since the force of sliding friction $\mu_k N$ is less than the maximum static force $\mu_s N$ needed to "break away" the force of static friction, it is clear that for a given interface, μ_k will always be less than μ_s. Also, since the force of kinetic friction between an object and the surface upon which it slides is practically independent of its velocity, the coefficient of kinetic friction is essentially independent of the velocity of the body with respect to the surface.

A description, such as the one given above stated in terms of experimental observation rather than fundamental principles is called an *empirical* description. The coefficients of static and kinetic friction, μ_s and μ_k, that enter into the description cannot be calculated in any way except by resorting to the very difficult arguments involving intermolecular forces outlined previously. But they *can be measured experimentally* very easily for all conceivable pairs of substances which might form a friction interface, and these measured values can be tabulated and referred to when needed. Since the coefficients of friction are the ratios of two forces, they are dimensionless.

The laws governing frictional forces stated above are approximate rather than exact. In particular, the coefficient of kinetic friction may actually vary with velocity if a wide range of velocities is involved, although the assumption that it does not is usually quite good over a moderate velocity range. The coefficient of static friction μ_s is always greater than the coefficient of kinetic friction μ_k, because it is invariably true that for any system a greater force is required to "break away" than to maintain a constant sliding or rolling motion.

Let us now return to the systems shown in Fig. 2.8 and examine in detail what happens in each instance illustrated there, using the isolated free body technique described previously. In Fig. 2.9b, in which the weight W_1 is illustrated as an isolated free body, it is evident that if this object is in equilibrium, then, according to (2.4.2), the sum of all the x-, y-, and z-compounds of the forces acting on it must vanish. As it happens, all the forces acting on the weight have only y-components, and so the only equation of the set (2.4.2) that gives us any information is the second, which states that

$$\sum F_{iy} = T' - W_1 = 0 \qquad (2.6.3)$$

whereby

$$W_1 = T' = T \qquad (2.6.4)$$

since, as mentioned previously, T' and T are equal in magnitude. The tension in the cord is thus equal to W_1, as one might expect at equilibrium. Considering the rectangular object now as an isolated body, as shown in Fig. 2.9a, and writing the equations for equilibrium of forces applied to this object, we get

$$\sum F_{ix} = T - F_{f1} = 0$$

and $\qquad (2.6.5)$

$$\sum F_{iy} = N - W = 0$$

In these equations, W may be regarded as given, and T is known to be equal to W_1, from (2.6.4). We are thus confronted with a set of two simultaneous equations which may be solved for the two unknown quantities F_{f1} and N, to give

$$F_{f1} = T = W_1$$

and $\qquad (2.6.6)$

$$N = W$$

The force of friction is equal to the suspended weight W_1, and the size of the normal force N which the supporting plane exerts on the supported weight is simply W, the weight of the supported object. In writing the top equation of (2.6.5), note that the force of friction is assumed to lie in the plane of the friction interface. In this example, it is assumed that the system is in equilibrium under the action of all these forces, and hence it is clear that W_1 must be less than $\mu_s N$ (that is, less than $\mu_s W$), for according to the laws describing friction forces, the force of friction may have any value *less* than $\mu_s N$ but cannot *exceed* $\mu_s N$.

The situation that prevails when the value of the suspended weight is increased to some value W_2, at which the friction force reaches the value $F_{f2} = F_{f(max)} = \mu_s N$, shown in Fig. 2.8b. The system is still in equilibrium, although it is at the extreme limit at which equilibrium can be maintained by the frictional

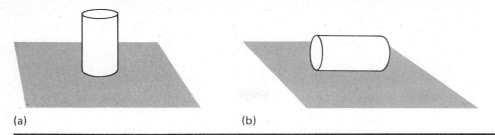

(a)   (b)

FIGURE 2.10. A "bistable" system, which can be in equilibrium in either of two different states (a) or (b).

force, and motion is impending. The equations of equilibrium are the same as before, the value $F_{f2} = \mu_s N$ replacing F_{f1} and W_2 replacing W_1. We shall find, then, as before, that $T = W_2$, while the two simultaneous equations which arise in place of (2.6.5) yield

$$N = W$$

and

$$F_{f2} = \mu_s N = T = W_2$$

But, since $N = W$, it is evident that

$$W_2 = \mu_s W \tag{2.6.7}$$

If the suspended weight exceeds this amount, equilibrium can no longer be maintained and the object "breaks away" and slides *with acceleration* along the y-direction.

There is yet another way in which this system can be in equilibrium. If some proper value W_3 for the suspended weight is chosen, the tension in the flexible cord will be just sufficient to balance the force of *kinetic* friction which arises when the supported object slides *at constant velocity* along the supporting surface, as illustrated in Fig. 2.8c. Since the system is still in equilibrium under these circumstances, the equations of equilibrium (2.6.5) are still applicable, provided that W_3 is substituted for W_1, and $F_{f3} = \mu_k N$ for F_{f1}. Again the isolated body approach applied to the suspended weight brings us to conclude that $T = W_3$, while the equations for equilibrium of the supported body yield

$$N = W$$

and

$$F_{f3} = \mu_k N = T = W_3$$

Also, since $N = W$, it is easily seen that

$$W_3 = \mu_k W \tag{2.6.8}$$

Since μ_k is less than μ_s, W_3 will be *less* than the force W_2, as given by (2.6.7), required to overcome the maximum possible static friction force and move the system from rest. It is also possible, therefore, for the system to be at equilibrium *at rest* when the suspended weight is W_3.

There are, thus, two possible equilibrium states, one in which the system is at rest, arising as a special case of the situation shown in Fig. 2.8a and discussed in connection with Eqs. (2.6.5) and in which W_1 has the value $\mu_k W$, and another in which the system slides to the right at constant velocity, as discussed immediately above. To effect a transition between these two states, an external force must be applied. For example, if the system is initially at rest, it can be set in motion by a touch of the hand, and will then persist in motion at constant velocity until stopped by another externally applied force. Such a system having two equilibrium states is often called a *bistable* system. Systems having three or, indeed, many equilibrium states are not at all uncommon. Another familiar example of a bistable system is shown in both of its equilibrium states in Fig. 2.10.

If the weight of the suspended body exceeds the value $F_{f(\max)} = \mu_s W$, then the force T exerted by the flexible cord will exceed the maximum force of friction that can exist at the friction interface. Under these circumstances, the sum of the x-components of forces acting upon the supported object cannot possibly be zero, but must instead sum to give a *net resultant force* along the x-direction, which, insofar as the object as an isolated free body is concerned, is an external force. The body is no longer in equilibrium and, according to Newton's second law, must experience an *acceleration* along the x-axis in response to the resultant force. If the weight of the suspended object is less than $\mu_s W$ but greater than $\mu_k W$, the supported body will be in equilibrium if it is set at rest, because a *static* friction force greater than $\mu_k W$ can then be maintained to balance the tension in the cord. If the object is set in motion along the positive x-direction, however, the maximum force of *kinetic* friction that can be maintained by the sliding contact is $\mu_k W$, and this is insufficient to balance the tension in the cord to make the resultant force zero. Instead, a net resultant force along the x-direction acts on the supported body, which again causes an acceleration in that direction. This situation is illustrated in Fig. 2.8d. For the moment, since we are concerned primarily with understanding how systems in equilibrium behave, we shall not

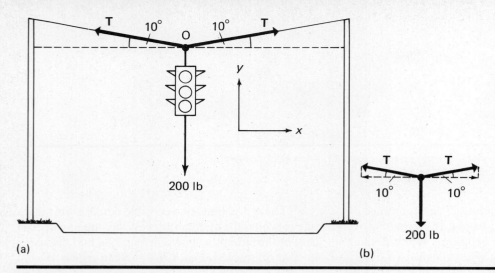

FIGURE 2.11

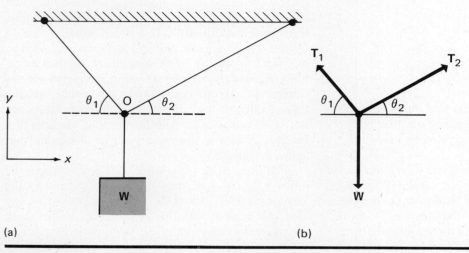

FIGURE 2.12

pursue the problem of discussing precisely what happens when the system is no longer in equilibrium, but we shall return to this subject in Chapter 4.

EXAMPLE 2.6.1

A traffic signal weighing 200 pounds is suspended from two light poles by cables, as shown in Fig. 2.11a. The stretched cables make an angle of 10° with the horizontal. Find the tension in the cables.

From the symmetry of the system, it is obvious that the tension in both cables is the same. Consider the junction at O as an isolated free body, as shown in Fig. 2.11b. Since this junction is in equilibrium, we may write

$$\sum F_{ix} = T \cos 10° - T \cos 10° = 0$$

This gives us no information of any value, but if the equation for the y-components is written, one obtains

$$\sum F_{iy} = 2T \sin 10° - 200 = 0$$

from which

$$T = \frac{100}{\sin 10°} = \frac{100}{0.1736} = 576 \text{ lb}$$

EXAMPLE 2.6.2

A weight W is suspended by two cables from an overhead beam. The cables make angles θ_1 and θ_2 with the horizontal as illustrated in Fig. 2.12. What is the tensions in each cable and what forces act on the overhead beam?

In this case, the tension in the two cables is not necessarily the same. Again, regard the junction at O as an isolated free body, as shown in Fig. 2.12b, and write the equations for equilibrium for each force component. From (2.4.2), then,

$$\sum F_{ix} = T_2 \cos \theta_2 - T_1 \cos \theta_1 = 0$$

and

$$\sum F_{iy} = T_1 \sin \theta_1 + T_2 \sin \theta_2 - W = 0$$

(2.6.9)

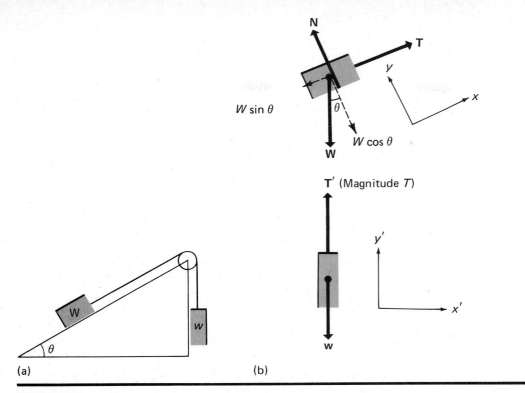

(a)   (b)

FIGURE 2.13

In these equations, θ_1, θ_2, and W are given. There are two unknowns, T_1 and T_2, for the pair of simultaneous equations. Solving the equations for these two quantities, obtaining from the first equation

$$T_2 = T_1 \frac{\cos \theta_1}{\cos \theta_2} \tag{2.6.10}$$

and substituting this value into the other,

$$T_1 = \frac{W \cos \theta_2}{\sin \theta_1 \cos \theta_2 + \cos \theta_1 \sin \theta_2}$$

$$= \frac{W \cos \theta_2}{\sin(\theta_1 + \theta_2)} \tag{2.6.11}$$

Substituting this back into (2.6.10), we find

$$T_2 = \frac{W \cos \theta_1}{\sin(\theta_1 + \theta_2)} \tag{2.6.12}$$

The forces acting on the overhead support are equal in magnitude but opposite in direction to the forces $\mathbf{T}_1$ and $\mathbf{T}_2$.

Clearly, Example 2.6.1 is a special case of this more general problem, where $W = 200$ lb and $\theta_1 = \theta_2 = 10°$. If we recall from trigonometry that $\sin 2x = 2 \sin x \cos x$, then, obviously, $\sin 20° = 2 \sin 10° \cos 10°$. Substituting this into (2.6.11) and (2.6.12), the answer obtained previously follows at once. It is often advantageous to solve a specific problem, such as the one treated in Example (2.6.1) by working a more general example such as the one explained directly above, in which unspecified values for the quantities that enter into the problem are used. Then general for-

mulas, such as (2.6.11) and (2.6.12), for the unknowns are derived, from which the required numerical answers may be obtained by inserting the given numerical data. In the first example, it is true that the correct answer was calculated, but the more general method used in the second example allows us to find the answers very easily, no matter what the weight of the suspended object is or what the angles of the cables are. In addition, the formulas for the tensions that were derived enable one to understand how those tensions will vary as functions of the suspended weight and the cable angles. In either case, the basic method of attack, through the equations for equilibrium, is the same. We shall often adopt the more general method of approach in the examples which are discussed in the text.[16]

EXAMPLE 2.6.3

An object of weight W is supported in equilibrium on a frictionless plane making an angle θ with the horizontal by a suspended weight w and a flexible cord passing over a frictionless pulley, as shown in Fig. 2.13a. Find (a) the normal force $\mathbf{N}$ that the plane exerts against the weight W and (b) the value of the suspended weight w required to keep the system in

[16] Although the method of attack used in this example is usually best inasmuch as it provides more information about the system being studied, it is not always quickest. Quite frequently, the initial substitution of numerical data in the equilibrium equations simplifies the arithmetic and may, therefore, be preferred when time is of the essence, as in an examination.

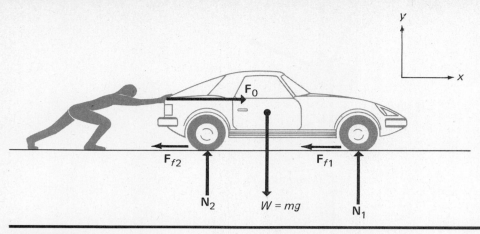

FIGURE 2.14

equilibrium. If $W = 100$ lb and $\theta = 30°$, what are the values of N and w?

Again, consider both weights as isolated free bodies and specify all the forces acting on each, as shown in Fig. 2.13b. The only forces acting on the weight w are vertical, and in order for this object to be in equilibrium,

$$\sum F_{iy'} = T' - w = T - w = 0 \qquad (2.6.13)$$

Since the only effect of the pulley in this case is to alter the direction in which the tension in the cord acts, the magnitudes of vectors T' and T are equal. The best way of dealing with the forces acting on the weight which is resting on the plane is to *resolve them into components parallel and perpendicular to the plane*. This approach is almost always advantageous in any system involving objects supported upon an inclined plane. In this case, writing the equations for equilibrium in the x- and y-directions (parallel and perpendicular to the plane), we find

$$\sum F_{ix} = T - W \sin \theta = 0 \qquad (2.6.14)$$

$$\sum F_{iy} = N - W \cos \theta = 0 \qquad (2.6.15)$$

The values of θ and W may be regarded as known, while w, N, and T are to be determined. Equations (2.6.13), (2.6.14), and (2.6.15) may be looked upon as three equations which can be solved as simultaneous equations for these three quantities. The solution, however, is very easy in this instance. From (2.6.13), it is clear that $T = w$; while from (2.6.14), we see that $T = W \sin \theta$. Therefore,

$$w = T = W \sin \theta$$

while from (2.6.15),

$$N = W \cos \theta$$

If $W = 100$ lb and $\theta = 30°$, then $w = T = 100 \sin 30° = 50$ lb, and $N = 100 \cos 30° = 86.6$ lb. In this example, the normal force N is the total force exerted by the

supporting plane against the weight W; there can be no component of force parallel to the plane, since the interface is *frictionless*. If this were not so, then forces of friction might act in the interface parallel to the plane, and the total force exerted by the plane on the weight would be the resultant of the force of friction and the normal force N.

EXAMPLE 2.6.4

A 1500-kilogram automobile stands on a level road, as shown in Fig. 2.14. The coefficient of (rolling) static friction is 0.05, and the coefficient of (rolling) kinetic friction is 0.03. How much force must be exerted horizontally to set the car in motion? How much horizontal force must be exerted to keep the car rolling at a constant velocity once it is in motion? Suppose now that the brakes are applied so that the wheels cannot rotate. The coefficient of (sliding) static friction is now 1.0 and the coefficient of (sliding) kinetic friction is 0.7. What are the answers to the above questions under these circumstances?

In this example, let us represent the weight of the car by $\mathbf{W}$, the normal force the pavement exerts against the wheels by $\mathbf{N}_1$ and $\mathbf{N}_2$ at the front and rear, respectively, the applied force by $\mathbf{F}_0$, and the frictional retarding forces at the front and rear wheels by $\mathbf{F}_{f1}$ and $\mathbf{F}_{f2}$.[17] The *total* normal force $\mathbf{N}$ is then equal to $\mathbf{N}_1 + \mathbf{N}_2$ and the total force of friction $\mathbf{F}_f$ is $\mathbf{F}_{f1} + \mathbf{F}_{f2}$. Since the car is in equilibrium we may set the sum of all the y-components of forces on it equal to zero, whereby

$$\sum F_{iy} = N_1 + N_2 - W = N - W = 0$$

or, recalling (2.3.5),

$$N = W = mg \qquad (2.6.16)$$

[17] Actually, of course, since there are two front wheels, $\mathbf{N}_1$ represents the sum of two equal normal forces at each front wheel and $\mathbf{N}_2$, the sum of two equal forces at the rear wheels. Similar remarks can be made concerning $\mathbf{F}_{f1}$ and $\mathbf{F}_{f2}$. It is not necessary in this problem to consider these forces individually.

Also,

$$\sum F_{ix} = F_0 - (F_{f1} + F_{f2}) = F_0 - F_f = 0$$

and

$$F_f = F_0 \qquad (2.6.17)$$

To set the car in motion, F_f must equal $\mu_s N$. But from (2.6.16), $N = W = mg$; and from (2.6.17), $F_0 = F_f$. Therefore, the required value of applied force F_0 is

$$F_0 = \mu_s mg = (0.05)(1500)(9.8) = 735 \text{ N}$$

To maintain the automobile's motion at constant velocity, F_f must equal $\mu_k N$, whence

$$F_0 = F_f = \mu_k mg = (0.03)(1500)(9.8) = 441 \text{ N}$$

When the brakes are locked, the set of forces which acts is essentially the same, except now the much higher coefficients of *sliding* friction $\mu_s = 1.0$ and $\mu_k = 0.7$ must be used, and it is easily seen that 14,700 and 10,290 newtons of force must be applied in order to initiate and maintain motion, respectively. In this example, wherein mks units are given, it is important to note that it is the *mass* of the car that is stated rather than the weight. The weight force must be calculated from the relation $W = mg$, in units of newtons. The other forces that are determined will then be expressed also in newtons.

EXAMPLE 2.6.5

A packing case weighing 100 pounds rests upon a horizontal floor, as shown in Fig. 2.15. A force $\mathbf{F}_0$ acting at an angle $\theta = 30°$ below the horizontal is applied to it, as illustrated. The coefficient of static friction between the packing case and the floor is 0.35, and the coefficient of kinetic friction is 0.25. How large an applied force is required to set the case in motion, and how much force is required to maintain it in motion at constant velocity?

The forces acting on the object, considered as an isolated free body, are shown in Fig. 2.15. Since the

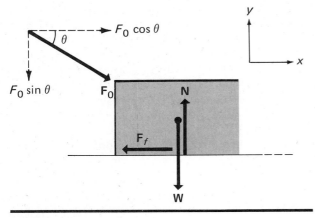

FIGURE 2.15

body is in equilibrium, we may write

$$\sum F_{iy} = N - W - F_0 \sin \theta = 0$$
$$\sum F_{ix} = F_0 \cos \theta - F_f = 0 \qquad (2.6.18)$$

In this problem, W and θ are specified, while under the conditions described above, F_f must equal $\mu_s N$ in order that the case be set in motion. We may then substitute this value for F_f in Eq. (2.6.18) and rewrite these equations in the form

$$N - F_0 \sin \theta = W$$
$$-\mu_s N + F_0 \cos \theta = 0$$

These can now be solved as a set of simultaneous equations for the two unknowns F_0 and N. The easiest way of doing this is to multiply the top equation by μ_s, add the two resulting equations together, and solve for F_0 to obtain

$$F_0 = \frac{\mu_s W}{\cos \theta - \mu_s \sin \theta} \qquad (2.6.19)$$

The normal force N may be determined by substituting this value back into either of the two original equations; in this way, we may show that

$$N = W\left(1 + \frac{\mu_s \sin \theta}{\cos \theta - \mu_s \sin \theta}\right) = \frac{W \cos \theta}{\cos \theta - \mu_s \sin \theta} \qquad (2.6.20)$$

Inserting the values $\theta = 30°$, $W = 100$ lb, $\mu_s = 0.35$ into (2.6.19), one may easily show that $F_0 = 50.7$ lb; and from (2.6.20), the corresponding normal force turns out to be 125.3 lb. If the case is sliding at uniform speed, the problem may be worked in the same way, except that F_f must now have the value $\mu_k N$ rather than $\mu_s N$. The answers are the same as those shown above in (2.6.19) and (2.6.20), except that μ_k replaces μ_s everywhere. If the values $\theta = 30°$, $W = 100$ lb, and $\mu_k = 0.25$ are substituted, Eq. (2.6.19) then yields $F_0 = 33.7$ lb, and (2.6.20) gives $N = 116.9$ lb.

In this problem, it is interesting to note that if the applied force $\mathbf{F}_0$ acts at a sufficiently steep angle (that is, if θ is sufficiently large), the force required to set the object in motion may become *infinitely large*. This is most easily seen by examining the expression $\cos \theta - \mu_s \sin \theta$ in the denominator of (2.6.19). If $\theta = 0$ (force applied horizontally), this quantity has the value unity and decreases as θ increases. At some value of θ (let us call it θ_m), it becomes zero, corresponding to an infinite value for F_0. The critical angle where this takes place can be found by setting $\cos \theta - \mu_s \sin \theta$ equal to zero and solving for θ. In this way, we find

$$\cos \theta_m - \mu_s \sin \theta_m = 0 \qquad (2.6.21)$$
$$\cot \theta_m = \mu_s$$

If θ is increased until its cotangent equals μ_s, no force F_0, however great, applied in that direction will set the object in motion. Physically, the reason for this is that at this angle, the vertical component of the applied force $\mathbf{F}_0$ acts to increase the normal force (and hence the force of friction $\mu_s N$) just as rapidly as it increases the horizontal component of $\mathbf{F}_0$ which opposes the force of friction. The same effect is observed in connection with sliding friction: there exists some value of θ (call it θ_k) beyond which no force will suffice to maintain the system in sliding equilibrium. Since the equations of sliding equilibrium are the same as (2.6.19) and (2.6.20) with μ_k replacing μ_s, the critical value θ_k where this effect takes place is that value for which

$$\cot \theta_k = \mu_k \qquad (2.6.22)$$

Since μ_k is *smaller* than μ_s, and since the value of $\cot \theta$ decreases as θ gets larger, this critical angle is *larger* than the one defined by (2.6.21). There is, therefore, a range of values of θ wherein motion cannot be initiated by any force, however large, but for which sliding equilibrium can be maintained provided that motion is *externally* initiated, for example by the temporary application of a horizontal force.

For values of θ in excess of the critical value determined by (2.6.21) or (2.6.22), the values of the applied force and normal force as given by (2.6.19) and (2.6.20) are negative, since the denominator in those expressions then has a value less than zero. This does not mean that the applied forces and normal force actually can ever become negative under these circumstances. Instead, it reflects the fact that it is impossible for the force of friction ever to attain the values $\mu_s N$ or $\mu_k N$ which it is assumed to have in arriving at (2.6.19) and (2.6.20), and serves to warn us that we are trying to use these equations in a region where they cannot possibly apply.

This example illustrates the advantage of using general algebraic symbols initially, rather than numerical values, in working out problems. Had we used only the given numerical values in the equilibrium equations, we would never have been aware of the interesting and important effects discussed in the two preceding paragraphs.

EXAMPLE 2.6.6

A rectangular object of weight W is placed upon a smooth inclined plane, as shown in Fig. 2.16a. The coefficient of static friction between the object and plane is μ_s, the coefficient of kinetic friction is μ_k. The angle θ which the plane makes with the horizontal is adjustable. Find the largest value θ may have while in a state of static equilibrium. Also, find the value for θ just sufficient to maintain the system in equilibrium with the object sliding to the left at constant speed.

In this case, as usual, we must regard the object

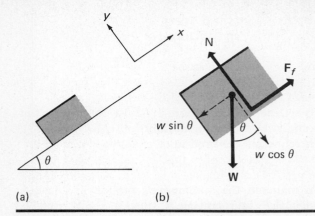

(a)   **(b)**

FIGURE 2.16

as an isolated free body in equilibrium and examine all the forces acting upon it, as shown by Fig. 2.16b. As always, when dealing with objects that are supported by an inclined plane, we shall resolve these forces into x-components parallel with the plane and y-components perpendicular thereto. Then, for equilibrium,

$$\sum F_{yi} = N - W \cos \theta = 0$$
$$\sum F_{xi} = F_f - W \sin \theta = 0 \qquad (2.6.23)$$

From the top equation, clearly,

$$N = W \cos \theta \qquad (2.6.24)$$

The maximum value that F_f may attain while at equilibrium is $\mu_s N$; this occurs at an extreme angle of tilt which we shall call θ_m. Putting these values into the lower equation of (2.6.23) and expressing N in terms of W by (2.6.24), we obtain

$$\mu_s \cos \theta_m - \sin \theta_m = 0$$

from which

$$\tan \theta_m = \mu_s$$

or

$$\theta_m = \tan^{-1} \mu_s \qquad (2.6.25)$$

In the case where the system is in sliding equilibrium, the force of friction must equal $\mu_k N$, and this value is attained at an angle of tilt which we shall call θ_k. Putting these values into the lower equation of (2.6.24) and expressing N in terms of W by (2.6.24), we may arrive by a similar series of steps at

$$\theta_k = \tan^{-1} \mu_k \qquad (2.6.26)$$

From (2.6.25) and (2.6.26), it is apparent that the coefficients of friction μ_s and μ_k can be measured by measuring the angles θ_m and θ_k associated with "breakaway" and with a constant sliding motion. This is, in fact, a very good way of measuring coefficients of friction in actual practice.

2.7 Moments and Torques: Rotational Equilibrium of Rigid Bodies

In section 2.4, we discussed the possibility that a rigid body might be in translational equilibrium without being in rotational equilibrium. We shall now examine more closely what is meant by this statement and explain what conditions must be fulfilled in order that a rigid body be in a state of rotational equilibrium. In order to accomplish this, we must first introduce the concept of the *moment of a force*, which is also referred to as *torque*. Consider a force **F** which acts at a point of application P whose position, relative to an arbitrarily chosen origin O, is specified by a vector **r**. Then the torque τ about the origin O, or the moment of the force **F** about the point O, is a vector defined by

$$\tau = \mathbf{r} \times \mathbf{F} \tag{2.7.1}$$

as illustrated in Fig. 2.17. The line of action of the moment vector τ is ordinarily taken to pass through the point O, in reference to which the moment is defined. According to the definition of the cross product, as discussed in section 1.6, then, the torque vector τ is perpendicular to both **F** and **r** and thus points along the direction of the normal to the plane determined by **F** and **r**, in the sense in which a right-handed screw would advance in turning the direction of **r** into the direction of **F** through the smallest angle. Also, its magnitude must be

$$\tau = rF \sin \theta \tag{2.7.2}$$

where θ is the angle between the directions of **r** and **F**.

Intuitively, we can understand why the torque

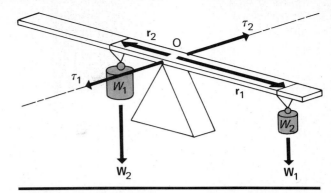

FIGURE 2.18. Effect of weight W_1 in producing a counterclockwise rotation about O can be counteracted by a much smaller weight W_2, placed at a proportionally larger distance from the axis.

is defined in this way by remarking that the quantity τ is supposed to represent *the effect of a force in producing a rotation* about O. It is easy to demonstrate experimentally that the rotational effect of a large force applied close to the axis of rotation can be balanced by a smaller, opposing force applied proportionally further from the axis, in such a way that the products of the forces and perpendicular distances between the application point and the rotation axis are equal in magnitude, as in Fig. 2.18. The *torques* about the point O associated with the two forces are then equal and oppositely directed and add to a total *net torque of zero*, producing rotational equilibrium. Thus, the magnitude of the rotational effect of a force **F** must be proportional to the product of F and the distance r to the origin. Also, since obviously any component of **F** along the direction of **r** cannot act to produce rotation about O, only the component of **F** perpendicular to **r** has any rotational effect. But the component of **F** in the direction normal to **r** is just $F \sin \theta$, and hence the magnitude of the rotational effect of the force about O will be proportional to the product of F, r, and $\sin \theta$. This product is just the magnitude of the torque vector τ defined by (2.7.1).

The *direction* of the torque vector is *along the axis about which rotation will occur;* but, although this fact must be clearly understood, many of its implications will become apparent only in later chapters. If a number of individual forces act on the same rigid body, they will give rise to individual torques, which may be summed *vectorially* to obtain the total or *resultant* torque acting upon the system. According to (2.7.2), torque must have the dimensions of force times length; it may be expressed in units of newton-meters in the mks system, dyne-centimeters in the cgs system, and pound-feet in the English system.

The condition for a rigid body to be in rotational equilibrium, then, is that the resultant torque upon it

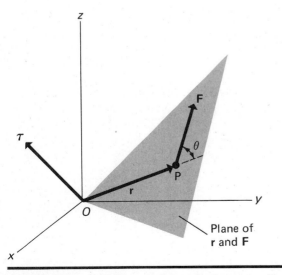

FIGURE 2.17. Definition of the vector torque as the cross product **r** × **F**.

(referred to any arbitrary point as an origin of co-ordinates) be zero; this can be stated in the form of a vector equation as

$$\sum_{i=1}^{n} \tau_i = 0 \qquad (2.7.3)$$

where $\tau_1, \tau_2, \tau_3 \ldots, \tau_i \ldots, \tau_n$ are individual torques acting on the body, and the sum above is interpreted as a *vector* sum. Since no vector can equal zero unless all three components vanish, this equation, like (2.4.1), can be written as a set of three equations, one for each component, having the form

$$\sum_{i=1}^{n} \tau_{ix} = 0$$

$$\sum_{i=1}^{n} \tau_{iy} = 0 \qquad (2.7.4)$$

$$\sum_{i=1}^{n} \tau_{iz} = 0$$

where the summations now simply represent *algebraic* sums.

The foregoing results have been justified only by remarking that they are in accord with all experimental observations. Although, intuitively, it would appear that they are in some way associated with Newton's laws as stated in section 2.2, it is not at all clear what the connection is. For example, we might reasonably ask why it is that when torque is defined in just such a way that its magnitude is $Fr \sin \theta$, the condition for rotational equilibrium for systems such as that shown in Fig. 2.18 is that the sum of those torques be zero. Why does it not turn out that a magnitude $Fr^2 \sin \theta$, or some more complex expression, is required to define torque in formulating this equilibrium condition? The correct answer to these questions requires detailed calculation involving the application of Newton's second law to rotational motion. We shall postpone performing this calculation explicitly until Chapter 7, since we wish to confine our discussion at this point to unaccelerated motions related primarily to the first and third laws. We may note, however, that the result of this investigation is that when a net torque, as defined by (2.7.1) *and in no other way*, is applied to a body, that body experiences a rotational acceleration proportional to the torque both in magnitude and direction. Therefore, when the net torque, as defined by (2.7.1), is zero, there is *no* rotational acceleration, which means that the body either does not rotate or is in a state of constant rotational motion, and hence is in equilibrium. The conditions for rotational equilibrium (2.7.3) and (2.7.4), therefore, follow directly as a consequence of

Newton's laws and do not really involve any new physical principles.

So far, we have discussed torques applied to rigid bodies in the most general way. In most of the actual examples with which we shall be confronted, however, only rotation of a rigid body *about a single externally fixed axis of rotation* is involved. Under these circumstances, matters become much simpler than they are in general. First of all, it is necessary to consider only components of torque parallel to the fixed rotation axis in investigating rotational motion of systems of this type. The reason for this can be understood by referring to Fig. 2.19, wherein a wheel mounted in two fixed bearings is illustrated. Suppose an external torque τ is applied to this system. This torque can always be resolved into two component torques $\tau_{\parallel}$ acting parallel to the rotation axis (corresponding to an applied force couple such as AB) and $\tau_{\perp}$ acting perpendicular to the rotation axis (corresponding to an applied couple such as CD). The torque $\tau_{\parallel}$ can act to change the rotational motion of the wheel about the axis of rotation, or to combine with other applied torque components acting parallel to the rotation axis in such a way as to produce rotational equilibrium about this axis; but the torque $\tau_{\perp}$ can act only against the fixed bearing supports and can in no way affect the rotational motion about the axis. It is opposed by an equal and opposite torque exerted *by the bearing supports on the bearing* as the couple EF, which maintains equilibrium and keeps the axis fixed in accord with the requirements of Newton's first law.

Likewise, in dealing with systems of this type, we need consider *only force components that lie in a plane perpendicular to the rotation axis*, since only

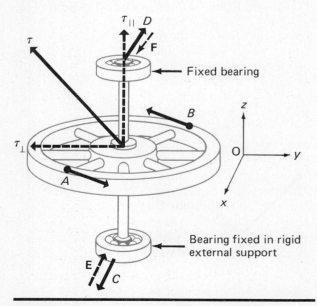

FIGURE 2.19. System of torques acting on a wheel mounted in frictionless fixed bearings.

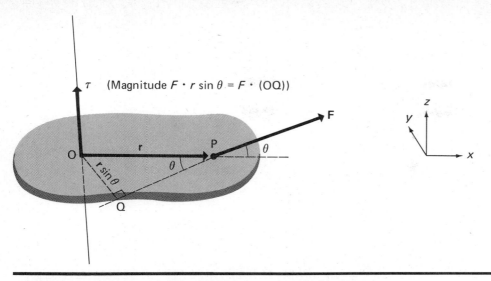

FIGURE 2.20. Definition of torque in terms of force and "moment arm" OQ.

such components can give rise to torques parallel to the rotation axis. We shall ordinarily discuss in examples only such force components, regarding the components parallel to the axis as having been rejected beforehand. Also, in such cases, the equations for equilibrium of torques about axes like the x- and y-axis in Fig. 2.19 express only the equality of force couples such as CD and EF, which is usually not of very much interest or importance. Therefore, in any case of rotation about a fixed axis like that shown in the figure, it is only the third equation of the set (2.7.4), expressing the equilibrium of torques about the actual axis of rotation (the z-axis in this case), that need be considered in detail. It should be emphasized, however, that this equation must be satisfied for any possible choice of the point O about which moments may be calculated. Since all the torque components which might affect the equilibrium of such a system point along the $+z$- or $-z$-directions, the picture of forces and torques relevant to this situation is as illustrated in Fig. 2.20. It is evident from this diagram that the magnitude of the torque τ arising from the application of the force $\mathbf{F}$ at point P to a rigid body constrained to rotate about a fixed axis through O can be represented by the product of the magnitude F and the *lever arm* or *moment arm* $OQ = r \sin \theta$, which is the perpendicular distance between the axis of rotation and the line of action of $\mathbf{F}$. The rule for finding the magnitude of a torque in a system with a fixed rotation axis, where all applied forces lie in a plane perpendicular to this axis, is therefore often stated in the following form:

The magnitude of the moment of a force about a given axis is the product of the magnitude of the force and the perpendicular distance between the line of action of the force and the axis.

EXAMPLE 2.7.1

In the pendulum shown in Fig. 2.21, an object of mass m swings on the end of a string of length r. Find the torques about the support point O due to the forces $m\mathbf{g}$ and $\mathbf{T}$ that act upon the mass.

The mass can be located with respect to the support point O by the dashed position vector $\mathbf{r}$ shown in the figure. According to (2.7.1), the torque associated with the tension force $\mathbf{T}$ should be

$$\tau_1 = \mathbf{r} \times \mathbf{T} = 0$$

the cross product vanishing since the angle between

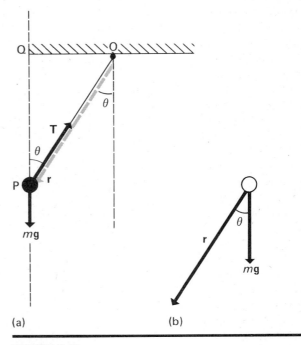

FIGURE 2.21

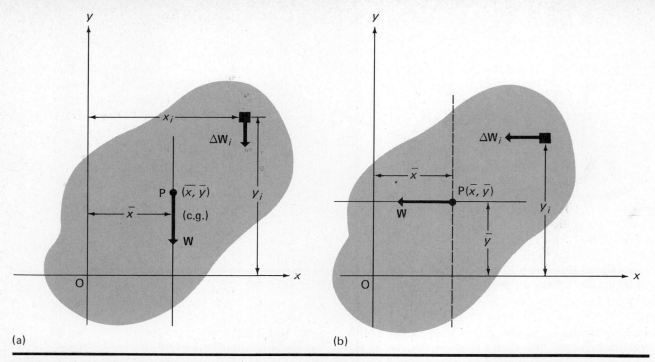

FIGURE 2.22. Action of weight forces on an irregular rigid object. At (a), force of gravity acts in the *y*-direction, while at (b), it acts along the *x*-axis.

vectors **T** and **r** is 180°. In the case of the weight force, the direction of **r** makes an angle θ with the direction of $m\mathbf{g}$, and therefore the torque τ_2 arising from the force $m\mathbf{g}$ will be

$$\tau_2 = \mathbf{r} \times m\mathbf{g}$$

The magnitude of this torque is given by

$$\tau_2 = mgr \sin \theta$$

The direction of τ_2 points out of the page, normal to the paper. This is most easily seen by translating the vector **r** along its line of action OP until its tail rests at P, as in Fig. 2.21b. It is then evident that a right-handed screw advances out of the page as **r** is turned into $m\mathbf{g}$ through the angle θ.

The same results are obtained in a slightly different but equivalent way by using the rule that the torque associated with a force can be expressed as the product of the force times the perpendicular distance between its line of action and the axis. The tension force **T** acts along the line OP. Since the line of action OP passes through the axis O about which we are computing torques, the perpendicular distance between the line along which **T** acts and the axis (referred to as the moment arm of the force **T**) is in this case zero. The tension force, therefore, causes zero torque about this axis. The weight force, however, acts vertically, along the line PQ. The perpendicular distance between this line and the axis O is simply the distance OQ $= r \sin \theta$. This distance represents the

moment arm of the force $m\mathbf{g}$ with respect to O. The torque τ_2 associated with the weight force is then the product of the force $m\mathbf{g}$ and the moment arm $r \sin \theta$, which gives $\tau_2 = mgr \sin \theta$, in agreement with our previous result.

EXAMPLE 2.7.2

At a certain instant of time, a particle's position with respect to an origin O is given by the vector $\mathbf{r} = 3\mathbf{i}_x + 4\mathbf{i}_y$ (component magnitudes in meters). It experiences a force $\mathbf{F} = 16\mathbf{i}_x + 32\mathbf{i}_y$ (component magnitudes in newtons). Find the torque associated with the force **F** acting on the particle, referred to the origin O.

In this instance, we may proceed most easily by directly evaluating the cross product in terms of cartesian components, using the rules for multiplying the unit vectors $\mathbf{i}_x$, $\mathbf{i}_y$, and $\mathbf{i}_z$ as expressed by (1.6.9), (1.6.10), and (1.6.11). We find, therefore, that

$$\begin{aligned}\tau = \mathbf{r} \times \mathbf{F} &= (3\mathbf{i}_x + 4\mathbf{i}_y) \times (16\mathbf{i}_x + 32\mathbf{i}_y) \\ &= (3)(32)\mathbf{i}_z - (4)(16)\mathbf{i}_z \\ &= 32\mathbf{i}_z \text{ N-m}\end{aligned}$$

2.8 Moments of Weight Forces and the Center of Gravity

There is one more point we must understand before the picture of rotational equilibrium is complete, having to do with the moments associated with weight forces. When a force is applied to a system via the tension in a cord or some other external mechanical agency, it is usually quite clear where the point of application is and where the line of action lies. But in the case of a rigid body of finite extent, how is the weight force to be represented? Where is the point of application and the line of action? Before answering these questions in detail, we should note that so far as *translational* equilibrium is concerned, the answer makes no difference so long as the direction of the weight force is specified as being vertically downward. This can be understood by referring to the example discussed in connection with Fig. 2.6 or to Examples 2.6.3, 2.6.4, 2.6.5, and 2.6.6. In none of the cases discussed there is it at all necessary to know where the point of application or the line of action of the weight force is in order to write the equations for translational equilibrium. All that is required is the assumption that the force acts vertically downward. As a matter of convenience, and in anticipation of results to be developed later, in drawing the illustrations which accompany these (and other) examples we have shown the weight force as though it acted downward from some more or less centrally located point within the body.

As soon as we have to discuss and calculate the *moments* of weight forces, however, we are at once faced with the problem of determining where those forces act. It is evident that they must act somewhere, that there is some appropriate line of action and some precise point of application along that line. But there is no obvious way of determining where that point is. A more precise understanding of the situation can be obtained by referring to Fig. 2.22a. In this figure we see a thin, irregularly shaped plate of total weight W. But the plate is a rigid body of finite extent, and the effect of the total weight force (shown in the figure as acting at some point P whose coordinates are $\bar{x}$ and $\bar{y}$) is in some sense a *summation* of the action of weight forces exerted by all parts of the body at their respective locations. The question now is seen to be, where must we apply the weight force **W** so that its effect is the same as that of all the individual weight forces exerted by every individual part of the body acting at its respective location?

Part of the answer is obtained by observing that the weight forces exerted by the individual weight elements of the body give rise to a torque about the point O and that the line of action of the total weight force *must be chosen so that the total weight force* **W**, *acting along this line, generates precisely that same torque.*

We shall, therefore, calculate the total moment of all the individual weight forces and equate them to the moment of the total weight force **W**. Let us begin by visualizing the body to be composed of a large number (let us call it N) of very tiny square elements, as shown in Fig. 2.22a. We may identify each of these elements by numbers 1, 2, 3, ..., N and their individual weights as $\Delta W_1, \Delta W_2, \Delta W_3, \ldots, \Delta W_N$. These weights may not all be the same, since the plate's density or thickness might not be uniform over its entire area. Consider now a typical weight element, let us say the ith one, whose weight is $\Delta \mathbf{W}_i$ and which is located at the point (x_i, y_i). The magnitude of the torque $\Delta \tau_i$ associated with the force $\Delta \mathbf{W}_i$ is simply the product of the force and the moment arm, which in this case is the perpendicular distance x_i between the weight element and the line of action of the weight force. Therefore,

$$\Delta \tau_i = x_i \, \Delta W_i \tag{2.8.1}$$

The total moment τ is found by summing these individual moments over all the elements (1, 2, 3, ... $i, \ldots, N$) comprising the body. Since the individual vector torques $\Delta \tau_i$ are all in the same direction (normal to the page), this sum need be only an algebraic sum of the individual magnitudes. We may, therefore, write

$$\tau = x_1 \, \Delta W_1 + x_2 \, \Delta W_2 + x_3 \, \Delta W_3 + \cdots$$
$$+ x_i \, \Delta W_i + \cdots + x_N \, \Delta W_N$$

or, using the usual summation notation,

$$\tau = \sum_{i=1}^{N} x_i \, \Delta W_i \tag{2.8.2}$$

The point of application P of the total weight force **W** must be chosen so that the total weight force acting at P$(\bar{x}, \bar{y})$ generates this same total torque about O. But the moment of the total weight force about O is simply $W\bar{x}$. Therefore,

$$W\bar{x} = \tau = \sum_{i=1}^{N} x_i \, \Delta W_i \tag{2.8.3}$$

from which the x-coordinate of P may be obtained as

$$\boxed{\bar{x} = \frac{1}{W} \sum_{i=1}^{N} x_i \, \Delta W_i} \tag{2.8.4}$$

The distance $\bar{x}$ has the property that the moment of the weight force is obtained by multiplying $\bar{x}$ by the total weight. Accordingly, $\bar{x}$ is the distance from O to the *effective line of action* of the weight force.

It is important to note that the sum in (2.8.4) is to be evaluated in the *limit of infinitesimal subdivision*, that is, the limit where the number of elements N approaches infinity, while the individual weights ΔW_i, and the dimensions Δx and Δy in Fig. 2.22, approach zero. This exact limiting value can be obtained by using the methods of integral calculus, for in the limit as $N \to \infty$, $\Delta W_i \to 0$, the summation in (2.8.4) can be represented as a definite integral. The evaluation of this integral is not very difficult for objects of simple and regular geometric form and uniform density.

At this point, however, it is more important to understand how the line of action of the weight force and the coordinates $\bar{x}$ and $\bar{y}$ of its point of application are defined physically than to go through the mathematics of calculating their exact locations. We shall, therefore, postpone discussing the calculation of these quantities by integration until the concept of center of mass is discussed in Chapter 4. It will suffice here simply to say that we can consistently identify a line of action and a point of application for the total weight force by the requirement that the moment of the total weight force about any axis be equal to the sum of the individual moments arising from every individual weight element of the body acting at its respective location in the object.

Now let us assume that the xy-coordinate system of Fig. 2.22 is fixed with respect to the body and that the object is rotated through an angle of 90° about point O, so that the force of gravity acts along the $-y$-direction rather than along $-x$. The situation will be as represented in Fig. 2.22b, where, for convenience, we have kept the same orientation for the body and coordinate axes and rotated the gravitational force to obtain the same effect. Let us again calculate the moments of the gravitational forces about O. For the element ΔW_i, we obtain a moment $\Delta\tau_i$ of magnitude

$$\Delta\tau_i = y_i\,\Delta W_i \tag{2.8.5}$$

The sum of all such moments contributed by the elements composing the body is

$$\tau = y_1\,\Delta W_1 + y_2\,\Delta W_2 + \cdots + y_i\,\Delta W_i + \cdots + y_N\,\Delta W_N$$

or

$$\tau = \sum_{i=1}^{N} y_i\,\Delta W_i \tag{2.8.6}$$

As before, this total moment can be considered as the product of the entire weight of the object times an *effective moment arm* $\bar{y}$. Then,

$$W\bar{y} = \tau = \sum_{i=1}^{N} y_i\,\Delta W_i \tag{2.8.7}$$

from which

$$\boxed{\bar{y} = \frac{1}{W}\sum_{i=1}^{N} y_i\,\Delta W_i} \tag{2.8.8}$$

Again, $\bar{y}$ has the property that the total moment τ about O can be expressed as the total weight times $\bar{y}$, and therefore $\bar{y}$ plays the role of an effective or average moment arm when the body is in this second orientation with respect to the gravitational force. The quantity $\bar{y}$ may thus be regarded as the distance from O to the effective line of action of the weight force when the

body is in this position. The two lines of action defined by $\bar{x}$ and $\bar{y}$ intersect at point P whose coordinates are $\bar{x}$ and $\bar{y}$ and which defines the *center of gravity* of the body. The effective line of action of the weight force, in fact, *always* passes through the center of gravity, whatever orientation the object may have with respect to the gravitational force. This assertion should be intuitively evident in the light of the above discussion, but it can be proved by a straightforward but slightly tedious extension of the arguments used above. The proof is assigned as an exercise for the student. It follows from all this that *we may correctly regard the weight force as always acting vertically downward from the center of gravity* of the body, since the point of application of a force may be taken to be anywhere on its line of action.

Figure 2.23a shows an irregularly shaped object suspended at point O and free to rotate about an axis passing through that point. The center of gravity of the body is at P, and we may therefore represent the weight force **W** as acting vertically downward from that point. For the object to be in equilibrium, the resultant torque about O must be zero. This can occur only when the line of action of the weight force passes through the point O, for then, and only then, will the moment arm (and therefore the torque) associated with the weight force **W** be zero. Since the weight force acts vertically through P, it is evident that the condition of equilibrium for torques is satisfied when the center of gravity lies directly below the point of suspension, as shown at (a) in Fig. 2.23.

If the location of the center of gravity is unknown, we may conclude from this argument that it must lie somewhere on a line drawn vertically through the suspension point O. If we now suspend the body

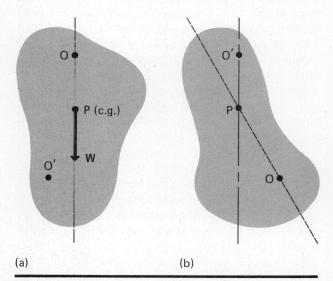

(a)   (b)

FIGURE 2.23. An irregular rigid body free to rotate about a point of suspension O (a) or O′ (b).

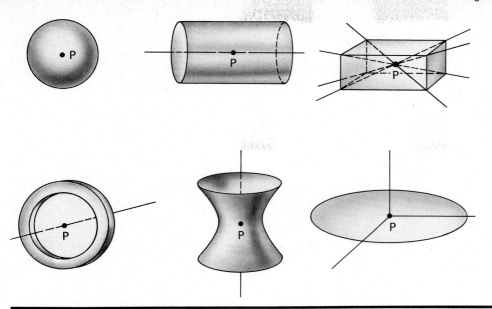

FIGURE 2.24. Location of the center of gravity for a number of symmetric objects of uniform density.

about a new suspension point O′, it will assume a new equilibrium position in which the center of gravity is on a line drawn vertically through O′, as in Fig. 2.23b. The intersection of these lines through O and O′ marks the body's center of gravity. This simple technique affords a practical experimental method of locating the center of gravity of an irregularly shaped body.

If the suspension point *coincides* with the center of gravity, the line of action of the weight force passes through the suspension axis irrespective of the orientation of the body. The moment arm associated with the weight force is always zero under these circumstances, as is the resultant torque. An object suspended about an axis passing through its center of gravity will, therefore, remain in equilibrium whatever its orientation with respect to that axis may be.

The above development, for the sake of simplicity, was restricted to a two-dimensional flat object, but the methods used and the results obtained are equally applicable to a three-dimensional body. Under these circumstances, the z-coordinate of the center of gravity will have the value

$$\bar{z} = \frac{1}{W} \sum_{i=1}^{N} z_i \, \Delta W_i \qquad (2.8.9)$$

It is intuitively obvious (and generally demonstrable) that, for solid objects of uniform density that are symmetric about some point P, the center of gravity coincides with P, since then for every element ΔW contributing a moment about this point, there is another which contributes an equal and opposite moment. Some common illustrations of this general state are shown in Fig. 2.24. For bodies lacking a

center of symmetry, or which are of nonuniform density, the center of gravity must be calculated mathematically using Eqs. (2.8.4), (2.8.8), and (2.8.9) or determined experimentally by methods similar to those outlined above.

It is frequently necessary to find the center of gravity of a composite system consisting of a finite number of individual bodies, for each of which the weight and center of gravity are specified. Suppose that there are N such objects, having weights W_1, W_2, W_3, ..., W_N, whose centers of gravity have the coordinates (x_1, y_1, z_1), (x_2, y_2, z_2), (x_3, y_3, z_3), ..., (x_N, y_N, z_N) as illustrated[18] in Fig. 2.25. If we denote by W the sum of all the individual weights, that is,

$$W = W_1 + W_2 + W_3 + \cdots + W_N = \sum_{i=1}^{N} W_i \qquad (2.8.10)$$

we may set the sum of all the individual moments of weight forces about O equal to this total weight times an effective moment arm $\bar{x}$, which represents the x-coordinate of the center of gravity of the whole system. In this way, we can write the torque about O due to the individual weight forces as

$$\tau = W_1 x_1 + W_2 x_2 + \cdots + W_N x_N = \sum_{i=1}^{N} W_i x_i = W\bar{x}$$
$$(2.8.11)$$

[18] In this drawing, the symbol $\odot$ is used to represent the fact that an arrow representing the z-direction stands vertically upward from the page. It is a frequently used convention that the symbols $\odot$ and $\otimes$ refer to arrows (which may represent either vectors or coordinate axes) pointing normally out of and into the page, respectively. The dot and cross are symbolic representations of the point and "tail feathers" of an arrow.

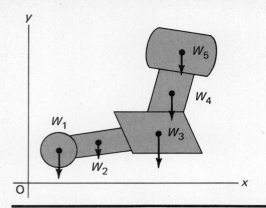

FIGURE 2.25. A system composed of a number of bodies whose individual centers of gravity are specified.

from which

$$\bar{x} = \frac{1}{W} \sum_{i=1}^{N} W_i x_i = \frac{\sum_{i=1}^{N} W_i x_i}{\sum_{i=1}^{N} W_i} \qquad (2.8.12)$$

If, now, the weight forces are regarded as acting in the $-y$-direction, the same procedure yields

$$\tau = W_1 y_1 + W_2 y_2 + \cdots + W_N y_N = \sum_{i=1}^{N} W_i y_i = W\bar{y} \qquad (2.8.13)$$

or

$$\bar{y} = \frac{1}{W} \sum_{i=1}^{N} W_i y_i = \frac{\sum_{i=1}^{N} W_i y_i}{\sum_{i=1}^{N} W_i} \qquad (2.8.14)$$

A similar expression for the z-coordinate of the center of gravity can be derived.

These expressions are somewhat similar to those given by Eqs. (2.8.4), (2.8.8), and (2.8.9) for the

center of gravity of a simple object of arbitrary irregular shape. There are, however, some important differences. First of all, there is no infinitesimal subdivision of the sample. The number N refers simply to the number of individual parts of the composite system—two, three, or five, etc., as the case may be. This number is not regarded as being infinite or even large. Also, the individual weights W_i are simply the *finite* weights of the separate parts of the system and need not be particularly small, let alone infinitesimal. Since we know where the center of gravity of each part is located, there is no question as to where the line of action of each weight force W_i is located. In evaluating the sums in (2.8.12) and (2.8.14), therefore, there are only a few finite terms to be added up, and the limiting processes of calculus required in the problem of finding the center of gravity of a single irregular object are irrelevant. These distinctions will be further clarified by the series of examples worked out below.

In Chapter 4, we shall have occasion to introduce the concept of *center of mass*, which has much in common with the idea of center of gravity. In many instances, we shall see that the center of gravity and the center of mass are the same. We shall also return then to the problem of using the methods of integral calculus to find the center of gravity of objects of irregular shape.

EXAMPLE 2.8.1

The centers of three uniform spherical bodies of mass 1 kg, 2 kg, and 3 kg are located at the vertices of an equilateral triangle whose sides are 1 m in length, as shown in Fig. 2.26a. Find the position of the center of gravity of this system.

In this case, we have three spherical objects whose masses are distributed symmetrically about their geometric centers. The center of gravity, therefore, obviously coincides with the geometric center in all instances. The centers of the spherical bodies are located at the vertices of the equilateral triangle shown in Fig. 2.26a. The situation is, therefore,

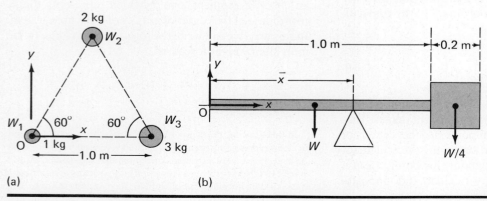

(a)

(b)

FIGURE 2.26

exactly that described above in connection with Eqs. (2.8.12) and (2.8.14). This is a system comprising three masses in which the individual weights and centers of gravity are given. In (2.8.12) and (2.8.14), therefore, $N = 3$. The xy-coordinates of the three centers of gravity, from Fig. 2.26a, are $(x_1, y_1) = (0, 0)$ for the 1-kg mass, $(x_2, y_2) = (\frac{1}{2} \text{ m}, \sqrt{3}/2 \text{ m})$ for the 2-kg mass, and $(x_3, y_3) = (1.0 \text{ m}, 0)$ for the 3-kg mass, using the coordinate system shown in the diagram. From (2.8.12), then, the x-coordinate of the center of gravity is

$$\bar{x} = \frac{\sum\limits_{i=1}^{3} W_i x_i}{\sum\limits_{i=1}^{3} W_i} = \frac{W_1 x_1 + W_2 x_2 + W_3 x_3}{W_1 + W_2 + W_3}$$

or

$$\bar{x} = \frac{(1.0)(9.8)(0) + (2.0)(9.8)(0.5) + (3.0)(9.8)(1.0)}{(1.0)(9.8) + (2.0)(9.8) + (3.0)(9.8)}$$

$$\bar{x} = \tfrac{2}{3} \text{ m}$$

In the same way, (2.8.14) gives us for the y-coordinate of the center of gravity

$$\bar{y} = \frac{\sum\limits_{i=1}^{3} W_i y_i}{\sum\limits_{i=1}^{3} W_i}$$

$$= \frac{(1.0)(9.8)(0) + (2.0)(9.8)(\sqrt{3}/2) + (3.0)(9.8)(0)}{(1.0)(9.8) + (2.0)(9.8) + (3.0)(9.8)}$$

$$\bar{y} = \sqrt{3}/6 \text{ m} = 0.289 \text{ m}$$

Since the centers of gravity of all three bodies lie in the xy-plane, the z-coordinate of the center of gravity of the system is clearly zero. Note that the factors of g needed to express the weight forces exerted by the respective masses cancel out completely, a circumstance that might have been forseen in view of the fact that the acceleration of gravity experienced by each mass is exactly the same. Whenever this is the case, as it is here, the center of gravity and the center of mass coincide.

EXAMPLE 2.8.2

A uniform rod 1.0 m long is of weight W. A uniform cubic block of wood of weight $W/4$, whose edge is 0.2 m, is fastened to one end of the rod as shown in Fig. 2.26b. At what point can the system be balanced on a knife-edge?

The point of balance must be at the center of gravity. The system again consists of individual weights each of whose center of mass is known. Since both the rod and block are of uniform composition, the center of gravity is in each case at the geometric center of the body. Therefore, using the coordinate system shown in Fig. 2.26b, the xy-coordinates of the center of gravity of the rod are $(x_1, y_1) = (0.5 \text{ m}, 0)$ and those of the block are $(x_2, y_2) = (1.1 \text{ m}, 0)$. In this case, since there are two individual objects, $N = 2$ and (2.8.12) becomes

$$\bar{x} = \frac{\sum\limits_{i=1}^{2} W_i x_i}{\sum\limits_{i=1}^{2} W_i} = \frac{W_1 x_1 + W_2 x_2}{W_1 + W_2}$$

$$= \frac{(W)(0.5) + (W/4)(1.1)}{W + (W/4)} = \frac{0.775W}{1.25W} = 0.62 \text{ m}$$

Since the y-coordinate of the center of gravity of each separate body is zero, it is evident that $\bar{y}$ will also be zero. The point of balance will thus be located along the rod 0.62 m from the left end.

2.9 Examples of Systems in Equilibrium

We have now laid the foundation for studying the mechanics of systems that are in equilibrium. We shall present in this section a number of examples that illustrate the application of the equations defining the conditions under which equilibrium of forces and torques prevail. Before doing this, however, we shall briefly reiterate the basic steps involved in the use of the equations of equilibrium:

1. First of all, we must decide exactly what constitutes the system in equilibrium and what is external to it. The system is regarded as an isolated entity upon which certain forces and torques act. All forces acting on the system are identified and their points of application indicated. Weight forces are considered to act downward from the centers of gravity of the objects involved. In most of the cases we shall study, the location of the centers of gravity will be clearly evident.

2. Newton's first law, stating that the vector sum of all forces acting on the system must vanish, is applied, usually in the form of component equations stating that $\sum F_x = \sum F_y = \sum F_z = 0$.

3. In the case of rigid bodies in equilibrium, the sum of torques about any axis is also zero. Therefore, an appropriate torque axis is chosen and the sum of all force moments about this axis equated to zero.

4. The equations expressing equilibrium of forces and torques are solved for the required unknowns. These unknowns may be unknown force components or the moment arms or points of application of individual forces on the system. The examples that follow illustrate the use of these basic procedures.

EXAMPLE 2.9.1

A uniform meter stick is supported at its center by a knife-edge as shown in Fig. 2.27a. A 1000-g mass is suspended from one end (at a distance of 50 cm from the knife edge), and a 1200-g mass is suspended at a distance of 20 cm from the knife-edge on the other side. How large a mass must be attached to the left end of the stick so that the system be balanced on the

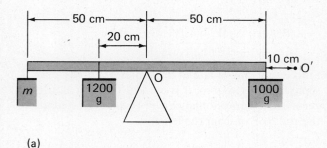

(a)

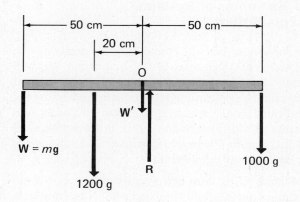

(b)

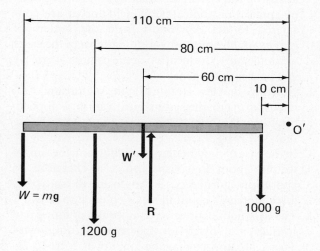

(c)

FIGURE 2.27

knife-edge? If the meter stick has a mass of 200 g, what is the total force exerted by the stick on the knife-edge support?

If the system is to be in rotational equilibrium (that is, in balance on the knife-edge), the sum of the torques about any point must be zero. Let us take the torques about point O, at the knife-edge. Representing the weight of the meter stick by W', we may observe that since the meter stick is uniform, its center of gravity is in the center directly over the knife-edge, and a downward force $\mathbf{W}'$ due to the weight of the stick will be exerted there. Downward forces due to the various weights will act as shown in Fig. 2.27b, and a force $\mathbf{R}$ will act upward at the knife-edge. The torques arising from $\mathbf{R}$ and $\mathbf{W}'$ about O are zero, however, since their moment arms are zero. The downward force exerted by the 1000-g mass will cause a torque about O which will be directed *out of* the page and which we shall regard to be positive. The forces exerted by the 1200-g mass and the weight W will cause torques about O which are directed *into* the page and which we shall view as negative. We shall always observe these conventions regarding the signs of torques about a common axis arising from coplanar forces throughout this book. Because the individual torque vectors are all parallel, in equilibrium the algebraic sum of their magnitudes must be zero. Accordingly, we may write (recalling that weight and mass are related by $W = mg$)

$$(1000)(980)(50) - (1200)(980)(20) - 50W = 0$$
$$50W = 2.548 \times 10^7$$
$$W = mg = 5.096 \times 10^5 \text{ dynes}$$

Dividing by g ($=980$ cm/sec^2), we find that the required mass is 520 grams. The value of the reaction force R may be obtained by observing that if the system is also in *translational* equilibrium, the algebraic sum of y-components of forces must be zero. Then,

$$R - W - W' - (1200)(980) - (1000)(980) = 0$$
$$R = W + W' + (2200)(980)$$
$$= (520)(980) + (200)(980) + (2200)(980)$$
$$R = (2920)(980) = 2.861 \times 10^6 \text{ dynes}$$

since, from above, $W = (520)(980)$ dynes and we are given the fact that the weight of the stick, W', is equal to $(200)(980)$ dynes, in view of the fact that its mass is 200 grams.

It is important to note that one may choose any point about which to calculate moments and arrive at the same results. For example, suppose we choose to compute moments about the point O', 10 cm to the left of the left-hand end of the stick. Then, referring to Fig. 2.27c, we may write the moments of all the forces about this point as

$$60(R - W') - (1000)(980)(10)$$
$$- (1200)(980)(80) - 110W = 0$$
$$60(R - W') - (106,000)(980) - 110W = 0$$

The equation for translational equilibrium of y-components can be written as

$$(R - W') - (2200)(980) - W = 0$$

The term $R - W'$ can be eliminated by multiplying the equation directly above by -60 and adding it to the other; the answer $W = (520)(980)$ dynes is then obtained from the resulting expression, and R can be determined as before by substituting this value and the value $W' = (200)(980)$ dynes into either equation. The algebraic work is simplified, of course, by taking moments about O, because then the moments of the forces **R** and **W'** vanish. From this point of view it is more *convenient* to use moments about O, but the same results are obtained by taking moments about any other point. It is true, in general, that taking moments about a point through which the lines of action of one or more forces pass results in a simplification of the ensuing algebraic work.

EXAMPLE 2.9.2
A 3200-pound automobile, wheelbase 113 inches, has 57 percent of its total weight supported by the front wheels. Where is the horizontal coordinate of its center of gravity?

Referring to Fig. 2.28, the weight force **W** acts downward from the center of gravity located, let us say, a distance x behind the rotation axis of the front wheels. There are normal forces $\mathbf{N}_f$ and $\mathbf{N}_r$ exerted by the road on the front and rear wheels, respectively. The equations for translational equilibrium tell us that the sum of y-components of forces must be zero, whereby

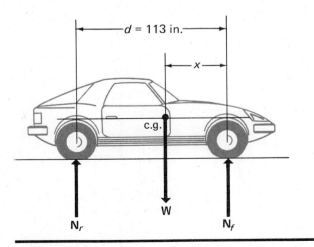

FIGURE 2.28

$$N_f + N_r - W = 0 \tag{2.9.1}$$

Also, for rotational equilibrium, the sum of moments of all forces about any point must be zero. Let us take moments about the rotation axis of the front wheels. Then, calling the wheelbase d, we may write

$$N_r d - Wx = 0 \tag{2.9.2}$$

Solving this equation for N_r and substituting the value so obtained into (2.9.1), we find

$$N_f + \frac{Wx}{d} - W = 0$$

from which

$$x = d\left(1 - \frac{N_f}{W}\right) \tag{2.9.3}$$

But $d = 113$ inches and N_f is given as $0.57W$. Substituting these values into (2.9.3), we obtain $x = 48.6$ inches.

EXAMPLE 2.9.3
A bridge 100 feet long weighs 100 tons and is supported by two concrete piers, one at either end, as illustrated in Fig. 2.29. A truck weighing 20 tons is driven onto the bridge and parked so that its center of gravity is 25 feet from one of the end supports. What are the upward forces exerted on the ends of the bridge by its piers?

Since the system is in equilibrium, the sum of the moments about any point is zero. Taking moments about point A of Fig. 2.29, we find

$$(100)(50) + (25)(75) - 100N_2 = 0$$
$$N_2 = 68.75 \text{ tons}$$

Taking moments about B in the same way,

$$100N_1 - (100)(50) - (25)(25) = 0$$
$$N_1 = 56.25 \text{ tons}$$

Note that the sum of the two forces exerted by the piers exactly balances the 125-ton weight of bridge and truck. This problem could also have been worked by writing moments about the center of the bridge and an equation of translational equilibrium for vertical force components.

EXAMPLE 2.9.4
A uniform ladder of length L and weight W leans at an angle θ against a vertical wall. A man of weight w stands on the ladder at a distance d from the end resting on the floor, as shown in Fig. 2.30. There is no friction between ladder and wall, but the floor supplies enough frictional force to prevent slipping and thus maintains the system in equilibrium. Find the normal force **N** and the friction force $\mathbf{F}_f$ at the floor,

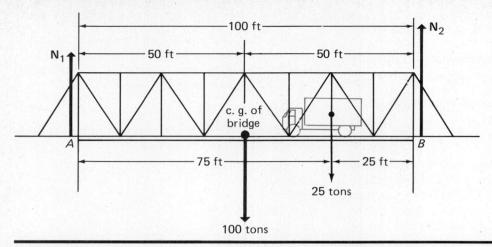

FIGURE 2.29

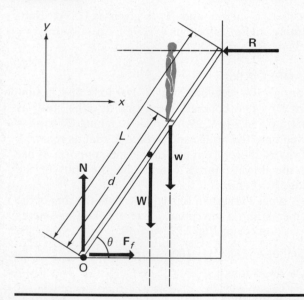

FIGURE 2.30

and the force **R** exerted by the wall on the ladder. If the coefficient of static friction between ladder and floor is μ_s, find the maximum value that w can attain before the ladder slips.

Referring to Fig. 2.30, it is clear that the force **R** at the wall must act normal to the wall, since no tangential frictional forces can arise there. We may write the following equations for the translational equilibrium of x- and y-components of force:

$$F_f = R \tag{2.9.4}$$

$$N - (w + W) = 0 \tag{2.9.5}$$

Also, taking moments about the point O, we obtain (recalling that the force **W** acts downward from the center of gravity of the ladder),

$$\frac{WL}{2} \cos \theta + wd \cos \theta - RL \sin \theta = 0 \tag{2.9.6}$$

From (2.9.5), we see at once that $N = w + W$, and from (2.9.6) and (2.9.4),

$$F_f = R = \left(\frac{W}{2} + \frac{wd}{L} \right) \cot \theta \tag{2.9.7}$$

When w is so large that the ladder is about to slip, $F_f = \mu_s N = \mu_s(w + W)$. Substituting this value for F_f into (2.9.7) and solving for w, we find that the value of w which brings this state of affairs to pass is

$$w = \frac{\frac{1}{2} \cot \theta - \mu_s}{\mu_s - \frac{d}{L} \cot \theta} W = \frac{\frac{1}{2} - \mu_s \tan \theta}{\mu_s \tan \theta - \frac{d}{L}} W \tag{2.9.8}$$

If μ_s is equal to or greater than $(d \cot \theta)/L$, the ladder will never slip, no matter how heavy the man is. If μ_s is equal to or less than $\frac{1}{2} \cot \theta$, the ladder will slip even without the man's added weight.

EXAMPLE 2.9.5

An object of the size and shape illustrated in Fig. 2.31 is supported by two "feet" at either end and rests upon a level floor. A force **T** is applied as shown. If the floor

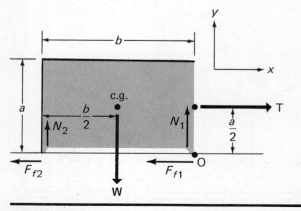

FIGURE 2.31

is very slippery, clearly the object will slide along in the x-direction if **T** is sufficient to overcome the frictional forces at the "feet." If the floor is very rough, however, and exerts large frictional forces on the feet, the object will tip over about an axis passing through the point O rather than slide. Find the maximum value for the coefficient of static friction μ_s between floor and feet that will permit the object to slide rather than tip over.

Writing the conditions for translational equilibrium for x- and y-components for forces, we may obtain

$$\sum_i F_{iy} = N_1 + N_2 - W = 0$$

whereby

$$N_1 + N_2 = W \tag{2.9.9}$$

and

$$\sum_i F_{ix} = T - F_{f1} - F_{f2} = 0$$

from which

$$T = F_{f1} + F_{f2} \tag{2.9.10}$$

If T is so large that sliding motion is imminent, then

$$\frac{F_{f1}}{N_1} = \frac{F_{f2}}{N_2} = \mu_s \tag{2.9.11}$$

Substituting this in (2.9.10) and recalling (2.9.9),

$$T = F_{f1} + F_{f2} = \mu_s N_1 + \mu_s N_2$$
$$= \mu_s(N_1 + N_2) = \mu_s W \tag{2.9.12}$$

Taking moments of forces about the point O, it is apparent that

$$N_2 b = W\left(\frac{b}{2}\right) - T\left(\frac{a}{2}\right)$$
$$N_2 = \frac{W}{2} - T\left(\frac{a}{2b}\right) \tag{2.9.13}$$

Substituting this result in (2.9.9) and solving for N_1, one finds

$$N_1 = \frac{W}{2} + T\left(\frac{a}{2b}\right) \tag{2.9.14}$$

If conditions are such that the body is about to tip over rather than slide, then N_2 (and hence F_{f2}) is equal to zero. Letting $N_2 = 0$ and substituting $\mu_s W$ for T from (2.9.12), in (2.9.13), one obtains

$$\frac{W}{2} = \mu_s W\left(\frac{a}{2b}\right)$$

Then, solving for μ_s, we find

$$\mu_s = \frac{b}{a} \tag{2.9.15}$$

If μ_s is smaller than this value, N_2 remains greater than zero, and the object slides. If μ_s is greater than this, N_2 vanishes before the total friction force reaches the critical value $\mu_s W$ and the body will tip over about the point O.

EXAMPLE 2.9.6

A boom of mass 200 kg makes an angle $\theta = 53°$ with a vertical support. Its upper end is fastened to this support by a horizontal cable. A load of 500 kilograms is supported by another cable hanging vertically from the end of the boom, as shown in Fig. 2.32. What is the tension in the horizontal supporting cable? What is the force exerted by the vertical supporting wall on

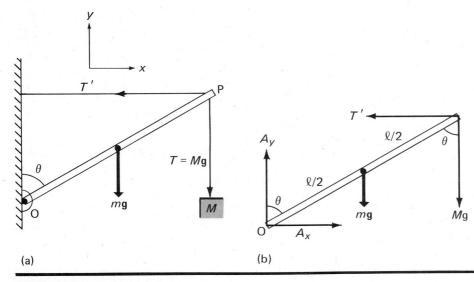

(a) (b)

FIGURE 2.32

the point where the boom is pivoted to the vertical support?

If we consider the boom as an isolated free body, it is clear that the forces it experiences will be those illustrated in Fig. 2.32b. Since the system is in equilibrium, the tension T in the vertical cable will be equal to the weight force Mg; the tension T' in the horizontal cable is, of course, unknown. The force due to the boom's own weight will be mg, where m is the mass of the boom, and it will act downward through the center of gravity of the boom, which we shall assume is at its center. The boom will also experience a force $\mathbf{A}$ exerted on it by the vertical support at the left. We know neither the magnitude nor the direction of this force, but we can specify it in terms of its horizontal and vertical components, A_x and A_y, as shown.

The equations for translational equilibrium then can be written

$$\sum F_{ix} = A_x - T' = 0 \tag{2.9.16}$$

and

$$\sum F_{iy} = A_y - mg - Mg = 0 \tag{2.9.17}$$

In these equations, we know mg and Mg, while A_x, A_y, and T' are to be determined. There is clearly no way in which these three quantities can all be found from only two equations. It is evident, then, that another relation between them must be found. This third relation can be obtained by observing that in spite of the fact that the forces acting on the boom are not all concurrent, the system is, nevertheless, in equilibrium with respect to rotation. This means that the resultant *torque* about any given axis must be zero. In this example, since the unknown forces A_x and A_y both pass through the point O, it is convenient to take torques about that point. Writing the sum of the torques about O arising from all the individual forces and equating it to zero, we find

$$\sum \tau_0 = T'(l \cos \theta) - Mg(l \sin \theta) - mg(\tfrac{1}{2}l \sin \theta) = 0 \tag{2.9.18}$$

where l is the length of the boom. This equation can be solved at once for T':

$$\begin{aligned} T' &= (M + \tfrac{1}{2}m)g \tan \theta \\ &= ((500) + (0.5)(200))(9.8)(\tan 53°) \\ &= 7800 \text{ N} \end{aligned} \tag{2.9.19}$$

From (2.9.16), now, it is apparent that

$$A_x = T' = 7800 \text{ N}$$

while from (2.9.17)

$$A_y = (m + M)g = (200 + 500)(9.8) = 6860 \text{ N}$$

The magnitude and direction of the total force $\mathbf{A}$

exerted by the support on the boom can now be determined from the components A_x and A_y, because

$$A = \sqrt{A_x{}^2 + A_y{}^2} = \sqrt{(7800)^2 + (6860)^2} = 10{,}390 \text{ N}$$

and because the angle ϕ between the direction of $\mathbf{A}$ and the x-axis is

$$\phi = \tan^{-1} \frac{A_y}{A_x} = \tan^{-1} \frac{6860}{7800} = 41.3°$$

2.10 Stable, Unstable, and Neutral Equilibrium

When an object is in equilibrium, the sum of all the external forces and external torques acting upon it is zero. In many instances, these external forces and torques are functions of the position or orientation of the object. When the object is in its equilibrium position, the forces and torques sum to zero; but if the body is *displaced* from its equilibrium position or orientation by some external means, some of the forces or torques may *change* as a result of this displacement, and they will add to some finite resultant torque or force rather than to zero. The direction of this resultant torque or force may be such as to cause the system to move back toward the position of equilibrium, or it may be such as to cause the system to move still further away from equilibrium. In the former case, the system will move back to the equilibrium state, overshoot it, and subsequently oscillate back and forth about that point. In the latter, it will move more and more rapidly *away* from the equilibrium point, coming to rest only after additional external forces are brought into play by some constraining influence such as a rigid container or support.

In the first instance above, the system is said to possess a point of *stable* equilibrium; in the second, the equilibrium point is said to be *unstable*. Examples of systems exhibiting stable equilibrium include a rocking chair, wherein a displacement from equilibrium sets up forces tending to restore the chair to its equilibrium state and make it oscillate about that point; a pendulum, wherein a displacement from the bottom of the swing sets up forces acting to restore equilibrium and eventually to make the system oscillate back and forth through the equilibrium position; and a ship, wherein displacement from equilibrium sets up torques that act to return the ship to an upright position and thereafter result in a rolling motion back and forth through that position. The most common examples of unstable equilibrium are objects balanced upon edges or corners, or objects which are in equilibrium at the top of a hill but slide or roll downward if displaced from the equilibrium point in any direction.

If all the external forces or torques that act are

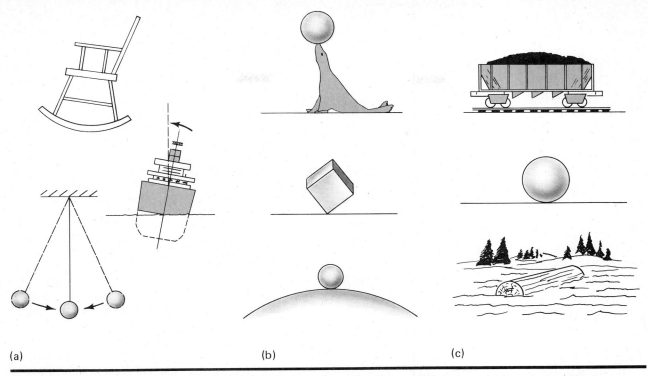

FIGURE 2.33. Illustrations exhibiting (a) stable equilibrium, (b) unstable equilibrium, and (c) neutral equilibrium.

independent of the position or orientation of the body, then, when the body is displaced in position or orientation, no changes in the forces and torques will result and the body will remain in equilibrium. A system of this sort is said to be in *neutral* equilibrium. Examples of neutral equilibrium are objects which may roll or slide upon flat surfaces. Some illustrations of stable, unstable, and neutral equilibrium are shown in Fig. 2.33.

SUMMARY

Classical mechanics is based upon Newton's laws of motion, which can be stated as follows:

First Law When the resultant or vector sum of all the forces acting on a body is zero, the body's acceleration is zero, and it remains either at rest, or in uniform motion with a velocity that is constant in magnitude and direction.

Second Law When the resultant or vector sum of all the forces acting on a body is not zero, the body acquires an acceleration in the direction of the resultant force. The magnitude of this acceleration is directly proportional to that of the resultant force, and inversely proportional to the body's mass.

Third Law For every action there is an equal and opposite reaction; thus, the force exerted by one body on another is equal in magnitude and opposite in direction to the force exerted by the second body on the first.

The "action" and "reaction" forces of Newton's third law always act on different bodies.

The mass of an object is a measure of its inertia and is an inherent physical property of the matter of which the object is made. The weight of an object is the gravitational force it experiences due to the gravitational attraction of the earth or any other body. The relationship between a body's weight W and its mass m is $W = mg$, where g is the acceleration acquired by a freely falling object under the influence of the local gravitational field.

Objects can experience torques as well as forces. If the vector $\mathbf{r}$ locates the point of application of a force $\mathbf{F}$ with respect to a given axis O, then $\mathbf{r} \times \mathbf{F}$ is the torque about that axis due to the force $\mathbf{F}$. The magnitude of the torque may also be expressed as the product of the force and its moment arm, defined as the perpendicular distance between the line of action of the force and the axis.

The condition that a body be in equilibrium is satisfied if the vector sum of all forces and all torques

53

acting on it is zero. This can also be expressed by the equations of equilibrium

$$\sum_i F_{ix} = \sum_i F_{iy} = \sum_i F_{iz} = \sum_i \tau_i = 0$$

In an extended rigid body, the weight force may be regarded as acting vertically downward from the center of gravity when writing the equations of translational and rotational equilibrium.

The kinetic friction force acts in the plane of a frictional interface to oppose the motion of the sliding object. Its magnitude is $\mu_k N$, where N is the normal force and μ_k is the coefficient of kinetic friction.

The static friction force acts also in the plane of the frictional interface, but its value adjusts itself to that which is necessary to maintain equilibrium. Its magnitude, however, can never exceed $\mu_s N$, where N is the normal force and μ_s is the coefficient of static friction.

QUESTIONS

1. Under some circumstances, a body of mass m is found to have no weight. Can the converse of this statement ever be true?
2. Are there any circumstances under which the action and reaction forces of Newton's third law act on the same body?
3. Two blocks of masses m_1 and m_2 sit on a table, block m_1 resting on block m_2. Explain very clearly why the force which block 1 exerts on block 2 is $m_1 g$.
4. Explain physically why we find no objects having negative mass.
5. Two objects accelerate as a result of the mutual forces exerted upon each other. If their masses are m_1 and m_2, show that $m_1/m_2 = a_2/a_1$.
6. Find your weight in newtons and also your mass in slugs.
7. When you open the door of a closet, you exert a force on the doorknob, and, according to Newton's third law, the doorknob exerts a force on you. Why, then, don't you accelerate and crash into the door?
8. A block is placed on a very steep inclined plane whose angle with the horizontal direction is very close to 90°. Show that even if there is a large coefficient of friction, the acceleration will be close to g.
9. When you walk forward, it must be the friction force that brings about your motion. Explain how it is possible for the friction force to be in the direction of your motion.
10. When a rocket ship gets further and further from its launch pad, it becomes easier to accelerate. Why is this true?
11. At a certain instant, a body known to be experiencing a force is not moving. Does this violate any of the laws of motion? Explain.
12. An ice cream cone has stable, unstable, and neutral equilibrium positions. Describe each of these with respect to a horizontal table.

PROBLEMS

1. Show that whenever three nonparallel coplanar forces maintain a rigid body in equilibrium, their lines of action must be concurrent, as illustrated in Fig. 2.5b.
2. Prove that a body acted upon by three distinct *noncoplanar* forces cannot be in equilibrium.
3. Show that 1 newton = 10^5 dynes; show that 1 pound = 4.445 newtons.
4. An object of mass 12 kg rests upon a frictionless horizontal surface. How large an acceleration will it undergo if a horizontal force of 25 newtons acts on it?
5. A man weighing 180 pounds jumps out of a plane with a parachute that weighs 25 pounds. After the parachute opens and things have settled down a bit, he is observed to be falling at a constant speed of 20 ft/sec. (a) Is the system of man plus parachute in equilibrium under these circumstances? (b) What is the upward force exerted on the man's body by the parachute harness, neglecting any force of air resistance acting directly on the man's body? (c) What is the downward force exerted on the parachute harness by the man's body? (d) What is the total upward force of air resistance that acts on the system of man plus parachute? (e) What is the total downward force of gravity acting on the system of man plus parachute? (f) Are the forces in (b) and (c) forces of action and reaction? (g) Are the forces in (d) and (e) an action-reaction pair?
6. A 5 kilogram block of steel rests upon a smooth plane level wooden surface. It is observed that a horizontal force of 26.5 newtons must be applied to the block in order to set it in motion, but that once it is in motion only 21.6 newtons of force is required to maintain it in motion at constant velocity. What are the static and kinetic coefficients of friction between the steel block and the plane?
7. The plane in the previous problem is tilted by raising one end. At what angle of tilt will the block start to slide? What angle of tilt is necessary to allow the block to slide down the plane at constant velocity?
8. A force of 1200 pounds is necessary to pull an empty railroad car weighing 40,000 pounds along a level track at constant speed. How much force is required to pull the same car along a level track when it is loaded with 100,000 pounds of coal?
9. How large a force, applied in a direction parallel to the track, is required to pull the railroad car in the preceding problem at constant speed along a track inclined at an angle of 1.0° to the horizontal? Along a track inclined at an angle of 30° to the horizontal?
10. A locomotive can exert a maximum tractive force of 140,000 pounds. How steep an incline can it climb at constant speed, neglecting any effect of rolling friction, while hauling a train of a total weight consisting of 3000 tons, including locomotive? Express your answer in terms of the angle of inclination of the track with the horizontal.
11. Eight men are engaged in a tug of war, as illustrated in the diagram. Each of them exerts a horizontal force of 200 pounds on the rope. (a) What is the tension T_1 in the central part of the rope? (b) What is the tension T_2 in

the segment of rope between the second and third members of the team on the right?

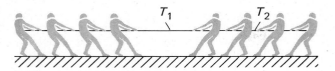

12. In the accompanying diagram, a 16-pound load of snow is seen resting on a snow shovel. Describe the magnitude, direction, and point of application of the following vectors: **(a)** the vector representing the force exerted by the earth on the snow; **(b)** the vector representing the force exerted by the snow on the earth; **(c)** the vector giving the force exerted by the shovel on the snow; **(d)** the vector showing the force exerted by the snow on the shovel. **(e)** If the force in **(c)** is increased in magnitude to 20 pounds, which, if any, of the other forces change? **(f)** Which of these forces can be identified as action–reaction pairs?

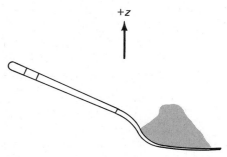

13. A weight W is supported by the arrangement of pulleys shown in the figure. What force F must be applied as shown to keep the system in equilibrium? What is the tension T in the rope securing the upper pulley to the overhead support under these circumstances?

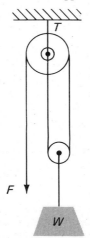

14. In the figure, block A has a mass of 8 kg and block B, a mass of 16 kg. The coefficient of kinetic friction between A and B, and also between B and the plane on which it slides, is 0.20. Find the force necessary to keep block B moving to the right at constant speed **(a)** when A rests on B and moves along with it and **(b)** when A is tied to the wall and slides on top of B.

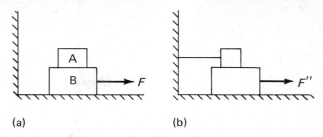

(a)   (b)

15. In the diagram, an 8-kg mass is pulled up an inclined plane at constant speed by a suspended 5-kg mass. The inclined plane is tilted at an angle of 30° to the horizontal, and there is no friction in the pulley. **(a)** What is the tension in the connecting cord? **(b)** What is the coefficient of kinetic friction between the 8-kg mass and the plane?

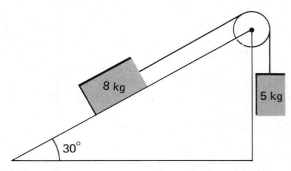

16. A 100-pound weight W_1 and a second weight W_2 are connected by a light flexible cord passing over a light frictionless pulley, and supported in equilibrium on a frictionless set of surfaces such as that shown in the figure with $\theta_1 = 30°$ and $\theta_2 = 60°$. What is the weight W_2?

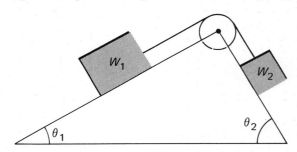

17. Suppose that W_1, θ_1 and θ_2 in the accompanying diagram have arbitrary values. Find the value of W_2 required to maintain the system in equilibrium. What are the normal forces exerted by the plane on W_1 and W_2?

18. Two weights are connected by a light, flexible cord passing over a light, frictionless pulley. One of the weights, which

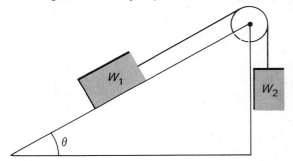

we shall refer to as W_1, is supported by a plane inclined at an angle θ to the horizontal, as shown in the figure. The static and kinetic coefficients of friction between the plane and the weight W_1 are 0.5 and 0.4, respectively. If $W_1 = 100$ lb and $\theta = 37°$, **(a)** what is the smallest value of W_2 which will prevent the weight W_1 from sliding down the plane? **(b)** What value must W_2 have in order that W_1 may slide down the plane with constant velocity? **(c)** What is the largest value which W_2 may have before W_1 is set in motion up the plane? **(d)** What value must W_2 have in order that W_1 may slide up the plane with constant velocity?

*19. What are the answers asked for in Problem 18 if W_1, μ_s, μ_k, and θ have arbitrary values?

20. In Problem 16, suppose that $W_1 = 100$ pounds, $\theta_1 = 37°$, and $\theta_2 = 53°$ and that the static and kinetic friction coefficients between both weights and the supporting surfaces are 0.5 and 0.4, respectively. Find **(a)** the smallest value of W_2 that will prevent W_1 from sliding down the plane; **(b)** the value for W_2 that will permit W_1 to slide down the plane with uniform speed; **(c)** the largest value that W_2 may have before W_1 is set in motion up the plane; **(d)** the value for W_2 that will permit W_1 to slide up the plane with uniform speed; and **(e)** the tension in the cord in each instance.

*21. What are the answers to the questions asked in Problem 20 if W_1, μ_s, μ_k, θ_1, and θ_2 have arbitrary values?

22. In the accompanying figure, find the tensions T_1 and T_2 in the supporting ropes.

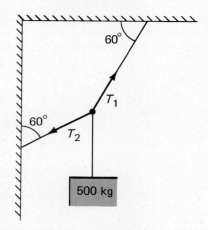

23. In the diagram, find the tensions in the three supporting ropes and the value of W required to maintain the system in equilibrium in this configuration.

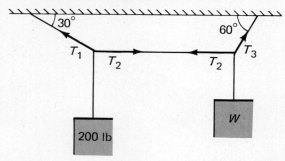

24. Suppose a 100-lb weight were substituted for the weight

W in the preceding problem. Discuss qualitatively what would happen to the system.

*25. A heavy, flexible chain of weight W is suspended between two vertical supports in such a way that the ends of the chain make an angle θ with the horizontal, as shown in the figure. Find **(a)** the magnitude of the forces exerted by the ends of the chain upon the supports and **(b)** the tension in the chain midway between the supports. **(c)** Is the tension in the chain the same at all points?

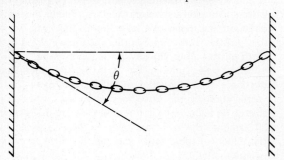

26. Find the horizontal force required to hold a suspended mass m at an angle θ to the vertical, as illustrated in the diagram. What is the tension in the rope connected to the overhead support? What are the magnitudes of the horizontal force and the tension if $m = 100$ kg and $\theta = 37°$?

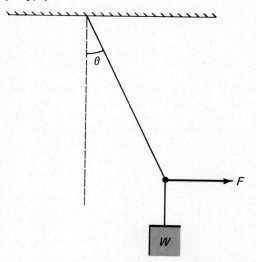

27. A uniform horizontal plank 6 m long is balanced at its center on a knife-edge support. A 50-kg mass is placed at the extreme left-hand end, and a 100-kg mass is placed halfway between the knife-edge and the 50-kg mass. Where must a 125-kg mass be located in order that the system be balanced on the knife-edge?

28. A uniform horizontal plank 6 m long carries a 285-kg mass attached to one end and a 100-kg mass attached to the other end. It is found to balance on a knife-edge support when the support is located 1.8 m from the end which carries the 285-kg mass. What is the mass of the plank itself?

29. A 4000-pound automobile has a wheelbase of 120 inches. Its center of gravity lies 50 inches behind the rotation axis of the front wheels. How much force do the front wheels exert on the road and how much force do the rear wheels exert when the car is at rest or in motion at constant speed?

30. A 50-pound door whose dimensions are given in the drawing is suspended at two hinges. The center of gravity of the door is at its geometric center P. **(a)** What are the horizontal forces that act at each hinge? **(b)** What is the sum of the vertical forces that act at the two hinges? **(c)** Is it possible, with the data given here, to find the individual vertical forces at the hinges? Explain. **(d)** A small child weighing 40 pounds swings on the doorknob; what are the horizontal forces that act at each hinge now?

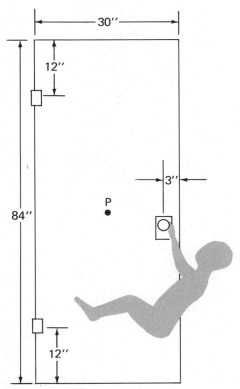

31. A uniform boom, 10.0 ft long shown in the diagram, weights 200 pounds and is hinged at A. It is held in place by a cable fixed at C, 6.0 ft from A, which makes an angle of 37° with the boom. A 500-pound weight hangs from the boom at a point 2.0 ft from A. Find **(a)** the tension in the supporting cable and **(b)** the magnitude and direction

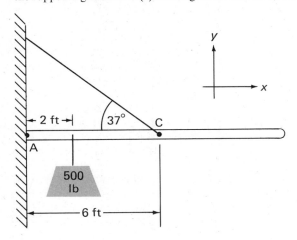

of the force exerted on the boom by the supporting wall at A.

32. A box of mass m is at rest on a rough inclined plane, which slopes at an angle θ to the horizontal. Attached to it by a light flexible cord is a cart of mass m' having frictionless wheels, as illustrated in the drawing. The coefficient of static friction between the box and the plane is μ_s. Find **(a)** the tension in the cord, **(b)** the friction force between the box and the plane, and **(c)** the smallest possible value for the mass m that will allow static equilibrium to be maintained.

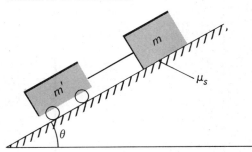

33. In the diagram, a uniform rigid bar in equilibrium makes an angle of 20° to a rough concrete floor. The bar is 3 meters long and has a mass of 90 kg. A force **P** acts at right angles to the bar at a point 0.6 meter from the left end. Find **(a)** the magnitude of force **P**, **(b)** the magnitude and direction of the force exerted by the floor on the bar, and **(c)** the least possible value of the coefficient of static friction between floor and bar that will permit equilibrium to be maintained.

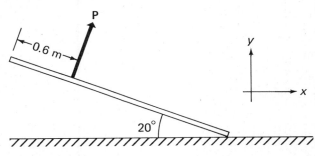

34. The crane shown in the diagram is pivoted at point A. Its boom has a mass of 100 kg and makes an angle of 37° with the horizontal. The rope BC is horizontal. A mass of 600 kg is suspended at D. The distance AD is

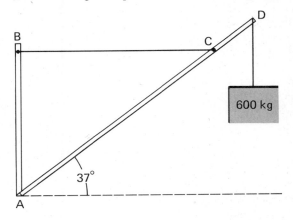

3.0 meters and distance CD is 1.0 meter. Find **(a)** the tension in the rope BC and **(b)** the horizontal and vertical components of the force exerted on the boom by the support at A.

35. A small, uniform drawbridge 10 meters long has a mass of 1200 kg. It is hinged at one side by a strong steel pin, as shown in the figure. The bridge is undergoing temporary repairs and is supported in equilibrium at an angle of 15° to the horizontal by a horizontal cable. Three men, each of mass 80 kg, are working on the end of the bridge. Find **(a)** the tension in the cable and **(b)** the horizontal and vertical components of the force exerted on the end of the bridge by the hinge pin.

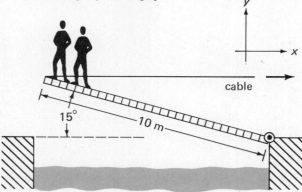

*36. A crane consists of a vertical mast of length l and a uniform boom of mass m and length $2l$, as shown in the figure. The angle θ can be varied by adjusting the length of the cable AB. Neglecting the weight of the cable, find the tension in the cable as a function of m, θ, and the load M.

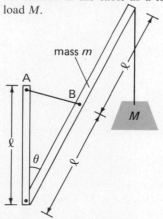

37. A hoist consists of a vertical center post of height a and a boom of length $3a$, as shown in the diagram. The boom

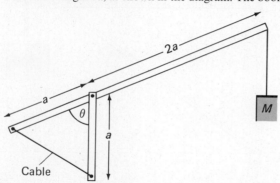

is raised and lowered by means of a cable of negligible mass, as shown. Neglecting the mass of the boom, find an expression for the tension in the cable as a function of the angle θ and the mass M which is suspended from the boom.

38. A rough inclined plane supports two blocks of mass m and m', as shown in the diagram. The coefficient of sliding friction between the mass m and the plane is μ_k, and the coefficient of sliding friction between mass m' and the plane is μ_k', where $\mu_k' > \mu_k$. At what angle θ must the plane be inclined so that the two blocks will slide down the plane, in contact with each other, at constant speed? What is the contact force between the two blocks in this situation?

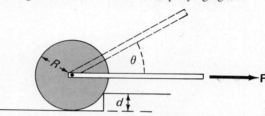

39. What is the minimum horizontal force needed to pull a lawn roller of radius R and mass m over a vertical step of height d, as shown in the accompanying figure?

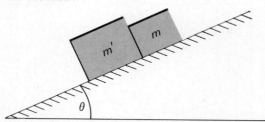

40. In the preceding problem, suppose the force is applied at an angle θ with the horizontal. Find **(a)** the force needed to pull the roller over the step, **(b)** the angle θ for which the force found in part **(a)** is a minimum, and **(c)** the angle between the roller handle and the line joining the axle and the step edge when the conditions of part **(b)** are satisfied.

*41. A stepladder is constructed as shown in the diagram. A 200-pound man stands on the ladder at point P. Ne-

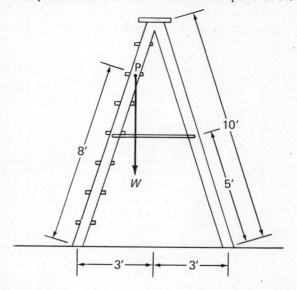

glecting the weight of the ladder itself, and assuming that it is resting upon a frictionless surface, find (a) the normal forces N_1 and N_2 at the left- and right-hand legs; (b) the tension T in the connecting rod which holds the two halves together; (c) the force components R_x and R_y which the hinge at the top exerts on the left half of the ladder. *Hint:* Write equations of equilibrium for each half of the ladder separately.

42. A 100-kilogram spherical ball lies in a groove as shown in the figure. Assuming that the contact between the groove wall and the ball is frictionless, calculate the forces which the walls exert upon the ball at A and B.

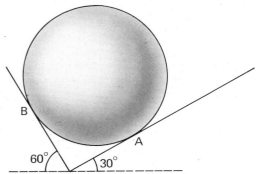

*43. An automobile of weight W rests on a road which is banked at an angle θ, as shown in the diagram. The center of gravity of the car is at point P, at a height h above the road, and the distance between the wheels is d. (a) Find the normal forces and frictional forces at both wheels. (b) Show that if the coefficient of static friction μ_s is less than $d/2h$, the car will slip before overturning if the angle θ is slowly increased, while the reverse is true if μ_s is greater than $d/2h$. *Hint:* With the information given above, it is not possible to determine unambiguously the distribution of frictional forces between the wheels on either side of the car! To overcome this difficulty make the physically reasonable assumption that $F_{f1}/N_1 = F_{f2}/N_2$.

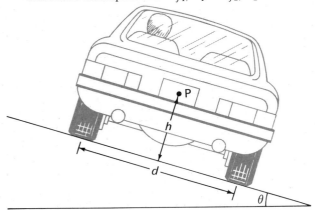

*44. A derrick boom of length l is supported by a cable perpendicular to it which is attached a distance d from the foot, as shown in the drawing. The boom is of weight W_0 and is of uniform diameter; a weight W is suspended from its end. It makes an angle θ with the horizontal. (a) Find the tension in the supporting cable. (b) Find the reaction force exerted on the foot of the boom by its support. (c) If $W = 3000$ lb, $W_0 = 1200$ lb, $l = 20$ ft, $d = 15$ ft, and $\theta = 30°$, what values do the quantities asked for in (a) and (b) assume?

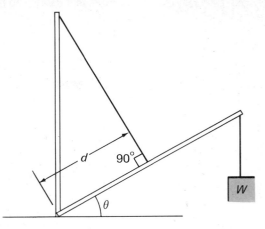

45. A wheelbarrow is illustrated in the accompanying diagram. The center of gravity of the wheelbarrow and its load is located at a point P, a distance d behind the axle. The total length of the handles is l. Let W represent the combined weight of the wheelbarrow and load. (a) What upward force must be applied to the handles to lift them? (b) What is the normal force exerted on the wheel by the road? (c) If the wheelbarrow weighs 50 lb and it carries a 400-lb load, and if $d = 2$ ft and $l = 6$ ft, what values do these quantities assume?

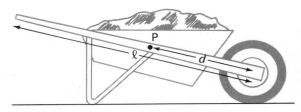

46. A rectangular block of weight W, a units high and b units long, is pulled at constant speed along a level surface by a horizontal force $\mathbf{T}$ acting at a height h as illustrated in the figure. The coefficient of sliding friction between the block and the supporting surface is μ_k. Find (a) the value of T required, (b) the distance x between the line of action of the normal force and the front of the block, (c) the maximum value for μ_k in order that the block may slide rather than tip over. (d) What are the values which the quantities referred to in (a) and (b) assume when $W = 100$ lb, $a = 3$ ft, $b = 4$ ft, $h = 2$ ft, $\mu_k = 0.2$? What is the maximum value for μ_k under these circumstances? Compare these results with those obtained in example 2.9.5.

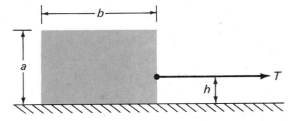

47. A 20-lb sign is attached to a 10-ft boom that rests against the side of a building as shown in the drawing. The end of the boom is supported by a rope making an angle of 30° with the horizontal. The weight of the boom is 10 lb and acts downward at its center. The sign is attached to the

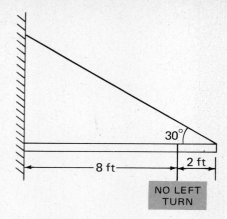

NO LEFT
TURN

boom at a point 8 ft from the building. Find **(a)** the tension in the supporting rope and **(b)** the horizontal and vertical components of the force exerted on the heel of the boom by the building.

48. Prove analytically that the center of gravity of a rectangular solid of uniform density is at its geometric center.

*49. A half-cylinder of uniform density and radius a is placed upon an inclined plane that makes an angle θ with the horizontal, as illustrated in the drawing. Find the angle ϕ that the plane upper surface of the half-cylinder makes with the horizontal in equilibrium, assuming that no slipping occurs at the line of contact between the inclined plane and the half-cylinder. You may assume that the center of gravity of the half-cylinder has been calculated to be on its axis of symmetry a distance of $4a/3\pi$ from the center, as shown.

50. A bench of mass 25 kg is 3.2 meters long and is supported by two "feet," one at each end. A person of mass 75 kg is sitting at a point 1.0 meter from the left end. Find the normal forces exerted by the ground on both of the feet.

51. A physicist owns a very large dog, whom he wishes to weigh, and a single bathroom scale. The dog is far too big to put on the single scale, so the physicist has the bright idea of weighing first the front feet, then the hind feet, and adding the two readings together to get the total weight of the dog. Can you state **(a)** whether this procedure will yield results that are exactly right, only approximately correct (say, within 5 or 10 percent), or altogether wrong? **(b)** the name of the dog.

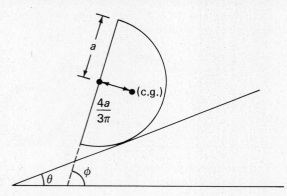

Bathroom scale

60

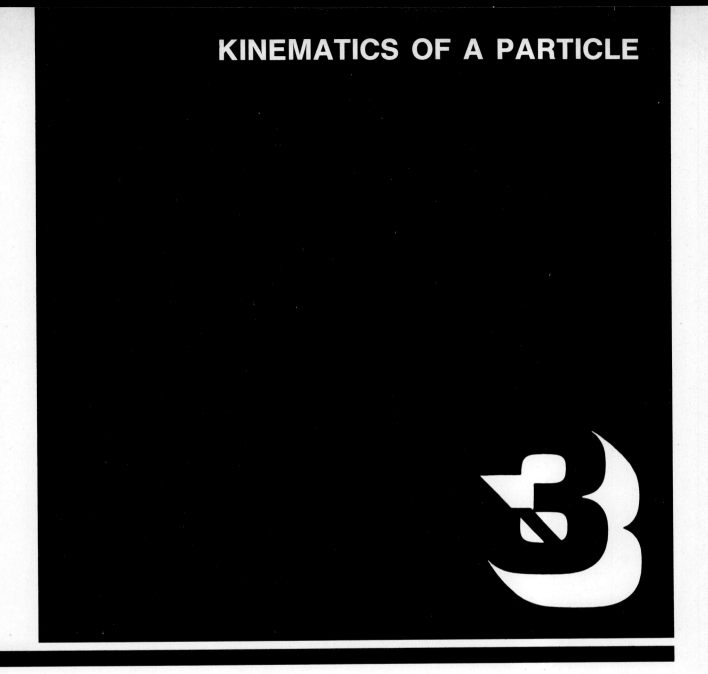

KINEMATICS OF A PARTICLE

3.1 Displacement, Velocity, and Acceleration

To understand how material objects move when acted upon by unbalanced external forces and torques, it is important to formulate exact physical and mathematical pictures of *displacement, velocity,* and *acceleration* and to understand the relationships between these three quantities. In doing this, we shall envision a system comprising three mutually orthogonal coordinate axes and a tiny point in motion which, in the course of time, traces out some sort of path in the coordinate space. We shall not concern ourselves at first with the physical forces that cause this motion nor with the relationship between these physical causes

and the resulting trajectory. We instead assume that an *equation of motion* is known, which can be solved to give explicit information about the particle's position, velocity, and acceleration at all times. We are concerned for the time being with the *geometric* aspects of the motion, the study of which is called *kinematics.* The relationship between the net forces and the ensuing motion—the branch of mechanics referred to as *dynamics*—will be considered in the next chapter.

Let us assume, initially, that the particle we are discussing is constrained in some way so as to move only along the x-axis. Its position at any time t can then be described by the distance x between the origin and the particle, and since a distinct value of x is associated with each value of time t, x is a function of t.

61

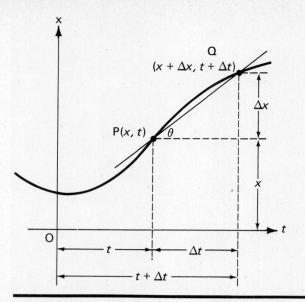

FIGURE 3.1. Displacement of an object moving along the x-axis plotted as a function of time. The quantity $\Delta x/\Delta t$ represents the average velocity over the time interval Δt, while the limit of this quantity as Δt approaches zero, which is the derivative dx/dt, represents the instantaneous velocity at time t.

Under these circumstances one may plot the displacement x versus time and obtain a graph such as that shown in Fig. 3.1. The *average velocity* $\overline{v}_x$ during some time interval Δt may be determined by finding the distance Δx through which the particle moves in that time interval and observing that

$$\overline{v}_x = \frac{\text{displacement}}{\text{time interval}} = \frac{\Delta x}{\Delta t} \qquad (3.1.1)$$

From Fig. 3.1, it is clear that $\overline{v}_x$ is the tangent of angle θ and thus represents the slope of the secant line PQ connecting the two points on the curve corresponding to time t and displacement x, and to time $t + \Delta t$ and displacement $x + \Delta x$. The *instantaneous* velocity v_x associated with some instant of time t and the corresponding displacement x may now be defined as the *limit* of $\overline{v}_x$ as the time interval Δt approaches zero. But this is just the definition of the derivative of x with respect to t, therefore,

$$v_x = \lim_{\Delta t \to 0} \frac{\Delta x}{\Delta t} = \frac{dx}{dt} \qquad (3.1.2)$$

The instantaneous velocity may be regarded as the slope of the tangent line to the curve of Fig. 3.1 at P. Clearly, as Δt and Δx approach zero in the limit, the slope of the secant line PQ will approach the slope of the tangent to the curve at P. In view of Eq. (3.1.2), we may regard the instantaneous velocity v_x as the

time rate of change of displacement. It is easily shown that if the instantaneous velocity is *constant*, then the average velocity over any interval of time is equal to the instantaneous velocity; the proof of this statement is assigned as an exercise. If the instantaneous velocity is not constant, then the average velocity will depend on the time interval chosen and will not, in general, equal the instantaneous velocity at the beginning or end of the interval.

We may also speak of the *average acceleration* $\overline{a}_x$ during some time interval as the change in instantaneous velocity Δv_x the particle experiences during the interval, divided by duration of the interval Δt, from which

$$\overline{a}_x = \frac{\text{change in velocity}}{\text{time interval}} = \frac{\Delta v_x}{\Delta t} \qquad (3.1.3)$$

As before, the *instantaneous acceleration* a_x associated with time t is regarded as the limit of $\overline{a}_x$ as the time interval Δt approaches zero, that is, as the derivative of v_x with respect to t or, in view of (3.1.2), as the second derivative of x with respect to t:

$$a_x = \lim_{\Delta t \to 0} \frac{\Delta v_x}{\Delta t} = \frac{dv_x}{dt} = \frac{d^2 x}{dt^2} \qquad (3.1.4)$$

We may, therefore, speak of the instantaneous acceleration as the *time rate of change of the instantaneous velocity.* If we were to plot the velocity v_x as a function of time (instead of the displacement as in Fig. 3.1), we would find that the slope dv_x/dt at any point would be equal to the instantaneous acceleration at the corresponding time. It is possible by using (3.1.2) and (3.1.4) to express the acceleration a_x in a slightly different way, which we shall often find most useful. By writing $dv_x/dt = (dv_x/dx)(dx/dt)$, which amounts to multiplying dv_x/dt by dx/dx (that is, unity), we may obtain

$$a_x = \frac{dv_x}{dt} = \frac{dv_x}{dx}\frac{dx}{dt} = v_x \frac{dv_x}{dx} \qquad (3.1.5)$$

We shall find this relation helpful whenever we are required to find displacement in terms of velocity, or vice versa. From this point onward, we shall have little use for average velocity or average acceleration, and, unless otherwise specified, the terms velocity or acceleration will refer to the instantaneous values of these quantities.

If the motion of the particle is no longer confined to the x-axis, it will trace out some curve or trajectory in space, as shown in Fig. 3.2. At time t, the particle will be at some point P whose space coordinates are (x, y, z); and at this time, its *displacement* with respect to the origin can be described by a *position vector* **r**

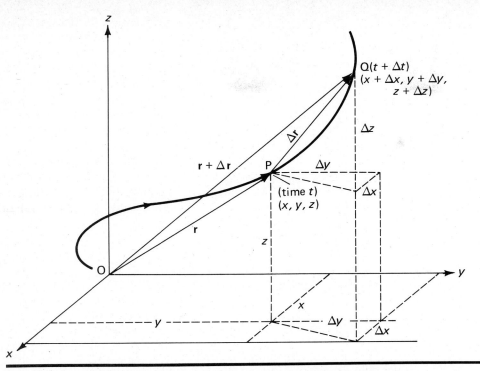

FIGURE 3.2. Displacement of a point moving along an arbitrary path in three-dimensional coordinate space. The vector $\Delta\mathbf{r}/\Delta t$ is the average velocity during the interval Δt, while the derivative $d\mathbf{r}/dt$, obtained in the limit $\Delta t \to 0$, represents the instantaneous vector velocity at time t.

whose x-, y-, and z-components are x, y, and z, respectively, as mentioned previously in connection with (1.5.14). The position vector $\mathbf{r}$ at time t is then

$$\mathbf{r} = x\mathbf{i}_x + y\mathbf{i}_y + z\mathbf{i}_z \qquad (3.1.6)$$

At a later time $t + \Delta t$, the particle will have moved along its path to some point Q, whose coordinates are $(x + \Delta x, \ y + \Delta y,$ and $z + \Delta z)$. The position vector $\mathbf{r} + \Delta\mathbf{r}$ associated with Q is

$$\mathbf{r} + \Delta\mathbf{r} = (x + \Delta x)\mathbf{i}_x + (y + \Delta y)\mathbf{i}_y + (z + \Delta z)\mathbf{i}_z \quad (3.1.7)$$

In analogy with (3.1.1), the average velocity can be written as the *vector* $\bar{\mathbf{v}} = \Delta\mathbf{r}/\Delta t$. Therefore, using (3.1.7) and (3.1.6), we may write

$$\bar{\mathbf{v}} = \frac{\Delta\mathbf{r}}{\Delta t} = \frac{(\mathbf{r} + \Delta\mathbf{r}) - \mathbf{r}}{\Delta t}$$

$$= \left(\frac{\Delta x}{\Delta t}\right)\mathbf{i}_x + \left(\frac{\Delta y}{\Delta t}\right)\mathbf{i}_y + \left(\frac{\Delta z}{\Delta t}\right)\mathbf{i}_z \qquad (3.1.8)$$

The instantaneous velocity $\mathbf{v}$ is now defined as a vector expressing the limiting value of $\bar{\mathbf{v}}$ as Δt approaches zero, whereby

$$\mathbf{v} = \lim_{\Delta t \to 0} \frac{\Delta\mathbf{r}}{\Delta t}$$

$$= \mathbf{i}_x \left(\lim_{\Delta t \to 0} \frac{\Delta x}{\Delta t}\right) + \mathbf{i}_y \left(\lim_{\Delta t \to 0} \frac{\Delta v}{\Delta t}\right) + \mathbf{i}_z \left(\lim_{\Delta t \to 0} \frac{\Delta z}{\Delta t}\right)$$

or

$$\mathbf{v} = \frac{d\mathbf{r}}{dt} = \left(\frac{dx}{dt}\right)\mathbf{i}_x + \left(\frac{dy}{dt}\right)\mathbf{i}_y + \left(\frac{dz}{dt}\right)\mathbf{i}_z \qquad (3.1.9)$$

The instantaneous velocity, then, is a vector whose x-, y-, and z-components are

$$
\begin{aligned}
v_x &= \frac{dx}{dt} \\[1mm]
v_y &= \frac{dy}{dt} \\[1mm]
v_z &= \frac{dz}{dt}
\end{aligned}
\qquad (3.1.10)
$$

The direction of this vector is the limiting direction of the vector $\Delta\mathbf{r}$ as $\Delta t \to 0$, hence, as Q slides along the curve toward P. It is apparent from Fig. 3.2 that in this limit, the direction $\Delta\mathbf{r}$ is the direction of the tangent to the path at P. Hence, the direction of $\mathbf{v}$ is also the direction of the tangent to the path at P. The magnitude of $\mathbf{v}$ is, of course, given by

$$v = \sqrt{v_x{}^2 + v_y{}^2 + v_z{}^2} \qquad (3.1.11)$$

We may now use precisely the same method in discussing the acceleration. The velocity vector **v** at time t is

$$\mathbf{v} = v_x\mathbf{i}_x + v_y\mathbf{i}_y + v_z\mathbf{i}_z \qquad (3.1.12)$$

with v_x, v_y, and v_z given by (3.1.10), while at time $t + \Delta t$, the velocity will be

$$\mathbf{v} + \Delta\mathbf{v} = (v_x + \Delta v_x)\mathbf{i}_x + (v_y + \Delta v_y)\mathbf{i}_y + (v_z + \Delta v_z)\mathbf{i}_z \qquad (3.1.13)$$

The average acceleration $\bar{\mathbf{a}}$ over the time interval Δt is $\Delta\mathbf{v}/\Delta t$, whereby

$$\bar{\mathbf{a}} = \frac{\Delta\mathbf{v}}{\Delta t} = \frac{(\mathbf{v} + \Delta\mathbf{v}) - \mathbf{v}}{\Delta t}$$

$$= \left(\frac{\Delta v_x}{\Delta t}\right)\mathbf{i}_x + \left(\frac{\Delta v_y}{\Delta t}\right)\mathbf{i}_y + \left(\frac{\Delta v_z}{\Delta t}\right)\mathbf{i}_z \qquad (3.1.14)$$

The instantaneous acceleration at time t is obtained by evaluating the average acceleration in the limit as $\Delta t \to 0$. As in (3.1.9), the quantities $\Delta v_x/\Delta t$, $\Delta v_y/\Delta t$, etc., become derivatives in this limit, the final result being

$$\mathbf{a} = \frac{d\mathbf{v}}{dt} = \left(\frac{dv_x}{dt}\right)\mathbf{i}_x + \left(\frac{dv_y}{dt}\right)\mathbf{i}_y + \left(\frac{dv_z}{dt}\right)\mathbf{i}_z \qquad (3.1.15)$$

The instantaneous acceleration **a** is a *vector* whose components are

$$
\begin{aligned}
a_x &= \frac{dv_x}{dt} = \frac{d^2x}{dt^2} \\[4pt]
a_y &= \frac{dv_y}{dt} = \frac{d^2y}{dt^2} \\[4pt]
a_z &= \frac{dv_z}{dt} = \frac{d^2z}{dt^2}
\end{aligned}
\qquad (3.1.16)
$$

The direction of the acceleration vector is the direction of the vector $d\mathbf{v}$ representing the change in velocity in an infinitesimal time interval. This vector *need not be in the same direction as the velocity vector* **v** and, in fact, is generally not in that direction. We shall soon run into a very simple example wherein the acceleration vector is *perpendicular* to the velocity! The magnitude of the acceleration vector is, as always, given by

$$a = \sqrt{a_x{}^2 + a_y{}^2 + a_z{}^2} \qquad (3.1.17)$$

As before, using the same algebraic argument as we used in deriving (3.1.5), we may show that the components of acceleration may be written in the alternative

form

$$
\begin{aligned}
a_x &= v_x\left(\frac{dv_x}{dx}\right) \\[4pt]
a_y &= v_y\left(\frac{dv_y}{dy}\right) \\[4pt]
a_z &= v_z\left(\frac{dv_z}{dz}\right)
\end{aligned}
\qquad (3.1.18)
$$

which we shall find very useful later on.

In working out actual problems in the dynamics of physical systems, we shall ordinarily be able to state from the laws of motion, in the form of a set of *equations of motion*, what the values of the *accleration* components a_x, a_y, and a_z must be. The velocity components may then be obtained by integration, since from (3.1.16)

$$
\begin{array}{lll}
dv_x = a_x\,dt & & v_x = \displaystyle\int a_x\,dt \\[6pt]
dv_y = a_y\,dt & \text{from which} & v_y = \displaystyle\int a_y\,dt \\[6pt]
dv_z = a_z\,dt & & v_z = \displaystyle\int a_z\,dt
\end{array}
\qquad (3.1.19)
$$

In evaluating the integrals, a constant of integration is obtained, which cannot be evaluated *unless the value of the velocity at some specific time is known beforehand.* Often, we shall find that the initial velocity (at $t = 0$) is known or can be specified. In order to work out the precise values of the velocity at all times, then, it is necessary to know (in addition to the equations of motion giving the acceleration) something about the velocity at some particular time or place. This subsidiary information is called a *boundary condition*. When the velocity components have been evaluated from the given acceleration, with the aid of a suitable boundary condition, the displacement of the particle can be found by integrating once again. From (3.1.10), we may write

$$
\begin{array}{lll}
dx = v_x\,dt & & x = \displaystyle\int v_x\,dt \\[6pt]
dy = v_y\,dt & \text{whence} & y = \displaystyle\int v_y\,dt \\[6pt]
dz = v_z\,dt & & z = \displaystyle\int v_z\,dt
\end{array}
\qquad (3.1.20)
$$

Once more, a constant of integration arises which cannot be evaluated without some additional knowledge—another boundary condition, this time specifying the *location* of the particle at some particular instant of time. The techniques involved in the integration of equations of motion and use of boundary conditions will be explored in detail in discussing the examples presented in the next section.

3.2 Motion in One Dimension with Constant Acceleration: Freely Falling Bodies

The simplest conceivable motion is one in which no acceleration at all is involved. In this case, $a_x = a_y = a_z = 0$, which, according to (3.1.10), means that v_x, v_y, and v_z are all *constant* in time. Performing the integrations in (3.1.20) regarding v_x, v_y, and v_z as constants, we may easily see that

$$x = v_x t + c_1$$
$$y = v_y t + c_2 \qquad (3.2.1)$$
$$z = v_z t + c_3$$

where c_1, c_2, and c_3 are constants of integration. The correctness of these equations is easily demonstrated by differentiating them with respect to time, regarding the velocity components as constants, to obtain once again the expressions (3.1.20). The integration constants cannot be evaluated unless we are given some boundary condition that tells us where the particle is at some given time. Let us suppose for simplicity that at $t = 0$, the particle starts from the origin. Substituting $t = 0$, $x = 0$, $y = 0$, and $z = 0$ into (3.2.1), we see that in this case $c_1 = c_2 = c_3 = 0$, from which

$$x = v_x t$$
$$y = v_y t \qquad (3.2.2)$$
$$z = v_z t$$

These are the equations of a straight line passing through the origin. The distance between the particle and the origin, where it started at $t = 0$, is given by

$$r = \sqrt{x^2 + y^2 + z^2} = t\sqrt{v_x^2 + v_y^2 + v_z^2} = vt \qquad (3.2.3)$$

where $v = \sqrt{v_x^2 + v_y^2 + v_z^2}$ is the magnitude of the velocity vector, sometimes referred to as the *speed*. These results are in accord with our intuitive expectations. This case is so simple that it is really unnecessary to use the formal methods of the previous section to treat it, but it is discussed here to assure the reader that these methods do provide what he expects in this elementary situation.

Let us now turn our attention to a more difficult example, wherein a particle constrained to move along the x-axis is subjected to a *constant acceleration* a_x. Since the particle is confined to the x-axis, $y = z = 0$ at all times, so that $v_y = v_z = 0$ and $a_y = a_z = 0$. We need, therefore, concern ourselves only with the equations involving the x-components of velocity and acceleration. We may obtain the velocity v_x by integration according to (3.1.19), regarding a_x as constant. This can be accomplished in two slightly different ways, regarding the integrals to be evaluated either as *indefinite integrals* with accompanying constants of integration or as *definite integrals* to be evaluated between appropriately chosen lower and upper limits.

In the first method, we are asked in (3.1.19) to evaluate the integral

$$v_x = \int a_x \, dt$$

where a_x is a constant quantity independent of time. Since a constant can always be written outside the integral sign, this can be expressed as the *indefinite integral*

$$v_x = a_x \int dt = a_x t + c_1 \qquad (3.2.4)$$

where c_1 is the constant of integration that inevitably arises when an indefinite integral is calculated. The correctness of this result is easily established by differentiating, which gives us once again $dv_x = a_x \, dt$, the equation we started with. In doing this, we are simply establishing that $a_x t + (\text{const.})$ is the "antiderivative" of the constant quantity a_x.

We are now left with the task of evaluating the integration constant c_1. This can be done only if we are given a boundary condition that specifies the value of v_x at some particular time. Let us suppose that we are told that the particle started off from the origin at $t = 0$ with an initial velocity v_{0x}. Then substituting $v_x = v_{0x}$ and $t = 0$ into (3.2.4), we find $c_1 = v_{0x}$, which allows us to write (3.2.4) as

$$v_x = a_x t + v_{0x} \qquad (3.2.5)$$

The displacement x may be found by integrating once more, according to (3.1.20), whence (noting that both a_x and v_{0x} are constants)

$$x = \int v_x \, dt = \int (a_x t + v_{0x}) \, dt = \tfrac{1}{2}a_x t^2 + v_{0x} t + c_2 \qquad (3.2.6)$$

In this expression, the constant of integration c_2 may be found by using the boundary condition, mentioned above in connection with (3.2.4), that at $t = 0$ the particle is at the origin. Substituting $t = 0$ and $x = 0$ in (3.2.6), it is readily seen that $c_2 = 0$ and that (3.2.6) can then be written as

$$x = \tfrac{1}{2}a_x t^2 + v_{0x} t \qquad (3.2.7)$$

As before, the quantity $(a_x t^2/2) + v_{0x}t + c_2$ is easily recognized as the "antiderivative" of $a_x t + v_{0x}$.

In these calculations, the integrals that have arisen have been treated as indefinite integrals, upon whose integration a constant of integration has appeared. It is also possible to treat the integrals as *definite integrals*, and if the limits of these definite integrals are assigned properly in agreement with the given boundary conditions, exactly the same results are obtained, without necessity for introducing integration constants.[1] For example, we may begin by noting from (3.1.19) that

$$dv_x = a_x \, dt \qquad (3.2.8)$$

where a_x is constant. Now let us integrate both sides of the equation, between the limits $t = 0$ (corresponding to the time when the particle has velocity v_{0x}) and time t (when the corresponding velocity is v_x). Accordingly, we find

$$\int_{v_{0x}}^{v_x} dv_x = a_x \int_0^t dt \qquad (3.2.9)$$

Integrating, evaluating the definite integrals, and solving for v_x, we may obtain (3.2.5). Furthermore, one may continue as follows by writing from Eqs. (3.1.20) and (3.2.5):

$$dx = v_x \, dt = (a_x t + v_{0x}) \, dt$$

Again, integrating both sides between the limits $t = 0$ (corresponding to the time when, according to the boundary conditions, the particle is at $x = 0$) and time t (when the particle has some displacement x), we obtain

$$\int_0^x dx = \int_0^t (a_x t + v_{0x}) \, dt$$
$$= a_x \int_0^t t \, dt + v_{0x} \int_0^t dt \qquad (3.2.10)$$
$$x = \tfrac{1}{2} a_x t^2 + v_{0x} t$$

in agreement with (3.2.7). When using definite integrals in this way, one must be sure that the values chosen for the lower and upper limits are *corresponding* values assumed by the respective variables. For example, in (3.2.9), the lower limit on the left-hand side must be the value that v acquires when t has the value given by the lower limit on the right-hand side, and likewise for the upper limits.

Equations (3.2.5) and (3.2.7) allow us to determine the velocity and displacement at any given time t. But suppose we are asked what the velocity is when the displacement x attains some given value? It is, of course, possible to find the answer by substituting that value of x into Eq. (3.2.5), solving for t, and substituting the value of t thus obtained into (3.2.7). A much neater and more general way of handling this situation can be found by noting from (3.1.18) that

$$v_x \, dv_x = a_x \, dx \qquad (3.2.11)$$

and integrating both sides of this equation between limits $x = 0$ (corresponding to $v_x = v_{0x}$) and a displacement x (corresponding to velocity v_x), regarding

a_x as constant. We find

$$\int_{v_{0x}}^{v_x} v_x \, dv_x = a_x \int_0^x dx$$

The integrals are easily evaluated to give

$$\left[\tfrac{1}{2} v_x{}^2\right]_{v_{0x}}^{v_x} = a_x [x]_0^x \qquad (3.2.12)$$

or

$$\boxed{v_x{}^2 - v_{0x}{}^2 = 2a_x x} \qquad (3.2.13)$$

Equation (3.2.13) relates the velocity v_x to the displacement x and is, therefore, just what is needed. This equation may also be obtained, incidentally, by eliminating the time between Eqs. (3.2.5) and (3.2.7).

Equations (3.2.5), (3.2.7), and (3.2.13) can be regarded as the "three golden rules of uniformly accelerated linear motion." *Any problem or example involving this type of motion can be worked out by the systematic use of these equations, without the necessity of writing out and integrating the equations of motion.* The techniques for using these rules will be illustrated by the series of examples that follows this section and the following one. Although these "golden rules" allow us to work problems involving uniformly accelerated motion very easily, it is important to realize that *not all motion is uniformly accelerated* and that cases in which the acceleration varies throughout the motion are very common also. Such systems cannot be treated by the use of the "golden rules"; we must instead write down the equations of motion and integrate them, regarding the acceleration as a *variable* quantity. So before you "write down the formula and plug in the numbers," ask yourself the question, "Is this an example of uniformly accelerated motion, where the golden rules are applicable, or can the acceleration vary during the course of the motion?"

The motion of a body falling freely under the influence of gravity is an example of one-dimensional motion under constant acceleration. As shown in the preceding chapter, we may represent the acceleration of gravity for all bodies near the earth's surface as a constant vector **g** pointing vertically downward toward the center of the earth. Its magnitude g has the value 32.16 ft/sec², 980 cm/sec², or 9.80 m/sec² in the three most commonly used systems of units.[2]

Suppose we orient a system of coordinate axes in such a way that the x- and y-axes are parallel to the earth's surface and the z-axis is pointing vertically upward. Then the motion of a freely falling object released somewhere along the z-axis will be a one-dimensional motion with constant acceleration; the body's motion will be confined to the z-axis so long

---

[1] In the definite integral, the integration constant, if included, has the same value at the upper and lower limits of the integral and vanishes when the upper and lower limit values of the integral are subtracted.

[2] In most of the examples that are worked out in this book, the approximate value $g = 32$ ft/sec² is used rather than the more precise value of 32.16 ft/sec² in order to simplify the numerical work.

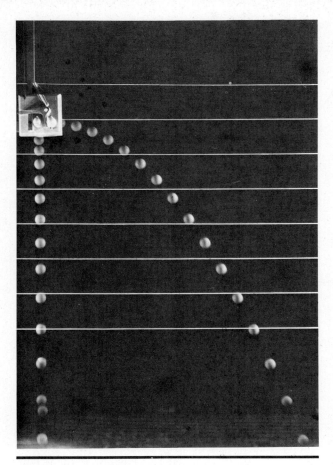

FIGURE 3.3. Multiflash photograph illustrating uniformly accelerated motion. The object on the left falls freely from rest, while the object on the right is given an initial horizontally directed velocity. In this case, the body follows the parabolic path characteristic of the "projectile motion" discussed in section 3.3. (From PSSC Physics, D. C. Heath and Company, Lexington, Mass., 1965)

as its initial velocity has no x- or y-component. Equations (3.2.5), (3.2.7), and (3.2.13) may then be used to describe its motion, provided that it is understood that the coordinate x (including subscripts) is replaced throughout by z. The acceleration vector has magnitude g and points downward along the $-z$-direction. The components of acceleration are, therefore, $a_x = a_y = 0$, $a_z = -g$. Substituting z for x in (3.2.5), (3.2.7), and (3.2.13) and letting $a_z = -g$, we may write these equations as

$$v_z = v_{0z} - gt \tag{3.2.14}$$

$$z = v_{0z}t - \tfrac{1}{2}gt^2 \tag{3.2.15}$$

$$v_{0z}{}^2 - v_z{}^2 = 2gz \tag{3.2.16}$$

In these expressions, v_{0z} refers to the initial velocity imparted to the body, along the z-direction. It is assumed that no initial x- or y-components of velocity

exist. One should also remember in using these equations that the system of coordinates is always arranged such that the object is at the origin when $t = 0$. The examples discussed below will serve to illustrate the way in which these expressions may be applied to specific problems, while the physical character of the motion is illustrated in Fig. 3.3. Equations (3.2.14), (3.2.15), and (3.2.16) will be recognized as the "golden rules" of uniformly accelerated motion for the special case in which $a_z = -g$.

EXAMPLE 3.2.1

A body, initially at rest, is dropped and allowed to fall freely. Describe its motion.

In this example, when the body is released, its velocity is zero. If we assume that the initial position of the object is at the origin, we may use the results expressed in Eq. (3.2.15) to discuss this problem simply by setting the initial velocity v_{0z} equal to zero. We then obtain

$$v_z = -gt$$
$$z = -\tfrac{1}{2}gt^2 \tag{3.2.17}$$
$$v_z{}^2 = -2gz$$

These equations allow one to express the position z and velocity v_z at the end of each second by setting $t = 0, 1, 2, 3, \ldots$ seconds. The resulting data expressed in tabular form, using $g = 32$ ft/sec^2, are

t, sec	z, ft	v_z, ft/sec
0	0	0
1	-16	-32
2	-64	-64
3	-144	-96
4	-256	-128
5	-400	-160
6	-576	-192

The minus signs before the values of z and v_z given above reflect the fact that the velocity vector is in the $-z$ direction and that the particle falls from the origin through positions whose z-coordinates are negative if the positive z-axis is pointing upward. Graphs of v_z and z versus time and v_z versus z are shown in Fig. 3.4. Another representation of the motion is given by the multiple-flash photograph shown in Fig. 3.3 above.

EXAMPLE 3.2.2

A baseball pitcher can impart a maximum initial velocity v_0 to a baseball. How high can he throw the ball vertically? If $v_0 = 100$ ft/sec, what maximum height does the ball attain?

In this case $v_{0z} = v_0$ in (3.2.16), while at the top of its path, the velocity v_z of the ball is zero. Substituting these values into the third equation of (3.2.17),

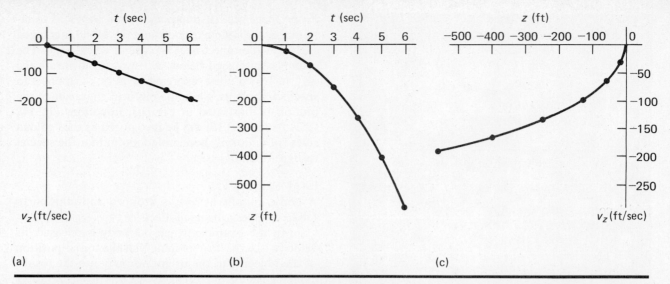

FIGURE 3.4. Plots of (a) velocity vs. time, (b) displacement vs. time, and (c) velocity vs. displacement in uniformly accelerated motion.

we obtain

$$v_{0z}^2 = 2gz$$

from which

$$z = v_0^2/2g \qquad (3.2.18)$$

This is the value of the z-coordinate of the ball at the time when its velocity v_z is zero, and hence its maximum height, assuming, of course, that the ball starts out at the origin at $t = 0$. If $v_0 = 100$ ft/sec and $g = 32$ ft/sec^2, $z = 156$ ft.

EXAMPLE 3.2.3

A truck moves at a constant speed of 40 mph along a two-lane highway. It is followed by a car, which is trying to pass. At full power, the car is capable of maintaining a constant acceleration of 5.0 ft/sec^2. The truck is 60 ft long, and 40 ft of additional "clearance" is needed for a safe passing maneuver. How long will it take the car to pass the truck? How far does the car travel in passing the truck and returning to its own lane? A second car, traveling at a constant speed of 55 mph, is approaching the first car and the truck from the opposite direction. As the first car moves out to pass the truck, it is 800 feet from the second car. Assuming that the second car does not brake, will the two cars collide, or will the passing maneuver be safely completed? If a collision takes place, find when and where it occurs. (Note that 60 mph = 88 ft/sec.)

Let us choose as an origin the position of the first car as it starts to pass the truck and measure all displacements with respect to that point. We shall represent the displacement, velocity, and acceleration of the first car by x, v, and a, respectively, while the displacement, velocity, and acceleration of the truck will be denoted by x', v', and a'. If the car begins to pass

the truck at $t = 0$, its displacement at time t is given by (3.2.7) as

$$x = v_0 t + \tfrac{1}{2} a t^2 \qquad (3.2.19)$$

where v_0 represents the initial speed of 40 mph. The truck's displacement, x', can be described by a similar equation. In writing this equation, however, it must be observed that the truck, at $t = 0$, is not at the origin but at a certain distance x_0' ahead of it. In writing Eq. (3.2.7), it is assumed that the object starts out at the origin. If it starts at some point whose x-coordinate is x_0', its displacement will be greater than that indicated by (3.2.7) by the initial displacement x_0'. Therefore, we must write for the truck

$$x' = x_0' + v_0' t + \tfrac{1}{2} a' t^2 \qquad (3.2.20)$$

Since the truck moves at constant speed, however, its acceleration a' is zero, and since the initial speeds of the car and truck are the same, the initial velocity v_0' in the above equation can be set equal to v_0. Equation (3.2.20) then becomes

$$x' = x_0' + v_0 t \qquad (3.2.21)$$

In passing the truck, the car will travel $x - 0$ feet, while the truck will travel $x' - x_0'$ feet. If the passing maneuver is to be safely executed, the car must travel the same distance as the truck, plus the truck's length (60 ft), plus 40 ft of "clearance" allowing it to get safely in and out of the passing lane. Therefore, for a safe pass, the distance x given by Eq. (3.2.19) must exceed the distance $x' - x_0'$ given by (3.2.21) by 100 ft, or

$$x = (x' - x_0') + 100 \text{ ft}$$
$$v_0 t + \tfrac{1}{2} a t^2 = v_0 t + 100 \text{ ft}$$
$$\tfrac{1}{2} a t^2 = 100 \text{ ft}$$

From this, we obtain immediately $t = 6.32$ sec as the time required for a safe pass, using the given value $a = 5.0$ ft/sec². The distance the car must cover in the process is found from (3.2.19) as

$$x = v_0 t + \tfrac{1}{2}at^2$$
$$= (58.67)(6.32) + (\tfrac{1}{2})(5.0)(6.32)^2$$
$$= 471 \text{ ft}$$

using $v_0 = 58.67$ ft/sec $= 40$ mph. The car's final speed v can be determined from (3.2.5) to be

$$v = v_0 + at = (58.67) + (5.0)(6.32) = 90.27 \text{ ft/sec}$$

which corresponds to 61.6 mph. This answer can also be obtained from (3.2.13) using $x = 471$ ft as determined above.

The second car, at $t = 0$, is 800 ft from the first one and approaches at a constant speed of 55 mph, or 80.67 ft/sec. The first car, in passing the truck, has to cover 471 ft in 6.32 sec. If, during that time, the second car travels further than $800 - 471 = 329$ ft, there will be a collision. But since, at 80.67 ft/sec, the second car will go some 510 ft in 6.32 sec, there is no doubt that a collision is in the offing!

The question as to where and when it will occur is easily settled. In analogy with (3.2.20), we can write the displacement x'' of the second vehicle from the origin as

$$x'' = x_0'' + v_0''t + \tfrac{1}{2}a''t^2 \qquad (3.2.22)$$

in which $x_0'' = 800$ ft is its initial displacement, $v_0'' = -80.67$ ft/sec its initial velocity, and a'' its acceleration. Since the initial velocity is directed along the $-x$-axis, the quantity v_0'' has a negative numerical value. Also, since the second car moves with constant speed, the acceleration a'' in (3.2.22) will be zero. At the instant of collision, the displacements x and x'' will be the same. Equating (3.2.20) and (3.2.22) and inserting the proper numerical values for the known quantities, we obtain

$$x_0'' + v_0''t + \tfrac{1}{2}a''t^2 = v_0 t + \tfrac{1}{2}at^2$$
$$800 + (-80.67)t + (\tfrac{1}{2})(0)t^2 = (58.67)t + (\tfrac{1}{2})(5.0)t^2$$
$$\qquad (3.2.23)$$

or, finally

$$2.50t^2 + 139.34t - 800 = 0$$

This quadratic equation can be solved for t using the quadratic formula to obtain

$$t = \frac{-139.34 \pm \sqrt{19{,}416 + 8{,}000}}{5.00} = \frac{-139.34 \pm 165.58}{5.00}$$

Like all quadratic equations, this one has two solutions. One of them corresponds to the physical conditions of the problem and, therefore, represents the answer we are seeking. The other satisfies the equation mathematically but violates the physical conditions of the problem. We must choose the one that is physically reasonable and reject the other as an extraneous mathematical possibility.

In this case, we know that any collision that takes place must occur at some time *after* $t = 0$, since the first vehicle does not move out to pass until then. Therefore, the solution that satisfies the physical circumstances of the problem must correspond to some positive value of the time. In this case, our mathematics gives us one positive answer and one that is negative, corresponding to $t < 0$. The positive solution, corresponding to $t = (-139.34 + 165.58)/5.00 = 5.247$ sec must, therefore, be chosen and the other rejected as an extraneous "unphysical" solution.

Using this value for t in (3.2.19), it is ascertained that the collision occurs at $x = 376.6$ ft, nearly 100 ft short of the 471-ft distance needed for safe passing. It is also a simple matter to show from (3.2.5) that the speed of the first car at the time of collision is 84.9 ft/sec, or 57.9 mph. The student should be able to show that, under the circumstances of this problem, a safe pass can be made only when the second vehicle is initially at a distance greater than 981 ft from the first.

3.3 Projectile Motion

In projectile motion, the moving object is acted upon by the constant force of gravity, as in the preceding examples. But, in addition, the object may be given an additional *horizontal* component of velocity, normal to the direction of gravitational acceleration. Let us consider a projectile fired at the origin at time $t = 0$ from a gun resting upon a horizontal plane surface, whose barrel is elevated to make an angle θ_0 with the horizontal plane, and which imparts an initial velocity $\mathbf{v}_0$ to the projectile. This situation is illustrated in Fig. 3.5.

Because there is an initial horizontal component of velocity $v_{0x} = v_0 \cos\theta_0$, the motion of the shell is not in a straight line; instead, the projectile traces out a curved path lying in the xz-plane. In Fig. 3.5, the x-axis has been chosen so as to point in the direction of the horizontal component of the initial velocity vector $\mathbf{v}_0$. Under these circumstances, the y-component of velocity, in a direction perpendicular to the page, is zero; and since there are no forces acting that have y-components, it will always remain zero, according to Newton's first law. Any motion that takes place must, therefore, be confined to the xz-plane.

As in the cases considered in the previous sections, the force of gravity will cause a constant acceleration $\mathbf{g}$ directed vertically downward. The vector $\mathbf{g}$, therefore, has the components $g_x = 0$, $g_y = 0$, and $g_z = -g$. The equations of motion for the x- and z-components may then be written

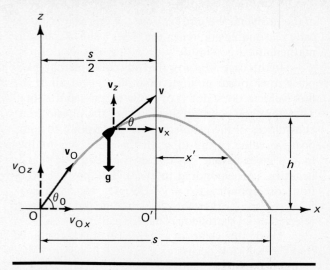

FIGURE 3.5. Motion of a projectile subject to a constant, downwardly directed gravitational force.

$$a_x = \frac{dv_x}{dt} = 0 \qquad (3.3.1)$$

$$a_z = \frac{dv_z}{dt} = -g \qquad (3.3.2)$$

From (3.3.1), it is evident that v_x must be constant, and since the initial value at time $t = 0$ is given to be v_{0x}, we must have

$$v_x = v_{0x} = v_0 \cos \theta_0 \qquad (3.3.3)$$

From (3.3.2) and from the given condition that for $t = 0$, $v_z = v_{0z} = v_0 \sin \theta_0$, we may write, integrating from time $t = 0$ (when $v_z = v_{0z}$) to some later time t when the vertical component of velocity has some value v_z,

$$\int_{v_{0x}}^{v_z} dv_z = -\int_0^t g\, dt \qquad (3.3.4)$$

Since g is a constant, these integrals can easily be evaluated to give

$$v_z - v_{0z} = -gt$$

and since $v_z = v_0 \sin \theta_0$, we finally obtain

$$v_z = v_{0z} - gt = v_0 \sin \theta_0 - gt \qquad (3.3.5)$$

We may now substitute $v_x = dx/dt$ in (3.3.3), multiply both sides of the resulting equation by dt, and integrate once more from time $t = 0$ (when $x = 0$) to a later time t corresponding to some horizontal displacement x, obtaining

$$\int_0^x dx = v_{0x} \int_0^t dt$$

from which

$$x = v_{0x}t = v_0 t \cos \theta_0 \qquad (3.3.6)$$

In this integration, the quantity v_{0x} is a constant and is, therefore, brought outside the integral sign, as shown above. The same procedure may be followed in connection with Eq. (3.3.5), giving

$$\int_0^z dz = \int_0^t (v_{0z} - gt)\, dt$$

and

$$z = v_{0z}t - \tfrac{1}{2}gt^2 = v_0 t \sin \theta_0 - \tfrac{1}{2}gt^2 \qquad (3.3.7)$$

Equations (3.3.6) and (3.3.7) define the path of the projectile by giving the coordinates x and z at each instant of time. They form a set of *parametric equations* for the path of the object which are connected by the parameter t. A single equation for the path may be obtained by eliminating t between the two equations, for example, by solving (3.3.6) for t and inserting the value so obtained into (3.3.7). The result is

$$z = x \tan \theta_0 - \frac{gx^2}{2v_0^2 \cos^2 \theta_0} \qquad (3.3.8)$$

This equation describes a parabola, although, since the central axis of the parabolic path does not coincide with the z-axis, the equation does not have the simple form associated with that more common case. By making a transformation to a new set of coordinates (x', z') about a new origin O', as shown in Fig. 3.5, Eq. (3.3.8) may be made to assume a more familiar form. The details of this calculation are assigned as an exercise.

The maximum height h is attained at the point where dz/dx is zero. Differentiating Eq. (3.3.8) with respect to x and setting the result equal to zero, we obtain

$$\frac{dz}{dx} = \tan \theta_0 - \frac{gx}{v_0^2 \cos^2 \theta} = 0 \qquad (3.3.9)$$

This equation may be solved for x as follows:

$$x = \frac{v_0^2}{g} \sin \theta_0 \cos \theta_0. \qquad (3.3.10)$$

The maximum height h is attained when x has this value, and is simply the value z takes on for this value of x. Substituting the value of x given by (3.3.10) into equation (3.3.8) giving z as a function of x, it is easy to show that

$$h = \frac{v_0^2}{2g} \sin^2 \theta_0 \qquad (3.3.11)$$

The projectile strikes the ground once more at some distance s from the origin along the x-axis. At this point, the z-coordinate of the path is zero. The distance

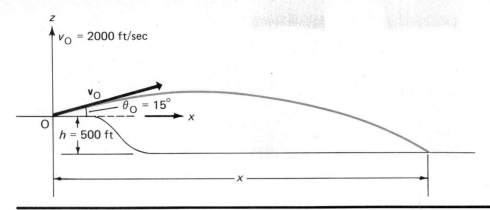

FIGURE 3.6

s can therefore be found by setting z equal to zero in (3.3.8) and solving for x. There are two solutions, one of which is $x = 0$, corresponding to the *initial* departure point. The second corresponds to the final impact point at $x = s$ and has the form

$$s = \frac{2v_0^2}{g} \sin \theta_0 \cos \theta_0 = \frac{v_0^2}{g} \sin 2\theta_0 \qquad (3.3.12)$$

It can be shown that for any given value of initial velocity v_0, the range s of the projectile attains the largest possible value when $\theta_0 = 45°$. Thus, for maximum range (neglecting air resistance, of course), a gun should be elevated $45°$ above the horizontal plane. The proof of this statement is assigned as an exercise.

The equations above tell us nearly all we would ever want to know about the motion of the projectile but provide little physical insight into its behavior. The student should, therefore, avoid the practice of memorizing the results expressed by Eqs. (3.3.3) through (3.3.12). One should instead note from the equations of motion, (3.3.1) and (3.3.2), and from the subsequent development that the x- and z-components of acceleration, velocity, and displacement are independent of one another, and that the z-components, according to (3.3.2), (3.3.5), and (3.3.7), are just those which would be obtained for a *freely falling* object with initial velocity v_{0z}; while the x-components, according to (3.3.1), (3.3.3), and (3.3.6), are simply those which are appropriate for an *unaccelerated* body of constant velocity v_{0x}. These features are characteristic of projectile motion in general. They may be utilized to arrive at the answers to problems in which the circumstances are different from those assumed in the system treated above and to which the formulas developed in this section may not be directly applicable, such as the one discussed in the example below.

EXAMPLE 3.3.1

A gun which imparts an initial muzzle velocity of 2000 ft/sec is fired from the top of a hill 500 ft above

a level plane, as shown in Fig. 3.6. The gun barrel is elevated at an angle of $15°$ with the horizontal. How far from the gun, in a horizontal direction, does the shell land? What is its velocity on impact? Neglect the curvature of the earth's surface and all effects arising from air resistance.

The angle of elevation of the gun barrel is $15°$. This means that the initial velocity vector $\mathbf{v}_0$ can be resolved into x- and z-components such that

$$v_{0x} = v_0 \cos 15° = (2000)(0.9659) = 1932 \text{ ft/sec}$$
$$v_{0z} = v_0 \sin 15° = (2000)(0.2588) = 517.6 \text{ ft/sec}$$

Now, if we choose the origin at the gun barrel and assume that the gun is fired at $t = 0$, we may write from (3.3.7)

$$z = v_{0z}t - \tfrac{1}{2}gt^2 = 517.6t - 16t^2 \qquad (3.3.13)$$

At the point of impact, whose z-coordinate is -500 ft, we have

$$16t^2 - 517.6t - 500 = 0$$

This quadratic equation can be solved for t to give

$$t = \frac{517.6 \pm \sqrt{(517.6)^2 + (4)(16)(500)}}{32}$$

$$= \frac{517.6 \pm 547.7}{32} \text{ sec}$$

Clearly, the time at which the shell arrives at the point of impact must be after the gun is fired, hence positive. We must, therefore, choose the positive sign in the above solution, since the negative sign leads to a negative time of impact which is impossible physically. Rejecting this extraneous solution, then, the time of impact is given by

$$t = \frac{517.6 + 547.7}{32} = 33.29 \text{ sec}$$

The x-coordinate of the shell at impact, which is what we are seeking, can be obtained, once t is known, from (3.3.6):

71

$$x = v_{0x}t = (1932)(33.29) = 64{,}316 \text{ ft} \qquad (3.3.14)$$

The velocity on impact is obtained by noting from (3.2.15) that

$$v_{0z}{}^2 - v_z{}^2 = 2gz$$

whence, on impact,

$$(517.6)^2 - v_z{}^2 = (2)(32)(-500)$$
$$v_z{}^2 = 299{,}900 \text{ ft/sec}$$

Also, from (3.3.3),

$$v_x = v_{0x} = 1932 \text{ ft/sec}$$

The impact speed v is then obtained from

$$v^2 = v_x{}^2 + v_z{}^2 = (1932)^2 + 299{,}900$$
$$v = 2008 \text{ ft/sec}$$

Though the solution discussed above is the most straightforward way of working the problem, there is another good way of approaching it that avoids the necessity of solving a quadratic equation for the time t. This method starts with the relation (3.2.16):

$$v_z{}^2 - v_{0z}{}^2 = -2gz$$

From the given data, we know that $v_{0z} = 517.6$ ft/sec and that on impact $z = -500$ ft, since the shell lands 500 ft below the point of firing. Substituting these values into the above equation, we find at once that $v_z{}^2 = 299{,}900$, $v_z = -547.6$ ft/sec. In obtaining these numerical values, we must take the negative sign of the square root of $v_z{}^2$, since we know that the z-component of velocity is negative when the shell hits the earth. But from (3.2.14), or by differentiating (3.3.13) with respect to time, we know that

$$v_z - v_{0z} = gt$$

Substituting $v_z = -547.6$ ft/sec and $v_{0z} = 517.6$ ft/sec into this equation and solving for t, we arrive at our previous answer, $t = 33.29$ sec. The other answers that are called for in the statement of the example are obtained as before.

EXAMPLE 3.3.2

Work Example 3.3.1 using generalized symbols h for the height of the hill, v_0 for the muzzle velocity, and θ_0 for the elevation.

As before, we may begin with Eq. (3.3.7), noting that at the moment of impact, the z-coordinate of the shell is $-h$. We obtain then

$$-h = v_{0z}t - \tfrac{1}{2}gt^2$$
$$\tfrac{1}{2}gt^2 - v_{0z}t - h = 0 \qquad (3.3.15)$$

This may again be solved for t using the quadratic formula:

$$t = \frac{v_{0z} + \sqrt{v_{0z}{}^2 + 2gh}}{g} \qquad (3.3.16)$$

Again, unless the positive sign of the radical is chosen, the time t will be negative, which is clearly impossible on physical grounds. Also, as before,

$$x = v_{0x}t = v_{0x}\left[\frac{v_{0z} + \sqrt{v_{0z}{}^2 + 2gh}}{g}\right]$$
$$= v_0 \cos \theta_0 \left[\frac{v_0 \sin \theta_0 + \sqrt{v_0{}^2 \sin^2 \theta_0 + 2gh}}{g}\right] \qquad (3.3.17)$$

noting that $v_{0x} = v_0 \cos \theta_0$ and $v_{0z} = v_0 \sin \theta_0$.

The z-component of the impact velocity is obtained from (3.2.16) by setting $z = -h$, which gives

$$v_z{}^2 = v_{0z}{}^2 + 2gh \qquad (3.3.18)$$

while

$$v_x = v_{0x}$$

as always. The impact velocity is then given by

$$v^2 = v_x{}^2 + v_z{}^2 = v_{0x}{}^2 + v_{0z}{}^2 + 2gh \qquad (3.3.19)$$

In this equation, however, $v_{0x}{}^2 + v_{0z}{}^2 = v_0{}^2$, which permits us to write (3.3.19) as

$$v^2 = v_0{}^2 + 2gh \qquad (3.3.20)$$

These results can also be obtained without using the quadratic formula by the alternate approach outlined at the end of the preceding example. The reader should verify this by actually obtaining (3.3.16) starting with (3.2.16) and obtaining the time of flight subsequently from (3.2.14).

3.4 Rotational Motion: Angular Displacement, Velocity, and Acceleration

In the rotational motion of rigid bodies about a fixed axis of rotation, the principles developed in the first section of this chapter apply to the motion of every infinitesimal element of the body. Unfortunately, however, the instantaneous linear velocity and acceleration of each element of the body is different from that of other elements, as illustrated by Fig. 3.7. For example, for a given rate of rotation, points close to the rotation axis move more slowly than points which are far from the rotation axis. Also, two different elements of the body, even though located at the same distance from the rotation axis and moving with the same instantaneous linear speed, do not have the same instantaneous direction associated with their motion. For this reason, it is difficult to describe the motion of such a body in terms of the instantaneous linear velocity and acceleration of each of its elements. We are thus led to seek some simpler way to represent such a motion, and we may begin by defining quantities referred to as *angular displacement*, *angular*

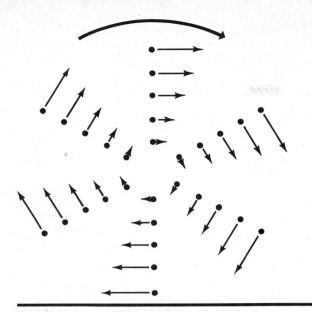

FIGURE 3.7. Instantaneous linear velocity vectors associated with various points within a rigid body rotating about a fixed axis normal to the page.

velocity, and *angular acceleration* and discussing the motion in terms of these quantities.

The case of projectile motion, discussed in the preceding section, offers an important and practical example of motion restricted to a plane. Its distinguishing feature is the constancy of the acceleration vector. Circular motion is another important example of motion in a plane. In this instance, however, the acceleration vector is no longer constant. In this section we shall consider circular motion involving both constant and time-varying rotational speeds.

Figure 3.8 shows the instantaneous position of a point particle moving counterclockwise in a circle of radius r lying in the xy-*plane*. At time t, the particle's instantaneous position is described by the displacement vector $\mathbf{r}$, whose Cartesian components are x and y. In terms of these components, we may write the vector $\mathbf{r}$ as

$$\mathbf{r} = x\mathbf{i}_x + y\mathbf{i}_y = (r\cos\theta)\mathbf{i}_x + (r\sin\theta)\mathbf{i}_y \qquad (3.4.1)$$

where $\mathbf{i}_x$ and $\mathbf{i}_y$ are the usual Cartesian unit vectors.

In discussing circular motion, we shall find it very useful to introduce another pair of unit vectors, called radial and azimuthal unit vectors, $\mathbf{i}_r$ and $\mathbf{i}_\theta$, as illustrated in Fig. 3.8. The radial unit vector $\mathbf{i}_r$ is a vector of unit magnitude along the direction of $\mathbf{r}$ and can, therefore, be expressed as the vector $\mathbf{r}$ divided by its own magnitude. We may write

$$\mathbf{i}_r = \frac{\mathbf{r}}{r} = \left(\frac{x}{r}\right)\mathbf{i}_x + \left(\frac{y}{r}\right)\mathbf{i}_r \qquad (3.4.2)$$

or

$$\mathbf{i}_r = (\cos\theta)\mathbf{i}_x + (\sin\theta)\mathbf{i}_y \qquad (3.4.3)$$

The azimuthal unit vector $\mathbf{i}_\theta$ shown in Fig. 3.8 is a vector of unit magnitude which is normal to $\mathbf{i}_r$ and points in the direction of increasing θ. It is evident from the geometry of Fig. 3.8 that the x- and y-components of $\mathbf{i}_\theta$ are, respectively, $-\sin\theta$ and $\cos\theta$. We may, therefore, write

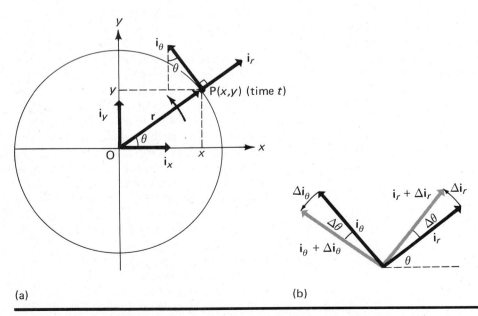

(a) (b)

FIGURE 3.8. (a) Point particle moving in a circular path of radius r whose center is at O.
(b) Unit radial vector $\mathbf{i}_r$ and unit tangential vector $\mathbf{i}_\theta$, and changes $\Delta\mathbf{i}_r$ and $\Delta\mathbf{i}_\theta$ that occur in time Δt during which rotation through angle $\Delta\theta$ occurs.

73

$$\mathbf{i}_\theta = (-\sin\theta)\mathbf{i}_x + (\cos\theta)\mathbf{i}_y \qquad (3.4.4)$$

or, recalling (3.4.1),

$$\mathbf{i}_\theta = \left(-\frac{y}{r}\right)\mathbf{i}_x + \left(\frac{x}{r}\right)\mathbf{i}_y \qquad (3.4.5)$$

Though the use of the unit vectors $\mathbf{i}_r$ and $\mathbf{i}_\theta$ is very convenient, there is one complicating factor which must be clearly understood from the beginning. While the Cartesian vectors $\mathbf{i}_x$ and $\mathbf{i}_y$ are truly constant vectors that change neither in magnitude nor in direction throughout the motion of the particle, this is not true of the unit vectors $\mathbf{i}_r$ and $\mathbf{i}_\theta$. Though the *magnitudes* of $\mathbf{i}_r$ and $\mathbf{i}_\theta$ have the fixed value of unity, their *directions* may continually change as the particle moves. The vectors $\mathbf{i}_r$ and $\mathbf{i}_\theta$ *are not constant vectors* with respect to time, and *their time derivatives are not zero*. This can easily be seen from Eqs. (3.4.3) and (3.4.4), where it is evident that the x- and y-components of the radial and azimuthal unit vectors vary whenever the angle θ changes. It is also apparent from a purely geometric point of view from Fig. 3.8.

To compute the linear velocity of the particle, we must evaluate the time derivative of the displacement vector $\mathbf{r}$. In doing this, we must take account of the time variation of the vectors $\mathbf{i}_r$ and $\mathbf{i}_\theta$. According to (3.4.2), the displacement vector $\mathbf{r}$ can be written also as $r\mathbf{i}_r$; therefore

$$\mathbf{v} = \frac{d\mathbf{r}}{dt} = \frac{d(r\mathbf{i}_r)}{dt} = \left(\frac{dr}{dt}\right)\mathbf{i}_r + r\left(\frac{d(\mathbf{i}_r)}{dt}\right) \qquad (3.4.6)$$

Now, dr/dt is the time rate of change of the *magnitude* of vector $\mathbf{r}$. In circular motion, where the point P describes a circle, the position vector $\mathbf{r}$ then has a *constant* magnitude r, and the quantity dr/dt is zero. The time rate of change of the vector $\mathbf{i}_r$ can be calculated by differentiating Eq. (3.4.3), observing that the time derivative of the vectors $\mathbf{i}_x$ and $\mathbf{i}_y$ is zero. In this way, we obtain

$$\frac{d\mathbf{i}_r}{dt} = \left(-\sin\theta\,\frac{d\theta}{dt}\right)\mathbf{i}_x + \left(\cos\theta\,\frac{d\theta}{dt}\right)\mathbf{i}_y$$

$$= [(-\sin\theta)\mathbf{i}_x + (\cos\theta)\mathbf{i}_y]\frac{d\theta}{dt} = \mathbf{i}_\theta\,\frac{d\theta}{dt}$$

or

$$\frac{d\mathbf{i}_r}{dt} = \omega\mathbf{i}_\theta \qquad (3.4.7)$$

where

$$\omega = \frac{d\theta}{dt} \qquad (3.4.8)$$

According to (3.4.7), $d\mathbf{i}_r/dt$ is a vector of magnitude $\omega = d\theta/dt$ in the direction of $\mathbf{i}_\theta$. That the vector $d\mathbf{i}_r$ can reasonably be expected to point in the direction $\mathbf{i}_\theta$ is evident geometrically from Fig. 3.8b. It is also apparent from this figure that the vector $d\mathbf{i}_\theta$ is in the direction of $-\mathbf{i}_r$. Substituting $dr/dt = 0$ and $d\mathbf{i}_r/dt = \omega\mathbf{i}_\theta$ into (3.4.6), we get the simple result

$$\mathbf{v} = \left(r\,\frac{d\theta}{dt}\right)\mathbf{i}_\theta = (r\omega)\mathbf{i}_\theta \qquad (3.4.9)$$

From this equation, we see that the magnitude of the linear velocity is given by

$$v = r\omega \qquad (3.4.10)$$

and its direction is that of the vector $\mathbf{i}_\theta$, tangential to the circular path.

The quantity $\omega = d\theta/dt$ represents the instantaneous time rate of change of the angle θ expressing the *angular displacement* of the particle from its starting point. It is referred to as the *angular velocity* of the particle, and it is always expressed in units of radians per second, since θ itself is measured in radians. It should be noted that one revolution per second equals 2π rad/sec, since $360° = 2\pi$ rad. If the angular displacement θ increases linearly with time, then the angular velocity ω is constant. In the present discussion, however, this not assumed to be the case.

The linear acceleration $\mathbf{a}$ may now be computed using the methods developed above. Since the linear acceleration is the time derivative of the linear velocity, we may write

$$\mathbf{a} = \frac{d\mathbf{v}}{dt} = \frac{d}{dt}(r\omega\mathbf{i}_\theta) = r\,\frac{d}{dt}(\omega\mathbf{i}_\theta) \qquad (3.4.11)$$

observing that in the case of circular motion, the magnitude r does not vary with respect to time. Taking the derivative as before, we find

$$\mathbf{a} = r\left[\left(\frac{d\omega}{dt}\right)\mathbf{i}_\theta + \omega\,\frac{d(\mathbf{i}_\theta)}{dt}\right] \qquad (3.4.12)$$

The time derivative of $\mathbf{i}_\theta$ is computed from (3.4.4) as

$$\frac{d\mathbf{i}_\theta}{dt} = \left(-\cos\theta\,\frac{d\theta}{dt}\right)\mathbf{i}_x + \left(-\sin\theta\,\frac{d\theta}{dt}\right)\mathbf{i}_y$$

$$= -\omega[(\cos\theta)\mathbf{i}_x + (\sin\theta)\mathbf{i}_y]$$

or

$$\frac{d\mathbf{i}_\theta}{dt} = -\omega\mathbf{i}_r \qquad (3.4.13)$$

The time derivative of $\mathbf{i}_\theta$ is seen to be in the direction of the vector $-\mathbf{i}_r$, as mentioned before in connection with Fig. 3.8b. Substituting this result into (3.4.12),

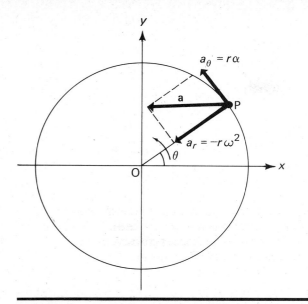

FIGURE 3.9. Instantaneous acceleration vector associated with circular motion, represented as sum of tangential and radial (centripetal) components.

we obtain

$$\mathbf{a} = (r\alpha)\mathbf{i}_\theta - (r\omega^2)\mathbf{i}_r \qquad (3.4.14)$$

where

$$\alpha = \frac{d\omega}{dt} = \frac{d^2\theta}{dt^2} \qquad (3.4.15)$$

The quantity α represents the time rate of change of angular velocity and is, therefore, referred to as the *angular acceleration*. It is expressed in units of rad/sec^2.

The linear acceleration has both a tangential component and a radial component, as shown in Fig. 3.9. The magnitude of the tangential component a_θ is, from (3.4.14),

$$a_\theta = r\alpha \qquad (3.4.16)$$

while the radial component a_r is of magnitude

$$a_r = -r\omega^2 = -\frac{v^2}{r} \qquad (3.4.17)$$

remembering from (3.4.10) that $v = r\omega$. The radial component of acceleration, in view of its negative sign, is directed inward, toward the center of the circular path, as illustrated in Fig. 3.9. It is frequently referred to as the *centripetal acceleration*.

The simplest case of circular motion is that in which the angular velocity ω is constant and the angular acceleration α is, therefore, zero. Under these circumstances, according to (3.4.14) or (3.4.16), the tangential component of acceleration vanishes, and

the *centripetal acceleration* is all that remains. The total acceleration vector is directed radially inward toward the center of the circular path and is of magnitude $-r\omega^2$ or $-v^2/r$. The reason that the acceleration is inwardly directed in this case can be understood geometrically from Fig. 3.10. A racing car traveling at constant speed around a circular track or a speck of dust on a rotating phonograph record is a simple example of *uniform circular motion*.

We have gone to considerable trouble, involving some tricky vector algebra, to introduce the concepts of angular displacement, angular velocity, and angular acceleration. Why? Simply to torture the student with mathematics? No, there is a better reason. The linear displacement, linear velocity, and linear acceleration of every point in any rotating object differ in magnitude and direction. It is hopeless to try to describe the geometry of rotation of such a body in terms of these parameters. But the rotational motion of all points within the object can be characterized by a single value of *angular* displacement, a single value of *angular* velocity, and a single *angular* acceleration. For example, it is obvious that every point within the body has *the same* rotational speed, in units of rad/sec, about the rotation axis, even though the linear speed of points near the axis are very different from the linear speed of those far from it. The angular displacement, angular velocity, and angular acceleration are, therefore, a set of useful quantities that enable us to describe the rotational motion of solid bodies in a simple way. They also afford us a set of easily observed and measured parameters for *calculating*, via (3.4.9) and (3.4.14), the linear velocity and acceleration of every point within any rotating object.

EXAMPLE 3.4.1

A speck of dust on a phonograph record rotates at a rotational speed of 45 revolutions per minute (rpm). If the dust particle is 10 cm from the rotation axis, find (a) its linear velocity and (b) its linear acceleration.

In deriving the expressions given above for linear velocity and linear acceleration, it was assumed that *angles are measured in radians throughout*. In working with these expressions, we must, therefore, always express angles or angular displacements in rad, angular velocities in rad/sec, and angular accelerations in rad/sec^2. For a record rotating at 45 rpm (or $\frac{45}{60}$ rps), the angular velocity in rads will be

$$\omega = \tfrac{45}{60}(2\pi) = 1.5\pi \text{ rad/sec} = 4.71 \text{ rad/sec}$$

The linear speed, according to (3.4.10), will be

$$v = r\omega = (10)(1.5\pi) = 15\pi \text{ cm/sec} = 47.1 \text{ cm/sec}$$

From (3.4.9), it is apparent that the linear velocity is tangentially directed.

In this case, the angular velocity is constant, and

75

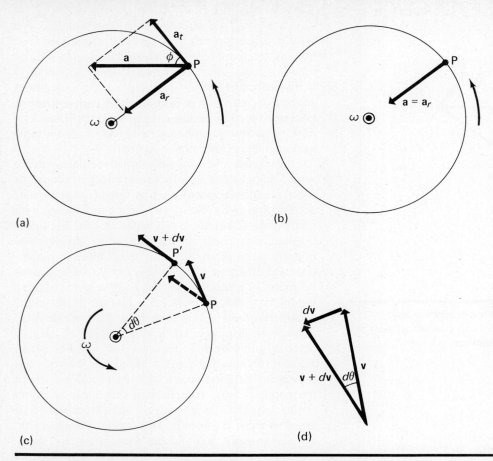

FIGURE 3.10. Geometry of rotational motion. (a) Disposition of instantaneous velocity and acceleration vectors. (b) Instantaneous velocity and acceleration in case of uniform circular motion. (c) and (d) Geometric demonstration of the fact that the acceleration is directed radially inward in uniform circular motion.

the angular acceleration α, equal to $d\omega/dt$, is zero. We are, therefore, dealing with an example of *uniform circular motion*. Since α is zero, there will be no tangential component of acceleration in (3.4.14). The only acceleration left is the centripetal acceleration, directed radially inward. According to (3.4.14), its magnitude will be

$$a_r = -r\omega^2 = -(10)(1.5\pi)^2$$
$$= -22.5\pi^2 = -221.8 \text{ cm/sec}^2$$

The negative sign signifies that the acceleration vector points radially inward, in the direction of decreasing values of r. The same numerical result could have been obtained by use of the alternative expression $-v^2/r$ for the centripetal acceleration.

EXAMPLE 3.4.2

Suppose, in the preceding example, that the speck of dust starts from rest at $t = 0$ and experiences a *linearly increasing angular velocity*. After 2.5 sec, its instantaneous angular velocity is 18.0 rad/sec. Find (a) the

angular acceleration α during this period and (b) the magnitude and direction of the linear acceleration vector $\mathbf{a}$ (i) at $t = 0$, (ii) at $t = 1.0$ sec, and (iii) at $t = 2.5$ sec.

Since ω changes with time, we are no longer dealing with uniform circular motion. In this case, at $t = 0$ the angular velocity ω is zero. Thereafter, it increases linearly with time, its value at any instant being, therefore, directly proportional to the time t elapsed after start of the motion at $t = 0$. We may thus express ω as a function of t by

$$\omega = \omega(t) = kt$$

where k is a constant of proportionality. The angular acceleration α is, from (3.4.15), the time derivative of the angular velocity, from which we obtain

$$\alpha = \frac{d\omega}{dt} = k$$

The value of the constant of proportionality k is found by noting that at $t = 2.5$ sec, $\omega = 18.0$ rad/sec.

Substituting these values into the first equation above, we see at once that $k = 7.2$ rad/sec^2, and the angular acceleration, according to the second equation, must then be

$$\alpha = k = 7.2 \text{ rad/sec}^2$$

In this instance, even though ω varies with time, the angular acceleration α is constant. Such motion is called *uniformly accelerated circular motion* and will be considered in more detail in section 3.6.

According to (3.4.14), the tangential component of the linear acceleration vector is $r\alpha$, and the radial component is $-r\omega^2$. Since the radius r is 10 cm, the tangential component will be, in all cases, (i), (ii), and (iii)

$$a_\theta = r\alpha = (10)(7.2) = 72 \text{ cm/sec}^2$$

Since we have shown above that $\omega(t) = kt = 7.2t$, it is evident that the radial component of acceleration, for $t = 0$, 1.0 sec, and 2.5 sec, will be

(i)
$$a_r = -r\omega^2 = -(10)(0) = 0 \qquad (t = 0)$$

(ii)
$$a_r = -r\omega^2 = -(10)(7.2)^2$$
$$= -518.4 \text{ cm/sec}^2 \qquad (t = 1.0 \text{ sec})$$

(iii)
$$a_r = -r\omega^2 = -(10)(18.0)^2$$
$$= -3240 \text{ cm/sec}^2 \qquad (t = 2.5 \text{ sec}).$$

In case (i) above, the radial component of acceleration is zero and the acceleration vector is of magnitude $a = 72$ cm/sec^2, directed tangentially. In cases (ii) and (iii), the magnitude of a and the angle ϕ between the acceleration vector and the radial direction are given by

(ii)
$$a = \sqrt{a_r^2 + a_\theta^2} = \sqrt{(-518.4)^2 + (72)^2} = 523.4 \text{ cm/sec}^2$$

$$\tan \phi = \frac{a_\theta}{a_r} = \frac{72}{518.4} = 0.1389 \qquad \phi = 7.91°$$

(iii)
$$a = \sqrt{a_r^2 + a_\theta^2} = \sqrt{(-3240)^2 + (72)^2} = 3240.8 \text{ cm/sec}^2$$

$$\tan \phi = \frac{a_\theta}{a_r} = \frac{72}{3240} = 0.0222 \qquad \phi = 1.27°.$$

In cases (ii) and (iii), the situation is similar to that illustrated in Fig. 3.10a. The acceleration vector now has both radial and tangential components, and the angle between the acceleration vector and the radial direction is defined by $\tan \phi = a_\theta / a_r$. In this case, it is useful to observe that the angular acceleration and the tangential component of linear acceleration are both constant, while the centripetal acceleration and the linear acceleration vector both vary with time.

EXAMPLE 3.4.3

Assuming that the earth's radius is 3963 miles, find the linear speed and the centripetal acceleration of an object at the equator that is fixed with respect to the earth's surface.

This is another case of uniform circular motion, in which the angular velocity is constant. Since the earth rotates through 2π radians in 24 hours (86,400 sec), its angular velocity is

$$\omega = \frac{2\pi}{86,400} = 7.27 \times 10^{-5} \text{ rad/sec}$$

The linear speed of an object at the equator, 3963 miles from the rotation axis, is

$$v = r\omega = (3963)(5280)(7.27 \times 10^{-5}) = 1522 \text{ ft/sec}$$

while the centripetal acceleration is

$$a_\theta = -r\omega^2 = -(3963)(5280)(7.27 \times 10^{-5})^2$$
$$= -0.111 \text{ ft/sec}^2$$

3.5 Angular Velocity and Angular Acceleration as Vectors: Rotation of Rigid Bodies

In the preceding section, the concepts of angular velocity and angular acceleration were introduced using algebraic techniques involving the time-varying unit vectors $\mathbf{i}_r$ and $\mathbf{i}_\theta$. This discussion was based on the examination of the motion of a single particle moving along a circular path. In this treatment, the angular velocity and angular acceleration appear as *scalar* quantities.

One of the most important instances involving circular motion is to be found in the rotational motion of rigid bodies. In this case, every point within the object simultaneously executes circular motion about the rotation axis, although the radii of different points with respect to the axis may be different and, thus, different points within the body may have different linear velocities and accelerations. In such situations, it is important to distinguish between angular velocities and angular accelerations that may take place about different axes of rotation. This can be done by associating directions as well as magnitudes with angular velocities and angular accelerations, and thereby defining these quantities as *vectors* rather than scalars.

In this section, therefore, we shall show how angular velocity and acceleration can be thought of as vector quantities. In so doing, we shall arrive at

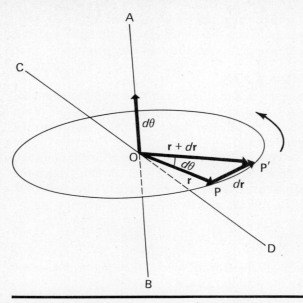

FIGURE 3.11. Linear and angular displacement vectors in circular motion.

many of the results of the preceding section using geometric rather than algebraic arguments. These arguments are important in developing a broad and comprehensive understanding of rotational kinematics.

Consider a point P rotating at a fixed distance r about a fixed axis AB which intersects the plane of rotation at O, as illustrated in Fig. 3.11. Suppose that in some time interval dt the point rotates through an angle $d\theta$. Let the vector $\mathbf{r}$ represent the displacement at time t, and the vector $\mathbf{r} + d\mathbf{r}$ the displacement after a time dt has elapsed. The vector $\mathbf{r}$, despite the fact that its magnitude is constant, is not a constant vector, since its *direction* is continually changing. Nevertheless, since the point is constrained to move at a fixed distance from the axis, the *lengths* of the vectors $\mathbf{r}$ and $\mathbf{r} + d\mathbf{r}$ are equal. The vector $d\mathbf{r}$, which represents their difference, is as shown in Fig. 3.11.

The *angular* displacement, from Fig. 3.11, must evidently be of magnitude $d\theta$. But it is also apparent that such an angular displacement has a *direction* associated with it because, obviously, infinitesimal rotations about different axes are not physically equivalent. For example, in Fig. 3.11, if the axis of rotation (and the plane normal to it, containing the vectors $\mathbf{r}$, $d\mathbf{r}$, and $\mathbf{r} + d\mathbf{r}$) were tilted so that the rotation axis assumed the orientation CD, and then a rotation of magnitude $d\theta$ were performed about CD, the vector $d\mathbf{r}$ would have a direction different from that shown in the figure, and the rotational displacement would clearly be physically nonequivalent to that shown in Fig. 3.11 about axis AB. It is reasonable, then, to associate with the infinitesimal angular displacement $d\theta$ the *direction* of the axis about which the infinitesimal rotation has been performed and, indeed, to

regard the infinitesimal angular displacement as a *vector $d\boldsymbol{\theta}$*, having magnitude $d\theta$ and *pointing along the rotation axis in the direction in which a right-handed screw would advance when turned in the direction of rotation*. The vector $d\boldsymbol{\theta}$ is shown in Fig. 3.11. This way of associating a vector with an infinitesimal rotation and of choosing the way in which this vector points is only a definition, but it is one which has been universally adopted, and which we shall conform to at all times. It simplifies matters greatly in that *it associates the same angular displacement with all elements of a rigid rotating body.*

Unfortunately, it is not at all clear that this way of associating a direction with an infinitesimal rotation really defines a vector at all. True enough, a quantity having both magnitude and direction has been described, but it is not intuitively obvious that quantities so defined obey the algebraic laws for vector quantities set down in Chapter 1. For instance, is the net rotational displacement corresponding to the "vector sum" of two infinitesimal rotations independent of the *order* in which the two infinitesimal rotations are added, (as it must be for vectors, according to Eq. (1.4.1)? And, assuming that it is, does the net rotational displacement so obtained correspond to what would be predicted by adding the two infinitesimal rotations according to the laws of vector addition? There is no offhand reason why the answers to these questions should be affirmative. We shall simply state at this point that the answers are "yes" to both questions and that the representation of infinitesimal angular displacements as vectors by the scheme outlined above is in accord with the laws of vector algebra. It must be noted, however, that only *infinitesimal* rotations can be represented as vectors in this way. Finite rotations cannot, since the net effect of two successive *finite* rotational displacements depends upon the *order* in which those rotations are performed, contradicting the commutative law of vector addition as expressed by Eq. (1.4.1). This is illustrated in Fig. 3.12.

Referring once more to Fig. 3.11, it is easily seen that the vector $d\boldsymbol{\theta} \times \mathbf{r}$ is a vector in the direction of $d\mathbf{r}$ whose magnitude is $r\,d\theta$, since the angle between the directions of $d\boldsymbol{\theta}$ and $\mathbf{r}$ is 90° and sin 90° = 1. But the magnitude of $d\mathbf{r}$ is simply PP', which, in the limit $d\theta \to 0$ approaches $r\,d\theta$. The vectors $d\boldsymbol{\theta} \times \mathbf{r}$ and $d\mathbf{r}$ are thus identical, or

$$d\mathbf{r} = d\boldsymbol{\theta} \times \mathbf{r} \qquad (3.5.1)$$

If each side of this equation is divided by dt, we may obtain

$$\frac{d\mathbf{r}}{dt} = \frac{d\boldsymbol{\theta}}{dt} \times \mathbf{r} \qquad (3.5.2)$$

In this expression, $d\mathbf{r}/dt$ is, by definition, the instantaneous linear velocity $\mathbf{v}$, while $d\boldsymbol{\theta}/dt$ is a vector whose

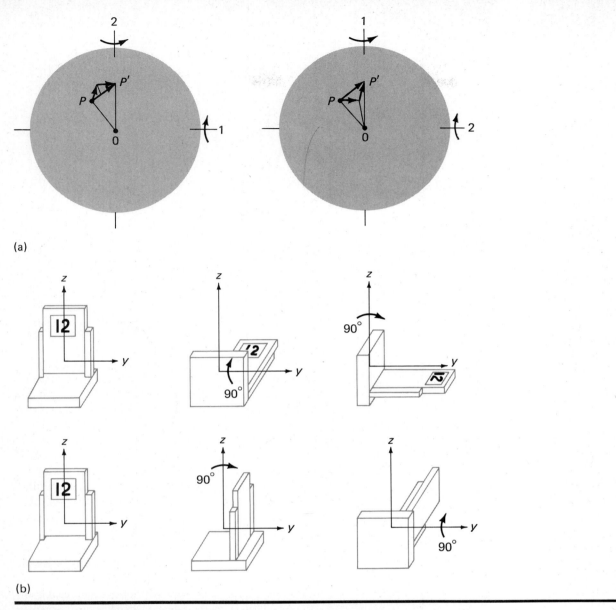

FIGURE 3.12. (a) Geometric argument leading to the conclusion that the resultant of *infinitesimal rotations* is the same irrespective of the order in which they occur. On the left, an infinitesimal rotation (1) occurs about a horizontal axis, carrying P into Q, and a second infinitesimal rotation (2) about a vertical axis carries Q into P'. On the right, the order of the rotations is reversed, carrying P into Q' and Q' into P', but the resultant infinitesimal displacement PP' is the same in either case. (b) Geometric demonstration that the sum of *finite* rotations depends upon the order in which they occur. In the top row, an object is first rotated clockwise 90° about the y-axis, then clockwise 90° about the z-axis. In the bottom row, the same rotation operations are performed in the reverse order. The final orientation of the object is clearly different in the two cases.

magnitude is equal to the time rate of change of angular displacement and which points along the rotation axis in the direction of advance of a right-handed screw. The latter vector is commonly designated the *angular velocity vector* and is represented by the symbol ω. Using this notation, (3.5.2) may be

written in the form

$$\mathbf{v} = \boldsymbol{\omega} \times \mathbf{r} \tag{3.5.3}$$

Since the angle between the directions of ω and $\mathbf{r}$ is 90°, the relation between the magnitudes of the linear and

79

angular velocities is, according to the above equation,

$$v = r\omega \tag{3.5.4}$$

This agrees with our previous result expressed in Eq. (3.4.10).

Now let us try to describe the instantaneous linear acceleration of point P. Since the linear acceleration is by definition $d\mathbf{v}/dt$, we may proceed by differentiating (3.5.3) to obtain

$$\mathbf{a} = \frac{d\mathbf{v}}{dt} = \frac{d}{dt}(\boldsymbol{\omega} \times \mathbf{r}) = \boldsymbol{\omega} \times \frac{d\mathbf{r}}{dt} + \frac{d\boldsymbol{\omega}}{dt} \times \mathbf{r} \tag{3.5.5}$$

The second step above follows from the rule for the differentiation of a product of two functions[3] $u(t)$ and $v(t)$: $d(uv)/dt = u(dv/dt) + v(du/dt)$. But $d\mathbf{r}/dt$ is simply the linear velocity $\mathbf{v}$, which according to (3.5.3) equals $\boldsymbol{\omega} \times \mathbf{r}$, while $d\boldsymbol{\omega}/dt$ is the time rate of change of angular velocity, which we may refer to as the *angular acceleration* and represented by the symbol $\boldsymbol{\alpha}$. We may write, then,

$$\mathbf{a} = \frac{d\mathbf{v}}{dt} = (\boldsymbol{\omega} \times \mathbf{v}) + \frac{d\boldsymbol{\omega}}{dt} \times \mathbf{r}$$

or

$$\mathbf{a} = \boldsymbol{\omega} \times (\boldsymbol{\omega} \times \mathbf{r}) + (\boldsymbol{\alpha} \times \mathbf{r}) \tag{3.5.6}$$

where

$$\boldsymbol{\alpha} = d\boldsymbol{\omega}/dt \tag{3.5.7}$$

Now suppose that the angular velocity vector $\boldsymbol{\omega}$ changes by some amount $d\boldsymbol{\omega}$ in time dt. In general, the change $d\boldsymbol{\omega}$ will not have the same direction as $\boldsymbol{\omega}$, and hence the direction (as well as the magnitude) of $\boldsymbol{\omega}$ will change with time, as illustrated by Fig. 3.13a. Under these circumstances the direction of the rotation axis changes in the course of time, and the angular acceleration vector $\boldsymbol{\alpha}$, which has the same direction as the velocity change $d\boldsymbol{\omega}$, will not point in the same direction as $\boldsymbol{\omega}$. If, however, $d\boldsymbol{\omega}$ should have the same direction as $\boldsymbol{\omega}$, the magnitude of $\boldsymbol{\omega}$ will change with time while the direction of $\boldsymbol{\omega}$ remains fixed, as shown by Fig. 3.13b. The direction of the rotation axis is then immobile, and the angular acceleration vector which, as always, has the direction of $d\boldsymbol{\omega}$ is *parallel to the angular velocity vector* (or antiparallel to it, if

---

[3] This is readily seen from Eq. (1.6.11). If one considers A_x, A_y, A_z, B_x, B_y, and B_z in that equation to depend upon time and then differentiates the right-hand side of the equation with respect to time, one may group separately all terms containing derivatives of components of $\mathbf{A}$ and all terms containing derivatives of components of $\mathbf{B}$. It is then possible to identify one group of terms as being identical to $\mathbf{A} \times (d\mathbf{B}/dt)$ and the other as $(d\mathbf{A}/dt) \times \mathbf{B}$, and hence to prove that

$$d(\mathbf{A} \times \mathbf{B})/dt = [\mathbf{A} \times (d\mathbf{B}/dt)] + [(d\mathbf{A}/dt) \times \mathbf{B}]$$

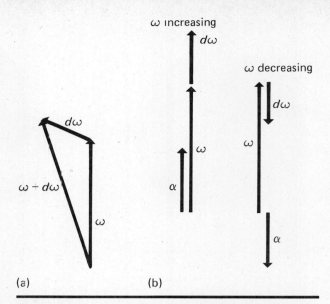

FIGURE 3.13. Change in angular velocity vector occurring (a) when the direction of the rotation axis is changing and (b) when the direction of the rotation axis is fixed.

the angular acceleration is negative). This special case, in which the rotation axis is fixed, is particularly simple and important and, except for a few special examples, is the only one we shall discuss in detail. Unless otherwise specified, then, we shall ordinarily assume in any situation where rotational motion is involved *that the rotation axis is fixed*, hence, that the angular velocity and angular acceleration vectors are parallel (or antiparallel). It should be noted, however, that expressions (3.5.3) and (3.5.6) are correct even when the rotation axis is not fixed.

Returning now to Eq. (3.5.6), one may easily see that the first term of that equation represents the component of acceleration along the radial direction while the second term represents the tangential component. This is evident because the vector $\boldsymbol{\omega} \times \mathbf{v}$ is perpendicular to both $\boldsymbol{\omega}$ and $\mathbf{v}$ and, therefore, must lie along the *radial* direction. Likewise, the vector $\boldsymbol{\alpha} \times \mathbf{r}$ must be perpendicular to $\boldsymbol{\alpha}$ (hence to $\boldsymbol{\omega}$, if the rotation axis is fixed) and also to $\mathbf{r}$; its direction must necessarily be *tangential*. The situation is illustrated in Fig. 3.14 and also in Fig. 3.10a, where the acceleration vector $\mathbf{a}$ is shown as the sum of radial and tangential components, such that

$$\mathbf{a} = \mathbf{a}_r + \mathbf{a}_t \tag{3.5.8}$$

From (3.5.6) it is apparent that

$$\mathbf{a}_r = \boldsymbol{\omega} \times (\boldsymbol{\omega} \times \mathbf{r}) = \boldsymbol{\omega} \times \mathbf{v} \tag{3.5.9}$$

$$\mathbf{a}_t = \boldsymbol{\alpha} \times \mathbf{r} \tag{3.5.10}$$

Since $\boldsymbol{\omega}$ and $\mathbf{v}$ are perpendicular, the magnitude of $\boldsymbol{\omega} \times \mathbf{v}$ is $v\omega$, the sine of the angle involved being unity.

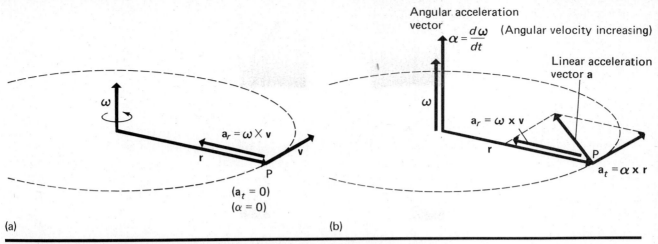

FIGURE 3.14. (a) Linear velocity, angular velocity, and linear acceleration vectors in uniform circular motion. In this case, the angular acceleration α is zero. (b) Angular velocity, angular acceleration, and linear acceleration vectors in nonuniform circular motion. The linear velocity vector is still tangential to the path, in the same direction as a_t, but is omitted from the diagram in the interest of clarity.

But, according to (3.5.4), $v = r\omega$, and hence the magnitude of $\boldsymbol{\omega} \times \mathbf{v}$, which is also the magnitude of the radial acceleration $\mathbf{a}_r$, is given by

$$a_r = r\omega^2 = \frac{v^2}{r} \qquad (3.5.11)$$

Likewise, for the case of a fixed axis of rotation, α is in the direction of $\boldsymbol{\omega}$ and is, therefore, perpendicular to $\mathbf{r}$, and the magnitude of the tangential component of acceleration is

$$a_t = r\alpha \qquad (3.5.12)$$

These equations are both in agreement with results obtained in section 3.4.

If the rate of rotation about the axis is *constant*, then the angular velocity ω is constant, and the *angular acceleration* $\alpha = d\omega/dt$ is zero. Under these circumstances, and according to (3.5.12), the *tangential* component of the instantaneous linear acceleration also vanishes. The *radial* component of linear acceleration, a_r, however, does not vanish, but, from (3.5.11), has the constant value $r\omega^2$. The *linear* acceleration vector is directed radially toward the axis of rotation, as shown in Fig. 3.10b. This constant linear acceleration $r\omega^2 = v^2/r$ directed toward the axis under conditions of constant rotational motion is the *centripetal* acceleration.

Figures 3.10c and 3.10d are intended to show on an intuitive level why the linear acceleration associated with rotational motion at constant angular velocity is directed centrally toward the axis. In Fig.

3.10c, the velocity vectors $\mathbf{v}$ and $\mathbf{v} + d\mathbf{v}$ associated with points P and P′ are shown. In order to obtain their vector difference $d\mathbf{v}$, we may translate the vector $\mathbf{v} + d\mathbf{v}$ parallel to itself until its tail rests on point P, as shown by the dotted arrow. Figure 3.10d shows the difference of vectors $\mathbf{v}$ and $\mathbf{v} + d\mathbf{v}$ as represented by the vector $d\mathbf{v}$. The direction of the linear acceleration vector $\mathbf{a} = d\mathbf{v}/dt$ is in the direction of $d\mathbf{v}$. But clearly, from the figure, in the limit where $d\theta \to 0$, the direction of $d\mathbf{v}$ will be perpendicular to $\mathbf{v}$, thus along the radial direction. This situation has been referred to before in connection with eq. (3.1.16). If the angular velocity is not constant, the magnitudes of the vectors $\mathbf{v}$ and $\mathbf{v} + d\mathbf{v}$ will be different and the vector $d\mathbf{v}$ will not be directed radially. There will then be a tangential component of linear acceleration, as well as a radial one, in accord with the predictions of Eq. (3.5.6), (3.5.10), and (3.5.14).

3.6 Uniformly Accelerated Rotational Motion

In the preceding section, we discussed the case of a particle moving in a circular path about a fixed central axis. A particle moving in this way is said to execute *circular motion*. The simplest case of circular motion occurs when the angular velocity is constant and the angular acceleration is, therefore, zero. This is uniform circular motion.

The next simplest example is one in which *constant* angular acceleration exists. This situation is one of great practical importance and is referred to

81

as *uniformly accelerated circular motion*. It is analogous to the constant linear acceleration discussed in section 3.2 and leads to a set of equations relating angular displacement, angular velocity, and time similar to the "golden rules" developed for linear motion in that section. For uniformly accelerated rotational motion, then, if the angular acceleration α $(=d\omega/dt)$ is constant,

$$dω = α \, dt \tag{3.6.1}$$

Then, integrating and assuming that ω has the initial velocity ω_0 at $t = 0$,

$$\int_{\omega_0}^{\omega} d\omega = \alpha \int_0^t dt \tag{3.6.2}$$

which leads to

$$[\omega]_{\omega_0}^{\omega} = \alpha[t]_0^t$$

or

$$\omega = \omega_0 + \alpha t \tag{3.6.3}$$

This equation expresses the fact that the angular velocity ω at time t equals the initial value ω_0 plus the product of the angular acceleration and the time, and it is in every way analogous to Eq. (3.2.5) setting forth a similar relation between linear velocity, linear acceleration, and time.

Similarly, setting $\omega = d\theta/dt$ in (3.6.3), multiplying both sides of the equation by dt, and integrating once more, we may obtain

$$\int_0^\theta d\theta = \int_0^t (\omega_0 + \alpha t) \, dt \tag{3.6.4}$$

$$[\theta]_0^\theta = [\omega_0 t + \tfrac{1}{2}\alpha t^2]_0^t$$

or

$$\theta = \omega_0 t + \tfrac{1}{2}\alpha t^2 \tag{3.6.5}$$

In (3.6.4), for assigning the limits of integration it is assumed that the value of θ at time $t = 0$ is zero. Equation (3.6.5) clearly bears a strong resemblance to (3.2.7). Equations (3.6.3) and (3.6.5) allow us to express the angular displacement and angular velocity for any given value of t. If it is desired to find the angular displacement as a function of angular velocity rather than time, we may proceed by noting that

$$\alpha = \frac{d\omega}{dt} = \frac{d\omega}{d\theta}\frac{d\theta}{dt} = \omega\frac{d\omega}{d\theta} \tag{3.6.6}$$

Multiplying both sides of this equation by $d\theta$ and integrating from $\theta = 0$, when the angular velocity has the initial value ω_0, up to a final angular displacement θ, for which the angular velocity is ω, we find that

$$\alpha \int_0^\theta d\theta = \int_{\omega_0}^{\omega} \omega \, d\omega \tag{3.6.7}$$

$$\alpha[\theta]_0^\theta = [\tfrac{1}{2}\omega^2]_{\omega_0}^{\omega}$$

or

$$\omega^2 = \omega_0{}^2 + 2\alpha\theta \tag{3.6.8}$$

This result may also be obtained by eliminating t between Eqs. (3.6.3) and (3.6.5). It is analogous to Eq. (3.2.13), which expresses a similar relationship between the corresponding linear quantities. Equations (3.6.3), (3.6.5), and (3.6.8) form a set of "golden rules" of uniformly accelerated rotational motion completely analogous to the corresponding rules for uniformly accelerated linear motion and which can be used in exactly the same way. As before, however, it is important to ascertain in every instance that the angular acceleration is really constant before trying to use these rules.

EXAMPLE 3.6.1

A flywheel of radius r rotates with constant angular velocity ω. Investigate the relations between the linear displacement, velocity, and acceleration of a point on the outer rim of the wheel and the angular displacement and velocity of the wheel.

Referring to Fig. 3.15, the relation between the linear displacement component and the angular displacement θ is simply

$$x = r \cos \theta$$
$$y = r \sin \theta \tag{3.6.9}$$

Since the angular velocity ω is constant, we may write, from the definition of ω,

$$\frac{d\theta}{dt} = \omega = \text{constant} \tag{3.6.10}$$

from which

$$\theta = \omega t + \theta_0 \tag{3.6.11}$$

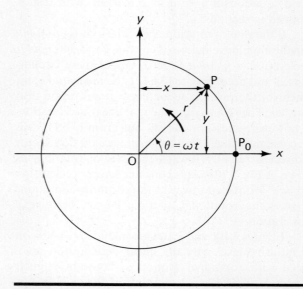

FIGURE 3.15

where θ_0 is the value of θ at $t = 0$. For simplicity, let us assume that the point starts out at the point P_0; under these circumstances θ_0 is zero and

$$\theta(t) = \omega t \tag{3.6.12}$$

We may, therefore, write Eqs. (3.6.9) as

$$\begin{aligned} x &= r \cos \omega t \\ y &= r \sin \omega t \end{aligned} \tag{3.6.13}$$

This set of circumstances is one which will arise frequently in discussing the motion of rotating systems. It will generally be assumed under these conditions that the angular displacement associated with time zero is zero and that the angular displacement θ for a body rotating at a constant angular speed can be expressed as ωt, as indicated by (3.6.12) above and as illustrated in Fig. 3.15. It should be noted, however, that this can be done only *if there is no angular acceleration*. Observe also that if α is set equal to zero in (3.6.5), we obtain the result obtained above as (3.6.12), because then the angular velocity always has the constant value $\omega = \omega_0$.

The velocities v_x and v_y can be obtained by differentiating (3.6.13) with respect to time. Thus, observing in this example that r is constant with respect to time,

$$\begin{aligned} v_x &= \frac{dx}{dt} = -\omega r \sin \omega t \\ v_y &= \frac{dy}{dt} = \omega r \cos \omega t \end{aligned} \tag{3.6.14}$$

The accelerations a_x and a_y may be found by differentiating once more with respect to t, whereby

$$\begin{aligned} a_x &= \frac{dv_x}{dt} = -\omega^2 r \cos \omega t \\ a_y &= \frac{dv_y}{dt} = -\omega^2 r \sin \omega t \end{aligned} \tag{3.6.15}$$

Sets of equations such as (3.6.13), (3.6.14), or (3.6.15), which describe the individual components of vectors in terms of some common parameter, such as time, are referred to as *parametric equations*. It is always possible to eliminate the parameter from the equations of such a set by solving one of the equations for this parameter and inserting the value thus obtained into the others. In the above examples, however, this may be accomplished more easily by squaring both sides of each equation and adding the two resulting equations, recalling that $\sin^2 \theta + \cos^2 \theta = 1$. Applying this procedure to eliminate the time between the two equations (3.6.14), we find

$$x^2 + y^2 = r^2 \tag{3.6.16}$$

This is the equation of the *path* of the point—a circle of radius r, according to the above expression. It is clearly correct in view of the original statement of the problem. The time may be eliminated in the same way between the two equations (3.6.14) to give

$$v_x{}^2 + v_y{}^2 = v^2 = (r\omega)^2 \tag{3.6.17}$$

in agreement with (3.4.10). Eliminating t between the pair of equations (3.6.15) yields

$$a_x{}^2 + a_y{}^2 = a^2 = (r\omega^2)^2 \tag{3.6.18}$$

in accord with (3.4.17).

EXAMPLE 3.6.2

A bicycle wheel 27 in. in diameter is mounted on a fixed axle connected to an electric motor which can be used to rotate the wheel. During a 10-sec interval, the motor imparts a constant angular acceleration α to the wheel. Initially, the wheel is at rest, but at the end of 10 sec the linear velocity of a point on its rim is 88 ft/sec. Find the angular velocity of the wheel at the end of the 10-sec period of acceleration, the angular acceleration during the period, and the number of revolutions that the wheel turns during the period.

If the linear speed of a point on the wheel's rim is 88 ft/sec, then the angular velocity ω must be

$$\omega = \frac{v}{r} = \frac{88}{1.125} = 78.2 \text{ rad/sec}$$

The angular acceleration follows from (3.6.3), in which the initial angular velocity ω_0 is zero, the final angular velocity is 78.2 rad/sec, and the time t is 10 sec. Accordingly,

$$\alpha = \frac{\omega - \omega_0}{t} = \frac{78.2 - 0}{10} = 7.82 \text{ rad/sec}^2$$

The angular displacement of the wheel can now be obtained from (3.6.5):

$$\theta = \omega_0 t + \tfrac{1}{2}\alpha t^2 = (0)(10) + (\tfrac{1}{2})(7.82)(10)^2 = 391 \text{ rad}$$

The number n of revolutions is found by dividing this figure by 2π rad/rev, which gives $n = 391/2\pi = 62.3$ rev. It should also be noted that the angular displacement can be obtained from (3.6.8), which gives

$$\theta = \frac{\omega^2 - \omega_0{}^2}{2\alpha} = \frac{78.2^2 - 0}{(2)(7.82)} = 391 \text{ rad}$$

in agreement with our former result.

3.7 Radial and Azimuthal Components of Motion for a Particle Moving in a Plane

In section 3.4, we developed expressions for the radial and azimuthal components of velocity and acceleration for a particle traveling in a circular path in the xy-plane. In this section, we shall extend that discus-

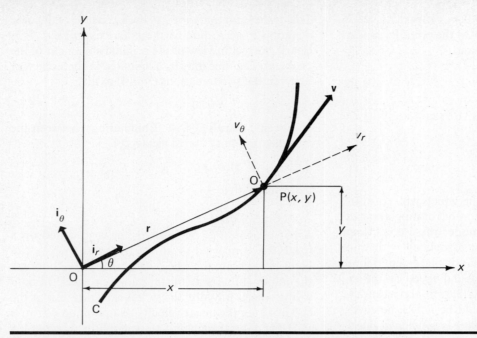

FIGURE 3.16. Radial and azimuthal components of the velocity vector. The velocity vector **v** is shown resolved into radial and azimuthal components. It is evident from this diagram that the velocity can be represented as $\mathbf{v} = v_r \mathbf{i}_r + v_\theta \mathbf{i}_\theta$, where $\mathbf{i}_r$ and $\mathbf{i}_\theta$ are the unit vectors in the radial and azimuthal directions.

sion to the calculation of the radial and azimuthal velocity and acceleration components for the case of *particle motion of any kind* occurring in the xy-plane.

The student may well wonder why we insist on subjecting him—or her—to calculations of this sort. After all, we have already exhibited the Cartesian velocity and acceleration components in section 3.1. Why not stick to Cartesian coordinates all the time and work out every problem using only x-, y-, and z-components? The answer is that doing this in certain problems, particularly those involving rotational motion, introduces algebraic complexities that have to be endured to be believed! In addition, although the use of Cartesian coordinate systems in these problems leads to results that are correct, they are expressed in a complex way that obscures most of the simple, beautiful, and important *physical* aspects of the motion. It is much better, in the long run, to fight your way through the mathematics at this point than to have to use Cartesian coordinates in every problem.

Suppose, now, that a particle P moves with velocity **v** along an arbitrary curve C in the xy-plane, as shown in Fig. 3.16. It will be possible, in general, to resolve the velocity into radial and azimuthal components v_r and v_θ, as shown.[4] The same radial and

---

[4] The reader should note the distinction between the terms *tangential* and *azimuthal* components of motion. The former refers to a component in the direction of the tangent to the path at a given point, while the latter refers to the component normal to the radius vector, thus the component in the θ-direction. Only if the path is circular are the two components identical.

azimuthal resolution can be applied to the acceleration vector **a**, which need not, however, be directed tangentially to the curve, like **v**. Now let us recall the work we did at the beginning of section 3.4. First, we should note that Eqs. (3.4.1) to (3.4.5) involve only relationships between the (x, y) and (r, θ) coordinates of the point P and in no way depend upon the path traced out by that point in the course of time. These equations simply relate the Cartesian and polar components of the unit vectors $\mathbf{i}_x$, $\mathbf{i}_y$, $\mathbf{i}_r$, and $\mathbf{i}_\theta$. They are, therefore, independent of any assumption as to whether or not the motion is circular and apply here equally well as in section 3.4. The same statement can be made regarding Eq. (3.4.6). In this case, as well as that treated in section 3.4, there is a radial displacement vector **r** whose Cartesian components are x (or $r \cos \theta$) and y (or $r \sin \theta$). Once more, this displacement vector can be written as the magnitude r times the radial unit vector $\mathbf{i}_r$, from which

$$\mathbf{v} = \frac{d\mathbf{r}}{dt} = \frac{d(r\mathbf{i}_r)}{dt} = \frac{dr}{dt}\mathbf{i}_r + r\frac{d\mathbf{i}_r}{dt} \tag{3.7.1}$$

We may now calculate $d\mathbf{i}_r/dt$ as before by differentiating (3.4.3), obtaining, as in section 3.4,

$$\frac{d\mathbf{i}_r}{dt} = \mathbf{i}_\theta \frac{d\theta}{dt} = \omega \mathbf{i}_\theta \tag{3.7.2}$$

where, as usual, $\omega = d\theta/dt$. Substituting this into Eq. (3.7.1), we get

$$\mathbf{v} = \frac{d\mathbf{r}}{dt} = \frac{dr}{dt}\mathbf{i}_r + (r\omega)\mathbf{i}_\theta \tag{3.7.3}$$

In the previous example involving a circular path, the magnitude r representing the radius of the circle is a constant, and we were justified in setting dr/dt equal to zero. In the present case, however, no such assumption is made, and both terms must be retained. But since any vector $\mathbf{A}$ can be expressed in terms of radial and azimuthal components by $\mathbf{A} = A_r\mathbf{i}_r + A_\theta\mathbf{i}_\theta$, (3.7.3) can be written as

$$\mathbf{v} = v_r\mathbf{i}_r + v_\theta\mathbf{i}_\theta = \frac{dr}{dt}\mathbf{i}_r + (r\omega)\mathbf{i}_\theta \qquad (3.7.4)$$

where v_r and v_θ are the radial and azimuthal components of $\mathbf{v}$. From this, however, it is apparent that

$$v_r = dr/dt \qquad (3.7.5)$$

$$v_\theta = r\omega \qquad (3.7.6)$$

The acceleration $\mathbf{a}$ may now be obtained by differentiating Eq. (3.7.3) to obtain

$$\mathbf{a} = \frac{d\mathbf{v}}{dt} = \frac{d^2r}{dt^2}\mathbf{i}_r + \frac{dr}{dt}\frac{d\mathbf{i}_r}{dt} + \left(r\frac{d\omega}{dt} + \omega\frac{dr}{dt}\right)\mathbf{i}_\theta + r\omega\frac{d\mathbf{i}_\theta}{dt} \qquad (3.7.7)$$

But $d\mathbf{i}_r/dt$ is given by (3.7.2), while $d\mathbf{i}_\theta/dt$, evaluated as before by differentiating (3.4.4), can be expressed as

$$\frac{d\mathbf{i}_\theta}{dt} = -\omega\mathbf{i}_r \qquad (3.7.8)$$

Substituting these values into (3.7.7), we find

$$\mathbf{a} = \left(\frac{d^2r}{dt^2} - r\omega^2\right)\mathbf{i}_r + \left(2\omega\frac{dr}{dt} + r\frac{d\omega}{dt}\right)\mathbf{i}_\theta = a_r\mathbf{i}_r + a_\theta\mathbf{i}_\theta \qquad (3.7.9)$$

or

$$a_r = \frac{d^2r}{dt^2} - r\omega^2 \qquad (3.7.10)$$

$$a_\theta = r\frac{d\omega}{dt} + 2\omega\frac{dr}{dt} \qquad (3.7.11)$$

Note that these results reduce to those obtained previously as (3.4.16) and (3.4.17) for the case of circular motion, where $dr/dt = 0$.

It is important to observe that a_r is not simply dv_r/dt and that a_θ is not equal to dv_θ/dt! The reason for this seemingly strange state of affairs is that the unit vectors $\mathbf{i}_r$ and $\mathbf{i}_\theta$ are not constant but may change direction as the particle moves. For example, if we write

$$\mathbf{v} = v_r\mathbf{i}_r + v_\theta\mathbf{i}_\theta \qquad (3.7.12)$$

and differentiate with respect to time to find the acceleration, we obtain not only the radial component $(dv_r/dt)\mathbf{i}_r$ but other radial components that arise from the fact that the unit vectors as well as the components

have time derivatives. Only when using Cartesian coordinate systems in which the unit vectors are constant both in magnitude and direction can we write the components of acceleration as the time derivative of the corresponding velocity components. In the case of uniform circular motion, $v_r = dr/dt = 0$ and, therefore, $dv_r/dt = 0$. The radial component of acceleration, however, is not zero, as we should now know very well, but $-r\omega^2$. The reality of this radial acceleration is well established not only by the calculations above and in section 3.4, but by Example 3.6.1 and by the geometric arguments illustrated in Fig. 3.14.

EXAMPLE 3.7.1

A man standing initially at the center of a merry-go-round rotating with constant angular velocity ω walks outward to the edge of the rotating platform along a radius of the turntable with uniform speed v_r with respect to the turntable. Find the radial and tangential components of his velocity and acceleration at any point.

The equations of the man's path, with respect to the earth, can be written as

$$r = v_r t \qquad (3.7.13)$$

$$\theta = \omega t \qquad (3.7.14)$$

where v_r and ω are constants. This is a set of parametric equations which tell what values r and θ have at any time t. If we eliminate t between these equations by solving for t in (3.7.14) and substituting the value so obtained in (3.7.13), we may obtain a single equation which tells us what r is as a function of θ, that is, the polar equation for the path described by the man as he walks outward. We may easily show in this manner that the polar equation of the path is

$$r = \frac{v_r}{\omega}\theta \qquad (3.7.15)$$

which is easily recognized as the equation of a spiral, as shown in Fig. 3.17. This is intuitively in agreement with what one could probably expect under the circumstances.

According to (3.7.5), the radial component of velocity is simply given by

$$\frac{dr}{dt} = v_r \qquad (3.7.16)$$

Likewise, according to Eqs. (3.7.6), (3.7.13), and (3.7.14), the θ-component of velocity is

$$v_\theta = r\frac{d\theta}{dt} = r\omega = v_r\omega t \qquad (3.7.17)$$

From (3.7.10), the radial acceleration is

$$a_r = \frac{d^2r}{dt^2} - r\left(\frac{d\theta}{dt}\right)^2 = -r\omega^2 = -v_r\omega^2 t \qquad (3.7.18)$$

85

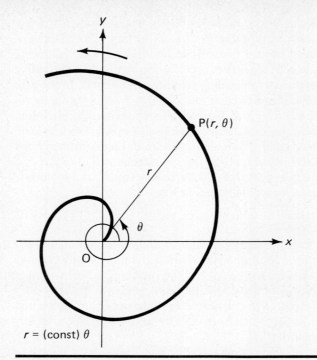

$r = (\text{const})\,\theta$

FIGURE 3.17. Spiral path traced out by a man who walks radially outward on a platform which rotates with constant angular velocity.

while the azimuthal acceleration, according to (3.7.11), is

$$a_\theta = 2\left(\frac{dr}{dt}\right)\left(\frac{d\theta}{dt}\right) + r\left(\frac{d^2\theta}{dt^2}\right) = 2\omega v_r \qquad (3.7.19)$$

These answers are in agreement with previously obtained results arising from uniform circular motion, in that an azimuthal velocity $r\omega$ and a centripetal acceleration $-r\omega^2$ are obtained; but, in addition, there is an *azimuthal acceleration* $2\omega v_r$ that was not present before. Such an acceleration can arise in any rotating system whenever the radial velocity is other than zero. In the case of uniform circular motion, of course, the radial velocity dr/dt is zero and there is no azimuthal acceleration. The present example does not fall into this category on account of the radial motion, which gives rise to a spiral path in place of the circular one associated with uniform rotation.

The azimuthal acceleration gives rise to certain curious dynamic effects that can be observed in any rotating system when radial motion is present and is responsible for certain observable terrestrial effects which occur in connection with the earth's axial rotation.

The magnitudes of the velocity and acceleration vectors may be calculated from Eqs. (3.7.16), (3.7.17), (3.7.18), and (3.7.19) by observing that

$$v = \sqrt{v_r^2 + v_\theta^2} = v_r\sqrt{1 + \omega^2 t^2} \qquad (3.7.20)$$

and

$$a = \sqrt{a_r^2 + a_\theta^2} = \omega v_r\sqrt{4 + \omega^2 t^2} \qquad (3.7.21)$$

3.8 Velocity and Acceleration in Different Reference Frames

It is often of great importance to be able to relate the motion of an object observed in a given coordinate system, or *frame of reference*, to the motion of the same object that would be observed in another coordinate system moving with respect to the original one. Such problems arise, for example, in interpreting the predictions of Newton's laws aboard moving, especially *accelerating*, vehicles or rotating systems, and in navigating ships and aircraft when air or water currents are present. Beyond this, some of the more elementary concepts related to the *special theory of relativity* are most easily understood by investigating how the results of certain measurement processes would be interpreted by two observers who are in motion with respect to one another. We shall not go into detail about this particular aspect of moving reference frames here but shall, nevertheless, build a foundation for understanding certain aspects of special relativity which will be useful later. In this section we shall emphasize unaccelerated reference frames, reserving discussion of accelerated reference systems for the section that follows.

Consider a point P, which traces out some path in the course of time, with a given velocity and acceleration at each point. We can easily describe the position of P by a vector $\mathbf{r} = \mathbf{i}_x x + \mathbf{i}_y y + \mathbf{i}_z z$ referred to an origin O about which a Cartesian (x, y, z)-coordinate system is erected. Likewise, the velocity vector $\mathbf{v} = \mathbf{i}_x v_x + \mathbf{i}_y v_y + \mathbf{i}_z v_z$ and the acceleration vector $\mathbf{a} = \mathbf{i}_x a_x + \mathbf{i}_y a_y + \mathbf{i}_z a_z$ can be defined with reference to the same coordinate system, the velocity and acceleration components being related to x, y, and z and the time t by (3.1.10) and (3.1.16). Suppose now that a second frame of reference exists, with axes x', y', and z' (which, for simplicity, we shall assume to be parallel to the x-, y-, and z-axes of the original system) intersecting in an origin O' which is in motion, but not necessarily constant motion, with respect to O. This situation is illustrated in Fig. 3.18. The coordinates of the origin of the second system are x_0, y_0, and z_0, and this point can be represented by a position vector $\mathbf{r}_0 = \mathbf{i}_x x_0 + \mathbf{i}_y y_0 + \mathbf{i}_z z_0$. The motion of the point P, as viewed from this second frame of reference, can be described by a position vector $\mathbf{r}' = \mathbf{i}_x x' + \mathbf{i}_y y' + \mathbf{i}_z z'$, where x', y', and z' are the coordinates of P in the new system.[5] There will be an apparent velocity vector

---

[5] Note that because the coordinate axes of the two systems are parallel at all times, a single set of unit vectors $(\mathbf{i}_x, \mathbf{i}_y, \mathbf{i}_z)$ suffices for both reference frames.

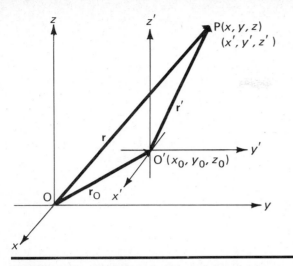

FIGURE 3.18. Relationship between the displacement vectors of a point particle referred to two different reference systems.

$\mathbf{v}' = \mathbf{i}_x v_x' + \mathbf{i}_y v_y' + \mathbf{i}_z v_z'$ and an apparent acceleration vector $\mathbf{a}' = \mathbf{i}_x a_x' + \mathbf{i}_y a_y' + \mathbf{i}_z a_z'$ associated with the motion of the point P as seen from the primed reference frame, with expressions of the same form as (3.1.10) and (3.1.16) relating the components of velocity and acceleration in this system to the variables x', y', z', and t'.

In doing this, it is assumed that an *absolute time scale* can be established applicable to both reference frames, so that $t' = t$. This assumption would be justified *if the velocity of light were infinitely large* but due to the finite (though large) speed of light, it is not quite correct.[6] A detailed examination of the reasoning upon which this assertion is based will be postponed until later, but it is founded upon the easily demonstrated fact that (again due to the finite speed of light) two events which appear to be simultaneous in the fixed system may not seem to be simultaneous in the moving system. The recognition that the time scales associated with the two reference frames are not the same if relative motion exists between them is one of the fundamental principles of the special theory of relativity proposed by Einstein in 1905. This effect results in certain deviations of the dynamic behavior of physical systems from what is predicted by Newton's laws. Although these predicted relativistic deviations have been observed and confirmed by experiment, unless relative velocities amounting to a significant fraction of the speed of light are involved, they are extremely small and are undetectable by ordinary experimental methods. Until we are ready to consider relativistic behavior explicitly in a later chapter, we shall always confine ourselves to the study of systems

in which all velocities are much less than the speed of light and in which, therefore Newton's laws in their usual form are essentially correct. This is not a very serious restriction, since most macroscopic terrestrial and even celestial systems that can be studied experimentally fall into this category.

From the vector diagram in Fig. 3.18 it is evident that the relation between $\mathbf{r}$ and $\mathbf{r}'$, the position vectors of P in the two reference frames, is

$$\mathbf{r}' = \mathbf{r} - \mathbf{r}_0 \qquad (3.8.1)$$

This equation may be differentiated with respect to time, to obtain

$$\mathbf{v}' = \mathbf{v} - \mathbf{v}_0 \qquad (3.8.2)$$

where $\mathbf{v}' = d\mathbf{r}'/dt$ is the velocity of P relative to the primed coordinate system, $\mathbf{v} = d\mathbf{r}/dt$ is the velocity relative to the unprimed system, and $\mathbf{v}_0 = d\mathbf{r}_0/dt$ is the velocity of the primed with respect to the unprimed system. This equation may be differentiated once more to get a relation between acceleration vectors of the form

$$\mathbf{a}' = \mathbf{a} - \mathbf{a}_0 \qquad (3.8.3)$$

where $\mathbf{a}' = d\mathbf{v}'/dt$, $\mathbf{a} = d\mathbf{v}/dt$, and $\mathbf{a}_0 = d\mathbf{v}_0/dt$. Equations (3.8.2) and (3.8.3) enable us to find the velocities and accelerations in the moving system in terms of those in the fixed system and the velocity and acceleration of the primed system with respect to the unprimed. If the velocity $\mathbf{v}_0$ of the primed system with respect to the unprimed is constant, then (3.8.3) reduces to $\mathbf{a}' = \mathbf{a}$, since under these circumstances $\mathbf{a}_0$ is zero. The accelerations in the two frames are then the same. We shall assume that this is the case for the rest of the discussion presented in this section. If the two origins coincide at $t = 0$, the relative position vector can be written

$$\mathbf{r}_0 = \mathbf{v}_0 t \qquad (3.8.4)$$

since then $d\mathbf{r}_0/dt = \mathbf{v}_0 = $ const., as assumed. Equation (3.8.1) now becomes

$$\mathbf{r}' = \mathbf{r} - \mathbf{v}_0 t \qquad (3.8.5)$$

or, taking components,

$x' = x - v_{0x}t$
$y' = y - v_{0y}t$
$z' = z - v_{0z}t$

This kind of transformation, relating coordinates measured in a frame moving with constant velocity to those in a fixed frame, assuming that t' and t are the same, is sometimes called a *Galilean transformation*. The corresponding relativistic transformation, in which the difference in time scales between the two systems is accounted for, is referred to as a *Lorentz transformation* and will be introduced in a later chapter.

EXAMPLE 3.8.1

A jet plane flies at a constant "air speed" of 560 mph between Dallas, Texas, and Birmingham, Alabama, 600 miles due east of Dallas. Find the total flight time from Dallas to Birmingham, assuming that a westerly wind blows at a constant speed of 60 mph. What is the flight time for the return trip under these conditions? If the wind speed increases to 90 mph, what will be the percent change in flying time for the round trip?

Let us assume that the unprimed reference system is fixed with respect to the earth and the primed system is fixed relative to the air, and that the x- and x'-axes both point due east. The wind comes from the west and, therefore, blows due east with velocity $\mathbf{v}_0 = \mathbf{i}_x v_{0x}$, with $v_{0x} = 60$ mph. This velocity represents the relative velocity of the primed system (air) with respect to the unprimed frame (earth). From Eq. (3.8.2), then,

$$v_x' = v_x - v_{0x}$$

In this equation, v_x represents the speed of the plane with respect to the ground, which is what we are seeking, while v_x' is its speed relative to the air, given as 560 mph. Therefore,

$$560 = v_x - 60$$
$$v_x = 620 \text{ mph}$$

This corresponds to a flight time of 600/620 hr, or 58.06 min. If the wind were in the other direction (along the $-x$-axis) we would have $v_{0x} = -60$ mph, and the above equation would read

$$560 = v_x - (-60)$$
$$v_x = 500 \text{ mph}$$

The flight time would now be 600/500 hr, or 72.0 min. If the wind blows from west to east at 90 mph, $v_{0x} = 90$ mph, the initial journey now taking place at $v_x = 650$ mph and requiring 600/650 hr, or 55.38 min. On the return trip, $v_x = 470$ mph, the time required being 76.60 min.

For the whole trip, representing the 600-mile distance as d, we can represent the flying time as

$$t = t_1 + t_2 = \frac{d}{v_x' + v_{0x}} + \frac{d}{v_x' - v_{0x}}$$
$$= \frac{d}{v_x'} \left[\frac{1}{1 + (v_{0x}/v_x')} + \frac{1}{1 - (v_{0x}/v_x')} \right]$$
$$= \frac{2d}{v_x'} \frac{1}{1 - (v_{0x}^2/v_x'^2)}$$

$$= \frac{(2)(600)}{560(1 - ((60)^2/(560)^2))}$$
$$= 2.168 \text{ hr}$$

using $v_{0x} = 60$ mph. If v_{0x} increases to 90 mph, the same calculation gives $t = 2.200$ hr, an increase of 1.48 percent. It is interesting to note that even though the wind shortens the time required to make one of the legs of the journey, its net effect is always to *increase* the time required for the round trip.

EXAMPLE 3.8.2

Show how to navigate a ship in a situation where the magnitude and direction of the ship's velocity with respect to the water are known and where there is an ocean current of known velocity and direction. Show also how to find what heading and velocity the ship must have to make good a predetermined speed along a predetermined course.

These results can be obtained from Eq. (3.8.2). Let us associate the unprimed fixed system with the earth's surface and the primed system with the water, which moves with velocity v_0 with respect to the surface of the earth. Orient the fixed system so that the x-axis is in the direction of $\mathbf{v}_0$. The situation is then as illustrated in Fig. 3.19a. The ship heads in direction $\mathbf{v}'$, making speed v' relative to the sea. The direction of $\mathbf{v}'$ makes an angle ϕ with the current, whose velocity vector is $\mathbf{v}_0$. The velocity vector $\mathbf{v}$ of the ship in the frame fixed with respect to the ocean bottom is, according to (3.8.2), the *vector sum* of $\mathbf{v}'$ and $\mathbf{v}_0$. Let θ be the angle between $\mathbf{v}$ and the current vector $\mathbf{v}_0$. Then, from Fig. 3.19a,

$$v_x' = v' \cos \phi \qquad \text{and} \qquad v_y = v_y' = v' \sin \phi$$

$$\tan \theta = \frac{v_y}{v_0 + v_x'} = \frac{v_y'}{v_0 + v_x'} = \frac{v' \sin \phi}{v_0 + v' \cos \phi} \qquad (3.8.7)$$

Also,

$$v^2 = (v_0 + v_x')^2 + v_y'^2 = v_0^2 + 2v_0 v_x' + v_x'^2 + v_y'^2$$

But since $v_x'^2 + v_y'^2 = v'^2$,

$$v^2 = v_0^2 + v'^2 + 2v_0 v_x'$$

or

$$v^2 = v_0^2 + v'^2 + 2v_0 v' \cos \phi \qquad (3.8.8)$$

Suppose that the ship's speed through the water is 18 knots and it steers a course making an angle of 60° with the x-axis. Then, $v' = 18$ knots, $v_0 = 4$ knots, and $\phi = 60°$; according to (3.8.7) and (3.8.8) or Fig. 3.19b,

$$\tan \theta = \frac{18(\sin 60°)}{4 + 18(\cos 60°)} = \frac{(18)(0.866)}{4 + (18)(0.500)}$$

$$\tan \theta = 1.200 \qquad \theta = 50.2°$$
$$v^2 = (4^2) + (18^2) + (2)(4)(18)(\cos 60°) = 412$$

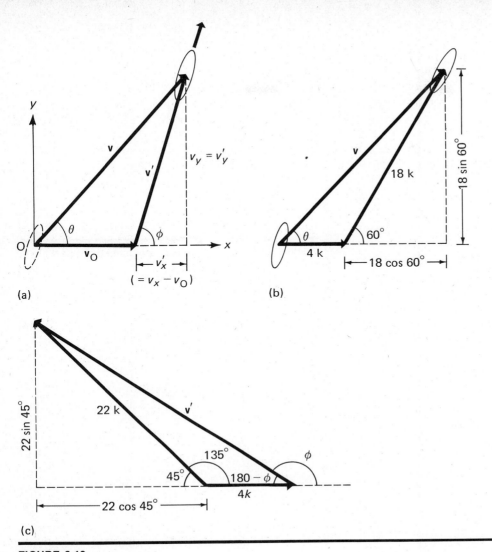

FIGURE 3.19

$v = 20.3$ knots

The ship will make an effective speed of 20.3 knots along a direction making an angle of 50.2° with respect to the current.

The other problem arises when $\mathbf{v_0}$ is known and it is desired to make a given speed along a given direction. In effect, $\mathbf{v_0}$ and $\mathbf{v}$ are given, and $\mathbf{v'}$ must be found in order to determine how the vessel must be headed and how fast she must go with respect to the water in order to make the required speed along the specified course. In this case, the angle θ is known and ϕ must be found. Referring to Fig. 3.19a, we may write

$$v_x = v\cos\theta \quad \text{and} \quad v_y = v\sin\theta$$

Then,

$$\tan\theta = \frac{v_y}{v_x - v_0} = \frac{v\sin\theta}{v\cos\theta - v_0} \tag{3.8.9}$$

$$v'^2 = (v_x - v_0)^2 + v_y^2 = v_x^2 - 2v_xv_0 + v_0^2 + v_y^2$$

However, since $v_x^2 + v_y^2 = v^2$, this may be written

$$v'^2 = v^2 + v_0^2 - 2v_xv_0 = v^2 + v_0^2 - 2vv_0\cos\theta \tag{3.8.10}$$

For example, suppose the ship is required to make an effective speed of 22 knots over the ocean bottom on a course making an angle of 135° with a 4-knot current. Then, from (3.8.9) and (3.8.10) or from Fig. 3.19c,

$$\tan\phi = \frac{22(\sin 135°)}{22(\cos 135°) - 4} = \frac{(22)(0.7071)}{(22)(-0.7071) - 4}$$

$$\tan\phi = 0.795 \qquad \phi = 141.5°$$

$$v'^2 = (22^2) + (4^2) - (2)(22)(4)(\cos 135°) = 624.4$$

$$v' = 25.0 \text{ knots}$$

The ship must steer a course making an angle of 141.5° with the current and make a speed of 25.0 knots with respect to the water to maintain the desired course and speed with respect to the ocean bottom.

3.9 Velocity and Acceleration in Accelerated Reference Systems

In working mechanics problems, we often regard the earth's surface as an unaccelerated frame of reference. We know, however, that the earth rotates about its own axis and also describes a more or less circular orbit around the sun. Both these motions are accelerated ones, and it is, therefore, evident that any coordinate system fixed with respect to the earth's surface must in reality undergo accelerations. In many instances, the effect of these accelerations is so small that it can be observed only with difficulty, but situations do exist in which they may be quite significant. It is, therefore, important to be able to discuss the laws of mechanics from the point of view of accelerated reference frames. The first step in doing this is in understanding the relationships between the velocity and acceleration of a body measured in an unaccelerated coordinate system and the corresponding quantities observed from a frame of reference having a known acceleration.

In an unaccelerated coordinate system, which is often referred to as an *inertial* reference frame, the acceleration of any object is directly attributable to real forces acting on it. These forces and accelerations are, of course, directly related by Newton's second law. In an accelerated or *noninertial* reference frame, however, the situation is somewhat different. In such a system, part of the observed acceleration may be due to the action of real forces, but *part of the observed acceleration may arise as a result of the acceleration of the coordinate system itself*.

As an illustration, consider the case where two men on a rotating merry-go-round, one near the center, the other near the outside, wish to toss a baseball back and forth. The man near the center throws the ball directly toward his friend. To an observer stationed directly above the merry-go-round, looking down on the rotating platform, the ball appears to move outward in a straight line, with no tangential acceleration whatsoever. During the time the ball is in the air, however, the platform's rotation carries the prospective catcher sideways relative to the ball's trajectory, to a point where he can no longer catch it. Everything is very clear to the outside observer in the unaccelerated reference system. But the man near the outside rim of the rotating platform, who is in an accelerated system of reference, sees the ball deflected sideways more and more rapidly as it moves toward him, until at length it is far beyond his reach. Actually, of course, it is his own acceleration that prevents him from catching the ball. But in his accelerated frame of reference, it seems as though it is the ball that is undergoing acceleration. In seeking to explain the ball's apparent acceleration, he looks for forces that could have been responsible for it and

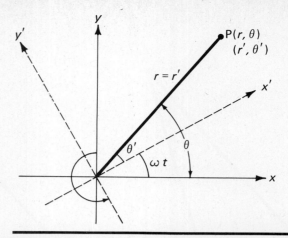

FIGURE 3.20. Coordinates of a moving point particle P, referred to two coordinate systems one of which rotates with respect to the other.

ascribes a certain reality to these forces. We often refer to "forces" of this sort as *pseudoforces* or *inertial forces*, and we shall discuss them in some detail in the next chapter.

To understand these effects more clearly, let us consider the situation illustrated in Fig. 3.20, in which a point P moves in some arbitrary fashion in the *xy*-plane. The coordinates of P can be expressed as (r, θ) in a fixed system of reference (x, y) or as (r', θ') in a rotating system (x', y') whose origin coincides with that of the fixed system at all times. It is assumed that the rate of rotation of the rotating system has the *constant* value ω_0 and that the (x, y)- and (x', y')-axes coincided at $t = 0$, so that the angle between the two systems at time t is ω_0, as illustrated. The point P has radial and azimuthal velocity and acceleration components v_r, v_θ, a_r, a_θ and v_r', v_θ', a_r', a_θ', as seen respectively from the fixed and rotating systems. The problem is to relate the true velocity and acceleration components v_r, v_θ, a_r, a_θ to the apparent values r', θ', v_r', v_θ', a_r', a_θ' seen in the rotating frame.

From Fig. 3.20, it is easily established that

$$\boxed{\begin{aligned} r &= r' \\ \theta &= \theta' + \omega_0 t \end{aligned}} \tag{3.9.1}$$

whence, differentiating,

$$\frac{dr}{dt} = \frac{dr'}{dt}$$

$$\frac{d\theta}{dt} = \frac{d\theta'}{dt} + \omega_0 \tag{3.9.2}$$

$$\frac{d^2r}{dt^2} = \frac{d^2r'}{dt^2}$$

$$\frac{d^2\theta}{dt^2} = \frac{d^2\theta'}{dt^2} \tag{3.9.3}$$

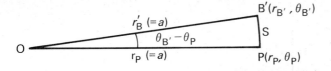

FIGURE 3.22

is small, the arc length S and the chord $B'P$ are very nearly equal. We see, then, that to a high degree of approximation,

$$B'P = S = a(\theta_{B'} - \theta_P) = \omega_0 v_{0x} t_0^2 \qquad (3.9.18)$$

According to this, the object falls to the right of point B' (as seen from A'), the displacement increasing as the square of the time of flight t_0. In the rotating system of reference, it is as if the object were subjected to a constant tangential acceleration of magnitude $-2\omega_0 v_{0x}$, since then its displacement after time t_0 would be

$$y' = \tfrac{1}{2}a_{y'}t_0^2 = \tfrac{1}{2}(-2\omega_0 v_{0x})t_0^2 = -\omega v_{0x}t_0^2$$

measured in an (x', y')-system fixed in the rotating turntable with the x'-axis along $A'B'$. This is in agreement with the result expressed by (3.9.18).

A much easier and less painful way to arrive at this result is to note that since, during flight, the object is subjected to no acceleration components in the plane of the turntable, the true tangential acceleration a_θ in (3.9.7) may be set equal to zero, whereupon the tangential acceleration in the rotating frame, a_θ', must be equal to $-2\omega_0 v_r$, in agreement with the results obtained previously. The detailed argument above is presented to allow the reader to obtain a clear physical picture of the origin of the accelerations which may be observed in rotating reference frames for objects that are in reality completely unaccelerated.

SUMMARY

The study of the purely geometric aspects of motion is known as *kinematics*. The study of forces and the motion that results from their action is referred to as *dynamics*.

Average velocity is defined as the vector displacement divided by the time interval during which it occurs, that is, $\overline{\mathbf{v}} = \Delta\mathbf{r}/\Delta t$. The instantaneous velocity at a given instant of time is obtained by taking the limit of the average velocity as the length of the interval Δt approaches zero. Therefore, the instantaneous vector velocity is the time derivative of the vector displacement; $\mathbf{v} = d\mathbf{r}/dt = \mathbf{i}_x(dx/dt) + \mathbf{i}_y(dy/dt) + \mathbf{i}_z(dz/dt)$.

Average acceleration is defined as the change in the (instantaneous) velocity vector divided by the time interval during which it takes place. Instantaneous acceleration is obtained in the limit as the length of the time interval approaches zero and can, therefore, be defined as the derivative of the vector velocity with respect to time.

In the case of motion in one dimension with *constant acceleration*, the "golden rules"

$$\begin{aligned} v_x &= v_{0x} + a_x t \\ x &= v_{0x}t + \tfrac{1}{2}a_x t^2 \\ v_x^2 &= v_{0x}^2 + 2a_x x \end{aligned}$$

are applicable. These rules are not valid when the acceleration varies with time. An example of motion with constant acceleration is provided by the motion of objects under the influence of the earth's gravitational attraction near the earth's surface.

Projectile motion is motion in two dimensions in which the acceleration vector is constant in magnitude and direction. Under such circumstances, the components of motion parallel to the acceleration vector are described by the "golden rules," while the motion perpendicular to that direction is unaccelerated.

Circular motion is another example of two-dimensional motion. In *uniform circular motion*, the rotational speed is constant and the acceleration vector is directed radially inward with magnitude v^2/r, or $r\omega^2$. In *accelerated circular motion*, the acceleration vector has, in addition, a tangential component whose magnitude is dv/dt, or $r(d\omega/dt)$.

Motion in a plane can be described in terms of Cartesian components or radial and azimuthal components. The latter choice frequently allows us to understand the important physical characteristics of the motion more easily.

Angular velocity is a vector $\boldsymbol{\omega}$ directed along the axis of rotation whose magnitude is given by $d\theta/dt = \omega$. The angular acceleration vector $\boldsymbol{\alpha}$ is defined as the time rate of change of the angular velocity vector, that is, $\boldsymbol{\alpha} = d\boldsymbol{\omega}/dt$. When rotation occurs about a single fixed axis, the vector $\boldsymbol{\alpha}$ (as well as $\boldsymbol{\omega}$) points in the direction of the rotation axis.

In *uniformly accelerated rotational motion*, the vectors $\boldsymbol{\omega}$ and $\boldsymbol{\alpha}$ are in the direction of the axis of rotation, and the magnitude of the angular acceleration is constant in time. Under these circumstances, the motion is described by a set of "golden rules" very similar to those pertaining to uniformly accelerated linear motion:

$$\begin{aligned} \omega &= \omega_0 + \alpha t \\ \theta &= \omega_0 t + \tfrac{1}{2}\alpha t^2 \\ \omega^2 &= \omega_0^2 + 2\alpha\theta \end{aligned}$$

93

QUESTIONS

1. The average velocity components of a particle can be either positive or negative. Can the same be said of the average speed?
2. An automobile is moving east, but its acceleration is west. What cay you infer about the way in which its speed depends on time?
3. At a recent track meet, one of the runners was timed at 4 minutes for the mile run. The start and finish lines are at the same place. What is the runner's average speed? What is his average velocity?
4. When the acceleration is constant, the average velocity is $(\mathbf{v}_1 + \mathbf{v}_2)/2$. Show that, in the absence of air resistance, a ball thrown up with a certain initial speed will return to the same location with the same speed.
5. A ball is thrown vertically, and air resistance is taken into account. Is the time of rise shorter or longer than the time of fall?
6. The instantaneous velocity of a body is known to be constant over a certain time interval. Show that its average velocity must be equal to the instantaneous velocity.
7. Suppose the earth were rotating about its axis at twice the present rate. Describe the change in the apparent weight of an object at the equator and at the north pole.
8. Do you think projectiles move on parabolic paths when air resistance is taken into account?
9. If air resistance is taken into account, do you think the horizontal range of a projectile is a maximum for an angle less than or greater than 45°?
10. Is it possible for an object to undergo circular motion and yet not be accelerating?
11. If an angular velocity is given in revolutions per minute, by what factor must this be multiplied to convert to radians per second?
12. A certain object is in motion in a circle. Can it have an angular acceleration but zero radial acceleration?
13. In planetary motion, the force on each planet is directed radially inward toward the sun. Does that mean there can be no azimuthal acceleration? Does this imply zero angular acceleration?
14. You can give an automobile a greater forwardly directed acceleration by stepping on the brakes than by stepping on the accelerator. Explain how this can be accomplished.

PROBLEMS

1. Prove, for motion in one dimension, that if the instantaneous velocity v_x is constant, the average velocity over any time interval must be equal to this instantaneous velocity.
2. The following is a passage from John D. MacDonald's *The Scarlet Ruse*, a detective novel starring Travis McGee. (Greenwich, Conn.: Fawcett Publications, 1973), quoted with the author's permission:

   "She grinned, threw the ski over the side, and dived after it. When she had the ski and had worked her feet into the slots, I pulled the bar past her at dead slow, and she grabbed it and turned to the right angle and nodded. I pushed both throttles, and she popped up out of the water and the *Munequita* jumped up onto the step in perfect unison. After she made a few swings, I knew she was not going to have any trouble. She gave me the pumping sign for more speed and then the circle of thumb and finger when it was where she wanted it. It translated to *thirty-two miles an hour*.

   "She was not tricky. There were no embellishments. All she did was get into the swooping rhythm of cutting back and forth across the almost-flat wake, out there in the expanse of Biscayne Bay, far from land, far from any other water craft. She edged the slalom ski as deeply as the men do, laying herself back at a steep angle, almost flat against the water, throwing a broad, thin, curved curtain of water at least ten feet into the air at the maximum point of strain. At that point before she came around and then came hurtling back across, ski flat, to go out onto the other wing, the strain would sag her mouth, wipe her face clean of expression, and pull all the musculature and tendons and tissues of her body so taut she looked like a blackboard drawing in medical school.

   "She took it each time to the edge of what she could endure. It was hypnotic and so determined that it had a slightly unpleasant undertaste, like watching a circus girl high under the canvas, going over and over and over, dislocating her shoulders with each spin, while the drums roll and the people count.

   "I put it into autopilot so I could watch her. From time to time I glanced forward to make certain no other boat was angling toward us. I knew she would have to tire soon. I tried to calculate her speed. She was going perhaps twenty-five feet outside the wake on one side and then the other. Call it fifty feet. I timed her from her portside turn back to her portside turn. Ten seconds. For a hundred feet. Miles per hour equals roughly two thirds of the feet per second. Ten feet per second. Add seven miles an hour then to the boat speed. Very close to forty miles an hour. At that speed, if she fell, the first bounce would feel like hitting concrete. Water is not compressible."

   What is wrong with private eye McGee's mathematics, and what is the correct answer for the girl's speed, in miles per hour?
3. The displacement of a point moving along the x-axis is given by $x(t) = v_0 t + \frac{1}{2}at^2$, where v_0 and a are constants. Show that during any time interval Δt of finite duration, beginning at time t, the average velocity $\bar{v}_x$ is given by $\bar{v}_x = v_0 + at + \frac{1}{2}a\,\Delta t$. Show from this result that the instantaneous velocity at any time is given by $v_x = v_0 + at$. *Hint:* The displacement Δx occurring during the time interval Δt is $x(t + \Delta t) - x(t)$.
4. An automobile drives north along a straight level road, covering a distance of 18.0 km in 15 minutes. It then turns around and returns along the same road to the starting point, the return trip taking 12 minutes. Find (a) the average speed for the outward journey, (b) the average speed for the return trip, (c) the average speed for the entire journey, (d) the average velocity for the outward journey, (e) the average velocity for the return trip, and (f) the average velocity for the whole journey.
5. Why can't you obtain the answer to part (c) of the preceding problem by finding the average of the answers to parts (a) and (b)?
6. An airplane flies 300 km due east in a time interval of 20 minutes. It then flies due north for 24 minutes, during which time a distance of 400 km is covered. Find (a) the average speed of the plane during the first leg, (b) the

average speed during the second leg, **(c)** the average speed for the whole trip, **(d)** the average velocity for the first leg, **(e)** the average velocity for the second leg, and **(f)** the average velocity for the whole trip.

7. An automobile is driven counterclockwise around a circular track whose circumference is 2000 meters at a constant speed of 20 meters/sec. Assuming that it starts out at time $t = 0$ headed due north, find **(a)** its displacement vector with respect to the center of the track at time $t = 0$, **(b)** its displacement vector at $t = 50$ sec, **(c)** its displacement vector at time $t = 25$ sec, **(d)** its instantaneous velocity at $t = 0$, **(e)** its instantaneous velocity at $t = 50$ sec, **(f)** its instantaneous velocity at $t = 12.5$ sec, **(g)** its average velocity during the interval $0 < t < 25$ sec, and **(h)** its average acceleration during the interval $0 < t < 25$ sec.

8. Using the data given in the preceding problem, find the average velocity **(a)** during the time interval $0 < t < 5$ sec, **(b)** during the time interval $0 < t < 1$ sec, and **(c)** during the time interval $0 < t < 0.1$ sec.

9. The accompanying figure shows the displacement x as a function of time for a particle moving along the x-axis. Draw schematic diagrams showing the instantaneous velocity v_x and the instantaneous acceleration a_x as functions of time.

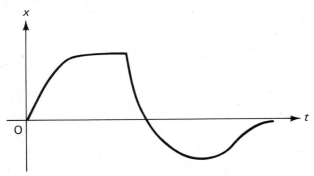

10. The accompanying diagram shows the acceleration a_x as a function of time for a particle that moves along the x-axis. Assuming that the particle starts from rest at $t = 0$, draw schematic diagrams showing the instantaneous velocity v_x and the displacement x as functions of time.

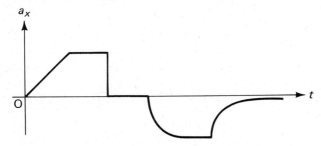

11. The displacement x of a particle moving along the x-axis is given as a function of time by $x(t) = at^2 - bt^4$, with x in meters and t in seconds. Find **(a)** the dimensions of the constants a and b. Assuming that a has the numerical value 16.0 units and b the numerical value 1.0 units, find **(b)** the instantaneous velocity as a function of time, **(c)** the instantaneous acceleration as a function of time, **(d)** the maximum value of the displacement for time $t > 0$, **(e)** the

maximum value of instantaneous velocity for $t > 0$, **(f)** the average velocity during the interval 1 sec $< t < 2$ sec, and **(g)** the average acceleration during the interval 1 sec $< t < 2$ sec.

12. The acceleration of a particle moving along the x-axis is given as a function of time by $a_x(t) = 36t - 24t^3$, when a_x is given in m/sec² and t in sec. **(a)** What are the dimensions of the numerical constants in this equation? Assuming that the particle starts from rest at the origin, find **(b)** the instantaneous velocity and **(c)** the instantaneous displacement as functions of the time. **(d)** What is the maximum value of displacement for $t > 0$? **(e)** What is the maximum value of velocity for $t > 0$? **(f)** What is the average velocity during the interval 1 sec $< t < 2$ sec? **(g)** What is the average acceleration during this same time interval?

13. A point moves along the x-axis in such a way that its displacement is given by $x(t) = x_0 \cos \omega t$, where x_0 and ω are constants. Find its instantaneous velocity and acceleration. If $x_0 = 5$ cm and $\omega = 25$ rad/sec, what are the values of the instantaneous velocity at $t = 0.05$ sec?

14. For the motion described in Problem 13, show that $a_x = -\omega^2 x$ and $v_x = -\omega\sqrt{x_0^2 - x^2}$. Plot $x(t)$, $v_x(t)$, and $a_x(t)$ as functions of time on a suitable coordinate scale.

15. A particle moves in the xy-plane, its position coordinates being given as functions of time by $x(t) = 3t$, $y(t) = t^3 - 12t$. Find **(a)** the displacement vector of the particle at $t = 3$ sec, **(b)** the velocity vector at $t = 3$ sec, **(c)** the acceleration vector at $t = 3$ sec, **(d)** the displacement vector, **(e)** the velocity vector, and **(f)** the acceleration vector when the particle is at the point P(6, −16). Assume that distances are measured in meters and time in seconds.

16. In the preceding problem, **(a)** what is the equation of the path along which the particle moves? Find the magnitude and direction of **(b)** the displacement, **(c)** the velocity, and **(d)** the acceleration at $t = 3$ sec.

17. A particle moves in the xy-plane, its position coordinates being given by $x(t) = 12 \cos 3\pi t$, $y(t) = 6 \sin 3\pi t$. Find **(a)** its displacement, **(b)** its velocity components, and **(c)** its acceleration components at time $t = 0.25$ sec. **(d)** What is the equation of the path along which the particle travels? Assume that distances are measured in meters, time is given in seconds, and that the angular quantity $3\pi t$ is expressed in radians.

18. A point executes a three-dimensional motion described by the following equations of motion: $x(t) = r \cos \omega t$, $y(t) = r \sin \omega t$, and $z(t) = bt$, where r, b, and ω are constants. Find **(a)** the x-, y-, and z-components of velocity at any time t; **(b)** the corresponding acceleration components; **(c)** the magnitude of the velocity vector; **(d)** the magnitude of the acceleration vector; **(e)** the numerical values of these quantities when $r = 5$ cm, $b = 1$ cm/sec, $\omega = 10$ rad/sec, and $t = 0.1$ sec. **(f)** What is the geometric form of the path followed by the point?

19. A point moves according to the following equations of motion: $x(t) = 5t^2 + 2t^3$, $y(t) = 5t^2 - t^4$, and $z(t) = 25t - t^3$, where x, y, z are expressed in centimeters and t is expressed in seconds. Find **(a)** the x-, y-, and z-components of velocity at any time t; **(b)** the corresponding acceleration components; **(c)** the magnitude of the velocity at any time t; **(d)** the magnitude of the acceleration at any time t; **(e)** the numerical values of all these

95

quantities when $t = 2$ sec; **(f)** the distance between the point and the origin when $t = 2$ sec.

20. A particle starts from the origin at $t = 0$, and thereafter its motion is described by the equations of motion $d^2x/dt^2 = \alpha t$ and $d^2y/dt^2 = d^2z/dt^2 = 0$, where α is a constant. Its initial velocity vector has a component v_0 along the x-direction and has no y- or z-component. Find the x-, y-, and z-components of its velocity and displacement for any time $t > 0$. What are its displacement, velocity, and acceleration components at $t = 3$ sec if α has the value 4.0 cm/sec^3 and $v_0 = 24$ cm/sec?

21. Find the equation corresponding to (3.2.13) which relates x and v_x under the conditions set forth in Problem 20.

22. A 1-lb object is dropped from the top of a building 720 ft high. Neglecting air resistance, find **(a)** the time required for it to fall to the earth and **(b)** the velocity on impact. **(c)** What are the answers to these questions if the object weighs 10 lb?

23. What are the answers to parts **(a)** and **(b)** of Problem 22 above if the object is thrown downward with an initial velocity of 40 ft/sec?

24. A driver passes an unmarked police car parked at the side of the road, his car traveling at a constant speed of 110 ft/sec (75 mph). At that instant, the policeman starts to chase the speeder, his car attaining a constant acceleration of 12.0 ft/sec^2. **(a)** How long does it take the police car to catch up with the speeding motorist? **(b)** How far does the police car travel before overtaking the other car? **(c)** How fast is the police car traveling when the other car is overtaken?

25. An object is dropped from rest from a window on the fortieth floor of an office building, 480 ft above the street. At the instant it is released, a second object is thrown downward from the roof of the building, which is 720 ft above street level. With what initial downward velocity does this second object have to be thrown in order that it may hit the street at the same time as the first body? Neglect air resistance.

26. In the situation described in the preceding problem, suppose that both objects are released from rest but at different times. Neglecting air resistance, how long after the object on the roof is dropped would the one on the fortieth floor be released in order that they hit the ground together?

27. Again referring to the situation considered in the two preceding problems, suppose that both objects are dropped from the roof of the building. The first object is released from rest, and the second is thrown downward 2.0 sec after the first one is dropped. Neglecting air resistance, what initial downward velocity does the second object have to be given in order that it may hit the street at the same instant as the first?

28. Two trains, initially 90 miles apart, are traveling toward one another on the same track. One of the trains is moving at a speed of 25 mph, the other at 35 mph. A bird, traveling at a constant speed of 60 mph, starts out at the first train, flies to the second, then reverses direction, flying back to the first, reversing direction again, and in this fashion flying back and forth between trains until they collide. What total distance has the bird flown at the instant the trains collide? *Hint:* If you're adding

up the length of the individual paths, you're barking up the wrong tree!

29. Two automobiles are traveling initially at the same speed along a straight road. The first car is 100 meters ahead of the second. The second car has a constant acceleration of 8 ft/sec^2, the first a constant acceleration of 6 ft/sec^2. **(a)** How long does it take the second car to overtake the first? **(b)** By how much does the speed of the second car exceed that of the first as they pass?

30. A high-speed electric subway train undergoes constant acceleration that increases its speed from zero to 30 meters/sec in 24 sec. **(a)** How far does it travel during this acceleration? **(b)** Provided the same constant acceleration is maintained, how fast will it be traveling if it starts from rest and travels 400 meters? **(c)** How long will it take to accelerate from 20 meters/sec to 30 meters/sec? **(d)** How much distance will be covered in accelerating from 15 meters/sec to 25 meters/sec?

31. An automobile traveling at an initial speed of 30 meters/sec on a foggy day suddenly sights a truck 50 meters ahead, traveling in the same direction at a constant speed of 12 meters/sec. It takes the driver 0.6 sec to react to the situation and apply the brakes. When he does, the car attains a constant deceleration of 4.0 meters/sec^2. **(a)** Will the car and truck collide, assuming neither swerves? If so, find **(b)** when and **(c)** where the collision occurs and **(d)** the relative speed of the vehicles on impact. **(e)** What is the minimum deceleration the car would have to have under these circumstances to prevent collision?

32. A ball is thrown vertically upward from the level of the street alongside a house. It is caught by a person standing at an open window 20 ft above the ground. The initial velocity of the ball is 48 ft/sec, and it is caught on the way down. Find **(a)** how high above the ground it rose, **(b)** how long it was in the air, and **(c)** its speed when caught.

33. A bomber is in level flight 10,000 ft above a stretch of flat terrain. It is flying at a constant speed of 800 ft/sec (545 mph). At a given point, a bomb is dropped. **(a)** Neglecting effects of air resistance, how far from the point vertically below the plane at the instant of release will the bomb reach the ground? **(b)** At what angle from the vertical must the bombsight telescope be adjusted so that the correct target is in the sights under these circumstances?

34. A warship's guns impart an initial speed of 2500 ft/sec to their shells. **(a)** To what angle above the horizontal must they be elevated in order to hit a target 10 miles away, neglecting effects of air resistance? **(b)** What maximum height do the shells reach during their flight? **(c)** What is the speed of the shells on impact?

35. **(a)** At what angle above the horizontal plane should a gun be elevated for maximum range? **(b)** If the initial velocity it imparts to a projectile is v_0, what would be the maximum range? Neglect air resistance.

36. A baseball pitcher can give a baseball a maximum initial speed of 100 ft/sec. How far can he throw the ball?

37. A ball is projected from the origin with an initial upward velocity component of 20 meters/sec and a horizontal x-component of velocity of 25 meters/sec. Neglecting air resistance, find **(a)** the position (x, y) of the ball after

2 sec, 3 sec, and 6 sec; **(b)** the velocity components (v_x, v_y) after 2 sec, 3 sec, and 6 sec; **(c)** the acceleration components (a_x, a_y) after 2 sec, 3 sec, and 6 sec; **(d)** the time required to reach the highest point of the path; **(e)** the height attained at the highest point of the trajectory; **(f)** the time required for the ball to return to the level of the x-axis; and **(g)** the horizontal distance covered when the ball returns to the level of the x-axis.

38. A golfer hits the ball giving it an initial velocity of 80 ft/sec at an angle of 37° with the horizontal. The ball hits a tree 160 ft from the tee. Neglecting air resistance, find **(a)** the time of flight, **(b)** at what height the ball strikes the tree, and **(c)** the magnitude and direction of the ball's velocity at the time of impact. **(d)** Does the ball hit the tree on the way up or on the way down?

39. A golf ball is hit from a level fairway at an angle of 30° with the horizontal. It is in flight for 4.50 sec before returning to the ground. Find **(a)** the magnitude of its initial velocity and **(b)** the horizontal range of the ball on the fairway. Neglect air resistance.

40. A batter hits a home run over the fence. The ball is hit 4 ft above ground level with an initial velocity of 150 ft/sec at an angle of 37° to the horizontal. If the fence is 400 ft from home plate and 16 ft high, find **(a)** how long it takes the ball to clear the fence and **(b)** by what vertical distance the ball clears the fence. **(c)** Assuming the same angle of projection, what is the minimum speed the batter would have to impart to the ball to make it clear the fence? Neglect air resistance throughout.

41. A batter hits a line drive in a direction making an angle of 82° with the first base line. Its initial velocity is 100 ft/sec, with an initial upward direction 37° above the horizontal. Find **(a)** the maximum height attained by the ball, **(b)** the total time it remains in flight, **(c)** the distance between home plate and the point where the ball lands, and **(d)** the x- and y-coordinates of the point where it hits the ground (x-axis along first base line, y-axis along third base line). At the instant the ball is hit, the left fielder, who is on the third base line 320 ft from home plate, starts to run to make the catch. Find **(e)** how fast he should run and **(f)** the direction in which he should run, as defined by an angle α with the third base line. Neglect air resistance and the height of the batter and fielder.

42. A ball is thrown from the top of a building 100 ft high at an angle of 37° above the horizontal, with an initial speed of 80 ft/sec. Find **(a)** how long it remains in the air, **(b)** the maximum height above ground it attains, **(c)** the horizontal distance between where it is thrown and where it hits the ground, and **(d)** the magnitude and direction of its velocity as it hits the ground. Neglect air resistance.

43. A hunter, armed with a blowgun and a supply of poisoned darts, sees a monkey in a tree munching a banana. Knowing nothing about the intricacies of projectile motion, he aims his weapon directly at the monkey and prepares to fire one of his darts. The monkey sees him and, also knowing nothing about projectile motion, reasons as follows: "The instant he fires that dart, I'll just drop off this limb of the tree and fall freely; that way the dart will certainly hit my former position, but by the time it arrives there, I'll have fallen some distance to safety!"

Is the monkey's reasoning correct, or will he end up as a stuffed trophy? Explain.

*44. By referring Eq. (3.3.8) to an origin O' as shown in Fig. 3.5, transform this equation to one more readily identifiable as a standard parabola. *Hint:* Note that if x' and z' are coordinates referred to O', $x' = x - \frac{1}{2}s$, $z' = z$.

45. An electric motor starts from rest and attains its normal rotational velocity of 1740 rpm in 1.0 seconds, running at constant speed thereafter. Assuming that a constant angular acceleration takes place during this period, find **(a)** the angular acceleration, **(b)** the angular velocity at the end of 0.5 seconds after the motor starts, **(c)** the number of rotations through which the shaft turns during the 1-second acceleration period, **(d)** the angular velocity of the shaft after 10 revolutions.

46. The shaft of the motor described in Problem 45 is attached to a pulley 10 inches in diameter. Find **(a)** the linear velocity of a point on the rim of the pulley 1.0 second after the motor starts, **(b)** the radial and **(c)** the tangential components of acceleration of a point on the rim of the pulley 0.8 second after the motor starts, and **(d)** the radial and **(e)** the tangential components of acceleration of this same point 1.2 second after the motor starts.

47. An automobile stops with constant deceleration from an initial velocity of 60 mph (88 ft/sec) in a distance of 180 ft. Its wheels are 28 inches in diameter. Find **(a)** the initial angular velocity of the wheels, and **(b)** the angular acceleration of the wheels during the stop.

48. A toy gyroscope is set into motion by pulling a string wrapped around the axle. The string is 3 ft long, the axle 0.2 inches in diameter, and the string is pulled free from the axle in 2.0 seconds in such a way as to cause a constant angular acceleration of the rotor. Find **(a)** the angular acceleration of the rotor, **(b)** the final angular velocity, and **(c)** the number of revolutions through which the rotor turns during the acceleration process.

49. A golfer hits a ball from a tee located a height h above a level fairway as shown in the figure. The ball lands on the fairway 4.0 seconds later, at a point whose horizontal distance from the tee is 332.6 ft. Find **(a)** the height h of the tee above the fairway, **(b)** the maximum height reached by the ball in its flight, and **(c)** the vertical and horizontal components of velocity as the ball hits the fairway.

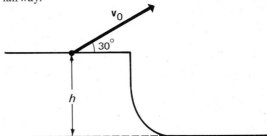

50. A car is driven around a circular track having a circumference of 2000 meters at a constant speed of 20 meters/sec. Assuming that it is headed due north at time $t = 0$, find the magnitude and direction of its instantaneous acceleration **(a)** at time $t = 0$ and **(b)** at time $t = 25$ sec.

51. In the preceding problem, assume that the car starts from rest at $t = 0$, at which time it is headed due north, and

that its speed increases linearly from zero to 40 meters/sec in 50 sec. Find the magnitude and direction of its instantaneous acceleration (a) at time $t = 0$, (b) at time $t = 25$ sec, and (c) at time $t = 50$ sec.

52. A particle moves along a curve whose (r, θ) coordinates are given as functions of time by $r = 2a \cos \theta$ and $\theta = \omega t$ ($\omega =$ const.). Show that the path of the particle is a circle of radius a intersecting the origin and whose center is at the point $x = a$, $y = 0$.

53. Find (a) the radial and (b) the azimuthal components of velocity and acceleration for a particle moving as described in Problem 52. Find also the components of velocity and acceleration tangential and normal to the path and show that the motion corresponds to uniform circular motion of angular velocity 2ω about the center of the circle.

54. Show by purely *geometric* means, without reference to the equations of section 3.5, that the motion of the particle described in connection with Problem 52 is uniform circular motion of angular velocity 2ω about the center of the circle.

*55. Consider a particle which moves in such a way that its polar coordinates (r, θ) are given as functions of time by $r = a \cos 3\theta$ and $\theta = \omega t$ ($\omega =$ const.). (a) Plot the path of the particle, assuming some convenient value for a. Find the radial and azimuthal components of (b) velocity and (c) acceleration. Find (d) the magnitude of the velocity vector and (e) the magnitude of the acceleration vector. Find the normal and tangential components (f) of the velocity and (g) of the acceleration.

56. A man walks radially outward on a circular platform which revolves with constant angular velocity ω. An attempt to calculate the azimuthal acceleration he experiences is made by noting that as he walks outward a distance dr in time dt, the change in tangential velocity of the turntable is $dv_\theta = d(r\omega) = \omega \, dr$. The azimuthal acceleration would then be $dv_\theta/dt = \omega \, dr/dt = \omega v_r$. But this is incorrect since, according to Eq. (3.7.19), the azimuthal acceleration is $2\omega v_r$. What is the fallacy in the reasoning leading to the incorrect value ωv_r?

57. A boy is on a merry-go-round which revolves in a counterclockwise direction (viewed from above) at a rotational speed of 0.1 revolution per second. (a) Standing 15 ft from the axis of rotation, he jumps vertically upward, attaining a maximum height of 3 ft above the platform. Where with respect to the platform does he land? (b) He runs radially outward with a velocity of 10 ft/sec, and when he is 15 ft from the axis of rotation, he jumps, remaining in the air 0.5 second before returning to the platform. Where does he land now?

58. A stone is dropped down a vertical mine shaft 1024 ft deep at the equator. Where, in respect to the point on the bottom of the shaft vertically below that from which it was dropped, does it land? *Hint:* Note that the earth rotates counterclockwise on its axis when viewed from a point directly above the north pole.

*59. A rowboat heads directly north across a stream flowing from west to east with a velocity v_0. The rowboat can make a velocity v' in still water. The width of the stream is d. Find (a) how long the rowboat takes to reach the other side, (b) the distance between the point at which the boat touches the opposite shore and the point directly opposite its starting point, (c) the angle between the boat's course and the line normal to the banks of the stream, and (d) the velocity of the boat with respect to the bottom of the stream.

60. The rowboat in Problem 59 is headed into the current, its heading making an angle θ with a line normal to the banks of the stream so that its course is normal to the banks. Find (a) the velocity of the boat with respect to the stream bottom, (b) the time taken to reach the opposite shore, (c) the angle θ at which the boat must be headed to ensure that its course be perpendicular to the current. (d) What happens when the current velocity v_0 exceeds the still-water speed of the boat, v'?

61. An airplane flies on course 60°, air speed 300 mph. The wind is from the southeast (135°) at 50 mph. Find the effective ground speed and the true course of the plane. Do not work this by merely substituting numbers into Eqs. (3.8.7) and (3.8.8). Work from vector diagrams and perform the proper vector sums numerically, using the component method.

62. A plane wishes to fly on course 45° at a ground speed of 300 mph. The wind is from the east at 50 mph. Find the airspeed and heading the plane must adopt to maintain this true course and speed. Do not work this by merely substituting numbers into Eqs. (3.8.9) and (3.8.10). Work instead from vector diagrams and perform the proper vector sums numerically by the method of components.

63. Verify the results obtained in Problems 61 and 62 by using Eqs. (3.8.7), (3.8.8.), (3.8.9), and (3.8.10).

NEWTON'S SECOND LAW AND TRANSLATIONAL DYNAMICS

4.1 Introduction

In Chapter 2, we investigated the conditions under which a system of forces can be in equilibrium. Under such conditions, the vector sum of all the forces and all the torques is zero, and the particle or object upon which they act is either at rest or moving at constant velocity. Newton's second law tells us that if the vector sum of all the forces acting on a body is not zero, the body, instead of remaining at rest or in a state of constant velocity, will undergo an acceleration proportional to the resultant force. The problem of finding the exact state of motion of the body may be solved by essentially the same methods used to solve problems of force equilibrium, except that the vector sum of all the forces is no longer zero, but, instead, must be equated to the mass of the object times its accelera-

tion. In much the same way, if the vector sum of all the torques acting on a rigid body is not zero, an angular acceleration is produced which is proportional to the resultant torque. In this chapter, we shall confine ourselves to the investigation of linear accelerations arising from net forces, leaving the problem of angular accelerations caused by torques to a later chapter, and restricting ourselves to systems in which there are no net torques.

4.2 The Application of Newton's Second Law to Systems of Forces

The foundations for applying Newton's second law to systems wherein several forces may act in such a way that the system is not in equilibrium have already been laid in section 2.4. We need only note that in the case where an object is in equilibrium, the vector sum of all the forces is zero, while if it is not, the resultant force on the body must equal the mass times the acceleration. This requirement may be met simply by replacing the zero on the right side of Eq. (2.4.1) by $m\mathbf{a}$ to obtain

$$\mathbf{F}_1 + \mathbf{F}_2 + \mathbf{F}_3 + \cdots + \mathbf{F}_i + \cdots + \mathbf{F}_n = m\mathbf{a}$$

or

$$\sum_{i=1}^{n} \mathbf{F}_i = m\mathbf{a} \tag{4.2.1}$$

Again, since two vectors are equal only if all their components are equal, we may equate the x-, y-, and z-components of the vectors on both sides of Eq. (4.2.1), the result being

$$\sum_{i=1}^{n} F_{ix} = ma_x$$

$$\sum_{i=1}^{n} F_{iy} = ma_y \tag{4.2.2}$$

$$\sum_{i=1}^{n} F_{iz} = ma_z$$

Equation (4.2.1) or its equivalent, the set of equations (4.2.2), represent the *equations of motion* of the system and can be integrated subject to the proper initial conditions to obtain values of velocity and displacement at any time, using the methods developed in Chapter 3. Often, however, we are satisfied to find only the acceleration and certain contact forces such as the tension in strings connecting separate parts of the system. Problems of this sort can usually be solved by straightforward application of the procedures already developed in Chapter 2 in connection with systems in equilibrium, as illustrated by the series of examples which follow.

EXAMPLE 4.2.1

A 100-pound weight is supported by a rope of negligible mass and given an upward acceleration of 12 ft/sec², as illustrated in Fig. 4.1. What is the tension in the rope? Suppose there is a downward acceleration of 12 ft/sec². What is the tension then?

Clearly, the tension T must exceed 100 pounds, otherwise there would be no net vertical component of force and the object would be in equilibrium, hence at rest or moving with constant velocity. As a matter of fact, the difference between T and 100 pounds is the net vertical component of force, which must equal mass times acceleration according to Newton's second law, as expressed by Eq. (2.3.2). But the mass of a 100-pound-weight is 100/32 slugs, according to (2.3.5).

FIGURE 4.1

We may therefore write

$$\sum F_{iy} = T - W = ma_y$$

or

$$T - 100 = ma_y = \frac{W}{g} a_y = \left(\frac{100}{32}\right)(12)$$

$$T = 100 + \frac{1200}{32} = 137.5 \text{ lb}$$

In the second instance, the y-component of acceleration is negative, the vector $\mathbf{a}$ pointing along the y-direction. Then, $a_y = -12$ ft/sec², whereby

$$\sum F_{iy} = T - W = ma_y = \left(\frac{100}{32}\right)(-12)$$

$$T = 100 - \frac{1200}{32} = 62.5 \text{ lb}$$

EXAMPLE 4.2.2

Suppose that the object on the end of the rope in Example 4.2.1 is a 100-kg mass and that it is accelerated with a positive (upward) acceleration of 3 m/sec². What is the tension in the rope under these circumstances?

The problem is done as before, except that now we may use the figure 100 kg directly as the mass, but we must express the weight force in proper mks force units, that is, newtons. The weight force W associated with a mass of 100 kg is, according to (2.3.5), given by $W = mg = (100)(9.8) = 980$ newtons. Accordingly,

$$\sum F_{iy} = T - (100)(9.8) = ma_y = (100)(3)$$

whence

$$T = 980 + 300 = 1280 \text{ N}$$

EXAMPLE 4.2.3

Suppose that the object on the end of the rope in Example 4.2.1 is a mass m and that it is given an acceleration a_y. What is the tension in the rope now?

The exercise is done as before, except that the general symbols m and a_y are used. Thus, since according to (2.3.5) the weight force is mg,

$$\sum F_{iy} = T - mg = ma_y$$
$$T = m(g + a_y) \qquad (4.2.3)$$

The answer to Example 4.2.2 is readily obtained by substituting $m = 100$ kg, $g = 9.8$ m/sec^2, and $a_y = 3$ m/sec^2 into (4.2.3). Equation (4.2.3) can be stated in terms of the weight W rather than the mass m, by use of (2.3.5), from which

$$T = W\left(\frac{g + a_y}{g}\right) = W\left(1 + \frac{a_y}{g}\right) \qquad (4.2.4)$$

The answers to Example 4.2.1 can now be obtained by substituting appropriate numerical values for W, a_y, and g.

One of the points to be noted from these examples is that the fact that a weight W is suspended from a rope does not necessarily imply that the tension in the rope is equal to W. This is true, when the weight is at rest or in uniform motion, but *not if it is undergoing acceleration*. In the latter instance, it is the *difference* between the force W and the actual tension that provides the force necessary to maintain the acceleration.

EXAMPLE 4.2.4

A mass M is suspended from a heavy flexible but inextensible rope of length L whose mass per unit length is μ and whose total mass, therefore, is μL. A force is applied to the free end of the rope in such a way as to impart a *constant* acceleration to the mass. Find (a) the tension at every point along the length of the rope, and (b) the force required to maintain the acceleration.

Referring to Fig. 4.2, consider a small element of the rope, of length dz, extending from a distance z to a distance $z + dz$ above the mass M. The mass for this element is its mass per unit length μ times the length dz. Denoting this mass by dm, we have

$$dm = \mu \, dz \qquad (4.2.5)$$

The forces acting on this element of the rope are shown in Fig. 4.2b. By $T(z)$ we mean the tension at a point a distance z from the lower end of the rope. The weight force due to the weight of the element is its mass dm times the acceleration of gravity, according to (2.3.5), and it acts vertically downward. Setting the sum of the forces equal to mass times acceleration, we obtain

$$T(z + dz) - T(z) - \mu g \, dz = (dm)a = \mu a \, dz \qquad (4.2.6)$$

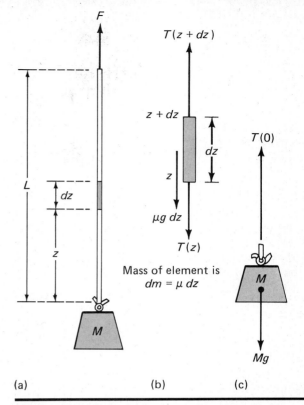

FIGURE 4.2. System of forces acting on a mass suspended from a heavy flexible rope.

But the quantity $T(z + dz) - T(z)$ is simply the difference in tension dT associated with the distance dz along the rope. Substituting this into (4.2.6), we may write

$$dT = \mu(g + a) \, dz \qquad (4.2.7)$$

Both sides of this equation may now be integrated from $z = 0$ at the bottom, where $T = T(0)$, to some point a distance z above where $T = T(z)$:

$$\int_{T(0)}^{T(z)} dT = \int_0^z \mu(g + a) \, dz = \mu(g + a) \int_0^z dz$$

In this equation, the constant quantities μ, g, and a may be brought outside the integral, as shown. We may now integrate both sides to obtain

$$[T]_{T(0)}^{T(z)} = \mu(g + a)[z]_0^z$$

or

$$T(z) - T(0) = \mu(g + a)z \qquad (4.2.8)$$

Now, $T(0)$ is the tension at the bottom, where the rope is attached to the mass M. Referring to Fig. 4.2c, where the forces acting on the lower fastening are shown, we may again set the sum of forces equal to mass times acceleration:

$$T(0) - Mg = Ma$$

101

or

$$T(0) = M(g + a) \qquad (4.2.9)$$

Substituting this value for $T(0)$ into (4.2.8) and solving for $T(z)$, we may finally obtain

$$T(z) = (g + a)(M + \mu z) \qquad (4.2.10)$$

as the tension at any point z. At the top of the rope, where the force F is applied, $z = L$, and the tension there is

$$T(L) = (g + a)(M + \mu L) \qquad (4.2.11)$$

The force F may be found by applying Newton's second law once more at the top of the rope. Again setting the sum of forces equal to mass times acceleration, noting now that the weight force is the *sum* of the weight forces associated with the mass M and the rope, hence $(M + \mu L)g$, we find

$$F - (M + \mu L)g = (M + \mu L)a$$
$$F = (g + a)(M + \mu L) \qquad (4.2.12)$$

The fact that F and $T(L)$ are equal, according to (4.2.11) and (4.2.12), could have been inferred from Newton's third law. In the same way, one may show that the force exerted by the mass M upon the lower end of the rope is equal to $T(0)$ as given by (4.2.9), as required by Newton's third law. Although the action and reaction forces are equal both at the top and bottom, their common magnitude is not the same at the top and bottom, because of the contribution of the rope's mass.

If the mass of the rope, μL, is *much less* than M, then, from (4.2.9), (4.2.10), and (4.2.11), it is evident (recalling that z must always be less than or equal to L) that *the tension in the rope is practically constant over its entire length*, its value then being essentially $M(g + a)$. Likewise, μL may be neglected in comparison with M in (4.2.12), and the force necessary to

produce the acceleration has this same value. You may recall that the same results were obtained in section 2.5 in discussing the difference between Newton's first and third laws. In the examples worked herein and in the problems which are assigned, it is assumed, unless otherwise stated, that the masses of the ropes or cords used in the system being investigated are much smaller than the masses they connect. This allows us to assume that the tensions in the connecting cords are everywhere the same and generally simplifies analysis of the situation. The example discussed above illustrates in a general way what happens and how to proceed when this assumption may no longer be justified.

EXAMPLE 4.2.5

A block of mass m_1, which slides along a frictionless surface, is connected to a vertically hanging mass m_2 by a flexible, inextensible cord of negligible mass which passes over a pulley, as shown in Fig. 4.3a. The inertia of the pulley and any frictional effects associated therewith may be neglected (we shall always assume this to be true unless otherwise specified). Find the common acceleration of the masses and the tension in the cord.

The free body diagrams for m_1 and m_2, as shown in Figs. 4.3b and 4.3c, serve as a convenient starting point for the solution. The signs of the force components, displacements, and accelerations involved depend upon which direction is assumed to be the positive direction of displacement for each mass. A reasonable way of making such a choice is to assume initially that the system is displaced in one direction or the other, and to regard the direction in which each mass moves under these circumstances as the positive direction of displacement for that mass. This initial assumption does not have to agree with the direction in which the system actually will move in response to the forces acting on it. Such a choice is indicated in

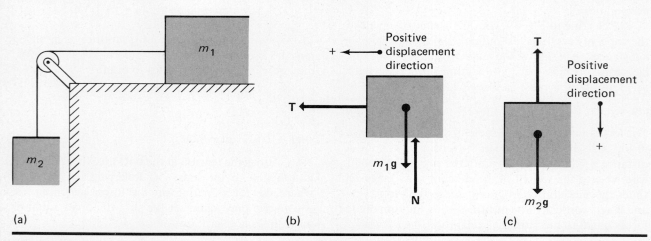

(a) (b) (c)

FIGURE 4.3

Figs. 4.3b and 4.3c. We shall always follow this procedure for choosing positive displacement directions unless some specific statement to the contrary is made. In Fig. 4.3b, there is no vertical component of acceleration, hence

$$N - m_1 g = 0$$

which, though true enough, tells us nothing of immediate interest in this case. Applying Newton's second law to the horizontal force components in the diagram, however, we may write

$$T = m_1 a \qquad (4.2.13)$$

where a is the acceleration. From Fig. 4.3c, in the same way, Newton's second law tells us that

$$m_2 g - T = m_2 a \qquad (4.2.14)$$

The acceleration magnitude a is the same in (4.2.13) and (4.2.14) because the cord does not stretch. Equations (4.2.13) and (4.2.14) may be solved as simultaneous equations for the two unknowns T and a; this is accomplished very simply in this case by substituting (4.2.13) into (4.2.14) and solving for a:

$$a = \frac{m_2}{m_1 + m_2} g \qquad (4.2.15)$$

The tension T is found by substituting this back into (4.2.13):

$$T = \frac{m_1 m_2 g}{m_1 + m_2} \qquad (4.2.16)$$

EXAMPLE 4.2.6

Suppose that the weight of the block sliding on the horizontal surface in the previous example is 100 pounds, while that of the vertically suspended object is 50 pounds, and suppose that, instead of being frictionless, the contact between the horizontal block and the surface on which it rests exhibits a kinetic friction coefficient $\mu_k = 0.2$. What is the acceleration and the tension in the cord now?

The situation is similar to that encountered in the previous example, except for the fact that there is now a force of friction acting on the sliding object in a direction opposite to the direction of motion, thus opposite the direction of T in Fig. 4.3b. The magnitude of this force (let us call it F_f) is

$$F_f = \mu_k N = \mu_k m_1 g = (0.2)(100) = 20 \text{ lb}$$

since we saw in Example 4.2.5 that $N = m_1 g$. Accordingly, applying Newton's second law to the sliding mass,

$$T - F_f = T - 20 = \frac{100}{32} a \qquad (4.2.17)$$

since the mass m_1 corresponding to the 100-pound

weight is (100/32) slugs. Applied to the suspended object, Newton's second law requires that

$$50 - T = \frac{50}{32} a \qquad (4.2.18)$$

the mass of the suspended body being (50/32) slugs. Adding Eqs. (4.2.17) and (4.2.18) to eliminate T, and then solving for a, we obtain

$$a = 6.4 \text{ ft/sec}^2$$

Substituting this value into either (4.2.17) or (4.2.18) and solving for T, we find

$$T = 40 \text{ lb}$$

EXAMPLE 4.2.7
ATWOOD'S MACHINE

Two masses M and m are connected by a cord passing over a pulley as shown in Fig. 4.4. Let us assume that M is larger than m, that the cord is essentially massless and inextensible, and that we may neglect the friction and inertia of the pulley. Describe the motion of the system and calculate the tension T in the cord.

Let us designate the upward direction as the positive direction of displacement for m and the downward direction as the positive direction of displacement for M; this is in accord with our previously stated convention. Since the cord is inextensible, m must go up as fast as M goes down, otherwise the cord would stretch or contract. But if the velocities of m and M are equal at all times, so are their accelerations, since, if $v_M = v_m$, $dv_M/dt = dv_m/dt$. Applying Newton's second law to the free body diagram in Fig. 4.4b,

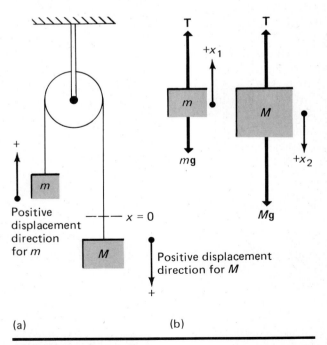

(a)   (b)

FIGURE 4.4. Atwood's machine.

$$T - mg = ma \qquad (4.2.19)$$

and

$$-T + Mg = Ma \qquad (4.2.20)$$

where a is the common acceleration of m and M. Adding these two equations to eliminate T,

$$(M - m)g = (M + m)a$$

or

$$a = \frac{M - m}{M + m} g \qquad (4.2.21)$$

Substituting this value for a into either (4.2.19) or (4.2.20), the value of T, is seen to be

$$T = \left(\frac{2mM}{m + M}\right) g \qquad (4.2.22)$$

The apparatus described above is sometimes referred to as *Atwood's machine* and can be used to make accurate determinations of the acceleration of gravity. Freely falling objects fall through distances conveniently accessible within the laboratory in very short time intervals, which are difficult to measure accurately. In order to make precise measurements of g, it is therefore desirable to have a device which in some way "dilutes" the effect of gravity and allows motion over distances of a meter or two to occupy a time span accurately measurable with a clock or stop watch. Atwood's machine is just such a device. If m and M are nearly equal, it is evident from (4.2.21) that the acceleration a may be made as small as desired; in fact, when m and M are precisely equal, the acceleration is zero. In practice, the masses m and M are selected to give a conveniently small value of acceleration and the ratio $(M - m)/(M + m)$ is determined carefully by weighing.[1] The acceleration is carefully determined by observing the time required for one of the masses (starting from rest) to ascend or descend through a measured distance, then using the "golden rules" of uniformly accelerated motion to infer the value of a. The magnitude of g is finally obtained from the measured quantities $(M - m)/(M + m)$ and a using Eq. (4.2.21).

4.3 Radial and Azimuthal Forces

In the preceding chapter, it was demonstrated that a body describing uniform circular motion is subject to a radial acceleration $-r\omega^2$, sometimes called centripetal acceleration. Likewise, a body having a radial velocity component v_r with respect to a rotating

[1] In this connection, it should be noted that according to (2.3.5), the quantities $(M - m)/(M + m)$ and $(W - w)/(W + w)$ are equal irrespective of what the value of g may be.

reference frame undergoes an *azimuthal acceleration* $2\omega v_r$ in relation to a fixed system of reference. According to Newton's first law, objects free from external forces simply do not experience any acceleration. Therefore, for any body to undergo such centripetal or azimuthal accelerations, it *must be acted upon by forces*. The force which produces centripetal acceleration $-r\omega^2$ is just *that force required to keep the body moving in a circular path*, and the force which produces the azimuthal acceleration $2\omega v_r$ is *that force necessary to keep the body moving in a purely radial path* with respect to the rotating reference system. The magnitudes of these forces are related to the corresponding accelerations by Newton's second law. Therefore, for the *centripetal force*,

$$F_{\text{cent}} = ma_{\text{cent}} = -mr\omega^2 = -\frac{mv_t^2}{r} \qquad (4.3.1)$$

while for the *azimuthal force*,

$$F_{\text{az}} = ma_{\text{az}} = 2m\omega v_r \qquad (4.3.2)$$

where v_t and v_r are the azimuthal and radial components of velocity, respectively. The minus sign in (4.3.1) simply means that the centripetal force is directed opposite the direction in which r increases—toward the axis of rotation.

These forces usually manifest themselves as forces exerted upon objects in rotating reference systems by other bodies fixed with respect to the rotating system. For example, when an automobile rounds a circular curve, a passenger has the impression that a force is pushing him outward, away from the center of the circular path. What happens, in reality, is that when the car enters the curve and starts changing direction, the passenger's body tends to move straight ahead, along a tangent to the circular path, in accord with Newton's first law, until the seat of the car exerts the *inward centripetal force* necessary (according to Newton's second law) to change the direction of the passenger's body to coincide with the direction of the car. To the passenger, however, it seems as though a force of unknown origin is pushing him outward. This force is often referred to as the *centrifugal force*. The inwardly directed force the car exerts upon him is interpreted as a reaction force that opposes this mysterious "primary" force to a degree sufficient to keep him in his seat. But in reality, more or less the opposite is true. The inwardly directed force exerted by the car seat on the passenger is the true "primary" force that gives him the acceleration he must have to travel in a circular path, that is, to keep him "at rest" in his accelerated reference system. The only outward "centrifugal" force present is the contact force his body exerts against the seat of the car. Its character is that of a reaction force that opposes the "primary"

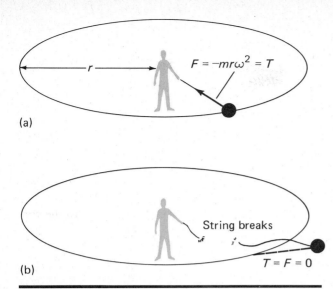

(a)

(b)

FIGURE 4.5. (a) The tension in the string, equal to $m r \omega^2$, provides the centripetal force necessary to keep the ball moving in a circular path with constant angular velocity. (b) When the string breaks, there is no longer any horizontal force component, and the ball moves off along a line tangent to the circular path, in accordance with Newton's first law.

centripetal force, but it acts on the car, not on the passenger, and therefore contributes exactly nothing to keeping the passenger moving in a circle.[2]

A similar situation is illustrated in Fig. 4.5. At (a), a boy swings a ball around him in a circular path on the end of a string with uniform angular velocity ω. For the ball to move in a circular path, it must undergo a centripetal acceleration $-r\omega^2$, and according to Newton's second law a force $-m r \omega^2$ is required to produce this acceleration. This force is supplied by the tension T in the string, which attains precisely this value. If the string suddenly breaks, the tension drops immediately to zero. There is now no force (except that of gravity, which will be discussed later) acting on the ball. It must now, according to Newton's first law, *continue in a straight line* along the tangent to the circular path at the point where the string broke. It does not fly outward along a radial path, although this would appear to be its fate to an observer stationed in a reference frame rotating with angular velocity ω! The effect of gravity is not particularly important in this example, since the gravitational force $m\mathbf{g}$ has no component in the plane in which the circular path lies. After the string breaks, its effect is to alter the path of the ball from a straight line tangent to the circular path to the parabolic path

[2] The term *centrifugal force* is often used incorrectly to refer to the *centripetal* force necessary to sustain a body in uniform circular motion.

characteristic of projectile motion along which it eventually falls to the ground.

EXAMPLE 4.3.1

A 160-pound man rounds a curve whose radius is 600 ft in an automobile traveling 60 mph (88 ft/sec). Find the inward force exerted on him by the car seat.

The centripetal force is related to the linear velocity v_t by Eq. (4.3.1). Recalling that the man's mass is $w/g = 160/32 = 5$ slugs, we may write

$$F_{\text{cent}} = -\frac{m v_t^2}{r} = -\frac{(5)(88)^2}{600} = 64.5 \text{ lb, inward}$$

EXAMPLE 4.3.2

An elevator ascends with a constant speed of 50 ft/sec in a vertical shaft in a building at the equator. Find the magnitude and direction of the azimuthal force exerted on a 160-pound man by the elevator cage.

The azimuthal force is given by (4.3.2). To use this result, however, we must calculate the angular velocity of the earth's rotation. Since the earth rotates through an angle of 2π radians in 24 hours (86,400 seconds),

$$\omega = \frac{2\pi}{86,400} = 7.27 \times 10^{-5} \text{ rad/sec}$$

This angular velocity is positive, since the earth rotates in a counterlockwise (positive) sense when viewed from above the north pole.[3] From (4.3.2), then, since the radial velocity v_r in the rotating terrestrial frame is 50 ft/sec,

$$F_{\text{az}} = 2m\omega v_r = 2\left(\frac{160}{32}\right)(7.27 \times 10^{-5})(50) = 0.0363 \text{ lb}$$

The positive sign of the result means that the force is in the positive sense of rotation, hence to the east. Such a small force would be imperceptible to a casual observer. If the elevator's velocity were increased tenfold, the force would be perceptible, but then the upward velocity would be nearly 350 miles per hour, which is beyond the capability of most elevators.

EXAMPLE 4.3.3

A mass m is swung in a circular path on the end of a light string of length l about a vertical rotation axis. The angular velocity is constant and has the value ω. The string makes an angle θ with the vertical, as shown in Fig. 4.6. Find the tension in the string and the angle θ. For a 1-meter string, how large would ω have to be for θ to have the value 89°? If $m = 0.1$ kg, what would the tension be under these conditions?

The only forces acting on the mass are the weight

[3] We shall find this numerical value useful and shall refer to it in working out subsequent examples.

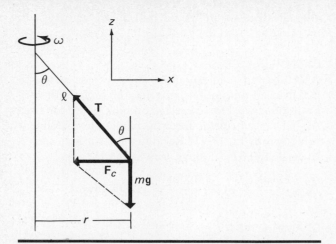

FIGURE 4.6

force $m\mathbf{g}$, which is vertically downward in direction, and the tension $\mathbf{T}$, as shown in Fig. 4.6. The mass executes a uniform circular motion in a horizontal plane, and it must, therefore, experience a centripetal force $\mathbf{F}_c$ to sustain it in its path. But this centripetal force must somehow be provided by the weight force and the tension in the cord; in fact, the resultant of these two forces must be exactly $\mathbf{F}_c$ if the mass is to describe uniform circular motion. If it were anything else, the mass would execute some other kind of motion! Because the centripetal force $\mathbf{F}_c$ must lie in the horizontal plane in which the circular motion takes place, the vertical force components must sum to zero (there being naturally no vertical acceleration) and the horizontal force components must sum to the magnitude of $\mathbf{F}_c$, which, according to (4.3.1), is $-mr\omega^2$. In other words, if r is the radius of the circular path, then

$$F_x = -T \sin \theta = -mr\omega^2 = F_c \qquad (4.3.3)$$

and

$$F_z = T \cos \theta - mg = 0 \qquad (4.3.4)$$

These equations may also be interpreted as following directly from Newton's second law. In (4.3.3), the net horizontal component of force, $-T \sin \theta$, is simply equated to the mass times the horizontal component of acceleration, which here is just the centripetal acceleration $-r\omega^2$. In (4.3.4), the net vertical force component is again equated to the mass times the vertical component of acceleration; but since the latter must vanish in uniform circular motion taking place in a horizontal plane, the product is zero.

Equations (4.3.3) and (4.3.4) contain three unknown quantities: T, θ, and r. We are unable to solve them unless at least one of these can be expressed in terms of the other two. From Fig. 4.6, however, it is obvious that r can be written in terms of l (which is known) and θ by

$$r = l \sin \theta \qquad (4.3.5)$$

If this is substituted into the original set of equations, only the variables T and θ remain. The solution is most easily obtained by observing that when (4.3.5) is substituted into (4.3.3), the value of T is obtained at once:

$$T = ml\omega^2 \qquad (4.3.6)$$

Substituting this into (4.3.4), we can solve for $\cos \theta$:

$$\cos \theta = \frac{g}{l\omega^2} \qquad (4.3.7)$$

In trying to understand these answers, it is best to begin by considering the case where ω is very large. Then, according to (4.3.6), T must also be very large; and from (4.3.7), $\cos \theta$ must be extremely small and θ itself must be close to $90°$. Under these circumstances, the string is nearly horizontal. As ω decreases, T decreases and $\cos \theta$ increases, the latter quantity attaining the value of unity, at which point $\theta = 0$, when $g/(l\omega^2) = 1$, corresponding to $\omega = \sqrt{g/l}$. When ω approaches this value, the string hangs almost vertically, the tension then being practically equal to mg, according to (4.3.6) or (4.3.4). At the same time, the radius r as given by (4.3.5) approaches zero. A value of ω *less* than $\sqrt{g/l}$ leads to a value of $\cos \theta$ *greater* than unity in (4.3.7), which means that no real value of θ can be found to satisfy the requirements of the problem under these conditions. In effect, it is *impossible to excite uniform circular motion* in a system such as the one shown in Fig. 4.6 with angular velocity smaller than $\sqrt{g/l}$. Physically, the reason for this is that the tension in the string in this case cannot possibly be smaller than the weight force mg.

Equation (4.3.7) expresses θ in terms of l and ω. If two of these are known, the third is easily found. We are required to find what value of ω is required in order that $\theta = 89°$ for a string 1 meter long. Solving (4.3.7) for ω and substituting these values for l and θ, we find (using mks units) that

$$\omega = \sqrt{\frac{g}{l \cos \theta}} = \sqrt{\frac{9.8}{(1.0)(0.0175)}}$$
$$= 23.7 \text{ rad/sec, or } 226 \text{ rpm}$$

The tension in the cord can now be calculated using (4.3.6). If $m = 0.1$ kg,

$$T = ml\omega^2 = (0.1)(1.0)(23.7)^2 = 56 \text{ N}$$

EXAMPLE 4.3.4

A man is weighed at the north pole using a very accurate spring balance. His weight is determined by this method to be 160.000 pounds. What result would be obtained if he were weighed on the same balance at the equator? Assume that $g = 32.000$ ft/sec^2 in both locations.

$$= 5.000(32.000 - 0.111) = 159.445 \text{ lb}$$

The spring tension and thus the *indicated* weight would be 159.445 pounds. Note that the force of gravity on the man's body and therefore his *actual* weight is still 160.000 pounds. The discrepancy in the reading of the spring balance comes about simply because at the equator a centripetal force $-mr\omega^2$ must be acting on the man's body to keep him in the circular path of the earth's surface as the earth rotates on its axis. The tension in the spring must, therefore, be less than the actual weight force acting on him by just this amount; if it were not, he would fall down or fly upward!

It is clear from these calculations that in accurate determinations of weight by spring balances, corrections for the centripetal forces arising from the axial rotation of the earth must be made. Corrections such as these should be applied in the scheme outlined in section 2.1 for measuring forces in terms of standard weight forces whenever the use of spring balances or equivalent devices is envisioned.

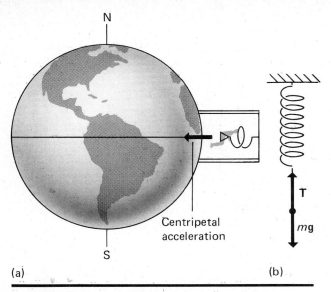

FIGURE 4.7. Effect of centripetal forces associated with the earth's rotation in determining weight forces at the equator.

Referring to Fig. 4.7, the forces acting on the spring balance are $m\mathbf{g}$ downward and $\mathbf{T}$, the spring tension, is in the opposite direction. At the north pole, the mass of the man's body is unaccelerated, and, therefore, according to Newton's second law,

$$F = T - mg = ma = 0$$

$$T = mg = \left(\frac{160.000}{32.000}\right)(32.000) = 160.000 \text{ lb} \qquad (4.3.8)$$

The spring balance, which is calibrated to read the spring tension, will indicate 160.000 pounds, in agreement with the data given above.

At the equator, precisely the same forces act, but now the man's body has a centripetal acceleration whose magnitude is $r\omega^2$ and which points toward the axis of rotation of the earth; its direction is, therefore, downward at the equator. We may again equate the sum of the vertical force components to mass times acceleration, as in (4.3.8), but now the equation will have the form

$$F = T - mg = ma_{\text{cent}} = m(-r\omega^2) \qquad (4.3.9)$$

whence

$$T = m(g - r\omega^2) \qquad (4.3.10)$$

In this equation, r refers to the earth's radius (3963 miles) and ω to the angular velocity of axial rotation, which was determined to be 7.27×10^{-5} rad/sec in Example 4.3.2. Substituting these values and those given above for m and g into (4.3.10), we find

$$T = \frac{160.000}{32.000}\left[32.000 - (5280)(3963)(7.27 \times 10^{-5})^2\right]$$

*EXAMPLE 4.3.5

A plumb bob of mass m is suspended from an overhead support by a light, flexible cord at a point on the earth's surface whose latitude is θ. What angle does the cord make with the vertical direction and what is the tension in the cord?

Due to the earth's rotation, the plumb bob will not hang quite vertically but will be displaced outward somewhat like the mass shown in Fig. 4.6. The only forces acting on the mass are the weight force $m\mathbf{g}$ directed toward the center of the earth, hence vertically downward, and the tension $\mathbf{T}$ acting along the cord, which we shall assume makes an unknown angle ϕ with the vertical direction, as shown in Fig. 4.8. As the earth rotates, the mass executes uniform circular motion in a plane parallel to that of the equator, about the point P in Fig. 4.8a. If the latitude is θ, then the vector $m\mathbf{g}$ makes an angle θ with this plane. Under these circumstances, the mass has a centripetal acceleration of magnitude $r\omega^2$ which lies in the plane of rotation and must experience a net resultant force equal to the mass times this centripetal acceleration. Since $\mathbf{T}$ and $m\mathbf{g}$ are the only forces acting on the mass, and since the total resultant force must be $m\mathbf{a}$ in order that uniform circular motion be sustained, the resultant of $\mathbf{T}$ and $m\mathbf{g}$ must be just this centripetal force $\mathbf{F}_c = m\mathbf{a}_c$. We may then equate the x- and z- components of the forces $\mathbf{T}$ and $m\mathbf{g}$ in Fig. 4.8b to the mass times the corresponding components of the centripetal acceleration. Accordingly, noting that $a_c = r\omega^2$,

$$T \cos \phi - mg = -ma_c \cos \theta = -mr\omega^2 \cos \theta \quad (4.3.11)$$

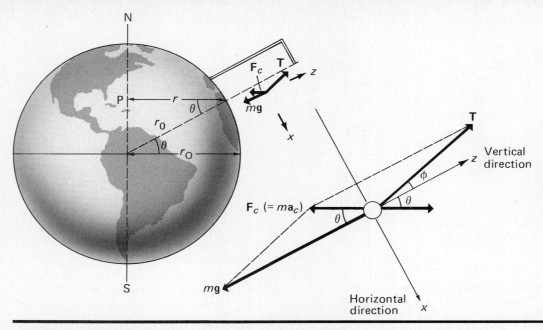

FIGURE 4.8. Effect of centripetal forces arising from the earth's rotation at an arbitrary point whose latitude is θ.

and

$$-T \sin \phi = -ma_c \sin \theta = -mr\omega^2 \sin \theta \qquad (4.3.12)$$

We must now solve these two equations for the two unknowns ϕ and T. From (4.3.12),

$$T = mr\omega^2 \frac{\sin \theta}{\sin \phi} \qquad (4.3.13)$$

Substituting this into (4.3.11) and rearranging, we may obtain

$$r\omega^2 \sin \theta \cot \phi = g - r\omega^2 \cos \theta$$

from which

$$\tan \phi = \frac{1}{\cot \phi} = \frac{r\omega^2 \sin \theta}{g - r\omega^2 \cos \theta} \qquad (4.3.14)$$

or, dividing numerator and denominator of the fraction on the right by g,

$$\tan \phi = \frac{\dfrac{r\omega^2}{g} \sin \theta}{1 - \dfrac{r\omega^2}{g} \cos \theta} \qquad (4.3.15)$$

This result will be more easily understood if we express the path radius r in terms of the earth's radius r_0 and the latitude θ. From Fig. 4.8a, it is obvious that r is given in terms of these other two quantities by

$$r = r_0 \cos \theta \qquad (4.3.16)$$

Substituting this into (4.3.15), we finally have,

$$\tan \phi = \frac{\dfrac{r_0\omega^2}{g} \sin \theta \cos \theta}{1 - \dfrac{r_0\omega^2}{g} \cos^2 \theta} \qquad (4.3.17)$$

To find the tension, let us write (4.3.13) in the form

$$T = mr\omega^2 \sin \theta \csc \phi = mr\omega^2 \sin \theta \sqrt{1 + \cot^2 \phi}$$

$$= mr\omega^2 \sin \theta \sqrt{1 + \frac{1}{\tan^2 \phi}} \qquad (4.3.18)$$

recalling that $\csc \phi = 1/\sin \phi$, $\csc \phi = \sqrt{1 + \cot^2 \phi}$, and $\cot \phi = 1/\tan \phi$. Substituting the expression given by (4.3.14) for $\tan \phi$ into (4.3.18), we may obtain

$$T = mr\omega^2 \sin \theta \sqrt{1 + \left(\frac{g - r\omega^2 \cos \theta}{r\omega^2 \sin \theta}\right)^2}$$

$$= m\sqrt{r^2\omega^4 \sin^2 \theta + (g - r\omega^2 \cos \theta)^2}$$

$$= m\sqrt{r^2\omega^4(\sin^2 \theta + \cos^2 \theta) - 2gr\omega^2 \cos \theta + g^2}$$

$$= m\sqrt{r^2\omega^4 - 2gr\omega^2 \cos \theta + g^2}$$

or

$$T = mg\sqrt{1 - 2\left(\frac{r\omega^2}{g}\right)\cos \theta + \left(\frac{r\omega^2}{g}\right)^2} \qquad (4.3.19)$$

Again expressing r in terms of r_0 and θ by (4.3.16),

$$T = mg\sqrt{1 - 2\left(\frac{r_0\omega^2}{g}\right)\cos^2 \theta + \left(\frac{r_0\omega^2}{g}\right)^2 \cos^2 \theta}$$

$$(4.3.20)$$

Equations (4.3.17) and (4.3.20) provide the exact answers to the questions which were asked in the original statement of the problem. Note that according to (4.3.17), $\phi = 0$ for $\theta = 0$ or $\theta = 90°$, which means that a plumb bob will hang vertically either at the equator or the north pole. This is in agreement with what we might expect on the basis of intuition. At the equator, where $\theta = 0$, Eq. (4.3.20) reduces to $T = m(g - r_0\omega^2)$, in agreement with the results of Example 4.3.4, while at the poles ($\theta = \pm 90°$), $T = mg$, as expected intuitively. In both (4.3.17) and (4.3.20), the quantity $r_0\omega^2/g$, which expresses the ratio of the centripetal acceleration to the acceleration of gravity, enters in an important way. Let us compute this ratio numerically: r_0 is the earth's radius, equal to 6380 km; ω was found to have the value 7.27×10^{-5} rad/sec for the earth's axial rotation in Example 4.3.2; and $g = 9.80$ m/sec^2. Therefore,

$$\frac{r_0\omega^2}{g} = \frac{(6.38 \times 10^6)(7.27 \times 10^{-5})^2}{9.8} = 0.00345$$

This means that on the earth's surface, the centripetal acceleration amounts to only 0.345 percent of the acceleration due to gravity, and this is the reason why centripetal effects due to the earth's axial rotation are small and not readily observed.

The fact that the terrestrial value of $r_0\omega^2/g$ is so small makes it desirable for us to obtain *approximate* expressions which are much simpler than (4.3.17) and (4.3.20) and which may be used in their place without incurring significant error so long as $r_0\omega^2/g$ is small compared with unity. For instance, since $\cos^2\theta$ is always less than unity, the second term in the denominator of the expression for $\tan\phi$ given by (4.3.17) will always be much less than the first in any situation (such as our terrestrial environment) where $r_0\omega^2/g$ is much less than unity. Under such circumstances, the second term in the denominator can be ignored altogether, since, for example, almost the same result is obtained if the quantity in the numerator is divided by 1.000 or by 0.999. Accordingly, we can write as a good approximation for (4.3.17)

$$\tan\phi \cong \frac{r_0\omega^2}{g}\sin\theta\cos\theta \quad \text{whenever } r_0\omega^2/g \ll 1 \quad (4.3.21)$$

The symbol $\ll$ is used here to signify "is much less than." We shall also sometimes use the symbol $\gg$ to mean "is much greater than."

At latitude 45°, $\sin\theta = \cos\theta = \sqrt{2}/2$, while on the earth's surface, the value of $r_0\omega^2/g$ has already been found to be 0.00345. According to (4.3.21), the angle ϕ between a plumb line and the vertical is given by

$$\tan\phi = (0.00345)(0.5000) = 0.00172 \qquad \phi = 0.099°$$

This is about the maximum error that can occur in terrestrial surroundings with a plumb line. It is so small that for most purposes no corrections to the indications of a plumb bob need be made.

An approximate formula for the tension in the cord can be obtained from (4.3.20) when $r_0\omega^2/g \ll 1$ by first observing that under those conditions the third term under the radical, containing the factor $(r_0\omega^2/g)^2$, is always much smaller than the other two, since the square of a number much less than unity is inevitably much less than the number itself. This term may then be safely neglected altogether. Of the remaining terms, the second is, of course, much less than the first, the resulting expression having the form

$$T \cong mg\sqrt{1 - y} \tag{4.3.22}$$

with

$$y = 2\left(\frac{r_0\omega^2}{g}\right)\cos^2\theta \tag{4.3.23}$$

Now it is easily shown that if $y \ll 1$, $\sqrt{1-y}$ is very nearly equal to $1 - \frac{1}{2}y$. This is most simply understood by starting with the algebraic identity

$$\left(1 - \frac{y}{2}\right)^2 = 1 - y + \frac{y^2}{4}$$

and observing that if $y \ll 1$, $y^2/4$ is much smaller than either unity or y. Accordingly, we may neglect the third term on the right-hand side whenever y is much less than unit and write

$$\left(1 - \frac{y}{2}\right)^2 \cong 1 - y \qquad \text{whenever } y \ll 1$$

Taking the square root of both sides of this equation, we see that, to a high degree of approximation,

$$1 - \frac{y}{2} \cong \sqrt{1 - y} \qquad \text{whenever } y \ll 1 \tag{4.3.24}$$

If we now substitute $1 - \frac{1}{2}y$ for $\sqrt{1-y}$ in (4.3.22), we obtain

$$T \cong mg\left(1 - \frac{y}{2}\right) \qquad \text{for} \quad y \ll 1$$

Substituting the value of y from (4.3.23) into this, we finally get

$$T \cong mg\left(1 - \frac{r_0\omega^2}{g}\right)\cos^2\theta \qquad \text{whenever} \quad r_0\omega^2/g \ll 1$$

$$(4.3.25)$$

At latitude 45°, $\cos^2\theta = \frac{1}{2}$, and, as before, $r_0\omega^2/g = 0.00345$. It is clear, then, that in such a situation

$$\frac{r_0\omega^2}{g}\cos^2\theta = (0.00345)(0.5000) = 0.00172$$

Thus, from (4.3.25)

$$T = (0.99828)mg$$

Under terrestrial conditions, then, the tension differs from the weight force mg by only a bit more than one part in a thousand due to centripetal effects arising from the earth's axial rotation. This difference is so small that it is difficult to detect without the use of sensitive instruments, and can safely be neglected in most practical calculations concerned with terrestrial surroundings. It has, for example, been neglected in all the calculations involving static force equilibria in Chapter 2, where it was always assumed that the downward forces upon surfaces supporting weights and the tensions in cords supporting unaccelerated weights were equal to the weight forces. We shall continue to make these assumptions in all future calculations, unless some specific assertion to the contrary is made.

It should be emphasized that (4.3.21) and (4.3.25) are not exactly correct but are approximations of (4.3.17) and (4.3.20), which nevertheless are nearly correct whenever $r_0\omega^2/g$ is much less than unity. They are much simpler than the exact expressions and save much time whenever numerical values must be computed. The technique of simplifying complex expressions by the use of good approximations is widely applied and, indeed, is one of the most useful devices in the physicist's arsenal of mathematical weapons. We shall encounter frequent uses of this technique and similar ones in subsequent discussions.

EXAMPLE 4.3.6

An automobile of mass m goes around a circular curve of radius r at speed v. Its center of mass, which in this case coincides with the center of gravity, is at a height h above the roadway and is located transversely halfway between the wheels, as shown in Fig. 4.9. The distance between the wheels is d. Assuming that the coefficient of friction between the tires and the roadway is sufficiently large to prevent the car from slipping sideways, find (a) the distribution of the normal forces and frictional forces between the wheels on the left side of the car and those on the right, and (b) the maximum speed at which the car can negotiate the curve without overturning.

The forces that act are shown in Fig. 4.9. The normal forces N_1 and N_2 and the frictional forces F_{f1} and F_{f2} represent, of course, the sum of the respective forces acting at the front and rear wheels of the car. Since the car is in a state of uniform circular motion, it must experience a centripetal force along the x-direction whose magnitude is m times the centripetal acceleration $r\omega^2$, where $\omega = v/r$. Equating x- and z-components of force to mass times the respec-

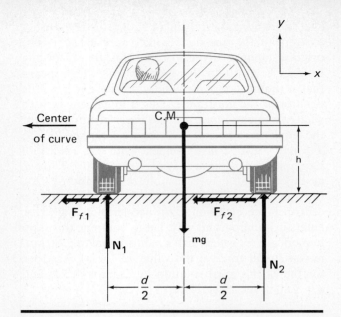

FIGURE 4.9. System of forces acting on a car rounding an unbanked curve.

tive acceleration components, we find

$$\sum F_z = mg - (N_1 + N_2) = 0 \qquad (4.3.26)$$

and

$$\sum F_x = -(F_{f1} + F_{f2}) = -mr\omega^2 \qquad (4.3.27)$$

The minus signs in (4.3.27) are needed because the centripetal force vector, as well as the vectors $\mathbf{F}_{f1}$ and $\mathbf{F}_{f2}$, are in the negative x-direction and hence have negative x-components. This equation simply expresses the fact that the entire centripetal force needed to keep the car in its circular path is supplied by the frictional forces between tires and roadway.

Since four quantities N_1, N_2, F_{f1}, and F_{f2} are unknown, we need four equations to determine all of them. Another equation may be obtained by taking *moments* of forces about some suitable point. In problems involving *accelerated bodies*, certain complications arise, which are beyond the scope of this book to treat in detail, unless *moments of forces are taken either about a point which is unaccelerated or about the center of mass.* It is very important, therefore, to observe these precautions in calculating moments of forces acting on accelerated objects. In this case, and in most others of a similar nature, it is most simple to take moments about the *center of mass.* Accordingly,

$$N_2 \frac{d}{2} - N_1 \frac{d}{2} - (F_{f1} + F_{f2})h = 0$$

the sum of the moments being zero, since the car does not have any angular acceleration about an axis through the center of mass (that is, it does not tip over). Since $F_{f1} + F_{f2} = mr\omega^2$, from (4.3.27), this can be written as

$$(N_2 - N_1)\frac{d}{2} - mr\omega^2 h = 0 \tag{4.3.28}$$

Strictly speaking, the three equations, (4.3.26), (4.3.27), and (4.3.28), are all that can be obtained from the laws of motion. In particular, the laws of motion cannot tell us how the frictional forces F_{f1} and F_{f2} are divided between the wheels on the right side of the car and those on the left. If both sets of wheels were traveling on dry concrete, for example, both F_{f1} and F_{f2} might be quite large. On the other hand, if the wheels on the left side were traveling on a very slippery surface such as ice while those on the right were on dry concrete, no appreciable frictional force could arise on the left while the frictional force at the right might be very great, indeed, even though neither set of wheels were actually slipping! Under these circumstances, F_{f1} would be small while F_{f2} would be much larger. The problem, then, as originally stated is *indeterminate* unless some further assumption is made regarding the division of frictional forces. A physically reasonable assumption (which nevertheless has nothing to do with the laws of motion) would be that the nature of the frictional contact between tires and roadway is the same on both sides of the car, and hence that the *ratio of frictional force to normal force is the same on both sides*. This means that

$$\frac{F_{f1}}{N_1} = \frac{F_{f2}}{N_2} \tag{4.3.29}$$

This equation may be regarded as a reasonable *ad hoc* assumption that completes the set of four equations necessary to solve for the four unknowns.

The set of Eqs. (4.3.26), (4.3.27), (4.3.28), and (4.3.29) may be solved most easily by observing, from (4.3.26), that $N_2 = mg - N_1$ and substituting this value for N_2 into (4.3.28), which may then be solved for N_1: the result being

$$N_1 = \frac{mg}{2} - mr\omega^2 \frac{h}{d} = mg\left(\frac{1}{2} - \frac{r\omega^2}{g}\cdot\frac{h}{d}\right) \tag{4.3.30}$$

Then, from (4.3.26)

$$N_2 = mg - N_1 = mg\left(\frac{1}{2} + \frac{r\omega^2}{g}\cdot\frac{h}{d}\right) \tag{4.3.31}$$

According to (4.3.29), $F_{f1} = F_{f2}(N_1/N_2)$; substituting this into (4.3.27), we find

$$F_{f2} = \frac{mr\omega^2}{1 + \dfrac{N_1}{N_2}} = mr\omega^2\left(\frac{1}{2} + \frac{r\omega^2}{g}\cdot\frac{h}{d}\right) \tag{4.3.32}$$

while, finally, from (4.3.29),

$$F_{f1} = F_{f2}\frac{N_1}{N_2} = mr\omega^2\left(\frac{1}{2} - \frac{r\omega^2}{g}\cdot\frac{h}{d}\right) \tag{4.3.33}$$

From these equations, it is easily seen that when $\omega = 0$, in which case the car is at rest, $N_1 = N_2 = mg$ and $F_{f1} = F_{f2} = 0$, as expected. As the velocity of the car, hence ω, increases, the normal force N_2 and the frictional force F_{f2} at the wheels on the right side increase while the corresponding forces N_1 and F_{f1} on the left side decrease. If ω is sufficiently large, the force N_1 as given by (4.3.30) will become zero. At this point, the wheels on the left side of the car can leave the roadway, the force exerted by those wheels on the road and the equal and opposite reaction force N_1 having now vanished, and the car will tip over. The value of ω for which N_1 vanishes is obtained by setting N_1 equal to zero in (4.3.30) and by solving for ω; from which

$$\omega = \frac{v}{r} = \sqrt{\frac{gd}{2hr}} \tag{4.3.34}$$

or solving for v,

$$v = \sqrt{\frac{g\,dr}{2h}} \tag{4.3.35}$$

According to this result, in order that v be as large as possible, the car should be designed so that the distance d between the wheels is as large as practicable and the height h of the center of mass is as small as possible. For an automobile for which $d = 5$ ft and $h = 2$ ft and for a curve of radius 600 ft, (4.3.35) predicts that a maximum velocity

$$v = \sqrt{\frac{(32)(5)(600)}{(2)(2)}} = 154.9 \text{ ft/sec}, \quad \text{or } 105.5 \text{ mph}$$

can be attained before the vehicle tips over. The ratio of friction force to normal force at the wheels on the right side, which remain on the road, can be determined from (4.3.31) and (4.3.32), which, recalling that $\omega = v/r$, predict that

$$\frac{F_{f2}}{N_2} = \frac{r\omega^2}{g} = \frac{v^2}{gr} \tag{4.3.36}$$

In the particular case considered above, this amounts to

$$\frac{F_{f2}}{N_2} = \frac{(154.9)^2}{(32)(600)} = 1.25$$

In order that the car turn over before the tires slip outward on the roadway, the coefficient of friction between tires and road must in this instance exceed 1.25. If it does not, then the ratio of force of friction to normal force will exceed the coefficient of friction before the car reaches the velocity required for it to overturn, and the car will slide outward.

111

4.4 Center of Mass

In Example 4.3.6 above, we found it necessary to introduce the concept of *center of mass*. We have already defined and discussed what is meant by the term *center of gravity* in section 2.8. The concept of *center of mass* is essentially similar, and in many instances, such as the one treated in the example above, the two points coincide. There are, nevertheless, certain differences between the two terms which should be clearly understood.

The center of mass is defined mathematically simply by replacing the weight W by the mass m in Eqs. (2.8.4), (2.8.8), and (2.8.9), which define the coordinates of the center of gravity. Thus, the coordinates x^*, y^*, and z^* of the center of mass of an arbitrary irregularly shaped body of total mass m are defined by

$$x^* = \frac{1}{m} \sum_{i=1}^{N} x_i \, \Delta m_i \qquad (4.4.1)$$

$$y^* = \frac{1}{m} \sum_{i=1}^{N} y_i \, \Delta m_i \qquad (4.4.2)$$

$$z^* = \frac{1}{m} \sum_{i=1}^{N} z_i \, \Delta m_i \qquad (4.4.3)$$

In these equations, as in (2.8.4), (2.8.8), and (2.8.9), we visualize the body as having been subdivided into a very large number N of tiny, individual mass elements $\Delta m_1, \Delta m_2, \Delta m_3, \ldots, \Delta m_i, \ldots, \Delta m_N$. These equations are regarded as being exact in the limit where this subdivision becomes truly infinitesimal, that is, in the limit as $N \to \infty$ and $\Delta m_i \to dm_i \to 0$. In this limit, then, these equations define the coordinates of the center of mass just as (2.8.4), (2.8.8), and (2.8.9) in the same limits define the coordinates of the center of gravity. In fact, if we note that the total mass m is simply the sum of the masses of the individual elements, thus that

$$m = \Delta m_1 + \Delta m_2 + \Delta m_3 + \cdots \Delta m_i \cdots + \Delta m_N$$

$$= \sum_{i=1}^{N} \Delta m_i \qquad (4.4.4)$$

and if we explicitly include the limiting conditions referred to above, we can write (4.4.1), (4.4.2), and (4.4.3) as

$$x^* = \lim_{\substack{N \to \infty \\ \Delta m_i \to 0}} \frac{\sum_{i=1}^{N} x_i \, \Delta m_i}{\sum_{i=1}^{N} \Delta m_i}$$

$$y^* = \lim_{\substack{N \to \infty \\ \Delta m_i \to 0}} \frac{\sum_{i=1}^{N} y_i \, \Delta m_i}{\sum_{i=1}^{N} \Delta m_i}$$

and

$$z^* = \lim_{\substack{N \to \infty \\ \Delta m_i \to 0}} \frac{\sum_{i=1}^{N} z_i \, \Delta m_i}{\sum_{i=1}^{N} \Delta m_i}$$

The limiting process in these equations is exactly that involved in defining the integral of a mathematical function. Using the definition of the integral, therefore, we can write these equations as

$$x^* = \frac{\int x \, dm}{\int dm} = \frac{1}{m} \int x \, dm \qquad (4.4.5)$$

$$y^* = \frac{\int y \, dm}{\int dm} = \frac{1}{m} \int y \, dm \qquad (4.4.6)$$

$$z^* = \frac{\int z \, dm}{\int dm} = \frac{1}{m} \int z \, dm \qquad (4.4.7)$$

the final form arising from the fact that, according to (4.4.4),

$$\int dm = m \qquad (4.4.8)$$

Equations (2.4.4), (2.4.8), and (2.4.9) defining the coordinates of the center of gravity can be written as integrals, simply by replacing m by W and dm by dW in these expressions. The result is

$$\bar{x} = \frac{\int x \, dW}{\int dW} = \frac{1}{W} \int x \, dW \qquad (4.4.9)$$

$$\bar{y} = \frac{\int y \, dW}{\int dW} = \frac{1}{W} \int y \, dW \qquad (4.4.10)$$

$$\bar{z} = \frac{\int z \, dW}{\int dW} = \frac{1}{W} \int z \, dW \qquad (4.4.11)$$

In the case of a finite collection of point masses m_1, m_2, $m_3, \ldots, m_N$ whose coordinates are (x_1, y_1, z_1), (x_2, y_2, z_2), $(x_3, y_3, z_3), \ldots, (x_N, y_N, z_N)$, or of a collection of irregular objects whose masses are m_1, m_2, $m_3, \ldots, m_N$ and whose individual centers of mass have those same coordinates, the center of mass can be defined by replacing w_i by m_i in Eqs. (2.8.12) and (2.8.14). In this case, therefore, we obtain

$$x^* = \frac{\sum_{i=1}^{N} m_i x_i}{\sum_{i=1}^{N} m_i} = \frac{1}{m} \sum_{i=1}^{N} m_i x_i \qquad (4.4.12)$$

$$y^* = \frac{\sum\limits_{i=1}^{N} m_i y_i}{\sum\limits_{i=1}^{N} m_i} = \frac{1}{m} \sum\limits_{i=1}^{N} m_i y_i \qquad (4.4.13)$$

$$z^* = \frac{\sum\limits_{i=1}^{N} m_i z_i}{\sum\limits_{i=1}^{N} m_i} = \frac{1}{m} \sum\limits_{i=1}^{N} m_i z_i \qquad (4.4.14)$$

where, naturally,

$$m_1 + m_2 + m_3 + \cdots + m_N = \sum_{i=1}^{N} m_i = m \qquad (4.4.15)$$

We have already shown in connection with Eq. (2.3.5) that mass and weight are related by $W = mg$. If we substitute mg for W in Eqs. (4.4.9), (4.4.10), and (4.4.11), we may show that the coordinates of the center of gravity can be written as

$$\bar{x} = \frac{\int x\, d(mg)}{\int d(mg)} \qquad \bar{y} = \frac{\int y\, d(mg)}{\int d(mg)} \qquad \bar{z} = \frac{\int z\, d(mg)}{\int d(mg)}$$

$$(4.4.16)$$

Now, if g *has the same value* for all mass elements within the system, then it may be treated as a constant in the integrations indicated above, in which case $d(mg) = g\, dm$. Furthermore, if g is constant, it may be brought outside the above integrals and canceled in numerator and denominator. The expressions (4.4.9) then reduce precisely to those given in (4.4.5), (4.4.6), and (4.4.7), which define the coordinates of the center of mass. In instances such as this, then, $x^* = \bar{x}$, $y^* = \bar{y}$, and $z^* = \bar{z}$ and the center of mass coincides with the center of gravity. The same conclusion may be reached for the case of point masses or separate bodies whose center of gravity is defined by equations such as (2.8.12) and (2.8.14). In those equations, $m_i g_i$ may be substituted for w_i. If the gravitational accelerations g_i have the common value g for all the masses of the system, then this quantity can be factored out of both numerator and denominator and canceled, leaving expressions identical to (4.4.12), (4.4.13), and (4.4.14). Again, the center of gravity and the center of mass are the same.

If, on the other hand, the value of g *varies* among the mass elements in an irregular body, or from one mass to another in a system of point masses or extended separate bodies, then the operations described above are no longer possible. For example, now, $d(mg) = g\, dm + m\, dg$; furthermore, g may not be brought outside the integrals. Under these conditions the center of mass of a system and its center of gravity may not necessarily coincide. Although such situations are not uncommon on the astronomical scale, in terrestrial situations the value of g is practically constant everywhere, and the center of mass and the center of gravity coincide. We shall encounter numerous references to the center of mass in subsequent work, especially in Chapters 6 and 7. The following examples illustrate the use of integration in calculating the coordinates of the center of mass or the center of gravity.

EXAMPLE 4.4.1

Find the center of mass of the right triangular plate of uniform density and thickness shown in Fig. 4.10a.

According to the figure, the legs of the triangle are a and b. The slope of the line representing the hypotenuse of the triangle is then $\mu = b/a$. From elementary analytic geometry, we know that the equation of a straight line whose slope is μ and whose x-intercept is x_1 is $y = \mu(x - x_1)$. Since the slope of the hypotenuse of the plate is b/a and since its x-intercept is zero, its equation must therefore be

$$y = (b/a)x \qquad (4.4.17)$$

The x-coordinate of the center of mass, according to (4.4.5), is

$$x^* = \frac{\int x\, dm}{\int dm}$$

We are now faced with the problem of *how to choose* the mass elements dm and *how to relate the mass of each of them to the size and shape of the plate*. Since we wish to sum $x\, dm$ for each element, it is best to choose elementary bits of the plate, each of which is at a given distance x from the x-axis, as shown by the shaded element of Fig. 4.10a. If the thickness of the plate is h and if its density (mass per unit volume) is ρ, then the mass of the shaded element will be the product of the density ρ and the volume dV, which is itself the product of the thickness h and the area $dA = y\, dx$. We may, therefore, write

$$dm = \rho\, dV = \rho h\, dA = \rho h y\, dx \qquad (4.4.18)$$

From (4.4.5), the mass is obtained by summing over mass elements such as those shown by dotted lines in Fig. 4.10a by integration. According to (4.4.5) and (4.4.18), then, since ρ and h are constants,

$$x^* = \frac{\int x\, dm}{\int dm} = \frac{\rho h \int xy\, dx}{\rho h \int y\, dx} \qquad (4.4.19)$$

The length y of each element is given in terms of x by (4.4.17) as $(b/a)x$. Substituting this expression for y in (4.4.19) and integrating from $x = 0$ to $x = a$, we find

113

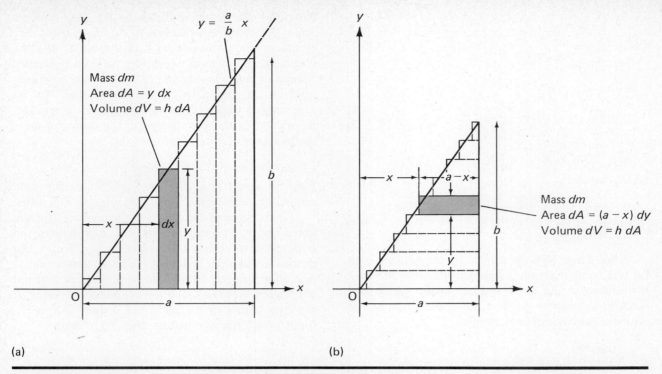

FIGURE 4.10. Subdivision of a triangular plate into mass elements appropriate for the determination of (a) the x-coordinate and (b) the y-coordinate of the center of mass by integration.

$$x^* = \frac{\rho h(b/a) \int_0^a x^2 \, dx}{\rho h(b/a) \int_0^a x \, dx} = \frac{[x^3/3]_0^a}{[x^2/2]_0^a} = \frac{2a}{3} \qquad (4.4.20)$$

This result can also be obtained without actually performing the integration in the denominator, since it is evident from simple geometry that the plate's area is $\frac{1}{2}ab$ and that its mass can be calculated as density times volume, from which

$$m = \rho V = \rho h A = \tfrac{1}{2}\rho hab \qquad (4.4.21)$$

Then,

$$x^* = \frac{1}{m} \int x \, dm = \frac{\rho h(b/a) \int_0^a x^2 \, dx}{\rho hab/2} = \frac{2a}{3} \qquad (4.4.22)$$

The y-coordinate of the center of mass is found in a similar way by dividing the plate into dm elements each of which is at a constant height y above the y-axis, as shown in Fig. 4.10b. For these elements,

$$dm = \rho \, dV = \rho h \, dA = \rho h(a - x) \, dy \qquad (4.4.23)$$

But from (4.4.17), $x = (a/b)y$. Substituting this into the above equation, we find

$$dm = \rho ha\left(1 - \frac{y}{b}\right) dy \qquad (4.4.24)$$

From (4.4.6), (4.4.21), and (4.4.24), then, integrating from $y = 0$ to $y = b$ over the dotted mass elements shown in Figure 4.10b, we obtain

$$y^* = \frac{1}{m} \int y \, dm = \frac{\rho ha \int_0^b \left(y - \frac{y^2}{b}\right) dy}{\rho hab/2} = \frac{\left[\dfrac{y^2}{2} - \dfrac{y^3}{3b}\right]_0^b}{b/2}$$

or

$$y^* = b/3 \qquad (4.4.25)$$

The (x, y)-coordinates of the center of mass are found to be $2a/3$ and $b/3$. Since the plate is flat and of uniform thickness and density, the z-coordinate of the center of mass will be such as to locate the center of mass halfway between the two large triangular faces that are normal to the z-axis.

EXAMPLE 4.4.2

Find the center of mass of a uniform semicircular plate of uniform density ρ, radius a, and thickness h, as shown in Fig. 4.11.

If the coordinate axes are oriented as shown in the figure, it is evident from the fact that the plate is symmetric about the horizontal axis that $y^* = 0$. Because of the uniform density and thickness of the plate, the z-coordinate of the center of mass will be such that this point is located midway between the two parallel semicircular faces. The x-coordinate may be obtained by integrating over appropriately chosen mass elements dm, as illustrated in Fig. 4.11. These elements are long, thin strips, parallel to the y-axis, of length $2y$ and width dx. The shaded element shown in the figure is at a distance x from the vertical

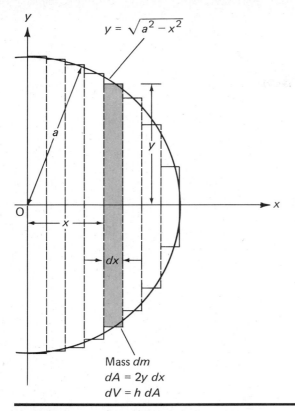

FIGURE 4.11. Mass elements appropriate for determining the *x*-coordinate of the center of mass of a semicircular plate by integration.

coordinate axis. For a plate of thickness h and density ρ, the mass dm will be

$$dm = \rho \, dV = \rho h \, dA = 2\rho h y \, dx \qquad (4.4.26)$$

The half-length y is given as a function of x by the equation of the circle,

$$y = \sqrt{a^2 - x^2} \qquad (4.4.27)$$

The plate's total mass m can be expressed in terms of its dimensions by

$$m = \rho V = \rho h A = \rho h \pi a^2 / 2 \qquad (4.4.28)$$

Then, from Eqs. (4.4.26), (4.4.27), and (4.4.28), integrating over the area elements of the semicircle from $x = 0$ to $x = a$,

$$x^* = \frac{1}{m} \int x \, dm = \frac{2\rho h \int xy \, dx}{\pi \rho h a^2 / 2}$$

$$= \frac{4}{\pi a^2} \int_0^a x \sqrt{a^2 - x^2} \, dx$$

or

$$x^* = -\frac{2}{\pi a^2} \left[\frac{2}{3} (a^2 - x^2)^{3/2} \right]_0^a = \frac{4a}{3\pi} \qquad (4.4.29)$$

The integral involved in the above equations may be evaluated by using a table of integrals such as the one included as an appendix at the end of this book. Alternatively, if we let $v = a^2 - x^2$ and $dv = -2x \, dx$, the integral is reduced to one involving only the integrand $v^{1/2} \, dv$, which is easy to evaluate.

EXAMPLE 4.4.3

Find the center of mass of a solid object of uniform density having the form of a right circular cone of base radius a and height h.

Referring to Fig. 4.12, the symmetry of the object about the x-axis is such as to ensure that $y^* = z^* = 0$. The coordinate x^* can be found by integration over coin-shaped elements of mass dm whose faces are parallel to the base of the cone. In the case of the shaded mass element dm shown in the figure, located at a distance x from the cone's vertex, we can write

$$dm = \rho \, dV = \rho A \, dx = \rho \pi y^2 \, dx \qquad (4.4.30)$$

The radius y is related to x by the equation of the line OA in the xy-plane, as in Example 4.4.1 above, from which

$$y = (a/h)x \qquad (4.4.31)$$

From (4.4.5), (4.4.31), and (4.4.30), integrating from $x = 0$ to $x = h$, we find

$$x^* = \frac{\int x \, dm}{\int dm} = \frac{\pi \rho \int xy^2 \, dx}{\pi \rho \int y^2 \, dx} = \frac{\dfrac{a^2}{h^2} \displaystyle\int_0^h x^3 \, dx}{\dfrac{a^2}{h^2} \displaystyle\int_0^h x^2 \, dx}$$

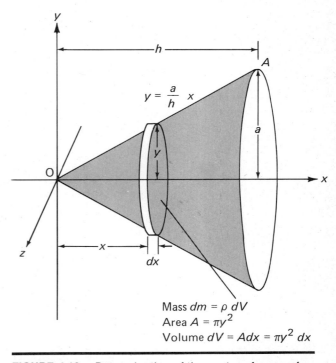

FIGURE 4.12. Determination of the center of mass of a cone, using circular lamina as mass elements.

or
$$x^* = \frac{[x^4/4]_0^h}{[x^3/3]_0^h} = \frac{3a}{4} \qquad (4.4.32)$$

Had we been able to recall that the volume of a cone is one third the area of the base times the height, we could have written the total mass as a function of the dimensions as $m = \int dm = \rho V = \rho a^2 h/3$ directly and thus avoided having to evaluate the integral in the denominator above. In this example and in all the others above, the center of gravity can be found in a similar way using (4.4.9), (4.4.10), and (4.4.11). It is obvious, of course, that if g is the same for all mass elements in the body, the calculations are essentially the same as those displayed above and the center of gravity and the center of mass are the same.

EXAMPLE 4.4.4

A solid object of uniform density has the form of a half-cylinder whose diameter and height are equal joined to a cube as shown in Fig. 4.13. Find the center of mass.

This example is most easily discussed as a system of two objects, a half-cylinder and a cube, whose individual centers of mass are known. We can, therefore, use Eqs. (4.4.12), (4.4.13), and (4.4.14), which allow us to find the desired answer without integrating. This example is very similar to Examples 2.8.1 and 2.8.2 involving the center of gravity of composite objects. Let us call the cube object 1, whose center of mass $C_1(x_1, y_1, z_1)$ is obviously located at its center. The center of mass of the half-cylinder, which we refer to as object 2, has been found in Example 4.4.2 above. If we choose our coordinate system as shown in Fig. 4.13, then C_1 is at the origin and $x_1 = y_1 = z_1 = 0$. From (4.4.29), the coordinates of the center of mass $C_2(x_2, y_2, z_2)$ of the second object will be $x_2 = y_2 = 0$, and $z_2 = (a/2) + (4a/3\pi)$. The masses of the two ob-

jects, m_1 and m_2, can be expressed in terms of their dimensions as

$$m_1 = \rho V_1 = \rho a^3 \qquad (4.4.33)$$

$$m_2 = \rho V_2 = \tfrac{1}{2}\pi\rho\left(\frac{a}{2}\right)^2 a = \pi\rho a^3/8 \qquad (4.4.34)$$

where ρ is the density. According to (4.4.14), then, since $N = 2$,

$$z^* = \frac{\displaystyle\sum_{i=1}^{2} m_i z_i}{\displaystyle\sum_{i=1}^{2} m_i} = \frac{m_1 z_1 + m_2 z_2}{m_1 + m_2}$$

$$= \frac{(\rho a^3)(0) + \dfrac{\pi\rho a^3}{8}\left(\dfrac{a}{2} + \dfrac{4a}{3\pi}\right)}{\rho a^3 + \dfrac{\pi\rho a^3}{8}}$$

Canceling factors of ρa^3 in numerator and denominator,

$$z^* = \frac{3\pi + 8}{6(\pi + 8)}\, a = 0.2607a \qquad (4.4.35)$$

From the symmetry of the object and from the fact that $x_1 = x_2 = y_1 = y_2 = 0$, it is easy to see that $x^* = y^* = 0$.

4.5 Inertial and Accelerated Reference Frames

It is abundantly evident that in our terrestrial environment Newton's laws predict the results of dynamic experiments quite well, even when the earth's centripetal and azimuthal accelerations are neglected. This has been established in Examples 4.3.2, 4.3.4, and 4.3.5, where the magnitude of these effects is calculated and found to be rather small. It is clear, nevertheless, that in a terrestrial reference frame, Newton's laws do not give us precisely correct answers *unless we provide complete information about the accelerations to which the reference system itself is subjected* and incorporate this into our calculation.[4] We are at the same time

[4] In view of the fact that two thousand years of effort on the part of philosophers who were highly skilled by the standards of their own time were required before the laws of motion were understood by Newton, it is interesting to speculate on the degree of understanding we would have achieved, and the way in which those laws would have been formulated, if the earth's axial rotation were, say, ten times as rapid as it actually is. Under those circumstances, the ratio $r_0\omega^2/g$ would have the value 0.345 and centripetal effects would be 100 times as large as they are under present conditions; coriolis effects would be larger by a factor of 10. A 5-slug mass would appear to weigh only about 105 pounds at the equator, and perceptible azimuthal forces would be encountered in high-speed elevators, while a plumb bob would hang at an angle of 10° from the vertical at latitude 45°N or S!

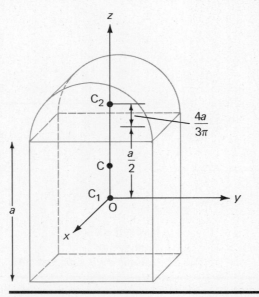

FIGURE 4.13

led to suspect that there may exist reference frames in which Newton's laws are obeyed exactly, without any necessity for introducing corrections arising from the motion of the reference frame. Such reference systems are referred to as *inertial* reference systems, and, if we are to believe implicitly in Newton's laws, we should be tempted to conclude that an inertial system is one which is unaccelerated. But unaccelerated with respect to what?

Certainly, we can construct a frame of reference which is fixed with respect to the sun rather than the earth and in which the somewhat disturbing effects associated with the earth's axial and orbital rotations would be absent. This would be a nearly perfect inertial frame so far as happenings on the scale of the earth or even the solar system are concerned. One may ask, however, whether the sun is really unaccelerated with respect to the center of mass of our galaxy. If not, this reference frame might not serve as a good inertial frame for events on a galactic scale. One might then have to go to a frame which is fixed with respect to the center of mass of our galaxy in order to obtain an inertial frame for phenomena on that scale. But is our galaxy unaccelerated with respect to others? Certainly not—and the game may go on endlessly. It is often said that an inertial frame is one which is stationary or moving with constant velocity with respect to "the fixed stars." But the stars are not really fixed; they have velocities and accelerations as well, so that this is an empty statement.

The quest for an inertial frame of reference is one in which the goal always eludes us. We cannot be sure that any reference system is a perfect inertial frame. We may, indeed, regard Newton's laws as *defining* what is meant by an inertial system, and *asserting that such systems actually exist.* The difficulties associated with finding such a reference system in practice illustrate one of the most serious logical inadequacies of Newtonian mechanics. They are responsible, in part, for having inspired the theory of relativity, in which all reference frames are equivalent, none having the preferred status of inertial frames in the Newtonian scheme.

Nevertheless, it is certainly not difficult to find reference frames that approximate inertial systems and in which Newton's laws work to a high degree of accuracy, and this is, after all, the main requirement of any physical theory. While recognizing that they are not free of logical inconsistency, we may confidently use Newton's laws to describe the physical behavior of systems built on the scale of ordinary terrestrial objects and, in many instances, involving microscopic or celestial systems. We must understand, however, that in these latter situations we cannot have quite so much confidence in Newtonian mechanics, and we must not be surprised under such circumstances to find that classical mechanics may sometimes lead to erroneous results. On the microscopic scale, an entirely different scheme, called *quantum mechanics*, must ordinarily be used to describe the behavior of atomic systems, while considerations arising from the theory of relativity are often important for both atomic and astronomic systems.

4.6 True Forces and Inertial Forces

There are only four fundamental forces known to exist in nature. The strongest is the *strong nuclear interaction*, which binds the elementary particles of atomic nuclei together. In addition, there is a *weak nuclear interaction*, which plays an important role in the process, sometimes referred to as beta decay, by which a given radioactive nucleus may be transmuted into a nucleus of a different kind by the emission of an electron. These forces, however, though very strong within atomic nuclei, fall off very rapidly with increasing distance and are, therefore, of little effect outside the nuclei of atoms. Of intermediate strength is the *electromagnetic force*, which exists between electric charges and currents. Weakest of all is the *gravitational force*, which is a force of mutual attraction between any two masses. The gravitational force is, indeed, so weak that mutual gravitational attraction between masses of the size we are accustomed to observing in everyday life is quite negligible, and only bodies of tremendous mass (such as the earth, for example) generate gravitational fields that are at all perceptible to us. We shall investigate the gravitational force in some detail in Chapter 8.

The forces that exist between solid objects in contact are forces of repulsion between the atoms of the respective bodies, and these interatomic forces are essentially electromagnetic in nature. The same general remarks apply to frictional forces. We shall refer to forces that fall into any of the four general categories described above as *true* forces.

In addition to these true forces, an observer in a reference frame accelerated with respect to an inertial system will experience forces arising from the acceleration of his system which, though real enough to him, are quite unreal from the point of view of an observer in the inertial system. Forces of this kind are called *inertial* forces. For example, consider two observers, one in an inertial system, the other riding in a train which travels at constant speed around a circular track. The observer on the train argues as follows:

"The objects surrounding me on this train certainly appear to be at rest and in a state of static equilibrium. Nevertheless, some very odd forces seem to be acting on me and on other objects on the train. For example, I feel a 'centrifugal force' pushing me outward, away from the center of the circular track, and if I jump off the floor of the car, I accelerate outward with an acceleration $r\omega^2$ with respect to the car while I am in the air. Also, if I move back and forth across the car, with radial speed v_r, I feel a strange sidewise azimuthal thrust, parallel to the track, of magnitude $2m\omega v_r$. I have to regard this force as a real one, and I refer to it as a 'coriolis force.' Moreover, if I set up experiments involving force systems that are static in my reference frame, I cannot explain the results I obtain using Newton's laws unless I assume that in addition to the gravitational and contact forces acting, there is a centrifugal force $mr\omega^2$ directed radially outward acting on all bodies in my reference frame. In the same way, I have to assume that an azimuthal coriolis force of magnitude $2m\omega v_r$ always acts in my reference system, if I wish to obtain experimentally correct answers in applying Newton's laws to situations involving bodies whose radial velocity component is not zero. Both these forces are easily perceived, and I cannot make any sense out of mechanics unless I introduce them. I, therefore, conclude that they are real forces."

The observer in the inertial system, however, takes a very different view of the matter. He states it this way:

"I am quite certain that Newton's laws agree extremely well with experiment in my frame of reference. The train, and all objects that travel with it, however, must be acted upon somehow by a resultant *centripetal* force $-mr\omega^2$, directed radially *inward*. If this were not the case, the train and any objects that accompany it could not be executing uniform circular motion, but would be following some other path. This centripetal force originates in the forces the track exerts on the wheel flanges and is transmitted by the rigid materials of which the cars are constructed to all bodies aboard the train. Likewise, a certain azimuthal force has to act on bodies on the train having a radial component of velocity to sustain them in a radial path. Again, the train itself, and ultimately the track, supplies this azimuthal force. The observer who fancies himself to be pushed outward by a 'centrifugal force' feels only the reaction his body exerts against the inward centripetal force, which is the only real force present. When he jumps off the floor of the car, he is mistaken in concluding he acquires an outward acceleration $mr\omega^2$. What really happens is that as soon as he is free of the floor, there is *no force whatever* acting on him, and he is in free flight as a projectile. But the train still has its *inward* acceleration $-mr\omega^2$ and leaves him behind, so to speak, producing the impression to an observer on the train that he has acquired an *outwardly* directed acceleration of that same magnitude.

"In much the same way, when the man on the train tries to analyze the results of a static force equilibrium experiment, in which the objects are unaccelerated in his reference frame, he has forgotten that his experimental bodies are subject to the same centripetal acceleration as the train itself. He finds that equating the sum of x-, y-, and z-force components to zero does not predict results that agree with experiment unless he

assumes that a centrifugal force $mr\omega^2$ acts radially outward from the center of mass of every body, and concludes from this that the centrifugal force is real. What he should have done is to note that all the objects in his experiment have a centripetal acceleration $-mr\omega^2$, directed radially inward, and set the force components equal not to zero but to the mass times the respective component of this centripetal acceleration vector! He would then find that Newton's laws give experimentally correct answers even though the only forces acting are real gravitational and contact forces. The 'centrifugal force' he introduces is superfluous, therefore, and the need for it arises only because he has not properly accounted for the acceleration of his reference system. Similar remarks can be made about the 'coriolis force' he believes to exist."

Both points of view are valid, and, provided that the proper inertial forces are introduced by the observer in the accelerated system, both lead to precisely the same solutions for dynamical problems. This is illustrated in Examples 4.6.1 and 4.6.2. In subsequent discussions, we shall for the most part adopt the methods of the observer in the inertial system, avoiding the necessity for introducing inertial forces, because his methods are more straightforward and proceed more directly from basic physical principles.[5]

EXAMPLE 4.6.1

A mass m is suspended on a string from the ceiling of one of the cars in the train discussed above, as illustrated in Fig. 4.14a. Find the tension in the cord and the angle θ the cord makes with the vertical direction, using both approaches outlined above.

The observer in the inertial reference frame would reason that the only two real forces acting are the tension $\mathbf{T}$ and the weight force $m\mathbf{g}$. Since the mass m is executing uniform circular motion, it must experience a net inwardly directed centripetal force $\mathbf{F}_c$, whose magnitude is $mr\omega^2$. This net force must then be the resultant of the real forces $\mathbf{T}$ and $m\mathbf{g}$ as shown in Fig. 4.14b. Since there is no upward acceleration, the net z-component of force must be zero, whence

$$T \cos \theta - mg = 0 \qquad (4.6.1)$$

The net x-component of force, on the other hand, must equal the mass times the x-component of acceleration, whereby

$$-T \sin \theta = mr\omega^2 \qquad (4.6.2)$$

[5] One should note, however, that in Example 4.3.6 an observer in the reference system of the automobile, who introduces a centrifugal force acting radially outward from the center of mass, may obtain the solution by taking moments of forces about any desired point other than the center of mass. An observer in the inertial frame, it will be recalled, runs into certain difficulties unless he takes moments about the center of mass. In this instance, the methods of the observer in the rotating frame, who is compelled to introduce an inertial force, have something to recommend them.

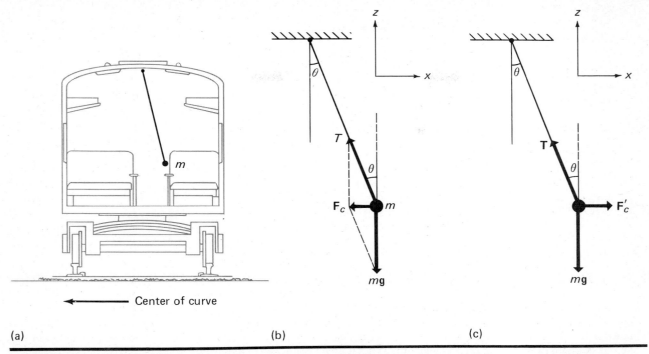

FIGURE 4.14. (a) Suspended mass in a train rounding a circular curve. (b) System of forces postulated by an observer in an inertial reference frame. (c) System of forces arrived at by an observer on the train.

From (4.6.1),

$$T = \frac{mg}{\cos \theta} \qquad (4.6.3)$$

Substituting this into (4.6.2) and solving for $\tan \theta$, we find

$$\tan \theta = \frac{r\omega^2}{g} \qquad (4.6.4)$$

The observer on the train takes the view that in his reference frame, the mass is unaccelerated, so that the problem is one involving only the *static equilibrium* of forces. However, since he concludes that every object in his reference frame experiences an outwardly directed centrifugal force, there must be such an inertial force $\mathbf{F}'_c$ of magnitude $mr\omega^2$ acting on the suspended mass. Accordingly he draws a force diagram such as that shown in Fig. 4.14c and sets both x- and z-components equal to zero, as required for static equilibrium. He then writes

$$F'_c - T \sin \theta = mr\omega^2 - T \sin \theta = 0 \qquad (4.6.5)$$

$$T \cos \theta - mg = 0 \qquad (4.6.6)$$

But these two equations are identical with (4.6.1) and (4.6.2), so the solution is the same. If the man on the train did not know that the correct magnitude for the inertial force $\mathbf{F}'_c$ was $mr\omega^2$, he would first have to solve the problem using the methods of the inertial frame

observer and note that if he gave his inertial force $\mathbf{F}'_c$ this magnitude, his equations would agree with the inertial frame solution. He would then be confident in using the magnitude $mr\omega^2$ for the centrifugal force in all dynamical problems in his reference frame.

EXAMPLE 4.6.2

The train discussed in connection with Example 4.6.1 proceeds along a straight level track with constant acceleration a as shown in Fig. 4.15a. Find the angle θ between the string supporting a suspended mass m and the vertical, and the tension in the string. What inertial force would have to be introduced by a passenger on the train to enable him to account for the acceleration of his reference system in solving mechanics problems in that frame?

To an observer in the inertial frame, the only real forces acting on the mass are the weight force $m\mathbf{g}$ and the tension $\mathbf{T}$, as shown in Fig. 4.15b. The mass m is undergoing constant linear acceleration, however, and thus, according to Newton's second law must experience a constant resultant force $\mathbf{F}_r$, equal to $m\mathbf{a}$. This total force $\mathbf{F}_r$ must be the vector sum of $\mathbf{T}$ and $m\mathbf{g}$. Since there is no vertical component of acceleration, we may equate the sum of the z-components of the real forces to zero, from which

$$T \cos \theta - mg = 0 \qquad (4.6.7)$$

The sum of the x-components, however, must equal

119

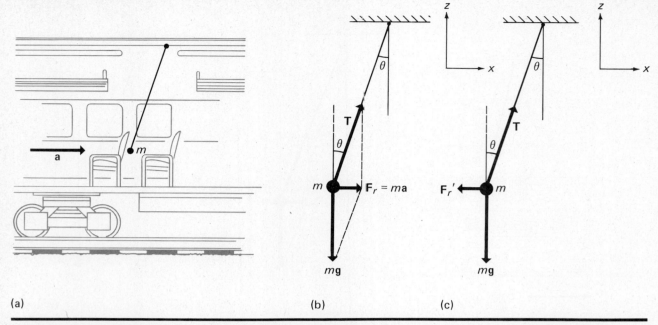

(a)   (b)   (c)

FIGURE 4.15. (a) Suspended mass in a train accelerating along a straight track. (b) The situation as viewed by an observer in an inertial system. (c) The situation as seen by an observer on the train.

the mass times the acceleration, and therefore

$$F_r = T \sin \theta = ma \qquad (4.6.8)$$

From (4.6.7),

$$T = \frac{mg}{\cos \theta} \qquad (4.6.9)$$

Substituting this value for T into (4.6.9) and solving for $\tan \theta$, we obtain

$$\tan \theta = \frac{a}{g} \qquad (4.6.10)$$

    An observer on the train would conclude that the mass is unaccelerated, hence in a state of equilibrium, but subject to an *inertial force* due to the acceleration of the reference system. His force diagram would look like the one shown in Fig. 4.15c, where $\mathbf{F}_I$ is the inertial force. Since to him the system is in equilibrium, he would equate both x- and z-components of force to zero, obtaining

$$T \cos \theta - mg = 0 \qquad (4.6.11)$$

and

$$T \sin \theta - F_I = 0 \qquad (4.6.12)$$

Equations (4.6.7) and (4.6.11) are clearly identical; Eqs. (4.6.8) and (4.6.12) are not the same, however, *unless* the inertial force $\mathbf{F}_I$ has the magnitude ma. Accordingly, the solution cannot agree with the one obtained in the inertial system unless

$$\mathbf{F}_I = -m\mathbf{a} \qquad (4.6.13)$$

The minus sign appears because the inertial force $\mathbf{F}_I$ is directed opposite the acceleration vector, as shown in Figs. 4.14b and c. Another way of seeing this is to note that in the inertial frame

$$\mathbf{T} + m\mathbf{g} = \mathbf{F}_r = m\mathbf{a} \qquad (4.6.14)$$

while in the accelerated frame of the train

$$\mathbf{T} + m\mathbf{g} + \mathbf{F}_I = 0 \qquad (4.6.15)$$

These vector equations will agree only if $\mathbf{F}_I = -m\mathbf{a}$, as given by (4.6.13). The inclusion of inertial forces $-m\mathbf{a}$ for all objects would be necessary to allow a passenger in the train to solve mechanics problems correctly using the "first observer's approach" in the accelerated reference frame.

    It should be noted in connection with both examples that it is quite easy for an observer in the accelerated system to find the inertial forces appropriate to his reference frame so long as he knows its acceleration relative to an inertial system. If he does not, and this may very well be the case, then the only way of determining the proper inertial forces is by experiment.

SUMMARY

Newton's second law may be used to describe systems acted upon by a resultant force in much the same way as the first law is applied to systems in equilibrium. In both cases, the bodies comprising the system are viewed as *isolated free bodies* acted upon by a number

120

of individual forces that are summed vectorially to obtain a resultant. In the case of a body in equilibrium, the components of this resultant are all zero. If the system is not in equilibrium, the components of the resultant force on each body, according to Newton's second law, are to be equated to the mass times the respective components of its acceleration. Therefore,

$$\sum_i F_{ix} = ma_x \qquad \sum_i F_{iy} = ma_y \qquad \sum_i F_{iz} = ma_z$$

Any object that moves in a circular path has a radial or centripetal acceleration $-v^2/r$ or $-r\omega^2$. To move in such a path, it must be acted upon by a resultant centripetal force component, directed radially inward, of magnitude mv^2/r or $mr\omega^2$. Likewise, anybody having a radial velocity component v_r in a system rotating with angular velocity ω with respect to an unaccelerated system has an azimuthal acceleration $2\omega v_r$; to move in such a way, it must be acted upon by a resultant azimuthal force component $2m\omega v_r$. These centripetal and azimuthal force components are frequently furnished by forces the surroundings exert on the body in question, such as normal forces, tension forces, and frictional forces.

The center of mass of a body is defined by mathematical relations like those that define the center of gravity, except that the weight is replaced throughout by the mass. The center of mass and the center of gravity coincide whenever the acceleration of gravity is the same for all mass elements of the system.

Newton's laws are stated with reference to inertial, or unaccelerated, coordinate systems. In accelerated reference systems, they are still valid, but effects arising from the acceleration of the coordinate system must be taken into account. The most simple and plausible way of doing this is to analyze the motion with reference to an inertial coordinate system and then transform it mathematically into the reference frame of the accelerated observer. Alternatively, certain "inertial forces" peculiar to the accelerated reference system may be assumed to act on all bodies within it, in such a way as to account for the acceleration of the reference frame, and allow Newton's laws in their usual form to be used therein.

QUESTIONS

1. An object at the end of a string is spun around in a vertical circle. In what position is the string most likely to break?
2. A circular room of radius R spins with constant angular velocity ω about an axis passing through its center. There are a number of paintings inside the room, the coefficient of static friction between the wall and the paintings being μ_s. Is it possible to hang the paintings simply by placing them against the wall? If so, formulate the conditions under which this will be possible.

3. A boy is standing on a scale inside an elevator which is descending at a constant but high speed. Will the reading of the scale be modified by the motion of the elevator?
4. A stone attached to a rope is whirled in a horizontal circle of constant radius. The rope makes an angle of 30° with the vertical. Is the tension in the rope greater or less than the weight of the stone?
5. You are riding in an open railroad car moving with constant speed v_0. You throw a ball up in the air. Describe its motion as seen by you and by an observer sitting under a nearby tree.
6. You are on a merry-go-round which rotates with constant angular frequency ω_0. If you throw a ball straight up in the air, will you be able to catch it when it comes down?
7. You are inside a vehicle which has no windows and is totally isolated from its surroundings. Propose an experiment to determine whether or not you are accelerating.
8. A heavy weight is supported on a horizontal plane by frictionless rollers. It is tied to the head of a heavy hammer by a length of stout clothesline. A man then "swings" the hammer in such a way that the clothesline tied to the weight restrains its motion about halfway through the swing. It is found that the clothesline invariably breaks under these circumstances before imparting any appreciable motion to the weight on the rollers. Yet if a fine thread is tied to the heavy weight and given a gentle but steady pull, the weight can be set in motion, and after a time made to move as fast as desired. Explain.
9. It is apparent from Example 4.2.1 that the indications of a spring balance used to weigh an object in an elevator that accelerates upward or downward will be incorrect. Will you obtain correct weights using a bathroom scale? How about an equal-arm balance and a set of standard masses?

PROBLEMS

1. What resultant force is required to accelerate a 3800-pound automobile at a constant rate from rest to a speed of 88 ft/sec in 12.0 sec?
2. A locomotive capable of exerting a tractive effort of 120,000 pounds pulls a train weighing 3000 tons up a 1% grade. The coefficient of rolling friction is 0.003. Find (a) the maximum acceleration attainable, (b) the time and distance required to attain a speed of 22 ft/sec, and (c) the maximum weight which could be drawn up the grade.
3. A box of mass 100 kg is drawn along a flat level surface by a horizontal force of 480 newtons. The coefficient of friction between the box and the surface upon which it slides is 0.20. What is its acceleration?
4. A 192-pound man rides in an elevator which accelerates vertically upward with an acceleration of 6.4 ft/sec². Find (a) the normal force exerted by the floor of the elevator upon the man's feet and (b) the value of this normal force if the acceleration were reversed in direction.
5. Suppose the man in the elevator in Problem 4 were suspended from a spring balance. (a, b) What would the balance reading be under the circumstances outlined in Problems 4a and 4b? (c, d) What would the corresponding indications of an equal-arm beam balance be?
6. A 2-kg block is placed on a plane inclined at an angle of

37° to the horizontal. It is drawn up the plane with a constant acceleration of 4.0 m/sec² by a force **T** directed parallel to the plane. The coefficient of sliding friction between block and plane is 0.1. Find **(a)** the component of the weight force parallel to the plane, **(b)** the force of friction, and **(c)** the magnitude of the force **T**.

7. Two rectangular blocks are in contact on a frictionless horizontal plane. The left-hand block is of mass m_1, the right-hand one of mass m_2. The left-hand block is pushed to the right by a horizontal force **F** acting against its left side. Find **(a)** the acceleration of the system and **(b)** the force of contact exerted on mass m_1 by the mass m_2. Now suppose that the direction of **F** is reversed and that it acts against the right side of the right-hand block. Find under these conditions **(c)** the acceleration of the system and **(d)** the force of contact exerted by the mass m_1 on mass m_2. **(e)** Why are the answers to **(b)** and **(d)** different?

8. Work out the answers to parts **(a)**, **(b)**, **(c)**, and **(d)** of the preceding problem, assuming that $F = 5.0$ newtons, $m_1 = 2.0$ kg, $m_2 = 3.0$ kg, and that the coefficient of sliding friction between both blocks and the plane is 0.05.

9. Three rectangular blocks are in contact on a frictionless plane. The left one is of mass 2.0 kg, the center one of mass 3.0 kg, the right one of mass 4.0 kg. A horizontal force **F** of magnitude 12 newtons is applied to the left side of the leftmost block, pushing the system to the right. Find **(a)** the acceleration of the system, **(b)** the force of contact between the left block and the middle block, and **(c)** the force of contact between the middle block and the right block.

10. Two rectangular blocks in contact with one another are pushed up a frictionless plane inclined at 37° to the horizontal by a force **F** of magnitude 60 newtons, which acts parallel to the plane. The first block, on which the force **F** acts, is of mass 3.0 kg; the other is of mass 5.0 kg. Find **(a)** the acceleration of the system and **(b)** the force of contact exerted by the first block on the second.

11. A boy and a girl on ice skates weigh 160 and 80 pounds, respectively. They are holding opposite ends of a horizontal rope. The boy is pulled along the ice by a third person who exerts a horizontal force of 40 pounds on him. There is a force of friction of 10 pounds acting on the boy's skates and a force of friction of 8 pounds on the girl's skates. Find **(a)** the acceleration of the boy, **(b)** the acceleration of the girl, and **(c)** the tension in the rope between them. Suppose now that the 40-pound force acting on the boy is removed. Find **(d)** the acceleration of boy and girl and **(e)** the tension in the rope under these circumstances.

12. Two rectangular blocks, one resting on top of the other, are accelerated along a frictionless horizontal surface by a horizontal force **F** applied to the lower block. The coefficient of static friction at the interface between the two blocks is 0.2, the mass of the lower block is 16.0 kg, and the mass of the upper block is 8.0 kg. How large does the force **F** have to be to make the upper block slide backward on the lower one?

13. A 6.0-kg block is accelerated along a rough horizontal surface by a force of 40 newtons acting downward at an angle of 37° below the horizontal. A force of friction of 20 newtons acts at the interface between block and plane.

Find **(a)** the acceleration of the block and **(b)** the coefficient of sliding friction between block and plane.

14. A 12.0-kg block is accelerated up a rough plane inclined at an angle of 30° with the horizontal by a horizontally directed force of 120 newtons. The coefficient of friction between block and plane is 0.20. Find the magnitude of the block's acceleration.

15. In Example 4.2.6, suppose that the kinetic friction coefficient had the value 0.6 rather than 0.2. The value of $\mu_k N$ would then be 60 pounds, and, from (4.2.17) and (4.2.18), we would find $a = -2.1$ ft/sec². But this would mean that the suspended mass has an acceleration directed *upward* rather than downward. This is clearly incorrect, but where is the error?

16. Find **(a)** the acceleration and **(b)** the tension in the string for the system illustrated in Fig. 2.8d. It is assumed that the suspended weight is greater than $\mu_k N$.

17. A packing case is given an initial velocity of 5 m/sec and allowed to slide to a stop on a level floor. The case slides to a stop in a distance of 8 meters. What is the coefficient of friction μ_k?

18. A mass of 25 kg slides on a plane inclined at an angle of 30° to the horizontal. It is attached by means of a string which runs over a pulley to a freely suspended mass of 40 kg, as shown in the figure. Find **(a)** the tension in the cord and **(b)** the acceleration of the system, assuming that the coefficient of kinetic friction μ_k is 0.2.

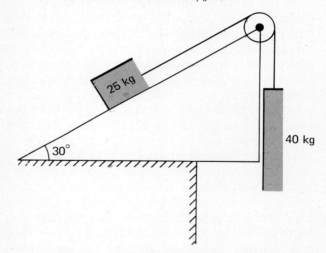

19. Work Problem 18 using the algebraic symbols m for the sliding mass, M for the suspended mass, θ for the angle of inclination of the plane, and μ_k for the kinetic friction coefficient.

20. A 40-pound weight rests upon a platform, as shown in the accompanying diagram, and is connected to a weight

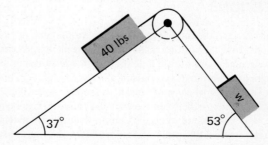

W by a light flexible cord passing over a pulley. The coefficient of kinetic friction between the 40-pound weight and the plane is 0.25, while the contact between the weight W and the surface it touches is essentially frictionless. Find (a) the acceleration and (b) the tension in the cord if $W = 10$ pounds. (c, d) Find the same quantities when $W = 60$ pounds.

21. (a) What values of W would be required, under the conditions set forth in Problem 20, to allow the system to move without acceleration? (b) What values of W would suffice to "break away" and start the system into motion if the coefficient of *static* friction between the 40-pound weight and the plane is 0.3?

22. Describe what happens physically in the situation discussed in Problem 21 when W is between 18 and 20 pounds or between 40 and 42 pounds.

23. Work Problem 20 assuming that both the weight W and the 40-pound weight are acted upon by frictional forces, the kinetic friction coefficient being 0.25 for each object.

24. Assuming that both the 40-pound weight and the weight W are acted upon by frictional forces, as described in Problem 23 (a) what values of W would be required to allow the system to move without acceleration, and (b) what values of W would suffice to "break away" and start the system into motion from rest if the *static* friction coefficient for each body is 0.3?

25. Two masses m_1 and m_2, resting on inclined planes as shown in the diagram, are connected by a light string passing over a pulley. Assuming that both masses slide on their respective planes without friction, calculate (a) the acceleration of the system and (b) the tension in the cord.

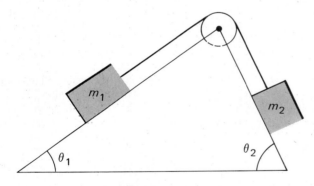

26. An 8.0-kg mass is accelerated along a horizontal friction-less plane by the arrangement shown in the accompanying figure. Neglecting friction in the pulleys, find (a) the acceleration of the 8.0-kg mass and the tension in the cords to the left (b) and right (c) of the 8.0-kg mass.

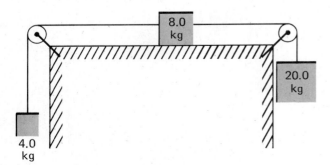

27. In the system shown in the accompanying diagram, find (a) the acceleration of the three masses, (b) the tension in the string connecting the 2.0-kg and 4.0-kg masses, and (c) the tension in the string connecting the 2.0-kg and 3.0-kg masses.

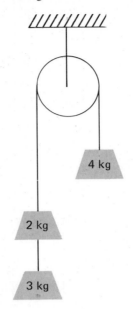

*28. Two objects having masses m and M $(M > m)$ are connected by a light, flexible cord passing over a frictionless pulley of negligible mass, as shown in the diagram. A known force F is applied vertically upward on the pulley, as shown. Show that for $0 < F < 2mg$ the system remains in equilibrium, as shown at (a); for $2mg < F < 2Mg$ the mass M remains in equilibrium while m accelerates upward, as in (b); while for $F > 2Mg$ both masses are

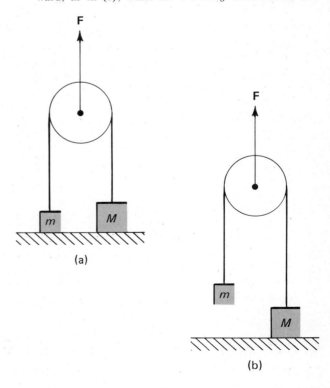

123

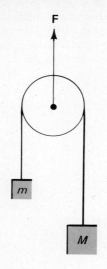

(c)

accelerated upward, as in **(c)**. Find for all three cases the tension T in the cord, the accelerations a and A of both masses, and the normal forces n and N exerted by the supporting surface on each mass.

29. Show that Eqs. (4.2.21) and (4.2.22) giving the tension and acceleration for Atwood's machine can be obtained from the results of the preceding problem.

*30. A car weighing 4000 pounds is braked so as to undergo a constant deceleration of -16 ft/sec^2. The car's wheelbase d is 10 ft, and the center of mass is located midway between front and rear wheels at a height h of 2 ft above the road, as shown in the diagram. Find **(a)** the normal force acting on the front wheels, **(b)** the normal force acting on the rear wheels, **(c)** the friction force acting at the front wheels, **(d)** the friction force acting at the rear wheels, and **(e)** the magnitude and direction of the total force on the rear wheels. *Hint:* Make the same assumption about the ratio of friction forces to normal forces as was made in Example 4.3.6, expressed by Eq. (4.3.29).

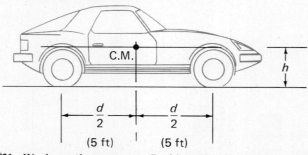

*31. Work out the answers to Problem 30 in the general case where the mass of the car is m, the acceleration is a, the wheelbase is d, and the height of the center of mass (which is still midway between the front and rear wheels) is h. Find also **(f)** the maximum deceleration which can occur before the rear wheels leave the ground, assuming, of course, that the front wheels do not slip.

32. Obtain the relevent equations of Problems 30 and 31 by introducing a proper inertial force and taking moments about the point where the front wheels touch the road.

Show that the equations so obtained are fully equivalent to those obtained using the methods described in Example 4.3.6.

33. An automobile is involved in an accident and suffers substantially constant deceleration from a speed of 60 mph (88 ft/sec) in a distance of 10 ft. The driver, who weighs 160 pounds, is restrained by seat belts. Find the force exerted by the belts on the driver.

34. A 2.0-kg mass moves in a circular path on a frictionless horizontal surface at a constant linear speed of 4.0 meters/sec. If the centripetal force acting on it is 12.0 newtons, find the radius of the path.

35. A small body of mass 0.2 kg is attached to a string 0.8 meters long and allowed to rotate in a circular path whose plane is horizontal. The string makes an angle of 50° with the vertical at all times. Find **(a)** the linear velocity of the mass, **(b)** the tension in the string, and **(c)** the time required for a complete revolution. **(d)** Does the tension in the string increase or decrease as the angle the string makes with the vertical increases?

36. A small body of mass 2.5 kg is attached to the end of a rigid rod of negligible mass and length 0.8 meters. The other end of the rod is pivoted to a horizontal arm of length 0.6 meters, which can rotate about a vertical axis. The arm rotates about the vertical axis at a constant rate of rotation in such a way that the rod supporting the mass makes an angle of 30° with the vertical. Find **(a)** the tension in the rod to which the mass is attached, **(b)** the linear speed of the mass, and **(c)** the time required for the mass to make one complete revolution.

37. The moon's mean distance from the earth is 382,000 km, and its mass is 0.01228 times the earth's mass. Find the mean position of the center of mass of the earth–moon system with respect to the center of the earth.

38. A mass m is rotated in a vertical plane with constant angular velocity ω on the end of a thin but rigid rod of length r. The mass of the rod is negligible in comparison with the mass m. Find **(a)** the tension in the rod when the mass is at the highest point in its path and **(b)** the tension in the rod when the mass is at the lowest point in its path.

39. A "space station" 100 ft in diameter, such as the one shown in the figure, is orbited about the earth in a circular orbit at an altitude of 200 miles. Since all the objects

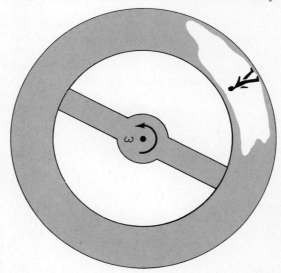

inside the station orbit with the same orbital velocity and orbital trajectory as the station itself, the inhabitants would have the sensation of weightlessness if some means of creating "artificial gravity" were not provided. Accordingly, the station is given an axial rotational velocity ω which sets up a centripetal acceleration to give the sensation of a gravitational field. **(a)** What value of ω is required to produce the effect of the earth's surface gravity, and **(b)** are the inhabitants of the station really weightless even with $\omega = 0$? Explain.

40. The earth's mean distance from the sun is 149,400,000 km. Assuming that its orbit is circular, find the centripetal acceleration associated with the earth's *orbital* motion around the sun, and compare the magnitude of orbital centripetal effects with *axial* centripetal effects due to the earth's rotation on its axis.

41. A man is positioned against the inside surface of a cylinder which rotates with constant angular velocity ω about a vertical axis, as shown in the diagram. The coefficient of friction between him and the surface of the cylinder is 0.35, and the inside radius of the cylinder is 10 ft. What is the smallest value that ω may have in order that he may not slip downward off the cylinder's wall?

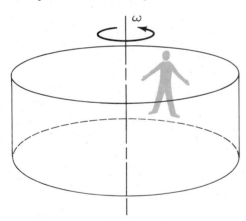

42. An automobile rounds an unbanked circular curve of radius 600 ft. The coefficient of friction between tires and roadway is 1.00. Assuming that the car will not overturn, how fast may it go before slipping outward?

43. Work Problem 42 using the general symbols m for the car's mass, v for its velocity, r for the radius of the curve, and μ for the coefficient of friction.

44. **(a)** At what angle θ should a curve of radius r be banked so that, for a car traveling with velocity v, the tires need provide no sideward frictional forces to keep the car on its course? **(b)** What would be the normal force N exerted by the roadway on the car under these circumstances? **(c)** What would the values of θ and N be for a 4000-pound car, a curve of 600-ft radius, and a speed of 88 ft/sec?

45. **(a)** How fast could a 4000-pound car round the curve described in Problem 44 without slipping outward if the coefficient of friction is 1.0, assuming that it will never overturn? **(b)** What is the normal force N under these conditions?

*46. A curve of radius r is banked at an angle θ, and a car of mass m rounds the curve at speed v. Assuming that the car will never overturn, find **(a)** the normal force exerted by the road on the wheels, **(b)** the total frictional force

acting perpendicular to the wheels, **(c)** the maximum velocity at which the car can negotiate the curve without sliding outward, and **(d)** the normal force prevailing at this maximum velocity.

47. A 10-kg mass is connected to a 5-kg mass with a cord, and the system is then pulled to the right along a level surface with another cord, as shown in the figure. The tension in the latter cord is 50 newtons, while the coefficient of kinetic friction between each mass and the surface upon which it rests is 0.25. Find **(a)** the acceleration of the system, **(b)** the tension in the cord connecting the 10-kg and 5-kg masses, and **(c)** the tension in the right-hand cord which would be required if the system were to move with constant velocity.

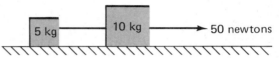

48. Suppose that the left-hand mass in the preceding problem is m_1, the right-hand mass m_2, the tension in the left-hand cord T_1, the tension in the right-hand cord T_2, and the coefficient of kinetic friction for each mass μ_k. Find the answers to the questions asked above under these conditions.

*49. Three masses connected by cords and arranged on an inclined plane with a pulley are shown in the figure. If $m_1 = 10$ kg, $m_2 = 25$ kg, $m_3 = 40$ kg, $\theta = 37°$ ($\sin \theta = 0.6$, $\cos \theta = 0.8$), and if the coefficient of kinetic friction between m_1 and m_2 and the plane is 0.2, find **(a)** the acceleration of the system, **(b)** the tension T_1, and **(c)** the tension T_2.

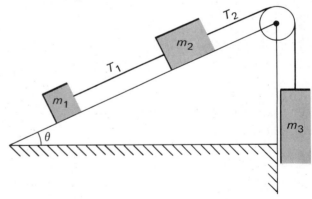

*50. Work Problem 49 using the general symbols m_1, m_2, m_3, T_1, T_2, and θ to represent the masses, tensions and angle of inclination, as shown in the figure. Assume that friction is *absent*.

51. A warship, at latitude 45°N, fires its guns due north. The guns (assumed to be at water level) are elevated 10° above the horizontal and impart to their shells an initial speed of 2500 ft/sec. Find **(a)** the range at which the shells will hit, based on the projectile theory of section 3.3, neglecting air resistance; **(b)** the actual point where the shells will hit, referred to the point predicted by the projectile theory of part **(a)** as an origin; and **(c)** the angular azimuthal correction to the aiming of the guns in order that the point predicted in **(a)** may actually be hit. *Hint*: What is the average *radial* component of velocity of the projectiles?

52. According to Bohr's theory of the hydrogen atom, a hydrogen atom is composed of an electron of mass 9.11×10^{-28} g revolving in a circular orbit of radius 5.28×10^{-9} cm about a much more massive proton. The proton exerts an attractive force of 8.27×10^{-3} dynes on the electron. Find (a) the centripetal acceleration of the electron and (b) its angular velocity about the proton.

*53. Two objects of mass m_1 and m_2 slide down a frictionless plane inclined at an angle θ with the horizontal as shown in the diagram. There is a force of friction F_f at the interface between the two bodies sufficient to prevent them from sliding over one another. Find (a) the acceleration of the system, (b) the normal force between the plane and the mass m_2, (c) the normal force between the two masses, and (d) the force of friction at the interface between the two bodies.

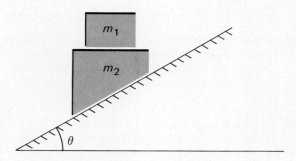

54. A 2-kg block is placed on a 37° inclined plane and drawn up the incline by a force **T**, parallel to the plane. Its acceleration up the plane is 4 m/sec². The coefficient of sliding friction is 0.1. Find the magnitude of the force **T**.

55. In Fig. 4.3a, let the pulley be massless and frictionless, but let there be a force of friction between the sliding mass and the plane. Suppose also that $m_1 = 10$ kg and $m_2 = 5$ kg. It is observed that mass m_2 descends with constant acceleration from rest through a distance of 7.66 m in 2.5 sec. Find (a) the tension in the string connecting the masses and (b) the coefficient of sliding friction between mass m_1 and the plane.

56. A 240-kg mass is pulled vertically upward as shown in the diagram. At the same time, an 80-kg mass is drawn up a 20° incline as illustrated. The coefficient of friction

between the latter mass and the surface upon which it moves is 0.50. A constant force T_1 exerted parallel to the incline is found to displace the system from rest a distance of 60 m in 10 sec. Find (a) the acceleration of masses, (b) the force of friction between the 80-kg mass and the inclined plane, and (c) the tensions T_1 and T_2 in both ropes.

57. Find the x- and y-coordinates of the center of mass of a plate of uniform thickness and density having the shape of the region bounded by the x- and y-axes and the parabola $y = 16 - x^2$.

58. Find the center of mass of a hemisphere of uniform density and radius a.

59. Using the answer to the preceding problem, find the center of mass of the object in the accompanying diagram, assuming it to be of uniform density.

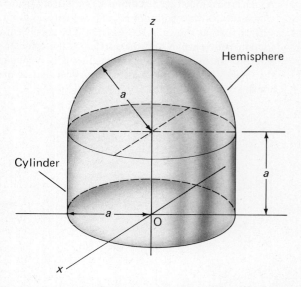

60. The upper hemisphere of the composite sphere shown in the figure is of uniform density ρ_1 and the lower hemisphere of uniform density ρ_2. Using the answer to Problem 58, find the center of mass of this composite body.

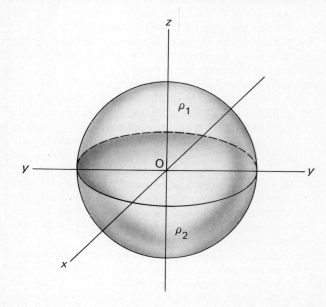

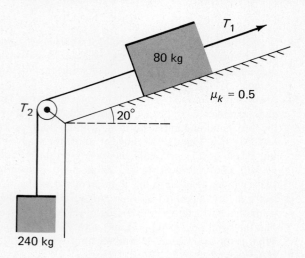

61. Find the center of mass of the composite cylinder shown in the diagram if the part above the xy-plane is of uniform density ρ_1 and the part below the xy-plane is of uniform density ρ_2.

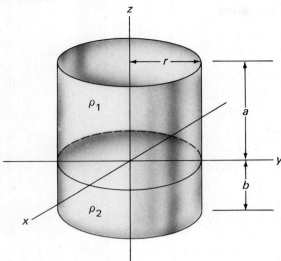

WORK AND ENERGY

5.1 Introduction

Today, everyone is talking about *energy*. Energy shortages, the high cost of energy, and future sources of energy are subjects of intense national and international concern. World energy needs and projected needs are expressed in numbers of astronomical size, plotted as exponentially rising lines on graphs, and discussed endlessly by politicians, economists, lawyers, journalists, businessmen, sociologists, urban planners, technologists, and, occasionally, scientists. Energy turns the wheels of our technological society. But just what is energy? How is it defined? What are the laws that govern its conversion from one form into another? Very well, there is an easy definition: the energy of a system is its capacity to do *work*. But then, what is work? How is it defined, and what are its properties? Now we are back to physics—mechanics, in fact—and must rely on that science to provide the answers to these questions that are obviously so important.

We have now finished the task of stating Newton's laws and illustrating their direct application to mechanics problems. The methods we have developed are useful in analyzing mechanical systems and solving practical examples. It is time now to examine some of the broader implications between mechanics and other areas of physics. These further implications are extremely interesting and important and lead ultimately to the formulation of *conservation laws* for energy and momentum. In this chapter, we shall concentrate on understanding the conservation of energy; momentum and the conservation of momentum will be discussed in the following chapter. The conservation laws transcend the limits of mechanics and serve as valid fundamental principles for the other branches of physics as well.

In order to understand these topics, we need to introduce the concept of *work*. In a colloquial sense, work is the end product of physical activity or muscular exertion. But such a vague definition would hardly satisfy Lord Kelvin; we must find a way of defining work quantitatively in precise numerical terms if it is to be of any use to us in physics. Once a satisfactory quantitative definition of work has been formulated, the concept of *energy* can be understood. In particular, the simple and useful concepts of *kinetic energy* and *potential energy* as different forms of mechanical energy can be introduced. When we have acquired an understanding of these topics, we are in a position not only to state the law of conservation of energy, but also to formulate an elegant alternative way of working mechanics problems from a quite simple, but very different point of view. We must begin, however, by saying precisely what we mean by *work*.

5.2 The Concept of Work

We shall first discuss the work and power associated with *constant* forces and then generalize the definitions which are proposed to cases in which the force varies with time or position. Mechanical *work* may be done by any force in motion; the amount of work done by a *constant* force **F** moving through a displacement **r** is a *scalar* quantity W defined by

$$W = Fr \cos \theta \qquad (5.2.1)$$

where θ is the angle between the directions of **F** and **r**. In view of the definition of the scalar product of two vectors, as given by Eq. (1.6.1), work can be expressed equally well as the dot product

$$W = \mathbf{F} \cdot \mathbf{r} \qquad (5.2.2)$$

This definition of work expresses precisely what is meant by that term in a scientific context, and the student should avoid confusing the conventional usage of the word (as encompassing any activity involving physical exertion) with the physical definition given above. For example, the man in Fig. 5.1a is doing work; but, according to our usage of the term, the man in Fig. 5.1b is not, because the force he exerts *produces no displacement*. The units of work, according to the above equations, are units of force times

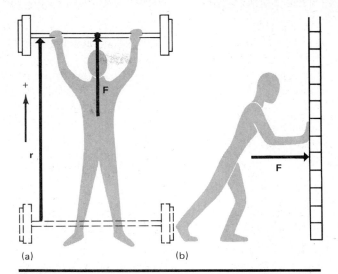

FIGURE 5.1. The concept of work. At (a), work is done when a force causes a displacement; but in (b), even though a force is exerted, no work is done, because there is no displacement.

units of distance. In the English system, accordingly, the proper unit is the foot-pound (ft-lb), which is the amount of work done by a force of one pound moving in its own direction a distance of 1 foot. Similarly, the dyne-centimeter, usually referred to as the *erg*, and the newton-meter, often called the *joule* (J), are the units of work in the cgs- and mks-metric systems. Since $1\ N = 10^5$ dyn and $1\ m = 10^2$ cm, it is evident that $1\ J = 10^7$ ergs. The reader is asked as an exercise to show that 1 ft-lb = 1.356 J.

EXAMPLE 5.2.1
How much work is done on a set of weights weighing 180 pounds, such as those shown in Fig. 5.1a, in lifting them from the floor to a height of 7 feet? How much work is done by the weights when they are set down again?

In computing the amount of work done in a given process, it is important to focus on a given object or set of objects, referred to as "the system." One may then reckon either the work done *on* the system by "its surroundings" or the work done *by* the system on its surroundings. Each of these quantities is the negative of the other, and it is always important to specify which of the two is being considered.

For example, let us designate as "the system" the set of weights which is being lifted and set down. The man is then what is referred to above as "the surroundings." In Fig. 5.1a, the weights are lifted; in order to do this, the man must exert a 180-pound upwardly directed force **F** on the system (the weights) and move them upward a distance of 7 feet. The work done by the *surroundings on the system* is then, according to (5.2.1),

$W = Fr \cos \theta = (180)(7)(1) = 1260$ ft-lb

the value of θ, the angle between **F** and **r**, being zero, $\cos \theta = 1$. At the same time, as required by Newton's third law, the system exerts an equal and opposite force (let us call it **F**′) of 180 pounds directed downward on its surroundings, the man. The work W' done *by the system on the surroundings* is

$W' = F'r \cos \theta' = (180)(7)(-1) = -1260$ ft-lb

the value of θ', the angle between **F**′ and **r**, being 180°, implying $\cos \theta' = -1$.

When the weights are set down, the displacement vector is directed downward (let us call it **r**′ = −**r**). The work W'' done *by the surroundings on the system* is now

$W'' = Fr' \cos \theta'' = (180)(7)(1) = -1260$ ft-lb

since the angle θ'' between **F** and **r**′ is 180°. The work W''' done *by the system on the surroundings* in setting the weights down is

$W''' = F'r' \cos \theta''' = (180)(7)(1) = 1260$ ft-lb

since the angle θ''' between **F**′ and **r**′ is zero.

EXAMPLE 5.2.2

A packing case weighing 80 pounds is moved at constant speed over a level floor a distance of 12 feet by a force **F** which is directed 30° below the horizontal, as shown in Fig. 5.2. The coefficient kinetic friction is 0.4. How much work is done by the force **F** on the packing case?

Since the object is moved at constant speed, it is in equilibrium. Accordingly,

$$\sum F_x = F \cos 30° - 0.4N = 0 \qquad (5.2.3)$$

and

$$\sum F_y = -F \sin 30° - 80 + N = 0 \qquad (5.2.4)$$

Solving these for F and N, we find $F = 48$ lb, $N =$

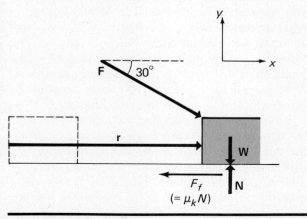

FIGURE 5.2. System of forces acting on a box moved at constant speed by force **F**.

130

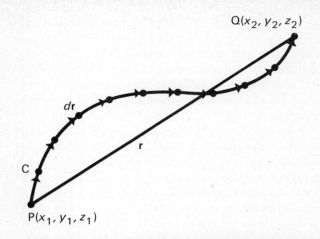

FIGURE 5.3. A resultant displacement **r** viewed as the vector sum of infinitesimal displacement elements $d\mathbf{r}$ along the actual path traversed.

104 lb. The work done *by the force* **F** *on the box*, which is what we were asked to find, is

$W = Fr \cos \theta = (48)(12) \cos 30° = 499$ ft-lb

Note that the force of friction is given by $F_f = 0.4N = 41.6$ lb. The work done by the force of friction on the box is, therefore,

$W_f = F_f r \cos \theta = (41.6)(12) \cos 180° = -499$ ft-lb

The work done by the forces **W** and **N** is zero, because both these vectors are perpendicular to **r**, hence $\theta = 90°$ and $\cos \theta = 0$. The total work done on the box by *all* forces acting on it is, therefore, $W + W_f = 0$. The same conclusion is reached by observing that since the resultant force on the box is zero, the work it does must also be zero.

Let us now examine the more general case where the force may not be constant throughout the displacement and where the displacement itself may take place along some curved path C extending from a starting point $P(x_1, y_1, z_1)$ to an endpoint $Q(x_2, y_2, z_2)$, rather than along the straight line joining those points. This situation is illustrated in Fig. 5.3. Suppose that the total displacement **r** is regarded as the vector sum of many tiny displacement elements $d\mathbf{r}$ along the curve C, as shown in the diagram. We may regard each element $d\mathbf{r}$ to be so small that the force **F** is substantially constant over that element, even though it may vary greatly over the total displacement. The work element dW associated with the displacement $d\mathbf{r}$ may now be written as

$$\boxed{dW = F \cos \theta \, dr = \mathbf{F} \cdot d\mathbf{r}} \qquad (5.2.5)$$

from Eqs. (5.2.1) and (5.2.2), because **F** is now essentially constant throughout the tiny displacement $d\mathbf{r}$.

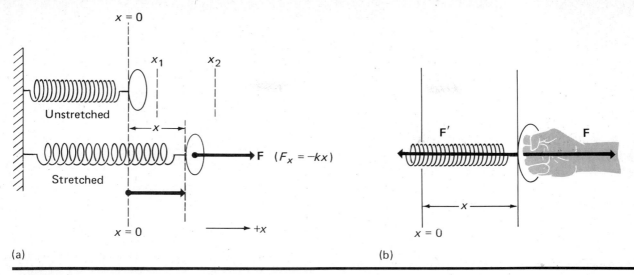

FIGURE 5.4. Forces exerted by an "ideal" spring.

The elements dW may then be summed by integration along the path C:

$$W = \int_c dW = \int_c F \cos \theta \, dr = \int_c \mathbf{F} \cdot d\mathbf{r} \qquad (5.2.6)$$

If the dot product $\mathbf{F} \cdot d\mathbf{r}$ is expressed in terms of the components of the vectors $\mathbf{F}$ and $d\mathbf{r}$ by Eq. (1.6.5), this integral can be expressed as three separate integrals involving the three force components:

$$W = \int_c \mathbf{F} \cdot d\mathbf{r} = \int_P^Q (F_x \, dx + F_y \, dy + F_z \, dz)$$
$$= \int_{x_1}^{x_2} F_x \, dx + \int_{y_1}^{y_2} F_y \, dy + \int_{z_1}^{z_2} F_z \, dz \qquad (5.2.7)$$

EXAMPLE 5.2.3

The force exerted by an *ideal spring* is directly proportional to the distance through which it is stretched or compressed. The force exerted by the spring can, therefore, be expressed in terms of the distance x through which it is extended or compressed from its equilibrium length as

$$F = -kx \qquad (5.2.8)$$

where k is a positive proportionality constant whose magnitude depends upon the characteristics of the particular spring at hand.[1] If the spring is stretched by

[1] The force law expressed by (5.2.7), in which the deformation is directly proportional to the force, is often referred to as *Hooke's law* of elasticity, and a spring whose force–displacement relation is adequately described by this equation is sometimes called a *Hooke's law spring*, or a *linear spring*. The terminology is derived from the name of Robert Hooke (1635–1703), the British scientist who first stated it. It is generally quite well obeyed by actual springs as long as the forces and displacements involved are relatively small.

unit distance ($x = 1$), then $F = -k$; we may, therefore, interpret k as the *magnitude of the force required to stretch the spring through a unit distance*. The dimensions of the spring constant k, according to (5.2.8), must be force/distance, thus, lb/ft, dyn/cm, or N/m, in the three most commonly used systems of units. The minus sign in Eq. (5.2.8) is used because the force exerted by the spring is always in the direction opposite to the displacement vector $\mathbf{r}$, as shown in Fig. 5.4a. Let us now calculate the work done on the spring in stretching it from x_1 to x_2.

For the purposes of this calculation, since we are trying to find the amount of work done *on the spring*, the spring is the "system" and the arm of the person stretching it is the "surroundings." When the spring is stretched through a distance x from the equilibrium point, it exerts a force $\mathbf{F}$, whose magnitude is given by (5.2.8), on whatever is stretching it. By Newton's third law, the agent that stretches the spring must exert an equal but opposite force $\mathbf{F}' = -\mathbf{F}$, as illustrated in Fig. 5.4b. The force acting *on the system* is, therefore, this force $\mathbf{F}'$, whose magnitude is kx. According to (5.2.8), the work done on the system in extending the end of the spring from x_1 to x_2 is

$$W = \int_c \mathbf{F} \cdot d\mathbf{r} = \int_{x_1}^{x_2} F_x \, dx = k \int_{x_1}^{x_2} x \, dx = k \left[\frac{x^2}{2} \right]_{x_1}^{x_2}$$

or

$$W = \tfrac{1}{2}k(x_2{}^2 - x_1{}^2) \qquad (5.2.9)$$

Since F_y and F_z (as well as dy and dz) are zero in this example, the integrals over dy and dz in (5.2.7) vanish, leaving only the integral over dx, which is evaluated in (5.2.9) above.

Suppose that we are told that the spring is of such construction that a force of 10 newtons will

131

stretch it through a distance of 0.1 meter, and we are asked how much work is done on it in extending it from the equilibrium position through a distance of 0.25 meter.

We do not know the spring constant k, but it can be calculated from the given data. It is known that a force of 10 newtons stretches the spring through 0.1 meter. We known also that Eq. (5.2.8) relates force and stretch for *any* value of those quantities. If, now, $F = 10$ newtons and the corresponding displacement, $x = 0.1$ meter, are subsituted into (5.2.8), we may solve for k, ignoring the minus sign in (5.2.8) and dealing only with magnitudes:

$$k = \frac{F}{x} = \frac{10}{0.1} = 100 \text{ N/m}$$

The amount of work required to stretch the spring can now be found from (5.2.9) using $x_2 = 0.25$ meter, $x_1 = 0$, and $k = 100$ newtons/meter, the result being

$$W = \tfrac{1}{2}k(x_2{}^2 - x_1{}^2) = (\tfrac{1}{2})(100)(0.25)^2 = 3.125 \text{ J}$$

The process of integration required to arrive at (5.2.9) can be visualized by referring to Fig. 5.5, in which the force F' required to produce a given extension x, as given by (5.2.8), is plotted against x. This graph of $F'\,(=kx)$ versus x is just a straight line of slope k. The quantity

$$\int_{x_2}^{x_1} F'\,dx$$

which is the work required to extend the spring from x_1 to x_2, is also the *area under the curve* $F'(x)$ plotted as a function of x between the ordinates x_1 and x_2. This is the shaded area in Fig. 5.5. It is easily verified, by computing the areas of the two triangles shown in that diagram as one half base times altitude and subtracting the smaller from the larger, that the shaded area is indeed $\tfrac{1}{2}kx_2{}^2 - \tfrac{1}{2}kx_1{}^2$, in agreement with the result obtained by integration.

It will be noted in this example, as well as in Examples 5.2.1 and 5.2.2, that the forces required to accelerate and decelerate the objects that are set in motion, and upon which work is done, are entirely neglected. For the moment, we may justify this neglect by regarding the displacement of the objects upon which work is done to take place very slowly and with application of the smallest possible forces, which permit only the tiniest accelerations or decelerations. The forces needed to produce accelerations will then be negligible in comparison with the other forces acting on the system. For example, in order for the man illustrated in Fig. 5.1a to lift the set of weights at all, he must exert an upward force not merely equal to their weight but exceeding it somewhat; the weights then acquire an upward acceleration. As the weights reach their maximum height, the lifting force exerted

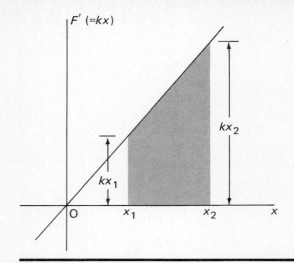

FIGURE 5.5. Work done in stretching an ideal spring represented as the area under the curve of force vs. displacement.

falls somewhat *below* the weight, resulting in a net force downward, which decelerates the weights until they stop. The force exerted by the man who lifts them is, therefore, not always exactly equal to 180 pounds but exceeds this value as the weights are picked up and falls short of that value as they are brought to rest above his head. Nevertheless, if he is very careful to exert only the smallest possible force in excess of 180 pounds as he picks the weights up, so as to produce the smallest possible acceleration, and to slow them down as gently as possible near the top of their path, the work done in accelerating and decelerating the system will be negligible compared to that done in moving the system through its prescribed path. The results obtained in all the above examples will then accurately represent the work done on the system.

We shall in due course consider in detail the work done by forces causing acceleration and deceleration, and we shall show that *if the initial and final velocities of the system are equal, the total work done by these forces is precisely zero.* This is brought about by the fact that the work done *on* the system in accelerating it is exactly recovered as work done *by* the system in deceleration back to the initial velocity.

EXAMPLE 5.2.4

A body moves along the path OABC in the xy-plane shown in Fig. 5.6, where the x- and y-coordinates are measured in meters. A force

$$\mathbf{F} = c(\mathbf{i}_x x + \mathbf{i}_y y + \mathbf{i}_z 0) \text{ N} \tag{5.2.10}$$

where the constant c has the value 1 newton/meter, acts on the body during its journey. Find the work done on it as it moves from O to C.

The work done on the body in moving along

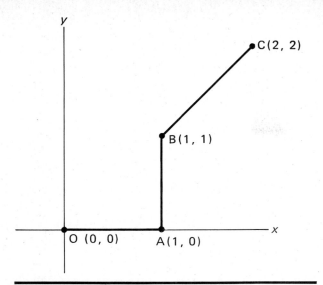

FIGURE 5.6

each segment of the path can be calculated from Eq. (5.2.8):

$$W_{OA} = \int_0^1 F_x\, dx + \int_0^0 F_y\, dy = \int_0^1 x\, dx = \left[\frac{x^2}{2}\right]_0^1 = \tfrac{1}{2}\,\text{J}$$

The integral over dy vanishes because the upper and lower limits are the same. There is no contribution from an integral over dz since $F_z = 0$. In the same way,

$$W_{AB} = \int_0^0 F_x\, dx + \int_0^1 F_y\, dy = \int_0^1 y\, dy = \left[\frac{y^2}{2}\right]_0^1 = \tfrac{1}{2}\,\text{J}$$

and

$$W_{BC} = \int_1^2 F_x\, dx + \int_1^2 F_y\, dy = \int_1^2 x\, dx + \int_1^2 y\, dy$$

$$= \left[\frac{x^2}{2}\right]_1^2 + \left[\frac{y^2}{2}\right]_1^2 = 3\,\text{J}$$

The total work is the sum of these three contributions:

$$W = W_{OA} + W_{AB} + W_{BC} = 3 + \tfrac{1}{2} + \tfrac{1}{2} = 4\,\text{J}$$

Suppose the force had been given as

$$\mathbf{F} = k(\mathbf{i}_x x y^3 + \mathbf{i}_y x^2 y^2 + \mathbf{i}_z 0) \qquad (5.2.11)$$

where the constant k has the value 1 newton/m^4. We should now proceed as before, calculating the work done on the three separate segments of the path:

$$W_{OA} = \int_0^1 F_x\, dx + \int_0^0 F_y\, dy = \int_0^1 x y^3\, dx \qquad (5.2.12)$$

To evaluate this integral along the path OA, we must observe that the line OA is a segment of the line $y = 0$. The y-coordinate of every point on this part of the path is zero, hence $xy^3 = 0$ for every point on the path along which the elements $F_x\, dx$ are summed; the integrand in Eq. (5.2.12) is, therefore, zero and hence

$$W_{OA} = 0$$

On the segment AB, we may write

$$W_{AB} = \int_0^0 F_x\, dx + \int_0^1 F_y\, dy = \int_0^1 x^2 y^2\, dy \qquad (5.2.13)$$

The segment AB is part of the line whose equation is $x = 1$. Substituting this value for x, which gives the x-coordinate of each point on this part of the path, (5.2.13) becomes

$$W_{AB} = \int_0^1 1 \cdot y^2\, dy = \left[\frac{y^3}{3}\right]_0^1 = \tfrac{1}{3}\,\text{J}$$

On segment BC, we may write

$$W_{BC} = \int_1^2 F_x\, dx + \int_1^2 F_y\, dy$$

$$= \int_1^2 x y^3\, dx + \int_1^2 x^2 y^2\, dy \qquad (5.2.14)$$

Segment BC is part of a curve whose equation is $y = x$; hence, the x- and y-coordinates of each point on this part of the path are related by this equation. Setting y equal to x in the first integral above and x equal to y in the second, we find

$$W_{BC} = \int_1^2 x^4\, dx + \int_1^2 y^4\, dy = \left[\frac{x^5}{5}\right]_1^2 + \left[\frac{y^5}{5}\right]_1^2$$

$$= \frac{62}{5} = 12.4\,\text{J}$$

Now, as before,

$$W = W_{OA} + W_{AB} + W_{BC} = 0 + \frac{1}{3} + \frac{62}{5} = 12.733\,\text{J}$$

EXAMPLE 5.2.5

Find the work done on the rotating object discussed in Example 4.3.3 by the forces which act on it.

This problem, while it may appear complex at first sight, is really extremely simple. The forces acting on the mass m are shown in Fig. 5.7 along with the vector $d\mathbf{r}$, which expresses the displacement of the object in an infinitesimal time interval and whose direction is tangential to the circular path. It is clear from this diagram that both the weight force $m\mathbf{g}$ and the tension $\mathbf{T}$ are *perpendicular* to $d\mathbf{r}$. The dot product of both these forces with $d\mathbf{r}$ must then be zero; and hence, according to (5.2.5) and (5.2.6), the work done on the mass by both forces is zero. Another way of arriving at the same result is to note that the *resultant* of $\mathbf{T}$ and $m\mathbf{g}$ must be a centripetal force whose magnitude is $mr\omega^2$ and which is directed radially inward. But this resultant force $\mathbf{F}_c$ is perpendicular to the vector $d\mathbf{r}$, hence the dot product $\mathbf{F}_c \cdot d\mathbf{r}$ and also its integral over the circular path must vanish. The work is, therefore, zero. This example illustrates the important point that not every force which moves through a given displacement need necessarily do work. Clearly, there is no work done by centripetal or centrifugal forces in circular motion.

133

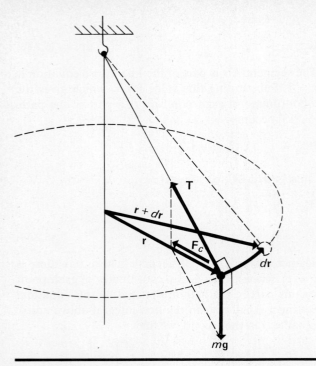

FIGURE 5.7

5.3 Power

The term *power* is used to express the *rate* at which work is done. A given amount of work done over a very long time interval may, therefore, require very little power, while if the same amount of work is to be done in an extremely short time, a great deal of power may be needed.

Suppose a very small amount of work dW is done in an infinitesimal time interval dt extending from time t to time $t + dt$. Then, the *instantaneous* power $P(t)$ required at time t to do this work is defined as dW/dt, from which

$$P(t) = \frac{\text{infinitesimal work done}}{\text{infinitesimal time required}} = \frac{dW}{dt} \qquad (5.3.1)$$

This can be expressed in a slightly different way by multiplying both sides of the equation by dt to obtain

$$dW = P(t)\,dt \qquad (5.3.2)$$

It is evident from the above discussion that power is related to work in much the same way as velocity is related to displacement, as expressed by Eqs. (3.1.1), (3.1.2), and (3.1.20).

Suppose that $P(t)$ is independent of time, having the *constant value* P at every time. Then, according to (5.3.2),

$$dW = P\,dt \qquad (5.3.3)$$

with P *constant*. This is easily integrated from time $t = 0$ to some later time t; let us suppose that the work done at $t = 0$ is zero, and that the work done at time t has the value W. Then, integrating both sides of (5.3.3) from time zero (when the work done is also zero) to time t, when work W has been accomplished, we find

$$\int_0^W dW = P \int_0^t dt \qquad (5.3.4)$$

or

$$[W]_0^W = P[t]_0^t \qquad (5.3.5)$$

whence

$$W = Pt \qquad (5.3.6)$$

or

$$\boxed{P = W/t} \qquad (5.3.7)$$

When the rate at which work is done is *constant* with respect to time, the power may be obtained from (5.3.7) or the work from (5.3.6). Under these conditions, the power is simply the work done per unit time. If the rate at which work is done *varies* with time, however, the power must be calculated from (5.3.1). If the power is known as a function of time and it is the work that is required, it must be obtained by integration of (5.3.2), from which

$$\boxed{W = \int dW = \int P(t)\,dt} \qquad (5.3.8)$$

The procedure for doing this is somewhat similar to that employed in deriving (5.3.6), except that, if $P(t)$ is not constant in time, it may not be brought outside the integral as above, but must instead remain inside and be integrated according to the established rules of calculus. This will be illustrated by the examples below.

Another very useful way of expressing the power consumed in a process wherein work is done may be arrived at by dividing both sides of Eq. (5.2.5) by dt:

$$\frac{dW}{dt} = \frac{\mathbf{F} \cdot d\mathbf{r}}{dt} = \mathbf{F} \cdot \frac{d\mathbf{r}}{dt} \qquad (5.3.9)$$

But dW/dt is, by definition, the power $P(t)$, and $d\mathbf{r}/dt$ is the velocity $\mathbf{v}$. Therefore, (5.3.9) may be written in the form

$$\boxed{P(t) = \mathbf{F} \cdot \mathbf{v}} \qquad (5.3.10)$$

This is an alternative expression for power which we shall find to be very useful in application to practical examples.

According to Eq. (5.3.1), the units of power must

be work per unit time. In the mks-metric system, then, the unit of power is the newton-meter/second or joule/second. This unit is often referred to as the *watt* (abbreviated W); it is clear that a watt is the power required to do 1 joule of work per second. Similarly, the dyn-cm/sec, or erg/sec, which is the corresponding unit of power in the cgs-metric system, is the power required to 1 erg of work per second. Since 1 erg = 10^{-7} joule, this unit corresponds to 10^{-7} watt. In the English system, the fundamental power unit is the ft-lb/sec, but the *horsepower* (abbreviated hp), an arbitrary unit equal to 550 ft-lb/sec (33,000 ft-lb/min), is also very frequently used. The reader, as an exercise, will kindly show that 1 kilowatt (KW) = 1.341 hp, while 1 kilowatt-hour (kWh) = 0.252 ton-mile.

EXAMPLE 5.3.1

The man discussed in Example 5.2.1 and illustrated in Fig. 5.1a is able to lift a set of weights weighing 180 pounds through a vertical distance of 7 feet at uniform velocity in 2.0 seconds. How much power does he develop?

This is a situation in which work is done at a constant rate, because the force involved in doing the work moves at constant velocity. We may, therefore, conveniently use (5.3.8) to compute the power:

$$P = \frac{W}{t} = \frac{1260}{2} = 630 \text{ ft-lb/sec}$$

$$= \frac{630}{550} = 1.146 \text{ hp}$$

The reader should not be surprised that a man can generate power in excess of 1 horsepower. So can a horse—for a short time!

Another way of doing this problem would be to note that the weights are lifted at constant speed through 7 feet in 2.0 seconds by a constant force of 180 pounds. The velocity with which the weights ascend is, therefore, given by

$$v = \frac{d}{t} = \frac{7}{2} = 3.5 \text{ ft/sec}$$

Then, according to Eq. (5.3.10),

$$P = \mathbf{F} \cdot \mathbf{v} = Fv = (180)(3.5) = 630 \text{ ft-lb/sec}$$

in agreement with the previous result. Note in this case that the vectors $\mathbf{F}$ and $\mathbf{v}$ are in the same direction, hence $\cos \theta = 1$ in (1.6.1) and $\mathbf{F} \cdot \mathbf{v} = Fv$.

EXAMPLE 5.3.2

An automobile's engine can deliver 85 horsepower to the rear wheels of the car. The car's maximum speed is 160 ft/sec (109 mph). What is the total magnitude of the frictional forces acting to resist its motion when it is moving without acceleration at its top speed?

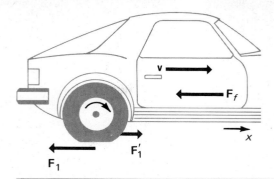

FIGURE 5.8. System of forces acting on a car accelerating along a level road.

If the car is moving at constant speed, the *resultant* force acting upon it must be zero. The car's engine and transmission produce forces which act to turn the rear wheels and which act *on the roadway* in a backward direction. The force $\mathbf{F}_1$ in Fig. 5.8 is just such a force. In view of Newton's third law, however, force $\mathbf{F}_1$ must be accompanied by an equal and opposite reaction force $\mathbf{F}_1'$ exerted *by the road on the car*. It is this force that serves to accelerate the vehicle when the driver steps on the accelerator. There are, in addition, frictional forces on the car acting opposite to the direction of the motive force $\mathbf{F}_1'$. These forces are associated with frictional resistances in the wheel bearings and tires and with air resistance, which is also frictional in nature. If the car moves at constant speed, it is in equilibrium, and the sum of the force components in any direction are then zero. If the sum of all the frictional forces just mentioned is $\mathbf{F}_f$, then, summing x-components of force,

$$F_1' - F_f = 0 \tag{5.3.11}$$

since neither F_1' nor F_f has y- or z-components.

Regarding the car as "the system," it is now clear that work is done *on* the system by $\mathbf{F}_1'$, which moves with the car's velocity $\mathbf{v}$, and *by* the system on the surroundings (the atmosphere and the roadway), by $\mathbf{F}_f$, which moves with this same velocity. Since $\mathbf{F}_1'$ and $\mathbf{F}_f$ are equal and opposite, it is evident that the work done on the car in a given distance by the propulsion system is equal in magnitude to the work done by the car on its surroundings. In other words, the rate at which the propulsion system does work on the car equals the rate at which work is done by the car on its surroundings via frictional forces. The latter quantity may then be inferred simply by calculating the former, by means of (5.3.10):

$$P = \mathbf{F}_1' \cdot \mathbf{v} = F_1'v$$

the cosine factor in the scalar product being unity, since $\mathbf{F}_1'$ and $\mathbf{v}$ are parallel. Then, recalling that 1 hp = 550 ft-lb/sec,

135

$$F_1' = \frac{P}{v} = \frac{(85)(550)}{160} = 292 \text{ lb}$$

But since $\mathbf{F}_1'$ and $\mathbf{F}_f$ have equal magnitudes, we may conclude that

$$F_f = F_1' = 292 \text{ lb}$$

EXAMPLE 5.3.3

An object of mass m executes a sinusoidal motion along the x-axis of an orthogonal coordinate system, so that its displacement $x(t)$ at time t is given by

$$x(t) = x_0 \sin \omega t \qquad (5.3.12)$$

In this expression x_0 is a constant having units of distance and expresses the maximum amplitude of the motion, and ω is a constant which expresses the frequency of the periodic displacement. This type of motion is usually referred to as *simple harmonic motion*, and we shall investigate its characteristics in great detail in a later chapter. For the moment, however, let us set for ourselves the objective of determining the *power* required to maintain this sort of motion at any time t and, in particular, to determine the power required when $x_0 = 1.5$ m, $m = 5$ kg, $\omega = 12.57$ rad/sec, and $t = 0.0625$ sec.

In this example, we may again conveniently use (5.3.10) as a basis for finding the power. Since the mass moves only along the x-direction, the velocity $\mathbf{v}$ can have no y- or z-components. Likewise, the force $\mathbf{F}$, which from Newton's second law must equal $m\mathbf{a}$, must have only an x-component. The dot product $\mathbf{F} \cdot \mathbf{v}$ then reduces to $F_x v_x$, and (5.3.10) may be written as

$$P(t) = F_x v_x \qquad (5.3.13)$$

Now, from (5.3.12),

$$v_x = dx/dt = x_0 \omega \cos \omega t \qquad (5.3.14)$$

and

$$a_x = dv_x/dt = -x_0 \omega^2 \sin \omega t \qquad (5.3.15)$$

Newton's second law tells us that the force F_x must be related to this acceleration by

$$F_x = ma_x = -x_0 m \omega^2 \sin \omega t \qquad (5.3.16)$$

Substituting these values into (5.3.13), we obtain

$$P(t) = -x_0{}^2 m \omega^3 \sin \omega t \cos \omega t$$

and since $2 \sin \theta \cos \theta = \sin 2\theta$, this may be written

$$P(t) = \tfrac{1}{2} x_0{}^2 m \omega^3 \sin 2\omega t \qquad (5.3.17)$$

This is the desired result. Observe that the power is proportional to the *square* of the amplitude and the *cube* of the frequency and that its time variation occurs at twice the frequency associated with the displacement. For $x_0 = 1.5$ m, $m = 5$ kg, $\omega = 12.57$ rad/sec,

and $t = 0.0625$ sec, we have

$$P = -(1/2)(1.5)^2(5)(12.57)^3 \sin (1.571 \text{ rad})$$

But 1.571 radians $= \pi/2$ radians, hence $\sin(1.571) = \sin(\pi/2) = 1$. Then,

$$P = -(1/2)(2.25)(5)(1980) = -11{,}200 \text{ watts}$$
$$= -11.2 \text{ kilowatts}$$

The negative sign of the result means simply that the *system* (the mass) *is doing work on its surroundings* rather than the other way around at this particular instant of time.

5.4 Kinetic Energy and the Work–Energy Theorem

In most of our previous discussions we have considered only the work done by the individual forces acting on unaccelerated bodies. We must now turn our attention to the problem of finding the work done by a *resultant* force $\mathbf{F}$ which acts to *accelerate* an object. Suppose that a body is initially at some position $P(x_1, y_1, z_1)$ with initial velocity $\mathbf{v}_1$ (components v_{1x}, v_{1y}, v_{1z}) and is then *accelerated* by the action of a resultant force $\mathbf{F}$, which need not necessarily be constant, achieving at some later time the velocity $\mathbf{v}_2$ (components v_{2x}, v_{2y}, v_{2z}) when it is located at position $R(x_2, y_2, z_2)$. This process is illustrated in Fig. 5.9 where, in addition, the body is shown at some arbitrary intermediate point $Q(x, y, z)$, where its velocity is $\mathbf{v}$. The question is, how much work is done by the resultant force $\mathbf{F}$ acting on the body as it moves from P to R?

According to Eq. (5.2.7), the work can be expressed as

$$W_{12} = \int_{x_1}^{x_2} F_x \, dx + \int_{y_1}^{y_2} F_y \, dy + \int_{z_1}^{z_2} F_z \, dz \qquad (5.4.1)$$

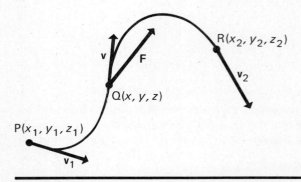

FIGURE 5.9. Path of the body discussed in Example 5.4. The body is initially at P and undergoes acceleration along the curved path finally arriving at R. Point Q illustrates an arbitrary intermediate stage at which the resultant force is **F** and the velocity **v**.

However, from Newton's second law,

$$F_x = m\frac{dv_x}{dt} \qquad F_y = m\frac{dv_y}{dt} \qquad F_z = m\frac{dv_z}{dt} \qquad (5.4.2)$$

Substituting these expressions for the force components in (5.4.1), we may write

$$W_{12} = m\int_{x_1}^{x_2} \frac{dv_x}{dt}\,dx + m\int_{y_1}^{y_2} \frac{dv_y}{dt}\,dy + m\int_{z_1}^{z_2} \frac{dv_z}{dt}\,dz$$

$$(5.4.3)$$

But

$$\frac{dv_x}{dt}\,dx = \left(\frac{dv_x}{dx}\frac{dx}{dt}\right)dx = v_x\frac{dv_x}{dx}\,dx = v_x\,dv_x$$

$$\frac{dv_y}{dt}\,dy = \left(\frac{dv_y}{dy}\frac{dy}{dt}\right)dy = v_y\frac{dv_y}{dy}\,dy = v_y\,dv_y \qquad (5.4.4)$$

$$\frac{dv_z}{dt}\,dz = \left(\frac{dv_z}{dz}\frac{dz}{dt}\right)dz = v_z\frac{dv_z}{dz}\,dz = v_z\,dv_z$$

Inserting these results into Eq. (5.4.3),

$$W_{12} = m\int_{v_{1x}}^{v_{2x}} v_x\,dv_x + m\int_{v_{1y}}^{v_{2y}} v_y\,dv_y + m\int_{v_{1z}}^{v_{2z}} v_z\,dv_z$$

$$(5.4.5)$$

In writing this equation, we have changed the variables of integration from x, y, and z to v_x, v_y, and v_z. In evaluating the definite integrals, therefore, we must choose the limits of integration for the velocity integrals from v_{1x} to v_{2x}, v_{1y} to v_{2y}, and v_{1z} to v_{2z} to correspond with the previous limits x_1 to x_2, y_1 to y_2, and z_1 to z_2 for the original integration shown in (5.4.3). We can now easily integrate (5.4.5), the result finally being

$$W_{12} = m\left[\tfrac{1}{2}v_x^2\right]_{v_{1x}}^{v_{2x}} + m\left[\tfrac{1}{2}v_y^2\right]_{v_{1y}}^{v_{2y}} + m\left[\tfrac{1}{2}v_z^2\right]_{v_{1z}}^{v_{2z}}$$
$$= \tfrac{1}{2}m(v_{2x}^2 + v_{2y}^2 + v_{2z}^2) - \tfrac{1}{2}m(v_{1x}^2 + v_{1y}^2 + v_{1z}^2)$$

or

$$W_{12} = \tfrac{1}{2}mv_2^2 - \tfrac{1}{2}mv_1^2 \qquad (5.4.6)$$

This expression is remarkably simple. It tells us that the work done in accelerating a body from velocity $\mathbf{v}_1$ to velocity $\mathbf{v}_2$ depends *only on the mass and the magnitudes of the initial and final velocities*. It is quite *independent* of the path taken by the body during the accelerations and of the time taken to accomplish the acceleration. It is likewise independent of the detailed way in which forces act during the acceleration. So long as the initial and final velocities are the same, the same result is obtained whether the force is constant or variable during the process, and the work done in the latter case is completely independent of the *way* in which the force and velocity may vary while the acceleration is taking place. The result embodied in Eq. (5.4.6) is often referred to as the *work–energy theorem*.

The very simple form of these results leads us to define a quantity U_k which we shall call the *kinetic energy* of a body, such that

$$U_k = \tfrac{1}{2}mv^2 \qquad (5.4.7)$$

In the case discussed in detail above, the initial kinetic energy of the object in question is

$$U_{k1} = \tfrac{1}{2}mv_1^2 \qquad (5.4.8)$$

while its final kinetic energy is

$$U_{k2} = \tfrac{1}{2}mv_2^2 \qquad (5.4.9)$$

The work done in accelerating a body from an initial velocity v_1 to a final velocity v_2, according to (5.4.6), then, can be written as

$$W_{12} = U_{k2} - U_{k1} = \Delta U_k \qquad (5.4.10)$$

In accomplishing the acceleration, we may say that the work W_{12} done on the object has been transformed into an *equivalent* amount of kinetic energy, or *energy of motion* ΔU_k. If the body were to be decelerated from velocity v_2 back to the original velocity v_1, the amount of work done *on* the body, according to the work–energy theorem, is

$$W_{21} = \Delta U_k = U_k(\text{final}) - U_k(\text{initial})$$
$$= \tfrac{1}{2}mv_1^2 - \tfrac{1}{2}mv_2^2 = -W_{12} \qquad (5.4.11)$$

The fact that this quantity is the *negative* of the work W_{12} required to accelerate it originally expresses the fact that work is done *by the body on the surroundings* as it undergoes deceleration and, moreover, that the work done by the body in being decelerated to its original speed is exactly equal in magnitude to the work done on it in accelerating it initially. Viewed in this way, we see that the kinetic energy of a body can be regarded as *the amount of work the body is capable of doing on its surroundings in being brought to rest*. The total work expended in both processes is zero, as it must be in any instance where initial and final kinetic energies are the same. It is obvious from Eq. (5.4.6) that the units of kinetic energy are the same as those of work. The result embodied in Eqs. (5.4.6) or (5.4.10) is one of the most important and general of all physical laws. We may state the work–energy theorem in words as follows:

The work done by the resultant force acting on any system is equal to the change in the system's kinetic energy.

It is important to note that the work–energy theorem relates the work done by the *resultant* force acting on a body to its change in kinetic energy. It is, therefore, incorrect to apply it to the individual forces that add up to the resultant force. For example, consider the case of an object which is dragged at constant speed along a level surface against forces of friction by an external force, as illustrated in Fig. 5.10. In this case, the individual forces $\mathbf{F}_{\text{ext}}$ and $\mathbf{F}_f$ must sum to a resultant force of zero, since the body is not accelerated. Accordingly,

$$F_f + F_{\text{ext}} = F_{\text{res}} = 0$$

or

$$F_f = -F_{\text{ext}} \qquad (5.4.12)$$

Now suppose the object moves through a given horizontal distance. From the work–energy theorem, the work done by the resultant force is zero, since the final and initial velocities are equal. This is correct, because the resultant force is itself zero. There is work done, however, by the individual forces $\mathbf{F}_{\text{ext}}$ and $\mathbf{F}_f$. As a matter of fact, the work done by the external force *on* the body is just equal to the work done *by* the body against the frictional force in view of the fact that $F_f = -F_{\text{ext}}$. It is incorrect, however, to try to apply the work–energy theorem to either of the individual forces $\mathbf{F}_f$ or $\mathbf{F}_{\text{ext}}$, because in the derivation of that theorem it is the resultant force that is set equal to $m\mathbf{a}$, as in (5.4.2), rather than any individual force. The indiscriminate application of this theorem to either of the two individual forces $\mathbf{F}_f$ or $\mathbf{F}_{\text{ext}}$ will lead to the conclusion that $F_f = F_{\text{ext}} = 0$, which is absurd. The examples below illustrate the application of the work–energy theorem to systems in which frictional or other effects that dissipate energy are absent. Although we shall postpone discussing systems in which friction is important until later in the chapter, it is evident from the example discussed immediately above that the work–energy theorem is applicable in either case.

EXAMPLE 5.4.1

A 4000-pound car can accelerate from rest to a speed of 88 ft/sec in 12.0 sec. How much work is required to accomplish this acceleration and what is the average power required? Neglect any effects of friction or air resistance.

According to the work–energy theorem, the amount of work required to produce a given velocity change is given by

$$W = U_{kf} - U_{ki} = \tfrac{1}{2}mv_f^2 - \tfrac{1}{2}mv_i^2$$

where the subscripts f and i refer to the final and initial states of motion. But $v_f = 88$ ft/sec, while $v_i = 0$. Therefore,

$$W = \frac{1}{2}\left(\frac{4000}{32}\right)(88)^2 - 0 = 484,000 \text{ ft-lb}$$

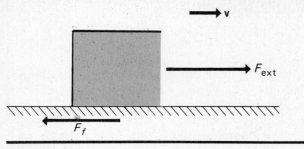

FIGURE 5.10

If this work is done in 12.0 sec, the average power is simply

$$P_{\text{av}} = \frac{W}{t} = \frac{484,000}{12} = 40,300 \text{ ft-lb/sec}$$

$$= \frac{40,300}{550} = 73.3 \text{ hp}$$

EXAMPLE 5.4.2

An object falls vertically from height h_1 to height h_2. Its initial velocity is $\mathbf{v}_1$, directed downward. Using the work–energy theorem, find the velocity at height h_2. Neglect the effect of air resistance.

The resultant force on the object is the weight force $m\mathbf{g}$, directed vertically downward. The displacement vector is also vertically downward in direction and is of magnitude $h_1 - h_2$. The work done on the object by the resultant force during this displacement, according to (5.2.1), is

$$W = mg(h_1 - h_2) \qquad (5.4.13)$$

But from the work–energy theorem, this must equal the final kinetic energy minus the initial kinetic energy. If the final velocity is $\mathbf{v}_2$, then

$$W = mg(h_1 - h_2) = U_{k2} - U_{k1} = \tfrac{1}{2}mv_2^2 - \tfrac{1}{2}mv_1^2$$

Solving for v_2^2, we can easily obtain

$$v_2^2 = v_1^2 + 2g(h_1 - h_2) \qquad (5.4.14)$$

5.5 Kinetic and Potential Energy: Conservative and Nonconservative Forces

In the previous section, the term *kinetic energy* was defined, and it was shown that the kinetic energy of a body in motion can be regarded as the amount of work the body is capable of doing before being brought to rest. The word *energy* in general is used by scientists to describe the *capacity of an object or a physical system to do work*. The total energy of a system is then the total amount of work that it can be made to do on its surroundings, and is naturally measured in the same units as work.

Energy can take many forms. For example, it is possible to make heat, light, electricity, or sound do mechanical work, and to extract mechanical work from chemical and nuclear reactions as well. We, therefore, recognize the existence of *thermal, light, electrical, acoustical, chemical,* and *nuclear* energy as well as purely mechanical energy. Each of these forms of energy can be converted into any other, and if each form is properly defined, quantitative equivalence can be established between all forms of energy. We shall have more to say about this subject later in connection with the law of conservation of energy. For the moment, however, let us confine ourselves to mechanical energy and inquire whether kinetic energy is the only form of mechanical energy.

The answer to this question is that there must be another form of mechanical energy since a body is capable of doing work not only because it may be in motion but *also by virtue of its position.* For example, an object which is held in an elevated position above the earth's surface is capable of doing work on its surroundings as it is lowered to the ground, even though it may be lowered so slowly that it never acquires any appreciable kinetic energy. The amount of work such a body can do on its surroundings by virtue of its position is referred to as its *potential energy.* Accordingly, the potential energy of a mass m which is raised to a height z above the earth's surface is mgz, since it is capable of doing this amount of work as it descends.

The potential energy as thus defined is clearly a function of the object's position. Another example of potential energy is afforded by a spring which is stretched or compressed. Any such spring is capable of doing work as its tension or compression is relaxed and hence serves to store potential energy. In Example 5.2.3, we calculated that the work required to stretch an ideal spring through a distance x from its equilibrium position ($x = 0$) is given by $W = \frac{1}{2}kx^2$, where k is the spring constant expressing the force required to produce unit extension. The stretched spring is capable of doing this same amount of work on its surroundings as it relaxes to its equilibrium condition. We may, therefore, infer that its potential energy is $\frac{1}{2}kx^2$. Again the potential energy of the spring, or any object attached thereto, is a function of the *position variable* x that expresses how far the spring is stretched.

It is evident that one may associate a potential energy with an object acted upon by the earth's gravitational force or with an object subjected to the force of an ideal spring merely by calculating the work the system is capable of doing by virtue of its position. In both cases the potential energy is a function only of the body's position. The resulting expressions for the potential energy, while directly giving the energy of the object, are also closely associated with the forces acting on the object, since it is the forces that ultimately determine how much work can be performed on the surroundings. There are many other simple examples of forces that permit the direct determination of the potential energy in precisely this same way. Such forces are often called *conservative* forces.

There exist, however, certain types of forces which cannot be associated with a uniquely defined potential energy in this simple and straightforward way. The most frequently encountered example of a force of this kind is the force of friction. While it is possible to do work on a friction interface *against* forces of friction, it is impossible to make the forces of friction do work on their surroundings. If work is done in stretching a spring, for example, all the work done *on the spring* to stretch it is recovered as work done *by the spring* on its surroundings when it is permitted to relax to its equilibrium position, as shown in Figs. 5.11a and b. The stored potential energy of the spring is thus uniquely determined by its extension.

On the other hand, suppose an object lies on a level horizontal surface, and suppose that frictional forces act between the object and the surface upon which it rests. Then, work must be done to move the object from one position to another; but this work is not stored as potential energy since it is *not recoverable as work done by the object* when it is moved back to its original position. As a matter of fact, *additional* work must be done *on the body* to return it to its original position because the direction in which the force of friction acts is reversed when the velocity vector reverses, as illustrated in Figs. 5.11c and d. The work done against the forces of friction, instead of being stored as potential energy recoverable as work, is instead transformed into *heat energy* at the friction interface and is dissipated in the body and the surface on which it rests by conduction and radiation. It is likewise impossible to associate any unique value with the work done against friction in moving the object between two given points because the work done depends upon the path taken. It is clear, for example, that more work is done against friction if an indirect, circuitous route is followed than if a direct path is taken, simply because the applied force necessary to overcome friction must move through a greater distance in the former case than in the latter. For these reasons, it is not possible to associate a potential energy with frictional forces. Forces such as frictional forces for which no unambiguous potential energy can be specified are called *nonconservative forces.*

139

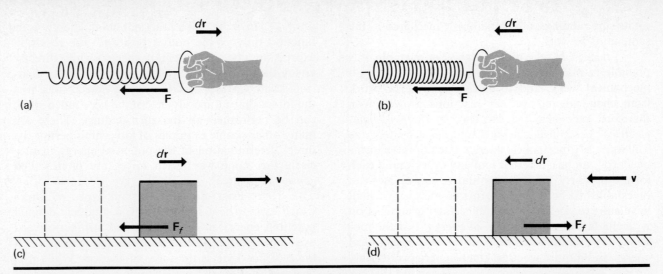

FIGURE 5.11. Conservative (a), (b) and nonconservative forces (c), (d). Work done in stretching an ideal spring is fully recovered as work done by the spring on its surroundings as the spring is allowed to return to its equilibrium position. Frictional work expended in moving an object from one place to another can never be recovered as work done by the object as it is returned to its initial position.

5.6 Conservation of Energy: Conservative Systems

Let us now consider a system (an object of given mass, for example) whose potential energy is a known function $U_p(x, y, z)$ of its position coordinates. We are hereby specifying that the *forces acting upon it are conservative*, since if they were not we would not be able to specify the potential energy in this unambiguous way. Let us also assume that the origin of our coordinate system is located at a point where the potential energy of the object is zero.[2] Since the *potential energy* is defined as *the amount of work done by the force in returning the body to the state of zero potential energy*, in view of (5.2.7) we may write

$$U_p(x, y, z) = \int_x^0 F_x\, dx + \int_y^0 F_y\, dy + \int_z^0 F_z\, dz \quad (5.6.1)$$

or, reversing the limits of integration and the signs of the integrals,

$$U_p(x, y, z) = -\int_0^x F_x\, dx - \int_0^y F_y\, dy - \int_0^z F_z\, dz \quad (5.6.2)$$

Now, suppose the body moves from a point (x_2, y_2, z_2) where its potential energy is U_{p2} to a point (x_1, y_1, z_1) of lower potential energy U_{p1}. Then, from (5.6.2), the *change* in potential energy, $U_{p2} - U_{p1}$, must be

$$U_{p2} - U_{p1} = \Delta U_p$$
$$= -\int_0^{x_2} F_x\, dx + \int_0^{x_1} F_x\, dx - \int_0^{y_2} F_y\, dy$$
$$+ \int_0^{y_1} F_y\, dy - \int_0^{z_2} F_z\, dz + \int_0^{z_1} F_z\, dz$$
$$(5.6.3)$$

However,

$$\int_0^{x_2} f(x)\, dx - \int_0^{x_1} f(x)\, dx = \int_{x_1}^{x_2} f(x)\, dx \quad (5.6.4)$$

and similarly for the y- and z-integrals. Therefore, (5.6.3) may be written

$$U_{p2} - U_{p1} = \Delta U_p$$
$$= -\int_{x_1}^{x_2} F_x\, dx - \int_{y_1}^{y_2} F_y\, dy$$
$$- \int_{z_1}^{z_2} F_z\, dz \quad (5.6.5)$$

This expression, however, is just the negative of the one evaluated in (5.4.1) in connection with the work–energy theorem and whose value was found to be $\frac{1}{2}mv_2^2 - \frac{1}{2}mv_1^2$. Therefore, (5.6.5) becomes

[2] The choice of a point where the potential energy is zero is arbitrary. Ordinarily, the potential energy of an object in the earth's gravitational field is taken to be zero at the earth's surface, but it is not necessary to make this choice, and the zero point can be chosen at any desired level. Different choices for the point at which the zero of potential energy is located, of course, result in the assignment of different values of potential energy and total energy to an object in a given position, but since it is energy *differences* between different states of a system that are physically meaningful, rather than absolute values of energy, this is not important. Similarly, the potential energy of an ideal spring is usually assumed to be zero in the equilibrium state of the spring, but other choices can be made, if desired. The additive constant which is introduced into the potential energy by the choice of the zero level does not affect energy differences between different states of the system nor the amount of work required to go from one state to another.

$$U_{p2} - U_{p1} = \Delta U_p = \tfrac{1}{2}mv_1^2 - \tfrac{1}{2}mv_2^2 = U_{k1} - U_{k2}$$

But the change in kinetic energy ΔU_k is $U_{k2} - U_{k1}$ or $-(U_{k1} - U_{k2})$. Therefore,

$$\Delta U_p + \Delta U_k = (U_{p2} - U_{p1}) + (U_{k2} - U_{k1}) = 0 \quad (5.6.6)$$

If we express the total mechanical energy, U_m as the sum of the kinetic and potential energies, we may write

$$U_m = U_k + U_p$$

By rearranging (5.6.6) slightly and by use of the equation above, we may now obtain

$$U_{k2} + U_{p2} - (U_{k1} + U_{p1}) = U_{m2} - U_{m1} = 0$$

or

$$U_{m2} = U_{k2} + U_{p2} = U_{k1} + U_{p1} = U_{m1} \quad (5.6.7)$$

which expresses the fact that the sum of kinetic energy and potential energy is the same at the end as it was at the beginning.

This is the *energy conservation theorem* for a system which is acted upon by conservative forces only. Stated in words, it tells us that *the total mechanical energy of a system which is acted upon only by conservative forces is constant.*

*   A subsidiary but nevertheless quite important question arising at this point is that concerning the relation between force and the potential energy. The potential energy is fundamentally defined in terms of work, but it is clear from Eq. (5.6.2) that it is also intimately connected with the force on the body. Suppose we calculate the potential energy U_p at point (x, y, z) and also at point $(x + dx, y + dy, z + dz)$ and subtract to obtain dU_p, the potential energy difference between these two points. Accordingly,

$$U_p(x, y, z) = -\int_0^x F_x\, dx - \int_0^y F_y\, dy - \int_0^z F_z\, dz$$

and

$$U_p(x+dx, y+dy, z+dz) = -\int_0^{x+dx} F_x\, dx - \int_0^{y+dy} F_y\, dy$$
$$- \int_0^{z+dz} F_z\, dz$$

Then,

$$dU_p = U_p(x + dx, y + dy, z + dz) - U_p(x, y, z)$$
$$= -\int_0^{x+dx} F_x\, dx + \int_0^x F_x\, dx - \int_0^{y+dy} F_y\, dy$$
$$+ \int_0^y F_y\, dy - \int_0^{z+dz} F_z\, dz + \int_0^z F_z\, dz$$

or

$$dU_p = -\int_x^{x+dx} F_x\, dx - \int_y^{y+dy} F_y\, dy - \int_z^{z+dz} F_z\, dz$$
$$(5.6.8)$$

But if F_x, F_y, and F_z are reasonably well-behaved

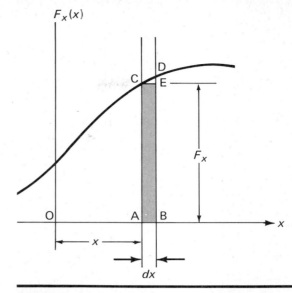

FIGURE 5.12. Work done by a force during an infinitesimal displacement, represented as the area under the curve, showing force as a function of displacement.

functions of x, y, and z in the neighborhood of the point (x, y, z), their values will not be significantly different at $(x + dx, y + dy, z + dz)$ from what they are at (x, y, z), nor will they be much different anywhere within the region of space between x and $x + dx$, y and $y + dy$, and z and $z + dz$. The value of the first integral in (5.6.8) can then be evaluated, for example, as the area of the region ABCD in Fig. 5.12. Clearly, the area ABCE is nearly the same as area ABCD and, in fact, approaches it in the limit where $dx \to 0$. We may then write, using a similar line of reasoning for the second and third integrals,

$$\int_x^{x+dx} F_x\, dx = F_x\, dx$$
$$\int_0^{y+dy} F_y\, dy = F_y\, dy \qquad (5.6.9)$$
$$\int_0^{z+dz} F_z\, dz = F_z\, dz$$

It should be noted that this method of evaluating integrals can be used only when the region of integration is infinitesimally thin, as it is in this case. If it were not, the areas ABCE and ABCD in Fig. 5.12 would not be essentially equal and the method would fail. Substituting the values given in (5.6.9) for the integrals in (5.6.8), we can easily show that

$$dU_p = -F_x\, dx - F_y\, dy - F_z\, dz \qquad (5.6.10)$$

But the total differential dU_p, which states the difference in the values of U_p at (x, y, z) and $(x + dx, y + dy, z + dz)$ can be expressed in terms of the *partial* derivatives[3] $\partial U_p/\partial x$, $\partial U_p/\partial y$, and $\partial U_p/\partial z$, as

[3] See Appendix D for an outline of important facts concerning partial derivatives, along with a derivation of (5.6.11).

141

$$dU_p = \frac{\partial U_p}{\partial x}\,dx + \frac{\partial U_p}{\partial y}\,dy + \frac{\partial U_p}{\partial z}\,dz \qquad (5.6.11)$$

Comparing Eqs. (5.6.10) and (5.6.11), we see[4] that

$$
\begin{aligned}
F_x &= -\frac{\partial U_p}{\partial x} \\[2mm]
F_y &= -\frac{\partial U_p}{\partial y} \\[2mm]
F_z &= -\frac{\partial U_p}{\partial y}
\end{aligned}
\qquad (5.6.12)
$$

$$
\begin{aligned}
\mathbf{F} &= \mathbf{i}_x F_x + \mathbf{i}_y F_y + \mathbf{i}_z F_z \\[2mm]
&= -\left[\mathbf{i}_x \frac{\partial U_p}{\partial x} + \mathbf{i}_y \frac{\partial U_p}{\partial y} + \mathbf{i}_z \frac{\partial U_p}{\partial z} \right]
\end{aligned}
\qquad (5.6.13)
$$

The vector $\mathbf{i}_x(\partial U_p/\partial x) + \mathbf{i}_y(\partial U_p/\partial y) + \mathbf{i}_z(\partial U_p/\partial z)$ is sometimes referred to as the *gradient* of the scalar function $U_p(x, y, z)$ and is written ∇U_p. Using this notation, (5.6.13) takes the form

$$\mathbf{F} = -\nabla U_p \qquad (5.6.14)$$

where

$$\nabla U_p = \mathbf{i}_x \frac{\partial U_p}{\partial x} + \mathbf{i}_y \frac{\partial U_p}{\partial y} + \mathbf{i}_z \frac{\partial U_p}{\partial z} \qquad (5.6.15)$$

We have shown that if a force is conservative, its components are related to the potential energy by (5.6.12), or, in other words, it can be expressed as minus the gradient of the potential energy $U_p(x, y, z)$. It is important to note that the converse of this statement is also true. If a force can be expressed as the negative gradient of any scalar function of position $U_p(x, y, z)$, then it is conservative and the scalar function U_p represents the potential energy. We shall not, however, attempt a proof of this statement.

It remains now to discuss some of the more important consequences of energy conservation and some of the important properties of conservative forces. Suppose a conservative force $\mathbf{F}$ acts upon a particle in a system such as that illustrated in Fig. 5.13. Let us assume that the potential energy of the particle at point $P(x_x, y_y, z_z)$ is U_{p2}, at point $Q(x_1, y_1, z_1)$ is U_{p1}, and at the origin is zero. From the definition of the potential energy as the work the particle can do in

---

[4] This inference may be rigorously justified by noting that both (5.6.10) and (5.6.11) must hold for *any* arbitrary infinitesimal choice of values for dx, dy, and dz. Suppose now that $dy = dz = 0$; then, equating the right-hand sides of (5.6.10) and (5.6.11), the first of Eqs. (5.6.12) follows. Likewise, supposing that $dx = dz = 0$, and equating the right-hand sides of (5.6.10) and (5.6.11), one may obtain the second equation of the set (5.6.12). Assuming that $dx = dy = 0$ and equating those same two expressions, the third equation of (5.6.12) is obtained.

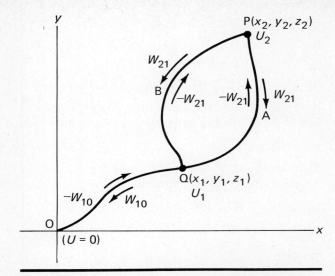

FIGURE 5.13. Different routes that may be followed by a particle moving from P to O. If the force acting is conservative, the work the force does is the same for every possible path.

going from the given point to a point where the potential energy is zero, and from Eq. (5.6.2), we may write

$$
\begin{aligned}
U_{p2} &= U_p(x_2, y_2, z_2) \\[2mm]
&= -\int_0^{x_2} F_x\,dx - \int_0^{y_2} F_y\,dy - \int_0^{z_2} F_z\,dz = W_{20}
\end{aligned}
$$
$$(5.6.16)$$

This is the work W_{20} done by the force as the particle moves from P to O along any path whatever. Let us assume, initially, that the path actually taken is PAQO. The amount of work done by the force as this same path (OQAP) is retraced is, from (5.2.7),

$$\int_0^{x_2} F_x\,dx + \int_0^{y_2} F_y\,dy + \int_0^{z_2} F_z\,dz = -W_{20} \qquad (5.6.17)$$

as would be expected. Now suppose that the two parts of the path PAQ and QO are considered separately. The work W_{10} done by the force in going from Q to O, which is by definition equal to U_{p1}, is, by the same token,

$$
\begin{aligned}
U_{p1} &= U_p(x_1, y_1, z_1) \\[2mm]
&= -\int_0^{x_1} F_x\,dx - \int_0^{y_1} F_y\,dy - \int_0^{z_1} F_z\,dz = W_{10}
\end{aligned}
$$
$$(5.6.18)$$

while the work done in retracing this same path would be

$$\int_0^{x_1} F_x\,dx + \int_0^{y_1} F_y\,dy + \int_0^{z_1} F_z\,dz = -W_{10} \qquad (5.6.19)$$

The work W_{21} done by the force then, in going from P to Q along PAQ is just the difference between

W_{20} and W_{10}, which, in turn, is equal *to the difference in potential energy between the two points.* In other words,

$$W_{21} = W_{20} - W_{10} = U_{p2} - U_{p1} \quad \text{along PAQ} \quad (5.6.20)$$

Now suppose the body goes from P to O along path PBQO. The work done by the force will again be given by (5.6.16), the same value W_{20} being obtained. If this were not so, the potential energy at P could not be assigned a unique value and the force would be nonconservative! The work done by the force, then, in going from P to Q along path PBQ would be that done in going along PBQO (namely, W_{20}) plus that done in retracing the curve OQ ($-W_{10}$). The result is again

$$W_{21} = W_{20} - W_{10} = U_{p2} - U_{p1} \quad \text{along PBQ}$$

The same result would be obtained along any path connecting P and Q. We may therefore conclude that

the work done by any conservative force in moving from one point to another is independent of the path followed and is equal to the initial potential energy of the system minus its final potential energy.

Another interesting property of conservative systems can be inferred from this discussion. The work done in going from P to Q along either PAQ or PBQ is W_{21}. Obviously, from the results obtained in (5.6.16) and (5.6.17) or in (5.6.18) and (5.6.19), the work done in returning from Q to P along either path must be $-W_{21}$. But suppose we go from P to Q along PAW and return along PBQ; how much work is done now? The answer is

$$W = W_{21} - W_{21} = 0$$

The same result would be obtained for any outward path and any other return path. This, however, means that the work done by the force is zero around any closed path on which both P and Q lie. But there is nothing special about the points P and Q; they could have been chosen to lie anywhere! The conclusion is that

the work done by any conservative force around any closed path is zero.

In all these calculations, the action of all the forces on the system is taken care of by properly defining the potential energy function, which, as we have seen, is closely related to these forces. Under such circumstances, the total mechanical energy of the system always remains constant. This result does not imply that there is no way of changing the system's energy but only that the energy cannot be changed unless *additional* "external" forces, not accounted for

in the potential energy function, are exerted on the system by its surroundings. If these forces do an amount of work W_{ext} on the system, then, according to the work–energy theorem, the system's energy will simply be increased by that amount, and, therefore, in such a situation its total energy becomes

$$U_{mf} = U_{mi} + W_{ext} \qquad (5.6.21)$$

It is important to distinguish between these two sets of circumstances. Most of our conclusions about energy conservation apply to "isolated" systems of the first type mentioned above, but it is important to recognize that the energy of any system can be changed by the introduction of new "external" forces. If these external forces are conservative, the energy of the surroundings will change by an amount equal in magnitude but opposite in sign to the system's energy change, the *total energy of system and surroundings* still remaining constant.

EXAMPLE 5.6.1

A baseball of mass m is projected vertically upward with initial velocity $\mathbf{v}_0$. Discuss its subsequent motion from the viewpoint of energy conservation. Neglect effects arising from air resistance.

Referring to Fig. 5.14, five separate stages can be identified. Initially, at (1), the ball is thrown upward with velocity $\mathbf{v}_0 = \mathbf{i}_z v_0$. Let us make the usual assumption that the potential energy is zero at the earth's surface. Then, since the potential energy is defined as the work done by gravity in returning the ball to the point where the potential energy is zero, if the

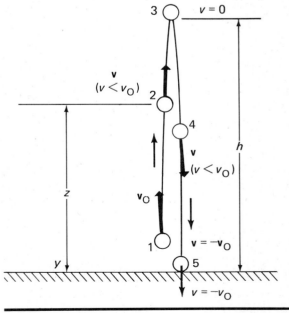

FIGURE 5.14

ball is at height z above the earth, its potential energy is given by the product of the weight force mg and the distance z through which this force moves as the ball returns to the ground. Since the angle between the force and displacement vectors in this case is zero, the factor $\cos\theta$ in (5.2.1) is unity. We then have

$$U_p = mgz \qquad (5.6.22)$$

Note that the force components, as predicted by (5.6.12), would be

$$F_x = -\frac{\partial U_p}{\partial x} = 0$$

$$F_y = -\frac{\partial U_p}{\partial y} = 0$$

$$F_z = -\frac{\partial U_p}{\partial z} = -mg$$

This is in agreement with the force originally described. The set of equations (5.6.12) is useful in checking the correctness of the potential energy which has been calculated, since it must always lead to the correct force components if the potential energy is correctly represented. Note also that the same expression for these force components is obtained even if a point other than the earth's surface is chosen as the zero level of potential energy. The potential energy would then be $U_p = mg(z - z_0)$, where z_0 is the elevation at which the potential energy is taken to be zero. But the partial derivatives $\partial U_p/\partial x$, $\partial U_p/\partial y$, and $\partial U_p/\partial z$ are the same as those listed above, since all derivatives of the constant mgz_0 are zero.

At (1) in Fig. 5.14, the kinetic energy is $\frac{1}{2}mv_0^2$, according to (5.4.7), while the potential energy is zero, from (5.6.22). The total energy U_m is the sum of these two quantities:

$$U_m = U_k + U_p = \tfrac{1}{2}mv_0^2 + 0 = \tfrac{1}{2}mv_0^2 \qquad (5.6.23)$$

In its journey upward and down, if we neglect air resistance, we may regard the ball as subject only to the force of gravity. Since a uniquely defined potential energy can be found, this force must be conservative. The total energy U_m is, therefore, constant throughout the ball's journey. As it rises, the potential energy mgz *increases* with the height z. Since the total energy U_m must have the constant value $\frac{1}{2}mv_0^2$, according to (5.6.23), the kinetic energy, and hence the velocity v, must decrease. In fact, the energy conservation theorem can be used to find the precise value of v at any given time, since at any given height z we must have, according to (5.4.7), (5.6.22), and (5.6.23),

$$U_m = U_k + U_p = \tfrac{1}{2}mv^2 + mgz = \tfrac{1}{2}mv_0^2 \qquad (5.6.24)$$

This may be solved for v:

$$v^2 = v_0^2 - 2gz$$
$$v = \sqrt{v_0^2 - 2gz} \qquad (5.6.25)$$

The reader will recognize this as one of the "golden rules" of uniformly accelerated motion previously obtained in Chapter 3 by integrating equations of motion derived from Newton's second law. It will be obvious that the result is obtained much more simply from the energy conservation theorem. It is also easy, letting $v = dz/dt$, to integrate the expression given above for v between appropriate limits to obtain the familiar result $z = v_0 t - \frac{1}{2}gt^2$. It is clear from this that the work–energy theorem and the energy conservation principle and Newton's second law are alternative ways of expressing the same physical principles.

As the height of the ball increases, a point is reached where all the initial kinetic energy has been converted to potential energy; at this point, $U_k = \frac{1}{2}mv^2 = 0$, hence $v = 0$. The ball has now reached the top of its path. If we denote the value of z at this point as $z = h$, we may write the total energy (which is conserved and still has the initial value $\frac{1}{2}mv_0^2$) as

$$U_m = U_k + U_p = 0 + mgh = \tfrac{1}{2}mv_0^2$$

or, solving for h,

$$h = v_0^2/2g \qquad (5.6.26)$$

in agreement with the results previously obtained in Example 3.2.2. The ball now begins to drop, its potential energy decreasing as z decreases and its kinetic energy increasing in such a way that the total energy remains constant. At some point of height z on the way down, if the expression for the total energy as the sum of kinetic and potential energies is written, Eq. (5.6.24) is once more obtained, leading again to (5.6.25). The velocity v must now be taken as the *negative* square root of the right-hand side:

$$v = -\sqrt{v_0^2 - 2gz} \qquad (5.6.27)$$

As the ball reaches the earth's surface, once again its potential energy is zero and its kinetic energy must have the original value $\frac{1}{2}mv_0^2$. The only difference is that now $v = -v_0$ rather than $+v_0$. These changes in kinetic and potential energy are plotted as a function of time throughout the flight of the ball in Fig. 5.15.

When the ball strikes the earth, it may bounce several times, then come to rest. What has now happened to the initial energy $\frac{1}{2}mv_0^2$? The answer is, in general, that some of it has been transferred to the earth as kinetic energy, altering the earth's orbital path around the sun ever so slightly, and some has been converted into heat since the forces acting during its collisions with the earth are nonconservative. It should be noted, also, that the force exerted on the ball by the arm of the man who threw it originally was

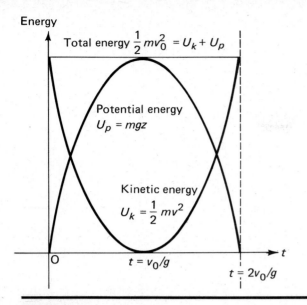

Energy

Total energy $\frac{1}{2}mv_0^2 = U_k + U_p$

Potential energy
$U_p = mgz$

Kinetic energy
$U_k = \frac{1}{2}mv^2$

$t = v_0/g$

$t = 2v_0/g$

FIGURE 5.15

nonconservative, having been generated by the conversion of chemical to mechanical energy in the muscles.

EXAMPLE 5.6.2

A mass m resting on a frictionless horizontal surface is attached to a rigid support by means of an ideal spring of spring constant k, as shown in Fig. 5.16. The spring is stretched a distance x_0 from its equilibrium position and released. Discuss the subsequent motion from the viewpoint of energy conservation and find, in particular, the maximum velocity attained by the mass.

In Fig. 5.16a, the mass is shown in its equilibrium position $x = 0$, at the limit of its excursion $x = x_0$, and at some intermediate position where the extension of the spring has an intermediate value x.

We have already seen in Example 5.2.3 that an amount of work equal to $\frac{1}{2}kx^2$ must be done in order to stretch an ideal spring through a distance x. Since, if there are no frictional energy losses, all this work is recoverable as work done by the system when the spring is released, the amount of work the system can do in returning to the equilibrium position is $\frac{1}{2}kx^2$, and, therefore, the potential energy of the spring-mass system is given by

$$U_p = \frac{1}{2}kx^2 \tag{5.6.28}$$

It is assumed that the potential energy associated with the equilibrium position is zero. This assumption is not necessary but is convenient in this example.

When the spring is stretched initially through the distance x_0 and released, at the instant of release its potential energy, from (5.6.28), is $\frac{1}{2}kx_0^2$; and since its initial velocity is zero, its kinetic energy at this moment is also zero. The total energy U_m which is *conserved* throughout the subsequent motion of the system is

$$U_m = U_k + U_p = 0 + \frac{1}{2}kx_0^2 = \frac{1}{2}kx_0^2 \tag{5.6.29}$$

As the spring relaxes, it does work on the mass, accelerating it and at the same time, of course, increasing its kinetic energy. In the process, its own stored potential energy decreases, the total energy remaining constant according to (5.6.7). At an intermediate point where the extension of the spring is x, we may write the total energy as the sum of the potential energy, which is then $\frac{1}{2}kx^2$, and the kinetic energy as $\frac{1}{2}mv^2$, where v is the velocity, setting the total equal to $\frac{1}{2}kx_0^2$ as required by (5.6.29) to obtain

$$U_m = U_k + U_p = \frac{1}{2}mv^2 + \frac{1}{2}kx^2 = \frac{1}{2}kx_0^2 \tag{5.6.30}$$

This equation may be solved for the velocity v:

$$v = \sqrt{\frac{k}{m}(x_0^2 - x^2)} \tag{5.6.31}$$

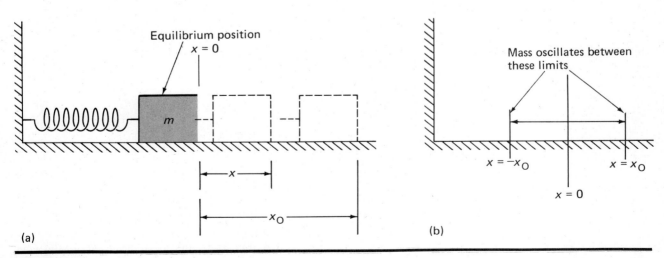

Equilibrium position
$x = 0$

m

x

x_0

(a)

Mass oscillates between these limits

$x = -x_0$

$x = x_0$

$x = 0$

(b)

FIGURE 5.16

The maximum velocity is reached when the spring has returned to its equilibrium position. Its potential energy is then zero, according to (5.6.28), and its total energy $\frac{1}{2}kx_0^2$ appears in the form of kinetic energy of motion. At this point, we may find the maximum velocity by again setting the sum of kinetic and potential energies equal to the total energy and solving fo v:

$$U_m = U_k + U_p = \tfrac{1}{2}mv^2 + 0 = \tfrac{1}{2}kx_0^2$$
$$v = x_0\sqrt{k/m} \qquad \text{at } x = 0 \tag{5.6.32}$$

The same result follows from (5.6.31) simply by setting $x = 0$. Beyond this point, the mass continues to the right, doing work on the spring now in *compressing* it, its kinetic energy decreasing and the potential energy stored in the spring increasing. When $x = -x_0$, the amount of stored potential energy, according to (5.6.28), is $\frac{1}{2}kx_0^2$, which equals the total energy. The kinetic energy and hence the velocity are zero, and the mass comes to rest at this point. The compressed spring now starts to expand, decreasing its potential energy and correspondingly increasing the kinetic energy of the mass, which is now moving to the right in Fig. 5.16. The mass passes the equilibrium position once again with maximum (positive) velocity and once more expends its kinetic energy in stretching the spring, increasing its extension until all its kinetic energy is gone and the spring is again stretched a distance x_0. The system is now once more in the *same state whence it began*; the spring is extended a distance x_0 and the velocity is zero. The same cycle of events described above, therefore, begins again and is repeated over and over, the mass oscillating between the limits of excursion x_0 and $-x_0$, as shown in Fig. 5.16b. The time taken for a complete oscillation can be shown to be $2\pi\sqrt{m/k}$. Oscillatory motion of this type is called *simple harmonic motion* and will be investigated in detail in Chapter 9.

5.7 One-Dimensional Motion

Consider a situation in which a body of mass m moves along the x-axis only. The forces acting on it are conservative and are expressible in terms of a given potential energy $U_p(x)$. The total energy of the system is U_m. Let us discuss the resulting motion from the viewpoint of energy conservation.

According to the energy conservation theorem, the total energy is constant and is equal to the sum of the kinetic and potential energies. Therefore,

$$U_m = \tfrac{1}{2}mv^2 + U_p(x) = \text{const.} \tag{5.7.1}$$

Solving for v, we find

$$v = \frac{dx}{dt} = \sqrt{\frac{2}{m}\left[U_m - U_p(x)\right]} \tag{5.7.2}$$

If we multiply both sides of this equation by dt and divide both sides by $\sqrt{2[U_m - U_p(x)]/m}$, we shall have an equation of the form

$$\frac{dx}{\sqrt{\dfrac{2}{m}\left[U_m - U_p(x)\right]}} = dt \tag{5.7.3}$$

in which the only variable on the left side is x and the only variable on the right side is t. This equation can be integrated from the initial time $t = 0$, at which an initial position x_0 is specified, to an arbitrary later time t, at which the position of the body is given by x. Accordingly,

$$\int_0^t dt = t = \int_{x_0}^x \frac{dx}{\sqrt{\dfrac{2}{m}\left[U_m - U_p(x)\right]}} \tag{5.7.4}$$

The velocity of the body is given by (5.7.2). Its position at any time can be found by substituting the given value for $U_p(x)$ in terms of x into (5.7.4), evaluating the integral, and solving the resulting equation for x. Unfortunately, this is often more easily said than done, because the integral is difficult to evaluate in all but the simplest cases. Nevertheless the procedure is straightforward and will always yield the desired result if the required integral can be calculated. The *force* acting on the body may be inferred from $U_p(x)$ by considering a displacement dx which is accomplished in time dt. During the displacement, the potential energy undergoes a change dU_p and the kinetic energy, a change dU_k; but the sum of these two must be zero, since the total energy cannot change. Then,

$$dU_p + dU_k = dU_p + d\left[\tfrac{1}{2}mv^2\right] = dU_p + mv\,dv = 0$$

or

$$dU_p = -mv\,dv = -mv\frac{dv}{dt}dt = -m\frac{dx}{dt}\frac{dv}{dt}dt \tag{5.7.5}$$

since $dx/dt = v$. But since dv/dt is the acceleration a,

$$dU_p = -ma\,dx = -F\,dx$$

or

$$\boxed{F = -\frac{dU_p}{dx}} \tag{5.7.6}$$

Note that Eq. (5.6.12) agrees with this whenever U_p depends *only* on x, because then the partial derivative $\partial U_p/\partial x$ and the total derivative dU_p/dx are the same.

Even when it is difficult or impossible to evaluate the integral in (5.7.4), a great deal of qualitative understanding can be obtained about the motion of the system by examining a *potential energy diagram* such as that shown in Fig. 5.17, where the potential energy U_p is plotted as a function of x. If the total

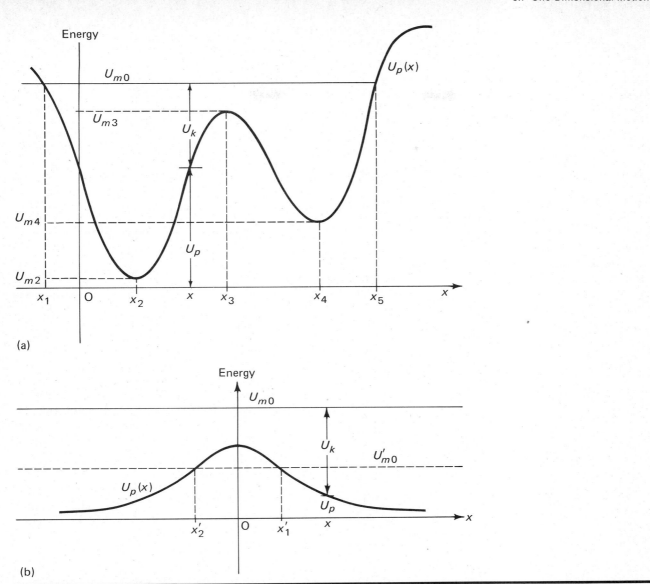

FIGURE 5.17. (a) Potential "well" in which a particle of total energy U_{m0} will execute periodic oscillatory motion along the x-axis. (b) Potential energy function that does not give rise to oscillatory motion.

energy is U_m, then at any point x it is easy to display graphically the values of both U_k and U_p as illustrated. Also, it is possible to determine whether or not the motion is oscillatory. For example, in Fig. 5.17a, a body whose total energy is U_m would be "trapped" in the "potential well" extending from x_1 to x_5, because its potential energy can never exceed its total energy, and the region of the potential diagram where the potential energy is either less than or equal to U_m is just the "potential well" between x_1 and x_5. The motion of such a body is of necessity oscillatory, because when it arrives at x_5, the force on it is negative (since the slope dU_p/dx there is positive and since $F = -dU_p/dx$); this moves it to the left, toward x_1. When it arrives at x_1, the force on it is positive (because the slope dU_p/dx, equal to $-F$, is now

negative), which sets it in motion to the right, again toward x_5. Upon arriving at x_5, the cycle is then repeated, leading to oscillatory motion.

On the other hand, if the potential energy diagram is of the form shown in Fig. 5.17b and contains no potential wells, then, for a body of total energy U_{m0}, the kinetic energy is positive everywhere. If the kinetic energy is never zero, the velocity can never be zero and no reversal of the direction of motion can ever take place. The motion must therefore proceed from left to right (or vice versa) continuously with no reversal of direction and hence cannot be oscillatory. For a body of energy U_{m0}' starting at the far right and moving leftward, a *single* reversal of direction occurs at $x = x_1'$, where the kinetic energy, hence the velocity, is zero. The force at this point

147

$(= -dU_p/dx)$ is positive, the force vector pointing toward the right, and the particle is set in motion again in the direction whence it came. It may undergo no further reversal of motion, however, since there is no point to the right of x_1' where the kinetic energy is zero and where a reversal may take place. The motion is therefore nonoscillatory. Similar observations may be made about a particle starting at the far left and approaching x_2'.

The potential energy diagram also illustrates the possible points of equilibrium of the system. For example, in Fig. 5.17a, a particle at x can *never be in equilibrium* under the influence of the forces acting on it, no matter what its total energy may be. There is always a net force equal to $-dU_p/dx$ acting on it at this point, and, therefore, it must always have an acceleration when it is there. However, at the minima x_2 and x_4, and at the maximum x_3, dU_p/dx is zero and there is *no net force* on a particle at these points. Particles of total energy U_{m2}, U_{m4}, and U_{m3}, respectively, may remain stationary at these *equilibrium points*. In the case of x_2 and x_4, a particle given a small displacement to the left experiences a force to the right (because $-dU_p/dx$ is positive), tending to return it toward the equilibrium point. Likewise, a small displacement to the right produces a force to the left, which again tends to return the body toward the equilibrium point. A small oscillatory motion is thus set up, the particle sliding back and forth in the potential minimum about the point of equilibrium. The condition is then said to be one of *stable equilibrium*. On the other hand, in the case of x_3, where the potential is a maximum, a small displacement to the left causes a force to the left to act, moving the particle rapidly away from the equilibrium point, while a small displacement to the right sets up a force to the right which behaves similarly. The particle does not tend to remain in the neighborhood of such an equilibrium point, and the condition is referred to as one of *unstable equilibrium*. These definitions of stable and unstable equilibrium are in complete agreement with the explanation of the terms as given from a somewhat different point of view in section 2.10.

It is apparent from these discussions that a particle moves somewhat as if it were sliding upon a frictionless surface having the contours of the potential energy curve. While this description of the motion is not precisely correct, it is sufficiently accurate in many cases to enable the main physical characteristics of the motion to be visualized. The idea that the points of stable equilibrium are those that correspond to minimum values of potential energy is one that is very useful and which we shall have occasion to use in the future.

Let us now consider again, in somewhat more detail, the one-dimensional system illustrated in Fig. 5.15. The potential energy $U_p(x)$ is given in this case by

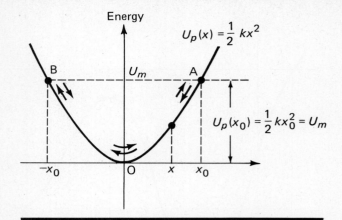

FIGURE 5.18. Potential "well" associated with an ideal spring.

$$U_p(x) = \tfrac{1}{2}kx^2 \tag{5.7.7}$$

and the corresponding parabolic potential energy curve is shown in Fig. 5.18. The force, according to (5.7.6), must be

$$F = -dU_p/dx = -kx \, dx \tag{5.7.8}$$

which, of course, agrees with (5.2.7). From Fig. 5.18 and from the discussion above, it is apparent that for any given total energy U_m, the motion must be oscillatory; let us call the excursion limits $\pm x_0$ to agree with our former notation. From Fig. 5.18, it is easily seen that the total energy is equal to the value assumed by $U_p(x)$ at the excursion limits $x = \pm x_0$, that is,

$$U_m = U_p(x_0) = \tfrac{1}{2}kx_0^2 \tag{5.7.9}$$

This is also in agreement with the results of Example 5.6.2. Substituting (5.7.7) and (5.7.9) into (5.7.2),

$$v = \sqrt{\frac{k}{m}(x_0^2 - x^2)} \tag{5.7.10}$$

which agrees with (5.6.31).

Substituting (5.7.7) and (5.7.9) into (5.7.4) and making the assumption that the particle is at $x = 0$ when $t = 0$, we obtain

$$t = \sqrt{\frac{m}{k}} \int_0^x \frac{dx}{\sqrt{x_0^2 - x^2}} \tag{5.7.11}$$

This may be integrated by reference to tables or simply by noting that

$$\frac{d}{dx}\left(\frac{1}{\sqrt{x_0^2 - x^2}}\right) = \sin^{-1}\left(\frac{x}{x_0}\right) \tag{5.7.12}$$

By either method, we obtain

$$t\sqrt{\frac{k}{m}} = \left[\sin^{-1}\left(\frac{x}{x_0}\right)\right]_0^x = \sin^{-1}\left(\frac{x}{x_0}\right) - \sin^{-1}0$$

$$= \sin^{-1}\left(\frac{x}{x_0}\right) \tag{5.7.13}$$

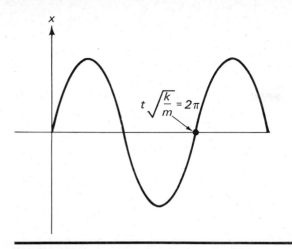

FIGURE 5.19. Oscillatory motion of the system illustrated in Fig. 5.16, consisting of a mass connected to an ideal spring.

since the angle whose sine is zero is zero. Now let us take the sine of both sides of this equation. Since

$$\sin\left[\sin^{-1}\left(\frac{x}{x_0}\right)\right] = \frac{x}{x_0}$$

Eq. (5.7.13) can be written[5] in the form

$$x = x_0 \sin\left(t\sqrt{\frac{k}{m}}\right) \tag{5.7.14}$$

This motion is an oscillatory function of time. A plot of the displacement x versus time is shown in Fig. 5.19. A complete cycle of the motion takes place during the time required for the argument of the sine function to increase by 2π radians. When $t = 0$, of course, the argument of the sine is zero. When

$$t\sqrt{\frac{k}{m}} = 2\pi$$

the argument will have increased by 2π and a new cycle will begin. The *period* associated with one cycle may be found simply by solving the above equation for t:

$$t = 2\pi\sqrt{\frac{m}{k}} \tag{5.7.15}$$

as mentioned at the end of the previous example.

The velocity can be obtained as a function of time simply by differentiating (5.7.14):

$$v = \frac{dx}{dt} = x_0\sqrt{\frac{k}{m}}\cos\left(t\sqrt{\frac{k}{m}}\right) \tag{5.7.16}$$

It is easy to verify by direct substitution that (5.7.14) and (5.7.16) are consistent with (5.7.10). The motion can be approximately visualized in terms of the x-component of the motion of a particle sliding back

[5] The sine of the angle whose sine is x/x_0 is simply x/x_0! Right?

and forth in a frictionless parabolic valley like that shown in Fig. 5.18. The minimum point of the valley at $x = 0$ is clearly a point of stable equilibrium. A body of total energy zero can thus remain in static equilibrium. A body of total energy zero can thus remain in static equilibrium at $x = 0$. This corresponds to the case where the mass is at rest with the spring being neither stretched nor compressed.

*EXAMPLE 5.7.1

A particle of mass m slides on a frictionless track whose height above the earth's surface is given by some known function $z(x)$, as illustrated in Fig. 5.20. It is released from rest at a given point $P(x_0, z_0)$ and starts to slide downhill toward Q. What is its velocity when it reaches the point Q, whose coordinates are (x, z)?

In this example there is no friction and the normal force $\mathbf{N}$, which is always perpendicular to the direction of motion, does no work. Therefore, the mechanical energy of the particle is conserved. The potential energy of the particle at $z(x)$ is equal to the amount of work done by gravity in returning the particle to the earth's surface, where the potential energy is assumed to be zero. Hence, as in Example 5.6.1, the potential energy at $z(x)$ is given by

$$U_p(z) = mgz = mgz(x) \tag{5.7.17}$$

Note that in this case for any point on the curve, z is given as the value of a known function $z(x)$, which is, the equation of the curve along which the body slides. At the starting point P, the initial velocity is zero. At that point there is no kinetic energy, so the potential energy there must equal the total energy. Therefore, the total energy U_m is

$$U_m = U_p(z_0) \doteq mgz_0 \tag{5.7.18}$$

At any point such as Q, whose coordinates are (x, z), we may, therefore, write the total energy U_m as

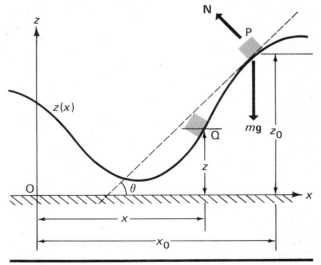

FIGURE 5.20

$$U_m = U_k + U_p = \tfrac{1}{2}mv^2 + mgz(x) = \tfrac{1}{2}mgz_0 \qquad (5.7.19)$$

from which

$$v = \sqrt{2g[z_0 - z(x)]} \qquad (5.7.20)$$

This result is arrived at very easily from conservation of energy. We could also obtain the same information directly from Newton's laws by equating force components to mass times respective acceleration components and integrating the accelerations to get the velocities. But this is much more difficult, since the forces and thus the accelerations vary from point to point, which vastly complicates the integration of the equations of motion.

Let us examine in detail the form these results take when the curve along which the particle descends is a parabola whose equation is

$$z(x) = \frac{1}{2}\frac{kx^2}{mg} \qquad (5.7.21)$$

This potential energy function leads to the same expression for potential energy as a function of x encountered in the previous example of a particle moving in one dimension along the x-axis, for now

$$U(z) = mgz(x) = \tfrac{1}{2}kx^2 \qquad (5.7.22)$$

while

$$U_m = U_p(z_0) = mgz_0 = \tfrac{1}{2}kx_0^2 \qquad (5.7.23)$$

Equation (5.7.20) now becomes

$$v = \sqrt{\frac{k}{m}}\,(x_0^2 - x^2) \qquad (5.7.24)$$

This looks very much the same as Eq. (5.7.10), which was derived for a particle moving in one dimension with the same potential energy as for the two-dimensional motion described immediately above. There is one difference, however; in this case, the velocity has both a z- and x-component, while in the former one, only an x-component was possible. This can be seen from Fig. 5.21. In this instance, v is no longer equal to

dx/dt but rather to $(v_x^2 + v_y^2)^{1/2}$ or $[(dx/dt)^2 + (dy/dt)^2]^{1/2}$. We may, however, proceed by noting from Fig. 5.21 that

$$\frac{dx}{dt} = v_x = v\cos\theta$$

Then, from (5.7.24),

$$dt = \frac{dx}{v\cos\theta} = \sqrt{\frac{m}{k}}\,\frac{dx}{\sqrt{x_0^2 - x^2}\,\cos\theta}$$

This may now be integrated from $x = x_0$ at $t = 0$ to some point x at time t to give

$$\sqrt{\frac{m}{k}}\int_{x_0}^{x}\frac{dx}{\sqrt{x_0^2 - x^2}\,\cos\theta} = \int_0^t dt = t \qquad (5.7.25)$$

Now, apart from the factor $\cos\theta$, which depends on the variable of integration x and must, therefore, remain within the integral, this expression is the same as (5.7.11), the result obtained for the one-dimensional situation with the same potential energy. If θ remains *always so small that $\cos\theta$ is practically equal to unity*, then the two results are essentially the same, and the physical character of the x-component of motion in this system will be substantially identical with the one-dimensional case where the potential energy function has the same form. If, however, θ becomes so large that this is no longer true, then significant differences will occur. The discussion of this special case outlines the precautions to be taken in visualizing the one-dimensional motion of a particle with a given potential energy in terms of the x-component of the two-dimensional motion of an object sliding on a frictionless contour representing the same potential energy function. The two motions are essentially the same only if the "hills" in the two-dimensional contour are not too steep.

EXAMPLE 5.7.2

An object of mass m slides along a frictionless track containing a vertical circular loop of radius r as shown in Fig. 5.22a. The object is released from rest initially at a point A on the track at height h above the bottom of the loop, traverses the loop, and leaves the loop at point B. What is the minimum value the initial height h must have to ensure that the object does not fall away from the track at point P?

At point P, the forces acting on the body are as shown in Fig. 5.22b. We may sum the z-components of these forces and set them equal to $-mr\omega^2$ or $-mv^2/r$ since, if the object is on the track, it is executing circular motion and, whether this motion is uniform or not, the radial component of acceleration, according to (3.4.17), must be $-r\omega^2$ or $-v^2/r$. This means that

$$-N - mg = -mv^2/r$$

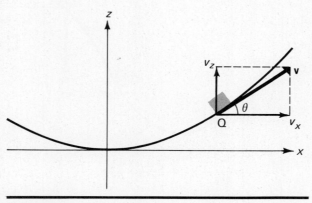

FIGURE 5.21

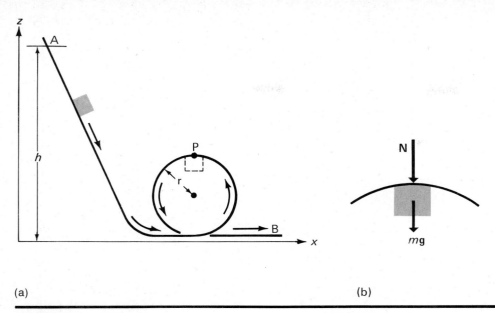

(a)

(b)

FIGURE 5.22

or

$$N = (mv^2/r) - mg \qquad (5.7.26)$$

When v has such a small value that the body is about to leave the track, the normal force N is zero. Under those circumstances, setting $N = 0$ in (5.7.26) and solving for v^2, we see that

$$v^2 = gr \qquad (5.7.27)$$

Since there is no frictional force or other dissipative force, the system is conservative. According to the energy conservation theorem (5.6.7), the sum of the initial kinetic and potential energies $U_{k1} + U_{p1}$ must equal the sum $U_{k2} + U_{p2}$ when the object reaches the point P. We may, therefore, write

$$U_{k1} + U_{p1} = U_{k2} + U_{p2}$$
$$0 + mgh = \tfrac{1}{2}mv^2 + 2mgr$$

or, solving for v^2,

$$v^2 = 2g(h - 2r) \qquad (5.7.28)$$

An object starting at height h attains the velocity given by this expression at the point P. But if h is the minimum height required to ensure that the object remain on the track, this velocity, according to (5.7.27), must also have the value $\sqrt{gr}$. Accordingly, equating (5.7.27) and (5.7.28) and solving for h,

$$h = 5r/2 \qquad (5.7.29)$$

If h is greater than this, the body will speed around the loop in contact with the track and the normal force exerted by the track on the object will never drop to zero. If h is less than $5r/2$, the body will fall from the track before reaching the top of the loop.

EXAMPLE 5.7.3

Work Example 4.2.5 using the energy conservation theorem rather than Newton's second law.

Since there are no frictional forces or other nonconservative forces, the system is conservative, and the sum of kinetic and potential energies remains the same at all times. Referring to Fig. 5.23, let us assume that at time $t = 0$, the mass m_2 is a distance z_0 above the horizontal reference plane $z = 0$, while the mass m_1 is at a height z_1 above this plane. Let us also suppose that both masses are moving initially with velocity v_0 such that m_1 travels to the left while m_2 is descending. The initial kinetic and potential energies U_{ki} and U_{pi} for the entire system are then given by

$$U_{ki} = \tfrac{1}{2}m_1 v_0{}^2 + \tfrac{1}{2}m_2 v_0{}^2$$

and

$$U_{pi} = m_1 g z_1 + m_2 g z_0 \qquad (5.7.30)$$

At a later time t, the system will have moved until the mass m_2 is some distance z above the reference plane, at which time the velocity will attain a corresponding value v. The final values of the kinetic and potential energies U_{kf} and U_{pf} are then

$$U_{kf} = \tfrac{1}{2}m_1 v^2 + \tfrac{1}{2}m_2 v^2$$

and

$$U_{pf} = m_1 g z_1 + m_2 g z \qquad (5.7.31)$$

Note that the potential energy of mass m_1 does not change, because the mass moves horizontally.

The energy conservation theorem (5.6.7) states that

$$U_{ki} + U_{pi} = U_{kf} + U_{pf} \qquad (5.7.32)$$

151

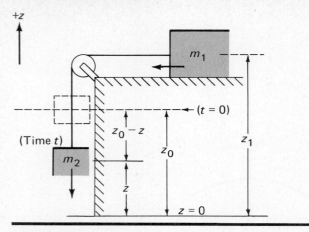

FIGURE 5.23

or, substituting from (5.7.30) and (5.7.31),

$$\tfrac{1}{2}m_1v_0^2 + \tfrac{1}{2}m_2v_0^2 + m_2gz_0 + m_1gz_1$$
$$= \tfrac{1}{2}m_1v^2 + \tfrac{1}{2}m_2v^2 + m_2gz + m_1gz_1 \quad (5.7.33)$$

Subtracting m_1gz_1 from both sides of this equation and rearranging slightly, this may be reduced to

$$\tfrac{1}{2}m_1(v^2 - v_0^2) + \tfrac{1}{2}m_2(v^2 - v_0^2) + m_2g(z - z_0) = 0 \quad (5.7.34)$$

But since the forces acting upon both masses of the system are constant in time, the acceleration of the system must be *constant*. Under such circumstances, according to the "golden rules" that describe this type of motion, the displacement $z - z_0$ is related to the initial and final velocities and the acceleration a by

$$v^2 - v_0^2 = 2a(z - z_0) \quad (5.7.35)$$

from which

$$z - z_0 = \frac{v^2 - v_0^2}{2a} \quad (5.7.36)$$

Substituting this result into (5.7.34), dividing both sides of the resulting equation by $v^2 - v_0^2$, and solving for a, we may find

$$a = -\frac{m_2g}{m_1 + m_2} \quad (5.7.37)$$

This is in agreement with (4.2.15) if it is recalled that in section 4.2 the positive displacement direction was assumed to be *downward*, while we have assumed it to be *upward* in the derivation immediately above. This accounts for the presence of the minus sign in (5.7.37).

The tension in the connecting cord can be inferred from the work–energy theorem, equating the work done on mass m_2, for example, during its journey from z_0 to z to its increase in kinetic energy. The work done by the resultant force is the product of that force times the distance through which it has moved, or $(m_2g - T)(z_0 - z)$. Equating this to the gain in kinetic

energy, we find

$$(m_2g - T)(z_0 - z) = \tfrac{1}{2}m_2v^2 - \tfrac{1}{2}m_2v_0^2$$
$$= \tfrac{1}{2}m_2(v^2 - v_0^2) \quad (5.7.38)$$

But according to (5.7.36), $v^2 - v_0^2 = -2a(z_0 - z)$. Substituting this into the above equation,

$$m_2g - T = -m_2a$$
$$T = m_2(g + a) \quad (5.7.39)$$

However, a is known from (5.7.37) to have the value $-m_2g/(m_1 + m_2)$. Therefore,

$$T = m_2g - \frac{m_2^2g}{m_1 + m_2} = \frac{m_1m_2g}{m_1 + m_2} \quad (5.7.40)$$

in agreement with (4.2.16).

It is evident from all this that any problem involving conservative forces may be solved by use of the energy conservation principle and the work–energy theorem, as well as by direct use of Newton's second law. This is hardly surprising if it is recalled that both the work–energy theorem and the energy conservation theory are direct consequences of Newton's second law. In general, it will be found that the energy conservation method is advantageous in working problems where initial or final velocity, or the distance through which a system moves, is sought, while the direct application of Newton's law is simpler when acceleration, force, or time is asked for. In the next section, we shall see that energy conservation methods can also be used when frictional forces are present, provided that proper account is taken of the dissipation of mechanical energy by these forces.

EXAMPLE 5.7.4

An ideal spring can be compressed 2.5 cm by the application of a force of 125 newtons. An object of mass 1.5 kg is placed against the spring, which is then compressed 10.0 cm and released. The mass is thus projected along a horizontal frictionless surface terminated by a frictionless inclined plane making an angle of 37° with the horizontal, as illustrated in Fig. 5.24. What is the velocity of the mass as it travels along the horizontal surface? What is its velocity after it has traveled along the inclined plane for 2.0 meters? How far along the plane will it go before coming to rest?

In this example, there are no frictional forces. The only forces are those exerted by gravity and by the spring, which are conservative. In order to discuss the energy stored by the spring, we must first find the spring constant k. This can be accomplished by using Eq. (5.2.8), since we know that a force of 125 newtons compresses the spring by 0.025 meter. Substituting these data into (5.2.8),

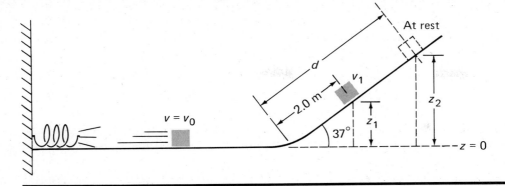

FIGURE 5.24

$$k = \frac{F}{x} = \frac{(125)}{(0.025)} = 5000 \text{ N/m}$$

Since all forces acting are conservative, the total mechanical energy as represented by the sum of the kinetic and potential energies is always the same. Initially, as the mass rests against the spring prior to "firing," its kinetic energy is zero, as is its potential energy. The spring, however, is compressed by 0.1 meter and, therefore, has stored a certain amount of potential energy, which in this initial state represents the total energy U_m of the system. This total energy can be calculated from (5.6.28) as

$$U_m = \tfrac{1}{2}kx^2 = (\tfrac{1}{2})(5000)(0.1)^2 = 25 \text{ J}$$

When the spring is "fired," it expends all its stored potential energy in accelerating the mass. Since the mass travels horizontally at first, its potential energy in this phase of the motion is zero. Therefore, writing the total energy U_m as the sum of the initial kinetic and potential energies U_{k0}, and U_{p0} of the mass, we find from the energy conservation principle that

$$U_m = U_{k0} + U_{p0} \tag{5.7.41}$$

But

$$U_{k0} = \tfrac{1}{2}mv_0^2$$

while

$$U_{p0} = mgz = 0$$

since $z = 0$ initially. Substituting these values into (5.7.41),

$$\tfrac{1}{2}mv_0^2 + mg(0) = U_m$$

from which

$$(\tfrac{1}{2})(1.5)\, v_0^2 = 25$$
$$v_0 = 5.77 \text{ m/sec}$$

When the object has traveled 2.0 meters along the plane, it has risen to a height z_1 above the reference

plane $z = 0$. This height is given by

$$\frac{z_1}{(2.0 \text{ m})} = \sin 37° = 0.60$$
$$z_1 = 1.20 \text{ m}$$

At this point, the sum of the kinetic and potential energies, U_{k1} and U_{p1}, is still U_m. Therefore, from (5.7.41),

$$U_m = U_{k0} + U_{p0} = U_{k1} + U_{p1} = 25 \text{ J} \tag{5.7.42}$$

In this case, however,

$$U_{k1} = \tfrac{1}{2}mv_1^2$$
$$U_{p1} = mgz_1$$

From (5.7.42), then, using $z_1 = 1.20$ meters,

$$(\tfrac{1}{2})(1.5)v_1^2 + (1.5)(9.80)(1.20) = 25$$
$$v_1 = 3.13 \text{ m/sec}$$

Finally, at some point of height z_2 above the plane $z = 0$, the kinetic energy is reduced to zero, and the object comes to rest. Its total energy as given by the sum of kinetic and potential energies U_{k2} and U_{p2} still has the initial value U_m. Therefore,

$$U_m = U_{k0} + U_{p0} = U_{k1} + U_{p1}$$
$$= U_{k2} + U_{p2} = 25 \text{ J} \tag{5.7.43}$$

In this stage of the motion, we have $v_2 = 0$, from which

$$U_{k2} = 0$$
$$U_{p2} = mgz_2$$

From (5.7.43), then,

$$U_{k2} + U_{p2} = 0 + (1.5)(9.8)z_2 = 25$$
$$z_2 = 1.70 \text{ m}$$

This corresponds to a distance d along the plane given by

$$\frac{z_2}{d} = \sin 37° = 0.6 \qquad d = \frac{z_2}{\sin 37°} = \frac{1.70}{0.6} = 2.83 \text{ m}$$

153

5.8 Conservation of Energy: Nonconservative Systems

In nonconservative systems, it is usually possible to identify the individual forces that act on the system as conservative and nonconservative forces and to treat each class separately. For example, in a system where there are applied forces for which suitable potential energy functions can be found, and also frictional forces that convert mechanical kinetic energy into heat, we may identify the former as conservative forces, which can be related directly to their respective potential energies, and the latter as nonconservative forces for which no potential energy exists. In a system of this sort it is still possible to write the kinetic energy of every body in the system as one half the mass times the square of the velocity. Also, we may still express the mechanical potential energy of every body as the amount of work done by conservative forces in returning to the position of zero potential energy.

Consider now a process in which the state of motion of the system changes. Let U_{ki} and U_{pi} represent the initial kinetic and potential energies, respectively, defined as above and let U_{kf} and U_{pf} be their final values. If there were no nonconservative forces, the sums $U_{ki} + U_{pi}$ and $U_{kf} + U_{pf}$ would be equal, according to the energy conservation theorem for isolated conservative systems. In nonconservative systems, this is no longer true. It may be that some of the system's initial mechanical energy is dissipated by frictional forces and converted into heat. The work done in these processes is not recoverable as mechanical kinetic or potential energy, but appears instead as energy that is *nonmechanical*. The final value of mechanical energy is then less than the initial value by an amount W_f which represents this nonrecoverable work lost in frictional processes. This statement can be expressed in the form of an equation as

$$U_{kf} + U_{pf} + W_f = U_k + U_{pi} \qquad (5.8.1)$$

or

$$U_{mf} + W_f = U_{mi} \qquad (5.8.2)$$

where U_{mf} and U_{mi} are the final and initial values of the system's total mechanical energy, kinetic plus potential.

It may also happen that some of the system's initial mechanical energy is consumed in doing work in nonconservative processes, resulting in the appearance of energy in a guise other than that of heat. A weight-driven generator may, for example, convert a weight's potential energy into electrical energy, which may in turn charge a battery, resulting finally in the production of chemical energy. In any and all cases, however, it is possible to show that the work done by the system in these nonconservative processes inevitably results in the production of a precisely equivalent amount of nonmechanical energy. This energy may appear in the form of chemical, electrical, or other energy whose quantity can be defined in a simple and physically reasonable way and measured accurately be experiment.[6]

If we are willing to admit the existence of these other forms of energy and to define them properly, we may extend the energy conservation law to nonconservative as well as conservative systems. In isolated systems wherein nonconservative forces act, then, *mechanical energy is not conserved*, but may be converted into other forms of energy or, for that matter, created from other forms of energy. The mechanical work done by the system in nonconservative processes, W_{nc}, is, however, always equal to the difference between the initial and final mechanical energies and also equals the amount of nonmechanical energy U' that appears in the process. We may, therefore, write

$$U_{mi} = U_{mf} + W_{nc} \qquad (5.8.3)$$

and also

$$U' = W_{nc} \qquad (5.8.4)$$

This, in turn, enables us to write

$$U_{mi} = U_{mf} + U' \qquad (5.8.5)$$

which is a statement that the *total energy*—embracing both mechanical energy and energy of other forms as represented by U'—is conserved even in systems in which nonconservative forces are present.[7] When $W_{nc} = U' = 0$, Eq. (5.8.5) reduces to the energy conservation principle for conservative systems, as given by (5.6.7).

To summarize all this, we can state that for *conservative* systems isolated from their surroundings, mechanical energy is conserved, the sum of potential and kinetic energy being the same at all times. For such systems, the external work required to take the system from a given initial state to a given final state is simply equal to the difference in total mechanical energy between the two states.

---

[6] We shall explain exactly how this may be done as the need for it arises in subsequent chapters.

[7] It may happen that mechanical energy is produced rather than consumed by nonconservative processes. This happens, for example, when a man throws a ball or when an electric motor run from a battery does mechanical work. In the first instance, chemical energy is converted to mechanical energy, while in the second, chemical energy is converted to electrical energy which is in turn converted to mechanical energy. Under circumstances such as these, nonconservative processes do work *on* the system, and *increase* its mechanical energy. When this happens, W_{nc} and U' are *negative* in the above equations.

In isolated *nonconservative* systems, mechanical energy is not conserved, but may be converted into other forms of energy, or created from other forms of energy. But the mechanical work done by nonconservative forces is equal to the nonmechanical energy that is created and, therefore, as expressed by (5.8.4) and (5.8.5), total energy is conserved.

Just as no purely mathematical argument for the validity of Newton's laws exist, no theoretical justifications can be given for the conservation of total energy in isolated nonconservative systems. Its acceptance as a general law of nature rests solely on the fact that it is vindicated by all the experiments that have ever been performed to test it.

EXAMPLE 5.8.1

Suppose, in Example 5.7.3, that the frictionless surface upon which the mass m_1 slides (Fig. 5.23) is replaced by one for which the kinetic coefficient of friction is μ_k. How is the solution now obtained from the viewpoint of work and energy?

Due to the introduction of frictional forces, this system is no longer a conservative one. Mechanical energy will no longer be conserved, although the *total* energy of the system, which includes energy converted into nonmechanical forms by the nonconservative forces, will still remain constant. We must, therefore, begin with the energy conservation principle as stated by (5.8.1). Let us assume, as before, that the system starts with mass m_2 at height z_0 above the reference plane, with initial velocity v_0, and arrives at a final state in which m has descended to distance z above the reference plane, with final velocity v. Under these circumstances, the expressions for K_i, U_i, K_f, and U_f are the same as those used previously[8] in Example 5.7.3, as given by Eqs. (5.7.30) and (5.7.31). The work done by nonconservative forces is, in this case, simply the work done in frictional processes during the motion, which can be expressed as the force of friction, $\mu_k m_1 g$, times the distance $z_0 - z$ through which it moves between the initial and final states of the system. Therefore,

$$W_f = \mu_k m_1 g(z_0 - z) \qquad (5.8.6)$$

Substituting the initial and final kinetic energies from (5.7.30) and (5.7.31), along with W_f from (5.8.6) into the energy conservation equation (5.8.1), we find

$$\tfrac{1}{2}m_1 v_0^2 + \tfrac{1}{2}m_2 v_0^2 + m_2 g z_0 + m_1 g z_1$$
$$= \tfrac{1}{2}m_1 v^2 + \tfrac{1}{2}m_2 v^2 + m_2 g z + m_1 g z_1 - m_1 g \mu_k (z - z_0) \qquad (5.8.7)$$

[8] It will be recalled that the nonconservative forces have no role in determining the potential energy of either body and are to be entirely neglected for the purpose of assigning potential energies to both masses in the system.

Subtracting $m_1 g z_1$ from both sides of this equation, transposing all terms to the right-hand side and factoring, we obtain

$$\tfrac{1}{2}(m_1 + m_2)(v^2 - v_0^2) + g(m_2 - \mu_k m_1)(z - z_0) = 0 \qquad (5.8.8)$$

As before, the acceleration of the system must be constant, in which case Eq. (5.7.36) is applicable. Substituting the value of $z - z_0$ from there into the above expression and dividing through by $v^2 - v_0^2$, we find

$$m_1 + m_2 + (m_2 - \mu_k m_1)\frac{g}{a} = 0 \qquad (5.8.9)$$

from which

$$a = -\left(\frac{m_2 - \mu_k m_1}{m_1 + m_2}\right)g \qquad (5.8.10)$$

Note that this reduces to the previously obtained expression (5.7.31) when $\mu_k = 0$.

The tension may be obtained as before, applying the work–energy theorem to mass m_2. Accordingly, we may equate the work done by the resultant force $m_2 g - T$ to the gain in kinetic energy $\tfrac{1}{2}m_2 v^2 - \tfrac{1}{2}m_2 v_0^2$, giving

$$(m_2 g - T)(z_0 - z) = \tfrac{1}{2}m_2(v^2 - v_0^2)$$

Again, however, from (5.7.35), $v^2 - v_0^2 = -2a(z_0 - z)$. Substituting this into the above equation, we find

$$m_2 g - T = -m_2 a$$
$$T = m_2(g + a) \qquad (5.8.11)$$

But the value of a is given by (5.8.10); substituting it into (5.8.11) and simplifying, we finally obtain

$$T = \frac{m_1 m_2(1 + \mu_k)g}{m_1 + m_2} \qquad (5.8.12)$$

Again, this reduces to the expression previously obtained as (5.7.40), as it should, when $\mu_k = 0$. Equation (5.8.12) may also be obtained by applying the work–energy theorem to mass m_1 in precisely the same way, equating the work done by the net resultant force on that body to its gain in kinetic energy.

It is important to observe that the work–energy theorem can be applied equally well to conservative systems and to nonconservative systems. This point is illustrated very well by the calculation of tension in this example and in Example 5.6.5, which deals with a conservative system. It is only necessary in either case to equate the work done by the *net resultant force* to the change in kinetic energy taking place. This point was discussed, though not quite so explicitly, in the treatment of the work–energy theorem in section 5.4. It should be contrasted with the energy conservation theorem, which must be used in a *different* form in conservative and nonconservative systems.

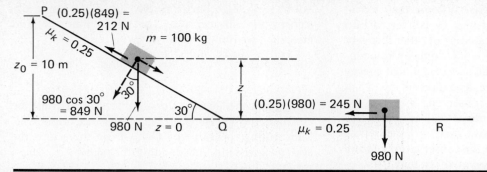

FIGURE 5.25

EXAMPLE 5.8.2

A packing case whose mass is 100 kilograms slides down a ramp whose vertical height is 10 meters and which is inclined at an angle of 30° to the horizontal onto a level floor, coming to rest at a point R some distance from the foot of the ramp, as shown in Fig. 5.25. The object starts from point P at the top of the ramp with an initial velocity $v_0 = 3$ meters/second directed along the ramp. The coefficient of kinetic friction between the case and the ramp and between the case and the level floor is 0.25. Find the velocity the packing case attains at the bottom of the ramp, the acceleration it has while descending the ramp, the distance it slides along the floor before coming to rest, and its acceleration while coming to rest along the level surface. Use work–energy methods.

Since frictional forces are present, the system is nonconservative. The energy conservation law must therefore be written in the form of Eq. (5.8.1). The kinetic energy of the object is $mv^2/2$, where v is the velocity. The potential energy is mgz, where z is the *vertical* height above the floor. The frictional work W_f is found by multiplying the frictional force by the distance d through which it moves. It is apparent from Fig. 5.25 that $d = PQ = 20$ meters. Regarding P as the initial point and Q the final point, and representing the final velocity at Q as v_Q, Eq. (5.8.1) gives

$$\tfrac{1}{2}mv_0^2 + mgz_0 = \tfrac{1}{2}mv_Q^2 + mgz_Q + \mu_k(mg\cos 30°)d$$
$$(\tfrac{1}{2})(100)(9) + (100)(9.8)(10)$$
$$= (\tfrac{1}{2})(100)v_Q^2 + 0 + (0.25)(849)(20)$$

Note that the final potential energy, at point Q, is zero. This equation can be simplified to give

$$450 + 9800 = 50v_Q^2 + 4240$$
$$v_Q = 10.96 \text{ m/sec}$$

The acceleration (which is constant, of course) may be obtained from

$$v^2 - v_0^2 = 2as \qquad (5.8.13)$$

where s is the distance measured along the ramp from the starting point P. As the packing case reaches the

bottom of the ramp, $v = v_Q = 10.96$ meters/second and $s = 20$ meters. Substituting these values into (5.8.13), along with the given value $v_0 = 3$ meters/second, and solving for a, it is easily determined that

$$a = \frac{v_Q^2 - v_0^2}{2s_{PQ}} = \frac{120.2 - 9.0}{40} = 2.78 \text{ m/sec}$$

As the body leaves point Q and slides along the level floor, work is done against friction, the initial kinetic energy being gradually used up until the object comes to rest at point R. Let us again use the conservation of energy equation (5.8.1), but now regarding the initial point as Q and the final point as R, and let us denote the distance QR as x. The initial and final potential energy is zero, the initial kinetic energy is $\tfrac{1}{2}mv_Q^2$, and the final kinetic energy is zero. The work done by the force of friction is the coefficient of friction times the normal force (which is now 980 newtons, as illustrated in Fig. 5.24) times the distance x. Substituting all this in the above equation, we may write

$$(\tfrac{1}{2})(100)(10.96)^2 + 0 = 0 + 0 + (0.25)(980)x$$

from which

$$245x = 6010$$
$$x = 24.5 \text{ m}$$

Again, the acceleration can be obtained from (5.8.13) regarding the initial velocity as v_Q rather than v_0 and interpreting the quantity s as the distance from the initial point Q measured along the level floor. When v becomes zero, s has attained the value x given directly above. In this way, we obtain

$$a = \frac{v^2 - v_Q^2}{2s} = \frac{0 - (10.96)^2}{(2)(24.5)} = -\frac{120.2}{49.0} = -2.45 \text{ m/sec}^2$$

as the acceleration along the level portion of the path. The total work done against friction is simply the sum of the initial kinetic and potential energy at point P $(450 + 9800 = 10,250 \text{ joules})$, since all of this energy is eventually used up in doing work against frictional forces as the case slides down the ramp and comes to rest on the floor.

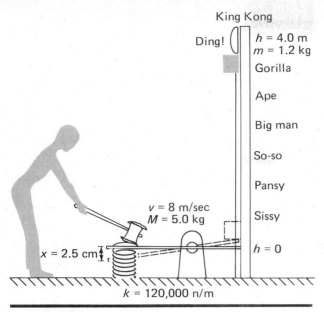

King Kong

Ding!

$h = 4.0$ m
$m = 1.2$ kg

Gorilla

Ape

Big man

So-so

Pansy

$v = 8$ m/sec
$M = 5.0$ kg

Sissy

$h = 0$

$x = 2.5$ cm

$k = 120,000$ n/m

FIGURE 5.26

EXAMPLE 5.8.3

A device familiar to those who while their time away at amusement parks instead of studying physics is shown in Fig. 5.26. The contestant must impart to a 5.0-kg hammer an impact velocity of 8.0 m/sec in order to ring the bell, claim the title of "King Kong," and be awarded a Kewpie doll. The anvil is restrained by an ideal spring whose spring constant is 120,000 N/m and is deflected downward a distance of 2.5 cm by a prize-winning blow. The weight, which is driven upward a distance of 4.0 m to ring the bell, has a mass of 1.2 kg. Assuming that the rising weight strikes the bell with zero velocity, what is its velocity as it leaves the anvil? What is its potential energy as it collides with the bell? How much potential energy is stored in the spring that restrains the anvil, just after the hammer blow? Are the forces that act during the hammer blow conservative? If not, how much mechanical energy is converted to heat and other nonmechanical forms of energy in the process?

The small weight that rings the bell starts from $h = 0$ where its potential energy is zero and rises to a final height of 4.0 m to ring the bell. Its final velocity and, therefore, its final kinetic energy are zero. The only force acting on it during its upward journey is the force of gravity, which is indisputably conservative. Therefore, in this part of the example, we can use the energy conservation law in the form

$$U_{ki} + U_{pi} = U_{kf} + U_{pf} \qquad (5.8.14)$$

identifying the initial state with the moment the weight rises from the anvil and the final state with the instant it arrives at the bell. Under these circumstances, (5.8.15) becomes

$$\tfrac{1}{2}mv^2 + mg(0) = \tfrac{1}{2}m(0)^2 + mgh$$
$$v^2 = 2gh = (2)(9.8)(4.0) = 78.4$$
$$v = 8.85 \text{ m/sec}$$

When the weight strikes the bell, its kinetic energy is zero, and its potential energy, which represents its total energy, will be

$$U_{pf} = mgh = (1.2)(9.8)(4.0) = 47.04 \text{ J}$$

The spring that restrains the anvil is deflected 2.5 cm = 0.025 m by the blow. Just after the blow, its stored potential energy must be

$$U_{pf}' = \tfrac{1}{2}kx^2 = (\tfrac{1}{2})(120,000)(0.025)^2 = 37.5 \text{ J}$$

As the hammer (of mass $M = 5.0$ kg) hits the anvil, its velocity v_i is 8.0 m/sec, and its kinetic energy U_{ki} will be

$$U_{ki} = \tfrac{1}{2}Mv_i^2 = (\tfrac{1}{2})(5.0)(8.0)^2 = 160 \text{ J}$$

Let us now identify an initial state with the instant the hammer arrives at the anvil and a final state in which the restraining spring is fully compressed and the weight has just arrived at the bell. Then, the initial mechanical energy is the total kinetic energy of the hammer, since the initial potential energy of spring, hammer, and rising weight are all zero. This initial mechanical energy amounts to 160 joules, according to the above calculation. In the final state, the hammer, spring, and weight are all at rest, so there is no final kinetic energy. Since the hammer is at height $h = -0.025$ m, its final potential energy is

$$U_{pf}'' = Mgh = (5.0)(9.8)(-0.025) = -1.22 \text{ J}$$

The final potential energy of the spring and the rising weight are, as calculated above, 37.5 joules and 47.04 joules, respectively. It is evident that the total final mechanical energy is $U_{pf} + U_{pf}' + U_{pf}'' = 83.32$ joules, which is a good deal less than the initial mechanical energy of 160 joules. Therefore, nonconservative forces must have been present during the collision of hammer and anvil. The amount of mechanical energy converted to energy of other types can be found by using the law of conservation of energy for nonconservative systems (5.8.3) and (5.8.4) from which

$$U_{mi} = U_{mf} + W_{nc}$$
$$160 = 83.32 + W_{nc}$$
$$W_{nc} = 160 - 83.32 = 76.68 \text{ J}$$

SUMMARY

The work dW done by a force $\mathbf{F}$ in an infinitesimal displacement $d\mathbf{r}$ is

$$dW = \mathbf{F} \cdot d\mathbf{r} = F \cos \theta \, dr$$

where θ is the angle between the force and displacement vectors. In a finite displacement, the work is obtained by integrating this expression to obtain

$$W_{12} = \int_1^2 \mathbf{F} \cdot d\mathbf{r} = \int_{x_1}^{x_2} F_x \, dx + \int_{y_1}^{y_2} F_y \, dy + \int_{z_1}^{z_2} F_z \, dz$$

If the force is constant and if the displacement takes place along a straight line, this can be written

$$W_{12} = \mathbf{F} \cdot \mathbf{r} = Fr \cos \theta$$

Power is the work done per unit time; it is defined by

$$P = dW/dt$$

or

$$P = \mathbf{F} \cdot \mathbf{v}$$

The kinetic energy of a body of mass m moving with speed v is

$$U_k = \tfrac{1}{2}mv^2$$

The work–energy theorem states that the work done by the *resultant* force acting on a system is equal to the change in the kinetic energy of the system. If the "system" is a body of mass m, this means that

$$W_{12} = \Delta U_k = U_{k2} - U_{k1} = \tfrac{1}{2}mv_2^2 - \tfrac{1}{2}mv_1^2$$

From the work–energy theorem, it is clear that the kinetic energy of a body is equal to the work the body can do on its surroundings in being brought to rest.

The term *energy* refers in general to the capacity of a system to do work.

The potential energy of a system is the energy a system may have by virtue of its position or displacement. Gravitational energy stored in an elevated weight or strain energy stored in a compressed or stretched spring are examples of mechanical potential energy. The total mechanical energy of a system is the sum of its kinetic and potential energies.

There are certain forces that do not store potential energy and from which work cannot be recovered by the release of stored potential energy. These forces are referred to as nonconservative forces. The most common nonconservative forces are frictional or dissipative forces. Nonconservative forces act to change mechanical energy into nonmechanical forms of energy. A conservative force is one that has the following property: work done against the force in altering the configuration of the system is fully recoverable as work done by the system as it returns to its initial state. The work that the system can do in returning to an initial state of zero potential energy defines the potential energy associated with a conservative force. There is no way of associating a potential energy with a nonconservative force. Systems in which only conservative forces act are called conservative systems.

The work done by a conservative force in moving from point A to point B along a given path is the negative of the work done by the force in retracing the path from B to A. The work done by a conservative force in moving from point A to point B is the same, irrespective of the path followed between these points. The work done by a conservative force in moving around a closed path is zero. The work done by a conservative force in moving from A to B is $U_p(B) - U_p(A)$, where $U_p(x, y, z)$ is the potential energy associated with the force. Any force that does not have all these properties is nonconservative.

The potential energy associated with a given conservative force $\mathbf{F} = \mathbf{i}_x F_x + \mathbf{i}_y F_y + \mathbf{i}_z F_z$ is given by

$$dU_p = -F_x \, dx - F_y \, dy - F_z \, dz$$

or

$$U_p(x, y, z) = -\int_0^x F_x \, dx - \int_0^y F_y \, dy - \int_0^z F_z \, dz$$

assuming that $U_p = 0$ at the origin. The force components F_x, F_y, and F_z are related to the potential energy function by

$$F_x = -\frac{\partial U_p}{\partial x} \qquad F_y = -\frac{\partial U_p}{\partial y} \qquad F_z = -\frac{\partial U_p}{\partial z}$$

Law of Conservation of Energy

Conservative Systems: The mechanical energy of a conservative system, expressed as the sum of its kinetic and potential energies, remains the same at all times. This can be written as

$$U_{mf} = U_{mi}$$

or

$$U_{kf} + U_{pf} = U_{ki} + U_{pi}$$

Nonconservative Systems: The final mechanical energy of the system differs from its initial mechanical energy by the amount of mechanical energy converted into nonmechanical energy (such as heat) by the action of nonconservative forces. The total energy, including nonmechanical as well as mechanical forms, remains the same at all times. This may be written as

$$U_{kf} + U_{pf} + W_{nc} = U_{ki} + U_{pi}$$

where W_{nc} represents the energy transformed into nonmechanical forms by the nonconservative forces.

QUESTIONS

1. In a tug of war, does the losing team do a positive or negative amount of work? How do you reconcile this result with the effort they have expended?
2. A pendulum consisting of a mass suspended by a light, flexible string is made to swing back and forth. Does the

weight force acting on the mass do work as the pendulum moves from the center of its swing to its outermost position? Does the tension in the string do work in this process? Explain your answers.

3. A boy is riding a bicycle at constant speed along a level road. Does the road do any work on the bicycle? If the boy stops pedaling, will the road do work and, if so, is it positive or negative?

4. A 105-pound diver does a double summersault off a 10-meter diving board. What is the total work done by gravity?

5. A constant force acting on an object does not result in a constant power output. Explain.

6. The power input to a light bulb is 150 watts. Express this in horsepower.

7. Invent a force law which can be used to describe the motion of a point object in two dimensions. Determine whether the work due to this force is independent of the path.

8. The power output of a certain device is known to increase linearly with time, starting from $t = 0$. Show that the work done increases quadratically with t.

9. The mass of any body is an intrinsic property of the body. Can you say that kinetic energy is also an intrinsic property?

10. A moving object has an initial kinetic energy U_{k0}. As a result of applied forces, the speed of the object doubles. How much work has been done by the resultant force on the body?

11. If the force on a body at a given point is known to be zero, does this necessarily imply that the potential energy is zero at that point?

12. A rubber ball is dropped from a platform onto the ground below. It bounces a number of times and finally comes to rest. Describe the various transformations of energy that have taken place.

13. A pendulum initially set in motion ultimately comes to rest. Explain this observation in terms of the work–energy theorem.

14. As a pendulum swings, its energy is transformed from kinetic energy to potential energy and vice versa. At what points is the energy all potential energy? When the kinetic energy is three quarters of its maximum value, what is the value of the potential energy?

15. When an object slides on a rough surface, friction does a negative amount of work. Explain in terms of the work–energy theorem.

16. A spring that is tightly compressed by means of a clamp is dissolved in acid. What becomes of the potential energy that was stored in the spring when it was originally compressed?

PROBLEMS

1. Show that (a) 1 ft-lb = 1.356 joules, (b) 1 kilowatt = 1.341 horsepower, (c) 1 kWh = 0.252 ton-mile = 3.600×10^7 joules. Note that the kilowatt-hour is a unit of work or energy rather than a unit of power.

2. Why is it necessary to introduce the constants c and k which are assigned unit value into the expressions for force, Eqs. (5.2.10) and (5.2.11), in Example 5.2.4?

3. In Eq. (5.7.25), express the left-hand member as a function of x alone, eliminating the variable θ.

4. A 180-pound man climbs a rope to a height of 25 ft in 15 seconds. (a) How much work is done? (b) Assuming he climbs at constant speed, what power is developed?

5. A 50-pound block is dragged at constant speed along a horizontal surface for a distance of 30 ft by a force **F** acting at an angle of 37° above the horizontal. The coefficient of sliding friction between block and surface is 0.20. Find (a) the magnitude of the force **F**, (b) the work done by **F**, (c) the work done by the force of friction, and (d) the work done by the weight force.

6. A man pushes a box of mass 40 kg up an inclined plane tilted at an angle of 15° with the horizontal with a force of 200 newtons, directed parallel to the plane. The box moves along the plane at constant speed a distance of 20 meters. The coefficient of kinetic friction between box and plane is 0.10. Find (a) the work done by the man on the box, (b) the work done by the force of friction, and (c) the work done by the weight force.

7. An ideal spring is compressed initially a distance of 0.05 meter from its equilibrium length. An additional force of 600 newtons is exerted upon it, increasing the distance by which it is compressed by 0.15 meter. Find (a) the spring constant, (b) the work done in compressing it initially a distance of 0.05 meter, (c) the additional work required to increase the compression by 0.15 meter.

8. A certain spring does not obey Hooke's law, but instead exerts a force given by $F_x = -kx - bx^3$, where k and b are constants and x is the distance through which it is stretched or compressed from its equilibrium length. Show that the work needed to stretch such a spring from an initial extension x_1 to a final extension x_2 is given by $W = (k/2)(x_2{}^2 - x_1{}^2) + (b/4)(x_2{}^4 - x_1{}^4)$.

9. The spring discussed in the preceding problem is found to exhibit an extension of 0.05 meter from its equilibrium length when a stretching force of 600 newtons is exerted and an extension of 0.200 meter when a force of 3600 newtons is applied. Find (a) the spring constants k and b, (b) the force exerted by the spring when it is extended a distance of 0.300 meter from its equilibrium length, and (c) the work required to increase the extension from $x_1 = 0.200$ meter to $x_2 = 0.300$ meter.

10. A uniform rope 100 ft long weighing 0.25 lb/ft is coiled on the ground. How much work must be done to lift one end of the rope vertically to a height of 100 ft so that the other end hangs just clear of the ground?

*11. Work Problem 10 for an arbitrary uniform rope weighing μ pounds per unit length, one end of which is lifted to a height h which is smaller than its total length.

12. How much work is required to push a crate weighing 250 pounds up a ramp 10 ft high which is inclined at an angle of 30° to the horizontal if the coefficient of friction between crate and ramp is 0.35? Assume that a horizontally directed external force is applied to the crate.

13. An ocean liner is powered by engines developing 180,000 hp. At full power, the speed of the ship is 32 knots. How much force must the propellers exert against the hull to propel the ship at this speed? (Recall that 1 nautical

159

mile = 1.152 statute miles = 6080 ft and that 1 knot is 1 nautical mile/hr.)

14. A jet engine develops 10,000 pounds of thrust as it propels an airplane at a speed of 600 mph. How much power is developed under these conditions?

15. In Problem 12, assuming that it takes 10.0 sec to push the crate up the ramp at constant speed, how much power is required in the process? How much power is consumed by frictional processes?

16. A swimmer can swim 50 yards in 23.0 sec, developing an average of 2.10 hp in the process. What is the average force he exerts against the water during his swim?

17. An object of mass 24.0 kg is accelerated up a frictionless plane inclined at an angle of 37° with the horizontal by a constant force. Starting at the bottom from rest, it covers a distance of 18 meters along the plane in 3.0 sec. (a) What is the average power required to accomplish this process? (b) What is the instantaneous power required at the end of the 3.0-sec interval?

18. A ship, formerly having a maximum speed of 18 knots with 12,000 hp, is undergoing reconstruction. It is desired to equip her with new engines capable of propelling her at a maximum speed of 24 knots. Assuming that the force of resistance exerted by the water on the ship's hull is directly proportional to her velocity, how much power will the new engines have to develop to give her the required 24-knot speed?

*19. A particle is acted upon by a force given (in dynes) by $\mathbf{F} = c(x^3\mathbf{i}_x + x^2y\mathbf{i}_y + 0\mathbf{i}_z)$, where x, y, and z are its position coordinates measured in centimeters from the origin along a particular set of axes and c is a constant having the numerical value 1.000. (a) What are the dimensions of the constant c? (b) How much work must be done on the particle to move it around the closed path OABCO in the xy-plane in the figure? (The equation of curve OCB is $y = \frac{1}{2}x^2$ and that of curve ODB is $y = 2x$.) (c) How much work must be done on the particle to move it around the closed path OABDO? (d) Is the force $\mathbf{F}$ conservative? Explain.

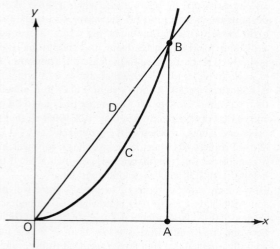

20. A particle of mass 10 kg starts from rest at the origin and moves along the x-direction under the influence of a net force whose magnitude in newtons is given by $F_x = 4/(1 + x)$, where x is the distance from the origin in

meters. Find (a) The velocity v_x when the mass has reached a point 12 meters from the origin, (b) the kinetic energy at this point, (c) the potential energy of the system at this point, (d) the amount of work done in bringing the mass from its starting point to this point, and (e) the instantaneous power required to sustain the motion described above at this time. Assume the potential energy to be zero at the origin.

21. A body of mass m experiences a time-varying force F_x which causes it to move along the x-axis in such a way that its position is given as a function of time by $x(t) = ae^{-\alpha t}$, where α is a positive constant. Find (a) the resultant force acting on the body when its distance from the origin has the value x; (b) the potential energy of the system as a function of the distance x from the origin, assuming that the potential energy is zero at the origin; and (c) the kinetic energy at any point x; (d) the kinetic energy at any time t.

*22. An object of mass m initially at rest at the origin is subjected to a variable net force given by $F_x = A/(1 + x)$, $F_y = F_z = 0$, where A is a constant. Find (a) the work done by the force in moving the body through a distance x, (b) the velocity of the body after it has traveled a distance x, (c) the kinetic energy of the body when it has moved through a distance x, (d) the potential energy of the body when it has moved through a distance x, assuming the initial potential energy to be zero, (e) the sum of the kinetic and potential energy at any point x, (f) the power required to sustain the motion at any point x, and (g) the definite integral which must be evaluated to find x as a function of time.

23. Suppose that, in addition to the force given in Problem 13, a frictional force $F_x = -\mu mg$ were also present. Using the work–energy theorem derive an equation for the distance x the object moves before coming to rest. Do not attempt to solve explicitly for x.

24. A particle of mass m initially at rest at the origin is subjected to a time-varying force given by $F_x = At^2$, $F_y = F_z = 0$, where A is a constant. Find (a) the velocity at time t, (b) the velocity at distance x, (c) the force at distance x, (d) the kinetic energy at time t, (e) the kinetic energy at distance x, (f) the potential energy of the body, assuming that the initial potential energy is zero, and (g) the total energy of the particle.

25. What would be the maximum speed a 150-horsepower automobile weighing 4000 pounds could attain up a road which rises uniformly 9 ft in every 100 ft along the horizontal, provided all frictional forces, including air resistance, are neglected?

26. In Problem 24, you are given the additional information that $m = 10$ kg and that at $t = 4$ sec, $x = 4$ m. Find (a) the value of x, v_x, F_x, a_x, and U_k when (a) $t = 8$ sec, (b) $x = 10$ m.

27. An automobile of mass m attains a constant speed v_0 along a level road when its engine develops a given power output P_0. (a) What speed will it attain when climbing a road inclined at an angle θ to the horizontal, with the same power output, assuming that all the forces of friction and air resistance it encounters are independent of velocity? (b) A 3200-pound car attains a constant speed of 110 ft/sec along a level road with a power output of

48 hp. What is its speed up a road inclined at an angle of 3° to the horizontal, with the same power output?

28. In the preceding problem, find **(a)** the kinetic energy of the car when it is traveling along a level road at a speed of 110 ft/sec and **(b)** the kinetic energy when climbing the 3° incline at a speed of 64.8 ft/sec. **(c), (d)** Through what vertical distances would the car have to fall to acquire the same kinetic energy calculated in parts **(a)** and **(b)**, respectively?

29. A body's kinetic energy is initially U_{k0}. As a result of forces acting on it, its velocity increases by an amount Δv. How much work, expressed in terms of U_{k0}, Δv, and the mass m, has been done by the resultant force on the object?

30. A body of mass m rests on a horizontal frictionless surface against a spring having spring constant k, which has been compressed through a distance x from its equilibrium length. The spring is suddenly released, setting the mass in motion along the surface. What is its velocity after leaving the spring, assuming that the mass parts company with the spring when the latter attains its equilibrium length?

31. An object of mass m is suspended from an overhead support by a light, flexible cord of length l. Initially, it hangs vertically, in equilibrium. It is then moved very slowly by a horizontally directed external force **F** until it is again in equilibrium, but now with the cord making an angle θ with the vertical. Find, in terms of the known quantities m, l, and θ, **(a)** the work done by the force of tension in the cord during the process, **(b)** the work done by the gravitational force acting on the mass, and **(c)** the work done by the external force **F**. *Hint:* Use the work–energy theorem!

32. In the preceding problem, the force **F** is suddenly removed, allowing the mass to swing back and forth like a pendulum. What is its velocity when it reaches the lowest point in its swing?

33. In the accompanying figure, a roller coaster car starts from rest at point A, 30 meters above ground level. Find **(a)** how fast it is going when it reaches point B, assuming that the track is frictionless; **(b)** the height h of the track at point C, assuming that its speed there is 20.0 meters/sec. **(c)** Finally, the car arrives at point D, where the brakes are applied. They grab, locking the wheels, which slide along the track, bringing the car to a stop at a point E 24 meters beyond D. What is the coefficient of kinetic friction between the wheels and the track?

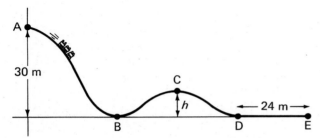

34. A rubber ball is dropped from rest onto a concrete surface from a height of 16 ft. On the first rebound, it reaches a maximum height of 12 ft. Find **(a)** the fraction of the ball's initial energy lost in the collision and **(b)** the height

attained by the ball on the second rebound, assuming that the same type of collision occurs in the second bounce.

35. A projectile of mass m is projected at an angle θ_0 with the horizontal, with initial velocity v_0. Show, using the law of conservation of energy, that the maximum height h to which it rises is given by $h = v_0^2 \sin^2 \theta_0 / 2g$. Neglect frictional effects due to air resistance.

36. A block of mass 25.0 kg is forced against an ideal horizontal spring, a horizontal force of 600 newtons acting to compress the spring through a distance of 0.2 meter. The horizontal force is removed and the spring expands, setting the block in motion. It slides through a total distance of 0.6 meter along the table top before coming to rest. What is the coefficient of sliding friction between block and table?

37. A thin, uniform rod of mass m and length l is pivoted at one end and is allowed to hang vertically in equilibrium. It is then displaced in such a way that it makes an angle θ with the vertical. By how much has its potential energy been increased?

38. A body of mass m is attached to a vertically suspended ideal spring of negligible mass whose spring constant is k. It is then lowered very slowly until it hangs freely suspended, in equilibrium, on the spring. Find **(a)** through what distance the spring has stretched during the process. **(b)** Suppose the mass had simply been released from rest on the unstretched spring. How far would it fall before reversing direction and starting upwards once again?

39. A 100-kg object is pulled through a distance of 6.0 meters along an inclined plane, which makes an angle of 30° with the horizontal, by a 700-newton force directed parallel to the plane. Its speed changes from 2.0 meters/sec to 4.0 meters/sec during this process. Find **(a)** the work done on the body by the 700-newton force, **(b)** the change in potential energy of the object, **(c)** the change in kinetic energy of the object, and **(d)** the work done by frictional forces.

40. In the preceding problem, what would the answers have been if the 700-newton force had been directed horizontally rather than parallel to the plane, assuming all the other given data were exactly the same?

41. An object of mass m is tied to an overhead support by a light, flexible string of length l. It is displaced until the string is stretched horizontally, as shown in the diagram, and released from rest. **(a)** Find the velocity attained by

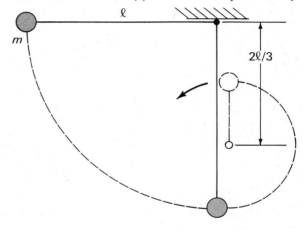

the mass when it is directly below the point of suspension at the bottom of its swing and **(b)** the tension in the string at this point. **(c)** The string is intercepted by a pin located at a distance $2l/3$ below the point of suspension, as illustrated in the figure. The object then moves in a circular path about the pin. What is its speed when it reaches a point directly above the pin? **(d)** What is the tension in the string at this point?

42. In the situation described in the preceding problem, how far below the overhead support does the pin have to be located to ensure that the string will remain taut at all points in the circular path described by the object after the string hits the pin?

43. A simple way of measuring the coefficient of kinetic friction between two surfaces is shown in the accompanying figure. A block of mass m slides on a level surface; the interface between the two is the friction interface to be studied. This block is accelerated through a distance h by the falling mass m'. After mass m' strikes the floor, m continues to move along the surface until stopped by the frictional force after having traveled an additional distance d. Using energy conservation principles, find **(a)** an expression for the coefficient of kinetic friction in terms of the measurable quantities m, m', h, and d, and **(b)** the coefficient of friction in the case where $m = 0.200$ kg, $m' = 20.0$ kg, $h = 0.200$ meter, and $d = 0.500$ meter.

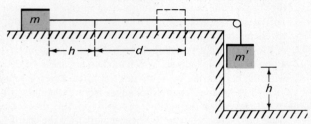

44. A 20-pound weight rests upon a spring at the bottom of a smooth, frictionless plane inclined at an angle of $37°$ to the horizontal, as shown in the figure. The spring is designed such that a force of 120 pounds is necessary to compress it a distance of 1 inch. The spring is compressed through a distance of 6 inches (the weight resting upon it during the process) and suddenly released, so that the weight is projected up the plane. Find **(a)** the distance, measured along the plane, through which the weight travels before it stops and begins to descend and **(b)** the velocity of the weight after it has traveled 10 ft along the plane.

45. Work Problem 44 assuming a kinetic friction coefficient $\mu_k = 0.2$ between the weight and the plane.

46. Work Problem 44 for an arbitrary weight w, an arbitrary angle θ, an arbitrary spring constant k, and initial spring compression ξ. Find **(a)** the maximum distance s_m through which the weight travels along the plane and **(b)** the velocity at any specified distance $d(<s_m)$ along the plane.

47. Work Problem 46 assuming a kinetic friction coefficient μ_k between the weight and the plane.

48. An automobile engine and transmission is capable of delivering maximum power P to the driving wheels. Assuming that this constant maximum power can be developed at any car speed and that the vehicle starts from rest at $t = 0$, and neglecting all frictional effects, find **(a)** the velocity of the vehicle (of mass m) as a function of time, **(b)** the distance covered as a function of time, and **(c)** the velocity as a function of the distance traveled. **(d)** For a 4000-pound automobile capable of delivering a maximum of 150 horsepowers to the drive wheels, what would be the minimum time required to attain a speed of 88 ft/sec (60 mph)? **(e)** Can this level of performance be attained in practice and, if not, why not?

49. In Problem 48, find **(a)** the acceleration of the car as a function of time and **(b)** the force required to maintain this acceleration. Do these results suggest additional reasons why the set of conditions assumed in the statement of Problem 46 cannot be precisely realized?

50. Two automobiles, each weighing 4000 pounds and going in opposite directions at a speed of 88 ft/sec (60 mph) meet in a head-on collision and are brought to a stop in 0.05 second. What is the average power associated with the collision?

51. **(a)** Using the work–energy theorem, find the shortest distance in which an automobile whose initial velocity is v can be stopped on a level surface where the coefficient of static friction between tires and roadway is μ_s. **(b)** What does this minimum stopping distance amount to if $v = 88$ ft/sec (60 mph) and $\mu_s = 1.0$? **(c)** Find the answer to part **(a)** assuming that a "reaction time" t_r elapses between the instant the driver is signaled to stop and the moment the brakes are applied. **(d)** What does the answer to part **(b)** become if the driver's reaction is 0.6 second?

52. A 4000-pound automobile accelerates from rest to a speed of 88 ft/sec (60 mph) in 12.0 seconds. What is the average power required to effect this acceleration?

53. An object of mass m is suspended from a pulley system

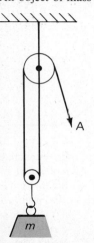

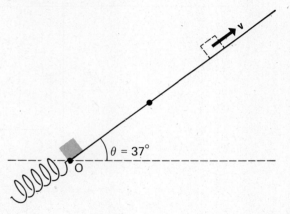

$\theta = 37°$

such as that shown in the diagram. From the energy conservation principle, determine how much force must be exerted at point A to keep the object in equilibrium, neglecting the mass of the pulleys and frictional losses in the system.

54. A body of mass m rests upon a frictionless plane inclined at an angle θ to the horizontal. Use the energy conservation principle to find the minimum force required to push the object up the plane at constant speed.

55. In Problem 54, suppose that the coefficient of kinetic friction between object and plane is μ_k. Using energy conservation, find (a) the minimum force required to push the object up the plane at constant speed and (b) the *efficiency* of the system, expressed as the ratio of the potential energy gained in a given displacement of the object to the total work done in effecting this displacement.

56. In the diagram, assume that mass M weighs 50 pounds, mass m weighs 10 pounds, h = 10 ft, and that the mass

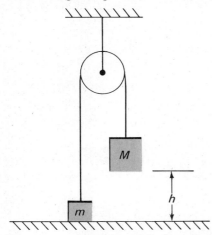

and frictional resistance associated with the pulley are negligible. If the system is initially at rest and mass m is released, find, using the energy conservation principle, the velocity with which mass M strikes the floor.

57. Work Problem 56 using arbitrary values for m, M, and h.

58. Work Problem 4.17 using energy conservation methods.

59. Work Problem 4.18 using the energy conservation principle.

60. Work Problem 4.25 using the energy conservation principle.

61. Work Problem 4.19 using the energy conservation principle.

LINEAR MOMENTUM AND THE CONSERVATION OF MOMENTUM

6.1 Introduction

In the preceding chapter, we found that the total mechanical energy of a system is or is not conserved, depending on the conservative or nonconservative character of the forces that act. While it is true that if the definition of energy is broadened to embrace nonmechanical forms, then total energy is always conserved, it is also pertinent to inquire whether there is any purely mechanical quantity which is conserved irrespective of the conservative or nonconservative nature of the system.

It turns out, as we shall soon see, that a property called the *linear momentum* may be defined for any system, which is conserved in all events, whether or not the system is conservative, provided that the sys-

tem is isolated in the sense that there are no external forces acting upon it. It may also be shown that if any external force is brought to bear on such a system, the linear momentum changes, the time rate of change of the linear momentum of the center of mass being precisely equal to the resultant external force. These characteristic properties of the linear momentum allow a restatement of Newton's second law in terms of momentum, enabling us to describe in very simple terms the internal dynamics of systems containing bodies that may collide with one another and allowing us to understand the dynamical behavior of bodies whose mass may vary with time. The analysis of collision processes from the point of view of momentum conservation has, in fact, provided much of what we know about the mutual interactions of atoms, molecules, and elementary particles. We shall, therefore, be amply rewarded by investigating in some detail how all this is accomplished.

6.2 Linear Momentum and Newton's Second Law

The linear momentum of a single point particle of mass m and velocity $\mathbf{v}$ is a *vector quantity* $\mathbf{p}$ defined by

$$\mathbf{p} = m\mathbf{v} \tag{6.2.1}$$

If m is constant (and for the time being, we shall restrict our discussions to situations where this is true), we may, upon differentiating both sides of (6.2.1) with respect to time and recalling Newton's second law, obtain

$$\frac{d\mathbf{p}}{dt} = m\frac{d\mathbf{v}}{dt} = \mathbf{F} \tag{6.2.2}$$

where $\mathbf{F}$ is the resultant force acting on the particle. If the resultant force on the particle is zero,

$$\frac{d\mathbf{p}}{dt} = 0 \tag{6.2.3}$$

and if the derivative of the momentum with respect to time is zero, then the momentum itself must be constant. We may, therefore, state (as a consequence of the definition of momentum and Newton's second law) that the momentum of a single particle acted upon by no external force is *conserved*. Also, from (6.2.2), it is clear that if a force acts on a particle, *the time rate of change of momentum $d\mathbf{p}/dt$ is equal to the resultant force.*

The momentum of a system of particles can readily be defined as the *vector sum* of the momenta of all the individual particles of the system. Thus, for a system of n particles having masses $m_1, m_2, m_3, \ldots, m_i, \ldots, m_n$, we may write for the total momentum

$$\mathbf{p} = \mathbf{p}_1 + \mathbf{p}_2 + \cdots + \mathbf{p}_i + \cdots + \mathbf{p}_n = \sum_{i=1}^{n} \mathbf{p}_i \tag{6.2.4}$$

or

$$\mathbf{p} = m_1\mathbf{v}_1 + m_2\mathbf{v}_2 + \cdots + m_i\mathbf{v}_i + \cdots + m_n\mathbf{v}_n$$

$$= \sum_{i=1}^{n} m_i\mathbf{v}_i \tag{6.2.5}$$

6.3 Conservation of Linear Momentum

The law of conservation of momentum for a single particle was observed in the preceding section to be a fairly obvious consequence of Newton's second law.

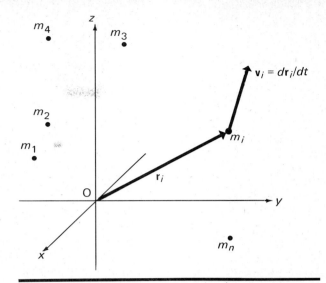

FIGURE 6.1. System of particles acted upon by external forces and by internal forces exerted by the particles on one another.

We shall now turn our attention to situations involving the momentum of systems of many particles or of extended rigid bodies.

Let us begin by considering a system of n separate mass particles having masses $m_1, m_2, m_3, \ldots, m_i, \ldots, m_n$. The positions of each particle may be described by a radius vector from the origin, and the velocity of each may be described by an appropriate velocity vector. The position vector $\mathbf{r}_i$ (components x_i, y_i, z_i) and the velocity vector $\mathbf{v}_i$, equal to $d\mathbf{r}_i/dt$ (components $dx_i/dt, dy_i/dt, dz_i/dt$) for the ith particle, which represents a typical particle of the system, are shown in Fig. 6.1. In general, the velocity vectors of each particle will be different; only in the case of a rigid body having no rotational motion will each particle's velocity vector be the same at all times. Suppose the various particles of the system are subjected to forces $\mathbf{F}_1, \mathbf{F}_2, \ldots, \mathbf{F}_i, \ldots, \mathbf{F}_n$. Then, according to Newton's second law, the masses will undergo accelerations $\mathbf{a}_1, \mathbf{a}_2, \ldots, \mathbf{a}_i, \ldots, \mathbf{a}_n$ related to the forces by

$$\mathbf{F}_1 = m_1\mathbf{a}_1 = m_1\frac{d\mathbf{v}_1}{dt}$$

$$\mathbf{F}_2 = m_2\mathbf{a}_2 = m_2\frac{d\mathbf{v}_2}{dt}$$

$$\vdots \quad \vdots \qquad \vdots$$

$$\mathbf{F}_i = m_i\mathbf{a}_i = m_i\frac{d\mathbf{v}_i}{dt} \tag{6.3.1}$$

$$\vdots \quad \vdots \qquad \vdots$$

$$\mathbf{F}_n = m_n\mathbf{a}_n = m_n\frac{d\mathbf{v}_n}{dt}$$

Let us now find the vector sum $\mathbf{F}$ of all these individual forces. From (6.3.1),

$$\mathbf{F} = \mathbf{F}_1 + \mathbf{F}_2 + \cdots + \mathbf{F}_i + \cdots + \mathbf{F}_n$$

$$= m_1 \frac{d\mathbf{v}_1}{dt} + m_2 \frac{d\mathbf{v}_2}{dt} + \cdots + m_i \frac{d\mathbf{v}_i}{dt} + \cdots + m_n \frac{d\mathbf{v}_n}{dt}$$

$$= \frac{d}{dt}(m_1 \mathbf{v}_1 + m_2 \mathbf{v}_2 + \cdots + m_i \mathbf{v}_i + \cdots + m_n \mathbf{v}_n) \quad (6.3.2)$$

But the expression in parentheses above is simply the total momentum of the system, and therefore, denoting this quantity by $\mathbf{p}$,

$$\mathbf{F} = \frac{d\mathbf{p}}{dt} \quad (6.3.3)$$

The resultant of the forces acting on the individual particles of the system is therefore equal to the time rate of change of the total momentum of the system.

Now, the forces $\mathbf{F}_i$ acting on a typical particle (the ith, for example) may be of two types: *external* forces which are exerted by agencies outside the system, and *internal* forces which are exerted by one particle of the system upon another, for instance, during collisions between particles. The latter forces always occur as pairs of forces which are equal in magnitude and opposite in direction, in accord with Newton's third law concerning action and reaction. This is illustrated in Fig. 6.2, where the forces arising from the collision of two particles m and n are considered. The force $\mathbf{F}_n$ exerted by particle m upon particle n is, according to Newton's third law, equal in magnitude and opposite in direction to the force $\mathbf{F}_m$ exerted by particle n upon particle m. Therefore,

$$\mathbf{F}_m + \mathbf{F}_n = \mathbf{F}_m + (-\mathbf{F}_m) = 0 \quad (6.3.4)$$

and the internal forces arising from the interaction of these two particles contribute exactly nothing to the vector sum of individual forces in (6.3.2). This same statement is true of all internal forces between particles belonging to the system, regardless of whether the

forces are contact forces arising from collision or longer-range forces such as gravitational attractions, because Newton's third law must hold in any event. The vector sum $\mathbf{F}$ of individual forces in (6.3.2), therefore, is identical with the resultant of the *external* forces acting on individual particles, inasmuch as the internal forces contribute nothing to the vector sum.

This result can be understood in even simpler terms by introducing the motion of the center of mass of the system. Let us briefly restate the calculations above using the summation notation wherein, for example, $\mathbf{F}_1 + \mathbf{F}_2 + \mathbf{F}_3 + \cdots + \mathbf{F}_i + \cdots + \mathbf{F}_n$ is written as

$$\sum_{i=1}^{n} \mathbf{F}_i$$

Equation (6.3.2) then becomes

$$\mathbf{F} = \sum_{i=1}^{n} \mathbf{F}_i = \sum_{i=1}^{n} m_i \frac{d\mathbf{v}_i}{dt} = \sum_{i=1}^{n} \frac{d}{dt}(m_i \mathbf{v}_i)$$

$$= \frac{d}{dt} \sum_{i=1}^{n} m_i \mathbf{v}_i = \frac{d\mathbf{p}}{dt} \quad (6.3.5)$$

Let us equate the x-component of the resultant external force $\mathbf{F}$ and the quantity $d(\sum m_i \mathbf{v}_i)/dt$, which, according to the above equation, must be the same. Then, since the x-component of $\mathbf{v}_i$ is dx_i/dt,

$$F_x = \frac{d}{dt} \sum_{i=1}^{n} m_i v_{ix} = \frac{d}{dt} \sum_{i=1}^{n} m_i \frac{dx_i}{dt} = \frac{d}{dt} \sum_{i=1}^{n} \frac{d}{dt}(m_i x_i)$$

or

$$F_x = \frac{d}{dt}\left(\frac{d}{dt} \sum_{i=1}^{n} m_i x_i\right) \quad (6.3.6)$$

In writing (6.3.6) in its final form, it is to be recalled that the derivative of a sum of individual terms is equal to the sum of the derivatives of the individual terms themselves. The center of mass, according to (4.4.5), is defined in such a way that if M is the total mass of all particles in the system and $\bar{x}$ the x-coordinate of the center of mass, then

$$M\bar{x} = m_1 x_1 + m_2 x_2 + \cdots + m_i x_i + \cdots + m_n x_n$$

$$= \sum_{i=1}^{n} m_i x_i \quad (6.3.7)$$

Accordingly, (6.3.6) becomes

$$F_x = \frac{d}{dt}\left(\frac{d}{dt} M\bar{x}\right) = \frac{d}{dt}\left(M \frac{d\bar{x}}{dt}\right) \quad (6.3.8)$$

But $d\bar{x}/dt$ is the time rate of change of the x-coordinate of the center of mass, in other words, the x-component of the velocity of the center of mass of the system. If we denote the vector velocity $\bar{\mathbf{v}}$ with which the center of mass moves by

$$\bar{\mathbf{v}} = \mathbf{i}_x \bar{v}_x + \mathbf{i}_y \bar{v}_y + \mathbf{i}_z \bar{v}_z \quad (6.3.9)$$

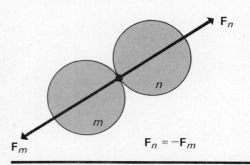

FIGURE 6.2. Physical character of forces exerted by particles on one another.

then (6.3.8) becomes

$$F_x = \frac{d(M\bar{v}_x)}{dt} \qquad (6.3.10)$$

Precisely analogous considerations apply to the motion of the y- and z-components of the center of mass; hence

$$F_y = \frac{d(M\bar{v}_y)}{dt} \qquad (6.3.11)$$

$$F_z = \frac{d(M\bar{v}_z)}{dt} \qquad (6.3.12)$$

If we regard the total mass of the system to be concentrated at the center of mass, then $M\bar{v}_x$, $M\bar{v}_y$, and $M\bar{v}_z$ represent the components of a vector $\bar{p}$ which we may refer to as *the momentum of the center of mass*. We find, therefore, that

$$\mathbf{F} = \frac{d\bar{\mathbf{p}}}{dt} \qquad (6.3.13)$$

where

$$\bar{\mathbf{p}} = M\bar{\mathbf{v}} \qquad (6.3.14)$$

We may summarize these results by stating that

the time rate of change of the momentum of the center of mass of any system of particles is equal to the resultant of all the external forces acting on the individual particles.

It follows also that

if the resultant of the external forces acting on individual particles of any system is zero, then the momentum of the center of mass of the system is conserved.

These statements describe the law of conservation of momentum for general systems of particles.

Since rigid bodies may be thought of as collections of mass particles (that is, individual atoms or molecules) whose locations relative to one another are firmly fixed in space, the above remarks relating to general particle systems refer also to rigid bodies. If the rigid body has no rotational motion (and we shall postpone the discussion of rotating rigid bodies to a later chapter), then the individual velocity vectors of all the particles in the body are the same and are equal to the velocity $\bar{v}$ of the center of mass.

One application of the principles just derived is illustrated in Fig. 6.3. In this diagram, the path of a shell fired from a gun into a medium in which there is no frictional resistance is represented. At point A the shell explodes, the fragments being ejected in all directions, and information regarding the subsequent path of the center of mass (c.m.) of the shell

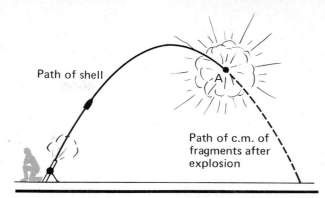

FIGURE 6.3. The motion of the center of mass of any system is affected only by the action of *external* forces. In this illustration, the internal forces acting on the shell fragments during the explosion have no effect on the motion of the center of mass (c.m.).

fragments is sought. The answer is obtained very simply in view of the results outlined above. The only *external* force which ever acts on the shell or the fragments resulting from the explosion is the constant downward force of gravity. The explosive forces that rupture the shell and scatter the fragments are purely *internal*. Therefore, the motion of the center of mass of the system is just that resulting from the external force of gravity alone, both before and after the explosion, and the motion of the center of mass of the fragments after the explosion takes place along the dotted extrapolation of the parabolic path of the shell up to point A. The velocity of the center of mass along this path, moreover, is just that which the shell would have attained at any given point had it not exploded. Further examples of momentum conservation and momentum changes resulting from applied forces are discussed in the examples which follow.

EXAMPLE 6.3.1

A rifle weighing 12 pounds imparts to a half-ounce ($\frac{1}{32}$ pound) bullet an initial muzzle velocity of 2400 ft/sec. What is the recoil velocity of this weapon when the bullet leaves the barrel, and why is it less dangerous to the user than to the victim?

Assuming that the rifle is at rest when fired and that its recoil is unchecked, the situation is as illustrated in Fig. 6.4. The total momentum of the system consisting of rifle and bullet before firing is zero. When the rifle is fired, no external forces act upon this system; it is true that the bullet is forced from the rifle barrel by the expansion of gases resulting from the explosion of the propellant, but these gases also exert an oppositely directed recoil force upon the rifle breech. These internal forces sum to zero just as do the collision forces illustrated in Fig. 6.2. Since no external forces are ever brought to bear on the system, its momentum after firing must be the same

FIGURE 6.4

as before. The sum of the momentum of the rifle plus that of the bullet after firing is, therefore, zero, which is to say:

initial momentum of system (rifle + bullet)

= final momentum of (rifle + bullet)

or

$$0 = m_r \mathbf{v}_r + m_b \mathbf{v}_b \qquad (6.3.15)$$

where the subscript r refers to the rifle and b to the bullet. Since the momentum vectors $\mathbf{v}_r$ and $\mathbf{v}_b$ are both parallel to the x-axis in Fig. 6.4, the x-component scalar equation corresponding to vector equation (6.3.15) is, therefore,

$$0 = m_r v_r + m_b v_b$$

whence

$$v_r = -\frac{m_b}{m_r} v_b \qquad (6.3.16)$$

Substituting the given values for v_b and m_b/m_r into this equation, it is easy to obtain the result

$$v_r = -6.25 \text{ ft/sec}$$

We observe that the recoil velocity of the rifle is much less than the muzzle velocity of the bullet. This is so because the momenta of rifle and bullet are equal and opposite while the mass of the rifle is much larger than that of the bullet, which requires that the opposite be true of the respective velocities. While this is part of the answer regarding the relative hazards in store for hunter and quarry, it is by no means the entire story. After all, 6.25 ft/sec is about 4.3 mph, and it is quite possible to sustain serious injury by being hit by a truck moving at that speed. In any collision, however, the severity of any resulting damage may be reasonably assessed in terms of the energy dissipated and the forces experienced by the objects involved. The kinetic energy U_{kb} imparted to the bullet is $\frac{1}{2} m_b v_b^2$, while the kinetic energy of recoil of the rifle is $\frac{1}{2} m_r v_r^2$. The ratio of these quantities is given by

$$\frac{U_{kr}}{U_{kb}} = \frac{m_r v_r^2}{m_b v_b^2} = \frac{m_b}{m_r} \qquad (6.3.17)$$

the final expression being obtained using (6.3.16) to express v_r in terms of v_b. It is evident from this that since m_b is ordinarily much less than m_r, the recoil kinetic energy imparted to the rifle is much less than

that given to the bullet. Substituting the numerical data given in the initial statement of the problem, we find, in fact, that in this case $U_{kr}/U_{kb} = 1/384$, the shoulder of the rifleman absorbing only 1/384 the energy imparted to the bullet. The kinetic energy of the bullet is in this case 2815 ft-lb, while the kinetic energy of recoil of the rifle is 7.33 ft-lb. Suppose the bullet were brought to rest in a distance of 1 ft. The constant force necessary to accomplish this would be 2815 pounds, according to the work–energy theorem, and this would be the force exerted by the bullet on its target. But the rifle could be brought to rest in 2 inches by the application of a constant force of $7.33/(1/6) = 44$ pounds, which would represent the "kick" of the rifle against the shoulder of the marksman if the weapon were stopped in that distance.

EXAMPLE 6.3.2
A 300-pound boat in which a 160-pound man and a 60-pound boy are standing is initially at rest in still water. The boy dives into the water from the stern of the boat, giving himself an initial horizontal velocity component of 8 ft/sec with respect to the water. The man then dives into the water over the stern of the boat, giving himself an initial horizontal velocity component of 8 ft/sec with respect to the boat's motion before his dive. Find (a) the velocity of boat and man after the boy dives and (b) the velocity of the boat after the man's dive. (c) Suppose, assuming the boat is initially at rest, that boy and man dive over the stern simultaneously with horizontal velocity 8 ft/sec with respect to the water. What is the final velocity of the boat in this case?

Let us designate the boy's mass by m, the man's mass by M, and the boat's mass by M_b. The initial momentum of the system is zero, and according to the law of conservation of momentum, the momentum of the boy plus the momentum of man and boat must be zero after the boy's body leaves the boat. Therefore, equating initial and final horizontal momenta for the boy's dive, we may write

$$0 = (M + M_b)v_b + m v_x \qquad (6.3.18)$$

where v_b is the boat's final velocity and v_x is the boy's final velocity. Since the boat is assumed to move off in the positive x-direction, and since the boy jumps over the stern in the negative x-direction, the magnitude of the quantity v_x, which represents the x-component of his velocity, is negative. Solving for v_b, it is easy to see that

$$v_b = -\frac{m}{M + M_b} v_x = -\frac{mg}{Mg + M_b g} v_x = -\frac{(60)(8)}{160 + 300}$$

$$= -1.044 \text{ ft/sec}$$

When the man dives, the final velocity of the boat (call it v_b') may be ascertained by once more equating initial and final x-components of momentum. It must be noted, however, that the initial velocity of the man with respect to the water (denoted by v_x') now has the magnitude $-8 + 1.044 = -6.956$ ft/sec. Equating initial and final x-momenta, we obtain

$$(M + M_b)v_b = M_b v_b' + M v_x' \qquad (6.3.19)$$

$$v_b' = \frac{(M + M_b)v_b - M v_x'}{M_b} = \left(1 + \frac{M}{M_b}\right)v_b - \frac{M}{M_b}v_x'$$

$$= (1.535)(1.044) - (0.535)(-6.956) = 5.32 \text{ ft/sec}$$

When the man and boy dive at the same time, momentum is again conserved; and since the initial momentum is now zero,

$$0 = M_b v_b'' + m v_x + M v_x \qquad (6.3.20)$$

where v_b'' refers to the final velocity of the boat and where v_x again denotes the common velocity of the divers with respect to the water. Solving for v_b'',

$$v_b'' = -\frac{m + M}{M_b}v_x = -\frac{mg + Mg}{M_b g}v_x$$

$$= -\frac{60 + 160}{300}(-8) = 5.87 \text{ ft/sec}$$

EXAMPLE 6.3.3

A half-ounce ($\frac{1}{32}$ pound) bullet is fired into a 5 pound block of wood suspended as a pendulum, as shown in Fig. 6.5. The block rises to a maximum height h of 2.5 ft above the equilibrium position. What is the initial velocity of the bullet? Is mechanical energy conserved in the collision of bullet and block?

When the bullet collides with the block of wood, the momentum of the system must be conserved, and if a velocity v_f is imparted to the block in the process, then momentum conservation requires that

$$mv_0 = (m + M)v_f \qquad (6.3.21)$$

where m and M are the respective masses of bullet and block and where v_0 is the initial velocity of the bullet. The kinetic energy of the system (consisting of block and imbedded bullet) moving just after im-

pact with speed v_f is gradually converted to potential energy by the pendulum suspension; and when the maximum height h is attained, it has all been converted to potential energy. According to the energy conservation principle, then, neglecting the mass of the pendulum suspension,

$$\tfrac{1}{2}(m + M)v_f{}^2 = (m + M)gh \qquad (6.3.22)$$

whence

$$v_f = \sqrt{2gh} \qquad (6.3.23)$$

Inserting this value for v_f into (6.3.21) and solving for v_0, it is easy to see that

$$v_0 = \left(1 + \frac{M}{m}\right)\sqrt{2gh} \qquad (6.3.24)$$

In this case, $M/m = 160$ and $h = 2.5$ ft. Correspondingly, from (6.3.24), we may obtain $v_0 = 2035$ ft/sec.

The initial kinetic energy of the bullet is $\frac{1}{2}mv_0{}^2$, or 2030 ft-lb. The kinetic energy of the system just after impact is $\frac{1}{2}(m + M)v_f{}^2$, which in view of (6.3.22) is equal to $(m + M)gh$, or 12.6 ft-lb. The difference is converted into heat during impact by frictional processes within the block. The velocity v_f just after impact is, according to (6.3.23), 12.62 ft/sec.

EXAMPLE 6.3.4

A bomb which is at rest at the outset explodes into three fragments, as shown in Fig. 6.6. The first fragment, with a mass of 2 kg, moves initially along the x-axis with a velocity of 300 m/sec, while the second, which has a mass of 3 kg, moves initially in the y-direction with a velocity of 100 m/sec. The third fragment has a mass of 5 kg. Find the magnitude and direction of the initial velocity of this third fragment. Describe the motion of the center of mass

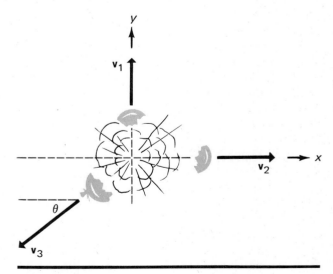

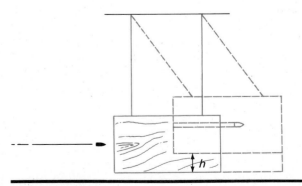

FIGURE 6.5

FIGURE 6.6

of the bomb before, during, and after the explosion, and find the amount of mechanical energy generated by the explosion. Where does this energy come from?

Since the bomb was initially at rest, and since it is acted upon by no external forces during the explosion, its center of mass remains at rest, and its final momentum is zero. Equating initial and final momenta, then, we may write

$$0 = m_1 \mathbf{v}_1 + m_2 \mathbf{v}_2 + m_3 \mathbf{v}_3 \qquad (6.3.25)$$

where m_1, m_2, and m_3 represent the masses of the three fragments and $\mathbf{v}_1$, $\mathbf{v}_2$, and $\mathbf{v}_3$, their final velocities. This can equally well be written as a set of three equations for the x-, y-, and z-momentum components:

$$m_1 v_{1x} + m_2 v_{2x} + m_3 v_{3x} = 0$$
$$m_1 v_{1y} + m_2 v_{2y} + m_3 v_{3y} = 0 \qquad (6.3.26)$$
$$m_1 v_{1z} + m_2 v_{2z} + m_3 v_{3z} = 0$$

In the last of these equations, since the velocity vectors $\mathbf{v}_1$ and $\mathbf{v}_2$ of fragments 1 and 2 lie in the xy-plane, $v_{1z} = v_{2z} = 0$; hence, if this equation is to be satisfied, v_{3z} must also be zero. The motion of the third fragment is likewise confined to the xy-plane.

Substituting the given values for m_1, m_2, v_{1x}, v_{1y}, v_{2x}, and v_{2y} into the first two equations, we obtain

$$(2)(300) + (3)(0) + (5)v_{3x} = 0$$
$$(2)(0) + (3)(100) + (5)v_{3y} = 0$$

which give

$$v_{3x} = -120 \text{ m/sec}$$
$$v_{3y} = -60 \text{ m/sec}$$

These are the velocity components of the third fragment. The magnitude of its velocity is given by

$$v_3^2 = v_{3x}^2 + v_{3y}^2 + v_{3z}^2 = 18000 \text{ m}^2/\text{sec}^2$$
$$v_3 = 134 \text{ m/sec}$$

while the angle θ between $\mathbf{v}_3$ and the $-x$-direction is described by

$$\tan \theta = \frac{v_{3y}}{v_{3x}} = \frac{-60}{-120} = \frac{1}{2}$$

from which

$$\theta = 26.6°$$

The center of mass of the bomb and of the fragments into which it bursts is initially at rest and *remains* at rest during and after the explosion. This statement should perhaps be qualified somewhat by remarking that it *neglects* the mass of the explosive charge itself and the terms in the momentum balance equation arising from the mass and momentum of the explosive charge and the gaseous products into which it is transformed. A more accurate description of the situation would be that the center of mass of the bomb casing plus the explosive charge and of the fragments and gaseous products which subsequently arise from the explosion remains at rest. In our calculations, the contributions from the explosive charge and the combustion products have been neglected.

The mechanical energy generated by the explosion is

$$U_m = \tfrac{1}{2}m_1 v_1^2 + \tfrac{1}{2}m_2 v_2^2 + \tfrac{1}{2}m_3 v_3^2$$
$$= \tfrac{1}{2}(2)(300)^2 + \tfrac{1}{2}(3)(100)^2 + \tfrac{1}{2}(5)(134)^2 = 150{,}000 \text{ J}$$

Again, any mechanical energy possessed by the gaseous products arising from the explosion is neglected in this calculation. The mechanical energy arises from the conversion of stored chemical energy in the explosive charge into other forms during the explosion. This mechanical energy is by no means the only energy which is produced from stored chemical energy, since heat and light energy also accompany the explosion.

6.4 Collisions

We may define a *collision* as a situation in which the bodies involved are initially (and finally, as well) free bodies acted upon by no resultant forces and in which they undergo a mutual interaction where large forces are exerted by the bodies upon one another during a very short time interval. Usually, only two bodies are involved in a collision, although strictly speaking this need not be so. Again, the mechanical energy of the bodies *may* be conserved during the collision, but frequently it is not, as anyone knows who has ever observed an automobile accident or an airplane crash. In the latter instance, some of the initial mechanical energy will be converted to heat, light, or sound energy.

Usually, one has little or no interest in knowing the exact details of what the forces one body exerts on another during a collision are or how they vary with time during the short interaction period. Indeed, to obtain information such as this, it is necessary to have precise knowledge of the geometric form of the bodies involved and the elastic behavior of the materials of which they are made, which is often not easily available. It is fortunately more often of interest and utility to be able to ascertain the final state of motion of the system from a knowledge of the initial dynamical state and a few simple facts about the degree to which energy is conserved in the interaction. The law of conservation of momentum enables us to solve problems of this type and leads us to conclude that the results obtained hold irrespective of the precise nature of the interaction or of the details about just how big the forces are, in which directions they act, and how they vary with time during the period of interaction!

The interactions of molecules in a gas, the interaction of elementary particles with one another and with atomic nuclei, the interaction of free electrons with the atoms of a crystal lattice or with impurity atoms in a metal or semiconductor, and even the interaction of light with electrons and other elementary particles are all examples of situations which may be regarded as collisions and whose basic mechanics may best be analyzed starting with the law of conservation of momentum. Indeed, momentum conservation in collision-type interactions is the basic fact of life in kinetic theory, nuclear physics, particle physics, and solid-state physics and has provided us with much of the knowledge of particle interactions that we now have in all those areas. Collision analysis from the viewpoint of momentum conservation is therefore one of the most important aspects of elementary mechanics.

In any two-body collision, the total initial and final momentum must be the same, which means that

$$m_1\mathbf{u}_1 + m_2\mathbf{u}_2 = m_1\mathbf{v}_1 + m_2\mathbf{v}_2 \qquad (6.4.1)$$

where $\mathbf{u}_1$ and $\mathbf{u}_2$ are the initial velocities of the two bodies and $\mathbf{v}_1$ and $\mathbf{v}_2$ are their final velocities after the collision. Suppose, for example, that the initial velocities $\mathbf{u}_1$ and $\mathbf{u}_2$ are known, as well as one of the two final velocities $\mathbf{v}_1$ or $\mathbf{v}_2$. Then, obviously, it is possible to solve for the missing final velocity. It is equally clear, however, that if only the initial velocities are specified, there is not enough information known to determine both final velocities, and that some other data must be produced in order to solve for both these quantities. These additional data often take the form of a statement of how much kinetic energy is lost during the collision or, perhaps, a statement that the bodies stick together after impact, if that happens to be true.

6.5 Collisions in One Dimension: Elastic and Inelastic Collisions

Let us now consider one-dimensional collisions involving two bodies moving along the same straight line. We shall assume that no external forces act on these bodies, the only forces being those which the objects exert on each other during the collision. The situation is that which arises when two billiard balls undergo a "head-on" collision, where the centers of mass are moving along the same straight line, as illustrated in Fig. 6.7a. The state of affairs shown in Fig. 6.7b is not representative of a one-dimensional collision, even though the initial velocities have only x-components. In this case, y-components of velocity will arise during the collision and the problem will be essentially two-dimensional, though the y-components of momentum of the two bodies will be equal and opposite after the collision.

Suppose we call the initial velocity components along the x-direction u_1 and u_2 and the final velocities, after the collision has taken place, v_1 and v_2. Often, the masses and the initial velocities will be known, and the final velocities v_1 and v_2 will be desired; but, obviously, in some cases other sets of known and unknown quantities may be specified. If both v_1 and v_2 must be found, however, it is evident that some consideration other than momentum conservation must enter the picture, since momentum conservation affords only one equation relating these quantities. It is certain that momentum will be conserved during collision, but *not* that kinetic energy will be conserved, since some kinetic energy may be dissipated as heat if the forces which act during collision are nonconservative. For simplicity, however, we may begin by assuming that the collision is one in which kinetic energy as well as momentum is conserved; such a collision is said to be a *perfectly elastic collision*. Under these circumstances, conservation of momentum requires that

$$m_1u_1 + m_2u_2 = m_1v_1 + m_2v_2 \qquad (6.5.1)$$

while, if kinetic energy is conserved,

$$m_1u_1^2 + m_2u_2^2 = m_1v_1^2 + m_2v_2^2 \qquad (6.5.2)$$

These equations may be rearranged by transposing all quantities containing m_1 to one side and those containing m_2 to the other side. Then

$$m_1(u_1 - v_1) = m_2(v_2 - u_2) \qquad (6.5.3)$$

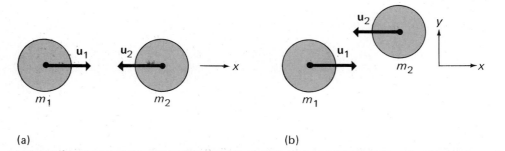

(a)   (b)

FIGURE 6.7.   (a) One-dimensional "head-on" collision. (b) Oblique or glancing collision, which is essentially two dimensional.

171

and

$$m_1(u_1{}^2 - v_1{}^2) = m_2(v_2{}^2 - u_2{}^2) \qquad (6.5.4)$$

Equation (6.5.4), however, may be written as

$$m_1(u_1 - v_1)(u_1 + v_1) = m_2(v_2 - u_2)(v_2 + u_2) \qquad (6.5.5)$$

If we divide this equation by Eq. (6.5.3), we obtain

$$u_1 + v_1 = v_2 + u_2$$

or

$$v_2 - v_1 = -(u_2 - u_1) \qquad (6.5.6)$$

In this expression, $u_2 - u_1$ represents the velocity of body 2 relative to body 1 before the collision, and $v_2 - v_1$ represents the corresponding relative velocity afterwards. A perfectly elastic collision preserves the magnitude of this relative velocity but reverses its sign.

Now let us proceed from the opposite extreme and consider what happens in a collision in which a maximum amount of kinetic energy is dissipated. Such a collision may occur, for example, between two balls of putty or some other deformable substance, and in this type of collision the two objects stick together and move along after collision with the same common velocity. This type of collision may be referred to as a *completely inelastic collision*. An automobile accident is usually a collision of this type, because the vehicles involved usually stick together and skid along the highway with the same speed after the impact. In any such collision, then, v_2 and v_1 are equal, and the equation corresponding to (6.5.6) would read

$$v_2 - v_1 = 0 \qquad (6.5.7)$$

In general, collisions are neither perfectly elastic nor completely inelastic but may be somewhere between these extremes. The bodies involved do not generally stick together after collision, but, on the other hand, there is usually some kinetic energy lost during the process. It is often possible, however, to characterize the energy dissipation of a collision by specifying a single number e, called the *restitution coefficient*, which represents the *ratio of relative velocities after and before collision*. Stated in mathematical terms, the definition of the restitution coefficient is

$$e = \frac{v_2 - v_1}{u_1 - u_2} \qquad (6.5.8)$$

From (6.5.6) it is evident that for a perfectly elastic collision, $e = 1$; and from (6.5.7), for a completely inelastic collision, $e = 0$. For collisions between these extremes, e will have a value between zero and unity. The restitution coefficient can never exceed unity; a value of e greater than 1 would require the total kinetic energy after collision to be greater than before, which is clearly impossible.

The value of the restitution coefficient depends upon the materials of which the colliding bodies are made, and, like the coefficients of friction, is not readily calculated but must be determined by experiment for each pair of colliding substances. A simple way of determining the restitution coefficient will be discussed in an example further on in this chapter. Once determined, however, it is found to be quite independent of initial and final velocities over a wide range of values and is therefore quite useful in the analysis of collision dynamics. In our treatment, we shall regard the masses m_1 and m_2, the initial velocities u_1 and u_2, and the restitution coefficient to be given, and we shall solve for the final velocities using the law of conservation of momentum.

Accordingly, we must begin with

$$m_1 u_1 + m_2 u_2 = m_1 v_1 + m_2 v_2$$

and

$$v_2 - v_1 = e(u_1 - u_2)$$

and solve this system of equations for v_1 and v_2 in terms of u_1, u_2, the masses and the restitution coefficient. The result is

$$v_1 = \frac{m_1 - em_2}{m_1 + m_2} u_1 + \frac{(1 + e)m_2}{m_1 + m_2} u_2 \qquad (6.5.9)$$

and

$$v_2 = \frac{(1 + e)m_1}{m_1 + m_2} u_1 + \frac{m_2 - em_1}{m_1 + m_2} u_2 \qquad (6.5.10)$$

For a completely inelastic collision, $e = 0$, and these results reduce to

$$v_1 = v_2 = \frac{m_1}{m_1 + m_2} u_1 + \frac{m_2}{m_1 + m_2} u_2 \qquad (6.5.11)$$

In the case of a perfectly elastic collision, $e = 1$, in which case

$$v_1 = \frac{m_1 - m_2}{m_1 + m_2} u_1 + \frac{2m_2}{m_1 + m_2} u_2 \qquad (6.5.12)$$

$$v_2 = \frac{2m_1}{m_1 + m_2} u_1 + \frac{m_2 - m_1}{m_1 + m_2} u_2 \qquad (6.5.13)$$

In order to obtain an intuitive picture of what happens in some particularly simple cases, let us investigate the special situation where the collision is perfectly elastic and the mass m_2 is initially *at rest*, whereby u_2 is zero. Then, (6.5.12) and (6.5.13) become

$$v_1 = \frac{m_1 - m_2}{m_1 + m_2} u_1 = \frac{(m_1/m_2) - 1}{(m_1/m_2) + 1} u_1 \qquad (6.5.14)$$

$$v_2 = \frac{2m_1}{m_1 + m_2} u_1 = \frac{2m_1/m_2}{(m_1/m_2) + 1} u_1 \qquad (6.5.15)$$

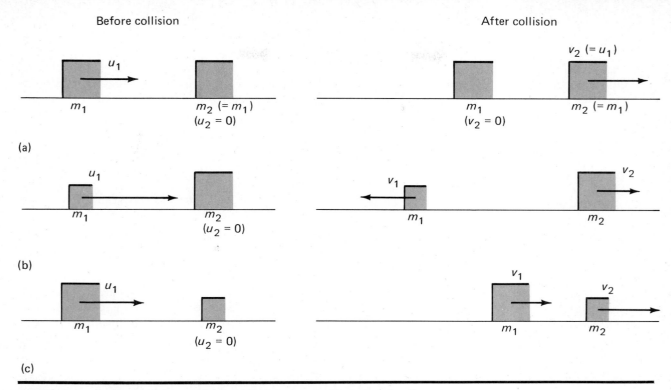

FIGURE 6.8. Some examples of elastic collisions in one dimension. In all cases, mass m_2 is initially at rest. In (a), m_1 and m_2 are equal; in (b), m_1 is less than m_2; in (c), m_1 is greater than m_2.

FIGURE 6.9. Linear air track, showing two "riders" that are supported by a cushion of air streaming through small holes in the track. Photo by Professor Howard Grotch, Pennsylvania State University.

In the case where the masses are equal, these further reduce to

$$v_1 = 0 \quad \text{and} \quad v_2 = u_1$$

which means that the mass m_1 which hits the initially stationary mass m_2 is itself brought to rest, while mass m_2 flies off with velocity equal to the inital velocity of m_1 in the same direction.

This result, illustrated in Fig. 6.8a, is easily demonstrated by an apparatus shown in Fig. 6.9 known as a *linear air track*, on which riders are supported by a thin cushion of air emitted from holes in the track. In this way, frictional effects are practically eliminated, and collisions, either elastic or inelastic, between riders may be enacted and quantitatively analyzed to provide an accurate experimental verification of the law of momentum conservation as well as of other laws of dynamics. If the mass ratio m_1/m_2 is less than unity (that is, m_1 is less than m_2), according to (6.5.14) and (6.5.15), after the collision v_1 is *negative* while v_2 is positive. This means that mass m_1 reverses direction while m_2 moves in the direction in which m_1 was traveling originally. If the ratio m_1/m_2 is greater than unity (m_1 greater than m_2), Eqs. (6.5.14) and (6.5.15) predict that after collision both v_1 and v_2 will be positive and that v_2 will be greater than v_1. These results, illustrated in Figs. 6.8b and 6.8c, are easily

173

verified with the linear air track or any other equivalent device.

6.6 One-Dimensional Collisions in the Center-of-Mass System

Some unique physical insights into collision dynamics may be obtained by viewing what happens in a coordinate system fixed with respect to the center of mass of the colliding bodies. Since no external forces act during the collision, the motion of the center of mass is the same before and after the collision, according to the results developed in section 6.3. This is easily seen to be true also from the results of the preceding section. If the position of mass m_1 at some time t before the collision is x_1 and that of mass m_2 at the same time is x_2, then, according to the definition of the center of mass,

$$(m_1 + m_2)\bar{x} = m_1 x_1 + m_2 x_2 \tag{6.6.1}$$

where $\bar{x}$ is the x-coordinate of the center of mass of the two objects. If we differentiate both sides of this equation with respect to time, we find

$$(m_1 + m_2)\frac{d\bar{x}}{dt} = m_1 \frac{dx_1}{dt} + m_2 \frac{dx_2}{dt} \tag{6.6.2}$$

or

$$(m_1 + m_2)\bar{u} = m_1 u_1 + m_2 u_2 \tag{6.6.3}$$

where $\bar{u}$ is the velocity of the center of mass before collision. It is clear from this that

$$\boxed{\bar{u} = \frac{m_1 u_1 + m_2 u_2}{m_1 + m_2}} \tag{6.6.4}$$

After collision, the velocities of the two bodies become v_1 and v_2, given in terms of the initial velocities u_1 and u_2 by (6.5.9) and (6.5.10). The velocity of the center of mass after collision will be, by the same line of reasoning that led to (6.6.4),

$$\boxed{\bar{v} = \frac{m_1 v_1 + m_2 v_2}{m_1 + m_2}} \tag{6.6.5}$$

But if the values of v_1 and v_2 as expressed in terms of u_1 and u_2 by (6.5.9) and (6.5.10) are substituted into this, one may obtain

$$\boxed{\bar{v} = \frac{m_1 v_1 + m_2 v_2}{m_1 + m_2} = \frac{m_1 u_1 + m_2 u_2}{m_1 + m_2} = \bar{u}} \tag{6.6.6}$$

the velocity of the center of mass of the system thus being *precisely the same before and after collision*.

Let us now try to work out the dynamics of collision in a coordinate system in which the center of mass is at rest. In this system, which moves with constant velocity $\bar{u}$ relative to the one that is fixed with respect to the earth, we shall designate the positions and velocities as primed quantities. Accordingly, referring to Fig. 6.10 which illustrates the relations between the two systems, if the two origins coincide at time $t = 0$, then

$$x' = x - \bar{u}t \tag{6.6.7}$$

where x is the position coordinate of point P referred to the fixed system and x' is the position coordinate of the same point referred to the system that moves with the center of mass of the two bodies. Then, differentiating (6.6.7) with respect to time,

$$\frac{dx'}{dt} = \frac{dx}{dt} - \bar{u} \tag{6.6.8}$$

Before the collision, we denote the velocities as (u_1, u_2) or (u_1', u_2') according to the system they are referred to; after the collision, we may write them as (v_1, v_2) or (v_1', v_2'). From (6.6.8),

$$u_1' = u_1 - \bar{u} \qquad v_1' = v_1 - \bar{u}$$
$$\text{and}$$
$$u_2' = u_2 - \bar{u} \qquad v_2' = v_2 - \bar{u} \tag{6.6.9}$$

where $\bar{u}$ is given by (6.6.6).

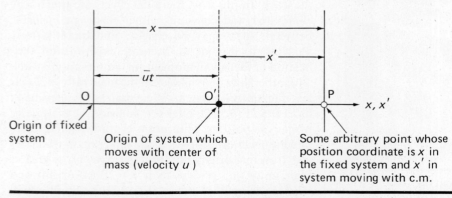

O — Origin of fixed system

O' — Origin of system which moves with center of mass (velocity u)

P — Some arbitrary point whose position coordinate is x in the fixed system and x' in system moving with c.m.

FIGURE 6.10. Coordinate system (origin O) fixed relative to the earth and a system (origin O'), moving with velocity $\bar{u}$, fixed relative to the center of mass (c.m.) of the system.

The law of conservation of momentum, which must hold in *either* system, allows us to write

$$m_1 u_1' + m_2 u_2' = m_1 v_1' + m_2 v_2' \qquad (6.6.10)$$

but, in addition, from (6.6.9) and (6.6.6),

$$u_1' = u_1 - \bar{u} = u_1 - \frac{m_1 u_1 + m_2 u_2}{m_1 + m_2} = \frac{m_2(u_1 - u_2)}{m_1 + m_2}$$

$$(6.6.11)$$

$$u_2' = u_2 - \bar{u} = u_2 - \frac{m_1 u_1 + m_2 u_2}{m_1 + m_2} = -\frac{m_1(u_1 - u_2)}{m_1 + m_2}$$

$$(6.6.12)$$

from which

$$m_1 u_1' + m_2 u_2' = 0 \qquad (6.6.13)$$

Equation (6.6.10) now becomes

$$\boxed{m_1 u_1' + m_2 u_2' = m_1 v_1' + m_2 v_2' = 0} \qquad (6.6.14)$$

This is to be expected and, indeed, need hardly have been calculated, since the momentum of the center of mass in the system at rest with respect to the center of mass certainly should be zero.

Let us again assume that the initial velocities u_1' and u_2' are known and that we also know the value of the restitution coefficient. We shall, as before, solve for the final velocities v_1 and v_2. We may start with

$$m_1 v_1' + m_2 v_2' = 0 \qquad (6.6.15)$$

and

$$v_2' - v_1' = e(u_1' - u_2') \qquad (6.6.16)$$

Solving for v_2' in the first of these, and substituting the value so obtained into the other, we find that

$$v_1' = -e(u_1' - u_2') \frac{m_2}{m_1 + m_2} \qquad (6.6.17)$$

and

$$v_2' = -\frac{m_1}{m_2} v_1' = e(u_1' - u_2') \frac{m_1}{m_1 + m_2} \qquad (6.6.18)$$

But now, we may use (6.6.13) to express u_2' in terms of u_1' in (6.6.17) and to express u_1' in terms of u_2' in (6.6.18). This allows us, finally, to write these two equations simply as

$$\frac{v_1'}{u_1'} = \frac{v_2'}{u_2'} = -e \qquad (6.6.19)$$

The picture of the collision, therefore, must be that illustrated in Fig. 6.11. Equation (6.6.19) should be compared with Eq. (6.5.8), which applies to a system that is at rest. It is apparent that the transformation to a system that is fixed with respect to the center of mass rather than the surroundings enables us to express the ratio of *absolute* final and initial velocities

of each particle in a very simple way. In the system fixed with respect to the surroundings, it is only the ratio of *relative* velocities that has this simple relationship.

It is also particularly interesting to calculate the initial and final kinetic energies for this system. The initial kinetic energy U_{ki} is, of course,

$$U_{ki} = \tfrac{1}{2} m_1 u_1'^2 + \tfrac{1}{2} m_2 u_2'^2$$

The final kinetic energy can be found directly from (6.6.19) from which

$$v_1'^2 = e^2 u_1'^2 \qquad (6.6.20)$$

$$v_2'^2 = e^2 u_2'^2 \qquad (6.6.21)$$

Then the final kinetic energy U_{kf} is

$$U_{kf} = \tfrac{1}{2} m_1 v_1'^2 + \tfrac{1}{2} m_2 v_2'^2 \qquad (6.6.22)$$

which, using (6.6.20) and (6.6.21), can be expressed as

$$U_{kf} = e^2(\tfrac{1}{2} m_1 u_1'^2 + \tfrac{1}{2} m_2 u_2'^2) = e^2 U_{ki} \qquad (6.6.23)$$

The *ratio* of final to initial energy is, therefore,

$$\boxed{\frac{U_{kf}}{U_{ki}} = e^2} \qquad (6.6.24)$$

We see that the restitution coefficient is very simply related to the fractional energy loss during collision, in that the square of the restitution coefficient represents the ratio of final to initial kinetic energy *calculated in the coordinate system of the center of mass*.

The restitution coefficient can be related to the initial and final energies so simply only in the center of mass coordinate system. Both the initial and final kinetic energy calculated in the system fixed with respect to the earth will be different from those calculated in the center of mass system, and the ratio of final to initial energy will then be a complex expression involving not only the restitution coefficient but also the velocities themselves. Of course, if energy is conserved in the collisions, it will be conserved in both coordinate systems, and then $U_{kf}/U_{ki} = 1$ in both cases.

The center of mass system is particularly useful in allowing us to understand certain important physical features of collision dynamics, as illustrated in this example. The center of mass coordinate system invariably provides a reference frame in which the essential physics of collision processes is much more easily expressed and understood than in any other reference system. It is nevertheless not always necessary, nor desirable, to make a coordinate transformation to the center of mass system to obtain the answers to ordinary, uncomplicated collision problems. Such problems can usually be solved easily enough in the fixed system without resorting to any coordinate transformation.

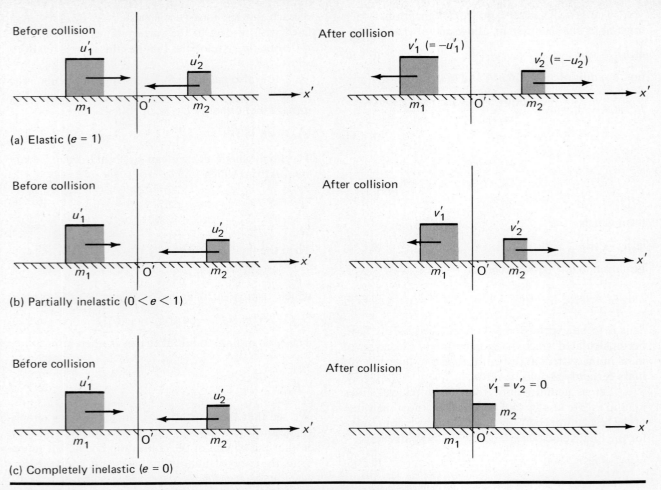

FIGURE 6.11. Some examples of collisions in one dimension, as observed in the coordinate system of the center of mass; (a) elastic; (b) partially inelastic; (c) completely inelastic.

EXAMPLE 6.6.1

A small ball is dropped from height h_0 upon a flat surface fixed with respect to the earth and rebounds to height h before starting downward once more. Find the restitution coefficient in terms of h and h_0.

The earth is so massive compared with the object that is dropped, that, even though momentum is conserved during the collision, the change in the earth's velocity caused by the collision may be neglected. In (6.5.8), then, we may assume $v_2 = u_2 = 0$, whence

$$e = -\frac{v_1}{u_1} \tag{6.6.25}$$

where u_1 is the ball's velocity immediately before collision and u_1 its velocity immediately afterward. Since it was dropped from height h_0, its initial potential energy mgh_0 must have been converted into an equivalent amount of kinetic energy just before impact. Therefore,

$$mgh_0 = \tfrac{1}{2}mu_1{}^2 \tag{6.6.26}$$

By the same sort of argument, all the initial kinetic energy on rebound is converted to potential energy by the time height h is attained and

$$mgh = \tfrac{1}{2}mv_1{}^2 \tag{6.6.27}$$

From these two equations, it is evident that $v_1{}^2/u_1{}^2 = h/h_0$. But, in view of (6.6.25), this means that

$$e^2 = \frac{h}{h_0} \tag{6.6.28}$$

This simple result affords an easy way of measuring the restitution coefficient e for various pairs of substances simply by dropping a sphere made of one substance on a surface made from another and noting the height of rebound. If numerator and denominator of (6.6.28) are multiplied by mg, the ratio of final to initial energy is just e^2. This is in agreement with the results obtained earlier in this section, since in this calculation the fixed system *is* for all practical purposes the center of mass system.

EXAMPLE 6.6.2

A body of mass m_1 collides with a body of mass m_2 which is initially at rest. The final velocities are observed to be v_1 and v_2. Find (a) the restitution coefficient e and (b) the initial velocity of m_1, in terms of the known final velocities v_1 and v_2. All velocities have only x-components, so that the collision is one-dimensional.

From conservation of momentum and the definition of the restitution coefficient,

$$m_1u_1 + m_2u_2 = m_1v_1 + m_2v_2 \qquad (6.6.29)$$

$$v_2 - v_1 = e(u_1 - u_2) \qquad (6.6.30)$$

But because the initial velocity u_2 of mass m_2 is zero,

$$m_1u_1 = m_1v_1 + m_2v_2 \qquad (6.6.31)$$

$$v_2 - v_1 = eu_1 \qquad (6.6.32)$$

We need now merely solve these two equations for e and u_1. We may easily eliminate u_1 between the two by multiplying (6.6.31) by e and (6.6.32) by m_1 and then subtracting one equation from the other. Solving the resulting equation for e, we may obtain

$$e = \frac{v_2 - v_1}{v_1 + (m_2/m_1)\,v_1} \qquad (6.6.33)$$

This value may then be substituted into (6.6.32), whereupon, solving for u_1, it is clear that

$$u_1 = v_1 + \frac{m_2}{m_1}\,v_2 \qquad (6.6.34)$$

In this example, the center of mass is in motion and, therefore, the fixed system and center of mass system are in motion with respect to one another. In the fixed frame, therefore, the ratio of final to initial kinetic energies is no longer e^2 but is more complex. Indeed, it can be shown—the details being assigned as an exercise—that in the fixed frame

$$\frac{U_{kf}}{U_{ki}} = \frac{m_1{}^2 + (1 + e^2)m_1m_2 + e^2m_2{}^2}{(m_1 + m_2)^2} \qquad (6.6.35)$$

Note, however, that for an *elastic* collision ($e = 1$), one obtains $U_{kf}/U_{ki} = 1$ in the fixed frame as well as in the center of mass frame. If energy is conserved in one reference system, it must be conserved in all!

EXAMPLE 6.6.3

A set of identical steel balls, each of mass m, is suspended by a set of strings whose spacing is equal to the diameter of the spheres, as shown in Fig. 6.12. If one ball is displaced and released, it collides with the row of stationary balls, and immediately one ball flies off the other end of the row with equal velocity. If two balls are displaced and released, two balls will fly off the other end; if three balls are displaced and let go, three balls fly off the other end, and so forth. In each case, the final velocity acquired by the balls leaving the far end of the row is essentially equal to the initial velocity with which the first set of balls collide with the rest. Explain why these results are observed, in particular, why, when two balls are released initially, one ball does not fly off the other end with velocity greater than that associated with the initial set of two balls before collision. Begin by assuming that the collision is perfectly elastic, but discuss what happens when this is no longer true.

Suppose, initially, that n balls are held aside and released, striking the rest of the stack with initial velocity u. The question of what forces act to transmit the momentum of the colliding balls from one end of the row of stationary balls to the other is quite complex; but in any case, whatever the mechanism of momentum transfer, momentum must be conserved in the collision interaction, and, if the collision is elastic, so must kinetic energy. If some number n' balls are expelled from the other end of the stack with velocity v immediately after collision, we may write

$$p_i = nmu = n'mv = p_f \qquad (6.6.36)$$

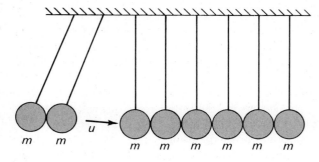

(a)

(b)

FIGURE 6.12

and

$$U_{ki} = \tfrac{1}{2}nmu^2 = \tfrac{1}{2}n'mv^2 = U_{kf} \qquad (6.6.37)$$

If we square both sides of (6.6.36) and divide the resulting equation by (6.6.37), we may obtain

$$\frac{n^2m^2u^2}{nmu^2} = \frac{n'^2m^2v^2}{n'mv^2}$$

or

$$n' = n \qquad (6.6.38)$$

Inserting this into (6.6.36), one finds also that

$$v = u \qquad (6.6.39)$$

The results actually observed are seen to follow at once from conservation of momentum and energy in the collision. There is no other way in which this can be accomplished; for example, if n' and n are not equal, (6.6.36) requires that

$$\frac{v}{u} = \frac{n}{n'} \qquad (6.6.40)$$

while (6.6.37) demands that

$$\frac{v}{u} = \sqrt{\frac{n}{n'}} \qquad (6.6.41)$$

But these two equations are inconsistent if n and n' are not equal, since the only number (other than zero) whose square root is equal to itself is 1.

The behavior described initially is obtained only if the collision is elastic, otherwise, what happens may be quite complex. For example, if the collision is perfectly inelastic ($e = 0$), the colliding balls will stick to the stationary ones, and all of them will move off after the collision with a velocity *less* than u, which, by application of the principle of momentum conservation, is given by

$$v = \frac{n}{N} u \qquad (6.6.42)$$

where N is the total number of balls in the entire system. The proof of this assertion is left for the reader as an exercise.

6.7 Collisions in Two and Three Dimensions

The principles of momemtum conservation for two- and three-dimensional systems are not essentially different from those which were applied in the one-dimensional system. We shall find, however, that the general mathematical analysis of a multibody collision in three dimensions, where all the known and unknown parameters are permitted to retain arbitrary

values throughout is hopelessly complex. While, for example, it is still possible to define a restitution coefficient, it does not help much to do so in terms of simplifying the mathematical work, and therefore this parameter is much less useful in two- and three-dimensional situations than it is for the one-dimensional case. We shall, therefore, be content to outline the general principles that are applicable and to examine several instances illustrating how these principles are used in certain particularly interesting and important examples. We shall focus our attention upon collision problems involving *two bodies only* wherein the conditions of the problem and the symmetry of the colliding bodies are sufficient to guarantee that the motion will be confined to a plane at all times, the *xy*-plane in particular. Such *two-dimensional collision* problems illustrate in the most general way the conservation of momentum *as a vector* and are at the same time simple enough so that the physical principles to be illustrated are not obscured by mathematical complexity.

If the vector momentum of the system is to be conserved, then, for a two-body collision,

$$m_1\mathbf{u}_1 + m_2\mathbf{u}_2 = m_1\mathbf{v}_1 + m_2\mathbf{v}_2 \qquad (6.7.1)$$

where $\mathbf{u}_1$ and $\mathbf{u}_2$ are the initial velocities and $\mathbf{v}_1$ and $\mathbf{v}_2$ the final velocities. This can be written as a set of three equations expressing the fact that if the vector momentum is conserved, each component must be conserved:

$$m_1u_{1x} + m_2u_{2x} = m_1v_{1x} + m_2v_{2x} \qquad (6.7.2)$$

$$m_1u_{1y} + m_2u_{2y} = m_1v_{1y} + m_2v_{2y} \qquad (6.7.3)$$

and

$$m_1u_{1z} + m_2u_{2z} = m_1v_{1z} + m_2v_{2z} \qquad (6.7.4)$$

In any situation where the motion is confined entirely to the *xy*-plane, of course, the *z*-components of velocity for both bodies, in both initial and final states, are all zero, and Eq. (6.7.4) then simply expresses the uninteresting fact that zero equals zero. In any such instance, its contribution is nil, and it need not be considered any further. Such *two-dimensional collision* situations may arise, for example, in the collision of billiard balls on the surface of a horizontal table or in the collision of cylindrical objects (not necessarily *circular* cylinders) on the surface of an air table.

Ordinarily, the initial velocities will be known, and the final velocities will be sought. There will then be four unknown quantities (v_{1x}, v_{1y}, v_{2x}, and v_{2y}) for a two-dimensional collision. But momentum conservation, by itself, furnishes only two equations relating them, and a statement regarding the degree to which energy is conserved (for example, specifying the ratio

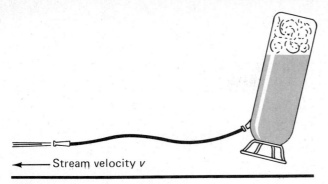

←――― Stream velocity v

FIGURE 6.22

mass of the column of water and v its velocity. According to the impulse–momentum theorem, this requires that a force F be exerted on the column of water whose relation to the initial and final momentum will be given by (6.8.22):

$$F(t_2 - t_1) = p_f - p_i = mv - 0$$

This equation assumes that F is constant, as it will be if the pressure inside the tank is constant.

The length of the column of water emitted during this time interval will be

$$l = v(t_2 - t_1) \qquad (6.8.26)$$

Its mass will be given by the product of its volume (expressed as the product of the cross-sectional area A times the length) and the density ρ of the fluid in slugs per cubic foot. Therefore,

$$m = (lA)\rho = v(t_2 - t_1) \cdot \pi r^2 \cdot \rho \qquad (6.8.27)$$

where r is the radius of the nozzle, hence of the stream of water. Substituting this into the impulse–momentum theorem,

$$F(t_2 - t_1) = v^2(t_2 - t_1)\pi r^2 \rho$$
$$F = \pi r^2 \rho v^2 \qquad (6.8.28)$$

Substituting the numerical values given above, we find

$$F = (3.14)\left(\frac{0.125}{12}\right)^2 \left(\frac{62.4}{32}\right)(40)^2 = 1.064 \text{ lb}$$

This is the force required to set the column of fluid in motion; and, according to Newton's third law, the reaction force exerted by the nozzle upon whoever is holding it is equal in magnitude and oppositely directed.

*EXAMPLE 6.8.3

A pile driver works by repeatedly dropping a 640-pound weight from a height of 12 ft upon a piling which is being driven into the earth. The force exerted on the piling, as a function of time after the initial impact, is given by

$$F(t) = F_{max}\left[1 - \left(\frac{2t}{t_2} - 1\right)^2\right] \qquad (6.8.29)$$

A plot of this relation is shown in Fig. 6.23. The time t_2 required for the weight to come to rest after the initial impact is 0.02 sec. Find F_{max} and calculate the time-average force $\bar{F}$ in terms of F_{max}.

The impulse–momentum theorem states that

$$p_f - p_i = \int_{t_1}^{t_2} F(t)\, dt$$

Applied to the weight of the pile driver, this means that

$$mv - 0 = \int_0^{t_2} F(t)\, dt$$
$$= F_{max} \int_0^{t_2}\left[1 - \left(\frac{2t}{t_2} - 1\right)^2\right] dt \qquad (6.8.30)$$

where v is the velocity with which the falling weight hits the piling. Squaring the quantity in parenthesis within the integral and performing the integration, we obtain

$$mv = 4F_{max}\int_0^{t_2}\left(\frac{t}{t_2} - \frac{t^2}{t_2^2}\right) dt$$
$$= 4F_{max}\left[\frac{t^2}{2t_2} - \frac{t^3}{3t_2^2}\right]_0^{t_2} = \frac{2}{3}F_{max}t_2 \qquad (6.8.31)$$

$$F_{max} = \frac{3}{2}\frac{mv}{t_2}$$

But

$$v^2 - v_0^2 = 2ax$$

Substituting $v_0 = 0$, $a = g = 32 \text{ ft/sec}^2$, and $x = 12 \text{ ft}$ into this, we obtain

$$v = 27.7 \text{ ft/sec}$$

whereby, since $m = 640/32$ slugs and $t_2 = 0.02$ sec,

$$F_{max} = \left(\frac{3}{2}\right)\left(\frac{640}{32}\right)\left(\frac{27.7}{0.02}\right) = 41,500 \text{ lb}$$

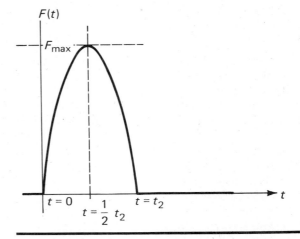

FIGURE 6.23

The impulse–momentum theorem written in the form of Eq. (6.8.18) can also be applied to this system, the result being

$$mv = \bar{F}t_2$$

$$\bar{F} = \frac{mv}{t_2} \tag{6.8.32}$$

Comparing this result with (6.8.31), which is stated in terms of F_{max}, it is seen that

$$\bar{F} = \frac{2}{3}F_{max}$$

6.9 Motion of Systems with Variable Mass

Up to this, point, our application of Newton's laws has been restricted to closed systems, containing constant mass. There are, however, a number of interesting examples in which mass varies continuously with time, and we would like to formulate a treatment for such systems. For example, the mass of a rocket ship decreases as the hot exhaust gases are ejected, and the total mass hauled by a locomotive increases during a rainstorm as water accumulates in the hopper cars.

According to the discussion given in section 6.3, the equation describing the motion of a system of constant mass is obtained by equating the net sum of all external forces to the time rate of change of the *total* momentum:

$$\mathbf{F}_{ext} = \frac{d\mathbf{p}}{dt} \tag{6.9.1}$$

We have emphasized the word *total* above to call attention to the fact that the momentum must include not only the instantaneous momentum of the system of interest, whose mass is given by $m(t)$, but also the momentum associated with any small mass $dm(t)$ which has either entered or left the system in the time interval dt.

It is important to stress that the equations $\mathbf{F}_{ext} = m\mathbf{a}(t)$ and $\mathbf{F}_{ext} = d(m\mathbf{v}(t))/dt$ that are valid for systems of constant mass do not properly describe the dynamics of systems with time-varying mass, since they do not correctly account for reaction forces such as those exerted by the escaping mass on the rocket in the case of rocket propulsion. It is, therefore, important to remember that in systems where there may be time-varying mass components, the solutions of dynamical problems must be approached from the point of view of overall momentum conservation of the entire system including all masses and mass increments.

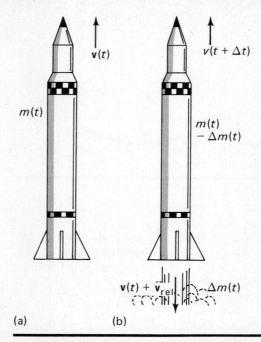

FIGURE 6.24. A rocket propelled by the expulsion of mass with given relative velocity $\mathbf{v}_{rel}$. (a) Situation at time t. (b) Situation at time $t + \Delta t$, after the expulsion of mass Δm.

As a specific example consider the principle of rocket propulsion as illustrated in Fig. 6.24. The rocket ship has an instantaneous velocity $\mathbf{v}(t)$ and a mass $m(t)$ at time t. The time variation of the mass, $m(t)$, is usually known, and the gases are ejected at a known velocity with respect to the rocket through an orifice of known size. On the other hand, $\mathbf{v}(t)$ can be ascertained only through the application of Newton's second law as stated in Eq. (6.9.1) above. Figure 6.24a shows the gases of mass $\Delta m(t)$ which will be ejected in the time interval Δt. Figure 6.24b describes the situation at the later time $t + \Delta t$. The mass of the rocket has been reduced by $\Delta m(t)$, this being the mass of the gases ejected during dt. In this example, we assume for simplicity that our rocket is in outer space and does not experience any gravitational (that is, weight) force.

To apply Eq. (6.9.1), we compute the *total* momentum of the system at time t and at time $t + \Delta t$. The exhaust gases have been ejected with a relative velocity $\mathbf{v}_{rel}$ and, therefore, have a velocity $\mathbf{v}(t) + \mathbf{v}_{rel}$ with respect to a stationary observer. The momenta at t and at $t + \Delta t$ are, therefore,

$$\mathbf{p}(t) = m(t)\mathbf{v}(t)$$
$$\mathbf{p}(t + \Delta t) = [m(t) - \Delta m(t)]\mathbf{v}(t + \Delta t) \tag{6.9.2}$$
$$+ \Delta m(t)[\mathbf{v}(t) + \mathbf{v}_{rel}]$$

In (6.9.2), $\Delta m(t)$ is a positive quantity which may be obtained once dm/dt, the time rate of change of the mass of the rocket, is known. The relationship needed is

$$\lim_{\Delta t \to 0} \Delta m(t) = \lim_{\Delta t \to 0} \left(-\frac{dm(t)}{dt} \right) \Delta t \qquad (6.9.3)$$

The negative sign on the right-hand side is required because dm/dt is negative for a system whose mass is decreasing with time.

Allowing Δt to approach zero, we obtain from Eqs. (6.9.1), (6.9.2), and (6.9.3)

$$\mathbf{F}_{ext} = m(t)\frac{d\mathbf{v}(t)}{dt} - \mathbf{v}_{rel}\frac{dm(t)}{dt} \qquad (6.9.4)$$

It is helpful to rewrite (6.9.4) in a slightly different form in order to compare it to Newton's second law for a system of constant mass. We may write this equation

$$\mathbf{F}_{ext} + \mathbf{F}_r = m(t)\frac{d\mathbf{v}(t)}{dt} \qquad (6.9.5)$$

where

$$\mathbf{F}_r = \mathbf{v}_{rel}\frac{dm(t)}{dt} \qquad (6.9.6)$$

is the reaction force which the ejected fuel exerts back on the rocket. This reaction force provides the propulsion necessary to accelerate the rocket in a controlled manner.

Equation (6.9.5) may also be derived by an alternative method in which our system consists only of the rocket and does not include the ejected gases. Here, we may use the impulse–momentum theorem in the form

$$\mathbf{p}(t + \Delta t) - \mathbf{p}(t) = (\mathbf{F}_{ext} + \mathbf{F}_r)\,\Delta t \qquad (6.9.7)$$

Equation (6.9.7) states that the change of momentum of the rocket in the interval Δt is equal to the total impulse, this being separated into a part due to external forces and a part due to the reaction force $\mathbf{F}_r$ which the escaping gases exert back on the rocket. In view of Newton's third law, this reaction force must be the negative of the force $\mathbf{F}_i$ which acted to eject the gas. The force $\mathbf{F}_i$ may be obtained by using

$$\mathbf{F}_i\,\Delta t = [\mathbf{v}(t) + \mathbf{v}_{rel}]\,\Delta m(t) - \mathbf{v}(t)\,\Delta m(t)$$
$$= \mathbf{v}_{rel}\,\Delta m(t) \qquad (6.9.8)$$

or

$$\mathbf{F}_r\,\Delta t = -\mathbf{v}_{rel}\,\Delta m(t) \qquad (6.9.9)$$

Using (6.9.3) for small Δt, we have

$$\mathbf{F}_r\,\Delta t = \mathbf{v}_{rel}\frac{dm(t)}{dt}\,\Delta t \qquad (6.9.10)$$

This gives a result for $\mathbf{F}_r$ which agrees with (6.9.6). Expressing $\mathbf{p}(t)$ and $\mathbf{p}(t + \Delta t)$ in terms of masses and

velocities at time t and $t + \Delta t$, and utilizing (6.9.10), we again arrive at (6.9.5).

Clearly, the two different points of view lead to the same result. In the first case we considered Newton's second law for the total system of the rocket and the ejected gases, whereas in the second case we applied the impulse–momentum theorem to the rocket and had to consider impulses due to $\mathbf{F}_{ext}$ and $\mathbf{F}_r$. The possibility of studying a given physical problem from several equivalent points of view is a common occurrence in physics and illustrates the maturity and intellectual consistency of the subject. It also furnishes certain insights not ordinarily arising from a more single-minded approach.

In outer space, where gravitational forces are negligible, it is this reaction force which is solely responsible for any acceleration or course alterations of the rocket. The dramatic Apollo moon voyages would not have been possible without a detailed understanding of this force. Of course, as a rocket approaches the moon, the earth, or any other body, gravitational forces also come into play. These forces could also have been included as part of $\mathbf{F}_{ext}$ in the equations developed above.

EXAMPLE 6.9.1

A rocket ship in outer space ejects exhaust gases at a constant velocity $\mathbf{v}_{rel}$ with respect to the rocket at a constant rate $R = dm/dt$ when fuel is being burned. Assuming that $\mathbf{v}_{rel}$ is directed opposite to the initial velocity $\mathbf{v}_0$ of the rocket, find the velocity $\mathbf{v}(t)$ after time t has elapsed and the distance traveled during the period of acceleration.

According to Eq. (6.9.5), during the acceleration period

$$\mathbf{v}_{rel}R = m(t)\frac{d\mathbf{v}(t)}{dt}$$

where $m(t) = m_0 + Rt$. This equation may also be written

$$d\mathbf{v}(t) = \frac{\mathbf{v}_{rel}R\,dt}{m_0 + Rt}$$

and is easily integrated to give

$$\int_{\mathbf{v}_0}^{\mathbf{v}(t)} d\mathbf{v} = \int_0^t \frac{\mathbf{v}_{rel}R}{m_0 + Rt}\,dt$$

or

$$\mathbf{v}(t) - \mathbf{v}_0 = \mathbf{v}_{rel}\ln\left(1 + \frac{R}{m_0}t\right) \qquad (6.9.11)$$

In this integration, we have assumed that at $t = 0$, the initial velocity is $\mathbf{v}_0$, in agreement with the conditions set forth in the original statement of the problem. If the time t during which fuel is burned is small so that $Rt \ll m_0$, the logarithm above may be approximated by using the well-known approximation, true for $x \ll 1$, that $e^x \cong 1 + x$, whence, taking the logarithm of both sides,

$$\ln(1 + x) \cong x \; (|x| \ll 1) \tag{6.9.12}$$

If we set $x = Rt/m_0$, we obtain from (6.9.11) the approximate result

$$\mathbf{v}(t) - \mathbf{v}_0 \cong \frac{R}{m_0} t \mathbf{v}_{\text{rel}} \tag{6.9.13}$$

which is nearly correct so long as $Rt/m_0 \ll 1$, hence $Rt \ll m_0$. Note that since $\mathbf{v}_{\text{rel}}$ is antiparallel to $\mathbf{v}_0$ and since R is negative, the velocity change is in the direction of $\mathbf{v}_0$.

To determine the distance traveled in time t, it is necessary to perform a further integration. We write (6.9.11) in the form

$$dx = v_0 \, dt + v_{\text{rel}} \ln\left(1 + \frac{R}{m} t\right) dt \tag{6.9.14}$$

This may readily be integrated between the initial displacement $x = 0$ and a final value $x(t)$. The details are left to the reader. The result is

$$x(t) = v_0 t - v_{\text{rel}} \left[t - \left(\frac{m_0}{R} + t\right) \ln\left(1 + \frac{R}{m_0} t\right) \right] \tag{6.9.15}$$

For small t, satisfying $Rt \ll m_0$, we may integrate (6.9.13) in exactly the same way to obtain the approximate expression

$$x(t) \cong v_0 t + \frac{1}{2} v_{\text{rel}} \frac{R}{m_0} t^2 \tag{6.9.16}$$

The example of rocket propulsion involves a system whose mass is diminishing with time. If the exhaust gases are ejected directly opposite to the vector velocity of the rocket, a positive acceleration in the direction of the velocity vector is achieved and the motion is confined to a straight line. Other motions are possible by aiming the velocity of the ejected gases in some other direction. In principle any motion is possible by controlled use of the reaction forces.

EXAMPLE 6.9.2

Let us consider now a system whose mass is increasing with time. A train with open-hopper cars may receive rainwater during a thunderstorm, and its motion may be affected by the continuous increase of mass. For simplicity, let us assume the rain is falling vertically and $R = dm/dt$ is the constant rate at which the total mass of the system is changing. If the train is initially coasting at velocity v_0 on a straight, level track, discuss its subsequent motion by using Eq. (6.9.1).

At $t = 0$, the total momentum in the direction of the track is

$$p_0 = m_0 v_0 \tag{6.9.17}$$

and at the later time t, it is

$$p(t) = (m_0 + Rt)v(t) \tag{6.9.18}$$

If we assume there are no *net* external forces acting in the direction of the track, then, since the total momentum component in the direction of the track is conserved, $p(t) = p_0$. Therefore,

$$m_0 v_0 = (m_0 + Rt)v(t) \tag{6.9.19}$$

To obtain the position of the train as a function of time, we may set $v(t) = dx/dt$ and integrate this equation to find

$$\int_0^x dx = m_0 v_0 \int_0^t \frac{dt}{m_0 + Rt}$$

from which

$$x(t) = \frac{m_0 v_0}{R} \ln\left(1 + \frac{Rt}{m}\right) \tag{6.9.20}$$

In establishing the limits of integration, it is assumed that $x = 0$ at $t = 0$.

EXAMPLE 6.9.3

A long train with open cars is moving along a railway track. The coefficient of friction between the wheels and the track is μ. During a rainstorm, water increases the total mass of the train at a rate $R = dm/dt$, which is constant. The initial mass of the train is m_0. How much power must the locomotive deliver in order to keep the train moving at a constant velocity v_0 if frictional forces are ignored? What will the result be if frictional forces are included?

The external force which must be supplied is related to the time rate of change of the linear momentum according to

$$\mathbf{F}_{\text{ext}} = \frac{d\mathbf{p}}{dt}$$

The train is assumed to be moving at constant velocity and, therefore, the momentum increase is due only to the increase in mass. Therefore,

$$F_{\text{ext}} = \frac{dp}{dt} = \frac{d}{dt}\left[(m_0 + Rt)v_0\right] = Rv_0 \tag{6.9.21}$$

If it is assumed that friction may be ignored, the locomotive must deliver the power

$$F_{\text{ext}} v_0 = Rv_0{}^2 \tag{6.9.22}$$

If frictional forces are taken into account, the total external force must still be given by $v_0 R$, but now it consists of two parts, F_{ext_1}, representing the force exerted by the locomotive, and F_{ext_2}, the friction force which *retards* the motion. If we write

$$F_{\text{ext}} = F_{\text{ext}_1} + F_{\text{ext}_2} = Rv_0 \tag{6.9.23}$$

and set the frictional force equal to the friction coefficient times the total normal force $m(t)g$, whence

$$F_{ext_2} = -\mu(m_0 + Rt)g \qquad (6.9.24)$$

we may obtain an expression for F_{ext_1} given by

$$F_{ext_1} = v_0 R + \mu(m + Rt)g \qquad (6.9.25)$$

The power delivered by the locomotive is, therefore,

$$F_{ext_1}v_0 = Rv_0{}^2 + \mu(m_0 + Rt)gv_0 \qquad (6.9.26)$$

This result shows that the power delivered by the locomotive must increase with time in order to maintain constant speed. This increase is due to the continuous increase of the retarding frictional force which, in turn, is caused by the increased normal force the tracks exert on the train.

SUMMARY

The linear momentum of a point mass m that moves with velocity $\mathbf{v}$ is defined as the vector quantity $\mathbf{p} = m\mathbf{v}$. Newton's second law for such a particle can be written in the form $\mathbf{F} = d\mathbf{p}/dt$, where $\mathbf{F}$ is the resultant force.

For a system of particles, the momentum of the system is the vector sum of the momenta of the individual particles. In this case, we must consider separately *external* forces, applied from outside the system, and *internal* forces, exerted by the particles on one another. When this is done, it turns out that the internal forces act in such a way as to have no effect on the system's total momentum. Therefore, in the absence of external forces, the momentum of a system of particles remains constant, or is *conserved*.

If external forces are present, the system's total momentum $\mathbf{p}$ changes, the time rate of change of momentum being equal to the resultant of all external forces acting on the particles of the system. Accordingly, we may write

$$\mathbf{F}_{ext} = d\mathbf{p}/dt$$

where $\mathbf{F}_{ext}$ is the vector sum of all external forces acting on the particles. We can also show that the system's total momentum can be represented as its total mass times the velocity with which its center of mass moves. The total momentum of the system, therefore, can be expressed as the "momentum of the center of mass."

In collision processes, momentum is always conserved, though kinetic energy may or may not be conserved. If it is, the collision is said to be *elastic*; if not, it is referred to as an *inelastic collision*. An inelastic collision in which the two colliding bodies stick together dissipates a maximum amount of kinetic energy, and is sometimes referred to as a *completely inelastic process*. For a two-body collision in one dimension, the *restitution coefficient*

$$e = \frac{v_2 - v_1}{u_1 - u_2}$$

gives a measure of the degree to which a collision is elastic, since for a completely elastic collision $e = 1$, and for one that is completely inelastic, $e = 0$.

In the reference frame of the center of mass, the dynamics of collisions are particularly simple, since in this coordinate system the total momentum is zero. For example, the ratio of final and initial kinetic energies in this system is the square of the restitution coefficient which, in turn, has the simple form given by Eq. (6.6.19).

The impulse of a force $\mathbf{F}$ acting over a time interval that extends from t_1 to t_2 is defined as

$$\int_{t_1}^{t_2} \mathbf{F}\, dt$$

According to the *impulse–momentum theorem*, this quantity must be equal to the change of momentum of the body on which the force acts during this time interval. The time-average force acting during the given time integral can be expressed as the impulse divided by the length of the time interval $t_2 - t_1$.

In the case of systems in which the mass varies with time, it is incorrect to use Newton's second law in the form $\mathbf{F} = m\mathbf{a}$. We must, instead, approach problems involving such systems from the point of view of momentum conservation, using Newton's law in the form $\mathbf{F} = d\mathbf{p}/dt$, with the momentum of every body written as $\mathbf{p}_i = m_i(t)\mathbf{v}_i$.

QUESTIONS

1. Two point masses m and $2m$ have exactly the same kinetic energies. Find the ratio of the magnitudes of their momenta. Is there enough information to determine the sum of their momenta?

2. A solid sphere of mass M rotates about a fixed axis with angular velocity ω in a laboratory demonstration. What is the total linear momentum of the sphere?

3. For a system of point masses, the total momentum is equal to the momentum of the center of mass. Explain this statement. Is it still true if there is no particle at the center of mass?

4. In many problems involving collisions, the analysis is carried out in the center-of-mass reference frame. This reference frame is sometimes also called the zero momentum frame. Explain the meaning of this statement.

5. Estimate the magnitude of your momentum when you are running at a speed of 8 miles per hour.

6. A tennis ball is dropped to the ground from a height of about 4 ft, and it rebounds somewhat. Is the momentum of the ball just before collision equal to the momentum just after? If you enlarge the system so that it includes the ball and the earth, will momentum be conserved?

7. If a collision is completely inelastic, does that mean that all the kinetic energy is lost? Give examples to support your reasoning. Is there a coordinate system in which all the kinetic energy is lost?

8. Two objects collide in mid air. Are all three components of momentum conserved? What happens to the momentum after some time has elapsed?

9. A cue stick collides with a billiard ball and remains in contact for a time interval Δt. If the final momentum of the ball is measured, can the force $F(t)$ be determined? Can the time average of the force be found? How does the impulse received by the cue stick compare with that received by the ball? Does the cue stick acquire any momentum?

10. Show that in any two-particle collision the speed of the center of mass cannot exceed the speed of both of the bodies.

11. A child is at the center of a large, frozen pond. Assuming the coefficient of friction to be zero, describe a method by which the child can move on the ice.

12. Is it true that internal forces do not affect the motion of the center of mass of a system?

13. In rocket propulsion, why is it necessary to eject mass in order to accelerate the rocket?

PROBLEMS

1. (a) What is the momentum of a 4500-pound car moving at a constant speed of 44 ft/sec? (b) How fast would a 2700-pound car have to go to attain the same momentum? (c) How fast would the smaller car have to go to attain the same kinetic energy as the car in (a)?

2. A 120-kilogram box slides along a level floor. Its velocity is initially 12 meters/sec, and it is brought to rest by a constant frictional force in 6 sec. Find (a) the initial momentum of the box, (b) the final momentum, (c) the rate of change of momentum, (d) the resultant force on the box, and (e) the force of friction.

3. (a) Is momentum conserved in the situation described in the preceding problem? (b) If not, what becomes of the initial momentum of the box?

4. Two children, each of mass m, sit at opposite ends of a rowboat of mass M. The boat is initially at rest in still water. The child in the stern throws a ball of mass m_0 forward with velocity v_0, and at the same time the child in the bow throws a ball of equal mass backward with velocity $2v_0$. The balls are projected horizontally. (a) Find the speed of the rowboat while the two balls are in the air. Neglect any effect of water resistance on the boat. (b) The child in the bow catches the ball thrown forward, but the other child misses, and the other ball falls in the water. What is the final velocity of the boat? (c) What would the final velocity of the boat have been had the second child caught the ball?

5. Two masses, $m_A = 2$ kg and $m_B = 3$ kg are separated by an ideal spring whose spring constant is 1200 newtons/meter and which is compressed a distance of 0.20 meter from its equilibrium length. Suddenly the spring is released, setting both masses in motion. The masses move

on a frictionless horizontal surface, and there are no frictional or dissipative processes that act during the expansion of the spring. The spring's mass is negligible. Find the final velocities v_A and v_B of both masses.

6. A switching locomotive weighing 200,000 pounds coasts along a level track at a constant speed of 11 ft/sec into a line of four freight cars each weighing 140,000 pounds and couples onto the line of cars. (a) What is the speed of the train after they are coupled together? (b) How much kinetic energy is dissipated during the collision?

7. A woman in a Volkswagen waiting for the light to change is struck directly from the rear by a man in a Cadillac moving at a speed of 40 ft/sec. The Volkswagen is propelled into the intersection with an initial speed of 40 ft/sec, the Cadillac following behind at some speed v. The total mass of the Cadillac is 3.0 times the mass of the VW and its driver. Find (a) the speed v of the Cadillac immediately after the collision, (b) the speed of the center of mass of the two cars, before and after collision, and (c) the kinetic energy loss during the collision, assuming that the Cadillac and its driver weigh 5400 pounds.

8. Two objects of equal mass undergo a perfectly elastic collision. Initially, one of the objects is at rest, and after collision its velocity vector makes an angle θ_1 of 30° with the initial velocity vector of the other object. Find the direction of motion of the other object after collision and the speeds of both, expressed in terms of the initial velocity u_2 of the body which is in motion before impact.

9. Two carts of mass 5 kg and 3 kg, respectively, are fastened together by an ideal spring whose spring constant is 500 N/m, as shown in the figure. The spring is stretched a distance of 0.25 m beyond its equilibrium length, and both carts are released simultaneously from rest with the spring so extended. Neglecting friction, find (a) the velocity v_1 of the 5-kg cart and the velocity v_2 of the 3-kg cart when the extension of the spring is 0.10 m beyond the equilibrium length and (b) the velocities v_1 and v_2 when the extension of the spring is 0.20 m beyond equilibrium. (c) What are the initial accelerations of the two carts?

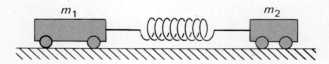

10. Work out the answers to Problem 9 above using arbitrary values $m_1, m_2, v_1, v_2, a_1, a_2$ for the masses, velocities, and accelerations of the two carts. Denote the extension of the spring beyond equilibrium by x_0 initially and by x at an arbitrary subsequent stage of the motion, and find v_1 and v_2 as functions of the masses m_1 and m_2, x and x_0, and the spring constant k.

11. A 14-inch naval gun fires a shell weighing 2000 pounds, imparting to it a muzzle velocity of 2800 ft/sec. The gun barrel weighs 100,000 pounds. The recoil energy of the gun barrel is absorbed in compressing a spring, which subsequently returns the gun barrel to its original position. (a) What is the initial recoil velocity acquired by the gun barrel as the gun is fired? (b) What must the spring constant be in order that the gun barrel shall recoil 2.5 ft

on firing? **(c)** What is the ratio of energy imparted to the shell to that acquired by the gun barrel on firing? **(d)** A 40,000-ton ship carries nine guns of this type. The guns are fired simultaneously at 0° elevation while aimed perpendicular to the ship's keel. Find the sideward velocity imparted to the vessel as the guns fire.

12. A body of mass m_1 and initial velocity u_1 moving along the $+x$-direction undergoes a completely inelastic collision with another body of mass m_2 which is moving initially along the $+y$-direction with velocity u_2. Find **(a)** the common velocity v of the two bodies after impact and **(b)** the angle between the final velocity vector and the x-direction.

13. A 3200-pound automobile moving initially due east with velocity 40 ft/sec collides with a 4800-pound car moving southwest with initial velocity 60 ft/sec. After collision, the vehicles stick together. Find the magnitude and direction of the common velocity of the cars immediately after impact. Is energy conserved in this collision?

14. Assuming that the impact phase of the collision described in Problem 13 above lasts 0.1 sec, find the magnitude and direction of the time-average force acting on the 3200-pound car during the collision. What is the magnitude and direction of the time-average force acting on the other vehicle during impact?

15. Object A, of mass 10 kg, moving initially in the $+x$-direction at a speed of 15 m/sec, collides with object B, of mass 25 kg, moving in the $+y$-direction at a speed of 10 m/sec. After collision, object A's velocity vector makes an angle of 50° with the $+x$-axis. Find the final velocities of both objects. Is this a perfectly elastic collision?

16. A variant of the device illustrated in Fig. 6.12 is shown in the accompanying diagram. Show that if N balls of mass M are drawn aside and released, n balls of mass m will fly off the other end of the stack, where $n = NM/m$, provided that the collisions between balls are elastic and that M/m is an integer. Discuss qualitatively what will happen when M/m is not an integer.

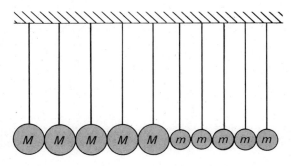

17. An object of mass 10 kg and initial velocity of 16 m/sec due east collides with a second object of mass 15 kg and initial velocity 8 m/sec southwest. After the impact, the second object has a velocity of 7 m/s in a direction 37° below the $+x$-axis. Find the magnitude and direction of the velocity vector of the first object after collision. Is this collision elastic?

18. A ball of mass m, suspended as shown in the diagram, is released from height h and allowed to strike a stationary mass M resting on a horizontal frictionless surface when

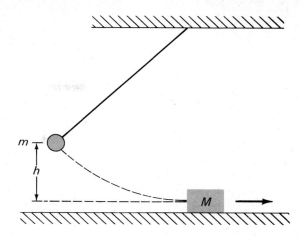

it reaches the lowest point in its path. Find the velocity V of mass M and the velocity v of mass m immediately after impact, assuming that the collision is perfectly elastic.

*19. A stream of liquid of cross-sectional area A and constant initial velocity v_0 is directed against the end of a railroad car of mass M, which is initially at rest on a level track. The situation is illustrated in the figure. Neglecting all frictional effects, show that after time t has elapsed, the car's velocity is

$$v(t) = v_0(1 - e^{-\rho A v_0 t/M})$$

where ρ is the mass per unit volume of the liquid. Show also that after time t the car will have covered a distance given by

$$x(t) = v_0 t - \frac{M}{\rho A}(1 - e^{-\rho A v_0 t/M})$$

and that the velocity v corresponding to a given position x may be found from

$$x = \frac{M}{\rho A}\left[\ln\left(\frac{v_0}{v_0 - v}\right) - \frac{v}{v_0}\right]$$

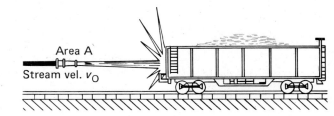

Area A
Stream vel. v_O

*20. A stream of fluid of cross-sectional area A, mass per unit volume ρ, and constant initial velocity v_0 is directed *into* a railroad car of mass M_0 which is initially at rest, as shown in the diagram. Again neglecting frictional effects,

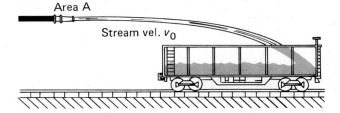

Area A
Stream vel. v_O

show that in this case the answers asked for in Problem 19 will be given by

$$v(t) = \frac{v_0 t}{t + (M_0/\rho A v_0)}$$

$$x(t) = v_0 t - \frac{M_0}{\rho A} \ln\left(1 + \frac{\rho A v_0 t}{M_0}\right)$$

$$x = \frac{M_0}{\rho A}\left[\frac{v}{v_0 - v} - \ln\left(\frac{v_0}{v_0 - v}\right)\right]$$

21. In the most general two-dimensional inelastic collision, a body of mass m_1 and velocity $\mathbf{u}_1$ collides with a body of mass m_2 and velocity $\mathbf{u}_2$. After the collision, the bodies have the same final velocity $\mathbf{v}$ which lies in the same plane as $\mathbf{u}_1$ and $\mathbf{u}_2$ but has, of course, a different direction. Show that

$$v_x = \frac{m_1 u_{1x} + m_2 u_{2x}}{m_1 + m_2}$$

and

$$v_y = \frac{m_1 u_{1y} + m_2 u_{2y}}{m_1 + m_2}$$

22. An object of mass m_1 moves initially along the $-y$-direction with velocity u and collides with a second body of mass m_2 which is moving in the $+x$-direction with twice that velocity. The collision is perfectly inelastic, the two bodies clinging together after collision; their final mutual velocity is $\frac{3}{2}$ the initial velocity of mass m_1. Find (a) the ratio of masses m_1/m_2 and (b) the angle the final velocity vector of the bodies makes with the x-axis

23. An object of mass m_1 moves initially with velocity u along a path which makes an angle of $53°$ with the x-axis, as shown in the figure. It collides with a second body having twice its mass which has been moving along the $+x$-direction with three times its velocity. After collision, the first mass is found to be traveling in the $-y$-direction, while the second continues to move along the $+x$-axis at a reduced speed. Find the speeds of both objects after the collision in terms of the initial velocity u of mass m_1.

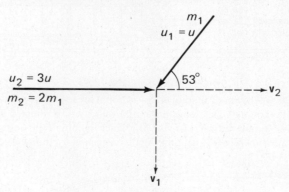

24. Suppose, in Problem 23 above, the second mass had been deflected to a final path whose direction was $5°$ below the x-axis. What would the answers to the problem be in this case?

*25. Two billiard balls of equal mass m and equal radii r_0 collide, one of them initially being at rest. (a) Show that

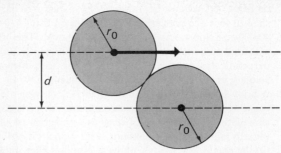

after the collision their velocities are perpendicular. (b) Find a relation between the angle of deflection of the incident ball and the impact parameter d shown in the diagram as the distance between centers of the billiard balls in a direction perpendicular to the incident velocity. *Hint:* Use conservation of momentum along the lines of centers and perpendicular to the line of centers.

26. An electron collides elastically with a deuterium atom. After the collision, the electron and the deuterium atom move in the same direction. What fraction of the electron energy has been transferred to the deuterium?

27. On a frictionless table, a 6-kg block moving at a speed of 1 m/sec collides with a second 6-kg block which was initially at rest. The collision is completely inelastic, and the two blocks stick together after the collision. (a) What is the velocity of the center-of-mass frame with respect to the table? (b) Find the velocity of each of the blocks before and after the collision in the center-of-mass frame (c) What is the final kinetic energy of the system in the table's reference frame? (d) What fraction of the initial kinetic energy has been dissipated?

28. A strange 200-pound man is standing on a completely frictionless ice-covered pond with his 10-pound Siamese cat in his arms. He navigates by periodically throwing his belongings. (a) He throws his 2-pound lunch box with a speed of 40 ft/sec in a horizontal direction. What speed does he acquire as a result? (b) He's tired of coasting with the speed obtained in (a) and wishes to come to rest. Unfortunately, he has already squandered most of his possessions and finds that he must throw his beloved cat in order to stop his motion. With what *relative* speed should he toss the cat in order to come to rest?

29. Two protons collide elastically, one of them initially being at rest in the laboratory frame. The diagram shows the collision as viewed in the center-of-mass frame. The protons come together, each having a speed of 10^8 cm/sec. The straight line connecting the particles has rotated by $90°$ as a consequence of the collision. Find the magnitude

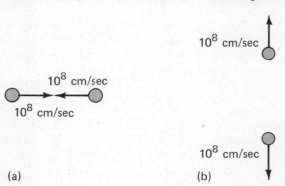

and direction of the initial and the *final* velocities in the laboratory reference frame.

30. Two pendulums, each having a length of 4 ft, are suspended from the same point, as shown in the diagram. The one on the left has a mass of 2 slugs, while the one on the right has a mass of 1 slug. The coefficient of restitution for a collision between the attached weights is 0.5. If the left pendulum is released from rest from the horizontal position as shown in the diagram, find the final maximum heights reached by each of the weights.

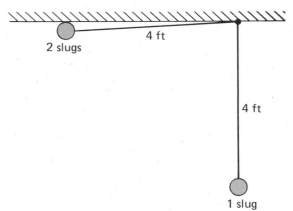

31. Two steel spheres, suspended as shown in the diagram, are displaced sideways and released from rest so as to collide as they reach the lowest point of their respective paths. The initial velocity of the left-hand sphere is 10 cm/sec and its mass is 90 grams. The right-hand sphere, whose mass is 20 grams, has a speed of 30 cm/sec just before collision. Immediately after the collision, the speed of the right-hand mass is observed to be 33.0 cm/sec, to the right. Find **(a)** the velocity of the left-hand mass immediately after collision and **(b)** the amount of kinetic energy dissipated in the collision. **(c)** If the spheres had been made of putty and had stuck together after collision, what would the magnitude and direction of their velocity have been just after the collision?

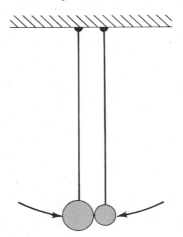

32. A student finds himself on the surface of a frozen river 100 ft wide, 50 ft from shore. We are not told how he got there initially, but he finds it impossible to get to shore because he is at rest and the surface is absolutely frictionless. Another student, on the river bank, observes his plight and tosses his 5.0-pound physics text, which the first student catches. The book is thrown at an angle of 37° to the horizontal with an initial speed of 40.8 ft/sec. **(a)** Assuming that the first student, who catches the book, weighs 160 pounds, how long will it take after the catch for him to reach the opposite bank of the river? **(b)** Is this "collision" elastic or inelastic?

33. A 10-kg block moving east with an initial speed of 0.2 meter/sec along a frictionless horizontal plane is struck by a 5-gram bullet moving horizontally due north. The bullet embeds itself in the block, and they move off together at an angle of 37° north of east. Find **(a)** their common speed after impact and **(b)** the bullet's speed upon impact.

34. A bullet of mass m is fired horizontally into a block of mass M which is resting on a level horizontal surface. The coefficient of kinetic friction between block and surface is μ_k. The bullet becomes imbedded in the block, which moves a distance d along the plane before coming to rest once again. Find the initial speed of the bullet in terms of the measurable quantities m, M, μ_k, and d.

35. A bullet of mass 10.0 grams is fired horizontally into a block of wood of mass 1.20 kg, which is suspended as a pendulum by a light string of length 2.0 meters as illustrated in the figure. It quickly penetrates this block and, upon emerging, travels horizontally a short distance, then embeds itself in a block of mass 0.5 kg resting on a horizontal level surface in the manner described in the preceding problem. The pendulum is observed to rise to a maximum height of 0.707 meter. The second block slides a distance of 0.24 meter along the plane before coming to rest again. The coefficient of friction between the second block and the plane on which it rests is 0.20. Find **(a)** The initial velocity of the bullet and **(b)** its velocity after emerging from the pendulum.

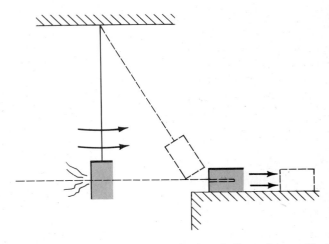

36. A stream of water from a garden hose is directed normally onto a vertical wall. The rate of flow is 45 kg/minute, and the velocity of the stream as it hits the wall is 6.0 meters/sec. Assuming that the velocity of the stream is reduced to zero after it hits the wall, find **(a)** the force exerted by the stream on the wall and **(b)** the force required to hold the nozzle steady and prevent it from "kicking back."

37. An object weighing 8 pounds experiences a force that increases linearly from zero to a maximum value of

18 pounds in a time interval of 2.5 sec. Assuming the body to be at rest initially, what is its speed at the end of the interval?

38. An object of mass 2.0 kg has an initial x-component of velocity of 2.4 meters/sec and an initial y-component of velocity of -3.6 meters/sec. It receives an impulse given by $3.2\mathbf{i}_x + 4.8\mathbf{i}_y$, the numerical coefficients having dimensions of newton-sec. Find (a) the final velocity of the object, (b) its final kinetic energy, and (c) the work done on it by the forces acting during the impulse.

39. A body of mass 3.6 kg and initial velocity 6.0 meters/sec makes a "head on" elastic collision with another body, which is initially at rest. After the collision, the first body continues in its original direction of motion at a speed of 2.4 meters/sec. What is the mass of the second body? What is its speed after the collision?

40. In the preceding problem, suppose that the first body has mass M and initial velocity u_0 and the second is of mass m. The speed of the first body after the collision is reduced to the value αu_0 (where $\alpha < 1$). Show that the mass of the second body is given by $M(1 - \alpha)/(1 + \alpha)$ and its speed after collision is $u_0(1 + \alpha)$.

41. A truck weighing 6400 pounds passes beneath an overpass at an initial constant speed of 30 ft/sec. A bale of cotton weighing 480 pounds is dropped into it from the overpass. What is the final velocity of the truck and bale of cotton?

42. An ore boat weighing 8000 tons and having an initial speed of 8.0 ft/sec passes underneath a loading chute which discharges 200,000 pounds of ore per second into its hold for a period of 100 sec, as illustrated in the figure. Find (a) its initial rate of deceleration, (b) its velocity and acceleration after 50 sec, and (c) its velocity and acceleration after 100 sec.

43. In the situation described in the preceding problem, let the mass of the boat be m_0, its initial velocity be v_0, and the mass of ore entering the hold every second be R. Show that the boat's velocity as a function of time during the loading process is $v(t) = m_0 v_0/(m_0 + Rt)$ and that its acceleration is given by $a(t) = -(Rv_0/m_0)/[1 + (Rt/m_0)]^2$.

44. (a) An object is dropped from rest onto a platform that is moving upward with a uniform speed of 10 ft/sec. The object's mass is negligible compared to that of the platform. Assuming that the collision with the platform is elastic and that at the time the object is released the platform is 100 ft below the object, find how far above its initial position the object rises at the highest point of its rebound. (b) What is the answer to this problem if the platform is moving downward at 10 ft/sec rather than upward?

45. An object is dropped from rest onto a platform that is

moving upward with a uniform speed v_0. The object's mass is negligible in comparison with that of the platform, as in the preceding problem. Assuming that the object collides elastically with the platform and that its speed at the instant of collision is v, show that the maximum height it attains above the site of the collision is given by $h = (v + 2v_0)^2/2g$.

46. A railroad car is traveling on a frictionless track with speed v_0 carrying a huge load of bananas. The total weight of car and bananas is 40 tons. Unfortunately, 5 tons of bananas are rotten, and it is decided that these should be dumped from the rear end of the car. The crates are placed on rollers and are given a speed of $0.1v_0$ with respect to the railroad car just before they fall off. If there are five crates dumped, one at a time, find the final percentage increase in the speed of the railroad car.

*47. A rocket with initial mass of 10^5 kg is fired vertically upward. Every second it ejects 200 kg of fuel at a velocity of 10^5 m/sec with respect to the rocket. (a) What is the thrust which the fuel exerts on the rocket? (b) How does the total mass of the rocket vary with time? (c) Assuming that the gravitational acceleration g does not vary with time, how does the acceleration of the rocket vary with time?

48. A steel ball falls from rest from a height h_0 and strikes a concrete floor. The coefficient of restitution between the ball and the floor is 0.5. How many bounces must the ball make in order that its energy will fall below 1% of its initial value?

49. A golfer, jealous of his opponent's 300-yard drive, decides to swing as hard as he possibly can when his turn comes. He strikes the ball, giving it an initial velocity directed 30° above the horizontal and a net impulse of 0.5 lb-sec. Unfortunately, shortly after impact, his ball splits into two equal halves both of which continue in the same direction, one with twice the velocity of the other. The weight of the golf ball is 0.101 pound. (a) Find the initial velocity of each part of the golf ball. (b) How far down the fairway does each part go, assuming no air resistance? (c) Since neither part of the golf ball has reached 300 yards, the golfer is disappointed; but he argues that his drive would have exceeded 300 yards if the ball had not split apart. Is he right or wrong?

50. A 10-kg block moving east at 0.2 m/sec along a frictionless horizontal surface is struck by a bullet of mass 0.005 kg moving straight north. The bullet imbeds itself in the block and both move 37° north of east together. Find (a) their common velocity after impact, (b) the bullet's velocity before impact, and (c) the percent loss of kinetic energy.

51. A 0.01-kg bullet is fired horizontally into a 10-kg block of wood which was initially at rest on a table. The bullet comes to rest in the block immediately (inelastic collision), and this block, after traveling a negligible distance, collides with a second stationary 10-kg block, as shown in the diagram. The coefficient of restitution for the second collision is 0.5, and the coefficient of sliding friction

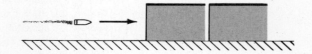

between the blocks and the table is 0.2. The first block travels 4 cm before coming to rest, while the second travels 36 cm. Find the initial speed of the bullet.

*52. Two carts of mass m_1 and m_2 are placed on a frictionless air track of length L. The lengths of the carts are small compared to L. The cart of mass m_1 traveling with velocity v makes an elastic collision with the mass m_2, which is at rest at a distance x from one end of the track, as shown in the diagram. (a) Prove that the velocity of m_1 after impact

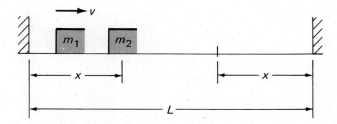

is positive, zero, or negative assuming respectively that m_1 is greater than, equal to, or less than m_2. (b) Assuming that $m_1 > m_2$ and that m_2 makes a subsequent partly inelastic impact with the bumper at the end of the track and meets m_1 again at a distance x from the far end of the track, show that $m_1/m_2 = L/(L - 4x)$.

53. A cart of mass m is traveling with speed v on a frictionless air track. It collides with an identical mass traveling with the same velocity in the opposite direction. Show that the loss of energy is $mv^2(1 - e^2)$, where e is the coefficient of restitution.

54. Two carts having masses m_1 and m_2 are placed on a linear air track. The cart of mass m_1 is initially at rest, and the other collides with it inelastically with initial velocity v. If e is the restitution coefficient, show that the amount of energy lost in the collision, measured in the reference frame of the air track, is $m_1 m_2 v^2(1 - e^2)/[2(m_1 + m_2)]$.

55. Derive Eq. (6.6.35) in Example 6.6.2.

56. Derive the result given as Eq. (6.6.42) in Example 6.6.3.

ROTATIONAL DYNAMICS

7.1 Introduction

Since we have already investigated the most important features of translational dynamics, we may now begin to apply the principles we have established to the study of rotational motion. There is nothing essentially different about the rotational dynamics of a rigid body from the dynamics of a particle in translational motion, and both cases can be completely understood in terms of Newton's laws as we have been using them all along. We could, in principle, calculate the net force acting on every mass element of a rotating body and arrive at a description of the motion of that element by equating this to mass times linear acceleration. But since the velocity and direction of the *linear* motion of elements of an object vary from point to point even in the simplest case of uniform rotational motion about a fixed rotation axis, it is an extremely, and unnecessarily, complex task to approach the problem that way.

It makes our work much easier to observe that a rigid body free to rotate about a given axis remains in rotational equilibrium (at rest or in rotational motion at constant angular velocity) in the absence of externally applied torques. This has been discussed in connection with the static equilibrium of rigid bodies with respect to torques in section 2.5. One might be led to suspect, then, that when a net torque is present, an angular acceleration about the axis might arise and to postulate that the magnitude of the angular acceleration should be directly proportional to the torque. It is experimentally observed that this is, indeed, true. It can be shown moreover, that these results follow directly from Newton's laws for translational motion, provided that torque, angular displacement, angular velocity, and angular acceleration are defined as we have already defined them. The point of doing all this is that while various points within the rotating body experience different forces, linear accelerations, and linear velocities, there is at any given instant only *one single value* of torque, angular acceleration, and angular velocity for the body as a whole. This leads to a much simpler description of the motion of the body and to a much easier and, at the same time, more profound appreciation of the fundamental characteristics of its rotational motion.

A complete treatment of the rotational motion of a dynamical system of the most general type is very complex and will be found to lie far beyond the scope of an introductory text. We shall, therefore, restrict our discussion largely to the rotation of *rigid bodies* about a *fixed axis of rotation*, though some simple yet important examples where the rotation axis is in motion will also be introduced. We shall also be somewhat less insistent on the rigorous derivation of our results than in the preceding chapters, and shall in many instances appeal to physical intuition rather than mathematical rigor in presenting them.

7.2 Newton's Second Law for Rotation: Moment of Inertia

Consider the simplest possible case of a single concentrated mass particle of mass m rotating at a fixed distance r from a fixed rotation axis, as illustrated in Fig. 7.1. Suppose that a tangentially directed net force $\mathbf{F}_t$ is acting on the object. According to Newton's second law, there will be a tangential acceleration $\mathbf{a}_t$ such that

$$\mathbf{F}_t = m\mathbf{a}_t \qquad (7.2.1)$$

But since the angular acceleration α is related to the linear acceleration a_t by

$$a_t = r\alpha \qquad (7.2.2)$$

and since the net force F_t is related to the torque τ about the axis at 0 by

$$\tau = F_t r \qquad (7.2.3)$$

we may write Eq. (7.2.1) in terms of torque and angular acceleration as

$$\tau = (mr^2)\alpha \qquad (7.2.4)$$

Since m and r are both constants, we may conclude that the angular acceleration is directly pro-

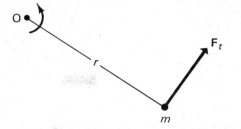

FIGURE 7.1. A point mass rotating at a fixed distance r about point O.

portional to the net resultant torque acting on the body, as a direct consequence of Newton's laws for linear motion. It is very important to observe, however, that the constant of proportionality between torque and angular acceleration is not the mass of the object but rather the quantity mr^2. This quantity is referred to as the *moment of inertia* of the rotating mass particle; and if it is represented by the symbol I in Eq. (7.2.4), the equation has the form

$$\boxed{\tau = I\alpha} \quad \text{where} \quad \boxed{I = mr^2} \qquad (7.2.5)$$

This is the usual way of writing the equation relating torque and angular acceleration for a rotating body and is, in fact, the *rotational analogue of the equation* $F = ma$. There is one important difference, however. In the expression $F = ma$, the factor of proportionality m between force and linear acceleration is a fixed and invariant property of the body in question. In the equation $\tau = I\alpha$, on the other hand, the corresponding factor I which relates torque and angular acceleration involves not only the mass of the object but also the distance r separating it from the axis of rotation. In the example we are discussing at present, it is apparent that if the distance between the mass and the axis of rotation is increased, the moment of inertia will be increased likewise, and a greater torque will then be required to produce a given angular acceleration, and vice versa. In this simple case, the moment of inertia is equal to mr^2. In more complex systems involving more than one mass particle or large, extended bodies, the moment of inertia will be given by more complex expressions, but it will always be true that the moment of inertia of a system will depend not only on the mass but how the mass of the system is situated with respect to the rotation axis.

This may be seen more clearly by an examination of the next simplest configuration, shown in Fig. 7.2. Here, there are n discrete, concentrated mass particles which are rigidly fixed with respect to one another and are disposed at constant distances from a fixed axis of rotation. Certain tangential forces are assumed to act on the particles, whose magnitudes are such that the same angular acceleration α is

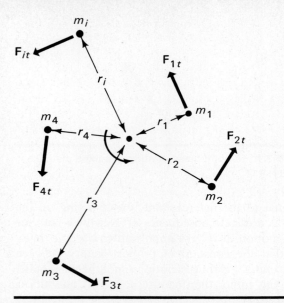

FIGURE 7.2. System of point masses rotating with a common angular velocity about a fixed axis.

produced for each mass (otherwise the system would not have a common angular acceleration or angular velocity!). The angular acceleration of each mass is related to the magnitude of the tangential force it experiences by a relation like that shown in (7.2.1). If the separate masses are numbered $1, 2, 3, \ldots, i, \ldots, n$, the equation of motion for the ith mass would be

$$F_i = m_i a_{ti} \tag{7.2.6}$$

and there would be just such an equation for each mass in the system. If Eq. (7.2.6) is multiplied on both sides by the distance r_i,

$$\tau_i = F_{ti} r_i = m_i r_i a_{ti} = (m_i r_i^2)\alpha \tag{7.2.7}$$

recalling that $a_{ti} = r_i \alpha$, each mass having the same angular acceleration. The quantity τ_i is the torque or moment about the axis associated with the force F_{ti}. The total torque τ acting on the system is the sum of the individual torques acting on each mass:

$$\tau = \tau_1 + \tau_2 + \tau_3 + \cdots + \tau_i + \cdots + \tau_n = \sum_{i=1}^{n} \tau_i \tag{7.2.8}$$

or

$$\tau = \sum_{i=1}^{n} \tau_i = \sum_{i=1}^{n} (m_i r_i^2)\alpha \tag{7.2.9}$$

Since α is the same for every mass in the system, it can be factored out of every term in (7.2.9) and written outside the summation sign, if desired, since it multiplies every term within. The result is

$$\tau = \alpha \sum_{i=1}^{n} m_i r_i^2 \tag{7.2.10}$$

which expresses once again the fact that the angular acceleration is directly proportional to the torque, every term m_i or r_i within the parentheses being constant throughout the motion of the system. The factor of proportionality between τ and α may be identified as the *moment of inertia* for this type of system, which allows us once again to write

$$\boxed{\tau = I\alpha} \tag{7.2.11}$$

with the moment of inertia for a system such as this defined as

$$\boxed{\begin{aligned} I &= \sum_{i=1}^{n} m_i r_i^2 \\ &= m_1 r_1^2 + m_2 r_2^2 + \cdots + m_i r_i^2 + \cdots m_n r_n^2 \end{aligned}} \tag{7.2.12}$$

It is clear that the moment of inertia will depend not only on the masses m_i but also their respective radii r_i, and hence not only on the total mass of the system but also upon how the individual mass particles are placed relative to the axis of rotation.

In performing the summation indicated in going from (7.2.7) to (7.2.8), it might seem suspect to add force magnitudes which are, after all, the magnitudes of vectors all of which have different directions. But if we recall that the *vector torque* is a vector which is normal both to the force vector and to the radius vector connecting its point of application to the point about which moments are taken, it is apparent that the vector torques arising from the individual tangential forces in Fig. 7.2 *all point in the same direction*, namely, directly out of the page, normal to its plane. Hence, the magnitudes of all these individual torques may be added algebraically to obtain the magnitude of the total torque τ.

The final example of this section is the most complex and, at the same time, the most interesting and important one. Consider now the case of an extended irregular rigid body rotating about a fixed rotation axis, as shown in Fig. 7.3. In the illustration, the axis of rotation passes through the body, although the results obtained are true whether it does or not. Any rigid body may be thought to be built up of mass elements of infinitesimal size, such as dm, which may be regarded as representative of any such element. A small net tangential force of magnitude dF_t acts on such an element, causing linear acceleration a_t, which may be different for different mass elements but which must lead to an *angular* acceleration α which must be the same for all mass elements, otherwise the body would be anything but rigid! The tangential force magnitude dF_t is related to the tangential linear acceleration a_t by

$$dF_t = a_t \, dm \tag{7.2.13}$$

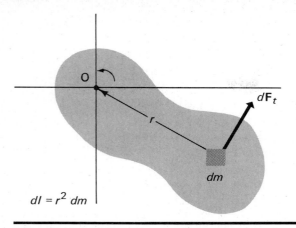

FIGURE 7.3. A rigid body rotating about a fixed rotation axis at O.

If both sides of this equation are multiplied by the constant distance r which separates the element dm from the rotation axis, the magnitude of the small moment $d\tau$ associated with the force dF_t acting about the rotation axis at O is seen to be

$$d\tau = r\, dF_t = (r\, dm)a_t = (r^2\, dm)\alpha \qquad (7.2.14)$$

since once more, $a_t = r\alpha$ for every point in the object. The total torque τ may now be obtained by integrating the above equation over every tiny mass element in the body. This leads to

$$\tau = \int_v d\tau = \int_v (r^2\, dm)\alpha = \alpha \int_v r^2\, dm \qquad (7.2.15)$$

The subscript v on the integral is used simply to remind us that the integration is to be performed over the entire volume of the object. The factor α can be taken outside the integral because it is the same for every mass element within the body and hence is a *constant with respect to integration over all such mass elements* (though not necessarily constant with respect to time). Equation (7.2.15) once more expresses the fact that the angular acceleration and the resultant torque are proportional. In this instance, however, the proportionality factor appears in the form of an integral. This integral is a constant factor for any given body and any given rotation axis, though its value will depend upon the size and shape of the object and also, for a given body, upon the location and orientation of the axis of rotation. As before, however, we may identify it as the moment of inertia of the object and, for a rigid body, write the equation

$$\boxed{\tau = I\alpha} \qquad (7.2.16)$$

with

$$\boxed{I = \int_v dI = \int_v r^2\, dm} \qquad (7.2.17)$$

In this expression, $dI = r^2\, dm$ may be regarded as the contribution of a representative mass element dm to the total moment of inertia of the body.

The details of determining the moment of inertia associated with mass systems of various types will be discussed in detail in the next section. It should be noted that the dimensions of the moment of inertia are in every instance mass times (distance)^2, hence, in the English system, lb-ft^2; in the cgs-metric system, g-cm^2; and in the mks-metric system, kg-m^2. We shall see later in this chapter that in rotational motion the moment of inertia plays the role assumed by mass in translational motion, not only in Newton's second law for rotation but also in the expressions we shall derive for rotational momentum and kinetic energy. Indeed, the moment of inertia of a body is occasionally referred to as its *rotational inertia*.

The reader may wonder why the forces acting on mass elements were in all cases assumed to have no radial components. This was done purely as a matter of convenience. But since the line of action of any radial force in all instances passes through the rotation axis about which torques are taken, it is clear that no torque or angular acceleration will ever arise on their account even when they are present. Such force components may, however, create forces on the bearings in which the rotation axis is mounted and would have to be considered in any example in which such forces are sought.

In many important cases, masses that execute translational motion may be *coupled* to the rotational motion of other elements which may turn about either fixed or moving axes of rotation. In working out the dynamics of systems of this type, one may apply Newton's second law for translational motion ($F = ma$) to the elements undergoing translation and Newton's second law for rotation ($\tau = I\alpha$) to the rotating bodies. The coupling between the two can generally be expressed by the previously developed relations between angular displacement, velocity, and acceleration and their linear analogues ($a = r\alpha$, for example). A system of equations may thus be obtained which can be solved for all angular and linear accelerations. The examples discussed below will illustrate how all this is accomplished in actual practice.

EXAMPLE 7.2.1

A small 5-kg mass rotates at a constant radius of 1.5 m about a fixed frictionless axis of rotation as shown in Fig. 7.1. Assuming the object is so small that its behavior is that of a mass particle concentrated at the center of mass and that the mass of the rod attaching it to the axis of rotation is so small as to be negligible, find (a) the angular acceleration produced by a constant tangentially directed force of 8 newtons, (b) the time required to accelerate the system from rest to a

201

ROTATIONAL DYNAMICS

rotational speed of 10 revolutions per second (rps), and (c) the number of revolutions executed before this speed is reached.

The torque and angular acceleration are related by Newton's second law for rotating systems:

$$\tau = I\alpha \qquad (7.2.18)$$

In this case, the torque is simply the product of the tangential force and the radius, while if we may consider the object as a point mass concentrated at its center of mass, the situation corresponds to that discussed in the previous material leading up to Eq. (7.2.4). In this simple case, the moment of inertia is given by the point-mass expression

$$I = mr^2 \qquad (7.2.19)$$

Using these concepts, it is easily seen that (7.2.18) becomes

$$(8)(1.5) = (5)(1.5)^2 \alpha$$

$$\alpha = \frac{(8)(1.5)}{(5)(1.5)^2} = 1.067 \text{ rad/sec}^2$$

which is the answer to part (a).

Since the torque is constant, so is the angular acceleration. We have already examined the characteristic features of uniformly accelerated rotational motion, and we have shown that the angular velocity may be expressed as

$$\omega = \omega_0 + \alpha t \qquad (7.2.20)$$

In this example, $\omega_0 = 0$, $\omega = 10 \text{ rps} = (10)(2\pi) \text{ rad/sec}$, while $\alpha = 1.067 \text{ rad/sec}^2$. Substituting these values into (7.2.20) and solving for t, we obtain

$$62.8 = 0 + 1.067t$$

$$t = \frac{62.8}{1.067} = 58.8 \text{ sec}$$

as the answer to part (b).

Finally, another familiar result of the theory of uniformly accelerated rotational motion is that

$$\omega^2 - \omega_0^2 = 2\alpha\theta \qquad (7.2.21)$$

where θ is the angular displacement measured from the starting point, where the angular velocity has the value ω_0. In our example, $\omega_0 = 0$, $\alpha = 1.067 \text{ rad/sec}^2$, and $\omega = 10 \text{ rps} = 62.8 \text{ rad/sec}$. Substituting these values into (7.2.21), it is evident that

$$(62.8)^2 - 0 = (2)(1.067)\theta$$

and

$$\theta = \frac{62.8^2}{2.134} = 1848 \text{ rad}$$

or

$$\theta = \frac{1848}{2\pi} = 294 \text{ rev}$$

This result could also have been obtained by substituting $t = 58.8 \text{ sec}$, $\alpha = 1.067 \text{ rad/sec}^2$, and $\omega_0 = 0$ into the equation $\theta = (1/2)\alpha t^2 + \omega_0 t$, which should now be familiar to the reader.

One may reasonably inquire as to when the assumption of a point mass for a rotating object is a good one and when it is not. In general, this approximation will be a very good one, allowing us to write the moment of inertia simply as mr^2 (or $\sum_i m_i r_i^2$ if there is a system of a number of such objects rigidly connected), provided that the size of the object itself is small compared to the distance between its center of mass and the rotation axis. In this example, had the object been a 5-kg sphere of lead (density 11.3 g/cm³), its radius, of about 4.7 cm, would have been small compared to the distance of 150 cm between the center of mass and the axis of rotation. Had the body been 5-kg sphere of styrofoam (density 0.025 g/cm³), however, it would have had a radius of over 36 cm, which is not much smaller than the 150 cm separating its center and the axis of rotation. In the former case, the idealization of the object as a point mass concentrated at the center of mass would be an excellent approximation; but in the latter case, this picture would certainly be a good deal less accurate. This point-mass assumption will be examined in a more quantitative way in section 7.4 in connection with Example 7.4.3.

EXAMPLE 7.2.2

Three 6-kg masses (regarded as point objects) are attached rigidly to a fixed frictionless rotation axis by light spokes, whose inertia may be neglected. The masses form an equilateral triangle when viewed from above, and each is located 2.5 m from the axis of rotation. A light drum of radius 0.5 m, whose inertia is negligible, is attached to the system, and a tangential force **F** is exerted on a rope wrapped around it, as illustrated in Fig. 7.4. How large must this force be for the system to have an angular acceleration of 5 rad/sec²?

In this case, there are a number of rigidly connected objects which behave as a system of point masses, as discussed in the second case considered in this section leading to Eq. (7.2.10). In this instance, the moment of inertia is

$$I = \sum_i m_i r_i^2 = m_1 r_1^2 + m_2 r_2^2 + m_3 r_3^2 \qquad (7.2.22)$$

In our example, $m_1 = m_2 = m_3 = 6 \text{ kg}$ and $r_1 = r_2 = r_3 = 2.5 \text{ m}$, so that

$$I = (6)(2.5)^2 + (6)(2.5)^2 + (6)(2.5)^2 = 112.5 \text{ kg-m}^2$$

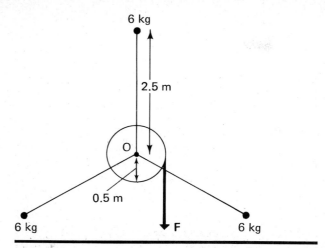

FIGURE 7.4

Then, according to Newton's second law for rotational motion, if r_d denotes the radius of the drum upon which the rope is wrapped,

$$\tau = F r_d = I\alpha$$

$$\tau = F(0.5) = (112.5)(5)$$

$$F = \frac{(112.5)(5)}{0.5} = 1125 \text{ N}$$

EXAMPLE 7.2.3

A toy gyroscope rotor has a moment of inertia of 3000 g-cm². A string 80 cm long is wrapped around the shaft, which is 0.5 cm in diameter, and pulled off with a constant force of 25 pounds. What is the angular acceleration of the rotor and what is its final rotational speed, assuming that frictional effects are negligible?

In this problem, we must first note that a mixed set of units has been employed in setting forth the data. Specifically, the force of 25 pounds must be converted to dynes. Since a mass of 454 grams exerts the same weight force as that associated with a 1-pound weight, 1 pound of force corresponds to (454)(980) dynes, and 25 pounds corresponds to $(25)(454)(980) = 1.112 \times 10^7$ dynes. Denoting the shaft radius by r,

$$\tau = F r = I\alpha$$

$$\tau = (1.112 \times 10^7)(0.25) = 3000\alpha$$

$$\alpha = \frac{(1.112 \times 10^7)(0.25)}{3000} = 927 \text{ rad/sec}^2$$

corresponding to $(927/2\pi) = 147.6$ rev/sec²

Now, when the string is pulled through a distance of one shaft radius, the shaft turns through an angle of 1 radian. The total angular displacement of the shaft of the rotor, therefore, is the length of the

string divided by the radius. In other words,

$$\theta = \frac{80}{0.25} = 320 \text{ rad}$$

The rotational speed is related to the angular displacement by

$$\omega^2 - \omega_0{}^2 = 2\alpha\theta$$

Since $\omega_0 = 0$,

$$\omega^2 = (2)(927)(320) = 593,000$$

$$\omega = 770 \text{ rad/sec}$$

This rotational speed can be expressed as $(770/2\pi) = 122.4$ rps, or 7350 rpm.

EXAMPLE 7.2.4

A mass m hangs from the end of a rope which is wrapped around a drum of radius R and moment of inertia I_0, as shown in Fig. 7.5a. Assuming that the drum rotates about its axis without friction, find (a) the angular acceleration of the drum, (b) the linear acceleration of the mass, and (c) the tension in the rope when the mass is allowed to descend under its own weight. Find numerical values for these quantities, assuming $m = 64$ pounds, $I_0 = 3$ slug-ft², and $R = 1.5$ ft.

It is useful in this case, as in many other mechanics problems, to consider the forces (or torques) acting on each isolated element of the system and to arrive thereby at a system of algebraic equations which may be solved for the desired set of unknowns. The forces acting on the drum are shown in Fig. 7.5b. The tension T acts to produce a vector torque about the rotation axis whose direction is into the page and, hence, whose magnitude may be regarded as *negative* if the conventions we adopted regarding this matter in Chapter 2 (section 2.7 and Example 2.7.1) are observed. The resulting net torque, $-TR$, can be related to the angular acceleration of the drum:

$$-TR = I_0\alpha \qquad (7.2.23)$$

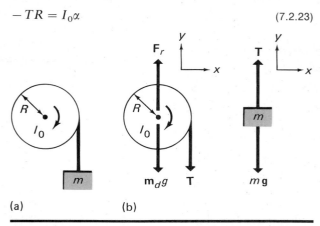

(a)　　　　(b)

FIGURE 7.5

The force arising from the weight of the drum, $m_d g$, acting downward through the center of mass of the drum, which is assumed to coincide with the axis of rotation, causes no contribution to the sum of the moments about that axis. Likewise, the force $\mathbf{F}_r$, which represents the upward force exerted by the shaft mountings on the rotation axis, has a line of action passing through the rotation axis and does not contribute to the sum of moments about that axis. These two forces, which in this example are equal and opposite, serve simply to maintain the system in translational equilibrium in the y-direction. They play no role in determining the acceleration of either drum or attached mass, and need not even be explicitly included if that information is all that is required, though they should always be considered for the sake of completeness.

The forces acting on the suspended mass are shown in Fig. 7.5c. Assuming that the positive y-axis points upward, we may write

$$F_y = T - mg = ma_y \qquad (7.2.24)$$

In Eqs. (7.2.23) and (7.2.24) there are three unknown quantities, namely, α, a_y, and T. To determine all three, another equation is needed. This is provided by the relation connecting a_y and α, which in this example is

$$a_y = R\alpha \qquad (7.2.25)$$

We have now three equations, (7.2.23), (7.2.24), and (7.2.25), which must be solved for a_y, α, and T. This is most easily done by multiplying Eq. (7.2.24) on both sides by R, then adding (7.2.23) and (7.2.24) to eliminate T, and finally expressing a_y in terms of α throughout in the resulting expression by using (7.2.25). We obtain

$$-mgR = I_0\alpha + mRa_y = I_0\alpha + mR^2\alpha$$

which may be solved for α to give

$$\alpha = \frac{-mgR}{I_0 + mR^2} \qquad (7.2.26)$$

From (7.2.25) it follows at once that

$$a_y = R\alpha = \frac{-mgR^2}{I_0 + mR^2} \qquad (7.2.27)$$

and from (7.2.23),

$$T = -\frac{I_0\alpha}{R} = \frac{mgI_0}{I_0 + mR^2} \qquad (7.2.28)$$

Had we been asked to find the upward force F_r exerted on the shaft of the drum by its bearings, we could have done so by noting that the drum *remains in equilibrium* insofar as its *translational* motion is concerned and, hence, that the sum of the

y-components of all forces acting on it must be zero, giving

$$\sum F_y = F_r - m_d g - T = 0$$

or

$$F_r = m_d g + T = m_d g + \frac{mgI_0}{I_0 + mR^2} \qquad (7.2.29)$$

Note that if $I_0 = 0$, the above equations give $T = 0$ and $a_y = -g$ corresponding to free fall of the mass m under the influence of gravity.

If we assume that the suspended object weighs 64 pounds ($m = 2$ slugs), that $I_0 = 3$ slug-ft^2, and that $R = 1.5$ ft, we obtain from (7.2.26), (7.2.27), and (7.2.28)

$$\alpha = \frac{-(64)(1.5)}{3 + (2)(1.5)^2} = 12.8 \text{ rad/sec}^2$$

$$a_y = R\alpha = (1.5)(12.8) = 19.2 \text{ ft/sec}^2$$

and

$$T = \frac{(64)(3)}{3 + (2)(1.5)^2} = 25.6 \text{ lb}$$

Questions regarding the angular displacement or velocity of the drum as a function of time, or the angular velocity as a function of angular displacement, may, of course, be answered by using the familiar "golden rules" for uniformly accelerated rotational motion, since the angular acceleration α is constant. Likewise, similar questions having to do with the linear motion of the suspended mass may be resolved by using the "golden rules" for uniformly accelerated translational motion, in view of the fact that the linear acceleration of that body is also constant.

In this example, and the ones considered previously, it is assumed that the bearings that support the rotating parts of the system are frictionless. Although this assumption is a very good one in many cases, there are instances in which frictional effects hinder the rotational motion and must be accounted for in the calculations. In most cases, the effect of friction manifests itself as a constant *frictional torque* τ_f whose magnitude is independent of angular velocity and whose direction is opposite that of the angular velocity vector. The magnitude of the frictional torque must be specified, of course, to allow us to completely determine the motion of the system when rotational friction is of any importance.

In the above example, had there been a frictional torque τ_f associated with the bearings, we should have had to write (7.2.23) in the form

$$\tau_f - TR = I_0\alpha$$

In this expression, the sign of frictional torque τ_f is dictated by the fact that the angular velocity vector ω points into the page, which means that its magnitude is *negative*, hence that τ_f, whose sign must be opposite that of ω, is *positive*. Had the system been rotating in the opposite direction, then τ_f would have been prefaced by a minus sign. The student will find it instructive to work this example with the friction torque τ_f present; this task is assigned as one of the problems at the end of the chapter.

EXAMPLE 7.2.5
ATWOOD'S MACHINE WITH A PULLEY
HAVING SIGNIFICANT ROTATIONAL INERTIA
In this example, we shall reconsider the apparatus described in Example 4.3.7, called Atwood's machine, which is used to determine accurately the acceleration of gravity. In our previous analysis of this device, it was assumed that the pulley had neither friction nor rotational inertia in any significant amount. We shall now rework this problem, continuing to neglect friction but including the effect of the rotational inertia of the pulley, comparing the results now obtained with those derived previously to see to what extent this effect might be important in practice.

Atwood's machine is illustrated in Fig. 7.6a. Two masses, m_1 and m_2 ($m_2 > m_1$), are suspended from a frictionless pulley which now has a moment of inertia I. The weights are released, and by observing their accelerations a determination of g is effected. For the moment, let us suppose that the unknown quantities are the accelerations of the weights, the angular acceleration of the pulley, and the tensions in the strings. If we can express these quantities in terms of the masses, the moment of inertia, the radius of the pulley, and the value of g, it is an easy matter to turn the situation around and assume that the accelerations are known and the value of g to be determined.

The reader may be puzzled by the fact that in the previous analysis of this device, given in Example 4.3.7, the tensions T and T' shown in Fig. 7.6b were assumed to be equal, while now they are regarded as having different values. The explanation of this lies in the fact that previously the pulley was assumed (tacitly, at least) to have *no rotational inertia*. Under this assumption, no net torque would be required to change the angular velocity of the pulley, and the angular acceleration imparted to the pulley by the linear motion of the weights could be imparted to it even though the equal tensions in the two strings give rise to no net torque on the pulley. This inertialess pulley is regarded simply as changing the direction of a tension force without affecting its magnitude. The situation may be understood more clearly by taking the opposite point of view and inquiring why, after all, the two tensions *should* be equal.

Clearly, any actual pulley must have some rotational inertia and, therefore, cannot possibly undergo angular acceleration unless some net torque is present. But if the two tensions T and T' are precisely equal, the positive torque due to T is exactly balanced by a negative torque of equal magnitude arising from T' and the net torque, and hence the angular acceleration must be zero. Therefore, no angular acceleration of any actual pulley can occur *unless the two tensions are different*. If the pulley is large and its mass is large compared to the two suspended masses, this difference between the two tensions may be very great. On the other hand, if the pulley is small and light compared to the weights of the suspended objects, the small amount of force required to effect any angular acceleration it may attain can be supplied by a difference in T and T' which may be very small, indeed, compared to either of those two quantities themselves. In this instance, it simplifies the work of analyzing the situation to *neglect* the rotational inertia of the

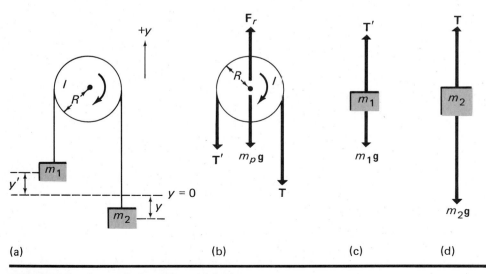

(a) (b) (c) (d)

FIGURE 7.6. Atwood's machine revisited.

pulley altogether, which, in effect, amounts to assuming that T and T' are equal. We are then back to the "ideal" inertialess pulley envisioned in our previous treatment of this and many other systems. There are many instances of practical importance in which the assumption of the "ideal," inertialess pulley is very good, indeed, but there are also some in which the inertia of the pulley is of importance, and in such cases the tensions on either side cannot be assumed to be equal in magnitude. The results of the calculations to be made below will illustrate more clearly when a given pulley can be assumed to be "ideal" and when it cannot.

In describing the motion of the system shown in Fig. 7.6, we may again proceed by isolating the different elements of the systems and considering the various forces and torques acting on each of them. The individual forces and torques acting on the pulley and on each of the masses are illustrated at (b), (c), and (d) in the diagram. The tensions T and T', give rise to torques on the pulley, the former corresponding to a torque vector pointing into the page (hence to a torque of negative magnitude), and the latter corresponding to a torque vector out of the page (hence positive). The *resultant* torque, equal to the algebraic sum of these two individuals torques, may be equated to the moment of inertia I of the pulley times its angular acceleration α, giving

$$T'R - TR = I\alpha \qquad (7.2.30)$$

The force corresponding to the weight of the pulley, $m_p\mathbf{g}$, and the upward force exerted on the pulley by its bearings, $\mathbf{F}_r$, are shown in Fig. 7.6b. These forces cause neither linear nor angular acceleration, since the pulley is in translational equilibrium and since the moments of both these forces about the rotation axis are zero. We therefore need not consider them any further, though we would have to consider them had we been asked[1] to find F_r. The net force in the y-direction acting on each suspended mass may be related to the respective acceleration by

$$\sum F_y = T - m_2 g = m_2 a_y \qquad (7.2.31)$$

and

$$\sum F_y = T' - m_1 g = m_1 a_y' \qquad (7.2.32)$$

where a_y and a_y' are the accelerations of m_2 and m_1, respectively. Also, the linear acceleration of m_2 and the angular acceleration of the pulley are related by

$$a_y = R\alpha \qquad (7.2.33)$$

We now have four equations to determine the five

[1] When you're done with the calculations below, you should be able to come back and find F_r very easily. Try it!

unknowns a_y, a_y', α, T, and T'. Another relation between these quantities must, therefore, be found. This is easily accomplished by noting that if the string is inextensible, any upward velocity v_y' attained by m_1 must be equal in magnitude and opposite in sign to the corresponding downward velocity v_y acquired by m_2; were it not so, the string would have to stretch. Then,

$$v_y' = -v_y$$

and differentiating with respect to time

$$\frac{dv_y'}{dt} = -\frac{dv_y}{dt}$$

or

$$a_y' = -a_y \qquad (7.2.34)$$

We must now solve the five equations (7.2.30) to (7.2.34) for the five required quantities. This can best be done by first substituting (7.2.34) into the first three of these equations, then substituting (7.2.33) into the three resulting expressions to obtain

$$(T' - T)R = I\alpha$$

$$T' - m_1 g = -m_1 R\alpha$$

$$T - m_2 g = m_2 R\alpha$$

The second of these equations may now be subtracted from the third and the quantity $T - T'$ in the expression so obtained replaced by $I\alpha/R$ to which, according to the top equation, it is equal. Solving for angular acceleration, one may show that

$$\alpha = -\frac{(m_2 - m_1)gR}{I + (m_1 + m_2)R^2} \qquad (7.2.35)$$

Then,

$$a_y = R\alpha = -\frac{(m_2 - m_1)gR^2}{I + (m_1 + m_2)R^2} = -a_y' \qquad (7.2.36)$$

while from (7.2.31) and (7.2.33) and the above equations,

$$T = m_2(a_y + g) = \frac{m_2 g(I + 2m_1 R^2)}{I + (m_1 + m_2)R^2} \qquad (7.2.37)$$

$$T' = m_1(a_y' + g) = \frac{m_1 g(I + 2m_2 R^2)}{I + (m_1 + m_2)R^2} \qquad (7.2.38)$$

The difference between the two tensions on either side of the pulley, from the two equations directly above, is

$$T - T' = \frac{(m_2 - m_1)gI}{I + (m_1 + m_2)R^2} \qquad (7.2.39)$$

First of all, it should be observed that if the moment of inertia I is set equal to zero, Eq. (7.2.36) reduces to the "ideal" pulley value obtained previously as Eqs. (4.2.21), where the difference in sign in the case of a_y simply reflects a different choice of direction for the positive y-axis in the two calculations. Also, if I is set equal to zero in (7.2.37) and (7.2.38), both of these reduce to the previously obtained expression for the common tension on both sides of the "ideal" inertialess pulley of section 4.2. In both instances, it can easily be seen that the expressions given above *will differ very little from those obtained in section 4.2 whenever the moment of inertia I is much less than* $(m_1 + m_2)R^2$. This statement tells us when it is permissible to use the simplified "ideal" pulley and when it is not. Also, from (7.2.39) it may be observed that the difference in tensions reduces to zero when I is set equal to zero; and from (7.2.37) and (7.2.38) it can be seen that the values of T and T' both reduce to that found previously in Eq. (4.2.22) in this case.

In this example, it is assumed that there are no frictional torques present in the pulley's bearings. It should be obvious, however, that frictional torques, when present, can also lead to differences in the tensions T and T', even in situations where the pulley's rotational inertia may be negligible.

In trying to determine the value of g by using Atwood's machine, we would determine experimentally the value of the acceleration a_y or $a_y{}'$ using accurately known masses and a pulley whose radius and moment of inertia were accurately known. We would then determine g from Eq. (7.2.34), which, when solved for g, reads

$$g = a_y{}' \frac{I + (m_1 + m_2)R^2}{(m_2 - m_1)R^2} \qquad (7.2.40)$$

Now, for a pulley fashioned in the form of a uniform *circular disk* of mass m_p and radius R, it can be proved (as we shall see in the next section) that

$$I = \tfrac{1}{2}m_p R^2 \qquad (7.2.41)$$

Under these circumstances, (7.2.40) becomes

$$g = a_y{}' \frac{m_1 + m_2 + \tfrac{1}{2}m_p}{m_2 - m_1} \qquad (7.2.42)$$

This result is the same as our previous value from Eq. (4.3.21) which neglects the rotational inertia of the pulley, *except for the presence of the term* $\tfrac{1}{2}m_p$ *in the numerator.* If $\tfrac{1}{2}m_p$ is much smaller than $m_1 + m_2$, then the two expressions will lead to numerical results for g which differ only very slightly. But if it is not, significant errors may be incurred if (4.3.21) is used. In any real experimental situation, the actual values of m_1, m_2, and m_p, as well as the inherent experimental uncertaities in measuring the distances and times required in ascertaining the acceleration $a_y{}'$, must all be taken into account in determining whether it is safe to use (4.2.21) or whether the more accurate and correct expression (7.2.42) should be employed.

Suppose, for example, we were required to determine the value of g with an accuracy of 1 part in 1000 (0.1 percent). We might find it convenient to uses masses $m_2 = 1000$ g and $m_1 = 980$ g, with a pulley of mass 75 g. It is easy enough to arrange the apparatus so that one of the masses descends 2 meters, the other rising the same distance. The time of descent is then about 6.5 seconds, as determined from (7.2.36) and the relations for uniformly accelerated motion. With care, these distances and the time interval can be measured with the necessary precision. This allows the acceleration $a_y{}'$ to be determined with the required degree of accuracy.

In this instance, $m_1 + m_2 = 1980$ g and $\tfrac{1}{2}m_p = 37.5$ g. The term $\tfrac{1}{2}m_p$ is, therefore, $37.5/1980 = 0.01885$, or 1.8 percent of the quantity $m_1 + m_2$. An error of 1.8 percent will therefore occur in the numerator of (7.2.42) and, therefore, in the calculated value of g if the term $\tfrac{1}{2}m_p$ arising from the rotational inertia of the pulley is ignored. This is much larger than the 0.1 percent precision we are trying to attain in the experiment, and it is, therefore, necessary to retain the rotational inertia term in (7.2.42) in calculating the results. If, on the other hand, an accuracy of 1 part in 20 (5 percent) in the measured value of g was all that was required and if experimental uncertainties of that order were inherent in the apparatus which was available, the 1.8 percent discrepancy between (7.2.42) and (4.3.21) would not have mattered much, and the simpler expression might equally well have been used.

7.3 Calculation of Moments of Inertia

The moment of inertia of a rigid body about a given axis is an important concept in the development of rigid body dynamics. As previously mentioned, it plays a role in rotational motion analogous to that assumed by mass in translational motion. It is important, therefore, to understand something of the technical aspects of determining moments of inertia for bodies of various shapes and sizes.

In order to obtain a moment of inertia I for a rigid body of irregular shape, such as the one illustrated in Fig. 7.3, it is usually necessary to use Eq. (7.2.12) and to actually subdivide the body into small masses, each of which contributes a term to the total moment of inertia I. Each of these terms represents a contribution from a point mass; the sum of them gives an approximation to the moment of inertia. The accuracy of this procedure is determined by the degree of subdivision, since for a continuous distribution of mass the "true" moment of inertia must be evaluated by means of the integral (7.2.17). The difficulties encountered in evaluating (7.2.17) are substantial when the body is irregular. However, if the body is regular or symmetric, it is often possible to evaluate I exactly.

Before proceeding to the task of evaluating moments of inertia for continuous bodies, we consider first the more elementary problem of determining I for a body consisting of several point constituents.

EXAMPLE 7.3.1

Three point masses such as those shown in Fig. 7.7 are located at the three vertices of an equilateral triangle whose sides have length l. The masses are m, $2m$, and $3m$, respectively. Find the moment of inertia about an axis perpendicular to the triangle and passing through point A. Repeat for an axis through D.

Equation (7.2.12) tells us that

$$I_A = (m)(0^2) + (2m)(l^2) + (3m)(l^2) = 5ml^2$$

In the same way, the moment of inertia I_D is given by

$$I_D = md^2 + 2md^2 + 3md^2 = 6md^2$$

where d is the distance from the vertex to the center of the triangle. It is an elementary exercise in trigonometry to show that $d^2 = l^2/3$, and, therefore, the final result, expressed in terms of l, is

$$I_D = 2ml^2$$

The two results, I_A and I_D, are quite different since in each case a different axis was chosen. This emphasizes the point made earlier that *the moment of inertia depends on both the form of the object and the orientation of the axis* and that it is meaningless to speak of a moment of inertia without simultaneously specifying the axis about which it is computed.

Calculation of the moment of inertia for any object consisting of discrete point masses proceeds in a manner similar to that used in the above calculation. On the other hand, for a continuous body (7.2.17) must be employed. This requires the use of calculus, as the following example will illustrate.

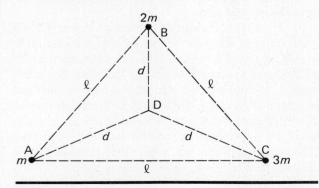

FIGURE 7.7

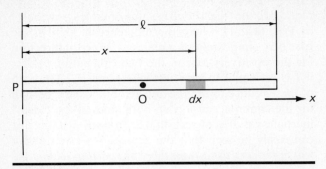

FIGURE 7.8

EXAMPLE 7.3.2

A thin, rigid rod of length l and mass m is free to rotate about an axis passing through the point P at its left end, as shown in Fig. 7.8. Compute the moment of inertia about this axis.

The moment of inertia about P is given by (7.2.17) as

$$I_p = \int x^2 \, dm$$

The rod may be subdivided into infinitesimal point masses dm each located a distance x from P, as shown in Fig. 7.8. As in all calculations of moments of inertia, it is necessary to express the mass element dm in terms of its position coordinates. In the present case, under the assumption of uniform mass density, the mass per unit length of the rod is m/l, and the mass dm contained in length dx is given by mass per unit length times length:

$$dm = \frac{m}{l} \, dx$$

Therefore,

$$I_p = \int_0^l x^2 \frac{m}{l} \, dx = \frac{m}{l} \int_0^l x^2 \, dx = \frac{m}{l} \left[\frac{x^3}{3} \right]_0^l$$

or

$$I_p = \tfrac{1}{3} ml^2 \tag{7.3.1}$$

In the event that the density is nonuniform, the constant linear density m/l is replaced by a prescribed function of x. The result would, of course, differ from (7.3.1).

EXAMPLE 7.3.3

Find the moment of inertia of a thin hoop of mass m and radius r about an axis perpendicular to its plane which passes through its center, as illustrated in Fig. 7.9.

As before, we may begin by breaking the hoop up into small elements of mass dm that may be regarded as point masses. It is assumed that the radial thickness of the hoop is so small as to be completely negligible

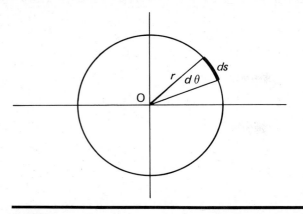

FIGURE 7.9

in comparison to the radius r. For a single mass element dm of length ds, which may be regarded as a point mass, the contribution to the moment of inertia is given by (7.2.17):

$$dI = r^2 \, dm \qquad (7.3.2)$$

Again, however, the mass dm can be expressed as mass per unit length (m/l) times the length ds. Therefore,

$$dm = \frac{m}{l} \, ds \qquad (7.3.3)$$

where l represents the total length (circumference) of the hoop. Substituting this into Eq. (7.3.2), we find, since $ds = r \, d\theta$,

$$dI = r^2 \frac{m}{l} \, ds = \frac{mr^2}{l} (r \, d\theta) \qquad (7.3.4)$$

We may now integrate this over all the mass elements in the hoop, thus over its entire length. In doing so, we note that for every mass element the radius r is the same. We may, therefore, regard r as a constant in the integration over the mass elements and write it outside the integral, along with the other constants m and l. We find, therefore, that

$$I = \frac{mr^3}{l} \int_0^{2\pi} d\theta = \frac{2\pi mr^3}{l} \qquad (7.3.5)$$

Since the circumference l can be written as $2\pi r$, however, this reduces to

$$I = mr^2 \qquad (7.3.6)$$

We might have expected this result on purely intuitive grounds, since the entire mass of the hoop is at a distance r from the axis. The above calculation confirms this expectation.

EXAMPLE 7.3.4

Calculate the moment of inertia of a thin, flat circular plate of concentric inner and outer radii R_i and R_0 about an axis normal to its plane that passes through the plate's center, as illustrated in Fig. 7.10a.

In this example, we must again evaluate the integral (7.3.2) over all mass elements of the plate. Now, however, we may simplify our work by regarding the plate as being made up of infinitesimally thin hoops, each of mass dm, radius r, and radial thickness dr, as illustrated. The results of the preceding example, as expressed by (7.3.6), enable us to write for the contribution dI from each such hoop element

$$dI = r^2 \, dm \qquad (7.3.7)$$

But the mass of each element can be expressed as the mass per unit area of the plate (call it σ) times the element's surface area dA. For the shaded element in Fig. 7.10a, which is shown slit open and laid out in a flat strip in Fig. 7.10b, it is evident that

$$dm = \sigma \, dA = \sigma 2\pi r \, dr \qquad (7.3.8)$$

We may now integrate this over all mass elements of the plate, from the inner radius $r = R_i$ to the outer edge $r = R_0$. Since the mass per unit area σ is the same for every mass element, it may be written outside the integral. We obtain, therefore, substituting (7.3.8) into (7.3.7),

$$I = 2\pi\sigma \int_{R_i}^{R_0} r^3 \, dr = 2\pi\sigma \left[\frac{r^4}{4} \right]_{R_i}^{R_0} = \tfrac{1}{2}\pi\sigma(R_0{}^4 - R_i{}^4) \qquad (7.3.9)$$

This can be expressed more simply by noting that σ can be written as the plate's total mass m divided by its total area:

$$\sigma = \frac{m}{\pi R_0{}^2 - \pi R_i{}^2} \qquad (7.3.10)$$

Substituting this into (7.3.9) and noting that $(R_0{}^4 - R_i{}^4) = (R_0{}^2 + R_i{}^2)(R_0{}^2 - R_i{}^2)$

$$I = \tfrac{1}{2}m(R_0{}^2 + R_i{}^2) \qquad (7.3.11)$$

This result was derived for a plate of infinitesimal thickness, but since a thick plate can be thought of as being built up of a stack of many thin ones, and since the moment of inertia of each of the thin constituents is given by the same expression (7.3.11), this result is also true for a plate of finite uniform thickness, in other words, for a hollow circular cylinder of inner radius R_i and outer radius R_0. In the case of a thin plate without the central hole, or in the case of a solid circular cylinder, the inner radius R_i is zero, and for such a body, according to (7.3.11)

$$I = \tfrac{1}{2}mR_0{}^2 \qquad (7.3.12)$$

Also, if the inner radius R_i approaches the outer radius R_0, the body we are dealing with approaches the form of the thin, circular hoop of the previous example; and under these circumstances, (7.3.11) approaches the value $mR_0{}^2$ given by (7.3.6).

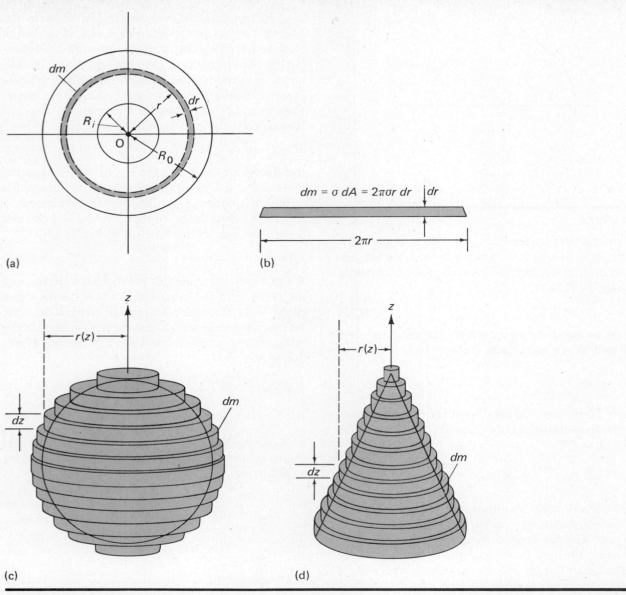

(a)

(b)

(c)

(d)

FIGURE 7.10

The moment of inertia of more complex solids of revolution can be computed by considering them to be constructed of a stack of thin circular plates of thickness dz and mass dm. The contribution of each such plate element to the total moment of inertia of the body is now, however, no longer expressed as $dI = r^2\,dm$ but, instead, in view of (7.3.12), by

$$dI = \tfrac{1}{2}r^2\,dm \qquad (7.3.13)$$

The total moment of inertia may be found by expressing the mass dm and the radius r in terms of the vertical distance z and integrating over dz. A sphere and a right circular cone are shown subdivided into cylindrical platelike elements in this way in Figs. 7.10c and 7.10d. The calculation of the moments of inertia

for these bodies is assigned in the problems at the end of this chapter.

EXAMPLE 7.3.5

A flat (laminar) rigid body lying in the xy-plane has moments of inertia I_x, I_y, and I_z about the x-, y-, and z-axes, respectively, as shown in Fig. 7.11. The z-axis is orthogonal to the plane of the body. Show that $I_x + I_y = I_z$.

Referring to Fig. 7.11, it is clear that

$$I_z = \int r^2\,dm = \int (x^2 + y^2)\,dm = \int x^2\,dm + \int y^2\,dm$$

But

$$I_x = \int y^2\,dm \qquad \text{and} \qquad I_y = \int x^2\,dm$$

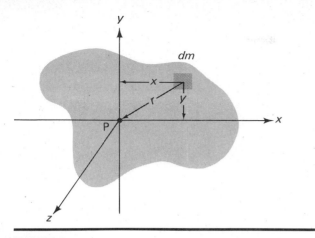

FIGURE 7.11

so that

$$I_z = I_x + I_y \qquad (7.3.14)$$

This easily proved theorem is extremely useful in some instances, as we shall see in the next example. In using it, however, we must remember that it is true only for thin, platelike objects.

EXAMPLE 7.3.6
Calculate the moment of inertia of a thin, circular hoop of mass m and radius R about any diameter (see Fig. 7.12).

Let y be the axis about which the moment of inertia is taken. Call the corresponding result I_y. Explicit calculation of the moment about this axis is fairly involved although it could certainly be done. However, we note that due to the symmetry of the ring, a calculation about the x-axis would yield I_x identical to I_y. Using (7.3.10), we find that

$$I_x + I_y = 2I_y = I_z \qquad (7.3.15)$$

But I_z has already been evaluated in Example 7.3.3 as mR^2. Thus,

$$I_y = \tfrac{1}{2}I_z = \tfrac{1}{2}mR^2 \qquad (7.3.16)$$

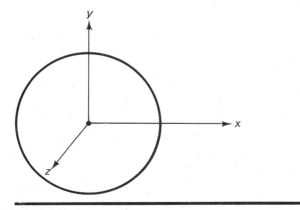

FIGURE 7.12

We have worked out several examples to indicate the basic techniques utilized in computation of moments of inertia. In Fig. 7.13, a tabulation of moments of inertia for various bodies and selected axes is given. These will be useful for reference in working some of the problems at the end of the chapter.

In addition to the methods outlined in this section, there are experimental methods for determining I for bodies in which explicit computation would be prohibitively difficult. In Chapter 9, we shall examine the connection between the period of oscillation of a rigid body, used as a physical pendulum, and the moment of inertia.

7.4 Parallel Axis Theorem

The *parallel axis theorem* provided a means of determining the moment of inertia of a rigid body about any axis parallel to a given axis through the center of mass, provided the moment of inertia about the latter axis is already known.

Let I_O denote the moment of inertia of a rigid body about any axis through its center of mass O and let $I_{O'}$ denote the moment of inertia with respect to an axis through O' *parallel* to the preceding axis. Then, according to the parallel axis theorem,

$$\boxed{I_{O'} = I_O + mR^2} \qquad (7.4.1)$$

where m is the total mass of the body and R is the distance between the two axes. The proof of this theorem is easily carried out by considering the situation shown in Fig. 7.14, which illustrates an irregular rigid body whose center of mass is located at O. We can assume, without loss of generality, both axes to be parallel to the z-direction as shown. An infinitesimal mass element dm is located by position vectors $\mathbf{r}'$ and $\mathbf{r}$ with respect to two coordinate systems centered on the respective axes, as shown in the figure. The position vector $\mathbf{R}$ in Fig. 7.14 locating the center of mass with respect to the axis O' is taken to lie in the $x'y'$-plane.

Now with respect to O',

$$I_{O'} = \int r'^2 \, dm = \int (x'^2 + y'^2) \, dm \qquad (7.4.2)$$

and with respect to O,

$$I_O = \int r^2 \, dm = \int (x^2 + y^2) \, dm \qquad (7.4.3)$$

But the vectors $\mathbf{r}'$ and $\mathbf{r}$ are related by

$$\mathbf{r}' = \mathbf{r} + \mathbf{R}$$

which implies

$$x' = x + X \qquad \text{and} \qquad y' = y + Y \qquad (7.4.4)$$

211

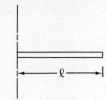

thin rod, axis normal to rod: $I = \frac{1}{3} m\ell^2$

thin annulus or hoop, about its axis of symmetry: $I = mR^2$

cylinder, about its axis of symmetry: $I = \frac{1}{2} mR^2$

thin spherical shell, about any axis passing through center $= I = \frac{2}{3} mR^2$

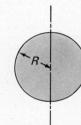

flat circular plate, about any axis in its own plane passing through center: $I = \frac{1}{4} mR^2$

cone about its axis of symmetry: $I = \frac{3}{10} mR^2$

FIGURE 7.13. Moment of inertia for several common symmetrically shaped bodies.

where X and Y refer to the horizontal and vertical components of **R**. Substituting these expressions into (7.4.2) and noting that $X^2 + Y^2 = R^2$, we may write

$$I_{O'} = \int (x^2 + 2xX + X^2 + y^2 + 2yY + Y^2) \, dm$$

or

$$I_{O'} = I_0 + \int R^2 \, dm + 2 \int (xX + yY) \, dm$$

$$= I_0 + \int R^2 \, dm + 2X \int x \, dm + 2Y \int y \, dm \quad (7.4.5)$$

In this last expression, the quantities X and Y are written outside the integral sign. This is possible because these quantities express the location of one axis with respect to the other and are, therefore, the same for all the mass elements in the body. They may be treated as constants with regard to the integration over mass elements. But in any situation in which the center of mass is at the origin and in which, therefore, the coordinates x and y measure distances with refer-

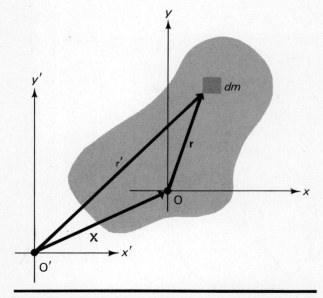

FIGURE 7.14. Geometric situation referred to in the proof of the parallel axis theorem.

$$\tau = m \frac{d}{dt}[\mathbf{r} \times (\boldsymbol{\omega} \times \mathbf{r})] \tag{7.5.5}$$

Now, since $\boldsymbol{\omega}$ points along the z-axis, it has only a z-component and hence can be written

$$\boldsymbol{\omega} = \omega \mathbf{i}_z \tag{7.5.6}$$

Also, since the position vector $\mathbf{r}$ lies in the xy-plane, its components must be x and y, whereby

$$\mathbf{r} = x\mathbf{i}_x + y\mathbf{i}_y$$
$$\boldsymbol{\omega} \times \mathbf{r} = \omega \mathbf{i}_z \times (x\mathbf{i}_x + y\mathbf{i}_y) \tag{7.5.7}$$
$$= \omega(-y\mathbf{i}_x + x\mathbf{i}_y)$$

and

$$\mathbf{r} \times (\boldsymbol{\omega} \times \mathbf{r}) = (x\mathbf{i}_x + y\mathbf{i}_y) \times \omega(-y\mathbf{i}_x + x\mathbf{i}_y)$$
$$= \omega(x^2 + y^2)\mathbf{i}_z = r^2\omega \mathbf{i}_z$$
$$= r^2\boldsymbol{\omega} \tag{7.5.8}$$

Substituting this value into (7.5.5), and noting that the magnitude r is constant for this motion, one may finally obtain

$$\tau = (mr^2)\frac{d\omega}{dt} = I\alpha \tag{7.5.9}$$

It is possible to extend this result, which has been obtained for a point mass, to rigidly connected collections of point masses and to extended rigid bodies rotating about fixed axes. In all these cases, the result may be expressed by Eq. (7.5.9). The orientation of the vectors representing the quantities $\mathbf{r}$, $\boldsymbol{\tau}$, $\boldsymbol{\omega}$, $d\boldsymbol{\omega}$, and $\boldsymbol{\alpha}$ is shown in Fig. 7.16b.

7.6 Work and Energy in Rotational Motion

Using the ideas presented earlier in this chapter in conjunction with those discussed in Chapter 5, it is possible to develop the concept of work, kinetic energy, and potential energy and the work–energy theorem and energy conservation principle for rotational motion.

It is reasonable to begin by asking how much work is done by a torque moving through a given angular displacement. The result is obtained by a calculation similar to that used to calculate the work done in a translation. In Fig. 7.17, a tangential force F_t applied at point P on a body creates a torque which results in an angular displacement θ about a fixed axis O. This force may be regarded as the tangential component of a more general vector force having a radial component as well; but since the line of action of a radial force component must necessarily pass through the point O, its moment arm will be zero and

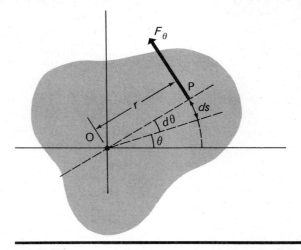

FIGURE 7.17. Calculation of work done by a torque during an infinitesimal angular displacement.

it will not contribute to the torque about that axis. Furthermore, since the body is assumed to be rigid and the axis O to be fixed, the distance r (OP) is constant and there will be no radial component of displacement, hence no work attributable to radial forces.

The work done, then, by the force F_t during the displacement of the point P through the distance ds will be

$$dW = F_t \, ds$$

or

$$dW = F_t r \, d\theta \tag{7.6.1}$$

Since $F_t r$ is simply the torque associated with the force, therefore (7.6.1) can be written as

$$dW = \tau \, d\theta \tag{7.6.2}$$

for the differential displacement $d\theta$. For a finite angular displacement from an initial angular displacement θ_i to a final value θ_f, this equation may be integrated with respect to θ, the result being

$$W = \int_{\theta_i}^{\theta_f} \tau \, d\theta \tag{7.6.3}$$

In general, the torque τ may vary with the displacement θ, and therefore the integral can be evaluated only when this variation is known, that is, when τ is specified as some particular function of θ. In many cases of practical importance, the torque may be constant throughout the rotational displacement; and if this is true, in (7.6.3) τ may be written outside the integral, the work done by the torque then being expressed by

$$W = \tau \int_{\theta_i}^{\theta_f} d\theta = \tau[\theta]_{\theta_i}^{\theta_f} = \tau(\theta_f - \theta_i) \tag{7.6.4}$$

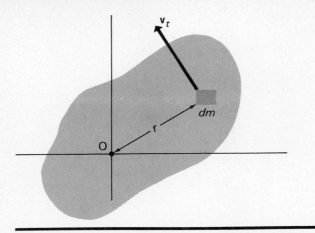

FIGURE 7.18. Calculation of rotational kinetic energy.

Equations (7.6.3) and (7.6.4) are the rotational analogues of Eqs. (5.2.6) and (5.2.2) which define work in translational displacements. The similarity in form between the two sets of expressions is easily observed.

When a rigid body is set into rotation about a fixed axis, each mass element of the body acquires a tangential velocity and, therefore, a kinetic energy. The sum of all the kinetic energies of all the mass elements is called the *rotational kinetic energy* of the body. Such a body, rotating about a fixed axis O, is shown in Fig. 7.18. The differential amount of rotational kinetic energy $dU_k^{(r)}$ associated with the mass element dm will be given by

$$dU_k^{(r)} = \tfrac{1}{2}(dm)v_t^2 \qquad (7.6.5)$$

and since the linear velocity v_t is related to the angular velocity ω by $v_t = r\omega$,

$$dU_k^{(r)} = \tfrac{1}{2}(dm)r^2\omega^2 \qquad (7.6.6)$$

The total rotational kinetic energy may be obtained by integrating over all the mass elements in the body:

$$U_k^{(r)} = \int_v \tfrac{1}{2}r^2\omega^2 \, dm = \tfrac{1}{2}\omega^2 \int_v r^2 \, dm \qquad (7.6.7)$$

The quantity ω may be written outside the integral since it is the same for all mass elements of the body and, therefore, may be regarded as a constant with regard to integration over those elements. (Note, however, that since ω may vary with time, this could not possibly be done if the integration were being carried out with respect to time.) In Eq. (7.6.7), the integral on the right is clearly identical with the moment of inertia as defined by (7.2.17), and hence we may finally write

$$\boxed{U_k^{(r)} = \tfrac{1}{2}I\omega^2} \qquad (7.6.8)$$

We should be careful to remember that the axis of rotation itself need not necessarily be fixed; in particular, it may itself have a translational motion. The total kinetic energy of the body will then be partially rotational (arising from rotation about the axis) and partially translational (arising from the translation of the axis). In this sort of system, the *total* kinetic energy may be expressed as a sum of rotational and translational contributions. We shall see later exactly how to do this.

It is possible also to envision situations where the work done by an externally applied torque may be stored and later totally recovered as kinetic energy of rotation. The circumstances are similar to those arising when work done in stretching a spring is stored as potential energy (of translation), which may later be completely recovered as translational kinetic energy if the spring is released to accelerate an attached mass, for example.

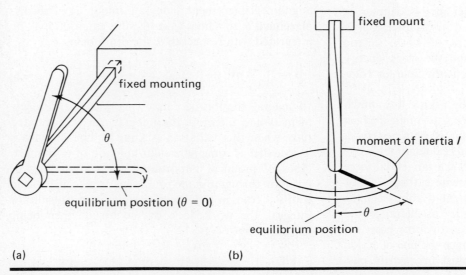

(a)

(b)

FIGURE 7.19. Rotational potential energy and the torsion pendulum.

The simplest example of such a situation is provided by a *torsion bar*, in which a torque twists a long uniform rod through a certain angle. Mechanisms of this type are sometimes used to support automobile bodies and to provide isolation from road irregularities. An illustration of the torsion spring is given in Fig. 7.19a. In this system, work may be done to twist the rod through a certain angle θ and stored as *potential energy of rotation* in the rod, which may then be converted into rotational kinetic energy if the system is released. Another illustration of such a system is afforded by the *torsional pendulum* shown in Fig. 7.19b, in which a disc or other object having a large moment of inertia may rotate back and forth on the end of a suspension rod, twisting and untwisting the rod as it rotates. Let us consider the torsional pendulum as a specific example of how rotational potential energy may be stored and converted into rotational kinetic energy.

It is not difficult to construct torsion suspensions which obey Hooke's law of elasticity, in that the torque required to produce a given angular displacement is directly proportional to the angular displacement, particularly if we do not insist on twisting the suspension too far. We shall assume, therefore, that Hooke's law is obeyed, and that

$$\tau = k'\theta \qquad (7.6.9)$$

where τ is the torque and θ the corresponding angular displacement. The proportionality constant k' expresses the torque necessary to effect unit angular displacement, and has the dimensions N-m/rad or lb-ft/rad. It is analogous to the spring constant for a linear spring although its dimensions are different. We shall begin by asking how much work must be done by an *external* torque τ to twist the disc very slowly from its equilibrium position, corresponding to zero angular displacement, until a final state is reached in which the angular displacement is θ_f.

Since according to Eq. (7.6.9) the torque is not constant throughout the displacement but instead depends on the value of θ, we cannot use (7.6.4) but must utilize the more general expression (7.6.3) to find the work done. Substituting the value of the torque τ as given by (7.6.9) into this equation and integrating from the initial state ($\theta = 0$) to the final condition of the system where the angular displacement has the value θ_f,

$$W = \int_{\theta_i}^{\theta_f} \tau \, d\theta = k' \int_0^{\theta_f} \theta \, d\theta = \tfrac{1}{2}k'\theta_f^2 \qquad (7.6.10)$$

This work is stored as potential energy in the torsion rod, and the rod can be made to do precisely this same amount of external work if it is allowed to untwist until the initial state of the system is again attained. We may, therefore, say that the work done

by an external torque in twisting the rod is stored as potential energy of rotation in the rod and that the stored potential energy in the rod, if it is twisted through an angle θ_f from the equilibrium position, is given by

$$U_p^{(r)} = \tfrac{1}{2}k'\theta_f^2 \qquad (7.6.11)$$

This expression is analogous to the equation $U_p^{(t)} = \tfrac{1}{2}kx^2$ which gives the translational potential energy stored in a linear spring that is stretched through a linear displacement x from the equilibrium position.

It is important to realize that in the process described above the torque τ is only one of the individual torques acting and not the resultant torque on the system. In particular, there is an essentially equal and opposite torque τ' exerted by the torsion rod on the pendulum disc. The total resultant torque acting on the pendulum disc during the twisting process is the sum of τ and τ', which is essentially zero. The work done by the torque τ is stored as potential energy in the torsion rod. The work done by the torque τ is accomplished by the expenditures of an equivalent amount of energy of some sort in whatever external agency is twisting the pendulum. In any event, it is clear that there is no work done by the *resultant* torque. The situation in this regard is similar to that described in section 5.2 in connection with the calculation of work done by a person who lifts a set of weights, and it may be of value at this point to reread that section carefully.

Now let us investigate what happens if the pendulum disc is twisted through some angle θ_0 and quickly released. The situation, then, is that the torque τ suddenly becomes zero, and only the torque $\tau' = -\tau = k'\theta$ acts. There is immediately a *resultant* torque $-k'\theta$; the situation ceases to be one of equilibrium, and an angular acceleration ensues.

In the case of linear motion, we found it possible to show that the work done by a *resultant* force under circumstances such as these was equal to the body's increase in kinetic energy of translation, and we referred to this result as the *work–energy theorem*. It is very easy to show that a similar theorem can be proved for rotational motion. Suppose a *resultant* torque τ_{res} acts upon a body, of moment of inertia I, which is free to rotate about fixed axis. Then, according to Newton's second law for rotation,

$$\tau_{res} = I\alpha = I\frac{d\omega}{dt} = I\frac{d\omega}{d\theta}\frac{d\theta}{dt} = I\omega\frac{d\omega}{d\theta} \qquad (7.6.12)$$

since $d\theta/dt = \omega$. If a rotation takes place, then the work done by the resultant torque in an infinitesimal angular displacement $d\theta$ will be

$$dW = \tau_{res} \, d\theta = I\omega\frac{d\omega}{d\theta} \, d\theta = I\omega \, d\omega \qquad (7.6.13)$$

217

Integrating this from an initial state wherein $\theta = \theta_i$, $\omega = \omega_i$ to a final condition in which $\theta = \theta_f$, we may easily show that

$$W = \int_{\theta_i}^{\theta_f} \tau_{\text{res}}\, d\theta = I \int_{\omega_i}^{\omega_f} \omega\, d\omega = \tfrac{1}{2}I\omega_f^2 - \tfrac{1}{2}I\omega_i^2$$

(7.6.14)

or, referring to (7.6.8), that

$$\boxed{W = U_{kf}^{(r)} - U_{ki}^{(r)} = \Delta U_k^{(r)}}$$

(7.6.15)

This equation tells us that if a resultant torque acts to produce an angular acceleration,

the work done by the resultant torque equals the increase in kinetic energy of rotation. This is the work–energy theorem for rotational motion.

In the case of the torsional pendulum, the resultant torque is given by

$$\tau_{\text{res}} = \tau' = -k'\theta$$

(7.6.16)

and if this is substituted into (7.6.14), it is apparent that

$$W = \int_{\theta_0}^{\theta} \tau'\, d\theta = -k' \int_{\theta_0}^{\theta} \theta\, d\theta = \tfrac{1}{2}k'\theta_0^2 - \tfrac{1}{2}k'\theta^2$$

(7.6.17)

where θ_0 is the initial value of angular displacement and θ the final value. But this result is equal to minus the amount of potential energy stored in the system when it was wound up from angular displacement θ_0 to angular displacement θ in the first place. Now, since the potential energy $U_{pi}^{(r)}$ corresponding to angular displacement θ_0 is $\tfrac{1}{2}k'\theta_0^2$ and the potential energy $U_{pf}^{(r)}$ at angular displacement θ is $\tfrac{1}{2}k'\theta^2$, Eqs. (7.6.14) and (7.6.17) can be combined to give

$$\boxed{\tfrac{1}{2}I\omega^2 - \tfrac{1}{2}I\omega_0^2 = \tfrac{1}{2}k'\theta_0^2 - \tfrac{1}{2}k'\theta^2}$$

(7.6.18)

where ω_0 is the initial angular velocity of the system and ω its final angular velocity. This, however, can be stated in the form

$$U_{kf}^{(r)} - U_{ki}^{(r)} = U_{pi}^{(r)} - U_{pf}^{(r)}$$

(7.6.19)

which can equally well be written

$$\boxed{U_{kf}^{(r)} + U_{pf}^{(r)} = U_{ki}^{(r)} + U_{pi}^{(r)}}$$

(7.6.20)

Equation (7.6.20) expresses *the law of conservation of energy for rotation.* It is evident from this that, provided no frictional losses are present, the sum of rotational kinetic energy and rotational potential energy is always constant in systems whose motion is purely rotational. Equation (7.6.18) allows us to understand all details of the motion of the pendulum after it is released. A complete discussion of these details will be given in an example at the end of this section.

It must be added that there are conservative torques, leading to the storage of potential energy which is totally recoverable as work, and nonconservative torques, for which this is not so. The most common example of a nonconservative torque is a frictional torque such as that encountered when a screw is driven into a block of wood. In this case, the work done by the externally applied torque in driving the screw is dissipated as heat and cannot be recovered to do work on the surroundings. It is not possible to associate a potential energy with a nonconservative torque. The situation is analogous to that encountered in regard to conservative and nonconservative forces in Chapter 5.

Though the work–energy theorem is applicable whether the system is conservative or nonconservative, the energy conservation law, which was discussed above for a conservative system, must be written for a nonconservative system in such a way as to account for the energy dissipated by nonconservative torques. This can be done by noting simply that the sum of the final kinetic and potential energy will be less than the initial energy by the amount of work done against any nonconservative torques which may be present. In this way, Eq. (7.6.20) should be modified to read

$$U_{kf}^{(r)} + U_{pf}^{(r)} + W_{nc} = U_{ki}^{(r)} + U_{pi}^{(r)}$$

(7.6.21)

for systems in which nonconservative torques are present.

One frequently encounters dynamical systems in which both rotating and translating bodies are present and which, therefore, have both rotational and translational kinetic energy. In such systems, the total kinetic energy is simply the sum of the translational and rotational parts. The same statement can be made regarding translational, rotational, and total potential energies for systems in which both translational and rotational potential energies are present; the total potential energy is the sum of the translational and rotational components.

It often happens in such systems that translational energy is *converted* into rotational energy within the system, or vice versa. Under these circumstances, it is the *total* energy of the system rather than the individual translational or rotational parts which is conserved. Written as an equation, this states that

$$U_{ki}^{(\text{tot})} + U_{pi}^{(\text{tot})} = U_{kf}^{(\text{tot})} + U_{pf}^{(\text{tot})}$$

(7.6.22)

or, writing each term as a sum of translational and rotational parts,

$$U_{ki}^{(t)} + U_{ki}^{(r)} + U_{pi}^{(t)} + U_{pi}^{(r)}$$
$$= U_{kf}^{(t)} + U_{kf}^{(r)} + U_{pf}^{(t)} + U_{pf}^{(r)} + W_f^{(r+t)}$$

(7.6.23)

We shall not try to prove this result in a rigorous mathematical way but shall note that the rotational kinetic energy is merely the sum of all the translational energies $\frac{1}{2}(dm)v_t^2$ for each mass element of the system. It is clearly this *total* energy of the system which must remain the same, irrespective of any internal exchange between rotational and translational kinetic energy. Similar remarks pertain to rotational and translational parts of the potential energy. The situation is not essentially different from the purely translational case wherein kinetic and potential energies may be converted into one another while the total energy remains constant. The following set of examples will provide additional insight into the treatment of systems in which both translational motions and rotations about fixed axes are present.

EXAMPLE 7.6.1

In Example 7.2.3, find the final kinetic energy of the gyroscope rotor and show that the increase in kinetic energy of the rotor equals the work done on the string, as predicted by the work–energy theorem.

It was found in working Example 7.2.3 that the final angular velocity of the rotor was 771 rad/sec. According to (7.6.8), then, the kinetic energy must be

$$U_k^{(r)} = \tfrac{1}{2}I\omega^2 = (\tfrac{1}{2})(3000)(770)^2 = 8.89 \times 10^8 \text{ ergs}$$

Since the rotor started from rest, the increase in its kinetic energy is equal to this final value. The work done on the string is given by

$$W = Fd \cos\theta = (1.112 \times 10^7)(80)(1)$$
$$= 8.89 \times 10^8 \text{ ergs}$$

The two quantities are found to be equal, as predicted by the work–energy theorem.

EXAMPLE 7.6.2

Obtain the results exhibited in Example 7.2.5 in connection with Atwood's machine, using work–energy methods.

The apparatus is shown in Fig. 7.6a. Suppose initially that the masses are at rest at the same level, at which point $y = 0$, and that they are then released. At some later time, the mass M will have descended a distance d, the mass m rising through the same distance. The masses will then be traveling with speed v, and the pulley will have an angular velocity ω. There is no friction, so the energy conservation equation may be written as

$$\boxed{\begin{aligned} U_{ki}^{(t)} + U_{ki}^{(r)} &+ U_{pi}^{(t)} + U_{pi}^{(r)} \\ &= U_{kf}^{(t)} + U_{kf}^{(r)} + U_{pf}^{(t)} + U_{pf}^{(r)} \end{aligned}} \quad (7.6.24)$$

Substituting in the appropriate initial and final energies,

$$0 + 0 + 0 + 0 = \tfrac{1}{2}mv^2 + \tfrac{1}{2}Mv^2 + \tfrac{1}{2}I\omega^2 + mgd$$
$$- Mgd + 0 \quad (7.6.25)$$

But, since $v = R\omega$, we may write

$$\tfrac{1}{2}\omega^2[(m + M)R^2 + I] = (M - m)gd \quad (7.6.26)$$

or

$$\omega^2 = \frac{2(M - m)gd}{I + (m + M)R^2} = 2\alpha(\theta - \theta_0) \quad (7.6.27)$$

the latter equality arising from the uniformly accelerated character of the motion. But since the pulley has been moved through an angle θ by the string on its circumference, moving through a linear distance d, we may write

$$d = -R\theta \quad (7.6.28)$$

In this equation, the minus sign must be used because d and R are clearly positive quantities, while the angular displacement as illustrated in Fig. 7.6a is obviously negative. Substituting this into (7.6.27), setting θ_0 equal to zero, and canceling θ from either side of the equation, the final result is

$$\alpha = -\frac{(M - m)gR}{I + (m + M)R^2} \quad (7.6.29)$$

This is identical to Eq. (7.2.35). The linear acceleration a_y of mass M and a_y' of mass m can be obtained from

$$a_y = R\alpha = -a_y' \quad (7.6.30)$$

the results being in each case the same as those obtained previously.

The tensions T and T' can be calculated using the work–energy theorem, which tells us that the work done by the *resultant* force (or torque) acting on any object is equal to its change in kinetic energy of translation (or rotation). Consider, for example, the pulley in Fig. 7.6a. The resultant torque acting on the body is $(T' - T)R$ as illustrated by Fig. 7.6b. During an angular displacement θ, the work done by this resultant torque is $(T' - T)R\theta$, which, assuming the body started from rest, must, according to the work–energy theorem, be equal to the final rotational kinetic energy of the object. Therefore, we may conclude that

$$(T' - T)R\theta = \tfrac{1}{2}I\omega^2 \quad (7.6.31)$$

and since $\omega^2 = 2\alpha\theta$, this becomes

$$(T' - T)R = I\alpha$$

whence

$$T' - T = \frac{I\alpha}{R} = -\frac{(M - m)gI}{I + (M - m)R^2} \quad (7.6.32)$$

This is identical with Eq. (7.2.39). Either T or T' may similarly be calculated by applying the work–energy theorem to the translational motion of the masses *m* and *M*. The details of doing this are left to the reader.

It should be observed that although the final velocities and accelerations are easily obtained from energy conservation, the procedure necessary to arrive at the forces or torques acting on the various objects in the system is somewhat roundabout. The procedure of summing individual forces or torques cannot be circumvented, even though the language of work and energy is used throughout. We might just as easily have summed the individual forces or torques and, instead of using the work–energy theorem, simply equated the resultants to ma_y', Ma_y, or $I\alpha$, using the known values of the various accelerations obtained by energy conservation considerations to arrive at the desired forces or torques.

EXAMPLE 7.6.3

A mass *m* resting on a plane inclined at an angle θ to the horizontal is fastened to a roller mounted on frictionless bearings by a massless string which is wrapped several times around the roller. This arrangement is shown in Fig. 7.20. The radius of the roller is *R*, its moment of inertia *I*. Assuming the plane to be frictionless, find, using energy conservation, the linear velocity of the mass after it has descended from rest through a vertical height *h*, corresponding to a distance *d* along the plane. What is its linear acceleration at this point? Repeat the calculation assuming now that the coefficient of sliding friction between mass and plane has some value μ_k.

We may begin by writing the energy conservation equation in the form

$$U_{ki}^{(t)} + U_{ki}^{(r)} + U_{pi}^{(t)} + U_{pi}^{(r)}$$
$$= U_{kf}^{(t)} + U_{kf}^{(r)} + U_{pf}^{(t)} + U_{pf}^{(r)} + W_f \quad (7.6.33)$$

In the initial state, the kinetic energy of the system is zero, the potential energy being *mgh*; the potential energy of rotation remains zero throughout. In the final state, when the mass has reached the bottom of the plane, the potential energy is zero, the initial amount having been converted into translational kinetic energy of the mass and rotational kinetic energy of the roller. Supplying the values of these energies in (7.6.33), we obtain

$$0 + 0 + mgh + 0 = \tfrac{1}{2}mv^2 + \tfrac{1}{2}I\omega^2 + 0 + 0 + 0$$
$$(7.6.34)$$

Since $v = R\omega$ ($\omega = v/R$), we may eliminate ω, obtaining

$$mgh = \tfrac{1}{2}mv^2 + \tfrac{1}{2}I\frac{v^2}{R^2}$$

whence, solving for *v*, $\qquad (7.6.35)$

$$v^2 = 2gh\frac{mR^2}{I + mR^2}$$

Since all the forces and torques which act are constant, the acceleration of the mass *m* must be uniform, in which case $v^2 = 2ad$. But in (7.6.35), *h* can be written in terms of *d* by virtue of the fact that $h = d \sin\theta$. Accordingly, (7.6.3) becomes

$$v^2 = 2gd \sin\theta \frac{mR^2}{I + mR^2} = 2ad$$
$$(7.6.36)$$

$$a = (g \sin\theta)\frac{mR^2}{I + mR^2}$$

In the case where a nonconservative frictional force acts, the situation is not so beautifully simple. The work done by the force of friction in moving through a distance *d* must be calculated and included on the right-hand side of Eq. (7.6.33). Unfortunately, this cannot be done unless the force of friction is known, which, in turn, requires a knowledge of the normal force. We, therefore, cannot avoid a certain amount of work involving a consideration of the individual forces acting on the mass shown in Fig. 7.20b. From the diagram, it is apparent that the normal force

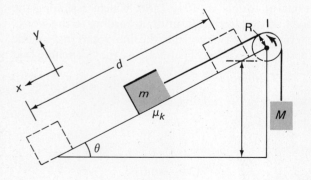

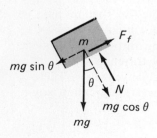

(a)   (b)

FIGURE 7.20

can be obtained in the usual way by the requirement that the resultant force component normal to the plane is zero; in other words,

$$\sum F_y = N - mg \cos \theta = 0$$

or

$$N = mg \cos \theta \qquad (7.6.37)$$

whence

$$F_f = \mu_k N = \mu_k mg \cos \theta \qquad (7.6.38)$$

and finally

$$W_f = F_f d = \mu_k mgd \cos \theta \qquad (7.6.39)$$

Including this term on the right-hand side of (7.6.33) and repeating the same calculations which led to (7.6.35), the equation now takes the form

$$v^2 = 2gd \frac{mR^2(\sin \theta - \mu_k \cos \theta)}{I + mR^2} = 2ad \qquad (7.6.40)$$

and, just as before, the acceleration must be

$$a = g \frac{mR^2(\sin \theta - \mu_k \cos \theta)}{I + mR^2} \qquad (7.6.41)$$

EXAMPLE 7.6.4

A torsional pendulum consists of a uniform circular disc of mass 5 kg and radius 0.12 meter suspended by a thin, vertical rod of negligible mass which may be twisted by rotating the disc about its axis, as shown in Fig. 7.19b. A tangential force of 5 newtons applied to the rim of the disc will twist the disc through an angle of 45° from its equilibrium position. The disc is initially turned through an angle of 180° from equilibrium and then released from rest. What is its rotational speed when it reaches the equilibrium position once more? How much time elapses from the instant of release until the equilibrium position is attained?

We may calculate the torsional spring constant k' from (7.6.16), since we know that a tangential force of 5 newtons at the disc's rim will turn it through 45°, or $\pi/4$ radians. In this way, we find that

$$\tau_{res} = Fr = k'\theta$$

$$k' = \frac{Fr}{\theta} = \frac{(5)(0.12)}{\pi/4} = \frac{2.4}{\pi} \text{ N-m/rad}$$

From (7.6.18), noting that $\theta_0 = \pi$ rad, $\omega_0 = 0$, we may then conclude that

$$\tfrac{1}{2}k'(\theta_0{}^2 - \theta^2) = \tfrac{1}{2}I(\omega^2 - \omega_0{}^2)$$
$$I\omega^2 = k'(\pi^2 - \theta^2) \qquad (7.6.42)$$

The moment of inertia I for a circular disc about a perpendicular axis passing through its center is given by

$$I = \tfrac{1}{2}mR^2 = (\tfrac{1}{2})(5)(0.12)^2 = 0.036 \text{ kg-m}^2$$

Substituting this and the previously found value for k' into (7.6.42) and noting that the equilibrium position corresponds to $\theta = 0$, it is evident that

$$(0.036)\omega^2 = \frac{2.4}{\pi}(\pi^2) = 2.4\pi$$

whence

$$\omega = 14.49 \text{ rad/sec} = 2.31 \text{ rev/sec}$$

Equation (7.6.18) tells us nothing directly about the time elapsed. However, since $\omega = d\theta/dt$, this equation can be written in the form

$$\frac{d\theta}{dt} = \sqrt{\frac{k'}{I}} \sqrt{\pi^2 - \theta^2}$$

or

$$\frac{d\theta}{\sqrt{\pi^2 - \theta^2}} = \sqrt{\frac{k'}{I}} \, dt \qquad (7.6.43)$$

This can be integrated between $t = 0$ (at which time $\theta = \theta_0 = \pi$ radians) and some later time t at which the angular displacement has some corresponding value θ. In this way, (7.6.43) becomes

$$\sqrt{\frac{k'}{I}} \int_0^t dt = \int_\pi^\theta \frac{d\theta}{\sqrt{\pi^2 - \theta^2}} \qquad (7.6.44)$$

Integrating, recalling that

$$\int dx/(a^2 - x^2)^{1/2} = \sin^{-1}(x/a)$$

we find

$$\sqrt{\frac{k'}{I}} \, [t]_0^t = \left[\sin^{-1}\left(\frac{\theta}{\pi}\right) \right]_\pi^\theta$$

or

$$\sqrt{\frac{k'}{I}} \cdot t = \sin^{-1}\left(\frac{\theta}{\pi}\right) - \sin^{-1}(1) = \sin^{-1}\left(\frac{\theta}{\pi}\right) - \frac{\pi}{2} \qquad (7.6.45)$$

since $\sin^{-1}(1) = \pi/2$. This can be rearranged to read

$$\sin^{-1}\left(\frac{\theta}{\pi}\right) = t\sqrt{\frac{k'}{I}} + \frac{\pi}{2} \qquad (7.6.46)$$

Now, let us take the sine of the quantities on either side of this equation. Since the quantities are equal, their sines will also be equal. We can write

$$\sin\left[\sin^{-1}\left(\frac{\theta}{\pi}\right)\right] = \sin\left(t\sqrt{\frac{k'}{I}} + \frac{\pi}{2}\right)$$

221

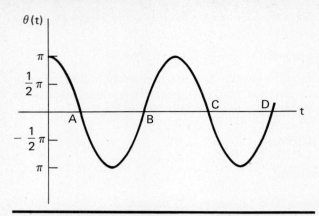

FIGURE 7.21. Sinusoidally varying angular displacement of the torsion pendulum.

or

$$\frac{\theta}{\pi} = \sin\left(t\sqrt{\frac{k'}{I}} + \frac{\pi}{2}\right)$$

since the sine of the quantity whose sine is θ/π is simply θ/π itself. We see, at last, that the angular displacement θ can be written as a function of time:

$$\theta(t) = \pi \sin\left(t\sqrt{\frac{k'}{I}} + \frac{\pi}{2}\right) \qquad (7.6.47)$$

A plot of this is shown in Fig. 7.21. It is evident that the angle θ varies sinusoidally with time, swinging periodically between the extreme values π radians and $-\pi$ radians; this is fully in accord with what is observed experimentally and what might have been foreseen intuitively. Motion of this type is often referred to as *angular harmonic motion*, and we shall return to a detailed discussion of its characteristics in a later chapter.

Returning belatedly to the problem of how much time elapses before the equilibrium position $\theta = 0$ is reached, it is apparent that this can be calculated by setting θ equal to zero in (7.6.45) and solving for t:

$$\sqrt{\frac{2.4}{0.036\pi}} \cdot t = \sin^{-1}(0) - \frac{\pi}{2}$$

Now the angle whose sine is zero can be chosen to be 0, $\pm\pi$, $\pm 2\pi$, $\pm 3\pi$, ..., etc. If we choose it to be zero, we shall arrive at a negative value for t; the same applies if we choose it to be $-\pi$, -2π, -3π, ..., etc. If we choose $\sin^{-1}(0)$ to have the value π, we obtain

$$\sqrt{\frac{2.4}{0.036\pi}} \cdot t = \pi - \frac{\pi}{2} = \frac{\pi}{2}$$

or

$$t = 0.341 \text{ sec}$$

This represents the smallest positive value of t corresponding to the equilibrium position, shown at

A in Fig. 7.21; it is the answer we were asked to find. Values of t corresponding to $\sin^{-1}(0) = 2\pi, 3\pi, 4\pi, \ldots$, etc. represent subsequent crossings of the equilibrium position as shown at B, C, D, ... in Fig. 7.21.

7.7 Rotational and Translational Energy in Systems Involving Simultaneous Translational and Rotational Motion

We must now attend to the sticky problem of how to handle dynamical systems in which bodies execute *simultaneous* translational and rotational motion. We shall begin by discussing how to view systems of this sort in terms of separate rotations and translations, and then show how the kinetic energy of these systems can be cleanly separated into rotational and translational contributions. Once this is accomplished, we shall see that the same general approach to problems of this type may be used whether we wish to apply work–energy methods or to use Newton's second law directly.

First of all, we must understand the important fact that

the most general possible displacement of a rigid object may be viewed as a translation of any point fixed in the object, plus a rotation of the object about an axis passing through this point.

Although no rigorous mathematical proof of this statement will be attempted, its truth will become intuitively evident from the following physical argument based upon Fig. 7.22a. In this diagram, a general irregularly shaped object lying in the plane of the paper at (a) undergoes an arbitrary displacement ending at (c). It is easily seen that any such displacement can be accomplished by choosing any point P fixed with respect to the object, translating the object through displacement r until the point P coincides with the location shown at (c), and then rotating the object about an axis perpendicular to the plane of the paper passing through point P until the object is moved from the position shown at (b) into the final position (c). The displacement is clearly the sum of a translation of point P and a rotation about an axis passing through P.

The same result can be shown to be true for a general displacement in three dimensions by the same sort of argument, though the final rotation is not so easily visualized. In Fig. 7.22b, a hammer, which is representative of a general irregular object, undergoes a displacement from (a) to (d). This can be accomplished by choosing a point P fixed within the hammer, translating the object through the translational displacement **r** until the point has the location shown at (d) (the hammer then being in position (b)), and then

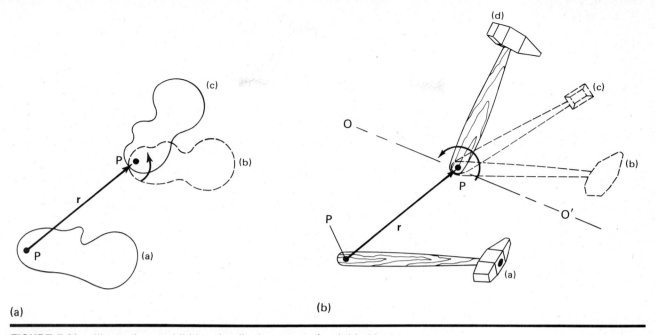

FIGURE 7.22. Illustrations exhibiting the displacement of a rigid object, as represented by a translation of a point fixed within the body and a rotation about an axis through that point.

rotating it about the axis OO', which passes through P, until the final position (d) is reached.

An intermediate stage in the rotation is shown at (c) of Fig. 7.22b for added clarity: in the diagram, the point O is closer to the observer's eye than O'. Again, the total displacement is the sum of a translation of a point within the object plus a rotation about an axis passing through that point, but this time it is not quite so simple to see how to choose the axis OO' about which the rotation is made. The student may find it helpful to try out the procedure outlined above with an actual three-dimensional object to assure himself that it works every time! The fixed point P can be any point within the object, in particular, the center of mass. We shall see that there are very substantial advantages to choosing this point to be the center of mass in all instances, for then the kinetic energy of the object splits very clearly into translational and rotational parts and the description of its motion becomes very simple. If this choice is strictly adhered to—and we shall always do so—we may state that

the most general displacement of a rigid body may be viewed as a translation of the center of mass of the body plus a rotation about an axis passing through the center of mass.

If we separate the rotational and translational motion of the body in this way during every infinitesimal time interval of its motion, we may write its total kinetic energy as

$$U_k = \tfrac{1}{2}mv_0^2 + \tfrac{1}{2}I_0\omega_0^2 \qquad (7.7.1)$$

where v_0 is the instantaneous translational velocity of the center of mass, ω_0 is the instantaneous angular velocity about the rotation axis passing through the center of mass, and I_0 is the moment of inertia of the body with respect to that axis. This remarkable result states that

the energy of motion of a rigid body may always be viewed as the translational kinetic energy of the center of mass of the body plus a rotational kinetic energy about an axis of rotation passing through the center of mass.

This result emphasizes the importance of the center of mass in the dynamics of rigid objects, since the same clear separation of rotational and translational energies *does not occur for any other choice of reference point* for the analysis of the body's motion. In that case, the above expression for the kinetic energy would contain a complex term involving both the linear velocity of the reference point and the angular velocity of the body with respect to that point. Equation (7.7.1) may be derived by expressing the linear velocity $\mathbf{v}_i$ of the ith mass element of the body as the vector sum of the translational velocity of the center of mass and the linear velocity of the mass element with respect to the center of mass, and then summing the quantity $\tfrac{1}{2}m_iv_i^2$ by integration over all the mass elements of the body. The details are com-

223

plex and somewhat beyond the scope of this book; we shall, therefore, merely state the result.

There is nothing in this development that indicates how one may in every case find the orientation of the proper rotation axis through the center of mass needed to make the clean separation of rotational and translational energies expressed by Eq. (7.7.1). The question of how this is accomplished in general is again somewhat beyond the scope of this book and will not be discussed here. It should be stated, however, that the orientation of this axis is not necessarily fixed but may vary with time. In the specific examples we shall consider, the way in which the rotation axis should be chosen will be immediately obvious from the geometry of the situation. We shall, in fact, with the single exception of the gyroscope, restrict ourselves to cases where the axis of rotation executes translational but not rotational motion, that is, to situations wherein the *direction* of the rotation axis does not change. Under such circumstances, the separation of rotational and translational motion can be made in a simple and natural way.

The result obtained above suggests that any dynamical problem involving simultaneous translation and rotation can be worked using the work–energy approach in connection with (7.7.1), provided one remembers that it is the *sum* of the rotational and translational energies which is conserved rather than each individual component. We are then led to the question of how to approach problems of this sort from the point of view of the direct application of Newton's laws. From what we have learned so far, it is natural to suppose that it is correct again to separate the motion into the translational motion of the center of mass and rotation about an axis through that point, and to apply $\mathbf{F} = m\mathbf{a}_0$ to the former and $\mathbf{M} = I_0\boldsymbol{\alpha}_0$ to the latter. This is indeed the case, as we shall soon see in detail; but once more it is important to understand that this can be done with absolute safety only if we separate the motion into the translation of the *center of mass* (not some other point) and rotation about an axis *through the center of mass* (not somewhere else).

We may now complete our discussion of work and energy in rotational motion by stating the conservation of energy theorem for systems executing simultaneous rotational and translational motion. This law is expressed by Eqs. (7.6.22) and (7.6.23) for systems involving only rotation about a *fixed* axis. It is easily extended to the more general case simply by observing that the total energy must be expressed as the sum of purely rotational and translational components with reference to the center of mass. The conservation of total energy in any such instance can, therefore, be expressed as before by (7.6.22) or (7.6.23).

EXAMPLE 7.7.1

An automobile weighs 3600 pounds, including the wheels. Each wheel weighs 48 pounds, and its dynamical behavior is essentially that of a uniform circular cylinder of radius 1.2 ft. What fraction of the total kinetic energy of the car is rotational kinetic energy when the car is in motion?

The moment of inertia of each wheel about its axis of rotation is given by

$$I = \tfrac{1}{2}mR^2 = \left(\frac{1}{2}\right)\left(\frac{48}{32}\right)(1.2)^2 = 1.08 \text{ slug-ft}^2$$

The kinetic energy of the body and chassis, minus wheels, is

$$U_{k(b+c)}^{(t)} = \tfrac{1}{2}m_b v^2 = \left(\frac{1}{2}\right)\frac{3600 - (4)(48)}{32}v^2 = 53.25v^2$$

where m_b represents the combined mass of body and chassis. The kinetic energy of the wheels is partly translational and partly rotational. It may be written as the sum of the translational kinetic energy of the center of mass plus the rotational kinetic energy about the axis of the wheel, which passes through its center of mass. For each wheel, then, letting m_w represent the mass of the wheel,

$$U_{kw} = U_{kw}^{(t)} + U_{kw}^{(r)} = \tfrac{1}{2}m_w v^2 + \tfrac{1}{2}I\omega^2$$

$$= \tfrac{1}{2}m_w v^2 + \tfrac{1}{2}I\frac{v^2}{R^2}$$

since $v = R\omega$. Substituting numerical values into this expression, it is evident that

$$U_{kw}^{(t)} = \left(\frac{1}{2}\right)\left(\frac{48}{32}\right)v^2 = 0.75v^2$$

$$U_{kw}^{(r)} = (\tfrac{1}{2})(1.08)\frac{v^2}{1.2^2} = 0.375v^2$$

The total kinetic energy is

$$U_k^{(tot)} = U_{k(b+c)}^{(t)} + 4U_{kw}^{(t)} + 4U_{kw}^{(r)}$$
$$= (53.25 + 3.00 + 1.50)v^2 = 57.75v^2$$

The total *rotational* kinetic energy is

$$U_k^{(r)} = 4U_{kw}^{(r)} = 1.50v^2$$

The fractional part f of the total energy represented by rotational kinetic energy is

$$f = \frac{U_k^{(r)}}{U_k^{(tot)}} = \frac{1.50v^2}{57.75v^2} = 0.0260$$

or, in other words, 2.6 percent. This calculation, of course, neglects rotational kinetic energy contributed by rotating parts in the engine and drive line, but these are so small as to be negligible in any actual vehicle.

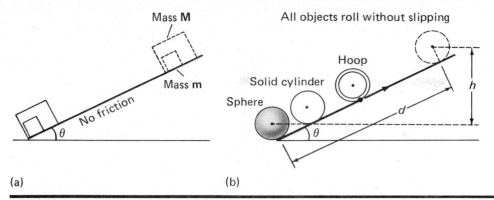

FIGURE 7.23

EXAMPLE 7.7.2

If two objects of different mass are allowed to slide down a *frictionless* inclined plane, having been released from rest at the top simultaneously, they will always arrive at the bottom at exactly the same time. This follows from the fact that for a mass m, the component of the weight force along the plane is $mg \sin \theta$, where θ is the angle of inclination. If there is no friction, this will be the only force component acting along the plane; and then, if the y-axis is assumed to point along the plane,

$$F_y = ma_y = mg \sin \theta$$
$$a_y = g \sin \theta \qquad (7.7.2)$$

The acceleration, according to (7.7.19), is *independent* of the object's mass, and is, therefore, the same for a light object or a heavy one. It follows, therefore, that since the two objects undergo equal acceleration, it takes them exactly the same time to cover the distance they slide along the plane, and they arrive in a dead heat at the bottom. The result is, of course, obvious and amounts to hardly more than Galileo's observation that different masses dropped simultaneously from the top of a tower hit the ground together.

Suppose, however, that a number of circularly symmetric objects, for example, a uniform solid cylinder, a rather thin circular hoop, and a uniform sphere are fashioned in such a way that each object has *exactly the same mass m* and *the same external radius R*. This can be accomplished in practice by using substances of different density to make each object. Now what happens if the objects are allowed to roll without slipping down an inclined plane, being released from rest together at the top, just as before?

The similarity of the circumstances in the two examples might tempt us to conclude that all three objects would arrive at the bottom at the same time, as before; in fact it might be assumed that this would happen even if their masses were not the same. But when the experiment is actually performed, it is found

that the sphere wins the race every time, beating the solid cylinder to the bottom by a small margin, while the hoop always comes in a poor third! The result is the basis of a demonstration experiment which has long been popular with physics lecturers, if not with students. It is illustrated schematically in Fig. 7.23.

This apparent contradiction to Newton's second law can easily be understood if both rotational and translational dynamic effects are considered. We may analyze the situation from the viewpoint of energy conservation, writing the total energy of the object in question as the sum of the translational kinetic energy of its center of mass, the rotational kinetic energy of the body about its axis of rotation (which in all cases passes through the center of mass, as required), and the potential energy. If we identify the initial state as one of rest at the top of the plane and the final state as that obtaining when the object reaches the bottom,

$$U_{ki}{}^{(t)} + U_{ki}{}^{(r)} + U_{pi} = U_{kf}{}^{(t)} + U_{kf}{}^{(r)} + U_{pf} + W_f$$
$$(7.7.3)$$

Both components of the initial kinetic energy are zero since the object starts from rest. The initial potential energy U_{pi} is equal to mgh, where h is the vertical distance through which the center of mass descends. The final kinetic energy of translation is $\frac{1}{2}mv^2$, where v is the final translational velocity of the center of mass, while the final kinetic energy of rotation is $\frac{1}{2}I\omega^2$, where I is the moment of inertia and ω the angular velocity of rotation about the center of mass. The final potential energy is zero.

There are also forces of friction in this system; indeed, it is the force of friction that gives rise to the angular accelerations that occur. In this example, however, the objects roll rather than slide, and therefore no mechanical energy is dissipated as heat in nonconservative frictional processes. The term W_f, therefore, in Eq. (7.7.3), which represents this nonconservative work, is in this case precisely *zero*. We have not

225

provided a proof of this statement, but it follows from the fact that when rolling occurs, the point of contact is instantaneously at rest. Therefore, the conditions that prevail at that point are those of static friction rather than sliding friction. A more precise discussion of rolling motion will be provided in the following section.

If we put all this information into Eq. (7.7.3), we write,

$$mgh = \tfrac{1}{2}mv^2 + \tfrac{1}{2}I\omega^2 \tag{7.7.4}$$

or, since $\omega = v/R$ and $h = d \sin \theta$,

$$2mgd \sin \theta = mv^2 + \frac{Iv^2}{R^2} = mv^2\left(1 + \frac{I}{mR^2}\right)$$

whence

$$v^2 = \frac{2gd \sin \theta}{1 + \dfrac{I}{mR^2}} = 2ad \tag{7.7.5}$$

The acceleration, obtained by equating v^2 to $2ad$ in (7.7.5) in accord with the laws of uniformly accelerated motion, is

$$a = \frac{g \sin \theta}{1 + \dfrac{I}{mR^2}} \tag{7.7.6}$$

If the moment of inertia is zero, this reduces to $a = g \sin \theta$ as Eq. (7.7.2) obtained for objects which slide down a frictionless surface. But if the object has an appreciable moment of inertia, as is nearly always the case, the acceleration is less than $g \sin \theta$, and the greater the moment of inertia, the smaller the acceleration. For a sphere, $I = \tfrac{2}{5}mR^2$; for a uniform solid cylinder, $I = \tfrac{1}{2}mR^2$; while for a thin hoop, $I = mR^2$. Substituting these values into (7.7.6),

$$a = \tfrac{5}{7}g \sin \theta = 0.714g \sin \theta \qquad \text{sphere}$$
$$a = \tfrac{2}{3}g \sin \theta = 0.667g \sin \theta \qquad \text{cylinder}$$
$$a = \tfrac{1}{2}g \sin \theta = 0.500g \sin \theta \qquad \text{hoop}$$

From this, it is apparent that the sphere should accelerate fastest, thus winning the race. The cylinder accelerates more slowly and comes in a bit later, while the hoop's acceleration is much less, and we should, therefore, expect it to be last by a wide margin. This is exactly what is observed. Newton's laws, instead of leading to an apparent disagreement with experiment, give results that are precisely correct when they are applied properly so as to account for both translational and rotational motions of the bodies involved.

These effects may be understood intuitively by recalling that each of the three objects begins with an equal amount of gravitational potential energy, which is finally entirely converted to kinetic energy of rota-

tion and translation. An object whose moment of inertia is large will, for a given translational speed, acquire more rotational kinetic energy than a body with a small moment of inertia. A greater *fraction* of its total kinetic energy will be rotational kinetic energy because of this. When its initial gravitational potential energy has been all converted to kinetic energy, therefore, a relatively large fraction of this total energy will be energy of rotation, and only a relatively small amount will be left over to appear as translational kinetic energy $\tfrac{1}{2}mv^2$. Its linear velocity will, therefore, be smaller at any given time than that of a body whose moment of inertia is less, and on this basis the results exhibited above are to be expected.

This example is an important one, embodying many of the most important aspects of combined rotation and translation. It will later be reworked using the direct application of Newton's second law for the translational motion of the center of mass and the rotation of the object about its axis. A thorough consideration of both these treatments will be amply repaid in terms of the student's comprehension of combined rotational and translational motion.

EXAMPLE 7.7.3

The construction of a Yo-Yo is illustrated in Fig. 7.24. Suppose it starts from rest and descends through a vertical height h. Find its final translational and rotational velocities and its linear acceleration. What is the tension in the string?

From the energy conservation principle, we may write

$$U_{ki}{}^{(t)} + U_{ki}{}^{(r)} + U_{pi} = U_{kf}{}^{(t)} + U_{kf}{}^{(r)} + U_{pf} \tag{7.7.7}$$

since there is no work done against friction. Since the initial kinetic energy is zero, the initial potential energy

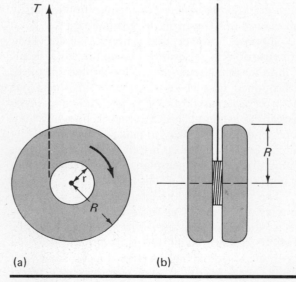

(a)   (b)

FIGURE 7.24.   The Yo-Yo.

is mgh, and the final potential energy is zero, this can be stated as

$$mgh = \tfrac{1}{2}mv^2 + \tfrac{1}{2}I\omega^2 \qquad (7.7.8)$$

where v is the linear velocity of the center of mass, ω is the rotational velocity about the rotation axis, which in this case clearly passes through the center of mass, and I is the moment of inertia of the Yo-Yo about this axis. The linear and angular velocities are related by

$$v = r\omega \qquad (7.7.9)$$

If we assume that the slit within which the string is wound is small in comparison with the total thickness of the Yo-Yo, as shown in the front view of Fig. 7.24b, the moment of inertia will be nearly that of a uniform cylinder of outside radius R, that is, $I = \tfrac{1}{2}mR^2$. In making this approximation, the rotational inertia of the shaft on which the string is wound is neglected in comparison to that of the rest of the device. Substituting these values for v and I into (7.7.8),

$$2mgh = mv^2 + \tfrac{1}{2}mR^2 \cdot \frac{v^2}{r^2} = mv^2\left(1 + \frac{R^2}{2r^2}\right)$$

whence

$$v^2 = \frac{2gh}{1 + (R^2/2r^2)} \qquad (7.7.10)$$

In this case, since all the forces and torques that act are constant, all the linear and angular accelerations will likewise be constant. Therefore, from the equations of uniformly accelerated motion, v^2 may be equated with $2ah$, the initial conditions being as they are. In this way, we obtain

$$a = \frac{g}{1 + (R^2/2r^2)} \qquad (7.7.11)$$

The angular velocity ω equals v/r, according to (7.7.9). Accordingly, (7.7.10) can be written in terms of ω to read

$$\omega^2 = \frac{2gh}{r^2 + \tfrac{1}{2}R^2} \qquad (7.7.12)$$

The tension in the string, T, can be found from Newton's second law by equating the resultant of all vertical forces to ma, as given by (7.7.11):

$$\sum F_y = T - mg = ma = \frac{-mg}{1 + (R^2/2r^2)}$$

$$T = mg\left[1 - \frac{1}{1 + (R^2/2r^2)}\right] = mg\,\frac{\tfrac{1}{2}R^2}{r^2 + \tfrac{1}{2}R^2} \qquad (7.7.13)$$

In performing this calculation, if we take the y-axis to point vertically upward, we should note that the acceleration, directed *downward*, becomes negative and must be prefaced with a minus sign.

In an effective Yo-Yo, the shaft radius r is much smaller than the external radius R; the ratio $R^2/2r^2$ is, therefore, quite large. Let us suppose, as a typical example, that $R = 5r$. Then we should find $R^2/2r^2 = \tfrac{1}{2}(R/r)^2 = 12.5$, and

$$v^2 = \frac{2gh}{13.5} = (0.0741)(2gh)$$

$$a = \frac{g}{13.5} = 0.0741g$$

while

$$T = \frac{12.5}{13.5}\,mg = 0.927mg$$

It is evident that the translational velocity v of the Yo-Yo is much less than the value $\sqrt{2gh}$ found for an object that has undergone free fall for a distance h. Also, the linear acceleration is found to be much less than the value g attained by a freely falling object. Indeed, this is exactly how a Yo-Yo really behaves. The reason for this rather strange action is that most of the Yo-Yo's initial gravitational potential energy goes into exciting rotational energy about the axis rather than translation of the axis itself.

All well and good, one might say, but what actually keeps the Yo-Yo up? What keeps it from falling like any other object let go to fall freely? The answer is, of course, the upward tension in the string, which is nearly equal to the Yo-Yo's total weight mg. It is interesting to note that if the string were wound very loosely on the shaft, so that it could slip around the shaft instead of gripping it tightly, then the Yo-Yo *would* fall freely without rotating when released, the acceleration would be g, and the tension in the string would be zero.

The reader should by now have acquired sufficient insight into the dynamics of this familiar device to explain why the Yo-Yo climbs up the string after the string has unwound to its fullest extent; the explicit details of this argument will be left for the reader to fill in as an exercise.

7.8 Equations of Motion for Rigid Bodies in Simultaneous Translational and Rotational Motion

The most general motion of a rigid body is one in which the body rotates about its center of mass while the center of mass undergoes translational motion. In the previous section, an extensive discussion of energy conservation for systems which are translating and rotating was given, and it was shown that the energy may be separated into translational energy of the center of mass plus energy of rotation about the center of mass.

In this section, we shall study the dynamics of bodies which are both translating and rotating from the point of view of Newton's laws of motion. We will be concerned with simultaneous application of the equations of motion which govern pure translation and pure rotation.

According to the work–energy theorem, the work done by the resultant of all forces and torques on a body is equal to the change in its kinetic energy. If the system's kinetic energy undergoes a change dU_k in time dt as a result of the performance of work dW by external forces and torques, then

$$dU_k = dW = F_{ext}\, dr + \tau_{ext}\, d\theta \tag{7.8.1}$$

where F_{ext} and τ_{ext} are the resultant external force and torque. But, since U_k is given by Eq. (7.7.1), we can write

$$dU_k = mv_0\, dv_0 + I_0\omega_0\, d\omega_0 \tag{7.8.2}$$

Now, v_0 and ω_0 represent, in general, the magnitudes of three-dimensional vectors $\mathbf{v}_0$ and $\boldsymbol{\omega}_0$ that may have components v_{0x}, v_{0y}, v_{0z} and $\omega_{0x}, \omega_{0y}, \omega_{0z}$:

$$
\begin{aligned}
v_0 &= \sqrt{v_{0x}^2 + v_{0y}^2 + v_{0z}^2}\\
\omega_0 &= \sqrt{\omega_{0x}^2 + \omega_{0y}^2 + \omega_{0z}^2}
\end{aligned} \tag{7.8.3}
$$

while

$$dv_0 = \frac{v_{0x}\, dv_{0x} + v_{0y}\, dv_{0y} + v_{0z}\, dv_{0z}}{\sqrt{v_{0x}^2 + v_{0y}^2 + v_{0z}^2}}$$

$$d\omega_0 = \frac{\omega_{0x}\, d\omega_{0x} + \omega_{0y}\, d\omega_{0y} + \omega_{0z}\, d\omega_{0z}}{\sqrt{\omega_{0x}^2 + \omega_{0y}^2 + \omega_{0z}^2}}$$

Substituting these expressions into (7.8.2), we find

$$
\begin{aligned}
dU_k = {}&m(v_{0x}\, dv_{0x} + v_{0y}\, dv_{0y} + v_{0z}\, dv_{0z})\\
&+ I_0(\omega_{0x}\, d\omega_{0x} + \omega_{0y}\, d\omega_{0y} + \omega_{0z}\, d\omega_{0z})
\end{aligned} \tag{7.8.4}
$$

Now, the first expression in parentheses above is simply $\mathbf{v}_0 \cdot d\mathbf{v}_0$, while the second is just $\boldsymbol{\omega}_0 \cdot d\boldsymbol{\omega}_0$. Therefore, (7.8.4) can be written as

$$
\begin{aligned}
dU_k &= m(\mathbf{v}_0 \cdot d\mathbf{v}_0) + I_0(\boldsymbol{\omega}_0 \cdot d\boldsymbol{\omega}_0)\\
&= m\mathbf{v}_0 \cdot \frac{d\mathbf{v}_0}{dt}\, dt + I_0\boldsymbol{\omega}_0 \cdot \frac{d\boldsymbol{\omega}_0}{dt}\, dt\\
&= m\frac{d\mathbf{v}_0}{dt} \cdot (\mathbf{v}_0\, dt) + I_0\frac{d\boldsymbol{\omega}_0}{dt} \cdot (\boldsymbol{\omega}_0\, dt)
\end{aligned} \tag{7.8.5}
$$

But $\mathbf{v}_0\, dt$ is the displacement of the center of mass $d\mathbf{r}_0$, while $d\mathbf{v}_0/dt$ is the acceleration $\mathbf{a}_0$ of the center of mass. Likewise, $\boldsymbol{\omega}_0\, dt$ is the angular displacement $d\boldsymbol{\theta}_0$, and $d\boldsymbol{\omega}_0/dt$ is the angular acceleration $\boldsymbol{\alpha}_0$ of the body relative to an axis through the center of mass. Therefore, (7.8.5) may be expressed as

$$dU_k = m\mathbf{a}_0 \cdot d\mathbf{r}_0 + I_0\boldsymbol{\alpha}_0 \cdot d\boldsymbol{\theta}_0 \tag{7.8.6}$$

This equation will be the same as (7.8.1) if

$$\mathbf{F}_{ext} = m\mathbf{a}_0 \tag{7.8.7}$$

and

$$\tau_{ext} = I_0\boldsymbol{\alpha}_0 \tag{7.8.8}$$

We see as a result of all this that (7.7.1) and the work–energy theorem lead us to the relations (7.8.7) and (7.8.8) between the resultant force and the acceleration *of the center of mass* and between the resultant torque and the angular acceleration *referred to an axis through the center of mass*.

We may summarize these results by stating that:

The equations of motion describing the motion of a rigid body undergoing simultaneous translation and rotation are properly formulated by applying Newton's second law for translation, $\mathbf{F}_{ext} = m\mathbf{a}_0$, to describe the linear translation of the center of mass and Newton's second law for rotation, $\tau_{ext} = I_0\boldsymbol{\alpha}_0$, with torques and angular accelerations referred to the center of mass, to describe the rotation of the body about the center of mass.

In particular, serious errors may result if torques or angular accelerations are taken about points (particularly points undergoing acceleration) other then the center of mass of the object. On the other hand, we shall assert, without proof, that it is generally possible to take moments and compute accelerations about points which are *unaccelerated*. These statements will be illustrated by the examples which follow.

EXAMPLE 7.8.1

A truck is moving rapidly down a highway carrying several crates containing valuable merchandise. The driver suddenly applies his brakes hard to avoid hitting a deer which is crossing the road. Assuming his (negative) acceleration is a_x and the coefficient of static friction between the truck floor and the crate is μ_s, determine the conditions under which crates of uniform density, height H, and width h will tip over rather than slide forward on the truck bed.

We expect intuitively that the crates will have a tendency to slide forward in the truck and also to tip over. We wish to examine the circumstances under which sliding or tipping will actually occur. According to Fig. 7.25, the only horizontal force the crate experiences is a friction force acting toward the rear of the truck. We also see that there is a vertical weight force $\mathbf{W}$ and a normal force $\mathbf{N}$. Now if the crate is not slipping, this means that the friction force is sufficient to equal the mass times acceleration. In other words,

$$F_f = ma_x \tag{7.8.9}$$

If the acceleration becomes too large, eventually ma_x will exceed the maximum possible friction force $\mu_s mg$ and the crate will slide. Thus, the condition that the crate does not slide can be expressed as

$$|a_x| < \mu_s g \tag{7.8.10}$$

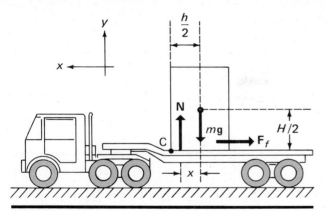

FIGURE 7.25

Let us consider now the application of Eq. (7.8.8). This equation is somewhat more difficult to apply since the normal force is really distributed along the whole surface of the crate. However, we may replace it by a single equivalent normal force **N** whose line of action (as described by the distance x) is determined so as to produce the same net torque about the center of mass. The condition for rotational equilibrium is that the sum of the torques about the center of mass must vanish. This may be expressed as

$$\sum \tau_0 = F_f\left(\frac{H}{2}\right) - Nx = 0$$

or since $N = mg$ and $F_f = ma_x$,

$$mgx = ma_x\left(\frac{H}{2}\right) \qquad (7.8.11)$$

Now, when the crate is about to topple, $x = h/2$, in which event the normal force acts at the left end of the crate to produce the largest torque it possibly can. Equation (7.8.11) then becomes

$$gh = a_x H$$

$$a_x = g\frac{h}{H} \qquad (7.8.12)$$

Thus, the condition that the crate does not topple is

$$|a_x| < g\frac{h}{H} \qquad (7.8.13)$$

As the acceleration increases, it reaches either $g(h/H)$ or $\mu_s g$ first, depending upon which is smaller. If $g(h/H) < \mu_s g$, then $g(h/H)$ is attained first, and the crate topples over. If $g(h/H) > \mu_s g$, then $\mu_s g$ is attained first and the crate slides. This result agrees with our intuitive perception of the situation in telling us that a low crate with a large base is less likely to topple than a tall narrow crate.

Now suppose we had made the mistake of taking torques about some other point such as C. Intuition tells us that just as the crate tips, the normal force passes through C. If we would take torques about

this point, we would find only a *clockwise* torque due to the weight force. This would tend to make the crate fall over in a manner completely at odds with our intuition and experience. The mistake we made was in using C as a point about which to apply Eq. (7.8.8). Point C is not the center of mass; it is an accelerated point, and (7.8.6) is simply inapplicable with respect to any accelerated point other than the center of mass. The only possible way of taking torques about C and of analyzing the problem with respect to a noninertial axis would involve introduction of fictitious inertial forces and thereby inertial torques. Introduction of fictitious forces in accelerated frames is a notion we encountered earlier in discussing circular motion. Though sometimes useful in particular contexts, it is generally wise to avoid the use of inertial forces, since they generally tend to obscure our physical picture of what is really happening.

As a second example illustrating the simultaneous use of Eqs. (7.8.5) and (7.8.6), we shall consider the general problem of rigid bodies rolling down an inclined plane. In this class of objects are included solid and hollow spheres, solid and hollow cylinders, and other objects that may roll. This same example was discussed using work–energy methods as Example 7.7.2 in the preceding section.

Before discussing this general example, however, let us formulate in a careful way the condition for *rolling*. We have discussed rolling motion and used the condition discussed here from a purely intuitive point of view in Example 7.6.2 and indeed in many other contexts. There are, however, certain features of rolling motion that have not been brought to light in our previous discussions and that may well be introduced at this point. In section 7.2, we studied the Atwood machine. We there encountered the equation $a = R\alpha$ which expressed a relationship between the linear acceleration of a rope and the angular acceleration of a pulley. When this relation is satisfied, the rope is not slipping with respect to the pulley surface.

A similar situation arises in the motion of a cylindrical object on a flat surface. Consider Fig. 7.26 which illustrates a tire moving on such a surface. If the tire is not rotating about its center but is in pure translation as shown in Fig. 7.26a, then the tire must be sliding on the surface. If *in addition* to translation there is some degree of rotation about the central axis, the overall motion is a *superposition* of these two motions.

The condition for rolling is met when the distance s through which the axis of rotation moves is related to the total angle of rotation θ about the axis by the relation $s = R\theta$, where R is the radius of the object. This, in turn, implies (through differentiation) that the relations $v = R\omega$ and $a = R\alpha$ must also be satisfied during rolling.

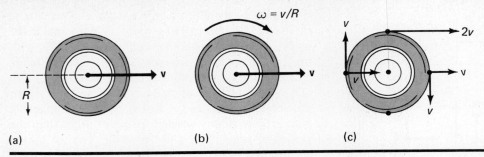

$$\omega = v/R$$

(a)   (b)   (c)

FIGURE 7.26. Geometry of rolling motion.

In Fig. 7.26b the condition for rolling motion is illustrated.

We note that at any instant of time, each point on the tire will have a different *vector velocity*, which may be obtained by adding the translational velocity **v** of the rotation axis to the translational velocity of each point with respect to the axis. As illustrated in Fig. 7.26c the uppermost point has the largest linear speed, while the point of contact with the surface has zero speed. Each point will, however, have also a linear *acceleration* and, therefore, the point which has zero speed at a given instant will have a finite speed shortly thereafter.

Let us now turn our attention to a specific problem involving the condition of rolling, which is, of course, only one special case of combined translation and rotation.

EXAMPLE 7.8.2

A rigid body of mass m, radius R, and moment of inertia I rolls down a plane inclined at an angle θ to the horizontal without slipping. It starts from rest at the top and rolls to the bottom, covering a total distance d (see Fig. 7.27). Write the equations of motion of the body, find the linear and angular acceleration, determine the linear speed at the bottom of the plane, and find the time it takes to reach the bottom.

The equations of motion are given by (7.8.7) and (7.8.8). Let us consider first the translational motion of the body. The forces which act on the body

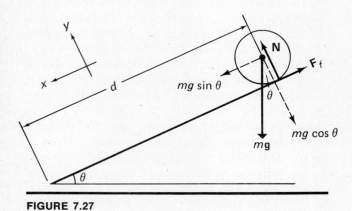

FIGURE 7.27

are the weight **W**, the normal force **N**, and the friction force $\mathbf{F}_f$. These act in the directions indicated in Fig. 7.27. Note that in the statement of the problem, no information has been given concerning any coefficients of friction. In fact, the statement that the object is rolling will completely specify the frictional force. It is assumed only that this actual force of friction is smaller than the *maximum possible* frictional force available. If this turned out not to be the case, we would then infer that the object could not be rolling but would slip instead.

The equations for translational motion then give

$$W \sin \theta - F_f = ma_0 \qquad (7.8.14)$$

and

$$N - mg \cos \theta = 0 \qquad (7.8.15)$$

To obtain the torque equation, we note that only the friction force exerts a torque about the center of mass. This torque *alone* tends to make the object roll down the plane. In the absence of a friction force, the rigid body would simply slide down the plane. We have, therefore,

$$F_f R = I \alpha_0 \qquad (7.8.16)$$

The *condition of rolling* provides a link between (7.8.14) and (7.8.15) in that the condition gives the relation

$$a_0 = R \alpha_0 \qquad (7.8.17)$$

which is necessary to provide enough equations to determine all the unknown quantities. Equations (7.8.14), (7.8.15), and (7.8.16) constitute a set of three equations in the three unknowns a_0, α_0, and F_f. This set of equations may now be solved for these quantities. Eliminating α_0 from (7.8.16) by use of (7.8.17) and subsequently eliminating F_f from (7.8.14) by using (7.8.16), we obtain

$$mg \sin \theta - \left(\frac{I}{R^2} \right) a_0 = ma_0$$

which may be solved for a_0 to obtain

$$a_0 = \frac{g \sin \theta}{1 + I/mR^2} \qquad (7.8.18)$$

in agreement with the result obtained previously in Example 7.7.2. We note that as $I \to 0$, we recover the well-known result for the acceleration of a point mass sliding down an inclined plane. The *angular* acceleration, is, of course, just a_0/R.

The linear speed at the bottom is readily obtained by using the equations developed earlier to discuss motion with constant acceleration. Thus, since our object started from rest, its speed at the bottom is found by recalling that $v^2 - v_0{}^2 = 2a_0 d$, whence, since $v_0 = 0$,

$$v^2 = \frac{2gd \sin \theta}{1 + I/mR^2} \tag{7.8.19}$$

and the time to travel the distance d is given by

$$t^2 = \frac{2d}{a_0} = \frac{2d}{g \sin \theta} \left(1 + \frac{I}{mR^2}\right) \tag{7.8.20}$$

EXAMPLE 7.8.3

We have discussed motions in which both translation of the center of mass as well as rotation about the center of mass occur. For such motions, separate equations may be written for translation and for rotation. It is important to note, however, that such systems may be treated by using *only the torque equation* if the system happens to be *in pure rotation about a particular point which is not accelerated*. The following example, solved in two different ways, will illustrate this important special case.

A meter stick stands vertically on a horizontal table. It is displaced slightly from the vertical and therefore falls. Assume that frictional forces are sufficient to ensure that the bottom of the meter stick does not slip. Set up the equations which determine the motion of the meter stick.

The position of the meter stick at an arbitrary angle θ is illustrated in Fig. 7.28. Since the motion

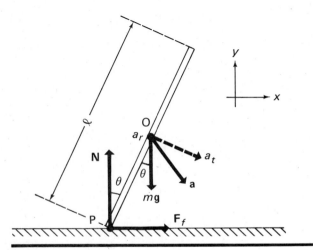

FIGURE 7.28

of any rigid body may be represented as a translation of the center of mass plus a rotation about the center of mass, we may on this basis set up equations describing the resulting motion. Consider first the forces acting on the meter stick. In the horizontal direction, there is a static friction force $\mathbf{F}_f$ which is responsible for the horizontal acceleration of the center of mass. In the vertical direction, the weight $\mathbf{W} = m\mathbf{g}$ acts, as does a normal force $\mathbf{N}$. The equations of translational motion are therefore given by

$$F_f = ma_x \tag{7.8.21}$$

and

$$mg - N = ma_y \tag{7.8.22}$$

The rotational motion about the center of mass is determined by the net torque:

$$N \frac{l}{2} \sin \theta - F_f \frac{l}{2} \cos \theta = I_0 \alpha = I_0 \frac{d^2\theta}{dt^2} \tag{7.8.23}$$

We should be careful to note at this point that the motion described by these equations is not uniformly accelerated, since the forces and torques vary as the stick tips over. It is somewhat more convenient to rewrite (7.8.21) and (7.8.22) in terms of radial and tangential components rather than Cartesian components since the center of mass is a fixed distance from the point of contact. Equations (7.8.21) and (7.8.22) are then replaced by

$$mg \sin \theta - N \sin \theta + F_f \cos \theta = \frac{ml}{2} \frac{d^2\theta}{dt^2} \tag{7.8.24}$$

$$mg \cos \theta - N \cos \theta - F_f \sin \theta = \frac{ml}{2} \left(\frac{d\theta}{dt}\right)^2 \tag{7.8.25}$$

in which, you will recall from our study of circular motion, $a_t = r\alpha = (d^2\theta/dt^2)$, and $a_r = -r\omega^2 = -r(d\theta/dt)^2$. Equations (7.8.23), (7.8.24), and (7.8.25) now form a set of three simultaneous equations for the three variables F_f, N, and θ. It is possible to eliminate F_f and N from these three to obtain a single differential equation describing how the angle θ varies as a function of time. The nasty details of doing this are left as an exercise for the reader.

Now, having gone through all this labor, let us see if perhaps there's a simpler way. We know that the point of contact is fixed. Therefore, the entire meter stick is in a state of pure rotation with respect to the point of contact. Since this point is fixed and therefore not accelerated, we may apply the torque equation *with respect to this point*. The result is

$$\frac{mgl}{2} \sin \theta = I_p \alpha \tag{7.8.26}$$

where I_p represents the moment of inertia of the stick about this fixed point. This is exactly the equation you would arrive at ultimately by using the more difficult method illustrated previously. The lesson to be learned from this example is that before plunging into a problem, some thought should be given to determine the easiest possible approach. The determination of moments with respect to an unaccelerated axis about which a state of pure rotation exists, if one can be found, usually makes the work involved in solving rotation–translation problems much easier.

7.9 Conservation of Angular Momentum

In our previous treatment of translational motion, we found it useful to introduce the idea of the linear momentum of a particle. Initially, one might have questioned the value of defining a new quantity $\mathbf{p}$ as a simple multiple of the velocity vector $\mathbf{v}$. The full virtue of doing this could not be appreciated until the motion of *systems* of particles was considered. We found then that in the absence of any net external force, the total linear momentum is always conserved and that when a net external force is applied, the time rate of change of linear momentum is precisely equal to this net external force. These results having to do with the linear momentum are found to be of great value in situations where a lack of detailed information regarding the forces acting between different bodies of the system makes a direct application of Newton's laws difficult or impossible. We shall now see that a somewhat similar situation prevails with respect to rotational motion in which it is advantageous to introduce a quantity called *angular momentum* and to investigate its relation to externally applied torques.

You will recall that the torque τ associated with a force $\mathbf{F}$ is given by $\tau = \mathbf{r} \times \mathbf{F}$, where $\mathbf{r}$ is the position vector that locates the point of application of $\mathbf{F}$ with respect to a given point O. You are also aware of the fact that torque plays much the same role in rotational motion that force does in translation. It is, therefore, reasonable to define the *angular momentum* $\mathbf{L}$ of a particle having linear momentum $\mathbf{p}$ as

$$\mathbf{L} = \mathbf{r} \times \mathbf{p} \tag{7.9.1}$$

where, again, $\mathbf{r}$ is a position vector that locates the particle with respect to an arbitrarily chosen origin. It is apparent from this definition that the angular momentum is a vector normal to the plane in which $\mathbf{r}$ and $\mathbf{p}$ lie, whose magnitude is $rp \sin \theta_{rp}$, as shown in Fig. 7.29. The dimensions of angular momentum are those of distance times linear momentum, hence, kg-m^2/sec or g-cm^2/sec in metric units, or slug-ft^2/sec in British units. While the linear momentum of a particle is the same no matter where the origin is

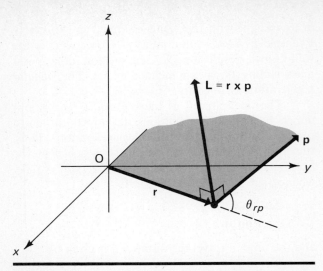

FIGURE 7.29. Definition of vector angular momentum as the cross product of displacement and linear momentum.

chosen, this is no longer true of the angular momentum. The angular momentum will always be found to depend upon the point O from which the position vector $\mathbf{r}$ originates. The same situation exists, of course, in relation to force and torque.

The generalization of the definition (7.9.1) to a system of particles is obvious. We may simply sum the individual angular momenta of all the particles of the system, remembering to add the various contributions *vectorially*, and to refer them all to the same given point O. If there are N particles in the system, then at a given time t, the angular momentum of the system will be

$$\mathbf{L} = \sum_{i=1}^{N} \mathbf{L}_i = \sum_{i=1}^{N} \mathbf{r}_i \times \mathbf{p}_i \tag{7.9.2}$$

where $\mathbf{r}_i$ and $\mathbf{p}_i$ characterize the position and momentum of the ith particle at time t and where $\mathbf{L}_i$ is its individual angular momentum. In many cases, the vector sum in (7.9.2) is quite difficult to evaluate, particularly when the various individual contributions $\mathbf{L}_i$ point in different directions. Let us set this difficulty aside for a moment, however, and consider first a few examples in which the angular momentum may be readily obtained.

EXAMPLE 7.9.1

Two trains having masses m_1 and m_2 travel along a straight railroad track with constant velocities v_1 and v_2, respectively ($v_2 > v_1$). At time $t = 0$, they are at positions d_1 and d_2 with respect to the station platform, as shown in Fig. 7.30. The platform is located by the vector $\mathbf{D}$ with respect to the ticket booth. Find the total angular momentum of the two trains for any time t, first with respect to the station platform, then with respect to the ticket booth.

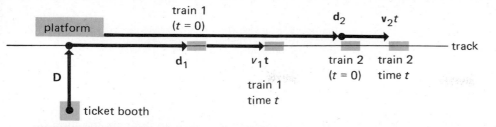

FIGURE 7.30

With respect to the *station platform*, the position vectors of the two trains are $\mathbf{r}_1 = \mathbf{d}_1 + \mathbf{v}_1 t$ and $\mathbf{r}_2 = \mathbf{d}_2 + \mathbf{v}_2 t$, and therefore their respective angular momenta are

$$\mathbf{L}_1 = \mathbf{r}_1 \times \mathbf{p}_1 = (\mathbf{d}_1 + \mathbf{v}_1 t) \times m_1 \mathbf{v}_1$$
$$= (\mathbf{d}_1 \times m_1 \mathbf{v}_1) + (\mathbf{v}_1 t \times m_1 \mathbf{v}_1) \qquad (7.9.3)$$

$$\mathbf{L}_2 = \mathbf{r}_2 \times \mathbf{p}_2 = (\mathbf{d}_2 + \mathbf{v}_2 t) \times m_2 \mathbf{v}_2$$
$$= (\mathbf{d}_2 \times m_2 \mathbf{v}_2) + (\mathbf{v}_2 t \times m_2 \mathbf{v}_2) \qquad (7.9.4)$$

These contributions, however, are both zero since they involve only vector cross products wherein both vectors are parallel. The total angular momentum with respect to the station platform is therefore zero.

With respect to the *ticket booth*, the position vectors are given by $\mathbf{r}_1' = \mathbf{D} + \mathbf{d}_1 + \mathbf{v}_1 t$ and $\mathbf{r}_2' = \mathbf{D} + \mathbf{d}_2 + \mathbf{v}_2 t$; the angular momenta of the two trains then become

$$\mathbf{L}_1' = \mathbf{r}_1' \times \mathbf{p}_1 = (\mathbf{D} + \mathbf{d}_1 + \mathbf{v}_1 t) \times m_1 \mathbf{v}_1$$
$$= (\mathbf{D} \times m_1 \mathbf{v}_1) + (\mathbf{d}_1 \times m_1 \mathbf{v}_1) + (\mathbf{v}_1 t \times m_1 \mathbf{v}_1) \quad (7.9.5)$$

$$\mathbf{L}_2' = \mathbf{r}_2' \times \mathbf{p}_2 = (\mathbf{D} + \mathbf{d}_2 + \mathbf{v}_2 t) \times m_2 \mathbf{v}_2$$
$$= (\mathbf{D} \times m_2 \mathbf{v}_2) + (\mathbf{d}_2 \times m_2 \mathbf{v}_2) + (\mathbf{v}_2 t \times m_2 \mathbf{v}_2) \quad (7.9.6)$$

Again, the cross products involving parallel vectors contribute nothing, and the second and third cross products in both equations are zero. Now, the magnitudes of $\mathbf{L}_1'$ and $\mathbf{L}_2'$ are, respectively, $L_1' = m_1 D v_1$ and $L_2' = m_2 D v_2$, since the vector $\mathbf{D}$ is perpendicular both to $\mathbf{v}_1$ and $\mathbf{v}_2$. The direction of $\mathbf{L}_1'$ and $\mathbf{L}_2'$ is in both cases into the plane of the page, or vertically downward with respect to the station platform. The resultant angular momentum $\mathbf{L}'$, equal to $\mathbf{L}_1 + \mathbf{L}_2$, therefore points in the same direction, its magnitude being given by

$$L' = L_1' + L_2' = D(m_1 v_1 + m_2 v_2) \qquad (7.9.7)$$

EXAMPLE 7.9.2

A flywheel of radius R and mass m is rotating about a *fixed* axis through its center, as shown in Fig. 7.31. It has constant angular velocity ω. Find its angular momentum.

At first sight, this problem seems formidable since the body is continuous and, therefore, seemingly involves a complicated sum of constituent angular momenta. To evaluate this sum, let us first look at the angular momentum of a single mass element dm located a distance r from the center, as shown in the figure. Since its linear momentum is tangentially directed and has magnitude $v\,dm$, its infinitesimal contribution to the angular momentum about the rotation axis will be

$$d\mathbf{L} = \mathbf{r} \times \mathbf{v}\,dm \qquad (7.9.8)$$

which points along the $+z$-axis, thus along the axis of rotation in the sense of advance of a right-handed screw rotating with the flywheel. Now, it is clear that any other mass element in the object will also contribute an angular momentum about the same axis *which points in the very same direction*. The fact that all constituent mass elements have angular momenta in the same direction leads to an enormous simplification, since it reduces the complicated vector summation of individual angular momentum vectors to a scalar summation.

To obtain the total angular momentum, then, it is necessary to add up contributions having magnitude

$$dL = |\mathbf{r} \times \mathbf{v}|\,dm = rv\,dm \qquad (7.9.9)$$

for all elements of the body. A further simplification results if we recall that each mass dm rotates about the center of a circle of radius r with the *same* angular velocity ω, related to the linear velocity v by $v = r\omega$.

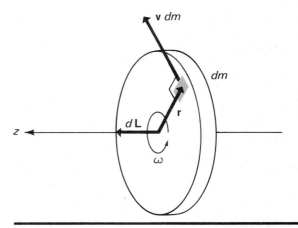

FIGURE 7.31. Calculation of the angular momentum of a rigid body rotating about a fixed axis.

233

Using this relation, we may replace the *variable* linear velocity v by the *constant* angular velocity ω, the result being

$$dL = rv\,dm = \omega r^2\,dm \qquad (7.9.10)$$

The total angular momentum may now be found by integrating over all the mass elements, and in doing this the constant[2] angular velocity ω can be brought outside the integral:

$$L = \int \omega r^2\,dm = \omega \int r^2\,dm$$

or, recalling (7.2.17),

$$L = I\omega \qquad (7.9.11)$$

The direction of **L** is, of course, along the axis of rotation, the same as all the constituent angular momenta $d\mathbf{L}$.

We have in this way obtained a very important result which we shall have occasion to use quite frequently later. The relation (7.9.11) provides us with a *general* way of calculating the angular momentum of *any symmetric body* which rotates about a *fixed* axis of rotation, whether or not the angular velocity is constant in time. The fact that we referred to a flywheel for the purpose of illustration is unimportant—the "flywheel" could equally well have been spherical, or conical, or square, so long as it rotated about a fixed axis of symmetry. In any such event, we should have found that $\mathbf{L} = I\omega$.

We have seen previously that Newton's second and third laws lead to an equation relating the time rate of change of linear momentum with the external forces which effect such a change. It is perhaps not surprising that a rotational analogue of this law exists involving external torques and the time rate of change of angular momentum. To see this more clearly, let us evaluate directly the time derivative of the angular momentum vector for a *single particle* of mass m and velocity **v**. This can be done by differentiating (7.9.1) with respect to time:

$$\frac{d\mathbf{L}}{dt} = \frac{d}{dt}(\mathbf{r} \times \mathbf{p}) = \frac{d}{dt}(\mathbf{r} \times m\mathbf{v})$$

$$= \frac{d\mathbf{r}}{dt} \times m\mathbf{v} + \mathbf{r} \times m\frac{d\mathbf{v}}{dt}$$

$$= (\mathbf{v} \times m\mathbf{v}) + \mathbf{r} \times m\frac{d\mathbf{v}}{dt} \qquad (7.9.12)$$

The first term vanishes since it involves parallel vectors, while in the second, by virtue of Newton's

[2] The angular velocity ω is constant in the sense that it is the same for all mass elements of the body over whose coordinates the integration is performed. There is no assumption whatever that ω is constant with respect to time.

234

second law, $m(d\mathbf{v}/dt)$ can be replaced by **F**. The resulting equation is

$$\mathbf{r} \times \mathbf{F} = \tau = \frac{d\mathbf{L}}{dt} \qquad (7.9.13)$$

This equation tells us that the

time rate of change of the angular momentum of a particle is equal to the resultant torque it experiences.

If the resultant torque is zero, then $d\mathbf{L}/dt = 0$ and the angular momentum is a constant of the motion—in other words, it is *conserved*.

There are a number of important special cases in which the torque vanishes, leading to conservation of angular momentum. First of all, if the angular momentum is referred to the location of the particle itself, then both torque and angular momentum will be zero; this case is not of very great interest, since in many instances a coordinate system attached to the particle will be an accelerated, hence *noninertial*, system, which may not be a very satisfactory reference frame. The torque τ also vanishes if the force **F** is parallel to the position vector **r** and has, therefore, only a radial component. Such a force is referred to as a *central force*, and central forces play a very important role in mechanics and other branches of physics; for example, the gravitational forces exerted by the sun and the planets upon one another are forces of this type, as we shall soon see. Another case in which the torque is zero and in which angular momentum is, therefore, conserved, is that in which F itself vanishes; under these circumstances the motion is simply a linear one with constant velocity.

So far, we have dealt only with a single particle, and we have found in Eq. (7.9.13) a rotational analog of Eq. (6.2.2) which expresses the time rate of change of linear momentum for a particle. In a similar way there exists a rotational analog of Eq. (6.3.5) which applies to a complete *system* of interacting particles. From the preceding result as embodied in (7.9.13), it is clear that the time derivative of the total angular momentum of a system of many particles is given by the sum of the torques acting on every particle of the system, or, for an N-particle system,

$$\frac{d\mathbf{L}}{dt} = \sum_{i=1}^{N} \frac{d\mathbf{L}_i}{dt} = \sum_{i=1}^{N} \tau_i \qquad (7.9.14)$$

The torques on each particle may be classified as being either internal or external. Internal torques are those due to forces exerted by particles of the system *on one another*, while external torques are those which are due to forces originating *outside* the system. We shall now assert, without formal proof, that the *sum* of all internal torques is zero if the forces between particles are always directed along the lines joining them. The proof of this statement, which depends in part upon Newton's third law, will not be given here. We shall always assume hereafter that this condition is fulfilled. The situation is basically similar to that encountered in the discussion of the time rate of change of linear momentum of a system of particles developed in section 6.3, where it was found that internal forces between particles of the system canceled one another and contributed nothing to the resultant force acting on the system. The reader may find it profitable at this point to reexamine that section. Under these circumstances, (7.9.14) may be written as

$$\tau_{ext} = \frac{d\mathbf{L}}{dt} \qquad (7.9.15)$$

where τ_{ext} is the resultant of all the external torques acting on the system. If the system is a rigid body, the same result is obtained, since a rigid body can always be regarded as a collection of mass particle elements whose number approaches infinity as a limit while their individual masses approach zero, such that the total mass remains constant.

We may summarize the result expressed by (7.9.15) by stating that:

The time rate of change of angular momentum of any system of particles is equal in magnitude and direction to the resultant of all the external torques acting on the system.

It follows also that:

If the resultant of the external torques acting on any system of particles is zero, then the angular momentum of the system remains constant in magnitude and direction.

The latter statement, which follows directly upon setting τ_{ext} equal to zero in (7.9.15), is referred to as the *law of conservation of angular momentum*, and constitutes one of the most important conservation laws of physics. The full impact of these important laws will be demonstrated by the following series of examples.

EXAMPLE 7.9.3

A flywheel rotating about a fixed axis through its center experiences a constant applied force, as illustrated in Fig. 7.32. Find the angular momentum of the flywheel as a function of time, assuming that it starts from rest at $t = 0$, and find the angular velocity ω at any time.

In this problem, there are really three external forces acting on the flywheel. First, there is a downward force at point O due to the flywheel's own

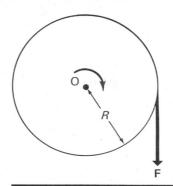

FIGURE 7.32

weight; there is also an upward force at O exerted by the bearings on which the wheel is mounted upon the wheel itself; finally, there is the force **F**, as shown.

The first two of these forces pass through the point O, about which we shall compute torques and angular momenta; therefore, neither force contributes any torque about this axis. The situation is very much the same as that discussed in detail in Example 7.2.4. The third force, however, does exert a torque about O, of magnitude RF and directed downward into the plane of the paper. If the flywheel is mounted on a rigid axle, then the angular momentum will also point into the plane of the page, and we shall find from (7.9.15) that

$$RF = \frac{dL}{dt} \qquad (7.9.16)$$

$$dL = RF \, dt$$

where R and F are both constant. If the angular momentum is zero at time $t = 0$ and has the value L at some later time t, (7.9.16) may be integrated to give

$$\int_0^L dL = RF \int_0^t dt$$

or

$$L = RFt \qquad (7.9.17)$$

We have already seen from Example 7.9.2 that the angular momentum of a symmetric rigid body such as this, rotating about a fixed axis, is given by $L = I\omega$. We may, therefore, write (7.9.16) as

$$\tau = RF = \frac{d(I\omega)}{dt} = I\frac{d\omega}{dt} = I\alpha \qquad (7.9.18)$$

a result previously obtained by direct application of Newton's laws in section 7.2. Equation (7.9.18) serves merely to illustrate that Newton's laws and conservation of angular momentum amount to the same thing insofar as rigid bodies rotating about fixed axes are concerned. Since R and F are constant, it is evident that the acceleration α in (7.9.18) is also constant, whereby $\omega - \omega_0 = \alpha t$, or since the initial angular velocity ω_0 is zero,

$$\omega = \alpha t = \frac{RF}{I} t \qquad (7.9.19)$$

This exercise does not exhibit any particular advantage in using (7.9.15) rather than the methods of section 7.2. However, as we shall soon see, Eq. (7.9.15) provides a more generally useful framework for working examples more difficult than those involving only rigid bodies rotating about fixed axes. For example, we may use (7.9.15) to approach problems in which the external torque is *not parallel* to the angular momentum as in the motion of a gyroscope or spinning top. Furthermore, the methods developed in this section lead directly to the concept of conservation of angular momentum, a most important result which is not so obvious from the outcome of section 7.2.

Conservation laws such as this are extremely important and useful in physics. Our earlier study of conservation of energy and linear momentum permitted us to analyze certain complicated problems without reference to the complex details of their internal dynamics. Conservation of angular momentum is a law of vast usefulness in situations of the same general type. Finally, the conservation laws of physics provide us with a guide to making the first steps into unexplored territory on the frontiers of scientific knowledge, where there are no generally accepted and reliably tested procedures to enable us to predict the outcome of physical processes. Some of the subsequent examples should indicate the value of these methods in certain situations of great fundamental importance.

EXAMPLE 7.9.4
A rigid, circular turntable of radius 2 meters and mass 100 kg is initially rotating about its fixed, vertical central axis with initial angular velocity $\omega_i = 2\pi$ rad/sec. The turntable's bearings are frictionless. A 50-pound boy who is initially at rest jumps onto the turntable, landing at a spot 1 meter from the center. Find the final angular velocity of the boy and turntable.

This is a problem in which a detailed consideration of forces is unnecessary; in fact, such a consideration would make the solution more difficult. The situation is really one involving an inelastic "rotational collision"; therefore, we proceed by attempting to use conservation of angular momentum, just as we would use conservation of linear momentum in a linear collision problem. Since there are no torques *external* to the system, but only the internal torques which arise between the two objects which comprise it, the angular momentum of boy plus turntable is conserved.

The initial angular momentum about the axis of the turntable is given by the sum of the initial angular momentum of the boy (which is zero, since he is at rest in the beginning) and that of the turntable, whose moment of inertia I, in view of its cylindrical shape, is $\frac{1}{2}m_t r^2$, where m_t is its mass and r its radius. The initial angular momentum is then

$$L_i = 0 + I\omega_i = \frac{1}{2}m_t r^2 \omega_i$$
$$= (\tfrac{1}{2})(100)(2)^2(2\pi) = 400\pi \text{ kg-m}^2/\text{sec} \qquad (7.9.20)$$

In the final state of the system, boy and turntable have the same final angular velocity. We shall assume that the boy's size is sufficiently small in comparison with his final distance from the axis that the point mass approximation can be used to express his moment of inertia; his moment of inertia I_b after landing on the turntable will then be $m_b r_b^2$, where m_b is his mass and r_b the distance from his center of mass to the axis. The final angular momentum ω_f then is

$$L_f = I\omega_f + I_b \omega_f = \frac{1}{2}m_t r^2 \omega_f + m_b r_b^2 \omega_f$$
$$= (\tfrac{1}{2})(100)(2)^2 + (50)(0.454)(1)^2 \omega_f = 222.7\omega_f$$
$$\qquad (7.9.21)$$

Equating initial and final angular momenta and solving for ω_f,

$$\omega_f = \frac{400\pi}{222.7} = 5.64 \text{ rad/sec}$$

EXAMPLE 7.9.5
A man stands on a platform mounted on frictionless bearings which is free to rotate about a vertical axis, as shown in Fig. 7.33a. The moment of inertia of the man and turntable is 5 slug-ft². The man is holding an 8-pound weight in each hand. He extends his arms so that the weights are 2.5 ft from the axis of the turntable

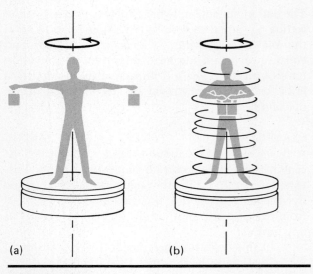

(a)   (b)

FIGURE 7.33

and is given an initial angular velocity of 2 revolutions per second by an assistant. He subsequently moves his arms inward until the weights are on the axis of rotation as in Fig. 7.33b. Assuming that the weights can be regarded as point masses, find his final angular velocity. Also, find the initial and final kinetic energy of the system, and explain why energy is *not* conserved.

There are no external torques acting on the system, and therefore angular momentum must be conserved. The initial angular momentum L_i can be written as

$$L_i = I_t\omega_i + I_w\omega_i = (I_t + 2mR^2)\omega_i \qquad (7.9.22)$$

where I_t represents the moment of inertia of the turntable plus the man, I_w represents the moment of inertia of the two weights, and ω_i is the initial angular velocity. Since the weights are treated as point masses, their combined moment of inertia will be $2mR^2$, where m is the mass of each and R the distance from the rotation axis. Substituting numerical values into this expression, it is apparent that

$$L_i = [5 + (2)(\tfrac{1}{4})(2.5)^2](2)(2\pi) = 102.0 \text{ slug-ft}^2/\text{sec}$$

The final angular momentum is given by

$$L_f = I_t\omega_f = 5\omega_f \qquad (7.9.23)$$

where ω_f is the final angular velocity, because the weights are now on the axis of rotation ($R = 0$) where their contribution to the rotational inertia of the system is negligible. Equating initial and final angular momenta and solving for ω_f, one finds

$$5\omega_f = 102.0$$
$$\omega_f = 20.4 \text{ rad/sec} = 3.25 \text{ rev/sec}$$

It is apparent that as a consequence of conservation of angular momentum, the turntable's angular velocity increases markedly as the weights are moved in near to the axis of rotation. This example is the basis of a frequently performed demonstration designed to verify the law of conservation of angular momentum. The same effect can also be observed when a figure skater causes her spin angular velocity to increase by moving her arms from an extended position in near to the body, or when a diver makes his body rotate more rapidly by tucking his arms and legs in close to his body.

Let us now investigate the question of energy conservation—or, to be more accurate, nonconservation. The initial energy of platform, man, and weights is

$$U_{ki} = \tfrac{1}{2}(I_t + 2mR^2)\omega_i^2$$
$$= (\tfrac{1}{2})[(5) + (2)(\tfrac{1}{4})(2.5)^2](4\pi)^2 = 641 \text{ ft-lb} \qquad (7.9.24)$$

while the final kinetic energy is

$$U_{kf} = \tfrac{1}{2}I_t\omega_f^2 = (\tfrac{1}{2})(5)(21.0)^2 = 1102 \text{ ft-lb} \qquad (7.9.25)$$

From this we see that, far from having been conserved, the total energy of the system has actually *increased*. We may now reasonably inquire where all this extra kinetic energy came from.

The answer lies in the fact that the man *must do work on the weights in order to move them in close to the axis of rotation.* It is this amount of work that eventually appears as increased kinetic energy of rotation. If the platform were not revolving, of course, no work would be required, since then the only force exerted by the weights on the man is mg, which is directed downward and is, therefore, normal to the displacement occurring as the weights are moved inward. But when the platform revolves, the man must exert forces on the weights to keep them in their circular path. He therefore finds it necessary to do work to make the weights move inward. The forces he exerts and the work he does generate torques that increase his angular velocity and that of the weights as he draws them inward. But all these torques are internal ones, exerted by one part of the system on another; the resultant external torque on the system is zero. The situation is similar in this respect to the example of the rifle and bullet discussed in Example 6.3.1. The reader may find it interesting at this point to examine the initial and final energies of the system discussed in Example 7.9.4 and see if he can explain why the final energy is less than the initial and what happened to the difference.

EXAMPLE 7.9.6

A circular turntable of radius R and moment of inertia I_0 is rotating with initial angular velocity ω_0 about a fixed vertical axis through its center. A man of mass m is riding the turntable at its outer rim and walks toward the center, along a radius of the turntable, with constant radial speed v_r with respect to the turntable. Find the angular velocity of man and turntable at any time, assuming that the turntable rotates in frictionless bearings.

The most straightforward approach to this problem involves the use of angular momentum conservation. No external torques act on the system of man plus turntable, and therefore the angular momentum of the system when the man is some distance r from the center is exactly the same as it was initially, when the man was located at a distance R from the axis. Assuming that the man is small enough with respect to the turntable that he can be regarded as a point mass whose moment of inertia can be expressed by mr^2, the initial angular momentum L_i can be written as the sum of the turntable's angular momentum, $I_0\omega_0$, and that of the man, $mR^2\omega_0$, as follows:

$$L_i = (I_0 + mR^2)\omega_0 \qquad (7.9.26)$$

while at a later time, when he is a distance r from the

center, the angular momentum is

$$L_f = (I_0 + mr^2)\omega(r) \qquad (7.9.27)$$

where $\omega(r)$ is the angular velocity at that time. We may now equate these two values of angular momentum and solve for $\omega(r)$ to obtain

$$\omega(r) = \frac{I_0 + mR^2}{I_0 + mr^2}\,\omega_0 \qquad (7.9.28)$$

Since r is always less than R, the numerator of the fraction in (7.9.28) is always larger than the denominator; in fact, as r becomes less, $\omega(r)$ becomes greater. The angular velocity of the system, therefore, increases steadily as the man walks inward. One may understand this more clearly by noting that the total moment of inertia of the system decreases as the man walks inward; and in order that the angular momentum (which is, after all, the product of the total angular momentum and the angular velocity) remain constant, the angular velocity must increase.

The man's resultant velocity has a constant radial component v_r and a variable tangential component given by

$$v_t(r) = r\omega(r) = \frac{I_0 + mR^2}{I_0 + mr^2}\,r\omega_0 \qquad (7.9.29)$$

His path, when viewed from a nonrotating reference frame, such as the stationary base on which the turntable is mounted, will be a *spiral*.

This result was easily obtained using a conservation law applied to the total system of man plus turntable. But suppose we had decided instead to do the problem by examining the motion of the man alone; clearly he experiences forces and torques which produce an acceleration from which the motion presumably could be completely determined. But, although the problem can be done in this way, the detailed consideration of all the forces that act and the calculation of the acceleration that results is a difficult and tedious task, indeed. The work of doing this is left as one of the more difficult problems at the end of the chapter.

Along the same line, it is not difficult to compute the initial and final kinetic energies of the system and to show that the total energy of the system always increases as the man walks inward; the details of doing this are again assigned as an exercise. This means that the man must do work on the system in order to go from the outside to the center; from his point of view, this work is done against the various force components exerted *on him by the turntable*. It is a difficult, but not quite impossible, task to compute the work done by the man on the turntable and to assure oneself via the work–energy theorem that this is equal to the turntable's gain in kinetic energy. One may also, with difficulty, compute the forces exerted *on the man by*

the turntable—which are, of course, equal and opposite to those exerted on the turntable by him—and show that the work done by these forces equals the man's gain (a negative gain in this case) in kinetic energy. The total change in kinetic energy of both man and turntable can thus be accounted for in detail and shown to agree with the result obtained much more easily from conservation of angular momentum.

7.10 Gyroscopes, Tops, and Precession

So far, we have concerned ourselves exclusively with rotational motions where the rotation axis is either fixed or, at worst, undergoing translational motion without changing its direction. While most of the simple rotational motions with which most of us are familiar fall into these categories, there is one particularly well-known example—that of the spinning top or gyroscope—which does not. If we cause the rotor of a gyroscope, for example, to spin rapidly and then mount one end of the rotation axis on a fixed support, as shown in Fig. 7.34a, the gyroscope does not simply fall downward off the support (as it would if the rotor were not spinning), but instead remains in a roughly horizontal position while the axis of the rotor turns slowly in a horizontal plane, thus performing a rotational motion of its own about a vertical axis coinciding with that of the pin or pylon supporting the end of the gyroscope. This peculiar slow rotation of the axis itself is referred to as *precession*.

The observed precession of a gyroscope or spinning top appears at first sight to defy all intuitive notions of how such a device should behave. What, after all, keeps the thing from falling over? We shall see that the rotational form of Newton's laws of motion, properly applied to a system such as this, easily leads to an explanation of the precessional motion. Unfortunately, the explanation, though perfectly correct, is somewhat mathematical in character and leaves our intuitive sense still unsatisfied to a degree. It is, indeed, difficult to trace the paths of all the individual mass elements in the spinning rotor, to describe the forces acting on them, and to assure oneself finally that the precessional motion arises simply from the requirement that $\mathbf{F} = m\mathbf{a}$ for all such elements, even though this is indisputably true.

We shall begin by trying to see how the precession arises from Newton's laws for rotating bodies. First of all, consider the symmetric spinning top shown in Fig. 7.34b. The top's spin angular momentum at time $t = 0$ is a vector $\mathbf{L}_0$, which may make an angle θ with the vertical, as shown in the diagram. The center of mass of the top is assumed to lie on the rotation axis, since the top is symmetric, at a distance

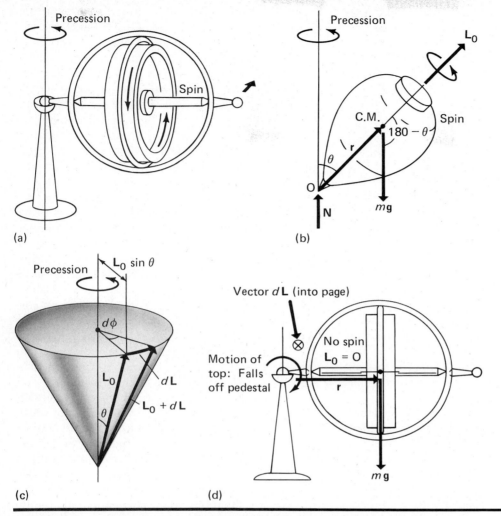

FIGURE 7.34. The gyroscope. (a) Precession of a toy gyroscope. (b) Geometry of a symmetric top. (c) Vector diagram for precession of a symmetric top. (d) Gyroscope with no spin angular momentum.

r from the point O, and its location is described by the position vector **r**.

The top experiences a torque about the point O due to its own weight; there is, of course, an upward normal force **N** acting on the top at O, but since it passes through that point, it contributes no torque. The total resultant torque, equal to the time rate of change of total angular momentum, is that due to the weight force $m\mathbf{g}$, whence

$$\tau = \frac{d\mathbf{L}}{dt} = \mathbf{r} \times m\mathbf{g} \qquad (7.10.1)$$

or

$$d\mathbf{L} = (\mathbf{r} \times m\mathbf{g})\, dt \qquad (7.10.2)$$

The vector $d\mathbf{L}$, representing the change of angular momentum during the time dt, must, according to this, be perpendicular both to **r** and $m\mathbf{g}$. Since $m\mathbf{g}$ points vertically downward, $d\mathbf{L}$ must lie in a horizontal plane; and since **r** points in the same direction as $\mathbf{L}_0$, the orientation of $d\mathbf{L}$ in this horizontal plane must be *normal* to $\mathbf{L}_0$.

The orientation of $d\mathbf{L}$ with respect to $\mathbf{L}_0$ is, therefore, as shown in Fig. 7.34c, the final angular momentum after time dt being given by the vector $\mathbf{L}_0 + d\mathbf{L}$. It is apparent that the angular momentum vector *precesses* through an angle $d\phi$ about the vertical axis during this time interval. Since we could now perform the same calculation starting with the vector $\mathbf{L} + d\mathbf{L}$ at time dt instead of with $\mathbf{L}_0$ at time zero, it is evident that this motion of precession is a continuous one and that in time the angular momentum vector will trace out a cone whose axis is vertical and whose apex angle is θ. This is actually the way a spinning top behaves. The angular velocity of precession (let us call it Ω) is given by the rate of change of the horizontal angle ϕ with respect to time, thus by $d\phi/dt$. Since in the limit of short time intervals $d\phi$ also becomes very small, and since in the limit of small angles the tangent of the angle approaches the angle itself, expressed in radians, it is easily seen from Fig. 7.34c that

$$\tan (d\phi) \simeq d\phi = \frac{dL}{L_0 \sin \theta}$$

whence

$$\Omega = \frac{d\phi}{dt} = \frac{dL/dt}{L_0 \sin \theta} \qquad (7.10.3)$$

But since dL/dt is equal to the magnitude of the resultant torque, this may be written as

$$\Omega = \frac{\tau}{L_0 \sin \theta} = \frac{mgr \sin(180^\circ - \theta)}{L_0 \sin \theta}$$

and since $\sin(180^\circ - \theta) = \sin \theta$,

$$\Omega = \frac{mgr}{L_0} = \frac{mgr}{I_0 \omega_0} \qquad (7.10.4)$$

where I_0 is the moment of inertia of the top or rotor and ω_0 its spin angular velocity. It is evident from this that the precession angular velocity increases as the quantity mgr (proportional to the resultant torque) increases and decreases with increasing spin angular velocity ω_0.

In all these calculations, it has been assumed that the angular momentum change dL is much less than L_0. This will always be true if the spin angular momentum is much larger than the angular momentum of precession, a condition which is generally satisfied whenever the spin angular velocity is much larger than the angular velocity associated with precession. However, suppose the spin angular velocity of the top or gyroscope is zero. In this case, L_0 also becomes zero, and the *only* angular momentum the system possesses is that represented by $d\mathbf{L}$, arising from the torque $\mathbf{r} \times m\mathbf{g}$. A total angular momentum in the direction of this vector corresponds to that which would arise, for example, in Fig. 7.34b if the top were to rotate in the plane of the vector $\mathbf{r}$ and the vertical axis in such a way as to increase the angle θ. This is just a complicated way of saying that the top falls down in a situation such as this, which is indeed what is observed to happen. The same thing is illustrated for a gyroscope with no spin in Fig. 7.34d. The only angular momentum present is that created by the torque $\mathbf{r} \times m\mathbf{g}$; it is an ever-increasing vector quantity whose direction is into the plane of the page corresponding to a clockwise rotation of the gyro about its pedestal as it topples over.

Looking back now toward the question of what keeps the gyroscope or top from falling over, we may perhaps answer that, to begin with, we don't really let a top or gyroscope engage in free fall, but instead *insist on supporting one end* so that it must change the direction of its angular momentum vector as it tips! Since the angular momentum of the rotor is large, this change could be brought about only by a very large torque, much larger than the small torque available from the weight force. All this torque can do is make the rotor slowly precess, and that about a vertical rather than horizontal axis. If we really allow such a device to engage in free fall by simply dropping it, it falls just like anything else, attaining acceleration $a_y = -g$, since in this case it can fall without changing the magnitude or direction of its very large angular momentum. It is, therefore, more accurate to say that a top or gyroscope will fall, *but not tip over.*

Falling, in the most general sense, refers to the behavior of objects acted upon only by gravitational forces or torques. The most conventional way of falling in our immediate habitat is to obtain a translational acceleration $a_y = -g$, but less common situations exist *in which other responses to unbalanced gravitational forces or torques arise as a natural consequence of Newton's laws.* There are, therefore, other ways of falling which satisfy all the requirements of Newton's laws even though they may not satisfy our intuitive expectations quite so well. Taking this view, we may say that precession is the gyroscope's own peculiar way of falling. The behavior of the Yo-Yo, discussed earlier in section 7.6, is another case which illustrates this point. We shall encounter still another interesting and particularly important way of "falling" in the next chapter. It is apparent from all this that a profound understanding of the laws of nature can *extend* our simple intuitive perception of the universe to embrace situations where an *untrained* intuition would fail. That is one of the reasons why people find it useful to study physics, even though it is a difficult, and sometimes not particularly entertaining, subject.

When a rapidly spinning gyroscope is mounted so that it is free from external torques (for example by allowing it to float in a pool of mercury), its axis of rotation points always in the same direction even though the base upon which it is mounted—a ship or airplane—may change its orientation continually. This property, which follows directly from the result that the angular momentum remains constant if there are no external torques, has led to the widespread use of gyroscopes as the essential elements of gyrocompasses and other inertial navigation systems for marine, aeronautical, or aerospace applications.

The phenomenon of precession, though utilized also in certain engineering applications, is an important feature of the behavior of atoms and fundamental subatomic particles under certain conditions. Since atoms and elementary particles often possess intrinsic angular momenta, they can be made to precess if subjected to external torques. These precessions, as we shall see later, can lead to important, easily observable magnetic effects and can contribute valuable information relating to our physical picture of atoms, molecules, and other microscopic systems.

SUMMARY

Newton's law for rotational motion can be stated in the form

$$\tau = I\alpha$$

where τ is the resultant torque about the axis of rotation and α is the angular acceleration. The *moment of inertia* I, for a simple point mass m at a distance r from the axis of rotation, is given by

$$I = mr^2$$

For a system of point masses m_i that rotate with a common angular velocity about a fixed axis, the moment of inertia is

$$I = \sum_{i=1}^{N} m_i r_i^2 = m_1 r_1^2 + m_1 r_2^2 + \cdots + m_N r_N^2$$

For an extended rigid body, the moment of inertia can be arrived at by integrating point mass contributions $dI = (dm)r^2$ over all mass elements of the body, from which

$$dI = r^2\, dm$$

and

$$I = \int dI = \int_v r^2\, dm$$

The moment of inertia in rotational motion plays a role analogous to that of mass in translational motion. It is important to remember, however, that the moment of inertia depends not only on the mass of the system but on how the mass is distributed with respect to the rotation axis and where the rotation axis itself is located.

If we know the moment of inertia I_0 of a body about an axis passing through the center of mass, we can find the moment I about any axis parallel to this one by using the parallel axis theorem, which states that

$$I = I_0 + mR^2$$

where R is the distance between the two axes.

A torque that gives rise to rotational motion does work that is given by

$$W = \int_{\theta_i}^{\theta_f} \tau\, d\theta$$

This work can be converted to kinetic energy of rotation, the definition of which can be stated as

$$U_k^{(r)} = \tfrac{1}{2} I \omega^2$$

where ω is the angular velocity. The *work–energy theorem for rotational motion* states that the work done by the *resultant* torque acting on a system equals the increase in the system's kinetic energy of rotation.

A system may possess potential energy of rotation, which represents the rotational work the system can do in going from its initial state to a standard final condition in which the rotational potential energy is, by definition, zero.

In systems in which there are no nonconservative forces or torques, the sum of translational and rotational kinetic and potential energies is conserved; if frictional or other nonconservative forces or torques act, the difference between initial and final mechanical energies will be accounted for by the work done by these nonconservative processes. These laws are expressed by Eqs. (7.6.24) and (7.6.23).

For bodies which simultaneously execute both translational and rotational motion, the total kinetic energy can be expressed as the sum of the translational kinetic energy of the center of mass plus the rotational kinetic energy about an axis through the center of mass.

The equations of motion of a rigid body undergoing simultaneous translation and rotation are formulated by applying Newton's second law for translation, $\mathbf{F} = m\mathbf{a}_0$, to describe the linear translation of the center of mass, and Newton's second law for rotation, $\tau = I_0\alpha_0$, with torques and angular accelerations referred to an axis through the center of mass, to describe the rotation of the body about the center of mass.

The angular momentum $\mathbf{L}$ of a body of mass m and linear momentum $\mathbf{p}$ is a vector defined as

$$\mathbf{L} = \mathbf{r} \times \mathbf{p} = \mathbf{r} \times m\mathbf{v}$$

The angular momentum of a system of particles is the vector sum of their individual angular momenta. For a rigid body of moment of inertia I rotating about a fixed axis with angular velocity ω, the angular momentum can be expressed as

$$\mathbf{L} = I\omega$$

For any object or system of objects, Newton's second law can be stated as

$$\tau_{\text{ext}} = d\mathbf{L}/dt,$$

where τ_{ext} represents the resultant of all the *external* torques acting. In this case, as in the corresponding case involving linear momentum, *internal* torques exerted by one object of the system on another cancel out and thus do not appear in the above equation at all. For systems in which the resultant external torque is zero, $dL/dt = 0$ and *angular momentum is conserved*.

In systems in which bodies revolve or roll without slipping, their translational and rotational motions are related by the equations

$$x = R\theta \qquad v = R\omega \qquad a = R\alpha$$

where R is the radius of the object. These equations are, therefore, useful in expressing the *coupling* between rotation and translation in systems where both types of motion take place.

QUESTIONS

1. Is the second hand of an electric clock in rotational equilibrium? In translational equilibrium?
2. To calculate the moment of inertia of the earth about an axis passing through the sun perpendicular to the plane of the orbit, we may treat the earth as a point object. However, to compute the moment about a diameter through the earth, we must consider the earth as an extended body. How do you reconcile these two statements?
3. In studying the translational motion of a rigid body, all its mass may be assumed to be concentrated at the center of mass. Give a reason why the same statement cannot hold with regard to rotational motion.
4. A certain rigid body has the same moment of inertia about three perpendicular axes passing through the center of mass of the body. What can you say about the mass distribution of the body?
5. A student calculates the kinetic energy of a rigid body solely in terms of rotational energy. Does this imply that the center of mass is not in motion?
6. A certain force acts upon a rigid body. Can it cause a change in both the translational and rotational energies? Give an example.
7. Using the parallel axis theorem, compute the moment of inertia of a point mass about any axis a distance R from the mass. Show that this is exactly the expected result.
8. On an examination in elementary physics, the class is asked to state the condition for rolling. One student responds by asserting only that the body should be round. How would you have graded his answer? What is the correct answer?
9. A meter stick stands vertically on a frictionless table and is then slightly displaced so that it falls. Describe the motion of the meter stick. What is the kinetic energy per unit mass when the stick strikes the ground?
10. When a diver wishes to make several turns in the air, he/she doubles up in a ball. Explain briefly why this is done.
11. It is rather difficult to maintain your balance on a stationary unicycle, but it is much easier if the unicycle is in motion. Explain.
12. A billiard ball is struck well above its center so that the condition $R\omega_0 > v_0$ holds initially. In which direction does the friction force act until rolling sets in?
13. A billiard ball is struck exactly in the center. Explain why the ball cannot roll initially.
14. What is your reaction to the following statement? The moon is more strongly attracted to the sun than it is to the earth and, therefore, it should rotate about the sun instead of about the earth.
15. What factors might bring about changes in the earth's rate of rotation about its axis?
16. You are given a sphere and an inclined plane. Explain how you can determine whether or not the sphere is hollow.
17. Estimate the amount of energy the earth has due to its rotation about its own axis.
18. If the polar ice caps were to melt, what effect would this have on the angular velocity of the earth about its axis?
19. During World War II, instrument technicians in the Navy became practical jokers. They would place a rapidly spinning gyroscope in a sealed box and ask someone to deliver it to a destination which required making at least one right-angle turn during the journey. Describe what happens during the turn.
20. Whenever the torque acting on a body is perpendicular to its angular momentum, the change in angular momentum must be perpendicular to the original angular momentum. Briefly explain why this is true.
21. A cylindrical can of tuna fish and a can of soup having the same size and shape have exactly the same mass and moment of inertia about the axis of symmetry. The labels have been lost and therefore it is difficult to identify the cans. Can you suggest a method of determining which is which, without opening the cans?

PROBLEMS

1. A toy gyroscope rotor has a moment of inertia of 2500 g-cm². A string 1 meter long is wrapped around the shaft which is 0.5 cm in diameter. If the string is pulled with a constant force and the rotor reaches an angular speed of $\omega = 750$ rad/sec when the string is fully unwound, find the angular acceleration and the constant force.
2. Four 3-kg point masses occupy the corners of a square 1 meter on each side as illustrated in the accompanying figure. Find the moment of inertia of the system about an axis perpendicular to the square and passing through its center.

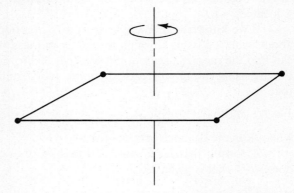

3. For the preceding problem, find the moment of inertia about an axis perpendicular to the square but passing through one of the masses.
4. The four 3-kg masses of Problem 2 are connected to a light drum as shown in the diagram. The drum has a

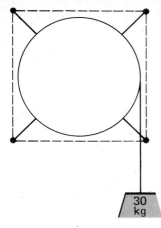

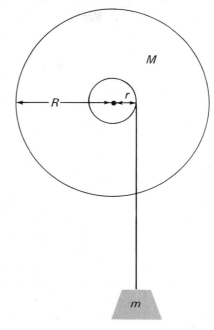

wrapped around the shaft and acts to produce an angular acceleration of the system. Find **(a)** the angular acceleration of the system, **(b)** the linear acceleration of mass m, and **(c)** the tension in the cord, in terms of m, M, r, and R.

radius of 50 cm and is assumed to have a negligible moment of inertia. A 30-kg mass is attached to a long string which is wrapped around the drum. Find the angular acceleration of the system and the angular velocity ω after 8 meters of string have unwrapped. Assume that the initial angular velocity is zero.

5. A grindstone consists of a cylindrical disc of radius 0.30 meter and mass 36.0 kg. Find **(a)** its moment of inertia and **(b)** the constant external torque required to accelerate the wheel from rest to an angular speed of 1800 rpm in a time interval of 5.0 sec, assuming that the bearings are frictionless. **(c)** When the wheel has reached an angular speed of 1800 rpm, the source of power is removed, and a piece of metal is held against the rim of the wheel with a radially directed force of 12.0 newtons. The coefficient of kinetic friction between the metal and the wheel is 0.8. How long will it take, under these circumstances for the wheel to come to rest? **(d)** Through how many revolutions will it have turned during the process described in **(c)**?

6. What are the answers to the preceding problem if the bearings in which the grindstone are mounted exert a constant retarding frictional torque of 0.6 newton-meter?

7. Work out the results presented in Example 7.2.4 for the case in which a constant frictional torque τ_f is present.

8. A cord is wrapped around the rim of a wheel of radius 0.5 meter and moment of inertia 2.4 kg-m². The wheel is mounted in bearings which exert a constant frictional torque of 8.0 newton-meters when the wheel is revolving. A mass of 24 kg is fastened to the end of the cord and allowed to descend vertically, giving the wheel an angular acceleration. Find **(a)** the angular acceleration of the wheel and **(b)** the tension in the cord.

9. A cord is wrapped around the periphery of a cylindrical drum of radius 0.4 meter, which is free to rotate about its axis in frictionless bearings. A mass of 12 kg is fastened to the cord and released from rest. It is observed to descend a distance of 3.6 meters in 1.80 sec. **(a)** What is the moment of inertia of the drum? **(b)** What is the tension in the cord? **(c)** What is the velocity of the mass at the end of the 1.80-sec interval? **(d)** What is the angular velocity of the drum at the end of the 1.80-sec interval?

10. A cylindrical disc of radius R, mass M, and moment of inertia $MR^2/2$ is supported in frictionless bearings by a shaft of radius r and negligible rotational inertia, as shown in the figure. A mass m is attached to a cord

11. In the accompanying figure, assuming that frictional forces and torques are completely absent and that $R = 0.2$ meter, $m = 12.0$ kg, $m' = 18.0$ kg, and $I = 1.60$ kg-m², find **(a)** the angular acceleration of the heavy roller of radius R, **(b)** the linear acceleration of the masses, **(c)** the tension T, and **(d)** the tension T'.

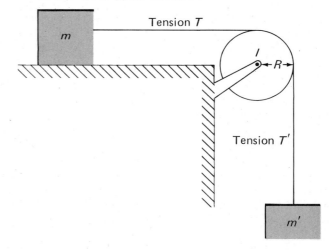

12. Find the answers to the preceding problem using the general symbols m, m', I, and R instead of the specific numerical values given previously.

13. A rigid, uniform rod of mass m and length l has attached to each of its ends a small sphere, likewise of mass m, to form a dumbbell. The spheres are so small that they can be regarded as point masses. What is the moment of inertia of this object about an axis perpendicular to the rod passing through its center?

14. What is the moment of inertia of the system described in

the preceding problem about an axis perpendicular to the rod passing through one of the small point masses attached to its end?

15. Find the moment of inertia of a thin, uniform rectangular plate of length a and width b about an axis perpendicular to the plate and passing through its center.

16. Find the moment of inertia of a thin, uniform plate of mass m having the form of an equilateral triangle of side a, about an axis lying in the plane of the plate, passing through one of the vertices and perpendicular to the side opposite that vertex.

17. Find the moment of inertia of a right circular cone of mass m and base radius R about its axis of symmetry. Use volume elements of the sort illustrated in Fig. 7.10d.

*18. Find the moment of inertia of a uniform sphere of mass m and radius R about an axis passing through its center. Use volume elements like those shown in Fig. 2.10c.

*19. Find the moment of inertia of a thin, spherical shell of mass m and radius R about an axis passing through its center. Assume that the thickness of the shell is negligible in comparison with its radius.

20. (a) What is the moment of inertia of a thin, uniform square plate of mass m having sides of length l about an axis which coincides with one of its sides? (b) What is the moment of inertia of such a plate about an axis in the plane of the plate passing through the center of the square and parallel to one of its sides? (c) What is its moment of inertia about an axis perpendicular to its plane passing through its center? (d) What is its moment of inertia about an axis parallel to the one described in (c) passing through the midpoint of one of the sides? (e) What is the moment of inertia of the plate about an axis parallel to the one described in (c) passing through one of the corners?

*21. Find the moment of inertia of a section of a uniform right circular cone intercepted between two planes perpendicular to the cone's axis, about the body's axis of symmetry.

22. Find the moment of inertia of a "nose cone" having the form of a uniform solid paraboloid of revolution of maximum radius R and height h, about its own axis.

23. A flywheel having a mass of 24.0 kg has the form of a circular disc of radius 0.3 meter which is free to rotate about its own axis. (a) What is the moment of inertia of the flywheel about its axis of rotation? (b) How much work must be done to accelerate the flywheel from rest to a final angular speed of 3000 rpm? (c) What is the flywheel's final rotational kinetic energy?

24. The flywheel in the preceding problem is slowed by a friction brake from 3000 rpm to a final angular speed of 12,000 rpm. How much kinetic energy is dissipated in the form of heat within the friction brake?

25. A speed-governing device consists of two small spheres each of mass 0.8 kg pivoted to a vertical shaft by two light rods, as shown in the figure. When the vertical shaft rotates at a rate of 800 rpm, the masses are found to move in a horizontal circular path of radius 0.18 meter. How much work was done on the governor in bringing it from rest to its final rotational speed? Neglect friction and regard the small spheres as point masses.

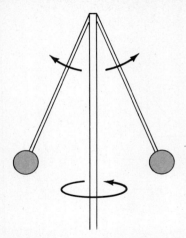

26. A torsion pendulum is formed by attaching a cylindrical disc of radius 0.25 meter whose moment of inertia is 0.72 kg-m^2 to an axial rod embedded in a fixed support. It is found that a tangential force of 36.0 newtons applied to the outer edge of the disc twists it through an angle of 90°. The disc is twisted through an angle of 150° from the equilibrium state, and released from rest, so that it rotates back and forth. Find (a) the torsional spring constant of the torsion rod, assuming that Hooke's law is obeyed; (b) the work needed to twist the disc through the initial 150° angular displacement; (c) the angular speed of the disc as it rotates through the equilibrium position; and (d) the angular speed of the disc when it has turned through an angle of 90° from the equilibrium position.

27. Find the ratio of rotational kinetic energy to total kinetic energy for the following bodies, which are considered to roll without slipping along a flat surface: (a) a thin, circular hoop, (b) a thin, uniform spherical shell ($I = 2MR^2/3$), (c) a uniform right circular cylinder, and (d) a uniform solid sphere.

28. A mass of 18.0 kg, resting on a horizontal frictionless surface, is connected to an ideal spring whose spring constant k is 2400 newtons/meter and to a cord wrapped around a circular flywheel of radius 0.36 meter and moment of inertia $I = 1.20$ kg-m^2, as illustrated in the diagram. The mass is displaced 0.2 meter from the equilibrium position and released from rest. What is its speed when it passes through the equilibrium position?

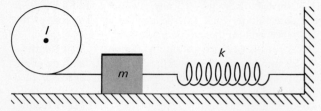

29. In the figure, a mass of 12.0 kg is shown suspended by two cords, one wrapped around a drum of radius 0.10 meter and moment of inertia 0.05 kg-m^2, the other wrapped around a drum of radius 0.15 meter and moment of inertia 0.12 kg-m^2. The mass is released from rest and descends a distance h of 6.0 meters. Neglect friction. Using energy conservation methods, find (a) the final

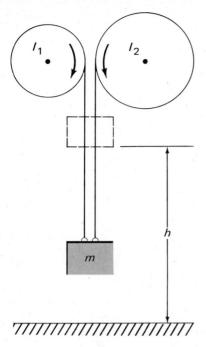

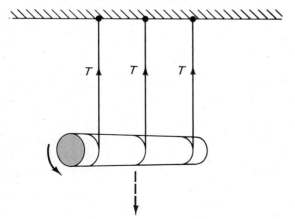

velocity of the descending mass, **(b)** its linear acceleration, **(c)** the angular acceleration of both drums, and **(d)** the tension in both cords.

30. A uniform cylindrical lawn roller is drawn along a level surface at a constant speed of 1.8 meter/sec. The roller's mass is 80 kg, its radius is 0.3 meter, and its moment of inertia is 3.6 kg-m^2. Find **(a)** the translational kinetic energy of the center of mass, **(b)** the rotational kinetic energy about the roller's axle, and **(c)** the total kinetic energy. Assume that the roller rolls without slipping.

31. The lawn roller in the preceding problem is drawn along a level surface by a constant horizontal force of 240 newtons. Assuming that it rolls without slipping and that axle friction is negligible, find **(a)** the linear acceleration of its center of mass and **(b)** the angular acceleration of the roller about its axle. Use a method based on work and energy.

32. Work the preceding problem using a method based directly on Newton's second law.

33. A uniform cylinder of radius 2 cm has a mass of 4 kg. It has three cords wound around it. The ends of the cords are

attached to the ceiling, as shown in the diagram. If the rod is held horizontally and released from rest with the cords in a vertical position, find the translational acceleration of the rod as it falls, and also the tension in each of the cords. Note $I = \frac{1}{2}mr^2$.

34. A meter stick is held upright and then allowed to fall. Assuming the end touching the table does not slip, find the velocity with which the far end strikes the table.

35. A uniform solid cylinder rolls down a plane without slipping. If the plane makes an angle of θ with the horizontal, show that the coefficient of friction between the plane and the cylinder must be at least $\frac{1}{3} \tan \theta$.

36. A thin circular ring of mass M has an inner radius R_1 and outer radius R_2. Show that the moment of inertia about an axis perpendicular to the plane of the ring passing through the ring's outer edge is given by $M(3R_2{}^2 - R_1{}^2)/2$.

37. For the ring of the preceding problem, show that the moment of inertia about any diameter is $M(R_2{}^2 - R_1{}^2)/4$.

38. A solid sphere rolls up an inclined plane having an angle of inclination of 45°. If the translational velocity of the center of mass at the bottom of the plane is 5 m/sec, find how far the sphere travels up the plane and also how long it takes.

39. A thin, square metallic plate of uniform density, shown in the figure, has a mass of 5 kg. It has an area of 0.25 m. Break the plate into segments as shown and obtain the moment of inertia about an axis through P and perpendicular to the plate. *Hint:* Use the parallel axis theorem.

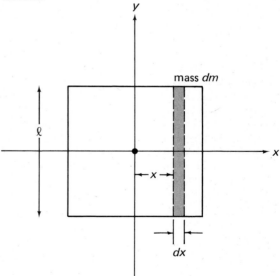

40. For the preceding problem, obtain an *approximate* answer by considering the sum of the contribution from ten strips. Use your calculator if you own one.

* 41. A flat circular disc of radius R and mass m has a nonuniform mass density which varies linearly with distance from the center. Find the radius of gyration of the disc with respect to an axis perpendicular to the disc and passing through its center.

42. A uniform circular disc is connected to a horizontal shaft, as illustrated in the diagram. The moment of inertia of

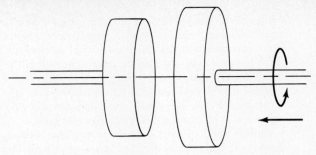

disc and shaft is 6.4 kg-m², and disc and shaft rotate initially at a uniform angular speed of 960 rpm. The shaft is then moved horizontally to the left until the face of the disc contacts that of a similar coaxially mounted disc that is initially at rest. The moment of inertia of this second disc and its shaft is 4.8 kg-m². After a short time, the two discs are coupled together and rotate with a common angular speed. **(a)** What is this final angular velocity? **(b)** Is rotational kinetic energy conserved in this process? **(c)** If not, how much is lost, and what becomes of it?

43. An empty cylindrical bucket of radius 0.60 ft is placed coaxially upon a horizontal rotating platform with frictionless bearings. The moment of inertia of the platform and bucket is 1.20 slug-ft². The platform and bucket are given an initial angular velocity of 5.0 revolutions per second. A load of sand weighing 48 pounds is then dumped into the bucket. What is the final rotational speed of the system?

44. A long-playing record of mass 100 grams and radius 15.0 cm is dropped from rest onto a turntable rotating at an initial speed of $33\frac{1}{3}$ rpm. The moment of inertia of the turntable is 0.0180 kg-m². **(a)** What is the rotational speed of the turntable just after the record is dropped onto it? **(b)** How much torque must the turntable motor supply if the initial speed of $33\frac{1}{3}$ rpm is to be regained in 1.5 sec? Assume that the record is a uniform circular disc.

45. A boy weighing 96 pounds stands 4 ft from the axis of a horizontal rotating platform whose moment of inertia is 80 slug-ft². A second boy, standing on the ground nearby, throws a 0.50-pound baseball, which is caught by the first boy. The baseball is traveling horizontally at a speed of 72 ft/sec as it is caught; its trajectory, viewed from above, is perpendicular to the radial line joining the boy on the turntable with the turntable axis. What is the angular speed of the turntable and the first boy immediately after the ball has been caught?

46. A uniform meter stick is projected into the air at time $t = 0$ in such a way that its center has an initial velocity of 10.0 meters/sec, directed at an angle of 37° above the horizontal, and an angular velocity about a horizontal axis passing through its center, of 12.0 rad/sec. If the meter stick is horizontal at the instant of projection, describe its position and orientation **(a)** at time $t = 0.5$ sec and **(b)** at $t = 1.0$ sec.

47. A child of mass 30 kg is on a merry-go-round which has a moment of inertia $I = 200$ kg-m² when not occupied. It rotates at a constant rate of 0.6 rev/s while the child sits at a distance of 1.5 m from the axis of rotation. The child then slowly walks toward the axis and sits down at a distance of 0.5 m from the axis. What is now the new

constant speed (rev/sec) of the merry-go-round? Treat the child as a point mass and neglect any slowdown due to friction.

48. A string is wrapped around a cylinder of length l and radius R, consisting of a material having uniform density ρ. A mass m hangs from the string. If the mass falls from rest through a height h and strikes the top of a table, obtain an expression for its speed when it hits.

49. A spool of mass m and moment of inertia I about P rests on a horizontal plane and experiences an applied force in the direction shown in the figure as well as a friction force F_f. **(a)** If $\theta = 0$ and the spool rolls without slipping, which way does it roll and what acceleration does it have? **(b)** Repeat part **(a)** if θ is $\pi/2$. **(c)** For what angle θ does the spool slide without rolling?

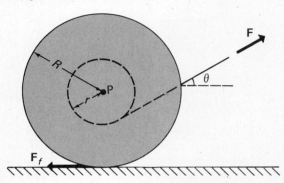

50. In the accompanying diagram, a light, frictionless pulley of negligible mass over which hangs a body C of mass m is shown. The other end of the string is wound around a cylinder B of mass M and radius of gyration R_0. Prove that if R is the radius of the cylinder, the acceleration of its center of mass is given by $a = g/[(2 + (M/2m)(1 + R_0{}^2 R^{-2})]$.

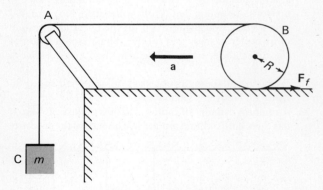

51. A 10-pound solid sphere of radius 6 inches rolls without slipping down an inclined plane making an angle of 37° with the horizontal. The equivalent vertical displacement of the sphere from the top to the bottom is 35 ft. When the sphere reaches the bottom of the incline, **(a)** what is the linear velocity of the center of mass? **(b)** What is the angular velocity about the center of mass? **(c)** How much potential energy has been lost? **(d)** What fraction of the kinetic energy gained is rotational? **(e)** Use Newton's laws to determine the linear acceleration of the center of mass and the force of friction.

52. A cylinder of mass M, radius R, and radius of gyration

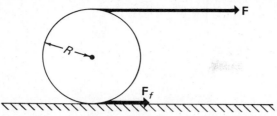

R_0 is pulled on a horizontal surface by a team of horses with a force F, as shown in the diagram. It rolls without slipping. **(a)** Derive an expression for the friction force F_f in terms of F, R_0, and R. Assuming that $R_0 \leq R$, show that F_f must be either zero or in the same direction as F. **(b)** Assume that the body is a solid cylinder ($R_0^2 = R^2/2$); find F_f in terms of F and also obtain the acceleration a in terms of F and M. **(c)** Repeat **(b)** for a hollow thin cylinder ($R_0 = R$).

53. A uniform circular disc with a mass of 300 kg and a radius of 0.6 meter is rotating at $\omega = 3000$ rpm. What constant tangential force acting on the rim will bring the disc to rest in 2 minutes?

54. Two ice skaters each of mass 75 kg skate toward each other at a speed of 4.5 meters/sec in straight lines 1.3 meter apart. Find their total angular momentum with respect to an axis passing through their center of mass. They join hands when they reach each other and together go into a circular path of diameter 1 meter. Find their common angular velocity. Assume their moments of inertia are obtained using $I = mr^2$.

55. For the preceding problem, by how much does the kinetic energy increase in the process of coupling the two skaters?

56. A beautiful woman and her handsome husband, respectively 105 pounds and 170 pounds in weight, find themselves at opposite ends of the diameter of a rotating turntable of weight 500 pounds and radius 4 ft. The turntable makes one revolution every 3 seconds. Realizing the foolishness of the situation, they immediately walk directly to the center and hug each other. Find the final angular velocity of the turntable and the change in kinetic energy. Account for the difference in kinetic energy.

57. A billiards expert explains to his novice nephew that if you strike the ball at a certain height h above center, the ball will begin to roll immediately without slipping. Find h if the billiard ball has a radius R.

58. A monkey stands on a frictionless rotating platform going around at $\frac{1}{2}$ rev/sec with his arms outstretched. He has a large bunch of bananas in each hand, and in this position the total moment of inertia is 4 kg-m². By drawing in his arms, the monkey reduces this to 3 kg-m². **(a)** Find the new angular velocity of the platform. **(b)** What is the kinetic energy of the system now?

59. An airplane moves in a horizontal circle of radius 300 yards with a speed of 200 mph. The propellor has an angular velocity of 4000 rpm, and the airplane has a moment of inertia of 25 slug-ft² about an axis perpendicular to the plane of rotation of the propellor and passing through its center. Find the torque exerted on the airplane (both magnitude and direction).

60. A mass M is in circular motion of radius r, and velocity v, as shown in the diagram. It is connected to a string which passes through a vertical tube. The radius can be varied by pulling on the string. If the string is shortened to a radius $r/2$, find the new speed and the change in kinetic energy. Calculate the work done to pull the mass in and show that it is equal to the change in kinetic energy.

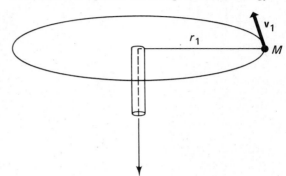

61. A gyroscope has a mass of 2.5 kg and a rotor of moment of inertia 0.005 kg-m². Its rotor has a constant angular speed of 12,000 rpm. It is set on a pylon with its axis of rotation horizontal. The center of mass of the system under these circumstances is 6.0 cm from the pylon. Find **(a)** the magnitude and direction of the initial angular momentum of the rotor, **(b)** the magnitude and direction of the torque about the pylon arising from the weight of the gyroscope, and **(c)** the magnitude and direction of the angular velocity of precession.

62. In the preceding problem, describe how the angular velocity of precession would change if **(a)** the angular speed of the rotor were doubled, **(b)** if the moment of inertia of the rotor were doubled, **(c)** if the weight of the top were doubled, and **(d)** if the gyro were set on its pylon with its rotation axis making an angle of 30° rather than 90° with the vertical.

63. A man is given a locked suitcase to deliver from one office to another. The delivery involves a trip down a corridor and a 90° turn into a perpendicular hallway. All goes well during the first part of the trip; but as soon as the man rounds the corner at the junction of the two corridors, the suitcase swivels through an angle of 90° about its handle hinge until, finally, it is sticking out at an angle of 90° to the body of the man carrying it. How do you explain this strange behavior?

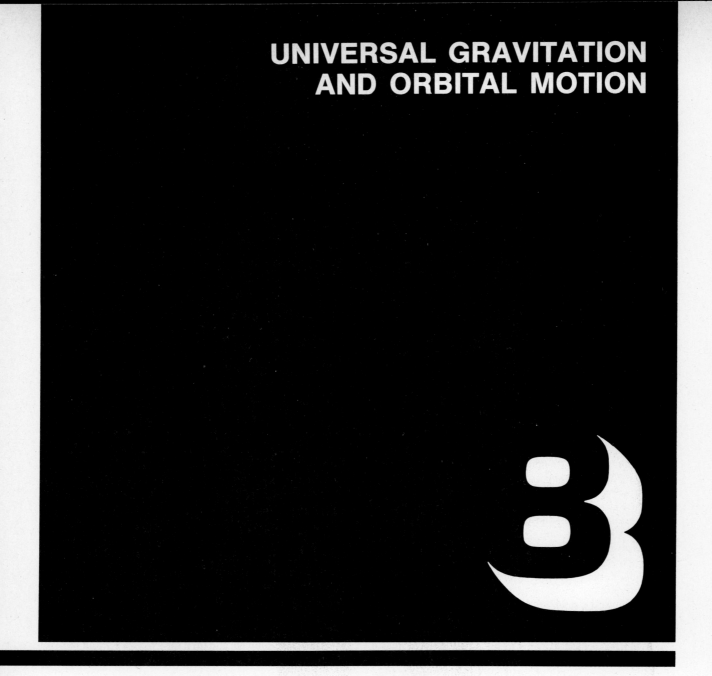

UNIVERSAL GRAVITATION AND ORBITAL MOTION

8.1 Introduction

Our development of the principles of mechanics is now practically complete. It is nevertheless important to have some knowledge of the *practice* of mechanics as applied to several particularly important areas such as gravitation, harmonic motion, wave motion, and the statics and dynamics of fluids. We shall begin by trying to understand gravitational forces and the motion of objects under the influence of these forces.

From a historical point of view, it would have been appropriate to have begun this book with a consideration of gravity and orbital motion. The motions of the sun, the moon, the stars, and the planets have excited—and puzzled—mankind's imag-

ination since the beginning of recorded history. Today, unfortunately, most of us are watching TV at night rather than contemplating the icy beauty of the heavens. Indeed, for those of us in urban locales, the regular and yet irregular panorama of the skies is discernible at best with difficulty, and the motions of stars and planets are hardly observed much less understood in detail. In earlier times, however, the explanation of the apparent motions of sun, moon, stars, and planets was one of the most important problems confronting scientists and philosophers. That the answers eluded mankind for five millennia of recorded history is one of the less flattering aspects of man's sojourn on this planet. The explanations are very simple as soon as man stops insisting that he

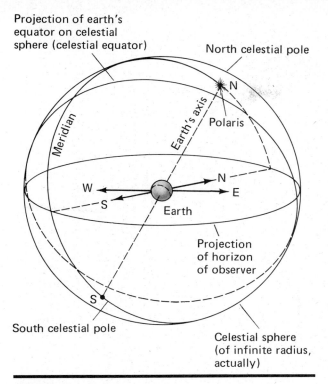

Projection of earth's equator on celestial sphere (celestial equator)

North celestial pole

Meridian

Earth's axis

Polaris

N

W

S

N

E

Earth

Projection of horizon of observer

South celestial pole

Celestial sphere (of infinite radius, actually)

FIGURE 8.1. Concept of the "celestial sphere."

is the center and focus of the universe. This idea, unfortunately, was not widely accepted until the time of Copernicus (1473–1543).

Once the idea of a geocentric universe was discarded, however, rapid progress was made in understanding the dynamics of celestial bodies, culminating, in 1689, in Newton's formulation of the *law of universal gravitation*, describing the precise nature of the gravitational forces exerted by any body on any other. This law, in conjunction with the three laws of motion previously stated, enables astronomers to predict with accuracy the orbital motion of all the planets and their satellites and forms the basis for our understanding of the dynamics of the cosmos to this day. Although we shall not try to present a historically complete account of this fascinating story,[1] we shall try to explain briefly the main features of the apparent motions of important celestial objects and try to understand in a general way how these motions were regarded before Copernicus and Newton came along. Initially, we shall try to concentrate on presenting a reasonably complete description of the observed motions, without concerning ourselves with detailed explanations. We shall then take a look at the views

[1] For a more complete accounting of the history of astronomy, the reader may find it interesting to consult Leon N. Cooper, *An Introduction to the Meaning and Structure of Physics* (New York: Harper & Row, 1968), chapter 5, or Gerald Holton and Duane H. D. Roller, *Foundations of Modern Physical Science* (Reading, Mass.: Addison-Wesley, 1958), chapters 6–12.

of ancient Greek astronomers, and finally examine the more modern theories of Copernicus, Kepler, and Newton.

8.2 The Apparent Motions of Celestial Bodies: The Geocentric Universe

If we were to make a practice of watching the motion of celestial objects day after day and night after night, certain patterns would arise over a period of time that would allow us to identify four different sorts of motion, characterizing, respectively, the apparent behavior of sun, moon, stars, and planets.

The sun is observed to rise every day in the east and set every day in the west, magically reappearing the next morning, not in the direction whence it disappeared at sunset but on the opposite part of the horizon. Its path in the sky during the course of the day is essentially circular; the location of the center of the circle depends upon the latitude of the observer and, in effect, marks the projection of the earth's polar axis on the celestial sphere (Fig. 8.1).[2] The sun's path does not remain the same day after day during the year but moves north of the celestial equator during the summer and south during the winter. In fact, the length of the solar day itself, as defined by the time between two successive passages of the sun across the meridian,[3] is not quite constant, but may vary by a few seconds during the course of a year. The solar day is shorter during the summer in the northern hemisphere and longer during the winter—a small variation, but one we shall be able eventually to understand.

The moon also rises in the east and sets in the west, but its apparent velocity is not quite the same as that of the sun, since it executes one revolution with respect to the sun's position on the celestial sphere roughly every 27 days. This change in the moon's position in the sky with respect to that of the sun is responsible for the moon's *phases*; therefore the moon goes through the full cycle of phases—full moon to

[2] The celestial sphere may be regarded as a sphere of infinite radius with ourselves at the center. When we look outward toward the sky, therefore, we see celestial objects as if they were attached to the inside of this celestial sphere, even though they actually may be at greatly varying distances from us. The concept of the celestial sphere is illustrated by Fig. 8.1.

[3] The *meridian* is an imaginary line on the celestial sphere in the form of a great circle intersecting the projections of the earth's north and south polar axis on the celestial sphere. It therefore forms a north–south axis in the sky. The meridian is illustrated in Fig. 8.1. In the same way, the projection of the earth's equator outward onto the celestial sphere forms a great circle on that sphere which intersects the east–west points of the observer's horizon (but *not* the zenith). This great circle is called the *celestial equator*, or *ecliptic*. The celestial equator is also shown in Fig. 8.1.

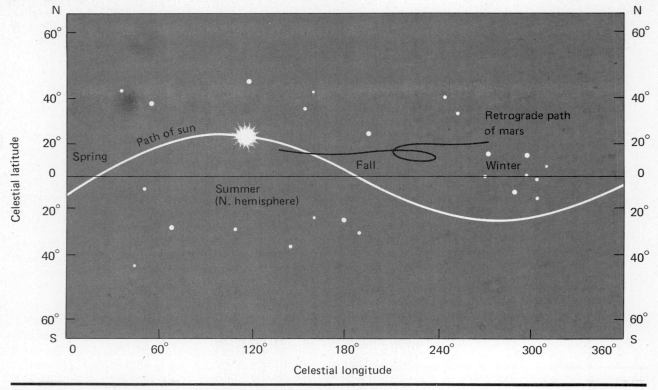

FIGURE 8.2. Map of the constellations, showing the path of the sun, and (schematically) illustrating the irregular retrograde motion of a planet like Mars.

full moon—also in about 27 days. (It is interesting to note in this regard that moon and month are the same word!) The moon also appears to wander north and south of the celestial equator, but its motion is less regular than that of the sun.

The motion of the stars is the simplest of all. The stars appear to execute uniform circular motions centered on the celestial pole, rotating in a counter-clockwise sense to an observer facing north, like the sun and moon. But the stars *do not alter their apparent positions with respect to one another*. It is as if they were *fixed* on the celestial sphere and the celestial sphere itself were rotating. Neither do they wander north and south of the celestial equator as the sun and moon do.[4] Their period of rotation, between successive transits across the meridian, is nearly, but not quite, the same as that of the sun. The *siderial* day, defined as the time between successive transits of the same star across the meridian, is, on the average, almost exactly 4 minutes shorter than the solar day and is, furthermore, very constant in duration. It does not exhibit the small variations associated with the length of the solar day during the course of the year.

[4] These statements are oversimplified to a degree. The stars do exhibit very slight motions on the celestial sphere, detectable only by very refined observations over long periods of time. These motions, though well understood, are negligible on the scale of our discussion, and we shall not need to consider them further.

Any celestial object may be assigned latitude and longitude coordinates on the celestial sphere. Those of the fixed stars are constant, though the sphere itself appears to rotate through 360° during the course of the siderial day, which is 23 hours, 56 minutes. The motion of the sun, referred to the stars, which we now regard as being fixed on the rotating celestial sphere, is periodic, retracing the same path every year. Since the sun appears to rotate through 360° in 24 hours 00 minutes while the celestial sphere's apparent rotation through the same angle takes 23 hours 56 minutes, the sun "falls behind" the stars by about 4 minutes, that is, 4/(24)(60), or 1/360, revolution each solar day. More accurately stated, this figure really should be 1/365.26 revolution per day, since the "4 minutes" is actually about 3 minutes 56 seconds. This amounts, of course, to *one revolution every year*. In other words, 365.26 solar days (or one solar year) amount to 365.26 + 1.00 siderial days. This is no accident—indeed, it's a dead giveaway of the fact that the earth revolves about the sun rather than the other way around, as we shall see more clearly later on. The motion of the sun with respect to the stars fixed on the celestial sphere is illustrated in Fig. 8.2.

While most of the starlike objects in the sky behave in the manner just described, there are a few that do not. These bodies, called the planets, exhibit roughly the same daily rotation as the sun and stars, but superimposed on that are most erratic wanderings back and forth among the stars which are fixed with respect to the celestial sphere. These strange motions are different for each planet and do not appear to be cyclical or periodic. In certain instances, notably Mars, the motion actually reverses direction every now and then, as illustrated in Fig. 8.2. This phenomenon is referred to as *retrograde* motion. Two of the planets, Mercury and Venus, never seem to get very far from the sun. In fact, most of us have never seen Mercury simply because it is always so close to the sun that it's not easily observed with the naked eye unless we know just when and where to look. The other planets may appear almost anywhere with respect to the sun, though they never stray very far from the celestial equator. There are five planets, known since prehistoric times, that are visible to the unaided eye. These are Mercury, Venus, Mars, Jupiter, and Saturn. Three others—Uranus, Neptune, and Pluto—have been discovered since the invention of the astronomical telescope by Galileo in 1610, the latter as recently as 1930. They can be seen only with the aid of powerful telescopes. The apparent motions of the planets against the celestial sphere are so complex and irregular that they had defied rational explanation until Copernicus arrived on the scene in the sixteenth century.

Before Copernicus, most of our ideas about the structure of the universe were founded on earth-centered models. The earliest of these were based in pure mythology and superstition. But around 400 B.C., at the height of the Greek civilization, much serious philosophical effort had been expended in trying to arrive at a rational explanation of the motion of celestial bodies. The Greek philosophers and mathematicians were in some ways well prepared to attack this problem. They were excellent logicians, good observers, and skilled geometers. They understood that the earth is spherical and had actually measured its radius quite accurately. Unfortunately, their efforts were of limited success since they were always constructed around the fundamental idea that the earth was fixed and that all heavenly bodies revolved about it in some fashion. These theories accounted quite satisfactorily for the observed motion of the stars, since, after all, they do appear to rotate about the earth just as if they were attached to the inside of a rotating sphere with us at the center. The sun's motion, which is not much more complex, could also be accommodated (with a bit of fudging to account for its annual trips north and south of the equator) by the assumption that it traveled around the earth independently of the sphere of stars and at a slightly different rate of rotation. But the motions of the planets could never be very satisfactorily integrated into a scheme such as this.

Actually, a few attempts were made during this period to construct heliocentric models of the universe, but they foundered on the Greeks' faulty understanding of dynamics. If the earth were rotating about the sun, or even revolving about its own axis, should we not somehow be able to sense this rapid motion, just as we sense that are are in motion in a rapidly moving carriage? Also, would not birds in flight or objects thrown into the air be rapidly left behind if the earth possessed such a rapid motion of its own? Furthermore, would not such motion give rise to violent winds and tides? Finally, whence comes the force necessary to *maintain* this motion? The student may profitably test his understanding of Newtonian dynamics by trying to answer all these objections. Anyway, the Greeks bought them, and it wasn't until nearly two thousand years later that we understood the laws governing the motion of material objects well enough to refute their objections.

In the second century A.D., Ptolemy of Alexandria made a sophisticated attempt to incorporate planetary motions into a geocentric universe. He assumed that the planets rotated about some point in space, which was itself rotating about the earth. This combination of two separate rotational motions does lead to an explanation of the observed retrograde orbits of the planets, and did indeed suffice to predict their future (as well as past) positions reasonably well, provided that proper values for the radius and angular velocity of both circular motions were assumed. The resulting *epicyclic* motion is illustrated in Fig. 8.3. But why this particular combination of two uniform circular motions? Well, everyone knows that circular motion is the most perfect form of motion, and therefore the most suitable for describing the paths of these most perfect bodies!

Although the Ptolemaic system of epicycles did account, after a fashion, for the observed motion of

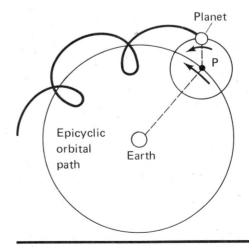

FIGURE 8.3. Explanation of planetary motion by Ptolemy's epicycles.

the known planets, it was not very satisfying for many reasons. For one thing, by choosing proper values for the radii and angular speeds which enter into the theory, almost any motion can be arrived at. At the same time, there is no rational way of arriving at the choices made to reproduce the observed orbits other than trial and error, working backward from actual observations. The theory affords no interrelations whatsoever between the motions of the different planets. Finally, if a new planet were to be discovered at a certain point in the heavens, going initially in a given direction with a given speed, there would be no basis afforded by the Ptolemaic theory for predicting where it would be a month or a year hence. Its course could be predicted only by observing its position night after night, week after week, year after year, for many years, and then laboriously guessing what kind of epicycle could be constructed to fit its observed behavior. Anyway, by the sixteenth century, techniques of astronomical observation were becoming much better than they had been previously, and more accurate determinations of the actual motions of the planets were being made. These observations uncovered significant disagreements in the observed positions of the planets and those expected on the basis of the most carefully constructed epicyclic orbits.

This was the state of affairs about 1500 A.D. In describing it, we have concentrated on explaining in detail the *apparent* motions of celestial bodies and the then existing ideas about those motions. The currently accepted explanations of many of the phenomena that were discussed will be familiar to many readers. We have nonetheless deliberately tried to avoid introducing modern explanations, in order to acquaint the student with the contemporary problems confronting astronomers and the existing theories which had previously formed the basis for their thought. We shall now turn our attention to the criticism of the geocentric picture of the universe, and try to understand the main features of the heliocentric theory which over a period of time came to be accepted in its place.

8.3 The Heliocentric Picture of the Solar System: Nicolaus Copernicus, Tycho Brahe, Galileo, and Johannes Kepler

Nicolaus Copernicus (1473–1543), a Polish astronomer and philosopher, is responsible for the revival of the sun-centered concept of the solar system, which had dwelt in obscurity and disrepute for nearly two thousand years. Based on the concept of a solar system with the sun at the center, surrounded by the planets then known in the order; Mercury, Venus,

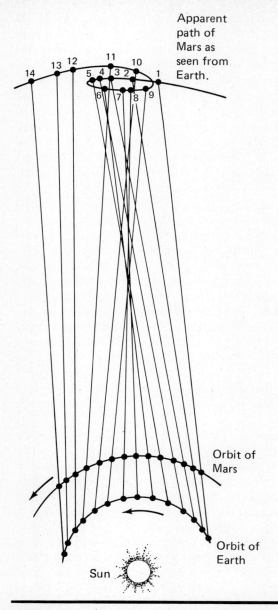

FIGURE 8.4. The apparent retrograde motion of the planets on the basis of the Copernican heliocentric model.

Earth, Mars, Jupiter, and Saturn, *all traversing circular orbits centered on the sun* with uniform but different angular velocities, he was able to reproduce the details of the observed motions of the planets, including the odd retrograde motion of some. The explanation of the retrograde motion of Mars is shown schematically in Fig. 8.4, where it is seen to arise naturally from the fact that the orbit of Mars is outside that of the earth and the angular velocity of the earth in its circular orbit is greater than that of Mars in its orbit.

The "sphere of the stars" was also envisioned as being centered on the sun, but of vastly greater radius than any of the planetary orbits. This assumption was necessary because, if the stars were not at a distance much greater than the size of the earth's orbit about the sun, they would be seen to change position on the celestial sphere as the earth traverses its orbit during the course of a year, just as the apparent position of the top of a relatively nearby flagpole might change relative to a background of very distant clouds as one moves his head back and forth. This effect of *stellar parallax* had never been observed[5] and had provided, and continued to provide, much ammunition for those whose views were opposed to the heliocentric picture of the universe.

The daily circular motions of the sun, stars, and planets were ascribed by Copernicus to the *rotation of the earth about its own polar axis* once every siderial day (23 hours 56 minutes). This automatically explained why the sun "falls behind" the motion of the stars to the extent of precisely one full revolution per year, the change in position of the sun against the background of stars being, naturally enough, caused by the earth's progress along its orbit around the sun from day to day, making up exactly 360° during the course of one year.

Though the sun-centered theory of Copernicus accounted satisfactorily for the motions of the sun, stars, and planets within the accuracy of the then available observations, the ideas of Copernicus did not gain rapid acceptance, even among astronomers. The reasons for this are to be found partially in the imperfect understanding of mechanics extant at that time (some of the dynamical objections based on contemporary ideas about mechanics have already been mentioned) and partially in the opposition of conservative religious and political authority. Copernicus was regarded as a heretic not only by the Pope but also by Martin Luther.

Within a half-century after the introduction of the sun-centered model of Copernicus, a very great advance in the experimental observation of stellar and planetary motions was accomplished by the Danish astronomer Tycho Brahe (1546–1601). Brahe's careful and painstaking observations, extending over all his mature life, improved the accuracy of the determinations of the positions of stars and planets by a factor of about 20 over those previously available which had formed the experimental basis of both the Ptolemaic and Copernican theories. Brahe's observations showed that there were *small but perceptible errors* in the positions of the planets predicted not only on

the basis of the Ptolemaic epicycles then in use, but *also in the positions arrived at from the Copernicus theory of circular orbits.*

During this same period, Galileo Galilei (1564–1632) constructed the first successful astronomical telescope, which he immediately put to use observing the moon and the planets. It became evident from his investigations that the planets all had the spherical form of the earth and that some of them were accompanied by satellites which could be observed to rotate about them, in much the same way Copernicus thought that the earth and the other planets rotated about the sun. Galileo was a firm adherent to the heliocentric model of the solar system. His beliefs got him into serious trouble with religious and political authorities, and he was ordered to stop teaching the theories of Copernicus. Nevertheless, his astronomical observations and his clever and sagacious arguments in favor of Copernicus' idea were very important contributions to its final acceptance.

Tycho Brahe did not believe in the Copernican theory because he, too, embraced some of the misconceptions about mechanics which were then so common. But his observations became the basis for our present understanding of planetary orbits and, indirectly, for Newton's development of the law of gravitation. Very little success resulted from initial efforts to reconcile Brahe's accurate data with epicyclic pictures of planetary motion. In 1596, however, Johannes Kepler (1571–1630) published an exhaustive analysis of Brahe's measurements, based on a slight modification of the Copernican theory, in which the orbits of the planets were envisioned as being slightly *elliptic* rather than circular. The basic conclusions of Kepler's analysis survive to this day.

The results of Kepler's investigations can be summarized in three laws which, taken together, reproduced Brahe's observations very accurately and satisfactorily and which could be used to make extremely good predictions of planetary positions far into the future. They may be stated as follows:

1. Each planet's motion is periodic, in a closed elliptic orbit with the sun at one of the foci.
2. The line joining the sun with any planet sweeps out equal areas in equal time intervals in all parts of the orbit.
3. The square of the time required for any planet to traverse its orbit about the sun is proportional to the cube of the mean radius of the orbit.

Kepler's first law reaffirms the Copernican picture of the solar system and at the same time makes a slight but significant modification. Actually, the elliptical orbits of the planets in the solar system are nearly but not quite circular. Nevertheless, the ellipticities of their orbits must be taken into consideration

---

[5] The effect of stellar parallax can be observed for the stars closest to our solar system with the aid of the very large telescopes available to us now. The effect is very small, though measurable accurately enough for stars not too distant, and is one of the minor stellar motions referred to in Footnote 4. It provides us with a very important way of making actual measurements of stellar distances, since the magnitude of the observed parallax depends on the ratio of the star's distance to the diameter of the earth's orbit.

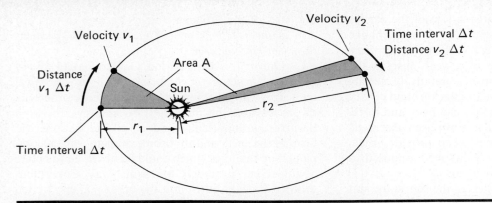

FIGURE 8.5. Geometric representation of Kepler's second law.

to account for the planetary positions observed by Tycho Brahe within the experimental accuracy of his data. Also, Kepler's evidence that the orbits of the planets were elliptical rather than circular wrought a subtle but nonetheless telling effect on contemporary ideas about dynamics. For a long time, the motions of the planets had been thought of exclusively in terms of circular motions, because the circle was the "most perfect" of curves and circular motion the "most natural" or "most perfect" form of motion. But if their paths were less than perfect, then why should they be elliptic rather than some other "less than perfect" shape, and, for that matter, why should the planets choose to travel in less perfect rather than more perfect paths at all? Such questions had the important ultimate effect of diverting the train of thought of astronomers from philosophical arguments to dynamical ones and, therefore, from purely speculative reasoning to rational or scientific thought.

Kepler's second law would predict that if the orbit of a planet were circular, its velocity would remain exactly the same at all times. Since the orbits are in fact elliptical, the result is that according to this law the planet must move faster when it is near the sun than it does when it is farther away. This may be understood more easily by reference to Fig. 8.5, in which a planet is shown at two different phases of its elliptic orbit whose eccentricity is very much exaggerated in the illustration. In each case, the position is shown after the lapse of time Δt, and according to Kepler's second law the two shaded areas should be the same. Since the two sectors can be regarded as triangles in the limit of small values of Δt, the two areas can be written, in this limit, as

$$A = \tfrac{1}{2}r_1 v_1 \,\Delta t = \tfrac{1}{2}r_2 v_2 \,\Delta t \qquad (8.3.1)$$

which means that

$$r_1 v_1 = r_2 v_2 \qquad (8.3.2)$$

In the illustration, since r_2 is larger than r_1, it is evident that v_2 must be smaller than v_1.

Since the orbit of the earth is slightly elliptical, the earth moves more rapidly along its orbit when it is nearest the sun (in winter in the northern hemisphere) than when it is farthest away (summer). The difference between the siderial and solar day is, therefore, slightly greater than the mean value of 3 minutes 55.9 seconds in winter than in summer. This variation in the length of the solar day, which was referred to previously when the apparent motion of the sun was discussed, can actually be observed, the observed values being in agreement with those calculated from Kepler's second law.

The third law relates the mean radii of the orbits of the various planets with their periods of revolution. The reader is invited to use the values of the mean radii and periods given in Table 8.1 to ascertain that Kepler's third law, which predicts that

$$\frac{T^2}{r^3} = \text{constant} \qquad (8.3.3)$$

where r is the mean radius and T the period of revolution, is indeed obeyed very accurately.

Kepler's laws serve to describe and predict the orbits of the planets very accurately, although in themselves they give us no explicit clue as to the underlying mechanics to which they must somehow be related. They nevertheless provided a very exciting challenge to the natural philosophers of the seventeenth century to propound a scheme of mechanics in which they would be encompassed as an integral part of the overall design. Sir Isaac Newton was finally able to do this first by stating the three laws of motion with which we are now familiar and, in addition, formulating the *law of universal gravitation*, which describes in precise mathematical terms the *forces* exerted by the sun on the planets, and by the planets on the sun and on each other. With these elements at hand, he was able to derive the results of Copernicus and Kepler from the fundamental principles of mechanics, and to show that the motions of heavenly bodies were understandable in terms of

TABLE 8.1 The Solar System

Body	Mean diameter, km	Mean density, gm/cm³	Surface gravity, earth g's	Period of axial rotation	Orbital period, solar days	Mean distance from sun, km	Mass, kg	Eccentricity of orbit (c/a)
Sun	1,391,000	1.41	28.	25d.			1.99×10^{30}	
Mercury	5,000	3.8	0.27	88.0d.	88.0	58×10^6	0.32×10^{24}	0.21
Venus	12,400	4.86	0.86	224.7d	224.7	108×10^6	4.87×10^{24}	0.01
Earth	12,740	5.52	1.00	23h.56m.	365.26	149×10^6	5.98×10^{24}	0.02
Mars	6,870	3.96	0.37	24h.37m.	687.0	228×10^6	0.64×10^{24}	0.09
Jupiter	139,800	1.33	2.64	9h.50m.	4332.6	778×10^6	1900×10^{24}	0.05
Saturn	115,100	0.71	1.17	10h.14m.	10759.2	1426×10^6	569×10^{24}	0.06
Uranus	51,000	1.26	0.92	10h.49m.	30687.	2869×10^6	87×10^{24}	0.05
Neptune	50,000	1.6	1.44	15h.40m.	60184.	4495×10^6	103×10^{24}	0.01
Pluto	14,000	4.	0.74	16h.	90700.	5900×10^6	5×10^{24}	0.25
Moon	3,476	3.36	0.17	27d.7h.43m.	27d.7h.43m.	0.384×10^{6}*	$.0734 \times 10^{24}$	0.07

* Mean distance from earth's center.
** The eccentricity of an ellipse is defined as the distance between the foci divided by the major axis. For the case of a circle, the two foci coincide at the center, and the eccentricity is zero.

the same laws that govern the motion of familiar everyday objects in terrestrial surroundings. In order to accomplish all this, he found it necessary to invent the calculus as well, thus, incidentally, giving birth to the powerful mathematical methods which have been so useful to us in developing not only mechanics but all other branches of science.

This was a stupendous intellectual achievement—perhaps the single greatest intellectual accomplishment of mankind to date. In a very real sense it marked the beginning of science as we understand it today. This is not to say that those who preceded Newton, such as Copernicus, Brahe, and Kepler, did not make fine and lasting contributions to science. It suggests instead that a certain watershed, marking the division between earlier thinking based upon religion, speculation, and philosophical argumentation and more modern scientific methods based on objective, analytical and mathematical understanding of natural phenomena in terms of rational fundamental principles had been crossed, for good.

8.4 Newton's Law of Universal Gravitation

Sir Isaac Newton was born in 1642 and died in 1727. His investigations of planetary motion embodying the law of universal gravitation are to be found in his celebrated *Principia Mathematica*, published in 1687. The story of how the fall of an apple from the branches of a tree provided the inspiration for the law of gravitation is so widely quoted that it has achieved the status of a cliche.

Newton's law of universal gravitation may be stated as follows:

A force of attraction exists between any two mass particles, directed along the line joining them, whose magnitude is directly proportional to each of the two masses and inversely proportional to the square of the distance separating them.

The magnitude of this gravitational force can be written in the form of an equation as

$$F = G \frac{m_1 m_2}{r^2} \tag{8.4.1}$$

where m_1 and m_2 are the masses, r is the distance between them, and G is a proportionality constant having the dimensions dyne-cm²/g² (cgs) or newton-m²/kg² (mks). This proportionality constant is sometimes referred to as the universal gravitation constant and has the value

$$G = 6.673 \times 10^{-11} \text{ N m}^2/\text{kg}^2$$
$$= 6.673 \times 10^{-8} \text{ dyne cm}^2/\text{g}^2$$

The methods by which this value was determined experimentally will be described later.

The force of gravitational attraction is illustrated in Fig. 8.6 by the forces **F** and **F'**. The force **F** represents the gravitational attraction experienced by the point mass m_1 at the origin from the presence of the nearby mass m_2 at the point P(x, y, z). This force can be written, as a vector, by multiplying the magnitude given by Eq. (8.4.1) times a unit vector $\mathbf{i}_r$ directed radially outward from the origin. We have seen previously, also, that this unit vector can be represented by the radial vector **r** (equal to $\mathbf{i}_x x + \mathbf{i}_y y + \mathbf{i}_z z$) divided by its own magnitude. Equation (8.4.1) may, therefore, be expressed as

$$\mathbf{F} = G \frac{m_1 m_2}{r^2} \mathbf{i}_r = F \frac{m_1 m_2 \mathbf{r}}{r^3} \tag{8.4.2}$$

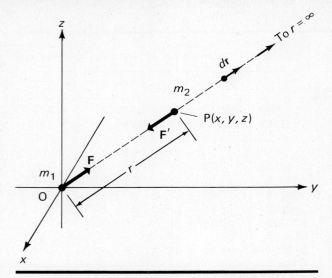

FIGURE 8.6. Gravitational forces between two point masses according to Newton's law of universal gravitation.

where **r** is a vector directed along the line joining m_1 and m_2.

The force **F'** felt by mass m_2 due to the presence of m_1, according to the law of universal gravitation, must have the same magnitude as **F** and must be directed along the line joining the two masses, but in the direction opposite to that of **F**. It is evident, then, that

$$\mathbf{F'} = -\mathbf{F} \qquad (8.4.3)$$

and, indeed, that **F** and **F'** are an *action–reaction pair*. This is in accord with our previously stated idea that, for example, the reaction force to a body's weight is an equal and oppositely directed force acting on the earth's center of mass.

It seems quite natural that any gravitational force existing between two masses might be proportional to the mass of the body experiencing the force and as well to the mass of the body acting as its source. Also, it does not strain our intuitive faculties much to imagine that such a force might be directed along the line joining the particles in question. Hypotheses such as these would occur to most persons trying to understand how bodies might influence one another. The inverse square dependence of force upon distance is less straightforward, but can be arrived at quite simply if Kepler's laws are kept in mind. Suppose, for example, we consider the case of perfectly circular planetary orbits; the actual orbits of the planets are elliptical, but only slightly so, and for the purposes of arriving at the broad and general features of their motion we can at least begin by picturing them as circular. If we believe Newton's laws of motion, a body traveling in a circular path with uniform angular velocity must have an acceleration directed toward the center of

magnitude $r\omega^2$. If the gravitational force is the only one which acts, then it alone must account for the centripetal acceleration, and we must have

$$F_g = ma_{\text{cent}} = mr\omega^2 \qquad (8.4.4)$$

where F_g is the gravitational attraction between sun and planet and m is the mass of the planet. It is well to note that the idea of the gravitational force being directed along the line joining the two bodies is already forced upon us by the character of the centripetal force itself.

Consider now how a gravitational force might conceivably behave. It is not unreasonable to suppose that it is proportional to each mass involved, and we have already had to accept that it is directed along the line joining them. We might also suppose that it is large when the masses are close together and becomes smaller and smaller as they are moved farther and farther apart. This requires that the force depend somehow on the distance r, and the simplest possible mathematical dependence of force on distance for which the force is large at small distances and small at large distances is one in which the force varies as r^{-n}, where n is a positive number. In this way we might be led to the tentative hypothesis that

$$F_g = G \frac{mM}{r^n} \qquad (8.4.5)$$

where M is the sun's mass, which, in conjunction with (8.4.4), leads us to

$$mr\omega^2 = G \frac{mM}{r^n}$$

or

$$\omega^2 = \frac{GM}{r^{n+1}} \qquad (8.4.6)$$

The angle θ through which the planet rotates in time t will be given by $\theta = \theta_0 + \omega t$, since we are dealing with uniform circular motion (zero angular acceleration). Therefore, according to (8.4.6), letting $\theta_0 = 0$, we may write

$$\theta = \omega t = \sqrt{GM} \frac{t}{r^{(n+1)/2}} \qquad (8.4.7)$$

When the angle θ is equal to 2π radians, the elapsed time is the period T, associated with one complete orbital revolution; hence, setting $\theta = 2\pi$ and $t = T$, we obtain

$$2\pi = \frac{T\sqrt{GM}}{r^{(n+1)/2}}$$

or

$$T = \frac{2\pi}{\sqrt{GM}} r^{(n+1)/2} \qquad (8.4.8)$$

Now, there is nothing available from theory alone to enable us to decide that one value of n might be better than another. But according to Kepler's third law, expressing the result of the *observed* behavior of the planets, the square of the periodic time T is found to be proportional to the cube of the radius of the orbit; in other words, T^2 should be proportional to r^3 if the actually observed motion of the planets is ever to be explained. Squaring both sides of (8.4.8), we find that

$$T^2 = \frac{4\pi^2}{GM} r^{n+1} \qquad (8.4.9)$$

If we choose $n = 2$ in this equation, the expression reduces to

$$T^2 = \frac{4\pi^2}{GM} r^3 \qquad (8.4.10)$$

and Kepler's third law is obtained at once. But no other choice of the exponent n in the assumed force law (8.4.5) will lead to agreement with Kepler's laws. Newton was, therefore, led to postulate the *inverse square* dependence of gravitational force with distance because that dependence, and no other, afforded agreement the Kepler's laws of planetary motion, which were derived directly from the experimental observations of Tycho Brahe.

8.5 Gravitational Potential Energy, the Gravitational Field, and the Superposition Principle

Let us now try to calculate the work done by the gravitational force $\mathbf{F}'$ acting on the mass m_2 in Fig. 8.6 as it moves from the position P, at a distance r from mass m_1, to a position infinitely distant from m_1 along the extension of the line OP. This is the prescription, developed in sections 5.5 and 5.6, for finding the *potential energy* of m_2. Actually, of course, work must be done *against* the gravitational force on m_2 in order to move m_2 to a point more distant from m_1, so we shall expect the calculated potential energy to be negative.

The work done by the force $\mathbf{F}'$ in moving through the displacement $d\mathbf{r}$ shown in Fig. 8.6 will be

$$dW = dU_p = \mathbf{F}' \cdot d\mathbf{r} = F' \cos \theta \, dr \qquad (8.5.1)$$

where θ is the angle between $\mathbf{F}'$ and dr, in this case $180°$. Accordingly, $\cos \theta = -1$ and

$$dW = dU_p = -F' \, dr \qquad (8.5.2)$$

But F' may be expressed by (8.4.1), giving

$$dW = dU_p = -Gm_1m_2 \frac{dr}{r^2} \qquad (8.5.3)$$

We must now integrate this from the point P, at a distance r from m_1, to infinity. We then obtain

$$W = U_p = -Gm_1m_2 \int_r^\infty \frac{dr}{r^2} = Gm_1m_2 \left[\frac{1}{r}\right]_r^\infty$$

$$= Gm_1m_2 \left(0 - \frac{1}{r}\right) \qquad (8.5.4)$$

or finally

$$U_p = -\frac{Gm_1m_2}{r} \qquad (8.5.5)$$

It is important to note that even though m_2 has been moved to a point infinitely distant from m_1, only a *finite amount* of work is required to do this. The reason is, of course, that the gravitational force itself approaches zero so rapidly with increasing r that the integral of $F' \, dr$, representing the limit of the individual work elements along the path, remains finite.

We may also recall from section 5.6 that if a force can be expressed as the negative gradient of any scalar function whatsoever, it is a conservative one, the scalar function representing the potential energy associated with the force. The scalar function U_p in (8.5.5) is just such a function, because if we write

$$U_p = -\frac{Gm_1m_2}{r} = -\frac{Gm_1m_2}{\sqrt{x^2 + y^2 + z^2}} \qquad (8.5.6)$$

then

$$\frac{\partial U_p}{\partial x} = Gm_1m_2 \frac{x}{(x^2 + y^2 + z^2)^{3/2}} = Gm_1m_2 \frac{x}{r^3}$$

$$\frac{\partial U_p}{\partial y} = Gm_1m_2 \frac{y}{(x^2 + y^2 + z^2)^{3/2}} = Gm_1m_2 \frac{y}{r^3} \qquad (8.5.7)$$

$$\frac{\partial U_p}{\partial z} = Gm_1m_2 \frac{z}{(x^2 + y^2 + z^2)^{3/2}} = Gm_1m_2 \frac{z}{r^3}$$

and

$$-\nabla U_p = -\left(\mathbf{i}_x \frac{\partial U_p}{\partial x} + \mathbf{i}_y \frac{\partial U_p}{\partial y} + \mathbf{i}_z \frac{\partial U_p}{\partial z}\right)$$

$$= -Gm_1m_2 \frac{(\mathbf{i}_x x + \mathbf{i}_y y + \mathbf{i}_z z)}{r^3} \qquad (8.5.8)$$

But $\mathbf{i}_x x + \mathbf{i}_y y + \mathbf{i}_z z$ is just the vector $\mathbf{r}$, which enables us to write

$$-\nabla U_p = -Gm_1m_2 \frac{\mathbf{r}}{r^3} = \mathbf{F}' \qquad (8.5.9)$$

since, according to (8.4.3), $\mathbf{F}'$ is simply minus $\mathbf{F}$, as expressed by (8.4.2).

The gravitational force is, therefore, conservative, its potential energy function being given by (8.5.5), and it possesses all the important properties of conservative forces described in sections 5.5 and 5.6. In particular, the work done by the force in any displacement depends only on the difference in potential energy between the initial and final points and *not in any way upon the path followed between them.* Also, the work done by the force in any displacement around a *closed* path is *zero.*

Although one may understand the nature of the gravitational force as simply a force acting through a given distance between two or more bodies, it is very useful to picture any mass particle as representing the source of a *gravitational force field* whose influence extends through all space, causing other masses to experience forces whose magnitudes and directions are given by (8.5.9). The field, according to this picture, is an inherent property of the mass particle that is its source and may be regarded as existing independently of any other masses which may happen to be around to experience its influence. The gravitational interaction of two masses may be understood in two separate stages as the creation of the field by the mass particle and the influence of that field on other bodies. Newton's law of universal gravitation, as expressed by (8.4.2), allows a vector **A**, having components

$$A_x = \frac{F_x}{m_2} = \frac{Gm_1 x}{r^3}$$

$$A_y = \frac{F_y}{m_2} = \frac{Gm_1 y}{r^3} \qquad (8.5.10)$$

$$A_z = \frac{F_z}{m_2} = \frac{Gm_1 z}{r^3}$$

to be defined at every point in space around mass m_1. Such a vector expresses the *force per unit mass* experienced by a body of mass m_2 located at $P(x, y, z)$; it is referred to as the *gravitational field intensity* associated with the field of m_1. Likewise, a scalar function representing the *potential energy per unit mass* of a body such as m_2 at every point P can be written as

$$\phi = \frac{U_p}{m_2} = -\frac{Gm_1}{r} \qquad (8.5.11)$$

This function, which, as a result of (8.5.9), has the property that

$$-\nabla\phi = \mathbf{A} \qquad (8.5.12)$$

whence

$$A_x = -\frac{\partial\phi}{\partial x} \qquad A_y = -\frac{\partial\phi}{\partial y} \qquad A_z = -\frac{\partial\phi}{\partial z} \qquad (8.5.13)$$

is referred to as the *potential function* associated with the gravitational field.

Although we shall frequently use the concept of the gravitational field, we shall not make much further use of the quantities **A** and ϕ but will instead work directly with the quantities **F** and U_p. It is

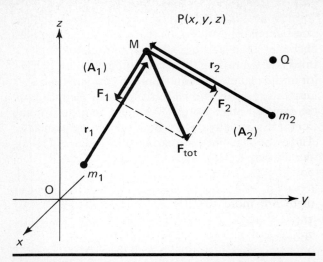

FIGURE 8.7. Addition, or superposition, of gravitational forces by the laws of vector addition.

important, nevertheless, to understand the field concept and to have an understanding of how gravitational intensity, field, and potential are defined, since ideas and quantities strictly analogous to these are used as the basis of the theory of electricity and magnetism, which we shall study at great length in later chapters.

It is also very important to understand that the gravitational forces, fields, and potentials at any point arising from more than one point mass source may be *added* to arrive at a total value of force, field intensity, potential energy, or gravitational potential at that point. In the case of the gravitational force or field intensity, which are vector quantities, this addition must be carried out *vectorially*, as illustrated in Fig. 8.7. In this diagram, the gravitational forces acting on the mass M at the point P(x, y, z) arising from the nearby masses m_1 and m_2 are shown. The forces $\mathbf{F}_1$ and $\mathbf{F}_2$ can be written as

$$\mathbf{F}_1 = -Gm_1 M \frac{\mathbf{r}_1}{r_1{}^3} \qquad \text{and} \qquad \mathbf{F}_2 = -Gm_2 M \frac{\mathbf{r}_2}{r_2{}^3}$$

$$(8.5.14)$$

the minus signs signifying that the vectors $\mathbf{F}_1$ and $\mathbf{F}_2$ are antiparallel to the vectors $\mathbf{r}_1$ and $\mathbf{r}_2$ which locate M with respect to the other two masses. The total force experienced by M is simply the *vector sum* of these two individual forces:

$$\mathbf{F}_{tot} = \mathbf{F}_1 + \mathbf{F}_2 = -GM\left(m_1 \frac{\mathbf{r}_1}{r_1{}^3} + m_2 \frac{\mathbf{r}_2}{r_2{}^3}\right) \qquad (8.5.15)$$

Similar considerations apply to the vector addition of the gravitational field intensity contributions $\mathbf{A}_1 = \mathbf{F}_1/M$ and $\mathbf{A}_2 = \mathbf{F}_2/M$ at P to obtain a total field intensity equal to $\mathbf{F}_{tot}/M$. In doing this, it is necessary to keep in mind the fact that while a mass M located at P may contribute to a total field intensity somewhere else (at Q, for instance), *the force, and thus the field intensity it experiences itself arises solely from the fields set up by the other masses m_1 and m_2.* In other words, no mass experiences any force or resultant field intensity from its own gravitational field. While this statement is fairly obvious, endless confusion can result if it is not clearly understood. The same general statement also applies in regard to the potential energy of M and the gravitational potential at P arising from masses m_1 and m_2.

The situation is not essentially different when there are more than two bodies that may set up gravitational forces at a point such as P, or when the various differential mass elements of a large, extended object may all participate in contributing to a gravitational field at that point. In every such instance, the total gravitational force or field intensity may be arrived at by summing all the individual contributions from each mass element *as vectors.*

The calculation of the total potential energy of a mass such as M in Fig. 8.7 or the total gravitational potential at the point P is even easier, because both these quantities are scalars, and it is necessary only to find the *scalar sum* of all the contributions of the other masses in the system. In the case illustrated in Fig. 8.7, the individual contributions to the potential energy of M arising from the presence of m_1 and m_2 are

$$U_{p1} = -\frac{Gm_1 M}{r_1} \quad \text{and} \quad U_{p2} = -\frac{Gm_2 M}{r_2} \qquad (8.5.16)$$

The total potential energy of M in the field of the two other masses is, therefore, given by

$$U_p = U_{p1} + U_{p2} = -GM\left(\frac{m_1}{r_1} + \frac{m_2}{r_2}\right) \qquad (8.5.17)$$

Since the gravitational potential ϕ_1 at P is simply U_{p1}/M and ϕ_2 is likewise given by U_{p2}/M, the total gravitational potential may be written as

$$\phi = \phi_1 + \phi_2 = \frac{U_{p1}}{M} + \frac{U_{p2}}{M} = \frac{U_p}{M} = -G\left(\frac{m_1}{r_1} + \frac{m_2}{r_2}\right) \qquad (8.5.18)$$

The fact that the total potential energy or gravitational potential arising from a number of masses or a distribution of mass elements can be arrived at so easily by a process of scalar addition is one of the reasons why these quantities are so important and useful. In calculating the gravitational field arising from a large number of individual masses or from the mass elements of an extended body, it is almost always much simpler to sum the scalar potential energy contributions from the elements to find the total potential energy $U_p(x, y, z)$ and then to find the gravitational force or field vector from the relation

$$
\begin{array}{lcl}
F_x = -\partial U_p/\partial x & & A_x = -\partial\phi/\partial x \\
F_y = -\partial U_p/\partial y & \text{or} & A_y = -\partial\phi/\partial y \\
F_z = -\partial U_p/\partial z & & A_z = -\partial\phi/\partial z
\end{array}
$$

as a final step than to try to take the vector sum of individual force or field contributions. This will be amply illustrated in the calculation of gravitational forces from spherical objects undertaken in the next section.

8.6 Forces and Fields from Spherical Objects: The Cavendish Experiment and the Earth's Mass

Newton's law of universal gravitation is stated in terms of forces between point mass particles, which are idealized objects. In order to understand the forces exerted by real bodies upon one another, it is necessary to extend our understanding of the law of gravitation to cases involving extended objects having finite dimensions. In principle, this can be done easily enough simply by superposing the forces arising from every mass element of the bodies in question, as described in the preceding section. In practice, however, the mathematical difficulties encountered in calculating gravitational forces between objects of arbitrary shape in this way are formidable. For this reason, we shall be content merely to develop methods for finding the gravitational forces exerted by spherical masses upon one another. This in itself is extremely useful, since it allows us to understand the forces exerted by the sun and planets (all of which are practically spherical) on one another and, in addition, permits us to understand how the law of gravitation was first tested experimentally and how Newton's universal gravitation constant may be measured in the laboratory.

Let us consider initially the gravitational interaction between a point mass m and a thin, spherical shell of radius R and thickness dR, as illustrated in Fig. 8.8. For this purpose, it is much easier to calculate the potential energy of the point mass in the gravitational field of the spherical sheel than to try to calculate directly the force it experiences. We shall, therefore, proceed by writing the potential energy contribution from a ring-shaped element of the spherical shell, such as the shaded element in the figure, and sum over all such elements of the spherical shell to obtain the total gravitational potential energy of the point mass. The potential energy of mass m in the gravitational field of the ring-shaped element, according to (8.5.5), is

$$dU_p = -\frac{Gm(dm_s)}{r'} \qquad (8.6.1)$$

where dm_s represents the mass of the ring element and r' is the distance between all the mass particles of this

259

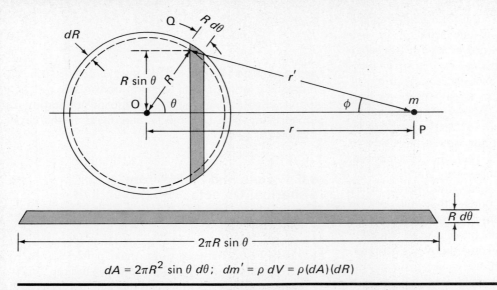

$$dA = 2\pi R^2 \sin \theta \, d\theta; \quad dm' = \rho \, dV = \rho(dA)(dR)$$

FIGURE 8.8. Calculation of the gravitational potential energy of a spherical shell.

mass element and the mass m. If we assume that the ring-shaped element subtends an angle $d\theta$ at the center of the sphere and that the element has constant density (mass per unit volume) ρ, we may write

$$dm_s = \rho \, dV = \rho \, dA \, dR = \rho(2\pi R^2 \sin \theta \, d\theta) \, dR$$

$$(8.6.2)$$

This may be easily understood by referring to the lower drawing in Fig. 8.8, where the ring element is shown as if it were removed from the sphere and flattened out into a single strip of length $2\pi R \sin \theta$, width $R \, d\theta$, and thickness dR. Substituting this expression for dm' into (8.6.1), we may write

$$dU_p = -\frac{2\pi G\rho mR^2 \sin \theta \, dR \, d\theta}{r'} \qquad (8.6.3)$$

But according to the law of cosines (remember your trigonometry?), the distance r' can be expressed in terms of the distances r and R in triangle OQP by

$$r'^2 = r^2 + R^2 - 2rR \cos \theta \qquad (8.6.4)$$

In integrating (8.6.3) over the ring-shaped elements shown in the diagram, the distance r, the radius R, and the thickness dR remain *fixed*, while the distance r' and the angle θ *vary* as we sum over the elements. We may therefore take the differential of both sides of Eq. (8.6.4), regarding r, R, and dR as constants, and r' and θ as variables, to obtain

$$r' \, dr' = rR \sin \theta \, d\theta \qquad (8.6.5)$$

from which

$$\sin \theta \, d\theta = \frac{r' \, dr'}{rR} \qquad (8.6.6)$$

We may now substitute this into (8.6.3), finally, to get

$$dU_p = -\frac{2\pi G\rho mR \, dR}{r} \, dr' \qquad (8.6.7$$

This must now be integrated over all ring elements of the spherical shell, which extends from $r' = r - R$ to $r' = r + R$ along the line joining the point mass and the center of the shell. We may therefore integrate over the variable dr', regarding r, R, and dR as constants, as follows:

$$U_p = \int dU_p = -\frac{2\pi G\rho mR \, dR}{r} \int_{r-R}^{r+R} dr' \qquad (8.6.8)$$

$$U_p = -\frac{2\pi G\rho mR \, dR}{r} \left[r' \right]_{r-R}^{r+R}$$

$$= -\frac{2\pi G\rho mR \, dR}{r}(r + R - r + R)$$

$$= -\frac{4\pi \rho R^2 Gm \, dR}{r} \qquad (8.6.9)$$

In this equation, the quantity $4\pi\rho R^2 \, dR$ represents simply the total mass of the spherical shell since, after all, its volume is simply the surface area $4\pi R^2$ times the thickness dR, and density times volume is mass. Accordingly, setting this quantity equal to the shell mass m_s, we may finally write

$$U_p = -\frac{Gmm_s}{r} \qquad (8.6.10)$$

which is identical to (8.5.5). The force experienced by mass m can easily be obtained by the procedure illustrated in the previous section leading to Eq. (8.5.9). Since the expression for U_p is identical to (8.5.5), it is clear even without going through the algebra that the force experienced by m will be given by

$$\mathbf{F} = -Gmm_s \frac{\mathbf{r}}{r^3} \qquad (8.6.11)$$

which is the same as (8.5.9).

In doing all this, we have proved that the gravitational attraction between a spherical shell and an exterior point mass is the same as that which would exist between the same point mass and another having the mass of the spherical shell and located *at its center*. Since any solid sphere can be regarded as being composed of concentric spherical shell mass elements, it is evident that the same result holds true for the gravitational attraction between a point mass and a *solid sphere* of uniform density.[6]

The case of *two extended spherical masses* leads to a similar conclusion, though the justification is more complex. Suppose we have two uniform spheres each exerting gravitational forces upon the other. Each mass element of the first experiences forces from all mass elements of the second, the total effect in each case being as if the second sphere were actually a point mass located at the position of its center. The net force acting on the first sphere is thus that which would arise from a point mass at the center of the second sphere. The same argument can be made regarding the forces acting on the individual mass elements of the second sphere from all the mass elements of the first, leading to the conclusion that the net force on the first sphere is that which would arise from the total mass of the second concentrated into a mass point at its center. Two uniform rigid spheres then exert gravitational forces upon one another *as though their entire masses were concentrated at their respective centers of mass.*

But haven't we known all along the gravitational forces always act at the center of mass of any object? Aren't we elaborating the obvious, proving the self-evident, reinventing the wheel, so to speak? The answer is that we are not; in fact, it has been essential to execute the elaborate maneuvers which were performed above in obtaining a seemingly obvious set of results.

The explanation of this lies in the fact that the definition of the center of mass, as given in section 4.4, is conceived from the outset in terms of an object in a *uniform* gravitational field, in which the acceleration of gravity is the same everywhere. Under these circumstances, the gravitational (weight) force acting upon it can indeed be regarded as acting upon its center of mass. In all the examples considered so far, only uniform gravitational fields have been involved, and our practice of regarding the entire gravitational force as acting at the center of mass has been justified.

[6] Or, for that matter, for a solid sphere of nonuniform density, so long as the density is a function only of the radial distance from the center of the sphere.

Now we are trying to deal with situations in which there are *nonuniform* gravitational forces. In such instances, the gravitational forces on each element of a rigid body add up differently from the uniform-field case, and it is no longer true in general that the entire gravitational force can be considered to act at the center of mass. *The exception is the case where the attracting masses are spherical*, as we have just finished demonstrating!

Another point worthy of note is that the results embodied in Eqs. (8.6.10) and (8.6.11) are valid only when the point mass m is *outside* the spherical shell, that is, where $r > R$. If the point mass is located *within* the spherical shell ($r < R$), one may show, using methods essentially identical to those used in the other case, that the gravitational potential energy of this mass is constant and that the gravitational force of attraction it experiences from the spherical shell is, therefore, zero. The explicit proof of this result is assigned as an exercise for the student.

We are now in a position (provided we are given the value of the gravitational constant G) to make actual numerical estimates of the gravitational forces between spherical masses of various size and density. Suppose, for example, that we consider two equal spheres of lead ($\rho = 11.3$ g/cm^3) 30 cm in diameter. The mass of such a sphere will be

$$m = \rho V = \tfrac{4}{3}\pi R^3 \rho = \tfrac{4}{3}(\pi)(15)^3(11.3)$$
$$= 159{,}700 \text{ g}, \qquad \text{or} \qquad 159.7 \text{ kg}$$

When two such spheres are in contact, the distance between their centers is about 30 cm. The force of gravitational attraction between the two under such circumstances will then be, in view of the results proved earlier in this section,

$$F = \frac{Gm^2}{r^2} = \frac{(6.673 \times 10^{-8})(1.597 \times 10^5)^2}{(30)^2}$$
$$= 1.891 \text{ dyne}$$

using $G = 6.673 \times 10^{-8}$ dyne-cm^2/g^2 and $r = 30$ cm. Now this is not a very large force, but neither is it completely undetectable provided that sensitive apparatus is used. An appreciation of the size of such a force can be gained by observing that an ordinary house fly weighs about 1 dyne. If the radii of the spheres were doubled, their mass would increase by a factor of 8, and their mutual gravitational attraction, as given by the above equation, would be 16 times greater, which is still not very big.

Turning this situation around and looking at it from another point of view, it is evident that if the masses and radii of the spheres were known and if their force of attraction could be *measured* by some sufficiently sensitive device, then the value of the constant G could be obtained experimentally. This was first

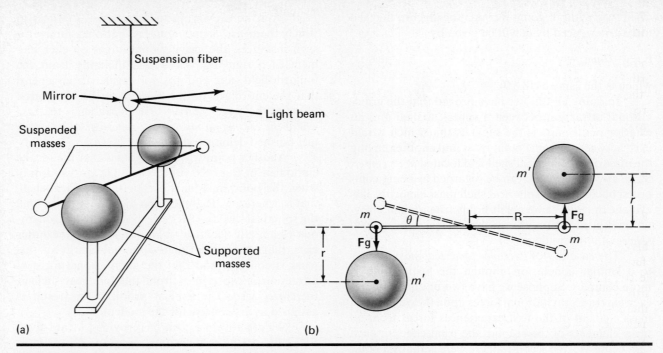

FIGURE 8.9. The Cavendish experiment, in which the direct determination of the universal gravitation constant was accomplished.

accomplished by Henry Cavendish (1731–1810), an English scientist, in 1790. The apparatus he designed to make this delicate and difficult experimental determination is illustrated in Fig. 8.9. It consists basically of two spherical masses connected by a light, rigid rod suspended at its center by a thin, elastic torsion fiber to which a mirror is clamped. There are, in addition, two much longer and heavier spherical masses which can be positioned near the suspended ones, as shown, or removed to a much greater distance. When the large masses are positioned far away, so that the gravitational forces exerted by them on the smaller ones are negligible, the suspended masses assume the equilibrium position shown by the dotted lines in Fig. 8.9b. When the large masses are positioned at some small distance from the smaller ones, a small attractive gravitational force F_g acts on each of the suspended masses, setting up torques which twist the torsion fiber, with its attached mirror and masses, through an angle θ into a new equilibrium position.

In this new equilibrium state, the gravitational torque on each mass is $F_g R$, and the restoring torque set up by the torsion fiber is $-k'\theta$, where k' is the torsional Hooke's law constant of the suspending fiber. Since the system is in equilibrium, these torques must add to zero, giving

$$2F_g R - k'\theta = 0$$

$$F_g = \frac{k'\theta}{2R} \qquad (8.6.12)$$

Since the quantities k', R, and θ can be easily measured experimentally (the latter in terms of the deflection of a light beam incident on the mirror and reflected by it to a distant screen), the Cavendish experiment leads to a direct measurement of the force of gravity. It was, therefore, possible to use it to verify the predictions of Newton's theory that this force should be proportional to the two masses m and m' and inversely proportional to the square of the distance r shown in Fig. 8.9b. Equally important, it was also used to ascertain that the gravitational force was dependent only on the masses of the two objects involved and completely independent of their density, size, and chemical composition.

Once having been assured by the experimental results that the form of Newton's postulated expression (8.4.1) is correct, one may substitute that relation into (8.6.12) to obtain

$$F_g = \frac{k'\theta}{2R} = G\frac{mm'}{r^2}$$

$$\qquad (8.6.13)$$

$$G = \frac{k'\theta r^2}{2mm'R}$$

This allows us to determine the value of G directly in terms of easily measured quantities related to the Cavendish apparatus. Cavendish's original value for G determined nearly two centuries ago differed by only about 1 percent from the currently accepted value given in section 8.4.

8.7 The Earth's Gravitational Field and the Earth's Mass

An object of mass m located at a distance r from the earth's center experiences a weight force mg as a result of the earth's attraction. At the same time, this weight force can be written using Newton's law of universal gravitation (8.4.1), setting m' equal to the earth's mass m_e. Equating these two expressions, we may write

$$mg = G \frac{mm_e}{r^2}$$

$$g = g(r) = \frac{Gm_e}{r^2} \qquad (8.7.1)$$

This equation expresses the acceleration of gravity as a function of the distance r between the mass m and the center of the earth. When r is set equal to the earth's radius r_e, g attains the familiar "surface gravity" value $g_0 = 9.8$ m/sec^2 = 32 ft/sec^2, which we have had much occasion to use in problems where the objects involved are always quite close to the earth's surface. Setting r equal to r_e in this equation, we obtain

$$g_0 = g(r_e) = \frac{Gm_e}{r_e^2} \qquad (8.7.2)$$

Also, dividing Eq. (8.7.1) by Eq. (8.7.2), it is easily seen that

$$g(r) = g_0 \frac{r_e^2}{r^2} \qquad (8.7.3)$$

Equation (8.7.3) expresses $g(r)$ as a function of r in a particularly simple way and allows us to see the justification for using the constant value g_0 for the acceleration of gravity in problems in which the bodies involved never get very far from the earth's surface. The earth's radius is about 6400 kilometers, and this is the value of r_e to be inserted in (8.7.3) in any such problem. But if the objects involved are always within a few kilometers, say, of the earth's surface, the ratio r_e^2/r^2 in (8.7.2) will always by very close to unity and the local value of g at their positions very close to g_0. The approximation of $g(r)$ by the constant value g_0 will, therefore, be a very accurate one in such cases.

Equation (8.7.2) can be solved for m_e to give

$$m_e = \frac{g_0 r_e^2}{G} \qquad (8.7.4)$$

From this, it is evident that if the earth's radius is known, if the acceleration of gravity at the surface can be measured by some device such as Atwood's machine (Example 4.2.7), and if the value of Newton's constant G can be determined experimentally by the Cavendish experiment, then the earth's mass follows

directly. Substituting the measured values for these quantities into (8.7.4), we find

$$m_e = \frac{(9.8)(6.37 \times 10^6)^2}{6.673 \times 10^{-11}} = 5.959 \times 10^{24} \text{ kg}$$

which is equivalent to about 6.6×10^{21} tons. From this, and from a knowledge of the earth's radius, it is easy to ascertain that the earth's average density must be about 5.52 g/cm^3.

Since Cavendish's measurement of the universal gravitation constant G was the final piece of information needed to carry out this calculation, it is often said that in performing his experiment Cavendish succeeded in "weighing the earth." In an indirect sense, this is quite correct, though it must be remembered that it was Newton who provided the means for calculating the earth's mass from the known data.

EXAMPLE 8.7.1

A body of mass m is projected vertically upward from the earth's surface with initial velocity v_0, which is large enough to send it so far that the earth's gravitational attraction may no longer be assumed constant. Find an expression relating its velocity and its height, measured from the earth's center. Is it possible to give it an initial velocity so large that it will never return? Neglect any effects arising from the earth's axial or orbital rotation, and also any effect of atmospheric friction.

The most straightforward way of approaching this problem is to equate the gravitational force Gmm_e/r^2 with ma and integrate once to find the velocity. This approach is perfectly feasible, as the student may easily verify, but we shall find it more productive of physical insight to proceed from the law of conservation of energy, which states that

$$U_{ki} + U_{pi} = U_{kf} + U_{pf} \qquad (8.7.5)$$

Initially, the velocity is v_0 and the distance from the earth's center is r_e. We may take the final state of the system to be one in which the body is at a distance r from the earth's center, with velocity v. Since the body is projected vertically upward (i.e., radially outward), there is no rotational energy, and, therefore, $U_{ki} = \frac{1}{2}mv_0^2$, $U_{kf} = \frac{1}{2}mv^2$, $U_{pi} = -Gmm_e/r_e$, and $U_{pf} = -Gmm_e/r$. Equation (8.7.5) may therefore be written as

$$\frac{1}{2}mv_0^2 - \frac{Gmm_e}{r_e} = \frac{1}{2}mv^2 - \frac{Gmm_e}{r}$$

or

$$v^2 - v_0^2 = 2Gm_e\left(\frac{1}{r} - \frac{1}{r_e}\right) \qquad (8.7.6)$$

But according to (8.7.2), $Gm_e = g_0 r_e^2$, which permits

us to write (8.7.6) in the form

$$v^2 = v_0^2 - 2g_0 r_e^2 \left(\frac{1}{r_e} - \frac{1}{r} \right) \qquad (8.7.7)$$

or, solving for $1/r$,

$$\frac{1}{r} = \frac{1}{r_e} - \frac{v_0^2 - v^2}{2g_0 r_e^2} \qquad (8.7.8)$$

The object reaches its maximum height r_m when its velocity is reduced to zero, all its initial kinetic energy then having been expended in increasing its gravitational potential energy. Accordingly, setting $v = 0$ in (8.7.8),

$$\frac{1}{r_m} = \frac{1}{r_e} - \frac{v_0^2}{2g_0 r_e^2} \qquad (8.7.9)$$

Thus, when v_0 is zero, $r_m = r_e$, reasonably enough. As v_0 increases, $1/r_m$ becomes less than $1/r_e$, implying that r_m is greater than r_e, which is also very plausible. Finally, if v_0 becomes so large that the second term on the right becomes as large as the first, r_m becomes infinite. Under these circumstances, the object has been given an amount of kinetic energy initially which is just equal in magnitude, but opposite in sign, to its (negative) initial gravitational potential energy. It then always has sufficient energy of motion to completely overcome the attractive gravitational potential energy of the earth's gravitational field and can altogether escape the earth's attraction, its velocity approaching zero as its height increases toward infinity. The initial velocity required to bring about this state of affairs is obtained by setting the right side of Eq. (8.7.9) equal to zero and solving for v_0:

$$v_0 = v_{\text{esc.}} = \sqrt{2g_0 r_e} = \frac{\sqrt{2GM_e}}{r_e} \qquad (8.7.10)$$

This critical value of initial velocity is referred to as the *escape velocity*. Its value for the earth is easily calculated to be

$$v_{\text{esc}} = \sqrt{(2)(9.8)(6.37 \times 10^6)} = 11,200 \text{ m/sec}$$

which is about 25,000 miles per hour. This result demonstrates the possibility of space travel, since it shows that a body may be projected to any desired distance by a finite expenditure of energy.

For initial velocities exceeding the escape velocity, the object also escapes the earth's gravitational attraction, its velocity approaching some finite positive value as r approaches infinity, as obtained by setting $1/r = 0$ in Eq. (8.7.7). The body's motion naturally falls into one of three physically different categories, according to whether the initial velocity is less than, equal to, or greater than the escape velocity. We shall see that these three same categories are characteristic of planetary or orbital motion in general.

To obtain an explicit expression relating the distance r with time, we must set $v = dr/dt$ in (8.7.7) and integrate once again. The resulting integral can be evaluated in a straightforward way, but the algebraic details are troublesome, except when v_0 is equal to the escape velocity, and the results are not sufficiently evocative of physical insight to bother with. The simple case where the initial velocity is equal to the escape velocity should be worked out by the reader.

EXAMPLE 8.7.2

An elevator shaft is sunk vertically downward through the center of the earth and to the surface again, as illustrated in Fig. 8.10. The shaft is evacuated and fitted with a frictionless car, which is released from rest at the earth's surface at point A. Discuss the subsequent motion of the elevator car, finding, in particular, its maximum velocity and the time required for it to make a complete trip through the earth to point B. Assume the density of the earth to be uniform throughout its volume.

When the car is beneath the earth's surface at a distance r from the center, it experiences forces of attraction from the part of the earth's mass that is closer to the center of the earth than the car, as represented by the shaded region, and also from the exterior, unshaded part of the earth. As mentioned in section 8.6, the forces exerted by various parts of the spherical mass shell exterior to the car cancel one another and sum to zero. The total effective gravitational force on the car is then simply that arising from the *interior* spherical shaded region of radius r. Let us denote the mass of the shaded region by $M(r)$ and that of the car by m. Then, according to the law of universal gravitation,

$$F_g = -\frac{GmM(r)}{r^2} \qquad (8.7.11)$$

But if r_e is the radius of the earth and ρ its density,

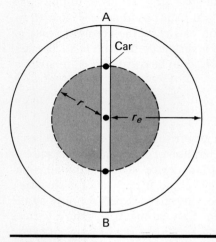

FIGURE 8.10

and if $V(r)$ represents the volume of the shaded region, then

$$M(r) = \rho V(r) = \rho \tfrac{4}{3}\pi r^3 \qquad (8.7.12)$$

Also, if V_e is the total volume of the earth and m_e its total mass,

$$\rho = \frac{m_e}{V_e} = \frac{m_e}{\tfrac{4}{3}\pi r_e^3} \qquad (8.7.13)$$

Substituting this value for ρ into (8.7.12), we obtain

$$M(r) = m_e \frac{r^3}{r_e^3} \qquad (8.7.14)$$

Inserting this into (8.7.11), we find, using Newton's second law to relate force to acceleration, that

$$F_g = -\frac{Gmm_e}{r_e^3}r = m\frac{d^2r}{dt^2}$$
$$\qquad (8.7.15)$$
$$\frac{d^2r}{dt^2} = -\frac{Gm_e}{r_e^3}r$$

or

$$\frac{d^2r}{dt^2} = -\omega^2 r \qquad (8.7.16)$$

where

$$\omega = \sqrt{\frac{Gm_e}{r_e^3}} = \sqrt{\frac{g_0}{r_e}} \qquad (8.7.17)$$

the final expression in (8.7.17) being obtained by noting from (8.7.2) that $Gm_e = g_0 r_e^2$.

Equation (8.7.16) relates the acceleration d^2r/dt^2 to the displacement r. This equation can be solved by writing the acceleration a in the form

$$a = \frac{d^2r}{dt^2} = \frac{dv}{dt} = \frac{dv}{dr}\frac{dr}{dt} = v\frac{dv}{dr} \qquad (8.7.18)$$

recalling that $v = dr/dt$. Substituting $v(dv/dr)$ for d^2r/dt^2 in (8.7.16), we may write

$$v\,dv = -\omega^2 r\,dr \qquad (8.7.19)$$

which may be integrated from the initial state in which $v = 0$ at the earth's surface (where $r = r_e$) to some subsequent stage of the journey at which the velocity is v and the distance from the earth's center is r. In this way, since ω is a constant, we obtain

$$\int_0^v v\,dv = -\omega^2 \int_{r_e}^r r\,dr$$

from which

$$v^2 = \omega^2(r_e^2 - r^2)$$
$$\qquad (8.7.20)$$
$$v = \frac{dr}{dt} = \omega\sqrt{r_e^2 - r^2}$$

This equation, in turn, can be rewritten in the form

$$\frac{dr}{\sqrt{r_e^2 - r^2}} = \omega\,dt \qquad (8.7.21)$$

and integrated from the initial state in which $r = r_e$ at $t = 0$ to a later phase of the trip, at time t, when the distance from the earth's center is r. Since

$$\int (r_e^2 - r^2)^{-1/2}\,dr = \sin^{-1}(r/r_e)$$

this becomes

$$\left[\sin^{-1}\left(\frac{r}{r_e}\right)\right]_{r_e}^r = \omega[t]_0^t$$

or

$$\sin^{-1}\left(\frac{r}{r_e}\right) = \omega t + \frac{\pi}{2} \qquad (8.7.22)$$

since $\sin^{-1} 1 = \pi/2$. Taking the sine of both sides of this equation, we find at last that

$$\frac{r}{r_e} = \sin\left(\omega t + \frac{\pi}{2}\right) = \cos\omega t$$

or

$$r = r_e \cos\omega t \qquad (8.7.23)$$

The velocity v can be found by differentiation, giving

$$v = \frac{dr}{dt} = \omega r_e \sin\omega t \qquad (8.7.24)$$

From Eq. (8.7.23) we see that the motion of the elevator car is a *periodic* one, the car moving up and down in its shaft sinusoidally with time between the limits $r = r_e$ (point A in Fig. 8.10) and $r = -r_e$ (point B). The motion has the same mathematical form as the motion of a mass attached to an ideal spring, as illustrated in Fig. 5.16 and discussed, from a different point of view, in Example 5.6.2. It is easily seen, in fact, that if we equate the force in Eq. (5.2.8), which describes the motion of the mass–spring system, to $m(d^2x/dt^2)$, we obtain an equation of motion identical to (8.7.16), except for the fact that $\sqrt{k/m}$ appears in place of the constant ω. We may also note that (8.7.23) is essentially the same motion as that described for the mass–spring system by (5.7.14), the cosine function appearing in place of the sine only because we assume here that at $t = 0$ the elevator car is at the limit of its excursion rather than at the center of its path at $r = 0$. This very common and quite important type of motion is referred to as *simple harmonic motion* and will be discussed extensively in the next chapter.

In any event, from (8.7.23) it is apparent that as the quantity ωt changes by 2π radians, an entire cycle of the motion will take place and the system will return to its initial state. At $t = 0$, for example, we find that $r = r_e$, corresponding to point A in

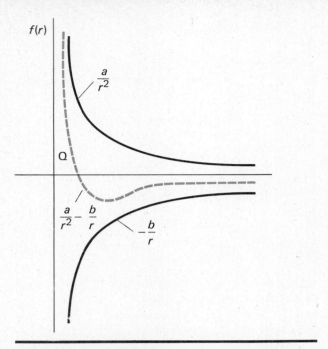

FIGURE 8.11. The quantity $U - \frac{1}{2}mv_r^2$ of Eq. (8.8.7), illustrated as the sum of two quantities that vary proportional to $-1/r$ and $+1/r^2$, respectively.

Fig. 8.10. When $\omega t = 2\pi$, that is, when $t = 2\pi/\omega$, the elevator will have descended to B, reversed direction, and returned to A, ready to begin another cycle. The time required is

$$t = \frac{2\pi}{\omega} = 2\pi \sqrt{\frac{r_e}{g_0}} = 2\pi \sqrt{\frac{6.37 \times 10^6}{9.80}}$$

$$= 5066 \text{ sec, or } 84.4 \text{ min}$$

The time for a one-way trip from A to B is, of course, just half of this, or 42.2 min. Since the velocity is given by (8.7.24), it is clear that the maximum speed will be attained when $\omega t = \pi/2$, $\sin \omega t = 1$, in which case Eqs. (8.7.24) and (8.7.17) inform us that

$$v_{\max} = \omega r_e = \sqrt{g_0 r_e} \qquad (8.7.25)$$

which is the same as the escape velocity given by (8.7.10) divided by $\sqrt{2}$, amounting to 7920 m/sec. This velocity is attained at the earth's center, where $r = 0$, as we may easily see by substituting $\omega t = \pi/2$ into Eq. (8.7.23).

8.8 Planetary and Satellite Orbits

The general problem of determining the orbital paths of planets and satellites is mathematically very difficult and is in all its generality beyond the scope of this book. There are, however, two special cases that are rather simple and that can be understood without recourse to complex mathematical detail. One of these

is the case of the vertically projected object which was treated in the previous section; the other is the case of circular orbits which we have already encountered in section 8.4.

It is, nevertheless, possible to understand the general properties of planetary motion in qualitative or semiquantitative terms, and in this section we shall undertake the development of that understanding, starting with *the energy conservation principle* and the *law of conservation of angular momentum*. Both these principles are, of course, direct consequences of Newton's second law, and all the results we shall obtain might very well have been derived from the second law. But we shall find the course we are going to follow simpler from a mathematical point of view and much more productive in terms of the understanding of the *physical* principles involved. We shall also return to the case of circular orbits and work out some of the important physical results pertaining to this type of motion in an example at the end of this section.

Let us consider the case of a spherical body of mass m moving in an orbital path around a very much more massive spherical object[7] of mass M. If the origin of coordinates is taken at the center of mass of the body of mass M, the situation is as illustrated in Fig. 8.10. The total energy of the system will be given as the sum of the kinetic and potential energies by

$$U = \frac{1}{2}mv^2 - \frac{GmM}{r} = \text{const.} \qquad (8.8.1)$$

Since there are no nonconservative forces acting, the total energy of the system will always remain the same. We may express the kinetic energy in terms of the polar coordinates r and θ by noting that if a vector displacement dr occurs in time dt, then the radial and tangential components of the velocity vector $d\mathbf{r}/dt$ will be given by $v_r = dr/dt$ and $v_\theta = r(d\theta/dt) = r\omega$, as discussed in section 3.7. Accordingly, Eq. (8.8.1) can be written in the form

$$U = \frac{1}{2}m(v_r^2 + r^2\omega^2) - \frac{GmM}{r} \qquad (8.8.2)$$

The gravitational force is a *central* force whose direction is always opposite that of the vector $\mathbf{r}$ in Fig. 8.10. The torque $\mathbf{r} \times \mathbf{F}$ experienced by the mass m is, therefore, zero, and this means that the time rate of change of angular momentum is zero, hence that the angular momentum of the mass m is constant.

---

[7] In doing this, we account only for the gravitational attraction of the sun on the planets and neglect the attraction of the other planets for the one in question. The neglect of these interplanetary attractions is justified by the fact that the masses of the other planets, and thus their gravitational attractions, are very much less than that of the sun.

But the angular momentum of mass m with respect to the origin in Fig. 8.10 is, by definition,

$$\mathbf{L} = \mathbf{r} \times \mathbf{p} = \mathbf{r} \times m\mathbf{v} = \mathbf{r} \times m(\mathbf{i}_r v_r + \mathbf{i}_\theta r\omega) \qquad (8.8.3)$$

However, since the vector $\mathbf{i}_r$ is in the same direction as the vector $\mathbf{r}$, their cross product is zero; also, the cross product of the perpendicular vectors $\mathbf{r}$ and $\mathbf{i}_\theta$ is a vector whose magnitude is $\mathbf{r}$ and whose direction is normal to the plane in which the vectors $\mathbf{r}$ and $\mathbf{i}_\theta$ lie. Such a vector points upward from the plane of the page, along the z-axis, and can be expressed as $r\mathbf{i}_z$. Under these circumstances, (8.8.3) can be written

$$\mathbf{L} = (mr^2\omega)\mathbf{i}_z \qquad (8.8.4)$$

This vector remains constant in magnitude and direction at all times. From (8.8.4), the angular velocity ω, which, it must be emphasized, is *not* necessarily constant, can be expressed in terms of the magnitude of the angular momentum as

$$\omega = \frac{L}{mr^2} \qquad (8.8.5)$$

Substituting this into the energy conservation relation (8.8.2), we obtain

$$U = \tfrac{1}{2}m\left(v_r{}^2 + \frac{L^2}{m^2 r^2}\right) - \frac{GmM}{r} \qquad (8.8.6)$$

or, finally,

$$U - \tfrac{1}{2}mv_r{}^2 = \frac{L^2}{2mr^2} - \frac{GmM}{r} \qquad (8.8.7)$$

In this equation, the only variable quantities are r and v_r, since U and L^2 are constant. The right-hand side of (8.8.7) contains only the variable r and has the form

$$f(r) = \frac{a}{r^2} - \frac{b}{r} \qquad (8.8.8)$$

where a and b are constants. The functions a/r^2 and $-b/r$ are plotted versus r in Fig. 8.11, and the function $f(r)$ in (8.8.8) is clearly representable as the sum of the ordinates of the two functions plotted in Fig. 8.11 for every value of r. Now, whatever the values of a and b may be, for sufficiently *small* values of r, a/r^2 will be much larger in magnitude than b/r, simply because $1/r^2$ increases faster than $1/r$ as r approaches zero. Likewise, for sufficiently *large* values of r, b/r will be much larger than a/r^2, because $1/r^2$ falls off faster than $1/r$ as r becomes very large. The function in (8.8.8), and therefore the function on the right-hand side of (8.8.7), will thus resemble a/r^2 for small values of r and $-b/r$ for large values of r, and, in fact, must have the general form shown in Fig. 8.12, where the quantity $U - \tfrac{1}{2}mv_r{}^2$ is shown plotted against r.

Now let us examine the expression $U - \tfrac{1}{2}mv_r{}^2$. The quantity U is a constant, which may be positive, zero, or negative. Since the kinetic energy $mv^2/2$ is always positive and the potential energy $-GmM/r$ is always negative, we may conclude that when U is positive, the kinetic energy always exceeds the potential energy. When U is negative, the potential energy always exceeds the kinetic energy; and when U is zero, the kinetic and potential energies are always equal in magnitude. Also, since $mv_r{}^2/2$ is always positive, irrespective of the sign of v_r, the quantity $U - \tfrac{1}{2}mv_r{}^2$ is always *less* than U itself, except when $v_r = 0$, in which case it reaches its maximum possible value, becoming equal to U.

Referring to Fig. 8.12, let us first consider the case where the total energy U of mass m has a fixed negative value U_1. At some arbitrary time, the

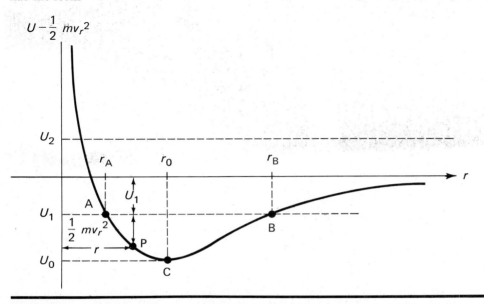

FIGURE 8.12. Schematic diagram illustrating bounded and unbounded orbital paths.

quantity $U_1 - \frac{1}{2}mv_r^2$ will have a value such as that associated with point P, which in any event must be more negative than U_1. At this time, the length of the radius vector will be the r-coordinate of P in the diagram. When v_r ($= dr/dt$) is zero, the distance r will have either a maximum or a minimum value, and at the same time the quantity $U_1 - \frac{1}{2}mv_r^2$ will attain the value U_1. This situation corresponds to the points A and B in Fig. 8.12. During the course of motion, therefore, the point P, whose coordinates are determined by the values of r and $U_1 - \frac{1}{2}mv_r^2$, moves back and forth along the curve, oscillating between the extremum points A and B which define its minimum and maximum distance from the origin. The radial velocity v_r at an arbitrary stage of the motion associated with point P can be determined from the quantity $\frac{1}{2}mv_r^2$, which is shown in the diagram, or from Eq. (8.8.7). The angular velocity ω can be found from (8.8.5). The resultant velocity v can then be found from

$$v^2 = v_r^2 + v_\theta^2 = v_r^2 + r^2\omega^2 \tag{8.8.9}$$

The actual equation of the orbital path can be obtained by writing $d\theta/dt$ for ω and dr/dt for v_r in Eqs. (8.8.5) and (8.8.6) and then eliminating the quantity dt between the two equations. One is then left with a differential equation relating $dr/d\theta$ to the variables r and θ. Unfortunately, the solution of this equation is a complicated and difficult task, and we shall not attempt to reproduce it here. Nevertheless, the final results obtained are quite simple, and their principal physical characteristics are clearly understandable from an analysis of Fig. 8.12, without going through the mathematical trauma of actually obtaining the equation of the orbital path.

Suppose we start with a body of mass m which has a certain *fixed* value of angular momentum L. The minimum possible energy such a body can have is U_0, corresponding to point C on the curve shown in Fig. 8.12. In this case, r can have only the fixed value r_0; the orbit is, therefore, circular and the radial velocity v_r is zero.

Also, since $v_r = 0$, and since point C corresponds to the minimum of the curve in Fig. 8.12, using (8.8.7) we may write

$$\left(\frac{dU}{dr}\right)_{r=r_0} = 0 = \frac{d}{dr}\left(\frac{L^2}{2mr^2} - \frac{GMm}{r}\right)_{r=r_0}$$

$$= \frac{GMm}{r_0^2} - \frac{L^2}{mr_0^3} \tag{8.8.10}$$

from which

$$r_0 = \frac{L^2}{GMm^2} \tag{8.8.11}$$

expresses the radius of the orbit. Finally, since $L = mr^2\omega$, this may be written as

$$\omega^2 = \frac{GM}{r_0^3} \tag{8.8.12}$$

or, since $\omega = 2\pi/T$, where T is the orbital period,

$$T^2 = \frac{4\pi^2 r_0^3}{GM} \tag{8.8.13}$$

This equation is identical with (8.4.9) and expresses Kepler's third law.

Some important information about the respective values of kinetic and potential energies of bodies in circular orbits can also be obtained from this treatment. By definition, the body's potential energy is

$$U_{0p} = -\frac{GMm}{r_0} \tag{8.8.14}$$

Also, since for a circular orbit $v_r = 0$, the total energy, according to (8.8.7), must be given by

$$U_0 = \frac{L^2}{2mr_0^2} - \frac{GMm}{r_0} \tag{8.8.15}$$

But since $L^2 = GMm^2 r_0$, from (8.8.11), this can also be written

$$U_0 = \frac{GmM}{2r_0} - \frac{GmM}{r_0} = -\frac{GmM}{2r_0} \tag{8.8.16}$$

It is now clear that the potential energy can be expressed as

$$U_{0p} = 2U_0 \tag{8.8.17}$$

and that the kinetic energy must be given by

$$U_{0k} = U_0 - U_{0p} = U_0 - 2U_0 = -U_0 = \frac{GmM}{2r_0} \tag{8.8.18}$$

When the total energy exceeds the minimum value U_0, as shown, for example, for the case $U = U_1$ referred to previously, the radial velocity is no longer zero, and the radius vector varies in magnitude between the values r_A and r_B in Fig. 8.12, as discussed earlier. In this case, one may show that the actual orbit is an *ellipse* with the mass M at one of the foci, as illustrated in Fig. 8.13. Kepler's second law follows also from the fact that since the gravitational force is always directed along the radius vector r, it can give rise to no torque. Angular momentum is, therefore, conserved and

$$L = mr^2 \frac{d\theta}{dt} = \text{const.} \tag{8.8.19}$$

But the element of area swept out by the radius vector during a small angular displacement $d\theta$ is given (in

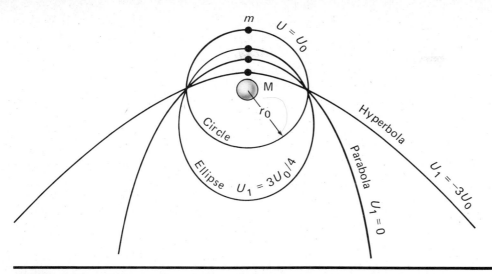

FIGURE 8.13. Different orbitals paths corresponding to the same value of angular momentum.

polar coordinates) by

$$dA = \tfrac{1}{2}r^2\, d\theta \qquad\qquad (8.8.20)$$

whence

$$\frac{dA}{dt} = \tfrac{1}{2}r^2\frac{d\theta}{dt} = \frac{L}{2m} \qquad\qquad (8.8.21)$$

which is also constant, according to (8.8.19).

Projectile motion, arising when an object is projected into the air near the earth's surface, is an important example of this type of orbital motion. The flight path of the projectile is an ellipse one focus of which is *at the center of the earth*. The path we observe is a portion of the ellipse near the other end. This "orbit" is, of course, intersected by the earth's surface, and the orbital motion of the object is interrupted permanently when it strikes the earth. The situation is illustrated in Fig. 8.14. If the earth's total mass were concentrated in a tiny region at its center instead of being distributed throughout a large spherical volume, an object projected upward from point A in the figure would trace out the complete elliptic path illustrated.

But haven't we already treated the case of projectile motion in section 3.3, and didn't we see there that projectile paths are *parabolic* rather than elliptical? So we did, but in section 3.3 we made the simplifying assumption that the acceleration of gravity was constant and did not vary either in magnitude or in direction during the projectile's path. In this section, we no longer make this assumption but regard the force of gravity always as being described by Newton's inverse square law. If the projectile never gets very far from the earth's surface, the assumption of constant g is a good one, and the parabolic path predicted by the calculations of section 3.3 and the elliptical one obtained from exact calculations where this assumption is no longer made are very nearly the same. Under these circumstances the orbital ellipse is very long and thin, and only the very end (the part AOB in Fig. 8.14) projects beyond the earth's surface. But it can be shown that the very end portion of any such ellipse is nearly parabolic in shape. The two treatments, therefore, lead to essentially the same results for the part of the orbital path near the earth's surface. But if the orbital path of the projectile extends upward so far that the variation of the force of gravity over the path is appreciable, then this part of the orbit is no longer approximately parabolic, the treatment of section 3.3 is no longer

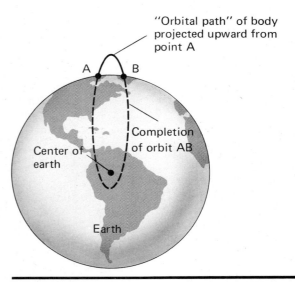

FIGURE 8.14. "Orbital path" of a body projected upward near the earth's surface.

valid, and exact calculations based upon Newton's law of universal gravitation must be used instead.

Returning to the curve shown in Fig. 8.12, consider now the case where the total energy of the system approaches zero. In this case, the minimum radial distance r_A remains finite while r_B becomes infinite. In effect, the major axis of our elliptical orbit has become infinitely large. Now we may show that if we start with an ellipse one end of which always remains at the origin and allow the major axis to become infinitely large, the elliptical curve approaches a parabola. The orbit in this limit is, therefore, parabolic. The motion is no longer periodic but rather one in which the mass m approaches the larger object along a parabolic path, reaches the closest distance of approach r_A, and recedes once more toward infinity, as illustrated in Fig. 8.13. Since the total energy is always zero in this orbit, the kinetic energy is always equal in magnitude, but opposite in sign, to the potential energy. The velocity is, therefore, related to the radial distance r by

$$\tfrac{1}{2}mv^2 = \frac{GmM}{r}$$

$$v = \sqrt{\frac{2GM}{r}}$$

It is important to note that this is an *entirely different kind of parabolic orbit* from those which were

discussed in section 3.3 for the case where g is constant. For this orbit, g is not constant either in magnitude or direction, and the fact that the orbit is parabolic in spite of all this is a rather remarkable kind of mathematical accident!

Finally, it is possible to have orbital paths for which the total energy is positive, corresponding to the case $U = U_2$ in Fig. 8.12. For these orbits, the positive kinetic energy always exceeds the negative potential energy in magnitude. Once more, the nearest distance of approach, r_A, is finite, while the object is able to escape to infinity just as in the parabolic case. It may be shown mathematically that these orbital paths are *hyperbolic* in form, as illustrated in Fig. 8.13.

The situation is precisely analogous to the one discussed in Example 8.7.1 for a body projected vertically upward. Here, also, for small values of kinetic energy, the mass m never escapes from the gravitational attraction of the larger one but traces out a periodic orbit having an elliptic form. When the kinetic energy reaches a certain critical value, the mass m "escapes" to infinity via a parabolic orbit, while for still more energetic orbits the mass may still escape to infinity, but now in hyperbolic orbits.

The various orbital possibilities are illustrated in Fig. 8.15, where the trajectories A, B, C, D, E, F, and G represent the paths taken by an object projected horizontally from a very high tower with initial tangential velocities which undergo successive in-

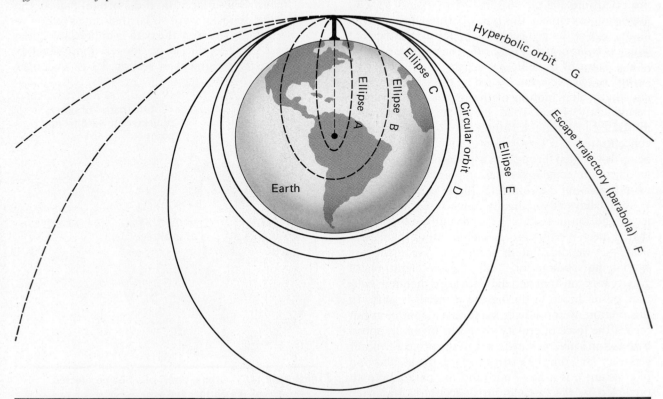

FIGURE 8.15. Orbital paths of bodies projected horizontally with different velocities' All these orbits have different angular momenta.

creases from A to G. The difference between this case and the one shown in Fig. 8.13 is that here, the minimum distance of approach is the same for all orbits and the angular momentum undergoes a systematic increase from A to G, while in Fig. 8.13, all orbits have the same angular momentum but exhibit different minimum radial distances, since the product $mr^2\omega$ (i.e., mvr) has to remain constant.

Finally, we may remark that these different orbital paths all arise from the behavior of mass particles acted upon by gravitational forces alone. Although they differ in many respects from our limited intuitive notions of how freely falling objects behave, gleaned from our circumscribed terrestrial experience, they provide a much more general picture of how bodies may behave when allowed to "fall." It may be useful at this point to recall the discussion of the "falling" of the gyroscope given at the end of section 7.10 and to interpret planetary and satellite motions also from this same point of view.

EXAMPLE 8.8.1

A satellite weighing 1000 pounds at the earth's surface is projected into a circular orbit whose height above the earth's surface is 100 miles. Find its period of rotation about the earth, its linear and angular velocities, and the amount of work required to put it in orbit.

The period can be found at once from Eq. (8.8.13), using the earth's mass for M and the value of G determined from the Cavendish experiment. From a purely calculational point of view, however, it is easier to express the quantity Gm_e in (8.8.13) in terms of the surface gravitational acceleration g_0 and the earth's radius r_e by using (8.7.2); in this way, replacing Gm_e in (8.8.13) by $g_0 r_e^2$, we may write (8.8.13) as

$$T^2 = \frac{4\pi^2 r_0^3}{g_0 r_e^2} = \frac{(4)(3.1416)^2(4060)^3(5280)^3}{(32.2)(3960)^2(5280)^2}$$

$$= 27.6 \times 10^6 \text{ sec}^2$$

$$T = 5256 \text{ sec, or } 87.6 \text{ min}$$

In these calculations, the earth's radius r_e has been taken to be 3960 miles, the orbit radius r_0 then being $3960 + 100 = 4060$ miles.

The angular velocity can be obtained from (8.8.12), again substituting $g_0 r_e^2$ for Gm_e, or even more easily by recalling that in any event

$$\omega = \frac{2\pi}{T} = \frac{6.2832}{5.256 \times 10^3} = 1.195 \times 10^{-3} \text{ rad/sec}$$

The linear velocity will be

$$v = r_0\omega = (4060)(5280)(1.195 \times 10^{-3})$$

$$= 2.562 \times 10^4 \text{ ft/sec}$$

which is about 17,500 mph, or 4.85 miles per second. These figures should be familiar to anyone who has witnessed orbital satellite flights via television or other news media.

According to the law of conservation of energy (and the work–energy theorem), the amount of work done in setting such a satellite into its orbit will be equal to its increase of kinetic energy, plus any increase in potential energy it may experience between its initial state of rest on the earth's surface and its final orbital condition. We may, therefore, write

$$W = \Delta U_k + \Delta U_p$$
$$= (U_{kf} - U_{ki}) + (U_{pf} - U_{pi}) \qquad (8.8.24)$$

But $U_{kf} = \frac{1}{2}mv^2$, $U_{ki} = 0$, $U_{pf} = -Gm_e m/r_0$, and $U_{pi} = -Gm_e m/r_e$; under these circumstances, (8.8.24) becomes

$$W = \frac{1}{2}mv^2 + Gm_e m\left(\frac{1}{r_e} - \frac{1}{r_0}\right)$$

$$= \frac{1}{2}mv^2 + g_0 r_e^2 m\left(\frac{1}{r_e} - \frac{1}{r_0}\right)$$

$$= \frac{1}{2}mv^2 + g_0 r_e m\left(\frac{r_0 - r_e}{r_0}\right)$$

$$= \left(\frac{1}{2}\right)\left(\frac{1000}{32.2}\right)(2.562 \times 10^4)^2$$

$$+ (32.2)(3960)(5280)\left(\frac{1000}{32.2}\right)\left(\frac{4060 - 3960}{4060}\right)$$

$$= (1.019 \times 10^{10}) + (5.15 \times 10^8)$$

$$= 1.070 \times 10^{10} \text{ ft-lb}$$

It is evident, in this case, that most of the work done on the satellite is expended in accelerating it to its final orbital velocity and that only a relatively small amount (5.15×10^8 ft-lb) is expended in increasing its potential energy.

EXAMPLE 8.8.2

Calculate the radius of the moon's orbit around the earth (assuming it to be circular) from Newton's law of universal gravitation, and from the observed fact that the period is 27.32 days. What justification have we for assuming that the moon's orbit is approximately circular?

From (8.8.13) and (8.7.2), we know that

$$T^2 = \frac{4\pi^2 r_0^3}{Gm_e} = \frac{4\pi^2 r_0^3}{g_0 r_e^2} \qquad (8.8.25)$$

Solving for the orbital radius r_0, this gives

$$r_0 = \left(\frac{g_0 r_e^2 T^2}{4\pi^2}\right)^{1/3}$$

$$= \left(\frac{(32.2)(3960)^2(5280)^2(27.32)^2(86400)^2}{4(3.1416)^2}\right)^{1/3}$$

$$= 1.257 \times 10^9 \text{ ft, or } 238,000 \text{ miles}$$

It is important to note that this calculation, like those of the previous example, was first accomplished by Newton without explicit use of measured values of G or m_e, neither of which Newton, or anyone else, had until Cavendish came along. The physical justification for the assumption that the orbit is circular rests upon two observations. First of all, if the moon were appreciably further from the earth at some phases of its orbit than at others, the *apparent size* of the moon's disc in the sky would vary from time to time, and no such variation is observed. Secondly, according to Kepler's second law, if the moon's orbit were not circular, its orbital velocity would be greater when it is close to the earth than when it is further away—but no large variation in the moon's steady progress through the sky can be detected throughout its period.

*EXAMPLE 8.8.3

Find the relations expressing the semimajor and semiminor axes of an elliptic orbit in terms of the minimum and maximum radial distances r_A and r_B of Fig. 8.12. Express the distance c between the ellipse center and the foci and the eccentricity e (equal to c/a) in terms of r_A and r_B. Find expressions for r_A and r_B in terms of the actual orbital energy U_1 and the energy U_0 and radius r_0 associated with a circular orbit of equal angular momentum. Finally, express a, b, c, and e in terms of U_0, U_1, and r_0.

The essential geometry of the ellipse is shown in Fig. 8.16. The geometric property which defines the ellipse is that the sum of the distances between any point on the curve and two fixed points f and f' is always the same. The points f and f' are called the foci of the ellipse. Let the semimajor axis be designated by a, the semiminor axis by b, and the distance between the origin and either focus by c. Then, since the distances f'P + Pf (equal to $2a$) and f'Q + Qf must be the same, it is clear that f'Q = Qf = a. But, applying the Pythagorean theorem to the right triangle OQf,

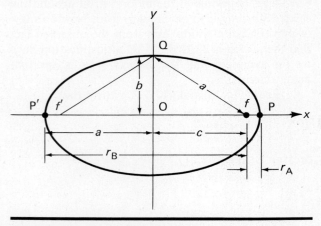

FIGURE 8.16. Geometry of the ellipse.

it follows that

$$c^2 = a^2 - b^2 \qquad (8.8.26)$$

From Fig. 8.16, it is immediately apparent that

$$r_A = a - c \qquad (8.8.27)$$

$$r_B = a + c \qquad (8.8.28)$$

These two equations can be solved for the quantities a and c:

$$a = \tfrac{1}{2}(r_B + r_A) \qquad (8.8.29)$$

$$c = \tfrac{1}{2}(r_B - r_A) \qquad (8.8.30)$$

From (8.8.26), the semiminor axis b can be expressed as

$$b^2 = a^2 - c^2$$
$$= \tfrac{1}{4}[(r_B^2 + 2r_A r_B + r_A^2) - (r_B^2 - 2r_A r_B + r_A^2)]$$
$$b = r_A r_B \qquad (8.8.31)$$

The eccentricity e is defined by the relation

$$e = c/a \qquad (8.8.32)$$

From Eqs. (8.8.29) and (8.8.30), this can be written as

$$e = \frac{r_B - r_A}{r_B + r_A} \qquad (8.8.33)$$

At either extremity of the elliptical orbit, when r has the value r_A or r_B, the radial velocity v_r is zero. But, therefore, from (8.8.7) the energy U_1 may be written

$$U_1 = \frac{L^2}{2mr^2} - \frac{GmM}{r} \qquad r = r_A, r_B \qquad (8.8.34)$$

The angular momentum L can be expressed by (8.8.11):

$$L^2 = GMm^2 r_0 \qquad (8.8.35)$$

where r_0 is the radius of the *circular* orbit having this same angular momentum. If we multiply the right-hand side of this above and below by $2r_0$, we find

$$L^2 = \frac{GMm}{2r_0} 2mr_0^2 = -2mr_0^2 U_0 \qquad (8.8.36)$$

where $U_0 = -GmM/2r_0$ is, from (8.8.16), the total *energy* associated with the circular orbit of radius r_0 and angular momentum L.

In (8.8.34), in the first term on the right, we shall replace L^2 by the equivalent value $-2mr_0^2 U_0$ given by (8.8.36); in the second term, we shall multiply numerator and denominator by $2r_0$ and in the resulting expression replace the quantity $-GmM/2r_0$ by U_0.

Equation (8.8.34) then becomes

$$U_1 = \frac{2r_0 U_0}{r} - \frac{r_0{}^2 U_0}{r^2} \qquad r = r_A, r_B \qquad (8.8.37)$$

Multiplying both sides of this by $r^2/r_0{}^2$ and rearranging slightly, this may be written as

$$\frac{r^2}{r_0{}^2} - \frac{2U_0}{U_1}\left(\frac{r}{r_0}\right) + \frac{U_0}{U_1} = 0 \qquad r = r_A, r_B \qquad (8.8.38)$$

This is a quadratic equation which can be solved for r/r_0; the two values of r so obtained obviously represent r_A and r_B. Using the quadratic formula to solve (8.8.38), it is apparent that

$$\frac{r}{r_0} = \frac{U_0}{U_1} \pm \sqrt{\left(\frac{U_0}{U_1}\right)^2 - \frac{U_0}{U_1}} = \frac{U_0}{U_1}\left[1 \pm \sqrt{1 - \frac{U_1}{U_0}}\right] \qquad (8.8.39)$$

Since r_B is larger than r_A, the root corresponding to the $+$ sign must represent r_B, the one corresponding to the $-$ sign, r_A. Therefore,

$$\frac{r_A}{r_0} = \frac{U_0}{U_1}\left[1 - \sqrt{1 - \frac{U_1}{U_0}}\right] \qquad (8.8.40)$$

$$\frac{r_B}{r_0} = \frac{U_0}{U_1}\left[1 + \sqrt{1 - \frac{U_1}{U_0}}\right] \qquad (8.8.41)$$

For elliptic orbits, both U_0 and U_1 are negative, as illustrated by Fig. 8.12, but U_1 is always smaller in magnitude. Therefore, the ratio U_1/U_0 is less than, or at best equal to, unity, and the quantity within the radical cannot be negative. Of course, where U_1 equals U_0, both (8.8.40) and (8.8.41) give $r_A = r_B = r_0$; this is the case of the circular orbit.

The semimajor and the semiminor axes a and b can be obtained by substituting the values of r_A and r_B from Eqs. (8.8.40) and (8.8.41) into (8.8.29) and (8.8.31); this can be done in a perfectly straightforward way, the result being

$$a = \tfrac{1}{2}(r_A + r_B) = \frac{U_0}{U_1} r_0 \qquad (8.8.42)$$

$$b = \sqrt{r_A r_B} = \sqrt{\frac{U_0}{U_1}} \cdot r_0 \qquad (8.8.43)$$

The eccentricity e can be obtained in the same way from (8.8.33), this equation taking the form

$$e = \frac{r_B - r_A}{r_B + r_A} = \sqrt{1 - \frac{U_1}{U_0}} \qquad (8.8.44)$$

For the circular orbit $(U_1 = U_0)$, e becomes zero, corresponding to a circle.

When U_1 approaches *zero*, the quantity $[1 - (U_1/U_0)]^{1/2}$ approaches unity. This quantity can be expanded in a series using the binomial theorem, giving

$$\left(1 - \frac{U_1}{U_0}\right)^{1/2} = 1 - \frac{1}{2}\frac{U_1}{U_0}$$
$$+ \left(\frac{1}{2}\right)\left(-\frac{1}{2}\right)\left(\frac{1}{2!}\right)\left(\frac{U_1}{U_0}\right)^2 + \cdots \qquad (8.8.45)$$

In the limit as $U_1 \to 0$, the terms containing $(U_1/U_0)^2$, $(U_1/U_0)^3$, etc., become much smaller than the term containing U_1/U_0, and in this limit may be neglected altogether. Therefore, as $U_1 \to 0$, the radical in (8.8.40) may be replaced by the quantity $1 - (U_1/2U_0)$, according to (8.8.45), the result being

$$\lim_{U_1 \to 0} \frac{r_A}{r_0} = \lim_{U_1 \to 0} \frac{U_0}{U_1}\left[1 - \sqrt{1 - \frac{U_1}{U_0}}\right]$$
$$= \lim_{U_1 \to 0} \frac{U_0}{U_1}\left[1 - \left(1 - \frac{U_1}{2U_0}\right)\right]$$

or

$$\lim_{U_1 \to 0} \frac{r_A}{r_0} = \frac{U_0}{U_1}\frac{U_1}{2U_0} = \frac{1}{2} \qquad (8.8.46)$$

In this limit, therefore, $r_A \to \tfrac{1}{2}r_0$ and, clearly from (3.8.41), $r_B \to \infty$. This is the case of the *parabolic* orbit. From (8.8.43), it is evident that the eccentricity of such an orbit is unity, and, indeed, this is the value of eccentricity associated with a parabola.

If U_1 becomes positive, r_A becomes less than $\tfrac{1}{2}r_0$ and the eccentricity e becomes greater than unity, corresponding to a *hyperbolic* orbit.

SUMMARY

It is important to understand the concept of the celestial sphere, shown in Fig. 8.1, and to know the important facts about the apparent motion of sun, moon, stars, and planets discussed in section 8.2. It is also necessary to understand the main features of the geocentric and heliocentric concepts of the universe.

Kepler's laws of planetary motion are:

1. Each planet's motion is periodic, in a closed elliptic orbit with the sun at one focus.
2. The line joining the sun with any planet sweeps out equal areas in equal time intervals in all parts of the orbit.
3. The square of the time required for any planet to traverse its orbit about the sun is proportional to the cube of the orbit's mean radius.

Newton's law of universal gravitation states:

A force of attraction exists between any two mass particles, directed along the line joining them, whose magnitude is directly proportional to each of the two masses and inversely proportional to the square of the distance between them.

This may be put in the form of an equation having the form

$$F = \frac{Gm_1m_2}{r^2}$$

The constant of proportionality G in this equation is referred to as the universal gravitation constant; its value in the mks system is 6.673×10^{-11} N m^2/kg. Kepler's laws follow from Newton's laws of motion when the law of universal gravitation is used to express the forces between the sun and the planets.

The potential energy of a point mass m_2 at a distance r from a fixed point mass m_1 at the origin, associated with their gravitational interaction, is

$$U_p(r) = -\frac{Gm_1m_2}{r}$$

In writing this expression, it is assumed that the potential energy of mass m_2 is zero when it is infinitely distant from m_1.

The gravitational forces on a point mass from several nearby mass particles is the *vector* sum of the individual forces caused by each of the particles. The potential energy is the *scalar* sum of the individual potential energies attributable to each of the nearby mass particles.

The gravitational force between two spherical masses is exactly that which would exist between two point particles having the same masses located at the centers of the spherical objects. The gravitational force inside a *hollow* spherical mass is zero.

The acceleration of gravity due to the earth's gravitational field, as a function of the distance r from the earth's center, is given by

$$g(r) = Gm_e/r^2 = g_0 r_e{}^2/r^2$$

where m_e is the mass of the earth and r_e its radius. It is evident that at $r = r_e$, the acceleration of gravity reduces to $g_0 = 9.8$ m/sec^2 = 32 ft/sec^2.

The characteristics of *circular* planetary and satellite orbits are particularly simple. They can most easily be arrived at by equating the centripetal force $mr\omega^2$ to the gravitational force GmM/r^2, from which the angular velocity

$$\omega^2 = GM/r^3$$

follows at once. In this expression, M refers to the large central mass about which the planet or satellite revolves. The orbital period and linear velocity then follow at once from $T = 2\pi/\omega$ and $v = r\omega$.

Orbital paths can be elliptical (which includes circular as a special case), parabolic, or hyperbolic.

Elliptical orbits are closed paths, and the orbiting object retraces the elliptical orbit again and again, never escaping the gravitational field in which it finds itself. Such orbits are characteristic of objects whose kinetic energy is always insufficient to overcome the attractive potential energy of the gravitational field and allow the body to "escape" to infinity. The parabolic orbit is characteristic of bodies with just sufficient energy to overcome the attractive potential energy, while objects having more than enough energy to "escape to infinity" do so along hyperbolic paths.

QUESTIONS

1. What is the gravitational acceleration g at a distance of $1.1r_e$ from the center of the earth? What value does it have at $0.5r_e$?
2. The temperature at the surface of a planet as well as the gravitational field determine whether the planet can retain an atmosphere. Comment on this statement.
3. Can you explain briefly why the mutual gravitational potential energy of two objects must always be negative?
4. Is there any point between the earth and the moon where a body would be weightless? Would it be massless also?
5. Explain why the gravitational effect of the moon is more important than that of the sun in producing tides, even though the gravitational force between the sun and the earth is much larger than that between the moon and the earth.
6. Two physics students observe a coconut which falls from a tree and strikes the earth. One says that the coconut falls downward because of the force of gravity. The other comments that the force of gravity the coconut exerts on the earth is just as strong as that which the earth exerts on the coconut, and that the earth, therefore, moves upward to meet the coconut. Discuss the reasoning of the two students and decide which explanation is correct.
7. The gravitational force law is exactly an inverse square law. If the exponent of r differs slightly from $-2.000\ldots$, would Kepler's laws remain valid? Would the force still be conservative?
8. Kepler's second law emerges from the fact that a certain dynamic quantity is conserved. What is this quantity?
9. If a spacecraft in a circular orbit about the earth loses some of its kinetic energy, what happens to the orbit?
10. Describe what happens to the period of a pendulum as it is transported from the surface of the earth to the surface of the moon.
11. What is the escape velocity from the surface of Saturn?
12. If a new planet with a period of ten years were to be discovered, what would be its mean distance from the sun?

PROBLEMS

1. What is the gravitational force of attraction between two 50-kg masses separated by 10 m? Compare this to the weight of one 50-kg mass at the surface of the earth.

2. Find the force, in newtons, of gravitational attraction between **(a)** the sun and the earth, **(b)** the earth and the moon, **(c)** the sun and the moon, and **(d)** the sun and the planet Pluto. Use the numerical data from Table 8.1.

3. What is the value of Newton's gravitation constant G in the British system of units?

4. What is the force of gravitational attraction between two spherical masses weighing 300,000 tons whose centers are separated by a distance of 40 meters? Assume that the mass can be considered in this case to be concentrated at the centers of the spheres.

5. What initial acceleration would the masses in the preceding problem experience if they were released from rest and allowed to move under the influence of their mutual gravitational attraction alone?

6. What mass should two equal point masses have to experience a mutual gravitational attraction of 1.0 newton at a separation of 1.0 meter?

7. At what separation would two 1.0-kg point masses experience a gravitational attraction of 1.0 newton?

8. **(a)** What is the gravitational potential energy of two 50.0-kg masses separated by a distance of 1.0 meter? **(b)** What is the gravitational potential energy at a separation of 10.0 meters? **(c)** How much work must be done to increase the separation of the masses from 1.0 meter to 10.0 meters?

9. Find the gravitational potential energy of two protons separated by a distance of 10^{-13} cm.

10. **(a)** What is the gravitational potential energy of a person of mass 75 kg at the earth's surface? **(b)** What is the corresponding potential energy at a distance of 1 km above the earth's surface? **(c)** What is the potential energy at a distance of 2000 km above the earth's surface? **(d)** At a distance of 20,000 km?

11. What is the gravitational potential **(a)** at the earth's surface, **(b)** at an altitude of 1.0 km above the earth's surface, **(c)** at an altitude of 2000 km, and **(d)** at an altitude of 20,000 km? **(e), (f), (g), (h)** What is the magnitude of gravitational field intensity at the four altitudes referred to in **(a), (b), (c), (d)**?

12. Show that the gravitational field intensity is really just the local acceleration of gravity and that its units of newtons/kg can equally well be written as meters/sec².

13. **(a)** What is the magnitude and direction of the gravitational force on the small mass m in the figure if $M =$

50 kg, $d = 0.5$ meter, $z = 0.3$ meter, and $m = 20$ kg. The earth's gravitational field is assumed to be absent. **(b)** What is the gravitational potential energy of the small mass?

14. Work the preceding problem using the general symbols m, M, d, and z for the physical quantities involved, finding algebraic expressions for **(a)** the gravitational force on mass m and **(b)** its potential energy. **(c)** How much work must be done on mass m to move it from the point $(x = 0, z = 0)$ to infinity?

15. In the preceding problem, show that $F_z = -\partial U_p / \partial z$.

16. Three 40-kg masses are located at the corners of a square of side 1.20 meters. **(a)** Find the magnitude and direction of the gravitational force on a 24-kg mass placed at the fourth corner of the square. **(b)** What is the potential energy of the 24-kg mass? Assume the earth's gravitational field to be absent.

17. How much energy is required to allow a 1000-kg satellite to escape from the earth's gravitational field?

18. Show that the gravitational force on a point mass within a thin spherical shell of uniform density is zero. *Hint:* Use the same procedure employed in section 8.6 to derive Eqs. (8.6.10) and (8.6.11).

19. **(a)** Find the gravitational attraction between two uniform spheres of radii r_1 and r_2 and densities ρ_1 and ρ_2 when they are in contact. **(b)** Find the potential energy under these circumstances. **(c), (d)** By what factors do the attractive force and potential energy increase when the radii of the spheres are doubled?

20. **(a)** How large would two lead spheres of equal radius have to be to attract one another, when in contact, with a force of 1.00 newton? **(b)** What would the mass of each sphere be under these circumstances? Note that the density of lead is 11.3 g/cm³.

21. Show that the weight of an object of mass m at a height h above the surface of the earth is approximately $mg(1 - 2h/R_e)$ if $h/R_e \ll 1$.

22. A satellite weighing 1500 pounds at the earth's surface is put into a circular orbit 120 miles above the surface. Find its weight at this altitude. What is its period of rotation about the earth, its angular velocity, and its potential energy? How much work was required to put it into this orbit?

23. Our sun is losing mass due to radiation at the rate of 4 million tons per second. What is the fractional change of the gravitational force on the earth during a period of 1000 years? During this same period of time, how does the length of the year change? Assume as a first approximation that the orbit radius does not change.

24. At what distance from the earth, as measured along a line from the earth to the sun, would a body be weightless?

25. Because of the rotation of the earth, a scale reads a person's true weight only at the north and south poles. Find the fractional error if a person is weighed at the equator. Is the true weight larger or smaller than the apparent weight?

26. A satellite is in a circular orbit 12 km above the moon's surface. What is its orbital speed and its period of revolution?

27. At what altitude above the surface of the earth will a freely falling body have an acceleration of 4.9 m/sec²?

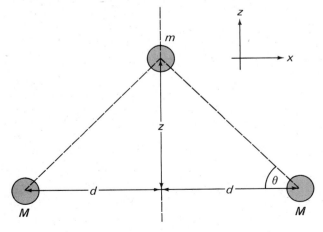

28. A satellite is sent up to record weather conditions at a certain location above the equator. If it remains stationary (or *synchronous*) with respect to this location, how high above the earth's surface must it be placed?

29. The first man-made satellite Sputnik I was put into orbit in 1957. Its mean distance from the center of the earth was 6950 km. Find the period of its rotation.

30. Three synchronous satellites used for communication are symmetrically placed above the equator. What is the distance between any two of them? Use the result of Problem 28.

31. Estimate the mass of the sun by using the period of revolution of the earth about the sun and the mean earth-to-sun distance.

32. A body is to be launched from the surface of the earth with a velocity large enough to escape from the gravitational field of the sun as well as the earth. Find the minimum velocity required to carry out this escape from the solar system.

33. How much work is required to move a 2-kg object from the surface of the earth to a point 5×10^4 km from the center of the earth? What is the gravitational potential energy at this separation?

34. Using the data given in Table 8.1 and the expression for the escape velocity, find the bodies from which the escape velocity is the largest and the smallest.

35. On a small planet, the gravitational acceleration is $g_{\text{planet}} = 0.1$ m/sec^2. A high jumper who can jump to a height $h_e = 7$ ft on earth tries the same jump on the small planet. How high will he jump and for how long are his feet above ground? Assume that the high jumper's initial velocity is independent of g.

36. M_A and M_B represent the masses of two planets whose centers are d meters apart and whose radii are R_A and R_B. Assume the planets are uniform spheres and $d > R_A + R_B$. A spaceship of mass m is traveling along the line of centers. Find the position x at which the net force on the spaceship due to the planets would be zero.

37. The residents of a spherical homogeneous asteroid of radius 4.0×10^4 m and mass 5.0×10^{16} kg have placed a 20-kg satellite in a circular orbit at 6.0×10^4 m above the surface of the asteroid. (a) Find the linear speed of the satellite in its orbit. (b) How much energy would be required to move the satellite into a higher circular orbit 1.1×10^5 m from the center of the asteroid?

38. Find the force in newtons with which a 100-kg mass would be attracted by Mars at the surface of that planet. Assume that the mass of Mars is 0.1 times the mass of Earth and the radius of Mars is 0.5 times the radius of Earth.

39. Assume a 10-kg mass weighs exactly 98 N at the surface of the earth. What would it weigh at a distance of 7 earth radii from the center of the earth?

40. A hypothetical planet has a mass $M_x = 5M_e$ and a radius $R_x = 2R_e$, where M_e and R_e are the mass and radius of the earth, respectively. (a) Find the acceleration of gravity at the surface of the planet and also at a distance R_x from the surface. (b) Find a formula for the speed of a satellite of the planet in a circular orbit at a distance of $3R_x$ from the center of the planet.

41. A pendulum of length 1 m makes an angle of $10°$ with the vertical due to the presence of a hypothetical point object of mass M placed a horizontal distance of 1 m from the equilibrium position of the pendulum in the absence of M. Find the mass of this object. Is it possible to carry out such an observation?

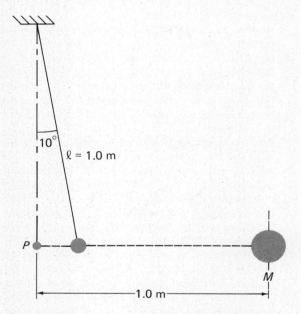

42. A satellite orbits the earth in an elliptic orbit. When closest to the earth, the satellite is at a height of 330 km above the earth's surface and has a velocity of 7740 meters/sec. Find (a) its height above the earth's surface and (b) its velocity when it is farthest from the earth. *Hint:* Use conservation of energy and angular momentum.

43. In the case of elliptic planetary orbits, show that, in general, the minimum radial distance r_0, the maximum radial distance r, and the velocities v_0 (at r_0) and v (at r) are related by the equations $r = r_0 v_0{}^2 / [(2GM/r_0) - v_0{}^2]$ and $v = (2GM/r_0 v_0) - v_0$.

OSCILLATORY MOTION

9.1 Introduction

In the preceding chapter, we studied the properties of planetary orbits and found that one of their most important characteristics is that they are *periodic*. By this we mean that after a certain lapse of time, referred to as the *period* of motion, the system regains its initial position and velocity, after which the motion begins to repeat itself. The motion, therefore, occurs in repetitive cycles each of which is exactly the same as any other. The earth, for example, traces out essentially the same elliptical orbit around the sun, year after year. Periodic motions occur frequently in nature and occupy a position of great importance in physics. Periodic motions that can be described in

terms of a single distance coordinate, such as the up-and-down motion of a mass suspended from a vertical spring, are referred to as *oscillatory* motions. The beat of a heart, the motion of a violin string, the swing of a pendulum, and the motion of the atoms in a crystal are all examples of oscillatory motion. Motion of this type will be the primary subject of interest in this chapter.

There are many different types of oscillatory motion, some of which are extremely complex. There is one form, however, that is very frequently encountered and is also very simple, called *simple harmonic motion*. We have already seen some examples of this type of motion, first in Chapter 5 in studying the motion of a mass resting on a frictionless horizontal table subject to forces exerted by an ideal spring, and again in Chapter 8 in investigating the motion of an elevator car in a vertical shaft drilled through the earth's center. In both cases, the displacement of the mass is a *sinusoidally varying function of time*, and this is the most important characteristic of simple harmonic motion. We shall direct most of our efforts in this chapter toward studying simple harmonic motion. Complex oscillatory motions that do not fall within this particularly simple category are referred to as *anharmonic* oscillations.

9.2 Simple Harmonic Motion

Simple harmonic motion arises whenever the resultant force on an object is *oppositely directed and directly proportional to its displacement* from a position of equilibrium in which the resultant force is zero. This may be stated mathematically as

$$F_x = -kx \qquad (9.2.1)$$

where k is the constant of proportionality between the force F_x and the displacement x from the origin, which in this case corresponds to the position of equilibrium, where the force is zero. The minus sign in Eq. (9.2.1) tells us that the force is in a direction *opposite* that of the displacement x, hence acts to the left when the displacement x is to the right and vice versa. This equation will be recognized as that which expresses the force exerted by an ideal spring as a function of its displacement from the equilibrium position, as discussed in Chapter 5, Example 5.6.2. We have already found in that example that sinusoidal simple harmonic motion is characteristic of such a system.

If we set F_x equal to mass times acceleration in (9.2.1), we can write that equation in the form

$$a_x = \frac{d^2x}{dt^2} = -\omega^2 x \qquad (9.2.2)$$

where ω^2, which in this case has the value

$$\omega^2 = k/m \qquad (9.2.3)$$

is a constant expressing the proportionality between acceleration a_x and displacement x. We shall soon see why it is advantageous to write this proportionality constant in this form. At this juncture, however, the important point is that *whenever the equation of motion of a body can be expressed mathematically in the form given by (9.2.2) above, the motion of the body will be simple harmonic motion.*

The truth of this assertion is evident from the results of Example 8.7.2, in which the equation of

motion of the elevator car is precisely of this form, the constant of proportionality ω having the value $\omega = \sqrt{g_0/r_e}$ given by (8.7.17). It can also be exhibited by *assuming* a solution which corresponds to a displacement that varies sinusoidally with time, of the form

$$x(t) = A \sin(\omega t + \delta) \qquad (9.2.4)$$

where A and δ are constants that express the maximum limits of displacement and the point in the cycle where the motion begins. This assumed solution can easily be shown to satisfy the equation of motion (9.2.2) by differentiating twice to obtain

$$v_x = \frac{dx}{dt} = A\omega \cos(\omega t + \delta) \qquad (9.2.5)$$

and

$$a_x = \frac{d^2x}{dt^2} = -A\omega^2 \sin(\omega t + \delta) \qquad (9.2.6)$$

and then substituting (9.2.4) and (9.2.6) into the equation of motion to obtain the identity

$$-A\omega^2 \sin(\omega t + \delta) = -\omega^2 [A \sin(\omega t + \delta)]$$

It is evident from all this that the sinusoidal motion described by (9.2.4) and illustrated in Fig. 9.1 *always* satisfies an equation of motion of the form of (9.2.2). It is equally true that whenever the equation of motion of a system can be expressed in the standard form given by (9.2.2), the motion must be simple harmonic.[1]

In Fig. 9.1a, we see the displacement of a particle that executes simple harmonic motion along the x-axis, as given by (9.2.4), plotted as a function of time. The motion can be thought of as that associated with the mass–spring system shown in Fig. 9.1c; as shown, the object's equilibrium point is chosen as the origin of the displacement coordinates measured in the x-direction. The body's velocity v_x ($=dx/dt$), as expressed by (9.2.5), is plotted as a function of time in Fig. 9.1b. The time $t = 0$ is *not* specified as the moment at which the object is at $x = 0$, which would admittedly be simpler, but rather at some odd instant when somebody started the stop watch.

Since the maximum and minimum values of the sine function are $+1$ and -1, respectively, the body oscillates between the displacement limits $+A$ and $-A$. The quantity A in (9.2.4), therefore, specifies these extreme limits of displacement and is referred to as the *amplitude* of motion.

The quantity ω in (9.2.4) specifies the rapidity

[1] It is important also to observe that in simple harmonic motion the acceleration is not constant but varies with time and displacement throughout the cycle. The "golden rules" of Chapter 2 that are so useful in describing uniformly accelerated motion are, therefore, definitely not applicable to simple harmonic motion!

$$x(t) = A \sin(\omega t + \delta) \quad \text{or} \quad \phi(t) = \phi_0 \sin(\omega t + \delta)$$

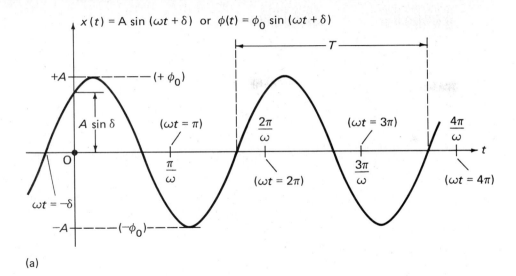

(a)

$$v(t) = A\omega \sin(\omega t + \delta)$$

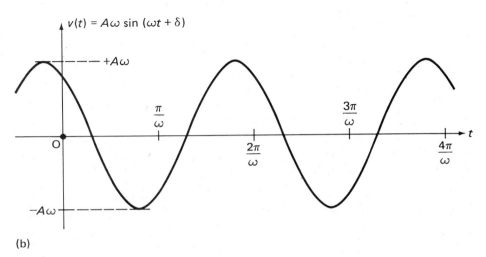

(b)

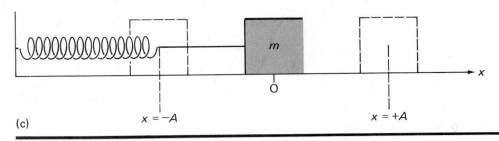

(c)

FIGURE 9.1. Displacement (a) and velocity (b) as functions of time in simple harmonic motion. (c) Mass–spring system that executes simple harmonic motion with angular frequency $\omega = \sqrt{k/m}$.

with which a complete cycle is executed or, more accurately, the number of cycles of motion that are completed per unit time. At time $t = 0$, for example, the angle $(\omega t + \delta)$ which forms the argument of the sine function in (9.2.4) has the value δ; at time $t = 1$ sec, it has the value $\omega + \delta$. In 1 sec, therefore, this angle increases by ω radians, and since one cycle of motion is completed as the angle increases by 2π radians,

$\omega/2\pi$ cycles will be executed in 1 sec. The *number of cycles of motion completed per second* is referred to as the *frequency* f, and from the above argument it is apparent that the frequency is related to ω by

$$f = \frac{\omega}{2\pi} \quad \text{or} \quad \omega = 2\pi f \tag{9.2.7}$$

279

The quantity ω itself is called the *angular frequency* of the motion. The units of ω are rad/sec, while those of f are cycles/sec (cps), which are often referred to by the not very descriptive term *hertz*, abbreviated Hz.[2]

The angle $(\omega t + \delta)$ of Eq. (9.2.4) is termed the *phase angle;* its value specifies the point in the cycle that is at hand at any given time. From (9.2.4) and (9.2.5), it is evident that whenever the phase angles associated with two different times differ by an integral multiple of 2π radians, the displacement and velocity of the particle will be the same and the two times will correspond to the same stage of motion within the cycle.

The *period* of the motion is the time required to complete one entire cycle. During this time, it is evident from what has been said above that the phase angle must increase by 2π radians to bring the system around to the same stage of the cycle. At time t_0, the phase angle will be $(\omega t_0 + \delta)$, while T seconds later, the phase angle will be $[\omega(t_0 + T) + \delta]$. But if the time T corresponds to one period, then these two phase angles must differ by exactly 2π radians, which means that

$$[\omega(t_0 + T) + \delta] - (\omega t_0 + \delta) = 2\pi$$

or

$$\boxed{T = \frac{2\pi}{\omega}} \tag{9.2.8}$$

From the way in which the period is defined, it is evident that the period of the motion is simply the *reciprocal of the frequency*. It is easy to see that (9.2.8) follows directly from (9.2.7) from this point of view.

The quantity δ in (9.2.4) is referred to as the *phase constant* of the motion; its value is determined by the choice of the point within the cycle at which we choose to "start the clock" or, in other words, to locate the point at which $t = 0$. At time $t = 0$, according to (9.2.4), (9.2.5), and (9.2.6),

$$x = x_0 = A \sin \delta \tag{9.2.9}$$

$$v_x = v_{x0} = A\omega \cos \delta \tag{9.2.10}$$

$$a_x = a_{x0} = -A\omega^2 \sin \delta = -A\omega^2 x_0 \tag{9.2.11}$$

The value of δ, then, can be determined by using the following criteria:

$$\sin \delta = \frac{x_0}{A} = -\frac{a_{x0}}{A\omega^2} \tag{9.2.12}$$

$$\cos \delta = \frac{v_{x0}}{A\omega} \tag{9.2.13}$$

or, dividing (9.2.12) by (9.2.13),

$$\tan \delta = \frac{\omega x_0}{v_{x0}} \tag{9.2.14}$$

If we know the value of the displacement x_0 and velocity v_{x0} at $t = 0$ and also A and ω, (9.2.12), (9.2.13), and (9.2.14) enable us to find the phase constant. If the phase constant is known together with A and ω, (9.2.9), (9.2.10), and (9.2.11) allow x_0 and v_{x0} to be found. We shall frequently find it advantageous to "start the clock" when the particle passes through the equilibrium position $x = 0$. This corresponds to choosing the value $\delta = 0$ for the phase constant, in which case (9.2.4) reduces to

$$x(t) = A \sin \omega t \tag{9.2.15}$$

The phase constant was chosen this way in discussing the motion of the mass–spring system in Example 5.6.2, though we did not say so specifically at the time. It is also often advantageous to start the clock when the particle has just come to rest at the extreme limit of its displacement at $x = +A$ and is poised to begin its return motion, so to speak. Under these circumstances, $x_0 = +A$ and $v_{x0} = 0$; and from (9.2.12) and (9.2.13), $\sin \delta = 1$ and $\cos \delta = 0$, corresponding to $\delta = 90° = \pi/2$ rad. Equation (9.2.4) now becomes

$$x(t) = A \sin\left(\omega t + \frac{\pi}{2}\right)$$

or

$$x(t) = A \cos \omega t \tag{9.2.16}$$

since $\sin(\theta + 90°) = \cos \theta$. The phase constant was chosen this way in discussing the motion of the car in the elevator shaft through the earth's center in Example 8.7.2.

EXAMPLE 9.2.1

A particle moves along the x-axis with simple harmonic motion, its displacement being given as a function of time by

$$x(t) = 3.5 \sin(8.25t + 0.75) \tag{9.2.17}$$

where distances are measured in meters and times in seconds. Find the amplitude of the motion, the angular frequency, the frequency, the period, and the phase constant. What is the displacement when $t = 0.5$ sec? What are the velocity and acceleration at that time? What is the maximum velocity? What is the maximum acceleration? What is the velocity when the particle is 2.8 meters from the equilibrium point? What is the acceleration at that point?

Equation (9.2.17) has the same standard mathematical form as (9.2.4). Comparing these two equations, it is easy to see that the amplitude must be $A = 3.5$ m, while the angular frequency is $\omega =$

8.25 rad/sec and the phase constant is $\delta = 0.75$ rad $= 43.0°$. The frequency, from (9.27), is

$$f = \frac{\omega}{2\pi} = \frac{8.25}{2\pi} = 1.313 \text{ Hz, or cps}$$

while the period is

$$T = \frac{2\pi}{\omega} = \frac{2\pi}{8.25} = 0.762 \text{ sec}$$

Since the displacement is given as a function of time by (9.2.17) the velocity and acceleration can be obtained by differentiating:

$$v_x(t) = dx/dt = (3.5)(8.25)\cos(8.25t + 0.75)$$
$$= (28.875)\cos(8.25t + 0.75) \text{ m/sec}$$
$$\text{(9.2.18)}$$

$$a_x(t) = dv_x/dt = -(28.875)(8.25)\sin(8.25t + 0.75)$$
$$= -(238.2)\sin(8.25t + 0.75) \text{ m/sec}^2$$
$$\text{(9.2.19)}$$

Substituting $t = 0.5$ sec into the above equations, we find

$$x(0.5 \text{ sec}) = (3.5)\sin(4.875 \text{ rad})$$
$$= (3.5)\sin 279.3° = -3.454 \text{ m}$$
$$v_x(0.5 \text{ sec}) = (28.875)\cos(4.875 \text{ rad})$$
$$= (28.875)\cos 279.3° = 4.675 \text{ m/sec}$$
$$a_x(0.5 \text{ sec}) = -(238.2)\sin(4.875 \text{ rad})$$
$$= -(238.2)\sin 279.3° = 235.1 \text{ m/sec}^2$$

The acceleration can be obtained even more simply from (9.2.2), which tells us that at all times

$$a_x = -\omega^2 x = -(8.25)^2(-3.454) = 235.1 \text{ m/sec}^2$$

The maximum velocity is attained when the cosine function in (9.2.18) reaches its maximum value of unity. It is easy to see, therefore, that this maximum velocity must be 28.875 m/sec. This value of velocity occurs when the phase angle $(8.25t + 0.75)$ equals zero or π radians, or when

$$8.25t = -0.75$$
$$t = -0.0909 \text{ sec}$$

and at intervals of $\frac{1}{2}T = 0.381$ sec thereafter. In the same way, the acceleration is maximum when the sine function in Eq. (9.2.19) attains its maximum value of unity, and therefore the maximum value of acceleration must be 238.2 m/sec².

To find the velocity when the displacement is 2.8 m, we may first substitute $x = 2.8$ m into (9.2.17) and solve for the time. In this way we find

$$\sin(8.25t + 0.75) = 0.8000$$

from which

$$8.25t + 0.75 = \sin^{-1}(0.800) = 0.9273 \text{ rad}$$
$$t = 0.0215 \text{ sec}$$

This value for the time is now substituted into Eq. (9.2.18), which gives us the velocity:

$$v_x = (28.875)\cos[(8.25)(0.0215) + 0.75]$$
$$= (28.875)(0.6000) = 17.33 \text{ m/sec}$$

In doing this, it is important to note that the arguments of all the sine and cosine functions must be expressed in *radians* rather than degrees.

The same result can be obtained more simply by squaring both sides of (9.2.5) to obtain

$$v_x^2 = A^2\omega^2 \cos^2(\omega t + \delta) = A^2\omega^2(1 - \sin^2(\omega t + \delta))$$

But, according to (9.2.4), $\sin^2(\omega t + \delta) = x^2/A^2$. Substituting this into the above equation, we find

$$\boxed{\begin{array}{l} v_x^2 = \omega^2(A^2 - x^2) \\ \text{or} \\ v_x = \pm\omega\sqrt{A^2 - x^2} \end{array}}$$
$$\text{(9.2.20)}$$

This equation allows us to express v_x directly as a function of the displacement x rather than the time. Substituting $\omega = 8.25$ rad/sec, $A = 3.5$ m, and $x = 2.8$ m into (9.2.20), we find at once that $v_x = 17.33$ m/sec, as before.

The acceleration when $x = 2.8$ m could be found by substituting $t = 0.0215$ sec into Eq. (9.2.19) and calculating a_x from that expression. It is simpler, however, to observe that from (9.2.2)

$$a_x = -\omega^2 x = -(8.25)^2(2.8) = -190.6 \text{ m/sec}^2$$

EXAMPLE 9.2.2

A frictionless mass–spring system such as that shown in Fig. 9.1c consists of a mass of 12 kg and an ideal spring of spring constant 4800 N/m. The mass is drawn aside from the equilibrium position a distance of 0.4 m and released. Show that the resulting motion is simple harmonic and find the amplitude, the angular frequency, the frequency, and the period. Write expressions for the displacement, velocity, and acceleration as functions of time and for the velocity and acceleration as functions of the displacement.

In this case, the equation of motion can be expressed as

$$F_x = -kx = m\frac{d^2x}{dt^2}$$
$$\text{(9.2.21)}$$

where x is the displacement from the equilibrium

position. This equation can be written in the *standard form*

$$\frac{d^2x}{dt^2} = -\omega^2 x \qquad (9.2.22)$$

if we identify ω as

$$\omega = \sqrt{\frac{k}{m}} \qquad (9.2.23)$$

The motion must, therefore, be simple harmonic motion whose angular frequency is given by the above expression. In this case, the angular frequency will be

$$\omega = \sqrt{\frac{k}{m}} = \sqrt{\frac{4800}{12}} = 20 \text{ rad/sec}$$

The frequency itself is

$$f = \frac{\omega}{2\pi} = 3.183 \text{ Hz, or cps}$$

while the period is

$$T = \frac{1}{f} = 0.3142 \text{ sec}$$

We sense intuitively from the initial conditions that the amplitude of the motion has to be 0.4 m, because whenever the spring's extension becomes that great, all the kinetic energy of motion the mass acquires as a result of the forces exerted on it by the spring is stored up again as potential energy. If the displacement were ever greater than 0.4 m, there would be more potential energy in the spring than was stored there initially; and since the system is a conservative one, there is, in the absence of external forces, no way in which this can occur. The validity of these statements is attested to mathematically by noting that at the instant the mass is released, $x = 0.4$ m and $v_x = 0$. But, according to (9.2.10), v_x can be zero only when $x = A$. Therefore, the amplitude has to be 0.4 m.

Knowing all these facts, we are almost in a position to write down an expression of the form of (9.2.4) relating displacement and time. The only thing we lack is the phase constant δ. If we start the clock at the moment the mass is released, then, at $t = 0$, $x_0 = A$ and $v_{x0} = 0$. According to (9.2.12) and (9.2.13), then, $\sin \delta = 1$ and $\cos \delta = 0$, which means that $\delta = \pi/2$. We may now write the expression relating displacement and time by substituting $A = 0.4$ m, $\omega = 20$ rad/sec, and $\delta = \pi/2$ into (9.2.4):

$$x = (0.4) \sin\left(20t + \frac{\pi}{2}\right) \text{ m} \qquad (9.2.24)$$

Since $\sin(\theta + 90°) = \cos\theta$, this could also be written very simply as

$$x = (0.4) \cos(20t) \text{ m} \qquad (9.2.25)$$

Equations relating velocity and acceleration with time can now be obtained, as before, by differentiating (9.2.24) or (9.2.25); for example,

$$v_x = dx/dt = (0.4)(20) \cos\left(20t + \frac{\pi}{2}\right) \text{ m/sec} \qquad (9.2.26)$$

$$a_x = dv_x/dt = -(0.4)(20)^2 \sin\left(20t + \frac{\pi}{2}\right) \text{ m/sec}^2 \qquad (9.2.27)$$

The equations relating acceleration and velocity to displacement are obtained from (9.2.2) and (9.2.20), from which we find

$$a_x = -\omega^2 x = -400x \text{ m} \qquad (9.2.28)$$

$$v_x = \pm\omega\sqrt{A^2 - x^2} = \pm 20\sqrt{0.16 - x^2} \text{ m/sec} \qquad (9.2.29)$$

Had we chosen to set the system in motion before starting the clock and then to start the clock as the mass passes the equilibrium position, we would then have required that $x_0 = 0$ when $t = 0$. Under these circumstances, (9.2.12) requires that $\sin \delta = 0$, or $\delta = 0$, and

$$x(t) = (0.4) \sin(20t) \text{ m} \qquad (9.2.30)$$

Had we started the clock when the mass was 0.15 m to the right of the equilibrium position on the outward-bound leg of its journey, we would have $x_0 = 0.15$ m and $A = 0.4$ m. In this case, (9.2.12) would tell us that $\sin \delta = 0.15/0.4 = 0.375$, for which $\delta = 22.02° = 0.3844$ rad, or $\delta = 157.98° = 2.7572$ rad. Since the time $t = 0$ corresponds to the outward leg of the cycle, we must choose the first of these values; the second corresponds to the later phase of the cycle when the mass is returning to the equilibrium position. In this case, then,

$$x(t) = (0.4) \sin(20t + 0.3844) \text{ m}$$

Finally, had we been told that the clock had been started, let us say, 0.032 sec after the mass moved to the right past the equilibrium position, the time at which the mass is at the equilibrium position would be $t = -0.032$ sec. But at that instant the phase angle $\omega t + \delta$ is zero. Therefore, substituting $\omega = 20$ rad/sec and $t = -0.032$ sec, we find

$$\omega t + \delta = (20)(-0.032) + \delta = 0$$
$$\delta = 0.64 \text{ rad}$$

which allows us to express the displacement as

$$x(t) = (0.4) \sin(20t + 0.64) \text{ m}$$

From all this, it is apparent that the phase constant δ can be determined *only when the displacement and velocity are specified at a given time or when some equivalent information is provided*, as in the example directly above. Usually, this information refers to the initial state of motion at $t = 0$, but information about the displacement or velocity at any other time will serve equally well. In this regard, the student is invited to show that for the system discussed in this example, if we are given that the displacement is $x = 0.12$ m at time $t = 0.027$ sec, the velocity then being 7.632 m/sec, the phase constant δ must have the value of -0.2352 rad. In doing this, it is important to recognize that Eqs. (9.2.12), (9.2.13), and (9.2.14) cannot be used, since they relate δ to the *initial* values of position and velocity at $t = 0$. One should proceed instead by substituting the given information into (9.2.4) and solving for δ.

9.3 Kinetic and Potential Energy in Simple Harmonic Motion

This subject has been extensively treated in Chapter 5, Example 5.6.2, and the reader should carefully review the detailed explanations set forth there, particularly in regard to conservation of total energy and the interchange of kinetic and potential energy during the cycle of motion. At this point, therefore, we need only add a few remarks to reinforce what has already been said and to relate it to what we are doing now.

If we multiply (9.2.2) on both sides by the mass m and identify ma_x as the resultant force F_x, we can express that equation as

$$F_x = m\frac{d^2x}{dt^2} = -m\omega^2 x \tag{9.3.1}$$

for any system that executes simple harmonic motion. For the case of the mass–spring system, for which $\omega^2 = k/m$, this reduces to the familiar equation $F = -kx$. The potential energy associated with such a force can be calculated as the work done by the force as it moves from some initial position x to a final position where the potential energy is defined to be zero. Taking this final position to be the equilibrium position $x = 0$, using (9.3.1) we may write,

$$U_p = \int_x^0 F_x\,dx = -\int_0^x F_x\,dx = m\omega^2 \int_0^x x\,dx$$

or

$$U_p = \tfrac{1}{2}m\omega^2 x^2 \tag{9.3.2}$$

This represents the potential energy stored in any system that executes simple harmonic motion. Again,

for the mass–spring system wherein $\omega^2 = k/m$, this reduces to the familiar result $U_p = \tfrac{1}{2}kx^2$.

The total energy U_t can now be written as the sum of kinetic and potential energies:

$$U_t = \tfrac{1}{2}mv_x^2 + \tfrac{1}{2}m\omega^2 x^2 = \text{const.} \tag{9.3.3}$$

which expresses the energy conservation principle for simple harmonic motion. This, however, can be written more conveniently in terms of the amplitude A by recognizing that at the extreme limits of displacement, when $x = \pm A$, the velocity is zero. If we substitute this information into (9.3.3), we see that $U_t = \tfrac{1}{2}m\omega^2 A^2$, which permits us to write the energy conservation equation for simple harmonic motion as

$$\tfrac{1}{2}m\omega^2 A^2 = \tfrac{1}{2}mv_x^2 + \tfrac{1}{2}m\omega^2 x^2 \tag{9.3.4}$$

This is solved for v_x to obtain

$$v_x = \pm\omega\sqrt{A^2 - x^2} \tag{9.3.5}$$

in agreement with our previous calculations based upon the sinusoidal expression (9.2.4) for the displacement. For a given value of y, there are thus two possible velocities of opposite sign, corresponding to motion toward or away from the equilibrium point. This relation clearly shows that the speed is a maximum at the equilibrium position and is zero whenever $x = \pm A$.

Starting from the other end, so to speak, we can obtain the equation of motion (9.2.2) by differentiating the energy conservation equation with respect to time, noting that the derivative of the constant total energy is zero:

$$mv_x\frac{dv_x}{dt} + m\omega^2 x\frac{dx}{dt} = 0 \tag{9.3.6}$$

In this equation, however, $dx/dt = v_x$, while $dv_x/dt = d^2x/dt^2$. Dividing both sides by mv_x, therefore,

$$\frac{d^2x}{dt^2} = -\omega^2 x \tag{9.3.7}$$

which is the same as (9.2.2).

EXAMPLE 9.3.1
MASS SUSPENDED VERTICALLY
FROM AN IDEAL SPRING
In Fig. 9.2, a mass m is shown suspended vertically from a spring whose upper end is attached to a fixed overhead support. The spring, which we assume to be of negligible mass, has a force constant k and obeys Hooke's law $F = -k(y - y_0)$, where y_0 is the spring's unstretched length. For this case, the spring has a potential energy given by

$$U_{ps} = \tfrac{1}{2}k(y - y_0)^2 \tag{9.3.8}$$

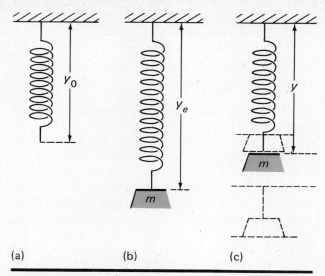

(a) (b) (c)

FIGURE 9.2. Vertical mass–spring system. (a) Spring alone, equilibrium position at $y = y_0$. (b) Mass attached to spring at equilibrium at $y = y_e$. (c) System oscillating about equilibrium position $y = y_e$; instantaneous displacement is $y(t)$.

This is not the entire potential energy of the system since the mass m also has gravitational potential energy.

Let us choose the gravitational potential energy U_{pg} to be zero when $y = y_0$. With this convention, the expression for U_{pg} is

$$U_{pg} = mg(y - y_0) \qquad (9.3.9)$$

It is the total potential energy $U_p = U_{ps} + U_{pg}$ which determines the motion of the system. This is given by

$$U_p = \tfrac{1}{2}k(y - y_0)^2 + mg(y - y_0) \qquad (9.3.10)$$

The functions given by (9.3.8) and (9.3.9) are plotted in Fig. 9.3 using solid lines, and their resultant is shown using a dotted curve. The total force F_y acting on the mass may be determined by

$$F_y = -\frac{dU_p}{dy} = -k(y - y_0) - mg \qquad (9.3.11)$$

This force is the slope of the dotted curve. When $y = y_0$, the force is not zero and, therefore, y_0 is not an equilibrium point. Setting F_y equal to zero in (9.3.11), we see that the equilibrium point is y_e, where

$$y_e = y_0 - \frac{mg}{k} \qquad (9.3.12)$$

The gravitational force has thus caused a *shift* of the equilibrium position by an amount mg/k below the unstretched position of the spring. It is convenient to rewrite (9.3.10) by substituting for y_0 the expression given by (9.3.12). The potential energy then becomes

$$U_p = \tfrac{1}{2}k\left(y - y_e - \frac{mg}{k}\right)^2 + mg\left(y - y_e - \frac{mg}{k}\right)$$

$$= \tfrac{1}{2}k(y - y_e)^2 - \tfrac{1}{2}k \cdot 2(y - y_e)\cdot\frac{mg}{k} + \frac{(mg)^2}{2k}$$

$$\quad + mg\left(y - y_e - \frac{mg}{k}\right)$$

$$= \tfrac{1}{2}k(y - y_e)^2 - \frac{(mg)^2}{2k} \qquad (9.3.13)$$

At this, point, we recall that we may add to a potential energy any constant without affecting the equations of motion. In the present case, addition of the constant $(mg)^2/2k$ leads to a new potential energy given by

$$U_p{}' = \tfrac{1}{2}k(y - y_e)^2 = \tfrac{1}{2}ky'^2 \qquad (9.3.14)$$

where

$$y' = y - y_e \qquad (9.3.15)$$

The difference between (9.3.14) and (9.3.13) arises from our choice of the point at which the potential energy has the value zero. In (9.3.14), it is zero at the equilibrium position y_e, while in (9.3.13), the zero occurs at y_0. According to (9.3.14), it is as though the entire potential energy were due solely to the stretch of the spring from the equilibrium position y_e, in the *absence* of any gravitational potential energy. The reason for this seemingly strange result is evident from Fig. 9.3, where it is clear that the sum of the parabolic potential energy U_{ps} and the linear potential energy U_{pg} is simply another parabolic curve having the same shape as U_{ps} but displaced slightly to the right.

From (9.3.14) it is easy to see that the motion must be simple harmonic motion, because the potential energy has the same form as that exhibited previously in (9.3.2), with $\omega = \sqrt{k/m}$. Alternatively, it is evident that the resultant force F_y can be obtained from (9.3.14) as

$$F_y = -\frac{dU_p{}'}{dy} = -k(y - y_e) = -ky' \qquad (9.3.16)$$

from which

$$a_y = \frac{d^2y}{dt^2} = \frac{d^2y'}{dt^2} = -\frac{k}{m}y' \qquad (9.3.17)$$

which is the standard equation (9.2.2.) with

$$\omega = \sqrt{k/m} \qquad (9.3.18)$$

It is important to note that when the force is expressed in the form (9.3.16), as $-k(y - y_e)$, it includes the *sum* of two forces: (a) the force $-k(y - y_0)$, which a spring stretched from its unstretched position exerts on the mass m; and (b) the force $-mg$, which the earth exerts on the mass m due to gravitational attraction. We may

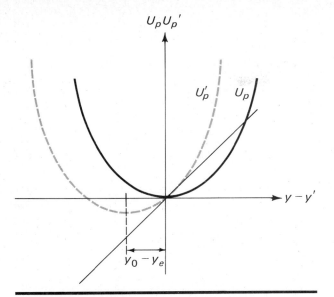

FIGURE 9.3. Resultant potential energy of vertical mass-spring system, viewed as the sum of linear (gravitational) and parabolic (spring) potential energies. Their sum is seen to be a parabola of the same shape as the original potential energy curve of the spring, but displaced to a new equilibrium position.

write the displacement as a function of time in the usual way:

$$y' = y - y_e = A \sin(\omega t + \delta) \qquad (9.3.19)$$

The amplitude A and the phase constant δ may then be determined as before from given data relating to the initial conditions.

EXAMPLE 9.3.2

A spring of force constant 10 N/m has a mass of 1 kg attached to its free end. The spring hangs from the ceiling. Find (a) the period of oscillation if the system is vibrating and (b) the equilibrium position if $y = 0$ corresponds to the unstretched position of the spring. (c) The mass is set in motion by pulling it down 0.5 m from the equilibrium position and then simply releasing it; find the equation which governs the motion.

To determine the period of oscillation, we recall that

$$\omega = \frac{2\pi}{T} = \sqrt{\frac{k}{m}} \qquad (9.3.20)$$

In the present case, this gives

$$\frac{2\pi}{T} = \sqrt{10/1} = 3.16 \ \text{sec}^{-1} \qquad (9.3.21)$$

$$T = 1.987 \ \text{sec} \qquad (9.3.22)$$

This gives the result for Part (a). To find the equilib-

rium position for Part (b), we equate the gravitational force and the spring force:

$$y_e = -\frac{mg}{k} = -\frac{(1)(9.8)}{10} = -0.98 \ \text{m} \qquad (9.3.23)$$

Since the motion is simple harmonic, the general solution takes the form

$$y - y_e = A \sin(\omega t + \delta) = A \sin(3.16t + \delta) \qquad (9.3.24)$$

$$v_y = dy/dt = A\omega \cos(\omega t + \delta)$$

$$= 3.16A \cos(3.16t + \delta) \qquad (9.3.25)$$

where t is expressed in seconds. A and δ may now be determined by conditions at $t = 0$. These conditions state that $y - y_e$ has the value -0.5 m and dy/dt has the value zero at the initial time $t = 0$. These two equations then give for $t = 0$

$$(y - y_e) = A \sin(0 + \delta) = -0.5 \ \text{m} \qquad (9.3.26)$$

$$v_x = 3.16A \cos \delta = 0 \qquad (9.3.27)$$

The second equation can be satisfied if A is zero or if δ is $\pi/2$. The first possibility is in conflict with (9.3.26) and must be rejected. Therefore, the initial phase δ must be $\pi/2$. When this is substituted into (9.3.24), we obtain

$$A \sin \delta = A \sin \pi/2 = A = -0.5 \ \text{m} \qquad (9.3.28)$$

The complete solution of Part (c) is then given by

$$y = -0.98 - 0.5 \sin(3.16t + \pi/2) \ \text{m} \qquad (9.3.29)$$

This may also be written as

$$y = -0.98 - 0.5 \cos(3.16t) \ \text{m} \qquad (9.3.30)$$

by using the identity $\sin(\theta + \tfrac{1}{2}\pi) = \cos \theta$

EXAMPLE 9.3.3

The spring and mass of the preceding example are in the equilibrium position. A bullet of mass $m_b = 10$ g is fired from below at a speed $v_0 = 500$ m/s into the mass m, which in this example we assume to be a wooden block. The bullet comes to rest quickly within the block. Find the equation which governs the subsequent motion of the system. Determine the frequency, amplitude, and initial phase, first assuming that the mass m_b is much smaller than m, and then assuming that it is not negligible. Assume the collision of bullet with block to be instantaneous.

Figure 9.4 illustrates the situation just before and just after the bullet strikes the block. Let $t = 0$ correspond to the situation just after the bullet has come to rest in the block. The initial conditions are specified at $t = 0$. In the situation in which the approximation $m_b \ll m$ is utilized, the general solution is of the same general form as in the preceding example, with

$$y = -0.98 + A \sin(3.16t + \delta) \ \text{(m)} \qquad (9.3.31)$$

285

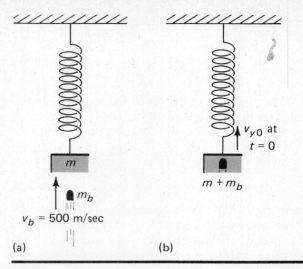

FIGURE 9.4

However, the initial conditions are now different. The new conditions must be used to specify new values of A and δ. One of the initial conditions is that at $t = 0$ the mass is at the equilibrium position and, therefore,

$$-0.98 = -0.98 + A \sin(0 + \delta) \text{ (m)} \qquad (9.3.32)$$

This implies that δ is zero. The second condition needed is the value of the velocity at $t = 0$. Using momentum conservation during the instantaneous collision, we obtain, for $t = 0$,

$$m_b v_b = (m_b + m)v_{y0} \qquad (9.3.33)$$

and therefore the initial velocity v_{y0} must be, neglecting m_b in comparison to m,

$$v_{y0} \simeq \frac{m_b v_b}{m} = \frac{(0.01)(500)}{(1.0)} = 5 \text{ m/sec} \qquad (9.3.34)$$

The velocity of the mass m at any later time t can be found by differentiating (9.3.31) to obtain

$$v_y = dy/dt = 3.16A \cos(3.16t + \delta) \text{ (m)} \qquad (9.3.35)$$

Combining (9.3.35) with the results $\delta = 0$ and $v_{y0} = 5$ m/sec, we can determine A:

$$A = \frac{5.0}{3.16} = 1.58 \text{ m} \qquad (9.3.36)$$

The complete solution is, therefore,

$$y = -0.98 + 1.58 \sin(3.16t) \text{ (m)} \qquad (9.3.37)$$

How is this result changed if the condition $m_b \ll m$ is not utilized? In the first place, the oscillatory motion would now have a frequency ω' given by

$$\omega' = \sqrt{\frac{k}{m_b + m}} \qquad (9.3.38)$$

The frequency of oscillation is slightly diminished or the period increased. Secondly, the equilibrium position no longer occurs at $y_e = 0.98$ m but is slightly lower due to the added weight. The new equilibrium position is at a value $y_E = -(m_b + m)g/k$. The general solution is now going to have the form

$$y = y_E + A' \sin(\omega' t + \delta) \qquad (9.3.39)$$

where again the arbitrary constants are determined from the initial conditions, which are now, for $t = 0$,

$$y = y_e = -0.98 \text{ m}$$

and

$$v_{y0}' = \frac{m_b v_b}{m_b + m} = \left(\frac{0.01}{1.01}\right) \times 500 = 4.95 \text{ m/sec} \qquad (9.3.40)$$

With these values, the initial phase is no longer zero, and the value of the amplitude also differs slightly from the previous result.

The preceding examples illustrate the importance of initial conditions in determining the exact motion. The idealization of a massless spring has been made throughout in the interest of simplicity and is justifiable whenever the attached mass has a value much larger than the spring mass.

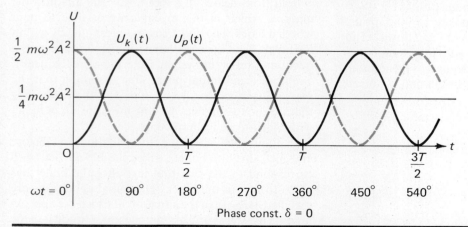

FIGURE 9.5. Time variation of kinetic and potential energy of mass–spring system in simple harmonic motion.

As the motion of the mass m proceeds, the energy of the system, which remains constant at all times, is given by the sum of a time-varying potential energy and a time-varying kinetic energy. The expressions for U_k and U_p may be obtained from (9.3.24) and (9.3.25):

$$U_k = \tfrac{1}{2}mv_y^2 = \tfrac{1}{2}mA^2\omega^2 \cos^2(\omega t + \delta) \qquad (9.3.41)$$

$$U_p = \tfrac{1}{2}k(y - y_e)^2 = \tfrac{1}{2}kA^2 \sin^2(\omega t + \delta)$$
$$= \tfrac{1}{2}mA^2\omega^2 \sin^2(\omega t + \delta) \qquad (9.3.42)$$

These functions are plotted in Fig. 9.5. Note that when the potential energy is a maximum, the kinetic energy is a minimum, and vice versa. Throughout the motion, energy is converted back and forth from one form to another, the conversion occurring in an oscillatory manner. This time-varying exchange of one form of energy to another is characteristic of all types of oscillatory motion.

From the form of the curves plotted in Fig. 9.5, it is apparent that the *time-average* value of both U_k and U_p is simply one half the maximum value $\tfrac{1}{2}m\omega^2 A^2$. We may therefore write these average values as

$$\boxed{\bar{U}_k = \bar{U}_p = \tfrac{1}{4}m\omega^2 A^2 = \tfrac{1}{2}U_t} \qquad (9.3.43)$$

This result is true for any system that executes simple harmonic motion and is an important characteristic of any simple harmonic oscillator.

9.4 Simple Harmonic Motion and Uniform Circular Motion: Superposition of Harmonic Oscillations of Equal Frequency

In the expressions we have derived for the displacement, velocity, and acceleration of an object that executes simple harmonic motion, there is a phase angle $(\omega t + \delta)$ and an angular frequency ω whose dimensions are the same as those of angular velocity. These quantities resemble those used in the discussion of uniform circular motion to describe angular displacement and angular velocity. All this suggests, quite correctly, that there is a close connection between simple harmonic motion and uniform circular motion.

This relation can be understood with reference to Fig. 9.6, in which a point $P(x, y)$ executes uniform circular motion in the xy-plane. The displacement of this point P, referred to the origin O, is described by the displacement vector $\mathbf{r}$, whose coordinates are x and y. If the uniform circular motion proceeds with constant angular velocity ω, then, assuming that the angular displacement θ is zero at $t = 0$, the angular

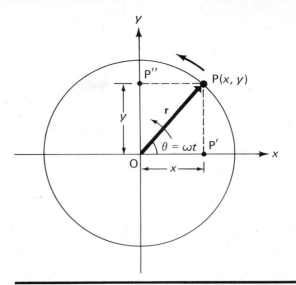

FIGURE 9.6. Simple harmonic motion viewed as the projection of uniform circular motion onto the x- or y-coordinate axis.

displacement θ can be written as

$$\theta = \omega t \qquad (9.4.1)$$

from which the x- and y-components of the displacement $\mathbf{r}$ can be written

$$x = r \cos \omega t = r \sin\left(\omega t + \frac{\pi}{2}\right) \qquad (9.4.2)$$

$$y = r \sin \omega t \qquad (9.4.3)$$

These are both simple harmonic motions of amplitude r, the phase constant δ being zero for the y-component and $\pi/2$ for the x-component.

The situation may be summarized by stating that if a vector $\mathbf{r}$ executes uniform circular motion with angular velocity ω and angular displacement $\theta = \omega t$, the projection of its motion on the x-axis (represented by the motion of point P' along the x-axis) is simple harmonic motion whose displacement is $r \cos \omega t$. Similarly, the projection of its motion on the y-axis (represented by the motion of point P'' along the y-axis) is simple harmonic motion whose displacement is $r \sin \omega t$. Simple harmonic motion of angular frequency ω can, therefore, be regarded as the rectangular resolution of uniform circular motion of angular velocity ω. From the opposite point of view, we can also think of uniform circular motion as being *the vector sum of two mutually perpendicular simple harmonic oscillations of equal amplitude and frequency but differing in phase by $\pi/2$ radians.*

This picture of simple harmonic motion is useful in understanding what happens when we must superpose the displacements corresponding to two different simple harmonic oscillations which are of *equal frequency but which differ in amplitude and phase.* This can, of course, be accomplished in a purely

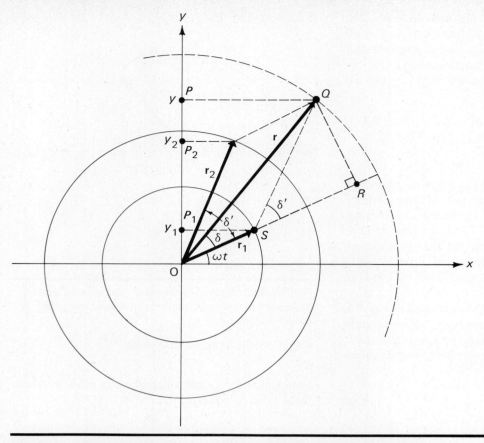

FIGURE 9.7. Superposition of two simple harmonic motions of the same frequency, achieved by adding the projections of two displacement vectors defining uniform circular motions of different radii and phase angle but the same angular velocity.

mathematical way simply by adding two simple harmonic displacements y_1 and y_2, expressed by

$$y_1 = r_1 \sin \omega t \qquad (9.4.4)$$

$$y_2 = r_2 \sin(\omega t + \delta') \qquad (9.4.5)$$

to obtain

$$y = y_1 + y_2 = r_1 \sin \omega t + r_2 \sin(\omega t + \delta') \qquad (9.4.6)$$

In writing (9.4.4) and (9.4.5), we have chosen to "start the clock" when the displacement y is zero. The effect of this is simply to ensure that the phase constant δ associated with this motion is zero and that the phase constant related to the motion y_2, denoted by δ' in (9.4.5), represents simply the *difference* in phase between the two separate oscillations. While this procedure is correct, it is not very informative and offers little in the way of physical insight into the character of the resulting motion. In particular, it does not tell us whether the resultant motion is simple harmonic and, if so, what its amplitude and phase are.

These questions are answered simply by regarding each of the individual motions as projections of uniform circular motions of equal angular velocity

whose radius vectors differ in magnitude and direction, as illustrated in Fig. 9.7. In this figure, displacement vectors $\mathbf{r}_1$ and $\mathbf{r}_2$, whose directions always differ by an angle δ', rotate with uniform angular velocity ω. Clearly, the projections of these vectors on the y-axis, designated by points P_1 and P_2, execute simple harmonic motions in which the displacements y_1 and y_2 are given precisely by (9.4.4) and (9.4.5). The *resultant* of $\mathbf{r}_1$ and $\mathbf{r}_2$ is the vector $\mathbf{r}$, which also rotates with uniform angular velocity ω. The projection of this vector on the y-axis, labelled P, must, according to the law of vector addition, be the sum of y_1 and y_2, hence the quantity y as represented by (9.4.6). But since this quantity is the projection of a vector $\mathbf{r}$ that rotates with uniform circular motion, its motion, also, has to be simple harmonic! We must, therefore, be able to express y in the form

$$y = r \sin(\omega t + \delta) \qquad (9.4.7)$$

where r is the magnitude of the vector $\mathbf{r}$ and $\omega t + \delta$ the angle between $\mathbf{r}$ and the x-axis. But since the vector $\mathbf{r}_1$ makes an angle ωt with the x-axis, δ represents the angle between $\mathbf{r}$ and $\mathbf{r}_1$, as shown in Fig. 9.7. In the drawing, the distance QS is equal to $\mathbf{r}_2$; therefore, QR and RS can be written

$$QR = r_2 \sin \delta' \quad \text{and} \quad RS = r_2 \cos \delta' \quad (9.4.8)$$

The magnitude of $\mathbf{r}$, however, is given by

$$r^2 = (OR)^2 + (QR)^2 = (r_1 + r_2 \cos \delta')^2 + r_2{}^2 \sin^2 \delta'$$

from which

$$r^2 = r_1{}^2 + r_2{}^2 + 2r_1 r_2 \cos \delta' \qquad (9.4.9)$$

The tangent of δ can be expressed as

$$\tan \delta = \frac{QR}{OR} = \frac{r_2 \sin \delta'}{r_1 + r_2 \cos \delta'} \qquad (9.4.10)$$

The result of all this is that the two simple harmonic motions (9.4.4) and (9.4.5), when added or superposed, result in a third simple harmonic motion whose amplitude and phase are given in terms of the amplitudes and phases of the original motions by (9.4.9) and (9.4.10). Another way of saying it is that when you add two sine waves of equal frequency but different amplitude and phase, you get a third sine wave whose phase and amplitude are represented by (9.4.9) and (9.4.10).

When the two constituent harmonic motions have the *same phase*, then the angle δ' is zero and (9.4.9) and (9.4.10) reduce to

$$r = r_1 + r_2 \qquad (9.4.11)$$

and

$$\delta = 0 \qquad (9.4.12)$$

The resulting motion, according to (9.4.7), is then

$$y = (r_1 + r_2) \sin \omega t \qquad (9.4.13)$$

as one might expect from Fig. 9.7.

When the two constituent harmonic oscillations have the *same amplitude but different phases*, $r_2 = r_1$, and (9.4.9) and (9.4.10) give

$$r^2 = 2r_1{}^2(1 + \cos \delta') = 4r_1{}^2 \cos^2 \frac{\delta'}{2} \qquad (9.4.14)$$

$$r = 2r_1 \cos \frac{\delta'}{2}$$

and

$$\tan \delta = \frac{\sin \delta'}{1 + \cos \delta'} = \frac{2 \sin(\delta'/2) \cos(\delta'/2)}{2 \cos^2(\delta'/2)} = \tan(\delta'/2)$$

whence

$$\delta = \tfrac{1}{2}\delta' \qquad (9.4.15)$$

In obtaining these results, we have used the trigonometric identity $1 + \cos \theta = 2 \cos^2(\theta/2)$. Again, these conclusions could have been predicted geometrically from Fig. 9.7.

EXAMPLE 9.4.1

Express the result of superposing two harmonic oscillations whose displacements are given by $1.5 \sin(24t)$ and $2.7 \sin(24t + 2\pi/3)$, where distances are expressed in meters and time in seconds.

According to the ideas developed above, this motion can be expressed as a third simple harmonic oscillation whose amplitude can be found by substituting $r_1 = 1.5$ m, $r_2 = 2.7$ m, and $\delta' = 2\pi/3$ rad into (9.4.9), the result being

$$r = \sqrt{(1.5)^2 + (2.7)^2 + (2)(1.5)(2.7) \cos 120°} = 2.343 \text{ m}$$

The phase δ can be found in the same way from (9.4.10):

$$\tan \delta = \frac{(2.7)(\sin 120°)}{(1.5) + (2.7)(\cos 120°)} = 15.59$$

$$\delta = 86.33° = 1.507 \text{ rad}$$

The displacement can now be written as

$$x(t) = r \sin(\omega t + \delta) = 2.343 \sin(24t + 1.507) \text{ (m)}$$

These results can also be obtained very easily without reference to Eqs. (9.4.9) and (9.4.10) simply by finding the resultant of the two vectors $\mathbf{r}_1$ and $\mathbf{r}_2$ shown in Fig. 9.8. These vectors represent the amplitudes of the two separate motions at time $t = 0$, the angle $2\pi/3$ rad ($120°$) between them corresponding to the phase difference δ' between the two harmonic oscillations. By the usual rules of vector algebra,

$$r_1 = 1.5\mathbf{i}_x \text{ (m)}$$
$$r_2 = (2.7)(\cos 120°)\mathbf{i}_x + (2.7)(\sin 120°)\mathbf{i}_y$$
$$= (-1.35)\mathbf{i}_x + (2.338)\mathbf{i}_y \text{ (m)}$$

from which

$$\mathbf{r} = \mathbf{r}_1 + \mathbf{r}_2 = (0.150)\mathbf{i}_x + (2.338)\mathbf{i}_y$$

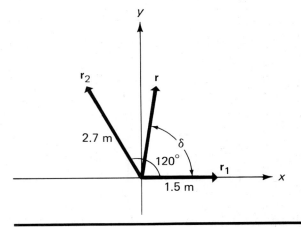

FIGURE 9.8

Then,

$$r = \sqrt{(0.150)^2 + (2.338)^2} = 2.343 \text{ m}$$

$$\tan \delta = \frac{2.338}{0.150} = 15.59$$

$$\delta = 86.33° = 1.507 \text{ rad}$$

as before. Knowing the amplitude and phase constant, the displacement as a function of time must be

$$x(t) = (2.343) \sin(24t + 1.507) \text{ (m)}$$

as obtained above.

9.5 The Simple Pendulum, the Torsion Pendulum, and Angular Harmonic Motion

The "simple" pendulum, illustrated in Fig. 9.9, is a very useful device that has regulated millions of clocks for the past eight centuries with reasonable accuracy. Though simple to conceive and construct, its motion, however, when discussed in the most general terms, is not "simple" at all.

We shall, therefore, begin by discussing a device that is really simple in all respects—the *torsion pendulum*. In fact, if you refer to Chapter 7, you will see that we have already analyzed the motion of this system in Example 7.6.4, from the viewpoint of energy conservation. Indeed, our original ideas about the conservation of energy in rotational motion were formed by a detailed consideration of this contrivance, which is illustrated in Fig. 7.19b. We showed in Chapter 7 that if the restoring torque τ exerted by the torsion rod is

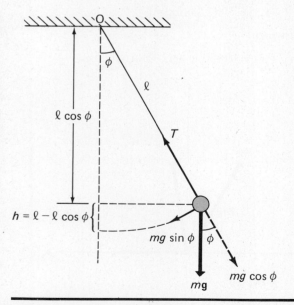

FIGURE 9.9. Geometry of the simple pendulum.

directly proportional to the angle ϕ through which it is twisted, then

$$\tau = -k'\phi \tag{9.5.1}$$

where k' is a torsion constant analogous to a spring constant.[3] Also, the potential energy stored in the twisted torsion rod was shown to be

$$U_p = \tfrac{1}{2}k'\phi^2 \tag{9.5.2}$$

In these equations, it is assumed that the equilibrium position where the torque vanishes is at $\phi = 0$. In our previous work, we saw in addition, using energy conservation, that the angular displacement ϕ is a sinusoidally varying function of time of the form

$$\phi(t) = \phi_0 \sin(\omega t + \delta) \tag{9.5.3}$$

where

$$\omega = \sqrt{\frac{k'}{I}} \tag{9.5.4}$$

This result can be obtained equally well by setting the resultant torque as given by (9.5.1) equal to the moment of inertia times the angular acceleration $d^2\phi/dt^2$, to obtain

$$\frac{d^2\phi}{dt^2} = -\omega^2\phi \tag{9.5.5}$$

where the constant ω is expressed by (9.5.4). It is easily seen that this equation is identical to (9.2.2), except that the angular displacement ϕ appears in place of the linear displacement x. The mathematical relation between ϕ and t arising from (9.5.5) must, therefore, be the same as the one relating x and t obtained from (9.2.2) and should follow from (9.2.4) if x is replaced by ϕ, and, clearly, except that the amplitude in (9.5.3) has been represented by the symbol ϕ_0 rather than A, it does.

The *angular* displacement of the torsion pendulum, therefore, has the same functional dependence on time as the *linear* displacement of the mass–spring system shown in Fig. 9.1c. Rotational motion of this type is referred to as *angular harmonic motion*. It results whenever the equation of motion of a rotating body can be stated in the form (9.5.5). As in the case of linear simple harmonic motion, the quantity ϕ_0 in (9.5.3) represents the maximum angular excursion of the system from the equilibrium position and is referred to as the *angular amplitude*. The constants ω

---

[3] In this chapter, we shall use the symbols ϕ, $d\phi/dt$, and $d^2\phi/dt^2$ to denote angular displacement, angular velocity, and angular acceleration rather than the previously used notation θ, ω, and α. The reason for this change in notation will become apparent later on.

and δ also represent the angular frequency of the motion and the phase constant. As in the case of linear simple harmonic motion, their values can be determined from information about the motion given for $t = 0$ or some other specific time. As before, the frequency and the period of the motion can be expressed in terms of ω by (9.2.7) and (9.2.8). The graph shown in Fig. 9.1a accurately represents the variation of angular displacement with time in angular harmonic motion, provided that the angular displacement ϕ rather than the linear displacement x is plotted on the vertical axis and that the linear amplitude limits $\pm A$ are replaced with the angular amplitude limits $\pm \phi_0$. In the same way, the graph shown in Fig. 9.1b represents the time variation of the angular velocity $d\phi/dt$, as given by

$$\frac{d\phi}{dt} = \omega \phi_0 \cos(\omega t + \delta) \qquad (9.5.6)$$

in angular harmonic motion if we visualize the vertical axis to be calibrated in units of angular velocity (rad/sec) rather than linear velocity. Finally, the angular acceleration in angular harmonic motion can be expressed by differentiating (9.5.3) twice to obtain

$$\frac{d^2\phi}{dt^2} = -\omega^2 \phi_0 \sin(\omega t + \delta) = -\omega^2 \phi \qquad (9.5.7)$$

In discussing angular harmonic motion, it is important, however, to avoid confusing the instantaneous *angular velocity* $d\phi/dt$ of the system with the constant quantity ω, which expresses the *angular frequency* with which the cyclical motion proceeds. The instantaneous angular velocity $d\phi/dt$, as given by (9.5.6), is a *time-dependent* quantity which varies between the limits $\pm \omega \phi_0$ as the cycle is traversed, as illustrated in Fig. 9.1b. The angular frequency ω, on the other hand, is a *constant* (represented by $\sqrt{k'/I}$ for the torsion pendulum) which, when divided by 2π, tells us how many cycles of motion will be completed in a second. In order to avoid confusing these two very different quantities, we have used ϕ (rather than θ) to represent the angular displacement and $d\phi/dt$ to represent angular velocity. Please note that in this section $d\phi/dt$ and ω are *not*, repeat, *not* equal!

The *simple pendulum*, illustrated in Fig. 9.9, appears at first sight to be an example of angular harmonic motion. Though under certain conditions its motion approximates angular harmonic motion, in general its behavior is *anharmonic*, as we shall soon see. In the diagram, it is assumed that a small mass m is suspended by a cord of length l and negligible mass from a fixed overhead support. The mass is assumed to be so small that its behavior is essentially that of a point mass. In the figure, we see the pendulum at some arbitrary point in its cycle where the string makes an angle ϕ with the vertical equilibrium position. The only forces acting on the mass are the weight force $m\mathbf{g}$ (shown resolved into components parallel and normal to the string) and the tension $\mathbf{T}$. The line of action of $\mathbf{T}$ passes through the axis at O and, therefore, gives rise to no torque about that point. The only force component that does generate such a torque is the component of the weight force $mg \sin \phi$ parallel to the path along which the mass moves. Setting the torque associated with this force component about O, $-mgl \sin \phi$, equal to the moment of inertia of the mass about point O times the angular acceleration $d^2\phi/dt^2$, we find

$$(ml^2)\frac{d^2\phi}{dt^2} = -mgl \sin \phi$$

or

$$\frac{d^2\phi}{dt^2} = -\frac{g}{l} \sin \phi \qquad (9.5.8)$$

Now this is not the same as (9.5.5), which represents the standard form of the equation of motion for angular harmonic motion. The motion, although clearly periodic, is *anharmonic*. It is not impossible to arrive at a mathematical solution of (9.5.8) that expresses the precise variation of the angular displacement with time, but it is nevertheless very difficult and involves some advanced and complex mathematics, which we shall not attempt here. Suffice it to say that the result of this mathematical analysis shows that the general motion of the simple pendulum differs from simple angular harmonic motion in that the displacement is *not* a simple sinusoidal function of time and in that the period of oscillation is dependent upon the amplitude of the motion rather than being independent of it as it is with angular simple harmonic motion.

On the other hand, we may note that (9.5.8) differs from the "standard" equation (9.5.5) only in that the function $\sin \phi$ appears on the right-hand side rather than ϕ itself. But we know that whenever the angular displacement ϕ, expressed in radians, is small in comparison to one radian ($57.3°$), $\sin \phi$ is very nearly the same as ϕ itself. If we restrict ourselves to angular displacements from the vertical that are small compared to one radian, then, under these circumstances, we shall find that

$$\sin \phi \cong \phi \qquad \phi \ll 1 \text{ rad} \qquad (9.5.9)$$

is always an excellent approximation which permits us to express the general equation of motion (9.5.8) in the form

$$\frac{d^2\phi}{dt^2} = -\omega^2 \phi \qquad \phi \ll 1 \text{ rad} \qquad (9.5.10)$$

291

where

$$\omega = \sqrt{\frac{g}{l}} \qquad (9.5.11)$$

This is the equation for angular harmonic motion, the angular frequency ω being now identified as $\sqrt{g/l}$.

We see, then, that as long as we restrict ourselves to the case where the pendulum never swings very far from the vertical, its motion is approximately angular harmonic motion in which the angular displacement is given by (9.5.3) and angular frequency by (9.5.11). Under these circumstances the period, given by

$$T = \frac{2\pi}{\omega} = 2\pi \sqrt{\frac{l}{g}} \qquad (9.5.12)$$

will be essentially independent of the amplitude ϕ_0. For an angle $\phi = 30°$, $\sin \phi$ and ϕ differ only by about 4.7 percent. We may, therefore, expect the motion of a simple pendulum to be represented accurately as angular simple harmonic motion to within a few percent so long as we do not allow it to swing more than about that far from the vertical. If we square both sides of Eq. (9.5.12) and solve for l, we may write

$$l = \frac{gT^2}{4\pi^2} \qquad (9.5.13)$$

Setting $T = 2.0$ sec in this expression, it is evident that a pendulum whose period is exactly 2 sec must, at sea level, have a length of 3.26 ft (39.1 in.), or 0.993 m. The correctness of this result is evident from the dimensions of the grandfather clock.[4] It is also evident from (9.5.12) that to lengthen the period and thus slow the clock down, you must increase l and thus lengthen the pendulum, and vice versa. Finally, and most remarkable of all, Eqs. (9.5.11) and (9.5.12) do not contain the mass m at all—nor does (9.5.8) for that matter—which tells us that the period of a pendulum is completely *independent* of the mass.

As the amplitude of a simple pendulum increases, eventually we reach a situation where the approximation (9.5.9) is no longer very good. Beyond that point, angular harmonic motion is no longer a valid representation of the motion, and under these conditions the motion is *anharmonic*. Since $\sin \phi$ is always *smaller* in magnitude than the angle ϕ itself, the acceleration of the real pendulum, as given by (9.5.8), will be significantly less than that given by (9.5.10) corresponding to simple angular harmonic motion. For this reason, once the motion becomes appreciably anharmonic, the period no longer remains constant

but becomes longer and longer as the amplitude increases.

The laws of motion were not stated until it was possible to clearly understand the concept of *time* and to make reasonably accurate measurements of time intervals. The simple pendulum and the torsion pendulum, as represented by the balance wheel of a watch, allowed us to make accurate clocks and time measurements for the first time around 1200 A.D. The development of these devices, therefore, paved the way for the discoveries of Galileo and Newton four hundred years later. Fine mechanical clocks and watches are still driven by these simple but accurate mechanisms, though electronic watches that utilize the vibrations of tiny quartz crystals or metal "tuning forks" as timekeeping elements are now also in widespread use. But even these modern and sophisticated electronic timepieces utilize a simple harmonic oscillation as their basic timing element.

A more general example of a pendulum is illustrated in Fig. 9.10, in which we see an irregularly shaped, rigid object that is free to rotate about a point of suspension O. Such an object is frequently referred to as a *physical pendulum*. The object has a mass m, and the center of mass is at point P, whose distance from the axis O is represented by h. The moment of inertia of the body about point O is I_0. In this case also, the only force that causes a torque about O is the component of the weight force perpendicular to the line OP. Setting this torque equal to the moment of inertia times the angular acceleration $d^2\phi/dt^2$, we obtain

$$I_0 \frac{d^2\phi}{dt^2} = -mgh \sin \phi \qquad (9.5.14)$$

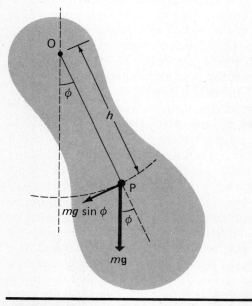

[4] The clock ticks once every half-period, thus twice during a full period. Therefore, if the full period is exactly 2 seconds, the clock will tick every second, which is what grandfather clocks usually do, when they work.

FIGURE 9.10. The "physical pendulum."

As in the case of the simple pendulum, this equation does not have the standard mathematical form (9.5.5), except when the angle ϕ is small and the approximation $\sin \phi \cong \phi$ is appropriate. When this approximation can be made, however, we may write (9.5.14) as

$$\frac{d^2\phi}{dt^2} = -\omega^2\phi \qquad \phi \ll 1 \text{ rad} \qquad (9.5.15)$$

where now

$$\boxed{\omega = \sqrt{\frac{mgh}{I_0}}} \qquad (9.5.16)$$

For small angular amplitudes, we see that the motion of the general physical pendulum, like that of a simple pendulum, is angular simple harmonic motion whose angular frequency is given by Eq. (9.5.16). For the simple pendulum, the moment of inertia of the point mass m about the axis is simply $I_0 = mh^2$; if this value is substituted into the above equation, we immediately obtain the previously derived expression (9.5.11).

EXAMPLE 9.5.1

The balance wheel of a watch is 1.00 cm in diameter and has a mass of 0.50 g. Assuming that the inertial properties of the balance wheel can be approximated by assuming that all its mass is concentrated in the rim, 0.50 cm from the rotation axis, what must be the torsion constant of the restraining "hairspring" for the watch to tick exactly five times per second, each tick corresponding to a half-period of torsional oscillation? If the amplitude of angular harmonic motion is π radians, write the equations for angular displacement, velocity, and acceleration of the balance wheel. What are the maximum values of these quantities?

If we know the angular frequency ω and the moment of inertia I, we can determine the torsion constant k' from (9.5.4). If we can consider all the mass of the wheel to be concentrated in the rim, at a distance of 0.50 cm from the axis, the moment of inertia will be

$$I = mr^2 = (0.50)(0.50)^2 = 0.125 \text{ g cm}^2$$

At the same time, if the watch ticks five times per second, then a half-period is 0.2 sec, and the full period of oscillation is $T = 0.40$ sec. Then, from (9.2.8),

$$\omega = \frac{2\pi}{T} = \frac{(2)(\pi)}{0.40} = 15.71 \text{ rad/sec}$$

Substituting these values into (9.5.4) gives

$$k' = I\omega^2 = (0.125)(15.71)^2 = 30.85 \text{ dyn-cm}$$

This means that when the balance wheel is rotated through an angle of 1 radian from the equilibrium position, the hairspring generates a restoring torque of 30.85 dyn-cm, in accord with (9.5.1).

Let us choose the time $t = 0$ at the instant the balance wheel passes through the equilibrium position $\phi = 0$. Then the phase constant δ in (9.5.3) will be zero. Since the amplitude is given as π radians and ω is known to have the value 15.71 rad/sec from the calculations above, we may write the displacement (9.5.3) in this case as

$$\phi(t) = \pi \sin(15.71t) \text{ rad}$$

Differentiating with respect to time, we obtain the angular velocity

$$\frac{d\phi}{dt} = 15.71\pi \cos(15.71t)$$

and differentiating this gives us the angular acceleration

$$\frac{d^2\phi}{dt^2} = -(15.71)^2\pi \sin(15.71t)$$

From these expressions it is easy to see that the maximum value of the angular velocity of the wheel is $15.71\pi = 49.35$ rad/sec, while the maximum angular acceleration will be $(15.71)^2\pi = 775.4$ rad/sec^2.

EXAMPLE 9.5.2

A grandfather clock is transported from Baltimore, Maryland, at sea level, to Denver, Colorado, which is at the same latitude but at an elevation of 1 mile above sea level. If the clock kept perfect time at Baltimore, how would its rate be affected by the move to Denver?

Since the latitude of the two cities is the same, there is no change in the effective value of g arising from any change in the centripetal force due to the earth's rotation, as calculated in section 4.3. But because Denver is 3961 miles distant from the earth's center while Baltimore is 3960 miles distant, the acceleration of gravity g' at Denver will be related to the value g at Baltimore according to (8.7.3) by

$$g' = g\left(\frac{3960}{3961}\right)^2 = 0.999495 \text{ g}$$

The period of a pendulum clock having a pendulum of length l is given by (4.5.12). The ratio of the period T of a clock at Baltimore to the period T' for the same clock at Denver will be

$$\frac{T'}{T} = \frac{2\pi\sqrt{l/g'}}{2\pi\sqrt{l/g}} = \sqrt{\frac{g}{g'}} = 1.000253$$

using the value of g/g' calculated previously. The period of the clock will be longer in Denver than in Baltimore by the factor 1.000253. Thus, if every tick represented exactly 1 sec in Baltimore, the interval between ticks in Denver would be 1.000253 sec, and since there are $(24)(3600) = 86400$ sec in a day, the clock when moved to Denver would be slow by $(0.000253)(86400) = 21.8$ sec per day, an amount which is by no means negligible.

This discrepancy could, of course, be corrected by decreasing the effective length of the pendulum to some appropriate length l'. If the periods T and T' are to be equal, we must choose l' so that

$$\frac{T}{T} = \frac{2\pi\sqrt{l'/g'}}{2\pi\sqrt{l/g}} = \sqrt{\frac{l'g}{lg'}} = 1.000\ldots \qquad \text{exactly}$$

This, however, requires that

$$\frac{l'}{l} = \frac{g'}{g} = 0.999495$$

If the pendulum was 1 m ($=1000$ mm long initially, the above correction could be made by shortening its length to 999.495 mm, a decrease of about 0.5 mm.

EXAMPLE 9.5.3

In the discussion of the simple pendulum, the oscillating mass was assumed to be a point mass at a distance l from the axis of rotation. Consider an actual pendulum in which the oscillating mass is a sphere of mass m and radius R whose center is at a distance l from the point of suspension. Find the true angular frequency of this actual pendulum and compare it with that of the "ideal" simple pendulum given by (9.5.11). Neglect the mass of the suspending cord.

The true angular frequency of the actual pendulum can be obtained by considering it to be an example of the *physical pendulum* discussed previously. Its angular frequency will, therefore, be expressed correctly by (9.5.16). The moment of inertia of a sphere about an axis through its own center is $I = \frac{2}{5}mR^2$, where m is its mass and R its radius. In this case, however, we need to know the moment of inertia I_0 about an axis parallel to one through the sphere's center but displaced by a distance l, so that it passes through the point of suspension; this corresponds to the distance h in Fig. 9.10. The moment of inertia about this axis can be calculated from the parallel axis theorem (7.4.1) which in this case tells us that

$$I_0 = I + ml^2 = \frac{2}{5}mR^2 + ml^2$$

or

$$I_0 = ml^2 \left(1 + \frac{2}{5}\frac{R^2}{l^2}\right) \qquad (9.5.17)$$

From (9.5.16), therefore, the angular frequency will be

$$\omega = \sqrt{\frac{mgh}{I_0}} = \sqrt{\frac{g}{l}\left(1 + \frac{2}{5}\frac{R^2}{l^2}\right)^{-1/2}} \qquad (9.5.18)$$

since in this case $h = l$. This expression differs from the one obtained as (9.5.11) for the angular frequency of the ideal point mass pendulum by the factor $(1 +$

$2R^2/5l^2)^{-1/2}$. It is evident that whenever the radius of the suspended mass is much less than the length l of the suspension, this factor will be very close to unity and the motion will be essentially that of an ideal pendulum. For the case where $l = 1.0$ m and $R = 2.5$ cm $= 0.025$ m, this factor amounts to 0.999875.

EXAMPLE 9.5.4

Discuss the motion of the ideal simple pendulum from the point of view of energy conservation.

Referring to Fig. 9.10, we see that the gravitational potential energy of the suspended mass is

$$U_p = mgh = mgl(1 - \cos\phi) \qquad (9.5.19)$$

provided we take the potential energy to be zero at the equilibrium point, where $h = 0$. Since the force T is always perpendicular to the direction of motion, it does no work on the mass and hence does not contribute to any change in its potential energy.

The kinetic energy will be

$$U_k = \frac{1}{2}mv^2 = \frac{1}{2}ml^2\left(\frac{d\phi}{dt}\right)^2 \qquad (9.5.20)$$

since the angular velocity $d\phi/dt$ is the linear velocity v times the radius. At the very end of the swing, when the angular displacement equals the amplitude ϕ_0, the velocity, and hence the kinetic energy, is zero and the potential energy $U_p(\phi_0)$ equals the total energy U_m. Therefore, by setting ϕ equal to ϕ_0 in (9.5.14), we may express the total energy as

$$U_m = U_p(\phi_0) = mgl(1 - \cos\phi_0) \qquad (9.5.21)$$

The sum of the kinetic and potential energies U_k and U_p, as given by (9.5.19) and (9.5.20), must at all times equal this constant total energy U_m. Therefore, we may write

$$U_m = mgl(1 - \cos\phi_0) = mgl(1 - \cos\phi) + \frac{1}{2}ml^2\left(\frac{d\phi}{dt}\right)^2$$

or

$$\frac{1}{2}\left(\frac{d\phi}{dt}\right)^2 = \frac{g}{l}(\cos\phi - \cos\phi_0) \qquad (9.5.22)$$

This expression gives the angular velocity $d\phi/dt$ as a function of the displacement ϕ. In principle, the angular displacement as a function of time could be calculated from this equation by taking the square root of both sides multiplying by $dt/(\cos\phi - \cos\phi_0)^{1/2}$, and integrating, but the integral is in general extremely difficult to evaluate. The resulting motion is not angular harmonic motion unless the amplitude ϕ_0 is a small angle.

If we differentiate both sides of (9.5.22), we may obtain

$$\frac{1}{2}\cdot 2\frac{d\phi}{dt}\frac{d^2\phi}{dt^2} = -\frac{g}{l}\sin\phi\frac{d\phi}{dt}$$

recalling that $d(z^2)/dt = 2z(dz/dt)$. This is easily seen to be exactly the same as (9.5.8), and we are thus led to precisely the same results as those obtained previously.

EXAMPLE 9.5.5

Show that in any system that executes angular harmonic motion, the instantaneous angular velocity $d\phi/dt$ can be expressed in terms of the instantaneous amplitude ϕ by the relation

$$\frac{d\phi}{dt} = \pm \omega \sqrt{\phi_0{}^2 - \phi^2} \qquad (9.5.23)$$

We may begin by multiplying both sides of (9.5.3) by ω and squaring both sides of the resulting expression to obtain

$$\omega^2 \phi^2 = \omega^2 \phi_0{}^2 \sin^2(\omega t + \delta)$$

Also, by squaring both sides of (9.5.6), we may write

$$\left(\frac{d\phi}{dt}\right)^2 = \omega^2 \phi_0{}^2 \cos^2(\omega t + \delta)$$

Adding these two equations, noting that $\sin^2 x + \cos^2 x = 1$ and solving for $d\phi/dt$, we may finally arrive at (9.5.23). Again, in (9.5.23) it is important to note that ω and $d\phi/dt$ are very different quantities, the latter representing the true instantaneous *angular velocity* of the system and the former giving us the constant *angular frequency* that gives the number of cycles of motion completed every second. Equation (9.5.23) is analogous to (9.2.19), which relates instantaneous velocity to instantaneous displacement in linear simple harmonic motion.

9.6 Damped Oscillations

Frequently, in actual practice, there are frictional, dissipative, or resistive forces that act to reduce the energy of systems in motion, eventually bringing them to a stop. We have encountered these forces previously as frictional forces opposing the motion of rigid objects that slide over one another, and as forces such as air resistance or fluid friction. Forces of the first kind are usually roughly independent of velocity, while forces of the latter type are generally found to increase with increasing velocity, being in many cases roughly proportional to velocity. Forces of this second kind, having the character of air resistance or fluid friction, act on objects that fall through fluid media, such as the atmosphere. Since they are proportional to the object's velocity, these forces increase in magnitude as the body's velocity increases, until eventually they are substantially equal to the body's weight. The resultant force then becomes essentially zero and no further increase in velocity occurs, the object having

now reached a *terminal* velocity. Resistive forces of this sort act in liquid media also; they are responsible, for example, in bringing a motorboat gradually to rest when the motor is shut off. Let us now briefly consider what occurs when, as frequently happens, forces of this sort act upon systems which in their absence would execute simple harmonic motion.

A resistive force that opposes the motion of the system will have a direction opposite that of the system's velocity. If the magnitude of the force is directly proportional to the velocity, then such a force can be represented mathematically as

$$F_r = -\alpha v_x = -\alpha \left(\frac{dx}{dt}\right) \qquad (9.6.1)$$

where α is a constant of proportionality. If such a force is present, the equation of motion (9.3.1) becomes

$$m \frac{d^2 x}{dt^2} = -m\omega^2 x + F_r \qquad (9.6.2)$$

or

$$\frac{d^2 x}{dt^2} = -\frac{\alpha}{m}\left(\frac{dx}{dt}\right) - \omega^2 x \qquad (9.6.3)$$

There are standard mathematical techniques for solving differential equations of this type; we shall not discuss them here, but simply quote the solution that is obtained when they are used. It is

$$x(t) = A e^{-(\alpha/2m)t} \sin\left[\sqrt{\omega^2 - \frac{\alpha^2}{4m^2}} \cdot t + \delta\right] \quad \alpha < 2m\omega$$

$$(9.6.4)$$

where A and δ are amplitude and phase constants that may be determined in the usual way from initial conditions. In writing this solution, it is assumed that the resistive force constant α is *less* than $2m\omega$, thus that the resistive term is relatively small. The correctness of (9.6.4) can be verified by differentiating the solution and substituting the resulting expressions into (9.6.3) to show that the solution satisfies the equation of motion. The algebraic details are assigned as an exercise. It is evident from (9.6.4) that the introduction of the "damping force" changes the character of simple harmonic motion in two ways. First of all, the frequency is decreased from the value ω to a value ω' such that

$$\omega' = \sqrt{\omega^2 - \frac{\alpha^2}{4m^2}} \qquad (9.6.5)$$

Also, the harmonic motion $\sin(\omega' t + \delta)$ is multiplied by an exponential factor $e^{-(\alpha/2m)t}$ that continually decreases the amplitude of the motion, making it approach zero for large values of time. A plot of the displacement $x(t)$ versus time is shown in Fig. 9.11. It is apparent from the figure that the total energy of the system also decays exponentially with time as the

295

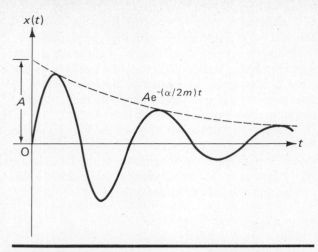

FIGURE 9.11. Displacement as a function of time for damped harmonic motion.

system's amplitude falls off. This energy is converted into heat via frictional work done by the damping force F_r.

The type of motion described by Eq. (9.6.3) and (9.6.4) is referred to as *damped harmonic motion*. We shall encounter it again in electrical circuits that contain resistance losses.

EXAMPLE 9.6.1

A steel ball having a mass of 16 g is dropped into a tall cylinder of lubricating oil. At first, it accelerates rapidly, as we might expect for a freely falling body, but then the acceleration becomes less and less and vanishes altogether. Finally, a constant *terminal velocity* $v_t = 60$ cm/sec is attained.

The same ball is then attached to a spring whose

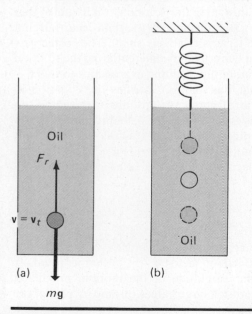

(a)   (b)

FIGURE 9.12

force constant is $k = 7500$ dynes/cm, lowered into a vessel containing the same type of oil, and set into oscillation, as shown in Fig. 9.12b. Find the frequency ω when the system executes simple harmonic motion in the absence of the oil. Express the damping constant α in terms of the terminal velocity v_t. Find the frequency ω' when the system oscillates with the ball immersed in oil. Write the equation expressing the displacement as a function of time in this case, assuming that the amplitude $A = 5.0$ cm. Assume that the resistive force exerted by the oil on the ball is directly proportional to its velocity.

When the terminal velocity v_t is reached, there is no longer any acceleration; therefore, the resultant force on the ball is zero. But the resultant force is the vector sum of the weight force $m\mathbf{g}$ and the resistive force $\mathbf{F}_r$, as illustrated in Fig. 9.12a. Therefore, when $v = v_t$,

$$F_r = mg \tag{9.6.6}$$

But since the resistive force is directly proportional to the velocity, its magnitude can be expressed, according to (9.6.1), as αv_t. Substituting this into (9.6.6) and solving for the constant α, we obtain

$$\alpha = \frac{mg}{v_t} = \frac{(16)(980)}{60} = 261.3 \text{ g/sec} \tag{9.6.7}$$

The frequency of the system in the absence of the damping fluid will be

$$\omega = \sqrt{\frac{k}{m}} = \sqrt{\frac{7500}{16}} = 21.65 \text{ rad/sec}$$

When the ball is immersed in the damping fluid, the frequency becomes, according to (9.6.5),

$$\omega' = \sqrt{\omega^2 - \frac{\alpha^2}{4m^2}} = \sqrt{(21.65)^2 - \frac{(261.3)^2}{(4)(16)^2}}$$

$$= 20.05 \text{ rad/sec}$$

Equation (9.6.4) for the displacement in this case has the form

$$x(t) = (5.0)e^{-8.16t} \sin(20.05t)$$

assuming that the displacement is zero at $t = 0$ so that the phase constant δ is zero.

9.7 Forced Oscillations and Resonance

Every system that can exhibit simple harmonic motion is characterized by a *natural angular frequency* ω. This natural frequency will be $\sqrt{k/m}$ for the mass–spring system, $\sqrt{g/l}$ for a simple pendulum, and $\sqrt{k'/I}$ for a torsion pendulum, but whatever the system, there will be some particular frequency at which it will oscillate when set into motion. Now let us ask ourselves what the behavior of such a system will be when it is subjected to an *external* sinusoidally varying force whose frequency may be the same as, or different from, the system's natural frequency. It the angular frequency of this externally imposed force is ω_0, we may express the force as a function of time as

$$F_{ext}(t) = F_0 \sin \omega_0 t \qquad (9.7.1)$$

To determine the effect of such a force on the system's motion, we must insert this additional external force on the right side of the equation of motion (9.3.1), which then becomes

$$m \frac{d^2 x}{dt^2} = -m\omega^2 x + F_0 \sin \omega_0 t \qquad (9.7.2)$$

or

$$\frac{d^2 x}{dt^2} = -\omega^2 x + \frac{F_0}{m} \sin \omega_0 t \qquad (9.7.3)$$

where ω is the system's natural frequency and ω_0 is the frequency of the externally imposed force.

Again, we may use standard mathematical techniques that have been developed to solve equations such as this. Since our primary objective is to understand the physical characteristics of the system rather than to teach mathematics, we shall rely on our intuition to bypass these techniques. To do this, we shall have to ask ourselves what might occur when we subject a system having a natural vibrational frequency to an external vibrational force whose frequency is, in general, different from this natural frequency. We know that the system will vibrate, but at what frequency? Its own or the externally imposed one? Now, nothing is going to change the externally imposed frequency; once it is there, it is there for good. The system, therefore, has to adjust its motion to the frequency of the external force. Its motion is, therefore, a *forced oscillation* at a frequency imposed from outside. These arguments lead us to suspect that the system's displacement may be written as

$$x(t) = A \sin(\omega_0 t + \delta) \qquad (9.7.4)$$

where A and δ are amplitude and phase constants. The correctness of this guess can be demonstrated by establishing the fact that the proposed solution does, in fact, satisfy the equation of motion (9.7.3). Differentiating (9.7.4) twice to find $d^2 x/dt^2$ and substituting $d^2 x/dt^2$ and $x(t)$ as given above into (9.7.3), we obtain upon rearranging terms

$$(\omega^2 - \omega_0^2) A \sin(\omega_0 t + \delta) = \frac{F_0}{m} \sin \omega_0 t \qquad (9.7.5)$$

These two sinusoidal functions can be equal at all times only if they have the same amplitude and phase.

This requires us to choose the phase constant δ to have the value *zero*, and the amplitude A must then be

$$A = \frac{F_0}{m(\omega^2 - \omega_0^2)} \qquad (9.7.6)$$

Substituting this into (9.7.4), we finally obtain

$$x(t) = \frac{F_0}{m(\omega^2 - \omega_0^2)} \sin \omega_0 t \qquad (9.7.7)$$

which satisfies the equation of motion exactly.

From all this, we see that system vibrates at the external forcing frequency ω_0 with an amplitude given by (9.7.6). The remarkable feature of this result is that *when the external forcing frequency ω_0 approaches the system's natural frequency, the amplitude of vibration approaches infinity!*

This phenomenon is referred to as *resonance*, the large-amplitude vibrations that are set up at the system's natural frequency are called *resonance vibrations*, and the system's natural frequency is often called its *resonant frequency*. We know that we can "pump" a swing to large amplitudes of oscillation by moving our body back and forth at the natural frequency of the swing. We are familiar with the fact that soldiers marching across a narrow suspension bridge ordinarily break their step so as to avoid exciting a resonant swaying motion of the bridge. We may be familiar with the fact that in 1940, a large suspension bridge across the Tacoma Narrows in the state of Washington was destroyed when a high wind set up fluctuating forces on the structure whose frequency was very close to its resonant frequency. The resulting motion is illustrated in Fig. 9.13. Finally, we may have experienced the annoying "shimmy" that can be felt in the front suspension of a car whose front wheels are unbalanced. This shimmy occurs at the speed for which the wheels' rotational frequency equals the natural frequency of the car's springs. These are all important examples of *mechanical resonance*.

Resonance phenomena also exist in electrical curcuits; the "tuning" of a radio to a certain broadcast frequency is an example. In the fields of atomic, nuclear, and solid-state physics, resonance effects occur on an atomic or subatomic level and are frequently utilized to study the structure of atoms, molecules, crystals, and atomic nuclei. Resonance phenomena are important in nearly all branches of science, and we shall encounter them on a number of occasions in our future work in elementary physics.

In the case considered above, of course, there is no frictional resistance or "damping" on our system, and it is this feature that leads to a prediction of infinite amplitude when $\omega_0 = \omega$ in (9.7.5). Actually, no system is entirely free of damping or other dissipative frictional resistance, and the effect of this is to limit the amplitude of the system to a finite—

FIGURE 9.13. Photograph of the destruction of the Tacoma Narrows bridge by resonant oscillations excited by high winds. More recent designs have been careful to avoid the possibility of exciting resonant oscillations or to provide mechanisms for damping when they occur.

though perhaps very large—value at the resonant frequency, as shown by the dotted curve in Fig. 9.14. The exact size of this maximum amplitude will depend, of course, on the strength of the frictional or dissipative effects that are present. In many cases, resonance vibrations can become large enough to cause the system to self-destruct, as in the case of the Tacoma Narrows bridge. Frequently, damping is intentionally introduced into oscillatory systems to limit the amplitudes that can be attained and, therefore, to counteract their tendency toward self-an-

nihilation. The use of hydraulic shock absorbers in the suspension systems of automobiles reflects this motivation, a fact that anyone who has driven a car without shock absorbers or with weak "shocks" will readily verify.

Physically, the large amplitudes associated with resonance arise because when the externally imposed force has the same frequency as the natural frequency of the system, the external force can act to increase the system's kinetic energy by the same small amount during each cycle. These increases of kinetic energy

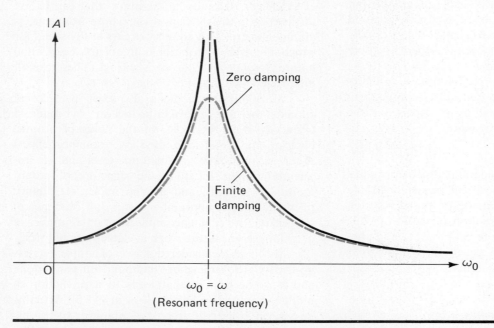

FIGURE 9.14. Amplitude of forced harmonic oscillations as a function of the forcing frequency, in systems with and without damping.

are cumulative and build up until the total energy of the system, and thus its amplitude, becomes very large. Far from resonance, the system tends to send its own kinetic energy back into the external system that provides the forcing oscillation, as well as to accept energy from that source. As a result, a balance is set up under these conditions in which the energy accepted by the system and rejected by it to the surroundings is equalized at amplitudes which are reasonably small.

SUMMARY

Periodic motion is motion that occurs in repetitive cycles all of which are exactly alike. Oscillatory motion is one-dimensional, periodic motion. Simple harmonic motion is an oscillatory motion in which linear or angular displacement varies sinusoidally as a function of time. Oscillatory motions that are not simple harmonic motion are sometimes referred to as anharmonic oscillations.

Simple harmonic motion occurs *whenever the resultant force (or torque) on a system is directly proportional to the linear (or angular) displacement from the equilibrium position and when, in addition, the direction of the force is opposite that of the displacement.* The equation of motion for a system of this type can always be written in the standard form

$$\frac{d^2x}{dt^2} = -\omega^2 x$$

where ω is a proportionality constant referred to as the angular frequency. The angular frequency divided by 2π is the number of cycles of motion executed during any one second time interval. The displacement $x(t)$ that satisfies this equation is

$$x(t) = A \sin(\omega t + \delta)$$

where A and δ are constants. The constant A represents the maximum possible displacement and is referred to as the amplitude. The angle $(\omega t + \delta)$ is called the phase angle. The quantity δ, usually called the phase constant, represents the value of the phase angle at time $t = 0$. The constants A and δ can be evaluated when the initial displacement and velocity, or other equivalent information, are given. The velocity and acceleration associated with simple harmonic motion are

$$v_x = A\omega \cos(\omega t + \delta)$$
$$a_x = -A\omega^2 \sin(\omega t + \delta) = -\omega^2 x$$

The frequency f refers to the number of cycles of motion that take place each second and is equal to $\omega/2\pi$. The period T is the length of time required for the execution of one cycle and is equal to the recip-

rocal of the frequency, hence to $2\pi/\omega$. Also, velocity and displacement are related by

$$v_x = \pm\omega\sqrt{A^2 - x^2}$$

The force that gives rise to simple harmonic motion is that described earlier in connection with Hooke's law, which describes the behavior of an ideal spring. We have already shown that this is a conservative force, and, therefore, the total energy of systems that execute simple harmonic motion is conserved.

Simple harmonic motion can be regarded as the projection of uniform circular motion onto the x- and y-axes; these two projections can be viewed as the x- and y-components of a vector of length A that rotates about the origin with angular velocity ω. Simple harmonic oscillations of the same angular frequency but of different amplitudes and phases can, therefore, be combined by using the laws of vector addition, as applied to two or more such rotating vectors, as described in section 9.4 and Example 9.4.1.

Angular harmonic motion, such as that exhibited by the torsion pendulum, occurs when the equation of motion for the angular displacement ϕ of a system about a given axis can be stated in the form

$$\frac{d^2\phi}{dt^2} = -\omega^2\phi$$

The quantity ω refers here to the number of cycles of motion that occur during one second and is *not* equal to the instantaneous angular velocity of the system. The latter quantity, designated in this chapter as $d\phi/dt$, is the time rate of change of the angular displacement. Since the angular displacement is given by

$$\phi(t) = \phi_0 \sin(\omega t + \delta)$$

where ϕ_0 is the angular amplitude, expressing the maximum possible angular excursion from equilibrium, we must have

$$d\phi/dt = \omega\phi_0 \cos(\omega t + \delta)$$

and

$$d^2\phi/dt^2 = -\omega^2\phi_0 \sin(\omega t + \delta) = -\omega^2\phi$$

as the instantaneous angular velocity and acceleration.

The angular frequency ω associated with a mass attached to a single ideal spring of spring constant k is given by

$$\omega = \sqrt{k/m}$$

For a torsion pendulum of moment of inertia I and torsion constant k', the angular frequency is

$$\omega = \sqrt{k'/I}$$

The motion of the simple pendulum is in general anharmonic, but for small angular amplitudes, it is very closely approximated by angular simple harmonic motion of angular frequency

$$\omega = \sqrt{g/l}$$

where l is the length of the pendulum. For the general physical pendulum, consisting of an irregular object suspended at a point O which is at a distance h from the center of mass, the motion is again anharmonic in general; but for small amplitudes, it approximates simple harmonic motion of angular frequency

$$\omega = \sqrt{mgh/I_0}$$

where I_0 is the moment of inertia of the object about the rotation axis.

Damped oscillations occur when frictional forces directly proportional to velocity are present. The effect of these forces is to make the amplitude of oscillation decrease exponentially with time and to decrease somewhat the frequency of oscillation from the value it would have in the absence of dissipative forces. The situation is illustrated by Example 9.6.1.

Forced oscillations occur when a mechanical system is forced to vibrate at a frequency imposed upon it by a sinusoidally varying external driving force. When the frequency of this external driving force approaches the natural frequency of vibration of the system, a condition of resonance is approached in which the amplitude of the system's oscillations may become very large. At the resonant frequency, the amplitude will be limited only by the degree to which dissipative forces may be able to get rid of energy or by the self-destruction of the system itself!

simple harmonic. Formulate several criteria to represent the degree to which an approximate simple harmonic motion approaches a true simple harmonic motion.

9. The frequencies ω and f can both be given in units of $\sec^{-1}$. Why does one usually say that ω is in rad/sec?

10. What is the significance of the two possible signs which give the velocity as a function of position in (9.2.20)?

11. Does the balance wheel of a watch execute exact simple harmonic motion? Explain your reasoning.

12. How can you determine the value of an unknown mass by using a watch and a spring of known force constant?

13. Describe what happens to a system at resonance if the damping is very small.

14. An elevator is accelerating downward. Describe how this affects the period of a pendulum suspended from the ceiling of the elevator.

15. Does the piston of a reciprocating automobile engine execute exact simple harmonic motion within the cylinder or is its motion only approximately simple harmonic?

16. A hollow sphere is attached to a long string. Would you expect the period of the pendulum to be increased or decreased if the sphere were partially filled with water?

17. When you have a force directly proportional to the velocity, is it possible that this is a conservative force?

18. In the presence of damping, the frequency of oscillation diminishes. Can you explain this qualitatively?

19. Clemson University's Foucault pendulum is a simple pendulum set up in a tower that extends from the top of the physics building to the ground floor four stories below. An undergraduate is given a stopwatch and told to find the height of the building. He has no way of climbing onto the roof and no other measuring instrument. How should he proceed in carrying out his assignment?

20. By the way, what *is* a Foucault pendulum, what is it used to demonstrate, and how does it work?

QUESTIONS

1. On the basis of dimensional arguments, show that the frequency of oscillation of a mass attached to a spring must be proportional to $\sqrt{k/m}$.

2. If a pendulum clock is transported from the earth to the moon, the period of motion changes. Does it increase or decrease?

3. If the potential energy as a function of position is given, describe how you can determine points of equilibrium. How do you determine whether the equilibrium is stable or unstable?

4. For a certain force law, the potential energy function is proportional to x^4. Where is the equilibrium position? Will motion about this position be simple harmonic?

5. A mass is executing simple harmonic motion. If the amplitude is doubled, what happens to the frequency? To the total energy?

6. A mass attached to a string swings as a pendulum with simple harmonic motion. At what position is the string most likely to break? Explain your reasoning.

7. It is somewhat rare to find any system which executes exact simple harmonic motion. Why is this true?

8. Give some examples of motions which are approximately

PROBLEMS

1. A mass m attached to a spring of force constant k executes simple harmonic motion given by the equation $x = 0.5 \cos(0.8t - 0.4)$, where x is in meters. For this motion, find (a) the amplitude, (b) the frequency, (c) the period, and (d) the ratio k/m.

2. For the preceding problem, find the velocity and acceleration of the mass at $t = 2$ sec.

3. A mass of 200 grams is suspended from a fixed overhead support by an ideal spring. It is displaced from its equilibrium position and released. It is subsequently found that the displacement $y(t)$, measured from the equilibrium position in centimeters, is $y(t) = 4.8 \sin(7.5t + \pi/4)$. The time t is measured in seconds. Find (a) the amplitude of the oscillatory motion, (b) its angular frequency, (c) the frequency, (d) the phase constant, (e) the period, and (f) the spring constant.

4. In the preceding problem, find (a) the initial displacement, (b) the initial velocity, and (c) the initial acceleration of the mass. (d), (e), (f) What are the values of displacement, velocity, and acceleration at $t = 0.18$ sec? (g) What is the velocity with which the mass passes through the equilib-

rium position? **(h), (i)** What are the velocity and acceleration when the displacement is 3.6 cm?

5. A simple harmonic oscillator consists of a mass of 1.2 kg resting on a horizontal frictionless surface attached to a fixed support by a horizontal ideal spring whose spring constant is 180 newtons/meter. The system is set into oscillation, and at time $t = 0.16$ sec it is observed that the displacement from the equilibrium position is 0.15 meter and the velocity is -0.80 meter/sec. Find **(a)** the angular frequency, **(b)** the period, **(c)** the amplitude, and **(d)** the phase constant. **(e)** Write the equation for displacement as a function of time.

6. A 2.4-kg mass oscillates on a frictionless horizontal surface. Its oscillations are sinusoidal, with amplitude 0.40 meter and frequency 1.8 cps. Find **(a)** its total energy, **(b)** its average kinetic energy, **(c)** its average potential energy, **(d)** its maximum kinetic energy, and **(e)** its maximum potential energy.

7. Find **(a)** the displacement and **(b)** the velocity of the oscillator in the preceding problem when the potential energy and kinetic energy of the system are equal. **(c)** For what values of the phase angle are the kinetic and potential energies equal?

8. A simple harmonic oscillator consists of a 1.60-kg mass connected to a fixed support by a horizontal ideal spring whose spring constant is 72.0 newtons/meter. The mass rests upon a horizontal frictionless surface. The system is set into vibration, and it is observed that at time $t = 0.08$ sec the mass is 0.180 meter from the equilibrium position, while at $t = 0.40$ sec it is 0.060 meter from the equilibrium point. Find **(a)** the angular frequency of oscillation, **(b)** the frequency, **(c)** the amplitude of vibration, and **(d)** the phase constant. **(e)** Write the equation giving displacement as a function of time.

9. The oscillator in the preceding problem is set into oscillation with an amplitude of 0.200 meter. Just as the mass passes through the equilibrium position, a large blob of clay having a mass of 1.20 kg is dropped onto the oscillating 1.60-kg mass. The clay sticks to the oscillating mass and thereafter vibrates back and forth with it. Find **(a)** the frequency and **(b)** the amplitude of oscillation of the system in its final state.

10. In the preceding problem, suppose that the ball of clay is dropped onto the oscillating 1.6-kg mass at the extreme limit of its excursion from equilibrium, when its velocity is zero. What would the answers be in this case?

11. Is energy conserved in the situations considered in the two preceding problems? If not, how much is lost in each case?

12. What are the amplitude and phase of the sum of two simple harmonic oscillations of equal amplitude A which differ in phase by **(a)** 0°, **(b)** 30°, **(c)** 45°, **(d)** 60°, **(e)** 90°, **(f)** 120°, **(g)** 135°, **(h)** 150°, and **(i)** 180°?

13. What are the amplitude and phase of the sum of the two harmonic oscillations A (amplitude 2.40 meters, phase constant zero) and B (amplitude 3.60 meters, phase constant 120°)?

14. Three sinusoidally varying ac voltages are superposed in an electrical circuit. All three have the same frequency of 60 Hz. The first is of amplitude 117.0 volts and phase constant 0°; the second is of amplitude 88.0 volts and phase constant 225°; and the third is of amplitude 240.0

volts and phase constant 120°. What are the amplitude, frequency, and phase of their sum?

15. A torsion pendulum consists of a horizontal plate supported by a thin, elastic rod whose upper end is rigidly clamped in a fixed support. It is observed that a torque of 32.0 newton-meters is needed to twist the plate through an angle of 60°. The plate is then released and is observed to execute angular simple harmonic motion whose period is 1.25 sec. **(a)** What is the torsion constant of the suspension rod? **(b)** What is the moment of inertia of the plate? **(c)** What is the angular displacement of the plate as a function of time, assuming that it is released from rest with initial angular displacement of 60°?

16. In the preceding problem, the angular displacement at time t was found to be $\phi = (\pi/3) \sin[5.027t + (\pi/2)]$ radians. Find **(a)** the angular amplitude of this motion, **(b)** its angular frequency, **(c), (d), (e)** the angular displacement, velocity and acceleration of the system at $t = 0.50$ sec, and **(f), (g)** the angular velocity and acceleration when the angular displacement is 45° from the equilibrium position.

17. A certain pendulum ordinarily has a period of 2 sec. What would the period be if it is suspended from the ceiling of an elevator which accelerates downward at 10 ft/sec²?

18. An 80-kg man finds that when he gets into his 1500-kg car, the center of gravity of the car is lowered by 0.50 cm. What is the value of the natural oscillation frequency of the car on its springs? Neglect damping effects.

19. A piston moves up and down with frequency f and amplitude A. A coin is placed at the top of the piston. For what value of f does the coin lose contact with the piston at the top of the stroke?

20. A mass attached to a spring oscillates with an amplitude A. When the displacement from equilibrium is $A/3$, what fraction of the energy is kinetic and what fraction is potential energy?

21. A 6-kg block extends a spring 18 cm from its equilibrium position. The 6-kg block is removed and replaced by a 4-kg block which is set into oscillation. Find the frequency and period of oscillation.

22. A spring has a constant $k = 40$ N/m. If a 3-kg mass is attached and set into oscillation, find the period of the motion.

23. Electrons in an oscilloscope experience simple harmonic motion in two directions, given respectively by $x = A \cos \omega t$ and $y = B \sin \omega t$. Show that the electrons follow elliptical paths.

24. A pendulum is released from rest with the string in a horizontal position, as shown in the accompanying

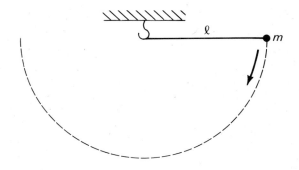

figure. Show that the time it takes for the mass to reach its lowest point is given by the expression

$$t = \sqrt{\frac{l}{2g}} \int_0^{\pi/2} \frac{d\phi}{\sqrt{\cos \phi}}$$

Do not try to work out the integral; it is an extremely difficult one.

25. The bob of a pendulum consists of a small 1-kg mass. The string has a length of 1 m. If the pendulum is released at $t = 0$ at an angle of 0.1 rad with the vertical, with an initial angular velocity of 0.5 rad/sec, obtain an expression which gives the angular displacement as a function of time. Assume the initial angular velocity is as shown in the diagram.

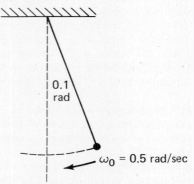

26. A particle whose mass is 20 g is suspended from an ideal spring and oscillates vertically with undamped simple harmonic motion with a frequency of 8 cycles per sec. If the particle starts from rest 6 cm below its equilibrium position, find **(a)** the displacement, velocity, and acceleration of the particle at an arbitrary time t (plot these quantities on a graph over one time period); **(b)** the maximum values of the potential and kinetic energies; and **(c)** the velocity and acceleration when the displacement is 3 cm.

27. A particle executes simple harmonic motion with a frequency of 5 cycles per second. Obtain an expression giving the displacement x as a function of t, using the following initial conditions: **(a)** $t = 0$, $x = A$, $v_x = 0$; **(b)** $t = 0$, $x = 0$, $v_x = v_0$; **(c)** $t = 0$, $x = 2$, $v_x = 5$.

28. A body weighing 3 pounds is hung on the lower end of a spring and produces an extension of 0.5 in. Find the period of the simple harmonic motion when the body is set in motion.

29. The accompanying figure illustrates three rigid bodies each of mass M suspended from a point O. Find the frequencies and periods of oscillation for each of the objects.

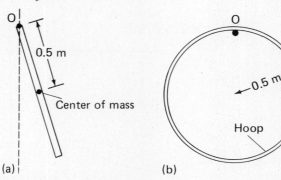

(a)   (b)

30. A point mass m is located on a frictionless horizontal table, halfway between two fixed supports a distance $2d$ apart, as shown at **(a)** in the accompanying diagram. Two ideal springs each of equilibrium length l_0 ($l_0 < d$) are attached to the supports. These springs are now stretched through the distance B and attached to the mass, as shown at **(b)** in the diagram. Find the angular frequency of the simple harmonic motion that takes place when the mass m is displaced slightly along the x-direction and released.

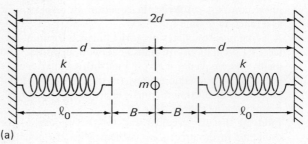

(a)

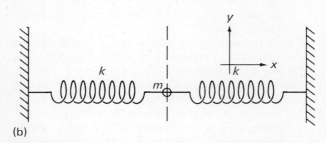

(b)

*31. Show that when the mass m in the preceding problem is given a *small* initial displacement (much less than B, to be exact) in the y-direction, its motion is *approximately* a simple harmonic oscillation of angular frequency given by $\omega^2 = (2k/m)(B/l)$.

*32. Show that when the equilibrium length of the springs in the preceding problem is exactly equal to d, an initial displacement along the x-direction results in simple harmonic motion as in Problem 31, but an initial displacement along the y-direction, *no matter how small*, always excites an oscillatory motion that is *anharmonic*.

33. Prove that when $\alpha < 2m\omega$, Eq. (9.6.4) is a solution of (9.6.3).

34. Using the solution (9.6.4) for damped harmonic motion, obtain **(a)** the kinetic energy as a function of t and **(b)** the potential energy as a function of t, and **(c)** calculate the sum and explain why this varies with t.

35. In the accompanying diagram, we see two cases illustrating an object of mass m suspended in different ways

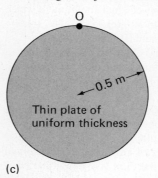

(c)

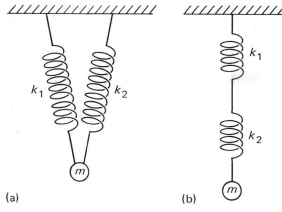

(a)   (b)

by two ideal springs having spring constants k_1 and k_2. In each case, it is desired to replace the two springs by a single spring in such a way that the frequency of oscillation remains unchanged. How should the spring constant k' of the equivalent single spring be chosen in each case?

36. A circular plate of radius R and moment of inertia I is suspended by a torsion rod whose torsion constant is k'. Its rim is connected to a fixed support via an ideal spring whose spring constant is k, as shown in the accompanying diagram. Show that when the plate is given an initial angular displacement and then released, it will oscillate with angular simple harmonic motion whose frequency is $\omega = ((k' + kR^2)/I)^{1/2}$.

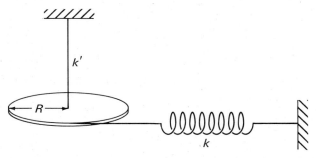

37. A mass m is suspended from a massive frictionless pulley of radius R and moment of inertia I and attached via an ideal massless spring of spring constant k to a fixed support, as shown in the diagram. Show that when the system

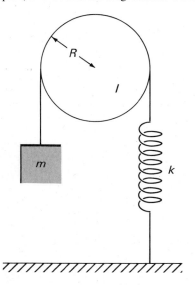

is given a small displacement from the equilibrium position and released, it executes simple harmonic motion whose angular frequency is given by $\omega = (kR^2/(I + mR^2))^{1/2}$.

38. In the situation described in the preceding problem, what is the maximum initial displacement that can be imparted without reducing the tension in the string supporting the mass to zero at some point in the cycle?

39. A uniform thin rod of length L and mass m is suspended freely in a vertical position at a point O, which is at a distance d (where $d < L/2$) from its upper end. The rod is displaced slightly from equilibrium and allowed to swing back and forth about an axis through O. Show that the angular frequency of the motion is given by $\omega^2 = 12gh/(L^2 + 12h^2)$.

40. A small ball of mass m and radius r rolls along a track having the form of a circular arc of radius R, as shown in the accompanying diagram. Assuming that the ball's moment of inertia is I, that it rolls without slipping, and that its radius is very much less than that of the circular track, show that when displaced slightly from its equilibrium position and allowed to roll back and forth, it executes simple harmonic motion of angular frequency given by $\omega^2 = (g/R)[mr^2/(I + mr^2)]$.

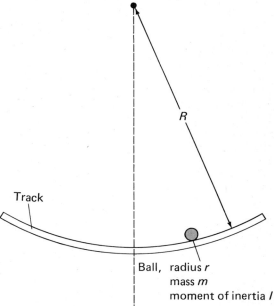

Track

Ball, radius r
mass m
moment of inertia I

41. A solid cylinder of uniform mass density and total mass M rolls without slipping on a circular surface of radius R,

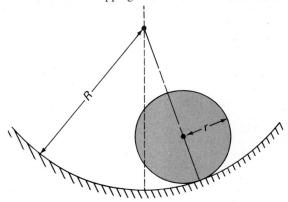

as shown in the diagram. If the cylinder's radius is r, show that its motion is oscillatory and that for small amplitudes the motion is approximately angular harmonic motion of angular frequency $\omega = \sqrt{2g/3(R - r)}$.

42. A uniform rod of length 1 m is suspended from a wire as shown in the figure. A torque of 0.2 N m causes an angular displacement of 0.1 rad. If the rod has a mass of 0.5 kg, find the frequency of torsional oscillations.

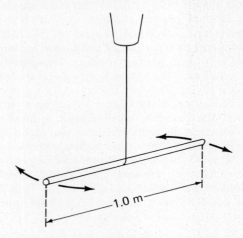

43. The period of damped oscillations of a 1-kg mass attached to a spring is 0.5 sec. If the 1-kg mass stretches the spring by 5 cm, find the constant α which expresses the damping.

*44. An automobile manufacturer wishes to fit his cars with shock absorbers that reduce the amplitude of up-and-down body oscillations to one tenth the initial value after one cycle of oscillation. The car's springs are ideal coil springs whose spring constant is 24,000 newtons/meter. Its mass is 1500 kg, which is equally distributed among the four wheels each of which is suspended by one of the springs. Assuming that the spring and shock absorber on each wheel behave as an ideal damped harmonic oscillator, find (a) the frequency of oscillation of the car on its springs before the shock absorbers are attached, (b) the damping parameter α required for each of the four shock absorbers if the manufacturers design objective is to be attained, and (c) the frequency of the damped oscillations that occur when the shock absorbers are mounted.

*45. In the situation described in the preceding problem, let m be the car's mass, $\omega = (k/m)^{1/2}$ the frequency of oscillation of the springs, α the combined damping parameter of the four shock absorbers, and β the ratio of initial amplitude to residual amplitude after one complete cycle of oscillation. Show that (a) $\alpha^2 = 4m^2\omega^2/[1 + (2\pi/\ln \beta^{-1})^2]$ and (b) $\omega'^2 = \omega^2/[1 + (\ln \beta^{-1}/2\pi)^2]$.

*46. (a) Show that the velocity of an ideal damped harmonic oscillator whose displacement is given by $x(t) = Ae^{-(\alpha/2m)t} \sin(\omega't + \delta)$ can be written as $v_x = A\omega e^{-(\alpha/2m)t} \cos(\omega't + \delta + \delta')$, where $\omega'^2 = \omega^2 - (\alpha/2m)^2$ and $\delta' = \tan^{-1}(\alpha/2m\omega')$. (b) Show that the acceleration of such a damped harmonic oscillator can be written as $a_x = -A\omega^2 e^{-(\alpha/2m)t} \sin(\omega' + \delta + 2\delta')$. Hint: Use the ideas developed in section 9.4 for combining sinusoidal oscillations of the same frequency.

47. An ideal harmonic oscillator is comprised of a mass of 4.0 kg attached to an ideal spring whose spring constant is $57600\pi^2$ newtons/meter. The mass is subjected to a sinusoidally time-varying external force of variable frequency whose magnitude is given as a function of time by $F(t) = 60.0 \sin 2\pi ft$, in units of newtons. (a) Find the resonant frequency of the spring–mass system. Find the amplitude of the vibrations of the mass when the variable frequency f of the external force has the value (b) 30 cps, (c) 40 cps, (d) 50 cps, (e) 55 cps, (f) 50 cps, (g) 59.5 cps, and (h) 59.9 cps.

10.1 Introduction

Man has always been fascinated by the rhythmic undulations of the surface of the sea. The complex breakers which regularly approach our shores have been observed and studied for thousands of years, and yet our understanding of ocean waves is even now far from complete. Ocean waves may originate very far from shore and may travel at high speed toward their final destiny, but during their long journey they neither transport the debris in the ocean nor do they move any water over large distances. This statement contains the very essence of wave motion, for a wave is a disturbance of a medium, a disturbance which itself moves from place to place but which does not carry the medium with it as it goes. The wave is capable of transporting *energy* over large distances but in doing so it does not transport any *matter*.

The world we live in is full of waves. Without the electromagnetic waves that light the sky and warm our planet, our world would be ice cold and devoid of life. Without the waves we detect as sound, we could not hear the music of Beethoven or Brahms nor could we communicate by speech. Without waves, we would have no radios, no television sets, nor most of our modern technology. Without waves, we could never have gone to the moon, or even have been aware of its existence. Wave phenomena are one of the most basic aspects of physical reality and are fundamental to any description of matter on an atomic or subatomic scale.

Not all waves work for the benefit of man. Tidal

waves (*tsunamis*) have devastated coastal cities and caused great loss of life in the past and will probably continue to do so in the future. Shock waves from supersonic aircraft cause a sonic boom which can shatter glass and damage the structure of buildings. Overdoses of energy carried by waves in the form of radiation can be harmful to life.

Physicists have studied waves at the macroscopic and microscopic levels, and these studies have proved to be of enormous value to civilization.

10.2 Types of Waves

Let us begin our study of wave motion by discussing the waves that are somewhat familiar to us. We will first consider waves that move in a physical medium such as water, air, or a string. Later, in a subsequent chapter, we will take up the very important special case of electromagnetic waves, which can travel, or *propagate*, in empty space or vacuum.

If someone were to ask you to describe in words the concept of a wave, you would probably have some difficulty doing this. Perhaps the best you could do would be to say that a wave is a *self-propagating disturbance* which travels in a medium and which owes its existence to a source that initially created the disturbance. Your questioner is likely to be confused by your choice of words, and perhaps the best thing to do is to take him into a laboratory and show him some examples of waves. This is likely to be the soundest way to approach the subject, and therefore this is the procedure we will adopt. After a number of examples have been discussed, the principal features of wave motion will become apparent.

One of the simplest examples of wave motion can be demonstrated by using a long rope tied to a post. If the rope is taut and the free end is jiggled, a disturbance propagates along the rope by traveling toward the post. Figure 10.1 schematically illustrates a number of snapshots of the rope taken at equal time intervals. This type of wave, which moves in a one-dimensional system, is called a *one-dimensional wave*. The rope is a one-dimensional medium since each element of it may be located by means of a single coordinate x when the medium is in equilibrium.[1] The surface of water is two-dimensional, and, therefore, surface waves in a fluid are called two-dimensional waves.

---

[1] In this example, since the motion of the rope that is initiated by the passage of the wave is in the y-direction, this is no longer strictly true when the equilibrium of the system is disturbed by the propagation of the wave. Now, the coordinate x locates the part of the system we are discussing, while a displacement along the y-direction describes the magnitude of the disturbance.

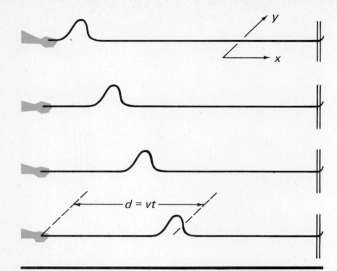

FIGURE 10.1. Propagation of a wave in the form of a transverse pulse along a rope.

As the wave of Fig. 10.1 progresses toward the post, those parts of the rope which are moving experience a transverse displacement along the y-direction. This transverse displacement will vary as a function of time and also as a function of the distance x along the rope. We must, therefore, regard the displacement $y(x, t)$ as a function of two variables, x and t. The displacements of various elements of the rope in this type of wave motion are always perpendicular, or transverse, to the direction of the wave propagation. We say that we have a *transverse wave* whenever particle velocity and wave velocity are perpendicular in this manner. Note that in this wave motion there is no net movement of any part of the rope in the direction of the wave.

There are many possible ways of initiating disturbances of different shapes on the rope of Fig. 10.1. All of these produce transverse waves, and all of these waves move at the same speed, provided the tension in the rope is the same in each instance.

Not all waves are transverse. Another class, known as *longitudinal waves*, are present when the displacements of particles undergoing disturbance are *parallel* to the direction of wave propagation. An example of this type of wave is illustrated in Fig. 10.2. The left end of a spring which is in a horizontal position on a frictionless table top is given a sudden momentary displacement by quickly pushing it to the right (causing a compression) and then pulling it back to the left (creating a rarefaction). This disturbance then propagates to the right with a speed governed by the characteristics of the spring. If we focus our attention on a small segment of the spring which is in motion, we notice that its motion is either *parallel or antiparallel* to the direction of motion of the wave. This wave is another example of a one-dimensional

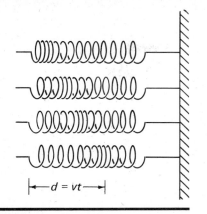

FIGURE 10.2. Propagation of a wave in the form of a longitudinal pulse along a spring.

wave. The displacements can again be characterized by means of a variable, u, which depends on the single variable x and also on t. In this case, however, it is important to realize that the displacement $u(x, t)$ is *parallel* to the x-axis. Any such wave is referred to as a longitudinal wave.

An essential characteristic of one-dimensional wave motion is that the initial disturbance moves with constant speed. Photographs of the medium taken at equal intervals of time all look alike in the sense that the disturbance maintains its initial shape but merely moves to a different location. The speed and direction of motion of the disturbance define a velocity vector known as the *wave velocity*. In the next section, some of the qualitative ideas discussed above will be made more quantitative.

We have mentioned two examples of waves, both of which are one-dimensional. There are also examples of waves which move in two dimensions and in three dimensions. An example of the former is provided by the surface waves which spread on a pond after a stone has been tossed into the water. An example of the latter is provided by sound waves coming from a train whistle. The type of wave created and transmitted depends on the medium that carries the wave as well as on the geometry of the source that produced the wave. For example, a point source such as a whistle will send out sound waves in all directions. At a given time, any displacement of the medium that exists at any point on a sphere, centered at the source, is the same at any other point on this sphere. In this case we say that *the wavefronts are spherical*. The stone falling into a pond sends out circular surface waves which are disturbances with *circular wavefronts*. Waves with plane and cylindrical wavefronts are also very commonly encountered. In Fig. 10.3, a number of

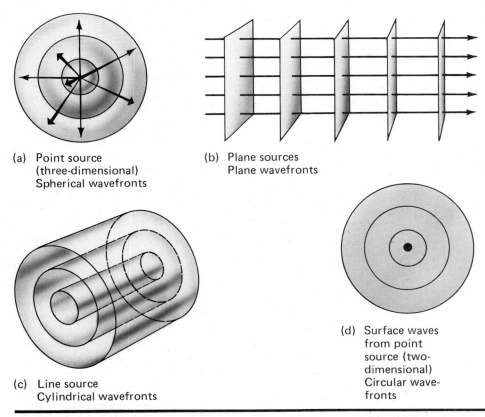

(a) Point source
(three-dimensional)
Spherical wavefronts

(b) Plane sources
Plane wavefronts

(c) Line source
Cylindrical wavefronts

(d) Surface waves from point source (two-dimensional) Circular wave-fronts

FIGURE 10.3. Spherical (a), plane (b), and cylindrical (c) wave fronts. (d) Circular waves on the surface of a liquid, generated by a point source.

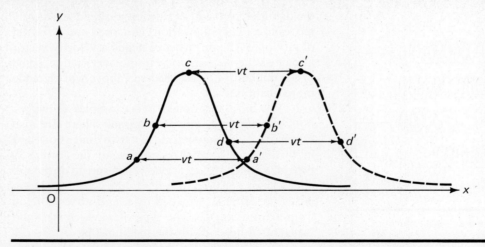

FIGURE 10.4. Constant velocity associated with different points in a disturbance propagating as a wave.

different waves are illustrated, with several wavefronts for each case. These wavefronts have the property that the magnitude of the disturbance of material particles (displacement from equilibrium) is everywhere the same on a given wavefront. Furthermore, the wave always propagates along a direction perpendicular to the wavefront at any point.

10.3 Mathematics of One-Dimensional Waves

There are two considerations which must be incorporated into the mathematical description of one-dimensional waves. One of these is the constant speed of propagation of the initial disturbance, while the second is the maintenance of the original form of the disturbance.

The physics can best be described with reference to the rope of Fig. 10.1. At $t = 0$, the rope has experienced an initial disturbance. This initial disturbance at $t = 0$ may then be characterized by means of the equation

$$y(x, 0) = F(x) \tag{10.3.1}$$

which gives the transverse displacement of each point x on the rope at the initial time. How do we now determine the dependence of y on t for all subsequent times for each value of x, subject to the considerations mentioned above? In order for the disturbance to maintain its initial shape, it is essential that after a time interval t has elapsed, each and every displacement y should have moved to a *new location* some constant distance d from the original location. The constancy of speed is guaranteed by requiring that $d = vt$, where the constant v is called the *wave velocity*, or the *phase velocity*. These considerations are demonstrated in

Fig. 10.4. The conditions mentioned above are fulfilled if the time dependence of the disturbance is given by the equation[2]

$$y(x, t) = F(x - vt) \tag{10.3.2}$$

If we set $t = 0$, this describes the initial disturbance as given by (10.3.1), but it also describes the situation at any later time t. Equation (10.3.2) gives a one-dimensional traveling wave moving to the right. For a wave traveling to the left, (10.3.2) would be modified by changing the sign in front of the velocity v. We shall see later that the magnitude of the phase velocity is in all instances determined by the physical properties of the medium in which the wave propagates.

The most important feature of (10.3.2) is the occurrence of the combination $x - vt$ as the *argument* of the function F. For example, if the maximum of the disturbance defined by the function $F(x - vt)$ should occur where the argument of the function is zero, it is clear that the location of that point as a function of time will be defined by the condition that $x - vt = 0$ or $x = vt$; in other words, the point of the disturbance moves to the right, along the direction of propagation, with constant wave velocity v. The same general remarks clearly apply to all other points defined by $F(x - vt)$, and it is evident, therefore, that writing the argument of such a function in the form $x - vt$ (rather than simply as x) causes the function to "travel" in the x-direction with velocity v just the way a wave would! The function F itself must be determined from initial conditions; and although *any* function F leads to a wave motion in the mathematical sense, in practice,

---

[2] In this equation, the quantity $x - vt$ represents the *argument* of the function F described (for $t = 0$) in Eq. (10.3.1). It is decidedly *not* a factor that multiplies F!

physical considerations will limit the range of possibilities, as the following example demonstrates.

EXAMPLE 10.1

Which of the following mathematical expressions can represent one-dimensional waves that are physically reasonable?

a. $y(x, t) = y_0 e^{-\lambda(x-vt)^2}$ (10.3.3)

b. $y(x, t) = \beta(x + vt)^4$ (10.3.4)

c. $y(x, t) = y_0 e^{-\lambda x^2(1+t/t_0)}$ (10.3.5)

d. $y(x, t) = y_0 \sin k(x - vt)$ (10.3.6)

Let us first consider which of these expressions represents a wave motion in one dimension. We recall that a one-dimensional traveling wave must be in all cases a function of $x - vt$ or of $x + vt$. The functions given by **(a)**, **(b)**, and **(d)**, therefore, meet the mathematical criteria, but **(c)** does not. The "wave" given by **(b)** is not physically sensible since it predicts that at each value of x the displacements become arbitrarily large as t increases. There are no physical systems which actually display this behavior and, therefore, (10.3.4) is not likely to be a wave of any physical interest. In Fig. 10.5, the above functions are illustrated for various equally spaced time intervals. Notice that **(a)**, **(b)**, and **(d)** maintain their *shape* as t increases, whereas **(c)** does not. Equation **(d)** represents a *sinusoidal* wave, which turns out to be the simplest and most important type of wave motion.

Equation (10.3.2) is a solution of a well-known differential equation called the *wave equation*. This equation relates the second partial derivative of y with respect to the variable x to the second partial derivative with respect to t. Partial derivatives[3] are needed because y depends on two variables, x and t. In calculating the partial derivatives of functions of x and t, it is important to note that in taking the partial derivative with respect to x, the quantity t is held constant, while in taking the partial derivative with respect to t, the quantity x is held constant. These derivatives may be obtained from (10.3.2) with the help of the "chain rule" of calculus. Letting $u(x, t) = x - vt$, we obtain

$$\frac{\partial y}{\partial x} = \frac{\partial}{\partial x} F(u) = \frac{dF(u)}{du}\frac{\partial u}{\partial x} = \frac{dF(u)}{du}\frac{\partial}{\partial x}(x - vt) = \frac{dF(u)}{du}$$

$$\frac{\partial^2 y}{\partial x^2} = \frac{\partial}{\partial x}\left(\frac{\partial y}{\partial x}\right) = \frac{\partial}{\partial x}\left(\frac{dF(u)}{du}\right) = \left[\frac{d}{du}\left(\frac{dF(u)}{du}\right)\right]\frac{\partial u}{\partial x}$$

$$= \frac{d^2F(u)}{du^2}\frac{\partial}{\partial x}(x - vt) = \frac{d^2F(u)}{du^2}$$ (10.3.7)

$$\frac{\partial y}{\partial t} = \frac{\partial}{\partial t} F(u) = \frac{dF(u)}{du}\frac{\partial u}{\partial t} = \frac{dF(u)}{du}\frac{\partial}{\partial t}(x - vt)$$

$$= -v\frac{dF(u)}{du}$$

$$\frac{\partial^2 y}{\partial t^2} = \frac{\partial}{\partial t}\left(\frac{\partial y}{\partial t}\right) = \frac{\partial}{\partial t}\left(-v\frac{dF(u)}{du}\right) = -v\left[\frac{\partial}{\partial u}\left(\frac{dF(u)}{du}\right)\right]\frac{\partial u}{\partial t}$$

$$= -v\frac{d^2F(u)}{du^2}\frac{\partial}{\partial t}(x - vt) = v^2\frac{d^2F(u)}{du^2}$$ (10.3.8)

Dividing (10.3.8) by v^2 and comparing it to (10.3.7), we see that $y(x, t)$ satisfies the *partial differential equation*

$$\frac{\partial^2 y}{\partial x^2} - \frac{1}{v^2}\frac{\partial^2 y}{\partial t^2} = 0$$ (10.3.9)

This equation is a *one-dimensional wave equation*. Any function $y(x, t)$ that has the properties of a wave as outlined above will satisfy this equation.

There are, of course, many systems in which waves do not propagate; their motion is not described by an equation of this form. In systems in which waves do propagate, we shall see that the wave equation arises directly from the physical principles that govern the system's motion. Thus, Newton's laws of motion should reveal that a string must satisfy a wave equation of the form given above rather than some other type of equation. Let us investigate this further by discussing the application of Newton's laws to a moving string.

Consider a small segment of a string which has been disturbed as in Fig. 10.1; this segment is shown in Fig. 10.6. The left end has been displaced an amount $y(x, t)$, while the right end has been displaced $y(x + \Delta x, t)$ at time t. The angles which the ends of this segment make with the x-axis are $\theta(x)$ and $\theta(x + \Delta x)$, respectively. These angles are assumed to be quite small, so that the piece of string has been displaced from equilibrium by only a small amount. It is also assumed that the tension in the entire string is uniform and has the value T_0. Now, according to Newton's second law, the resultant of all forces acting on this segment is equal to its mass multiplied by its acceleration. If the segment is taken to be very small (infinitesimal), then the concept of a single acceleration is meaningful since the segment approaches a point mass. Accordingly, we have

$$T_0 \sin \theta(x + \Delta x) - T_0 \sin \theta(x) = ma_y = (\mu \Delta x)a_y$$

$$= (\mu \Delta x)\frac{\partial^2 y}{\partial t^2}$$ (10.3.10)

where μ is the linear mass density, that is, the mass per unit length. The acceleration a_y is written as a *partial* derivative since the quantity y is a function not only of time but *also* of position x along the string.

309

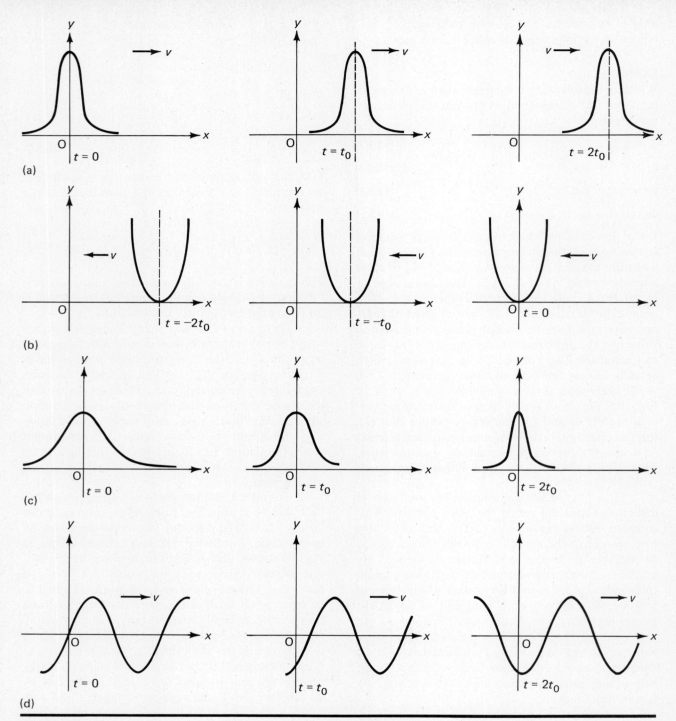

FIGURE 10.5

Since we have assumed that the angles are very small and that they are expressed in radians, it is legitimate to use the approximations

$$\sin \theta(x) \cong \theta(x) \cong \tan \theta(x) = \left(\frac{\partial y}{\partial x}\right)_x$$

recalling that the slope $\partial y/\partial x$ is the tangent of the angle θ. Also, at $x + \Delta x$,

$$\sin \theta(x + \Delta x) \cong \theta(x + \Delta x) \cong \tan \theta(x + \Delta x)$$

$$= \left(\frac{\partial y}{\partial x}\right)_{x+\Delta x} \quad (10.3.11)$$

Substituting these approximations in (10.3.10) and dividing both sides of the equation by Δx,

$$T_0 \frac{\left(\dfrac{\partial y}{\partial x}\right)_{x+\Delta x} - \left(\dfrac{\partial y}{\partial x}\right)_x}{\Delta x} = \mu \frac{\partial^2 y}{\partial t^2} \quad (10.3.12)$$

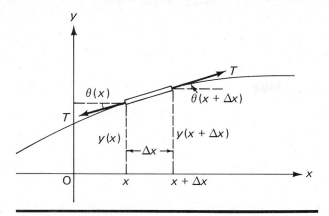

FIGURE 10.6. Forces and displacements associated with an infinitesimal element of a string along which a wave propagates.

Finally, if we allow Δx to become arbitrarily small, we obtain

$$T_0 \frac{\partial^2 y}{\partial x^2} = \mu \frac{\partial^2 y}{\partial t^2} \qquad (10.3.13)$$

or

$$\frac{\partial^2 y}{\partial x^2} - \frac{1}{(T_0/\mu)} \frac{\partial^2 y}{\partial t^2} = 0 \qquad (10.3.14)$$

since in the limit as $x \to 0$ the quantity on the left side of equation (10.3.13) becomes,[4] by definition, $T_0(\partial/\partial x)(\partial y/\partial x)$, or $T_0(\partial^2 y/\partial x^2)$.

This equation may now be compared to (10.3.8), the general equation for a one-dimensional wave. This comparison reveals that the string does, indeed, satisfy the wave equation and can therefore propagate waves. It also tells us that the waves which propagate must have a *wave velocity* such that $1/v^2 = \mu/T_0$, or

$$v = \sqrt{\frac{T_0}{\mu}} \qquad (10.3.15)$$

Thus, we see that the emergence of wave motion is a direct consequence of Newton's laws applied to the string as well as some approximations which are quite valid in many instances. We see also that the velocity of the waves is related directly to the physical characteristics of the propagating medium, which in this case is the string. From (10.3.14), it is apparent that the wave velocity is directly proportional to the square root of the tension and inversely proportional to the mass per unit length.

The wave equation, (10.3.14), has an enormous number of possible solutions. The simplest familiar

examples of these are *sinusoidal* solutions, having the form

$$y(x, t) = y_0 \sin[k(x - vt) + \delta] \qquad (10.3.16)$$

The fundamental importance of sinusoidal solutions will become more apparent later when we discuss more complex solutions. It may be shown directly by differentiating (10.3.16) and substituting the resulting expressions into (10.3.14) that the sine wave (10.3.16) will be a solution of the wave equation, but it is not the only solution, since we have already demonstrated that *any* function of the variable $x - vt$ will be a solution, and (10.3.15) is such a function.

Let us now describe some of the properties of the sinusoidal waves described by (10.3.16). At a fixed time t_0, (10.3.16) specifies the displacement of every point of the string from its equilibrium position; we shall continue to use the string as an example although many of the remarks we make are applicable to other systems as well. In Fig. 10.7, we illustrate schematically the dependence of y on x at some time t_0. This figure shows a number of locations where y assumes a maximum value of y_0 or a minimum value of $-y_0$. These values occur whenever sine takes on a value of $+1$ or -1, respectively. Therefore, the quantity y_0, which is called the *amplitude* of the wave, represents the *maximum possible displacement from equilibrium*.

The entire argument of the sine function in (10.3.16) is $[k(x - vt) + \delta]$. This is called the *phase* of the wave. The constant δ appearing in the phase is determined from initial conditions. For example, it may be regarded as the phase angle at $x = 0$ and at the initial time $t = 0$.

Referring to Fig. 10.7 again, we see that the distance λ represents the distance between successive maximum positive displacements. This distance is called the *wavelength*. Now suppose a maximum of displacement occurs at $x = x_1$. If we substitute $y = y_0$, $x = x_1$, into (10.3.16), this tells us that

$$y_0 = y_0 \sin[k(x_1 - vt_0) + \delta]$$

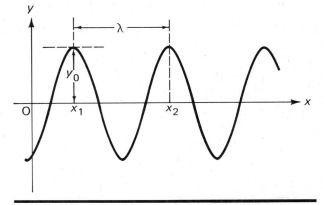

FIGURE 10.7. Sinusoidal wave, illustrating the wavelength and amplitude.

---

[4] See Appendix D for the definition of the partial derivative in terms of fundamental limiting processes.

or

$$k(x_1 - vt_0) + \delta = \frac{\pi}{2} \qquad (10.3.17)$$

If λ represents the wavelength, then the next maximum will occur at $x = x_1 + \lambda$, and in the interval between these two successive maxima, the phase angle $[k(x - vt) + \delta]$ will have advanced by precisely 2π radians. Equation (10.3.17) then becomes

$$k(x_1 + \lambda - vt_0) + \delta = \frac{\pi}{2} + 2\pi \qquad (10.3.18)$$

Subtracting (10.3.17) from (10.3.18), we now obtain

$$\boxed{k = \frac{2\pi}{\lambda}} \qquad (10.3.19)$$

The constant k, called the *propagation constant*, can, therefore, be expressed in terms of the wavelength by means of the above relation.

We have examined several features of a "snapshot" of the wave at time t_0 and have, therefore, studied the dependence of y on x for a *fixed value of t*. Let us now study the dependence of y on t for a *fixed value of x*, for example, $x = x_0$. Here, we are looking at only a tiny point on the string and we are watching how that point moves as time progresses. From (10.3.16) it is evident, setting $x = x_0$, that the motion is *simple harmonic motion*, with frequency f and period T given by the relation

$$\omega = kv = 2\pi f = \frac{2\pi}{T} \qquad (10.3.20)$$

But since $k = 2\pi/\lambda$ and $\omega = 2\pi f$, we may express the velocity v by

$$\boxed{v = \frac{\omega}{k} = \lambda f} \qquad (10.3.21)$$

which relates the frequency, the wavelength, and the wave velocity.

Using the various relations above, we can write the original wave equation (10.3.16) in a number of different equivalent forms. These include

$$\boxed{\begin{aligned} y(x, t) &= y_0 \sin[k(x - vt) + \delta] \\ &= y_0 \sin\left[\frac{2\pi}{\lambda}(x - vt) + \delta\right] \\ &= y_0 \sin[(kx - \omega t) + \delta)] \\ &= y_0 \sin\left(2\pi \frac{x}{\lambda} - ft + \delta\right) \\ &= y_0 \sin\left(2\pi \frac{x}{\lambda} - \frac{t}{T} + \delta\right) \end{aligned}} \qquad (10.3.22)$$

All of these expressions contain the same information. The student should be able to derive one form from any other. To reinforce some of these concepts, let us discuss a few examples.

EXAMPLE 10.3.1

A long string of mass density 0.001 slug/ft is stretched so that the tension is 0.5 pound. The left end is moved up and down with simple harmonic motion having a period of 0.5 sec and an amplitude of 1.5 ft. Assume the tension is constant throughout the motion. (a) Find the speed of the wave generated in the string. (b) What is the frequency of the wave? (c) Determine the wavelength. (d) Obtain a mathematical expression for the displacement $y(x, t)$ at any point.

The expression for the velocity of propagation has already been derived and is given by (10.3.15). Substituting the numbers given above, we obtain

$$v = \sqrt{\frac{T_0}{\mu}} = \sqrt{\frac{0.5}{0.001}} = 22.4 \text{ ft/sec} \qquad (10.3.23)$$

If we look at a small segment of the string and observe its motion, we will find that it is simple harmonic motion with a frequency given by the frequency of the source of the wave. Therefore, the frequency of the wave is

$$f = \frac{1}{T} = 2.0 \text{ Hz} \qquad (10.3.24)$$

Now that we know both the frequency and the speed of the wave, we can use (10.3.21) to determine the wavelength. In the present example, we obtain

$$\lambda = \frac{v}{f} = \frac{22.4}{2.0} = 11.2 \text{ ft} \qquad (10.3.25)$$

Finally, we may try to determine the mathematical expression for the wave. This can be done by using (10.3.23), (10.3.24), and (10.3.25) in conjunction with the expressions given by (10.3.22). We find that

$$y(x, t) = y_0 \sin[(kx - \omega t) + \delta]$$
$$= 1.5 \sin\left(\frac{2\pi}{11.2}x - 2\pi(2.0)t + \delta\right)$$

or

$$y(x, t) = 1.5 \sin(0.561x - 12.57t + \delta) \text{ ft} \qquad (10.3.26)$$

Now, this cannot be a complete solution since the phase angle δ has not been specified, nor can it be determined from the conditions stated in the problem. We need to know something else. The reader should show that if it is specified that $y = 0$ and $\partial y/\partial t < 0$ at $x = 5$ ft and $t = 0.4$ sec, then this information is enough to give $\delta = 2.221$ rad. It is clear from this that a given wave motion is completely specified by amplitude, frequency or wavelength, velocity, and initial phase. The velocity may be obtained by knowing the physical properties of the medium (tension, mass per unit length, etc.); but the others can be ascertained only by determining, somehow, the position, velocity, and acceleration of any point of the system at some stage of the motion.

EXAMPLE 10.3.2

The equation for a transverse wave on a string is given by

$$y = 2 \cos[\pi(x - 100t)] \text{ m} \qquad (10.3.27)$$

where x and y are in meters and t is in seconds. (a) What is the amplitude of this wave? (b) What is the phase angle δ? (c) Determine the frequency of vibration of the string. (d) Find the wave velocity. (e) At $t = 1$ sec, find the displacement, velocity, and acceleration of a small segment of the string located at $x = 2$ m.

The amplitude of a simple wave such as that expressed by (10.3.17) is the maximum possible displacement y_0. Now in (10.3.17) the wave is specified by means of a sine function, whereas in the present example we have used a cosine function. The simple trigonometric identity $\cos \theta = \sin(\theta + \frac{1}{2}\pi)$ may be used to convert one form to another. Thus, (10.3.27) may also be written as

$$y = 2 \sin[\pi(x - 100t) + \pi/2] \qquad (10.3.28)$$

Let us now compare this to (10.3.22). We find that the amplitude is $y_0 = 2$ m, the phase angle δ is $\pi/2$, the frequency of vibration is $f = 50$ Hz, and the wave velocity is $v = 100$ m/sec. Using (10.3.28), we can determine directly the displacement of any part of the string at any time t. We may also determine the velocity and acceleration of any portion of the string by evaluating appropriate derivatives of Eq. (10.3.28). Thus, the up-and-down velocity of a segment located at a given distance x from the origin will be given by the partial derivative $\partial y / \partial t$ evaluated at x, from which

$$v_y = \frac{\partial y}{\partial t} = -200\pi \cos[\pi(x - 100t) + \pi/2] \qquad (10.3.29)$$

The acceleration is given by $\partial v_y / \partial t$ evaluated at x, or

$$a_y = \frac{\partial v_y}{\partial t} = \frac{\partial^2 y}{\partial t^2} = -20{,}000\pi^2 \sin\left[\pi(x - 100t) + \frac{\pi}{2}\right] \qquad (10.3.30)$$

The values at $t = 1$ sec and $x = 2$ m are, therefore,

$$y = 2 \sin\left[\pi(2 - 100) + \frac{\pi}{2}\right] = 2 \sin\left[-98\pi + \frac{\pi}{2}\right]$$

$$= 2 \sin \frac{\pi}{2} = 2 \text{ m}$$

$$\frac{\partial y}{\partial t} = -200\pi \cos\left[-98\pi + \left(\frac{\pi}{2}\right)\right] = -200\pi \cos\left(\frac{\pi}{2}\right) = 0$$

and

$$\frac{\partial^2 y}{\partial t^2} = -20{,}000\pi^2 \sin\left[-98\pi + \left(\frac{\pi}{2}\right)\right]$$

$$= -20{,}000\pi^2 \text{ m/sec}^2 = -1.97 \times 10^5 \text{ m/sec}^2$$

This is a very large acceleration. It is evident from the above calculation that the maximum magnitude of the acceleration is always given by

$$(a_y)_{\max} = \omega^2 y_0 \qquad (10.3.31)$$

Therefore, waves of large amplitude and high frequency can possess a very large acceleration, as in the present example.

So far, we have been discussing the physics of one-dimensional transverse waves, using a string as our primary example. Transverse waves can be produced and propagated in other ways. For example, if a pebble is dropped into a pond, a familiar circular pattern of waves emanates from the source. A stretched membrane also provides a medium in which two-dimensional waves can propagate. For two-dimensional waves, the displacements z from equilibrium depend on the time t as well as the segment of the medium undergoing displacements. Now each part of the undisturbed medium must be located by means of two spatial coordinates x and y. If z represents the displacement of water molecules located at (x, y) when a wave is present, then z can be shown to satisfy the equation

$$\frac{\partial^2 z}{\partial x^2} + \frac{\partial^2 z}{\partial y^2} - \frac{1}{v^2}\frac{\partial^2 z}{\partial t^2} = 0 \qquad (10.3.32)$$

The above equation is a natural generalization of (10.3.9) and is a *two-dimensional wave equation*.

In the case of one-dimensional wave motion, a disturbance propagates *undiminished* (in the absence of dissipation), and the amplitude of the simple harmonic motion described by (10.3.16) is unchanged as the wave progresses. This is a consequence of the fact that the initial energy given to the system must be transferred through every point x. The situation is quite different for a two-dimensional circular wave. The initial disturbance created by a stone falling into a pond imparts a certain energy to the system. As this energy propagates outward from its source, more and more particles must share that energy. As a consequence, each particle will have an amplitude of vibration that *decreases* with increasing distance from the source.

Waves also propagate in three-dimensional media and are governed by the wave equation

$$\frac{\partial^2 u}{\partial x^2} + \frac{\partial^2 u}{\partial y^2} + \frac{\partial^2 u}{\partial z^2} - \frac{1}{v^2}\frac{\partial^2 u}{\partial t^2} = 0 \qquad (10.3.33)$$

313

where u represents the displacement from equilibrium. An example of a solution of this equation is provided by sound propagation in three dimensions. Sound waves propagate in all directions from a point source of sound in the atmosphere. Since the velocity is the same in all directions, the wavefronts emitted by such a source are spherical. Once again, since the energy emitted by the source is spread over a larger and larger area as the wavefronts propagate outward, the wave amplitude decreases as the distance r from the source increases. In this instance, the amplitude varies inversely with the distance from the source. For this reason, sound seems fainter and fainter to our ears as we move further away from the source. Light waves emitted from a point source behave in the same way.

If we are very far from a point source that emits spherical wavefronts, the curvature of the wavefronts becomes very small, and for practical purposes the wavefronts can be regarded as *planes* rather than spheres. For example, the sun emits light in the form of spherical electromagnetic waves. But when these wavefronts reach the earth, their radius is nearly 150,000,000 km. When we consider that the earth's radius is less than 6400 km, it is apparent that the earth intercepts only a very tiny portion of the spherical wavefronts emitted by the sun; this restricted portion of such an enormous sphere is for all intents and purposes flat, just as the small part of the earth's surface we see in our immediate vicinity is essentially flat. Plane waves are the simplest and most important examples of waves that propagate in three-dimensional systems. Indeed, even though they propagate in three dimensions, their amplitude can be described by the *one-dimensional* wave equation (10.3.9) or by the solutions given in Eq. (10.3.22). We shall have

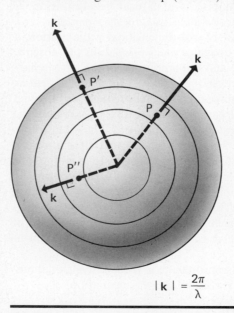

$$|\mathbf{k}| = \frac{2\pi}{\lambda}$$

FIGURE 10.8. The concept of the propagation vector.

occasion to study the characteristics of plane waves in great detail, not only in this chapter, but also in our later study of electromagnetic waves and optics.

In the case of a wave that travels in a three-dimensional medium, the characteristic features of its propagation may be conveniently described by a *propagation vector* $\mathbf{k}$. The propagation vector is a vector which is perpendicular to the wavefront at every point and whose magnitude is given by $k = 2\pi/\lambda$. The propagation vector at every part of the wavefront points in the direction of propagation at that point, as illustrated by Fig. 10.8. The propagation vector is a natural generalization of the propagation constant introduced previously for one-dimensional wave motion. It enables us to visualize the way in which the wavefronts are moving at every point. It is a concept that we shall have occasion to refer to often in our future work concerning sound waves, electromagnetic waves, and the propagation of light.

10.4 Longitudinal Waves: Sound Waves

The emphasis in the previous section has been placed on transverse wave motion even though much of the mathematics is also applicable to the case of longitudinal waves. In this section, we consider the propagation of longitudinal waves and, in particular, sound waves, which are the most important example of longitudinal waves. There is a profound difference between the waves previously discussed and the sound waves we will study. If you have seen waves traveling on water or moving along a string, you know that the displacements of particles from equilibrium are macroscopic ones that are usually visible to the human eye. On the other hand, the displacements occurring in sound waves are displacements of individual molecules or atoms from equilibrium positions and can be extremely minute, in some cases as small as 10^{-8} cm.

Newton's second law was used to show that a string under tension can support wave motion. The derivation of its one-dimensional wave equation also described the velocity of wave propagation. We would like to ask now whether Newton's second law can also be used to predict wave motion in a solid bar which has been disturbed by being struck with a hammer. If wave motion does result, what is the wave velocity and how does it depend on the properties of solids? To answer these questions, it is necessary first to know a little about elastic deformation of solids.

Let us consider a solid bar of length l_0 and cross-sectional area A, as shown in Fig. 10.9. If this rod experiences forces of magnitude F on both of its faces, the rod is said to be under longitudinal stress, and the amount of stress is defined by

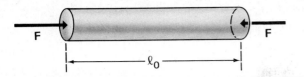

FIGURE 10.9. Solid elastic bar in which longitudinal waves are excited.

$$\text{stress} = \frac{F}{A} \qquad (10.4.1)$$

This stress can be due to compression, as in Fig. 10.9, or it can be due to tension. An idealized "rigid object" will not exhibit any deformation at all when under stress; but real substances, even those that are quite "rigid" in the conversational sense of the word, will undergo a slight deformation when stress is applied. Therefore, when longitudinal stress is applied to the bar shown in Fig. 10.9, its length will change by a small amount Δl. The change in length divided by the length l_0 is known as the *longitudinal strain:*

$$\text{strain} = \frac{\Delta l}{l_0} \qquad (10.4.2)$$

Now, if the stress is not too large, the relationship between stress and strain for most substances is a linear one. The stress is directly proportional to the strain and the proportionality constant is given by a number Y, which is called Young's modulus. The magnitude of Young's modulus will depend upon the mechanical properties of the substance and is, in fact, a quantitative measure of its rigidity. The equation relating stress and strain may thus be written

$$\frac{F}{A} = Y \frac{\Delta l}{l_0} \qquad (10.4.3)$$

This equation, in which the stress varies linearly with the strain, is simply another form of Hooke's law.

Equation (10.4.3) may be used to study the propagation of a disturbance in a bar of finite length. Let us consider a *very small* section of such a bar, having mass Δm and length Δx when in equilibrium, and located at position x. This section is shown in Fig. 10.10. When the bar is struck by a hammer, this section is no longer in equilibrium. Its left end, at some time t, has undergone a displacement $u(x, t)$ from its equilibrium position. Its right end has experienced a similar displacement $u + \Delta u$. Its length, therefore, due to applied stress, has changed by an amount Δu from Δx to $\Delta x + \Delta u(x)$. Accordingly, at the left end, where the force is $F(x)$ and the elastic elongation $\Delta u(x)$, (10.4.3) becomes

$$\frac{F(x)}{A} = Y \frac{\Delta u(x)}{\Delta x} \qquad (10.4.4)$$

At the right end, the force is $F(x + \Delta x)$ and the elastic elongation is $\Delta u(x + \Delta x)$. Under these circumstances, (10.4.3) becomes

$$\frac{F(x + \Delta x)}{A} = Y \frac{\Delta u(x + \Delta x)}{\Delta x} \qquad (10.4.5)$$

The total force acting on the element is, therefore,

$$F(x + \Delta x) - F(x) = AY \left[\frac{\Delta u(x + \Delta x)}{\Delta x} - \frac{\Delta u(x)}{\Delta x} \right] \qquad (10.4.6)$$

If Δx approaches zero, $\Delta u(x)/\Delta x$ represents the partial derivative of u with respect to x, evaluated at x. Using Newton's second law, we find, therefore, that

$$AY \left[\left(\frac{\partial u}{\partial x} \right)_{x + \Delta x} - \left(\frac{\partial u}{\partial x} \right)_{x} \right] = (\Delta m) \frac{\partial^2 u}{\partial t^2} = \rho A \, \Delta x \frac{\partial^2 u}{\partial t^2} \qquad (10.4.7)$$

where ρ is the density, or mass per unit volume, and $\partial^2 u/\partial t^2$ is the acceleration of the element. If we now

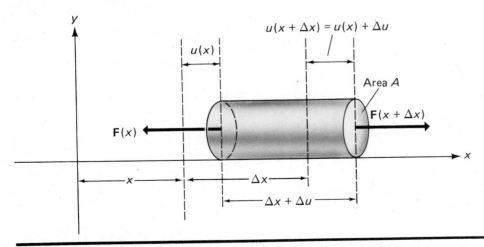

FIGURE 10.10. Forces and displacements associated with an infinitesimal element of an elastic bar along which longitudinal waves propagate.

divide both sides of (10.4.7) by $Y \Delta x$ and take the limit as $\Delta x \to 0$, we obtain

$$\frac{\partial^2 u}{\partial x^2} - \frac{1}{Y/\rho} \frac{\partial^2 u}{\partial t^2} = 0 \qquad (10.4.8)$$

Equation (10.4.8) is a one-dimensional wave equation describing the displacement u of a small segment located at x. The wave velocity is given by (10.3.9) as

$$\boxed{v = \sqrt{Y/\rho}} \qquad (10.4.9)$$

We find, therefore, that any initial displacement of molecules at the left end of the bar, the end struck by the hammer, can propagate as a *longitudinal wave* described by a solution of (10.4.8). Such a wave is nothing more nor less than a *sound wave*. A typical sinusoidal solution of Eq. (10.4.8) can be written as

$$u(x, t) = A \sin(kx - \omega t) \qquad (10.4.10)$$

with

$$v = \omega/k = \sqrt{Y/\rho}$$

It is important to realize that the displacement $u(x, t)$ of any part of the bar from its equilibrium position takes place *along the x-direction*. The particles of the bar execute simple harmonic motion about their equilibrium positions, vibrating back and forth along the x-axis, while the wave itself is, according to Eq. (10.4.10), propagating also along the x-direction. The situation is very similar to that illustrated in Fig. 10.2 by the passage of a compressional pulse along a spring. This type of wave motion, in which the particle displacements take place parallel to the direction in which the wave itself travels, is called *longitudinal wave motion* and is to be contrasted with the transverse type of wave motion studied previously, in which particle motions took place perpendicular to the direction of travel of the wave.

EXAMPLE 10.4.1

A metallic bar of length 1 m and cross-sectional area 2 cm^2 has a density of 3 g/cm^3. It is subjected to compression by forces of 10^3 newtons at its ends, and as a result of this compression, the bar contracts by 0.1 mm. (a) Find the longitudinal strain on the bar. (b) What is the longitudinal stress? (c) Determine Young's modulus. (d) What is the velocity of propagation of longitudinal waves traveling in the bar?

The strain is given by the expression

$$\text{strain} = \frac{\Delta l}{l_0} = \frac{0.1 \text{ mm}}{1000 \text{ mm}} = 10^{-4}$$

while the stress is given by

$$\text{stress} = \frac{F}{A} = \frac{10^8}{2} = 0.5 \times 10^8 \text{ dyn/cm}^2$$

Using these two results in conjunction with the definition of Young's modulus, we obtain

$$Y = \frac{\text{stress}}{\text{strain}} = 0.5 \times 10^{12} \text{ dyn/cm}^2$$

To find the speed of wave propagation, we use the formula given by Eq. (10.4.9) to find

$$v = \left(\frac{0.5 \times 10^{12}}{3} \right)^{1/2} = 0.41 \times 10^6 \text{ cm/sec} = 4100 \text{ m/sec}$$

This wave velocity is typical of the speed of longitudinal waves in metallic objects and gives an idea of what we might expect for the velocity of sound in solid substances.

When a longitudinal disturbance propagates in a solid as a wave, the form of the disturbance depends on its source. The blow of a hammer causes a sharp pulse to propagate from one end to the other. On the other hand, a periodic disturbance of a definite frequency causes a periodic or sinusoidal wave to propagate. Just as we found that the wave equation for a one-dimensional string could exhibit sinusoidal solutions of a definite frequency and wavelength, we also find that one-dimensional longitudinal waves possess these same characteristics.

Longitudinal waves propagating through a solid (or liquid or gas) are also called *sound waves*, although not all waves of this type are audible to the human ear. Those waves which have frequencies below 20 Hz are called infrasonic waves and are inaudible. These waves are usually generated by very big sources since their wavelength is generally very long. For example, a longitudinal wave of 1/20 Hz traveling through the earth, where the wave speed is about 5000 m/sec, would have a wavelength $\lambda = v/f = 100,000$ meters. Such waves can be produced inside the earth's mantle by seismic activity such as earthquakes. In fact, longitudinal waves with a wavelength as large as the diameter of the earth have been detected.

When longitudinal waves contain frequencies in the range of 20 to 20,000 Hz, they are in the audible range and can be heard as sound by the human ear.[5] Such waves may originate in vibrating strings, air columns, membranes, loudspeaker diaphragms, and other sources. Their wavelength depends on the sound velocity in the medium in which the waves are traveling as well as their frequency. In air, where the sound velocity is about 330 m/sec, 20-Hz waves have

---

[5] Other animal species can hear much higher frequencies. For example, bats can hear up to 120,000 Hz and porpoises, up to 240,000 Hz. However, at the low-frequency end, bats and porpoises are limited to 10,000 Hz and would, therefore, not be able to appreciate our music. Although many animals hear frequencies that are ultrasonic, few can match the human sensitivity in the low-frequency end of the spectrum.

a wavelength of about 16 m, while 20,000-Hz sound waves have a wavelength of around 1.6 cm.

High-frequency longitudinal waves (above 20,000 Hz) cannot be detected by the human ear. These waves may be produced by electrically induced high-frequency vibrations of crystals; they can also by produced and detected by bats and porpoises, and are extremely important for navigation and food gathering for both of these species. They are referred to as *ultrasonic waves.*

We have discussed to some extent the propagation of sound in a solid. Sound also travels through gases and liquids, although the speed is usually smaller than in solids. Liquids and gases are fluids rather than rigid substances, and, therefore, the description of sound propagation in them is different in one important respect. Instead of using Young's modulus to characterize their elastic properties, we must use instead the *bulk modulus B* of the substance, which is defined by

$$\Delta P = B \frac{\Delta V}{V_0} \qquad \text{or} \qquad B = \frac{\Delta P}{(\Delta V/V_0)} \qquad (10.4.11)$$

where $\Delta V/V_0$ is the fractional volume change a volume V_0 of the substance undergoes when subjected to an externally applied change in pressure (force per unit area) ΔP. The bulk modulus B or, more properly, its inverse, is a direct measure of the *compressibility* of the substance. The velocity of sound in a liquid or a gas is, therefore, expressed by

$$v = \sqrt{\frac{B}{\rho}} \qquad (10.4.12)$$

In water, this velocity is approximately 1500 m/sec, while in air at normal atmospheric pressure and at a temperature of 0°C, the sound velocity is 331.4 m/sec, or 1089 ft/sec.

For a gas, both the bulk modulus B and the density depend upon the temperature and pressure, and, therefore, the velocity of sound varies significantly if these parameters change to any large extent. In normal terrestrial surroundings, the atmospheric temperature and pressure variations that arise from changing climatic conditions are not sufficiently great to alter the sound velocity very much from the figure given above. However, we shall learn more about the variation of the sound velocity with temperature and pressure in Chapter 13.

10.5 Energy Transport by Waves

Energy may be transported from one place to another in two ways. In the first of these, the energy is carried between two places because of the motion of a body that possesses energy. For example, the energy of a thrown baseball is carried by the ball from the pitcher's hand to the catcher's glove. The second means of transporting energy is by wave motion, and it is this mechanism which is of primary interest in this section.

To illustrate the basic concepts, let us consider the very idealized problem of transferring energy between two peaks of a mountain range by using the transverse waves that can propagate along a rope strung between the two peaks. On one of the peaks, an energy source is available in the form of an oscillator which can vibrate up and down at a definite frequency. The end of the rope is attached to the oscillator, and, therefore, energy is transferred to the rope. Figure 10.11 illustrates a number of snapshots of the rope at various time intervals. In the absence

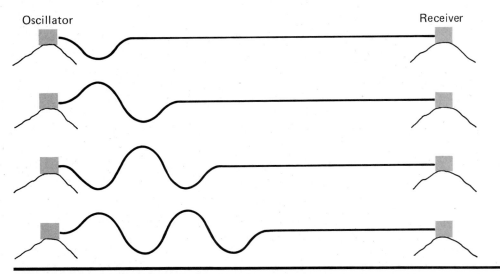

FIGURE 10.11. Successive stages in the transport of energy from an oscillator to a distant receiver via transverse waves in a rope.

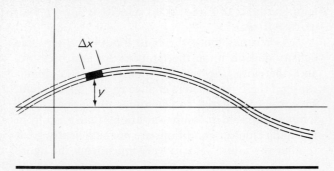

FIGURE 10.12. Typical mass element of the rope illustrated in Fig. 10.11.

of any mechanism for loss of energy, we have a rope which ultimately delivers energy to the second peak at the same rate that it receives it. If the rope oscillates in such a manner as to ultimately establish a wave of the form

$$y = y_0 \sin(kx - \omega t) \qquad (10.5.1)$$

then at what rate is energy being transferred between the two peaks?

To answer this question, let us consider a small element of the rope of length Δx whose mass is $\Delta m = \mu \, \Delta x$, where μ is the linear mass density, or mass per unit length. Such an element, as illustrated in Fig. 10.12, moves up and down with simple harmonic motion of amplitude y_0 and frequency ω. Its contribution to the kinetic energy of the rope will be

$$\Delta U_k = \tfrac{1}{2}(\Delta m)v_y{}^2 = \tfrac{1}{2}(\mu \, \Delta x)\left(\frac{\partial y}{\partial t}\right)^2 \qquad (10.5.2)$$

since $v_y = \partial y / \partial t$. But $y(t)$ is given by (10.5.1), from which

$$\frac{\partial y}{\partial t} = \omega y_0 \cos(kx - \omega t) \qquad (10.5.3)$$

Substituting this into (10.5.2), we obtain

$$\Delta U_k = \tfrac{1}{2}\mu\omega^2 y_0{}^2 \, \Delta x \cos^2(kx - \omega t) \qquad (10.5.4)$$

But we may use the well-known trigonometric identity

$$\cos^2 \theta = \tfrac{1}{2} + \tfrac{1}{2}\cos 2\theta \qquad (10.5.5)$$

to write (10.5.4) in the form

$$\Delta U_k = \tfrac{1}{4}\mu\omega^2 y_0{}^2 \, \Delta x + \tfrac{1}{4}\mu\omega^2 y_0{}^2 \, \Delta x \cos 2(kx - \omega t) \qquad (10.5.6)$$

If we average this over a length of time corresponding to many periods of the harmonic motion, we will obtain the time-average kinetic energy associated with this portion of the rope, $\Delta \bar{U}_k$. This is easily accomplished, since the first term in the above equation is constant with respect to time, and the time average of the second term is zero because the average value of $\cos 2\omega t$ is zero over many cycles. We then obtain

$$\boxed{\Delta \bar{U}_k = \tfrac{1}{4}\mu\omega^2 y_0{}^2 \, \Delta x} \qquad (10.5.7)$$

It is possible, though somewhat less straightforward, to calculate directly the average potential energy associated with this element of the system. To simplify matters, we shall recall from our previous work in section 9.3 that the average potential energy of a simple harmonic oscillator is equal to its average kinetic energy. Therefore,

$$\boxed{\Delta \bar{U}_p = \Delta \bar{U}_k} \qquad (10.5.8)$$

and the average total energy $\Delta \bar{U}$ can, therefore, be written

$$\Delta \bar{U} = \Delta \bar{U}_k + \Delta \bar{U}_p = 2\Delta \bar{U}_k = \tfrac{1}{2}\mu\omega^2 y_0{}^2 \, \Delta x \qquad (10.5.9)$$

In the limit where $\Delta x \to 0$, this can be expressed as

$$d\bar{U} = \tfrac{1}{2}\mu\omega^2 y_0{}^2 \, dx \qquad (10.5.10)$$

We may interpret this result in the following way. Equation (10.5.9) tells us that every segment of length dx has total energy $d\bar{U}$. Since the system executes a wave motion of constant amplitude and frequency, all the energy received by every such segment from its neighbor on the left in any given time interval must be balanced by an equal amount that is passed on to the neighboring segment on the right. At the extreme left end of the rope, energy is fed in by an externally powered oscillator. The rate at which this energy is absorbed from the external source by the rope is simply the power output of the oscillator. At the same time, if there is no internal energy dissipation in the rope, energy will be transferred *at the same rate* into whatever energy-absorbing system is fastened to the right-hand end. In any time interval dt, therefore, the first segment of the rope of length dx on the left-hand end—which, you will recall, receives *all* its energy from the externally powered source—will receive an amount of energy $d\bar{U}$ as given by (10.5.10), and the last segment of length dx on the right will give up an equal amount to whatever it is connected to. Clearly, then, the amount of energy transported by the system during this time interval must be

$$d\bar{U} = \tfrac{1}{2}\mu\omega^2 y_0{}^2 \, dx = \tfrac{1}{2}\mu\omega^2 y_0{}^2 \left(\frac{dx}{dt}\right) dt \qquad (10.5.11)$$

Since dx/dt corresponds to the velocity with which energy is transported by the system, hence the wave velocity v, this can be written

$$d\bar{U} = \tfrac{1}{2}\mu v\omega^2 y_0{}^2 \, dt \qquad (10.5.12)$$

or finally,

$$\boxed{\bar{P} = \frac{d\bar{U}}{dt} = \tfrac{1}{2}\mu v\omega^2 y_0{}^2} \qquad (10.5.13)$$

In this equation, $\bar{P}$ represents the *average power* transmitted by the wave motion that has been set up in the system.

Equation (10.5.13) exhibits some very important features of energy transfer in wave propagation. In particular, we note that the power depends on the *square of the amplitude and on the square of the frequency*. It also has a linear variation with the wave velocity. Although the above features were derived for a transverse wave traveling in one dimension, they are also valid for one-dimensional longitudinal waves.

In the case of one-dimensional waves, in which there is no dissipation of energy, any energy which flows past any given point in 1 sec must also flow past any other point in that same time interval. The situation is very much like water flowing through a pipe; the rate of flow is constant everywhere.

For waves in two or in three dimensions, the energy goes out in many directions, and, as a consequence, the energy flowing through a given unit area perpendicular to the direction of propagation depends crucially on the exact location of the area. For this reason it is useful to define a new term called the *intensity*. The intensity I at a given location is defined as the average energy which crosses a unit area perpendicular to the propagation direction in unit time, that is,

$$I = \frac{\Delta \bar{U}}{(\Delta A)(\Delta t)} = \frac{\bar{P}}{\Delta A} \qquad (10.5.14)$$

as illustrated by Fig. 10.13. The value of I depends upon the energy input by the source of the wave as well as the geometry which relates the given area to the source. Typical units used to measure I are watts/cm² or watts/m². In the case of a sound wave propagating in a gaseous medium, Eq. (10.5.13) is still correct, even though the wave is longitudinal rather than transverse. The average power crossing an area A per unit time, therefore, is

$$\frac{\bar{P}}{A} = \frac{\mu v \omega^2 y_0^2}{2A} \qquad (10.5.15)$$

In this expression, the quantity μ/A represents mass divided by (length × area), or mass per unit volume. It may, therefore, be replaced by the volume density ρ of the medium, which expresses its mass per unit volume. Making this substitution, (10.5.15) becomes

$$I = \tfrac{1}{2}\rho v \omega^2 y_0^2 \qquad (10.5.16)$$

EXAMPLE 10.5.1

A point source emits energy with a power of 10 watts in the form of spherical sound waves. Find the intensity 5 m from the source, and also find the energy

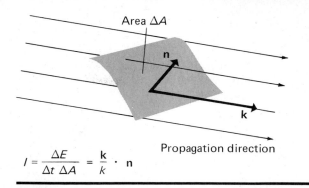

$$I = \frac{\Delta E}{\Delta t\, \Delta A} = \frac{\mathbf{k}}{k} \cdot \mathbf{n}$$

FIGURE 10.13. Transport of wave energy across an arbitrarily oriented area element.

which crosses an area 3 cm² (perpendicular to the direction of propagation of the wave) every 5 sec if the area is 5 m away from the source.

Let $\bar{P}$ be the average power emitted by the source. Thus, $\bar{P}$ joules are emitted each second, where $\bar{P}$ has the numerical value 10 in the present case. This amount of energy must pass through any sphere whose center is at the source (assuming no absorption); and therefore, from (10.5.13),

$$I = \frac{\Delta U}{(\Delta t)(\Delta A)} = \frac{\bar{P}}{\Delta A} = \frac{\bar{P}}{4\pi r^2} \qquad (10.5.17)$$

where I is the intensity a distance r from the source. It is evident from this equation that the intensity of spherical sound waves radiating from a point source falls off *inversely as the square of the distance* from the source. Substituting numbers, we obtain

$$I = \frac{10}{4\pi(5)^2} = 3.2 \times 10^{-2} \text{ watt/m}^2$$

$$= 3.2 \times 10^{-6} \text{ watt/cm}^2$$

This expression may now be utilized to find the energy passing through any area on a sphere of radius 5 m. In 5 sec, the energy passing through 3 cm² is

$$U = (3.2 \times 10^{-6})(3)(5) = 4.8 \times 10^{-5} \text{ J}$$

It should be remarked that $\bar{P} = 10$ watts is the *acoustic* power output, which is considerably smaller than the electrical power input needed to create this acoustic power. The conversion of electrical power in devices such as loudspeakers to useful acoustic power is usually quite inefficient, much of the electrical energy being wasted in the form of heat.

The faintest sound the human ear can detect has an intensity (at the ear) of 10^{-16} watts/cm². This represents an extremely small amount of energy received by the ear each second, and it reminds us that the ear is an enormously sensitive organ. The ear can respond effectively to intense sound waves but generally cannot tolerate intensities larger than about 10^{-4} watts/cm² without experiencing pain. Due to the large range of intensities we hear, it has become customary to characterize sound intensities by means of a logarithmic scale which gives the intensity of sound relative to the minimum discernible intensity $I_0 = 10^{-16}$ watts/cm². If I is the intensity of a sound wave at a given location, then

$$\beta = 10 \log \frac{I}{I_0} \qquad (10.5.18)$$

gives a measure of intensity in so-called *decibels*, or db. Thus, if $I = I_0$, $\beta = 0$; whereas if $I = 10^{-4}$ watts/cm², $\beta = 120$ db. A value of 10 db implies an intensity 10 times as large as I_0, while 20 db corresponds to 100 times I_0. Ordinary conversation occurs at about 65 db, while whispering corresponds to 20 db. It should be kept in mind that intensity is a quantitative measure of energy received and is not synonymous with the sensation of loudness, although there is, to be sure, a correlation. It could happen that sound waves of the same intensity appear to have different loudness since they may be of different frequency. The sensation of loudness is a qualitative physiological response which depends on many factors. Two people can disagree on the "loudness' of a TV set, and often do, but they will not disagree on its intensity if they have the proper instruments with which to measure it.

EXAMPLE 10.5.2

At a football stadium, the plays are announced from a single loudspeaker, since the university cannot afford a more elaborate system. The most distant spectator is 600 ft from the speaker. If an intensity level of 70 db is desired at this point, what is the minimum acoustic power required?

The intensity I may be determined from Eq. (10.5.18), according to which $70 = 10 \log I/I_0$, which implies (noting that I_0 represents the minimum audible intensity of 1.00×10^{-16} watt/cm²),

$$I = 10^7 I_0 = 10^7 \times 10^{-16} = 10^{-9} \text{ watt/cm}^2$$

Substituting this numerical value into Eq. (10.5.17), we obtain, for $r = 600$ ft $= 1.83 \times 10^4$ cm,

$$P = 4\pi r^2 I = (4\pi)(1.83 \times 10^4)^2(10^{-9}) = 4.21 \text{ watts}$$

We can compare the relative sound intensities I and I' at different distances r and r' from a point source that radiates at a constant power level $\bar{P}$ by noting from (10.5.17) that

$$\bar{P} = 4\pi r^2 I = 4\pi r'^2 I' \qquad (10.5.19)$$

from which

$$\frac{I'}{I} = \frac{r^2}{r'^2} \qquad (10.5.20)$$

This equation is simply another way of expressing the inverse square law for a point source. In this example,

we know that for $r = 600$ ft, $I = 1.0 \times 10^{-9}$ watt/cm². Using (10.5.20), it is easy to show that for $r = 30$ ft, $I' = 4.0 \times 10^{-7}$ watt/cm²; and for $r = 3$ ft, $I' = 4.0 \times 10^{-5}$ watt/cm². Both of these figures are still below the pain threshold of 10^{-4} watt/cm², although in the latter case the intensity level would certainly be uncomfortable.

EXAMPLE 10.5.3

At 10,000 cycles per second, the threshold of audibility occurs at about 11 db. Find the amplitude of molecular vibrations in air (at standard temperature and pressure) at this frequency. The density of air is $\rho = 1.29 \times 10^{-3}$ g/cm³ and the wave speed is $v = 3.31 \times 10^4$ cm/sec. Note that 1 watt = 1 joule/sec = 10^7 ergs/sec = 10^7 dyne-cm/sec = 10^7 g cm²/sec³.

From (10.5.18), we can determine the intensity at 11 db. We may write, therefore,

$$11 = 10 \log \frac{I}{I_0}$$

which implies $I/I_0 = 12.6$, or

$$I = 12.6 \times 10^{-16} \text{ watts/cm}^2$$
$$= 12.6 \times 10^{-16} \text{ g cm}^2/\text{sec}^3 \text{ cm}^2$$

since $I_0 = 1.00 \times 10^{-9}$ watt/cm². Then, from (10.5.16),

$$y_0{}^2 = \frac{2I}{\rho v \omega^2} = \frac{(2)(12.6 \times 10^{-9})}{(1.29 \times 10^{-3})(3.31 \times 10^4)(2\pi \times 10^4)^2}$$

$$= 14.9 \times 10^{-20} \text{ cm}^2$$

$$y_0 = 3.86 \times 10^{-10} \text{ cm}$$

This is an incredibly small displacement amplitude. In fact, the displacement of molecules or atoms from their equilibrium positions is considerably smaller than the molecules or atoms themselves, for they are generally larger than 10^{-8} cm. It is almost incredible, but nonetheless true, that the human ear can detect motions of such minute proportions.

10.6 Superposition of Waves

One of the most profound principles underlying wave motion is the so-called *superposition principle*. It states that two or more waves may travel through the same region of space in a completely independent way and that the displacement of particles in the medium is obtained by direct addition of the displacements which each of the separate waves would produce in the absence of all others. This principle is confirmed by experiment, provided the amplitudes of the various waves are not too large. For very large amplitudes, the restoring forces on the particles do not obey Hooke's law, and this simple result is no longer true.

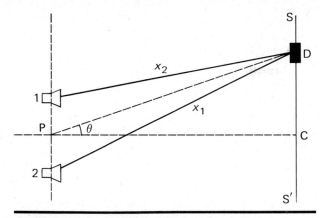

FIGURE 10.14. Spatial interference of sound waves emitted by two loudspeakers.

As a consequence of the superposition principle, waves exhibit an effect called interference. When two or more waves interfere, they do not in any way affect one another. It is unfortunate, therefore, that the word interference has become universally adopted, for in the present context it merely means that two or more waves add or superpose to provide a resultant which may produce effects that were not inherent in any of the individual waves. Thus, interference is really synonymous with superposition. There are two main classes of interference phenomena, since waves can interfere either spatially or temporally. The distinction between these classes will be made below.

Consider an example illustrated by Fig. 10.14 in which two loudspeakers emit sound waves of the same frequency and amplitude, and let these waves be denoted by

$$y_1 = y_0 \sin(kx_1 - \omega t)$$
$$y_2 = y_0 \sin(kx_2 - \omega t)$$

(10.6.1)

In these expressions, x_1 and x_2 are the distances, respectively, from speaker 1 and speaker 2 to a detector D, and y_1 and y_2 are the molecular displacements produced there at time t by each speaker acting *by itself*. The movable detector D, which is far from the speakers in comparison to their separation, is free to move along the line SS'. What is the resultant wave received by the detector and how does the intensity of sound vary as the detector is moved?

The geometry of the situation is illustrated in Fig. 10.14. According to the superposition principle, the molecular displacements at the detector are given by

$$y = y_1 + y_2 = y_0[\sin(kx_1 - \omega t) + \sin(kx_2 - \omega t)]$$

(10.6.2)

The above sum may be evaluated by utilizing the trigonometric identity

$$\sin a + \sin b = 2 \sin \tfrac{1}{2}(a + b) \cos \tfrac{1}{2}(a - b)$$

(10.6.3)

If this is used, we obtain

$$y = 2y_0 \sin[\tfrac{1}{2}k(x_1 + x_2) - \omega t] \cos \tfrac{1}{2}k(x_1 - x_2)$$

(10.6.4)

for the result. Since we have assumed that the speaker separation is small compared to the distance from the detector, the distance $\tfrac{1}{2}(x_1 + x_2)$ is approximately the distance from the detector to the point P midway between the speakers.

Let us now rewrite Eq. (10.6.4) as

$$y = Y_0 \sin(kX - \omega t)$$

(10.6.5)

where

$$X = \tfrac{1}{2}(x_1 + x_2) \quad \text{and} \quad Y_0 = 2y_0 \cos \tfrac{1}{2}k(x_1 - x_2)$$

(10.6.6)

Now, Eq. (10.6.5) looks like a single wave, except that its amplitude Y_0 *depends on the position of the detector*. If x_1 and x_2 are equal, which corresponds to having the detector midway between the speakers, the amplitude Y_0 assumes its maximum value of $2y_0$. As we have seen previously, the intensity of a wave is proportional to the square of the amplitude, and therefore an amplitude of $2y_0$ gives an intensity four times as large as the intensity which would have been present if the sound from only one speaker were received. Now, as the distance $x_2 - x_1$ increases, the amplitude and therefore the intensity decreases until, when $\tfrac{1}{2}k(x_1 - x_2) = \pi/2$, it has the value *zero*. At this point, the difference in path length $x_1 - x_2$ is π/k, which (since $k = 2\pi/\lambda$) amounts to *one half the wavelength* λ. There is, then, a difference in phase of π radians, or 180°, between the two waves. When the amplitude is a maximum, the waves reinforce one another and we have what is known as *constructive* interference. When the amplitude assumes its minimum value of zero, the two waves cancel one another and we have complete *destructive* interference. Whether the interference is constructive or destructive depends upon the path difference $x_1 - x_2$ and thus upon the spatial position of the detector. Therefore, this phenomenon is known as *spatial interference* of the waves.

Constructive interference occurs whenever

$$\tfrac{1}{2}k(x_1 - x_2) = 0, \pi, 2\pi, 3\pi, \text{etc.,}$$

thus, whenever

$$x_1 - x_2 = n\lambda \qquad n = 0, \pm 1, \pm 2, \pm 3, \ldots$$

(10.6.7)

The condition for destructive interference is

$$\tfrac{1}{2}k(x_1 - x_2) = \frac{\pi}{2}, \frac{3\pi}{2}, \frac{5\pi}{2}, \ldots$$

or, in other words,

$$x_1 - x_2 = (n + \tfrac{1}{2})\lambda \qquad n = 0, \pm 1, \pm 2, \pm 3, \dots$$

(10.6.8)

These conditions simply state that constructive interference occurs whenever the path difference $x_1 - x_2$ amounts to an integral number of wavelengths, or whenever the phase difference is an integral multiple of 2π radians. Destructive interference takes place whenever the path difference amounts to an odd number of half-wavelengths, or when the phase difference is an odd multiple of π radians.

There are a number of subtle features concerning the above derivation of interference which we have glossed over but which should at least be mentioned. We have assumed that the amplitude of each wave, y_0, does not depend upon position. This is actually not true; the amplitude is much larger near a speaker than far away. We may expect, however, that Eqs. (10.6.1) are valid expressions for giving the displacements along the line SS', provided this line is so far from the speakers that, as D is varied, the distance x_1 changes by an amount much smaller than x_1 itself. The amplitude y_0 then will not vary appreciably as the

detector is moved. The assumption that the detector is far away is needed for still another reason. Since the waves are longitudinal the displacements characterized by y_1 and y_2 are not really parallel, as we have assumed. If, however, the distance to the detector is large in comparison to the speaker separation, the displacements are practically parallel, and our assumption is a good one. Finally, we have been assuming, without specifically saying so, that the two waves are *coherent*. This means there is a definite constant phase relation between the two waves at every instant of time.

A second example of spatial interference of waves is illustrated by Fig. 10.15, a photograph of "double slit interference" of water waves. At the bottom of the picture, waves are generated by two mechanically driven oscillators excited at the same frequency and phase. Circular waves spread out from each of the two sources and superpose to produce a pattern in which light and dark spots correspond, respectively, to destructive and constructive interference. This example is very similar to the preceding one, except that now the waves are transverse rather than longitudinal.

EXAMPLE 10.6.1

The two sources in a "ripple tank" are separated by a distance d, as illustrated in Fig. 10.16. A wave of angular frequency ω is incident. Discuss the pattern produced by the superposition of waves. Assume that each wave has an amplitude which is a function of distance from the source given by $y_0(r_1)$ and $y_0(r_2)$.

Let P be the point of observation. The superposition principle gives a resultant wave at P as

$$y = y_0(r_1) \sin(kr_1 - \omega t) + y_0(r_2) \sin(kr_2 - \omega t)$$

(10.6.9)

FIGURE 10.15. Spatial interference of water waves excited by two point sources. (Ripple tank photo courtesy of Professor T. A. Wiggins, Pennsylvania State University)

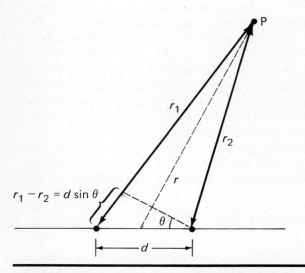

FIGURE 10.16. Geometry of spatial interference from two point sources.

The complete interference pattern is very difficult to calculate explicitly because $y_0(r_1)$ and $y_0(r_2)$ may be quite different at an arbitrary point. Let us, therefore, simplify the problem by assuming that r_1 and r_2 are very large in comparison to the distance d. Under this condition, r_1 and r_2 are almost equal. In fact, as an approximation, we may assume that $y_0(r_1) \cong y_0(r_2) \cong y_0(r)$, where $r = \frac{1}{2}(r_1 + r_2)$. Under this assumption, Eq. (10.6.9) may be transformed, once again using the identity (10.6.3), to yield the displacement

$$y = 2y_0(r) \cos k\left(\frac{r_1 - r_2}{2}\right) \sin(kr - \omega t) \qquad (10.6.10)$$

Here, we have a resultant wave with an effective amplitude

$$A = 2y_0(r) \cos \tfrac{1}{2}k(r_1 - r_2) \qquad (10.6.11)$$

The individual waves are transverse two-dimensional waves, and, therefore, there is no difficulty in numerically adding the displacements. This would be true even without the approximation made above.

The amplitude A will vary slowly with r. The reader should show that for large r it is reasonable to expect a dependence $y_0(r) \sim r^{-1/2}$. Figure 10.16 shows the geometry of the situation. When r_1 and r_2 are large, their difference may be expressed in terms of the separation of the sources and the angular variable θ. It is easy to see from the figure that

$$r_1 - r_2 = d \sin \theta \qquad (10.6.12)$$

The amplitude, therefore, can be expressed in terms of r and θ by means of the equation

$$A(r, \theta) = 2y_0(r) \cos \tfrac{1}{2}(kd \sin \theta). \qquad (10.6.13)$$

For a fixed value of r, this amplitude varies significantly with the angle θ. The amplitude is zero whenever

$$\boxed{\frac{k}{2} d \sin \theta = \frac{\pi}{2}, \frac{3\pi}{2}, \frac{5\pi}{5}, \ldots} \qquad (10.6.14)$$

and has its maximum value when

$$\boxed{\frac{k}{2} d \sin \theta = 0, \pi, 2\pi, 3\pi, \ldots} \qquad (10.6.15)$$

When the amplitude is zero, we have complete destructive interference. The points at which it vanishes are called *nodes*. Constructive interference is implied by the condition (10.6.15). These locations are called *antinodes*. At all other points, the waves superpose to produce neither a minimum nor a maximum in the amplitude.

Let us rewrite the condition for a *minimum amplitude* by using the relation $k = 2\pi/\lambda$ discussed earlier. This gives

$$\boxed{d \sin \theta = \frac{\lambda}{2}, \frac{3\lambda}{2}, \frac{5\lambda}{2}, \ldots} \qquad (10.6.16)$$

Now, the value of $\sin \theta$ lies between zero and unity. Therefore, if $d < \lambda/2$, it is impossible to obtain any minimum in the interference pattern. If d is very much larger than $\lambda/2$, the maxima and minima may be located too close to one another to reveal a discernible interference pattern. Thus, the condition for a nice, visible pattern is that d be larger than $\lambda/2$, perhaps five or ten times as large, but certainly not very much larger (for example, not 100 times as large).

In much the same way, the condition for *maximum amplitude* can be written as

$$\boxed{d \sin \theta = 0, \lambda, 2\lambda, 3\lambda, \ldots} \qquad (10.6.17)$$

We see that the variation of amplitude with θ is quite pronounced. For a fixed angle θ, the amplitude does vary with r but the variation is quite gradual. This variation is due to the fact that at large values of r, the available energy is distributed over a larger area and, therefore, the displacement amplitude at each point will be correspondingly reduced.

We have discussed two similar examples of superposition of waves to produce an interference pattern that exhibits a strong spatial variation of the displacement amplitude. The first case involved a longitudinal wave, while the second involved a transverse wave. Let us now consider a third example of spatial interference, this one resulting from waves traveling in opposite directions with the same speed and frequency. This situation is one which commonly occurs in systems in which a wave may be *reflected*. It is illustrated in Fig. 10.17. A string is given a pulse that travels to the right toward the support to which its right end is tied. If no energy is lost at the support, the pulse is reflected and *inverted* as shown in the figure. This inversion and reflection is a consequence of Newton's third law. As the pulse approaches the

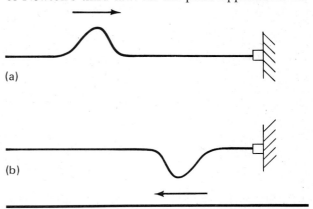

(a)

(b)

FIGURE 10.17. Reflection of a transverse wave in a string by a fixed boundary, illustrating the 180° phase change.

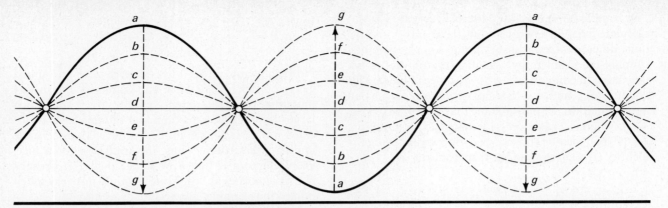

FIGURE 10.18. System of standing waves in a vibrating string.

support, the string exerts a force on the support while the support exerts an equal and opposite force on the string. The force the support exerts has a time variation and a magnitude exactly such as to cause the inversion. This same phenomenon of inversion also occurs for a continuous wave and in this case can be regarded as a phase change of 180° upon reflection.

Consider now the case of a string fixed at *both* of its ends. In this case, reflection occurs at both ends. Let us suppose, for simplicity, that the string is vibrating with a single frequency ω. The motion must then be a superposition of waves traveling to the right and waves traveling in the opposite direction, to the left. Mathematically, this superposition can be written as[6]

$$y = y_0 \sin(kx - \omega t) + y_0 \sin(kx + \omega t) \qquad (10.6.18)$$

Once again the trigonometric identity, (10.6.3), can be utilized to transform this expression into

$$y = 2y_0 \sin kx \cos \omega t \qquad (10.6.19)$$

The superposition of waves has produced a situation in which the string element at position x undergoes a simple harmonic vibration with amplitude $2y_0 \sin kx$. This amplitude varies with x but not with time. It attains the maximum value $2y_0$ whenever the following condition is fulfilled:

$$kx = \frac{\pi}{2}, \frac{3\pi}{2}, \frac{5\pi}{2}, \ldots \qquad (10.6.20)$$

These locations of maximum amplitude of vibration are known as *antinodes*. There are also positions for which the amplitude of vibration vanishes. These are called *nodes* and they occur whenever

$$kx = 0, \pi, 2\pi, \ldots \qquad (10.6.21)$$

If the string is attached to fixed supports at $x = 0$ and

[6] Recall that when we replace v by $-v$ in the expression $y = F(x - vt)$, we reverse the direction in which the wave travels.

at $x = L$, then these locations must be nodes since the string cannot move at these points. The expression (10.6.19) certainly guarantees the existence of a node at the left end. The condition to have a node at the right end requires that for $x = L$, $\sin kL = 0$. This implies, in turn, that

$$kL = \pi, 2\pi, \ldots$$

or

$$\frac{2\pi L}{\lambda} = \pi, 2\pi, \ldots \qquad (10.6.22)$$

or

$$\lambda = 2L, \frac{2L}{2}, \frac{2L}{3}, \frac{2L}{4}, \frac{2L}{5}, \ldots$$

The wavelength and the frequency are related by the condition $f\lambda = v$, where v is the velocity of the component traveling waves. Therefore, the equation above also implies that a string fixed at both ends cannot vibrate with any arbitrary frequency, but only those frequencies that result when condition (10.6.22) is satisfied. The only possible frequencies are, therefore, given by

$$f = \frac{v}{2L}, \frac{2v}{2L}, \frac{3v}{2L}, \frac{4v}{2L}, \ldots \qquad (10.6.23)$$

These frequencies are all multiples of the lowest frequency, $v/2L$, which is known as the *fundamental*.

The disturbance produced by means of this superposition is not a traveling wave at all but is referred to as a standing wave. The reason for this terminology is clear when one examines Fig. 10.18. Here, the whole string is viewed at a number of instants in time. Since the nodes are places which are at rest, energy cannot be transmitted across a node. Therefore, the energy contained in the region between two adjacent nodes "stands" there, although it alternates between kinetic and potential energy during the period of vibration. The standing wave pattern may also be understood from a slightly different viewpoint. Two component waves have been superposed, and each transports energy. The wave moving to the right transports energy in that direction, while the wave moving to the left carries energy to the left. The total effect is no energy transfer at all, since both waves are transferring equal energies each second but in opposite directions.

The various frequencies which are possible, as given by Eq. (10.6.22), are known as the *natural frequencies* of the string, or the harmonics of the string. In addition to the fundamental frequency $f_1 = v/2L$, also called the first harmonic, we may have all integral multiples of it. These are known as the overtones, or the higher harmonics. Thus, $f_2 = 2f_1$ and $f_3 = 3f_1$ are the second harmonic (first overtone) and third harmonic (second overtone), respectively. The series continues in the same manner for all other harmonics.

In the preceding chapter on simple harmonic motion, we encountered various systems which vibrate at only one natural frequency. Examples of this were provided by a mass attached to a simple spring and also by a pendulum. We have now seen that other systems can have many natural frequencies of vibration. It is very fortunate that this is the case, for otherwise the music emanating from a stringed instrument such as a violin would be incredibly dull. The *quality* of sound emanating from a violin or other musical instrument depends cirtically on the relative amplitudes of the various overtones as well as upon the fundamental vibration frequency. A typical sound spectrum from a violin is shown in Fig. 10.19. It should be mentioned that our ears differentiate the sound of an oboe, for example, from the sound of a violin playing the same fundamental notes completely on the basis of the different spectrum of harmonics or overtones emitted by the two instruments.

Let us now turn our attention to a very different type of interference, namely, interference in time. In the case of spatial interference of waves, the amplitude of the resultant disturbance is strongly dependent on *position*. The resultant displacement in any wave motion also has a time variation, of course, which is characterized by the frequency, but this variation is generally too rapid for our ears to detect. The frequency of most sound waves is simply too high to allow identification of the individual maxima and minima of the displacements. The ear responds, therefore, only to the time-averaged intensity. It is, however, true that different frequencies cause distinctly different physiologic responses which we identify as the *pitch*, and therefore the individual waves are not entirely without their effect.

By superposition of waves, it is nevertheless possible to create sounds in which distinct *time variations* of intensity are readily detected by the ear. The phenomenon leads to the identification of *beats*. Let us suppose, for example, that we have two tuning forks which when struck emit sound waves of frequency f_1 and f_2. At the position of an observer's ear, the net displacement of molecules of the medium is given by

$$y = y_0 \sin(k_1 x - \omega_1 t) + y_0 \sin(k_2 x - \omega_2 t) \qquad (10.6.24)$$

provided the amplitude of each of the waves is the same. The resultant wave, Eq. (10.6.24), can be re-expressed as

$$\begin{aligned} y &= 2y_0 \sin \tfrac{1}{2}[(k_1 + k_2)x - (\omega_1 + \omega_2)t] \\ &\quad \times \cos \tfrac{1}{2}[(k_1 - k_2)x - (\omega_1 - \omega_2)t] \\ &= 2y_0 \sin(\bar{k}x - \bar{\omega}t) \cos \tfrac{1}{2}[(\Delta k)x - (\Delta \omega)t] \qquad (10.6.25) \end{aligned}$$

where

$$\begin{aligned} \bar{k} &= \tfrac{1}{2}(k_1 + k_2) & \bar{\omega} &= \tfrac{1}{2}(\omega_1 + \omega_2) \\ \Delta k &= k_1 - k_2 & \Delta \omega &= \omega_1 - \omega_2 \end{aligned} \qquad (10.6.26)$$

The quantities $\bar{\omega}$ and $\bar{k}$ represent the average of the

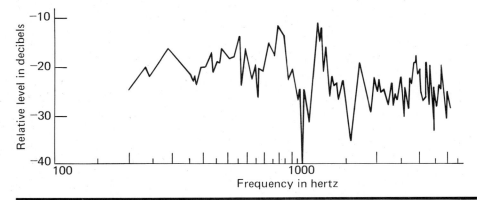

FIGURE 10.19. Frequency spectrum of sound emitted by a violin. Sound intensity level, in decibels, is plotted on the vertical axis, while frequency is plotted along the horizontal axis. (From M. V. Mathews and J. Kohut, *J. Acoust. Soc. Amer. 53*, 1620, 1973)

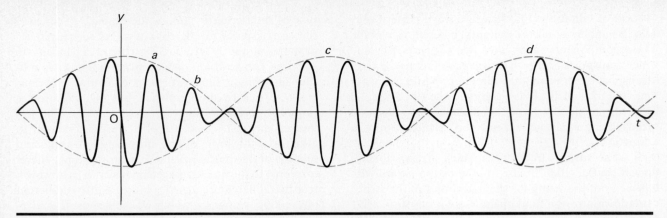

FIGURE 10.20. Temporal interference of two waves having slightly different frequencies, illustrating the phenomenon of beats.

individual frequencies and propagation constants, while $\Delta\omega$ and Δk represent their differences.

At a *fixed position x*, this expression displays a *variation in time* in both the sine factor as well as the cosine factor. Let us now assume that the two angular frequencies ω_1 and ω_2 are very nearly equal. Then, the first factor contains a very rapid time variation determined by the average of both frequencies. However, since $\Delta\omega$ is now small, the variation in time governed by the second factor may indeed by quite noticeable and may show up as *beats*, or successions of maximum amplitude spaced far enough in time to be distinguished by the ear. The situation is illustrated by means of Fig. 10.20, in which a graph of y as a function of t is given at the position $x = 0$. Any other value for x would, of course, illustrate the same general characteristics. This graph exhibits two distinct time scales. The time between any two relative maxima such as a and b is the period T, which is the reciprocal of the average frequency f. On the other hand, there is also a time T' between two absolute maxima such as c and d. These maxima occur whenever the cosine function becomes $+1$ or -1. This time interval between maxima corresponds to a change in the phase of the cosine of $180°$, or π radians, and therefore

$$\tfrac{1}{2}(\Delta\omega)T' = \pi$$

or

$$T' = \frac{2\pi}{\omega_1 - \omega_2} = \frac{1}{f_1 - f_2} = \frac{1}{f_b} \qquad (10.6.27)$$

There are, therefore, absolute maxima occurring at a "beat frequency" of $f_b = 1/T'$ given by the *difference between the frequencies f_1 and f_2*. This phenomenon is one which most of us have noticed while listening to musical instruments that are slightly "out of tune." The phenomenon is not restricted to sound waves but is also present for other forms of waves. In a

subsequent section, its application to radar will be discussed.

EXAMPLE 10.6.2

Two identical piano wires of equal length have a frequency of 386 Hz when kept under the same tension. What fractional increase in tension will lead to 2 beats per second when the wires vibrate at the same time?

A change in tension will have the effect of modifying the wave speed v, which is related to the tension T_0 by means of $v = \sqrt{T_0/\mu}$. The product of frequency and wavelength is the wave speed v, and, therefore, $f\lambda = \sqrt{T_0/\mu}$, or

$$f^2\lambda^2 = T_0/\mu \qquad (10.6.28)$$

This implies that frequency is proportional to the square root of the tension since the wavelength is fixed by geometry. Differentiating the above equation with respect to T_0, regarding λ as fixed, we find

$$2f\lambda^2 \frac{df}{dT_0} = \frac{1}{\mu} \qquad (10.6.29)$$

Multiplying this equation on both sides by T_0 and writing T_0/μ as $f^2\lambda^2$, this can be expressed as

$$\frac{df}{f} = \frac{1}{2}\frac{dT_0}{T_0} \qquad (10.6.30)$$

To obtain 2 beats per second, we require a fractional increase in frequency given by $df/f = 2/386 = 0.00518$. This can be obtained by a fractional increase in tension of $2\,df/f$, or 0.0104.

10.7 Fourier Series

In studying the vibrating string, we have tried to restrict the discussion to vibrations of a single frequency. The same restriction was also made for sound waves. The superposition principle tells us, however, that a very complicated wave may be built by means of the addition of many waves with differing amplitudes and frequencies. Now suppose we have a very complicated wave like that which is excited when a violin string is bowed or plucked. Is it possible to decompose this wave into a sum of simple sinusoidal waves each of which vibrates at its own definite frequency? Or, turning the question around, is it possible to synthesize a complex nonsinusoidal wave by superposing simple sinusoidal components? The answers to these questions were provided many years ago, in 1807, by the brilliant French mathematician J. B. J. Fourier in the course of his studies of heat conduction. Fourier showed that an "arbitrary" periodic function could be described as an infinite sum of sinusoidal functions.

Specifically, he discovered that if $f(x)$ is a function which is periodic over an interval $-L \leq x \leq L$, then $f(x)$ can be written as an infinite sum of sine and cosine functions of the form

$$f(x) = \frac{a_0}{2} + \sum_{n=1}^{\infty} \left(a_n \cos \frac{n\pi x}{L} + b_n \sin \frac{n\pi x}{L} \right) \qquad (10.7.1)$$

The coefficients a_n and b_n are called the Fourier coefficients, and their values are given by the formulas

$$a_n = \frac{1}{L} \int_0^L f(x) \cos \frac{n\pi x}{L} \, dx \qquad (10.7.2)$$

$$b_n = \frac{1}{L} \int_0^L f(x) \sin \frac{n\pi x}{L} \, dx \qquad (10.7.3)$$

For these integrals to exist, the function $f(x)$ must be reasonably well behaved.

In Fig. 10.21, a few illustrations of the above method are given for various simple periodic functions. In principle, an infinite number of terms are needed in the series to reproduce exactly the function $f(x)$. In practice, however, it usually happens that a very good approximation to the actual function may be obtained by keeping a limited finite number of terms. If a small number of terms suffices as an approximation, we say that the series converges rapidly, while if many terms are needed, we say that the convergence is slow.

If we have a function which is not periodic but is limited in its domain of definition to an interval from $-L$ to L, we may extend the function to make it periodic, and then the above formulas are valid for describing the function in the region between $-L$ to L. As an example, consider the human profile traced from a photograph in Playboy magazine and shown in Fig. 10.22. Professor G. Ward of Penn State University has Fourier analyzed this profile numerically. The sequence of pictures illustrates various approximations to the profile by taking one, two, three, five, ten, 20, 30, and 40 terms in the Fourier series. It is clear from these figures that 40 terms constitute a very good approximation to the actual profile. Even so, 40 is a lot less than the infinite number needed for its exact reconstruction.

The technique of Fourier analysis is extremely powerful and marks a significant advance in the development of mathematical physics. Since the method may be somewhat beyond the mathematical level of many of the students for whom this text is intended, we shall try to convey an appreciation of the basic ideas involved without a full presentation of the mathematics. To fix a framework for these ideas, let us attempt to describe the general motion of a string of length L which is fixed at both ends. We have previously shown in (10.6.19) that if waves of a definite

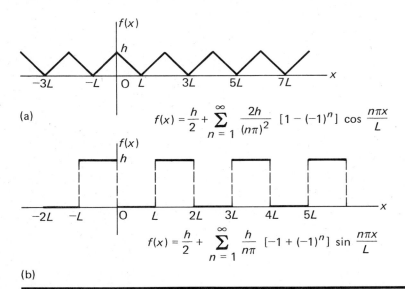

(a)

$$f(x) = \frac{h}{2} + \sum_{n=1}^{\infty} \frac{2h}{(n\pi)^2} \left[1 - (-1)^n \right] \cos \frac{n\pi x}{L}$$

$$f(x) = \frac{h}{2} + \sum_{n=1}^{\infty} \frac{h}{n\pi} \left[-1 + (-1)^n \right] \sin \frac{n\pi x}{L}$$

(b)

FIGURE 10.21. (a) Fourier series representing "sawtooth" wave of wavelength 2L. (b) Fourier series representing "square" wave of wavelength 2L.

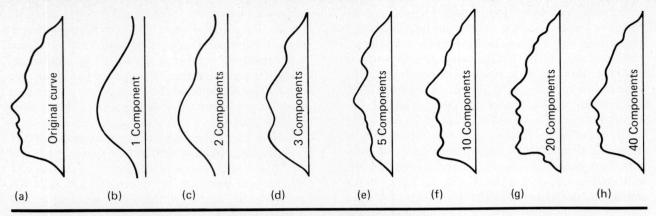

FIGURE 10.22. Synthesis of a human profile by a Fourier Series. (a) Original profile. (b)–(h) Fourier series representation, retaining 1, 2, 3, 5, 10, 20, and 40 components. (Courtesy of G. Ward, Pennsylvania State University)

frequency moving in opposite directions are superposed, a *standing wave* of a definite frequency is established. The possible frequencies were given by $f_n = nv/2L$. According to the superposition principle, the most general standing wave would, therefore, have the form of a sum of waves like those shown in (10.6.19), that is,

$$y(x, t) = \sum_{n=1}^{\infty} (b_n \sin k_n x) \cos \omega_n t$$

In this equation, according to (10.6.22) and (10.3.21), we must choose

$$k_n = \frac{\pi}{L}, \frac{2\pi}{L}, \frac{3\pi}{L}, \dots \quad \text{or} \quad k_n = \frac{n\pi}{L}$$

and

$$\omega_n = vk_n = \frac{n\pi v}{L}$$

The above equation may, therefore, be written as

$$y(x, t) = \sum_{n=1}^{\infty} \left(b_n \sin \frac{n\pi x}{L} \right) \cos \left(\frac{n\pi v}{L} t \right) \quad (10.7.4)$$

Suppose now that at $t = 0$, the string has the form shown in Fig. 10.23, corresponding to a violin string that is plucked and suddenly released. Then, at $t = 0$, since cos $(n\pi vt/L)$ has the value of unity, the function

$$y(x, 0) = \sum_{n=1}^{\infty} b_n \sin \frac{n\pi x}{L} \quad (10.7.5)$$

must somehow represent the triangular form of the plucked string shown in Fig. 10.23. This can be accomplished by making the proper choice of the Fourier coefficients in Eq. (10.7.1). Indeed, it is evident that if we choose a_0 and all the coefficients a_n of the cosine terms in that expression to be zero, (10.7.1) reduces to the equation above. The numerical values of the re-

maining coefficients b_n are calculated from (10.7.3), using for $f(x)$ the function

$$f(x) = \frac{2hx}{L} \quad (0 < x < L/2)$$

$$f(x) = \frac{2h}{L}(L - x) \quad (L/2 < x < L),$$

which represents the triangular shape of the plucked string at $t = 0$. We shall not go through the details of the calculation, but instead simply state that the coefficients b_n are found to have the form

$$b_n = \frac{8h}{n^2\pi^2} \sin \frac{n\pi}{2} \quad (10.7.6)$$

In this expression, we should note that $\sin(n\pi/2) = 1$, $0, -1, 0, \dots$ for $n = 1, 2, 3, 4, \dots$. According to (10.7.4), then, the form of the string at any time t can be represented as

$$y(x, t) = \sum_{n=1}^{\infty} \frac{8h}{n^2\pi^2} \sin \frac{n\pi}{2} \sin \frac{\pi n}{L} x \cos \frac{\pi n v}{L} t \quad (10.7.7)$$

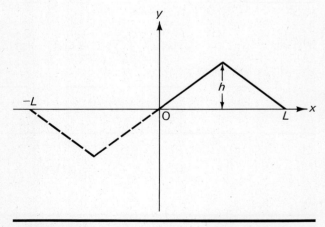

FIGURE 10.23. Geometry of the plucked string at time $t = 0$.

Now let us review what has been accomplished. We started with a snapshot of a plucked violin string at $t = 0$, at which time the string was not moving. We went through some fairly complicated mathematics to obtain the result shown above as Eq. (10.7.7). This equation tells us, however, exactly what a snapshot of the string would look like at any instant of time greater than zero. We were able to determine the entire time dependence of the motion of the string by knowing the time dependence of the component standing waves as well as the Fourier coefficients, which provide a measure of *how much of each standing wave is present in the entire sum*.

It should be noted from Eq. (10.7.6) that the Fourier coefficients contain the factor $1/n^2$ and, therefore, drop off rapidly with increasing n. This is an example of a series which has rapid convergence, and we might expect that the first ten terms already represent an excellent approximation to the actual function. We have carried out a numerical evaluation by keeping ten nonvanishing terms and have plotted a number of different snapshots of what the string would look like at different times. The results are given in Fig. 10.24. The idea that a complex nonsinusoidal wave can always be represented as a superposition of simple sinusoidal components is an important concept and one that we may on occasion ask you to recall.

10.8 The Doppler Effect and Some Related Phenomena

Whenever there is relative motion between the source of a wave and an observer, there are some very important and easily observable effects that occur. The siren of a speeding police car or ambulance appears to increase in pitch as the vehicle approaches and to decrease when it is moving away. This alteration in the frequency of detected sound is even more pronounced when the source is a train whistle or a jet plane. In the case of electromagnetic waves, the frequencies are also modified by relative motion. The frequency of light waves coming from distant galaxies is thus lowered as a result of the recession of the galaxies from our solar system.

The effect in which the emitted and detected frequencies differ due to relative motion is known as the *Doppler effect*. It was named after the Austrian physicist Christian Johann Doppler who discovered the effect in 1842 for light waves.

Consider the simple case of emission of a sound wave of definite frequency f from a fixed point source. The source may be considered to emit f disturbances or f wavefronts per second which then travel outward at the speed of sound. An observer who is stationary in the medium will detect f wavefronts per second. In other words, the number emitted and the number received are identical, as shown in Fig. 10.25a. In Fig. 10.25b, the observer is moving toward the source of sound, and in 1 second he now receives more wavefronts as a result of his motion. In t seconds, he moves a distance $d = v_0 t$, and as a consequence he detects $d/\lambda = v_0 t/\lambda$ additional wavefronts. Since the frequency f' detected by him is merely the number of wavefronts he receives per second,

$$f' = f + \frac{v_0}{\lambda} \tag{10.8.1}$$

where λ is the wavelength of the wave. But since $\lambda = v/f$, where v is the wave velocity, this equation can be written as

$$f' = f + f\frac{v_0}{v} = \left(1 + \frac{v_0}{v}\right)f \tag{10.8.2}$$

By the same reasoning, if the observer recedes from the source of sound as in Fig. 10.25c, he receives fewer disturbances per second and, therefore, the detected frequency would be

$$f' = \left(1 - \frac{v_0}{v}\right)f \tag{10.8.3}$$

We may combine both (10.8.2) and (10.8.3) by writing

$$f_\pm' = \left(1 \pm \frac{v_0}{v}\right)f \tag{10.8.4}$$

where the plus sign denotes a frequency increase (observer approaching source) and the minus sign denotes a decrease (observer receding from source).

For sound waves, Eq. (10.8.4) does not account for all frequency shifts due to relative motion since we must also consider the possibility of motion of the source of sound toward or away from the observer. When the source of sound moves toward the observer, there is a decrease in wavelength directly ahead of the source and an increase directly behind the source, as shown in Fig. 10.26. In the forward direction, the decrease in wavelength is easily calculated since it is equal to the distance traveled by the source in the time interval between two successive emissions of wavefronts or disturbances. This distance is equal to the velocity of the source v_s multiplied by the time interval between two successive emissions. But this time interval is the period T, which is, in turn, the reciprocal of the frequency f emitted by the source. Therefore, the distance referred to above is given by v_s/f. The wavelength as seen by the stationary observer is then

$$\lambda' = \lambda - \frac{v_s}{f} \tag{10.8.5}$$

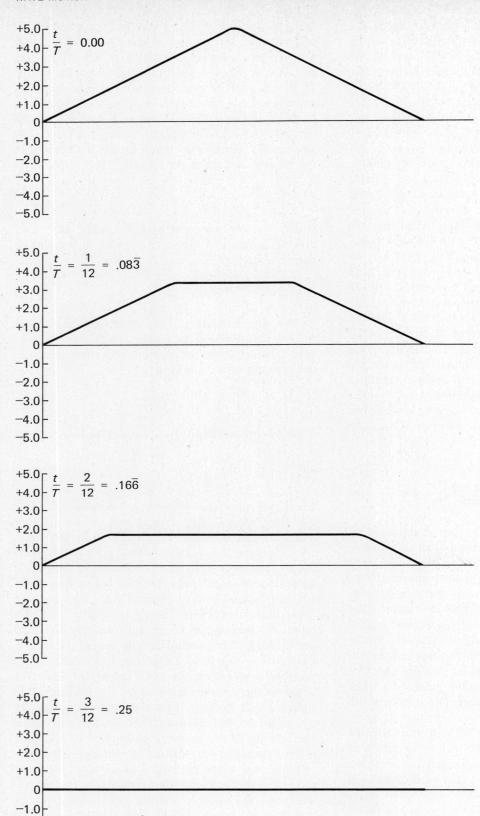

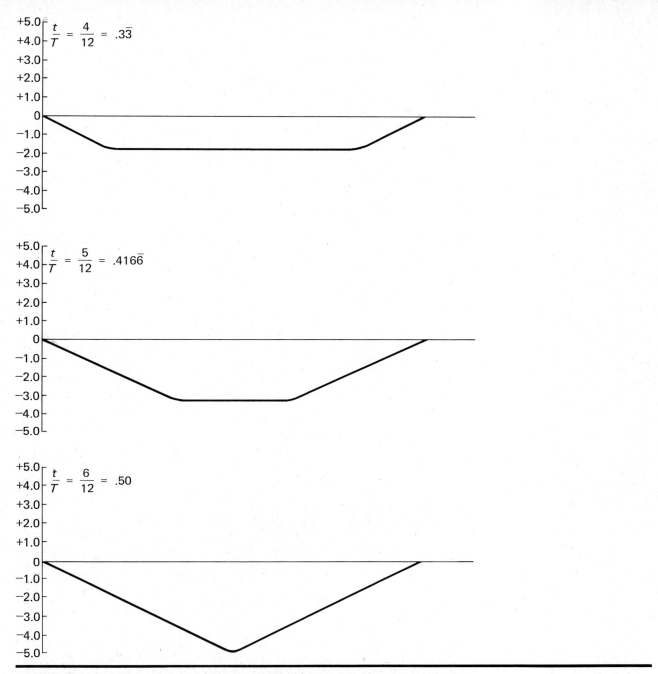

FIGURE 10.24. Fourier series solutions, according to (10.7.7), showing the plucked string in various stages of its vibrational cycle. Each succeeding illustration corresponds to a time interval of $\frac{1}{12}$ the period of vibration.

and hence the frequency of sound heard by the observer is

$$f'' = \frac{v}{\lambda'} = \frac{v}{\lambda - (v_s/f)} \tag{10.8.6}$$

But since λ is given by v/f, we may write

$$f'' = f\left(\frac{1}{1 - (v_s/v)}\right) \tag{10.8.7}$$

as the frequency detected by the observer at P. An observer located at Q will hear a smaller frequency, given by

$$f'' = f\left(\frac{1}{1 + (v_s/v)}\right) \tag{10.8.8}$$

since the wavelength he receives is greater than that emitted.

Let us summarize Eqs. (10.8.7) and (10.8.8) by

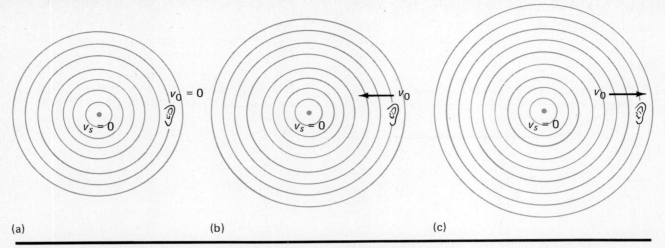

FIGURE 10.25. Schematic illustration of the Doppler effect involving motion of the observer.

writing, as before,

$$f_\pm'' = f\left(\frac{1}{1 \pm (v_s/v)}\right) \qquad (10.8.9)$$

for the situation in which the source is in motion.

From the arguments presented above, we see that there is a distinct difference between motion of the observer and motion of the source in determining the Doppler shift of sound waves. This difference is present because sound propagates through a physical medium, and there is a distinct difference in the way in which wavefronts are emitted and propagate when the source is at rest and when it is in motion. For light waves, which propagate through empty space,

however, there is no distinction to be made between the two cases discussed above. Light does not need a material medium in which to propagate, and, therefore, the effect depends *only on the relative motion* of the source and the detector of light.

We may state, without proof, that if this relative velocity u is much smaller than the speed of light c, the Doppler shift for light is given by

$$f_\pm' \cong f\left(1 \pm \frac{u}{c}\right) \qquad (10.8.10)$$

As a matter of fact, even for sound waves the distinction between $f_\pm'$ and $f_\pm''$ disappears if the relative velocity of source and observer is very much smaller than the wave speed. The following examples will illustrate some of these ideas.

EXAMPLE 10.8.1

An ambulance speeding down a street at 60 mph has a siren which emits sound at 440 Hz. At an intersection at which there is a red traffic light, a number of pedestrians hear the siren. What frequency do they hear as the ambulance is coming toward them? Away from them? How fast must the ambulance go in order for the red light to appear green to the driver? The numerical information required is $v(\text{sound}) = 331$ m/sec, $v(\text{light}) = c = 3 \times 10^8$ m/sec, $\lambda(\text{green}) = 540 \times 10^{-9}$ m, and $\lambda(\text{red}) = 660 \times 10^{-9}$ m, 1 mph = 0.45 m/sec.

In this example, we have a case in which the source of sound is in motion. When the ambulance is moving toward the pedestrians, the frequency they detect is

$$f_+'' = 440 \frac{1}{1 - \dfrac{60 \times 0.45}{331}} = 479.1 \text{ Hz} \qquad (10.8.11)$$

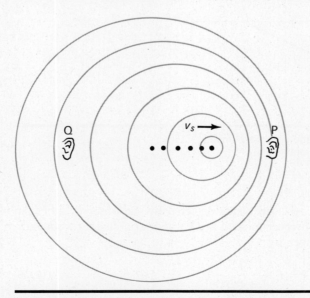

FIGURE 10.26. Schematic illustration of the Doppler effect involving motion of the source.

while when it is receding from them, they hear

$$f_-'' = 440 \frac{1}{1 + \dfrac{60 \times 0.45}{331}} = 406.8 \text{ Hz} \qquad (10.8.12)$$

These frequency shifts are quite substantial and are easily noticed.

When the ambulance is moving toward the red light, we have a Doppler shift of *light* due to the relative motion of source and observer. We expect a frequency increase so that the detected light frequency will be

$$f_+' = f\left(1 + \frac{u}{c}\right) \qquad (10.8.13)$$

where u is the speed of the ambulance and c is the speed of light. The frequencies are related to the wavelengths by $f\lambda = c$, and, therefore, (10.8.13) may be rewritten as

$$\frac{c}{540 \times 10^{-9}\,\text{m}} = \frac{c}{660 \times 10^{-9}\,\text{m}}\left(1 + \frac{u}{c}\right) \qquad (10.8.14)$$

This is easily solved for u/c:

$$\frac{u}{c} = \frac{66}{54} - 1 = 0.22 \qquad (10.8.15)$$

If we now substitute for c, we find that the ambulance must be going at a speed of 0.66×10^8 m/sec, or 1.47×10^8 mph. This is a pretty good speed for an ambulance. It certainly suggests that if you get a ticket for passing a red light and you argue that the light appeared green because of the Doppler effect, you are not likely to be convincing.[7] If the policeman wants to show off his knowledge of physics, he might very well say, "Perhaps you are right, Sir, about seeing a green light, but if you did, then you surely deserve a speeding ticket."

The numerical answer to the latter part of the above example may seem a bit absurd. It is undoubtedly true that if you can detect any light frequency shift at all with your eyes, your speed is extremely high. There are, however, instruments much more sensitive than the human eye, and these are capable of measuring very small frequency shifts and, therefore, low speeds. The radar sets that are used to measure auto speeds are examples of such refined instruments. Their operation is only possible because of the Doppler effect. The following example illustrates the method.

---

[7] Although R. W. Wood, formerly Professor of Physics at Johns Hopkins University, is reputed to have used this argument on a judge and gotten away with it!

EXAMPLE 10.8.2

On Interstate Route 80, a radar trap operates by sending out electromagnetic waves of frequency 2400×10^6 cycles per second. This frequency is in the so-called microwave region. A station wagon moving at 85 mph gets pulled over for speeding. Explain how the radar system based on the Doppler effect works.

The state trooper knows that the station wagon was speeding because some of the microwaves he sent out were reflected by the approaching vehicle and came back with a higher frequency. The station wagon first receives microwaves which are shifted in frequency by Δf, where Δf, according to (10.8.13) is given by

$$\frac{\Delta f}{f} = \frac{f_+' - f}{f} = \frac{u}{c} = \frac{85 \times 0.45}{3 \times 10^8} = 1.28 \times 10^{-7}$$

The station wagon, in reflecting the microwaves back to the radar set, acts once again as a moving source of microwave energy, introducing a *second* Doppler shift of the same magnitude. The total frequency shift in the waves received by the patrol car is, therefore,

$$2\,\Delta f = 2(u/c)f = (2)(1.28 \times 10^{-7})(2.4 \times 10^9) = 614.4 \text{ Hz}$$

The speed of light is so high that if the trooper's radar hits your car when you are 100 m from him, it takes only 0.66×10^{-6} sec for the reflected radar waves to return to his car. During this time interval, your vehicle, moving 85 mph, has only moved a distance of 2.5×10^{-5} m, or 0.025 mm. Fantastic, but true!

The actual *shift* in frequency between the emitted radar and the detected radar may be measured by detecting the beats between the two frequencies. The number of beats is directly proportional to the vehicle's speed, and in a typical situation the trooper in the radar car reads your speed directly on a digital display panel. The entire process is practically instantaneous.

There are many other important uses of the Doppler effect, especially for light. We will return to the subject again in the chapters on electromagnetic waves and will discuss more examples at that time. The Doppler effect provides an example of a physical phenomenon that arises from the relative motion between a source of waves and a receiver of waves. Other interesting phenomena also emerge as a consequence of this relative motion. As one example, we shall consider the sonic boom observed when the speed of an aircraft or any other object is in excess of the speed of sound.

EXAMPLE 10.8.3

In Fig. 10.27, a projectile is in motion with a constant velocity v_s in excess of the speed of sound v in the

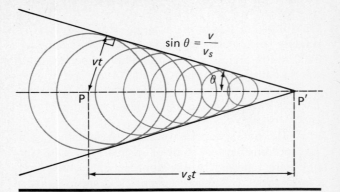

FIGURE 10.27. Wavefronts emitted by an object traveling at a speed in excess of the speed of sound, illustrating the formation of a shock front.

$$\sin \theta = \frac{v}{v_s}$$

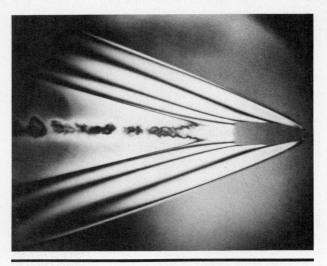

FIGURE 10.28. Photograph of shock wave associated with a bullet moving at Mach 2.5. Reconstructed image of a double-exposed holographic interferogram of a high-velocity 22-caliber bullet. The first exposure was made just before the firing, and the second exposure was made when the bullet was passing through the scene volume. A pulsed ruby laser was used. The interference fringes give the change in optical path which, in turn, gives the air density behind the shock wave. (Photo courtesy of L. O. Heflinger, R. F. Wuerker, and R. E. Brooks; TRW Systems)

medium. At $t = 0$, the projectile is at a point P, while at some later time t, it is at point P′. In this figure, the spherical wavefronts which have been emitted in the time of flight between P and P′ are shown. In the present case, in which $v_s > v$, the wavefront has a conical shape, with the angle θ at the apex given by $\sin \theta = v/v_s$. The reciprocal of this number, v_s/v, is called the *Mach number*. An actual photograph of this conical shape is illustrated in Fig. 10.28 for a bullet moving at Mach 2.5. When the wavefronts of a sound wave have this form as a result of supersonic speeds, we have what is known as a *shock wave*. This shock wave is responsible for the so-called sonic boom or explosion caused by supersonic aircraft.

Let us now refer to the situation shown in Fig. 10.29. At time t when an airplane (or other supersonic object) is at point P′, let us consider the sound which arrives at a point A on the conical front, which might represent some given point on the ground. Points C and D are former locations of the airplane separated by a distance $v_s \Delta t$, where Δt is the time of flight between C and D. Now let us calculate the difference in arrival times at A of a sound wave emitted at C and one at D. If t_C and t_D denote these times, we find

$$t_C = \frac{CA}{v}$$

and

$$t_D = \Delta t + \frac{DA}{v} \tag{10.8.16}$$

where CA and DA are the distances. Therefore,

$$t_D - t_C = \Delta t + \frac{1}{v}(DA - CA)$$

$$= \Delta t + \frac{1}{v}(DA - EA - CD) \tag{10.8.17}$$

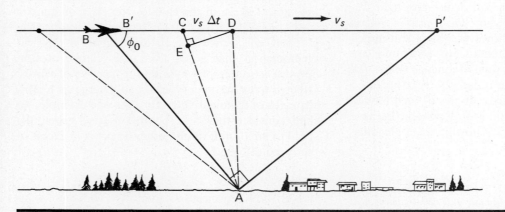

FIGURE 10.29. Geometry of shock wave excited by a supersonic aircraft.

Now, if the distance *CD* is much smaller than *CA*, it follows that $DA \cong EA$. We also note that $CE = CD \cos \phi = (v_s \Delta t) \cos \phi$, and, therefore, the difference in arrival times is

$$t_D - t_C = \Delta t \left(1 - \frac{v_s}{v} \cos \phi \right) \qquad (10.8.18)$$

Several conclusions can now be drawn from Eq. (10.8.18), an equation which would also be valid at subsonic speeds. If $v_s < v$ (subsonic), there is no possibility of simultaneous arrival at point A. Waves emitted at different times arrive at A at different times. However, if $v_s > v$, then if the angle ϕ is chosen to be ϕ_0, where $\cos \phi_0 = v/v_s$, the arrival times are *identical*. In fact, subject to the approximations made, the arrival time of *all* sound waves emitted between points B and B′ are the same. This implies that a very substantial constructive interference of sound occurs at point A, and as a result a *sonic boom* or loud explosion is heard at A. The only condition is that B and B′ should be points whose separation is substantially smaller than their distances to the detector (at A). It is seen that the angle ϕ_0 obtained above is the complement of the previously defined angle θ.

The very large intensity at A and the correspondingly large pressure variations which result pose a very serious problem. The sonic boom produced by such a shock wave can shatter glass and do considerable damage to the structure of buildings. It is also extremely unpleasant to hear the intense sound wave. Actually, supersonic aircraft produce a double boom. One is caused by the nose and the other by the tail of the airplane. This is illustrated in Fig. 10.30.

Sonic boom cannot be eliminated. Supersonic aircraft are designed so as to minimize the shock waves, but there seems to be no way in which they can be done away with altogether. The only solution to the problem seems to be to require aircraft to fly subsonically until they attain enough altitude so that the sonic boom will not pose any difficulty at ground level.

10.9 Ocean Waves

We began our discussion of wave motion by referring to the water waves that travel across the surface of the sea. Although these waves are familiar to all of us, their behavior is very complex. We shall, therefore, present only a brief qualitative survey of their more important properties.

Ocean waves originate in the action of the wind. Winds create small ripples on a calm sea. The profile of these ripples presents additional surfaces that absorb energy from the air currents and further increases their size. The ultimate amplitude of the waves that are formed depends upon the average wind velocity, the time during which the wind acts, and the distance over which it is effective.

Ocean waves can carry huge amounts of energy and can, therefore, cause severe damage to coastal installations when they come ashore. At Wick, Scotland, for example, in a great storm that occurred in 1872, the designer of the breakwater watched from a nearby cliff as the sea wrought its destruction on his handiwork. The end of the breakwater was capped by an 800-ton piece of concrete secured to the foundation by iron columns 3.5 inches in diameter. Both cap and foundation, weighing a total of 1350 tons were removed as a unit and swept into the sea. The designer, undaunted by this, rebuilt the installation, using a larger cap that weighed 2600 tons. The new breakwater, however, suffered a similar fate a few years

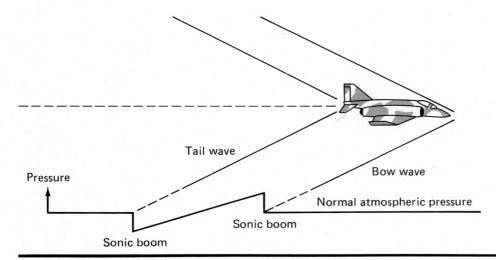

FIGURE 10.30. Double "sonic boom" excited by nose and tail of aircraft.

later. Whether he kept his job and made a third try is not documented.[8]

Although most ocean waves are generated by the wind, underwater earthquakes, landslides, and volcanic activity can also cause huge waves. Waves such as these are often referred to as *tidal waves*, or *tsunamis*, which is the Japanese equivalent. On August 27, 1883, an eruption of the volcano Krakatoa in the East Indies caused a tremendous tsunami in which waves 100 feet high swept away the town of Merak, killing 36,380 persons. Tidal waves can travel thousands of miles at speeds of 300 to 400 mph, yet are hardly visible in the open sea. Only when they approach the shore can their impact, which is usually deadly, be realized.

One of the main reasons for the complex behavior of water waves is that the wave velocity is not a single fixed number but is instead a function of the wavelength. A water wave of short wavelength travels slower than a wave of longer wavelength, the relation between velocity and wavelength being something like $v^2 = 5.12\lambda$, if λ is measured in feet and v in ft/sec. This behavior is distinctly different from the way in which waves propagate in stretched strings or in gases, where the wave velocity is completely independent of wavelength or frequency. Waves such as water waves, whose velocity varies with wavelength, are said to exhibit the phenomenon of *dispersion*.

Dispersion plays an important role in physics, particularly in the propagation of light and electromagnetic waves in dense transparent substances like glass. However, light waves traveling in a vacuum or in a rarified gaseous medium such as air do not exhibit dispersion; they travel with the same speed (2.997×10^8 m/sec) regardless of wavelength. In a dense, transparent substance such as glass or water, on the other hand, their speed is significantly reduced and depends on the wavelength. Because the phenomenon of dispersion is present when light propagates through glass or water, these substances are often referred to as *dispersive media* with respect to the propagation of light. Dispersion is responsible for the separation of white light into its constituent spectral colors in the rainbow or in a glass prism. We shall investigate the subject of dispersion in more detail in our later work on light and optics.

Another complexity in the study of water waves arises from the fact that they are neither purely transverse nor purely longitudinal, but rather a mixture of the two. For plane water waves in an infinitely large body of very deep water, the motion of individual volume elements within the medium is circular, the size of the circular paths decreasing with

increasing depth. But in shallow water, this picture is no longer valid and the situation becomes even more complex. As the water depth decreases, the wave amplitude increases, the wavelength decreases, and the wave velocity is reduced. This decrease in wave velocity accounts for the fact that wavefronts are nearly always parallel to the coast when they reach shore. When a wave approaches shallow water at an angle to the coastline, the part nearest the shore, where the water is shallowest, begins to travel more slowly than before. The part farther out, in deeper water, therefore, catches up, so to speak, until the entire wavefront is practically parallel with the shoreline. Ultimately, the effect of shallow water in reducing the wavelength and at the same time increasing the amplitude causes the waves to crest and break up in what we recognize as surf. In this process, the originally circular orbits of water elements first become elliptical, and finally extremely irregular as the waves break up.

Our understanding of the complex dynamics of water waves is still incomplete, and a better understanding of waves in fluid media is necessary to resolve existing problems in ship design, harbor planning, and the construction of coastal structures. Research in the behavior of water waves is therefore an active field of endeavor which utilizes sophisticated experimental tanks, testing basins, and computing equipment.

SUMMARY

Waves are self-propagating disturbances that travel through material media, transporting energy without transport of the medium itself. In the case of light and electromagnetic waves, transport of energy through empty space can be accomplished by interaction of electric and magnetic fields.

Periodic waves are characterized by amplitude, phase angle, wavelength, frequency, and propagation velocity: *Amplitude* refers to the maximum possible displacement of a point in the medium from its equilibrium position. The *phase angle* specifies the part of the periodic cycle that is at hand for any point in the medium at any given time. *Wavelength* λ refers to the distance between successive maxima or minima, or between any two neighboring points having equivalent phase angles.

Frequency f refers to the number of cycles of motion executed by a given particle per unit time *angular frequency* ω is simply 2π times the frequency. *Propagation velocity* v refers to the rate at which a point having a given phase (such as a maximum or minimum of displacement) travels through the me-

---

[8] See, for example, W. Bascom, "Ocean Waves," *Scientific American* (August 1959). The reader will find many other interesting facts about ocean waves in this comprehensive article.

dium. Propagation velocity, frequency, and wavelength are related by $v = f\lambda$.

The *propagation constant k* is a quantity defined by $k = 2\pi/\lambda$. The wave velocity can be expressed in terms of the angular frequency and the propagation constant as

$$v = f\lambda = \frac{\omega}{k}$$

Wavefronts represent the locus of points in the medium of equal phase, such as points corresponding to maximum displacement or zero displacement. Wavefronts propagate through the medium with wave velocity v. In general, the shape of the wavefront depends upon the nature of the source of wave motion and the distance of the wavefront from the source, but there are three particularly simple cases: *point source*, giving rise to *spherical* wavefronts; *line source*, giving rise to *cylindrical* wavefronts; and *plane source*, giving rise to *plane* wavefronts.

Plane wavefronts also approximate the behavior of cylindrical and spherical wavefronts at large distances from the source of the waves. The direction in which the wave disturbance propagates is everywhere perpendicular to the wavefront.

Mathematically, for one-dimensional waves or plane waves in three dimensions, the displacement $u(x, t)$ of the medium from equilibrium at any point x and any time t can be expressed as

$$u(x, t) = f(x - vt)$$

where v is the wave velocity and f represents any function of the argument $x - vt$. The most important special case of this situation is the sinusoidal plane waves, where

$$u(x, t) = A \sin[(kx - \omega t) + \delta]$$
$$= A \sin\left[2\pi\left(\frac{x}{\lambda} - ft\right) + \delta\right]$$

with $\omega/k = f\lambda = v$. In the case of *longitudinal waves*, the displacement of the medium is parallel to the direction of propagation, while for *transverse waves*, the displacement is normal to the propagation direction. Any of the three functions written above satisfies a partial differential *wave equation* of the form

$$\frac{\partial^2 u}{\partial x^2} = \frac{1}{v^2}\frac{\partial^2 u}{\partial t^2}$$

where v is the wave velocity.

The propagation velocity v may always be expressed in terms of the physical properties of the medium in which the waves travel. For example, for transverse waves in a stretched string where the tension is T_0 and the linear mass density is μ,

$$v = \sqrt{T_0/\mu}$$

while for longitudinal acoustic waves in an elastic solid substance

$$v = \sqrt{Y/\rho}$$

where Y is Young's modulus defined by (10.4.3) and ρ is the density. For longitudinal sound waves in gases, the velocity is

$$v = \sqrt{B/\rho}$$

where B is the bulk modulus defined by (10.4.11) and ρ is the density.

The average energy $\bar{U}$ transmitted by a one-dimensional wave, such as might propagate along a rope, in time Δt is given by

$$\Delta\bar{U} = \tfrac{1}{2}\mu v\omega^2 y_0{}^2 \Delta t$$

where y_0 is the amplitude and μ is the mass per unit length.

The intensity I of a wave is the average energy which crosses unit area normal to the propagation direction in unit time. For a rope, as discussed above,

$$I = \frac{\Delta\bar{U}}{(\Delta A)(\Delta t)} = \frac{\mu v\omega^2 y_0{}^2}{2\,\Delta A}$$

while for a wave in a three-dimensional medium of density ρ, the intensity is

$$I = \tfrac{1}{2}\rho v\omega^2 y_0{}^2$$

Waves can be superposed or added by adding the separate displacements caused by each at every point in the medium at every instant of time.

Spatial interference occurs when the phase relationships between two waves of equal frequency are such as to cause reinforcement or cancellation of wave amplitudes at points whose distances from the two sources that emit the waves are different. Reinforcement, referred to as *constructive interference*, occurs at points where the difference in path between the two sources amounts to an integral number of wavelengths; while cancellation, called *destructive interference*, occurs at points where the difference in path between the two sources is an odd-integral number of half wavelengths.

Temporal interference occurs when waves emitted from two different sources have slightly different frequencies. The intensity of the received wave then varies in amplitude periodically with a "beat" frequency equal to the difference in the two source frequencies.

The *Doppler effect* is an alteration in perceived frequency caused by motion of source or observer. For sound waves, in the case where the observer is moving, the perceived frequency f' is related to the source frequency f by

$$f' = \left(1 \pm \frac{v_0}{v}\right)f$$

where v_0 is the velocity of the observer. If the source is moving with velocity v_s and the observer is stationary, then the perceived frequency f'' will be

$$f'' = \left(\frac{1}{1 \pm (v_s/v)} \right) f$$

For light waves, the perceived frequency depends *only on the relative velocity u* between source and observer. It is given by

$$f' \cong f \left(1 \pm \frac{u}{c} \right)$$

in cases where the relative velocity u is much less than the speed c with which light propagates.

Dispersion occurs in situations where the wave velocity is not independent of wavelength but varies for waves of different wavelengths. It is absent in most of the simple examples discussed in this chapter, but occurs in the case of water waves, and in the case of light waves which travel through dense transparent substances such as water or glass.

QUESTIONS

1. Can you think of any way of propagating a one-dimensional longitudinal wave along a rope? Present your reasoning.
2. The speed of propagation of sound waves is temperature dependent while that of light waves is not. Explain why this is true.
3. What are some of the properties you can associate with a wavefront?
4. In writing the wave equation, it is necessary to use partial derivatives. Explain.
5. Show on the basis of dimensional arguments that the speed of propagation of a wave along a string must be proportional to $\sqrt{T/\mu}$.
6. What is the difference between wave frequency and wave velocity?
7. When light waves from a point source travel through empty space, their intensity obeys an exact inverse square law. Suggest a reason why the same statement does not apply for sound waves.
8. All waves have energy associated with them. Suggest some empirical ways of verifying this.
9. Two waves of different amplitude propagate in a certain medium. Is it possible for these waves to produce complete destructive interference?
10. When the temperature of air increases by 1 percent, how does the speed of sound in air change?
11. In the propagation of wave motion in a given medium, how would the power transmitted change if the frequency and amplitude were both doubled?
12. Describe the difference between loudness and intensity.
13. Is the principle of superposition of waves an obvious principle or must it be subject to empirical verification? Under what circumstances could it be false?

14. Beats occur when there is "temporal interference." Explain this statement.
15. In the propagation of sound through a gas, is it true that the wavelength is always inversely proportional to the frequency? Is it inversely proportional to the propagation constant?
16. Give a number of examples which illustrate the Doppler effect for sound waves. Would there be an analogous effect for water waves?
17. When a standing wave is established in a vibrating string, energy cannot be transported across a node. Explain.
18. Standing waves can only occur for specific values of the wavelength. What is the basic reason for this?
19. Explain clearly the concept of dispersion.

PROBLEMS

1. A traveling wave with a frequency of 20 cps and a wavelength of 8 ft propagates along a rope. The maximum displacement of any point from equilibrium is 0.5 ft. (a) Write an expression which gives the displacement of any point on the rope as a function of its position x and of the time t. (b) What is the velocity of propagation of the wave? (c) At $t = 3/160$ sec, find the velocity and acceleration of the point on the rope located at $x = 0$. The answer will depend on part (a). (d) If the tension in the rope is 64 pounds, what is the weight of 1 ft of rope?
2. Sound waves in air travel with a velocity of about 331 meters/sec, while light or radio waves have a velocity, in air or in free space, of about 3.00×10^8 meters/sec. Find the wavelength of (a) sound waves and (b) light or radio waves, of frequency 20,000 Hz. Find the frequency of (c) sound and (d) light waves of wavelength 1.00×10^{-6} meter.
3. What are the angular frequencies and the propagation constants associated with the waves in parts (a), (b), (c), (d) of the preceding problem?
4. A traveling wave having an angular frequency of 180 rad/sec and a velocity of 80 meters/sec propagates along a rope. (a) What is the propagation constant? (b) What is the frequency? (c) What is the wavelength? Assuming that the amplitude of the wave is 0.40 meter, write (d) an expression for the displacement of any point on the rope at any time, in terms of propagation constant and angular frequency, and (e) a similar expression in terms of the frequency and wavelength.
5. A transverse traveling wave is described by the equation $y(x, t) = 0.72 \sin(3.60x - 270t)$, giving the displacement in meters. Find (a) the amplitude, (b) the angular frequency, (c) the propagation constant, (d) the wavelength, (e) the frequency, and (f) the wave velocity.
6. For the wave described in the preceding problem, find (a) the displacement, (b) the velocity, and (c) the acceleration at the point $x = 0.80$ meter at time $t = 0.025$ sec.
7. A one-dimensional traveling wave is described by the equation

$$y(x, t) = 0.15 \sin \left(\frac{5\pi}{6} x - 24\pi t \right)$$

where x and y are in meters. (a) What is the amplitude,

frequency, and wavelength characterizing the wave? **(b)** Find the velocity of propagation. **(c)** Find the velocity and acceleration, $\partial y/\partial t$ and $\partial^2 y/\partial t^2$ at $x = 0$, at time $t = 1/72$ sec. **(d)** If this wave is traveling along a rope with a mass density of 0.1 kg/m, what is the tension in the rope?

* **8.** Which of the following expressions represent traveling waves:

(a) $y(x, t) = y_0 \sin^2(kx - \omega t)$

(b) $y(x, t) = y_0(t) \cos kx$

(c) $y(x, t) = y_0 e^{-\lambda(x^2 - v^2 t^2)}$

For those which do give traveling waves obtain expressions for the speed of propagation of the wave, the wavelength, the velocity $\partial y/\partial t$ and the acceleration $\partial^2 y/\partial t^2$.

9. A stretched rope 15 m long is fastened to a post at one end and is held by a physics professor at the other end. The tension in the rope is 5 N. The professor sends a pulse down the rope. The pulse is reflected at the post and returns to its point of origin, the round trip taking 6 sec. How much does the rope weigh?

10. Show that the fractional change in the speed of a wave on a rope is given by

$$\frac{dv}{v} = \frac{1}{2}\frac{dT_0}{T_0} - \frac{d\mu}{\mu}$$

where dT_0/T_0 and $d\mu/\mu$ are the fractional changes of tension and of mass density.

11. Show that for very large r, the amplitude of a circular wave emitted by a point source varies approximately as $r^{-1/2}$.

12. The audible frequency range for bats is between 10,000 and 120,000 Hz. Find the corresponding wavelengths for sound in air at standard conditions.

13. Show that Young's modulus is the strain needed to stretch a bar to twice its original length.

14. Consider a transverse wave whose displacement in meters is given by $y(x, t) = 0.072 \sin(3.60x - 270t)$, with t in seconds, propagating along a rope of mass density 0.080 kg/m. Find **(a)** the tension in the rope, **(b)** the velocity of the wave, **(c)** the average kinetic energy per unit length, **(d)** the average potential energy per unit length, **(e)** the average total energy per unit length, and **(f)** the average energy propagated per unit time.

15. The speed of a transverse wave progagating along a thin cylindrical copper wire is 200 meters/sec. Find the speed of such a wave along a second copper wire of half the diameter, assuming that the tension is the same in both cases.

16. An astronaut on the moon has a superpower amplifier and loudspeaker. Assuming no absorption of sound and that the speaker behaves as a point source, what power output would be needed at the speaker in order to produce sound that is barely audible on earth? Take the distance from earth to moon to be 384,000 km. Assume (unrealistically, of course) that the sound propagates in air at normal temperature and pressure.

17. Show that the average power transferred by transverse waves traveling in a rope past any given point x is

$$\bar{P} = -T\overline{\left(\frac{\partial y}{\partial x}\right)}\overline{\left(\frac{\partial y}{\partial t}\right)}$$

where T is the tension in the rope. Assume a sinusoidal waveform.

18. The total daily energy output of the sun is 3×10^{32} joules. **(a)** Find the power output in watts. **(b)** Assuming the atmosphere absorbs 50 percent of the energy, find the intensity at the surface of the earth 1.49×10^8 km away. **(c)** Estimate the total energy you would receive from the sun, during midday, in a 1-hour interval.

19. A single loudspeaker is located 4 m away from a person listening to music. The maximum desired sound intensity is 90 decibels. Find the required acoustic power output from the speaker, assuming that it acts as a point source.

20. Two waves traveling in a one-dimensional medium are superposed to produce a resultant wave. If the individual waves are represented by displacements $y_1 = A \sin(k_1 x - \omega_1 t)$ and $y_2 = A \sin(k_2 x - \omega_2 t)$, show that the resultant wave is given by

$$y = 2A \cos\left[\left(\frac{k_1 - k_2}{2}\right)x - \left(\frac{\omega_1 - \omega_2}{2}\right)t\right]$$

$$\times \sin\left[\left(\frac{k_1 + k_2}{2}\right)x - \left(\frac{\omega_1 + \omega_2}{2}\right)t\right]$$

$$= 2A \cos\left[\left(\frac{\Delta k}{2}\right)x - \left(\frac{\Delta \omega}{2}\right)t\right]\sin(\bar{k}x - \bar{\omega}t)$$

21. Two hi-fi speakers are placed in a room and are separated by 12 ft, as shown in the figure. The manufacturer suggests the following procedure to determine if the speakers have been connected in phase: Play a monophonic record with a good deal of bass. Turn the speaker selector switch first on one speaker and listen for the bass volume, and then turn the switch so both speakers are in use. Again listen for the bass volume. If the bass volume diminishes when both speakers are on, the speakers are out of phase and the wires to one of them should be reversed. Explain the manufacturer's reasoning.

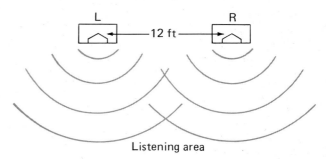

Listening area

22. A siren, which is at rest, emits sound whose frequency is 500 Hz. **(a)** What frequency would be perceived by an observer moving toward the siren at a speed of 30 meters/sec? **(b)** What frequency would he hear if he were moving away from the siren at this same speed? Suppose, now, that the observer is at rest and the siren is moving. **(c)** What frequency would be heard if it were approaching the observer at 30 meters/sec? **(d)** What frequency would be heard if it were receding from the observer at 30 meters/sec?

23. A passenger train, traveling at a constant speed of 30 meters/sec is approaching a freight train traveling along a parallel track in the opposite direction at a constant speed of 15 meters/sec. The freight train sounds its horn, which emits sound waves at a frequency of 256 Hz. **(a)** What frequency does a passenger abroad the passenger train hear? **(b)** What frequency would he hear after the trains have passed each other and are departing in opposite directions? *Hint:* What would a stationary observer alongside the track hear in each case?

24. An observer at a railroad station hears the horn of an approaching locomotive at a frequency of 320 Hz. After the locomotive has passed by, the frequency appears to drop to 256 Hz. Based on these observations, what frequency would be heard if the locomotive were at rest? What is the speed of the locomotive in miles per hour?

25. An approaching train sounds its whistle and appears to have a frequency of 280 Hz, according to passengers waiting at the station. The frequency experienced by passengers aboard the train is 262 Hz. The velocity of sound in air is 1080 ft/sec under prevailing conditions. Find the speed of the train.

26. A microwave source of wavelength 0.15 m is used at a radar trap to ticket speeding motorists. If the frequency change between the emitted and the reflected waves is 400 Hz, find the speed of the automobile.

27. A supersonic plane is cruising at a speed of 500 m/sec at an altitude of 30,000 ft. What is the value of the Mach angle? How much time elapses between the moment when the plane is directly overhead and when the sonic boom is heard?

28. Two identical guitar strings 50 cm long, each having a mass of 50 g, are under a tension of 360.0 N. They are plucked simultaneously. What is the fundamental frequency? If the tension in one of the strings is increased to 364.0 N, what beat frequency will be heard?

29. An angry musician, trying unsuccessfully to tune his instrument, throws his tuning fork at the wall at a speed of 20 meters/sec. The natural frequency of the tuning fork is 528 Hz. Find **(a)** the frequency the musician hears from the tuning fork, **(b)** the frequency he hears from the sound reflected from the wall, and **(c)** the frequency of the beats that result from these two sound waves.

30. A violin string having a length of $\frac{1}{3}$ m and a mass of 5.5×10^{-5} kg is under a tension of 90 N. Find the frequencies of the fundamental note (first harmonic) and the second and third harmonics.

31. An organ pipe of length L is open at one end and closed at the other. The open end is an antinode (maximum vibration amplitude), while the closed end is a node (zero vibration amplitude). Find the wavelengths and frequencies of the sound waves caused by the vibration of air in the organ pipe.

32. A string of length 1 m is fixed at both ends. It has a mass of 60 g and is under a tension of 300 N. If the string is vibrating, find the frequencies of the fundamental and of the overtones.

*33. A set of point masses m are separated by ideal springs whose force constant is β, as illustrated in the figure. In equilibrium, the separation of the masses is a. The system is assumed to be of infinite extent in both directions **(a)** Show that when longitudinal vibrations of this system are excited, the equation of motion describing the displacement u_n of the nth mass from its equilibrium position is $m(d^2u_n/dt^2) = \beta(u_{n+1} + u_{n-1} - 2u_n)$. **(b)** Show that for amplitudes u_n small compared to the spacing a, this equation is satisfied by $u_n = A \sin(kna - \omega t)$, where the frequency ω and the propagation constant k are related by $\omega = (4\beta/m)^{1/2} \sin(ka/2)$. **(c)** Does this system exhibit the phenomenon of dispersion? Note that the quantity na that appears in the solution for the amplitude represents the x-coordinate of the nth mass.

34. In the preceding problem, show that for very long wavelengths, that is, for very small values of k, the wave velocity has the value $a(\beta/m)^{1/2}$, independent of frequency. Note that in this limit, the effect of dispersion is negligible.

*35. In Problem 33, consider the limiting case where the masses become very small and closely spaced. In this limit, the system approaches one in which there is a continuous distribution of mass along the x-axis rather than a group of finite masses with finite spacing. Designating each mass as Δm, investigate the behavior of the system in the limit where Δm approaches zero and a approaches zero in such a way that the mass per unit length $\mu\,(=\Delta m/a)$ remains constant. Show, specifically, that in this limit the equation of motion of Problem 33,

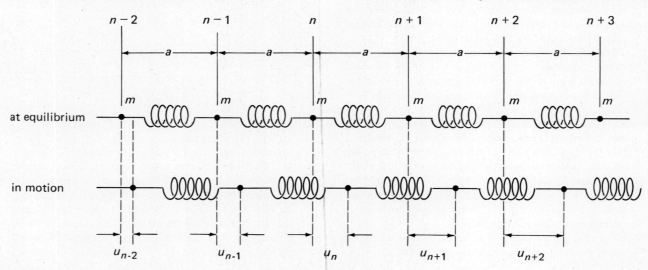

part **(a)**, becomes $\partial^2 u/\partial x^2 = (\mu/Y)\,\partial^2 u/\partial t^2$, where Young's modulus Y equals βa. In this limit, it is evident that the system's behavior is described by the wave equation for longitudinal elastic waves and that this procedure constitutes a way of deriving the wave equation from a system of many closely spaced "atomic particles" bound to one another by Hooke's law forces. The result suggests, therefore, that this picture of the behavior of atoms in a solid substance is a valid one. In conjunction with the result of Problem 33, it also suggests that *very short* sound waves in solids (whose wavelength is comparable to the interatomic spacing and whose frequency is far above even the usual ultrasonic range) exhibit the effect of dispersion, which has been found to be true.

36. A certain lecture room has flat parallel side walls 8.0 meters apart. They have a hard, smooth plaster surface. A lecturer connects an audio oscillator to a loudspeaker.

When the oscillator is tuned to a frequency of 165.5 Hz, a student sitting in the lecture room observes that at certain locations he hears its sound very loudly, but when he moves his head sideways from such a position a distance of about 10 inches, the sound is much less distinct. When the oscillator is tuned to a frequency of 186 Hz, however, no such effect occurs, the apparent sound intensity then being more or less independent of the listener's position. **(a)** Explain the odd effect described above for $f = 165.5$ Hz. **(b)** Why is this effect not observed for $f = 186$ Hz?

37. **(a)** In the situation described in the preceding problem, are there other frequencies for which the effect found at 165.5 Hz could be observed, and, if so, what are those frequencies? **(b)** How might the room be redecorated so as to minimize such effects?

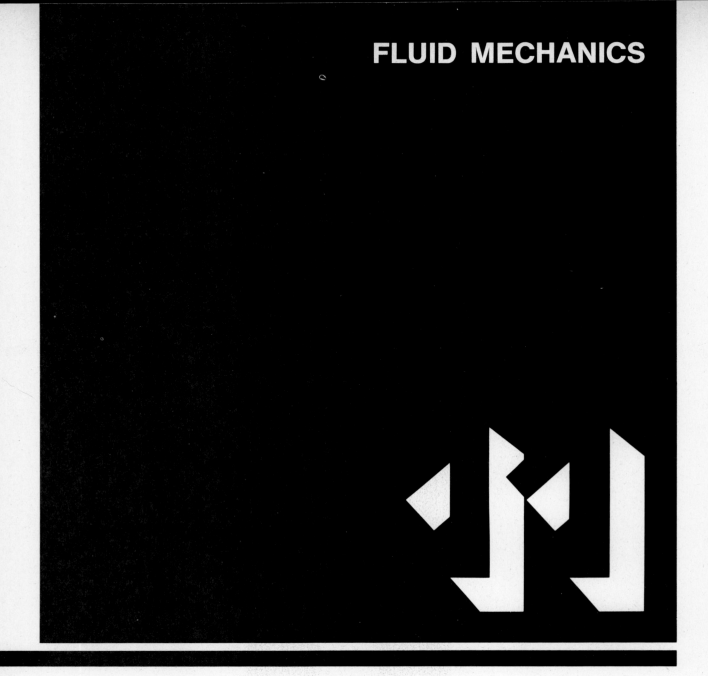

FLUID MECHANICS

11.1 Introduction

We shall now attempt to apply the laws of mechanics to fluids. No change needs to be made in Newton's laws themselves, but there is the problem of understanding in precise terms how fluids behave and how we may most effectively apply Newton's laws to them.

First of all, what do we mean by the term *fluids*? Fluids are substances that can be made to *flow* by the proper application of forces. Generally speaking, fluids can be classified as *liquids* or *gases*. Liquids are practically incompressible and, therefore, can be regarded as having a fixed volume, even though their shape may change as, for example, when they are poured from one container to another. Gases are

highly compressible (or, looking at it the other way round, highly expandable) and, therefore, have no characteristic volume; they simply expand to fill any container in which they may be placed.

Well, then, what makes a fluid flow, and under what conditions will it flow? The answer to this is that any fluid body will support *normal* forces that may act on its boundaries, without flowing, and can be in equilibrium under the action of a number of such normal forces. But a fluid cannot resist *tangential*, *shearing*, or *bending* forces; as soon as such forces are exerted on a fluid, it will flow in response to them. The situation, which is illustrated by Fig. 11.1, has a resemblance to that at a frictionless interface between an object and a plane which supports it; there can be

342

normal forces at such an interface, but no tangential ones. The difference is, of course, that in the case of a rigid body on a frictionless surface, this condition applies only at the surface, while for a fluid it holds for *every point* within the substance.

A necessary condition for a fluid body to be in equilibrium, therefore, is that its boundaries experience *only normal forces*. Water in a pond or in a pan or other container, or a compressed gas within a cylinder are examples of such situations.

On the atomic scale, the difference between solids, liquids, and gases is attributable entirely to the interplay between the attractive forces that exist between individual atoms, ions, or molecules and to the random motion these particles acquire as a result of their thermal energy. In a *solid*, at very low temperatures, the thermal motion of atoms or molecules is very slight and excites only small vibrations in the rigid crystalline lattice of atoms, molecules, or ions. With increasing temperature, the thermal motions increase in energy until they are strong enough occasionally to tear apart the bonds between neighboring atoms or molecules; atoms or molecules are now still held together by long-range cohesive forces, but can move more or less freely past one another. The solid has now melted and has become a *liquid*. Finally, as the temperature is increased still further, the average thermal kinetic energy of each atom or molecule becomes greater than the average attractive potential energy that binds it to the other particles. The atoms or molecules now fly apart and are constrained only by the walls of whatever container they may be in. The substance has now vaporized and exists as a *gas*.

It is clear that the *same* substance may exist at different temperatures as solid, liquid, or gas and that the transition temperatures between these states depend primarily on the strength of the cohesive forces between individual atoms, molecules, or ions. We shall postpone the detailed description of thermal relations between solids, liquids, and gases to a later chapter, but the student will perhaps more readily appreciate on an intuitive level the difference between the properties of these *states of aggregation* in the light of this simple discussion.

11.2 Basic Properties of Fluids: Pascal's Principle; Pressure, Volume, and Density

In studying the mechanics of particles and rigid bodies, we found it most useful to work with the masses of individual objects, the forces acting on individual bodies, and the linear dimensions of such bodies. In the case of fluids, due to their special properties, it is generally more convenient to speak instead in terms of density, pressure, and volume.

We are already familiar with volume. We have also occasionally worked with the density ρ, defined as mass per unit volume. For an incompressible fluid, the density is constant throughout, but this is not generally true for a compressible fluid, though it may be approximately true under certain restricted conditions. The earth's atmosphere is an example of variable density in a compressible fluid; in this instance, the density decreases with altitude, in response to decreasing atmospheric pressure. Over a restricted range of altitudes, however, this variation can often be neglected.

The concept of *pressure* may best be understood by considering an infinitesimal element of area da somewhere in the fluid, as illustrated in Fig. 11.2. It may in some cases be helpful to envision a tiny piece of paper or plastic film of zero thickness and mass and area da to represent such a surface area element, but this is not absolutely necessary. In any case, there may be forces exerted by the fluid on such an area element. Due to the properties of fluids, these forces, at least at equilibrium, will be normal to the element, since fluids cannot sustain tangential forces without flowing. The pressure on such an element of area is defined as the magnitude of the force acting on it divided by the area of the element. This may be written as

$$P = \frac{dF}{da} \qquad (11.2.1)$$

or as an expression for the force, in the form

$$dF = P \, da \qquad (11.2.2)$$

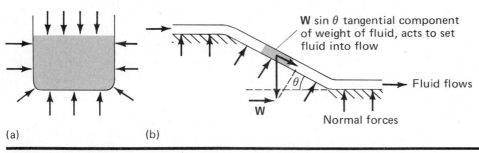

(a) (b)

W sin θ tangential component of weight of fluid, acts to set fluid into flow

θ

W

Fluid flows

Normal forces

FIGURE 11.1. (a) Fluids in equilibrium experience only forces normal to their boundaries. (b) The action of tangential forces is to cause fluid flow.

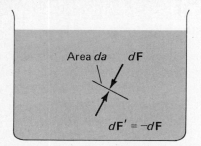

FIGURE 11.2. Equilibrium of forces within fluids, illustrating the definition of pressure as the ratio of force to area.

The total force on an area of finite size can be obtained by integrating this over the finite area:

$$F = \int_s P \, da \qquad (11.2.3)$$

It is important to understand that pressure, as ordinarily defined, is a *scalar* rather than a vector; it is the ratio of the *magnitude* of a force to an area. In Fig. 11.2, it is clear that if the element da is to be in equilibrium, there can be no resultant force on it. The force $d\mathbf{F}$ acting on one side of it must, therefore, always be accompanied by an equal and opposite force $d\mathbf{F}'$ on the other side. In defining the pressure, of course, since only the magnitude of the force is used to define P, it makes no difference whether the magnitude of $d\mathbf{F}$ or $d\mathbf{F}'$ is employed, since they are equal.

In Eq. (11.2.3), the pressure P over an area of finite size may vary from point to point. In general, therefore, the pressure P *cannot* be written outside the integral. This can be done *only if the pressure is constant over the entire surface in question.* This condition is often satisfied in cases of particular interest, and when it is, (11.2.3) becomes

$$F = P \int_s da = Pa \qquad (11.2.4)$$

or

$$P = \frac{F}{a} \qquad (11.2.5)$$

It is, however, only under the conditions set forth above that we may resort to the simple idea that "pressure is force divided by area" or "force is pressure times area."

In the metric system, density is expressed in units of g/cm^3 and kg/m^3, while pressure is expressed as $dynes/cm^2$ or $newtons/m^2$. It is easy to show that $1 \, g/cm^3 = 10^3 \, kg/m^3$ and that $1 \, dyne/cm^2 = 0.1 \, newton/m^2$. In the English system, the unit of pressure is the lb/ft^2 or, more commonly, $lb/in.^2$. Also in common use is the *bar*, equal to $10^6 \, dynes/cm^2$, and the *standard atmosphere*, equal to $1.013 \times 10^6 \, dynes/cm^2$, or

$14.7 \, lb/in.^2$. One atmosphere is the average pressure exerted by the earth's atmosphere at sea level at a temperature of 0°C.

The basic properties of a fluid in equilibrium, discussed in the previous section from a rather different point of view, can be expressed concisely in a statement often referred to as *Pascal's principle*, since it was first set forth by the French mathematician and philosopher Blaise Pascal (1623–1662). Pascal's principle states:

Pressure applied to a fluid is transmitted undiminished to all parts of the fluid and to the walls of the container enclosing it.

For example, when a fluid is enclosed in a vessel fitted with a piston, as illustrated in Fig. 11.3, certain pressures P_A, P_B, and P_C will exist at points A, B, and C. If now a sudden change in pressure ΔP is caused by the application of a force $\mathbf{F}$ on the piston, Pascal's principle tells us that the pressures at A, B, and C immediately take on the values $P_A + \Delta P$, $P_B + \Delta P$, and $P_C + \Delta P$. At point C, the fluid is kept in equilibrium by a normal force exerted by the wall of the container. This normal force per unit area also experiences an immediate increase of magnitude ΔP.

Another consequence of Pascal's principle is exhibited by Fig. 11.4. In this illustration, four different containers are shown, each filled to the same level with the same fluid. The arrows represent the forces exerted per unit area (thus the pressures exerted) by the fluid on the container walls at various points. These forces arise ultimately from the weight forces acting on each volume element of the fluid which transmits force, hence pressure, to all parts of the fluid beneath and thence to the container walls. We can easily understand how this occurs by imagining, in Fig. 11.3, that the force $\mathbf{F}$ could equally well arise from the weight force associated with a layer of fluid on top of that shown instead of from an external force applied to a piston. In the slightly different situation

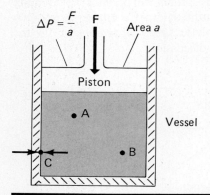

FIGURE 11.3. Pascal's principle. Forces acting on one part of a fluid are transmitted to the interior of the fluid and to its boundaries.

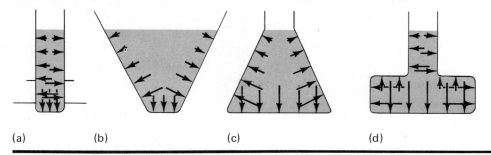

FIGURE 11.4. Weight forces exerted by fluids on containing vessels, as predicted by Pascal's principle.

shown in Fig. 11.4, it is apparent from Newton's third law that the forces exerted by the fluid on the container walls must be equal and opposite those exerted by the walls of the container on the fluid. If the fluid is in equilibrium, however, the latter forces must be everywhere normal to the walls of the vessel. Otherwise, they would have tangential components that would cause the fluid to flow. The forces the fluid exerts on the container, shown in Fig. 11.4, which are equal and opposite, then must likewise be normal to the container wall as shown in the drawing. Pascal's principle, along with Newton's laws, allows us in this way to understand how the pressure of higher layers of fluid are transmitted to those below, and also that these pressures generate *normal* forces on the container walls as illustrated in Fig. 11.4.

Though Pascal's principle applies in general only at equilibrium, it is frequently useful even in non-equilibrium situations. Though we shall not try to show why this is true in a rigorous mathematical way, the argument that follows will serve as a rough qualitative explanation. In the case of compressible fluids, forces are transmitted by longitudinal pressure waves in the fluid which propagate with the velocity of sound from the point of application of the force. Therefore, there is a time lag corresponding to the travel time of a longitudinal wave from the point of application of a force to the point where the pressure it ultimately creates is observed. Pascal's principle neglects this time lag and, therefore, can be applied rigorously only in situations of equilibrium, where the dynamic state of the fluid is independent of time. There are, nevertheless, many important cases where the effect of this time lag is negligible for practical purposes and where Pascal's principle can be used even in problems involving accelerated motion of compressible fluids to derive results which, though not exact, are very good approximations. For a perfectly incompressible fluid, the bulk modulus $(-V \Delta P/\Delta V)$ is infinite, since ΔV is identically zero. The wave velocity, equal to $\sqrt{B/\rho}$, becomes infinite in this limit, and hence, for such a substance there is no time lag and Pascal's principle is always valid.

11.3 The Variation of Fluid Pressure with Depth

Let us now try to calculate in detail how the pressure in a fluid in equilibrium changes as a function of depth. Consider the shaded element of fluid of thickness dy in Fig. 11.5. As usual in mechanics problems dealing with situations of equilibrium, we may proceed by isolating the fluid element, writing all the individual forces acting on the element, and setting their vector sum equal to zero. The forces **F**, **F′**, **A**, and **A′** are forces exerted by the rest of the fluid against this element; they arise from the pressure in the fluid. The forces **A** and **A′**, which are equal in magnitude but opposite in direction, serve merely to maintain the system in equilibrium along the *x*-direction and are of no particular interest otherwise. There are three forces which have *y*-components, and since the system is in equilibrium these *y*-components must sum to zero. Therefore,

$$\sum F_y = F - F' + mg = 0 \tag{11.3.1}$$

Now let the pressure at depth y be P, and at depth $y + dy$ let it be $P + dP$. Then, since the pressure over the top surface of the shaded element is constant, as is the pressure over the bottom surface, $F = Pa$ and $F' = (P + dP)a$, where a is the area of the surface of the element. Also, since the mass of the element can be written as the density ρ times the volume $a\,dy$, Eq.

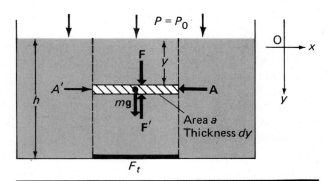

FIGURE 11.5. System of forces acting on a fluid element of infinitesimal thickness, in equilibrium.

345

(11.3.1) can be expressed as

$$Pa - (P + dP)a + \rho ga \, dy = 0 \qquad (11.3.2)$$

or

$$\boxed{dP = \rho g \, dy} \qquad (11.3.3)$$

This is the general expression relating a differential pressure variation dP to the depth change dy.

To understand what finite pressure change occurs over a finite distance y, we must integrate. If we are dealing with an *incompressible* fluid, ρ is *constant* and does not change as the pressure (or the depth) varies. Therefore, ρ may be written outside the integral. We may then integrate (11.3.3) from $y = 0$, at which point P has the value P_0 and expresses whatever atmospheric or other pressure exists at the surface of the fluid, to the depth y, at which point the pressure is P. Thus,

$$\int_{P_0}^{P} dP = \rho g \int_{0}^{y} dy \qquad (11.3.4)$$

or

$$P - P_0 = \rho g y \qquad (11.3.5)$$

$$\boxed{P = P_0 + \rho g y} \qquad (11.3.6)$$

This simple result tells us that the pressure change $P - P_0$ in an incompressible fluid increases linearly with depth. It is an important finding and one that we shall make use of constantly. The quantity P is often referred to as the *absolute* pressure, and the quantity $P - P_0$ is sometimes called the *gauge pressure*, since it is this pressure that is measured by any instrument calibrated to read zero at atmospheric pressure.

In a *compressible* fluid, such as a gas, the density ρ is no longer constant but changes as the pressure (or the depth) varies. In this case, the specific form of the dependence of density upon pressure must be known before (11.3.3) can be integrated; now, needless to say, the quantity ρ may no longer be written outside the integral. Under these circumstances, the variation of pressure with depth may be quite complex. Though we may examine one or more such instances in detail at a later point, at the present we shall confine our pressure-versus-depth investigations to incompressible substances.

Let us now inquire into the total force F_t exerted on that part of the bottom of the container directly below the area a. This can most easily be found by multiplying both sides of (11.3.6) by a and setting y equal to the total depth of fluid h in the resulting equation:

$$Pa = P_0 a + (\rho ah)g \qquad (11.3.7)$$

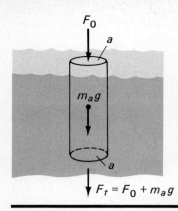

FIGURE 11.6. Forces acting on a column of fluid, in equilibrium.

Now, Pa is the force F_t on the bottom of the vessel, while $P_0 a$ is simply the force exerted by the atmosphere on the surface. The quantity ah is simply the volume of the fluid in a column having surface area a and height h (the boundaries of this column are the vertical dotted lines in Fig. 11.5). The quantity ρah is, therefore, the mass of the fluid within that column. We may write, then,

$$\boxed{F_t = F_0 + m_a g} \qquad (11.3.8)$$

where F_t is the force on the bottom, F_0 is the atmospheric force on the surface, and m_a is the mass of the fluid within the column. The force on any area of the bottom, then, is the atmospheric force on the corresponding area of the surface plus the *weight* of the fluid within the column standing over that area of the bottom. The situation is illustrated in Fig. 11.6. This result, by the way, computed only for incompressible fluids here, is true even for compressible substances. For example, the force exerted by the earth's atmosphere upon 1 cm^2 of the earth's surface at sea level is equal to the weight of atmosphere in a column of base area 1 cm^2 that extends upward as far as the atmosphere extends. Since a pressure of one standard atmosphere equals 1.013×10^6 dynes/cm^2, the weight of air in such a column is 1.013×10^6 dynes, corresponding to a mass of $(1.013 \times 10^6)/980$, or 1033 grams. This calculation, of course, neglects the variation of g over the upward extent of the atmosphere; but since most of the earth's atmosphere is within a few miles of the earth's surface, the error involved is quite small.

The discovery that gases such as air have appreciable weight and that the atmosphere exerts a definite and measureable pressure is usually attributed to the German scientist von Guericke, who in 1654 demonstrated that atmospheric pressure upon the two halves of an evacuated sphere a few feet in diameter gave rise to such strong forces that the two halves could not be separated by two teams of horses. Torricelli, an Italian scientist who in 1643 invented the barometer and first accurately measured the atmosphere's pressure, is also credited with this discovery. We shall investigate both these developments in the series of examples which follows. In these examples, we shall neglect the variation of the atmospheric pressure P_0 with height above the earth's surface. Near the earth's surface, the atmospheric pressure decreases by only 1 percent in an ascent of about 100 meters. Such a small variation can safely be ignored when the differences in height involved are on the scale of a few meters. But it should be understood that this variation must sometimes be accounted for in situations where height differences of hundreds of meters are encountered.

EXAMPLE 11.3.1

A swimming pool is rectangular in shape, having a length of 75 ft, a width of 40 ft, and a depth of 6 ft. Find (a) the gauge pressure at the bottom of the pool, (b) the total force on the bottom of the pool due to the water in it, and (c) the total force on one of the ends of the pool, whose dimensions are 40 ft by 6 ft. What is the absolute pressure on the bottom of the pool under normal sea level atmospheric conditions? The density of water is 62.4 lb/ft³, or 1.940 slugs/ft³.

The gauge pressure on the bottom follows directly from Eq. 11.3.5:

$$P_g = P - P_0 = \rho gy = (1.94)(32.2)(6)$$
$$= 375 \text{ lb/ft}^2, \text{ or } 2.604 \text{ lb/in.}^2$$

It is important to note that the quantity ρ is the *mass* per unit volume, while ρg represents the *weight* per unit volume. Since the pool is of constant depth, the pressure on the bottom is everywhere the same. Under these circumstances, we may use (11.2.4) to find

$$F = P_g a = (375)(75)(40) = 1,125,000 \text{ lb}$$

To find the force on one of the 40-ft by 6-ft pool ends we can no longer use (11.2.4), because the gauge pressure $P - P_0$ is not constant but varies from zero at the surface of the water to 375 lb/ft² at the bottom. We must now resort to (11.2.3) to express the gauge pressure as a function of depth and integrate over the area of the pool end. To see how this is done, let us refer to Fig. 11.7. In this diagram, the width of the pool

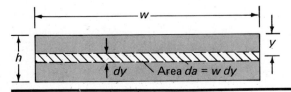

FIGURE 11.7

is represented by w and the depth by h. The shaded element of area is of width w and height dy, which approaches zero in the limit of integration so that the pressure variation across it is, in the limit, zero. The force dF on the shaded element of area is given by

$$dF = P_g \, da = \rho gyw \, dy \qquad (11.3.9)$$

This must be integrated from $y = 0$ to $y = h$, which gives

$$F = \int dF = \rho gw \int_0^h y \, dy \qquad (11.3.10)$$

or

$$F = \rho gw[\tfrac{1}{2}y^2]_0^h = \tfrac{1}{2}\rho gwh^2 \qquad (11.3.11)$$

Substituting the proper numerical values of the quantities in (11.3.11),

$$F = (\tfrac{1}{2})(1.94)(32.2)(40)(6)^2 = 45,000 \text{ lb}$$

The absolute pressure on the pool bottom is simply the sum of the gauge pressure and the atmospheric pressure P_0, which equals 14.7 lb/in.², or 2120 lb/ft². Then,

$$P = P_g + P_0 = 375 + 2120$$
$$= 2495 \text{ lb/ft}^2, \text{ or } 17.3 \text{ lb/in.}^2$$

Why is it that the bottom of the pool "feels" only the gauge pressure $P - P_0$ rather than the absolute pressure P? The atmospheric pressure P_0 is exerted not only *downward* on the pool bottom (transmitted there undiminished, according to Pascal), but also *upward* on the other side of the pool bottom, from below. These two pressures give rise to equal and opposite vertical forces on either side of the pool bottom, which add to zero. But there is water on only one side of the pool bottom, and the pressure it generates, which is $P - P_0$, or P_g, is *unbalanced*. Similar considerations apply to other systems on which atmospheric pressures act, including our own bodies. Since atmospheric pressure acts on us from within as well as without, we experience resultant pressure forces only when these two pressures become appreciably different, as in an ascending airplane or elevator. This point is illustrated also by the following example.

EXAMPLE 11.3.2

Two hollow hemispheres whose inside diameter is 2.0 ft can be fitted together and sealed, so that the air can be pumped out of the hollow sphere so formed with a vacuum pump, as illustrated in Fig. 11.8. How much force is required to separate the hemispheres when both the inside of the spherical enclosure and the outside surroundings are at atmospheric pressure (14.7 lb/in.², or 2120 lb/ft²)? How much force is required to separate the hemispheres when all the air

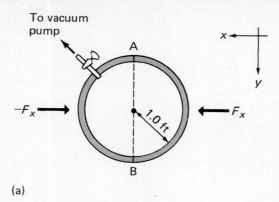

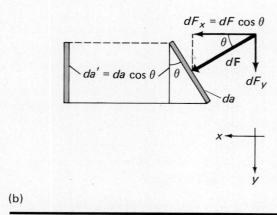

(b)

FIGURE 11.8. Magdeburg hemispheres, illustrating forces acting on an area element of the container wall.

is pumped out of the interior, reducing the absolute pressure there to zero, while the outside remains at atmospheric pressure?

A sectional view of the hemispheres is shown in Fig. 11.8a, and the forces acting on an area element da are illustrated in Fig. 11.8b. The external force on this element arising from atmospheric pressure, $d\mathbf{F}$, acts normally to the surface, as shown. If the inside of the sphere is at atmospheric pressure, there is an equal but oppositely directed force acting on the inside of the area element; the resultant of these two forces is zero. The situation is the same on all surface elements, so the net force is zero everywhere. Under these circumstances, no force is needed to separate the hemispheres.

Now suppose the inside of the sphere to be evacuated; the only forces then acting are those shown in Fig. 11.8b. It is only the x-component of the force $d\mathbf{F}$ which is effective in forcing the two halves of the sphere together. The x-component of $d\mathbf{F}$ can be written as

$$dF_x = dF \cos \theta = \frac{dF}{da}(da \cos \theta) = P(da \cos \theta)$$

(11.3.12)

since dF/da is the atmospheric pressure P. This can,

in turn, be stated in the form

$$dF_x = P\, da'$$

(11.3.13)

where da' is the area da *projected* onto the plane AB of Fig. 11.8a. This can be integrated over the right-hand hemisphere to find the total force F_x which holds it against the left-hand one; the force components dF_y, of course, sum to zero over the hemisphere in such an integration. Since the pressure of the atmosphere is essentially constant over the surface, the pressure P may be written outside the integral during the integration. Integrating (11.3.13), then,

$$F_x = P \int_s da' = Pa'$$

(11.3.14)

where a' is the area of the hemisphere projected on the plane AB. Since this is simply the area of a circle having the same radius as the sphere, or πr^2, F_x can be written as

$$F_x = \pi r^2 P = (3.1416)(1)^2(2120) = 6660 \text{ lb}$$

The same result can be arrived at more simply by noting that each hemisphere can be considered as supporting a circular column of atmosphere of area $a' = \pi r^2$ which exerts a pressure of 2120 lb/ft^2 on its base. We preferred, however, to do the calculation in a more complex but rigorous fashion to show precisely how the atmospheric forces act and why the projected area a' rather than the total area a must be used.

This example illustrates Otto von Guericke's experiment of the so-called *Magdeburg hemispheres* which demonstrated in 1654 the forces exerted by the earth's atmosphere and which we referred to earlier.

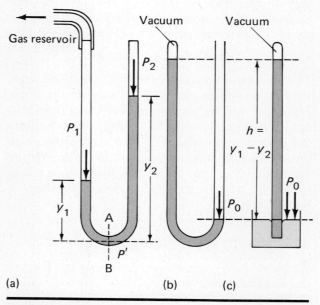

FIGURE 11.9. (a) Open U-tube manometer. (b) Closed U-tube manometer. (c) Torricelli's barometer.

Von Guericke was, at the time, mayor of the city of Magdeburg. Can you imagine, nowadays, the mayor of Pittsburgh or Baltimore concerning himself with such doings?

EXAMPLE 11.3.3

Consider the open U-tube *manometer* shown in Fig. 11.9. A gas at pressure P_1 is connected to the left side, while the right side is at another pressure P_2 (usually atmospheric, in which case $P_2 = P_0$, but not necessarily so). Find the relation between the liquid heights y_2 and y_1 and the pressure difference at equilibrium. What happens when the left side is evacuated? Assume the fluid in the U-tube to be incompressible.

Consider the plane AB which intersects the tube at its lowest point. Because the system is in equilibrium, the pressure on the left side of the plane from the column of fluid in the left side of the tube must be the same as that on the right side due to the fluid in that column. If it were not so, the fluid would move. Call this common pressure at the lowest point P'. Then, according to (11.3.5), for the column of fluid on the left,

$$P' - P_1 = \rho g y_1 \qquad (11.3.15)$$

while for the column on the right

$$P' - P_2 = \rho g y_2 \qquad (11.3.16)$$

If we now subtract (11.3.15) from (11.3.16), we will eliminate P', obtaining

$$\boxed{P_1 - P_2 = \rho g(y_2 - y_1)} \qquad (11.3.17)$$

which is the desired result. If the pressure on the surface of the liquid on the right is atmospheric pressure, then $P_2 = P_0$ and

$$P_1 - P_0 = \rho g(y_2 - y_1) \qquad (11.3.18)$$

If, now, the left side of the tube is evacuated, P_1 becomes zero, and (11.3.18) becomes

$$\boxed{P_0 = \rho g(y_1 - y_2) = \rho g h} \qquad (11.3.19)$$

where h is the difference between the levels y_1 and y_2. This gives a way of *measuring* atmospheric pressure in terms of the height of a fluid column, and the device so arranged is called a *liquid barometer*, as shown in Fig. 11.9b. In this device, which was first invented by Evangelista Torricelli in 1648, the pressure of the earth's atmosphere in the open tube or reservoir is balanced by that due to the column of liquid in the closed tube. In practice, instead of evacuating and sealing off one side of a U-tube, as shown in Fig. 11.9b, it is usually more convenient to entirely fill a barometer tube of sufficient length which is already sealed at one end, invert it, and release the open end while keeping it immersed in a reservoir of the fluid. This is the way in which the familiar *mercury barometer* shown in Fig. 11.9c is usually arranged.

There are many liquids which could be used to construct a barometer. One important requirement is that the *vapor pressure* of any such liquid be very small, so as to allow as good a vacuum as possible in the portion of the tube above the fluid. Mercury is particularly suitable in this respect, since its vapor pressure at room temperature is quite negligible in comparison to atmospheric pressure. Also, mercury is very dense and allows a column height which is quite convenient in the laboratory. A mercury barometer such as the one shown in Fig. 11.9c exhibits a column height of 76.0 cm at sea level under average conditions of atmospheric pressure. Since the density of mercury is 13.6 g/cm^3, Eq. (11.3.19) allows us to establish that the average atmospheric pressure should be

$$P_0 = \rho g h = (13.6)(980)(76) = 1.013 \times 10^6 \text{ dynes/cm}^2$$

Fluctuations in atmospheric pressure arising from meteorologic conditions are easily measured by noting the height of the mercury column in a barometer of this sort. If the barometric fluid were water instead of mercury, the column height would be 13.6 times greater because water is 13.6 times less dense than mercury. A water barometer would, therefore, have a column of height 10.34 meters under average atmospheric pressure. Water is not a very good barometric fluid, because, in addition to the inconvenient column height, its vapor pressure is too high.

The widespread use of the mercury barometer and manometer has led to the use of a unit of pressure related to the height of the mercury column in such a device. One may define a unit of pressure as the pressure exerted by a column of mercury 1 millimeter in height. This unit is called the *torr* (after Torricelli), or the *millimeter of mercury*. It is easy to see that 1 torr (1 mm Hg) = (13.6)(980)(0.1) = 1333 dynes/cm^2. The torr is not a very good unit in some respects, but its use has become widespread, particularly in the measurement of low pressures in vacuum systems and in measuring the barometric pressure of the atmosphere. It is, therefore, in line with common terminology to refer to the pressure of the atmosphere under standard conditions as 760 torr, or 760 mm Hg, or to refer, for example, to the tiny residual pressure inside an electronic tube as 10^{-6} torr, or 10^{-6} mm Hg. One even occasionally sees pressures expressed in *inches of mercury*, but this practice, fortunately, is becoming less common that it once was, despite the fact that there are millions of inexpensive household barometers calibrated this way.

EXAMPLE 11.3.4

A rowboat has a flat bottom whose area is 60 ft^2. The sides of the boat are perpendicular to the plane of the

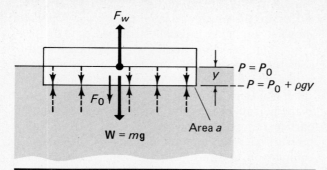

FIGURE 11.10. System of forces acting in equilibrium on a boat.

bottom and have a uniform height of 1.5 ft. The boat weighs 480 pounds. Find (a) the height of the water line of the unloaded boat above the bottom, (b) the distance the boat sinks when four passengers whose total weight is 640 pounds step aboard, and (c) the maximum weight the boat can carry before sinking. The loads are assumed to be uniformly distributed in all cases.

This example illustrates how a boat works, a subject not well understood even by many persons who have spent most of their lives afloat. Consider first the unloaded boat floating at equilibrium in the water, as shown in cross section in Fig. 11.10. The atmospheric pressure at the surface of the water is P_0; this same pressure is exerted by the atmosphere downward on the top surface of the boat's bottom. There is an upward force on the lower surface of the bottom of the boat, the surface that is in contact with the water, at a distance y beneath the surface. Since the pressure at this depth is greater than P_0, in view of the law of pressure versus depth expressed by Eq. (11.3.6), the upward force exerted by the water on the boat's bottom (which we shall learn to refer to in the next section as the *total buoyant force*) is greater than the downward force caused by atmospheric pressure acting inside. But there is also the boat's weight, which acts downward, too, and which, of course, in equilibrium, makes up the difference. If we call the downward force due to atmospheric pressure F_0, the total upward force of buoyancy due to water pressure on the boat's bottom F_w, and the weight force W, Newton's first law tells us that in equilibrium,

$$\sum F_y = F_w - F_0 - W = 0 \tag{11.3.20}$$

Since the pressures involved in generating F_w and F_0 are both uniform over the boat's bottom, we may write

$$F_0 = P_0 a \tag{11.3.21}$$

$$F_w = Pa = (P_0 + \rho g y)a \tag{11.3.22}$$

where a is the area of the bottom of the boat and y is its depth beneath the surface. Inserting these values

into (11.3.20),

$$\rho g y a = W$$

or

$$y = \frac{W}{\rho g a} \tag{11.3.23}$$

Inserting the values of the quantities on the right-hand side given in the initial statement of the problem, we find, using $\rho g = 62.4$ lb/ft^3, that

$$y = \frac{480}{(62.4)(60)} = 0.128 \text{ ft}$$

When the passengers step aboard, of course, the total weight becomes 640 pounds greater; otherwise the argument is the same. Now,

$$y' = \frac{480 + 640}{(62.4)(60)} = 0.299 \text{ ft}$$

The boat, then, has sunk a distance $y' - y = 0.171$ ft. The question of how much of a load the boat will carry before it sinks can be settled by setting y equal to 1.5 ft in (11.3.23) and solving for W. The total weight is then found to be

$$W = \rho g a y = (62.4)(60)(1.5) = 5616 \text{ lb}$$

This represents the weight not only of the cargo but also of the boat itself. The weight of cargo which can be carried is $5616 - 480$, or 5136 lb.

The important thing to realize in this example is that the buoyant force, which supports the boat and its cargo, arises because the upward pressure of the water on the bottom of the boat *is greater than the downward pressure of the atmosphere inside*. This is so because as one descends beneath the surface of an incompressible liquid, the pressure rises steadily in view of the pressure-versus-depth law (11.3.6); this law, in turn, you will recall, has its origins in Pascal's principle and Newton's first law. If there were no increase in pressure with depth, there would be no buoyant force, and no boat would float. This example illustrates clearly the origin of buoyant forces. In the next section, we shall examine in detail the action of such forces on objects immersed or floating in fluids.

11.4 Buoyant Forces and Archimedes' Principle

Anyone who has ever swum or owned a boat has experienced very directly the buoyant forces exerted by fluids on objects which are immersed in them or floating on them. In Example 11.3.4, we examined how these forces arise in a specific example involving a rowboat. It is now time to look at the situation in a more general way and try to arrive at some idea of how

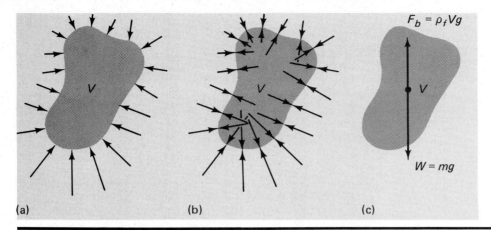

FIGURE 11.11. System of forces acting on an irregular, immersed object. (a) Forces exerted by the fluid on the object. (b) Forces exerted in equilibrium on a volume of fluid of the same size and shape as the immersed object, and forces exerted by this volume of fluid on its surroundings. (c) Equivalent system of forces acting on immersed body, with the forces shown at (a) replaced by the equivalent buoyant force $\mathbf{F}_b$.

to express buoyant forces acting on any object in a fluid.

Consider first a body of irregular shape which is totally immersed in a fluid, as shown in Fig. 11.11a. It is important to realize from the outset that such an object is *not necessarily in equilibrium*, for the net buoyant force exerted by the liquid may be greater than, less than, or equal to the body's mass. In the first instance, the body will sink; in the second, it will rise to the surface; and only in the final case will it remain in equilibrium. In any event, when immersed, any body will experience forces due to the pressures acting on it; a diagram showing the general appearance of these forces is shown in Fig. 11.11a. Since fluids can exert only forces normal to the boundaries, the forces must be perpendicular to the surface of the object at all points; and since the pressure within the fluid increases with increasing depth, the magnitude of the forces increases with depth. The fact that the upward forces acting on the bottom of the object are greater than the downward forces acting on the top is what causes the net buoyant force in the first place. The force vectors in Fig. 11.11a will be seen to display both these characteristics.

Now suppose that the object is removed and the volume it formerly occupied is filled with fluid identical to that surrounding the volume, as shown in Fig. 11.11b. The system of forces acting upon the dotted volume of fluid formerly occupied by the immersed body is no different from what it was when the body was there; the surrounding fluid does not know (nor care!) whether its pressure forces are exerted upon an immersed object or upon a volume of fluid put there in its place. But there is one difference; when the body is replaced with an identical chunk of fluid, the system now *has to be in equilibrium*.

If it were not, then there would be a flow of fluid, which clearly cannot take place under the conditions of Fig. 11.11b. Since the volume of fluid occupying the dotted volume previously filled by the body is in equilibrium, the vector sum of all forces acting on it must be zero. The forces acting on the fluid in this region are, first, the resultant of all the pressure forces exerted on it by the surrounding fluid (which, *in toto*, represents the net buoyant force itself) and, second, the weight of the fluid in the dotted volume. Since the sum of these two forces is zero, the net buoyant force must be equal in magnitude, but opposite in direction, to the weight force associated with the fluid in this volume.

Now let us put the immersed object back into the fluid in its former location. The net buoyant force, which is, after all, due to the surroundings, is again no different from what it always was. But now, we know that under the conditions of Fig. 11.11b it had to be an upward force equal to the weight of fluid occupying the same volume as the object itself. Therefore, that is what it is when the object itself is in place, as in Fig. 11.11a or 11.11c. The basic elements of the situation are shown in Fig. 11.11c, in which both the object's weight force and the resultant buoyant force are illustrated.

The result of all this can be stated in the following way:

A body immersed in a fluid experiences a resultant upward buoyant force equal in magnitude to the weight of displaced fluid.

This statement is called *Archimedes' principle*, and expresses essentially all there is to know about the buoyant force. It is important to note that in deriving Archimedes' principle, it was not necessary to assume that the fluid in question was incompressible. Archimedes' principle is, therefore, very general and applies not only to liquids but to gases as well. It is the buoyant force of the earth's atmosphere which allows a balloon or blimp to rise, for example.

In the case of an object only partially immersed, buoyant forces arise only from the part of the body below the surface of the liquid. By the same token, the volume of fluid displaced is no longer the total volume of the object but only the volume of that part which is immersed. Archimedes' principle, nevertheless, applies equally well to bodies totally immersed or partially immersed.

There are three ways for a body in a fluid to be in equilibrium. First, the density of the body may be precisely equal to that of the fluid. The weight of displaced fluid is then exactly equal to the body's own weight, and the buoyant force is precisely equal and opposite the weight force. The object will then remain in equilibrium while totally immersed, under the action of weight force and buoyant force alone; a submerged submarine is a good example.

Secondly, a body less dense than the fluid in which it is placed will come to equilibrium floating on the surface only partially immersed. Suppose, for example, the body shown in Fig. 11.11c were less dense than the surrounding fluid and it was released from rest while totally immersed. The upward buoyant force would be equal to the weight of displaced fluid, in this case $\rho_f V_g$, while the weight force would be $mg = \rho V_g$, where ρ is the density of the object and ρ_f is the density of the fluid. Since in this case ρ is less than ρ_f, the buoyant force will exceed the weight force and the body will undergo acceleration upward, eventually arriving at the surface. When it arrives there, part of its volume will rise *above* the surface, leaving the object only *partially* immersed. As this happens, the buoyant force will diminish, because the volume of fluid displaced by the body is no longer equal to the total volume of the object, but only to the volume of that part beneath the level of the fluid surface. The object will continue to rise out of the fluid until the weight of the volume of displaced fluid (which is now less than the total volume of the body) is equal to the body's own weight. At this point the weight force and the buoyant force will again be equal and opposite, and equilibrium will now be attained with the object floating on the surface only partially immersed, as shown in Fig. 11.12. In this condition, the buoyant force will be g times the mass of displaced fluid, hence $\rho_f V_{imm} g$, where V_{imm} is the volume of the part of the body beneath the level of the surface. Since the body is in equilibrium under the action of buoyant and weight forces alone,

$$F_b - W = \rho_f V_{imm} g - mg = 0 \qquad (11.4.1)$$

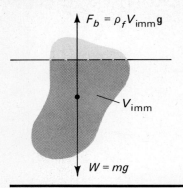

FIGURE 11.12. System of forces acting on a partially immersed object.

$$V_{imm} = \frac{m}{\rho_f} \qquad (11.4.2)$$

Finally, there is the possibility that *external* forces may be introduced to hold an object in equilibrium either partially or wholly immersed. The magnitude and direction of any such external force can always be determined by the condition of force equilibrium in conjunction with Archimedes' principle. These three possibilities will be discussed in detail in the series of examples which follow. Before beginning this series, the reader may find it instructive to return to Example 11.3.4 and rework it using Archimedes' principle.

EXAMPLE 11.4.1

Work out the results of Example 11.3.4 using Archimedes' principle.

This is easily enough accomplished, but first, referring to Fig. 11.10, it is important to realize that the total upward force F_w exerted by the water on the bottom of the boat is a *gross* buoyant force, attributable to the total pressure $P_0 + \rho g y$ instead of the pressure difference $\rho g y$, rather than the *resultant* buoyant force F_b referred to in Archimedes' principle. The force F_w is opposed by the downward force F_0 arising from atmospheric pressure, and it is the *difference* between these two quantities which represents the resultant buoyant force F_b of Archimedes' principle and which is to be equated to the weight of displaced fluid. The situation will be clearly illustrated by Figs. 11.10 and 11.13. From these drawings, it is evident that since the boat is in equilibrium,

$$\sum F_y = F_b - W = 0 \qquad (11.4.3)$$

This is the same as Eq. (11.3.20), except that $F_w - F_0$ is identified as the net buoyant force F_b. But now, according to Archimedes' principle, the net buoyant force equals the weight of water displaced, whence

$$F_b = \text{weight of displaced water} = \rho V_g = \rho a y g \qquad (11.4.4)$$

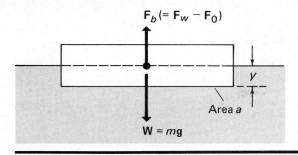

FIGURE 11.13

Therefore,

$$W = \rho a y g \qquad y = \frac{W}{\rho a g}$$

which is the same as Eq. (11.3.23). The numerical results asked for in the exercise all follow directly from this equation, just as before.

EXAMPLE 11.4.2

An engineer is assigned the task of designing a spherical balloon which will have a gross lifting capacity of 4900 newtons. This corresponds to a mass of 500 kg, which may include the mass of the balloon itself. The balloon is to be filled with hydrogen whose density is $\rho_{H_2} = 0.090$ kg/m^3, while the density of air is $\rho_{air} = 1.293$ kg/m^3. What is the minimum radius such a balloon can have to lift this total load? If the balloon itself is made of flexible plastic of density $\rho = 900$ kg/m^3 and thickness 0.100 cm, what will be the net lifting capacity or maximum payload? What will the payload be if the balloon is filled with helium of density $\rho_{He} = 0.180$ kg/m^3?

Let us first neglect the mass of the balloon and deal with the lifting capacity of the gas alone. As shown in Fig. 11.14, for the gas to lift a mass of 500 kg, corresponding to a weight force of (500)(9.8) newtons, the buoyant force must *exceed* the weight of the gas itself by that amount; in other words,

$$F_b - m_{H_2}g = F_l = 4900 \text{ N} \qquad (11.4.5)$$

where F_l is the lifting force of the gas and m_{H_2} is the mass of the hydrogen in the balloon, which is equal, in turn, to $\rho_{H_2}V$. The buoyant force is equal to the

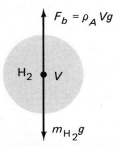

FIGURE 11.14

weight of the displaced air, which is $\rho_{air}Vg$. Substituting these values into (11.4.5), we find

$$Vg(\rho_{air} - \rho_{H_2}) = F_l \qquad (11.4.6)$$

Since the volume is $\frac{4}{3}\pi r^3$, where r is the radius, this may be written as

$$\tfrac{4}{3}\pi r^3 g(\rho_{air} - \rho_H) = F_l$$

from which

$$r = \left(\frac{F_l}{\frac{4}{3}\pi g(\rho_{air} - \rho_H)}\right)^{1/3}$$

$$= \left(\frac{(4900)}{(\frac{4}{3})(3.1416)(9.8)(1.293 - 0.090)}\right)^{1/3} = 4.63 \text{ m}$$
$$(11.4.7)$$

If the skin of the balloon is made of plastic 0.1 cm ($= 10^{-3}$ m) thick and of density $\rho = 900$ kg/m^3, its mass m_b will be given by

$$m_b = \rho V = \rho a d \qquad (11.4.8)$$

where a is the surface area and d the thickness. Since $a = 4\pi r^2$ for a sphere,

$$m_b = 4\pi r^2 \rho d = (4)(3.1416)(4.63)^2(900)(10^{-3}) = 242 \text{ kg}$$
$$(11.4.9)$$

The net lifting capacity, or payload, would be $500 - 242 = 258$ kg.

If the balloon were to be filled with helium, the buoyant force would still be given by

$$F_b = \rho_{air}Vg \qquad (11.4.10)$$

but now the weight of lifting gas would be

$$W_{He} = \rho_{He}Vg \qquad (11.4.11)$$

The available lifting force would now be

$$F_b - W_{He} = Vg(\rho_{air} - \rho_{He}) = \tfrac{4}{3}\pi r^3 g(\rho_{air} - \rho_{He})$$
$$= (\tfrac{4}{3})(3.1416)(4.63)^3(9.8)(1.293 - 0.180)$$
$$= 4535 \text{ N} \qquad (11.4.12)$$

Such a force could lift a mass of $(4535/9.8) = 463$ kg. Since the plastic skin has a mass of 242 kg, the net mass which could be lifted under these circumstances is $463 - 242 = 221$ kg.

EXAMPLE 11.4.3

A vessel contains a layer of water, $\rho_2 = 1.00$ g/cm^3, upon which is floating a layer of oil, $\rho_1 = 0.800$ g/cm^3. A cylindrical object of unknown density ρ whose base area is a and whose height is h is dropped into the vessel, finally coming to equilibrium floating on the oil–water interface, immersed in the water to a depth $\frac{2}{3}h$, as shown in Fig. 11.15. What is the density of the object?

In this case, the object is partly immersed in the

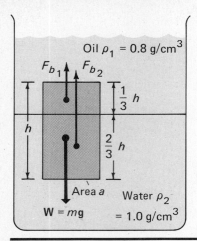

FIGURE 11.15

water and partly in the oil. There are two separate buoyant forces, one from the part of the body in the water, the other from the part in the oil, as shown in the diagram. Since the system is in equilibrium,

$$\sum F_y = F_{b1} + F_{b2} - mg = 0 \qquad (11.4.13)$$

The buoyant force F_{b1} due to the oil is equal to the weight of the displaced oil, namely, the weight density $\rho_1 g$ times the volume of the object immersed in the oil, which is one third the total volume. The force F_{b2} may be calculated in the same way:

$$F_{b1} = \tfrac{1}{3}\rho_1 gV = \tfrac{1}{3}\rho_1 gah \qquad (11.4.14)$$

$$F_{b2} = \tfrac{2}{3}\rho_2 gV = \tfrac{2}{3}\rho_2 gah \qquad (11.4.15)$$

Substituting these values into Eq. (11.4.13) and expressing the total mass m of the object as $\rho V (= \rho ah)$, we find

$$\tfrac{1}{3}\rho_1 gah + \tfrac{2}{3}\rho_2 gah - \rho ahg = 0 \qquad (11.4.16)$$

Canceling gah throughout and solving for ρ, we obtain

$$\rho = \tfrac{1}{3}\rho_1 + \tfrac{2}{3}\rho_2 = (\tfrac{1}{3})(0.800) + (\tfrac{2}{3})(1.00) = 0.933 \text{ g/cm}^3$$
$$(11.4.17)$$

Somewhat related to this example is the question of why we always neglect the buoyant force of the atmosphere in any problem involving an object floating partly immersed in a liquid and partly in the air, as we did in Example 11.4.1 and the explanatory material leading to Eq. (11.4.2). The answer lies in the fact that the atmosphere's density is so small that the buoyant force arising from it is usually quite negligible in comparison with that of the liquid. Suppose, for example, in this exercise, instead of oil ($\rho_1 = 0.800$ g/cm^3) we were dealing with an upper layer of air ($\rho_1 = 0.001293$ g/cm^3) and that the other conditions were the same, including the fact that the floating object was two thirds immersed in the water.

(This would mean that its density is no longer 0.933; can you figure out what it would be?) The buoyant forces due to the two fluids are still given by (11.4.14) and (11.4.15), and the *ratio* of the buoyant forces arising from the two fluids is

$$\frac{F_{b1}}{F_{b2}} = \frac{\tfrac{1}{3}\rho_1 gah}{\tfrac{2}{3}\rho_2 gah} = \frac{1}{2}\frac{\rho_1}{\rho_2} \qquad (11.4.18)$$

Now when ρ_1 and ρ_2 have the values 0.8 and 1.0, as in the original statement of the example, this ratio amounts to 0.4, which is quite large. But if $\rho_1 = 0.001293$, as for air, the ratio of the buoyant force of air to the buoyant force of the water is only 6.46×10^{-4}, which is so small as to be of no consequence in most problems.

EXAMPLE 11.4.4

A certain object has a mass of 250 g and a density of 2.70 gm/cm^3. When "weighed" while immersed in a liquid of unknown density by the arrangement shown in Fig. 11.16a, it is found that balance is obtained when 180 g is placed on the right-hand side of the laboratory balance. What is the density of the fluid in the container?

The forces acting on the object while immersed are shown in Fig. 11.16b. The tension T in the string is equal to the weight force associated with a mass of 180 g. Therefore, $T = (180)(980) = 176,400$ dynes. The net buoyant force F_b, according to Archimedes' principle, must be

$$F_b = \rho gV \qquad (11.4.19)$$

where ρ is the density of the fluid (*not* the density of the object, remember!). According to Newton's first law, then

$$\sum F_y = T + F_b - mg = 0 \qquad (11.4.20)$$

or, in view of Eq. (11.4.19),

$$\rho gV = mg - T \qquad (11.4.21)$$

Let us now divide both sides of this equation by mg,

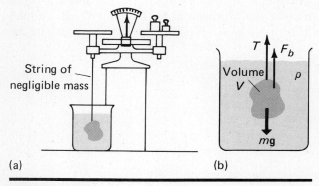

(a)　　　　　　　　　　　　　　　　(b)

FIGURE 11.16

which can also be written as $\rho_0 g V$, where ρ_0 is the object's density:

$$\frac{\rho g V}{mg} = \frac{\rho g V}{\rho_0 g V} = \frac{mg - T}{mg}$$

or

$$\rho = \rho_0 \frac{mg - T}{mg} \tag{11.4.22}$$

In this case, $\rho_0 = 2.70$, $mg = (250)(980)$ dynes, and $T = (180)(980)$ dynes. Substituting these values into (11.4.22) gives the answer $\rho = 0.756$.

EXAMPLE 11.4.5

An object of mass 180 g, but of *unknown* density, is "weighed" in water (density $\rho_w = 1.00$ g/cm^3), the weight so obtained corresponding to a balancing mass of 150 g, and "weighed" again in a liquid of unknown density ρ_f, a balancing mass of 144 g being needed this time. What is the density of the second fluid? What is the density of the object?

Let the density of the object be ρ and its volume V. Then, from Newton's first law, the two "weighings" give

$$T_1 + F_{b1} - mg = 0 \tag{11.4.23}$$

$$T_2 + F_{b2} - mg = 0 \tag{11.4.24}$$

where, according to Archimedes' principle,

$$F_{b1} = \rho_w g V \tag{11.4.25}$$

$$F_{b2} = \rho_f g V \tag{11.4.26}$$

Equations (11.4.23) and (11.4.24) may now be written

$$T_1 + \rho_w g V - mg = 0 \tag{11.4.27}$$

$$T_2 + \rho_f g V - mg = 0 \tag{11.4.28}$$

In these equations there are two unknowns, V and ρ_f. One may most easily eliminate V by multiplying the first equation through by ρ_f, the second by ρ_w, and then subtracting the second equation from the first. The resulting expression can be solved for ρ_f to give

$$\rho_f = \rho_w \frac{mg - T_2}{mg - T_1} \tag{11.4.29}$$

Also, one may solve (11.4.27) for V, the result being

$$V = \frac{mg - T_1}{\rho_w g} \tag{11.4.30}$$

Now, the density of the object is simply $\rho = m/V$, with V given above. Then,

$$\rho = \frac{m}{V} = \frac{\rho_w mg}{mg - T_1}$$

We are given that $\rho_w = 1.00$ g/cm^3,

$$mg = (180)(980) \text{ dynes}$$
$$T_1 = (150)(980) \text{ dynes}$$

and

$$T_2 = (144)(980) \text{ dynes}$$

Substituting these values into (11.4.29) and (11.4.31), it is easy to establish that $\rho_f = 1.20$ and $\rho = 6.00$.

11.5 Fluid Dynamics and Bernoulli's Equation

Up to this point, we have discussed only situations involving fluids in equilibrium. It is, of course, also important to understand the mechanics of fluids which may undergo acceleration. This branch of mechanics is referred to as *fluid dynamics*. The general study of fluid dynamics is a complex and difficult undertaking, involving much advanced mathematics; it is a task we shall not attempt here, except to sketch roughly some of the problems involved and some of the possible areas of application. It is possible, however, to understand many of the important features of fluid dynamics in certain restricted cases by the development and use of a simple relation based on the work–energy theorem and the energy conservation principle, called *Bernoulli's equation* after its originator the Swiss scientist Daniel Bernoulli (1700–1782).

In developing Bernoulli's equation, we shall restrict ourselves to the case of steady-state, incompressible, nonviscous fluid flow. By *steady-state flow* we mean that at every point within the flowing fluid medium the velocity of the fluid does not change with time. This does not mean that the fluid is unaccelerated; each droplet of liquid may undergo acceleration as it moves from point to point. At a fixed *location*, however, the velocity of the fluid flowing past remains at all times the same in magnitude and direction. A fountain or a waterfall are common examples of this type of flow. In the case of a fountain, illustrated in Fig. 11.17, in which a stream of water is projected upward and then falls back to the level of the nozzle, the velocity of the water droplets is zero when they reach their maximum height, and this state of affairs does not change with time. But the acceleration of each droplet is *not* zero here; indeed, it has the value **g** at this point and, for that matter, at all other points along the path taken by the stream.

Under conditions of steady-state flow, a steady pattern of *lines of flow* marking the paths taken by fluid particles within the stream is set up. These flow lines can be made visible by dropping tiny particles such as confetti into a liquid, or by injecting a series of thin streams of ink at various points. In a gas flow,

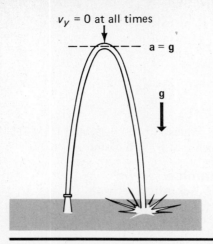

$v_y = 0$ at all times

a = g

g

FIGURE 11.17. Motion of a projected stream of liquid.

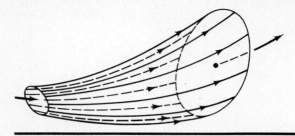

FIGURE 11.19. Tube of flow bounded by streamlines.

the lines of flow, or *streamlines* as they are often called, can be seen if tiny smoke jets are provided at a number of points. Some situations involving steady-state flows in which the streamlines have been rendered visible in this way are illustrated in Fig. 11.18. There are two important observations to make in regard to the lines of flow, or streamlines. First and most obvious, since the direction of fluid flow is along the streamlines, there is no fluid flow perpendicular to them. Secondly, families of streamlines can be used to define *tubes of flow* within the fluid, as shown in Fig. 11.19. Since there can be no flow across the surface of such a tube of flow, the fluid flow is entirely along and within the tube.

In deriving Bernoulli's equation, we shall confine ourselves to the case of *incompressible* fluids. It is quite possible to modify the derivation to allow for compressibility, though for our purposes this introduces unnecessary complications. It turns out also that the incompressible Bernoulli equation gives a good *approximate* description of the steady-state flow of compressible fluids in cases where the flow velocity is everywhere small compared to the velocity of sound. We shall, therefore, find ourselves using the incompressible Bernoulli equation even in gases, though in doing so we have to be careful to avoid cases of high-velocity flow.

Finally, we shall have to concern ourselves with fluid flow situations which are *nonviscous*. By *viscosity* we mean the property exhibited by fluids of generating internal forces of friction or resistance when one layer of fluid is made to move across another parallel layer with finite relative velocity. We are familiar with the fact that when a liquid such as water or ethyl alcohol is stirred with rapid rotation in a cylindrical can, it may remain in rotation for a very long time before all evidence of the original motion ceases. In the case of a heavy oil, however, such rotational motion, if excited, falls off very rapidly when stirring is ceased. It is the internal viscous forces between fluid layers moving at different velocities that are responsible for the damping of the motion in both cases. In both cases, it is evident that the kinetic energy of rotation which was excited by stirring is dissipated (actually transformed into heat energy) by the action of these forces.

Viscous forces, therefore, have the characteristics of *frictional* forces in fluids. These forces, which arise from interactions between the molecules of the substance itself, are quite weak in the case of water and alcohol. In a given time interval, say a second, they may transform only a small fraction of the initial kinetic energy into heat. For such substances, it may be quite reasonable to completely ignore the effect of viscosity in turning mechanical energy into heat, just as it is possible to neglect the effect of sliding friction in many mechanics problems of the types studied in earlier chapters. On the other hand, if we are dealing with a very viscous fluid such as lubricating oil or glycerine, a very large fraction of the kinetic energy may be dissipated in a similar time interval; now the viscous forces are large and cannot be neglected.

We shall not try to treat viscous flow in this chapter, but shall confine ourselves to situations in which viscous friction is unimportant. Finally, we shall also exclude from our treatment any situation in which internal rotations of the fluids, or *vortex* effects, are important. If these effects are absent, the flow pattern is said to be *irrotational*, and we shall consider only instances of this type in deriving Bernoulli's equation.

Now, having restricted ourselves to a fairly narrow and simple range of fluid flow conditions—steady-state, incompressible, nonviscous, and irrota-

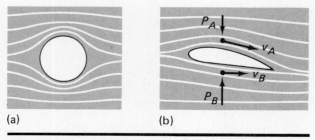

(a)   (b)

FIGURE 11.18. Streamlines illustrating steady fluid flow around a spherical obstacle (a) and an airfoil (b).

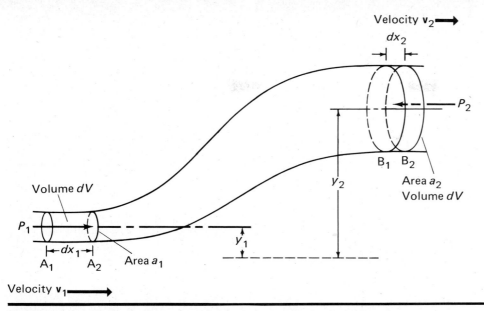

FIGURE 11.20. System of forces acting on fluid confined within a tube of variable cross section.

tional—we are ready to derive Bernoulli's equation. Consider the motion of a fluid in a pipe of variable cross-sectional area, as illustrated in Fig. 11.20. In particular, let us investigate the motion of an element of fluid which at time t extends from point A_1 along the pipe to B_1 and which, at a time dt later, extends from point A_2 to B_2, having moved a distance dx_1, at speed dx_1/dt, from A_1 to B_1, and, at the other end, a distance dx_2, at speed dx_2/dt, from A_2 to B_2. Let the cross-sectional area of the tube be a_1 from A_1 to B_1 and a_2 from A_2 to B_2. We shall assume that the section A_1B_1 is at height y_1 above the earth's surface and that the section A_2B_2 is at height y_2. Since the fluid is incompressible, the same volume dV is swept out at either end, and this means that the velocities v_1 and v_2 and the areas a_1 and a_2 are *related*. The connection between them stems from the relation

$$dV = a_1\, dx_1 = a_2\, dx_2 \qquad (11.5.1)$$

or, dividing by dt and noting that $v_1 = dx_1/dt$, $v_2 = dx_2/dt$,

$$\frac{dV}{dt} = a_1v_1 = a_2v_2 \qquad (11.5.2)$$

Bernoulli's equation is a consequence of energy conservation and the work–energy theorem. The work–energy theorem states that the work done on any body by the resultant force acting on it must equal its change in kinetic energy. We must, therefore, compute the work done by all forces acting on the element of fluid and also its change in kinetic energy. The forces acting are of two kinds: pressure forces P_1 and P_2 at the two ends and gravitational forces arising from the weight of the fluid in the element. As the fluid element moves from its initial to its final position, the work done by the forces attributable to pressures P_1 and P_2 are $P_1 a_1\, dx_1$ and $-P_2 a_2\, dx_2$, the minus sign arising from the fact that the pressure on the element acts in the $-x$ direction on the right-hand end of the fluid element A_1B_1 or A_2B_2. Recalling Eq. (11.5.1), the net work done by the pressure forces is then

$$dW_p = P_1 a_1\, dx_1 - P_2 a_2\, dx_2 = (P_1 - P_2)\, dV \qquad (11.5.3)$$

Work is also done on the element by gravitational forces; in effect, as the element of fluid moves from its initial position A_1B_1 to its final position the effect is to raise a volume of fluid dV, hence a mass of fluid $dm = \rho\, dV$, through a height $y_2 - y_1$. The amount of work done by gravity is therefore

$$dW_g = -(dm)g(y_2 - y_1) = -\rho g(y_2 - y_1)\, dV \qquad (11.5.4)$$

Finally, the change in kinetic energy of the element, dU_k, can be calculated by observing that the element A_1B_1 has lost its kinetic energy $\frac{1}{2}(dm)v_2{}^2$ and the element A_2B_2 has acquired kinetic energy $\frac{1}{2}(dm)v_2{}^2$. The velocities of all *other* points in the fluid column remain unchanged and, therefore, give no contribution to the change in kinetic energy. The total change in kinetic energy during the process may, therefore, be written as

$$dU_k = \tfrac{1}{2}(dm)v_2{}^2 - \tfrac{1}{2}(dm)v_1{}^2$$
$$= \tfrac{1}{2}\rho(v_2{}^2 - v_1{}^2)\, dV \qquad (11.5.5)$$

According to the work–energy theorem, the work done on the fluid by the resultant force, which can be expressed as the sum of the work contributions

357

of the individual forces, equals the change in kinetic energy of the element. Hence, we may write

$$dW_p + dW_g = dU_k \qquad (11.5.6)$$

or, from Eqs. (11.5.3), (11.5.4), and (11.5.5),

$$(P_1 - P_2)\,dV - \rho g(y_2 - y_1)\,dV = \tfrac{1}{2}\rho(v_2{}^2 - v_1{}^2)\,dV \qquad (11.5.7)$$

Canceling dV throughout and rearranging the terms, this can be written finally as

$$\boxed{P_1 + \rho g y_1 + \tfrac{1}{2}\rho v_1{}^2 = P_2 + \rho g y_2 + \tfrac{1}{2}\rho v_2{}^2} \qquad (11.5.8)$$

This is Bernoulli's equation. Note that if the fluid is at equilibrium ($v_1 = v_2 = 0$), it reduces to what should now be the familiar result

$$P_1 - P_2 = \rho g(y_2 - y_1) \qquad (11.5.9)$$

which was set forth previously as Eq. (11.3.7).

The derivation was carried out for a fluid within a pipe or tube, and it might, therefore, be thought that the results so obtained would not apply to freely flowing fluids which are not so constrained. *This is not so,* however, because the same derivation could also have been applied to the flow of a certain portion of an unconstrained fluid enclosed by one of the *tubes of flow* of the system! Since no fluid ever flows across the boundary of any tube of flow, the derivation would have been equally valid for such a situation. After all, it makes no difference whether a certain portion of a flowing fluid is constrained by the physical presence of the walls of a pipe or merely by the rest of a larger fluid system surrounding it and flowing along with it.

Bernoulli's equation, (11.5.8), tells us that the quantity $P + \rho g y + \tfrac{1}{2}\rho v^2$ is the same at any two points in the flow system, which implies that its value is the same everywhere in the system. The quantity $\rho g y$ represents the potential energy of the fluid per unit volume at a given point, and the quantity $\tfrac{1}{2}\rho v^2$ likewise represents the kinetic energy per unit volume at that point. Bernoulli's equation may then be interpreted as stating that the sum of the pressure, the potential energy per unit volume, and the kinetic energy per unit volume is everywhere the same—hence is *conserved.* Since the derivation of Bernoulli's equation is based on the principle of energy conservation, it is hardly surprising that it has the form of a conservation law.

One of the important physical effects predicted by Bernoulli's equation can be understood in simple terms by considering the case where there is no appreciable difference in gravitational potential energy throughout the flowing fluid. We may then set both y_1 and y_2 equal to zero, and (11.5.8) becomes

$$P_1 + \tfrac{1}{2}\rho v_1{}^2 = P_2 + \tfrac{1}{2}\rho v_2{}^2 \qquad (11.5.10)$$

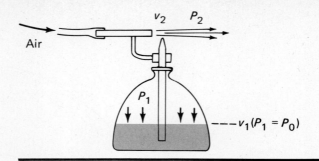

FIGURE 11.21. An atomizer.

$$P_2 - P_1 = \tfrac{1}{2}\rho(v_1{}^2 - v_2{}^2) \qquad (11.5.11)$$

In this equation, suppose that v_2 is greater than v_1; then $v_1{}^2 - v_2{}^2$ is negative, as is $P_2 - P_1$. This means than P_1 is greater than P_2 or, turning it around, that P_2 is smaller than P_1. In other words, when gravitational effects are unimportant, an *increase* in the flow velocity of a fluid is inevitably accompanied by a *decrease* in pressure. This is sometimes called the *Venturi effect* and is responsible for the action of atomizers, automotive carburetors, wind tunnels, and many other devices, as well as for the curvature of the path of a rapidly spinning baseball and for the aerodynamic lift of an airplane wing.

The example of an atomizer is shown in Fig. 11.21. In this device, a large flow velocity v_2 is created by a stream of air across the nozzle, the flow velocity being zero or very small elsewhere, in particular at the surface of the liquid in the reservoir. According to Bernoulli's equation, then, the pressure in the region of the nozzle must be less than that experienced in the reservoir, where atmospheric pressure prevails. Accordingly, the atmospheric pressure in the reservoir forces the liquid to rise in the tube leading to the nozzle (just as in a fluid barometer) and finally to emerge from the nozzle orifice to be broken up by and carried away in the air stream.

The case of the spinning baseball is shown in

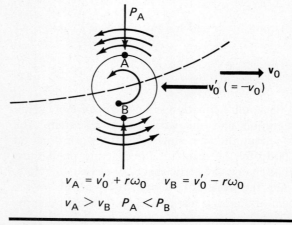

$$v_A = v_0' + r\omega_0 \qquad v_B = v_0' - r\omega_0$$
$$v_A > v_B \qquad P_A < P_B$$

FIGURE 11.22. Motion of a spinning baseball.

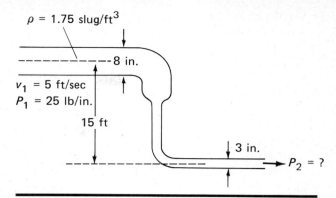

FIGURE 11.23

Fig. 11.22. In this illustration, we are looking down from above on a baseball thrown by a pitcher at the far right. The pitcher has imparted an initial spin angular velocity ω_0 to the ball and, at the same time, a linear velocity v_0 to the center of mass. As the ball spins, it drags along a rotating layer of air as shown by the arrows. Let us temporarily move with the center of mass of the baseball and try to account for the flow velocities in its vicinity. In doing this, it is as if the baseball is motionless (except for spin) in a wind tunnel which sends a stream of air at it from right to left with velocity v_0', equal to $-v_0$, as shown in the diagram. The layer of air dragged by the surface of the ball has linear velocity $r_0\omega_0$ at point B and $-r_0\omega_0$ at point A, where r_0 is the ball's radius. The total magnitude of flow velocity with respect to the center of the ball is then $v_0' + r_0\omega_0$ at A and $v_0' - r_0\omega_0$ at B. Since the velocity at B is less than at A, the pressure on the ball, according to Bernoulli's equation, is greater at B than at A. This means that there will be a net force which will deflect the ball upward in the diagram or sideways to a person in the position of the pitcher. It is evident from all this that atmospheric friction is responsible for the curvature of the ball's path. A pitcher, no matter how good he is, could never throw a curve in a vacuum!

Somewhat the same analysis applies to the problem of the dynamic lift of an airplane wing. The streamlines about the cross section of such an airfoil are shown in Fig. 11.18b. The airfoil is so shaped and positioned in the airstream, at an appropriate angle of attack, that a pattern of flow is produced in which the velocity v_A near the upper surface of the wing is necessarily greater than the velocity v_B near the lower surface. According to Bernoulli's equation, this means that the pressure P_B on the undersurface must be greater than the pressure P_A on the top surface and that a resultant upward aerodynamic lift is created. It is not entirely obvious how the airfoil is to be shaped or positioned so that this relation between the velocities is obtained; there are some flow velocities and attack angles for which it is not so, as, for example, in "stall" conditions. A more detailed discussion of aerodynamic lift in airfoils is somewhat beyond the scope of this work and will not be included here.

The following series of examples will illustrate more explictly how Bernoulli's equation may be applied to various instances of steady-state fluid flow problems.

EXAMPLE 11.5.1

An incompressible fluid is flowing from left to right through a cylindrical pipe such as the one shown in Fig. 11.23. The density of the fluid is 1.75 slug/ft^3. Its velocity at the input end is 5 ft/sec, and the pressure there is 25 lb/in.^{2} The output end is 15 ft below the input end. What is the pressure at the output end?

Bernoulli's equation tells us that

$$P_1 + \rho g y_1 + \tfrac{1}{2}\rho v_1^2 = P_2 + \rho g y_2 + \tfrac{1}{2}\rho v_2^2 \qquad (11.5.12)$$

We know P_1, ρ, y_1, y_2, and v_1; neither P_2 nor v_2 are given. However, since the fluid is incompressible, v_2 must be related to v_1 by (11.5.2):

$$v_2 = \frac{a_1}{a_2} v_1 = \frac{\pi r_1^2}{\pi r_2^2} v_1 = \frac{(\tfrac{1}{3})^2}{(\tfrac{1}{8})^2}(5) = 35.6 \text{ ft/sec}$$

Substituting this value along with $P_1 = 25$ lb/in.$^2 = 3600$ lb/ft^2, $v_1 = 5$ ft/sec, $\rho = 1.75$ slug/ft^3, $y_1 = 15$ ft, and $y_2 = 0$ into (11.5.12), we obtain

$$3600 + (1.75)(32.2)(15) + (0.5)(1.75)(25)$$
$$= P_2 + 0 + (0.5)(1.75)(35.6)^2$$

$P_2 = 3358$ lb/ft^2, or 23.3 lb/in.^{2}

EXAMPLE 11.5.2

A *Venturi flowmeter* is illustrated in Fig. 11.24. It consists of a constricted tubular section of frontal area

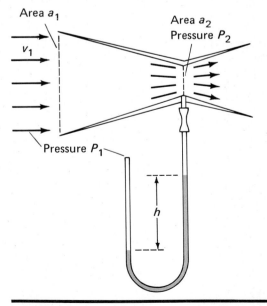

FIGURE 11.24. The Venturi flowmeter.

a_1 and throat area a_2. It is inserted into a stream of gas or liquid flowing at velocity v_1, which is to be determined, and a small orifice in the throat is connected to a pressure gauge, shown as an open U-tube manometer in the figure. A pressure P_2 is measured on the manometer. Find the equation relating the flow velocity v_1 which is to be measured to the known areas a_1 and a_2 and the measured pressure difference $P_1 - P_2$. What is the flow velocity of a stream of air ($\rho = 1.293 \times 10^{-3}$ g/cm^3) in a situation where the frontal area a_1 is 100 cm, the throat area a_2 is 20 cm^2, and the height h measured on an open tube mercury manometer 25 cm?

We begin again with Bernoulli's equation:

$$P_1 + \rho g y_1 + \tfrac{1}{2}\rho v_1{}^2 = P_2 + \rho g y_2 + \tfrac{1}{2}\rho v_2{}^2 \qquad (11.5.13)$$

In this situation, $y_1 = y_2 = 0$, and the throat velocity v_2 can be expressed in terms of the stream velocity v_1 by the relation (11.5.2), which states that $a_1 v_1 = a_2 v_2$, or

$$v_2 = \frac{a_1}{a_2} v_1 \qquad (11.5.14)$$

Bernoulli's equation then reduces to

$$P_1 + 0 + \tfrac{1}{2}\rho v_1{}^2 = P_2 + 0 + \tfrac{1}{2}\rho \frac{a_1{}^2}{a_2{}^2} v_1{}^2$$

which upon rearranging may be expressed as

$$P_1 - P_2 = \tfrac{1}{2}\rho v_1{}^2 \left(\frac{a_1{}^2}{a_2{}^2} - 1 \right)$$

from which

$$v_1{}^2 = \frac{P_1 - P_2}{\tfrac{1}{2}\rho \left(\dfrac{a_1{}^2}{a_2{}^2} - 1 \right)} \qquad (11.5.15)$$

For an open-tube mercury manometer, using the notation of this example, the pressure difference, according to (11.3.17), is

$$P_1 - P_2 = \rho_{Hg} g h \qquad (11.5.16)$$

where ρ_{Hg} is the density of mercury, equal to 13.6 g/cm^3. Equation (11.5.15) now becomes

$$v_1{}^2 = \frac{2\rho_{Hg} g h}{\rho \left(\dfrac{a_1{}^2}{a_2{}^2} - 1 \right)} \qquad (11.5.17)$$

For an air stream where $\rho = 1.293 \times 10^{-3}$ g/cm^3 and where $h = 25$ cm, $a_1 = 100$ cm^2, and $a_2 = 20$ cm^2, this gives

$$v_1{}^2 = \frac{(2)(13.6)(980)(25)}{(1.293 \times 10^{-3})(5^2 - 1)} = 21.47 \times 10^6 \text{ cm}^2/\text{sec}^2$$

$$v_1 = 4634 \text{ cm/sec, or } 46.34 \text{ m/sec}$$

EXAMPLE 11.5.3

A cylindrical tank 6 ft in diameter rests atop a platform 20 ft high, as shown in Fig. 11.25. Initially, the tank is filled with water ($\rho = 1.938$ slug/ft^3) to a depth h_0 equal to 10 ft. A plug whose area is 1.00 in.^{2} is removed from an orifice in the side of the tank at the very bottom. With what velocity does the water flow initially from this orifice? What is the velocity of the stream initially as it strikes the ground? How long will it take to empty the tank entirely?

Let us first of all assign dimensions and values of pressure, velocity, and height to the system as shown in Fig. 11.25. We shall work with gauge pressures throughout; this is permissible when using Bernoulli's equation, provided it is done consistently. What one *cannot* do is use gauge pressure on one side of the equation and absolute pressure on the other, so beware of doing that!

At the very top surface of the liquid, P, v, and y have the values P_1, v_1, and y_1, and in the stream which is just outside the orifice at the bottom of the tank, they have the values P_2, v_2, and y_2. According to Bernoulli's equation,

$$P_1 + \rho g y_1 + \tfrac{1}{2}\rho v_1{}^2 = P_2 + \rho g y_2 + \tfrac{1}{2}\rho v_2{}^2 \qquad (11.5.18)$$

Now, obviously, the gauge pressure P_1 is zero; but would you believe that P_2 is also zero? Well, it is, and the reasoning goes like this. Inside the tank, in the water *behind* the orifice, the pressure is not zero but rather $\rho g h_0$; but all this pressure is transformed to kinetic energy in the trip through the orifice. Outside the orifice, the only pressure on the now rapidly moving stream is that of the atmosphere, hence the gauge pressure there is once more zero! This argument neglects, as usual, the slight rise in atmospheric pressure encountered in descending through the distance h_0, which in this problem is only 10 ft.

The height y_1 is equal to $h_0 + H$, while y_2 equals H. The velocities v_1 and v_2 are again related by

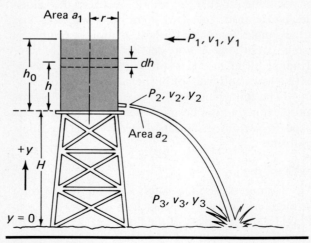

FIGURE 11.25

$a_1v_1 = a_2v_2$. In this case, however, the area a_1 is $\pi r^2 = (\pi)(3)^2 = 9\pi = 28.27$ ft$^2 = 4072$ in.^{2}, while $a_2 = 1.00$ in.^{2} This means that v_2 is some 4000 times larger than v_1, and under these circumstances the term $\frac{1}{2}\rho v_1{}^2$ will be some 1,600,000 times less than $\frac{1}{2}\rho v_2{}^2$. We might then just as well neglect it altogether and set v_1 equal to zero. Putting all this information into Eq. (11.5.18), we find

$$0 + \rho g(h_0 + H) + 0 = 0 + \rho g H + \tfrac{1}{2}\rho v_2{}^2$$
$$v_2 = \sqrt{2gh_0} = \sqrt{(2)(32.2)(10)} = 25.4 \text{ ft/sec} \quad (11.5.19)$$

Once more assigning values for P, v, and y as P_2, v_2, and y_2 just outside the tank orifice and P_3, v_3, and y_3 as the stream hits the ground, according to Bernoulli's equation,

$$P_2 + \rho g y_2 + \tfrac{1}{2}\rho v_2{}^2 = P_3 + \rho g y_3 + \tfrac{1}{2}\rho v_3{}^2 \quad (11.5.20)$$

But now, $P_2 = P_3 = 0$, $y_2 = H$, $y_3 = 0$, and, from Eq. (11.5.19), $v_2{}^2 = 2gh_0$. Putting all this into (11.5.20), we obtain

$$0 + \rho g H + \rho g h_0 = 0 + 0 + \tfrac{1}{2}\rho v_3{}^2$$
$$v_3 = \sqrt{2g(h_0 + H)} = \sqrt{(2)(32.2)(30)} = 44.0 \text{ ft/sec}$$
$$(11.5.21)$$

In finding how long it takes to empty the tank, the time required for the depth to change from h to zero is required, and this is closely related to the velocity v_1 which we found we could neglect for the purpose of determining v_2. We must now, however, start worrying about v_1. As always in the case of an incompressible fluid,

$$a_1v_1 = a_2v_2 \quad (11.5.22)$$

Now suppose the height h of the liquid in the tank decreases by the amount dh during a time interval dt. The velocity v_1 must now be given by

$$v_1 = -\frac{dh}{dt} \quad (11.5.23)$$

the minus sign being necessary because dh/dt is negative (h is decreasing with time) while the fluid speed v_1 is essentially positive, since it is the magnitude of a velocity vector. Substituting this value for v_1 into (11.5.22), we obtain

$$-\frac{dh}{dt} = \frac{a_2}{a_1}v_2 = \frac{a_2}{a_1}\sqrt{2gh} \quad (11.5.24)$$

recalling from (11.5.19) that $v_2 = \sqrt{2gh}$. We shall now have to integrate this equation to find the relationship between the depth h and the time. Multiplying the equation by dt and dividing by $\sqrt{h}$, we may express (11.5.24) in the form

$$-\frac{dh}{\sqrt{h}} = \frac{a_2}{a_1}\sqrt{2g}\, dt \quad (11.5.25)$$

This may now be integrated between time $t = 0$, when the depth has the initial value h_0 and a later time t, when the depth is h. Accordingly,

$$-\int_{h_0}^{h} h^{-1/2}\, dh = \frac{a_2}{a_1}\sqrt{2g}\int_0^t dt$$
$$(11.5.26)$$
$$-[2\sqrt{h}]_{h_0}^{h} = 2(\sqrt{h_0} - \sqrt{h}) = \frac{a_2}{a_1}\sqrt{2g}\, t$$

and finally, solving for t,

$$t = \frac{a_1}{a_2}\frac{\sqrt{2}(\sqrt{h_0} - \sqrt{h})}{\sqrt{g}} \quad (11.5.27)$$

This equation gives the elapsed time t as a function of the fluid depth h. Initially, the depth h_0 is 10 ft; when the tank is empty, the depth is zero. Setting $h = 0$ in (11.5.27), we find

$$t = \frac{a_1}{a_2}\sqrt{\frac{2h_0}{g}} = \frac{(4072)}{(1.00)}\frac{(2)(10)}{[(32.2)]^{1/2}} = 3209 \text{ sec, or } 53.5 \text{ min}$$

EXAMPLE 11.5.4

A small fixture is attached to the tank orifice of the previous example to direct the stream upward at an angle θ without affecting its speed or cross-sectional area. What is the maximum height h' attained by the stream?

The situation is now that illustrated in Fig. 11.26. From Bernoulli's equation, we know that

$$P_2 + \rho g y_2 + \tfrac{1}{2}\rho v_2{}^2 = P_4 + \rho g y_4 + \tfrac{1}{2}\rho v_4{}^2 \quad (11.5.28)$$

But $P_2 = P_4 = 0$, $y_2 = H$, and from the results of the preceding example, $v_2{}^2 = 2gh_0$. We may also write $v_2{}^2 = v_{2x}{}^2 + v_{2y}{}^2 = v_2{}^2 \cos^2\theta + v_2{}^2 \sin^2\theta$. Bernoulli's equation now reads

$$\rho g H + \tfrac{1}{2}\rho v_2{}^2 \cos^2\theta + \tfrac{1}{2}\rho v_2{}^2 \sin^2\theta = \rho g h' + \tfrac{1}{2}\rho v_4{}^2$$
$$(11.5.29)$$

In this equation, we know neither h' nor v_4. We do know, however, that since the pressure is everywhere

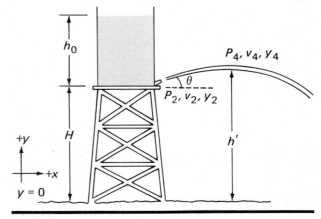

FIGURE 11.26

zero in the stream of fluid emitted by the orifice, the only force on the water in that stream is the force of gravity acting on each individual droplet. Under these circumstances, each droplet of the stream executes *projectile motion;* there is, therefore, no x-component of acceleration for the liquid in the stream because the resultant force has only a y-component. The x-component of velocity for the liquid in the stream is, therefore, *constant*, and this, in turn, tells us that

$$v_4 = v_{2x} = v_2 \cos \theta \qquad (11.5.30)$$

Substituting this value for v_4 into Eq. (11.5.29) along with the value $2gh_0$ for $v_2{}^2$ and solving for h', it is easy to obtain

$$h' = H + h_0 \sin^2 \theta \qquad (11.5.31)$$

It is evident that the jet of fluid emitted by the tank nozzle can never rise higher than the level of fluid in the tank, since the largest value $\sin^2 \theta$ can have is unity, attained when $\theta = 90°$, in which case $h' = H + h_0$, the stream then rising just to the inside fluid level.

EXAMPLE 11.5.5

A cylindrical tank 1.2 m in diameter is filled to a depth of 0.3 m with water ($\rho = 1000$ kg/m^3). The space above the water is occupied by air, compressed to a gauge pressure of 1.00×10^5 N/m^2. A plug is removed from an orifice in the bottom of the tank having an area of 2.5 cm^2. What is the initial velocity of the stream which flows through this orifice? What is the upward force experienced by the tank when the plug is removed?

The situation is illustrated in Fig. 11.27. From Bernoulli's equation,

$$P_1 + \rho g y_1 + \tfrac{1}{2}\rho v_1{}^2 = P_2 + \rho g y_2 + \tfrac{1}{2}\rho v_2{}^2 \qquad (11.5.32)$$

In this example, just as in Example 11.5.3, the velocity v_1 may be neglected because the area a_2 is so much smaller than a_1. The pressure P_1 is equal to 1.00×10^5 N/m^2, while P_2 is zero; also $y_1 = h$, while $y_2 = 0$. Putting all this into (11.5.32), we find

$$P_1 + \rho g h + 0 = 0 + 0 + \tfrac{1}{2}\rho v_2{}^2$$

$$v_2{}^2 = \frac{2(P_1 + \rho g h)}{\rho} = \frac{(2)[(10^5) + (10^3)(9.8)(0.3)]}{10^3}$$

$$= 205 \text{ m}^2/\text{sec}^2$$

$$v_2 = 14.35 \text{ m/sec} \qquad (11.5.33)$$

To find the initial upward thrust on the tank, we may note that before the plug is removed the initial momentum of the system is zero. Since there are no *external* forces acting on the system (tank plus water), the momentum remains zero after the plug is removed. But in time interval dt, the liquid acquires

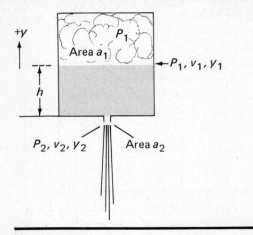

FIGURE 11.27

a negative y-component of momentum equal to

$$dp_{\text{liq}} = -v_2 \, dm \qquad (11.5.34)$$

where dm is the mass which squirts from the orifice in time dt. This quantity can be expressed as the product of the density times the volume emitted in time dt, the latter in turn being equal to the area a_2 times the distance the stream travels in time dt, which is $v_2 \, dt$. We may write, therefore,

$$dm = \rho a_2 v_2 \, dt \qquad (11.5.35)$$

Substituting this into Eq. (11.5.34), we find

$$dp_{\text{liq}} = -\rho a_2 v_2{}^2 \, dt \qquad (11.5.36)$$

The momentum imparted to the tank is just the negative of this, in view of the fact that the total final momentum must be zero. But the momentum imparted to the tank in time dt may also be equated to the *impulse* of the force acting on the tank, according to the impulse–momentum theorem. We may write, therefore,

$$dp_{\text{tank}} = \rho a_2 v_2{}^2 \, dt = F \, dt$$

whereupon it is evident that the force F must be given by

$$F = \rho a_2 v_2{}^2 = (10^3)(2.5 \times 10^{-4})(14.35)^2 = 51.5 \text{ N}$$

11.6 Surface Tension and Capillary Attraction

We have already had occasion to mention the phenomenon of viscosity, or internal friction, in fluids, and we have seen that this property has its origin in the forces between individual molecules of the substance. There are other unique properties of liquids attributable to these intermolecular forces, notably the phenomena of *surface tension* and *capillary attraction*. While these effects, like viscosity, do not alter the basic laws of fluid mechanics, they do alter the way they must be applied in certain particular situations. Therefore, though not treating these subjects in great detail, we shall offer a brief description of each, trying to indicate under what circumstances they may be important and when they may safely be neglected.

The phenomenon of surface tension arises from the fact that the intermolecular forces acting on molecules at or near a liquid surface differ from those which act on molecules deep in the interior of the liquid. Deep inside the liquid, the forces acting on each individual molecule are exerted equally in all directions, while a molecule at the surface experiences no force whatsoever from outside. A surface molecule, when displaced slightly from its equilibrium position toward the outside of the liquid, therefore, experiences strong resultant forces which tend to return it to its original position. This gives the surface of a liquid somewhat the character of a stressed elastic membrane, like the surface of an inflated balloon, except, of course, that it is self-healing when punctured!

The force necessary to rupture the surface "skin" of any liquid can be measured conveniently by determining experimentally how much upward force is required to pull a wire loop free of the surface, as illustrated in Fig. 11.28a. From the force diagram shown there, the force of surface tension T_s can be expressed in terms of the maximum force F required to free the loop from the liquid and the weight of the loop as

$$\sum F_y = F - mg - T_s = 0 \qquad (11.6.1)$$

or

$$T_s = F - mg \qquad (11.6.2)$$

Now, the force T_s is clearly proportional to the *length* of the loop that has to be pulled through the surface and also to the *strength of the intermolecular forces* that have to be overcome. The latter, of course, depends in detail upon just what forces are exerted by whatever molecules are present in the conditions of temperature, pressure, etc., which prevail, in other words, upon the substance itself and its ambient conditions. In any case, the force T_s can be written as a constant of proportionality times the *length* of fluid surface ruptured:

$$T_s = \gamma l \qquad (11.6.3)$$

All the information about the strength of the intermolecular forces is contained in the proportionality constant γ, called the *specific surface tension parameter*, which will be different for each liquid. This parameter is very difficult to calculate from the fundamental properties of the molecules, but it is very easy to *measure* by the arrangement shown in Fig. 11.28;

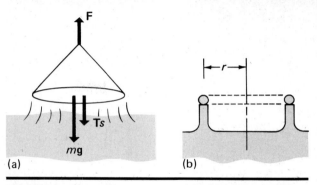

FIGURE 11.28. System of forces acting on a wire loop being pulled upward through the surface of a liquid.

in that diagram, the length of surface to be broken is given by $l = 2(2\pi r) = 4\pi r$, whence from Eqs. (11.6.2) and (11.6.3),

$$\gamma l = 4\pi r\gamma = F - mg$$

$$\gamma = \frac{F - mg}{4\pi r} \qquad (11.6.4)$$

It should be noted that the length of surface to be ruptured is in this case *twice* the circumference of the loop, since a surface film on the inside as well as the outside of the wire loop has to be broken. This is illustrated in Fig. 11.28b.

The surface tension parameter γ provides us with an easily measured index of the strength of intermolecular forces. From (11.6.3), it is evident that γ has the dimensions of force per unit length, dynes/cm or newtons/m in the metric system or pounds/ft in the English system. Because of surface tension, a pin or a razor blade can, with care, be supported by the surface of a liquid much less dense than the metal from which these objects are fashioned. In these cases, the surface tension force is much greater than the Archimedean buoyant force. Surface tension is also responsible for the fact that small quantities of liquids assume the form of spherical droplets, because the stressed surface "skin" tends to contract, molding the liquid into a form having minimum surface area for its volume, that is, into a *sphere*. Surface tension is also important in understanding the behavior of bubbles and soap films.

There are situations in which surface tension forces are quite important and others in which they are negligible. The surface tension of water will support a razor blade but not a battleship. Surface tension forces are wholly responsible for the shape of a raindrop but have nothing to do with the shape of Lake Erie. The difference lies in the different ways in which surface tension forces and weight forces vary as a function of the linear dimensions of the fluid mass in question. The perimeter of a fluid body varies as the first power of its linear extent, its area varies as the square of the linear extent, while the volume (or mass) changes as the cube. Forces that are directly proportional to an object's mass, therefore, fall off much more rapidly as the dimensions of the object decrease than do forces proportional to the surface area or linear dimensions. The surface tension force is an example of this latter type, while weight forces are of the first kind.

Let us consider, for example, a raindrop. For a spherical volume of water of radius, r, the surface tension force holding the two hemispheric halves together is the surface tension parameter γ times the circumference of the sphere:

$$F_s = \gamma l = 2\pi r \gamma \qquad (11.6.5)$$

The weight force, on the other hand, is given by

$$W = mg = \rho V g = \tfrac{4}{3}\pi r^3 \rho g \qquad (11.6.6)$$

It will be equal to the external normal force experienced by the drop when it rests on a rigid surface. For a spherical water droplet of radius 0.1 cm, for which $\rho = 1.00$ g/cm^3 and $\gamma = 81$ dynes/cm, we find from (11.6.5) and (11.6.6) that the surface tension force holding the droplet together is 50.9 dynes, while the external normal force exerted by the surface on which it rests, equal to its weight, is 4.1 dynes. In this case, the external normal force is practically negligible in comparison with the surface tension force; it is, therefore, the latter that determines the form of the droplet and renders it practically spherical.

Now consider a spherical volume of water whose radius is 10 cm. The surface tension force holding the two halves together is now, according to Eq. (11.6.5), 100 times larger than before, or 5090 dynes. But, from (11.6.6), the external normal force exerted on the volume by its surroundings is proportional to r^3, so it will be 10^6 times larger than before and thus have the value of 4.1×10^6 dynes. In this case, the force attributable to surface tension is almost 1000 times *less* than the normal force from the surface on which the volume of liquid rests. The surface tension force now is clearly insufficient to maintain the spherical form of the liquid volume. So the liquid spreads out to ultimately take the shape of whatever container it happens to be in. Now the container furnishes the normal force necessary to establish equilibrium with the fluid's weight force, and the relatively small surface tension force is of little or no importance.

If the liquid is supported only by a perfectly flat surface, a circular puddle will form which will spread until its circumference is so large that the surface tension force is again sufficient to allow the system to reach equilibrium. When this occurs, the surface tension force on the upper half of the puddle (which acts downward) will just balance the normal force exerted by the lower half of the puddle on the

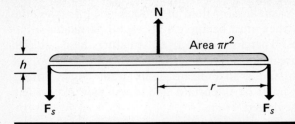

FIGURE 11.29. System of forces acting on fluid in a circular puddle.

upper half, as illustrated in Fig. 11.29. The normal force N is half the weight of a sphere of water of radius 10 cm and thus of volume $V = \tfrac{4}{3}\pi(10)^3 = 4189$ cm^3. The surface tension force F_s is γ times the circumference of the puddle, $2\pi r$, where r is the radius. Equating these quantities, we find

$$N = \tfrac{1}{2}\rho g V = \gamma 2\pi r$$

$$r = \frac{\rho g V}{4\pi\gamma} = \frac{(1.00)(980)(4189)}{(4)(3.1416)(81)} = 4033 \text{ cm, or } 40.33 \text{ m} \qquad (11.6.7)$$

The resulting puddle is, then, over 80 m (264 ft) in diameter! The corresponding depth of fluid, h, can be obtained from

$$v = \pi r^2 h = 4189 \text{ cm}^3$$

$$h = \frac{V}{\pi r^2} = \frac{(4189)}{(3.1416)(4033)^2} = 8.20 \times 10^{-5} \text{ cm} \qquad (11.6.8)$$

The puddle so formed is clearly very large and very shallow. The quantitative accuracy of these numbers should not be taken very seriously, since there are many practical factors that interfere with actually observing the formation of such a puddle in reality. The numbers, however, do illustrate the relative weakness of surface tension forces in situations where relatively large volumes or masses of fluids are involved.

Closely related to surface tension is the phenom-

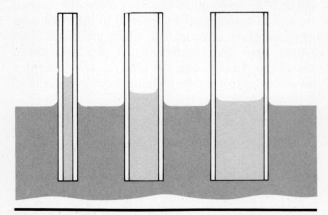

FIGURE 11.30. Capillary rise of a liquid in tubes of various radii.

enon of capillary attraction. When liquids confined within thin tubes, or capillaries, are allowed to reach equilibrium with a free liquid outside, as shown in Fig. 11.30, it is found that the level of the fluid within the tube is slightly higher than the free outside level, and the smaller the radius of the tube, the higher the level within. The difference in levels depends also upon the liquid used and the composition of the capillary tubing; in some cases, such as mercury within glass, the liquid surface within the capillary may be lower rather than higher. Associated with the capillary rise is the formation of a curved liquid surface (or "meniscus") within the tube, the level of the fluid rising toward the edge of the capillary where liquid and tube wall meet, though in the mercury–glass system the curvature is in the other direction.

This phenomenon is also caused by intermolecular attraction, but now between the molecules of the liquid and those of the capillary wall rather than between the molecules of the liquid alone. Consider the situation illustrated in Fig. 11.31. From this drawing, one may easily infer that the forces of intermolecular attraction responsible for the capillary rise are those which exist where the liquid surface meets the capillary wall, denoted by F_s. Elsewhere within the capillary, of course, there are also forces of intermolecular attraction between molecules of liquid and those in the capillary wall, such as F and F' in Fig. 11.31. But these forces act horizontally, as shown, and cannot support the liquid column within the capillary. We need, therefore, only consider the surface tension force F_s as supporting the liquid column. Indeed, it is only the vertical component of this force, $F_s \cos \theta$, where θ is the contact angle at which the liquid intersects the tube wall, that supports the column within the capillary.

Now, F_s is equal to the surface tension γ times the circumference of the tube:

$$F_s = 2\pi r \gamma \qquad (11.6.9)$$

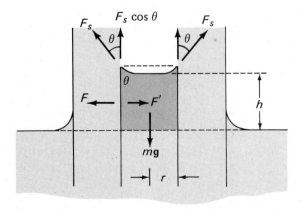

FIGURE 11.31. System of forces acting on the liquid in a capillary tube.

where r is the inside radius of the capillary. Also, since the system is in equilibrium,

$$\sum F_y = F_s \cos \theta - mg = 0 \qquad (11.6.10)$$

where m represents the mass of the shaded liquid column in Fig. 11.31. Neglecting the small amount of fluid above the level h at the edges of the column, this can be written as

$$m = \rho V = \rho \pi r^2 h \qquad (11.6.11)$$

Inserting the values of m and F_s as given by (11.6.11) and (11.6.9) into (11.6.10), we obtain

$$2\pi r \gamma \cos \theta - \pi \rho g r^2 h = 0$$

$$h = \frac{2\gamma \cos \theta}{\rho g r} \qquad (11.6.12)$$

Thus, we see that the capillary rise h is proportional to the surface tension γ and to the cosine of the contact angle and inversely proportional to the capillary radius and to the density of the fluid. For water in a glass capillary ($\rho = 1.00$ g/cm^3, $\gamma = 81$ dynes/cm, $\theta = 25.5°$) of radius 0.05 cm,

$$h = \frac{(2)(81)(0.9026)}{(1.00)(980)(0.05)} = 2.98 \text{ cm}$$

The strength of the intermolecular attraction between liquid molecules and those of the capillary wall enters into this calculation through the *contact angle* θ. This quantity has a definite, measurable value for each pair of substances. For example, for water in a glass capillary, $\theta = 25.5°$. If the force of attraction between molecules of liquid and capillary is the same as between the molecules of liquid themselves, then $\theta = 90°$ and $\cos \theta = 0$, and there will be no capillary effect at all. If the forces of attraction between liquid and wall molecules are stronger than the forces of attraction between liquid molecules, then $\theta < 90°$ and $\cos \theta$ is positive, resulting in a capillary rise. This is the situation encountered with water in a glass tube. If the forces of attraction between individual liquid molecules are stronger than those existing between fluid molecules and capillary molecules, the contact angle will be greater than 90° and $\cos \theta$ will be negative, leading to a capillary *depression*. This effect is observed with mercury in glass capillaries.

SUMMARY

Fluids such as liquids and gases are substances that flow under the action of applied forces. Gases are highly compressible, while liquids are practically incompressible.

The pressure exerted by a liquid or on a liquid is the force per unit area it exerts or experiences. The

365

density of a liquid is its mass per unit volume. Pascal's principle states that

pressure applied to a fluid is transmitted undiminished to all parts of the fluid and to the walls of its container.

For incompressible fluids, the variation of pressure with depth is given by

$$P = P_0 + \rho g y$$

where P_0 is the surface pressure and y is the depth.

When a body is partially or totally immersed in a fluid, the fluid exerts an upward force upon it which is referred to as buoyant force. Archimedes' principle states that the net buoyant force on a body is equal to the weight of the fluid displaced by the object.

The dynamics of steady-state, nonviscous, irrotational fluid flow are described by Bernoulli's equation, which states that, at any two points within a tube of flow,

$$P_1 + \rho g y_1 + \tfrac{1}{2}\rho v_1^2 = P_2 + \rho g y_2 + \tfrac{1}{2}\rho v_2^2$$

Surface tension and capillary attraction are effects that arise from the action of intermolecular attractive forces. Near the surface of a liquid, these forces tend to pull fluid molecules that are displaced from equilibrium toward the outside of the liquid back into the fluid; this gives the surface of the liquid the character of a self-healing stressed elastic membrane. The surface tension parameter γ is the force necessary to rupture a unit length of fluid surface. Surface tension is responsible for the spherical form of liquid droplets and for the forces that allow a pin or a razor blade to float on the surface of a fluid despite the fact that the weight exceeds the Archimedean buoyant force.

Capillary attraction is responsible for the rise of a fluid within a fine capillary with respect to the fluid level outside. It arises because the intermolecular attractive forces between fluid molecules and the molecules of the capillary tubing are stronger than those between the fluid molecules themselves.

QUESTIONS

1. Explain why a helium-filled balloon soars toward the sky while an air-filled one drops to the ground.
2. Can you think of any way of using Archimedes' principle to determine the weight of your head without removing it?
3. An ice cube is placed in a glass of water. The fluid now completely fills the glass. Describe what happens when the ice cube melts.
4. Pressure is expressed fundamentally in units of force per unit area. What is meant, then, when we describe pressures in units such as "millimeters of mercury"?
5. It is always easier to float in the ocean than in a swimming pool. Can you explain why?
6. Is it true that when an object floats, the fraction submerged depends on the ratio of densities of the fluid and the object?
7. Under what conditions does the pressure in a fluid increase linearly with depth?
8. A body sinking to the bottom of the ocean after a time achieves a certain constant terminal velocity. What forces act on it while it sinks?
9. What methods can be used to alter the depth at which a submarine cruises?
10. A plastic bag is weighed in air when empty and also when filled with air at atmospheric pressure. Are the two weights different? Explain. The two weighings are now repeated under vacuum. Are the weights equal to each other in this case?
11. When a truck passes your car on the highway, your car experiences a force. Explain this on the basis of Bernoulli's equation.
12. A fluid flows through a funnel the upper part of which is conical in shape. How does the fluid speed vary with distance from the upper surface of the fluid?
13. In steady streamline flow at any given location, the velocity vector **v** does not change with time. Does that mean that elements of fluid are unaccelerated?
14. A solid cube suspended from a spring balance under vacuum has a weight W. When it is completely submerged in liquid of density ρ, its weight is $W/2$. Find the volume of the cube.
15. The purity of gold can be determined by weighing it in air and also in water. Describe how this procedure works.
16. Why is it that a towel dries you off after a shower much more efficiently than a sheet of waxed paper having the same dimensions?

PROBLEMS

1. Show that (a) $1 \text{ g/cm}^3 = 10^3 \text{ kg/m}^3$, (b) $1 \text{ kg/m}^3 = 1.939 \times 10^{-3} \text{ slug/ft}^3$, (c) $1 \text{ dyne/cm}^2 = 0.1 \text{ N/m}^2$, (d) $1 \text{ lb/in.}^2 = 68,900 \text{ dynes/cm}^2$.
2. In Fig. 11.4, vessels (a) and (b) have the same base area, as do vessels (c) and (d). The pressure at the bottom of each vessel is the same, since they are all filled with the same fluid to the same level. This means that the force exerted by the liquid on the base of (a) and (b) is the same, and, therefore, each vessel should exert the same downward force on the surface supporting it. The same remarks apply to vessels (c) and (d). But this is saying that (a) and (b) should have the same weight, as should (c) and (d). However, isn't it clear from the drawing that (b) should weigh much more than (a), and (c) should weigh much more than (d)? What is the resolution of this apparent hydrostatic paradox?
3. All the air is evacuated from a spherical glass bulb of radius 0.12 m. (a) Taking normal atmospheric pressure to be $1.013 \times 10^6 \text{ dyne/cm}^2$ or $1.013 \times 10^5 \text{ newtons/meter}^2$, find the total force exerted by the atmosphere on its sur-

face. **(b)** What is the total force on its inner surface? **(c)** What is the net force on the bulb? Air is now permitted to reenter the bulb, until the interior is at normal atmospheric pressure. **(d), (e), (f)** What are the answers to (a), (b), and (c) under these circumstances?

4. A cylindrical aerosol spray container is filled with a gas the absolute pressure of which is 3.50 atmospheres. The can is 10.0 cm in diameter and 25.0 cm long. **(a)** What is the magnitude and direction of the force the gas exerts on an area of 1.00 cm^2 on the can's interior surface? **(b)** What is the magnitude and direction of the force exerted by the atmosphere on the other side of this surface? **(c)** What is the magnitude and direction of the resultant force on this area? **(d)** What is the total force exerted by the gas on the interior surface of the can? **(e)** What is the total force exerted by the atmosphere on the outside of the can? **(f)** What is the total net force acting on the surface of the can?

5. A cylindrical tank 2.5 meters in diameter and 1.5 meters deep is full of water (density 1.000 g/cm^3 or 1000 kg/m^3). **(a)** What is the pressure on the top surface of the water, under normal atmospheric conditions at sea level? **(b)** What is the additional pressure exerted by the water on the bottom of the tank? **(c)** What is the total, or absolute, pressure on the inner surface of the tank bottom? **(d)** What is the pressure exerted by the atmosphere on the outer surface of the tank bottom? **(e)** What is the net force on the bottom of the tank? **(f)** What is the additional pressure exerted by the water on the side of the tank at the point where it intersects the bottom?

6. **(a)** For the hydraulic lift illustrated in the diagram, assuming that the fluid is incompressible, show, using Pascal's principle, that the force F_1 required on the small piston to exert a force F_2 at the large piston is related to the areas A_1 and A_2 by the equation $F_1 = F_2(A_1/A_2)$. **(b)** Show that when the small piston is displaced downward a distance d_1, the large piston moves upward a distance $d_2 = (A_1/A_2)d_1$. **(c)** Show that the work F_1d_1 done by the small piston and the work F_2d_2 done by the large piston are equal.

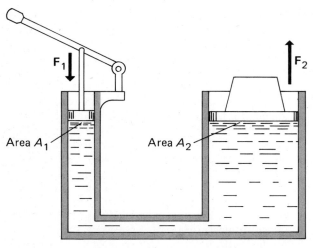

7. An automobile weighing 3200 pounds is driven onto a service station lift, the piston of which is 10.0 inches in diameter. The lift and piston weigh 1600 pounds. **(a)** What

pressure (in excess of atmospheric) must be exerted by the hydraulic fluid on the piston of the lift in order to raise the car at constant speed? **(b)** What pressure would be required to give the car an upward acceleration of 2.4 ft/sec^2?

8. The hull of a steamship extends to a depth of 27 ft below the surface of the ocean. The vessel's sides are perpendicular to the water and it has a flat bottom whose area is 34,000 ft^2. Find **(a)** the gauge pressure of the water on the bottom of the ship, **(b)** the absolute pressure of the water on the bottom of the ship, **(c)** the total weight of the vessel and its cargo, and **(d)** the amount of cargo that could be accommodated by increasing the allowed draft to 28 ft. One cubic foot of sea water weighs 64.3 lb.

9. A rectangular tank whose base is 6 ft^2 and whose height is 15 ft is completely filled with water, which weighs 62.4 lb/ft^3. Find **(a)** the gauge pressure at the bottom of the tank, **(b)** the net force on the bottom of the tank, and **(c)** the net force on one of the tank sides. **(d), (e),** and **(f)** What are the answers to these questions if the tank is half full?

10. The maximum depth of the oceans is about 35,000 ft. Assuming that sea water is incompressible and weighs 64.2 lb/ft^3, what would the gauge pressure be at this depth? Do you feel that the assumption of incompressibility is a very good one in this case?

11. The velocity of sound in sea water weighing 64.2 lb/ft^3 is 4760 ft/sec. What is the bulk modulus of sea water?

*12. Suppose that a compressible liquid substance of density ρ_0 at atmospheric pressure has a bulk modulus B which is constant with respect to pressure or density. What is the law relating pressure P with depth y in such a substance?

*13. Using the bulk modulus of sea water obtained in Problem 6, and regarding sea water as a substance of constant bulk modulus such as is described in Problem 12, find the gauge pressure at a depth of 35,000 ft. Can you now say more authoritatively whether the assumption of incompressibility used in Problem 10 is justified?

14. A U-tube contains water whose surface stands initially at a height of 15 cm from the bottom of the tube in each arm. An immiscible liquid is poured into one arm of the tube until a layer of the liquid 15 cm deep has been formed on top of the water. How high above the bottom of the tube is the water surface in each arm now?

15. A U-tube of uniform cross-sectional area 1.50 cm^2 contains initially 50.0 cm^3 of mercury (density 13.6 g/cm^3). An equal volume of an unknown liquid is added to one arm of the tube, and it is observed that the difference in height of the mercury surface in the two arms is now 2.75 cm. What is the density of the unknown liquid?

16. Gallium is a metallic element which is liquid above about 30°C under atmospheric pressure. Its density in the liquid state is 6.096 g/cm^3. What would be the height of the liquid column in a gallium barometer under normal sea level atmospheric pressure?

17. Gallium and indium form a system of alloys which are liquid at room temperature. A barometer filled with an alloy of this sort exhibits a column height of 157.0 cm under normal sea level atmospheric pressure. What is the density of the alloy?

18. The introduction of a small droplet of water into the space above the mercury in a mercury barometer results in a lowering of the column height by 1.80 cm. What is the pressure of the water vapor in the enclosed space above the mercury?

19. A glass cube of side length 2.0 cm and density 2.4 g/cm^3 is immersed in water (density 1.000 g/cm^3). Its bottom face is parallel to the water surface and is 10.0 cm below it. Find (a) the total force on the bottom of the cube (magnitude and direction) due to the gauge pressure of the fluid, (b) the total force on the top of the cube, (c) the total force on each of the other four sides, and (d) the resultant of all these forces. (e) Does this result give the net buoyant force predicted by Archimedes' principle?

20. Neglecting all effects due to surface tension, how much work would have to be done to lift the glass cube of Problem 19 from the original position described there until its entire volume just clears the surface of the water?

21. A cylindrical water tank 6 ft in diameter and 10 ft high is made, like a barrel, of vertical wooden staves held together with steel hoops which encircle them. The tank is entirely full of water ($\rho g = 62.4$ lb/ft^3). Find (a) the net force exerted by the water on the bottom of the tank and (b) the sum of the tensions in the steel hoops which run around the outside of the tank.

22. A piece of ore, having the appearance of gold, has a mass of 250.0 g. When it is "weighed in water" using the arrangement of Fig. 11.16, balance is obtained when 200.5 g is added to the right-hand pan of the balance. What is the density of the ore sample? Is it likely to be gold?

23. How much would the ore sample of Problem 22 (mass still 250 g) have "weighed" immersed in water had it been pure gold (density 19.3 g/cm^3).

24. The German battle cruiser *Seydlitz* was a ship whose normal displacement was 25,000 tons (sea water $\rho g = 64.2$ lb/ft^3). Her main deck had a minimum height of 14.0 ft above her normal water line, and her hull area above the normal water line was 41,360 ft^2. In June 1916, during the battle of Jutland, the ship was so severely damaged that she shipped 5300 tons of water; but she managed, nevertheless, to return to port safely. How much more water could have been taken on (assuming she could remain on an even keel) before her main deck was awash, in which event she most certainly would have sunk?

25. Prove that for an object floating on the surface of a liquid, the line of action of the buoyant force is through the center of mass of the volume of displaced fluid.

26. The density of ice is 0.92 g/cm^3. What fraction of the volume of an iceberg is submerged (a) when it is floating in fresh water (density 1.00 g/cm^3) and (b) when it is floating in sea water (density 1.03 g/cm^3)?

*27. A spherical balloon of mass m_b and radius r is attached to a very long string of mass γ per unit length. It is filled with a gas of density ρ_g and surrounded by an atmosphere of density ρ_0. The balloon is released, the string being laid out in a pile on the ground. It rises until it reaches equilibrium, the weight of the string (one end of which is still lying on the ground, of course) balancing the net buoyant force. Show that the height y_0 to which the bottom of the

balloon rises under these circumstances is given by

$$y_0 = [4\pi r^3(\rho_0 - \rho_g) - 3m_b]/3\gamma$$

What is y_0 for a 2.00-g balloon of 15 cm radius filled with hydrogen ($\rho_g = 0.90 \times 10^{-4}$ g/cm^3) surrounded by air ($\rho_0 = 1.293 \times 10^{-3}$ g/cm^3) and attached to a string whose mass is 0.0100 g per centimeter of length? Assume that the density of the atmosphere is constant with height.

*28. The string of the balloon of Problem 27 is pulled downward, displacing the floating balloon from its equilibrium position, and suddenly released. Neglecting all frictional or viscous resistance, show that the balloon executes vertical simple harmonic motion about the equilibrium position whose angular frequency is given by $\omega = [3\gamma g/(4\pi r^3\rho_0 + m_b)]^{1/2}$. What is the period of the motion for the balloon described in Problem 27? *Hint:* Assume that the displacement from the equilibrium position is always *small* compared to y_0 itself.

*29. A hydrometer is shown schematically in the figure. Its total mass is m, and it consists of a weighted bulb of volume V_B to which is attached a narrow tubular stem of cross-sectional area A. Show that when this device is allowed to float in a liquid of density ρ_f, it reaches equilibrium when the distance y_0 from the top of the bulb to the liquid surface is given by $y_0 = (m - \rho_f V_B)/\rho_f A$.

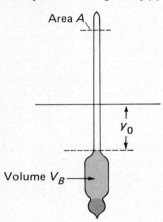

Area A

y_0

Volume V_B

*30. Show that when the hydrometer of Problem 29 is displaced slightly up or down from its equilibrium position and then released, it executes simple harmonic motion whose angular frequency is $\omega = (\rho_f gA/m)^{1/2}$. What is the period of oscillation for a hydrometer of stem area 0.4 cm^2 and mass 100 g in water (density 1.00 g/cm^3)?

31. A hollow concentric sphere of outer radius r is made of aluminum (density 2.70 g/cm^3). It is fashioned in such a way that it neither floats nor sinks when placed in water (density 1.00 g/cm^3). What is its inner radius?

32. How much pressure at the water mains is required so that a fire hose at street level can be used to fight a fire on the roof of a building 120 ft high? With what velocity does the stream leave the nozzle under these conditions? *Hint:* Assume that the water main is so large that the velocity of flow inside may be neglected.

33. A cylindrical tank 2.5 m in diameter contains three layers of liquids. The bottom layer, 1.5 m deep, is ethyl bromide whose density is 1470 kg/m^3. On top of this lies a layer of water 0.9 m thick and, finally, floating on the water layer,

is a layer of benzene whose density is 880 kg/m³ and which is 2.0 m thick. Find the gauge pressure at the bottom of the tank and the total force exerted by the liquid on the tank bottom.

34. An airtight cylindrical tank 5 ft in diameter and 30 ft high contains water to a height of 20 ft above the tank bottom. The space above the water contains air, compressed to a gauge pressure of 25 lb/in.² A plug is removed from an orifice in the side of the tank at the bottom, as in Fig. 11.25, allowing a stream of water of cross-sectional area 1 in.² to emerge horizontally. (a) Find the velocity of this stream as it emerges from the orifice. (b) A nozzle is connected to the orifice directing the stream vertically upward; find the maximum height it attains.

35. A 3.00-in.-diameter horizontal water main is connected to the bottom of the tank of Problem 34. The gauge pressure in this main is found to be 28 lb/in.² (a) What is the flow velocity in the water main? (You may neglect the velocity within the tank.) (b) After a certain horizontal distance, the diameter of the water main narrows to 2.00 in.; what is the flow velocity in this part of the main? (c) What is the gauge pressure there?

36. Water is allowed to flow unhindered from the end of the 2.00-in.-diameter main in Problem 35. Find the flow velocity and gauge pressure in the 3.00-in. section preceding it.

*37. A pyramidal tank 15 ft deep, constructed as shown in the diagram, is initially full of water. An orifice in the bottom of area 5 in.² is opened and water is allowed to drain through it. How long will it take to completely empty the tank? *Hint:* Can you use the approximations made in Example 11.5.3 in this problem?

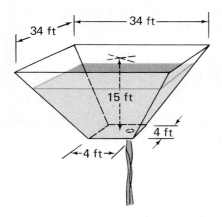

38. A semitrailer is rectangular in shape and is 40 ft long, 10 ft wide, and 15 ft high. It is driven through a tunnel of rectangular cross section at a speed of 60 mph (88 ft/sec). The tunnel wall is 3 ft from the side of the trailer. Assuming that the airstream flows symmetrically about the trailer, what sideward force is experienced by the trailer as it goes through the tunnel?

39. A razor blade of rectangular shape is 3.00 cm long and 2.00 cm wide and has a mass of 1.750 g. It is observed that this blade can, with care, be floated upon the surface of a dish of water and can, indeed, be loaded with additional mass until the plane of the blade is 0.180 cm below the

plane of the surrounding fluid surface before sinking, as illustrated in the figure at (a). When the blade is just on the point of sinking, the configuration of the fluid surface near the edge of the blade is as shown at (b). How much additional mass has been loaded onto the blade to bring it to the point of sinking? The surface tension of water is 81.0 dynes/cm.

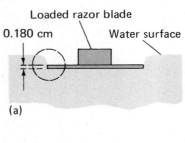

(a)

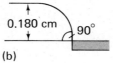

(b)

40. The battleship U.S.S. New Jersey weighs 45,000 tons and is 887 ft long and 108 ft wide. The perimeter of its hull at the water line is about 1820 ft. What would be the maximum force the surface tension of water could contribute to supporting the vessel?

41. A mercury barometer constructed of capillary tubing of inside diameter 1.00 mm reads 726.0 torr. The density of mercury is 13.6 g/cm³, its surface tension is 465 dyn/cm, and the contact angle between mercury and glass is 140°. What is the correct value of atmospheric pressure?

42. The density of pure ethyl alcohol is 0.8062 g/cm³, that of pure water 1.0000 g/cm³, and that of ice 0.9175 g/cm³, all at 0°C. Assuming no interaction between alcohol and water, show that an ice cube should sink in a drink stronger than 37.4 weight percent alcohol (about 76 proof) and float in any drink which is weaker.

43. Any serious drinker knows that the result of Problem 42 is incorrect, since he will be acquainted with the fact that an ice cube dropped into 100 proof whiskey (about 50 percent ethyl alcohol by weight) will float! Can you suggest why the calculation of the previous problem is in error?

44. (a) Show that the pressure within a spherical soap bubble of radius r, blown from a soap solution whose surface tension is γ, exceeds atmospheric pressure by an amount given by $P - P_0 = 4\gamma/r$. (b) How much would this excess pressure amount to for a bubble of 5.0 cm radius blown from a solution of surface tension 25 dyne/cm? (Hint: Consider the equilibrium of a single hemisphere of bubble).

45. (a) How much work must be done (in excess of that required to overcome normal atmospheric pressure) to blow a bubble like the one discussed in the preceding problem from an initial radius r_0 to a final radius r? (b) What would this amount to for a bubble of initial radius zero and final radius 5.0 cm, blown from a soap solution of surface tension 25 dyne/cm?

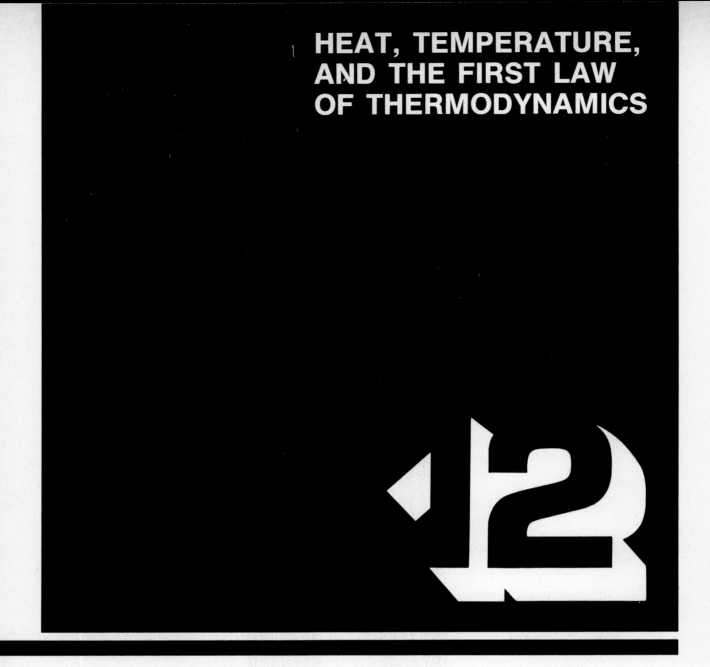

HEAT, TEMPERATURE, AND THE FIRST LAW OF THERMODYNAMICS

12.1 Introduction

In 1854, a great iron ship was laid down at Millwall, in England. She was 680 feet long, with an 83-foot beam and a 48.2-foot depth of hull, and her loaded displacement was 32,160 tons. Her name was *Great Eastern*. She was powered by steam engines working at a maximum steam pressure of 25 lb/in.^{2} which developed about 10,000 horsepower. These engines drove the ship by means of giant paddle wheels and a screw propeller. She was the mechanical marvel of the era, and her builders hoped to realize substantial profits on their investment by employing her in the Atlantic passenger service. But, alas, though she did manage to lay the first transatlantic telegraph cable,

she was a total failure economically. In the period following her completion, therefore, and for a very long time afterward, ships of much more modest size were the rule, and vessels of the scope of the *Great Eastern* were not again constructed for over half a century.

In 1906, however, two huge vessels were launched. The first of these was a warship, the revolutionary *HMS Dreadnought*, 520 feet long, which was to carry ten 12-inch guns, more than double the number mounted by any of her contemporaries. She was built at Portsmouth and displaced nearly 20,000 tons when fully loaded. The second, a passenger vessel, was even larger. It had a displacement of 42,000 tons, the first ship to exceed the size of the

Great Eastern. She was 762 feet long, 88 feet broad, and 57.1 feet deep. She was built at Newcastle-on-Tyne and was christened *Mauretania.*

Both these vessels were extremely successful. The *Dreadnought* served as the prototype of dozens of other similar ships, culminating during World War II in the Japanese-built *Yamato* class of 68,000 tons, bearing 18-inch guns. The *Dreadnought* served throughout World War I, emerging without a scratch. She achieved the distinction of being the only battleship ever to sink a U-boat single handed—a feat which was accomplished, believe it or not, by ramming!

The *Mauretania* was even better. She remained in the Cunard Lines' Atlantic passenger fleet until 1935, when she was replaced by the *Queen Mary,* having returned to her owners many times their original investment. She held the transatlantic speed record for over 22 years, and during the last year of her service made her fastest crossing, averaging over 27 knots (31 mph).

The reasons for the failure of the *Great Eastern* and for the success of *Mauretania* and *Dreadnought* are complex, and probably no single factor can be said to have made all the difference. Certainly the design of the *Great Eastern* was in advance of her time, but such simple things as deep harbor channels and sufficiently long docks were not commonly available when she was built. But one of the most important reasons for her failure lay in the weaknesses of her power plant. Though most of the important physical principles underlying the design and efficient operation of steam engines had been discovered at her inception, the usual lag of a few decades between the enunciation of scientific principles and their embodiment in engineering technology resulted in her being equipped with low-power, wasteful, inefficient, and unreliable engines. She could count on maintaining a sea speed of only about 10 knots, not bad for the time, but not sufficient to allow her to make enough voyages annually to recoup her operating costs. These expenses were truly enormous, for she consumed a veritable mountain of coal every day at cruising speed. Her bunkers accommodated no less than twelve thousand tons of coal, and not much remained after a transatlantic voyage.

The *Mauretania* and the *Dreadnought* were constructed at a time when heat and thermodynamics had been developed into a complete and well-understood branch of scientific endeavor whose principles were appreciated by engineers everywhere and incorporated into the design of efficient and reliable steam plants. Both ships were equipped with the powerful and economical quadruple screw steam turbine machinery which had been developed a few years earlier by Sir Charles Parsons (1854–1931). These were the first large-scale marine installations of this

(a)

(b)

(c)

FIGURE 12.1. (a) The 32,000-ton steamship *Great Eastern,* designed by Isambard K. Brunel and laid down in 1854. (b) The 18,000-ton battleship *HMS Dreadnought,* propelled by Parsons turbines, mounting ten 12-inch guns, was the brain child of Admiral Sir John Fisher and served as the prototype for dozens of similar vessels. (c) The 42,000-ton passenger steamer *RMS Mauretania,* launched in 1906, remained in service until 1935. She could exceed 27 knots after her original coal-fired boilers were replaced by oil-burning units. The sinking of her sister ship *Lusitania,* by a German submarine in 1915, was one of the factors that involved the United States in World War I.

371

(a)   (b)

FIGURE 12.2. (a) Sir Charles Parsons, developer of the marine steam turbine. (b) Dr. Rudolf Diesel, inventor of the diesel engine.

type of machinery and the precursors of innumerable marine power plants of this type. They also had high-pressure boilers, the *Mauretania* generating steam at 195 lb/in.² and the *Dreadnought*, at 250 lb/in.² The *Dreadnought* could make about 22 knots using the full power of her 23,000-horsepower machinery, while the larger *Mauretania*, which developed 70,000 horse-power, made 26 knots on her trials and later, after conversion to oil fuel, had no trouble averaging 27 knots. Both ships were far more economical than the *Great Eastern* and also far more reliable and trouble free.

The science of heat and thermodynamics made the success of these fine vessels possible, and its continuing application to technology has resulted in even larger and faster ships and, in more recent years, in gas turbines for powering conventional jet airplanes and supersonic aircraft.

During the years when the power plants of the *Mauretania* and the *Dreadnought* were being perfected, the German engineer, Dr. Rudolf Diesel (1858–1913) was developing the internal combustion engine which bears his name and which was later to power the underwater enemies of the big turbine-powered ships. Again, the success of his work, and of the other engineers who developed the internal combustion engines which now power everything from lawn mowers to freighters, was rendered possible only through the understanding and applications of the principles of heat and thermodynamics.

We have now finished our treatment of the mechanics of macroscopic bodies; it is time for us to turn our attention to the dynamics of molecular motion, and to the science of heat and thermal physics which is closely associated with that subject.

12.2 Heat as a Form of Energy: Temperature, Internal Energy, and the First Law of Thermodynamics

Our familiarity with heat energy in everyday life stems from the fact that our bodies can sense the temperature of objects in contact with them—we can "feel" that a body is hot or cold. But this is a very crude perception and does not permit us to determine with accuracy exactly how hot or cold a given object may be, or exactly how much heat may be absorbed by or extracted from a system by a given thermal process. It is, therefore, necessary to develop more objective and quantitative ways of defining and measuring temperature and thermal energy.

First, we must develop a clear understanding of what the terms *heat*, *temperature*, and *internal energy* mean. This is not as easy as it might appear. We can sense temperature and construct thermometers easily enough, but we have no direct sensation of heat flow or internal energy. We now know that heat is a form of energy which may be transferred from the molecules of one body to those of another when there is a temperature difference between the two, but this was not clearly understood until well after the beginning of the nineteenth century. Before that time, heat was generally thought to be a "subtle caloric fluid" which could somehow be accommodated within material bodies, whose presence rendered them hot, and whose absence made them cold.

Count Rumford (1753–1814) studied the frictional processes involved in boring out the barrels of artillery pieces and observed that much more work was needed to bore out a given cannon barrel when dull boring tools were used than was required when sharp ones were employed. In addition, much more heat was produced (as measured by the amount of cooling water boiled away) when dull tools were used than when sharp ones were substituted. Though these results could be explained (in a somewhat contrived way) on the basis of the caloric fluid theory, Rumford suggested that they could be much more simply understood on the hypothesis that heat is a form of energy into which mechanical work might be converted. He inferred also that the increased temperature resulting as a consequence of the conversion of mechanical work into heat might arise simply as a result of this external work having excited more energetic random agitation of the atoms of the metal in the cannon barrel. The equivalence of heat and work was also studied by Julius Mayer (1814–1878), and, above all, by James Prescott Joule (1818–1889), who established by a long series of precise experiments that the loss of mechanical energy in frictional or dissipative processes could always be accounted for by the appearance of a precisely equivalent amount of thermal energy.

In order to fully appreciate these findings, it is necessary to understand the differences between *heat*, *temperature*, and *internal energy*. By *heat* we mean the energy transferred to or from a system as a result of a difference in *temperature* between the system and its surroundings. The *internal energy* of a system refers to the random kinetic energy of translation, rotation, or vibration its atoms or molecules may possess and, in addition, to the potential energy of interaction between its atoms or molecules. It does *not* include any *external* kinetic energy a body as a whole may have due to translational motion of its center of mass or due to rotational motion of the object as a whole

about the center of mass, nor does it include any external potential energy a body may have due to the action of external forces. Generally speaking, the internal energy of a system refers to the random energy of motion associated with atomic translational or vibrational motions—the "molecular agitations" of the system, so to speak—and to the potential energies of interaction of the atoms or molecules with one another. The internal energy is closely related to the temperature, and a precise relationship between temperature and internal energy can be established for any system. But the relationship is not a universal one and depends, instead, upon the internal structure of the system at hand. It is not the same, for example, for an ideal gas, a dense vapor, and a solid crystalline substance.

When heat is allowed to flow into a system as a result of a temperature difference between the system and its surroundings, an equivalent increase in internal energy takes place provided—and this is an important qualification—that the system is allowed to perform *no mechanical work* on its surroundings. In general, of course, this will not be the case, and then we shall find that:

The increase in internal energy of the system plus the amount of external work done by the system is equivalent to the heat energy that is absorbed.

This observation constitutes *the first law of thermodynamics* and can be stated in the form of an equation by

$$\Delta Q = \Delta U + \Delta W \qquad (12.2.1)$$

where ΔQ represents the heat energy absorbed by the system, ΔU is the change in the system's internal energy, and ΔW is the work done by the system. Naturally, in a process where heat is extracted from the system, ΔQ is negative, as is ΔU in a situation where the internal energy decreases or ΔW in an instance where work is done on the system rather than by it.

The first law of thermodynamics as stated above is simply a statement of conservation of energy, in which it is postulated that an equivalence between heat energy and mechanical energy can be unambiguously arrived at. We have not yet discussed how this equivalence is to be established, nor have we even defined the concept of temperature. So, in a sense, our introduction of the first law of thermodynamics may seem to be a bit premature. Nevertheless, its importance is so great and its content so fundamental to the understanding of all thermal processes that it is best to have it in mind during all our subsequent discussions.

From the way in which internal energy is defined, it makes sense to discuss the *total internal energy content* of a system (U) as well as changes in internal energy (ΔU) which take place as a result of thermal or

FIGURE 12.3. James Prescott Joule, who first demonstrated the quantitative equivalence between heat and work.

mechanical processes. But the same statement *cannot be made with regard to the heat energy* (ΔQ). Since the heat energy absorbed by a system may act partly to increase its internal energy (and thus raise its temperature) and partly to make the system perform external work, it cannot be said, as the caloric fluid theory would have us believe, that a system possesses some precisely defined and conserved "amount of heat" (Q). It is only the increments of heat energy absorbed by or extracted from the system (denoted by ΔQ) that can be accounted for on the basis of the first law of thermodynamics. The same can be said in regard to external work; there is no way of discussing the "total work" (W) associated with a thermal system, but only the increments of work (ΔW) associated with individual thermal or mechanical processes influencing the system. We shall undertake a much more complete and detailed discussion of the first law of thermodynamics in a later section. At this point, however, we must return to our consideration of the definition of temperature and the equivalence of heat and work.

The *temperature* of a body can be determined in a very rough way simply by feeling it to determine whether it is hot or cold. But this provides no quantitative measure of temperature nor even any very definite clue to the physical meaning of temperature. A rigorous definition of temperature is not easy to formulate, and we shall not attempt to do so until later when we have described the properties of ideal gases. At this point, we shall have to be content to understand a few basic facts having to do with temperature and to establish for the time being a working definition in terms of which temperature scales can be constructed and temperature measurements made.

First of all, it should be evident that a temperature difference establishes a flow of heat energy and defines the direction in which such a flow takes place. We all know that when a body which feels hot is placed in thermal contact with one which feels cold, after a time the hot one feels cooler and the cold one warmer, and finally a situation of equilibrium is reached wherein no difference between the two bodies is evident to our senses. We interpret these facts by stating first of all, that the temperature of a body is related to the *average internal kinetic energy per molecule*, and in addition *that a difference in temperature gives rise to a flow of heat energy from the hotter to the colder region.* For the system comprising a hot object and a cold object in thermal contact, heat flows from the hot object to the cold one. Considering for the moment the hot object alone, the first law of thermodynamics tells us that, since no work is done by it,

$$\Delta Q = \Delta U \qquad (12.2.2)$$

where ΔQ is the heat energy absorbed by the hot body and ΔU is the change in its internal energy. But heat energy leaves the hotter object, so ΔQ is negative, and so is ΔU. Since the internal energy of the hot object decreases, its temperature decreases also. Now suppose we view the cold object as our "system." It again does no external work, so that

$$\Delta Q' = \Delta U' \qquad (12.2.3)$$

where $\Delta Q'$ is the heat absorbed (now positive) and $\Delta U'$ is the change in internal energy. In this case, the change in internal energy is positive, which means that the temperature has increased. Finally, let us assume that the "system" consists of *both* objects in thermal contact. Then, since no heat flow from the surroundings takes place, for this system

$$\Delta Q_{\text{tot}} = \Delta Q + \Delta Q' = 0$$

or

$$\Delta Q = -\Delta Q' \qquad (12.2.4)$$

which says simply that the amount of heat energy flowing out of the hot object equals that flowing into the cold one. Similarly, from (12.2.1), (12.2.3), and (12.2.4),

$$\Delta U' = -\Delta U \qquad (12.2.5)$$

which states that the increase of internal energy for the initially colder body equals the decrease of internal energy for the initially hotter one.

In the final state of the combined system, there are no temperature differences and, therefore, no further flow of heat. This final state is said to be one of *thermal equilibrium.* In general, we may say that

a system is in a state of thermal equilibrium if there is no flow of heat or any other form of energy between the various parts of the system, or between the system and its surroundings.

Since a flow of heat ensues only in response to a temperature difference, this means that all parts of the system are at the same temperature and that the system is either thermally insulated from its surroundings or that the surroundings are at the same temperature as the system itself.

It is important to note that although temperature and internal energy are related to one another, and that an increase in temperature always results in an increase in internal energy, heat flows from one object to another *only* in response to a *temperature* difference, and the condition of equilibrium is defined by an equality of *temperatures* rather than internal energies. The relation between temperature and internal energy per molecule is not always a direct proportionality and, furthermore, it differs from substance to substance and from system to system. While objects in thermal equilibrium with one another always have the same temperature, two substances may be in thermal equilibrium and yet have a *different* average internal energy per molecule. Conversely, the mere fact that the average internal energy per molecule may differ from one substance to another does not of itself guarantee that heat will flow from one to the other when they are brought into thermal contact.

The condition of thermal equilibrium occupies a position of importance in the study of thermodynamics, for thermodynamics attempts to describe the properties of systems in thermal equilibrium and the energy differences connecting different thermal equilibrium states.[1]

We can obtain some idea of an object's temperature by direct sensation, but this will not do if we have to obtain quantitative measurement of its temperature. Fortunately, there are a number of easily observable physical effects that are directly proportional to temperature, in terms of which quantitative temperature measurements can be made. For example, most substances expand when heated and contract when cooled. It is possible, therefore, by confining a small volume of a liquid such as mercury in a glass capsule to which a thin capillary tube is attached, to construct a *thermometer*. Because mercury expands to a greater extent than glass when heated, a measure of the temperature of a body can be obtained simply by bringing the bulb of this thermometer into thermal contact with the body and noting the extent to which the column of mercury in the capillary rises or falls. For convenience, the thermometer may be calibrated by marking off a given number of equally spaced *degrees of temperature* along the capillary tube between the observed mercury levels at two easily defined and simply reproducible *reference temperatures* such as the freezing and boiling points of water at normal atmospheric pressure. In this way, the *Celsius* (or *centigrade*) temperature scale was defined by dividing the thermometer scale into 100 equally spaced degrees between these two reference points, the value zero degrees Celsius (0°C) being assigned to the freezing point and 100 degrees Celsius (100°C) to the boiling point. In a similar way, the *Fahrenheit* scale may be defined, by dividing the same temperature interval into 180 equally spaced degrees, in this case assigning to the freezing point the value 32 degrees Fahrenheit

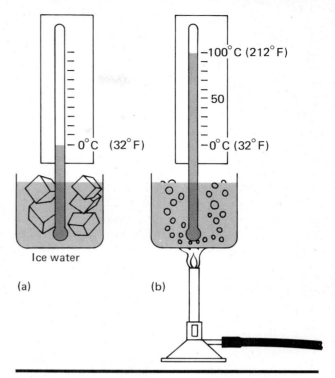

FIGURE 12.4. Calibration of temperature measuring instruments, using the melting and boiling points of water as standards of temperature.

(32°F), the boiling point then having the temperature 212°F. This calibration procedure is illustrated in Fig. 12.4. It is possible to use thermometric fluids other than mercury; alcohol, pentane, and other organic liquids are being used frequently in inexpensive household thermometers.

The relationship between the Celsius and Fahrenheit scales is easily arrived at from their respective definitions. Since the Celsius degree is 1.8 times as large as the Fahrenheit degree, and since the zero of the Celsius scale falls at 32°F, we may write

$$T_F = 1.8T_C + 32 \qquad (12.2.6)$$

whence

$$T_C = \tfrac{5}{9}(T_F - 32) \qquad (12.2.7)$$

where T_C is a given temperature expressed in Celsius degrees and T_F is its Fahrenheit counterpart.

12.3 The Equivalence of Mechanical Energy and Heat: Joule's Experiment

It is well known that the temperature and, therefore, the internal energy of an object may rise when frictional or dissipative processes act on it. Thus, as every Boy Scout knows, when two pieces of wood are

[1] In order to subject the statements made above to any kind of experimental test, one must make the rather reasonable assumption that two bodies, each of which is in thermal equilibrium with a given third body, are in thermal equilibrium with each other. This hypothesis, which we shall take for granted hereafter, is sometimes referred to as the *zeroth law of thermodynamics*. The necessity for such a hypothesis stems from the fact that one can compare the temperatures of two objects only by bringing a third object (a thermometer) into thermal equilibrium with each of them in succession.

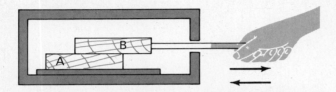

FIGURE 12.5. Conversion of mechanical work into internal energy by frictional processes.

vigorously rubbed together, they may become quite hot. Let us consider an example in which a block of wood is rubbed rapidly back and forth against an identical block, as shown in Fig. 12.5. After a time, both blocks are found to have undergone a noticeable increase in temperature. One may then ask whether any heat flowed into block A or block B. At first, this seems to be a ridiculously simple question, to be answered "yes" at once. After all, the temperature of both blocks has increased, so heat must be generated at the rubbing interface and must flow into blocks A and B on either side—how else would the temperature rise occur?

But, the matter is not quite so simple as this, and strictly speaking, the answer is that *there has been no flow of heat energy into either object!* The resolution of this apparent paradox lies in the fact that a flow of heat energy is *always* caused by a temperature difference between an object and its surroundings. Let us focus our attention on body A for a moment. The temperature at the surface of block A at the rubbing interface has increased, but unless the temperature of the corresponding surface of block B were even greater, no flow of heat, as such, into block A could ever be established. But if this were the case, how did the surface of body B get to be hotter than the surface of A? That could only have happened if, somehow, previously, heat had flowed into B from the friction interface, which would have meant that at that time, A had been warmer than B! This line of argument obviously adds up to sheer nonsense; we are driven to conclude that there can be no difference in temperature between the surfaces of the two blocks.

What has happened, in fact, is that the internal energy of both objects has increased, and, as a result, their temperatures have gone up. But the increase in internal energy arose not as a consequence of the introduction of heat energy at the friction interface but rather *as a result of the conversion of mechanical work performed in rubbing the blocks together into molecular internal energy* by frictional processes at the interface! The same temperature increases *could*, of course, have been produced by the introduction of heat energy into both bodies from an external source at a higher temperature, and in view of this, we could say that this amount of heat energy is *equivalent* to the mechanical work expended at the interface in the original rubbing process. But in the example at hand, there has been no introduction of heat energy at all into either object.

Looking at the situation from the point of view of the first law of thermodynamics, we might consider a system consisting of the two blocks thermally insulated from their surroundings so as to *exclude* the possibility of any heat entering or leaving. Then, since there can be no introduction of heat energy into the system, $\Delta Q = 0$, whence

$$\Delta Q = \Delta U_A + \Delta U_B - \Delta W_f = 0 \qquad (12.3.1)$$

where ΔU_A and ΔU_B represent the changes in internal energy experienced by A and B, respectively, and ΔW_f is the work done against friction in rubbing the objects together; the minus sign is needed because external work is done *on* the system rather than by it. Clearly, from (12.3.1),

$$\Delta U_A + \Delta U_B = \Delta W_f \qquad (12.3.2)$$

which expresses the fact that the total change in the internal energy of both bodies equals the mechanical work done on them. Any temperature changes which are observed come about as a result of these changes in internal energy. It is evident from all this that a flow of heat occurs only in response to a temperature difference but that a temperature change can be created without a heat flow.

We have mentioned that the thermal effects of mechanical work can be reproduced by the introduction of a certain amount of heat from an external source and that, in doing so, there is a definite relationship between the amount of heat needed and the amount of work involved. The first law of thermodynamics, indeed, postulates that such a relationship must exist. The experimental proof of this assertion rests upon the investigations of Rumford, Mayer, and Joule. It was Joule, in particular, who accurately established the relationship between mechanical energy and heat in a famous series of experiments performed between 1843 and 1849, whose general thrust is illustrated in Fig. 12.6.

In this drawing, an apparatus consisting of a tank containing a known mass of water is shown, along with a series of rotating paddles driven by descending masses, which may be used to stir the water vigorously. Work is done against internal viscous friction as the water is stirred, and this work is expended in exciting an increase in the internal energy of the water, exactly as was the work done against friction in rubbing the previously discussed wooden blocks together. This increase in the internal energy of the water, in turn, causes an increase in its temperature. The apparatus is thermally insulated from its surroundings so that little or no heat can flow in or out of the system. Joule found that when a total mass

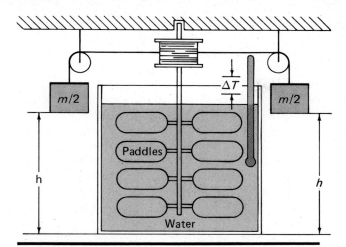

FIGURE 12.6. Joule's experiment, in which external work is converted into internal energy, causing a rise in the temperature of the liquid.

m descended through a known distance h, corresponding to a total loss of mechanical potential energy[2] $\Delta U_p^{(m)} = mgh$, the ensuing temperature rise ΔT could be described by the relation

$$\Delta U_p^{(m)} = mgh = Cm_w \, \Delta T \qquad (12.3.3)$$

$$\Delta T = \frac{\Delta U_p^{(m)}}{Cm_w} = \frac{mgh}{Cm_w} \qquad (12.3.4)$$

where m_w is the mass of water in the tank and C is a constant of proportionality. From this, it is evident that the observed temperature rise is strictly proportional to the loss of mechanical energy occasioned by the descent of the weights per unit mass of water. If we conclude that an equal amount of internal energy has appeared in place of the mechanical energy, we should expect a temperature rise proportional to the increase in internal energy per unit mass of water, which, in turn, would be equal to the loss of potential energy per unit mass of water. This is precisely what is observed according to (12.3.4). Furthermore, the same expenditure of external work per unit mass of water was always found to produce the same temperature increase, regardless of whether the external work was expended in stirring the water, rubbing together under water two bodies between which frictional forces acted, generating electrical power which was expended in an immersed electrical resistor, or in a number of other ways.

From the viewpoint of the first law of thermodynamics, Joule's experiment may be regarded, once

[2] The student should be careful not to confuse the symbols U and ΔU, which in the chapters on heat and thermodynamics are used to represent *internal energy*, with the symbols $U_p^{(m)}$ or $\Delta U_p^{(m)}$, representing *mechanical potential energy*. In these chapters, such symbols representing mechanical potential or kinetic energies will always bear the superscript (m) wherever there is the least chance of confusing them with an internal energy term.

again, as involving a system wherein *there is no introduction of heat energy as such*. Then, $\Delta Q = 0$, and the first law of thermodynamics tells us that

$$\Delta Q = \Delta U + \Delta W = \Delta U - mgh = 0 \qquad (12.3.5)$$

where ΔU is the change in the internal energy of the water. Solving for ΔU and dividing by the mass of the water m_w, this becomes

$$\frac{\Delta U}{m_w} = \frac{mgh}{m_w} \qquad (12.3.6)$$

and if we assume that the observed temperature rise should be proportional to the increase of internal energy per unit mass of water, we may write

$$\Delta T = \frac{1}{C} \frac{\Delta U}{m_w} = \frac{mgh}{Cm_w} \qquad (12.3.7)$$

where $1/C$ is a constant of proportionality. This is in agreement with (12.3.4) expressing Joule's experimental findings.

Joule's experiment may be regarded as establishing the equivalence of heat and mechanical work since, of course, it enables us to relate the amount of heat absorbed from an external heat source in producing a given rise in temperature to the amount of mechanical work required to produce the same effect. It may also be considered to have furnished one of the earliest and most important experimental tests of the first law of thermodynamics.

Quantitatively, Joule's experimental data showed that 772.5 ft-lb of mechanical energy is always expended in raising the temperature of 1 pound of water at 60°F by one degree Fahrenheit. Subsequent experimental refinements have established this numerical value to be 777.9 ft-lb. An equivalent statement, using metric units, is that 4186 joules of energy is expended in raising the temperature of 1 kg of water at 14.5°C by one degree Celsius. Though there is no reason why we should not simply use joules or foot-pounds as units of heat energy, it is common to express units of thermal energy in terms of the amount of energy required to raise the temperature of unit mass (or weight) of water by one degree Celsius (or Fahrenheit). The English thermal unit, defined in this way, represents the heat required to raise the temperature of 1 pound of water at 60°F by one degree Fahrenheit, and is referred to as the *British thermal unit* (Btu). The corresponding metric unit, the *kilocalorie* (kcal), represents the heat energy necessary to raise the temperature of 1 kilogram of water at 14.5°C by one degree Celsius. From the above numerical data,

1 kcal = 4186 J

while

1 Btu = 777.9 ft-lb

It is not difficult to establish from these data that

1 Btu = 0.2520 kcal

The *calorie*, equal to 1.000×10^{-3} kcal, is also frequently used; it is the amount of heat energy required to raise the temperature of 1 gram of water at 15.5°C by one degree Celsius. According to the above figures, 1 cal = 4.186 joules, and 1 Btu = 252.0 cal.

The constant C in Eqs. (12.3.6) and (12.3.7) represents the amount of energy required to raise the temperature of a unit mass (or weight) of substance (in this case water) by one degree of temperature. This quantity is called the *specific heat* of the substance. One might think offhand that this quantity might be the same for all materials, but it is not and, in fact, depends upon the composition and internal molecular arrangement of the substance in question, for reasons we shall discuss in detail later. It is also, in general, dependent upon the temperature, though frequently this temperature dependence is so small as to be negligible for most purposes. The specific heat of a substance is expressed in units of cal/g-°C, or in the British system, Btu/lb-°F. It is evident from the results of Joule's experiment that the specific heat of water is 1.000 cal/g-°C or 1.000 Btu/lb-°F at 14.5°C and 60°F, respectively. A compilation of specific heats of some common substances is given in the tables at the end of this chapter. The specific heat of a material is an important index of its internal molecular constitution and often furnishes valuable insights into the details of molecular arrangement and intermolecular forces. In this connection it is often valuable to speak of the *molar specific heat c*, defined as the quantity of energy required to raise the temperature of one mole of a substance by one degree.

In view of the result of Joule's experiment, we may relate the quantity of heat absorbed by a body, ΔQ, to the corresponding rise in temperature ΔT by the equation

$$\Delta Q = mC \, \Delta T \qquad (12.3.8)$$

where m is the mass and C the specific heat. In writing this equation, it is assumed that the temperature dependence of the specific heat is negligible over the temperature range ΔT. If this is not so, we must consider the specific heat as being a function of temperature, $C(T)$, and write, for an *infinitesimal* temperature increment dT, over which the variation of $C(T)$ is insignificant,

$$dQ = mC(T) \, dT \qquad (12.3.9)$$

This equation may now be integrated from initial temperature T_1 to final temperature T_2 to obtain

$$\Delta Q = \int dQ = m \int_{T_1}^{T_2} C(T) \, dT \qquad (12.3.10)$$

In actually evaluating the integral, of course, we must somehow know the specific functional dependence of C upon the temperature.

The above equations are written on the assumption that the quantity C is expressed in units of energy per unit *mass* per degree temperature. Since, in the British system, the Btu is a unit of energy per unit *weight* per degree temperature, specific heat values in that system are usually expressed in units of energy per unit *weight* per degree temperature. These equations, in the latter case, should all be written substituting weight for mass throughout, (12.3.8) being expressed using these units as $\Delta Q = wC \, \Delta T$, and so forth.

There is an additional important point to be made about the equivalence of heat and work, one which caused much confusion during the period when the science of heat and thermodynamics was in its infancy. The trouble is this: while it is always possible to convert a given amount of mechanical energy into a precisely equivalent amount of internal energy or heat, as was done in Joule's experiment, *the converse is not true*. In other words, when we try to convert a certain quantity of heat energy into its equivalent amount of mechanical work, we find that it is *impossible* to do so! It is easy enough to devise processes which convert *some* thermal energy into mechanical energy; such processes are in everyday use in steam engines, turbines, and internal combustion engines. But no such device *ever* succeeds in converting 1.000 kcal of heat into 4186 joules of mechanical work. The reason for this is not that heat and mechanical work are not equivalent in terms of energy, but rather that *no thermal process can be devised which is 100 percent efficient in turning heat into work!*

Any such process is by its very nature accompanied by an inevitable loss of thermal energy to the system's surroundings. The amount of mechanical work which is extracted may vary in each individual case and will depend upon the circumstances surrounding each process. This state of affairs is due to the fact that mechanical energy arises from an *ordered* motion of macroscopic bodies, while the internal energy of any system arises from the completely *random* motions of its molecules. It is accordingly easy enough to devise schemes (involving frictional processes and collisions) which dissipate ordered motion of large objects into random internal molecular motion, but it is never possible to reverse these schemes and create large-scale order from molecular chaos. One might ask, for instance, how you might go about *reversing Joule's experiment* by somehow making the random internal energy of the water molecules "unstir" the liquid in the tank, imparting all their energy as kinetic energy to the paddles, causing them to turn and raise the suspended masses, the temperature of the liquid itself being lowered in the process.

While not denying that some thermal energy can be extracted and turned into mechanical work in certain ways, our intuition may fairly easily guide us to the conclusion, correct in this case, that this is not one of them. The reason is that such a process would require the molecules of the liquid to go *of themselves* from a random state of high probability to an ordered state whose probability, on a purely statistical basis, is vanishingly small. It is as if we should require all the air molecules in the classroom to put themselves in the wastebasket. We can put them there by the expenditure of much mechanical work, but they will not congregate there themselves!

We shall examine this matter in a more thorough and systematic way a bit later; in the meantime, it is well to remember that much of the behavior of thermal systems can be understood in the light of the remarks made just above. These statements give an illustration of the content of the *second* law of thermodynamics, about which we shall have more to say in the next two chapters.

EXAMPLE 12.3.1

At a swim meet, there are six swimmers who swim 3000 ft in an average time of 11.0 minutes, their individual power output averaging 0.65 hp. The pool is 75 ft long, 50 ft wide, and has an average depth of 8 ft. Find (a) the swimmers' total work output in Btu and kcal and (b) the rise in water temperature in the pool, assuming that no heat is lost from the water. Assume that the specific heat of water is 1.000 Btu/lb-°F and neglect its temperature dependence.

The work output of each swimmer can be obtained directly from the given power output; thus, recalling that 1 hp = 550 ft-lb/sec,

$$\Delta W = Pt = (0.65)(550)(11.0)(60) = 236,000 \text{ ft-lb}$$

or, for six swimmers,

$$\Delta W = (236,000)(6) = 1,416,000 \text{ ft-lb}$$

This can be expressed in heat units by noting that the results of Joule's experiments show that 1 Btu = 779 ft-lb:

$$\Delta Q \text{ Btu} = \frac{\Delta W_{tot} \text{ ft-lb}}{778 \text{ ft-lb/Btu}} = \frac{1,416,000}{779} = 1817 \text{ Btu}$$

and since 1 Btu = 0.2520 kcal,

$$\Delta Q \text{ kcal} = \Delta Q \text{ Btu } 0.2520 \text{ kcal/Btu} = 458 \text{ kcal}$$

The rise in water temperature due to the stirring of the water by the swimmers is calculated from (12.3.8), which, since we shall use the British units of energy per unit weight per degree Fahrenheit, we write as

$$\Delta Q = wC \, \Delta T$$

where w is the weight of water in the pool. Therefore, since for water the specific heat C is 1.000 Btu/lb-°F,

$$\Delta T = \frac{\Delta Q}{wC} = \frac{\Delta Q}{\rho g V C} = \frac{1817}{(62.4)(75)(50)(8)(1.000)}$$

$$= 0.000971°F$$

EXAMPLE 12.3.2

4.0 Kilograms of water, initially at a temperature of 60°C, is poured into an aluminum kettle of mass 3.0 kg which is initially at a temperature of 10°C. Find the final temperature of water and kettle assuming that the specific heat of water is 1.00 cal/g-°C and that the specific heat of aluminum is 0.215 cal/g-°C in this range of temperature. Neglect any variation of specific heat with temperature. Find also the subsequent thermal energy input needed to bring the water to the boiling point in this container. Supposing that the energy were supplied by a 500-watt electric immersion heater, how long would it take to boil water in this container, assuming that no heat is lost to the surroundings?

When the water is added to the container, there is no conversion of heat into mechanical work or any other form of energy. Therefore, the amount of heat lost by the warmer water is equal to the amount gained by the colder container; in other words, the algebraic sum of these energies is zero, whence

$$\Delta Q_w + \Delta Q_{Al} = 0 = m_w C_w \, \Delta T_w + m_{Al} C_{Al} \, \Delta T_{Al}$$

(12.3.11)

where ΔQ_w represents the heat gained by a mass m_w of water of specific heat C_w whose temperature changes by an amount ΔT_w, the quantities m_{Al}, C_{Al}, and ΔT_{Al} being the corresponding values for the aluminum kettle. Since ΔT_w is negative, ΔQ_w will also be negative, which signifies only that heat is lost rather than gained by the water. We do not know yet what the final temperature will be; let us call it T_f. Then $\Delta T_w = T_f - 60$ and $\Delta T_{Al} = T_f - 10$, in units of degrees Celsius. Substituting these expressions into (12.3.11) along with the given values of the masses and specific heats, we obtain

$$0 = (4.0)(1.00)(T_f - 60) + (3.0)(0.215)(T_f - 10)$$

which gives $T_f = 53.1°C$

The additional heat input ΔQ necessary to raise water and container from the temperature T_f to the boiling point of water is given by

$$\Delta Q = m_{Al} C_{Al}(100 - 53.1) + m_w C_w(100 - 53.1)$$
$$= (3.0)(0.215)(46.9) + (4.0)(1.00)(46.9) = 218 \text{ kcal}$$

Note that in these calculations the specific heats in kcal/kg-°C are numerically the same as the values given in units of cal/g-°C; this is, of course, due to the fact that the ratios kcal/cal and kg/g are the same.

If the heat input is supplied by an electric immersion heater which consumes P watts of electrical power and turns it all into heat, then after t seconds, the amount of heat ΔQ produced is

$$\Delta Q = Pt \tag{12.3.12}$$

In this equation, however, if P is in watts (and remember that a watt is a joule per second), ΔQ must be expressed in joules. If we utilize the result that 1 kcal = 4186 joules, then 218 kcal of heat may be expressed as

$$\Delta Q = (218)(4186) = 913,000 \text{ J, or W-sec}$$

Substituting this value into (12.3.7) along with $P =$ 500 watts, we find that the time t required for a 500-watt electric heater to produce 913,000 joules, or 218 kcal, of heat is

$$\Delta Q = 913,000 = (500)t$$

$$t = \frac{913,000}{500} = 1825 \text{ sec} = 30.4 \text{ min}$$

EXAMPLE 12.3.3

An automobile weighing 3200 pounds, with driver, driven over level terrain at a constant speed of 60 mph (88 ft/sec) is found to consume 1 gallon of gasoline, weighing 7.00 lb, in 25.5 miles. The heat of combustion of gasoline is 21,400 Btu/lb. The same automobile, once more on level terrain, is found to coast from an initial speed of 65 mph to a final speed of 55 mph in 12.0 seconds when the transmission is in "neutral." What percent of the thermal energy obtained by burning the gasoline does the engine convert to mechanical energy when the car is driven over level ground at a constant speed of 60 mph?

In order to work this example, we must know how much heat energy is available from one gallon of gasoline, and also how much mechanical work the engine must do against forces of friction to keep the car moving at a constant speed of 60 mph for 25.5 miles. Let us call the former quantity ΔQ_c, the latter, W_f. The force of friction F_f can be calculated from the known change of velocity Δv in given time Δt during the coasting trial, since from Newton's second law

$$\sum F_x = F_f = ma = m \frac{dv}{dt} \tag{12.3.13}$$

$$m \, dv = F_f \, dt \tag{12.3.14}$$

Equation (12.3.14) may be integrated from time t_1, corresponding to velocity v_1, to time t_2, when the velocity is v_2, giving

$$m \int_{v_1}^{v_2} dv = F_f \int_{t_1}^{t_2} dt \tag{12.3.15}$$

$$m(v_2 - v_1) = F_f(t_2 - t_1)$$

or

$$m \, \Delta v = F_f \, \Delta t \tag{12.3.16}$$

where we have written Δv for $v_2 - v_1$ and Δt for

$t_2 - t_1$. In performing this integration we have assumed that the net frictional force F_f is constant over the velocity range from v_1 to v_2; this is not strictly true because the forces which arise from air resistance and which contribute significantly to F_f are larger at high speeds than at lower ones; but if the velocity interval $v_2 - v_1$ is not very large, the variation of F_f over this range will be quite small and the error incurred by assuming F_f to be constant will be negligible. In our example, of course, $\Delta v = 65$ mph $-$ 55 mph $= 10$ mph $= 14.7$ ft/sec, and $\Delta t = 12.0$ sec.

From (12.3.16), then, we may express F_f in the form

$$F_f = \frac{m \, \Delta v}{\Delta t} \tag{12.3.17}$$

The mechanical work done against friction as the vehicle travels a distance r during which 1 gallon of gasoline is consumed is then

$$W_f = F_f r = \frac{mr \, \Delta v}{\Delta t} \tag{12.3.18}$$

This is the total mechanical work output of the engine required to keep the car moving at constant speed against the net forces of friction during the consumption of 1 gallon of gasoline. The ratio of this quantity to the total heat of combustion of the gasoline, ΔQ_c (which must, of course, be expressed in ft-lb per gallon rather than Btu/lb) will give the fraction of heat energy available from the gasoline that ultimately is converted into mechanical energy. Let us call this ratio the *thermal efficiency ratio* η; then,

$$\eta = \frac{W_f}{\Delta Q_c} = \frac{mr \, \Delta v}{\Delta Q_c \, \Delta t} \tag{12.3.19}$$

In our example, $m = 100$ slugs, $r = 25.5$ mi $= 134,600$ ft, $\Delta v = 10$ mph $= 14.7$ ft/sec, $\Delta Q_c = 21,400$ Btu/lb $=$ 149,800 Btu/gal $= (149,800)(778)$ ft-lb/gal $= 1.167 \times 10^8$ ft-lb/gal; therefore,

$$\eta = \frac{(100)(134,600)(14.7)}{(1.167 \times 10^8)(12.0)} = 0.141, \text{ or } 14.1 \text{ percent}$$

These figures are typical for an overdrive-equipped Volvo 164E. You may wish to try this procedure out on your own car and see what you come up with.

12.4 Phase Equilibria: Energies of Fusion and Vaporization

So far, we have limited our discussion of thermal equilibrium to the case of two different solid bodies. It is obvious, of course, that a solid object can be in thermal equilibrium with a liquid or a gas at any temperature or that a liquid and a gas can be in thermal equilibrium with one another over a wide range of temperatures. What is less obvious is that two different *phases*, or states of aggregation, of a *single* substance, can be in thermal equilibrium with each other under the proper circumstances.

For example, it is possible to establish conditions under which ice (that is, the solid phase of water) and liquid water are in thermal equilibrium. Under these conditions, if ice crystals and liquid water are mixed, a condition of dynamic equilibrium on the molecular scale exists, in which, in a given time interval, the number of ice molecules set free from the surface of the ice crystals by the thermal agitation of the crystal lattice of the ice equals the number of liquid water molecules striking the surfaces of the ice crystals that stick and are incorporated into the ice phase. If the temperature of the system, under atmospheric pressure, is 0°C, these two mutually inverse processes proceed at precisely the same rate, and thermal equilibrium is established between the ice and liquid water since both substances are at the same temperature and no heat energy flows from one phase to the other.

Now, obviously, the number of ice molecules per unit time shaken loose from the surface of the ice crystals will increase as the temperature, and thus the intensity of thermal agitation in the crystal, increases. At the same time, the number of water molecules per unit time impinging on the ice surface, with kinetic energy small enough to permit them to be "captured" by intermolecular attraction and incorporated into the crystal, will decrease with increasing temperature. Therefore, if the temperature of the ice–water mixture is increased just a bit above 0°C, the former process will proceed at a greater rate than the latter, and the ice will start to melt. If the temperature is maintained slightly above 0°C for a sufficiently long time, the ice phase will disappear altogether, and nothing but liquid water will remain. By the same token, if the temperature of the ice–water system is maintained a bit below 0°C, the latter process will proceed at a greater rate than the former, and the water will begin to freeze and will disappear entirely, leaving only ice after a certain time.

The point of all this discussion is that a system consisting of the solid and liquid forms of a given substance, at a given constant pressure, may be in thermal equilibrium, *but only at a single, sharply defined temperature* which we usually refer to as the *melting temperature* of the substance. This melting temperature may change with pressure; but for most substances over normally accessible ranges of pressure, the change is not very large.

A final and most important aspect of systems such as these arises from the fact that *work must be done* against the strong intermolecular forces that act between the individual molecules of a crystal in order to separate a molecule from the crystal and free it into the liquid phase, where intermolecular attractions are significantly weaker. As a consequence, heat energy introduced into an ice–water system at the melting temperature may be absorbed in doing nothing more than supplying the work needed to transfer molecules from the solid phase to the liquid phase. This work is expended largely in overcoming the strong attractive potential energy of interaction between molecules in the crystal. When this has been accomplished, neither the average kinetic energy of the remaining ice molecules nor the average kinetic energy of the molecules now in the liquid phase are any different from what they were previously, in spite of the fact that heat energy has been added to the system. In the final state, the system is again in thermal equilibrium at the melting temperature, the only difference being that there is less ice and more water. The heat energy absorbed by the system has, of course, increased the internal energy of the ice–water mixture; the fact that there is more water and less ice reflects this increase. The process can be reversed, by removing from the system the thermal energy supplied by work done on water molecules as they are incorporated into ice crystals. The ice phase now grows at the expense of the liquid phase, the internal energy of the ice–water mixture decreasing correspondingly.

The situation can be summarized by observing that at the melting temperature, a given amount of heat is necessary to melt a given quantity of solid material, *without any significant or lasting change in temperature taking place*. Likewise, the same amount of heat energy is generated when this amount of liquid is frozen, even though no appreciable change in temperature may occur. This energy is usually referred to as the *latent heat of fusion;* it is different for different substances, because the intermolecular forces between different molecules are not the same. The heat of fusion of a substance, like the melting temperature, may vary slightly with pressure. The heat of fusion represents the energy necessary to tear a tightly bound solid phase apart and render it liquid. It may be expressed in units of calories per gram, calories per mole, or Btu per pound. The heat of fusion of water, for example, under normal atmospheric pressure is about 80 cal/g, 144 Btu/lb, or 1440 cal/mole.

Very similar principles govern the thermal relationships existing between a liquid and its vapor. Consider, for example, a system in which water and water vapor are in thermal contact and in which is maintained *a constant pressure of 1 atmosphere*. In such a system, once again, two competitive processes occur. The first is the expulsion of water molecules from the liquid into the vapor phase due to the thermal agitation of the liquid molecules. The second is the inverse process, in which a water molecule in the vapor strikes the surface of the liquid with insufficient kinetic energy to allow it to rebound into the vapor phase; it is then incorporated into the liquid. In this case also, the first process occurs more frequently as the temperature is increased, while the second process happens less frequently, simply because the average kinetic energy of the molecules in both phases increases as the temperature is raised.

There is, as a result, a single, sharply distinguished value of temperature (100°C for the water–steam system at atmospheric pressure) for which both processes occur at the same rate. At this temperature, the number of molecules per unit time leaving the liquid phase and entering the vapor phase equals the number per unit time going in the opposite direction; there is then no net flow of matter or thermal energy from one phase to the other. The system is, therefore, in thermal equilibrium at this temperature, which we call the *boiling point*. If the temperature is raised slightly, the vapor phase being allowed to expand so as to maintain the pressure at the constant value of 1 atmosphere, the first process will proceed more rapidly than the second, the liquid boils or evaporates, and we end up eventually with no liquid at all but only a certain volume of steam. If the temperature is decreased, the vapor phase volume again being adjusted to maintain constant pressure, the second process is now the more rapid one, the vapor now condensing into liquid, leaving after a while only the liquid phase.

In precise analogy with the solid–liquid system, work must be done against the relatively strong intermolecular forces in the liquid phase in order to separate a molecule from this phase and transfer it into the vapor phase. Once more, heat may be absorbed by the system with no effect other than the disappearance of some of the liquid and a corresponding increase in the volume of vapor, and vice

versa. This thermal energy is called the *latent heat of vaporization* of the substance and represents the energy necessary to separate liquid molecules from one another and transfer them to the vapor phase where intermolecular forces are very much smaller. For water at a constant pressure of 1 atmosphere, the latent heat of vaporization amounts to 539 cal/g = 9700 cal/mole = 970 Btu/lb. The heat of vaporization is a function of pressure, and, in fact, varies with pressure much more strongly than the heat of fusion; its pressure variation must often be taken into account if substantial pressure changes are involved. Once again, vaporization of liquid is accompanied by an increase in the internal energy of the liquid–vapor system, while condensation results in a decrease of internal energy.

Now, let us consider the behavior of a system containing a given substance, such as water, which is enclosed in a rigidly constructed, confining vessel, as illustrated in Fig. 12.7. Such a system is compelled always to occupy a fixed constant volume, and, as the temperature within the vessel is changed, the pressure will vary as well. In Fig. 12.7c, for instance, the liquid is in equilibrium with its vapor at a certain fixed temperature T_3. Under these circumstances, the pressure will assume a value just sufficient to ensure that the two competing processes of evaporation of liquid molecules and condensation of vapor molecules occur at the same rate. If the temperature is 100°C, then this *equilibrium vapor pressure* will be exactly 1 atmosphere. Now suppose the temperature is raised to some higher value T_4. At this temperature, the thermal agitation of the liquid molecules is increased, which means that more of them will be able to escape the intermolecular attraction of their neighbors and evaporate into the vapor phase. At the same time, since the average kinetic energy of the vapor molecules will be increased, a smaller fraction of them will have

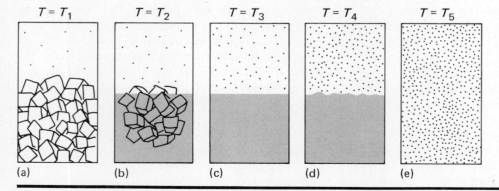

FIGURE 12.7. Various equilibrium configurations for an isolated system. (a) Solid–vapor equilibrium; (b) triple-point equilibrium involving solid, liquid, and vapor; (c) liquid–vapor equilibrium; (d) critical point equilibrium, in which the separate existence of liquid and vapor disappears; and (e) supercritical equilibrium of high-density vapor.

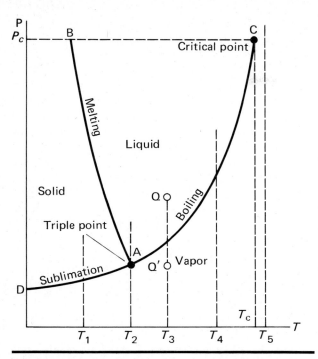

FIGURE 12.8. Equilibrium phase diagram illustrating the equilibrium configurations shown in Fig. 12.7.

sufficiently low kinetic energy to be "captured" by the intermolecular forces of the liquid phase. Evaporation will be speeded up and condensation slowed down, and, therefore, the vapor phase will grow at the expense of the liquid. As this happens, the concentration of molecules in the vapor increases, and, as we shall see more explicitly in a later section, this in turn causes an increase in the *pressure* of the vapor. Finally, the increase in vapor concentration provides a sufficient number of molecules with low enough kinetic energy to be captured by the liquid on striking the surface, to once more balance the now increased evaporation rate. Equilibrium is again established at the higher temperature T_4, but now at a significantly larger value of equilibrium vapor pressure. The situation is now that illustrated in Fig. 12.7d.

It is apparent, then, that the equilibrium vapor pressure of a substance increases substantially with increasing pressure. This is the reason why water may boil at a temperature considerably lower than 100°C at high altitudes and why water in a high-pressure steam boiler may reach temperatures much in excess of 100°C before boiling. The variation of vapor pressure with temperature is plotted in Fig. 12.8 for temperatures above the point labeled T_2. At any point on the line AC in this diagram, the liquid and its vapor can exist in equilibrium with one another at a vapor pressure corresponding to the prevailing temperature. At a given temperature, such as T_3 or T_4, if the pressure is increased to a value above the equilibrium vapor pressure and maintained there, the condensa-

tion process will always proceed at a greater rate than the evaporation process and the liquid phase will grow at the expense of the vapor. After a time, the vapor will disappear entirely and nothing but liquid will remain.[3] Under conditions such as these, then, the two phases *cannot exist* in equilibrium, the equilibrium state consisting solely of liquid at temperature and pressure corresponding, for example, to those associated with point Q. The region of the diagram above the line AC is one in which the liquid phase is the stable equilibrium state. In much the same way, should the pressure be reduced and maintained below the equilibrium vapor pressure at a given fixed temperature, evaporation will occur more rapidly than condensation, the vapor phase will grow at the expense of the liquid, and, after a while, nothing but vapor will remain.[3] Under such circumstances, the two phases again cannot coexist permanently in equilibrium, but now the equilibrium state consists wholly of vapor at a temperature and pressure such as might prevail at Q′ in Fig. 12.8. The region of this diagram below the line AC is, therefore, one in which the vapor phase is the stable equilibrium state. A diagram such as the one shown in Fig. 12.8, in which the regions corresponding to stable equilibrium states of a given substance are exhibited and in which the boundaries between these regions are outlined, is referred to as an *equilibrium phase diagram* of the substance.

Returning now to the confined liquid–vapor system illustrated in Fig. 12.7c and d, it is evident that as the temperature is increased, the density of the vapor continually increases, while as a result of thermal expansion the density of the liquid decreases. As the temperature, and thus the pressure, become greater and greater, the difference in density between the two phases becomes less and less and eventually, at a certain temperature and pressure, vanishes altogether. The liquid–vapor interface then disappears, and *liquid and vapor cease to exist as two separate and distinct phases!* Under these conditions, the substance is said to have reached the *critical point*, which is shown at C on the equilibrium phase diagram of Fig. 12.8. The temperature and pressure for which this occurs are referred to as the *critical temperature* and the *critical pressure*, respectively. For water, the critical temperature is 374°C and critical pressure is 218.5 atmospheres; its density at the critical point is 0.325 g/cm³. At temperatures and pressures above the critical point, separate liquid and vapor phases do not exist, the material then having

---

[3] Such a process can be carried out only by allowing the volume to change in such a way as to maintain the pressure at its new level. It could not, therefore, be accomplished, strictly speaking, within the system shown in Fig. 12.7.

somewhat the physical character of a vapor but with a density more like that of a liquid; the situation is illustrated by Fig. 12.7e. The latent heat of vaporization of the liquid decreases continually as the critical point is approached and becomes zero as the critical temperature is attained. The distinction between liquid and vapor having vanished, naturally, the energy required to effect the phase transition must vanish as well!

If the temperature of the system is reduced below the value T_3 in Fig. 12.8, the vapor pressure decreases correspondingly, and at some point, labeled T_2 in the drawing, the liquid will begin to freeze. If the temperature is reduced even further, the vapor pressure decreases still more. But now there is no equilibrium possible between liquid and vapor; the only state of equilibrium which may exist is one between solid and vapor. Under these circumstances, the atoms on the surface of the solid phase evaporate directly into the vapor without the formation of any intermediate liquid phase at all. This process is often referred to as *sublimation* of the solid. Conversely, the vapor molecules condense into the solid phase directly, without the formation of a liquid phase. At equilibrium at any given temperature, the pressure again adjusts itself so that the rates of sublimation and condensation are equal, and the relations between the two phases are in every way similar to those prevailing in the liquid–vapor system at higher temperatures. In particular, there is a *latent heat of sublimation* associated with the transfer of matter from the solid to the vapor phase, and vice versa. The sublimation phase boundary is shown in Fig. 12.8, and the solid–vapor equilibrium for a temperature such as $T = T_1$ is illustrated in Fig. 12.7a.

There must also be a line separating the solid and liquid phases in the equilibrium phase diagram. This line would be vertical if the melting temperature were independent of pressure; but since this is not usually the case, it will ordinarily slope one way or the other according to whether the melting point decreases or increases with rising pressure. In the case of water, the melting temperature decreases upon the application of high pressure; this is what makes ice skating so practical and enjoyable. The phase boundary between liquid and solid, therefore, slopes to the left as the pressure increases, as shown in Fig. 12.8. Many other substances, however, behave oppositely, and for them this phase boundary would slope to the right. At any point along this line, the molecular processes discussed in connection with the ice–water system occur at the same rate, and liquid–solid equilibrium is possible.

It is evident, also, that at a sufficiently low value of pressure, attainable in a system such as that shown in Fig. 12.8 by adjusting the temperature properly,

the solid–liquid phase boundary must *intersect* the solid–vapor and liquid–vapor boundaries. At the point of intersection, corresponding to temperature T_3 in Fig. 12.8, solid, liquid, and vapor are all in equilibrium with one another, and all the molecular processes involved in solid–liquid, solid–vapor, and liquid–vapor equilibrium proceed at precisely the same rate. This situation is illustrated schematically in Fig. 12.7b. This point in the equilibrium phase diagram is referred to as the *triple point* of the substance. For water, the triple point occurs at a temperature of 0.0075°C and a pressure of 0.000605 atmospheres. At the triple point, strange as it may seem, the substance is freezing and boiling at the same time.

The triple point affords a sharply defined set of conditions which may easily be set up to establish a very accurate *reference temperature* for the calibration of precision temperature measuring instruments. This reference temperature can, in fact, serve as an exact calibration temperature to define the internationally accepted standard Celsius temperature scale itself. Initially, of course, the Celsius scale was defined in terms of the equilibrium melting and boiling temperatures of water at normal atmospheric pressure, and this is quite satisfactory for scientific work not involving temperature measurements of extreme precision. But slight variations of atmospheric pressure affect the boiling temperature (and to a lesser degree the melting temperature) to a measurable extent, and in situations where reference temperatures must be standardized to $10^{-3°}$ or $10^{-4°}$C, it is difficult to make accurate corrections for these variations. But if a sealed triple-point cell is constructed, in which both the pressure and temperature are held at the triple-point value by the dynamic equilibrium of the molecular processes of the substance in the cell, the temperature within the cell will have to correspond very accurately to the triple-point temperature. Such a cell affords a very precise temperature reference. For this reason, the triple point temperature of water is used as one of the fixed reference temperatures in terms of which the international Celsius scale is now defined.

It is sometimes possible, also, for several different *allotropic forms* of a substance to exist in the solid phase. These are solid forms of the same substance having a different arrangement of atoms or molecules in the crystal lattice. The most familiar examples of these allotropic modifications are diamond and graphite, monoclinic and rhombic sulfur, and gray and white tin, but there are many others which are less well known. In general, only one of these allotropic forms is stable under a given set of temperature–pressure conditions. For example, at atmospheric pressure, gray tin is the stable form for $T < 18°$C and white (metallic) tin the stable form for $T > 18°C$. This transition temperature, however, is also a function of pressure. The solid-phase region of the equilibrium phase diagram may, therefore, be divided into two or more regions corresponding to different allotropic modifications of the solid phase. The equilibria existing between such forms are very much like those which prevail between solid and liquid or liquid and vapor; in particular, latent heats of transformation analogous to the latent heat of fusion or vaporization may exist between allotropic forms of the same substance.

EXAMPLE 12.4.1

How much heat would have to be supplied to 4.0 kg of ice (specific heat 0.50 cal/g-°C) in a 3.0-kg aluminum kettle (specific heat 0.215 cal/g-°C), both initially at a temperature of $-20°C$, to turn the ice into steam at 100°C under normal atmospheric pressure? How long would it take a 500-watt immersion heater to do this? Assume that all specific heats do not vary significantly with temperature.

First of all, the ice must be raised to the melting temperature, requiring the addition of an amount of thermal energy given by $\Delta Q_1 = mC_i \Delta T_1$, where m is the mass of ice, C_i the specific heat of ice, and ΔT_1 the difference between the initial temperature and the melting point. It is important to note that the specific heat of ice is *not* the same as that of liquid water. The reason is that the intermolecular forces are different in the two cases, so that different amounts of thermal energy per unit mass are required to produce an increase in molecular kinetic energy corresponding to a one-degree temperature rise.

Once the ice has been brought to the melting temperature, the heat of fusion $\Delta Q_2 = m \Delta q_f$ must be added to melt the ice; Δq_f represents the heat of fusion per unit mass, in this case 80 kcal/kg. Then, the water must be raised to a temperature of 100°C, requiring an additional heat input $\Delta Q_3 = mC_w \Delta T_2$, where ΔT_2 represents the 100-degree temperature increment. The water must next be vaporized into steam, which takes another heat input $\Delta Q_4 = m \Delta q_v$, where Δq_v is the heat of vaporization per unit mass, i.e., 539 kcal/kg. Finally, the aluminum kettle must be raised in temperature from $-15°$ to 100°C, requiring $\Delta Q_5 = m_{Al}C_{Al}(\Delta T_1 + \Delta T_2)$. The total heat input is, therefore,

$$\Delta Q = \Delta Q_1 + \Delta Q_2 + \Delta Q_3 + \Delta Q_4 + \Delta Q_5$$
$$= mC_i[0 - (-15)] + m \Delta q_f + mC_w(100 - 0)$$
$$+ m \Delta q_v + m_{Al}C_{Al}[100 - (-15)]$$
$$= (4.0)(0.50)(15) + (4.0)(80) + (4.0)(1.00)(100)$$
$$+ (4.0)(539) + (3.0)(0.215)(115)$$
$$= 2980 \text{ kcal} = (2980)(4186) = 12,474,000 \text{ J}$$

The time t required for the production of this amount of energy by a 500-watt heater can be found from

$$\Delta Q = 12,474,000 \text{ J} = 500t$$
$$t = 24,950 \text{ sec, or } 416 \text{ min}$$

This example is an extension of Example 12.3.2, to which it should be compared.

EXAMPLE 12.4.2

A 50-g lump of ice initially at a temperature of $-25°C$ is dropped into 300 g of water whose temperature is 20°C. The mixture is stirred until a final equilibrium state is reached. Assuming that there is no exchange of heat energy between the system and its container, describe the final equilibrium state. Repeat the exercise assuming that 150 g of ice at $-25°C$ is added initially. What is the final state of the system if 300 g of ice initially at a temperature of $-80°C$ is added to 50 g of water at 20°C? Assume that the specific heat of ice has the constant value of 0.50 cal/g-°C in all cases.

In all instances, no heat flows into or out of the ice-water mixture, nor is there conversion of heat into mechanical or any other form of energy. Therefore, the total amount of heat absorbed by the system is zero, which is the same as saying that the heat given up by the water must be equal to that absorbed by the ice.

There are three possibilities regarding the final state of the system. The first is that the water may melt all the ice, experiencing a temperature drop in the process, the final state being one in which there is nothing but liquid water at a temperature in excess of 0°C but lower than its initial temperature. Alternatively, the water may be cooled to a temperature of 0°C *before* all the ice is melted, the final state then consisting of a certain quantity of ice and a certain quantity of water in equilibrium at a temperature of 0°C. Finally, the quantity of ice may be so large, and its initial temperature so far below the melting point, that it may be able, before it even reaches the melting point, to absorb enough heat from the liquid not only to cool it to 0°C but to freeze it altogether! The final equilibrium condition is one in which there is nothing but ice at some final temperature below 0°C but above the initial temperature of the ice.

It is necessary at the beginning to decide which of these three possibilities applies to the system at hand. In the first calculation, it is evident that the amount of heat that the water can give up in being cooled from 20°C to the freezing point is $m_wC_w(20) = (300)(1.00)(20) = 6000$ cal. It is also a simple matter to ascertain that the amount of heat necessary to melt all the ice is $m_iC_i(25) + m_i \Delta q_f = (50)(0.50)(25) + (50)(80) = 4625$ cal. From this, it follows that the water can melt all the ice before being cooled to the freezing point itself, and the final system will consist wholly of water at some final temperature T_f between 0°C and 20°C. The final temperature may be calculated by noting that

$$\Delta Q_i + \Delta Q_w = 0 \qquad (12.4.1)$$

where ΔQ_i represents the heat absorbed by the ice (and the water into which it changes upon melting) and ΔQ_w is the heat given up by the water which was present initially. But these can be written as

$$\Delta Q_i = m_i C_i \, \Delta T_i + m_i \, \Delta q_f + m_i C_w \, \Delta T_w' \qquad (12.4.2)$$

$$\Delta Q_w = m_w C_w \, \Delta T_w \qquad (12.4.3)$$

The three terms in (12.4.2) represent, respectively, the heat required to raise the temperature of the ice through ΔT_i degrees to the melting point, the heat required to melt the ice once that temperature is attained, and the heat required to raise the temperature of the *water* formed when the ice melted through $\Delta T_w'$ degrees to the final equilibrium temperature. The right-hand side of (12.4.3) represents the heat given up by the water initially present as it cools through ΔT_w degrees to the final equilibrium temperature. In view of (12.4.1), then, we may write

$$m_i C_i \, \Delta T_i + m_i \, \Delta q_f + m_i C_w \, \Delta T_w' + m_w C_w \, \Delta T_w = 0 \qquad (12.4.4)$$

Now, since $\Delta T_i = 0 - (-25)$, $\Delta T_w' = T_f - 0$, $\Delta T_w = T_f - 20$ (in units of °C), using the given values for masses and specific heats,

$$(50)(0.50)(+25) + (50)(80) + (50)(1.00)T_f$$
$$+ (300)(1.00)(T_f - 20) = 0$$

$$T_f = 3.93°C$$

In the second calculation we are asked to make, an amount of heat equal to $(150)(0.5)(25) = 1875$ cal is needed to raise the ice to the melting point; this can be supplied from the 6000 cal which would be given up by the liquid as it is cooled to the freezing point. But the additional quantity of $(80)(150) = 12,000$ cal which would be needed to melt the ice once the melting temperature is reached is not available. Therefore, some of the ice will melt, but not all of it. Let us call the mass of ice melted when equilibrium is finally attained M. Then, since the final equilibrium termperature of both ice and water is 0°C, (12.4.1) becomes

$$m_i C_i \, \Delta T_i + M \, \Delta q_f + m_w C_w \, \Delta T_w = 0 \qquad (12.4.5)$$

where now $\Delta T_i = [0 - (-25)]°C$ and $\Delta T_w = [(0 - 20)]°C$. In this equation, the first term represents the heat absorbed by the ice in being brought to the melting temperature, the second the heat absorbed in melting the mass M of ice, and the third the heat given up by the water as its temperature is reduced to the

final equilibrium temperature. Substituting numerical values, we obtain

$$(150)(0.50)(25) + M(80) + (300)(1.00)(-20) = 0$$
$$M = 51.6 \text{ g}$$

The final equilibrium is reached, then, at a temperature of 0°C, after 51.6 g of the original 150 g of ice has melted.

The final example involves a situation wherein 300 g of ice initially at -80°C is added to 50 g of water at 20°C. In this case, $(300)(0.50)(80) = 12,000$ cal is needed to raise the temperature of the ice to the melting point. But only $(50)(1.0)(20) = 1000$ cal is given up in cooling the water to the freezing temperature, while a further withdrawal of $(50)(80) = 4000$ cal will result in solidification of the whole initial quantity of water. Since all this energy will be withdrawn from the initially present liquid by the ice long before it reaches the melting temperature, the final equilibrium state will be one in which nothing but ice is present. The final temperature T_f can be calculated from (12.4.1), which now takes the form

$$m_i C_i \, \Delta T_i + m_w C_w \, \Delta T_w - m_w \, \Delta q_f + m_w C_i \, \Delta T_i' = 0 \qquad (12.4.6)$$

In this equation, the first term is the heat acquired by the initial mass of ice in being warmed through ΔT_i degrees to the final equilibrium temperature. The second is the heat lost by the water as it experiences a temperature drop ΔT_w to the freezing temperature. The third represents the heat withdrawn from the water to freeze it; it must have a minus sign because it represents a *negative* flow of heat from the water in calculating ΔQ_w in (12.4.1). The fourth term is the heat withdrawn from the ice formed from the freezing of the initially present liquid as it undergoes a temperature change of $\Delta T_i'$ degrees from the freezing temperature to the final equilibrium temperature.

Once more denoting this final equilibrium temperature by T_f, it is clear that $\Delta T_i = T_f - (-80)$, $\Delta T_w = 0 - 20$, and $\Delta T_i' = T_f - 0$, in units of Celsius degrees. Equation (12.4.6) then becomes

$$(300)(0.50)(T_f + 80) + (50)(1.00)(-20)$$
$$- (50)(80) + (50)(0.50)T_f = 0$$

$$T_f = -40.0°C$$

12.5 Transfer of Energy by Conduction, Convection, and Radiation

We have already seen that heat may flow from one body to another when the two objects have different temperatures and when thermal contact between the two is established. In a similar way, heat may flow from one part of a single homogeneous body to another part if a temperature difference between the two parts is maintained. In such a case we say that energy has been transported by *thermal conduction*.

The transport of heat by thermal conduction occurs when the internal kinetic energy of the atoms or molecules in one part of a substance increases in response to a temperature rise and these atoms or molecules interact with other nearby atoms or molecules, passing on some of their newly acquired internal energy to other particles of the system. In this way, heat energy may flow from one part of a substance to another whenever a temperature difference exists. We say that heat is *conducted* from one region of the substance to another.

Thermal conduction takes place in solids, liquids, and gases, though the molecular processes which cause it differ somewhat in each case. In gases, internal kinetic energy of translation is transferred from one molecule to another by direct collisions between molecules, while in solids, and to a large extent in liquids, too, internal kinetic energy is transferred from one part of the substance to another by very short acoustical waves traveling through the substance. These acoustical waves are excited by the thermal vibration of atoms or molecules about equilibrium positions in the crystal lattice. It is beyond the scope of a general text such as this to treat the dynamics of these interactions in detail; we shall be content with merely describing the physical effects arising from them and relating them to the geometry of the system.

Consider the arrangement shown in Fig. 12.9. Here, a large, plane slab of thickness x separates two heat reservoirs maintained at constant temperatures T_1 and T_2 ($T_2 > T_1$). When the temperature difference between the two sides of the slab is first established, the temperatures at various locations within it will change with time as heat enters one side of the slab and starts to flow through it. But eventually, after a sufficiently long time, a *steady-state* heat flow is attained in which the temperatures everywhere within the slab thereafter remain the same. Once this steady-state condition is set up, the temperature within the substance varies linearly between the value T_2 at the left-hand side and T_1 at the right. A constant quantity of heat energy per unit time unit area of slab will flow through the material from left to right, and it is easily shown experimentally that the magnitude of this thermal energy flux is for most substances proportional to the temperature difference $T_2 - T_1$ and inversely proportional to the slab thickness Δx. These observed facts can be stated in the form

$$\frac{1}{A}\frac{\Delta Q}{\Delta t} = K\frac{T_2 - T_1}{\Delta x} \tag{12.5.1}$$

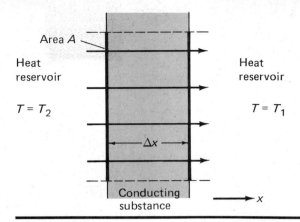

FIGURE 12.9. Flow of heat by conduction through a planar slab of thickness Δx.

where ΔQ is the heat flowing through area A in time Δt and the quantity K, called the *thermal conductivity* of the material, is a proportionality constant, different for different substances. Equation (12.5.1) can be rewritten in the more convenient form

$$\boxed{\frac{\Delta Q}{\Delta t} = KA\frac{\Delta T}{\Delta x}} \tag{12.5.2}$$

where ΔT stands for the temperature difference $T_2 - T_1$. This is the fundamental equation of thermal conduction. It summarizes the observed facts that in steady-state transport of heat by conduction, the amount of heat transported per unit time is proportional to the area through which it flows and to the temperature difference ΔT, and inversely proportional to the thickness Δx. The quantity $\Delta T/\Delta x$ in (12.5.2) is often referred to as the *temperature gradient*. Strictly speaking, this equation is applicable only in situations where the temperature difference ΔT is not extremely large, since for most real substances the specific thermal conductivity K is found to exhibit a slight temperature dependence. It is also important to remember that (12.5.2) applies only to heat transport through a planar slab of material when steady-state heat flow conditions have been attained. In objects of cylindrical, spherical, or irregular shape, or in instances where time-dependent temperature distributions are involved, the finite differences represented by the Δ symbols in (12.5.2) must be allowed to approach zero, thereby becoming *differential* quantities. The resulting differential equation may then be solved for the temperature distribution and thermal current, a mathematical task which is, in general, too complex to consider here. The thermal conductivities of a number of common substances are listed in one of the thermal tables at the end of this chapter. It should be observed that the thermal conductivities of metallic substances—in particular copper, silver and gold—are very much larger than those of nonmetals. Substances such as copper are said to be good thermal conductors, while materials such as glass or asbestos are often referred to as thermal insulators. From Eq. (12.5.2) it is apparent that the units of thermal conductivity are cal/sec-cm-°C or Btu/sec-ft-°F.

It is often said that "heat may flow by *conduction*, *convection*, and *radiation*." In a strict sense, this statement is incorrect, for, as we have already seen, a flow of *heat energy* exists only when there is a *temperature difference*. Therefore, if we insist on the strictest interpretation of the term "heat flow," we must say that it is accomplished only by thermal conduction. Nevertheless, *energy* may be transported by mechanisms other than thermal conduction, and it may happen that the *net effect* of such energy transport may be to transfer a certain amount of internal molecular energy from one body to another, despite the fact that in the transfer process certain intermediate steps involving the transport of energy by means other than heat flow may be present. This is, in fact, what happens in the case of *convective and radiative energy transport*.

In the case *of convection*, the internal energy of a hot substance is carried from one place to another by *mass transport*, that is, by actual translational motion of the substance as a whole from one place to another. The most common examples of convection are found in systems where hot gases or liquids experience an upward motion through a fluid medium under the influence of Archimedean buoyant forces which exist because the fluid medium is less dense when it is hot than when cold. Thus, because water expands when heated, the density of the hot water in the tubes of a boiler in the basement of a house is less dense than the cooler water in the upstairs radiators. According to Archimedes' principle, therefore, this heated water experiences an upward buoyant force like that acting on a helium-filled balloon in the atmosphere. A circulation in the pipes is, therefore, established in which hot water from the boiler rises into the radiators, while cooler water from the radiator outlet flows downward into the boiler, as illustrated in Fig. 12.10. Heat flows into the water by thermal *conduction* in the boiler tubes, since there is a temperature difference between the outside surfaces of the tubes (which are exposed to the flames within the boiler) and the water within the tubes. The heated water rises into the radiators, due to the fact that it is now less dense than any other fluid in the system; this constitutes *convective* transport of its internal energy. Once it is within the radiator, heat flows once more from the water through the radiator tubing into the room; this heat flow is by *conduction*, since there is a temperature difference between the water in the radiator and the air outside. The net effect has been a

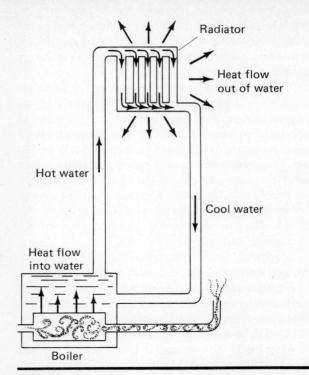

FIGURE 12.10. Convective transport of energy by mass transport in a hot-water heating system.

transfer of heat energy from the heated gases within the boiler to the air of the upstairs room, but there has been an *intermediate process of convection* in which, strictly speaking, there has been no heat flow as such at all but only a transport of matter from the boiler to the radiator. Convective energy transport is also observed in the "natural draft" associated with chimneys and smokestacks, in the thermal mixing of ocean strata, and in the motion of atmospheric currents.

In somewhat the same way, we may regard the transport of thermal energy by *radiation* as a process in which there is an intermediate step wherein energy is propagated in the form of electromagnetic waves at the speed of light. The properties of electromagnetic waves are treated in detail in a later chapter, and we shall not try to understand anything but a few basic facts about them at this point. In 1873, a British physicist, James Clerk Maxwell, found that whenever any *electrically charged particle* (such as an electron or an atomic nucleus, for example) *undergoes acceleration, it radiates energy in the form of electromagnetic waves*. These waves, of which light waves are one particular example, are transverse waves made up of oscillatory electric and magnetic fields. They may propagate not only through material media, such as air, water, or glass, but also through *empty space;* their velocity, in a vacuum, is given by $c = 2.998 \times 10^8$ m/sec, the speed of light.

Since the atoms and molecules of solid bodies are made up of charged particles and since their thermal vibration is an accelerated motion, the atoms of any substance radiate electromagnetic energy because of their thermal vibration. This radiation increases in intensity with increasing temperature; it is also found that though a wide range of frequencies are radiated, lower frequencies predominate at low temperatures, and more and more radiation of high frequency is emitted as the temperature rises. This thermal radiation is responsible for the fact that very hot objects emit visible light. The observed color of this light shifts from red to orange to yellow as the temperature is increased, simply because more radiation of high frequency (thus short wavelength) is emitted at the higher temperatures. The emitted radiation, however, is not all in the form of visible light; some of it may have a frequency less than that which our eyes can detect as visible light, and some of it may be radiated at a frequency greater than that of any visible light. The former radiation is referred to as infrared radiation, the latter, ultraviolet radiation. Even objects at room temperature, indeed, even those which may be far below room temperature, give off radiation of this kind, though most of it is emitted at infrared frequencies, the intensity emitted as visible light being far too small for us to detect with our eyes.

When a body is in thermal equilibrium with its surroundings, the amount of electromagnetic energy it radiates to the surroundings is exactly balanced by electromagnetic energy radiated by the surroundings and received by it. There is, therefore, no net transfer of internal energy between the body and the surroundings, in agreement with our previously developed ideas about thermal equilibrium.

But suppose the situation is as shown in Fig. 12.11, where an iron ball is heated to incandescence by a Bunsen burner. In this case, heat flows by con-

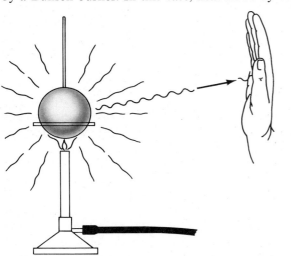

FIGURE 12.11. Radiative transport of energy, in which the emission and absorption of electromagnetic energy is a vital intermediate step.

duction into the iron since the Bunsen flame has a higher temperature than the metal. The internal energy of the iron, and thus its temperature, increases until the iron is very much hotter than its surroundings (as typified by the observer's hand). Since the amplitude of vibration of the atoms of the iron increases as the temperature rises, so also does the acceleration of the charges associated with the atoms. Therefore, the electromagnetic energy radiated by the object becomes more intense and, at a certain temperature, it begins to emit a perceptible amount of visible light as well as infrared radiation. Under these circumstances, it is radiating much more electromagnetic energy than it receives from its surroundings, the situation being no longer one of equilibrium. When this radiant energy strikes the hand of the observer, part of it may be reflected, but some of it is absorbed in the observer's hand. In being absorbed, it will increase the internal kinetic energy of atoms and molecules of the observer's hand, therefore raising the temperature near the surface of the skin. Heat may then flow by conduction from the surface of the skin to the interior of the observer's hand because there is now a temperature difference between these two regions.

Again, the net effect is that heat absorbed by the iron ball from the Bunsen flame has been transported to the interior of the observer's hand. But there has been an *intermediate step* in which energy is transported not by heat flow, but by the radiation and subsequent absorption of electromagnetic energy. This radiation process can occur even in a perfect vacuum—fortunately for us, because that is what happens when solar energy is transported from the sun to the earth through the near-perfect vacuum of outer space. The quantitative aspects of the radiation emitted by incandescent heated objects were what gave birth to the *quantum theory* of matter and energy, and they are discussed in a later chapter when that subject is introduced.

EXAMPLE 12.5.1
One face of a copper plate 50 cm long, 30 cm wide, and 2.5 cm thick is maintained at a temperature of 150°C, while the other is held at a constant temperature of 15°C. Assuming that the average thermal conductivity of copper in this temperature range is $K_{Cu} = 0.92$ cal/sec-cm-°C, find the quantity of heat which flows through the plate in a period of 10 minutes, after steady-state thermal flow has been established.

This problem can be worked by direct application of Eq. (12.5.2), which for our purposes is best rewritten as

$$\Delta Q = K_{Cu} A \frac{\Delta T}{\Delta x} \Delta t$$

In our example, $A = (50)(30) = 1500 \text{ cm}^2$, $\Delta T = 150 - 15 = 135°C$, $\Delta x = 2.5$ cm, and $\Delta t = 10$ min $= 600$ sec. Substituting these values, along with the given value of K_{Cu} in cgs units, we obtain

$$\Delta Q = (0.92)(1500)\left(\frac{135}{2.5}\right)(600)$$

$$= 4.47 \times 10^7 \text{ cal} = 44{,}700 \text{ kcal}$$

This very large amount of heat flows through the plate by virtue of the fact that copper is an excellent thermal conductor. Had the plate been made of an insulating substance, such as glass (for which $K = 0.002$ cal/cm-sec-°C), the result would have been about 460 times smaller.

EXAMPLE 12.5.2

The copper plate described in the above example is firmly bonded to an aluminum plate of the same length and width but having a thickness of 4.0 cm. The temperature of the copper face is now held at 150°C and that of the aluminum face, at 15°C. Assuming that the thermal conductivity of aluminum is $K_{Al} = 0.46$ cal/sec-cm-°C, find how much heat flows through the composite plate in 10 minutes under steady-state conditions. What is the temperature within the plate at the copper–aluminum interface?

In this case, neglecting any heat losses at the edges, the heat flowing in any given time interval through the copper and aluminum sections must be the same, since they are "in series," so to speak, with respect to the thermal energy flow. Then,

$$\Delta Q = K_{Cu} A \frac{\Delta T_{Cu}}{\Delta x_{Cu}} t = K_{Al} A \frac{\Delta T_{Al}}{\Delta x_{Al}} \Delta t \qquad (12.5.3)$$

If we let T_i represent the temperature at the copper–aluminum interface, then $\Delta T_{Cu} = 150 - T_i$ and $\Delta T_{Al} = T_i - 15$, expressed in units of centigrade degrees. Substituting these values into the above equation and canceling the common factors A and t, we obtain

$$K_{Cu} \frac{(150 - T_i)}{\Delta x_{Cu}} = K_{Al} \frac{(T_i - 15)}{\Delta x_{Al}}$$

or, solving for T_i,

$$T_i = \frac{(150)K_{Cu}\,\Delta x_{Al} + (15)K_{Al}\,\Delta x_{Cu}}{K_{Al}\,\Delta x_{Cu} + K_{Cu}\,\Delta x_{Al}}$$

$$= \frac{(150)(0.92)(4.0) + (15)(0.46)(2.5)}{(0.46)(2.5) + (0.92)(4.0)}$$

$$= 117.9°C$$

Once the interface temperature T_i is determined, it is easy to evaluate ΔT_{Cu} and ΔT_{Al} and then to find ΔQ by substituting into either of the expressions shown in (12.5.3). In our example, $\Delta T_{Cu} = 150 - T_i = 32.1°C$, and then

$$\Delta Q = K_{Cu} A \frac{\Delta T_{Cu}}{\Delta x_{Cu}} \Delta t = (0.92)(1500)\left(\frac{32.1}{2.5}\right)(600)$$

$$= 1.063 \times 10^7 \text{ cal} = 10{,}630 \text{ kcal}$$

12.6 Thermal Expansion

We are familiar with the fact that nearly all substances expand slightly when heated and contract somewhat when cooled. This effect, called *thermal expansion*, is exhibited by solids, liquids and gases and has been known, though not understood, since ancient times. We have already seen how the thermal expansion of liquids confined within glass capillary tubes may be utilized in the construction of thermometers and in defining quantitative temperature scales.

In the case of solid objects, thermal expansion results in a change in the *linear dimensions* of any body, while in the case of liquids and gases, which have no permanent shape, thermal expansion manifests itself in a change of *volume*. In nearly all cases of practical importance, thermal expansion measurements are made under conditions of *constant pressure*, and we shall, therefore, confine our discussion of this effect to this important case. Because of the way in which thermal expansion manifests itself, it is most useful to speak in terms of the *linear expansion* of a solid body and the *volume expansion* of a liquid or gas, though these two ways of describing thermal expansion are, as we shall see, closely related.

Let us first consider the linear expansion of a solid object. Such an object, one of whose linear dimensions we may represent by l_0, will expand by an amount Δl when the temperature is raised by an amount ΔT. It is found experimentally that for most substances in the normal range of temperature, the linear expansion Δl is directly proportional to the initial size l_0 and to the temperature change ΔT. Stated as an equation, this becomes

$$\boxed{\Delta l = \alpha l_0 \, \Delta T} \qquad (12.6.1)$$

where α is a coefficient of proportionality usually referred to as the *linear expansion coefficient*. Turning Eq. (12.6.1) around and solving for α, it becomes

$$\boxed{\alpha = \frac{1}{l_0} \frac{\Delta l}{\Delta T}} \qquad (12.6.2)$$

When written in this form, the equation makes clear that the linear expansion coefficient represents the fractional variation in length $\Delta l / l_0$ per unit temperature change. It is evident also from (12.6.2) that the units of the expansion coefficient are reciprocal degrees, that is, $°C^{-1}$ or $°F^{-1}$.

In a liquid or a gas, thermal expansion is

observed as a volume change ΔV in a quantity of substance having initial volume V_0 related to a temperature change ΔT. In this case, the volume change ΔV is directly proportional to the initial volume V_0 and to the temperature change ΔT for most substances in normally accessible ranges of temperature. In the language of mathematics, this means that

$$\Delta V = \beta V_0 \, \Delta T \qquad (12.6.3)$$

where β is a constant of proportionality called the *volume expansion coefficient*. Once more, if we solve this for β,

$$\beta = \frac{1}{V_0} \frac{\Delta V}{\Delta T} \qquad (12.6.4)$$

it is easily seen that the volume expansion coefficient expresses the fractional change in volume $\Delta V / V_0$ per unit temperature change.

In the case of solids, it is sometimes useful to discuss volume expansion as well as linear expansion, and, as shown in Example 12.6.3 below, it is not difficult to show that the linear coefficient α and the volume coefficient β are related by the simple expression

$$\beta = 3\alpha \qquad (12.6.5)$$

The values of the thermal expansion coefficient for a number of solid and liquid substances are tabulated in the thermal tables at the end of this chapter. It will be noted that the coefficients differ from one substance to another.

The reason for this is that in liquids and solids the thermal expansion coefficients are closely related to the forces that act between individual molecules to hold the substance together. In solids, the atoms or molecules are held together in a regularly spaced *crystal lattice* by intermolecular forces that act somewhat but not exactly like the Hooke's law forces for ideal springs. The crystal may thus be visualized as a three-dimensional bed spring such as the one illustrated in Fig. 12.12, where, for simplicity, only a two-dimensional section is shown. In equilibrium, in the absence of internal kinetic energy, each atom or

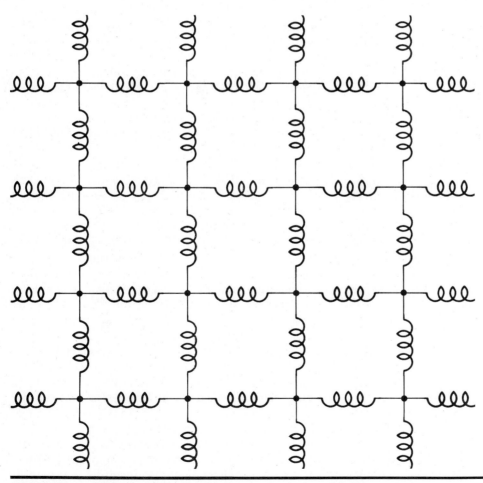

FIGURE 12.12. Conceptual two-dimensional model of an "ideal" crystal in which the molecules are represented as identical point masses bound to one another by Hooke's law springs.

molecule would occupy an equilibrium position on its respective lattice site. As the temperature is increased, internal kinetic energy is excited in the form of molecular vibrations about the equilibrium positions the amplitude of which increases with increasing temperature. For many purposes, it is sufficient to regard the "springs" (which represent the intermolecular forces) as ideal Hooke's law springs for which the potential energy of any molecule would be given by

$$U_p(x) = \tfrac{1}{2}kx^2 \qquad (12.6.6)$$

where x is the displacement of that molecule from its equilibrium position. Under these circumstances, the molecular vibrations would be those of a simple harmonic oscillator having the potential energy function (12.6.6). For such a system, however, the *average displacement* of the particle from its equilibrium position is zero, since the time average of the displacement $x = A\cos(\omega t + \delta)$ over a complete period is zero. For this reason, the thermal expansion coefficient of a crystal in which the intermolecular forces obey Hooke's law is zero.

At first sight, this seems to imply that a "Hooke's law" crystal is a poor molecular representation of a solid substance, since we know that every solid material exhibits thermal expansion. But wait—thermal expansion is, after all, a *small* effect. No crystal swells to twice its linear dimensions even when subjected to very large temperature changes. Indeed, a 1 percent change in linear dimensions can be obtained in most substances only by a temperature change of many hundreds of Celsius degrees. So, though the prediction of zero linear expansion by the Hooke's law lattice is not correct, the fact that actual expansion coefficients are so small tells us that it may for many other purposes serve as a good simple picture of a crystal. Also, we must conclude that thermal expansion is critically dependent upon the *deviation* of the intermolecular forces and their related potential energies from the form represented by Hooke's law.

In liquids, the picture is somewhat similar but is complicated by the fact that the internal energy of the substance not only excites vibrations of the molecules but also may sometimes actually create voids or vacant spaces between otherwise densely packed molecules. For this reason, the volume expansion coefficients of substances in the liquid state are generally greater than those exhibited by solid crystalline materials.

Gases, generally speaking, have larger expansion coefficients than either liquids or solids, but here it is the increased translational kinetic energy of the gas molecules resulting from an increase in temperature that causes thermal expansion. Since the molecules of gaseous substances are relatively free, the intermolecular forces between them are quite weak

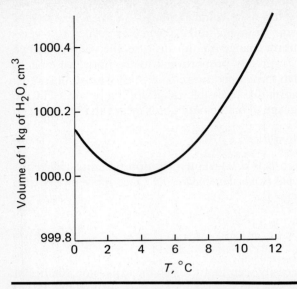

FIGURE 12.13. Nonlinear thermal expansion of water at temperatures just above the melting point.

and, therefore, have little or no effect upon the expansion of the gas. An increase in internal energy instead gives the molecules greater translational kinetic energy, which in turn causes them to exert larger forces upon the walls of the container when they collide with these walls. If the container were completely rigid, of course, there would be an increase in pressure, but if some means such as a movable piston is provided to relieve this pressure rise and thus maintain constant pressure, the gas will expand to an extent governed by the increase in translational kinetic energy of the molecules. The quantitative treatment of the expansion coefficient of "ideal" gases is quite simple but will be postponed until the following chapter, in which the thermal properties of these substances are treated in detail.

In a few substances, the complex interplay of intermolecular forces results in thermal contraction rather than thermal expansion in certain restricted temperature ranges. The most familiar example of this phenomenon is afforded by the behavior of water between 0°C and 4°C. In this temperature range, water contracts when heated and expands when cooled, as illustrated by Fig. 12.13. This strange behavior reveals the fact that water is in some respects a very complex substance. The physical behavior of water departs in a number of ways from that expected of simpler and more "ideal" substances. For example, the thermal expansion of water in the temperature range from 4°C to the boiling point does *not* exhibit the linear behavior shown by most liquids and embodied in Eq. (12.6.3)—there is no temperature-independent expansion coefficient. The volume expansion of water in this temperature range is illustrated in Fig. 12.14.

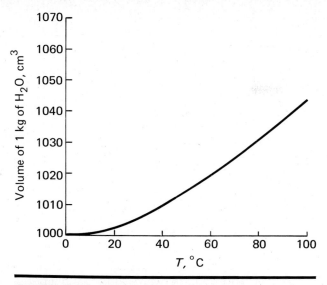

FIGURE 12.14. Thermal expansion of water in the temperature range from 0° to 100°C.

EXAMPLE 12.6.1

A steel bridge is 800 ft long. Find the change in length due to thermal expansion when the temperature changes from $-20°C$ to $+35°C$. The linear expansion coefficient for steel in this temperature range is $1.1 \times 10^{-5}°C^{-1}$.

According to (12.6.1), the length change is given by

$$\Delta l = \alpha l_0 \Delta T$$

In this case, $\alpha = 1.1 \times 10^{-5}°C^{-1}$, $l_0 = 800$ ft, and $\Delta T = 35 + 20 = 55°C$. Substituting these values into (12.6.1), we obtain $\Delta l = 0.484$ ft $= 5.8$ in. The expansion must be accounted for and expansion joints provided so that the roadway or structure is not damaged by the very large forces which might otherwise be exerted as this expansion takes place.

EXAMPLE 12.6.2

A circular copper plate 16.0 cm in diameter has a concentric hole 4.000 cm in diameter drilled in it. It is desired to fixed this disc onto a solid cylindrical steel rod 4.010 cm in diameter by heating the disc to a temperature high enough to make the hole expand to a diameter equal to that of the rod, allowing it to be fitted over the end of the rod and then cooling it. To what temperature change must the disc be subjected in order to accomplish this if the linear expansion coefficient of copper is $1.66 \times 10^{-5}°C^{-1}$? What temperature change could the rod (expansion coefficient $1.0 \times 10^{-5}°C^{-1}$) and the attached disc undergo before the disc becomes loose again?

First of all, let us dispel the fallacious argument that the hole would get smaller rather than larger as the disc is heated because the expansion of the surrounding metal would push its boundaries inward. This is best accomplished by considering the case of a thin circular hoop of diameter d_0 at temperature T_0. Suppose we were to cut the hoop apart and lay it out flat; we would then have a bar of length $l_0 = \pi d_0$. Now suppose that the temperature of the bar is increased by ΔT degrees. An increase in length given by $\Delta l = \alpha l_0 \Delta T$ occurs, if α is the linear expansion coefficient. The new length of the bar is

$$l = l_0 + \Delta l = l_0(1 + \alpha \Delta T) \tag{12.6.7}$$

Now, while keeping the temperature at the elevated value $T_0 + \Delta T$, suppose we were to bend the bar into a hoop once more, joining the two ends together at the point where they meet. The diameter of the hoop is

$$d = \frac{l}{\pi} = \frac{l_0}{\pi}(1 + \alpha \Delta T) = d_0(1 + \alpha \Delta T) \tag{12.6.8}$$

the hole having *expanded* by the amount $\alpha d_0 \Delta T$. Indeed, the hole has expanded *just as though it were made of solid material having the expansion coefficient* α! The circular copper plate of our example can, of course, be envisioned as being made up of a very large number of concentric hoop elements whose diameters increase uniformly from the inner diameter of the hole to the outer diameter of the disc, so clearly the same result will apply to it also. The general lesson to be learned is that all the dimensions of an object, internal ones as well as external, expand with the expansion coefficient of its material when the body is heated.

We now need to find what temperature change ΔT is needed to produce a diameter change Δd equal to 0.010 cm. From (12.6.1), this is given by

$$\Delta d = \alpha_{Cu} d_0 \Delta T \tag{12.6.9}$$

where α_{Cu} is the expansion coefficient of copper and d_0 is the internal diameter. Solving for ΔT and inserting the appropriate numerical values, we find

$$\Delta T = \frac{\Delta d}{\alpha_{Cu} d_0} = \frac{(0.010)}{(1.66 \times 10^{-5})(4.00)} = 150.6°C$$

When both the rod and the attached disc are heated, both will expand, the diameter of the hole in the copper disc being

$$d = d_0 + \Delta d = d_0 + \alpha_{Cu} d_0 \Delta T \tag{12.6.10}$$

and that of the steel rod being

$$d' = d_0' + \Delta d' = d_0' + \alpha_s d_0' \Delta T \tag{12.6.11}$$

In (12.6.11), d', d_0', and $\Delta d'$ represent the actual diameter, initial diameter, and thermal expansion of the steel rod, whose thermal expansion coefficient is

α_s. When d and d' are equal, the disc will become loose once more on the rod; therefore, setting the right-hand sides of these two equations equal and solving for ΔT, we find

$$\Delta T = \frac{d_0' - d_0}{\alpha_{Cu}d_0 - \alpha_s d_0'}$$

$$= \frac{4.010 - 4.000}{(4.000)(1.66 \times 10^{-5}) - (4.010)(1.10 \times 10^{-5})}$$

$$= 449°C$$

EXAMPLE 12.6.3

Show that the volume expansion coefficient β and the linear expansion coefficient α for any solid substance are related by the simple expression $\beta = 3\alpha$.

Consider the case of a body of cubic form having sides of length l_0 at temperature T_0. If the temperature is increased by an amount ΔT, the length of each side increases by an amount Δl equal, according to (12.6.1), to $\alpha l_0 \Delta T$. Under these circumstances, each side of the cube has length l given by

$$l = l_0 + \Delta l = l_0(1 + \alpha \Delta T) \qquad (12.6.12)$$

If we denote the original volume by V_0 ($= l_0^3$) and the expansion in volume by ΔV, then the volume V at temperature $T_0 + \Delta T$ will be

$$V = V_0 + \Delta V = l^3 = l_0^3(1 + \alpha \Delta T)^3$$
$$= l_0^3[1 + 3\alpha \Delta T + 3(\alpha \Delta T)^2 + (\alpha \Delta T)^3] \quad (12.6.13)$$

Since $V_0 = l_0^3$, it is evident from this equation that the volume change ΔV is simply

$$\Delta V = V_0[3\alpha \Delta T + 3(\alpha \Delta T)^2 + (\alpha \Delta T)^3] \qquad (12.6.14)$$

In most instances involving thermal expansion the change in length Δl will be very much less than the initial length l_0. According to (12.6.12), this means

that the quantity $\alpha \Delta T$ will ordinarily be very much less than unity. But if this is true, the second and third terms within the parenthesis in (12.6.14) will be negligible in comparison with the first, the equation then becoming

$$\Delta V = 3\alpha V_0 \Delta T \qquad (12.6.15)$$

This equation is precisely the same as (12.6.3), the equation defining volume expansion, except for the fact that the volume expansion coefficient is represented by 3α. From (12.6.3) and (12.6.15), then, we have

$$\Delta V = 3\alpha V_0 \Delta T = \beta V_0 \Delta T \qquad (12.6.16)$$

which means that α and β must be related by

$$\beta = 3\alpha \qquad (12.6.17)$$

The situation may be understood geometrically by referring to Fig. 12.15, where at (a) the quantities l_0 and Δl are marked off on the cubical body, and at (b) the separate volumes thus defined are shown as disassembled volume elements. The large, flat pieces represent a total volume increase

$$3l_0^2 \Delta l = 3l_0^2 \cdot \alpha l_0 \Delta T$$

corresponding to the first term in (12.6.14), which is retained. The smaller shaded pieces represent the volume increments

$$3l_0 \cdot (\Delta l)^2 = 3l_0 \cdot l_0^2(\alpha \Delta T)^2 \quad \text{and} \quad (\Delta l)^3 = l_0^3(\alpha \Delta T)^3$$

which are neglected. The relative size of the volumes which are retained and those which are neglected makes it evident that the approximation which was used is fully justified whenever Δl is much less than l_0.

This proof is based on an object of cubical form. It is nevertheless easily extended to the case of a body of arbitrary irregular shape simply by noting that any such object can be thought of as consisting

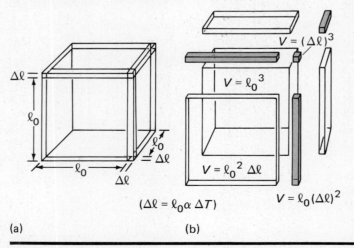

$$V = (\Delta l)^3$$
$$V = l_0^3$$
$$V = l_0^2 \Delta l$$
$$V = l_0(\Delta l)^2$$
$$(\Delta l = l_0 \alpha \Delta T)$$

(a)   (b)

FIGURE 12.15. Changes in volume occurring as a result of a length change from l_0 to $l_0 + \Delta l$, due to thermal expansion.

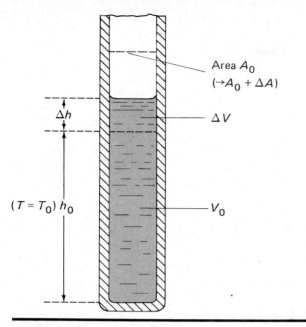

Area A_0
$(\rightarrow A_0 + \Delta A)$

ΔV

Δh

$(T = T_0)\, h_0$

V_0

FIGURE 12.16

approximately of a very large number of tiny cubic elements each of which exhibits the volume expansion characteristics discussed above. In the limit, the number of these elements becomes infinitely large, their dimensions at the same time approaching zero and the "approximate" object then approaching the "real" irregular shape more and more closely. The volume expansion of the irregular object is the sum of the volume expansion of the elements, for each of which (12.6.17) holds. Therefore, the relation must apply to the body as a whole.

EXAMPLE 12.6.4

A uniform cylindrical glass tube is filled with mercury to a height of 1.000 m at 20°C, as illustrated in Fig. 12.16. The temperature is then raised to 120°C. What is the height of the mercury column at 120°C? The coefficient of linear expansion of the glass is $8.5 \times 10^{-6}\,°\text{C}^{-1}$, and the volume coefficient of expansion of mercury is $1.82 \times 10^{-4}\,°\text{C}^{-1}$ in this temperature range.

Let us call the height of the mercury column h_0 at the initial temperature T_0, its cross-sectional area A_0, and its volume V_0, and let us assume that it undergoes dimensional changes Δh, ΔA, and ΔV when the temperature changes by an amount ΔT. The total volume of mercury is always given by the cross-sectional area times the height, so that at the final temperature $T + \Delta T$ we may write

$$V_0 + \Delta V = (h_0 + \Delta h)(A_0 + \Delta A)$$
$$= h_0 A_0 + A_0\, \Delta h + h_0\, \Delta A + \Delta A\, \Delta h$$
$$\text{(12.6.18)}$$

Now the quantity $A_0 h_0$ is equal to the initial volume V_0; this quantity may, therefore, be canceled on either side of the equation. Also, since the quantities ΔA and Δh are both small in comparison with A_0 and h_0, the product of these two small quantities $\Delta A\, \Delta h$ may safely be neglected in comparison with the other terms. Equation (12.6.18) may, therefore, be written as

$$\Delta V = A_0\, \Delta h + h_0\, \Delta A \qquad \text{(12.6.19)}$$

or, solving for Δh,

$$\Delta h = \frac{1}{A_0}(\Delta V - h_0\, \Delta A) \qquad \text{(12.6.20)}$$

But the volume expansion ΔV of the liquid is given by

$$\Delta V = \beta_l V_0\, \Delta T = \beta_l A_0 h_0\, \Delta T \qquad \text{(12.6.21)}$$

since $V_0 = A_0 h_0$. The quantity ΔA refers to the increase in cross-sectional area of the mercury column caused by the expansion of the hole in the glass tube enclosing it. Such an expansion can be described in terms of an area expansion coefficient γ defined in analogy with the linear expansion coefficient α and the volume coefficient β as the fractional increase in area $\Delta A/A_0$ per unit temperature change; this may be expressed as

$$\gamma = \frac{1}{\Delta T}\frac{\Delta A}{A_0} \qquad \text{(12.6.22)}$$

$$\Delta A = \gamma A_0\, \Delta T \qquad \text{(12.6.23)}$$

The coefficient γ expresses the area expansion of the glass tube; it is related to the linear expansion coefficient α_g and the volume coefficient β_g, since it describes only another aspect of the expansion of the same material. In the light of the discussions given in the previous example, it should not be too hard to believe that the area expansion coefficient will be related to α and β by

$$\gamma = 2\alpha_g \qquad \text{(12.6.24)}$$

whence

$$\gamma = 2\frac{\beta_g}{3} = \tfrac{2}{3}\beta_g \qquad \text{(12.6.25)}$$

This result can easily be proved using the same methods as those used in deriving the relation $\alpha = 3\beta$ in the previous example. The details are assigned as an exercise. Substituting (12.6.21) and (12.6.23) into (12.6.20), noting that $\gamma = 2\beta_g/3$, we find

$$\Delta h = h_0(\beta_l - \tfrac{2}{3}\beta_g)\, \Delta T \qquad \text{(12.6.26)}$$

Inserting the values $h_0 = 1.00$ m, $\beta_l = 1.82 \times 10^{-4}\,°\text{C}^{-1}$, $\beta_g = 3\alpha_g = 0.255 \times 10^{-4}\,°\text{C}^{-1}$, and $\Delta T = 100°\text{C}$, one may easily find that $\Delta h = 0.0165$ m $= 1.65$ cm.

It is interesting to note that if β_l and β_g are equal,

(12.6.26) predicts, reasonably enough, that $\Delta h = h_0 \Delta T(\beta/3) = \alpha h_0 \Delta T$. In this case, the liquid and the "hole" expand at the same rate, the liquid column lengthening as if it were a solid piece of glass with a linear expansion coefficient $\beta/3$. In the case where β_l is less than $2\beta_g/3$, Δh becomes negative, the liquid column then falling when the temperature is raised! This happens because the expansion of the liquid is insufficient to fill in the more rapidly expanding hole in the tube. In practice, it would be difficult to find a pair of substances which would satisfy the conditions necessary for this effect to be observed over an appreciable range of temperatures.

We could have worked this example somewhat differently, by substituting (12.6.21) and (12.6.23) directly into (12.6.18), to find

$$V_0(1 + \beta_l \Delta T) = (h_0 + \Delta h)A_0(1 + \tfrac{2}{3}\beta_g \Delta T) \qquad (12.6.27)$$

This equation is easily solved for Δh:

$$\Delta h = h_0 \left(\frac{\beta_l - \tfrac{2}{3}\beta_g}{1 + \tfrac{2}{3}\beta_g} \right) \Delta T \qquad (12.6.28)$$

This result differs from (12.6.26) by the factor $1 + \tfrac{2}{3}\beta_g$ in the denominator. Which equation is correct? The answer is, *both!* The two results differ only because the very small quantity $\Delta A\, \Delta h$ was neglected in deriving (12.6.26), but was retained in arriving at (12.6.28). Thermal expansion of most materials is a small effect, and, therefore, the quantity $2\beta_g/3$ will always be much less than unity; in the specific example discussed above, it is 1.7×10^{-5}. The inclusion of the term in the denominator of (12.6.28) thus has very little effect on the numerical result. It should not be thought that the retention of the term $\Delta A\, \Delta h$ in (12.6.28) makes that result in any way superior to (12.6.26), because the use of the relations $\beta = 3\alpha$ and $\gamma = 2\alpha$, which was resorted to in both cases, involves the neglect of terms of this sort in any event.

TABLE 12.1. Specific Heats of Solid and Liquid Substances at Atmospheric Pressure

Substance	Temperature, °C	C, cal/g-°C	c, cal/mole-°C
Water	0	1.0087	18.173
	15	1.0000	18.016
	40	0.9976	17.973
	100	1.0064	18.131
Ice	−21–0 (av)	0.505	9.10
Aluminum	0	0.2079	5.61
	20	0.214	5.77
	100	0.225	6.07
	300	0.248	6.69
	0–100 (av)	0.211	5.69
Copper	0–100 (av)	0.0930	5.91
Gold	0–100 (av)	0.0316	6.22
Iron	0–100 (av)	0.1097	6.13
Lead	0–100 (av)	0.0309	6.40
Silver	0–100 (av)	0.0561	6.05
Tin (metallic)	0–100 (av)	0.0556	6.59
Zinc	0–100 (av)	0.0935	6.11
Mercury	0–100 (av)	0.0331	6.64
Sodium chloride	0–100 (av)	0.210	12.27
Brass	0–100 (av)	0.0917	—
Steel	0–100 (av)	0.113	—
Glass	0–100 (av)	0.199	—
Quartz	0–100 (av)	0.188	11.30

TABLE 12.2. Specific Heat Data for Gases at Atmospheric Pressure

Substance	Temp., °C	c_p, cal/mole-°C	c_v, cal/mole-°C	$c_p - c_v$, cal/mole-°C	γ	c_v/R	c_p/R
He	18	4.964	2.992	1.972	1.659	1.506	2.499
A	15	5.005	3.001	2.004	1.668	1.510	2.520
Ne	15	4.965	3.027	1.938	1.64	1.524	2.500
Kr	15	4.944	2.942	2.002	1.68	1.481	2.489
Xe	15	4.963	2.990	1.973	1.66	1.505	2.499
H_2	15	6.832	4.846	1.986	1.410	2.439	3.440
	2000	8.241	6.253	1.988	1.318	3.148	4.149
N_2	15	6.939	4.942	1.997	1.404	2.488	3.494
O_2	15	6.976	4.979	1.997	1.401	2.507	3.512
	2000	8.541	6.555	1.986	1.303	3.300	4.300
NO	15	6.993	4.995	1.998	1.400	2.514	3.521
CO	15	6.944	4.946	1.998	1.404	2.490	3.496
Cl_2	15	8.17	6.03	2.14	1.355	3.03	4.11
Br_2	20–300	8.79	6.66	2.13	1.32	3.35	4.42
I_2	200–400	8.63	6.63	2.00	1.30	3.34	4.34
SO_2	15	9.73	7.55	2.18	1.29	3.80	4.90
NH_3	15	8.92	6.81	2.11	1.31	3.43	4.49

TABLE 12.3. Melting Temperatures, Boiling Temperatures, and Heats of Fusion and Vaporization Under Normal Atmospheric Pressure

Substance	Melting point, °C	Heat of fusion, cal/g	Boiling point, °C	Heat of vaporization, cal/g
He	—	—	-268.6	6.0
A	-190	8.94	-186	37.6
H_2	-259.25	13.8	-252.8	108.
N_2	-210	6.09	-195.55	47.6
O_2	-219	3.30	-182.9	50.9
NH_3	-75	108.1	-33.4	327.1
H_2O	0.00	79.71	100.0	539.6
Ethanol	-114.4	24.9	78.3	204.
Methanol	-97	16.4	64.7	262.8
Acetone	-95	19.6	56.1	124.5
Acetic acid	16.58	44.7	118.3	96.8
Benzene	5.42	30.3	80.2	94.3
NaCl	804.3	124	—	—
Aluminum	658	76.8	—	—
Copper	1083	42	—	—
Silver	961	21.07	—	—
Gold	1064	15.8	—	—
Lead	327	5.86	—	—
Tin	232	14.0	—	—
Zinc	419	28.13	—	—
Mercury	-39	2.82	—	—

TABLE 12.4. Thermal Conductivity of Solid Substances

Substance	Temperature, °C	Thermal conductivity,* cal/cm-sec-°C
Aluminum	18	0.480
	100	0.492
Copper	18	0.918
	100	0.908
Gold	18	0.700
Silver	18	0.974
Iron	18	0.161
Steel	18	0.115
Lead	18	0.083
Mercury (liq.)	17	0.0197
Tin	18	0.155
Zinc	18	0.265
Ice	0	0.0022
Glass	20	0.0025
Silica (fused)	20	0.00237
Asbestos	500	0.00019

* These values must be multiplied by 10^{-5} to obtain thermal conductivities in units of kcal/m-sec-°C.

TABLE 12.5. Thermal Expansion Coefficients of Solid and Liquid Substances

Substance	Temperature, °C	α, °C^{-1}	β, °C^{-1}
Aluminum	20–100	23.8×10^{-6}	
Brass	25–100	19.0×10^{-6}	
Copper	25–100	16.8×10^{-6}	
Gold	15–100	14.3×10^{-6}	
Silver	15–100	18.8×10^{-6}	
Iron	-18–100	11.4×10^{-6}	
Steel	0–100	10.5×10^{-6}	
Lead	18–100	29.4×10^{-6}	
Invar*	20	0.9×10^{-6}	
Tin	18–100	26.9×10^{-6}	
Zinc	10–100	26.3×10^{-6}	
Glass	0–100	8.9×10^{-6}	
Ice	-20–0	51.0×10^{-6}	
Quartz (fused)	0–100	0.5×10^{-6}	
Mercury	0–100		0.1818×10^{-3}
Acetic acid	16–107		1.06×10^{-3}
Acetone	0–50		1.32×10^{-3}
Ethanol	27–50		1.01×10^{-3}
Methanol	0–60		1.13×10^{-3}
Benzene	11–80		1.18×10^{-3}
Carbon tetrachloride	0–76		1.18×10^{-3}
Ether			1.51×10^{-3}
Pentane			1.464×10^{-3}

* Nickel–steel low-expansion alloy.

TABLE 12.6. Critical Temperatures, Pressures, and Densities

Substance	T_{crit}, °C	P_{crit}, atm	ρ_{crit}, g/cm³
Helium	−267.9	2.26	0.0693
Hydrogen	−239.9	12.8	0.0310
Oxygen	−118.8	49.7	0.430
Nitrogen	−141.1	33.5	0.3110
Argon	−122	48.	0.531
Neon	−228.7	25.9	0.484
Krypton	−63	54.	0.78
Xenon	16.6	58.2	1.155
Chlorine	144.0	76.1	0.573
Carbon dioxide	31.1	73.0	0.460
Sulfur dioxide	157.2	77.7	0.52
Ammonia	132.4	111.5	0.235
Methane	−82.5	45.8	0.162
Ethane	32.1	48.8	0.21
Propane	95.6	43.	—
Ethylene	9.7	50.9	0.22
Acetylene	36.	62.	0.231
Water	374.0	218.5	0.325
Methanol	240.0	78.7	0.272
Ethanol	243.1	63.1	0.275
Acetone	235.0	47.	0.268
Acetic acid	321.6	57.2	0.351
Benzene	288.5	47.7	0.304

SUMMARY

Heat is energy that flows from one body to another, or from one part of a body to another, as a result of a temperature difference between two objects or between different parts of the same object.

The internal energy of a system is the potential energy of interaction between its individual atoms or molecules plus the kinetic energy of translation, vibration, or rotation they may have as a result of their thermal agitation. It does not include energy arising as a result of the motion of the system as a whole or arising from *external* forces that act on the system.

The temperature of a system, roughly speaking, is a quantitative measure of the average thermal kinetic energy of its atoms or molecules. It can be measured by constructing thermometers in which the expansion or contraction of a liquid such as mercury provides a quantitative indication of temperature. On the Celsius, or centigrade, temperature scale, the freezing point of water is taken to be at 0°C and the boiling point under normal atmospheric pressure at 100°C. On the Fahrenheit scale, these same temperatures are 32°F and 212°F, respectively. The relations between Celsius and Fahrenheit temperatures are

$$T_C = \tfrac{5}{9}(T_F - 32) \quad \text{and} \quad T_F = 1.8 T_C + 32$$

The first law of thermodynamics is an expression of the law of conservation of energy. It states that for any system,

$$\Delta Q = \Delta U + \Delta W$$

where ΔQ is the amount of heat energy absorbed by the system, ΔU is the change in the system's internal energy, and ΔW is the work done by the system.

A system is in thermal equilibrium if there is no flow of heat or any other form of energy between the various parts of the system, or between the system and its surroundings. Since a flow of heat occurs only where there is a temperature difference, this means that all parts of the system are at the same temperature and that the system is either thermally isolated from the surroundings or has the same temperature as the surroundings.

Joule's experiment demonstrated the equivalence of heat and mechanical energy. It showed that

777.9 ft-lb = 1 Btu

4186 joules = 1 kcal

4.186 joules = 1 cal

where one calorie is the heat required to raise the temperature of 1 g of water at 14.5°C by 1°C, and 1 Btu is the heat needed to produce a 1°F rise in the temperature of 1 pound of water initially at 60°F.

The specific heat of a substance is defined as the heat required to raise the temperature of unit mass of the material, or unit weight, by one degree, Celsius or Fahrenheit, as the case may be. Molar specific heats are also frequently used. As a consequence of this definition, the amount of heat dQ absorbed by a system and the temperature rise dT it experiences are related by

$$dQ = mC\,dT$$

where C is the specific heat. If the specific heat is independent of temperature, this can be written as

$$\Delta Q = mC\,\Delta T$$

A liquid and a solid, at a given pressure, can be in equilibrium at a single temperature referred to as the melting temperature. Heat is absorbed in transforming solid to liquid under these circumstances, without effecting any lasting change in the temperature of the system. The amount of heat required to transform a unit mass of a substance from solid to liquid at the melting temperature is referred to as its latent heat of fusion. Similar circumstances of equilibrium exist between liquid and vapor phases and between solid and vapor phases. In the case of liquid–vapor equilibrium at the boiling temperature, the amount of heat required to transform unit mass of substance from liquid to vapor is called the latent heat of vaporization. The corresponding quantity for solid–vapor equilibrium is referred to as latent heat of sublimation. Solid, liquid, and vapor can all exist together in equilibrium at a particular temperature and a particular pressure referred to as the "triple point." At the triple-point temperature and pressure, the substance in question is boiling and freezing at the same time!

When a confined liquid–vapor system is heated, a *critical temperature* (and a corresponding *critical pressure*) at which the liquid–vapor interface disappears is eventually attained. At this point, the liquid and vapor phases cease to have separate identities.

Heat energy is transferred primarily by conduction, a process wherein a temperature gradient within a substance gives rise to a flow of heat. The heat that flows per unit area per unit time is proportional to the temperature gradient, from which we can write

$$\frac{\Delta Q}{\Delta t} = KA \frac{\Delta T}{\Delta x}$$

where the temperature gradient $\Delta T/\Delta x$ is represented as the ratio of the temperature change ΔT that occurs in a linear distance Δx within the material. The constant of proportionality K is a quantity characteristic of the particular substance at hand and is referred to as its *thermal conductivity*.

Energy may also be transported by convection, in which mass transport of material acts eventually to transport thermal energy, and by radiation, in which the emission and subsequent reabsorption of electromagnetic energy serves to accomplish the same end result.

The linear dimensions of solid objects and the volume of liquid and gaseous substances change in response to temperature changes. This effect is referred to as thermal expansion. The linear expansion Δl of a solid substance can be expressed as

$$\Delta l = \alpha l_0 \, \Delta T$$

where ΔT is the temperature change, l_0 is the initial length, and α is a proportionality constant characteristic of the substance, called the coefficient of linear expansion. In the case of fluids, the volume change ΔV is related to the temperature change by a similar equation of the form

$$\Delta V = \beta V_0 \, \Delta T$$

where the constant β is referred to as the volume expansion coefficient. It is possible, also, to show that the linear and volume expansion coefficients of any substance are related by

$$\beta = 3\alpha$$

QUESTIONS

1. Does it make sense to talk about the temperature in a region which is essentially a perfect vacuum, such as interstellar space?

2. Sometimes it is very difficult to open a bottle of ketchup. Heating the cap under the hot water tap generally helps. Explain.

3. A sheet of copper has a hole drilled in it. As the temperature increases, what happens to the diameter of the hole?

4. When a body is in thermal equilibrium, what does this imply in terms of temperature? What does it mean to say it is in thermal equilibrium with its surroundings?

5. If you hold a piece of metal in your hand and touch the metal to a block of ice, your hand soon feels cold. Can you explain why?

6. A thermos bottle consists of double-walled glass with a layer of vacuum between. The two inner glass surfaces are silvered. Explain how this arrangement reduces heat loss due to conduction, convection, and radiation.

7. On a chilly morning, when the temperature is just above freezing, the grass in the back yard is covered with frost but the stones on the driveway are not. Explain this in terms of the conductivity of grass and stone.

8. Standing barefoot on a slate floor feels colder than standing on a rug even though both are at the same temperature. Explain.

9. The process of boiling, which occurs near the surface of a liquid, serves to cool the remaining liquid. Explain this statement. What prevents the liquid's temperature from decreasing?

10. Explain how a pressure cooker can be used to shorten the time required to cook a meal.

11. On a certain day, the temperature outdoors is 102°F and yet we are able to maintain a body temperature of 98.6°F. How is this accomplished?

12. Liquid contained in a beaker experiences a rise in temperature as a result of rapid stirring. Has heat been added to the liquid?

13. We know that 1 calorie is equivalent to 4.186 joules. Can you describe some features of a *hypothetical* world in which 1 joule would be 4.186 calories?

14. Comment on the following statement: "Any calibrated instrument which responds to temperature changes is a good thermometer."

15. When two systems are brought into physical contact, they adjust so that the combined system tends toward a uniform value of temperature. Can you think of other quantities which behave in the same manner?

16. Can you think of some examples in which the internal energy of a system increases without the addition of heat?

17. Are there situations in which heat is added to a system without causing a corresponding temperature increase?

18. In a foreign country, a doctor measures a tourist's temperature as 40°. What scale is he using and how sick is his patient?

PROBLEMS

1. The boiling point of benzene is 80.2°C at 1 atmosphere. Convert this to a temperature in Fahrenheit.

2. Make a graph of T_F vs T_C. Is there a temperature at which $T_F = T_C$? If so, find its value.

3. Mercury solidifies at -38.9°C and boils at 357°C at 1 atmosphere. Convert these temperatures to degrees Fahrenheit.

4. Normal body temperature is 98.6°F. On a cold January day, the outdoor temperature is 3°F. Find the temperature difference between your body and the air in degrees Celsius. Determine both temperatures in degrees Celsius.

5. At what temperature is the Fahrenheit reading four times as large as the Celsius reading?

6. A motor of 0.5 hp is used to stir 8 gallons of water (1 gallon = 8.3 lb). If the motor is used for 30 minutes and all the work goes into heating the water, by how many degrees Fahrenheit does the water temperature rise?

7. The hot water tank in the basement of a house has been at 65°F for several days while repairs were being done on a ruptured natural gas line. If the tank has a capacity of 42 gallons and the natural gas can deliver 1000 Btu per cubic foot, how much gas is needed to heat the water to 145°F? Assume that all the natural gas is used to heat the water and there are no losses of energy.

8. A college professor owns a 350 w immersion heater. He wishes to make a pot of tea and must therefore bring $\frac{1}{2}$ kg of water from 65°F to 212°F. How long does it take to boil the water?

9. A saucepan with three quarts of water in it is put on the stove in preparation for a spaghetti dinner. It is found that the water is brought from 70°F to boiling in 10 minutes. Assuming no energy loss, find the power output of the stove burner. Unfortunately, just as the water begins to boil, a neighbor calls to say that Craig has just destroyed all her tomatoes and that she's very angry. The conversation lasts a long time and soon the smell of a burning pot is evident. How long did the phone call last if it terminated just as the pot began to burn?

10. Steam at 100°C is added to a 50-g piece of ice at −10°C. How much steam is needed to just melt the ice?

11. A lead bullet of mass 25 g is fired at 350 m/sec into a wood block and is brought to rest. If the specific heat of lead is 0.031 cal/g-°C, find the temperature increase of the bullet assuming all the energy is used to heat the bullet.

12. The specific heat of a certain substance is temperature dependent; $C(T)$ of Eq. (12.3.9) is given by $C(T) = a + bT$. If the temperature is increased from T_1 to T_2, show that the heat required per unit mass is given by

$$\frac{\Delta Q}{m} = (T_2 - T_1)\left[a + \frac{b}{2}(T_2 + T_1)\right]$$

13. A mother tells her 10-year-old daughter to let the water run so that she can take a bath. The child turns on only the hot water faucet, and 25 gallons of water at 140°F is placed in the tub. How many gallons of cold water at 50°F will be needed to lower the temperature to 105°F?

14. A calorimeter of mass 75 g is made of aluminum which has a heat capacity of 0.22 cal/°C-g. It contains 100 g of a mixture of water and ice. A 90-g chunk of aluminum at a temperature of 100°C is dropped into the calorimeter, which then increases in temperature to 6°C. Find the mass of ice originally present in the calorimeter.

15. Assume the heat of fusion of snow is 340 joules/g. A skier of mass 90 kg descends a 32° slope at a speed of 16 m/sec. How many grams of snow melts beneath his skis per second if all the energy available goes into melting the snow?

16. Water, 150 g, is placed in an insulated vessel of very small heat capacity. An immersion heater brings the water from 18°C to the boiling point in 9 minutes. Find the power output of the heater. How long will it take to vaporize this amount of water assuming the full power of the heater is utilized?

17. Three liquids, A, B, and C, are maintained at temperatures T_A, T_B, and T_C. When equal masses of A and B are mixed, the temperature of the mixture is T_{AB}, and when equal masses of B and C are mixed, the temperature becomes T_{BC}. What temperature T_{AC} (expressed in terms of T_{AB} and T_{BC}) would be obtained by mixing equal amounts of A and C?

18. A plate of copper has dimensions of 100 cm by 100 cm by 3 cm. One side is maintained at temperature 0°C, while the other side is held at 100°C. If the average thermal conductivity is $K_{Cu} = 0.92$ cal/sec-cm-°C, how much heat flows through the plate under steady-state conditions during a 15-minute period?

19. An automobile radiator contains 20 quarts of water. If 300,000 calories of heat is given to the cooling system and it is all used to raise the water temperature, find the rise in temperature.

20. The linear expansion coefficient of a lead plug is 29.4×10^{-6}°C^{-1}. If the plug has a diameter of 11.000 cm at 40°C, at what temperature will it fit exactly into a hole of uniform diameter 10.996 cm?

21. A copper rod has a length of 50.00 cm when measured with a steel tape measure at a temperature of 10°C. What would the length be when measured at 30°C?

22. A section of copper wire has a length of 400 m. What is the fractional change in length due to a temperature change from 10°C to 30°C?

23. At 20°C, the density of gold is 19.30 g/cm^3. What is its density at 100°C?

24. The specific heat of a certain substance is found to be $C(T) = C_1 + C_2 T^2$, where T is in °C and C_1 and C_2 are constants. Find the amount of heat ΔQ per unit mass required to raise the temperature of the substance from 0°C to 10°C. If one assumes instead that the specific heat at the temperature 5°C will provide a good approximation in this temperature range, obtain an expression for the error introduced in ΔQ by using this approximation.

25. A physics student decides to measure the mass of a copper kettle in a rather unconventional way. He pours 5 kg of water at 70°C into the kettle which is initially at 10°C. He then finds that the final temperature of water and kettle is 66°C. From this information, find the mass of the copper kettle. Comment on the sources of error involved in this measurement.

26. A 100-g lump of ice at −20°C is placed in a 300-g copper calorimeter and 250 g of water at 30°C is poured in. Determine the final temperature of the system. Is the system ice, water, or a mixture?

27. A 300-g copper calorimeter contains 100 g of ice. The system is initially at 0°C. If 50 g of steam at 100°C (1 atm of pressure) is introduced into the calorimeter, what is the final temperature of the contents?

28. Find the amount of heat needed to convert 10 g of ice at −15°C to steam at 100°C. Compare this heat with the

energy available when a 3-kg object falls through a vertical distance of 10 meters.

29. A storm window shown in the figure consists of a layer of air sandwiched between two plates of glass. If the thermal conductivities of glass and air are, respectively, K_g and K_a, show that the thermal conductivity of the storm window is

$$K = \frac{2L_1 + L_2}{\dfrac{L_1}{K_g} + \dfrac{L_2}{K_a} + \dfrac{L_1}{K_g}}$$

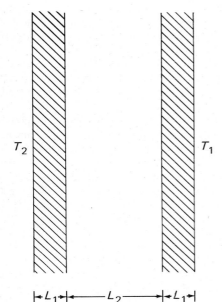

T_2 T_1

$\vdash L_1 \dashv \vdash \!\!\longleftarrow L_2 \longrightarrow\!\! \vdash L_1 \dashv$

30. The heat conductivity of air is $K_a = 6 \times 10^{-6}$ kcal/sec-m-°C while that of glass is $K_g = 2 \times 10^{-4}$ kcal/sec-m-°C A window of area 2 m^2 and thickness 0.006 m conducts heat from the interior to the exterior of a house. If the inside temperature is 20°C and the outside temperature is 3°C, find the heat loss over a 24-hour period. If the price of fuel is \$0.015 per 6500 Btu, find the heating cost per day just due to losses through this window. Explain why this result is not an accurate estimate of the actual cost.

31. For the preceding problem, the glass window is replaced by a storm window in which a layer of air of thickness 0.06 m is between two layers of glass, each 0.04 m thick. Find the heat loss over 24 hours and compare this with the result obtained with ordinary glass.

32. The differential form of Eq. (12.5.2) is $dQ/dt = KA\, dT/dx$. Although this was given for one-dimensional conduction through a plane area A, it is also applicable to other cases. For example, heat can flow radially outward from a small hot spherical object. In this case, x is replaced by r and A by $4\pi r^2$. Show that the temperature difference corresponding to distances r_1 and r_2 from the object is given by

$$T_1 - T_2 = \left(\frac{\Delta Q}{\Delta t}\right)\frac{1}{4\pi K}\left(\frac{1}{r_1} - \frac{1}{r_2}\right)$$

33. Consider the case where heat flows radially outward from a very long cylindrical object of radius r_0 and surface temperature T_0 through a medium of thermal conduc-

tivity K. Find the temperature T at a radial distance r (where $r > r_0$) from the center of the system.

*34. A very long cylindrical pipe of inner radius r_0 and outer radius r_1 carries steam at temperature T_0. It is covered with a concentric cylindrical layer of insulating material of inner radius r_1 and outer radius r_2. The outside of this layer is maintained at a known constant temperature T_2. The thermal conductivity of the pipe is K_1 and that of the insulating material is K_2. Utilizing the results of the preceding problem, find the temperature T_1 at $r = r_1$, where the outer surface of the pipe is in contact with the inner surface of the insulating substance.

*35. The outer surface of a long cylindrical wire of radius R_1 and length l is maintained at a constant temperature T_1, which is greater than that of the surrounding air. Assuming it requires an energy per unit time of H to maintain this surface temperature, show that the temperature at a distance R_2 from the center of the wire is given by

$$\left(\frac{H}{2\pi K_a l}\right)\ln\frac{R_2}{R_1} = T_1 - T_2$$

where K is the thermal conductivity.

36. The density of ice is 0.92 g/cm^3 and its thermal conductivity is 0.0022 cal/cm-sec-°C. Its heat of fusion is 79.7 cal/g. A lake contains a layer of ice x cm thick when the air temperature is -15°C. Show that the rate of increase of this layer, in cm/sec, is given by

$$\frac{dx}{dt} = \frac{(0.0022)(15)}{(79.7)(0.92)}\frac{1}{x}$$

and find this rate of increase when $x = 10$ cm. If at $t = 0$, $x = 0$, how long does it take for the thickness to increase to 10 cm?

37. A pan of water at 0°C is placed outdoors when the temperature is -10°C. If the pan has an area of 500 cm^2 and the water is 5 cm in depth, how long does it take for all the water to turn to ice? Neglect any effects due to the heat capacity of the pan.

38. (a) How much work is needed to transfer a molecule of water at 0°C and 1 atm pressure from an ice crystal into the liquid phase? (b) How much work is needed to transfer a molecule of liquid water at 100°C and 1 atm pressure into the vapor phase? (c) Do processes such as these result in a change in temperature? (d) Do they result in a change in internal energy? (Hint: There are 6.023×10^{23} molecules in one "gram mole" of water, which has a mass of 18.0 g. The density of ice at 0°C and 1 atm pressure is 0.92 g/cm^3, while the density of steam at 100°C and 1 atm pressure is 0.5974 g/l.) Neglect the work done against the atmosphere by the expansion or contraction associated with the phase changes.

39. Show that the change of density with temperature that occurs as a result of thermal expansion can be expressed by $\Delta\rho/\Delta T = -\beta\rho$.

40. A steel bar, whose Young's modules is 3.0×10^7 pounds/in^2, and whose linear expansion coefficient is 1.20×10^{-5} °C^{-1}, is held by its ends in a rigid vise. The cross-sectional area of the bar is 1.00 in^2. How much force will be exerted on the jaws of the vise if the temperature of the bar is increased by 100°C?

41. In the situation described in the preceding problem, show

401

that if the bar's Young's modules is Y, its linear expansion coefficient α, its cross-sectional area A, and the temperature rise ΔT, the force exerted on the vise is given by $F = \alpha A Y (\Delta T)$.

42. A very rigid confining vessel has a volume expansion coefficient β_c, and is completely filled with a liquid whose bulk modulus is B_l and whose volume expansion coefficient is β_l. (a) What is the pressure increase that would result from a temperature rise ΔT? (b) What would this pressure rise amount to for alcohol ($B_l = 1.28 \times 10^5$ pounds/in^2, $\beta_l = 7.5 \times 10^{-4}$ °C^{-1}) confined in a steel vessel ($\beta_c = 3.6 \times 10^{-5}$ °C^{-1}) for a temperature rise of 100°C?

43. Show that the coefficient γ defined by Eq. (12.6.22), expressing the thermal expansion of the area of a thin plate of a material whose linear expression coefficient is α, is given by $\gamma = 2\alpha$.

44. An automobile manufacturer utilizes in his engines a cast iron cylinder block, with cylinders bored to a diameter of 3.500 inches at 20°C. Aluminum pistons are to be fitted to these cylinders. Unfortunately, the coefficient of linear expansion of cast iron is 1.2×10^{-5} °C^{-1}, while that of aluminum is 2.4×10^{-5} °C^{-1}, which means that unless the pistons are deliberately designed slightly smaller than the cylinder bore when the engine is cold, they will expand more rapidly than the cylinders as the engine warms up and will eventually "seize" to the cylinder walls, causing the engine to self-destruct. The manufacturer, regarding this behavior as undesirable, specifies that the pistons shall not "seize" until their temperature rises to 420°C. (a) What is the maximum diameter the pistons may have at 20°C in order that this requirement be satisfied? What will be the clearance between pistons and cylinder walls at the normal operating temperature of 220°C? Assume in all cases that the surface temperature of the cylinders and the pistons is the same.

45. A metal bar of length l_1 is welded, end to end, to a bar of a different metal, whose length is l_2. The coefficient of linear expansion of the first metal is α_1, that of the second α_2. What is the effective linear expansion coefficient of the composite welded bar?

46. A metal bar of length L at temperature 0°C is heated nonuniformly in such a way that the temperature is given as a function of the distance x along its length, measured from one end, as $T(x) = T_0 \sin(\pi x/L)$. According to this, the ends at $x = 0$ and $x = L$ are still at zero temperature, while at $x = L/2$, where the argument of the sine function is $\pi/2$, the temperature has the maximum value T_0. The coefficient of linear expansion of the bar is α. Find the increase in the length of the bar as a function of α and T_0. (Hint: What is the average temperature of the bar?)

THERMAL PROPERTIES OF IDEAL SYSTEMS AND THE ABSOLUTE TEMPERATURE SCALE

13.1 Introduction

In the last chapter, we tried to develop the concept of heat as a form of energy, to accurately define the terms *heat, internal energy,* and *temperature,* to state the first law of thermodynamics, and to introduce and examine briefly the more important thermal properties of matter. We shall now attack the problem of describing the thermal properties of the very simplest types of systems—"ideal" gases and "perfect" crystals—from the basic dynamics of their individual molecules. In doing this, we shall find it convenient, indeed necessary, to introduce the concept of the *absolute temperature scale.* These systems contain astronomical numbers of atoms or molecules, and

we shall soon find that the only reasonable way of understanding their thermal properties on the basis of molecular mechanics lies in finding certain *average* dynamical quantities and relating the observed physical properties of the system to these average molecular dynamical properties. The science of the development of thermal properties from molecular mechanics is referred to as *kinetic theory,* and the techniques of relating the overall macroscopic behavior of material systems to the average behavior of their molecular constituents are known as *statistical mechanics.*

We shall also be concerned with further developing our understanding of thermodynamics, particularly in the direction of examining the application of the first law of thermodynamics to ideal gases and perfect crystals. Finally, we shall try to develop general methods of treating "nonideal" or "imperfect" systems, even though we shall not be able to go very far toward demonstrating their properties in explicit mathematical terms.

13.2 The Equation of State

Consider the case of a given quantity of gas, 1 mole,[1] for example, which is maintained at a *constant temperature* T_0. Under such conditions, there is a unique relation between the volume and the pressure; once the pressure is specified, the volume is uniquely determined, and vice versa. If the temperature is changed to another temperature T_1 and thereafter maintained at that value, there will again be a unique relation between volume and pressure for this new temperature, though it will not be the same as the one prevailing at the old temperature T_0. In the same way, if the gas is confined to a given *constant volume* V_0, there will be some definite relationship between pressure and temperature in that, once one of these quantities is specified, the other is also determined. If the volume is changed to some other value V_1 and then fixed at that value, there is again a definite relationship connecting pressure and temperature, though not the same one which held for the original fixed volume V_0. Finally, if the *pressure* is held constant, there is a unique relation between volume and temperature, which may be different at different values of the pressure.

The most general form of these relationships may differ considerably from one gas to another, but for *all* gases at temperatures high enough and pressures small enough to ensure that the *density of the gas is very much less than its density at the critical point*, these relations may be expressed quite accurately by the following equations:

$$PV = P_0V_0 \qquad \begin{matrix} \text{constant temperature} \\ T = T_0 \end{matrix} \qquad (13.2.1)$$

$$\frac{P}{T - \lambda} = \frac{P_0}{T_0 - \lambda} \qquad \begin{matrix} \text{constant volume} \\ V = V_0 \end{matrix} \qquad (13.2.2)$$

$$\frac{V}{T - \lambda} = \frac{V_0}{T_0 - \lambda} \qquad \begin{matrix} \text{constant pressure} \\ P = P_0 \end{matrix} \qquad (13.2.3)$$

where P_0 and V_0 represent, respectively, the pressure

---

[1] A gram-mole is that quantity of a substance whose mass, in grams, numerically equals the mass of one molecule of the substance, expressed in atomic mass units (amu) wherein the mass of the carbon atom is defined to be 12.000 ... amu. Thus, since on this scale the mass of the oxygen molecule is 32.00 amu, a gram-mole of oxygen would contain 32.00 g of gaseous oxygen. The kilogram-mole is defined in a similar way.

and volume of a given quantity of the gas at temperature T_0, while P, V, and T are the corresponding values of pressure, volume, and temperature under other conditions. The quantity λ is a constant having the *same value for all gases;* if temperatures are measured in Celsius degrees, its value is $\lambda = -273.16°C$, while if the Fahrenheit scale is used, it is given by $\lambda = -459.69°F$. Equations (13.2.1), (13.2.2), and (13.2.3) are presented here simply as representing the results of experimental observations on the behavior of actual gases at high temperatures and low pressures—we have not derived them in any way at all. They define the behavior of what we usually refer to as an *ideal gas*, and the fact that the *observed* behavior of all gases at temperatures sufficiently high and pressures sufficiently low is in accord with these equations implies that *all gases behave very much like ideal gases at temperatures high enough and pressures low enough to render the density far less than the critical density.* Equation (13.2.1) is a mathematical statement of what is sometimes called Boyle's law, while (13.2.3) is often referred to as Charles's law.

These three equations can be combined into a single one fairly simply. Consider, first of all, Eq. (13.2.1); it tells us that for an ideal gas at constant temperature, the product PV has some constant value, at the temperature T_0, represented by the product P_0V_0 of some initial reference volume V_0 and some initial reference pressure P_0. This relationship between pressure and volume is plotted as the curve marked $T = T_0$ in Fig. 13.1. If the temperature has the constant value $T_0 + \Delta T$, then the product PV will still be constant, but the value of this constant is no longer P_0V_0 but will be different in view of the change in temperature. For temperature $T_0 + \Delta T$, therefore, the pressure–volume relation will still be a curve of the form $PV = \text{const.}$, but with a different value of the constant. Such a curve is plotted for temperature

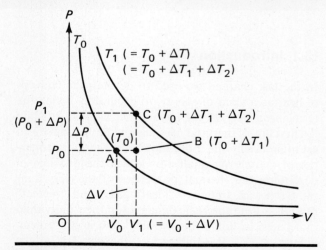

FIGURE 13.1. Pressure–volume relationship for an ideal gas at two fixed temperatures.

$T_0 + \Delta T$ in Fig. 13.1, and its equation can be written as $PV = P_1 V_1$ inasmuch as the point C for which $P = P_1 = P_0 + \Delta P$ and $V = V_1 = V_0 + \Delta V$ lies on the curve.

Now suppose that we start with the gas at point A, where the temperature is T_0, the pressure P_0, and the volume V_0, and raise the temperature by some amount ΔT, the final state of the gas then being represented by point C, at temperature $T_1 = T_0 + \Delta T$, pressure $P_1 = P_0 + \Delta P$, and volume $V_1 = V_0 + \Delta V$. The route of intermediate stages of pressure and volume between A and C could be anything we might choose; it could be taken along a straight diagonal line joining A and C by adjusting the experimental conditions properly at each stage of the process, or it could be made to follow more complicated curved paths, if necessary. Since we know the relations between P, T, and V for constant temperature, constant pressure, and constant volume processes, however, it is advantageous for us to choose a route which first goes along a horizontal path AB at *constant* pressure P_0 until volume V_1 is reached; at this point, an intermediate temperature change ΔT_1 will have occurred. Our path then proceeds along the vertical line BC at *constant volume* V_1 until the pressure P_1 is reached at C. This involves a second intermediate temperature change ΔT_2, and it must be true that

$$\Delta T_1 + \Delta T_2 = \Delta T \qquad (13.2.4)$$

Along path **AB**, since the process occurs at constant pressure, we may conclude from (13.2.3) that the final and initial volumes V_1 and V_0 and the corresponding temperatures $T_0 + \Delta T_1$ and T_0 are related by

$$\frac{V_1}{T_0 + \Delta T_1 - \lambda} = \frac{V_0}{T_0 - \lambda} \qquad (13.2.5)$$

In the same way, along path **BC**, since the process takes place at constant volume, we can infer from (13.2.2) that

$$\frac{P_1}{T_0 + \Delta T_1 + \Delta T_2 - \lambda} = \frac{P_0}{T_0 + \Delta T_1 - \lambda} \qquad (13.2.6)$$

Solving (13.2.5) for V_1 and (13.2.6) for P_1 and multiplying the resulting expressions to find the product $P_1 V_1$, it is evident that

$$P_1 V_1 = P_0 V_0 \left(\frac{T_0 + \Delta T_1 + \Delta T_2 - \lambda}{T_0 + \Delta T_1 - \lambda} \right) \left(\frac{T_0 + \Delta T_1 - \lambda}{T_0 - \lambda} \right) \qquad (13.2.7)$$

or, recalling that $\Delta T_1 + \Delta T_2 = \Delta T$ and that $T_1 = T_0 + \Delta T$,

$$\frac{P_1 V_1}{T_1 - \lambda} = \frac{P_0 V_0}{T_0 - \lambda} = \text{const.} \qquad (13.2.8)$$

It is important at this point to note that nothing we have done assumes that ΔT, ΔV, and ΔP are small quantities or must approach differentials as a limit. Indeed, these quantities may be as large as we please, and, therefore, T_1, V_1, and P_1 may be as different from T_0, V_0, and P_0 as we choose—the delta (Δ) notation was adopted simply for convenience in doing the mathematics. We may then just as well label the variables associated with the final state T, V, and P as in (13.2.1), (13.2.2), and (13.2.3) and write (13.2.8) as

$$\frac{PV}{T - \lambda} = \frac{P_0 V_0}{T_0 - \lambda} = \text{const.} \qquad (13.2.9)$$

This single equation describes the relationship between the pressure, temperature, and volume of an ideal gas and is referred to as the *equation of state* of the ideal gas. It embodies all the information contained in the three equations (13.2.1), (13.2.2), and (13.2.3) since it reduces to (13.2.1) for a process in which the temperature has the fixed value T_0, to (13.2.2) for a process conducted at constant volume V_0, and to (13.2.3) for a process carried out at constant pressure P_0.

The equation of state of an ideal gas may be written in a slightly different, and somewhat more convenient, form by noting that (13.2.9) states that

$$PV = c(T - \lambda) \qquad (13.2.10)$$

where c is a constant. Now, for any given pressure P and temperature T, we must certainly expect that the volume of gas must be directly porportional to the mass of gas in the system; if we double the amount of gas, we must clearly double its volume if P and T are to remain fixed. Therefore, the constant c must be proportional to the amount of gas in the system, and we can represent it as

$$c = nR \qquad (13.2.11)$$

where n is the number of moles of gas present and R is another constant usually referred to as the *gas constant*. Equation (13.2.10) then becomes

$$PV = nR(T - \lambda) \qquad (13.2.12)$$

This is the form in which the equation of state of an ideal gas is usually expressed, though the form (13.2.9) is also useful at times. The ideal gas constant R is found, by experiment, to be *the same for all gases* that exhibit the "ideal" behavior of Eq. (13.2.9) and has the value

$R = 8314$ joules/mole-°C

when mole = kilogram-mole

$= 8.314$ joules/mole-°C

when mole = gram-mole

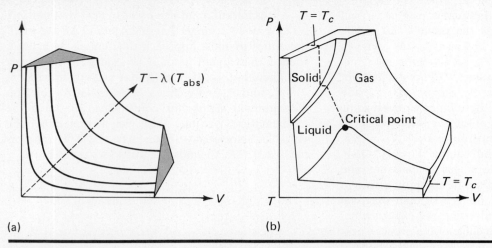

FIGURE 13.2. (a) PVT surface representing the equation of state of an ideal gas. (b) PVT surface for a real substance, exhibiting solid, liquid, and gas phases.

$$= 8.314 \times 10^7 \text{ ergs/mole-°C}$$
$$\text{when mole} = \text{gram-mole}$$

$$= 0.08206 \text{ liter-atm/mole-°C}$$
$$\text{when mole} = \text{gram-mole}$$

$$= 1.986 \text{ kcal/mole-°C}$$
$$\text{when mole} = \text{kilogram-mole}$$

$$= 1.986 \text{ cal/mole-°C}$$
$$\text{when mole} = \text{gram-mole}$$

It should be noted that the product PV has the dimensions of work, since $P = F/A$ and $V = Al$. The unit liter-atm/mole-°C, much beloved by chemists, is a bastard unit if there ever was one, but even good physicists admit that it is convenient for some purposes. By setting $P = 1$ atm, $n = 1$ mole, $R = 0.08206$ liter-atm/mole-°C, and $\lambda = -273.16$°C in Eq. (13.2.12), it is evident that this equation states that *the volume of one gram-mole of any ideal gas at 1 atm pressure and 0°C is 22.415 liters.*

The equation of state of a substance is a mathematical relationship between volume, pressure, and temperature for that substance. We have investigated the explicit form of this relationship only for the case of the "ideal gas," but such a relationship exists (even though it may be difficult to express it in mathematical form) for gases in the "nonideal" range of temperature and pressure and even for substances in the liquid or solid form. Much useful information about the thermal properties of any substance is embodied in its equation of state. Unfortunately, the equation of state cannot be deduced from the laws of thermodynamics alone. It may be arrived at, of course, from the results of experimental measurements; our understanding of the ideal gas is at this point derived from just such data. But it may also be shown to follow from an understanding of the dynamics of the molecules of the system, provided that a correct and detailed expres-

sion for the forces acting between individual molecules can be found. We shall see in a later section how the ideal gas equation of state can be derived in this fashion. The important thing to understand here is that the derivation of the equation of state from molecular mechanics rests upon the discovery of a reasonable "model" for representing the intermolecular forces.

The equation of state of any substance can be plotted as a surface referred to three mutually perpendicular axes representing P, V, and T. Such a surface, representing a plot of (13.2.12) for an ideal gas, is shown in Fig. 13.2a. Another, for a real substance which condenses into solid and liquid phases in certain ranges of temperature and pressure, is illustrated in Fig. 13.2b. The reader will note that the equilibrium phase diagrams discussed in Chapter 12 are two-dimensional plane sections of this three-dimensional plot. The equation of state is usually expressed as a relationship between volume, temperature, and pressure but can be expressed in terms of any equivalent set of state variables such as density, temperature, and pressure.

13.3 The Kelvin or Absolute Temperature Scale

One of the most striking features of the ideal gas law, in the forms presented in the preceding section, is the fact that while the pressure and volume enter the equations simply as P and V, the temperature always appears as the expression $T - \lambda$. It is not difficult to understand why this is so; the origin of volume and pressure has been chosen, logically enough, as the point where those quantities are zero, but the zero of temperature has been chosen *arbitrarily* for both

Celsius and Fahrenheit scales. Thus, though we never encounter negative values of volume nor, ordinarily, negative pressures, it is quite possible to achieve negative Celsius or Fahrenheit temperatures. The appearance of the quantity $T - \lambda$ in our ideal gas law reflects this offhand choice of the zero of temperature. Moreover, it suggests the possibility of making a more natural choice of the origin of temperature—the definition of an *absolute zero* of temperature—and the construction of a more fundamental temperature scale based on this natural minimum temperature. This possibility should come as no surprise, for if our understanding of temperature as a measure of the internal kinetic energy of a system is correct, we should be able in principle to reduce this quantity to zero, or at least to an irreducible minimum, whereupon no further removal of heat energy could ever be accomplished and no further reduction in temperature would be possible.

Let us proceed in the simplest way—which happens to be correct—and define the *absolute*, or *Kelvin*, temperature T_K by

$$T_K = T_C - \lambda \qquad (13.3.1)$$

where T_C represents the Celsius temperature; since the experimental results having to do with the behavior of ideal gases show that $\lambda = -273.16°C$, this can be written as

$$\boxed{T_K = T_C + 273.16°} \qquad (13.3.2)$$

From this, it is clear that the absolute zero of temperature corresponds to $-273.16°C$ and that $0°C$ corresponds to $+273.16°K$. Room temperature (about $27°C$) falls at about $300°K$ on the absolute scale.

Using T now to represent the *absolute* temperature, the ideal gas laws can be written using T to replace $T - \lambda$; (13.2.9) and (13.2.12) now become

$$\boxed{\frac{PV}{T} = \frac{P_0 V_0}{T_0}} \qquad (13.3.3)$$

$$\boxed{PV = nRT} \qquad (13.3.4)$$

These are the simplest possible forms of the equation of state for an ideal gas; in using these equations, however, it is very important to remember that T *must be stated in Kelvin degrees*.

Now let us consider the case of an ideal gas maintained at *constant pressure*. Solving (13.3.4) for V, it is clear that the relationship between volume and temperature must be

$$V = \left(\frac{nR}{P}\right) T = (\text{const.})T \qquad T \text{ in } °K \qquad (13.3.5)$$

From (13.2.9), the corresponding relation, using temperatures expressed in Celsius degrees, would be

$$V = \left(\frac{nR}{P}\right)(T - \lambda) = (\text{const.})(T - \lambda) \qquad T \text{ in } °C$$
$$(13.3.6)$$

In either case, the prediction of the equation of state is that the volume of the gas *vanishes* at $T = 0°K = -273.16°C$.

Actually, of course, this does not happen, because as the temperature is lowered, a point is reached where the substance departs from ideal gas behavior and ultimately turns into a liquid or a solid whose equation of state is very different from that of an ideal gas. Equations (13.3.5) and (13.3.6) realistically represent the behavior of actual substances only at rather high temperatures and rather low pressures. It is the *extrapolation* of the ideal gas equation of state which predicts the vanishing of the volume at $T = 0°K$, rather than the true equation of state for the substance in the applicable range of temperature and pressure. Nevertheless, it is this extrapolated ideal gas behavior that defines the absolute temperature scale and enables us to measure the value of the absolute zero point in Celsius degrees. This can be done, for example, by finding the intercept of the extrapolation of the volume–temperature relation of an ideal gas on the temperature axis, as illustrated in Fig. 13.3. In this diagram, a solid curve representing experimental data relating the volume of a real gas as a function of temperature (at constant pressure P) is plotted as the solid curve. Above a certain temperature, in this instance about $25°C$, the curve is a straight line, in agreement with the prediction of the ideal gas equation of state (13.3.6). Below this temperature, the equation of state of the gas departs significantly from what is expected from the ideal gas laws; in this region the density of the gas is presumably not vastly greater than the critical point density. At about $-70°C$, the gas condenses into a liquid at the given pressure; and at lower temperatures, the volume is that of the substance in the liquid or solid states under the prevailing conditions of temperature and pressure.

In the ideal gas region, the asymptotic dotted straight line can be fitted to the solid experimental curve; its equation will be given by (13.3.6). Accordingly, if we determine the point on the temperature axis where the extrapolated dotted line reaches zero volume, in this case $T = -273.16°C$, we should expect this to correspond to $T - \lambda = 0$, thus to the zero of the absolute scale. In every case where such a plot is made, irrespective of the gas used or the pressure at which the experiment is conducted, *this same value for the intercept on the temperature axis is obtained*. "Absolute zero" turns out to have the same value in every experiment.

In much the same way, if the amount of gas is accurately known, and if the pressure is carefully

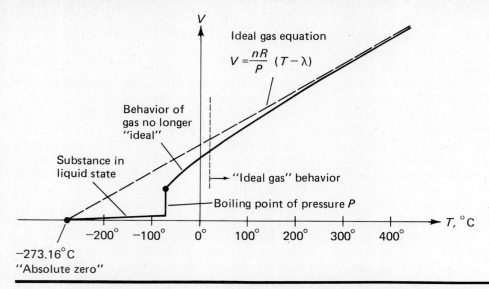

FIGURE 13.3. Volume–temperature relation for a real substance and for an ideal gas. Extrapolation of the linear behavior of the real substance in the "ideal" region always intercepts the temperature axis at $-273.16°C$, or "absolute zero."

determined, the value of the gas constant R can be arrived at by measuring the *slope* of the dotted line and equating it to the value predicted by the ideal gas equation of state, which is nR/P. Once again, the experimentally measured value of R is the same regardless of the chemical composition of the gas or of the pressure under which the measurements are carried out.

The properties of ideal gases suggest that they might very profitably be used as thermometric fluids and that their properties could be utilized *to define and standardize the temperature scale*. For example, suppose that a known quantity of an actual gas were enclosed at a known constant pressure P_0 in a glass bulb at initial absolute temperature T_0 by a mercury column connected to an external reservoir by flexible tubing, as shown in Fig. 13.4a. The volume V_0 of the enclosed gas can be determined by calibration markings along the mercury column. If the temperature of the bulb were raised to T_1, the gas would expand, depressing the mercury column. The pressure in the system would also increase since now the mercury levels in the thermometer bulb and the external reservoir would be different, but by lowering the reservoir until they were again the same, it could be brought back to the initial value. The new volume V_1 could be determined from the new height of the mercury column. If the gas in the bulb were truly "ideal," both the initial temperature T_0 and the final temperature T_1 could be found in terms of the known constant pressure P_0, the measured volumes V_0 and V_1, the known molar number n and the gas constant:

$$T_0 = \frac{P_0 V_0}{nR} \qquad (13.3.7)$$

$$T_1 = \frac{P_0 V_1}{nR} \qquad (13.3.8)$$

This would afford a convenient way of defining the absolute temperature scale operationally and of calibrating conventional mercury thermometers and other practical temperature-measuring instruments against such a scale.

The one hitch in the construction and operation of the *ideal gas thermometer* in such a simple way is that there are no truly ideal gases around. Some come rather close, of course; but under realizable experimental conditions, none is good enough for

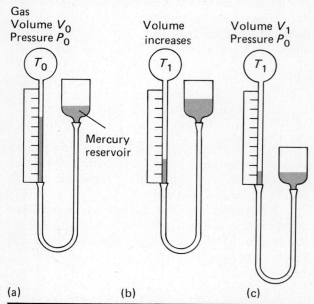

FIGURE 13.4. Constant pressure ideal gas thermometer.

ultraprecise temperature measurement, scale standardization, and calibration purposes. Once more, we may accomplish the desired results by the technique of *extrapolating to zero density or pressure*. For this purpose, the measured values of T, as given by (13.3.7) or (13.3.8), are noted for the initial bulb pressure P_0; these values, though perhaps not far off, are incorrect, since the behavior of the gas is not quite ideal. Then, the gas in the bulb is made less dense by lowering the bulb pressure, and the measurements are repeated, giving new and more accurate values of temperature.[2] Finally, a whole series of measurements is made, using lower and lower gas densities in the bulb enclosure. These temperature measurements for T_0 and T_1 are plotted against pressure and extrapolated to zero pressure in analogy to the volume measurements of Fig. 13.3. In this way, a true measurement of the initial absolute temperature T_0 and any desired temperature increment $T_1 - T_0$ can be accomplished using the properties of the ideal gas.

The possibility of using an ideal gas thermometer in this way leads, in turn, to the idea of *defining* the temperature scale in terms of the behavior of gases extrapolated to the limit of zero pressure. Accordingly, we might reasonably define the absolute temperature T by the equation

$$T = \lim_{P \to 0} \frac{PV}{nR} \qquad (13.3.9)$$

where the limiting process is the process of extrapolation to zero pressure described in connection with the ideal gas thermometer. We shall hereafter adhere to this definition of temperature and, unless specific indications to the contrary are made, shall always assume that temperatures are absolute temperatures measured in Kelvin degrees.

Finally, a few words are in order about why gases behave more "ideally" under some conditions than under others. The key to the simple "ideal" behavior of gases at low pressures and densities is the realization that *such a gas consists mostly of empty space*. For example, a mole of ideal gas at $0°C$ and 1 atmosphere pressure, containing some 6×10^{23} molecules, occupies a volume of 22.4 liters, while a mole of liquid water, containing the same number of molecules, occupies only 0.018 liters under the same conditions. It is clear, therefore, that most of the 22.4 liters occupied by the gas is nothing but empty space.

The simplest conceivable system, of course, is

empty space itself. The equation of state of empty space is $P = 0$, irrespective of volume, for there are no molecules in the space to exert pressure. Since the internal kinetic energy is zero, and so is the number of molecules, the internal kinetic energy per molecule, and thus the temperature, is undefined.

The next simplest kind of thermal system imaginable is one of empty space occupied by point particles each having a given mass and possessing a certain amount of kinetic energy. Though enclosed in a container of definite volume, the particles are otherwise completely free in the sense that they exert no forces upon one another, except, possibly, in instantaneous elastic collisions. In such a system, *all the internal energy of the system resides in the kinetic energy of the individual particles*. There is no potential energy of interaction between the molecules, because they do not exert forces upon one another. Since the particles have momentum and since they collide with the walls of the enclosure, the system does exert pressure on its containing vessel. A system having these dynamical characteristics is one with the thermodynamic properties of an ideal gas, and its equation of state is $PV = nRT$. We shall discuss the molecular mechanics of an ideal gas in more detail in the next section.

In an actual gas, the molecules are not point particles, and, moreover, they exert attractive forces upon one another. These forces are particularly strong when the average distance between molecules is small, as in the solid or liquid state. In the gaseous state, they are much weaker in view of the much larger average distance between particles. If the density of the gas is very small, the average distance between molecules is so large that the intermolecular attractive forces are insignificant and the fact that the molecules have a small but finite size does not matter; the system then behaves very much like an ideal gas, obeying the ideal gas equation of state quite closely. As the pressure, and thus the density, is increased, the average intermolecular distances become less, and effects of finite molecular size and stronger intermolecular attractions become important. In particular, the entire internal energy of the gas is no longer simply the kinetic energy of the molecules but consists partly of the *potential energy of interaction between molecules*. The average molecular kinetic energy is, therefore, reduced, which leads to a reduction in the average momentum of the molecules striking the container walls and hence to a decrease in the pressure in comparison to that which an ideal gas would exert. The gas no longer accurately obeys the ideal gas equation of state, particularly when the density approaches the critical density.

Numerous attempts to construct detailed dynamic models of nonideal gases and to deduce their equations of state have been made. Though some have

---

[2] Constant pressure conditions throughout the measurement may be assured by confining the mercury in the external reservoir in a sealed container, lowering the pressure over the mercury to about that in the thermometer bulb, and manipulating the reservoir as before so that the mercury levels in the reservoir and the thermometer stem are the same when the volume is read off.

been successful in describing certain restricted aspects of the physical behavior of nonideal gases, none has been completely successful in reproducing all the observed thermal phenomena associated with real gases. The reason for this lies primarily in the difficulty of describing the action of intermolecular forces in a sufficiently detailed and realistic manner. One of the simplest and most successful of the attempts at formulating an equation of state for real gases is due to van der Waals, who proposed an equation of state of the form

$$\left(P + \frac{a}{V^2}\right)(V - b) = RT \tag{13.3.10}$$

for 1 mole of gas, where a and b are constants. This equation can be more readily understood by solving for P and writing it

$$P = \frac{RT}{V - b} - \frac{a}{V^2} \tag{13.3.11}$$

The quantity b may be thought of as representing the volume of the gas molecules themselves, and thus $V - b$ represents the "free volume" or "empty space volume" in the gas. It is this volume which changes in response to pressure, since the molecules themselves are nearly incompressible. Hence, it makes sense to replace the total volume V in the ideal gas equation by $V - b$. This gives us (13.3.11), except for the term $-a/V^2$. This term is included to account for the reduction in pressure caused by the intermolecular attractive forces, which, naturally enough, becomes larger as the volume, and thus the average intermolecular distance decreases.

The van der Waals equation predicts the existence of a critical point and describes the behavior of real gases in the neighborhood of the critical point moderately well. Below the critical temperature, with a little fudging which is reasonably well justified on physical grounds, it can even describe the liquid–vapor transition and the phase boundaries of the liquid and vapor states. The general features of real gas behavior are fairly well described, then, though quantitative agreement with experimental data for many substances is not always very good.

EXAMPLE 13.3.1
Express the ideal gas equation of state in the form of a relationship between density, temperature, and pressure.

Although this example is a very simple one, it is also extremely important, since it is often advantageous to use the ideal gas law in this form. Substituting $V = m/\rho$ into the ideal gas law (13.3.4), we obtain

$$\frac{P}{\rho} = \frac{n}{m} RT \tag{13.3.12}$$

But m/n is the mass of 1 mole of gas, that is, its molecular mass (or molecular weight, as the chemists continually and incorrectly refer to it). Calling this quantity M, we may write (13.3.12) as

$$P = \frac{RT}{M} \rho \tag{13.3.13}$$

where ρ is the density. This is the desired result, but in using it we must be careful about units. If the units liter-atm/mole-°K are used for R, °K for T, g/mole for M, and g/liter for ρ, P will be obtained in units of atmospheres. If units of joules/mole-°K are to be used for R, one must use units of °K for T, kg/mole for M, and kg/m³ for ρ in order to obtain P in newtons/m².[3]

EXAMPLE 13.3.2
The atmosphere is composed of 78.09 mole percent N_2 ($M = 28.016$ g/mole), 20.95 mole percent O_2 ($M = 32.00$ g/mole), and 0.96 mole percent A ($M = 39.94$ g/mole). Assuming all these substances to be ideal gases, calculate the density of the atmosphere at 0°C and 1.000 atmospheres pressure and compare the result so obtained with the observed value of 1.2923 g/liter. Is the assumption of ideal gas behavior a good one for the atmosphere?

Consider a system consisting of $n_{N_2} = 0.7809$ moles N_2 at 1 atmosphere pressure, occupying a volume V_{N_2}, $n_{O_2} = 0.2095$ mole O_2, occupying a volume of V_{O_2}, and 0.0096 mole A, occupying a volume V_A. If all these gases were mixed together, we would have what might well be called a "mole of air"; and if all these substances behaved as ideal gases, its total volume would be $V_{tot} = V_{N_2} + V_{O_2} + V_A$. According to the ideal gas law, $V_{N_2} = n_{N_2}RT/P$, $V_{O_2} = n_{O_2}RT/P$, $V_A = n_A RT/P$. The density of the gas can then be expressed as

$$\rho = \frac{m_{tot}}{V_{tot}} = \frac{n_{N_2}M_{N_2} + n_{O_2}M_{O_2} + n_A M_A}{V_{N_2} + V_{O_2} + V_A}$$
$$= \frac{P}{RT}\left(\frac{n_{N_2}M_{N_2} + n_{O_2}M_{O_2} + n_A M_A}{n_{N_2} + n_{O_2} + n_A}\right) \tag{13.3.14}$$

Inserting the values $P = 1.000$ atm, $R = 0.08206$ liter-atm/mole-°K, $T = 273.16$°K, $n_{N_2} + n_{O_2} + n_A = 1.000$ along with the given values of molar quantities and molar masses into this expression, we may easily obtain the result $\rho = 1.29222$ g/liter. Since this is very

---

[3] In this book, the term *mole* usually refers to the *gram*-mole. Occasionally, as in this example, when using mks metric units, it is necessary to interpret the molar mass as the kilogram-molecular mass, that is, the number of kilograms of the substance equal numerically to its molecular mass expressed in atomic mass units. In either instance, of course, the numerical value of the quantity M is the same; in the case of O_2, for example, $M = 32.00$ g/g-mole or 32.00 kg/kg-mole!

close to the observed value of 1.2923 g/liter, we may safely conclude that air under normal atmospheric temperatures and pressures behaves very much like an ideal gas. By equating the calculated density of air to the quantity $(M/RT)P$ given by (13.3.13), inserting the corresponding values of T and P, and solving for M, one may find the "equivalent gram-molecular mass" $M = 28.97$ g/mole. This means simply that air acts very much like an ideal gas of gram-molecular mass 28.97 g/mole, a fact which simplifies ideal gas law problems involving air.

EXAMPLE 13.3.3

The air pressure in the tires of a 3200-pound car is adjusted at a temperature of 0°C to 32 lb/in.², as read on a tire pressure gauge at normal atmospheric pressure. The temperature then rises to 20°C. What would the tire pressure gauge read now if the tire pressures were checked? What is the total area of the tires' "footprints" at 0°C and at 20°C? Assume that air behaves as an ideal gas and that the volume of the tires does not change as a function of pressure.

In this example, since we are not given the value of R in English units, it is probably best to use the ideal gas law in the form (13.3.3). Since the volume does not change, the final volume and the initial volume both have the value V_0 and, therefore, cancel from both sides of the equation, leaving

$$\frac{P}{T} = \frac{P_0}{T_0} \tag{13.3.15}$$

It goes without saying that the final and initial temperatures must be expressed on the absolute scale, so that $T_0 = 273°$K and $T = 293°$K. But the same is true in regard to the *pressures*; the pressure terms in the ideal gas equation are expressed as *absolute* pressures rather than gauge pressures. Accordingly, we must write $P_0 = 32 + 14.7 = 46.7$ lb/in.² Inserting these numerical data and solving for the final pressure P, $P = 50.12$ lb/in.² (absolute) $= 35.42$ lb/in.² (gauge). Since the force exerted by the tires against the road is equal to the car's total weight, the "footprint area" of the tires is $3200/32 = 100$ in.² at 0°C and $3200/35.42 = 90.3$ in.² at 20°C.

EXAMPLE 13.3.4

Assuming that the atmosphere has a uniform temperature of 300°K, what is the normal atmospheric pressure at an elevation of 4000 meters?

In Chapter 11, we found that the atmosphere could be treated as an incompressible fluid of constant density whenever height changes of only a few meters or tens of meters are involved. In this instance, however, we are dealing with a vertical distance of 4000 meters, and a more careful treatment taking into account the compressibility of the atmosphere is re-

quired. We may begin with the general equation relating the change of pressure with height, which we have met previously as equation (11.3.3),

$$\frac{dP}{dz} = -\rho g \tag{13.3.16}$$

where z is the height above the earth's surface.[4] When ρ is constant, as for an incompressible fluid, this is easily integrated to give the familiar relation $P = P_0 + \rho gh$. In this case, ρ is no longer constant but is a function of pressure, expressed by the ideal gas equation of state in the form (13.3.13). Substituting this value of ρ into (13.3.16),

$$\frac{dP}{dz} = -\frac{Mg}{RT} P \tag{13.3.17}$$

This can be put into a form which is readily integrable by collecting everything that varies with P on the left and everything that varies with z on the right. This is accomplished by dividing both sides by P and multiplying through by dz. The result is

$$\frac{dP}{P} = -\frac{Mg}{RT} dz \tag{13.3.18}$$

In this equation, Mg/RT is constant because the temperature is uniform and because the variation of g over the extent of the earth's atmosphere is negligibly small. We may now integrate (13.3.18) from $z = 0$, at which point $P = P_0$ to any desired height z at which the pressure has the value P. In this way, we obtain

$$\int_{P_0}^{P} \frac{dP}{P} = -\frac{Mg}{RT} \int_0^z dz \tag{13.3.19}$$

or

$$[\ln P]_{P_0}^P = \ln P - \ln P_0 = \ln \frac{P}{P_0} = -\frac{Mgz}{RT} \tag{13.3.20}$$

Exponentializing both sides of the final equality shown above, we find

$$P = P_0 e^{-Mgz/RT} \tag{13.3.21}$$

The value of the exponent is obtained, using mks units, as

$$-\frac{Mgz}{RT} = -\frac{(28.97)(9.80)(4000)}{(8314)(300)} = -0.4553$$

whence

$$P = P_0 e^{-0.4553} = 0.6343 P_0$$

Since P_0 represents normal atmospheric sea level pressure, the pressure at $z = 4000$ meters is 0.6343

[4] In this equation, the minus sign appears because the positive z-direction is straight up, while in the development leading to Eq. (11.3.3), the positive y-axis was assumed to point vertically downward.

atmospheres. Though the assumption of uniform temperature throughout the atmosphere is not completely realistic, the answer given above is probably quite close to what is observed under actual conditions. The pressure-versus-height law (13.3.21) can be expressed as a density-versus-height law by using (13.3.13) to express P in terms of ρ:

$$\rho = \rho_0 e^{-Mgz/RT} \qquad (13.3.22)$$

where $\rho_0 = (M/RT)P_0$ is the density at the earth's surface. This example is of particular importance in that it exhibits the exponential dependence of the number of molecules having a given total energy upon the energy. We shall refer to the results obtained here in a later section when we shall examine this dependence in detail.

13.4 The Molecular Mechanics of the Ideal Monatomic Gas

Up to this point, our study of physics has largely been confined to the mechanics of objects and masses of fluid substances that are of appreciable size on the scale of our own bodies and that are, in any event, large enough for us to see and touch. We have had to refer to the ultimate atomic and molecular constitution of matter on occasion, to justify this point or that, and though these occasions have arisen with increasing frequency since we began the study of heat and thermodynamics, we have so far not had to apply the laws of physics to individual atoms or molecules in any systematic way. We are now ready to tackle some problems in which the mechanics of individual atoms and molecules is of importance, and we shall begin, naturally, by applying the laws of *macroscopic* mechanics that have served us so well in the past to *microscopic* systems in which the behavior of individual atoms or molecules is important.

In doing this, we shall find that in some ways the laws of Newtonian mechanics work very well in describing the behavior of atomic systems but that sometimes they are incomplete and even inaccurate in predicting certain aspects of atomic or molecular behavior. These areas of incompleteness and inaccuracy have been investigated and thoroughly understood during the twentieth century, but only by the development of a new and more general system of mechanics referred to as *quantum mechanics*. Quantum mechanics is a comprehensive scheme of mechanics which is applicable to atomic, molecular, and nuclear systems yet which reduces to the more familar Newtonian mechanics for macroscopic bodies and for energies perceptible to the human senses. In this chapter, we shall not try to apply quantum mechanics

to molecular problems, but shall only indicate the limits of applicability of Newtonian mechanics and the results obtained from quantum theory beyond those limits. The basic fundamentals of quantum theory and some elementary examples of how it may be applied to the mechanics of atomic systems are discussed in a later chapter.

Let us now try to apply the laws of mechanics— ordinary Newtonian mechanics—to the atoms of a monatomic ideal gas. We shall make several simplifying assumptions which, as we shall see more clearly at the end, define the dynamic properties of the ideal gas. The assumptions are these:

1. We assume that there are N atoms occupying a fixed volume V and having a fixed total internal energy U.
2. The atoms are regarded as mass particles having negligible radius but carrying energy and momentum and obeying Newton's laws.
3. The atoms are independent of one another; they exert no long-range forces upon one another and interact with one another, and with the walls of the container, only through *instantaneous elastic collisions*. This assumption means that there is no potential energy of interaction between molecules and that, therefore, *all the internal energy of the system resides in the translational kinetic energy of the molecules.*
4. The pressure of the gas is assumed to arise from elastic collisions between the gas atoms and the walls of the container.
5. The number of atoms is assumed to be very large and their thermal motion is regarded as completely random. The macroscopic properties of the gas are assumed to arise essentially from the *average* properties of the randomly moving atoms.

Let us now turn our attention to the collisions between the gas atoms and the walls of the container, as illustrated in Fig. 13.5. Consider a unit area on the wall of the container, 1 cm², let us say. During each collision of a gas atom with the wall, a certain amount of momentum is transferred to the container wall. The force on this unit area, hence the pressure, can be calculated by finding the total change of momentum of all particles striking the wall, per unit time. In any elastic collision, a molecule having an initial component of velocity v_x along the direction normal to the wall ends up with a final velocity component $-v_x$, as illustrated by Fig. 13.5. The other momentum components are unchanged. The total change in the x-component of momentum for any such particle thus has the magnitude $2mv_x$.

Now, to begin, let us consider *only* those particles having some specific value of the x-component of velocity, v_{x0}, forgetting all the rest for the moment.

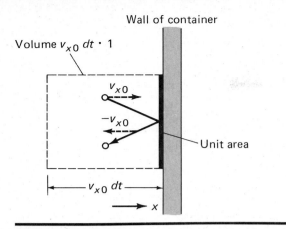

Volume $v_{x0} \, dt \cdot 1$

Wall of container

v_{x0}

$-v_{x0}$

Unit area

$v_{x0} \, dt$

x

FIGURE 13.5. Dynamic model of an ideal monatomic gas.

During a time interval dt, the number of such particles that strike the wall are just those that are within a distance $v_{x0} \, dt$ of the wall. These particles can reach the wall during time dt, but any particle further from the wall will not be able to reach it during that interval and will collide with it later. If there are $N(v_{x0})$ particles having this specific value of v_x, the number of such particles per unit volume is $N(v_{x0})/V$, and the number within the volume indicated in Fig. 13.5 that will strike the wall during the interval dt is this quantity times the volume marked off there, which in turn is $v_{x0} \, dt$, since the base of the indicated volume has unit area. There are then $(N(v_{x0})/V)(v_{x0} \, dt)$ particles striking the wall, each imparting momentum $2mv_{x0}$ to it. The total momentum change $(dp)_{v_{x0}}$ for all such particles is then

$$(dp)_{v_{x0}} = \frac{2m}{V} v_{x0}^2 N(v_{x0}) \, dt \qquad (13.4.1)$$

Finally, we must sum this result over all possible values of v_x, that is, over all *positive* values of v_x. Particles having negative values of v_x, amounting to half the total number, are not going to hit the wall at all, since they're headed in the wrong direction at the outset. The total change in momentum for all particles is thus

$$(dp)_{tot} = \frac{2m}{V} \left[\sum_n v^2_{xn} N(v_{xn}) \right] dt \qquad (13.4.2)$$

where we represent the different possible values of initial velocity that particles can have by $v_{x0}, v_{x1}, v_{x2}, v_{x3}, \ldots, v_{xn} \ldots$. This sum can be expressed in a simpler form by recognizing the fact that it is related to the *average* of the quantity v_x^2 over the different possible values of v_x. Calling this average $\overline{v_x^2}$ and writing out how it might be found, we see that

$$\overline{v_x^2} = \frac{\left(\begin{array}{c} v_{x0}^2 N(v_{x0}) + v_{x1}^2 N(v_{x1}) \\ + v_{x2}^2 N(v_{x2}) + \cdots + v_{xn}^2 N(v_{xn}) + \cdots \end{array} \right)}{N(v_{x0}) + N(v_{x1}) + N(v_{x2}) + \cdots + N(v_{xn}) + \cdots} \qquad (13.4.3)$$

This can be understood most simply by regarding it as the problem of finding the average grade on an exam, where there are $N(v_{xn})$ exams each having a grade of v_{xn}^2 points. The total number of points earned on all exam papers is, therefore, the number in the numerator and the total number of papers is the number in the denominator. In our case, the numerator is just the sum in brackets in (13.4.2). The denominator, since we're summing only over particles having positive values of v_x, adds up to *half* the total number of particles in the container, or $N/2$. In this way, we can write (13.4.3) as

$$\overline{v_x^2} = \frac{\sum\limits_n v_{xn}^2 N(v_{xn})}{N/2} \qquad (13.4.4)$$

whence

$$\sum_n v_{xn}^2 N(v_{xn}) = \frac{N \overline{v_x^2}}{2} \qquad (13.4.5)$$

Equation (13.4.2) then becomes

$$(dp)_{tot} = \frac{mN \overline{v_x^2}}{V} \, dt \qquad (13.4.6)$$

But now, the velocity components v_x, v_y, and v_z are related to the speed v of the particle by

$$v_x^2 + v_y^2 + v_z^2 = v^2 \qquad (13.4.7)$$

and if both sides are averaged over all the particles, then

$$\overline{v_x^2} + \overline{v_y^2} + \overline{v_z^2} = \overline{v^2} \qquad (13.4.8)$$

However, since the motion of the particles is random, there can be nothing which would distinguish the x-, y-, and z-components from one another; therefore, $\overline{v_y^2} = \overline{v_z^2} = \overline{v_x^2}$ and (13.4.8) can be written as $3\overline{v_x^2} = \overline{v^2}$, or

$$\overline{v_x^2} = \tfrac{1}{3} \overline{v^2} \qquad (13.4.9)$$

We can, therefore, write $(dp)_{tot}/dt$, from (13.4.6), finally as

$$\frac{(dp)_{tot}}{dt} = \text{force on unit area} = P = \frac{N}{V} \tfrac{1}{3} m \overline{v^2} \qquad (13.4.10)$$

$$PV = \tfrac{1}{3} N m \overline{v^2} \qquad (13.4.11)$$

Now this is *beginning* to look like the ideal gas law (13.3.4). It tells us that the product PV is constant for a gas having a given internal energy, for after all, $\overline{v^2}$ is just a constant expressing the average of the velocity squared taken over all the molecules. Furthermore, it states that this constant value is proportional to the number of molecules in the system, as does (13.3.4), since the number of moles n is proportional to the number of molecules N. What seems to be lacking in (13.4.11) is information regarding the temperature. But this is not surprising, since temperature is *not* one of the concepts of Newtonian mechanics but is instead a thermal quantity related to the average internal kinetic energy of translation of the molecules.

Now, the total internal energy of the gas can be expressed as the total kinetic energy of all the molecules, which can equally well be written as the total number of molecules times their *average* kinetic energy. Thus, the internal energy U of the gas is

$$U = N(\tfrac{1}{2}m\overline{v^2}) = \tfrac{1}{2}Nm\overline{v^2} \qquad (13.4.12)$$

and the average internal energy per particle is

$$\frac{U}{N} = \tfrac{1}{2}m\overline{v^2} \qquad (13.4.13)$$

Using this information, (13.4.11) may be expressed in terms of the average kinetic energy per particle as

$$PV = N\,\frac{2}{3}\,\frac{U}{N} \qquad$$

We are now faced with the problem of relating the right side of this equation to the temperature. Classical mechanics will not help us; we must adopt some rational definition of the temperature ourselves. Clearly enough, the quantity in parentheses above depends only upon the *translational kinetic energy per particle* and, therefore, satisfies our intuitive notions of temperature in all respects. Also, we have *already* settled upon a definition of temperature from the observed behavior of an ideal gas in which the right side of (13.4.14) had to equal nRT. The way we define what we mean by temperature here must agree with that previous definition. We are, therefore, forced to write

$$\frac{2}{3}\frac{U}{N} = kT \qquad (13.4.15)$$

where k is a constant, called *Boltzmann's constant*, which, in order that (13.4.14) agree with (13.3.4), has to have a numerical value such that

$$Nk = nR \qquad (13.4.16)$$

that is,

$$R = \frac{N}{n}\,k = N_{A}k \qquad (13.4.17)$$

where N_A $(=N/n)$ is *Avogadro's number*, expressing the *number of molecules in a mole*, equal to 6.023×10^{23} molecules per gram-mole. This means that Boltzmann's constant has to have the value

$$k = 1.381 \times 10^{-16} \text{ erg/molecule-}^\circ\text{K}$$
$$= 1.381 \times 10^{-23} \text{ joule/molecule-}^\circ\text{K}$$

Since from (13.4.17) $k = R/N_A$, we may interpret Boltzmann's constant as the *gas constant per molecule*.

With the definition of temperature established by (13.4.15), we see that the ideal gas law (13.4.14) may be written in terms of the number of molecules N and Boltzmann's constant k as

$$PV = NkT \qquad (13.4.18)$$

In view of the relation (13.4.16), this is fully in agreement with our former way of stating it as Eq. (13.3.4). It is perhaps not wholly facetious to regard (13.4.18) as the physicist's way of writing the ideal gas law and (13.3.4) as the chemist's way, for the former emphasizes the number of molecules and the fundamental molecular constant k, while the latter accentuates the importance of the number of moles and the molar gas constant.

Looking back over what we have done, a number of important facts can be observed. First of all, Eq. (13.4.15) allows us to write the total internal energy U as

$$U = \tfrac{3}{2}NkT \qquad (13.4.19)$$

and the internal energy per particle as

$$\frac{U}{N} = \tfrac{3}{2}kT \qquad (13.4.20)$$

We could never have found these results from the equation of state alone. *It was the adoption of a reasonable molecular model and the application of the laws of motion to that model that allowed us to calculate them.*

From (13.4.19) and (13.4.20) it is clear that *the internal energy of an ideal gas is a function of temperature alone* and is independent of the volume or pressure. This result is of great importance and allows us to calculate quite simply a number of other important properties of ideal gases. This result applies only to ideal gases and a few other very simple dynamical systems; in general, the internal energy will be a function of volume or pressure as well as temperature.

Now suppose that 1 mole of gas containing N_A molecules is enclosed in a rigid container so that its volume remains constant. Under these circumstances, it cannot expand and, therefore, *cannot do any work on its surroundings.* If a quantity of heat dQ is absorbed by the gas through the container walls, then, according to the first law of thermodynamics,

$$dQ = dU + dW = dU \qquad (13.4.21)$$

414

since $dW = 0$. Dividing both sides of (13.4.21) by dT, we find that the *molar specific heat* at constant volume, c_v, is given by

$$c_v = \frac{dQ}{dT} = \frac{dU}{dT} = \frac{d}{dT}\left(\tfrac{3}{2}N_A kT\right)$$

or, recalling (13.4.17),

$$\boxed{c_v = \tfrac{3}{2}N_A k = \tfrac{3}{2}R}$$

$$= 2.979 \text{ cal/mole-}°\text{K} \qquad (13.4.22)$$

This predicts that the molar specific heat of every monatomic ideal gas is independent of temperature and has the same numerical value given above. This is in good agreement with experiment, for the measured specific heats of the monatomic gases He, Ne, A, Kr, and Xe all have very nearly this constant specific heat in the range of temperatures and pressures where the ideal gas equation of state is obeyed. The specific heats of diatomic and polyatomic gases, however, are larger than the value given in (13.4.22), because part of the heat absorbed goes into exciting motions *other* than those which contribute to translational kinetic energy of the molecules, such as rotational motion about the center of mass and internal molecular vibrations. We shall examine these rotational and vibrational contributions in some detail in a later section, where a more detailed discussion of the specific heats of polyatomic ideal gases will be given.

Finally, comparing Eqs. (13.4.13) and (13.4.20), we see that

$$\tfrac{1}{2}m\overline{v^2} = \tfrac{3}{2}kT \qquad (13.4.23)$$

whence

$$\boxed{\overline{v^2} = \frac{3kT}{m}} \qquad (13.4.24)$$

This relation allows us to calculate the average value of v^2 for the gas molecules. The square root of this quantity represents an average velocity called the *root-mean-square velocity* or, as it is often abbreviated, the *rms velocity* of the molecules. This can be expressed in a form better suited to numerical calculation by multiplying numerator and denominator in (13.4.24) by N_A and noting that $N_A k = R$, while $nN_A = M$, the gram-molecular mass. Accordingly,

$$v_{\text{rms}} = \sqrt{\overline{v^2}} = \sqrt{\frac{3kT}{m}} = \sqrt{\frac{3RT}{M}} \qquad (13.4.25)$$

For helium, $M = 4.00$, and for $T = 300°\text{K}$,

$$v_{\text{rms}} = \sqrt{\frac{(3)(8314)(300)}{(4)}} = 1368 \text{ m/sec}$$

For nitrogen, $M = 28.01$, at $300°\text{K}$, we find that $v_{\text{rms}} = 517$ m/sec. It is clear from (13.4.25) that the molecular velocities are proportional to the square root of the absolute temperature and inversely proportional to the square root of the molecular mass. The values of the molecular average velocities predicted by (13.4.25) have been compared with experimentally measured values and found to be in good agreement in the range of conditions where the ideal gas model is valid.

Although our calculations were based on the assumption of a gas composed of monatomic point molecules, they would be equally valid for polyatomic gases, except that the quantity U in that case would represent only the part of the internal energy contributed by the translational kinetic energy of the molecules. This is the only part that gives rise to the linear momentum of the molecules and hence it is the only part that contributes to the pressure. It is also just this part of the total internal energy that goes into our reckoning of the temperature, and, as a result, *the equation of state derived as Eq. (13.4.18) is valid whether the gas is monatomic or not.* The same remarks apply to Eq. (13.4.25), which is correct for polyatomic as well as monatomic ideal gases. On the other hand, as we mentioned in connection with specific heat, in the case of a polyatomic gas, there is more to the internal energy than just the translational kinetic energy of the center of mass. In addition, there may be kinetic energy of *rotation* about the center of mass and also kinetic and potential energy associated with possible *harmonic oscillations* these complex molecules may perform.

In the simplest case of a diatomic molecule, the forces connecting the two molecules can be represented, at least for small amplitudes of motion, as behaving like Hooke's law springs, the molecule then being like a dumbbell in which the two masses are connected by a spring. The molecule, in addition to rotating, can then execute simple harmonic vibrational motion in which the masses move alternately toward and away from one another. The potential and kinetic energy associated with this vibrational or oscillatory motion also constitutes part of the total internal energy.

In more complex polyatomic molecules, there may be several different and independent modes of vibrational motion each of which contributes to the total internal energy. The total internal energy of polyatomic gases, and thus their specific heat, is greater than that associated with the monatomic ideal gas. At the same time, since in the dynamic model of the ideal gas the molecules are essentially independent of one another, their total internal energy, even though no longer equal to $3NkT/2$, is still independent of the

415

average distance between molecules—and hence independent of the volume and pressure. It is, therefore, true *even for polyatomic ideal gases* that the total internal energy can always be represented as a function of temperature only.

Our application of classical mechanics to the molecular dynamics of an ideal gas has been generally quite satisfactory. For the *monatomic* ideal gas, it is very good indeed, leading to answers that are in agreement with experiment throughout. We might have expected that some discrepancy would arise because we used classical mechanics rather than quantum theory to treat the mechanics of an atomic system. But none did, and in fact, the quantum mechanical treatment of the monatomic ideal gas leads to precisely the same results as the classical calculation. In the case of polyatomic gases, there are certain discrepancies between the classical results and what is observed experimentally, particularly in the way in which rotational and vibrational motions contribute to the total internal energy. In this case, quantum theory leads to somewhat different predictions in those areas, which are in every case in accord with what is obtained experimentally.

EXAMPLE 13.4.1

Find the coefficient of volume expansion β for an ideal gas under constant pressure.

The volume thermal expansion coefficient β can be defined by

$$\beta = \frac{1}{V}\frac{dV}{dT} \tag{13.4.26}$$

But from (13.4.18), $V = NkT/P$, and if the pressure is held constant,

$$\frac{dV}{dT} = \frac{Nk}{P} \tag{13.4.27}$$

Equation (13.4.26) then becomes

$$\beta = \frac{1}{V}\frac{Nk}{P} = \frac{Nk}{NkT} = \frac{1}{T} \tag{13.4.28}$$

since $PV = NkT$.

13.5 Work of Expansion: Isothermal and Adiabatic Processes

When a gas expands, it may do work on its surroundings; and, conversely, in order to compress a gas into a smaller volume, external work must be done on it. The actual amount of work done in these processes depends not only upon the equation of state of the substance but *also upon the conditions under which the expansion or compression takes place*, that is, whether

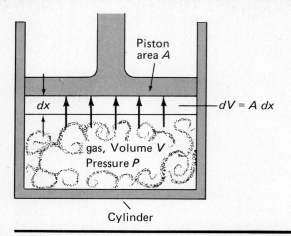

FIGURE 13.6. Expansion of a substance confined within a cylinder having a movable piston.

it is carried out at constant temperature or at constant pressure, or with no flow of heat in or out of the system or in some other way. In all such cases, the actual computation of the work done can be accomplished by a procedure illustrated in Fig. 13.6, in which a cylinder containing a volume V of gas at pressure P, fitted with a piston of area A is shown. The gas exerts a total force $F = PA$ on the piston, and, if the piston moves a small distance dx, the work done by this force is

$$dW = \mathbf{F} \cdot d\mathbf{r} = PA\,dx \tag{13.5.1}$$

But $A\,dx$ represents dV, the increase in volume of the gas during this small expansion. The work done by the gas on the surroundings as a result of the expansion is, therefore,

$$\boxed{dW = P\,dV} \tag{13.5.2}$$

In general, the pressure may not remain the same during the course of a large expansion in volume but may change as the volume changes. Under such circumstances, the total work done must be computed by expressing P as a function of V and integrating (13.5.2) from the initial volume V_1 to the final volume V_2:

$$\boxed{\Delta W = \int dW = \int_{V_1}^{V_2} P\,dV} \tag{13.5.3}$$

If the expansion is carried out in such a way that the pressure *remains constant* throughout the process, P may be written outside the integral, and (13.5.3) becomes

$$\boxed{\Delta W = P\int_{V_1}^{V_2} dV = P(V_2 - V_1) \qquad P = \text{const.}}$$

$$\tag{13.5.4}$$

These results are quite general and hold even though the expansion may occur in a manner different from that shown in Fig. 13.6. Also, it should be clear that even though derived by considering the example of a confined gas, these same equations are valid for determining the work done in the expansion of a solid or liquid substance as well.

Geometrically, the integral in (13.5.3) represents the *area under the curve* representing pressure as a function of volume in the expansion process between the ordinates V_1 and V_2, as shown in Fig. 13.7a. If the process is an expansion from the smaller volume V_1 to the larger volume V_2, the integration proceeds in the positive direction along the V-axis, the resulting area, and the work, being positive. This is in agreement with our previous convention that work done *by* the system is positive. If the process goes in the reverse direction, as a compression from the larger volume V_2 to the smaller volume V_1, the limits of the integral are reversed, the integration now proceeding in the negative direction along the V-axis, and the area under the curve is negative, as is the work. This again agrees with our earlier idea that work done *on* the system must be negative.

Now let us look at the situation shown in Fig. 13.7b. Here, we have a system (let us think of it as a confined gas, just to be specific) initially at point A in the PV-plane. The gas is then allowed to expand along path AB, its pressure decreasing at the same time, until point B is reached. During this expansion, positive work W_1, represented by the area DABC, is done by the gas. From point B, the gas is compressed until its volume has been reduced to the initial value at E, in such a way that the pressure stays constant during the process. This could be accomplished, for example, by cooling the gas as its volume is decreased. During this step, negative work W_2 is done on the system, this being represented on the diagram by *minus* the area DEBC. The gas is then increased in pressure, keeping the volume constant, until the

initial state at point A is regained. This could be done by increasing the temperature of the gas. No work is done during this process, since the volume does not change, and this is true for any process carried out at constant volume, because in (13.5.2) $dV = 0$ in any such process. The *net* work done by the system during the cyclic process ABEA is $W_1 - W_2$, which is represented on the diagram by area DABC minus area DEBC, in other words, by the shaded area ABE. This result is characteristic of any cyclic process which returns the system to the initial conditions of pressure and volume (and temperature as well, since that is determined by the equation of state once P and V are specified). Such a process is always described by a closed path in the PV-plane, and the net work done by the system during any process of this type is represented by the area within the closed curve. This work will be positive if the curve is traversed clockwise, as in the example above, and negative if it is traversed counterclockwise.

In any process where all the work done is work of expansion or compression, the entire amount of work accomplished can be accounted for by (13.5.2), and the first law of thermodynamics for differential changes in heat, internal energy, and work, can then be written as

$$dQ = dU + P\,dV \qquad (13.5.5)$$

If the first law is expressed as a relation between finite changes in these quantities, then it must take the form

$$\Delta Q = \Delta U + \int_{V_1}^{V_2} P\,dV \qquad (13.5.6)$$

because then the finite quantity ΔW is represented by (13.5.3).

Let us now examine in detail some of the simpler and more important processes by which the thermal state of a gas or other substance may be altered. These are as follows:

1. *Isothermal Process.* Such a process is one in which the temperature remains constant, that is, one in which $dT = 0$. In the case of an ideal gas, since then the internal energy depends solely upon temperature, $dU = 0$.

2. *Constant Volume Process.* In such a process, the volume remains the same throughout; therefore, $dV = 0$, and *no external work is done*. From (13.5.5), then,

$$dQ = nc_v\,dT = dU \qquad V = \text{const.} \qquad (13.5.7)$$

where c_v is the molar specific heat measured at constant volume. For an ideal monatomic gas, we found in the previous section that $c_v = \frac{3}{2}R$.

3. *Constant Pressure Process.* In this kind of process,

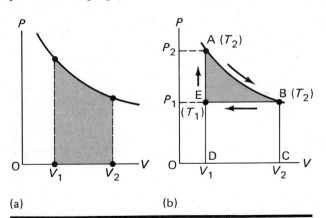

(a)   (b)

FIGURE 13.7. (a) Work done in an isothermal expansion process, illustrated as the area under the isothermal PV curve. (b) Work done in a cyclic process, represented as the area enclosed by the cycle in the PV plane.

the pressure is always the same. For this reason, the work done is always very easily figured from (13.5.4). The specific heat of most substances measured at constant pressure is greater than the specific heat at constant volume, because at constant pressure the expansion of the substance does external work. This means that at constant pressure the heat input goes only partially into increasing the internal energy and hence the temperature, the rest being used to perform external work against the surroundings—the atmosphere, for example. At constant pressure, therefore, more heat must be added to raise the temperature a given amount than would be necessary at constant volume when no external work is done. In one of the following examples, we shall see that for an ideal gas, $c_p - c_v = R$, where c_p and c_v represent the molar specific heats measured, respectively, at constant pressure and constant volume.

4. *Adiabatic Process.* In an adiabatic process, the system is thermally insulated from its surroundings, so that no heat flow into or out of the system can occur. Under these circumstances, $dQ = 0$, and from the first law of thermodynamics

$$dU = -dW = -P\,dV \qquad \text{adiabatic} \qquad (13.5.8)$$

Adiabatic processes play an important role in the theory of heat engines, as we shall see in the next chapter. One of the examples which follow is devoted to the problem of the adiabatic expansion of an ideal gas.

EXAMPLE 13.5.1
Find the work done by n moles of an ideal gas in an isothermal expansion from an initial volume V_0 to a final volume V.

According to (13.5.1), the work done in expansion will be given by

$$\Delta W = \int_{V_1}^{V_2} P\,dV \qquad (13.5.9)$$

In order actually to carry out the integration, we must know how P varies with volume during the process in question. In this case, the desired relation follows at once from the ideal gas law, $PV = nRT$, in which T is *constant* since the expansion occurs isothermally. This allows us to write (13.5.9) as

$$\Delta W = \int_{V_0}^{V} \frac{nRT}{V}\,dV = nRT \int_{V_0}^{V} \frac{dV}{V} = nRT\big[\ln V\big]_{V_0}^{V}$$

$$= nRT\big[\ln V - \ln V_0\big] = nRT \ln \frac{V}{V_0} \qquad (13.5.10)$$

EXAMPLE 13.5.2
Show that for any *ideal gas*, the difference between the molar specific heats at constant pressure and at constant volume is equal to the gas constant R.

If the temperature of n moles of an ideal gas is increased by an amount ΔT at *constant volume*, then $\Delta W = 0$; and from the first law of thermodynamics and the definition of specific heat, a heat input ΔQ_1 and an internal energy increase ΔU_1 given by

$$\Delta Q_1 = \Delta U_1 = nc_v\,\Delta T \qquad (13.5.11)$$

are required. Now suppose we increase the temperature of the gas *by the same amount* at constant pressure. Since the internal energy of an ideal gas depends only on the temperature, the same increase of internal energy, ΔU_1, will result in this case as well. However, external work $P(V_2 - V_1) = P\,\Delta V$ will be done, according to (13.5.4), since this is a constant pressure process. Therefore,

$$\Delta Q_2 = \Delta U_1 + P\,\Delta V \qquad (13.5.12)$$

But $\Delta Q_2 = nc_p\,\Delta T$, and from (13.5.11), ΔU_1 is equal to $nc_v\,\Delta T$. Therefore, (13.5.12) can be written

$$n(c_p - c_v)\,\Delta T = P\,\Delta V \qquad (13.5.13)$$

or, in the limit of small increments dT and dV,

$$P\frac{dV}{dT} = n(c_p - c_v) \qquad (13.5.14)$$

For an ideal gas, however, $V = nRT/P$, and for a constant pressure process, $dV/dT = nR/P$. Substituting this value into (13.5.14), we find

$$c_p - c_v = R \qquad (13.5.15)$$

the desired result. For a *monatomic* ideal gas, according to (13.4.22), $c_v = \frac{3}{2}R$. Therefore, for such a gas

$$c_p = \frac{5}{2}R \qquad (13.5.16)$$

It is frequently convenient to express the relation between the constant pressure and constant volume properties of gases in terms of the ratio

$$\gamma = \frac{c_p}{c_v} \qquad (13.5.17)$$

For monatomic ideal gases, we have

$$\gamma = \frac{\frac{5}{2}R}{\frac{3}{2}R} = \frac{5}{3} \simeq 1.67 \qquad (13.5.18)$$

For polyatomic ideal gases, the value of γ differs from this, because even though (13.5.15) holds for such gases, it is no longer true that $c_v = \frac{3}{2}R$.

Equation (13.5.15) holds only for ideal gases because the internal energy change ΔU, connected with the temperature increase ΔT, has been assumed to be the same for both constant pressure and constant volume processes. This is true if the internal energy depends only upon temperature, as it does in the case of the ideal gas. But if the internal energy were depen-

418

dent upon *volume* as well as temperature, the internal energy change in a constant volume process would be *different* from what it would be in a process in which a change in volume were involved, even though the same temperature change occurred in both cases.

EXAMPLE 13.5.3

The diagram in Fig. 13.7b can be regarded as a sort of cyclical *heat engine* in which heat is absorbed from an outside source (boiler) in processes EA and AB and rejected to an external cooling system (condenser) during process BE. These processes are repeated, in successive cycles of the operation of the engine, the path EABEABEAB... being traced out repeatedly. Assuming that the substance undergoing these processes is an *ideal gas*, which begins the cycle at point E, where its temperature is T_1, and assuming that the process AB is an isothermal expansion taking place at temperature T_2 $(T_2 > T_1)$, find the values of ΔQ, ΔU, and ΔW for all three phases of the cycle and for the cycle as a whole. Find the *efficiency*, defined as the net work output divided by the heat input from the boiler. You may regard the molar specific heats c_p and c_v as known quantities.

The engine begins its cycle at point E, where the pressure is P_1, the volume V_1 and the temperature of the working substance T_1. The process EA is a pressure increase at *constant volume*, and if we are dealing with a fixed amount of substance, this can only be accomplished by increasing the temperature to a higher value T_2. In a constant volume process, no external work is done, and hence, if we write the first law of thermodynamics in the form

$$\Delta Q_{EA} = \Delta U_{EA} + \Delta W_{EA} \qquad (13.5.19)$$

we see at once that

$$\Delta W_{EA} = 0 \qquad (13.5.20)$$

Then,

$$\Delta Q_{EA} = \Delta U_{EA} = nc_v \, \Delta T = nc_v(T_2 - T_1) \qquad (13.5.21)$$

assuming that we have n moles of working substance of constant-volume molar specific heat c_v.

In process AB, which is isothermal, the temperatures at A and B have the same value T_2. Since the internal energy of an ideal gas depends *only* on the temperature and since the temperature has not changed, no change in internal energy takes place, and

$$\Delta U_{AB} = 0 \qquad (13.5.22)$$

We know that the gas must have done an amount of work against the surroundings during this isothermal expansion given by (13.5.10); therefore,

$$\Delta W_{AB} = nRT_2 \ln \frac{V_2}{V_1} \qquad (13.5.23)$$

This quantity corresponds to the area DEABC in

Fig. 13.7b. Since we are dealing with an ideal gas, the equation of state tells us that at point E,

$$P_1 V_1 = nRT_1 \qquad (13.5.24)$$

while at B,

$$P_1 V_2 = nRT_2 \qquad (13.5.25)$$

Dividing one of these equations by the other, we see that

$$\frac{V_2}{V_1} = \frac{T_2}{T_1} \qquad (13.5.26)$$

which allows us to express (13.5.23) in the form

$$\Delta W_{AB} = nRT_2 \ln \frac{T_2}{T_1} \qquad (13.5.27)$$

Now, from the first law of thermodynamics, recalling that $\Delta U_{AB} = 0$, we obtain

$$\Delta Q_{AB} = \Delta U_{AB} + \Delta W_{AB} = nRT_2 \ln \frac{T_2}{T_1} \qquad (13.5.28)$$

In the process BE, the temperature is reduced from T_2 to T_1, and, therefore, an amount of heat

$$\Delta Q_{BE} = -nc_p(T_2 - T_1) \qquad (13.5.29)$$

flows out of the working substance, this process taking place at *constant pressure*. We must use the minus sign in (13.5.29) since heat is given up by the system. At the same time, the work done on the system (which is again negative because the path BE is traversed in the negative direction by the V-axis) is, for a constant pressure process, according to (13.5.4),

$$\Delta W_{BE} = -P_1(V_2 - V_1) \qquad (13.5.30)$$

which, by virtue of (13.5.24), (13.5.25), and (13.5.15), can also be written as

$$\Delta W_{BE} = -nR(T_2 - T_1) = -n(c_p - c_v)(T_2 - T_1) \qquad (13.5.31)$$

Since ΔQ_{BE} and ΔW_{BE} are known, ΔU_{BE} can be found from the first law, which states that

$$\Delta U_{BE} = \Delta Q_{BE} - \Delta W_{BE} = -nc_v(T_2 - T_1) \qquad (13.5.32)$$

For the whole cycle, the change in internal energy will be, from (13.5.21), (13.5.22), and (13.5.32),

$$\Delta U_{tot} = \Delta U_{EA} + \Delta U_{AB} + \Delta U_{BE} = 0 \qquad (13.5.33)$$

This is to be expected, because as we pointed out in the preceding chapter, the internal energy, as contrasted with the heat and work, is a *well-defined function* of temperature and volume (temperature alone in this particular case); and if the temperature and volume are returned to their initial values, the internal energy also returns to its initial value. As a result, $\Delta U = 0$ in *any* cyclic process.

The net work done by the system can be obtained in the same way from (13.5.20), (13.5.23), and (13.5.27) as

$$\Delta W_{\text{tot}} = \Delta W_{\text{EA}} + \Delta W_{\text{AB}} + \Delta W_{\text{BE}}$$

$$= nR\left[T_2 \ln \frac{T_2}{T_1} - (T_2 - T_1) \right] \qquad (13.5.34)$$

This corresponds to the area EAB in Fig. 13.7b. The net heat absorbed by the system, finally, is, from (13.5.21), (13.5.28), and (13.5.29), using $c_p - c_v = R$,

$$\Delta Q_{\text{tot}} = \Delta Q_{\text{EA}} + \Delta Q_{\text{AB}} + \Delta Q_{\text{BE}}$$

$$= nR\left[T_2 \ln \frac{T_2}{T_1} - (T_2 - T_1) \right] \qquad (13.5.35)$$

As one must expect from the first law of thermodynamics, since $\Delta U_{\text{tot}} = 0$,

$$\Delta Q_{\text{tot}} = \Delta W_{\text{tot}}$$

To determine the efficiency of the engine, one might be tempted to calculate the ratio $\Delta Q_{\text{tot}}/\Delta W_{\text{tot}}$, which gives the answer unity (that is, 100 percent). This is incorrect, the reason being that there is a fundamental difference between the contributions ΔQ_{EA} and ΔQ_{AB} as distinguished from ΔQ_{BE}. The quantity $\Delta Q_{\text{EA}} + \Delta Q_{\text{AB}}$ represents heat received from the boiler at a high temperature. The quantity ΔQ_{BE}, on the other hand, represents heat rejected into the condenser at a *lower* absolute temperature. This heat is *lost forever* so far as contributing to the work output of this particular engine is concerned. The reason for this is that this heat is given up by the working substance at the minimum temperature T_1, and to utilize it in a continuation of the operating cycle, it somehow has to be returned to the boiler *at the higher temperature* T_2. But *heat does not flow of its own accord from a region of low temperature to a region of higher temperature*, and such a spontaneous transfer of heat energy cannot occur. This limitation is a particular example of the *second law of thermodynamics*, which we shall investigate in more detail in the next chapter.

It is evident from this discussion, in any event, that the proper definition of efficiency for this engine is the ratio of the net work output per cycle to the total heat absorbed from the high temperature source. Therefore, if the efficiency is denoted by η,

$$\eta = \frac{\Delta W_{\text{tot}}}{\Delta Q_{\text{EA}} + \Delta Q_{\text{AB}}} = \frac{T_2 \ln \dfrac{T_2}{T_1} - (T_2 - T_1)}{T_2 \ln \dfrac{T_2}{T_1} + \dfrac{c_v}{R}(T_2 - T_1)}$$

$$(13.5.36)$$

This can be simplified by noting that since $R = c_p - c_v$,

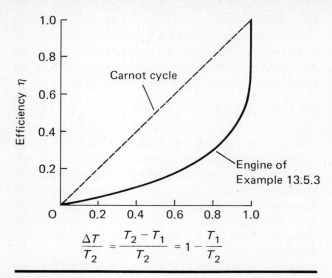

FIGURE 13.8

$$\frac{c_v}{R} = \frac{c_v}{c_p - c_v} = \frac{1}{\dfrac{c_p}{c_v} - 1} = \frac{1}{\gamma - 1} \qquad (13.5.37)$$

where $\gamma = c_p/c_v$. Making this substitution and dividing numerator and denominator by T_2,

$$\eta = \frac{\ln \dfrac{T_2}{T_1} - \left(1 - \dfrac{T_1}{T_2}\right)}{\ln \dfrac{T_2}{T_1} + \dfrac{1}{\gamma - 1}\left(1 - \dfrac{T_1}{T_2}\right)} \qquad (13.5.38)$$

From this expression, it is clear that the efficiency depends only upon the ratio of the absolute temperatures T_2 and T_1. A graph of the efficiency plotted versus the quantity $1 - (T_1/T_2) = (T_2 - T_1)/T_2 = \Delta T/T_2$, using the value $\frac{5}{3}$ (representative of a monatomic ideal gas) for γ, is shown in Fig. 13.8. For $\Delta T = 0$, T_1 and T_2 are equal; in this case, there is no net work done and the efficiency is zero. As the ratio $\Delta T/T_2$ increases, the fraction of input heat converted to net work output increases, reaching unity at $\Delta T/T_2 = 1$. This condition is attained only when $T_2 - T_1 = T_2$, that is, when $T_1 = 0$, the condenser then operating *at absolute zero*. Such a state of affairs, though it may be approached, cannot ever be realized in practice, and so the *efficiency remains always less than unity*. A cyclically operating heat engine using this cycle of processes, therefore, can never convert all the heat input received into mechanical work. This is a specific example of a completely general principle, true for all heat engines and any cycle of thermal processes, the general principle being referred to as the *second law of thermodynamics*.

EXAMPLE 13.5.4

Find the relations between pressure and volume, pressure and temperature, and volume and temperature for n moles of an ideal gas which is allowed to undergo adiabatic expansion from some initial state P_0, V_0, and T_0 to a final state where these state variables have the values P, V, and T. Find the work done by the gas in such an adiabatic expansion. Assume that the value of the specific heat ratio γ is known. Suppose that 0.8 liters of air ($\gamma = \frac{7}{5}$) initially at 300°K and 1.00 atmosphere pressure are compressed adiabatically in one of the cylinders of a diesel engine until the volume is reduced to $\frac{1}{24}$ of the original value. What are the final values of temperature and pressure? How much work must be done on the air to effect this compression? Assume that air behaves as an ideal gas in this process.

The equation of state governing the relationship between pressure, volume, and temperature in an adiabatic process is the same as for an isothermal process or for any other process; in the case of an ideal gas, $PV = nRT$. The trouble is this: starting with an initial volume V_0 and pressure P_0, one may easily find the temperature T_0 from the relation

$$P_0 V_0 = nRT_0 \tag{13.5.39}$$

When the gas is allowed to expand (or contract) adiabatically until it has some new volume V, both temperature and pressure will change. Their final values T and P are related to the volume, as always, by

$$PV = nRT \tag{13.5.40}$$

Unfortunately, however, neither P nor T are known, and at the moment, Eq. (13.5.40) gives us only one relation between them. We must, therefore, uncover another relation between these variables. This is afforded by the first law of thermodynamics, which tells us that, since in an adiabatic process $dQ = 0$,

$$0 = dU + P\,dV \tag{13.5.41}$$

Now, in any process involving a given temperature change dT in an ideal gas, *the change in internal energy will be the same*, whether the process is adiabatic, constant volume, constant pressure, or what have you. The reason is that for an ideal gas the internal energy depends upon the temperature and upon nothing else. We saw previously, in connection with Eq. (13.5.11), that, in a constant volume process resulting in a temperature change dT, the change in internal energy must be given by $dU = nc_v\,dT$. This value of dU must then accompany any process resulting in temperature change dT, and, therefore, (13.5.41) can be written as

$$0 = nc_v\,dT + P\,dV \tag{13.5.42}$$

$$dT = -\frac{P}{nc_v}\,dV \tag{13.5.43}$$

If we differentiate both sides of (13.5.9), we see that

$$P\,dV + V\,dP = nR\,dT = n(c_p - c_v)\,dT \tag{13.5.44}$$

recalling that $R = c_p - c_v$. Substituting into this the value of dT given by (13.5.43), this reduces to

$$P\,dV + V\,dP = -\left(\frac{c_p - c_v}{c_v}\right)P\,dV = (1 - \gamma)P\,dV \tag{13.4.45}$$

where $\gamma = c_p/c_v$, as usual. Subtracting $P\,dV$ from either side, this can finally be expressed as

$$\gamma P\,dV = -V\,dP \tag{13.5.46}$$

We could integrate this if we could arrange to have only functions of V on one side and functions of P on the other. This can easily be brought about by dividing both sides by PV, which gives us

$$\gamma\,\frac{dV}{V} = -\frac{dP}{P} \tag{13.5.47}$$

Integrating both sides of this equation from the initial volume V_0, when the pressure has the value P_0, to the final state in which the volume is V and the pressure P, we have

$$\gamma \int_{V_0}^{V} \frac{dV}{V} = -\int_{P_0}^{P} \frac{dP}{P}$$

or

$$\gamma[\ln V]_{V_0}^{V} = \gamma[\ln V - \ln V_0] = -[\ln P]_{P_0}^{P}$$
$$= \ln P_0 - \ln P$$

But $\ln V - \ln V_0 = \ln(V/V_0)$ and $\ln P_0 - \ln P = \ln(P_0/P)$; so this can be written

$$\gamma \ln \frac{V}{V_0} = \ln\left(\frac{V}{V_0}\right)^{\gamma} = \ln \frac{P_0}{P} \tag{13.5.48}$$

Now if the logarithms of two real quantities are equal, the quantities themselves must be equal, and, therefore, (13.5.48) implies that

$$\frac{V^{\gamma}}{V_0^{\gamma}} = \frac{P_0}{P} \tag{13.5.49}$$

or

$$PV^{\gamma} = P_0 V_0^{\gamma} = \text{const.} \tag{13.5.50}$$

This is the desired result. If P_0, V_0, and V are known, then P can be determined from (13.5.50), and this allows the final temperature T to be found using the ideal gas law (13.5.40). Equation (13.5.50) is often referred to as the *adiabatic equation of state* for an ideal gas. This equation can be stated in two other useful ways, as a relation between V and T and as a relation between P and T. The first of these can be obtained by substituting, from the ideal gas law, the values $P = nRT/V$ and $P_0 = nRT_0/V_0$ into (13.5.50). The resulting expression has the form

$$TV^{\gamma-1} = T_0 V_0^{\gamma-1} = \text{const.} \tag{13.5.51}$$

Finally, we may note that (13.5.51) may be written

$$\frac{V^{\gamma-1}}{V_0{}^{\gamma-1}} = \frac{T_0}{T} \tag{13.5.52}$$

Raising both sides of this equation to the power $\gamma/(\gamma - 1)$ we get, in view of (13.5.49),

$$\frac{V^{\gamma}}{V_0{}^{\gamma}} = \frac{T_0{}^{\gamma/(\gamma-1)}}{T^{\gamma/(\gamma-1)}} = \frac{P_0}{P}$$

or

$$\boxed{PT^{(\gamma-1)/\gamma} = P_0 T_0{}^{(\gamma-1)/\gamma} = \text{const.}} \tag{13.5.53}$$

The work done in an adiabatic process can be obtained from (13.5.3), using (13.5.49) to express P as a function of V. Thus,

$$\Delta W = \int_{V_0}^{V} P \, dV = \int_{V_0}^{V} \frac{P_0 V_0{}^{\gamma}}{V^{\gamma}} \, dV = P_0 V_0{}^{\gamma} \int_{V_0}^{V} V^{-\gamma} \, dV$$

$$= P_0 V_0{}^{\gamma} \left[\frac{V^{-\gamma+1}}{-\gamma + 1} \right]_{V_0}^{V} = \frac{P_0 V_0{}^{\gamma}}{\gamma - 1} \left[\frac{1}{V_0{}^{\gamma-1}} - \frac{1}{V^{\gamma-1}} \right] \tag{13.5.54}$$

But $P_0 = nRT_0/V_0$; substituting this into (13.5.54), we find

$$\Delta W = \frac{nRT_0}{\gamma - 1} \left[1 - \left(\frac{V_0}{V} \right)^{\gamma-1} \right] \tag{13.5.55}$$

In the numerical example we were asked to work, $V = V_0/24$ and $\gamma = 7/5$. Therefore, from (13.5.50),

$$PV^{\gamma} = P \frac{V_0{}^{\gamma}}{(24)^{\gamma}} = P_0 V_0{}^{\gamma}$$

from which

$$P = P_0(24)^{\gamma} = (1)(24)^{1.4} = 85.6 \text{ atm, or} = 1258 \text{ lb/in.}^2$$

The final temperature T is then, from (13.5.39) and (13.5.40), given by

$$\frac{T}{T_0} = \frac{PV}{P_0 V_0} = (85.6) \left(\frac{1}{24} \right) = 3.57$$

or, since $T_0 = 300°\text{K}$,

$$T = (3.57)(300) = 1070°\text{K} = 797°\text{C}.$$

This temperature is sufficient to ignite diesel fuel as it is injected into the combustion chamber; but in order to attain it, a compression ratio of 24 : 1 must be adopted. This entails pressures of 85 atmospheres even before the fuel ignites. The high combustion chamber pressure requires that the connecting rods, bearings, and engine block of a diesel engine be more strongly constructed than the corresponding parts of a gasoline engine.

The work done during compression is obtained directly from (13.5.54) as

$$W = \frac{P_0 V_0}{\gamma - 1} \left[1 - \left(\frac{V_0}{V} \right)^{\gamma-1} \right] = \frac{(1)(0.8)}{(0.4)} \left(1 - (24)^{0.4} \right)$$

$$= (2.0)(1 - 3.565) = -5.13 \text{ liter-atm} = 520 \text{ J}$$

It is useful to remember that 1 liter-atmosphere = 101.3 joules; this conversion factor is easily inferred from the values of R given in various units in section 13.2.

EXAMPLE 13.5.5

A quantity of a diatomic ideal gas ($c_v = \frac{5}{2}R$) having an initial volume $V_0 = 25.0$ liters, an initial pressure $P_0 = 1.5$ atmosphere, and an initial temperature $T_0 = 300°\text{K}$ is subjected to the cyclical process ABC shown in Fig. 13.9. Starting at A, the pressure of the gas is raised to the value $P_0 = 4.5$ atmospheres at constant volume. The pressure and volume are then changed, following the straight-line path BC until at point C the initial pressure $P_0 = 1.5$ atmosphere is reached. At this point, the temperature is the same as it was at point B. Finally, the volume of the gas is reduced at the constant pressure P_0 until the gas reaches the initial state at A. Find the changes in heat, work, and internal energy in each leg of the process and in the complete cycle. What is the net work done by the system per cycle?

The number of moles of gas in the system can be found by substituting the initial conditions of pressure, temperature, and volume into the ideal gas law. Thus,

$$n = \frac{P_0 V_0}{RT_0} = \frac{(1.5)(25.0)}{(0.08206)(300)} = 1.523 \text{ mole}$$

At point B, the pressure has been tripled, while the volume remained constant at the value V_0. According to the ideal gas law, this can only take place if the absolute temperature is increased by a factor of 3. Therefore, the values of pressure, volume, and tem-

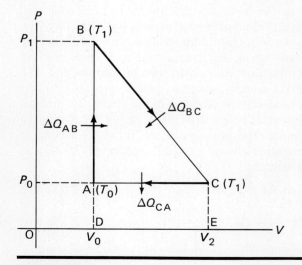

FIGURE 13.9

perature at B must be $P_1 = 4.5$ atmospheres, $V_1 = V_0 = 25$ liters, and $T_1 = 3T_0 = 900°K$. At point C, the pressure is back at the initial value $P_0 = 1.5$ atmosphere, while the temperature still has the value $T_2 = T_1 = 900°K$, three times the initial value. According to the ideal gas law, this means that the volume at this point must be three times the initial value, thus $V_2 = 3V_0 = 75$ liters.

Along AB, since there is no change in volume, $\Delta W = 0$, hence

$$\Delta W_{AB} = 0$$

$$\Delta Q_{AB} = \Delta U_{AB} = nc_v \, \Delta T$$
$$= (1.523)(\tfrac{5}{2})(8.314)(900 - 300) = 18,990 \text{ J}$$

Note that since n is expressed in gram-moles rather than kilogram-moles, we must use the value 8.314 joules/(g-mole)-°K rather than 8314 joules/(kg-mole)-°K for R.

Along BC, since the endpoints are at the same temperature, $\Delta U_{BC} = 0$, and, therefore,

$$\Delta Q_{BC} = \Delta W_{BC} = \text{area DABCE}$$
$$= \text{area DACE} + \text{area ABC}$$
$$\Delta Q_{BC} = \Delta W_{BC} = P_0(V_2 - V_0) + \tfrac{1}{2}(V_2 - V_0)(P_1 - P_0)$$
$$= \tfrac{1}{2}(V_2 - V_0)(P_0 + P_1)$$
$$= \tfrac{1}{2}(75 - 25)(1.5 + 4.5)$$
$$= 150 \text{ liter-atm} = 15195 \text{ J}$$

We could, of course, have determined the work by expressing P as a function of volume along the straight line BC and then finding the integral of $P(V) \, dV$ between the limits $V_2 = 75$ liters and $V_1 = 25$ liters. This is, however, complex and unnecessary since we can figure the areas involved using simple geometric formulas for the areas of rectangles and triangles.

Along CA, since the process takes place at constant pressure and since, for an ideal gas, $c_p = c_v + R \, (= \tfrac{7}{2}R$ in this case),

$$\Delta Q_{CA} = nc_p \, \Delta T = (1.523)(\tfrac{7}{2})(8.314)(300 - 900)$$
$$= -26590 \text{ J}$$

Also,

$$\Delta W_{CA} = -\text{area PACE} = -(V_2 - V_0)P_0$$
$$= -(75 - 25)(1.5)$$
$$= -75 \text{ liter-atm} = -7600 \text{ J}$$

Then,

$$\Delta U_{CA} = \Delta Q_{CA} - \Delta W_{CA} = -26,590 - (-7600)$$
$$= -18,990 \text{ J}$$

For the complete cycle,

$$\Delta Q = \Delta Q_{AB} + \Delta Q_{BC} + \Delta Q_{CA}$$
$$= 18,990 + 15,195 - 26,590 = 7595 \text{ J}$$

$$\Delta U = \Delta U_{AB} + \Delta U_{BC} + \Delta U_{CA}$$
$$= 18,990 + 0 - 18990 = 0$$
$$\Delta W = \Delta W_{AB} + \Delta W_{BC} + \Delta W_{CA}$$
$$= 0 + 15,195 - 7600 = 7595 \text{ J}$$

The sum $\Delta Q_{AB} + \Delta Q_{BC}$ represents heat absorbed by the system from external sources at temperatures higher than that of the system, while the negative quantity ΔQ_{CA} represents heat rejected by the system to external surroundings at temperatures lower than that of the system.

The results relating to the changes of *internal energy* should be compared with those obtained in Example 13.5.3, which, though similar to this one, differs from it in some respects. We shall emphasize two important facts about the internal energy which, although pointed out in the last chapter, will be restated here in the light of the examples just worked out, in which they are clearly evident and play an important role:

1. The internal energy U represents the random kinetic energies and the mutual potential energies of all the molecules of the system. It is, therefore, a perfectly well-defined function of the temperature and the volume of the system, and in any cyclic process, whenever these variables are returned to their initial values, the internal energy returns also to its initial value. Therefore, in *any cyclic process, ΔU is zero*, irrespective of the closed path laid out in the $P-V$ diagram by the way the process is carried out. This is *not* true for ΔQ and ΔW, for they have values which cannot be defined until the details of how the cyclic path is traversed are revealed. This means that, unlike the internal energy, the "total" heat and "total" work Q and W are not well-defined functions of P, T, and V. Indeed, it makes no sense at all to speak of these "total" quantities, but only the path-dependent changes ΔW and ΔQ. In the same way, the quantity dU is the *differential* of a perfectly well-defined function of P, T, and V, while dQ and dW, though denoting quantities which may approach zero as a limit, do not have this property since their values depend not only on the values of P, T, and V but also on what thermal processes are adopted to bring about changes in those quantities.

2. The change in internal energy between two states, P_1, V_1, and T_1 and P_2, V_2, and T_2, is simply given by the difference $\Delta U = U_2 - U_1$ in the internal energy associated with the states and is altogether *independent* of the path taken to go from one to the other. This again is not true of the changes ΔQ and ΔW which depend upon thermal processes and areas associated with these processes in the $P-V$ diagram. The reason, once more, is that there are

no well-defined functions Q and W that play a role similar to that of U.

EXAMPLE 13.5.6

Find the isothermal bulk modulus of an ideal gas and use the result in connection with Eq. (10.4.12) of Chapter 10 to derive an expression for the velocity of sound in an ideal gas. Comment on the variation of sound velocity with the physical characteristics of the medium predicted by this equation.

In Chapter 10, we showed that the velocity of sound in a fluid medium could be expressed as

$$v = \sqrt{\frac{B}{\rho}} \qquad (13.5.56)$$

where B is the bulk modulus of the substance, expressing the ratio of the stress dP to the fractional volume change dV/V. Accordingly, we may write the bulk modulus as

$$B = -\frac{dP}{dV/V} = -V\frac{dP}{dV} \qquad (13.5.57)$$

In this equation, the minus sign is used since a positive pressure increment dP results in a negative volume change, while at the same time the bulk modulus is inherently positive. For an ideal gas, of course, $PV = nRT$, and, therefore, in an *isothermal process* in which T remains fixed,

$$B = -V\frac{d}{dV}\left(\frac{nRT}{V}\right) = -V\left(-\frac{nRT}{V^2}\right) = \frac{nRT}{V} \qquad (13.5.58)$$

We find, therefore, that we can express the bulk modulus of an ideal gas in the very simple form

$$B = \frac{nRT}{V} = P \qquad (13.5.59)$$

The density is the ratio of mass to volume; for n moles of gas, the mass is nM, where M is the gram-molecular mass. Therefore,

$$\rho = \frac{nM}{V} \qquad (13.5.60)$$

Substituting the above expressions for B and ρ into (13.5.56),

$$v = \sqrt{\frac{nRT}{V}\frac{V}{nM}} = \sqrt{\frac{RT}{M}} \qquad (13.5.61)$$

This equation gives the speed of sound in an ideal gas. It is apparent that the sound velocity varies as the square root of the absolute temperature and inversely as the square root of the molecular mass.

Inserting the numerical data $R = 8314$ joules/mole-°K, $T = 300$°K, and $M = 28.97$ g/mole for air at normal ambient temperature, we obtain

$$v = \sqrt{\frac{(8314)(300)}{28.97}} = 293.4 \text{ m/sec}$$

This is about 11.4 percent less than the observed speed of 331 m/sec. The discrepancy is accounted for by the fact that the rapid compression and rarefaction processes that occur in the propagation of sound are not really isothermal. In fact, they occur so rapidly that heat has no opportunity to flow into or out of the regions of compression or rarefaction as they propagate. The situation is, therefore, more accurately represented in terms of adiabatic processes rather than isothermal ones. The assumption of adiabatic conditions can be shown simply to introduce a factor γ multiplying the quantity RT/M under the radical in (13.5.61). For air, for which $\gamma = 1.40$, this predicts a sound velocity of 347 m/sec under the conditions assumed above, in much better agreement with what is observed. The predicted variation of sound velocity with temperature and molecular mass is the same in either case.

13.6 Distribution of Energies and Velocities: The Ideal Crystal

In a system such as an ideal gas, the individual gas molecules may have a large range of velocities. Some molecules may move very slowly, while others will have much greater velocities and hence much greater kinetic energies. We have already derived an expression for the root-mean-square velocity, (13.4.25), which gives us a good idea of the average velocity of the molecules, and have also determined the closely related fact that the average translational kinetic energy is $\frac{3}{2}kT$.

It is important for many purposes, particularly in determining average values of molecular dynamic quantities, to know how the molecules of a system are distributed in velocity or energy, that is, to know what fraction may have speeds in a range Δv centered on a given speed v or what fraction may have total energy in a range Δu centered on a given energy u. The velocity and energy distributions of an ideal gas are closely related because all the energy resides in the kinetic energy (hence velocity) of the molecules. In more complex systems, the relationship is not so direct because only part of the energy stems from molecular velocities, the rest being potential energy of interaction between molecules and within individual molecules. The distribution in energy is the more fundamental quantity; when this is known, the velocity distribution can always be found. Furthermore, it can be shown that all systems, at least insofar as classical mechanics can be regarded as describing their behavior correctly, have essentially the *same* distribution of total energies, even though their velocity distributions may differ as a result of different molecular potential energy contributions. This distribution of energies is called the *Maxwell–Boltzmann distribution*.

The distribution law arises primarily from the workings of probability and statistics rather than from classical dynamics. The assumptions upon which it rests are that any position coordinate and any value of a particle momentum component is as likely (on the basis of pure chance) as any other, subject, however, to the important restriction that the total number of particles in the system and the sum total of their energies remain fixed. Now, clearly, there are some distributions of energy among particles of smaller probability than others on the basis of pure chance. One distribution that is possible, but extremely improbable, is one in which a single particle has all the internal energy of the system while the others have none and are motionless. Naturally, there are other distributions of energy among particles that occur with higher probability than this; in particular, there must be one whose probability of occurrence is greater than any other. It is this distribution of maximum probability that is assumed to represent the actual physical state of the system. *Assumed* is the right word, for it is an assumption, but it is justified by the good agreement between predictions based upon it and experimental observations.

The calculation of the Maxwell–Boltzmann energy distribution based on these concepts can be accomplished using the laws of probability and statistics, along with a generous helping of advanced mathematics. We shall not attempt a complete presentation of the calculation but shall, instead, indicate how the result may be inferred in a simple, though not very general or rigorous, manner from the results derived in Example 13.3.4. In that example, we found that the density of an ideal gas at a uniform temperature, subject to a constant gravitational force, could be expressed by Eq. (13.3.22). Using the relation $R = N_A k$ and noting that the mass m of a single molecule is the molar mass M divided by the number of molecules per mole, N_A, this can be written in the form

$$\rho = \rho_0 e^{-mgz/kT} \tag{13.6.1}$$

The density ρ_1 at height z_1 and the density ρ_2 at height z_2 can be calculated directly from this expression. The ratio of these two densities would then be

$$\frac{\rho_2}{\rho_1} = e^{-(mgz_2 - mgz_1)/kT} \tag{13.6.2}$$

The quantity $mgz_2 - mgz_1$ represents the difference in potential energy of molecules at the two levels. But since the temperature of the gas is the same at both levels, there can be no difference in the average *kinetic*

energy of the molecules nor in the distribution of kinetic energies between the two levels z_2 and z_1. The quantity $mgz_2 - mgz_1$, therefore, represents the difference in *total energy* of molecules at height z_2 and those at height z_1. Let us call this difference $u_2 - u_1$, using the small letter u to represent the energy of an individual particle so as to avoid confusion with the capital letter U, which in Chapters 12, 13, and 14 represents the total internal energy of *all* the particles in the system. We may then write (13.6.2) as

$$\frac{\rho_2}{\rho_1} = e^{-(u_2 - u_1)/kT} \tag{13.6.3}$$

This result suggests that the number of particles in the system, per unit volume, having total energy u is an exponential function of energy of the form

$$f(u) = A e^{-u/kT} \tag{13.6.4}$$

where A is a constant, and, therefore, that the fundamental probability that a particle will have an energy which lies in a small range du centered on u will also have this basic form. This supposition is generally correct, though the final probability and particle density are sometimes modified by another factor, as we shall soon see. Equation (13.6.4) exhibits the basic mathematical form of the Maxwell–Boltzmann energy distribution.

Now let us turn our attention to an ideal gas consisting of N molecules confined in a container of volume V, having a fixed internal energy U. The latter condition implies that the temperature is also fixed. If we are asked to find the distribution of molecular velocities, we might conclude that the distribution would once more involve the same simple exponential law, the particle energy u now being expressed by

$$u = \tfrac{1}{2}mv^2 = \tfrac{1}{2}m(v_x^2 + v_y^2 + v_z^2) \tag{13.6.5}$$

This is correct, provided the proper interpretation is made. The way of arriving at the proper interpretation, however, involves the clear recognition of the two basic assumptions upon which our understanding of statistical mechanics rests. The first assumption is a purely *statistical* one, which states that in the absence of any other consideration, a molecule's momentum components are just as likely to have one value as another. This is often stated in another way by visualizing the momentum components (p_x, p_y, p_z) as coordinates which locate a point with respect to three mutually perpendicular axes. The location of such a point in this "momentum space" specifies completely the momentum of the particle. The statistical assumption we are making specifies that in the absence of other considerations, the probability associated with all points in momentum space is the same.

This is ridiculous, of course. We know that the

gas molecules have some average speed $\bar{v}$ and that there are very few with speeds many times greater than that and very few with speeds vastly smaller. Anyway, if all momenta—hence all velocities—were equally likely, then there would be many molecules with astronomical speeds, and the total energy of the system would be astronomical, too, *infinite*, in fact! But this is just the point. This contradiction can be resolved by recalling that there is *another* assumption we made about the system, a dynamical one, to the effect that the total energy U was *fixed*. You may now correctly conclude that this is why those words "in the absence of other considerations" were so carefully included in stating assumption number one. The other consideration (that is, the fixed total energy U) does not modify the *statistical* facts about the equality of probability for all points in momentum space, but it does impose a constraint on how many of these points can be occupied at the same time. Indeed, it is this dynamical constraint on the total energy that leads to the exponential Maxwell–Boltzmann factor; without it the energy distribution function would simply be a constant, independent of the energy.

Returning now to the question of stating the proper form for the distribution of particle velocities in an ideal gas, it is clear that the probability associated with all points in velocity space (v_x, v_y, v_z) is the same, because in this case the particle momentum coordinates are just $p_x = mv_x$, $p_y = mv_y$, and $p_z = mv_z$. Momentum and velocity space are exactly alike, apart from a constant scale factor m. The probability that a molecule would be located in a little brick-shaped region in velocity space having dimensions dv_x, dv_y, dv_z centered upon some velocity (v_x, v_y, v_z) is, apart from the dynamical constraint of fixed total energy, just a constant times the volume of velocity space in the region, that is, $C(dv_x\, dv_y\, dv_z)$. The constraint on the total energy, however, introduces the Maxwell–Boltzmann factor

$$f(u) = Ae^{-u/kT} = Ae^{-m(v_x{}^2 + v_y{}^2 + v_z{}^2)/2kT} \qquad (13.6.6)$$

The *actual* probability associated with the region, and thus the number of particles of the system associated with it, will be proportional to the product of these factors:

$$
\begin{aligned}
N(v_x, v_y, v_z)\, & dv_x\, dv_y\, dv_z \\
&= Be^{-m(v_x{}^2 + v_y{}^2 + v_z{}^2)/2kT}\, dv_x\, dv_y\, dv_z \\
&\doteq Be^{-mv_x{}^2/2kT}e^{-mv_y{}^2/2kT}e^{-mv_z{}^2/2kT}\, dv_x\, dv_y\, dv_z
\end{aligned}
\qquad (13.6.7)
$$

where $B\,(=AC)$ is a constant. The quantity

$$N(v_x, v_y, v_z)\, dv_x\, dv_y\, dv_z$$

represents the number of particles of the gas in the

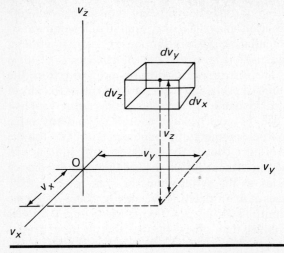

FIGURE 13.10. Brick-shaped volume element of dimensions dv_x, dv_y, dv_z about the point v_x, v_y, v_z in velocity space. Equation (13.6.7) gives the number of molecules of a Maxwell-Boltzmann gas in such an element.

brick-shaped element shown in Fig. 13.10. It is written in this fashion to exhibit the fact that it is proportional to the size of the infinitesimal velocity–space volume element $dv_x\, dv_y\, dv_z$.

It is possible to visualize (13.6.7) as the product of three factors, which we shall call $N(v_x)\, dv_x$, $N(v_y)\, dv_y$, and $N(v_z)\, dv_z$ as follows:

$$
\begin{aligned}
N(v_x, v_y, v_z)\, & dv_x\, dv_y\, dv_z \\
&= (B^{1/3}e^{-mv_x{}^2/2kT}\, dv_x)(B^{1/3}e^{-mv_y{}^2/2kT}\, dv_y) \\
&\quad \times (B^{1/3}e^{-mv_z{}^2/2kT}\, dv_z) \\
&= N(v_x)\, dv_x\, N(v_y)\, dv_y\, N(v_z)\, dv_z
\end{aligned}
$$

$$(13.6.8)$$

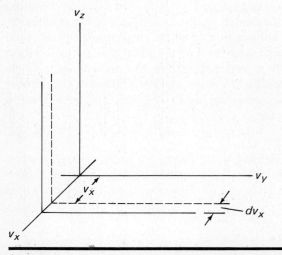

FIGURE 13.11. Plane-parallel slab of velocity space of thickness dv_x about velocity v_x. Equation (13.6.8) expresses the number of molecules of a Maxwell-Boltzmann gas in such a region.

426

ceding example. We are told that the energies of the oscillators are distributed according to the Maxwell–Boltzmann law, hence that the number of oscillators with total energy in a range du about the value u is given by

$$N(u)\, du = Ae^{-u/kT}\, du \qquad (13.6.16)$$

The average energy, computed according to what should by now be regarded as the standard technique, will be

$$\bar{u} = \frac{\int_0^\infty uN(u)\, du}{N} = \frac{\int_0^\infty uN(u)\, du}{\int_0^\infty N(u)\, du}$$

$$= \frac{\int_0^\infty ue^{-u/kT}\, du}{\int_0^\infty e^{-u/kT}\, du} = \frac{-(kT)^2 \left[\dfrac{u}{kT} e^{-u/kT} + e^{-u/kT}\right]_0^\infty}{(-kT)\left[e^{-u/kT}\right]_0^\infty}$$

$$(13.6.17)$$

giving

$$\boxed{\bar{u} = kT} \qquad (13.6.18)$$

The integral in the numerator may be evaluated by the technique of integration by parts or, more conveniently, by consulting a standard table of integrals. In evaluating this integral at the upper limit, it is necessary to recognize that the quantity xe^{-x} approaches *zero* in the limit $x \to \infty$.

We have constructed and examined in detail a dynamical model of the ideal gas based upon a system that is mostly "empty space." We shall now go to the opposite extreme and try to construct a model of an "ideal crystal" that is based upon a system of atoms or molecules that are packed into a regular, closely spaced array. In this array, it is supposed that strong forces of attraction exist between neighboring atoms to limit their movement to small vibrations about their equilibrium positions.

The simplest model of a perfect cubic crystal, therefore, envisions such a substance as a collection of point masses, representing the atoms, arranged in a regularly spaced lattice one unit of which is a cubic cell of side a, as shown in Fig. 13.15. The atoms are

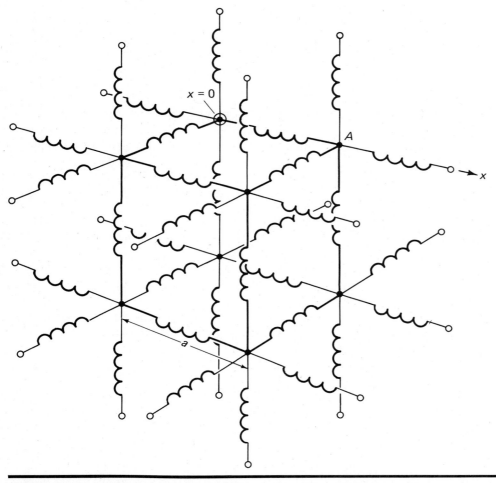

FIGURE 13.15. Model of the lattice of an "ideal crystal" in which point molecules of mass m are bound to one another by attractive forces which act like Hooke's law springs.

429

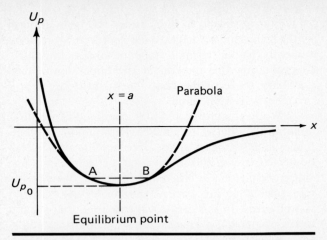

FIGURE 13.16. Potential energy of an atom in an "ideal" crystal, represented by a parabolic potential well, and the form of the potential well in a real substance.

bound together by attractive forces which behave in the first approximation like *ideal Hooke's law springs*. This behavior arises from the fact that at equilibrium the resultant force on any atom is zero. But the resultant force is related to the potential energy of the atom, U_p, by

$$F_x = -\frac{\partial U_p}{\partial x} = 0$$

$$F_y = -\frac{\partial U_p}{\partial y} = 0 \qquad (13.6.19)$$

$$F_z = -\frac{\partial U_p}{\partial z} = 0$$

The fact that $\partial U_p/\partial x$, $\partial U_p/\partial y$, and $\partial U_p/\partial z$ are all zero means that at equilibrium U_p will be a minimum. Plotted as a function of the coordinate x in Fig. 13.15, then, the potential energy of the atom A will have to look something like what is shown in Fig. 13.16. If the atom has no kinetic energy, it remains at rest in equilibrium at the point $x = a$, which represents the equilibrium nearest-neighbor interatomic spacing. If the atom is somehow displaced along the x-direction from the equilibrium point, it will oscillate back and forth in the "potential well" of Fig. 13.16, for example, between the limits A and B. This motion will correspond to that of a simple harmonic oscillator *only if the potential well is parabolic*, that is, only if $U_p(x) - U_{p0} = \frac{1}{2}ks^2$, where s is the displacement from the equilibrium point $x = a$. But if we consider only the case where the oscillations have relatively small amplitudes, it will always be possible to represent the rounded minimum in the potential energy curve quite accurately by a section of a parabola, as shown by the dotted curve of Fig. 13.16. Therefore, if the amplitude of oscillation does not exceed the limits A and B in the diagram, the oscillatory motion of the atoms will be

very nearly simple harmonic motion, and the picture of the lattice held together with Hooke's law springs will be justified. For larger excursions from the equilibrium position, however, this picture may break down, because then the portion of the potential minimum in which the atom is confined will no longer be well represented by a parabola. Fortunately, the harmonic approximation is usually quite good except near the melting temperature of the substance.

Anyway, assuming that the atoms behave like ideal harmonic oscillators, let us now try to find the total *vibrational* internal energy of a simple cubic crystal consisting of N identical point atoms. From this, as we shall soon see, it is a simple matter to find also the specific heat of such a crystal under conditions of constant volume.

We may regard the crystal as a collection of N harmonic oscillators, each of which is free to vibrate along any of the three coordinate axes. But since the x-, y-, and z-components of the motion are executed independently of one another, as a result of the fact that the x-component of force on an atom depends only on the x-component of its displacement and not on the y- and z-components, we may regard the crystal to be made up of $3N$ independent *one-dimensional* classical simple harmonic oscillators. The energy of these oscillators will be distributed according to the familiar Maxwell–Boltzmann law.

In that event, the problem is nearly worked, for in Example 13.6.2 we found that the average energy of any oscillator such as this is kT. We need now only multiply this by the equivalent number of oscillators in the crystal to obtain the internal energy:

$$\boxed{U = 3N\bar{u} = 3NkT} \qquad (13.6.20)$$

There is another component of the internal energy, of course, which represents the total potential energy of interaction of the molecules with one another; at equilibrium, this amounts to $-NU_{p0}$ and is constant, independent of temperature. It, therefore, contributes nothing to the specific heat.

The specific heat at constant volume is simply the rate of change of total internal energy with temperature, and thus, from (13.6.20),

$$c_v = \frac{dU}{dT} = 3Nk \qquad (13.6.21)$$

For one mole of substance, $N = N_A$, and the molar specific heat of an "ideal crystal" is, therefore,

$$\boxed{c_v = 3N_A k = 3R = 5.96 \text{ cal/mole-}°\text{K}} \qquad (13.6.22)$$

We find that the molar specific heat of any "ideal" crystal has this constant, temperature-independent value *irrespective* of the atomic species from which it

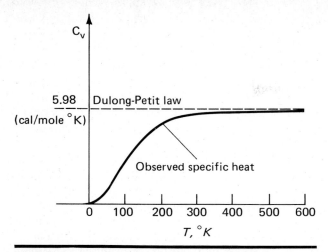

FIGURE 13.17. Specific heat of an ideal crystal according to the laws of classical physics and the observed specific heat of a typical crystalline solid.

is formed. This result is known as the *law of Dulong and Petit*. It is generally in good agreement with experimental observations at and above room temperature; but at low temperatures, measured values of specific heat for crystalline substances drop below the Dulong–Petit value of 5.96 cal/mole-°K and, indeed, approach zero as the absolute temperature approaches zero. This state of affairs is illustrated in Fig. 13.17. The discrepancy between the predictions of the classical ideal crystal model and these experimental findings arises from the inadequacy of Newtonian mechanics in describing all details of the mechanics of atomic and molecular systems. Classical mechanics assumes that a dynamical system—a harmonic oscillator in this case—may have any value of total energy in the averaging process expressed by (13.6.17). Quantum theory reveals, however, that many systems, including this one, may have only *certain discrete allowed values of total energy*, which are referred to as allowed *energy levels*, or allowed *energy states* of the system. In the case of the molecular harmonic oscillators of the crystal lattice, these allowed energies are evenly spaced in energy, the energy between adjacent levels being very small on the scale of energies we can observe macroscopically. At high temperatures, the average energy of the oscillators is large compared to the energy difference between adjacent levels, and in this case the discrete spacing of the levels does not really matter—particle energies are so large that it makes little difference whether they are distributed continuously or in closely spaced discrete states. Classical mechanics and quantum theory then lead to the same result. But at low temperatures, average molecular energies are so small as to be comparable to the energy difference between neighboring allowed energy states, or even smaller than this difference, and in this case the discreteness of the

allowed energies plays an important role. Quantum mechanics, therefore, predicts a specific heat which differs from the classical result at low temperatures and which agrees very well in most instances with the results of experimental measurements.

In the high-temperature range, the constant value $3R$ for the molar specific heat of a crystal is exactly twice that associated with a monatomic ideal gas. The reason for this is that when heat energy flows into an ideal crystal, it excites not only kinetic energy of vibration but also acts to increase the molecular *potential* energy above the equilibrium value U_{p0} shown in Fig. 13.16. This increase in potential energy represents energy stored in the intermolecular forces—in the "Hooke's law springs" which hold the crystal together, so to speak. We have already seen that the average kinetic and potential energies of a simple harmonic oscillator are *equal*, each component accounting for half the total energy. But it is only the molecular *kinetic* energies that affect the temperature; half the total energy input is stored as potential energy and therefore has no effect on the temperature. In a monatomic ideal gas, on the other hand, all the heat energy added to the system acts to increase the kinetic energy of the molecules and thus to raise the temperature. In the case of the ideal crystal, twice as much heat must be supplied per mole of substance to produce a given temperature increase than is needed in the case of an ideal gas, and, therefore, we should expect the molar specific heat of the crystal to be twice as large, which is precisely what is found.

13.7 Equipartition of Energy and the Specific Heats of Polyatomic Gases

We found in section 13.4 that the internal energy of an ideal monatomic gas amounts to $\frac{3}{2}kT$ per atom. Since the x-, y-, and z-components of velocity are entirely independent of one another, we had no trouble in swallowing the assertion that

$$\overline{v_x^2} = \overline{v_y^2} = \overline{v_z^2} = \tfrac{1}{3}\overline{v^2} \tag{13.7.1}$$

If we multiply this equation through by half the particle mass, we see that

$$\tfrac{1}{2}m\overline{v_x^2} = \tfrac{1}{2}m\overline{v_y^2} = \tfrac{1}{2}m\overline{v_z^2} = \tfrac{1}{3}\tfrac{1}{2}m\overline{v^2} = \tfrac{1}{3}\tfrac{3}{2}kT = \tfrac{1}{2}kT \tag{13.7.2}$$

since the average kinetic energy per particle, according to (13.4.20), is $\frac{3}{2}kT$. What (13.7.2) tells us is that the average energy residing in the x-, y-, and z-velocity components is the same for each component and has the value $\frac{1}{2}kT$ in each case. This the most elementary example of the principle of *equipartition of energy*. Each particle can be regarded as having three independent components of motion along the three co-

431

ordinate axes; it is said that the particle has three *degrees of freedom*. The principle of equipartition of energy states simply that:

The average internal energy per molecule for any dynamical system obeying the laws of classical mechanics amounts to $\frac{1}{2}kT$ per degree of freedom.

To find the internal energy of a polyatomic gas or other more complex system, all we need to do is to determine the number of degrees of freedom for a molecule and multiply that number by $\frac{1}{2}kT$. This is simple enough in principle, but there are one or two pitfalls to be avoided which we shall examine by considering a few simple examples.

First, let us undertake the example of an ideal gas consisting of *rigidly connected* diatomic molecules. The molecules are assumed to be point masses rigidly held together like a dumbbell. If there were no rigid connections between pairs of atoms, we would have a monatomic gas and there would be six degrees of freedom per *pair* of atoms. When we rigidly connect a pair of atoms into a molecule, however, we remove one degree of freedom by removing the possibility of one of the atoms executing independent motion along one direction. We are, therefore, left with five degrees of freedom per pair of atoms (that is, per molecule). There will then be, according to the principle of equipartition of energy, an internal energy per molecule amounting to $\frac{5}{2}kT$, or a molar internal energy of

$$U = \tfrac{5}{2}N_A kT = \tfrac{5}{2}RT \qquad (13.7.3)$$

The specific heat at constant volume is then given by

$$c_v = \frac{dU}{dT} = \tfrac{5}{2}R = 4.965 \text{ cal/mole-}°\text{K} \qquad (13.7.4)$$

and since, according to (13.5.15), $c_p = c_v + R$, the constant pressure specific heat c_p will be

$$c_p = \tfrac{7}{2}R = 6.951 \text{ cal/mole-}°\text{K} \qquad (13.7.5)$$

The five degrees of freedom for a rigid diatomic molecule may be regarded physically as three *translational* degrees associated with the translational motion of the center of mass and two *rotational* degrees of freedom corresponding to rotation about two independent axes through the center of mass lying in a plane perpendicular to the line joining that atoms. A third rotational degree of freedom corresponding to rotation about an axis *coinciding* with that line cannot be excited since the moment of inertia about that axis is zero—remember, these are *point* masses! In the case of rigid, nonlinear, triatomic molecules, three independent rotational degrees of freedom can be excited, the internal energy per molecule being then

$3kT$ and the constant volume molar specific heat, $3R$.

But are diatomic molecules really rigid? Isn't it more nearly correct to envision the two atoms as being connected by interatomic attractive forces which might, to a first approximation, act like ideal springs just as they do in an ideal crystal? Such a picture is, indeed, more truly representative of the actual physical behavior of the molecule, and it requires us to introduce, in addition to the translational and rotational degrees of freedom, the possibility of vibrational motion in which the molecules oscillate alternately toward and away from one another along the line joining them. If the interatomic forces can be approximated by Hooke's law, then the molecule may vibrate as a one-dimensional simple harmonic oscillator, which, according to the results of Example 13.6.2, will acquire an average energy of kT per oscillator, or $N_A kT = RT$ per mole. The vibrational motion is essentially independent of the translational and rotational motions and, from the point of view of the equipartition principle, can be regarded as introducing *two* additional degrees of freedom into the system, each of which contribute $\frac{1}{2}kT$ toward the internal energy. Why two degrees of freedom? Because a harmonic oscillator can store *potential* energy as well as kinetic energy, and $\frac{1}{2}kT$ must be assigned to each possible way of absorbing energy from an external heat source. The situation is much the same as that we encountered in the analysis of the ideal crystal. Indeed, the high-temperature specific heat of an ideal crystal can be arrived at by simply assigning $\frac{1}{2}kT$ to each of the two degrees of freedom associated with the $3N$ one-dimensional oscillators of the lattice. If we include the possibility of vibrational motion of the diatomic molecule, the internal energy will be increased by kT to $\frac{7}{2}kT$ per molecule, or to $\frac{7}{2}N_A kT = \frac{7}{2}RT$ per mole. The molar specific heat, including this vibrational contribution, will then be

$$c_v = \frac{dU}{dT} = \tfrac{7}{2}R = 6.951 \text{ cal/mole-}°\text{K} \qquad (13.7.6)$$

while

$$c_p = c_v + R = \tfrac{9}{2}R = 8.937 \text{ cal/mole-}°\text{K} \qquad (13.7.7)$$

How does all this compare with what is actually observed? A look at the thermal data tables at the end of Chapter 12 shows that at 300°K, for diatomic gases the observed value of c_v/R is about 2.5, but only in the case of iodine vapor at about 550°K is the value $c_v/R = 3.5$ approached. It would appear from these data that the rotational contribution is observed, but that the vibrational component is missing, or possibly, vice versa.

In fact, the vibrational component can be observed, but it can usually only be excited at temperatures substantially above 300°K, and the reason has once more to do with quantum theory. According to quantum theory, molecules can have essentially any *translational* energy, but both the rotational and vibrational energy components are *quantized*, that is, they can have only certain discrete values. In the case of the allowed rotational energy values, the spacing between allowed energy levels is so small that even at room temperature there is sufficient thermal energy in the system to excite rotational energy values of many times this spacing in energy. The system then behaves very much as if quantization were unimportant, and, therefore, acts very much according to the predictions of classical mechanics.

The equipartition theorem, stemming as it does from classical mechanics, thus gives us correct information regarding the internal energy. This might not be the case at much lower temperatures, when the thermal energy per molecule becomes of the order of or less than the spacing between allowed energy states. The allowed vibrational energies, on the other hand, are more widely spaced in energy, and it is not until quite high temperatures (usually of the order of many hundreds or even thousands of degrees Kelvin) are attained that average molecular energies become many times larger than the spacing between the allowed vibrational energy levels. The value $c_v/R = 3.50$ predicted for diatomic gases by the classical equipartition principle is not observed until such temperatures are reached. In the case of iodine vapor, the rotational levels are unusually closely spaced, so that this value is approached at much lower temperatures. A plot of the specific heat of a typical diatomic gas is shown in Fig. 13.18; in this diagram, the onset of rotational and vibrational contributions to the specific heat is illustrated.

The equipartition principle, then, arising as it does from Newtonian mechanics, can be trusted to give correct answers only in the limiting case of high temperatures. Nevertheless, it provides a simple and easily used guide to the prediction of thermal properties of simple molecular systems which would be difficult to calculate in any other way. The system is visualized as a collection of independent translatory, rotatory, and vibratory motions. An internal energy $\frac{1}{2}kT$ per molecule is then assigned to each translatory and rotatory motion and an internal energy kT to each vibratory mode. The constant volume specific heat may easily be calculated by finding the temperature derivative of the total internal energy per mole.

Turning it around, the equipartition principle, in connection with *measured* values of specific heats, is also valuable for ascertaining the *number* of translational, rotational, and vibrational degrees of freedom associated with a given molecule. This information is sometimes extremely useful in determining the *structural arrangement of atoms* in the molecule.

SUMMARY

The equation of state of a substance is an expression that states the relationship between pressure, volume, and temperature for that substance. The equation of state can also be stated in terms of other sets of "state variables" such as density, temperature, and volume. For all substances in the gaseous state, when the density is very much less than the density at the critical point, the equation of state is represented by the ideal gas law, which has the form

$$PV = nRT$$

where n is the number of moles and R is a constant referred to as the universal gas constant. The ideal gas law can also be stated as

$$\frac{PV}{T} = \frac{P_0 V_0}{T_0}$$

where P_0, V_0, and T_0 are given values of pressure, volume, and temperature in some particular reference state. In these expressions, the temperatures are in all cases *absolute* temperatures, or *Kelvin* temperatures, measured in degrees having the same size as Celsius degrees, but referred to a zero point at $-273.16°C$, which is called the absolute zero of temperature. The experimental reality of ideal gas behavior allows us to use "ideal" gases as thermometric fluids and to define the absolute temperature scale in terms of the relation $T = PV/nR$ for a substance that behaves as an ideal gas. One gram-mole of any ideal gas at $273.16°K$ and 1 atmosphere pressure occupies a volume of 22.415 liters.

The behavior of ideal gases can be understood on the basis of molecular dynamics by applying the laws of classical mechanics to a system containing N

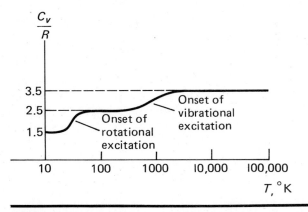

FIGURE 13.18. Specific heat of a diatomic gas, showing the excitation of rotational and vibrational quantum levels at high temperatures.

molecules, which are assumed to behave as point particles of given molecular mass m and which obey the laws of conservation of energy and momentum. The assumptions regarding this system are discussed in detail in section 13.4. It is found that the system exhibits the behavior of a monatomic ideal gas, provided that the temperature is defined so that RT represents two thirds the internal kinetic energy per mole of gas. This definition is equivalent to the temperature scale measured by an ideal gas thermometer. Accordingly, we infer that the molar internal energy U of an ideal monatomic gas is

$$U = \tfrac{3}{2}RT$$

from which

$$c_v = \frac{dU}{dT} = \tfrac{3}{2}R$$

represents the constant volume specific heat. Another result of the dynamic calculation of the properties of a monatomic ideal gas is that the root-mean-square average molecular velocity is

$$v_{rms} = \sqrt{3RT/M}$$

where M is the gram-molecular mass. It is frequently useful to speak in terms of the gas constant per molecule, k. This quantity is referred to as Boltzmann's constant and has the numerical value 1.381×10^{-23} joule/molecule-°K. It is evident from the definition that

$$k = \frac{R}{N_A}$$

where $N_A = 6.023 \times 10^{23}$ is the number of molecules in a mole, which we sometimes refer to as Avogadro's number. One of the most important facts to emerge from the dynamical analysis of the ideal gas is that *the internal energy of any ideal gas is a function only of the temperature.*

When any substance is allowed to expand against an external pressure P, the work it does can be expressed in differential form as

$$dW = P \, dV$$

or, upon integration over a finite volume change, as

$$\Delta W = \int dW = \int_{V_1}^{V_2} P \, dV$$

Accordingly, the work done in any thermal process can be represented as the *area under the curve* representing the process in a pressure–volume diagram. In the case of an expansion at *constant* pressure, this amounts to $P(V_2 - V_1)$, or $P \, \Delta V$. Also, in any process in which there is no volume change, $dV = 0$ and, therefore, there can be no work done by the system.

Adiabatic processes play an important role in thermodynamics; such a process is one in which the system is isolated thermally from its surroundings and in which no heat can enter or leave. In any such process, $dQ = 0$, and, therefore, $dU = dW = P \, dV$.

In using the first law of thermodynamics, it is important to understand the fact that the internal energy of the system is always a well-defined function of pressure, volume, and temperature and that, whenever these variables are restored to their initial values, the internal energy also assumes its former value. This is not true of the heat absorbed or the work done; these quantities depend not only on P, V, and T but also on the processes that are employed to take the system from one state to another. Thus, while it makes perfect sense to speak of the internal energy U, there is no such well-defined quantity as Q or W; we can speak only of changes ΔQ and ΔW that take place when the state of a system changes. Moreover, when the state of a system changes, the changes ΔQ and ΔW depend not only on the initial and final states but also on the route followed on the P–V diagram between them. The internal energy, however, depends only on the initial and final states, and the change ΔU is simply the difference in the internal energies associated with these states, irrespective of the path connecting them. It follows that the internal energy change in any cyclic process is *zero*, though, again, this is not true of ΔQ and ΔW.

The distribution of total molecular energies is given, in systems wherein the laws of classical physics can be used, by

$$f(u) = Ae^{-u/kT}$$

where u is the particle energy and $f(u) \, du$ is the fraction of particles having energies in a range du about u. This equation expresses the Maxwell–Boltzmann energy distribution law. This fundamental law of energy distribution leads to the velocity distribution laws of section 13.6.

The principle of equipartition of energy allows us to understand how molecular internal energies are distributed among possible translational, rotational, and vibrational motions of the molecule. It states that the average internal energy per molecule for any dynamic system obeying the laws of classical mechanics amounts to $kT/2$ per degree of freedom. One degree of freedom is assigned to each independent component of translational or rotational motion, and two degrees of freedom to each possible vibrational motion of the system, since vibrational modes absorb both kinetic and potential energy.

QUESTIONS

1. The equation of state of an ideal gas is not applicable to any substance at high pressures and low temperatures. Give several reasons.

2. Why is the absolute or Kelvin temperature scale more fundamental than the Celsius or Fahrenheit scale?

3. How does the temperature of an ideal gas change when any of the following quantities is tripled, assuming all other variables are fixed: (a) the pressure, (b) the volume, and (c) the number of moles?

4. What is the constant-volume specific heat of any substance in the limit of infinite temperature?

5. Two different gases have exactly the same temperature. Does this mean that their molecules have the same rms molecular speed?

6. When a system changes from state A to state B, is it true that the change in internal energy is independent of the processes involved in the transition from A to B?

7. A gas expands adiabatically doing work on its surroundings. Where does the energy which is converted to work come from?

8. The ideal gas law can be written using the gas constant R or Boltzmann's constant k. Discuss both forms.

9. What happens to the internal energy per particle as the temperature approaches absolute zero?

10. Explain qualitatively why the specific heat of a gas at constant pressure is greater than the specific heat at constant temperature.

11. If a substance has a large specific heat, does this mean that it has the capacity to hold a lot of heat?

12. Heat is "energy in transit" that flows when a temperature difference exists. How does it differ from internal energy?

13. Give an example in which the internal energy of a system increases even though no heat flows into the system.

14. Is it possible to construct an engine with unit efficiency? What laws of physics bear on this question?

15. The atmosphere does not contain any hydrogen (H_2). Can you explain this? Why does the moon have no atmosphere?

16. Is it conceivable that some molecules in a gas may have speeds in excess of $10v_{rms}$?

17. The velocity of sound in an ideal gas is found to be

$$\sqrt{\gamma RT/M}$$

while the average molecular speed is $\sqrt{2.546RT/M}$. Aside from a rather small numerical factor, these velocities are the same. We conclude, therefore, that sound propagates with a velocity roughly equal to the average molecular speed. Can you explain, using qualitative physical arguments, why this is true?

PROBLEMS

1. One gram-mole of oxygen at 1 atm and 0°C occupies a macroscopic volume of 22.415 liters. If each oxygen molecule is assumed to be a sphere of radius 10^{-8} cm, what is the volume actually occupied by the molecules?

2. Find the value of the ideal gas constant R in the British units (a) ft-lb/mole °F, and (b) btu/mole °F. Interpret the mole as a "pound-mole", that is, a weight in pounds numerically equal to the molecular weight in units wherein the atomic weight of the abundant carbon isotope is assigned the integer value 12. (c) How many molecules are there in such a pound-mole?

3. A given quantity of an ideal gas, whose mass is 12.0 g,

occupies a volume of 12.30 liters at 26.8°C and normal atmospheric pressure. Find (a) its volume at a temperature of 120°C and a pressure of 2.5 atm, (b) its pressure when the temperature is 65°C and the volume is 18.0 liters, (c) the temperature when the volume is 24.0 liters and the pressure 0.85 atm, (d) the gram molecular mass of the gas.

4. An oxygen tank has a volume of 40 liters. Initially, the absolute pressure in the tank is 20 atm. As oxygen is removed from the tank, the temperature drops from 35°C to 12°C while the pressure drops to 10 atm. (a) How many moles of oxygen are originally in the tank? (b) How many grams have been removed?

5. Two liters of gas at 1 atm at a temperature of -25°C is to be compressed to $\frac{1}{2}$ liter. If the final temperature is 50°C, find the pressure. How many molecules of gas are present?

6. The three temperature scales in widespread use are Fahrenheit, Celsius, and Kelvin. Convert each of the following temperatures to the other two scales: 50°F, 212°F, 42°F, 90°C, 3°K, 100°K.

7. A large steel tank contains carbon dioxide at 20°F and a pressure of 10 atm. If the CO_2 is heated to 200°F, what is the new gas pressure in the tank?

8. The molecular mass of methane is 16 g/mole. Using the ideal gas law, find its density in g/liter at 20°C and 4 atm. Also, find its density at S.T.P. (0°C and 1 atm).

9. The equivalent gram-molecular mass of air is 28.97 g/mole. How many grams of air are required to raise the gauge pressure of a tire with a 6-liter volume from 0 to 32 lb/in.^{2} if the temperature is 0°C?

*10. In Example 13.3.4, it was assumed that the temperature of the atmosphere is uniform. A more realistic assumption is that T decreases linearly with distance in the lower part of the atmosphere, according to $T = T_0 - \lambda z$, where $\lambda \simeq 0.006$°K/m characterizes the average decrease in temperature. Show that

$$P = P_0\left(1 - \frac{\lambda z}{T_0}\right)^{Mg/\lambda R}$$

Calculate P/P_0 at 4000 m if $T_0 = 300$°K and compare with the result of Example 13.3.4.

*11. If x is small compared to 1, the series expansion of $\ln(1 - x)$ is $\ln(1 - x) = -x - \frac{1}{2}x^2 - \frac{1}{3}x^3 - \cdots$. Show, using this result, that the answer to the preceding problem can be expressed in the approximate form

$$P = P_0 e^{-(Mg/RT_0)z(1 + (\lambda z/2T_0))}$$

Hint: Use only the first two terms of the series expansion.

12. At atmospheric pressure and at a temperature of 20°C, a helium-filled balloon has a volume of 12 liters. The balloon rises to a height of 4 km where the temperature is -20°C and the pressure is 0.63 atm. Find the new volume of the balloon. How many moles of He are in the balloon?

13. A plastic container confines 12 g of oxygen at 10 atm pressure and a temperature of 25°C. Due to a leak in the container, the pressure drops to 6 atm and the temperature becomes 21°C. How many grams of oxygen remain in the container? What is the volume of the container?

14. A cubical box is divided into two equal portions by means of a partition. Each half contains n moles of an ideal gas, and the temperature initially has the uniform

value T_0. The gas on one side is now heated to temperature T_1 while the other side is maintained at T_0. No gas leaks across the partition. If the total volume of the box is L^3, show that the total force on the partition is $2nR(T_1 - T_0)/L$.

15. The bottom of a lake 35 m deep is at a temperature of 5°C while the surface is at 25°C. If an air bubble has a volume of 15 cm^3 at the bottom of the lake, find its volume when it rises to the surface.

16. Atmospheric pressure corresponds to 760 mm Hg. A well-evacuated container has a pressure of 5×10^{-10} mm Hg. If the container is at a temperature of 25°C, find the number of molecules per cm^3 in the evacuated system.

17. A helium-filled blimp has a volume of 90,000 liters. If the pressure and temperature are 1 atm and 20°C, respectively, find the total mass of helium present. One mole of helium has a mass of 4.003 g.

18. The root-mean-square thermal velocity of molecular oxygen can be obtained from Eq. (13.4.25) with $M = 32$ g/mole. Obtain v_{rms} for oxygen at the surface of the earth when the temperature is 300°K. Compare this to the escape velocity of oxygen from the gravitational field of the earth.

19. A hypothetical planet has a mean density of 5000 kg/m^3 and a surface temperature of 375°C. If the planet is smaller than r_{min}, the root-mean-square thermal velocity of molecular oxygen will exceed the escape velocity and the planet will not be able to retain an oxygen atmosphere. Find the value of r_{min}.

20. The gram-molecular mass of CO_2 is 44 g/mole while that of O_2 is 32 g/mole. A tank maintained at a constant temperature originally contains 8 kg of CO_2 at a pressure of 6 atm. If the CO_2 is removed and replaced by 7 kg of O_2, find the new pressure.

21. A storage tank for methane (CH_4) has a volume of 100 liters. If the temperature is 50°C and the pressure is 12 atm, find (a) the number of molecules in the tank and (b) the mass of this gas in grams.

22. Two monatomic gases A and B are contained in the same container at a temperature T. If the molecular masses are M_A and M_B, find the ratio of their rms thermal velocities.

23. A volume of 5 liters of an ideal gas, initially at a pressure of 2 atm, is heated from a temperature of 27°C to 57°C. Seventy five calories of heat is required. (a) What is the final pressure? (b) How many moles of gas are present? (c) What is the heat capacity per mole of gas?

24. How many moles of O_2 are stored in a 70-liter cylinder filled to an absolute pressure of 1470 lb/in.^{2} at 27°C?

25. The coefficient of volume expansion of an ideal gas at constant pressure depends only on temperature and is given by $1/T_0$. Show that a gas which obeys the van der Waals equation of state,

$$\left(P + \frac{a}{V^2}\right)(V - b) = RT \qquad \text{1 mole of gas}$$

has a coefficient of expansion β (at constant P) given by

$$\beta = \frac{RV^2(V - b)}{V^3RT - 2a(V - b)^2}$$

26. A certain nonideal gas obeys an equation of state

$$PV = c_1 T + c_2 T^2$$

(a) Obtain an expression for the coefficient of expansion β. (b) If the temperature changes from T_1 to T_2 at constant pressure, how much work is done by the gas?

27. Three moles of an ideal gas undergoes an isothermal expansion at a temperature of 30°C. If the volume increases from 5 liters to 20 liters, find (a) the initial and final pressures of the gas and (b) the amount of work done by the gas on its surroundings.

28. One mole of a diatomic gas expands adiabatically from a volume V_0 to volume $2V_0$. Find the final pressure in terms of the initial pressure P_0.

29. One hundred calories of heat is added to 2 moles of an ideal monatomic gas. Find the temperature change if the gas is kept at constant volume. What will the temperature change be if the pressure is kept constant?

30. Ten moles of an ideal gas kept at temperature 100°C expands isothermally and does 400 joules of work on its surroundings. If the gas initially occupied a volume of 10 liters, find (a) the final volume occupied by the gas and (b) the initial and final gas pressures. (c) If the 400 joules can be fully utilized to raise the temperature of 5 moles of an ideal monatomic gas at constant volume, what temperature increase will be brought about? Repeat for a constant pressure process (isobaric).

31. The efficiency of a cyclical heat engine is given by Eq. (13.5.38). If the lower temperature is 0°C and the higher temperature is 100°C, find the efficiency assuming $\gamma = 1.4$.

32. Show that the work done in an adiabatic compression of an ideal gas can be written in the form $W = (P_0V_0 - PV)/(\gamma - 1)$, where V_0 is the initial volume and V is the final volume. For a gas for which $\gamma = 1.4$ and for which the initial pressure and volume are 10 atm and 1.5 liters, respectively, find the work done if the final pressure and volume are 4 atm and 3 liters, respectively.

33. Find the fractional change in the velocity of sound in air between 6 A.M. and 6 P.M. if the temperatures at these times are 40°F and 60°F, respectively.

34. Two liters of an ideal gas originally under a pressure of 1 atm is expanded at constant temperature until the volume triples. The gas is then compressed to its original volume at constant pressure and finally brought back to its original pressure at constant temperature. Make a plot of the various stages on a $P-V$ diagram and find the total work done on the gas during this process.

35. Find v_{rms} for a hydrogen molecule at 30°C. At what temperature will a molecule of oxygen (O_2) have the same rms average velocity?

36. One cubic meter of air is initially at a temperature of 120°F at 2 atm pressure. It is allowed to expand at constant pressure to a volume of 5 m^3. It further expands adiabatically until the volume is 10 m^3, the final pressure being 0.5 atm. Sketch the entire process on a $P-V$ diagram and find the work done by the gas at each stage.

37. Ten liters of oxygen at 100°C is under a pressure of 5 atm. Find the number of moles of oxygen, the number of molecules, and the total internal energy.

38. The $P-V$ diagram for a complete cycle of an engine is shown in the accompanying figure. The working substance is n moles of an ideal monatomic gas. Calculate the heat input and change of internal energy during each

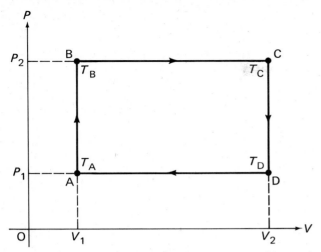

process. Show that the sum of the changes of internal energy adds up to zero for the complete cycle ABCA.

39. You are given a $P-V$ diagram for a cyclic process involving 3 moles of an ideal monatomic gas, as shown in the diagram, as well as the data stated with the diagram, $\Delta Q_{AB} = 200$ cal. Find **(a)** the pressure at point A, **(b)** the temperature and pressure at point B, **(c)** the work done by the gas in going from B to C, **(d)** the temperature at C, **(e)** the heat input and change of internal energy from B to C, **(f)** the temperature at D, and **(g)** the total work done by the gas.

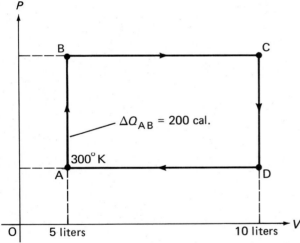

40. Four moles of an ideal gas expands isothermally at a temperature of 300°K. If the work output is 50 joules, find the ratio of the final volume to the initial volume and the ratio of the final pressure to the initial pressure.

41. Show using $N(v)$ as given by Eq. (13.6.10) that the most probable molecular speed is given by

$$v = \sqrt{\frac{2kT}{m}}$$

*42. Find the probability that a molecule in an ideal gas will have a speed greater than the most probable speed obtained in the preceding exercise. (*Hint:* In this problem and the succeeding one you may find it useful to expand the exponential, to integrate in that form, and to use your calculator to find a numerical result.)

*43. Find the root mean square velocity v_{rms} of O_2 at 300°K. At this temperature, 1 mole of oxygen is confined in a cylinder. Approximately how many molecules will have speeds between zero and $v_{rms}/2$?

44. The figure shows the $P-V$ diagram of an ideal gas. You are given the following data: $T_a = 350°K$, $P_3 = 2$ atm, $V_1 = 10$ liters, and $V_2 = 25$ liters. The curve from AE is an adiabatic curve $PV^{1.4} = $ const. There are two curves connecting A and D; the curved path is an isothermal process, the straight line is simply one in which the temperature is adjusted during the process so that the path on the $P-V$ diagram is linear. Find **(a)** the pressures P_1 and P_2, **(b)** the temperature. at B, C, and E, and **(c)** the work done by the gas in the processes ABD, ACD, CAE, AD (linear path), and AD (isotherm).

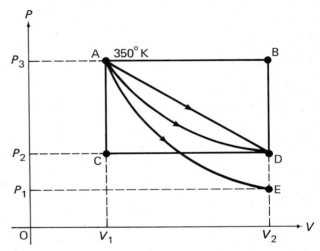

45. The Clausius equation of state is obtained from the van der Waals equation by setting the coefficient a equal to zero. Thus, $P(V - b) = RT$. Show that at constant pressure the coefficient of expansion β is given by $\beta = R/PV$.

46. An aspiring young engineer says he has invented a new heat engine which delivers 40 kWh (kilowatt-hours) of work by taking in 120,000 Btu from the fuel supply and rejecting 20,000 Btu to the condenser. His boss asks your opinion of the engineer's talents. What would you say?

47. **(a)** What is the root-mean-square velocity of an atom in a sample of crystalline copper at a temperature of 300°K? **(b)** What would the corresponding figure be for crystalline silver? **(c)** for crystalline gold? (Note that the gram molecular mass is 63.54 for Cu, 107.87 for Ag, and 196.97 for Au.)

48. The specific heat of a certain solid metallic element is observed to be 0.0308 cal/g °K at temperatures above 300°K. What would you expect its gram molecular mass to be?

49. The molecule of a gaseous substance consists of three atoms arranged at the vertices of an equilateral triangle. Neglecting vibrational contributions, what would you expect the constant volume specific heat of such a substance to be, in the range of temperatures and pressures where it behaves as an ideal gas? What would its constant pressure specific heat be under these conditions? What is the value of γ for this substance?

*50. For an ideal gas, the internal energy depends only on

437

the temperature and is completely independent of the volume. In this instance, therefore, $dU/dV = 0$. In a nonideal gas, however, the internal energy depends on volume as well as temperature, and then the change in internal energy between two states at slightly different temperatures will be $\Delta U = (\partial U/\partial V)_T \Delta V + (\partial U/\partial T)_V \Delta T$, where $(\partial U/\partial V)_T$ represents the partial derivative of U with respect to V in a situation where the temperature is held constant, while $(\partial U/\partial T)_V$ is a partial derivative with respect to T in a situation where the volume is held fixed. This relation is derived in Appendix D. Show, using this relation, that for a nonideal substance

$$\left(\frac{\partial U}{\partial V}\right)_T = \frac{nC_v(\gamma - 1)}{\beta V} - P$$

where β is the expansion coefficient $(1/V)(\partial V/\partial T)_p$ measured at constant pressure. (Hint: Proceed along the line of the derivation of Eq. (13.5.15), noting in this case that the internal energy changes in the constant-volume and constant-pressure processes are *different*.)

51. Work problem 38 at the end of Chapter 12 taking into account work done against the atmosphere by the expansion or contraction which takes place in the phase changes.

DISORDER, REVERSIBILITY, ENTROPY, AND THE SECOND LAW OF THERMODYNAMICS

14.1 Introduction

The first law of thermodynamics amounts to a statement of the law of conservation of energy and a definition of the internal energy function. By itself, it can lead to a great deal of insight into the thermal properties of matter. However, there are many important questions that cannot be resolved by the first law alone. These questions generally have to do with why some thermal processes are easily reversed while others are essentially irreversible, and with why the irreversible ones go the way they do and not in the opposite direction.

For example, in a system consisting of ice and water in equilibrium at 0°C under atmospheric pres-

sure, if 80 calories of heat is slowly and carefully introduced in such a way that thermal conditions in the system never depart far from those prevailing at equilibrium, exactly 1 gram of ice will be transformed to liquid water at 0°C. Subsequently, 80 calories of heat may be slowly returned to the surroundings, and the melting process will be reversed, resulting in the freezing of 1 gram of water to form ice. Both the system and the surroundings are back to where they started, thermally; no *net* absorption or rejection of heat energy has taken place nor was any *net* work done by either system or surroundings in the whole cycle.

On the other hand, suppose we have two identical blocks of copper, one at a uniform temperature of 0°C, the other at a uniform temperature of 100°C, and suppose that they are suddenly brought into thermal contact. Heat will flow from the hot object into the cold one until both arrive at a common final temperature of some 50°C. But this process is inherently *irreversible*, since heat will flow spontaneously from a hot body into a colder one but not the other way. Of course, we could restore the initial state of the system by putting one of the blocks into the freezing compartment of a refrigerator and the other onto the warm outside heat radiating fins of the icebox, plugging the refrigerator in, and letting it run until the inside block had cooled to 0°C and the outside one had warmed up to a temperature of 100°C. But this requires *work input* to the refrigerator. In other words, the surroundings has had to do work to restore the initial temperatures of the bodies and, therefore, must have less internal energy than before. The thermal state of the surroundings has, therefore, *changed*. Nobody has ever found a way to restore the initial temperatures of two such bodies without changing the thermal state of the surroundings. In fact, it is *impossible* to do so, and the statement of this impossibility is one way of stating *the second law of thermodynamics*. Note that there is nothing in the *first* law of thermodynamics to forbid the spontaneous transfer of heat from one body to the other until their temperatures differ by 100°C; after all, energy is indubitably conserved in such a process, and this is all the first law requires.

Then there is the closely related question of why heat flows spontaneously from hot to cold in the first place and not the other way. After all, there is once more nothing in the *first* law of thermodynamics that contradicts the possibility of heat energy flowing from a cold object into a hotter one, the cold one getting colder in the process and the hot one getting even hotter. Energy is conserved, which is all the first law requires. Why is it, then, that heat flows one way but not the other? Again, many schemes have been devised to try to exhibit spontaneous heat flow from cold to hot, but all have failed utterly. Once more, we are led to postulate that nature has doomed all such attempts to failure, and this would be another way of stating the second law of thermodynamics.

The second law of thermodynamics enables us to understand and answer questions such as those discussed above, about which the first law tells us nothing. There are literally dozens of ways of stating the second law of thermodynamics, some of which have absolutely no resemblance to others. We shall have occasion to examine and understand a number of these equivalent statements in this chapter, and a comprehensive list of equivalent statements of the second law is set forth in section 14.5. To begin, the simplest way of expressing the essential physical content of the second law of thermodynamics is to state it in the following way:

I. Systems evolve spontaneously with the passage of time from states of low probability to states of higher probability, and not in the opposite direction.

We shall sometimes refer to this statement of the second law as Statement I.

The second law, stated in this form, can be understood by thinking of an example drawn from the game of bridge. Suppose that the cards are shuffled and dealt, and suppose that the North player receives 11 spades and two other cards, East receives 11 hearts and two other cards, South ends up with 11 diamonds and two odd cards, while West gets 11 clubs and two other cards. Now this is a very unusual distribution of cards, it must be admitted; indeed, it is one of very low probability in a random deal. It is much more probable that each player might receive three or four cards in each of the suits, because there are many more independent choices of individual cards that will result in such even distributions of suits than is the case for very lopsided distributions like the one mentioned first. But then, there are distributions that are even more lopsided and have even lower probabilities in a random deal; one player may receive 13 spades, another 13 hearts, the third 13 diamonds, and the fourth 13 clubs.

The hand that was originally discussed above is played out and the cards reshuffled and redealt for a new hand. Now, before the cards are picked up, one of the players suggests a side bet on how the suits are distributed between hands this time. How would *you* bet—on the distribution 13 spades in one hand, 13 hearts in another, 13 diamonds in another, 13 clubs in the last one, or on a normal distribution in which three or four cards of each suit were found in all four hands? Naturally, you would make the latter choice, since the even distribution is one of greater probability. The low probability distribution *could* occur—there is no law that forbids it—but it is very unlikely that it will actually do so. The system, which started in a configuration of rather low probability on the first deal, will spontaneously evolve toward a configuration of higher probability on the next deal, and not the other way.

Another way of viewing the situation, which emphasizes the spontaneous evolution of the system from states of lower probability to higher probability is to envision a game in which North starts with 13 spades, East with 13 hearts, South with 13 diamonds, and West with 13 clubs. A fifth person draws a card from the North hand—it has to be a spade, at the beginning—and replaces it with a card drawn from a second well-shuffled deck. The replacement might, of course, be a spade, but the probability is only one in four, or 0.25. The probability that the replacement will be a card of another suit is three in four, or

0.75. The card originally drawn from the North hand is then incorporated at random into the second deck. The fifth person then draws a card from the East hand, replacing it with one drawn from the second deck and incorporating East's card into the second deck at random. The process is then repeated with South and West, and then again and again around the table in the sequence NESWNESWNESW . . . .

On the second cycle, it is quite likely that North now has 12 spades and one card of another suit. Let us assume that this is the case. On the second draw, the probability that a spade will be drawn is 12/13, while the probability that the other card will be drawn is only 1/13. The probability that the replacement will be a spade is 1/4 and that the replacement will be a card of another suit is 3/4. It is evident that there is a high probability that after the second round, North's hand will contain only 11 spades, two cards of other suits now having replaced them. Without going into the mathematics of the probabilities at each stage, it is clear that as round after round is completed, the North hand will be further infiltrated by cards of suits other than spades, while at the same time the solid initial line of spades is depleted. After a long time, the hand will tend to assume the configuration of highest inherent probability, in which the cards are nearly equally apportioned among suits. After all, the successive rounds of drawing and replacing cards amounts to nothing more than a very long and drawn-out way of shuffling the deck! The configurations of the East, South, and West hands, will, of course, change along similar lines.

The system evolves over a period of time from a highly ordered state of inherently low probability to a random or disordered state of inherently higher probability. The second law of thermodynamics asserts simply that the properties of physical systems evolve spontaneously in the same way. The second law has far-reaching consequences which affect the thermal behavior of solids, liquids, and gases, and which are sometimes rather hard to exhibit from Statement I. We shall, therefore, have to examine and understand several other equivalent statements of this law to understand completely how it bears on the behavior of thermal systems. The most important of these statements has to do with the efficiency of heat engines in converting heat energy into mechanical work, and it is this subject which will soon occupy our attention.

There is one important aspect of the second law that must be clearly understood, which is illustrated by Statement I and the card game examples discussed above. It has to do with the fact that the second law of thermodynamics is a statement about *probabilities* rather than a definitive statement of what will or must happen under given circumstances. The second law states the most probable way in which certain physical

systems develop, rather than how they *have* to develop. In the examples discussed above, there is a *chance* that successive rounds of drawing and replacing cards might make an evenly distributed hand into one containing 13 spades. The probability of this coming to pass in, say, 1000 rounds, is exceedingly small, while the probability associated with the reverse development is very large—but it could happen, just as one might roll "snake eyes" 1000 times in succession with dice. Usually, though, the dice and the cards do not cooperate, and neither do the molecules in any physical system. The air molecules in the room might all congregate in the wastebasket, but nobody would bet on it!

In playing bridge with a deck of 52 cards, part of the fascination of the game is that the laws of probability cannot always be counted upon to have exact predictive value. When we know that the opposing side has six trumps distributed among the two defenders, it is usually fatal to count on their being split three and three. In the same way, we must be cautious about applying statistical laws such as the second law of thermodynamics to systems containing only 52 molecules. Fortunately, most of the physical systems we are concerned with contain huge numbers of molecules, and in this situation the predictive value of probabilistic laws is very much better. If we were playing bridge with a deck containing *52 million* cards and were concerned with how *6 million* outstanding trumps were split between our two opponents, we would be very much better off than before, since we would find that the likelihood of a split worse than, say, 3,020,000 to 2,980,000 would be very small, indeed. But, then, this state of affairs would probably make bridge as dull as physics, particular inasmuch as the probability of your side ever getting 7 million trumps out of a total of 13 million would be quite negligible.

The final point to remember is that there is no mathematical proof of the second law of thermodynamics. It is a reasonable enough assertion but one whose final justification rests solely upon the fact that it is everywhere in agreement with experiment and that its predictions, to date, have always been fulfilled.

14.2 Reversible and Irreversible Processes

In the preceding section, we mentioned the fact that certain processes which act to change the thermal state of a system are reversible while others are not. We discussed examples of both types of processes and examined briefly some of the consequences of each. We now wish to examine the two types of processes more carefully and to understand in a precise way what is meant and implied by each.

In a reversible process, the state of a system is

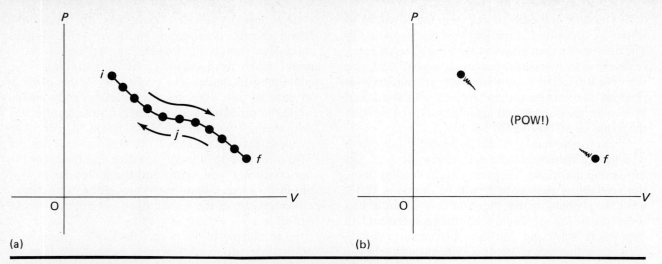

FIGURE 14.1. Schematic representation of how (a) reversible and (b) irreversible processes take place, using the standard *PV* diagram.

changed from an initial equilibrium state *i* to a final equilibrium state *f* so slowly and gradually that at each state of the process, the state of the system can be represented by an intermediate equilibrium state such as *j*, shown in Fig. 14.1a. The system is always close enough to equilibrium that its state can be described physically to a very high degree of approximation by the characteristics of an equilibrium state like *j*. The system is said then to proceed from an initial state to a final state through a *succession of equilibrium states.* The system of ice and water discussed in the preceding section is a good example. If heat is introduced slowly and the system stirred throughout the course of the process, ice may be melted in such a way that the temperature of the system hardly differs from the equilibrium value of 0°C. The system's state at all times is well approximated by an equilibrium state, and the process can easily be reversed at any time simply by withdrawing heat instead of adding it. The system can, therefore, be brought from *i* to *f* and back to *i* again, with no net change having taken place either in the system or in its surroundings. Both system and surroundings regain their initial values of pressure, temperature, and volume, with no *net* interchange of heat or work having been made between the system and its surroundings. This property may, in fact, be said to define a reversible process.

There are many other processes which can be carried out reversibly, among them the isothermal, adiabatic, constant volume, and constant pressure processes relating to gases—and not only ideal ones—discussed in the preceding chapter. In fact, in doing the examples and deriving many of the results brought to light in that chapter, we made the tacit assumption that these processes were, in fact, carried out that way. It is, of course, possible to carry out processes that are inherently reversible so suddenly and abruptly that they are in actuality *not* reversible when carried out that way. Reversible changes imply vanishingly small departures from equilibrium at all stages; this is a condition which may be approached, but never quite attained. Therefore, a process which is truly reversible is something like an ideal gas. It is a representation of a process which, though never precisely capable of realization, can be closely approached in many practical situations and which allows us to visualize and calculate results related to actual processes that would be difficult to obtain in any other way.

Irreversible processes are those in which substantial departures from equilibrium states are inherent in the nature of the process and that, therefore, cannot be carried out in such a way that the former state of the system is recovered with no net exchange of heat or work between system and surroundings. Ordinarily, we have no control over the rate at which such processes take place, as in the example of the bodies of different temperature in contact discussed in the preceding section. One very important class of inherently irreversible processes involves frictional or dissipative systems in which an ordered form of energy (such as the macroscopic kinetic or potential energy of a large body, the ordered motion of charge in an electric current, or the chemical energy of the molecules of a compound) is turned into random internal energy of molecules, and thence into heat. Joule's experiment involves this kind of process; its inherent irreversibility is obvious; we have no way of "unstirring" the liquid, sorting out the random internal thermal energy of its molecules and making them push the paddles. The situation has been discussed in detail in Chapter 12.

Any process in which random internal energy is generated by friction has the same inherent irreversibility for much the same reason. At the molecular interface, no matter how gently we may rub two materials together, very violent disruption of normal molecular motions occur. Locally, then, very large temperatures and temperature gradients are found and, locally, vast departures from the equilibrium state are taking place. State changes do not occur through a succession of equilibrium states, but rather through a series of conditions so far removed from equilibrium that it may be impossible even to define accurately the temperature near the friction interface. The system can be returned to its original configuration, but only by withdrawing the heat generated by friction. In a cyclic process, therefore, net work against friction is done on the system by the forces which have acted upon it, and a net transfer of heat from system to surroundings has to take place to return it to its original temperature. The process cannot take place, as cyclic reversible processes do, with no net interchange of work or heat between system and surroundings. The thermodynamic properties of any system undergoing irreversible change cannot be represented by a succession of intermediate equilibrium states; the transition between the initial equilibrium configuration i and a final equilibrium state f, therefore, has *no well-defined path* on a state diagram, but may be visualized, roughly, in the way illustrated by Fig. 14.1b.

We mentioned a bit earlier that even a reversible process may become irreversible if we are not careful about the way in which it is carried out. Consider, for example, the adiabatic compression of an ideal diatomic gas in an insulated cylinder fitted with a frictionless piston, from an initial equilibrium state (P_i, T_i, V_i) to a final equilibrium state (P_f, T_f, V_f). If the compression is carried out smoothly and reasonably slowly, the quantities P, T, and V will change gradually enough that the gas molecules can maintain among themselves what is essentially a state of equilibrium. Under these circumstances, the relation between P, T, and V is always represented by the equilibrium equation of state, $PV = nRT$, and the work done is represented by the area under the adiabatic path connecting the initial and final states on the $P-V$ diagram, expressed in this case by Eq. (13.5.55). On the other hand, if we were to hit the piston as hard as possible with a sledge hammer, reducing the volume very suddenly from V_i to V_f, intense shock waves would be generated by the blow and would propagate back and forth through the gas. These shock waves would excite internal molecular motions not normally excited in a reversibly conducted process. In particular, they might excite vibrational motions of the diatomic molecules that

would occur in equilibrium only at very high temperatures. Indeed, in a sense, the fact that these excitations occur is an indication that the temperature of the gas might in some sense be much higher than the equilibrium value given by (13.5.52). The true state of the system, however, is not accurately described by *any* value of temperature, because our definitions of temperature are all derived from the properties of equilibrium states or, at any rate, states that are hardly distinguishable from equilibrium states. The distribution of energies of the molecules, for example, may be nothing like the Maxwell–Boltzmann distribution under these conditions. All we can say is that the molecules are excited to higher energies than would ever be the case in a reversible adiabatic compression.

It takes work to generate the system of shock waves that causes all this excitation of molecular vibrational motion, and, therefore, we shall find that more work is required to effect the same volume change than was required when the compression was carried out slowly. This excess work is stored as molecular vibrational energy when the final volume V_f is attained. Once that happens, the compression stops, and the system after a time once more reaches equilibrium at volume V_f. In reaching equilibrium, however, the excess vibrational energy stored in the diatomic molecules is dissipated in exciting the other normal molecular motions that prevail in the equilibrium state—molecular kinetic energy of translation and rotation, for the most part. As this happens, and as an equilibrium state once more reappears, the temperature is greater than Eq. (13.5.51) would predict, since, after all, the work originally done to generate shock waves has now been channeled into the random internal energy of the gas. Heat must be removed from the gas to the surroundings, somehow, if the final equilibrium state is to be characterized by the values P_f, T_f, and V_f chosen previously. The system might then be expanded reversibly (and adiabatically) until its initial state is reached once again, but *not* without that removal of heat from the system to the surroundings! How much heat would have to be withdrawn? That would depend upon how much excess energy of vibration has been stored by the molecules, which would depend upon the amount of energy absorbed from the shock waves, which is in turn related to how much shock wave energy was created in the first place, which, finally, would depend upon how hard and how suddenly the hammer blow was struck.

The importance of the distinction between reversible and irreversible processes lies in the fact that the degree of disorder, and hence the probability, associated with a given state of a system plus its surroundings *increases* in an irreversible process but *remains the same* during one that is reversible. This is a profound and far-reaching statement, one that we shall not comprehend in all its aspects immediately, but shall return to again and again to try to understand more clearly. Roughly speaking, however, it is the conversion of highly ordered forms of energy—mechanical, electrical, chemical—into random thermal energy of molecules that is responsible for the inevitable increase in the disorder of system plus surroundings in irreversible processes.

At this point, we must recognize, first of all, that the *probability* associated with a given state and the *randomness*, or *degree of disorder*, associated with that state go hand in hand. The reason for this is that what we mean by a random or disordered state is the kind of state we expect to find just by chance, and hence this kind of state is by definition the one of highest probability. Highly structured or ordered states are those that can be achieved only by a few independent chains of events which are not likely to occur just by chance. Therefore, highly ordered systems, in other words, those of low disorder, are those whose probability of occurrence is low. Also, we should note that the words "system plus surroundings" in the above context is synonymous with "universe."

If these arguments are accepted, then the statement made at the beginning of the next to last paragraph can be taken to mean that *the degree of disorder of the universe increases in an irreversible process and remains the same in a reversible process.*

Finally, since every process is either reversible or irreversible, this could be rephrased to state that *the degree of disorder of the universe either increases or remains the same, but never decreases.* These statements will be seen to resemble strongly Statement I of the second law of thermodynamics. Indeed, they are simply slightly different statements of the second law, and can be shown to mean exactly what Statement I means. If systems evolve spontaneously from states of low probability to states of high probability, then the universe, as a system, must evolve spontaneously from states of low disorder to states of higher disorder, because the degree of disorder of a state and the probability with which it occurs are closely related. Ultimately, then, the universe approaches a state of maximum probability, or maximum disorder, in which all its matter is distributed uniformly and in which no highly ordered forms of energy nor temperature differences exist. Fortunately, none of us will be around to witness this unhappy state of affairs when it comes about.

Once more, there is no way of proving these statements; they are expressions of how nature is observed to operate, and their justification lies in the fact that, though many attempts to disprove them have been made, none has succeeded. Nevertheless, their plausibility can clearly be established by considering one or two examples.

In the case of the adiabatic compression we discussed earlier, when conducted reversibly, any mechanical work done is fully and completely recoverable by reversing the process and letting the gas expand to its original state. There is no increase in the disorder of the system or its surroundings under these circumstances. But when the compression is effected irreversibly, by hitting the piston with a hammer, the kinetic energy of the hammer is changed to internal thermal energy, which has to be allowed to flow out of the system as heat in order to attain the final equilibrium state f. The kinetic energy of the hammer arises from an ordered motion of all its molecules, and this energy ultimately is converted into random heat energy. Clearly, the degree of disorder of the system and surroundings has increased, the net increase in this example being associated in the end exclusively with the surroundings.

Now let us consider the case of the two identical copper blocks that have initial temperatures of $0°C$ and $100°C$ and are suddenly placed in contact with one another. We may regard the system as the two blocks and specify that they are thermally isolated from their surroundings. In this case, there is no exchange of heat or work with the surroundings at all, so there can be no net change in its state of disorder. In the system itself, the initial state is one in which there is a significant difference in the average internal energies of the molecules in the two blocks. If we assume that the copper blocks behave as ideal crystals, which in this case is largely justified, their internal energies are given by $U = 3NkT$ as found in section 13.6, and the average energy per molecule is given by $U/N = 3kT$, where N is the number of molecules in each block. Since $T = 373°K$ in the hot object and $T = 273°K$ in the cold one, it is evident that the average internal energy per molecule is greater in the hot body than in the cold one by about 40 percent. In the final state, after thermal equilibrium is established, the average internal energy per molecule in the two blocks is equal. It is obvious that the final state, in which the average internal energy per molecule is the same in all parts of the system, is the one having the greatest degree of disorder. The condition prevailing when the two blocks are just brought into contact is not unlike the situation in which all the gas molecules in the room are congregated in the wastebasket—a highly ordered condition whose probability of occurring by chance is very small. The chance of such an initial state appearing as a result of random processes in a system wherein there are 2.4×10^{22} atoms is something like the chance that in 2.4×10^{22} tosses of a coin, 1.4×10^{22} heads and 1.0×10^{22} tails will result. This chance, you must admit, has to be much less than that associated with most probable result, that in which the number of heads and tails is equal.

In this process, which is, of course, irreversible,

the degree of disorder associated with the system increases, while the surroundings is unaffected in any way. The total disorder of system plus surroundings, therefore, is increased. This example clearly illustrates the spontaneous evolution of a system from a state of low probability to one of higher probability. It is evident that if heat were to flow in the reverse direction, the cold block getting colder and the hot block getting hotter, the evolution of the system would be in the opposite direction, from a state of admittedly low probability to one whose probability is even less. This is in contradiction to the second law of thermodynamics. It is, in fact, possible to state the second law of thermodynamics in the form:

Heat will not flow spontaneously from a body at a given temperature to another whose temperature is higher.

Finally, let us discuss the reversible process in which heat is introduced into, or withdrawn from, a system in which a mixture of ice and water is present at 0°C and 1.0 atmosphere pressure. When heat is introduced, ice melts and liquid water appears in its place. But a crystal is a more strongly ordered system than a liquid, just as a liquid is a more highly ordered system than a gas. As ice melts, therefore, the ice–water system becomes more disordered. At the same time, however, heat is being withdrawn from the surroundings. When heat is introduced into any system, its effect is always to increase the randomness or degree of disorder of the system, and when it is withdrawn, the effect is the opposite—the system becomes less disordered. In this case, the increase of disorder of the ice–water system that occurs as heat is introduced and ice is melted is exactly balanced by a corresponding decrease in the disorder of the surroundings as it gives up heat to the system. There is no *net* change in degree of disorder when we consider *both* the system and its surroundings, as we must always do. The process we are discussing is, therefore, reversible.

The reader has been subjected to a great deal of discussion on a qualitative, intuitive, and physical level about disorder, probability, and the spontaneous evolution of systems from one state to another. This has had its uncomfortable aspects, the most conspicuous one being the lack of a quantitative description for the term "degree of disorder" and the lack of any way of associating this concept quantitatively with changes in heat and work. We shall soon see that a simply definable thermal quantity called the *entropy* provides this needed quantitative measure of "degree of disorder."

Nevertheless, our seemingly clumsy and argumentative approach was not without a surreptitious motive. Entropy is a subtle, difficult, and confusing concept to grasp, particularly when it is approached by the mathematical rather than the physical route.

We shall be exploring the other route shortly, and when we do, it will be useful to remember that "entropy" and "degree of disorder" are synonymous and to return to this section for help in relating the abstract mathematics of entropy and entropy changes to what is actually going on in the system at hand. We shall at that time return to the same examples we considered in this section and rework them from a more quantitative point of view.

14.3 Heat Engines, Real and Ideal: The Carnot Cycle

We have already encountered a heat engine of sorts in Example 13.5.3. In that example are demonstrated the techniques of analyzing all aspects of the cycle and calculating the efficiency. We shall now use those techniques to analyze the behavior of a cyclical heat engine that, though utterly without importance as an operating device, has been particularly important in the development of thermodynamics. This engine, which we shall refer to as the *ideal Carnot engine*, was first discussed by the French engineer Sadi Carnot in 1828.

First, we shall consider the case in which the working substance is an *ideal gas* that is subjected to a series of *reversible* isothermal and adiabatic processes, as shown in Fig. 14.2. In the figure, the solid curves are two isothermal curves corresponding to temperatures T_1 and T_2. Their equations are, of course, $PV = nRT_1$ and $PV = nRT_2$. The temperature T_2 can be regarded as that of a high-temperature heat reservoir—a boiler in practical terms—from which heat is withdrawn to effect an isothermal expansion of the working fluid. Temperature T_1 corresponds to that of a second heat reservoir, such as a condenser, into which heat is rejected as the working fluid undergoes an isothermal compression at the lower temperature T_1. The dashed curves are adiabatic curves whose equations are $PV^\gamma = P_A V_A^\gamma$ and $PV^\gamma = P_B P_B^\gamma$, in accord with Eq. (13.5.49). It will be noted that the adiabatic curve passing through a given point on the $P-V$ diagram has a steeper negative slope than the isothermal curve through the same point. It is easy to show, in fact, that the slope of the adiabatic curve is greater than that of the isothermal by a factor γ. The proof of this statement is assigned as a problem.

The Carnot engine may be regarded as a cylinder in which n moles of an ideal gas are confined by means of a frictionless piston. The cylinder can be brought into thermal contact with either heat reservoir, at temperatures T_1 and T_2, or it can be thermally insulated so that no heat can flow in or out. The cycle proceeds around the path ABCD in four steps, as follows:

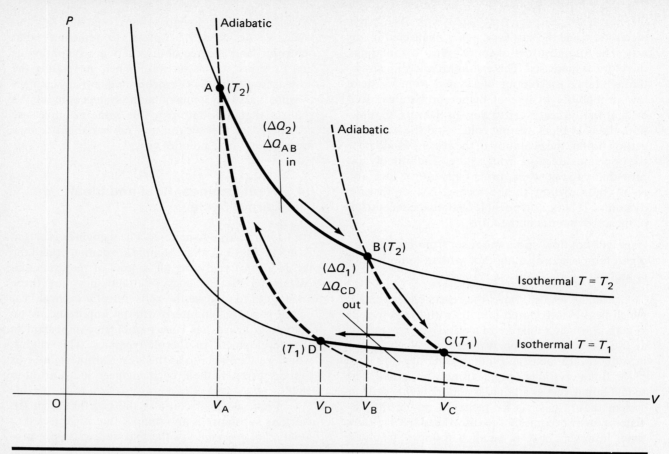

FIGURE 14.2. The ideal Carnot cycle. Processes AB and CD are reversible isothermal processes, while BC and DA are reversible adiabatic processes. The cycle operates as a heat engine when conducted in the sequence ABCDA, as shown, and as a refrigerator or heat pump when reversed.

Step AB. The system, initially at point A, is subjected to a *reversible isothermal expansion* at temperature T_2 along the path AB. This may be accomplished by bringing the cylinder into contact with the "boiler" at temperature T_2 and allowing heat to flow into the working substance slowly (that is, reversibly), decreasing the pressure during the process so that the isothermal path AB is followed, until point B has been reached. During this part of the cycle, the system absorbs heat ΔQ_{AB} (which we shall later call ΔQ_2) at constant temperature T_2, and a certain amount of work ΔW_{AB} is done by the system.

Step BC. When point B is reached, the cylinder is removed from thermal contact with the high-temperature reservoir and thermally insulated from the surroundings. Then the gas is slowly allowed to expand adiabatically along path BC until point C is reached. During this *reversible adiabatic expansion*,

work ΔW_{BC} is done by the system on the surroundings, the internal energy and the temperature decreasing until, at point B, $T = T_1$.

Step CD. At point C, the cylinder is placed in thermal contact with the low-temperature reservoir, whose temperature is T_1, and slowly compressed along the isothermal path CD. As the pressure increases, heat is rejected from the gas into the low-temperature reservoir so that the temperature remains constant at $T = T_1$ during the process. During this reversible isothermal compression, heat ΔQ_{CD} (which will later be referred to as ΔQ_1) is given off by the system while work ΔW_{CD} is done on it.

Step DA. When point D is reached, the cylinder is removed from thermal contact with the "condenser" and thermally insulated once more. Then, the gas is slowly compressed along the adiabatic path DA while its temperature increases as the process continues until finally the value $T = T_2$ is reached at point A, the system now having completed one cycle and returned to its initial state. During the *reversible adiabatic compression*, work ΔW_{DA} is done on the system by the surroundings. The engine can be run through any desired number of cycles, a certain net amount of heat being taken from the "boiler" and a certain net amount of work being done in each cycle. It is, of course, necessary to consider only one such cycle to learn all that can be learned about the engine.

Now, processes AB and CD are isothermal, and if the working substance is an ideal gas, then $\Delta U_{AB} = \Delta U_{CD} = 0$. According to the first law of thermodynamics, then, in these processes

$$\Delta Q_{AB} = \Delta U_{AB} + \Delta W_{AB} = \Delta W_{AB} = nRT_2 \ln \frac{V_B}{V_A}$$

$$(14.3.1)$$

$$\Delta Q_{CD} = \Delta U_{CD} + \Delta W_{CD} = \Delta W_{CD} = -nRT_1 \ln \frac{V_C}{V_D}$$

$$(14.3.2)$$

the expressions for the work done following directly from (13.5.10). Processes BC and DA, on the other hand, are adiabatic, and, therefore, by definition,

$$\Delta Q_{BC} = \Delta Q_{DA} = 0 \qquad (14.3.3)$$

Also, for an ideal gas, the internal energy depends only upon the temperature. Therefore, the gain in the system's internal energy in going from T_1 to T_2 along DA is equal to the loss in going from T_2 to T_1 along BC. Therefore,

$$\Delta U_{BC} = -\Delta U_{DA} \qquad (14.3.4)$$

and from the first law of thermodynamics, recalling (14.3.3), we find

$$\Delta W_{BC} = -\Delta U_{BC} = \Delta U_{DA} = -\Delta W_{DA} \qquad (14.3.5)$$

The net work done by the system is the sum of the ΔW terms around the cyclic path, whence, using (14.3.5),

$$\Delta W_{tot} = \Delta W_{AB} + \Delta W_{BC} + \Delta W_{CD} + \Delta W_{DA}$$
$$\Delta W_{tot} = \Delta W_{AB} - \Delta W_{DA} + \Delta W_{CD} + \Delta W_{DA}$$
$$= \Delta W_{AB} + \Delta W_{CD} \qquad (14.3.6)$$

But, recalling (14.3.1) and (14.3.2), this can in turn be written

$$\Delta W_{tot} = \Delta Q_{AB} + \Delta Q_{CD} \qquad (14.3.7)$$

This equation states that the work done by the engine is equal to the difference between the amount of heat ΔQ_{AB} received from the boiler and the amount ΔQ_{CD}

rejected to the condenser; ΔQ_{CD} is, of course, a negative quantity. This is an inevitable consequence of the first law of thermodynamics, since the net change in internal energy in a complete cycle is zero.

The efficiency of the cycle is defined as the ratio of the net work output to the amount of heat absorbed from the high-temperature source per cycle. Therefore, calling the efficiency η, we find

$$\eta = \frac{\Delta W_{tot}}{\Delta Q_{AB}} = \frac{\Delta Q_{AB} + \Delta Q_{CD}}{\Delta Q_{AB}} \qquad (14.3.8)$$

This is what we are after, but it looks a little simpler if we relabel our symbols, letting

$$\Delta Q_{AB} = \Delta Q_2 \qquad \Delta Q_{CD} = \Delta Q_1 \qquad (14.3.9)$$

because, after all, ΔQ_{AB} is associated with a process occurring at temperature T_2 while ΔQ_{CD} is associated with a process at temperature T_1. We may then write (14.3.8) as

$$\boxed{\eta = \frac{\Delta Q_1 + \Delta Q_2}{\Delta Q_2} = 1 + \frac{\Delta Q_1}{\Delta Q_2}} \qquad (14.3.10)$$

It appears from this that the efficiency must be greater than unity, but just the opposite is true, since ΔQ_1 represents heat *given up* by the system to the low-temperature reservoir. It is, therefore, a *negative* quantity, while ΔQ_2, representing heat *absorbed* from the high-temperature reservoir, has to be positive.

This is a useful relation in its own right, but it can be expressed even more simply by substituting, from (14.3.1) and (14.3.2), the values for ΔQ_1 and ΔQ_2, giving

$$\eta = 1 - \frac{nRT_1 \ln(V_C/V_D)}{nRT_2 \ln(V_B/V_A)} \qquad (14.3.11)$$

Now, in Fig. 14.2, points B and C lie on the same adiabatic curve, the temperature T_1 prevailing at point C and T_2 at B. From (13.5.51), therefore, the temperatures and volumes must be related by

$$T_1 V_C^{\gamma-1} = T_2 V_B^{\gamma-1} \qquad (14.3.12)$$

The same kind of relation can be written for points A and D, which also lie on the same adiabatic curve, so that

$$T_1 V_D^{\gamma-1} = T_2 V_A^{\gamma-1} \qquad (14.3.13)$$

Dividing Eq. (14.3.12) by (14.3.13),

$$\left(\frac{V_C}{V_D}\right)^{\gamma-1} = \left(\frac{V_B}{V_A}\right)^{\gamma-1} \qquad (14.3.14)$$

and therefore, taking the $(\gamma - 1)$th root of both sides,

$$\boxed{\frac{V_C}{V_D} = \frac{V_B}{V_A}} \qquad (14.3.15)$$

447

Substituting this result into (14.3.11) and canceling factors in the numerator and denominator, we obtain finally the very simple result

$$\eta = 1 - \frac{T_1}{T_2} = \frac{T_2 - T_1}{T_2} \qquad (14.3.16)$$

This is very reasonable; it tells us that, since T_2 is always greater than T_1, the efficiency must be less than unity. Physically, this is so because a certain part of the heat initially received from the "boiler" must always be wasted in being rejected to the "condenser" during process CD. There is no possibility, therefore, of constructing an engine of this type that converts *all* the heat input ΔQ_2 into work. The best we can do is make the ratio T_1/T_2 in (14.3.16) as small as possible by operating the "boiler" at the highest possible temperature and the "condenser" at the lowest.

Since all the processes in the ideal Carnot cycle are *reversible*, the engine can be run backward around the cycle as well as foward. In this mode of operation, the sequence of events in Fig. 14.2 is best thought of as CBADCBAD . . . . Net work is done *on* the system rather than by it, since the closed cycle is traversed in the opposite direction. Also, the signs of ΔQ_1 and ΔQ_2 are reversed; the process CD is now an isothermal expansion in which heat ΔQ_1 is *absorbed* by the system at the lower temperature T_1, while process BA is an isothermal compression in which heat is *given off* by the system at the higher temperature T_2. The device now works like a refrigerator. As work is done on the engine to make it traverse the cycle, heat is absorbed from a reservoir at a low temperature T_1 (the freezing comparment) and rejected into a reservoir at a high temperature (the kitchen). All the work of analysis we have done for the Carnot engine is still valid; we need only remember that the signs of ΔQ_1, ΔQ_2, and ΔW_{tot} are reversed. The Carnot refrigerator is best understood by solving (14.3.10) for $\Delta Q_1/\Delta Q_2$ and substituting into the resulting expression the value of η given by (14.3.16). The result is

$$\frac{\Delta Q_1}{\Delta Q_2} = -\frac{T_1}{T_2} \qquad (14.3.17)$$

By definition, T_2 is greater than T_1. Therefore, the quantity T_1/T_2 is less than unity, which means that the magnitude of ΔQ_1 must be less than that of ΔQ_2. But ΔQ_1 is the heat removed from the "freezing compartment" and ΔQ_2 is the heat rejected to the "kitchen." It is clear that the latter must be greater than the former. Less heat is removed from the ice trays, then, than is given off to the kitchen. *You can heat the room, but never cool it, by letting the refrigerator door remain open!* Strangely enough, as we shall see later,

this statement is a valid expression of the second law of thermodynamics.[1]

The work required to transport a given quantity of heat ΔQ_1 from the low-temperature reservoir to the high-temperature reservoir is easily calculated from the results already obtained. From (14.3.8), (14.3.16), and (14.3.17), it is seen that

$$\Delta W_{\text{tot}} = \eta \, \Delta Q_2 = \left(1 - \frac{T_1}{T_2}\right)\left(-\frac{T_2}{T_1}\Delta Q_1\right)$$

$$= -\Delta Q_1 \left(\frac{T_2}{T_1} - 1\right) \qquad (14.3.18)$$

The minus sign reflects the fact that work is done on the system when it is used as a refrigerator. If the ratio T_2/T_1 is expressed as a function of η by (14.3.16),

$$\Delta W_{\text{tot}} = -\frac{\eta}{1 - \eta}\Delta Q_1 \qquad (14.3.19)$$

where η is the Carnot efficiency defined in the usual way.

A schematic drawing of a heat engine operating between two fixed reservoirs at temperatures T_1 and T_2 is shown in Fig. 14.3a, and a refrigerator operating between two similar fixed-temperature reservoirs is illustrated in Fig. 14.3b. These need not be Carnot cycle devices, nor even reversible engines; the relations shown there are true for any device working between fixed-temperature reservoirs. In a complete cycle of operation, the initial and final states are the same, in which case the net change in internal energy must be zero; hence, from the first law, the net heat added, let us call it ΔQ_{tot}, and the net work done are related *in all instances* by

$$\Delta Q_{\text{tot}} = \Delta Q_1 + \Delta Q_2 = \Delta W_{\text{tot}} \qquad (14.3.20)$$

Also, if the efficiency η is defined as the ratio of net work output to heat input from high-temperature reservoir per cycle, then

$$\Delta W_{\text{tot}} = \eta \, \Delta Q_2 \qquad (14.3.21)$$

and, from (14.3.20),

$$\Delta Q_1 = -(1 - \eta)\Delta Q_2 \qquad (14.3.22)$$

irrespective of whether we are dealing with a Carnot cycle device or whether the engine is reversible or irreversible.

Now, we know what the efficiency of a reversible Carnot engine using an ideal gas is. We have, however,

---

[1] This also enables us to understand why an air conditioner will not cool a room unless it is mounted in a window, with the cooling coils on the inside and the heat radiating fins outside.

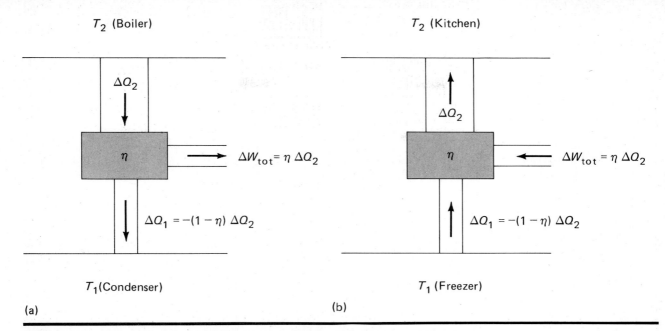

T_2 (Boiler)

T_2 (Kitchen)

T_1 (Condenser)

T_1 (Freezer)

(a)

(b)

FIGURE 14.3. Energy flow in (a) a heat engine and (b) a refrigerator, operating between two heat reservoirs at fixed temperatures T_1 and T_2.

no way of knowing what the efficiency of any other type of engine which operates between the same two fixed temperatures is. There is no initially apparent reason, for example, to suppose that an engine using a nonideal gas or a condensable vapor would have the same efficiency as an ideal Carnot engine. Let us, therefore, suppose that we have such an engine, that it is *reversible*, and that its efficiency η' is *less* than the Carnot efficiency (14.3.16) for the same operating temperatures T_1 and T_2.

Let us now provide a frictionless mechanical linkage that will couple this engine to an *ideal* Carnot engine of efficiency η, operating between the same temperatures, in such a way that the work output of the ideal Carnot engine is used to drive this new engine in reverse, as a refrigerator. This scheme is illustrated in Fig. 14.4. In the equations we must write, the quantity ΔW_{tot} must represent both the work output of the Carnot engine, equal to $\eta \Delta Q_2$ according to Eq. (14.3.8) and Fig. 14.3a, and the work done *on* the refrigerator, equal according to Fig. 14.3b to $\eta' \Delta Q_2'$. Unfortunately, our previously adopted sign convention requires us to regard work done by a system as positive and work done on a system negative. In the system we are dealing with, ΔQ_2 is obviously positive, as are η and η', while $\Delta Q_2'$ is clearly negative; hence $\eta \Delta Q_2$ and $\eta' \Delta Q_2'$ cannot be equal. We must instead denote the magnitude of ΔW_{tot} by $|\Delta W_{\text{tot}}|$ and write, from Figs. 14.3 and 14.4,

$$\eta \Delta Q_2 = |\Delta W_{\text{tot}}| \tag{14.3.23}$$

$$-\eta' \Delta Q_2' = |\Delta W_{\text{tot}}| \tag{14.3.24}$$

which means that

$$\boxed{\eta \Delta Q_2 = -\eta' \Delta Q_2'} \tag{14.3.25}$$

Let us now regard our "system" as everything enclosed by the dashed lines in Fig. 14.4. It is as if we had a "black box" with four pipes sprouting from it to be connected to the heat reservoirs, as shown in the figure. What would the properties of this strange contraption be if we hooked it up and set it in motion? The answer to this question can be found by calculating how much heat is received by the system from the high-temperature reservoir and how much is received from the low-temperature reservoir.

The heat absorbed by the system from the upper reservoir at temperature T_1 per cycle is

$$\Delta Q_{T_2} = \Delta Q_2 + \Delta Q_2' \tag{14.3.26}$$

However, using (14.3.25) to express $\Delta Q_2'$ in terms of ΔQ_2, this can be written as

$$\Delta Q_{T_2} = -\Delta Q_2 \left(\frac{\eta}{\eta'} - 1 \right) \tag{14.3.27}$$

Since ΔQ_2 is positive and since η is larger than η', the quantity ΔQ_{T_2} has to be negative, in other words, the system rejects heat into the high-temperature reservoir, like a refrigerator.

The heat absorbed by the system per cycle from the lower reservoir may be calculated the same way. Thus, calling this quantity ΔQ_{T_1}, we may write

$$\Delta Q_{T_1} = \Delta Q_1 + \Delta Q_1' \tag{14.3.28}$$

449

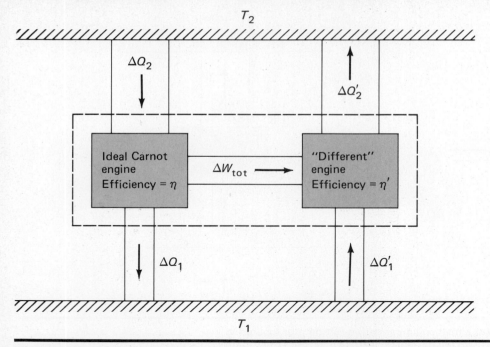

FIGURE 14.4. Ideal Carnot engine used to drive a different heat engine in reverse, as a refrigerator.

But from (14.3.22) or Fig. 14.3,

$$\Delta Q_1 = -(1 - \eta)\,\Delta Q_2 \qquad (14.3.29)$$

$$\Delta Q_1' = -(1 - \eta')\,\Delta Q_2' \qquad (14.3.30)$$

Substituting these values into (14.3.28) and expressing $\Delta Q_2'$ in terms of ΔQ_2 by (14.3.25), we obtain

$$\Delta Q_{T_1} = \Delta Q_2\left(\frac{\eta}{\eta'} - 1\right) = -\Delta Q_{T_2} \qquad (14.3.31)$$

Since ΔQ_{T_2} was of necessity negative, ΔQ_{T_1} is the same in magnitude but clearly positive. Heat is, therefore, absorbed by the system from the low-temperature reservoir, again like a refrigerator.

But our black box is unlike the usual sort of refrigerator shown in Fig. 14.3 in one vital respect. What it does, in effect, is to transfer heat from the low-temperature reservoir to the high-temperature reservoir *without the necessity of any external work input.* The refrigerator section on the right gets all the work it needs to keep running from the Carnot engine on the left. All we have to do is crank it up initially and let it run. If the heat capacities of the thermal reservoirs are not truly infinite, the hot one will soon start getting hotter and the cold one will get colder until at length all the heat in the universe will wind up in the high-temperature reservoir. Also, you could run a steam engine or something of the sort from the high-temperature reservoir and keep replacing the heat it used with this wonderful device. No fuel would

ever be needed.[2] Surely this is the way out of the energy crisis!

Alas, this is not the way out of the energy crisis, but the way into the funny farm. Thousands of basement inventors have gone bankrupt trying to make such schemes work; nobody has ever had the least success. The reason is that troublesome second law of thermodynamics, which we once saw stated in the form: *heat cannot flow spontaneously from a colder body to one that is hotter.* Now, of course, one may question what exactly is meant by the word *spontaneously,* but in this context a good translation is "without the performance of external work."

We are now able to state the second law in a well-known way, which was first accomplished by Rudolf Clausius (1822–1888), as follows:

It is impossible for any cyclical process to result in no effect other than the transfer of heat from a body at a lower temperature to another whose temperature is higher.

Related to this statement of the second law, and fully equivalent, is the so-called Kelvin-Planck statement originated by Lord Kelvin (1824–1907) and Max Planck (1858–1947), which denies the possibility of

[2] A machine such as this, which violates the second law of thermodynamics (though not the first), is sometimes referred to as a perpetual motion device of the second kind. A perpetual motion device of the first kind is one which violates the law of conservation of energy.

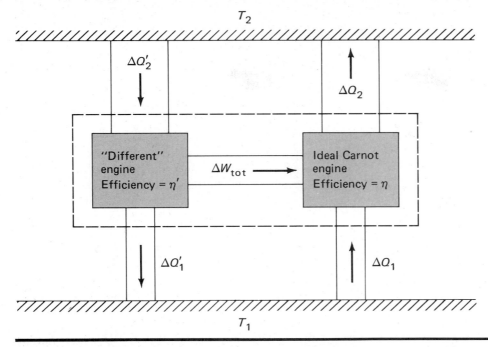

FIGURE 14.5. Ideal Carnot refrigerator powered by the work output of a heat engine using a different thermal cycle.

extracting work from a system in the manner suggested above or by any equivalent scheme. It states that:

> *A process whose only net effect is to transform heat extracted from a reservoir of uniform temperature into work is impossible.*

All very well, then, but where did our calculations go wrong? After all, the properties of ideal gases are well known, and there is no law which prevents us from constructing and operating a Carnot engine. What is wrong, and what is prevented by the second law, is our supposition that a reversible engine whose efficiency is *less* than the Carnot efficiency can exist! But what about one whose efficiency is *greater* than the Carnot efficiency? This can be shown to be impossible in a similar way. In this case, the other engine, of efficiency η', is made to function as a heat engine, driving the Carnot engine of lesser efficiency η in reverse, as in Fig. 14.5.

Once again, the relation between $\Delta Q_2'$ and ΔQ_2 will be given by (14.3.25). Therefore, the heat absorbed by the system from the reservoir at the higher temperature will be

$$\Delta Q_{T_2} = \Delta Q_2' + \Delta Q_2 = -\Delta Q_2'\left(\frac{\eta'}{\eta} - 1\right) \qquad (14.3.32)$$

where now ΔQ_2 has been expressed in terms of $\Delta Q_2'$ by (14.3.25). Since, from Fig. 14.5, $\Delta Q_2'$ is positive and since $\eta' > \eta$, ΔQ_{T_2} must be negative, as before. Heat is,

therefore, rejected by the system enclosed by the dashed lines in the figure to the high-temperature reservoir, as in the former case.

The heat absorbed by the system from the cooler reservoir is

$$\Delta Q_{T_1} = \Delta Q_1' + \Delta Q_1 \qquad (14.3.33)$$

Expressing $\Delta Q_1'$ and ΔQ_1 in terms of $\Delta Q_2'$ and ΔQ_2 by (14.3.29) and (14.3.30), which are valid in this case just as before, and then using (14.3.25) to express ΔQ_2 in terms of $\Delta Q_2'$, this becomes

$$\boxed{\Delta Q_{T_1} = \Delta Q_2'\left(\frac{\eta'}{\eta} - 1\right) = -\Delta Q_{T_2}} \qquad (14.3.34)$$

This result is the same as the earlier one. An amount of heat ΔQ_1, which is necessarily positive, since $\eta' > \eta$, is absorbed in each cycle from the cooler reservoir, and an equal quantity of heat per cycle is rejected to the high-temperature reservoir, in violation of the second law of thermodynamics. So it appears that there can be no reversible heat engine whose efficiency is *greater* than the Carnot efficiency.

This leaves very little room for speculation other than to suppose that

all reversible heat engines operating between the same two constant temperatures have the same efficiency, given by the Carnot expression (14.3.16).

This conclusion does not violate the second law, because then $\eta' = \eta$ in (14.3.27), (14.3.31), (14.3.32), and

451

(14.3.34); and in both cases considered, $\Delta Q_{T_1} = \Delta Q_{T_2} = 0$, there then being no net transfer of heat between the system and either reservoir.

In the case of an *irreversible* heat engine of efficiency η', it is evident that if η' were greater than the Carnot value, the irreversible engine might be set up just as before to drive the Carnot engine as a refrigerator, resulting in a net transfer of heat from the cooler to the hotter reservoir with no net work input, in violation of the second law. Therefore, η' cannot be greater than the Carnot value η given by (14.3.16). But the opposite argument cannot be made, since there is no assurance whatever that when the irreversible engine is run backward it will have the same efficiency as in the forward direction. Therefore, irreversible heat engines may have efficiencies lower than the Carnot efficiency associated with the same reservoir temperatures. The Carnot efficiency, thus, sets an upper limit to the efficiency of actual heat engines, which often exhibit efficiencies significantly less than the Carnot value because of the inherent irreversibility in the operating cycle and heat losses from exhaust gases and imperfectly insulated boilers, steam pipes, and other parts.

The second law of thermodynamics enables us to understand the operating characteristics of reversible and irreversible heat engines in a very precise and, at the same time, general way. Although this understanding of heat engines was developed in connection with the study of reciprocating steam engines and employs in its derivations certain analogies and terminology closely related to them, it is much more general and encompasses any device in which heat is converted into work or in which work is used to transfer heat from one body to another. Such devices include not only internal combustion engines but also steam and gas turbines in which the "cycle" is going on continuously and is, therefore, not very easy to visualize as a cycle at all. Indeed, this theory applies even to such devices as thermocouples and thermoelectric power generators, whose efficiency, just as that of any other contrivance designed to convert heat into work, is limited by the Carnot factor.

EXAMPLE 14.3.1

If all reversible engines working between temperatures T_1 and T_2 have the same efficiency, as given by Eq. (14.3.16), how is it that the reversible heat engine discussed in Example 13.5.3, has an efficiency, given by Eq. (13.5.38), *different* from the Carnot value?

The point is that the engine of Example 13.5.3 involves a cycle in which the maximum temperature of the working substance is T_2 and the minimum temperature T_1, but not one in which all the heat received by the substance is received at $T = T_2$ and

all the heat given up by the substance is given up at $T = T_1$. In Fig. 13.7b, for example, heat is received by the working substance along path EA at temperatures varying from a minimum of T_1 at point E to a maximum of T_2 at point A. Similarly, along path BE, heat is given up by the substances at values of temperature varying between T_2 at point B to T_1 at point E. Since, in contrast to the Carnot engine, much of the heat received by the working substance enters at temperatures below the maximum external source temperature T_2 and since much of the heat given up is rejected at temperatures above the minimum external sink temperature T_1, it is clear that the efficiency of such a cycle must be *less* than the Carnot efficiency. That this is true is demonstrated in Fig. 13.8, where the dashed line represents the Carnot efficiency.

We have tried here to explain the situation with qualitative physical arguments, but more rigorous and mathematical proofs that the Carnot cycle is the most efficient of all reversible cycles can be constructed. In regard to the engine of Example 13.5.3, one can prove mathematically starting with Eq. (13.5.35) that the efficiency of this cycle is always less than the Carnot value for any positive value of γ. These proofs are assigned as exercises for the student.[3]

We have tried to establish the following main points: (i) The statement at the beginning of this example, and that set off below Eq. (14.3.34), refer to engines in which all the heat is received at the *constant* maximum temperature T_2 and rejected at the *constant* minimum temperature T_1. (ii) Engines in which this requirement is not satisfied must have efficiencies *smaller* than that given by the Carnot expression. The Carnot efficiency, therefore, represents the maximum attainable by any cyclical heat engine, reversible or irreversible, working between maximum temperature T_2 and minimum temperature T_1.

EXAMPLE 14.3.2

The steam turbines of the German World War II battleship *Bismarck* took in steam from the boilers at 450°C, exhausting it into condensers at about 27°C. At full speed (30 knots), the engines developed 138,000 hp. The ship's fuel oil consumption was 44,800 kg/hr. The heat of combustion of the oil was about 10,500 kcal/kg. Find (a) the maximum possible efficiency for such a plant, (b) the minimum possible thermal output from the boilers, (c) the minimum possible thermal input into the condensers, (d) the minimum possible consumption of fuel oil under

---

[3] It is interesting to note in this connection that, while the efficiency of the cycle of Example 13.5.3, and most other imaginable cycles as well, depends upon the value of γ, the efficiency of the Carnot engine is *independent* of this quantity.

"ideal" conditions, and (e) the actual operating efficiency of the plant.

The greatest possible efficiency for any heat engine is the Carnot efficiency, given by (14.3.16). Here, we have a maximum hot-reservoir temperature $T_2 = 450°C = 723°K$ and a minimum cold-reservoir temperature $T_1 = 27°C = 300°K$. The efficiency will then be

$$\eta = \frac{T_2 - T_1}{T_2} = \frac{423}{723} = 0.5851, \text{ or } 58.5\%$$

The power output is known to be 138,000 hp, or $(138,000)(550) = 7.59 \times 10^7$ ft-lb/sec. Since 1 ft-lb = 1.356 joules, so in each second,

$$\Delta W_{tot} = (7.59 \times 10^7)(1.356) = 10.29 \times 10^7 \text{ J}$$

$$= \frac{10.29 \times 10^7}{4186} = 24,590 \text{ kcal}$$

recalling that 1 kcal = 4186 joules. The heat input ΔQ_2 is related to ΔW_{tot} by (14.3.21):

$$\Delta Q_2 = \frac{\Delta W_{tot}}{\eta} = \frac{24590}{0.5851} = 42,030 \text{ kcal per sec}$$

This is the least possible rate at which heat could be supplied by boilers to produce the required rate of work output. The minimum rate at which heat might be rejected into the low-temperature reservoirs is found from (14.3.22) to be

$$\Delta Q_1 = -(1 - \eta) \Delta Q_2 = -(0.4149)(42030)$$

$$= -17,440 \text{ kcal per sec}$$

The input heat ΔQ_2 is generated by burning fuel oil whose heat of combustion is 10,500 kcal/kg. The actual rate at which fuel is used is 44,800 kg/hr, or 44,800/3,600 = 12.44 kg/sec. To produce the 42,030 kcal/sec heat input required by the reversible Carnot cycle, the fuel consumption rate would be only 42,030/10,500 = 4.00 kg/sec. The actual efficiency is related to the Carnot efficiency by the ratio of these two values representing the ratio of actual to ideal rates of energy consumption:

$$\eta_{real} = \eta_{Carnot} \cdot \frac{4.00}{12.44} = (0.5851)\left(\frac{4.00}{12.44}\right)$$

$$= 0.1883, \text{ or } 18.8\%$$

The main reasons for the wide disparity between actual and ideal efficiencies are (i) irreversibility of the actual processes occurring in the thermal cycle, (ii) heat losses by conduction and radiation from boilers, pipes, and equipment and by the passage of hot gases out through the stack, (iii) mechanical losses in frictional effects in bearings and reduction gears, and (iv) the fact that some of the power output is used not for propelling the ship but for driving auxiliary devices such as pumps, electrical generators, and heating systems.[4]

EXAMPLE 14.3.3

How long should it take to freeze 1 liter of water at 0°C into ice cubes in a refrigerator operating as an ideal reversible Carnot device, assuming that the unit is driven by a motor developing 0.250 hp (0.186 kilowatt), that the temperature of the freezer compartment is −20°C, and that the temperature of the kitchen is 30°C? How much heat is rejected into the kitchen in the process?

The efficiency may be calculated from the Carnot formula as usual, whence, since the hot reservoir is at $T_2 = 30°C = 303°K$ and the cold one at $T_1 = -20°C = 253°K$,

$$\eta = \frac{T_2 - T_1}{T_2} = \frac{50}{303} = 0.165 \qquad (14.3.35)$$

In time t, the work done on the system by the motor will be

$$\Delta W_{tot} = -186t \text{ J} \qquad t \text{ in sec} \qquad (14.3.36)$$

since its power output is 0.186 kilowatt = 186 watts = 186 joules/sec. This may be related to the amount of heat ΔQ_1 removed from the low-temperature reservoir (the freezing compartment in this case) most easily by (14.3.18) or (14.3.19). Since we must remove 80,000 calories = (80,000)(4.186) = 334,900 joules of energy to freeze 1 liter of water at 0°C, the heat of fusion being 80 cal/g, we must have $\Delta Q_1 = 334,900$ joules. This quantity is positive, since it represents heat absorbed by the system. According to (14.3.19), then,

$$\Delta W_{tot} = -\frac{\eta}{1 - \eta} \Delta Q_1 \qquad (14.3.37)$$

Substituting $\Delta Q_1 = 334,900$ joules and the values of ΔW_{tot} and η given by (14.4.35) and (14.3.36) into this equation,

$$-186t = -\frac{(0.165)}{(1 - 0.165)}(334,900)$$

---

[4] From the given rate of fuel consumption, which is easily expressed as about 1150 tons per day, it is apparent that in a four-day transatlantic voyage at an average speed of 30 knots, about 4600 tons of fuel is consumed. At a more economical speed—perhaps 25 knots, which is still over twice the speed of the *Great Eastern*—the *Bismarck*'s fuel consumption would be only about half as great, perhaps 2200 tons for the crossing. This should be compared with the fuel consumption of the *Great Eastern* for such a trip, which, at less than half the speed of the *Bismarck*, would be at least 10,000 tons! Figures such as these show at once the vast gains in efficiency achieved by the application of scientific principles to the technology of ship propulsion and demonstrate at the same time why the *Great Eastern* was not a commercial success.

$$186t = (0.1976)(334,900) = 66,180$$

$$t = 356 \text{ sec}$$

Actually, we know that it usually takes much longer than this to freeze a liter of water in an ice cube tray, but that is because the heat must flow to the freezing interface through an inch or so of water, which is not a very good conductor of heat. The above answer becomes much more credible if you visualize the water as spread out in a thin pool covering a metal plate in which the cooling coils are embedded.

The heat ΔQ_2 given off to the kitchen may be obtained from (14.3.17), which states that

$$\Delta Q_2 = -\frac{T_2}{T_1}\Delta Q_1 = -\frac{(303)}{(253)}(334,900) = -401,100 \text{ J}$$

$$= -95,800 \text{ cal}$$

the minus sign signifying, of course, that ΔQ_2 is heat given off by the refrigerator to the surroundings. There are other equally valid ways of approaching this example. For instance, we could have found ΔQ_2 first, using (14.3.17), then determining the time required by noting that, according to (14.3.21), $\Delta W_{\text{tot}} = \eta \, \Delta Q_2 = -186t$. There is little to recommend one scheme over the other.

14.4 Entropy, Probability, and Disorder

We have discussed at length the natural tendency of thermal systems to evolve from ordered to disordered states. We also have the general understanding that when heat flows into a system it tends to increase the degree of disorder, and vice versa. But we have had difficulty in arriving at a quantitative measure of disorder and, in particular, have not found any way of associating the amount of disorder created in a system with the amount of heat added. We are now faced with the problem of trying to find some simple thermal quantity that will express precisely the change in probability of the state of a system when heat is introduced.

What would we expect the important properties of such a quantity to be? One important aspect of its behavior can be arrived at simply by noting that it is a function of a well-defined mathematical parameter describing the system, that is, the probability that the system be in the state it is in. If the state of the system is somehow changed and later returned to its initial condition, this probability will clearly return to its original value, since it is determined only by the number and kind of particles in the system, their total energy, and the forces acting on them. The system's disorder parameter (let us call it S), therefore, must

have the same kind of mathematical behavior as the internal energy. In other words, it must be a well-defined function of P, V, and T; the change in S incurred between a given initial and final state must be a function only of the initial and final values of pressure, volume, and temperature and must be *independent* of the process path by which the system was conducted from the initial to the final state. For the same reason, the total change in S around a closed path in which the system is finally returned to its initial state is *zero*, which, expressed mathematically means that around such a path,

$$\Delta S = \oint dS = 0 \tag{14.4.1}$$

The simplest possible hypothesis might be that the internal energy itself is the disorder parameter of the system. But this is not so, because there are many situations where the disorder clearly increases while the internal energy remains the same. One of the simplest is the case of the hot and cold blocks of copper that are brought into thermal contact. Assuming that the law of Dulong and Petit is correct, their initial internal energy will be $3nRT_1 + 3nRT_2$, while when the final temperature $\frac{1}{2}(T_1 + T_2)$ is attained, it is $3nR\frac{1}{2}(T_1 + T_2) + 3nR\frac{1}{2}(T_1 + T_2)$, which is precisely the same as the initial value. Yet it is amply evident, as we have already seen in section 14.2, that the degree of disorder associated with the system has increased. So we must look further.

The clue we are looking for is provided by the Carnot cycle. We know, from (14.3.1), (14.3.2), (14.3.3), and (14.3.15) that the net heat input per cycle is

$$\Delta Q_{\text{tot}} = \Delta Q_{\text{AB}} + \Delta Q_{\text{BC}} + \Delta Q_{\text{CD}} + \Delta Q_{\text{DA}}$$

$$= nRT_2 \ln \frac{V_{\text{B}}}{V_{\text{A}}} - nRT_1 \ln \frac{V_{\text{B}}}{V_{\text{A}}} \tag{14.4.2}$$

This is not zero. This is an example of a familiar situation, one in which

$$\oint dQ \neq 0 \tag{14.4.3}$$

around a closed path. But now, let us find the integral of the quantity dQ/T around the closed path; along the adiabatic parts, $dQ = 0$, and no contribution will result there, as before. Along the isothermal paths, for an ideal gas, at any rate, $dU = 0$, and

$$dQ = dW = P \, dV = nRT \frac{dV}{V} \tag{14.4.4}$$

The integral of dQ/T around the closed path is

$$\oint \frac{dQ}{T} = \int_{\text{A}}^{\text{B}} \frac{dQ}{T} + \int_{\text{B}}^{\text{C}} \frac{dQ}{T} + \int_{\text{C}}^{\text{D}} \frac{dQ}{T} + \int_{\text{D}}^{\text{A}} \frac{dQ}{T} \tag{14.4.5}$$

As mentioned above, the second and fourth integrals along the adiabatic sections are zero, since dQ along these paths is zero. We are left with the integrals along the isothermal paths, which, according to (14.4.4) and (14.4.5), have the form

$$\oint \frac{dQ}{T} = \int_{V_A}^{V_B} \frac{1}{T_2} nRT_2 \frac{dV}{V} + \int_{V_C}^{V_D} \frac{1}{T_1} nRT_1 \frac{dV}{V}$$

$$(14.4.6)$$

In the first of these integrals, we may substitute the constant value T_2 for T, because T_2 has that value everywhere along the isothermal path of integration between A and B in Fig. 14.2. Likewise, the constant value $T = T_1$ may be used in the second integral. Since the temperatures cancel, we are left with

$$\oint \frac{dQ}{T} = nR \int_{V_A}^{V_B} \frac{dV}{V} + \int_{V_C}^{V_D} \frac{dV}{V} = nR[\ln V]_{V_A}^{V_B}$$

$$+ nR[\ln V]_{V_C}^{V_D}$$

$$= nR\left(\ln \frac{V_B}{V_A} + \ln \frac{V_D}{V_C} \right) = nR\left(\ln \frac{V_B}{V_A} - \ln \frac{V_C}{V_D} \right)$$

$$(14.4.7)$$

recalling in writing the last expression that $\ln(1/x) = -\ln x$. But from (14.3.15), $V_C/V_D = V_B/V_A$, and, therefore, the integral is *zero*. We have, then, tentatively identifying the differential of the disorder parameter dS with dQ/T,

$$\Delta S = \oint dS = \oint \frac{dQ}{T} = 0 \qquad (14.4.8)$$

as required mathematically by Eq. (14.4.1). Note that the crucial point at which this result was ensured came in Eq. (14.4.6) with the cancellation of T_1 and T_2 in the integrands. Had we, for example, tried to compute the integral of dQ/T^n around the path, we would have obtained this cancellation only for $n = 1$; the result obtained, therefore, is not one that could have been obtained for many proposed disorder parameters. It will be observed, also, that the quantity dQ/T has another most important property expected of a disorder parameter: when dQ is positive (heat added to system), there is an increase in the disorder parameter, and when dQ is negative (heat withdrawn from system), it decreases, in agreement with our intuitive notion that adding heat to a system increases its internal disorder, and vice versa.

Despite all these encouraging developments, however, our identification of S as a disorder parameter is still only tentative, because, although we have shown that its value remains unchanged for an ideal gas about a closed Carnot path, we are not yet sure that it possesses this property for *any* closed reversible path and for any conceivable working substance. The proof that it does for any closed reversible path may be understood by referring to Fig. 14.6a, in which an arbitrary reversible cyclic process is shown. Figure 14.6b shows that an approximation to the path represented by such a process may be constructed of small isothermal and adiabatic sections. In the limit where the number of these sections becomes very large and the length of the individual parts very small, such an approximation approaches the actual path arbitrarily closely. Any actual closed reversible cycle

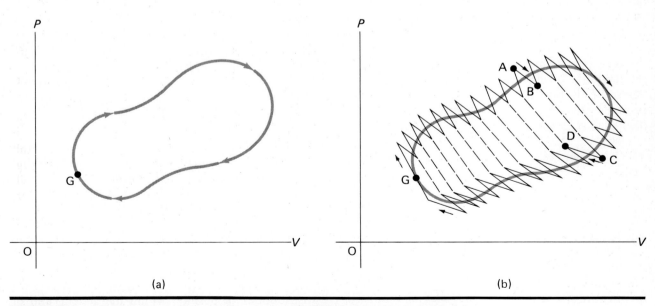

(a)   (b)

FIGURE 14.6. (a) Arbitrary cyclic reversible process. (b) The process illustrated in (a) represented as the sum of an arbitrarily large number of reversible Carnot cycles.

455

can thus be regarded as a superposition of a very large number of very small reversible Carnot cycles, such as ABCD in Fig. 14.6b. The change ΔS associated with the actual path may be calculated by summing the changes associated with all the individual sections of the approximate path in the limit as the number of these sections becomes infinitely large. There can be no contribution to the quantity ΔS as given by (14.4.8) along any of the adiabatic sections, because along any adiabatic path the quantity dQ is identically *zero*. The contributions to ΔS from isothermal sections, such as AB and CD can be calculated by the same methods used before in obtaining (14.4.7). But for each isothermal section such as AB there is another, such as CD, which forms a part of the same elementary Carnot cycle, and the contributions to ΔS from these sections inevitably add to zero, just as they do in (14.4.7). The total change ΔS is zero, therefore, whether the reversible cycle is a Carnot cycle or not.

Finally, if the quantity S is truly representative of the degree of disorder of the system—and by now we have good reason to think that it is—the degree of disorder will always return to the given initial value when a cyclic process is brought back reversibly to its starting point *irrespective of whether the working substance is an ideal gas or not*. The reason for this is that the thermal state of a system, including its degree of disorder, is completely determined by the state variables P, V, and T, and when they return to their initial values, the degree of disorder must return to its original value, whether the working substance is an ideal gas, a condensable vapor, 100-proof bourbon, or Chanel No. 5! Therefore, if S is really the disorder parameter, and if it can be demonstrated that $\Delta S = 0$ for a cyclic process involving an ideal gas, then ΔS will be zero for a cyclic process involving *any* working substance, solid, liquid, or gas.

This, of course, is not a rigorous derivation of this result, since it hinges on the *belief* rather than upon the established fact that S is related uniquely to the degree of disorder of the system. We can, however, prove conclusively, from the second law of thermodynamics, that $\Delta S = 0$ for any reversible cyclic process in any system, whether it behaves like an ideal gas or not. The proof hinges upon Eq. (14.3.10), which states that

$$\frac{\Delta Q_1}{\Delta Q_2} = \eta - 1 = -\frac{T_1}{T_2} \qquad (14.4.9)$$

The result is obtained in view of the fact that $\eta = 1 - (T_1/T_2)$, an expression which was first obtained from a consideration of the operating cycle of an ideal-gas Carnot engine but which was then shown to be true for any reversible engine using *any working substance*, as a direct consequence of the second law of thermodynamics. We find, therefore, for a general reversible

Carnot cycle, whether the working substance is an ideal gas or not, that

$$\boxed{\frac{\Delta Q_1}{T_1} = -\frac{\Delta Q_2}{T_2}} \qquad (14.4.10)$$

In Eq. (14.4.5), the second and fourth integrals will still be zero, irrespective of the working fluid, since they are along adiabatic paths for which dQ is zero, no heat flowing in or out. The first and third integrals are along isothermal paths, the temperature having the *constant* values T_2 in the first and T_1 in the third. Since the temperatures are constant throughout the integration, they can be written outside the integrals, which leaves

$$\Delta S = \oint dS = \oint \frac{dQ}{T} = \frac{1}{T_2} \oint_A^B dQ + \frac{1}{T_1} \oint_C^D dQ \qquad (14.4.11)$$

But the first integral here is just ΔQ_2 and the second is ΔQ_1, as illustrated in Fig. 14.2. Therefore,

$$\Delta S = \oint dS = \oint \frac{dQ}{T} = \frac{\Delta Q_2}{T_2} + \frac{\Delta Q_1}{T_1} \qquad (14.4.12)$$

This is *zero*, according to (14.4.10), irrespective of whether we are dealing with an ideal gas or not.

The quantity S, thus, has all the mathematical properties expected of the disorder parameter in the most general of systems involving any possible working substance. It has been given the name *entropy*, and an understanding of its definition and basic properties is fundamental to a comprehension of a great many common thermal processes. We have had occasion previously to apologize for the loosely defined term "degree of disorder," which we have nevertheless used frequently because of its conceptual simplicity. We are, therefore, concerned about giving this term a certain degree of legitimacy, which, to a physicist, means establishing a quantitative definition. We have now established that legitimacy in regard to the term "degree of disorder." In fact, we have established it twice, having previously mentioned that the *probability* of a system being in its given state constitutes a quantitative measure of disorder, and now having just identified the *entropy* of a system as also being a quantitative measure of its disorder from the point of view of macroscopic thermal properties.

This situation leads us to suspect that there is a relation between the entropy of a system in a given state and the probability that the state of the system is what it is. The "probability of the state of the system" refers to the probability that the individual particles are distributed in position and velocity corresponding to the given state, among all other possible distributions of position within the system, and of velocity, that they might possibly have. The calculation of these probabilities is a complex and bewildering affair and one that in this book we shall try to avoid. There is, however, one simple fact about probability and entropy which lets us deduce the general form of the relation between them.

If we have two systems in given states, there is a probability w_1 that system 1 is in its prescribed state and another probability w_2 that system 2 is in its given state. Viewing the two systems as a whole, however, the probability that system 1 is in its given state while *at the same time* system 2 is in its prescribed state is the *product* $w_1 w_2$ of the separate probabilities associated with the individual systems. It is as though we had two sets of dice. If we roll set 1, we know that the probability of "snake eyes" (that is, one spot showing on both dice) is one in 36 or 1/36. If we roll set 2, it is equally clear that the probability of rolling a total of 12 is again 1/36. But if we roll the two sets simultaneously, what is the probability of set 1 giving us "snake eyes" and set 2 at the same time coming up with a total of 12? On the average, set 1 will come up snake eyes once in every 36 tries. But set 2 will usually not total 12 in this event; as a matter of fact, you will find that, on the average, only once in 36 rolls for which snake eyes appear on set 1 will, in addition, the total of 12 be registered on set 2. Since it took 36 rolls of the dice for every appearance of snake eyes in set 1 to begin with, you will have to roll the dice on the average $36 \times 36 = 1296$ times for every time you come up with snake eyes on set 1 and a total of 12 on set 2 *simultaneously*. Considering the two sets as a whole, the probability of this simultaneous occurrence is the *product* of the probabilities of the individual occurrences, that is, $(1/36) \times (1/36) = 1/1296$.

The entropy, however, depending as it does on dQ or ΔQ, is clearly an *additive* property. To find the entropy or change in entropy of system 1, we evaluate the integral of dQ_1/T_1, while to make the corresponding calculation for system 2, we evaluate the integral of dQ_2/T_2. The total entropy is the *sum* of the individual entropies, just as the total heat added would be the sum of the heat inputs to the individual systems. Adding the entropies turns out, then, to multiply the probabilities. But then the entropy must be proportional to the *logarithm* of the probability, since this is the characteristic property of logarithms. We may therefore write the relation between entropy and probability as

$$S = c \ln w \qquad (14.4.13)$$

where c is a constant of proportionality. Then, $S_1 = c \ln w_1$ and $S_2 = c \ln w_2$, whereby

$$S_{tot} = S_1 + S_2 = c(\ln w_1 + \ln w_2) = c \ln w_1 w_2$$
$$= c \ln w_{tot} \qquad (14.4.14)$$

in agreement with (14.4.13) and the arguments preceding it. Actually, it can be shown that the proportionality constant c can be expressed as the product of the number of particles N belonging to the system and Boltzmann's constant k, (14.4.13) then becoming

$$S = Nk \ln w \qquad (14.4.15)$$

14.5 Entropy Changes in Reversible and Irreversible Processes

Having now defined entropy and established it as a valid thermodynamic disorder parameter, the next step is to examine the entropy changes associated with thermal systems in reversible and irreversible processes.

In reversible processes, the entropy of a system may increase or decrease, according to the sign of dQ, that is, according to whether heat flows in, increasing the molecular disorder, or flows out, decreasing it. But since in a reversible process, every bit of heat that flows into a system is accompanied by an equal flow of heat out of the surroundings, we must have

$$dQ' = -dQ \qquad (14.5.1)$$

where the unprimed quantity refers to the change experienced by the *system* and the primed quantity the change associated with the *surroundings*. The same statement can be made regarding dU, dU', and dW, dW' in a reversible process; if it were not so, the system would not return to its original state when the process was reversed. If the infinitesimal reversible flow of heat occurs at temperature T, then, since $dS = dQ/T$ and $dS' = dQ'/T$, Eq. (14.5.1) informs us that

$$T \, dS' = -T \, dS \qquad (14.5.2)$$

or

$$dS + dS' = 0 \qquad (14.5.3)$$

What this tells us is that in any reversible process, the change in entropy of the system plus that of its surroundings is zero. Since the entropy represents disorder, this means that in a reversible process any increase in the molecular disorder of a system is accompanied by a corresponding decrease of the disorder of the surroundings. Another way of saying it is that in any reversible process, the entropy, or disorder, of the universe remains the same, since, after all, system plus surroundings encompasses everything. The reader will recall this statement having been made and discussed qualitatively in section 14.2; our understanding of it is now, hopefully, more precise and quantitative.

But wait; no heat flow ever occurs without a

temperature difference, and if there's a temperature difference, shouldn't Eq. (14.5.2) read

$$T' \, dS' = -T \, dS \qquad (14.5.4)$$

where now T and T' are not equal? And doesn't this, then, invalidate Eq. (14.5.3)?

The answer to this lies in our understanding of an ideal reversible process. A reversible process, like an ideal gas, never occurs in reality but is only approached as a limit. As the flow of heat dQ or dQ' occurs more and more slowly, the temperature difference $T' - T$ which causes it will become smaller and smaller. In every case where the given heat flow occurs in a finite time, $T' - T$ will not be zero, and in every such case the process will not really be reversible but in the limit where the rate at which heat flows approaches zero, the temperature $T - T'$ also approaches zero, and it is only in this limit that the process is truly reversible. Therefore, if we are talking about an ideal reversible process understood in this way, Eqs. (14.5.2) and (14.5.3) are correct as written. The usefulness of the concept of the reversible process, unattainable though it may be, is justified in the insight and understanding of real processes it affords and, of course, by the fact that many real processes of great importance are closely approximated by ideal reversible processes.

Before considering irreversible processes, it will be worthwhile to state a few rather obvious but nevertheless important facts about entropy changes in certain particularly simple types of reversible processes.

In any *adiabatic reversible process*, no heat can ever flow between system and surroundings, since the system is thermally insulated from the surroundings, and, therefore, $dQ = dQ' = 0$. In this case, not only is the sum $dS + dS'$ equal to zero but also

$$\boxed{dS = 0 \qquad \text{and} \qquad dS' = 0} \qquad (14.5.5)$$

In an *isothermal reversible process*, we may write from the definition of entropy

$$dQ = T \, dS \qquad (14.5.6)$$

This may be integrated very easily to find the entropy change associated with a finite heat flow ΔQ, because in the integration since T is a constant in any isothermal process, it may remain outside the integral. Therefore,

$$\Delta Q = T \int_i^f dS = T(S_f - S_i) \qquad (14.5.7)$$

or

$$\Delta Q = T \, \Delta S$$

$$\boxed{\Delta S = \frac{\Delta Q}{T}} \qquad (14.5.8)$$

In these equations, the i and f subscripts refer to the initial and final states of the system, respectively. For the surroundings, we will find in much the same way that

$$\Delta Q' = T \, \Delta S' = -\Delta Q \qquad (14.5.9)$$

We see that in reversible processes the entropy of system plus surroundings is always conserved. In *irreversible* processes, however, highly ordered forms of energy are converted to random thermal energy, and an accompanying increase in the state of disorder of the universe occurs. This corresponds to an *increase* in the entropy of system plus surroundings. We may say, therefore, that

in irreversible processes entropy is created and the total entropy of system plus surroundings increases.

The increase in entropy associated with irreversible processes is mandated by the second law of thermodynamics, which tells us that systems—the universe in this case—evolve spontaneously from states of lesser disorder to states of higher disorder and not the opposite way. Indeed, the natural direction for any irreversible process can be *predicted* by finding the direction in which an increase rather than decrease of entropy takes place. In many of the irreversible processes we have considered as examples, this might seem to be a trivial exercise, the natural direction for the process being clearly evident to begin with. But there are others, chemical reactions, for example, in which it is not at all obvious in which direction the process will proceed spontaneously.

The amount of entropy created in an irreversible process can be calculated by making use of the fact that although the process does not proceed through any definable succession of equilibrium or near-equilibrium states, and thus is not easily represented as a path on a P–V diagram (see Fig. 14.1), it nevertheless begins with an equilibrium state i and, when all the dust has settled, ends up in another equilibrium state f. Since the entropy associated with both these states is a well-defined quantity corresponding to the degree of disorder associated with them, the change in entropy of the system can be found simply by calculating the entropy difference between the final and initial states, much as if the process had been reversible. In other words, one may write

$$\boxed{\Delta S_{\text{irr}} = S_f{}^{(\text{eq})} - S_i{}^{(\text{eq})}} \qquad (14.5.10)$$

where $S^{(\text{eq})}$ represents the entropy of any equilibrium state of the system as a function of pressure, temperature, and volume.

But how does this differ from what happens in the case of a reversible process? Simply in this way:

had the process been reversible, there would inevitably have been an accompanying change in entropy, equal in size but opposite in sign, associated with the surroundings. But now there is no such requirement, in fact, the entropy of the surroundings may not change at all. This is the case in the example of the two bodies at temperatures of 100°C and 0°C that are brought into thermal contact with one another. Both bodies are thermally isolated from the surroundings, so that there is no heat exchanged between either of the two bodies comprising the system and the surroundings. The entropy of the surroundings, therefore, does not change. But the flow of heat from one body to the other within the system, as we have already seen in section 14.2, takes the system itself from a comparatively ordered state to one that is more highly disordered, thereby generating an *increase* in entropy. The entropy change of system plus surroundings in this irreversible process is, therefore, a net increase, as required by the second law of thermodynamics. In order for the entropy of system plus surroundings to decrease, heat would have to flow from the colder body to the hotter one, contradicting the second law.

It should be evident at this point that the second law of thermodynamics may be stated in the following way:

In any reversible process, the entropy of any system plus that of its surroundings remains constant, whereby

$$dS + dS' = 0$$

while in any irreversible process, the entropy of any system plus that of its surroundings must increase, in which case

$$dS + dS' > 0$$

The equivalence of this statement to our initial postulate, expressed by Statement I, is clearly established by the definitions of entropy and reversibility and by Eq. (14.4.15).

The details of calculating entropy changes associated with reversible and irreversible processes will be spelled out more clearly in the following series of examples, some of which are designed to provide a final and authoritative discussion of effects that have been used continually within the text to illustrate phenomena related to reversible and irreversible processes and the second law. They therefore form an integral part of the chapter and should not be omitted by the reader.

EXAMPLE 14.5.1

Find the entropy change of a system containing ice and water in equilibrium at 0°C when 1 gram of ice is melted by the very slow introduction of 80 calories of heat from the surroundings. What is the entropy change of the surroundings? What happens when 80 calories of heat is subsequently slowly withdrawn to the surroundings, returning the system to its initial state?

We have considered this process before, in section 14.2, and found that it is reversible, provided, of course, that heat is introduced and withdrawn sufficiently slowly. It is also isothermal, the temperature of both ice and water remaining essentially at 0°C throughout the process. The change in entropy of the system will, of course, be positive, since heat is absorbed by the system. Its magnitude will be given by (14.5.8):

$$\Delta S = \frac{\Delta Q}{T} = \frac{(80)}{(273)} = 0.293 \text{ cal/°K}$$

Since the process is reversible, $\Delta Q' = -\Delta Q$ and the entropy change of the surroundings, according to (14.5.9) is given by

$$\Delta S' = -\frac{\Delta Q}{T} = -0.293 \text{ cal/°K}$$

The entropy change of the system plus that of the surroundings according to these results is *zero*, as we must expect for any reversible process. When the heat is withdrawn and the system returned reversibly to its initial state, the sign of ΔQ now becomes negative, giving $\Delta S = -0.293$ cal/°K. At the same time, since heat is absorbed by the surroundings, $\Delta Q'$ becomes positive, whence $\Delta S' = +0.293$ cal/°K.

EXAMPLE 14.5.2

Two cubes of copper of equal mass are enclosed within a perfectly insulating container. Initially, one of the bodies is at a uniform temperature T_2, the other at a lower uniform temperature T_1. The blocks are initially separate but are then brought into thermal contact with one another, so that heat flows from the warmer object into the colder one. After a long time, both bodies have the final temperature T_f. Find the change in entropy of the system in this process and the change in entropy of the surroundings. Is the process reversible or irreversible? Assume that the specific heat of copper has the constant value C in the range of temperatures considered. Find the numerical values of these entropy changes if $T_1 = 0°C$, $T_2 = 100°C$, and $C = 0.0925$ cal/g-°K, assuming that the mass of each block is 1.000 kg.

We have discussed this system before and concluded that the process is irreversible. But let us pretend for the time being that we don't know that. The methods used in calculating entropy changes will be the same in any event. The situation is illustrated in Fig. 14.7, the initial and final equilibrium states being

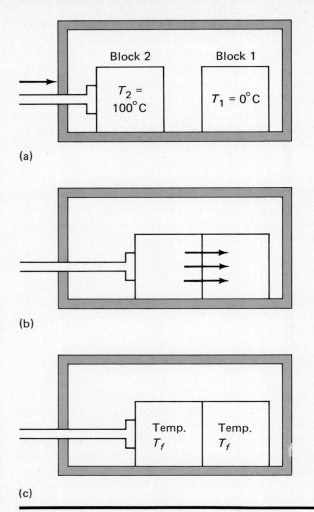

(a)

(b)

(c)

FIGURE 14.7. Isolated system containing two bodies initially at different temperatures. The bodies are brought into contact, and after a time a final equilibrium state is reached in which both bodies are at the same final temperature.

shown as (a) and (c), respectively. At (b), an intermediate stage is shown during which heat is flowing from block 2 into block 1. In this intermediate stage, the *average* temperature of block 1 has some value between T_1 and T_f, while block 2 is at an average temperature between T_2 and T_f. Unfortunately, however, the temperature is no longer *uniform* throughout each body; points in body 1, for example, that are close to the boundary between the two objects are clearly warmer than points on the opposite side, far from this interface. While this situation might seem to present formidable obstacles to the calculation of entropy changes, we can get around it by calculating what would happen if, instead of having the two blocks in contact all the time and letting the heat flow from 2 to 1 as it will, we conduct the process in stages letting only a small amount of heat flow each time before interrupting the heat flow process. We might

allow the two blocks to remain in thermal contact for a small fraction of a second initially, separating them after a small amount of heat has been transferred from 2 to 1 and allowing a condition of equilibrium to be established. We might then find that object 1 has a temperature of 1.0°C while object 2 has a temperature of 99.0°C. But, since equilibrium is established in this condition, these temperatures are now *uniform* throughout the respective objects, and the entropy changes of the two blocks can be calculated (to a high degree of approximation, at least) as $\Delta Q_1/T$ and $\Delta Q_2/T^*$, where ΔQ_1 is the amount of heat absorbed by object 1, ΔQ_2 $(= -\Delta Q_1)$ is the amount of heat given up by object 2, T has the average value 0.5°C (273.66°K), and T^* has the average value 99.5°C (372.66°K). The two blocks can then be allowed to touch one another for another fraction of a second where another small heat transfer takes place, and are then separated until another condition of equilibrium is reached in which body 1 is a bit warmer than 1.0°C and body 2 a bit cooler than 99°C. The entropy change of each body is then calculated on the same basis and added to what was obtained in the first process. In this way, a large number of small entropy changes—associated in each instance with *equilibrium* states in which the bodies are separated and at a uniform temperature throughout—can be added up to obtain the total entropy difference between the given initial state $T_1 = 0°C$, $T_2 = 100°C$ and the given final state $T_1 = T_2 = T_f$.

In this way, we may proceed from the initial to the final condition through a closely spaced sequence of equilibrium states, somewhat like those designated by the points in Fig. 14.1a, instead of in one fell swoop as in Fig. 14.1b. This is perfectly legitimate, since the entropy at equilibrium is a unique function of the variables P, V, and T defining the state of the system. Both initial and final states are indeed equilibrium states in this case, and the entropy difference between them has to be the same no matter how we choose to go from one to the other. In this regard, the entropy behaves just like the internal energy. By taking the route we are taking, however, we can calculate what we wish to know; by any other, we cannot.

Let us now look at Fig. 14.8, in which the results of one of these intermediate elementary heat transfer steps is represented. Object 1, at temperature T $(T_1 < T < T_f)$, absorbs heat dQ_1 and undergoes a temperature change dT, while object 2, at temperature T^* $(T_2 > T^* > T_f)$, gives up heat dQ_2 $(= dQ_1)$ and undergoes a temperature change dT^*. Since both objects have mass m and specific heat C, we may write

$$dQ_1 = mC\,dT \qquad (14.5.11)$$

and

$$dQ_2 = mC\,dT^* = -dQ_1 = -mC\,dT \qquad (14.5.12)$$

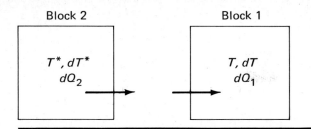

FIGURE 14.8. Intermediate stage in the heat flow process illustrated in Fig. 14.7.

From (14.5.12), of course, it is clear that

$$dT^* = -dT \qquad (14.5.13)$$

The entropy change of object 1 (let us call it dS_1) will then be

$$dS_1 = \frac{dQ_1}{T} = mC \frac{dT}{T} \qquad (14.5.14)$$

while that associated with object 2 is

$$dS_2 = \frac{dQ_2}{T^*} = mC \frac{dT^*}{T^*} \qquad (14.5.15)$$

The finite entropy change resulting from summing up the elementary changes associated with all the intermediate elementary heat transfer steps can be found by integrating Eqs. (14.5.14) and (14.5.15):

$$\Delta S_1 = mC \int_{T_1}^{T_f} \frac{dT}{T} = mC[\ln T]_{T_1}^{T_f} \qquad (14.5.16)$$

$$\Delta S_1 = mC \ln \frac{T_f}{T_1}$$

In much the same way,

$$\Delta S_2 = mC \int_{T_2}^{T_f} \frac{dT^*}{T^*} = mC[\ln T^*]_{T_2}^{T_f} \qquad (14.5.17)$$

$$\Delta S_2 = mC \ln \frac{T_f}{T_2}$$

Observe that since $T_1 < T_f < T_2$, $T_f/T_1 > 1$ while $T_f/T_2 < 1$. This means that Eqs. (14.5.16) and (14.5.17) give a positive value for ΔS_1 and a negative value for ΔS_2, in agreement with what we expect on the basis that ΔQ (representing heat absorbed) is positive and ΔQ_2 (representing heat given off) is negative.

The total entropy change of the system, ΔS, is the sum of ΔS_1 and ΔS_2; therefore,

$$\Delta S = mC \left[\ln \frac{T_f}{T_1} + \ln \frac{T_f}{T_2} \right] = mC \ln \frac{T_f^2}{T_1 T_2} \qquad (14.5.18)$$

Since both objects have the same mass and the same constant specific heat, however, the final temperature

will be just halfway between T_1 and T_2, that is, it will be the average of those two quantities. Hence,

$$T_f = \tfrac{1}{2}(T_2 + T_1) \qquad (14.5.19)$$

Substituting this into (14.5.18), we then obtain

$$\Delta S = mC \ln \frac{(T_2 + T_1)^2}{4T_1 T_2} \qquad (14.5.20)$$

This can be stated in a more understandable form by observing that

$$(T_2 + T_1)^2 - (T_2 - T_1)^2 = 4T_1 T_2 \qquad (14.5.21)$$

and replacing the quantity $4T_1 T_2$ in (14.5.20) with its equivalent on the left side of (14.5.21). After dividing numerator and denominator of the argument of the logarithm by $(T_2 + T_1)^2$, the final result obtained is

$$\Delta S = mC \ln \frac{1}{1 - \left(\dfrac{T_2 - T_1}{T_2 + T_1} \right)^2} \qquad (14.5.22)$$

Now in this equation, both T_1 and T_2 are positive quantities, but T_2 is bigger than T_1. Therefore, $T_2 - T_1$ is a positive quantity, but one that is necessarily smaller than $T_2 + T_1$. The fraction $(T_2 - T_1)^2/(T_2 + T_1)^2$ is, therefore, positive but less than unity, since its denominator is bigger than its numerator. This means that the argument of the logarithm in (14.5.22) is a positive number *greater* than unity. And since the logarithm of any positive number greater than unity is positive, we find that the entropy change of the system has to be positive. There is, thus, always an *increase* in the entropy of the system, whose magnitude will be given by (14.5.22). This is to be expected, since we have already found that the degree of disorder of the system increases in this process.

Since the system is thermally insulated from the surroundings and since no heat ever flows into or out of the insulated enclosure shown in Fig. 14.7, we have $dQ' = 0$; and the entropy change of the surroundings, given by

$$dS' = \int \frac{dQ'}{T} = 0 \qquad (14.5.23)$$

must also be zero.

The entropy change of the system plus the entropy change of the surroundings is given by

$$\Delta S + \Delta S' = mC \ln \frac{1}{1 - \left(\dfrac{T_2 - T_1}{T_2 + T_1} \right)^2} \qquad (14.5.24)$$

which, as we know, is always positive. The entropy of system plus surroundings always increases in a process such as this, and, therefore, the process is *irreversible*, as we suspected all along.

Substituting the values $m = 1.000$ kg, $C =$

0.0925 kcal/kg-°K, $T_1 = 273°K$, and $T_2 = 373°K$ into (14.5.23) or (14.5.24), we find

$$\Delta S = \Delta S + \Delta S' = (1.000)(0.0925) \ln(1.183)$$
$$= 0.01556 \text{ kcal/°K}$$

EXAMPLE 14.5.3

An apparatus consisting of two enclosures having volumes V_1 and V_2 connected by a tube in which a valve is installed is shown in Fig. 14.9. The whole apparatus is insulated thermally from the surroundings. Initially, the volume V_1 contains n moles of an ideal gas in equilibrium at temperature T, while volume V_2 is evacuated. The valve between the two enclosures is then opened, the gas expanding into volume V_2 until the pressure on both sides is the same, so that an equilibrium condition again prevails. What is the final temperature of the gas? Is this process reversible or irreversible?

During the expansion, no heat flows in or out of the system, nor is there any external work done. Therefore, $\Delta Q = \Delta W = 0$, and from the first law of thermodynamics

$$\Delta U = \Delta Q - \Delta W = 0 \qquad (14.5.25)$$

the internal energy of the gas also remaining the same. But, in the case of an ideal gas, the internal energy depends only on the temperature, and vice versa. Since there is no change in internal energy, there can be no change in temperature, the final temperature, therefore, having the initial value T.

The question of reversibility can be settled by finding the change in entropy of system plus surround-

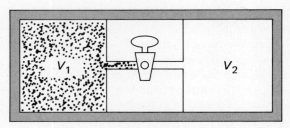

(a)

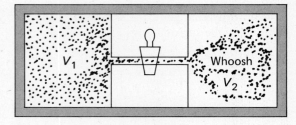

(b)

FIGURE 14.9. Free expansion of a gas into an evacuated reservoir.

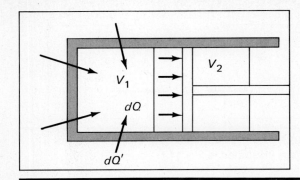

FIGURE 14.10. Reversible process connecting the same initial and final states as the free expansion process of Fig. 14.9.

ings in the process. The system goes from an initial equilibrium state of volume V_1 and temperature T to a final equilibrium state of volume $V_1 + V_2$ and temperature T. We cannot conveniently calculate the entropy change of the system as it undergoes this particular free expansion process, but we can easily enough compute it for other processes connecting the same initial and final states, and, as we have already seen, the entropy difference is a function only of the initial and final equilibrium states, independent of the process that takes the system from one to the other. In particular, the system can proceed from the initial to the final state by a process of *reversible isothermal expansion*, in which the heat absorbed, and therefore the entropy change, can easily be calculated.

In such an expansion, the temperature remains constant but the gas does external work. It is as if the gas were enclosed in a cylinder with a frictionless piston, as shown in Fig. 14.10. As the gas is allowed to expand from volume V_1 to $V_1 + V_2$, it does work against the piston, and in order to keep the temperature constant, heat must be absorbed by the gas from the surroundings. Since the temperature is, in fact, maintained constant throughout the process, the internal energy will also be constant, and hence $dU = 0$. The first law of thermodynamics then tells us that, for an infinitesimal expansion of volume dV,

$$dQ = dU + P\,dV = 0 + nRT\frac{dV}{V} \qquad (14.5.26)$$

since for an ideal gas, $P = nRT/V$. The entropy change dS is, then, by definition

$$dS = \frac{dQ}{T} = nR\frac{dV}{V} \qquad (14.5.27)$$

This can be integrated between the initial volume V_1 and the final volume $V_1 + V_2$ to give the finite entropy change ΔS, the result being

$$\Delta S = \int \frac{dQ}{T} = nR \int_{V_1}^{V_1+V_2} \frac{dV}{V} = nR[\ln V]_{V_1}^{V_1+V_2}$$

or

$$\Delta S = nR \ln\left(1 + \frac{V_2}{V_1}\right) \qquad (14.5.28)$$

Since $1 + (V_2/V_1)$ is always a positive quantity greater than unity, its logarithm must be positive. The entropy of the system, therefore, always increases.

In the reversible isothermal expansion, the heat absorbed by the system dQ is always accompanied by an equal loss of heat $dQ' = -dQ$ by the *surroundings*. This means that accompanying any entropy increase of the system dS will be a corresponding entropy *decrease* $dS' = -dS$ of the surroundings. The entropy change of the system plus that of the surroundings will, therefore, be *zero* for the reversible isothermal expansion, as it must for any reversible process.

But in the free expansion process of Fig. 14.9, the system is thermally isolated so that no heat is ever withdrawn from the surroundings. Therefore, $dQ' = 0$ and, as a result, $dS' = dQ'/T = 0$ as well. There is therefore, *no* change whatever in the entropy of the surroundings. The change in the entropy of the system, ΔS, plus that of the surroundings, $\Delta S'$, is accordingly

$$\Delta S + \Delta S' = nR \ln\left(1 + \frac{V_2}{V_1}\right) \qquad (14.5.29)$$

which, as we know, is always positive. There is, therefore, always an increase in the entropy of system plus surroundings associated with the free expansion process, and hence it is *irreversible*. We might have suspected this on the basis of the fact that once the valve is opened, we lose control over the rate of the process; it goes at its *own* rate, not one that we assign to it. In particular, there is no way of making it proceed at an arbitrarily slow pace.

RESUMÉ

Starting with an initial statement of the second law of thermodynamics to the effect that systems evolve spontaneously from states of low probability to states of higher probability, we have tried to examine the consequences of this proposition with regard to thermal processes. Along the way, we have seen that the essential content of the initial statement of the second law can be expressed in many equivalent ways, some of which bear little or no superficial resemblance to the initial form. Although the equivalence of all these statements is rather clearly evident in the context of the discussion, we have made no rigorous attempt to establish their equivalence on a formal "necessary and sufficient conditions" basis.

We have thus collected a substantial number of ways of expressing the wisdom embodied in the second law of thermodynamics. There is no way of deciding which is to be preferred. In one situation, one of the given statements may very aptly explain the behavior of the system, while in another, it may be relatively useless and one of the alternative statements may apply more directly. In any event, we shall at this point list all the ways we have so far found to express the physical content of the second law.

  I. Systems evolve spontaneously with the passage of time from states of low probability to states of higher probability, and not in the opposite direction.

  II. Systems evolve spontaneously with the passage of time from more highly ordered configurations to states which are less ordered.

  III. The degree of disorder of the universe either increases or remains the same but may never decrease.

  IV. The degree of disorder of the universe increases in irreversible processes and remains constant in reversible processes.

  V. Heat will not flow spontaneously from a body at a given temperature to one whose temperature is higher.

  VI. Clausius' Statement. It is impossible for any cyclical process to result in no effect other than the transfer of heat from a body at a lower temperature to another whose temperature is higher.

  VII. Kelvin-Planck Statement. A process whose net effect is to transform heat extracted from a reservoir of uniform temperature into work is impossible.

  VIII. McKelvey's Statement. You may be able to heat the room by letting the refrigerator door remain open, but you will never succeed in cooling it that way.

  IX. It is impossible to construct a perpetual motion machine of the second kind. (See footnote 2.)

  X. All reversible heat engines operating between the same two constant temperatures have the same efficiency, given by the Carnot expression, and any irreversible engine operated under the same conditions must be less efficient than this.

  XI. In any reversible process, the entropy of the system plus that of its surroundings remains constant, while in any irreversible process the entropy of the system plus that of its surroundings must increase.

  XII. The entropy of the universe may either increase or remain constant in any process, but it may never decrease.

Statement I is, of course, the initial way in which the second law was presented in section 14.1. Statement II is derived from this simply by replacing the concept of probability with that of disorder and by assuming that the state of maximum probability is

463

the one in which the molecular randomness is greatest. Statements III and IV are clearly derived from II; their connection with I and II is discussed in section 14.2. Statements II, III, and IV are less than completely satisfactory due to the fact that the term "degree of disorder" has no precise or quantitative meaning. Statement V is much better, even though it stems directly from II; there is no element of vagueness in it at all. Clausius' statement VI resembles V but is phrased in the language of cyclical heat engines. The Kelvin-Planck statement VII is, in effect, a statement that perpetual motion machines of the second kind—those in which a compressor drives an engine whose output is harnessed to drive the compressor, for example—cannot work. Statement IX is a more explicit denial of this possibility. The equivalence of both VII and IX with V was established in section 14.3. Statement X was also shown to be equivalent to V or VI in that section. Statement VIII is, on the surface, a facetious one, but it is every bit as good as the Kelvin-Planck or Clausius statement. It is equivalent to VII or IX; if there are no perpetual motion machines then you cannot run one backward as a refrigerator. Statements XI and XII are formulated in an attempt to deal with the ambiguity of Statements II, III, and IV by replacing the term "degree of disorder" by the more precise and quantitative measure of disorder represented by *entropy*.

14.6 The Third Law of Thermodynamics

While it may be true that nobody may succeed in constructing a perpetual motion machine of the second kind that will actually do work, we might still be able to construct reversible engines that would convert 100 percent of their heat input into mechanical work if we could operate their low-temperature reservoirs at absolute zero. In this event, the temperature T_1 in the Carnot efficiency factor (14.3.16) is zero and η becomes unity! Unfortunately, nobody has succeeded in reducing the temperature of any system to absolute zero, though attempts to produce lower and lower absolute temperatures have been going on since the heyday of the caloric fluid theory and, in fact, are still going on. The current record low temperature, by the way, is of the order of 10^{-6}°K. The difficulty of attaining a temperature of absolute zero can be appreciated by referring to Eq. (14.3.19), which gives the amount of work we must do on a reversible refrigerator to transfer a certain amount of heat, ΔQ_1, from a low-temperature reservoir at temperature T_1 to a reservoir at a higher temperature T_2. Clearly, if the "freezing compartment" temperature T_1 is absolute zero, the refrigerator's Carnot

factor η must be unity. But then, according to (14.3.19), the amount of work done to effect the transfer of any finite amount of heat must be *infinite*. We are thus led to suspect that though we may be able to approach a temperature of absolute zero, we will never attain it.

What is often referred to as the *third law of thermodynamics* is simply a denial of the attainability of absolute zero, based upon the considerations discussed above and upon the fact that all experimental attempts to reduce the temperature of any system to absolute zero have failed. We may state it in the following way:

> *It is impossible to reduce the temperature of any system to absolute zero in a finite number of processes involving the expenditure of a finite amount of work.*

SUMMARY

In a reversible process, a system goes from an initial state to a final state through a succession of intermediate states that approximate equilibrium states. The system can always be returned from the final state to the initial state by reversing the heat and work inputs that produced the original state change. The system, therefore, can be carried from an initial state to a final state and back again to the initial condition, with no net interchange of heat or work having been made between system and surroundings.

Irreversible processes are those for which the system passes through states in which physical conditions are far from those prevailing in any equilibrium state and which, therefore, cannot be visualized as an orderly succession of intermediate equilibrium states connecting the initial and final conditions. There is no way, therefore, of regaining the initial state by simply reversing the heat and work inputs that caused the original state change. The original state of the system can be reestablished, of course, but only at the expense of net work done by, or net heat exchanged with, the surroundings. Dissipative or frictional processes in which the energy of ordered systems is expended in generating internal energy are inevitably irreversible.

A Carnot cycle is a cyclical combination of reversible isothermal and adiabatic processes that can serve either as a heat engine or, in reverse, as a refrigerator. The efficiency of the Carnot cycle is defined as the ratio of the net work done to the heat absorbed from the high-temperature heat source per cycle. It is found that any Carnot cycle operating between temperatures T_2 and T_1 ($T_2 > T_1$) has an efficiency given by

$$\eta = \frac{\Delta W_{tot}}{\Delta Q_2} = \frac{T_2 - T_1}{T_2}$$

It is also true that

$$\frac{\Delta Q_1}{\Delta Q_2} = -\frac{T_1}{T_2}$$

where ΔQ_1 is the heat rejected to the low-temperature reservoir. Every reversible engine operating between the constant temperatures T_2 and T_1 must have the same efficiency as a reversible Carnot engine, and the effect of irreversibility is inevitably to reduce the efficiency below that given by the Carnot expression.

The second law of thermodynamics is an expression of the fact that systems spontaneously evolve from ordered states of low probability to more random configurations of higher statistical probability. In the language of heat engines, the second law states that: (a) it is impossible for any cyclic process to result in no effect other than the transfer of heat from a body at a lower temperature to one whose temperature is higher (Clausius statement) or (b) a process whose net effect is to transform heat extracted from a reservoir at uniform temperature into work is impossible (Kelvin–Planck statement).

The entropy of a system is a quantitative measure of its disorder. The change in entropy dS associated with a quantity of absorbed heat dQ is defined by

$$dS = dQ/T$$

from which, in a process involving the flow of a finite amount of heat and a finite entropy change,

$$\Delta S = \int \frac{dQ}{T}$$

The entropy of any system, like the internal energy, is a well-defined quantity that depends entirely on the state variables P, V, and T. The change in the entropy of a system involved in *any* process that connects the same initial and final equilibrium states is the same, and the change in its entropy in any *cyclic* process is zero.

In reversible processes, the total entropy change of system and surroundings is zero, and entropy is conserved. In irreversible processes, entropy is created, and the total entropy of system and surroundings *increases*. The direction in which irreversible processes proceed, therefore, is that corresponding to an increase in entropy of system plus surroundings.

In terms of entropy, the second law of thermodynamics states that the entropy of the universe remains the same in reversible processes and always increases in ones that are irreversible.

The third law of thermodynamics can be stated as follows: it is impossible to reduce the temperature of any system to absolute zero in a finite number of processes involving the expenditure of a finite amount of work.

QUESTIONS

1. List several thermodynamic processes and classify each as either reversible or irreversible.
2. An ideal spring is compressed, storing a certain amount of potential energy. Is this process reversible?
3. Milk is poured into a cup of coffee and is mixed with a spoon. Is this an example of a reversible process?
4. Is it possible to characterize the various stages of an irreversible process by means of an equation of state?
5. Give several examples to show that mechanical energy can be completely converted to heat.
6. A room can be heated by leaving the refrigerator door open. Comment on the efficiency of this method in comparison to more conventional methods of heating.
7. During an isothermal expansion all the heat added is converted to work. Does this violate the second law of thermodynamics?
8. In a power plant, the temperature of the steam should be as high as possible. Why?
9. Can you explain very briefly why two different isotherms cannot intersect?
10. The efficiency of a real Carnot engine is always less than that of an ideal Carnot engine. Can you suggest factors which lead to this reduced efficiency?
11. A gas undergoes an adiabatic compression from volume V_1 to V_2. By how much does its entropy change?
12. A block slides down an inclined plane. What happens to the total entropy during this process?
13. At one time, a serious proposal was made to use mercury vapor rather than steam as the working fluid in commercial turbine installations used to generate electricity. Mercury boils at 357°C at atmospheric pressure. Can you suggest some advantages and disadvantages of this scheme?
14. Can you suggest some other possible substitutes for steam as a working fluid in the application discussed in the preceding question?
15. In example 14.5.3 an ideal gas undergoes a free expansion into a vacuum. Can you suggest any way in which it might be made to undergo a spontaneous "free compression" process? Is such a process possible? Explain your answers.

PROBLEMS

1. In playing bridge, the 52 cards are dealt among four players. After the bidding the "South" player is obligated to make a contract of four spades, that is, to take six plus four tricks in a game wherein spades are the trump suit. After his partner, who is "dummy" puts his cards on the table, face up, South knows that his hand and that of his partner contain eight of the thirteen spades in the deck. **(a)** What is the probability that all five of the remaining spades belong to one of the two opposing

players? **(b)** What is the probability that one of the two opponents has one spade and the other has four? **(c)** What is the probability that one of the two opponents has two spades, while the other has three?

2. Identify the following processes as reversible or irreversible. **(a)** A pan of water evaporates slowly at a constant temperature of 27°C. **(b)** A teaspoonful of cream is slowly stirred into a cup of coffee at the same temperature. **(c)** A teaspoonful of water at 80°C is slowly stirred into a glass of water at 20°C. **(d)** An automobile is lifted on a hydraulic lift which employs a frictionless piston and a nonviscous fluid. **(e)** A match is lit by focusing the sun's rays on its head with a magnifying lens.

3. Identify the following processes as reversible or irreversible. **(a)** A set of "racked" billiard balls is broken up by a cue ball aimed at the set; the collision is elastic and there is no friction between the balls and the surface of the table. **(b)** A slice of bread is slowly toasted in an electric toaster. **(c)** An object is set into motion on a frictionless horizontal surface by a spring gun employing an ideal spring. **(d)** A lead-acid "storage battery" provides electric current to run a frictionless electric motor; no heat is generated anywhere in the circuit. **(e)** A salt solution and an equal volume of pure water occupy the two halves of a rectangular tank; they are prevented from mixing by a thin plastic panel which separates the two halves. The plastic panel is slowly and carefully withdrawn so that the two liquids are now in contact. After several days the salt solution and water are observed to have formed a single homogeneous solution.

4. Calculate the slope $\partial P/\partial V$ for an adiabatic curve and also for an isothermal curve. Show that at a point where two such curves intersect, the slope on the adiabatic curve is greater than that of the isothermal by a factor γ. Using this result, show that these two curves can intersect only once.

5. A Carnot engine absorbs 50 calories of heat from the "boiler" process. It rejects 30 calories to the "condenser." What is the efficiency η of this engine?

6. The efficiency of a certain Carnot engine is 0.36. If the "boiler" is at a temperature of 600°K, what is the temperature of the "condenser"?

7. A refrigerator operates between reservoirs of temperature at 210°K and 360°K. It absorbs 600 joules at the lower temperature. If its efficiency is $\frac{1}{2}$ that of a Carnot refrigerator, how much heat is given up to the high-temperature reservoir?

8. What is the thermal efficiency of a Carnot cycle operating between 160°F and 350°F? If the efficiency is to be doubled by raising the upper temperature, what value must this new temperature have?

9. One mole of a monatomic ideal gas performs a Carnot cycle between 300°K and 600°K. For the upper isothermal process, the volume increases from 2 liters to 5 liters. Find the work done by the gas during one cycle, the heat exchanged in the two isothermal processes, and the thermal efficiency.

10. What is the minimum amount of work needed to extract 10 calories of heat from a body at 0°F when the environment is at 70°F?

11. A refrigerator has an efficiency $\frac{1}{3}$ that of an ideal Carnot refrigerator. Two hundred kilograms of water at 0°C is to be converted to ice. If the room temperature is 29°C, how many calories of heat are transferred to the room and how much work is required (in joules)?

12. Air is carried through a Carnot cycle starting from a volume of 4 liters at 2 atm and a temperature of 320°K. The processes are as follows: **(a)** adiabatic compression to a temperature 500°K, **(b)** isothermal expansion to a volume of 8 liters, **(c)** adiabatic expansion to a volume of 12 liters, and **(d)** isothermal compression to close the cycle. Plot each of these processes on a $P-V$ diagram and calculate the net work done by the gas. Assume that air behaves as an ideal gas.

13. A steam engine has a thermal efficiency which is 25 percent that of an ideal Carnot engine. The boiler temperature is 450°F and the condenser temperature is 130°F. If the engine delivers 10 hp, how many Btu/sec are absorbed from the boiler?

14. A gas obeying the Clausius equation of state $P(V - b) = RT$ is used as the working substance of a Carnot engine. Show that the efficiency is exactly the same as that obtained when the working substance is an ideal gas.

15. A heat pump or Carnot refrigerator operates between a low-temperature reservoir at 0°C and a high-temperature reservoir at 80°C. What is the efficiency of the pump? For every 10,000 joules absorbed from the low-temperature reservoir, how many joules are rejected to the high-temperature reservoir and how much work is done on the system?

16. It is possible to heat a building by using refrigeration cycles. Pipes buried underground carry fluid which absorbs heat at a low temperature and delivers it to the building interior at a higher temperature. Assume the low temperature is 0°C and the high temperature is 27°C. For each kilowatt-hour of electrical energy needed to operate the system, how many kilowatt-hours of heat are delivered to the building? Assume the most ideal conditions.

17. A Carnot engine operates between temperatures T_1 and T_2 ($T_1 < T_2$). The efficiency can be increased by lowering T_1 by ΔT or raising T_2 by ΔT. Which of these alternatives leads to the greater final efficiency?

18. Assume that an adiabatic curve and an isothermal curve intersect twice as shown in the figure. Show that this would constitute a violation of the second law of thermodynamics.

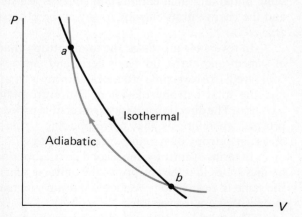

19. Assume that two different adiabatic curves intersect as shown in the diagram. They can then be connected by a

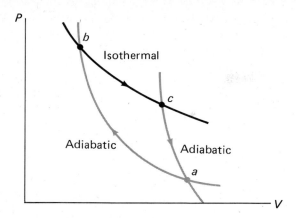

single isotherm. Show that this cannot happen since it violates the second law of thermodynamics.

20. A refrigerator kept at a temperature of 0°C is located in the kitchen where the temperature is 27°C. If the refrigerator is to stay cold, any heat that enters it from the room must be pumped out. Assuming that 10^7 joules of heat enter the refrigerator each day, how much mechanical power is needed to operate the refrigerator if it is assumed to be an ideal Carnot refrigerator? At 2.5¢ per kilowatt-hour, what is the daily cost?

21. A reversible Carnot engine receives 12,000 joule/sec from a boiler at 400°C, and rejects heat to a condenser at 40°C. Find (a) the efficiency of the cycle and (b) the amount of heat rejected to the condenser per second.

22. The Carnot engine in the preceding problem is reversed and run as a refrigerator. The temperatures of the heat reservoirs remain the same as before. Find (a) how much work must be done on the system per second to extract 3600 joules of heat per second from the reservoir at 40°C, (b) how much heat per second is rejected to the high-temperature reservoir, (c) why the answer to the preceding question is different from 3600 joule/sec, (d) the change in entropy of the working substance for one complete cycle.

23. When two wires made from different metals are joined to form an electrical cricuit, as shown in the diagram, a current will flow in the circuit whenever the junctions between the two materials are maintained at different temperatures. Such a device is called a thermocouple. In the diagram, a motor is run by a thermocouple, one of whose junctions is held at $T_2 = 600°C$, the other at

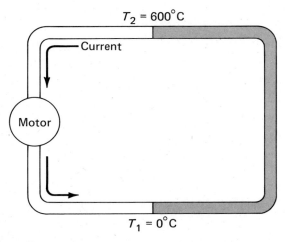

$T_1 = 0°C$. The motor develops 12 watts of power. The heat input to the hot junction is 100 joule/sec. (a) Is this "heat engine" a reversible one? (b) How much of the input energy, if any, is lost every second in exciting irreversible processes within the system? (c) What is the maximum power obtainable from such a system, if it were possible for it to operate as a reversible heat engine?

24. Find the change of entropy of water when 200 g at 0°C is mixed with 50 g at 50°C.

25. One kilogram of water is raised from freezing to boiling temperature. Find the entropy change of the water.

26. One kilogram of ice at −20°C is heated at atmospheric pressure until a temperature of 120°C (superheated steam) is attained. Using the information c_p(ice) = 2.1 × 10^3 joules/kg-°K, c_p(water) = 4.2 × 10^3 joules/kg-°K, c_p(steam) = 2.1 × 10^3 joules/kg-°K, heat of fusion = 3.3 × 10^5 joules/kg, and heat of vaporization = 22.6 × 10^5 joules/kg, obtain the change of entropy. Ignore variations of the above quantities with temperature.

27. Show (a) that in any constant-volume process, $dS/dU = 1/T$, and (b) that in a process in which the internal energy of the system remains unchanged, $dS/dV = P/T$.

28. Show that the entropy change of an ideal monatomic gas between an initial state (P_0, V_0, T_0) in which the entropy is S_0, and a final state (P, V, T) in which the entropy is S is given by

$$S - S_0 = (3nR/2) \ln(T/T_0) + nR \ln(V/V_0)$$

where n is the number of moles of gas.

29. Five moles of an ideal monatomic gas $(\gamma = 5/3)$ undergo a reversible adiabatic expansion from an initial volume of 24.0 liters. Find (a) the change in internal energy of the gas, (b) the change in internal energy of the surroundings, (c) the entropy change of the gas, (d) the entropy change of the surroundings, (e) the change in the entropy of the universe.

30. Two pieces of ice, whose total mass amounts to 25 g, are melted within a perfectly insulating enclosure by an electrically powered device which rubs the pieces of ice together vigorously, doing frictional work at the surface where they are in contact. The temperature within the enclosure remains at 0°K throughout the process. Find (a) the change in the internal energy of the system, (b) the change in the internal energy of the surroundings, (c) the entropy change of the system, (d) the entropy change of the surroundings, (e) the entropy change of the universe. (f) How would you classify this process?

31. In Joule's experiment, 20.0 kg of water are stirred within a thermally insulated enclosure, and it is found that 83,660 joules of external work must be expended in order to raise the temperature of the water from 15.00°C to 16.00°C. Find (a) the change in internal energy of the system, (b) the change in internal energy of the surroundings, (c) the entropy change of the system, (d) the entropy change of the surroundings, (e) the change in the entropy of the universe. You may assume, for simplicity, that the change in absolute temperature is so small that the process can be characterized by a constant average temperature of 15.50°K for the purpose of calculating entropy changes.

32. Five moles of a monatomic ideal gas $(C_v = 3R/2)$ undergo a reversible isothermal expansion from an initial volume

of 24.0 liters to a final volume of 120.0 liters at a temperature of 300°K. Find (a) the change in internal energy of the gas, (b) the change in the internal energy of the surroundings, (c) the entropy change of the gas, (d) the entropy change of the surroundings, (e) the entropy change of the universe.

33. A 75-kg bag of sand is dropped from a platform 4 meters high and lands on the pavement. The temperature outside is 30°C. Assuming no energy transfer to the pavement, find the increase of entropy of the bag of sand.

34. At low temperatures, many crystalline substances have a specific heat given by the Debye T^3 law

$$c_v = AT^3$$

Assuming such a substance would have an entropy of zero at $T = 0°K$, find the entropy per mole at a finite temperature T_1.

35. (a) Make a plot of temperature vs. entropy for the four processes (two isothermal, two adiabatic) of a Carnot cycle. (b) Demonstrate that the area contained within the closed curve is the net heat absorbed by the system.

36. The average specific heat of copper is 0.093 cal/g-°C in the temperature range of 0°–100°C. Calculate the total change of entropy for the following cases: (a) A 500-g copper block at a temperature of 90°C is put into a lake in which the temperature is 10°C. (b) The block is dropped into the lake from a helicopter 200 meters up. (c) Two 500-g copper blocks, one at 10°C, the other at 90°C, are joined together in a container which is isolated from its surroundings.

37. A cylinder of cross-sectional area A has insulating walls and is fitted with an insulating piston, as shown in the diagram. The end of the cylinder opposite the piston is thermally conducting and allows heat to be introduced from the surroundings. There is a constant force of friction F_f that acts between the piston and the cylinder wall. The cylinder contains n moles of an ideal gas with initial volume V_0 and temperature T_0. The piston is initially at a distance x_0 from the end of the cylinder. An isothermal expansion is then carried out, until the piston is at a distance x from the end of the cylinder, the final volume being $V = Ax$. Find (a) the changes ΔQ, ΔU, and ΔW for the gas within the cylinder, (b) the changes $\Delta Q'$, $\Delta U'$, and $\Delta W'$ for the surroundings, (c) the entropy change ΔS for the ideal gas, (d) the entropy change $\Delta S'$ of the surroundings, (e) the entropy change of the universe.

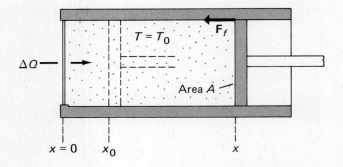

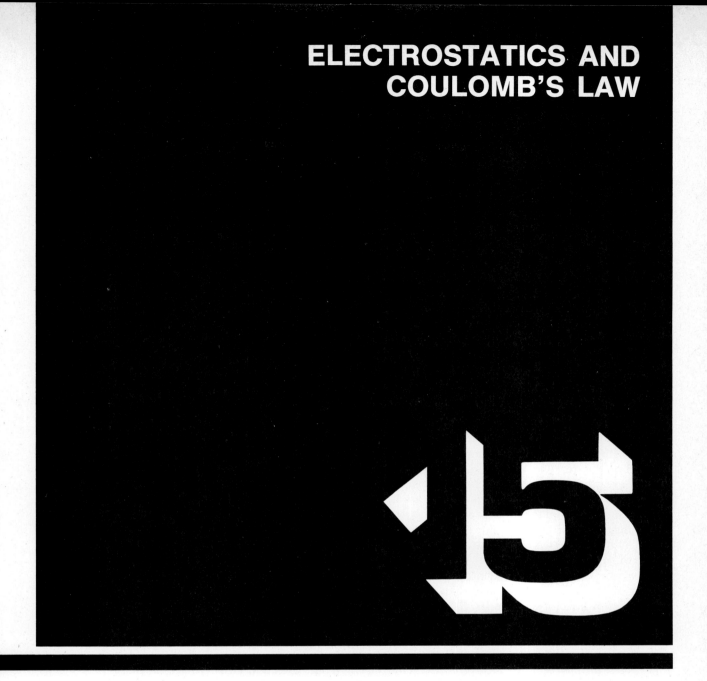

ELECTROSTATICS AND COULOMB'S LAW

15.1 Introduction

The present chapter, as well as a number of succeeding ones, is devoted to the subject of electromagnetism, the study of the laws which govern electric and magnetic phenomena. These laws constitute a well-established part of the theoretical framework of physics and have been of enormous value in the progress of all the sciences. Before we embark on the study of electromagnetism, it is perhaps worthwhile to spend a little time discussing the fundamental force laws of physics which are now known to play a vital role in physical processes.

We have already mentioned that really only four known forces exist in nature. Presumably, if the origins and relationships between these forces were clearly understood, they would comprise a basic foundation for all of physics. This does not mean that if and when such understanding is achieved, physics will grind to a halt and physicists will have to seek employment elsewhere, for there will always be a gap between the understanding of these fundamental laws and their application to explain observed phenomena and to make new predictions. The task of applying basic laws of nature to explain observed effects will always exist and will always be a great challenge, as it is today.

The four force interactions now known to exist are gravitational, electromagnetic, and the so-called "weak" and "strong" interactions between elementary particles. The latter two forces fall off very rapidly with increasing distance between particles and so are quite imperceptible on a macroscopic scale. But at distances of the order of the size of atomic nuclei (of the order of 10^{-13} cm) they are very important. The strong interaction is the strongest of all known forces; within the nuclei of atoms it is hundreds of times stronger than the electromagnetic force. It is this force that holds the nucleus together and prevents it from flying apart because of the mutual electrostatic repulsion of the protons within the nucleus. Next strongest is the electromagnetic force, which, in the guise of an electrostatic force between charges, we shall discuss very soon. The weak interaction between elementary particles is next. Like the strong interaction, it falls off very rapidly with distance and is, therefore, appreciable only when particles are very close together, as in the nucleus of the atom. Within atomic nuclei, however, this force is quite significant in the processes responsible for the radioactive transformation of radioactive nuclei that decay by emitting an electron. It also plays an important role in our theories of nuclear fission and fusion.

Electromagnetic and gravitational forces fall off much less rapidly with distance than the nuclear forces. For this reason, they are perceptible on the scale of everyday experience and are responsible for most of the directly observable effects we associate with the interaction of tangible macroscopic systems. Electromagnetic forces are quite strong, being exceeded in that respect only by the strong nuclear interaction, while gravitational forces are the weakest of all, being weaker than the "weak" interaction force by a factor of about 10^{-27} and weaker than electromagnetic forces by a factor of some 10^{-39}. This assertion about the relative strength of electric and gravitational forces leads to the question of why the effects of gravitational forces are experienced all the time, while we usually have to go to a certain amount of trouble to observe electrical forces.

The answer to this is simple enough and has to do with the basic properties of matter. In the case of the gravitational force, which arises from the mutual interaction of masses, the force is *always* attractive. So far as we can determine, there is no such thing as negative mass or antigravity, except in the imaginative fancies of the science fiction writers. Electric forces arise from another fundamental property of matter, called *electric charge*. But there are two types of electric charge, referred to as positive and negative, which interact in such a way that charges of the same sign *repel* one another, while charges of opposite sign *attract*. Ordinary matter, consisting of "neutral" atoms, contains equal amounts of positive and negative charges and, therefore, gives rise to no observable electrical forces. It is only when we go to some trouble

to add or remove charge of a given sign, thus endowing the body with a "net electric charge," that observable electrical forces are associated with it.

At the same time, as we saw in Chapter 8, the gravitational force between objects (even those that are quite massive on an ordinary scale) is very small indeed. The gravitational attraction between two adults of average mass separated by a distance of 1 meter has the incredibly small magnitude of about 10^{-30} pounds. It takes bodies of literally astronomical size to generate appreciable gravitational fields. But, of course, we live on the surface of the earth, which is just such a body, and we are continually reminded of gravitational forces every time we walk up or down the stairs, or get in or out of bed! So gravitational forces are part of everyday life, while forces that are obviously of electrical origin manifest themselves only when we interfere with the natural tendency of material systems to contain equal numbers of positive and negative charges.

The understanding we have achieved of electrical, or electromagnetic, forces has been used to great advantage for technological as well as purely scientific purposes. This knowledge has enabled us to understand the behavior of atoms and molecules, to develop a comprehensive theory of radiation, to study the properties of conducting and insulating substances, and to understand many other phenomena. At the same time, it has been put to use in the development of electrical power networks and electrically operated transportation and communication systems. Without an understanding of electromagnetism, the state of the world would be primitive indeed, and the economy of the United States, Western Europe, and Japan, for better or worse, would bear little resemblance to its present form.

15.2 Electric Charge

In Chapter 8, we found that mass can create a gravitational field **g**, which in turn can exert a force $m\mathbf{g}$ on a body of mass m. *Electric charge*, like mass, is another important inherent property of matter which can be present in both large and small bodies. Electric charge can also create force fields in space, and these fields in turn transmit forces to other charged bodies and thereby affect their motion.

All of us have at one time or another experienced or seen effects due to *electrified* or *charged* bodies. We may have walked on a new carpet and as a result experienced a minor shock upon touching a metal object. We may have rubbed a balloon on the rug and then placed it on a wall or ceiling, where it stays, apparently attracted by some curious force. If you place a piece of paper against a blackboard and rub the palm of your hand over it quickly four or five times, the paper stays there for a while, attracted to the blackboard. These are a few of many simple examples which demonstrate that in the process of rubbing objects together, there can be a net transfer of some entity between the bodies. The entity transferred is called *electric charge;* and in the process of rubbing, bodies become *electrified,* or *electrically charged.*

The existence of electric charge, and for that matter the entire science of electromagnetism, dates back to 600 B.C. when the Greeks observed that amber rubbed with wool could attract light objects such as bits of paper or straw. The phenomena of magnetism were also known to the Greeks because the attraction of iron to magnetite, also known as lodestone, had been observed. It took many years, however, until these qualitative observations were made quantitative and the close connection between electricity and magnetism firmly established. James Clerk Maxwell (1831–1879), building on the important work of others, finally laid the complete foundation of classical electromagnetism by summarizing the known laws of electricity and magnetism in an elegant set of four equations which were thereafter universally referred to as Maxwell's equations. His contribution has been of monumental significance to the development of all of science.

Why do bodies become electrified as a result of rubbing? The reason for this can be partially understood on the basis of the atomic theory of matter. The fundamental elementary building blocks of matter are *electrons, protons,* and *neutrons.* In any atom, the protons and neutrons share a very small volume of space called the *nucleus* of the atom. For light nuclei, those which contain only a few protons and neutrons, the atomic nucleus has a radius of about 10^{-13} cm, while for heavy nuclei, the radius is somewhat larger. The electrons in the atom are typically about 10^{-8} cm from the nucleus and are attracted to the nucleus by a force called the *electrostatic force,* or the *Coulomb force.* This force exists by virtue of the fact that electrons and nuclei have electric charges of *opposite* sign, the electron being negatively charged and the nucleus being positively charged. The positive charge of the nucleus is entirely due to the charges of protons, since it is known that neutrons have no net electric charge.

Ordinary matter is made up of enormous numbers of atoms and molecules and is ordinarily electrically neutral. This implies that *equal* numbers of electrons and protons are present. The atomic electrons are bound to their nuclei by forces that, though reasonably strong, are not insurmountable. These electrons can, therefore, be transferred from one body to another when two substances are brought into intimate contact. Thus, in the process of rubbing two

bodies together, many electrons may be transferred from one object to another. When this happens, one of the bodies has an *excess* of electrons while the other has a *deficiency.* The one with the excess is *negatively charged,* while the one which is deficient is *positively charged.*

The charge of an object can therefore be regarded as the summation of all of the elementary atomic charges which make up the object. The mass of any *elementary particle,* such as the electron, is an intrinsic characteristic of the particle. Likewise, the *charge* of any such particle is also an intrinsic property. Ultimately, the existence of mass and charge (as well as their particular numerical values) must be ascertained by experiment, for there is no theory at present which can predict the mass or charge of elementary particles. The magnitude of a given particle's mass can be determined from the ratio of force to acceleration, and in a similar manner its electric charge is measured by the strength of the electric forces acting on the particle under carefully standardized experimental conditions.

As we have just seen, bodies may be negatively or positively charged, whenever there is an excess or a deficiency of electrons with respect to protons. Originally, negative charge was defined by Benjamin Franklin as the type of charge residing on hard rubber which had been rubbed with cat's fur, while positive charge was the type acquired by glass electrified by rubbing with silk. These conventions, which are entirely arbitrary, have persisted to this day, although as we shall see later, it might perhaps have been more appropriate to have adopted the reverse convention. An equivalent and more modern definition of the sign of electric charge would be to say that negatively charged objects have a charge of the same sign as that of the electron while positively charged objects have a charge of the same sign as that of the proton. It is readily established by experiment that charges having the same sign repel one another, while charges of opposite sign exhibit a mutual attraction. This leads to the familiar assertion that "like charges repel, unlike charges attract."

Consider, for example, the situation illustrated in Fig. 15.1, where we see an apparatus consisting of two rods, one of glass and one of hard rubber, suspended from an overhead support by a thread, so that they are free to rotate in a horizontal plane. There are also two similar rods that can be electrified and held in the vicinity of the suspended ones and pieces of silk and cat fur with which to electrify the rods by rubbing, as described previously. If the suspended glass rod is charged by rubbing it with a silk cloth and if a similarly charged rod is held in its vicinity, we find that the suspended rod swings away from the one that has been brought near it, the electric

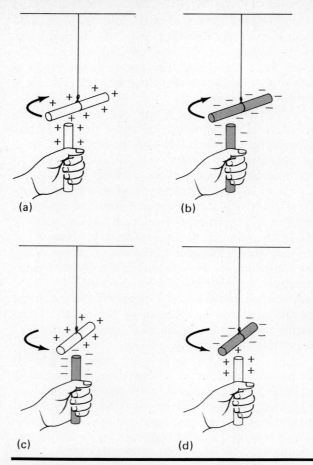

(a)   (b)

(c)   (d)

FIGURE 15.1. Experimental observations leading to the conclusion that "like charges repel, unlike charges attract."

force between the two similarly charged objects being clearly repulsive. Both rods have been positively charged, and it is evident that there is a repulsive force between positive electric charges. This is illustrated in Fig. 15.1a.

In Fig. 15.1b, we see the existence of a similar repulsive force between two hard rubber rods that have been negatively charged by rubbing with cat fur. We conclude, therefore, that there is a similar repulsive force between negative charges. But now, if one of the rods is charged positively and one charged negatively, as illustrated in Figs. 15.1c and d, it is found that the suspended rod swings *toward* the oppositely charged one that is held nearby. Thus, there is an attractive force between electric charges of opposite sign. Let us now examine some of the other important facts about electric charge.

One of the most important facts about electric charge is that it appears as an integral number of electronic charges. The charge on the electron, therefore, so far as we now know, is the smallest possible quantity of negative charge that can be found in nature. Likewise, the charge on the proton, which is

precisely equal in magnitude but opposite in sign to that on the electron, is the smallest unit of positive charge to be found in the universe. If we denote the electronic charge by $-e$, then the charge on the proton can be written as $+e$, while the neutron, which is electrically neutral, has charge zero. The charges on other elementary particles are either zero or some integral multiple of the charge of the electron. In the same way, the charges exhibited by ions or atomic nuclei are inevitably exact integral multiples of the charge on the electron or proton. This characteristic occurrence of electric charge in units of an indivisible elementary charge is referred to as *charge quantization*, and we say that electric charge is *quantized* in units of the electron charge.

The experimentally determined masses and charges of some of the more common elementary particles are listed in Table 15.1. From these data, it is apparent that electric charge is quantized. On the other hand, careful examination of the figures listed for the masses of the particles reveals no evidence that mass is quantized in the same way as electric charge.

The reason why electric charge is quantized is indeed a mystery. It is not understood theoretically but is simply an experimental fact, which as far as we know has no basic justification. At the present time, both theoretical and experimental high-energy physicists are actively looking for fractionally charged particles known as *quarks*. If these as yet hypothetical particles, for which there is at the moment no direct experimental evidence, can be shown to exist, they might constitute fundamental building blocks from which particles like the proton, neutron, pion, and others could be explained. In the theoretical schemes invented thus far, the various quarks would have

TABLE 15.1. Some Elementary Particles and Their Charges and Masses

Particle name	Symbol	Charge	Mass in units of the proton mass
Proton	p	e	1.000
Neutron	n	0	1.001
Electron	e^-	$-e$	0.000545
Positron	e^+	e	0.000545
Muon	μ^-, μ^+	$-e, +e$	0.1126
Pi meson	π^+, π^-	$+e, -e$	0.1488
	π^0	0	0.1438
Photon	γ	0	0
Neutrino	ν	0	0
Antineutrino	$\bar{\nu}$	0	0
Lambda	Λ^0	0	1.189
Rho meson	ρ^+, ρ^0, ρ^-	$+e, 0, -e$	0.82
Omega meson	ω	0	0.836

TABLE 15.2. Some Reactions Illustrating Charge Conservation

Beta decay of the neutron*

$n \rightarrow p + e^- + \bar{\nu}_e$

0   e   −e   0

Electron and positron annihilation

$e^+ + e^- \rightarrow \gamma + \gamma$

e   −e   0   0

Decay of unstable boron to stable carbon

$^{12}_{5}B \rightarrow {}^{12}_{6}C + e^- + \bar{\nu}_e$.

5e   6e   −e   0

Production of carbon 14 by collision of neutrons with nitrogen nuclei

$^{14}_{7}N + n \rightarrow {}^{14}_{6}C + p$

7e   0   6e   e

---

* The symbol $\bar{\nu}_e$ denotes a particle associated with the electron called an *antineutrino*. Neutrinos and antineutrinos are massless particles which were theoretically predicted by W. Pauli in 1930. They were first detected experimentally in 1956 by C. L. Cowan and F. Reines.

fractional charge; that is, charges less than that of the electron. This would violate our present ideas about charge quantization, but since charge quantization has not been explained in any fundamental way by theory, the existence of fractional charge *per se* would not be objectionable.

Another extremely important feature of electric charge is that *electric charge is always conserved*. This means that in any interaction or reaction, the initial and final values of the total electric charge must be the same. Thus, *total electric charge* is neither created nor is it destroyed. In Table 15.2, we show a number of processes or reactions between particles or nuclei which occur in nature. In each case, the total charge before the interaction occurs is identical to that which exists afterward. The validity of charge conservation has been established by the results of a vast number of experiments. It is possible, however, also to develop a *theory* in which this conservation law is universally valid. Thus, charge conservation can be explained on the basis of a comprehensive theory of electromagnetism, while charge quantization cannot, at present, be understood in any fundamental way.

15.3 Coulomb's Law

We have discussed qualitatively several aspects of electric charge and have found that like charges repel and unlike charges attract. Physics is a quantitative science, however, and it is, therefore, desirable to establish an exact force law between charged objects.

FIGURE 15.2. Charles Augustin de Coulomb (1736–1806), discoverer of the law expressing the force between electrostatic charges.

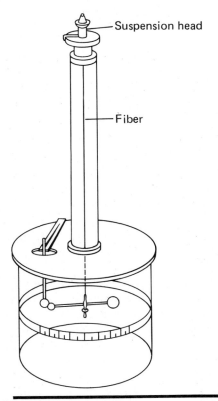

FIGURE 15.3. Coulomb's torsion balance, used in early experimental studies of electrostatics.

Let us discuss the development of this force law, for the case of two charged bodies, in detail.

In 1784, the French physicist Charles Augustin de Coulomb (1736–1806) discovered the quantitative force law between two "point charges" by measuring the forces of attraction and repulsion using a torsion balance (Fig. 15.3). The operation of this balance is

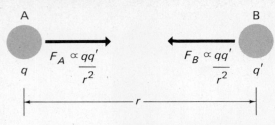

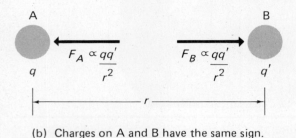

(a) Charges on A and B have opposite signs.

(b) Charges on A and B have the same sign.

FIGURE 15.4. Schematic illustration of forces between electric charges, according to Coulomb's law.

similar to that of the apparatus used by Cavendish to investigate the action of gravitational forces. Coulomb found that the force between charges q and q'

is directly proportional to the magnitude of each of the charges and inversely proportional to the square of the distance between them. The force is found to be directed along the straight line drawn between the charges.

In Fig. 15.4, the findings of Coulomb are illustrated for both attractive and repulsive forces.

At the time Coulomb made these observations, he did not have an accurate way of measuring the exact magnitude of a given charge, but was nonetheless able to vary in a precise and quantitative manner the relative charges on his "point objects." This was accomplished by bringing a small charged sphere having some charge q into contact with a second identical but uncharged sphere; the initial charge will then be equally shared by both spheres. In this manner, Coulomb was able to subdivide a given unknown charge q into parts and thus obtain spheres having charges $q/2$, $q/4$, $q/8$, and so forth. Of course, the charged spheres used are not really point objects and, therefore, the observations do not directly give the force law between infinitesimally small point charges. However, if the distance of separation between charges is much larger than the radius of the spheres, then the point charge concept is valid to a high degree of approximation. It must be stressed that the statement of Coulomb's law applies rigorously only to *point charges*. We shall see later how to determine electrical forces between large or extended *distributions* of charge.

The findings of Coulomb may be summarized by writing

$$F \propto \frac{qq'}{r^2}$$

or

$$F = k \frac{qq'}{r^2} \qquad (15.3.1)$$

which says that the magnitude of the force between two point charges whose magnitudes are q and q' is directly proportional to the product of the charges and inversely proportional to the square of the distance of separation. To make this relation precise, we need an equation rather than a statement of a proportionality. At this point, it becomes necessary to define a unit of charge. Once the unit of charge is specified, the proportionality constant k shown in Eq. (15.3.1) will be uniquely determined. There is a good deal of freedom in the choice of the unit of charge, and therefore in the choice of k, and different books may use different conventions. For example, if we require that the proportionality constant k be equal to unity, we must measure the charge in terms of a unit called the *statcoulomb*. One statcoulomb is defined as the amount of charge which, when separated by 1 cm from an equal charge, leads to a repulsive electrical force of 1 dyne. The system of electrical units so defined is called the electrostatic, or esu, system. In this text, however, *we shall adopt the more commonly used mks system of electrical units and use them consistently in all our work involving electricity and magnetism.* In the mks system, the *unit of charge* is called the *coulomb*, and 1 coulomb (1 C) is that amount of charge which when separated by 1 meter from a similar charge leads to a force 8.98742×10^9 newtons. This definition of the unit of electric charge appears at this point to be an arbitrary one, somewhat unrelated to any simple physical picture. This is actually not so, as we shall be able to see more clearly later, after we are familiar with magnetic forces. For the moment, we shall simply regard the coulomb as an arbitrarily chosen unit of charge in accord with the above definition. In Eq. (15.3.1), setting $q = q' = 1$ coulomb, $r = 1$ meter, and $F = 8.98742 \times 10^9$ newtons, it is evident that in the mks system of electrical units, the proportionality constant k must have the value

$$k = 8.98742 \times 10^9 \text{ N-m}^2/\text{C}^2$$

The coulomb is quite a large amount of charge, especially when compared to the charge of elementary particles such as the electron or the proton. In fact, the electron charge $-e$ is -1.6022×10^{-19} coulomb, and it therefore takes about 6.25×10^{18} electrons to comprise 1 coulomb of charge. In Coulomb's experiments, the electrical forces are large enough to detect using a torsion balance, and therefore the charges involved are enormous by comparison to the electron charge. The statement of Coulomb's law is nevertheless universal and applies also to the force between two electrons or between an electron and a proton. Certainly, a direct measurement of the attractive electrical force between a single electron and a single proton is impossible, but there is an enormous wealth of indirect evidence that it is correctly represented by Coulomb's law. Much of atomic physics is, in fact, based on this force law, and therefore the successful experimental verification of the predictions of atomic physics serve as a confirmation of the validity of Coulomb's law as applied to extremely small elementary charges.

In the period that followed Coulomb's original work, and in more recent times, much more sensitive instruments for detecting and measuring electric charges and forces have been designed. One of the most familiar of these is the *gold leaf electroscope*, illustrated in Fig. 15.5. In this instrument, two very thin pieces of gold leaf are suspended from a metal rod which extends outside the instrument to form an external electrode. This electrode is carefully insulated from the instrument's case, which is equipped with glass windows to allow the user to observe the gold leaves inside. When an electric charge is transferred to the electroscope's external electrode, for example, by touching it with a body that has been frictionally electrified, the charges are eventually conducted to the gold leaves. These, since they now bear charges of like sign, repel one another and diverge, as illustrated in Fig. 15.5b. The magnitude of the charge on the electroscope can be measured quantitatively by measuring the extent to which the leaves diverge. In more modern times, electronic charge-sensing instruments that are vastly more sen-

sitive than the electroscope have been developed; these instruments are referred to as *electrometers*.

We have already mentioned the fact that, even though electrostatic and gravitational forces both obey an inverse square force law, electrostatic forces are very much stronger than gravitational forces. The example which follows will serve to illustrate the correctness of this statement.

EXAMPLE 15.3.1

In a hydrogen atom, the average distance between the electron and the proton is about 0.5×10^{-8} cm. Find the ratio of the electrical attraction to the gravitational attraction between the electron and the proton.

The magnitude of the electric force, as given by Coulomb's law, is

$$F_e = \frac{kq_1q_2}{r^2}$$

$$= \frac{(9 \times 10^9 \text{ N-m}^2/\text{C}^2)(1.6 \times 10^{-19} \text{ C})^2}{(0.5 \times 10^{-10} \text{ m})^2}$$

which gives

$$F_e = 9.2 \times 10^{-8} \text{ N} \qquad (15.3.2)$$

The gravitational force between the electron and the proton is given by

$$F_g = \frac{GmM}{r^2}$$

$$= \frac{\left(\begin{array}{c}(6.67 \times 10^{-11} \text{ N-m/kg}^2) \\ \times (9.11 \times 10^{-31} \text{ kg})(1.67 \times 10^{-27} \text{ kg})\end{array}\right)}{(0.5 \times 10^{-10} \text{ m})^2}$$

$$F_g = 4.06 \times 10^{-47} \text{ N} \qquad (15.3.3)$$

Therefore, the ratio of electric to gravitational forces is

$$\frac{F_e}{F_g} = 2.27 \times 10^{39} \qquad (15.3.4)$$

We can easily see that this result is actually independent of the separation r of the charges since both Coulomb's law and Newton's law of gravitation have *exactly* the same dependence on r. They are both *inverse square laws*.

One of the most remarkable (and fortunate) characteristics of ordinary matter is its *overall electrical neutrality*. Macroscopic bodies tend toward electrical neutrality because atoms are themselves ordinarily neutral. If large bodies are not electrically neutral, the electrical forces between them can be enormous, as the following example illustrates.

EXAMPLE 15.3.2

What is the approximate magnitude of the electrical force between two people 10 meters apart if each water molecule in their bodies is assumed to have

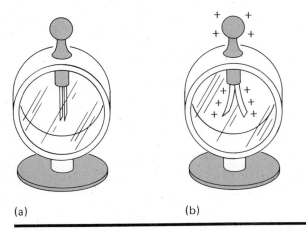

(a)   (b)

FIGURE 15.5. The gold leaf electroscope (a) uncharged and (b) charged.

1 unit of negative charge ($q = 1.6 \times 10^{-19}$ C)? Assume that the bodies of the individuals concerned contain 45 kg ($= 2500$ gram-moles) of water.

Since Avogadro's number is about 6×10^{23}, each of these persons contains $(2500)(6)(10^{23}) = 1.5 \times 10^{27}$ water molecules. If each of these is assumed to have a charge of -1.6×10^{-19} coulomb, the net charge on each person is approximately $Q = -2.4 \times 10^8$ coulombs. Therefore, at 10 meters of separation, there would be a repulsive force of roughly

$$F = (9 \times 10^9 \text{ N-m}^2/\text{C}^2)\left(\frac{(2.4)^2 \times 10^{16} \text{ C}^2}{100 \text{ m}^2}\right)$$

$$= 5.18 \times 10^{24} \text{ N}$$

This electrical repulsion is truly enormous (the weight of the earth, for comparison, is about 6×10^{25} newtons). In fact, even if only one molecule in a trillion (10^{12}) had such a negative charge, the electrical repulsion would still amount to 5.18 newtons at a distance of 1 meter, which would still be quite noticeable. Thus, the electrical neutrality of matter is a very well-established fact.

The electrical force between charged particles is a *vector quantity*. In the case of only two charged particles, the force each one experiences is directed along the line connecting the two particles. If, however, there are many charges present, the net force experienced by a *given charge* is the *vector sum of all the Coulomb forces* the other charges exert on the given charge. This method of *superposing* electrostatic forces due to different point charges, you will note, is exactly the same as that used in section 8.5 for superposing gravitational forces on a given mass arising from other point masses in its vicinity.

The principle is illustrated in Fig. 15.6, where we see the force vectors $\mathbf{F}_{ab}$, $\mathbf{F}_{ac}$, and $\mathbf{F}_{ad}$ that represent the individual electrostatic forces experienced by charge q_a arising from the nearby charges q_b, q_c, and q_d. The total force on charge a is the *resultant or vector sum* of these individual forces. We may write these forces in vector form by noting that the magnitude of $\mathbf{F}_{ab}$, the force on q_a due to q_b, must be, according to Coulomb's law, kq_aq_b/r_{ab}^2, where r_{ab} is the distance between the charges, and must point in the direction of the vector $\mathbf{r}_{ab}$ shown in Fig. 15.6. It can, therefore, be written by multiplying this magnitude by a vector of unit magnitude whose direction is that of $\mathbf{r}_{ab}$. Such a vector can be obtained by dividing $\mathbf{r}_{ab}$ by its own magnitude r_{ab}. In this way, we obtain

$$\mathbf{F}_{ab} = \frac{kq_aq_b}{r_{ab}^2}\left(\frac{\mathbf{r}_{ab}}{r_{ab}}\right) = \frac{kq_aq_b\mathbf{r}_{ab}}{r_{ab}^3} \qquad (15.3.5)$$

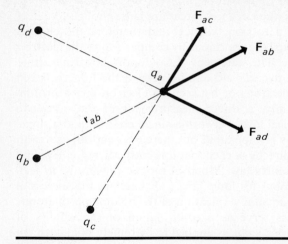

FIGURE 15.6. Superposition of electrostatic forces by vector addition.

In this expression, if q_a and q_b have the same sign, the magnitude kq_aq_b/r_{ab}^2 is positive and the vector $\mathbf{r}_{ab}$ has the direction shown in Fig. 15.6, its head on q_a and its tail on q_b. The force $\mathbf{F}_{ab}$ is in this direction also, as we must expect from Coulomb's force law. If q_a and q_b are opposite in sign, the quantity kq_aq_b/r_{ab}^2 becomes *negative*, which reverses the direction of the force $\mathbf{F}_{ab}$, which now has the character of an attractive force, as required by Coulomb's law.

The total force on charge q_a due to all other charges may be written as the *vector* sum of contributions having the form of (15.3.5) from each charge. Therefore, the total force $\mathbf{F}_a$ experienced by charge q_a is

$$\mathbf{F}_a = \mathbf{F}_{ab} + \mathbf{F}_{ac} + \mathbf{F}_{ad} + \cdots = \sum_i \frac{kq_aq_i}{r_{ai}^2}\frac{\mathbf{r}_{ai}}{r_{ai}} \qquad (15.3.6)$$

where the summation index i represents a, b, c, ..., and where the vector $\mathbf{r}_{ai}$ is a vector extending from charge q_i to q_a. We shall now illustrate the principle of superposition of Coulomb forces by a series of examples.

EXAMPLE 15.3.3

Three positive charges q are located at the vertices of an equilateral triangle whose sides have length l, as shown in Fig. 15.7. Find the magnitude and direction of the resultant force $\mathbf{F}$ acting on the charge located at a.

The charge located at position a experiences repulsive forces $\mathbf{F}_{ab}$ and $\mathbf{F}_{ac}$ as shown in the figure. These force vectors are given, respectively, by

$$\mathbf{F}_{ab} = \frac{kq^2}{l^2}\frac{\mathbf{r}_{ab}}{l} \qquad (15.3.7)$$

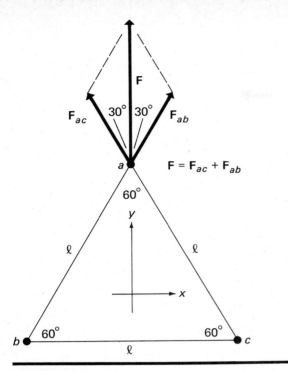

FIGURE 15.7

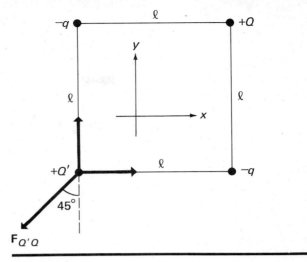

FIGURE 15.8

and

$$\mathbf{F}_{ac} = \frac{kq^2}{l^2} \frac{\mathbf{r}_{ac}}{l} \qquad (15.3.8)$$

The resultant force is, therefore, $\mathbf{F} = \mathbf{F}_{ab} + \mathbf{F}_{ac}$. In this problem, it is convenient to *resolve* both these forces into x- and y-components. The sum of the x-components vanishes, and the sum of the y-components is *twice* the y-component of one of the forces, say, $\mathbf{F}_{ab}$.

Therefore, the direction of the resultant force is along the positive y-axis, and its magnitude is

$$F_a = \frac{2kq^2}{l^3}(l \cos 30°) = \frac{1.732kq^2}{l^2} \qquad (15.3.9)$$

EXAMPLE 15.3.4
Four charges, shown in Fig. 15.8, are located at the corners of a square each of whose sides has length l. The magnitude and sign of the charges are indicated. What must be the value of the charge $+Q'$ in order that the net force on $+Q$ be zero?

The force on charge Q is the resultant of three forces, two of which are attractive forces toward the equal negative charges $-q$ and one of which is a repulsion away from $+Q'$. If the resultant force is to be zero, then the x- and y-components of the resultant, F_x and F_y, must both vanish. These components, in this case, have equal magnitudes given by

$$F_x = F_y = \frac{k(-q)Q}{l^2} + \frac{kQ'Q}{(\sqrt{2}l)^2} \cdot \frac{\sqrt{2}}{2} \qquad (15.3.10)$$

Setting this expression equal to zero,

$$Q' = 2\sqrt{2}q$$

EXAMPLE 15.3.5
A charge $+Q$ is placed on the axis of a narrow ring of radius R which carries a total charge Q' that is uniformly distributed along its circumference. If $+Q$ is at a distance x from the center of the ring, find the Coulomb repulsion experienced by the charge $+Q$.

This problem is illustrated in Fig. 15.9. When there is a continuous distribution of charge, as in the present case, the charge distribution must be broken up into very small or infinitesimal parts each of which may be considered to be a point charge. In the illustration, a small segment with charge dQ' is shown. Since the charge is uniformly distributed around the circumference, the part dQ' must represent a fraction $d\theta/2\pi$ of the total charge Q' on the ring. Therefore,

$$dQ' = \frac{Q'}{2\pi} d\theta \qquad (15.3.11)$$

The force $d\mathbf{F}$ on charge Q arising from the element dQ' of the ring's charge has a component parallel to the axis $dF_{\parallel}$ and a component $dF_{\perp}$ perpendicular to the axis. The magnitudes of these forces are given, respectively, by

$$dF_{\parallel} = \frac{kQ\,dQ'}{r^2}\cos\phi = \frac{kQQ'}{r^2}\frac{d\theta}{2\pi}\frac{x}{r}$$

$$\qquad (15.3.12)$$

$$dF_{\perp} = \frac{kQ\,dQ'}{r^2}\sin\phi = \frac{kQQ'}{r^2}\frac{d\theta}{2\pi}\frac{R}{r}$$

irrespective of the location of the charge dQ'. The directions of each of the forces $dF_{\parallel}$ are the same and, therefore, add to produce a final force $F_{\parallel}$, while the various components $dF_{\perp}$ *cancel* upon addition. This

477

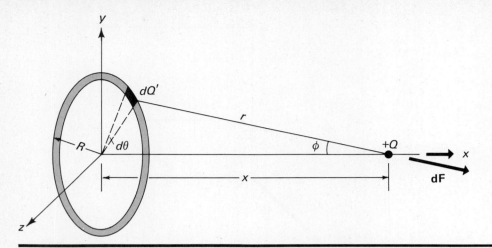

FIGURE 15.9

cancellation occurs because elements dQ' that are diametrically opposite lead to forces $dF_\perp$ which are equal in magnitude but opposite in direction. The total force $F_{\parallel}$ can be obtained by integrating the top equation in (15.3.12) with respect to $d\theta$. In performing the integration, it is important to note that k, Q, Q', r, and R are all the same for every element of charge on the ring and can, therefore, be written outside the integral. Under these circumstances, integrating $d\theta$ elements around the entire circumference of the ring, we may obtain

$$F_{\parallel} = \frac{kQQ'x}{2\pi r^3} \int_0^{2\pi} d\theta = \frac{kQQ'x}{r^3}$$

or, since $r^2 = R^2 + x^2$,

$$F_{\parallel} = kQQ' \frac{x}{(R^2 + x^2)^{3/2}} \qquad (15.3.13)$$

In Fig. 15.10, a plot of $F_{\parallel}$ versus x is given. At the center of the ring, the force on Q is exactly zero, the forces from the charge elements of the ring canceling one another completely. As Q is moved out-

ward, the Coulomb repulsion increases to a maximum value at $x = R/\sqrt{2}$ and thereafter decreases. For values of x that are much larger than R, the force varies essentially as $1/x^2$. The $1/x^2$ dependence eventually sets in for $x \gg R$, because then the ring of charge looks very much like a point charge from the point of view of the charge Q!

The procedure, illustrated in the previous examples, for calculating the Coulomb force on a point charge involves summing Coulomb forces between the given point charge and all other charges. This summation procedure is, in principle, straightforward; but if the charge distribution is complicated or asymmetric, it is generally very difficult to add up all the forces as vectors. Alternative methods which are somewhat easier exist and will be discussed in a subsequent chapter.

Let us consider another example of the vector addition of Coulomb forces.

EXAMPLE 15.3.6
An infinitely long, thin wire contains a uniformly distributed positive charge density (charge per unit

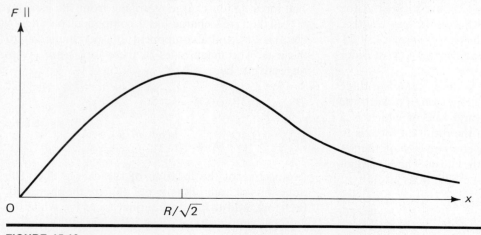

FIGURE 15.10

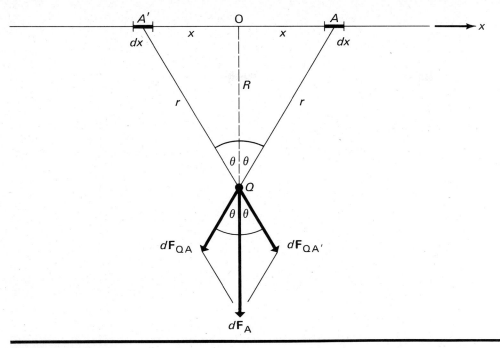

FIGURE 15.11

length) λ. Find the Coulomb repulsion experienced by a positive point charge Q a distance R from the wire.

In Fig. 15.11, a long section of the wire and two small elements of charge located at positions x and $-x$ are shown. The dotted line drawn from Q perpendicular to the line of charge provides an origin of coordinates at O. The force on Q due to the element at A whose total charge is $\lambda\,dx$ has magnitude

$$dF_{QA} = \frac{kQ\lambda\,dx}{r^2} \tag{15.3.14}$$

and a direction as indicated. Now, the force $dF_{QA'}$ also has the same magnitude as dF_{QA} and a direction such that its x-component is *opposite* to the x-component of dF_{QA}, while its y-component is the same as that of $dF_{QA'}$. Thus, the vector addition of these two forces gives an x-component of zero and a y-component equal to twice the y-component of dF_{QA}. Hence, we may write the total y-component of force as

$$dF_y = \frac{2kQ\lambda\,dx}{r^2}\cos\theta \tag{15.3.15}$$

To obtain the complete force on Q, we have to add up all contributions from all parts of the wire. This involves either an integration over the variable x from zero to infinity (remember, our method of addition already includes elements of charge at $-x$) or an integration over *angles* from $\theta = 0$ to $\theta = \pi/2$. We shall find it somewhat easier to integrate over angles. From Fig. 15.11, it is easily seen that

$$x = R\tan\theta$$

$$dx = R\sec^2\theta\,d\theta$$

$$r^2 = \frac{\cos^2\theta}{R^2}$$

Therefore, we may write (15.3.15) as

$$dF_y = \frac{2kQ\lambda R\sec^2\theta\,d\theta}{R^2}\cos\theta\cos^2\theta$$

$$= \frac{2kQ\lambda}{R}\cos\theta\,d\theta \tag{15.3.16}$$

From this, by integration, we may express the total force on Q as

$$F_y = \int dF_y = \frac{2kQ\lambda}{R}\int_0^{\pi/2}\cos\theta\,d\theta$$

$$= \frac{2kQ\lambda}{R}\left[\sin\theta\right]_0^{\pi/2} = \frac{2kQ\lambda}{R} \tag{15.3.17}$$

This calculation is illustrative of the general technique used in adding up Coulomb forces vectorially. In Chapter 16, we shall learn another method which enables us to obtain the above result with much less labor.

15.4 Conductors and Insulators

One of the most striking aspects of the behavior of matter with respect to electric charge is provided by the very different properties of electrical conductors and insulating substances. The basic features of both types of substance can be understood by observing that electrons are free to move about in, or flow through, a conducting substance, while in an insulator they remain bound to atoms and do not ordinarily move about within the material. The word *ordinarily* is used intentionally in this statement about insulators, for it is true that the frictional processes involving silk cloths and cat fur do succeed in tearing electrons away from atoms in insulating substances. Once this is done, however, and once an insulating body acquires a distribution of charge on or within it, the charge distribution remains fixed and does not move under the influence of any electrical force that may be experienced by it. Insulating substances are often referred to as *dielectrics*.

In the case of a conductor, the picture is different. Not only can a conductor acquire an electric charge as a result of an excess or a deficiency of electrons within the material, but the electrons within the substance are free to move about in the conductor in response to any electric or other forces they may experience. Therefore, if the electrons within a conductor are subjected to electrostatic forces from external charges, they may flow within the conductor and will in a very short time reach a condition of *equilibrium* in which the resultant electrical force on every electron within the substance is zero. In this respect, their behavior resembles that of a mechanical system in which particles or objects will execute certain motions until a condition of equilibrium is reached wherein the resultant force on all of them is zero.

Metallic substances are always good electrical conductors. Indeed, many of the characteristic properties of metals, such as their large thermal conductivity and their high optical reflectivity, are directly related to the presence of mobile free electrons within them. Glass, wood, rubber, stone, and most plastics, waxes, oils and other organic substances exhibit the behavior of typical electrical insulators.

There are, in addition, numerous substances that have a few free electrons but in concentrations much lower than those typical of metallic substances. Such substances conduct electricity, but only rather poorly, their electrical behavior being intermediate between that of metallic conductors and good insulators. Substances such as graphite, germanium, and silicon, which are referred to as semiconductors, belong in this intermediate category.

There is an enormous range in the degree of electrical conductivity associated with good insulators and metallic conductors, that of quartz (which is an excellent insulator) being lower by a factor of about 10^{25} than that of copper. We shall develop this simple picture of conductors and insulators into a more precise and quantitative description in subsequent chapters.

15.5 The Atomic View of Matter

The importance of Coulomb's law of electrical attraction and repulsion is indisputable, and its role in our present understanding of the behavior of matter is a vital one. The internal structure of matter and all the properties which are derived as a consequence of that structure are ultimately explained by using Coulomb's law to describe forces between charged particles. Since this law is so vital to our understanding, let us digress momentarily to discuss the origins of our present knowledge of matter on the atomic scale.

The view that matter is discontinuous or discrete when looked at on a small enough scale is a difficult concept to grasp, for it is at variance with the visual appearance of ordinary substances. Everything we look at appears to be continuous, smooth, and homogeneous in composition. Aristotle, a man whose thoughts were greatly valued by his contemporaries, was a proponent of this continuous view of matter, as advocated by Empedocles. According to his view, all substances were homogenous mixtures of earth, air, fire, and water. The opposite viewpoint, with its notion of discontinuous or discrete matter, was advanced by the "atomic" school of Democritus, Leucippus, and Lucretius. However, the former point of view prevailed for a long time, since it was supported not only by Aristotle's status as a philosopher, but also by medieval Christian theology. It was not until the sixteenth or seventeenth century, when experimental science began to gain its proper role and the testing of ideas in the laboratory became widely accepted, that the atomic point of view emerged once again.

The most basic early experiments in support of the discrete character of matter were carried out by chemists. In the latter part of the eighteenth century, men such as Joseph Black and Henry Cavendish performed quantitative measurements which demonstrated conservation of mass in chemical reactions, and Antoine Lavoisier studied and identified a number of gaseous elements. Shortly thereafter, it was discovered that elements such as hydrogen or carbon would combine with oxygen only in certain definite proportions. It was extremely difficult to reconcile the observed reactions of these substances in *definite ratios* with a continuous view of matter. All the evidence gleaned from a study of chemical reactions supported the atomic theory. The chemists discovered numerous elements and were able to determine their atomic weights based on a scale in which oxygen was chosen as the standard. It was chemists who first provided firm evidence for the atomic view of matter, but the problem of understanding the structure of individual atoms and of providing a unified theory of all atoms was ultimately attacked and resolved by physicists.

By the end of the nineteenth century, dozens of elements had already been discovered. In 1872, the Russian chemist Mendeleev succeeded in classifying the elements according to their valences, chemical similarities, and atomic masses. Families of elements such as the alkali metals and the halogens had been identified. The overall structure of the periodic table began to emerge and exhibited vacancies which led to the prediction and discovery of hitherto unknown chemical elements.

By the end of the nineteenth century, it was known that the negative charges carried by atoms are due to electrons. The celebrated experiments of J. J. Thomson in 1897 on cathode rays lead to a determination of the ratio of the electron charge to its mass (e/m). In 1917, R. A. Millikan succeeded in directly measuring the charge of individual electrons. By combining his measurement with Thomson's determination of e/m, Millikan was able also to ascertain the mass of a single electron. It was apparent from these and from certain other observations that the electrons in an atom could only account for a very small fraction of the total atomic mass and that most of the mass must, therefore, reside elsewhere in the atom.

In 1911, Ernest Rutherford proposed a theory according to which the mass of the atom was primarily concentrated in a very small region of space (about 10^{-13} cm in radius) which he referred to as the *atomic nucleus*. He was led to this brilliant discovery by the results of an experiment done by Geiger and Marsden on the scattering of alpha particles[1] by matter. Geiger and Marsden observed that some of the scattered alpha particles were deflected through *very large angles*, often in excess of 90°. If the charge in a given atom were uniformly distributed throughout the atom, we should expect the deflection of alpha particles to be very small, and large-angle scatterings would be highly improbable. Rutherford interpreted these data by asserting that most of the mass of the atom, as well as all the positive charge, resides in the tiny atomic nucleus. The large-angle scattering is then explained as due to the strong Coulomb repulsion between the alpha particle and the nucleus. Rutherford's nuclear model of the atom was discussed quantitatively by using Coulomb's law, together with the principles of classical mechanics, to predict the angular distribution of alpha particles scattered from atoms of gold, silver, copper, and aluminum. The nuclear theory of matter was verified experimentally in a beautiful series of observations carried out in Rutherford's laboratory in Cambridge.

---

[1] Alpha particles are helium nuclei; their mass is very nearly four times that of the proton, and they bear a positive charge whose magnitude is exactly twice that of the charge on the electron. Particles such as these are emitted when certain heavy radioactive elements such as radium and uranium undergo radioactive decay.

Subsequently, Niels Bohr used Rutherford's picture of the atom as a basis for the introduction of quantized electron orbits, in which the electrons in an atom were permitted to have only a discrete set of energies rather than the continuous energy range allowed by the laws of classical mechanics. This was one of the earliest applications of quantum theory. In the years that followed, quantum mechanics was developed by Schrodinger, Heisenberg, Dirac, and others. We shall discuss the history and the basic ideas of this theory in greater detail in subsequent chapters.

SUMMARY

The electric charge of a fundamental particle such as an electron or proton is, like the rest mass, one of its inherent physical properties. The charges of all fundamental particles that have been observed experimentally are exact integral multiples of the charge on the electron. Charges are of two fundamentally different types, positive and negative. Charges of like sign repel one another, while unlike charges attract. If we denote the magnitude of the charge on the electron by e, then fundamental particles are all observed to have charges $0, \pm e, \pm 2e, \ldots$. We summarize this situation by saying that charge is *quantized* in multiples of e. Ordinary matter is, as a rule, electrically neutral, containing the same number of positive and negative charges. It may be electrified, however, by rubbing processes that tear electrons away from atoms and act to effect a partial separation of positive and negative charge.

Coulomb's law expresses the way in which forces between electrically charged bodies are observed to act. It states that the force between two point charges is directly proportional to the magnitude of each charge and inversely proportional to the square of the distance between them. The force is directed along the straight line joining the charges. As an equation, this can be written

$$F = k \frac{q_a q_b}{r_{ab}^2} \qquad \text{or} \qquad \mathbf{F} = \frac{k q_a q_b}{r_{ab}^2} \left(\frac{\mathbf{r}_{ab}}{r_{ab}} \right)$$

In the mks system, the unit of charge is the coulomb. One coulomb is that amount of charge which, when placed at a distance of 1 meter from a similar amount, experiences a force of 8.98742×10^9 newtons. When mks units are used, the proportionality constant k in Coulomb's law, therefore, has the value 8.98742×10^9 N-m^2/C^2.

Electrical forces arising from a number of separate charges can be superposed like gravitational forces, adding the force contributions on a given

481

charge from all the other charges in the system, as *vectors*.

In an insulating substance, electric charges are tightly bound to atoms within the substance and do not move under the influence of electrostatic or other forces. Once a charge distribution is established on or within an insulator, it remains there until it is altered by some process, such as frictional electrification, that can effect charge separation in materials of this sort. Substances such as glass, rubber, wax, and most plastic materials are good insulators.

In a conductor, electrons are free to move about within the substance in response to electrostatic or other forces they may experience. Therefore, if the electrons in a conductor are subjected to electrostatic forces from an external charge distribution, they can flow within the conductor and quickly reach a condition of electrostatic equilibrium in which the resultant force on every electron in the conductor is zero. Substances such as copper, silver, aluminum, and other metals are good conductors of electricity.

QUESTIONS

1. Can a neutral point object experience an electrical force? Can a neutral, extended body feel a net electrical force?
2. After a balloon is rubbed on a rug, it can cling to the ceiling without falling. Explain why this is so. Why does the balloon eventually fall?
3. A fixed charge A exerts a certain Coulomb force on charge B. Is this force which A exerts on B altered if other charges are brought near B?
4. What features do Coulomb's law and Newton's law of gravitation have in common? In what ways do they differ?
5. Electric charge is always conserved. Is the same statement true regarding mass? Can you suggest an example in which mass is not conserved?
6. Coulomb's law is a radial, inverse square law. Explain the meaning of these terms.
7. The electrical force an object experiences does not depend on its mass. Suggest an experiment to prove this assertion.
8. Is it more likely for you to experience a shock due to electrostatic effects when the weather is dry or when it is humid? Explain.
9. How would you explain to a layman what it means for an object to have an electric charge?
10. Is it possible to subdivide matter in such a way as to obtain charges smaller than the charge of the electron?
11. In what way does the law of charge conservation differ from the law of conservation of momentum?
12. A very tiny, charged ball hangs vertically from a thread. Suggest a method for determining the sign of the charge.
13. The electrons in a conductor are free to move under the influence of any force that acts on them, including the force of gravity. Why is it then, despite the fact that every electron in a conductor experiences a downward weight force, that all the electrons do not fall to the lowest part of the conducting object?

PROBLEMS

1. Two positive charges, each of magnitude one coulomb, are placed at opposite ends of a football field 100 yards long. Find the force of repulsion between them.
2. An electron is projected horizontally with a speed of 10^8 cm/sec. It passes into a region between a pair of oppositely charged horizontal plates 5 cm long and experiences a total downward acceleration of 10^{16} cm/sec^2 in this region. (a) Find the force on the electron in newtons. (b) Find the horizontal and vertical components of velocity as functions of time. (c) By what angle has the electron velocity changed in the time interval during which it was between the plates?
3. Find the equation of the path along which the particle in the preceding problem moves. Take the origin to be at the point of projection and the x-axis to be parallel to the initial velocity.
4. What equal positive charges would have to be placed at the center of the earth and on a person weighing 150 pounds in order for the person to appear weightless at the surface of the earth?
5. Two point charges each experience a force of 0.05 newton when they are 0.2 meter apart. What force do they feel when they are (a) 1 meter apart? (b) 0.1 meter apart? (c) 50 meters apart?
6. An aluminum paper clip has a mass of 0.5 g. Find the number of electrons it contains and the total charge due to electrons. It also contains an equal number of protons and is, therefore, electrically neutral. If, however, we could separate the positive and negative charges into two groups, what force of attraction would they have if the positive and negative charges were 10 cm apart?
7. Find the magnitude of the minimum electrostatic repulsion force between two protons in a carbon nucleus. The radius of the nucleus is 3.8×10^{-13} cm. Why don't the protons fly out of the nucleus?
8. Two charges q_1 and q_2 satisfy the condition $q_1 + q_2 = Q$, where Q is fixed. If the force between them is to be as large as possible for a fixed separation, find the charges q_1 and q_2.
9. An electric current of 1 ampere flowing in a wire implies a flow of charge of 1 coulomb per second. In a wire which is carrying 3×10^{-3} amperes, how many electrons pass a given point each second?
10. Two pith balls, one with mass m, the other of mass $2m$, are suspended from silk threads of length l, as illustrated in the accompanying diagram. They each have charge q. Show that their equilibrium separation d is given by

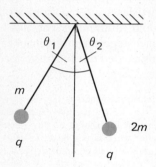

$d = (3kq^2l/2mg)^{1/3}$, assuming the angles θ_1 and θ_2 are small.

11. A muon is an elementary particle whose charge is the same as that of the electron but whose mass is 206 times greater. If we assume that the electron in a hydrogen atom travels in a circle about the proton, attracted by a force of 8.1×10^{-8} newtons, and that Newton's second law applies, find (a) the radius of the electron in its orbit, (b) the radius of the muon orbit having the same angular momentum in an atom consisting of a muon orbiting about a proton, (c) the kinetic energy of each in their respective orbits, and (d) the electron velocity and muon velocity in their respective orbits.

12. A small body of mass 10.0 g bearing a charge of 1.00×10^{-6} coulombs is at rest at a distance of 1.00 cm from a second fixed charge of -1.00×10^{-6} coulombs. What initial speed, directed along the line joining the charges, would have to be given to the first charge to insure that it would "escape" from the second and never return?

13. Two equal charges of magnitude $Q = 2.40 \times 10^{-6}$ coulomb are located on the x-axis at $x = +10.0$ cm and $x = -10.0$ cm respectively. What is the magnitude and direction of the force experienced by a third charge of magnitude $q = 1.20 \times 10^{-6}$ coulomb located on the y-axis at the point $(x = 0, y = 10\sqrt{3}$ cm$)$?

14. (a) In the preceding problem, what are the coulomb forces acting on the two charges lying on the x-axis? (b) What is the vector sum of the coulomb forces acting on all three charges?

15. Two equal charges of magnitude $+Q$ are located on the x-axis at distances $\pm a$ from the origin. Find the magnitude and direction of the force on a third charge $+q$ located on the y-axis at a distance y from the origin.

16. What is the answer to the preceding problem when the charges on the x-axis have equal and opposite charges $\pm Q$?

17. Two equal charges $+Q$ are located on the x-axis at distances $\pm a$ from the origin. Find the magnitude and direction of the force on a third charge $+q$ located on the x-axis at a distance x from the origin.

18. What is the answer to the preceding problem when the charges on the x-axis have equal and opposite charges $\pm Q$?

19. (a) What equal charges would two spheres each of mass 10.0 kg have to bear in order that their electrostatic repulsion exactly balance their gravitational attractions? (b) How many electronic charges does this amount to?

20. How much charge is needed to electroplate one gram of metallic silver from a silver nitrate solution in which the silver is present as Ag^+ ions?

21. It is found that 289,500 coulombs of charge are needed to deposit one mole of a certain metallic substance from an ionized solution of one of its salts. What electric charge does each of the metallic ions bear?

22. It is found that when 10.0 g of a certain zinc–indium alloy is put into solution in which Zn^{++} and In^{+++} ions are present, exactly 26,222 coulombs of electric charge are needed to deposit electrolytically all of the indium and zinc from the solution. What are the atomic percentages of indium and zinc in the alloy?

23. It is often convenient to ignore the fact that charges come in quantized units and to consider a continuous "charge density" $\rho(x, y, z)$ which represents the charge per unit volume in an infinitesimal volume element around the point (x, y, z). Suppose the charge density everywhere inside a sphere of radius R is given by $\rho(r) = \lambda r$, where r is the distance from the center of the sphere and λ is a proportionality constant. Show that the total charge Q contained in the sphere is given by $Q = \pi \lambda R^4$.

24. The nucleus of a lead (Pb) atom contains 82 protons and has a radius of approximately 7.2×10^{-13} cm. If an additional proton approaches this nucleus, find the force of repulsion when the approaching proton is at the surface of the nucleus. Assume for purposes of applying Coulomb's law that the distributed charge can be replaced by a single charge at the center of the nucleus.

25. The lead nucleus in the above problem has 82 protons and 124 neutrons. Assuming it is spherical and has uniform mass density and charge density, obtain expressions for the mass and charge densities.

26. A point particle with charge q_1 and mass m is projected directly toward the center of a nucleus of charge q_2; both charges are positive. If the projectile has an initial velocity v_0 very far from the nucleus, as illustrated in the diagram, show that the minimum distance of approach is given by $D = 2kq_1q_2/mv_0^2$.

27. Three charges are placed in a straight line as shown in the figure. Show that the charge in the middle, which is in equilibrium, is in stable equilibrium with respect to displacements in the y-direction but in unstable equilibrium with respect to displacements in the x-direction.

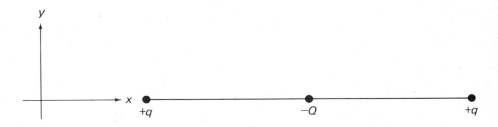

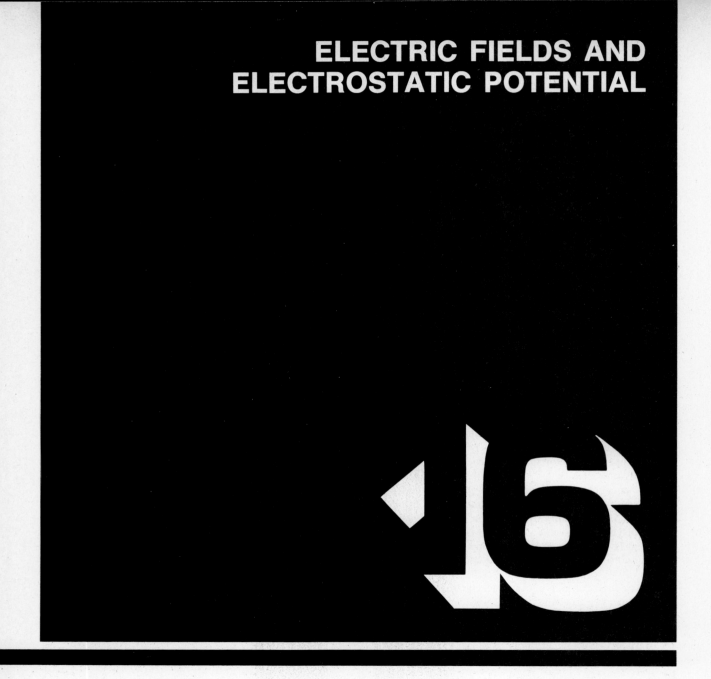

ELECTRIC FIELDS AND ELECTROSTATIC POTENTIAL

16.1 Introduction to Fields

The *point particle* concept has been used extensively in our discussions of classical mechanics. For example, the earth has been treated as a point object in discussing motion with respect to the sun, and astrophysicists in some instances regard entire galaxies as point particles. Electrons, protons, and neutrons are likewise considered point particles for most purposes. One might wonder how it is possible to regard both the proton, of radius 10^{-13} cm, and the earth, 10^9 cm in diameter, as point objects. In the strictest possible sense such an object occupies at any instant only one point of space and has no volume. However, in physics, an object can be idealized and assumed to be a point object *if its size is irrelevant in the context*

of the problem at hand. An automobile may be thought of as a point particle when viewed from an airplane, but it certainly cannot be viewed as a point particle by the people who are inside it. An electron may be thought of as a point object when we are interested in its motion about the nucleus, and the earth may be treated as a point object when studying its orbit about the sun. Thus, although there is a strict mathematical definition of a point, the point particle concept is an *idealization* used extensively in physics, and there is usually no difficulty in deciding when this idealization can be used in any given situation.

Another idea widely used in physics, and equally important, is the concept of a *field*. The field, which we have discussed already to some extent in connection with gravitation, is a more difficult concept than the particle. The reader may, at this point, profitably review what has already been said about the field concept in connection with gravitation in the discussion following Eq. (8.5.9) in Chapter 8.

A field is a function or a set of functions representing, for example, each component of a vector, defined at all points in a given coordinate space, that associates a certain physical quantity with every point in that space. In physics, the space is usually three-dimensional real space defined by Cartesian coordinates (x, y, z), or, in some cases, a four-dimensional space in which the time is included as a fourth "coordinate."

The fields encountered in physics are many and varied. A very simple example of a field is provided by temperature. At every point in space there is a unique temperature at a given time t and, in principle, we can write a function $T(x, y, z, t)$, of four variables, which specifies this *temperature field*. This type of field is known as a *scalar* field since only a single number, the temperature, is assigned to each point in space. Any temperature field can be shown to satisfy certain mathematical equations and also to obey certain physical restrictions which, taken together, determine the temperature at every point and at every given time. It is quite evident in the case of the temperature field that no one has ever *calculated* $T(x, y, z, t)$ at all points on the earth's surface since there are an enormous number of factors which would have to be included in the equations. However, the *physical existence* of the temperature field is apparent, for we know that a temperature does exist at each point in space, and it can be *measured* experimentally, subject to the usual experimental errors involved in any observation. Indeed, the morning paper usually gives us a map of the experimentally determined temperature field over the United States every day.

Another example of a scalar field is the mass density $\rho(x, y, z)$ of a fluid substance at any fixed time. In principle, this may also be measured at any point in space. Pressure $P(x, y, z)$ is yet another example of a scalar field of physical interest. Can you think of or invent other examples of scalar fields?

In addition to scalar fields, there are fields known as vector fields. A *vector field* is defined by associating a *unique vector* with every point of space. One example of a vector field, namely, gravitation, has already been discussed in Chapter 8. With respect to a coordinate system fixed on earth, to each point(x, y, z) we can in this case assign a *vector* $\mathbf{g}(x, y, z)$. The totality of all such vectors defines the vector field. Ultimately, this field can be calculated by using Newton's law of universal gravitation, which defines the vector field in terms of the mass distribution that acts as its source.

When a fluid such as water is undergoing steady or streamline flow, every location in the fluid can be assigned a unique velocity vector which gives the magnitude and direction of the fluid velocity at that location. The totality of all such vectors $\mathbf{v}(x, y, z)$ comprises a vector field. When sound waves are propagating through a gas, these waves cause time-dependent molecular displacements; these displacements can be characterized by vectors and, therefore, form a vector field. Two very important vector fields, the *electric field* and the *magnetic field*, will be discussed extensively in this chapter and in succeeding ones. The laws of electricity and magnetism are best formulated using these vector fields as fundamental entities. At the beginning, the advantages of this point of view will not be very apparent, but they will become more visible as our understanding of electricity and magnetism develops. The field concept is illustrated in Fig. 16.1.

In addition to scalar and vector fields, there are other more complicated fields, which we shall not discuss here. Some of these fields are very important in quantum physics and cosmology and are studied by physicists engaged in research in quantum field theory and astrophysics.

How do we know that fields exist? To be convinced, we need to carry out an experiment to demonstrate the existence of the field. This implies detecting some physical effect caused by the field. In the case of gravity, you can hold an object at a point(x, y, z) and then release it. The object accelerates downward, and this acceleration substantiates the existence of the field. If you are more ambitious, you can actually measure the magnitude and direction of the field $\mathbf{g}(x, y, z)$ at any point. A field of temperature can be demonstrated by measuring the temperature locally. It is sometimes difficult or impossible to determine fields by carrying out measurements, since the field may have to be determined at hundreds or thousands of different points in space. In such cases, the mathematical equations that define the field frequently provide a convenient recipe for their direct calculation. The fields themselves, or the effect they produce, must, of course, ultimately be in accord with empirical observations.

Let us try to understand the utility of the field concept by using gravity as an example. A distribution of mass such as the earth establishes a field $\mathbf{g}(x, y, z)$ in all of space. If we then wish to find the gravitational force experienced by an object of mass m located at a point(x, y, z), this force is given by $m\mathbf{g}(x, y, z)$. A different mass will experience a different force. We have, therefore, a situation in which there is a mass distribution which causes a force of gravity to act on a mass m. In studying this interaction, it is extremely useful to view the distribution of mass within the earth as a *source distribution* that sets up a gravitational field which then interacts with the object in question to produce a force $m\mathbf{g}(x, y, z)$ upon

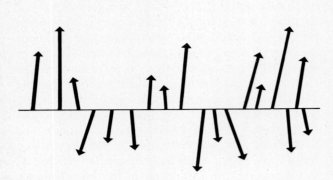

(a) To each point x along a line we can associate a vector **A**(x). Hence a vector field is obtained.

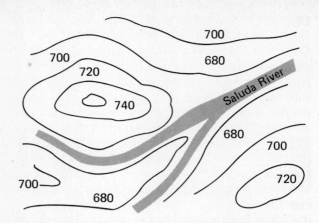

(b) To each location on a plot of land we can assign an altitude. This is a scalar field.

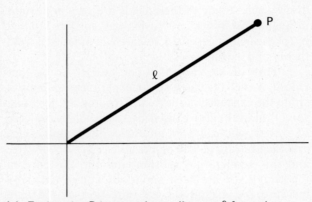

(c) Each point P in space has a distance ℓ from the origin. This is another scalar field.

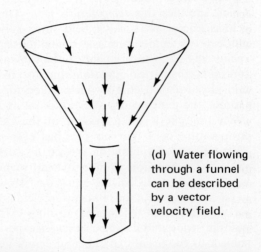

(d) Water flowing through a funnel can be described by a vector velocity field.

FIGURE 16.1. Several illustrations of the "field" concept. (a) An arbitrarily constructed vector field in which a vector **A**(x) is associated with every point on the x-axis. (b) Scalar field of altitudes above sea level, as illustrated by a contour map. (c) Scalar field in which the scalar distance r between the origin and any point in space is associated with that point. (d) Vector field representing the magnitude and direction of fluid velocity at every point for water flowing through a funnel.

it. This concept of a field as an intermediary between a source distribution and a system ultimately influenced by it is not only very useful for calculational purposes, but also, in the case of gravitation and electromagnetism, appears to coincide with physical reality.

In the case of electric fields, the *electric charges constitute the sources of the field.* Once the field is known, the force on an individual charged object can be determined. *Moving electric charges* can also establish *magnetic fields* in space, and these in turn can exert forces on charged objects in motion. We will spend considerable time studying how to cal-

culate electric and magnetic fields and the effects charged objects experience in the presence of these fields. At the outset, however, we shall consider only electric fields created by charges that are at rest. Such fields are referred to as *electrostatic fields*, and the description of how they are generated and calculated and how they act upon charges is referred to as the science of *electrostatics*.

It is often convenient, although not necessary, to have geometric pictures of fields. When we discuss electric and magnetic fields, these geometric constructs will be introduced and used wherever they help to clarify the physics.

16.2 The Electric Field

One of the most useful concepts in the theory of electromagnetism is that of the *electric field vector* **E**. At any given location, characterized by Cartesian coordinates (x, y, z), the electric field **E** is defined as the *electric force per unit charge*, namely,

$$\mathbf{E}(x, y, z) = \frac{\mathbf{F}(x, y, z)}{q} \qquad q \to 0 \tag{16.2.1}$$

and is given in units of newtons/coulomb (N/C) in the mks system of units. In this equation, **F** is the sum of all Coulomb forces exerted on the charge q by other charges or distributions of charge; and since each of these forces, according to Coulomb's law, *is itself proportional to q*, the electric field **E** is *independent* of q. In fact, it is presumed to exist independent of any attempt to measure it by observing the force which acts on an actual "test charge" q.

If we wanted to make an accurate measurement of **E** by determining the ratio of force to charge, it would frequently be necessary to use a test body having an infinitesimally small charge. If the charge were not infinitesimally small, it might itself influence the distribution of those source charges which are responsible for the original force **F**. This distortion might change the value of **F**, thus rendering the measurement unacceptable or inaccurate. One of the most important tasks of the experimentalist is to make sure that the act of measurement does not appreciably alter the quantity he is trying to measure. For example, you would not try to measure the temperature of a tiny cup of tea using a large thermometer that has been stored in the freezing compartment of a refrigerator, for you surely would get an incorrect result. Exactly the same reasoning applies to the act of measuring an electric field. The force **F** must, therefore, represent the force on an infinitesimally small "test charge" q and, in fact, should be evaluated in the limit as q approaches zero. In this limit, of course, the force **F** also approaches zero, but the quantity F/q remains finite, just as the derivative dy/dx does in the limit where Δy and Δx both become infinitesimally small.

The usefulness of the electric field stems from its action as an *intermediary* in the transmission of electrical forces from one or more charged bodies to others. Thus, a given set of charges can act as the *source* of an electric field; this electric field, which exists everywhere, then exerts a force on the charges that are located within it. We shall devote a great deal of effort to the task of learning how to calculate electric fields in a variety of circumstances. These will be illustrated initially in the examples which follow.

EXAMPLE 16.2.1

A positive point charge Q is located at the origin of a coordinate system. Find the electric field it produces.

This calculation can be done by first obtaining the force which a positively charged test body q would experience due to the charge Q. According to Coulomb's law, the force **F** acting on q will be given by

$$\mathbf{F} = \frac{kqQ}{r^2}\mathbf{i}_r = \frac{kqQ}{r^2}\frac{\mathbf{r}}{r} \tag{16.2.2}$$

where $\mathbf{i}_r$, which can also be written as $\mathbf{r}/r$, is a unit vector pointing radially away from Q. This is shown in Fig. 16.2. To obtain the electric field that exists at the location of the charge q, we use (16.2.1) to obtain

$$\mathbf{E} = \frac{\mathbf{F}}{q} = \frac{kQ}{r^2}\mathbf{i}_r = \frac{kQ}{r^2}\frac{\mathbf{r}}{r} \tag{16.2.3}$$

Since $\mathbf{i}_r = \mathbf{r}/r$ is a unit vector pointing from the charge Q toward the point of observation, the electric field points radially outward for this case. Moreover, it is spherically symmetric since the magnitude of the field depends *only* on the distance r from the origin. Thus, we see that a single point charge produces a spherically symmetric field pointing radially away from the charge, whose magnitude varies inversely as the square of the distance r. In this example, and in many others that are similar, the requirement that the "test charge" q be vanishingly small is extraneous; the same result for the field is obtained whatever the size of the test charge. There are other instances, however, in which this is not true, the correct result for the field being obtained only in the limit where q becomes vanishingly small.

It is very useful also to have a *geometric* way of representing electric fields. Electric fields can be visualized by drawing so-called *field lines*, or *lines of force*. Michael Faraday, a nineteenth-century physicist with great insight into electromagnetic phe-

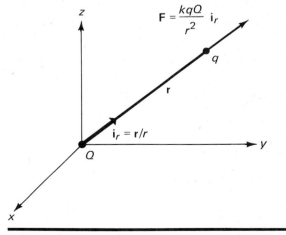

FIGURE 16.2. Electric field of a point charge.

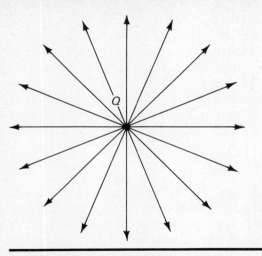

FIGURE 16.3. Radial field lines emanating from a point charge.

nomena, liked to use these lines to visualize pictorially how electric fields behave. Their construction is stated in a few simple rules:

1. Lines of force are drawn in the direction of the force a positive test charge would experience at every point in the field.

2. The density of the lines of force, as expressed by the number of lines crossing unit area perpendicular to their direction, is a measure of the magnitude of the electric field.

It should be noted that lines of force can never intersect. If they did, in what direction would the force on a charge be at the point of intersection? Since the force at any point can have only one direction, it is evident that field lines must never intersect. The lines of force due to a point charge are illustrated

in Fig. 16.3. These lines extend radially outward from the charge Q located at the origin of coordinates. The number of lines per unit area varies inversely as the square of the distance from the origin.

Once we know the electric field produced in all of space due to a single charge, as in the preceding example, or due to some distribution of charge, we can always calculate the Coulomb force exerted on any point charge q by using the equation $\mathbf{F} = q\mathbf{E}$.

Let us now consider the case where there are a number of point charges q_i located at various places, as shown in Fig. 16.4. We wish to calculate the electric field at observation point a. This is carried out by summing *as vectors* the electric fields which each individual charge would itself produce at point a. Thus the total electric field will be

$$\mathbf{E}(\mathrm{a}) = \sum_i \frac{kq_i}{r_{ai}^2} \frac{\mathbf{r}_{ai}}{r_{ai}} \qquad (16.2.4)$$

It is evident from this that individual point charge fields may be superposed in exactly the way in which forces on a point charge arising from the presence of other charges are superposed, as expressed by Eq. (15.3.6). In general, it may be very difficult to evaluate this vector sum in closed form since there could be many vectors to add. In special circumstances, in which symmetry can be exploited, the task of evaluating $\mathbf{E}$ can be much easier. Although (16.2.4) is written for a discrete set of charges $q_1, q_2, \ldots, q_n$, it can easily be extended to a continuous distribution of charge. In that case, the various charges become infinitesimal quantities dq whose size can be expressed as the product of the charge density, or charge per unit volume, which we shall write as $\rho(x, y, z)$, times the volume dV occupied by charge dq. The sum

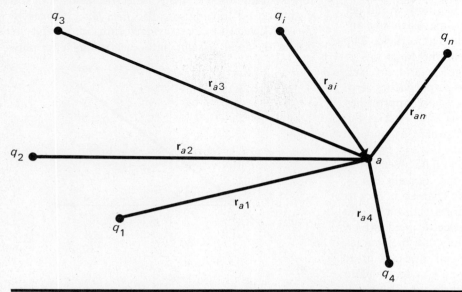

FIGURE 16.4. Calculation of the field of several point charges by vector addition.

appearing in (16.2.4) then becomes a sum of infinitely many infinitesimal contributions and can be written as an integral of the form

$$\boxed{\begin{aligned}
\mathbf{E}(a) &= \int \frac{k}{r_{ai}^2} \frac{\mathbf{r}_{ai}}{r_{ai}} \, dq \\
&= \int \frac{k\mathbf{r}_{ai}}{r_{ai}^3} \rho(x, y, z) \, dV
\end{aligned}} \tag{16.2.5}$$

Actually, (16.2.5) embodies three separate equations for the three separate components of the electric field. In many cases, however, it will not be necessary to carry out three separate integrations to determine **E**.

EXAMPLE 16.2.2

A thin, hollow spherical shell of radius R and total charge Q has a uniform density of charge on its surface. Obtain a mathematical expression for the electric field in terms of the distance r from the center of the sphere and make a plot of E versus r.

Problems such as this one involve the use of Eq. (16.2.5). The most important initial task is the selection of an *appropriate* infinitesimal charge element dq. Let us first do the calculation for the case of the observation point outside the charge distribution, as shown in Fig. 16.5a. The charge distribution is broken up into annular rings of charge such that the plane of each ring is perpendicular to the axis passing through the center of the sphere and the observation point a.

The electric field $d\mathbf{E}$ produced at point a due to the charge dQ is parallel to the line Oa and points away from the charge dQ if dQ is positive. The calculation of the magnitude dE is readily accomplished in exactly the same manner as the calculation of $F_{\parallel}$ in Chapter 15. In this case, in (15.3.13), we may replace Q', the charge on the ring, by dQ and the radius R by the quantity y illustrated in Fig. 16.5a, writing (15.3.13) in the form

$$dE = \frac{F_{\parallel}}{Q} = \frac{kx \, dQ}{(x^2 + y^2)^{3/2}} \tag{16.2.6}$$

It is now convenient to reexpress dE in terms of the angle θ shown in Fig. 16.5a. The variables y and x are easily expressed as

$$y = R \sin \theta \qquad x = r - R \cos \theta \tag{16.2.7}$$

while the charge dQ divided by the total charge Q is the same as the area of the charge element divided by the area of the entire sphere. That is,

$$\frac{dQ}{Q} = \frac{(2\pi R \sin \theta)R \, d\theta}{4\pi R^2} = \frac{\sin \theta \, d\theta}{2} \tag{16.2.8}$$

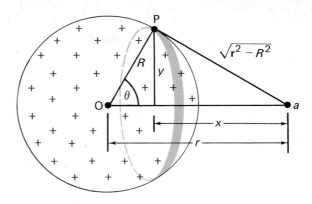

(a)

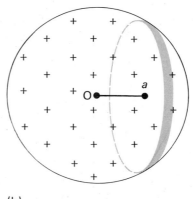

(b)

FIGURE 16.5

Equations (16.2.7) and (16.2.8) enable us to write (16.2.6) as

$$dE = \frac{k(r - R \cos \theta)Q \dfrac{\sin \theta \, d\theta}{2}}{(r^2 + R^2 - 2rR \cos \theta)^{3/2}} \tag{16.2.9}$$

To obtain the complete electric field, we now have to integrate over all the charged rings. With this subdivision into charged rings, it is evident that all the infinitesimal contributions $d\mathbf{E}$ must be parallel, and therefore the total electric field may be obtained by integrating (16.2.9) between $\theta = 0$ and $\theta = \pi$, since these angles encompass the entire charge distribution. To evaluate the above integral, we note first that since

$$2r(r - R \cos \theta) = (r^2 + R^2 - 2rR \cos \theta) + (r^2 - R^2)$$

we can write (16.2.9) as

$$\begin{aligned}
dE &= \frac{kQ}{4r} \frac{\left[(r^2 + R^2 - 2rR \cos \theta) + (r^2 - R^2)\right] \sin \theta \, d\theta}{(r^2 + R^2 - 2rR \cos \theta)^{3/2}} \\
&= \frac{kQ}{4r} \frac{\sin \theta \, d\theta}{\sqrt{r^2 + R^2 - 2rR \cos \theta}} \\
&\quad + (r^2 - R^2) \frac{\sin \theta \, d\theta}{(r^2 + R^2 - 2rR \cos \theta)^{3/2}}
\end{aligned} \tag{16.2.10}$$

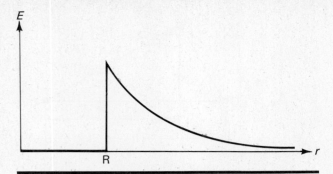

FIGURE 16.6. Electric field magnitude plotted as a function of the distance r from the center of the system, for a thin, uniformly charged spherical shell.

By setting $v = r^2 + R^2 - 2rR \cos \theta$ and $dv = 2rR \sin \theta$, these expressions can be integrated in a straightforward way, the result being

$$E = \frac{kQ}{4r} \left[\frac{1}{rR} \sqrt{r^2 + R^2 - 2rR \cos \theta} \right.$$

$$\left. - \frac{(r^2 - R^2)}{rR} \frac{1}{\sqrt{r^2 + R^2 - 2rR \cos \theta}} \right]_0^\pi$$

$$= \frac{kQ}{4r^2 R} \left[\sqrt{r^2 + R^2 - 2rR \cos \theta} \right.$$

$$\left. \times \left(1 - \frac{r^2 - R^2}{r^2 + R^2 - 2rR \cos \theta} \right) \right]_0^\pi$$

$$= \frac{kQ}{4r^2 R} (2R) \left[\frac{R - r \cos \theta}{\sqrt{r^2 + R^2 - 2rR \cos \theta}} \right]_0^\pi$$

$$= \frac{kQ}{2r^2} \left(\frac{R + r}{R + r} - \frac{R - r}{r - R} \right)$$

or finally

$$E = \frac{kQ}{r^2} \tag{16.2.11}$$

We see from all this that the electric field produced *outside* a spherical shell of charge turns out to be the same as the electric field which would have been produced by placing the entire charge Q at the origin! Although we have only proved this result for the special case of a charged spherical shell, it is also true more generally. *Any* spherically symmetric distribution of charge produces an electric field *outside* of the charge distribution identical with the field which would be produced if all the charge were placed at the center of the distribution.[1]

We have deliberately stressed the word *outside*, since the above conclusion is *not true* when the ob-

[1] The correctness of this assertion is easily understood by noting that any spherically symmetric distribution of charge can be regarded as being composed of a large number of thin, concentric spherical shells each of which bears a uniform density of charge.

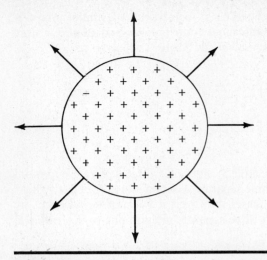

FIGURE 16.7. Radial field lines outside a thin, uniformly charged spherical shell.

servation point is in the *interior* of the charge distribution. To see this, we must examine the second part of our problem in which we suppose that $r < R$ as in Fig. 16.5b. The calculation is identical to the one carried out above, except for the very important difference that $(r^2 + R^2 - 2rR)^{1/2}$ is now equal to $R - r$, and therefore the result is

$$E = \frac{kQ}{2r^2} \left(\frac{R + r}{R + r} - \frac{R - r}{R - r} \right) = 0 \tag{16.2.12}$$

According to this, the electric field anywhere *inside* a charged spherical shell with uniform charge density is *identically zero*. The electric field is, therefore, *discontinuous* at the boundary.[2] A plot of electric field versus distance r is shown in Fig. 16.6.

This calculation is a long and arduous one. We will see later, after having studied Gauss's law, that the very same result can be arrived at in a much simpler way by exploiting the symmetry inherent in the problem. In the present problem, the electric field may easily be visualized in terms of lines of force. There are no lines of force present inside the sphere since the electric field is zero there. The lines originate at the surface of the sphere and extend radially outward, as illustrated in Fig. 16.7.

*EXAMPLE 16.2.3
THE FIELD OF AN ELECTRIC DIPOLE
Consider two point charges $+Q$ and $-Q$ separated by a distance d. Such a configuration of equal and

[2] These results are analogous to those obtained in Chapter 8, where we found that the gravitational field outside a spherically symmetric mass shell is the same as that produced by a point mass located at the center of the shell, while the field anywhere inside the shell is zero. The similarity between the gravitational and electrostatic cases is accounted for by the fact that an inverse square force law is involved in both instances.

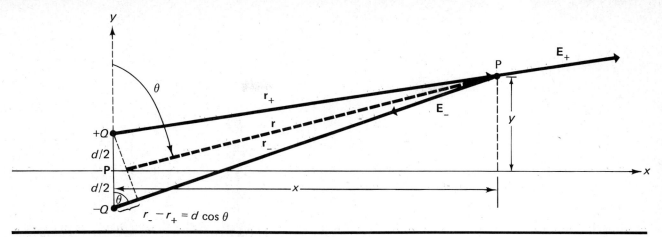

FIGURE 16.8. Calculation of the electric field of a dipole.

opposite charges is called an *electric dipole*. Compute the electric field produced by the dipole at points distant in comparison to the size of the dipole. Consider only the field in two dimensions.

The situation is illustrated in Fig. 16.8. The center of the dipole is at the origin, while point P is some arbitrary observation point at which we would like to compute the electric field **E**. The vector from $+Q$ to P is denoted by $\mathbf{r}_+$, while that from $-Q$ to P is $\mathbf{r}_-$. The vector from the center of the dipole to P is **r**. The electric field at P is a superposition of the electric fields which each of the charges produce at P. It is, therefore, given by

$$\mathbf{E} = \mathbf{E}_+ + \mathbf{E}_- = \frac{kQ\mathbf{r}_+}{r_+{}^3} + \frac{k(-Q)\mathbf{r}_-}{r_-{}^3} \qquad (16.2.13)$$

Let us examine this sum in greater detail by writing the total field in terms of its cartesian components E_x and E_y. We use the relations

$$\mathbf{r}_+ = x\mathbf{i}_x + \left(y - \frac{d}{2}\right)\mathbf{i}_y$$

$$\mathbf{r}_- = x\mathbf{i}_x + \left(y + \frac{d}{2}\right)\mathbf{i}_y$$

$$r_+{}^3 = \left[\left(y - \frac{d}{2}\right)^2 + x^2\right]^{3/2} \qquad (16.2.14)$$

$$r_-{}^3 = \left[\left(y + \frac{d}{2}\right)^2 + x^2\right]^{3/2}$$

to obtain the components

$$E_x = kQ \frac{x}{\left[\left(y - \frac{d}{2}\right)^2 + x^2\right]^{3/2}} - \frac{x}{\left[\left(y + \frac{d}{2}\right)^2 + x^2\right]^{3/2}}$$

and

$$E_y = kQ \frac{y - \dfrac{d}{2}}{\left[\left(y - \dfrac{d}{2}\right)^2 + x^2\right]^{3/2}} - \frac{y + \dfrac{d}{2}}{\left[\left(y + \dfrac{d}{2}\right)^2 + x^2\right]^{3/2}}$$

$$(16.2.15)$$

If we imagine that d is very small in comparison to the distance of the dipole from the point P, then we see that the components E_x and E_y are both very nearly zero. In other words, from very far away, there is almost complete cancellation of the electric fields produced by each of the charges. This is reasonable, in view of the fact that at a large distance the dipole presents the appearance of equal and opposite charges located at nearly (but not quite) the same point. Since Eqs. (16.2.15) actually approach zero as d vanishes, it is customary to make a series expansion of E_x and E_y and to retain only those terms which are linear in d. We make use of the following relations, which are valid for small d:

$$\frac{1}{\left[\left(y \pm \dfrac{d}{2}\right)^2 + x^2\right]^{3/2}} \cong \frac{1}{[x^2 + y^2 \pm yd]^{3/2}}$$

$$= \frac{1}{(x^2 + y^2)^{3/2}\left(1 \pm \dfrac{yd}{x^2 + y^2}\right)^{3/2}}$$

$$\cong \frac{1}{(x^2 + y^2)^{3/2}}\left[1 \mp \frac{3}{2}\frac{yd}{x^2 + y^2}\right].$$

$$(16.2.16)$$

In this expression, the quantity $d^2/4$ is neglected in expanding $(y \pm \frac{1}{2}d)^2$, while the final form results by the binomial expansion

$$(1 \pm b)^{-3/2} = 1 \pm \left(-\frac{3}{2}b\right) + \frac{1}{2!}\left(-\frac{3}{2}\right)\left(-\frac{5}{2}\right)b^2 + \cdots$$

in which, since b is much less than unity if d is small, terms in b^2, b^3, etc., are neglected. Let us now sub-

491

stitute the relation (16.2.16) into Eq. (16.2.15), dropping terms that contain d^2 whenever they arise:

$$E_x = kQ\left[\frac{x}{(x^2+y^2)^{3/2}}\left(1+\frac{3}{2}\frac{yd}{x^2+y^2}\right)\right.$$

$$\left.-\frac{x}{(x^2+y^2)^{3/2}}\left(1-\frac{3}{2}\frac{yd}{x^2+y^2}\right)\right]$$

$$=\frac{kQd}{(x^2+y^2)^{3/2}}\left[\frac{3xy}{x^2+y^2}\right]$$

$$=\frac{kQd}{(x^2+y^2)^{3/2}}\left[\frac{3xy}{x^2+y^2}\right]$$

$$E_y = kQ\left[\left(y-\frac{d}{2}\right)\frac{1}{(x^2+y^2)^{3/2}}\left(1+\frac{3}{2}\frac{yd}{x^2+y^2}\right)\right.$$

$$\left.-\left(y+\frac{d}{2}\right)\frac{1}{(x^2+y^2)^{3/2}}\left(1-\frac{3}{2}\frac{yd}{x^2+y^2}\right)\right]$$

$$=\frac{-kQd}{(x^2+y^2)^{3/2}}\left[1-\frac{3y^2}{x^2+y^2}\right] \qquad (16.2.17)$$

It is often convenient to rewrite the Cartesian components in terms of the polar coordinates r and θ as shown in Fig. 16.8. Furthermore, it is customary to define an *electric dipole vector* which points from the negative to the positive charge; it is denoted by **p**. The magnitude of this vector is defined to be the charge Q multiplied by the charge separation d; that is, $p = |\mathbf{p}| = Qd$. This vector is frequently referred to as the *dipole moment* of the dipole. In terms of these definitions, we rewrite the field produced when $r \gg d$ as

$$E_y = -\frac{kp}{r^3}(1-3\cos^2\theta)$$

$$E_x = \frac{kp}{r^3}(3\cos\theta\sin\theta)$$

A rough sketch of the field lines which are produced by an electric dipole is shown in Fig. 16.9.

One might ask whether there are any real systems in nature that act like electric dipoles and therefore produce the type of field calculated above. It turns out that there are many examples of atoms and molecules in which there is a spatial separation of the positive and negative charges and whose electric field is essentially that of the point dipole discussed in this example. The hydrogen molecule provides one such example. Although the separation of charges is more complicated than that of the simple dipole discussed above, at large enough distances the electric fields are very nearly approximated by those given in (16.2.17). In addition to naturally occurring dipoles, there are also electric dipoles which can be created when atomic systems are subjected to the influence of an external electric field. For example, the presence of an electric

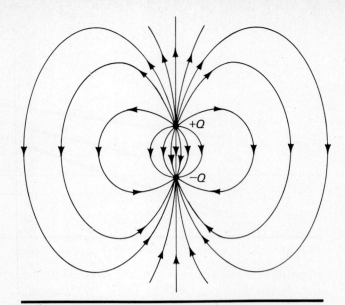

FIGURE 16.9. Electric field of an electric dipole.

field can bring about a separation of positive and negative charges in an insulating material. This manifests itself by establishing within the material many microscopic atomic dipoles which in turn produce their own electric fields. This phenomenon is particularly important in the study of devices known as capacitors (a subject which we will return to later) and is, therefore, of interest in the theory of electric circuits.

It is particularly important to note that the electric field produced by an electric dipole has a $1/r^3$ dependence. This should be contrasted with the $1/r^2$ dependence of the field produced by a single point charge. The strength of the field due to a dipole falls off more rapidly than that of a point charge because of the near-cancellation of the fields produced by the individual charges that constitute the dipole.

16.3 Gauss's Law

One of the most useful and important laws in electricity is known as Gauss's law. This law is not an independent principle but can be shown to be a consequence of Coulomb's law. Before we discuss the law and its many applications in detail, we must first introduce a few basic preliminaries.

The first of these is the important mathematical idea of a *surface integral*. Let us suppose that we have a function $f(x, y, z)$ that we can evaluate at every point on some given surface S. Let us now subdivide the surface into very small area elements. For every such element, we can calculate the product of the function f and the area Δa; for the ith area element, Δa_i, this product will be $f_i \Delta a_i$, where f_i is the value

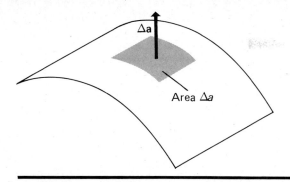

FIGURE 16.10. Definition of the vector area element associated with an infinitesimal surface area.

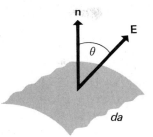

FIGURE 16.11. Geometry of field vector and surface normal used in defining electric flux.

of the function f at this ith element. The sum of products of f_i and Δa_i defines an integral when we allow the number of points to become infinite and also allow the areas to approach zero. Thus, the integral defined by

$$\oint_S f \, da = \lim_{\substack{N \to \infty \\ \Delta a_i \to 0}} \sum_{i=1}^{N} f_i \, \Delta a_i \qquad (16.3.1)$$

is called a *surface integral*, of the function f over the surface S, the symbol $\oint$ denoting integration over a closed surface.[3] If the function f has the value unity everywhere, then the surface integral is simply the total area of the surface. If, on the other hand, the function f is a variable such as the electric charge density (charge per unit surface area), then the surface integral gives the total charge on the surface. Surface integrals of this type occur frequently in physics, and it is important to learn how to evaluate them. This is a subject which is usually studied in elementary calculus; we assume that the student already has some familiarity with the basic idea of surface integration.

If we have some given vector field defined in all space, then at *each point on any surface we draw*, we can find a definite field vector. A small area element on the surface can also be thought of as an infinitesimal vector, as illustrated in Fig. 16.10. Its magnitude is given by the area Δa and its direction is *that of the normal* to the surface. Since the normal to an infinitesimal surface can point in two possible directions, we must resolve any possible ambiguity by adopting an appropriate convention. In the cases of interest to us, we are going to be calculating surface integrals over *closed* surfaces that define the boundary of some volume. We shall always take the normal to be the *outward normal*, that is, the normal that points away from the volume enclosed by the surface.

We may, therefore, write the vector characteriz-

ing an infinitesimal area da as $d\mathbf{a} = \mathbf{n} \, da$, where $\mathbf{n}$ is the *outward normal unit vector* and da is the infinitesimal area. Now at each element, suppose there exists a field such as an electric field $\mathbf{E}$. Then, to each area element we can associate a scalar quantity obtained by evaluating the scalar, or dot, product of $\mathbf{E}$ and $d\mathbf{a}$. In this way, as illustrated in Fig. 16.11, we can define a quantity $d\Phi$ known as the *differential electric flux*, given by

$$d\Phi = \mathbf{E} \cdot d\mathbf{a} = \mathbf{E} \cdot \mathbf{n} \, da \qquad (16.3.2)$$

The entire flux Φ through a closed surface S may then be obtained by adding up the contributions from all the infinitesimal areas of the surface. Thus, the flux is given by

$$\Phi = \oint_S \mathbf{E} \cdot \mathbf{n} \, da \qquad (16.3.3)$$

In this expression, it is important to note that $\mathbf{n}$ represents a vector *of unit magnitude* normal to the surface.

The flux of a vector field through a given surface is a concept that has found widespread use in all of physics. It is used in gravitation, in which case the field $\mathbf{g}$ replaces $\mathbf{E}$. It is used in fluid flow,[4] where the fluid velocity is used in place of $\mathbf{E}$. In heat conduction, the vector which characterizes the magnitude and direction of heat flow would be used in place of $\mathbf{E}$ in (16.3.3).

Let us consider now the calculation of the electric flux through any imaginary closed surface that surrounds a point charge Q. We assume that the electric field $\mathbf{E}$ is due *only* to the charge Q. The hypothetical closed surfaces we use are usually called

---

[3] Instead of the symbol $\oint$, which denotes integration over a closed surface, we could also have $\int$, which implies integration over a surface that is not necessarily closed.

[4] It is the case of fluid flow that prompted the terminology *flux* in the first place. In this instance, the flux integral (16.3.2) represents the volume of fluid per unit time that flows through the surface S. In *electric* flux (or *magnetic* flux, which we shall encounter later), there is nothing that is actually *flowing* across the surface in the usual sense of the word, and in these instances the terminology is somewhat misleading.

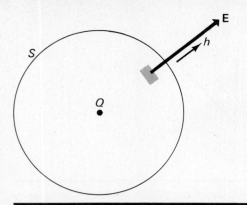

FIGURE 16.12. Field vector and surface normal for a point charge surrounded by a concentric spherical Gaussian surface.

Gaussian surfaces; it must be kept in mind that they are not real surfaces but merely mathematical constructs.[5] Thus, we may surround the charge Q by any Gaussian surface we wish; but, for the present, let us choose a spherical surface such that Q is at the center of a sphere of radius r. We break up the spherical surface into infinitesimal areas, as shown in Fig. 16.12. At each of these surfaces, the unit normal **n** points radially outward from the center of the sphere. Under these circumstances, the normal vector **n** is the unit radial vector $\mathbf{i}_r$. Moreover, we have already seen that the electric field **E** produced by a single point charge Q is given by

$$\mathbf{E} = \frac{kQ\mathbf{i}_r}{r^2} \qquad (16.3.4)$$

In other words, **E** also points radially outward. Therefore, the dot product of **E** and da is simple to evaluate:

$$\mathbf{E} \cdot \mathbf{n}\, da = \frac{kQ}{r^2} da(\mathbf{i}_r \cdot \mathbf{i}_r) = \frac{kQ}{r^2} da \qquad (16.3.5)$$

To obtain the complete flux, we now have to integrate over the entire Gaussian surface. Since this Gaussian surface is spherical, with a fixed radius r, we find a total flux

$$\Phi = \oint_S \mathbf{E} \cdot \mathbf{n}\, da = \oint_S \frac{kQ}{r^2} da = \frac{kQ}{r^2} \oint_S da$$

$$\Phi = \frac{kQ}{r^2} 4\pi r^2 = 4\pi kQ \qquad (16.3.6)$$

In this integration, the constant radius r can be written outside the integral. The factor $4\pi k$ appears very frequently in electric flux calculations and elsewhere in electrostatic theory. It is convenient, therefore, to

[5] The solution of many problems in physics is made much easier by taking full advantage of any symmetry which may be present. This is true in classical mechanics, quantum mechanics, electromagnetism, and almost every other subfield of physics.

replace the factor $4\pi k$ with a new constant ε_0 given by

$$4\pi k = \frac{1}{\varepsilon_0} \qquad (16.3.7)$$

The constant ε_0 has the value

$$\varepsilon_0 = 8.85418 \times 10^{-12}\ \text{C}^2/\text{N-m}^2$$

and is often called the *permittivity constant of free space.*

We have found that the net electric flux through a sphere of radius r with a positive charge Q at the center is

$$\Phi = \frac{Q}{\varepsilon_0} \qquad (16.3.8)$$

It is remarkable that this result turns out to be *independent of the radius* of the Gaussian sphere. This independence arises only because of the precise inverse square dependence of the electric field as given by Coulomb's law. As r increases, E decreases as $1/r^2$, but the area increases as r^2. The combined decrease of E and increase of area leads to a flux which is independent of the size of the sphere.

To understand this result intuitively, it is useful to think of the flux through a closed surface as a net outflow of "something." We recall that the strength or magnitude of the electric field can be represented geometrically in terms of the number of field lines per unit area. In the calculation of flux, we are multiplying the number of field lines per unit area by an area, and therefore the electric flux through a surface is proportional to the total number of field lines passing through the surface. Now suppose we surround the point charge Q by some arbitrary *nonspherical* Gaussian surface S', as shown in Fig. 16.13. A mathematical proof can be given to show that the flux $\Phi_{S'}$ passing

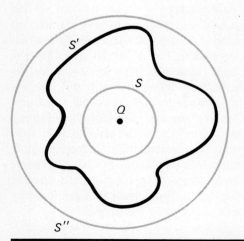

FIGURE 16.13. Field lines that intersect both spherical surfaces S and S'' must also pass through an arbitrary nonspherical surface S' that lies between them.

through an arbitrary surface S' is exactly the same as the flux passing through a sphere surrounding the charge, that is, it also has the value Q/ε_0. The strict mathematical proof of this statement is somewhat involved, and we shall postpone its proof to a brief section at the end of the chapter. In the meantime, we shall justify this result on the basis of our intuitive picture of field lines. Referring to Fig. 16.13 once again, we see that the number of field lines passing through spheres S and S'' as well as through the surface S is the same, and therefore we expect the flux to be identical through *all* of these surfaces. We have already seen mathematically that the flux through sphere S is the same as the flux through sphere S''. Since flux represents a "flow" of field lines, it is reasonable to suppose that the flux through any closed surface such as S, which is between S and S', is also the same. This is in fact true, and we conclude that *the flux through any closed surface surrounding a single point charge Q is Q/ε_0.*

Now, suppose we have a single point charge which is *outside* some closed Gaussian surface. What electric net flux does this charge, by virtue of the electric field it creates, produce through the surface? The answer is, exactly *zero*, because *the number of field lines entering the volume surrounded by the surface is identical to the number of field lines emerging*, as shown in Fig. 16.14. Thus, the net flow of field lines is zero.

At this point the preceding statements can easily be generalized to the case of many point charges or to a continuous charge distribution. This follows because the total electric field produced by many point charges is equal to the sum of the electric fields produced by the individual charges. Consider then the total electric flux through some closed surface, as shown in Fig. 16.15. Assume that the closed Gaussian surface en-

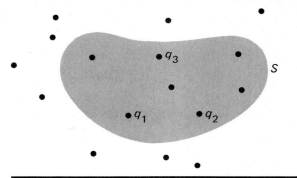

FIGURE 16.15. Only charges within the closed surface S contribute to the electric flux through it.

closes charges $q_1, q_2, \ldots, q_n$ but that charges $q_{n+1}, q_{n+2}, \ldots, q_m$ are *outside* the Gaussian surface. The total flux through S is then given by

$$\Phi_S = \oint_S \mathbf{E} \cdot \mathbf{n} \, da \qquad (16.3.9)$$

and the total electric field $\mathbf{E}$ at any point on the surface is simply the sum of the electric fields produced by the individual charges. Thus,

$$\Phi_S = \oint_S \mathbf{E}_1 \cdot \mathbf{n} \, da + \oint_S \mathbf{E}_2 \cdot \mathbf{n} \, da$$
$$+ \cdots \oint_S \mathbf{E}_i \cdot \mathbf{n} \, da + \cdots \qquad (16.3.10)$$

Any given term in this sum represents the electric flux through the Gaussian surface S due to the presence of a single charge. We have already seen that if a given charge q_a is *within* the Gaussian surface, the electric flux it produces is given by q_a/ε_0, while if the charge is *outside* the Gaussian surface, the contribution is zero. Thus we find, referring to Fig. 16.15, that

$$\Phi_S = \oint_S \mathbf{E} \cdot \mathbf{n} \, da = \frac{1}{\varepsilon_0}(q_1 + q_2 + q_3 + \cdots + q_n)$$

$$(16.3.11)$$

Let us now formulate in words the above statement, which is *Gauss's law:*

Gauss's Law Given any distribution of charges, discrete or continuous, the total electric flux produced by these charges through *any closed Gaussian surface S* is related to the total charge *inside* the Gaussian surface by the equation

$$\oint_S \mathbf{E} \cdot \mathbf{n} \, da = \frac{q_0}{\varepsilon_0} \qquad (16.3.12)$$

where $\mathbf{E}$ is the electric field produced by *all the charges*, those inside as well as those outside, and q_0 is the total charge contained *within* the Gaussian surface.

It should be kept firmly in mind that while the right-hand side of Eq. (16.3.12) is determined *only* by the *charges inside* the Gaussian surface, the left-hand

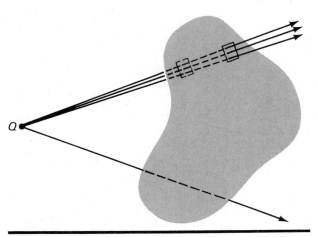

FIGURE 16.14. Flux through a closed surface from a charge outside the surface is zero, since all field lines entering the surface must eventually leave it.

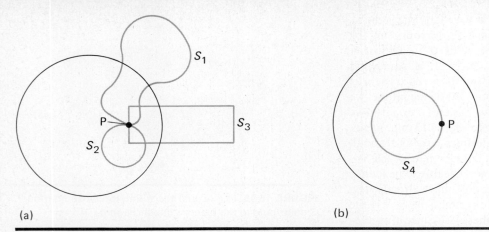

(a)

(b)

FIGURE 16.16

side involves the *total electric field* and, therefore, includes the electric fields contributed by charges outside the Gaussian surface as well as the electric fields produced by charges inside. It is often easy to misunderstand the statement or the application of Gauss's law by failing to appreciate the logic of the above statement. It is also important to realize that Gauss's law is a statement about the electric flux through a *closed* surface S. Endless confusion and mistakes can arise by trying to use Gauss's law with Gaussian surfaces that are not closed!

The reader may be wondering at this point why we have introduced Gauss's law and what role it plays in the solution of problems. In the arguments which lead to Eq. (16.3.12), no new physics was introduced. Thus, the statement of Gauss's law is only a mathematical consequence of the application of Coulomb's law for many charges. Now, if we have some distribution of discrete charges, then the electric field at any given point is *always* given by (16.2.4) as

$$\mathbf{E}(x, y, z) = \sum_{i=1}^{n} \frac{Q_i}{4\pi\varepsilon_0} \frac{\mathbf{r}_{ai}}{r_{ai}^3} \qquad (16.3.13)$$

In principle, (16.3.13) can always be used to calculate the electric field, even in cases of a continuous charge distribution. In practice, however, the calculation may be *very difficult* since it involves a summation of many vectors. Using these same charges and an arbitrary closed Gaussian surface, we have obtained (16.3.12), which involves the electric field at many points on this arbitrary surface. Can this equation be used instead of (16.3.13) to obtain the electric field at a single point (x, y, z), and, if so, how can this be accomplished?

The answer to this question depends upon the specific problem under consideration, for although (16.3.12) is always true in electrostatics, it is only useful in a limited number of cases. The problem *must possess enough symmetry* so that by a judicious choice of a Gaussian surface we can, in a sense, carry out the integration implied in (16.3.12) without knowing the

value of $\mathbf{E}$. In fact, we shall soon see that it is only *after* we have carried out this integration that we learn the value of $\mathbf{E}$ at any given point. We shall now discuss a number of examples which will hopefully serve to clarify some of the preceding remarks and to demonstrate the basic technique for calculating electric fields by using Gauss's law.

EXAMPLE 16.3.1

Calculate the electric field of Example 16.2.2 using Gauss's law.

In Fig. 16.16a, three Gaussian surfaces are drawn which pass through a point P at which we would like to calculate the electric field. Now according to Gauss's law,

$$\oint_{S_2} \mathbf{E} \cdot \mathbf{n} \, da = 0$$

$$\oint_{S_1 \text{ or } S_2} \mathbf{E} \cdot \mathbf{n} \, da = \frac{1}{\varepsilon_0} (q_0)_{1 \text{ or } 3} \qquad (16.3.14)$$

for each of these cases. In this expression, $(q_0)_{1 \text{ or } 3}$ represents the total charge included inside Gaussian surfaces 1 or 3, which, in either case, we do not know. The three surfaces chosen are very different geometrically, but they have one common characteristic— *none of them is useful for working out the electric field at point P.*

Now, let us look at Fig. 16.16b, in which we have chosen a *spherical* Gaussian surface passing through P, which is concentric with the charge distribution. Two facts emerge from consideration of the *symmetry* of this charge distribution: (a) the *magnitude* of the electric field must depend only on r, the distance from the center of the sphere, and (b) the *direction* of the electric field must be radial.

According to Gauss's law, for the interior surface S_4 we have

$$\oint_{S_4} \mathbf{E} \cdot \mathbf{n} \, da = 0 \qquad (16.3.15)$$

since there is no charge inside this Gaussian surface. Now, at any point on this surface, $\mathbf{E} \cdot \mathbf{n}$ is simply E, since the vector $\mathbf{E}$ is *always parallel* to $\mathbf{n}$. Moreover, since the magnitude E has the same value at all points on the sphere,

$$\oint_{S_4} \mathbf{E} \cdot \mathbf{n} \, da = \oint_{S_4} E \, da = E \oint_{S_4} da = 4\pi r^2 E = 0$$

Thus, $4\pi r^2 E$ vanishes, which implies that $E = 0$. We find, therefore, that the electric field at point P within the shell of charge is zero. Since the same calculation could have been carried out for *any* point inside the hollow shell, it follows that the electric field vanishes everywhere within the shell. This result agrees with that obtained earlier in Example 16.2.2. We see that in order to apply Gauss's law successfully to this problem, in which there is spherical symmetry, we had to exploit that symmetry by selecting an appropriate Gaussian surface. Arbitrarily chosen Gaussian surfaces are of no use in this problem.

To continue the solution of this example, let us calculate the electric field outside the distribution of charge. We again choose a spherical Gaussian surface passing through point P', as shown in Fig. 16.17. According to Gauss's law, the total flux is now Q/ε_0, where Q is the total charge on the shell. Therefore,

$$\oint \mathbf{E} \cdot \mathbf{n} \, da = 4\pi r^2 E = \frac{Q}{\varepsilon_0}$$

$$E = \frac{Q}{4\pi \varepsilon_0 r^2} \tag{16.3.16}$$

This result also agrees with that previously obtained. The reader should contrast the two methods of calculating the electric field, both of which lead to exactly the same result. The calculation by means of Gauss's law is much simpler, although to use it we had to invoke the symmetry inherent in the problem's geometry.[5]

In Example 16.2.3, the electric field produced by a point dipole was calculated by using (16.3.13). Is it possible to obtain the same result by using Gauss's

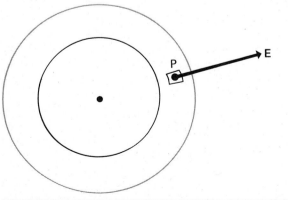

FIGURE 16.17

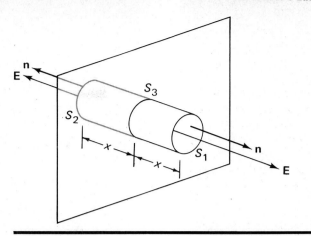

FIGURE 16.18

law? The answer is unequivocally no, because it just isn't possible to select a Gaussian surface which will permit a useful evaluation of the surface integral and thereby provide the answer we need. The reader should verify this by choosing any Gaussian surface he wishes. None will work. There is not enough inherent geometric symmetry in this situation to allow us to make any meaningful use of Gauss's law, although, to be sure, the law is certainly valid for any Gaussian surface we might choose.

EXAMPLE 16.3.2

A thin infinite sheet of positive charge bears a uniformly distributed charge density σ per unit area. Find the electric field as a function of distance from the sheet.

This type of problem could be done by using (16.3.13) for the case of a continuous distribution of charge, but it is much easier to solve it by Gauss's law. Due to the symmetry of the charge distribution, the electric field on both sides of the sheet must surely be perpendicular to the sheet, and, moreover, at a given distance its magnitude should be the same on the right and on the left. As a matter of fact, since the sheet is presumed to be truly infinite in extent, the electric field should not depend on any variable other than possibly the distance from the sheet. Now, let us choose a cylindrical Gaussian surface as shown in Fig. 16.18. This surface has circular faces S_1 and S_2 and a cylindrical portion S_3. The surfaces S_1 and S_2 are deliberately chosen to be the same distance x from the sheet. The particular choice of a circular cylindrical Gaussian surface is not crucial to the arguments which follow, as the reader can readily verify.

Let us now calculate the total electric flux through our Gaussian surface. We can write the flux as a sum of three terms:

$$\oint_S \mathbf{E} \cdot \mathbf{n} \, da = \oint_{S_1} \mathbf{E} \cdot \mathbf{n} \, da + \oint_{S_2} \mathbf{E} \cdot \mathbf{n} \, da + \oint_{S_3} \mathbf{E} \cdot \mathbf{n} \, da \tag{16.3.17}$$

This simply expresses the fact that the flux through the entire closed surface can be written as a sum of contributions from the various portions of the surface. The contribution from S_2 is zero; on that surface, the scalar product of $\mathbf{E}$ and $\mathbf{n}$ vanishes because $\mathbf{E}$ is perpendicular to the plane of the charge distribution and is, therefore, also perpendicular to the normal anywhere on S_2. On the faces S_1 and S_3, the electric field is constant in magnitude and parallel to the normal. By symmetry, E on S_1 is the same as E on S_3 because both surfaces are the same distance from the charge distribution. The total flux is therefore $Ea + 0 + Ea$, or $2Ea$, where a is the area of surface S_1 or S_3. Now, according to Gauss's law, the net flux is also obtained as q_0/ε_0, where q_0 is the net charge contained within the Gaussian surface. In the present case, q_0 is given by σa, where σ is the surface charge density. Thus, according to Gauss's law,

$$2Ea = \frac{\sigma a}{\varepsilon_0}$$

or

$$E = \frac{\sigma}{2\varepsilon_0} \qquad (16.3.18)$$

We thus obtain the rather interesting result that the electric field does not depend at all on the distance from the sheet but only on the charge density on the sheet.

In realistic situations, of course, there cannot be any truly infinite sheet of charge and one might think, therefore, that the result is purely academic. However, the conclusion reached above certainly suggests that (16.3.18) should be a reasonable approximation to the electric field prevailing close to a large but finite sheet of charge. Some of the problems at the end of the chapter will illustrate this point in greater detail.

EXAMPLE 16.3.3
A long, straight wire bears a uniformly distributed positive charge of density per unit length λ. Calculate the magnitude of the electric field in terms of the distance r from the wire.

In this example, the perfect cylindrical symmetry suggests that the electric field should be perpendicular to the wire and that it should depend only on the distance r from the wire. The geometry is illustrated in Fig. 16.19, in which an appropriate *cylindrical* Gaussian surface has also been drawn. This Gaussian surface consists of circular end faces S_1 and S_3 and a curved cylindrical portion S_2. Now S_1 and S_3 are surfaces in which the normal is perpendicular to the direction of the $\mathbf{E}$ field; therefore, there is no contribution to the flux from either S_1 or S_3. On the other hand, for any point on S_2, the electric field, which is

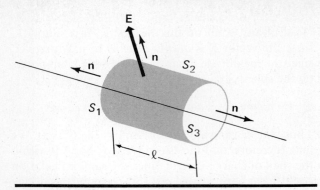

FIGURE 16.19

constant in magnitude on this entire surface, is *parallel* to the normal vector $\mathbf{n}$. The total flux, therefore, comes from surface S_2 and is readily seen to be

$$\Phi = \oint_{S_2} \mathbf{E} \cdot \mathbf{n}\, da = \int E\, da = E \int da = E(2\pi rl) \qquad (16.3.19)$$

According to Gauss's law, this flux is related to the net charge inside the Gaussian surface. Therefore,

$$E(2\pi rl) = \frac{q_0}{\varepsilon_0} = \frac{\lambda l}{\varepsilon_0}$$

or

$$E = \frac{\lambda}{2\pi\varepsilon_0 r} \qquad (16.3.20)$$

The electric field produced by a line of charge, therefore, falls off as the inverse *first* power of the distance from the charge.

16.4 Applications of Gauss's Law to Conductors and Insulators

Materials are classified as conductors or insulators depending on the ease with which electrons can move in response to an applied electric field. The electrons in a good conductor such as copper are essentially free and can migrate easily whenever fields are present. On the other hand, in an insulator such as quartz, the application of an electric field does not set charges in motion. In such a material, the electrons are strongly bound to atomic nuclei and, therefore, the applied field only causes some stretching or distortion of chemical bonds. Many materials are neither good conductors of electricity nor good insulators, but rather fall somewhere between the two extremes.

In this section, we shall discuss the application of Gauss's law to conductors and insulators by studying a number of examples in which electric fields are calculated for a variety of geometric arrangements.

In the case of any *conducting* object that is in equilibrium in an external electrostatic field, *the electric field everywhere within the object must be zero.* If this were not the case, free charges within the conductor would experience forces which would set them in motion, causing currents to flow within the substance, which contradicts the assertion that the substance is in a condition of equilibrium. This statement, used in conjunction with Gauss's law, allows us to learn much about electric fields and charge distributions on conductors. The conclusions that are to be drawn can be summarized in three general statements about the behavior of charge distributions and electric fields associated with conducting bodies. They are: (a) any net charge possessed by a conducting object resides on the outside surface of the conductor; (b) the electric field at the surface of any conducting body is normal to the surface at every point; and (c) the electric field just outside the surface of a conductor is related to the surface charge density (charge per unit area) $\sigma(x, y, z)$ at every point by the relation $E = \sigma/\varepsilon_0$.

In Fig. 16.20, a solid conductor is shown. To prove statement (a), we construct a Gaussian surface within the conductor, arbitrarily close to the surface, as shown. Now, according to Gauss's law, the net flux passing through this Gaussian surface is proportional to the total charge contained within it. The total flux passing through this Gaussian surface must, however, be zero. The reason for this is that the electric field within the conductor must be identically zero when all the charges are at rest. If this were not so, the electrons within the conductor would experience accelerations and, therefore, we would not be dealing with a system whose charges are stationary. Since the electric field is zero within the conductor, it follows that no net charge can be contained within any Gaussian surface embedded in the conductor. Therefore, any excess charge must reside on the surface of the conductor. In the present case of a solid conductor, the only boundary is the single surface

which bounds the conducting volume; therefore, the total charge Q must be on that surface. Shortly, we shall consider conductors that have interior as well as exterior boundaries; under these conditions, the complete geometry must be analyzed before any conclusions about the charge distribution can be made.

On the outer boundary of the above conductor, there is a total charge Q. In general, this charge is not distributed uniformly over the surface and, therefore, a variable surface charge density $\sigma(x, y, z)$ describes the distribution of charge. This surface density may vary according to the geometry of the sample and the external field. It must, however, always be set up *in such a way as to guarantee the vanishing of the electric field within the conductor.* On the surface of the conductor, the electric field at any point must be *normal* to the surface. The reason is fairly simple. If the electric field had any component parallel to the surface, then the electric charges, which are free to move, would migrate along the surface. But in any situation of electrostatic equilibrium, all charges are *stationary* and, therefore, no such component of the electric field is possible. Hence, we can state that when the charges are at rest, *electric field lines are always perpendicular to conducting surfaces.*

Let us now calculate the electric field very close to the surface of a conductor. In Fig. 16.21, a small cylindrical Gaussian surface is drawn very near the point(x, y, z). The axis of the cylinder is perpendicular to the surface at the point(x, y, z). This Gaussian surface has ends S_1 and S_3 and a curved portion S_2. The direct computation of the flux out of this surface is elementary. On S_1, the electric field is exactly zero, since S_1 is within the conductor. Through S_2, the electric flux must be zero since the electric field, where it does exist, is perpendicular to the normal to S_2. Finally, on S_3, the electric field $\mathbf{E}$ is parallel to the normal and a contribution EA is obtained, where A is the area of surface S_3. The total charge contained

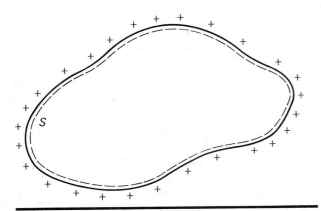

FIGURE 16.20. Excess charge on a solid conducting body must reside on its surface.

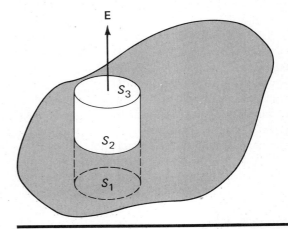

FIGURE 16.21. Gaussian surface used in calculating the electric field just outside a charged conducting body.

within this Gaussian surface is given by the charge density $\sigma(x, y, z)$ multiplied by the area A. Then, according to Gauss's law, we find $EA = \sigma A/\varepsilon_0$, or $E = \sigma/\varepsilon_0$. We conclude, therefore, that right at or slightly above the point(x, y, z) on a conducting surface, the electric field is given by

$$E(x, y, z) = \frac{\sigma(x, y, z)}{\varepsilon_0} \qquad (16.4.1)$$

If the local charge density is positive, the electric field points outward, away from the conductor; while if the charge density is negative, the field points inward.

One might wonder why the charges on the surface do not leave the conducting object altogether, since, after all, they experience an electric field which causes an electric force to be exerted on them. The answer to this question can be understood by observing that charges on the surface of a conductor encounter *nonelectrostatic* forces that maintain equilibrium in the direction perpendicular to the surface and that ultimately keep them from leaking off into the region outside the conductor.

EXAMPLE 16.4.1

A hollow conductor in the form of a thin spherical shell has a metallic lid through which a charged metallic ball suspended by an insulating thread can be introduced. This ball is used to charge the sphere. In Fig. 16.22, four stages of the charging process are shown. Describe what happens at each stage if the conductor is originally uncharged.

In Fig. 16.22a, a charged metallic ball bearing total charge $+Q$ has been inserted into the center of the spherical cavity. This positive charge exerts attractive electrostatic forces on the free electrons in the outer conducting shell which attract them to the inner wall of the spherical cavity, where they appear as a distribution of surface charges representing a total charge $-Q$ on the interior surface of the cavity. At the same time, since the outer conductor is initially uncharged and must remain so, a deficiency of electrons in the outer part of the conducting shell is created, which finally has the effect of creating a positive distribution of charges (due to atoms which do not have enough electrons to render them neutral) on the outer surface of the conductor. Since the total charge on the outer conducting shell must be zero, the total charge on its outer surface must be $+Q$.

These results may also be verified by the application of Gauss's law. Consider the Gaussian surface S shown in Fig. 16.22a, which lies entirely within the metallic outer shell, where the field is everywhere zero. Since the electric field E is zero everywhere on S,

$$\oint_S \mathbf{E} \cdot \mathbf{n} \, da = \frac{1}{\varepsilon_0}(Q + Q') = 0 \qquad (16.4.2)$$

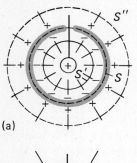

(a)

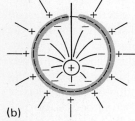

(b)

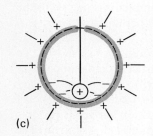

(c)

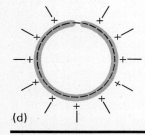

(d)

FIGURE 16.22. Transfer of charge from a charged conducting ball to a hollow conducting sphere.

where $+Q$ is the charge on the metallic ball and Q' represents the total charge on the inner surface of the shell. But from (16.4.2) it is evident that

$$Q' = -Q \qquad (16.4.3)$$

as stated above. The charge distribution that appears on the outer conducting shell as a result of the nearby charged metallic object is referred to as a distribution of *induced charges*. In this case, the induced charge distribution consists of the uniformly distributed spherically symmetric charges $-Q$ on the inner surface and $+Q$ on the outer surface of the shell.

When the center of the charged metallic ball coincides with the center of the spherical outer shell, as in Fig. 16.22a, the system has perfect spherical symmetry, and it is evident that the fields must be radial at every point, with no tangential components whatever. Under these circumstances, the reader should have no difficulty, using a concentric spherical Gaussian surface such as S', which encloses total charge $+Q$, in showing that the electric field within the interior cavity has the magnitude

$$E = \frac{Q}{4\pi\varepsilon_0 r^2} \qquad (16.4.4)$$

In the same way, using a spherical Gaussian surface such as S'', which encloses total charge $+Q - Q + Q = +Q$, it is easy to show that the electric field outside the outer conducting shell must also have the value given by (16.4.4). It is also important to observe that the electric fields arising from the three separate charge distributions $+Q, -Q,$ and $+Q$ on the charged ball and the inner and outer surfaces of the conducting shell add up in such a way as to render the field inside the conducting material of the shell *precisely zero*.

In Fig. 16.22b, the charged ball has been lowered so that it approaches, but does not yet touch, the lower interior surface of the cavity. Under these conditions, the attractive Coulomb forces on the free electrons in the outer conducting shell are stronger near the bottom of the interior cavity, where there is only a short distance between the charged ball and the wall of the cavity, than on those further away near the top of the interior space. These changed forces cause the surface charged distribution on the interior wall of the cavity to rearrange itself in such a way that there is a larger surface charge density near the bottom than near the top. There is now no longer perfect spherical symmetry, and the fields are no longer radial at all points. Under these circumstances, we should not be able to use Gauss's law to find the electric fields, since we would have no clue for constructing a Gaussian surface that is everywhere perpendicular—or parallel—to the field! Nevertheless, the surface charges, and the electric fields they create, still rearrange themselves in exactly such a way as to add up *within the conductor* to give an electric field of zero magnitude. The surface charges on the *outside* surface of the spherical shell do not suffer any lasting rearrangement and are thus essentially unaffected by the rearrangement of interior fields and surface charges. The reason is simple; in electrostatic equilibrium, the field within the conducting substance is always zero, both before and after the rearrangement of interior fields and surface charges. The forces experienced by the outside surface charge distribution are, therefore, the same after the interior rearrangement of surface charge has taken place as they were before, and, so is the configuration of outer surface charges themselves!

Finally, as illustrated in Fig. 16.22c, the charged ball is allowed to touch the inner surface of the spherical shell. When this occurs, charges can flow freely between the outer surface of the ball and the inner surface of the shell. The attractive Coulomb forces between the free electrons on the interior of the shell and the positive charges on the outer surface of the charged ball now cause free electrons to *flow* from the interior wall of the cavity onto the metal ball until the charge distributions completely neutralize one another, leaving no charges whatsoever within the spherical shell. All this happens much more quickly than it takes to describe it; it is all over in a tiny fraction of a second and is perhaps accompanied by a small spark that marks the spot where large Coulomb forces between closely spaced charges of opposite sign have torn electrons from atoms of oxygen or nitrogen within the gap and ionized the air in that region.

But the *positive* surface charge distribution on the *outer* surface of the spherical shell is essentially unaffected by all this, for reasons similar to those discussed previously. It remains there quite undisturbed by all that has gone on within the interior cavity since the charged metal ball was put there to induce it in the first place. Indeed, the lid can be lifted, the metal ball, now uncharged, removed, and the lid replaced, and the positive surface charge $+Q$ will still remain on the outer surface of the spherical shell. The *net effect* of all these processes has been to *transfer the positive charge $+Q$ from the charged ball to the outer surface of the shell*, though, when examined in detail, it is easily seen that this transfer is not accomplished directly.

It is important also to notice that this transfer could never have taken place at all had the inner metal ball *not* been allowed to touch the inner surface of the spherical shell. Had the ball been *withdrawn* from the inner cavity, for example, when the stage illustrated in Fig. 16.22b was reached, there would have been nothing to prevent electrons on the inner wall of the shell from flowing through the conducting substance of the shell itself to neutralize the excess positive charge on the outside. This can be seen by applying Gauss's law to the surface S within the metal, assuming now that the conducting ball bearing charge Q is gone. Under these circumstances, Gauss's law tells us that

$$\oint_S \mathbf{E} \cdot \mathbf{n} \, da = 4\pi R^2 E = \frac{Q'}{\varepsilon_0}$$

$$(16.4.5)$$

$$E = \frac{Q'}{4\pi\varepsilon_0 R^2}$$

where Q' is the total charge on the inside surface of the shell and R is the radius of the surface S. Now, if Q' has any value other than zero, it is evident from the above expression that the field E within the metal will not be zero and, hence, that any charges there will experience forces and move *within* the conductor. Under these circumstances, the situation cannot be one of equilibrium; the charges that make up the distribution Q' must now flow within the metallic conducting material *until the value of Q' is, in fact, zero* and a condition of electrostatic equilibrium in which E is zero everywhere within the conductor is at length reestablished. In doing this, the charge $Q' = -Q$ on the inner surface simply migrates *through* the shell to neutralize the positive charge $+Q$ on the outer surface and to establish, finally, an equilibrium state in which there is no charge on either surface of the shell.

EXAMPLE 16.4.2

A spherical insulator having radius R contains a total charge Q uniformly distributed throughout its volume. It is surrounded by a concentric conducting shell which has a net charge q. The shell has an inner radius of $5R$ and an outer radius of $6R$. (a) Obtain expressions for the electric field for all values of r, where r is the distance from the center of the insulator. (b) Find the charge and charge density on the inside and outside surfaces of the conductor. (c) Plot E versus r in the various regions illustrated in Fig. 16.23.

This example can be worked using Gauss's law. Regardless of the value of r, the perfect spherical symmetry of the system assures us that the field will be radial everywhere. This allows us to construct a spherical Gaussian surface, and in each case the flux through the surface is given by

$$\Phi = \oint_S \mathbf{E} \cdot \mathbf{n} \, da = 4\pi r^2 E \qquad (16.4.6)$$

According to Gauss's law this is equal to q_0/ε_0. It follows, therefore, that

$$E(r) = \frac{q_0(r)}{4\pi\varepsilon_0 r^2} \qquad (16.4.7)$$

where $q_0(r)$ is the total charge contained *within the*

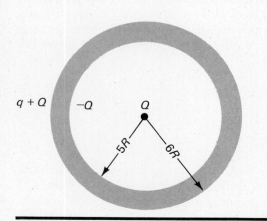

FIGURE 16.23

Gaussian sphere of radius r. We now consider various cases:

Case 1

$0 < r < R$

In this region, the net charge is the charge within the insulator contained in a sphere of radius r. If the charge within the insulator is uniformly distributed throughout its volume, we may express the constant charge density or charge per unit volume ρ as

$$\rho = \frac{Q}{v} = \frac{Q}{(4\pi R^3/3)} \qquad (16.4.8)$$

The charge $q_0(r)$ contained within a sphere of radius r ($r < R$) and volume $v = 4\pi r^3/3$ will then be

$$q_0(r) = \rho v = \frac{Q}{(4\pi R^3/3)} \frac{4\pi r^3}{3} = \frac{Qr^3}{R^3} \qquad (16.4.9)$$

We may now substitute (16.4.9) into (16.4.7) to obtain

$$E(r) = \frac{Qr}{4\pi\varepsilon_0 R^3} \qquad 0 < r < R \qquad (16.4.10)$$

in this region. The electric field thus increases linearly with r within the charged insulating substance.

Case 2

$R < r < 5R$

For values of r restricted to this range, the net charge within the Gaussian surface is given by Q, and, therefore, the electric field falls off as the inverse square of the distance. Substituting Q for $q_0(r)$ ($R < r < 5R$) into (16.4.7), we find

$$E(r) = \frac{Q}{4\pi\varepsilon_0 r^2} \qquad R < r < 5R \qquad (16.4.11)$$

Case 3

$5R < r < 6R$

This is the region within the conductor; therefore, the electric field is necessarily *zero*.

Case 4

$r > 6R$

In the final region, the total charge within the Gaussian surface is $q + Q$. Therefore, setting $q_0(r) = q + Q$ for $r > 6R$ in (16.4.7), we obtain

$$E(r) = \frac{q + Q}{4\pi\varepsilon_0 r^2} \qquad (16.4.12)$$

In order to obtain a vanishing electric field inside the conductor, it is necessary to have *zero* net charge within any Gaussian sphere drawn *inside* the conductor. This implies that the interior surface of the conductor must have a total charge $-Q$, and the exterior surface must, therefore, have a charge of $q + Q$. The charge densities on the inside and outside

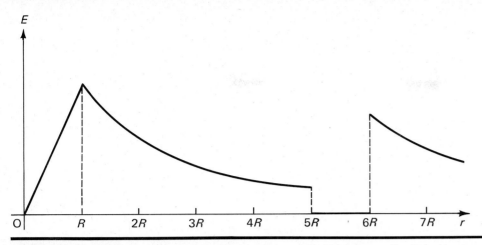

FIGURE 16.24. Field magnitude E vs. radial distance r for the system discussed in Example 16.4.2.

surfaces of the conductor are then given

$$\sigma_{\text{inside}} = \frac{-Q}{4\pi(5R)^2}$$

$$\sigma_{\text{outside}} = \frac{q+Q}{4\pi(6R)^2}$$

(16.4.13)

respectively.

As a check on the previous computation of the electric field, we may use (16.4.1) together with the above equations. It is then readily verified that the magnitudes of the fields agree with the results obtained from (16.4.11) and (16.4.12), with r given by $5R$ and $6R$, respectively. It should be noted that a corresponding expression (such as σ/ε_0) cannot be used outside the insulator since the charge is distributed *throughout the volume* rather than on a surface. A typical plot of E versus r is shown in Fig. 16.24. It should be noted that the electric field is discontinuous at the conducting surfaces.

16.5 Electrostatic Potential

The concept of potential energy has been discussed extensively in Chapter 5 in connection with our earlier study of work and energy, and the concept of the gravitational potential associated with the gravitational field has been introduced in Chapter 8. We may recall that a conservative force is one in which the work done by the force, on an object moving under its influence between two points, is independent of the path. The work depends, therefore, only on the initial and final positions of the body and not on the details of how it got from its initial position to its final location. Thus, for conservative forces, it is possible to define a *potential energy* as the negative of the work done by the force in moving from the initial

position to the endpoint. The Coulomb force law, like the law of gravitation, is an inverse square law, in which the force between two charges depends only on their separation. The direction of the force is parallel to the straight line connecting the two charges; there are two possible directions depending on whether the charges have the same or opposite signs. Except for the possibility of these two directions of force, the Coulomb law has a mathematical form identical to the law of gravitation and may readily be shown to be conservative. The arguments presented to demonstrate its conservative character are identical to those discussed in Chapter 8 and will, therefore, not be repeated here.

Let us now follow the motion of an individual charge q which experiences a *total* Coulomb force **F**. The force **F**, which depends on the location of q, is assumed to be due to the presence of one, several, or perhaps very many charges whose positions are fixed. In fact, **F** could also be due to a continuous distribution of charge. Initially, q is at a location specified by i, and it moves to a final position specified by f. We may, therefore, equate the net work done by the force **F** to the negative of the change in potential energy. Accordingly, we have

$$U_{pf} - U_{pi} = -\int_i^f \mathbf{F} \cdot d\mathbf{r}$$

(16.5.1)

The total Coulomb force **F** may be expressed as the product of the charge q and the total electric field **E** produced by all other charges. Thus, we can also write

$$U_{pf} - U_{pi} = -\int_i^f q\mathbf{E} \cdot d\mathbf{r}$$

(16.5.2)

An alternative way of writing (16.5.2) is obtained by dividing both sides of the equation by the charge q and then defining the *electrostatic potential energy per unit charge* as the *electrostatic potential V*, while referring to the *potential energy change per unit charge* as the *difference of potential* ΔV. The equation, then, takes the form

$$\Delta V = V_f - V_i = \frac{U_{pf} - U_{pi}}{q} = -\int_i^f \mathbf{E} \cdot d\mathbf{r} \qquad (16.5.3)$$

Since potential energy U is expressed in joules and q is a charge measured in coulombs, the ratio $\Delta U/q$ has units of joules/coulomb. These units of electrostatic potential and potential difference are more commonly referred to as *volts*. It is evident from their definition that 1 coulomb of charge does 1 joule of work on its surroundings in undergoing a potential decrease of 1 volt. Also, the units of the electric field E, which are newtons/coulomb, can equally well be expressed as volt/meter, since

$$\frac{\mathrm{V}}{\mathrm{m}} = \frac{\mathrm{J}}{\mathrm{C}\text{-}\mathrm{m}} = \frac{\mathrm{N}\text{-}\mathrm{m}}{\mathrm{C}\text{-}\mathrm{m}} = \frac{\mathrm{N}}{\mathrm{C}}$$

It is, in fact, much more common to use units of volt/meter for the electric field than units of newtons/coulomb, and from this point on, we shall always do so.

The usefulness of Eq. (16.5.3) follows from the fact that it is an equation which involves only the source charges which produce the field $\mathbf{E}$; it does not involve the charge q which is acted upon by the field. The electric field $\mathbf{E}$ is a *vector field* which a certain charge distribution produces at every point in space. According to (16.5.3), the electrostatic potential represents a *scalar field* produced at each point in space. Differences in the value of this field at any two points may be obtained by evaluating the integral above along any path connecting the two points. The absolute value of potential is not defined by (16.5.3); only differences are specified. We may, therefore, choose the point where the potential is taken to be zero anywhere we please. In most cases, it is convenient to define the potential V_i to be zero when the initial point i is *infinitely far* from the charges which produce the field.

There are several very important advantages that stem from knowing the electrostatic potential associated with a given distribution of charges at all points in space. Among them are the following:

1. If we know the potential difference between any two points, we can easily obtain the change in potential energy and the work done when a charge placed in the field moves between these two points.
2. Knowledge of the potential function V makes it possible to calculate the electric field $\mathbf{E}$ by taking certain derivatives of V. The exact method will be discussed later in this section.
3. The construction of surfaces upon which the po-

tential is the same at all points is an aid to visualizing the electric field pattern.

We shall appreciate the significance of these comments better after discussing in detail a number of applications which illustrate the usefulness of the concept of potential. Since it is important to learn how to calculate the potential produced by a distribution of charge, we consider first a few simple examples of potential calculations.

EXAMPLE 16.5.1

Calculate the potential function V produced by a single positive charge q which is located at the origin of the coordinate system.

To solve this example, we make direct use of (16.5.3). We already know that a single point charge q establishes a spherically symmetric electric field $\mathbf{E}$ which points radially outward. This field can, therefore, be written in vector form as

$$\mathbf{E} = \frac{q}{4\pi\varepsilon_0} \frac{\mathbf{i}_r}{r^2} \qquad (16.5.4)$$

where $\mathbf{i}_r$ is a unit vector which always points *radially outward*. In Figs. 16.25a and 16.25b, we draw two possible paths going from point i to point f. Either of these might be used to evaluate the integral that appears in (16.5.3) because the result is independent of path. Let us demonstrate this independence, however, by discussing the calculation for any *arbitrary* path, such as that illustrated in Fig. 16.25c.

The small displacement vector $d\mathbf{r}$ can be written in terms of the unit vectors $\mathbf{i}_r$ and $\mathbf{i}_\theta$. The vector $\mathbf{i}_r$ points radially outward while $\mathbf{i}_\theta$ points azimuthally in a direction perpendicular to $\mathbf{i}_r$, as shown in Fig. 16.25c. Using these unit vectors, we can write

$$d\mathbf{r} = (dr)_r \mathbf{i}_r + (dr)_\theta \mathbf{i}_\theta \qquad (16.5.5)$$

where $(dr)_r$ and $(dr)_\theta$ are the components[6] of the vector $d\mathbf{r}$. The component $(dr)_r$ can also be written as dr, while $(dr)_\theta$ is equal to $r\,d\theta$, as illustrated in Fig. 16.26. The scalar product of $\mathbf{E}$ and $d\mathbf{r}$ is, therefore,

$$\mathbf{E} \cdot d\mathbf{r} = \frac{q}{4\pi\varepsilon_0 r^2}\, \mathbf{i}_r \cdot ((dr)\mathbf{i}_r + (r\,d\theta)\mathbf{i}_\theta)$$

$$= \frac{q\,dr}{4\pi\varepsilon_0 r^2} \qquad (16.5.6)$$

According to (16.5.3), we must now integrate this expression between the initial point i, which is a distance r_i from q, and the final point f, which is a

[6] It is important to note that the displacement or position vector $\mathbf{r}$ from the origin to any point P has, by definition, only a radial component. But a vector $d\mathbf{r}$ representing an infinitesimal change in $\mathbf{r}$ does not have to be in the same direction as $\mathbf{r}$ and, therefore, may have an azimuthal component as well as a radial one. This is illustrated in Fig. 16.26.

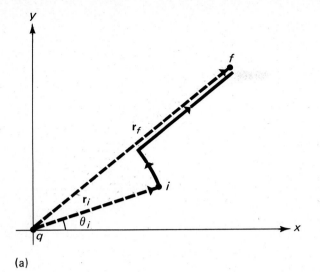

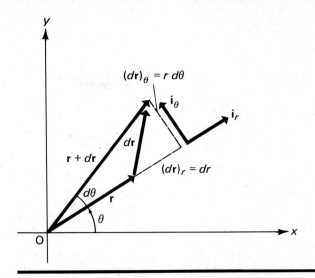

FIGURE 16.26. Radial and azimuthal components of vector $d\mathbf{r}$.

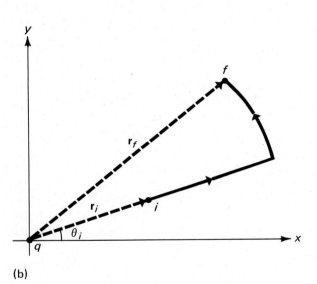

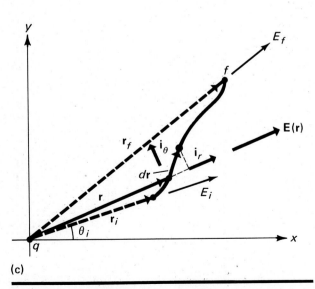

FIGURE 16.25. Calculation of electrostatic potential of a point charge.

distance r_f from q. Therefore,

$$\int_i^f \mathbf{E} \cdot d\mathbf{r} = \int_{r_i}^{r_f} \frac{q}{4\pi\varepsilon_0 r^2} \, dr = \left[\frac{-q}{4\pi\varepsilon_0 r} \right]_{r_i}^{r_f}$$

$$= -\left(\frac{q}{4\pi\varepsilon_0 r_f} - \frac{q}{4\pi\varepsilon_0 r_i} \right) \qquad (16.5.7)$$

When (16.5.3) is used, the result can be stated as

$$V(r_f) - V(r_i) = \frac{q}{4\pi\varepsilon_0} \left(\frac{1}{r_f} - \frac{1}{r_i} \right) \qquad (16.5.8)$$

As we mentioned earlier, the potential infinitely far from the charge distribution is usually defined to be zero. Therefore, if r_i is chosen to be infinity, and if this value is substituted into (16.5.8), writing r in place of r_i and $V(r)$ in place of V_f, we obtain

$$V(r) = \frac{q}{4\pi\varepsilon_0 r} \qquad (16.5.9)$$

as the potential produced by a point charge. In this expression, r is simply the distance of the point of observation from the point charge. The reader should note that *nowhere* in the above calculation did we find it necessary to state which specific path is being used in the integration. In the end, the only integration required was a radial integral carried out between the initial and final radial distances. The above calculation was done assuming q to be a positive charge. The result, however, is also true if the charge is negative. A positive point charge, therefore, creates a positive potential everywhere, whereas a negative point charge establishes a negative potential.

505

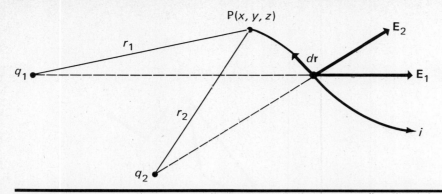

FIGURE 16.27. Calculation of electrostatic potential of two point charges.

EXAMPLE 16.5.2

Two point charges q_1 and q_2 produce an electric field in space. Obtain an expression for the electric potential which they produce at some arbitrary point P as shown in Fig. 16.27.

Let P represent the final point f and let the initial point i be infinitely far from the charges. Then, if P is the point (x, y, z), we have

$$V(x, y, z) - 0 = -\int_i^P \mathbf{E} \cdot d\mathbf{r} \qquad (16.5.10)$$

At this stage, it looks as though we have a difficult integration to perform since $\mathbf{E}$ is the vector sum of the electric fields produced by both charges and is, in general, quite complicated. It would indeed be difficult to first add the electric fields $\mathbf{E}_1$ and $\mathbf{E}_2$ shown in Fig. 16.27 and then to integrate. A much more ingenious procedure involves separating the single difficult integration into two easier integrations, according to which

$$-V(x, y, z) = \int_i^P \mathbf{E} \cdot d\mathbf{r} = \int_i^P (\mathbf{E}_1 + \mathbf{E}_2) \cdot d\mathbf{r}$$
$$= \int_i^P \mathbf{E}_1 \cdot d\mathbf{r} + \int_i^P \mathbf{E}_2 \cdot d\mathbf{r} \qquad (16.5.11)$$

where $\mathbf{E}_1$ and $\mathbf{E}_2$ are, respectively, the electric fields produced by charges q_1 and q_2. The two integrals which now appear are actually identical to the one already worked out in the preceding example. The first integral above is, in fact, the potential produced at point P by charge q_1, while the second integral is the potential produced at P by q_2. Thus, we find that

$$V(x, y, z) = \frac{q_1}{4\pi\varepsilon_0 r_1} + \frac{q_2}{4\pi\varepsilon_0 r_2} \qquad (16.5.12)$$

where r_1 and r_2 are the distances from charges q_1 and q_2 to the observation point P.

Thus, we see that *potentials* are *additive*. The potential at any point due to an arbitrary number of charges is the *scalar* sum of the potentials produced by each of the charges. This additivity is extremely useful. Electric fields produced by many charges are also additive, of course, but it is, in general, more difficult to add electric fields because of their vector character than to add potentials, which are scalars. In fact, we will show later that if one succeeds in adding up all the potentials to find the total electrostatic potential, then the electric field can be obtained *from the potential*. This allows us to circumvent totally the task of directly adding many electric fields to obtain the total $\mathbf{E}$ field.

The generalization of (16.5.12) to the case of n charges is readily seen to be

$$V(x, y, z) = \frac{1}{4\pi\varepsilon_0} \sum_{i=1}^{n} \frac{q_i}{r_i} \qquad (16.5.13)$$

where r_i is the distance from the ith charge to the point (x, y, z). Equation (16.5.13) is also valid even if n tends to infinity and the charge distribution becomes one that is continuously distributed. In this event, however, we must first find out how much charge is contained in any given small volume element surrounding an arbitrary point (x', y', z') within the charge distribution. This involves *knowing* the volume charge density $\rho(x', y', z')$, since

$$dq' = \rho \, dx' \, dy' \, dz' = \rho \, d(\text{vol}) \qquad (16.5.14)$$

provides the relation needed to obtain the amount of charge within the volume element. Next, we must know the distance between this element of charge and the observation point. If the charge element is located by means of a vector $\mathbf{r}'$ and the point at which the potential is to be evaluated is located by $\mathbf{r}$, this distance will be given by the magnitude of the vector $\mathbf{r} - \mathbf{r}'$ illustrated in Fig. 16.28 and is denoted by $|\mathbf{r} - \mathbf{r}'|$. The charge element, therefore, produces a small potential given by

$$dV(x, y, z) = \frac{1}{4\pi\varepsilon_0} \frac{dq'}{|\mathbf{r} - \mathbf{r}'|}$$
$$= \frac{1}{4\pi\varepsilon_0} \frac{\rho(x', y', z') \, dx' \, dy' \, dz'}{|\mathbf{r} - \mathbf{r}'|} \qquad (16.5.15)$$

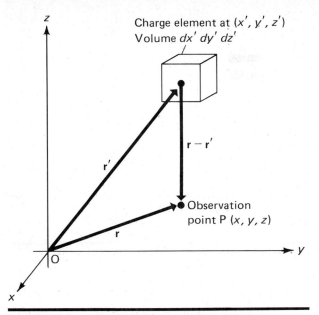

FIGURE 16.28. Relationship between vectors **r**, **r'**, and **r–r'** for an arbitrary observation point P.

The total potential produced is obtained by summing over the entire charge distribution. For a continuous distribution of charge, this means carrying out an integration over the coordinates x', y', z', which describe the distribution. The final expression is therefore

$$V(x, y, z) = \frac{1}{4\pi\varepsilon_0} \int \frac{\rho(x', y', z')\, dx'\, dy'\, dz'}{|\mathbf{r} - \mathbf{r}'|} \qquad (16.5.16)$$

For charge distributions that have some symmetry, it is frequently possible to carry out the above integration, but it requires careful choice of the volume element and the integration variables. When it is possible to evaluate (16.5.16) directly one may obtain the potential at an *arbitrary* point whose coordinates are (x, y, z). If, however, the integration cannot be directly accomplished, one can integrate (16.5.16) numerically, either on a big computer or with a pocket calculator. The disadvantage of numerical integration is that a separate calculation must be performed for each point (x, y, z). With modern high-speed computers, one may, however, obtain accurate values of potentials at many observation points even when analytic calculations are not possible.

EXAMPLE 16.5.3

A thin cylindrical disc of radius R and thickness a contains a total charge Q which is uniformly distributed. Obtain an expression for the electrostatic potential for points on the axis of the disc.

The charge distribution and also a typical point P on the axis are illustrated in Fig. 16.29. Since

the charge distribution is continuous, we use (16.5.16) to calculate the potential. To apply (16.5.16), we need to obtain expressions for the charge density, volume element, and spatial separation $|\mathbf{r} - \mathbf{r}'|$. The first step is to decide what volume element is appropriate. In the present case, we can take this to be the volume contained between r' and $r' + dr'$, as shown in the figure. Since d is very small, all the charge within this volume element is at the same distance $(r'^2 + r^2)^{1/2}$ from the point P. The volume element consists of the area of the ring $2\pi r'\, dr'$ multiplied by the thickness a of the disc. Thus,

$$d(\text{vol}) = 2\pi r' a\, dr' \qquad (16.5.17)$$

is the differential volume element. The density ρ is uniform and is, therefore, the total charge Q divided by the volume $\pi R^2 a$, that is,

$$\rho = \frac{Q}{\pi R^2 a} \qquad (16.5.18)$$

Using (16.5.16), we obtain

$$V_P = \frac{1}{4\pi\varepsilon_0} \int_0^R \frac{\dfrac{Q}{\pi R^2 a} 2\pi r' a\, dr'}{\sqrt{r'^2 + r^2}}$$

$$= \frac{Q}{2\pi\varepsilon_0 R^2} \int_0^R \frac{r'\, dr'}{\sqrt{r'^2 + r^2}} \qquad (16.5.19)$$

The integral appearing in (16.5.19) can readily be carried out, setting $u = r'^2 + r^2$ and $du = 2r'\, dr'$, to obtain

$$V_P = \frac{Q}{2\pi\varepsilon_0 R^2} \left[\sqrt{r'^2 + r^2} \right]_0^R$$

$$= \frac{Q}{2\pi\varepsilon_0 R^2} \left[\sqrt{R^2 + r^2} - r \right] \qquad (16.5.20)$$

It is important to observe that the distance r remains *fixed* during integration. It is instructive to examine the dependence of this result on a single variable

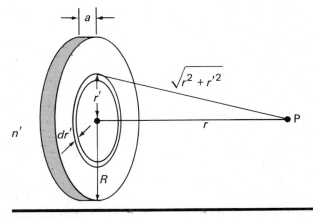

FIGURE 16.29

while the others are held constant. We shall consider several cases:

Case 1 Q, R Fixed but r Variable

Here, we are studying the dependence of V on distance from a disc of fixed size and charge. For r equal to zero, we obtain $Q/2\pi\varepsilon_0 R$. As r is increased, V decreases; and for values of r much larger than R, V becomes essentially $Q/4\pi\varepsilon_0 r$. This simply means that at large distances, the disc looks like a point charge.

Case 2 Q, r Fixed but R Variable

In this case, we are a fixed distance from a disc of fixed charge Q and we are considering the variation of V as the size of the disc is allowed to vary. When R is extremely small, we again recover the point potential result of $Q/4\pi\varepsilon_0 r$. As R is increased, the potential decreases and, in fact, approaches zero as R becomes very large, for all values of r. As R increases, the charge distribution approaches an infinite sheet of charge and, therefore, we expect the electric field to approach the constant value $\sigma/2\varepsilon_0$ (as demonstrated from Gauss's law in Example 16.3.2). However, with fixed Q and increasing R, the surface charge density σ approaches zero and, therefore, we expect the electric field to approach zero likewise. This means that V approaches a constant value, and in the present case, this value turns out to be zero.

Case 3 Q/R² and r Fixed but R Variable

In this last situation for small values of R the potential approaches zero. Clearly, as R gets close to zero so does Q, since the quantity Q/R^2 is held constant. Therefore, small R means we have a point charge with very small Q, and therefore we expect V to go to zero. As R is increased, Q increases and V likewise increases. Finally, as R approaches infinity, V also approaches infinity for all R. In this limit we have an infinite sheet of charge with finite charge density. This implies the existence of a constant electric field $\sigma/2\varepsilon_0$. In the presence of such a constant field the potential must be infinite as one may readily see by evaluating the integral of (16.5.3), using the lower limit $r_i = \infty$.

EXAMPLE 16.5.4

Three point charges having values of 2×10^{-10} coulomb, 4×10^{-10} coulomb, and -5×10^{-10} coulomb are placed at the vertices of an equilateral triangle whose sides are each 10 cm. Obtain the electrostatic potential at the center of the triangle.

In this example, we use (16.5.13) to superpose the potentials produced by each of the charges. These charges are shown in Fig. 16.30. The charges are equidistant from the center of the triangle. Using elementary trigonometry, we find that the distance r

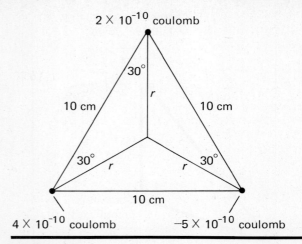

FIGURE 16.30

to the center is $10/\sqrt{3}$ cm. Therefore, from (16.5.13),

$$V_P = 9 \times 10^9 \left(\frac{4 \times 10^{-10}}{0.1/\sqrt{3}} + \frac{2 \times 10^{-10}}{0.1/\sqrt{3}} - \frac{5 \times 10^{-10}}{0.1/\sqrt{3}} \right)$$

$$= 15.6 \text{ N-m/C} = 15.6 \text{ J/C} = 15.6 \text{ volts} \qquad (16.5.21)$$

Our definition of electrostatic potential, as embodied in (16.5.3), allows us to calculate the potential in situations in which the total electric field is known. We have found, however, that the potential can also be calculated in a somewhat more direct manner by using either (16.5.13) for the case of discrete charges or (16.5.16) for a continuous charge distribution. These *methods do not require prior knowledge of the electric field*, but rather involve a *summation of the individual potentials* produced by small point charges. It is interesting to inquire whether or not we can somehow invert (16.5.3) to find the total **E** field from the potential V rather than vice versa. It turns out that this is quite practical, and it provides a very useful method for evaluating electric fields.

To see how this works, we write the differential form of (16.5.3) as

$$dV = -\mathbf{E} \cdot d\mathbf{r} \qquad (16.5.22)$$

This gives the change in voltage over an infinitesimally small displacement $d\mathbf{r}$. We know that the complete potential function V produced by all charges can be expressed in terms of Cartesian coordinates as a function $V(x, y, z)$. Therefore, if dV is viewed as the change of V in going from point(x, y, z) to point$(x + dx, y + dy, z + dz)$, then dV can be expressed as

$$dV = \frac{\partial V}{\partial x}\, dx + \frac{\partial V}{\partial y}\, dy + \frac{\partial V}{\partial z}\, dz \qquad (16.5.23)$$

according to the rules of calculus. The derivatives above are *partial* derivatives. For example, $\partial V/\partial x$ is the ordinary derivative of V with respect to x, regarding y and z to have *fixed constant values*. On the other hand, the right-hand side of (16.5.22) can be written as follows:

$$\mathbf{E} \cdot d\mathbf{r} = E_x \, dx + E_y \, dy + E_z \, dz \qquad (16.5.24)$$

This form of the scalar product was discussed extensively in Chapter 1. Let us now substitute Eqs. (16.5.23) and (16.5.24) into (16.5.22) and put both terms on the same side of the equation to obtain

$$\left(E_x + \frac{\partial V}{\partial x}\right)dx + \left(E_y + \frac{\partial V}{\partial y}\right)dy + \left(E_x + \frac{\partial V}{\partial z}\right)dz = 0$$

$$(16.5.25)$$

The displacements dx, dy, and dz may be chosen completely independently of one another and are *not* generally zero. For example, we could choose dx and dy to be zero but dz to be nonzero, in which case we would find $E_z + \partial V/\partial z = 0$. Likewise, if we chose $dx = dz = 0$ and $dy \neq 0$, we would have to conclude that $E_y + \partial V/\partial y = 0$. And if we had $dy = dz = 0$ and $dx \neq 0$, we would have to admit that $E_x + \partial V/\partial x = 0$. Under the circumstances, the only way we can *guarantee* that the left side of this equation will be zero for all values of dx, dy, and dz is to have

$$
\begin{aligned}
E_x &= -\frac{\partial V}{\partial x} \\[2mm]
E_y &= -\frac{\partial V}{\partial y} \qquad\qquad (16.5.26)\\[2mm]
E_z &= -\frac{\partial V}{\partial z}
\end{aligned}
$$

at any arbitrary point in space.

Thus, we see that if the potential function $V(x, y, z)$ is explicitly known, then the electric field may readily be obtained by evaluating the partial derivatives of $V(x, y, z)$. The total electric field vector $\mathbf{E}$ may be written in terms of the unit vectors $\mathbf{i}_x$, $\mathbf{i}_y$, and $\mathbf{i}_z$ as

$$
\begin{aligned}
\mathbf{E} &= E_x\mathbf{i}_x + E_y\mathbf{i}_y + E_z\mathbf{i}_z \\[2mm]
&= -\frac{\partial V}{\partial x}\mathbf{i}_x - \frac{\partial V}{\partial y}\mathbf{i}_y - \frac{\partial V}{\partial z}\mathbf{i}_z \qquad (16.5.27)
\end{aligned}
$$

or

$$\mathbf{E} = -\nabla V \qquad (16.5.28)$$

where ∇ denotes the differential operator defined by

$$\nabla = \mathbf{i}_x\frac{\partial}{\partial x} + \mathbf{i}_y\frac{\partial}{\partial y} + \mathbf{i}_z\frac{\partial}{\partial z} \qquad (16.5.29)$$

which was introduced previously in Chapter 5. This operator ∇ is referred to as the *gradient operator*, and its use in mathematics and physics is very widespread.

The use of this symbol allows us to write equations in a more compact form, which is an advantage in problems involving much algebraic manipulation before the derivatives are actually evaluated.

Let us now try to understand the importance of the relation between the electric field $\mathbf{E}$ and the potential V. Suppose we are given a certain distribution of charges and wish to obtain the electric field. We can sometimes, though not always, calculate $\mathbf{E}$ by using Gauss's law; when that is possible, it is usually the best method. If Gauss's law cannot be applied, then we can obtain $\mathbf{E}$ in either of two ways:

Method 1

Add vectorially the electric fields which each of the charges or charge elements produces at an arbitrary observation point. Frequently, this may be difficult since the addition of a large number of vectors is usually hard and time consuming.

Method 2

Calculate the electrostatic potential, at an arbitrary point (x, y, z), by adding the potentials produced by each of the charges and obtain the components of the electric field vector by evaluating the partial derivatives given by (16.5.26). This method involves the addition of *scalars* and is, therefore, much easier to carry out. There is, however, a penalty one must pay; although the summation is easier, the electric field is not obtained until the appropriate derivatives of V are calculated. Generally, though, the procedure of taking derivatives is simpler than adding individual fields vectorially.

We have studied earlier the construction of lines of force due to electric fields in order to help us visualize the field. It is also very helpful to construct surfaces whose points are all at the same potential and which are, therefore, called *equipotential surfaces*. It turns out that if we can construct the equipotential surfaces for a given problem, it is then a simple task to draw the field lines, since we will see shortly that these field lines must always be *perpendicular* to the equipotential surfaces. Just as it is impossible to have intersecting field lines, it is also impossible for two different equipotential surfaces, corresponding to different values of the potential, to cross one another.

Let us now prove that field lines are always perpendicular to equipotential surfaces. Let A and B be two points on a given equipotential surface, as shown in Fig. 16.31. Assume A is fixed but allow the position of B, which is infinitesimally close to A, to vary. Then, since A and B are at the same potential,

$$dV_{AB} = -\mathbf{E} \cdot d\mathbf{r}_{AB} = 0 \qquad (16.5.30)$$

where $d\mathbf{r}_{AB}$ is the vector from A to B. But if the scalar product vanishes, the vectors $\mathbf{E}$ and $d\mathbf{r}_{AB}$ must be

509

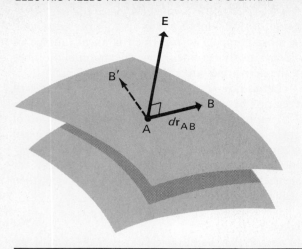

FIGURE 16.31. Relationship of field vectors and equipotential surfaces.

mutually perpendicular. Equation (16.5.30) holds for every point B on the equipotential surface in the neighborhood of A. This means that the vector **E** is perpendicular to *any* vector $\mathbf{r}_{AB}$ lying in the equipotential surface, for example the one joining A and B' in Fig. 16.31. It therefore follows that **E** must be perpendicular to the surface itself. The vector **E** points in the direction of *decreasing* potential in view of the minus signs in (16.5.26). Since the electric field is always perpendicular to the surface of any conducting object, it follows that *the surface of any conducting body in electrostatic equilibrium is an equipotential surface.* Every point on the surface of a conductor under such circumstances must be at the *same* potential.

The problem of calculating the electric field produced by a dipole was discussed earlier in the chapter by superposing the electric fields produced by each of the two charges. Let us now consider the calculation of the potential and field of a dipole by Method 2 discussed above.

EXAMPLE 16.5.5
Calculate the electrostatic potential produced by the electric dipole of Example 16.2.3 at an arbitrary point in the xy-plane. Determine the x- and y-components of the electric field by differentiating the potential with respect to the variables x and y.

The reader should refer back to Fig. 16.8 for an illustration of the dipole. According to (16.5.3), the electrostatic potential at the field point (x, y) is given by

$$V(x, y) = \frac{1}{4\pi\varepsilon_0}\left(\frac{Q}{r_+} + \frac{-Q}{r_-}\right) = \frac{Q}{4\pi\varepsilon_0}\frac{r_- - r_+}{r_+ r_-} \quad (16.5.31)$$

or

$$V(x, y) = \frac{1}{4\pi\varepsilon_0} \times$$

$$\left\{\frac{Q}{\left[\left(y - \dfrac{d}{2}\right)^2 + x^2\right]^{1/2}} + \frac{-Q}{\left[\left(y + \dfrac{d}{2}\right)^2 + x\right]^{1/2}}\right\} \quad (16.5.32)$$

To obtain the electric field components, we make use of (16.5.26). Then

$$E_x = -\frac{\partial V}{\partial x} = -\frac{Q}{4\pi\varepsilon_0}\left\{\frac{\partial}{\partial x}\frac{1}{\left[\left(y - \dfrac{d}{2}\right)^2 + x^2\right]^{1/2}}\right.$$

$$\left. -\frac{\partial}{\partial x}\frac{1}{\left[\left(y + \dfrac{d}{2}\right)^2 + x^2\right]^{1/2}}\right\}$$

$$= \frac{Q}{4\pi\varepsilon_0}\left\{\frac{x}{\left[\left(y - \dfrac{d}{2}\right)^2 + x^2\right]^{3/2}}\right.$$

$$\left. -\frac{x}{\left[\left(y + \dfrac{d}{2}\right)^2 + x^2\right]^{3/2}}\right\} \quad (16.5.33)$$

This is identical to the result obtained earlier in (16.2.15). The y-component of electric field is calculated in a similar manner and is found to be

$$E_y = -\frac{\partial V}{\partial y} = \frac{Q}{4\pi\varepsilon_0}$$

$$\times \left\{\frac{y - \dfrac{d}{2}}{\left[\left(y - \dfrac{d}{2}\right)^2 + x^2\right]^{3/2}} - \frac{y + \dfrac{d}{2}}{\left[\left(y + \dfrac{d}{2}\right)^2 + x^2\right]^{3/2}}\right\}$$

$$(16.5.34)$$

From this, we see that the calculation of electric field in which we use the intermediate step of first obtaining the potential is, in principal, straightforward and quite easy. Figure 16.32 gives a sketch of the equipotential surfaces and field lines of the electric dipole.

A particularly simple expression for the potential of a dipole at large distances follows from (16.5.31), by observing from Fig. 16.8 that under such circumstances $r_- - r_+ \cong d\cos\theta$, while $r_+ r_- \cong r^2$. Equation (16.5.31) can then be expressed as

$$\boxed{V(r, \theta) = \frac{p\cos\theta}{4\pi\varepsilon_0 r^2}} \quad (16.5.35)$$

where p represents the dipole moment Qd.

EXAMPLE 16.5.6
Find the electrostatic potential both inside and outside a thin hollow spherical shell of charge Q and radius R and sketch the equipotential surfaces and the field lines (see Example 16.2.2).

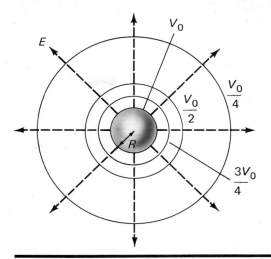

FIGURE 16.33

FIGURE 16.32. Equipotentials and field lines of a point dipole.

The electrostatic potential may be calculated by using (16.5.16). However, in the present case, the total electric field both inside and outside the sphere is easily obtained by using Gauss's law (see Example 16.3.1), so it appears to be simpler to obtain the potential by using (16.5.3) instead. The potential at any point a distance r from the center of the sphere can be written as

$$V(r) = -\int_\infty^r \mathbf{E} \cdot d\mathbf{r} \tag{16.5.36}$$

If r is greater than the radius of the sphere, the electric field points radially outward; therefore,

$$\mathbf{E} = \frac{Q\mathbf{i}_r}{4\pi\varepsilon_0 r^2} \tag{16.5.37}$$

In this case, as in the work leading to (16.5.6), the scalar product $\mathbf{E} \cdot d\mathbf{r}$ can be written

$$\mathbf{E} \cdot d\mathbf{r} = \frac{Q\,dr}{4\pi\varepsilon_0 r^2} \tag{16.5.38}$$

We then find that

$$V(r) = -\int_\infty^r \frac{Q\,dr}{4\pi\varepsilon_0 r^2} = \left[\frac{Q}{4\pi\varepsilon_0 r}\right]_\infty^r = \frac{Q}{4\pi\varepsilon_0 r} \tag{16.5.39}$$

outside the sphere. This formula is also valid at the surface of the sphere, where we may set r equal to R. If we continue to integrate to any point *within* the sphere, we obtain no potential change in the interior region because $\mathbf{E}$ *is zero inside*. Therefore, the entire

volume within the sphere is at the same potential as the spherical surface, namely, $V(R) = Q/4\pi\varepsilon_0 R$.

In Fig. 16.33, equipotential surfaces and field lines are sketched. The equipotentials plotted differ from one another by the constant potential difference $V_0/4$, where V_0 is the potential at the surface of the shell of charge. Thus, the potentials at various values of r are given by $V(R) = V_0$, $V(\tfrac{4}{3}R) = \tfrac{3}{4}V_0$, $V(2R) = V_0/2$, etc.

EXAMPLE 16.5.7

A solid sphere of radius R has a uniform charge density ρ and a total charge Q. Find the electrostatic potential as a function of r and make a graph of V versus r. Calculate the electric field both inside and outside the sphere using the gradient operator. Also compute the electric field by using Gauss's law, and show that the two answers agree.

We shall make use of the result obtained in Example 16.5.6. Let us restate that result in a way which will be useful. A shell of charge having radius r_0 and charge q produces a potential given by

$$V(r) = \frac{q}{4\pi\varepsilon_0 r} \qquad \text{for } r > r_0$$

and

$$V(r) = \frac{q}{4\pi\varepsilon_0 r_0} \qquad \text{for } r \leq r_0 \tag{16.5.40}$$

in the exterior and interior regions, respectively. We note that the potential in the *exterior region* $r > r_0$ is exactly the same as the potential that would be produced by a point charge q located at the origin.

Now, to handle the present problem of a solid sphere, we break up the charge distribution into shells of charge having radius r_0 and thickness dr_0, where r_0

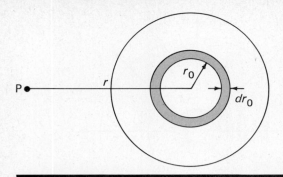

FIGURE 16.34. Spherical shell element of radius r_0 and thickness dr_0 in a spherically symmetric charge distribution of outside radius R.

is now a variable quantity depending upon the particular shell (see Fig. 16.34). Let r denote the position of the observation point P at which the potential is to be calculated. There are two cases to be distinguished, depending on whether r is greater than or less than the radius R of the sphere.

Case 1

$r > R$

In this case, the point of observation is exterior to *all* the shells of charge and, therefore, each shell of charge produces a potential identical to that produced if all of its charge were at the origin. Thus, if a given shell carries charge dq, it produces a potential

$$dV = \frac{1}{4\pi\varepsilon_0} \frac{dq}{r} \qquad (16.5.41)$$

The total potential is given by

$$V(r) = \int \frac{1}{4\pi\varepsilon_0} \frac{dq}{r} = \frac{1}{4\pi\varepsilon_0} \frac{1}{r} \int dq$$

$$\qquad (16.5.42)$$

$$V(r) = \frac{Q}{4\pi\varepsilon_0 r}$$

In this integration over charge shells, the distance r to the observation point is the same for all charge elements and may, therefore, be written outside the integral. In other words, the entire potential at point P turns out to be identical to the potential produced at P by putting all the charge at the origin.

Case 2

$r \leq R$

This part of the problem is more subtle because the observation point is exterior to those shells for which $r > r_0$ and interior to those for which $r < r_0$. Thus, we can state that the contribution to the potential consists of two terms, V_1 $(r_0 < r)$ and V_2 $(r_0 > r)$, and we can write the total potential V as $V_1 + V_2$. The calculation of V_1 may be carried out, for shells of radius *less* than r_0, by assuming their total charge to be concen-

trated at the origin. Therefore,

$$V_1 = \frac{1}{4\pi\varepsilon_0} \frac{q_1}{r} \qquad (16.5.43)$$

where q_1 consists of all the charge within a sphere of radius r. Since the charge density is uniform, we find, just as in Example 16.4.2, that

$$q_1 = \frac{r^3}{R^3} Q$$

and, therefore,

$$V_1 = \frac{q_1}{4\pi\varepsilon_0 r} = \frac{Q}{4\pi\varepsilon_0} \frac{r^2}{R^3} \qquad (16.5.44)$$

In the case of the contribution V_2, the potential arising from a shell of radius r_0 bearing a charge dq is given by

$$dV_2 = \frac{1}{4\pi\varepsilon_0} \frac{dq}{r_0} \qquad (16.5.45)$$

for any point such as P within its interior. The amount of charge dq depends on the volume of the shell. It can be found by multiplying the charge density $Q/\frac{4}{3}\pi R^3$ by the volume of the shell, which is $4\pi r_0^2 \, dr_0$, the result being

$$dq = \frac{3Q r_0^2}{R^3} dr_0 \qquad (16.5.46)$$

Substituting this into (16.5.43), we find

$$dV_2 = \frac{3Q}{4\pi\varepsilon_0} \frac{r_0 \, dr_0}{R^3} \qquad (16.5.47)$$

The total contribution V_2 is obtained by integrating (16.5.47) between the limits of integration r and R. Thus,

$$V_2 = \frac{3Q}{4\pi\varepsilon_0 R^3} \int_r^R r_0 \, dr_0 = \frac{3Q}{4\pi\varepsilon_0 R^3} \left(\frac{R^2}{2} - \frac{r^2}{2} \right) \quad (16.5.48)$$

The entire potential in Case 2 is, therefore, given as the sum of Eqs. (16.5.44) and (16.5.48). We find, therefore,

$$V(r) = \frac{Q}{4\pi\varepsilon_0 R^3} \left(r^2 + \tfrac{3}{2}R^2 - \tfrac{3}{2}r^2 \right)$$

$$= \frac{Q}{4\pi\varepsilon_0 R^3} \left(\tfrac{3}{2}R^2 - \tfrac{1}{2}r^2 \right) \qquad (16.5.49)$$

Notice that as r approaches the radial coordinate R of the charge, this result becomes identical to (16.5.42) in the same limit. Thus, the potential is continuous at the boundary. The continuity of the potential is a physical requirement, for if the potential were discontinuous, its derivative would be infinite and this would imply an infinitely large value for the electric field, and thus for the forces experienced by charges.

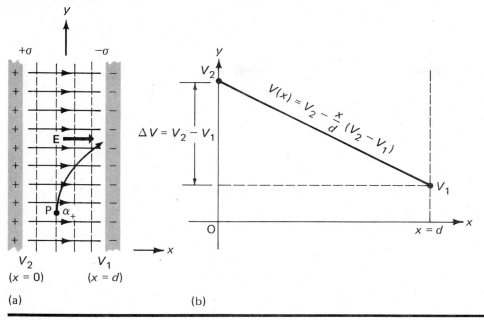

FIGURE 16.35

16.6 Electrostatic Potential Energy

When a particle of charge q moves throughout a region of space in which there exists an electrostatic potential $V(x, y, z)$ which varies from point to point, its potential energy changes as it moves. If we know the electrostatic potential V, then we may obtain the potential energy $U_p(x, y, z)$ by using the expression

$$U_p = qV \tag{16.6.1}$$

which relates potential energy to electrostatic potential. Equation (16.6.1) follows from (16.5.3) when we assume that both V and U vanish when q is infinitely far from any of the charges which produce the field. The above equation is particularly useful in several ways. If the gradient of U is evaluated, we find the relation between the force $\mathbf{F}$ on the charge and the potential $V(x, y, z)$. From (16.6.1) and our previously established relationship between force and potential energy, we may, therefore, write

$$\mathbf{F} = -\frac{\partial U_p}{\partial x}\,\mathbf{i}_x - \frac{\partial U_p}{\partial y}\,\mathbf{i}_y - \frac{\partial U_p}{\partial z}\,\mathbf{i}_z$$

$$= -q\left(\frac{\partial V}{\partial x}\,\mathbf{i}_x + \frac{\partial V}{\partial y}\,\mathbf{i}_y + \frac{\partial V}{\partial z}\,\mathbf{i}_z\right) \tag{16.6.2}$$

According to (16.5.27), this is in complete agreement with the idea that $\mathbf{F} = q\mathbf{E}$. Once the force is known, the Newtonian equations of motion may be utilized to study the motion of the particle. If we are not interested in the detailed trajectory followed by the charged particle, but rather only the kinetic energy or speed at a given location, energy conservation methods can be used to calculate these quantities in a particularly simple way.

There are actually rather few examples in electromagnetism in which we can solve the equations of motion and thereby study exact trajectories.[7] Fortunately, approximation techniques or energy conservation methods can be used to great advantage in many problems. Let us consider, at this point, a simple example in which the exact motion can be studied.

EXAMPLE 16.6.1

Two very large parallel planar conducting plates are separated by a distance d which is small compared to their length and width, as shown in Fig. 16.35a. The plates bear equal and opposite charge densities $\pm\sigma$, as shown, distributed uniformly over their surfaces.

Calculate the electrostatic potential function $V(x)$ in the region between the plates. From it, find the potential energy of a particle of charge q in the

---

[7] We must emphasize at this point that it is *not* generally true that "charged particles move along the lines of force," however logical this statement may sound. The reason is that, as we have seen many times before, a body's acceleration vector, which is in the direction of the resultant force, and its velocity vector, which defines the direction in which it moves, are not necessarily parallel. It is true that the lines of force give the direction of the *acceleration vector* of a charged body, but this is not the same as the direction in which the particle moves. In our future studies, we shall see several examples in which this situation arises. The electric field pattern exhibits lines of force, or lines of acceleration, but *not necessarily lines of motion!*

field, the electric field **E**, and the force on a particle of charge q. Show that the difference in potential between the plates, $\Delta V = V_2 - V_1$, is related to the charge density by $\Delta V = \sigma d/\varepsilon_0$. Show also that $E = \Delta V/d = \sigma/\varepsilon_0$ in this example.

An alpha particle (charge $+2e$, mass approximately $4M_p$, where $M_p = 1.67 \times 10^{-27}$ kg is the proton mass) enters the region between the plates at a point exactly halfway between them, with an initial velocity $\mathbf{v}_0$ parallel to the plates. Show that the path of the alpha particle is parabolic, and obtain an expression for its speed at the instant it strikes the right-hand plate in Fig. 16.35a.

From the planar symmetry of the geometry of Fig. 16.35a, it is evident that the electric field must be everywhere perpendicular to the conducting plates. The situation in this regard is very similar to that discussed in Example 16.3.2, since the charge distributions on the plates have the form of planar sheets that are essentially infinite in extent. Therefore, the field components E_y and E_z will be zero. According to (16.5.26), then, the electrostatic potential V will not vary in the y- or z-directions parallel to the plates, the derivatives $\partial V/\partial y$ and $\partial V/\partial z$ being zero. The electrostatic potential will vary only along the x-direction and can be expressed as a function of the single variable x. Under these circumstances, the partial derivative $\partial V/\partial x$ and the total derivative dV/dx are identical, and the third equation of (16.5.26) gives us

$$E_x = -\frac{dV(x)}{dx} \tag{16.6.3}$$

$$dV = -E_x \, dx \tag{16.6.4}$$

Recalling once more the results of our former calculations in Example 16.3.2, we found that the electric field produced by an infinite planar sheet of charge, uniformly distributed, is *constant* and does not vary with the distance between the charge sheet and the observation point. In this case, the total field between the plates can be regarded as a superposition of two such fields from the individual plates and must, thus, also be constant. Therefore, we may integrate Eq. (16.6.4) to find the potential function $V(x)$ in terms of the electric field, and in doing so may regard E_x to be constant. Integrating (16.6.4) from $x = 0$, at which point the potential has some value V_2, to a point whose distance from the left-hand plate is x and where the potential has the value $V(x)$, we find

$$\int_{V_2}^{V(x)} dV = -E_x \int_0^x dx \tag{16.6.5}$$

$$V(x) - V_2 = -E_x x$$
$$V(x) = V_2 - E_x x \tag{16.6.6}$$

Since it is evident from Fig. 16.35a that E_x is

positive, the potential $V(x)$ as a function of x must have the form shown in Fig. 16.35b, starting out at $x = 0$ with the value V_2 and decreasing linearly with slope E_x until, at the other plate, where $x = d$, it has the value V_1 which represents the potential of the plate on the right. According to Eq. (16.6.6), setting $x = d$ and $V(x) = V_1$, this quantity V_1 must have the value

$$V_1 = V_2 - E_x d \tag{16.6.7}$$

This, in turn, tells us that the field E_x is related to the *difference in potential*—referred to by the electricians as the "voltage difference"—between the plates, ΔV, by

$$\Delta V = V_2 - V_1 = E_x d \tag{16.6.8}$$

or

$$E_x = \frac{\Delta V}{d} = \frac{V_2 - V_1}{d} \tag{16.6.9}$$

Since the electric field has only an x-component, the field lines must always be parallel to the x-axis, as shown in Fig. 16.35a. The equipotential surfaces, which are always perpendicular to the lines of the electric field, must, therefore, have the form shown by the dotted lines parallel to the plates in that diagram.

We know now how the electric field and the potential difference between the plates are related. But how are these quantities themselves related to the surface charge density σ? This connection follows at once if we recall that the electric field at the surface of a conductor is always related to the surface charge density at every point on the surface by (16.4.1). In this example, the electric field has the same value *everywhere* within the region between the plates, so that its value at the surface as obtained from (16.4.1) is the same as its value at any other point. From (16.4.1), therefore, we find that

$$E_x = E_{x(\text{surf})} = \frac{\sigma}{\varepsilon_0} \tag{16.6.10}$$

which allows us to write (16.6.8) as

$$\Delta V = \frac{\sigma d}{\varepsilon_0} \tag{16.6.11}$$

Also, using Eq. (16.6.9), we can write the potential function $V(x)$ in the slightly different and more useful form

$$V(x) = V_2 - (V_2 - V_1)\frac{x}{d} = V_2 - \frac{\sigma x}{\varepsilon_0} \tag{16.6.12}$$

The force on a charge q in the field may, of course, be obtained as $\mathbf{F} = q\mathbf{E}$, where E is given by (16.6.9) or (16.6.10). It may also be obtained from the potential function by writing the potential energy of the charge, which, from (16.6.1), is

$$U_p(x, y, z) = qV(x, y, z) = qV_2 - q(V_2 - V_1)\frac{x}{d}$$

$$(16.6.13)$$

The force components F_x, F_y, and F_z can then be found as the partial derivatives of the potential energy, $\partial U_p/\partial x$, $\partial U_p/\partial y$, and $\partial U_p/\partial z$. In this case, of course, the second and third of these are zero, and

$$F_x = -\frac{dU_p}{dx} = q(V_2 - V_1)/d \qquad (16.6.14)$$

It is apparent from (16.6.9) that this agrees with $F_x = qE_x$.

We now know the force on the charge and are, therefore, in a position to determine its motion from ordinary Newtonian mechanics. We write, therefore, equating (16.6.14) with ma_x,

$$a_x = \frac{F_x}{m} = \frac{q(\Delta V)}{md}$$

$$(16.6.15)$$

$$a_y = a_z = 0$$

There is a *constant* acceleration a_x along the x-direction, while the other components of acceleration are zero. There is a constant initial velocity v_0 along the y-direction. The problem is thus one of projectile motion. From the "golden rules" we can write

$$x - x_0 = v_{0x}t + \tfrac{1}{2}a_xt^2 = \tfrac{1}{2}a_xt^2 \qquad (16.6.16)$$

since $v_{0x} = 0$. Since $v_{0y} = v_0$, the golden rules also allow us to write

$$y - y_0 = v_{0y}t = v_0t$$

$$t = \frac{y - y_0}{v_0} \qquad (16.6.17)$$

If we now substitute the value of t given above into (16.6.16), we eliminate the time between the two equations, obtaining $x - x_0$ as a function of y, which is the equation of the path taken by the charged particle. In this way, using (16.6.15) to express a_x, we find

$$x - x_0 = \frac{q\,\Delta V}{2mv_0^2 d}(y - y_0)^2 \qquad (16.6.18)$$

which is clearly the equation of a *parabola*. We should have been able to infer this, in any event, as soon as we saw that the motion was that of a projectile. It is important to observe that this trajectory, as illustrated in Fig. 16.35a, is *not* along the lines of force. Initially, at least, it is directly *across* them!

We could also find, by setting $y - y_0 = d/2$ in (16.6.17), how long it takes the particle to go from its initial position midway between the plates to the surface of the plate on the right. We could then determine

v_x for this time from $v_x = v_{0x} + a_xt = a_xt$. Since we know also that $v_y = v_{0y} = v_0$ at all times, we could finally arrive at the particle's speed upon impact. Since we are only asked for the speed, it is much easier, however, to find its final kinetic energy using the law of conservation of energy, which tells us that

$$U_{kf} + U_{pf} = U_{ki} + U_{pi} \qquad (16.6.19)$$

In this instance, $U_{ki} = \tfrac{1}{2}mv_0^2$ while $U_{kf} = \tfrac{1}{2}mv_f^2$, where v_f is the final velocity. From the potential energy function (16.6.18), it is easy to show that the initial potential energy, at $x = d/2$, midway between the plates is given by $U_{pi} = q(V_1 + V_2)/2$, while the final potential energy, when $x = d$, is $U_{pf} = qV_1$. Substituting all these values into (16.6.19), we find

$$\tfrac{1}{2}mv_f^2 + qV_1 = \tfrac{1}{2}mv_0^2 + \tfrac{1}{2}qV_1 + \tfrac{1}{2}qV_2 \qquad (16.6.20)$$

from which

$$v_f^2 = v_0^2 + \frac{q\,\Delta V}{m} \qquad (16.6.21)$$

For an alpha particle,

$$m = 4M_p = (4)(1.673 \times 10^{-27})\text{ kg}$$

and $q = 2e = (2)(1.602 \times 10^{-19})$ coulomb. If the initial velocity were given by $v_0 = 1.00 \times 10^6$ m/sec, and if there were a potential difference of 20,000 volts across the plates, which are spaced 2.5 cm apart, the velocity on impact will be

$$v_f^2 = (1.00 \times 10^6)^2 + \frac{(3.204 \times 10^{-19})(2 \times 10^4)}{6.69 \times 10^{-27}}$$

$$= 1.958 \times 10^{12}\text{ m}^2/\text{sec}^2$$

$$v_f = 1.400 \times 10^6\text{ m/sec}$$

The initial kinetic energy of the particle is

$$U_{ki} = \tfrac{1}{2}mv_0^2 = (\tfrac{1}{2})(6.69 \times 10^{-27})(1.00 \times 10^6)^2$$

$$= 3.346 \times 10^{-15}\text{ J}$$

while its final kinetic energy will be

$$U_{kf} = \tfrac{1}{2}mv_f^2 = (\tfrac{1}{2})(6.69 \times 10^{-27})(1.400 + 10^6)^2$$

$$= 6.551 \times 10^{-15}\text{ J}$$

The potential energy it loses in traveling from its starting point to the plate on the right is

$$\Delta U_p = U_{pf} - U_{pi} = -\tfrac{1}{2}q\,\Delta V$$

$$= -(\tfrac{1}{2})(3.204 \times 10^{-19})(2 \times 10^4)$$

$$= -3.204 \times 10^{-15}\text{ J}$$

These numerical values are inconveniently small. In expressing the energies associated with elementary particles, it is therefore customary to use energy units of *electron volts* (eV) rather than joules.

The electron volt is defined as the *change in potential energy experienced by a particle of charge e in moving through a potential difference of 1 volt.* In joules, this energy is

$$1.000 \text{ eV} = (e)(1.000 \text{ V}) = (1.6022 \times 10^{-19} \text{ C})(1.000 \text{ J/C})$$
$$= 1.6022 \times 10^{-19} \text{ J}$$

Expressed in these units, the initial kinetic energy of the alpha particle is

$$U_{ki} = \frac{3.346 \times 10^{-15}}{1.6022 \times 10^{-19}} = 20,880 \text{ eV}$$

while its final kinetic energy U_{kf} would be 40,880 eV and the loss in potential energy would be exactly 20,000 eV. This figure can most easily be arrived at by observing that the alpha particle has charge $2e$ and falls in potential by half the total potential difference ΔV, that is, 10,000 volts. Its change in potential energy is, therefore, $-q \, \Delta V = -(2e)(10,000) = -20,000e$ joules, or, in view of the definition of the electron volt, $-20,000$ eV.

The student should be able to verify that the numerical data assumed above imply that the electric field between the plates is 8.00×10^5 volts/meters, and that their surface charge density is 7.08 coulombs/meter^2.

EXAMPLE 16.6.2

In the lowest atomic state (ground state) of atomic hydrogen, the electron has an energy of -13.6 eV. This is called the binding energy. The magnitude of the potential energy is twice the magnitude of the kinetic energy. Find the average distance between the electron and the proton in the ground state of hydrogen.

In the absence of any external forces, the hydrogen atom has a constant total energy of -13.6 eV, which is equal to the *sum* of the kinetic and potential energies. Therefore,

$$-13.6 \text{ eV} = -2.18 \times 10^{-18} \text{ J} = \tfrac{1}{2}mv^2 - \frac{e^2}{4\pi\varepsilon_0 r} \quad (16.6.22)$$

The potential energy is negative and is obtained by using (16.6.1) with Q given by $-e$ (the electron charge) and V the potential produced by the proton, whose charge is $+e$. Now, since the *average value* of the kinetic energy is one half the average value of the magnitude of the potential energy, (16.6.22) can be rewritten as

$$-2.18 \times 10^{-18} = \frac{1}{2}\left(\frac{e^2}{4\pi\varepsilon_0 r}\right)_{av} - \left(\frac{e^2}{4\pi\varepsilon_0 r}\right)_{av}$$

$$= -\frac{1}{2}\left(\frac{e^2}{4\pi\varepsilon_0 r}\right)_{av}$$

$$= -(\tfrac{1}{2})(9 \times 10^9)\frac{(1.6 \times 10^{-19})^2}{r_{av}}$$

$$r_{av} = 0.53 \times 10^{-10} \text{ m} \quad (16.6.23)$$

This average distance between the electron and proton in the ground state of hydrogen is known as the *first Bohr radius*, and it gives us the approximate size of the hydrogen atom in its lowest state. We have deliberately emphasized the need for the averaging process above since it is believed that the kinetic energy and the potential energy are perpetually changing in the atom. This point of view is supported by modern treatments of atomic physics, which ultimately involve the use of quantum mechanics.

EXAMPLE 16.6.3

A thin spherical shell of radius R has a total charge Q which is uniformly distributed throughout the shell. Obtain an expression for the potential energy of the charges on the sphere in terms of R and Q.

The charge distribution on the sphere can be thought of as having been assembled there from very many tiny elements of charge, each of magnitude dq, which are transported from infinity, where their potential energy is zero, up to the sphere. Each of these acquires potential energy as it is placed on the sphere because it has been moved from infinity, where the electrostatic potential is zero, to the surface of the sphere, where the electrostatic potential has the value $q/4\pi\varepsilon_0 R$ as determined in Example 16.5.6, where q is the amount of charge *already* present on the sphere.

Figure 16.36 illustrates an intermediate stage of this process in which charge q is uniformly distributed on the sphere and charge dq is brought from infinity to the sphere. The change in the potential energy due to the movement of dq from infinity to the sphere will be called dU. This quantity is given by dq multiplied by the difference of potential experienced by the charge. Since the potential at the surface of the sphere is $q/4\pi\varepsilon_0 R$, the change in potential energy is

$$dU = V \, dq = \frac{q \, dq}{4\pi\varepsilon_0 R} \quad (16.6.24)$$

In this example, we integrate to find the potential energy U after the total charge Q has been brought in. Upon integration between the initial limits $q = 0$ when $U = 0$ to the final stage in which $q = Q$ and the total potential energy is U, we have

$$U = \int_0^U dU = \int_0^Q \frac{q \, dq}{4\pi\varepsilon_0 R} = \frac{1}{4\pi\varepsilon_0 R} \int_0^Q q \, dq$$

$$= \frac{1}{4\pi\varepsilon_0 R}\left[\frac{q^2}{2}\right]_0^Q = \frac{Q^2}{8\pi\varepsilon_0 R} \quad (16.6.25)$$

It is interesting to note that in this example, the potential energy of the final charge distribution is proportional to the *square* of the total charge. It should also be evident that the work that must be done to assemble the charge distribution is equal to the final potential energy stored in its field.

FIGURE 16.36

The electrostatic potential energy of the protons in atomic nuclei can be estimated using this result. For example, the nuclei of lead atoms, whose atomic number Z is 82, have a radius R of 7.8×10^{-13} cm. We may assume for this purpose that the nucleus behaves like a thin spherical shell of radius R having total charge Ze. This is probably not a very realistic model of the nucleus, but it does result in a spherically symmetric distribution of charge of about the right size, which is the most important aspect involved in determining the final potential energy. Anyway, since all we are after is an order-of-magnitude estimate, it will serve adequately for the purpose at hand.

Equation (16.6.25) represents the potential energy of a distribution of charge having this form. If we set $Q = Ze$ in that expression, we obtain

$$U = \frac{(Ze)^2}{8\pi\varepsilon_0 R} \tag{16.6.26}$$

Now, using $Z = 82$ and the value of the nuclear radius R given above, we obtain

$$U = \frac{[(82)(1.602 \times 10^{-19})]^2}{(8)(\pi)(8.85 \times 10^{-12})(7.8 \times 10^{-15})}$$

$$= 9.95 \times 10^{-11} \text{ J} = 6.20 \times 10^8 \text{ eV} = 620 \text{ MeV}$$

where MeV = million electron volts. This energy is much larger than typical electronic energies in atoms, which are of the order of a few electron volts (you may recall that the energy of the electron in the hydrogen atom is 13.6 eV).

The potential energy above is a purely repulsive electrostatic energy and is not the only type of energy of importance in the nucleus. Clearly, the nucleus would fly apart as a result of Coulomb repulsion unless there were also attractive forces to hold it together. These attractive forces are due to the *strong interaction* between particles within the nucleus. They also contribute significantly to the potential energy and are, in fact, responsible for changing the *sign* of the total energy of the nucleus, thus resulting in a total energy that is, on balance, attractive rather than repulsive. The above calculation reveals, nevertheless, that nuclear binding energies may be on the scale of *billions of electron volts*—thus billions of times greater than the energies involved in electronic and atomic processes—

and that huge amounts of energy may be involved in fission reactions in which atomic nuclei are broken up. As a general rule, the energy contained within a nucleus, including both kinetic and potential energy, may be estimated by assuming a total energy of about 8 million electron volts for each proton and neutron within the nucleus.

EXAMPLE 16.6.4

Find the torque that a dipole of moment **p** experiences in the presence of a uniform external electric field **E**. Find the potential energy of such a dipole as a function of its orientation with respect to such an external field.

A dipole consisting of charges $+q$ and $-q$ separated by a distance d is shown in Fig. 16.37. It is placed in a uniform electric field **E**, and in the figure it is shown with its axis oriented at an angle θ to the field. Due to the field, there are forces of magnitude qE exerted in opposite directions on the individual charges as shown, and these forces generate a torque that tends to align the dipole so that its axis is parallel to the direction of the field. Work must be done, therefore, to turn the dipole so that its axis makes an angle θ with the field. This work is stored as *potential energy* of the dipole, and by calculating the amount of work done in turning the dipole through a given

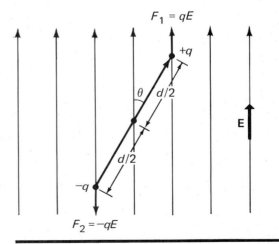

FIGURE 16.37. Forces and torques on a dipole in a uniform external field.

angle, the potential energy of the dipole in the external field may be ascertained.

Let us assume that the dipole is oriented initially at an angle θ_0 and then turned to some other angle θ. The torque about the center of the dipole caused by the two forces at the ends is

$$\tau = qE \frac{d}{2} \sin \theta + qE \frac{d}{2} \sin \theta = qEd \sin \theta$$

or, since the dipole moment has magnitude $p = qd$,

$$\boxed{\tau = pE \sin \theta} \qquad (16.6.27)$$

It is evident that this is the same as the magnitude of the vector $\mathbf{p} \times \mathbf{E}$ and that the torque vector τ also has the same direction as the vector $\mathbf{p} \times \mathbf{E}$. We may, therefore, write Eq. (16.6.27) as

$$\boxed{\tau = \mathbf{p} \times \mathbf{E}} \qquad (16.6.28)$$

to express the torque experienced by a dipole in an external electric field.

Since this torque varies with the angle θ, we must write the work done in an infinitesimal displacement $d\theta$ as a function of θ and integrate over the angular displacement to find the potential energy of the dipole:

$$dW = dU = \int_{\theta_0}^{\theta} \tau \, d\theta = pE \int_{\theta_0}^{\theta} \sin \theta \, d\theta$$

or

$$U(\theta) = pE[-\cos \theta]_{\theta_0}^{\theta} = pE(\cos \theta_0 - \cos \theta) \qquad (16.6.29)$$

But it makes things simpler if we choose θ_0, the angle at which the potential energy is taken to be zero, to be $90°$, so that $\cos \theta_0 = \cos 90° = 0$. Under these circumstances, (16.6.29) becomes

$$\boxed{U(\theta) = -pE \cos \theta} \qquad (16.6.30)$$

16.7 Poisson's Equation

There are many different ways to solve problems in electrostatics. The electric field may be calculated by superposing point charge fields or by the use of Gauss's law, or by finding the electrostatic potential and calculating its partial derivatives with respect to x, y, and z. This range of possibilities may seem to be a bit confusing at first sight, but all of these possible alternatives stem directly from Coulomb's law, coupled with the concepts of conservative forces, electrostatic potential, and electric flux. It is apparent at this stage that mathematics plays a rather crucial role in electrostatics, and the serious student should, therefore be sure to master the mathematical concepts that are important in this area of physics.

In addition to all these methods, there is still another approach that is very useful in solving electrostatic problems, particularly those in which a specified continuous three-dimensional distribution of charge is concerned. This involves the use of an expression called *Poisson's equation*, which relates the given charge density at any point in space with the electrostatic potential at that point.

Any continuous three-dimensional distribution of charge can be characterized by a volume charge density $\rho(x, y, z)$ which gives the amount of charge per unit volume in the neighborhood of any point. For simplicity, we shall consider only a charge density distribution $\rho(x)$ that depends on the single coordinate x and is completely independent of y and z. Since such a charge distribution is completely symmetric about the x-axis, it can give rise to no resultant electric fields with y- or z-components; any electric field that it creates will have only an x-component E_x. For this reason, also, $\partial V/\partial y = \partial V/\partial z = 0$; the electrostatic potential will then depend only upon x and will be related to the field E_x by

$$E_x = -\frac{dV(x)}{dx} \qquad (16.7.1)$$

This situation is illustrated in Fig. 16.38.

Now let us draw a Gaussian surface having the form of a thin, rectangular box of thickness dx and surface area A, such as the surface S in the figure. We shall now apply Gauss's law to the surface. We see at once that on the parts of the surface that are normal to the y- and z-directions, the field $\mathbf{E}$ and the normal vectors (such as $\mathbf{n}'' = \mathbf{i}_y$) to these surfaces are mutually perpendicular. There can thus be no contribution to the electric flux from these surfaces, since $\mathbf{E} \cdot \mathbf{n}$ is zero there.

On the two surfaces that are normal to the x-direction, at x and $x + dx$, the field has the values $\mathbf{i}_x E_x(x)$ and $\mathbf{i}_x E_x(x + dx)$, respectively. The outward unit vector normal to the surface at x is $\mathbf{n}' = -\mathbf{i}_x$, while at $x + dx$ it is $\mathbf{n} = \mathbf{i}_x$. Gauss's law then tells us that

$$\oint_s \mathbf{E} \cdot \mathbf{n} \, da = \int_{s_1} E_x(x + dx) \, da - \int_{s_2} E_x(x) \, da = \frac{q_0}{\varepsilon_0}$$

$$(16.7.2)$$

But since the field E_x depends only on x and is independent of y and z, its value is the same at all points on both of the surfaces S_1 and S_2, and therefore $E_x(x)$ and $E_x(x + dx)$ can be written outside the integrals in the above equation. This gives us

$$\oint_s \mathbf{E} \cdot \mathbf{n} \, da = [E_x(x + dx) - E_x(x)]A = \frac{q_0}{\varepsilon_0} \qquad (16.7.3)$$

The charge inside the Gaussian surface, q_0, can now

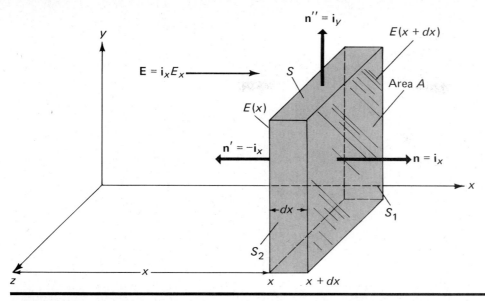

FIGURE 16.38. Gaussian surface used in the derivation of Poisson's equation.

be written as the product of the charge density $\rho(x)$ and the volume $A\,dx$. Dividing by dx on both sides, we now obtain,

$$\frac{[E_x(x + dx) - E_x(x)]A}{dx} = \frac{\rho(x)A}{\varepsilon_0}$$

or, in view of the definition of the derivative,

$$\boxed{\frac{dE_x}{dx} = \frac{\rho(x)}{\varepsilon_0}} \qquad (16.7.4)$$

Since E_x can be written as $-dV(x)/dx$, however, this can also be written in the form

$$\boxed{\frac{d^2V}{dx^2} = -\frac{\rho(x)}{\varepsilon_0}} \qquad (16.7.5)$$

Equations (16.7.4) and (16.7.5) are referred to as the one-dimensional forms of *Poisson's equation*. They are *differential* equations, which can be solved, much like those that arise from Newton's second law in mechanics, whenever the charge density $\rho(x)$ is known, to give the electrostatic potential and the electric field. Alternatively, of course, if E_x or $V(x)$ is known, the charge density at any point follows at once from (16.7.4) or (16.7.5).

In studying the application of Gauss's law to electrostatic problems, we are always faced with two very troublesome problems. First of all, we are assured of success only in situations that are highly symmetric, in which we can be sure that the magnitude of the field is constant and either perpendicular or parallel to an appropriate surface of integration. Also, there is no well-defined way of choosing such an appropriate surface, even when one exists; it is

a matter of trial and error. Therefore, even though Gauss's law is extremely important and useful in many instances, it has certain inherent limitations that stem chiefly from the fact that it involves *the evaluation of a surface integral.*

You will observe that the derivation given above simply *restates* Gauss's law in the form of a differential equation involving the *derivative* of the field (or potential) evaluated at any point rather than an integral evaluated over a surface. Thus, Poisson's equation can be thought of as the differential form of Gauss's law. This restatement of Gauss's law eliminates the two objections referred to above in connection with the integral form and gives us an equation with which we can accomplish the same purposes in situations where the use of Gauss's law in its integral form would be completely out of the question. It is, of course, not always easy to solve Poisson's equation either, and in situations that are highly symmetric it is still much easier to use Gauss's law in its original form. But, where that high degree of symmetry is absent and where simple procedures involving direct superposition of point charge fields or potentials are not possible, Poisson's equation provides a powerful and rational starting point for solving extremely difficult problems that could not be worked in any other way.

The extension of Poisson's equation to three-dimensional systems in which the potential and charge density depend upon all three Cartesian coordinates is tedious but relatively straightforward, and we shall not attempt it here. As in the one-dimensional case, it involves nothing more than the transformation of Gauss's law to the form of a differential equation. The three-dimensional Poisson equation has the form

519

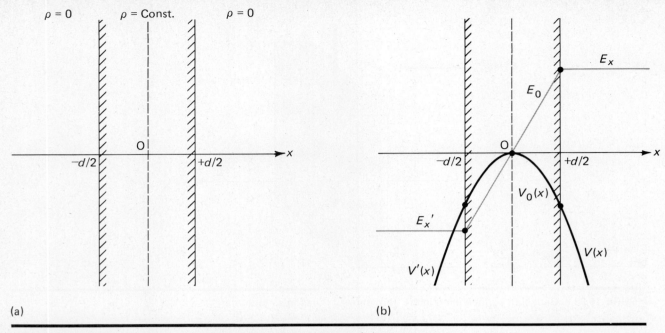

$\rho = 0$ $\rho = \text{Const.}$ $\rho = 0$

(a)

(b)

FIGURE 16.39. (a) Geometry of the charge distribution of Example 16.7.1. (b) Plot of field and potential distributions as given by Eqs. (16.7.14).

$$\frac{\partial E_x}{\partial x} + \frac{\partial E_y}{\partial y} + \frac{\partial E_z}{\partial z} = \frac{\rho}{\varepsilon_0} \qquad (16.7.6)$$

or, since $E_x = -\partial V(x, y, z)/\partial x$, etc.,

$$\frac{\partial^2 V}{\partial x^2} + \frac{\partial^2 V}{\partial y^2} + \frac{\partial^2 V}{\partial z^2} = -\frac{\rho}{\varepsilon_0} \qquad (16.7.7)$$

Since these are partial differential equations, their solution is a more complex business than that involved in the one-dimensional case.

The use of Poisson's equation is by no means restricted to systems in which there is a continuously distributed density of charge throughout all space. In situations where the charge distributions are on conducting surfaces, for example, the volume charge density is zero in the surrounding space. Under these circumstances, we may simply set $\rho = 0$ in Eqs. (16.7.4) to (16.7.7) and solve the resulting differential equation—now referred to, incidentally, as *Laplace's equation*—using as a boundary condition the requirement that the component of **E** *tangential* to those surfaces vanish at every point on them. In the solution that is obtained, the component of **E** *normal* to the conducting surfaces tells us, by way of (16.4.1), just what the surface charge density at every point must be! This would be the only way of obtaining the electric field, electrostatic potential, and surface charge density distribution in a situation such as that illustrated in Fig. 16.22c. The reader, in this way, should now be able to use (16.7.5) and the boundary condition

(16.4.1) to obtain all the results of the potential calculation of Example 16.6.1.

Poisson's equation can be transformed for use in cylindrical and spherical coordinate systems. The cylindrical and spherical forms of Poisson's equation are fairly simple and very useful in situations whose geometry makes cylindrical or spherical coordinates particularly appropriate. We shall not, however, attempt to discuss these cases any further at this point. The usefulness of Poisson's equation will be illustrated by the following example.

EXAMPLE 16.7.1

A plane-parallel plate of thickness d, infinite in extent along the y- and z-directions, carries a constant charge density ρ_0 distributed throughout its volume as illustrated in Fig. 16.39. Using Poisson's equation, calculate the electric field and electrostatic potential at every point, both inside and outside the plate.

Outside the plate, $\rho(x) = 0$, and (16.7.4) and (16.7.5) become

$$\frac{dE_x}{dx} = 0 \quad \text{and} \quad \frac{d^2 V}{dx^2} = 0 \qquad \text{outside} \qquad (16.7.8)$$

This tells us at once that E_x must be *constant* everywhere outside the plate, while $V(x)$ must vary linearly with distance. Therefore,

$$V(x) = \alpha x + \beta \qquad E_x = -\frac{dV}{dx} = -\alpha \qquad x > \frac{d}{2}$$

$$(16.7.9)$$

$$V'(x) = \alpha'x + \beta' \qquad E_x' = -\frac{dV}{dx} = -\alpha' \qquad x < -\frac{d}{2}$$

$$(16.7.10)$$

where α, β, α', β' are arbitrary constants. Inside the plate, $\rho(x) = \rho_0$. This means that the potential must be of the form

$$V_0(x) = -\frac{\rho_0 x^2}{2\varepsilon_0} + \alpha_0 x + \beta_0$$

while

$$E_0 = -\frac{dV_0}{dx} = \frac{\rho_0 x}{\varepsilon_0} - \alpha_0 \qquad -\frac{d}{2} < x < \frac{d}{2} \qquad (16.7.11)$$

where α_0 and β_0 are arbitrary constants.

Now, we are free to choose the point where the potential is zero anywhere we like. Let us require that it be zero at $x = 0$. This means that in (16.7.11), which is applicable in the range $(-d/2 < x < d/2)$, we may set β_0 equal to *zero*. Also, due to the symmetry of the charge distribution, we should expect the field on either side to have the same magnitude, though, since it will be in opposite directions in the two cases, it will have the opposite sign. Therefore, in (16.7.9) and (16.7.10) we must take $-\alpha = -(-\alpha')$, or $\alpha' = -\alpha$. Likewise, again on account of the symmetry of the charge distribution, we know that the change in potential energy experienced by a test charge in moving from $x = 0$ to a given distance l from the origin will be the same whether the charge moves to the left or to the right. Therefore, we must find that $V(l) = V'(-l)$, or, since $\alpha' = -\alpha$,

$$\alpha l + \beta = -\alpha(-l) + \beta'$$
$$\beta = \beta'$$

Putting all this information into (16.7.9), (16.7.10), and (16.7.11), we obtain

$$V(x) = \alpha x + \beta \qquad E_x = -\alpha \qquad x > \frac{d}{2}$$

$$V_0(x) = -\frac{\rho_0}{2\varepsilon_0}x^2 + \alpha_0 x \quad E_0 = \frac{\rho_0 x}{\varepsilon_0} - \alpha_0 \quad -\frac{d}{2} < x < \frac{d}{2}$$

$$V'(x) = -\alpha x + \beta \qquad E_x' = \alpha \qquad x < -\frac{d}{2}$$

$$(16.7.12)$$

However, both the electric fields, representing the force on a test charge, and the potentials, representing the work done in moving such a charge, must vary continuously everywhere within the field regions, including the regions near $x = \pm d/2$ which mark the boundaries of the plate. There is no reason why the force on any test charge should undergo a sharp discontinuous change there, though it may start increasing or decreasing more rapidly in crossing such

a boundary.[8] We may, therefore, equate the field E_x outside the plate and the field E_0 inside the plate at $x = d/2$, and the field E_x' outside and E_0 inside at $x = -d/2$. In this way, from (16.7.12) we obtain

$$\frac{\rho_0 d}{2\varepsilon_0} - \alpha_0 = -\alpha$$

and

$$-\frac{\rho_0 d}{2\varepsilon_0} - \alpha_0 = \alpha \qquad (16.7.13)$$

Adding these two equations, we find at once that $\alpha_0 = 0$. Subtracting one from the other informs us that $\alpha = -\rho_0 d/2\varepsilon_0$. In the same way, equating the expressions for the potential $V(x)$ outside the plate and $V_0(x)$ inside at $x = d/2$, we may find (using the values for α and α_0 given above) that $\beta = \rho_0 d^2/8\varepsilon_0$. Putting all these values back into (16.7.12), then, the final result is

$$V(x) = \frac{\rho_0 d^2}{8\varepsilon_0}\left(1 - \frac{4x}{d}\right) \qquad E_x = \frac{\rho_0 d}{2\varepsilon_0} \qquad x > \frac{d}{2}$$

$$V_0(x) = -\frac{\rho_0 x^2}{2\varepsilon_0} \qquad E_0 = \frac{\rho_0 x}{\varepsilon_0} \qquad -\frac{d}{2} < x < \frac{d}{2}$$

$$V'(x) = \frac{\rho_0 d^2}{8\varepsilon_0}\left(1 + \frac{4x}{d}\right) \qquad E_x' = -\frac{\rho_0 d}{2\varepsilon_0} \qquad x < -\frac{d}{2}$$

$$(16.7.14)$$

16.8 Applications of Electrostatics[9]

Electrostatic principles can be effectively utilized in a variety of ways to accomplish tasks which might otherwise be difficult or even impossible. Thus, the advent of xerography has revolutionized the copying industry, electrostatic coating and painting has saved huge amounts of money in paint alone, electrostatic separators have provided an excellent way of separating minerals from one another, and electrostatic precipitators are used at processing plants to reduce the level of atmospheric pollution. The van de Graaff generator is a device which has many applications in

---

[8] The electric field does change discontinuously, from zero inside to some finite value outside, at the surface of a conductor. In this case, however, there is a surface charge distribution consisting of a finite charge contained within a region of zero thickness, hence of zero volume. In this sense, therefore, such a surface charge distribution represents a singularity in a volume distribution, where the volume charge density becomes infinite in a region of vanishingly small spatial extent. In any situation such as this one, in which only finite volume charge densities are involved, fields and potentials will always be continuous.

[9] A more detailed account of some of the examples discussed in this section is given in an article by A. D. Moore, *Scientific American* (March 1972).

science and technology, especially as an accelerator of elementary particles. The principle of electrostatic shielding is extremely important, for it provides a method of shielding delicate instrumentation from unwanted spurious electrical phenomena which may be occurring elsewhere.

In this section, we shall discuss some examples to demonstrate the usefulness of electrostatics, thereby providing motivation for further study of this beautiful subject.

ELECTROSTATIC SHIELDING

Let us consider first the most elementary example of electrostatic shielding. A conductor completely encloses some region of space as shown in Fig. 16.40. Assume the interior region contains no electric charges at all. We can then show that the electric field inside this region is necessarily zero, regardless of the changes of charge distribution on the *outside* surface of the conductor or the electric fields that might be present outside the conductor. To prove this assertion, we use the fact that the interior wall of the conductor must be an *equipotential surface*. Assume the electric field within the cavity is not zero and that a typical field line starting at point A and terminating at point B is drawn, as shown in the diagram. To obtain the potential difference between points B and A, we must integrate the electric field along any path. Thus,

$$V_{\mathrm{B}} - V_{\mathrm{A}} = - \int_{\mathrm{A}}^{\mathrm{B}} \mathbf{E} \cdot d\mathbf{r} \qquad (16.8.1)$$

Let us choose a path which follows this field line. Then, along this path **E** and *d***r** are parallel and **E** · *d***r** is always a positive number. This implies that the integral on the right-hand side is positive. But since A and B lie on the same equipotential surface, $V_{\mathrm{B}} - V_{\mathrm{A}}$ is zero and, therefore, the integral must also be zero. This leads to a contradiction which can only

be resolved by assuming there are *no field lines* that commence at a point such as point A and terminate at point B. Thus, the electric field within the cavity must be zero when there are no charges within the cavity. We see, therefore, that by surrounding a cavity with conducting walls, it is possible to maintain a *field-free region*. This region might be a small region in which we want to have a zero field, or it might be an entire laboratory in which sensitive electrical measurements are being made.

There is a generalization of the above discussion for the case in which the cavity does contain electric fields produced by charges inside. Under these conditions, it is possible to show that the fields inside are produced only by the charges inside. Any **E** fields or charges outside have no effect on the **E** field inside. Thus, the conductor electrostatically shields the cavity inside. These same conclusions follow from the situation discussed in Example 16.4.1, which might profitably be reviewed at this point. It is well known that one of the safest places to be during a severe thunderstorm in which there is lightning is in the interior of an automobile. The automobile is not a perfectly conducting shell but it does function nevertheless quite well as an electrostatic shield because a large part of the body is made of sheet metal of high electrical conductivity.

ELECTROSTATIC PRECIPITATION

Electrostatic precipitation is a process by which solid particles or liquid droplets are removed from the gas in which they are suspended. The basic technique is illustrated in Fig. 16.41. The outer walls of the duct which carries the gas and suspended particles is *grounded*. This simply means that the electrostatic potential of the walls is exactly the same as that of the earth. If a person standing on the ground touches the walls, there is no potential difference between his hand and his foot and, therefore, no danger of current flow through his body.[10] At the center of the duct, a straight axial wire (called the corona wire) is kept at a very high potential of many kilovolts relative to ground. This voltage could be positive or negative. If it is *negative*, an electric field will exist within the duct and will point radially inward toward the corona wire. For this type of geometry, the electric field varies as $1/r$, as demonstrated in Example 16.3.3. Assume

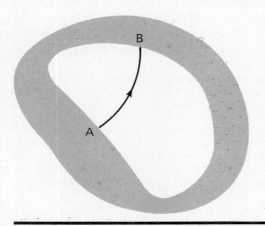

FIGURE 16.40. Calculation of the field inside a cavity in a conducting body.

[10] The electrical conductivity of the earth is sufficiently high that the earth's electrical behavior is in most situations much like that of a highly conducting body. The electrical conductivity of the earth is often utilized to provide a path through which electric charges sent from generating plants to commercial users of electricity can return to the generating station. The terminals of the generator, and of the user's appliance, that are led into the earth in such a situation are said to be grounded.

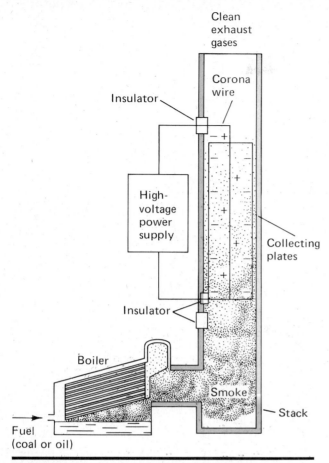

FIGURE 16.41. Precipitation of smoke and dust particles by electrostatic means in a coal- or oil-burning power plant.

$E = \gamma/r$, where γ is a constant to be determined from the potential difference between the corona wire and the outer walls. Assume a negative potential difference V between a corona wire of radius r_c and the walls a radial distance r_w from the center. Then,

$$V = -\int_{r_c}^{r_w} E \, dr = -\int_{r_c}^{r_w} \gamma \frac{dr}{r} = -\gamma [\ln r]_{r_c}^{r_w}$$

$$= -\gamma \ln \frac{r_w}{r_c} \tag{16.8.2}$$

from which it follows that

$$E = -\left(\frac{V}{\ln \dfrac{r_w}{r_c}} \right) \frac{1}{r} \tag{16.8.3}$$

From this, it is evident that the electric field increases with decreasing r and is, therefore, strongest close to the negative corona wire. If the field is strong enough, it can actually tear electrons out of gas atoms and in this way *ionize* the gas. In air, this begins to happen when the electric field intensity E reaches some 30,000 volts/cm. Under these circumstances, a *corona discharge* is set up in which currents of charged particles flow, the electrons being attracted to the positive outer wall of the chamber and the ions to the negative corona wire. The corona discharge is visible to the naked eye as a greenish glow emitted by the excited atoms.

As electrons move from the region of the corona discharge near the central wire toward the positive outer wall, they encounter the particles of dust or other matter that are to be precipitated from the gas. In doing so, the electrons stick to the particles, imparting to them a negative charge. The negatively charged particles are then attracted to the outer wall where they collect and may be periodically removed. *Electrostatic painting* may be accomplished if instead of dust or soot, paint spray droplets are present and the object to be painted is made the positive electrode of the system.

If Eq. (16.8.3) is solved for V and if we set $r = r_c$, we find that the potential difference between the corona wire and the outer electrode required to produce a given field E at the surface of the corona wire is

$$V = -Er_c \ln \frac{r_w}{r_c} \tag{16.8.4}$$

For a typical situation involving a chamber for which $r_w = 100$ cm and a corona wire of radius $r_c = 0.25$ cm $= 0.025$ m, this tells us that a potential difference of some 6200 volts will suffice to produce a field of 30,000 volts/cm at the surface of the corona wire. Such a potential difference can be established and maintained quite easily.

Electrostatic precipitators are in widespread use. There are many different designs to suit various purposes. Some of the uses involve the reduction of pollution by eliminating potentially harmful industrial effluent gases in cement plants, steel mills, power stations, and smelters. Electrostatic precipitators can also recover valuable materials such as oxides of copper, tin, and other metals from smelter stack gases.

XEROGRAPHY

In the past fifteen years, the imaging process known as *xerography* has come into widespread use and has proved to be of great benefit to those who own Xerox stock, as well as to those who are required to copy printed or handwritten documents. The basic process was conceived by Chester Carlson shortly before World War II. It depends upon a physical effect known as *photoconductivity*, in which light striking a highly insulating material gives electrons that are normally tightly bound to the atoms enough energy to set them free and to enable them to travel about in the substance very much like the free electrons in highly conducting metals.

523

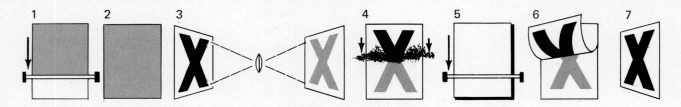

a. How xerography works

1. Surface of selenium-coated plate is electrically charged as it passes under wires. 2. Plus marks represent positively charged plate. 3. Original document is projected through camera lens. Plus marks here represent latent image retaining positive charge. Charge is drained away in areas that are exposed to light. 4. Negatively charged powder is cascaded over plate and adheres to positive image. Latent image now becomes visible. 5. Sheet of paper (or paper offset master) is placed over plate, and paper is given positive charge. 6. Positively charged paper attracts powder from plate, forming direct positive image. 7. Print or offset master is fused by heat for permanency.

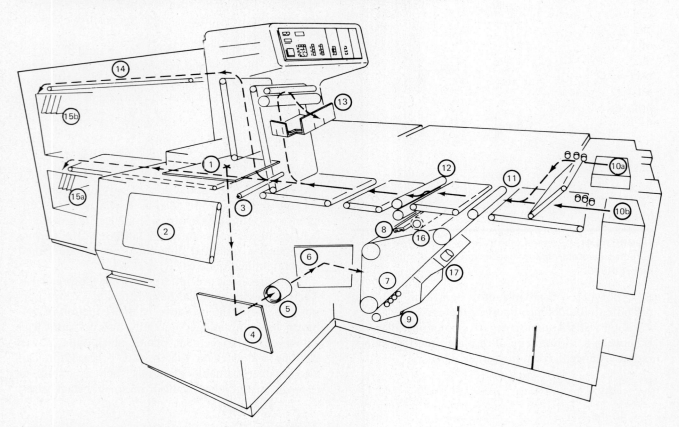

b. Step-by-step operation of the xerox 9200 duplicating system

Original document is moved to flat platen (1) by automatic document handler (2) and flashed by high intensity flash lamps (3). Image is reflected to object mirror (4) through stationary lens (5) and reflected by image mirror (6) to the selenium — alloy coated belt (7).

The charge corotron (8) places a positive charge of electricity onto the belt. The light reflected through the optical system removes the charge except for the desired image. Developer with negatively charged toner (9) is brushed onto the moving belt and the toner sticks to the charged portions of the belt.

Ordinary, uncoated copy paper is fed from the main paper tray (10a) or the auxiliary paper tray (10b). Image transfer (11) to the paper is achieved as it passes over the belt. The imaged copy is transported to the fuser (12) where heat and pressure fuse the toner image into the paper. The copy is then transported to the receiving tray (13).

When the sorter (14) is used, the copies will automatically feed into the lower bin module (15a) until full, then switch to the upper bin module (15b).

Any toner on the belt not transferred to the paper is removed by a brush (16) and vacuum system and recovered for re-use. More toner is added automatically by a toner dispenser (17).

A rotating metallic drum coated with a thin layer of a substance such as selenium, which is highly insulating in the dark but which becomes photoconductive when illuminated, is charged positively by a corona discharge like that discussed in the preceding section. A lens then focuses an image of the page to be copied on the selenium coating. Where light strikes the drum, the selenium film becomes conducting, and the positive charge in these areas leaks through the conducting film into the metallic drum. Where the film remains dark, however, the selenium is still highly insulating, and in these areas the charge distribution remains fixed, as in any normal insulating substance.

A negatively charged toner in the form of an intensely black powder is now dusted onto the drum. It adheres only to the places that retain their positive charge, corresponding to the dark areas of the image of the original document. It is subsequently transferred to the surface of a sheet of white paper that has been given a strong positive charge. Finally, the toner is fused onto the surface of the paper by the temporary application of heat to form a permanent positive copy of the original page.

The Xerox process is a beautiful demonstration of the utilization of the concepts of physics. Its success depends upon the application of the principles of electrostatics, optics, and photoconductivity. Since the initial introduction, significant advances in the speed with which copies are made and in the overall quality of reproduction have been achieved. With the most modern version of this process, Xerox copies can

be made in color. It has even been suggested that the Xerox process might be used to duplicate black-and-white and colored photographic material, but at present it is unable to approach the quality of photographic reproduction processes.

FIELD-ION MICROSCOPE

In section 16.4, it was pointed out that in the vicinity of sharp conducting points, the charge density and electric field may become very large. Indeed, fields as large as 50 million volts per centimeter are possible very close to the end of a sharp, needlelike point. This principle forms the basis of the field-ion microscope, a very high-resolution microscope invented in 1956 by Erwin W. Mueller, of The Pennsylvania State University.

Figure 16.43 shows the basic design of a field-ion microscope. A specimen is fashioned in the form of a wire with a very sharp point whose tip has a radius of about 10^{-15} cm. The sample chamber is evacuated, but then a small amount of helium gas is introduced into the chamber. A moderately small potential difference between the fluorescent viewing screen and the wire now produces an enormous electric field near the tip. The field is largest close to the individual atoms protruding from the tip. When helium atoms approach these high field regions, they are ionized, their electrons being torn away by the large electric field. The positively charged helium ions then move radially outward along the field lines and strike the fluorescent screen, producing a detailed image of the tip of the specimen, in which the individual atoms on the surface of the sample are clearly visible. A typical pattern is shown in Fig. 16.44. This field-ion image exhibits a characteristic ring pattern which represents the stereographic projection of the positions of protruding atoms that act as the centers at which field ionization takes place. In this way, the arrangement of atoms on the specimen surface and physical effects such as surface migration can be studied. With the field-ion microscope, magnification factors of about 2,000,000 are possible. The system must be operated at low temperature and high vacuum in order to eliminate effects that might reduce the resolution and blur the image. Also, the high mechanical stresses that accompany the enormous electric fields near the specimen tip restrict its use to samples of strong metallic substances such as tungsten and rhenium.

VAN DE GRAAFF ELECTROSTATIC GENERATOR

The electrostatic generator was put into practical use by R. J. van de Graaff in the early 1930's. Its primary purpose is to provide a potential difference of up to

FIGURE 16.42. Diagram of an electrostatic copier. Courtesy of Xerox Corporation. Original document is moved to flat platen (1) by automatic document handler (2) and flashed by high-intensity flash lamps (3). Image is reflected to object mirror (4) through stationary lens (5) and reflected by image mirror (6) to the selenium alloy-coated belt (7). The charge corotron (8) places a positive charge of electricity onto the belt. The light reflected through the optical system removes the charge except for the desired image. Developer with negatively charged toner (9) is brushed onto the moving belt and the toner sticks to the charged portions of the belt.

Ordinary, uncoated copy paper is fed from the main paper tray (10a) or the auxiliary paper tray (10b). Image transfer (11) to the paper is achieved as it passes over the belt. The imaged copy is transported to the fuser (12) where heat and pressure fuse the toner image into the paper. The copy is then transported to the receiving tray (13). When the sorter (14) is used, the copies will automatically feed into the lower bin module (15a) until full, then switch to the upper bin module (15b). Any toner on the belt not transferred to the paper is removed by a brush (16) and vacuum system and recovered for re-use. More toner is added automatically by a toner dispenser (17).

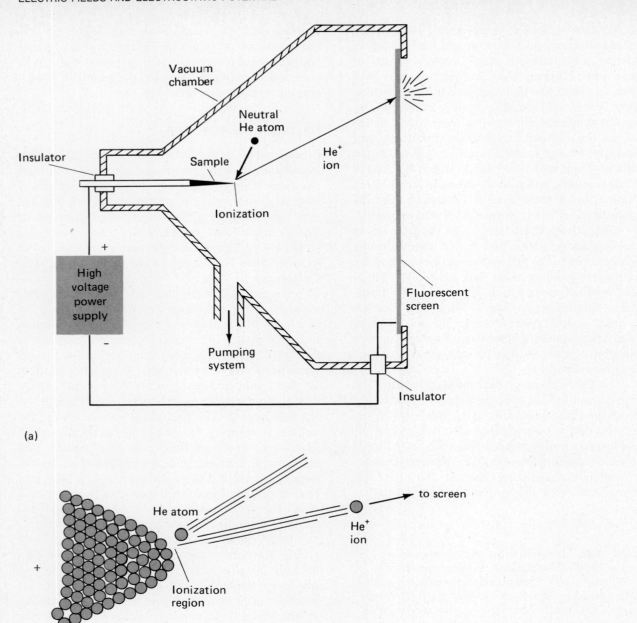

FIGURE 16.43. (a) Schematic diagram of field-ion microscopy. (b) Field ionization near surface atom and subsequent repulsion of positively charged helium ion toward viewing screen.

10 million electron volts (MeV), which can be used to accelerate particles such as protons. A schematic diagram is shown in Fig. 16.45. A continuous belt made of an insulating material such as rubber or silk rotates on two pulleys, one of which is at ground potential. The pulleys are driven by a motor. Positive charge is "sprayed" onto the belt near the lower pulley from a series of fine points at which a corona discharge is produced and is ultimately transferred to the exterior of a hollow metallic sphere which is essentially connected to the upper pulley. We may recall that charge transferred to the interior wall of a conductor migrates to the outside, keeping the interior of the conductor field free. As a consequence, in the van de Graaff generator a great deal of charge can be transferred continuously to the upper collect-

FIGURE 16.44.   Field-ion micrograph of rhenium tip. Photo courtesy of Professor Erwin W. Mueller, Pennsylvania State University.

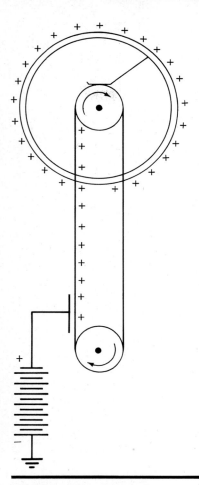

FIGURE 16.45.   Schematic diagram of van de Graaff electrostatic generator.

ing electrode, thus creating very large potential differences. There are, however, certain limits to how much charge can be placed on the conductor. We may observe that the exterior field near the conducting surface is σ/ε_0. If this gets too large, say, 30,000 volts/cm, the air becomes ionized and the charge can then leak off by causing a spark or corona discharge. This can be made to occur at higher fields by enclosing the machine in a pressurized container, since the breakdown field of air increases with pressure. Also, by using a collecting electrode whose outer radius is quite large, we can achieve a high electrostatic potential on the electrode without creating an unduly large surface field. Van de Graaff generators that produce potential differences of 6 or 8 MeV are quite common nowadays.

The primary function of this type of machine is to provide and maintain a high enough potential difference so that particles such as protons, deuterons, and alpha particles can be used as projectiles in "low energy" nuclear physics experiments, where energies up to about 10 MeV are needed.

16.9 The Mathematical Proof of Gauss's Law

Our acceptance of Gauss's law has been based so far upon qualitative physical arguments. While these are quite valid and very useful in imparting an intuitive understanding of this important principle, we shall now provide a more rigorous mathematical derivation for those who may not be easily persuaded by qualitative reasoning.

Let us begin by considering a single point charge q. Construct an arbitrary closed Gaussian surface S that surrounds such a charge and also a large sphere S'' of radius r_0, with q at its center, that completely encloses the surface S, as illustrated in Fig. 16.46a. A small element of area da of this surface is shown in cross section; the unit vector normal to this surface is $\mathbf{n}$. This area element is at a distance r from the charge. The electric flux $d\Phi$ through da is, by definition,

$$d\Phi = \mathbf{E} \cdot \mathbf{n} \, da = E \cos \theta \, da \qquad (16.9.1)$$

where θ is the angle between $\mathbf{n}$ and the field $\mathbf{E}$, which,

527

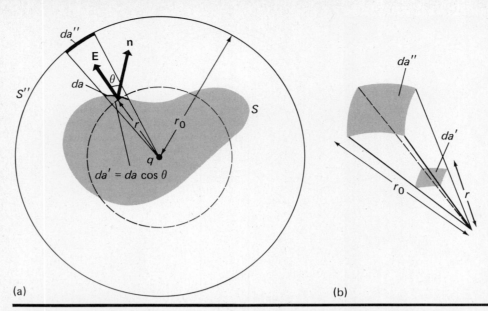

FIGURE 16.46. Calculation of the electric flux arising from a point charge inside an arbitrary closed surface S.

according to Coulomb's law must point radially outward, away from q. The angle θ is also the angle between the actual surface S and the surface of a sphere of radius r (shown as a dotted curve). The projection da' of the area da onto the dotted spherical surface is also shown in Fig. 16.46; the relation between these two areas, from the figure, is

$$da' = da \cos \theta \qquad (16.9.2)$$

which enables us to write (16.9.1) as

$$d\Phi = E \, da' \qquad (16.9.3)$$

Now, according to Coulomb's law, the field is given by

$$E = \frac{q}{4\pi\varepsilon_0 r^2} \qquad (16.9.4)$$

Also, the area da' and the area da'' that represents its projection onto the large sphere S'' that surrounds S must be related by

$$\frac{da'}{da''} = \frac{r^2}{r_0^2} \qquad (16.9.5)$$

because the dimensions of *both* sides of the elementary rectangle da'' in Fig. 16.46b are larger than those of da' by the ratio r_0/r. If we now substitute the value of E from (16.9.4) and the value of da' in terms of da'' given by (16.9.5) into (16.9.3), we find

$$d\Phi = \frac{q}{4\pi\varepsilon_0 r^2} \frac{r^2}{r_0^2} \, da'' = \frac{q}{4\pi\varepsilon_0 r_0^2} \, da'' \qquad (16.9.6)$$

This, in conjunction with (16.9.1), allows us to write

$$d\Phi = \mathbf{E} \cdot \mathbf{n} \, da = \frac{q}{4\pi\varepsilon_0 r_0^2} \, da'' \qquad (16.9.7)$$

Now let us integrate $\mathbf{E} \cdot \mathbf{n}$ over the area elements da that comprise dS, while integrating $q/4\pi\varepsilon_0 r_0^2$ over the area elements da'' of the sphere S''. Since the flux through each corresponding pair of elements is the same, we shall find, because r_0 is a constant in the integration over S'', that

$$\Phi = \oint_S \mathbf{E} \cdot \mathbf{n} \, da = \frac{q}{4\pi\varepsilon_0 r_0^2} \oint_{S''} da'' = \frac{q}{4\pi\varepsilon_0 r_0^2} (4\pi r_0^2)$$

or

$$\oint_S \mathbf{E} \cdot \mathbf{n} \, da = \frac{q}{\varepsilon_0} \qquad (16.9.8)$$

This looks very much like Gauss's law, but we have not quite finished our proof, since we have not considered what happens when the charge q is *outside* the Gaussian surface S, as illustrated in Fig. 16.47. In this case, it is evident from the figure, and from the above calculation that, first of all, the flux associated with areas da_1 and da_2 is

$$d\Phi = \mathbf{E}_1 \cdot \mathbf{n}_1 \, da_1 + \mathbf{E}_2 \cdot \mathbf{n}_2 \, da_2$$
$$= E_1 \cos \theta_1 \, da_1 + E_2 \cos(180° - \theta_2) \, da_2 \qquad (16.9.9)$$

where $\mathbf{n}_1$ and $\mathbf{n}_2$ are the *outwardly* directed normals that are shown. However, since $da_1' = da_1 \cos \theta_1$ and $da_2' = da_2 \cos \theta_2$ and since $\cos(180° - \theta_2) = -\cos \theta_2$, we may write (16.9.9) as

$$d\Phi = E_1 \, da_1' - E_2 \, da_2' \qquad (16.9.10)$$

But from our previous calculations, it is evident that, just as before,

$$E_1 \, da_1 = E_2 \, da_2 = \frac{q}{4\pi\varepsilon_0 r_0^2} \, da'' \qquad (16.9.11)$$

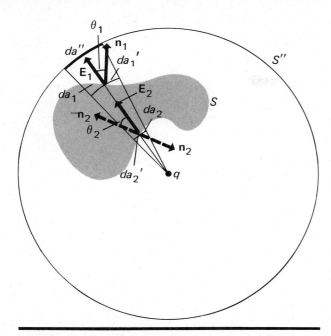

FIGURE 16.47. Calculation of the electric flux arising from a point charge outside an arbitrary closed surface S.

If this is true, then (16.9.10) reduces to

$$d\Phi = \mathbf{E}_1 \cdot \mathbf{n}_1 \, da_1 + \mathbf{E}_2 \cdot \mathbf{n}_2 \, da_2 = 0$$

And if this is integrated over all the surface elements (da_1, da_2) of S, we obtain

$$\oint_S \mathbf{E} \cdot \mathbf{n} \, da = 0 \tag{16.9.12}$$

This comes about, of course, because the outward flux across da_1 is equal in magnitude but *opposite in sign* to the *inward* flux across da_2.

Combining the results embodied in (16.9.8) and (16.9.12) and observing that every point charge or charge element in a system wherein many such charge elements are present contributes to flux across such a surface in exactly this way, we arrive at

$$\oint_S \mathbf{E} \cdot \mathbf{n} \, da = \frac{q_0}{\varepsilon_0} \tag{16.9.13}$$

where q_0 represents the total charge enclosed *within* the Gaussian surface S. This is Gauss's law as originally stated in section 16.3.

SUMMARY

A field is a function or a set of functions representing, for example, each component of a vector that associates a given physical quantity with each point in space. In electrostatics, the electric field **E** is defined as a vector representing the *electrostatic force per unit charge* experienced by a test charge placed at any desired observation point, evaluated in the limit as the size of the test charge approaches zero. Thus,

$$\mathbf{E} = \lim_{q \to 0} \frac{\mathbf{F}}{q}$$

The electric field of a single point charge q, located at the origin, is given by

$$\mathbf{E} = \frac{kq}{r^2} \mathbf{i}_r = \frac{q}{4\pi\varepsilon_0 r^2} \mathbf{i}_r$$

The constant ε_0, referred to as the *permittivity of free space*, is simply related to the force constant k in Coulomb's law by

$$\varepsilon_0 = \frac{1}{4\pi k} = 8.85418 \times 10^{-12} \text{ C}^2/\text{N-m}^2$$

For a number of point charges, the field is calculated by forming the vector sum of the individual point charge field contributions, from which the field at an observation point $P(x, y, z)$ can be expressed as

$$\mathbf{E}(P) = \mathbf{E}_1 + \mathbf{E}_2 + \mathbf{E}_3 + \cdots = \sum_{i=1}^{n} \frac{1}{4\pi\varepsilon_0} \frac{q_i}{r_i^2} \frac{\mathbf{r}_i}{r_i}$$

where $\mathbf{r}_i$ is the vector from the ith point charge to the point P where the field is being evaluated.

The electric field of a continuous charge distribution can be arrived at by superposition of fields of contributions from properly chosen infinitesimal charge elements by integration, as described in Example 16.2.2.

The electric flux across a given surface S is defined by

$$\Phi = \int_S \mathbf{E} \cdot \mathbf{n} \, da$$

Gauss's law is a theorem relating the electric flux over a closed surface S to the total charge enclosed by the surface. It states that

$$\oint \mathbf{E} \cdot \mathbf{n} \, da = \frac{q}{\varepsilon_0}$$

where q is the total charge within the closed surface S. Gauss's law can be used to calculate the electric fields arising from highly symmetric charge distributions, as illustrated by the examples worked out in section 16.3. It may also be used to infer certain useful facts about conductors and insulators, as follows: (a) The electric field within any conductor at equilibrium is zero. (b) Any net charge borne by a conductor resides *on its surfaces*. (c) The electric field just outside a conductor is normal to its surface at every point. (d) The magnitude of the field just outside the surface of a conductor is related to the surface charge density σ at every point on the conductor's surface by $E = \sigma/\varepsilon_0$.

The electrostatic potential associated with an electric field is defined as the potential energy per unit charge possessed by a test charge placed at any desired point of observation. It can also be regarded as the work per unit charge such a test charge can do in going from the observation point to another point where the potential is taken to be zero. The potential due to a single point charge q at the origin is

$$V = \frac{q}{4\pi\varepsilon_0 r}$$

if we take the potential at infinity to be zero. The potential arising from a number of point charges can be arrived at by superposing scalar contributions from each of the source charges. Thus, the potential at point P(x, y, z) from such a distribution of source charges is

$$V(x, y, z) = V_1 + V_2 + V_3 + \cdots = \frac{1}{4\pi\varepsilon_0} \sum_{i=1}^{n} \frac{q_i}{r_i}$$

where r_i is the distance between the ith charge and the observation point P.

The potential can be related to the electric field by the relation

$$dV = -\mathbf{E} \cdot d\mathbf{r}$$

from which the potential difference between points A and B can be obtained as

$$\Delta V_{AB} = V_B - V_A = -\int_A^B \mathbf{E} \cdot d\mathbf{r}$$

The electric field can be obtained from the potential $V(x, y, z)$ by the relations

$$E_x = -\frac{\partial V}{\partial x} \qquad E_y = -\frac{\partial V}{\partial y} \qquad E_z = -\frac{\partial V}{\partial z}$$

Since the electrostatic field has associated with it a unique potential energy, it is a conservative force field. Therefore, it has the following properties: (a) The work required to move a charge between any two points is the same whatever path is taken and is equal to $q\,\Delta V_{AB}$. (b) The work required to circulate a charge around any closed path is zero.

Surfaces upon which the electrostatic potential is everywhere the same are called *equipotential surfaces.* Equipotential surfaces are perpendicular at every point to the electric field lines. The surface of any conductor, in equilibrium, is an equipotential surface. The work required to move a charge between any two points on the same equipotential surface is zero.

The torque τ experienced by a dipole in a uniform external field $\mathbf{E}$ can be expressed by

$$\tau = \mathbf{p} \times \mathbf{E}$$

where $\mathbf{p}$ is the dipole moment. Its magnitude is $\tau = pE \sin \theta$, where θ is the angle between the field and the dipole axis. The potential energy of such a dipole is

$$U(\theta) = -pE \cos \theta$$

Another useful relation between the electric field at any point and the charge density $\rho(x, y, z)$ at that point is given by Poisson's equation, which can be derived directly from Gauss's law. For a situation in which the charge density and field vary only along a single direction in space, Poisson's equation states that

$$\frac{dE_x}{dx} = \frac{\rho(x)}{\varepsilon_0}$$

or, since the field E_x equals $-dV/dx$,

$$\frac{d^2 V}{dx^2} = -\frac{\rho(x)}{\varepsilon_0}$$

These equations are replaced by more complex partial differential equations in cases where the fields and charge density depend on all three spatial coordinates.

QUESTIONS

1. A student of elementary physics insists that the electric field pattern produced by two point charges will have lines of force which intersect. Would you agree with him, and why?

2. A point charge is in motion under the influence of an electric field. Does the charge follow a trajectory parallel to the field lines? Explain.

3. Are there any circumstances in electrostatics for which Gauss's law is invalid? Is this law always useful in finding the $\mathbf{E}$ field?

4. If the exponent of r in Coulomb's law were slightly different from $-2.000 \ldots$, how would the properties of the electric field, the electrostatic potential, and the lines of force be affected?

5. The lines of force produced by a point charge can be directed toward or away from the charge. In the case of the gravitational field, the lines of force produced by a point mass can have only one direction, namely, toward the mass. What is the reason for this difference?

6. There is no net charge in a certain region of space. Is it true that the electric field at any point on the surface that bounds the region is zero?

7. A point charge produces an infinitely large field at its own position. Does the charge experience a force due to this field?

8. The electric field everywhere on a closed surface is zero. How much charge is contained within the volume bounded by this surface? What is the electric flux through the surface?

9. In the application of Gauss's law, is the electric field due to all charges or only to those charges within the Gaussian surface?

10. The electric field is known at every point in space. Show how to determine from this field the amount of charge within any prescribed volume.

11. Discuss the differences between electrostatic potential

and electrostatic potential energy. Under what conditions are these quantities proportional to one another?

12. The electrostatic potential at a certain point is known. Is it possible to obtain the electric field at the point using only the given value of the potential?

13. Explain why lines of force are always perpendicular to equipotential surfaces.

14. Gauss's law is useful for computing electric fields only when the charge distribution is highly symmetric. Comment on the validity of this assertion.

15. It is very simple to draw the equipotential surfaces due to a point charge. Why is it difficult to do the same for an electric dipole?

16. Can two equipotential surfaces intersect? Why, or why not?

17. Two points, P_1 and P_2, are at the same electrostatic potential. Does this mean that no work is done in bringing a charge from P_1 to P_2? Is it true that the electric fields at P_1 and P_2 are the same?

18. A Gaussian surface encloses two electric dipoles. What is the net electric flux through this surface?

19. Equal charges are located at the corners of a cube. Can Gauss's law be used to compute the electric field?

20. A point charge is placed in the neighborhood of an uncharged spherical conductor that is insulated from the surroundings. How does its presence affect the distribution of free charge in the conducting sphere? Is there a force between the charge and the spherical conductor, and, if so, is it attractive or repulsive?

21. A point charge is placed near a "grounded" conducting plane, that is, one whose constant potential is the same as that of the earth, say, $V = 0$. The plane is assumed to be of infinite extent. How does the presence of the charge affect the distribution of mobile charges in the planar conductor? Is there a net charge on the plane, and, if so, how large is it? Does the point charge experience a force, and, if so, is it attracted or repelled by the plane?

22. Answer the preceding question in the case where the point charge is replaced by a dipole.

PROBLEMS

1. A spherical charged oil drop of mass 10^{-4} grams is stationary in a vertical electric field of strength 200 newtons per coulomb. Find the net charge of the drop.

2. A thin conducting sheet is inserted midway between the two charges in the preceding problem. The angle θ is observed to decrease. Explain qualitatively why the electrical force on each charge is diminished.

3. A thin circular ring of radius 20 cm has a linear charge density given by $\lambda = 10^{-6} \cos \theta$ coulomb/cm, as illus-

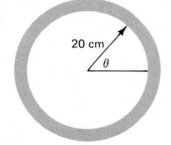

trated in the diagram. Find the total charge contained on the ring.

4. An electric quadrupole consists of four charges as shown in the figure. Calculate the electric field produced at the axial point P whose distance r from the point O is very large compared to a or b.

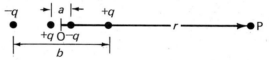

5. A circular disc of radius 10 cm contains a total charge of 10^{-6} coulomb. The surface charge density σ is directly proportional to the distance r from the center of the disc. If r is expressed in cm, find the value of the proportionality constant. How much charge is contained within the circle of radius 5 cm?

6. The electric field in a certain region of space is constant. Show by direct calculation that the flux through the hemispherical surface shown in the accompanying figure is zero.

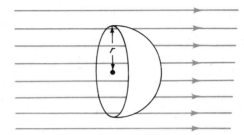

7. A large, thin metallic plate of area 10 m² is placed in a uniform field perpendicular to its surface. If the field is 50,000 newtons/coulomb find the charge induced on the surface.

8. At the surface of the earth the electric field is 100 newtons/coulomb, pointing away from the earth. If we assume the earth's surface is conducting, estimate the total charge on its surface.

9. Two point charges $q_1 = 10^{-8}$ coulomb and $q_2 = 2 \times 10^{-8}$ coulomb are separated by a distance of 1 meter. How much work is done in moving a third charge $q_3 = 3 \times 10^{-8}$ coulomb from a point on the line joining the charges 60 cm from q_1 and 40 cm from q_2 to the point midway between the charges?

10. Three concentric conducting spheres are shown in the diagram. The interior sphere, which is solid, has a charge

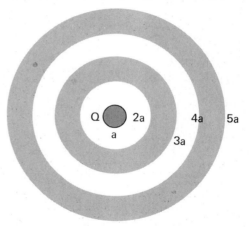

Q. **(a)** Find the distribution of charge on the outer two spheres and the charge density on their surfaces. **(b)** Use Gauss's law to obtain the electric field in each region of space. **(c)** Make a schematic plot of E versus r for values from $r = 0$ to $r = \infty$.

11. A semicircle of radius R bears a uniform linear charge density λ along its entire length. Show that the electric field at the center of the semicircle is $E = \lambda/2\pi\varepsilon_0 R$.

12. Show that if the electric field produced by a point charge were given by $E = Q/4\pi\varepsilon_0 r^{2+\varepsilon}$, where ε is a small number, then Gauss's law could not be valid. *Hint:* Evaluate the flux through a spherical surface surrounding the charge Q.

13. The hydrogen atom in its lowest allowed energy state contains an electron whose average distance from the proton is 0.5×10^{-8} cm. Find the electric field which the proton produces at the position of the electron. If we assume that the electron is in a circular orbit, find its speed in miles per hour.

14. An insulating rod of length 1 meter bears a charge of $+10^{-6}$ coulomb, uniformly distributed along its length. Find the magnitude and direction of the electric field within the rod 75 cm from the left end.

15. A conducting sphere has a radius of 1.0 cm. How much charge can it be given without causing electrical breakdown of the surrounding air?

16. Two point charges $q_1 = 5 \times 10^{-6}$ coulomb and $q_2 = -2 \times 10^{-6}$ coulomb are 1 meter apart. Find the electric field and the electrostatic potential at a point P, which is 75 cm from q_1 and 25 cm from q_2.

17. The spherical conducting electrode of a van de Graaff generator is charged to 2×10^6 volts. What is the minimum radius the shell must have to avoid electrical breakdown of the air?

18. A metal sphere of radius 50 cm is charged by transferring 10^{17} electrons to it. Find the electrostatic potential of the sphere. What is the magnitude and direction of the force on a proton 1 meter from the center of the sphere?

19. A proton is accelerated from rest in a uniform field E of strength 10^6 newtons/coulomb. Find the distance it goes in 10^{-7} seconds. Through what potential difference has it been accelerated and how much energy does it have in joules? In electron volts?

20. A particle of charge $+Q$ and mass m hangs vertically from a massless thread of length l, as shown in the diagram. A second charge $+\delta q$ is used as a test charge to measure the electric field; it is placed at position x. Find the electric field at x. Show that the fractional error in the measurement is given by

$$\frac{\delta E}{E} = -\frac{Q\,\delta q l}{2\pi\varepsilon_0 m g x^3}$$

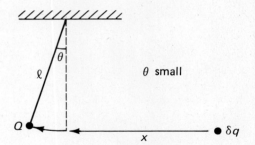

θ small

(Assume in the second case that the thread remains almost vertical.)

21. Two charged balls, each having a very small radius and carrying a charge Q, are suspended from threads of length l. Each ball has the same mass m. They repel each other and in equilibrium make an angle θ with the vertical. **(a)** Show that the angle θ is given by a solution of the equation

$$mg \tan \theta = \frac{Q^2}{4\pi\varepsilon_0} \frac{1}{4l^2 \sin^2 \theta}$$

(b) Find an expression for the potential energy (gravitation plus electrical) in terms of θ. **(c)** Show that the expression in (a) can be obtained by minimizing the potential energy with respect to the angle θ.

22. Assuming the potential at infinity to be zero, find **(a)** the electrostatic potential at a distance of 1.0 meter from a point charge of 1.00×10^{-6} coulombs, **(b)** the electrostatic potential at a distance of 0.10 meters from this same charge, **(c)** the amount of work needed to move a charge of 2.5×10^{-7} coulombs from a distance of 1.0 meter from the first charge to a distance of 0.1 meter from it.

23. How much work must be done to move an electron from an initial distance of 0.528×10^{-10} meters from a proton to an infinite separation?

24. A charge $q_1 = +8.00 \times 10^{-7}$ coulombs is at the point $x = 0.12$ meter, $y = 0.08$ meter, $z = 0$. What is the difference in potential between the points $x = 0.18$ meter, $y = z = 0$, and $x = 0.36$ meter, $y = z = 0$?

25. A charge Q is located at the origin, while two equal charges of magnitude $-Q/2$ are located on the z-axis at $z = \pm a$. Show that the electrostatic potential at any point in the xy-plane due to this distribution of charges is $V = (Q/4\pi\varepsilon_0)((1/r_0) - (a^2 + r_0^2)^{-1/2})$, where r_0 is the distance between that point and the origin.

26. A dipole consists of two point charges of $\pm 1.00 \times 10^{-6}$ coulombs separated by a fixed distance of 0.0100 meter. Find **(a)** how much external force must act on the charge to keep the dipole from collapsing as a result of the coulomb attraction of its charges, **(b)** the dipole moment of the charge configuration. **(c)** Assuming that the dipole moment vector is oriented along the z-axis, and that the center of the dipole is at the origin, how much work is needed to move a point charge of 1.00×10^{-6} coulombs from the point $x = 0.0100$ meter, $y = z = 0$ to infinity, along the x-axis? **(d)** How much work is needed to move the point charge of part (c) from its original location to infinity along a line parallel to the z-axis?

27. **(a)** How much work must be done to assemble the dipole discussed in the preceding problem? **(b)** Assuming that the dipole moment vector is oriented along the z-direction, how much work is needed to move a charge of magnitude -1.00×10^{-6} coulombs from the point $x = 0.0050$ meter, $y = 0$, $z = 0.0050$ meter to infinity, along the line $x = z$? **(c)** How much work is needed to move a charge of magnitude -1.00×10^{-6} coulombs from the point $x = 0.050$ meter, $y = 0$, $z = 0.050$ meter to infinity along the line $x = z$? **(d, e)** What would the answers to (b, c) be if the charge were moved to infinity along a line parallel to the z-axis?

28. A *point* dipole whose dipole moment is of magnitude 1.00×10^{-8} coulomb-meters, oriented along the z-direction, is located at the origin. Find **(a)** the work needed to move a charge of magnitude -1.00×10^{-6} coulombs from the point $x = 0.0050$ meter, $y = 0$, $z = 0.0050$ meter to infinity, **(b)** the work needed to move a charge of magnitude -1.00×10^{-6} coulombs from the point $x = 0.050$ meter, $y = 0$, $z = 0.050$ meter to infinity. Could this point dipole serve as a good approximation to the finite dipole of the preceding problem **(c)** for the process involving removing the charge from (0.005 meter, 0, 0.005 meter) to infinity? **(d)** for the process of removing the charge from (0.05 meter, 0, 0.05 meter) to infinity?

29. A point dipole, whose dipole moment is 2.4×10^{-8} coulomb-meter, is oriented at an angle of $60°$ to a constant externally applied electric field of 3.0×10^{4} volts/meter. Find **(a)** the torque exerted on the dipole by the external field, **(b)** the work that can be done by the dipole as its dipole vector aligns parallel to the external field.

30. Show that the potential energy of a particle of charge q in a constant electric field $\mathbf{E}$ is given by $U = -q\mathbf{r} \cdot \mathbf{E}$, where $\mathbf{r}$ is the position vector of the charged particle. *Hint:* Recall that $F_x = -\partial U/\partial x$, etc.

31. Four charges, each of 10^{-8} coulomb, occupy the four corners of a square of sides 1 meter. Let x measure the distance from one corner to a point along the diagonal line connecting this corner with another, as shown in the diagram. **(a)** Make a sketch of the potential $V(x)$ as a function of x. **(b)** Use this graph as a basis for drawing a second sketch which shows the variation of the electric field along the same line.

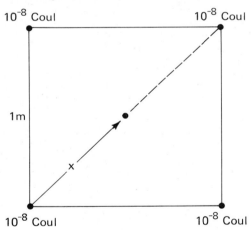

32. For the preceding problem, how much work was done in putting the charges at the four corners of this square?

33. Show that the electric field between two parallel plates cannot drop abruptly to zero at the edge but that some

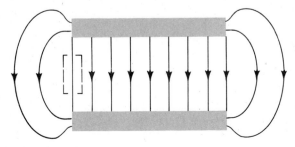

fringing must occur. See the accompanying figure, using the suggested loop to evaluate $\oint \mathbf{E} \cdot d\mathbf{l}$.

34. Two infinitely large parallel conducting plates have the uniform charge densities shown in the diagram. Calculate the electric fields in regions (a), (b), and (c) by using Gauss's law. Show that these results can also be obtained by superposing the electric fields produced by the four sheets of charge. Find the potential difference between the conductors.

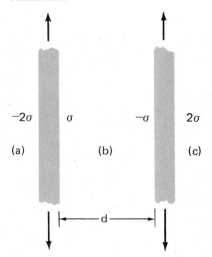

35. An isolated spherical conductor of radius r_1 is charged until its surface is at potential V_0. It then bears a total charge q_1. It is then desired to put charge q_2 on a second spherical conductor of radius r_2, likewise isolated and independent of the first, such that its surface potential will attain the same value V_0. Show that in order that this be accomplished, $q_1/q_2 = r_1/r_2$. Show also that this implies that the surface charge densities σ_1 and σ_2 are related by $\sigma_1/\sigma_2 = r_2/r_1$. This result, which shows that for a given potential the surface charge density (and thus the surface field) varies inversely with the radius of curvature, explains why electric fields are very large near sharply curved or pointed conducting objects.

36. Four point objects each bearing charge Q are located at the four corners of a square whose sides have length l. Find the potential energy of the system. Obtain the electrostatic potential at the center of the square.

* 37. A very large circular conducting plate of radius R bears a nonuniform surface charge density σ given by $\sigma = \sigma_0[1 - (R/r)]$, where r is the distance from the center of the plate. Find the potential at any point along an axis perpendicular to the plate passing through its center.

*38. Using the result of the preceding problem, find the electric field at any point on the axis.

39. A region of space is characterized by a potential $V(x) = -30e^{-0.1x}$ volts, where x is in meters. The potential is independent of y and z. A proton is initially at rest at the point $x = 10$ meters, $y = z = 0$. What will its velocity be when it reaches the point $x = 1.0$ meter, $y = z = 0$?

40. A straight line containing charge per unit length λ produces a potential proportional to $\ln r$, where r is the distance from the line. Find the proportionality constant, using the relation $E = -dV/dr$ and the previously determined expression for E.

*41. In a given region of space there exists a potential given by $V(r) = V_0(e^{-\mu r}/r)$, where μ is a constant. Find the electric field at any point within the region. Find the charge density distribution that gives rise to a potential of this form. *Hint*: Use

$$E_x = -\frac{\partial V}{\partial x} \qquad E_y = -\frac{\partial V}{\partial y} \qquad E_z = -\frac{\partial V}{\partial z}$$

where

$$r = \sqrt{x^2 + y^2 + z^2}$$

42. We have shown that a spherical shell of charge containing a uniform charge density produces a potential outside the shell identical to that produced by an equal point charge located at its center. This follows because according to Coulomb's law, the potential of a point charge varies as $1/r$. Suppose, contrary to fact, that the potential of a point charge were given by $(4\pi\varepsilon_0)^{-1}qe^{-\beta r}/r$, where β is a constant. Under these conditions show that the potential outside a charged spherical shell is no longer given by this expression but is instead

$$V(r) = \frac{q}{4\pi\varepsilon_0} \frac{e^{\beta R} - e^{-\beta R}}{2\beta R}$$

where R is the radius of the shell. Show also that the potential inside the shell is no longer zero.

43. A continuous, spherically symmetric distribution of charge produces a potential which varies as $\ln r$. How would the electric field vary? What kind of charge distribution would produce this field?

44. A wire of length L carries a uniform linear charge density λ. Find the electric field as a function of the distance r from the center of the wire (measured perpendicular to the wire).

45. Calculate the electrostatic potential at a distance x from a finite but very large charged sheet bearing a uniform charge density σ. Using this result, calculate the electric field. Now take the limit of an infinite sheet and show that the field agrees with that obtained from Gauss's law. Assume, for simplicity, that the sheet is circular and the field point is on an axis perpendicular to the center of the sheet.

46. The electric field in a small region of space is given by $E_x = 0$; $E_y = 0$; $E_z = 10z$ volts/meter. (a) Find the flux passing through a cube whose side is 30 cm and which is centered on the origin; the faces of the cube are normal to the coordinate axes. (b) How much charge is within the cube? (c) Find an expression for the charge density ρ.

47. A balloon of radius r contains a charge Q uniformly distributed on its surface. Show that the balloon experiences an outward electrostatic force per unit surface area given by

$$\frac{F}{A} = \frac{Q^2}{32\pi^2\varepsilon_0 r^4}$$

Hint: Find how the potential energy varies with r.

48. The electric field in a given region is expressed by $E_x = 3x^2$ volts/meter, $E_y = 0$, $E_z = 0$. Find the charge contained in a cube of side 1 meter if one of the faces lies in the xy-plane and the opposite face is at $x = 1$ meter.

49. Let us assume for a moment that the proton is a point charge. How much work is done in bringing an electron from infinity to within 10^{-16} cm of the center of the proton? If the proton is assumed to be a sphere of radius 10^{-13} cm, bearing a uniform charge distribution throughout its volume, how much work would be required to move an electron from infinity to within 10^{-16} cm of the center?

50. Let us naively assume that in a helium atom the two electrons are in circular orbits of radii 0.5×10^{-8} cm and 1.0×10^{-8} cm about a point nucleus of charge $+2e$. Find the minimum and maximum values of the potential energy of the system.

51. A distribution of charge density is uniform along the y- and z-directions, but varies along the x-direction according to the formula $\rho(x) = \alpha x^2$, where α is a constant. The electric field at $x = 0$ is known to be zero. Find (a) the electric field at any distance x from the plane $x = 0$, (b) the difference in potential between any point and the plane $x = 0$.

CAPACITANCE, DIELECTRICS, AND POLARIZATION

17.1 Capacitance

There are many devices, or *circuit elements*, that are in everyday use in ordinary electrical circuits. We shall study some of these in detail in the remaining chapters of the book. One of the most important of these is an *electrostatic* element referred to as a *capacitor*, or *condenser*. Consider two conductors A_1 and A_2 which carry charges of equal magnitude but opposite sign, as shown in Fig. 17.1. The charges $-q$ and $+q$ contained on the respective conductors are distributed on the surfaces. As we have seen in the preceding chapter, each of the conducting surfaces must be an equipotential. Therefore, all of conductor A_1 is at a potential V_1, while the entire conductor A_2 is at a potential V_2. Any such arrangement of con-

ductors carrying equal and opposite charges is called a *capacitor*.

The magnitude of the potential difference $V_2 - V_1$ between these conductors will be denoted by ΔV. This potential difference is ordinarily *directly proportional* to the magnitude of the charge q. The proportionality constant can be written as $1/C$, where C is called the *capacitance* of the two conductors. Thus, we may write

$$\Delta V = \frac{1}{C} q$$

or

$$C = \frac{q}{\Delta V}$$

(17.1.1)

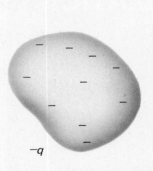 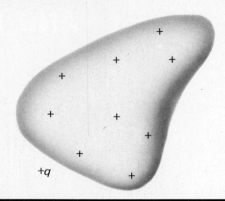

FIGURE 17.1. Capacitor consisting of a pair of conductors having equal and opposite charges.

Although this equation contains q in the numerator, we must remember that ΔV is usually directly proportional to q, and, therefore, the capacitance C is actually *independent of the charge* contained on the conductors or of the difference in potential between them. The capacitance C is measured in units of *coulombs per volt*, which are more frequently referred to as *farads* (abbreviated F), after the British physicist Michael Faraday. The farad is an extremely large unit of capacitance in practical terms; therefore, the microfarad (μF, equal to 10^{-6} farad) or the picofarad (pF, equal to 10^{-12} farad) are usually more appropriate units.

The capacitance of any given system of conductors depends on two important factors:

1. The *geometric arrangement* of the conductors. This includes the size, shape, and spacing of the conductors as well as their geometric relation to one another.
2. The properties of the *medium* in which the conductors are placed (air, vacuum, dielectric, etc.).

We shall study the influence of the medium later in this chapter after we have discussed some important properties of dielectrics. Let us now take up the problem of how the capacitance depends on geometry. Perhaps the simplest example of a capacitor is that of the *parallel-plate capacitor* shown in Fig. 17.2. Two conducting plates each having an area A are separated by a distance d, which is usually very small compared to the linear dimensions of the plates. A typical value of d might be less than 1 mm, while the plate dimensions might be hundreds or thousands of times larger. For such a configuration of conducting plates, the electric field between the plates is quite uniform except near the edges, where some fringing might occur.

We have already seen in Example 16.6.1 that in such a situation, the electric field is uniform between the plates, the magnitude of the potential difference being given by

$$\Delta V = Ed \tag{17.1.2}$$

where d is the plate separation. We found also that the magnitude of the electric field between the plates can be expressed as

$$E = \frac{\sigma}{\varepsilon_0} \tag{17.1.3}$$

where σ is the surface charge density on the plates. Using the expression $\sigma = q/A$ in conjunction with Eqs. (17.1.1) through (17.1.3), we find

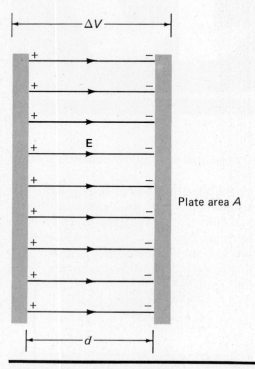

FIGURE 17.2. Parallel-plate capacitor.

$$C = \frac{q}{\Delta V} = \frac{q}{Ed} = \frac{q}{(\sigma/\varepsilon_0)d} = \frac{qA\varepsilon_0}{qd}$$

or

$$C = \frac{\varepsilon_0 A}{d} \qquad (17.1.4)$$

for the *capacitance of a parallel-plate capacitor in vacuum*. In the above calculation, the permittivity ε_0 of the vacuum, which is not very different from that of air, was used. We may anticipate some modification of this result if the space between the conducting plates is filled with a dielectric material, in which case the permittivity will be modified. We shall discuss later how and why permittivity is modified by the presence of various insulating substances.

EXAMPLE 17.1.1

A parallel-plate capacitor is made from aluminum foil. The plates are separated by 1 mm of air. What area should the plates have to produce a capacitance of 1 pF? 1 μF? 1 F?

According to (17.1.4), the area of the plates should be

$$A = \frac{Cd}{\varepsilon_0}$$

where d is the distance in meters, C is the capacitance in farads, and ε_0 is the permittivity of air. Substituting the above values, we obtain

$$A_1 = \frac{(10^{-12})(10^{-3})}{8.85 \times 10^{-12}} = 1.13 \times 10^{-4} \text{ m}^2$$

$$A_2 = \frac{(10^{-6})(10^{-3})}{8.85 \times 10^{-12}} = 1.13 \times 10^2 \text{ m}^2$$

$$A_3 = \frac{(1)(10^{-3})}{8.85 \times 10^{-12}} = 1.13 \times 10^8 \text{ m}^2$$

Clearly, the area A_1 is reasonable in size and could very well represent an actual capacitor. On the other hand, A_2 is quite large while A_3 is only slightly smaller than the state of Connecticut! It would, needless to say, be totally impractical to construct a 1 F capacitor using the above design.

EXAMPLE 17.1.2

The electrical breakdown of air takes place whenever the electric field exceeds about 30,000 volts/cm. What is the maximum charge a parallel-plate capacitor of 0.002 μF can hold if the plates have an area of 100 cm²?

The maximum charge q_{max} is proportional to the maximum voltage V_{max} that can be applied between the plates:

$$q_{max} = C \, \Delta V_{max} \qquad (17.1.5)$$

but

$$E_{max} = \frac{\Delta V_{max}}{d}$$

where d is the separation of the plates. Since

$$C = \frac{\varepsilon_0 A}{d}$$

we obtain, noting that 30,000 volts/cm = 3×10^6 volts/m and 100 cm² = 0.01 m²,

$$q_{max} = C E_{max} d = E_{max} \varepsilon_0 A$$
$$= (3 \times 10^6)(8.85 \times 10^{-12})(1.0 \times 10^{-2})$$
$$= 2.66 \times 10^{-7} \text{ C}$$

If the plates are charged in excess of this amount, the air between the plates becomes conducting and the charges on the plates are neutralized.

In addition to the parallel-plate capacitor, there are several other types of simple capacitors. In the next two examples, we discuss the cases of cylindrical and spherical capacitors.

EXAMPLE 17.1.3
CYLINDRICAL CAPACITOR

Two long, conducting hollow cylinders of radii r_a and r_b are coaxial, as shown in Fig. 17.3. Obtain an expression for the capacitance per unit length if the region between the cylinders contains air.

Assume that the inner cylinder bears a charge per unit length of $+\lambda$. This charge must reside on the outer wall of the cylindrical shell in order for the field inside the conductor to be zero. The outer cylinder contains a charge per unit length of $-\lambda$; this charge is located on the inner boundary. According to Gauss's law, using a cylindrical Gaussian surface of radius r, the electric field points radially outward and is obtained from

$$\oint_s \mathbf{E} \cdot \mathbf{n} \, da = (E)(2\pi r l) = \frac{q}{\varepsilon_0} = \frac{\lambda l}{\varepsilon_0}$$

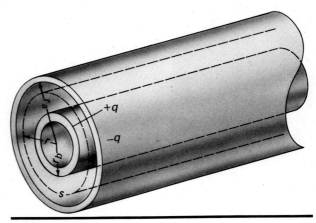

FIGURE 17.3. Long cylindrical capacitor.

or

$$E = \frac{\lambda}{2\pi\varepsilon_0 r} \qquad (17.1.6)$$

The potential difference between the cylinders, according to (16.5.22), then has a magnitude

$$\Delta V = \int_{r_b}^{r_a} E \, dr = \frac{\lambda}{2\pi\varepsilon_0} \int_{r_b}^{r_a} \frac{dr}{r}$$

or

$$\Delta V = \frac{\lambda}{2\pi\varepsilon_0} \ln \frac{r_a}{r_b} \qquad (17.1.7)$$

A length l of the cylinder contains a charge $q = \lambda l$, and, therefore, this length has a capacitance

$$C = \frac{\lambda l}{\Delta V} = \frac{\lambda l}{\dfrac{\lambda}{2\pi\varepsilon_0} \ln \dfrac{r_a}{r_b}}$$

The capacitance per unit length is, therefore,

$$\frac{C}{l} = \frac{2\pi\varepsilon_0}{\ln \dfrac{r_a}{r_b}} \qquad (17.1.8)$$

EXAMPLE 17.1.4
SPHERICAL CAPACITOR

Two conducting concentric spheres have radii r_a and r_b, as shown in Fig. 17.4. Find the capacitance C of this device.

The electric field can be obtained by using Gauss's law in conjunction with the Gaussian surface

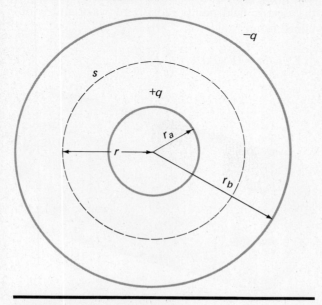

FIGURE 17.4. Spherical capacitor.

shown in Fig. 17.4. In this way we can write

$$\frac{q}{\varepsilon_0} = \oint_s \mathbf{E} \cdot \mathbf{n} \, da = 4\pi r^2 E \qquad \text{or} \qquad E = \frac{q}{4\pi\varepsilon_0 r^2} \qquad (17.1.9)$$

in the region between the spheres. The potential difference between the spheres is, therefore, found once again from (16.5.22):

$$\Delta V = \int_{r_a}^{r_b} \frac{q}{4\pi\varepsilon_0 r^2} \, dr = \frac{q}{4\pi\varepsilon_0} \left(\frac{1}{r_a} - \frac{1}{r_b} \right)$$

Therefore,

$$C = \frac{q}{\Delta V} = \frac{q}{\dfrac{q}{4\pi\varepsilon_0} \left(\dfrac{r_b - r_a}{r_a r_b} \right)} = \frac{4\pi\varepsilon_0 r_a r_b}{r_b - r_a} \qquad (17.1.10)$$

It is often convenient to describe the capacitance of a single conductor, in which case it is assumed that the second conductor of opposite charge is located at infinity. With this definition, we can use Eq. (17.1.10) to calculate the capacitance of a single spherical conductor of radius r_a. The second conductor is then taken as a sphere of very large radius r_b (i.e., $r_b \to \infty$). In this way we may obtain

$$C = 4\pi\varepsilon_0 r_a$$

as the capacitance of a sphere of radius r_a.

17.2 Capacitors in Series and in Parallel

Capacitors in electrical circuits may be connected to one another and to other current elements in a variety of ways. In a circuit diagram, we use the symbol ―||― to represent a capacitor with a fixed capacitance as an element of an electric circuit, while the symbol ―||― is used to represent a capacitor whose capacitance can be varied. In a variable capacitor, provision is made for adjusting the spacing between the parallel plates of the capacitor, by turning a dial, for instance. There are, of course, other ways of building capacitors whose capacitance can be adjusted.

Let us consider the effect of connecting two different, initially uncharged, capacitors in the manner shown in Fig. 17.5a. If a voltage is applied between the points A and C, charges appear on the four conducting surfaces. For simplicity, assume we are dealing with parallel-plate capacitors whose separation is much larger than the separation of plates within any one of the capacitors. When the voltage is applied, charge $+q$ appears on plate 1, and an opposite charge $-q$ must appear on plate 4. The plates 2 and 3 as well as the external device which supplies the voltage were

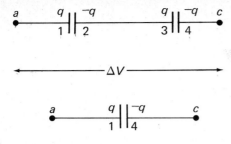

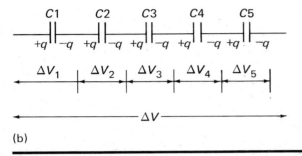

(a)

(b)

FIGURE 17.5. (a) Two capacitors in series (above) and a single equivalent capacitant (below). (b) Charges and potential differences for a series array of several capacitors.

originally electrically neutral and must *remain* in that condition since they are assumed to be insulated from their surroundings. Therefore, if charge $+q$ appears on plate 1, a compensating charge $-q$ must appear on plate 2. But this leaves plate 3 with a charge $+q$, since the sum of the charges on plates 2 and 3 must be zero. Fortunately, however, the charge $+q$ on plate 3 is exactly what is needed to balance the charge $-q$ on plate 4.

Since plates 1 and 4 are conductors carrying opposite charges, they constitute a capacitor of some sort. However, this is not a *simple* parallel plate capacitor because the plates may be quite far apart and, moreover, there are additional conductors in between. One may regard plates 2 and 3 as well as their connecting wire as a *single conductor* in which a separation of positive and negative charges has taken place. As explained above, then, the charge on plate 2 must be $-q$, while that on plate 3 must be $+q$. When capacitors are connected as in Fig. 17.5a, we say that the capacitors are in *series*. The statement that capacitors are in series implies that any increment of charge appearing on one of the capacitors necessarily appears on all the others. In the present case of two initially uncharged capacitors, the first had its charge increased from 0 to q and, as we have seen, the second one necessarily experienced the same increment. As mentioned before, we may regard this combination of capacitors as a *single* capacitor having some *equivalent capacitance* C, as shown in Fig. 17.5b. Then, C is

related to ΔV_{AC} by means of

$$\Delta V_{AC} = \frac{q}{C} \qquad (17.2.1)$$

But clearly

$$\Delta V_{AC} = \Delta V_{AB} + \Delta V_{BC}$$

or

$$\Delta V_{AC} = \frac{q}{C} = \frac{q}{C_1} + \frac{q}{C_2} \qquad (17.2.2)$$

The equivalent capacitance C must, therefore, be given by

$$\frac{1}{C} = \frac{1}{C_1} + \frac{1}{C_2} \qquad \text{or} \qquad C = \frac{C_1 C_2}{C_1 + C_2} \qquad (17.2.3)$$

Thus, if we wanted to replace the two capacitors C_1 and C_2 by a single one which would contain the *same* charge q when the *same* voltage ΔV_{AC} is applied, its capacitance would be given by Eq. (17.2.3). The generalization of (17.2.3) to the case of N capacitors connected in series is very straightforward and results in

$$\boxed{\frac{1}{C} = \frac{1}{C_1} + \frac{1}{C_2} + \cdots + \frac{1}{C_N}} \qquad N \text{ capacitors in series}$$

$$(17.2.4)$$

A group of capacitors may also be connected in *parallel*. When hooked up in this way, as shown in Fig. 17.6, the voltage across each capacitor has exactly the same value ΔV_{AB}. Each of the capacitors contains a charge which depends on its individual capacitance. Thus, for Fig. 17.6, we have

$$\Delta V_{AB} = \frac{q_1}{C_1} = \frac{q_2}{C_2} \qquad (17.2.5)$$

Referring to the same diagram, we see that the conductors carrying charges $+q_1$ and $+q_2$ may be regarded as a single conductor, while the conductors with charges $-q_1$ and $-q_2$ may be regarded as a second conductor. Thus, the combination may be replaced by a single *equivalent* capacitor of charge $q_1 + q_2$ which also has a voltage ΔV_{AB} across it. Now, since the combined or equivalent capacitor has the capacitance $C = (q_1 + q_2)/\Delta V_{AB}$, we find that

$$C = \frac{q_1 + q_2}{\Delta V_{AB}} = \frac{q_1}{\Delta V_{AB}} = \frac{q_2}{\Delta V_{AB}} = C_1 + C_2 \qquad (17.2.6)$$

Thus, the equivalent capacitance of a group of *parallel* capacitors is obtained by summing the values of the individual capacitances. For N capacitors in parallel, the equivalent capacitance will be

$$\boxed{C = C_1 + C_2 + \cdots + C_N} \qquad N \text{ capacitors in parallel}$$

$$(17.2.7)$$

539

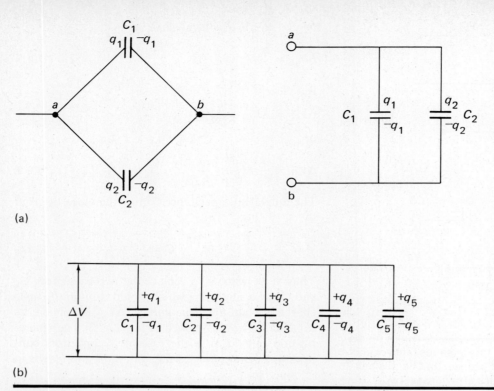

(a)

(b)

FIGURE 17.6. (a) Equivalent representations for a parallel combination of two capacitors. (b) Charges and potential differences for a parallel array of several capacitors.

Some complex networks of capacitors, such as the one shown in Fig. 17.7, can be reduced, step by step, using the laws of series and parallel combination derived above to a single equivalent capacitor. But there are many networks for which this is not possible. An illustration of such a situation is shown by the arrangement of capacitors of Fig. 17.8. For the time being, let us restrict our attention to problems which can be handled in terms of series and parallel combinations.

EXAMPLE 17.2.1

Two parallel-plate capacitors having capacitances of 4 μF and 6 μF are available in a laboratory. What are the possible values of capacitance that can be obtained by combining them? If a voltage of 6 V is applied across each of the combinations of capacitor, find the charge on each capacitor, assuming no initial charges are present.

There are four possible values of capacitance:

a. $C_1 = 4 \ \mu$F   capacitor 1 alone
b. $C_2 = 6 \ \mu$F   capacitor 2 alone
c. $C_1 + C_2 = 10 \ \mu$F   1 and 2 in parallel

d. $\dfrac{C_1 C_2}{C_1 + C_2} = 2.4 \ \mu$F   1 and 2 in series

In case (a), a potential difference of 6 V charges

the capacitor so that

$$q_1 = C_1 \, \Delta V = 2.4 \times 10^{-5} \ \text{C}$$

In a similar manner, in case (b), capacitor 2 develops a charge

$$q_2 = C_2 \, \Delta V = 3.6 \times 10^{-5} \ \text{C}$$

For the parallel connection (c), the total charge on the combination is $q_1 + q_2$, where again $q_1 = 2.4 \times 10^{-5}$ C and $q_2 = 3.6 \times 10^{-5}$ C.

Finally, for the series connection (d), the charge on the combined capacitor is

$$q = (2.4 \times 10^{-6})(6) = 1.44 \times 10^{-5} \ \text{C}$$

EXAMPLE 17.2.2

Three identical capacitors, each having a capacitance of 1 μF, are respectively charged by applying 1 V, 2 V, and 4 V across them. The sources of voltage are then removed leaving the three capacitors charged. Next, they are connected to one another in parallel by connecting the negatively charged plates to one another and likewise connecting the positively charged plates. Find the charges and voltages across each of the capacitors after this connection is made. What is the capacitance of the combination?

Initially, the individual capacitors have charges

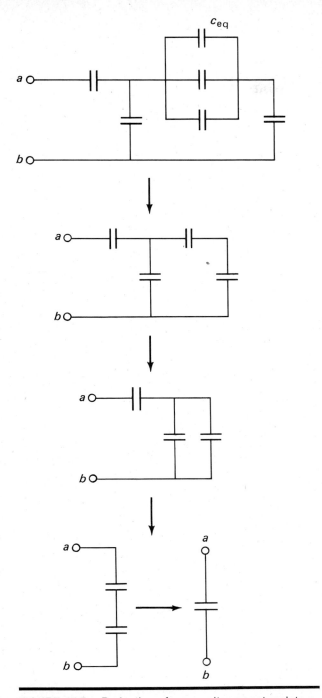

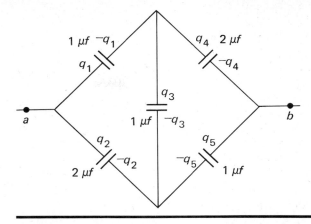

FIGURE 17.8. Network of capacitors which cannot be represented as a simple series on parallel combination.

each of the capacitors and q_1', q_2', and q_4' their respective charges. Then, from (17.1.1),

$$\Delta V = \frac{q_1'}{1\ \mu F} = \frac{q_2'}{1\ \mu F} = \frac{q_4'}{1\ \mu F}$$

and, therefore, each of the capacitors, being identical, will carry the same charge. Thus, we have

$$q_1 + q_2 + q_4 = 7 \times 10^{-6}\ C = 3q_1'$$

and

$$q_1' = q_2' = q_4' = \tfrac{7}{3} \times 10^{-6}\ C = 2.33 \times 10^{-6}\ C$$

The voltage ΔV is, therefore,

$$\Delta V = \frac{2.33 \times 10^{-6}}{1 \times 10^{-6}} = 2.33\ V$$

The capacitance of the combination is clearly

$$C_1 + C_2 + C_4 = 3\ \mu F$$

since the capacitors have been connected in parallel.

EXAMPLE 17.2.3

Two capacitors of capacitance 3 μF and 6 μF are connected in series across a 12-V battery. What are the charges and potential differences on each of them? They are then disconnected from the battery and connected together in parallel. Find the resulting potential difference for both polarities of connection.

The charges acquired during the series connection are most easily found by using the formula

$$q = C\ \Delta V$$

where C, the equivalent capacitance, has the value $C_1 C_2 / (C_1 + C_2)$, from which

$$C = \frac{(3)(6)}{(3 + 6)} = 2\ \mu F$$

Therefore, the charge on the combination, and also on

FIGURE 17.7. Reduction of a capacitance network to a single equivalent capacitor by successive application of laws derived for series and parallel equivalent capacitance.

$$q_1 = C_1\ \Delta V_1 = (1 \times 10^{-6})(1) = 10^{-6}\ C$$
$$q_2 = C_2\ \Delta V_2 = (1 \times 10^{-6})(2) = 2 \times 10^{-6}\ C$$
$$q_4 = C_4\ \Delta V_4 = (1 \times 10^{-6})(4) = 4 \times 10^{-6}\ C$$

After the connection is made, the voltage across each of the capacitors is identical since they are connected in parallel. Let ΔV denote the unknown voltage across

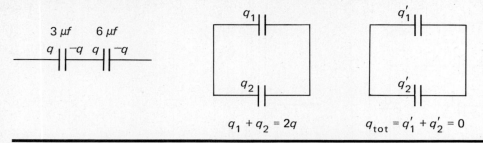

FIGURE 17.9

each of the capacitors, is

$$q = C \, \Delta V = (2 \times 10^{-6})(12) = 2.4 \times 10^{-5} \text{ C}$$

After the battery is removed, we can separate the capacitors, and we shall assume that each retains its charge until they are again connected together. In Fig. 17.9, the two possible parallel connections are illustrated. If the positive plates are connected, the combined capacitor has a charge of $2q$, or 4.8×10^{-5} coulomb. Since the equivalent capacitance is $C_1 + C_2 = 9 \ \mu\text{F}$, the voltage across the combination is

$$\Delta V = \frac{2q}{C_1 + C_2} = \frac{4.8 \times 10^{-5} \text{ C}}{9 \times 10^{-6} \text{ F}} = 5.33 \text{ V}$$

and the charges on the respective capacitors are

$$q_1 = C_1 \, \Delta V = 1.6 \times 10^{-5} \text{ C}$$
$$q_2 = C_2 \, \Delta V = 3.2 \times 10^{-5} \text{ C}$$

When the connection is made with the polarities opposite, the net charge is zero and, therefore, the voltage is zero. In this case, the separate capacitors are restored to an uncharged condition.

17.3 Energy Storage in a Capacitor

The usefulness of capacitors as circuit elements stems from their ability to *store energy*. To understand how energy can be stored, consider a simple parallel-plate capacitor which is initially uncharged as shown in Fig. 17.10a. The capacitor is connected across a source of potential difference represented by a battery of voltage ΔV_f which, when the switch is closed, as in Fig. 17.10b, transfers charge from one plate to the other until the potential difference across the capacitor is *equal* (though opposite in polarity) to that provided by the battery. When this stage, illustrated in Fig. 17.10c, is reached, the flow of charge from battery to capacitor stops. The capacitor is now charged to a potential difference ΔV_f and bears a charge $\pm Q$ on its plates. We will discuss the details of this charging process at greater length after we have studied such concepts as resistance and Kirchhoff's circuit equations, which are introduced in the next chapter.

At some intermediate stage of the charging process, such as the one shown in Fig. 17.10b, one plate has charge $-q$ while the second plate has charge $+q$. At this moment, the potential difference across the capacitor is given by

$$\Delta V = \frac{q}{C} \qquad (17.3.1)$$

A very short time later, after a time interval dt, the charge will have increased by an amount dq. This charge increment has experienced a change in potential energy of dq times the difference of potential ΔV across which it has been transferred, according to (16.6.1). Therefore, during the time dt, the potential energy of the capacitor has increased by an amount

$$dU = \Delta V \, dq = \frac{q}{C} \, dq \qquad (17.3.2)$$

This increase in energy occurs at the expense of the battery which is doing *work* on the system by increasing the potential energy of the charges. A mechanical analogy, in which a man does work by lifting crates onto a platform supported by ropes, is shown in Fig. 17.10d. Here, the man does work whenever he lifts up a crate, and he therefore increases the gravitational potential energy of the system. There is, however, an important difference between the two different cases illustrated in Fig. 17.10. The potential energy difference between the capacitor plates is *variable*, and therefore the amount of work done by the battery depends not only on dq but also *on the potential difference* ΔV between the plates, which *changes* as the charge builds up. In the mechanical analogy, the man does the same amount of work every time he moves up another crate, since the height through which he moves it is always the same. Can you think of any modification of the mechanical analogy which more closely resembles the result expressed in Eq. (17.3.2)?

Now, using Eq. (17.3.2), we can obtain by integration the total energy stored by the capacitor. In integrating, we may write the capacitance C outside the integral, since it is *independent* of the charge q on the capacitor:

$$U = \int_0^Q \frac{q}{C} \, dq = \frac{1}{C} \int_0^Q q \, dq = \frac{Q^2}{2C} \qquad (17.3.3)$$

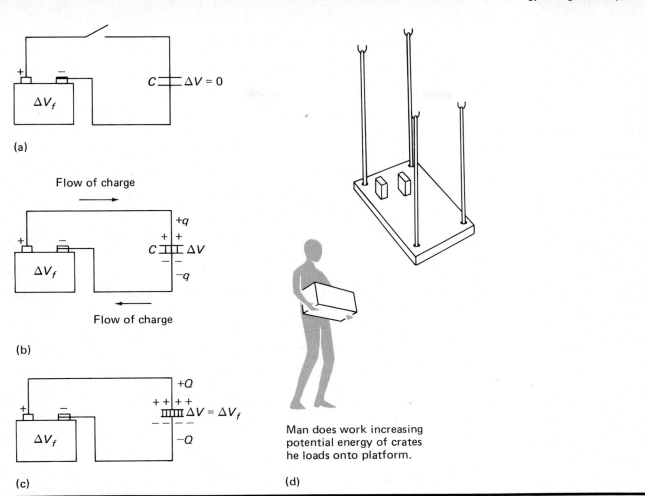

FIGURE 17.10. (a), (b), and (c). Successive stages in the charging of a condenser with a battery; at (c), a condition of equilibrium is reached, in which a certain amount of electrostatic energy is stored in the capacitor. (d) A mechanical process that is analogous to the storage of electrostatic potential energy illustrated in (a), (b), and (c). In this process, mechanical potential energy is stored by the man loading bricks onto the raised platform. The original source of the stored potential energy is, in both instances, chemical energy which is transformed into electrical energy by the battery in (a), (b), and (c) and converted into mechanical energy in the man's muscle in (d).

where Q is the final charge imparted to the capacitor. Equation (17.3.3) may be written in several different ways by using the relation (17.1.1) between Q, C, and the final potential difference ΔV. Thus,[1]

$$U = \frac{Q^2}{2C} = \tfrac{1}{2} Q \, \Delta V = \tfrac{1}{2} C \, (\Delta V)^2 \qquad (17.3.4)$$

[1] The situation here is very similar to that discussed in Example 16.6.4. The process discussed there amounts to charging a single conducting sphere of radius R to a final charge Q. As we saw in Example 17.1, such a sphere can be thought of as a spherical *capacitor* in which one of the plates has an infinite radius, the capacitance of the sphere being $4\pi\varepsilon_0 R$. It will become evident that Eq. (16.6.25), which represents the potential energy stored in this spherical capacitor, can be written as $U = Q^2/2C$, where $C = 4\pi\varepsilon_0 R$; in this form, it is identical with the result given above.

The electrostatic potential energy stored in the capacitor can be used later by connecting the capacitor to some other electrical circuit. The use of this energy is analogous to the possibility of using the potential energy of the stored crates in Fig. 17.10d. If the ropes in that diagram were cut, for example, the crates would lose their potential energy and could do useful work.

EXAMPLE 17.3.1

A 5-μF capacitor and a 3-μF capacitor are connected in series and are initially uncharged. If a potential difference of 10 volts is applied across the series combination, find the total energy stored as well as the energy stored by each capacitor. They are disconnected from the 10-volt source and from each other and then reconnected in parallel with positive plates

543

connected together and negative plates likewise joined. Find the new energy of the system and explain why it is *different* from that of the original series connection.

The equivalent capacitance of the series connection is

$$C = \frac{C_1 C_2}{C_1 + C_2} = \frac{15}{8} \, \mu F$$

When this series combination is connected across 10 volts, each of the capacitors acquires a charge

$$q = C \, \Delta V = \left(\frac{15}{8} \times 10^{-6}\right)(10) = 1.88 \times 10^{-5} \, C$$

According to (17.3.3), the energy of the combination is then

$$U = \tfrac{1}{2}C \, (\Delta V)^2 = \left(\frac{1}{2}\right)\left(\frac{15}{8} \times 10^{-6}\right)(10)^2 = 9.4 \times 10^{-5} \, J$$

The individual capacitors have energies

$$U_1 = \frac{q^2}{2C_1} = \frac{(1.88 \times 10^{-5})^2}{(2)(5 \times 10^{-6})} = 3.5 \times 10^{-5} \, J$$

$$U_2 = \frac{q^2}{2C_2} = \frac{(1.88 \times 10^{-5})^2}{(2)(3 \times 10^{-6})} = 5.9 \times 10^{-5} \, J$$

When the plates are reconnected in parallel, we have an equivalent capacitance of

$$C_1 + C_2 = 8 \, \mu F$$

carrying a charge of

$$2q = 3.76 \times 10^{-5} \, C$$

and therefore an energy

$$U' = \frac{(3.76 \times 10^{-5})^2}{(2)(8 \times 10^{-6})} = 8.8 \times 10^{-5} \, J$$

We see that the final energy U' is less than the initial energy U. The difference in electrostatic energy is accounted for as energy lost as heat in the process of transferring charge from one of the conductors to another.[2] This process of energy loss, known as joule heating, will be discussed in the following chapter.

EXAMPLE 17.3.2

N identical capacitors each of capacitance C are to be connected across a voltage V. Derive expressions for the stored energy for both series and parallel connections.

For the series connection, the stored energy is

---

[2] Initially, of course, the electrostatic energy appears as *kinetic energy* of the mobile free electrons that move through the conductors when the plates are reconnected, but this is ultimately dissipated as heat as the moving electrons undergo inelastic collisions within the conductors.

$U_s = \tfrac{1}{2}C_s \, (\Delta V)^2$, where

$$\frac{1}{C_s} = \frac{1}{C} + \cdots + \frac{1}{C} = \frac{N}{C}$$

or

$$C_s = \frac{C}{N} \qquad (17.3.5)$$

Hence,

$$U_s = \tfrac{1}{2}C \, (\Delta V)^2 \, \frac{1}{N} \qquad (17.3.6)$$

When the capacitors are connected in parallel, each one has an energy of $\tfrac{1}{2}C \, (\Delta V)^2$; therefore, the energy of all of them is U_p, where

$$U_p = \tfrac{1}{2}C \, (\Delta V)^2 N \qquad (17.3.7)$$

Thus, for a parallel connection, it is clear that greater energy storage is possible. For a fixed potential difference, the energy stored is linearly proportional to the capacitance. This is *much greater for N capacitors in parallel than for N in series*.

It is worth noting that the plates of a parallel-plate capacitor attract each other electrically since they are oppositely charged. This attractive force can readily be calculated from the potential energy U if U is expressed in terms of the separation x of the capacitor plates. The potential energy is given by

$$U = \frac{Q^2}{2C} = \frac{Q^2 x}{2\varepsilon_0 A} \qquad (17.3.8)$$

The force can now be expressed as

$$F = -\frac{dU}{dx} = -\frac{Q^2}{2\varepsilon_0 A} \qquad (17.3.9)$$

The plates, therefore, experience a mutual attraction that varies quadratically with the charge on the capacitor.

We have found that the electrostatic energy stored by a capacitor is equal to the work required to charge the capacitor. In this view of the situation, the stored energy resides in the electrostatic potential energy of the charges on the plates of the device. An alternative view can be formulated in terms of the *electric fields* established between the plates of the capacitor. According to this view, it is possible to define an *energy density u_e* as the *stored electrostatic field energy per unit volume*. This quantity is analogous to other densities such as mass density or charge density. It turns out that at any location the electric energy density u_e depends quadratically on the electric field E at that location. To be more precise, we can show that u_e is given by

$$\boxed{u_e = \tfrac{1}{2}\varepsilon_0 E^2} \qquad (17.3.10)$$

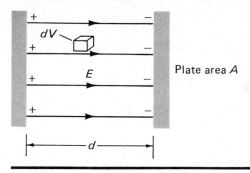

FIGURE 17.11. Volume element dV in an electric field **E** contains electrostatic energy $(\varepsilon_0 E^2/2)\, dV$.

This relation is very general, and although we shall not give a general proof, we shall demonstrate the compatibility of this equation with the expressions for the energy of various capacitors. Thus, as we shall see, we may regard the energy of a capacitor as being somehow *distributed throughout the volume of space containing the electric field created by the capacitor.* From this point of view, the energy is the *work needed to establish the entire electric field.*

The parallel-plate capacitor, for example, has a constant electric field E within the volume between its plates, as illustrated in Fig. 17.11. The magnitude of this field is given by Eq. (17.1.3), and the relationship between the field and the potential difference ΔV is expressed by (17.1.2). These relations allow us to calculate the capacitance as shown in Eq. (17.1.4). The total electrostatic energy of the charge distribution, according to (17.3.4), is $U = C\,(\Delta V)^2/2$. But, for the parallel-plate capacitor, (17.1.3) and (17.1.1) allow us to write

$$U = \tfrac{1}{2}C\,(\Delta V)^2 = \frac{1}{2}\frac{\varepsilon_0 A}{d}\,(E^2 d^2) = \tfrac{1}{2}\varepsilon_0 E^2 A d \qquad (17.3.11)$$

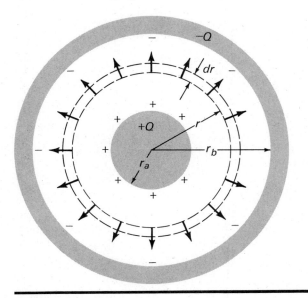

FIGURE 17.12

But Ad is simply the volume of the region between the plates of the condenser. Dividing both sides of (17.3.11) by this quantity, it is evident that in the case of the parallel-plate condenser, at least, the energy per unit volume within the field region is $\varepsilon_0 E^2/2$, as predicted by (17.3.10). Equation (17.3.10) is very general in applicability and provides a way of calculating the electrostatic energy associated with any field configuration arising from any charge distribution, irrespective of whether or not it is created by a capacitor.

EXAMPLE 17.3.3

A spherical capacitor consists of a sphere of outer radius r_a concentric with a larger sphere of inner radius r_b. If the capacitor bears a total charge Q, find the total energy of the system (a) by expressing the energy using Eq. (17.3.4) and (b) by integrating u_e over all space (see Fig. 17.12). According to (17.1.10), the capacitance of this spherical capacitor is

$$C = \frac{4\pi\varepsilon_0 r_a r_b}{r_b - r_a}$$

and, therefore, the stored energy is

$$U = \frac{Q^2}{2C} = \frac{Q^2(r_b - r_a)}{8\pi\varepsilon_0 r_a r_b} \qquad (17.3.12)$$

To do part (b) of the problem, we need to know the electric field at each point of space. Using Gauss's law, one finds that for $r_a < r < r_b$,

$$E = \frac{1}{4\pi\varepsilon_0}\frac{Q}{r^2} \qquad (17.3.13)$$

while $E = 0$ for all other values of r. The energy is, therefore, all contained between r_a and r_b. In this region, the energy density is $\varepsilon_0 E^2/2$, or

$$u_e = \varepsilon_0 \frac{Q^2}{32\pi^2 \varepsilon_0^2 r^4} \qquad (17.3.14)$$

The energy dU contained between r and $r + dr$ is the energy contained between spheres of these radii. Since the volume between the spheres is $4\pi r^2\, dr$, we find

$$dU = \frac{Q^2}{32\pi^2 \varepsilon_0 r^4}\,(4\pi r^2\, dr)$$

$$= \frac{Q^2}{8\pi\varepsilon_0 r^2}\, dr \qquad (17.3.15)$$

The total energy is, therefore,

$$U = \int_{r_a}^{r_b} \frac{Q^2}{8\pi\varepsilon_0 r^2}\, dr = -\left[\frac{Q^2}{8\pi\varepsilon_0 r}\right]_{r_a}^{r_b}$$

$$= \frac{Q^2(r_b - r_a)}{8\pi\varepsilon_0 r_a r_b} \qquad (17.3.16)$$

as previously obtained.

17.4 Dielectrics Increase Capacitance

As we have seen in the preceding sections, the capacitance of a pair of conductors depends on their geometry. In previous sections, it has been assumed that the region between the conductors of the capacitor is filled with air (more precisely, vacuum). It was discovered by Michael Faraday that for a fixed geometry, the capacitance of a condenser can be increased by replacing the vacuum by a *dielectric*, that is, by an insulating substance.

Before we study what happens when a dielectric is used in a capacitor, let us consider how the parallel-plate capacitor is altered if *part* of the space between the plates is replaced by a conductor, as shown in Fig. 17.13. Initially, the capacitance has the value

$$C_0 = \frac{\varepsilon_0 A}{d} \qquad (17.4.1)$$

Assume that the inserted conductor is a distance a_1 from one plate and a_2 from the other. Since the electric field within the conductor vanishes, the conducting material must itself somehow establish an electric field in its interior which exactly cancels the field produced by the capacitor. In a conductor, this is accomplished by a physical separation of free charges, which produces a positive charge density at one surface and a negative charge density at the other. From the fact that we now have two capacitors in series, with capacitances

$$C_1 = \frac{\varepsilon_0 A}{a_1} \quad \text{and} \quad C_2 = \frac{\varepsilon_0 A}{a_2} \qquad (17.4.2)$$

it follows that the equivalent capacitance is

$$C = \left(\frac{1}{C_1} + \frac{1}{C_2}\right)^{-1} = \frac{\varepsilon_0 A}{a_1 + a_2} \qquad (17.4.3)$$

But since $a_1 + a_2$ must necessarily be less than d, this means that C must be *larger* than before.

The increased capacitance due to insertion of a conducting material is easily understood. Any conductor contains charges that are free to migrate in response to applied electric fields. The separated charges produce an electric field which cancels the applied field in the interior of the conductor. The same amount of charge is stored on the plates, but by an external potential difference *smaller* than was there originally. This means that the capacitance, which is the ratio of the stored charge to the potential difference, has increased.

Now let us repeat the above observations, inserting now an insulating substance or *dielectric* material such as glass or mica. We find, just as Faraday did, that even though there are no free charges which move in response to applied electric fields, a dielectric can also increase the capacitance. Faraday showed conclusively that for any geometry, the effect of *completely filling* the space between the conductors with a dielectric is to multiply the capacitance by a constant K, which depends *only* on the choice of the dielectric. This constant is independent of the geometry of the capacitor. It is also practically independent of the potential difference across the device, so long as the interior field does not become excessively large.

The constant K is called the *dielectric constant* of the insulator. For vacuum it assumes the value 1, while for air it is only slightly greater, 1.0006 under standard conditions. Glass has a dielectric constant of about 6, while water has the value 81. We shall try to understand why a dielectric increases the capacitance and why this increase depends only on the properties of the medium. We shall attempt, at first, to develop a reasonable macroscopic model of why the capacitance is altered and will later discuss some aspects of the microscopic theory of dielectrics.

We have seen that when a conductor is inserted between the capacitor plates, the macroscopic charge separation creates an electric field which completely cancels the electric field produced by the capacitor plates in its interior. The fact that a dielectric also leads to an enhancement of capacitance suggests that the dielectric material is also capable of producing an electric field. What mechanism is responsible for this electric field? We have learned in the preceding chapter that *electric dipoles*, even though they are electrically neutral, can produce electric fields. The atoms and molecules of most substances do not have inherent dipole moments in the absence of electric fields. But when an external electric field is present, it has the effect of *inducing* a dipole moment in each atom or molecule, which is directly proportional to the strength of the external field. We will have more to say about this later, but for now let us assume that an applied electric field causes a rather small spatial separation of charge within each atom or molecule, and, as a consequence, each of them produces a dipole electric field whose direction is such as to oppose the applied field $\mathbf{E}$ (see Fig. 17.14).

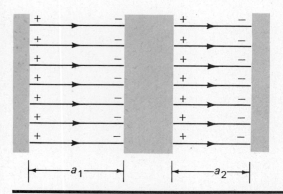

FIGURE 17.13. Parallel-plate capacitor into which an uncharged conducting plate has been inserted.

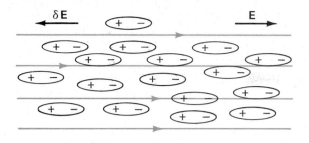

(a)

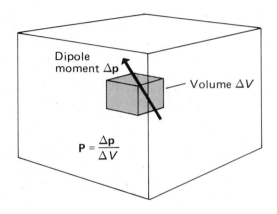

(b)

FIGURE 17.14. Induced elementary dipole moments arising from the application of an external electric field. The total electric field within the substance is the sum of the external field and the fields arising from the induced dipoles.

Assume that the ith molecule acquires a small molecular dipole moment $\mathbf{p}_i$. A macroscopic volume element $\Delta\mathcal{V}$ containing m such dipoles, therefore, has a dipole moment[3]

$$\Delta\mathbf{p} = \sum_{i=1}^{m} \mathbf{p}_i \qquad (17.4.4)$$

Even though $\Delta\mathcal{V}$ may be an extremely small volume, m can be huge since the number of molecules in any macroscopic volume element is very large. The ratio of $\Delta\mathbf{p}$ to $\Delta\mathcal{V}$ is the average dipole moment per unit volume of dielectric. We may now define a *macroscope polarization vector* $\mathbf{P}$ at every point within the dielectric according to the prescription

$$\mathbf{P} = \lim_{\Delta\mathcal{V}\to 0} \frac{\Delta\mathbf{p}}{\Delta\mathcal{V}} = \frac{d\mathbf{p}}{d\mathcal{V}} \qquad (17.4.5)$$

This defines a new vector field inside the dielectric. According to this definition, the polarization is a vector that represents the dipole moment acquired by

[3] In this section, we shall use the script symbol $\mathcal{V}$ to represent volume, so as to avoid confusion with velocity (v) and potential (V).

every volume element $\Delta\mathcal{V}$ divided by the volume of that element. The limiting procedure above states that $\Delta\mathcal{V}$ approaches zero, but this limit need not be taken too literally. In effect, $\Delta\mathcal{V}$ must be very small compared to the macroscopic dimensions of the sample but should be large enough to contain a very large number of molecules.

The existence of the polarization $\mathbf{P}$ implies the presence of an additional electric field which partially cancels the original field that produced $\mathbf{P}$ to begin with. It is difficult, however, to calculate the new electric field by directly superposing the fields produced by each small volume element within the dielectric. There is a better method, one in which Gauss's law and certain relations between polarization and charge density are both utilized. Consider a dielectric material in which $\mathbf{P}$ is uniform throughout, as shown in Fig. 17.15. The total charge density in any small interior volume will be zero, for there are just as many positive as negative charges there. On the other hand, as we can see from Fig. 17.15, a *surface charge density* develops on the left and right exterior surfaces of the dielectric. Suppose that within any molecule or atom there is an average charge separation Δ between charges $+q$ and $-q$. Let the dielectric plate area be A.

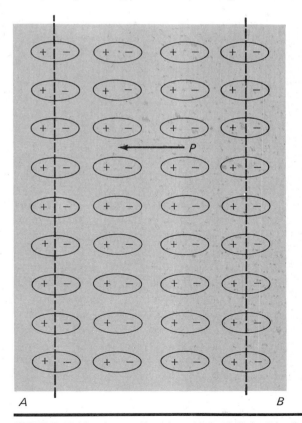

FIGURE 17.15. In a uniformly polarized dielectric, the charges of adjacent interior dipoles cancel each other; the net field produced is solely due to the unbalanced surface charge distribution that appears on the exterior surfaces of the material.

547

Then, within the volume $A\Delta$ there are $NA\Delta$ molecules, where N is the number per unit volume.

Therefore, at the left face we have a total charge, referred to as *bound charge*, or *polarization charge*,

$$q_b = +qNA\Delta$$

or a positive surface charge density σ_b given by

$$\sigma_b = qN\Delta \qquad (17.4.6)$$

The magnitude of the polarization vector $\mathbf{P}$ is the dipole moment per unit volume and must be given by the product of the dipole moment $q\Delta$ of an individual molecule and the number of molecules per unit volume:

$$P = q\,\Delta N \qquad (17.4.7)$$

We see, therefore, that for the special case treated here, in which the polarization vector is perpendicular to the surface boundary, the magnitude of the polarization vector is equal to the bound charge density:

$$P = \sigma_b \qquad (17.4.8)$$

More generally, if the local polarization vector $\mathbf{P}$ makes an angle θ with the boundary, the bound charge density is given by

$$\boxed{\sigma_b = \mathbf{P} \cdot \mathbf{n} = P\cos\theta} \qquad (17.4.9)$$

where $\mathbf{n}$ is the unit normal to the boundary.[4]

The charge density σ_b is as real as any other charge density we have encountered. The use of descriptive adjectives such as "bound" and "free" merely distinguishes between the charge density arising from the polarization of a dielectric and the charge density present on or within a conductor. A distribution of free charge is, of course, completely free to move when electric fields are present, which is not the case for a distribution of bound charge or polarization charge.

Using the charge density σ_b, let us calculate the electric field within the dielectric by using Gauss's law. We choose a Gaussian surface, as shown in Fig. 17.16. The charge contained within this surface now consists of negative *free* charges on the conducting surface as well as positive bound charges. We therefore obtain from Gauss's law

$$\oint_s \mathbf{E} \cdot \mathbf{n}\, da = EA = \left(\frac{\sigma_f - \sigma_b}{\varepsilon_0}\right) A$$

or

$$E = \frac{\sigma_f - \sigma_b}{\varepsilon_0} \qquad (17.4.10)$$

---

[4] It is easy to see why this is so by noting that if the polarization vector is *parallel* to the boundary, there is no surface charge there at all. It is, therefore, only the component of the polarization vector *normal* to the surface (that is, $\mathbf{p} \cdot \mathbf{n}$) that is effective in generating a surface distribution of bound charge

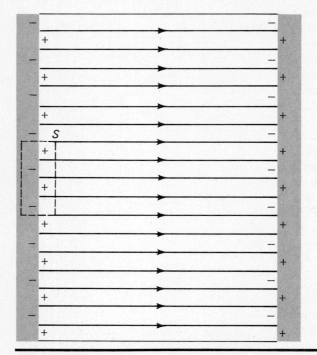

FIGURE 17.16. Gaussian surface, containing both free and bound polarization charge, used in the calculation of the field within a dielectric.

Thus, for a given free surface charge density on the plates of the capacitor, the presence of the dielectric brings about a *reduction* of the electric field. Substituting $\sigma_b = \mathbf{P} \cdot \mathbf{n} = P$, we can rewrite the above equation as

$$\sigma_f = \varepsilon_0 E + P \qquad (17.4.11)$$

The vector combination $\varepsilon_0 \mathbf{E} + \mathbf{P}$ is called the *displacement vector* $\mathbf{D}$. As we see from the simple case above, this vector is directly related to the free charge density. In fact, since we can regard $\sigma_b A/\varepsilon_0 = (\mathbf{P} \cdot \mathbf{n})A/\varepsilon_0$ as the integral of $(\mathbf{P} \cdot \mathbf{n})/\varepsilon_0$ over the Gaussian surface, we can *restate* Gauss's law in terms of $\mathbf{D}$ in the simple form

$$\boxed{\oint_s (\varepsilon_0 \mathbf{E} + \mathbf{P}) \cdot \mathbf{n}\, da = \oint_s \mathbf{D} \cdot \mathbf{n} = q_f} \qquad (17.4.12)$$

where q_f is the total amount of *free* charge within the Gaussian surface. This is the form in which we ordinarily use Gauss's law when dielectric materials are present. It should be noted that the charge on the left side represents only the free charge within the Gaussian surface; effects due to the distribution of bound charges are completely accounted for by the inclusion of the polarization term in the integral. We shall return to a somewhat more detailed discussion of the displacement vector $\mathbf{D}$ in a subsequent section.

From the above discussion, we see that in a

dielectric a mechanism exists for reducing an applied external electric field. The presence of bound charges accounts for this reduction. This same mechanism, therefore, also reduces the potential difference between the conducting plates of a capacitor, for a given fixed charge, and thereby leads to an increase in its capacitance.

When the electric field within a dielectric increases, the spatial separation of positive and negative charge centers within each atom or molecule also increases. A reasonable approximation, valid for electric fields that are not enormously large and for most dielectric materials, is to assume that the polarization **P** varies *linearly* with the applied electric field. In such a case,

$$\mathbf{P} = \chi \varepsilon_0 \mathbf{E} \tag{17.4.13}$$

relates the polarization **P** and the electric field **E** inside the dielectric. In this relation, the proportionality constant χ is called the *electric susceptibility* of the dielectric. Equation (17.4.13), in which χ is constant, defines what we refer to as *linear* dielectric behavior.

Let us substitute Eq. (17.4.13) into (17.4.11). We then find

$$\sigma_f = \varepsilon_0 E + \chi \varepsilon_0 E = (1 + \chi)\varepsilon_0 E$$

or

$$E = \frac{1}{1+\chi}\frac{\sigma_f}{\varepsilon_0} = \frac{\sigma_f}{\varepsilon} \tag{17.4.14}$$

where ε is defined by the relation

$$\varepsilon = (1 + \chi)\varepsilon_0 \tag{17.4.15}$$

The quantity ε defined in this way is called the *permittivity of the dielectric*. Since χ is always a positive number, the permittivity of a dielectric always exceeds the value ε_0 associated with free space.

Since we now know that the field within the dielectric is σ_f/ε, as given by (17.4.14), we can proceed now to recalculate the capacitance of a parallel-plate condenser, assuming that the space between the plates is entirely filled with a uniform dielectric medium of permittivity ε. Since now the field within the dielectric is given by (17.4.14), we may write

$$\sigma_f = q_f/A = \varepsilon E = \frac{\varepsilon \, \Delta V}{d}$$

from which

$$C = \frac{q_f}{\Delta V} = \frac{\varepsilon A}{d} \tag{17.4.16}$$

The capacitance has clearly increased by the factor $\varepsilon/\varepsilon_0$ with respect to the case where there is no dielectric present. It is just this factor, however, that was used

to define the *dielectric constant K* referred to in the initial phases of our discussion. Therefore, we see that the dielectric constant must be given by

$$K = \frac{\varepsilon}{\varepsilon_0} = 1 + \chi \tag{17.4.17}$$

EXAMPLE 17.4.1

A parallel-plate capacitor has a capacitance $C_0 = 10^{-6}$ F in air. It is charged to 200 volts and has a plate separation of 2 mm. A dielectric with an electric susceptibility of 49 is now inserted so as to fill the space between the plates. (a) Find the new value of capacitance. (b) How much bound charge is contained on one of the dielectric boundaries? (c) Find the polarization of the dielectric. (d) What value does the electric field and the displacement field have inside the dielectric?

According to (17.4.16), insertion of a dielectric with $\chi = 49$ (corresponding to $K = 50$) increases the capacitance to

$$C = (1 + 49)C_0 = 50 C_0 = 50 \times 10^{-6} \text{ F}$$

To obtain the amount of bound charge, let us first compute the polarization vector inside the dielectric. From (17.4.13),

$$P = 49 \varepsilon_0 E \tag{17.4.18}$$

According to (17.4.14), this may also be written as

$$P = 49 \varepsilon_0 \cdot \frac{1}{50}\frac{\sigma_f}{\varepsilon_0} = 0.98 \sigma_f$$

Since the polarization P is equal to the bound charge density,

$$\sigma_b = 0.98 \sigma_f$$

or

$$q_b = 0.98 q_f = 0.98 C_0 \, \Delta V$$
$$= (0.98)(10^{-6})(200) = 1.96 \times 10^{-4} \text{ C}$$

The polarization is, therefore,

$$P = \sigma_b = \frac{q_b}{A} = \frac{q_b \varepsilon_0}{C_0 d}$$

$$= \frac{(1.96 \times 10^{-4})(8.85 \times 10^{-12})}{(10^{-6})(2 \times 10^{-3})} = 8.7 \times 10^{-7} \text{ C/m}^2 \tag{17.4.19}$$

The electric field inside the dielectric is

$$E = \frac{1}{50}\frac{\sigma_f}{\varepsilon_0} = 0.02 \frac{q_f}{A\varepsilon_0} = 0.02 \times \frac{C_0 \, \Delta V}{A\varepsilon_0}$$

$$= 0.02 \frac{\varepsilon_0 A}{A d \varepsilon_0} \Delta V = 0.02 \frac{\Delta V}{d} \quad \text{(as expected)}$$

549

Thus,

$$E = 0.02 \frac{200}{2 \times 10^{-3}} = 2 \times 10^3 \text{ V/m} \qquad (17.4.20)$$

Finally, the magnitude of the displacement vector **D** is the free charge density, see (17.4.11):

$$D = \frac{q_f}{A} = \varepsilon_0 \frac{\Delta V}{d} = 8.85 \times 10^{-7} \text{ C/m}^2$$

As a check, we note that

$$\varepsilon_0 \frac{\Delta V}{d} = \varepsilon_0 \frac{1}{1+\chi} \frac{\Delta V}{d} + \varepsilon_0 \frac{\chi}{1+\chi} \frac{\Delta V}{d}$$

$$D = \varepsilon_0 E + P$$

EXAMPLE 17.4.2

Capacitors of 3 μF and 6 μF are connected in series across a potential difference of 100 volts. It is possible to alter the equivalent capacitance by inserting a dielectric material between the plates of each capacitor separately, or between both simultaneously. If the dielectric constant of the dielectric is 15, find the possible values of capacitance that can be obtained. For each case, find the potential difference across the combination. Assume in all cases that the dielectric entirely fills the space between the plates.

Originally, we have a 3-μF capacitor with a potential difference ΔV_3 across it and a 6-μF capacitor with potential difference ΔV_6. Clearly,

$$Q_f = C_3 \Delta V_3 = C_6 \Delta V_6$$

and

$$V_3 + V_6 = 100 \text{ V}$$

This implies that

$$\Delta V_3 = 66.7 \text{ V}$$
$$\Delta V_6 = 33.3 \text{ V}$$

When the 3-μF capacitor is altered by inserting the dielectric, its capacitance is increased to $(3)(15) = 45$ μF and the voltage across it is reduced to $66.7/15 = 4.44$ volts. Thus, the capacitance of the equivalent combination is

$$C = \frac{(45)(6)}{45 + 6} = 5.29 \text{ } \mu\text{F}$$

and the voltage across it is

$$\Delta V = 4.44 + 33.3 = 37.7 \text{ V}$$

If the 6-μF capacitor has the dielectric inserted, its capacitance becomes $6 \times 15 = 90$ μF and the voltage across it becomes $33.3/15 = 2.22$ volts. The equivalent capacitor has a capacitance

$$C = \frac{(3)(90)}{3 + 90} = 2.90 \text{ } \mu\text{F}$$

The voltage across the combination is

$$\Delta V = 66.7 + 2.22 = 68.9 \text{ V}$$

Finally, if both capacitors are simultaneously altered, the equivalent capacitance is

$$C = \frac{(45)(90)}{45 + 90} = (15) \frac{(3)(6)}{3 + 6} = 30 \text{ } \mu\text{F}$$

The voltage is

$$\Delta V = \frac{100}{15} = 6.66 \text{ V}$$

We note that in each case, the product of the equivalent capacitance C and the voltage ΔV has the constant value of 200×10^{-6} coulomb, corresponding to the fact that the free charge on the capacitor is unaltered.

17.5 Boundary Conditions on E, P, and D

The three vectors, **E**, **P**, and **D**, play a central role in dielectric theory. The statement of Gauss's law can be given in terms of the electric vector **E**, in which case the charge appearing in the equation is the total (free plus bound) charge, or it may be stated using the displacement vector **D**, in which case only the free charge appears. There is a fundamental difference between the vectors **E** and **D**, which should be emphasized. While **E** represents a sum of microscopic fields produced by individual atoms or molecules, **D** is necessarily a *macroscopic* field arising from the polarization of a macroscopic volume. Put another way, it would be entirely reasonable to consider the average electric field produced by one or two molecules, but it would not be meaningful to discuss the displacement vector **D** for a system containing so few molecules.

The solution of problems involving dielectrics requires an understanding of how the vectors **E**, **P**, and **D** change at the boundary between two different dielectrics. Let us consider a small portion of an interface between materials with dielectric constants K_1 and K_2, respectively, as shown in Fig. 17.17. At the interface, there will be a bound charge density σ_b if the dielectrics are polarized, and also (possibly) a free charge density σ_f. Let us consider the integral of $\mathbf{E} \cdot d\mathbf{l}$ around the rectangular path consisting of sides ab, bc, cd, and da, as shown in Fig. 17.17a. This integral must vanish in electrostatics as a consequence of the conservative character of the field.[5] The contributions from segments bc and da vanish because these distances are chosen to be infinitesimally

---

[5] This is so because the work done by a conservative force in traversing any closed path must be zero.

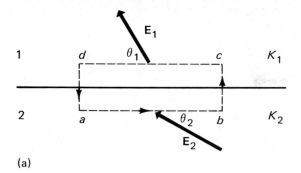

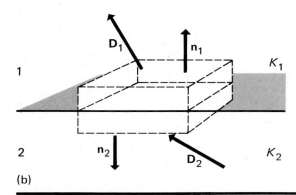

FIGURE 17.17. Boundary conditions on the vectors **E** and **D** at the interface between two dielectric media of different permittivity. (a) The component of **E** parallel to the interface is the same in both media. (b) The component of **D** normal to the interface is the same in both media.

small. Thus, with $l = cd = ab$, we find

$$\oint \mathbf{E} \cdot d\mathbf{l} = 0 = (E_1 \cos \theta_1)l - (E_2 \cos \theta_2)l = 0$$

or

$$\boxed{E_{1p} = E_{2p}} \tag{17.5.1}$$

where the subscript p denotes the component of a vector parallel to the boundary. Equation (17.5.1), therefore, states that the parallel component of the electric field is continuous across the boundary.

Consider next the condition on the displacement vector **D** which follows from the application of Gauss's law in the form of Eq. (17.4.12). In Figure 17.17b, a Gaussian surface is drawn. It consists of rectangular faces of equal area A parallel to the boundary, as well as four other faces each of infinitesimal area. The flux of the vector **D** comes only from the faces parallel to the boundary. It may be written

$$\oint \mathbf{D} \cdot \mathbf{n} \, da = (D_{1n} - D_{2n})A \tag{17.5.2}$$

where D_{1n} and D_{2n} are, respectively, the normal com-

ponents of the displacement vector in the two media. According to Eq. (17.4.12), this flux is exactly equal to $q_f = \sigma_f A$, the free charge contained within the Gaussian surface. Thus, we obtain

$$\boxed{D_{1n} - D_{2n} = \sigma_f} \tag{17.5.3}$$

which implies a *discontinuity* of the normal component of **D** when there is free charge on the boundary. In the absence of free charge, the normal component of **D** is *continuous* across the interface, even though bound polarization charges may be present there.

For a linear, isotropic dielectric, Eqs. (17.5.1) and (17.5.3) are sufficient to determine the boundary conditions placed on the normal component of **E**, the parallel (tangential) component of **D**, and also the components of **P**. From (17.4.13) and the relation

$$\mathbf{D} = \varepsilon_0 \mathbf{E} + \mathbf{P} \tag{17.5.4}$$

we find

$$\mathbf{P} = \chi \varepsilon_0 \mathbf{E} \tag{17.5.5}$$

$$\mathbf{D} = (1 + \chi)\varepsilon_0 \mathbf{E} = K\varepsilon_0 \mathbf{E} = \varepsilon \mathbf{E} \tag{17.5.6}$$

The continuity of the parallel component of **E** implies, therefore, that

$$\frac{P_{1p}}{\chi_1} = \frac{P_{2p}}{\chi_2} \tag{17.5.7}$$

and

$$\frac{D_{1p}}{K_1} = \frac{D_{2p}}{K_2} \tag{17.5.8}$$

On the other hand, Eq. (17.5.3) along with (17.5.4), (17.5.5), and (17.5.6) tells us that

$$K_1 E_{1n} - K_2 E_{2n} = \frac{\sigma_f}{\varepsilon_0} \tag{17.5.9}$$

and

$$\frac{K_1 P_{1n}}{\chi_1} - \frac{K_2 P_{2n}}{\chi_2} = \sigma_f$$

$$\left(1 + \frac{1}{\chi_1}\right) P_{1n} - \left(1 + \frac{1}{\chi_2}\right) P_{2n} = \sigma_f \tag{17.5.10}$$

We see, therefore, that all of the boundary conditions can be expressed in terms of the free charge density. The application of Gauss's law for the electric field, as expressed by (16.3.12), yields the result

$$E_{1n} - E_{2n} = \frac{1}{\varepsilon_0}(\sigma_f + \sigma_b)$$

or

$$E_{1n} - E_{2n} = \frac{1}{\varepsilon_0}(D_{1n} - D_{2n} + \sigma_b) \tag{17.5.11}$$

When this is combined with (17.5.4), we find that the bound charge density at an interface is given by

$$P_{1n} - P_{2n} = -\sigma_b \qquad (17.5.12)$$

If region 1 turns out to be vacuum while region 2 is a dielectric, this reduces to the previously obtained special case $\mathbf{P} \cdot \mathbf{n} = \sigma_b$, as discussed earlier.

These boundary conditions are important in determining how capacitors function in the presence of dielectrics. They are also very valuable in the theory of geometric optics and can be used to understand the behavior of light at the interface of two materials. Let us now consider several examples that illustrate how dielectrics modify capacitance and how the various vectors discussed above can be calculated in a variety of circumstances.

EXAMPLE 17.5.1

A parallel-plate capacitor in vacuum has a capacitance $C_0 = 2 \times 10^{-6}$ farad and is charged to 50 volts. The plates are separated by 5 mm. A very thin, uncharged plate of dielectric material 2 mm wide is centrally placed between the plates of the capacitor, as shown in Fig. 17.18. If this dielectric has a susceptibility of 15, find (a) the electric field inside and outside the dielectric, that is, in regions 1, 2, and 3; (b) the displacement vector in each of these regions; (c) the polarization vector; (d) the bound surface charge at the dielectric boundaries; and (e) the capacitance of the device.

Let us first compute the charge density on the capacitor plates. The total charge on the positive plate is $Q = C_0 \, \Delta V = (\varepsilon_0 A/d) \, \Delta V$; therefore, the charge density is given by

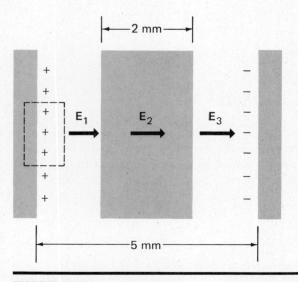

FIGURE 17.18

$$\sigma_f = \frac{\varepsilon_0 \Delta V}{d} = \frac{(8.85 \times 10^{-12})(50)}{5 \times 10^{-3}}$$

$$= 8.85 \times 10^{-8} \; \text{C/m}^2 \qquad (17.5.13)$$

By choosing a Gaussian surface with one side within the conductor, we may easily show that the electric field in regions 1 and 3 points to the right and has the value

$$E_1 = E_3 = \frac{\sigma_f}{\varepsilon_0} = 10^4 \; \text{V/m} \qquad (17.5.14)$$

At the interface between regions 1 and 2, there is no free charge density and, therefore, with $K_1 = 1$ and $K_2 = 1 + 15 = 16$, we find from equation (17.5.9) that

$$E_1 - 16E_2 = 0$$
$$E_2 = 0.063E_1 = 6.25 \times 10^2 \; \text{V/m} \qquad (17.5.15)$$

Note that for this geometry all field vectors are already normal to the conducting and dielectric surfaces.

The displacement vector in each of the regions is obtained from (17.5.6). Clearly, $\mathbf{D}$ must be continuous since there is no *free* charge contained at the boundaries. Thus,

$$D_1 = D_2 = D_3 = 8.85 \times 10^{-8} \; \text{C/m}^2 \qquad (17.5.16)$$

The polarization vector is obtained from (17.5.5). In vacuum, where the susceptibility is zero, we have

$$P_1 = P_3 = 0$$

while in region 2, we have

$$P_2 = 15\varepsilon_0 E_2 = 8.30 \times 10^{-8} \; \text{C/m}^2 \qquad (17.5.17)$$

At the left boundary, we have a bound charge density of

$$\sigma_b = -8.30 \times 10^{-8} \; \text{C/m}^2$$

and a total charge of

$$Q_b = \sigma_b A = \sigma_b \frac{C_0 d}{\varepsilon_0}$$

$$= -(8.30 \times 10^{-8}) \frac{(2 \times 10^{-6})(5 \times 10^{-3})}{8.85 \times 10^{-12}}$$

$$= -9.38 \times 10^{-5} \; \text{C} \qquad (17.5.18)$$

At the right boundary of the dielectric, the bound charge is $+9.38 \times 10^{-5}$ coulomb.

The potential difference between the capacitor plates is found by integrating $\mathbf{E} \cdot d\mathbf{l}$ from the left-hand plate to the one on the right through all three regions within the intervening space.

$$\Delta V = \int \mathbf{E} \cdot d\mathbf{l} = (10^4)(1.5 \times 10^{-3})$$
$$+ (0.0625 \times 10^4)(2 \times 10^{-3})$$
$$+ (10^4)(1.5 \times 10^{-3})$$
$$= 31.25 \; \text{V}$$

The new value of capacitance is, therefore,

$$C = \frac{Q}{\Delta V} = \frac{50}{31.25} C_0 = 3.2 \times 10^{-6} \text{ F} \qquad (17.5.19)$$

Note that, although we originally introduced dielectrics by completely *filling* the space with a dielectric material, we are now in a position to determine how the capacitance varies when any amount of dielectric is used.

EXAMPLE 17.5.2

A point charge of 5×10^{-6} coulomb is surrounded by a neutral, spherical Pyrex glass dielectric of inner radius 10 cm and outer radius 15 cm. The region between the point charge and the glass contains gaseous hydrogen, and the region outside the glass is air. (a) Find how the electric field varies with distance from the charge. (b) Obtain the bound charge density at the inner and outer glass surfaces. (c) Obtain an expression for the average polarization of hydrogen atoms at 5 cm from the point charge. What is the average separation of positive and negative charge in each of these atoms? See Table 17.1 for dielectric constants and dielectric strengths (maximum electric field to cause electrical breakdown). The density of atomic hydrogen can be given as 2.69×10^{19} atoms/cm^3 at 1 atmosphere and 0°C.

Let us use Gauss's law for the displacement vector $\mathbf{D}$ by choosing a spherical Gaussian surface, as shown in Fig. 17.19. The only free charge is situated at the origin. Therefore,

$$\oint \mathbf{D} \cdot \mathbf{n} \, da = D \times 4\pi r^2 = q$$

or

$$D = \frac{q}{4\pi r^2} \qquad (17.5.20)$$

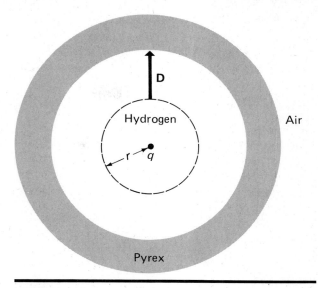

FIGURE 17.19

This quantity is continuous throughout all three regions since there is no free charge on any of the boundaries. The electric field can be obtained in each region by using the relation $\mathbf{D} = K\varepsilon_0\mathbf{E}$. Thus, we find $\mathbf{E}$ in the three regions to be

$$E_1 = \frac{q}{4\pi\varepsilon_0 r^2(1.00026)}$$

$$E_2 = \frac{q}{4\pi\varepsilon_0 r^2(4.5)} \qquad (17.5.21)$$

$$E_3 = \frac{q}{4\pi\varepsilon_0 r^2(1.00054)}$$

We now calculate the polarization vector on the inner and outer portions of the glass surface, using the expressions $\mathbf{P} = \chi\varepsilon_0\mathbf{E} = (K-1)\varepsilon_0\mathbf{E}$. Thus

$$P_1 \text{ (at 0.1 m)} = (1.00026 - 1)\varepsilon_0 E_1(0.1 \text{ m})$$

$$= (0.00026)\frac{5 \times 10^{-6}}{(4\pi)(1.00026)(0.1)^2}$$

$$= 1.03 \times 10^{-8} \text{ C/m}^2$$

$$P_2 \text{ (at 0.1 m)} = (4.5 - 1)\varepsilon_0 E_2(0.1 \text{ m})$$

$$= (3.5)\frac{5 \times 10^{-6}}{(4\pi)(4.5)(0.1)^2}$$

$$= 3.09 \times 10^{-5} \text{ C/m}^2$$

$$\qquad (17.5.22)$$

$$P_2 \text{ (at 0.15 m)} = (4.5 - 1)\varepsilon_0 E_2(0.15 \text{ m})$$

$$= (3.5)\frac{5 \times 10^{-6}}{(4\pi)(4.5)(0.15)^2}$$

$$= 1.37 \times 10^{-5} \text{ C/m}^2$$

TABLE 17.1. Approximate Dielectric Constants of Some Materials

Material	Dielectric constant K	T, °C
Vacuum	1	
Air	1.00059 (1 atm)	20
Hydrogen	1.00026	100
Water	80.4	20
Mica	3–6	25
Plexiglas	3.12	27
Quartz (fused)	3.75–4.1	20
Glass (Pyrex)	4.5	20
Lucite	2.84	23
Hevea rubber	2.94	27
Paraffin	2–2.5	20

P_3 (at 0.15 m) $= (1.00054)\varepsilon_0 E_3(0.15\ \text{m})$

$$= (0.00054)\frac{5 \times 10^{-6}}{(4\pi)(1.00054)(0.15)^2}$$

$$= 0.95 \times 10^{-8}\ \text{C/m}^2$$

On the interface at $r = 0.1$ m, there is a bound charge density

$$P_1\ (0.1\ \text{m}) - P_2\ (0.1\ \text{m}) = -3.089 \times 10^{-5}\ \text{C/m}^2$$
$$(17.5.23)$$

while at the outer surface, the bound charge density is

$$P_2\ (0.15\ \text{m}) - P_3\ (0.15\ \text{m}) = 1.369 \times 10^{-5}\ \text{C/m}^2$$
$$(17.5.24)$$

At $r = 0.05$ m, the polarization vector has a magnitude

$$P_1\ (0.05\ \text{m}) = 4.12 \times 10^{-8}\ \text{C/m}^2 \qquad (17.5.25)$$

The polarization vector is the average dipole moment per unit volume and is, therefore, equal to the number of atoms per unit volume multiplied by the average dipole moment per atom. Therefore,

$$4.12 \times 10^{-8}\ \text{C/m}^2 = (2.69 \times 10^{25})p$$

where p is the average atomic dipole moment. From this, we find

$$p = 1.53 \times 10^{-33}\ \text{C-m} \qquad (17.5.26)$$

Since

$$p = q\Delta = (1.6 \times 10^{-19})\Delta = 1.53 \times 10^{-33}\ \text{C-m}$$

we find an average charge separation

$$\Delta = 0.96 \times 10^{-14}\ \text{m} = 0.96 \times 10^{-12}\ \text{cm} \qquad (17.5.27)$$

The average size of the hydrogen atom is about 0.5×10^{-8} cm, but under ordinary conditions, the center of the average electronic charge distribution coincides exactly with the proton charge. We now see that, due to the application of an external field, a certain separation of positive and negative charges can be achieved. Although this separation is extremely small, even on the atomic scale, a large number of atoms can collectively bring about a macroscopic polarization, and thereby alter the electric field within a dielectric substance.

17.6 Microscopic Theory of Polarization

It has been assumed that a linear, isotropic dielectric develops a polarization **P** which is *proportional* to the applied field **E** if the applied field is not too large. There are, of course, some substances in which these conditions are violated, but a large class of materials do, in fact, satisfy the simple criteria we have assumed. Let us consider how a macroscopic polarization is produced for a few simple cases. To be completely accurate, we would have to study the phenomena of polarization by using quantum mechanics together with statistical physics; a treatment of this type is beyond the scope of this text. We shall instead try to use some elementary physical concepts to describe the polarization of atoms and molecules.

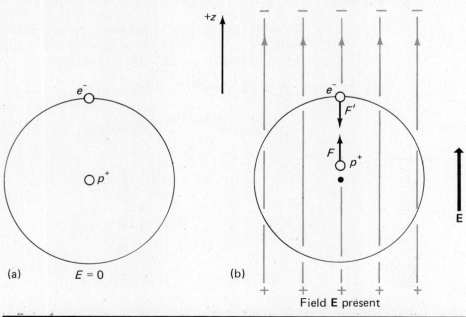

FIGURE 17.20. Polarization of a hydrogen atom by an externally applied electric field.

Suppose we have a gas of hydrogen atoms. In a crude way, we can think of the electron as being in a circular orbit around the proton at the center, as shown in Fig. 17.20a. This situation is assumed to exist in the absence of any applied fields. Under these conditions, the average electronic charge distribution coincides precisely with the proton charge, since the electronic charge is symmetric with respect to the proton. Thus, the hydrogen atom has no *permanent* dipole moment and is therefore said to be *nonpolar*. An electric field **E** is now applied, say, in the z-direction. In the absence of this field, the circulating electron has a z-coordinate which executes simple harmonic motion. It therefore satisfies an equation of motion of the form

$$m \frac{d^2 z}{dt^2} + m\omega_0^2 z = 0 \qquad (17.6.1)$$

where ω_0 is the angular frequency of the electron in its orbit. When a constant electric field **E** is applied in the z-direction, an additional force $-e\mathbf{E}$ acts on the electron, and a force $+e\mathbf{E}$ acts on the proton. In comparison to the electron, the proton is hardly affected, however, since it is much more massive. In the presence of this field, Eq. (17.6.2) is altered to give

$$m \frac{d^2 z}{dt^2} + m\omega_0^2 z = -eE \qquad (17.6.2)$$

Letting $z_0 = -eE/m\omega_0^2$, and noting that since z_0 is constant $d(z - z_0)/dt = dz/dt$, this equation can be rewritten in the form

$$\frac{m d^2 (z - z_0)}{dt^2} + m\omega_0^2 (z - z_0) = 0 \qquad (17.6.3)$$

Note that for a constant **E** field, z_0 is independent of t. Equation (17.6.1) describes simple harmonic motion in the z-direction with an average z-value of zero, while Eq. (17.6.3) gives a similar motion where the average z-value has been *shifted* to z_0. Thus, we see that the atom develops a microscopic dipole moment oriented in the direction of the **E** field, as shown in Fig. 17.20b. This dipole is created because, in the presence of the applied field, the average position of the electron charge *no longer coincides* with the average position of the proton charge.

This microscopic dipole has a dipole moment

$$p = e|z_0| = \frac{e^2 E}{m\omega_0^2} \qquad (17.6.4)$$

The angular frequency ω_0 may be obtained from the ionization energy of the lowest quantum energy level state of hydrogen as $\omega_0 = 2.06 \times 10^{16}$ rad/sec. Since each atom has a dipole moment in the z-direction, the macroscopic polarization is

$$P = \frac{Ne^2}{m\omega_0^2 \varepsilon_0} \varepsilon_0 E = \chi \varepsilon_0 E \qquad (17.6.5)$$

where $N = 2.69 \times 10^{25}$ atoms/m³ at standard temperature and pressure. Thus, the susceptibility of hydrogen gas is found to be

$$\chi = \frac{Ne^2}{m\omega_0^2 \varepsilon_0}$$

$$= \frac{(2.69 \times 10^{25})(1.6 \times 10^{-19})^2}{(9.11 \times 10^{-31})(2.07 \times 10^{16})^2(8.85 \times 10^{-12})}$$

$$= 0.00020$$

The experimentally observed susceptibility is 0.00026; therefore, this classical calculation agrees quite well with the actual value. To do better, we would have to use quantum mechanics instead of classical mechanics. The calculation of the susceptibility for other nonpolar gases is somewhat more complicated but can still be carried out to reasonable accuracy.

Let us now consider the calculation for a system in which each atom or molecule contains a *permanent* dipole. Such materials are called *polar substances* in view of the presence of these permanent microscopic dipoles. Here, the calculation is quite different. For nonpolar materials, the electric field actually produces microscopic dipoles, each very small in size and all in alignment with each other. A polar substance usually has a rather large permanent dipole moment associated with each molecule or atom. In the absence of an applied field, the average dipole moment per unit volume vanishes as a result of cancellations of the various dipoles, which point in random directions. When an electric field is applied, the dipoles tend to align in the direction of the field, this being the configuration which leads to the smallest energy. Of course, there are competing collision processes which tend to randomize the moments. Therefore, only a partial alignment is possible, as shown in Fig. 17.21, which illustrates the alignment process.

The degree of alignment that occurs in any given situation will depend on the strength of the applied field and also, inversely, on the effectiveness with which thermal processes operate to randomize the moments. We should, therefore, expect the actual dipole moment per unit volume to be directly proportional to the strength of the aligning field E and inversely proportional to the temperature, since the effectiveness of thermal agitation in randomizing the distribution will depend directly upon the temperature. Though we shall not go through the details of calculating the polarization in this case,[6] it can, in fact, be shown that for moderate values of field strength and for temperatures in the room-tempera-

---

[6] We shall consider details of this calculation in Chapter 21 when we derive results that are completely analogous to these for permanent *magnetic* dipoles in an externally applied magnetic aligning field.

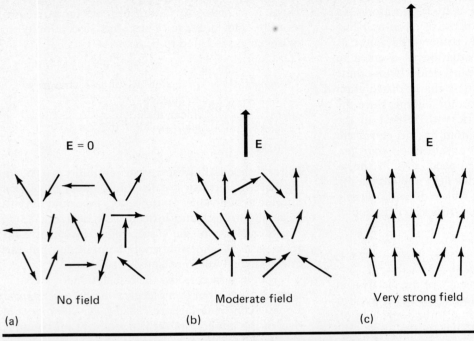

FIGURE 17.21. Polarization of randomly oriented dipolar molecules by an externally applied field.

ture range, the polarization P will be given by

$$P = \frac{Np^2E}{3kT} = \chi\varepsilon_0 E \qquad (17.6.6)$$

from which

$$\chi = \frac{Np^2}{3\varepsilon_0 kT} \qquad (17.6.7)$$

In these expressions, N represents the number of molecules per unit volume and p, the strength of the individual permanent molecular dipoles. If these results are applied, for example, to water vapor at normal temperatures, they agree very well with experiment. It is clear, of course, that Eq. (17.6.7) cannot be valid at very low temperatures for it predicts an infinite susceptibility as T approaches zero.

The treatment of liquid dielectrics such as water and solid dielectrics such as quartz is beyond the scope of this text. The two cases discussed above, however, serve to illustrate the origin of the small susceptibility which arises for simple polar and nonpolar materials. We have shown that in each case the polarization varies linearly with the applied **E** field. However, in the case of nonpolar material, the **E** field actually causes each molecule or atom to develop a microscopic dipole, all of which are aligned. For polar materials, the individual molecules all have permanent dipoles; the application of an electric field then produces a partial alignment, which leads to a macroscopic polarization.

SUMMARY

A capacitor, or condenser, is an electrical device composed of two conductors carrying equal and opposite charges. The capacitance of such a device is defined by the stored charge divided by the potential difference between the conductors:

$$C = \frac{q}{\Delta V}$$

The units of capacitance are coulombs/volt; these units are more commonly referred to as *farads*. The capacitance of any given condenser depends upon the geometry and spacing of the electrodes. An ideal parallel-plate capacitor has a capacitance

$$C = \varepsilon_0 A/d$$

where A is the plate area and d is the spacing between the plates. The electric field in such a device is related to the potential difference ΔV by

$$E = \frac{\Delta V}{d} = \frac{\sigma}{\varepsilon_0}$$

For capacitors in series, the equivalent capacitance is given by

$$\frac{1}{C} = \frac{1}{C_1} + \frac{1}{C_2} + \cdots + \frac{1}{C_n}$$

while for capacitors connected in parallel,

$$C = C_1 + C_2 + \cdots + C_n$$

556

For the series connection, the charge on every capacitor has the same value, while in the parallel connection, the total charge is the sum of the charges on each individual condenser.

A capacitor stores electrostatic energy; the amount is given by

$$U = \frac{q^2}{2C} = \tfrac{1}{2}q\,\Delta V = \tfrac{1}{2}C\,(\Delta V)^2$$

The energy density, or electrostatic energy per unit volume, within the capacitor is given by

$$u_e = \tfrac{1}{2}\varepsilon_0 E^2$$

When an insulating substance, or *dielectric* material, fills the region between the plates of a capacitor, the capacitance increases by a factor of the *dielectric constant K* associated with this particular substance. When a material dielectric is present, the field either induces atomic dipole moments within each atom or aligns permanent atomic dipoles. In either case the result is to induce a dipole moment directly proportional to the field, ideally, at least, in every volume element of the dielectric. The *polarization vector* $\mathbf{P}$ at any point is the ratio of the dipole moment $\Delta\mathbf{p}$ acquired by a small volume element about the point to its volume $\Delta\mathscr{V}$, from which

$$\mathbf{P} = \frac{\Delta\mathbf{p}}{\Delta\mathscr{V}}$$

In uniformly polarized dielectrics, the polarization can be viewed as arising from a distribution of bound surface charges on the surfaces of the dielectric. The surface density of bound polarization charge is given by

$$\sigma_b = \mathbf{P}\cdot\mathbf{n} = P\cos\theta$$

where $\mathbf{n}$ is the unit vector normal to the surface and θ is the angle between the vectors $\mathbf{n}$ and $\mathbf{P}$. It is important to note that surface polarization charge is fixed on the sample and is *not free to move about* like the charges on or within a conductor.

The polarization of an ideal linear dielectric is related to the field within it by

$$\mathbf{P} = \chi\varepsilon_0\mathbf{E}$$

where the proportionality constant χ is referred to as the electric susceptibility. The free charge on the metal plates of a condenser in which a dielectric is present is

$$\sigma_f = \varepsilon_0 E + P$$

where $\mathbf{E}$ is the field within the dielectric at the surface of the plate. The vector quantity $\varepsilon_0\mathbf{E} + \mathbf{P}$ is referred to as the *electric displacement* $\mathbf{D}$. Therefore,

$$\mathbf{D} = \varepsilon_0\mathbf{E} + \mathbf{P}$$

Gauss's law in systems wherein dielectric substances are present can be stated as

$$\oint_s (\varepsilon_0\mathbf{E} + \mathbf{P})\cdot\mathbf{n}\,da = \oint_s \mathbf{D}\cdot\mathbf{n}\,da = q_f$$

where q_f is the total amount of *free* charge within the closed Gaussian surface.

In an ideal linear dielectric, $\mathbf{P} = \chi\varepsilon_0\mathbf{E} = \varepsilon\mathbf{E}$; in this case, $\mathbf{D}$ can be written

$$\mathbf{D} = (1 + \chi)\varepsilon_0\mathbf{E} = \varepsilon\mathbf{E}$$

where

$$\varepsilon = \varepsilon_0(1 + \chi)$$

The quantity ε is referred to as the *permittivity* of the dielectric. The dielectric constant K is the ratio of the permittivity of the dielectric to ε_0, the permittivity of free space. Thus,

$$K = \frac{\varepsilon}{\varepsilon_0} = 1 + \chi$$

At any interface between two dielectrics having dielectric constants K_1 and K_2, the following relations exist between the field, displacement, and polarization vectors on either side:

$$E_{1p} = E_{2p}$$
$$D_{1n} - D_{2n} = \sigma_f$$
$$\frac{P_{1p}}{\chi_1} = \frac{P_{2p}}{\chi_2}$$
$$P_{1n} - P_{2n} = \sigma_b$$

In these expressions, the subscripts n and p refer respectively to components normal and parallel to the interface, while σ_f and σ_b refer to free charge density and polarization charge density on the interface. In the simplest cases involving electrostatic equilibrium of ordinary dielectric substances, σ_f will ordinarily be zero.

QUESTIONS

1. Must the plates of a capacitor always carry equal but opposite amounts of charge? If not, cite an example.
2. The potential difference across a parallel-plate capacitor is halved, whereupon the amount of stored energy decreases. By what factor does it change?
3. Does the capacitance of a capacitor depend on the potential difference? Describe what might happen when the potential difference across a capacitor becomes very, very large.
4. When two capacitors C_1 and C_2 are connected in series, the equivalent capacitance is always less than either C_1 or C_2. On the other hand, if they are connected in parallel, the equivalent capacitance is greater than C_1 or C_2. Can you explain these facts in a qualitative way?

5. In the space between the plates of a parallel-plate capacitor, energy is stored. Is this still possible if there is a perfect vacuum between the plates?

6. The maximum voltage available for storing 1.2×10^{-5} coulomb of charge on a capacitor is 1500 volts. What is the minimum capacitance needed?

7. Discuss and distinguish clearly between the electrostatic potential and potential energy of a capacitor.

8. Discuss two useful functions performed by a dielectric when it is used within a capacitor.

9. What is the difference between the dielectric strength of a material and its dielectric constant?

10. A capacitor is charged by placing it across a battery which is then disconnected. It is then immersed in a liquid dielectric material of dielectric constant K. Is the energy content of the capacitor altered in this process? If so, by what factor?

11. The knob of a tuning capacitor within a radio is turned, reducing the effective plate area from 50 cm² to 20 cm². Find the ratio of final to initial capacitance.

12. Why would a liquid dielectric such as water be unsatisfactory for use as the dielectric medium of a capacitor?

13. If a polar material is used as the dielectric for a capacitor, would you expect the capacitance to vary with temperature?

14. Explain the distinction between free and bound charge.

15. What are the dimensions of energy density? Demonstrate by working out the dimensions of $\varepsilon_0 E^2/2$.

16. The capacitance of a single sphere is $4\pi\varepsilon_0 R$. How large must R be to obtain a capacitance of 1 farad?

PROBLEMS

1. A charge of 30×10^{-6} coulomb is placed on a 200-μF parallel-plate capacitor. If the plate separation is 5 mm, find the electric field between the plates.

2. A capacitor is observed to have a charge of 2.5×10^{-6} Coulomb when a potential difference of 125 volts is maintained across its terminals. What is its capacitance?

3. The charge on a capacitor increases by 6.0×10^{-6} Coulomb when the potential difference across it increases from 100 to 120 volts. What is its capacitance?

4. An air-filled parallel-plate capacitor of capacitance 0.0025 μf has an area of 0.80 m². (a) What is the spacing between the plates? (b) How large a potential difference can be applied to the capacitor, assuming that the air between the plates can sustain a maximum field of 3.0×10^6 volt/meter before electrical breakdown or spark discharge occurs.

5. A 0.1-μf parallel-plate capacitor is to be designed so as to have a plate area not exceeding 0.10 m². (a) What is the maximum potential difference that such a device can sustain without undergoing "breakdown", assuming that the space between the plates is filled with air. (b) What is the maximum charge density on the plates under these circumstances.

6. A cylindrical capacitor consists of two very long concentric conducting cylinders. The outer radius of the inner cylinder is 9.50 cm and the inner radius of the outer one is 10.0 cm. (a) What is the capacitance of this structure,

per unit length? (b) How large a potential difference could this structure sustain without breakdown, assuming that the electrical breakdown of the air in the region between the plates occurs whenever an electric field of 3.0×10^6 volt/meter occurs at any point?

7. What would the answers to the preceding problem be for a concentric spherical capacitor having the same inner and outer radii?

8. A 0.0014 μF air-filled parallel-plate capacitor is made of circular plates of radius r and separation d. What is the smallest value r can have without producing electrical breakdown of the air when 500 volts is applied?

9. Find the capacitance of a metallic sphere with a radius equal to that of the earth. If it is charged to 1000 volts, find the charge on the sphere, the charge density, and the electric field at the surface.

10. Four capacitors of capacitance 1, 2, 3, and 4 μF are connected in series. Find their equivalent capacitance. If they are connected in parallel, what is their capacitance?

11. A long, cylindrical capacitor consists of an inner cylinder of radius 1 meter and an outer one of radius 2 meters. If a potential difference of 200 volts is applied, find the charge contained on a 5-meter length of the outer conductor.

12. A cylindrical capacitor consists of concentric cylindrical shells of radii r_1, r_2, r_3, and r_4, as shown in the figure. Show that the capacitance per unit length is given by

$$\frac{C}{l} = \frac{2\pi\varepsilon_0}{\ln\dfrac{r_2 r_4}{r_1 r_3}}$$

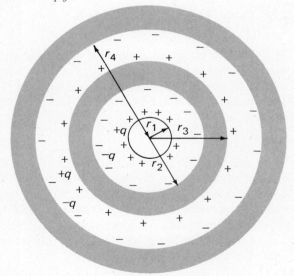

13. Two uncharged capacitors of respective capacitance 1 and 3 μF are connected in parallel. A potential difference of 12 volts is supplied by a battery. Find the amount of charge on each plate. They are then disconnected from the battery without losing their charge and connected to each other in series so that the positively charged plate of the 1-μF capacitor is connected to the negatively charged plate of the 3-μF capacitor. Find the potential difference across the equivalent capacitor. What is the potential difference across each capacitor?

14. Several capacitors are connected in the circuit shown in the diagram. The capacitances are 1, 1, 2, and 4 μF,

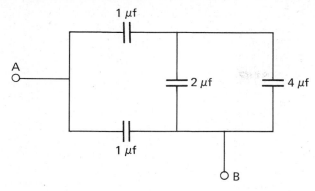

respectively. Find the equivalent capacitance of the combination and also the amount of charge on each capacitor when $\Delta V_{ab} = 50$ volts.

15. Two capacitors C_1 and C_2 can be connected in series and in parallel. When they are in series, the equivalent capacitance is $C_1/3$. When they are in parallel, it is $3 \, \mu F$. Find C_1 and C_2.

16. Five capacitors, each of capacitance $1 \, \mu F$, are connected as shown in the figure. Find the effective capacitance.

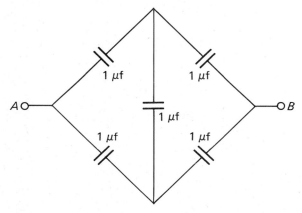

17. A variable air capacitor consists of four movable plates and three stationary ones, as shown in the diagram. Show that the *maximum* capacitance is $C_{max} = 3\varepsilon_0 A/d$, where A is the area of the plates and d is their separation.

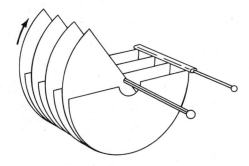

18. Two capacitors of capacitance 2 and $9 \, \mu F$, respectively, are connected in series across a 1000-volt potential difference. Find the charges and potential differences across each. The capacitors are then disconnected from the source and connected to each other. For both polarities of the connection, find the charge and potential difference across each capacitor.

19. When a conducting plate of thickness b is inserted between the plates of a parallel-plate capacitor, show that the capacitance becomes $C = C_0/(1 - b/d)$, where C_0 is the original capacitance and d is the original plate separation. Suppose now a spherical shell of thickness b is placed between a spherical capacitor whose plates are separated by d. Find the expression corresponding to the one above that applies to this geometry. Assume the conducting plates have radii r_1 and r_4 and the radii of the inserted shell are r_2 and r_3.

20. Two capacitors, each of capacitance C, are connected in series to a battery of voltage ΔV. Find the charge and energy stored in the equivalent capacitor. The capacitors are disconnected from the battery and from each other; they are reconnected positive plate to positive plate and negative plate to negative plate. Find the voltage across the combination and the stored energy. Account for the difference in energy in the two cases.

21. Two hundred identical capacitors each of capacitance $10 \, \mu F$ are connected in parallel. They are charged to 30,000 volts. At the rate of 3¢ per kilowatt-hour, how much money is the stored energy worth? If the same capacitors are instead charged while in series, how much is it worth now?

22. A metal sphere of radius 50 cm carries a charge of 10^{-6} coulomb on its surface. Obtain the electric energy density u_e for all values of r. Find the radius R such that half the stored energy is within a sphere of radius R.

23. A uniform charge density is distributed within a small sphere of radius 10^{-13} cm. If the amount of charge is that of a proton, 1.6×10^{-19} coulomb, find the energy density in all of space and also the total energy in all space. How much energy would be contained in a sphere of radius 5 cm if the energy density is uniform and has the value given by the previously obtained energy density at $r = 0.5 \times 10^{-13}$ cm?

24. Starting with $u_e = \varepsilon_0 E^2/2$ and expression (17.1.9), verify by integration over the field region between the electrodes, that the total energy stored in a cylindrical capacitor can be written as $q(\Delta V)/2$ with ΔV given by the expression directly above (17.1.10).

25. Two thin, long, coaxial cylindrical conductors having radii r_1 and $r_1 + x$, respectively, carry charges q_1 and $-2q_1$, respectively, on a length l. **(a)** Find the electric field in each region of space. **(b)** Obtain the energy density everywhere. **(c)** How much energy per unit length is contained in the space between the conductors? **(d)** Find the force of attraction between the conductors for length l.

26. A wire of length L is bent to form a semicircle. If it carries a total charge Q, find the energy density at the center of the circle assuming the charge to be distributed uniformly along its length.

27. A spherical conductor of radius r and charge Q consists of two separable hemispheres. Show that the force required to hold the halves together is $Q^2/32\pi\varepsilon_0 r^2$. *Hint:* First calculate the net outward force from $-dU/dr$, where U is the potential energy of the sphere.

28. A positive point charge Q sets up a spherically symmetric electric field. A dipole $\mathbf{p}$ is oriented in the direction of a field line, with the negative charge closer to Q. Show the

system has a potential energy of approximately

$$U = -\frac{pQ}{4\pi\varepsilon_0 r^2} - \frac{q^2}{4\pi\varepsilon_0 d}$$

where r is the distance to the center of the dipole and d is the separation between its charges $\pm q$.

29. A parallel-plate capacitor of capacitance 6 μF is charged by a 25-volt battery. A linear dielectric of dielectric constant 3 is inserted between the plates, completely filling the whole region between them. (a) Find the new potential difference between the plates. (b) What is the bound charge density on the dielectric surfaces? (c) Find the energy of the capacitor before and after the dielectric is inserted. Account for the difference.

30. A parallel-plate condenser of capacitance 0.003 μF employs paper ($K = 3.5$) as a dielectric between the plates. The capacitor is connected to a 12-volt battery. (a) How much charge does it acquire? (b) If the paper is removed and a sheet of mica ($K = 5.4$) put in its place, the battery being connected during the process, what is the charge on the condenser now? (c) What is its capacitance under the circumstances of part (b)?

31. A piece of quartz having a dielectric constant of 3.8 is placed in an electric field of 20,000 volts/meter, as shown in the figure. The electric field vector makes an angle of 45° with the top and bottom faces and is parallel to the front and back faces. Obtain the charge density on each of the faces.

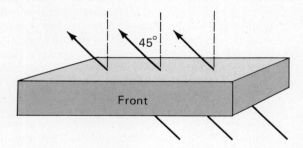

32. A parallel-plate capacitor of area 0.12 m^2 and plate separation 1.0 mm is immersed in a uniform dielectric of dielectric constant $K = 3.6$. A potential difference of 360 volts is maintained across the capacitor. Find (a) the capacitance of the device, (b) the free charge on the plates, (c) the polarization charge on the dielectric, (d) the magnitude of the electric field in the dielectric, (e) the magnitude of the displacement in the dielectric, (f) the magnitude of the polarization in the dielectric.

33. A certain "fictitious" dielectric contains permanent atomic electric dipoles of magnitude 3×10^{-22} coulomb-meter. The density of atoms is 10^{26} atoms/cubic meter. If an electric field of 10^4 volts/meter produces an effective polarization corresponding to the alignment of 25 percent of the atomic dipoles in the direction of the field, find the susceptibility of the dielectric.

*34. A plane-parallel slab of dielectric of thickness a and dielectric constant K is inserted into a parallel plate condenser of area A and electrode spacing d ($d > a$), so as to partially fill the space between the condenser plates with dielectric material. The dielectric slab is in contact with

the left-hand plate of the capacitor, leaving a gap of thickness $d - a$ between the other surface of the dielectric and the right-hand capacitor plate. The left-hand plate is at zero potential and the right-hand plate at a higher potential ΔV. Find (a) the electric field within the dielectric, (b) the electric field in the rest of the region between the plates, (c, d) the displacement in the two regions referred to, respectively, in (a) and (b), (e) the polarization within the dielectric, (f) the bound charge on the surfaces of the dielectric slab, (g) the free charge on the condenser plates, (h) the capacitance of the condenser, (i) the potential at the air–dielectric interface within the capacitor.

35. A parallel-plate capacitor is filled with dielectric material, as shown in the diagram. Show that the capacitance is given by

$$\frac{1}{C} = \frac{1}{C_0}\left(\frac{1}{K_1} + \frac{1}{K_2} + \frac{1}{K_3}\right)$$

where C_0 is the capacitance in the absence of any dielectric and K_1, K_2, and K_3 are dielectric constants of the three dielectric materials illustrated.

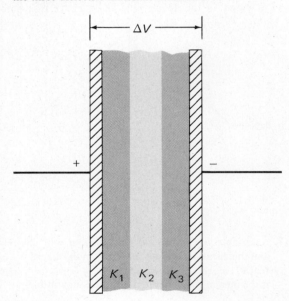

36. A point charge of 10^{-6} coulomb is embedded at the center of a dielectric sphere of radius 10 cm and susceptibility $\chi = 5$. Find the polarization and displacement vectors at the surface of the sphere. How much bound charge is present on the surface?

37. A cylindrical capacitor consists of a long conducting shell of inner radius 2 cm and outer radius 3 cm, and a second shell of inner radius 5 cm and outer radius 6 cm. The regions $r < 2$ cm and 3 cm $< r < 5$ cm contain a dielectric with $K = 3$. The region $r > 6$ cm is air. The capacitor is charged to 300 volts. Find (a) the capacitance per unit length, (b) the displacement vector in each region, (c) the polarization vector in each region, (d) the bound charge density at each interface, and (e) the stored energy per unit length.

38. A long glass prism of square cross section contains uniform electric field components of 2000 and 3000 volts/meter in the x- and y-directions, respectively, within its

interior. Find the polarization and the displacement vector at the inner boundaries of the glass. Determine the electric field at points just outside the glass. Assume that $K = 4.0$ for the glass.

39. An interface between glass and air contains no free charge but may contain some bound charge. The electric field at a point P just outside the glass is 20,000 volts/meter and makes a 30° angle with the normal to the surface, as shown in the diagram. Find the magnitude and direction of the electric field just inside the glass. Assume $K = 4.0$.

40. A parallel-plate capacitor of area A and plate separation d is charged to a voltage ΔV by a battery. While the capacitor is still connected to the battery, a slab of dielectric constant K and thickness $d/3$ is inserted parallel to the plates. Find the capacitance, the charge on the plates, and the energy stored after the dielectric slab is in place.

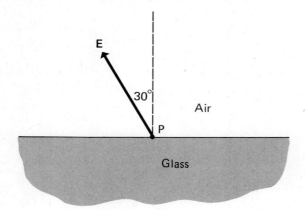

STEADY CURRENTS AND DIRECT CURRENT CIRCUITS

18.1 Introduction

In contemporary society, the word *electricity* is immediately associated with devices such as electric lights, electric motors, electric stoves, radios, and television sets. On a historical scale, all this is of comparatively recent origin. For perhaps two thousand years before 1800, our knowledge of electricity was strictly confined to *electrostatic* phenomena, which had been observed by the ancient Greeks. But the science of electrostatics did not develop very significantly until the Renaissance, and by 1800, the work of scientists such as Coulomb, Gauss, Gilbert, and Franklin had brought our understanding of electrostatics nearly to its present state.

After 1800, an explosive development in our knowledge of electricity took place, beginning with the discovery by Oersted of the magnetic fields accompanying electric currents and culminating in 1864 with Maxwell's brilliant theoretical work which predicted the existence of electromagnetic waves and the actual generation and detection of these "radio waves" by Hertz in 1888. During this period, the electric motor and generator were invented, electrical communication by telegraph and telephone was developed, and numerous other electrical and electromechanical devices were brought into everyday use. By 1888, practically all the scientific knowledge necessary for the development of our sophisticated communications technology, including radio, television, and microwave systems, was available. The actual realization of these devices was by then simply a matter of finding practical ways in which this knowledge could be applied to the construction of operating devices and systems.

The fact that this breakthrough in our understanding of electricity began shortly after 1800 can be traced to certain events that took place between 1786 and 1800, which do not seem to be accorded their proper significance in many contemporary physics texts. Up to 1786, there was no practical way of establishing a steady, continuous flow or *current* of electric charge. This restricted our experimental acquaintance with electricity to systems in which a situation of equilibrium of charges prevails, in other words, to *electrostatic* systems. Of course, it was possible to set up experiments in which a flow of charge might take place, for example, in the charging or discharge of a condenser or in the sudden production of a spark between two electrodes at different potentials. But in these instances, the flow of charge occurs so rapidly and is completed so quickly that it was impossible to observe any of its physical effects with the relatively unsophisticated detection apparatus available at the time.

In 1786, the Italian scientist Aloisio Galvani observed that continuous muscular contractions of frog's legs took place when electrical contact was established in a circuit which placed nerve endings in series with a conducting path including two different metals. Galvani's research was continued by Alessandro Volta (1745–1827) who attributed this "galvanic action" to the presence of the dissimilar metals and then constructed "voltaic piles" consisting of strips of dissimilar metals such as zinc and copper, separated by strips of cloth moistened with salt solution. These voltaic cells—which we now refer to as *batteries*—were capable of generating chemically differences in electric potential that could cause rather large steady-state flows of charge through conducting circuits.

The invention of the battery by Volta, therefore, paved the way for subsequent investigators such as Oersted, Ampere, Faraday, Henry, Maxwell, and Hertz, who quickly and systematically discovered and analyzed the magnetic, thermal, and chemical effects associated with *continuous currents* of electricity. That Oersted's discovery that a compass needle is deflected in the presence of a wire carrying a current of electricity occurred in 1820 was not an accident. This could not have happened much earlier, since before the voltaic cell had come into being nobody could produce substantial continuous current flow. Once that device was available, it was not long before all sorts of exciting and important discoveries about electricity were made.

In our study of electricity, we have now arrived at the state of understanding that prevailed in 1800. We shall now begin to acquaint ourselves with the adventurous discoveries of the nineteenth century on which our modern era of electrical technology and instant communication is founded.

18.2 Charge Flow in Conductors: Current and Current Density

When charge flow occurs within conductors, the conditions within the conducting substance are no longer those of electrostatic equilibrium. In particular, it is no longer true that the electric field is zero everywhere in the conductor. Indeed, the electric field *has to be different from zero* to establish and maintain the flow of charge.

In metallic substances, which represent the most important and familiar class of conductors, the valence electrons of the atoms are not bound strongly to individual atoms but are, instead, free to move within the conductor. This state of affairs arises on account of the interaction between the atoms that constitute the crystal lattice of the metal. Whenever an external force acts on these *free electrons* in a conducting substance, as it will, for example, when an electric field is imposed, the free charge distribution will move, setting up a flow of electric *current*. It is important to note that since the number of free valence electrons is balanced by an equal number of positive charges on the metal ions in the conductor, the conductor as a whole is, in general, electrically neutral and bears no *net* charge. The positive charges on the metal ions, however, are *fixed* within the crystal lattice of the metal and cannot move like the mobile free electrons. When an electric field exists within the conductor, then, it is only the free electrons that contribute to the current flow, while the positively charged metallic ions remain fixed and play no role save that of maintaining the overall electrical neutrality of the substance. In general, the flow of current in a conductor need not be constant with respect to time; but when it is, we say that a *steady current flow*, or a *direct current* (dc) flow, has been established. In this chapter, we shall confine ourselves, with one exception, to situations in which this is the case. We shall have occasion in subsequent chapters to investigate non-steady-state currents.

Consider first the case of a uniform conductor carrying a steady current of charge that is uniform and constant at all points within the conductor. The *current I* carried by the conductor is then defined as the total mobile charge passing a fixed plane normal to the conductor per unit time. According to this definition, if a quantity of mobile charge Δq crosses

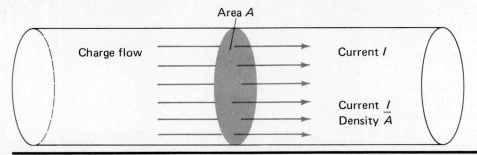

FIGURE 18.1. Concept of current as a flow of charge and of current density as charge flow per unit area.

the shaded area shown in Fig. 18.1 in time Δt, then

$$I = \frac{\Delta q}{\Delta t} \tag{18.2.1}$$

For currents that vary with time, the current at time t is defined as the limit of $\Delta q/\Delta t$ as the time interval becomes infinitesimally small; hence,

$$I(t) = \lim_{\Delta t \to 0} \frac{\Delta q}{\Delta t} = \frac{dq}{dt} \tag{18.2.2}$$

From these equations, it is obvious that current is a *scalar* quantity and that the units of current in the mks system are coulombs per second, more frequently referred to as *amperes* (abbreviated amp, or A).[1]

Another quantity related to the current is the *current density*, which expresses the strength or concentration of charge flow at any point in a conducting medium. The current density is a *vector in the direction of charge flow* at the given point. Its magnitude is determined by taking the limit of the charge flow— or current—per unit area through a small area $\Delta a'$ oriented perpendicular to the direction of the current at the point as the area $\Delta a'$ approaches zero, as shown in Fig. 18.2. Mathematically, this means that the magnitude of the current density vector $\mathbf{j}$ is given by

$$j = \lim_{\Delta a' \to 0} \frac{\Delta I}{\Delta a'} = \frac{dI}{da'} \tag{18.2.3}$$

whence, of course,

$$dI = j\,da' \tag{18.2.4}$$

In the case of a conductor within which the rate of flow of free charge is the same at all points, as shown in Fig. 18.1, the current density $\mathbf{j}$ is the same throughout the conductor. The relation between the total current I and the current density $\mathbf{j}$ may be found by

[1] After André Marie Ampère (1775–1836), a French physicist who investigated the magnetic effects of electric currents and who first stated the general law governing these effects.

integrating (18.2.4) over the shaded cross-sectional area of the conductor, regarding $\mathbf{j}$ as a *constant*. In this way, we obtain

$$I = \int j\,da' = j \int_A da' = jA \tag{18.2.5}$$

from which

$$j = \frac{I}{A} \tag{18.2.6}$$

When the current density *varies* from point to point within the conducting substance, as it may, for example, in a gas discharge tube or in the semiconducting crystal material in a transistor or solid-state rectifier, the relation between the current density and the total current is more complex. In Fig. 18.3, a situation of this sort is shown. A total current I flows through an irregularly shaped conducting substance, with magnitude and direction of the charge flow (and therefore the current density) changing from point to point. Let us first construct a surface S spanning the conductor—any surface will do, in principle—and

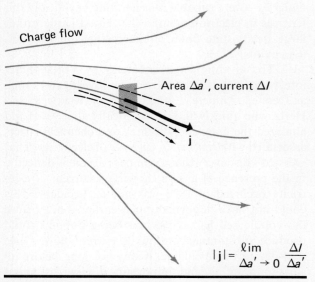

FIGURE 18.2. Current density in a nonuniform current.

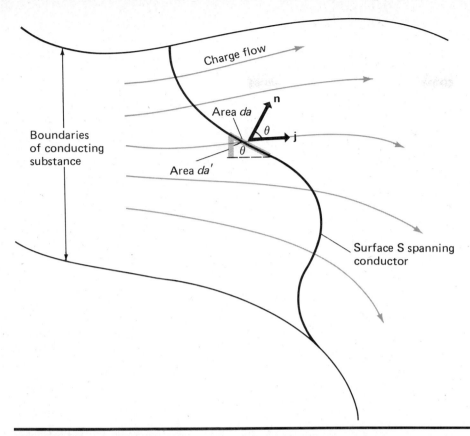

FIGURE 18.3. Calculation of the general relationship between current and current density.

let us focus our attention on an infinitesimal area element of the surface, *da*, whose orientation is specified by the *unit vector* **n** normal to the element. In general, the surface S is not normal to the current flow direction, which is defined by the direction of the current density vector **j**. The magnitude of the current density vector is defined by the current flowing through the element of area *da'* normal to the charge flow direction, as expressed by Eq. (18.2.4). But, from Fig. 18.3, the relation between the sizes of these two elements is

$$da' = da \cos \theta \tag{18.2.7}$$

where θ is the angle between the current density **j** and the unit vector **n** that specifies the orientation of the surface S at the point in question. From (18.2.4), then,

$$dI = j \cos \theta \, da \tag{18.2.8}$$

Now, $j \cos \theta$ is the same as $\mathbf{j} \cdot \mathbf{n}$, in view of the definition of the scalar product and of the fact that the vector **n** has unit magnitude. Equation (18.2.5) then becomes

$$dI = \mathbf{j} \cdot \mathbf{n} \, da \tag{18.2.9}$$

which can be integrated over the whole surface S to give

$$I = \int_S \mathbf{j} \cdot \mathbf{n} \, da \tag{18.2.10}$$

This is the relation connecting the total current and the current density in the most general case. The form of this relation, and the fact that it expresses the total flow of charge across the surface, may help us to understand more clearly why the electric "flux" Φ across a surface is defined as it is in Eq. (16.3.3).

The pattern of current density vectors throughout a medium in which current is flowing can be used to construct maps of *current flow lines* in much the same way in which the pattern of electric field vectors lead to maps of the lines of force for an electrostatic field. The lines marked *charge flow* in Figs. 18.2 and 18.3 are current flow lines of just this sort. The current flow lines are also very similar to the streamlines in a fluid flow field. It is evident, therefore, that the current density may be regarded as a *vector field*.

It is important to understand, nevertheless, that even though an electric field may be what establishes a given current flow, the current flow lines are not necessarily the same as the lines of force acting on the charges. The reason for this is that the current flow lines point in the direction of the *velocity* of individual charges, while the lines of force associated with the electric field point in the direction of electric *force* experienced by them—hence, according to Newton, in the direction of their *acceleration*. But, as we have seen many times in our study of mechanics, the velocity and acceleration vectors of a particle do not necessarily have the same direction. The simplest example of this is found in uniform circular motion, where the velocity vector is tangential to the circular path, while the force and acceleration vectors are normal to it, being directed radially inward. So the current flow lines and the electric field lines may in general have different directions, in which event the maps of the lines of force and current flow have quite different appearances.

Another important relation is the one connecting the current density and the *velocity* of the charges that contribute to the current flow at any point. This can be understood with reference to Fig. 18.4, which illustrates a part of the current flow pattern confined to a "tube of current flow" of area da', in analogy with a tube of fluid flow as discussed in Chapter 11. Since the current flow lines are parallel to the sides of the tube of flow, no current flows across these sides. In a time interval dt, any charge within the substance will move with the flow of current through some distance $dx = v\, dt$, where v is the average charge velocity. Within this time interval, a total charge dq flows past the shaded area da'. Suppose that the electric charge *density* is denoted by ρ, representing the electric charge per unit volume in the volume element shown. The amount of charge dq within the volume dV can then be written

$$dq = \rho\, dV = (\rho\, da')\, dx \qquad (18.2.11)$$

But since $dx = v\, dt$, this can be stated as

$$dq = (\rho v\, da')\, dt \qquad (18.2.12)$$

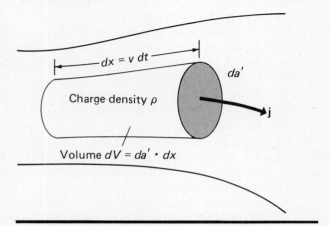

FIGURE 18.4. Calculation of the relation between current density and mobile charge density.

or

$$\frac{dq}{dt} = \rho v\, da' \qquad (18.2.13)$$

However, according to (18.1.2), dq/dt represents the total current in the tube of flow and is related to the current density there by (18.2.4). Substituting $j\, da'$ for dq/dt in (18.2.13) and canceling the area elements, the current density, charge density, and charge velocity are shown to be related by

$$j = \rho v \qquad (18.2.14)$$

Indeed, since the current density vector and the velocity vector have the same direction, this relation also holds as a vector relation of the form

$$\boxed{\mathbf{j} = \rho \mathbf{v}} \qquad (18.2.15)$$

Current density, therefore, equals charge density times charge velocity at all points within the conducting substance.

If the charges within the substance exist, as they always do, in the form of very large numbers of elementary charged particles, then the charge per unit volume will be given by

$$\rho = nq_0 \qquad (18.2.16)$$

where n is the number of elementary charged particles per unit volume and q_0 is the charge associated with each of them. Equation (18.2.15) then takes the form

$$\boxed{\mathbf{j} = nq_0\overline{\mathbf{v}}} \qquad (18.2.17)$$

where $\overline{\mathbf{v}}$ represents the average velocity of the charged particles. Finally if the elementary charges are electrons, their charge q_0 is $-e$, where e represents the size of the charge on the electron, which is 1.6022×10^{-19} coulomb. In this case,

$$\boxed{\mathbf{j} = -ne\overline{\mathbf{v}}} \qquad (18.2.18)$$

EXAMPLE 18.2.1

A conductor of uniform cross section carries a steady current of 1.00 ampere. How many electrons per second flow past a given point on the conductor?

If the current is 1.00 ampere, then the total rate of charge flow past any point is 1.00 coulomb per second. But since the size of the electronic charge is given by $e = 1.6022 \times 10^{-19}$ coulomb and since the number of electrons in a coulomb is the *reciprocal* of this quantity, or 6.241×10^{18} electrons/coulomb, the number of electrons passing a given point in the conductor carrying a current of 1.00 ampere ($= 1.00$ coulomb per second) is just the number of electronic charges in a coulomb, or 6.241×10^{18}. Our most sensitive current measuring instruments are capable of measuring currents of the order of 10^{-16} ampere, which corresponds to $(6.241 \times 10^{18}) \cdot (10^{-16}) \cong 600$ electrons per second.

EXAMPLE 18.2.2

A silver conductor (density 10.50 g/cm^3 and atomic mass 107.9 g/mole) has a uniform circular cross section 0.1 cm in diameter. This conductor is made to carry a uniform current of 100 amperes. Find the current density and the mean velocity with which charges move within the conductor, assuming that each silver atom contributes one mobile free electron to the conductor.

In this case, since the conductor is homogeneous and of uniform cross section, the current density is the same everywhere and can be found by dividing the current by the area, as expressed by (18.2.6). The cross-sectional area A is

$$A = \pi r^2 = (3.1416)(0.05^2) = 0.00785 \text{ cm}^2$$

The current density would then be

$$j = \frac{I}{A} = \frac{100}{0.00785} = 12{,}740 \text{ amp/cm}^2$$

$$= 1.274 \times 10^8 \text{ amp/m}^2$$

The average velocity of the charges may be obtained from Eq. (18.2.18). In a mole of silver, containing 107.9 grams of the substance, there are $N_A = 6.023 \times 10^{23}$ atoms and, therefore, 6.023×10^{23} mobile free electrons. On the other hand, since the density is 10.50 g/cm^3, a mole of silver occupies a volume of $107.9/10.50 = 10.28$ cm^3. Therefore, in each cubic centimeter of silver, there are $N_A/10.28 = (6.023 \times 10^{23})/10.28 = 5.86 \times 10^{22}$ atoms, hence, $n = 5.86 \times 10^{22}$ electrons/cm$^3 = 5.86 \times 10^{28}$ electrons/m^3. Then, according to (18.2.18),

$$\bar{v} = \frac{j}{ne} = \frac{1.274 \times 10^8}{(5.86 \times 10^{28})(1.602 \times 10^{-19})}$$

$$= 1.357 \times 10^{-2} \text{ m/sec} = 1.357 \text{ cm/sec}$$

In using Eqs. (18.2.15) and (18.2.18), it is important to *adhere strictly to mks units*, expressing j in amperes/m^2, ρ in coulombs/m^3, n in electrons/m^3, and v in meters/sec. Though in some contexts it is useful to express current density in amperes/cm^2 and electron density in units of electrons/cm^3, these are *not* mks units and will lead to erroneous results when used indiscriminately in the equations derived above.

EXAMPLE 18.2.3

A bar of *semiconductor* crystal such as germanium is of uniform, square cross section with side length 0.1 cm, as shown in Fig. 18.5. Due to the semiconducting properties of the substance, the current density is not uniform over the cross section of the sample but varies from point to point according to the equation

$$\mathbf{j} = j_0 \left(\sin \frac{\pi x}{h} \sin \frac{\pi y}{h} \right) \mathbf{i}_z \tag{18.2.19}$$

where h represents the 0.1-cm side of the square sample. This equation simply states that the current density falls to zero on the surfaces of the bar at $x = 0$, $x = h$ and $y = 0$, $y = h$; has the maximum value j_0 in the center at $x = y = h/2$; and varies sinusoidally as a function of x and y elsewhere. The direction of $\mathbf{j}$, according to (18.2.19), is everywhere parallel to the z-axis, that is, to the length of the bar. Assuming that j_0 has the value 800 amperes/m^2, find the total current I through the sample.

In this example, the current density varies from point to point in the sample, and therefore we must begin with Eq. (18.2.10), integrating the quantity $\mathbf{j} \cdot \mathbf{n}$ over the sample cross section. The sample cross section can be regarded as being composed of rectangular area elements having dimensions dx and dy, as shown in Fig. 18.5; since the cross section plane is normal to the z-direction, the unit vector $\mathbf{n}$ normal to the area element has unit magnitude and points along the z-axis. Because this represents the magnitude and direction of the vector $\mathbf{i}_z$, we may in this case write

$$\mathbf{n} = \mathbf{i}_z \tag{18.2.20}$$

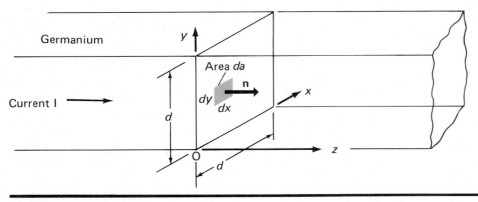

FIGURE 18.5

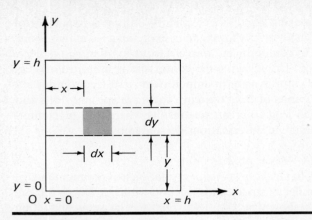

FIGURE 18.6

Substituting (18.2.19) and (18.2.20) into (18.2.10) and noting that $\mathbf{i}_z \cdot \mathbf{i}_z = 1$, we obtain

$$I = \int_S \mathbf{j} \cdot \mathbf{n} \, da = \int j_0 \left(\sin \frac{\pi x}{h} \sin \frac{\pi y}{h} \right) dx \, dy \quad (18.2.21)$$

Now, in integrating over the whole cross-sectional area of the sample, we must integrate *twice*, summing once over dx from $x = 0$ to $x = h$ to produce the strip element of width dy shown in Fig. 18.6, and then over dy from $y = 0$ to $y = h$ to cover the whole area. In doing the first integral, over dx, we must regard y as *constant*, since all the area elements in the strip have the same height above the x-axis. In this way, we may write

$$I = j_0 \int_0^h \int_0^h \sin \frac{\pi x}{h} \sin \frac{\pi y}{h} \, dx \, dy \quad (18.2.22)$$

Integrating over x, this becomes

$$I = j_0 \int_0^h \frac{h}{\pi} \left[-\cos \frac{\pi x}{h} \right]_0^h \sin \frac{\pi y}{h} \, dy$$

$$= \frac{2j_0 h}{\pi} \int_0^h \sin \frac{\pi y}{h} \, dy = \frac{2j_0 h^2}{\pi^2} \left[-\cos \frac{\pi y}{h} \right]_0^h = \frac{4j_0 h^2}{\pi^2}$$

$$(18.2.23)$$

Since h^2 represents the total cross-sectional area A of the sample, this may finally be written as

$$I = \frac{4j_0 A}{\pi^2} = \frac{(4)(800)(10^{-6})}{(3.1416)^2} = 3.24 \times 10^{-4} \text{ amp}$$

$$(18.2.24)$$

18.3 Electromotive Force and Potential Difference

Consider the situation shown in Fig. 18.7, where a battery, represented by the symbol —|I|I|I|—, is connected to a closed conducting circuit containing a motor that is doing mechanical work by lifting a weight. It is evident that certain forces must act on the mobile charges within the conducting circuit and within the wires that comprise the coils inside the motor. If this were not so, no current would ever be established in the circuit. It is also evident that *the work done by the motor is caused by the flow of current through its windings and, therefore, ultimately by the forces that cause the current to flow.*

We know that charges exert electrostatic forces upon one another, whose action is described by Coulomb's law and Gauss's law. The distribution of charge creates an *electrostatic field*, let us call it $\mathbf{E}_0$, whose intensity at any point is defined by

$$\mathbf{E}_0 = \lim_{q \to 0} \frac{\mathbf{F}_0}{q} \quad (18.3.1)$$

where $\mathbf{F}_0$ represents the vector sum of all the Coulomb forces on the "test charge" q arising from the *other* charges present. If these other charges are designated $q_1, q_2, \ldots, q_n$, then the force $\mathbf{F}_0$, according to Coulomb's law, will always be of the form

$$\mathbf{F}_0 = \sum_{k=1}^n \frac{q q_k}{r_k^2} \cdot \frac{\mathbf{r}_k}{r_k} \quad (18.3.2)$$

where $\mathbf{r}_k$ is the position vector describing the location of charge q_k with respect to the test charge q.

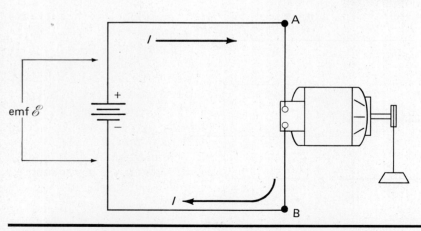

FIGURE 18.7. An electric current that performs mechanical work.

We know from our study of electrostatics that we can associate a potential energy with electrostatic forces, just as with gravitational forces. We can conclude, therefore, that electrostatic forces are *conservative*. But we learned in Chapter 5 that *the work done by any conservative force in traversing any closed path must be zero*. Since the electrostatic force acting on charge q is $q\mathbf{E}_0$, this means that the work it does can be written

$$W = \oint_c \mathbf{F} \cdot d\mathbf{l} = q \oint_c \mathbf{E}_0 \cdot d\mathbf{l} = 0$$

from which

$$\oint_c \mathbf{E}_0 \cdot d\mathbf{l} = 0 \qquad (18.3.3)$$

We see from this that the net work done by electrostatic forces on any charge in going around the circuit is *zero*. As a consequence, we must conclude that *electrostatic forces* cannot be responsible for the performance of work by electric currents or even, for that matter, for establishing or maintaining the flow of current in a conducting circuit, since that requires work also.

We are in this way forced to conclude that steady currents can flow only as a result of *nonelectrostatic* forces acting on the charges. At this point, it is necessary to understand that while charges exert electrostatic forces on one another, they may also be acted upon by forces other than Coulomb forces from other charges. In general, then, the total *electric* field $\mathbf{E}$, representing the *total* force per unit charge experienced by a given charge, *may not be entirely of electrostatic origin*. An *electrostatic field* is thus only a particularly simple special case of an *electric field*. We may, however, always write the total electric field intensity $\mathbf{E}$ in terms of the vector sum of electrostatic and nonelectrostatic forces acting on a test charge q, from which

$$\mathbf{E} = \lim_{q \to 0} \frac{\mathbf{F}_0}{q} + \lim_{q \to 0} \frac{\mathbf{F}'}{q} = \mathbf{E}_0 + \mathbf{E}' \qquad (18.3.4)$$

where $\mathbf{F}_0$ is the net electrostatic force experienced by the charge q and $\mathbf{F}'$ is the resultant nonelectrostatic force. The total field may in this way be expressed as the sum of an *electrostatic* field $\mathbf{E}_0$ and a *nonelectrostatic* field $\mathbf{E}'$. *Only fields of the latter type, which may arise from magnetic, chemical, thermal, or other sources, can cause steady current to flow in an electrical circuit.*

Referring again to Fig. 18.7, it is evident that in any given time interval, during which the motor performs external mechanical work ΔW_m, an equivalent amount of work must somehow be done on the mobile charges that comprise the current. This must be true as a result of the law of conservation of energy;

the motor does work because current is made to flow through its windings, and therefore an equivalent amount of work must be done on the charges comprising the current to maintain the current flow. If it were not so, we would be getting our work free—we could use the motor to run a generator to generate more current to do more work to generate more current, and so on—a perpetual motion machine. As we have already seen, it is *nonelectrostatic* forces acting upon the charges that must do this work. The effect of a current source, such as a battery, a generator, or a photocell, then, is to do work on the charges in the circuit, the effect of which is *to increase their potential energy* as they flow through such a device. The amount of potential energy *per unit charge* such a source may impart is referred to as its *electromotive force*, or *emf*. Thus, if a current source is capable of increasing the potential energy of a charge q by an amount ΔU_p, its emf $\mathscr{E}$ is given by

$$\mathscr{E} = \frac{\Delta U_p}{q} \qquad (18.3.5)$$

Since the potential energy ΔU_p of the charge q is related to a difference in *electrical potential* ΔV by the relation $\Delta U_p = q\,\Delta V$, it is evident that (18.3.5) can be written very simply as

$$\mathscr{E} = \frac{q\,\Delta V}{q} = \Delta V \qquad (18.3.6)$$

This equation allows us to interpret the emf $\mathscr{E}$ as *a difference in potential, attributable to nonelectrostatic forces, that is capable of establishing a steady current in a closed conducting circuit.*

From (18.2.6), it is clear that the emf of a current source has the dimensions of potential difference, which is *volts* in the mks system. A battery whose emf is 12 volts, therefore, is capable of maintaining, by chemical processes, a difference in potential of 12 volts between its terminals. When charge flows through such a battery, each coulomb of charge entering the negative terminal of the battery receives an increase in potential energy of 12 joules in the battery before it emerges from the positive terminal. In going through the external circuit, represented by the motor in Fig. 18.7, each coulomb of charge does 12 joules of work against external forces (represented by the weight in the diagram), whereby *its potential energy decreases by 12 joules as it returns to the negative terminal of the battery.* Assuming that charge Δq flows in time Δt, then, since $I = \Delta q / \Delta t$, the power supplied by the emf is

$$P = \frac{\Delta U_p}{\Delta t} = \frac{\Delta q\,\Delta V}{\Delta t} = I\,\Delta V \qquad (18.3.7)$$

569

or recalling (18.3.6),

$$P = \mathscr{E}I \qquad\qquad (18.3.8)$$

We should note that the terminology *electromotive force* is really a misnomer, since the emf is not a force at all but a difference in potential. It would have been more accurately denoted electromotive potential difference, but the term electromotive force has been sanctified by such long usage that it is hopeless to try to change it now. We shall thus continue to use the term emf in the customary way, remembering, hopefully, that it is a voltage difference and not a force.

Sources of emf other than batteries, while relying on different internal processes to produce and maintain their terminal potential differences, behave in essentially the same way when incorporated into conducting circuits. In all instances, a steady nonelectrostatic potential difference is maintained between the source terminals. Ideally, this potential difference or emf is *independent of the current* flowing and of any electric fields of external origin that may be present. Actual sources of electromotive force depart somewhat from this ideal behavior for reasons that are simple and well understood, though we shall defer their discussion until later in the chapter.

The electric fields—and now we mean *electric* and not *electrostatic* and intend to distinguish carefully between these terms henceforth—around circuits in which sources of emf are present are illustrated schematically in Figs. 18.8 and 18.9. In Fig. 18.8, there is a gap in the rectangular conducting loop, the circuit remains open, and no current flows. In Fig. 18.9, the gap has been closed (by throwing a switch, for example), the circuit has been completed, and a steady current I is flowing through the conducting loop. In Fig. 18.8, allowing that the emf of the battery is present, the case is essentially one of electrostatic equilibrium. Since the emf is present, the lower part of the rectangular conductor is (let us say) at a uniform potential $V = 0$, and the upper part at a potential of $+\mathscr{E}$ volts. Since in equilibrium there can be no field within a conductor, the entire lower part of the circuit is an equipotential region having zero potential. There is a certain distribution of surface charge on the surface of the conductor that forms in just such a way as to guarantee that this happens. Likewise, the entire upper part of the conducting circuit is an equipotential region of potential $+\mathscr{E}$ volts. A part of the surface charge distribution consists of a collection of positive charges on the upper part of the loop at the point where the circuit is interrupted and of an equal distribution of negative charges on the end surface of the lower loop. The gap in the conducting circuit strongly resembles a parallel-plate condenser charged to a potential corresponding to the emf $\mathscr{E}$ of the potential source. There is a strong electrostatic field in this gap whose source is these positive and negative charges on the conducting surfaces. The drop in electric potential of $-\mathscr{E}$ volts between A and B just balances the rise in potential of $+\mathscr{E}$ volts from C to D occasioned by the emf of the potential source. Since there is a strong drop in potential between A and B, a series of closely spaced equipotential contours will intersect the line connecting A and B as shown.

The same thing can be said of the battery in Fig. 18.8. Since its terminals C and D are at different potentials, a similar series of equipotential countours must be crossed in going from C to D. Away from the gap in the circuit and outside the battery, there will be fringing fields as shown, terminating on the equipotential surfaces of the circuit. In the same way, outside the circuit, but close by, there are equipotential lines that follow the contour of the circuit quite closely, becoming much more rounded at larger distances. These general features all add up to produce the picture of field lines and equipotential contours shown in Fig. 18.8.

When the circuit is completed and a steady current I allowed to flow, the picture changes drastically to that shown in Fig. 18.9. The field within the conductor is now no longer zero but has the constant magnitude $\mathscr{E}/l$, where l is the length of the conductor; the field is directed *along* the conductor at all points. This pattern of fields sets up a similar pattern of current flow lines in the loop—this is one case in which the electric field intensity pattern and the current flow pattern are the same. The potential fall of $-\mathscr{E}$ volts in the external conducting circuit between A and B is now no longer concentrated in an external gap but occurs gradually and uniformly along the length of the conductor. The surface of the conductor is no longer an equipotential surface. On the contrary, there are now plane equipotential surfaces that *intersect* the conductor normally at equal intervals of distance along its length, perpendicular to the field lines inside. In addition, the difference in potential between the battery terminals sets up a series of equipotential lines *within* the source of emf, as shown. The rise in potential of $+\mathscr{E}$ volts inside the battery from C to D is balanced by the gradual fall in potential of $-\mathscr{E}$ volts along the length of the conducting loop from A to B. The difference in potential between the battery terminals also gives rise to a fringing field outside the circuit as illustrated in the diagram. These features, taken together, add up to the picture of fields and potential contours shown in Fig. 18.9.

The discussion given above is a very important one. The investment of time and effort required to achieve a clear understanding of it will be amply repaid by the insight it affords into the physical behavior of conducting circuits in which emfs are present.

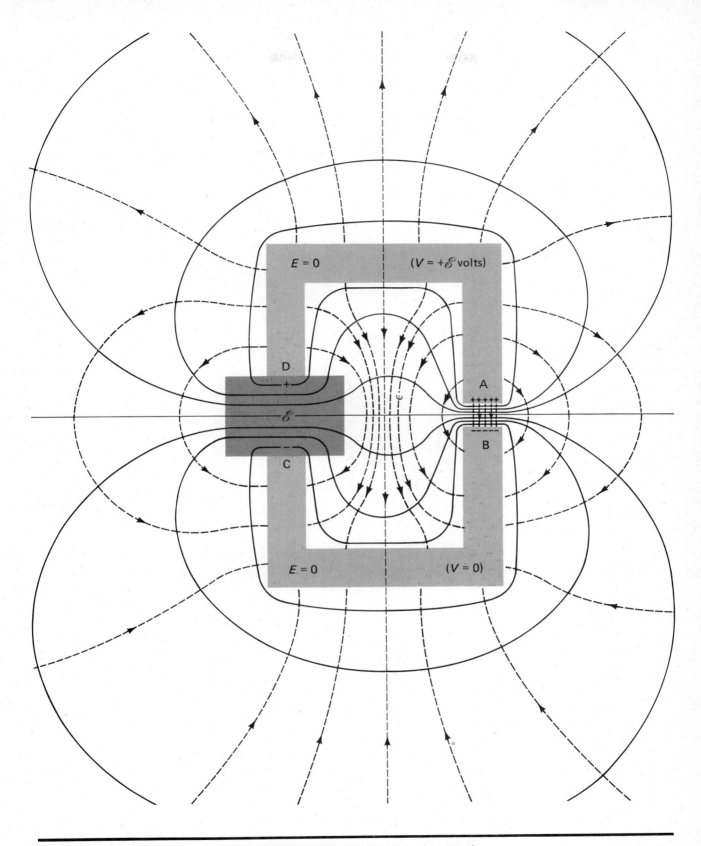

FIGURE 18.8. Electric field surrounding a circuit in which a gap prevents current from flowing.

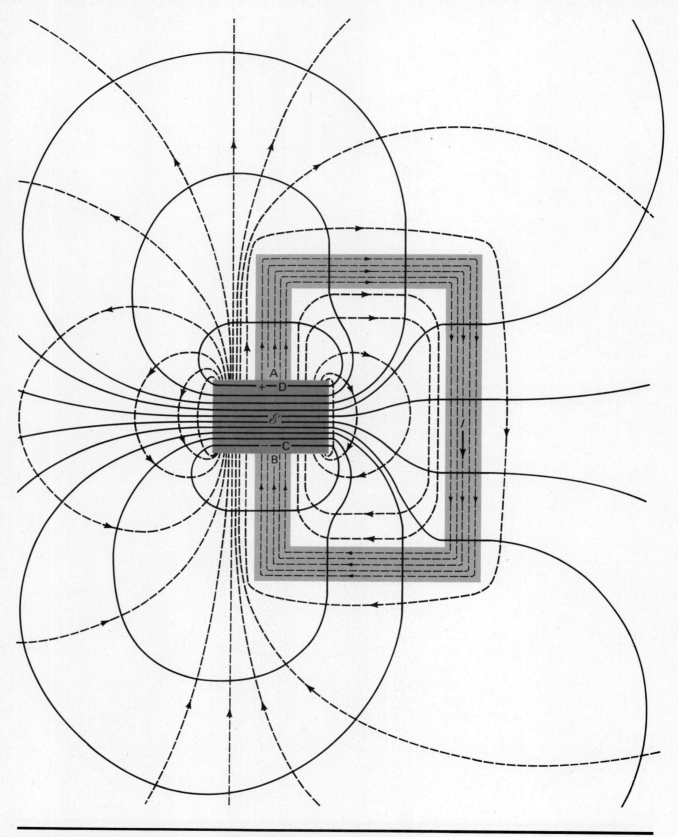

FIGURE 18.9. Electric field surrounding a circuit in which a steady current is circulating.

18.4 Ohm's Law and the Conduction of Electricity by Free Electrons

When a source of emf such as a battery is connected to an external conducting circuit, as illustrated in Fig. 18.9, it is observed experimentally that a steady current flows through the conducting substance. It has been further established from experiments that for most conducting materials, the current density **j** at any point is directly proportional to the electric field **E** within the conductor at that point, provided that the magnitude of the field is not excessively large. These experimental observations can be summarized in the form of an equation:

$$\mathbf{j} = \sigma \mathbf{E} \tag{18.4.1}$$

where σ is a constant of proportionality that expresses the current density obtained per unit electric field intensity. The value of the constant σ depends on the particular conducting substance in the circuit and also, usually to a lesser extent, on the temperature. The quantity σ is referred to as the *electrical conductivity* of the material comprising the circuit. Its units are those of current density divided by field intensity, amp volt^{-1} m^{-1}. The electrical conductivity of familiar substances varies a great deal from one material to another; metallic conductors exhibit conductivity of the order of 10^8 amp volt^{-1} m^{-1}, while good insulators may have conductivity of the order of 10^{-13} amp volt^{-1} m^{-1} or even less. The experimental law expressed by Eq. (18.4.1) was first established in 1826 by the German physicist Georg Simon Ohm (1789–1854) and is universally referred to as *Ohm's law*. Ohm's law is sometimes written in the form

$$\mathbf{E} = \frac{1}{\sigma} \mathbf{j} = \rho \mathbf{j} \tag{18.4.2}$$

where

$$\rho = \frac{1}{\sigma} \tag{18.4.3}$$

The reciprocal of the electrical conductivity, denoted above by ρ, is usually called the *electrical resistivity*; its dimensions are volt-meter/ampere.

We are usually told that steady currents flow in conducting materials when an electric field is present by virtue of the fact that mobile electrons in the conductor are set into motion by the field. If that is so, then we should be able to express the forces on the free electrons in terms of the applied field, write their equation of motion using Newton's second law, and finally find their velocity, from which the current density may be obtained. In this way, we should expect to *derive* Ohm's law from first principles using a free electron model as a basis for our calculations. We shall soon undertake to do just that, but we shall find that certain other aspects of the free electron model must be understood before the derivation can be successfully completed.

First, it is important to recognize that the drift velocity of an electron imparted by the applied emf, as calculated for a typical case in Example 18.3.2, is usually only an *extremely small part* of its total instantaneous velocity. The reason for this is that at ordinary temperatures electrons have *thermal energy* just like that associated with the molecules of a gas. Indeed, the distribution of free electrons in a metallic conductor or in a semiconducting substance is frequently referred to as an *electron gas* and behaves in many respects like a gas of charged particles confined by the physical boundaries of the substance, just as a gas is confined by a container. In Chapter 13, Eq. (13.6.15), we found that the average speed of a gas molecule of mass m in an ideal gas having a Maxwell-Boltzmann distribution of velocities is given by

$$\bar{v} = \sqrt{\frac{8kT}{\pi m}} \tag{18.4.4}$$

If we substitute $k = 1.381 \times 10^{-23}$ joule/molecule-°K, $T = 300$°K, and $m = 9.11 \times 10^{-31}$ kg, the mass of the electron, into (18.4.4), we find $\bar{v} = 1.07 \times 10^5$ meters/sec. Actually, since electrons obey the laws of quantum mechanics rather than classical mechanics, in metals, at least, their energy distribution differs from the Maxwell–Boltzmann distribution, resulting in even *larger* values of average thermal speed. The average electron speed arising from its thermal energy is clearly very much larger than the drift velocity it would ordinarily acquire from any applied emf; the drift velocity calculated for a typical situation of steady current flow in Example 18.3.2 is only a small fraction of a meter per second.

On the other hand, the random thermal motion of the electrons contribute exactly *nothing* to the current, because the number of electrons having given positive x-components of velocity $+v_x$ arising from thermal motion and the number having equal negative components $-v_x$ are equal. The average x-component of velocity associated with the thermal equilibrium motion is therefore *zero*. This is evident (for the Maxwell–Boltzmann distribution) from Eq. (13.6.8) and from Fig. 13.12, but it must be true whatever the form of the equilibrium distribution of velocities. The same remarks apply, of course, to the y- and z-velocity components. Since the average values of v_x, v_y, and v_z associated with the thermal motion are zero and since current density is simply charge density times velocity, this means that the average values of j_x, j_y, and j_z associated with the thermal motion of the electrons are *all* zero and that no current is ever contributed by

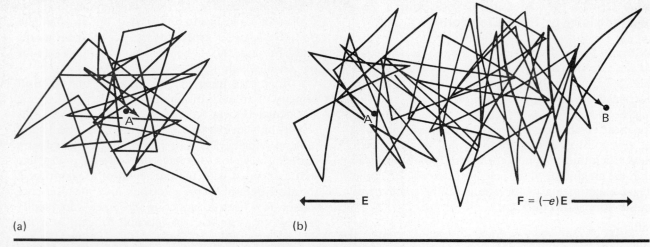

(a)   (b)

FIGURE 18.10. (a) Thermal motion of free charges in a conductor in which no current flows. (b) Motion of free charges in a conductor in which an externally created field is present and in which a current is flowing.

the thermal motion of free electrons. Any current flowing in a conducting substance, therefore, arises entirely from the velocity that individual electrons acquire from the applied electric field, even though that velocity may be very much smaller than the average velocity the electron has by virtue of its thermal energy. This velocity contributed by the applied emf is frequently referred to as the *drift velocity* of the electron, and the quantity $\bar{v}$ used in the expressions in section 18.1 should be interpreted as the average of the drift velocities of the free electrons involved in the current flow.

The situation may be understood easily by referring to Fig. 18.10. At (a), we see the random path of a single free electron in a conducting substance at equilibrium. There is no external force on the electron, and its path resembles that of a gas molecule in a monatomic ideal gas. The individual segments of the path are straight lines since between collisions that change the electron's course and speed, the force on it is zero. The electron can be thought of as being at point A initially. Although its random wandering may sometimes carry it a considerable distance from that point, it is just as likely to be found to the left of A as to the right of A. Its average displacement, and therefore its average velocity, is zero. Over a long period of time, on the average, it gets nowhere. In Fig. 18.10b, we see what happens when an electric field due to an applied emf is present. In this case, the electric field vector points to the left and gives rise to a constant force $-e\mathbf{E}$, to the right, on the electron. During each segment of the electron's path, this force imparts to the electron a small drift velocity, which, added to its random thermal velocity, results in the electron's path ending up just slightly to the right of where it would have been had there been no field. The electron thus "drifts"

systematically to the right during each path segment, and the ends of these segments are no longer quite randomly distributed. There is, instead, a slight tendency for the electron to end up a bit further "downstream" on each successive leg of its journey than would be the case if its motion were purely random. Its motion can thus be regarded as a purely random thermal motion, like that of a molecule in a gas on which a systematic drift in the direction of current flow is superposed. When the field is present, the individual path segments are no longer quite straight but become curved slightly into parabolic paths, this curvature being what finally results in the electron ending up a bit to the right of where it otherwise would have been. The problem, then, is that of a particle moving in a plane subject to a constant force—thus to a constant acceleration—which, we may recall from previous acquaintance with projectile motion, results in a parabolic trajectory rather than a straight one.

Using this physical picture as a basis, we may now begin to calculate quantitatively the relation between the current density and the electric field within the conductor. Let us first consider the motion of a single electron in a uniform conducting substance in which a constant electric field $\mathbf{E}$ is suddenly applied at time $t = 0$. Assuming that the electric field $\mathbf{E}$ is in the x-direction, then, since the field intensity represents the force per unit charge, we may write the force on the electron as

$$F_x = -eE \qquad\qquad (18.4.5)$$

Using Newton's second law to relate force and acceleration, this may be written as an equation of motion having the form

$$a_x = \frac{dv_x}{dt} = -\frac{eE}{m} \qquad\qquad (18.4.6)$$

Now, assuming that the field E is constant, this represents a uniformly accelerated motion, and the velocity v_x must, therefore, be given by

$$v_x = v_{0x} + a_x t = v_{0x} - \frac{eEt}{m} \qquad (18.4.7)$$

Naturally, since the electric field vector has no y- and z-components, a_y and a_z are both zero, and, therefore, the velocity components along the y- and z-directions have the equilibrium values v_{0y} and v_{0z}. In Eq. (18.4.7), if the field were absent, the second term on the right would be zero, leaving only the field-independent quantity v_{0x}. This quantity, therefore, must represent the effect of the thermal motion of the electrons, since, after all, it is all that remains if there is no field. In the same way, the second term, which is the only one that depends on the field at all, clearly expresses the *drift velocity* acquired by the electron due to the action of the field. If we now average v_x over all the electrons in the conductor, we shall find that since the average value of v_{0x} for the random thermal motion is zero, and since each electron in the substance acquires the same drift velocity $-eEt/m$ in the same time interval, the average value of v_x will be

$$\bar{v}_x = -\frac{eEt}{m} \qquad (18.4.8)$$

From (18.1.18), this means that the current density will be

$$j_x = -ne\bar{v}_x = \frac{ne^2}{m} Et \qquad (18.4.9)$$

with $j_y = j_z = 0$ since $\bar{v}_y = \bar{v}_z = 0$.

Now, this result as it stands is not Ohm's law; in fact, it is absurd since it states that the current density, rather than being constant, *must increase linearly with time*. The current density will become as large as we wish if we wait long enough! The origin of this paradoxical result can be traced back to Eq. (18.4.7) which states that the drift velocity of an electron in the direction of the applied field increases linearly with time. This is true enough as far as it goes, but what is omitted is the fact that this increase in drift velocity does not go on indefinitely but instead *ceases as soon as the electron undergoes a collision*, which radically alters its course and speed. After such a collision, the electron is headed in a different direction with a different speed. The collision wipes out all the drift motion of the electron and serves simply to return the characteristics of its motion to those of thermal equilibrium. The effect of the collisions, in other words, is to transform the kinetic energy the electron has acquired by virtue of its drift velocity into random thermal energy, that is, into *heat*. These arguments emphasize the critical importance of *collisions* in the conduction process. We shall soon see

that when we incorporate their effect into our calculations, our results at once become very much more reasonable.

Since the effect of collisions is to return the electrons to a state of thermal equilibrium, we may assume that *on the average, a collision reduces the drift velocity to zero*. If this is so, then during the time between successive collisions, the drift velocity may be pictured as having the form shown in Fig. 18.11. The electron starts at $t = 0$, at which time the average initial value of v_x will be zero, as shown. Of course, most of the time the velocity v_x will not be returned to zero by collisions; rather there will be a wide distribution of final velocities. The point is that since such a distribution corresponds essentially to thermal equilibrium, the *average* value of v_x after a collision is zero. We may as well make our work as easy as possible, though, by assuming that this is always so, since we shall later have to average over all electrons anyway.

In any event, if the electron starts from rest at $t = 0$, its subsequent drift speed in the x-direction increases linearly with time as predicted by (18.4.8), *until the electron undergoes a collision* at time $t = \tau_c$, at which time its x-component of velocity drops, on the average, to zero. This dependence of drift speed upon time is illustrated by Fig. 18.11, where, for simplicity, only the magnitude of v_x is plotted, its negative sign being ignored for the moment. After the collision at $t = \tau_c$, the process is repeated and the drift speed rises linearly until it is again abruptly reduced to zero by a second collision at some later time. This cycle of events then goes on indefinitely, time after time. During the interval from $t = 0$ to $t = \tau_c$, since the drift speed increases *linearly* with time from zero to a maximum value of $v_m = -eE\tau_c/m$ at $t = \tau_c$, its average

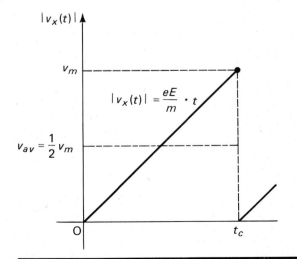

FIGURE 18.11. Average effect of collisions on the motion of charges within a conductor.

value during this particular cycle will be just half the maximum value, or

$$\bar{v}_x = -\frac{eE\tau_c}{2m} \qquad (18.4.10)$$

In this equation, the quantity τ_c represents the free drift time between one collision and the next. This time interval between successive collisions will not always be the same since, because of the random nature of the electron's thermal motion, there will be instances where successive collisions occur very quickly and those in which successive collisions occur only after much longer time intervals. These free time intervals between collisions have a statistical distribution of lengths, but over a long time there will be an *average* free time τ_c between collisions, and the time-average electron drift speed over such a long time interval will be

$$\boxed{\bar{v}_x = -\frac{eE\bar{\tau}_c}{2m}} \qquad (18.4.11)$$

From (18.2.18), we may now write the current density as

$$j_x = -ne\bar{v}_x = \frac{ne^2\bar{\tau}_c}{2m} E \qquad (18.4.12)$$

while, as before, $j_y = j_z = 0$.

If the average time $\bar{\tau}_c$ between collisions is independent of the field E, this equation tells us that the average current density is proportional to the electric field, that is, that Ohm's law is obeyed. Since the time between collisions depends primarily on the thermal speed of the electrons, and since this speed is hardly altered by the small drift speed superimposed by the field, it is easy to understand on physical grounds that the interval $\bar{\tau}_c$ should, indeed, be essentially unaffected by the value of E. This being the case, (18.4.11) may be written as

$$\boxed{j_x = \sigma E_x} \qquad (18.4.13)$$

in agreement with Ohm's law, where the conductivity is now expressed by

$$\boxed{\sigma = \frac{ne^2\bar{\tau}_c}{2m}} \qquad (18.4.14)$$

For a substance for which the conductivity is known, the average free time $\bar{\tau}_c$ can be calculated using this equation.[2] For copper at room temperature, for exam-

ple, the measured value of conductivity is $\sigma = 5.91 \times 10^7$ amp volt^{-1} m^{-1}. Substituting this value into (18.4.14), along with $e = 1.602 \times 10^{-19}$ coulomb, $m = 9.11 \times 10^{-31}$ kg, and $n = 8.47 \times 10^{28}$ electrons/m^3 (calculated assuming density 8.93 g/cm^3, molecular mass 63.54 g/mole, and one free electron per atom, using the methods of Example 18.2.2), we find

$$\bar{\tau}_c = 4.95 \times 10^{-14} \text{ sec}$$

This is a very small time, indeed; it suggests that electrons undergo collisions extremely frequently. For most conducting substances under realizable experimental conditions, values of $\bar{\tau}_c$ in the range of 10^{-12} to 10^{-14} seconds are typical.

Because the average time between successive electron collisions is so short, it is impossible for us to observe certain "start-up" and "shut-down" effects that might otherwise be expected when circuits are opened and closed. For example, suppose that electron collisions occurred on the average at the rate of one collision per minute for every electron. Assuming the field to be switched on at $t = 0$, the number of electrons stopped by collisions during the first few seconds would be negligible, and the current density should increase linearly with time as predicted by (18.4.9). This increase in current density with time would continue until nearly all the electrons had experienced at least one collision—this would take a time interval of the order of $\bar{\tau}_c$, or minute in this case—after which the current density would level off at the constant value given by (18.4.11). The rate at which electrons acquire kinetic energy from the field would then exactly balance the rate at which it is lost as heat in collisions.

In the same way, if the field is suddenly reduced to zero, by shortcircuiting the battery, for example, the current density would not fall to zero at once. Instead, the electrons would "coast" for a while with undiminished drift velocity along the field direction, as required by Newton's first law. This "coasting" would persist until the electrons experienced collisions, which would occur, however, only after an average time interval of the order of $\bar{\tau}_c$ *after* the field drops to zero! The current density would fall off gradually to zero, then, with a characteristic decay time of about $\bar{\tau}_c$ seconds or, in this particular example, 1 minute. These effects are illustrated in Fig. 18.12. That they are *not* observed is due to the fact that the average time $\bar{\tau}_c$ between collisions is not 1 minute but instead the incredibly small value of some 10^{-13} seconds. Such short time intervals are beyond the capability of even the fastest oscilloscopes and other electronic detectors to resolve.

So far, we have emphasized the importance of electron collisions in the conduction process but we have, deliberately, not answered the question, collisions with *what*? The impression may have been con-

---

[2] In certain other texts and references, Eq. (18.4.14) is often written the form $\sigma = ne^2\tau/m$, the factor of 2 in the denominator being absent, and the quantity τ in the numerator being referred to as the *relaxation time*. The difference in these two representations is due simply to the fact that relaxation time τ is defined in such a way that it represents one half of our average free time $\bar{\tau}_c$.

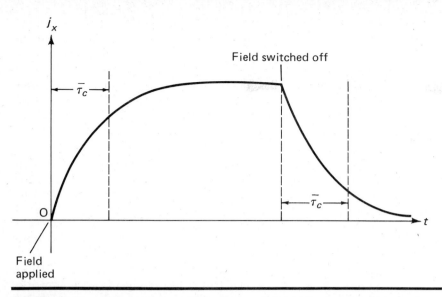

FIGURE 18.12. Delayed rise and fall of current due to the inertia of conducting electrons.

veyed that we have been talking about collisions of electrons with one another; if so, let us dispel it at once. It is true, of course, that electron–electron collisions occur very frequently, but such interactions have essentially no effect in reducing the average drift velocity! This may be understood easily by remembering that linear momentum must be conserved in electron–electron collisions and, in fact, in all others as well. In the collision of two electrons, the equality of the initial and the final momentum requires that for the x-component of the vector momentum,

$$mu_{1x} + mu_{2x} = mv_{1x} + mv_{2x} \qquad (18.4.15)$$

where u_1 and u_2 are the initial velocities and v_1 and v_2 are the final velocities. Multiplying this equation by $-e/m$, we obtain

$$-eu_{1x} - eu_{2x} = -ev_{1x} - ev_{2x} \qquad (18.4.16)$$

But now, $-eu_{1x}$ represents the initial x-component of current flow carried by the first electron—you will recall that charge times velocity equals current. Similarly, $-eu_{2x}$ is the initial x-component of current carried by the second electron, while $-ev_{1x}$ and $-ev_{2x}$ are the final x-components of current carried by the two electrons. What Eq. (18.4.16) tells us, then, is that a collision between two electrons *does not change* the x-component of the current they carry collectively! Similar remarks can be made with regard to the y- and z-components, since all components of the vector momentum are conserved. Far from reducing the current to zero, an electron–electron collision leaves the current the two electrons carry unchanged.

Very well, then, how about the possibility of electrons undergoing collisions with the atoms of the crystal lattice? So far as *classical mechanics* is concerned, there is no reason at all why such collisions

should not occur and should not act to limit the drift velocity of the electron distribution. But strangely enough, the laws of *quantum mechanics*, which, after all, are the final arbiters of how elementary particles such as electrons behave, predict that an electron in a *perfectly periodic crystal* does not collide with the atoms of the lattice at all but moves very much like a free particle in a vacuum! At this point, it appears that there isn't anything around that an electron can possibly collide with—but this is not quite the case.

The crux of the matter resides in the words "perfectly periodic crystal." Although an electron does not undergo collision interactions in a perfectly periodic crystal lattice, such collisions may occur *in a crystal lattice in which there is any departure from perfect periodicity*. In actuality, departures from perfect periodicity can occur in two important ways. First of all, in a crystal, the atoms *vibrate* about their equilibrium positions as a result of their thermal energy. The thermal vibrations of the lattice result in deviations from perfect periodicity. If we were to take a snapshot of the atoms at some given time, we would find them not at the perfectly periodically spaced equilibrium positions but randomly displaced just a bit from these positions as a result of their thermally impelled vibrations. The higher the temperature, the larger these vibrational displacements will be, on the average, and the more the lattice of the substance will deviate from perfect periodicity. These thermal deviations from periodicity give rise to the possibility of occasional collisions of the free electrons with the crystal lattice, and these electron–lattice interactions are the "collisions" that play the dominant role in limiting the drift velocity in many instances, particularly in pure crystals and at higher values of temperature. In pure metallic substances, the fact that electron–lattice collisions occur more frequently as the temperature increases means that the average free time $\bar{\tau}_c$ will become smaller with increasing temperatures. As a result, the electrical conductivity, as given by Eq. (18.4.14), decreases as the temperature is increased.

In addition, the presence of foreign impurity atoms in an otherwise perfect crystal will also result in a deviation from perfect periodicity in the neighborhood of such an atom. This also gives rise to the possibility of electron–impurity atom collisions which may limit the drift velocity of the electron distribution. The average time between such collisions varies inversely with the impurity concentration, high concentrations of impurity atoms producing small values of $\bar{\tau}_c$ and low electrical conductivity, and vice versa. Electron–impurity interactions are the dominant ones in impure crystals, particularly at low temperatures when thermal vibrations are weak. Such interactions are responsible for the fact that alloys usually have much lower electrical conductivity than the pure metallic elements of which they are composed. Copper manufacturers, for example, must take great care to produce extremely pure copper for use in electrical conductors, because the presence of even small concentrations of impurities promotes electron–impurity interactions that lower the electrical conductivity.

We have emphasized Ohm's law and described in detail the physical basis of electrical conductivity in conductors in which Ohm's law is obeyed because this type of electrical conduction is most frequently observed in common conducting materials. There are, however, certain situations in which Ohm's law is not obeyed and in which, therefore, the relation between current density and electric field intensity is not linear. The simplest of these are found in electronic vacuum tubes, semiconductor rectifiers, and transistors. A less commonly encountered example of *non-Ohmic* behavior is represented by *superconductivity*, which is a property exhibited by many metals and compounds at very low temperatures. A superconducting substance exhibits an electrical conductivity that, so far as we now know, is truly infinite, in that a steady electrical current may be made to flow in a superconducting loop even though there are no emfs in the circuit and the electric field is zero. Such a current, once established, will persist for days, weeks, or months—in fact, until the refrigerating system gives out! Unfortunately, however, the highest temperature at which superconductivity has been observed is about $23°K$.

We shall not try to explain the physical reasons for the non-Ohmic behavior of substances and systems such as those described above; they are somewhat beyond the scope of this book and incidental to its objectives. It is important, however, to understand that Ohm's law is not universally obeyed but is restricted to a particular, though familiar, range of substances and conditions.

18.5 Electrical Resistance: Ohm's Law for Circuits

Consider the situation illustrated in Fig. 18.13, where a steady direct current flows through a conductor of uniform cross-sectional area and constant conductivity σ. Under these circumstances, there will be a constant electric field $\mathbf{E}$ in the conductor. Assuming that the conduction process within the substance is adequately described by Ohm's law, the relation between current density and electric field is given by

$$\mathbf{j} = \sigma \mathbf{E} \tag{18.5.1}$$

According to (16.5.22), the potential difference dV across any infinitesimally short segment of length dl, which we can describe vectorially as a vector $d\mathbf{l}$ in the direction of the conductor, will be

$$dV = -\mathbf{E} \cdot d\mathbf{l} = -E\,dl \tag{18.5.2}$$

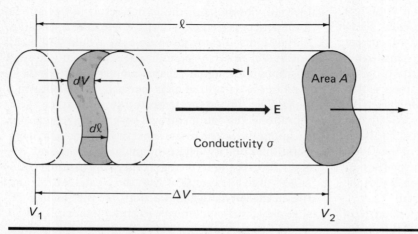

FIGURE 18.13. Flow of current and charge in a conductor of uniform cross section.

In this equation, the dot product has the value $-E\,dl$ because the vectors $\mathbf{E}$ and $d\mathbf{l}$ are parallel. Equation (18.5.2) may be integrated along the finite segment of conductor shown in Fig. 18.13. In writing the integral, the field, which is constant, can be written outside the integral sign:

$$\int_{V_1}^{V_2} dV = V_2 - V_1 = -E \int dl \tag{18.5.3}$$

or

$$\Delta V = -El \tag{18.5.4}$$

where the difference in potential between the ends of the segment, $V_2 - V_1$, is written as ΔV. The minus sign in (18.5.4) reflects the fact that the potential change from V_1 to V_2 is a potential drop rather than a potential rise. In the material which follows, we shall be interested only in the relative magnitudes of the current and potential difference and not in their signs. We shall, therefore, ignore the minus sign, simply taking it for granted that any potential change along the current flow direction through any conductor is a *drop*. In this way, substituting the value of field $E = \Delta V/l$ given by (18.5.1), we may write Ohm's law for a uniform conductor as

$$j = \frac{I}{A} = \sigma \frac{\Delta V}{l} \tag{18.5.5}$$

or, solving for I,

$$I = \frac{\sigma A}{l} \Delta V \tag{18.5.6}$$

This can be written in a slightly more familiar form by substituting for σ the value $1/\rho$, where ρ is the electrical resistivity,[3] and stating (18.5.6) as

$$I = \frac{A}{\rho l} \Delta V = \frac{\Delta V}{\frac{\rho l}{A}} \tag{18.5.7}$$

or

$$I = \frac{\Delta V}{R} \tag{18.5.8}$$

[3] It is important to observe that the *same symbol* ρ is used for *electric charge density* and *resistivity*. This is unfortunate, but since this notation is used almost universally, we have had to adopt it in this text. Usually, the meaning of the symbol ρ is clear enough in the context in which it is used; but to avoid confusion, we have attempted to explain what it means whenever there is the slightest possibility of misunderstanding. In this chapter, ρ is used for charge density in section 18.2 and for resistivity in all subsequent sections It is particularly important to understand that the symbol ρ in Eqs. (18.2.15) and (18.4.2) refers to different—repeat, *different*—quantities. In the former, it is charge density, in the latter, resistivity!

where

$$R = \frac{\rho l}{A} = \frac{l}{\sigma A} \tag{18.5.9}$$

The quantity R is called the *electrical resistance* of the segment of conducting material in question. In view of (18.5.9), the resistance of any conductor of uniform cross section and constitution will be directly proportional to its length and inversely proportional to its area. Equation (18.5.8), which expresses Ohm's law in terms of the current, resistance, and potential drop, is a relation that will be familiar to electricians and electronics technicians. While we shall have frequent occasion to use Ohm's law in this form, we should recognize that the form (18.5.1) is superior in some ways, particularly in that the conductivity is expressed as a fundamental atomic property of the conducting material and is independent of the size or shape of any sample that may be fashioned from it.

In view of the fact that (18.5.8) may be written as $R = \Delta V/I$, the electrical resistance of a conductor has the units of volts/ampere, or volt-seconds/coulomb. These units are universally referred to as *ohms* (abbreviated Ω), and resistance values are always given in such units. Furthermore, since from (18.5.9) $\rho = AR/l$ and $\sigma = 1/\rho$, the units of resistivity can be (and usually are) expressed as ohm-meters or ohm-cm, and those of conductivity as $\text{ohm}^{-1}\,\text{meter}^{-1}$ or $\text{ohm}^{-1}\,\text{cm}^{-1}$. From (18.5.8), it is clear that in a conductor whose resistance is 1 ohm, a potential difference of 1 volt causes a current of 1 ampere to flow. The same reasoning applied to (18.5.9) should convince us that the resistivity of a material in ohm-meters is numerically equal to the resistance in ohms between opposite faces of a sample in the shape of a cube 1 meter on a side; in the same way, the resistivity in ohm-cm is numerically equal to the resistance in ohms between opposite faces of a 1-cm cube.

In a conducting object of irregular shape, such as the one shown in Fig. 18.14, it is not usually possible even to define a unique cross-sectional area or length nor calculate its resistance from Eq. (18.5.9). Nevertheless, if the object is fashioned from a substance in which Ohm's law is obeyed, the current density at all points within the substance will be strictly proportional to the local value of the electric field. In the sample shown in Fig. 18.14, this means that if the externally applied potential difference ΔV is doubled, the electric field inside becomes twice as large as it was before, the current density is everywhere double its former magnitude, and as a result the total current through the sample is doubled. The current flowing through such a substance, then, is still strictly proportional to the applied potential difference, even though Eq. (18.5.9), which was developed for conductors of uniform cross

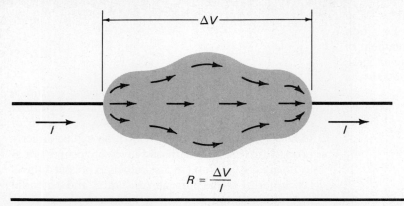

$$R = \frac{\Delta V}{I}$$

FIGURE 18.14. Current flow in a conductor of irregular shape.

section, is no longer applicable in calculating the resistance. In such instances, the resistance may be defined, and measured, as the *constant ratio* of the externally applied potential difference to the current. In other words, we *define* the resistance as the value of the constant ratio

$$R = \frac{\Delta V}{I} \qquad (18.5.10)$$

in accord with (18.5.8), which, strictly speaking, was developed for cases of uniform cross section.

EXAMPLE 18.5.1
The lump of Ohmic conducting material shown in Fig. 18.14 carries a current of 3.55 amperes when a potential difference of 75 volts is established across the two terminals. Find the electrical resistance of the sample.

 This is the simplest possible example of the application of Ohm's law. It is necessary only to insert the given values of current and voltage into (18.5.10):

$$R = \frac{\Delta V}{I} = \frac{75}{3.55} = 21.1 \text{ ohms}$$

In this example, it is evident that it would be very difficult to calculate the resistance of the sample even if its conductivity and dimensions are accurately known. In spite of this, we see that the resistance can be simply and accurately *measured* merely by determining the ratio of current to potential difference.

EXAMPLE 18.5.2
A sample of seawater is poured into a conductivity measuring cell such as that shown in Fig. 18.15. The square end surfaces of the cell, of dimensions 5.00 cm × 5.00 cm, are highly conducting plates, but the rest of the cell is made of glass. The length of the cell is 30.0 cm. A battery is connected to the terminals of the cell as shown, and a voltmeter and ammeter are in-

serted into the circuit to measure the current in the circuit and the potential difference across the sea water. It is found that the ammeter indicates a current of 0.730 ampere, while the voltmeter indicates a potential difference of 12.00 volts. Assuming that the sea water is an Ohmic conducting substance, find the resistivity of sea water. Express your answer in both cgs and mks units.

 The resistance of the seawater is obtained as in the preceding example:

$$R = \frac{\Delta V}{I} = \frac{12.00}{0.730} = 16.43 \text{ ohms}$$

For an Ohmic conductor of uniform cross section, such as the column of seawater in the cell, the resistivity is related to the resistance, length, and cross-sectional area by (18.5.9):

$$\rho = \frac{A}{l} R = \frac{(5.00)^2}{(30.0)} (16.43) = 13.69 \text{ ohm-cm}$$

To express this result in mks units of ohm-m, we note that

$$1 \text{ ohm-cm} = 1 \text{ ohm-cm} \left(\frac{1}{100}\right) \frac{m}{cm} = 0.01 \text{ ohm-m}$$

whence

$$\rho = (0.01)(13.69) = 0.1369 \text{ ohm-m}$$

EXAMPLE 18.5.3
A coil of copper wire 0.010 inch in diameter consisting of 300 turns has a mean diameter of 0.50 inch. The resistivity of copper at 20°C is 1.692×10^{-6} ohm-cm. Find the resistance of the coil at 20°C.

 The resistivity and resistance are related by

$$R = \frac{\rho l}{A}$$

In this case, if r is the radius of the wire's circular cross section, the area is

$$A = \pi r^2 = (3.1416)(0.0127)^2 = 5.067 \times 10^{-4} \text{ cm}^2$$

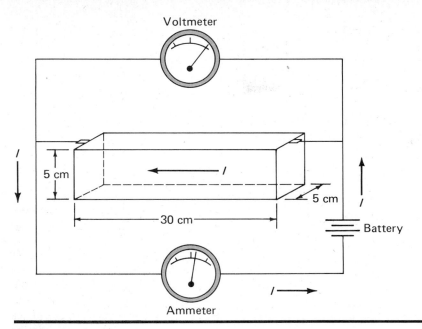

FIGURE 18.15

while, if r' is the mean radius of the coil and n the number of turns, the length will be

$$l = 2\pi r'n = (2)(3.1416)(0.25)(2.54)(300) = 1197 \text{ cm}$$

Substituting these values along with the given resistivity into the above expression for the resistance, one obtains $R = 3.997$ ohms.

EXAMPLE 18.5.4

The temperature of the coil described in the preceding example is raised to 100°C, and its resistance at that temperature is found to be 5.536 ohms. Assuming that the resistivity of copper varies linearly with the absolute temperature between 0°C and 100°C, find the *temperature coefficient of resistivity* that expresses the fractional change of resistivity per unit temperature change in this range of temperatures.

The fractional change in resistivity, caused by a temperature change ΔT, will be $\Delta\rho/\rho$; therefore, the temperature coefficient of resistivity δ will be the ratio of $\Delta\rho/\rho$ to ΔT, or

$$\delta = \frac{1}{\rho}\frac{\Delta\rho}{\Delta T} \qquad (18.5.11)$$

If the change in resistivity with temperature is linear, then $\Delta\rho$ will be simply a constant times ΔT, and δ will then be constant. In the statement of the example, we are given the change in resistance rather than the change in resistivity. From (18.5.9), however, since l and A are constant, $\Delta R = (l/A)\,\Delta\rho$; therefore,

$$\frac{\Delta R}{R} = \frac{\Delta\rho}{\rho} \qquad (18.5.12)$$

Substituting this result into (18.5.11), we obtain

$$\delta = \frac{1}{R}\frac{\Delta R}{\Delta T} \qquad (18.5.13)$$

Inserting the numerical values $\Delta R = 5.536 - 3.997 = 1.539$ ohm, $R = 3.997$ ohms, $\Delta T = 373 - 273 = 100°\text{K}$ we find that $\delta = 0.00385°\text{K}^{-1}$.

18.6 Resistances and EMFs in Series and in Parallel: Equivalent Resistance of Networks

When a number of ideal emfs are connected in *series*, as shown in Fig. 18.16, terminals such as those designated A and B in the figure must, when joined, have the same final potential. Since a charge, in passing through the cells in series, will experience a gain in potential energy equal to the sum of the potential energy gain associated with each individual cell, the potential rise associated with the array, ΔV, must be equal to the sum of the individual cell potential rises. Then, for emfs in series,

$$\Delta V = \Delta V_1 + \Delta V_2 + \Delta V_3 + \cdots + \Delta V_n \qquad (18.6.1)$$

from which, according to (18.3.6),

$$\boxed{\mathscr{E} = \mathscr{E}_1 + \mathscr{E}_2 + \mathscr{E}_3 + \cdots + \mathscr{E}_n} \qquad (18.6.2)$$

When *identical* emfs are connected in *parallel*, as shown in Fig. 18.17 the negative terminals are all at a common potential, and the positive terminals are likewise at a common potential that is higher by the

581

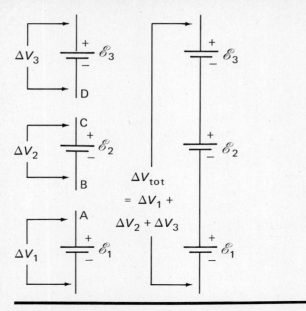

FIGURE 18.16. Sources of emf in series.

cell potential rise ΔV_0, whence, for *identical* cells in parallel,

$$\mathscr{E} = \Delta V_0 \qquad (18.6.3)$$

When cells whose emfs are not identical (or nearly identical) are connected in parallel, very large circulating currents will flow, and the observed potential differences will depend critically upon the resistance of the circuit conductors and the internal resistances of the cells, in a way that is not always simple to calculate. This is a good way to ruin batteries, as we shall

see in one of the examples at the end of this section.

Resistances, which we denote in circuit diagrams by the symbol ⌇⌇⌇, may also be combined in series and parallel combinations. When arranged in series, as shown in Fig. 18.18, the same current I flows through all the resistors. If this current is caused by an externally applied potential difference ΔV across the array of resistors, then it is clear from the figure that for n resistors in series,

$$\Delta V = \Delta V_1 + \Delta V_2 + \Delta V_3 + \cdots + \Delta V_n \qquad (18.6.4)$$

But if the resistances are ohmic in behavior, then, since the current through each of them has the common value I,

$$\Delta V_1 = IR_1$$
$$\Delta V_2 = IR_2$$
$$\Delta V_3 = IR_3 \qquad (18.6.5)$$
$$\vdots \qquad \vdots$$
$$\Delta V_n = IR_n$$

Adding these equations together, using (18.6.4), and noting that the equivalent resistance R is equal to the product of the current and the total potential drop across the array, we may write

$$\Delta V = I(R_1 + R_2 + R_3 + \cdots + R_n) = IR \qquad (18.6.6)$$

from which,

$$R = R_1 + R_2 + R_3 + \cdots + R_n \qquad (18.6.7)$$

The equivalent resistance of a series array of resistances is equal, then, to the sum of the individual resistances.

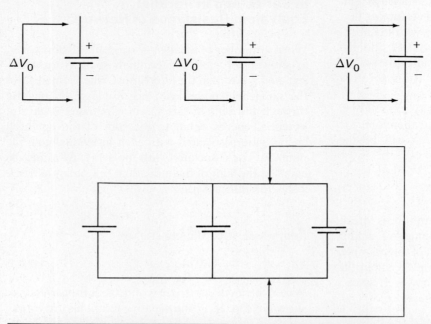

FIGURE 18.17. Sources of emf in parallel.

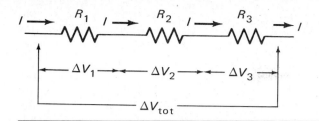

FIGURE 18.18. Resistances in series

In the case of a number of resistors connected in parallel, as shown in Fig. 18.19, the potential difference across all the resistors has the same value ΔV_0. The total current I which flows through the array, however, divides at point A into a number of smaller currents each of which flows through one of the resistors and which reunites with the others at B. For an array of n resistors in parallel, the total current I must always equal the sum of these individual currents:

$$I = I_1 + I_2 + I_3 + \cdots + I_n \qquad (18.6.8)$$

If the resistances obey Ohm's law, however, the current and potential drop associated with each can be written as

$$I_1 = \Delta V_0/R_1$$
$$I_2 = \Delta V_0/R_2$$
$$I_3 = \Delta V_0/R_3 \qquad (18.6.9)$$
$$\vdots \qquad \vdots$$
$$I_n = \Delta V_0/R_n$$

since the potential drop across each has the same value ΔV_0. If we add all these equations together, using (18.6.4) and noting that the total current must equal the potential drop ΔV_0 across the array divided by its equivalent resistance R, we obtain

$$I = \frac{\Delta V_0}{R} = \Delta V_0 \left(\frac{1}{R_1} + \frac{1}{R_2} + \frac{1}{R_3} + \cdots + \frac{1}{R_n} \right) \qquad (18.6.10)$$

from which

$$\boxed{\frac{1}{R} = \frac{1}{R_1} + \frac{1}{R_2} + \frac{1}{R_3} + \cdots + \frac{1}{R_n}} \qquad (18.6.11)$$

The reciprocal of the equivalent resistance is, therefore, equal to the sum of the reciprocals of the individual resistances.

For resistances in series, the equivalent resistance is always greater than that of any of the individual resistors, while for resistances in parallel, the equivalent resistance will always be smaller than any of the individual resistances. In the latter instance, alternative current paths allow more current to flow through the array than through any one resistor. Resistances may, of course, be connected into arrays that are more complex than simple series or parallel com-

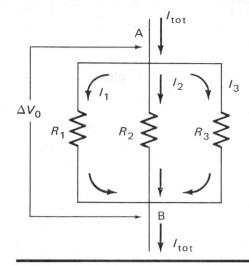

FIGURE 18.19. Resistances in parallel.

binations. Frequently, though not always, these may be reduced to a simple series or parallel arrangement by a step-by-step process, in which individual parts of the array are identified as series or parallel combinations and replaced by their equivalent resistances as given by (18.6.7) or (18.6.11). A repetition of this process is then possible, and eventually an equivalent resistance for the entire combination can be found. Some examples of this technique are illustrated in the series of examples which follows. For more complex networks, we must resort to a more general scheme of circuit analysis which will be developed in a later section.

In many of our discussions of the behavior of electrical circuits, we shall find it convenient to frame our calculations in terms of *ideal* circuit elements. By this term we mean resistances that are pure resistance and have no capacitance or other property that might alter their behavior in any way, and sources of emf that do *nothing* but maintain a given constant potential difference across their terminals. Such elements are usually regarded as being connected by conducting wires that are, likewise, assumed to contribute no resistance to the passage of current. The symbols included in the circuit diagrams shown here are ordinarily understood to represent "ideal" circuit elements of this kind, unless some specification to the contrary is made.

Actually, there are no truly ideal circuit elements. Connecting leads may have very little resistance compared to other resistive elements in the circuit, but their actual resistance is not zero. Resistances invariably have some capacitance associated with them, even though its effects may be for most purposes negligible. Finally, and most important at this point, emfs may not be ideal devices that do nothing but maintain a given difference in potential between their terminals, but may exhibit terminal potential differences that depend upon the physical and chemical processes that occur within them and, in particular, upon the amount of current they are required to deliver to the external circuit. If a battery, for instance, were an "ideal emf," we should expect it to maintain its stated terminal potential difference regardless of the resistance of the external circuit it is connected to. In principle, then, if we were to short-circuit its terminals with an Ohmic conductor of negligible resistance, we should expect it to deliver arbitrarily large currents. This, we know, does not happen in practice, because of dissipative processes that occur *within the source of emf itself*, which convert electrical energy into heat. After all, in a battery, current must flow through electrolytes and through other substances in which the same collision processes operate that act in Ohmic conductors to terminate the free path of electrons and to convert their kinetic energy to heat. As a result, the terminal potential difference of any real source of emf will be found to decrease as the amount of current furnished by the source increases and, in fact, drop to zero at some finite value of current. This latter condition is realized when the terminals of the source of emf are short-circuited, and the limiting current is sometimes referred to as the *short-circuit current* of the source. It is as if the real source of emf, as well as generating a potential rise offers an *internal resistance* to the passage of the current through it. The real source of emf can, indeed, be said to behave very much like an ideal emf connected in series with an ideal resistance within the device itself.

Such an equivalent representation of a real emf, in the form of a battery, is illustrated in Fig. 18.20. In the diagram, R_i represents the internal resistance of the battery and $\mathscr{E}$ its "ideal" terminal potential difference when no current is being drawn. The dotted lines represent schematically the boundaries of the

source—the battery case, so to speak. When the current I is zero, of course, the difference in potential between the terminals A and B is equal to the emf $\mathscr{E}$. When a current I flows, the potential energy of a charge decreases by the amount $q \, \Delta V_i$ as it traverses the "internal resistance" R_i. The difference in potential across R_i in the direction of current flow is the change in potential energy per unit charge, or $-\Delta V_i$. The net rise in potential between the terminals A and B is then given by

$$\Delta V_{AB} = \mathscr{E} - \Delta V_i \tag{18.6.12}$$

If the internal resistance behaves Ohmically, then $\Delta V_i = IR_i$, and (18.6.12) can be written

$$\Delta V_{AB} = \mathscr{E} - IR_i \tag{18.6.13}$$

When the current is so large that IR_i equals $\mathscr{E}$, the terminal potential difference ΔV_{AB} is reduced to zero. This is the condition established when an external short circuit between A and B exists. Now the current that flows is the short-circuit current whose value is given by

$$I_{SC} = \frac{\mathscr{E}}{R_i} \tag{18.6.14}$$

This is the largest current the source can ever deliver to an external circuit.

The physical model suggested by Eq. (18.6.12) and by Fig. 18.20 is oversimplified to the extent that the internal resistance of real sources is sometimes not Ohmic in character. In the case of batteries, the approximation of an Ohmic internal resistance is frequently quite a good one, but there are other important sources of electromotive potential difference, solid-state photocells, for example, where it is not. In these instances, a more complex physical model of the source of emf may have to be constructed.

EXAMPLE 18.6.1

A prominent and influential attorney, who is a member of a number of public commissions charged with formulating legislation on environmental pollution and on electrical energy generation, conversion, and distribution, has two automobiles, a new Cadillac and an old Volkswagen. The VW is used primarily by his wife for local errands. One morning, the VW refuses to start, its lights having been inadvertently left on overnight and its battery (of emf 6.3 volts and internal resistance 0.02 ohm) completely discharged. The lawyer's technical expertise informs him that though the 6-volt battery of the VW could not possibly be used to start the Cadillac, the Cadillac's 12.6 volt battery (internal resistance 0.03 ohm), should be just the thing to start the VW.

So he drives the Cadillac alongside the VW, extricates a set of connecting cables from the trunk,

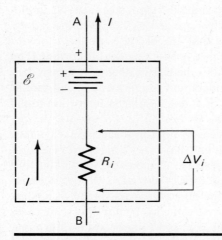

FIGURE 18.20.   Effect of internal resistance in a source of emf.

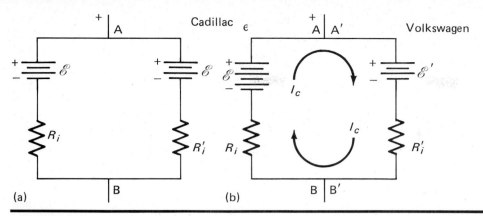

FIGURE 18.21

connects the two batteries positive to positive, negative to negative (just as the instructions indicate), gets into the VW, reaches for the ignition key—and is interrupted to take a long-distance telephone call, which keeps him occupied for half an hour. Upon returning, he starts the VW with no difficulty, puts the cables away, and gets into the Cadillac, congratulating himself surreptitiously on his mastery of elementary electrotechnology. But turning the key of the Cadillac, he finds that *it* now refuses to start! Puzzled and irritated, he calls the local garage to come and start the Cadillac, for, as everyone knows, you can't start a 12-volt car with a 6-volt battery! That evening, climaxing the events of the day, it was discovered that the poor VW's headlights were burned out. We are required to explain the odd circumstances of that episode involving the two cars.

There is no particular reason to avoid combining batteries or other sources of emf in parallel, as shown in Fig. 18.21a, so long as their emfs are identical or nearly identical. In that case, the zero-current terminal potential difference ΔV_{AB} will be equal to the common value of emf, and the internal resistance of the parallel combination will be lower than that of either cell, since the two internal resistances are now in parallel. When the emfs of the two cells, $\mathscr{E}$ and $\mathscr{E}'$, are significantly different, however, as illustrated in Fig. 18.21b, the larger emf $\mathscr{E}$ forces a *circulating current I* to flow around the loop and through the smaller emf $\mathscr{E}'$ in the *reverse* direction, from negative to positive! This is what happens in the case where the 12-volt battery of the Cadillac is connected in parallel with the 6-volt battery of the Volkswagen, and it is this current that is responsible for charging the battery of the VW and discharging the battery of the Cadillac.

We may calculate the size of this circulating current by reasoning as follows: since the points A and A' are really identical, as are the points B and B', the potential differences ΔV_{AB} associated with the left side of the loop and $\Delta V_{A'B'}$ associated with the right side must be equal. From Eq. (18.6.13), however, we know that ΔV_{AB} must be given by

$$\Delta V_{AB} = \mathscr{E} - I_c R_i \qquad (18.6.15)$$

In the case of $\Delta V_{A'B'}$, however, the current flows through the internal resistance in the direction *opposite* to that shown in Fig. 18.20, and, therefore, the potential difference ΔV_i of (18.6.12) across this resistance is *opposite in sign* to the potential differences in (18.6.12) and (18.6.13). Going from B' to A', the potential difference across R_i is a potential rise rather than a potential drop. The difference in potential between B' and A' in the right side of the loop must, therefore, be

$$\Delta V_{A'B'} = \mathscr{E}' + I_c R_i' \qquad (18.6.16)$$

But, as we pointed out previously, ΔV_{AB} and $\Delta V_{A'B'}$ must be equal. Equating the right sides of (18.6.15) and (18.6.16), therefore, and solving for I_c,

$$I_c = \frac{\mathscr{E} - \mathscr{E}'}{R_i + R_i'} \qquad (18.6.17)$$

In the case involving our prominent friend, $\mathscr{E} = 12.6$ volts, $\mathscr{E}' = 6.3$ volts, $R_i' = 0.02$ ohm, and $R_i = 0.03$ ohm. Inserting these numerical values into (18.6.17), we find the circulating current to be 126 amperes. This current, furnished by the emf of the Cadillac's battery, is sufficient, in a half-hour, to discharge the battery to the point where it will no longer start the car.[4] At the same time, lead–acid cells, after all, being reversible, the circulating current, flowing in reverse through the battery of the Volkswagon, is sufficient to charge it to the point where it will start the car. Had the cells been irreversible, like dry-cell flashlight batteries, the only observable effect would

---

[4] In time, of course, the resistances R_i and R_i' change owing to the discharge of one battery and the charging of the other. As a result, the circulating current I_c will decrease as the Cadillac's battery goes flat; this effect is, however, incidental to the point of the example.

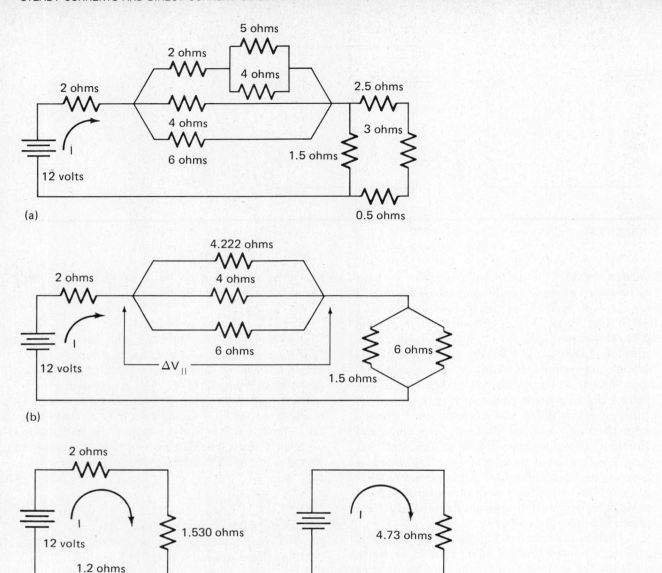

FIGURE 18.22

have been the discharge of the battery of higher emf.

The common value of ΔV_{AB} and $\Delta V_{A'B'}$ can be calculated by inserting the given numerical values of the emfs and resistances, along with the value $I_c = 126$ amperes, into (18.6.15) or (18.6.16). In this manner, it is easily ascertained that $\Delta V_{AB} = \Delta V_{A'B'} = 8.82$ volts. This potential difference, applied across the headlights of the VW, our poor victim having forgotten to turn off the headlight switch before connecting up his cables, is what is responsible for burning out the 6.3-volt headlamps of the VW.[5] The moral of the story has been stated by Alexander Pope as follows:

[5] Actually, what happened is that the headlight fuse blew, but the garage man charged our hero with a new pair of headlights anyway.

A little learning is a dangerous thing.
Drink deep, or taste not the Pierian spring.
There, shallow drafts intoxicate the brain,
While drinking largely sobers us again.

Essay on Man.

EXAMPLE 18.6.2

Find the current that flows in the circuit shown in Fig. 18.22a.

This circuit consists of a complex network of resistances, but, by using the equations developed above for the equivalent resistance of series and parallel arrays, it can be reduced to a single equivalent resistance, as illustrated in Figs. 18.22b, c, and d. At Fig. 18.22b, for example, the parallel combination of a 4-ohm and a 5-ohm resistance in the topmost branch of the original circuit has been replaced by its equivalent resistance, which, according to (18.6.11), is

$$\frac{1}{R} = \frac{1}{R_1} + \frac{1}{R_2} = \frac{1}{4} + \frac{1}{5} = \frac{9}{20} \text{ ohm}^{-1}$$

$$R = \frac{20}{9} = 2.222 \text{ ohms}$$

This resistance is, of course, in series with a 2-ohm resistor, and, therefore, according to (18.6.7), the total resistance in the top branch of that circuit is 2.222 + 2 = 4.222 ohms, as shown. At the same time, it is recognized that the 2.5-ohm, 3-ohm, and 0.5-ohm resistors in the right-hand branch are in series and can, according to (18.6.7), be replaced with a single resistance of 6 ohms, which then is in parallel with the 1.5-ohm resistor. In Fig. 18.22a, the equivalent resistance of the three parallel resistors in Fig. 18.22b has been replaced with an equivalent resistance of 1.530 ohms, calculated according to

$$\frac{1}{R} = \frac{1}{R_1} + \frac{1}{R_2} + \frac{1}{R_3} = \frac{1}{4.222} + \frac{1}{4} + \frac{1}{6} = 0.6536 \text{ ohm}^{-1}$$

from which

$$R = \frac{1}{0.6536} = 1.530 \text{ ohm}$$

In the same way, the two parallel resistances in Fig. 18.22b are replaced with an equivalent resistance of 1.2 ohms, calculated in the same way. There are now just three series resistances, of 2 ohms, 1.530 ohms, and 1.2 ohms. These are, by (18.6.7), equivalent to a single resistance of 4.73 ohms. The potential drop ΔV across this resistance is the emf of 12 volts; and, therefore, according to Ohm's law, the current must be

$$I = \frac{\Delta V}{R} = \frac{12}{4.73} = 2.54 \text{ amp}$$

The currents flowing through the individual resistances in the parallel arrays are found by first finding the potential drop across the array and then the individual currents by the use of Ohm's law. Thus, for the three branches of resistance 4.222 ohms, 4 ohms, and 6 ohms in Fig. 18.22b, the equivalent resistance of the array is 1.530 ohms. The total current of 2.54 amperes flows through the array, so the potential drop across it must be

$$\Delta V = IR = (2.54)(1.530) = 3.886 \text{ volts}$$

The current in the 4.222-ohm, 4-ohm, and 6-ohm branches must then be

$$I_1 = \frac{\Delta V}{R_1} = \frac{3.886}{4.222} = 0.920 \text{ amp}$$

$$I_2 = \frac{\Delta V}{R_2} = \frac{3.886}{4.00} = 0.972 \text{ amp}$$

$$I_3 = \frac{\Delta V}{R_3} = \frac{3.886}{6.00} = 0.648 \text{ amp}$$

since the potential drop ΔV appears across all three resistances. The currents through the other branches of the circuit can be calculated in a precisely similar fashion, the details being left to the student as an exercise.

There are many networks whose equivalent resistance can be determined in this way, simply by replacing series and parallel combinations by their equivalent resistances successively until only a single equivalent resistance is left. There are others, however, that cannot be resolved by this technique. In such instances, we must resort to a more general scheme of circuit analysis that will be presented in section 18.8.

18.7 Energy and Power in DC Circuits

We have already seen, in section 18.2, that an emf $\mathscr{E}$ that furnishes a current I to an external circuit supplies power to that circuit at a rate given by

$$P = \mathscr{E}I \tag{18.7.1}$$

In the external circuit, whenever a charge Δq flows through a resistance R, it undergoes a loss of potential energy $\Delta q\,\Delta V$, where ΔV is the potential drop across the resistance. This potential energy is expended in creating and maintaining the drift kinetic energy of the free electrons in the substance of which the resistor is composed. The drift kinetic energy, in turn, is lost in collisions with the crystal lattice that turn the kinetic energy of the electrons into random thermal vibrations—that is, into heat.[6] If potential energy ΔU_p is consumed in a time interval Δt, then the electrical power dissipated by the resistance in the form of internal energy will be

$$P = \frac{\Delta U_p}{\Delta t} = \frac{\Delta q\,\Delta V}{\Delta t} = I\,\Delta V \tag{18.7.2}$$

since $I = \Delta q/\Delta t$. The total amount of electrical energy consumed, thus the total internal energy generated, will be

$$\Delta U_p = \Delta Q = P\,\Delta t = I\,\Delta V\,\Delta t \tag{18.7.3}$$

The conversion of electrical energy supplied by a source of emf into heat by the passage of current

---

[6] Or, to be really precise, into *internal energy*. You may find it instructive in this example to refer back to Chapter 12, section 2.

587

through a resistance is, of course, the basis upon which most familiar electrical heating elements operate. If the resistance is of an Ohmic character, then $\Delta V = IR$, in which case (18.7.2) may be written as

$$P = I^2 R \qquad (18.7.4)$$

Alternatively, in (18.7.2) we may use Ohm's law to express I in terms of ΔV as $I = \Delta V/R$ and write (18.7.2) in still another way as

$$P = \frac{(\Delta V)^2}{R} \qquad (18.7.5)$$

It is important to note, from (18.7.3), that the rate at which electrical energy is converted to heat in an Ohmic resistance is proportional to the *square* of the current flowing through it. This conversion of electrical energy into heat by the passage of a current through an electrical resistance is frequently referred to as *Joule heating*.

EXAMPLE 18.7.1

What is the resistance of a light bulb that consumes 60 watts of electrical power at 117 volts? How much current flows through the bulb under these conditions?

The light bulb consumes 60 watts of power when the potential drop across it is 117 volts. Substituting these values into (18.7.5), we find that

$$60 = \frac{117^2}{R}$$

$$R = \frac{117^2}{60} = 228 \text{ ohms}$$

The current flowing then follows from Ohm's law:

$$I = \frac{\Delta V}{R} = \frac{117}{228} = 0.513 \text{ amp}$$

We could equally well have found I first from (18.7.2) and then computed R from (18.7.3) or (18.7.5).

EXAMPLE 18.7.2

A battery of emf $\mathscr{E}$ and internal resistance R_i is connected as illustrated in Fig. 18.23 to an external circuit of resistance R, which is adjustable. How should the external resistance be adjusted so that the largest possible amount of power be dissipated in the external circuit? What is this value of maximum power dissipation?

In Fig. 18.23, the emf $\mathscr{E}$ establishes a potential rise of $\mathscr{E}$ volts between A' and B. This sets up a current I which flows through the external circuit producing a potential drop $\Delta V = IR$ across the external resistance R and an internal potential drop $\Delta V_i = IR_i$ due to the

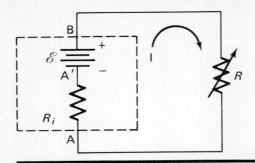

FIGURE 18.23

internal resistance of the emf. Clearly, the sum of these two potential drops around the path BAA' is exactly equal to the potential rise from A' to B. Were it not so, a charge passing around the circuit would experience a net rise or fall in potential every time it went around, and the system could not possibly be in a steady state. Under these circumstances, we may write

$$\mathscr{E} = \Delta V + \Delta V_i = I(R + R_i) \qquad (18.7.6)$$

from which

$$I = \frac{\mathscr{E}}{R + R_i} \qquad (18.7.7)$$

The power dissipated as heat in the external resistance R, according to (18.7.3), will be

$$P_{\text{ext}} = I^2 R = \frac{\mathscr{E}^2 R}{(R + R_i)^2} \qquad (18.7.8)$$

To find the value of R for which P_{ext} is a maximum we must find the derivative of P_{ext} with respect to R and set it equal to zero. In this way, we obtain

$$\frac{dP_{\text{ext}}}{dR} = \mathscr{E}^2 \frac{d}{dR} \left[R(R + R_i)^{-2} \right]$$

$$= \mathscr{E}^2 \left[(R + R_i)^{-2} - 2R(R + R_i)^{-3} \right]$$

$$= \mathscr{E}^2 \left(\frac{1}{(R + R_i)^2} - \frac{2R}{(R + R_i)^3} \right)$$

$$= \mathscr{E}^2 \frac{R + R_i - 2R}{(R + R_i)^3}$$

or

$$\frac{dP_{\text{ext}}}{dR} = \mathscr{E}^2 \frac{R_i - R}{(R + R_i)^3} \qquad (18.7.9)$$

The above expression for dP_{ext}/dR can be zero only if the numerator vanishes, that is, only if

$$R = R_i \qquad (18.7.10)$$

Maximum power is, therefore, dissipated in the external circuit *when the external resistance R equals the internal resistance of the source.*

This is the simplest possible example of quite a general principle called *impedance matching*. We are required to attach our 8-ohm stereo speakers to the "8-ohm" output terminal of the amplifier because the effective internal resistance of the amplifier between those terminals is equal to the speaker resistance, ensuring that a maximum amount of signal power is dissipated in the external circuit. The power dissipated under these conditions can be found by setting R equal to R_i in (18.7.8), from which

$$P_{ext}^{(max)} = \frac{\mathscr{E}^2}{4R_i} \qquad (18.7.11)$$

The power dissipated within the source of emf is $I^2 R_i$, and since I is given by (18.7.7),

$$P_i = I^2 R_i = \frac{\mathscr{E}^2 R_i}{(R + R_i)^2} \qquad (18.7.12)$$

When $R = R_i$, this tells us that the internal power dissipation is

$$P_i = \frac{\mathscr{E}^2}{4R_i} = P_{ext}^{(max)} \qquad (18.7.13)$$

The potential drop across the external resistance is IR, or, substituting the value of I from (18.7.7),

$$\Delta V = \frac{\mathscr{E} R}{R + R_i} \qquad (18.7.14)$$

Under conditions of maximum external power dissipation, when R is equal to R_i, this becomes

$$\Delta V = \mathscr{E}/2 \qquad (18.7.15)$$

the situation then being that half the potential difference created by the emf appears across the external resistance and half across the internal resistance of the emf itself. A plot of the external power dissipation as a function of R, as given by (18.7.7), is shown in Fig.

18.24. This plot clearly illustrates the maximum at $R = R_i$.

18.8 The Analysis of DC Circuits by Kirchhoff's Laws

So far, we have needed only the definitions of potential, emf, current, and Ohm's law to describe the behavior of the simple electrical circuits we have examined. These simple methods are, however, insufficient for the analysis of more complex networks, for which we shall have to resort to a more general set of principles referred to as *Kirchhoff's laws for dc circuits*. Kirchhoff's laws, named for the German physicist Gustav Kirchhoff (1824–1887), are based on the conservation of charge and current and on the fact that the potential always returns to its original value after a complete traversal of any closed path. Kirchhoff's laws may be stated as follows:

I. The algebraic sum of the currents flowing into any junction of conductors in a circuit is zero.

II. The algebraic sum of potential increases and decreases around any closed loop in the circuit is zero.

Law I may be restated to say that the sum of the currents flowing into any junction point in the circuit equals the sum of the currents flowing out of that point. Stated in this way, it clearly arises from the conservation of charge; if more current enters a junction point than is leaving, there will obviously be a buildup of charge. This cannot happen in a steady-state dc circuit—otherwise there would be no steady state! Kirchhoff's first law may be readily understood by referring to Fig. 18.25a.

Law II is simply a statement of the fact that the electric potential can be uniquely stated at any point in a steady-state circuit. In traversing a closed path,

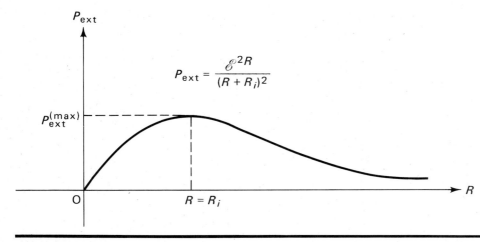

$$P_{ext} = \frac{\mathscr{E}^2 R}{(R + R_i)^2}$$

FIGURE 18.24

589

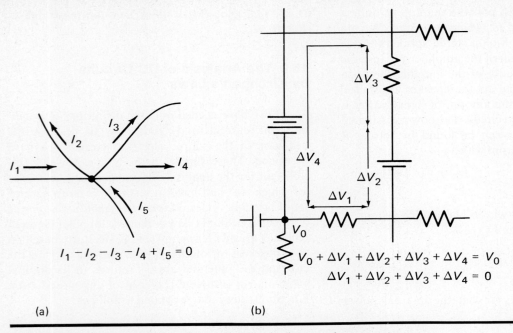

$$I_1 - I_2 - I_3 - I_4 + I_5 = 0$$

$$V_0 + \Delta V_1 + \Delta V_2 + \Delta V_3 + \Delta V_4 = V_0$$

$$\Delta V_1 + \Delta V_2 + \Delta V_3 + \Delta V_4 = 0$$

(a)   (b)

FIGURE 18.25. Kirchhoff's circuit laws. (a) Algebraic sum of currents into any junction of conductors is zero. (b) Algebraic sum of potential differences around any circuit loop is zero.

then, the value of the potential must always return to its initial value upon returning to the starting point. This means that the sum $\Delta V_1 + \Delta V_2 + \Delta V_3 + \Delta V_4$ in Fig. 18.25b must be zero.

The application of Kirchhoff's laws to actual electrical circuits is illustrated by the series of examples that follows. In these and in other problems as well, the directions of current flow in the individual circuit branches will not usually be known with certainty. One must, therefore, assume certain arbitrary directions for current flow in each branch. It is not necessary that the guesses be correct; if they are not, the final solutions will give negative values of current in the instances where the current flow is opposite to that which was assumed.

EXAMPLE 18.8.1

In the case of the Cadillac and Volkswagen outlined in Example 18.6.1, and illustrated in Fig. 18.21b, we neglected the fact that the Volkswagen's lights were on during the events described there. This means that there is an additional resistance R_0 of 0.2 ohm connected across the positive and negative terminals of the two batteries. Find the effect of this resistance upon the quantities calculated in Example 18.6.1, assuming, for the purposes of this calculation, that the light filaments or the lighting fuse do not burn out.

The circuit now looks like the one illustrated in Fig. 18.26. The lighting current I_0 flows through the resistance R_0 that represents the lighting circuit of the Volkswagen. Our previous methods of analysis are

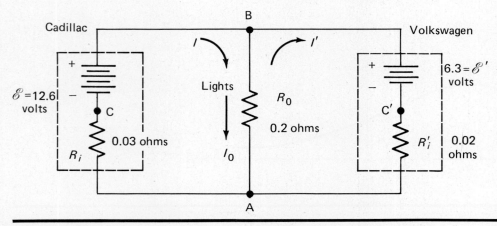

FIGURE 18.26

not directly applicable in this case, but we may now proceed by applying Kirchhoff's laws. Law I applied to the currents flowing into and out of point B tells us that

$$I - I' - I_0 = 0 \qquad (18.8.1)$$

In writing this equation, we assume, arbitrarily, that the currents I, I', and I_0 have the directions shown in Fig. 18.26 and that currents into the junction point B are positive and currents out of the junction are negative. The opposite assumption could equally well have been made, but one or the other must be selected and adhered to consistently throughout.

We may now apply law II to the left-hand loop of the circuit; we shall start at point C and go clockwise around the loop CBAC. In doing so, we sum every rise and fall in potential around the loop and equate the result to zero. From C to B, there is a potential *rise* $\Delta V_{CB} = +\mathscr{E}$ volts. From B to A, there is a potential *fall* due to the passage of the current I_0 through the resistance R_0, of amount $\Delta V_{AB} = -I_0 R_0$; and from A to C, there is a potential fall $\Delta V_{AC} = -IR_i$ due to the passage of the current I through the internal resistance R_i. Adding these three potential changes and equation their sum to zero, we get

$$\Delta V_{CB} + \Delta V_{BA} + \Delta V_{AC} = +\mathscr{E} - I_0 R_0 - IR_i = 0$$
$$(18.8.2)$$

In writing this equation, it is important to note that the current through the branch ACB is I, while that through branch BA is I_0.

In much the same way, we may apply law II to the right-hand loop, starting from point B and going around clockwise. The first potential change we meet, between B and C', is a *fall* in potential as we go through the emf $\mathscr{E}'$ from positive to negative—the "wrong way," so to speak, since the emf tries, unsuccessfully, in this case, to make current flow in the opposite direction. We write, therefore, $\Delta V_{BC'} = -\mathscr{E}'$. From C' to A, there is another *fall* in potential $\Delta V_{C'A} = -I'R_i'$ as the current I' flows through R_i'. From A to B, there is a *rise* in potential across resistance R_0 from the flow of current I_0. Had we been going around the loop counterclockwise, from B to A, the potential change would have been a *fall* associated with the potential *drop* across R_0. As it is, however, we are traversing this part of the loop from A to B, *opposite* the direction of the current I_0. Since the passage of I_0 through R_0 renders the end of R_0 nearest A lower in potential than the end nearest B, the change in potential from A to B is positive, and, therefore, $\Delta V_{AB} = +I_0 R_0$. Adding the potential changes and equating their sum to zero, we get

$$\Delta V_{BC'} + \Delta V_{C'A} + \Delta V_{AB} = -\mathscr{E}' + I_0 R_0 - I'R_i = 0$$
$$(18.8.3)$$

None of the three currents I, I', and I_0 is known,

but Eqs. (18.8.1), (18.8.2), and (18.8.3) give us three equations that can be solved simultaneously for these three quantities. Adding (18.8.2) and (18.8.3), we obtain

$$\mathscr{E} - \mathscr{E}' - IR_i - I'R_i' = 0 \qquad (18.8.4)$$

Substituting the value $I_0 = I - I'$ from (18.8.1) into (18.8.2), we find

$$\mathscr{E} - I(R_0 + R_i) + I'R_0 = 0 \qquad (18.8.5)$$

We may now multiply (18.8.4) by R_0, (18.8.5) by R_i', add the two equations, and solve for I:

$$I = \frac{R_0(\mathscr{E} - \mathscr{E}') + \mathscr{E}R_i'}{R_0 R_i + R_0 R_i' + R_i R_i'} \qquad (18.8.6)$$

Inserting this expression for I into (18.8.4) and solving for I', we may write, after much algebra,

$$I' = \frac{(\mathscr{E} - \mathscr{E}')R_0 - \mathscr{E}'R_i}{R_0 R_i + R_0 R_i' + R_i R_i'} \qquad (18.8.7)$$

Finally, from (18.8.1), (18.8.6), and (18.8.7),

$$I_0 = I - I' = \frac{\mathscr{E}R_i' + \mathscr{E}'R_i}{R_0 R_i + R_0 R_i' + R_i R_i'} \qquad (18.8.8)$$

The potential difference ΔV_{AB} between points B and A, as we saw earlier, is

$$\Delta V_{AB} = I_0 R_0 = \frac{(\mathscr{E}R_i' + \mathscr{E}'R_i)R_0}{R_0 R_i + R_0 R_i' + R_i R_i'} \qquad (18.8.9)$$

Substituting the numerical values $R_i = 0.03$ ohm, $R_i' = 0.02$ ohm, $R_0 = 0.2$ ohm, $\mathscr{E} = 12.6$ volts, and $\mathscr{E}' = 6.3$ volts, we find from (18.8.6) through (18.8.9) that $I = 142.6$ amperes, $I' = 101.0$ amperes, $I_0 = 41.6$ amperes, and $\Delta V_{AB} = 8.32$ volts. These answers should be compared with those found for Example 18.6.1. The reason that the currents I and I' are not vastly different from the circulating current I_c of Example 18.6.1 and that the values obtained for the potential difference ΔV_{AB} are nearly the same in both cases stems from the fact that the lighting circuit resistance R_0 is much larger than the internal resistance of either battery.

EXAMPLE 18.8.2

In the dc circuit shown in Fig. 18.27, find the currents flowing in each branch of the circuit and the potential difference ΔV_{AB} between the points A and B.

This circuit problem may be solved by a straightforward application of Kirchhoff's laws. At the beginning, we do not know with certainty the direction of all the currents in the various circuit branches. We begin, therefore, simply by assuming that the currents I_1, I_2, I_3, I_4, I_5, and I_6 in Fig. 18.27 have the directions shown there. From law I, at points A, C, and D, we may write

$$I_1 - I_2 - I_4 = 0 \qquad (18.8.10)$$

$$I_2 + I_6 - I_3 = 0 \qquad (18.8.11)$$

$$I_4 + I_5 - I_6 = 0 \qquad (18.8.12)$$

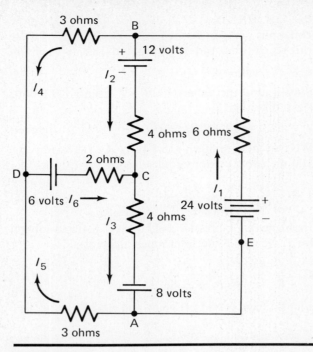

FIGURE 18.27

To determine completely the six currents, six equations are needed; the remaining three may be obtained by applying law II to the loops EBCAE, CADC, and BCDB. For loop EBCAE, starting at E, we may write $24 - 6I_1 - 12 - 4I_2 - 4I_2 - 4I_3 + 8 = 0$, the potential change associated with the 12-volt battery being a drop rather than a rise since we traverse it from positive to negative as we go around the loop. This equation may be simplified to

$$6I_1 + 4I_2 + 4I_3 = 20 \qquad (18.8.13)$$

For loop CADC, starting at C, we get in exactly the same way the equation $-4I_3 + 8 - 3I_5 - 6 - 2I_6 = 0$, which may be simplified to

$$4I_3 + 3I_5 + 2I_6 = 2 \qquad (18.8.14)$$

For loop BCDB, starting at B, we may write $-12 - 4I_2 + 2I_6 + 6 + 3I_4 = 0$, the signs of the third and fifth terms being positive because in each case we traverse the resistance in the direction opposite that of the assumed current flow, which means that the potential will rise rather than fall as we go through the resistance. This equation may be written more simply as

$$-4I_2 + 3I_4 + 2I_6 = 6 \qquad (18.8.15)$$

Equations (18.8.10) through (18.8.15) are a set of six simultaneous equations which may be solved for the six currents. Probably the easiest way to do this is to solve (18.8.10) for I_1, (18.8.11) for I_3, and (18.8.12) for I_5, inserting the values so obtained for I_1, I_3, and I_5 into (18.8.13), (18.8.14), and (18.8.15), respectively.

These latter equations then reduce to

$$-4I_2 + 3I_4 + 2I_6 = 6 \qquad (18.8.16)$$

$$4I_2 - 3I_4 + 9I_6 = 2 \qquad (18.8.17)$$

$$7I_2 + 3I_4 + 4I_6 = 20 \qquad (18.8.18)$$

The current I_6 is obtained immediately by adding (18.8.16) and (18.8.17), thereby eliminating I_2 and I_4, and giving $I_6 = 8/11$ ampere. Substituting this value into (18.8.17) and (18.8.18) and then adding those two equations eliminates I_4 and leads directly to the result $I_2 = 4/11$ ampere. If these values in turn are substituted into any of the above equations, $I_4 = 2$ amperes is obtained. Substituting these values for I_2, I_4, and I_6 into (18.8.10), (18.8.11), and (18.8.12), respectively, then leads directly to the results $I_1 = 26/11$ ampere, $I_3 = 12/11$ ampere, and $I_5 = -14/11$ ampere. The negative sign of I_5 tells us that we guessed the wrong direction for that current and that it actually flows opposite the direction indicated in Fig. 18.27. The potential difference ΔV_{AB} between A and B can be arrived at by starting at A, where the potential is V_A and going to B through point C, keeping track of all the potential rises and falls along the way. In this way, we may write

$$V_A - 8 + 4I_3 + 4I_2 + 12 = V_B \qquad (18.8.19)$$

which may be rearranged to give

$$V_B - V_A = \Delta V_{AB} = -8 + 4I_3 + 4I_2 + 12$$

Substituting the known numerical values for the currents finally yields $\Delta V_{AB} = 108/11$ volt.

The choice of loops and branch points we made is somewhat arbitrary. We could equally well have written a relation between currents according to law I for point A, and we could have written loop equations around paths such as AEBDA and ADCBEA instead of the ones we actually used. Such choices lead, of course, always to the same numerical answers for the currents.

EXAMPLE 18.8.3

A frequently used device for comparing the values of unknown resistors with precisely calibrated standard resistors is shown in Fig. 18.28. This apparatus, known as the *Wheatstone bridge*, was developed by the British physicist Sir Charles Wheatstone (1802–1875) in 1843. Current is made to flow in the four arms of the bridge through precisely calibrated resistances R_1, R_2, and R_3 one of which (R_3) is adjustable. In practice, R_3 may be a "substitution" box in which a precision resistance of any value desired may be inserted into the circuit simply by setting a series of dial switches. In the fourth arm of the bridge, the unknown resistance R_x is inserted. A galvanometer, which is an instrument used to detect very small currents—a very sensitive ammeter, really—is connected between A

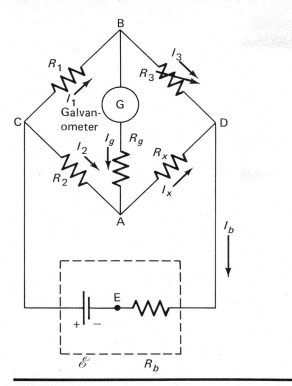

FIGURE 18.28

and B. Its resistance is represented by R_g in Fig. 18.28. In general, current will be flowing through the galvanometer, and in this condition the bridge is said to be *unbalanced*. By adjusting the calibrated resistance R_3, the current through the galvanometer can be reduced to zero, the bridge then being *balanced*.

Using Kirchhoff's laws, write a set of equations from which all the currents in the bridge can be determined even though the bridge may be unbalanced. Show that when the bridge is balanced, the unknown resistance is related to the known resistances R_1, R_2, and R_3 by $R_x = R_2 R_3 / R_1$.

Law I can be applied at points A, B, and C to give

$$I_b - I_1 - I_2 = 0 \tag{18.8.20}$$

$$I_1 - I_g - I_3 = 0 \tag{18.8.21}$$

$$I_2 + I_g - I_x = 0 \tag{18.8.22}$$

Law II, applied to loops CBA, ADB, and ECADE, gives

$$-I_1 R_1 - I_g R_g + I_2 R_2 = 0 \tag{18.8.23}$$

$$-I_x R_x + I_3 R_3 - I_g R_g = 0 \tag{18.8.24}$$

$$-I_2 R_2 - I_x R_x - I_b R_b = 0 \tag{18.8.25}$$

These six simultaneous equations can now be solved for the six different currents. The mathematics of the solution is tedious and difficult and will not be attempted here.

If the bridge is balanced, however, we know at

once that the current I_g is *zero*. This means that (18.8.21) through (18.8.24) reduce to

$$I_1 = I_3 \tag{18.8.26}$$

$$I_2 = I_x \tag{18.8.27}$$

$$I_1 R_1 = I_2 R_2 \tag{18.8.28}$$

$$I_x R_x = I_3 R_3 \tag{18.8.29}$$

The last of these equations states that

$$R_x = \frac{I_3}{I_x} R_3 \tag{18.8.30}$$

But from (18.8.27), I_x may be replaced by I_2; from (18.8.26), I_3 may be replaced by I_1. Then, from (18.8.28), the ratio I_1/I_2 is equal to R_2/R_1. The result of these successive substitutions is to transform (18.8.30) into the desired result

$$R_x = \frac{R_3 R_2}{R_1} \tag{18.8.31}$$

EXAMPLE 18.8.4

An array of resistors all having the same resistance R is connected as shown in Fig. 18.29. Find the current through each resistor as a fraction of the circuit current I_0 and find the equivalent resistance R_{tot} of the array.

In the circuit shown in Fig. 18.29, there are 12 individual resistances and 12 separate branch currents. At first sight, therefore, the problem would seem to be a very difficult one. But the resistances are all identical and are connected in a particularly symmetric array.

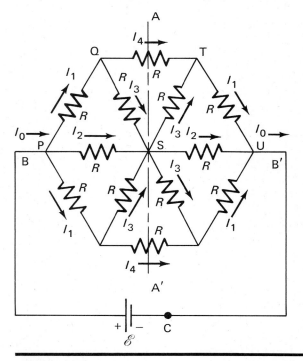

FIGURE 18.29

593

We shall use the *symmetry properties* of the array to argue that there are not 12 but only four currents that need to be determined, since certain branch currents must be the same. This technique of simplification by resort to symmetry properties is one that is very useful in many physical problems, particularly in the quantum physics of atoms, molecules, and crystal lattices.

If we note that the circuit is perfectly symmetric about the lines AA' and BB', we may conclude that all the currents in the part of the circuit above BB' must be the same as those below that line and that all the currents to the right of AA' must be the same in magnitude as those to the left but in the opposite direction, because the current I_0 flowing into the line AA' from the left flows out of that line to the right. It is clear, therefore, that there can be only four distinctly different currents, as shown in the diagram. Using law 1, we may write for the functions at P and Q the equations

$$2I_1 + I_2 = I_0 \qquad (18.8.32)$$

$$I_1 - I_3 - I_4 = 0 \qquad (18.8.33)$$

Around the circuit loops QPS and TQS, law II gives $I_1 R - I_2 R + I_3 R = 0$ and $I_4 R - I_3 R - I_3 R = 0$, or

$$I_1 - I_2 + I_3 = 0 \qquad (18.8.34)$$

$$-2I_3 + I_4 = 0 \qquad (18.8.35)$$

Equations (18.8.32) through (18.8.35) form a set of simultaneous equations for I_1, I_2, I_3, and I_4. From (18.8.35), we may obtain $I_4 = 2I_3$; substituting this into (18.8.33) yields $I_1 = 3I_3$. Substituting these values in turn into (18.8.32) and (18.8.34) gives

$$6I_3 + I_2 = I_0 \qquad (18.8.36)$$

and

$$4I_3 - I_2 = 0 \qquad (18.8.37)$$

Adding these two equations eliminates I_2 and allows us to express I_3 in terms of I_0 as

$$I_3 = I_0/10 \qquad (18.8.38)$$

Then, from (18.8.37),

$$I_2 = 4I_3 = 4I_0/10 \qquad (18.8.39)$$

$$I_1 = 3I_3 = 3I_0/10 \qquad (18.8.40)$$

$$I_4 = 2I_3 = 2I_0/10 \qquad (18.8.41)$$

Using law II around the path CPQTUC, we obtain

$$\mathscr{E} - I_1 R - I_4 R - I_1 R = 0 \qquad (18.8.42)$$

which we can write in the form

$$\mathscr{E} = R(2I_1 + I_4) = I_0 R_{\text{tot}} \qquad (18.8.43)$$

where R_{tot} represents the equivalent network resistance. But I_1 and I_4 can be expressed in terms of I_0 from (18.8.40) and (18.8.41). We may, therefore, write (18.8.43) in the form

$$\mathscr{E} = I_0 R_{\text{tot}} = R\left(\frac{6I_0}{10} + \frac{2I_0}{10}\right) \qquad (18.8.44)$$

from which

$$R_{\text{tot}} = \tfrac{4}{5}R \qquad (18.8.45)$$

18.9 The Charge and Discharge of Capacitors: Simple R–C Circuits

So far, we have confined ourselves to the discussion of purely dc or steady-state situations. We shall now consider a particularly important example in which the currents and potential differences in the circuit are time dependent. This example involves the charging of a capacitor C by a source of emf $\mathscr{E}$ through a resistance R and the subsequent discharge of the capacitor through a second resistance R', as illustrated in Figs. 18.30a and b, respectively.

Suppose that the capacitor C is initially uncharged and that the switch S is thrown to the left to complete the circuit shown in Fig. 18.30a at $t = 0$. A current I flows in the direction shown but decreases

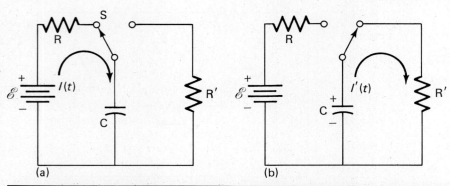

FIGURE 18.30. A simple series R–C (Resistance–Capacitance) circuit. At (a), the emf $\mathscr{E}$ charges the condenser through resistor R. At (b), the capacitor discharges through resistor R'.

From (18.9.12) and (18.9.16), we may express the charge q on the capacitor as

$$dq = -I'(t)\, dt = -\frac{\mathscr{E}}{R'} e^{-t/R'C}\, dt' \qquad (18.9.17)$$

which we may integrate from $t = 0$ when $q = q_0 = \mathscr{E}C$ to a later time t when q has the value $q(t)$. We find, therefore, that

$$\int_{\mathscr{E}C}^{q} dq = -\frac{\mathscr{E}}{R'} \int_{0}^{t} e^{-t/R'C}\, dt \qquad (18.9.18)$$

whence

$$[q]_{\mathscr{E}C}^{q} = -\frac{\mathscr{E}}{R'} R'C[-e^{-t/R'C}]_{0}^{t}$$

$$q - \mathscr{E}C = -\mathscr{E}C(-e^{-t/R'C} + 1)$$

or finally

$$\boxed{q(t) = \mathscr{E}Ce^{-t/R'C}}$$

In this case, both the current and the charge on the capacitor die out exponentially with the time constant $R'C$, as illustrated by Fig. 18.32. The potential difference across the resistance R' at any time t, which is equal to $I'R'$, can now be obtained easily from (18.9.16). In the same way, the potential difference across the capacitor at any time, which is equal to q/C, can be found from (18.9.19).

EXAMPLE 18.9.1

In the circuit of Fig. 18.30, a capacitor of capacitance $C = 2.0\ \mu\text{F}$ is charged by an emf of 100 volts through a resistance $R = 5000$ ohms. How long does it take to charge the capacitor to a potential difference equal to 99 percent of that created by the emf? If the condenser is subsequently discharged through a resistance of 20,000 ohms, what is the initial current and what would the current be after 0.15 second? Assume an initial potential difference of 100 volts across the condenser.

The potential difference across the capacitor is related to the charge it carries by $\Delta V_c = q/C$. But during the charging process, the charge is expressed as a function of time by (18.9.10). Therefore, the potential difference can be written

$$\Delta V_c = \mathscr{E}(1 - e^{-t/RC})$$

Setting ΔV_c equal to $0.99\mathscr{E}$, this becomes

$$0.99 = 1 - e^{-t/RC}$$

or

$$e^{-t/RC} = 0.01$$

Taking the natural logarithm of both sides and solving for T, we find

$$t = -RC \ln 0.01 = RC \ln 100$$

since $\ln(1/x) = -\ln x$. Substituting the given numerical values of R and C, we may obtain, finally,

$$t = (5 \times 10^3)(2 \times 10^{-6})(4.605) = 0.0461 \text{ sec}$$

If the capacitor is charged to the potential difference $\mathscr{E} = 100$ volts and then discharged through a resistance $R' = 20,000$ ohms, the current is given as a function of time by (18.9.16). For $t = 0$, the initial current is

$$I'(0) = \frac{\mathscr{E}}{R'} = \frac{100}{20000} = 0.005 \text{ amp}$$

After 0.15 second has elapsed, the current will be

$$I = \frac{\mathscr{E}}{R'} e^{-t/R'C} = 0.005e^{-0.15/(2 \times 10^4)(2 \times 10^{-6})}$$

$$= 0.005e^{-3.75} = (0.005)(0.02352) = 0.0001176 \text{ amp}$$

EXAMPLE 18.9.2

We have mentioned that a capacitor, since it stores electrical energy, behaves in some respects like a

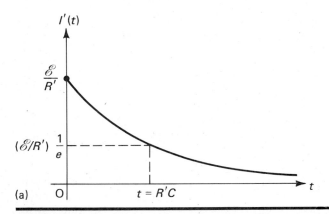

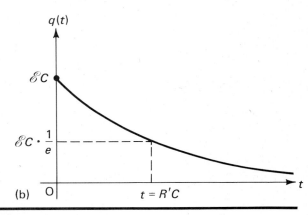

FIGURE 18.32. Plots of (a) current vs. time and (b) charge on the capacitor vs. time for the circuit of Figure 18.30 (b).

source of emf. Why is it, then, that we use batteries instead of capacitors as current sources for purposes such as starting cars?

To start an automobile, a current of about 200 amperes at 12 volts is required, and it should be possible to supply such a current for perhaps 3 minutes or so before significantly discharging the source. The starting circuit resistance R will be, under these circumstances,

$$R = \frac{\Delta V}{I} = \frac{12}{200} = 0.06 \text{ ohm}$$

Now, in order to supply a 200-ampere current from the discharge of a capacitor for 3 minutes, we should need a time constant of at least that much, whereby

$$RC = 180 \text{ sec}$$

or

$$C = \frac{180}{R} = \frac{180}{0.06} = 3000 \text{ F}$$

This is a very large capacitance. Under favorable circumstances, the dielectrics used in capacitors can withstand an electric field of about 10^7 volts/meter before breakdown. If ΔV is the potential difference between the plates and d their separation, and if the field within has the value 10^7 volts/meter, then, assuming parallel-plate geometry,

$$E = \frac{\Delta V}{d} = \frac{12}{d} = 10^7 \text{ V/m}$$

and the plate separation d must be at least

$$d = 1.2 \times 10^{-6} \text{ m}$$

At the same time, the capacitance of a parallel-plate capacitor is given by

$$C = \frac{K\varepsilon_0 A}{d}$$

where K is the dielectric constant, which in a good dielectric may be about 4.0. If we now set $C = 3000$ farads and $d = 1.2 \times 10^{-6}$ meter in this equation, we find, using $K = 4.0$,

$$3000 = \frac{(4.0)(8.854 \times 10^{-12})}{1.2 \times 10^{-6}} A$$

which tells us that we must have

$$A = 1.016 \times 10^8 \text{ m}^2$$

The volume of the capacitor must then be

$$\text{volume} = Ad = (1.016 \times 10^8)(1.2 \times 10^{-6}) = 122 \text{ m}^3$$

This volume could barely be contained within a cube 5 meters (16.5 feet) on a side! The storage capacitor would therefore have to be far larger than the auto-

mobile itself. This should be compared with the size of a lead-acid storage battery that can store as much or more *chemical* energy in a volume of perhaps 0.02 m^3. The lead-acid battery has a stored energy density about 6000 times greater than that of the capacitor. Beyond this, there are other problems, too, the most important being that no dielectric is a perfect insulator; therefore, even the best dielectrics permit leakage currents to flow which discharge the capacitor internally over a period of time that is undesirably short for the application envisioned above.

The use of lead-acid storage batteries as a primary source of energy to run automobiles has been carefully considered and, indeed, from time to time put into practice. But even though batteries provide much higher energy storage densities than capacitors, they still do not store nearly as much energy per unit volume as that which results from the combustion of hydrocarbon fuels such as gasoline or diesel oil. Thus, though providing sufficient energy to start internal combustion engines and run their auxiliary equipment, their use as *prime energy storage units* for automotive applications is marginal at best. A great deal of scientific research directed toward the development of batteries with improved storage densities and toward the development of *fuel cells* that convert the chemical energy of hydrocarbon fuels directly to electrical energy is being done at the present time. The availability of such units at reasonable cost would permit the production of practical electric automobiles and would eliminate some of the noise and atmospheric pollution now caused by the extensive use of internal combustion vehicles.

SUMMARY

Whenever an electric field exists within a conductor, a flow of charge occurs, which we refer to as an *electric current*. The current I in a conductor refers to the amount of charge dq that passes a given point in the conductor in time dt. The current is thus a *scalar* quantity defined by

$$I = dq/dt$$

The units of current are coulombs/second, more frequently referred to as amperes.

The *current density* is a *vector* quantity. Its direction is the direction of charge flow at any point within the conductor, and its magnitude is the current per unit area across an infinitesimal area da oriented normal to the charge flow at that point. This means that the current dI through a surface da that may not be normal to the charge flow can be written

$$dI = \mathbf{j} \cdot \mathbf{n} \, da$$

where **n** is the unit vector normal to *da*. The total current *I* that flows across a surface *S* can be obtained by integrating the current density over the surface, from which

$$I = \int_S \mathbf{j} \cdot \mathbf{n}\, da$$

For the case of a conductor of cross-sectional area *A* carrying a current of uniform current density *j*, this amounts to

$$j = \frac{I}{a}$$

The dimensions of current density are amperes/m^2 or amperes/cm^2. The current density is related to the charge density ρ and its average flow velocity **v** by

$$\mathbf{j} = \rho \overline{\mathbf{v}}$$

In the case of a conductor in which current flow occurs due to the motion of mobile free electrons, $\rho = -ne$, where *n* is the number of valence electrons per unit volume. The above equation then becomes

$$\mathbf{j} = -ne\overline{\mathbf{v}}$$

In order to maintain a steady current flow within a conductor, a source of *emf*, or *electromotive force*, must be present. An emf acts to increase the potential energy of charges by nonelectrostatic means. We may define the emf as a difference in potential ΔV, attributable to nonelectrostatic forces, that is capable of establishing a steady current in a closed conducting circuit. From this definition, it is evident that the magnitude $\mathscr{E}$ of the emf and the nonelectrostatic potential difference it creates is

$$\mathscr{E} = \Delta V$$

The power supplied to a circuit by an emf $\mathscr{E}$ is

$$P = \mathscr{E}I = I\,\Delta V$$

In many conducting substances, the current density is directly proportional to the electric field at every point. Such substances are said to be Ohmic conductors, or to obey Ohm's law. Stated mathematically, Ohm's law tells us that

$$\mathbf{j} = \sigma\mathbf{E} \quad\text{or}\quad \mathbf{E} = \frac{1}{\sigma}\mathbf{j} = \rho\mathbf{j}$$

where the proportionality factors σ and ρ are referred to, respectively, as the *conductivity* and *resistivity* of the conducting material. Note that the symbol ρ is used to refer *both* to electric charge density and to resistivity; this is unfortunate but is very common practice. The Ohmic behavior of conductors can be understood physically by assuming that free electrons undergo acceleration according to Newton's laws but undergo frequent collisions whose effect is, on the average, to reduce the drift velocity given them by the electric field to zero, thus returning them to the thermal equilibrium condition and dissipating their drift energy as heat.

In an Ohmic conductor of uniform cross section *A* and length *l*, the current and potential drop are related by

$$I = \frac{\Delta V}{R}$$

where the *resistance R* is defined by

$$R = \frac{l}{\sigma A} = \frac{\rho l}{A}$$

where ρ is the electrical resistivity. In other Ohmic conductors, the relation $I = \Delta V/R$ also holds, the resistance in all cases being the proportionality factor relating *I* and ΔV.

Resistances *in series* add to an equivalent resistance equal to the sum of the individual resistances, whereby

$$R = R_1 + R_2 + \cdots + R_n$$

For resistances *in parallel*, the equivalent resistance is given by

$$\frac{1}{R} = \frac{1}{R_1} + \frac{1}{R_2} + \cdots + \frac{1}{R_n}$$

Resistors in series all carry the same current, while resistors in parallel all experience the same potential difference.

The power dissipated in *Joule heating* in a resistance can be expressed by

$$P = I^2 R = I\,\Delta V = (\Delta V)^2/R$$

Circuits that cannot be decomposed into simple series and parallel combinations of resistances can be analyzed by the application of *Kirchhoff's laws*. These can be stated as follows:

I. The algebraic sum of the currents into any junction of conductors is zero.
II. The algebraic sum of potential increases and decreases around any closed circuit loop is zero.

Time-dependent voltage and current relations in simple resistance–capacitance (*R–C*) circuits can be investigated by the use of Kirchhoff's laws. It is found that the charge and discharge of series *R–C* combinations occur as exponential functions of time and that the time scale on which the charge and discharge of the capacitor, and the growth and decay of current, take place is described by a time constant t_0 given by

$$t_0 = RC$$

QUESTIONS

1. Describe briefly the main differences between metallic and nonmetallic materials.
2. How much charge passes through an x-ray tube during a 0.1-second exposure when the current is 150 milliamperes?
3. It it true that a conductor has free charge only if it has an excess of charge?
4. In electrostatics, we were always using the fact that the electric field within a conductor is zero. How do you reconcile this with the need to have a nonvanishing field within a conductor in order to produce a current flow?
5. Is it true that the relation $R = \Delta V/I$ is a statement of Ohm's law, or must some other condition also be satisfied?
6. Describe the main difference between an emf and a "voltage." Give some examples in which the existence of a voltage (potential difference) does not imply the presence of an emf.
7. If a current is established in a wire and then suddenly interrupted by removing the emf, how long does it take for the current to drop to zero?
8. What is a superconductor? Discuss its electrical behavior.
9. Two wires have the same resistance and the same length but one of them has a cross-sectional area twice that of the other. How are the conductivities of the two wires related?
10. Can you think of examples of flow, other than that of electric charge, in which the resistance is proportional to length and inversely proportional to area?
11. Describe in a careful way how you know when resistors are in series with one another. Do the same for resistors in parallel.
12. Explain qualitatively why the equivalent resistors of a series combination of resistors is always greater than any one of the individual resistances. Explain why, for a parallel combination, the equivalent resistance is less than any of the resistances.
13. Why is it more likely for a light bulb to burn out when the switch is turned on than when it is already in operation?
14. A fuse is rated at 10 amperes. What does this mean? Why is it dangerous to replace a 10-ampere fuse with one that is rated at 20 amperes?
15. Describe what happens to the light intensity of each lamp in a series circuit as more lamps are added.
16. A number of lamps are in parallel. As more lamps are added, describe what happens to the intensity of each lamp.
17. Electrical power is transmitted at high voltage. Explain why.
18. It would be very dangerous to shave with a plug-in electric razor while taking a bath. The danger is diminished, however, if the razor is battery operated. Explain.
19. Explain, on an atomic level, the differences between the processes involved in the conduction of electricity in a metal and in the ionized gas in a neon tube.
20. Can you suggest a possible design for a "dc transformer" that would increase the voltage of a battery by a large factor, using resistances, capacitors, and switching elements?

PROBLEMS

1. A conductor of uniform cross section carries a steady current of 5.00 amperes. How many electrons flow past a given point during a 1-minute time interval?
2. In an electroplating process, 40,000 coulombs are transferred at a current of 10 amperes. What is the time required?
3. The density of copper is 9.0 g/cm^3 and it has one conduction electron per atom. A steady current of 50 amperes is established in a wire which has a uniform circular cross section 0.1 cm in diameter. (a) Find the current density j. (b) What is the mean velocity of the electrons?
4. The resistivity of copper at 20°C is 1.7×10^{-8} ohm-meter. At this temperature, find the resistance of a copper wire 5 meters long if it has a circular cross section of radius 0.1 cm.
5. An aluminum wire of length l and circular cross section of radius r has a resistance R. By what factor is the resistance multiplied in each of the following cases. (a) The length of the wire is doubled. (b) The length of the wire is tripled and the radius is diminished to $r/3$. (c) Both l and r are doubled.
6. The current density j in a long, straight wire having a circular cross section of radius R varies with the distance from the center of the wire according to the relation $j = \lambda r$, where λ is a proportionality constant. Show that the current flowing through the wire is given by $i = 2\pi\lambda R^3/3$.
7. A steel rail has a cross-sectional area of 30 cm^2. If the resistivity of steel is 6.0×10^{-7} ohm-meter, find the resistance of a rail 6 km long.
8. A block of iron in the form of a rectangular solid has dimensions 2 cm × 3 cm × 100 cm. The resistivity at 20°C is 1.0×10^{-7} ohm-meter. Find the resistance between the three pairs of opposite faces, that is, between the two ends, the sides, and between the top and bottom.
9. A copper bar has dimensions 0.1 meter × 0.3 meter × 5 meter. The resistivity of copper is 1.7×10^{-8} ohm-meter. Find the resistance of the bar if the current flows along its length.
10. A metallic bar of length 12 meters contains 6×10^{25} mobile electrons. If a current of 3 amperes is flowing in the bar, what is the drift velocity of the electrons?
11. The resistivity of silver is 1.6×10^{-8} ohm-meter at 20°C. A coil is to be made with 25 km of 1-mm-diameter silver wire. What will be its resistance? If the wire is uniformly stretched to a length of 50 km, what would the new resistance be?
12. The density of aluminum is 2.7 g/cm^3 and its atomic mass is 27 g/mole. There are three conduction electrons per atom. Find the number of conduction electrons per cm^3. If a current of 10^{-3} ampere flows in an aluminum wire of 1 mm^2 cross section, what is the drift velocity of the electrons?
13. A wire of uniform cross section and of length 5 meters has a resistance of 2 ohms. If the resistivity is 1.6×10^{-6} ohm-meter, what is the area of the wire's cross section?
14. The earth's atmosphere carries positive charges toward the earth and negative charges away from it. The total current is 1800 amperes. Assuming the charge flow is

symmetric with respect to the earth, find the magnitude of the current density at the surface. The electric field at the surface has a value of 100 volts/meter, directed towards the earth. Find the electrical conductivity of the air near the earth's surface.

15. A van de Graaff generator has a belt 75 cm wide which moves at a speed of 30 meter/second. If a current of 2×10^{-4} ampere is carried to the collecting sphere, find the surface charge density on the belt.

16. Over a restricted temperature range, the resistivity of a material varies as $\rho(T) = \rho(T_0)[1 + \delta(T - T_0)]$, where T_0 is some reference temperature and T is the temperature at which the resistivity is required. The constant δ is called the temperature coefficient of resistivity. For copper and silver, we have

$$\rho_{Cu}(20°C) = 1.7 \times 10^{-8}\ \Omega\text{-m} \qquad \delta_{Cu} = 3.9 \times 10^{-3}°C^{-1}$$

$$\rho_{Ag}(20°C) = 1.6 \times 10^{-8}\ \Omega\text{-m} \qquad \delta_{Ag} = 3.8 \times 10^{-3}°C^{-1}$$

(a) At what temperature is the resistivity of silver equal to that of copper at 20°C? (b) A copper bar has a resistance of 12.00 ohms at 40°C. What would its resistance be if it is heated to 100°C? Neglect any possible expansion of the bar.

17. A copper bar is heated from 20°C to 200°C and is stretched uniformly to twice its original length, its volume remaining unchanged. If the bar was initially 1 meter along with a resistance of 0.02 ohms, find (a) the cross-sectional area of the bar before it was stretched, and (b) the resistance of the bar after it was stretched and its temperature increased.

18. At 20°C, a copper wire 5 meter long with a cross-sectional area of 0.1 cm² carries a current of 2 amperes. Find the potential difference between the ends of the wire.

19. A circular coil of aluminum wire 0.010 inch in diameter has 400 turns and has a mean diameter of 0.7 inch. The resistivity of aluminum at 20°C is 2.8×10^{-8} ohm-meter. If a voltage of 12 volts is applied between the ends of the coil, find the current which flows and also the heat dissipated during a 5-minute time interval.

20. A 350-watt immersion heater operates on a 120-volt line and is used to bring 250 cm³ of water from 27°C to the boiling point. (a) Find the current passing through the heater. (b) What is the rate at which energy is transferred to the water? (c) How long does it take to boil this amount of water? (d) How much longer does it take to convert all the water to steam?

21. A 100-watt light bulb operates on a 110-volt line. Find the current passing through the filament and the filament resistance.

22. A given amount of power is to be sent from a power station to a distant point. Find the ratio of the Joule heating losses for voltages of 50,000 and 1,000 volts assuming the same line is used in both cases.

23. An electric toaster operates on a 110-volt line. If it draws 6 amperes, find the resistance of the heating element as well as the energy consumed during a 30-second time span during which it is operative. At 3¢ per kilowatt-hour, how much does it cost to toast a piece of bread?

24. A 300-watt lamp operates on a 220-volt line. The bulb is immersed in 8 kg of water at 27°C. What current flows through the lamp? What is the water temperature after 5 minutes?

25. A motor which is 85 percent efficient draws 6 amperes from a 220-volt line. What is the horsepower of the motor?

26. A 110-volt 20-ampere motor lifts a 4000-pound weight at 7 ft/min. Find the horsepower of the motor and the motor's efficiency.

27. Find the equivalent resistance of the network of resistors shown in the diagram.

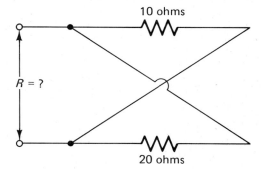

28. Show that if resistors R_1 and R_2 are in parallel, their effective or total resistance is $R_1R_2/(R_2 + R_2)$. Show also that for three resistors in parallel, the total resistance is $(R_1R_3 + R_1R_2 + R_2R_3)/(R_1 + R_2 + R_3)$.

29. A physics student has three identical resistors each with resistance R. How many different resistances can he obtain by using one, two, or three resistors in all possible combinations? What values of equivalent resistance are possible?

30. A 4-ohm and 8-ohm resistor are to be connected to a 6-volt battery. The resistors can be in series or in parallel. Find the power delivered by the battery for each of these possible connections as well as the Joule heat generated in each resistor.

31. Find the equivalent resistance of the combination of resistors shown in the accompanying figure.

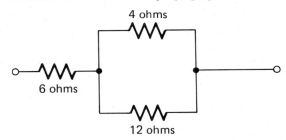

32. Find the equivalent resistance of the combination of resistors shown in the diagram.

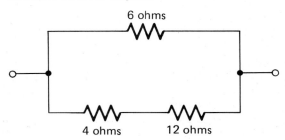

33. The combination of resistors shown in the figure is equivalent to a single resistor of what value?

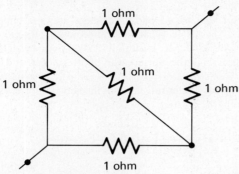

34. For the electrical circuit shown in the accompanying diagram, find the following: **(a)** the potential difference $V_b - V_a$; **(b)** the net power delivered by the 3-volt battery; **(c)** the heat generated in the 8-ohm resistor during a 2-minute interval.

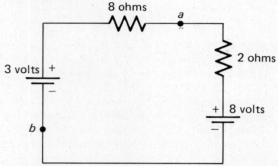

35. Find the currents I_1, I_2, which I_3 which flow in the circuit shown in the diagram.

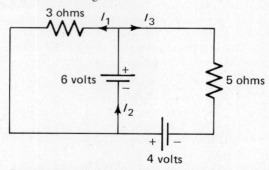

36. Consider the electric circuit illustrated in the figure. The

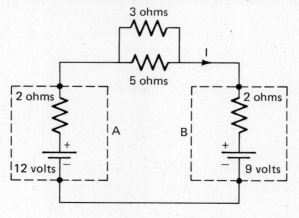

2-ohm resistors shown inside the batteries are internal resistances. **(a)** Find the combined resistance of the 3-ohm and 5-ohm resistors. **(b)** What is the current I? **(c)** What is the potential difference across battery B? **(d)** At what rate is battery B converting chemical energy to electrical energy? **(e)** Find the current through the 3-ohm resistor. **(f)** At what rate is heat developed in the 5-ohm resistor?

37. Given the circuit shown in the diagram, **(a)** write the loop equation for loop ABDEA. **(b)** Write the loop equation for the upper loop BCDB. **(c)** Find the currents I_1 and I_2. **(d)** At what rate does the 2-volt battery deliver energy? **(e)** At what rate does the 3-ohm resistor dissipate energy?

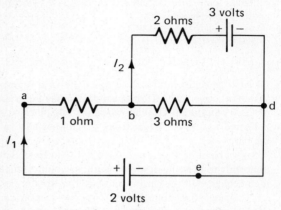

38. A 6-volt storage battery has an internal resistance of 0.03 ohm and delivers a current of 25 amperes. **(a)** What is the terminal voltage of the battery? **(b)** At what rate is heat generated within the battery? **(c)** What power is delivered to the load?

39. Two batteries of emf $\mathscr{E}_1$ and $\mathscr{E}_2$ have internal resistances R_1 and R_2. They are connected in parallel with each other and with a resistor R as shown in the figure. Show that the current I through resistance R is given by

$$I = \frac{\mathscr{E}_1 R_2 + \mathscr{E}_2 R_1}{R_1 R + R_2 R + R_1 R_2}.$$

40. Two dry cells are connected together, positive to positive and negative to negative. The newer of the two cells has an emf of 1.53 volt and an internal resistance of 0.05 ohm, while the older cell has an emf of 1.45 volt and a resistance of 0.15 ohm. **(a)** Find the current that flows in the circuit.

(b) At what rate is heat dissipated in each of the cells? **(c)** Find the potential difference (terminal voltage) across the terminals of each cell.

41. In the figure, n identical cells constituting a battery are connected in parallel to each other and to a resistance R. Find expressions for the current and potential difference across R. Show that for large n the potential difference of the battery approaches $\mathscr{E}$. Thus, by connecting many cells of emf $\mathscr{E}$, we may produce a battery with very low internal resistance.

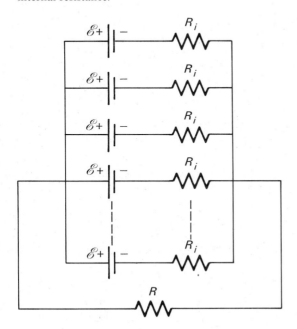

42. Two cells, one with emf 1.5 volt and internal resistance 0.10 ohm, the other with emf 2.0 volts and internal resistance 0.05 ohm, are connected in parallel to each other and to a 10-ohm resistor as shown in the diagram. **(a)** Find the currents I_1, I_2, and I_3. **(b)** At what rate do the cells

convert chemical energy to electrical energy? **(c)** Find the power dissipated in each resistor. **(d)** What is the terminal voltage of each cell?

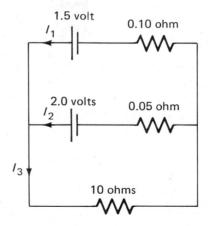

43. The cells in the above problem are connected in series to each other and to the 10-ohm resistor. Repeat parts **(a)**, **(b)**, **(c)**, and **(d)** in the preceding problem.

44. The Wheatstone bridge shown in Fig. 18.28 is balanced with $R_1 = 10$ ohm, $R_2 = 30$ ohm, and $R_3 = 5$ ohm. Find the unknown resistance R_x. If the battery has an emf of 12 volts and a negligible resistance, find the current which flows through R_x.

45. An electrician who is a novice at his profession has improperly designed the home circuit shown below. What is wrong with his design and what suggestion would you have for the distraught homeowners?

46. Twelve wires, each of resistance R, are arranged as the sides of a cube as shown in the diagram on top of the next page. Show that the equivalent resistance between points X and Y is $5R/6$. *Hint:* Convince yourself by symmetry arguments that the three corners marked A are at the same potential, and likewise the corners marked B are at the same potential. Use this to rearrange the

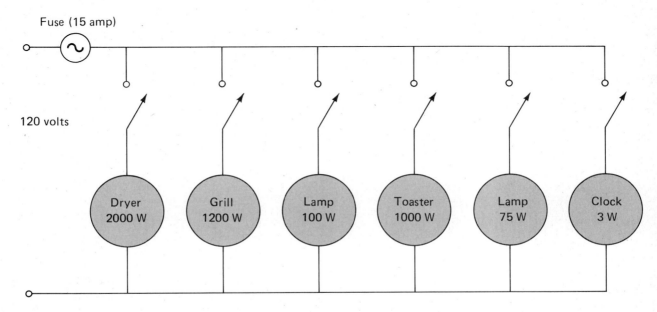

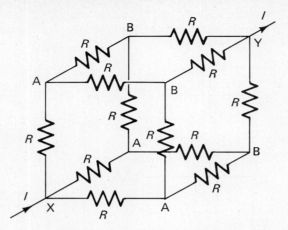

resistors so they may be combined as parallel and series combinations.

47. Twelve wires each of resistance R are arranged as the sides of a cube as shown in the accompanying figure. Show that the equivalent resistance between X and Y is $7R/12$. *Hint:* By symmetry, the two corners marked A have the same potential and the two corners marked B have the same potential.

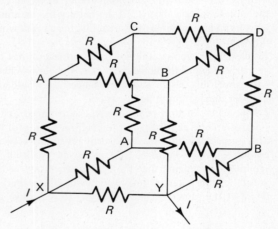

48. A galvanometer is a scientific instrument containing a coil whose angular deflection is proportional to the current passing through the coil. If it is to be used as an ammeter, a very small shunt resistance R_{sh} is connected in parallel to the coil which itself has a resistance R_c, as shown in the diagram. **(a)** Why is it necessary to have $R_{sh} \ll R_c$? **(b)** A certain galvanometer has a full-scale deflection when the current passing through it is 0.5×10^{-6} ampere. If the coil resistance is 300 ohm what should the shunt resistance be if the galvanometer is to be converted to an ammeter with a full-scale deflection of 10 amperes?

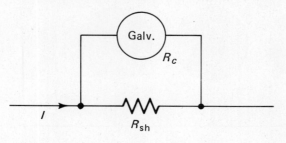

49. A galvanometer may be used to measure the potential difference between two points in an electrical circuit. To convert a galvanometer to a voltmeter, a high resistance R_s must be put in series with the coil, as shown in the diagram. **(a)** Explain why it is necessary to insert a large series resistor R_s. **(b)** A galvanometer with $R_c = 300$ ohms is fully deflected when a current of 0.5×10^{-6} ampere flows. It is to be used as a voltmeter with a full-scale deflection of 20 volts. What resistance R_s should be used? What is the potential drop across the galvanometer when it exhibits full-scale deflection?

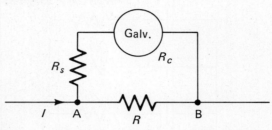

50. Explain, with relevant calculations, how a 0 to 50-micro-ampere galvanometer of 400-ohm resistance can be converted to an instrument reading **(a)** 0 to 3 amperes, **(b)** 0 to 10 volts, **(c)** 0 to 5 amperes, **(d)** 0 to 30 volts.

51. A capacitor C is charged to a potential difference ΔV. At $t = 0$, a switch is closed, thus completing a circuit with a series resistor R. The charge on the capacitor is then given by

$$q = q_0 e^{-t/RC}$$

Find the total energy dissipated in the resistor and show that this is equal to the energy initially stored by the capacitor.

52. A 5-μF capacitor is charged to 300 volts and then discharged through a 75,000-ohm resistor. After 3 seconds of discharge, find the remaining charge on the capacitor, the current through the resistor, and the rate at which Joule heat is being produced in the resistance.

53. The capacitor in an R–C circuit is charged to within 0.1 percent of its maximum charge. How many time constants have elapsed since the charging began? In the diagram, a capacitor of 2×10^{-6} farad carries a charge of 3×10^{-6} coulomb. The resistor has a resistance of 4×10^{6} ohms. What is the charge on the capacitor 24 seconds after the switch is closed? Find the current through the resistor immediately after the switch is closed.

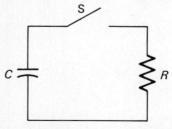

54. A 12-volt battery is connected in series with a resistor of resistance $R = 3.0 \times 10^{6}$ ohms and a capacitor of capacitance $C = 2.0 \ \mu$F. Find the time constant of the circuit. When the capacitor has a charge of one half the maximum value, what charge does it have?

55. In the circuit shown in the diagram, the switch is closed at $t = 0$. **(a)** Make a plot of the voltage ΔV_c across the capacitor as a function of time. **(b)** Find the charge on the capacitor after a very long time has elapsed. **(c)** How much energy is supplied by the battery in the charging process? **(d)** How much is stored in the capacitor when it is fully charged? **(e)** Where did the remainder of the energy go?

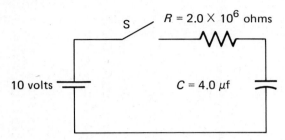

MAGNETIC FIELDS FROM STEADY CURRENTS

19.1 Introduction

The phenomenon of magnetism has been known for at least as long as static electricity. The magnetic forces exerted by permanently magnetized substances, such as magnetite, upon objects made of iron was known to the ancient Greeks, and the Chinese were using magnetic compasses by about 1000 A.D. By the time of the American revolution, the magnetic compass had been developed into a dependable instrument good enough to navigate ships all over the world, and the earth's magnetism had been accurately charted. It was by then well known that the earth's magnetic poles do not coincide with the polar axis, and accurate maps of magnetic declination had been made in which the deviation of "magnetic north" from true north is plotted. During the latter part of the nineteenth century, there was considerable speculation regarding the possible connections between electric and magnetic phenomena.

It was not until 1819, however, that the Danish scientist Hans Christian Oersted (1777–1851), using the newly invented "voltaic pile" to set up a steady current, discovered that a compass needle is deflected in the neighborhood of a conductor in which an electric current flows. The implication of this experiment, namely, that magnetism and electricity are related phenomena, is profound and gave rise to a period of rapid development in our understanding of magnetism and its relationship to electricity. Oersted's experiment is illustrated in Fig. 19.1.

That our initial acquaintance with magnetic

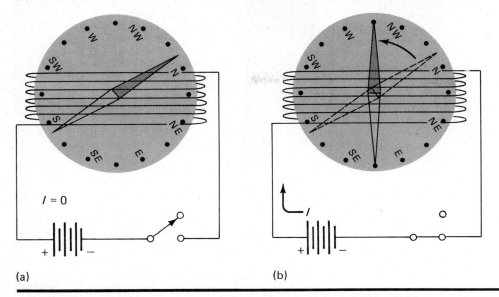

FIGURE 19.1. Oersted's experiment, illustrating the magnetic field produced by an electric current.

forces came by way of naturally occurring permanent magnets tended to obscure rather than clarify the connection between magnetism and electricity. It was only after a comparatively long time that it was generally understood that all magnetic fields, *even those associated with permanent magnets,* arise from the flow of electric currents. The magnetic fields of permanently magnetized substances arise from currents that circulate within the atoms of the substance and which add up on the surface of the magnet to an *equivalent surface current,* as illustrated by Fig. 19.2. It is this surface current that represents the source of the magnet's field. Permanent magnetism, more fre-

quently referred to as *ferromagnetism,* is a complex subject, which we shall not attempt to master in detail, though we shall learn more about it in Chapter 21. For the time being, our objective will be to define and understand magnetic fields and magnetic forces, and to examine how they arise from electric currents.

The approach we shall take is to describe the behavior of charges and currents that are placed in *externally produced* magnetic fields, without concerning ourselves at first with how the magnetic field is created. This procedure is advantageous in enabling us to formulate a reasonable definition of magnetic field strength in the beginning. Once that is accomplished,

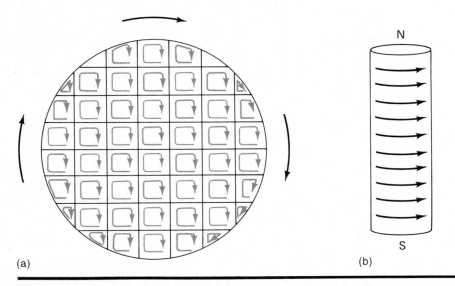

FIGURE 19.2. Schematic diagram illustrating, at (a), circulating atomic currents. On the interior boundaries shown in the diagram, these currents cancel one another, but at the surface add up to produce an equivalent surface magnetization current, shown at (b).

we shall turn our attention toward describing and understanding the magnetic fields produced by moving charges and by currents flowing in conductors, and toward explaining forces between current-carrying conductors.

19.2 Magnetic Forces and Fields: Magnetic Dipoles

Certain basic facts regarding the forces between the poles of magnets and the behavior of magnetic substances had been established long before Oersted's discovery of electromagnetism. In outlining these facts, we shall designate the poles of a magnet as north (N) or south (S). The north pole of a magnet is the one that points northward when the magnet is allowed to rotate freely like a compass needle, and the south pole is the one that points south under such circumstances. It was known long before Oersted's time that *unlike magnetic poles attract one another and that like poles repel one another.* By making quantitative measurements of the forces between the poles of bar magnets that are long and thin (so that the north end of the magnet is essentially free of interference from effects created by its own south pole) it had further been ascertained that these forces of attraction and repulsion are inversely proportional to the square of the distance between the magnet poles. The forces between magnet poles in many respects resemble the Coulomb forces between electric charges, particularly in that they exhibit the same *inverse square* dependence on distance. It would be tempting to assume that there are positive (N) and negative (S) *magnetic charges* within the magnets that act, in accord with Coulomb's force law, as the source of the magnetic field in exactly the same way that electric charges act as the source of electric fields. Unfortunately, this does not appear to be true, since every experimental effort that has been made to demonstrate the existence of these magnetic charges has failed.

In electrostatics, there is no difficulty in producing net positively and negatively charged objects. One needs only to bring out the amber and cat fur, or the glass rod and silk cloth, to accomplish this. It is also quite possible to break an electric dipole in two and thereby obtain a positively charged end and a negatively charged end. Magnets, on the other hand, *invariably have north and south poles of precisely equal strengths.* They are, therefore, essentially *dipoles.* Furthermore, when a bar magnet is cut in half, instead of getting an isolated north pole and an isolated south pole, we obtain two halves each of which have north and south poles of exactly equal strength, as illustrated by Fig. 19.3. Other more sophisticated schemes for isolating magnetic north or south charges have been

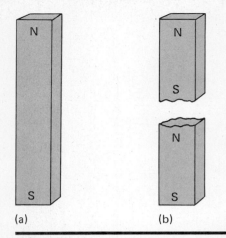

(a) (b)

FIGURE 19.3. When a magnet is cut in half, two magnets each having north and south poles of equal strength are produced. This suggests that magnetic fields are essentially dipolar in character, rather than arising from free magnetic charges.

thought of, but all have resulted in utter failure, and we are left with the conclusion that though we may have magnetic dipoles, there is no such thing as an isolated north or south pole. This remarkable behavior may be explained by the assertion that *the only source of magnetism, even in permanent magnets, is an electric current.* In permanently magnetized substances, these currents circulate within the atoms of the substance, as mentioned previously and as illustrated by Fig. 19.2.

When an electric current flows through a coil of wire, such as the one illustrated in Fig. 19.4, it establishes a magnetic field similar in every respect to the field of a bar magnet. The field that arises has all the characteristics of a *dipolar* Coulomb field and has properties analogous in all respects to the electric field associated with electric dipoles as discussed in Chapter 16. Under these circumstances, of course, it makes no sense to speak of separating the north pole

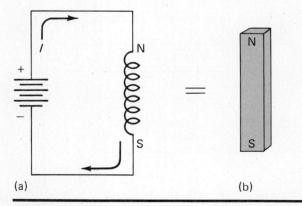

(a) (b)

FIGURE 19.4. A steady electric current in a coil produces exactly the same kind of magnetic field as a bar magnet.

from the south pole. The fields that arise are, nevertheless, identical in every respect to those produced by the permanent magnets with which we are all familiar. In addition, if the current-carrying coil is subjected to an *externally produced* magnetic field, such as that of a bar magnet, it will be acted upon by magnetic forces of attraction or repulsion just as if it were a permanent magnet. The north pole of the coil will thus be attracted by the south pole of a permanent magnet and repelled by its north pole. In fact, since any conductor in which a current flows has associated with it a magnetic field, *there will be a magnetic force on any current-carrying conductor that lies in an externally produced magnetic field.* Finally, since a current flow arises from the motion of any charged particle, *there must be a magnetic force on any moving charge in an externally produced magnetic field.*

We must now face the problem of how to set up a quantitative definition of magnetic field strength. The most obvious way to do this is to proceed in analogy to the electrostatic case and express the magnetic field strength in terms of the force per unit magnetic charge experienced by a magnetic charge placed in the field. Since, so far as we know, there is no such thing as a magnetic charge, however, this procedure could never be implemented experimentally. It is therefore logically inconsistent and operationally impossible. We must try, then, to find some other means that accomplishes essentially the same end in a way that is, in principle at least, experimentally feasible.

We may begin by recalling from our study of electrostatics that the torque on an electric dipole whose dipole moment is $\mathbf{p}$ is given by Eq. (16.6.28) as

$$\tau = \mathbf{p} \times \mathbf{E} \tag{19.2.1}$$

According to this expression, the magnitude of the torque is

$$\tau = pE \sin \theta \tag{19.2.2}$$

where θ is the angle between the field vector $\mathbf{E}$ and the direction of the dipole moment. The torque vanishes when $\theta = 0$, which means that when the dipole is aligned along the direction of the field, there is no net torque and the dipole is in *equilibrium*. We could, therefore, map out the direction of the field by noting the direction in which a freely suspended test dipole points at equilibrium.

The same approach can be taken to ascertain the direction of the magnetic field. We observe that electromagnets and permanent magnets both behave like *magnetic dipoles*. Let us denote the *magnetic dipole moment* of our magnet by $\mathbf{p}_m$ and the vector representing the magnetic field, which we shall hereafter refer to as the *magnetic induction*, by $\mathbf{B}$. In the case of an electric dipole, the moment can usually be expressed as the product of the electric charge on the ends of the

dipole times the length of the dipole. In the magnetic case, this is no longer possible, since there is no magnetic charge as such. The *magnetic moment* $\mathbf{p}_m$, therefore, is all that we have now to describe the magnet and serves as the sole fundamental parameter that establishes its strength. We shall see later, however, that it is related to the current that acts as the source of the field and to the size of the circuit through which it flows. We would like to define the magnetic induction $\mathbf{B}$ as a field that has the same properties with regard to magnetic charge that $\mathbf{E}$ has with regard to electric charge—or would have, if magnetic charge as such existed. In this event, the torque on a *magnetic dipole* of strength $\mathbf{p}_m$ at any point in an externally produced magnetic field $\mathbf{B}$ must be

$$\tau = \mathbf{p}_m \times \mathbf{B} \tag{19.2.3}$$

Once more, the magnitude of this torque is

$$\tau = p_m B \sin \theta \tag{19.2.4}$$

where θ is the angle between the axis of the magnetic dipole and the direction of the magnetic induction field $\mathbf{B}$ at the point where the dipole is located. As before, this torque vanishes, and a freely suspended dipole comes to equilibrium, when the axis of the dipole is aligned parallel to the magnetic induction vector $\mathbf{B}$.

Using these ideas as a basis, the magnetic induction field $\mathbf{B}$ can be mapped by noting the direction that a small freely suspended magnetic dipole, such as a compass needle, assumes at various points within the field. This technique is illustrated by Fig. 19.5. Patterns illustrating magnetic lines of force analogous to those discussed previously exhibiting electric lines of force are easily drawn up in this way. Magnetic field patterns

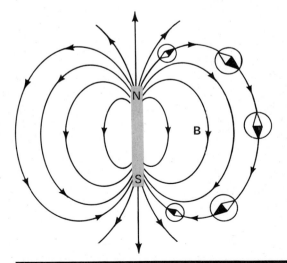

FIGURE 19.5. Magnetic lines of force, mapped with the aid of a compass.

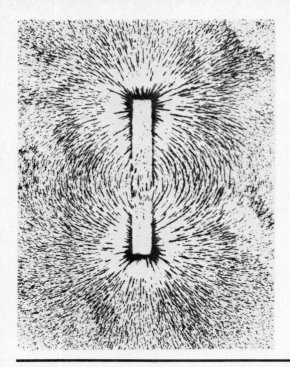

FIGURE 19.6. Pattern formed by iron filings in the field of a bar magnet.

can be made visible by sprinkling iron filings on a sheet of paper placed in a magnetic field. In these patterns, the individual atomic magnetic moments in each of the filings are aligned parallel to the external **B** field when placed in the field. This produces an *induced magnetic dipole moment* in each of the filings, which then aligns with the external field, producing patterns such as the one shown in Fig. 19.6, where the field lines are clearly visible. This phenomenon of *induced mag-*

netism in initially unmagnetized specimens of permanently magnetizable materials is responsible for the attraction of even unmagnetized bits of iron by magnets. The effect, illustrated in Fig. 19.7, resembles the production of an induced electric dipole moment in an uncharged conducting body brought into an electric field. The physical mechanisms that are responsible for induced magnetism differ from those upon which induced electric moments in conductors depend inasmuch as in the electric case the motion of mobile free charges is involved. Induced magnetism and related effects having to do with the behavior of magnetizable materials in magnetic fields will be discussed in more detail in Chapter 21.

A measure of the strength of the **B** field may be obtained by ascertaining the torque required to displace a test dipole through some given angle from the equilibrium position. This torque, according to (19.2.4), is proportional to the magnitude of **B**. Using these methods, we may systematically plot the lines of magnetic induction and determine the relative magnitude of **B** under any experimental conditions. In particular, we shall find certain arrangements of current sources or magnet poles that produce a **B** field *constant both in magnitude and direction* and, play an especially important role in experimental investigations of magnetic phenomena.

Two of these arrangements are of particular importance. The first may be obtained by bending a bar magnet into a circular shape, leaving only a narrow plane-parallel air gap between the two magnet poles. The **B** field will then be practically constant in magnitude and direction within this gap, as illustrated in Fig. 19.8. The analogy between this case and the case of the electric field of a parallel-plate capacitor is

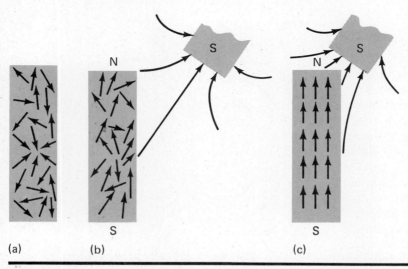

(a) (b) (c)

FIGURE 19.7. Alignment of atomic magnets in an initially unmagnetized piece of iron by the field of a permanent magnet, illustrating the phenomenon of induced magnetism.

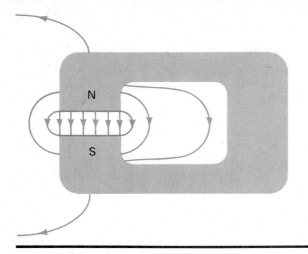

FIGURE 19.8. Uniform magnetic field in a narrow gap separating the north and south poles of a "horseshoe" magnet.

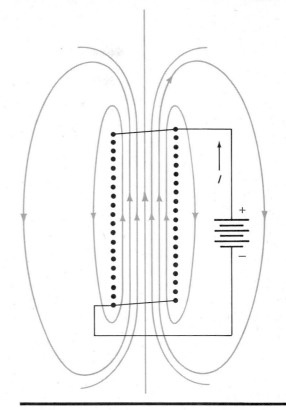

FIGURE 19.9. Uniform magnetic field within a long, closely wound coil carrying a steady current.

evident. The second arrangement by which a constant magnetic induction field may be produced is obtained in the interior of a long, closely wound, current-carrying cylindrical coil. In this instance the strength of the magnetic field is found to be strictly proportional to the *current* in the windings of the coil and may be varied at will by varying the current. This arrangement is illustrated in Fig. 19.9. We shall see later why this particular configuration leads to a constant interior field and what the relation between field and current is. In any event, these easily realizable experimental configurations allow one to perform simple and accurate experiments to determine the nature of the magnetic forces that act upon moving charges in magnetic fields and upon current-carrying conductors in magnetic fields.

In the case of a charged particle, the results of experiments such as these lead to the following conclusions: (i) The magnetic force on a charged particle is perpendicular both to the vector **B** and to the velocity vector of the particle. (ii) The force is proportional in magnitude to the size of the **B** field, the speed of the particle, and the charge it bears. (iii) The force is proportional in magnitude to the sine of the angle between the velocity vector of the charge and the vector **B**. Observations (ii) and (iii) allow us to conclude that the magnitude of the magnetic force on a charged particle is proportional to $qBv \sin \theta$, where θ is the angle between the velocity vector **v** and the field vector **B**. Observation (i) tells us that the force is perpendicular to both **v** and **B**. From the definition of the vector cross product, we can express the magnetic force $\mathbf{F}_m$ as

$$\mathbf{F}_m = \lambda q(\mathbf{v} \times \mathbf{B}) \tag{19.2.5}$$

where λ is a proportionality constant whose value will depend upon the size of the units chosen to express the magnitude of **B**. Naturally, it is best to choose these units in such a way that the constant λ has the value unity. In the mks system, this means that if the magnetic induction has unit magnitude, a point charge of 1 coulomb moving perpendicular to **B** with a speed of 1 meter per second will experience a force (perpendicular to both **B** and **v**) of magnitude 1 newton. From (19.2.5), we may conclude that the mks units of **B** are newton-sec/coulomb-meter. However, since coulombs/sec may be expressed as amperes, we may equally well write them as newtons/amp-meter. The geometric relation between **B**, **v**, and $\mathbf{F}_m$ is illustrated by Fig. 19.10. The size of the mks unit of magnetic induction may be visualized by noting that the earth's magnetic field is typically about 10^{-4} newton/amp-meter, while the induction near the pole face of a powerful permanent magnet is of the order of 0.2 newton/amp-meter. The most powerful electromagnets now available may create fields of about 10 newtons/amp-meter.[1] Using mks units for **B**, Eq.

---

[1] Since the mks unit for magnetic induction is inconveniently large for some purposes, the *oersted*, defined as 10^{-4} newton/ampere-meter, is frequently used instead. If B is measured in oersteds, of course, the constant λ in (19.2.5) is no longer unity but has the value 1.000×10^{-4}. In this book, we shall consistently use the mks units of newtons/ampere-meter throughout.

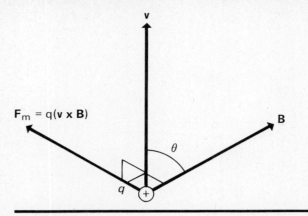

FIGURE 19.10. The relationship between the vectors representing magnetic field, charge velocity, and magnetic force.

(19.2.5) may be expressed as

$$\boxed{\mathbf{F}_m = q(\mathbf{v} \times \mathbf{B})} \tag{19.2.6}$$

If *both* an electric field **E** and a magnetic field **B** are present, there will be an *electrical* force $\mathbf{F}_e = q\mathbf{E}$ as well, in addition to the force given by (19.2.6). The total force on the charge will then be the vector sum of the electrical and magnetic forces, or

$$\mathbf{F} = \mathbf{F}_e + \mathbf{F}_m = q\mathbf{E} + q(\mathbf{v} \times \mathbf{B}) \tag{19.2.7}$$

This is the force experienced by a charged particle under the most general circumstances when both electric and magnetic fields act upon it. It is sometimes referred to as the *Lorentz force expression*, in honor of the Dutch physicist H. L. A. Lorentz (1853–1928). It is important to understand that there is *no magnetic force on a charge at rest*, because no current is then associated with it. Under these circumstances it creates no magnetic field of its own that might allow it to experience forces from external magnetic fields.

EXAMPLE 19.2.1

An electron moves along the x-axis with a speed of 1.00×10^7 meters/sec. There is a constant magnetic induction of 2.50 newtons/amp-meter along the z-direction. Find the magnitude and direction of the magnetic force experienced by the electron. How large an electrostatic field would be needed to produce an electrostatic force of the same magnitude?

According to (19.2.6), the magnetic force is

$$\mathbf{F}_m = q(\mathbf{v} \times \mathbf{B}) = q(v_x\mathbf{i}_x \times B_z\mathbf{i}_z) = qv_xB_z(\mathbf{i}_x \times \mathbf{i}_z)$$

or, since $\mathbf{i}_x \times \mathbf{i}_z = -\mathbf{i}_y$,

$$\mathbf{F}_m = -qv_xB_z\mathbf{i}_y$$
$$= -(-1.602 \times 10^{-19})(1.00 \times 10^7)(2.5)\mathbf{i}_y$$
$$= (4.00 \times 10^{-12} \text{ N})(\mathbf{i}_y)$$

The force is 4.00×10^{-12} newton and is directed along the positive y-direction. For an electrostatic field **E** to produce a force of this same magnitude, we should require that

$$E = \frac{F}{q} = \frac{4.00 \times 10^{-12}}{1.602 \times 10^{-19}} = 2.50 \times 10^7 \text{ V/m}$$

EXAMPLE 19.2.2

A body bearing a charge $q = 3.00 \times 10^{-8}$ coulomb is observed to experience a force **F** whose components are $F_x = 37.5 \times 10^{-8}$ newton, $F_y = -15.0 \times 10^{-8}$ newton, and $F_z = -60.0 \times 10^{-8}$ newton when it moves through a given constant magnetic field with velocity **v** having components $v_x = 20$ meters/sec, $v_y = 50$ meters/sec, and $v_z = 0$. The same charge is acted upon by a force **F'** having components $F_x' = 22.5 \times 10^{-8}$ newton, $F_y' = 7.5 \times 10^{-8}$ newton, and $F_z' = -36.0 \times 10^{-8}$ newton when it is given a velocity **v'** with components $v_x' = 10$ meters/sec, $v_y' = 30$ meters/sec, and $v_z' = 0$ in the same field. Find the magnitude and direction of the constant magnetic induction **B** from these data.

From (19.2.6), we may write

$$\frac{\mathbf{F}}{q} = \mathbf{v} \times \mathbf{B} \tag{19.2.8}$$

or, in terms of components,

$$\frac{1}{q}(F_x\mathbf{i}_x + F_y\mathbf{i}_y + F_z\mathbf{i}_z)$$
$$= (v_x\mathbf{i}_x + v_y\mathbf{i}_y + 0 \cdot \mathbf{i}_z) \times (B_x\mathbf{i}_x + B_y\mathbf{i}_y + B_z\mathbf{i}_z) \tag{19.2.9}$$

noting that $v_z = 0$. Evaluating the cross product and equating the x-, y-, and z-components of the vectors on either side of the resulting equation, it is apparent that

$$\frac{F_x}{q} = v_yB_z \tag{19.2.10}$$

$$\frac{F_y}{q} = -v_xB_z \tag{19.2.11}$$

$$\frac{F_z}{q} = v_xB_y - v_yB_x \tag{19.2.12}$$

Now we know v_x, v_y, F_x, F_y, F_z, and q. We may obtain B_z, therefore, from either (19.2.10) or (19.2.11). On the other hand, we have only one equation, (19.2.12), for B_y and B_x, and there is no way of solving it for both of these quantities. This is a reflection of the fact that there are *different* possible **B** fields that may lead to the *same* magnetic force on a charge with a given velocity. In order to ascertain B_x and B_y uniquely, therefore, we need a second set of data in which the force components are measured with some other known velocity; that is why the force $\mathbf{F}'$ which results when the velocity of the charge is $\mathbf{v}'$ is specified in the statement of the problem. From (19.2.6), $\mathbf{F}'$ must be related to $\mathbf{v}'$ by $\mathbf{F}'/q = \mathbf{v}' \times \mathbf{B}$, the magnetic field being the same in both instances. Once more, we may write out the components, as in (19.2.9), evaluate the cross product, and equate x-, y-, and z-components to get

$$\frac{F_x'}{q} = v_y'B_z \tag{19.2.13}$$

$$\frac{F_y'}{q} = -v_x'B_z \tag{19.2.14}$$

$$\frac{F_z'}{q} = v_x'B_y - v_y'B_x \tag{19.2.15}$$

Again, B_z may be obtained from either (19.2.13) or (19.2.14); we now have no less than *four* ways of evaluating this component. Fortunately, they all agree! From (19.2.10), for example, we find

$$B_z = \frac{F_x}{qv_y} = \frac{37.5 \times 10^{-8}}{(3.00 \times 10^{-8})(50)} = 0.25 \text{ N/amp-m}$$

Equations (19.2.12) and (19.2.15) now give us two independent equations from which B_x and B_y can be obtained; using the given values for the force and velocity components, they can be written as

$$20B_y - 50B_x = -20$$
$$-10B_y - 30B_x = -12$$

Multiplying the bottom equation by 2 and adding it to the top one, we find that $B_x = 0.4$ newton/amp-meter. Substituting this value into either of the two equations above then tells us that $B_y = 0$. It is evident that the vector $\mathbf{B}$ lies in the xz-plane; its magnitude is

$$B = \sqrt{B_x^2 + B_z^2} = \sqrt{(0.4)^2 + (0.25)^2}$$
$$= 0.472 \text{ N/amp-m}$$

The tangent of the angle θ between $\mathbf{B}$ and the z-axis will be given by

$$\tan\theta = \frac{B_x}{B_z} = \frac{0.4}{0.25} = 1.6$$

from which $\theta = 58.0°$.

EXAMPLE 19.2.3

A proton is introduced into a region in which there is a constant magnetic field in the z-direction, with an initial velocity vector lying in the xy-plane and having a magnitude of 5.00×10^6 meters/sec. It is found to move in a circular orbit of radius 12.5 cm in the xy-plane. Explain why the proton follows a circular path, find its angular velocity, and find the magnitude of the magnetic field $\mathbf{B}$ from the given data.

In the constant magnetic field, the proton experi-

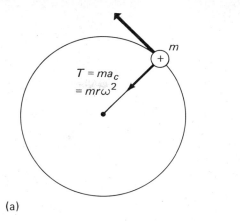

(a)

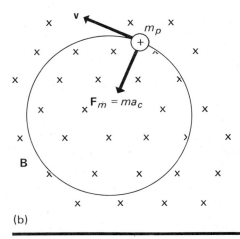

(b)

FIGURE 19.11. Schematic diagram illustrating the role of the magnetic force as a centripetal force for a charged particle moving with constant speed in a plane perpendicular to a uniform magnetic field.

ences a magnetic force of constant magnitude that is always directed perpendicular to the direction of motion. This force, therefore, has all the physical characteristics of a *centripetal* force as mentioned previously, it is similar in all respects to the tension in a string that holds a whirling object executing uniform circular motion in its path. The situation is illustrated in Fig. 19.11. Since the force $\mathbf{F}_m$ is perpendicular to both $\mathbf{B}$ and $\mathbf{v}$, it must play the role of the centripetal acceleration $r\omega^2$. In this way we obtain

$$q|\mathbf{v} \times \mathbf{B}| = qvB = mr\omega^2 \tag{19.2.16}$$

But v and ω are related by the expression $v = r\omega$. Substituting the value $r\omega$ for v in (19.2.16) and solving for ω,

$$\omega = \frac{v}{r} = \frac{qB}{m} \tag{19.2.17}$$

Both v and r are given, so that

$$\omega = \frac{v}{r} = \frac{5.00 \times 10^6}{12.5} = 4 \times 10^5 \text{ rad/sec}$$

Substituting this along with the charge $q = 1.602 \times 10^{-19}$ coulomb and the proton mass $m = 1.67 \times 10^{-27}$ kg into (19.2.17), we may solve for B to obtain 0.00417 newton/amp-meter.

19.3 Magnetic Flux and Gauss's Law for the Magnetic Field

You may recall, from Chapter 16, the way in which the *electric flux* across a given surface was defined as the integral of the component of **E** normal to the surface over a given area. In exact analogy to this, we shall define the *magnetic flux* Φ_m across a given surface S as the integral of the component of **B** normal to the surface over the specified area. For a surface such as that shown in Fig. 19.12, at point P there is a small element of surface area da, and a vector **n** of unit magnitude may be drawn normal to the surface area element at that point. The magnetic induction **B** will not, in general, be constant, either in magnitude or direction over the entire surface, but the vector **B** depicting the local value of the magnetic field at point P may be something like that shown in Fig. 19.12. The component of **B** perpendicular to the surface at this point is simply the component of **B** along the direction of the normal vector **n**, which is $B \cos \theta$. But since **n** has unit magnitude, the scalar product **B · n** also has the value $B \cos \theta$, and the component of **B** normal to the surface can be expressed as

$$B_n = B \cos \theta = \mathbf{B} \cdot \mathbf{n} \tag{19.3.1}$$

The element of magnetic flux across the area da is defined as the component of magnetic induction normal to the surface times the area da, that is,

$$d\Phi_m = B_n \, da = \mathbf{B} \cdot \mathbf{n} \, da \tag{19.3.2}$$

This says that the magnetic flux across da is the component of induction B_n normal to the surface times the area da. The reader may note the precise analogy between this definition of magnetic flux and the relationship between current and current density

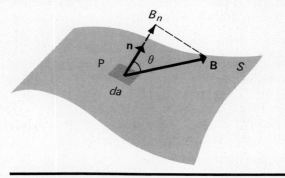

FIGURE 19.12. Vector quantities involved in the definition of magnetic flux.

described in Chapter 18 as Eq. (18.2.9). There, the current dI through the area element da was the normal component of the current density **j** times da—thus, $dI = \mathbf{j} \cdot \mathbf{n} \, da$. The idea here (and also in the case of electric flux discussed in Chapter 16) is the same. The idea of charge flux related to a current intensity vector **j** is one that is quite natural. For the magnetic field, the analogy of a similar "flux" related in the same way to a field intensity vector **B** seems somewhat strained, inasmuch as there is no actual flow of matter or energy to constitute what we ordinarily call a flux. The nineteenth-century scientists who developed the ideas of electric and magnetic fields thought otherwise, however, and imparted a reality to this "flux" of the electric or magnetic field that seems somewhat superfluous to modern physicists. Nevertheless, the terms have stuck with us, and the mathematical relationship between flux and field vectors is the same for electric and magnetic fields as it is for current flow. Therefore, we continue to use the terms although we understand that their literal interpretation in the case of electric and magnetic fluxes is somewhat misleading. We shall see in the next chapter that the concept of magnetic flux is extremely important in discussing emfs induced in electrical circuits by changing magnetic fields.

In any event, Eq. (19.3.2) describes the element of flux $d\Phi_m$ associated with area element da. To find the total magnetic flux Φ_m across the entire surface S in Fig. 19.12, we need only integrate over all the area elements that constitute the surface. In this way, we may write the magnetic flux as the *integral of the component of* **B** *normal to the surface at every point over the entire surface*, or

$$\boxed{\Phi_m = \int_S \mathbf{B} \cdot \mathbf{n} \, da} \tag{19.3.3}$$

In general, it may be a lengthy and difficult task to evaluate this integral. The work is much simpler, however, if the magnetic induction **B** is *constant in magnitude* and *direction* at all points and if the surface S through which the flux is to be calculated is *planar*, so that the normal vector **n** is likewise constant everywhere. Under these circumstances, the quantity **B · n** is the same for all area elements on the surface and may, therefore, be written outside the integral:

$$\Phi_m = \mathbf{B} \cdot \mathbf{n} \int_S da \tag{19.3.4}$$

or

$$\boxed{\Phi_m = (\mathbf{B} \cdot \mathbf{n}) A = BA \cos \theta} \tag{19.3.5}$$

where A denotes the total area of the surface S. If, in addition, the direction of **B** is normal to the surface, the angle θ between **B** and the normal vector **n** is zero and (19.3.5) becomes

$$\Phi_m = BA \qquad (19.3.6)$$

The units of magnetic flux are those of magnetic induction times area, which can be written as newton-meters/ampere, or joules/ampere. These mks units of magnetic flux are very frequently referred to as *webers* (abbreviated Wb), after the German physicist Wilhelm Weber (1804–1891). In fact, the units of magnetic induction itself are more often seen stated as webers/m^2 than as newtons/ampere-meter, which is what we have called them so far. From now on, as a matter of fact, we shall bow to convention and use the units webers/m^2 for B. It is also quite common to refer to the magnetic induction **B** as the *flux density* in view of the fact that its units can be expressed as flux per unit area. The student should be aware of this terminology even though we shall not adopt it here.

We have mentioned previously that magnetic

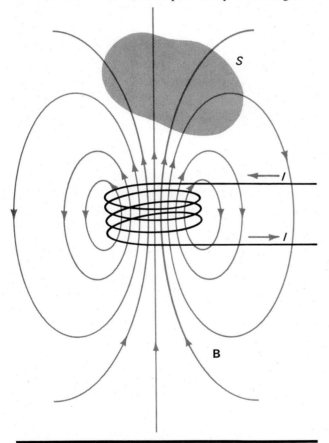

FIGURE 19.13. Magnetic lines of force do not begin or end on free charges, since magnetic charge does not exist. They must, therefore, close upon themselves, forming closed loops as illustrated. In such circumstances, any flux line entering a closed surface must also leave it, leading to Gauss's law for magnetism as given by Eq. (19.3.8).

forces between the poles of long, thin bar magnets (where the effect of poles at the far ends of the magnets are negligible) are found to obey Coulomb's law in that they fall off inversely as the square of the distance between poles. We shall see this inverse square law stated more precisely in section 19.5. You may recall from our previous work that it is precisely this dependence of field intensity on distance that enables us to prove Gauss's law for electric fields. Since the "magnetic charges" that may be regarded as the sources of magnetic dipoles give rise to inverse square magnetic fields, we may prove Gauss's law for magnetism, temporarily visualizing the magnetic field (for the purpose of the proof) to originate from magnetic charge. The proof is carried out in the same way as for the electric field. The result is

$$\oint_S \mathbf{B} \cdot \mathbf{n} \, da = 4\pi k q_m \qquad (19.3.7)$$

The integral of magnetic flux is carried out over *any closed surface S*, and the quantity q_m refers to the total magnetic charge *enclosed within* this "Gaussian surface." The constant k is the proportionality constant in the Coulomb's law expression relating the field B to the magnetic charge q_m and the distance, that is, $B = k(q_m/r^2)$.

Of course, there really is no such thing as "magnetic charge." The only source of any magnetic field is electric currents, and *magnetic fields are invariably dipolar in character* corresponding to zero total equivalent magnetic charge! Therefore, the quantity q_m on the right side of Eq. (19.3.7) *always has to be zero*, and for any realizable magnetic field configuration, Gauss's law for the magnetic field must be

$$\oint_S \mathbf{B} \cdot \mathbf{n} \, da = 0 \qquad (19.3.8)$$

where S is *any closed surface*.

Geometrically, the significance of this result can be understood by recalling that electrostatic lines of force always emanate from, or end upon, electric charges. Since there are no magnetic charges, however, the lines of magnetic induction cannot begin or end anywhere, but instead have to form *closed loops*, as shown in Fig. 19.13. In the case of a permanent magnet, as illustrated in Fig. 19.5, the loops are closed *within* the magnet itself. The requirement that the right side of (19.3.8) be zero is then satisfied inasmuch as all the field lines that go into any closed surface such as S must also come out of that surface somewhere; none can ever begin or end within.

EXAMPLE 19.3.1

Find the magnetic flux through a circular loop 30 cm in diameter whose normal makes an angle of 30°

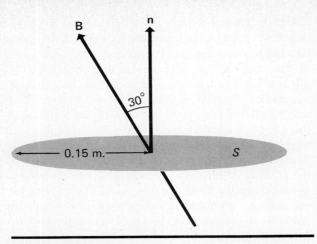

FIGURE 19.14

with a constant magnetic induction of 0.8 weber/meter², as illustrated in Fig. 19.14. What would the flux be if the plane of the loop were perpendicular to the magnetic field?

In this case we must determine the flux through a plane, circular area 30 cm in diameter. The magnetic induction **B** is constant in magnitude and direction, the angle between the vectors **B** and **n** being 30°. We may, therefore, use the simple expression (19.3.5) to calculate the flux:

$$\Phi_m = BA \cos \theta = (0.8)(\pi)(0.15)^2(0.866) = 0.0490 \text{ Wb}$$

If the field were perpendicular to the plane of the loop, then the angle θ would be zero, $\cos \theta = 1$, and

$$\Phi_m = BA = (0.8)(\pi)(0.15)^2 = 0.0565 \text{ Wb}$$

19.4 Forces on Currents and Torques on Magnetic Dipoles

Now that we know what force acts on a moving charge in an externally produced magnetic field, it is not difficult to calculate the magnetic force on a current-carrying conductor in an external field, since such a conductor behaves like a conduit for moving charges. Once that is done, it is in turn fairly easy to inquire into the forces and torques on closed current

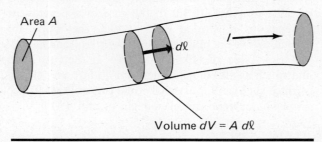

FIGURE 19.15. Flow of steady current in a uniform conductor.

loops and to calculate the strength of the dipole moment of a current-carrying loop.

Let us consider first the case of a conductor of uniform cross section A carrying a steady current I, as shown in Fig. 19.15. The charges within the conductor move with velocity **v** and, in time dt, sweep out a volume $dV = A \, dl$, as shown in the diagram. The displacement vector for such a charge is $d\mathbf{l}$, a vector which points along the conductor at this point. From the definition of instantaneous velocity, we may write the velocity **v** with which charge flows along the conductor as

$$\mathbf{v} = \frac{d\mathbf{l}}{dt} \tag{19.4.1}$$

From Eqs. (19.2.6) and (19.4.1), the element of force $d\mathbf{F}$ acting on the charge element dq is

$$d\mathbf{F} = dq \frac{d\mathbf{l}}{dt} \times \mathbf{B} \tag{19.4.2}$$

But if we denote the charge density per unit volume as ρ,

$$dq = \rho \, dV = \rho A \, dl \tag{19.4.3}$$

Substituting this into (19.4.2), we then find that

$$d\mathbf{F} = \rho A \frac{dl}{dt} (d\mathbf{l} \times \mathbf{B}) \tag{19.4.4}$$

From (19.4.1), we see that dl/dt is simply the magnitude of the velocity v. Furthermore, since charge density times velocity represents the current density, and since current density times area is the total current I, we may replace $\rho A \, (dl/dt)$ in (19.4.4) with I, the final result being

$$\boxed{d\mathbf{F} = I(d\mathbf{l} \times \mathbf{B})} \tag{19.4.5}$$

This equation gives the element of force $d\mathbf{F}$ that acts upon the charge element dq within a segment of the conductor of length $d\mathbf{l}$. The vector $d\mathbf{l}$ in this equation is a vector of length $dl \ (=v \, dt)$ that points along the conductor in the region where the charge

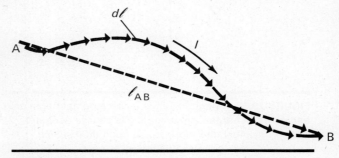

FIGURE 19.16. Vector $\mathbf{l}_{AB}$ connecting two given circuit points, illustrated as the sum of infinitesimal $d\mathbf{l}$ elements along the actual current path.

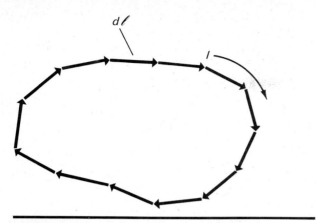

FIGURE 19.17. The vector sum of infinitesimal elements around a closed loop is zero, by the laws of vector addition.

element dq is flowing, as shown in Fig. 19.15. The total force on a finite segment of conductor can be found by integrating (19.4.5) over the successive elements $d\mathbf{l}$ that define such a finite segment, as shown in Fig. 19.16. Equation (19.4.5) is then written

$$\mathbf{F} = \int_c d\mathbf{F} = I \int_A^B d\mathbf{l} \times \mathbf{B} \qquad (19.4.6)$$

The current I is brought outside the integral because the same current flows through each element of the conducting segment. For a complete circuit, the integral is evaluated around the closed loop formed by the conductor, the integral now being written as

$$\mathbf{F} = I \oint_c d\mathbf{l} \times \mathbf{B} \qquad (19.4.7)$$

There are two special cases that deserve particular emphasis. In both of these, the externally produced magnetic field is *uniform* in magnitude and direction over all parts of the conductor. If this is the case, then **B** also can be written outside the integral as a constant, and (19.4.6) becomes

$$\mathbf{F} = \left[I \int_A^B d\mathbf{l} \right] \times \mathbf{B} \qquad (19.4.8)$$

The integral in the brackets above represents simply the *vector sum* of all the tiny $d\mathbf{l}$ vectors shown in Fig. 19.16, and according to the geometric picture of vector addition, this is just the vector $\mathbf{l}_{AB}$ whose tail rests on A and whose head is at B. In this case, (19.4.8) may be written

$$\mathbf{F} = I(\mathbf{l}_{AB} \times \mathbf{B}) \qquad (19.4.9)$$

Our second special case is that where **B** is again uniform and where the current path is a closed loop. Now, the $d\mathbf{l}$ elements form a *closed polygon* as illustrated in Fig. 19.17, their vector sum being zero according to the polygon method for vector addition. The integral in (19.4.8) vanishes, the result now becoming

$$F = 0 \qquad \text{closed loop, } \mathbf{B} \text{ constant} \qquad (19.4.10)$$

In this case, it is important to note that even though the resultant *force* on a current loop may vanish, the magnetic field may, nevertheless, give rise to a net *torque* on the loop, as we shall soon see.

Consider the rectangular current loop shown in Fig. 19.18. In this example, a current I flows around a plane, rectangular loop ABCD as shown. The loop is in a region where there is a constant magnetic field **B** which is directed vertically. The loop is free to rotate about the axis OO' which in the drawing is shown nearly, but not quite, perpendicular to the plane of the page. In the diagram, the plane of the

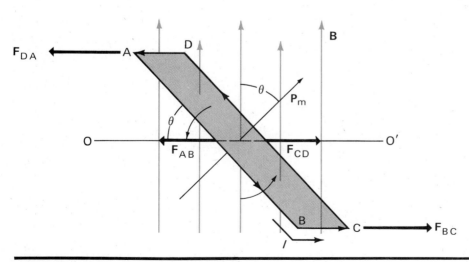

FIGURE 19.18. Rectangular loop carrying a steady current in a magnetic field. The plane of the loop is nearly, but not quite, normal to the page, segment AB being closer to the reader than segment CD.

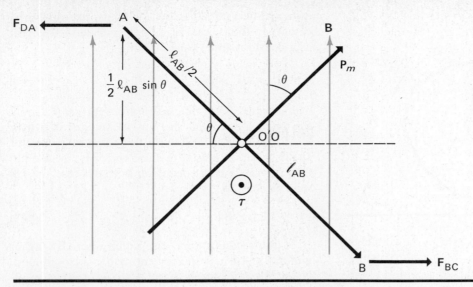

FIGURE 19.19. Another view of the current loop of Fig. 19.18, in which the lines AD, BC, and OO′ are normal to the page.

loop makes an angle θ with the direction of the magnetic induction **B**. Since the current flows around a closed circuit, the resultant force on the loop, according to (19.4.10), must be zero. The forces on the four sides of the rectangle, however, may be computed separately from (19.4.9) to be

$$\mathbf{F}_{AB} = I(\mathbf{l}_{AB} \times \mathbf{B}) \qquad (19.4.11)$$

$$\mathbf{F}_{BC} = I(\mathbf{l}_{BC} \times \mathbf{B}) \qquad (19.4.12)$$

$$\mathbf{F}_{CD} = I(\mathbf{l}_{CD} \times \mathbf{B}) \qquad (19.4.13)$$

$$\mathbf{F}_{DA} = I(\mathbf{l}_{DA} \times \mathbf{B}) \qquad (19.4.14)$$

where vector $\mathbf{l}_{AB}$ extends in the direction of the loop A to B, $\mathbf{l}_{BC}$ from B to C, etc. From the rules for vector multiplication, it is readily apparent that the forces $\mathbf{F}_{AB}$ and $\mathbf{F}_{CD}$ are equal in magnitude ($l_{AB} = l_{CD}$, the loop being rectangular in shape) and opposite in direction, their lines of action being along the axis OO′. These forces clearly give rise to no net torque. The forces $\mathbf{F}_{BC}$ and $\mathbf{F}_{DA}$ are also equal in magnitude, since $l_{BC} = l_{DA}$, and opposite in direction. Because the plane of the loop is tilted at an angle θ to the vertical, however, the lines of action of these two forces do not coincide and therefore give rise to a torque about OO′ as shown, which tends to turn the loop until its plane is normal to the field **B**. These forces are shown more clearly in Fig. 19.19, which is drawn with the axis OO′ exactly perpendicular to the page. From this diagram it is clear that the moment arm associated with $\mathbf{F}_{DA}$ and $\mathbf{F}_{BC}$ is $\frac{1}{2}l_{AB} \sin \theta$ and that the magnitude of the torque τ will then be

$$\tau = \tfrac{1}{2}F_{DA}l_{AB} \sin \theta + \tfrac{1}{2}F_{BC}l_{AB} \sin \theta \qquad (19.4.15)$$

But since $\mathbf{l}_{DA}$ and $\mathbf{l}_{BC}$ are perpendicular to **B**, from

(19.4.12) and (19.4.14) $F_{DA} = Il_{DA}B$ and $F_{BC} = Il_{BC}B$. Using these results and remembering that l_{DA} and l_{BC} are equal, (19.4.15) becomes

$$\tau = Il_{BC}l_{AB}B \sin \theta = IAB \sin \theta \qquad (19.4.16)$$

where A is the *area* of the loop. The direction of the torque *vector* τ is vertically upward out of the page in Fig. 19.19.

This result may be expressed more simply by defining the vector $\mathbf{p}_m$ to be *a vector normal to the loop whose magnitude is given by* IA, the product of the current times the area of the loop. If we do this, then it is readily apparent that the torque vector τ may be expressed as

$$\boxed{\tau = \mathbf{p}_m \times \mathbf{B}} \qquad (19.4.17)$$

because then the magnitude of τ will be that given by (19.4.16) and its direction will be vertically upward out of the page in Fig. 19.19.

It is easily seen that Eqs. (19.4.7) and (19.2.3) are identical. This means simply that the Lorentz force law (19.2.6) and Eq. (19.4.5), which follows directly from it, lead to the conclusion that the torque on a current loop in a magnetic field is exactly that which we should expect to act on a *magnetic dipole* in a magnetic field. This is in accord with how current loops and other electromagnets are actually found to behave when placed in externally produced magnetic fields and can be regarded as one of the important experimental verifications of the Lorentz force law. It is found, moreover, that the *size* of the magnetic dipole moment IA obtained from the Lorentz force law is exactly right in predicting the magnitude of the torques observed on current loops and solenoid electromagnets in magnetic fields. Current loops are found, in fact, to behave in all respects like magnetic dipoles; we shall see later that the magnetic fields they themselves generate are dipole fields.

The results given above are derived rigorously only for rectangular current loops, but they are correct whatever shape the loop may have, provided the dipole moment is expressed as the product of the current in the loop and its area, In relating the direction of the magnetic dipole moment associated with a current loop, it is useful to remember that if your right hand is oriented in such a way that the fingers curl in the direction of current flow in the loop, the thumb points in the direction of the magnetic moment vector $\mathbf{p}_m$ associated with the current. This is evident from Figs. 19.18 and 19.19, but is illustrated more explicitly in Fig. 19.20.

EXAMPLE 19.4.1

A straight segment of a horizontal conductor 15 cm long carrying a current of 75 amperes is placed in a constant vertically directed magnetic induction of magnitude 0.25 weber/m². Find the magnitude and direction of the force on this segment of conductor.

Let us suppose that the conductor is placed so that the current flows along the x-direction, the field $\mathbf{B}$ being in the z-direction. Since the $\mathbf{B}$ field is constant, we may express the force on the conductor by Eq. (19.4.9). Since $l_{AB} = 0.15$ meter and since the current flows along the x-axis, we may write $\mathbf{l}_{AB} = 0.15\mathbf{i}_x$; and since the magnetic induction vector is in the z-direction and is of magnitude 0.25 weber/m², we may express it by $\mathbf{B} = 0.25\mathbf{i}_z$. Equation (19.4.9) then gives

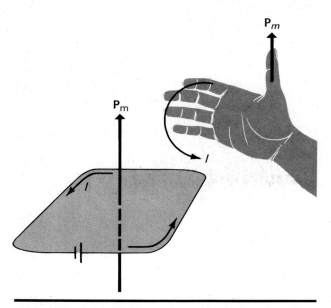

FIGURE 19.20. "Right hand rule" for the direction of the dipole moment (and the axial magnetic field) of a current loop.

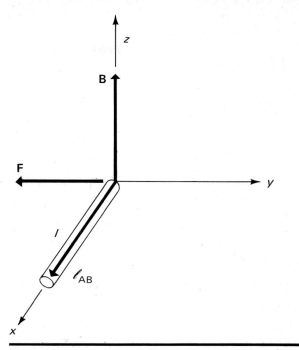

FIGURE 19.21

$$\mathbf{F} = I(\mathbf{l}_{AB} \times \mathbf{B}) = (75)(0.15\mathbf{i}_x \times 0.25\mathbf{i}_z)$$
$$= (75)(0.15)(0.25)(\mathbf{i}_x \times \mathbf{i}_z) = 2.81(-\mathbf{i}_y)$$

The force is, therefore, 2.81 newtons along the $-y$-direction, as shown in Fig. 19.21. Alternatively, one may note that the magnitude of $\mathbf{F}$ is given by

$$F = Il_{AB}B \sin \theta = (75)(0.15)(0.25)(1) = 2.81 \text{ N}$$

since the angle θ between $\mathbf{l}_{AB}$ and $\mathbf{B}$ is 90°, and that the force vector $\mathbf{F}$ must be in the direction shown in Fig. 19.21 as a consequence of the definition of the vector cross product.

EXAMPLE 19.4.2

A *d'Arsonval galvanometer*, commonly used as an accurate current-measuring instrument, is shown schematically in Fig. 19.22. A coil consisting of several hundred turns of very fine wire is pivoted in a uniform magnetic field generated by a permanent magnet. The coil's rotation is restrained by a fine hairspring, so that when current is allowed to flow through the coil, it experiences a torque from the magnetic field and rotates until the magnetic torque is balanced against an equal and opposite restoring torque from the hairspring, and equilibrium is once more attained. In this new state of equilibrium, the coil and its attached pointer have now rotated through an angle ϕ that is related to the current through the coil. Find the general relation between the angle ϕ and the current, assuming that the restoring torque of the hairspring is directly proportional to ϕ. For a magnet giving a uniform field of 0.25 weber/m², a rectangular coil 3 cm

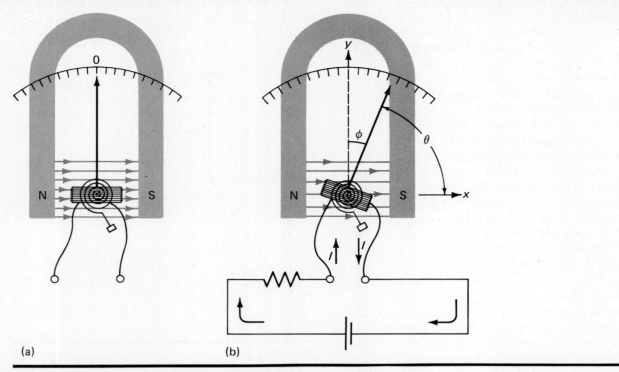

FIGURE 19.22. The d'Arsonval galvanometer.

long and 2 cm wide wound with 300 turns of wire, and a hairspring with a torque constant of 1.60×10^{-7} newton-meter/radian, how much current must flow through the coil to produce a deflection of 30°?

According to (19.4.16), the torque on a single current loop is $IAB \sin \theta$, where θ is the angle between the normal to the plane of the loop and the direction of **B**. In this case, the angle θ is equal to $90° - \phi$, as shown in Fig. 19.22b. Also, since there are many turns of wire in the coil, there are, in effect, N current loops, where N denotes the number of turns. Since the torques on each of these loops are additive, the torque on the coil will be N times the torque on a single loop. Therefore, according to (19.4.16), the torque on a coil of this type will be

$$\tau = NIAB \sin(90° - \phi) = NIAB \cos \phi \qquad (19.4.18)$$

The hairspring will exert a negative restoring torque τ_r on the coil. Assuming that the spring obeys Hooke's law, this torque will be given by

$$\tau_r = -k'\phi \qquad (19.4.19)$$

where k' is the torque constant expressing the torque necessary to produce unit angular displacement. Equilibrium will be reached, and the pointer will come to rest when the resultant torque, equal to the sum of these individual torques, is zero. Under these circumstances,

$$\sum \tau = NIAB \cos \phi - k'\phi = 0 \qquad (19.4.20)$$

Solving this equation for the current, we obtain

$$I = \frac{k'\phi}{NAB \cos \phi} \qquad (19.4.21)$$

Substituting the numerical values given in the statement of the problem into this equation, we find $I = 2.15 \times 10^{-6}$ ampere.

EXAMPLE 19.4.3

Suppose the coil described in the previous example is pivoted so as to allow it to rotate freely about a horizontal axis in a constant vertical magnetic field of 1.5 weber/m². There is no hairspring restraint in this example, but a 1-gram mass is hung from one side of the coil, as shown in Fig. 19.23. Find the angle θ between the vector $\mathbf{p}_m$, which is normal to the plane of the coil, and the magnetic field when a current $I = 2 \times 10^{-4}$ ampere flows through the coil.

Once more the problem is one involving equilibrium of torques. Again the magnetic torque given by (19.4.16) is present, but now there is a torque of opposite sign caused by the weight force of the suspended mass. This torque can be written

$$\tau_m = -mgl = -mga \cos \theta \qquad (19.4.22)$$

while the magnetic torque, from (19.4.16), can be expressed by

$$\tau = NIAB \sin \theta \qquad (19.4.23)$$

620

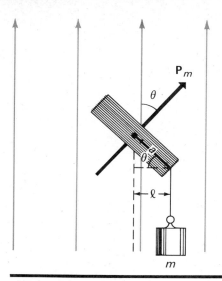

FIGURE 19.23

If the system is in equilibrium, the sum of the torques must be zero; hence,

$$\sum \tau = NIAB \sin \theta - mga \cos \theta = 0 \qquad (19.4.24)$$

Dividing both sides of this equation by $\cos \theta$, noting that $\sin \theta / \cos \theta = \tan \theta$ and solving for $\tan \theta$, we find

$$\tan \theta = \frac{mga}{NIAB} \qquad (19.4.25)$$

In our case, $a = 1.0$ cm $= 0.01$ meter, $m = 0.001$ kg, $N = 300$, $I = 2 \times 10^{-4}$ ampere, $A = 0.0006$ m^2, $B = 1.5$ weber/m^2. These values, substituted into (19.4.25), yield $\tan \theta = 1.815$, $\theta = 61.1°$.

19.5 The Magnetic Field of a Current-Carrying Conductor: Law of Biot and Savart

So far, we have concerned ourselves with trying to describe the forces upon charges and currents that are placed in externally produced magnetic fields. In doing this, we have not considered the question of what sort of magnetic field is produced by currents or moving charges themselves and thus have not yet attacked the problem of describing and explaining the results of Oersted's experiments. Our progress was guided by the necessity of defining and understanding the magnetic induction vector **B**. Having accomplished this objective, we now wish to return to Oersted's experiments and describe the magnetic field produced by a steady current in a uniform conductor.

In our study of electrostatics, we observed that Coulomb's law describing the electric field of point charges was simply the way in which *experimental* observations regarding electrostatic forces on charged

bodies could best be summarized. The situation is the same with regard to the magnetic fields produced by steady currents. There is no way of deriving an expression for these fields; all we can do is observe the magnetic forces created by actual currents experimentally and then try to find a mathematical expression for the magnetic field that agrees with the results of all the observations. It was in just this manner that the *law of Biot and Savart*, which gives the magnetic field created by the current flow in a conductor, was arrived at. The law of Biot and Savart tells us that the element of magnetic induction $d\mathbf{B}$ associated with a current I in a segment of conductor described by the vector $d\mathbf{l}$ is:

1. in a direction perpendicular both to $d\mathbf{l}$ and to the position vector **r** from the segment of conductor to the point P at which the field is being measured, as illustrated in Fig. 19.24;
2. directly proportional to the length dl of the segment and to the current I it carries;
3. inversely proportional in magnitude to the square of the distance r between the current element and the point P;
4. proportional to the sine of the angle θ between the vectors $d\mathbf{l}$ and **r**.

In mathematical form we can write this law as

$$d\mathbf{B} = \text{const} \cdot \frac{I}{r^2} d\mathbf{l} \times \mathbf{i}_r \qquad (19.5.1)$$

where $\mathbf{i}_r$ is a *unit* vector in the direction of the vector **r** in Fig. 19.24. It is evident that Eq. (19.5.1) embodies all the results set down above, for it tells us that $d\mathbf{B}$ is perpendicular both to $d\mathbf{l}$ and **r** and has a magnitude proportional to $I \, dl \sin \theta / r^2$, which is just what is observed. We have previously referred to the fact that magnetic forces exhibit an inverse square dependence on distance, like the Coulomb forces between electric

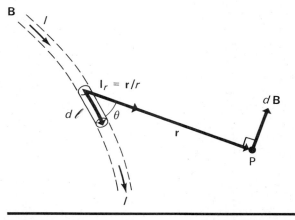

FIGURE 19.24. Magnetic field $d\mathbf{B}$ produced at a point P by a current element $d\mathbf{l}$ in a conductor carrying a steady current I, according to the law of Biot and Savart.

621

charges. This is clearly exhibited by (19.5.1). The constant of proportionality in Eq. (19.5.1) is usually written in the form $\mu_0/4\pi$, for reasons somewhat the same as those stated for writing the proportionality constant for Coulomb's law for electric fields in the form $1/4\pi\varepsilon_0$. Writing the proportionality constant this way, (19.5.1) becomes

$$dB = \frac{\mu_0 I}{4\pi} \frac{dl \times i_r}{r^2} \qquad (19.5.2)$$

Since the unit vector i_r in the direction of $\mathbf{r}$ can be arrived at by taking the vector $\mathbf{r}$ (of magnitude r) and dividing it by that magnitude, we may write

$$i_r = \frac{\mathbf{r}}{r} \qquad (19.5.3)$$

Equation (19.5.2) then becomes

$$dB = \frac{\mu_0 I}{4\pi} \frac{dl \times \mathbf{r}}{r^3} \qquad (19.5.4)$$

The constant μ_0 is a fundamental measure of the strength of the magnetism associated with the flow of electric charge. It is sometimes referred to as the *permeability of free space*. Its numerical value depends, of course, upon the system of units we use to describe the other quantities in the equation. In the mks-metric system of units, this constant has the value

$$\mu_0 = 4\pi \times 10^{-7} \text{ Wb/amp-m}$$

It seems strange that the value of this constant should be an exact multiple of 4π; and in fact this is no mere accident. It stems from the fact that the coulomb is not simply an arbitrary hunk of charge, nor the ampere (coulomb/sec) an arbitrary slug of current. The size of the ampere—and therefore the coulomb—has been chosen in just such a way as to *require* the value of μ_0 to be what it is. We shall see how this happens more clearly later when we discuss the forces that act between two straight parallel conductors carrying steady currents.

Unfortunately, the law of Biot and Savart as stated above tells us only the *differential element* of magnetic induction at a given point arising from a short differential segment of condutor. In order to determine the total magnetic induction arising from a segment of conductor of finite size, we must integrate (19.5.4) over the differential elements dl that make up the finite segment. We may write the total magnetic induction $\mathbf{B}$ in the form of the integral

$$\mathbf{B} = \frac{\mu_0 I}{4\pi} \int_c \frac{dl \times \mathbf{r}}{r^3} \qquad (19.5.5)$$

the integral being taken over the whole conducting segment involved.

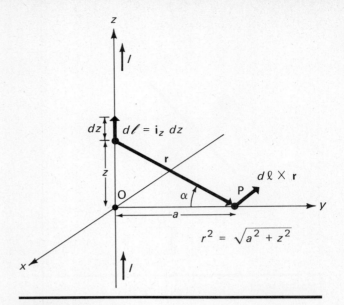

FIGURE 19.25

One simple case in which this may be done is that of an infinitely long, straight conductor of negligible diameter. Let us consider this case in detail. Suppose the conductor coincides with the z-axis, the current I flowing from $z = -\infty$ to $z = +\infty$, as in Fig. 19.25. Let us try to find the magnetic induction at point P a distance a from the conductor along the y-axis. Consider first the part of the induction arising from a current element of length dl extending from z to $z + dz$ along the z-axis. Since the vector dl can be written $dl = i_z\, dz$, and since the vector $\mathbf{r}$, whose y-component is a and whose z-component is $-z$, can be written as $\mathbf{r} = ai_y - zi_z$, Eq. (19.5.4) can be expressed as

$$dB = \frac{\mu_0 I}{4\pi} \frac{i_z\, dz \times (ai_y - zi_z)}{(a^2 + z^2)^{3/2}} = -\frac{\mu_0 I}{4\pi} \frac{a\, dz}{(a^2 + z^2)^{3/2}} i_x \qquad (19.5.6)$$

The expression on the right above results from the expansion of the cross product using the rules for vector multiplication of the unit vectors. It is apparent that the vector dB is in the $-x$-direction; this will be true also for the total magnetic induction $\mathbf{B}$, because the vectors dl and $\mathbf{r}$ always lie in the yz-plane, and their cross product must, therefore, have only an x-component.

We may now evaluate $\mathbf{B}$ by integrating (19.5.6) from $z = -\infty$ to $z = +\infty$ to obtain

$$\mathbf{B} = -\frac{\mu_0 I a}{4\pi} i_x \int_{-\infty}^{\infty} \frac{dz}{(a^2 + z^2)^{3/2}} \qquad (19.5.7)$$

In this expression, the unit vector i_x can be written outside the integral, since it is constant in magnitude and direction. This integral can be evaluated most easily by expressing all functions of z in terms of the angle α shown in Fig. 19.25. Thus,

$$r = \sqrt{a^2 + z^2} = \frac{a}{\cos \alpha} \tag{19.5.8}$$

$$z = a \tan \alpha \tag{19.5.9}$$

$$dz = a \sec^2 \alpha \, d\alpha = \frac{a \, d\alpha}{\cos^2 \alpha} \tag{19.5.10}$$

Substituting all this into (19.5.7), we may write that equation in the form

$$\mathbf{B} = -\frac{\mu_0 I}{4\pi a} \, \mathbf{i}_x \int_{-\pi/2}^{\pi/2} \cos \alpha \, d\alpha = -\frac{\mu_0 I}{4\pi a} \, \mathbf{i}_x [\sin \alpha]_{-\pi/2}^{\pi/2} \tag{19.5.11}$$

In this expression, the integration over the angle α is taken between the limits $-\pi/2$ to $+\pi/2$ because as we integrate over dz from $-\infty$ to $+\infty$, as required by (19.5.7), the angle α in Fig. 19.25 varies between these two values. Evaluating (19.5.11) between these limits, we finally obtain

$$\boxed{\mathbf{B} = \frac{\mu_0 I}{2\pi a} (-\mathbf{i}_x)} \tag{19.5.12}$$

This tells us that the magnitude of $\mathbf{B}$ is $\mu_0 I / 2\pi a$, where a is the distance between the wire and the point P, and that the direction of $\mathbf{B}$ is that of the $-x$-axis. There is nothing particularly important about our having chosen the point P to lie on the y-axis. This was merely a matter of convenience in doing the calculation. To look at it another way, we could first have chosen P and then oriented the y-axis so as to coincide with it. The magnitude of $\mathbf{B}$ is, therefore, the same for *any* point having the distance a from the wire, and its direction will always be perpendicular both to the direction of the current and to the line OP between the point and the wire, since this is the direction of $d\mathbf{l} \times \mathbf{r}$. These results are summarized in the picture of the field shown in Fig. 19.26. The lines of $\mathbf{B}$ are circular, the direction of the $\mathbf{B}$ vector being that assumed by the fingers of the right hand when the thumb points in the current. *A clear appreciation of the characteristics of the field surrounding a long, straight conductor is extremely important in understanding more complex fields and in dealing with the forces that current-carrying conductors exert on one another.* Historically, the result (19.5.12) for a long, straight conductor preceded the more fundamental differential form (19.5.4) and is itself frequently referred to as the Biot–Savart law.

Magnetic fields from different current sources can be *superposed* or added to obtain a total magnetic field, just as the separate fields from a number of electric point charges can be superposed to arrive at a

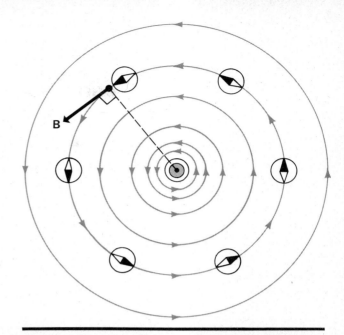

FIGURE 19.26. Magnetic field surrounding a long, straight conductor carrying a steady current.

total electric field. Thus, if there are a number of conductors present carrying currents $I_1, I_2, I_3, \ldots$, each of which gives rise to magnetic fields $\mathbf{B}_1, \mathbf{B}_2, \mathbf{B}_3, \ldots$ according to the law of Biot and Savart as expressed by (19.5.5), the total magnetic induction $\mathbf{B}$ can be represented as the *vector sum* of the individual fields $\mathbf{B}_1, \mathbf{B}_2, \mathbf{B}_3, \ldots$. This allows us to write

$$\boxed{\mathbf{B} = \mathbf{B}_1 + \mathbf{B}_2 + \mathbf{B}_3 + \cdots + \mathbf{B}_n = \sum_{i=1}^{n} \mathbf{B}_i} \tag{19.5.13}$$

where each of the individual fields $\mathbf{B}_i$ is related to its own current source I_i by

$$\mathbf{B}_i = \frac{\mu_0 I_i}{4\pi} \int_{c_i} \frac{d\mathbf{l}_i \times \mathbf{r}_i}{r_i^3} \tag{19.5.14}$$

In this equation, $d\mathbf{l}_i$ are the differential $d\mathbf{l}$ elements that go to make up the conductor c_i through which the current I_i flows, while $\mathbf{r}_i$ is the vector from the differential element $d\mathbf{l}_i$ to the point P where the field is sought. In superposing individual magnetic fields in this way, it is important that they be added as *vectors*, the summation indicated in (19.5.13) being carried out by the laws of *vector* addition.

The simplest example of the superposition of magnetic fields is afforded by the case of two straight parallel conductors, where the resultant field $\mathbf{B}$ is the vector sum of the contributions $\mathbf{B}_1$ and $\mathbf{B}_2$ of the individual fields ascribable to the two currents in the separate wires. This is illustrated in Fig. 19.27.

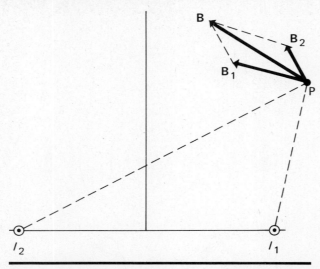

FIGURE 19.27. Superposition of magnetic fields by the law of vector addition.

EXAMPLE 19.5.1

Find the magnetic induction at any point on the z-axis created by a current I flowing in a circular loop of radius a in the xy-plane, whose center is at the origin.

The situation is illustrated in Fig. 19.28. It is important in this example to realize that the total field **B** at all points along the z-axis is in the z-direction and has *only* a z-component. This can be understood by considering the separate contributions to the total **B** field from the element $d\mathbf{l}$ shown in Fig. 19.28 and the element $d\mathbf{l}'$ that is of equal length but diametrically opposed across the current loop. These two contributions $d\mathbf{B}$ and $d\mathbf{B}'$ are in the direction of $d\mathbf{l} \times \mathbf{r}$ and $d\mathbf{l}' \times \mathbf{r}'$ and thus lie in the plane defined by OQP, perpendicular to both the $d\mathbf{l}$ element and the position vector **r** or **r**'. The horizontal components of $d\mathbf{B}$ and $d\mathbf{B}'$ are clearly of opposite sign and thus will add to zero when all the $d\mathbf{l}$ contributions are summed. The vertical components, however, are of like sign and will sum to a definite contribution in integrating over the $d\mathbf{l}$ elements. Since each element of the loop has its diametrically opposed companion $d\mathbf{l}'$ on the other side of the loop, there can be no horizontal component of the resultant **B** field, and hence **B** will have only a vertical z-component. In summing up the contributions from the elements of the current loop, therefore, we need sum only the z-components of the $d\mathbf{B}$ vectors, since we know that the horizontal components cancel out.

Since the vectors on either side of Eq. (19.5.4) are equal, their z-components must be equal. We may then equate the z-components of these vectors to obtain

$$dB_z = \frac{\mu_0 I}{4\pi} \frac{(d\mathbf{l} \times \mathbf{r})_z}{r^3} \qquad (19.5.15)$$

From Fig. 19.28 and the definition of the cross pro-

duct, the z-component of $d\mathbf{l} \times \mathbf{r}$ is given by

$$(d\mathbf{l} \times \mathbf{r})_z = |d\mathbf{l} \times \mathbf{r}| \cos \theta = r\, dl\, \frac{a}{r} = a\, dl \qquad (19.5.16)$$

Also, the arc length dl can be expressed as the product of the radius a of the circular loop and the central angle $d\phi$. Therefore, (19.5.16) becomes

$$(d\mathbf{l} \times \mathbf{r})_z = a^2\, d\phi \qquad (19.5.17)$$

which upon substitution into (19.5.15) gives

$$dB_z = \frac{\mu_0 I a^2}{4\pi r^3}\, d\phi \qquad (19.5.18)$$

This equation may now be integrated over $d\phi$, and in doing so the entire coefficient multiplying $d\phi$ may be written outside the integral, since in summing over all the $d\mathbf{l}$ elements of the loop the quantity r^3 is *the same for each element* and is, therefore, a constant in this particular integration. In this way, we may obtain, integrating around the loop from $\phi = 0$ to $\phi = 2\pi$,

$$B_z = \frac{\mu_0 I a^2}{4\pi r^3} \int_0^{2\pi} d\phi = \frac{\mu_0 I a^2}{2r^3} \qquad (19.5.19)$$

From the figure, it is evident that $r^2 = a^2 + z^2$ and, therefore, that (19.5.19) can be written finally as

$$B_z = \frac{\mu_0 I a^2}{2(a^2 + z^2)^{3/2}} \qquad (19.5.20)$$

This result can be expressed in terms of the magnetic moment p_m of the current loop by recalling that $p_m =$

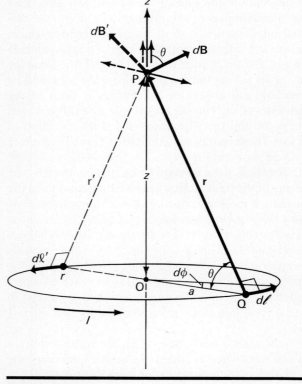

FIGURE 19.28

624

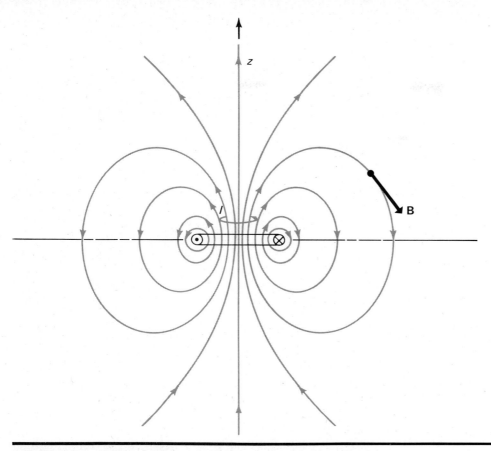

FIGURE 19.29. Magnetic field surrounding a circular current loop.

$\pi a^2 I$. In this way, (19.5.20) becomes

$$B_z = \frac{\mu_0 p_m}{2\pi(a^2 + z^2)^{3/2}} \qquad (19.5.21)$$

It is more difficult to calculate the magnetic induction at points that are not on the axis of the current loop, and we shall not attempt to do so here. Suffice it to say that at large distances from the loop, the field **B** can be shown to be the field of a *dipole* whose moment is $p_m \mathbf{i}_z$. For values of z that are much larger than the radius a, we may neglect the value of a in the denominator of (19.5.21); the expression for the field then becomes

$$B_z = \frac{\mu_0}{2\pi} \cdot \frac{p_m}{z^3}$$

This may be compared with the expression for the axial electric field of an electric dipole derived in Chapter 16; the similarity in form is evident. The lines of **B** for this field are shown in Fig. 19.29; we shall have frequent cause to refer to this field pattern in subquent work.

EXAMPLE 19.5.2

Find the magnetic induction at the center of a circular coil of 100 turns whose diameter is 10 cm and which carries a steady current of 10 amperes.

This example can be regarded as a special case of the preceding one. The field on the axis of a circular current loop is given by (19.5.20). For a circular coil of N turns, there are N current loops rather than one, and the fields of these superpose to give a field N times that expressed by (19.5.19). At the same time, if we are to find the field in the center of the coil, we must set z equal to zero in (19.5.20). Multiplying the right side of (19.5.20) by N and letting $z = 0$ in (19.5.20), we obtain

$$B_z = \frac{N\mu_0 I}{2a} \qquad (19.5.22)$$

In our example, $N = 100$, $I = 10$ amperes and, $a = 5$ cm $= 0.05$ meter, in which case (19.5.22) tells us that

$$B_z = \frac{(100)(4\pi)(10^{-7})(10)}{(2)(0.05)} = 0.0126 \text{ Wb/m}^2$$

EXAMPLE 19.5.3

Find the field at point P in the diagram shown in Fig. 19.30. What is the numerical value of the induction B for a circuit in which $a = 5$ cm, $b = 15$ cm, $l = 10$ cm, and $I = 100$ amperes?

In this example, we must first compute the fields from the straight sections AB and CD and then those from the circular sections BC and DA and finally

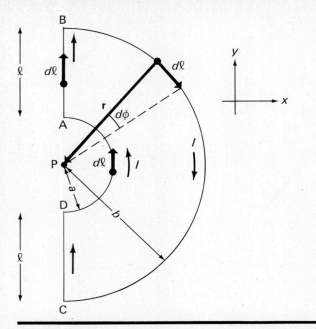

FIGURE 19.30

outer circular section, it is clear that $d\mathbf{l} \times \mathbf{B}$ will have no x- or y-component and that

$$(d\mathbf{l} \times \mathbf{r})_z = -r \, dl = -(b)(b \, d\phi) = -b^2 \, d\phi \qquad (19.5.23)$$

The minus sign is necessary because the direction of the cross product vector $d\mathbf{l} \times \mathbf{r}$ is *into* the page, that is, in the minus z-direction in the set of coordinates we are using. From (19.5.4), then, we may write

$$dB_z = \frac{\mu_0 I}{4\pi} \frac{-b^2 \, d\phi}{b^3} = -\frac{\mu_0 I}{4\pi b} \, d\phi \qquad (19.5.24)$$

noting that $r = b$ at every point on this segment of the circuit. This may be integrated over the semicircle BC from $\phi = 0$ to $\phi = \pi$ to give

$$B_z = -\frac{\mu_0 I}{4\pi b} \int_0^\pi d\phi = -\frac{\mu_0 I}{4b} \qquad (19.5.25)$$

The field $B_z{}'$ from the circular segment DA is calculated in exactly the same way. The only difference is that now the cross product vector $d\mathbf{l} \times \mathbf{r}$ is out of the page, in the positive z-direction. The result of this calculation is of the same form as (19.5.25), except that the minus sign is absent and, b is replaced by a. Therefore, we may write

$$B_z{}' = \frac{\mu_0 I}{4a} \qquad (19.5.26)$$

The total field at P is obtained by vector addition of the two fields given by Eqs. (19.5.25) and (19.5.26).

superpose all of them by vector addition. For the straight segments AB and CD, the $d\mathbf{l}$ elements are *parallel* to the vector $\mathbf{r}$ connecting the current element and the point P at which the field is to be determined. For these segments, $d\mathbf{l} \times \mathbf{r}$ is *zero*, and, therefore, according to (19.5.4), there can be no contribution to the field from these sections of the circuit. On the

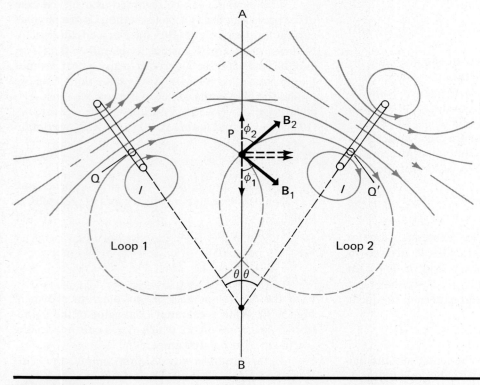

FIGURE 19.31

Since both these fields have only z-components, all that we need do in this case is add these components to give

$$B_{z(\text{tot})} = \frac{\mu_0 I}{4}\left(\frac{1}{a} - \frac{1}{b}\right) \qquad (19.5.27)$$

For $a = 0.05$ meter, $b = 0.15$ meter, and $I = 100$ amperes, this expression tells us that $B_{z(\text{tot})} = 4.19 \times 10^{-4}$ weber/m^2.

EXAMPLE 19.5.4

Figure 19.31 shows two identical circular current loops that carry equal currents I. Their planes are tilted with respect to each other by an angle shown as 2θ in the diagram. Show that anywhere on the plane AB bisecting this angle, the total magnetic induction must be normal to the plane.

This result can be understood by noting that the current loops are identical and carry equal currents and that the fields $\mathbf{B}_1$ and $\mathbf{B}_2$ associated with the circular loops, as shown in Fig. 19.29, are symmetric with respect to the plane in which the loop lies. In the plane AB, all points are *equidistant* from the centers of the two loops. Let us consider one representative line of magnetic induction in the field $\mathbf{B}_1$ produced by the loop on the left, for example, the line QP, and its counterpart PQ' in the field $\mathbf{B}_2$ produced by the right-hand loop. From the symmetry of the field about the planes of the loops and from the symmetric disposition of the plane AB with respect to the loops, it is evident that the arc lengths QP and $Q'P$ along the corresponding lines of $\mathbf{B}$ are equal, as are their angles of intersection with plane AB, ϕ_1 and ϕ_2. If the currents in the two loops are equal, then the magnitudes B_1 and B_2 will also have to be the same. Under these circumstances, it is clear from Fig. 19.31 that the transverse components $B_1 \cos \phi_1$ and $B_2 \cos \phi_2$ are equal in magnitude but opposite in sign and therefore cancel, leaving only the normal components $B_1 \sin \phi_1$ and $B_2 \sin \phi_2$. The resultant $\mathbf{B}$ field at point P, therefore, is normal to the plane AB, and its magnitude is given by

$$B = B_1 \sin \phi_1 + B_2 \sin \phi_2 = 2B_1 \sin \phi_1 \qquad (19.5.28)$$

Since P is arbitrarily chosen and is representative of any point lying in the plane AB, these results will be true for all points in the plane. The results of this example are important, and we shall have occasion to refer back to them when we are investigating the magnetic fields inside toroidal coils and straight solenoids.

19.6 Ampère's Law

In our study of electrostatics, we found that while the electric field set up by a given charge distribution can frequently be determined by superposing the field

contribution of the individual charge elements, the introduction of *Gauss's law* provides us with a most useful and important alternative way of accomplishing the same objective. The same general situation prevails with respect to the determination of magnetic field from their source currents, but with some aspects that are rather different. We have now covered the main features of the determination of magnetic fields by superposition of fields arising from individual current source elements. We might now expect Gauss's law for magnetism to be a useful alternative here just as it is in electrostatics. This is unfortunately not so, however, because in the case of magnetic fields the surface integral of the magnetic flux through a closed surface is invariably *zero* in view of the nonexistence of magnetic charge. Gauss's law for magnetism is informative and has important fundamental significance in relation to the properties of magnetic fields, but it is of very little practical use in helping us to calculate magnetic fields from current distributions.

We must, therefore, seek a different method that will do for us essentially what Gauss's law does in electrostatics. The desired alternative is provided by a theorem called *Ampère's law*, which relates the integral of $\mathbf{B} \cdot d\mathbf{l}$ around any closed path to the total current that flows through the path of integration. Ampère's law is reminiscent of Gauss's law for electric charges in that an integral over an arbitrary closed path is related to the source strength within. The practical application of Ampère's law in calculating magnetic fields, therefore, proceeds along somewhat the same lines as that of Gauss's law, except that Ampère's law involves an integration of *line* elements about a closed curve, while Gauss's law is related to the integration of *surface* elements over a closed surface.

We have shown how Gauss's law for electrostatic fields follows as a direct consequence of Coulomb's force law. It is true also that Ampère's law can be shown to follow directly from the differential form of the law of Biot and Savart expressed by Eq. (19.5.4). Unfortunately, this is a formidable mathematical task, even with the help of mathematics advanced far beyond the level we are constrained to use in this book. We shall, therefore, have to present a proof of Ampère's law that is less comprehensive than we should like it to be, but we shall be careful to note that the results obtained apply in much more general circumstances.

Consider the situation illustrated in Fig. 19.32, where a steady current I flows out of the page along a long, straight conductor normal to the paper. For convenience, we have chosen the coordinate system so that the current flows along the z-axis through the origin in the xy-plane. An arbitrary closed curve C lying in the xy-plane and surrounding the conductor is shown; we shall proceed by finding the integral of $\mathbf{B} \cdot d\mathbf{l}$ over all the $d\mathbf{l}$ elements that make up the curve C.

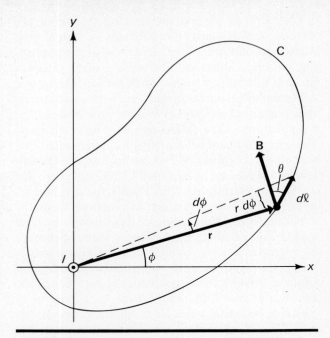

FIGURE 19.32. Calculation of the integral of $\mathbf{B} \cdot d\mathbf{l}$ around a closed path C surrounding a long, straight conductor which carries a steady current.

From the definition of the dot product, we know that

$$\mathbf{B} \cdot d\mathbf{l} = B \cos \theta \, dl \qquad (19.6.1)$$

where θ is the angle between the vectors $\mathbf{B}$ and $d\mathbf{l}$. But it is apparent from Fig. 19.32 that

$$dl \cos \theta = r \, d\phi \qquad (19.6.2)$$

We have also seen that the law of Biot and Savart for a long, straight conductor, as represented by (19.5.12), tells us that the magnitude of $\mathbf{B}$ must be

$$B = \frac{\mu_0 I}{2\pi r} \qquad (19.6.3)$$

Substituting the values given for B and $dl \cos \theta$ by (19.6.2) and (19.6.3) into (19.6.1), we get

$$\mathbf{B} \cdot d\mathbf{l} = \frac{\mu_0 I}{2\pi r} r \, d\phi = \frac{\mu_0 I}{2\pi} d\phi \qquad (19.6.4)$$

We may now integrate over all the dl elements of the curve C; in doing so, the angle ϕ goes from 0 to 2π. Therefore,

$$\oint_C \mathbf{B} \cdot d\mathbf{l} = \frac{\mu_0 I}{2\pi} \int_0^{2\pi} d\phi \qquad (19.6.5)$$

or

$$\oint_C \mathbf{B} \cdot d\mathbf{l} = \mu_0 I \qquad (19.6.6)$$

If the current I does not pass through the closed path C, the situation is as shown in Fig. 19.33. In integrating along the curve from A to B (in a counter-

clockwise sense as shown), the angle ϕ goes from ϕ_1 to ϕ_2, while in returning from B to A, it goes from ϕ_2 to ϕ_1. In this event, the integral over $d\phi$ on the right side of (19.6.5) would be

$$\frac{\mu_0 I}{2\pi} \left(\int_{\phi_1}^{\phi_2} d\phi + \int_{\phi_2}^{\phi_1} d\phi \right) = \frac{\mu_0 I}{2\pi} (\phi_2 - \phi_1 + \phi_1 - \phi_2) = 0$$

$$(A \to B) \quad (B \to A) \qquad (19.6.7)$$

We see thus that the integral of $\mathbf{B} \cdot d\mathbf{l}$ around a closed curve through which a current I passes is equal to $\mu_0 I$, but that if the current does not flow through the closed curve, the integral is zero. This is the essential content of Ampère's law.

Our proof is restricted to the case of a straight conductor of infinite length and a curve C that lies in a plane normal to it. It is easily enough extended to the case where there are a number of long, straight parallel conductors carrying currents $I_1, I_2, I_3, \ldots, I_n$, giving rise to individual fields $\mathbf{B}_1, \mathbf{B}_2, \mathbf{B}_3, \ldots, \mathbf{B}_n$, the vector sum of which is the total field $\mathbf{B}$. In this case, we calculate the integral of $\mathbf{B}_1 \cdot dl$, $\mathbf{B}_2 \cdot dl$, $\mathbf{B}_3 \cdot dl$, etc., around the curve C, getting $\mu_0 I_1, \mu_0 I_2, \ldots$, so long as the currents pass through C, but *getting zero if they do not*. The integrals are then added up, the result being

$$\oint_C (\mathbf{B}_1 + \mathbf{B}_2 + \mathbf{B}_3 + \cdots + \mathbf{B}_n) \cdot d\mathbf{l} = \mu_0 I_c \qquad (19.6.8)$$

or

$$\oint_C \mathbf{B} \cdot d\mathbf{l} = \mu_0 I_c \qquad (19.6.9)$$

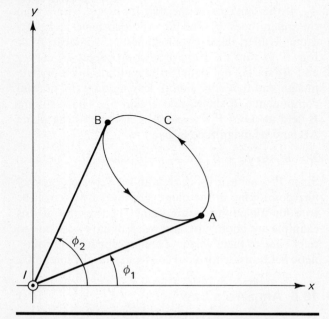

FIGURE 19.33. Calculation of Fig. 19.32 in the case where the path of integration does not surround the conductor.

where I_C represents the *total amount of current flowing through the arbitrary closed path of integration C*. In general, this will be less than the sum of all the currents $I_1, I_2, I_3, \ldots, I_n$ by the sum of those currents that do not pass through C and, therefore, give zero contribution to the integral.

However, Ampère's law is much more general than this. It matters not whether the conductors are straight or curved, or long or short, or parallel or not. Indeed, it makes no difference whether there are individual conductors or a continuous distribution of current density. The path of integration C does not have to lie in a plane normal to the current nor, in fact, does it have to lie in a plane at all. The relation (19.6.9) expressing Ampère's law is valid in every case.

The application of Ampère's law proceeds, to a degree, along lines similar to those followed in using Gauss's law in electrostatics. It is usually desirable to choose a contour C along which the *symmetry* of the current distribution permits simple evaluation of the integral of $\mathbf{B} \cdot d\mathbf{l}$. In situations where this symmetry is absent, it may be difficult or impossible to use Ampère's law in the form given above to relate $\mathbf{B}$ to the current distribution. The applications of Ampère's law will be illustrated by the following series of examples.

EXAMPLE 19.6.1

Use Ampère's law to find the magnetic induction $\mathbf{B}$ in the neighborhood of an infinitely long cylindrical conductor of radius a that carries a current I. Assume that the current density is *uniform* over the cross section of the conductor.

The symmetry of the situation allows us to guess that the lines of $\mathbf{B}$ are everywhere tangential, as shown in Fig. 19.34. So far as symmetry is concerned, they might, of course, be radial, but we know from the form of Biot and Savart's law (19.5.4) and from our experience in deriving (19.5.12) that this is unrealistic. Under the given conditions, it is difficult to see how a component of $\mathbf{B}$ other than a tangential one might exist. If it does, why should it have the magnitude it has, and why is its direction what it is? The unanswerability of these questions leads us to conclude that there are, in fact, no components of $\mathbf{B}$ in the radial direction or in the z-direction.

This being the case, it makes sense to choose *circular* paths such as C and C' to carry out the integration specified in Ampère's law. The vectors $\mathbf{B}$ and $d\mathbf{l}$ are always parallel on such a contour, so that for these paths

$$\mathbf{B} \cdot d\mathbf{l} = B \, dl \qquad (19.6.10)$$

Furthermore, the complete circular symmetry of conductor and current flow about point O leads us to conclude that the *magnitude* of $\mathbf{B}$ will be the same at

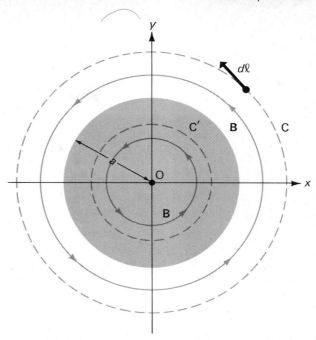

FIGURE 19.34

all points of such a path. Why would it be different at one point on C, for example, from what it is somewhere else? Again, the unanswerability of the question leads to the inevitable conclusion that in fact it isn't! Under these circumstances, B is the same for all the $d\mathbf{l}$ elements that make up path C and, therefore, can be regarded as constant in the evaluation of the integral of $\mathbf{B} \cdot d\mathbf{l}$. In view of these considerations and of (19.6.10), Ampère's law (19.6.6) can be written

$$\oint_C \mathbf{B} \cdot d\mathbf{l} = \oint_C B \, dl = B \oint_C dl = \mu_0 I_C \qquad (19.6.11)$$

Now the integral of dl around the path C is simply the length of the path, which in this case is the circumference of the circle C. Denoting the radius of the circular path C by r, (19.6.11) then can be written as

$$Bl = 2\pi r B = \mu_0 I_C \qquad (19.6.12)$$

Now, I_C is the total current flowing through the closed curve C. If the radius r of the path C is greater than the radius of the conductor, I_C is simply the total current I carried by the conductor. In this instance,

$$2\pi r B = \mu_0 I$$

from which

$$B = \frac{\mu_0 I}{2\pi r} \qquad r > a \qquad (19.6.13)$$

This expression gives the value of B for any value of r greater than a, that is, everywhere *outside* the conductor. You will note that the expression has the same form as (19.5.12), which is perhaps not surprising, though this is not quite the same situation

629

as that considered in deriving (19.5.12), because there, the current was confined to a line along the z-axis rather than a conductor of finite radius. For points *inside* the conductor, we must proceed in a slightly different way.[2] Suppose we wish to apply Ampère's law around the circular path C' which lies within the conductor and whose radius r is less than the radius of the conductor. The integral of $\mathbf{B} \cdot d\mathbf{l}$ is evaluated in exactly the same way and still has the value $2\pi rB$ given by (19.6.12). But now it must be equated to μ_0 times $I_{C'}$, which represents the current that flows through the path C', and this is now only a *fraction* of the total current I. If we assume that the current density j is uniform over the cross-sectional area of the conductor, the current which passes through C' is simply the current density times the area within the closed path C'. Accordingly,

$$I_{C'} = j\pi r^2 \qquad (19.6.14)$$

Because the current density is the total current I divided by the total area of the conductor, which is πa^2, (19.6.14) can be written

$$I_{C'} = I\frac{\pi r^2}{\pi a^2} = I\frac{r^2}{a^2} \qquad (19.6.15)$$

If we now equate the integral of $\mathbf{B} \cdot d\mathbf{l}$, which has the value $2\pi rB$ as before, to $\mu_0 I_{C'}$, according to Ampère's law we may now write

$$2\pi rB = \mu_0 I_{C'} = \mu_0 I \frac{r^2}{a^2} \qquad (19.6.16)$$

We can then solve for B to get, for r *less* than a,

$$B = \frac{\mu_0 I r}{2\pi a^2} \qquad r < a \qquad (19.6.17)$$

So, outside the conductor, B is given by (19.6.13) and falls off with increasing distance as $1/r$, while inside the conductor it is expressed by (19.6.17) and increases linearly as a function of r from the center to the outside surface. At the surface, where $r = a$, both (19.6.13) and (19.6.17) lead to the same value for the field, $B = \mu_0 I/2\pi a$. Since the $\mathbf{B}$ vector is everywhere tangential to circular paths such as C or C', the lines of $\mathbf{B}$ are circular, as illustrated in Fig. 19.34. The variation of $\mathbf{B}$ as a function of r is illustrated in Fig. 19.35. For a conductor of radius 1 mm carrying a current of 100 amperes, the maximum field, which occurs at the surface of the conductor, will be

$$B = \frac{\mu_0 I}{2\pi a} = \frac{(4\pi)(10^{-7})(100)}{(2\pi)(10^{-3})} = 0.0200 \text{ Wb/m}^2$$

---

[2] While it is true that in electrostatic equilibrium there can be no *electric* field within a conducting body, there is no law against there being a magnetic field within a conductor. As we shall soon see, the magnetic fields within conductors can be as strong or stronger than those outside.

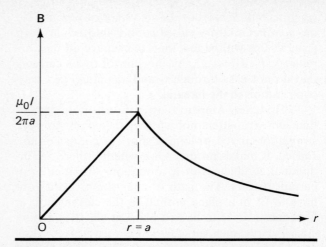

FIGURE 19.35. Plot of the field magnitude *vs* the radial distance r from the center of the conductor shown in Fig. 19.34.

EXAMPLE 19.6.2

Use Ampère's law to find the magnetic field of an infinitely long, hollow concentric cylindrical conductor of outer radius a and inner radius b carrying a total current I. Determine the field outside the conductor, within the conducting substance, and within the hollow tubular cavity near the center. Assume that the current density is constant throughout the conductor.

A cross-sectional diagram of this conductor is shown in Fig. 19.36. Once more, the symmetry of the conductor and the current flow is such that we may safely assume the lines of $\mathbf{B}$ to be circular, and thus the $\mathbf{B}$ vector always to be tangential, as illustrated. For this reason, if we choose circular paths such as C, C', and C'' to evaluate the Amperian integral, we shall once more find that $\mathbf{B}$ and the $d\mathbf{l}$ elements of

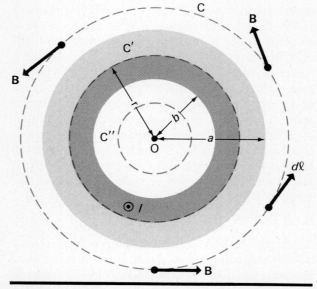

FIGURE 19.36

the path are always parallel and that $\mathbf{B} \cdot d\mathbf{l}$ reduces simply to $B \, dl$. In this instance also, the symmetry is such that the magnitude of $\mathbf{B}$ will be the same everywhere on the circular contours and can, therefore, be written outside the integral. As in the previous example, we may write Ampère's law as

$$\oint_C \mathbf{B} \cdot d\mathbf{l} = B \oint_C dl = \mu_0 I_C \qquad (19.6.18)$$

Again, the integral of dl around the path represents simply the total arc length of the path, in this case the circumference of the circular contour. Therefore, for the contour C, which is outside the conductor altogether, we have

$$2\pi r B = \mu_0 I_C = \mu_0 I \qquad r > a \qquad (19.6.19)$$

where r is the radius of the circular path C and I is the total current carried by the conductor. Solving for B,

$$B = \frac{\mu_0 I}{2\pi r} \qquad r > a \qquad (19.6.20)$$

This is the same result found in the previous example for a solid cylindrical conductor. By simply observing the magnetic field outside the conductor, therefore, we cannot tell whether it is solid or hollow.

Inside the conductor, on a path of integration such as C', we must equate the Amperian integral to that part of the total current $I_{C'}$ that passes through the contour. The current within the contour can be written as the product of the constant current density j and the area of the conductor's cross section that falls inside the contour C' of radius r, shown heavily shaded in Fig. 19.36. Since this area is $\pi r^2 - \pi b^2$, we have

$$I_{C'} = j\pi(r^2 - b^2) \qquad (19.6.21)$$

But since we can express the current density j as the total current I divided by the total cross-sectional area of the conductor, which in this case is $\pi a^2 - \pi b^2$, we may write (19.6.21) as

$$I_{C'} = \frac{I(r^2 - b^2)}{a^2 - b^2} \qquad (19.6.22)$$

As before, the integral of $\mathbf{B} \cdot d\mathbf{l}$ around C' is still $2\pi r B$, so that Ampère's law gives us

$$2\pi r B = \mu_0 I_{C'} = \frac{\mu_0 I(r^2 - b^2)}{a^2 - b^2} \qquad (19.6.23)$$

or

$$B = \frac{\mu_0 I(r^2 - b^2)}{2\pi r(a^2 - b^2)} \qquad b < r < a \qquad (19.6.24)$$

When the radius of the path of integration is *less* than the inner radius of the conductor, as is the case for the contour C'' in Fig. 19.36, there is no current flowing within the closed path of integration. For any

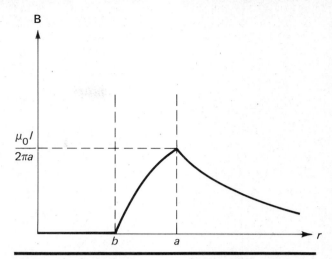

FIGURE 19.37. Plot of the field magnified *vs* the radial distance from the center of the system, for the hollow conductor shown in Fig. 19.36.

such path, therefore, $I_{C''} = 0$ and Ampère's law gives

$$2\pi r B = \mu_0 I_{C''} = 0$$

whence

$$B = 0 \qquad r < b \qquad (19.6.25)$$

The magnetic induction within the hollow interior of the conductor is, therefore, zero. This result is somewhat similar to that obtained from Gauss's law in electrostatics for the case of a hollow spherical charged body, where it is found that the electric field in the interior spherical cavity is zero. It is also reminiscent of the vanishing of the gravitational field in the interior of a hollow spherical planet. In the case of the magnetic field, however, we find that the field vanishes within hollow *cylinders* rather than spheres because the Biot–Savart law is somewhat different in form from Coulomb's law. It will be noted once more that when $r = a$, both (19.6.20) and (19.6.24) reduce to $B = \mu_0 I/2\pi a$, while for $r = b$, (19.6.24) gives $B = 0$, in agreement with (19.6.25). A plot of the field B as a function of r as given by (19.6.20), (19.6.24), and (19.6.25) is shown in Fig. 19.37.

EXAMPLE 19.6.3

Figure 19.38 shows the cross section of an infinitely long cylindrical conductor in which an off-center hole has been drilled. The conductor carries a current I that is uniformly distributed over the conducting cross-sectional area. The outer radius of the conductor is a, the radius of the hole is b, and the axis of the conductor and the hole are separated by a distance c, as shown in the figure. Find the magnitude and direction of the magnetic induction at the axis O' of the hole.

This is a difficult example which at first sight seems to defy all the standard methods of attack.

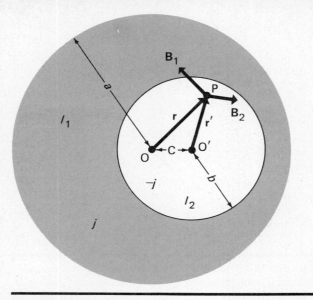

FIGURE 19.38

Since the hole is off center, we can neither conclude that the **B** field is zero within it nor evaluate the Amperian integral easily around a circular contour centered on either O or O′, since we are now no longer justified in assuming that the **B** field is tangential with respect to either point. This means that on such a circular path, **B** and $d\mathbf{l}$ are no longer parallel and $\mathbf{B} \cdot d\mathbf{l}$ is no longer equal to $B\,dl$ but rather to $B\,dl\cos\theta$, where θ is the unknown angle between the two vectors. Nor is it at all clear how to choose a noncircular path on which **B** and $d\mathbf{l}$ *are* parallel. The answer can be obtained by superposing field contributions from all the current elements in the conductor according to the law of Biot and Savart (19.5.4), but the calculation is a difficult one, definitely not recommended for those who are the least bit slow at mathematics. Nevertheless, it can be solved fairly easily by just the right combination of Ampère's law and superposition if one proceeds from the starting point of physical insight rather than mathematical formalism. Indeed, it provides a valuable object lesson in demonstrating the value of physical insight over straight mathematical calculation. It is an end run, so to speak, rather than a plunge into the line.

The trick is to realize that the current distribution specified by the statement of the example can be realized by superposing a uniform current density j (out of the page) that extends over the *entire* interior of the conductor, *including* the hole, and an equal but oppositely directed current density $-j$ (into the page) that extends only over the region where the hole is situated. These two distributions sum to the uniform current density j in the conductor and to zero in the region where the hole is supposed to be, which is just what we are given. We may then calculate the mag-

netic field $\mathbf{B}_1$ arising from the first of the distributions mentioned, using Ampère's law. In doing this, the symmetry of current flow and conductor that enables us to make the usual simplifying assumptions regarding $\mathbf{B}_1$ and $d\mathbf{l}$ is now perfectly reasonable. In the same way, we can calculate the magnetic field $\mathbf{B}_2$ arising from the second current distribution in the region where the hole is supposed to be, again using these same simplifying assumptions. Then, we simply superpose the fields $\mathbf{B}_1$ and $\mathbf{B}_2$ by vector addition to obtain the total field **B** associated with the total current distribution, which is that specified in the statement of the example!

Actually, we do not need to go through the Ampère's law calculations, since we already know the results from Example 19.6.1. Referring to Eqs. (19.6.14) and (19.6.16), we can easily see that the field $\mathbf{B}_1$ in the interior of the cylindrical conductor of radius a with uniform current density j will be

$$B_1 = \frac{\mu_0 j \pi r^2}{2\pi r} = \frac{\mu_0 j r}{2} \qquad (19.6.26)$$

The lines of $\mathbf{B}_1$ are circles centered upon O and the field $\mathbf{B}_1$ at any point such as P will always be tangential to the vector **r** connecting it with the origin. In the case of $\mathbf{B}_2$, we have a field arising from a current density $-j$ (into the page) spread over a circular region centered on O′. According to Eqs. (19.6.14) and (19.6.16), the magnitude of this field will be

$$B_2 = \frac{\mu_0 (-j) \pi r'^2}{2\pi r'} = -\frac{\mu_0 j r'}{2} \qquad (19.6.27)$$

where r' is the distance from the point P, where the field is to be found, to the center of the current density distribution at O′. The lines of $\mathbf{B}_2$ are circles centered on O′ and the direction of $\mathbf{B}_2$ at a point such as P will naturally be tangential to the vector **r′** connecting the point with O′. From this, it is clear that $\mathbf{B}_1$ and $\mathbf{B}_2$ are, in general, in different directions, and that in expressing the total field **B** as the sum $\mathbf{B}_1 + \mathbf{B}_2$, the laws of vector addition must be carefully taken into account. Even so, the directions of the two fields are specified rather simply by the vectors **r**, **r′**, and OO′; and though some fairly tedious calculation has to be done, it is not too difficult to arrive at mathematical expressions for the components of **B** at any point.

But we are not asked to do this. We are required only to find the magnitude and direction of $\mathbf{B}_1 + \mathbf{B}_2$ at the point O′. If the point P were to move to O′, we would have $r' = 0$; and from (19.6.27), $B_2 = 0$. Under these circumstances, the vector **r** would extend from O to O′, and its magnitude would be $r = c$. At this point, therefore, $\mathbf{B} = \mathbf{B}_1$, and from (19.6.26),

$$B = B_1 = \frac{\mu_0 j c}{2} \qquad (19.6.28)$$

It is evident from Fig. 19.38 that the direction of **B** under these conditions would be vertically upward in the plane of the page.

All we need do now is find the relationship between the current density j and the total current I carried by the conductor. The total current I_1 that was assumed to serve as a source for $\mathbf{B}_1$ is given by the current density j times the total area within the outer boundary of the conductor, from which

$$I_1 = \pi a^2 j \qquad (19.6.29)$$

In the same way, the total current I_2 that acts as the source of $\mathbf{B}_2$ is the current density $-j$ times the area of the hole:

$$I_2 = -\pi b^2 j \qquad (19.6.30)$$

The algebraic sum of these two currents is the net current I actually carried by the conductor:

$$I = I_1 + I_2 = \pi j (a^2 - b^2)$$

or

$$j = \frac{I}{\pi(a^2 - b^2)} \qquad (19.6.31)$$

Substituting this into (19.6.28), we get finally

$$B = \frac{\mu_0 I c}{2\pi(a^2 - b^2)} \qquad (19.6.32)$$

You may note that when $c = 0$, the hole is concentric. In this case, (19.6.32) gives $B = 0$, which agrees with the result obtained in Example 19.6.2.

19.7 Magnetic Fields Within Toroidal Coils and Straight Solenoids

Toroidal coils and straight solenoids are of particular importance in experimental work involving electromagnetism in the laboratory and in technology as well. We shall find, also, that we can add to our understanding of Ampère's law in describing their properties. We shall, therefore, now undertake an investigation of these structures from the starting point of Ampère's law.

One might naturally conclude that the best way to begin would be with the infinitely long, closely wound, straight solenoidal coil, which is to all appearances the simplest kind of electromagnet. Indeed, many textbooks do begin by using Ampere's law in a very simple derivation of the magnetic induction within such a coil, the sole virtue of which is that it gets the right answer. The application of Ampère's law to a straight solenoid is not a straightforward matter, however, if one insists upon understanding clearly the answers to all the questions that suggest themselves in the process.

We shall, therefore, proceed by first investigating the properties of closely wound *toroidal* coils, which, oddly enough, are really simpler and easier to understand than straight solenoids. We have already laid the foundation for the work we are about to do in Example 19.5.4 and in Fig. 19.31, which is related to it. The motivation of this example was eventually to show that the **B** field within a toroidal coil such as the one illustrated in Fig. 19.39 is purely tangential and has no component whatsoever in a plane perpendicular to the axis of the solenoid, that is, in a plane such as AB. The resemblance between the situations depicted in Figs. 19.31 and 19.39 is immediately evident. In the case of Fig. 19.31 and Example 19.5.4, we showed that two circular loops disposed as shown give rise to a field that is normal to the midplane AB at every point.

In Fig. 19.39, the point is to show that a toroidal coil can be regarded as being made up of pairs of current loops like those shown in Fig. 19.31, each of which is disposed similarly with respect to the plane AB, although at different angles θ for each such pair. Since the field on plane AB is simply the superposition of all the fields associated with the individual pairs of current loops, the total field **B** everywhere on the plane is likewise normal to the plane in direction. Since there is nothing special about where we chose to put AB to begin with, the same statement can be made about the direction of **B** anywhere in the toroid. The conclusion is that the **B** field in a toroidal coil must always be tangential, and the lines of **B** must, therefore, be circular, as shown in Fig. 19.39. From symmetry alone, we may also conclude that the *mag-*

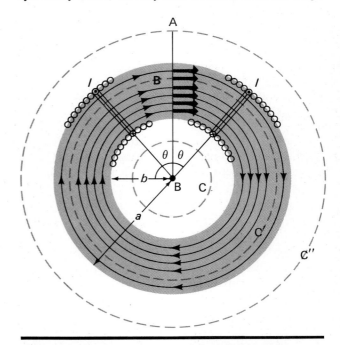

FIGURE 19.39. Toroidal coil carrying a steady current I.

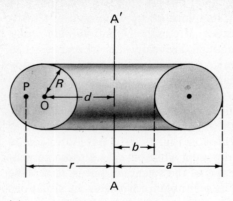

(a)

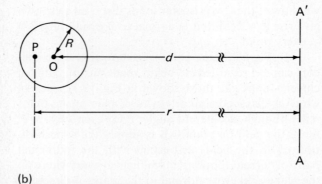

(b)

FIGURE 19.40. Geometric parameters of (a) "fat" and (b) "thin" toroids.

nitude of **B** is the same everywhere on any circular path centered on A.

Now, suppose that the outer radius of the coil is a and the inner radius b. This means that the *average* radius is given by

$$d = \frac{a + b}{2} \tag{19.7.1}$$

The average circumference c, then, will be

$$c = 2\pi d = \pi(a + b) \tag{19.7.2}$$

These dimensional parameters are illustrated in Fig. 19.40a. Let us assume that the total number of turns of wire, or current loops, is N. Then the number of turns per unit length are

$$n = \frac{N}{c} = \frac{N}{\pi(a + b)} \tag{19.7.3}$$

Let us now apply Ampère's law to the coil shown in Fig. 19.39, using circular paths such as C, C', and C'' as contours along which to evaluate the integral of $\mathbf{B} \cdot d\mathbf{l}$. Along a path such as C, we may write Ampère's law as

$$\oint_C \mathbf{B} \cdot d\mathbf{l} = B \oint_C dl = 2\pi rB = \mu_0 I_C \tag{19.7.4}$$

where r is the radius of the circular path C. But for a path such as C whose radius is less than the inner radius b, $I_C = 0$, and therefore (19.7.4) tells us that

$$B = 0 \qquad r < b \tag{19.7.5}$$

Along a curve such as C', the current $I_{C'}$ passing through the closed path of integration is simply NI, since there are N current loops each of which carries current I. Ampère's law evaluated along such a path gives

$$\oint_{C'} \mathbf{B} \cdot d\mathbf{l} = B \oint_{C'} dl = 2\pi rB = \mu_0 NI \tag{19.7.6}$$

Therefore, inside the toroid, for $b < r < a$, we may write the field as

$$B = \frac{\mu_0 NI}{2\pi r} \qquad b < r < a \tag{19.7.7}$$

Since from (19.7.3), $N = n\pi(a + b)$, where n is the average number of turns per unit length of coil, (19.7.7) can also be written

$$B = \frac{\mu_0 n(a + b)I}{2r} = \mu_0 nI \frac{d}{r} \tag{19.7.8}$$

In the very center of the toroid, where $r = d = \frac{1}{2}(a + b)$, the field B, will be given by

$$B_c = \mu_0 nI \tag{19.7.9}$$

This expression can be regarded as giving the *mean* field within the toroidal coil.

For circular paths such as C'', whose radius is greater than the maximum radius of the toroid, $I_{C''}$ again becomes zero since for each turn there is a current I going into the plane of the page and an equal but opposite current $-I$ coming out. These currents add to zero for every loop, giving $I_{C''} = 0$. Ampère's law therefore gives

$$\oint_{C''} \mathbf{B} \cdot d\mathbf{l} = 2\pi rB = \mu_0 I_{C''} = 0 \tag{19.7.10}$$

from which

$$B = 0 \qquad r > a \tag{19.7.11}$$

These results are remarkable in that inside the toroid there may be a very strong magnetic field as given by (19.7.8) or (19.7.9), while outside the toroid there is no field at all! This property of toroidal coils leads to their widespread use in electrical circuits where it is necessary to generate a strong magnetic field but undesirable to have stray exterior fields that may interfere with the operation of other circuit elements. Actually, the windings of toroidal coils, or any coil for that matter, are not really circular current loops but *spirals* whose pitch is so small that they are very nearly circular turns. For this reason, in addition to flowing around the circular cross section, the current also *advances along the toroid*, eventually traversing its entire length. This means that the external field of an actual toroidally wound coil is never really zero, but has instead a small residual value something like that which would be associated with the current I flowing in a single circular loop of radius d. This may be thousands of times smaller than the interior field, nevertheless.

The case of the very long straight solenoid can be treated as a limiting case of the toroidal coil in which the mean radius d is allowed to become infinitely large, while the cross-sectional radius R is held constant, as illustrated in Fig. 19.40. It is true, after all, that a straight line can be regarded as a segment of a circle of infinite radius. From Eqs. (19.7.8) and (19.7.9), it is immediately clear that the field at the center of the toroid always has the value $B_c = \mu_0 nI$, which is altogether *independent* of the radius d. This equation, then, will give us the field in the center of the coil, at point O in Fig. 19.40, however large the radius may become, even in the limit $d \to \infty$.

The field at other points in the interior of the coil is given by (19.7.8), where r is the distance of the point from the central axis AA'. In the case where the mean radius d is fairly small, as in Fig. 19.40a, the ratio d/r of Eq. (19.7.8) may be significantly larger or smaller than unity for points within the coil that may be near the inner or outer radius. But when d becomes very large (the cross section of the coil remaining the same) as illustrated in Fig. 19.40b, the ratio d/r for any point within the coil can never differ significantly from unity. As d increases without bound, in fact, the ratio d/r for all interior points approaches unity, in which case we find from Eq. (19.7.8) that *everywhere* in the interior of the solenoid, the field will be given by

$$B = B_c = \mu_0 nI \tag{19.7.12}$$

where n is the number of turns per unit length. At the same time, from (19.7.5) and (19.7.11), the field everywhere outside the coil will be zero. The fields associated with toroidal coils and infinite straight solenoids, therefore, have the configurations shown in Fig. 19.41.

Unfortunately, all the straight solenoids we must deal with in real life are finite in length. Their properties depart to a greater or lesser degree from those of the ideal infinite solenoid described above. In no event is the external magnetic field of a finite solenoid zero as it is for one of infinite length. Instead, for a long, slim solenoid, such as the one shown in Fig. 19.4, the external field is similar to that of a long bar magnet, while for one that is short and fat, the field is very much like that of the circular coil illustrated by Fig. 19.29. In the case of a long, slim solenoid, however, the field inside the coil is quite uniform over a large part of the interior and has approximately the constant value given by (19.7.12). The field pattern of such a solenoid is shown in Fig. 19.42; such coils can be used to advantage to generate a constant magnetic field in the laboratory and are also employed in the construction of electromagnets, relays, and other electromechanical devices.

EXAMPLE 19.7.1

A toroidally wound coil has an inner diameter $a = 5$ cm, and an outer diameter $b = 10$ cm and is wound with 1500 turns of wire. It carries a current of 5 amperes. Find (a) the magnetic field at the center of the circular cross section, (b) the interior field just beyond the inner radius, and (c) the interior field just short of the outer radius. (d) What would be the field within an infinite straight solenoid having the same number of turns per unit length?

In this example, the inner radius is 0.05 meter, the outer radius is 0.10 meter, and the central radius d is given by

$$d = \tfrac{1}{2}(a + b) = (\tfrac{1}{2})(0.15) = 0.075 \text{ m}$$

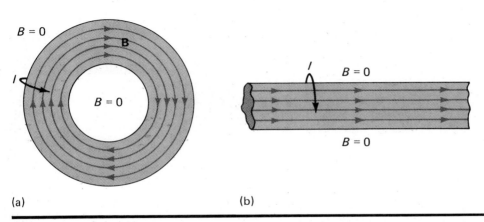

(a) (b)

FIGURE 19.41. Field configuration for (a) toroidal coil and (b) long, straight solenoid.

635

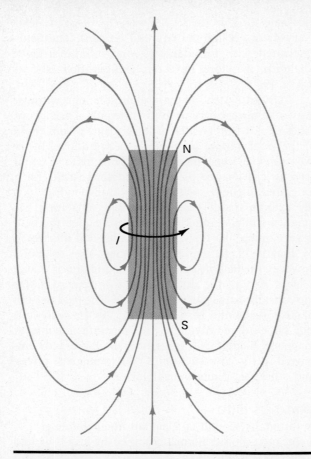

FIGURE 19.42. Field configuration of long, but finite, solenoid.

The average circumference is, therefore,

$$c = 2\pi d = (2\pi)(0.075) = 0.4712 \text{ m}$$

The number of turns per unit length is

$$n = \frac{N}{c} = \frac{1500}{0.4712} = 3183 \text{ turns/m}$$

The central field B_c, according to (19.7.9), will be

$$B_c = \mu_0 nI = (4\pi)(10^{-7})(3183)(5) = 0.0200 \text{ Wb/m}^2$$

Just beyond the inner radius, the distance r to the origin is 0.05 meter. The field there, according to (19.7.8), is

$$B = \mu_0 nI \frac{d}{r} = \frac{(0.0200)(0.075)}{(0.05)} = 0.0300 \text{ Wb/m}^2$$

Just short of the outer radius, $r = 0.10$ meter, and B will be given by

$$B = \mu_0 nI \frac{d}{r} = \frac{(0.020)(0.075)}{(0.10)} = 0.0150 \text{ Wb/m}^2$$

For a straight solenoid of infinite length with the same number of turns per unit length, the field every-

where in the interior is

$$B = \mu_0 nI = (4\pi)(10^{-7})(3183)(5) = 0.0200 \text{ Wb/m}^2$$

19.8 Forces Between Currents and the Definition of the Ampère

We have previously discussed the fact that currents in conductors give rise to magnetic fields and also that conductors in which currents flow experience forces when they are subjected to external magnetic fields. Putting these two observations together, it is easy to understand that currents are capable of exerting forces upon one another via their magnetic fields.

It is easy enough to write a mathematical expression for such a force of the form

$$\mathbf{F}_{12} = \frac{\mu_0 I_1 I_2}{4\pi} \oint_{C_1} \oint_{C_2} \frac{d\mathbf{l}_2 \times (d\mathbf{l}_1 \times \mathbf{r}_{12})}{r_{12}^3} \tag{19.8.1}$$

In this expression, $\mathbf{F}_{12}$ represents the force on current I_2 flowing in a conductor C_2 composed of the elements $d\mathbf{l}_2$ arising from current I_1 flowing in another conductor C_1 described similarly by the elements $d\mathbf{l}_1$. The vector $\mathbf{r}_{12}$ is a vector from element $d\mathbf{l}_1$ of the latter conductor to $d\mathbf{l}_2$ of the former. The integrals are evaluated with respect to the elements $d\mathbf{l}_1$ and $d\mathbf{l}_2$ of both circuits and must be taken around the whole circuit in each case. While it is simple to derive (19.8.1), starting with (19.4.5) and (19.5.4), there is not much use in doing so, since the evaluation of the integral above is so tedious as to make the expression practically useless for our purposes. Its derivation, however, is somewhat instructive and is assigned as an exercise.

We shall be satisfied at this point simply to consider the case of two infinitely long, parallel conductors separated by a distance a, carrying steady currents I_1 and I_2 as illustrated in Fig. 19.43. We know from the law of Biot and Savart, or from Ampère's law, that the magnetic induction $\mathbf{B}_1$ due to the current I_1 in the conductor on the left is

$$B_1 = \frac{\mu_0 I_1}{2\pi r} \tag{19.8.2}$$

where r is the distance from the conductor. The direction of $\mathbf{B}_1$ will be into the page, as shown in Fig. 19.43, according to the right-hand rule. At the second conductor, where $r = a$, the magnitude of $\mathbf{B}_1$ is clearly

$$B_1 = \frac{\mu_0 I_1}{2\pi a} \tag{19.8.3}$$

Now, the force on an element $d\mathbf{l}$ of the conductor on the right, which carries current I_2, from this magnetic field will be, according to (19.4.5),

$$d\mathbf{F} = I_2(d\mathbf{l} \times \mathbf{B}_1) \tag{19.8.4}$$

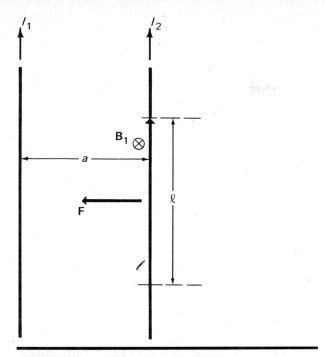

FIGURE 19.43. Calculation of magnetic forces between long parallel conductors carrying steady currents.

In fact, since $\mathbf{B}_1$ is constant over all parts of the right-hand conductor, we may write the total force on a segment of that conductor of length l as

$$\mathbf{F} = I_2(\mathbf{l} \times \mathbf{B}_1) \tag{19.8.5}$$

where the vector $\mathbf{l}$ is illustrated in Fig. 19.43. Clearly, from the figure, the vector $\mathbf{F}$ has the direction shown, the current on the right, therefore, experiences a force of *attraction* from the parallel current on the left. The magnitude of the force $\mathbf{F}$, according to (19.8.5) and (19.8.3), must be

$$\boxed{F = I_2 l B_1 = \frac{\mu_0 I_1 I_2 l}{2\pi a}} \tag{19.8.6}$$

We may summarize these results by stating that the force per unit length experienced by the right-hand conductor is attractive and is given by

$$\boxed{\frac{F}{l} = \frac{\mu_0 I_1 I_2}{2\pi a}} \tag{19.8.7}$$

A similar calculation of the force experienced by the conductor on the left due to the magnetic field of the right-hand current tells us that the force acting on that conductor is equal in magnitude and opposite in direction to that given by (19.8.7). This is what we must expect if Newton's third law is to be obeyed, for these forces qualify in every way as an action–reaction pair.

If the current I_2 is reversed in direction, the

vector $\mathbf{l}$ reverses direction also, which in turn reverses the direction of the force. The two currents are then found to repel one another. The law, therefore, must be that

parallel currents attract one another while anti-parallel currents repel.

Let us now calculate the force on 1 meter of conductor carrying a current of exactly 1 ampere, which is situated parallel to a second long conductor precisely 1 meter away carrying exactly the same current. In this case, $l = 1.0$ meter, $I_1 = I_2 = 1.000$ ampere exactly, and $a = 1.000$ meter, and (19.8.7) gives

$$F = \frac{(4\pi)(10^{-7})(1.0)(1.0)}{(2\pi)(1.0)} = 2.00\ldots \times 10^{-7} \text{ N}$$

exactly. That exactly $2.00\ldots \times 10^{-7}$ newton is the result of this calculation is not accidental but rather stems from the fact that *the ampere is defined as that current which, when flowing through a straight conductor parallel to an equal current 1 meter away, generates an attractive force of precisely $2.00\ldots \times 10^{-7}$ newton per meter.* This definition, in turn, *fixes the size of the coulomb,* since the coulomb can be regarded as the amount of charge transported past a given point in a conductor when a current of 1 ampere flows for 1 second

This definition of the ampere and the coulomb *also* determines the value of the permeability constant μ_0. Suppose, for example, we had not known the value of μ_0 but insisted, nevertheless, on defining the ampere in the way described above. Then we should have insisted that when a force of exactly $2.00\ldots \times 10^{-7}$ newton was produced in such a situation, the current *had* to be $1.00\ldots$ ampere, that is, $1.00\ldots$ coulomb/second. Putting these values into (19.8.7), we should have found that

$$2.00\ldots \times 10^{-7} = \frac{(\mu_0)(1.0)(1.0)}{(2\pi)(1.0)}$$

from which

$$\mu_0 = 4\pi \times 10^{-7} \text{ Wb/amp-m}$$

The way in which the ampere and hence the coulomb are defined, therefore, of itself *determines* that the value of μ_0 be $4\pi \times 10^{-7}$ weber/ampere-meter. If we take our coulomb of charge as given to us by this definition and now ask what *electrostatic* force would be exerted by two 1-coulomb point charges 1 meter apart, we find by experiment that this force just *happens* to have the odd value of 8.988×10^9 newtons. However, in writing Coulomb's law for the electrostatic forces between point charges, we have

$$F = k\frac{qq'}{r^2} \tag{19.8.8}$$

637

where k is the proportionality constant. For $q = q' = 1.00...$ coulomb and $r = 1.00...$ meter,

$$8.988 \times 10^9 = k \frac{(1.00...)(1.00...)}{(1.00...)^2} = k$$

which tells us that the proportionality constant we must use in Coulomb's law of electrostatic forces has to be

$$k = \frac{1}{4\pi\varepsilon_0} = 8.988 \times 10^9 \text{ N-m}^2/\text{C}^2$$

or

$$\varepsilon_0 = 8.854 \times 10^{-12} \text{ C}^2/\text{n-m}^2$$

It is reasonable to ask, of course, why it is that μ_0 rather than ε_0 is given such a simple assigned value. The answer is that because of the availability of steady current sources such as storage batteries and accurate current measuring instruments such as the d'Arsonval galvanometer, it is much easier to measure forces between currents and to standardize currents than it is to measure electrostatic charges. Historically, therefore, when the need for accurately defined electrical standards arrived, it was much simpler to exhibit a reproducible and accurately measurable ampere of current than a precise coulomb of charge. Actually, in the Gaussian system of electrical units, which is occasionally used, the force constant in Coulomb's law is, in fact, assigned the value of unity, which in turn fixes the size of the electrostatic unit of charge (esu) in that system. The constant μ_0 then no longer has the simple numerical value it has in the mks system.

While we are on the subject of the numerology of μ_0 and ε_0, it is of interest to point out that the quantity $1/\sqrt{\varepsilon_0\mu_0}$ has the dimensions of velocity. This can be seen by recalling that the weber is a unit of flux, which in turn is equal to magnetic induction (N/amp-m) times area (m^2). Then, dimensionally,

$$\frac{1}{\varepsilon_0\mu_0} = \frac{\text{amp-m}}{\text{weber}} \cdot \frac{\text{N-m}^2}{\text{C}^2} = \frac{\text{amp}^2\text{-m}^2}{\text{N-m}^2} \cdot \frac{\text{N-m}^2}{\text{C}^2}$$

$$= \frac{\text{amp}^2\text{-m}^2}{\text{C}^2} = \frac{\text{C}^2\text{-m}^2}{\text{C}^2\text{-sec}^2} = \left(\frac{\text{m}}{\text{sec}}\right)^2 \qquad (19.8.9)$$

Let us now inquire into the magnitude of the velocity $1/\sqrt{\varepsilon_0\mu_0}$. Substituting our known numerical values μ_0 and ε_0, which we obtain, remember, from measuring the forces between currents and electric charges, we get

$$\frac{1}{\sqrt{\varepsilon_0\mu_0}} = \frac{1}{\sqrt{(1.256637 \times 10^{-6})(8.854185 \times 10^{-12})}}$$

$$= 2.997925 \times 10^8 \frac{\text{m}}{\text{sec}}$$

or

$$\frac{1}{\sqrt{\varepsilon_0\mu_0}} = 299{,}792.5 \text{ km/sec} \qquad (19.8.10)$$

When we compare this with the known *velocity of light*, which has been independently and very accurately determined to be 299,792 kilometers per second, we find that the two numbers agree so closely that it must be more than pure accident they are so nearly the same. We are thus led to suspect that light is somehow intimately connected with electric and magnetic fields.

While we are not yet in a position to make this connection very clear, we should keep in mind the fact that it is one of the things we are working toward. We shall see in Chapter 23 that the properties of electric and magnetic fields lead directly to the possibility of *electromagnetic waves* composed of time-varying electric and magnetic fields that propagate with velocity $1/\sqrt{\varepsilon_0\mu_0}$ and exhibit all the experimentally observed characteristics of light waves.

SUMMARY

An electric current in a conductor, or a moving charge in free space, produces a magnetic field. Likewise, any current or any charge that is in motion in an externally produced magnetic field experiences a magnetic force. This magnetic force is perpendicular both to the magnetic field and the charge velocity vector, or current flow direction. It can, in fact, be expressed by

$$\mathbf{F}_m = q(\mathbf{v} \times \mathbf{B}) \qquad \text{for a moving charge}$$

or

$$d\mathbf{F}_m = I(d\mathbf{l} \times \mathbf{B}) \qquad \text{for an element } d\mathbf{l} \text{ of a current-carrying circuit}$$

These expressions can be regarded as defining the magnitude and direction of the magnetic induction vector $\mathbf{B}$. If both electric and magnetic forces act on a moving charge q, the total force the charge experiences is

$$\mathbf{F} = q\mathbf{E} + q(\mathbf{v} \times \mathbf{B})$$

Magnetic flux through a given surface S, in analogy with electric flux, is defined by

$$\Phi_m = \int_S \mathbf{B} \cdot \mathbf{n} \, da$$

For a constant magnetic field and a plane surface, this reduces to

$$\Phi_m = BA \cos \theta$$

where θ is the angle between the direction of $\mathbf{B}$ and the normal to the surface. Gauss's law for the magnetic induction has the form

$$\oint_S \mathbf{B} \cdot \mathbf{n} \, da = 0$$

the zero on the right reflecting the presumed nonexistence of magnetic charge or the fact that electric currents represent the source of all magnetic fields.

The magnetic field of an electric current loop is dipolar in character. The magnetic dipole moment $\mathbf{p}_m$ of any current loop can be expressed as

$$p_m = IA$$

where A is the area of the loop. The direction of the dipole moment vector is normal to the plane of the circuit, in the direction of the thumb of the right hand when the fingers curl in the direction of the current. The torque on a current loop in an external magnetic field is

$$\tau = \mathbf{p}_m \times \mathbf{B}$$

where $\mathbf{p}_m$ is the magnetic dipole moment of the loop.

The magnetic field created by a given current can be found in two principal ways. First, according to the law of Biot and Savart, the field element $d\mathbf{B}$ resulting from the flow of current I along the path element $d\mathbf{l}$ of a conductor can be written

$$d\mathbf{B} = \frac{\mu_0 I}{4\pi} \frac{d\mathbf{l} \times \mathbf{r}}{r^3}$$

where $\mathbf{r}$ is the vector from the path element $d\mathbf{l}$ to the point P at which the field element $d\mathbf{B}$ is to be calculated. The total field can then be obtained by integrating over all the path elements $d\mathbf{l}$ along which the current flows. In performing this summation, we must be careful to use the rules of *vector* addition. Thus,

$$\mathbf{B} = \frac{\mu_0 I}{4\pi} \int_C \frac{d\mathbf{l} \times \mathbf{r}}{r^3}$$

In these expressions, the constant μ_0 has the value $4\pi \times 10^{-7}$ weber/ampere-meter in mks units. It is referred to as the *permeability of free space*. In the case of a long, straight conductor carrying current I, the integration of the differential form of the Biot–Savart law over the $d\mathbf{l}$ elements of the wire for the field at a distance r from the conductor yields the simple result

$$B = \frac{\mu_0 I}{2\pi r}$$

The lines of $\mathbf{B}$ in this case are circles surrounding the wire whose direction is that of the fingers of the right hand when the thumb points in the direction of the current.

The second way of obtaining the magnetic field associated with a given distribution of current is by the use of Ampère's law, which states that for any *closed* path C,

$$\oint_C \mathbf{B} \cdot d\mathbf{l} = \mu_0 I_C$$

where I_C is the current flowing through the path C

around which the integral is calculated. Ampère's law is particularly useful in situations in which symmetry allows the above integral to be expressed as the magnitude of B times the length of the closed path C. The law of Biot and Savart is a statement somewhat similar to Coulomb's law, while Ampère's law has a strong resemblance to Gauss's law in electrostatics.

The magnetic field outside an ideal toroidal coil is zero. Within the coil, along the central axis of the cross section, the magnetic field is parallel to the axis of the toroid and has the magnitude

$$B = \mu_0 nI$$

where I is the current in the winding and n is the average number of turns per unit length along the toroid. In a long, straight solenoid, the field within the coil is uniform, points along the axis of the coil, and has magnitude

$$B = \mu_0 nI$$

where n is the number of turns per unit length and I is the current in the coil.

The force per unit length of a straight conductor carrying current I_1 a distance a from a second parallel straight conductor carrying current I_2 is

$$\frac{F}{l} = \frac{\mu_0 I_1 I_2}{2\pi a}$$

The force is attractive if the two currents are in the same direction, repulsive if they are in opposite directions.

The ampere is defined as that current which, when flowing through a straight conductor parallel to an equal current 1 meter distant, experiences an attractive force of exactly $2.000\ldots \times 10^{-7}$ newton per meter. This definition fixes the size of the coulomb, which is 1 ampere-second.

QUESTIONS

1. The electric field is defined so that the electric force is parallel to the field. Would it have been possible to define the magnetic field in such a way that the magnetic force is parallel to the magnetic field?

2. The maximum speed of a particle of charge q is v_0. What are the maximum and minimum magnitudes of the magnetic force on it?

3. A particle of charge q and velocity $\mathbf{v}$ is undeflected as it moves through a region of space in which there are uniform $\mathbf{E}$ and $\mathbf{B}$ fields. Obtain an expression for $\mathbf{E}$ in terms of the other quantities.

4. Magnetic fields arise from the motion of electric charges. Are there any other ways in which a $\mathbf{B}$ field can be produced?

5. In a cathode ray tube, electrons are initially projected along the positive z-axis and the beam is deflected toward negative x-values in the xy-plane. If the deflection is due to a magnetic field, find the direction of the field.

6. A wire carrying a current I experiences no magnetic force even though it is in a uniform magnetic field. How can you account for this?

7. In the northern parts of Canada, the cosmic ray flux is greater than it is at locations closer to the equator. How would you explain this fact?

8. A child watching TV notices a distortion of the picture whenever he brings his toy magnet close to the picture tube. Explain.

9. Discuss the various factors which determine the magnetic flux through a given surface. Is it always true that the magnetic flux through a closed surface is zero?

10. Can you suggest a method for measuring the dipole moment of a compass needle?

11. Compare and contrast the Biot–Savart law and Coulomb's law for electric charges.

12. Is Ampère's law valid for all possible closed paths which might be chosen? Is it useful for all such paths?

13. Ampère's law and the Biot–Savart law may both be used to compute magnetic fields. Which of these is more general in terms of applicability?

14. The magnetic field due to a toroid is completely confined to its interior, while the field due to a long solenoid is nonvanishing outside. Explain this in terms of the necessity for magnetic field lines to close.

15. Explain in simple terms why the force between two current-carrying wires is proportional to the product of their currents.

16. A proton is moving from west to east near the earth's equator. What is the direction of the magnetic force it experiences?

17. Two particles of equal charge are confined to move parallel to one another. Analyze the magnetic fields and forces produced and experienced by each of the charges.

18. Much fuss is being made in the United States and Britain over conversion from British to metric units. But, insofar as *electrical* units are concerned, both countries have been on the metric system for over a century since, after all, volts, amperes, coulombs, ohms, and watts are all mks units! Just to be perverse, and to annoy the potentates of the Système Internationale commission, can you suggest a way of defining electrical units on the basis of the *British* system? What are the conversion factors between your Britcoulombs, Britvolts, Bramperes, Britwatts, and Brohms and the conventional mks units? What are the values of ε_0 and μ_0?

PROBLEMS

1. A point charge of 1.8×10^{-6} Coulomb moves along the $+x$-axis with a speed of 800 meter/sec. There is a constant magnetic induction B whose magnitude is 0.72 newton/amp-meter. The B-field lies in the yz-plane and makes an angle of $45°$ with the $+y$-axis. Find the magnitude and direction of the magnetic force on the charge.

2. A point charge of 2.7×10^{-5} Coulomb moves along the $+y$-axis with a speed of 600 meters/sec. There is a constant B-field of magnitude 0.96 newton/amp-meter. Its direction lies in the xy-plane, making equal angles with the $+x$- and $+y$-axis. Find the magnitude and direction of the magnetic force on the charge.

3. An electric field of 200 volts/meter and a magnetic field of 0.80 newtons/amp-m act on an electron to produce zero net force. If the electric and magnetic fields are perpendicular, what is the magnitude and direction of the electron's velocity?

4. A deuteron (mass $2M_p$, charge $+e$) is accelerated from rest across a potential difference of 500 volts, thus acquiring a speed v. For this speed, what would be the radius of its orbit in a magnetic field of 0.5 weber/m²?

5. Show that the kinetic energy of a particle of charge q and mass m in a magnetic field B is given by $q^2B^2r^2/2m$, where r is the radius of its circular orbital path.

6. At $t = 0$, a particle of charge q and mass m is projected into a region of space where a uniform magnetic field $\mathbf{B} = B\mathbf{i}_z$ acts upon it. If the initial velocity is $\mathbf{v}_0 = v_{0y}\mathbf{i}_y + v_{0z}\mathbf{i}_z$, describe its subsequent motion. Obtain the velocity $\mathbf{v}$ at an arbitrary time t.

7. A charge $q_0 = 6 \times 10^{-5}$ coulomb moves with a velocity of 1500 meters/sec along the x-axis, in the presence of a magnetic field $\mathbf{B}$ of 8.0 newtons/amp-m which lies in the xy-plane at an angle of $30°$ to the x-axis. Find the magnitude and direction of the force on the charge.

8. A proton circles the earth in an orbit above the equator. The radius of the orbit (measured from the center of the earth) is 1.2×10^4 kilometers. Assume that the earth's magnetic field at the orbit is 1.5×10^{-5} newtons/amp-m in a direction parallel to its axis. Neglect all gravitational effects. Using $M_p = 1.67 \times 10^{-27}$ kg and $e = 1.6 \times 10^{-19}$ coulomb, find (a) the magnitude and direction of the magnetic force on the proton and (b) the period of the orbital motion of the proton.

9. If the speed of an electron is 10^7 meters/sec and its velocity is perpendicular to a magnetic field, what strength must the field have if the electron orbit is to have a diameter of 1 meter?

10. A particle of charge q and mass m enters a region of uniform magnetic field $\mathbf{B}$ with an initial velocity $\mathbf{v}_0$. By calculating the work done by the magnetic force, show that the kinetic energy of the particle is constant.

11. An ion with a charge $+3e$ is projected into a uniform magnetic field of magnitude 1.5 newtons/amp-m. It has a velocity of 10^7 meters/sec at an angle of $45°$ with the field direction. Compute the magnitude and direction of the force on the ion.

12. A potential difference ΔV exists between the plates of a parallel plate capacitor with plate separation d. A uniform magnetic field $\mathbf{B}$, perpendicular to the electric field, is present between the plates. Electrons are projected into the region, and for an appropriate value of ΔV they are observed to be undeflected. The electric field is then removed, ΔV drops to zero, and the electrons are now found to travel in a circular path of radius r. (a) Obtain an expression for e/m in terms of ΔV, B, r, and d. (b) If the voltage ΔV is such as to produce an electric field of 10^4 volts/meter and if B and r are, respectively, 10^{-3} newtons/amp-m and 5.71 cm, find the charge-to-mass ratio e/m.

13. In a given region of space, a uniform magnetic field of 0.2 newtons/amp-m is present. A charged particle having a velocity of 10^5 meters/sec enters the region, its velocity

vector making an angle of 30° with the direction of **B**. Find the magnitude and direction of the electric field which would result in no deflection of the particle's trajectory.

14. A magnetic field **B** points in the z-direction and has a magnitude of 2 webers/m². Find the magnetic flux Φ_m passing through a square loop of area 2 m² if the plane of the loop makes an angle of 15° with the z-axis.

15. A rectangular coil whose dimensions are 18 by 36 cm is placed in a constant magnetic induction field of magnitude 0.24 weber/m². The **B** vector makes an angle of 53° with the plane of the coil. What is the magnetic flux through the coil?

16. A magnetic induction field whose direction is parallel to the z-axis has a magnitude that varies along the x-direction and is given by $B = B_0(1 + (x/x_0)^2)$ where B_0 and x_0 are constants. There is no variation of B along the y- and z-directions. Show that the magnetic flux through a rectangular loop lying in the xy-plane of length a and width b whose center is at the point $x = c, y = 0$ can be expressed as $\Phi_m = AB(c) + AB_0b^2/12x_0^2$, where $B(c)$ is the value of B at the center of the loop. Assume that the side of the loop of length b is parallel to the x-axis.

17. Show, using Eq. (19.3.8), that it is impossible to have a spherically symmetric radial magnetic field, that is, a magnetic field such as $B = (k/r^2)\mathbf{i}_r$, like the Coulomb field of a point charge.

18. Two isotopes of a given element have atomic masses m_1 and m_2. They are both singly ionized and are accelerated through the same potential difference. Following this acceleration from rest, they both enter a uniform magnetic field, travel in a semicircle, and are detected on a photographic plate. If they follow paths of radii r_1 and r_2, respectively, find the ratio of their masses.

19. A long, straight wire carries a current of 50 amperes. It is placed in a uniform magnetic field of 0.0200 weber/m², which makes an angle of 30° with the wire. Find the magnetic force per unit length of the wire.

20. At the equator, the magnetic field of the earth is horizontal, points north, and has a strength of about 1.0×10^{-4} weber/m². A 100-meter length of transmission line carries a current of 700 amperes from east to west. Find the magnitude and direction of the force exerted on the transmission line due to the earth's magnetic field.

21. A rectangular coil of length 0.12 meter and width 0.18 meter consists of 240 turns of fine wire. (a) What is the dipole moment of this coil when it is carrying a current of 6.0 amp? (b) What torque would it experience in a uniform magnetic induction field of magnitude 0.24 weber/m² whose direction is at an angle of 30° with the plane of the coil?

22. A thin insulating circular loop of radius 0.16 meter bears a uniformly distributed charge of 1.8×10^{-5} Coulomb. It is rotated about its own axis at an angular speed of 3000 revolutions/minute. Find its magnetic moment.

23. A thin, insulating disc of radius R bearing a uniform charge density σ rotates with angular frequency ω about its axis. Find the magnetic moment of the disc.

24. A galvanometer consists of a coil 2.5 cm long by 2 cm wide wound with 250 turns of wire, placed in a uniform magnetic field of 0.12 weber/m². The plane of the coil is normal to the field when no current passes through the instrument. If a current of 5×10^{-6} ampere causes a 45° deflection of the coil, find the torque constant of the hair spring of the galvanometer.

25. A magnetic moment $\mathbf{p}_m$ in a uniform magnetic field **B** experiences a torque given by Eq. (19.4.17). This torque must be equal to the time rate of change of angular momentum $d\mathbf{L}/dt$. If the magnetic moment is proportional to the angular momentum, that is, $\mathbf{p}_m = \lambda\mathbf{L}$, discuss the motion of the angular moment vector **L** in the magnetic field **B**.

26. A rectangular coil of wire of size 6 cm × 12 cm contains 500 turns of wire and carries a current of 10^{-6} ampere. Find the maximum torque on the coil in a uniform magnetic field of magnitude 0.2 weber/m².

27. In the accompanying diagram, there are two very long, parallel wires AB and CD. They lie in a horizontal plane and are connected by the fixed wire BC and also by a frictionless sliding conductor AD whose mass is m. A current I flows in the circuit, as shown. Neglecting the effect of the current in conductor BC, which is assumed to be very distant from AD, find an expression for the acceleration of the sliding contact AD.

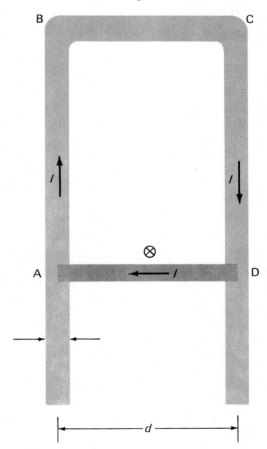

28. A long, straight wire of length L carries a current I. Using the Biot–Savart law, calculate the magnetic field produced at a distance r from the center of the wire in the plane perpendicular to the wire. Now, let L approach infinity and show that you obtain the magnetic field $\mu_0 I/2\pi r$ of an infinitely long wire.

29. A current I flows clockwise around a loop of wire in the form of an equilateral triangle of side L. Find the magnetic field at the center of the triangle.

30. Compute the magnetic flux passing through the path shown in the accompanying figure.

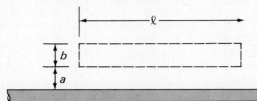

31. A closed electric circuit carries a current I, as shown in the diagram. Find the magnetic field produced at point P.

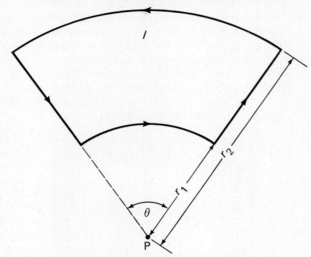

32. What is the magnitude and direction of magnetic field at a distance of 1.5 cm from a long straight conductor carrying a current of 400 amp?

33. Using the Biot–Savart law, obtain an expression for the magnetic field produced at the center of a circular loop of radius a carrying a current I.

34. Two very long, parallel wires separated by a distance d carry current I in opposite directions, as shown in the diagram. Find the magnitude and direction of the magnetic field at point P.

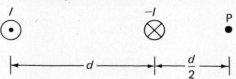

35. Two long, parallel wires separated by 0.2 meter carry

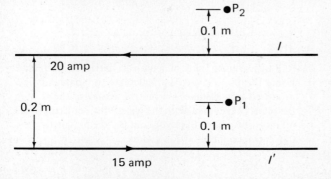

currents of 15 and 20 amperes, respectively, as shown in the diagram. Calculate the magnetic field **B** at points P_1 and P_2.

36. A particle of charge q and mass m enters a uniform magnetic field **B** and traverses a circular orbit of radius R. Show that due to the circular motion of the charge, an additional field of magnitude

$$B' = \frac{\mu_0}{4\pi} \frac{q^2 B}{mR}$$

will exist at the center of the particle's orbit.

37. Two parallel, coaxial circular coils each of radius b are separated by the same distance b, as shown in the diagram. They both carry the same current I. Show that the magnetic field at a point P on the axis is given by

$$B = \tfrac{1}{2}\mu_0 Ib^2\{[b^2 + (x + \tfrac{1}{2}b)^2]^{-3/2} + [b^2 + (x - \tfrac{1}{2}b)^2]^{-3/2}\}$$

This arrangement is known as Helmholtz coils, and it produces a very uniform field at the midpoint of the line joining the centers of the two coils. Show that the first and second derivatives of B with respect to x vanish at this point ($x = 0$).

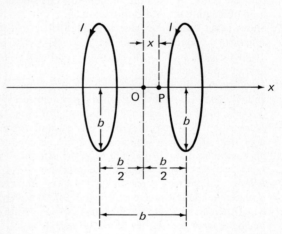

38. At a certain location, the magnetic field of the earth has a magnitude of 2.0×10^{-4} weber/m^2. A set of Helmholtz coils, such as those shown in the diagram, consists of two groups of circular coils each of which contains 40 turns. If the radius of the coils is 15 cm, what is the current needed to cancel the earth's magnetic field, thus producing a small field-free region at the point $x = 0$ along the axis of the coils? Assume that the axis of the coils is in the direction of the earth's magnetic field.

39. A square coil whose sides have length L carries a current I. Find the magnitude of the magnetic field at a distance z above the center of the coil. *Hint:* Make use of the result of Problem 28.

40. A circular ring of radius R bears a uniformly distributed charge Q. The ring rotates about a diameter with angular velocity ω. Obtain an expression for the magnetic field at the center of the ring.

41. A loop of flexible wire of length L carries a current I. The wire can be shaped to form a plane circular loop, a square loop, or a rectangular loop. For which of these possibilities is the magnetic field at the center of the loop largest?

42. A very long coaxial cable consists of an inner conductor of radius r_1 and an outer conductor of inner radius r_2

and outer radius r_3, as shown in the diagram. The inner and outer conductors carry equal but oppositely directed currents I. The current density within both conductors is uniform. Obtain expressions for the magnetic field as a function of r, the distance from the axis, for values of r between zero and infinity. Plot B versus r.

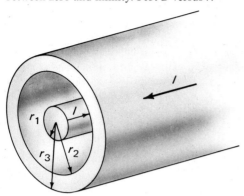

43. A long, hollow cylindrical conductor has inner radius r_1 and outer radius r_2, as shown in the figure. The conductor carries a total current I, but the current density j within the conductor is nonuniform. Calculate the magnetic field B for $r < r_1$, $r_1 < r < r_2$, and $r > r_2$ for each of the following cases: **(a)** j varies linearly with the distance from the central axis in the region between r_1 and r_2, that is, $j(r) = \alpha r$, where α is a constant, for $r_1 < r < r_2$, and **(b)** j varies quadratically in the same region, that is, $j(r) = \beta r^2$, where β is a constant, for $r_1 < r < r_2$. *Hint:* How do you find values for α and β in terms of the *total* current I in the cases above?

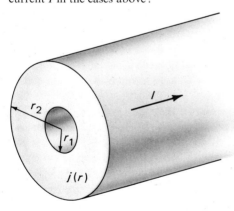

44. A long, straight solenoid has a length of 50 cm and carries a current of 0.4 ampere. It contains 300 turns of wire each of radius 4 cm. Find **(a)** the magnetic field **B** near the center of the solenoid and **(b)** the total magnetic flux passing through each turn. **(c)** If the current is increased by 1 percent, what is the fractional increase of the magnetic field?

45. A long, straight coil consisting of closely wound circular turns of wire has a radius of 3.0 cm and carries a current of 1.5 ampere. The magnetic field at the center of the coil is 1.4×10^{-4} weber/m². Find the number of turns in the coil.

46. A uniformly wound toroidal coil contains 1000 turns of wire. The inner radius is 15 cm and the outer radius is 20 cm. What is the value of the magnetic field at the center of the coil when the current in the winding is 10 amperes?

47. What is the magnetic field at the center of the "hole" in the doughnut-shaped coil of the preceding problem?

48. A very long conducting cylinder contains two cylindrical cavities, as shown in the diagram. Find the magnetic field produced at the point P just above the surface of the cylinder if the total current in the conductor is I and the current density is uniform over its cross section.

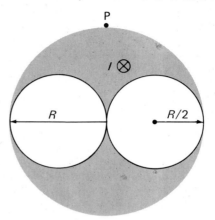

49. Two very long, parallel wires carry respectively currents of 5 and 10 amperes in the same direction. If the wires are 9 cm apart, find the force per unit length exerted on one of the wires.

50. A current loop of rectangular shape is placed near a very long, straight conductor, as illustrated in the diagram. A current $I = 3$ amperes flows in the straight conductor, and a current $I' = 1$ ampere flows in the loop. The dimensions of the loop are $b = 0.5$ meter, $c = 0.3$ meter, while the distance a is 0.2 meter. Find the magnitude and direction of the magnetic force on the current loop.

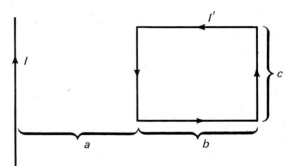

51. A long, horizontal wire AB, shown in the figure, carries a current of 50 amperes. The wire CD is free to move up and down while making electrical contact at points C and D. The mass per unit length of the wire CD is 5×10^{-3} kg/meter. What is the equilibrium height h to which CD will rise when the currents flow as illustrated?

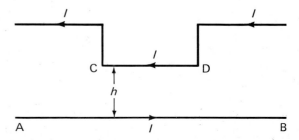

52. A mass spectrograph is an instrument which is used to measure the masses of charged ions and to determine the relative abundance of the different isotopes of any given element. In the figure, we see a schematic diagram of a Bainbridge-type mass spectrograph. Positive ions from a source are collimated by passing through slits S_1 and S_2.

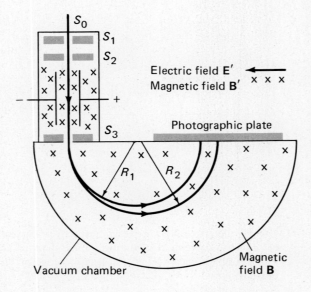

They then pass into a region of perpendicular electric and magnetic fields, which selects a unique velocity for passage through slit S_3. The ions passing through S_3 enter a region in which there is a constant magnetic field $\mathbf{B}$, perpendicular to the page. They then travel in a semicircular path and are finally recorded on a photographic plate. The density of the photographic image measures their relative abundance, while the radius of the path is used to measure the mass. (a) In terms of $\mathbf{E}'$ and $\mathbf{B}'$, find the magnitude of the velocity of the ions which pass through S_3. (b) Find the radius of the path traversed by an ion of charge e and mass m. (c) Show that for two equally charged ions of mass m_1 and m_2, $R_1/R_2 = m_1/m_2$.

53. Potassium has two isotopes with atomic masses of 38.976 and 40.975 atomic mass units (amu), respectively (1 amu $= 1.660 \times 10^{-24}$ g). If the heavier isotope travels in a circular path of radius 8.0 cm in the mass spectrograph discussed in the preceding problem, find the radius of the path of the lighter isotope. If the potassium ions have charge e, find the magnetic field B if you know that $E' = 70,000$ volts/meter and $B' = 0.5$ weber/m². Also, find the speed of the potassium ions.

54. The two isotopes of chlorine have masses of 34.980 and 36.978 amu, respectively. If the heavier isotope travels in a circular path of radius 8.0 cm, find the separation between the isotopes on the photographic plate in the Bainbridge mass spectrograph.

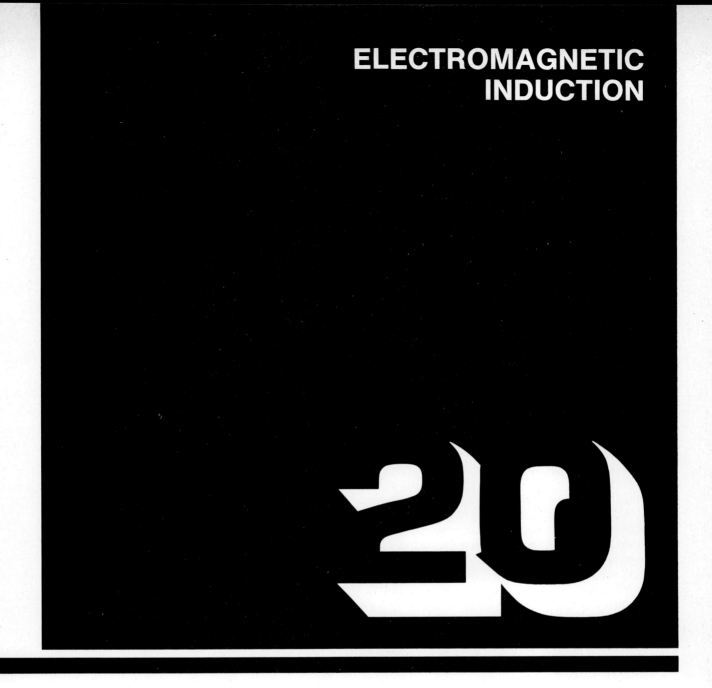

ELECTROMAGNETIC INDUCTION

20.1 Introduction

In the preceding chapter, we saw that the passage of an electric current creates a magnetic field around the conductor through which it flows. Oersted's discovery of this fact in 1819 led scientists to wonder whether it might equally well be possible, somehow, to reverse the process and to excite the flow of current in a circuit by means of a magnetic field. Initial experiments to exhibit such an effect were unsuccessful, because it was not at first realized that *steady* magnetic fluxes do not induce any emf or current flow in a circuit. It was not until about 1831 that it was discovered that an electric current could be generated magnetically but that such an effect is observed *only* *when the magnetic flux through the circuit changes with time*. This effect is referred to as *electromagnetic induction*, and the currents and emfs that are generated this way are called *induced currents and induced emfs*. Electromagnetic induction was discovered quite independently and practically simultaneously by the British physicist Michael Faraday (1791–1867) and by Joseph Henry (1797–1878), who was the first in a long line of distinguished American physicists.

The discovery of electromagnetic induction led at once to the possibility of constructing machines that would convert mechanical work into electrical energy simply by rotating coils of wire in a strong magnetic field. Such machines, which we now refer to as *generators*, were first called *dynamos*, and though many engineers and technicians were involved in the technical development of the dynamo from a laboratory curiosity into an efficient commercial machine, we may regard Henry and Faraday as the inventors of the generator. Since simple dynamos are reversible and can be run as motors that produce mechanical work when fed electric current from an external source, we may equally well credit Faraday and Henry with having invented the electric motor.

Both Henry and Faraday observed that when a time-varying current flows in a given circuit, the circuit's own magnetic field acts to induce an emf in the circuit itself whose effects are such as to oppose the external emf that causes the current to vary in the first place. This effect is usually referred to as *self-induction*. They also studied the emfs and currents induced in a coil by time-varying currents flowing in another nearby coil and found that very large induced emfs could be excited in a coil having a large number of turns of wire by a smaller time-varying emf in a coil consisting of relatively few turns. They thereby built the first *induction coils*, the precise counterparts of the ones used in every gasoline-driven automobile to excite the spark plugs, and invented the principles upon which the *transformer* operates.

Although Michael Faraday and Joseph Henry were both interested primarily in understanding and explaining the scientific principles underlying the behavior of electric currents and magnetic fields, rather than inventing and developing commercial products, it is safe to say that the social and commercial values that resulted from their work were sufficient to have paid their modest salaries many times over.

In this chapter we shall study how emfs and induced currents are produced from changing magnetic fluxes and how changing fluxes act to affect currents flowing in the circuits from which they originate and in other circuits. In doing this, we shall have to master the fundamental concepts underlying the phenomena of induced emfs and currents and to understand self-induction, mutual induction, and the behavior of resistive-inductive circuits.

20.2 Motional EMFs, Induced Currents, and Faraday's Law of Induction

The basic physics that underlies the production of induced electromotive forces and currents by changing magnetic fluxes can be understood quite well by considering examples where circuits are subjected to time-varying magnetic fluxes. At this point, the reader may find it expedient to review the definition of the term *magnetic flux* given in section 19.3, for the material in this chapter cannot be understood unless the student understands this concept.

The experimental results of Faraday and Henry regarding the production of induced emfs and currents may be summed up in the observation that

whenever there is a time-varying magnetic flux through a circuit, an emf is induced in the circuit the magnitude of which is directly proportional to the rate of change of magnetic flux with respect to time.

This statement is known as *Faraday's law of induction*. Mathematically, referring to the definition of the magnetic flux set forth in section 19.3, this can be written in the form of an equation as

$$\mathscr{E}_m = C \frac{d\Phi_m}{dt} = C \frac{d}{dt} \int_S \mathbf{B} \cdot \mathbf{n} \, da \qquad (20.2.1)$$

where $\mathscr{E}_m$ refers to the magnetically induced emf in the circuit and C is a constant of proportionality whose magnitude depends upon the units in which $\mathscr{E}_m$, Φ_m, and t are expressed. The integral in (20.2.1) is taken over the area enclosed by the circuit. In the mks system of units, the constant C has the exact value -1, for reasons similar to those which determine the value $4\pi \times 10^{-7}$ for the permeability of free space. In the mks system, then, Faraday's law can be expressed as

$$\mathscr{E}_m = -\frac{d\Phi_m}{dt} = -\frac{d}{dt} \int_S \mathbf{B} \cdot \mathbf{n} \, da \qquad (20.2.2)$$

Under circumstances in which the integral in (20.2.2) defining the magnetic flux can be expressed, according to (19.3.5), as $BA \cos \theta$, where A is the area enclosed by the circuit and θ the angle between the magnetic field and the normal to the plane of the circuit, (20.2.2) can be simplified to

$$\mathscr{E}_m = -\frac{d\Phi_m}{dt} = -\frac{d}{dt} BA \cos \theta \qquad (20.2.3)$$

This equation can be used whenever $\mathbf{B}$ is uniform in magnitude over the area enclosed by the circuit, provided that the circuit lies in a plane. These conditions will be satisfied in most of the examples considered in this text, but there are some cases of practical interest where they are not. In those cases, the induced emf must be calculated using the more general expression (20.2.2).

The magnetic flux through a circuit can be varied, according to (20.2.3), in a number of ways, specifically by varying the magnitude of $\mathbf{B}$ with respect to time, by varying the area of the circuit with respect to time, or by varying the angle θ with respect to time. An induced emf described by (20.2.3) or (20.2.2) is produced in *all cases*. These different possibilities can be seen more clearly by differentiating the product

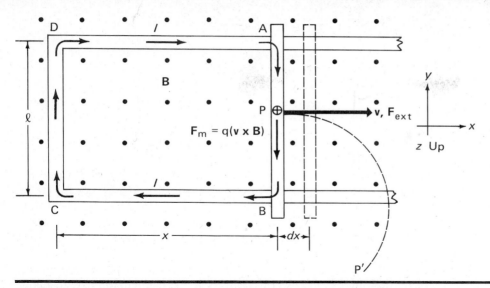

FIGURE 20.1. Magnetic forces and currents in a U-shaped conductor with a sliding contact spanning its arms.

$BA \cos \theta$ in (20.2.3), using the standard rule for differentiating the product of three functions:

$$\frac{d}{dt}(BA \cos \theta) = A \cos \theta \frac{dB}{dt} + B \cos \theta \frac{dA}{dt} + BA \frac{d(\cos \theta)}{dt}$$

(20.2.4)

Since $d(\cos \theta)/dt = -\sin \theta (d\theta/dt)$, this enables us to write (20.2.3) in the form

$$\mathscr{E}_m = -A \cos \theta \frac{dB}{dt} - B \cos \theta \frac{dA}{dt} + BA \omega \sin \theta$$

(20.2.5)

where ω is the angular velocity $d\theta/dt$. The negative sign in (20.2.2) and (20.2.3) is of considerable physical significance also, but we shall defer a detailed discussion until later in this chapter.

The physical principles underlying Faraday's law of induction can be illustrated by detailed consideration of a few specific instances in which induced emfs are generated in particularly simple ways. The first of these is illustrated in Fig. 20.1. In that diagram, we see a rigid U-shaped conductor whose arms are connected by a straight conducting segment AB that can slide or roll upon the arms of the U. A constant magnetic field **B** directed upward, out of the page, surrounds the whole system. The sliding segment of conductor is caused to move with constant velocity **v** to the right by the application of a force $\mathbf{F}_{ext}$ in the same direction. A mobile, free charge within the conductor AB will experience a magnetic force, since it moves with velocity **v**. According to the Lorentz force expression (19.2.7), this force is given by

$$\mathbf{F}_m = q(\mathbf{v} \times \mathbf{B}) = q(v\mathbf{i}_x \times B\mathbf{i}_z)$$

(20.2.6)

because in Fig. 20.1 the vector **v** is in the x-direction while **B** is in the z-direction. For a positive charge, the force $\mathbf{F}_m$ has the direction shown in Fig. 20.1; evaluating the cross product in Eq. (20.2.6), it is clear that

$$\mathbf{F}_m = (-qvB)\mathbf{i}_y$$

(20.2.7)

the minus sign signifying that the vector $\mathbf{F}_m$ points along the negative y-direction. Since the charges in the conductor are mobile, this force sets up a current I that flows around the loop ABCD in a clockwise direction. In moving a charge q from A to B a distance l in the figure, an amount of work given by the product of force times distance, that is, $-qvBl$, must be done on the charge, corresponding to an amount of work per unit charge $-qvBl/q = vBl$. But this, by the definition of electromotive force given in Chapter 18, section 18.3, is the emf that acts to make current flow through the segment AB and thence around the loop ABCD. Adopting this physical picture of the induced emf, we may express this *induced motional emf* $\mathscr{E}_m$ by

$$\mathscr{E}_m = -vBl$$

(20.2.8)

In the other arms of the circuit BC, CD, and DA, there are $I \, d\mathbf{l} \times \mathbf{B}$ forces acting on the charges that constitute the current I, but they are in all cases normal to the direction of current flow and, therefore, do no work. The emf given by (20.2.8) is, therefore, the only emf that causes current to flow in the circuit. The magnitude of the current I arising from the induced emf will be

$$I = \frac{|\mathscr{E}_m|}{R} = \frac{vBl}{R}$$

(20.2.9)

where R is the resistance of the path ABCD, assuming that all the conductors are Ohmic.

When current flows around the path ABCD, electrical energy will be dissipated in the form of heat at the rate of $I^2 R$ joules/sec as the current encounters the circuit resistance R. This electrical energy stems originally from the kinetic energy associated with the ordered drift velocity of charges along the conductor. This drift motion of the free charge distribution within the conductor is, in turn, excited by the Lorentz force arising from the sideward velocity $\mathbf{v}$ that is imparted to the conducting segment AB by the external force. It is, therefore, this external force that is responsible for doing the work on free charges within the conductor that accelerates them to their average drift velocity in the direction of the current—or in the opposite direction if they are negative. The law of conservation of energy would lead us to expect, therefore, that the work done on the conductor by the external force during a given time interval should be equal to the electrical energy produced by the induced emf during that period. This, in turn, should be equal to the amount of electrical energy dissipated as heat by the circuit resistance during that time.

According to this view, in which energy is conserved, the work done by the force $\mathbf{F}_{ext}$ in moving the segment AB through the distance dx in time dt is simply $F_{ext}\, dx$, while the amount of electrical energy dissipated during that time interval must be given by

$$dW = I^2\, R\, dt = (I)(IR)\, dt \qquad (20.2.10)$$

or, recalling (20.2.9),

$$dW = IlBv\, dt \qquad (20.2.11)$$

Equating these two quantities, we find

$$F_{ext}\, dx = IlBv\, dt \qquad (20.2.12)$$

$$F_{ext}\, \frac{dx}{dt} = F_{ext}\, v = IlBv \qquad (20.2.13)$$

whence, recalling (20.2.9),

$$F_{ext} = IlB = \frac{vB^2 l^2}{R} \qquad (20.2.14)$$

We have already seen in section 19.4 that a straight conducting segment defined by a vector $\mathbf{l}$ carrying a current I experiences a force $I(\mathbf{l} \times \mathbf{B})$ when placed in a uniform magnetic field $\mathbf{B}$. In the case of the segment AB of Fig. 20.1, this means that the conductor AB experiences a force of magnitude IlB in the *negative* x-direction from the field $\mathbf{B}$, since the vector $\mathbf{l}_{AB} \times \mathbf{B}$ points in that direction. This, however, according to (20.2.14) is equal and opposite the externally applied force $\mathbf{F}_{ext}$. The *resultant* force on the conducting segment AB is zero, which means that the conducting segment is in equilibrium under the action

of the forces that act upon it. The force $\mathbf{F}_{ext}$, therefore, does *not* act to accelerate the conductor AB, but only to overcome the magnetic force $I(\mathbf{l}_{AB} \times \mathbf{B})$ in the opposite direction and, therefore, to keep it in motion at the constant[1] velocity $\mathbf{v}$. A person who pulls a conducting segment such as AB along with his hand, indeed, experiences a resisting force that can be quite appreciable under the proper circumstances.

In this case, it is apparent that all the work done by the externally applied force goes into exciting the internal thermal energy of the atoms in the conductors. If we inserted an electric motor in the circuit ABCD of Fig. 20.1, however, some of that work might appear as mechanical work performed by the motor. The arrangement shown in the figure may be regarded as a crude form of dynamo, or electric generator.

We have gone to some trouble to explore the physical basis of the induced emf in this case to enable the student to understand its origins in a purely qualitative and intuitive way, and to show how it fits into the picture of currents, fields and forces developed in the preceding chapter. We could have derived its value much more easily, and we shall soon fall into the habit of doing so, simply by starting out with Faraday's law of induction as expressed by (20.2.3):

$$\mathscr{E}_m = -\frac{d\Phi_m}{dt} = -\frac{d}{dt}(BA \cos \theta) \qquad (20.2.15)$$

Now, in this example, B does not change with time, and the normal to the plane of the circuit is parallel to $\mathbf{B}$, so that $\theta = 0$ and $\cos \theta = 1$. Under these circumstances the only thing that now varies with time is the area ABCD, and (20.2.15) becomes

$$\mathscr{E}_m = -B\frac{dA}{dt} = -B\frac{d}{dt}(lx) = -Bl\frac{dx}{dt} \qquad (20.2.16)$$

or

$$\mathscr{E}_m = -Blv \qquad (20.2.17)$$

This answer is in exact agreement with our previous result (20.2.8). We could, of course, have used instead Eq. (20.2.5), in which the first and third terms vanish because both B and θ are constant, to obtain the same result. It may also be noted that the agreement between this result and that expressed as (20.2.8) are obtained only if the constant C in (20.2.1) is assigned the value -1. We may regard this agreement as justifying the choice $C = -1$ in that equation.

Another example which is somewhat similar but

---

[1] This statement is strictly true only if the resistance R of the circuit ABCD remains constant. Actually, because the length of the segments AD and BC varies as the conductor AB moves to the left, the resistance of the circuit changes also. This is a subsidiary point, however, and does not affect the general validity of the discussion given in this example.

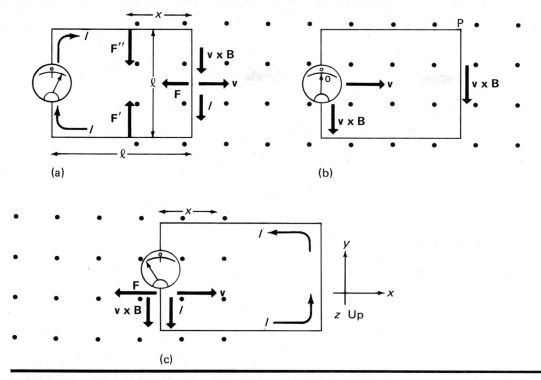

FIGURE 20.2. Currents and "motional emfs" in a circuit (a) entering, (b) traversing, and (c) leaving a region in which a uniform magnetic field has been created.

in which certain other aspects of electromagnetic induction are illustrated is shown in Fig. 20.2. In this example, a square loop of wire having sides of length l, originally in a field-free region, is thrust with constant velocity $\mathbf{v}$ into a region where there is a constant magnetic field $\mathbf{B}$, through this region, and out on the other side until it is once more clear of the magnetic field. The magnetic field is assumed to be normal to the plane of the loop. We wish to examine the currents induced in the loop and the forces acting upon it throughout this series of events.

As long as the loop is in a field-free region, there can be no motional emfs, but as soon as the right-hand side of the square enters the region where the constant field $\mathbf{B}$ exists, charges in this part of the loop experience magnetic forces given by

$$\mathbf{F}_m = q(\mathbf{v} \times \mathbf{B}) \tag{20.2.18}$$

which set up a motional emf of magnitude

$$\mathscr{E}_m = -vBl \tag{20.2.19}$$

in much the same way as in the previous example. A clockwise current I whose magnitude is

$$I = \frac{\mathscr{E}_m}{R} = \frac{vBl}{R} \tag{20.2.20}$$

is set up in the loop by this emf, as shown in Fig. 20.2a. This current in the right side of the loop experiences a force given by $\mathbf{F} = I(\mathbf{l} \times \mathbf{B})$. The force points in the

negative x-direction, as shown in the diagram, and its magnitude, of course, is

$$F_x = -IlB = -\frac{vB^2l^2}{R} \tag{20.2.21}$$

To counteract this force and to keep the loop in equilibrium so that it moves with constant velocity, an external force equal in magnitude and opposite in direction must be applied. In the sections of the upper and lower parts of the square that are inside the region where the field is set up, there are also $I(\mathbf{l} \times \mathbf{B})$ forces $\mathbf{F}'$ and $\mathbf{F}''$ as shown, but these are equal and opposite and give rise to no net force or torque.

Once the left side of the loop enters the field region, as shown in Fig. 20.2b, charges in both the left-hand and right-hand arms of the circuit experience equal magnetic forces of the form (20.2.18). But these tend to set up currents that would flow in opposite senses around the circuit and therefore cancel one another. Looking at it another way, a positive charge q that starts out at point P and traverses the circuit in a clockwise sense first experiences an emf that increases its potential energy by an amount $qBlv$ in the left side of the loop, but then experiences an emf in the right side that decreases its potential energy by a like amount. The net emf around the circuit is, therefore, *zero*, and no induced current is observed. Since there is no current in the loop, there are no forces acting on it or on any of its parts.

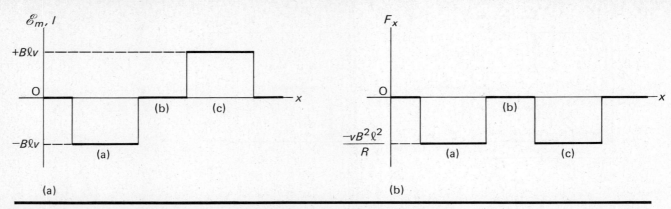

FIGURE 20.3. (a) Electromotive force and (b) forces experienced by the circuit of Fig. 20.2 during phases (a), (b), and (c) shown previously.

In Fig. 20.2c, the right side of the loop has emerged from the field region, and the motional emf in that part of the circuit has fallen to zero. But the emf in the left-hand segment, of magnitude vBl, is still there, and it causes a current vBl/R to flow around the circuit, this time in a counterclockwise sense. The current in this arm of the circuit experiences a force F whose magnitude is given by (20.2.21) and which is in the negative x-direction, since this is the direction of $I(\mathbf{l} \times \mathbf{B})$. Once more, to keep the loop moving at constant speed, an equal and opposite external force, directed to the right, must be applied. It is this force that ultimately supplies the energy dissipated as heat as the induced current passes through the Ohmic resistance of the current loop.

Again, using Faraday's law, these results can be obtained very simply by relating the net induced emf to the change in flux through the circuit. In Fig. 20.2a, therefore, we need only note that since $\cos \theta = 1$, the magnetic flux through the circuit is simply the product of the magnetic induction B and the area of that part of the field region within the loop, which is lx. Then, since B and l remain constant,

$$\mathscr{E}_m = -\frac{d\Phi_m}{dt} = -\frac{d(Blx)}{dt} = -Bl\frac{dx}{dt} \qquad (20.2.22)$$

or

$$\mathscr{E}_m = -Blv \qquad (20.2.23)$$

in agreement with (20.2.17). In Fig. 20.2b, the flux through the circuit clearly is not changing, whence $\mathscr{E}_m = -d\Phi_m/dt = 0$, while in Fig. 20.2c, the flux is again equal to the product of the induction and the area of the field region that is within the loop, which enables us to write

$$\mathscr{E}_m = -\frac{d\Phi_m}{dt} = -Bl\frac{dx}{dt} = -Bl(-v) = Blv \qquad (20.3.24)$$

since dx/dt is negative and has magnitude v.

The net emf, the current, and the force experienced by the loop are shown plotted against distance in Fig. 20.3. It is important to remember that, as shown in Fig. 20.3b, the mere fact that a circuit moves within a region where there is a magnetic field does *not* guarantee that there will be an induced emf or current in the circuit. The induced emf or current appears only when there is a *change in the magnetic flux* through the circuit. In the preceding example, this flux change was produced by increasing the area of the circuit, while in this one, the change is produced by moving the circuit as a whole in and out of a region where a magnetic field is present. The means by which the flux change is brought about is unimportant; the induced emf will always be given by Faraday's law irrespective of how the flux change occurs.

As a final illustration, let us consider the case of a small circular coil of N turns and area A, placed inside a very long straight solenoid of n turns per unit length in which a current I is maintained that increases linearly with time. The axis of the small coil is assumed to be parallel to that of the solenoid. Find the current and emf induced in the small coil. The situation is illustrated in Fig. 20.4.

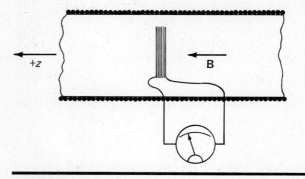

FIGURE 20.4. Induction in a stationary coil through which an increasing current produces an increasing flux.

Let us denote the current in the solenoid by I_0. Since this current increases linearly with time, we may write, assuming that $I_0 = 0$ at time $t = 0$,

$$I_0(t) = \alpha t \qquad (20.2.25)$$

where α is a constant. In the preceding chapter, we found that the magnetic induction within a long straight solenoid could be written according to (19.7.10) as

$$B = \mu_0 n I_0 = \mu_0 n \alpha t \qquad (20.2.26)$$

where n is the number of turns per unit length. The magnetic flux intercepted by each turn of the small coil will be

$$\Phi_m = BA = \mu_0 n I_0 A = \mu_0 n A \alpha t \qquad (20.2.27)$$

while

$$\frac{d\Phi_m}{dt} = \mu_0 n A \alpha \qquad (20.2.28)$$

Now, a certain emf given by $-d\Phi_m/dt$ will be generated in *each turn* of the small coil, because each of the N turns experiences the flux change. These individual emfs clearly appear in series with one another, so that it is evident that for a coil of N turns, we must write Faraday's law of induction as

$$\mathscr{E}_m = -N\frac{d\Phi_m}{dt} = -\mu_0 n N A \alpha \qquad (20.2.29)$$

If we note that, according to (20.2.25), $\alpha = dI_0/dt$, we can write this finally as

$$\mathscr{E}_m = -\mu_0 n N A \frac{dI_0}{dt} \qquad (20.2.30)$$

The current in the small coil will then be

$$I = \frac{\mathscr{E}_m}{R} = \frac{\mu_0 n N A}{R}\frac{dI_0}{dt} \qquad (20.2.31)$$

where R is its resistance.

It is apparent that when the current in the solenoid increases at a *constant rate*, an induced current that is *constant in magnitude* and proportional to the *rate of increase* of the solenoid current is produced. It is not easy to understand why this should be so on the basis of motional emfs, as in the preceding illustrations of induced emfs and currents. There is no relative motion between the two coils, nor is there any change in their area. It is true, of course, that the magnetic flux of the solenoid propagates somehow through the turns of the small coil, but there is no clearly evident way of associating a motional emf with it. Nevertheless, it is observed experimentally that a current whose magnitude is correctly predicted by (20.2.31) does flow in the small coil. Similarly, as both Faraday and Henry discovered, it is easy to

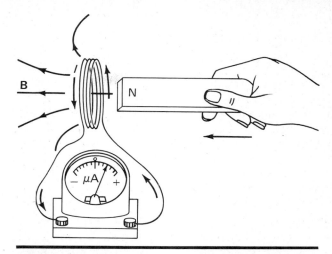

FIGURE 20.5. Another example of an emf induced by a flux change.

show that when a permanent magnet is inserted into a coil connected to a galvanometer, a momentary current is observed as the magnetic flux through the turns of the coil increases. As the magnet is withdrawn, a momentary current in the opposite direction flows as the flux through the coil decreases, as illustrated in Fig. 20.5. The emf associated with these currents, though not a motional emf in the conventional sense, is a real one and is correctly described by Faraday's law in terms of the time rate of change of flux through the coil.

We were able to analyze the first two examples starting from the Lorentz force expression and using it to find an appropriate motional emf, quite independent of Faraday's law. This was no longer possible in the third example, since there was no readily apparent way to find any motional emf. We found it necessary to use Faraday's law to describe what is observed; of course, we could equally well have discussed the other examples as well from the viewpoint of Faraday's law. It appears, then, that Faraday's law encompasses the information regarding the magnetic forces on moving charges contained in the Lorentz force law; indeed, the observed correctness of Faraday's law provides adequate experimental justification of the Lorentz force, although there is other experimental confirmation also. Faraday's law, however, goes beyond this to generalize the behavior of induced emfs and changing magnetic fluxes in situations where obvious "motional emfs" are not clearly definable. It is, therefore, Faraday's law that summarizes in a single statement all that is known and has been observed about electromagnetically induced emfs and currents. We shall see later that it is one of the four basic laws of electromagnetism from which Maxwell formulated the theory of electromagnetic radiation.

Accompanying the electromagnetically induced

emfs and currents is an electromagnetically *induced electric field*. Indeed, it is this induced electric field that can be regarded as the source of the emfs that are predicted by Faraday's law. It is important, therefore, to be able to relate the change in magnetic flux to the induced electric field as well as to induced circuit emfs. We shall now try to show how this may be done. The student will recall from the discussion of the concept of electromotive force in section 18.3 that any emf can be related to an electric field that is at least partially *nonelectrostatic* in origin (electric fields are not always electrostatic fields, you may remember). If we write the total electric field **E** at any point as the sum of an *electrostatic* field $\mathbf{E}_0$ and a *nonelectrostatic* field $\mathbf{E}'$ that expresses the force per unit charge not ascribable to electrostatic interactions with other charges, we may write, as we did in Chapter 18,

$$\mathbf{E} = \mathbf{E}_0 + \mathbf{E}' \qquad (20.2.32)$$

The emf $\mathscr{E}$, you may recall, was defined as the work done, per unit charge, on a charge making a complete traversal of a closed circuit. But since the work done on a unit charge in any displacement $d\mathbf{l}$ is $dW = \mathbf{E} \cdot d\mathbf{l}$, we may write

$$\mathscr{E} = \oint_C \mathbf{E} \cdot d\mathbf{l} = \oint_C \mathbf{E}_0 \cdot d\mathbf{l} + \oint_C \mathbf{E}' \cdot d\mathbf{l} \qquad (20.2.33)$$

the integrals being taken around the closed circuit path C, which may, *or may not*, follow the path of a metallic conductor. We saw, however, in Chapter 16 that, since the electrostatic field is *conservative* in character, the work done on a charge in moving it around any closed path is zero. The first integral on the right side of (20.2.33) is, therefore, zero, and

$$\mathscr{E} = \oint_C \mathbf{E} \cdot d\mathbf{l} = \oint_C \mathbf{E}' \cdot d\mathbf{l} \qquad (20.2.34)$$

the emf thus representing the *nonelectrostatic* work per unit charge done in moving a charge around the closed path C. Since the integral of $\mathbf{E}' \cdot d\mathbf{l}$ cannot be zero if there is an induced emf, the nonelectrostatic part of the electric field, $\mathbf{E}'$, must necessarily be *nonconservative*.

In any event, we may now replace the emf in the statement of Faraday's law given previously as Eq. (20.2.2) by the expression above, which enables us to write Faraday's law as follows:

$$\oint_C \mathbf{E} \cdot d\mathbf{l} = -\frac{d}{dt} \int \mathbf{B} \cdot \mathbf{n} \, da \qquad (20.2.35)$$

In this equation, $\mathbf{E} \cdot d\mathbf{l} = E \cos \theta \, dl$, where θ is the angle between the induced electric field **E** and the $d\mathbf{l}$ element of the path C at that point. The quantity

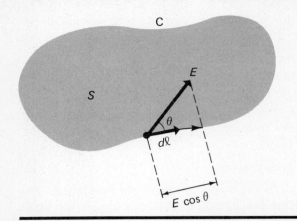

FIGURE 20.6. Calculation of the integral of $\mathbf{E} \cdot d\mathbf{l}$ around a closed circuit.

$E \cos \theta$, therefore, represents the component of **E** that is tangential to the closed path C at any given point, as shown in Fig. 20.6. Faraday's law written in this form tells us that *wherever there is a changing magnetic flux, there is an accompanying induced electric field* and that the integral of the tangential component of this field around any closed path C bears a relationship to the time rate of change of magnetic flux through the surface S enclosed by C, as given by (20.2.35). We shall find this statement of Faraday's law to be useful in a later chapter in developing the properties of electromagnetic waves. The application of Faraday's law in this form and in the originally given form will be further illustrated by the series of examples that follows.

EXAMPLE 20.2.1
A straight conductor of length l is moved with constant velocity **v** normal to a constant magnetic field **B**, as shown in Fig. 20.7. What is the net force on a charge within the conductor? What is the electrostatic field? What is the nonelectrostatic "motional" field? What is the difference in electrostatic potential between the two ends of the conductor? What happens

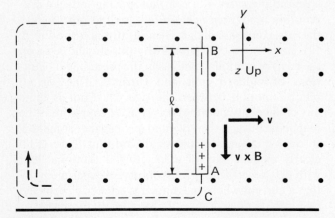

FIGURE 20.7

if the two ends of the conductor are connected by a wire that completes a circuit through some region outside the field?

Since there is no circuit around which current can flow, the mobile charges within the conductor have to be at rest with respect to the conductor, which means that they move with the constant velocity $\mathbf{v}$ with respect to a stationary observer. In any case, they are in equilibrium, and therefore the resultant force on any charge must be zero. But the force on a moving charge can be expressed as the sum of electrostatic and magnetic forces by

$$\mathbf{F} = q\mathbf{E}_0 + q(\mathbf{v} \times \mathbf{B}) = 0 \qquad (20.2.36)$$

where $\mathbf{E}_0$ is the electrostatic field within the conductor. In this example, the vector $\mathbf{v} \times \mathbf{B}$ is in the negative y-direction, and we can write the relation between the y-components of the forces in (20.2.36) as

$$F_y = qE_{0y} - qvB = 0 \qquad (20.2.37)$$

which tells us that the electrostatic field E_{0y} *within* the conductor has the constant value

$$E_{0y} = vB \qquad (20.2.38)$$

In (20.2.37), we can identify the term $-qvB$ with a nonelectrostatic "motional" field qE', so that

$$E_y{}' = -vB \qquad (20.2.39)$$

The emf $\mathscr{E}$ around any closed path, such as the dotted line C, will be the integral of $\mathbf{E} \cdot d\mathbf{l}$ around the path, that is,

$$\mathscr{E} = \oint_C \mathbf{E} \cdot d\mathbf{l} = \oint_C \mathbf{E}_0 \cdot d\mathbf{l} + \oint_C \mathbf{E}' \cdot d\mathbf{l} \qquad (20.2.40)$$

Looking only at (20.2.38) and (20.2.39), it is hard to see how the value of the integral in (20.2.40) could be anything but zero, since $\mathbf{E}_0$ and $\mathbf{E}'$ are equal and opposite within the conductor. The resolution of this apparent paradox lies in the fact that, though it is true that $\mathbf{E}_0$ and $\mathbf{E}'$ are equal and opposite within the conducting segment AB, they are not equal and opposite *outside* the conductor along the rest of path C. The fact is that outside the conductor, the motional field $\mathbf{E}'$ is *zero*. The electrostatic field $\mathbf{E}_0$, on the other hand, is set up by a distribution of charge that has been established on the top and bottom ends of the conductor by the force $\mathbf{v} \times \mathbf{B}$ that pushes positive charges toward A and negative charges toward B. Within the conductor, this charge distribution has to set up a field given by (20.2.38) if no current is to flow. But the electrostatic field from this charge distribution also extends *outside* the conductor throughout the territory around which the dotted path C is drawn. Therefore, there is a contribution to the integral of $\mathbf{E}_0 \cdot d\mathbf{l}$ around the part of C outside the conductor that exactly cancels the contribution from within.

Why does it cancel exactly? Because, as you will remember from Chapters 5 and 16, the integral $\mathbf{F} \cdot d\mathbf{l}$ is zero around any closed path for any force that is *conservative* in nature (that is, with which we can associate a unique potential energy), and the electrostatic Coulomb force is of this type.

The total emf $\mathscr{E}$ around the path C is, therefore, given by

$$\mathscr{E} = \oint_C \mathbf{E}' \cdot d\mathbf{l} \qquad (20.2.41)$$

and since the field $\mathbf{E}'$ is given by (20.2.39) within the conductor and is zero outside, this can be written

$$\mathscr{E} = \int_{y_A}^{y_B} E_y{}' \, dy = -\int_{y_A}^{y_B} vB \, dy = -vB \int_{y_A}^{y_B} dy$$

$$\mathscr{E} = -vB(y_B - y_A) = -vBl \qquad (20.2.42)$$

The difference in electrostatic potential between the two ends of the conductor can be calculated in the usual way from the relation $dV = -\mathbf{E}_0 \cdot d\mathbf{l}$. In this way, we obtain

$$\Delta V_{AB} = V_B - V_A = -\int_A^B \mathbf{E} \cdot d\mathbf{l}$$

$$= -\int_{y_A}^{y_B} E_{0y} \, dy = -vB \int_{y_A}^{y_B} dy$$

or

$$\Delta V_{AB} = -vB(y_B - y_A) = -vBl \qquad (20.2.43)$$

When an external circuit is completed, by connecting a wire around the dotted path C, for example, the accumulated excess charges that represent the source of the electrostatic field are free to flow out of the conductor through the external circuit. When this happens, the electrostatic field falls to zero, leaving only the motional emf $-vBl$, which now maintains an induced current flow in the external circuit, as in the example considered previously in connection with Fig. 20.2a.

EXAMPLE 20.2.2

A straight conductor of length R rotates about one end with a constant angular velocity ω in a constant magnetic field B that is parallel to the axis of rotation, as illustrated in Fig. 20.8a. Find the electrostatic field within the conductor and the difference in electrostatic potential between its ends. If a circuit is established, as in Fig. 20.8b, by connecting the end of the conductor at the rotation axis with a circular rail of radius R upon which the other end slides, find the emf in the circuit.

The resultant force on a charge q in the conductor is once again

$$\mathbf{F} = q\mathbf{E}_0 + q(\mathbf{v} \times \mathbf{B}) \qquad (20.2.44)$$

In this case, we cannot immediately set this force equal to zero as in the preceding example because the

653

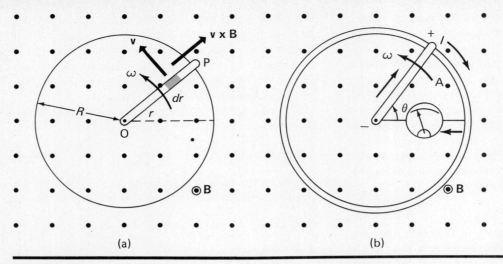

FIGURE 20.8

charges are not unaccelerated and, therefore, not in equilibrium under the action of the forces. We must instead equate the resultant force **F** as given by (20.2.44) to the mass of the charge times its centripetal acceleration, which is $-r\omega^2$, where r is the distance from the axis of rotation. Since, according to Fig. 20.8a, the magnitude of the force $q(\mathbf{v} \times \mathbf{B})$ is $qvB = qr\omega B$, we may write the following equation for the radial force components:

$$qE_0 + qr\omega B = -mr\omega^2 \qquad (20.2.45)$$

This may be solved for E_0 to give

$$E_0 = -r\omega\left(B + \frac{m\omega}{q}\right) \qquad (20.2.46)$$

The difference in potential between points O $(r = 0)$ and P $(r = R)$ can now be found, as usual, from $dV = -\mathbf{E}_0 \cdot d\mathbf{l}$. In this case, since $d\mathbf{l}$ and $\mathbf{E}_0$ are both radial vectors, $\mathbf{E}_0 \cdot d\mathbf{l} = E_0\,dr$, and we may write

$$\Delta V_{OP} = V_P - V_O = -\int_O^P \mathbf{E}_0 \cdot d\mathbf{l}$$

$$= -\int_0^R E_0\,dr = \int_0^R r\omega\left(B + \frac{m\omega}{q}\right)dr$$

$$\Delta V_{OP} = \omega\left(B + \frac{m\omega}{q}\right)\int_0^R r\,dr \qquad (20.2.47)$$

$$\Delta V_{OP} = \left(B + \frac{m\omega}{q}\right)\frac{\omega R^2}{2}$$

In the above integration, since B, m, ω, and q are all constant, the quantity $\omega B + (m\omega^2/q)$ may be written outside the integral.

Now, in (20.2.46) and (20.2.47) there are two separate terms. The first, the one that contains B, arises from the motion of the conductor in the magnetic field and the consequent $\mathbf{v} \times \mathbf{B}$ force, in the same

way that the field in the preceding example expressed by (20.2.38) comes about. If the magnetic field were zero, this contribution would vanish altogether. The second term, in which the factor $m\omega/q$ appears, arises simply because the right-hand side of Eq. (20.2.45) must have the value $-mr\omega^2$ because of the centripetal acceleration. This term, therefore, represents an electric field of *inertial* origin that serves only to keep the charges in the conductor moving in circular paths with angular velocity ω. It is there because the mobile charges move outward in the conductor, leaving uncompensated charge of opposite sign on the fixed metallic ions of the crystal lattice until there is an inward force of electrostatic attraction between the two that is just sufficient to furnish the centripetal force necessary to hold the mobile charges in their circular orbits. If there were no centripetal acceleration, this term would be absent, as it is in the previous example.

Let us calculate how much of a potential difference this inertial field can generate in practice. We shall for the moment set B equal to zero and use a value for ω that would lead to the *largest* possible potential difference that could be realized in normal experimental procedures. Suppose, then, we take $R = 0.3$ meter and $\omega = 6280$ radians/second; this corresponds to a conductor about 1 foot long that revolves at 60,000 rpm. We could hardly use a conductor that rotates more rapidly than that in any feasible experiment. In this event, using appropriate numerical values for the mass and charge of the electron in mks units, we find

$$\Delta V = \frac{m\omega^2 R^2}{2q} = \frac{(9.11 \times 10^{-31})(6.28 \times 10^3)^2(9 \times 10^{-2})}{(2)(-1.602 \times 10^{-19})}$$

$$= -1.01 \times 10^{-5} \text{ volt}$$

This is a very small voltage difference, indeed, one

that we might have to go to a great deal of trouble even to detect. Suppose, for example, that the conductor were rotating in a plane normal to the earth's magnetic field, which is, in fact, very weak, having a strength of about 0.5×10^{-4} weber/m^2 in most places. Under these circumstances, the contribution from the first term in (20.2.47), arising from the motion of the conductor in the earth's magnetic field, would be

$$\Delta V = \tfrac{1}{2} B \omega R^2 = (0.5)(0.5 \times 10^{-4})(6.28 \times 10^3)(9 \times 10^{-2})$$
$$= 1.41 \times 10^{-2} \text{ volt}$$

This is 1400 times as large as the inertially produced voltage $m\omega^2 R^2 / 2q$. In order to have a fair chance of detecting the inertial potential difference, we would have to somehow reduce the earth's field by magnetic shielding until it is less by a factor 1000 to 10,000 than it is normally. This is not an impossible task, but neither is it one that can be done easily.

The situation can be summed up by remarking that the inertial voltage $m\omega^2 R^2 / 2q$ is under normally realizable conditions so small as to be completely negligible in comparison with the contribution due to the conductor's motion, even when only stray fields due to terrestrial magnetism are present. Under these circumstances, we may *neglect* the inertial contribution, and write as very good approximations for (20.2.46) and (20.2.47),

$$E_0 = -r\omega B \tag{20.2.48}$$

and

$$\Delta V_{\text{OP}} = \tfrac{1}{2} B \omega R^2 \tag{20.2.49}$$

We shall occasionally encounter situations in future work where these electronic inertial potential differences exist, and, on the basis of the work we have done here, we shall consistently *neglect* them in comparison with the much larger electromagnetically induced voltages. It should be noted, however, that there are situations, notably in atomic physics, where the centripetal forces acting on electrons can be very important indeed. The hydrogen atom, in which the electrostatic force of attraction between a proton and an electron affords the centripetal force that keeps the electron in its path around the nucleus, is one simple instance where this is true.

Returning now to the situation shown in Fig. 20.8b, where a complete circuit that includes our previous rotating conductor has been formed, we may use Faraday's law to find the emf in the circuit very easily. The flux through the shaded area in this case can be written

$$\Phi_m = BA \tag{20.2.50}$$

and since B is constant,

$$\mathscr{E}_m = -\frac{d\Phi_m}{dt} = -B\frac{dA}{dt} \tag{20.2.51}$$

The area A of the shaded sector is a fraction $\theta/2\pi$ of the total area πR^2 within the outer circular conductor. Therefore, since R is constant,

$$\mathscr{E}_m = -B\frac{d}{dt}\left(\frac{\theta}{2\pi}\pi R^2\right) = -\frac{BR^2}{2}\frac{d\theta}{dt} = -\tfrac{1}{2}BR^2\omega \tag{20.2.52}$$

the minus sign simply reflecting that the current in the external circuit flows in the negative (clockwise) sense rather than the positive (counterclockwise) sense. This, in turn, stems from the fact that we measure counterclockwise angular displacements as positive and clockwise ones as negative. Since Faraday's law tells us only about *electromagnetically* induced emfs and currents, it does not give us any indication that an *inertially* produced emf or current might also be there.

It should also be clear from this and from several preceding examples that the potential difference in a conducting segment that moves in a magnetic field can sometimes be found very simply *by using Faraday's law to find the emf in a complete circuit in which the moving conductor forms one branch.* The validity of this procedure is reflected by the agreement between (20.2.52) and (20.2.48) and between (20.2.17) and (20.2.43), and elsewhere in the examples discussed so far. It stems from the fact that the integral of $\mathbf{E} \cdot d\mathbf{l}$ along the conducting segment in question represents both the emf in the isolated conducting segment and, *so long as the magnetic field does not vary with time,* also in the circuit in which that conductor forms the moving branch. The italicized words in the last sentence define the limitations of this procedure, since when there are time-varying fields, there are emfs induced in the circuit that are associated with the time variation of $\mathbf{B}$ as well as the motion of the conducting segment.

EXAMPLE 20.2.3

An alternating current (ac) generator, sometimes referred to as an *alternator*, consists, in its simplest form, of a coil of N turns rotating with constant angular velocity ω in a uniform magnetic field B, as illustrated in Fig. 20.9. For a coil of 150 turns with an area of 200 cm^2 and a resistance of 2.5 ohms rotating with a constant angular speed of 60 revolutions per second in a magnetic field of 0.15 weber/m^2, find the emf as a function of time, the amplitude of the emf, the amplitude of the maximum current that can be delivered to an external circuit, and the power delivered to an external circuit of resistance R as a function of time.

The instantaneous emf is calculated from Faraday's law in the usual way, remembering that since there are N turns, the emf in each appears in series. Therefore,

$$\mathscr{E} = -N\frac{d\Phi_m}{dt} = -N\frac{d}{dt}BA\cos\theta \tag{20.2.53}$$

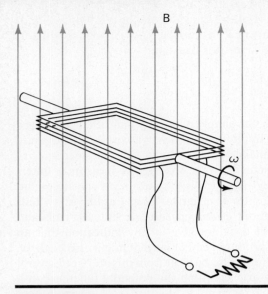

FIGURE 20.9

where θ is the angle between the magnetic field and the normal to the plane in which the coil lies. In this example, both B and A are constant, and it is the angle θ that changes as the coil rotates. As the coil turns, the intercepted flux changes from the maximum value BAN when $\theta = 0$, in which case the plane of the coil is normal to **B**, to zero when $\theta = 90°$ when the plane of the coil is parallel to the lines of **B**. Under these circumstances, (20.2.52) can be written as

$$\mathscr{E} = -BAN \frac{d \cos \theta}{dt} = BAN \sin \theta \frac{d\theta}{dt} \qquad (20.2.54)$$

But $d\theta/dt$ is the constant angular velocity ω; therefore, this can be expressed as

$$\mathscr{E} = BAN\omega \sin \theta \qquad (20.2.55)$$

This could also have been obtained from (20.2.5), in which, because $dB/dt = dA/dt = 0$, only the third term would contribute anything to the result. Since in this example the coil rotates continuously about a fixed axis, the angle θ will increase continuously with time according to the relation $\theta = \omega t$, and the emf can finally be expressed in the form

$$\mathscr{E}(t) = BAN\omega \sin \omega t \qquad (20.2.56)$$

Since $N = 150$, $A = 200 \text{ cm}^2 = 0.02 \text{ m}^2$, $B = 0.15$ weber/m^2, and $\omega = 60$ rev/sec $= 120\pi$ rad/sec, (20.2.56) becomes

$$\mathscr{E} = (0.15)(0.02)(150)(120\pi) \sin(120\pi t)$$
$$= 169.6 \sin(120\pi t) \text{ volts}$$

From these equations we observe that the emf varies with time in the fashion of a *simple harmonic oscillator* of angular frequency ω and amplitude $BAN\omega$. In the given example, the amplitude of the emf is clearly

169.6 volts, a value attained when $\omega t = \pi/2$, $3\pi/2$ $5\pi/2, \ldots$ rad. The output emf plotted as a function of time is shown in Fig. 20.10. An emf that varies with time in the same manner as a simple harmonic oscillator is referred to as a sinusoidally *alternating emf.*

If the internal resistance of the rotating coil is R_i and the resistance of the external circuit to which the device is connected is R, a current whose instantaneous value is

$$I(t) = \frac{\mathscr{E}}{R_i + R} = \frac{BAN\omega}{R_i + R} \sin \omega t \qquad (20.2.57)$$

will flow in the circuit. The amplitude of this current will be

$$I_{\max} = \frac{BAN\omega}{R_i + R} = \frac{169.6}{2.5} = 67.8 \text{ amp}$$

in the case where $R_i = 2.5$ ohms and the external resistance is zero, in which event the current in the external circuit is as large as it can ever be. The instantaneous power dissipated in the external resistance R will be I^2R, where I is given by (20.2.57). We may thus write

$$P_{\text{ext}} = I^2R = \left(\frac{BAN\omega}{R_i + R}\right)^2 R \sin^2 \omega t \qquad (20.2.58)$$

EXAMPLE 20.2.4

A uniform magnetic field **B** is confined to a cylindrical volume of radius R, the direction of **B** being parallel to the axis of the volume; outside this volume, B is zero. This geometry is illustrated in Fig. 20.11. The magnitude of the field **B** varies linearly with time, so that B is described as a function of time by

$$B(t) = \alpha t \qquad (20.2.59)$$

where the constant α represents the time rate of change dB/dt. The situation is seen to resemble that discussed in connection with the third of the illustrations dis-

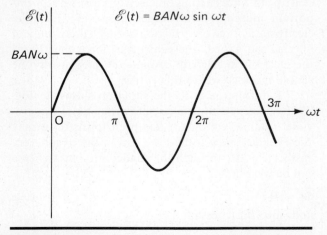

FIGURE 20.10. Harmonically varying emf of an ac generator, plotted as a function of time.

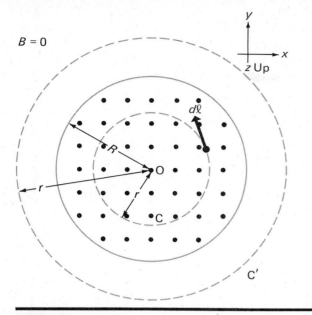

$B = 0$

FIGURE 20.11

cussed in the text preceding this series of examples. Find the magnitude and direction of the nonelectrostatic induced electric field **E** that exists within the field region and external to it as a result of the changing magnetic flux.

The relation between the field **E** and the time rate of change of the magnetic flux is best displayed by Faraday's law in the form

$$\oint_C \mathbf{E} \cdot d\mathbf{l} = -\frac{d}{dt} \int_S \mathbf{B} \cdot \mathbf{n}\, da \qquad (20.2.60)$$

In this case, the cylindrical symmetry of the field **B** is exactly the same as the symmetry of the current distribution in straight conductors, we discussed in Chapter 19, section 19.6, in connection with the derivation of the magnetic fields around cylindrical conductors using Ampere's law. The same symmetry arguments used there also apply here to tell us that the induced field **E** must be either purely radial or purely tangential. But if the field were purely radial, then **E** and $d\mathbf{l}$ would always be mutually perpendicular and $\mathbf{E} \cdot d\mathbf{l}$ zero at every point around a circular path such as C or C' in Fig. 20.11. Under these circumstances, the integral of $\mathbf{E} \cdot d\mathbf{l}$ around any path, and hence the emf around such a path, would have to vanish. Experimentally, we know very well that there *are* emfs induced around such paths, and, in any event, we have already seen from the development leading to (20.2.30) that we can calculate what they are from Faraday's law in the form (20.2.3). The field **E** thus cannot be purely radial; therefore it must be purely tangential, like the magnetic field around a straight current-carrying conductor, the field lines lying in the plane of the page in Fig. 20.11.

If this is the case, then, around a circular path such as C or C', the field **E** and the $d\mathbf{l}$ elements of the path are parallel, and E is the same for all parts of the path. We may, therefore, write

$$\oint_C \mathbf{E} \cdot d\mathbf{l} = \oint_C E\, dl = E \oint_C dl = 2\pi r E \qquad (20.2.61)$$

At the same time, if we evaluate the surface integral on the left side of (20.2.60) over the surface enclosed by C or C', we see that the lines of **B** and the normal **n** to this surface are parallel and, in addition, that the magnitude of **B** is the same everywhere within the field region. We may write, therefore, for path C

$$\int_S \mathbf{B} \cdot \mathbf{n}\, da = \int_S B\, da = B \int_S da = \pi r^2 B \qquad (20.2.62)$$

while for path C', noting carefully that the **B** field extends out only to $r = R$ and is zero beyond,

$$\int_S \mathbf{B} \cdot \mathbf{n}\, da = \int_S B\, da = B \int_{r<R} da + 0 \int_{r>R} da = \pi R^2 B$$
$$(20.2.63)$$

For paths such as C for which $r < R$, then, we may substitute (20.2.61) and (20.2.62) into Faraday's law (20.2.60) to obtain

$$2\pi r E = -\frac{d}{dt}(\pi r^2 B) = -\pi r^2 \frac{dB}{dt} \qquad (20.2.64)$$

from which

$$E = -\frac{r}{2}\frac{dB}{dt} = -\frac{\alpha r}{2} \qquad r < R \qquad (20.2.65)$$

For paths such as C' for which $r > R$, we may substitute (20.2.62) and (20.2.63) into (20.2.60) to find

$$2\pi r E = -\frac{d}{dt}(\pi R^2 B) = -\pi R^2 \frac{dB}{dt} \qquad (20.2.66)$$

or

$$E = -\frac{R^2}{2r}\frac{dB}{dt} = -\frac{\alpha R^2}{2r} \qquad r > R \qquad (20.2.67)$$

It will be noted that for $r = R$, both (20.2.65) and (20.2.67) lead to the same result, $E = -\alpha R/2$. The magnitude of the induced electric field, plotted as a function of r, is shown in Fig. 20.12. It is very much like the plot of the magnetic field of a uniform, straight current-carrying conductor as illustrated by Fig. 19.35, except for the negative value of E. A plot of the lines of **E** is shown in Fig. 20.13; it is apparent that the field resembles the magnetic field around a straight wire. It is to be emphasized that this is an *induced* field and not an electrostatic field. The integral of $\mathbf{E} \cdot d\mathbf{l}$ around closed paths *does not vanish*, as it always does for an electrostatic field. For this reason, also, these induced fields are *nonconservative*. The field lines do not begin and end on charges as is invariably the case for electrostatic fields. Instead

657

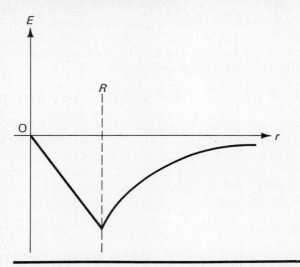

FIGURE 20.12. Induced electric field magnitude plotted as a function of the radial distance from the center of the system shown in Fig. 20.11.

they close upon themselves in a manner reminiscent of magnetic field lines. In the case of magnetic fields, the source of the field is always a distribution of current, therefore a distribution of moving charge. In the case of the induced electric fields, the source of the field is always a distribution of magnetic flux that is *changing with time*. A steady magnetic field induces no electric field, just as a stationary set of charges gives rise to no magnetic field. We shall see later that induced electric fields play an important role in the description of electromagnetic waves.

20.3 Lenz's Law and Eddy Currents

What is usually referred to as *Lenz's law* is simply an aspect of Faraday's law that we have not yet examined closely. It concerns the *direction* of the currents that are induced by flux changes and their own associated magnetic fields. In the preceding section, we emphasized mainly how to calculate induced emfs, electric fields, and currents, using Faraday's law, without worrying very much about their direction. Of course, the direction of any current that is induced by a change in the magnetic flux through a circuit is uniquely determined by Faraday's law; there is no ambiguity about that. But there is the question of the minus sign before the rate of change of magnetic flux in Eq. (20.2.2), (20.2.3), or (20.2.35), and what we call Lenz's law is simply a statement of the physical effects that require that that algebraic sign be minus rather than plus.

Lenz's law may be stated as follows:

Induced currents flow in such a way that their own magnetic effects oppose the flux change that created them originally.

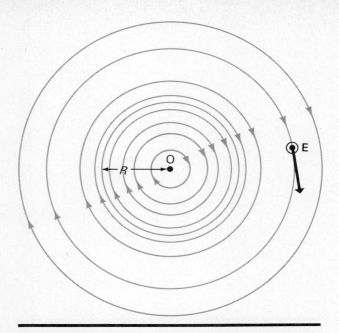

FIGURE 20.13. Field lines for the electric field induced in the system shown in Fig. 20.11.

The examples that were worked out in the preceding section afford numerous illustrations of how Lenz's law operates. For instance, referring to Fig. 20.1, we see that the induced current I is created by an increase in flux arising from a **B** field that points along the positive z-direction as the length of the circuit increases from x to $x + dx$. The flux change $d\Phi_m = B[l(x + dx) - lx] = Bl\,dx$ is, therefore, *positive*. According to Lenz's law, the induced current will flow around the loop ABCD in such a way that its own magnetic field will oppose this flux change. The magnetic field arising from the current must, therefore, create a magnetic field that is in the *negative* z-direction, and, therefore, according to the right-hand rule relating currents and their magnetic fields, must flow *counterclockwise* around ABCD. This is in agreement with our previous calculations based on the direction of the Lorentz force and on Faraday's law itself. We see that although Lenz's law does not give us the exact magnitude of the induced emf or current, it does tell us very quickly and easily the direction of the induced current, from which, of course, it is easy to infer the sign of the emf.

In the example illustrated in Fig. 20.2, we see that in (a) the magnetic field is in the positive z-direction; the flux through the loop is, therefore, increasing. According to Lenz's law, the magnetic effect of the current induced in the loop must be opposite in direction; hence, it must produce a magnetic field that opposes this increase in flux and which, therefore, must be in the negative z-direction. According to the right hand rule, the current must flow clockwise to produce such a magnetic effect. In Fig. 20.2c, we observe that although the magnetic field is still in the positive z-direction, the area of the field region enclosed by the loop decreases as the loop is withdrawn from the field. The flux through the loop is, therefore, decreasing also. The magnetic effect of the induced current must be such as to oppose this decrease in flux through the loop; therefore, it must give rise to a magnetic field in the positive z-direction. According to the right-hand rule for currents and fields, the induced current as a result must pass counterclockwise around the loop. These conclusions are again in agreement with those reached previously in a different way.

In the example shown in Fig. 20.4, the **B** field of the solenoid is along the positive z-direction and is increasing in magnitude. The increasing flux through the small coil must create a magnetic field that opposes this increase of flux along the +z-direction, and therefore this magnetic field must be in the −z-direction. But this can only occur if the current flow in the small coil is out of the page at the top and into the page at the bottom of the coil, as illustrated in Fig. 20.4.

Finally, in Fig. 20.5, we see a magnet whose north end is approaching a circular coil of wire. The flux through the coil is rapidly increasing as the magnet nears the coil. The direction of the induced current in the coil has to be such as to oppose the increase in flux, and, therefore, the induced current creates a magnetic field whose direction is opposite to that of the field of the magnet, that is, whose **B** vector points to the right in the figure. But such a field can result only if the induced current has the direction that is shown. In doing this, the magnetic field of the coil is such that the end nearest the magnet becomes a north pole, while the other end is a south pole, as shown, for example, in Fig. 19.4. There is, therefore, a force of *repulsion* set up that opposes the approach of the pole of the bar magnet toward the coil. If the motion of the magnet were reversed, the magnet being withdrawn to the right, the primary flux through the coil would decrease. The current in the coil would then have to flow in such a way as to oppose the decrease and therefore generate a field of its own that would have the same direction as that of the magnet, represented by a **B** vector to the left. The direction of the current in the coil would in this case be opposite that shown in the diagram. The magnetic field of the coil is such that the end nearest the bar magnet is a south pole, the other end being north. Under these circumstances, there is a force of *attraction* between the coil and the bar magnet that opposes the withdrawal of the magnet.

It is important in using Lenz's law to understand that the induced current's magnetic field is always in opposition not to the primary magnetic flux but to the *change* in primary magnetic flux that induces the current. This is illustrated in the very last example

considered above, in which the north pole of the magnet in Fig. 20.5 is withdrawn to the right. In this case, the primary field vector **B** within the coil still points to the left, but its magnitude is decreasing, so that the vector Δ**B** determining the *change* in flux in a given time interval would point in the opposite direction, to the *right*. It is this flux *change* the induced current must oppose by generating a magnetic field whose **B** vector points left.

The physical significance of Lenz's law can be understood by considering the consequences of its repeal. Suppose, for the moment, that induced currents were always in a sense *opposite* to that predicted by Lenz's law. The directions of the forces experienced by conductors carrying induced currents in external magnetic fields would then also be reversed. This would mean that, in Fig. 20.2a, the direction of the force **F** acting on the right side of the loop, given by Eq. (20.2.21), would be in the opposite direction to what is shown. The loop would now be attracted *into* the field region, and we would need to apply an equal and opposite force to hold it back and prevent it from being accelerated. Indeed, the loop would do work against this external force and would serve as a source of energy that could power whatever external device we might wish to run! The same thing would happen on the other side of the field region, as shown in Fig. 20.2c. The direction of the force **F** on the left side of the loop would be reversed and the loop would be expelled from the field region, during which process it could do external work. The process could then be repeated, the loop being reinserted from the right and returning, once more performing work, to its original position. The magnetic field is still there; it has given up no energy to the outside—for all we have said it might have been set up by a permanent magnet. The only net change has been in the performance of external work with no energy input to the system. Need we say more? If we are to repeal Lenz's law, we must also repeal the law of conservation of energy. And this nobody to date has discovered how to do.

A discussion of Lenz's law would not be quite complete without a brief description of *eddy currents*, for the generation and subsequent behavior of eddy currents is closely connected with the operation of Lenz's law. Eddy currents are currents that are induced, just as any other current, by changes in magnetic flux, but in *extended conducting objects* rather than in linear conductors. The flow of eddy currents, though of course limited by the boundaries of the conductor, is not restricted to a single closed, linear path the way the flow of current in an ordinary conducting circuit is. The paths of eddy currents within the conductor are defined primarily by the electric and magnetic forces that act on the mobile charge distribution in the conducting object, and the sense

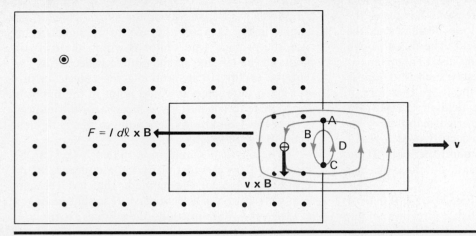

FIGURE 20.14. Eddy currents in a thin conducting plate being removed from a region in which a uniform magnetic field exists.

in which currents flow around these paths is determined by Lenz's law.

The action of eddy currents is illustrated in Fig. 20.14. A rectangular conducting sheet is pulled out of a region in which a uniform magnetic field has been established. Inside the field region, there are magnetic forces on mobile charges within the conductor given by $q(\mathbf{v} \times \mathbf{B})$, while outside, these forces vanish. The forces cause downward currents to flow in the part of the sheet inside the field, since there is an emf along any path such as ABC that lies in this region. The origin of this emf is the same as that associated with the linear conducting loop of Fig. 20.2c. Current flow along this path is established, the

current returning to point A along the path CDA that lies outside the field. As in the case of the loop shown in Fig. 20.2c, the current experiences a retarding force $\mathbf{F} = I\,\mathbf{l} \times \mathbf{B}$ that makes it necessary to exert an external force in the opposite direction to keep the conductor moving at constant speed. There are many such current loops in the sheet, the aggregate effect of which is to generate *heat* by Joule heating as the current passes through the resistance of the conducting paths and to create a retarding force that may be quite appreciable if one attempts to pull the conducting sheet out of the field very quickly.

Another demonstration of the effects of eddy currents is illustrated in Fig. 20.15. A pendulum whose

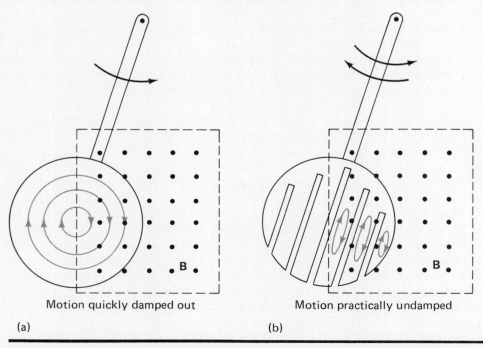

Motion quickly damped out   Motion practically undamped

(a)   (b)

FIGURE 20.15. Eddy current flow in (a) a highly conducting pendulum bob and (b) a slotted pendulum bob made from the same material.

bob is made of a flat conducting sheet is arranged to swing between the poles of a powerful magnet, as shown at (a). Owing to the generation of eddy currents and the retarding forces they experience as they enter and leave the field region, the motion of the pendulum is quickly damped and its kinetic energy appears as Joule heat within the conductor. At (b), the bob has been replaced with one that is identical except that a number of vertical slots are cut into it. These slots prevent the formation of large-scale eddy currents and limit those that form to long, narrow loops between the slots. As a result, the eddy currents are reduced in magnitude and generate much less Joule heat within the conductor, the retarding forces experienced by the pendulum as the bob enters and leaves the field region are much smaller, and the motion of the pendulum suffers very little damping. This method of breaking up eddy current flow and reducing eddy current heating is used also in transformers, where the iron cores are made of laminations that are electrically insulated from one another rather than from a single solid piece of metal.

20.4 Self-Induction and Self-Inductance: The Behavior of *L–R* Circuits

Up to this point, our concern has centered upon emfs generated by externally produced magnetic fluxes. Of course, *any* flux change experienced by a circuit, *even a change in the magnetic flux produced by the current flowing in the circuit itself*, will induce an emf in the circuit. Electromotive forces generated by a circuit's own current in this way are called *self-induced emfs*, and their generation is referred to as *self-induction*.

An understanding of self-induced emfs can best be gained by looking at a few specific examples. Let us first consider the case of a long, straight solenoid wound with n turns of wire per unit length. We know from our previous work, in Eq. (19.7.9), that the axial magnetic induction B within the turns of such a coil has the value

$$B = \mu_0 n I \tag{20.4.1}$$

where I is the current in the coil. In the ideal, long solenoid, all the magnetic flux associated with this field passes through every turn of the coil, as shown in Fig. 19.41, though this is no longer true for a solenoid of finite length, as illustrated by Fig. 19.42. Even so, if the length of the solenoid is much greater than its radius, as we shall assume, we shall not go far wrong in making this assumption anyway. Proceeding in this way, the flux through each turn can be written as the product of the magnetic induction and the cross-sectional area A of the coil as

$$\Phi_m = BA = \mu_0 n I A \tag{20.4.2}$$

If the current changes with time, the rate of change of flux within the solenoid is

$$\frac{d\Phi_m}{dt} = \mu_0 n A \frac{dI}{dt} \tag{20.4.3}$$

The self-induced emf can now be obtained from Faraday's law as

$$\mathscr{E} = -N \frac{d\Phi_m}{dt} = -\mu_0 N n A \frac{dI}{dt} \tag{20.4.4}$$

where N is the total number of turns equal to nl, where l is the length of the solenoid. We may finally express the self-induced emf as

$$\mathscr{E} = -\mu_0 n^2 l A \frac{dI}{dt} \tag{20.4.5}$$

Had our circuit been of some form other than a long solenoid, we could no longer have used (20.4.1) to express the magnetic field around and within it. But even if we had been unable to calculate the field surrounding the circuit, we could have concluded from the law of Biot and Savart, as expressed by (19.5.5), that the magnetic field at all points must be *directly proportional to the current* that creates it. The same is true of the flux Φ_m, since it is simply a summation of field values over the area of the circuit, in the simplest case equal to field times area. Accordingly, the time rate of change of flux $d\Phi_m/dt$ must be directly proportional to the time rate of change of current dI/dt, whatever form the circuit may have. We shall always find, therefore, that the self-induced emf $\mathscr{E}$ will be related to the current in the circuit by

$$\boxed{\mathscr{E} = -N \frac{d\Phi_m}{dt} = -L \frac{dI}{dt}} \tag{20.4.6}$$

where L is a proportionality constant referred to as the *self-inductance* of the circuit. The term self-inductance is frequently abbreviated to *inductance*. For a circuit in the form of a long, straight solenoid of length l and cross-sectional area A wound with n turns of wire per unit length, it is easy to see from (20.4.5) that the self-inductance must be

$$\boxed{L = \mu_0 n^2 l A} \tag{20.4.7}$$

For a circuit of another form, we should have to seek a relation other than (20.4.2) to give us the flux within the circuit. Also, as in the case of solenoids whose radius is not small compared to their length, all the flux that is generated may not be enclosed by every loop in the circuit; in this case, we would have to know just how much of the flux produced is effective in contributing to emfs in each turn. The self-inductance of the circuit, then, would clearly no longer be given by (20.4.7) and, indeed, might be difficult or impossible

661

to arrive at by mathematical calculation alone. Nevertheless, even in these instances, due to the strict proportionality between the current in a circuit and the magnetic flux it produces, Eq. (20.4.6) will still be satisfied, even though we may have no simple way of determining the self-inductance L other than by measuring it experimentally. Fortunately, this is not particularly difficult, and it is enough for most purposes to know simply that the self-induced emf is given by (20.4.6), in which L is a property of an individual circuit or collection of circuit loops, to be determined by measurement.

The situation is similar to that which arises in the case of Ohm's law in electrical circuits. It is quite often difficult to *calculate* the electrical resistance of a resistor even if the electrical resistivity of its substance is accurately known, which is usually not the case. But it is practically child's play to pass an accurately known current through it, measure the potential difference across it, and then use $R = \Delta V/I$ to arrive at the desired result. The important matter is not to know how to calculate the resistance but to be able to determine its value in some practical fashion and to know that it expresses the proportionality between current and potential difference according to Ohm's law.

It is likewise not terribly important to know how to calculate the self-inductance of a circuit or coil. It is much more important to be able to determine the self-inductance of a circuit in some practical experimental fashion and to understand that the self-induced emf in any circuit is related to the time rate of change of current in the circuit by

$$\mathscr{E}_L = -L\frac{dI}{dt} = \Delta V_L \qquad (20.4.8)$$

where the proportionality factor L is the self-inductance. Like any other emf, the effect of this one is to produce a potential difference ΔV_L across the terminals of the device by *nonelectrostatic* means. In this case, the emf is of electromagnetic origin; and, as we shall soon see, although the potential difference it contributes to the circuit can always be represented mathematically as $\Delta V_L = -L(dI/dt)$, its *direction* depends upon whether dI/dt is positive (current increasing) or negative (current decreasing).

It is also important to note that the inductance in a circuit can be concentrated in a single circuit element such as a solenoid or toroidal coil purposely designed to exhibit a large value of self-inductance, or distributed around the circuit in the event no such element is present. In either event, the emf $-L(dI/dt)$ is present, but in the first case, it is clearly associated with the inductive element, more or less like a battery emf, while in the second, it is distributed around the

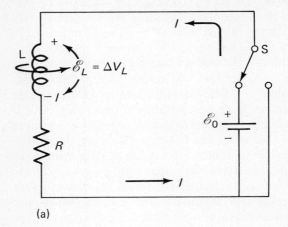

(a)

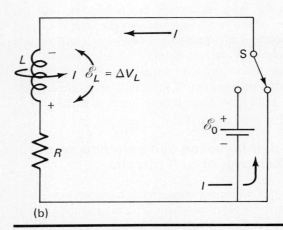

(b)

FIGURE 20.16. Self-induced emfs in a series $R-L$ (resistance–inductance) circuit (a) when current is turned on and (b) when current is switched off.

entire loop. We shall deal mainly with circuits containing concentrated inductive elements, referred to as *inductors* or, less frequently, as *chokes*. The presence of inductors in schematic diagrams of electrical circuits is illustrated by the symbol ⎓⎓⎓⎓. We shall discuss a method of measuring the self-inductance of circuits in an example to be presented a bit later. From (20.4.6), it is apparent that the units of self-inductance can be expressed as units of flux divided by units of current. In the mks system, then, the unit of self-inductance is weber/ampere. This unit is usually referred to as the *henry* (abbreviated H).

The *direction* of self-induced emfs in inductive circuits is easily determined by the application of Lenz's law. For example, in Fig. 20.16a, a coil having self-inductance L and resistance R is shown in series with a battery of emf $\mathscr{E}_0$ and a switch S. The resistance R, representing the resistance of the wire in the windings of the coil, is shown as a separate element in the circuit, so that the effects associated with resistance and self-inductance can be discussed separately, just as we might show the internal resistance of a battery

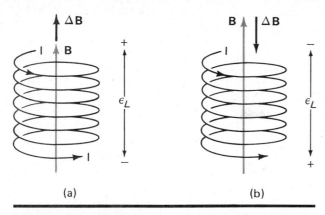

FIGURE 20.17. Flux changes and induced emfs (a) in Fig. 20.16a and (b) in Fig. 20.16b.

as a separate circuit element. We shall assume that *all* the circuit resistance is accounted for by the resistance R and *all* the circuit self-inductance by L.

When the switch S is closed, the current I, initially zero, starts to flow through the circuit. As it flows through the coil, a magnetic field builds up, and the time-varying flux through the turns of the coil, and the rest of the circuit as well, immediately generates a self-induced emf $\mathscr{E}_L = -L(dI/dt)$. The sign of this emf is such as to *oppose* the increase of magnetic flux through the coil caused by the current I. The current I produces a **B** field within the coil that points upward and continually increases, since the current is increasing with time. In a time interval Δt, there is a flux change caused by a field change $\Delta \mathbf{B}$, as shown in Fig. 20.17a. The self-induced emf in this case must be such that if it were to determine the direction of the current in the coil, it would generate an opposite flux change. The self-induced emf $\mathscr{E}_L$, therefore, tends to set up a current that *opposes* the one that actually flows; current flows the way it actually does only because the battery emf $\mathscr{E}_0$ is larger than $\mathscr{E}_L$. Acting by itself it would create a circuit potential difference that would make the top end of the coil positive with respect to the lower end, as shown in Fig. 20.17a. The induced $\mathscr{E}_L$, therefore, is opposite in sign to the battery emf. Instead of causing a potential rise in the circuit as the circuit is traversed in the direction of current flow, it causes a potential drop. This is already reflected in the minus sign in Eq. (20.4.8), which can be traced back to Lenz's law. The application of Kirchhoff's law II, which tells us that the sum of all emfs and potential changes around any closed circuit loop must be zero, therefore shows that with the switch S positioned as shown in Fig. 20.16a,

$$\mathscr{E}_0 - L\frac{dI}{dt} - RI = 0 \tag{20.4.9}$$

This is a differential equation that can be solved to

give the current I in the circuit as a function of time and thus exhibit the way in which current grows when the switch is closed.

The equation can be solved most easily by first differentiating both sides of (20.4.9) with respect to time:

$$L\frac{d^2I}{dt^2} + R\frac{dI}{dt} = 0 \tag{20.4.10}$$

Now let us refer to the quantity dI/dt as w. Then, if $w = dI/dt$, we must have $dw/dt = d^2I/dt^2$, which enables us to write (20.4.10) as

$$L\frac{dw}{dt} = -Rw$$

from which

$$\frac{dw}{w} = -\frac{R}{L}dt \tag{20.4.11}$$

We may now integrate both sides of this equation from time $t = 0$, when the current is also zero, to some later time t, when the current has the value I and its time derivative the value w. But if the current I is zero initially, from (20.4.9) we must have in the initial state of the system

$$\left(\frac{dI}{dt}\right)_{t=0} = \frac{\mathscr{E}_0}{L} \tag{20.4.12}$$

Since dI/dt is w in (20.4.11), when $t = 0$, w must have the initial value $\mathscr{E}_0/L$. We therefore must integrate (20.4.11) from $t = 0$ to the later time t, and, on the left side, from $w = \mathscr{E}_0/L$ corresponding to time $t = 0$, to the value w, corresponding to time t. In this way, remembering that R and L are constants that do not vary with time, we obtain

$$\int_{\mathscr{E}_0/L}^{w} \frac{dw}{w} = -\frac{R}{L}\int_0^t dt$$

or

$$[\ln w]_{\mathscr{E}_0/L}^{w} = \ln w - \ln \frac{\mathscr{E}_0}{L} = -\frac{Rt}{L} \tag{20.4.13}$$

Since $\ln A - \ln B = \ln(A/B)$, however, this can be written as

$$\ln \frac{wL}{\mathscr{E}_0} = -\frac{Rt}{L} \tag{20.4.14}$$

and, if we exponentialize both sides of this,

$$\frac{wL}{\mathscr{E}_0} = e^{-Rt/L} \tag{20.4.15}$$

or

$$w(t) = \frac{\mathscr{E}_0}{L}e^{-Rt/L} = \frac{dI}{dt} \tag{20.4.16}$$

663

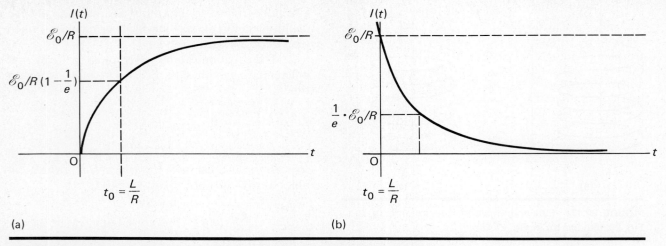

FIGURE 20.18. Current plotted as a function of time (a) for the situation shown in Fig. 20.16a, and (b) for the case of Fig. 20.16b.

recalling that all along w has been used as shorthand for dI/dt.

Now, (20.4.16), multiplying through by dt, can be written as

$$dI = \frac{\mathscr{E}_0}{L} e^{-Rt/L} dt \qquad (20.4.17)$$

We may now integrate this once more from $t = 0$, when the current I is also zero, to time t, when the current has the value I:

$$\int_0^I dI = \frac{\mathscr{E}_0}{L} \int_0^t e^{-Rt/L} dt$$

or

$$I = \frac{\mathscr{E}_0}{L} \left[-\frac{L}{R} e^{-Rt/L} \right]_0^t$$

$$\boxed{I(t) = \frac{\mathscr{E}_0}{R} (1 - e^{-Rt/L})} \qquad (20.4.18)$$

Equation (20.4.18), obtained at last after all this maneuvering, is what we are after. It describes how the current in the circuit of Fig. 20.16a builds up with time when the switch is closed.

From Eq. (20.4.8), the potential difference across the inductance is

$$\Delta V_L = \mathscr{E}_L = -L \frac{dI}{dt} \qquad (20.4.19)$$

and since dI/dt is given by (20.4.16), we find that

$$\boxed{\Delta V_L = -\mathscr{E}_0 e^{-Rt/L}} \qquad (20.4.20)$$

the minus sign reflecting the fact that the current experiences a fall in potential in traversing the inductance. The potential drop across the resistance is IR, with I given by (20.4.18); it is a simple matter to verify

that the sum of these two potential changes given by Eqs. (20.4.18) and (20.4.19) equals the applied emf $\mathscr{E}_0$.

The variation of the circuit current as a function of time is shown in Fig. 20.18a. The current is seen to rise from its initial value of zero and, after a time, approach the value $\mathscr{E}_0/R$ that would be expected in simple resistive circuit in which the effects of self-induction are negligible. The effect of self-induction is clearly to *delay* the rise of current in the circuit for a time until the magnetic field of the inductor is established. In Eq. (20.4.18), at $t = L/R$ seconds, the exponent of the exponential term is unity; and after that length of time, the current has risen to a fraction $1 - (1/e)$ of its final value—in other words, it *differs* from that final value by a fraction $1/e$ of the final amount. The quantity L/R, which has the dimensions of time and expresses this property of inductive–resistive circuits, is referred to as the *inductive time constant* of the circuit. For circuits in which the inductance is large and the resistance small, this time constant may be quite appreciable. It is not too difficult, in practice, to wind an inductor that has an inductance of perhaps 10 henrys and a resistance of no more than 1 ohm. Such an inductor would have a time constant of 10 seconds. More frequently, we shall encounter circuit inductances whose time constants are a very small fraction of a second. Even so, the effects of self-induction in circuits such as these may be of great importance.

Now, let us suppose that the current has after a long time attained the asymptotic value $\mathscr{E}_0/R$, and then the switch S is suddenly thrown to the right, as in Fig. 20.16b. The battery emf is now suddenly removed from the circuit. There is no external energy source to sustain the current in the circuit or the magnetic field it produces, so that the current starts to decrease, and the magnetic flux enclosed by the turns of the inductor (or by the circuit itself) starts to decrease.

According to Faraday's law, an emf must be induced as a result. The sign of this emf can again be ascertained by Lenz's law. As illustrated in Fig. 20.17b, the magnetic field is now decreasing; the vector $\Delta \mathbf{B}$ representing the change in the field $\mathbf{B}$ in a time interval Δt, has the direction shown. It is this field change that produces the flux change in the inductor that generates the induced emf. According to Lenz's law, the emf must be induced in such a way as to *oppose* the flux change that produces it. It must, therefore, be induced so as to contribute to a current flow that would produce a field opposing $\Delta \mathbf{B}$. The emf appears, then, as illustrated in Fig. 20.17b, in such a way that the lower end of the coil is positive and the upper end negative. An emf of this polarity tends to *sustain* the current flow through the inductor and through the rest of the circuit, so long as the current continues to vary with time, in much the same way as a battery of similar polarity. The emf appears, of course, only so long as the current is changing, and once the current and its associated magnetic field have subsided, the induced emf disappears also. Again, the emf contributes a potential difference $\Delta V_L = -L(dI/dt)$ across the terminals of the inductive element; but since dI/dt is now negative, it is opposite in polarity to what it was in the case discussed previously, illustrated by Fig. 20.16a.

Once again, we may equate the algebraic sum of potential increases and decreases around the circuit of Fig. 20.16a to zero, according to Kirchhoff's second law, to obtain

$$-L\frac{dI}{dt} - RI = 0 \tag{20.4.21}$$

Multiplying through by dt and rearranging algebraically, this can be put into the form

$$\frac{dI}{I} = -\frac{R}{L}dt \tag{20.4.22}$$

in which the left side depends only on I and the right side only on t. We may now integrate both sides from $t = 0$, at which time the current has the steady-state value $\mathscr{E}_0/R$ given by (20.4.18) for large t, to time t, when the current has the value I. Therefore,

$$\int_{\mathscr{E}_0/R}^{I} \frac{dI}{I} = -\frac{R}{L}\int_0^t dt \tag{20.4.23}$$

or

$$[\ln I]_{\mathscr{E}_0/R}^{I} = \ln I - \ln(\mathscr{E}_0/R) = -\frac{Rt}{L} \tag{20.4.24}$$

$$\ln\left(\frac{IR}{\mathscr{E}_0}\right) = -\frac{Rt}{L}$$

Exponentializing both sides of this equation, we fi-

nally obtain

$$I(t) = \frac{\mathscr{E}_0}{R}e^{-Rt/L} \tag{20.4.25}$$

The potential difference ΔV_L across the terminals of the inductor, according to (20.4.19), will be

$$\Delta V_L = -L\frac{dI}{dt} = -\frac{\mathscr{E}_0 L}{R}\left(-\frac{R}{L}e^{-Rt/L}\right)$$

or

$$\Delta V_L = \mathscr{E}_0 e^{-Rt/L} \tag{20.4.26}$$

The positive sign of this potential difference confirms our previous reasoning that led us to expect a voltage rise across the inductor under these conditions, rather like that associated with a battery. Indeed, in the circuit of Fig. 20.16b, the inductor is the only source of emf in the circuit and it is the inductor's emf, generated as the magnetic field of the coil decays, and this emf alone that sustains the current in the circuit.

Again, the potential drop across the circuit is IR, where I is given by Eq. (20.4.25). It is once more easy to verify that the algebraic sum of the potential differences across inductor and resistor is zero as required by (20.4.20). The current as given by (20.4.25) is plotted as a function of time in Fig. 20.18b. It will be observed that the current decays exponentially with time; after a time $t = L/R$ seconds, equal to the circuit's time constant, the current has decreased to a fraction $1/e$, about 37 percent, of its initial value. The similarity between the behavior of the current shown in Fig. 20.18 to that occurring during the charge and discharge of a capacitor, as discussed in Chapter 18, section 18.9, should not be overlooked. There is, in fact, a close analogy between the behavior of R–C circuits and inductive–resistive circuits, which are sometimes referred to as L–R circuits. In the case of R–C circuits, the capacitor stores and releases the energy of an electric field that forms within the plates; in the case of the L–R circuit the inductor stores and releases the *magnetic* energy associated with its magnetic field.

Finally, the reader should be careful not to confuse the delayed rise in current in L–R circuits as an external emf is applied, and the prolonged decay of current when it is removed, with the much shorter delays in the rise and fall of current due to the inertia of the mobile conducting electrons in purely resistive circuits discussed in Chapter 18, section 18.4, under the development of Ohm's law. These latter effects occur for entirely different reasons and on a much shorter time scale, so short, in fact, that they are not even observable directly. The inductive phenomena we are discussing here are exclusively of magnetic origin and produce effects on the rise and fall of currents in circuits that can be observed even with the most rudimentary apparatus.

EXAMPLE 20.4.1

Calculate the self-inductance L of a closely wound toroidal coil of inner radius a, outer radius b, and mean radius d that is closely wound with n turns per unit length along the mean circumference. Assume that the toroid is a "thin" one, that is, one for which the difference $b - a$ between the outer and inner radii is much less than the mean radius d.

We have already seen in (19.7.8) that the **B** field within a toroidal coil is given by

$$B = \mu_0 nI \left(\frac{d}{r}\right) \qquad (20.4.27)$$

and is zero outside the coil. In a "thin" toroid, the ratio d/r is very close to unity everywhere inside the toroid, as illustrated by Fig. 19.40b. In this case, (20.4.27) reduces to the approximate expression

$$B \simeq \mu_0 nI \qquad (20.4.28)$$

Since in this type of toroidal coil, B is practically constant over the cross section, the magnetic flux within the coil will be

$$\Phi_m = BA = \mu_0 nIA \qquad (20.4.29)$$

where A is the cross-sectional area. The induced emf is then

$$\mathscr{E} = -N\frac{d\Phi_m}{dt} = -\mu_0 nNA\frac{dI}{dt} = -L\frac{dI}{dt} \qquad (20.4.30)$$

the final equality following from (20.4.6). From this equation, it is apparent that

$$\boxed{L = \mu_0 nNA} \qquad (20.4.31)$$

But the total number of turns N is related to the number n per unit length by $N = nc$, where c is the mean circumference equal to $2\pi d$. Substituting this into (20.4.31),

$$L = 2\pi\mu_0 n^2 \, dA \qquad (20.4.32)$$

For a coil of radius 10 cm, cross-sectional area 3 cm², wound with 50 turns/cm, this gives

$$L = (6.283)(12.57 \times 10^{-7})(5.0 \times 10^3)^2(0.1)(3 \times 10^{-4})$$
$$= 5.92 \times 10^{-3} \text{ H}$$

EXAMPLE 20.4.2

The coil described in the previous example is wound with copper wire 0.2 mm in diameter, whose specific resistivity ρ is 1.75×10^{-8} ohm-meter. Find the time constant of the coil.

The time constant, according to the discussion

of L–R circuits given above, is simply L/R, where R is the resistance. To find the resistance we need to know the length of wire used in the winding. This is given by

$$l = 2\pi r_c N \qquad (20.4.33)$$

where r_c is the radius of the coil's circular cross section. We know that the cross-sectional area is 3 cm² = 3×10^{-4} m². Therefore,

$$A = \pi r_c^2 = 3 \times 10^{-4} \text{ m}^2$$

from which

$$r_c = 9.772 \times 10^{-3} \text{ m}$$

The total number of turns is the number per unit length, in this case 50 cm⁻¹, or 5000 m⁻¹, times the mean circumference of the toroid, which is $2\pi d = (2\pi)(0.1)$ m. Therefore, $N = (5000)(2\pi)(0.1) = 3142$. The length of wire in the coil is, therefore, according to (20.4.33),

$$l = (2\pi)(9.772 \times 10^{-3})(3142) = 192.9 \text{ m}$$

The cross-sectional area of the wire is

$$A_w = \pi r_w^2 = (\pi)(10^{-4})^2 = 3.142 \times 10^{-8} \text{ m}^2$$

The wire's resistance is

$$R = \frac{\rho l}{A_w} = \frac{(1.75 \times 10^{-8})(192.9)}{3.142 \times 10^{-8}} = 107.5 \text{ ohms}$$

The time constant t_0 of the coil is

$$t_0 = \frac{L}{R} = \frac{5.92 \times 10^{-3}}{107.5} = 5.51 \times 10^{-5} \text{ sec}$$

EXAMPLE 20.4.3

A coil of unknown inductance is found to carry a steady-state dc current of 3.50 amperes when a potential difference of 2.80 volts is maintained across it. When connected in a circuit such as that shown in Fig. 20.19, it is observed, with the aid of an oscilloscope, that the potential difference across a 1.0-ohm resistor placed in series with the coil rises to 90 percent of its asymptotic value in 4.2×10^{-3} seconds. What is the inductance of the coil? What is the instantaneous value of the current at this instant? What is the potential difference across the terminals of the inductor at this moment? Assume that the battery shown in the diagram exhibits negligible internal resistance and has an emf of 6.3 volts.

This circuit is similar to the one shown in Fig. 20.16a. The only difference is that part of the total resistance of the circuit is the *internal* resistance R_i of the coil and part is an *external* resistance R_e, equal to 1.0 ohm, the potential difference across which can be observed on an oscilloscope. We can see from the data given above that the value of the internal resis-

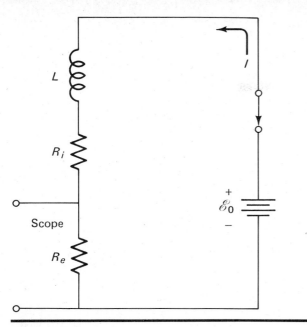

FIGURE 20.19

tance of the inductor is given by the ratio of potential difference to current measured under steady dc conditions:

$$R_i = \frac{\Delta V_L}{I_{dc}} = \frac{2.80}{3.50} = 0.80 \text{ ohm}$$

Indeed, we could replace the series resistances R_i and R_e in Fig. 20.19 with a *single* equivalent total resistance R given by

$$R = R_i + R_e = 0.80 + 1.00 = 1.80 \text{ ohm}$$

in which case it is apparent that the circuit is identical to that of Fig. 20.16a. The results we derived for that circuit are, therefore, applicable to this one, provided that we identify R as the *sum* of the internal and external resistances.

Accordingly, (20.4.18) gives the variation of the current with time after the switch is closed. We know that for $t = 4.2 \times 10^{-3}$ second, I equals 0.90 times the final asymptotic value $\mathscr{E}_0/R$. Therefore, substituting these values into (20.4.18), along with $R = 1.80$ ohm,

$$(0.90)(\mathscr{E}_0/R) = (\mathscr{E}_0/R)(1 - e^{-(1.80)(4.2 \times 10^{-3})/L})$$

We can now cancel $\mathscr{E}_0/R$ on either side to find

$$e^{-7.56 \times 10^{-3}/L} = 1 - 0.90 = 0.10$$

and taking the natural logarithm of both sides,

$$-\frac{7.56 \times 10^{-3}}{L} = \ln 0.10 = -2.303$$

$$L = 3.28 \times 10^{-3} \text{ H}$$

The instantaneous current at this instant can be found

by substituting the values $\mathscr{E}_0 = 6.3$ volts, $R = 1.80$ ohm, $t = 4.2 \times 10^{-3}$ second, and $L = 3.28 \times 10^{-3}$ henry into (20.4.18) to find $I = 3.15$ amperes. It is simpler, however, to observe that for $t = 4.2 \times 10^{-3}$ second,

$$I = (0.90)(\mathscr{E}_0/R) = (0.90)\left(\frac{6.3}{1.8}\right) = 3.15 \text{ amp}$$

The potential difference across the terminals of the inductor is the sum of the induced potential difference $-L(dI/dt)$ across the inductance and the potential difference $-IR_i$ due to the internal resistance of the coil; we write each of these terms with a minus sign to indicate that a potential *drop* is involved in both cases. Therefore, using (20.4.20), we may write

$$\Delta V = \Delta V_L + \Delta V_{R_i} = -\mathscr{E}_0 e^{-Rt/L} - IR_i$$
$$= -(6.3)(0.10) - (3.15)(0.80) = -3.15 \text{ volts}$$

In doing this calculation, we have observed that we already know from our previous work that $e^{-Rt/L} = 0.10$ for the given values of R, t, and L. More simply, we could have added the potential differences around the circuit to find

$$\mathscr{E}_0 + \Delta V_L + \Delta V_{R_i} + \Delta V_{R_e} = 0$$

from which

$$\Delta V_L + \Delta V_{R_i} = -\mathscr{E}_0 - \Delta V_{R_e} = -\mathscr{E}_0 + IR_e$$
$$= -6.3 + (3.15)(1.0) = -3.15 \text{ volts}$$

The *time constant* exhibited by the circuit of Fig. 20.19 in which both internal and external resistances are present would be

$$t_0 = \frac{L}{R} = \frac{L}{R_i + R_e} = \frac{3.28 \times 10^{-3}}{1.80} = 1.82 \times 10^{-3} \text{ sec}$$

If the external resistance were removed from the circuit, then the time constant associated with the coil and its internal resistance alone would become

$$t_0' = \frac{L}{R_i} = \frac{3.28 \times 10^{-3}}{0.8} = 4.10 \times 10^{-3} \text{ sec}$$

20.5 Energy in Inductive Circuits and the Energy Density of Magnetic Fields

Let us now return to the situation illustrated in Fig. 20.16b, where the magnetic field initially associated with the inductor sustains the flow of current in the circuit for a time as the magnetic flux decays. We have already noted that the current that flows during this time produces a certain amount of thermal energy in Joule heating as it passes through the circuit resistance R. Where did this heat come from? It was certainly not created from nothing, unless we are willing to abandon our faith in the law of conservation of energy.

The thermal energy was caused by the passage of current through a resistance, and the emf that caused the current to flow was generated by the change in magnetic flux through the turns of the inductance as the magnetic field in its neighborhood died away. It is evident, therefore, that the thermal energy created by Joule heating *must originally have been magnetic energy stored in the inductor's magnetic field.* If we are to believe the law of conservation of energy, the sum of the magnetic and thermal energies must always be equal. Therefore, if we denote the magnetic field energy by U_m and the thermal energy by Q and equate magnetic plus thermal energy at times t and $t + dt$, between which changes dU_m and dQ take place, we have

$$U_m + Q = U_m + dU_m + Q + dQ \qquad (20.5.1)$$

or

$$-dU_m = dQ \qquad (20.5.2)$$

We may divide both sides of this equation by dt to obtain

$$-\frac{dU_m}{dt} = \frac{dQ}{dt} = I^2R \qquad (20.5.3)$$

recalling that the rate at which electrical energy is converted to heat is I^2R. But now, let us look at Eq. (20.4.20), which tells us that, for this system,

$$-L\frac{dI}{dt} = RI \qquad (20.5.4)$$

Multiplying both sides of this equation by I and noting that $I(dI/dt) = d(\frac{1}{2}I^2)/dt$, this becomes

$$-LI\frac{dI}{dt} = -\frac{d}{dt}(\tfrac{1}{2}LI^2) = I^2R \qquad (20.5.5)$$

or

$$-\frac{d}{dt}(\tfrac{1}{2}LI^2) = \frac{dQ}{dt} \qquad (20.5.6)$$

Comparing this with (20.5.3), we see that we must express *the energy in the magnetic field of an inductor as*

$$\boxed{U_m = \tfrac{1}{2}LI^2} \qquad (20.5.7)$$

Although this result was derived by consideration of a particular process, it is evident that it is one of general applicability.

It is of importance also to be able to calculate the *energy density* or energy per unit volume exhibited by the magnetic field. This can be done by referring to the example of the long, straight solenoid wound with n turns per unit length in the interior of which we have already shown there exists a constant field given by

$$B = \mu_0 nI \qquad (20.5.8)$$

and for which we recently calculated the self-inductance as

$$L = \mu_0 n^2 lA \qquad (20.5.9)$$

where l is the length and A the cross-sectional area. According to (20.5.7) and (20.5.8), the amount of magnetic energy associated with the field is

$$U_m = \tfrac{1}{2}LI^2 = \tfrac{1}{2}(\mu_0 n^2 lA)\left(\frac{B^2}{\mu_0{}^2 n^2}\right) = lA\frac{B^2}{2\mu_0} \qquad (20.5.10)$$

The product lA simply represents the volume $\mathscr{V}$ of the interior of the solenoid where the magnetic field resides. This means that the energy density, or *energy per unit volume associated with the magnetic field,* can be expressed as

$$\frac{U_m}{\mathscr{V}} = \frac{B^2}{2\mu_0} \qquad (20.5.11)$$

Equations (20.5.7) and (20.5.11) are analogous to the previously derived expressions $U_e = \tfrac{1}{2}C(\Delta V)^2$ and $U_e/\mathscr{V} = \tfrac{1}{2}\varepsilon_0 E^2$ that express the *electric* field energy within a capacitor and the energy per unit volume associated with an *electric* field.

It is clear, therefore, that inductors can store energy in their magnetic fields just as capacitors store electrostatic energy and, further, that any magnetic field has associated with it an energy density that is similar to that exhibited by the electrostatic field. In oscillatory circuits, and in electromagnetic waves, as we shall soon see in detail, energy can be continually exchanged back and forth between electric and magnetic fields.

EXAMPLE 20.5.1

A large research electromagnet has parallel circular pole faces that are 30 cm in diameter and are spaced 5 cm apart. The magnet can produce a maximum field of 1.2 weber/m^2 that is essentially constant within the cylindrical region between the poles. How much energy is stored in the magnetic field in this region under these conditions? How long could a 60-watt bulb be illuminated at rated power by this field energy?

The energy density in the field region is given by (20.5.11). The total energy in the volume $\mathscr{V}$ of the cylindrical region between the poles is, therefore,

$$U_m = \frac{B^2}{2\mu_0}\mathscr{V} = \frac{B^2\pi R^2 h}{2\mu_0} \qquad (20.5.12)$$

where R is the radius of the pole face and h the spacing between the faces. Substituting the given values of these quantities, we find

$$U_m = \frac{(1.2)^2(\pi)(0.15)^2(0.05)}{(2)(4\pi)(10^{-7})} = 2025 \text{ J}$$

A 60-watt light bulb at rated power consumes 60 joules

of energy per second. The amount of energy stated above can, therefore, illuminate such a lamp for a time given by

$$t = \frac{2025}{60} = 33.75 \text{ sec}$$

20.6 Inductors in Series and in Parallel

Inductors can be combined as circuit elements in series and in parallel in the same way as resistors and capacitors can. The question then arises as to what the total inductance of the circuit is in terms of the individual inductances that are combined in series or in parallel. This question can be answered by reference to Fig. 20.20. In the case of inductances in series, as shown in Fig. 20.20a, all of the inductances carry the *same current* $I(t)$. The potential difference across the array is the sum of the individual potential difference associated with each element, from which we may write

$$\Delta V = \Delta V_1 + \Delta V_2 + \Delta V_3 \qquad (20.6.1)$$

Now, from (20.4.8), since all inductors carry the same current, the magnitude of the potential drops across them will be $L_1(dI/dt)$, $L_2(dI/dt)$, and $L_3(dI/dt)$. Also, the magnitude of the potential difference ΔV across the whole array must be $L(dI/dt)$, where L is the equivalent inductance of the series combination. Sub-

stituting these values into (20.6.1), we obtain

$$L\left(\frac{dI}{dt}\right) = L_1\left(\frac{dI}{dt}\right) + L_2\left(\frac{dI}{dt}\right) + L_3\left(\frac{dI}{dt}\right) \qquad (20.6.2)$$

or, dividing by (dI/dt),

$$L = L_1 + L_2 + L_3 \qquad (20.6.3)$$

For n inductors, the same procedure will give us

$$\boxed{L = L_1 + L_2 + L_3 + \ldots + L_n} \qquad (20.6.4)$$

For several inductances in parallel, as shown in Fig. 20.20b, all the inductors experience the *same potential difference* ΔV, though they carry different currents I_1, I_2, and I_3. The total current I entering and leaving the array can be expressed in terms of these individual currents:

$$I = I_1 + I_2 + I_3 \qquad (20.6.5)$$

Differentiating this with respect to time, we find

$$\frac{dI}{dt} = \frac{dI_1}{dt} + \frac{dI_2}{dt} + \frac{dI_3}{dt} \qquad (20.6.6)$$

But from (20.4.8), the magnitude of the potential difference ΔV across each individual inductor has to be related to the time rate of change of current through it by

$$\Delta V = L_1\frac{dI_1}{dt} = L_2\frac{dI_2}{dt} = L_3\frac{dI_3}{dt} \qquad (20.6.7)$$

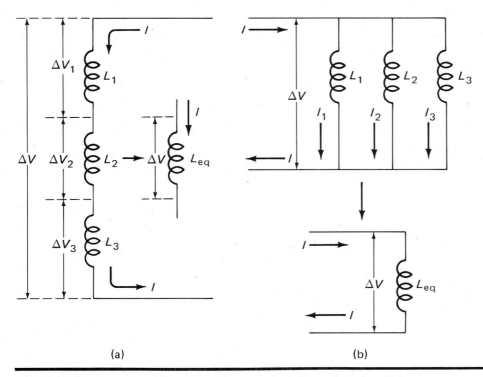

(a)　　　　　　　　　　(b)

FIGURE 20.20. (a) Inductances in series and (b) inductances in parallel.

This means that

$$\frac{dI_1}{dt} = \frac{\Delta V}{L_1} \qquad \frac{dI_2}{dt} = \frac{\Delta V}{L_2} \qquad \frac{dI_3}{dt} = \frac{\Delta V}{L_3} \qquad (20.6.8)$$

Substituting these values into (20.6.6) and noting that the relation between the potential difference ΔV and the time rate of change of the total current must be expressed in terms of the equivalent inductance L of the parallel array by $\Delta V = L(dI/dt)$, we find

$$\frac{\Delta V}{L} = \frac{\Delta V}{L_1} + \frac{\Delta V}{L_2} + \frac{\Delta V}{L_3} \qquad (20.6.9)$$

or

$$\frac{1}{L} = \frac{1}{L_1} + \frac{1}{L_2} + \frac{1}{L_3} \qquad (20.6.10)$$

Again, a repetition of this same derivation shows that an array of n inductors in parallel has an equivalent inductance given by

$$\boxed{\frac{1}{L} = \frac{1}{L_1} + \frac{1}{L_2} + \frac{1}{L_3} + \cdots + \frac{1}{L_n}} \qquad (20.6.11)$$

It is evident from these results that series and parallel arrays of inductors exhibit equivalent inductances that are of the same form as the equivalent resistances of series and parallel resistance arrays. This being the case, we need present no further specific examples of how calculations related to such series and parallel combinations are carried out.

A word of caution on another point is in order, nevertheless. In deriving Eqs. (20.6.4) and (20.6.11), we made the assumption that the potential drop across any given inductor is created only by the current change—thus the magnetic flux change—associated with that inductor and that this drop is completely *independent* of any changes in magnetic flux generated by the other inductances in the circuit. This will be true only if the inductors are well separated so that the magnetic field generated by each does not affect any of the others to a significant degree. The results quoted herein, therefore, relate only to the case of combinations of inductors that are *magnetically isolated* from one another. In the next section, we shall address ourselves to the problem of how inductors that are not magnetically isolated from one another behave.

20.7 Mutual Induction, Induction Coils, and Transformers

In sections 20.5 and 20.6, we considered how emfs were induced in circuits by the currents flowing in the circuits themselves; this phenomenon was given the

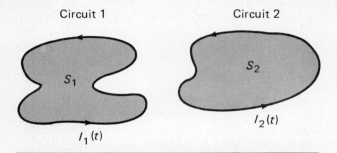

FIGURE 20.21. Two circuits whose magnetic fields may induce emfs in each other as well as in themselves.

name of self-induction. It is possible for emfs to be induced by flux changes arising from other circuits as well. This phenomenon, which we have already met in the circuit illustrated in Fig. 20.4, is referred to as *mutual induction*.

In order to understand the essentials of mutual induction, let us consider the situation illustrated in Fig. 20.21. There we see two circuits, numbered 1 and 2, carrying currents I_1 and I_2 that may vary with time. We now wish to know the emfs induced in each circuit by the magnetic flux changes associated with these time-varying currents. It is important to recognize that in this case each circuit encloses magnetic flux generated not only by its own current but also by the current in the other circuit. For example, recalling the definition of magnetic flux given in Chapter 19 as Eq. (19.3.3), the total flux Φ_1 enclosed by circuit 1 can be represented as the sum of fluxes generated by current I_1 flowing in circuit 1 and current I_2 flowing in circuit 2:

$$\Phi_1 = \int_{S_1} \mathbf{B}_1 \cdot \mathbf{n}_1 \, da + \int_{S_1} \mathbf{B}_2 \cdot \mathbf{n}_1 \, da = \Phi_{11} + \Phi_{12} \qquad (20.7.1)$$

Likewise, the flux Φ_2 enclosed by circuit 2 can be written as the sum of the fluxes generated by current I_1 in circuit 1 and I_2 in circuit 2 as

$$\Phi_2 = \int_{S_2} \mathbf{B}_1 \cdot \mathbf{n}_2 \, da + \int_{S_2} \mathbf{B}_2 \cdot \mathbf{n}_2 \, da = \Phi_{21} + \Phi_{22} \qquad (20.7.2)$$

In these equations, $\mathbf{B}_1$ and $\mathbf{B}_2$ are the separate magnetic induction fields generated by I_1 and I_2, S_1 and S_2 are the areas enclosed by the circuits, and $\mathbf{n}_1$ and $\mathbf{n}_2$ are unit vectors normal to S_1 and S_2, respectively.

From Faraday's law, we may write the emfs in the two circuits, $\mathscr{E}_1$ and $\mathscr{E}_2$, as

$$\mathscr{E}_1 = -\frac{d\Phi_1}{dt} = -\frac{d\Phi_{11}}{dt} - \frac{d\Phi_{12}}{dt}$$

$$= -\frac{d\Phi_{11}}{dI_1}\frac{dI_1}{dt} - \frac{d\Phi_{12}}{dI_2}\frac{dI_2}{dt} \qquad (20.7.3)$$

$$\mathscr{E}_2 = -\frac{d\Phi_2}{dt} = -\frac{d\Phi_{21}}{dt} - \frac{d\Phi_{22}}{dt}$$

$$= -\frac{d\Phi_{21}}{dI_1}\frac{dI_1}{dt} - \frac{d\Phi_{22}}{dI_2}\frac{dI_2}{dt} \qquad (20.7.4)$$

Now, since, according to (20.4.6), the self-induced emf in any circuit carrying current I and creating flux Φ can be written for a single loop as

$$\mathscr{E} = -\frac{d\Phi}{dt} = -\frac{d\Phi}{dI}\frac{dI}{dt} = -L\frac{dI}{dt} \qquad (20.7.5)$$

it is evident that the *self-inductance* L can be expressed as

$$\boxed{L = \frac{d\Phi}{dI}} \qquad (20.7.6)$$

We may, therefore, write (20.7.3) and (20.7.4) as

$$\mathscr{E}_1 = -L_1\frac{dI_1}{dt} - M_{12}\frac{dI_2}{dt} \qquad (20.7.7)$$

$$\mathscr{E}_2 = -M_{21}\frac{dI_1}{dt} - L_2\frac{dI_2}{dt} \qquad (20.7.8)$$

where L_1 and L_2 are the self-inductances of circuits 1 and 2 and M_{12} and M_{21} are defined by

$$\boxed{M_{12} = \frac{d\Phi_{12}}{dI_2} \qquad M_{21} = \frac{d\Phi_{21}}{dI_1}} \qquad (20.7.9)$$

The coefficient M_{12} expresses the rate at which the magnetic flux through circuit 1 changes with respect to the current in circuit 2, while M_{21} represents the rate at which the magnetic flux through circuit 2 changes as a function of the current in circuit 1. These coefficients are called the *coefficients of mutual induction* relating circuits 1 and 2.

It is, in principle, possible to calculate these coefficients, just as it is possible to calculate the coefficient of self-inductance for a single circuit. To do so, however, requires a knowledge of the fraction of flux generated by one of the circuits that is enclosed by the other, and this is usually very difficult to represent mathematically. We are, therefore, usually content to express our inability to do the required calculations by simply treating the coefficients of mutual induction as parameters to be determined not by calculation but by measurement. In this respect, they are treated in much the same way as the self-inductance coefficients L_1 and L_2. The magnitude of the self-induction coefficients is a measure of how strongly the magnetic flux generated by one of the circuits affects the emf induced in the other. For circuits so far apart that the magnetic field of one is very small in the neighborhood of the other, the mutual inductance coeffi-

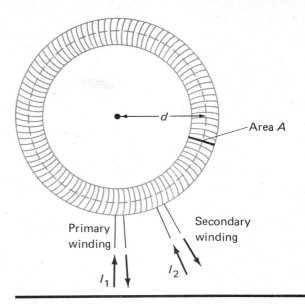

FIGURE 20.22. Toroidal induction coil having two separate windings carrying "primary" current I_1 and "secondary" current I_2.

cients will be negligible compared to the self-inductance coefficients, and (20.7.7) and (20.7.8) reduce to

$$\mathscr{E}_1 \cong -L_1\frac{dI_1}{dt} \qquad (20.7.10)$$

$$\mathscr{E}_2 \cong -L_2\frac{dI_2}{dt} \qquad (20.7.11)$$

In this case, the only important effect is the *self-induced* emf in each circuit, and the circuits are isolated, or magnetically independent of one another. On the other hand, if the circuits are so close together that one encloses a significant fraction of the flux generated by the other, the coefficients of mutual induction will be comparable in magnitude to the self-induction coefficients, and the circuits are said to be *magnetically coupled* to one another.

As a specific example, let us consider the case of a "thin" toroidal coil, like the one discussed in Example 20.4.1, except that there are two windings, one on top of the other, as illustrated in Fig. 20.22. The first, which we shall refer to as the *primary* winding, consists of n_1 turns per unit length along the mean circumference and carries a time-varying current I_1. The second, which we call the *secondary* winding, has n_2 turns per unit length and carries a time-varying current I_2. From the results of Example 20.4.1, we know that the current I_1 in the first winding produces a **B** field within the toroid given by

$$B_1 = \mu_0 n_1 I_1 \qquad (20.7.12)$$

In this case, all the magnetic flux that is generated is enclosed by every turn of each of the windings. Therefore, the fluxes Φ_{11} and Φ_{21} generated by the primary current I_1 and defined by (20.7.1) and (20.7.2) are

$$\Phi_{11} = \Phi_{21} = B_1 A = \mu_0 n_1 A I_1 \qquad (20.7.13)$$

where A is the cross-sectional area of the toroid. Likewise, the fluxes Φ_{12} and Φ_{22} generated by the secondary current I_2 are

$$\Phi_{12} = \Phi_{22} = B_2 A = \mu_0 n_2 A I_2 \qquad (20.7.14)$$

The emf induced in the primary, which we shall regard as having N_1 total turns of wire, is, according to (20.7.3),

$$\mathscr{E}_1 = -N_1 \frac{d\Phi_{11}}{dt} - N_1 \frac{d\Phi_{12}}{dt}$$

$$= -2\pi D n_1 \frac{d\Phi_{11}}{dt} - 2\pi D n_1 \frac{d\Phi_{12}}{dt} \qquad (20.7.15)$$

while the emf induced in the secondary winding is, from (20.7.4),

$$\mathscr{E}_2 = -N_2 \frac{d\Phi_{21}}{dt} - N_2 \frac{d\Phi_{22}}{dt}$$

$$= -2\pi D n_2 \frac{d\Phi_{21}}{dt} - 2\pi D n_2 \frac{d\Phi_{22}}{dt} \qquad (20.7.16)$$

where D is the mean radius of the coil and n_1 and n_2 are the numbers of turns per unit length on each winding. Substituting the values of the fluxes in terms of the currents from (20.7.13) and (20.7.14), these equations become

$$\mathscr{E}_1 = -2\pi \mu_0 D n_1{}^2 A \frac{dI_1}{dt} - 2\pi \mu_0 D n_1 n_2 A \frac{dI_2}{dt} \qquad (20.7.17)$$

$$\mathscr{E}_2 = -2\pi \mu_0 D n_1 n_2 A \frac{dI_1}{dt} - 2\pi \mu_0 D n_2{}^2 A \frac{dI_2}{dt} \qquad (20.7.18)$$

These equations are of the form of (20.7.7) and (20.7.8) with

$$L_1 = 2\pi \mu_0 D n_1{}^2 A \qquad L_2 = 2\pi \mu_0 D n_2{}^2 A \qquad (20.7.19)$$

and

$$M_{12} = M_{21} = 2\pi \mu_0 D n_1 n_2 A \qquad (20.7.20)$$

The calculated values of the self-inductance coefficients L_1 and L_2 agree with the result previously obtained for the thin toroid as Eq. (20.4.32). We see that the two mutual inductance coefficients M_{12} and M_{21} are equal, though this is not true for inductively coupled circuits in general. From these relations, the mutual induction coefficients can also be expressed in terms of the self-inductances L_1 and L_2:

$$M_{12} = M_{21} = \sqrt{L_1 L_2} \qquad (20.7.21)$$

though this again is a result that pertains only to this particular circuit configuration. We may, therefore,

write (20.7.17) and (20.7.18) as

$$\mathscr{E}_1 = -L_1 \frac{dI_1}{dt} - \sqrt{L_1 L_2} \frac{dI_2}{dt} \qquad (20.7.22)$$

$$\mathscr{E}_2 = -\sqrt{L_1 L_2} \frac{dI_1}{dt} - L_2 \frac{dI_2}{dt} \qquad (20.7.23)$$

These equations describe the emfs induced in both of the windings in terms of the currents that flow in each of them. While some fairly complex algebra may be involved in disentangling the currents and potential differences that arise when the windings are connected in circuits, as we shall see in the examples worked out below, the essential feature of the device we are dealing with can be exhibited by solving Eq. (20.7.22) for dI_1/dt, and substituting the value so obtained into (20.2.23). In this way, recalling (20.7.19), we can easily obtain

$$\frac{\mathscr{E}_2}{\mathscr{E}_1} = \sqrt{\frac{L_2}{L_1}} = \frac{n_2}{n_1} \qquad (20.7.24)$$

or, since $N_1 = 2\pi D n_1$ and $N_2 = 2\pi D n_2$,

$$\boxed{\frac{\mathscr{E}_2}{\mathscr{E}_1} = \frac{N_2}{N_1}} \qquad (20.7.25)$$

This equation tells us that the ratio of the emfs equals the ratio of the total number of turns in the two windings. If the secondary winding has more turns than the primary, therefore, and an emf $\mathscr{E}_1$ is induced in the primary by connecting that winding to an external current source, the emf $\mathscr{E}_2$ induced in the secondary winding will be larger than $\mathscr{E}_1$ by the factor N_2/N_1. Physically, of course, this occurs because the same flux changes are enclosed by both coils, and, therefore, the same emf, let us call it $\mathscr{E}'$, is induced in *each turn of each winding*. But since every turn of each winding is in series with every other turn, these individually induced emfs add up so that

$$\mathscr{E}_1 = N_1 \mathscr{E}' \qquad \text{and} \qquad \mathscr{E}_2 = N_2 \mathscr{E}' \qquad (20.7.26)$$

which leads immediately to Eq. (20.7.25).

If the secondary winding has many more turns than the primary, the emfs induced in the secondary coil may be much larger than the primary emf. This leads to the possibility of producing very large voltages from much smaller ones with such a circuit configuration. A device constructed in this way is referred to as an *induction coil*, if it is used in direct current (dc) circuits to produce large emfs as the circuit is closed or opened, or a *transformer*, if it is used in an alternating current (ac) circuit to step up emfs on a continuous basis. The former configuration is used to produce the large, transient emfs needed to excite the spark plugs in an automobile engine from the much smaller emf furnished by the battery, while the latter

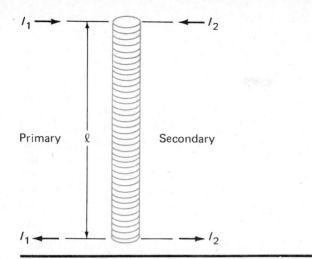

FIGURE 20.23. A straight, cylindrical induction coil whose important characteristics are similar to those of toroidal coil of Fig. 20.22.

is widely used to increase ac emfs in electric power transmission systems and electronic circuits. Naturally, if the secondary winding has *fewer* turns than the primary, the secondary emf will be smaller than the primary emf. This allows us to construct devices that *reduce* emfs with little or no resulting energy loss. Such step-down coils or transformers are also useful in many technical applications.

It should be observed that all the magnetic energy generated by the primary circuit is contained within the interior of the toroidal coil, since the external **B** field, at least in an ideal toroid, is zero. It is evident that, ideally at least, *all* this magnetic energy is coupled into the secondary circuit. There is, therefore, in principle, no loss of electromagnetic energy in an induction coil or transformer. This statement, however, is never quite true in practice, since there are always resistive losses incurred when the primary and secondary currents pass through their respective windings, which will never be of zero resistance. But by using sufficiently heavy wire for the windings, these losses can usually be reduced until they are negligible in practice.

If we neglect internal resistive losses, then, the power $\mathscr{E}_1 I_1$ delivered to the primary winding and the power $\mathscr{E}_2 I_2$ taken from the secondary winding will be equal, and, therefore, for an ideal transformer or induction coil,

$$\boxed{\frac{I_2}{I_1} = \frac{\mathscr{E}_1}{\mathscr{E}_2} = \frac{N_1}{N_2}} \tag{20.7.27}$$

As we have already seen, the long, straight solenoid can be considered as a thin toroidal coil of infinite radius. The results discussed for the thin toroidal transformer or induction coil are, therefore,

equally applicable to a transformer or induction coil made in the form of a long, straight cylinder closely wound with two separate windings, as illustrated in Fig. 20.23. All the equations written for the thin toroid are applicable to the long, straight solenoid, provided that we remember in Eqs. (20.7.15) to (20.7.20) that the quantity $2\pi D$ is to be replaced with the length l of the straight coil.

The operation of the toroidal coil with primary and secondary windings as an induction coil in a dc circuit is illustrated by Example 20.7.1, and as a transformer in an ac circuit, by Example 20.7.2. The second of these examples establishes some of the principles that we shall study in detail later, in Chapter 22, in connection with alternating current circuits. Though we shall make no strong attempt to consider them in detail here, the reader will probably find it of interest to refer back to Example 20.7.2 after having learned the main topics treated in Chapter 22.

EXAMPLE 20.7.1

An automotive induction coil that is electrically equivalent to the thin toroidal induction coil described above is used in a circuit as shown in Fig. 20.24. In the diagram, the conventional circuit symbol is used to depict the primary and secondary windings of the coil. Assuming that the currents in both windings are initially zero, find the induced emfs and currents in both windings as a function of time after the switch S is thrown to the left, as shown. Find the potential difference across the resistor R in the secondary circuit. Describe what happens when, the currents I_1 and I_2 having reached steady-state values after a long time, the switch S is suddenly thrown to the right, removing the battery emf $\mathscr{E}_0$ from the circuit.

In the primary circuit, using Kirchhoff's law to sum the emfs and potential drops around the circuit in the direction of the current I_1, we may write

$$\mathscr{E}_0 - I_1 R_1 + \mathscr{E}_1 = 0 \tag{20.7.28}$$

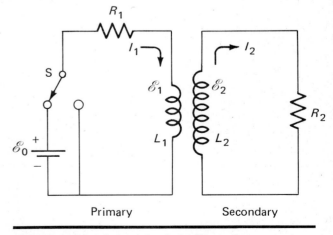

FIGURE 20.24. Automotive induction coil.

while in the secondary circuit, by the same token,

$$\mathscr{E}_2 - I_2 R_2 = 0 \tag{20.7.29}$$

In these equations, $\mathscr{E}_1$ and $\mathscr{E}_2$ are the mutually induced coupled emfs in the primary and secondary windings, as given by (20.7.22) and (20.7.23). Since these quantities are related as always by (20.7.24), we may replace $\mathscr{E}_2$ in (20.7.27) by $\mathscr{E}_1 \sqrt{L_2/L_1}$ and write

$$\mathscr{E}_1 - I_1 R_1 + \mathscr{E}_0 = 0 \tag{20.7.30}$$

$$\mathscr{E}_1 \sqrt{L_2/L_1} - I_2 R_2 = 0 \tag{20.7.31}$$

with $\mathscr{E}_1$ given by (20.7.22). These equations form a system of simultaneous differential equations that can be solved, with difficulty, for the currents I_1 and I_2. Equations (20.7.22) and (20.7.23) will then give the emfs $\mathscr{E}_1$ and $\mathscr{E}_2$. The simplest way of doing this is to substitute the value of $\mathscr{E}_1$ from (20.7.30) into (20.7.31) to obtain

$$I_1 R_1 \sqrt{\frac{L_2}{L_1}} = \mathscr{E}_0 \sqrt{\frac{L_2}{L_1}} + I_2 R_2 \tag{20.7.32}$$

This gives us immediately a relation between I_1 and I_2. A relation between dI_1/dt and dI_2/dt can be obtained simply by differentiating (20.7.30) with respect to time, observing that the first term on the right side is constant:

$$\frac{dI_1}{dt} = \frac{R_2}{R_1} \sqrt{\frac{L_1}{L_2}} \frac{dI_2}{dt} \tag{20.7.33}$$

If we substitute the value of $\mathscr{E}_1$ given by (20.7.22) into (20.7.31), we may write

$$\sqrt{L_1 L_2} \frac{dI_1}{dt} + L_2 \frac{dI_2}{dt} + I_2 R_2 = 0 \tag{20.7.34}$$

We can write an equation containing only I_2 and dI_2/dt by substituting the value of dI_1/dt given by (20.7.33) into this equation, to obtain

$$\left(\frac{L_1}{R_1} + \frac{L_2}{R_2} \right) \frac{dI_2}{dt} = -I_2 \tag{20.7.35}$$

This equation can easily be rearranged so as to include only the variable I_2 on the left and t on the right, by multiplying by dt and dividing by I_2:

$$\frac{dI_2}{I_2} = -\frac{dt}{\dfrac{L_1}{R_2} + \dfrac{L_2}{R_2}} \tag{20.7.36}$$

We must now integrate this equation from $t = 0$, at which time the current I_2 has the initial value I_{20} to time t when the current has the value I_2. In this way, we find

$$\int_{I_{20}}^{I_2} \frac{dI_2}{I_2} = -\frac{1}{\dfrac{L_1}{R_1} + \dfrac{L_2}{R_2}} \int_0^t dt \tag{20.7.37}$$

The only thing that prevents us from carrying out this prescription is that we do not know the value of the current I_{20} at time $t = 0$. Of course, for t less than zero, I_2 is zero. But when the switch is thrown, I_1 begins to increase at a finite rate which generates a finite flux change through the secondary circuit. This flux change at once makes the current undergo a *discontinuous* jump in magnitude from zero to I_{20} at $t = 0$. In the primary circuit, however, we do know that I_1 is zero at time $t = 0$. Now if we set $t = 0$ and $I_1 = 0$ in Eqs. (20.7.30) and (20.7.31), we get

$$\mathscr{E}_1(0) + \mathscr{E}_0 = 0 \tag{20.7.38}$$

$$\mathscr{E}_1(0) \sqrt{L_2/L_1} - I_{20} R_2 = 0 \tag{20.7.39}$$

where $\mathscr{E}_1(0)$ represents the value of the emf $\mathscr{E}_1$ at time $t = 0$. But, from (20.7.38), it is clear that $\mathscr{E}_1(0) = -\mathscr{E}_0$, and then, inserting this value into (20.7.39) and solving for I_{20},

$$I_{20} = -\frac{\mathscr{E}_0}{R_2} \sqrt{\frac{L_2}{L_1}} \tag{20.7.40}$$

We can now integrate both sides of (20.7.37) to find

$$[\ln I_2]_{I_{20}}^{I_2} = \ln I_2 - \ln I_{20} = \ln \frac{I_2}{I_{20}} = -\frac{t}{\dfrac{L_1}{R_1} + \dfrac{L_2}{R_2}} \tag{20.7.41}$$

Exponentializing the final equality in the above expression and replacing I_{20} with the value given by (20.7.40), we may finally show that

$$I_2(t) = I_{20} e^{-t/(L_1/R_1 + L_2/R_2)}$$

or

$$I_2(t) = -\frac{\mathscr{E}_0}{R_2} \sqrt{\frac{L_2}{L_1}} \, e^{-t/t_0} \tag{20.7.42}$$

where the time constant t_0 is given by

$$t_0 = \frac{L_1}{R_1} + \frac{L_2}{R_2} \tag{20.7.43}$$

It will be noted that the current I_2 falls off exponentially with time, the exponential time constant t_0 being the *sum* of the separate L–R time constants L_1/R_1 and L_2/R_2 that the primary and secondary circuits would have were they not coupled together by the mutual inductance of the device.

The current I_1 may be found by substituting (20.7.42) into (20.7.32) and solving for I_1 to obtain

$$I_1(t) = \frac{\mathscr{E}_0}{R_1} (1 - e^{-t/t_0}) \tag{20.7.44}$$

The emfs $\mathscr{E}_1$ and $\mathscr{E}_2$ are found by differentiating (20.7.42) and (20.7.44) to find dI_1/dt and dI_2/dt, then substituting these values into (20.7.22) and (20.7.23) to get

$$\mathscr{E}_1(t) = -\frac{\mathscr{E}_0}{t_0}\left[\frac{L_1}{R_1} + \frac{\sqrt{L_1 L_2}}{R_2}\right] e^{-t/t_0} \qquad (20.7.45)$$

$$\mathscr{E}_2(t) = -\frac{\mathscr{E}_0}{t_0}\left[\frac{\sqrt{L_1 L_2}}{R_1} + \frac{L_2}{R_2}\right] e^{-t/t_0} \qquad (20.7.46)$$

If $\mathscr{E}_1(t)$ as given above is multiplied by $\sqrt{L_2/L_1}$, one obtains $\mathscr{E}_2(t)$ as expressed by (20.7.46), in accord with what one might expect from (20.7.24). Finally, we may observe that the potential difference across the resistance R_2 is simply

$$\Delta V_2 = I_2(t)R_2 = -\mathscr{E}_0\sqrt{\frac{L_2}{L_1}}\, e^{-t/t_0} \qquad (20.7.47)$$

or, since from (20.7.24) and (20.7.25) the ratio $\sqrt{L_2/L_1}$ is equal to the secondary to primary turns ratio N_2/N_1,

$$\Delta V_2 = -\mathscr{E}_0\frac{N_2}{N_1}\, e^{-t/t_0} \qquad (20.7.48)$$

In the automotive ignition system, the switch S is the breaker points in the distributor, and the induction coil is the car's "ignition coil." The resistance R_1 represents the resistance of the coil's primary winding, while the resistance R_2 is the combined resistance of the secondary winding, the ignition wires, and the spark gap. It is clear that the output potential difference, essentially all of which appears across the spark gap since its resistance is ordinarily much larger than that of the windings or ignition wires, is initially *greater* than the battery emf by the turns ratio N_2/N_1. If a 12-volt battery is used, a turns ratio of $N_2/N_1 = 1000$ will give an initial firing voltage of 12000 volts.

The currents I_1 and I_2 are shown plotted as functions of time in Fig. 20.25. After a long time, if the switch S remains closed, the primary current I_1 reaches the steady-state dc value $\mathscr{E}_0/R_1$ and no longer

changes. In this condition, the magnetic flux it generates also no longer changes and the emf $\mathscr{E}_2$ in the secondary circuits falls to zero. The secondary current I_2 must under these conditions also fall off asymptotically to zero, as shown by Fig. 20.25b or by Eq. (20.7.42).

If the switch S is now suddenly thrown to the right, eliminating the battery emf $\mathscr{E}_0$, the primary current I_1 will start to decrease. But due to the fact that the decaying magnetic fields induce emfs $\mathscr{E}_1$ and $\mathscr{E}_2$ in both branches of the circuit, it will not fall to zero at once, but will be sustained from a time by the emf $\mathscr{E}_1$ which, by Lenz's law, will now be in the opposite direction. The current I_1 will, therefore, decay exponentially to zero with time constant t_0 as given by (20.7.43) in much the same way the self-induced current decays in the inductor discussed previously in section 20.4. At the same time, the suddenly changing magnetic flux through the secondary of the coil will induce another large exponential "spike" of emf and current in the secondary, similar to that shown in Fig. 20.25b, but again opposite in direction, as required by Lenz's law. We could, of course, verify these qualitative conclusions mathematically by methods similar to those used above, but we shall not bother to do so, since the methods are well established and the effects describable in qualitative terms by purely physical reasoning. If the switch S is then thrown back and forth, the primary current must behave as shown in Fig. 20.26a, while the secondary current and voltage exhibit a series of "spikes" corresponding to very large potential differences across the output resistance R, as illustrated in Figs. 20.26b and 20.26c.

In the automotive system, this repetitive make and break of the primary circuit is accomplished automatically by a distributor cam that repeatedly opens and closes the ignition points. In laboratory induction coils, this action is usually provided by a magnetically driven make-and-break circuit similar to that used in

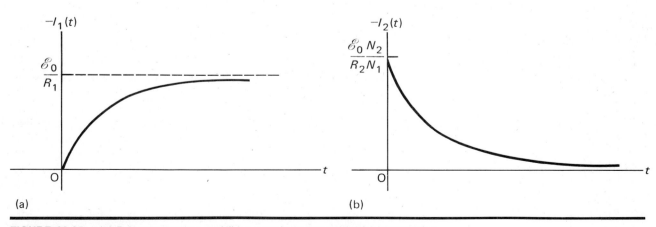

(a)

(b)

FIGURE 20.25. (a) Primary current and (b) secondary current in the automotive induction coil of Fig. 20.24, plotted as a function of time after switch S is closed.

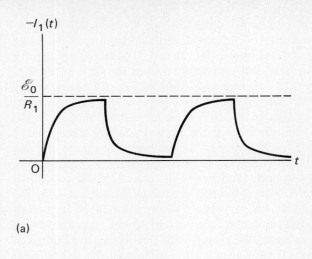

(a)

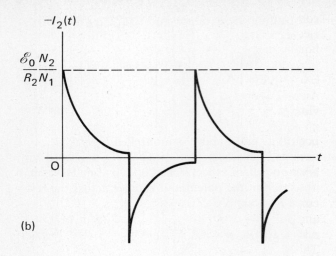

(b)

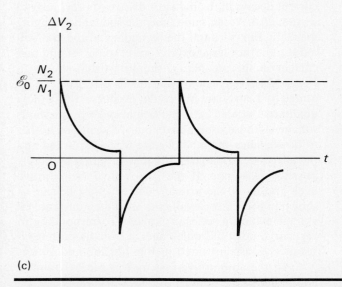

(c)

FIGURE 20.26. Growth and decay of (a) primary current, (b) secondary current, and (c) secondary emf in automotive coil as the "points" alternately make and break the primary circuit.

a buzzer or doorbell. The automotive ignition system has one complication that we have not taken into account. Before a spark is excited across the spark gap of the spark plug, the resistance R_2 of the secondary circuit is very high, since the only currents that can flow are leakage currents through, and on the surface of, insulating components. After the plug fires, however, current is conducted through the spark, which provides a relatively low-resistance conducting path through the ionized gases that are formed along the path of the discharge once it is initiated. The resistance R_2 then drops to a much smaller value, determined primarily by the resistance of the secondary winding and the ignition wiring, and remains there until the current falls to such an extent that the spark is extinguished. It then returns to the former higher value. This means that the time constant t_0

given by (20.7.40) does not remain the same during all phases of the induction process but varies as the resistance of the secondary circuit varies. While this changes the quantitative aspects of what we have presented to some extent, it does not essentially affect the important physical features of the results that are obtained.

Historically, induction coils have played an important part in the study of electrical phenomena, since they were the first really practical and reliable high-voltage generators. Before Joseph Henry built the first of these devices, the only way of obtaining high voltages was from fussy and unreliable electrostatic machines, which were no more than mechanical arrangements for electrification by rubbing one substance over another. The invention of the induction coil led to the systematic investigation of electrical conduction in ionized gases and the phenomenon of electrical "breakdown" which takes place in gases, liquids, and solids as arcs or sparks are initiated.

EXAMPLE 20.7.2

An induction coil like the one considered in the previous example is used as a *transformer* by connecting an ac emf $\mathscr{E}(t) = \mathscr{E}_0 \sin \omega t$ to the primary. Find the potential difference that appears across the output terminals AB in Fig. 20.27, assuming that no secondary current is being drawn by an external circuit.

In Fig. 20.27, we have essentially the same circuit as shown in Fig. 20.24, except that the battery and switch are replaced by an ac generator that furnishes an alternating emf given by $\mathscr{E}(t) = \mathscr{E}_0 \sin \omega t$. This is, in principle, the same sort of device as that described in Example 20.2.3. Also, to simplify our calculations somewhat, we have left the secondary circuit open, so that $I_2 = 0$, and we shall be content simply to find the potential difference ΔV_{AB} developed across the output terminals of the device. In Fig. 20.27, the resistance R_1 can be thought of as the combined internal resistance of the generator and the primary winding of the transformer, while R_2, which, as we shall see, plays little or no part in the calculations, represents the resistance of the secondary winding.

As always, (20.7.22) and (20.7.23) represent the induced emfs $\mathscr{E}_1$ and $\mathscr{E}_2$, though, since in this case $I_2 = 0$, the expressions are simplified to a large extent. In this case also, $\mathscr{E}_1$ and $\mathscr{E}_2$ are related by (20.7.24) and (20.7.25). If we write Kirchhoff's circuit law for the primary circuit, we see that

$$\mathscr{E}_0 \sin \omega t - I_1 R_1 + \mathscr{E}_1(t) = 0 \tag{20.7.49}$$

while for the secondary circuit,

$$\mathscr{E}_2(t) - I_2 R_2 - \Delta V_{AB} = 0$$

or, since $I_2 = 0$,

$$\mathscr{E}_2(t) - \Delta V_{AB} = 0 \tag{20.7.50}$$

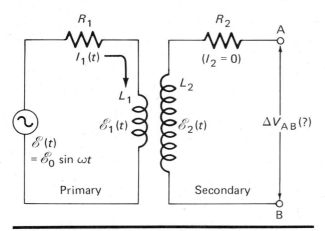

FIGURE 20.27. Schematic diagram of an induction coil used as a transformer in an alternating current circuit.

From (20.7.22), it is clear that in this case $I_2 = 0$ and $\mathscr{E}_1 = L_1 \, dI_1/dt$, which enables us to write for (20.7.49)

$$L_1 \frac{dI_1}{dt} + R_1 I_1 = \mathscr{E}_0 \sin \omega t \tag{20.7.51}$$

This is a differential equation that can be solved for I_1 as a function of time. It is of a slightly different-sort from those we have dealt with in our previous electrical examples because of the term $\mathscr{E}_0 \sin \omega t$ on the right. It is, in fact, related to equations that describe the forced vibrations of mechanical systems that are subject to external sinusoidally varying forces.

In any event, if there is an alternating emf applied to the circuit, it is reasonable to believe that it will excite an alternating current of the same frequency in the circuit. We are not quite certain that the current will be in *phase* with the emf; it might be—it is perhaps not unreasonable that it should be—but of course it is safer to assume that it is *not* if we're not quite sure. So we shall assume that an alternating current of frequency ω flows, such that

$$I_1(t) = I_{10} \sin(\omega t + \phi) \tag{20.7.52}$$

This current has peak amplitude I_{10} and differs in phase from the applied emf $\mathscr{E}_0 \sin \omega t$ by the phase angle ϕ. If the current and applied emf are in phase, $\phi = 0$; but in actuality, as we shall soon see, they are not. If we differentiate (20.7.52), we find

$$\frac{dI_1}{dt} = I_{10} \omega \cos(\omega t + \phi) \tag{20.7.53}$$

Substituting the values for I_1 and dI_1/dt given above into (20.7.51),

$$L_1 I_{10} \omega \cos(\omega t + \phi) + R_1 I_{10} \sin(\omega t + \phi) = \mathscr{E}_0 \sin \omega t \tag{20.7.54}$$

However, from elementary trigonometry we know that

$$\sin(\omega t + \phi) = \sin \omega t \cos \phi + \cos \omega t \sin \phi$$
$$\cos(\omega t + \phi) = \cos \omega t \cos \phi - \sin \omega t \sin \phi$$

Substituting these relations into (20.7.54) and grouping all terms containing $\sin \omega t$ and all terms containing $\cos \omega t$ separately, we find that (20.7.54) can be written in the form

$$I_{10}(\omega L_1 \cos \phi + R_1 \sin \phi) \cos \omega t$$
$$+ \left[I_{10}(R_1 \cos \phi - \omega L_1 \sin \phi) - \mathscr{E}_0 \right] \sin \omega t = 0 \tag{20.7.55}$$

The left side of this equation varies with time through the sinusoidal functions $\sin \omega t$ and $\cos \omega t$, while the right side is zero. In order that the two terms on the left side add to zero for all values of time, the coefficients before the terms $\sin \omega t$ and $\cos \omega t$ must both vanish. At $t = 0$, for example, $\sin \omega t$ is zero while

677

cos ωt is unity; in this event, it is evident that the coefficient of cos ωt must be zero in order that (20.7.51) be satisfied. On the other hand, at time $t = \pi/2\omega$, when $\omega t = 90°$, cos ωt is zero while sin $\omega t = 1$. Then, the only way in which (20.7.55) can be satisfied is that the coefficient of sin ωt vanish. But the equation can be satisfied for both these times only if *both* coefficients vanish. Therefore, setting both coefficients equal to zero, we can write

$$\omega L_1 \cos \phi + R_1 \sin \phi = 0 \qquad (20.7.56)$$

and

$$I_{10}(R_1 \cos \phi - \omega L_1 \sin \phi) = \mathscr{E}_0 \qquad (20.7.57)$$

From (20.7.56), dividing both sides by cos ϕ, we see at once that

$$\tan \phi = -\frac{\omega L_1}{R_1} \qquad (20.7.58)$$

so that there is a phase difference between the externally applied emf and the current, after all. How clever of us to have insisted on allowing for such a contingency back at equation (20.7.52)! From the minus sign in (20.7.58), we know that the phase angle will be negative and, therefore, that the current will lag behind the applied emf in phase. We shall see later, in Chapter 22, that this is a general characteristic of inductive ac circuits. If the tangent of the phase angle is $-\omega L_1/R$, then the sine and cosine can be figured out from Fig. 20.28, in which we see a right triangle wherein the side opposite the angle ϕ is $-\omega L_1$ and the side adjacent is R_1, which guarantees that the tangent will be as given by (20.7.58). From this triangle it is apparent that

$$\sin \phi = \frac{-\omega L_1}{\sqrt{R_1{}^2 + (\omega L_1)^2}}$$

and

$$\cos \phi = \frac{R_1}{\sqrt{R_1{}^2 + (\omega L_1)^2}} \qquad (20.7.59)$$

Substituting these values back into (20.7.57) and

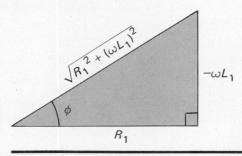

FIGURE 20.28. Phase relationships in the primary circuit of the transformer illustrated in Fig. 20.27.

solving for I_{10}, we find at last that

$$I_{10} = \frac{\mathscr{E}_0}{\sqrt{R_1{}^2 + (\omega L_1)^2}} \qquad (20.7.60)$$

Finally, putting this value into (20.7.52),

$$I_1(t) = \frac{\mathscr{E}_0}{\sqrt{R_1{}^2 + (\omega L_1)^2}} \sin(\omega t + \phi) \qquad (20.7.61)$$

An odd feature of this result is that the ratio of the current I_1 to applied emf $\mathscr{E}_0$—which we would identify as the resistance in a dc circuit—is not constant but *depends on the frequency*. This is another very general characteristic of ac circuits that contain either inductance or capacitance, and we shall learn more about it in Chapter 22.

At present, however, we are interested in obtaining the induced emf $\mathscr{E}_1(t)$, which is obtained from (20.7.22), noting that $I_2 = 0$, as

$$\mathscr{E}_1(t) = -L_1 \frac{dI_1}{dt} = -\frac{\omega L_1 \mathscr{E}_0}{\sqrt{R_1{}^2 + (\omega L_1)^2}} \cos(\omega t + \phi)$$
$$(20.7.62)$$

The secondary emf $\mathscr{E}_2$ may then be found from (20.7.24) as

$$\mathscr{E}_2(t) = \sqrt{\frac{L_2}{L_1}}\, \mathscr{E}_1(t) = -\frac{\omega \sqrt{L_1 L_2}\, \mathscr{E}_0}{\sqrt{R_1{}^2 + (\omega L_1)^2}} \cos(\omega t + \phi)$$
$$(20.7.63)$$

This enables us finally, from (20.7.50), to express the output potential difference ΔV_{AB} as

$$\Delta V_{AB} = -\mathscr{E}_2(t) = \frac{\omega \sqrt{L_1 L_2}\, \mathscr{E}_0}{\sqrt{R_1{}^2 + (\omega L_1)^2}} \cos(\omega t + \phi)$$
$$(20.7.64)$$

From this, it is obvious that if the frequency ω is zero, the output voltage ΔV_{AB} is also zero. This means that *a transformer simply will not work in a direct current circuit*, a fact that has bothered electrical engineers and electricians for decades but about which little can be done. The reason is, of course, that in a dc circuit there is no flux change and, therefore, no induced secondary emf to create the output voltage ΔV_{AB}. For ac frequencies such that $\omega L_1 \gg R_1$, however, we may neglect R_1 in the denominator of (20.7.64), which then becomes

$$\Delta V_{AB} = \mathscr{E}_0 \sqrt{\frac{L_2}{L_1}} \cos(\omega t + \phi) \qquad (20.7.65)$$

Also, in this case, the phase angle ϕ as given by (20.7.58) approaches $-\pi/2$, since tan ϕ approaches $-\infty$. But cos$(\omega t - \pi/2) = \sin \omega t$, so that we can write (20.7.65) as

$$\Delta V_{AB} = \mathscr{E}_0 \sqrt{\frac{L_2}{L_1}} \sin \omega t \qquad (20.7.66)$$

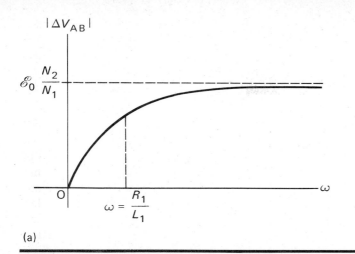

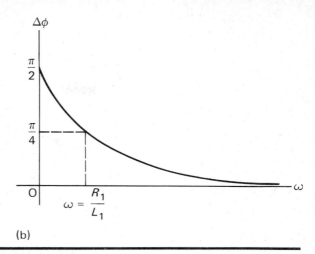

FIGURE 20.29. (a) Secondary output voltage and (b) phase angle ϕ between output and input voltages, plotted as a function of frequency.

or, recalling (20.7.24) and (20.7.25),

$$\Delta V_{AB} = \mathcal{E}_0 \frac{N_2}{N_1} \sin \omega t = \frac{N_2}{N_1} \mathcal{E}(t) \qquad (20.7.67)$$

The situation, then, is this: a transformer will not work at all on direct current. For low frequencies, there is an output voltage given by (20.7.64). This voltage differs in phase from the emf $\mathcal{E}(t) = \mathcal{E}_0 \sin \omega t$ applied to the primary circuit, and its amplitude depends on frequency, increasing as the frequency increases, as illustrated in Fig. 20.29. For frequencies so high that $\omega L_1 \gg R_1$, the output voltage amplitude is essentially *independent* of frequency and equals the input emf multiplied by the secondary-to-primary turns ratio N_2/N_1. In this limit, there is no longer any significant phase difference between the input and output voltages; this is evident from Eq. (20.7.67) and from Fig. 20.29b.

It will be noted that the condition $\omega L_1 \gg R_1$, under which the induction coil "operates the way a transformer should operate," can be stated as $\omega \gg R_1/L_1$, or, since L_1/R_1 is the inductive time constant t_1 of the primary circuit, as

$$\omega \gg \frac{1}{t_1} \qquad (20.7.68)$$

Therefore, so long as the frequency is much greater than the reciprocal of the primary circuit time constant, the device will behave like an "ideal" transformer in the sense described above.

If a current I_2 is allowed to flow in the secondary circuit, the essential features of the system are practically unaltered, although the mathematical analysis becomes much more complex, as do the expressions for the emfs and currents in the two circuits. We shall not attempt to tackle the details of this situation, except to mention that the condition (20.7.68) then

becomes

$$\omega \gg \frac{1}{t_1 + t_2} \qquad (20.7.69)$$

where t_2 represents the inductive time constant of the secondary circuit.

We have considered here only the case of transformers whose windings enclose no very strongly magnetizable substance, and transformers of this type are very useful in practice, particularly for high-frequency use, as in radio or television circuits. For power-line frequencies, however, such a device has to be very large and bulky and has to contain lots of expensive copper to satisfy condition (20.7.69). Under these circumstances, transformers are frequently wound around iron cores. As we shall see more clearly in the next chapter, the effect of the iron core is mainly to increase vastly the magnetic flux within the windings. This flux increase comes about because the flux arising from the current in the windings is augmented with magnetic flux arising from the magnetic moments of the iron atoms in the core, which align themselves willingly in whatever way they are impelled to line up by the currents in the windings. Under these circumstances, a small change in primary or secondary current can result in a huge change in magnetic flux. According to (20.7.6), this amounts to saying that the *inductances* of the primary and secondary circuits are greatly increased by an iron core. This, in turn, means that a given time constant, L/R, can be achieved with a much higher value of primary or secondary circuit resistance. The windings of the transformer can, therefore, be made using finer wire, the resulting device being much smaller, lighter, and less expensive than would otherwise be possible.

The transformer is a device of enormous technical importance. Its invention, in its day, was an event of the same order of importance as the invention of nuclear power or solid-state electronic devices in more recent years. The reason for this is that it is economical to transmit power over long-distance power lines *only at very high voltages*, but it is extremely dangerous and inconvenient to generate or use it at those voltage levels. Suppose we are faced with the problem of transmitting 10^8 watts of power over a power line whose resistance is 1 ohm. We could send a current of 10^6 amperes over the line using a potential difference of 10^2 volts. In that case, the power dissipated in the resistance of the power line itself would have to be $P = I^2R = (10^6)^2(1) = 10^{12}$ watts. But then, 10,000 times as much power is used in sending electrical energy to its destination as is delivered when it gets there! On the other hand, suppose that instead we sent a current of 10^3 amperes using a potential difference of 10^5 volts. The power dissipated within the power line itself is now $I^2R = (10^3)^2(1) = 10^6$ watts, only 1 percent of the power delivered to the final users. This is an acceptable level of power line loss, but it is difficult to design generators to generate current at 100,000 volts and even more difficult for consumers to use it under those conditions. It might be possible to lower power line resistances to a certain degree, but this requires huge conductors that represent vast investments in copper, which is not an inexpensive metal. The more reasonable alternative, one that is eminently practical and economical, is to generate the power at low voltages, use step-up power transformers to increase the voltage to what is required for economical power transmission, and lower the voltage on the other end for convenient distribution and use with step-down transformers. But, of course, transformers do not work on dc circuits, and that is why virtually all commercially generated electricity is used in the form of *alternating current*.

EXAMPLE 20.7.3

The toroidal coil described in Example 20.4.2 is used as the secondary of a toroidal transformer. A primary winding having 628 turns—one fifth the number of secondary turns—is wound on top of the secondary winding, using wire of the same kind as that used for the secondary winding. We then have a transformer with secondary resistance $R_2 = 10.75$ ohms, secondary inductance $L_2 = 5.92 \times 10^{-3}$ henry, primary resistance $R_1 = (0.2)(10.75) = 2.15$ ohms, and primary inductance (recalling from (20.4.32) that the inductance is proportional to the *square* of the number of turns) $L_1 = (0.04)(5.92 \times 10^{-3}) = 2.368 \times 10^{-4}$ henry. In what range of frequencies will this device behave as an "ideal" transformer? What will be the ratio of output potential difference to input emf in this range of operation?

For a device like this to operate as an ideal transformer, condition (20.7.69) must be satisfied. The two time constants are

$$t_2 = \frac{L_2}{R_2} = \frac{5.92 \times 10^{-3}}{10.75} = 5.51 \times 10^{-4} \text{ sec}$$

and

$$t_1 = \frac{L_1}{R_1} = \frac{2.368 \times 10^{-4}}{2.15} = 1.10 \times 10^{-4} \text{ sec}$$

Condition (20.7.69) then tells us that

$$\omega \gg \frac{1}{t_1 + t_2}$$

or

$$\omega \gg 1513 \text{ rad/sec, or 241 Hz}$$

We should expect the device to operate as an ideal transformer for frequencies much larger than 240 Hz—perhaps, as a practical matter, above 1000 Hz. Had we wound our primary and secondary coils on an iron core, we could have achieved primary and secondary inductances that are higher than those given above by a factor of about 1000, the time constant t_1 and t_2 thereby increasing by the same factor. Condition (20.7.68) would then give $\omega \gg 1.513$ radian/second (0.241 Hz) and above 1.0 Hz or so we should have a system that approximates the behavior of an ideal transformer. Such a device would make a very good transformer for 60 Hz ac power-line operation.

The ratio of output voltage to input emf for the ideal transformer is given by (20.7.67). Therefore, we can write in this case

$$\frac{\Delta V_{AB}}{\mathscr{E}(t)} = \frac{N_2}{N_1} = 5.0$$

EXAMPLE 20.7.4

A long, straight solenoid of cross-sectional area A_1 is wound with a primary winding of n_1 turns per unit length and carries a time-varying current I_1. This primary coil lies within a larger secondary coil of area A_2, with n_2 turns per unit length that carries a current I_2, as shown in Fig. 20.30. Calculate the coefficients of mutual inductance M_{12} and M_{21} for this structure.

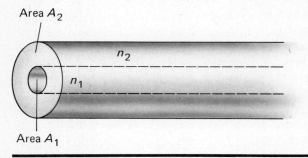

FIGURE 20.30

From (20.4.1), we know that the field B_2 inside the large secondary coil has the constant value

$$B_2 = \mu_0 n_2 I_2 \qquad (20.7.70)$$

In the case of the smaller primary coil, the field B_1 is given by

$$B_1^{(a)} = \mu_0 n_1 I_1 \qquad \text{inside coil 1}$$

and

$$B_1^{(b)} = 0 \qquad \text{outside coil 1} \qquad (20.7.71)$$

From the integrals shown in Eqs. (20.7.1) and (20.7.2), we can see that in this case

$$\Phi_1 = \Phi_{11} + \Phi_{12} = B_1 A_1 + B_2 A_1$$
$$= \mu_0 n_1 I_1 A_1 + \mu_0 n_2 I_2 A_1$$

$$\Phi_2 = \Phi_{21} + \Phi_{22} = B_1 A_2 + B_2 A_2$$
$$= B_1^{(a)} A_1 + B_1^{(b)}(A_2 - A_1) + B_2 A_2$$
$$= \mu_0 n_1 I_1 A_1 + \mu_0 n_2 I_2 A_2$$

recalling that $B_1^{(b)} = 0$. The emfs $\mathscr{E}_1$ and $\mathscr{E}_2$ in the two circuits are

$$\mathscr{E}_1 = -N_1 \frac{d\Phi_1}{dt} = -N_1 \frac{d\Phi_{11}}{dt} - N_1 \frac{d\Phi_{12}}{dt}$$
$$= -(n_1 l)\mu_0 n_1 A_1 \frac{dI_1}{dt} - (n_1 l)\mu_0 n_2 A_1 \frac{dI_2}{dt} \qquad (20.7.72)$$

$$\mathscr{E}_2 = -N_2 \frac{d\Phi_2}{dt} = -N_2 \frac{d\Phi_{21}}{dt} - N_2 \frac{d\Phi_{22}}{dt}$$
$$= -(n_2 l)\mu_0 n_1 A_1 \frac{dI_2}{dt} - (n_2 l)\mu_0 n_2 A_2 \frac{dI_2}{dt} \qquad (20.7.73)$$

Comparing these equations with (20.7.7) and (20.7.8), it is apparent that the coefficients of dI_1/dt and dI_2/dt in (20.7.72) are L_1 and M_{12}, respectively, while the coefficients of those same quantities in (20.7.73) are M_{21} and L_2, respectively. Therefore,

$$L_1 = \mu_0 n_1^2 l A_1 \qquad \text{and} \qquad L_2 = \mu_0 n_2^2 l A_2 \qquad (20.7.74)$$

in agreement with (20.4.7), while

$$M_{12} = M_{21} = \mu_0 n_1 n_2 l A_1 \qquad (20.7.75)$$

In this example, it is important to note that the relationship between the self-inductances L_1 and L_2 and the mutual inductance coefficients is no longer given by (20.7.21) but, instead, by

$$M_{12} = M_{21} = \sqrt{\frac{A_1}{A_2} L_1 L_2} \qquad (20.7.76)$$

SUMMARY

Whenever there is a time-varying magnetic flux through a circuit, an emf is induced in the circuit the magnitude of which is equal to the time rate of change of magnetic flux intercepted by the circuit. This statement is referred to as *Faraday's law of induction.* Mathematically, this law can be written as

$$\mathscr{E}_m = -\frac{d\Phi_m}{dt} = -\frac{d}{dt}\int_S \mathbf{B} \cdot \mathbf{n}\, da$$

If the magnetic induction is uniform over the area enclosed by the circuit, this becomes

$$\mathscr{E}_m = -\frac{d}{dt}(BA \cos\theta)$$

where A is the area of the circuit, and θ is the angle between the normal to the plane in which it lies and the induction vector $\mathbf{B}$. The induced emf appears in the circuit irrespective of how the flux change is brought about.

Faraday's law implies that a time-varying magnetic flux generates an electric, though nonelectrostatic, field $\mathbf{E}$ whose integral, taken around a closed circuit, represents the induced emf $\mathscr{E}_m$. Viewed in this way, Faraday's law can be written as

$$\mathscr{E}_m = \oint_C \mathbf{E} \cdot d\mathbf{l} = -\frac{d}{dt}\int_S \mathbf{B} \cdot \mathbf{n}\, da$$

Lenz's law informs us that induced currents flow in such a way that their own magnetic effects oppose the flux change that created them originally. Lenz's law is related to the law of conservation of energy and is responsible for the negative sign of the induced emf given by Faraday's law. Eddy currents are magnetically induced currents that flow in extended conducting bodies rather than linear circuits. Their direction and general flow pattern can be inferred from Lenz's and Faraday's laws.

Self-induced emf refers to an emf induced in a circuit by a change in its own magnetic flux or, in turn, by a change in the current that flows through the circuit itself. Since the circuit's magnetic flux is always proportional to the current, the self-induced emf will always be proportional to dI/dt. Therefore, the self-induced emf in a circuit of N turns is

$$\mathscr{E} = -N \frac{d\Phi_m}{dt} = -L\frac{dI}{dt}$$

where L is a proportionality constant referred to as the self-inductance of the circuit. The self-inductance of a circuit or circuit element depends on the number of turns of conductor, the area, and the shape of the circuit but is independent of the current. The magnetically self-induced emf in a coil can oppose the growth of current through the coil or can sustain the current through the coil after an external emf is removed; the self-induced emf gives rise to a potential difference across the terminals of an inductance whose instantaneous magnitude is

$$\Delta V_L = -L\frac{dI}{dt}$$

In the case of a long, straight solenoid, the inductance is easily calculated to be

$$L = \mu_0 n^2 l A$$

where l is the length of the coil, A is the cross-sectional area, and n is the number of turns per unit length. The direction of self-induced emfs is determined by Lenz's law.

Because of the self-induced emfs associated with inductances, series inductance–resistance (L–R) circuits exhibit exponential growth and decay of current, similar to that found in series R–C circuits. The time constant t_0 that gives the time scale on which these exponential changes occur is

$$t_0 = L/R$$

Like a capacitor, an inductance stores energy, though the energy resides in the *magnetic* field of the inductor rather than in an electrostatic field, as in a capacitor. It can be shown that the magnetic energy stored in an inductance that carries a current I is

$$U_m = \tfrac{1}{2}LI^2$$

and that the magnetic energy density (magnetic field energy per unit volume) anywhere in a magnetic field is

$$u_m = \frac{B^2}{2\mu_0}$$

Isolated inductances in series and in parallel combine like resistances, that is, for inductors in series,

$$L = L_1 + L_2 + \cdots + L_n$$

and for inductors in parallel,

$$\frac{1}{L} = \frac{1}{L_1} + \frac{1}{L_2} + \cdots + \frac{1}{L_n}$$

In the former case, the potential difference across the array is the sum of those associated with individual elements, the current through each being the same, while in the latter, the potential difference is the same across each element, the current in each element adding to produce the total current through the array.

Mutual induction between different circuits arises when a changing flux generated by one of them induces an emf in another. If there are two such circuits, the emf in circuit 1 will depend not only on dI_1/dt but also on dI_2/dt. Therefore,

$$\mathscr{E}_1 = -L_1\frac{dI_1}{dt} - M_{12}\frac{dI_2}{dt}$$

$$\mathscr{E}_2 = -M_{21}\frac{dI_1}{dt} - L_2\frac{dI_2}{dt}$$

In these equations, M_{12} and M_{21} are called coefficients of *mutual inductance*; they express the strength of the magnetic fields generated by the circuits and the degree to which magnetic flux generated by one circuit intercepts the area enclosed by the other. The laws of mutual induction govern the behavior of devices such as induction coils and transformers, in which emfs may be stepped up or down by mutual inductive coupling between coils with different numbers of turns. In such devices, the primary and secondary emfs, $\mathscr{E}_1$ and $\mathscr{E}_2$, are related to the number of turns N_1 and N_2 in the two windings by

$$\frac{\mathscr{E}_2}{\mathscr{E}_1} = \frac{N_2}{N_1}$$

QUESTIONS

1. The emf induced in a circuit depends upon a number of factors. State as many of these as you can.
2. In each of the cases shown in the diagram, indicate clearly the direction of the current induced in the dotted circuit.
3. A circular loop of wire is placed between the poles of an electromagnet. Which orientations lead to maximum and minimum magnetic flux?
4. Explain carefully the distinction between magnetic induction and magnetic flux. Give an example for which the magnetic induction is constant but the magnetic flux is not.
5. A circular loop is placed near a long wire in which the current is constantly increasing with time. In which direction will current flow around the loop?
6. Explain clearly why the angular frequency ω of rotation of a coil in a magnetic field determines the magnitude of the induced voltage as well as its time variation. What other factors affect the magnitude of this voltage?
7. Can you think of examples of phenomena which obey a law analogous to Lenz's law in the sense that an effect tends to oppose the cause which produced it?
8. Explain very briefly what is meant by a self-induced emf.
9. Can you explain why a self-induced emf is always proportional to the time rate of change of current passing through the circuit? On what factors does the self-inductance L depend?
10. How does the inductance per unit length of a given solenoid change if the length of wire used for the windings is doubled?
11. The emf of a generator produces an alternating current. Describe several ways of converting this to a direct, although pulsating, current.
12. An L–R circuit is connected to a battery. Does the time constant depend on the voltage of the battery? Is it possible for the voltage across the inductor to exceed the emf of the battery?
13. Give some convincing arguments to show that an inductor can store energy.

14. Suppose that the primary of a transformer were supplied with 250 volts dc instead of ac. Describe what would happen.

15. A transformer is constructed so that the secondary emf is one half the primary emf. What comments can you make concerning its construction?

PROBLEMS

1. A rectangular loop of wire 12.0 cm long and 8.0 cm wide is placed between the poles of an electromagnet that produces a uniform magnetic field in the region where the loop is located. Initially, the electromagnet is off and the magnetic field in this region is zero. A linearly increasing current is then made to flow in the coils of the magnet, producing a magnetic field in the region that rises at a constant rate from zero to 1.25 weber/m^2 in a time interval of 0.15 sec. **(a)** What is the magnitude of the emf induced in the loop during this period? **(b)** If the loop is replaced by a rectangular coil of wire consisting of 250 turns of wire, each of which has the same dimensions as the original loop, what is the emf induced in the coil?

2. A circular loop of flexible wire of radius 15.0 cm and resistance 0.01 ohm is placed in a uniform magnetic field of 0.80 weber/m^2. Equal and opposite forces are applied to the loop at opposite ends of a diameter, deforming it into a straight pair of parallel conductors of negligible separation in a time interval of 0.025 sec. Find **(a)** the average emf induced in the loop, **(b)** the average current that flows, **(c)** the work needed to effect the deformation.

3. A straight copper wire of length 2 meters moves at a speed of 1.5 meter/sec in a direction perpendicular to its length and perpendicular to a uniform magnetic field of magnitude 0.7 weber/m^2. Find the induced emf between the ends of the wire. If the ends are joined by completing a circuit through a 3-ohm resistor, at what rate must work be done to keep the wire moving at the constant velocity of 1.5 meter/sec?

4. A copper bar 20 cm long is perpendicular to a uniform magnetic field of 0.5 weber/m^2. As a result of its motion at right angles to its length, an emf of 0.10 volt exists between its ends. Find the speed with which the bar moves through the field.

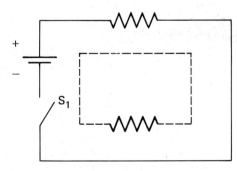

As switch S_1 is closed

(a)

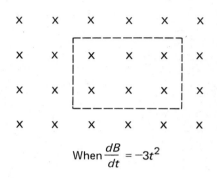

As switch S_2 is closed

(b)

When $\dfrac{dB}{dt} = -3t^2$

(c)

5. A circular loop of wire of radius 10 cm is oriented with its plane perpendicular to a uniform magnetic field of 0.2 weber/m^2. The loop is flipped over in a time interval of 0.5 sec. **(a)** Find the average induced emf during this process. **(b)** What is the average electric field in the loop? **(c)** If the loop has a resistance of 2.0 ohms, estimate how much work was required to flip the loop.

6. An electrical circuit contains a battery of emf $\mathcal{E}_0$ and a resistor R of length l and mass m which is free to move laterally, as shown in the figure. A uniform magnetic field B is perpendicular to the plane of the circuit, as illustrated. When switch S is closed, you may assume that the current through R is $\mathcal{E}/R$ at $t = 0$. Show that the induced current is given by $i_{ind} = \mathcal{E}l^2B^2t/(mR^2)$, provided t is not too large. What happens for large t?

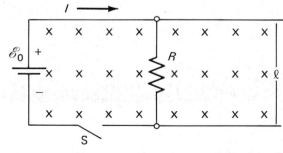

7. A long solenoid wound with n turns per unit length carries a current which varies with time according to $I = I_0 \sin \omega t$. A small, circular coil of N turns is placed inside the solenoid. The axes of the coil and solenoid are parallel and the turns of the coil each have area A' and resistance R'. Find **(a)** the emf induced in the small coil, **(b)** the induced current, and **(c)** the rate of power dissipation in the small coil.

683

8. A rectangular loop of wire is placed near a long, straight wire in which the current increases linearly with time according to $I = \alpha t$, as shown in the drawing. Find an expression for **(a)** the magnetic flux through the loop and **(b)** the emf induced in the loop.

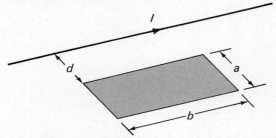

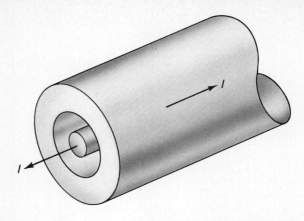

9. A single loop of wire of resistance R initially has a magnetic flux Φ_i passing through it. After a time interval, the flux has changed to Φ_f. Show that the net charge passing any point is given by $q = (\Phi_f - \Phi_i)/R$.

10. A circular coil wound with 100 turns of wire, each of radius 5 cm, is placed in a uniform magnetic field of 0.5 weber/m². The axis of the coil is parallel to the magnetic field. If the coil has a resistance of 200 ohms, how much net charge flows through the coil if its orientation with respect to the magnetic field is suddenly reversed?

11. A thin disc has a radius of 20 cm and rotates about its axis at 50 revolutions per second as shown in the diagram. A uniform magnetic field of 0.5 weber/m² is parallel to the axis of the disc. Calculate the potential difference between the center and the circumference. *Hint:* The potential difference along a small segment of the line AB is $d\mathscr{E} = Bv\,dl$.

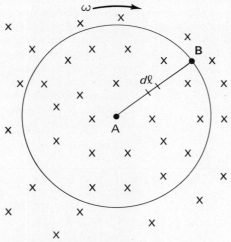

12. A long coaxial cable consists of an inner conductor of radius r_1 and an outer conductor of inner radius r_2 and outer radius r_3 as shown. The inner conductor carries a current I which varies with time in some prescribed way; the current density is uniform. The outer conductor carries the same current I but in the opposite direction. Show that there is an induced electric field parallel to the axis of the conductor and that for $r_2 < r < r_3$ its magnitude is given by

$$E(r) = E(r_1) + (\mu_0/2\pi)(dI/dt)\ln\left(\frac{r}{r_1}\right)$$

13. A straight conducting bar of mass 100 grams and length 15 cm rests on two parallel horizontal rails 15 cm apart. The bar, which carries a current of 30 amperes, is perpendicular to the rails. If the coefficient of static friction between the bar and rails is 0.25, what is the minimum uniform magnetic field required normal to the plane of the circuit to set the bar in motion?

14. A long solenoid has 500 turns per meter of cross-sectional area 0.05 m² and carries a current of 2 amperes. A secondary coil is wound closely over the primary. The secondary coil has a resistance of 4 ohms. It is found that if the current in the primary solenoid is reversed in a time interval of 0.05 second, an emf of 4 volts appears in the secondary during the reversal. **(a)** How many turns are contained in the secondary coil? **(b)** What is the net charge passing a given point in the secondary during the reversal of primary current?

15. A small, circular coil used to measure magnetic fields is known as a search coil. The plane of the coil is initially perpendicular to the magnetic field and the coil is attached to an instrument which measures the total charge that flows, for example, a ballistic galvanometer. The coil is quickly flipped over and the charge flow is measured. A particular search coil used in a student laboratory consists of 100 turns of wire, each turn having a cross-sectional area of 3 cm². The total coil resistance is 40 ohms. In a specific measurement, it is found that a charge of 2×10^{-5} coulomb is recorded. From the above information, how large is the magnetic field?

16. A circular coil of 300 turns and area 3.0×10^{-2} m² rotates at 60 rev/sec in a uniform field of 0.25 weber/m². What is the maximum emf generated and the maximum current flow if the coil has a resistance of 25 ohms?

17. A long straight solenoid of radius 5.0 cm is wound with 300 turns of wire per cm and carries initially a current of 2.40 amp. The current is reduced linearly to zero in 0.04 sec. Find **(a)** the electric field at a point 4.0 cm from the axis of the solenoid, **(b)** the electric field at a point 8.0 cm from the axis of the solenoid. **(c)** What is the maximum electric field strength and where does it occur?

18. A coil of N turns having cross-sectional area A is arranged as a simple pendulum, which is made to oscillate with simple harmonic motion $\theta = \theta_0 \sin \omega t$, as shown in the diagram. A uniform magnetic field B points in the horizontal direction, as illustrated. Obtain an expression for the induced emf in the coil as a function of the time.

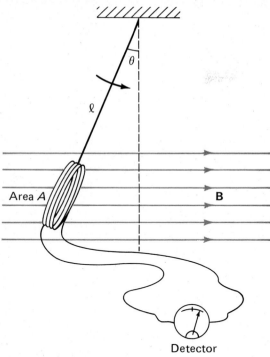

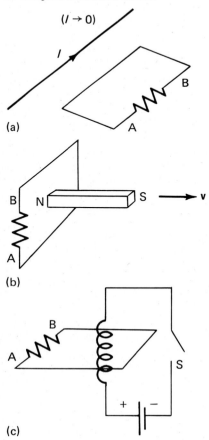

19. For each of the cases shown in the accompanying diagram, find the direction in which the induced current flows through the resistor ($A \to B$ or $B \to A$). **(a)** The current I suddenly drops to zero. **(b)** A bar magnet is pulled away from the loop of wire. **(c)** A switch is closed to complete an electric circuit.

20. In what direction does the induced current flow for the two cases shown in the figure?

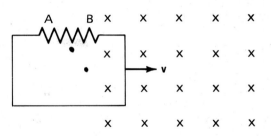

(a)

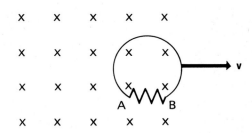

(b)

21. A long, straight solenoid 50 cm long has 500 turns of wire of cross-sectional area 10 cm^2. Find the inductance of the solenoid. If the current through the solenoid is reduced from 10 amperes to zero in 0.1 second, what is the average emf during this time?

*22. A coaxial cable contains an inner cylindrical conductor of radius r_1 and an outer cylindrical conductor of inner radius r_2 and outer radius r_3, as shown in the figure. Show that the cable has an inductance per unit length given by

$$\frac{L}{l} = \frac{\mu_0}{2\pi} \ln \frac{r_2}{r_1}$$

Hint: Find the total flux passing through the dotted rectangular loop in the diagram when the inner and outer conductors carry equal currents I in opposite directions.

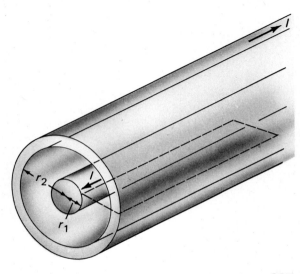

685

23. A "thin" toroid of radius 15 cm has 100 turns/cm and a cross-sectional area of 4 cm². Find the inductance of the toroid and the maximum self-induced emf when the current is reversed, from 15 amperes to −15 amperes, in 0.01 second.

24. The flux passing through each of the N turns of a coil is Φ_m. Show that the inductance of the coil is $L = N\Phi_m/I$, where I is the current flowing. If the inductance of a closely packed coil is 10 millihenrys and the coil has 300 turns, find the magnetic flux through each turn when a current of 2×10^{-3} ampere is flowing in the coil.

*25. A toroid has a rectangular cross section as shown in the diagram. The inner and outer radii are r_1 and r_2, respectively, and the width of each loop is a. The total number of turns is N. Find an expression for the inductance of the toroid. Calculate L if $N = 800$, $a = 2$ cm, $r_1 = 4$ cm, $r_2 = 8$ cm. *Hint:* Use the expression $B = \mu_0 IN/2\pi r$ obtained from Ampere's law for the magnetic field a distance r from the center of the toroid and integrate over the cross section to find the flux.

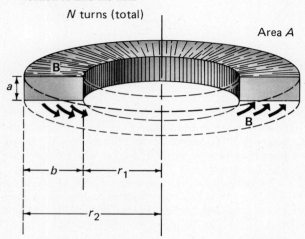

N turns (total)

Area A

26. A solenoid has a resistance of 40 ohms and a time constant of 0.08 second. What is its inductance?

27. A coil with a resistance of 65 ohms has an inductance of 25 henrys. By means of a switch, the coil is connected to a 50-volt battery. How long does it take for the current through the coil to reach one half its equilibrium value? For this value of t, what is the time rate of change of the current?

28. A 20-henry inductor has a resistance of 150 ohms. A switch is thrown at $t = 0$, applying a constant 30-volt potential difference across the inductor. **(a)** What is the time constant of the inductor? **(b)** Find the steady current which will flow in equilibrium after a long time has elapsed. **(c)** At what time will the current have reached 95 percent of its equilibrium value? **(d)** What is the rate of increase of the current when the current is one half the equilibrium value?

29. In an L–R circuit, what fraction of the equilibrium current is reached after five time constants have elapsed?

30. From Eq. (20.4.21), show that the charge q that has passed a given point in the circuit shown in Fig. 20.16b after the switch is thrown satisfies the equation $L(I - I(0)) + qR = 0$, and from this show that $q = (\mathscr{E}_0 L/R^2)(1 - e^{-Rt/L})$.

31. A 100-volt potential difference is suddenly applied across a coil with a resistance of 200 ohms and an inductance of 75 millihenrys. At $t = 0.002$ second, what part of the potential difference across the coil is due to the self-induced emf, and what part is due to current flow through the resistance? At $t = 0.004$ second, what current flows through the coil?

32. A coil has an inductance of 2.5 henrys and carries a steady current of 3 amperes. Find the energy stored in the magnetic field. If the volume within which the **B** field is significant is 200 cm³ and if the field therein is assumed to be uniform, find the strength of the magnetic field.

33. A certain solenoid has an inductance of 15 millihenrys and a resistance of 25 ohms. It is connected to a 6-volt battery of negligible internal resistance, with a switch in the circuit. Just after the switch is closed, when the current has reached 10 percent of its equilibrium value, find **(a)** the rate of change of the current, **(b)** the self-induced emf, **(c)** the energy stored in the solenoid, and **(d)** the rate of energy dissipation in the circuit.

34. How much energy is needed, in excess of that dissipated in Joule heating, to establish a 10-ampere current in a toroidal coil of volume 200 cm³ and inductance 1 henry? What is the average strength of the magnetic field within the coil?

35. A toroidal coil with a resistance of 20 ohms is connected to a battery. The current reaches one half its equilibrium value after 0.01 second. If the equilibrium current is 5 amperes, find the energy stored in the magnetic field and the emf of the battery, assuming its internal resistance is negligible.

36. A long, straight wire of radius a carries a current I. Find the energy stored in the magnetic field in a volume of length l that extends between $r = d$ and $r = 2d$, where $d > a$.

37. Find the magnetic energy per unit length for the coaxial cable of Problem 22 in the regions $r < r_1$ and $r_1 < r < r_2$.

38. A long, straight wire of radius a carries a total current I distributed uniformly over its cross-sectional area. Find the magnetic energy per unit length stored within the wire and show that it is independent of the radius.

39. In a certain region of space, a uniform magnetic field **B** is present. How large an electric field is needed to produce an electrical energy density of the same magnitude as the magnetic energy density?

40. Two solenoids have inductances of 5×10^{-6} henry and 8×10^{-6} henry, respectively. If they are well separated spatially, find the equivalent inductance when they are connected in series and when they are connected in parallel. If the resistances are 50 ohms and 100 ohms, respectively, find the time constant for both series and parallel connections.

41. A thin toroidal coil has a radius of 15 cm and a cross-sectional area of 4 cm². Its primary winding has 75 turns/cm while the secondary has 40 turns/cm. What is the value of the mutual induction coefficient M_{12}? Assume that the secondary is wound directly upon the primary winding.

42. Derive Eq. (20.7.24) by using Eqs. (20.7.22) and (20.7.23).

43. The primary and secondary coils of a transformer have 20 and 300 turns of wire, respectively. If the primary voltage is given, in volts, by $\mathscr{E}(t) = 10 \sin \omega t$, what is the maximum voltage in the secondary?

44. Assume that the primary inductance of the transformer of the preceding problem is 0.080 henries, and that the resistance of the primary coil is 3.6 ohms. If no current flows in the secondary coil, find the ratio of secondary output voltage amplitude to the amplitude of the source emf **(a)** for a source emf of infinitely large frequency, **(b)** for a source frequency of 400 Hz, **(c)** for a source frequency of 60 Hz, **(d)** for a source frequency of 10 Hz, **(e)** for a source frequency of 1.0 Hz. **(f)** Would this unit be acceptable for use in 60 Hz ac circuits?

45. In the situation described in the preceding problem, what would the phase difference between source emf and output emf be **(a)** for a source emf of infinitely large frequency, **(b)** for a source emf of frequency 400 Hz, **(c)** for a source frequency of 60 Hz, **(d)** for a source frequency of 10 Hz, **(e)** for a source frequency of 1.0 Hz.

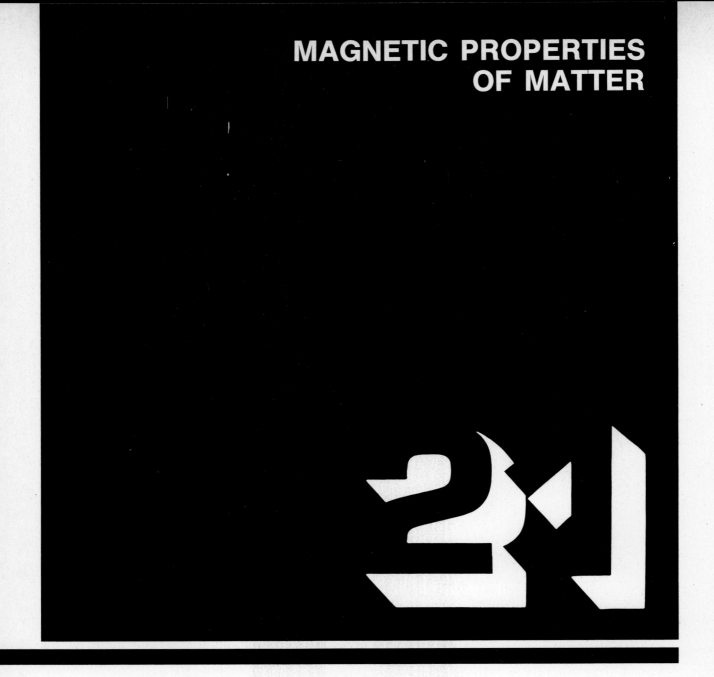

MAGNETIC PROPERTIES OF MATTER

21.1 Introduction

Magnetism was first discovered in ancient times as a result of the magnetic properties of the *lodestone*, which we now refer to as *magnetite*, or black iron oxide, Fe_3O_4. The permanent magnetism of iron was utilized in magnetic compasses for hundreds of years before the connection between magnetic fields and electric currents was discovered. In the preceding chapter, we have mentioned that the use of iron cores for inductors and transformers can result in significant changes in their characteristics that are extremely important in certain applications. The use of magnetic materials in microphones, loudspeakers, and other communications equipment is now very common.

Such materials are also widely used in computer memories, logic circuitry, and high-speed switching applications. It is, therefore, important to understand some of the basic principles that govern the interaction of magnetic fields with matter.

The general theory of the magnetic properties of matter is very complex, and we shall not attempt a detailed explanation of all its aspects. In permanently magnetizable substances particularly, a fundamental understanding of magnetic interactions depends upon a knowledge of the quantum-mechanical behavior of atoms and molecules, a subject that is too complex mathematically for us to go into at this time. On the other hand, we shall find that much of the observed behavior of magnetic substances can be understood by

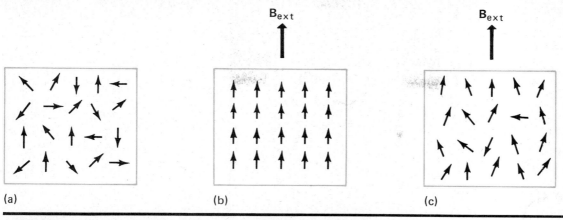

FIGURE 21.1. (a) Random alignment of atomic magnetic moments caused by thermal agitation. (b) Complete magnetic alignment in a system subjected to a strong external field, where thermal agitation is negligible. (c) Partial magnetic alignment in a system wherein the competing tendencies of thermal agitation and the external aligning field are both significant.

studying the effects of currents that circulate within atoms in creating atomic magnetic moments, and by considering the interaction of these atomic moments with one another and with externally applied magnetic fields. In the study of the electrical behavior of dielectrics, we found that it was very helpful to formulate Gauss's law for situations in which electrically polarizable materials are present. In the same way, we shall find that it is sensible for us to begin our study of magnetic substances by trying to formulate the fundamental theorem that relates magnetic fields to the currents from which they arise when magnetizable substances are present. That fundamental theorem, is, of course, Ampère's law, which, as we know, resembles Gauss's law for electric fields in certain important respects. Let us, therefore, see what Ampère's law tells us when magnetic materials are at hand.

21.2 Magnetization, Magnetic Intensity, and Ampère's Law

The fact that magnetizable materials can act as the source of magnetic fields stems from the fact that the individual atoms of the substance possess magnetic moments. These atomic magnetic moments arise from current loops associated with the motion of electrons within atoms and also from the fact that the inherent "spin" of the negatively charged electrons gives every electron an intrinsic spin magnetic moment.[1]

It is extremely important to realize that the net magnetic moment associated with any material arises as the outcome of a struggle between two competing

effects. These effects are the tendency of externally applied fields, or molecular fields originating within the material itself, to make the atomic magnetic moments line up in the same direction, and the tendency of random thermal agitation to break up aligned moments and produce a completely random distribution of atomic magnetization corresponding to zero total magnetic moment. These competing phenomena are illustrated in Fig. 21.1; we shall see frequent examples illustrating their interplay. In considering their behavior, it must be noted that the aligning effects of applied fields are proportional to the field but independent of temperature, while the randomizing influence of thermal agitation is proportional to temperature but independent of applied field.

Since all magnetic fields are caused by currents, it is sensible to try to relate the magnetic field caused by the atomic magnetic moments of magnetized substances to the circulating electronic currents that flow within the atoms of the substance. The way in which this may be done has already been suggested in a previous chapter in connection with Fig. 19.2 and is shown in somewhat more detail in Fig. 21.2. In this figure, each square represents schematically the volume occupied by a single atom; it is assumed that all the atoms are aligned magnetically so that their magnetic dipole moments are oriented vertically upward, out of the page. The arrows around the outer borders of each square show the circulation of electronic current within each atomic cell. The currents along each *internal* boundary are in opposite directions, and their effects in contributing to a total current distribution, therefore, cancel each other. But the atomic currents at the surface of the material are not canceled in this way and contribute to a net *surface magnetization current* I_m that flows around the outer surface of

---

[1] The intrinsic "spin" of the electron and the magnetic moment associated with it are discussed in greater detail in Chapter 28.

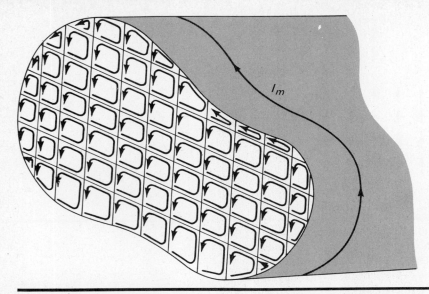

FIGURE 21.2. Cancellation of internal atomic circulating currents at internal boundaries, resulting in a net magnetic effect ascribable to a circulating surface magnetization current flowing around the outside surface of the sample.

the substance in the direction shown. This surface magnetization current can be regarded as the source of the sample's magnetic field, and that field can be regarded as a magnetic dipole field arising from a dipole moment p_m given by

$$p_m = I_m A \tag{21.2.1}$$

where A is the cross-sectional area. That this is true can be seen by noting from Fig. 21.2 that the magnitude of the surface magnetization current I_m is equal to that of the atomic current in each cell. But the individual dipole moment Δp_m associated with each cell is simply the circulating current, equal to I_m, times the area ΔA of the cell:

$$\Delta p_m = I_m \, \Delta A \tag{21.2.2}$$

But if we sum over all the cells, the individual dipole moments sum to the total dipole moment p_m, while the individual areas sum to the total area A, leading directly to (21.2.1).

We saw in Chapter 17 that the strength of the electric field of a polarized dielectric can be described by a polarization vector representing the electric dipole moment per unit volume. In the same way, we may express the magnetic field of a magnetized substance in terms of a *magnetization vector* $\mathbf{M}$ that is related to the magnetic dipole moment $\Delta \mathbf{p}_m$ arising from a volume ΔV of material, as shown in Fig. 21.3, by

$$\mathbf{M} = \frac{\Delta \mathbf{p}_m}{\Delta V} \tag{21.2.3}$$

We then say that $\mathbf{M}$ is the *magnetization* associated with the volume element ΔV. If the dipole moment

$\Delta \mathbf{p}_m$ is the same for all volume elements of size ΔV in the sample, the magnetization is constant everywhere within the material; this is the only case we shall consider in detail. If different volume elements give rise to different dipole moments as a result of more complete alignment of atomic dipoles in one region than in another, then we may allow the volume element ΔV surrounding any point P to become indefinitely small and express the magnetization vector at that point as

$$\mathbf{M} = \lim_{\Delta V \to 0} \frac{\Delta \mathbf{p}_m}{\Delta V} = \frac{d\mathbf{p}_m}{dV} \tag{21.2.4}$$

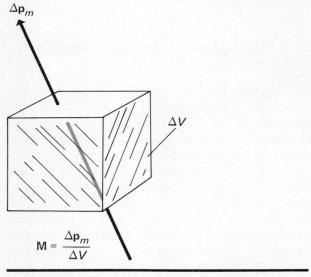

FIGURE 21.3. Diagram illustrating the definition of magnetization as the ratio of dipole moment to volume for every volume element of the sample.

The magnetization will then be a vector whose magnitude and direction may vary from point to point within the sample. Since from (21.2.4) we have

$$d\mathbf{p}_m = \mathbf{M}\,dV \qquad (21.2.5)$$

the total dipole moment of the sample will be

$$\mathbf{p}_m = \int_V \mathbf{M}\,dV \qquad (21.2.6)$$

the integral being taken over the total volume of the magnetized sample. When the magnetization $\mathbf{M}$ is constant, it may be written outside the integral:

$$\mathbf{p}_m = \mathbf{M}\int_V dV$$

or

$$\mathbf{p}_m = \mathbf{M}V \qquad (21.2.7)$$

If the dipole moment of the sample is expressed in terms of an effective atomic current distribution, for example, by (21.2.2), it is clear that the magnetization $\mathbf{M}$ is determined ultimately by this current distribution; and, conversely, if $\mathbf{M}$ is known, the source current distribution can be determined.

Now, with this background, let us see what happens when we apply Ampère's law to a magnetized substance. Because of the complexity of the mathematics required, we shall not be able to do this in the most general way. Instead, we shall develop our results by considering the behavior of a thin toroidal coil made by winding N total turns of wire on the surface of a uniform toroidal sample of magnetizable material. The results so obtained are of general applicability and illustrate very well the essential physical principles involved. It is best to visualize the magnetic substance on which the coil is wound not as a permanently magnetizable substance but rather as one that acquires a certain magnetization in proportion to the current in the windings, which subsequently disappears if the current is reduced to zero. In doing this, we do not exclude the possibility of using a permanently magnetizable core, but merely spare ourselves the aggravation of being bothered with questions best postponed until later, when we shall be in possession of enough basic understanding to answer them. The situation is illustrated in Fig. 21.4. We shall assume that a steady current I flows in the windings of the toroidal coil, and we shall apply Ampère's law (19.6.9) around a circular path C whose radius r is the mean radius of the toroid. The total current passing through the contour C consists of the current NI that flows in the windings and *also of the atomic magnetization current* I_m that flows around the cross section of the toroid, as shown in Fig. 21.4b.

Using this figure, also, we may relate the magnetization $\mathbf{M}$ to the dipole moment $d\mathbf{p}_m$ of the thin section of the toroidal core that is shown and to the part of the surface magnetization current that flows around this section and represents the source of its magnetic field. From (21.2.4) and (21.2.1), we may write the magnitude of the magnetization vector $\mathbf{M}$ as

$$M = \frac{dp_m}{dV} = \frac{A\,dI_m}{A(r\,d\theta)} = \frac{dI_m}{r\,d\theta} \qquad (21.2.8)$$

Now, dI_m is that portion of the total surface magnetization current that flows around the thin section of material subtending a central angle $d\theta$ shown in Fig. 21.4b. This portion dI_m is, therefore, equal to the total current I_m times the ratio $d\theta/2\pi$ that represents the fractional part of the toroid that is subtended by the

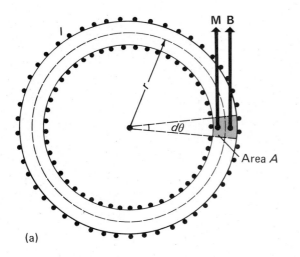

(a)

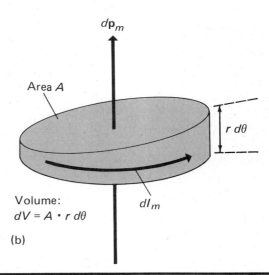

(b)

FIGURE 21.4. Relations between current, magnetic induction, and magnetization for a uniformly magnetized toroidal sample.

central angle $d\theta$. In the form of an equation, this means that

$$dI_m = I_m \frac{d\theta}{2\pi} \qquad (21.2.9)$$

Substituting this into (21.2.8), we find

$$M = I_m \frac{d\theta}{2\pi} \frac{1}{r\,d\theta} = \frac{I_m}{2\pi r}$$

or

$$\boxed{I_m = 2\pi r M} \qquad (21.2.10)$$

From (21.2.4), it is apparent that the vector $\mathbf{M}$ has the same direction as the vector $d\mathbf{p}_m$, and from Fig. 21.4b it is equally evident that the direction of this vector is everywhere *tangential* to the circular contour C. The vectors $\mathbf{M}$ and $\mathbf{B}$, therefore, are *always parallel*, because as we know from our previous work, the $\mathbf{B}$ field of a toroidal coil has this same tangential character. In addition, if we compute the integral of $\mathbf{M} \cdot d\mathbf{l}$ around the contour C, we find

$$\oint_C \mathbf{M} \cdot d\mathbf{l} = \oint_C M\,dl = M \oint_C dl = 2\pi r M \qquad (21.2.11)$$

But, using (21.2.10), we may now relate M and I_m by

$$\boxed{I_m = \oint_C \mathbf{M} \cdot d\mathbf{l}} \qquad (21.2.12)$$

We may now proceed to apply Ampère's law as given by (19.6.9) to the contour C, writing the total current I_t enclosed by C as the sum of the current NI that flows in the windings and the magnetization current I_m. Ampère's law in this form tells us that

$$\oint_C \mathbf{B} \cdot d\mathbf{l} = \mu_0 I_t = \mu_0 (NI + I_m) \qquad (21.2.13)$$

Using (21.2.12), this becomes

$$\oint_C \mathbf{B} \cdot d\mathbf{l} = \mu_0 NI + \mu_0 \oint_C \mathbf{M} \cdot d\mathbf{l} \qquad (21.2.14)$$

or

$$\boxed{\oint_C \left(\frac{\mathbf{B}}{\mu_0} - \mathbf{M}\right) \cdot d\mathbf{l} = NI} \qquad (21.2.15)$$

From this, it is evident that the true current NI that flows in the windings is related to the magnetic induction not simply as the integral of $\mathbf{B} \cdot d\mathbf{l}/\mu_0$, as it would be if no magnetic substance were present, but instead is similarly related to the integral of the quantity $[(\mathbf{B}/\mu_0) - \mathbf{M}] \cdot d\mathbf{l}$. The quantity $(\mathbf{B}/\mu_0) - \mathbf{M}$ is referred to as the *magnetic intensity* and is usually denoted by the symbol $\mathbf{H}$. Using this notation, we can write (21.2.15) as

$$\boxed{\oint_C \mathbf{H} \cdot d\mathbf{l} = I_C} \qquad (21.2.16)$$

where I_C ($= NI$) is the *true current* arising from the flow of mobile charges enclosed by the contour C and where the magnetic intensity $\mathbf{H}$ is given by

$$\boxed{\mathbf{H} = \frac{\mathbf{B}}{\mu_0} - \mathbf{M}} \qquad (21.2.17)$$

Solving this equation for $\mathbf{B}$, we find

$$\boxed{\mathbf{B} = \mu_0(\mathbf{H} + \mathbf{M})} \qquad (21.2.18)$$

Equations (21.2.14) through (21.2.18) are generally applicable to all situations where magnetic fields and magnetizable materials are present, even though they

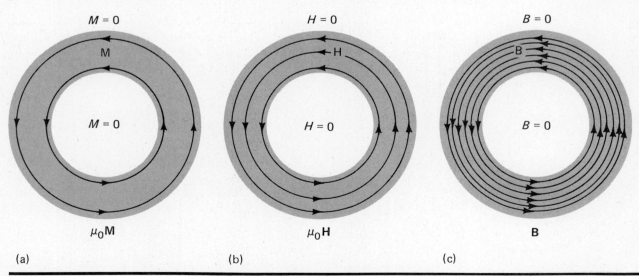

FIGURE 21.5. The (a) magnetization, (b) magnetic intensity, and (c) magnetic induction fields for a uniformly magnetized toroidal sample.

are derived only for the toroidal case. It is clear from (21.2.16) that the mks units for magnetic intensity H are amperes/meter. From (21.2.17), it is easily seen that M and H have the same dimensions. The relations between $\mathbf{H}$, $\mathbf{M}$, and $\mathbf{B}$ in a toroidal magnetized sample are illustrated in Fig. 21.5.

The situation resembles that which arose in the case of electrically polarizable materials in Chapter 17. There, we found that Gauss's law for cases in which polarizable materials are present state that the surface integral of $(\varepsilon_0 \mathbf{E} + \mathbf{P}) \cdot \mathbf{n}$—rather than $\varepsilon_0 \mathbf{E} \cdot \mathbf{n}$—is to be set equal to the mobile free charge inside the surface. In that case, we gave the label $\mathbf{D}$ to the quantity $\mathscr{E}_0 \mathbf{E} + \mathbf{P}$, referring to it as the electric displacement.

In the magnetic case, we see that it is really the contour integral of the quantity $[(\mathbf{B}/\mu_0) - \mathbf{M}] \cdot d\mathbf{l}$ that gives the *true* current enclosed by the contour, not simply the integral of $\mathbf{B} \cdot d\mathbf{l}$. We are, therefore, impelled to regard the quantity $(\mathbf{B}/\mu_0) - \mathbf{M}$ as a new field, which we call the magnetic intensity $\mathbf{H}$. By using Ampère's law in the form (21.2.16) in exactly the same way as we always have, evaluating the integral of $\mathbf{H} \cdot d\mathbf{l}$ around a given closed path, and setting its value equal to I_C, we may obtain an expression for the magnetic intensity $\mathbf{H}$ in terms of the true current that flows in the external circuits which excite the magnetization in the beginning. The induction $\mathbf{B}$ can then be found by a simple method involving (21.2.18) that we shall describe directly. The lines of $\mathbf{B}$ and $\mathbf{H}$ and the magnetization field $\mathbf{M}$ for a toroidal magnetized sample are illustrated in Fig. 21.5.

It is important to note that Ampère's law relates $\mathbf{B}$ to the *total* current, that is, the sum of the true current that arises from the macroscopic motion of charge in conducting media, *plus* any circulating atomic currents that may be present on or within magnetizable substances. The magnetic intensity $\mathbf{H}$, on the other hand, is related in essentially the same way by Ampère's law to the *true current only*. We may say, therefore, that the magnetic intensity $\mathbf{H}$ is a field arising from a source distribution consisting solely of true currents in conducting media, while the magnetic induction $\mathbf{B}$ is a field related similarly to a total current distribution consisting of true currents plus atomic magnetization currents.

In the electrostatic case, we found that the electric field $\mathbf{E}$ was related by Gauss's law to a source distribution consisting of all charges present, including polarization charges associated with polarizable dielectric substances, while the displacement $\mathbf{D}$ was related similarly to a source distribution consisting only of mobile charges that are free to move in response to electric forces acting on them. There is, therefore, something of an analogy to be drawn in one instance between $\mathbf{E}$ and $\mathbf{B}$ and, on the other hand, between $\mathbf{D}$ and $\mathbf{H}$.

Although the $\mathbf{H}$ field is related by Ampère's law to a source distribution consisting only of true currents, this does not mean that the magnetic intensity is completely unaffected by the magnetization currents and magnetic materials that may be present. Ampère's law guarantees only that the *integral of $\mathbf{H} \cdot d\mathbf{l}$ around a closed path* will be determined solely by true currents within the path. But the magnitude and direction of $\mathbf{H}$ can still be influenced in certain ways by magnetization currents even though the *integral* of $\mathbf{H} \cdot d\mathbf{l}$ is unaffected by them. In the case of a permanent magnet, for example, there are no true currents at all, and the integral of $\mathbf{H} \cdot d\mathbf{l}$ about any closed path is *zero*. This does not, however, guarantee that $\mathbf{H}$ itself is zero, as we shall see later when we examine these fields in detail. The same situation arises in electrostatics in connection with the electrostatic field $\mathbf{E}$. We saw in Chapter 16 that the integral of $\mathbf{E} \cdot d\mathbf{l}$ about a closed path is always zero, but clearly this does not imply that $\mathbf{E}$ itself must vanish.

In practice, we usually know the distribution of true currents that act as the sources of magnetic fields. The distribution of circulating magnetization currents associated with magnetic materials that may be present is, on the contrary, ordinarily unknown. It is, therefore, generally quite a simple matter, using the results derived above, to find the magnetic intensity $\mathbf{H}$ using Ampère's law. But there are many situations in which we may wish to determine the magnetic induction $\mathbf{B}$ as well as $\mathbf{H}$; in using Faraday's law of induction, for instance, we need to know $d\mathbf{B}/dt$ rather than $d\mathbf{H}/dt$. In this event, according to (21.2.18), there is no escaping the fact that in order to find $\mathbf{B}$, we must somehow find the magnetization $\mathbf{M}$.

Since, ordinarily, the net magnetization of the material is excited by the magnetic effect of a true current distribution to begin with, it is reasonable to assume that there is some simple relationship between the magnetization of the material and the magnetic intensity $\mathbf{H}$. In the simplest possible situation, this relation would be a direct proportionality of the form

$$\boxed{\mathbf{M} = \chi_m \mathbf{H}} \tag{21.2.19}$$

where χ_m is a constant quantity independent of $\mathbf{H}$. It is found, in fact, that many magnetizable substances behave this way, at least for values of magnetic intensity that are not excessively large. Such substances are said to exhibit *linear* magnetic behavior and are often called linear magnetic materials. The proportionality constant χ_m in (21.2.19) is usually referred to as the *magnetic susceptibility* of the material. In many instances, it is possible to calculate the magnetic susceptibility from the fundamental magnetic properties of the atoms of the substance and thus to predict its dependence on temperature, composition, and other physical parameters. We shall see how this is done in a few simple examples later in this chapter.

Even in the case of magnetizable substances that display linear magnetic behavior for moderate values of magnetic intensity, the simple linear relationship given by (21.2.19) is violated when the magnetic intensity becomes too large. In addition, permanently magnetizable substances, such as iron, nickel, and cobalt, hardly ever exhibit linear magnetic behavior as defined by (21.2.19). In such materials, the susceptibility is field dependent even for rather small values of H and, moreover, may depend upon *previous magnetic influences* the sample may have undergone. We shall investigate the physical reasons for the unique properties of these *ferromagnetic* substances later in this chapter.

For the time being, however, let us assume that the relation (21.2.19) is obeyed and limit ourselves first to describing the properties of linear magnetic materials. We shall assume that we can find appropriate ways of measuring χ_m experimentally and that it is, therefore, a known constant of the magnetic material present. We shall soon see one way in which this experimental measurement of susceptibility can be accomplished. Then, using (21.2.19) to express **M** in terms of **H**, we may write (21.2.18) as

$$\boxed{\mathbf{B} = \mu_0(1 + \chi_m)\mathbf{H} = \mu\mathbf{H}} \qquad (21.2.20)$$

where the quantity μ, defined by

$$\boxed{\mu = \mu_0(1 + \chi_m)} \qquad (21.2.21)$$

is called the *magnetic permeability* of the material. It is sometimes useful, also, to speak of the *relative permeability K_m*, which is defined as the ratio of the magnetic permeability μ to the permeability of free space. The relative permeability may, therefore, be expressed as

$$\boxed{K_m = \frac{\mu}{\mu_0} = 1 + \chi_m} \qquad (21.2.22)$$

The analogies between electric and magnetic susceptibility, magnetic permeability and dielectric permittivity, and relative permeability and the dielectric constant are readily apparent from these definitions.

If no magnetic material is present, then $M = 0$, and from (21.2.18), $\mathbf{B} = \mu_0\mathbf{H}$. If instead, without disturbing the true current distribution that forms the source of the **H** field, we substitute a material with a large positive magnetic susceptibility, the value of **H** is unchanged while the magnitude of **B**, as given by (21.2.20), will become very much larger. For this reason, the magnetic induction field **B** within materials of high magnetic permeability is much larger than

would be the case were such substances absent. We observe, therefore, that the lines of the induction field **B** tend to become highly *concentrated* within substances of high permeability. For this reason, a given change in true current may lead to a much larger flux change through magnetically permeable materials than through the same region of empty space. This is the essential reason for the large increase in inductance observed when an iron core is substituted for air or vacuum in an inductor. We must remember, however, that iron, though always exhibiting very large values of permeability, does not obey the linear relationship (21.2.19) between magnetic intensity and magnetization.

Table 21.1 gives the magnetic susceptibilities and permeabilities of a number of substances at room temperature. It will be observed that in most cases the numerical magnitudes of the magnetic susceptibilities are quite small, indeed, much *smaller* than the electric susceptibilities that are associated with common dielectric substances. The exception is iron, which goes to the other extreme and exhibits values of magnetic susceptibility much larger than the other materials listed and, in fact, much larger than the electric susceptibilities of most substances. No firm value can be quoted, since iron is not a linear magnetic substance, its susceptibility varying with applied magnetic intensity, previous magnetic history, and the metallurgical preparation of the sample. The huge values of permeability observed in iron and in other permanently magnetizable substances arise because in such ferromagnetic materials, the atomic magnetic dipoles tend not only to align in response to externally applied magnetic fields but *also to align one another as well*.

TABLE 21.1. Magnetic Susceptibilities

Substance	χ_m
Aluminum	2.3×10^{-5}
Bismuth	-1.7×10^{-4}
Copper	-1.0×10^{-5}
Gold	-3.6×10^{-5}
Lead	-1.7×10^{-5}
Magnesium	1.2×10^{-5}
Platinum	2.9×10^{-4}
Silver	-2.6×10^{-5}
Water	-0.88×10^{-5}
$CrK(SO_4)_2 \cdot 12H_2O$	2.32×10^{-5}
$Cu(SO_4) \cdot 5H_2O$	1.43×10^{-5}
$Gd_2(SO_4)_3 \cdot 8H_2O$	2.21×10^{-4}
MnF_2	4.59×10^{-4}
$CoCl_2$	3.38×10^{-4}
$FeCl_2$	3.10×10^{-4}
$FeCl_3$	2.40×10^{-4}
$NiCl_2$	1.71×10^{-4}
Iron (soft)	~ 5000

There are other substances which have positive magnetic susceptibilities that are much smaller and whose susceptibilities are, at least for reasonably small fields, independent of applied magnetic intensity. In such substances, individual atomic magnetic moments tend to align themselves parallel to the applied external field but do not significantly influence one another as in ferromagnetic materials. Such substances are are said to be *paramagnetic*.

Finally, we observe from the values in Table 21.1 that there are many materials whose magnetic susceptibility is *negative*, corresponding to relative permeability less than unity. In materials of this sort, the atomic magnetic moments, while practically independent of one another as in the paramagnetic materials, tend to align themselves in the direction opposite to that of the externally applied field. Such substances are referred to as *diamagnetic*.

We shall devote most of the rest of this chapter to the task of trying to explain, on the basis of the interaction of atomic and electronic moments with external fields and with one another, the magnitudes of the observed magnetic susceptibilities of various substances and why the susceptibility varies as it does with temperature and applied field. We shall first focus our attention on paramagnetic and diamagnetic substances, then go on to ferromagnetic materials, and finally say a few words about the commonest and most confusing magnetic systems, to wit, permanent magnets.

EXAMPLE 21.2.1

In a hydrogen atom, an electron may be regarded as rotating about a proton in a circular orbit of radius $r = 0.528 \times 10^{-10}$ meter with constant angular velocity ω. The centripetal force needed to hold the electron in its circular orbit is provided by the electrostatic attraction between the electron and the proton. Using the classical notion that the magnetic moment is given by the product of the current and the area enclosed by the path around which it flows, find the orbital magnetic moment of the hydrogen atom. What is the ratio of this magnetic moment to the orbital angular momentum of the electron?

The centripetal force $-mr\omega^2$ must be just the attractive force between electron and proton, which is $(+e)(-e)/4\pi\varepsilon_0 r^2$. Equating these two quantities, we find

$$\frac{e^2}{4\pi\varepsilon_0 r^2} = mr\omega^2$$

which we can solve for ω to obtain

$$\omega = \frac{e}{\sqrt{4\pi\varepsilon_0 mr^3}} \tag{21.2.23}$$

The system can be visualized as a circuit of radius r

around which a current flows as the negatively charged electron circles the proton. The current is the charge that flows, in this case the electronic charge, divided by the time required for one complete revolution, which is the period of revolution $2\pi/\omega$. We may, therefore, write

$$I = \frac{\omega e}{2\pi} = \frac{e^2}{2\pi\sqrt{4\pi\varepsilon_0 mr^3}} \tag{21.2.24}$$

The orbital magnetic moment of the atom is the current as expressed above times the area of the "circuit," which in this example is πr^2; this finally leads to

$$p_m = IA = \frac{e^2}{4}\sqrt{\frac{r}{\pi\varepsilon_0 m}} \tag{21.2.25}$$

Inserting the known values for the charge and mass of the electron and the value $r = 0.528 \times 10^{-10}$ meter for the radius of the orbit, we find

$$p_m = \frac{(1.602 \times 10^{-19})^2}{4}\sqrt{\frac{0.528 \times 10^{-10}}{\pi(8.853 \times 10^{-12})(9.11 \times 10^{-31})}}$$

$$= 9.27 \times 10^{-24} \text{ amp-m}^2$$

The orbital angular momentum of the electron with respect to the proton is given by

$$L = mr^2\omega \tag{21.2.26}$$

But from (21.2.24) and (21.2.25),

$$p_m = IA = \frac{\omega e}{2\pi}\pi r^2 = \frac{e\omega r^2}{2} \tag{21.2.27}$$

From these two equations, then, we see at last that

$$\frac{p_m}{L} = \frac{e}{2m} \tag{21.2.28}$$

It is apparent that the orbital magnetic moment is always directly proportional to the orbital angular momentum, the proportionality constant being simply expressed in terms of the ratio of the electronic charge to the electron mass. The proportionality of magnetic moment and angular momentum holds true for atomic systems in general, though the proportionality constant may vary from one case to another.

In addition to this orbital magnetic moment, there are also magnetic moments associated with what is called the *spin* of the particle. The spin of the electron, or the proton, is an angular momentum that is a *characteristic property* of the particle, as is its rest mass or charge. Any charged particle with such an angular momentum behaves like a rotating charge distribution, hence like a current loop, and, therefore, possesses an *inherent magnetic moment*. The characteristic spin of the electron and other elementary particles is discussed more fully in Chapter 28, and it is recommended that the reader refer briefly to the discussion given there, for we shall have more to say in

695

this chapter about electron spin. In this particular example, however, we are concerned only with the *orbital* angular momentum and the magnetic moment of the atom and shall, therefore, not consider the spin moments any further.

EXAMPLE 21.2.2

The thin, toroidal coil referred to in Examples 20.4.1 and 20.4.2 has a mean radius of 10 cm and a cross-sectional area of 3.0 cm^2 and is wound with 3142 turns of wire (50 turns/cm along the mean circumference). Let us assume that this coil is wound on the surface of a toroidal paramagnetic core of susceptibility $\chi_m = 4.59 \times 10^{-4}$. A steady current of 3.5 amperes is made to flow in the winding. Find (a) the magnetic intensity H within the coil, (b) the magnetization M, (c) the magnetic induction B within the coil, and (d) the total surface magnetization current I_m. What would the induction B be if there were no paramagnetic core? What would these answers be if a *ferromagnetic* core of relative permeability 1200, whose magnetic behavior is essentially linear, were substituted for the paramagnetic core?

The magnetic intensity $\mathbf{H}$ can be obtained from Ampère's law (21.2.16). Since the $\mathbf{H}$ field is tangential everywhere, as is $\mathbf{B}$, we can integrate around a circular path C having the mean radius of the toroid and write

$$\oint_C \mathbf{H} \cdot d\mathbf{l} = 2\pi r H = I_c = NI$$

or

$$H = \frac{NI}{2\pi r} = \frac{(3412)(3.5)}{(2\pi)(0.1)} = 19{,}000 \text{ amp/m}$$

The magnetization $\mathbf{M}$ is related to $\mathbf{H}$ by (21.2.19), from which

$$M = \chi_m H = (4.59 \times 10^{-4})(1.9 \times 10^4) = 8.721 \text{ amp/m}$$

The induction is obtained from (21.2.20), the permeability being given by

$$\mu = \mu_0(1 + \chi_m) = (4\pi \times (10^{-7})(1.000459)$$
$$= 12.572139 \times 10^{-7}$$

and the induction B by

$$B = \mu H = (12.572139 \times 10^{-7})(19{,}000)$$
$$= 0.0238871 \text{ W/m}^2$$

In these calculations, an ordinary unjustifiable number of significant figures are included simply because the permeability is hardly different from the free space value μ_0 since χ_m is so much less than unity. In this situation, the magnitude of $\mathbf{B}$ will differ very little from what it would be without the paramagnetic core. If a valid comparison is to be made between these values, therefore, they must be calculated to many more significant figures than would be justified by the accuracy with which the coil parameters and the current are specified.

According to (21.2.10), the magnetization current I_m will be

$$I_m = 2\pi r M = (2\pi)(0.1)(8.721) = 5.48 \text{ amp}$$

This is seen to be much less than the true current $I_c = NI = (3412)(3.5) = 11{,}940$ amperes due to the small value of the susceptibility. Had there been no paramagnetic core, the value of B would have been, according to Ampère's law (the magnetizing current I_m then being zero),

$$B = \frac{\mu_0 NI_{\text{tot}}}{2\pi r} = \frac{\mu_0 NI}{2\pi r} = 0.0238761 \text{ W/m}^2$$

We could have obtained this answer in an even simpler way by noting that without the paramagnetic core, the permeability of the region inside the coil is that of free space, μ_0. In this event, the relation between $\mathbf{B}$ and $\mathbf{H}$ is given according to (21.2.20) by $\mathbf{B} = \mu_0 \mathbf{H}$. Since we have already calculated $\mathbf{H}$, the value of $\mathbf{B}$ would then follow directly from this. It is evident from these results that the paramagnetic core acquires a magnetization current that, though small in comparison with the true current NI, is perceptible. This magnetization current, in turn, causes a small but detectable increase in the $\mathbf{B}$ field within the coil and, therefore, increases the magnetic flux through the coil and, in turn, the inductance of the coil. The relative permeability K_m of the core is given by (21.2.2) by

$$K_m = 1 + \chi_m = 1.000459$$

When a linear ferromagnetic core of relative permeability $K_m = 1200$ is substituted for the paramagnetic core, we find from (21.2.22) that $\chi_m = K_m - 1 = 1199$. The $\mathbf{H}$ field is determined as before, and since $I_c = NI = (3.5)(3412) = 11{,}942$ amperes as before, its value is unchanged. The magnetization is now given by $M = \chi_m H = (1199)(19000) = 2.28 \times 10^7$ amperes/meter. The magnetic permeability in this case becomes $\mu = \mu_0(1 + \chi_m) = (1200)(4\pi \times 10^{-7}) = 1.508 \times 10^{-3}$ weber/ampere-meter, while $B = \mu H = (1.508 \times 10^{-3})(19000) = 28.7$ weber/m^2. The magnetization current is now given by $I_m = 2\pi r M = (2\pi)(0.1)(2.28 \times 10^7) = 1.43 \times 10^7$ amperes, an amount of current that is truly formidable. This very large surface magnetization current causes a very large enhancement in the $\mathbf{B}$ field and in the magnetic flux, which in turn renders the inductance of the coil much larger that it would be without the ferromagnetic core. You cannot see it or feel it as it circulates around the surface of the magnetic core, but it is there all the same, and its effect on the inductance of the coil is easily observed.

It is to be noted, however, that ferromagnetic substances very often exhibit magnetic behavior that is not even approximately described by the linear law

(21.2.19) which is assumed to hold in this instance. In some cases, particularly in those in which relatively small values of **H** are involved, certain types of ferromagnetic substances, such as soft iron, behave in an essentially linear fashion, their magnetization being described fairly well by a constant susceptibility or permeability. For other ferromagnetic materials, however, particularly those from which permanent magnets are readily prepared, the picture is very different, and the assumption of linear magnetization or constant permeability can lead to results that are false and extremely misleading. We shall discuss this situation in detail in a later section.

21.3 Diamagnetic and Paramagnetic Substances: Larmor's Theory of Diamagnetism

In any given substance, there are a number of magnetic effects that are important in determining the overall magnetic susceptibility of the material. Some of these individual effects may in themselves lead to diamagnetism, while others may give paramagnetic contributions. The question of the total susceptibility of the material and whether the net susceptibility is paramagnetic or diamagnetic depends upon the relative importance of these separate and independent effects in the substance at hand. It is, therefore, reasonable to consider the most important of these effects under the same general heading.

In substances whose atoms possess permanent net magnetic dipole moments, it is usually these moments that make the dominant contribution to the magnetic susceptibility. Since these atomic moments tend to align themselves in the direction of the applied field, the net resulting magnetization is in that direction also, and, therefore, such materials are paramagnetic. In certain other materials, however, it may happen that the circulating currents due to individual atomic electrons give rise to *zero* total orbital magnetic moment. This occurs because, though individual electrons may have nonvanishing values of angular momentum, there are *pairs* of electrons whose angular momenta are equal and opposite, leading to zero total angular momentum and, therefore, according to (21.2.28) to zero net magnetic moment. It may also happen that the net spin magnetic moments may be paired and add to zero in the same way. Even though the atoms of such substances have zero magnetic moment in the absence of external fields, when a magnetic field is applied, the changing magnetic flux it produces may interact with the circulating current loops associated with atomic electrons in such a way that an *induced* magnetic moment may arise. According to Lenz's law, the direction of such an induced

magnetic moment must be *opposite* that of the applied magnetic field, and, therefore, this effect leads to a weak *diamagnetic* susceptibility. This diamagnetic effect is present, of course, even in materials without net atomic magnetic moments, but in these cases the effect is usually masked by the stronger paramagnetism due to the permanent atomic moments. The weak induced diamagnetism associated with circulating electron currents is frequently referred to as *Larmor diamagnetism*, after Sir Joseph Larmor, the British scientist who first discussed the effect.

In both diamagnetic and paramagnetic substances, it is important to note that although the alignment of atomic moments is influenced by the externally applied field, *the magnetic interaction of the individual atomic moments with one another is negligible*. This is true because the magnetic fields experienced by individual atoms due to the magnetic moments of other atoms are quite small and also because the magnetic field energy associated with the atomic moments is much less than the thermal energy of the atoms.

The way in which Larmor diamagnetism arises can be seen by referring to Fig. 21.6. This illustration shows two electrons that are constrained to move in opposite directions around a circular orbit of fixed radius r_0 about a positively charged nucleus bearing a total charge $+Ze$, where Z is an integer. When the external magnetic field is zero, the electrostatic force $-e(+Ze)/4\pi\varepsilon_0 r_0^2$ on each electron is the total centripetal force and may be equated to the centripetal acceleration $-mr_0\omega_0^2$. Therefore, for either electron we have

$$mr_0\omega_0^2 = \frac{Ze^2}{4\pi\varepsilon_0 r_0^2}$$

or

$$\omega_0^2 = \frac{Ze^2}{4\pi\varepsilon_0 mr_0^3} \tag{21.3.1}$$

In the absence of an external magnetic field, then, both electrons rotate with the same angular velocity ω_0, though in opposite directions. Their individual magnetic moments are given by the product of current and the area of the orbit, from which, as in Example 21.2.1,

$$p_{m1}(0) = \frac{(-e)\omega_0}{2\pi}(\pi r_0^2) = -\tfrac{1}{2}e\omega_0 r_0^2 \tag{21.3.2}$$

while

$$p_{m2}(0) = \frac{(-e)(-\omega_0)}{2\pi}(\pi r_0^2) = +\tfrac{1}{2}e\omega_0 r_0^2 \tag{21.3.3}$$

From this it is clear that for $B = 0$, the magnetic moments of the electrons cancel, leaving no net magnetic moment. It is assumed also that the spin mag-

697

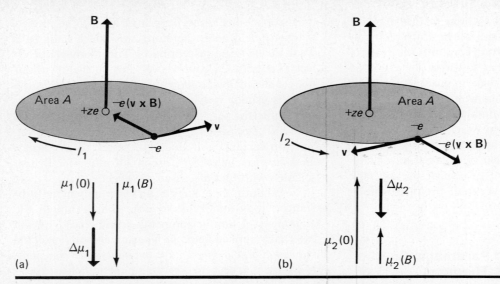

FIGURE 21.6. Magnetic forces on atomic electrons circulating (a) counterclockwise and (b) clockwise.

netic moments are similarly opposed and add to zero in the same way.

When the magnetic field **B** is present, there are, in addition to electrostatic attraction, magnetic forces $q(\mathbf{v} \times \mathbf{B})$ acting on the electrons in the directions shown in Fig. 21.6. For the first electron, shown at (a), this magnetic force increases the centripetal force and hence the centripetal acceleration $-mr\omega^2$. The angular velocity of this electron, therefore, becomes larger than before, changing the magnetic moment by an amount $\Delta\mathbf{p}_{m1}$. But for the second electron, shown at (b), the magnetic force decreases the centripetal force and, therefore, makes the angular velocity of this electron less than before. Its magnetic moment changes by an amount $\Delta\mathbf{p}_{m2}$, as shown in Fig. 21.6b. Since $\Delta\mathbf{p}_{m1}$ and $\Delta\mathbf{p}_{m2}$ are both opposite in direction to **B**, the material is diamagnetic.

We can calculate the magnetic moments, and ultimately the susceptibility, by writing the net radial force on each electron and setting it equal to the centripetal acceleration $-mr\omega^2$. For the first electron, in Fig. 21.6a, this gives

$$F_1 = -\frac{Ze^2}{4\pi\varepsilon_0 r_0^2} - evB = -mr_0\omega^2 \tag{21.3.4}$$

But, from (21.3.1), $Ze^2/(4\pi\varepsilon_0 r_0^2) = mr_0\omega_0^2$, where ω_0 is the initial angular velocity in the absence of the field **B**. Furthermore, since $v = r_0\omega$, we can write (21.3.4) as

$$F_1 = -mr_0\omega_0^2 - er_0\omega B = -mr_0\omega^2 \tag{21.3.5}$$

The corresponding equation for the second electron is the same, except that the term $er\omega_0 B$, representing the magnetic force $-e(\mathbf{v} \times \mathbf{B})$, is opposite in sign,

from which

$$F_2 = -mr_0\omega_0^2 + er_0\omega B = -mr_0\omega^2 \tag{21.3.6}$$

If we now divide both sides of (21.3.5) and (21.3.6) by mr_0^2, we can write these equations as

$$\omega^2 - 2\omega\omega_L - \omega_0^2 = 0 \tag{21.3.7}$$

and

$$\omega^2 + 2\omega\omega_L - \omega_0^2 = 0 \tag{21.3.8}$$

where the so-called *Larmor frequency* ω_L is defined by

$$\boxed{\omega_L = \frac{eB}{2m}} \tag{21.3.9}$$

If we add ω_L^2 to both sides of Eq. (21.3.7), we may write it in the form

$$\omega^2 - 2\omega\omega_L + \omega_L^2 = \omega_0^2 + \omega_L^2$$

or

$$(\omega - \omega_L)^2 = \omega_0^2 + \omega_L^2 \tag{21.3.10}$$

Now, from (21.3.1) and (21.3.9), it is evident that we can write the ratio of ω_L to ω_0 as

$$\frac{\omega_L}{\omega_0} = B\sqrt{\frac{\pi\varepsilon_0 r_0^3}{mZ}} \tag{21.3.11}$$

Let us see how large this ratio might possibly be. The largest value of B that can be realized experimentally is about 100 webers/m². In most solid substances, a mole occupies a volume of about 10 cm³, or 10^{-5} m². There are, therefore, typically about 6×10^{22} atoms per cm³, or 6×10^{28} atoms per m³, corresponding to a volume of about 1.7×10^{-29} m³ per atom. Taking the cube root of this, we get 2.6×10^{-10} meter, which must correspond to the atomic diameter $2r_0$. It is, therefore, reasonable to set r_0 equal to 1.3×10^{-10} meter. The ratio ω_L/ω_0 will be greatest, also, when Z is smallest, so we shall take $Z = 1$. Substituting these values into (21.3.11) along with $\varepsilon_0 = 8.85 \times 10^{-12}$ farad/meter and $m = 9.11 \times 10^{-31}$ kg, we find

$$\frac{\omega_L}{\omega_0} = (100)\left(\frac{(\pi)(8.85 \times 10^{-12})(1.3 \times 10^{-10})^3}{(9.11 \times 10^{-31})(1)}\right)^{1/2}$$

$$= 8.2 \times 10^{-4}$$

Clearly, then, $\omega_L \ll \omega_0$ under all experimentally realizable conditions. Therefore, we may safely *neglect* ω_L^2 on the right side of (21.3.10) in comparison with ω_0^2 and, taking the square root of both sides, write

$$\omega = \omega_0 + \omega_L = \omega_1 \qquad (21.3.12)$$

For the second electron, we may add ω_L^2 to both sides of (21.3.8):

$$(\omega + \omega_L)^2 = \omega_0^2 + \omega_L^2 \qquad (21.3.13)$$

Once again, we may neglect ω_L^2 in comparison with ω_0^2 on the right and, taking the square root of both sides, obtain for the angular frequency ω_2 of the second electron

$$\omega = \omega_0 - \omega_L = \omega_2 \qquad (21.3.14)$$

The magnetic moment due to the first electron is now

$$p_{m1} = I_1 A = \frac{-e\omega_1}{2\pi}\pi r_0^2 = -\frac{er_0^2}{2}(\omega_0 + \omega_L) \qquad (21.3.15)$$

while for the second,

$$p_{m2} = I_2 A = \frac{(-e)(-\omega_2)}{2\pi}\pi r_0^2 = \frac{er_0^2}{2}(\omega_0 - \omega_L) \qquad (21.3.16)$$

The total magnetic moment per atom is then

$$p_m = p_{m1} + p_{m2} = -er_0^2\omega_L$$

$$= -er_0^2\frac{eB}{2m} = -\frac{e^2r_0^2B}{2m} \qquad (21.3.17)$$

The magnetic moment per unit volume is the magnetic moment per atom, times the number of atoms per unit volume, n. By definition, however, this also represents the magnetization M. Also, since the susceptibility is the magnetization per unit magnetic intensity and since in the free space in which the electrons move about the nucleus $H = B/\mu_0$, we may finally write

$$\chi_m = \frac{M}{H} = \frac{\mu_0}{B}np_m$$

or

$$\chi_m = -\frac{\mu_0 ne^2 r_0^2}{2m} \qquad (21.3.18)$$

For a typical "atom" constructed on the admittedly crude model adopted here, we have already calculated that the radius r_0 is about 1.3×10^{-10} meter and that the number of atoms per unit volume, n, is about 6×10^{28} per m³. Inserting these along with the known values for the other quantities into (21.3.18), we find

$$\chi_m = -\frac{\left[\begin{array}{c}(4\pi \times 10^{-7})(6 \times 10^{28})\\ \times (1.6 \times 10^{-19})^2(1.3 \times 10^{-10})^2\end{array}\right]}{(2)(9.11 \times 10^{-31})}$$

or

$$\chi_m = -1.8 \times 10^{-5}$$

This is a weak but measurable diamagnetic susceptibility. It will be noted that it agrees, in order of magnitude at least, with the values of susceptibility listed in Table 21.1 for typical diamagnetic substances. Also, it is evident from (21.3.18) that the Larmor diamagnetic susceptibility should be essentially independent of temperature; this is also in general agreement with experiment.

More than this we should probably not expect. The model of the "atom" that we have adopted is a crude one, indeed. Actually, the electrons do not execute circular orbits, nor are they constrained to remain at a fixed distance from the nucleus. In many-electron atoms, there are electrostatic forces *between electrons* as well as between individual electrons and the nucleus, and these we have neglected altogether. Also, of course, we have used classical mechanics rather than quantum mechanics to describe the behavior of our atomic system, a procedure which, as we know, can sometimes lead to incorrect results. Finally, in any sample containing many atoms, we cannot expect the electrons in all of them to be moving in a plane perpendicular to the applied **B** field, as we assumed above. All the same, the Larmor theory successfully incorporates the physical mechanisms from which actual atomic diamagnetism arises.

21.4 Classical Paramagnetism of Substances Having Permanent Net Atomic Moments

In the case of substances whose atoms or molecules have *permanent* magnetic moments, these moments experience magnetic forces that tend to align them parallel to externally applied magnetic fields. Such materials are, therefore, paramagnetic. At the same

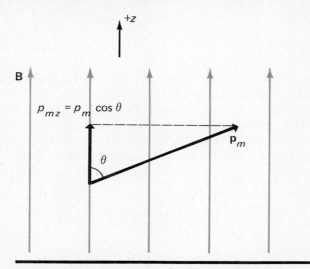

FIGURE 21.7. Magnetic moment contribution for a partially aligned magnetic dipole.

time, the random thermal energy of the atoms of the substance tends to destroy this alignment and produce a completely disordered arrangement of atomic moments that adds up to zero total magnetic moment. In gases, this thermal energy is mostly the kinetic energy of the molecules, and the disordering of atomic moments occurs in intermolecular collisions. In solid substances, the thermal energy manifests itself in the form of vibrational energy of the crystal lattice, and it is this vibrational energy that is responsible for the disordering of individual moments. In liquids, both effects are present to a significant extent. In all instances, however, there is a competition between the aligning influence of the externally applied field and the disordering effect associated with the thermal internal energy of the substance, as illustrated in Fig. 21.1. In general, the end result of this competition produces a *partial* alignment of moments, the extent of which depends upon the strength of the external field and also upon the temperature, as shown in Fig. 21.1c.[2]

If there is no external magnetic field, then all the atomic dipoles will be randomly oriented; the *average* vertical z-component of their magnetic moments will be zero, since positive and negative values of that component, or any other, occur with equal probability. If an external field is applied along a given direction, such as the z-axis as shown in Fig. 21.7, there will be more atomic moments having positive

[2] The results we shall obtain in this section pertaining to the magnetic susceptibility of substances whose atoms have permanent atomic magnetic moments are equally applicable to the discussion of *electrical* susceptibility of dielectrics whose atoms have permanent *electric* dipole moments. The calculations are made in exactly the same way, substituting the electric dipole moment of the atom for the magnetic moment. Equation (17.6.9) of section 17.6 is the exact analog of Eq. (21.4.20) in this section.

z-components than negative ones. There will then be some *positive* average z-component of the atomic moment. From Fig. 21.7, it is clear that, since

$$p_{mz} = p_m \cos \theta \tag{21.4.1}$$

this means that the average value of the z-component of all the atomic moments in the sample, $\bar{p}_{mz}$, is related to the average angle θ an atomic moment makes with the magnetic field by

$$\bar{p}_{mz} = p_m \overline{\cos \theta} \tag{21.4.2}$$

Since there is no component of magnetic field along the x- and y-directions, it is evident that the average value of the atomic magnetic moments along each of those directions must be zero. The total magnetic moment of the sample per unit volume, which is by definition the magnetization **M**, can then be obtained simply by multiplying the average atomic moment (21.4.2) by the number of atoms per unit volume to give

$$M = n\bar{p}_{mz} = np_m \overline{\cos \theta} \tag{21.4.3}$$

The problem of finding the magnetization of the sample is, therefore, tantamount to finding the average value of the cosine of the angle between an atomic moment and the external field.

In the presence of a magnetic field, we know from Eqs. (19.2.4) and (16.6.30) that the potential energy of a magnetic dipole making an angle θ with the field is

$$U_m = -p_m B \cos \theta \tag{21.4.4}$$

According to the Maxwell–Boltzman energy distribution law, the probability with which an atomic moment will be found oriented within the shaded area dA shown in Fig. 21.8 will be

$$Ce^{-U_m/kT} \, dA \tag{21.4.5}$$

where C is a constant. It should be carefully noted that the probability of orientation within the area dA is proportional to the *size* of that area, just as the probability of molecules having given ranges of velocity, as expressed by (13.6.7) or (13.6.10), is proportional to the size of the element in velocity space that defines the velocity range in question.

If we call the number of atomic dipoles oriented within the shaded area dA in Fig. 21.8 $N(\theta, \phi) \, dA$, then, from (21.4.5) and (21.4.4),

$$N(\theta, \phi) \, dA = Ce^{p_m B \cos \theta/kT} \, dA \tag{21.4.6}$$

The average value of $\cos \theta$ over all the atomic dipoles is then obtained by multiplying the number of dipoles within the area dA by $\cos \theta$, summing that contribution over all possible orientations by integration, and dividing by the total number of dipoles N_0, which from (21.4.6) is given by

$$N_0 = \int N(\theta, \phi)\, dA \qquad (21.4.7)$$

The calculation is in all respects equivalent to that carried out in Example 13.6.1 in connection with determining the average velocity of a molecule in an ideal gas. Since the area dA, from Fig. 21.8, can be expressed in terms of the angular coordinates θ and ϕ and the magnitude p_m by

$$dA = p_m{}^2 \sin\theta\, d\theta\, d\phi \qquad (21.4.8)$$

we may write

$$\overline{\cos\theta} = \frac{\int N(\theta, \phi)\cos\theta\, dA}{\int N(\theta, \phi)\, dA}$$

$$= \frac{\int_{\theta=0}^{\theta=\pi}\int_{\phi=0}^{\phi=2\pi} e^{p_m B\cos\theta/kT}\cos\theta\sin\theta\, d\phi\, d\theta}{\int_{\theta=0}^{\theta=\pi}\int_{\phi=0}^{\phi=2\pi} e^{p_m B\cos\theta/kT}\sin\theta\, d\phi\, d\theta} \qquad (21.4.9)$$

In this expression, the constant factors C and $p_m{}^2$ have been canceled in the numerator and denominator of the right-hand expression. In this expression, also, it is apparent that the integrands both above and below are completely *independent* of the azimuthal angle ϕ. Since this is true, those factors are *constants* with regard to integration over ϕ and can, in fact, be moved outside the integral sign having to do with that integration, the result being

$$\overline{\cos\theta} = \frac{\int_0^\pi e^{p_m B\cos\theta/kT}\cos\theta\sin\theta\, d\theta \int_0^{2\pi} d\phi}{\int_0^\pi e^{p_m B\cos\theta/kT}\sin\theta\, d\theta \int_0^{2\pi} d\phi}$$

$$= \frac{\int_0^\pi e^{p_m B\cos\theta/kT}\cos\theta\sin\theta\, d\theta}{\int_0^\pi e^{p_m B\cos\theta/kT}\sin\theta\, d\theta} \qquad (21.4.10)$$

The remaining integration over θ can be done quite easily by setting

$$\alpha = p_m B/kT \qquad (21.4.11)$$

and

$$w = \cos\theta \qquad dw = -\sin\theta\, d\theta \qquad (21.4.12)$$

Substituting this into (21.4.10) and noting, in the limits of the integrals, that $w = -1$ when $\theta = \pi$ and $w = +1$ when $\theta = 0$,

$$\overline{\cos\theta} = \frac{\int_{-1}^1 w e^{\alpha w}\, dw}{\int_{-1}^1 e^{\alpha w}\, dw} = \frac{\left[\dfrac{e^{\alpha w}}{\alpha^2}(\alpha w - 1)\right]_{-1}^1}{\left[\dfrac{e^{\alpha w}}{\alpha}\right]_{-1}^1} \qquad (21.4.13)$$

$$= \frac{1}{\alpha}\left(\frac{e^{\alpha}(\alpha - 1) + e^{-\alpha}(\alpha + 1)}{e^{\alpha} - e^{-\alpha}}\right)$$

$$= \frac{1}{\alpha}\left(\frac{\alpha(e^{\alpha} + e^{-\alpha}) - (e^{\alpha} - e^{-\alpha})}{e^{\alpha} - e^{-\alpha}}\right)$$

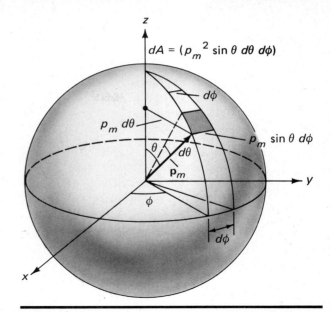

FIGURE 21.8. Calculation of the probability that a randomly oriented dipole of moment $\mathbf{p}_m$ will be within the shaded area dA illustrated above depends simply on finding the ratio of area dA to the total area of the sphere.

or

$$\overline{\cos\theta} = \frac{e^{\alpha} + e^{-\alpha}}{e^{\alpha} - e^{-\alpha}} - \frac{1}{\alpha} \qquad (21.4.14)$$

The magnetization M is now obtained from (21.4.3):

$$M = np_m\left(\frac{e^{\alpha} + e^{-\alpha}}{e^{\alpha} - e^{-\alpha}} - \frac{1}{\alpha}\right) = np_m f(\alpha) \qquad (21.4.15)$$

The magnetic susceptibility is determined from (21.2.19) as the ratio of the magnetization to the magnetic intensity H. But the permeability experienced by the individual magnetic moments of the atoms is essentially that of free space, since they are, after all, magnetically independent of one another. Therefore, we may write

$$\chi_m = \frac{M}{H} = \frac{\mu_0 M}{B} = \frac{\mu_0 np_m f(\alpha)}{B} \qquad (21.4.16)$$

where $f(\alpha)$ is defined by (21.4.15).

The function $f(\alpha)$ is shown plotted against α $(= p_m B/kT)$ in Fig. 21.9. For small values of α, corresponding to relatively small magnetic fields or relatively high temperatures, the function is a *linear* function of α having slope $\frac{1}{3}$. This is not easily seen from (21.4.15) but is not difficult to prove starting with the original integrals shown in (21.4.13). For $\alpha \ll 1$, we may write in those integrals, using the power series expansions for e^{α},

$$e^{\alpha} = 1 + \alpha + \frac{\alpha^2}{2!} + \frac{\alpha^3}{3!} + \cdots \qquad (21.4.17)$$

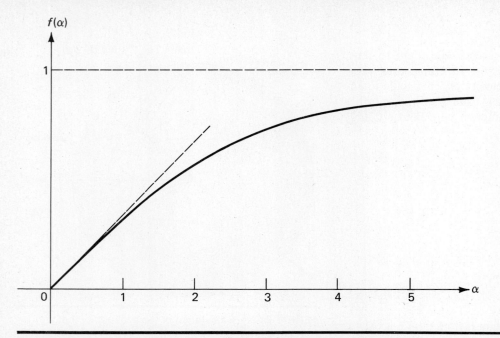

FIGURE 21.9. Function $f(\alpha)$ defined in Eq. (21.4.15) plotted as a function of α.

But for values of α that are small compared to unity, the terms in α^2, α^3, etc., are negligible compared with the first two. Therefore, for $\alpha \ll 1$, we may write

$$e^\alpha \cong 1 + \alpha$$

Substituting this into the integrals of (21.4.13), we find

$$f(\alpha) = \frac{\int_{-1}^{1} (w + \alpha w^2)\, dw}{\int_{-1}^{1} (1 + \alpha w)\, dw} = \frac{\left[\dfrac{w^2}{2} + \dfrac{\alpha w^3}{3}\right]_{-1}^{1}}{\left[w + \dfrac{\alpha w^2}{2}\right]_{-1}^{1}} = \frac{\alpha}{3}$$

$$(21.4.18)$$

which is the result referred to above. But from (21.4.11), $\alpha = p_m B / kT$. Therefore, according to (21.4.15) under conditions for which the value of α is *small* in comparison to unity, the magnetization will be

$$M = \frac{n p_m^2 B}{3kT} \qquad (21.4.19)$$

while from (21.4.16) the susceptibility is

$$\boxed{\chi_m = \frac{M}{H} = \frac{\mu_0 n p_m^2}{3kT}} \qquad (21.4.20)$$

Under these circumstances, the magnetic susceptibility is constant in the sense that it does not depend on the magnetic field. It is evident, however, that it does depend inversely on the temperature. This temperature dependence occurs because as the temperature is increased, the random internal thermal energy of the atoms or molecules also increases, and this tends to decrease the degree of alignment of atomic moments for any fixed value of external field.

For very large magnetic fields, or for very low temperatures, the value of α may exceed unity, and under extreme conditions may even become considerably larger than unity. The magnetization as expressed by Eq. (21.4.15) or Fig. 21.9 now reaches an asymptotic *saturation value* of $n p_m$, since under these conditions $f(\alpha)$ approaches unity. In this limit, we are approaching a condition of complete alignment of atomic dipoles along the direction of the external field. We should then expect the magnetic moment per unit volume to be equal to the product of the atomic moment p_m and the number of atoms per unit volume, which is just what is predicted by our calculations. At room temperature, the magnetic fields required to make α much greater than unity and thus saturate the magnetization are so large that they cannot be achieved experimentally; but at lower temperatures, proportionally lower fields are needed. This is clearly illustrated from the definition of α in (21.4.11) and also by Fig. 21.10.

In this calculation, once again, we assume that the atoms of the substance obey the laws of classical mechanics in all respects. Classically, according to (21.4.4), the potential energy of the atom in the magnetic field can have any value from $U_m = -p_m B$ corresponding to $\theta = 0$ to $U_m = +p_m B$ corresponding to $\theta = 180°$. But we know, of course, that sometimes the laws of classical mechanics do not correctly describe the behavior of atomic systems and that, in such instances, we must resort to quantum mechanics.

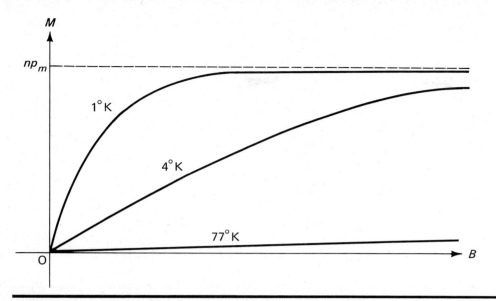

FIGURE 21.10. Magnetization plotted as a function of field, according to Eq. (21.4.15), for several temperature values.

This is one such situation. Quantum theory predicts that there are a number of *discrete* allowed values of energy lying between $-p_m B$ and $+p_m B$, corresponding to certain discrete values of $\cos\theta$ in (21.4.4). Therefore, the laws of quantum mechanics do not permit the atomic moment to have all possible orientations with respect to the magnetic field direction but only a discrete series of allowed orientations corresponding to certain discrete allowed values of the angle θ. This introduces certain added complications into the calculation of the susceptibility that have the effect of somewhat modifying the form of the function $f(\alpha)$. These modifications, however, are minor ones and do not significantly affect the general shape of the curve shown in Fig. 21.9 nor any of the physical conclusions we drew from the classical calculation, although, naturally, the quantum mechanical result is in better numerical agreement with experimental data than the classical one calculated above.

If in Eq. (21.4.20) we substitute the typical values $n = 6 \times 10^{28}$ m^{-3} and $p_m = 9.27 \times 10^{-24}$ ampere-m^2, we find that for room temperature (300°K), we obtain $\chi_m = 5.2 \times 10^{-4}$. This agrees fairly well with the values of χ_m listed in Table 21.1 for paramagnetic salts such as MnF$_2$ or CoCl$_2$ but is much higher than the susceptibilities observed experimentally in paramagnetic *metals* such as aluminum or magnesium. In substances such as these, the paramagnetism is due entirely to the conduction electrons, the metallic ions on the lattice sites having no net magnetic moments because all their orbital moments and spin moments are paired and add to zero. The paramagnetism exhibited by the conduction electrons is entirely due

to their spin magnetic moments since they have no orbital motion. But many of the electrons cannot align magnetically with an applied field because this would require them to occupy a quantum energy state that is already filled. This they cannot do, since the laws of quantum mechanics allow only one electron to occupy a given quantum state. Their paramagnetic susceptibility is strongly reduced by this effect, resulting in the comparatively low susceptibilities that are observed in most paramagnetic metallic substances.

EXAMPLE 21.4.1

What value of magnetic induction is needed to "saturate" the magnetic moment of a typical paramagnetic substance to the extent that the magnetization attains 80 percent of the asymptotic value np_m at a temperature of 1.0°K? What value of field is required to produce the same effect at 77°K, the boiling point of liquid nitrogen? At room temperature, 300°K? Assume that $p_m = 9.27 \times 10^{-24}$ ampere-m^2.

The magnetization is proportional to $f(\alpha)$. From Fig. 21.9, it is apparent that α must be quite large compared to unity for $f(\alpha)$ to approach the asymptotic value of 1. In this event, however, e^{α} is very large and $e^{-\alpha}$ is very small, so that the term containing these exponential quantities in (24.1.15) approaches unity very closely, and we may write

$$f(\alpha) \cong 1 - \frac{1}{\alpha} \qquad (\alpha \gg 1) \tag{21.4.21}$$

If the magnetization is to be 80 percent of the saturation value np_m, $f(\alpha)$ must have the value 0.80, from

703

which, according to (21.4.21), $\alpha = 5$. But from (21.4.11), this requires that

$$\frac{p_m B}{kT} = 5$$

$$B = 5 \frac{kT}{p_m} \qquad (21.4.22)$$

For $T = 1°K$, this gives

$$B = \frac{(5)(1.38 \times 10^{-23})(1.0)}{(9.27 \times 10^{-24})} = 7.44 \text{ W/m}^2$$

This is a very large magnetic field but one that could be produced in the laboratory without extreme difficulty. The corresponding field at $77°K$ would be 77 times larger, according to (21.4.22), and would thus amount to 573 webers/m². At $300°K$, by the same token, we would require a field of 2230 webers/m² to accomplish the same degree of saturation. Fields that are as large as these are far beyond what can be attained experimentally. Indeed, at temperatures of $77°K$ and above, it is quite impossible to observe anything but the linear portion of the magnetization curve corresponding to the condition $\alpha \ll 1$.

21.5 Ferromagnetic Materials

In *ferromagnetic* substances such as iron, nickel, and cobalt, there are interactions between the magnetic moments of neighboring atoms that are so strong that the atomic moments can align themselves with little or no assistance from externally applied fields. Naturally, if there are external fields present, the atomic moments do tend to line up in the direction of the external field; in fact, they align themselves almost completely, even for external fields that are not very strong. Ferromagnetic materials, therefore, exhibit very large magnetic permeabilities and can also, generally speaking, be permanently magnetized. Since the atomic moments are nearly completely aligned even by relatively weak external fields, the saturation value of the magnetization is attained readily at small values of magnetic intensity, and the magnetization is *not* a linear function of the applied **H** field. As a consequence, the magnetic susceptibility of ferromagnetic substances is not constant but varies with the strength of the external **H** field.

It would be tempting to jump to the conclusion that the forces that produce the magnetic alignment in ferromagnetic materials are the magnetic dipole forces that the individual atomic magnets exert on one another. But this is not true. These magnetic forces are not significantly stronger in ferromagnetic materials than they are in paramagnetic substances, where, as we have already seen, they are far too weak

to resist the randomizing effects that stem from thermally excited atomic or molecular motions. The atomic moments in ferromagnetic substances, as in substances that are paramagnetic, are too weak to align themselves!

In any solid crystalline substance, repulsive forces arise between neighboring atoms as the regions around the nuclei occupied by the atomic electrons begin to overlap. These forces are, in fact, what keeps the atoms apart and establishes a condition of equilibrium in which the interatomic spacing of the crystal is what it is. We feel these forces when we try to compress a solid crystalline substance; they enable the material to resist compression rather than collapsing when compressive stress is applied. The magnitude and direction of such forces is determined by the characteristics of the swarm of electrons, or *electron cloud*, that surrounds the nucleus of each atom. These characteristics, in turn, depend upon the number of electrons and upon how they interact through the laws of quantum mechanics. The electrostatic attraction of the electrons for the atomic nucleus and the electrostatic repulsion between one electron and another contribute to the resulting configuration of the electron swarm, but not in the simple manner that we might suppose from Coulomb's law and the laws of classical mechanics. The laws of quantum mechanics modify the manner in which these particles interact in certain ways, one of which is to forbid two electrons in any atom to execute the same form of orbital motion. This prohibition plays an important role in the quantum mechanics of electrons in atomic systems and in crystals and is referred to as the *Pauli exclusion principle*. We shall study this effect later and in more detail in Chapter 28. In any event, we may observe that the forces between atoms in crystals depend upon the atomic species involved and that even though they are not electrostatic forces in the usual sense of the term, they arise largely from electrostatic interactions between elementary particles. In particular, we find that the character of these forces is determined by the configuration of the electron cloud surrounding the nucleus of each atom.

But it is found that the configuration of the electron cloud, and hence the forces that act between neighboring atoms, *is critically influenced by both the orbital and spin angular momenta of the electrons in the atoms*. At the same time, we know from the results of Example 21.2.1 that the orbital and spin magnetic moments of an atom are always *proportional* to the corresponding angular momenta. This means that the magnetic moment of one atom can influence that of its neighbors—indirectly, to be sure—through these electronic interactions that occur when there is overlap between the electron swarms of neighboring atoms. It is said that there is a *coupling* between the atomic moments. Since this coupling is set up basically by electrostatic rather than magnetic interactions, it may be thousands of times stronger than the direct magnetic forces between neighboring atoms. On the other hand, it may be quite weak, or it may even act in such a way as to tend to align neighboring moments *antiparallel*[3] rather than parallel; it all depends upon the detailed configuration of the electron cloud about each atom.

There are a few elements, notably iron, cobalt, nickel, gadolinium, dysprosium, and manganese, in which the form of the electron distribution surrounding the atoms leads to extremely strong coupling between neighboring atoms which tends to align their magnetic moments parallel with one another. These elements are ferromagnetic substances and, under suitable conditions, are found to exhibit all the effects outlined previously that are characteristic of such materials.[4]

Even though the coupling between neighboring atoms in ferromagnetic substances may be very strong, the disordering effect of the crystal's thermal energy is also important and, at a sufficiently high temperature, can become so strong as to overcome the effect of ferromagnetic coupling altogether. Above this temperature, which varies from substance to substance depending upon the strength of the ferromagnetic coupling between atoms, ferromagnetic materials abruptly lose their spontaneous magnetization and exhibit magnetic behavior that is essentially paramagnetic in character. This transition temperature is referred to as the *Curie temperature* and for pure iron occurs at 1043°C. The Curie temperatures for some other ferromagnetic substances are listed in Table 21.2.

Ferromagnetic substances exhibit a diverse range of magnetic properties. These properties depend on the composition of the substance, on whether it is a single crystal or a polycrystalline substance (in the latter case upon the size and shape of the tiny crystallites of which it is composed), and on the previous thermal and magnetic treatment to which it has been subjected. Despite the wide variations in magnetic properties, there are two rather important general classes of ferromagnetic materials that can be identified. These are *hard* ferromagnetic substances, such as certain kinds of steel, that can be made to retain their magnetization even when the magnetizing current is switched off; and *soft* ferromagnetic mate-

TABLE 21.2. Curie Temperatures of Ferromagnetic Substances

Substance	Curie temperature, °C
Iron	770
Cobalt	1122
Nickel	358
Gadolinium	16
Dysprosium	−188

rials, such as soft iron, which, although easily magnetized, lose practically all of their magnetization when the magnetizing current is removed. Hard ferromagnetic substances can easily be made into strong permanent magnets, while for soft ferromagnets this is practically impossible. The terms "hard" and "soft," though in common use, are somewhat misleading, since there are some substances that exhibit the magnetic properties of hard ferromagnets but are rather soft and easily deformed, and vice versa. We shall, nevertheless, use these terms to categorize these two classes of magnetic materials, even though they sometimes do not correspond accurately to the mechanical properties of the substance at hand.

Let us now consider what happens when we magnetize a hard ferromagnetic substance by placing it in a magnetic field created by an electric current in the windings of a *toroidal coil*. The dependence of the magnetization M of the toroidal sample upon the magnetic intensity H is illustrated in Fig. 21.11. It is assumed here that the sample is unmagnetized initially and that the initial magnetic intensity is also zero. At the beginning, therefore, we are at O, where $H = 0$ and $M = 0$. Now let us suppose that the magnetic intensity is increased by increasing the current that flows through the coil. The atomic moments align with the field, causing an increase of M that is nearly linear with H at first but which tends to saturate as total magnetic alignment is approached. The magnetization of the sample proceeds, therefore, along the line OA. If the magnetization were carried further, of course, a *saturation magnetization* $M = np_m$ would be approached corresponding to the alignment of all atomic dipoles in the sample in the direction of **H**. Let us assume, however, that we stop at point A and then gradually reduce the **H** field again. When we do this, however, the magnetization does not retrace the curve OA, but instead follows path AB. When **H** is reduced to zero, point B is attained. At this point, even though the magnetic intensity is zero, a large proportion of the atomic dipoles retain their former alignment and there is, therefore, a substantial magnetization M_r, which is frequently called the *remanent magnetization* of the sample. We now have a *permanent magnet* that has a magnetization and produces

---

[3] Substances in which this type of coupling is present are referred to as *antiferromagnetic* materials.

[4] In the case of manganese, the pure metallic substance is not ferromagnetic, but ferromagnetic coupling between manganese atoms can be achieved by slightly altering the nearest-neighbor interatomic spacing. This can be accomplished by alloying manganese with other substances that may themselves be completely nonferromagnetic.

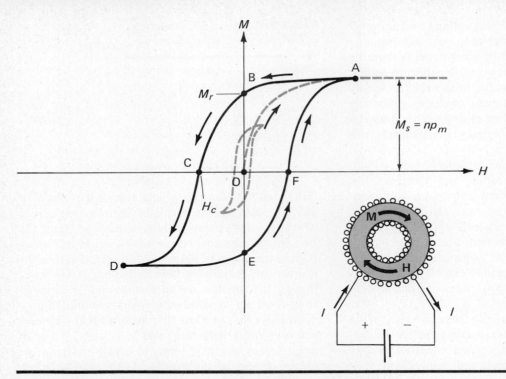

FIGURE 21.11. Magnetization as a function of magnetic intensity for a toroidal ferromagnetic sample, exhibiting the phenomenon of magnetic hysteresis.

its own **B** field even though there is no magnetizing current and, at least in the case of a toroidal sample, no **H** field at all.

If we now reverse the direction of the current, we reverse the direction of the magnetic intensity **H**. This tends to destroy the magnetic alignment of the sample; and at point C, when a certain negative value of H, referred to as the coercive field H_c, is reached, the magnetization is completely removed and the sample is once again unmagnetized. For larger negative values of H, the magnetization becomes negative and proceeds along the curve CD, reaching D when the **H** field is equal to and opposite from what it was at A. If, now, the value of H is once again allowed to increase, the path DEFA is traced out, until finally, when the **H** field returns to its former maximum value, the point A is regained. By retracing these steps, the sample can be cycled around the open loop shown in Fig. 21.11.

These processes clearly illustrate the fact that the magnetization depends not only upon the magnetic intensity *but also upon the previous magnetic history of the sample.* The fact that the path AB is followed rather than the initial path AO is a direct consequence of the fact that we are beginning with a magnetically aligned sample rather than one that is initially unmagnetized. The question of what value of H makes the magnetization zero depends likewise upon whether we are going from B toward C or from E toward F. This effect is called *magnetic hysteresis*

and is exhibited to some degree by all ferromagnetic substances, though it is most evident in hard ferromagnetic materials. The closed loop shown in Fig. 21.11 is referred to as a *hysteresis loop.* Its size and shape depends upon the composition and crystalline form of the sample and also upon the maximum values selected for the magnetic intensity at A and D. A smaller hysteresis loop that results when the sample is subjected to smaller maximum magnetic intensities is shown by dotted lines in Fig. 21.11.

In Fig. 21.11, the effect of hysteresis is illustrated by plotting the magnetization as a function of H, because it is quite easy to visualize what happens to the magnetization as the **H** field is increased or decreased. It is frequently of interest, however, to depict the effect of hysteresis in a plot of the magnetic induction B versus H. Such a plot looks very much like the original curve relating M and H, simply because, as we saw in Example 21.2.2, when a very easily magnetizable substance ($\mu \gg 1$) is involved the atomic magnetization current I_m is much larger than the true current I and the magnetization M, therefore, is much larger than the magnetic intensity H. From Eq. (21.2.18), therefore, in any such situation, $B \cong \mu_0 M$, and a plot of M versus H differs from a graph of B versus H only by the constant scale factor μ_0. A plot of B against H is shown in Fig. 21.12.

If we were to magnetize the sample and then suddenly replace the external emf that supplied the

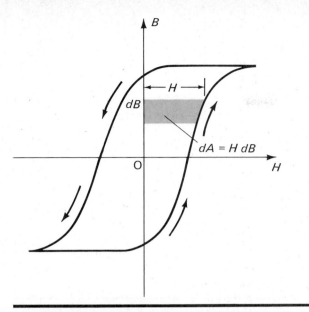

FIGURE 21.12. Magnetic induction as a function of magnetic intensity, illustrating the fact that the area within the hysteresis loop represents the magnetic energy dissipated during a cyclic magnetization process.

magnetizing current with a conducting path between the external terminals of the toroidal winding, the collapse of the magnetic field within the coil would generate a self-induced emf $\mathscr{E}$ $(= -N\, d\Phi_m/dt)$ that would tend to sustain the current in the external circuit, which would then decay exponentially with a time constant determined by the self-inductance of the coil and the resistance of the circuit. The magnetic energy that is dissipated in any time interval dt can now be calculated by writing, from energy conservation,

$$dU_m + \mathscr{E}I\, dt = 0 \tag{21.5.1}$$

from which

$$dU_m = -\mathscr{E}I\, dt = N\frac{d\Phi_m}{dt}\, I\, dt \tag{21.5.2}$$

$$dU_m = NI\, d\Phi_m$$

However, since $\Phi_m = BA$, and since the area is constant, we may write

$$d\Phi_m = A\, dB \tag{21.5.3}$$

Also, from (21.2.16) or from Example 21.2.2, we know that

$$H = \frac{NI}{2\pi r} = \frac{NI}{l} \tag{21.5.4}$$

or

$$NI = lH \tag{21.5.5}$$

where l is the mean circumference of the coil. Substituting (21.5.5) and (21.5.3) into (21.5.2), we find that

$$dU_m = AlH\, dB = VH\, dB \tag{21.5.6}$$

where $Al = V$ is the volume within the coil. But this can be written as

$$\frac{1}{V} dU_m = H\, dB \tag{21.5.7}$$

or, integrating,

$$\boxed{\frac{\Delta U_m}{V} = \int H\, dB} \tag{21.5.8}$$

It is clear that the integral in (21.5.8) represents the change in magnetic energy per unit volume within a magnetized substance that results from a change in magnetization. Also, from Fig. 21.12, it is apparent that this integral of (21.5.8), taken around the whole magnetization cycle, corresponding to the *area within the hysteresis loop* in a B-versus-H plot, represents the change in magnetic energy of the sample during the magnetization cycle. The situation is much the same as the one encountered previously in connection with cyclic heat engines, where the area within the closed loop in the P–V diagram representing the cyclic process corresponds to the work done by the system in the course of a cycle.

In this case, we find that the magnetic fields within the sample do *net work* during the magnetization cycle. This net work is done against certain dissipative processes within the substance that *resist* the action of the magnetic intensity in changing the magnetization. The action of these processes, like frictional effects, is to turn magnetic field energy into internal thermal energy which raises the temperature of the sample. At the same time, in order to repeat the magnetization cycle again and again, energy must be supplied by an external emf.

It is apparent that hysteresis has the effect of turning magnetic field energy into heat and that the amount of magnetic energy dissipated in this way per cycle is proportional to the area within the hysteresis loop. Hysteresis is primarily responsible for the rather well-known fact that iron-core transformers and inductors generate more heat than can be accounted for by resistive losses and eddy current generation. Hard ferromagnetic materials have rather wide hysteresis loops that enclose quite a large area in the B–H plot. This situation arises because of their large magnetic remanence, as shown by the distance OB in Fig. 21.11 and because rather strong coercive fields, as exhibited by the distance OC in the same figure, are required to destroy the magnetization once it is established. We find that such materials do not make very good transformer or inductor cores because since they waste much of the input power that is fed them in generating heat through hysteresis. On the

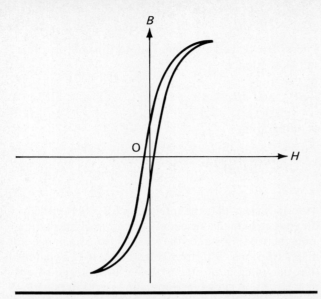

FIGURE 21.13. Hysteresis loop of a "good" ferromagnetic transformer core.

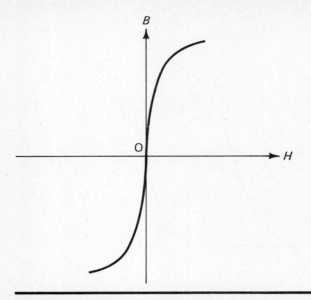

FIGURE 21.14. Hysteresis curve of an "ideal" soft ferromagnetic substance.

other hand, it is these large values of magnetic remanence and coercive field that render hard ferromagnetic substances eminently suitable for the fabrication of *permanent magnets* which—aside from their curiosity value—are much in demand for use in meter movements, loudspeakers, telephone equipment, and certain kinds of motors.

For iron core inductors and transformers, we would naturally seek ferromagnetic materials with as little magnetic remanence as possible and with coercive fields that are as small as can be had. Such materials have very narrow hysteresis curves, as illustrated by Fig. 21.13, which enclose a very small area in the B–H plot. Ideally, of course, what is desired is a substance of very high permeability that is completely *reversible*, exhibiting no hysteresis at all, such as that shown in Fig. 21.14. Soft iron and certain other *soft ferromagnetic substances* are found to have these characteristics, in certain instances approaching, but never quite reaching, the ideal case of Fig. 21.14. In such materials, the dissipative processes that resist magnetization changes are extremely weak and the magnetization cycle can be repeated frequently without much dissipation of energy. On the other hand, it is virtually impossible to magnetize them so that they retain any significant trace of permanent magnetism. In addition to the extreme cases of soft and hard ferromagnetic substances, of course, there is a wide range of materials whose magnetic properties are intermediate in character and do not closely resemble those of either hard or soft materials.

All ferromagnetic substances exhibit the phenomenon of *domain formation*, in which there are small discrete regions or *domains* in the material each of which exhibits complete magnetic alignment, even though the sample as a whole may be far from the saturation magnetization or may, in fact, have no net magnetization at all. The effect of applied magnetic fields is, therefore, principally either to change the direction of magnetic alignment of domains that are already totally aligned or to extend the boundaries of domains that are aligned parallel to the applied field at the expense of those that are aligned in other directions. These effects are illustrated in Fig. 21.15 and 21.16. The first of these effects, in which domains already aligned change their alignment direction to coincide with that of an applied field, occurs most frequently in substances that are polycrystalline aggregates containing many crystallites that are very small, so as to consitute single domains by themselves, or that contain impurities or imperfections that interfere with the motion of domain boundaries. The second, in which domains aligned with the external field grow at the expense of those aligned in other directions, occurs most frequently in large and pure single crystal specimens where there is nothing to interfere with the motion of domain boundaries. Of course, intermediate situations exist in which both effects occur at the same time. The size of magnetic domains may vary over a wide range, from dimensions of the order of 10^{-4} cm up to sizes of the order of millimeters or even centimeters.

Domain formation takes place because the creation of domains lowers the potential energy stored in the external magnetic field of a magnetizable substance and allows the system of atomic moments to approach a condition of *equilibrium* which cor-

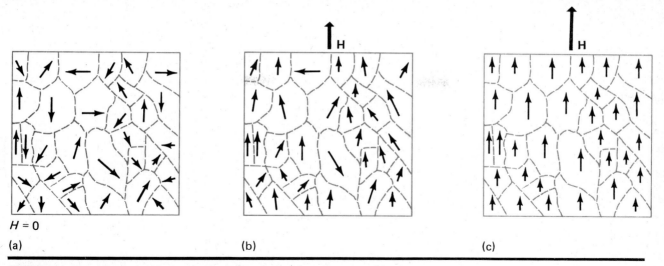

FIGURE 21.15. Magnetic alignment by alignment of fixed domains.

responds to minimum stored potential energy. The situation resembles in this respect that which we encounter when a spring is stretched. The spring is capable of storing potential energy and, therefore, of exerting forces on objects attached to it. The condition of equilibrium, in which the net force is zero, however, corresponds to the unstretched condition of the spring, in which the stored potential energy is a minimum. In the case of a single crystal which is totally aligned magnetically, a great deal of magnetic potential energy is stored in the magnetic field of the sample; the precise amount could be calculated by evaluating the integral of the energy per unit volume (given by $B^2/2\mu$) over the entire magnetic field. But if the crystal were to split into two domains of equal size but opposite alignment, the external field of the

sample, and hence the magnetic energy it contains, is much reduced.

This effect is illustrated in Fig. 21.17. Further subdivision into more domains can reduce this stored field energy even further and may, in fact, occur until there is hardly any detectable external field associated with the sample. The process of domain formation, however, cannot proceed indefinitely because there are, after all, strong forces *between neighboring atoms* that tend to align them ferromagnetically with parallel moments. An equilibrium condition is, therefore, reached when the energy changes associated with these two conflicting requirements just balance each other. The attainment of this equilibrium determines a certain average domain size characteristic of the particular crystal involved. The dissipative processes

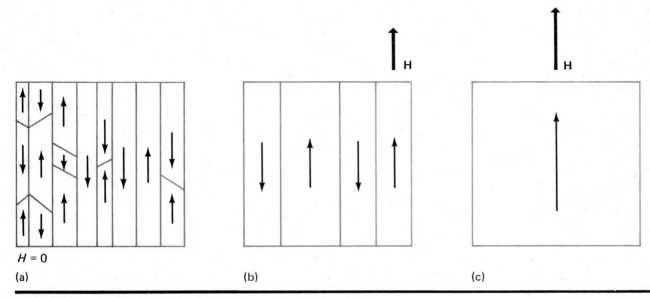

FIGURE 21.16. Magnetic alignment achieved by domain boundary motion.

709

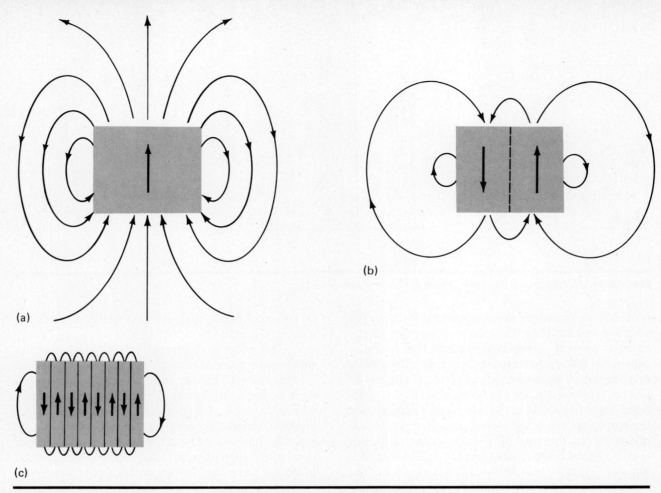

FIGURE 21.17. Reduction in external field energy as a result of domain formation.

that are associated with hysteresis losses in magnetic materials are largely related to the energies required to effect the magnetic alignment of individual magnetic domains and to accomplish domain boundary motion. Magnetic domain boundaries can be observed under the microscope by applying a liquid suspension of very finely powdered ferromagnetic material to the surface of the sample being examined. The fine particles, acting somewhat like iron filings sprinkled around a magnet, tend to congregate at the boundaries between domains, allowing them to be observed visually.

21.6 Permanent Magnets

Most of us derive our initial familiarity with magnetism from observing the curious behavior of permanent magnets. The forces they exert upon one another, and upon initially unmagnetized ferromagnetic substances, seem to stem from no visible source and thus are frequently accorded a certain magical or occult status. It is true enough that magnetism is most easily understood by firmly establishing the relationship between magnetic fields and electric currents at the very beginning, as we have done in our own study of the subject, and it is also true that in the case of permanent magnets this connection seems on the surface to have been severed somehow. But a closer examination of the situation shows that permanent magnetism can be described by the same laws that enable us to understand electromagnetism.

It is probably best to begin by returning to the now familiar toroidal geometry of section 21.1. In a permanently magnetized substance, the true current is *zero*, and the only current present is the equivalent magnetization current I_m, as given by Eqs. (21.2.10) and (21.2.12). This current can be visualized as a surface current that flows around the outside of the toroid as illustrated by Fig. 21.2 or Fig. 21.4. It is generated by electron currents that circulate within the atoms of the ferromagnetic material. These currents are not subject to the usual processes of scattering by collision that impede the flow of macroscopic currents of free electrons in ordinary conductors. Therefore, the magnetization currents, once established by aligning the atomic moments within the substance, flow unimpeded around the sample—forever or, at any rate, until something changes or destroys the magnetic alignment. The situation in this regard is reminiscent of the flow of true currents in *superconducting* materials at very low temperatures. There are, in fact, many similarities between the phenomena of ferromagnetism and superconductivity.

In a toroidal sample that has a uniform permanent magnetization field **M**, as shown in Fig. 21.18a, we may calculate **H** from Ampère's law, as expressed by (21.2.16), in the usual way. But in this case, the true current through any circular contour C is *zero*; therefore,

$$\oint \mathbf{H} \cdot d\mathbf{l} = 2\pi r H = 0$$

or

$$H = 0 \qquad (21.6.1)$$

This is true whether the contour C lies within the toroid or not, so there is no **H** field anywhere! This result is not unexpected in view of the fact that there are no true currents anywhere, although, as we shall soon see, this by itself does not guarantee that H must vanish at all points. Calculating **B** in the same way, we find from (21.2.13) that for a circular contour C within the toroid,

$$\oint_C \mathbf{B} \cdot d\mathbf{l} = 2\pi r B = \mu_0 I_m \qquad (21.6.2)$$

since the true current I is zero, while outside, where the total current within any contour is zero,

$$\oint_C \mathbf{B} \cdot d\mathbf{l} = 2\pi r B = 0 \qquad (21.6.3)$$

Since the magnetization current on the surface of the toroid is given by (21.2.10), (21.6.2) tells us that, *within* the sample,

$$\mathbf{B} = \mu_0 \mathbf{M} \qquad (21.6.4)$$

while *outside* the sample, from (21.6.3),

$$B = 0 \qquad (21.6.5)$$

These results can also be obtained by substituting $H = 0$, as given by (21.6.1), into (21.2.18), because the magnetization has the constant value M within the sample and is zero outside.

The situation is, therefore, as illustrated by Fig. 21.18. There is an **M** field and a **B** field that looks very much like it, but no **H** field. It is all very straightforward. Unfortunately, however, outside the sample, both B and H are everywhere zero, so it would be *impossible* for us to determine that we have a permanent magnet unless we could somehow dig inside to discover that the **B** field is there. This, you will admit, would be difficult.

To make the situation more realistic, then, let us take a saw and hack out a small piece of material, leaving a horseshoe magnet with a narrow air gap, as shown in Fig. 21.19. The magnetization field **M** is, of course, unaffected by this procedure and remains as illustrated in Fig. 21.19a. The magnetic induction **B** can be determined as above, using Ampère's law, but there are some complications involved in doing so, since, by removing the piece of magnetized material, we have destroyed the perfect toroidal symmetry of the sample. We can, therefore, no longer assert that the magnitude of **B** is the same everywhere along a circular contour nor that its direction is tangential to the circle. Nevertheless, since we have a permanent magnet, we are confident that there has to be a **B** field in the air gap, and since nobody has ever suc-

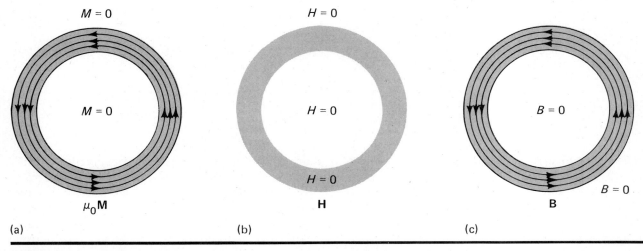

FIGURE 21.18. (a) Magnetization, (b) magnetic intensity, and (c) magnetic induction fields for a permanently magnetized toroidal sample.

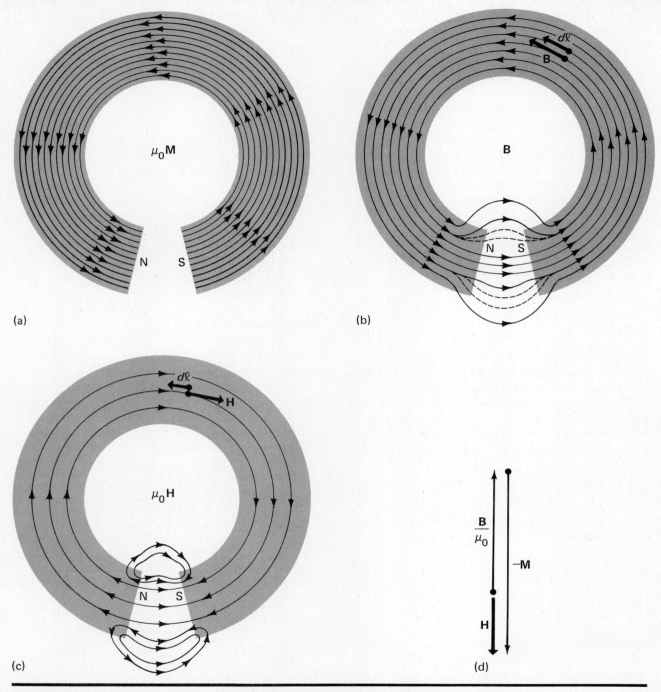

FIGURE 21.19. Lines of (a) magnetization, (b) magnetic induction, and (c) magnetic intensity for a permanent magnet in the form of a toroid having a small air gap. (d) Relationship between **B**, **M**, and **H** within the magnet.

ceeded in bringing an isolated magnetic charge into captivity, we are sure that the lines of this **B** field must close upon themselves and, therefore, continue into the magnetized material more or less as shown in Fig. 21.19b. Looking at it from a slightly different point of view, we may also note that the relationship of the **B** field with magnetization currents and true

currents is the same; there is no way of telling, inside or outside the sample, whether the **B** field arises from a magnetization current or from a true current of the same size flowing in a conductor.

The **B** field, therefore, must look like that which could be produced by a closely wound current-carrying toroidal coil without a magnetized core—but with a gap in the winding corresponding to the air gap of the permanent magnet. In either way, we are led to the picture of the magnetic induction field shown in Fig. 21.19b. The field is strong both within the material and in the air gap, though in the vicinity of the gap the lines may stray outside the confines of the gap and produce a weak magnetism in the surrounding volume. Within the magnetized material, also, the magnetic induction will be just a bit less than the former value $\mu_0 M$ because the total magnetization current I_m has been decreased by the amount carried by the missing chunk. This is also clear from the usual expression for the **B** field within a thin toroidal coil, $B = \mu_0 n I_{\text{tot}}$.

In the case of the magnetic intensity **H**, it is evident from (21.2.18) that outside the magnetized substance, in particular in and around the air gap, we must have

$$\mathbf{H} = \frac{\mathbf{B}}{\mu_0} \qquad \text{outside sample} \qquad (21.6.6)$$

because in this region, $M = 0$. The **H** field outside the sample is, therefore, the same as the **B** field, apart from the constant scale factor μ_0. Within the sample, however, from (21.2.18) we may infer that

$$\mu_0 \mathbf{H} = \mathbf{B} - \mu_0 \mathbf{M} \qquad \text{inside sample} \qquad (21.6.7)$$

But since we just reached the conclusion that, deep inside the toroidal magnet, the magnitude of **B** is less than the value $\mu_0 M$ in a toroid without the air gap, we are forced to admit that the direction of **H** must be *opposite* that of **M** and **B** in that region. This conclusion is reinforced by the argument that, since there are no true currents here, according to (21.2.16),

$$\oint_C \mathbf{H} \cdot d\mathbf{l} = 0 \qquad (21.6.8)$$

But along the mean circumference of the coil, in the air gap the integral of $\mathbf{H} \cdot d\mathbf{l}$ is clearly positive. Therefore, if (21.6.8) is to be satisfied, its value inside the magnetized material, along the rest of the circular contour, must be *of the opposite sign*. This can occur, however, only if **H** and $d\mathbf{l}$ are in opposite directions, thus only if the direction of **H** is opposite—or nearly opposite—that of **B**. The resulting field is shown schematically in Fig. 21.19c. Since **H** is in the opposite direction to **B** in the interior of the magnetized material and in the same direction as **B** in the gap, it is evident that the direction of the lines of **H** undergo a *reversal in direction* at the faces of the gap. This reversal of field is illustrated in Fig. 21.19c; it is quite characteristic of the behavior of the lines of magnetic intensity at the pole faces of permanent magnets. Clearly, without this reversal, the requirement of Ampère's law that the integral of $\mathbf{H} \cdot d\mathbf{l}$ around every

possible closed path be zero could not be satisfied, because without it **H** and $d\mathbf{l}$ would always be in the same direction around all of the closed loops formed by the lines of intensity, and the scalar product would then have to add up to a positive result around any of these paths.

In the case of the uninterrupted toroidal magnet discussed in connection with Fig. 21.18, we found that H was zero within the magnetic material as well as outside, while **B** and $\mu_0 \mathbf{M}$ were equal within the magnet. In the example of Fig. 21.19, we see that B is a bit less than $\mu_0 M$ deep within the magnet but that now there is an **H** field within the material, in the opposite direction to **B**. This **H** field accounts for the difference between $\mu_0 \mathbf{M}$ and **B** required by Eq. (21.2.18), which demands that at every point within the substance $\mathbf{H} = (\mathbf{B}/\mu_0) - \mathbf{M}$, as in Fig. 21.19d.

SUMMARY

All material substances exhibit magnetic effects when subjected to externally produced magnetic fields. These magnetic effects arise from the orientation of permanent or induced atomic or molecular magnetic moments. The effects create a magnetic dipole moment in every volume element of the substance. The *magnetization* of the material at a given point is defined as the local dipole moment per unit volume:

$$\mathbf{M} = \frac{\Delta \mathbf{p}_m}{\Delta V}$$

Since the atomic dipole moments can be expressed as circulating current times area, the magnetization can be related to a circulating surface current I_m, which in a uniformly magnetized substance can be written

$$I_m = M \cdot l$$

where l is the circumference of the cross section. The circulating surface current is shown in Fig. 21.2. Also, in such a substance the total dipole moment is

$$\mathbf{p}_m = \mathbf{M}V = \mathbf{M}lA = I_m A\mathbf{n}$$

where **n** is a unit vector normal to the plane of area A.

Ampère's law for systems containing magnetizable materials can be written in the form

$$\oint_C \mathbf{B} \cdot d\mathbf{l} = \mu_0 (I + I_m)$$

where I is the true current arising from the macroscopic motion of free charges and I_m is the circulating magnetization current. But the magnetization current can always be expressed in the form

$$I_m = \oint_C \mathbf{M} \cdot d\mathbf{l}$$

which allows Ampère's law to be written as

$$\oint_C \mathbf{H} \cdot d\mathbf{l} = I$$

where

$$\mathbf{H} = \frac{\mathbf{B}}{\mu_0} - \mathbf{M}$$

The vector **H** as defined above is referred to as the *magnetic intensity*.

In many magnetizable substances, particularly for small values of magnetic intensity, the magnetization is directly proportional to magnetic intensity, which allows us to write

$$\mathbf{M} = \chi_m \mathbf{H}$$

where the proportionality constant **M** is referred to as the *magnetic susceptibility*. This equation defines the behavior of a *linear* magnetic material. It then follows that

$$\mathbf{B} = \mu \mathbf{H}$$

where

$$\mu = \mu_0(1 + \chi_m)$$

The quantity μ is called the *magnetic permeability* of the substance. The relative permeability K_m is defined by

$$K_m = \frac{\mu}{\mu_0} = (1 + \chi_m)$$

Materials for which χ_m is positive are referred to as *paramagnetic substances*. In such materials, atomic moments align themselves parallel to the external field. There are also substances for which χ_m is negative. In these *diamagnetic materials*, atomic moments are aligned antiparallel to the external field. In ferromagnetic substances, the alignment of dipoles is accomplished not only by the direct influence of the external field but also by a strong coupling interaction between the individual atomic moments themselves. Above a certain temperature, referred to as the *Curie temperature*, ferromagnetic materials lose their ferromagnetic character and behave like paramagnetic substances. Ferromagnetic materials, below the Curie point, exhibit very large susceptibilities and magnetic behavior that is decidedly nonlinear. They can be permanently magnetized and exhibit magnetic *hysteresis* effects, in which the magnetization depends not only on the magnetic intensity but on the past history of magnetic influence the sample has undergone. Ferromagnetic materials also display the phenomenon of *domain formation*, in which small discrete regions of saturated magnetization are formed the effect of which is to reduce the magnetic energy stored in the sample's external field.

QUESTIONS

1. What two competitive effects determine the total magnetic moment of any material?
2. The magnetization vector at a point is the dipole moment per unit volume near the point. What difficulties arise if the volume element is infinitesimally small?
3. Describe the difference between magnetic induction and magnetic intensity in magnetizable materials and in free space.
4. Discuss the similarities between the magnetization vector **M** and the polarization vector **P**.
5. Explain briefly the difference between paramagnetism, diamagnetism, and ferromagnetism.
6. An unmagnetized steel paper clip is attracted to a bar magnet. Explain the origin of the attractive force.
7. What properties of iron cause it to be a ferromagnetic material?
8. Paramagnetic susceptibility is strongly temperature dependent while diamagnetic susceptibility is practically independent of temperature. Can you explain these observations?
9. Why do some materials have atoms or molecules with permanent magnetic dipole moments while others do not?
10. The magnetic susceptibility of ferromagnetic material varies with the strength of the external **M** field. Explain this in terms of the B-versus-H curve observed during the magnetization of a sample.
11. Of what significance is the Curie temperature?
12. Soft ferromagnetic materials are not ordinarily used to make strong permanent magnets. Explain.

PROBLEMS

1. A small sample of a magnetizable substance has the form of a cube of side 1.0 mm. The magnetic moment of the sample is 1.0×10^{-3} amp-m^2. Find **(a)** the magnetization of the sample, assuming that it is uniform, **(b)** the surface magnetization current. Assume that the magnetic moment vector is normal to one of the cube faces.
2. A uniformly magnetized sample has magnetic susceptibility of 2.4×10^{-4}. **(a)** What is its magnetic permeability? **(b)** What is the relative permeability of the substance?
3. The magnetization of a sample of iron is such that it contributes 1.9 weber/m^2 to a uniform magnetic induction **B**. What is the magnetic moment of a 1-cm^3 volume of this material?
4. A bar magnet 20 cm long and 6 mm in diameter is made of iron with a magnetic susceptibility of 5000. At the center of the magnet, the magnetic induction is 0.85 weber/m^2. Assuming uniform magnetization of the iron, find **(a)** the magnetic intensity at the center, **(b)** the magnetization within the material, and **(c)** the magnetic moment of the magnet.
5. For the preceding problem, how many electrons are needed to produce the magnetic moment observed if each electron contributes the magnetic moment given by Eq. (21.2.25)? Compare this result with the total number of

iron atoms. Iron has a density of 7.85 g/cm^3 and an atomic mass of 56 g/mole.

6. The uniform magnetic induction within a sample of aluminum is 1 weber/m^2 when it is placed between the poles of a magnet. Find the magnetic intensity and the magnetization within the aluminum. If the aluminum has a volume of 10 cm^3, what is the value of its magnetic moment?

7. A long solenoid is wound on an iron core of magnetic susceptibility 200. If the current in the solenoid is 2 amperes and the solenoid is wound with 750 turns of wire per meter, find (a) the magnetic intensity $\mathbf{H}$ inside the solenoid, (b) the magnetic induction $\mathbf{B}$, (c) the magnetization, and (d) the magnetic moment per unit length, assuming a cross-sectional area of 8 cm^2. In working out your answers, you may assume that the iron is uniformly magnetized and that the magnetic induction is uniform throughout the solenoid.

8. A toroidal coil has a mean circumference of 40 cm and a cross-sectional area of 5 cm^2. It is wound with 400 turns of wire on an iron core of magnetic susceptibility 550. A current of 1.5 ampere passes through the wire. (a) Using Eq. (21.2.16) find the magnetic intensity. (b) Determine the magnetic induction. (c) What is the total magnetic flux through the sample? Assume that the magnetic behavior of the iron is linear.

9. A toroidal sample of magnetizable material is of average radius 12.5 cm and has a cross-sectional area of 4.0 cm^2. It is wound with fine wire, 80 turns per cm along the mean circumference. Its magnetic susceptibility is 3.60×10^{-4}. The winding carries a steady current of 5.0 amp. What are (a) the magnitudes of the $\mathbf{H}$- and $\mathbf{B}$-fields within the substance, (b) the magnitude of the magnetization within the substance, and (c) the surface magnetization current? (d, e) What would the magnitude of the magnetic vectors $\mathbf{B}$ and $\mathbf{H}$ be if there were no paramagnetic core present?

10. What would the answers to the preceding problem be if a soft iron core having a constant susceptibility of 750 were substituted for the original paramagnetic core?

11. Find (a) the inductance of the toroidal coil of problem 9. (b) the inductance of the coil of problem 10.

12. How much magnetic energy is stored within the coils described (a) in problem 9 and (b) in problem 10?

13. The magnetic induction $\mathbf{B}$ inside a long solenoid wound on an iron core is found to be 0.28 weber/m^2 when a field intensity of 205 amperes/meter is applied. What is the permeability of the iron core? Assume linear magnetic behavior and uniform magnetization and magnetic induction throughout the core.

14. For an infinite solenoid of n turns per meter and current I, the magnetic induction on the axis is given by $B = \mu_0 nI$. (a) Will the induction field $\mathbf{B}$ on the axis increase, decrease, or remain the same if into the solenoid we insert a core of (i) ferromagnetic material, (ii) diamagnetic material, (iii) paramagnetic material? (b) If we insert a linear magnetic material of permeability μ, what is the magnetic intensity $\mathbf{H}$ on the axis of the solenoid?

15. Find the magnitude of the magnetization of a material in each of the following cases: (a) Each atom has a magnetic dipole moment of 1.7×10^{-33} ampere-m^2 and there are

9.4×10^{28} atoms/m^3; the moments are mutually parallel. (b) The magnetic susceptibility is $\chi_m = -3 \times 10^{-6}$ and $B = 3.6 \times 10^{-5}$ weber/m^2. (c) $H = 0.40$ ampere/meter and the relative permeability is 1.00041.

*16. In the analysis of diamagnetism in Section 21.3, it was assumed that application of a weak magnetic induction $\mathbf{B}$ does not bring about a change in the radius of the circular orbit. Show that the radius does change but that this change depends quadratically on $\mathbf{B}$ and can, therefore, be neglected for small fields.

17. The molecules of a paramagnetic substance have a magnetic dipole moment of 5.0×10^{-24} amp-m^2. The substance is placed in a magnetic field of 3.6 weber/m^2 at a temperature of 4.0°K. (a) What is the average angle between the molecular dipole moments of the substance and the magnetic induction vector. What would the answer be if the temperature were raised (b) to 40°K, (c) to 400°K?

18. Assuming that there are 7.2×10^{28} molecules per m^3, find the magnetic susceptibility corresponding to the conditions outlined in parts (a)–(c) of the preceding problem.

19. By using Eq. (21.4.5), obtain an expression which gives the ratio of the number of magnetic dipoles aligned parallel to the magnetic induction $\mathbf{B}$ to those which are antiparallel. Investigate and plot this ratio as a function of the temperature T from absolute zero to large temperatures.

20. A typical paramagnetic material experiences a 75 percent "saturation" of its magnetization at a temperature of 2°K. What value of magnetic induction is needed to achieve this saturation if the values of n and p_m are those given in Example 21.4.1?

21. The magnetic susceptibility of a paramagnetic substance is 4.5×10^{-4} at 300°K. If there are 5.5×10^{28} atoms/m^3, find the average magnetic moment per atom.

22. Sketch the field lines associated with the magnetic intensity $\mathbf{H}$, the magnetic induction $\mathbf{B}$, and the magnetization $\mathbf{M}$ in the neighborhood of a bar magnet of uniform cross section.

23. Sketch the field lines of $\mathbf{B}$, $\mathbf{H}$, and $\mathbf{M}$ for a uniformly magnetized spherical sample.

24. A toroidal permanent magnet of mean radius 6.0 cm is fashioned from a uniform ferromagnetic substance whose magnetization has the constant magnitude 2.40×10^7 amp/m. Find (a) the magnetic induction inside the toroid, (b) the magnetic intensity inside the toroid, (c) the magnetic induction outside the toroid, (d) the magnetic intensity outside the toroid, (e) the surface magnetization current.

*25. In the situation described in the preceding problem, suppose that a small sector of the toroid subtending a central angle of 10° is removed, as illustrated in Figure 21.19. Assuming that the surface current distribution is disturbed so little that the value of B on the mean circumference in the air gap is the same as it is within the magnetized material, and that the tangential symmetry of the $\mathbf{B}$- and $\mathbf{H}$-fields on the mean circumference is unaltered, find (a) the value of B on the mean circumference in the air gap and in the interior of the magnet, (b) the corresponding value of H in the air gap, and (c) the corresponding value of H inside the magnetized substance.

715

ALTERNATING CURRENT CIRCUITS AND RESONANCE

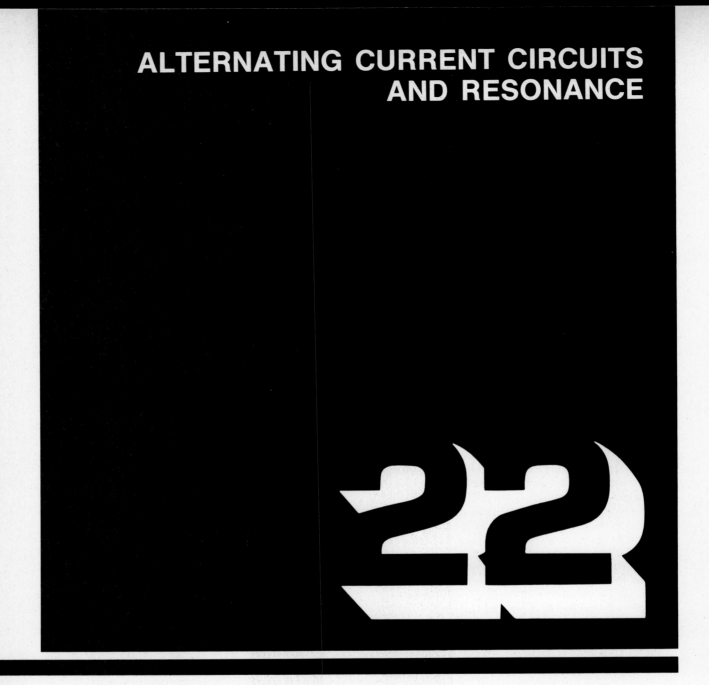

22.1 Introduction

Today, most of the electrical energy sold by power companies is transmitted as 60 cps alternating current (ac). The primary reason for the widespread use of ac rather than dc current is the relative ease and efficiency of converting one voltage to another by means of transformers. The necessity for this conversion arises from purely economic considerations. Power transmitted over large distances should be carried at low currents and consequently high voltages in order to minimize the Joule heating losses, which increase quadratically with the magnitude of the current.

Every time we close an electrical circuit by putting a plug into an outlet and turning the switch on, we are providing a potential difference between two points which varies with time according to $\mathscr{E} = \mathscr{E}_0 \sin(\omega t + \delta)$, where $\mathscr{E}_0$ is the voltage amplitude, δ is a phase angle, and ω is the angular frequency (120π rad/sec). Since most of the electrical energy consumed nowadays is supplied by ac emfs of this form, it is very important to study the behavior of circuits excited by such emfs. It should also be pointed out that such voltages are employed not only in commercial power lines but also in the reception of radio and TV signals. In the process of tuning a radio, for example, we adjust the knob to a given station and thereby receive only electromagnetic signals in a narrow band about a desired ac frequency. We shall discuss the physical principles involved in tuning a radio later in this chapter.

22.2 The Simple L–C Circuit

To begin our study of alternating current, let us examine a very simple series circuit consisting of an inductor L, a capacitor C, a battery of constant emf $\mathscr{E}$, and a switch S, as shown in Fig. 22.1. We are interested in finding out how the current in the circuit varies with time after the switch S has been closed at $t = 0$. The sum of all potential differences around the closed circuit loop must add up to zero, and, therefore, we may write

$$\mathscr{E} = L\frac{dI}{dt} + \frac{q}{C} \qquad (22.2.1)$$

for all values of $t \geq 0$.

We can obtain an equation which involves only the current I and its derivatives by differentiating both sides of (22.2.1) with respect to t. Since $\mathscr{E}$, L, and C are independent of t and dq/dt is the current I,

$$\frac{d\mathscr{E}}{dt} = 0 = L\frac{d^2I}{dt^2} + \frac{1}{C}\frac{dq}{dt}$$

or

$$\frac{d^2I}{dt^2} + \frac{1}{LC}I = 0 \qquad (22.2.2)$$

This equation has exactly the same form as (9.2.2), whose solution represented simple harmonic motion. Thus, by analogy to the solution of that equation, we find that the current I is

$$I = I_0 \cos(\omega t + \delta) \qquad (22.2.3)$$

where

$$\omega = \sqrt{\frac{1}{LC}} = 2\pi f = \frac{2\pi}{\tau} \qquad (22.2.4)$$

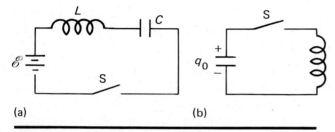

(a)   (b)

FIGURE 22.1. Series L–C (inductance–capacitance) circuit (a) with emf and (b) without emf but with initially charged capacitor.

where ω is the angular frequency of oscillation, in rad/sec, and f and τ represent the frequency and period, respectively; f is often called the *natural frequency* of the L–C circuit. We shall generally consider f to be given in cycles per second, or *hertz* (Hz), and τ in seconds. The *maximum current* I_0 is called the *current amplitude*, while the quantity δ is called the *phase angle*. The values of I_0 and δ can only be determined when an appropriate set of initial conditions have been specified.

EXAMPLE 22.2.1

A 2 μF capacitor and a 0.5 henry inductor are in series with a 12 volt battery. At $t = 0$, the capacitor is uncharged and a switch is closed, thus making a complete circuit. (a) How does the current in the circuit vary with time? (b) Find the charge q on the capacitor as a function of t. (c) Obtain an expression for the voltage across the inductor.

The angular frequency ω is given by

$$\omega = \sqrt{\frac{1}{LC}} = \sqrt{\frac{1}{(0.5)(2 \times 10^{-6})}}$$

$$= 10^3 \text{ rad/sec} \qquad (22.2.5)$$

Now, at $t = 0$, we know that q and I are both zero. From (22.2.3), therefore, we can write

$$I_0 \cos \delta = 0 \qquad (22.2.6)$$

from which

$$\cos \delta = 0 \qquad \delta = \pm\frac{\pi}{2} \qquad (22.2.7)$$

Also, if we set $q = 0$ in (22.2.1) it is evident that at $t = 0$, the entire potential difference $\mathscr{E}$ associated with the emf appears across the inductor. Therefore, for $t = 0$ we may write

$$\mathscr{E} = L\left(\frac{dI}{dt}\right)_{t=0} = -I_0\omega L \sin \delta$$

from which

$$I_0 = -\frac{\mathscr{E}}{\omega L} \sin \delta \qquad (22.2.8)$$

From (22.2.7), the phase angle δ must be either $+\pi/2$ or $-\pi/2$. If we choose it to be $+\pi/2$, then (22.2.8) will give a negative value for the current amplitude I_0, which must clearly be a positive quantity. We, therefore, have to assign the value

$$\delta = -\pi/2 \qquad (22.2.9)$$

to the phase angle, which then gives (22.2.8) the form

$$I_0 = \frac{\mathscr{E}}{\omega L} = \mathscr{E}\sqrt{\frac{C}{L}} \qquad (22.2.10)$$

717

since $\omega = 1/\sqrt{LC}$. The current, as expressed by (22.2.3) now can be written

$$I = \frac{\mathscr{E}}{\omega L} \cos\left(\omega t - \frac{\pi}{2}\right) = \frac{\mathscr{E}}{\omega L} \sin \omega t$$

$$= \frac{12}{(1000)(0.5)} \sin(1000t)$$

$$= (0.024) \sin(1000t) \qquad (22.2.11)$$

In this equation, of course, we must use t in seconds to obtain the correct numerical value for I in amperes.

Since

$$I = \frac{dq}{dt} \qquad \text{or} \qquad dq = I\,dt \qquad (22.2.12)$$

we can find the charge q on the capacitor at any time by substituting the current as a function of time into (22.2.12) and integrating from $t = 0$, when q is also zero, to a later time t, when the charge on the capacitor is q:

$$q = \int_0^q dq = \frac{\mathscr{E}}{\omega L} \int_0^t \sin \omega t\,dt$$

$$= \frac{\mathscr{E}}{\omega^2 L} \left[-\cos \omega t \right]_0^t$$

$$= \frac{\mathscr{E}}{\omega^2 L} (1 - \cos \omega t) = \mathscr{E}C(1 - \cos \omega t) \qquad (22.2.13)$$

The potential difference ΔV_C across the capacitor will be simply

$$\Delta V_C = \frac{q}{C} = \mathscr{E}(1 - \cos \omega t) \qquad (22.2.14)$$

while the potential difference across the inductor, according to (22.2.11), is

$$\Delta V_L = L\frac{dI}{dt} = \mathscr{E} \cos \omega t \qquad (22.2.15)$$

There are a number of features to be observed from these results. First, it is evident that the *instantaneous* potential differences ΔV_C and ΔV_L across the capacitor and inductor, respectively, add up to the emf $\mathscr{E}$, as required by Kirchhoff's law. The same statement *cannot* generally be made, however, about the *amplitudes* of the oscillating emfs across individual circuit elements because, generally speaking, these individual emfs *will not have the same phase*. This is readily apparent from (22.2.14) and (22.2.15), where it is seen that the oscillating emfs across the inductor and capacitor have opposite signs and are, therefore, 180° out of phase. It is evident, in addition, that the current as given by (22.2.11) does not have the same phase as either of the oscillating voltages across the inductor and capacitor. In fact, if we note that the oscillating part of the emf across the capacitor can

be written as $-\mathscr{E} \cos \omega t = \mathscr{E} \sin(\omega t - 90°)$, it is apparent that the phase of the current in the circuit is 90° ahead of that associated with the oscillating voltage across the capacitor, while it is 90° behind that exhibited by the potential difference ΔV_L across the inductor. We shall see that these phase relations between current and voltage across capacitive and inductive elements always occur in just this way. In the case of resistances, of course, the voltage drop is always in phase with the current since, from Ohm's law, $\Delta V_R = IR$, where R is constant.

Let us now turn our attention to some of the energy considerations involved in this oscillating circuit. If we multiply Eq. (22.2.1) by I, we obtain

$$\mathscr{E}I = LI\frac{dI}{dt} + \frac{Iq}{C}$$

or

$$\boxed{\mathscr{E}I = \frac{d}{dt}\left(\tfrac{1}{2}LI^2 + \frac{q^2}{2C}\right)} \qquad (22.2.16)$$

The left-hand side is readily recognized as the power delivered by the battery. Since I is an oscillating current, the power P also oscillates in time. For one half of each cycle, energy is converted from chemical to electrical, while for the other half, the conversion goes the other way. In making this statement, we are assuming that the battery is a completely reversible storage device which stores a certain amount of chemical energy on one half of the cycle and releases an equivalent amount of electrical energy into the circuit on the other half. The right-hand side of (22.2.16) is clearly the time rate of change of the sum of the energies stored in the inductor and the capacitor. We can also write the above equation as

$$P = \frac{d}{dt}\left(\int \tfrac{1}{2}\mathscr{E}_0 E^2\,d\mathscr{V} + \int \frac{1}{2\mu_0} B^2\,d\mathscr{V}\right) \qquad (22.2.17)$$

where E and B are the time-dependent electric and magnetic fields which exist within capacitor and inductor, respectively.

It is interesting to note that if $P = 0$, as is the case when the battery is not present, the energy of the system is conserved and, moreover, oscillates between energy stored in the capacitor and energy stored in the inductor.

To illustrate this more clearly, let us consider the circuit of Fig. 22.1b. Here, there is no emf but there is initially a charge Q_0 on the capacitor and, therefore, an initial potential difference $\Delta V_0 = Q_0/C$ across its plates at the moment the switch S is closed. Equation (22.1.1) now becomes

$$L\frac{di}{dt} + \frac{q}{C} = 0 \qquad (22.2.18)$$

or, since $I = dq/dt$,

$$\frac{d^2q}{dt^2} = -\frac{q}{LC} \tag{22.2.19}$$

This is once again the equation describing a simple harmonic oscillator of frequency $\omega = 1/\sqrt{LC}$; therefore, we may write

$$q = q_0 \cos(\omega t + \delta) \tag{22.2.20}$$

where $\omega = 1/\sqrt{LC}$ and

$$I = \frac{dq}{dt} = -\omega q_0 \sin(\omega t + \delta) \tag{22.2.21}$$

$$\frac{dI}{dt} = -\omega^2 q_0 \cos(\omega t + \delta) = -\frac{q_0}{LC} \cos(\omega t + \delta) \tag{22.2.22}$$

Now, since $q = Q_0$ when $t = 0$, and since $I = 0$ initially, Eqs. (22.2.20) and (22.2.21) give us, for $t = 0$,

$$Q_0 = q_0 \cos \delta \tag{22.2.23}$$

$$0 = \omega q_0 \sin \delta \tag{22.2.24}$$

From (22.2.24), it is easily seen that we must take $\delta = 0$, and then (22.2.23) tells us that $q_0 = Q_0$. Equations (22.2.20), (22.2.21), and (22.2.22) now become

$$q = Q_0 \cos \omega t$$
$$I = -\omega Q_0 \sin \omega t$$
$$\frac{dI}{dt} = -\frac{Q_0}{LC} \cos \omega t$$

From these, it is easy to see that

$$\Delta V_C = \frac{q}{C} = \frac{Q_0}{C} \cos \omega t = \Delta V_0 \cos \omega t \tag{22.2.25}$$

$$\Delta V_L = L \frac{dI}{dt} = -\Delta V_0 \cos \omega t = \Delta V_0 \cos(\omega t - \pi) \tag{22.2.26}$$

In this case, the sum of ΔV_C and ΔV_L is zero at all times, as required by (22.2.18). Again, it is easy to see that these two oscillating voltages are $180°$ out of phase at all times. What happens here is that, initially, the capacitor discharges into the inductor, the current flow establishing a magnetic field within its windings. When the capacitor is discharged, however, the magnetic field of the inductor starts to decay and produces a self-induced voltage that *sustains* the flow of current for a time until the capacitor once more becomes charged, this time with opposite polarity. As the current finally subsides, the capacitor's charge once more reaches the value Q_0, but with opposite polarity. Now the capacitor discharges into the inductor once more, but the current flows in the opposite direction, again exciting a magnetic field in and about the inductor, which is, of course, directed oppositely to what

it was before. Again, as the capacitor's charge is exhausted, the decay of current in the circuit and the associated decay of the inductor's field produce a self-induced voltage that sustains the current flow for a while until the capacitor is once more charged, this time with its original polarity. As the current flow drops to zero, this charge attains its initial value Q_0, at which point the cycle is ready to start all over again.

It follows from all this that the *energy* stored in the capacitor, or, more precisely, in its electric field, is given by

$$U_C = \frac{q^2}{2C} = \frac{Q_0{}^2}{2C} \cos^2 \omega t \tag{22.2.27}$$

The energy stored in the inductor's magnetic field is

$$U_L = \tfrac{1}{2} L I^2 = \tfrac{1}{2} L \omega^2 Q_0{}^2 \sin^2 \omega t$$

or, since $\omega^2 = 1/LC$,

$$U_L = \frac{Q_0{}^2}{2C} \sin^2 \omega t \tag{22.2.28}$$

The total energy, which is the sum of the two quantities represented by (22.2.27) and (22.2.28), is then

$$U = U_C + U_L = \frac{Q_0{}^2}{2C}$$

which is constant and equal to the initial energy stored in the capacitor as the switch was closed.

Under these *ideal* conditions, the energy initially given to the circuit is conserved and merely oscillates back and forth, from the capacitor to the inductor and back. This situation is illustrated in Fig. 22.2. It is reminiscent of the physics encountered in Chapter 9, in which we studied a number of examples of simple harmonic oscillatory motion. In those examples, the energy of the system changed back and forth from potential to kinetic in a periodic manner. In relating that situation to the one discussed here, one can say that the potential energy $kx^2/2$ of a spring is analogous to the energy $q^2/2C$ of the capacitor, while the kinetic energy $mv^2/2$ is analogous to the inductor's magnetic energy $LI^2/2$.

The analogy between electrical and mechanical systems can often be exploited to great advantage. It

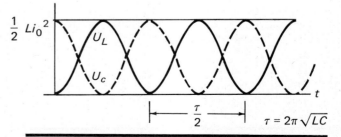

FIGURE 22.2. Electrostatic energy U_C and magnetic energy U_L as functions of time for the L–C circuit.

is purely a mathematical analogy, based on the fact that both systems satisfy the same kind of equation. Nevertheless, it is very often possible to solve mechanical problems by setting up the analogous electrical circuit and studying its properties instead. A simple pendulum executing approximate simple harmonic motion ultimately damps out as a result of resistive forces due to the surrounding air and the support. In a similar manner, we may expect that a *resistance* in an *L–C* circuit of this type will cause the current amplitude I_0 to diminish with time and eventually drop to zero, and this is exactly what is found to happen.

22.3 The *R–L–C* Circuit

We now consider a circuit like the one in Fig. 22.1a, except that a resistor R has now been added, as shown in Fig. 22.3. A resistor is a circuit element that always dissipates electrical energy as heat and is, therefore, quite different from a capacitor or inductor, which are both elements that can store energy. The presence of the resistor leads to an additional term in (22.2.1) that expresses the potential difference IR across its terminals. We then have, from Kirchhoff's law,

$$\mathscr{E} = L\frac{dI}{dt} + RI + \frac{q}{C} \tag{22.3.1}$$

as the basic circuit equation. By differentiating (22.3.1) with respect to the time, we again obtain an equation which involves only I and its derivatives

$$L\frac{d^2I}{dt^2} + R\frac{dI}{dt} + \frac{I}{C} = 0 \tag{22.3.2}$$

An equation having exactly this form was shown in Chapter 9 to describe the damped oscillatory motion that results when a simple harmonic oscillator is subject to a dissipative force that is directly proportional to velocity. That equation is written in section 9.6 as Eq. (9.6.3), and its solution has been shown to be of the form given by (9.6.4). Equation (22.3.2) is exactly the same as (9.6.3), except that different symbols are used to represent different quanti-

ties. For example, it is easily seen that the current in (22.3.2) corresponds to the displacement in (9.6.3), while the inductance corresponds to the mass, and so on. In fact, if we make the following substitutions, we can translate the electrical circuit equation above into the mechanical expression (9.6.3):

$$I \leftrightarrow x$$
$$L \leftrightarrow m$$
$$R \leftrightarrow \alpha \tag{22.3.3}$$
$$\frac{1}{C} \leftrightarrow k\,(=m\omega^2)$$

The quantities I, L, R, and $1/C$ are said to represent the *electrical analogs* of displacement, mass, dissipation constant, and spring constant, respectively. By using these substitutions, we can easily switch back and forth between the mechanical system and its electrical analog.

To write the solution for the electrical Eq. (22.3.2), all we have to do is replace every mechanical symbol in (9.6.4) with its electrical counterpart. In this way, we obtain

$$I(t) = I_0 e^{-(R/2L)t}\cos\left(\sqrt{\omega^2 - \frac{R^2}{4L^2}}\,t + \delta\right)$$

where ω (equal to $\sqrt{k/m}$ in the mechanical system) is given by

$$\omega = 1/\sqrt{LC} \tag{22.3.4}$$

This can be expressed even more simply as

$$I(t) = I_0 e^{-(R/2L)t}\cos(\omega' t + \delta)$$

where

$$\omega' = \sqrt{\frac{1}{LC} - \frac{R^2}{4L^2}} \tag{22.3.5}$$

These solutions are good only for resistances so small that $\alpha < 2m\omega$, or

$$R < 2\sqrt{\frac{L}{C}}$$

in other words,

$$\frac{R^2}{4L^2} < \frac{1}{LC} \tag{22.3.6}$$

As long as this inequality is satisfied, the quantity ω' in (22.3.5) will be real and the current will have the *damped oscillatory* form shown in Fig. 22.4 as curve (a). This has essentially the same form as the motion of the mechanical system shown in Fig. 9.12. It should be noted that an equally valid solution can be written

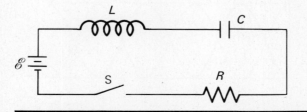

FIGURE 22.3. Series *R–L–C* (Resistance–Inductance–capacitance) circuit with emf.

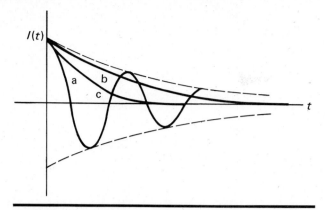

FIGURE 22.4. Current plotted as a function of time for the series *R–L–C* circuit: (a) damped oscillatory current for relatively small values of resistance, (b) critically damped current, and (c) overdamped current corresponding to very large values of resistance.

by replacing the cosine function in (22.3.5) with a sine function; the effect of this is only to change the value of the phase constant δ. The exponential factor that multiplies the cosine function in (22.3.5) will decrease by a factor $1/e\,(=0.368)$ in a time interval t_0 for which

$$\frac{Rt_0}{2L} = 1$$

or

$$t_0 = \frac{2L}{R} \qquad (22.3.7)$$

The quantity $2L/R$ is, therefore, said to express the *time constant* of an *R–L–C* circuit, in the same way that L/R and RC represent respectively the time constants of simple *L–R* and *R–C* circuits.

If the resistance is made so large that $R^2/4L^2 = 1/LC$, then, in (22.3.5), $\omega' = 0$ and the solution becomes

$$I(t) = I_0 e^{-(R/2L)t} \qquad (22.3.8)$$

The current now loses its oscillatory character altogether and simply dies off exponentially with a time constant given by (22.3.7), as illustrated by curve (b) in Fig. 22.4. Under these conditions, the circuit is said to be *critically damped*. If the resistance is increased still further, ω' becomes imaginary; in this case, the solution can no longer be expressed by (22.3.5) but must be written as a sum of two exponentially decaying terms. We shall not examine this solution in detail; its general form is indicated by curve (c) in Fig. 22.4. The circuit is now referred to as *overdamped*. The conditions of critical damping and overdamping, though not mentioned at the time when damped harmonic motion of mechanical systems was discussed, apply equally well to them also.

From the viewpoint of energy conservation, the decrease in current amplitude associated with damped oscillatory circuits is caused by the continual conversion of electrical energy stored in the electric and magnetic fields of the inductor and capacitor into heat. In any time interval dt, the amount of electrical energy dissipated as heat will be $I^2R\,dt$, and according to the law of conservation of energy, the sum of electric field and magnetic field energies must in the same interval decrease by exactly that amount.

The various solutions of the *R–L–C* circuit can be understood quantitatively by considering the analogy of a simple pendulum swinging in a viscous medium. In this case, the *viscosity* of the medium is analogous to the resistance *R* in the circuit. If the pendulum could swing in a vacuum, it would keep going forever (ignoring support friction); this is the analogy of the pure *L–C* circuit. An actual pendulum used, for example, in a typical lecture demonstration would swing to and fro in the air, executing an oscillatory motion which ultimately damps. This is damped oscillatory motion. The restoring force is much larger than the force which the air exerts on the pendulum bob. Therefore, the pendulum always overshoots its equilibrium position. In a similar manner, in the corresponding electrical circuit the current changes its direction many times as its amplitude decreases toward zero. Now, if the viscosity of the medium in which the pendulum swings is increased to a certain value, the pendulum just fails to overshoot its equilibrium point and no longer oscillates. This is the case of critically damped motion. It occurs in the electrical circuit when the resistance *R* has the value $R = 2\sqrt{L/C}$. Finally, if the viscosity of the medium is further increased, the motion is overdamped since the pendulum's movement is severely restrained by the medium; it then approaches its equilibrium position only very slowly. Whenever *R* exceeds $2\sqrt{L/C}$ in the *L–C–R* circuit, the current is overdamped and it approaches zero more slowly than in the critically damped circuit.

The three types of solutions discussed above are all called *transient solutions*, since the current and also the energy drop off essentially to zero after many time constants have elapsed. It should be understood, of course, that the current is never actually zero, although it may become extremely small after a short time interval, as the following example illustrates.

EXAMPLE 22.3.1
An electrical circuit contains a 5-μF capacitor, a 20-henry inductor, a 2000-ohm resistor, and a switch S. Initially, the capacitor is fully charged by a 200-volt battery. At $t = 0$, the switch is closed, making a complete circuit. (a) How does the current vary with time? (b) How long does it take for the current amplitude to drop to 1 percent of its maximum value?

Let us first compute the quantity ω' given by

$$\omega' = \sqrt{\frac{1}{LC} - \frac{R^2}{4L^2}} = \sqrt{\frac{1}{(20)(5 \times 10^{-6})} - \left(\frac{2000}{40}\right)^2}$$
$$= \sqrt{10^4 - 2500} = \sqrt{7500} \text{ (rad/sec)}^2$$

Since the quantity under the radical is positive, the solution is oscillatory, with an angular frequency

$$\omega' = \sqrt{7500} = 86.6 \text{ rad/sec}$$

The time constant is, therefore,

$$t_0 = \frac{2L}{R} = \frac{40}{2000} = 0.02 \text{ sec}$$

Thus, the solution has the form

$$I = I_0 e^{-t/0.02} \cos(86.6t + \delta) \tag{22.3.9}$$

We must now find I_0 and δ from the initial conditions. Since I is zero at the initial time $t = 0$, when the switch is closed, we must choose $\delta = \pi/2$. The determination of I_0 can then be made by using the equation

$$L\frac{dI}{dt} + RI + \frac{q}{C} = 0 \tag{22.3.10}$$

at time $t = 0$.

If we differentiate (22.3.9), we may write

$$\frac{dI}{dt} = -\frac{1}{0.02}I_0 e^{-t/0.02} \cos\left(86.6t + \frac{\pi}{2}\right)$$
$$- 86.6 I_0 e^{-t/0.02} \sin\left(86.6t + \frac{\pi}{2}\right)$$

or

$$\left(\frac{dI}{dt}\right)_{t=0} = -86.6 I_0 \tag{22.3.11}$$

Also, at time $t = 0$,

$$I(0) = 0$$

and

$$q(0) = C(\Delta V_C) = (5 \times 10^{-6})(200) = 10^{-3} \text{ C}$$

We now substitute these values into (22.3.10) to find

$$-(86.6)(20)I_0 + \frac{10^{-3}}{5 \times 10^{-6}} = 0$$

or

$$I_0 = 11.5 \times 10^{-2} \text{ amp} \tag{22.3.12}$$

The current amplitude is, therefore, given by

$$I = I_0 e^{-(R/2L)t} = (11.5 \times 10^{-2})e^{-t/0.02}$$

To find the time t at which $I/I_0 = 0.01$, we need only to solve the equation

$$0.01 = e^{-t/0.02}$$

Taking natural logarithms of both sides of this equation, we obtain

$$\ln 0.01 = \ln(e^{-t/0.02}) = -\frac{t}{0.02}$$
$$t = 0.092 \text{ sec}$$

22.4 The R–L–C Series Circuit with an Alternating EMF

In the preceding section, we studied the transient solutions of an R–L–C circuit. The energy stored in the inductor and the capacitor in such instances is ultimately dissipated, and the current finally falls off to zero.

In order to sustain a continuous oscillatory current in an R–L–C circuit, we must have a source of emf which is oscillatory also. An ac generator, for example, can supply an oscillating emf of the form

$$\mathscr{E}(t) = \mathscr{E}_0 \cos \omega t \tag{22.4.1}$$

where ω is the angular frequency. Thus, for a frequency of 60 Hz, the power line frequency of the voltage supplied to our homes, ω is 120π rad/sec. Let us now study R–L–C circuits which contain an emf of this form. We shall first study the mathematical solution of the series circuit, illustrated in Fig. 22.5.

According to Kirchhoff's circuit law, this series R–L–C series circuit now satisfies the differential equation

$$\mathscr{E}_0 \cos \omega t = L\frac{dI}{dt} + IR + \frac{q}{C} \tag{22.4.2}$$

instead of (22.3.1). By differentiating both sides of the equation with respect to t, we obtain, recalling that $I = dq/dt$,

$$-\mathscr{E}_0 \omega \sin \omega t = L\frac{d^2I}{dt^2} + R\frac{dI}{dt} + \frac{I}{C} \tag{22.4.3}$$

This equation is called an *inhomogeneous* differential equation, while (22.3.2) is called a *homogeneous* differential equation. The distinction arises because (22.4.3)

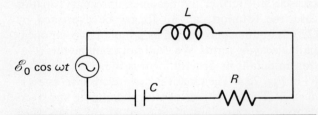

FIGURE 22.5. Series R–L–C circuit with sinusoidally alternating (ac) emf.

contains a specified function $f(t)$ on the left, whereas in (22.3.2) the function $f(t)$ is zero.

The most general solution of (22.4.3) can be obtained by writing I in the form

$$I = I_t + I_s \qquad (22.4.4)$$

where I_t is one of the *transient* solutions we discussed in the preceding section. The second term I_s is a particular solution of (22.4.3) which does *not* decay or diminish with time. It is, therefore, called the *steady-state solution*. Since I_t inevitably drops to zero after a short time, we shall concentrate now on studying only the steady-state solution. Hereafter, we shall simply use I to denote this solution.

Let us now attempt to find an oscillatory solution of (22.4.3) of the form

$$\boxed{I = I_0 \cos(\omega t - \phi)} \qquad (22.4.5)$$

which has the same frequency as the oscillating emf. In view of the fact that the emf $\mathscr{E}_0 \cos \omega t$ drives the current I, it is logical to expect that the current will oscillate at the same frequency as the driving emf. In writing the solution in this way, however, we assume that the current *need not necessarily be in phase with the emf*. Such a state of affairs is, in fact, found to occur in the ac inductive circuit that forms the primary of the transformer discussed in Example 20.7.2. The situation here is very similar; the reader is referred to the previous example for a more extended discussion. From (22.4.5), we find in this way that

$$\frac{dI}{dt} = -\omega I_0 \sin(\omega t - \phi) \qquad (22.4.6)$$

$$\frac{d^2 I}{dt^2} = -\omega^2 I_0 \cos(\omega t - \phi) \qquad (22.4.7)$$

Therefore, by substitution into (22.4.3),

$$-\mathscr{E}_0 \omega \sin \omega t = -L\omega^2 I_0 \cos(\omega t - \phi)$$
$$- R\omega I_0 \sin(\omega t - \phi)$$
$$+ \frac{I_0}{C} \cos(\omega t - \phi) \qquad (22.4.8)$$

At this point, we utilize the trigonometric identities

$$\sin(\omega t - \phi) = \sin \omega t \cos \phi - \cos \omega t \sin \phi$$
$$\cos(\omega t - \phi) = \cos \omega t \cos \phi + \cos \omega t \sin \phi \qquad (22.4.9)$$

to rewrite (22.4.8) in the form

$$-\mathscr{E}_0 \omega \sin \omega t$$
$$= I_0 \omega \left(\frac{1}{\omega C} - \omega L \right)(\cos \omega t \cos \phi + \sin \omega t \sin \phi)$$
$$- R\omega I_0 (\sin \omega t \cos \phi - \cos \omega t \sin \phi) \qquad (22.4.10)$$

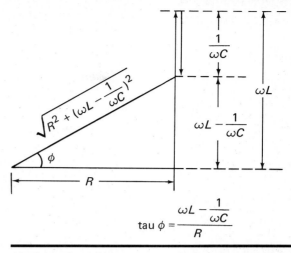

FIGURE 22.6. Impedance triangle showing geometric relationship between the quantities $\omega L - (1/\omega C)$ and R suggested by Eq. (22.4.11).

Now it turns out that this equation can only be satisfied for all times t if the coefficients of $\cos \omega t$ on both sides of the equation are identical. The coefficients of $\sin \omega t$ must likewise be the same. The situation in this respect is again very similar to the one we encountered in the discussion of the transformer in Example 20.7.2. Equating coefficients of $\cos \omega t$, therefore, we find

$$I_0 \omega \left(\frac{1}{\omega C} - \omega L \right) \cos \phi + R\omega I_0 \sin \phi = 0$$

or

$$\boxed{\tan \phi = \frac{\omega L - \dfrac{1}{\omega C}}{R}} \qquad (22.4.11)$$

The phase angle ϕ is shown in Fig. 22.6, which we shall refer to as the *impedance triangle*. From this diagram, we can read off the values of $\sin \phi$ and $\cos \phi$ needed in the relation obtained by equating the coefficients of $\sin \omega t$ on both sides of (22.4.10). In this way, we find

$$-\mathscr{E}_0 \omega = I_0 \omega \left(\frac{1}{\omega C} - \omega L \right) \sin \phi - R\omega I_0 \cos \phi$$

$$\mathscr{E}_0 = I_0 \left(\omega L - \frac{1}{\omega C} \right) \frac{\omega L - \dfrac{1}{\omega C}}{\sqrt{R^2 + \left(\omega L - \dfrac{1}{\omega C} \right)^2}}$$

$$+ I_0 R \frac{R}{\sqrt{R^2 + \left(\omega L - \dfrac{1}{\omega C} \right)^2}}$$

or finally

$$I_0 = \frac{\mathscr{E}_0}{\sqrt{R^2 + \left(\omega L - \dfrac{1}{\omega C}\right)^2}} = \frac{\mathscr{E}_0}{Z}$$

where

$$Z = \sqrt{R^2 + \left(\omega L - \frac{1}{\omega C}\right)^2}$$

(22.4.12)

The quantity Z is called the *impedance* of the R–L–C circuit; Z is easily seen from Fig. 22.6 to correspond to the *hypotenuse* of the impedance triangle. We may, therefore, write the current as

$$I = \frac{\mathscr{E}_0}{Z} \cos(\omega t - \phi)$$

(22.4.13)

The impedance of a series R–L–C circuit is somewhat analogous to the resistance R in a direct current circuit, since the current amplitude I_0 varies inversely with this quantity. It should be noted, however, that the quantity Z is a *function of the driving frequency ω* as well as of the quantities R, L, and C. As a consequence, the current amplitude I_0 is frequency dependent, and it attains a maximum for a certain value of ω, as the following example shows.

EXAMPLE 22.4.1
SERIES RESONANCE
An R–L–C circuit contains a resistance of 500 ohms and an inductance of 20 henrys. The capacitance C can be varied by turning a knob. If the applied ac voltage is at 60 hertz with an amplitude of 150 volts,

find (a) the value of C that gives the largest current amplitude I_0 and (b) the value of this maximum current. (c) Plot I_0 versus ω (considered now as a variable) for the above values of R, L, and C. Repeat this with values of the resistance of 50 ohms and 5 ohms.

The current amplitude I_0 is

$$I_0 = \frac{\mathscr{E}_0}{\sqrt{R^2 + \left(\omega L - \dfrac{1}{\omega C}\right)^2}} = \frac{\mathscr{E}_0}{Z}$$

and, therefore, for fixed values of $\mathscr{E}_0$, R, L, and ω, I_0 attains its maximum value when the impedance is as small as possible. This maximum value is reached at a frequency for which the quantity $\omega L - (1/\omega C)$ in the above equation becomes zero. This may be brought about in this case by adjusting the capacitance to satisfy the condition

$$\omega L = \frac{1}{\omega C} \quad \text{or} \quad \omega^2 = \frac{1}{LC}$$

(22.4.14)

Thus, the maximum value of I_0 occurs when the driving frequency ω is equal to the natural frequency $\omega_0 = \sqrt{1/LC}$ of the resistanceless L–C circuit. In the present example we find, accordingly, that

$$(120\pi)^2 = \frac{1}{20C} \qquad C = 0.35 \ \mu\text{F}$$

The maximum value of the current is, therefore,

$$(I_0)_{max} = \frac{\mathscr{E}_0}{R} = \frac{150}{500} = 0.3 \ \text{amp}$$

In Fig. 22.7, several graphs of I_0 versus ω are shown for resistances $R = 500$ ohms, 50 ohms, and

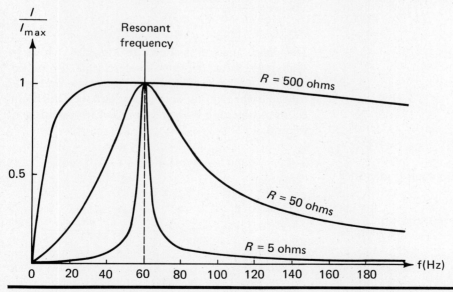

FIGURE 22.7. Variation of current with frequency in the R–L–C circuit of Example 22.4.1 for resistances of 500 ohms, 50 ohms, and 5 ohms, showing the condition of resonance at $f = 60$ Hz.

5 ohms. The feature clearly illustrated by these graphs is the very sudden variation of I_0 with ω as R decreases in value. In other words, the peak or maximum of I_0, which occurs at the *resonant* frequency $\omega_0 = \sqrt{1/LC}$, becomes much sharper or more pronounced for small values of R since the current amplitude then becomes very large. This phenomenon, in which I_0 assumes a sharp maximum at a particular frequency, is known as *resonance*. The sharpness of the resonance is measured by a quantity $\mathcal{Q}$, known as the quality factor, which is defined as the ratio of the inductive reactance to the resistance evaluated at the resonant frequency ω_0. Mathematically, we may write the definition of the quality factor as

$$\mathcal{Q} = \frac{\omega_0 L}{R} = \frac{1}{R}\sqrt{\frac{L}{C}} \qquad (22.4.15)$$

It is apparent from Fig. 22.7 that high $\mathcal{Q}$ implies a sharp resonance and low $\mathcal{Q}$, a resonance that is broad and ill defined.

When you tune your radio to a particular station, you are actually varying the value of the capacitance C to achieve a maximum current or voltage amplitude at the particular frequency at which the station broadcasts. Your radio is capable of amplifying signals at many different applied frequencies, received in the form of electromagnetic waves; but at resonance, it rejects most stations and amplifies only the one whose frequency matches the natural frequency of its tuning circuits. Though the circuits used in a radio employ resonance in a slightly different way from the one described above, the basic idea of varying a capacitance to tune to the resonant frequency is the same.

One might suppose that by reducing R to zero, one could draw an infinite amount of current. Although the mathematics suggests this, all circuits do have some resistance, and, therefore, the current is always limited to a finite value. This same resonance phenomenon also occurs in mechanical problems, as pointed out in Chapter 9. In mechanical systems, it is often important to *avoid* the occurrence of resonance, for if a system experiences a very large vibration amplitude, the consequences can often be disastrous, as in the case of the collapse of the Tacoma Narrows Bridge in 1940.

22.5 Rotating Vectors and Reactance

In the series R–L–C circuit of the preceding section, we had an alternating emf of the form

$$\mathscr{E}(t) = \mathscr{E}_0 \cos \omega t \qquad (22.5.1)$$

and we found that the circuit current is a sinusoidally varying current that differs in phase from the emf and

which, therefore, has the form

$$I(t) = I_0 \cos(\omega t - \phi) \qquad (22.5.2)$$

The voltages across the resistance, inductance, and capacitance are all sinusoidally varying quantities that are related to the current in different ways. In the case of the resistance, of course, this relation is simply

$$\Delta V_R = IR = I_0 R \cos(\omega t - \phi) \qquad (22.5.3)$$

the potential difference across the resistance being *in phase* with the current. The potential difference across the inductance is given by

$$\Delta V_L = L\frac{dI}{dt} = -\omega L I_0 \sin(\omega t - \phi)$$

$$= \omega L I_0 \cos\left(\omega t - \phi + \frac{\pi}{2}\right) \qquad (22.5.4)$$

since $\cos[\theta + (\pi/2)] = -\sin\theta$. From this, it is evident that the phase of the voltage across the inductor is 90° *ahead* of the current in the circuit. In the case of the capacitance, there will be a sinusoidally oscillating charge $q(t)$ on the condenser and a corresponding sinusoidal potential drop ΔV_c across it of the form

$$q(t) = q_0 \cos(\omega t - \delta) = C\,\Delta V_c \qquad (22.5.5)$$

where δ is a phase angle whose value we do not yet know. However, the instantaneous current in the circuit and the instantaneous charge on the condenser are related by

$$I(t) = \frac{dq}{dt} = -q_0\omega \sin(\omega t - \delta) = q_0\omega \cos\left(\omega t - \delta + \frac{\pi}{2}\right) \qquad (22.5.6)$$

since $\cos(90° + \theta) = -\sin\theta$. But now, from (22.5.2) and (22.5.6), $I(t)$ can be expressed in two different ways:

$$I(t) = I_0 \cos(\omega t - \phi) = q_0\omega \cos\left[\omega t - \left(\delta - \frac{\pi}{2}\right)\right] \qquad (22.5.7)$$

These two sinusoidally varying quantities can be equal only if they have the same amplitude and phase. Therefore,

$$q_0\omega = I_0 \qquad \text{and} \qquad \delta = \phi + \frac{\pi}{2} \qquad (22.5.8)$$

But this allows us to write the voltage ΔV_C across the capacitor, given by (22.5.5), as

$$\Delta V_C = \frac{I_0}{\omega C} \cos\left(\omega t - \phi - \frac{\pi}{2}\right) \qquad (22.5.9)$$

Comparing this with (22.5.2), it is easily seen that the phase of the potential difference across the capacitor is 90° behind that of the current in the circuit.

We may summarize these results as follows:

1. The circuit current I, the emf $\mathscr{E}$, and the voltage across the resistance, inductance, and capacitance are all sinusoidally varying quantities having the same frequency but different amplitudes and phases.
2. The amplitude of the voltage across a resistor is $I_0 R$ and it is in phase with the current.
3. The amplitude of the voltage across the inductor is $I_0 \omega L$, and the current through the inductor *lags behind* the voltage across it by 90°. We may express the ratio X_L of the amplitude of the voltage drop to the current amplitude for an inductor by

$$X_L = \frac{I_0 \omega L}{I_0} = \omega L \qquad (22.5.10)$$

4. The amplitude of the voltage across the capacitor is $I_0/\omega C$, and the current through the capacitor *leads* the voltage across it by 90°. The ratio X_C of the amplitude of the voltage drop to the current amplitude for a capacitor can be written

$$X_C = \frac{I_0}{\omega C I_0} = \frac{1}{\omega C} \qquad (22.5.11)$$

It is important, first of all to note that these results, though arising from the example of a series R–L–C combination, are true for *any ac circuit*. In deriving the relation (22.5.4) between the current and potential drop across the inductor, for example, all that is assumed is the presence of an ac voltage and a current through the element related at all times by $\Delta V_L = L(dI/dt)$. Similar considerations clearly apply to all the other conclusions stated above.

The quantities X_L and X_C defined above, expressing the ratio of voltage drop to current for inductors and capacitors in ac circuits, are referred to as the *inductive reactance* and *capacitive reactance* of the respective elements. Since they express the ratio of voltage to current, they have units of resistance—ohms in the mks system; and they play a role in ac circuits that resembles that of resistance in some ways but *differs* from it in other important respects. In particular, these reactances, in contrast to resistances, depend upon the frequency and, as we shall soon see, cannot be combined in quite the same way as resistances. Referring to (22.4.12), it is evident that the reactances X_L and X_C make up part of the circuit *impedance*.

Now, the voltages ΔV_R, ΔV_L, and ΔV_C are all simple harmonic oscillations that have *the same frequency*, though in general different amplitudes and different phases as well. The *instantaneous values* of these three quantities, in the case of the series R–L–C

circuit, must according to (22.4.2) add up at all times to the emf $\mathscr{E}_0 \cos \omega t$. What is involved here, therefore, is the addition of three different simple harmonic oscillations to give a fourth oscillation of the same sort. This addition can, of course, be done by complicated mathematics but, as we have shown in section 9.4 and, in particular, in Example 9.4.1, can also be accomplished by regarding the individual oscillations to be projections of the horizontal or vertical components of *vectors* that rotate about the origin with uniform angular velocity ω. For this purpose, the amplitudes of the individual oscillations are represented by the lengths of the vectors, and the phase differences between them by the angles between the directions of the individual vectors. The amplitude of the sum of the oscillations is then represented by the length of the resultant of the individual vectors, while its direction indicates the phase of the resultant oscillation. In adding the oscillatory disturbances this way, we take proper account of their *differences in phase*. We shall have further occasion in our study of optics to combine individual oscillatory disturbances in this way.

In the case of the series R–L–C circuit, then, we may represent the quantities ΔV_R, ΔV_L, and ΔV_C by vectors of magnitude $I_0 R$, $I_0 X_L$, and $I_0 X_C$, as illustrated in Fig. 22.8. The voltage drop ΔV_R has the same phase as the current, the angle between the vector representing this voltage and the x-axis being $\omega t - \phi$, as specified by Eq. (22.5.2). The vector is visualized as rotating about the origin with angular velocity ω, and its projection onto the x-axis is, therefore, the voltage $\Delta V_R = I_0 R \cos(\omega t - \phi)$ in agreement with (22.5.2). Since the voltage across the inductor is 90° ahead of the current in phase and the voltage across the capacitor is 90° behind the current, the vectors representing those voltages must be 90° ahead of and 90° behind

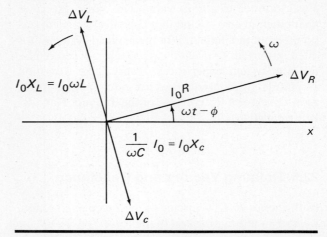

FIGURE 22.8. Magnitude and relative phase of voltage drops across resistive, capacitive, and inductive elements represented as vectors rotating with angular velocity ω.

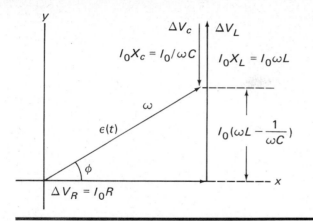

FIGURE 22.9. Addition of vectors representing the magnitude and relative phase of voltage drops across the resistance, inductance, and capacitance, exhibiting the "impedance triangle" of Fig. 22.6 whose hypotenuse has the magnitude and phase of the applied emf.

the first vector, respectively, as shown in the figure. All three vectors rotate together with angular velocity ω. The net effect of combining these oscillations, whose sum is represented by $\mathscr{E}(t)$, can now be exhibited by simple vector addition, as shown in Fig. 22.9. In this drawing, just to make things simple, we have "frozen" the system's rotation just as the vector representing ΔV_R (in phase with the current) is along the x-direction. The magnitude and phase of the resultant of the three vectors representing ΔV_R, ΔV_L, and ΔV_C represents the amplitude and phase angle of the applied emf. From this diagram, it is evident that

$$\mathscr{E}_0 = I_0 \sqrt{R^2 + \left(\omega L - \frac{1}{\omega C}\right)^2}$$

or

$$I_0 = \frac{\mathscr{E}_0}{Z} \qquad (22.5.12)$$

where

$$Z = \sqrt{R^2 + (X_L - X_C)^2}$$
$$= \sqrt{R^2 + \left(\omega L - \frac{1}{\omega C}\right)^2} \qquad (22.5.13)$$

This is in agreement with the results of the preceding section. It is also easily seen that the current differs in phase from the emf by a phase angle ϕ such that

$$\tan \phi = \frac{X_L - X_C}{R} = \frac{1}{R}\left(\omega L - \frac{1}{\omega C}\right) \qquad (22.5.14)$$

This also agrees with our previously obtained result.

It is evident from this and from our previous

calculations that the current in the circuit has the maximum value $I_0 R$ at resonance, when the capacitive and inductive reactances ωL and $1/\omega C$ are equal and when the quantity $\omega L - (1/\omega C)$ is zero. Under these circumstances, the current and applied emf are *in phase*. If the inductive reactance and capacitive reactance are not equal, the current will be less than this, and there will be a phase difference between the current and emf. When the inductive reactance exceeds the capacitive reactance, the phase angle ϕ will be positive and the current in the circuit will lag behind the emf, while if the capacitive reactance is larger, the phase angle is negative and the current is ahead of the emf in phase. The close relation between this method of summing ac voltage and currents and the geometry of the "impedance triangle" of the preceding section is readily apparent.

Those of our readers who are sensitive to the niceties of language will have noticed a certain indirectness in expression in the above discussion. We have said that the voltage ΔV_R, for example, is "described by the projection of a rotating vector whose length is the magnitude of ΔV_R and whose direction represents the phase angle of the current," rather than simply saying that "ΔV_R is a rotating vector." There is an excellent reason for all these excess words, and it is this: the quantity ΔV_R is *not a vector at all*, it is a *scalar quantity*; and the same goes for ΔV_L, ΔV_C, $\mathscr{E}(t)$, and $I(t)$. In the same way, the resistance R and the reactances X_L and X_C are all *scalars*. It simply happens that the amplitudes and phases of harmonically varying scalars combine mathematically in the same way as do the projections of rotating vectors of corresponding magnitudes and directions. The "rotating vectors" that "represent" harmonically varying scalar quantities are introduced only to provide us with a simple way of adding these scalar quantities using a rule that we already know as the law of vector addition!

EXAMPLE 22.5.1

A capacitor of 5-μF capacitance and a 100-ohm resistance are connected in series with an ac emf of variable frequency and negligible internal impedance. The amplitude of the emf is 150 volts. Find the current through the inductor for frequencies of 0, 10, 100, and 1000 hertz. Plot the current as a function of frequency and explain why it varies as it does. Find the phase angle between current and applied emf at the above frequencies and plot the phase angle as a function of frequency.

In this example, there is no inductive reactance at all—only capacitance. The circuit is illustrated in Fig. 22.10a. Using the picture developed in the preceding section, then, we can represent the harmonically varying quantities ΔV_R and ΔV_C as rotating

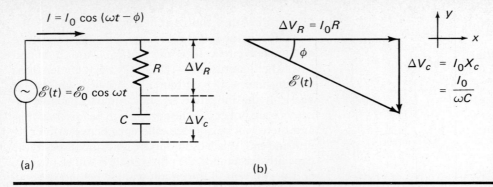

(a)

(b)

FIGURE 22.10. (a) Series R–C circuit with ac emf. (b) Vectors representing magnitude and relative phases of the voltages in (a).

vectors whose magnitudes are given by

$$\Delta V_R = I_0 R \quad \text{and} \quad \Delta V_C = I_0 X_C = \frac{I_0}{\omega C} \quad (22.5.15)$$

The vector representing ΔV_R is in phase with the current, and the vector representing ΔV_C is 90° behind that vector in phase, because the phase of the current through a capacitor is always 90° ahead of the voltage across it. These phase relations are shown in Fig. 22.10b, where, as usual, we have chosen to draw the vector diagram at the instant the "rotating vector" ΔV_R is horizontal. Since there is no inductance in the circuit, there is no vector to represent an inductive voltage drop. From Fig. 22.10, it is evident that the two harmonically varying voltages ΔV_R and ΔV_C must sum to the emf $\mathscr{E}(t)$ at all times, and, therefore, the sum of the vectors representing these quantities must give a vector whose magnitude and direction represent the amplitude and phase of the emf. Therefore,

$$\mathscr{E}_0 = I_0 \sqrt{R^2 + (1/\omega C)^2}$$

from which

$$I_0 = \frac{\mathscr{E}_0}{\sqrt{R^2 + (1/\omega C)^2}} = \frac{\mathscr{E}_0}{Z} \quad (22.5.16)$$

while

$$\tan \phi = \frac{1}{R\omega C} \quad (22.5.17)$$

This example is similar to the circuit representing the primary of a transformer considered in connection with Example 20.7.2, in which only inductance and resistance are present. The reader could profitably return to that example and apply the methods developed above to its solution. The situation here differs mainly in that the circuit *leads* the applied emf, rather than lagging behind as in the former example.

Inserting the numerical values $\mathscr{E}_0 = 100$ volts, $R = 100$ ohms, $C = 5 \times 10^{-6}$ faraday into these equations, we find that for $\omega = 0$, $I_0 = 0$, $\phi = 90°$. For $f = 10$ hertz, $\omega = 20\pi$ rad/sec, $I_0 = 0.0471$ ampere, $\phi = 88.2°$. For $f = 100$ hertz, $\omega = 200\pi$ rad/sec, $I_0 = 0.450$ ampere, $\phi = 72.6°$, etc. These results are shown in Fig. 22.11. For low frequencies, the capacitive reactance $1/\omega C$ is very high and the current amplitude, therefore, very low. As the frequency increases, the capacitive reactance (and thus the total impedance) decreases, resulting in an increase in current. At high frequencies, the capacitive reactance becomes much less than the resistance, and the current then approaches the limiting value $\mathscr{E}_0/R$. At low fre-

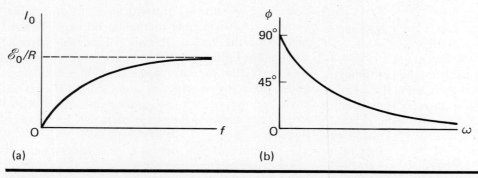

(a)

(b)

FIGURE 22.11. (a) Current amplitude and (b) phase angle between current and emf, plotted against frequency, for the circuit of Fig. 22.10.

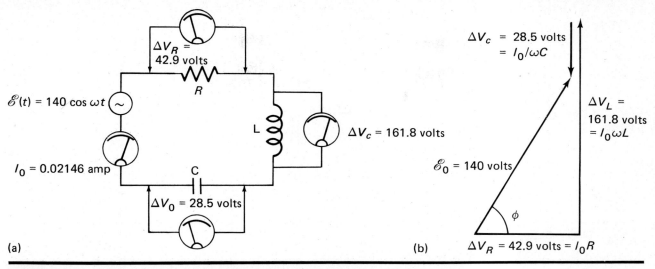

FIGURE 22.12. (a) Series *R–L–C* circuit of Example 22.5.2. (b) Vectors representing magnitude and relative phase of the voltages in (a).

quencies, there is a phase difference of nearly 90° between current and applied emf, which decreases with rising frequency and approaches zero at frequencies for which $R \gg 1/\omega C$.

EXAMPLE 22.5.2

An *R–L–C* circuit contains a resistance, an inductance, and a capacitance, none of whose values is known, in series with an emf of amplitude 140 volts and frequency 60 hertz, as shown in Fig. 22.12. An ac voltmeter is used to measure the voltage drops across the resistance, capacitance, and inductance, as illustrated in Fig. 22.12a. The meter indicates that the amplitude of the voltage across the resistor is 42.9 volts, that of the voltage across the inductor is 161.8 volts, and that of the voltage across the capacitor is 28.5 volts. An ac ammeter in the circuit indicates that the current amplitude is 0.02146 ampere.[1] Explain the rather curious fact that the sum of the indicated voltages across the resistance, inductance, and capacitance does not add up to the amplitude of the emf, and find the resistance, inductance, and capacitance from the given data.

It is found that the sum of the three measured voltage drops is $42.9 + 161.8 + 28.5 = 233.2$ volts, which is very different from the applied emf of 140 volts. This is not an indication that someone has repealed the laws of arithmetic, but simply reflects the fact that the voltmeter indicates only the *amplitude* of the voltage and knows nothing about its *phase*.

We must be careful, therefore, to properly take account of phase as well as amplitude when adding harmonically varying quantities. In the series *L–C* circuit with no emf considered in section 22.2, for example, the voltages across the inductance and capacitance were found to differ in phase by 180°. In that case, a voltmeter would have indicated a perfectly definite voltage across the inductor and a voltage of equal amplitude across the capacitor. But the sum of these is not really twice the indicated voltage, but *zero*, since the two are of opposite phase.

In this example, the voltages ΔV_R, ΔV_L, and ΔV_C must add up to the emf as illustrated in Fig. 22.12b. In the case of the voltage across the resistor, since the current is known to be 0.02146 ampere, we find

$$\Delta V_R = I_0 R$$

or

$$42.9 = (0.02146)R \qquad R = 2000 \text{ ohms}$$

For the inductor,

$$\Delta V_L = I_0 \omega L = (0.02146)(2\pi)(60)L = 161.8 \text{ V}$$
$$L = 20 \text{ H}$$

Finally, for the capacitor,

$$\Delta V_C = \frac{I_0}{\omega C} = \frac{(0.02146)}{(2\pi)(60)C} = 28.5 \text{ V}$$

$$C = 2.0 \times 10^{-6} \text{ F} = 2.0 \ \mu\text{F}$$

The phase angle ϕ between the current and emf, from

---

[1] Actually, most ac ammeters and voltmeters are calibrated to read root-mean-square (rms) average values of voltage and current rather than the amplitude itself. But the amplitude is easily found if the rms average is known. As we shall soon see, the amplitude is simply $\sqrt{2}$ times the rms average value.

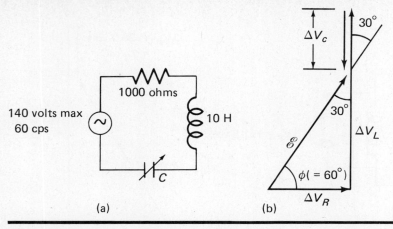

FIGURE 22.13. Series R–L–C circuit of Example 22.5.3.

Fig. 22.12b, is given by

$$\tan \phi = \frac{\Delta V_L - \Delta V_C}{\Delta V_R} = \frac{161.8 - 28.5}{42.9} = 3.107$$

$$\phi = 72.2°$$

In this case, it is clear from the figure that the current lags behind the emf. The reader should now be able to find the resonant frequency of the circuit and should be able to predict the voltages across all three circuit elements if the frequency of the emf were adjusted to the resonant frequency.

EXAMPLE 22.5.3

A series R–L–C circuit, shown in Fig. 22.13, contains an emf of magnitude 140 volts, oscillating at 60 cps. The resistance is 1000 ohms, the inductance is 10 henrys, and C can be varied. What value should C have in order for the voltage across the inductor, ΔV_L, to lead the applied emf by 30°? At this value of C, find the magnitude of the current I_0.

In this case, the phase relationships are as shown in Fig. 22.13b. The vector representing ΔV_L points vertically upward and must be 30° *ahead* of the vector representing the applied emf, as illustrated. But this requires that the phase angle ϕ between the current, which is in the same direction as the horizontal vector representing ΔV_R, and the emf must be 60°. Then,

$$\tan \phi = \tan 60° = \frac{X_L - X_C}{R}$$

$$X_C = X_L - R \tan 60° = \omega L - R \tan 60°$$

$$\frac{1}{\omega C} = (120\pi)(10) - (1000)(1.732) = 2037.9$$

$$= \frac{1}{(120\pi)C}$$

$$C = 1.3 \ \mu F$$

The impedance of the circuit is found to be

$$Z = \sqrt{R^2 + (X_L - X_C)^2} = R \sqrt{1 + \frac{(X_L - X_C)^2}{R^2}}$$

$$= R\sqrt{1 + \tan^2 60°} = (1000)(2) = 2000 \text{ ohms}$$

and hence the current amplitude is

$$I_0 = \frac{\mathscr{E}_0}{Z} = \frac{140}{2000} = 0.07 \text{ amp}$$

22.6 Some Other Examples of AC Circuits

In this section, a variety of simple ac networks will be studied. In each case, the circuit has two *input* terminals, at which a signal is supplied in the form of a sinusoidally varying voltage. There are also two *output* terminals which furnish a harmonically varying output voltage having the same frequency but different amplitude and phase. The situation is illustrated in Fig. 22.14, in which the actual circuit elements are schematically confined within a "black box." Such a black box could represent an amplifier, an attenuator, a filter circuit, or other useful signal processing device, but it is not necessary to specify its purpose in advance. The purpose of the box is simply to convert the input signal ΔV_{in} to the output signal ΔV_{out}. The exact contents of the box depend on what type of output signal is desired.

Let us consider first the two circuits shown in

FIGURE 22.14. Arbitrary four-terminal "black box" network.

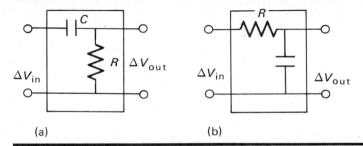

FIGURE 22.15. (a) $R-C$ high-pass filter and (b) $R-C$ low-pass filter.

Fig. 22.15. In Fig. 22.15a, the output signal is the voltage across the resistor, while in Fig. 22.15b, it is the voltage across the capacitor. In each of these cases, we assume we are dealing with a series $R-C$ circuit and that the solutions developed in sections 22.4 and 22.5 are applicable. In using these solutions, since there is no inductive element present, we shall consider the inductive reactance $X_L = \omega L$ to be zero throughout. In much of what follows, we shall be interested in computing the *time average* of the magnitude of $\Delta V_{out}/\Delta V_{in}$ for the case in which $\Delta V_{in} = \mathcal{E}_0 \cos \omega t$. This ratio is called the "gain" of the circuit at frequency ω, and its inverse is the "attenuation." In the series $R-C$ circuit, the current and the charge on the capacitor are respectively given by

$$I = \frac{\mathcal{E}_0}{\sqrt{R^2 + \left(\dfrac{1}{\omega C}\right)^2}} \cos(\omega t - \phi) \qquad (22.6.1)$$

and

$$q = \frac{\mathcal{E}_0}{\omega \sqrt{R^2 + \left(\dfrac{1}{\omega C}\right)^2}} \sin(\omega t - \phi) \qquad (22.6.2)$$

where

$$\tan \phi = -\frac{1}{\omega R C} \qquad (22.6.3)$$

For Fig. 22.15a, the ratio of the output to the input is, therefore,

$$\frac{\Delta V_{out}}{\Delta V_{in}} = \frac{IR}{\mathcal{E}_0 \cos \omega t} = \frac{R}{\sqrt{R^2 + \left(\dfrac{1}{\omega C}\right)^2}} \frac{\cos(\omega t - \phi)}{\cos \omega t}$$

$$= \frac{R}{\sqrt{R^2 + \left(\dfrac{1}{\omega C}\right)^2}} \frac{\cos \omega t \cos \phi + \sin \omega t \sin \phi}{\cos \omega t}$$

$$= \frac{R}{\sqrt{R^2 + \left(\dfrac{1}{\omega C}\right)^2}} (\cos \phi + \sin \phi \tan \omega t)$$

$$(22.6.4)$$

To obtain the gain at this frequency, we take the *average* value of the absolute magnitude of this quantity. The time average of $\tan \omega t$ is zero, however, and since in the "impedance triangle"

$$\cos \phi = \frac{R}{\sqrt{R^2 + \left(\dfrac{1}{\omega C}\right)^2}} \qquad (22.6.5)$$

we find the gain of the device to be

$$\left|\frac{\Delta V_{out}}{\Delta V_{in}}\right|_{av} = \frac{R^2}{R^2 + \left(\dfrac{1}{\omega C}\right)^2} = \cos^2 \phi \qquad (22.6.6)$$

It is important to observe that this ratio is a *function of the frequency*. For very large ω, the capacitive reactance approaches zero, and, therefore, the ratio approaches unity. At the opposite extreme, for very small ω, the ratio approaches zero. The above frequency dependence of the gain implies that any input signal which contains a *superposition* of different frequencies (such as a recording of a musical selection) can be processed in such a way as to have almost 100 percent transmission of the high-frequency part of the signal and very little transmission of the low-frequency part. Such a circuit is called a *high-pass filter*. In Fig. 22.16a, we plot $|\Delta V_{out}/\Delta V_{in}|_{av}$ versus ω to demonstrate that low frequencies are filtered out.

In Fig. 22.15b, the output signal is taken across the capacitor. For this case, the output-to-input ratio is

$$\frac{\Delta V_{out}}{\Delta V_{in}} = \frac{\mathcal{E}_0 \sin(\omega t - \phi)}{(\mathcal{E}_0 \cos \omega t)\omega C \sqrt{R^2 + \left(\dfrac{1}{\omega C}\right)^2}} \qquad (22.6.7)$$

Using the trigonometric identity

$$\sin(\omega t - \phi) = \sin \omega t \cos \phi - \cos \omega t \sin \phi$$

and averaging the absolute value over time, we now obtain

$$\left|\frac{\Delta V_{out}}{\Delta V_{in}}\right|_{av} = -\left(\frac{1}{\omega C}\right) \frac{1}{\sqrt{R^2 + \left(\dfrac{1}{\omega C}\right)^2}} = \sin \phi \qquad (22.6.8)$$

731

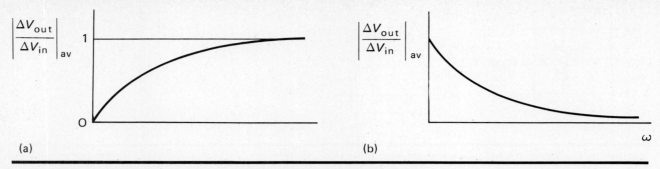

(a)

(b)

ω

FIGURE 22.16. Gain plotted as a function of frequency for (a) $R-C$ high-pass filter and (b) $R-C$ low-pass filter.

This expression, as a function of frequency, behaves quite differently from (22.6.6). In this case, the gain approaches unity at *low* frequencies and falls to zero at high frequency. Therefore, this arrangement serves as a *low-pass filter* since low-frequency components all tend to be transmitted to the output while the high-frequency voltages are suppressed. Figure 22.16b shows a plot of (22.6.8) in which the gain is plotted as a function of ω.

Let us now discuss several examples of resonant ac circuits, one of which we have already encountered in our studies of the series $R-L-C$ circuit shown in Fig. 22.5. We have seen that such a circuit exhibits resonant behavior when I is plotted versus ω. The resonance occurs when ω is equal to $\sqrt{1/LC}$. This implies that the *output voltage* across the resistor is very much larger near the resonance frequency than it is far from resonance. Now, if the input voltage contains many different frequencies, the output signal will be strong only at frequencies near the resonance, and, therefore, this type of circuit could be used in tuning a radio to a particular broadcast frequency. In practice, however, there are some technical disadvantages in the use of this circuit; as a result, a different circuit, sometimes called a "tank circuit," is more frequently used for tuning a radio. In this circuit, which is shown in Fig. 22.17, the capacitor and the inductor are connected in *parallel*, and this combination is in series with a resistor R. Although the inductor also usually has some resistance, in the following analysis we shall neglect its effect. The output voltage is taken across the parallel $L-C$ combination. In the series $R-L-C$ circuit, the resonance condition implies

that the current is a *maximum*. In the present case, however, a properly tuned circuit will be one in which the current at the resonant frequency is a *minimum*. In this event, the voltage across the resistor is a minimum, and, therefore, the *output voltage* ΔV_{out} will be a *maximum*. Circuits of this type can be solved by rather general and elegant methods, in which the various impedances are combined to find a total impedance for the circuit. We shall not develop these general methods in this text but shall instead apply Kirchhoff's laws directly to the circuit at hand.

Let the current through the resistor be denoted by I, while the currents through the inductor and capacitor are, respectively, I_1 and I_2. We shall again work out the case in which there is a single input frequency ω and then analyze the resonant behavior as a function of frequency. The circuit equations below follow from Kirchhoff's law, from conservation of current, and from the fact that the potential difference across the inductance is the same as that across the capacitance. Accordingly, we can write

$$I = I_1 + I_2 \qquad (22.6.9)$$

$$\mathscr{E}_0 \cos \omega t = \Delta V_R + \Delta V_L = IR + L\frac{dI_1}{dt}$$

$$= (I_1 + I_2)R + L\frac{dI_1}{dt} \qquad (22.6.10)$$

$$\Delta V_L = L\frac{dI_1}{dt} = \frac{q_2}{C} = \Delta V_C \qquad (22.6.11)$$

From (22.6.11), we obtain by differentiation

$$I_2 = LC\frac{d^2 I_1}{dt^2} \qquad (22.6.12)$$

Substituting this expression in (22.6.10), we obtain

$$\mathscr{E}_0 \cos \omega t = RLC\frac{d^2 I_1}{dt^2} + L\frac{dI_1}{dt} + RI_1 \qquad (22.6.13)$$

This equation is quite similar in form to (22.4.4), except that instead of sin ωt, we now have cos ωt. The solution can, therefore, again be written in the form

$$I_1 = I_{10} \cos(\omega t - \delta) \qquad (22.6.14)$$

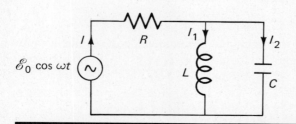

FIGURE 22.17. $R-L-C$ "tank circuit" of section 22.6.

where the current amplitude I_{10} and the phase angle δ are frequency dependent. Before we proceed to determine the values of I_{10} and δ, however, let us use (22.6.12) to find an expression for I_2. Carrying out the two required differentiations, we see that

$$I_2 = -\omega^2 LC I_{10} \cos(\omega t - \delta) \qquad (22.6.15)$$

Combining I_1 and I_2 to obtain the total current, we obtain

$$I = (1 - \omega^2 LC) I_{10} \cos(\omega t - \delta)$$
$$\equiv I_0 \cos(\omega t - \delta) \qquad (22.6.16)$$

where I_0 is the amplitude of the total current $I_1 + I_2$. This equation certainly suggests that at the frequency $\omega = \sqrt{1/LC}$, the current amplitude I_0 will vanish. This is true provided I_{10} does not become infinite at the same value of ω. As we shall show later, I_{10} is indeed finite, and, therefore, the current amplitude I_0 drops to zero when the applied frequency matches the natural frequency of the resistanceless L–C circuit.

Under these conditions, the output and input voltages are identical; moreover, there are no Joule heating losses at this frequency since the current through the resistor vanishes. All of this ordinarily implies the existence of a sharp output signal at the desired frequency. But, in general, the input signal to a radio contains the frequencies of all broadcasting stations near enough to produce a sizable input voltage. By tuning the radio, which is usually accomplished by varying the value of C, we process the input signal in such a manner as to virtually eliminate all frequencies *except those near the resonance*, for which the condition $\omega^2 = 1/LC$ is satisfied.

Let us now calculate the current I_1 through the inductor by substituting the assumed solution (22.6.14) into (22.6.13). We then find

$$\mathscr{E}_0 \cos \omega t = R(1 - \omega^2 LC) I_{10} \cos(\omega t - \delta)$$
$$- \omega L I_{10} \sin(\omega t - \delta) \qquad (22.6.17)$$

Using trigonometric identities to expand the sine and cosine functions, we can write

$$\mathscr{E}_0 \cos \omega t$$
$$= I_{10} R(1 - \omega^2 LC)(\cos \omega t \cos \delta + \sin \omega t \sin \delta)$$
$$- I_{10} \omega L (\sin \omega t \cos \delta - \cos \omega t \sin \delta) \quad (22.6.18)$$

This equation can only be satisfied if the coefficients of $\cos \omega t$ and of $\sin \omega t$ are identical on both sides of the equation. Equating coefficients of $\sin \omega t$, we obtain

$$\tan \delta = \frac{\omega L}{R(1 - \omega^2 LC)} \qquad (22.6.19)$$

Note that when the condition for resonance is satisfied, $\tan \delta$ is infinite and, therefore, the phase angle δ will be $\pi/2$. By equating coefficients of $\cos \omega t$, we find also

that

$$\mathscr{E}_0 = I_{10}[R(1 - \omega^2 LC) \cos \delta + \omega L \sin \delta]$$
$$= I_{10}\left[R(1 - \omega^2 LC) \frac{R(1 - \omega^2 LC)}{\sqrt{(\omega L)^2 + R^2(1 - \omega^2 LC)^2}} \right.$$
$$\left. + \frac{(\omega L)^2}{\sqrt{(\omega L)^2 + R^2(1 - \omega^2 LC)^2}} \right]$$
$$= I_{10}\sqrt{(\omega L)^2 + R^2(1 - \omega^2 LC)^2} \qquad (22.6.20)$$

Finally, upon solving for I_{10}, we obtain

$$I_{10} = \frac{\mathscr{E}_0}{\sqrt{(\omega L)^2 + R^2(1 - \omega^2 LC)^2}} \qquad (22.6.21)$$

for the amplitude of current I_1. We now see that this amplitude does not become infinite when the resonance condition is imposed. Using the above expression for I_{10}, I_0, which is defined by (22.6.16), can now be written as

$$I_0 = \frac{\mathscr{E}_0(1 - \omega^2 LC)}{\sqrt{(\omega L)^2 + R^2(1 - \omega^2 LC)^2}}$$
$$= \frac{\mathscr{E}_0}{\sqrt{R^2 + \dfrac{L^2/C^2}{\left(\omega L - \dfrac{1}{\omega C}\right)^2}}} = \frac{\mathscr{E}_0}{Z'} \qquad (22.6.22)$$

where

$$Z' = \sqrt{R^2 + \frac{L^2/C^2}{\left(\omega L - \dfrac{1}{\omega C}\right)^2}} \qquad (22.6.23)$$

In this equation, the impedance Z' is obviously *different* from that of an R–L–C series circuit. As we mentioned earlier, there are methods that can be used to determine the impedance without going through the intermediate steps of solving the differential equation.[2] These methods will also lead to the conclusions reached above. It is now clear from (22.6.23) that when the resonance condition is met, the impedance is *infinite*, and, therefore, the "tank circuit" draws no current. In Fig. 22.18 we show the impedance Z', the current I, and the ratio $|\Delta V_{\text{out}}/\Delta V_{\text{in}}|_{\text{av}}$ as functions of the frequency ω, assuming a single input frequency.

22.7 Power in AC Circuits

Let us consider now the energy transformations which occur in ac circuits. The instantaneous power delivered to any circuit element is given by $I(t)\Delta V(t)$, where $I(t)$

[2] See, for example, *Introductory Electronics for Scientists and Engineers* by Robert E. Simpson, published by Allyn and Bacon, Inc., 1974, for a discussion of these methods.

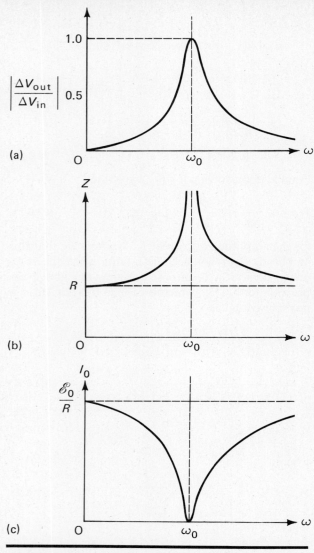

FIGURE 22.18. (a) Gain, (b) impedance, and (c) circuit current I_0 plotted as functions of frequency for "tank circuit."

and $\Delta V(t)$ are the instantaneous current through the element and the voltage across it, respectively. The power supplied by a source of emf such as a generator is also given by current multiplied by voltage. In general, even though the average values of $I(t)$ and $\Delta V(t)$ may be zero, the average of their product is *not necessarily zero*.[3] For example, when a light bulb is rated at 100 watts, this implies that the average energy transfer to the light bulb is 100 joules/second even though the average current and average voltage are both zero.

If $P(t)$ denotes an instantaneous power, the average power $\bar{P}$ is defined by

[3] The reason for this basically stems from the fact that though the average value of $\sin \omega t$ over a complete cycle is zero, the average value of $\sin^2 \omega t$ is not zero, since $\sin^2 \omega t$ can never have negative values.

$$\bar{P} = \frac{1}{T} \int_0^T P(t)\, dt \tag{22.7.1}$$

where T is the period of oscillation. This average, taken over one complete cycle, would be the same as the average taken over many cycles.

The value of $\bar{P}$ at any circuit element depends on two very important factors. The first of these is the *product of the magnitudes* of the voltage and current. The second is the *phase relation* between the voltage and the current. For example, consider a situation in which the voltage across a given circuit element is

$$\Delta V(t) = \Delta V_0 \cos(\omega t + \phi) \tag{22.7.2}$$

while the current is

$$I(t) = I_0 \cos(\omega t + \phi - \delta) \tag{22.7.3}$$

The current, therefore, lags behind the voltage by the phase angle δ. The instantaneous power $P(t)$ is now given by

$$\begin{aligned} P(t) &= I(t)\, \Delta V(t) \\ &= I_0\, \Delta V_0 \cos(\omega t + \phi) \cos(\omega t + \phi - \delta) \end{aligned} \tag{22.7.4}$$

This expression can be rewritten in a more convenient form by using the trigonometric identity

$$\cos a \cos b = \tfrac{1}{2}\left[\cos(a - b) + \cos(a + b)\right]$$

We then have

$$P(t) = \frac{I_0\, \Delta V_0}{2}\left[\cos \delta + \cos(2\omega t + 2\phi - \delta)\right] \tag{22.7.5}$$

We now substitute this expression in (22.7.1) to find the average power delivered to the circuit element:

$$\bar{P} = \frac{1}{T} \int_0^T \frac{I_0\, \Delta V_0}{2} \cos \delta\, dt$$

$$+ \frac{1}{T} \int_0^T \frac{I_0\, \Delta V_0}{2} \cos(2\omega t + 2\phi - \delta)\, dt$$

$$= \frac{I_0\, \Delta V_0}{2} \cos \delta$$

$$+ \frac{1}{T} \frac{I_0\, \Delta V_0}{2} \left[\frac{1}{2\omega} \sin(2\omega t + 2\phi - \delta)\right]_0^T$$

or

$$\boxed{\bar{P} = \frac{I_0\, \Delta V_0}{2} \cos \delta} \tag{22.7.6}$$

To obtain this result, we used

$$\sin(2\omega T + 2\phi - \delta) - \sin(2\phi - \delta)$$
$$= \sin(4\pi + 2\phi - \delta) - \sin(2\phi - \delta) = 0$$

The appearance of the factor of $\frac{1}{2}$ always results from the *averaging* process and suggests the following definition of *effective* values of current and voltage:

$$I_{\text{eff}} = \frac{I_0}{\sqrt{2}} \qquad \Delta V_{\text{eff}} = \frac{\Delta V_0}{\sqrt{2}} \qquad (22.7.7)$$

In practical usage, these effective values are nearly always specified in place of the amplitude or "peak" values. Thus, a "110 volt" ac line has 110 volts for its effective voltage, hence $110 \times \sqrt{2} = 156$ volts as its peak or maximum voltage.

The effective voltages and currents are frequently called the root-mean-square (abbreviated rms) values since they are given by the square root of the time average of the square, that is, $I_{\text{rms}} = \sqrt{\overline{I^2}}$. Hereafter, we will refer to these effective quantities by using the subscript rms. Ordinary ac voltage- and current-measuring instruments are nearly always calibrated to read rms average values of voltage or current rather than peak values.

We see from the above calculations that the average power delivered to a circuit element can be written as

$$\bar{P} = I_{\text{rms}} \Delta V_{\text{rms}} \cos \delta \qquad (22.7.8)$$

The quantity $\cos \delta$ is called the *power factor* of the circuit element. Thus, restating (22.7.8) in words,

average power = (rms current) × (rms voltage)
× (power factor)

Let us now consider the power factors of some simple ideal ac circuit elements.

1. Resistor: The voltage and current across a resistor are in phase since $\Delta V_R = IR$. This implies that $\delta = 0$ and, therefore, the power factor is unity.
2. Inductor: As we have seen earlier, the voltage across the inductor leads the current by 90° or $\pi/2$. Therefore, $\delta = \pi/2$ and the power factor is zero.
3. Capacitor: The voltage across the capacitor lags the current by 90°. This implies $\delta = -\pi/2$ and, hence, the power factor is again zero.
4. Generator in an *R–L–C* Circuit: The generator supplies the emf to the circuit. In this case, the current lags the voltage by a phase angle ϕ which is given by (22.4.11). The power factor thus turns out to be

$$\cos \phi = \frac{R}{Z}$$

where R is the resistance of the circuit and Z is the total impedance.

Let us now check the energy balance in a series *R–L–C* circuit by examining the average power for each element. The generator supplies an average power

$$\bar{P} = I_{\text{rms}} \mathscr{E}_{\text{rms}} \frac{R}{Z} = I_{\text{rms}}^2 R \qquad (22.7.9)$$

At the resistor, the power absorbed is

$$\bar{P} = I_{\text{rms}} \Delta V_{\text{rms}}(1) = I_{\text{rms}}^2 R \qquad (22.7.10)$$

while at the inductor and capacitor, there is no average energy transfer because the power factor is zero in both cases. Thus we see that the average energy input $\bar{P}$ is exactly equal to the amount of Joule heating in the circuit. A circuit of this type in which all the energy supplied is dissipated in Joule heating is called *passive*. An *active* circuit is one in which some of the energy supplied is utilized to do external work, for example, a circuit in which a generator is used to drive a motor.

In transmitting electric power, it is evident from (22.7.8) that for a given amount of power $\bar{P}$ transmitted at a fixed voltage, the least current is needed when the power factor is unity. Since excessive currents entail unnecessary power line losses in the form of Joule heating, the power company always likes the power factor to be as near to unity as possible. Indeed, they frequently penalize you for using low power factor circuits by charging you for the *volt-amperes* you consume rather than the watts that are dissipated according to (22.7.8).

It should also be pointed out that the power factors for the resistor, inductor, and capacitor were quoted for ideal circuit elements. In practice, a realistic inductor contains some resistance and capacitance and will, therefore, not have a power factor of zero. Similarly, the power factor of a realistic capacitor is not zero, and the power factor of an actual resistor is never exactly unity. Let us now consider several examples involving power in ac circuits.

EXAMPLE 22.7.1

A 100-watt incandescent light bulb is manufactured under the assumption that it will be used on a 110-volt ac line. The local average cost of electricity is 4¢ per kilowatt-hour. (a) Find the resistance of the bulb and the peak and rms currents which pass through it. (b) If the bulb is on for an average of 5 hours per day, what is the monthly cost? (c) What would be the fractional savings if the rms voltage were cut back to 105 volts?

The manufacturer designed the bulb so that

$$100 \text{ W} = I_{\text{rms}}^2 R = \frac{\Delta V_{\text{rms}}^2}{R}$$

Thus, the resistance R is 121 ohms and the rms current is 0.91 ampere. The peak current is therefore $I_0 = \sqrt{2} I_{\text{rms}} = 1.29$ ampere. In this calculation, we have assumed that the light bulb is an ideal resistor with unit power factor. If the bulb is used under the above conditions for 5 hours a day for 30 days, the amount of power is (0.1 kilowatt)(150 hours) = 15 kilowatt-hours. At 4¢ per kWh, this much energy would cost the consumer 60¢. If the line voltage is reduced to 105 volts, the bulb would not be operating at its rated output power, but would draw only $(105)^2/121 = 91$ watts. At this power, we would save some 9 percent of our original power cost, but obviously there would also be a reduction in light intensity.

EXAMPLE 22.7.2

A series R–L–C circuit has an impedance of 2000 ohms on a 60-cps 110-volt ac line. It contains an inductance of 5 henrys and the entire circuit has a power factor of $\frac{1}{2}$. (a) What is the average power supplied to the circuit? (b) Find the values of the capacitance and the resistance.

The average power taken from the line is

$$\overline{P} = \frac{\Delta V_{\text{rms}}^2}{Z}\cos\phi = \frac{(110)^2}{2000}\frac{1}{2} = 3.03 \text{ W}$$

For a series R–L–C circuit the equations

$$Z = \sqrt{R^2 + \left(\omega L - \frac{1}{\omega C}\right)^2} \qquad (22.7.11)$$

and

$$\cos\phi = \frac{R}{Z} \qquad (22.7.12)$$

must both be satisfied. Thus, with a power factor of $\frac{1}{2}$, we find $R = Z/2 = 1000$ ohms. To find the capacitance C, we use (22.7.8):

$$\frac{1}{\omega C} = \omega L - \sqrt{Z^2 - R^2}$$

(Note that $\omega L > 1/\omega C$ when the power factor is $\frac{1}{2}$.) We then have

$$\frac{1}{120\pi C} = (120\pi)(5) - \sqrt{(2000)^2 - (1000)^2}$$

which leads to $C = 17.3 \ \mu\text{F}$.

SUMMARY

The simplest case of electrical oscillations arises in the ideal L–C circuit, in which a harmonic oscillatory current

$$I(t) = I_0\cos(\omega_0 t + \delta)$$

with frequency

$$\omega_0 = \sqrt{1/LC}$$

can be excited. In this circuit, energy is transferred back and forth from the capacitor to the inductor by

the alternating current in the circuit, at the characteristic circuit frequency ω_0. The sum of the electrostatic energy in the capacitor and magnetic energy in the inductor remains constant at all times, in the absence of any resistance in the circuit, though in reality it will in time decrease and eventually be dissipated altogether by Joule heating in the circuit wires.

In the case of the series R–L–C circuit, in the absence of an emf, a damped oscillatory current given by

$$I(t) = I_0 e^{-Rt/2L}\cos(\omega' t + \delta)$$

with frequency

$$\omega' = \sqrt{\frac{1}{LC} - \frac{R^2}{4L^2}} \qquad \frac{R^2}{4L^2} < \frac{1}{LC}$$

can be excited. In this case, the current decays with a time constant $t_0 = 2L/R$, the energy being dissipated by Joule heating in the resistance. For $R^2/4L^2 \geq 1/LC$, the behavior is no longer oscillatory, the current then simply dropping off to zero exponentially with time.

When an oscillatory emf of the form $\mathscr{E}(t) = \mathscr{E}_0\cos\omega t$ is included in a series R–L–C circuit, an ac current of frequency ω given by

$$I(t) = \frac{\mathscr{E}_0}{Z}\cos(\omega t - \phi)$$

flows, with

$$Z = \sqrt{R^2 + \left(\omega L - \frac{1}{\omega C}\right)^2}$$

and

$$\tan\phi = \frac{\omega L - \dfrac{1}{\omega C}}{R}$$

The current and emf differ in phase by the angle ϕ. It is further observed that the voltages across the inductor and capacitor are not in phase with one another nor with the current. The results can be summarized as follows:

1. The current I, the emf $\mathscr{E}$, and the voltage across all the circuit elements are all sinusoidally varying quantities having the same frequency but different amplitudes and phases.
2. The amplitude of the voltage across the resistor is $I_0 R$, where I_0 is the current amplitude, and it is in phase with the current.
3. The amplitude of the voltage across the inductor is $I_0 X_L$, where

$$X_L = \omega L$$

is the inductive reactance, and the phase of the current through the inductor lags behind that of the voltage across it by 90°.

4. The amplitude of the voltage across the capacitor is $I_0 X_C$, where

$$X_C = \frac{1}{\omega C}$$

is the capacitive reactance, and the phase of the current through the capacitor leads that of the voltage across it by 90°.

The instantaneous voltages across all the circuit elements add up to that of the emf, but this is not true of the amplitudes or rms average values, because of the fact that the voltages across the individual circuit elements are not in phase with each other or with the emf. The situation can be visualized geometrically, as shown in Fig. 22.6, by regarding the voltages across the inductor, capacitor, and resistor as rotating vectors, each with its appropriate phase with respect to the current. These are then added vectorially to obtain a vector that represents the amplitude and phase of the emf.

The current amplitude is a maximum and the impedance is a minimum in a series R–L–C circuit, when $\omega L - (1/\omega C) = 0$, thus, when the frequency ω equals the circuit's *resonant frequency*

$$\omega = \omega_0 = \sqrt{1/LC}$$

For frequencies greater or small than this, the current is correspondingly less and the impedance is greater. At resonance, the capacitive and inductive reactances are equal, and the voltages across the capacitor and inductor are equal in magnitude but, of course, out of phase by 180°.

The average power dissipated in an ac circuit can be expressed by

$$\bar{P} = \tfrac{1}{2}\mathscr{E}_0 I_0 \cos \delta = \mathscr{E}_{rms} I_{rms} \cos \delta$$

where δ represents the phase angle between circuit current and emf and where

$$\mathscr{E}_{rms} = \mathscr{E}_0/\sqrt{2} \quad \text{and} \quad I_{rms} = I_0/\sqrt{2}$$

are the root-mean-square or effective values of emf and current. The quantity $\cos \delta$ is referred to as the power factor of the circuit.

QUESTIONS

1. What are the advantages in using alternating current rather than direct current for commercial power systems?
2. The current in an ac power line changes its direction 60 times each second. How is it possible to extract electrical energy from the circuit, in view of the fact that the average current and emf are both zero?
3. What is meant by a linear differential equation? Write down several examples of nonlinear differential equations.

4. Show that the time constant in a series R–L–C circuit has dimensions of seconds.
5. Describe the physical characteristics of the series R–L–C circuit with an alternating emf, and with no emf but with an initial charge on the capacitor.
6. In studies of alternating current circuits, the transient solution of the circuit equation is often discarded. Under what circumstances is this permissible?
7. In a simple R–L–C series circuit, what value does the impedance have when the frequency is tuned to resonance?
8. The voltage across a resistance R is given by $V_R = I_0 R \cos(\omega t - \phi)$. Express this using the sine function rather than a cosine function.
9. The voltage across the inductor leads the voltage across the capacitor in a series ac circuit by 180°. Explain what this statement means.
10. Explain qualitatively why the impedance of a capacitor decreases as the frequency increases while the impedance of an inductor increases with frequency.
11. In a series R–L–C circuit, are there any circumstances under which the rms voltages measured across the resistor, capacitor, and inductor add up algebraically to the rms value of the applied emf?

PROBLEMS

1. A generator provides an alternating emf with $\mathscr{E}_{rms} = 2000$ volts at 100 Hz. Find the peak voltage and the period. Write an expression for the time-varying emf.
2. A simple series L–C circuit contains a capacitance of 10 μF. If it oscillates at 100 Hz, find the inductance L and the inductive reactance X_L.
3. A 100-watt light bulb ordinarily operates with a 117-volt, 60-Hz voltage across it. If the emf is applied instead across a series combination of the light bulb and a 2-μF capacitor, find the voltage across the bulb. Assume that the bulb acts as an ideal resistance. Do you think the bulb will light up under these circumstances?
4. A capacitor C is initially given a charge q. If an inductor L is connected to it and a complete circuit is made, find the maximum current in the circuit.
5. A series R–L–C circuit consists of $R = 1000$ ohms, $L = 15$ henrys, and $C = 30$ μF. Find the frequency of oscillation in this circuit and determine the time constant t_0. There is no emf. If the current amplitude in the circuit is 10 amperes at $t = 0$, find its value when $t = 3t_0$.
6. Show by direct substitution that Eq. (22.3.4) is a solution of (22.3.2) when

$$\frac{1}{LC} - \left(\frac{R}{2L}\right)^2 > 0$$

7. Find the inductive reactance of a 0.02-henry inductor **(a)** at 60 Hz, **(b)** at 600 Hz, **(c)** at 6000 Hz. **(d)** What is the phase angle between the voltage across the inductor and the current that flows through it at these three frequencies?
8. Find the capacitive reactance of a 0.5-μf capacitor **(a)** at 6 Hz, **(b)** at 60 Hz, **(c)** at 600 Hz. **(d)** What is the phase

angle between the voltage across the capacitor and the current that flows through it?

9. A 0.036-henry inductor and a 12.0-ohm resistor are in series with an ac emf whose amplitude is 165 volts and whose frequency is 60 Hz. Find (a) the amplitude of the current in the circuit, (b) the amplitude of the voltage across the inductor, (c) the amplitude of the voltage across the resistor, (d) the phase angle between the emf and the current in the circuit.

10. A capacitor and a 3600-ohm resistor are in series with a 60-Hz ac emf whose amplitude is 165 volts. The current in the circuit is observed to have an amplitude of 0.0320 amp. Find (a) the capacitance of the capacitor, (b) the amplitude of the voltage across the capacitor, (c) the amplitude of the voltage across the resistor, (d) the phase angle between the emf and the current in the circuit.

11. A 0.24-henry inductor consists of a coil of copper wire having a dc resistance of 7.5 ohms. It is connected in series with a 60-ohm resistor and a 60-Hz ac emf of amplitude 312 volts. Find (a) the current in the circuit, (b) the phase angle between the current and the emf, (c) the amplitude of the voltage across the inductor, (d) the phase angle between the voltage across the inductor and the current through it, (e) the amplitude of the voltage across the 60-ohm resistor. (Hint: Represent the inductor as an "ideal" inductance in series with a resistance of 7.5 ohms.)

12. What would the answers to parts (c) and (d) of the preceding problem be if the frequency of the emf were changed to (a) to 6 Hz (b) to 600 Hz?

13. A series RLC circuit contains a 0.12-henry inductor, a 25-ohm resistor, and a 2.0-μf capacitor in series with a 400-Hz ac emf of amplitude 72 volts. Find (a) the impedance of the circuit, (b) the total reactance of the circuit, (c) the amplitude of the current in the circuit, (d) the phase angle between the current in the circuit and the emf, and the amplitude of the voltage across, (e) the resistor, (f) the inductor, and (g) the capacitor.

14. (a) What is the resonant frequency of the series RLC circuit of the preceding problem? Find (b) the inductive reactance, (c) the capacitive reactance, (d) the impedance of the circuit at the resonant frequency. Find (e) the current, and the voltages across (f) the resistor, (g) the inductor, and (h) the capacitor at resonance.

15. An emf of 2×10^{-3} volt (rms) and frequency 400 kHz is induced in a series R–L–C circuit. The circuit has a quality factor $\mathcal{Q} = 20$ and a resistance of 10^5 ohms. (a) Find the value of the inductance. (b) At resonance, what value does C have? (c) What is the voltage across the resistor at resonance? (d) What would this voltage be at 380 kHz assuming the same values for R, L, and C? (e) At this frequency, by what angle ϕ does the current lag the input voltage?

16. An experimental physicist has a coil with an inductance of 2 millihenrys (2×10^{-3} H). He wishes to construct a resonant L–C circuit with a frequency of 1500 Hz. What value of C should the capacitor have?

17. An R–L–C circuit contains an inductance of 25×10^{-3} henry, a capacitance of 3 μF, and a resistance of 100 ohms. Find the natural frequency of oscillation and the time constant associated with the current decay.

18. You are given three black boxes with input and output terminals. You know that one contains a resistor, another a capacitor, and the third an indictor, but you do not know which box contains each element. The terminals of the boxes are connected in series with each other and with a 60-Hz ac emf of amplitude 116 volts. It is observed that the amplitude of the voltage across box 1 is 320 volts, while a voltage of amplitude 84 volts is measured across box 2 and a voltage of amplitude 240 volts across box 3. It is found that the amplitude of the current in the circuit is 2.50 amp, and that the current lags the emf by a phase angle of 43.6°. Which box contains which element, and what are the values of the resistance, inductance, and capacitance?

19. The natural frequency of oscillation of an R–L–C circuit is given by Eq. (22.3.5). Show that as this natural frequency approaches zero, the time constant associated with current decay approaches the *reciprocal* of the natural frequency of the pure L–C circuit.

20. An emf of 105 volts causes a current of 10 amperes to flow in an ac circuit. The current lags the voltage by 15°. Find the power factor and the power delivered by the source.

21. A series R–L–C circuit contains elements $R = 10$ ohms, $\omega L = 10$ ohms, and $1/\omega C = 5$ ohms. If the rms voltage is 60 volts, find the rms current, the power factor, and the $\mathcal{Q}$ of the circuit.

22. A coil having both inductance and resistance has a $\mathcal{Q}$ value of 100 and an inductance of 200 μH at 600 kHz. At this frequency, find the resistance, the impedance, and the phase angle between the impedance and the resistance.

23. Consider a parallel L–C circuit as shown in the diagram. When an inductor and a capacitor are in parallel, with reactances X_L and X_C, respectively, the combined impedance is given by $X_L X_C/(X_L - X_C)$. For the L–C circuit shown, find the impedance of the circuit, the rms current ($\Delta V_{rms}/Z$) through the source, and the rms currents in the capacitor and inductor.

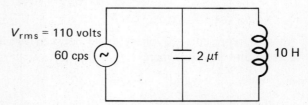

24. A 10-μF capacitor and an inductor L are connected in series with a 60-Hz ac emf. Using an ac voltmeter, a student measures 100 volts across the capacitor and 150 volts across the inductor. Find the inductance L and the rms current flowing in the circuit. What voltage would he measure across the combination of inductor and capacitor?

25. A resistor R and a 1-henry inductor are in series with an ac emf of 60 Hz. The voltage measured across the resistor is 30 volts, while that across the inductor is 60 volts. Find the resistance R and the voltage measured across the combination of resistor and inductor.

26. An inductor, resistor, and capacitor are connected in series with a 48-volt rms ac emf. If $\omega = 60$ rad/sec, $L = 0.05$ henry, $C = 0.011$ μF, and $R = 2.8$ ohms, find (a) the

inductive reactance, **(b)** the capacitive reactance, **(c)** the impedance of the circuit, and **(d)** the rms current through the circuit.

27. A circuit contains a coil with a resistance of 10 ohms and an inductance of 2 henrys. Find the impedance of the coil in a dc circuit. What is its impedance at 10^6 Hz?

28. A circuit contains a 2-henry coil with 200 ohms resistance. The current passing through the coil is $I = 0.5 \sin \omega t$, with I in amperes. If $\omega = 600$ rad/sec, obtain an expression for the time-varying voltage across the coil. By what angle does the current lag the voltage?

*29. Find the current as a function of time through the series R–L–C circuit of section 22.3 (no emf) for the "overdamped" case

$$\left(\frac{R}{2L}\right)^2 - \frac{1}{LC} > 0$$

30. A series R–L–C circuit contains an inductance $L = 60$ henrys, a resistor $R = 10^4$ ohms, and a 60-Hz ac emf of 110 volts (rms). A student measures the voltage across the inductor and finds it to be 200 volts. Find the capacitance C and the voltages he would obtain across the capacitor and the resistor. What is the natural frequency of oscillation for this circuit? What is the time constant?

31. A series R–L–C circuit has a time constant of 0.01 second and a natural frequency of $\omega = 200$ rad/sec. Make a graph of $\ln q$ versus t, where q is the charge on the capacitor, assuming a solution of the same form as (22.3.5) with $q_0 = 10^{-6}$ coulomb and $\delta = 0$. There is no emf in the circuit.

32. An R–C circuit containing a 100-volt (rms), 60-Hz power supply is shown in the diagram. If a student measures 80 volts across the capacitor, what voltage would he observe across the resistor?

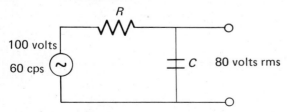

33. A series R–L–C circuit contains a 300-watt bulb, a 2-μF capacitor, and a variable inductor. The circuit is connected to a 117-volt, 60-Hz ac emf. The light bulb is an ideal resistance, designed to consume 300 watts of power at 117 volts. **(a)** What is the resistance of the bulb? **(b)** What value of L will make the bulb light up with maximum intensity? **(c)** When the inductance is now doubled to dim the light (for example, by inserting a permanent magnet into the coil of the inductor), how much power is consumed by the bulb?

34. A coil has a resistance of 10 ohms and an inductance of 0.2 henry. It is connected in series with a variable capacitor and a 110-volt, 60-Hz power source. Find the value of C which gives the largest current amplitude and also the quality factor $\mathcal{Q}$ of the circuit. At resonance, find the voltage across the coil and the capacitor, respectively.

35. An alternating current $I_1 = I_0 \cos \omega t$ and a direct current I_0' both pass through a resistor R. Show that the root-mean-square (rms) current is given by $I_{rms}^2 = I_0'^2 + I_0^2/2$. *Hint:* Calculate the average power dissipated in the resistor.

36. An alternating current passing through a 30-ohm resistor produces Joule heating at the rate of 3000 watts. Find the rms current and voltage through the resistor.

37. An R–L–C series circuit contains a resistor of 1000 ohms, an inductor of 10 henrys, and a capacitor of 1 μF. Draw the impedance diagrams for angular frequencies of $\omega = 2$ rad/sec, 10 rad/sec, and 100 rad/sec. Find the total impedance at each of these frequencies.

38. An ac motor has a power factor of 0.5. It draws 4 amperes from a 220-volt, 60-Hz line. Find the average power used by the motor.

39. A current $I = I_1 \sin \omega t + I_0 \sin 2\omega t$ flows through a resistor R. Find the average power dissipated during the time interval $2\pi/\omega$ and determine the rms current and voltage.

40. An R–C circuit contains a resistance of 10^3 ohms and capacitance of 10^{-8} faraday. If this is used as a high-pass filter, as in Fig. 22.15a, show that frequencies in the kilohertz range are effectively eliminated while those in the megahertz range remain.

MAXWELL'S EQUATIONS AND ELECTROMAGNETIC WAVES

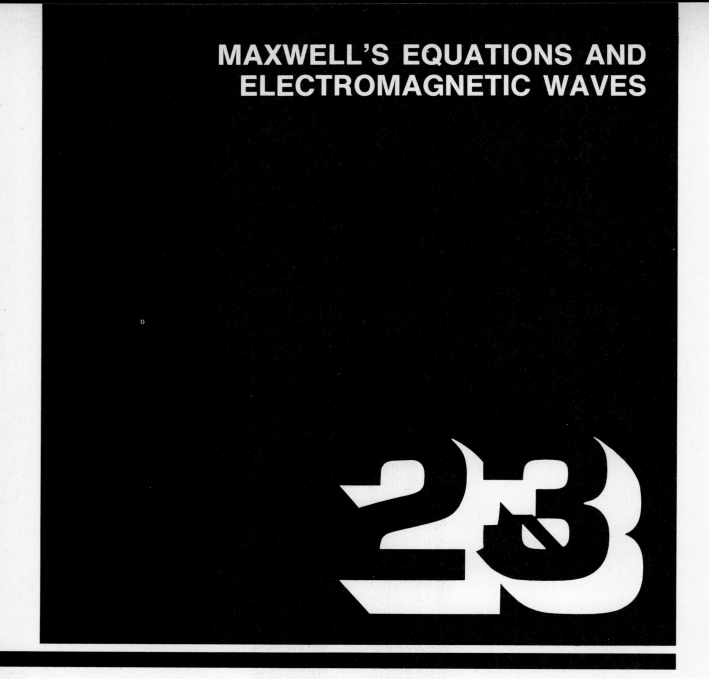

23.1 Introduction

James Clerk Maxwell (1831–1879), at the young age of twenty-four, set himself the task of placing the known laws of electricity and magnetism on a firm mathematical basis. Michael Faraday (1791–1867) had already discovered that a changing magnetic field would induce an electric field, but his explanations in terms of lines of force were not regarded as fully satisfactory by his contemporaries. It was Maxwell who succeeded in finding a correct mathematical formulation of Faraday's law of induction and who further predicted that a *time-varying electric field would induce a magnetic field*. This prediction originated from Maxwell's discovery of the *displacement current*, a new concept

that will be discussed in detail in this chapter. In addition to postulating this very essential missing link, which significantly modifies the laws of electricity and magnetism, Maxwell was also able to reformulate these laws in terms of four differential equations involving electric fields, magnetic fields, and distributions of charge and current density. These equations, which are universally referred to as *Maxwell's equations*, are the physical foundation for the classical theory of electromagnetism and also form the basis for building the quantum formulation known as quantum electrodynamics. On the basis of Maxwell's equations, Maxwell himself predicted theoretically the existence of electromagnetic waves and thus unified the subjects of light and electromagnetism. Eight years after Max-

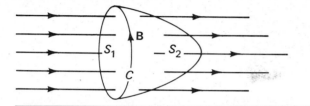

FIGURE 23.1. The current I_c in Ampère's law can be regarded as the total current passing through either surface S_1 or S_2, since both surfaces are bounded by the same Amperian contour C.

well's death, Heinrich Hertz was able to confirm these predictions experimentally by generating and detecting electromagnetic waves.

Maxwell also made major contributions in other areas of science, including color vision and statistical physics. In 1931, one hundred years after Maxwell's birth, there was a celebration in honor of the man believed to be the most outstanding theoretical physicist of the nineteenth century. At that time Max Planck said of Maxwell,[1]

> "It was his task to build and complete the classical [electromagnetic] theory and in so doing he achieved greatness unequalled. His name stands significantly over the portal of classical physics, and we can say of him: by his birth, James Clerk Maxwell belongs to Edinburgh, by his personality he belongs to Cambridge, by his work he belongs to the whole world."

In the present chapter, our main effort will be to understand the origin of the displacement current, to discuss the Maxwell equations in their integral form, and to study electromagnetic waves.

23.2 Displacement Current

In Chapter 19, we showed that when free charges move through a wire, a magnetic field is established in the surrounding space. The moving charges are the conduction electrons; the current that appears as a result of their motion is called the conduction current, hereafter denoted I_c. The Biot–Savart law provides the means by which the magnetic field can be calculated for a specified geometry and known conduction currents. If the problem has sufficient symmetry, Ampère's law

$$\oint \mathbf{B} \cdot d\mathbf{l} = \mu_0 I_c \qquad (23.2.1)$$

can be used to determine **B**. The integration above is carried out around a closed Amperian loop through which the conduction current I_c passes.

[1] James Clerk Maxwell, *A Collection of Commemorative Essays* (New York: Macmillan, 1931), p. 65.

It is important to stress that the conduction current appearing on the right-hand side of (23.2.1) is supposed to be the total current passing through *any surface bounded by the path used to evaluate the line integral.* This is illustrated in Fig. 23.1. We shall now show that a paradox exists if (23.2.1) is always valid and that this paradox suggests an important modification of Ampère's equation.

Consider a small portion of a closed electrical circuit in which there is a parallel-plate capacitor containing a dielectric of permittivity ε, as shown in Fig. 23.2a. A conduction current I_c is assumed to flow through the wires, charging the capacitor and producing an electric field in the region between the plates. As current flows into the capacitor, a charge builds up, and the electric field within it becomes increasingly strong. It is evident that the flow of cur-

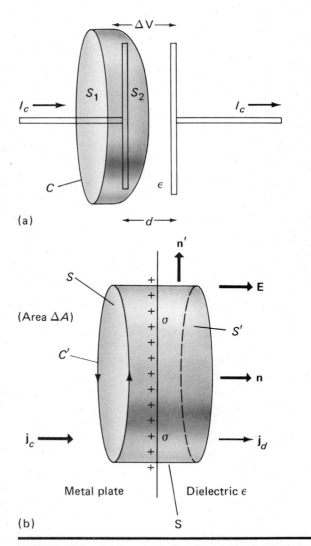

FIGURE 23.2. (a) Ampère's law for an ideal parallel-plate capacitor. (b) Evaluation of Amperian integral over a small element of the system of area ΔA.

rent into the capacitor causes a *time-varying* electric field between the plates.

Let us now apply Ampère's law to a contour C surrounding one of the conductors. According to Ampère's law, the integral of $\mathbf{B} \cdot d\mathbf{l}$ around the contour must equal μ_0 times the current that *passes through any surface bounded by the contour* C. If we choose the surface S_1, there is no problem, since it is evident that the conduction current goes directly through this surface. The Amperian integral is then clearly equal to $\mu_0 I_c$. On the other hand, we might equally well have chosen to set the integral equal to μ_0 times the current passing through the surface S_2, which also bounds the contour. But, now, there is no current, at least no *conduction* current, at all flowing through the surface, and we are led to equate the Amperian integral to *zero*. Obviously, there is something wrong here, for both these answers cannot be correct.

The paradox arises from the apparent discontinuity in the conducting current, which clearly flows in the wires connected to the capacitor but which does not exist in the region between the plates. The remedy was provided by Maxwell, who proposed a reinterpretation of the current to be used in Ampère's law. Maxwell suggested that this current should be regarded as the sum of two separate parts, one of which is the ordinary conduction current I_c, the other a so-called *displacement current*, which we shall write as I_d. In the present example, even though there is no conduction current of mobile free charge in the region inside the condenser, there is, according to Maxwell's view of the situation, an *equivalent* displacement current of such magnitude to assure that the total current is continuous throughout the system. Since the effect of the current that flows into the capacitor is to increase the electric field within the device, Maxwell reasoned that the displacement current must somehow be connected to the time rate of change of electric field or electric flux between the plates.

Consider now the situation shown in Fig. 23.2b. In this drawing, we see a small portion of the metal surface of one of the plates of the capacitor. Within the conductor, there is a current density $\mathbf{j}_c$ due to the motion of free charges. There is also a field $\mathbf{E}$ within the region outside the plate, whose strength increases as charge accumulates on the surface of the conductor. The changing electric field gives rise to a displacement current density $\mathbf{j}_d$ which must be equal to the conduction current density within the conductor to assure continuity of total current throughout the system. The conduction current and the charge q on the capacitor plate are related by

$$I_c = \frac{dq}{dt} \tag{23.2.2}$$

from which, dividing both sides of the area A of the capacitor plate,

$$j_c = \frac{d\sigma}{dt} \tag{23.2.3}$$

But as we saw previously in our study of electrostatics in Chapters 16 and 17, the charge density σ and the electric field within a parallel-plate capacitor are related by

$$E = \frac{\sigma}{\varepsilon} \tag{23.2.4}$$

from which

$$\sigma = \varepsilon E$$

Substituting this into (23.2.3), we obtain

$$j_c = \varepsilon \frac{dE}{dt} = j_d \tag{23.2.5}$$

which expresses the current density j_c in terms of the rate of change of electric field and which must represent the displacement current within the capacitor if continuity of total current is to be guaranteed.

Now, let us see what happens if we apply Ampère's law to the contour C' in Fig. 23.2b, using the *sum* of conduction current and displacement current in place of conduction current alone. If we state the current as the total current that passes through the surface S_1, we obtain

$$\oint_C \mathbf{B} \cdot d\mathbf{l} = \mu_0 j_c \, \Delta A = \mu_0 \varepsilon \, \Delta A \frac{dE}{dt}$$

If we interpret the current as that which passes through the surfaces S' and S'' which, taken together, also bound C', we find

$$\oint_C \mathbf{B} \cdot d\mathbf{l} = \mu_0 j_d \, \Delta A = \mu_0 \varepsilon \, \Delta A \frac{dE}{dt}$$

which is now exactly the same as our previous result. In writing the above equation, it is to be noted that neither conduction nor displacement current flows through the surface S'', since its orientation is normal to both current densities. It appears now that if we *define* the displacement current in terms of the time rate of change of the electric field as given by (23.2.4) and substitute the *sum* of conduction and displacement currents for the current used in evaluating the Amperian integral, we shall have no further trouble. If we integrate the current density over the entire plate of the device, we may express the total current as

$$I_0 = \int_S \mathbf{j} \cdot \mathbf{n} \, da = \varepsilon \int_S \frac{d\mathbf{E}}{dt} \cdot \mathbf{n} \, da$$

or

$$I_c = \varepsilon \frac{d}{dt} \int_S \mathbf{E} \cdot \mathbf{n} \, da = \varepsilon \frac{d\Phi_e}{dt}$$

where Φ_e is the total electric flux within the capacitor. In this way, we obtain

$$I_c = \varepsilon \frac{d\Phi_e}{dt} = I_d \qquad (23.2.6)$$

which relates the total displacement current to the electric flux. The dilemma originally encountered in Fig. 23.2a is now also resolved, since the Amperian integral associated with current through surface S_1 is simply $I_c = \varepsilon \, d\Phi_e/dt$, and that arising from passage of current through surface S_2 is $I_d = \varepsilon \, d\Phi_e/dt$, which is exactly the same.

We have now fixed up Ampère's law so as to forestall any further paradoxical behavior of the type described above. In doing so, however, we found that we must write the current through the Amperian contour as the sum of conduction and displacement currents. This requires us to write Ampère's law in a slightly, but very significantly, different form:

$$\oint_C \mathbf{B} \cdot d\mathbf{l} = \mu_0(I_c + I_d) \qquad (23.2.7)$$

or, using (23.2.6),

$$\oint_C \mathbf{B} \cdot d\mathbf{l} = \mu_0 I_c + \mu_0 \varepsilon \frac{d\Phi_e}{dt} \qquad (23.2.8)$$

Written in this way, it is evident that Ampère's law predicts a *new physical effect*. This is easily seen by considering the region between the capacitor plates in the example above, where I_c is zero and the term $\mu_0\varepsilon \, d\Phi_e/dt$ represents the entire Amperian integral. In such a region, if we are to believe Eq. (23.2.8), whenever there is a *changing electric flux*, the integral on the right will have a nonzero value, which means that a magnetic field must exist within the region! This may or may not be true, according to whether Maxwell's displacement current has any physical reality or is simply a mathematical pipe dream. It is easy to find out; experiments to detect the effect described above are not hard to perform. The results of these experiments are in all cases fully in accord with Maxwell's theory and attest to the reality of the displacement current. Indeed, it can be shown that without the displacement current there could be no electromagnetic waves! We can easily see that Ampère's law, as written in the amended form (23.2.8), is analogous to Faraday's law of induction, which states that an electric field is invariably associated with a changing magnetic flux; that is,

$$\oint_C \mathbf{E} \cdot d\mathbf{l} = -\frac{d\Phi_m}{dt} \qquad (23.2.9)$$

In using Ampère's law, it is important to understand the qualitative difference between a conduction current and a displacement current. When there is a conduction current, real charges are moving and, therefore, matter is in motion. On the other hand, a displacement current exists whenever there is a time-varying electric flux. If you try to attribute this current to matter in motion, you necessarily fail. The reason for the failure can be appreciated by understanding that all the preceding arguments remain intact even if the space between the capacitor plates has a permittivity ε_0. If that is the case, this space between the plates is a *vacuum*. It contains no charge and no mass, for there is nothing there. Yet there still exists a time-varying electric field and hence a displacement current $\varepsilon_0(d\Phi_e/dt)$.

The ideas expressed by (23.2.8) and (23.2.9), that a time-varying electric field causes a magnetic field and a time-varying magnetic field causes an electric field, led ultimately to the prediction of electromagnetic waves. We shall discuss how this was accomplished in detail later. Let us consider now several examples which will hopefully clarify some of these ideas.

EXAMPLE 23.2.1

A parallel-plate capacitor of capacitance 5 μF is directly connected to a 60-cps ac line with an rms voltage of 110 volts. Find the displacement current between its plates.

The total displacement current is

$$I_d = \varepsilon \frac{d\Phi_e}{dt} \qquad (23.2.10)$$

Now,

$$\Phi_e = \int_S \mathbf{E} \cdot \mathbf{n} \, da = EA \qquad (23.2.11)$$

since the electric field between the plates of a parallel-plate capacitor is uniform. The electric field is related to the potential difference across the plates by $E = \Delta V/d$, where d is the plate separation and ΔV is the instantaneous voltage. It follows that

$$I_d = \frac{\varepsilon A}{d} \frac{d(\Delta V)}{dt} = C \frac{d(\Delta V)}{dt} \qquad (23.2.12)$$

But if

$$\Delta V = 110\sqrt{2}\cos(120\pi t) = 154\cos(120\pi t) \qquad (23.2.13)$$

743

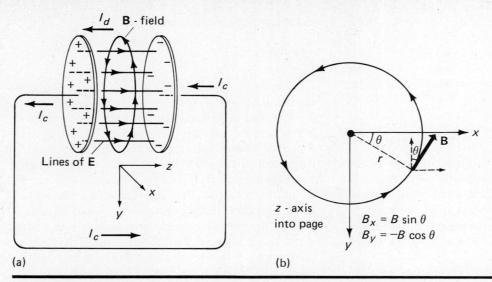

FIGURE 23.3. Geometry of magnetic field induced by the changing electric field within a parallel-plate capacitor.

characterizes the voltage across the plates, then

$$I_d = C \frac{d(\Delta V)}{dt}$$

$$= (5 \times 10^{-6})(-120\pi)(154 \sin 120\pi t)$$

$$= -0.290 \sin 120\pi t \text{ amp} \qquad (23.2.14)$$

gives the time-varying displacement current. As mentioned earlier, this current is equal to the conduction current dq/dt that flows in the circuit leads connected to the condenser plates. The rms current is $0.29/\sqrt{2} = 0.205$ ampere.

EXAMPLE 23.2.2
The charge q on a circular parallel-plate capacitor is given by

$$q = q_0 e^{-t/RC} \qquad (23.2.15)$$

where C is the capacitance and R is the resistance in the external circuit. The circular plates have radii r_0. (a) Find the total displacement current between the plates, assuming the field to be uniform in the region between them. (b) Determine the *magnetic* field between the plates as a function of the distance from the central axis connecting the plates. (c) Relate the *spatial variation* of the magnetic field to the *time variation* of the electric field.

Since, once again, the total displacement current is identical to the total conduction current in the leads, we have

$$I_{d0} = I_{c0} = \frac{dq}{dt} = -\frac{q_0}{RC} e^{-t/RC} \qquad (23.2.16)$$

where I_{d0} and I_{c0} are the total displacement and conduction currents, respectively. To find the magnetic field between the plates, we can use the equation

$$\oint_C \mathbf{B} \cdot d\mathbf{l} = \mu_0 \varepsilon_0 \frac{d\Phi_E}{dt} = \mu_0 I_d \qquad (23.2.17)$$

for an appropriate Amperian loop, as shown in Fig. 23.3. The Amperian loop is circular, with a radius r, and is concentric with the capacitor plates. Since the electric field is uniform between the capacitor plates, we may readily conclude by symmetry arguments such as those used previously in applying Ampère's law that B depends only on the distance from the axis and not on the distance from one plate to another. We may also reasonably conclude that the lines of $\mathbf{B}$ have to be tangential to the circular path of integration, by arguments similar to those used in the rather similar situation[2] discussed in Example 20.2.4.

To apply (23.2.16), we must integrate around the circular loop. Since $\mathbf{B}$ is uniform and tangential to the loop, we have

$$2\pi r B = \mu_0 \varepsilon_0 \frac{d}{dt} \int_S \mathbf{E} \cdot \mathbf{n} \, dA = \mu_0 I_d \qquad (23.2.18)$$

The amount of displacement current I_d that actually passes through the loop of radius r can be related to the total displacement current I_{d0} by

$$I_d = I_{d0} \frac{\pi r^2}{\pi r_0{}^2} = \frac{r^2}{r_0{}^2}\left(-\frac{q_0}{RC}\right)e^{-t/RC} \qquad (23.2.19)$$

[2] In Example 20.2.4, we calculated the electric field produced by a changing magnetic flux, using Faraday's law of induction. Here, we are using Ampère's law to find the magnetic field created by a changing electric flux. But the identical mathematical form of Eq. (23.2.17) above and (20.2.60) should lead us to expect that the **B** field we calculate here will have the same form as the **E** field calculated in the previous example.

Substituting this result into (23.2.17), we may then write

$$B = \frac{\mu_0}{2\pi} \frac{r}{r_0{}^2} \left(-\frac{q_0}{RC} \right) e^{-t/RC} \qquad (23.2.20)$$

The negative sign here simply characterizes the direction of **B**, which is indicated in Fig. 23.3. It is obtained by applying the rules developed earlier for finding the direction of a **B** field once the direction of the current is known.

If we assume that the z-axis is oriented along the direction of the electric field, the displacement current can *also* be written as

$$I_d = \varepsilon_0 \frac{d}{dt} \int_S \mathbf{E} \cdot \mathbf{n} \, dA = \varepsilon_0 \pi r^2 \frac{\partial E_z}{\partial t}$$

Combining this with our earlier expression, we see that

$$\varepsilon_0 \pi r^2 \frac{\partial E_z}{\partial t} = \frac{r^2}{r_0{}^2} \left(-\frac{q_0}{RC} \right) e^{-t/RC}$$

or

$$\frac{\partial E_z}{\partial t} = \frac{1}{\varepsilon_0 \pi r_0{}^2} \left(-\frac{q_0}{RC} \right) e^{-t/RC} \qquad (23.2.21)$$

The spatial variation of the magnitude of **B** can be obtained by taking its derivative with respect to r. Let us instead use (23.2.20) first to obtain the Cartesian components of the vector **B**. In Fig. 23.3, the Amperian loop is shown along with the directions of the **E** and **B** fields. When **E** is decreasing, **B** is counterclockwise, as shown. The *magnitude* of B is

$$B = \frac{\mu_0}{2\pi} \frac{r}{r_0{}^2} \frac{q_0}{RC} e^{-t/RC} \qquad (23.2.22)$$

Therefore, the components are

$$B_x = B \sin \theta = \left[\frac{\mu_0}{2\pi} \frac{r}{r_0{}^2} \frac{q_0}{RC} e^{-t/RC} \right] \frac{y}{r}$$

$$= \frac{\mu_0 y}{2\pi r_0{}^2} \frac{q_0}{RC} e^{-t/RC} \qquad (23.2.23)$$

and

$$B_y = -B \cos \theta = -\frac{\mu_0 x}{2\pi r_0{}^2} \frac{q_0}{RC} e^{-t/RC} \qquad (23.2.24)$$

From Eqs. (23.2.20), (23.2.22), and (23.2.23), we can write

$$B_x = -\frac{\mu_0 \varepsilon_0 y}{2} \frac{\partial E_z}{\partial t} \qquad (23.2.25)$$

$$B_y = \frac{\mu_0 \varepsilon_0 x}{2} \frac{\partial E_z}{\partial t} \qquad (23.2.26)$$

Since E_z and $\partial E_z / \partial t$ are independent of position, we

may also write

$$\frac{1}{\mu_0} \frac{\partial B_x}{\partial y} = -\frac{\varepsilon_0}{2} \frac{\partial E_z}{\partial t} \qquad (23.2.27)$$

$$\frac{1}{\mu_0} \frac{\partial B_y}{\partial x} = \frac{\varepsilon_0}{2} \frac{\partial E_z}{\partial t} \qquad (23.2.28)$$

and finally

$$\frac{1}{\mu_0} \left(\frac{\partial B_x}{\partial y} - \frac{\partial B_y}{\partial x} \right) = -\varepsilon_0 \frac{\partial E_z}{\partial t} \qquad (23.2.29)$$

This relation, which we have derived from a rather specific example, turns out to be very general. It is a special case of one of Maxwell's equations in which the conduction current density at the point in question is zero. Equation (23.2.28) provides a relation which is valid at each point in space and thus expresses the extremely important result that *if an electric field has a time variation at any point in space, then there must be a spatial variation of the magnetic field at that same* point.

23.3 The Integral Form of Maxwell's Equations

We shall consider in this section the statement of Maxwell's equations of electromagnetism. In the interest of simplicity, we shall state the equations which are applicable in free space, that is, in the absence of dielectric or magnetic material. All of the laws constituting the integral form of Maxwell's equations have already been discussed. This section will, therefore, consist mainly of a review of these relations.

In our study of electrostatics, we found a very important relation between the charge contained in a volume element and the electric flux Φ_e out of the element. That relation, known as *Gauss's law*, has the form

$$\oint_S \mathbf{E} \cdot \mathbf{n} \, da = \frac{q}{\varepsilon_0} \qquad (23.3.1)$$

For some types of problems in which the charge distribution is very symmetric, this law can be used to calculate the electric field **E**. However, even in the absence of symmetry, the statement provided by Gauss's law is extremely useful and it constitutes the integral form of one of Maxwell's equations. You will recall, also, that the proof of Gauss's law depends on the validity of Coulomb's force law. In a sense, then, (23.3.1) is another way of expressing Coulomb's law for the force between electric charges.

There is also a relationship that expresses *Gauss's law* for magnetism. This relationship was developed as Eq. (19.3.8) in Chapter 19. It has the simple form

$$\oint_S \mathbf{B} \cdot \mathbf{n}\, da = 0 \qquad (23.3.2)$$

and is the second of Maxwell's equations. A comparison of (23.2.1) and (23.3.2) reveals a rather fundamental difference between electricity and magnetism. In the former case, the *flux* of the *electric field* is equal to the *total electric charge* contained within the Gaussian surface. The same statement is true in magnetism, but now, since isolated magnetic charges do not appear to exist, the *magnetic* flux evaluated over a closed surface must be *zero*. It should be noted, by the way, that physicists are continuing to search for magnetic charges, also called *magnetic monopoles*, for there is no fundamental reason why they should not exist. If magnetic monopoles did exist, Maxwell's equations would have a more symmetric mathematical form, which would be appealing to many theoretical physicists. Until someone succeeds in catching one, however, we shall have to get along with Maxwell's equations as they are.

Faraday's law of induction, which relates a time-varying magnetic flux to the line integral of an induced electric field, is given by (20.2.35) in the form

$$\oint_C \mathbf{E} \cdot d\mathbf{l} = -\frac{d}{dt}\int_S \mathbf{B} \cdot \mathbf{n}\, dA \qquad (23.3.3)$$

This equation is the third of Maxwell's equations. In any steady-state situation, which includes all of electrostatics, the right side of this equation is zero. The integral of $\mathbf{E} \cdot d\mathbf{l}$ around any closed path vanishes and the electric field is then a conservative one. In general, however, this is not the case, and its line integral around any loop will then be related by Eq. (23.3.3) to the time rate of change of the magnetic flux through the loop.

Finally, the fourth of Maxwell's equations is obtained from *Ampère's law* and its generalization resulting from the introduction of the *displacement current*. This law can be written as

$$\oint_C \mathbf{B} \cdot d\mathbf{l} = \mu_0(I_c + I_d)$$

$$= \mu_0\left(I_c + \varepsilon_0 \frac{d}{dt}\int_S \mathbf{E} \cdot \mathbf{n}\, da\right) \qquad (23.3.4)$$

This equation states that a magnetic field can be produced by moving electric charges or by a time-varying electric field. Its mathematical form is similar to that of Faraday's law, except that Faraday's law, as given by (23.3.3), has no term that corresponds to the conduction current $\mu_0 I_c$ in Ampère's law. This difference

in the form of (23.3.3) and (23.3.4) can also be attributed to the absence of magnetic monopoles. If magnetic charges did exist, there would surely be a current associated with their motion, and this current would then presumably contribute an additional term to (23.2.2).

The four equations discussed above are all summarized in Fig. 23.4. Together, they form the basis of electromagnetism in free space. They imply, as we shall see later, the existence of electromagnetic waves. In fact, the *displacement current* introduced in the previous section provides the essential ingredient to establishing the presence of these waves.

23.4 Electromagnetic Waves

The most impressive prediction to emerge from Maxwell's formulation of the laws of electricity and magnetism is the existence of *electromagnetic waves*. The rigorous theoretical prediction of the presence of these waves proceeds most easily from a form of Maxwell's equations different from those given above. Maxwell's equations, as discussed in the preceding section, are stated in the form of relationships between integrals of the electric and magnetic fields around closed paths and over certain surface areas. But the essential properties of waves are expressed by the fact that the physical quantity that propagates wave motion must obey a wave equation of the form (10.3.9) at every point in space. This, however, is a *differential equation* rather than an integral relationship of the form in which we now know Maxwell's equations. In order to demonstrate that Maxwell's equations lead inevitably to electromagnetic waves, therefore, it is desirable to transform Maxwell's equations to a set of differential equations relating the electric and magnetic fields and their space and time derivatives at any given point.

We have already seen in section 16.7 that it is possible to express Gauss's law in the form of a differential equation (Poisson's equation) rather than the usual integral form. In a somewhat similar way, we can express the other three equations in differential form also. In deriving Eq. (23.2.28), we have already stated Ampère's law in the form of a differential equation. Needless to say, the same treatment can be applied to the other equations. Since the mathematics involved is rather tedious, we shall not attempt to display all these equations; but in exploring the relationships between Maxwell's equations and electromagnetic waves, we shall make implicit use of some of them.

In this section, we are going to make use of (23.3.3) and (23.3.4) in free space or vacuum, where I_c is zero. We shall show that these equations allow the

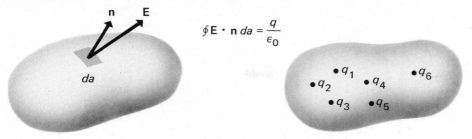

$$\oint \mathbf{E} \cdot \mathbf{n}\, da = \frac{q}{\epsilon_0}$$

(a) Electric flux out of arbitrary volume equals $(1/\epsilon_0)$ times total charge within volume.

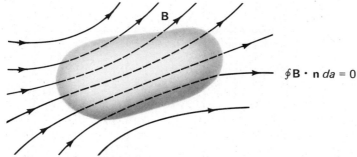

$$\oint \mathbf{B} \cdot \mathbf{n}\, da = 0$$

(b) Magnetic flux out of arbitrary volume equals net magnetic charge in volume. Flux must be zero due to nonexistence of magnetic charges. Therefore flux into volume and flux out of volume must always be equal.

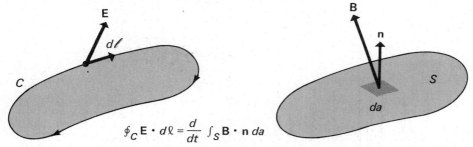

$$\oint_C \mathbf{E} \cdot d\boldsymbol{\ell} = \frac{d}{dt} \int_S \mathbf{B} \cdot \mathbf{n}\, da$$

(c) Line integral of $\mathbf{E} \cdot d\boldsymbol{\ell}$ around a closed path equals time rate of change of magnetic flux through area bounded by path.

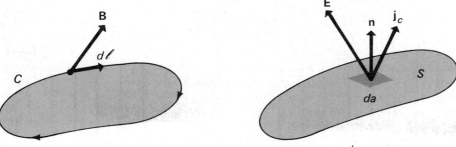

$$\oint_C \mathbf{B} \cdot d\boldsymbol{\ell} = \mu_0 (i_c + i_d) = \mu_0 \int_S \mathbf{j}_c \cdot \mathbf{n}\, da + \mu_0 \epsilon_0 \frac{d}{dt} \int_S \mathbf{E} \cdot \mathbf{n}\, da$$

(d) Line integral of $\mathbf{B}$ around a closed path equals μ_0 times the sum of conduction and displacement currents.

FIGURE 23.4. Schematic diagrams illustrating the physical content of Maxwell's equations. (a) Gauss's law, relating electric flux through a closed surface to enclosed charge. (b) Gauss's law for magnetism, denying the existence of magnetic charge. (c) Faraday's law relating the emf induced around a closed path to the time rate of change of magnetic flux through the path. (d) Ampère's law, as amended by Maxwell, relating the integral of the component of the magnetic induction tangential to a closed contour to the true current through the contour and to the time rate of change of electric flux through it.

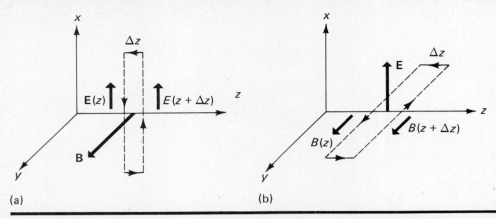

FIGURE 23.5. Geometry used in calculating Eqs. (23.4.4) and (23.4.5) from Maxwell's equations.

existence of certain waves characterized by appropriate temporal and spatial variations of the **E** and **B** fields. Moreover, we shall calculate explicitly the speed with which these waves propagate and show that this is exactly the speed of light.

To begin, let us assume that it is possible to establish electric and magnetic fields of a very special kind. The electric field has a *single component only in the x-direction and is uniform in the entire xy-plane.* The magnetic field *points only in the y-direction and is also uniform in the xy-plane.* Both **E** and **B**, therefore, depend only on the time t and the coordinate z. We can, therefore, write these fields as

$$\mathbf{E}(x, y, z, t) = E(z, t)\mathbf{i}_x$$
$$\mathbf{B}(x, y, z, t) = B(z, t)\mathbf{i}_y \qquad (23.4.1)$$

Fields characterized in this manner can be produced at large distances from a radio or radar antenna, but for now we are not concerned with how they are produced but rather with how they depend on z and t.

Figure 23.5a shows a rectangular path in the xz-plane. Two of the sides are of length l, while two have the length Δz, which is assumed to be very small. We now apply (23.3.3) to this loop. The line integral of the electric field is given by

$$\oint_C \mathbf{E} \cdot d\mathbf{l} = [E(z + \Delta z, t) - E(z, t)]l \qquad (23.4.2)$$

for small Δz. This, according to Faraday's law, is equal to the negative of the time rate of change of magnetic flux passing through the loop. Since $\mathbf{B}(z, t)$ is parallel to **n**, the magnetic flux is

$$\Phi_m = \int_S \mathbf{B} \cdot \mathbf{n} \, da = B(z, t)l \, \Delta z \qquad (23.4.3)$$

Thus,

$$-\frac{d\Phi_m}{dt} = -\frac{\partial}{\partial t} B(z, t)l \, \Delta z$$

Therefore, according to (23.3.3),

$$[E(z + \Delta z, t) - E(z, t)]l = -\frac{\partial}{\partial t} B(z, t)l \, \Delta z$$

Dividing by $l \, \Delta z$ and letting $\Delta z \to 0$, we find

$$\frac{\partial}{\partial z} E(z, t) = -\frac{\partial}{\partial t} B(z, t) \qquad (23.4.4)$$

Note that since **E** and **B** depend on two variables, the derivatives are partial derivatives. The preceding equation states that whenever there is a spatially varying electric field, there is also a time-varying magnetic field. It is, in effect, a differential form of Faraday's law.

If we now choose a similar rectangular path in the yz-plane as shown in Fig. 23.5b, we can use (23.3.4) with $I_c = 0$ to provide another relation between the electric and magnetic fields. By arguments similar to those leading to (23.4.4), it is readily shown that

$$\frac{\partial}{\partial z} B(z, t) = -\varepsilon_0 \mu_0 \frac{\partial}{\partial t} E(z, t) \qquad (23.4.5)$$

This equation is simply a slightly restated version of (23.2.28), in which the x-, y-, and z-directions are now relabeled y, z, and x, respectively, and in which the z-component of **B** is zero. By combining (23.4.4) and (23.4.5), we may now obtain a single equation for the electric field. To accomplish this we first take the z-derivative of (23.4.4) and then interchange the order of the derivatives of B to obtain

$$\frac{\partial}{\partial z} \frac{\partial}{\partial z} E(z, t) = -\frac{\partial}{\partial z} \frac{\partial}{\partial t} B(z, t) = -\frac{\partial}{\partial t} \frac{\partial}{\partial z} B(z, t)$$

Now we substitute for $\partial B/\partial z$ the value given by (23.4.5) to obtain

$$\frac{\partial^2}{\partial z^2} E(z, t) = \frac{\partial}{\partial t} \varepsilon_0 \mu_0 \frac{\partial}{\partial t} E(z, t)$$

or

$$\frac{\partial^2 E}{\partial z^2} = \varepsilon_0 \mu_0 \frac{\partial^2 E}{\partial t^2} \qquad (23.4.6)$$

748

This equation is a *one-dimensional wave equation* with exactly the same form as (10.3.9). It implies that the electric field $E(z, t)$ can propagate as a one-dimensional wave with a propagation speed of

$$v = c = \frac{1}{\sqrt{\mu_0 \varepsilon_0}} \qquad (23.4.7)$$

But we have already seen in connection with (19.8.10) that this velocity c is 299792.5 km/sec, which is exactly the measured *velocity of light!*

Also, by differentiating (23.4.5) which respect to z, we may obtain

$$\frac{\partial}{\partial z} \frac{\partial}{\partial z} B(z, t) = -\varepsilon_0 \mu_0 \frac{\partial}{\partial z} \frac{\partial}{\partial t} E(z, t) = -\varepsilon_0 \mu_0 \frac{\partial}{\partial t} \frac{\partial}{\partial z} E(z, t)$$

Substituting the value of $\partial E / \partial z$ from (23.4.4), however, this becomes

$$\frac{\partial^2 B}{\partial z^2} = \varepsilon_0 \mu_0 \frac{\partial^2 B}{\partial t^2} \qquad (23.4.8)$$

which is a wave equation for the magnetic field of exactly the same from as that obtained as (23.4.6) for the electric field. From this, it is apparent that the magnetic field $B(z, t)$ can propagate as a wave having the same velocity as that associated with the propagation of the electric field—the velocity of light. In the years preceding Maxwell's great discoveries, physicists considered the theory of light to be unrelated to the theory of electricity and magnetism. Maxwell's achievements included the realization that the propagation of light could be explained from his theory of electromagnetism and that light was, in fact, a form of electromagnetic radiation. Maxwell also showed that *electromagnetic waves are generated whenever electric charges are accelerated.* He was, therefore, able to predict that electromagnetic waves would be radiated by any circuit in which ac currents, particularly ac currents of very high frequency, are made to flow. This is a simple consequence of the fact that the free charges in the conductor in an ac circuit execute simple harmonic motion and, as a result, are continually undergoing acceleration. This prediction was verified experimentally in 1887, several years after Maxwell's death, by the German physicist Heinrich Hertz, who in effect, set up the first *radio transmitter and receiver.*[3]

Let us now examine in detail several properties of electromagnetic waves. We shall not concern our-

[3] The engineers are fond of attributing the invention of radio communication to Guglielmo Marconi. While it is true that Marconi, during the period 1895–1900, developed the first *commercially successful* radio transmitting and receiving systems, the honor of having first transmitted and received radio signals clearly belongs to Hertz, and it is he whom we should regard as the inventor of radio.

selves at the beginning with the question of how these waves originate, but will focus instead on some aspects of their propagation through space. When such a wave propagates, energy is transported from one location to another. Since these waves can move through *empty space*, there is no actual displacement of matter, as in the case of sound propagation. Nevertheless, the presence of the wave, which consists of oscillating electric and magnetic fields, would be readily detected by appropriate instruments. For example, a time-varying electric field could be measured at any point, in principle, by determining its influence on charged particles placed at that location, although usually more practical methods are available.

We have asserted at the outset that the electric and magnetic fields are given by (23.4.1). They are uniform in the entire xy-plane and depend only on z and t. When the electric and magnetic fields obey these assumed conditions, which lead to the wave Eqs. (23.4.6) and (23.4.8), we say that the electromagnetic wave is a *plane wave*. The electric and magnetic fields associated with plane waves are uniform in any plane such as $z = z_1$, as shown in Fig. 23.6, but vary from plane to plane. They are different, for example, in the planes $z = z_1$ and $z = z_2$. It is important to stress that electromagnetic waves are not *always* plane waves. We choose, however, to study only plane waves for two reasons. First, they are fairly simple to understand mathematically and physically. Also, very far from their point of origin, all electromagnetic waves are well approximated by plane waves over a limited region of space. Thus, although the light waves coming to us from the sun are actually spherical waves, the spheres have such large radii in the vicinity of the earth that their surfaces are practically planes.

When an electromagnetic wave travels through space, its electric and magnetic fields are always *perpendicular* or *transverse* to the direction of propagation. In the case presented here, the electric vector will be a function of $z - ct$. Therefore, whatever values the fields have at the instant t_1 in the plane $z = z_1$ will also be obtained at later times t_2 in the plane $z = z_2$, provided $z_1 - ct_1 = z_2 - ct_2$. Therefore, the configuration of fields advances in the z-direction with speed c. Since **E** and **B** are transverse to the direction of wave propagation, *electromagnetic waves are transverse waves.* In this respect, they are analogous to the transverse waves generated on a rope or stringed instrument but are to be contrasted with sound waves, which are longitudinal. Figure 23.7 illustrates the propagation of transverse electromagnetic waves.

The solution of the wave equation (23.4.6) can be written as

$$E(z, t) = E(z - ct) \qquad (23.4.9)$$

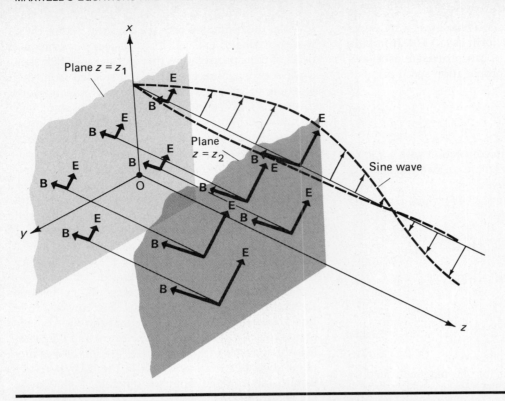

FIGURE 23.6. Geometry of electric and magnetic field vectors in a plane electromagnetic wave.

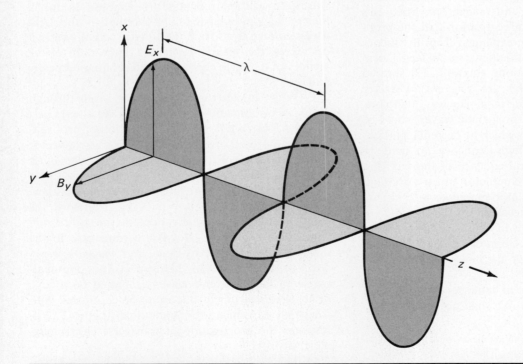

FIGURE 23.7. Another view of a plane electromagnetic wave propagating along the z-direction, showing the wavelength and the propagation vector **k**.

if the wave is propagating in the positive z-direction.[4] However, the exact dependence of the electric field on the quantity $z - ct$ does not emerge from the equation itself. This dependence is determined exclusively by the characteristics of the source of the wave. Thus, even for waves on a rope, we know that a simple pulse can travel along the rope or that, under other conditions, a sinusoidal wave can be established. In exactly the same way, the source of electromagnetic waves determines the exact form of the wave.

In this discussion, we have also assumed that both electric and magnetic fields have a single un-varying direction. It could happen, of course, that electromagnetic energy might be propagated in the form of a random superposition of waves whose electric and magnetic fields have all possible directions in the planes normal to the direction of propagation. In the former instance, which is the one considered here, we say that the wave is *linearly polarized*. In the latter case, the wave is said to be *unpolarized*.

Let us now consider the very special case of sinusoidal plane waves with a definite angular frequency of vibration ω. These are called *monochromatic waves* and can be expressed by

$$
\begin{aligned}
E(z, t) &= E_0 \cos \frac{2\pi}{\lambda} (z - ct) \\
&= E_0 \cos(kz - \omega t) \\
B(z, t) &= B_0 \cos \frac{2\pi}{\lambda} (z - ct) \\
&= B_0 \cos(kz - \omega t)
\end{aligned}
\tag{23.4.10}
$$

where

$$
c = \omega/k = \lambda f \quad \text{and} \quad k = 2\pi/\lambda \tag{23.4.11}
$$

In Fig. 23.7, a wave with the above characteristics is illustrated.

At any fixed position $z = z_0$, the electric and magnetic vectors oscillate harmonically with angular frequency ω given by

$$
\omega = ck = \frac{2\pi c}{\lambda}
$$

Since the angular frequency of oscillation is related the actual frequency f by $\omega = 2\pi f$, we readily obtain the important relation

$$
\lambda f = c \tag{23.4.12}
$$

The quantity λ, discussed in Chapter 10, is the wave-length. It determines the distance between successive maxima of the electric field, as shown in Fig. 23.7.

[4] In this equation, $z - ct$ represents the argument of the function $E(z, t)$, *not* a factor multiplying E!

The frequency f gives the number of oscillations of the **E** or **B** field per second at any location $z = z_0$.

The quantities E_0 and B_0 are the maximum possible values of the electric and magnetic fields and are called the amplitudes; E_0 and B_0 are *not independent*, for there is a constraint imposed by (23.4.4). Evaluating the required derivatives gives

$$
\frac{\partial E}{\partial z} = \frac{\partial}{\partial z} E_0 \cos \frac{2\pi}{\lambda} (z - ct) = -\frac{2\pi}{\lambda} E_0 \sin \frac{2\pi}{\lambda} (z - ct)
$$

$$
\frac{\partial B}{\partial t} = \frac{\partial}{\partial t} B_0 \cos \frac{2\pi}{\lambda} (z - ct) = \frac{2\pi c}{\lambda} B_0 \sin \frac{2\pi}{\lambda} (z - ct)
$$

Substituting into (23.4.4), we arrive at

$$
B_0 = \frac{E_0}{c} \tag{23.4.13}
$$

It is important to note that the electric and magnetic fields oscillate *in phase*.

In vacuum or empty space, electromagnetic waves of all frequencies travel at exactly the same speed. In a medium such as water or glass, the speed may vary with the frequency. This phenomenon, known as *dispersion*, accounts for the splitting of white light into its various color components following passage through a prism.

EXAMPLE 23.4.1

A plane electromagnetic wave with $\lambda = 5 \times 10^{-7}$ meter travels through empty space. Find the frequency and propagation constant for this wave whose wavelength corresponds to that of green light.

The frequency is given by applying the equation $\lambda f = c$. Thus,

$$
f = \frac{c}{\lambda} = \frac{3 \times 10^8}{5 \times 10^{-7}} = 6 \times 10^{14} \text{ Hz}
$$

The propagation constant is

$$
k = \frac{2\pi}{\lambda} = 1.26 \times 10^5 \text{ cm}^{-1}
$$

23.5 Energy Flow in Electromagnetic Waves

In this section, we take up the important topic of energy flow in electromagnetic waves. The great technologic import of electromagnetic waves stems from their capability to transfer energy from one place to another. As well as conveying TV commercials from sponsor to victim, they also carry energy from the sun to the earth, thus making life possible. This transfer of energy occurred long before man came into existence, and will continue to occur until the sun expends all of its vast energy. The primary mechanism for generating energy within the sun is nuclear fusion, a topic we shall take up later, but most of this energy is carried away from the sun in the form of electromagnetic waves.

In Chapters 17 and 20, we discussed the idea of electrostatic and magnetic energy density and pointed out that energy can be stored in space whenever electric and magnetic fields are present. In empty space, this energy density u (energy per unit volume) consists of the sum of electric and magnetic contributions

$$u = u_e + u_m = \tfrac{1}{2}\varepsilon_0 E^2 + \frac{1}{2\mu_0} B^2 \qquad (23.5.1)$$

where $\mathbf{E}$ and $\mathbf{B}$ are the electric and magnetic fields. Then, for any given volume, the energy U stored in the volume is

$$U = \int_v \left(\tfrac{1}{2}\varepsilon_0 E^2 + \frac{1}{2\mu_0} B^2 \right) dV \qquad (23.5.2)$$

When the field energy is present in the form of electromagnetic waves, energy can be carried into or out of the given region by the waves that propagate across its boundaries. Under these circumstances, the quantity U in (23.5.2) will decrease when wave propagation transports energy out of the volume, and vice versa. The flux of energy across the boundary of the region can be described in terms of a vector called the *Poynting vector*, defined by

$$\mathbf{S} = \frac{1}{\mu_0} (\mathbf{E} \times \mathbf{B}) \qquad (23.5.3)$$

The direction of the Poynting vector is that of the energy flux at any point, and its magnitude is the amount of energy per unit time that crosses unit area normal to the direction of the vector at that point. Its units, in the mks system, are watts/m^2. We shall not attempt to derive Eq. (23.5.3) in detail. Instead, we shall simply state that from Maxwell's equations it can be shown that the *flux* of the Poynting vector through any closed surface is equal to the change in the total electromagnetic field energy enclosed by the surface. It is, therefore, reasonable to interpret the Poynting vector in terms of the energy flow across the boundary, as described above. Mathematically, this property of the Poynting vector can be stated as

$$\Phi_s = \oint_\Sigma \mathbf{S} \cdot \mathbf{n} \, da = -\frac{\partial}{\partial t} \int_V \left(\frac{\varepsilon_0 E^2}{2} + \frac{B^2}{2\mu_0} \right) dV \qquad (23.5.4)$$

In this equation, $\mathbf{n}$ represents the unit outwardly directed vector normal to the closed surface Σ across

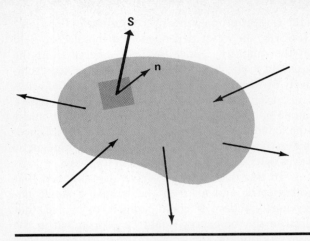

FIGURE 23.8. Schematic diagram illustrating the fact that the flux of the Poynting vector over a closed surface represents the rate of change of electromagnetic energy within the surface.

which the flux is calculated. Equation (23.5.4) states that the decrease in electromagnetic energy within any given volume is balanced by the energy propagated outward across its boundaries, or vice versa. The situation is illustrated in Fig. 23.8.

The following examples will illustrate the usefulness of the Poynting vector in calculating electromagnetic energy fluxes and will also serve to illustrate the validity of Eq. (23.5.4) in simply physical situations.

EXAMPLE 23.5.1

A plane wave traveling in a positive z-direction consists of oscillating electric and magnetic fields given by

$$E_x(z, t) = E_0 \cos(kz - \omega t) \qquad E_y = 0 \qquad (23.5.5)$$

$$B_y(z, t) = \frac{E_0}{c} \cos(kz - \omega t) \qquad B_x = 0 \qquad (23.5.6)$$

Obtain an expression for the Poynting vector in the plane $z = z_0$ and determine its time-average value.

According to (23.5.3), the Poynting vector is in the positive z-direction (see Fig. 23.9). Therefore, noting that $E_y = B_x = 0$ and $B_y = E_x/c$, we can write

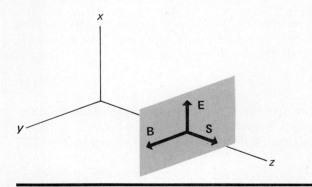

FIGURE 23.9

$$S_z = \frac{1}{\mu_0}(E_x B_y - E_y B_x) = \frac{E_x B_y}{\mu_0} = \frac{E_x{}^2}{c\mu_0}$$

$$= \frac{1}{\mu_0}\frac{E_0{}^2}{c}\cos^2(kz - \omega t) \qquad (23.5.7)$$

At $z = z_0$, then, we obtain

$$S_z = \frac{E_0{}^2}{\mu_0 c}\cos^2(kz_0 - \omega t)$$

$$= \frac{E_0{}^2}{\mu_0 c}\tfrac{1}{2}[1 + \cos 2(kz_0 - \omega t)] \qquad (23.5.8)$$

The time average of the cosine function vanishes, and thus the time average of the Poynting vector is

$$\overline{S}_z = E_0{}^2/2\mu_0 c \qquad (23.5.9)$$

EXAMPLE 23.5.2

Hydrogen atoms within the sun emit ultraviolet radiation of wavelength 1.216×10^{-7} meter. If the average Poynting vector due *only* to this component of the radiation has the magnitude 6×10^{-3} watts/m^2 at the earth, find the energy output of the sun at this wavelength and determine the electric and magnetic field amplitudes at the earth and at the surface of the sun due to this particular radiation.

If R denotes the distance from the sun to the earth (about 150,000,000 km) and $\overline{S}$ the magnitude of the Poynting vector at the earth's surface, the total energy propagated outward through a sphere of radius R per unit time will be

$$P = 4\pi R^2 \overline{S} = (6 \times 10^{-3})(4\pi)(1.50 \times 10^{11})^2$$

$$= 1.7 \times 10^{21} \text{ W}$$

This represents the total power radiated by the sun at this wavelength. To determine the electric field amplitude, we use (23.5.9) to find

$$\overline{S} = \frac{E_0{}^2}{2\mu_0 c} = 6 \times 10^{-3} \text{ W/m}^2$$

Substituting the values of μ_0 and c, we find $E_0 = 2.13$ volts/meter. Then recalling that $B_0 = E_0/c$, we obtain $B_0 = 7.1 \times 10^{-9}$ weber/m^2. At the surface of the sun, the same total power is radiated through a sphere having the sun's radius R_s, which is about 696,000 km. Therefore, writing the magnitude of the Poynting vector at the sun's surface as S_s,

$$P = 4\pi R^2 \overline{S} = 4\pi R_s{}^2 \overline{S}_s$$

from which

$$\overline{S}_s = \frac{R^2}{R_s{}^2}\overline{S} = \left(\frac{1.50 \times 10^{11}}{6.96 \times 10^8}\right)^2 (6.0 \times 10^{-3}) = 279 \text{ W/m}^2$$

Calculating E_0 and B_0 as before using (23.5.9) and (23.4.11), we find, at the surface of the sun, $E_0 = 459$ volts/meter, and $B_0 = 1.53 \times 10^{-6}$ weber/m^2. In

this example, we have assumed that the radiation is linearly polarized, which is not true. The calculated numerical values should, therefore, be regarded as illustrative rather than quantitatively correct.

EXAMPLE 23.5.3

A sinusoidal, plane electromagnetic wave propagates along the z-direction with velocity c. Starting with the energy density relations for electric and magnetic fields, show that the amount of energy propagated per unit area per unit time is given by $\overline{EB}/\mu_0 = \overline{E^2}/c\mu_0$, in agreement with the results predicted by the Poynting vector as discussed in Example 23.5.1.

Let us assume in this example that the plane wave is emitted by a source somewhere along the $-z$-axis, which is suddenly switched on at a certain time. When the source is switched on, a plane wave propagates along the z-direction with velocity c. At some later time, the wave will arrive at the plane surface S of area A shown in Fig. 23.10. At this instant, energy starts to flow across the surface. After a time interval t, the wave front has advanced to the surface S', a distance $l = ct$ further along the z-axis. The plane wave now fills the volume $V = Al = Act$ in Fig. 23.10. Electromagnetic energy now occupies this volume, and to get there, it must have crossed the surface S. The electrostatic energy density at every point is $\varepsilon_0 E^2/2$, and the *average* electrostatic energy density within the volume will be

$$u_e = \frac{\varepsilon_0 \overline{E^2}}{2} \qquad (23.5.10)$$

where $\overline{E^2}$ is the time average of E^2 at any point. Similarly, the magnetic energy density is

$$u_m = \frac{\overline{B^2}}{2\mu_0} \qquad (23.5.11)$$

where $\overline{B^2}$ is the time average of B^2. The total energy U within the volume is, therefore,

$$U = \left(\frac{\varepsilon_0 \overline{E^2}}{2} + \frac{\overline{B^2}}{2\mu_0}\right)V = \left(\frac{\varepsilon_0 \overline{E^2}}{2} + \frac{\overline{B^2}}{2\mu_0}\right)Act \qquad (23.5.12)$$

This energy has crossed the surface S in a time interval t. Therefore, the energy flux per unit area per unit time across the surface must be

$$\frac{U}{At} = c\left(\frac{\varepsilon_0 \overline{E^2}}{2} + \frac{\overline{B^2}}{2\mu_0}\right) \qquad (23.5.13)$$

But, according to (23.3.11) and (23.4.9), $B = E/c$. Therefore,

$$\frac{\varepsilon_0 \overline{E^2}}{2} = \frac{\varepsilon_0}{2}(c\overline{EB}) = \frac{c\varepsilon_0 \overline{(EB)}}{2}$$

and

$$\frac{\overline{B^2}}{2\mu_0} = \frac{1}{2\mu_0}\frac{\overline{(EB)}}{c} = \frac{\overline{(EB)}}{2c\mu_0} \qquad (23.5.14)$$

753

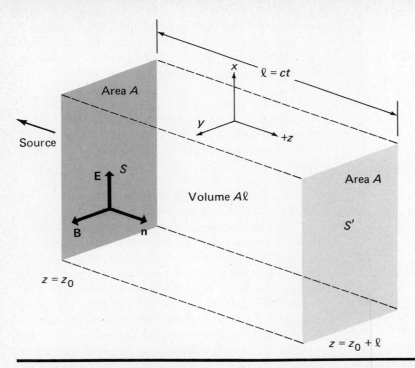

FIGURE 23.10. Geometry used for the calculations of energy flux in Example 23.5.3.

Substituting these results into (23.5.13) and noting that $c^2 = 1/(\varepsilon_0\mu_0)$, Eq. (23.5.13) can be written as

$$\frac{U}{At} = \frac{\overline{EB}}{\mu_0} \tag{23.5.15}$$

This is clearly the same result obtained by use of the Poynting vector defined by (23.5.3). From Fig. 23.10, it is evident that the magnitude of the Poynting vector is $S = EB/\mu_0$ and that its direction

is that of the z-axis. Therefore,

$$\mathbf{S} = \frac{E\overline{B}}{\mu_0}\,\mathbf{i}_z \tag{23.5.16}$$

and the time-average value of its magnitude is exactly that given by (23.5.15). Also, since $B = E/c$, it is easy to express (23.5.15) as

$$\frac{U}{At} = \frac{\overline{E^2}}{c\mu_0} = \frac{E_0{}^2}{2\mu_0 c} \tag{23.5.17}$$

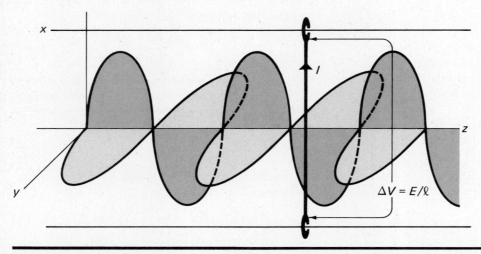

FIGURE 23.11. Transfer of momentum by an electromagnetic wave to a straight conductor.

754

since, as we have already ascertained in connection with (23.5.8), $E^2 = E_0{}^2/2$. This result agrees precisely with (23.5.9), which was obtained using the Poynting vector. Clearly, the concept of the Poynting vector leads to results that are in accord with what is to be expected from our previously derived expressions for electric and magnetic energy densities and from energy conservation.

23.6 Radiation Pressure

When a particle is in motion, its energy is transported from one location to another. It is also apparent that momentum is carried by the particle. In the previous section, we have seen that an electromagnetic wave carries energy, and, therefore, we might reasonably surmise that it carries momentum as well.

We can demonstrate that the wave transports momentum with the help of Fig. 23.11. A wire of length l and resistance R is oriented along the x-axis and is free to move in the z-direction. It is attached to a track made of an insulating material. A plane electromagnetic wave given by (23.4.13) and (23.4.14) is traveling in the z-direction. The electric field, which varies sinusoidally and which is parallel to the wire, leads to a current I given by

$$I = \frac{\Delta V}{R} = \frac{El}{R} \tag{23.6.1}$$

This implies that energy is dissipated within the wire by Joule heating at a rate given by $I^2 R$. If dU/dt is the rate at which the wave supplies energy to the wire, then energy conservation requires

$$\frac{dU}{dt} = I^2 R = \frac{E^2 l^2}{R} \tag{23.6.2}$$

Since a current flows through the wire, there is a magnetic force in the z-direction given by

$$F_m = IlB = \frac{EBl^2}{R} \tag{23.6.3}$$

Making use of the fact that $F_m = dp/dt$, where p is the z-component of momentum of the wire, and $B = E/c$, we find

$$\frac{dp}{dt} = \frac{1}{c}\frac{E^2 l^2}{R} = \frac{1}{c}\frac{dU}{dt} = \frac{d}{dt}\frac{U}{c} \tag{23.6.4}$$

By integrating both sides of this equation, we see that if the electromagnetic wave gives the wire an energy U, it also imparts a momentum p, given by

$$p = \frac{U}{c} \tag{23.6.5}$$

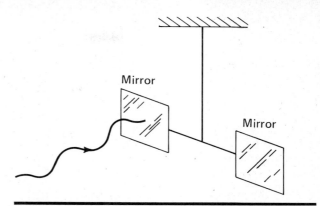

FIGURE 23.12. Schematic diagram of the experiment of Nichols and Hull designed to measure radiation pressure.

An electromagnetic wave, therefore, carries both energy and momentum.

When the sun shines on your hand, you are well aware that energy is being absorbed from the electromagnetic radiation. Momentum is also being absorbed, but because c is very large, the amount of momentum transferred is extremely small and difficult to detect. In 1903, the American scientists Nichols and Hull succeeded in verifying Eq. (23.6.5) by actually measuring *radiation pressure*. Their technique involved directing a light beam toward one of two mirrors suspended from a torsion fiber in a vacuum chamber, as shown in Fig. 23.12. They were able to measure the pressure by determining the angle through which the torsion balance twisted. The principal difficulty in doing this experiment is to eliminate spurious effects from residual gas atoms that can be much larger than the effect one wishes to observe unless great care is taken. The radiation pressures measured by Nichols and Hull were of the order of 7×10^{-6} newton/m^2. Thus, on a small mirror of area 10 cm^2, the force due to the radiation is only about 7×10^{-9} newton!

EXAMPLE 23.6.1

Light with an energy flux of 15 watts/cm^2 falls on a nonreflecting surface at normal incidence. If the surface has an area of 40 cm^2, find the average force exerted on the surface during a 30-minute time span.

The total energy falling on the surface is

$$\Delta U = (15 \text{ W/cm}^2)(1800 \text{ sec})(40 \text{ cm}^2) = 1.08 \times 10^6 \text{ J}$$

Thus, the total momentum delivered is

$$\Delta p = \frac{\Delta U}{c} = \frac{1.08 \times 10^6}{3 \times 10^8} = 0.36 \times 10^{-2} \text{ kg-m/sec}$$

The average force is

$$F = \frac{\Delta p}{\Delta t} = \frac{0.36 \times 10^{-2}}{0.18 \times 10^4} = 2 \times 10^{-6} \text{ N}$$

23.7 The Doppler Effect for Light

In Chapter 10, we discussed the origin of the Doppler effect and saw that changes in the detected frequency and wavelength of waves are present whenever there is relative motion between the source of the waves and the detector. Primary emphasis was placed on applications involving sound waves, although some discussion for light was also given. In the present section, we return to briefly take up the Doppler effect for electromagnetic waves.

When waves move through a material medium as sound waves do, there are two distinct effects, as discussed in Chapter 10. If the observer is in motion but the source is stationary, the frequency of the detected radiation is shifted. On the other hand, if the source is in motion and the observer is at rest in the medium, the wavelength of the radiation is changed. In the case of electromagnetic radiation traveling through *empty space* or vacuum, the Doppler effect can depend *only on the relative motion* between the source of radiation and the detector. Physically, there is no way to distinguish whether the source is moving with respect to the detector or the other way around.

The exact derivation of the Doppler frequency shift involves the use of the theory of relativity and will not be presented here. The result for the case in which the source is receding from the observer is

$$f' = f \frac{1 - (v/c)}{\sqrt{1 - (v/c)^2}} \qquad (23.7.1)$$

If the source and observer are moving toward one another, this equation is modified by replacing v by $-v$. In most cases involving motion of macroscopic sources, $v/c \ll 1$; therefore, the quantity $(v/c)^2$ in the denominator can usually be neglected.

One of the most important manifestations of the Doppler effect is the famous *red shift* of the light coming from distant galaxies. All measurements of known atomic spectral frequencies in the light emitted by these sources indicate shifts toward larger wavelengths or lower frequencies, leading to the conclusion that distant galaxies are receding or moving away from us. This provides direct evidence to support the theory of an expanding universe. In 1919, the American astronomer Hubble found that galaxies are moving away from us with varying speeds and that their speeds are directly proportional to their distance from us. This linear relation, which is extremely important in astronomy, is called Hubble's law.

EXAMPLE 23.7.1
A galaxy in the constellation Ursa Major is known to be receding at 1.4×10^4 km/sec. If the galaxy emits

light at wavelength λ, what wavelength will we detect?

Since $v/c = 0.047$ is much less than unity in this case, we may ignore the quantity v^2/c^2 in the denominator of Eq. (23.7.1). From this equation,

$$\frac{f' - f}{f} = \frac{\Delta f}{f} = -\frac{v}{c} = -0.047$$

But now, since Δf and $\Delta \lambda$ are small in comparison with f and λ and since

$$d\lambda = d(c/f) = -c \frac{df}{f^2} = -\frac{c}{f} \frac{df}{f} = -\lambda \frac{df}{f}$$

we may finally write

$$\frac{d\lambda}{\lambda} = -\frac{df}{f} = -\frac{v}{c} = +0.047$$

Thus, the detected wavelength λ' will be $\lambda + \Delta\lambda$, or

$$\lambda' = 1.047\lambda$$

EXAMPLE 23.7.2
A certain quasar is receding at $v = 0.8c$. If one of the spectral lines detected in its light has a wavelength of 1.25×10^{-7} meter when emitted from a source at rest, what wavelength is actually observed in the quasar's light?

Using the relation (23.7.1) and $f\lambda = c$, we obtain[5]

$$\frac{c}{\lambda'} = \frac{c}{\lambda} \left(\frac{1 - (v/c)}{\sqrt{1 - (v/c)^2}} \right)$$

from which

$$\lambda' = \lambda \frac{\sqrt{1 - (v/c)^2}}{1 - (v/c)} \qquad (23.7.2)$$

Substituting, we have

$$\lambda' = (1.25 \times 10^{-7}) \frac{\sqrt{1 - (0.8)^2}}{1 - 0.8} = 3.75 \times 10^{-7} \text{ m}$$

23.8 Generation and Detection of Electromagnetic Waves: The Electromagnetic Spectrum

The wave nature of light had already been exhibited and studied long before Maxwell's theory was developed. The theory of *optics* was fairly refined, but it only covered that portion of the electromagnetic spectrum which is referred to as visible light. This light, which spans wavelengths between 4.0×10^{-7}

---

[5] In this example, we cannot use the simple relation $d\lambda/\lambda = -df/f$ used in the previous equation because here, v/c is no longer small and the shifts Δf and $\Delta \lambda$ in frequency and wavelengths are no longer much smaller than the wavelengths and frequency themselves! We, therefore, cannot approximate them as differentials.

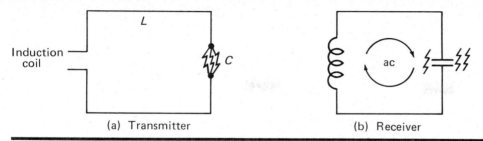

(a) Transmitter   (b) Receiver

FIGURE 23.13. Hertz's "radio transmitter" and "radio receiver."

and 7.0×10^{-7} meter or frequencies between 7.5×10^{14} and 4.3×10^{14} Hz, constitutes only a minute portion of the possible spectrum of electromagnetic frequencies or wavelengths. In this section we shall consider some qualitative aspects of the electromagnetic spectrum, including the generation and detection of electromagnetic waves.

The electromagnetic waves predicted by Maxwell were successfully generated and detected by Heinrich Hertz. His celebrated work, which was published in 1887, provided direct proof of Maxwell's theory. Hertz created these waves in the laboratory (see Fig. 23.13) by exciting high-frequency ac currents in a circuit consisting of a single conducting loop in which there is a narrow spark gap between two small, spherical electrodes. Such a circuit functions as an ordinary L–C circuit in which the capacitance is largely that of the spark gap and the inductance that of the single circuit loop. Since both the inductance and capacitance associated with the circuit are quite small, the frequency $\omega_0 \, (= 1/\sqrt{LC})$ of the ac current excited in the circuit by the high-voltage induction coil is very high. The rapid oscillation of current in the primary circuit stems from rapid harmonic oscillation of free charges in the conducting loop. This, in turn, means that these charges must undergo very large sinusoidally varying accelerations, which, according to the predictions of Maxwell's theory, requires that they radiate energy in the form of electromagnetic waves at their own vibrational frequency. In Hertz's experiments, a single loop of wire with a narrow spark gap was used as a detector, or *receiver*. This loop is constructed in such a way that its own capacitance and inductance give rise to a natural frequency equal to that of the transmitting circuit. Hertz found that when high-frequency ac oscillations were excited in the transmitting loop, as evidenced by the appearance of sparks in the primary spark gap, an accompanying spark discharge could be observed in the receiving loop, even though it might be several meters distant from the transmitting loop. He showed that these sparks were created by an ac current at the primary frequency that was induced in the secondary receiving circuit. He demonstrated that this current had all the characteristics expected on the assumption that it was in fact induced by electric and magnetic fields propagated as Maxwell's electromagnetic waves.

Hertz not only showed that the waves predicted by Maxwell exist, but he also demonstrated that his waves, which had wavelengths of ten million times the length of light waves, could be diffracted, reflected, and also polarized. They were shown to interfere with one another and also to propagate at the speed of light. Thus Hertz, in a series of careful experiments, clearly established the wave properties of this radiation and discovered that these *radiofrequency waves* behaved, in most respects, just like ordinary light waves. So far as we know, Hertz never attempted to use his strange apparatus to send Morse code messages from one point to another. It is clear, however, that he could easily have done so had he been so inclined. His interest, however, was to subject Maxwell's theory to experimental test and thus to confirm or deny its predictions rather than to develop a practical communications system. He was, nevertheless, the man who first sent and received radio signals and in this sense deserves to be honored as the inventor of radio communication.

Maxwell's theory allows for the existence of electromagnetic waves of any frequency or wavelength. Figure 23.14 shows the spectrum of waves which have been studied and utilized. Before 1800, the only waves which had been studied were the visible waves. During the nineteenth century, the spectrum was expanded in both directions toward higher and lower frequencies. The ultraviolet and infrared waves were studied first.

X rays were discovered by Roentgen in 1895 during a series of experiments in which he was studying the possible penetration of energetic electrons (then called cathode rays) through the glass wall of a cathode ray tube. He found, quite by accident, that radiation was emitted from the spot where the electron beam struck the glass. Roentgen had a screen coated with a material which was supposed to fluoresce when the electrons struck it. Instead, he discovered that it fluoresced as a result of a new and strange form of radiation emanating from the spot where the electrons hit the glass surface of the tube. He called this

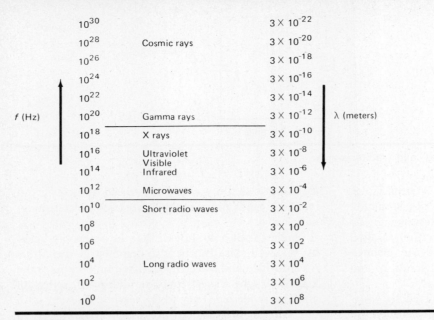

FIGURE 23.14. The electromagnetic spectrum.

new radiation X radiation. The wave character of this radiation was later studied by Max von Laue (in 1912) in a series of interference experiments. He discovered that X rays could be made to interfere by scattering them from the atoms in a crystal. In recent times, this interference phenomenon has been utilized extensively in crystallography, in which X rays serve as a tool for studying the geometric arrangements of atoms in crystals and other substances. Of course, other uses of X rays, for example, taking shadow pictures of parts of the body that are opaque to light and the treatment of cancerous growths, are well known to most of us.

Radiation at frequencies lower than that of visible light has also had a profound technologic impact, particularly in electrical engineering. In communications, both long-wavelength and short-wavelength radio waves are used extensively. Radar, which uses waves of a few centimeters in wavelength, was developed in the 1940s. By reflecting a radar beam from a moving plane, ship, or car, one can establish its location and also its speed. The extensive use of radio, television, microwaves, and radar today exhibits the vast technological importance of the scientific contributions of Maxwell and Hertz.

The electromagnetic spectrum has no boundaries in the theoretical sense, but there are practical limits on frequencies that can be produced and detected. For example, we do not have the means for producing electromagnetic radiation in the laboratory at the frequencies of high-energy cosmic rays. At the opposite extreme of very low frequency or long wavelength, the limitations primarily concern the efficiency of the radiation process. In a circuit which oscillates at very low frequency, radiation is emitted but in extremely small amounts and with very low efficiency. Under these circumstances, the energy needed to produce the radiation is very much greater then the energy radiated. As a practical matter, frequencies of a few kilohertz, corresponding to a wavelength of several hundred kilometers are sometimes used for long-range communications involving submerged underseas craft.

Let us now make some remarks concerning *sources* of electromagnetic radiation. We have already mentioned that, whenever charged particles accelerate, they can radiate energy. The wavelength of the radiation is often correlated with the characteristic size of the system that radiates. Thus, gamma radiation, having dimensions of 10^{-14} to 10^{-15} meter, typically originates from an atomic nucleus. X rays, ultraviolet rays, visible rays, infrared rays, and microwaves can all be emitted from atoms or molecules, which lose some of their energy in the process. Radio waves are produced by accelerating electrons in an ac circuit. A transmitting antenna can most efficiently radiate waves having a wavelength of about the same size as the antenna. This is achieved by producing between ground and antenna an oscillating current having the appropriate frequency.

In all these cases, electromagnetic radiation is emitted by accelerated charges. Since electromagnetic waves consist of oscillating electric and magnetic fields, the detection apparatus usually makes use of the fact that these oscillating fields can in turn accelerate charges and can produce oscillating currents.

Hertz's original "receiver" worked in exactly this way, and the same basic principle is utilized in practically all modern receiving devices. In modern communications equipment, however, there are no spark gaps.[6] Instead, in order to reproduce the original signal, the antenna current is processed through the elaborate detecting and amplifying circuits which inhabit radios or television sets. Since electromagnetic waves travel at 3×10^8 meters/sec, the entire process of converting sound or a picture to an electric signal, transmitting the waves over large distances, receiving the wave, and converting it back into the original signal occurs almost instantaneously.

Higher-frequency electromagnetic waves are detected by other means. For example, a photographic plate can be used to detect a range of frequencies from gamma rays to the infrared. Some waves can even be detected by means of fluorescence. For example, in a fluorescent lamp, invisible ultraviolet radiation of a mercury arc is absorbed by a fluorescent material, which reradiates energy in the visible range. Gamma rays, X rays, and ultraviolet rays can also be detected by means of an electroscope or an ionization chamber. This technique makes use of the fact that high-energy radiation can ionize atoms and, therefore, change the surrounding atmosphere into a conducting medium. The photoelectric effect is yet another means for detecting some forms of radiation. In this effect, electromagnetic radiation absorbed by a metal surface acts to eject some of the conduction electrons in the metal into the surrounding space, where they may be collected at an external electrode and detected as a dc current.

There are many ways to produce and detect electromagnetic radiation. The techniques that are utilized in all cases depend upon the frequency of the radiation. In every instance, however, electromagnetic waves originate from *accelerated charges*. Oscillating electric and magnetic fields then move away from these charges with the speed of light and are ultimately detected by observing the response of *other* charges. The underlying laws which allow a complete analysis of the radiation and detection of these waves are stated by the mathematically beautiful Maxwell equations.

23.9 A Postscript to the Study of Electromagnetic Fields

We have introduced the concept of a *field* in our original study of electrostatics in Chapter 15. At that time, the field concept seemed to introduce an un-

---

[6] Hertz used a spark gap in the receiving circuit simply because, in his day, there was no other way of detecting the presence of the high-frequency ac current there.

necessary complication into our study of the interaction between charges. It seemed then as though it would have been easier to have stated Coulomb's law as a force law in which the forces from one charge acted "at a distance" upon another. Instead, we insisted upon stating that the charges set up a "field" in the space surrounding them and that other charges experience forces as a result of the action of the field upon them. The role of the field as an intermediary in the process seems, at the beginning, extraneous and unnecessary.

We have now completed our formal study of electricity and magnetism and are about to start in on optics and modern physics. Before doing so, however, it is probably worth pausing for a moment to reconsider our understanding of fields, since we are now in a much better position to appreciate the usefulness of the field concept.

The trouble with the simple idea of "action of forces at a distance" is that the "action" does not occur instantaneously, but only after a *delay of time* sufficient to allow electromagnetic interactions to propagate between the current or charge source and the point where its effect is felt. When we switch on a current in a circuit or move an electric charge, the physical effects of these actions are manifested at another position only after an interval corresponding to the time required for light to travel between the two points.

But now, which concept is the simpler and more satisfying: the idea of a force that acts at a distance, but only after a time that depends upon how far the body it acts upon is from the body that exerts it; or the concept of a source that creates a field which propagates with the velocity of light and which acts upon other systems as it encounters them?

It is now the concept of action at a distance that seems to have a certain contrived aspect—to have been altered in a curious way to account for effects that it really does not cope with very well. Embarrassing questions difficult to answer satisfactorily have to be dealt with to defend it. How, for instance, does the second body know how far it is from the first in order to know how long it should wait before responding to the force? There is no good answer. We can patch up the theory to make it square with the facts if we work hard enough, but is it worth it? The situation with regard to action at a distance has taken on the same coloring as the theory of Ptolemaic epicycles in Chapter 8.

The concept of a field propagating outward from charge or current sources with the velocity of light is now seen to give all the required answers in a simple, clear way. Even the propagation velocity is obtained in terms of the constants ε_0 and μ_0 that tell us the strength of the basic interactions between charges and currents.[7] It appears to predict the results of whatever experiments are designed to test it in a very satisfactory way. It is at least as good as the action-at-a-distance concept and is indisputably simpler, less contrived, less troublesome, and therefore more beautiful and intellectually satisfying. So we accept it and believe it—until something even better comes along.

We could have made this argument bit by bit as we went along, because as the subject of electromagnetism is developed, the older action-at-a-distance idea becomes less and less appealing while the beauty and simplicity of the field concept becomes ever more apparent. We have chosen, however, to return to it at the very end because an understanding of electromagnetic wave propagation is needed to appreciate fully its usefulness.

SUMMARY

Maxwell showed that Ampère's law leads to certain contradictions unless the current enclosed by the Amperian loop is regarded as the sum of the "true" current arising from the motion of mobile charge and a *displacement current* given by

$$I_d = \varepsilon \frac{d\Phi_e}{dt}$$

This corresponds to a displacement current density

$$\mathbf{j}_d = \varepsilon \frac{d\mathbf{E}}{dt}$$

at any point. It is evident that a displacement current is present whenever a *time-varying electric field* exists and that its magnitude is proportional to the time rate of change of the field or electric flux. When the displacement current is included, Ampere's law has the form

$$\oint_c \mathbf{B} \cdot d\mathbf{l} = \mu_0(I_c + I_d) = \mu_0 I_c + \mu_0 \varepsilon \frac{d\Phi_e}{dt}$$

The laws of electromagnetism can be stated in the form of four equations, which are referred to as Maxwell's equations. In empty space they are

$$\oint_s \mathbf{E} \cdot \mathbf{n} \, da = \frac{q}{\varepsilon_0} \qquad \text{Gauss's law}$$

$$\oint_s \mathbf{B} \cdot \mathbf{n} \, da = 0 \qquad \text{Gauss's law for magnetism, plus the nonexistence of free magnetic charge}$$

$$\oint_c \mathbf{E} \cdot d\mathbf{l} = -\frac{d\Phi_m}{dt}$$

$$= -\frac{d}{dt} \oint_s \mathbf{B} \cdot \mathbf{n} \, da \qquad \text{Faraday's law}$$

$$\oint_c \mathbf{B} \cdot d\mathbf{l} = \mu_0 I_c + \mu_0 \varepsilon_0 \frac{d\Phi_e}{dt}$$

$$= \mu_0 I_c + \mu_0 \varepsilon_0 \frac{d}{dt} \int_s \mathbf{E} \cdot \mathbf{n} \, da \qquad \text{Ampère's law and displacement current}$$

When the displacement current term is included, it can be shown that Maxwell's equations can be expressed in the form of wave equations which predict the existence of electromagnetic waves that propagate with velocity $c = 1/\sqrt{\varepsilon_0 \mu_0} = 2.998 \times 10^8$ meters/sec. These waves are transverse waves, in which the electric and magnetic field vectors lie normal to each other in a plane that is itself normal to the direction of propagation. For the case of sinusoidal waves in free space, the electric and magnetic vectors at any given point oscillate sinusoidally with the frequency of the wave and with the same phase. For plane waves propagating in the z-direction, the electric and magnetic vectors can be expressed as

$$\mathbf{E}(z, t) = E_0 \cos(kz - \omega t)\mathbf{i}_x$$

$$\mathbf{B}(z, t) = B_0 \cos(kz - \omega t)\mathbf{i}_y$$

with

$$c = f\lambda = \frac{\omega}{k} = \frac{1}{\sqrt{\varepsilon_0 \mu_0}} \qquad \text{and} \qquad B_0 = \frac{E_0}{c}$$

Electromagnetic waves in free space are dispersionless, their velocity being independent of wavelength or frequency. The propagation constant k is related to the wavelength by

$$k = \frac{2\pi}{\lambda}$$

The propagation vector is a vector $\mathbf{k}$ of magnitude $2\pi/\lambda$ in the direction of propagation. For the sinusoidal wave discussed above,

$$\mathbf{k} = k\mathbf{i}_z = \frac{2\pi}{\lambda} \mathbf{i}_z$$

---

[7] It is interesting to note that if the velocity of light were truly infinite, the concept of action at a distance would regain some of its former attractiveness. But then, in order that $c = 1/\sqrt{\varepsilon_0\mu_0} = \infty$, we would have to have $\mu_0 = 0$, and there would be no magnetic fields at all. Nothing but electrostatics would remain.

The instantaneous energy density associated with an electromagnetic wave in free space is

$$u = u_e + u_m = \tfrac{1}{2}\varepsilon_0 E^2 + \frac{1}{2\mu_0} B^2$$

and the energy stored in any volume V is

$$U = \int_V \left(\tfrac{1}{2}\varepsilon_0 E^2 + \frac{1}{2\mu_0} B^2 \right) dV$$

The instantaneous flux of electromagnetic energy, per unit area normal to the propagation direction per unit time, is given by the Poynting vector,

$$\mathbf{S} = \frac{1}{\mu_0} (\mathbf{E} \times \mathbf{B})$$

The total flux of energy across any closed surface Σ is then

$$\Phi_S = \oint_\Sigma \mathbf{S} \cdot \mathbf{n} \, da$$

where $\mathbf{n}$ is the unit vector normal to the surface. It can be shown that for any closed surface Σ,

$$\Phi_S = -\frac{\partial U}{\partial t}$$

which expresses the fact that the flux of energy across the surface is equal to the rate at which the amount of energy within decreases.

For a plane electromagnetic wave in free space, the average amount of energy per unit time that flows through unit area normal to the propagation direction is

$$\overline{S} = \frac{E_0 B_0}{2\mu_0} = \frac{E_0{}^2}{2c\mu_0}$$

Electromagnetic radiation of energy U has an associated momentum p given by

$$p = \frac{U}{c}$$

When radiation is reflected or absorbed, this momentum is changed and, correspondingly, forces due to radiation pressure equal to the time rate of change of electromagnetic momentum are observed.

Electromagnetic radiation exhibits a Doppler effect, in which the observed frequency shift depends only on the relative velocity v between source and observer. The detected frequency f' is related to the emitted frequency f by

$$f' = f \frac{1 \pm (v/c)}{\sqrt{1 - (v/c)^2}}$$

The plus sign is used when source and observer are approaching one another, the minus sign when they are receding from one another.

Maxwell's theory shows that *electromagnetic waves are radiated whenever charges undergo acceleration*. Since charges are accelerated whenever ac currents flow in circuits, any ac circuit will radiate electromagnetic waves at the ac frequency of the flowing current. Furthermore, since in any harmonic oscillation of given amplitude the acceleration increases with increasing frequency, high-frequency ac circuits radiate more readily than circuits of low frequency.

Electromagnetic radiation of every conceivable wavelength has been generated and studied. These radiations, in order of decreasing wavelength, comprise radio waves, microwaves, infrared radiation, visible light, ultraviolet light, X rays, and gamma radiation, the wavelength spectrum covering a range of thousands of meters to less than 10^{-15} meter. Collectively, these waves constitute what is referred to as the *electromagnetic spectrum*.

QUESTIONS

1. Explain qualitatively the difference between conduction current and displacement current. Are both equally important in the application of Ampère's law?

2. Is it possible to have both a conduction current and a displacement current and yet have no magnetic field nearby?

3. It is often said that a time-varying electric field produces a magnetic field and a time-varying magnetic field produces an electric field. How does this come about?

4. The wavelength and frequency of a monochromatic wave are the reciprocals of one another. Explain qualitatively why this is so.

5. Show directly from the definition of the Poynting vector that its units are watts per square meter.

6. Is it possible for an object to absorb momentum from an electromagnetic wave and yet not absorb energy? Can a body absorb energy but not absorb momentum?

7. Is it possible to construct a system from parts of macroscopic size that will radiate monochromatic infrared radiation of wavelength 1.0×10^{-6} meter? Describe some of the problems you might encounter in trying to carry out such an assignment.

8. If a galaxy recedes from an observer at a speed very close to the speed of light, describe what happens to the frequency of the radiation that is received. Would it be possible even to detect such a galaxy?

9. One of Maxwell's equations is a statement of the nonexistence of isolated magnetic charges. Explain this by comparing the equation to Gauss's law.

10. Explain clearly the distinction between a plane wave, a spherical wave, and a cylindrical wave. Why do most waves look like plane waves far from the source which produced them?

11. An electromagnetic wave propagates through a certain region of space. How can this be verified experimentally?

12. The antenna of a broadcasting station radiates energy in

the form of radio-frequency electromagnetic waves. A receiver 50 km from the station picks up the program. Does a current flow from the broadcasting antenna to the receiver's antenna? If so, what are its physical characteristics?

13. A light bulb illuminates the interior of a room. The energy it radiates is absorbed by the walls. Does a current flow from the bulb to the walls? If so, what kind of a current? If not, why not?

14. A light bulb is located at the center of a cubical enclosure with *perfectly reflecting* interior surfaces. Describe the variation of light intensity within the enclosure as a function of time. What happens to the light intensity when the bulb is switched off?

15. Suppose you were told that the existence of isolated magnetic charges had been verified experimentally. How would you suggest altering Maxwell's equations in order to account for this new discovery? What new physical effects would you expect to occur?

PROBLEMS

1. A parallel-plate capacitor has circular plates of area 0.25 m^2. If the electric field between the plates is changing at the rate $dE/dt = 2 \times 10^{11}$ volts/meter-sec, find the displacement current, assuming that the region between the plates contains air.

2. A parallel-plate condenser has plates of area 0.6 m^2 separated by a distance of 0.12 mm. A sinusoidal emf of amplitude 360 volts and frequency 400 Hz is applied across the plates. Find (a) the amplitude of the conduction current into the plates, (b) the amplitude of the time rate of change of electrical field within the capacitor, (c) the amplitude of the displacement current density, (d) the amplitude of the total displacement current.

3. What is the magnitude of the magnetic induction **B** at a distance of 2 cm from the axis of the capacitor in the preceding problem?

4. A parallel-plate capacitor for which $C = 3$ μF is connected across a 60-cps ac line. If the displacement current has an amplitude of 0.5 ampere, find the rms voltage of the line.

5. Two small conductors are charged to a potential difference of 100,000 volts and are separated by a distance 3.33 cm. This potential difference is just large enough to cause electrical breakdown of the air between the conductors. If complete breakdown occurs in 10^{-6} second, find the average displacement current density during this time interval. *Hint:* When electrical breakdown has occurred, the electric field will have fallen essentially to zero.

6. A parallel-plate capacitor with circular plates is charged to a potential difference ΔV and is then disconnected from the battery. The capacitor plates are separated by a distance d and have area A, and the dielectric between them has permittivity ε. Assume that because the dielectric is not perfectly insulating but has a small leakage conductance, the capacitor discharges through the dielectric. Show that during this discharge, the total magnetic induction within the dielectric is zero.

7. There is an electric field parallel to the axis of an evacuated cylindrical volume of radius R. The field is spatially uni-

form but has a time variation given by $E = E_0 \cos \omega t$. Find the induced magnetic field **B** as a function of r and t, where r is the distance from the axis of the cavity.

8. A spherical capacitor is connected to an ac potential difference $\Delta V = \Delta V_0 \cos \omega t$. Show that there exists between its electrodes a displacement current given by $I_d = -C\omega V_0 \sin \omega t$, where C is the capacitance.

9. A circular parallel-plate capacitor contains a dielectric of permittivity ε and has a capacitance C. If the potential difference between the plates is $\Delta V = \Delta V_0 \cos \omega t$, obtain an expression for the displacement current. Assume that the field is uniform within the region between the plates.

10. Using the results of the preceding problem, find the magnetic induction **B** as a function of the time and the distance from the axis of the capacitor. The plates are separated by a distance d.

11. Using Eq. (23.3.4) with $I_c = 0$, prove Eq. (23.4.5). Do this by integrating $\mathbf{B} \cdot d\mathbf{l}$ along the rectangular path in the yz-plane as illustrated in Fig. 23.5b.

12. Show by direct differentiation that $E(z, t)$ given by (23.4.10) is a solution of the wave Eq. (23.4.6).

13. How long does it take for an electromagnetic signal to propagate from the earth to the moon and back?

14. Professors Carl Sagan and Frank Drake of Cornell University sent coded messages to a star cluster M13 in November 1974. If there is life anywhere in M13, an answer to the message might be expected in 48,000 years. How far away are these stars in miles and in meters?

15. (a) Show that the equation $\partial u/\partial x = v^{-1}(\partial u/\partial t)$ is a wave equation satisfied by any function having the form $u(x, t) = f(x + vt)$. (b) Show that the equation $\partial u/\partial x = -v^{-1}(\partial u/\partial t)$ is a wave equation whose solutions are $u(x, t) = f(x - vt)$. (c) Show that all solutions of both these equations also satisfy the more familiar wave equation $\partial^2 u/\partial x^2 = v^{-2}(\partial^2 u/\partial t^2)$. (d) Do all solutions of the second-order equation of part (c) satisfy the wave equation of part (a)? (e) Why is it that we always insist on using the wave equation in the form of the second-order partial differential equation shown in part (c)?

16. Show that $\mathbf{E}(z, t) = E(z + ct)\mathbf{i}_x$ is a solution of the one-dimensional wave equation in which the electric vector propagates in the negative z-direction.

17. The electric vector of an electromagnetic wave propagating in the positive z-direction is given by

$$E_0 \cos 5000[z - (3 \times 10^{10})t]\mathbf{i}_x$$

the numerical values being given in cgs units. Find the wavelength and the frequency of the wave. Obtain an expression for the magnetic induction $\mathbf{B}(z, t)$ by using Eq. (23.4.5).

18. What are the frequencies and propagation constants for electromagnetic waves of the following wavelengths if the waves propagate in vacuum: 10^{-8} cm (X rays), 100 cm (radio waves), 5.5×10^{-5} cm (visible light)?

19. A radio station is broadcasting at a frequency of 750,000 Hz. Find the wavelength of the electromagnetic waves it broadcasts.

20. A plane electromagnetic wave propagates in vacuum with a wavelength of 2.5×10^{-6} meter. It has an intensity of 4.24 watts/m^2. The electric vector is along the z-direction, while the magnetic vector is along the $-y$-direction.

(a) Calculate the magnitude of the electric field due to this wave. (b) Write a mathematical expression for the instantaneous electric field associated with the wave at any point in space and at any time. (c) What is the magnitude and direction of the Poynting vector?

21. The energy density associated with a certain electromagnetic wave of a single frequency is 10^{-7} joule/m^3. Find the amplitudes of the electric and magnetic fields.

22. Show that the electric and magnetic energy densities due to the propagation of a plane electromagnetic wave are equal.

23. Plane electromagnetic waves of a particular frequency are incident normally on the earth's surface. Assume the amplitude E_0 of the electric field to be 500 volts/meter. (a) What is the amplitude B_0 of the magnetic induction? (b) Find the average value of the Poynting vector. (c) Draw a diagram showing the electric field, the magnetic induction, and the Poynting vector.

24. A plane electromagnetic wave is observed to create a magnetic field of 2.5×10^{-8} weber/m^2 (rms). (a) How much energy per unit time is propagated across an area of 1 m^2 normal to the direction of propagation? (b) What is the rms electric field intensity? (c) What is the electromagnetic energy density?

25. Spherical electromagnetic waves proceed outward from a point source. At a distance of 300 meters from the source, it is found that the energy per unit time crossing unit area normal to the propagation direction is 0.0084 w/m^2. (a) What is the rms electric field strength at a distance of 300 meters from the source? (b) What is the rms magnetic field strength at this distance? (c) What is the rms electric field strength at a distance of 1800 meters from the source? (d) What is the total power emitted by the source?

26. Assuming that the transmitting antenna radiates like a point source in free space, what is the rms electric field strength (a) at a distance of 1 km from a 100,000 watt radio station, (b) at a distance of 100 km (c) at a distance of 1000 km?

27. In the preceding problem, a more realistic picture of the power radiated by the station would have to take into account the fact that the earth acts to reflect the emitted radiation, and that the radiation is also reflected by the ionosphere (an ionized atmospheric layer.) Assuming that all the radiated power is confined between the earth's surface and a reflecting layer at a height of 100 km, what is the rms electric field strength (a) at a distance of 100 km, (b) at a distance of 1000 km, (c) at a distance of 5000 km from the 100,000-watt radio station of the preceding problem. Neglect the earth's curvature in your calculation. (Hint: Assume, for simplicity, that the power is radiated from a line source normal to the earth and extending between earth and ionosphere.)

28. Calculate the frequency and propagation constant for electromagnetic radiation of wavelength 6.5×10^{-7} meter.

29. A point source emits 40 watts of electromagnetic energy in the form of spherical waves of a single frequency. Find the average electric field amplitude at a distance of 1.5 meter from the bulb.

30. Light from a bulb with a 100-watt output falls on a mirror with area 5 cm^2. If the mirror is 2 meters from the bulb, find the radiation pressure and the total force on the mirror assuming the light strikes it at normal incidence.

31. Sunlight is absorbed completely by the solar panels of a satellite. The intensity of the light is 100 watts/m^2 and the panels have a total area of 16 m^2. (a) Calculate the total momentum delivered to the panels over a 24-hour period. (b) Would the momentum delivered increase or decrease if the panels reflected some of the light?

32. In the laboratory, one of the spectral lines emitted by atomic hydrogen has a wavelength of 4.341×10^{-7} meter. The spectrum of a distant galaxy reveals that this line has a wavelength of 6.00×10^{-7} meter. Find the relative speed of the earth and the galaxy. Is the galaxy approaching or receding?

33. Microwaves of frequency 2.5×10^9 Hz are used to detect speeding on an interstate highway. If the speed limit is 55 mph, what frequency shift Δf between the transmitted and reflected waves indicates a speeding vehicle?

34. Due to the rotation of the earth, the frequency emitted by a laser can differ at the north pole and the equator. For a wavelength of 6.317×10^{-7} meter at the north pole, find the maximum and minimum wavelengths at the equator.

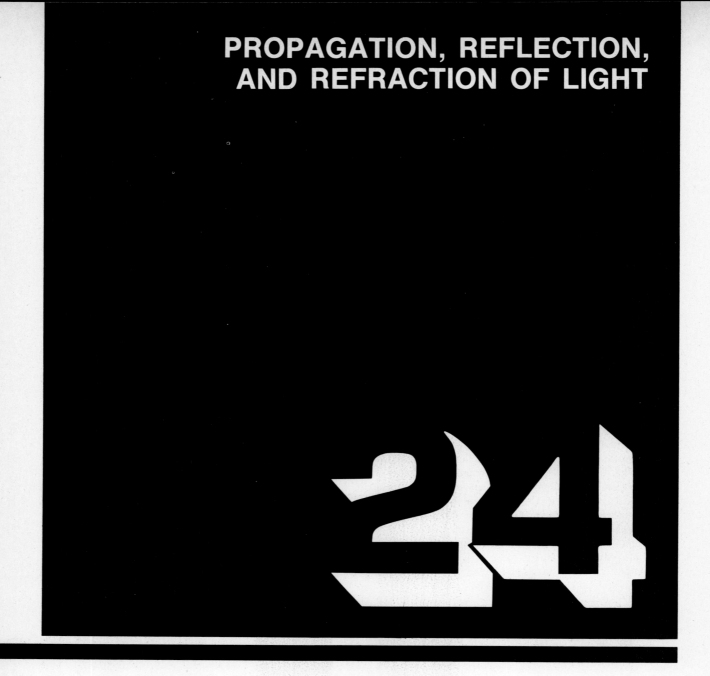

PROPAGATION, REFLECTION, AND REFRACTION OF LIGHT

24.1 Introduction

Like other physical effects that are directly evident to our senses, the properties of light have been the subject of speculative and experimental investigations since ancient times. The phenomena of reflection, refraction, and dispersion of white light into its constituent colors were all familiar to the Greeks and Romans, even though they were not very well understood by them. The refractive properties of glass were utilized in medieval times to correct imperfect vision, which is the most common of man's chronic physical imperfections.

The systematic study of the science of optics may be traced to Roger Bacon, Galileo, and Sir Isaac Newton, all of whom succeeded in constructing good astronomical telescopes. Newton's *Opticks* may be regarded as the source from which much of our subsequent understanding of the subject developed. The concept of light as a wave phenomenon was first proposed in 1678 by the great Dutch physicist Christian Huygens (1629–1695), but his hypothesis was not widely accepted for over a century. Then, in the early years of the nineteenth century, Thomas Young (1773–1829) in England and Augustin Jean Fresnel (1788–1827) in France clearly established the wave theory of light by showing that under proper conditions light exhibits the characteristic effects of *interference*, *diffraction*, and *polarization* that are uniquely associated with transverse waves. And finally, as we have now learned, Maxwell and Hertz demonstrated that light waves are really a form of high-frequency electromagnetic radiation.

Parallel to the development of our understanding of the physical nature of light, there is the growth of practical application of optics and the development of optical instruments. The understanding of the laws of reflection and refraction led to the development of telescopes, microscopes, and highly sophisticated lens systems for cameras, projectors, and other instruments. The discovery of the wave nature of light led immediately to the development of interferometers, polarimeters, diffraction gratings, and antireflection coating processes. The study of optics is still an active field of research in physics. The development of the stimulated emission coherent light amplifier, better known as the *laser*, was accomplished as recently as 1957. It has provided us not only with a large number of new and very useful optical instruments but has also enabled us to achieve a much better understanding of the interaction of intense electromagnetic radiation with matter.

The contribution of optics and optical instruments to the development of other areas of science and to man's general health and well-being should not be underestimated. Applied optics has provided us not only with pocket binoculars and single-lens reflex cameras, but also with sophisticated and powerful research instruments. The development of biological and medical science, it is safe to say, could not possibly have advanced very far had the microscope not been available. The reason for this lies in the simple fact that living cells are much too small to be visible to the naked eye. Indeed, it is safe to say that the much publicized technology of modern medicine and immunology would simply not be there had powerful optical instruments not been invented. At the other end of the scale, our knowledge of the sun, the planets, our galaxy, and the universe has been largely acquired through the use of huge but exquisitely designed and constructed astronomical telescopes.

Newton suggested that light consists of tiny particles or corpuscles emitted by a luminous object that travel outward from the source in straight lines. These particles, upon striking the retina of the eye, impart their energy to the retina allowing the reception of sensations that can be interpreted by the brain as a visual image. With the development of the wave theory, this particulate theory was discarded, since all the physical effects of light that were known at the time could be explained in terms of the behavior of electromagnetic waves, while the corpuscular theory could not satisfactorily account for the observed effects of interference, diffraction, and polarization.

By about 1905, however, a number of phenomena had been observed that could not be understood purely on the hypothesis that light exists only in the form of continuous electromagnetic waves. The most important of these effects are the emission of sharply defined light frequencies by atoms in ionized gases, the spectral characteristics of light radiated by incandescent hot objects, and the emission of electrons from metal surfaces resulting from the absorption of light by the substance. To describe these phenomena adequately, it was necessary to revive some of the features of the corpuscular theory and to postulate that light must exist in the form of discrete bundles or packets of electromagnetic wave energy that exhibit all the characteristics of electromagnetic waves while also possessing the discrete or corpuscular character of particles. These packets of electromagnetic energy are referred to as *light quanta* or *photons*. The initial successes achieved by the introduction of the idea that electromagnetic energy can exist in the form of discrete particles or quanta served to establish the foundations of the *quantum theory* as a new form of mechanics that is necessary to describe the behavior of matter and energy on the level of atomic systems and elementary particles.

And so it came about that we must regard light as having the characteristics of *both* waves and particles, exhibiting one or the other aspects of its dual nature according to the circumstances of how it interacts with physical systems and is observed experimentally. In this chapter and in the two that follow, we shall emphasize the behavior of light as an electromagnetic wave. In spite of this, however, we should not lose sight of the fact that it can display the particulate side of its dual personality under the proper circumstances.

At first, we shall rely heavily on the properties of waves in order to derive the fundamental laws of propagation, reflection, and refraction. We shall then try to describe how light rays, which travel in straight lines in free space, are bent upon refraction and reflection. Finally, we shall show how the laws of reflection and refraction can be used to describe how light rays are acted upon by mirrors, lenses, and prisms. In doing this, we shall find that once the laws of reflection and refraction are established, the wave nature of light is not manifested very obviously in the behavior of light rays, at any rate, in systems whose dimensions are large compared to the light wavelength. Under these circumstances, all we need to know is that light rays travel in straight lines in free space and are deflected according to simple geometrical laws at reflecting and refracting surfaces. These facts in themselves establish the geometry of light rays and determine their paths in optical systems. This aspect of optics which describes the geometry of light rays in systems whose dimensions greatly exceed the wavelength of light is referred to as *geometrical optics*. It is this subject that will largely occupy our attention in this chapter and in the one that follows.

In Chapter 26, however, we shall be concerned with the effects of interference and diffraction, which are produced when light waves of different amplitude and phase are combined. In these effects, it is the wave character of light that is of primary importance. The study of physical effects in which the amplitude and phase relationships of individual light beams are significant is called *physical optics*, or *wave optics*. Finally, there are many instances wherein the particulate or corpuscular behavior of individual light quanta is of importance. These situations broadly define the subject of *quantum optics*, about which we shall have something to say in Chapters 27 and 28.

24.2 Characteristics of Electromagnetic Wave Propagation: Intensity, Amplitude, Frequency, and Phase

We have already seen in connection with our study of electromagnetic radiation that what our eyes respond to as visible light is merely electromagnetic radiation that lies within a relatively small part of the electromagnetic spectrum between the frequency limits of about 4.2×10^{14} to 7.5×10^{14} Hz. This range of frequencies corresponds to free-space wave-lengths between 4.0×10^{-7} and 7.2×10^{-7} meter, though in dealing with light wavelengths it is customary to use either *nanometers* (nm) or *Ångström units* (Å) to express the wavelengths. Since a nanometer is 10^{-9} meter, we may say that the wavelength range of the visible spectrum extends from 400 to 720 nm, and since the Ångstrom unit is defined as 10^{-10} meter the same wavelength range may be expressed as 4000 Å to 7200 Å. In addition to these wavelengths, radiation in the parts of the electromagnetic spectrum adjacent to the visible light range, called infrared radiation on the long-wavelength side and ultraviolet radiation on the short-wavelength side, behaves very much like visible light. The laws of optics, generally speaking, apply equally well to visible light and in-frared and ultraviolet "light." The visible range of the electromagnetic spectrum, and the adjacent infrared and ultraviolet regions, are illustrated in Fig. 24.1.

This figure illustrates also that what we perceive as the *color* of a light beam is related to the wavelength or frequency of the light. It also shows how what we sense as *white* light is really light consisting of a super-position of waves of all visible frequencies. Finally, it demonstrates the fact that a beam of white light in passing through a refractive element such as a glass prism is dispersed into a band of light in which the component wavelengths are spatially separated, the long wavelengths being bent or *refracted* through a smaller angle than the short ones. Such a display is referred to as a *spectrum*, and we find that in the case of white light the colors displayed by the visible spec-

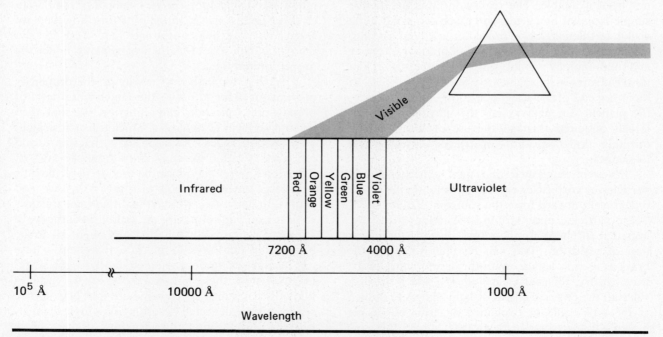

FIGURE 24.1. Visible, infrared, and ultraviolet regions of the electromagnetic spectrum, illustrating the production of the spectrum by the dispersion of white light in passing through a glass prism.

trum are, in order of decreasing wavelength, red, orange, yellow, blue, green, and violet. The separation of white light into its spectral components by a prism is based on the fact that in a dense transparent substance such as glass, the speed of light is less than in vacuum and, moreover, its velocity is different for different wavelengths. Different component wavelengths are thus deviated or refracted through different angles in passing through a prism and each color emerges in a slightly different direction, as illustrated in Fig. 24.1. This phenomenon is referred to as dispersion and has already been introduced, you may recall, in Chapter 10. We shall discuss the refraction and dispersion of light in some detail in a later section. Although the colors of the spectrum have been familiar from prehistoric times in the form of the rainbow, in which raindrops play the role of the glass prism, the phenomena of refraction, dispersion, and the resolution of white light into its spectral components were not clearly understood until Sir Isaac Newton studied them in 1672.

It is frequently important to distinguish between *monochromatic light*, which is light of a single pure color having a uniquely specified frequency and wavelength, and *white light*, which is a superposition of light waves of many different frequencies. In the case of a single monochromatic electromagnetic wave, the energy transported by the wave is given by the *Poynting vector* **S**, which according to (23.5.3) can be written

$$S = \frac{1}{\mu_0} (E \times B)$$ (24.2.1)

The vector **S** is oriented in the direction in which energy is transported by the wave, and its magnitude is the energy per unit area per unit time carried in that direction by the wave. Generally speaking, since the Poynting vector is normal to both **E** and **B**, as is the *propagation vector* **k** associated with the wave, the Poynting vector and the propagation vector are in the same direction.

For a *plane* electromagnetic wave propagating along the positive z-direction, the **E** and **B** fields lie in the xy-plane and have the magnitudes

$$E = E_0 \cos(kz - \omega t)$$ (24.2.2)

$$B = B_0 \cos(kz - \omega t)$$ (24.2.3)

where the propagation constant k is related to the wavelength λ by

$$k = \frac{2\pi}{\lambda}$$ (24.2.4)

For such a wave, since **E** and **B** are always perpendicular, the instantaneous magnitude of the Poynting vector will be

$$S = \frac{EB}{\mu_0} = \frac{E_0 B_0}{\mu_0} \cos^2(kz - \omega t)$$ (24.2.5)

and since its direction is the direction of the propagation vector **k**, it must point along the positive z-axis. At any given point, this quantity will fluctuate rapidly due to the harmonic variation of the **E** and **B** fields. For light waves, the frequency is so high that it is difficult to observe this fluctuation directly. In any event, what we usually wish to know is the *average* energy transported by the wave in a time interval much longer than the duration of a single cycle. This corresponds to the *time average* of the magnitude of the Poynting vector, which has already been calculated for this situation in section 23.5 of the preceding chapter. The result, as given by Eq. (23.5.9), is

$$\overline{S} = \frac{E_0^2}{2c\mu_0} = \frac{E_0 B_0}{2\mu_0}$$ (24.2.6)

The value of $\overline{S}$ so obtained, expressed in units of watts/m^2, is referred to as the *intensity* of electromagnetic radiation, or the *light intensity* in the xy-plane. For a single monochromatic wave, it is evident that the light intensity is proportional to the *square* of the electric vector amplitude E_0. For light consisting of a superposition of a number of components of different frequency, the total intensity is simply the sum of the intensities associated with each spectral component.

For a *point source* of light, as illustrated in Fig. 24.2, the wavefronts are no longer planes but *spherical* surfaces whose radii expand with velocity c. Expressions (24.2.2) and (24.2.3) are still valid for the magnitudes of the magnetic and electric fields, provided we realize that now the amplitudes E_0 and B_0 may vary with the radius of the spherical wavefront. This variation can be calculated by noting that the law of conservation of energy requires that for a source radiating a constant total amount of energy per second, the energy radiated across the surface of any sphere centered on the source has to be the same *irrespective* of the radius of the sphere. If this were not so, light energy would in time accumulate in some regions and be depleted from others, and this clearly does not happen. The light intensity $\overline{S}$ represents *the amount of energy per unit time per unit area* transported across any surface oriented normally to the propagation direction. The total rate of flow of electromagnetic energy dU/dt across any such surface may, therefore, be obtained by integrating the intensity $\overline{S}$ over the area of the surface. In the case of a point source, where the propagation vector **k** is everywhere radial and the wavefronts are spherical, we may write, integrating over a spherical surface of radius r,

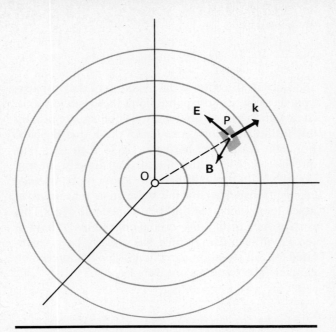

FIGURE 24.2. Spherical wavefronts produced by a point source of light at O, illustrating the propagation of a ray along the direction of the propagation vector **k**.

$$\frac{dU}{dt} = \oint_S \bar{S}\, da = \bar{S} \oint_S da = \bar{S}A \qquad (24.2.7)$$

or

$$\frac{dU}{dt} = 4\pi r^2 \bar{S} \qquad (24.2.8)$$

But since the total rate of flow of energy across any spherical surface is the same, for two spherical surfaces of radii r_1 and r_2 at which the intensity has the values $\bar{S}_1$ and $\bar{S}_2$ we may write

$$4\pi r_1^2 \bar{S}_1 = 4\pi r_2^2 \bar{S}_2 \qquad (24.2.9)$$

or

$$\frac{\bar{S}_2}{\bar{S}_1} = \frac{r_1^2}{r_2^2} \qquad (24.2.10)$$

From this, it is evident that for a point source of light, the intensity at any point is inversely proportional to the square of the distance from the point to the source. The intensity of illumination of extended light sources may be calculated by superposing the intensities contributed by infinitesimal areas of the source distributions (each being regarded as a point source) by integration.

Figure 24.2 illustrates some of the important features of light propagation from a point source. As we have already seen, the wavefronts—that is, surfaces of constant phase—are concentric spheres that expand radially outward with the speed of light. The

vectors **E** and **B** are, of course, tangent to the wavefronts, since electromagnetic waves are transverse. The propagation vector **k** is everywhere normal to **E** and **B** and, therefore, is a vector that points *radially outward* at any point on the wavefront. As the wavefront propagates, for example, from point O to point P, the propagation vector **k** associated with this point on the wavefront translates radially outward along the line OP. It is clear that the propagation vector anywhere on the wavefront always points in the same radial direction as the wavefront expands outward from the source. The most common way of summarizing this behavior is to say that *light travels in straight lines* through any uniform medium. The straight line OP along which the vector **k** associated with point P on the wavefront moves outward is referred to as a *light ray*. We may visualize the pattern of light radiated from a point source as a pattern of rays that go radially outward from the source as illustrated in Fig. 24.3 as well as by the pattern of spherical wavefronts shown in Fig. 24.2. It is important to note that *the rays are always normal to the wavefronts*.

The statement that light travels in straight lines must, however, be qualified in two ways. First of all, it must be understood that light rays may be bent by *reflection* at a reflecting surface or by *refraction* when the light ray enters or leaves a dense transparent medium such as water or glass. Also, we shall see in Chapter 26 that when light encounters obstacles or apertures that are *of the order of the light wavelength* in size, light rays may be *bent* when passing near such an obstacle or through such an aperture. This effect is referred to as *diffraction* of light, and it is said that light waves are *diffracted* by small obstacles or apertures.[1] Since the wavelength of visible light is very short, perceptible diffraction of visible light occurs only when light encounters apertures or obstacles that are very small indeed when measured on the scale of ordinary macroscopic systems. Therefore, unless such structures are present, we may neglect the effect of diffraction altogether and simply adopt the picture of light rays that go in straight lines whose direction is altered only by reflection or refraction.

---

[1] An easy way of observing the diffraction of light is to view a white light source reflected at grazing incidence from the playing surface of a $33\frac{1}{3}$ rpm disc recording. The bands of spectral colors that are observed arise as a consequence of diffraction of light reflected from the flat surfaces between the individual grooves. The distance between individual grooves is about 10^{-4} meter, which is about 50 times the wavelength of visible light (5×10^{-6} meter). Though the spacing between grooves in this case is considerably larger than the light wavelength, the effects of diffraction are enhanced by viewing the array of evenly spaced reflecting surfaces at a small angle; under these conditions, their apparent separation is much smaller. The array of reflectors on the record surface is an example of what is commonly referred to as a *diffraction grating*, a detailed analysis of which will be presented in Chapter 26.

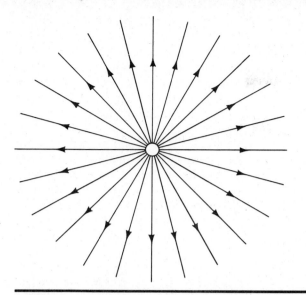

FIGURE 24.3. Pattern of rays emitted by a point source of light.

The description of the behavior of light rays and optical systems obtained in this way is referred to as *geometrical optics.*

EXAMPLE 24.2.1

The southern slope of the roof of a house is inclined at an angle of 30° to the horizontal; it forms a rectangular panel 12 meters long and 6 meters wide. The house is located in the southern part of the United States at latitude 35°N. This area of the roof is completely covered with silicon solar cells that convert 12 percent of the solar light energy incident upon them into electrical energy. Discuss the adequacy of the electrical output of this array in meeting household electrical needs of 50 kilowatt-hours per day in late June and late December. The intensity of solar radiation at normal incidence on the upper atmosphere is 1340 watts/m², but atmospheric absorption reduces this to about 750 watts/m² at the earth's surface in clear weather.

At the beginning of spring or fall, the sun's maximum altitude above the horizon is 90° minus the latitude. At latitude 35°N, this amounts to 55°. In late June, this figure is increased by the 23.5° inclination of the earth's axis to the plane of its orbit, and in late December it is decreased by the same amount. In summer, therefore, the sun's maximum altitude is 78.5°, while in winter it is 31.5°. In summer, the sun rises at about 5:00 A.M. and sets at 7:00 P.M. (standard time), while in winter, these times are more like 7:00 A.M. and 5:00 P.M. But during the first and last 2 hours of daylight in winter, the sun is so low in the sky that it is of negligible effectiveness in powering a solar array. In summer, its path is more nearly normal to the horizon as it rises and sets, and so it is only

during about the first hour after sunrise and before sunset that it is so low that it contributes negligible power. Roughly speaking, then, there are 6 hours of "generating time" in midwinter and 12 hours in midsummer. Not all this generating time is equally productive, however, since the angle that the sun's rays make with the normal to the roof panel varies throughout the day. Since the roof's normal is inclined 30° from the vertical toward the southern horizon, the sun's rays will be normal to the roof panel at midday when the sun is 60° above the southern horizon. In summer, of course, the sun gets as much as 78.5° above the southern horizon, thus actually *overshooting* the normal to the roof panel by 18.5° at times. In winter, however, the sun does not get more than 31.5° above the southern horizon and, therefore, never gets closer than 28.5° to the normal to the panel.

When sunlight falls on the array at other than normal incidence, as shown in Fig. 24.4, the effective area A' of the array is less than its actual area A. The relationship between these areas, from Fig. 24.4, is

$$A' = A \cos \phi = A \sin(90° - \phi) = A \sin \theta \qquad (24.2.11)$$

where ϕ is the angle between the sun's rays and the normal to the roof, and θ is the angle the rays themselves make with the panel.

It is an exceedingly complicated geometric problem to find the average value of sin θ over the productive generating time, but that is what we would have to do if we were to obtain a really precise answer to the question we are discussing. However, we shall be satisfied with a result that is approximate and take the average value of the angle θ during the productive

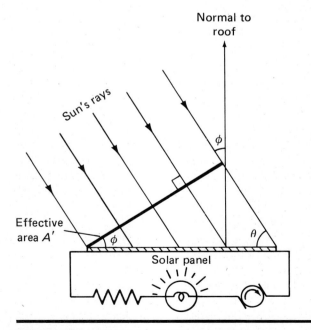

FIGURE 24.4. Geometry of solar energy collector.

generating time to be just *one half* the maximum value that is attained at midday. Since the roof slopes toward the southern horizon at an angle of 30° to the horizontal, this maximum value of θ will be just 30° greater than the sun's midday altitude above the southern horizon. In summer, we find the maximum value of θ to be 78.5° + 30° = 108.5° (in agreement with the previously mentioned fact that the sun overshoots the normal by 18.5°), while in winter, it is 31.5° + 30° = 61.5°. Dividing these values by 2, we obtain an average value $\overline{\theta}_s = 54.25$° for summer and $\overline{\theta}_w = 30.75$°.

The total daily energy output of the solar cell array will then be

$$U = \eta\overline{S}A't = \eta\overline{S}tA \sin\overline{\theta} \qquad (24.2.12)$$

where $\overline{S}$ is the solar energy per unit area, η is the fraction that is converted to electrical energy by the solar cells, A' is the effective area as expressed by (24.2.11), and t is the generating time. In summer, using the given values for all these quantities, we have

$$U = \eta\overline{S}tA \sin\overline{\theta}_s$$
$$= (0.12)(750)(12)(3600)(12)(6)(\sin 54.25°)$$
$$= 2.27 \times 10^8 \text{ J} = 63.1 \text{ kWh}$$

recalling that 1 joule = 1 watt-sec = $1/3.6 \times 10^6$ kilowatt-hours. In winter, we would have only 6 hours of generating time and an effective angle of only 30.75°, and in this case,

$$U = (0.12)(750)(6)(3600)(12)(6)(\sin 30.75°)$$
$$= 7.16 \times 10^7 \text{ J} = 19.9 \text{ kWh}$$

Evidently, then, such a scheme would provide adequate power in summer for a power requirement of 50 kilowatt-hours per day, but not in winter. Also, it neglects the possibility of cloudy weather, which would substantially reduce the amount of solar energy per unit area per unit time reaching the array. In any event, the price of silicon solar cells is about $0.15/cm^2$, or $1500/m^2$. For a roof installation of area 72 m^2 such as this one, the price would have to be $108,000, not counting storage batteries and charging equipment that would be needed to store energy collected in the daytime for use during the night. If the cost of the solar cells could be reduced by a factor of 10, however, the scheme would probably be quite attractive for southerly locations. Such cost reductions do not seem to be attainable in the near future, unfortunately.

EXAMPLE 24.2.2

An isotropic point source of light emits 1 watt of monochromatic light as a single spherical electromagnetic wave as described by Eqs. (24.2.2) and (24.2.3). What are the amplitudes of the **E** and **B** fields associated with such a wave at distances of 0.1 meter and 1.0 meter from the source? At what distance from the source would the intensity be the same as that of sunlight falling at normal incidence on the earth?

The source radiates power at the rate of 1 watt, and this power is radiated uniformly in all directions. Therefore, 1 joule of energy per second is transported across any spherical surface centered on the source. The area of a surface of radius 1 meter, is

$$A = 4\pi r^2 = 4\pi(1)^2 = 12.57 \text{ m}^2$$

The energy per unit area per unit time, represented by the quantity $\overline{S}$ of Eq. (24.2.6), will be

$$\overline{S} = \frac{E_0{}^2}{2c\mu_0} = \frac{1}{12.57} \text{ W/m}^2$$

Solving for $E_0{}^2$, we find

$$E_0{}^2 = \frac{(2)(3 \times 10^8)(4\pi \times 10^{-7})}{12.57} = 60 \text{ V}^2/\text{m}^2$$

$$E_0 = 7.74 \text{ V/m}$$

at a distance of 1 meter from the source. From (23.4.13), however,

$$B_0 = \frac{E_0}{c} = \frac{7.74}{3 \times 10^8} = 2.58 \times 10^{-8} \text{ Wb/m}^2$$

At a distance of 0.1 meter, the intensity $\overline{S}$ will be 100 times as great in view of the inverse square law (24.2.10), but since the amplitude E_0, from (24.2.6), varies as the square root of the intensity, E_0 will be only 10 times larger, giving $E_0 = 77.4$ volts/meter. Since $B_0 = E_0/c$, B_0 will also be 10 times larger, in which case $B_0 = 2.58 \times 10^{-7}$ weber/m^2.

From the previous example, we saw that the intensity of sunlight is $\overline{S} = 750$ watts/m^2 at normal incidence on the earth's surface. In this case, then, the total energy per second through any spherical surface of radius r will be given by intensity times area, as in (24.2.7) or (24.2.8). Therefore, since dU/dt is in this example 1 watt,

$$\frac{dU}{dt} = 4\pi r^2\overline{S}$$

or

$$r^2 = \frac{1}{4\pi\overline{S}}\frac{dU}{dt} = \frac{1}{(4\pi)(750)}(1.0) = 1.06 \times 10^{-4} \text{ m}^2$$

$$r = 0.0103 \text{ m}$$

24.3 The Principle of Least Time and Huygens's Principle: The Laws of Reflection and Image Formation

We shall now try to establish the law of reflection and on the basis of the behavior of light rays and wavefronts to understand why light is reflected by plane surfaces as it is. This might be regarded as a trivial exercise; after all, everyone knows that the angle of incidence and the angle of reflection are equal. It is important, however, to show how this follows from the properties of electromagnetic waves and to see what it implies about the behavior of rays and wavefronts. In doing so, we shall obtain certain insights that will be of value when we study the laws of refraction, where the situation is not so simple.

The laws of reflection and refraction, and a good deal more as well, can be derived from Maxwell's electromagnetic theory by writing mathematical expressions that represent incident, reflected, and refracted waves and insisting that the **E** and **B** fields associated with the waves obey the proper physical *boundary conditions* at the reflecting or refracting surface. For the **E** field, these are represented by Eqs. (17.5.1) and (17.5.3); there is a similar but somewhat different set for the **B** field. The mathematics involved in doing all this is for the most part beyond the scope of this book. When it is carried out, however, not only do we obtain the laws of reflection and refraction, but, in addition, equations that give information about the intensity and polarization of the reflected and refracted waves.

We shall approach the subject in a more elementary way by deriving the laws of reflection and refraction from simple physical notions regarding the behavior of rays and wavefronts. We shall proceed by two different methods, the first of which starts from a fundamental statement about the behavior of light rays, referred to as *Fermat's principle of least time*, the other proceeding from a basic postulate regarding the propagation of wavefronts, called *Huygens's principle*.

In the first case, we may begin with the fact that light travels in straight lines, or, more accurately, that light rays in free space or in homogeneous optical media are propagated as straight lines. This, as we have already seen, is in accord with the way in which wavefronts propagate and also with the observed behavior of light. It is hardly going too far to say that a straight line is defined by the way in which a light ray behaves! We also know that a straight line is the shortest distance between two given points. It might appear, then, that a light ray in going from one point to another simply takes the *shortest path*. On the other hand, it might equally well be that it takes the path that allows it to make the trip in the *shortest possible time*. These two possibilities frequently amount to the same thing, but there are some instances, one of which we shall encounter very shortly, where they do not. In these situations, experiments decisively tell us that it is the second possibility that is the correct one. The behavior of light rays may, therefore, be characterized by the statement that *a light ray in traveling between two points takes the path that allows it to make the trip in the shortest possible time*. This statement is called *Fermat's principle of least time*.

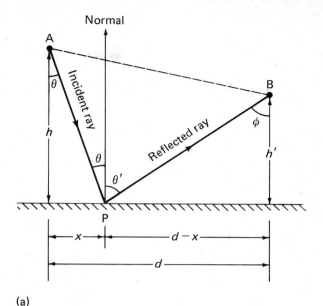

(a)

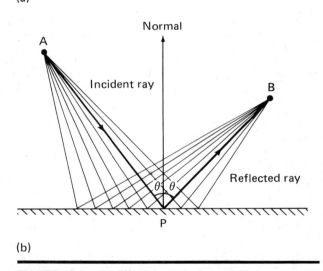

(b)

FIGURE 24.5. (a) Possible route for a light ray traveling between A and B. (b) Path of a light ray reflected from the plane surface that satisfies Fermat's principle of least time.

For the reflection of light from a plane reflecting surface, Fermat's principle can be understood with reference to Fig. 24.5a, where we see a ray going from point A to point B by the path APB, which involves a reflection at point P. The part of the path labeled AB is referred to as the incident ray, while part PB is called the reflected ray. The angle θ between the incident ray and the normal to the reflecting plane is referred to as the *angle of incidence*, while the corresponding angle θ' between the reflected ray and the normal is called the *angle of reflection*. In Fig. 24.5a, we have taken perverse pleasure in drawing these angles as very unequal. The vertical projections of A and B on the reflecting surface are separated by a distance d, the distances between the point of reflection and these two points being labeled x and $d - x$, respectively, as shown.

The path between A and B that can be traversed by a light ray in the very least possible time is the dotted direct path AB. Such a ray will, of course, not be reflected at all. For any ray that is reflected, the path of the ray must somehow consist of two straight segments that meet on the reflecting plane, because in free space light rays do travel in straight lines. Now, reflected rays may or may not follow path APB, but they will follow some similar path made up of two straight segments; and according to Fermat's principle, this path will be such as to allow the ray to proceed from A to the reflector and thence to B in the *minimum possible time.*

Mathematically, we can find this path by writing an equation for the time required for light to follow a path such as APB and use the usual technique of differentiation to find what specific path makes this time a minimum. The time t required for light to transverse path APB is given by

$$t = \frac{AP + PB}{c} = \frac{\sqrt{x^2 + h^2} + \sqrt{(d - x)^2 + h'^2}}{c}$$

or

$$ct = \sqrt{x^2 + h^2} + \sqrt{(d - x)^2 + h'^2} \qquad (24.3.1)$$

In this equation, h, h', and d are constants that express the location and spacing of the two fixed points A and B. The quantity x tells us where the incident light ray hits the reflecting plane. This we do not know beforehand but must *determine* from the condition that the time of transit be as small as possible. Accordingly, we must differentiate the time as given by (24.3.1) as a function of x with respect to x and set the resulting derivative equal to zero. Differentiating (24.3.1) with respect to x, we find

$$c\frac{dt}{dx} = \frac{x}{\sqrt{x^2 + h^2}} - \frac{d - x}{\sqrt{(d - x)^2 + h'^2}} = 0 \qquad (24.3.2)$$

By transposing one of the two terms to the opposite side of the equation and then squaring both sides, this can be written as

$$\frac{x^2}{x^2 + h^2} = \frac{(d - x)^2}{(d - x)^2 + h'^2} \qquad (24.3.3)$$

or, clearing of fractions,

$$x^2(d - x)^2 + x^2 h'^2 = x^2(d - x)^2 + h^2(d - x)^2$$

or, finally,

$$x^2 h'^2 = (d - x)^2 h^2$$

whence

$$\frac{d - x}{h'} = \frac{x}{h} \qquad (24.3.4)$$

From Fig. 24.5a, however, it is clear that $x/h = \tan\theta$ and $(d - x)/h = \tan\theta'$; therefore, (24.3.4) simply tells us that

$$\boxed{\theta' = \theta} \qquad (24.3.5)$$

This says that *the angle of incidence and the angle of reflection are equal.* Figure 25.4a should, therefore, be redrawn as Fig. 24.5b. In this drawing, the actual path of the ray, characterized by equal angles of incidence and reflection, is shown as the heavy black line APB, while other paths having unequal incidence and reflection angles, and therefore corresponding to transit times longer than the minimum, are drawn in with very light lines. Here, since both incident and reflected rays travel with constant speed c, the path corresponding to minimum time is also the *shortest* one a reflected ray can take between points A and B.

A point source such as that shown at A in Fig. 24.6a emits light rays in all directions. If there is a plane mirror nearby, those rays that hit the mirror are reflected in such a way that the angles the incident rays and the corresponding reflected rays make with the normal to the plane are equal. It is evident from the drawing that the effect of reflection is to generate a set of reflected light rays that diverge outwardly just as though they had been emitted from a fictitious, or *virtual*, point source behind the mirror at A'. A person whose eye is positioned at a point such as B, therefore, cannot tell visually that the reflected rays entering his eye really emanate from A rather than A'; the visual appearance presented is that of a point source such as A located behind the mirror at A'. We say that an *image* of the point source is formed by the mirror, and since there are no actual light rays that diverge from the point A', but only a set of reflected rays whose projections all intersect at that point, this image is said to be a *virtual* image rather than a real one.

To find the position of the image of a point source, it is necessary only to consider the paths of two rays emanating from the source in different directions and to find their point of intersection after reflection. In this case, since we have a virtual image rather than a real one, we must find the point of intersection of their projections in the space behind the mirror. This is illustrated in Fig. 24.6b, where the paths taken by the rays APB and AQC, which radiate in different directions from the source A, are shown. The projections PA' and QA' of the reflected rays PB and QC intersect at point A', and it is here that the image of the source point A is located. Because the law of reflection requires that the angles of incidence and reflection be the same, angles OAP and OA'P are both equal to θ. Then, in triangles OPA and OPA',

$$OA = OP \cot\theta = OA' \qquad (24.3.6)$$

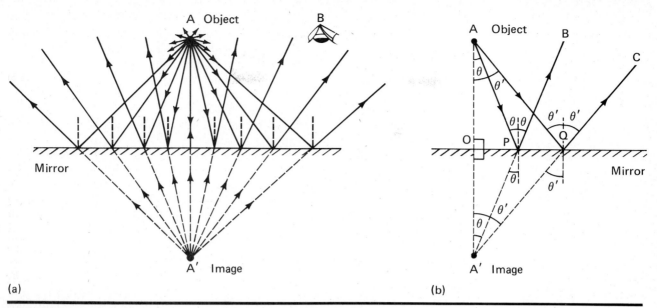

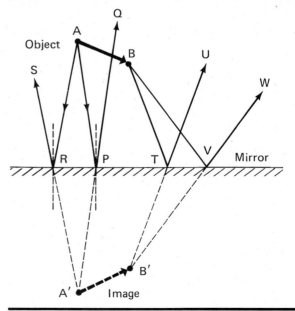

FIGURE 24.6. Geometry of light rays reflected from a plane mirror.

Clearly, then, the image of a point in a plane mirror is a virtual image located as far behind the mirror as the object point is in front of it, as illustrated in in Fig. 24.6b.

The size and orientation of the image of an extended object can be ascertained simply by locating the images of two or more separate points on or within the object, as shown in Fig. 24.7. In this illustration, the object has the form of an arrow; the image of the head is located at A′, where the projections of the reflected rays RS and PQ intersect, while the image of the tail is at B′ where the projections of the reflected rays TU and VW cross. The ray-tracing techniques developed in these examples for locating image points and establishing the size and orientation of images are important and are applicable to curved mirrors and lenses as well as to plane mirrors. It is important that the reader be able to construct similar ray diagrams and locate object and image points for himself.

The propagation of light can be understood from another point of view on the basis of a simple postulate regarding the behavior of wavefronts that was first enunciated by Christian Huygens in 1678. This postulate, which we shall refer to as *Huygens's principle*, states that the way in which wavefronts propagate can be ascertained by supposing that *every point on an existing wavefront serves as a point source from which spherical wavelets are emitted and that the newly propagated wavefront is a surface, or envelope, that is tangent to all these individual spherical wavelets.* The successive propagation of new wavefronts may then be described in terms of the emission of new spherical wavelets from the wavefronts that were previously obtained by the use of Huygens's principle. This is illustrated in

Fig. 24.8, where an initial wavefront AA′ is extended by imagining that simultaneously, at every point along the wavefront, a wavelet is emitted that spreads spherically outward as shown by the dotted lines. A new wavefront, BB′, can then be constructed at any later time by constructing a surface that is tangent to all the individual spherical wavelets, as shown. This new wavefront then serves as the source for a new series of wavelets that allows us to extend it further to the new position CC′ and, successively, to DD′ and thence to

FIGURE 24.7. Determination of the position and orientation of the image formed by a plane mirror, utilizing the ray geometry prescribed by the law of reflection.

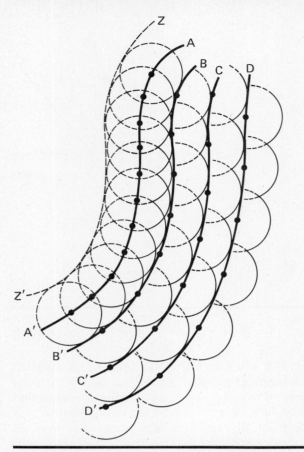

FIGURE 24.8. Construction of successive wavefronts using Huygens's principle.

any desired further extent. The propagation of spherical wavefronts from a point source at O, as constructed by Huygens's principle, is shown by Fig. 24.9.

Huygens was the first person to conceive of light as a *wave phenomenon*, and Huygens's principle was the first instance in which it was shown that the known facts regarding light propagation could be satisfactorily accounted for by the wave theory. Unfortunately, it was not generally accepted at the time of its introduction, and most scientists clung to the older particle emission concepts for over 130 years, until Young's interference experiments demonstrated unequivocally the wave nature of light.

Huygens's principle raises certain questions that are difficult to resolve in an elementary way. The most important and obvious of these has to do with why, for example, light is propagated forward from AA′ to BB′ in Fig. 24.8 rather than backward from AA′ to ZZ′. This difficulty also arises in Maxwell's electromagnetic theory, where solutions that propagate backward rather than forward can usually be found. The most reasonable explanation is that these situations correspond to allowed mathematical constructions or solutions that we reject because they do not correspond to

the real physical state of the system at hand. The same situation is frequently encountered in the mathematics of physical problems that lead to solutions in the form of quadratic equations. In these instances, we often find that one of the two possible solutions corresponds to the real physical state of the system, and the other corresponds to a fictitious state that is created by the mathematics but must be excluded on physical grounds.

In spite of these difficulties, Huygens's principle allows us to visualize in a simple and yet elegant geometric way how light waves get from one place to another and how they behave when they are reflected or refracted. We may derive the law of reflection, for example, quite easily by using Huygens's principle, as demonstrated by Fig. 24.10. We see an incident plane wavefront AA′ that advances through space, according to Huygens's principle (though the emitted wavelets are not shown) from AA′ to GG′. The wavefront AA′ strikes the mirror at A, at which time a spherical reflected wavelet is emitted as shown. By the time the incident wavefront has moved to B, the radius of the reflected wavelet has expanded to the distance A″B″, which is, of course, equal to A′B′ since the velocities of the incident wavefront and of the reflected wavelet are the same.

Let us now suppose that new reflected wavelets are emitted simultaneously at B and R. By the time the incident wavefront has advanced to CC′, the reflected wavelets have expanded in such a way that the reflected wavefront CC″ can be constructed by drawing lines tangent to them. In the same way, if spherical reflected wavelets are now emitted from CC″, by the time the incident wavefront has progressed to DD′, these will have grown to the point where, according to

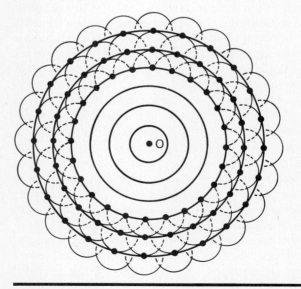

FIGURE 24.9. Construction of wavefronts emitted by a point source, by Huygens's principle.

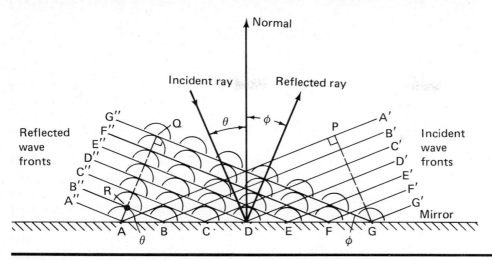

FIGURE 24.10. Construction of wavefronts reflected by a plane mirror, using Huygens's principle.

Huygens's principle, the reflected wavefront DD″ can be constructed. The process can be repeated *ad infinitum*. By this reasoning, we see that as the incident wave moves through the distance PG, the reflected wave set up by the growth of the reflected wavelets will have progressed through an *equal* distance AQ. But if these distances AQ and PG are equal, then, since in right triangle APG sin θ = PG/AG and in right triangle AQG sin θ' = AQ/AG, we must have

$$\sin \theta = \sin \theta'$$

whence

$$\boxed{\theta = \theta'} \tag{24.3.7}$$

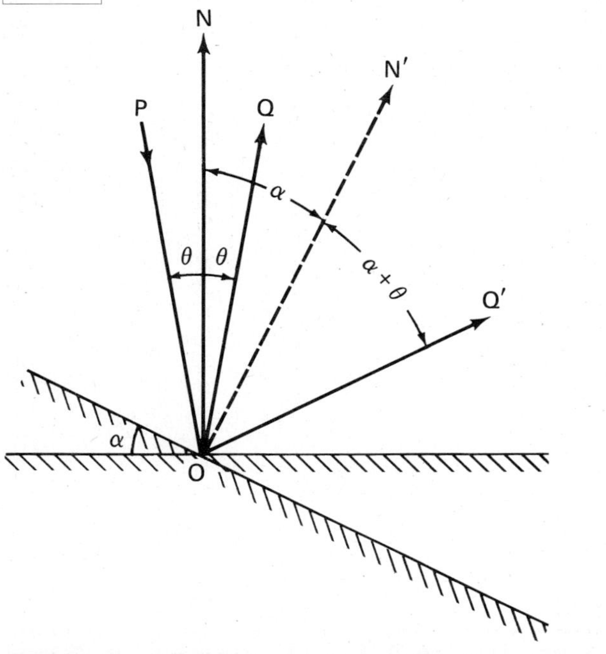

in agreement with (24.3.5). Evidently, Huygens's principle, in which the behavior of the wavefront is all-important, and Fermat's principle of least time, in which the behavior of the ray is of primary import, lead to exactly the same result in the case of reflection.

EXAMPLE 24.3.1

Show that when a plane mirror is turned through an angle α, the direction of a light ray reflected from it changes by twice that angle.

The situation is illustrated in Fig. 24.11, where an incident ray PO hits a plane mirror initially at an angle of incidence θ and is reflected as ray OQ. The mirror is then turned through an angle α, the angle between the incoming ray and the new normal N′ now becoming PON′, or $\alpha + \theta$. By the law of reflection, the new reflected ray OQ′ must also make an angle $\alpha + \theta$ with the normal N′. But now, what is the angle between the old reflected ray OQ and the new reflected ray OQ′? From the diagram, it is clearly given by

$$\text{angle QOQ}' = \text{angle NON}' + \text{angle N}'\text{OQ}'$$
$$= \text{angle NOQ} \tag{24.3.8}$$

or

$$\text{angle QOQ}' = \alpha + (\alpha + \theta) - \theta = 2\alpha \tag{24.3.9}$$

EXAMPLE 24.3.2

How high does a plane mirror have to be to enable you to see all of an object of height h located a distance d in front of the mirror if you are stationed a distance a in front of it?

The situation is as shown in Fig. 24.12. The mirror is at PQ, its height being designated by y. The object AB, of height h, is at a distance d in front of the mirror; and from the properties of plane reflecting surfaces we have already established, we know

FIGURE 24.11

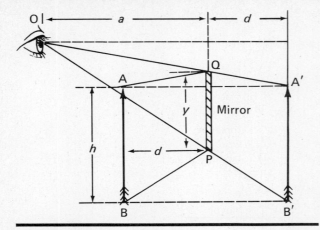

FIGURE 24.12

that its image A′B′ will be of the same size and located at a distance d behind the mirror. The observer's eye is at a distance a in front of the mirror, and if the entire object is to be seen in the mirror, the mirror must be at least as large as shown; if it is smaller, no mirror surface will be there to reflect the extreme rays AQO and BPO back to the observer's eye.

In the diagram, it is evident that triangles OPQ and OA′B′ are similar; therefore, their corresponding parts have to have the same ratios. We may thus write the ratios of the bases to the altitudes as

$$\frac{y}{a} = \frac{h}{a+d} \tag{24.3.10}$$

from which we obtain

$$y = h\frac{a}{a+d} = \frac{h}{1+\dfrac{d}{a}} \tag{24.3.11}$$

as the desired result. If the observer is looking at himself in the mirror, then d and a are equal and (24.3.11) reduces to $y = h/2$. To see all of yourself in a mirror, the mirror has to be half as tall as you are. Try it and see!

24.4 Light in Dense Transparent Media: Snell's Law of Refraction

When a light ray in free space encounters the surface of a dense transparent medium such as glass or water, it is *refracted* or bent toward the normal to the surface upon entering the medium, as illustrated in Fig. 24.13. This phenomenon is familiar to anyone who has taken note of the fact that a pencil or other straight object appears to be bent when part of it is immersed in water, as shown in Fig. 24.14. At the same time, the surface of the refracting medium also *reflects* a small portion of the incident light energy in a reflected beam

that obeys the usual law of reflection. For this reason, we can see the images of surrounding objects more or less faintly reflected from the surface of a pool of water or a flat sheet of glass. There are also changes in the state of polarization of the reflected and refracted rays as compared with that of the incident beam. These polarization changes are more subtle and less easily observed than the effects of refraction and reflection themselves, but nevertheless they can be detected and studied without great difficulty. They will be discussed in some detail in Chapter 26, but for the present we shall say no more about them. While all the details of these numerous effects are predicted accurately by Maxwell's electromagnetic theory, we shall for the most part adhere to the spirit of the previous section and discuss them from the point of view of Fermat's principle and Huygens's principle, which are less demanding mathematically.

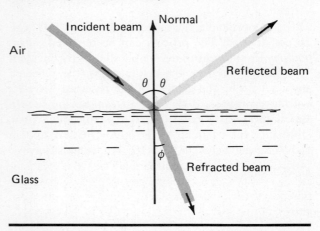

FIGURE 24.13. Geometry of refraction, illustrating incident, refracted, and reflected rays.

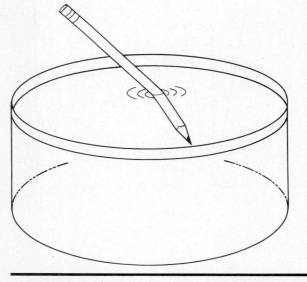

FIGURE 24.14. Image of a pencil partially immersed in water, showing the "bending" of light rays by refraction.

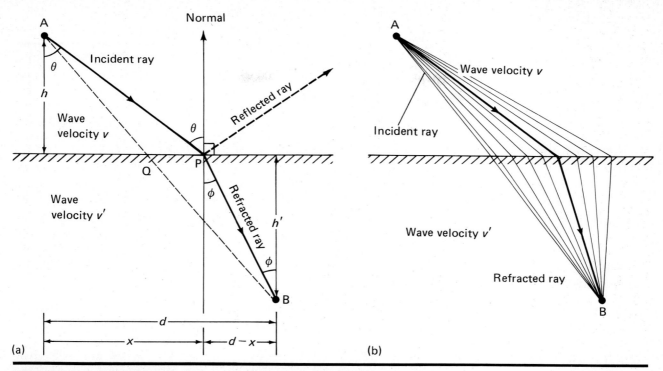

FIGURE 24.15. (a) Possible path for a light ray traveling from A to B. (b) The path satisfying Fermat's principle of least time (heavily accented line) and alternative paths (lightly drawn lines).

The geometry of refraction is shown in Fig. 24.15a, which is the counterpart of Fig. 24.5a, demonstrating reflection. The plane refracting boundary between two different transparent substances such as air and glass, or perhaps water and glass, is illustrated. We are to consider what path a light ray will take in going from a point A in the upper, less refractive medium, to a point B in the lower, more highly refracting substance. If light rays followed the path of shortest distance between two such points, they would take the straight path AQB, and in that case there would be no refraction. Since we do observe that refraction takes place, however, this cannot be how light rays behave. Also, if light rays follow the path which takes them from A to B in the shortest time, as Fermat's principle tells us they do, and if, at the same time, the speed of light were *the same* in the two substances, then light rays would still follow the straight path AQB.

On the other hand, if light rays follow the path that takes them from A to B in the least time, and if *the speed of light were less in the lower, more highly refracting substance than in the upper medium,* from the standpoint of time it would pay for the light ray to take a path that is longer in distance but which covers a greater proportion of the journey in the upper medium where light travels fast than in the lower one in which the velocity is smaller. The actually observed path, shown in the figure as APB, is one that has just

this characteristic form. It appears, then, that as Fermat's principle states, light does actually take the path that allows it to proceed between two given points in the least possible time and also that light travels more slowly in highly refracting substances than in less refractive ones.

Let us assume, now, that light travels with speed v in the upper, less refractive medium, and with a lesser speed v' in the lower one. If the upper medium were free space (or, as a practical matter, air, which has little effect on the speed of light), the velocity v would be given by $c = 1/\sqrt{\varepsilon_0 \mu_0} = 2.998 \times 10^8$ meters/second. Suppose we consider two points A and B, separated by a horizontal distance d, and inquire what the distance x in Fig. 24.15a would have to be in order to allow a light ray to go from A to B in the least possible time. We can find the answer by expressing the time t required for the journey as a function of the distance x to the point of refraction, differentiating this time with respect to x, and setting the derivative equal to zero so as to find the minimum value of the dependent variable t.

The time required for the trip between A and B will be

$$t = \frac{AP}{v} + \frac{PB}{v'} = \frac{\sqrt{x^2 + h^2}}{v} + \frac{\sqrt{(d - x)^2 + h'^2}}{v'} \tag{24.4.1}$$

where h and h' are the perpendicular distances between A and B and the refracting boundary. Differentiating

777

this with respect to t and setting the derivative equal to zero, we obtain

$$\frac{dt}{dx} = \frac{x}{v\sqrt{x^2 + h^2}} - \frac{d - x}{v'\sqrt{(d - x)^2 + h'^2}} = 0 \qquad (24.4.2)$$

Transposing one of these two terms to the opposite side of the equation,

$$\frac{v}{v'} \cdot \frac{d - x}{\sqrt{(d - x)^2 + h'^2}} = \frac{x}{\sqrt{x^2 + h^2}} \qquad (24.4.3)$$

But it is apparent from Fig. 24.15a that

$$\frac{x}{\sqrt{x^2 + h^2}} = \sin \theta \quad \text{and} \quad \frac{d - x}{\sqrt{(d - x)^2 + h'^2}} = \sin \phi \qquad (24.4.4)$$

from which (24.4.3) can be written as

$$\boxed{\frac{\sin \theta}{\sin \phi} = n_r} \qquad (24.4.5)$$

where

$$\boxed{n_r = \frac{v}{v'} = \frac{\text{velocity of incident ray}}{\text{velocity of refracted ray}}} \qquad (24.4.6)$$

Equation (24.4.5) tells us that *the sine of the angle of incidence θ divided by the sine of the angle of refraction ϕ has the constant value n_r*, referred to as the *relative index of refraction* between the two media. The relative index of refraction is the ratio of the velocity of the incident ray to that of the refracted ray.

If the upper medium were free space, we would have $v = c$, and then

$$n_r = \frac{c}{v'} = n \qquad (24.4.7)$$

where the symbol n is used to signify the *absolute index of refraction* of a single dense transparent substance with respect to free space or, in practice, air. It is important to observe the distinction between the relative index of refraction between two refracting media and the absolute index pertaining to the interface between a given substance and free space. When the simple expression "index of refraction" is used, it generally refers to the absolute index unless some specific indication to the contrary is made. A compilation of the absolute refractive indices of a number of familiar substances is provided in Table 24.1. Figure 24.15b illustrates the actual least-time path compared to other possible paths corresponding to different values of x and hence to greater transit times.

The law of refraction expressed by Eq. (24.4.5) is found to agree precisely with the observed behavior of refracted light rays in transparent media. This

TABLE 24.1. Refractive Indices of Common Transparent Substances

Substance	Refractive index*	Critical angle θ_c, degrees
Water	1.333	48.61
Acetone	1.359	47.38
Benzene	1.501	41.78
Carbon disulfide	1.625	37.98
Chloroform	1.446	43.75
Ethyl ether	1.351	47.75
Methyl alcohol	1.329	48.80
Ethyl alcohol	1.361	47.31
Glycerol	1.474	42.72
Diamond	2.4173	24.44
Fused quartz	1.4584	43.29
Sodium chloride	1.5442	40.36
Fluorite	1.4339	44.22
Light crown glass	1.5090	41.51
Dense crown glass	1.5691	39.59
Light flint glass	1.5734	39.46
Dense flint glass	1.6553	37.17
Heavy flint glass	1.7555	34.73
Heaviest flint glass	1.8900	31.95
Lanthanum flint	1.7950	33.85
Gallium phosphide	3.50	16.6

* $\lambda = 5890$ Å, $T = 20°$C.

experimental correspondence was first observed by the Hollander, Willebrord Snell in the early part of the seventeenth century, and the law of refraction expressed by Eq. (24.4.5) is frequently referred to as *Snell's law.*

You may recall that the velocity of light in free space as predicted by Maxwell's theory is given by

$$c = \frac{1}{\sqrt{\varepsilon_0 \mu_0}} = 2.998 \times 10^8 \text{ m/sec} \qquad (24.4.8)$$

If the permittivity and permeability of the medium in which light travels have respective values ε and μ that differ from those associated with free space, the velocity with which light is propagated will be

$$v = \frac{1}{\sqrt{\varepsilon \mu}} \qquad (24.4.9)$$

The ratio c/v which represents the absolute refractive index of the material medium of permittivity ε and permeability μ will then be

$$n = \frac{c}{v} = \sqrt{\frac{\varepsilon \mu}{\varepsilon_0 \mu_0}} \qquad (24.4.10)$$

Since most transparent substances have magnetic permeabilities that are extremely close to the free space value μ_0, this may be written to a very high degree of

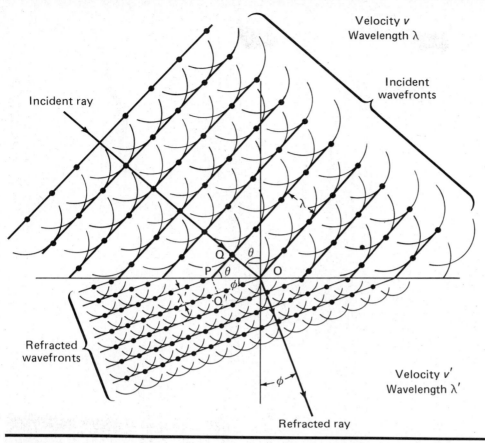

Velocity v
Wavelength λ

Incident wavefronts

Incident ray

Refracted wavefronts

Velocity v'
Wavelength λ'

Refracted ray

FIGURE 24.16. Snell's law of refraction, as obtained from Huygens's principle.

accuracy as

$$n = \frac{c}{v} = \sqrt{\frac{\varepsilon}{\varepsilon_0}} = \sqrt{K} \tag{24.4.11}$$

where K is the dielectric constant. It is evident that Maxwell's theory confirms our previous conjecture that light travels more slowly in dense material media than in free space and yields the additional information that the ratio of free space to medium propagation velocities, which is what we refer to as the refractive index, is equal to the *square root of the dielectric constant K*. This prediction has been adequately confirmed by experiment.

Snell's law of refraction follows also as a consequence of Huygens's principle. This is illustrated in Fig. 24.16, where we see the pattern of wavefronts and Huygens wavelets that results when light is refracted from a plane interface between two refracting media. The light *frequency* in both of the media is, of course, the same. Therefore, if the velocity of light in the upper medium is v and that in the lower medium is v', we must have

$$v' = \frac{\omega}{k'} \quad \text{and} \quad v = \frac{\omega}{k} \tag{24.4.12}$$

where k' and k are the propagation constants in the two media. But since $k = 2\pi/\lambda$ and $k' = 2\pi/\lambda'$ from (24.2.12),

$$\omega = \frac{2\pi v'}{\lambda'} = \frac{2\pi v}{\lambda} \tag{24.4.13}$$

or

$$\frac{\lambda}{\lambda'} = \frac{v}{v'} = n_r \tag{24.4.14}$$

In the situation shown in Fig. 24.16, the speed of light in the upper substance, v, is greater than its value v' in the lower medium. The ratio v/v' in (24.4.14) is, therefore, larger than unity. From this, it is apparent that the light wavelength λ' in the lower, more refractive substance is smaller than the wavelength λ in the upper medium where light travels faster. This fact is reflected in the closer spacing between the successive Huygens wavefronts in the lower medium; in fact, the distance between these wavefronts in the upper medium is λ, while in the lower medium it is $\lambda' = \lambda(v'/v) = \lambda/n_r$, where n_r is the relative refractive index.

In the upper substance, the distance OP in triangle OPQ can clearly be written as

$$OP = \frac{\lambda}{\sin \theta} \qquad (24.4.15)$$

while in the lower medium, the same distance in triangle OPQ' can be expressed as

$$OP = \frac{\lambda'}{\sin \phi} \qquad (24.4.16)$$

From these two equations, we see at once that

$$\frac{\lambda'}{\sin \phi} = \frac{\lambda}{\sin \theta} \qquad (24.4.17)$$

or

$$\boxed{\frac{\sin \theta}{\sin \phi} = \frac{\lambda}{\lambda'} = \frac{v}{v'} = n_r} \qquad (24.4.18)$$

which is the same as (24.4.5). Snell's law, therefore, follows directly from Huygens's principle.

Both Fermat's principle of least time and Huygen's principle are concerned exclusively with the relationship between incident and refracted rays or wavefronts, and neither predicts the existence of the *reflected* ray. Therefore, they give at best only partial information about certain aspects of the whole system of waves. This defect is not inherent in Maxwell's theory, which not only predicts the existence of the reflected ray but gives complete information about its intensity and polarization. An important characteristic of the reflected wave is the phase relation it exhibits with respect to the incident wave. It is found that when the incident wave proceeds from a less refractive medium into one of higher refractive index, the wave that is reflected back into the less refractive medium undergoes a *phase change* of π radians, or 180°. On the other hand, when the incident wave goes from the more refractive substance into one of smaller refractivity, the wave that is reflected back into the more highly refracting medium exhibits *no phase change at all* with respect to the phase of the incident wave. These phase changes must be carefully accounted for in order to understand correctly certain phenomena involving interference between incident and reflected light rays that will be discussed in Chapter 26. In Example 24.4.4 below, we shall see more clearly how these results are obtained.

EXAMPLE 24.4.1

Light enters a flat piece of glass whose absolute index of refraction is 1.520 at an angle of incidence of 53.0°. What is the angle of refraction? What is the velocity of light in the glass? If the free-space wavelength of the light is 589 nanometers, what is the wavelength in the glass? This piece of glass is immersed in water whose absolute refractive index is 1.333. What is the relative index of refraction for a ray entering the glass from the water? What is the relative index for a ray

leaving the glass and entering the water? What is the angle of refraction for a ray incident upon the glass from the water at an angle of incidence of 53.0°?

According to Snell's law, the angle the refracted ray makes with the normal is given by

$$\sin \phi = \frac{\sin \theta}{n_g} = \frac{\sin 53.0°}{1.520} = 0.5254$$

where n_g is the absolute refractive index of the glass. From this,

$$\phi = 31.7°$$

The velocity of light v_g within the glass will be, according to (24.4.7),

$$v_g = \frac{c}{n_g} = \frac{2.998 \times 10^8}{1.520} = 1.972 \times 10^8 \text{ m/sec}$$

From (24.4.14), it is evident that if the free-space wavelength is 589 nanometers, the wavelength within the glass must be

$$\lambda_g = \frac{v_g}{c} \lambda = \frac{\lambda}{n} = \frac{589}{1.520} = 387.5 \text{ nm}$$

When the plate of glass is immersed in water whose refractive index is $n_w = 1.333$, the relative index of refraction for a ray entering the glass from the water, according to (24.4.6), will be

$$n_r = \frac{v_w}{v_g} = \frac{\dfrac{c}{n_w}}{\dfrac{c}{n_g}} = \frac{n_g}{n_w} = \frac{1.520}{1.333} = 1.140$$

If the ray is incident from within the glass and goes into the water, the relative index n_r' is

$$n_r' = \frac{v_g}{v_w} = \frac{\dfrac{c}{n_g}}{\dfrac{c}{n_w}} = \frac{n_w}{n_g}$$

from which

$$n_r' = \frac{1}{n_r} = 0.877$$

For a ray entering the glass from the water at an angle of 53.0°, the angles of incidence and refraction are related by

$$\sin \phi = \frac{\sin \theta}{n_r} = \frac{\sin 53.0°}{1.140} = 0.700$$

$$\phi = 44.5°$$

EXAMPLE 24.4.2

A coin is thrown into a swimming pool 6.0 feet deep. When it has sunk to the bottom, it is viewed at an angle of 30° to the horizontal, as shown in Fig. 24.17a. How deep beneath the surface does it appear to be to

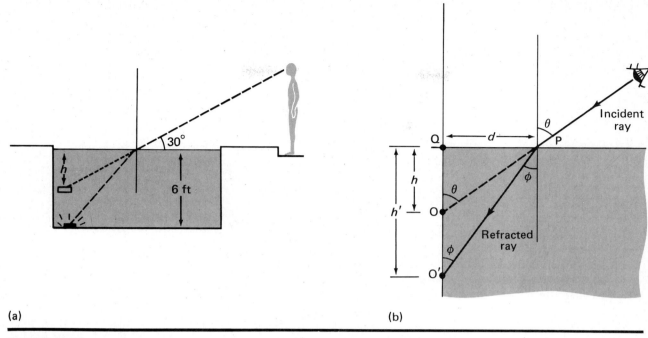

FIGURE 24.17

the observer? The index of refraction of the water is 1.333.

The essential features of the problem are shown in Fig. 24.17b. Light rays that leave the object at O′ are refracted at point P and appear to an external observer to be coming from a virtual image of the object at O, whose depth h is *less* than the actual depth h' of the object. From the drawing, it is apparent that

$$d = h \tan \theta = h' \tan \phi \qquad (24.4.19)$$

from which

$$\frac{h}{h'} = \frac{\tan \phi}{\tan \theta} \qquad (24.4.20)$$

But if we express the tangents of both angles as the ratio of sine to cosine, and recall that $\cos^2 \phi = 1 - \sin^2 \phi$, this can be written as

$$\frac{h}{h'} = \frac{\cos \theta}{\sin \theta} \frac{\sin \phi}{\sqrt{1 - \sin^2 \phi}} \qquad (24.4.21)$$

Also, according to Snell's law, $\sin \phi = (\sin \theta)/n$, from which finally

$$\frac{h}{h'} = \frac{\cos \theta}{\sqrt{n^2 - \sin^2 \theta}} \qquad (24.4.22)$$

In our example, $h' = 6.0$ feet, $\theta = 60°$, and $n = 1.333$. Therefore,

$$h = \frac{(6.0)(0.500)}{\sqrt{(1.333)^2 - (0.866)^2}} = 2.96 \text{ ft}$$

It is interesting to note that according to (24.4.22), the position of the image as given by h depends not only on the index of refraction but also on the angle of incidence at which it is observed. This is in sharp contrast with the case of the reflected image in a plane mirror, whose location is quite independent of the position of the observer. At nearly normal incidence, when θ is a relatively small angle, $\cos \theta \cong 1$ and $\sin^2 \theta \ll n^2$; Eq. (24.4.22) then becomes

$$\frac{h}{h'} \cong \frac{1}{n} \qquad (24.4.23)$$

EXAMPLE 24.4.3
Show that the direction of a ray passing through a plane-parallel sheet of glass is unaltered but is shifted parallel to its initial line of incidence through a distance that depends upon the index of refraction and the angle of incidence.

The geometry upon which this example is based is shown in Fig. 24.18. It is evident that the ray emerging from the glass at P is parallel to the incident ray, because the angle of refraction ϕ for the initial refraction at O is the same as the angle of incidence for the second refraction at P. Snell's law then tells us that in the refraction at P,

$$n_r' = \frac{\text{sine of angle of incidence}}{\text{sine of angle of refraction}} = \frac{\sin \phi}{\sin \theta'}$$

whence

$$\sin \theta' = \frac{\sin \phi}{n_r'} \qquad (24.4.24)$$

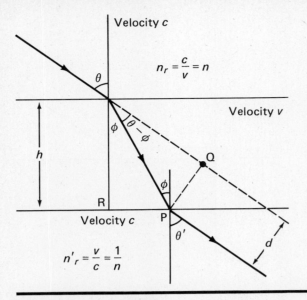

FIGURE 24.18. Deviation of a light ray by a plane-parallel refracting sheet.

where n_r' is the relative index of refraction *in going from the more highly refractive medium into the one of lower refractivity*. In this case, the index is given by the ratio of the velocities as

$$n_r' = \frac{\text{velocity of incident ray}}{\text{velocity of refracted ray}} = \frac{v}{c} = \frac{1}{n} \qquad (24.4.25)$$

the velocity of the incident ray having the smaller value v, while the refracted ray has the larger velocity c. Then,

$$\sin \theta' = n \sin \phi = \sin \theta \qquad (24.4.26)$$

or

$$\theta' = \theta \qquad (24.4.27)$$

It is important to note that the relative index n_r' for a ray incident upon the interface from within the more highly refracting medium is the reciprocal of the value n_r that is appropriate when the light is incident from the other side.

The distance d through which the ray emerging at P is deviated can be obtained by noting that in triangles OPR and OPQ, the distance OP can be expressed as

$$OP = \frac{h}{\cos \phi} = \frac{d}{\sin (\theta - \phi)} \qquad (24.4.28)$$

where h is the thickness of the sheet. Solving for d gives

$$d = \frac{h \sin(\theta - \phi)}{\cos \phi} = h\left[\frac{\sin \theta \cos \phi - \cos \theta \sin \phi}{\cos \phi}\right]$$

$$= h\left[\sin \theta - \frac{\cos \theta \sin \phi}{\sqrt{1 - \sin^2 \theta}}\right] \qquad (24.4.29)$$

But, from Snell's law, $\sin \phi = \sin \theta / n$, whereupon

$$d = h \sin \theta \left[1 - \frac{\cos \theta}{\sqrt{n^2 - \sin^2 \theta}}\right] \qquad (24.4.30)$$

In the case of a ray incident at an angle of 60° upon a sheet of glass of refractive index 1.52 and thickness 1.0 cm, this gives $d = 0.519$ cm. For near-normal incidence, when θ is a relatively small angle, $\cos \theta \cong 1$ and $\sin^2 \theta \ll n^2$, in which case (24.4.29) becomes

$$d \cong h\left(1 - \frac{1}{n}\right) \sin \theta \qquad (24.4.31)$$

*EXAMPLE 24.4.4

Show that for a plane electromagnetic wave normally incident upon a plane refracting interface of relative refractive index n_r, the intensity of the wave reflected from the surface is given by $(n_r - 1)^2/(n_r + 1)^2$ times the incident intensity.

A diagram representing the physical situation is shown in Fig. 24.19. A normally incident beam of light, represented by the propagation vector $\mathbf{k}_0$ proceeds along the negative z-direction and encounters the plane surface of a substance of different refractivity at $z = 0$. At this surface, it undergoes partial reflection, the reflected wave being represented by the propagation vector $\mathbf{k}''$, whose direction is that of the positive z-axis. It is important to note that since the wavelengths of the incident and reflected waves are the same, the magnitudes of these two propagation vectors must be equal; hence, if λ is the incident wavelength,

$$k_0 = \frac{2\pi}{\lambda} = k'' \qquad (24.4.32)$$

At the same time, a transmitted (or refracted) ray proceeds on into the lower substance, its propagation vector being represented by $\mathbf{k}'$ in the diagram. The magnitude of this vector is $2\pi/\lambda'$, where λ' is the wavelength in the second medium. Since

$$\frac{\lambda}{\lambda'} = \frac{v}{v'} = \frac{n'}{n} \qquad (24.4.33)$$

we find

$$k' = \frac{2\pi}{\lambda'} = \frac{2\pi}{\lambda(n/n')} = \frac{n'}{n} k_0 \qquad (24.4.34)$$

In Fig. 24.19, the incident and reflected rays are shown slightly separated for clarity, but in actuality they coincide.

The electric and magnetic field vectors are always perpendicular to the z-axis and to each other. If we assume that the electric field vector is in the x-direction and the magnetic field vector is in the y-direction, then the traveling waves that represent the electric and magnetic fields of the respective waves can be written

$$\mathbf{E}_0 = \mathbf{i}_x E_0 \sin(-kz - \omega t)$$
$$\mathbf{E}' = \mathbf{i}_x E' \sin(-k'z - \omega t) \qquad (24.4.35)$$
$$\mathbf{E}'' = \mathbf{i}_x E'' \sin(kz - \omega t)$$

and

$$\mathbf{B}_0 = \mathbf{i}_y B_0 \sin(-kz - \omega t)$$
$$\mathbf{B}' = \mathbf{i}_y B' \sin(-k'z - \omega t) \qquad (24.4.36)$$
$$\mathbf{B}'' = \mathbf{i}_y B'' \sin(kz - \omega t)$$

where $\mathbf{E}_0$ is the electric field of the incident wave, $\mathbf{E}'$ is that of the transmitted wave, and $\mathbf{E}''$ is that of the reflected wave. Similar notation is used for the magnetic field vectors. Note that the expressions for $\mathbf{E}_0$, $\mathbf{E}'$, $\mathbf{B}_0$, and $\mathbf{B}'$ represent waves that propagate along the negative z-direction, while those for $\mathbf{E}''$ and $\mathbf{B}''$ represent waves propagating along the positive z-axis, as shown in Fig. 24.19.

You may recall previous work regarding dielectric substances in Chapter 17, section 17.5, where it was shown that at an interface between two substances with different electric permittivities, *the components of* $\mathbf{E}$ *parallel to the interface are the same on both sides of the interface.* This condition is expressed mathematically by Eq. (17.5.1) and stems from the fact that the integral of $\mathbf{E} \cdot d\mathbf{l}$ around a small closed loop containing the interface vanishes.[2] If this is the case, then, since the electric fields (24.4.35) are all parallel to the interface, we may equate the sum of $\mathbf{E}_0$ and $\mathbf{E}''$ at the interface (where $z = 0$) to the value of $\mathbf{E}'$ at that location. Setting z equal to zero in (24.4.35) and equating the second expression to the sum of the first and third, we obtain

$$E_0 + E'' = E' \qquad (24.4.37)$$

We are dealing with an interface between two substances whose *dielectric permittivities* are different but whose *magnetic susceptibilities* differ hardly at all from that of free space. We may assume, then, that the magnetic permittivity has the free space value μ_0 everywhere in the system, including the region at the interface. Since there is no discontinuity in the magnetic permittivity at the interface, there should be no discontinuity in the $\mathbf{B}$ fields there either. Therefore, the total value of $\mathbf{B}$ as expressed by the sum of the first and third of Eqs. (24.4.36) should be equal, at

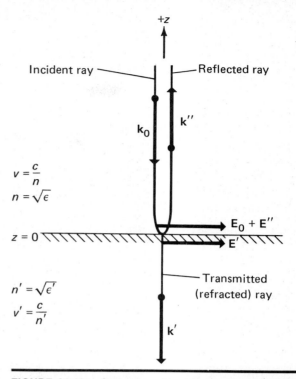

FIGURE 24.19. Geometry of incident, reflected, and refracted rays used in the calculations of Example 24.4.4.

$z = 0$, to that given by the second. Therefore, setting $z = 0$ in (24.4.36), we may write

$$B_0 + B'' = B' \qquad (24.4.38)$$

From our detailed study of electromagnetic waves in Chapter 23, we learned that the $\mathbf{E}$ and $\mathbf{B}$ fields of a plane electromagnetic wave that propagates in the z-direction are related by

$$\frac{\partial E_x}{\partial z} = -\frac{\partial B_y}{\partial t} \qquad (24.4.39)$$

as originally stated in Eq. (23.4.4). Taking the appropriate partial derivatives of the x-components of the electric fields and the y-components of the magnetic fields given by (24.4.35) and (24.4.36) and equating them in accord with (24.4.39), we find

$$-kE_0 \cos(-kz - \omega t) = \omega B_0 \cos(-kz - \omega t)$$
$$-k'E' \cos(-k'z - \omega t) = \omega B' \cos(-k'z - \omega t)$$
$$kE'' \cos(kz - \omega t) = \omega B'' \cos(kz - \omega t)$$
$$(24.4.40)$$

or, noting that $\omega/k = v$,

$$B_0 = -\frac{k}{\omega} E_0 = -\frac{E_0}{v}$$

$$B' = -\frac{k'}{\omega} E' = -\frac{E'}{v'} \qquad (24.4.41)$$

$$B'' = \frac{k}{\omega} E'' = \frac{E''}{v}$$

---

[2] This condition on the tangential components of the electric fields at the interface was derived in Chapter 17 for an electrostatic situation in which the integral of $\mathbf{E} \cdot d\mathbf{l}$ around *any* closed path is zero. For fields that vary with time, this integral, according to Faraday's law, is no longer zero but is instead equal to the time rate of change of the magnetic flux through the loop. However, the loop around which the integral is evaluated in Chapter 17, as illustrated by Fig. 17.17a, is chosen to be only infinitesimally wide, in which case its area and thus the magnetic flux through it are infinitesimally small. The induced emf is, therefore, negligible in this case and the result expressed by (17.5.1) is still correct, in spite of Faraday's law.

Substituting these values for the **B** fields into (24.4.38),

$$-E_0 + E'' = -n_r E' \qquad (24.4.42)$$

where

$$n_r = \frac{v}{v'} = \frac{n'}{n} \qquad (24.4.43)$$

is the relative index of refraction between the two media. Since we are ultimately seeking only the *ratios* of reflected and transmitted intensities to the incident light intensity, we may express Eqs. (24.4.37) and (24.4.42) in terms of such ratios by dividing both sides of those equations by E_0 to obtain

$$\frac{E'}{E_0} - \frac{E''}{E_0} = 1 \qquad (24.4.44)$$

and

$$n_r \frac{E'}{E_0} + \frac{E''}{E_0} = 1 \qquad (24.4.45)$$

Adding these two equations, we find that

$$\frac{E'}{E_0} = \frac{2}{1 + n_r} \qquad (24.4.46)$$

Substituting this into (24.4.44) we then obtain

$$\frac{E''}{E_0} = -\frac{n_r - 1}{n_r + 1} \qquad (24.4.47)$$

We have already seen from Eq. (24.2.8) that the *intensity* of any electromagnetic wave is expressed by the time average of the magnitude of the Poynting vector. For the incident wave, this is given by (24.2.6) as

$$\overline{S_0} = \frac{E_0 B_0}{2\mu_0} \qquad (24.4.48)$$

while for the transmitted and reflected waves it will be

$$\overline{S'} = \frac{E'B'}{2\mu_0} \quad \text{and} \quad \overline{S''} = \frac{E''B''}{2\mu_0} \qquad (24.4.49)$$

provided that the magnetic permeability has the free-space value μ_0 in both media. The fraction of incident energy that is reflected, which we shall call the *reflection coefficient R*, will, therefore, be given by

$$R = \left| \frac{S''}{S_0} \right| = \left| \frac{E''B''}{E_0 B_0} \right| \qquad (24.4.50)$$

while the fraction that is transmitted, which we shall identify as the *transmission coefficient T*, will be

$$T = \left| \frac{S'}{S_0} \right| = \left| \frac{E'B'}{E_0 B_0} \right| \qquad (24.4.51)$$

From (24.4.41), it is apparent that

$$\frac{B'}{B_0} = \frac{E'}{E_0} \cdot \frac{v}{v'} = n_r \frac{E'}{E_0} \qquad (24.4.52)$$

and

$$\frac{B''}{B_0} = -\frac{E''}{E_0} \qquad (24.4.53)$$

Inserting these values along with the ratios of the electric field amplitudes expressed by (24.4.46) and (24.4.47) into (24.4.50) and (24.4.51), the reflection coefficient becomes

$$R = \left(\frac{E''}{E_0} \right)^2 = \left(\frac{n_r - 1}{n_r + 1} \right)^2 \qquad (24.4.54)$$

and the transmission coefficient

$$T = n_r \left(\frac{E'}{E_0} \right)^2 = \frac{4n_r}{(n_r + 1)^2} \qquad (24.4.55)$$

Adding these equations, it is easy to see that

$$R + T = 1 \qquad (24.4.56)$$

which is in line with what we might expect on the basis of energy conservation.

These results confirm the prediction of Maxwell's theory about the existence of the reflected ray. While it could be correctly argued that we *assumed* from the beginning that the reflected ray was there, it is evident from the results we obtain that the conditions that the tangential components of both electric and magnetic fields match at the interface can be satisfied in its absence *only* when $n_r = 1$, that is, when $n' = n$. But in this situation, the interface vanishes altogether and the two substances involved form one continuous medium so far as optical properties are concerned. The inevitability of this conclusion is apparent in Eqs. (24.4.44) and (24.4.45), which stem directly from these conditions on the **E** and **B** fields at the interface. It is clear that if the reflected ray were missing, E'' would be zero and the two equations would be inconsistent unless n_r had the value unity.

In the case of an interface between air ($n = 1$) and glass ($n' \cong 1.52$), we have $n_r = 1.52$, and Eqs. (24.4.54) and (24.4.55) give

$$R = \left(\frac{0.52}{2.52} \right)^2 = 0.0426$$

and

$$T = \frac{(4)(1.52)}{(2.52)^2} = 0.9574$$

In this instance, a bit over 4 percent of the incident light energy is reflected, while slightly less than 96 percent is transmitted. For substances of higher refractive index, the reflected ray is more intense; in the case of leaded flint glass ($n = 1.96$), about 10.5 percent of the incident energy is reflected. It is important to note that these results hold *only* for light that is

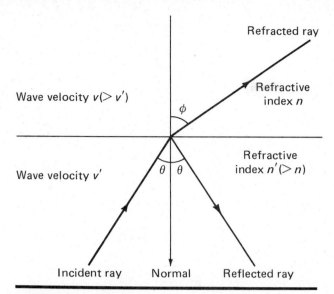

Wave velocity $v(> v')$

Wave velocity v'

Refracted ray

Refractive index n

Refractive index $n'(> n)$

Incident ray   Normal   Reflected ray

FIGURE 24.20. The case of "reverse refraction" in which light is incident on the refracting interface from within the more highly refractive medium. In this situation, the relative index of refraction is less than unity.

normally incident upon the refracting interface. For obliquely incident light, the reflectivity is found to vary with the angle of incidence as well as with the direction in which the incident beam is polarized, although for light that is randomly polarized and for indices of refraction that are not excessively large, Eqs. (24.4.54) and (24.4.55) are good approximations for angles of incidence up to about 45°.

The phase changes exhibited by the reflected ray, referred to previously in the text, can be understood with reference to (24.4.47). In this equation, so long as n_r is greater than unity, in which case $n' > n$, the incident ray goes from the medium of lower refractivity into the one of higher refractive index. Under these circumstances, (24.4.47) tells us that the reflected amplitude E'' has an algebraic sign *opposite* to that of the incident amplitude E_0. This reversal in sign corresponds to a phase change of 180°, since $\sin(\theta + 180°) = -\sin\theta$. When n_r is less than unity, and therefore $n' < n$, the incident ray goes from a more highly refracting medium into one that is less refractive. But then, the ratio of E'' to E_0 in (24.4.47) is positive, the algebraic signs of E'' and E_0 are the same, and no phase change is encountered on reflection.

This example, though admittedly fairly complex, is one of the simplest ones in which Maxwell's theory can be applied to a practical problem of optics. It is designed to give the reader a feel for how to apply electromagnetic theory to actual physical problems and also an appreciation of the mathematical difficulties that are encountered in working out their solutions.

24.5 "Reverse Refraction" and Total Internal Reflection

We have considered in some detail refraction phenomena in which the incident beam proceeds from the less refractive medium into one of higher refractive index in the preceding section. We have also encountered the reverse situation, once or twice, in which the light is incident upon the refracting interface from within the highly refracting substance and goes from it into a medium of smaller refractivity. We have seen that in this "reverse refraction" situation, the relative refractive index between the two media, given by the ratio of the velocity of the incident ray to that of the refracted one—or, alternatively, by the ratio of the absolute index of the substance in which the refracted ray propagates to that of the medium in which the incident ray travels—is *less* than unity. We may write, therefore, that the relative index n_r' pertaining to this case is, as in Example 24.4.1,

$$n_r' = \frac{v'}{v} = \frac{n}{n'} = \frac{1}{n_r} \qquad (24.5.1)$$

the symbols used referring to those shown in Fig. 24.20. Snell's law, which holds in either case, gives us

$$\frac{\sin\theta}{\sin\phi} = \frac{\text{velocity of incident ray}}{\text{velocity of refracted ray}} = \frac{v'}{v} = n_r' \qquad (24.5.2)$$

where in this case $n_r' < 1$. This equation tells us that the sine of the angle of incidence is smaller than the sine of the angle of refraction, and, therefore, that the angle of incidence is less than the angle of refraction, as illustrated in Fig. 24.20. If we rewrite (24.5.2) as

$$\sin\phi = \frac{\sin\theta}{n_r'} \qquad (24.5.3)$$

it is apparent that when $\sin\theta$ approaches the value n_r' (which is quite possible when $n_r' < 1$), $\sin\phi$ will approach unity and the angle of refraction will approach 90°. Under these conditions, the refracted ray just grazes the interface, barely escaping the more refractive medium. For angles of incidence that are even larger, $\sin\theta$ is greater than n_r' and Eq. (24.5.3) gives a value *greater than unity* for the sine of the angle of refraction. Clearly, there cannot be any real angle that satisfies this equation, and this suggests that *under these circumstances there can be no refracted ray.* This turns out to be true, and it is found that for angles of incidence larger than the critical value θ_c for which

$$\sin\theta_c = n_r' = \frac{n}{n'} \qquad (24.5.4)$$

the incident ray undergoes *total internal reflection* at the interface. This is physically reasonable in view of the fact that in the absence of any refracted ray in the

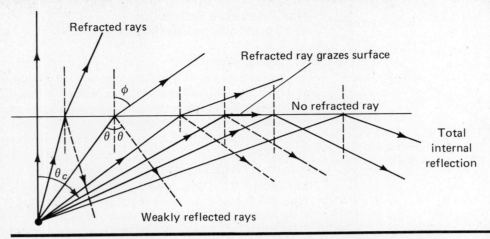

FIGURE 24.21. Light incident on a refracting interface from within the more highly refractive medium at various angles, exhibiting the phenomenon of reverse refraction, and for angles exceeding the critical angle θ_c, the phenomenon of *total internal reflection*.

less refractive medium, all the energy of the incident ray must remain within the more refractive medium.

The situation is illustrated in Fig. 24.21, where a number of rays emitted from a source at S in different directions are shown impinging on an interface beyond which lies a medium of smaller refractivity than the one which contains the source. For relatively small angles of incidence, there are refracted rays that escape into the less refractive substance, in the manner illustrated in Fig. 24.20, where the angles of refraction and incidence are related by (24.5.3). There are also weak rays that are reflected from the interface back into the medium whence they came, just as in the case when the light is incident from the less highly refracting substance into the medium of higher refractive index. As the angle of incidence increases, and as its sine approaches n_r', the angles of refraction become larger and larger, finally approaching 90°. Just below the critical angle of incidence θ_c, whose sine is equal to n_r', the angle of refraction is almost 90°, the refracted ray just grazing the interface between the two refracting media. Suddenly, when the angle of incidence reaches, or exceeds, this critical angle, *all incident light is reflected at the interface* by the usual law of reflection, back into the medium from which it originated.

The phenomenon of total internal reflection is easy to observe from the bottom of a swimming pool or, if you prefer to stay dry, by looking at the lower surface of the water in an aquarium at an angle beyond the critical angle. It can be put to useful purposes in devices such as totally reflecting prisms, discussed in Example 24.5.2, and light conduits such as the one illustrated in Fig. 24.22. A very useful method of determining the refractive index of liquids involves the determination of the critical angle at which total

reflection takes place, at an interface between the liquid and a glass prism whose refractive index is accurately known. The total internal reflection of laser beams within thin glass fibers, as shown in Fig. 24.22, allows such fibers to be used as *optical waveguides*. These waveguides can be used as the basis for communications networks that work at optical frequencies. The technology of *fiber optics* is, indeed, highly sophisticated and forms the basis for many useful optical devices.

EXAMPLE 24.5.1

A ray is incident upon an interface between water ($n' = 1.333$) and air ($n = 1.000$) from within the water at an angle of 30° to the normal. What is the angle between the normal and the refracted ray? What is the critical angle θ_c for this interface? What is the critical angle for an interface between glass ($n' = 1.520$) and air? For an interface between glass ($n' = 1.520$) and water?

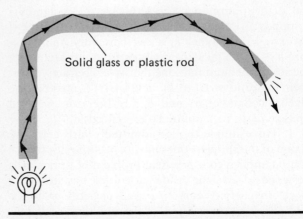

Solid glass or plastic rod

FIGURE 24.22. Propagation of light through a "light pipe" or optical fiber by total internal reflection.

For an interface between diamond ($n' = 2.418$) and air?

According to Snell's law (24.5.3), the angle of refraction is

$$\sin \phi = \frac{\sin \theta}{n_r'}$$

In the case where the incident ray proceeds from the more refractive medium into the less refractive one, the relative index n_r' is given by

$$n_r' = \frac{n}{n'} = \frac{1.000}{1.333} = 0.750$$

the numerical values pertaining to the air–water system. Therefore, for $\theta = 30°$, we may obtain

$$\sin \phi = \frac{\sin 30°}{0.750} = \frac{0.500}{0.750} = 0.667$$

$$\phi = 41.8°$$

The critical angle θ_c is, according to (24.5.4), given by

$$\sin \theta_c = n_r' = \frac{n}{n'}$$

For the air–water interface, $n = 1.000$ and $n' = 1.333$, from which

$$\sin \theta_c = \frac{1.000}{1.333} = 0.750 \qquad \theta_c = 48.6°$$

Similarly, for the interface between glass ($n' = 1.520$) and air,

$$\sin \theta_c = \frac{1.000}{1.520} = 0.658 \qquad \theta_c = 41.1°$$

In the case of the glass–water interface, since glass is more refractive than water, we must use $n' = n_g = 1.520$ and $n = n_w = 1.333$. Then, from (24.5.4),

$$\sin \theta_c = \frac{n}{n'} = \frac{1.333}{1.520} = 0.877 \qquad \theta_c = 61.3°$$

For the interface between diamond and air, $n = 1.000$ and $n' = n' = 2.418$, from which

$$\sin \theta_c = \frac{1.000}{2.418} = 0.414 \qquad \theta_c = 24.4°$$

In the case of a cut diamond, the small critical angle, resulting from the very large refractive index, ensures that most of the light entering the gem is totally reflected back toward the source. This explains, partially at least, the brilliant and lustrous appearance of gem-cut diamonds.

EXAMPLE 24.5.2

In binoculars of good quality, the internal light path is folded upon itself and an initially inverted image is

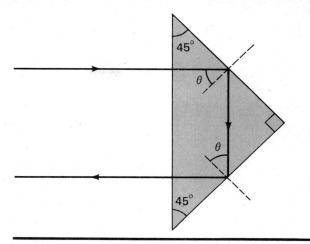

FIGURE 24.23. Binocular prism.

erected by total internal reflection from two prisms in the form of isosceles right triangles, as illustrated in Fig. 24.23. The folding of the light path renders binoculars much shorter and more compact than a straight telescope of the same magnification. What is the minimum index of refraction that these prisms must have to ensure that the light passing through will in fact be totally reflected?

Referring to the diagram shown in Fig. 24.23, it is apparent that the angle of incidence θ for both internal reflections is 45°. If the critical angle θ_c for the glass–air interface is greater than this, the light will not be totally reflected, but will be refracted instead and will leave the prism altogether. If θ_c, on the other hand, is less than 45°, total internal reflection will take place and the prism will work as intended. We must have, then, a minimum index of refraction n' for the glass such that, using $n = 1.000$ for air,

$$n_r' = \frac{1.000}{n'} = \sin \theta_c = \sin 45°$$

or

$$\frac{1}{n'} = \frac{\sqrt{2}}{2}$$

$$n' = \frac{\sqrt{2}}{2} = 1.4142$$

This presents no problem in practice, since most optical glass has a refractive index in excess of 1.50.

EXAMPLE 24.5.3

A small light bulb is immersed in a liquid of refractive index n', as shown in Fig. 24.24. What fraction of the light it emits will escape into the air beyond the surface of the liquid? Assume that the light bulb radiates light energy uniformly in all directions.

It is evident that light striking the surface at an angle of incidence less than θ_c will be refracted into

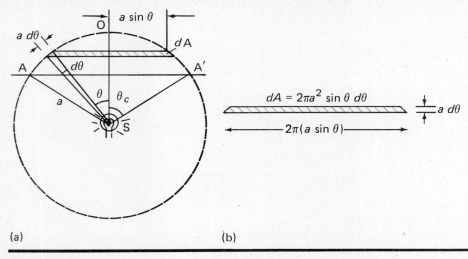

FIGURE 24.24

the air outside, while light striking the surface with a larger angle of incidence will be totally reflected back into the liquid. The light that escapes will, therefore, be emitted within a cone of apex angle θ_c. The fraction of the total energy represented by this light is the ratio of the area of the spherical cap AOA' of central angle θ_c to the total area of the dotted sphere shown in Fig. 24.24. The shaded area element dA of this cap in Fig. 24.24a is a ring of radius $a \sin \theta$ and width $a\, d\theta$; this element is shown as if it had been cut with scissors and flattened out onto the page in Fig. 24.24b. It is evident that the area element dA is

$$dA = 2\pi a^2 \sin \theta \, d\theta \qquad (24.5.5)$$

where a is the radius of the sphere. At the same time, the area of the dotted sphere is $4\pi a^2$. The fraction f of total energy escaping, as expressed by the ratio of areas referred to above, can then be written as

$$f = \frac{\int_0^{\theta_c} dA}{4\pi a^2} = \frac{2\pi a^2 \int_0^{\theta_c} \sin \theta \, d\theta}{4\pi a^2} = \tfrac{1}{2}[-\cos \theta]_0^{\theta_c}$$

or

$$f = \tfrac{1}{2}(1 - \cos \theta_c) \qquad (24.5.6)$$

However, from (24.5.4), we know that $\sin \theta = n_r'$. Therefore,

$$\cos \theta_c = \sqrt{1 - \sin^2 \theta_c} = \sqrt{1 - n_r'^2} = \sqrt{1 - \frac{1}{n^2}} \qquad (24.5.7)$$

since the relative index n_r' for light incident from the medium of absolute index n into air is just $1/n$. Substituting this into (24.5.7),

$$f = \frac{1}{2}\left[1 - \sqrt{1 - \frac{1}{n^2}}\right] \qquad (24.5.8)$$

For water, $n = 1.333$, and according to (24.5.8), $f = 0.1694$ for this case. It is interesting to note that the

fraction of light that escapes is completely *independent* of the radius a of the dotted sphere and hence of the depth at which the light is placed—so long as the medium is perfectly transparent and does not absorb any light.

24.6 Dispersion of Light by Refracting Substances

Our initial acquaintance with the subject of dielectrics was formed in the context of electrostatic systems. For this reason, our previous use of numerical values for the dielectric permittivity of such materials was founded upon measurements that were essentially electrostatic ones, though we have in some cases used them successfully in ac circuits at frequencies of hundreds, thousands, or even millions of cycles per second. If the dielectric permittivity of transparent substances were truly independent of frequency, their index of refraction would, of course, be independent of the light wavelength and would be numerically equal to the square root of the static dielectric constant. Unfortunately, this is never the case, and over the range of frequencies between zero and the frequency of ultraviolet light (about 3×10^{15} Hz), the dielectric constants of many substances undergo large changes. For water, which provides a rather extreme example, the observed value of the static dielectric constant is $K = \varepsilon/\varepsilon_0 = 81$, while for optical frequencies an index of refraction of about 1.333 is measured, corresponding to a dielectric constant of $K = n^2 = 1.77$. Most substances exhibit much less variation than this, it is true, and for some, the square root of the static dielectric constant provides a reasonably accurate figure for the optical refractive index, but an offhand assumption that this is the case for any given substance is generally unwarranted.

Physically, there are several different effects that contribute to this behavior, and to discuss all of them at this point would lead us far from the main point we are trying to make. In the range of optical frequencies, however, the variation of dielectric constant with frequency in most cases arises from the dynamics involved in the polarization of individual atoms by the rapidly varying sinusoidal electric fields of light waves. At these high frequencies, the polarization of the material is due almost entirely to atomic dipole moments arising from the displacement of the "electron clouds" of the atoms with respect to their nuclei by the electric field of the wave. This occurs because, in the presence of an electric field, the Coulomb force acts to displace the positive nucleus slightly in the direction of the field and the electron cloud of the atom in the opposite direction. As this displacement of charge occurs, an internal electric field is created that is opposite in sign to the external field and which tends to restore the atom to its original condition, much as the restoring force of a spring tends to return it to its equilibrium state. Indeed, this internal restoring force endows the atom with a natural frequency of vibration like that of a spring. If the external field were suddenly cut off, the internal Coulomb force would act to restore the equilibrium condition of the atom, reducing its dipole moment to zero. Once this is accomplished, however, the nucleus and the electron cloud, having accumulated kinetic energy in the process, continue to "coast" until an oppositely directed polarization and a concomitant oppositely directed restoring force stops them again. This cycle of events occurs over and over again, so that the atoms act just like harmonic oscillators with a well-defined *natural frequency* of vibration.

This means that when an external electric field of the natural frequency is applied, an atomic *resonance* is excited in which large vibration amplitudes and large values of polarization occur. At this frequency, which for most optically transparent materials is in the *ultraviolet* region of the spectrum, considerably beyond the visible range, the polarizability of the atoms and thus the dielectric constant and the optical refractivity are very large. Below this frequency, the dielectric constant and hence the refractive index will decrease with decreasing frequency, and this effect continues to be observed to some degree as we enter and pass through the visible region of the spectrum. For most substances, therefore, the refractive index for violet light at the high-frequency end of the spectrum is slightly larger than for red light of lower frequency. Typically, for substances such as glass and water, this variation, over the range of visible wavelengths, amounts only to a change of a few percent in the refractive index. This is large enough, nevertheless, to create easily observable effects of *dispersion* associated with the refraction of white light.

TABLE 24.2. Dispersion Characteristics of Optical Glasses

	Refractive index n				
λ, Å	Light crown	Dense crown	Light flint	Dense flint	Heavy flint
4000	1.5238	1.5854	1.5932	1.6912	1.8059
4500	1.5180	1.5801	1.5853	1.6771	1.7843
5000	1.5139	1.5751	1.5796	1.6670	1.7706
5500	1.5108	1.5732	1.5757	1.6591	1.7611
6000	1.5085	1.5679	1.5728	1.6542	1.7539
6500	1.5067	1.5651	1.5703	1.6503	1.7485
7000	1.5051	1.5640	1.5684	1.6473	1.7435

As a result of dispersion, the different wavelengths within a white light beam are refracted in slightly different directions to produce a visible separation of spectral colors. The amount of dispersion encountered over the visible spectrum in any individual substance depends critically on the value of the resonant frequency and on certain other properties of the atoms of the refracting substance. In particular, for substances in which the resonant frequency is relatively near the visible region of the spectrum, dispersion effects are particularly large. While materials having large refractive indices frequently exhibit relatively high dispersion, it is important to note that the dispersion exhibited by a refracting substance is *not* directly proportional to its refractive index. This fact makes it possible to construct superb binoculars and photographic lenses that create sharp and clearly defined images, free of visible color fringes, in spite of the fact that they rely upon the refraction of white light to form these images. We shall see later exactly how this is accomplished. The refractive indices of some common types of optical glass over the range of visible light wavelengths are compared in Table 24.2. From these data, it is apparent that the refractive index of light crown glass changes by some 1.38 percent over the visible spectrum, while the corresponding dispersive change in the refractive index of heavy flint glass is about 3.85 percent. The effect of dispersion in heavy flint glass is thus more than twice that in light crown glass. The ratio of refractive indices, however, is only $1.7611/1.5108 = 1.166$ at 5500 Å. The behavior of light rays in dispersive substances is illustrated in detail in the examples that follow.

EXAMPLE 24.6.1

A white light ray in air is incident upon a flat plate of heavy flint glass at an angle of incidence of 45.00°. What is the speed of light in the glass for the component of white light of wavelength 4000 Å? What is the angle of refraction for light of this wavelength? What is the speed in the medium and the angle of refraction for the component of wavelength 7000 Å? What are the critical angles for total internal reflection for these two wavelengths in glass of this sort? Use the numerical data in Table 24.2 for heavy flint glass.

Since the index of refraction for $\lambda = 4000$ Å is $n = 1.8059$, the speed of light in the medium for this wavelength is

$$v = \frac{c}{n} = \frac{2.998 \times 10^8}{1.8059} = 1.6601 \times 10^8 \text{ m/sec}$$

while for red light ($\lambda = 7000$ Å), the corresponding velocity is

$$v = \frac{c}{n} = \frac{2.998 \times 10^8}{1.7435} = 1.7195 \times 10^8 \text{ m/sec}$$

the refractive index being 1.7435 for $\lambda = 7000$ Å.

The angle of refraction for $\lambda = 4000$ Å is, according to Snell's law,

$$\sin \phi = \frac{\sin \theta}{n} = \frac{\sin 45.00°}{1.8059} = 0.3916$$

$$\phi = 23.05°$$

For $\lambda = 7000$Å, the corresponding results are

$$\sin \phi = \frac{\sin \theta}{n} = \frac{\sin 45.00°}{1.7435} = 0.4065$$

$$\phi = 23.93°$$

The angular dispersion over the range 4000–7000 Å is, therefore,

$$\Delta \phi = 23.93 - 23.05 = 0.88°$$

The critical angles for total internal reflection follow from (24.5.4), using $n_r' = 1/n$ for the respective wavelengths. For $\lambda = 4000$ Å,

$$\sin \theta_c = n_r' = \frac{1}{1.8059} = 0.5537 \qquad \theta_c = 33.62°$$

while for $\lambda = 7000$ Å,

$$\sin \theta_c = n_r' = \frac{1}{1.7435} = 0.5736 \qquad \theta_c = 35.00°$$

The change in critical angle over this range is

$$\Delta \theta_c = 35.00° - 33.62° = 1.38°$$

EXAMPLE 24.6.2

A prism made of heavy flint glass with an apex angle α refracts light as shown in Fig. 24.25, the light passing through symmetrically, entering and leaving at the same angle θ with respect to the normal to the prism face. Find the total angle Δ through which a light ray is deviated in passing through a prism in this way. Express the index of refraction of the prism in terms of this angle of deviation Δ. Find the angle of deviation Δ for a heavy flint glass prism of apex angle $\alpha = 60°$ for wavelengths of 4000 Å and 7000 Å, using the indices of refraction given in Table 24.2.

The path of the ray is shown in Fig. 24.25 as APQB. In triangle OPQ, since the sum of the three

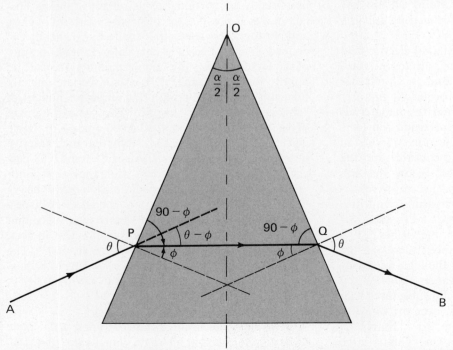

FIGURE 24.25. Symmetric deviation of a light ray by a prism of apex angle α.

angles has to be 180°, we may write

$$2(90° - \phi) + \alpha = 180° \tag{24.6.1}$$

from which

$$\phi = \frac{\alpha}{2} \tag{24.6.2}$$

At the same time, the ray is bent through an angle $\theta - \phi$ at P and through an equal angle at Q. Therefore, the entire angle Δ through which the ray is deviated—thus the angle between rays AP and QB—must be given by

$$\Delta = 2(\theta - \phi) \tag{24.6.3}$$

From Snell's law, however, and from (24.6.2), we can write

$$\sin \theta = n \sin \phi = n \sin \frac{\alpha}{2} \tag{24.6.4}$$

where n is the refractive index of the prism. Taking the inverse sine of this equation, we find

$$\theta = \sin^{-1}\left(n \sin \frac{\alpha}{2} \right) \tag{24.6.5}$$

Finally, inserting this value for θ and the value for ϕ given by (24.6.2) into (24.6.3), we obtain

$$\Delta = 2 \sin^{-1}\left(n \sin \frac{\alpha}{2} \right) - \alpha \tag{24.6.6}$$

This can be solved for n to obtain an expression for the index of refraction in terms of the deviation angle Δ by rearranging (24.6.6) to read

$$\frac{\Delta + \alpha}{2} = \sin^{-1}\left(n \sin \frac{\alpha}{2} \right) \tag{24.6.7}$$

and taking the sine of both sides of this equation to obtain

$$\sin\left(\frac{\Delta + \alpha}{2} \right) = n \sin \frac{\alpha}{2} \tag{24.6.8}$$

or

$$n = \frac{\sin\left(\dfrac{\Delta + \alpha}{2} \right)}{\sin \dfrac{\alpha}{2}} \tag{24.6.9}$$

This affords a good method for determining the index of refraction of a sample of glass in the form of a prism simply by measuring the apex angle α and the angle Δ through which a symmetrically disposed ray is deviated.

For a prism fashioned from heavy flint glass, using Table 24.2 and (24.6.6), with $\alpha = 60° = \pi/3$ rad and $n = 1.8059$, for light of wavelength 4000 Å we find that

$$\Delta = 2 \sin^{-1}[(1.8059)(0.5000)] - (\pi/3)$$
$$= 2 \sin^{-1}(0.90295) - (\pi/3)$$

or

$$\Delta = 2.25317 - 1.04720 = 1.20597 \text{ rad} = 69.10°$$

In making these calculations, it is important to express the inverse sine as an angle in *radians*. In the same way, for light of wavelength 7000 Å, for which $n = 1.7435$, we find

$$\Delta = 2 \sin^{-1}[(1.7435)(0.5000)] - (\pi/3)$$
$$= 1.07033 \text{ rad} = 61.33°$$

SUMMARY

Light is electromagnetic radiation of wavelength between 4000 and 7200 Ångströms (Å), or 400 and 720 nanometers (nm). Within this range of wavelengths, the *color of light* corresponds to the wavelength. In order of decreasing wavelengths, the colors of the spectrum are red, orange, yellow, green, blue, and violet. White light is light containing all spectral wavelengths with approximately equal intensities. In systems that are large in comparison with the light wavelength, light propagates along straight lines normal to the wavefronts, called *rays*. The electromagnetic energy per unit area normal to the propagation direction, per unit time, is referred to as the *light intensity*; it is equal to the time average of the magnitude of the Poynting vector. For a single linearly polarized monochromatic plane wave, the intensity is related to the electric and magnetic vector amplitudes E_0 and B_0 by

$$\overline{S} = \frac{E_0 B_0}{2\mu_0} = \frac{E_0{}^2}{2c\mu_0}$$

For a point source of light, the light intensity varies inversely with the square of the distance from the source:

$$\frac{\overline{S_2}}{\overline{S_1}} = \frac{r_1{}^2}{r_2{}^2}$$

The behavior of light rays can be understood completely in terms of Maxwell's electromagnetic theory. In simpler terms, they can be explained either on the basis of Fermat's principle of least time or by Huygens's principle. Fermat's principle states that a light ray travels between two points by that path which requires the least possible time. Huygens's principle states that the propagation of a wavefront occurs by the simultaneous emission of spherical wavelets at each point of an existing wavefront. The new wavefront can be represented at a later time as the envelope of the Huygens wavelets, and the path of the ray is defined by the normals to the successive wavefronts.

All these methods lead to the laws of reflection and refraction at plane surfaces. The law of reflection states that the angles between the normal to the reflecting plane and the incident and refracted rays are *equal*. When light travels through dense transparent media, such as glass or water, its speed is less than its speed in free space. Also, in different transparent media, the velocity of light will differ according to the properties of the individual substances. A light ray in passing through a plane interface separating two different transparent media experiences a change in velocity and, as a consequence, a change in direction. This effect is referred to as refraction. Fermat's principle or Huygens's principle can be used to derive Snell's law of refraction, which relates the direction of the incident and transmitted (or refracted) rays at a refracting interface. Snell's law states that

$$\frac{\sin \theta}{\sin \phi} = n_r$$

where θ and ϕ are the angles the incident and refracted rays make with the normal to the interface, and n_r is the relative index of refraction defined by

$$n_r = \frac{\text{velocity of incident ray}}{\text{velocity of refracted ray}}$$

Maxwell's theory of electromagnetism indicates that the refractive index can be expressed as the square root of the ratio of the permittivities of the two media. It also tells us that a certain amount of incident light is *reflected* by an interface between substances of different refractivity, the intensity of the incident and reflected beams, at normal incidence, being given by

$$R = \left(\frac{n_r - 1}{n_r + 1}\right)^2$$

When light passes from a highly refracting medium into one of smaller refractivity, the relative index, which we now designate by n_r', is smaller than unity. Under these circumstances, the angle of refraction ϕ is greater than the angle of incidence θ. When $\sin \theta$ approaches the value n_r', $\sin \phi$ approaches unity, and the angle of refraction is very close to 90°. The refracted ray is then essentially parallel to the interface. For angles of incidence beyond the critical angle θ_c for which

$$\sin \theta_c = n_r'$$

the refracted ray disappears and the incident light undergoes *total internal reflection*. The incident ray is now reflected back into the medium of incidence, and no energy is transmitted into the less dense substance.

Dispersion of light occurs within refracting substances because the refractive index ordinarily exhibits a slight but significant variation with wavelength.

This means that light of different colors is refracted through slightly different angles by a plane refracting interface or by a refracting element such as a prism or lens. Accordingly, a beam of white light is separated into its constituent spectral colors upon refraction at a plane interface or a prism. The effect of dispersion arises from the way in which the electric fields of light waves excite vibrational motions of the atomic charge distributions within the atoms of the refractive medium. Dispersion is responsible for the colors of the rainbow and affords a useful means of separating light beams into their separate spectral components for analysis in spectrometers and spectrographs.

QUESTIONS

1. Fermat's principle states that light rays follow the path that requires the shortest time. Give some examples to show that paths of minimum time and minimum distance are not necessarily equivalent.
2. The speed of light is less than c in all material media. What does this imply about the permittivity and permeability of materials?
3. When light goes from vacuum into a medium such as glass or water, the wavelength changes. Does the frequency also change? Why or why not?
4. Huygens's principle applies to visible light. Is it also applicable to radio waves? Do radio waves propagate in straight-line paths?
5. The direction of wave propagation is always perpendicular to the wavefront. Explain this using Huygens's principle.
6. A swimming pool always appears shallower than it really is. Explain, using Snell's law.
7. Different light wavelengths have the same speed in vacuum but different speeds in dense transparent materials. Cite some examples to demonstrate that this is the case.
8. Explain why red light is bent less than blue light when passing from air to water.
9. Does refraction of light by the atmosphere cause the days to be longer or shorter?
10. Describe how you can use total internal reflection to measure the index of refraction of a liquid.
11. Can you explain, using the law of refraction, why objects that are below the geometric horizon can still be visible?
12. Why is it that images observed in plane mirrors are reversed left to right but are not reversed top to bottom, that is, are not upside down?
13. It is sometimes said that the law of reflection can be regarded as a special case of Snell's law of refraction in which the relative index has the value $n_r = -1$. In what sense is this a justifiable statement?

PROBLEMS

1. **(a)** What is the wavelength, in meters, of light whose frequency is 6.4×10^{14} Hz? What would this wavelength be if expressed **(b)** in microns (1 micron $= 10^{-6}$ m),

(c) in nanometers, (d) in Ångstrom units? (e) How many wavelengths are there in a centimeter?

2. (a) What is the magnitude of the propagation constant k for light of frequency 6.40×10^{14} Hz. (b) What is the angular frequency? (c) What is the period?

3. A 100-watt incandescent light bulb is found to produce a total visible light intensity of 0.0458 w/m² at a distance of 2.5 meters when operated at its rated voltage. What fraction of the electrical energy it consumes is radiated as visible light?

4. A laser emits monochromatic light in the form of a thin cylindrical beam of diameter 1.00 mm. The light is emitted as a single electromagnetic wave, and the total radiated power is 1.0 watt. Find (a) the rms electric field magnitude of the emitted wave, (b) the rms magnetic induction, (c) the average intensity of the beam.

5. A ray of light incident on a plane mirror makes an angle of 30° with the normal to its surface. If the mirror is now rotated by 7°, thus increasing the angle to 37°, by what angle does the reflected ray rotate?

6. Suppose that the xz- and yz-planes of an orthogonal coordinate system are plane mirrors, which intersect along the z-axis. (a) What are the coordinates of the image of a point $P(x_0, y_0)$ in the xy-plane formed by reflection in both mirrors? (b) Find the angle between a ray from point P in the xy-plane incident on the first reflecting surface and the outgoing ray that results when this ray has undergone two reflections.

7. In the preceding problem, suppose the point P is moving in the xy-plane with velocity **v**. What is the velocity of the image formed after two reflections?

8. The rangefinder of a camera consists of two mirrors that are arranged as shown in the diagram. Mirror A is very lightly silvered in such a way that it is only partially reflecting, and remains partially transparent to an observer looking through it. Mirror B is an ordinary plane reflecting surface. In practice, the user views an object at P through mirror A and observes a double image resulting from the fact that light from P reaches his eye by two different routes; the direct path PA and the reflected path PBA. If mirror B is rotated about a vertical axis, through B, however, the two images can be made to coincide. Show that when this adjustment is made, the angle ϕ between the two mirrors will be given by $\tan(2\phi) = l/d$, where d is the distance AP between the mirrors and the object and l the distance AB between the mirrors. If l is accurately known, then the distance d to a distant object can be determined by observing the angle ϕ between the two reflecting surfaces when the images are superimposed.

9. A planar light source is parallel to and midway between two parallel plane mirrors. It emits light in the form of two plane waves of intensity $S_0/2$ which proceed outward to the left and to the right from the source. The mirrors reflect a fraction R of the light energy incident on them, and absorb the rest. Show that the total steady-state intensity, which includes reflected light as well as that emitted by the source, in the region between the mirrors is given by $\bar{S} = (\bar{S}_0/2)(1 - R)^{-1}$. Assume that reflected light can pass through the planar source unimpeded. (Hint: Be sure you include the effect of multiple reflections in one way or another.)

10. For a point source of light, the intensity falls off inversely with the square of the distance, as given by Eq. (24.2.17). Show that for a cylindrical source of light, such as a fluorescent tube, $S_2/S_1 = r_1/r_2$.

11. The optical path length of light is defined as the geometric path length multiplied by the index of refraction of the medium in which the light propagates. Show that the optical path length is the distance the light would have traveled in vacuum in the time interval during which it traverses the true geometric path length in the medium.

12. A charged particle such as an electron emits radiation (called Čerenkov radiation) when its speed in a material medium exceeds the speed of light in the medium. It is found that when an electron travels with $v > 1.85 \times 10^8$ meters/second in a certain medium, Čerenkov radiation is emitted. Find the index of refraction of the medium.

13. Light has a wavelength 4200 Å in light crown glass ($n = 1.51$). What are the wavelength and frequency of the light in vacuum? What are the frequency and wavelength in heavy flint glass ($n = 1.890$)?

14. What is the velocity of light (a) in water ($n = 1.333$), (b) in crown glass ($n = 1.51$), (c) in flint glass ($n = 1.890$), (d) in diamond ($n = 2.42$).

15. (a) What is the relative refractive index n_r for light incident from air into water ($n = 1.333$)? (b) What is the relative index for light incident from water into air? (c) What is

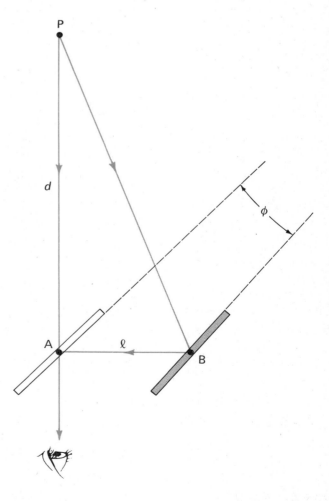

the relative refractive index for light incident from water into heavy flint glass ($n = 1.890$)? **(d)** What is the relative index for light incident from heavy flint glass into water?

16. A light ray in air incident on a planar air–water interface makes an angle of 45° with the normal to the surface. **(a)** What is the angle the refracted ray within the glass makes with the normal? **(b)** What would the angle of refraction be if the ray were incident on the same interface from within the glass at an incidence of 45°? The index of refraction of the water is 1.333.

17. Light is incident from water ($n = 1.333$) into a sheet of flint glass ($n = 1.860$) at an angle of 35° to the normal. **(a)** What angle does the refracted ray make with the normal? **(b)** What would the answer to (a) be if the light were incident from within the glass at an angle of 35° to the normal?

18. Light is incident from air into water ($n = 1.333$) at an angle of incidence of 60°. Upon traversing a layer of water it encounters a layer of flint glass ($n = 1.860$) whose surface is parallel to that of the water. **(a)** What angle does the ray make with the normal in the flint glass? **(b)** What effect does the presence of the water layer have upon the angle of refraction within the flint glass?

19. An athletic woman who can run twice as fast as she can row a boat wishes to go from point P to point P′ on opposite sides of a stream in the shortest possible time, as shown in the diagram below. What distance should she run along the shore and what distance should she row? Assume that the stream's flow velocity is zero. Show that the result you obtain can be exhibited also by using Snell's law.

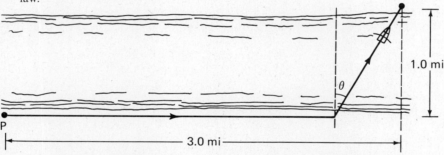

20. A swimming pool has a uniform depth of 8 feet. When viewed vertically through air, what is its apparent depth?

21. Yellow light of wavelength $\lambda = 5800$ Å is incident on a

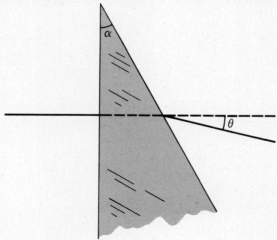

thin prism for which $n = 1.60$, as shown in the figure. The prism angle ϕ is 0.070 radians. **(a)** Find the angle of deviation θ for the emerging beam. **(b)** Find the wavelength of the light inside the prism. **(c)** Find the ratio of the frequency of the light inside the prism to that in air.

22. What is the reflection coefficient for light incident normally from air onto a plane surface of **(a)** water ($n = 1.333$), **(b)** light crown glass ($n = 1.520$), **(c)** heavy flint glass ($n = 1.880$), **(d)** diamond ($n = 2.518$).

23. A ray of light falls on a rectangular glass block ($n = 1.5$) which is almost completely submerged in water ($n = 1.33$), as shown in the diagram. **(a)** Find the angle θ when total internal reflection just occurs at P. **(b)** Would total internal reflection occur if the water were removed?

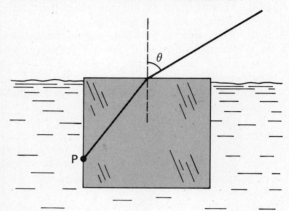

24. The binocular prism shown in Fig. 24.23 is immersed in a liquid of refractive index n_e. The refractive index of the prism is 1.700. What is the maximum value of n_e for which total internal reflection will take place as illustrated in Fig. 24.23?

25. A prism of refractive index n is used so that beams A, B, and C are internally reflected from the inner surface, as shown in the diagram on top of the next column. **(a)** Calculate the minimum value of n for which beam B is totally reflected. **(b)** Would the minimum index found in part (a) be higher or lower for beam A? **(c)** Would the minimum index found in part (a) be higher or lower for beam C?

*26. In the preceding problem, what would be the lowest possible value for the refractive index of the prism in the case where beams A, B, and C are all to be totally reflected, and where beams A and C are at an angle of 15° to the normal?

27. A light ray enters a glass plate at an angle of incidence of 30°. The glass plate has a refractive index of 1.500 and a

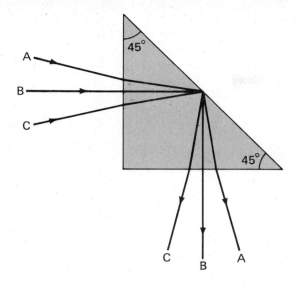

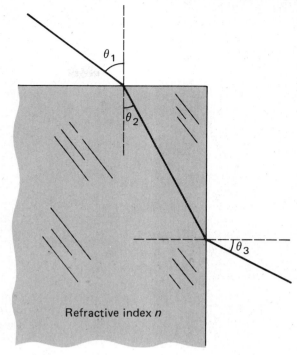

Refractive index n

uniform thickness of 1 cm. Show that the outgoing ray on the other side of the plate is parallel to the incoming ray. Find the distance through which the outgoing ray is displaced relative to the incident ray.

28. A Sunday afternoon golfer takes a tremendous swing at the ball but misses. He swings again but this time hits a terrible slice, and his brand-new ball lands in a nearby pool of clear water. He is determined to retrieve the ball from the water which he knows to have a uniform depth of 3.0 feet. As he approaches the edge of the pool, he does not see the ball at first; but when he is 24 feet from the edge of the pool, he is finally able to see it on the bottom. Assuming that his eye is 6.0 feet above the level of the water (for which $n = 1.333$), how far into the pool will he have to wade to retrieve the ball?

29. A silver dollar is at the bottom of a swimming pool at a point 3 meters from the pool wall. The pool is 2 meters deep. At what angle do light rays coming from the dollar leave the water surface at the edge of the pool?

30. A point source of light is located at the bottom of a large pool 2 meters deep. Find the area of the largest circle on the surface through which light coming from the source can emerge.

31. Light is normally incident at an interface between air and water. If the light rays are initially traveling in air, find the fraction of the incident energy reflected and the fraction transmitted into the water.

32. In water, the light from a certain source has a wavelength of 4100 Å. What is its wavelength in fused quartz? Find the critical angle for total internal reflection at a quartz–water interface. Use the refractive indices provided in Table 24.1.

33. A rectangular block of transparent material has an index of refraction n. A light ray is incident from the top at an angle θ_1 and emerges from the side at an angle θ_3 as shown in the diagram on top of the next column. (a) Show that

$$\sin \theta_3 = [n^2 - \sin^2 \theta_1]^{1/2}$$

(b) If the index of refraction of the block is 1.50, can the light ray ever be transmitted as illustrated? (c) For what index n will the emerging light ray be parallel to the side if the angle of incidence is $30°$?

34. At the interface between air and a certain transparent substance, a student is able to ascertain that the critical angle is $45°$. What is the velocity of light in this substance, and what is the wavelength of light for which $\lambda = 6000$ Å in air?

35. A light ray enters a glass prism of refractive index 1.50 at an angle of incidence of $30°$, as shown in the diagram. The incident ray enters the glass midway between A and B. Trace the progress of the light through the prism. Where does it emerge? What angle does the emerging ray make with the normal to the surface from which it emerges?

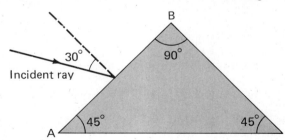

36. Light is incident normally on one face of a $45°$ prism, as shown in the figure. Find the angle between the incident and emerging beams if the index of the prism is (a) 1.30 and (b) 1.50.

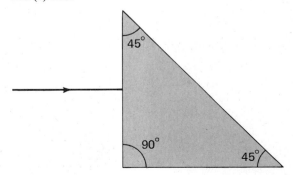

37. The optical fibers in a "light pipe" are thin, parallel cylindrical filaments of glass embedded in a plastic matrix of lesser refractivity. If the refractive index of such a fiber is n_g and the index of the surrounding plastic is n_p, find an expression for the largest angle θ for which light can enter the fiber and travel along its entire length without ever being transmitted into the plastic. The situation is illustrated in the diagram below.

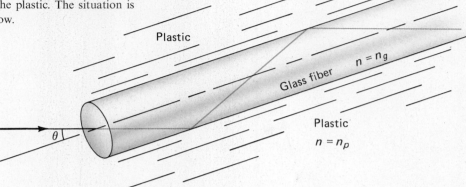

38. The index of refraction of a liquid can be measured by observing the onset of total internal reflection. In the figure, the liquid is poured onto a glass hemisphere for which the refractive index has been accurately determined to have the value $n = 1.9500$. **(a)** Describe how one can measure the refractivity of the liquid. **(b)** If total internal reflection occurs for $\theta > 80.00°$, what is the refractive index of the liquid?

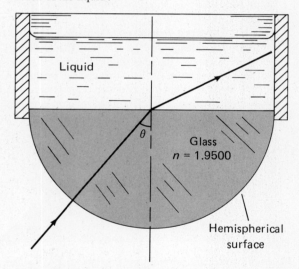

39. A little girl 3 feet tall is admiring her birthday dress by looking at her image in a vertical plane mirror 2 feet tall and 4 feet distant from her eyes. Her father wishes to photograph her image in the mirror. What is the minimum distance from the mirror from which it will be possible for him to photograph her entire image?

*40. A lifeguard stands at the edge of a swimming pool 6.0 feet deep. His eyes are 6.33 feet above the water. A straight pole 12.0 feet long rests on the bottom of the pool, perpendicular to the pool wall. One end of the pole touches the wall at the point above which the lifeguard is standing. The refractive index of the water is 1.333. **(a)** Find a mathematical expression giving the distance between the edge of the pool and the point where a ray traveling from the far end of the pole to the lifeguard's eye intersects the surface of the water. **(b)** Solve this equation for the required distance numerically, using a calculator if one is available. **(c)** Does the pole appear to the lifeguard to be straight or curved?

41. A ray of white light is incident at an angle of 60° upon a a plane layer of the "heavy flint glass" whose characteristics are given in Table 24.2. Find **(a)** the angle of refraction for light of wavelength 4000 Å, **(b)** the angle of refraction for light of wavelength 7000 Å, **(c)** the angular dispersion within the glass between wavelengths 4000 Å and 7000 Å. **(d)** The ray emerges from the other side of a plane-parallel sheet of glass; what is the angular dispersion now?

42. A ray of white light is incident upon one face of a prism whose apex angle is 60° at an angle of incidence of 57.0°. The prism is made of the "heavy flint" glass listed in Table 24.2. Find **(a)** the angle between the emerging ray and the normal to the prism face from which it emerges, for light of wavelength 4000 Å, **(b)** the corresponding angle for light of 7000 Å, **(c)** the angular dispersion of the prism between wavelengths 4000 Å and 7000 Å.

43. A ray of light of wavelength 4000 Å passes symmetrically through a prism of apex angle 45°, as in Example 24.6.2. The prism is made of the "dense flint" glass of Table 24.2. Find **(a)** the angle between the entering and exiting rays and the normals to the respective prism faces, **(b)** the total angle through which the ray is deviated. **(c), (d)** Repeat the calculation for a ray of wavelength 7000 Å.

44. A prism of apex angle 55.00° is placed symmetrically with respect to an incident monochromatic light ray. It is found that the ray is deviated through a total angle of 58.480° in passing through the prism. What is the refractive index of the prism for light of this wavelength?

PINHOLES, LENSES, AND MIRRORS

25.1 Introduction

The laws of geometrical optics and the imaging properties of lenses are beautifully exhibited by the 35-mm single-lens reflex cameras that have come to dominate the market in recent years. Those who are fortunate enough to own one of these beautifully fashioned instruments will be familiar with the way in which the viewfinder image—which is identical to the image that will eventually appear on the film—snaps in and out of focus as the focusing mount of the lens is turned and with the way in which the depth of the sharply defined zone can be made to vary by adjusting the lens diaphragm. A contraption such as this is a portable optics laboratory; and in one or another

aspect of its employment it exhibits nearly all the important laws of image formation that we shall have to study. At the other end of the line is the lowly "pinhole camera," which is the simplest possible photographic device. It consists of nothing more than a light-tight box with a small pinhole in one end and a film against the opposite end. The single-lens reflex camera and the pinhole camera are illustrated in Fig. 25.1.

It is our objective in this chapter to understand how images are formed by lenses and spherical mirrors and how these optical elements are used in devices such as telescopes, microscopes, cameras, and spectrometers. In discussing this material, we shall make frequent reference to the single-lens reflex camera and

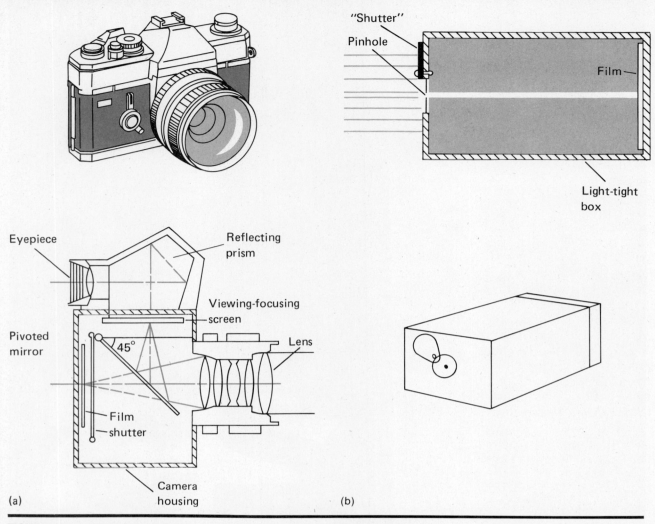

FIGURE 25.1. (a) Single-lens reflex camera. (b) A simple "pinhole" camera.

to the pinhole camera because they illustrate many of the principles to be learned. It will be particularly instructive to the student who owns a single-lens reflex camera, or can borrow one, to examine for himself the effects that are discussed. It is usually possible to accomplish the same end, though in a less elegant manner, with a light-tight box with a lens or pinhole at one end and a piece of ground glass or frosted plastic on which to observe the image at the other. The student who does not have access to a single-lens reflex camera will probably find it worthwhile to fashion a device of this sort, such as the one illustrated in Fig. 25.2.

In the preceding chapter, we saw how the image in a plane mirror is formed, and we also discussed the alteration of the apparent position of an object

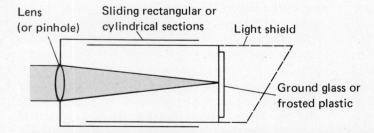

FIGURE 25.2. A "light box" that incorporates most of the illustrative features provided by a single-lens reflex camera.

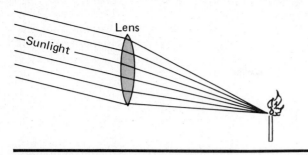

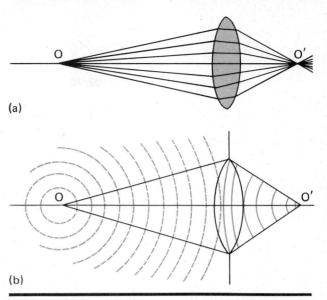

(a)

(b)

FIGURE 25.3. Use of a lens to concentrate light energy into a point.

lying at the bottom of a pool by the refraction of light at the surface of the water. It is evident, therefore, that optical *images* of light sources can be formed either by reflection or refraction, though up to this point only images formed by plane reflecting or refracting surfaces have been described.

Everyone is familiar with the use of a magnifying glass to produce an enlarged image of an object, or as a "burning glass" to concentrate enough solar energy on a tiny area to ignite combustible materials. It is evident that all the light from a distant point source that falls on the lens can be concentrated into an image point not far behind the lens, as illustrated in Fig. 25.3. The images produced by lenses are formed by *refraction* at the lens surfaces. These surfaces are almost always *spherical* in shape, although for special applications aspherical lenses are occasionally used.

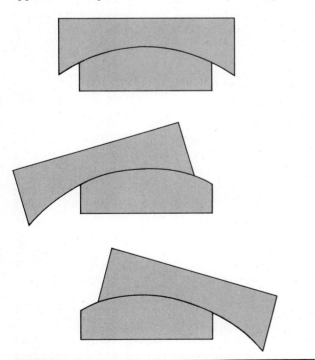

FIGURE 25.4. Production of a spherical refracting surface by grinding two pieces of glass in contact with one another.

FIGURE 25.5. Formation of an image of a point object by a lens, as illustrated (a) by the paths of rays refracted at the lens surfaces and (b) by the picture of spherical wavefronts traveling through the system.

The reason for the almost universal use of spherical lens surfaces is that spherical lenses form adequately good images and are very easy to produce. A convex spherical surface and a concave one of equal radius remain in close contact no matter how they are positioned with respect to one another, as shown in Fig. 25.4. No nonspherical surface has this property. Therefore, spherical surfaces are naturally generated when one piece of optical glass is ground against another with a fine, aqueous suspension of abrasive material, such as carborundum, between the two. Spherical lenses are easily and inexpensively manufactured in this way, and since the images they produce are reasonably good, they are always chosen for use in any commercially produced optical system, unless some very unusual requirement must be satisfied. Furthermore, the shortcomings or *aberrations* in the images produced by a single spherical lens can be drastically reduced or, in some cases, eliminated altogether by properly combining a number of spherical lenses into a compound lens, such as the lens of the single-lens reflex camera of Fig. 25.1a. The same objectives can be accomplished in some instances by the use of aspherical refracting surfaces, but since these are very difficult and expensive to produce, this is not often attempted.

The formation of an image of a point source of light by a lens with spherical surfaces is shown in terms of light rays and wavefronts in Fig. 25.5. It is evident, particularly from Fig. 25.5a, that a sheet of paper or, better, a ground-glass or frosted plastic screen placed at O' will display a bright point of light corresponding to the image of the source point at O. If the source point is part of a scene in front of the camera, the ambient light it reflects toward the lens is refracted in the same way as it would be if the source point were a luminous point object. In this case, the point O′ is a constituent of a larger image that is made up of light from every part of the scene refracted to image points in exactly the same way; a ground-glass screen or a sheet of paper placed behind the lens at O′ will then display a visible image of the scene before the lens. This is easily verified by removing the lens of a camera (or using a magnifying glass) and holding a sheet of white paper at an appropriate distance behind the lens to display the image. This is what you see when you look into the viewfinder of the single-lens reflex camera. As shown by Fig. 25.1a, the rays that converge to form the image are reflected through a right angle by a pivoted diagonal mirror onto a ground-glass screen on top of the camera body. This, in turn, is viewed through a reflecting prism, sometimes called a pentaprism because of its pentagonal shape, that enables the camera to be positioned in the conventional way at eye level and also renders the image right side up and free from the left–right reversal that would result in a single reflection from a plane mirror. When an exposure is made, the pivoted mirror flips up momentarily to cover the ground-glass screen obscuring the image for an instant, the shutter located just in front of the film opens and closes, and the mirror then returns to its former position restoring the finder image. The plane reflecting mirror is located just equidistant between the film plane and the ground-glass screen, so that the image received by the screen is precisely the same as that registered on the film during the exposure.

Images are formed by reflection from spherical mirrors as well as by refraction within spherical lenses, and optical instruments in which these elements are utilized have been used for centuries. Sir Isaac Newton invented a reflecting telescope in which the primary image is formed by a spherical mirror, and the largest astronomical telescopes even now are invariably reflecting instruments. Primary mirrors for astronomical telescopes over 5 meters (200 inches) in diameter have been made. The mirrors for modern reflecting telescopes are invariably paraboloidal rather than spherical in form since paraboloidal reflecting surfaces form more perfect images of objects on or near the telescope axis at very large distances. This represents one of the rare instances in which a non-spherical surface is the rule rather than the exception.

In this chapter, we shall study the formation of images by pinholes, spherical lenses, and spherical mirrors. We shall also show how individual optical elements may be combined to construct useful optical instruments such as camera lenses and telescopes. Finally, we shall discuss in elementary terms some of the common imperfections and aberrations that afflict lenses and mirrors. In doing this, we shall concentrate on plane and spherical refracting and reflecting surfaces and say little, if anything, about aspherical elements.

25.2 Images Formed by Pinholes

There are two optical elements of particular simplicity. One of these, the plane mirror, we have encountered in the preceding chapter. The plane mirror's most notable characteristic is that it is the only known optical element that forms a perfect image of any object in its vicinity, completely free of any imperfections or aberrations, irrespective of where the object is located with respect to the mirror. Unfortunately, the image formed by a plane mirror is always the same size as the object, and this severely restricts its usefulness in practical applications. Nevertheless, there are situations, the single-lens reflex camera being one of them, in which the plane mirror can be especially useful. The other extremely simple optical element is the pinhole, which consists of nothing more than a tiny hole in an otherwise opaque surface. In sharp contrast to the plane mirror, the images formed by pinholes are far from perfect. Nevertheless, pinholes illustrate some of the basic facts about optical imaging that apply also to more complex systems. In addition, they have some unique properties of their own, and comparing their properties with those of lenses and mirrors affords us some very entertaining insights into the behavior of these more complex elements.

For a number of years now, zoom lenses, which can be transformed continuously from telephoto lenses through the normal magnification range to wide-angle lenses, by moving a single control on the lens mount, have been the fashion among amateur photographers. These lenses are extremely useful for motion picture photography, though much less so for still cameras. Still, they are readily usable with single-lens reflex cameras, wherein the field of view is always displayed in the finder just as it will appear on the film. Their manufacturers, therefore, have succeeded in selling large numbers of them—at fancy prices—to amateur photographers as well as to their professional counterparts, some of whom actually find certain legitimate ways to use them.

Suppose, then, a salesman came to you with a lens that can provide any image size at all, from extreme telephoto to a wide-angle view embracing 180° and one, moreover, that works at a fixed aperture setting and requires no focusing whatever, at a price of less than one dollar. Would you buy it? If you did, you would be in for a severe disappointment because,

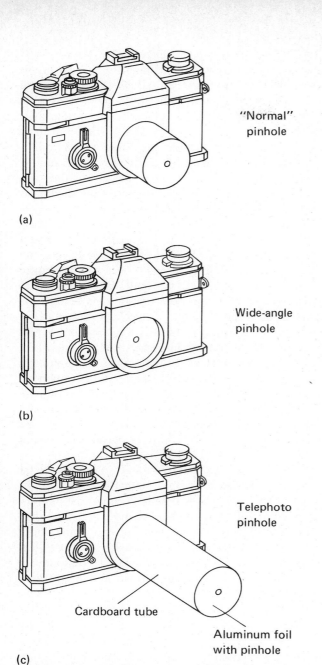

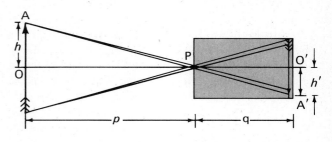

"Normal"
pinhole

(a)

Wide-angle
pinhole

(b)

Telephoto
pinhole

Cardboard tube

Aluminum foil
with pinhole

(c)

FIGURE 25.6. "Normal," wide-angle, and telephoto pinholes.

though it would do everything that is claimed, you would find it less than useless as a practical photographic tool.

You have probably guessed that the optic we have in mind is the pinhole, and you are right. That it will, in fact, do everything we were told it will do can easily be verified by replacing the lens of your single-lens reflex or light box with a piece of aluminum foil in which a tiny hole a few tenths of a millimeter in diameter has been made with a sharp needle and observing the image in the finder. The use of the wide-angle, normal, and telephoto pinhole is illus-

trated in Fig. 25.6. At the same time, you will observe why the pinhole is useless as a practical optical device, for the images you see will be very dim and also not very sharp. Alas, the pinhole is an excruciatingly "slow" lens—it admits so little light to the camera it takes practically all day to make an exposure. And then, what is even more exasperating, when the picture is developed it is not sharp and clear but blurred and fuzzy. You may also verify that the sharpness of a pinhole image can be improved—at the expense of even dimmer images and longer exposures—by using pinholes that are smaller and smaller. Unfortunately, however, this works only up to a certain point, at which the images are still not very sharp. Beyond this point, using a smaller pinhole makes the image fuzzier rather than sharper. The reason for this is that if the pinhole is sufficiently small, the effects of *diffraction* act to bend the rays as they enter the hole and spread them out to form a blurred spot rather than a sharply defined disc the exact size of the hole. We shall investigate this effect quantitatively in an example in the next chapter and determine just what size the pinhole should have for the sharpest possible image, but for the moment we shall not discuss this aspect of the pinhole's behavior any further.

The way in which pinholes form images can be understood with reference to Figs. 25.1b and 25.7. The case where the object is an infinitely distant point source on the axis of the pinhole is shown in Fig. 25.1b.

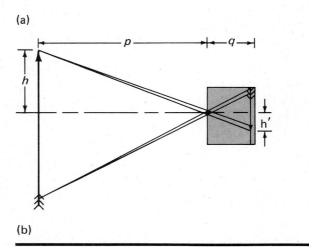

FIGURE 25.7. Geometry of pinhole images.

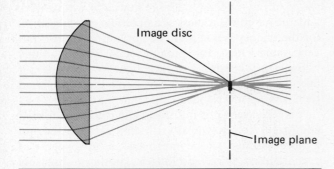

FIGURE 25.8. Path of rays traveling through different parts of a simple lens, showing the effect of spherical aberration in producing a small image disc of finite size rather than a perfect point image.

Since the object is infinitely far away, the light rays coming from it are parallel. Clearly, all the pinhole does is block off all but a tiny column of light whose diameter is that of the pinhole, which forms a spot of this same size on the ground glass or film. It is this spot that we recognize as the "image" of the source. Since the pinhole is of small but finite size, the image of the point source is not really a point at all but a small disc of the same diameter as the pinhole, and this is the reason why the images formed by pinholes are blurred and diffuse rather than sharp and clear. In the true sense of the word, these pinhole "images" *are not true images at all*, because *different light rays given off by the source object do not converge there*, as they do, for example, in the case of the images formed by plane mirrors as illustrated by Fig. 24.6 or by spherical lenses as illustrated by Fig. 25.5. *It is this convergence of different rays from the same source point, generally speaking, that defines what we mean by the term "image."*

So, the image formed by a pinhole, despite the fact that we can see it on a viewing screen, is strictly speaking, not an image at all. Nevertheless, we shall learn something about the behavior of rays in optical systems that do form true images by studying the pseudoimages formed by pinholes. The images formed by lenses, though they are much brighter and sharper

suffer from the same deficiency and, if we are really insistent on preserving the linguistic purity of the term "image," must be regarded as pseudoimages for the same reason, namely, that different rays from an object point are not focused by any lens onto a single point but, instead, into a tiny disc of small but finite size. In the case of well-designed and well-manufactured lenses, these image discs are very much smaller and brighter than any pinhole image, but they are, nevertheless, not mathematical points; and the image, beautiful as it may appear on the viewing screen of your single-lens reflex camera, is never *perfectly* sharp. One way in which such image imperfections arise is illustrated in Fig. 25.8. Collectively, the effects that lead to imaging imperfections of this sort are called *optical aberrations*, and we shall have occasion to describe the more important ones in detail later. You may recall that the only optical element that is free from these aberrations and produces a pattern of rays that converge upon (or, more accurately, whose projections diverge from) a mathematical image point is a plane reflecting surface. We shall not insist upon the fine semantic distinction regarding the term "image" brought out above but will, instead, use it in a broader sense for lenses, spherical mirrors, and pinholes, wherein rays are brought only approximately to focus at a point.

For an object point that is not infinitely far from the pinhole, it is clear from Fig. 25.7 that the size of the image spot will be larger than the pinhole, the picture, therefore, being even more indistinct than when the object is at "infinity." For a point object on the axis of the pinhole, it is easily seen from Fig. 25.9 that if the distance from object to pinhole is p and the distance between pinhole and image spot is q, then

$$\tan \theta = \frac{a}{p} = \frac{r}{p + q} \qquad (25.2.1)$$

where a is the pinhole radius and r the image radius. From this, it is easy to calculate the image spot size:

$$r = a\left(1 + \frac{q}{p}\right) \qquad (25.2.2)$$

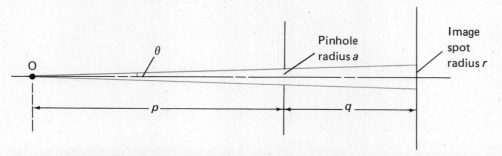

FIGURE 25.9. Formation of a pinhole image.

For an object point that is infinitely distant, $p = \infty$, and (25.2.2) reduces to $r = a$, in agreement with our previous assertion that the image spot size and pinhole size are the same for very distant object points. It is also apparent from Fig. 25.7 that images formed by pinholes are always *inverted*.

The use of a pinhole camera in the telephoto and wide-angle modes is illustrated by Figs. 25.7a and 25.7b. In the case where the angle between the incoming rays from the object make an appreciable angle with the pinhole axis, as the rays from the head and tail of the object at 25.7b do, the image spots will be elliptical rather than circular, and Eq. (25.2.2) will no longer accurately describe their size. The spot size in this case can be calculated easily enough by an extension of the methods used above, but we shall not bother to go through the calculations. The following examples will serve to illustrate some other important points about pinholes and optical images.

EXAMPLE 25.2.1

Explain the practical criteria which lead us to characterize an optically produced image as "sharp" or "fuzzy."

We may begin by noting that persons having normal vision can just distinguish points that have an angular separation of 1 to 3 minutes of arc as two separate entities rather than a single source. This in itself establishes the visual criterion on which the performance of telescopes or binoculars may be judged. For example, if we wish to design an astronomical telescope through which we wish to see a double star system whose two members have an apparent angular separation of 0.1 second, or 1/600 minutes, of arc, it is evident that it will have to achieve at least a 600-fold (600×) angular magnification, preferably two or three times this figure.

In the case of an image produced photographically, or by any other means that results in a picture of normal size on a sheet of paper, it is necessary to recognize that the normal distance between the picture and the observer's eye under ordinary viewing circumstances is about 25 cm. At this distance, image points on the picture, which to the observer's eye have an angular separation of 3 minutes of arc or less, generally cannot be seen as individual entities, while those separated by an angle greater than that will be seen as two separate points. Consider the case of a picture made up of image spots resulting from the use of imperfect optical elements, such as pinholes or lenses that are not free of aberrations. If these spots have an angular diameter to the viewer of less than about 3 minutes of arc, image points separated by *less* than that angle will be rendered on the photo as separate spots, though these will not be discernible as such to the viewer. Under these circumstances, the

image will have details that are finer than those the eyes of the viewer can distinguish, and he will conclude that the photograph is "sharp." On the other hand, if the image discs corresponding to source points on the object have an apparent diameter *larger* than about 3 minutes of arc, distinct image points separated by *more* than that angle will be rendered as a single spot. Since the observer's eyes would have resolved two such points as separate images had he been observing the actual object rather than a photograph of it, he concludes that the picture has less apparent fine detail than he associates with the object itself, and it, therefore, seems to be "fuzzy."

The actual diameter of an image spot that subtends an angle $\theta = 3.0$ minutes of arc to an observer whose eye is at a distance $r = 25.0$ cm away will be

$$d = r\theta = (25)\left(\frac{3}{60}\right)\left(\frac{1}{57.30}\right) = 0.0218 \text{ cm} = 0.218 \text{ mm}$$

since 1 radian = 57.30°. If an optical system is to produce a "sharp-looking" photograph, it must render point images as discs no larger than this in diameter. If the picture is a 20×25 cm (8 in. × 10 in.) enlargement from a 35-mm negative 2.4 cm wide, the negative must be enlarged at least $20/2.4 = 8.33$ times, and the image discs on the negative must be no more than $0.218/8.33 = 0.0262$ mm in diameter. The reciprocal of 0.0262 is about 38, so this would correspond to 38 separately resolved point images or parallel-line images per millimeter on the negative, a fact often conveyed in terminology such as "image resolution on the negative of 38 lines per millimeter." Optics, and camera-handling and processing techniques, that yield resolution of point images that are this good will produce prints that appear to be quite sharp, while those whose resolving power is less will give photographs that are noticeably blurred.

EXAMPLE 25.2.2

What is the *image magnification* of a pinhole camera such as that shown in Fig. 25.7?

The image magnification M is the ratio of the image size h' to the object size h shown in Fig. 25.7. Since the triangles OAP and O'A'P are similar, the ratios of their corresponding parts must be the same. Therefore, we may write

$$\frac{h}{p} = \frac{h'}{q} \tag{25.2.3}$$

or

$$M = \frac{h'}{h} = \frac{q}{p} \tag{25.2.4}$$

For a pinhole camera in which the film is 8 inches behind the pinhole and an object at a distance of 12 feet

803

from the front surface of the pinhole, the magnification will be

$$M = \frac{q}{p} = \frac{\frac{2}{3}}{12} = 0.0555$$

This means that the size of the image h' is 0.0555 times the size of the object. If the object is a person 6 feet tall, the image on the film will be of size

$$h' = Mh = 0.333 \text{ ft} = 4.00 \text{ in.}$$

25.3 Image Formation by Spherical Refracting Surfaces

In the previous section, we outlined the deficiencies of pinhole images in some detail. These deficiencies are serious ones, and we are fortunate that we do not have to rely upon pinhole optical systems to make photographs or to produce magnified images.

Simple spherical lenses, though far from perfect, form images that are vastly better in almost every way than those formed by pinholes. They are, in particular, very much brighter and very much sharper than pinhole images. We do not know who invented the lens, for the use of lenses goes back at least to the time of the Roman Empire and possibly even further. It is clear, however, that the lens is one of the greatest of mankind's early inventions and that its development marked the beginning of the science of optics. Lenses were invented and used long before anyone knew how they work, and, in fact, rather good spectacles and telescopes had been made before Snell's law of refraction had been discovered. The lens, then, provides an example in which technology outruns science and in which the invention is made and put into practice long before anyone really understands the fundamental principles involved in its operation. Another closely related example is the photographic process itself, in which light-sensitive silver salts suspended in a gelatin-based emulsion are exposed momentarily to produce an invisible *latent image* that may then be developed using exotic chemical reducing agents. This process was invented over a century ago, more or less by accident, and developed initially by cut-and-try methods. Later development, in which scientific understanding was systematically brought to bear on the subject, resulted in a highly sophisticated and very effective way of storing and reproducing extremely detailed optical images, either in black and white or in full color. But to this day, the precise details of the process by which light is absorbed by silver halide crystals and the reasons why these light-struck crystals, and they alone, are subsequently developed by chemical reducing agents are not clearly understood! Photography, then, like the invention of the lens itself, is an instance in which technological understanding was obtained and put into everyday practice long before anyone could understand what was happening on the basis of fundamental science.

Even though fundamental scientific knowledge has suggested countless practical technological applications, there are many examples like these where technology has arisen, initially at least, without too much help from science. Even in these instances, however, it is usually true that science can contribute a great deal to the subsequent development of the technology, and it is very clear that this is what has occurred in the case of lenses and the photographic process. Still, it is interesting to speculate about what kind of process might have been developed by the straightforward use of fundamental science had photography not have arisen as it did. Quite possibly, the physicists would have given us some sort of electrostatically based scheme like the familiar Xerox process, which would have done the job effectively enough but which might have required a lot of exotic, bulky, and expensive apparatus. The actual photographic process in everyday use is not pure optics nor pure chemistry,

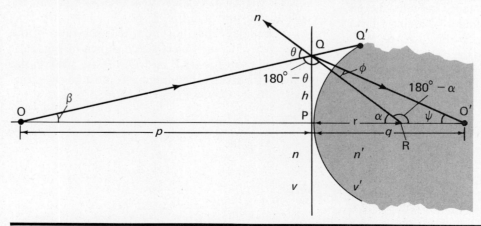

FIGURE 25.10. Geometry of refraction at a single spherical refracting surface.

but an extremely complex mixture of a number of different branches of science and technology. This is probably one of the reasons that its origins do not stem from basic science alone.

In any event, the discovery of Snell's law provided the key to understanding how lenses work and, in addition, furnished the understanding necessary to improve the relatively poor image quality that had been associated with lenses then in use. The application of Snell's law to lenses is best understood by first considering the example of a single *spherical* refracting interface separating two refracting media of refractive indices n and n', as illustrated in Fig. 25.10. The surface of the refracting interface has a radius of curvature r, and the substance having the higher refractive index n' is assumed to extend for some distance to the right of the interface, so that the image O' of the object at O is formed *within* this highly refractive substance. This is not a very practical or realistic case, but it is the simplest one imaginable involving a spherical refracting surface, and it is one in which the bare essentials are not unduly obscured by complicated geometric details. Even in this simple example, however, the mathematics involved in locating the exact path of the refracted ray and the exact position of the image point O' is not at all simple, and we shall find it necessary to make certain simplifying approximations to obtain a reasonably elementary, understandable result. Fortunately, as we shall soon see, the approximations are reasonable ones, and the results will be very accurate in many cases and will serve at least as a point of departure in those instances where they are not.

In Fig. 25.10, an incident ray OQ emitted from a point object at O, at an angle β to the axis OPR, that connects the object point and the center of the spherical refracting surface at R. This incident ray is actually refracted at the point Q' on the spherical surface. To simplify our calculations, however, we shall assume that the refraction takes place *not at* Q' *but at the point* Q on a plane normal to the axis OPR tangent to the surface at P. This is, of course, not a very good assumption unless the points Q and Q' are very close together. Indeed, for the ray OQQ' shown in Fig. 25.10, these two points are rather far apart, and the assumption that refraction occurs at Q rather than Q' would not be very good; the figure, in fact, is drawn to show a ray for which these two points are rather widely separated. On the other hand, if we consider only rays for which the angle α in Fig. 25.10 is a small angle, then, for these rays, the points Q and Q' will be very close together, and we can safely neglect the distance QQ' and consider the refraction to take place at Q. This means that our results will be good only for refracting surfaces whose height h is rather small compared to the radius of curvature r,

since the ratio h/r is, after all, the tangent of α. In Fig. 25.10, then, we might consider covering that part of the surface for which h exceeds perhaps 0.2 times r with black paint, just to be on the safe side. If the angle α is small, it is easy to see that the angles β, θ, ϕ, and ψ must also be small angles.

The line RQ is the normal to the spherical surface, and, therefore, the angle θ is the angle of incidence while ϕ is the angle of refraction. Since both these angles are small, and since for small angles the sine of the angle is very nearly equal to the angle itself, expressed in radians, we can write Snell's law as

$$\frac{\sin \theta}{\sin \phi} = \frac{\theta}{\phi} = n_r \qquad (25.3.1)$$

where

$$n_r = \frac{n'}{n} \qquad (25.3.2)$$

is the relative index of refraction for a ray incident on the interface from the left. Denoting the distance OP of the object from the refracting surface as the *object distance* p and the distance OQ from the refracting surface to the point where the refracted ray once more converges upon the axis OPRO' as the *image distance* q, we may also write

$$h = r \tan \alpha = p \tan \beta = q \tan \psi \qquad (25.3.3)$$

Since α, β, and ψ are small angles, however, and since the tangent of a small angle is very nearly equal to the angle itself, in radians, this can be expressed as

$$h = r\alpha = p\beta = q\psi \qquad (25.3.4)$$

Also, in triangle O'QR, since the sum of the three angles is 180°, we may write

$$\psi + \phi + (180° - \alpha) = 180° \qquad (25.3.5)$$

or

$$\psi = \alpha - \phi \qquad (25.3.6)$$

In the same way, in triangle OQR,

$$\alpha + \beta + (180° - \theta) = 180° \qquad (25.3.7)$$

or

$$\beta = \theta - \alpha \qquad (25.3.8)$$

Substituting the values so obtained for ψ and β into (25.3.4), we find

$$h = r\alpha = p(\theta - \alpha) = q(\alpha - \phi) \qquad (25.3.9)$$

From (25.3.9), we have

$$r\alpha = p(\theta - \alpha) \qquad (25.3.10)$$

Solving for θ, this becomes

$$\theta = \alpha\left(1 + \frac{r}{p}\right) \qquad (25.3.11)$$

805

Equation (25.3.9) also tells us that

$$q(\alpha - \phi) = r\alpha \qquad (25.3.12)$$

And since ϕ can be expressed in terms of θ by Snell's law (25.3.1), this can be written as

$$r\alpha = q\left(\alpha - \frac{\theta}{n_r}\right) \qquad (25.3.13)$$

Substituting the value of θ given by (25.3.11) into (25.3.13) and dividing both sides of the resulting equation by α, we find

$$r = q - \frac{q}{n_r} - \frac{qr}{pn_r} \qquad (25.3.14)$$

Finally, multiplying both sides of this equation by n_r/qr, we find

$$\boxed{\frac{1}{p} + \frac{n_r}{q} = \frac{n_r - 1}{r}} \qquad (25.3.15)$$

This is the characteristic equation relating the object distance p and the image distance q for a single, spherical refracting surface. It is important to note that (25.3.15) does *not* contain the angles α, β, ψ, θ, or ϕ at all. The result is, therefore, *independent* of all these angles, as long as the approximation that they are all small angles is valid and as long as distance QQ' is negligible. This means that any ray emerging from the image point O at *any small angle β* will cross the axis again at essentially the *same* distance q from the refracting surface. In other words, all such rays will come together, or converge, forming an *image* of the object point at a distance q from the refracting surface, as illustrated in Fig. 25.11. If the angle α becomes too large, the approximations we made break down, and under these circumstances, the image distance q for larger values of α depends on α. The outer rays, then, do not cross the axis at quite the same distance from the refracting surface as the inner ones that are nearer the axis. We then run into the situation illustrated in Fig. 25.8, wherein the image is no longer a point but a smeared out disc or spot of finite size. We say that the lens, operated in this way, has developed an imperfection, or *aberration*, that pre-

vents it from imaging a point object into an infinitesimally small point. This particular type of aberration is referred to as *spherical aberration*. It is clear that spherical aberration is more or less negligible so long as only rays for which the angle α in Fig. 25.10 is small are allowed to enter the refracting surface. When rays for which this angle is large are allowed to enter, however, spherical aberration may become large enough to seriously degrade the sharpness of the image that is formed.

You will now be dismayed to learn that we intend to inflict the preceding derivation on you all over again, but this time taking a different point of view. We do not do this just to be sadistic; on the contrary, we hope to be able to show you a fairly easy way of obtaining (25.3.15), this time starting with Fermat's principle of least time. This derivation will afford us certain insights not evident from the geometric calculation just completed and will leave us with a much clearer understanding of the approximations that are involved in obtaining Eq. (25.3.15).

Referring to Fig. 25.12, Fermat's principle tells us that a ray such as OQO' travels between the point O and O' by the path that corresponds to the minimum possible transit time t. For a path such as OQO', transit time t is

$$t = \frac{OQ}{v} + \frac{QO'}{v'} = \frac{1}{v}(OQ + n_r QO') \qquad (25.3.16)$$

where, as usual

$$n_r = \frac{v}{v'} \qquad (25.3.17)$$

From Fig. 25.12, it is easily seen that the distance BA can be written

$$BA = r - r\cos\alpha = r(1 - \cos\alpha) \qquad (25.3.18)$$

If the angle α is small, however, $\cos\alpha \simeq 1 - \frac{1}{2}\alpha^2$, and then (25.3.18) becomes

$$BA \simeq \frac{r\alpha^2}{2} \qquad (25.3.19)$$

Also,

$$OQ = \sqrt{r^2\sin^2\alpha + (p + BA)^2}$$

or

$$OQ = \sqrt{r^2\alpha^2 + \left(p + \frac{r\alpha^2}{2}\right)^2} \qquad (25.3.20)$$

since for small angles $\sin^2\alpha \cong \alpha^2$. In the same way,

$$QO' = \sqrt{r^2\sin^2\alpha + (q - BA)^2}$$

$$= \sqrt{r^2\alpha^2 + \left(q - \frac{r\alpha^2}{2}\right)^2} \qquad (25.3.21)$$

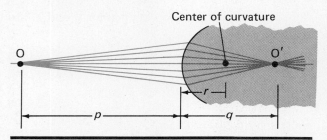

Center of curvature

FIGURE 25.11. Pattern of rays refracted by a single spherical refracting interface.

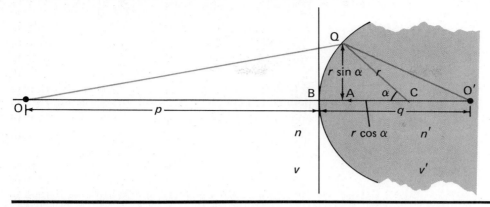

FIGURE 25.12. Calculation of the imaging equation of a spherical refracting surface using Fermat's principle of least time.

Substituting (25.3.20) and (25.3.21) into (25.3.16), we find that

$$vt = \sqrt{r^2\alpha^2 + \left(p + \frac{r\alpha^2}{2}\right)^2} + n_r\sqrt{r^2\alpha^2 + \left(q - \frac{r\alpha^2}{2}\right)^2}$$

(25.3.22)

If the ray OQO' is to be a minimum transit time path, as required by Fermat's principle, then the transit time for OQO' must be smaller than that of other paths, corresponding to different values of α. For a minimum transit time ray, then, the derivative of the transit time with respect to α must vanish. Differentiating (25.3.22) with respect to α (remembering that the velocity v is constant) and setting the result equal to zero, we get

$$v\frac{dt}{d\alpha} = \frac{r^2\alpha + \left(p + \frac{r\alpha^2}{2}\right)r\alpha}{\sqrt{r^2\alpha^2 + \left(p + \frac{r\alpha^2}{2}\right)^2}} + n_r\frac{r^2\alpha - \left(q - \frac{r\alpha^2}{2}\right)r\alpha}{\sqrt{r^2\alpha^2 + \left(q - \frac{r\alpha^2}{2}\right)^2}}$$

$$= 0 \qquad (25.3.23)$$

or

$$r\alpha\left(\frac{r + p + \frac{r\alpha^2}{2}}{\sqrt{r^2\alpha^2 + \left(p + \frac{r\alpha^2}{2}\right)^2}} + n_r\frac{r - q + \frac{r\alpha^2}{2}}{\sqrt{r^2\alpha^2 + \left(q - \frac{r\alpha^2}{2}\right)^2}}\right) = 0$$

(25.3.24)

One possible though uninformative solution to this equation is, of course, $r\alpha = 0$, from which $\alpha = 0$. Another is obtained when the other quantity in parentheses in the above equation is set equal to zero. When we do this, the resulting expression still looks complicated until we remember that we are restricting ourselves to use only that part of the refracting surface for which the angles α are reasonably small. If this is the case, then the terms containing α in (25.3.24) are much smaller than those that do not contain α

and can, therefore, as an approximation, be dropped altogether. Therefore, when we set this quantity equal to zero and neglect these terms, we find

$$\frac{r + p}{p} = -n_r\frac{r - q}{q} \qquad (25.3.25)$$

Dividing both sides of this equation by r and rearranging terms, however, this can be written

$$\boxed{\frac{1}{p} + \frac{n_r}{q} = \frac{n_r - 1}{r}} \qquad (25.3.26)$$

which is the same as Eq. (25.3.15).

This tells us, in effect, that when the object distance and the image distance are related by (25.3.26), any of the paths shown in Fig. 25.11 is a least-time path, irrespective of the angle α, *so long as that angle is small*. For larger angles, it is no longer possible to neglect the terms in α in (25.3.24) or, for that matter, to make the approximations that $\sin\alpha \cong \alpha$ and $(1 - \cos\alpha) \cong \frac{1}{2}\alpha^2$. Equation (25.3.26) will no longer be valid, and there will be instead a different relation between object and image distance, a relation that contains the angle α as well as the refractive index and the radius of curvature of the surface, For an object at a fixed distance p from the surface, this means that when α becomes large, the refracted ray will cross the axis not at O' but at another point whose exact location depends upon the angle. This leads us back once more to the situation shown in Fig. 25.8 and to the description of spherical aberration that we gave earlier.

A few important points are yet to be made about the imaging characteristics of a refracting surface. First, it is evident that *the object and image can always be interchanged without affecting the paths of the rays that traverse the system.* This is obvious from Fermat's principle, because the transit time associated with any ray, such as OQO' in Fig. 25.12, is exactly the same whether the ray starts at O and proceeds

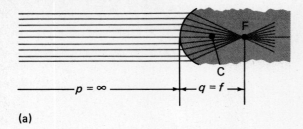

(a)

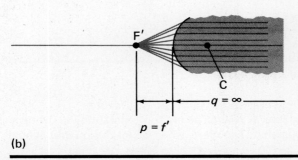

(b)

FIGURE 25.13. Focal points of a spherical refracting surface.

to O′ or starts at O′ and goes to O. In our discussion, we assumed that the rays that were refracted started at O and went from there to O′, but it would have made no difference had we regarded them as traveling in the opposite direction.

Also, it should be noted that the position assumed by the image for an object that is infinitely distant and the position the object must assume to produce an infinitely distant image play a role of special importance in the analysis of any optical system. These two points are referred to as the *focal points* of the system. In the first case, illustrated in Fig. 25.13, the object is infinitely far away along the axis OO′, and, therefore, all the rays emitted by it that reach the lens are parallel to one another and to the axis. In this case, according to (25.3.26), the location of the image is at the focal point F, whose distance f from the refracting surface may be ascertained, by setting $p = \infty$ in the equation, as

$$\frac{n_r}{q} = \frac{n_r - 1}{r}$$

or

$$\frac{1}{q} = \frac{1}{f} = \frac{1}{r}\frac{n_r - 1}{n_r} \qquad (25.3.27)$$

The second focal point, F′, is the point where an object must be placed to produce an image that is infinitely far beyond the refracting surface. In this case, for an object at F′ at a distance f' from the surface, the outgoing rays will converge only at infinity and hence may be regarded as parallel to the optical axis OO′. In this case, the object distance f' can be found

by setting $q = \infty$ in (25.3.26), which gives

$$\frac{1}{p} = \frac{1}{f'} = \frac{n_r - 1}{r} \qquad (25.3.28)$$

It is important to note that for this system, the focal points are *not* equidistant from the refracting surface and that neither one coincides with its center of curvature C.

EXAMPLE 25.3.1

Suppose we replace the image at O′ in Fig. 25.12 with a real object, denoting its distance from the refracting surface by p'. An image of this object will now be formed at the former position of the object at O. Call the distance of this image from the refracting surface q'. By replacing q with p' and p with q', show how the imaging equation is transformed in this case, and show that the new equation is physically equivalent to the old one.

In Eq. (25.3.26), if we replace q by p' and p by q', we obtain

$$\frac{1}{q'} - \frac{n_r}{p'} = \frac{n_r - 1}{r} \qquad (25.3.29)$$

Dividing both sides by n_r, we can write

$$\frac{1}{p'} + \frac{1}{n_r q'} = \frac{1}{r}\left(1 - \frac{1}{n_r}\right) \qquad (25.3.30)$$

In this situation, of course, light traveling from O′ is incident upon the refracting surface from *inside* the more refractive medium and is bent *away* from the normal as it emerges. The relative index that one must use in Snell's law in this event is, as we have seen on a number of previous occasions,

$$n_r' = \frac{n}{n'} = \frac{1}{n_r} \qquad (25.3.31)$$

Replacing $1/n_r$ in (25.3.30) by n_r', we obtain

$$\frac{1}{p'} + \frac{n_r'}{q'} = \frac{1 - n_r'}{r} = \frac{n_r' - 1}{r'} \qquad (25.3.32)$$

where

$$r' = -r \qquad (25.3.33)$$

This is the same as Eq. (25.3.26), except that (1) n_r' replaces n_r, which physically is correct, since the light from the object is incident from within the medium of higher refractive index, and (2) that the radius r is replaced by $-r$. But this also makes sense physically, since now the refracting surface looks concave to the incident light rather than convex and hence has a curvature of opposite sign. This is an important point which we shall refer to frequently in working problems and examples, We can put it most clearly by saying that the radius of curvature of a surface is to be considered as positive for surfaces that tend to make rays incident upon them converge and negative if they tend to make the incident rays less convergent. For the surface in Fig. 25.12, if rays are incident from an object outside at O, the surface tends to bend those rays toward the normal and hence to make them more convergent. In this case—in which we would use Eq. (25.3.26) as our imaging equation—we would take r to be positive. On the other hand, if rays are incident upon the surface from an object inside at O'—in which case (25.3.32) is the proper form for the imaging equation—we would observe that the incident rays are refracted away from the normal and, therefore, made less strongly convergent on passing through the surface. We would now regard the radius of curvature r' in (25.3.32) to be negative.

As a concrete example, suppose $n_r = 1.52$, $r = 12.0$ cm, and $p = 48.0$ cm. From (25.3.26), then,

$$\frac{1}{q} = \frac{n_r - 1}{n_r r} - \frac{1}{n_r p} = \frac{0.52}{(1.52)(12.0)} - \frac{1}{(1.52)(48.0)} = 0.0148$$

$$q = 67.6 \text{ cm}$$

On the other hand, if an object were located inside the refracting medium at $p' = 67.6$ cm, just where the image is in the above calculation, then $n_r' = 1/1.52 = 0.658$ and $r' = -r = -12.0$ cm, the light now being considered as going from O' to O. The imaging equation we must use now, however, is (25.3.32):

$$\frac{1}{q'} = \frac{n_r' - 1}{n_r' r'} - \frac{1}{n_r' p'} = \frac{0.658 - 1.000}{(0.658)(-12.0)} - \frac{1}{(0.658)(67.6)}$$

$$= 0.0208$$

$$q' = 48.0 \text{ cm}$$

which is just where the object was in the former calculation. One could also proceed simply by noting that the positions of object and image can always be interchanged and let it go at that, but this example serves as a formal proof that this observation is in agreement with the mathematical laws of image formation.

EXAMPLE 25.3.2
Find the focal points F and F' for the refracting interface ($n_r = 1.52$, $r = 12.0$ cm) considered in the previous example. Where would the image of a point on the axis 18.0 cm to the left of the refracting surface be formed?

The distance f between the right-hand focal point F and the refracting surface is found by setting $p = \infty$ in (25.3.26), corresponding to an object infinitely far away. This gives for the image distance

$$\frac{1}{q} = \frac{1}{f} = \frac{n_r - 1}{n_r r} = \frac{0.52}{(1.52)(12.0)} = 0.0285$$

$f = 35.1$ cm

The distance f' between the left focal point F' and the surface can be obtained by putting $q = \infty$ in (25.3.26), corresponding to an image at an infinite distance from the surface within the highly refracting medium. In this way, we find

$$\frac{1}{p} = \frac{1}{f'} = \frac{n_r - 1}{r} = 0.0433$$

or

$$f' = 23.1 \text{ cm}$$

For an object located on the axis 18.0 cm to the left of the refracting surface, we have $p = 18.0$ cm. Putting this into (25.3.26), we find

$$\frac{1}{q} = \frac{n_r - 1}{n_r r} - \frac{1}{n_r p} = \frac{0.52}{(1.52)(12.0)} - \frac{1}{(1.52)(18.0)}$$

$$= -0.00804$$

$$q = -124.4 \text{ cm}$$

What does this negative value of q mean? Simply that the image point is on the *same* side of the refracting surface as the object rather than the opposite side. This can be understood with reference to Fig. 25.13. From this diagram, it is evident that the effect of the refracting surface is to make rays that strike it more convergent than they were initially. If all the rays that strike the surface are parallel, then they converge to a point F behind the surface. If they are slightly divergent, as would happen if they came from an object quite far (though not infinitely far) from the surface, they would be converged to a point a bit to the right of F in Fig. 25.13a. For a closer object, the rays striking the refracting surface are more strongly divergent, and though the surface can still make them converge upon a point, they do so less strongly, and the point of focus is even further to the right of F. Finally, for an object at F', the rays from the object diverge so strongly that the refracting surface can do no more than straighten them out, as in Fig. 25.13b. This, of course, is the same as saying that they "converge" to form an image at infinity.

For an object closer to the surface than the point F', the refracting surface can no longer convert the divergent rays striking it into a convergent bundle. All it can do is make them *less divergent*, as shown in Fig. 25.14. This divergent set of rays, projected backward, appears to emanate from an image point O' on the same side of the refracting surface as the object. The situation resembles that which we encounter in the case of a plane mirror, where the reflected rays diverge from a point behind the mirror where the image appears to be. Just as for the plane reflecting surface, the image is highly visible and gives the

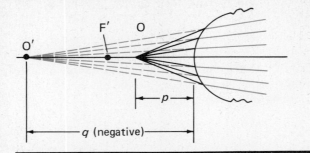

FIGURE 25.14. Pattern of incident and refracted rays for a point object very close to a spherical refracting surface.

illusion of actually being located at O', even though no light rays actually intersect there. The image, therefore, is a *virtual image*, in contrast to the *real images* that occurred in the previous examples in this section, in which light rays actually converged upon the image and in which the image could actually be displayed upon a sheet of white paper or ground glass placed at the image point.

That the image distance q is in this case negative is *not* directly related to the fact that the image is virtual rather than real, however. The negative value of q is a reflection only of the fact that the image is located in front of the refracting surface, on the same side of it as the object, rather than behind it. We shall see that whenever this situation occurs, the image distance q is to be regarded as negative.

25.4 The Thin-Lens Equation and the Imaging Properties of Thin Lenses

In the preceding section, we discussed the imaging properties of a single refracting surface. While this is the simplest possible example of imaging by refraction, it is not a very realistic one, since all refracting elements have to have two surfaces. We must, therefore, develop an imaging equation suitable for simple

lenses having two refracting surfaces. While it is possible to do this using Snell's law for two single refracting surfaces and combining the results of the two separate calculations, it is a very tedious and awkward procedure. We shall instead bypass Snell's law altogether and begin with Fermat's principle of least time. We shall find the calculation to be done is hardly more difficult or complex than that undertaken in the previous section. We shall begin by referring you to Fig. 25.15, in which a thin lens having spherical refracting surfaces of radii r and r' is illustrated.

The time of transit for the ray OQQ'O' can be written as

$$t = \frac{OQ}{v} + \frac{OQ'}{v'} + \frac{Q'O'}{v} \tag{25.4.1}$$

or

$$vt = OQ + n_r QQ' + Q'O' \tag{25.4.2}$$

where v and v' represent the velocity of light outside the lens and within the lens, respectively, and where, as usual, n_r is the relative index of refraction given by

$$n_r = \frac{n'}{n} = \frac{v}{v'} \tag{25.4.3}$$

From Fig. 25.15, it is easily seen that for *small* angles α and α' we can write

$$QA = r \sin \alpha \cong r\alpha \tag{25.4.4}$$

$$Q'A' = r' \sin \alpha' \cong r'\alpha' \tag{25.4.5}$$

while

$$BA = r(1 - \cos \alpha) \cong \frac{r\alpha^2}{2} \tag{25.4.6}$$

$$B'A' = r'(1 - \cos \alpha') \cong \frac{r'\alpha'^2}{2} \tag{25.4.7}$$

For a lens whose thickness d is small in comparison with the radii of curvature r and r', it is reasonable to neglect the difference in the height of

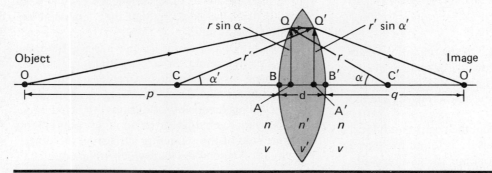

FIGURE 25.15. Derivation of the "thin-lens equation" using Fermat's principle of least time.

the ray at the point of entrance Q and at the exit point Q'. In doing this, we make what is frequently called the *thin lens approximation*, setting

$$QA = Q'A' \qquad (25.4.8)$$

or

$$r\alpha = r'\alpha'$$

$$\alpha' = \frac{r}{r'}\,\alpha \qquad (25.4.9)$$

If we substitute (25.4.9) into (25.4.7), the distance $B'A'$ can be expressed as

$$B'A' = \frac{r^2\alpha^2}{2r'} \qquad (25.4.10)$$

From the figure, it is evident that

$$OQ = \sqrt{(QA)^2 + (p + BA)^2} = \sqrt{r^2\alpha^2 + \left(p + \frac{r\alpha^2}{2}\right)^2} \qquad (25.4.11)$$

$$QQ' = d - BA - B'A' = d - \frac{r\alpha^2}{2} - \frac{r^2\alpha^2}{2r'} \qquad (25.4.12)$$

$$Q'O' = \sqrt{(Q'A')^2 + (q + B'A')^2}$$

$$= \sqrt{r^2\alpha^2 + \left(q + \frac{r^2\alpha^2}{2r'}\right)^2} \qquad (25.4.13)$$

If the ray OQQ'O' is to be a path of minimum transit time, then from Fermat's principle, $dt/d\alpha = 0$. Substituting (25.4.11), (25.4.12), and (25.4.13) into (25.4.2) and differentiating with respect to α, remembering that v is a constant, we find

$$v\frac{dt}{d\alpha} = \frac{d}{d\alpha}\sqrt{r^2\alpha^2 + \left(p + \frac{r\alpha^2}{2}\right)^2} + n_r \frac{d}{d\alpha}\left(d - \frac{r\alpha^2}{2} - \frac{r^2\alpha^2}{2r'}\right)$$

$$+ \frac{d}{d\alpha}\sqrt{r^2\alpha^2 + \left(q + \frac{r^2\alpha^2}{2r'}\right)^2}$$

$$= 0 \qquad (25.4.14)$$

or

$$\frac{r^2\alpha + \left(p + \dfrac{r\alpha^2}{2}\right)r\alpha}{\sqrt{r^2\alpha^2 + \left(p + \dfrac{r\alpha^2}{2}\right)^2}} - n_r r\alpha\left(1 + \frac{r}{r'}\right)$$

$$+ \frac{r^2\alpha + \left(q + \dfrac{r^2\alpha^2}{2r'}\right)\dfrac{r^2\alpha}{r'}}{\sqrt{r^2\alpha^2 + \left(q + \dfrac{r^2\alpha^2}{2r'}\right)^2}} = 0 \qquad (25.4.15)$$

Dividing both sides of this equation by $r\alpha$, we obtain

$$\frac{r + p + \dfrac{r\alpha^2}{2}}{\sqrt{r^2\alpha^2 + \left(p + \dfrac{r\alpha^2}{2}\right)^2}} - n_r\left(1 + \frac{r}{r'}\right)$$

$$+ \frac{r + \left(q + \dfrac{r^2\alpha^2}{2r'}\right)\dfrac{r}{r'}}{\sqrt{r^2\alpha^2 + \left(q + \dfrac{r^2\alpha^2}{2r'}\right)^2}} = 0$$

As before, we shall restrict ourselves to lenses for which the angle α is quite small, that is, lenses whose outer radius is small compared to the radii of curvature of their surfaces. In this case, the terms that contain α in the above equation will be negligible in comparison with the others, and (25.4.15) can then be written as

$$\frac{r + p}{p} - n_r - n_r\frac{r}{r'} + \frac{r + q(r/r')}{q} = 0 \qquad (25.4.16)$$

Dividing both sides of this by r, we get

$$\frac{1}{p} + \frac{1}{q} = \frac{n_r}{r} - \frac{1}{r} + \frac{n_r}{r'} - \frac{1}{r'}$$

or, finally,

$$\boxed{\frac{1}{p} + \frac{1}{q} = (n_r - 1)\left(\frac{1}{r} + \frac{1}{r'}\right)} \qquad (25.4.17)$$

This is the imaging equation of a thin lens, sometimes referred to as the *thin-lens equation*.

As before, the existence of small terms in α^2 in (25.4.15), neglected in arriving at (25.4.17), implies that there are really slight differences in the approximately equal transit times corresponding to different values of α and that these differences become important as α becomes large. This gives rise to spherical aberration of the same sort that is associated with single refracting surfaces and illustrated in Fig. 25.8.

The focal points F and F' associated with a thin lens can be found, as with a single refracting surface, by setting $p = \infty$ and finding the resulting image distance f, and by setting $q = \infty$ and finding the image distance f'. In *both* instances, we find the same focal distance given by

$$\boxed{\frac{1}{f} = \frac{1}{f'} = (n_r - 1)\left(\frac{1}{r} + \frac{1}{r'}\right)} \qquad (25.4.18)$$

The situation is illustrated in Fig. 25.16. For a thin lens, then (or, for that matter, any lens in a medium of uniform refractive index), the focal distances f and f' are *equal*. Their common value, denoted simply by f, is referred to as the *focal length* of the lens, and the

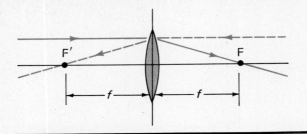

FIGURE 25.16. Focal points of a thin lens.

thin-lens equation can be written very simply, using the focal length, as

$$\frac{1}{p} + \frac{1}{q} = \frac{1}{f} \qquad (25.4.19)$$

In this equation, p is assumed to be inherently positive, while a positive image distance q is obtained when the image is on the opposite side of the lens from the object and a negative image distance when the image and object are on the same side. The focal length f is positive for any lens that is convergent, as a whole, and negative for one that, in toto, is divergent. By a convergent lens, in this sense, we mean one that converts parallel light rays into a set of rays that converge to a point, while a divergent one converts parallel light rays into a bundle that diverges outward away from a "virtual" image point on the opposite side of the lens.

EXAMPLE 25.4.1

A double convex lens, such as the one shown in Fig. 25.15, has radii of curvature $r = 15$ cm and $r' = 25$ cm. It is made of glass whose refractive index in air is 1.520. What is the focal length of this lens in air? What would the focal length be if it were immersed in water, refractive index 1.333? What would the focal length be if it were immersed in carbon disulfide, refractive index 1.628? For the lens in air, find the position of the image for an object at infinity, at a distance of 20 meters, at a distance of 2 meters, at a distance of 20 cm, and at a distance of 15 cm.

In this case, both surfaces of the lens act to bend incident rays *toward* the optical axis, and hence both surfaces act as *converging* refracting surfaces. Therefore, both radii of curvature are positive. When the lens is in air, the relative index is $n_r = 1.520$, and, therefore, from (25.4.18)

$$\frac{1}{f} = (1.520 - 1.000)\left(\frac{1}{15} + \frac{1}{25}\right) = (0.520)\left(\frac{8}{75}\right)$$

$$= 0.0554 \text{ cm}^{-1}$$

$$f = 18.03 \text{ cm}$$

If the lens is immersed in water, for which the refractive index is $n = 1.333$, the relative index is

$$n_r = \frac{n'}{n} = \frac{1.520}{1.333} = 1.140$$

The focal length as given by (25.4.18) is then

$$\frac{1}{f} = (1.140 - 1.000)\left(\frac{1}{15} + \frac{1}{25}\right) = 0.01496 \text{ cm}^{-1}$$

$$f = 66.8 \text{ cm}$$

If the lens is immersed in carbon disulfide, for which $n = 1.628$, the relative index is

$$n_r = \frac{n'}{n} = \frac{1.520}{1.628} = 0.9337$$

and (25.4.18) gives

$$\frac{1}{f} = (0.9337 - 1.000)\left(\frac{1}{15} + \frac{1}{25}\right) = -0.00708 \text{ cm}^{-1}$$

$$f = -141.3 \text{ cm}$$

That f is now negative means that the lens under these circumstances becomes divergent. The reason for this is that since the medium is now more highly refractive than the lens that is immersed in it, the rays incident upon it are bent *away* from the normal upon entering the lens and *toward* it upon leaving, thus exerting a *diverging* action upon these rays, as illustrated in Fig. 25.17a. A convex *air lens*, such as one might make by joining together and sealing the rims of two thin watch glasses, also acts as a divergent lens when immersed in water for the same reason. Conversely, a doubly concave air lens such as the one shown in Fig. 25.17b acts as a convergent lens, in contrast to the diverging action of a doubly concave glass lens immersed in air.

In the case where the lens discussed above is in air, the focal length is 18.03 cm. In this instance, for an object at infinity, $1/p$ is zero, and (25.4.19) gives $1/q = 1/f$, or

$$q = f = 18.03 \text{ cm}$$

For an object at a distance $p = 20$ meters $= 2000$ cm from the lens, we find from (25.4.19)

$$\frac{1}{q} = \frac{1}{f} - \frac{1}{p} = \frac{1}{18.03} - \frac{1}{2000} = 0.05496 \text{ cm}^{-1}$$

$$q = 18.19 \text{ cm}$$

As the object moves from infinity to a point 20 meters distant from the lens, the image recedes from the focal point 18.03 cm behind the lens to a point 18.19 cm behind the lens, moving through only the tiny distance of 0.16 cm. When the object is placed at a distance $p = 2.0$ meters $= 200$ cm from the lens, we have

$$\frac{1}{q} = \frac{1}{f} - \frac{1}{p} = \frac{1}{18.03} - \frac{1}{200} = 0.05046 \text{ cm}^{-1}$$

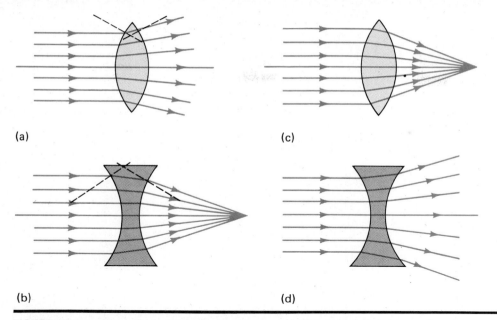

FIGURE 25.17. Refraction of light by (a) convex and (b) concave "air lenses," contrasted with refraction by (c) convex and (d) concave glass lenses in air.

$q = 19.82$ cm

The image is now formed at a distance of 1.79 cm behind the focal point. For an object distance $p = 20.0$ cm

$$\frac{1}{q} = \frac{1}{f} - \frac{1}{p} = \frac{1}{18.03} - \frac{1}{20.0} = 0.005463 \text{ cm}^{-1}$$

$q = 183.0$ cm

In this case, the image is far from the lens, at a distance of some 165 cm beyond the focal point. This comes about because now the object is nearly up to the focal point on its side of the lens, and the converging power of the lens is only barely sufficient to make the diverging beam of rays come together at all behind the lens. The successive stages of this situation are shown in Fig. 25.18. Finally, at the object distance $p = 15.0$ cm, the object has moved inside the focal point on its side of the lens. The rays that it sends toward the lens are so strongly divergent at the lens that, although the lens tries to converge them to an image point on the opposite side, all it can do is to make them *somewhat less divergent* than they were initially. These less divergent rays seem to emanate from a *virtual* image point on the object's side of the lens, as shown in Fig. 25.18e. In this case, the image distance will be

$$\frac{1}{q} = \frac{1}{f} - \frac{1}{p} = \frac{1}{18.03} - \frac{1}{15.0} = -0.01120 \text{ cm}^{-1}$$

$q = -89.3$ cm

the negative sign signifying that the image is on the same side of the lens as the object. In Fig. 25.18 at (a), (b), (c), and (d), the images that form are *real* ones. In these cases, actual light rays converge upon the image points, and it is possible to display such images upon a ground-glass screen or a sheet of paper placed at the image point. In all these cases, if the object were a white, hot incandescent body, the lens could be used as a burning glass. In the case of the virtual image at (e), there are no actual light rays that emanate from, or converge upon, the image point. Of course, you can *see* the image through the lens, just as you can see an image in a plane mirror. But you can never succeed in displaying it on a ground-glass screen placed at the image point because there are no converging light rays there to illuminate it. For the same reason, you could never succeed in using the lens as a burning glass if it were placed in this way with respect to an incandescent hot source object.

EXAMPLE 25.4.2

A number of lenses are shown in Fig. 25.19. Characterize each as converging (positive focal length) or diverging (negative focal length). Discuss the signs of the radii of curvature of each surface.

At (a), we see a double-convex lens. Since both surfaces tend to bend parallel rays from the left toward the axis of the lens, both radii of curvature are positive, and so, according to (25.4.18), is the focal length. The lens as a whole is convergent. In (b) a plano-convex lens is illustrated in which the radius r' is infinite, corresponding to the plane right surface of the lens. The left-hand surface still exerts a converging effect upon parallel rays from the left, while the plane sur-

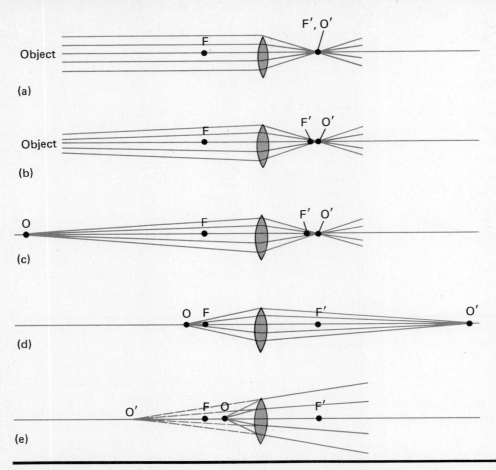

FIGURE 25.18. Image formation for objects at successively smaller distances from a thin converging lens.

face has little or no effect upon them. The radius of curvature r of the converging surface on the left is positive, while for the plane surface $1/r'$ is zero. As a whole, therefore, according to (25.4.18), the focal length is positive and the lens convergent. At (c), we see a double-concave lens. In this case, refraction at both surfaces tends to bend parallel rays from the left away from the axes and hence makes them diverge, as in Fig. 25.19d. Both radii of curvature are, therefore, negative, as is the focal length given by (25.4.18). The lens as a whole is, therefore, divergent. In (d), a plano-concave lens is shown in which the radius of curvature r is negative, while the radius r' is infinite for the plane right-hand surface. The reciprocal of this latter radius is zero, but due to the negative radius of the other surface, the focal length given by (25.4.18) is still negative, and the lens as a whole, therefore, is divergent. In (e), we see a lens having a convex surface of small (positive) radius r and a concave surface of larger (negative) radius r'. Refraction at the former surface tends to converge rays from the left, while refraction at the latter tends to diverge them. However, since $1/r$ is *greater* in magnitude than $1/r'$, the focal length given by (25.4.18) is positive and

the lens as a whole is thus convergent. Such a lens is sometimes referred to as a *converging meniscus* lens. Finally, at (f), we see the opposite case in which the converging surface has a positive radius of curvature r larger in magnitude than the negative radius r' asso-

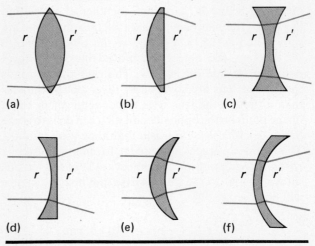

FIGURE 25.19. (a) Double convex, (b) plano-convex, (c) double concave, (d) plano-concave, (e) converging meniscus and (f) diverging meniscus lenses.

ciated with the diverging surface. In this case, the negative quantity $1/r'$ is of larger magnitude than the positive quantity $1/r$, and (25.4.18), therefore, indicates that the focal length of the lens will be negative and the lens as a whole divergent. A lens such as this is referred to as a *diverging meniscus* lens.

EXAMPLE 25.4.3

In Fig. 25.20, we see a photographic lens of 85-mm focal length. This lens is focused by turning a ring A which moves the whole lens out further from the film, on a set of screw threads, as the distance setting is varied from "infinity" to the closest setting. Explain what this accomplishes.

Also, note that the focusing scale is not linear but is constructed according to a peculiar but easily deciphered scheme. In this scheme, there are seven equally spaced marks corresponding to objects at "infinity," 30 meters, $30/2 = 15$ meters, $30/3 = 10$ meters, $30/4 = 7.5$ meters, $30/5 = 6$ meters, and $30/6 = 5$ meters. What is the basis of this scheme, and, in particular, why is it that equally spaced intervals on the focusing scale correspond to a given distance—30 meters in this case—and a series of closer object distances corresponding to $\frac{1}{2}, \frac{1}{3}, \frac{1}{4}, \frac{1}{5}, \frac{1}{6}, \dots$ of this distance?

The lens is mounted on the camera body so that when the lens is set at "infinity," its distance from the film plane is equal to the focal length f, in this case[1] 85 mm. Under these circumstances, according to (25.4.19), the image of an object for which $p = \infty$ coincides exactly with the plane of the film. Rays emitted from various points on the object are brought to precise convergence at corresponding points on the image plane, and a sharply defined image of the scene is formed on the film. In the viewfinder of a single-lens reflex camera, this image is reflected upon a ground-glass viewing screen whose distance from the lens is equal to that of the film plane, and it is evident to the viewer that when the lens is set on "∞," very distant objects are sharply imaged—"in focus," so to speak—upon this screen.

For an object that is closer to the camera, for example, at a distance $p = 30$ meters, the image is formed not at a distance of 85 mm behind the lens but at a slightly greater distance given by (25.4.19) as

---

[1] This statement is certainly true in the case of a "thin" lens. Actual photographic lenses, as illustrated by Fig. 25.1a, however, are frequently far from thin. In the case of "thick" lenses such as these, it is always possible to locate two points within the lens such that when the object distance is measured from the object to one of these points and the image distance from the image plane to the other point, the thin-lens Eq. (25.4.19) relating object and image distance is still obeyed. In this discussion, we assume that object and image distances are defined in this way and, therefore, that for an object at infinity, the image is formed 85 mm behind the point within the lens from which we must measure image distances according to this scheme.

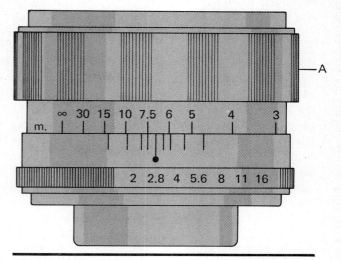

FIGURE 25.20. The "focusing scale" of a photographic lens.

$$\frac{1}{q} = \frac{1}{f} - \frac{1}{p} = \frac{1}{85} - \frac{1}{30,000} = 0.011731 \text{ mm}^{-1}$$

$$q = 85.241 \text{ mm}$$

This image is now focused sharply not on the film plane but on a plane that lies 0.241 mm behind it, as illustrated by Fig. 25.21. What appears on the film is now a small spot or disc of light corresponding to each point on the object rather than a sharp point, and the image is seen to be slightly blurred rather than sharply focused. For an object even closer to the lens, the image lies even further behind the film plane, and its appearance on the film or the ground-glass viewing screen is even more fuzzy and indistinct. This difficulty can easily be overcome by incorporating some arrangement by which the lens-to-film distance can be increased at will. In most cases, this is accomplished simply by mounting the lens barrel in a set of screw threads that enables us to screw the lens in and out simply by turning a milled collar. If the lens is moved

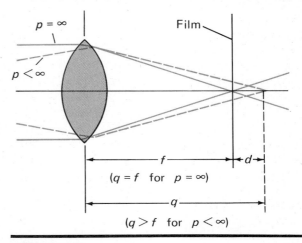

FIGURE 25.21

out a distance of 0.241 mm, by turning the focusing collar until the "30 m" mark is opposite the index, the rays emitted from various object points at object distances of 30 meters, as illustrated by the dotted lines in Fig. 25.21, will converge to a focus at the plane of the film, which, of course, is now 85.24 mm behind the lens. At the same time, rays from infinitely distant objects now converge at a point 0.241 mm in front of the film and diverge again beyond that point. These rays now form images of objects at infinity that are slightly blurred rather than sharply defined. Rays from objects closer than 30 meters still come to a focus behind the film plane, but since the lens is now further from the film than before, their point of focus is closer to the film, and they will appear to be sharper than before, even though not in good focus. As the lens is extended even more by turning the focusing collar further, images of objects at smaller and smaller distances come into sharp focus, while more distant ones get more and more blurred. These effects are easily observed through the finder of a single-lens reflex camera or with the viewing arrangement illustrated in Fig. 25.2.

Let us now try to find how far from the infinity setting the lens must be extended to bring an object at a given object distance q into precise focus on the film. At the "infinity" setting, the distance from the lens to the film plane equals the focal length f, as discussed above. For an object at a smaller object distance p, the image will be formed at a distance q behind the lens that is somewhat larger than f, and the lens must be extended through the distance $d = q - f$ shown in Fig. 25.21 to bring it into precise focus on the film. But since p and q are always related by the thin-lens equation (25.4.19),

$$\frac{1}{q} = \frac{1}{f} - \frac{1}{p} = \frac{p - f}{pf}$$

$$q = \frac{pf}{p - f} \tag{25.4.20}$$

Subtracting f from both sides of this equation, we find

$$d = q - f = \frac{pf}{p - f} - \frac{f(p - f)}{p - f}$$

$$d = \frac{f^2}{p - f} \tag{25.4.21}$$

This expression tells us how far we must extend a lens of focal length f beyond its infinity setting to focus it on an object at a distance p from the lens. Of course, for $p = \infty$, Eq. (25.4.21) gives $d = 0$, as we might expect. If the object distance is twice the focal length of the lens, then $p = 2f$ and, from (25.4.21), $d = f$. To focus a lens on an object at a distance of twice the

focal length, then, the lens must be moved out a distance equal to its focal length.

In the example we are asked to explain, all the object distances involved—∞ and 30, 15, 7.5, 6, and 5 meters—are very much larger than the 85-mm focal length. In all these cases, therefore, p is much larger than f, and in the denominator of the fraction in (25.4.21) we may replace $p - f$ with p without introducing any significant error, so that (25.4.21) reduces to

$$d \cong \frac{f^2}{p} \qquad \text{for } p \gg f \tag{25.4.22}$$

which can equally well be written

$$p \cong \frac{f^2}{d} \tag{25.4.23}$$

It is easy to verify that for $f = 85$ mm and $d = 0.2408$ mm, the object distance is 30,000 mm or 30 meters. This tells us that if we extend the lens a distance $d = 0.2408$ mm beyond the "infinity" setting, we are sharply focusing objects 30 meters distant on the film, a fact we already know. But suppose we make d two, three, four, five, or six times this figure. Then, according to (25.4.23), the corresponding object distances, for which objects are sharply imaged on the film, are $\frac{1}{2}$, $\frac{1}{3}$, $\frac{1}{4}$, $\frac{1}{5}$, and $\frac{1}{6}$ as great as the 30 meter distance obtained for the initial value $d = 0.2408$ mm. This corresponds to object distances of 30, 15, 10, 7.5, 6, and 5 meters, associated with lens extensions of $d = 0.2408$ mm, $2d$, $3d$, $4d$, $5d$, and $6d$. Equal angular rotations of the focusing collar that produce such a series of lens extensions, therefore, will sharply focus images on the film corresponding to object distances ∞, 30 meters, $30/2 = 15$ meters, $30/3 = 10$ meters, $30/4 = 7.5$ meters, $30/5 = 6$ meters, $30/6 = 5$ meters, etc. This is clearly in agreement with the way the focusing scale shown in Fig. 25.20 is constructed. Examination of any photographic lens will reveal that all focusing scales are constructed along similar lines, as required by Eq. (25.4.21) or (25.4.22).

We saw previously, in connection with our discussion of pinholes in section 25.2, that the sharpness of pinhole images is practically independent of the object distance, so long as it is large compared with the image distance. Pinhole images of distant and close objects are equally sharp, or equally unsharp. However, the same is not true in the case of lenses. The images formed by lenses are ideally sharp only for a single object distance, and objects at other distances are more or less fuzzy. To obtain a sharp image of an object at a given distance from a lens, we must adjust the image distance for maximum sharpness; in other words, we must *focus* the lens. This minor disadvantage, however, is vastly outweighed by the fact that properly focused lens images are vastly superior to pinhole images in brightness and sharpness.

25.5 Geometric Construction of Lens Images

In the preceding section, it was shown that the thin-lens equation allows us to locate the position of the image formed by a thin lens of known focal length for any given object distance. While the thin-lens equation is an extremely important relation, there are many practical situations in which it is of rather limited usefulness. For example, in a system in which two or more lenses must be used in combination, the use of the thin-lens equation frequently becomes extremely complex and difficult. In this instance, one may proceed by locating the image formed by the first lens using the thin-lens equation and then regarding this image as the *object* from which the second lens forms a second image whose position is calculated once again by using the thin-lens equation. It is possible to calculate the imaging properties of a multilens system this way, and we shall have occasion to do so in one or two instances, although it is not an easy method to use, especially when more than two lenses are involved. Furthermore, it is difficult to see what happens when the object or one of the lenses is moved from one position to another. You must start all over and repeat the whole calculation. Finally, although the use of the lens equation in this way will tell you where the image is formed, it gives no information about its size and hence about the magnification of the system.

It is, therefore, frequently advantageous to use geometric methods to show the position and size of the images produced by optical systems. These geometric methods are based on the construction of *ray diagrams* that show how various rays emitted from one or more points on the object intersect once more to produce an image at the appropriate position. In constructing such diagrams, there are *three rays* that proceed outward from any object point, whose sub-

sequent paths are particularly easy to trace. These are illustrated in Fig. 25.22, where an image formed by a converging lens is shown. An object AB having the form of an arrow is shown, the lens forming its image at A'B'. *All* rays that leave the point A on the image and pass through the lens must, of course, converge upon the point A' and go through that point. But if one does not know from the outset where the image actually is, it is not particularly simple to find the direction that any ray, chosen at random, will take after having been refracted by the lens. In the case of the three rays AP, AO, and AQ, however, it is easy to plot the paths taken and, therefore, to locate the image position at their intersection.

Ray AP is emitted from the object point A parallel to the optical axis of the lens. Such a ray might have come from an object *on* the axis at an infinite distance from the lens. In any event, we have already seen that any such parallel ray will, after refraction, pass *through the focal point* F. This is illustrated in Fig. 25.16 and Fig. 25.18a.

Ray AO passes through the center of the lens at O. Since the surfaces of the lens are vertical and parallel there, a thin lens, insofar as this ray is concerned, behaves just like a plate of glass with parallel surfaces and negligible thickness; the direction of ray AO is, therefore, not changed at all upon passing through the lens. The intersection of ray AOA', at point A', serves to locate the point A' and hence to define the position of the image of point A.

The third ray that can be used to locate the image point A' is the ray AF' that passes through the focal point F' on the same side of the lens as the object. This ray, after refraction, must be parallel to the optic axis, as illustrated in Fig. 25.16. From Fig. 25.22 it is clear that this ray also intersects the rays APFA' and AOA' at the image point A'. Figure 25.22 makes it clear also that the path of this ray must be as shown. Had the ray been emitted from A' and gone toward the lens along A'Q, it would have been refracted through the focal point F' and hence have passed through the point A. The fact that its actual path was

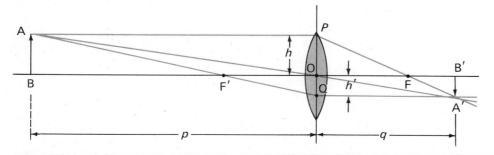

FIGURE 25.22. Construction of converging thin lens image by geometric tracing of rays.

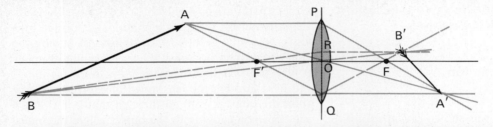

FIGURE 25.23. Size and orientation of lens images is easily exhibited by geometric ray construction.

in the opposite direction makes no difference; the location of the ray is the same in either case.

It is evident that point A′ could have been *constructed* by tracing—and producing—the rays APF, AOF, and AF′QA′. These rays, produced beyond the lens to a sufficient distance, *all intersect at the single point* A′, where the image of point A must be found. Indeed, we need not even trace the paths of all three of these rays since the intersection of any two of them is enough to establish the location of the image A′. The third will then automatically pass through this same point.

Figure 25.23 shows the construction of the image A′B′ of an object AB that is not perpendicular to the optical axis. This proceeds in exactly the same way. Point A′ is located as the intersection of rays APFA′, AOA′, and AF′QA′, while point B′ is found as the point where BQFB′, BOB′, and BF′RB′ intersect. Note that the images in Figs. 25.22 and 25.23 are both upside-down, or *inverted*, images.

The case where the object is closer to a converging lens than the focal point F′ is illustrated in Fig. 25.24. The ray construction is carried out by the rules that were developed above. A ray leaving the object AB at A traveling parallel to the axis of the lens must, after passing through the lens, pass through the focal point F. It, therefore, follows the path APF. The

ray AO that leaves the object at A and passes through the center of the lens goes through without any significant deviation, as discussed previously, following the path AOR. Now, the rays PF and OR that strike the eye of an observer looking through the lens from the right do not actually cross anywhere. However, if we project these rays backward, it is evident that their projections, shown as dashed lines, intersect at A′. The observer's eye has no way of knowing that the real rays PF and OR do not actually diverge from A′; to the observer it appears that these rays actually emanate from an image point A′ and that there is an image of the object, virtual though it may be, at A′B′. It will also be noted that in this case the image is right side up, or *erect*, and larger than the object. When used this way, the lens acts as a *simple magnifier* and affords a magnified, erect image of any object that is within the focal distance *f*. It is important to observe that geometric ray construction, though in these simple examples it does not furnish precise numerical answers about the image distances, does tell us whether the image is real or virtual, erect or inverted, or magnified or reduced in size. These additional details are not readily obtained from the thin-lens equation alone.

The methods of ray construction can, however, if properly used, afford quantitative as well as qualitative information about lenses and lens systems. In Example 25.5.1 below, we shall see that they provide yet another way of deriving the thin-lens equation. They can also give us a very simple expression for the magnification of a simple converging lens. This may be derived by noting from Fig. 25.22 that since triangles AOB and A′OB′ are similar ones, the ratios of their corresponding sides must be equal. This means that

$$\frac{AB}{OB} = \frac{A'B'}{OB'}$$

or

$$\frac{A'B'}{AB} = \frac{OB'}{OB} \qquad (25.5.1)$$

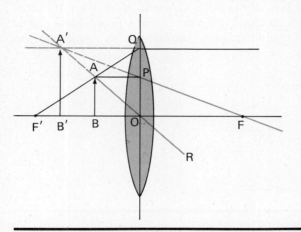

FIGURE 25.24. Ray diagram for an object very close to a thin converging lens.

But A′B′/AB is the ratio of image size to object size, which is by definition the magnification *M*. Also, OB is the object distance p while OB′ is the image distance q. Therefore, Eq. (25.5.1) can be rewritten as

$$M = \frac{q}{p} \tag{25.5.2}$$

From the thin-lens equation (25.4.19), the ratio q/p is equal to $(q/f) - 1$, which allows us to write the magnification in terms of the object distance and the focal length as

$$M = \frac{q}{f} - 1 \tag{25.5.3}$$

These equations are good for both real and virtual images and for converging and diverging lenses, provided that we recall that the focal length of a diverging lens is negative and that, whenever the image is formed on the same side of the lens as the object in Fig. 25.24, the image distance q is negative. A positive answer for M, as obtained, for example, in Fig. 25.22, corresponds to an *inverted* image, while one that is negative, resulting from a situation such as the one shown in Fig. 25.24, results whenever the image is *erect*. It is also interesting to note that this expression for the magnification, as given by (25.5.2), is exactly the same as the one obtained for the pinhole in Example 25.2.2.

Some additional insight into the magnification factor of lenses, particularly as it affects photographic lenses, can be obtained by eliminating q between (25.5.2) and (25.53). Writing (25.5.2) as $p = Mq$, substituting this into (25.5.3) and solving for M, we can easily obtain

$$M = \frac{1}{\dfrac{p}{f} - 1} \tag{25.5.4}$$

When written this way, the magnification factor is expressed in terms of the object distance p and the focal length. In the case of photographic lenses used for photographing relatively distant objects, the object distance p is normally very much larger than the focal length f, and under these circumstances, in the denominator of (25.5.4), the ratio p/f is much greater than unity. Neglecting unity as compared to this ratio, then, this equation can be written as a good approximation in the form

$$M = \frac{f}{p} \qquad p \gg f \tag{25.5.5}$$

What this tells us is that for an object at a given fixed distance p, the magnification is directly proportional to the focal length f and that the longer the focal length of the lens, the larger the image it produces on the film. Telephoto lenses, therefore, are lenses of

longer than normal focal length; the fact that they protrude further from the camera body than normal lenses simply reflects the fact that their focal length is greater than that of a normal lens. Wide-angle lenses, on the other hand, are lenses of shorter than normal focal length, whose magnification is correspondingly smaller and which, therefore, cover, on the same size negative, a wider field of view.

For the standard size 35-mm negative, whose dimensions are 24×36 millimeters, a lens of 50- to 55-mm focal length corresponding to an angular field of view of about 45° is most useful for general photography and is regarded as the "normal" lens for a 35-mm camera. A lens of 135-mm focal length, would according to (25.5.5) produce an image magnification 2.7 times that of a normal 50-mm lens, producing on the film a relatively mild telephoto effect. Lenses of from 15-mm to about 1000-mm focal length are commonly provided for use with 35-mm single-lens reflex cameras.

EXAMPLE 25.5.1

Given the facts that thin lenses divert rays parallel to their axes so as to pass through two focal points F and F′ located at a distance f from the lens on either side and that rays passing through the center of the lens are undeviated, use the methods of ray construction to derive the thin-lens equation.

Referring to Fig. 25.22, it is easily seen that the rays APFA′, AOA′, and AF′QA′ exhibit precisely the properties described above. Since triangles OPA and OQA′ are similar, the ratios of their corresponding sides must be equal. Therefore,

$$\frac{h'}{h} = \frac{A'Q}{AP} = \frac{q}{p} \tag{25.5.6}$$

from which

$$h' = \frac{hq}{p} \tag{25.5.7}$$

Also, since triangles OPF and QPA′ are similar, we may write

$$\frac{h + h'}{QA'} = \frac{h}{OF} \tag{25.5.8}$$

But since QA′ = q and OF = f, and since h' can be stated in terms of h from (25.5.7), (25.5.8) may be written as

$$\frac{h}{q} + \frac{h'}{q} = \frac{h}{q} + \frac{1}{q}\left(\frac{hq}{p}\right) = \frac{h}{f} \tag{25.5.9}$$

or

$$\frac{1}{q} + \frac{1}{p} = \frac{1}{f} \tag{25.5.10}$$

which is the thin-lens equation.

819

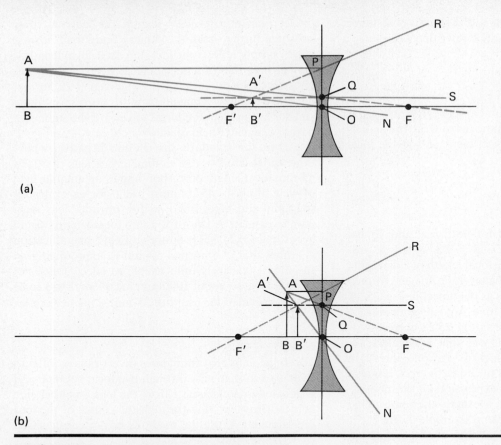

(a)

(b)

FIGURE 25.25. Ray diagrams for diverging lenses.

EXAMPLE 25.5.2

Discuss the imaging properties of a thin *diverging* lens using ray construction methods.

In Fig. 25.25a, we see an object AB beyond the focal point F′ of a divergent thin lens. A ray such as AP, emitted from the object point A parallel to the axis of the lens, is refracted so as to pass through a focal point. Since the lens is divergent, however, the ray APR is bent *away* from the axis. Instead of passing through the point F, as would be the case if the lens were convergent, it is refracted so that its *projection* PF′ passes through the focal point F′ on the *object* side of the lens. The ray AQ, whose projection would pass through the other focal point, by the same token is bent by the lens until its course is parallel with the axis, as ray AQS. The projections of refracted rays QS and PR meet in the virtual image point A′, and, therefore, the image A′B′ must be located as shown. The ray AON that passes undeviated through the center of the lens intersects these other two rays at A′ also.

The situation that arises when the object AB is closer to the lens than the focal point F′ is not essentially different; it is shown in Fig. 25.25b. The ray AP, initially parallel to the axis, is diverged by the lens so that its projection passes through F′, while a ray AQ

whose projection would intersect the focal point F on the other side of the lens is diverged to become parallel to the axis. Its backward projection QA′ intersects the projection of PR at the image point A′, the image A′B′, therefore, being situated as in the diagram. At the same time, the undeviated ray AON passing through the center of the lens intersects both RPA′ and SQA′ at the image point A′. In both cases, it is evident that the image is virtual, erect, and reduced in size.

EXAMPLE 25.5.3

A *telescope* is constructed by combining two simple *convergent* lenses as shown in Fig. 25.26. Explain the operation of such a device using ray construction methods.

The lens L_1 nearest the rather distant object AB is called the *objective* lens. Its focal length is f_1, and its focal points are F_1 and $F_1′$. It forms a real image A′B′ at the intersection of the rays P_1F_1 and AO_1 which can be located, as illustrated, by the usual ray construction. This image A′B′ in turn becomes the *object* for a convergent *eyepiece lens* L_2 that is used very much like the simple magnifier discussed in the text above in connection with Fig. 25.24. The eyepiece, whose focal length is f_2 and whose focal points are shown as F_2 and $F_2{}'$, is placed so that the image $A'B'$ is just inside the focal point $F_2{}'$. The eyepiece then acts as a simple magnifier like the one in Fig. 25.24, forming an enlarged virtual image $A''B''$ at the intersection of the projections of $A'O_2$ and P_2F_2. This image is formed at a large distance from the eyepiece when the image $A'B'$ is just inside the focal point $F_2{}'$. Indeed, its distance from the point O_2 may be adjusted by moving the eyepiece back and forth along the optical axis, thus varying the distance between the image $A'B'$ and the point $F_2{}'$. This is what happens when we focus a telescope or binoculars by moving the eyelenses slightly in and out. When we have the instrument "focused" for comfortable viewing of a distant object, we have, in fact, adjusted the eyelens so that the image $A''B''$ is at the same distance from the telescope as the object AB itself! In Fig. 25.26, the focus of the eyepiece is adjusted so that the image $A''B''$ is not quite as far away as the object, but this could be easily remedied by moving the eyelens L_2 just a trifle to the left so that $F_2{}'$ is just somewhat closer to $A'B'$. Ordinarily, in the case of a telescope, the object distance will be so large as to be, for practical purposes, infinite, though (to allow us to draw a reasonable diagram) this is not actually exhibited by Fig. 25.26.

Now, clearly, the magnification of the telescope is the product of the magnifications M_1 and M_2 of the objective and eyepiece lenses, which, according to (25.5.2) would be

$$M = M_1 M_2 = \frac{q_1 q_2}{p_1 p_2} \qquad (25.5.11)$$

In Fig. 25.26, $AP_1 = p_1$ and $O'B = q_1$, while $A'P_2 = p_2$ and $O_2B'' = q_2$. If the eyepiece is focused so that the image $A''B''$ is as distant as the object AB, then the image distance q_2 is given by

$$q_2 = AP_1 + O_1B' + A'P_2 = p_1 + q_1 + p_2 \qquad (25.5.12)$$

Ordinarily, of course, telescopes are used to view objects that are *very distant* compared to the size of the telescope itself. In such an instance, the object distance p_1 is very much larger than either $q_1\ (=O'B)$ or $p_2\ (=A'P_2)$, and hence (25.5.12) can be written essentially as

$$q_2 = p_1 \qquad (25.5.13)$$

Substituting this into (25.5.11), we find

$$M = \frac{q_1}{p_2} \qquad (25.5.14)$$

Now, if the telescope is aimed at a very distant object, the image $A'B'$ is formed at an image distance q_1 from the objective that is very nearly equal to its focal length f_1. At the same time, since the eyepiece is positioned so that its focal point $F_2{}'$ is very near the image $A'B'$, the eyelens object distance p_2 is very close to its own focal length f_2. Substituting these values for q_1 and p_2 into (25.5.14), we arrive finally at

$$\boxed{M = \frac{f_1}{f_2}} \qquad (25.5.15)$$

as the overall magnification of a telescope of this type. For a large overall magnification, therefore, the eyepiece lens must have a much smaller focal length than the objective lens. When you "look through the wrong end of the telescope", you reverse the function of eyepiece and objective, the ratio (25.5.15) then being replaced by its reciprocal, the "eyepiece" now having the long focal length and the "objective" the short one. Therefore, you see a *reduced* image whose size is reduced by the same factor by which it is magnified when you use the telescope in the customary way.

It will be observed that this kind of telescope produces an *inverted* image rather than an erect one.

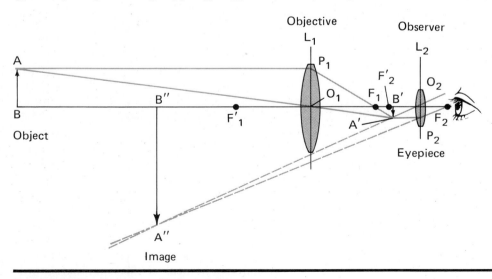

FIGURE 25.26. Ray geometry for a telescope.

This is not a very serious disadvantage for astronomical work, but is annoying for terrestrial use. The image can, however, be erected by use of an additional erecting lens or by a pair of erecting prisms in the light path. The latter scheme is very frequently used in binoculars, where, in addition to rendering the image right side up, it performs the useful function of folding the light path up upon itself and significantly reducing the overall size of the resulting instrument. It should be noted, in the case of the telescope shown in Fig. 25.26, that the observer's angular field of view is shown to be rather wider than what is normally attainable in simple telescopes and binoculars to allow the diagram to be drawn in an understandable way.

EXAMPLE 25.5.4

Explain the operation of the compound microscope illustrated schematically in Fig. 25.27.

The telescope and the microscope are in principle quite similar, but the rather different conditions governing their use introduce certain differences of detail. In the case of a telescope, we are ordinarily interested in viewing large objects that are very distant, while the microscope is generally used to observe very small objects close at hand. The telescope helps us when it is impossible for us to get very close to what we want to see. The microscope is used when our object is so small that we cannot get our eye close enough to examine it and, at the same time, keep it in sharp focus.

In the case of the microscope shown in Fig. 25.27, a small object AB is positioned fairly close to the focal

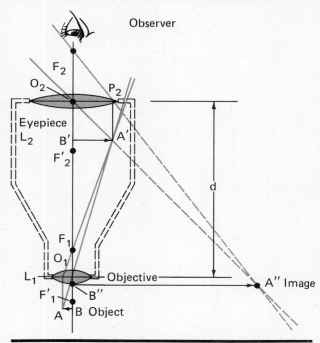

FIGURE 25.27. Ray diagram for a compound microscope.

point F′ of an objective lens L_1 whose focal length $f_1 = O_1F_1'$ is quite short. This produces an enlarged intermediate real image A′B′ at an image distance $q_1 = O_1B'$ that is much larger than the object distance $p_1 = BO_1$. The intermediate image A′B′ is then viewed through an eyepiece lens L_2 that acts as a simple magnifier, also of rather short focal length. The eyepiece, as in the case of the telescope, is shown positioned so that the intermediate image falls rather close to the focal point F_2', and, as a result, an enlarged inverted virtual image is formed at the intersection of the projected rays O_2A and F_2P_2 as shown. The exact position of this image can be varied by varying the position of the image A′B′, which in turn can be accomplished by slightly varying the original object distance p_1; this is what happens when we "focus" the microscope. In the case of the telescope, focusing was done by moving the eyepiece lens with respect to the objective, but in the microscope both the objective and eyelens are separated by a fixed distance d, as shown in the diagram, since they are mounted in a rigid cylindrical tube. The positioning of this tube, which bears both lenses, with respect to object AB then determines the exact location of the intermediate image A′B′. As the position of the image A′B′ changes with respect to the focal point F_2', of course, the position of the final image A″B″ also changes; in practice, the focus is adjusted so that it is at a comfortable distance for the viewer. If it is adjusted so that the final image is located at the same position as the object, as is nearly the case in Fig. 25.27, we can make a direct comparison of the object and image sizes and, therefore, obtain an expression for the magnification.

As before, the magnification of the objective, according to (25.5.2), is $M_1 = q_1/p_1$, while for the eyepiece the magnification is $M_2 = q_2/p_2$. The total magnification M can, therefore, be written

$$M = M_1M_2 = \frac{q_1q_2}{p_1p_2} \qquad (25.5.16)$$

In this expression, both p_1 and p_2 are not very different from f_1 and f_2 in view of the fact that the object and the intermediate image are positioned close to the focal points F_1' and F_2'. At the same time, provided that the focal lengths of both objective and eyepiece are short compared to the tube length d, the image distances q_1 $(=O_1B')$ and q_2 $(=O_2B'')$ are both approximately equal to this tube length. Inserting these values into (25.5.16), we find

$$\boxed{M \cong \frac{d^2}{f_1f_2}} \qquad (25.5.17)$$

Typically, we might use a tube length $d = 200$ mm. With an objective of focal length 5 mm and an eyepiece lens of focal length 15 mm, this gives

$$M = \frac{(200)^2}{(5)(15)} = 533$$

That the image is an inverted one ordinarily causes little or no difficulty in using a microscope, so no particular measures are ordinarily taken to erect it.

25.6 The Brightness and Sharpness of Lens Images

We have emphasized the importance of the lens as an optical element whose images are much sharper and brighter than those of pinholes. It is important, therefore, to know something of the factors that affect the sharpness and brightness of lens images and to be able to make certain quantitative calculations involving these factors for simple lenses and optical systems.

The brightness of the image formed by a lens is primarily influenced by two factors: (1) the amount of light that the lens collects from the object and (2) the area over which the lens distributes that light, which depends, in turn, on the magnification. The amount of light energy from the object that is collected by the lens is proportional to the area of the lens and hence to the square of its diameter d. The magnification of the lens, on the other hand, according to (25.5.2), is given by the ratio of image distance to object distance, q/p. Since the areas of the object and image are proportional to the squares of their linear dimensions, the ratio of image *area* to object *area* will be

$$\frac{A'}{A} = M^2 = \frac{q^2}{p^2} \qquad (25.6.1)$$

Light that leaves an area A corresponding to the object is concentrated by the lens on area A' comprising the image, as shown in Fig. 25.28. A certain total amount of light, proportional to the area A of the object, leaves the object and is focused by the lens into the area A' at the image. Clearly, the intensity of light at the image must be directly proportional to the area A and inversely proportional to the area A' and hence must be proportional to the ratio A/A'. Because the amount of light reaching the image is only that fraction of the light leaving the object that is collected by the lens, the image intensity must *also* be proportional to this fraction. The fraction of light collected by the lens, however, can be approximately expressed as the ratio of the lens area A_l to the area of a sphere of radius p centered on the object, which is $4\pi p^2$. This approximation is a good one so long as the lens diameter d is only a small fraction of the object distance p, which we shall assume is always the case. The total image intensity $\bar{S}_i$, being proportional to both these factors, can be written as

$$\bar{S}_i = K \frac{A}{A'} \frac{A_l}{4\pi p^2} \qquad (25.6.2)$$

where K is a proportionality constant. Since A/A' is given by (25.6.1) and since $A_l = \pi d^2/4$, where d is the lens diameter, this can be expressed as

$$\bar{S}_i = K \frac{p^2}{q^2} \frac{\pi d^2}{16\pi p^2} = K' \frac{d^2}{q^2} \qquad (25.6.3)$$

where $K' = K/16$ is simply another representation of the proportionality constant.

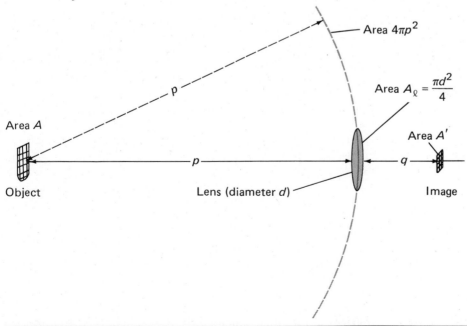

FIGURE 25.28. Diagram used for the calculation of image light intensity.

This equation illustrates how the brightness of lens images is related to image distance and lens diameter. It is very common, however, particularly with telescopes and photographic lenses, which are customarily used to form images of distant objects, to express this information in terms of a quantity called the *focal ratio*, or the f: *number* of the lens. In order to do this, let us substitute the value given by the thin-lens equation (25.4.19) for $1/q$ in terms of p and f into (25.6.3) to obtain

$$\bar{S}_i = K'd^2\left(\frac{1}{f} - \frac{1}{p}\right)^2 = \frac{K'd^2}{f^2}\left(1 - \frac{f}{p}\right)^2 \qquad (25.6.4)$$

When the object distance p is much larger than the focal length f, we may neglect the term f/p and write this as

$$\bar{S}_i = \frac{K'}{f:^2} \qquad (25.6.5)$$

where

$$f: = \frac{f}{d} \qquad (25.6.6)$$

The quantity represented by the rather odd symbol f: is the ratio of the focal length of the lens to its diameter. This ratio is called the *focal ratio* of the lens, or, more commonly, the f: *number* of the lens. The symbol f:, representing the ratio f/d, *should not be confused with the focal length f.* It is written in this way simply to conform to standard optical notation. A lens whose focal length is four times its diameter is said to have a focal ratio of 4, or an f: number that is commonly written as f:4.0 and pronounced "f-four." Likewise, a lens that is specified as a "135-mm f:2.5" can be counted upon to have a focal length of 135 mm and, the focal ratio being 2.5, a diameter of $135/2.5 = 54$ mm.

From (25.6.5), it is evident that the image brightness is proportional, *inversely*, *to the square of the f: number*. Low f: number lenses, whose diameter is an appreciable fraction of the focal length, produce relatively bright images, while those of high f: number, whose diameter is much less than the focal length, produce dimmer ones. At the same time, decreasing the f: number by a factor of 2 increases the image intensity by a factor of 4. An f:2.0 lens is, therefore, capable of producing image intensities that are four times greater than could be obtained with an f:4.0 lens. The f:2.0 could be used, for practical photographic purposes, in light four times less intense than could the f:4.0 lens; or, in light of a given intensity, it could expose the film properly four times faster. The latter property has led to the terminology "fast lenses"

for lenses of low f: number that can be used with relatively short exposure times, even in dim light. To double the image intensity, it is necessary to decrease the f: number by a factor of $\sqrt{2}$ ($=1.414$), while to halve the image intensity, one must increase the f: number by the same factor.

It is possible, by incorporating a circular aperture stop of adjustable diameter fashioned from thin interleaved metallic blades into an optical system, to vary the f: number from that of the wide-open lens to a much larger value corresponding to the smallest possible stop opening. Such a device works to accomplish the same purpose as the iris of the human eye and is referred to as an *iris diaphragm*. The control that enables the user to adjust the aperture of the diaphragm is usually calibrated in f: numbers corresponding to the effective focal ratio at any given setting. Such a series of lens aperture settings can be seen on the iris diaphragm control of the photographic lens shown in Fig. 25.20 as the series of numbers 2, 2.8, 4, 5.6, 8, 11, 16, 22. Each successive number differs from the previous one by a factor of approximately $\sqrt{2}$, and, therefore, each numbered setting increases or decreases the image intensity by a factor of 2 relative to the one that is next smallest or next largest.

Lenses of large effective aperture, or low f: number, are capable of producing brighter images than lenses of smaller aperture. Unfortunately, the effects of imaging imperfections or *lens aberrations* become much more serious as the effective aperture increases. Though these aberrations may be largely corrected by refining a single lens into a compound lens containing several elements with various refractive indices and radii of curvature, it is still generally true that image sharpness suffers as f: numbers decrease. Also, as we shall soon see, while lenses of large aperture may produce sharp images of objects that are at the proper distance to be in correct focus, images of objects at other distances are very much out of focus. On the other hand, lenses of smaller aperture, or higher f: number, render reasonably sharp images of objects that are not in perfect focus. Finally, lenses of large aperture are usually more complex and expensive than "slower" ones and, due to their large effective diameter, are larger, heavier, and less convenient to use.

Finally, we must remember that in all our discussions of image brightness based on the concept of focal ratio or f: numbers, we have assumed that the object distance is large compared with the focal length and, therefore, have neglected the f/p term in (25.6.4). When the object is rather close to the lens, this may no longer be a good approximation, and the effect of the f/p term may have to be taken into account. From (25.6.4), it is apparent that in cases where the f/p value is significantly larger than unity, its effect is to *decrease* the image intensity from what it would be

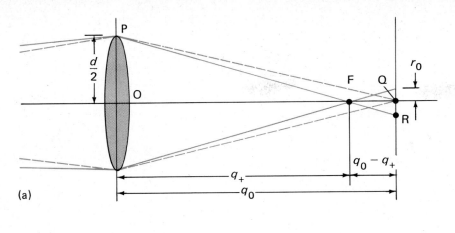

(a)

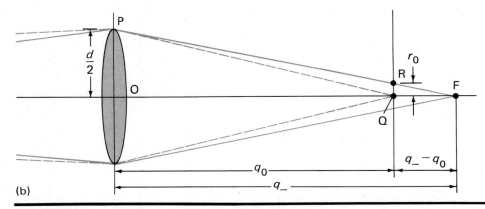

(b)

FIGURE 25.29. Geometry of out-of-focus image discs for thin lenses.

otherwise or, in other words, to produce an effect similar to increasing the f: number of the lens. We shall make some calculations regarding the magnitude of this effect in one of the examples at the end of this section.

Let us now turn our attention to the sharpness of lens images. Lens images may be less than perfectly sharp due to the presence of lens aberrations and also because the film or viewing screen may not coincide with the point of best focus for all of the image points that are in the picture. Aberrations do, of course, make significant contributions to the lack of sharpness of optical images, but we are not concerned with them here. Our problem at the moment is to determine the relative sharpness of images produced by an aberration-free lens that are not *precisely* in focus and the related problem of determining the range of object distances that are in *acceptable* focus for a given lens set to produce perfectly sharp images for a single given object distance.

Consider an aberration-free lens that produces a perfectly sharp image of an object at a distance p_0 on a film or focusing screen a distance q_0 from the lens. The diameter of the lens is d, and its focal length is f. Rays coming from object points at distances *other* than p_0 are brought to focus not at the distance q_0 but

at image distances that are slightly larger or smaller. On the film or screen that intercepts these rays, we see a small circular disc of light of radius r_0 rather than a perfect point image. The situation is illustrated in Fig. 25.29a for an object at a distance p_+ *greater* than p_0, in which case the rays converge to a point at an image distance q_+ less than q_0, and in Fig. 25.29b for an object at a distance p_- *less* than p_0, for which the image is formed at a distance q_- that is greater than q_0. In Fig. 25.29a, triangles OPF and QRF are similar and OP/OF = QR/QF; this can also be written as

$$\frac{d}{2q_+} = \frac{r_0}{q_0 - q_+} \tag{25.6.7}$$

This equation could be solved immediately for r_0 to obtain an expression for the radius of the image disc in terms of the image distances q_0 and q_+. But this is not very useful. What we are really after is an expression that relates the radius r_0 of the image spot to the object distances p_0 and p_+. This is obtained most easily by solving Eq. (25.6.7) for q_+, then inverting it to get

$$\frac{1}{q_+} = \frac{1}{q_0}\left(\frac{2r_0}{d} + 1\right) \tag{25.6.8}$$

825

But both q_0 and q_+ are related to their respective image distances p_0 and p_+ by the thin-lens equation, which tells us that

$$\frac{1}{p_0} + \frac{1}{q_0} = \frac{1}{f} = \frac{1}{p_+} + \frac{1}{q_+} \qquad (25.6.9)$$

Substituting the values of $1/q_0$ and $1/q_+$ in terms of $1/p_0$, $1/p_+$, and $1/f$ given by (25.6.9) into (25.6.8), we obtain

$$\frac{1}{p_+} = \frac{1}{p_0} - \frac{2r_0}{d}\left(\frac{1}{f} - \frac{1}{p_0}\right) \qquad (25.6.10)$$

from which

$$r_0 = \frac{\dfrac{d}{2}\left(\dfrac{1}{p_0} - \dfrac{1}{p_+}\right)}{\dfrac{1}{f} - \dfrac{1}{p_0}} \qquad (25.6.11)$$

Since from (25.6.6) $d = f/f:$, these results can be stated in terms of the $f:$ number of the lens as

$$\frac{1}{p_+} = \frac{1}{p_0} - \frac{2r_0 f:}{f}\left(\frac{1}{f} - \frac{1}{p_0}\right) \qquad (25.6.12)$$

and

$$r_0 = \frac{\dfrac{f}{2f:}\left(\dfrac{1}{p_0} - \dfrac{1}{p_+}\right)}{\dfrac{1}{f} - \dfrac{1}{p_0}} \qquad (25.6.13)$$

In the case where the object distance p_0 is much larger than the focal length, which is usual for telescopes and photographic lenses, we may neglect $1/p_0$ in comparison with $1/f$ to obtain the simple approximate equations

$$\boxed{\begin{array}{ll} \dfrac{1}{p_+} \cong \dfrac{1}{p_0} - \dfrac{2r_0 f:}{f^2} & p_0 \gg f \qquad (25.6.14) \\[2ex] \text{and} & \\[1ex] r_0 \cong \dfrac{f^2}{2f:}\left(\dfrac{1}{p_0} - \dfrac{1}{p_+}\right) & p_0 \gg f \qquad (25.6.15) \end{array}}$$

From these equations, we see that the radius of the out-of-focus image spot r_0 increases as the focal length is increased and as the $f:$ number is decreased. This means that in the case of lenses of long focal length—telephoto camera lenses, for example—and of low $f:$ number, objects that are not in the plane of correct focus appear to be much more fuzzy and indistinct than is the case with lenses of shorter focal length and higher $f:$ number. The viewfinder of the single-lens reflex camera or the device shown in Fig. 25.2 will verify these conclusions at once. These results clearly form the basis of the well-known photog-

rapher's maxim "stop the lens down (that is, make the $f:$ number higher) to improve the sharpness of out-of focus objects."

When the object is at a distance p_- that is less than p_0, its image will be formed behind the film or viewing screen at an image distance q_-, as shown in Fig. 25.29b. Once again, we see a circular disc of radius r_0 rather than a perfect point image. From the similarity of triangles OPF and QRF in that figure, OP/OF = QR/QF, or

$$\frac{d}{2q} = \frac{r_0}{q_- - q_0} \qquad (25.6.16)$$

As before, this equation relates the size of the image disc to the image distances rather than the object distances; but once again, using the thin-lens equation

$$\frac{1}{p_0} + \frac{1}{q_0} = \frac{1}{f} = \frac{1}{p_-} + \frac{1}{q_-} \qquad (25.6.17)$$

we can, by a series of algebraic steps exactly like those employed in deriving (25.6.12) and (25.6.13) from (25.6.7), obtain

$$\frac{1}{p_-} = \frac{1}{p_0} + \frac{2r_0 f:}{f}\left(\frac{1}{f} - \frac{1}{p_0}\right) \qquad (25.6.18)$$

from which

$$r_0 = \frac{\dfrac{f}{2f:}\left(\dfrac{1}{p_-} - \dfrac{1}{p_0}\right)}{\dfrac{1}{f} - \dfrac{1}{p_0}} \qquad (25.6.19)$$

Once more, whenever the object distance p_0 is much larger than the focal length, $1/p_0$ can be neglected in comparison with $1/f$ to give the simple approximate expressions

$$\boxed{\begin{array}{l} \dfrac{1}{p_-} = \dfrac{1}{p_0} + \dfrac{2r_0 f:}{f^2} \qquad\qquad (25.6.20) \\[2ex] \text{and} \\[1ex] r_0 \cong \dfrac{f^2}{2f:}\left(\dfrac{1}{p_-} - \dfrac{1}{p_0}\right) \qquad (25.6.21) \end{array}}$$

The same dependence of image spot size upon focal length and $f:$ number is obtained as in the previous case, and all the conclusions drawn previously apply here as well.

Related to the question of the size of out-of-focus image discs is the concept of *depth of field*. Strictly speaking, if we can ignore the effects of diffraction and lens aberrations, then, for a given lens-to-image plane distance q_0, only objects at the corresponding object distance p_0 will be perfectly sharp. Objects at other distances will be more or less fuzzy, because on the image plane rays coming from such objects are registered as small discs of light rather than sharp points. Since the eye's ability to resolve fine detail is limited, the eye is unable to distinguish between points and discs that are sufficiently small. Let us suppose that the eye can distinguish a disc of radius larger than R_0 as a distinct circular object, while a disc of radius less than R_0 is observed as an apparently sharp point image. We shall find, then, that objects whose distance from the lens is such that image discs of radii less than R_0 are formed will appear to be sharply defined, while those whose distances from the lens is such that image discs of radii greater than R_0 are produced will be unsharp and "out of focus." There will be a range of object distances on either side of the value p_0 for which the observed images have image disc sizes less than R_0 and within which the image appears to be sharp. Outside this range of object distances, images will appear to be out of focus. The range of apparently sharp object distances is said to define the *depth of field* of the lens. Assuming that the critical image size R_0—sometimes referred to as the *circle of confusion*[2]— can be unequivocally defined if we set the spot size r_0 equal to this critical value in Eq. (25.6.12), we shall find that p_+ then represents the largest object distance for which sharp focus will be obtained. Likewise, if we set $r_0 = R_0$ in (25.6.19), then p_- represents the smallest object distance that will be in satisfactory focus. The limits of the depth of field, p_+ and p_-, are, therefore, given by

$$\frac{1}{p_+} = \frac{1}{p_0} - \frac{2R_0 f:}{f}\left(\frac{1}{f} - \frac{1}{p_0}\right) \qquad (25.6.22)$$

and

$$\frac{1}{p_-} = \frac{1}{p_0} + \frac{2R_0 f:}{f}\left(\frac{1}{f} - \frac{1}{p_0}\right) \qquad (25.6.23)$$

In the case where the object distance is much greater than the focal length, these equations can be written approximately as

$$\frac{1}{p_+} = \frac{1}{p_0} - \frac{2R_0 f:}{f^2} \qquad (25.6.24)$$

and

$$\frac{1}{p_-} = \frac{1}{p_0} + \frac{2R_0 f:}{f^2} \qquad (25.6.25)$$

It will be observed that for lenses of large aperture, corresponding to small f: number, and of long focal length, the second term on the right side of these equations is small, and p_+ and p_- do not differ greatly from p_0. Under these circumstances, only a small range of distances about the chosen object distance p_0 will be in reasonably sharp focus, and objects

[2] No pun intended, this is really the terminology that is used.

elsewhere will be blurred. Conversely, if the lens is stopped down, thereby increasing the f: number, or if a lens of smaller focal length is substituted, p_+ and p_- will differ from p_0 to a much greater extent, and the range of distances that are in satisfactory focus will be very much greater. These considerations form the basis of the photographer's rules that a large depth of field may be obtained by stopping down to a suitably large f: number or by using a short focal-length, wide-angle lens. It is also to be noted that in the case of the pinhole, even though the image is dim and not very sharp, an object at practically any distance is reasonably well imaged. The penalty we pay for the sharp and bright images that lenses provide is that only a certain range of object distances is in satisfactory focus, and when we shift to an object at another distance we must adjust the image distance so that satisfactory focus is once more obtained.

EXAMPLE 25.6.1

A photographer uses a 35-mm single-lens reflex camera equipped with an f:1.4 lens of 50-mm focal length to photograph an object at a distance of 250 mm from the lens. Using an exposure meter[3] to determine the intensity of light reflected from the object, he concludes that the proper exposure is 1/60 second at f:4. Find the object distance and the linear magnification of the image on the film. What will happen if the exposure given is actually 1/60 second at f:4? What lens and shutter settings should actually be used for optimum exposure?

In this case, the lens equation tells us that the image distance will be

$$\frac{1}{q} = \frac{1}{f} - \frac{1}{p} = \frac{1}{50} - \frac{1}{250} = 0.016 \text{ mm}^{-1}$$

$$q = 62.5 \text{ mm}$$

When the lens is focused on "infinity," the image distance is equal to the focal length of 50 mm. It must, therefore, be extended 12.5 mm further from the film to bring the subject into focus. The magnification on the film will be given by

$$M = \frac{q}{p} = \frac{62.5}{250} = 0.25$$

The image on the film will, therefore, be just one quarter the actual size of the object.

Equation (25.6.4) expresses the image intensity as

$$\bar{S}_i = \frac{K'd^2}{f^2}\left(1 - \frac{f}{p}\right)^2 = \frac{K'}{f:^2}\left(1 - \frac{f}{p}\right)^2 \qquad (25.6.26)$$

[3] An external hand-held meter, that is, not one that is built into the camera and registers the intensity of light coming through the lens. The built-in meters automatically make the corrections we shall be discussing.

the second form of this expression arising from the fact that the f: number is defined as f: $= f/d$. So long as the object distance is very much larger than the focal length, f/p is much less than unity and the image intensity is simply represented in terms of the f: number alone by Eq. (25.6.5). In this case, the object distance is somewhat less than the focal length, but not vastly smaller. Therefore, the effect of the term in parentheses must be taken into account. This can be done simply by noting that in this case the ratio $f/p = 50/250 = 0.20$ and the term in parenthesis becomes $(1 - 0.20)^2 = (0.80)^2 = 0.64$. The image intensity is, therefore,

$$\bar{S}_i = 0.64\bar{S}_{io}$$

where

$$\bar{S}_{io} = K'/f:^2 \qquad (25.6.27)$$

is the image intensity that would result for an object at infinity. The energy that is absorbed by the film during the exposure can be restored to the required level by increasing the exposure time by a factor $1/0.64 = 1.56$. We might, therefore, allow an exposure time of $(1/60)(1.56) = 0.0260 \simeq 1/40$ second rather than the value prescribed by the external exposure meter.

Another way of making this correction is to use the original shutter speed and increase the image brightness to the level of what it would be if the object were at infinity, that is, $\bar{S}_{io}$. This requires that the f: number at the given object distance be changed to a smaller value f:*. In (25.6.26), therefore, when the f: number has the value f:*, the original intensity $\bar{S}_{io} = K'/f:^2$ is restored, whence

$$\bar{S}_{io} = \frac{K'}{f:^2} = \frac{K'}{f:^{*2}}\left(1 - \frac{f}{p}\right)^2 \qquad (25.6.28)$$

Solving this for the corrected diaphragm setting f:*, we obtain

$$f:^* = f:\left(1 - \frac{f}{p}\right) \qquad (25.6.29)$$

In our case, f: $= 4.0$ and $f/p = 0.2$. Therefore,

$$f:^* = (0.8)(4.0) = 3.2$$

and the camera should be set for 1/60 second at f:3.2 for correct exposure. In photographic terms, this is not an extremely large correction, but, of course, the smaller the object distance becomes, the larger the correction that must be made. For $p = 2f$ (in which $q = 2f = p$ and $M = 1$), according to (25.6.29), the f: number must be decreased by a factor of 2, which increases the image brightness by a factor of 4, to restore the image intensity to the level it would have for $p = \infty$.

Physically, the reason why this correction must

be made lies in the fact that for close object distances, the lens must be moved so far out from the film to bring the object into focus that it intercepts a significantly smaller angle when seen from a point on the film than is the case when the object is at infinity.

In this connection, Eq. (25.6.3) tells us that the image intensity for a given lens varies inversely as the square of the *image* distance, which, in turn, increases significantly as the object nears the lens.

EXAMPLE 25.6.2

A lens of 50-mm focal length is adjusted so that an object 5.0 meters distant is in perfect focus. What is the depth of field within which satisfactory focus results if the lens aperture is set at f:1.4? What is the depth of field at f:2.8, f:4, f:5.6, f:11, f:20, and f:32?

In this case, the object distance is 100 times the focal length. The approximate expressions (25.6.24) and (25.6.25) will, therefore, yield excellent results. For the diameter $2R_0$ of the smallest acceptable image spot, we shall refer to the calculations of Example 25.2.1, where we found that in order to produce "sharp" pictures, points about 0.026 mm apart had to be resolved as separate images. In that example a numerical figure of 1/38 mm was obtained. But visual acuity varies from one individual to another, and for practical purposes, if we assign a value of 0.025 mm to the diameter of the smallest possible image disc, we shall be close enough. Let us, therefore, take $2R_0 = 0.025$ mm $= 2.5 \times 10^{-5}$ meter. For an aperture of f:1.4, (25.6.24) and (25.6.25), using distances in meters, give

$$\frac{1}{p_+} = \frac{1}{5.0} - \frac{(2.5 \times 10^{-5})(1.4)}{(0.05)^2} = 0.20000 - 0.0140$$

$$= 0.1860 \text{ m}^{-1}$$

$$\frac{1}{p_-} = \frac{1}{5.0} + \frac{(2.5 \times 10^{-5})(1.4)}{(0.05)^2} = 0.20000 + 0.0140$$

$$= 0.2140 \text{ m}^{-1}$$

or

$$p_+ = 5.38 \text{ m} \qquad p_- = 4.67 \text{ m}$$

At an aperture of f:1.4, everything between 4.67 meters and 5.38 meters will appear sharp, while object distances outside this range will be more or less out of focus, the degree of fuzziness being dependent upon how far from this range the object is located. For an aperture of f:2.8,

$$p_+ = 5.81 \text{ m} \qquad p_- = 4.38 \text{ m}$$

and for

$$f\!:\!4, \ p_+ = 6.25 \text{ m} \qquad p_- = 4.17 \text{ m}$$
$$f\!:\!5.6, \ p_+ = 6.94 \text{ m} \qquad p_- = 3.91 \text{ m}$$
$$f\!:\!11, \ p_+ = 11.1 \text{ m} \qquad p_- = 3.23 \text{ m} \qquad \text{and so on}$$

At $f:20$, we find

$$\frac{1}{p_+} = \frac{1}{5.0} - \frac{(2.5 \times 10^{-5})(20)}{(0.05)^2} = 0.2000 - 0.2000 = 0$$

$$\frac{1}{p_-} = \frac{1}{5.0} + \frac{(2.5 \times 10^{-5})(20)}{(0.05)^2} = 0.2000 + 0.2000$$

$$= 0.4000$$

from which

$$p_+ = \infty \qquad p_- = 2.50 \text{ m}$$

At this aperture, p_+ has become infinite, and *all* objects further from the lens than $p_- = 2.50$ meters will appear sharply focused. This distance is referred to as the *hyperfocal distance* of the lens for this particular aperture. The hyperfocal distance can be found by setting $1/p_+$ equal to zero in (25.6.24), to obtain

$$\frac{1}{p_0} = \frac{2R_0 f:}{f^2} \qquad (25.6.30)$$

This value is then substituted into (25.6.25) to obtain

$$\frac{1}{p_-} = \frac{1}{p_0} + \frac{2R_0 f:}{f^2} = \frac{2R_0 f:}{f^2} + \frac{2R_0 f:}{f^2} = \frac{4R_0 f:}{f^2} \qquad (25.6.31)$$

from which the hyperfocal distance p_h can be found as

$$p_- = p_h = \frac{f^2}{4R_0 f:} \qquad (25.6.32)$$

When the lens is focused for the object distance p_0 given by (25.6.30), all objects beyond the hyperfocal distance p_h—which is half of p_0—to infinity will be sharply in focus. According to (25.6.33), the hyperfocal distance increases as the $f:$ number is decreased. Extreme wide-angle lenses of very short focal lengths have small hyperfocal distances and, therefore, are capable of forming images in which everything from infinity to a point quite close to the lens is in acceptably good focus. For example, for a lens of 18-mm focal length at an aperture of $f:4$, the hyperfocal distance is

$$p_h = \frac{f^2}{4R_0 f:} = \frac{(0.018^2)}{(5 \times 10^{-5})(4)} = 1.62 \text{ m}$$

Such a lens would produce pictures in which all objects at distances of greater than 1.62 meter are in focus, even at the large aperture of $f:4.0$. If it were stopped down to $f:16$, everything beyond 0.4 meter would be sharp!

At apertures below which p_+ has become infinite, it is meaningless to use (25.6.24) to compute p_+; in this event, p_+ *remains* infinite. The near field limit p_- may still be calculated from (25.6.25), however. For our original lens at $f:32$, for example, we would find $p_+ = \infty$, while $p_- = 1.92$ meter, according to (25.6.25).

EXAMPLE 25.6.3

We are required to produce a photograph of a scene in which all objects between the distance $p_- = 8.0$ meters and $p_+ = 16.0$ meters are in good focus with a lens of 85-mm focal length. What is the largest aperture (smallest $f:$ number) we can use, and upon what object distance p_0 should the focusing scale of the lens be set? What are the image spot sizes under these circumstances for objects at distances of 4.0 and 32.0 meters?

Once more, the object distances involved are much greater than the focal length of the lens, so we may use the simple approximate expressions (25.6.24) and (25.6.25). We know what p_+ and p_- are to be, but we do not know the object distance setting and the $f:$ stop required to arrive at these values. However, we may easily solve (25.6.4) and (25.6.5) for these quantities. If we add (25.6.4) to (25.6.5), we obtain

$$\frac{2}{p_0} = \frac{1}{p_+} + \frac{1}{p_-} \qquad (25.6.33)$$

from which

$$p_0 = \frac{2p_+ p_-}{p_+ + p_-} \qquad (25.6.34)$$

Likewise, subtracting (25.6.24) from (25.6.25), we find

$$\frac{1}{p_-} - \frac{1}{p_+} = \frac{4R_0 f:}{f^2} \qquad (25.6.35)$$

or, solving for $f:$,

$$f: = \frac{f^2}{4R_0}\left(\frac{1}{p_-} - \frac{1}{p_+}\right) \qquad (25.6.36)$$

In the example outlined above, $f = 85$ mm $= 0.085$ meter, $p_- = 8.0$ meters, and $p_+ = 16.0$ meters, we shall, as usual, set $2R_0 = 2.5 \times 10^{-5}$ meter. In this way, we get

$$p_0 = \frac{(2)(16.0)(8.0)}{16.0 + 8.0} = 10.67 \text{ m}$$

and

$$f: = \frac{(0.085)^2}{(5 \times 10^{-5})}\left(\frac{1}{8.0} - \frac{1}{16.0}\right) = 9.03$$

If we set our diaphragm at $f:9.0$ and our focusing scale at 10.7 meters, therefore, we shall obtain the desired depth of field.

The image spot size is given by (25.6.11) or (25.6.19) or, alternatively, by (25.6.15) and (25.6.21), when the object distance is much larger than f, as it is here. Using these latter two equations, then, for $p_+ = 32.0$ meters, (25.6.15) gives

$$r_0 = \frac{f^2}{2f:}\left(\frac{1}{p_0} - \frac{1}{p_+}\right) = \frac{(0.085)^2}{(2)(9.03)}\left(\frac{1}{10.67} - \frac{1}{32.0}\right)$$

$$= 2.50 \times 10^{-5} \text{ m}$$

829

corresponding to an image disc diameter of 5.00×10^{-5} meter $= 0.05$ mm. At this distance, separate image components are resolved only to the extent of "20 lines per mm" on the film. Using (25.6.21) for $p_- = 4.0$ meters, we obtain

$$r_0 = \frac{f^2}{2f} : \left(\frac{1}{p_+} - \frac{1}{p_0}\right) = \frac{(0.085)^2}{(2)(9.03)}\left(\frac{1}{4.0} - \frac{1}{10.67}\right)$$

$$= 6.25 \times 10^{-5} \text{ m}$$

corresponding to a spot diameter of 1.25×10^{-4} meter $= 0.25$ mm, or a resolution figure of 8 lines per mm on the film. On a print enlarged eightfold from the negative, these resolution figures would be worse by a factor of 8, or 2.5 lines per mm for objects at a distance of 32 meters and 1.0 lines per mm for objects 4.0 meters distant.

25.7 Mirrors, Spherical and Parabolic

Nobody who has ever looked into a reflecting telescope or even a curved shaving or make-up mirror will seriously dispute that mirrors can produce magnified or reduced images just as lenses can. Indeed, as we shall soon see, the whole business of the lens equation, including considerations of image brightness, magnification, aberrations, depth of field, and so forth can be applied practically unaltered to situations wherein spherical or even parabolic mirrors are the imaging elements.

In the case of the spherical mirror, we could proceed to derive the focal properties by applying the geometric law of reflection to the spherical surface, but we may more conveniently and simply accomplish the same results starting with Fermat's principle of least time.

In Fig. 25.30, we see a spherical reflecting surface of radius r whose center of curvature is at C. A ray OQ

emitted from the axial object point O strikes the mirror at Q and is reflected so as to return to the axis at the image point O'. Figure 25.31 shows the behavior of converging rays and wavefronts that corresponds to this picture. The time required for a reflected ray to go from the object point O to the image point O' via some path such as OQO' is

$$t = \frac{OQ + QO'}{c} \tag{25.7.1}$$

But, from Fig. 25.30, it is apparent that

$$OQ = \sqrt{[p - r(1 - \cos \alpha)]^2 + r^2 \sin^2 \alpha} \tag{25.7.2}$$

and

$$QO' = \sqrt{[q - r(1 - \cos \alpha)]^2 + r^2 \sin^2 \alpha} \tag{25.7.3}$$

where p and q are the object and image distances, respectively, as illustrated and α is the angle between the radius CQ and the optical axis. We shall restrict ourselves, as in the case of lenses, to situations where this angle α is a small one, for which the approximations

$$\sin \alpha \cong \alpha \tag{25.7.4}$$

and

$$\cos \alpha \cong 1 - \tfrac{1}{2}\alpha^2 \tag{25.7.5}$$

can be used. Using these approximations in (25.7.2) and (25.7.3) and substituting the resulting expressions into (25.7.1), we find

$$ct = \sqrt{\left(p - \frac{r\alpha^2}{2}\right)^2 + r^2\alpha^2} + \sqrt{\left(q - \frac{r\alpha^2}{2}\right)^2 + r^2\alpha^2} \tag{25.7.6}$$

As with lenses and refracting surfaces, if the path OQO' is to be traversed in the minimum possible time, then the value of t for that path must be smaller than for other possible paths corresponding to differ-

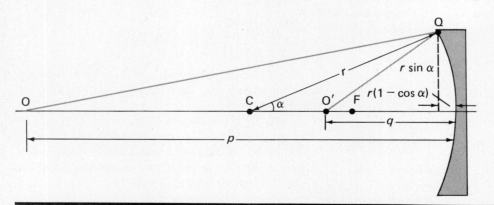

FIGURE 25.30. Calculation of the imaging equation for a spherical reflector using Fermat's principle.

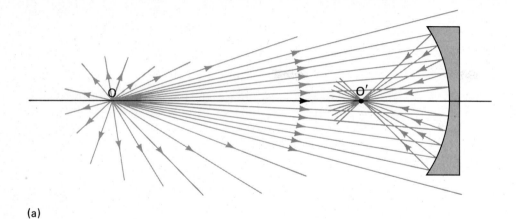

(a)

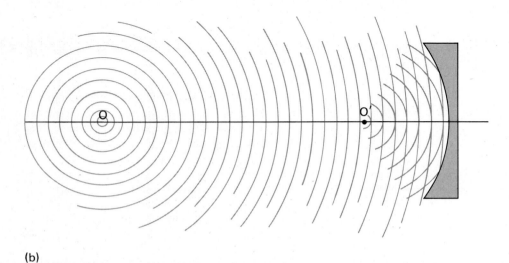

(b)

FIGURE 25.31. Pattern of (a) rays and (b) wavefronts associated with a spherical reflecting surface.

ent values of α. This means that the derivative of t with respect to α must be zero or, differentiating (25.7.6),

$$c \frac{dt}{d\alpha} = \frac{-\left(p - \dfrac{r\alpha^2}{2}\right)r\alpha + r^2\alpha}{\sqrt{\left(p - \dfrac{r\alpha^2}{2}\right)^2 + r^2\alpha^2}} + \frac{-\left(q - \dfrac{r\alpha^2}{2}\right)r\alpha + r^2\alpha}{\sqrt{\left(q - \dfrac{r\alpha^2}{2}\right)^2 + r^2\alpha^2}}$$

$$= 0 \tag{25.7.7}$$

or

$$r\alpha \left[\frac{r - p + \dfrac{r\alpha^2}{2}}{\sqrt{\left(p - \dfrac{r\alpha^2}{2}\right)^2 + r^2\alpha^2}} + \frac{r - q + \dfrac{r\alpha^2}{2}}{\sqrt{\left(q - \dfrac{r\alpha^2}{2}\right)^2 + r^2\alpha^2}} \right] = 0 \tag{25.7.8}$$

Once again, $r\alpha = 0$ is a possible solution but one that gives us little information. Another is obtained by

setting the quantity in brackets equal to zero. When we do this, we may greatly simplify the resulting expressions by recalling that we are, after all, restricting outselves to conditions under which the angle α is small and neglecting the terms containing α in comparison with the others. In this way, we obtain

$$\frac{r - p}{p} + \frac{r - q}{q} = 0 \tag{25.7.9}$$

Then, dividing both sides by r and simplifying, we obtain

$$\boxed{\frac{1}{p} + \frac{1}{q} = \frac{1}{f} = \frac{2}{r}} \tag{25.7.10}$$

where the focal length f is given by

$$\boxed{f = \frac{r}{2}} \tag{25.7.11}$$

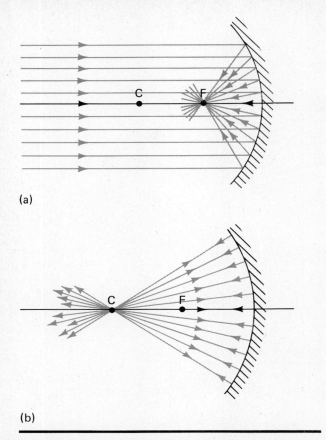

(a)

(b)

FIGURE 25.32. Reflection of (a) parallel incoming rays and (b) rays emanating from the center of curvature of a spherical reflector.

This is the form the imaging equation takes for a spherical mirror. It has the same form as the thin-lens equation, except that the focal length is equal to *one half the radius of curvature* of the spherical surface rather than the quantity shown in (25.4.18). For a mirror whose surface is concave as viewed from the object O, as shown in Fig. 25.30, the focal length f is positive, while for a surface that is convex when viewed from the object, it is negative. The image distance q is positive for images that are formed on the same side of the mirror as the object; but for virtual images that are formed behind the mirror, q is negative. In the case of a *plane* mirror, the radius of curva-

ture, and thus the focal length, is infinite. In this situation, $2/r = 1/f = 0$, and (25.7.10) gives

$$q = -p \qquad (25.7.12)$$

which tells us that the image is located behind the mirror at a distance equal to the distance between the object and the front surface of the mirror. This is in agreement with the ideas regarding the imaging properties of plane mirrors developed in the preceding chapter.

It is also possible to locate the images formed by spherical mirrors and to determine their size and orientation by ray construction methods very similar to those developed for thin lenses. These methods may be understood by noting from (25.7.10) that for an object at an infinite distance from the mirror, $1/p = 0$ and the image distance is $q = r/2 = f$. Under these circumstances, incoming rays from the object are *parallel* and come to a focus at a *focal point* F that is halfway between the center of curvature and the mirror, as illustrated in Fig. 25.32a. On the other hand, when the object is located at the center of curvature, $p = r$ and Eq. (25.7.10) then gives us $q = r$ also. In this case, the image coincides with the object at the center of curvature C, as illustrated in Fig. 25.32b. Rays emitted from the object at the center of curvature now are *radii* of the spherical mirror surface and are, therefore, perpendicular to this surface. According to the law of reflection, then, they will be reflected back upon themselves to reconverge to an image point coinciding with the object at C, as shown in Fig. 25.32c.

This can be summarized as follows: *Any ray that is parallel to the optical axis of the mirror will after reflection pass through the focal point F, while any ray passing through the center of curvature C will be reflected directly back through that point after striking the mirror.*

The use of these principles can be understood with reference to Fig. 25.33. An object is shown as the arrow at AB. A ray, such as AQ, that is emitted parallel to the axis BO passes through F after reflection, tracing out the path AQF. A ray, such as AC, passing through the center of curvature is reflected back upon itself along the path ACPC. These two

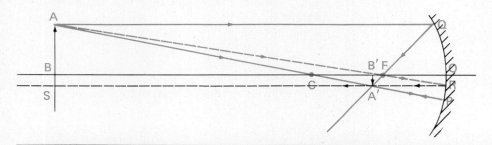

FIGURE 25.33. Ray diagram for a concave spherical mirror.

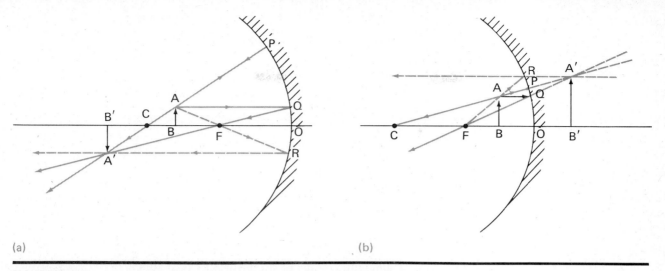

(a)   (b)

FIGURE 25.34.   Ray construction for objects very close to a spherical reflector.

rays intersect at A′, as do all other rays leaving A that are reflected from the mirror, and the image of point A is, therefore, formed at that point. It is also possible to see that the dotted ray AFR emitted from A and passing through F will after reflection travel parallel to the axis. The truth of this is clearly evident if the ray is retraced in the opposite sense. It is evident from the figure that this ray also passes through the image point A′. It is useful in some situations to utilize rays constructed like the dashed ray AFRA′ in addition to those illustrated by the solid lines in Fig. 25.33. In the case shown in this diagram, in which the object is beyond the center of curvature of the mirror, an inverted image of reduced size is formed, and since light rays such as AQFA′ and ACPA′ actually converge at the image point A′, the image is a real one.

The linear magnification M is the ratio of the image size A′B′ to the object size AB. But from the fact that triangles A′B′C and ABC are similar ones, whose corresponding parts are in the same ratio, we may write

$$M = \frac{A'B'}{AB} = \frac{B'C}{BC} = \frac{r-q}{p-r} = \frac{q}{p} \qquad (25.7.13)$$

since, after all, $r = OC$, $q = OB'$, and $p = OB$ in this example. The final form of this expression is obtained by substituting the value of r in terms of p and q obtained from (25.7.10). This equation can be stated in a somewhat more useful form by eliminating the image distance q. This may be done most easily by solving (25.7.10) for q to obtain

$$q = \frac{rp}{2p - r} \qquad (25.7.14)$$

and substituting this value into (25.7.13). It is then apparent that

$$M = \frac{r}{2p - r} \qquad (25.7.15)$$

The situation in which the object is closer to the mirror than the center of curvature is illustrated in Fig. 25.34, in which the images are located by the same ray construction rules that were developed above. In Fig. 25.34a, the object is within the center of curvature but still beyond the focal point. The image, located by the intersection of rays AQFA′ and APCA′ at point A′, is real, inverted, and enlarged in size. At (b), the object is closer to the mirror than the focal point, and the image, located by the intersections of the projections of rays APC and AQF, is a virtual one situated behind the mirror, upright, and enlarged in size.

The imaging properties of mirrors are frequently utilized in the design of astronomical reflecting telescopes. In the most common form, as first designed by Sir Isaac Newton, a large primary mirror forms a real image of a distant object that is very near the focal point F, as in Fig. 25.32a or Fig. 25.33. An enlarged secondary image is then obtained by positioning an eyepiece lens so that the primary mirror image lies just within the focal point of the eyelens, as in the refracting telescope of Fig. 25.26. A schematic diagram of a Newtonian reflecting telescope is shown in Fig. 25.35.

One of the most serious imaging deficiencies of simple lenses is referred to as *chromatic aberration*. Chromatic aberration stems from the dispersion of light within the optical glass of which the lens is fashioned, and its effect is to focus light of different colors at slightly different image points. As a result, images that are formed by simple lenses are surrounded by color fringes that seriously degrade their

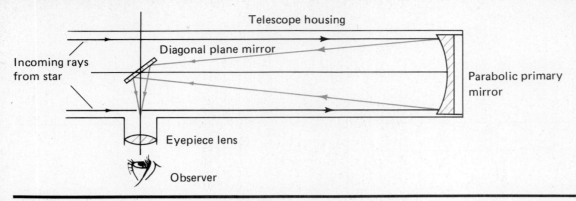

FIGURE 25.35. The Newtonian reflecting telescope.

sharpness and clarity. While it is possible to correct this condition by combining lenses having different refractive indices and dispersive powers, *mirrors* reflect light of all colors in exactly the same way and are, therefore, *completely free of chromatic aberration.* This is of great importance in the design of astronomical telescopes and is responsible for the fact that

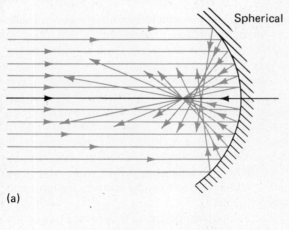

(a)

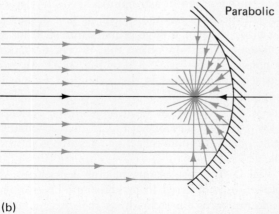

(b)

FIGURE 25.36. Ray pattern (a) for a concave spherical mirror, showing the effect of spherical aberration, and (b) for a parabolic reflector, which is free of spherical aberration for distant axial objects.

the largest and most powerful astronomical instruments are invariably reflectors.

At the same time, it is most desirable that spherical aberration of the kind shown for a simple lens in Fig. 25.8 be minimized or eliminated altogether. For the spherical mirror, if the object is at the center of curvature, all the rays that emanate from it are reflected so as to converge precisely once again at the center of curvature, no matter what part of the mirror they are reflected from. This is illustrated by Fig. 25.32a. A spherical mirror, therefore, has no spherical aberration for an object placed at the center of curvature. Unfortunately, it is difficult for us to place a distant star or planet at the center of curvature of a spherical mirror—objects such as stars are, whether we like it or not, at infinity for all practical purposes. Equally unfortunate is the fact that for objects at infinity, spherical mirrors have rather bad spherical aberration,[4] as illustrated in Fig. 25.36a. This is more or less evident in the derivation leading up to Eq. (25.7.10), wherein we ultimately assume the angle α to be so small that terms containing this angle can be dropped. If α is not a small angle, those terms will be important, and, as a result, the distance q at which a ray returns to the optical axis will differ for different values of α, resulting in spherical aberration of the type shown in Fig. 25.36a. For this reason, spherical mirrors are of limited usefulness in reflecting telescopes.

We are led to seek a substitute for a spherical mirror having no spherical aberration for objects at infinity giving rise to parallel incoming rays. As is proved in one of the examples below, *parabolic*[5] mir-

---

[4] This spherical aberration is suppressed for purposes of simplicity in Fig. 25.32a, but in actuality only the rays striking the mirror near the axis should pass through F. The outer rays reflected near the rim of the mirror should behave as in Fig. 25.36a.

[5] Strictly speaking, we should refer to parabolic mirrors as *paraboloids of revolution* whose axes coincide with the optical axis of the system. We shall always interpret the term *parabolic mirror* in this way.

FIGURE 25.37. The 200-in. Hale telescope at the Mt. Palomar observatory.

rors have *no spherical aberration* so long as the object distance is infinitely large. The reflecting properties of mirrors of this type are illustrated by Fig. 25.36b. Parabolic reflectors are, therefore, universally used as the primary mirrors in astronomical reflecting telescopes of large size. Mirrors of this type of over 5-meter (200-inch) diameter have been constructed. A photograph of the 200-inch Hale reflecting telescope at the Mt. Palomar Observatory in California is shown in Fig. 25.37. Parabolic reflectors are also used with a light source at the focus as headlight and searchlight reflectors to form accurately parallel beams from point sources.

The questions that arise about the sharpness and brightness of images formed by mirrors can be treated in exactly the same way used to discuss these matters in connection with lenses. Indeed, because the imaging equation is of the same form and the ray geometry essentially identical, the answers arrived at in section 25.6 related to lenses can be applied equally well to systems in which images are formed by reflection rather than refraction. The focal ratio, or f: number, of a mirror is defined as the ratio of its focal length to its diameter, as expressed by (25.6.6), just as for a lens. It is then not difficult to show that the brightness relations (25.6.4) and (25.6.5), the sharpness equations (25.6.13), (25.6.15), (25.6.19), and (25.6.21) and the depth of field equations (25.6.22), (25.6.23), (25.6.24), and (25.6.25) apply to mirrors as

well as lenses. Some additional information having to do with the properties of spherical and parabolic reflecting surfaces is brought to light by the examples that follow.

EXAMPLE 25.7.1

A concave spherical reflecting surface whose radius of curvature is 30.0 cm is used to form an image of an object at a distance of 120.0 cm from the mirror. What is the focal length of the mirror? Where is the image formed? What is the linear magnification? What are the answers to these questions for object distances of 60, 24, and 12 cm?

According to (25.7.11), the focal length is

$$f = \frac{r}{2} = \frac{30.0}{2} = 15.0 \text{ cm}$$

From (25.7.10), the image distance for an object distance of 120.0 cm is given by

$$\frac{1}{q} = \frac{1}{f} - \frac{1}{p} = \frac{1}{15.0} - \frac{1}{120.0} = 0.05833 \text{ cm}^{-1}$$

$$q = 17.14 \text{ cm}$$

The magnification can be found from either (25.7.13) or (25.7.15). Using the latter equation, for example, it is easy to see that

$$M = \frac{r}{2p - r} = \frac{30.0}{240.0 - 30.0} = \frac{30.0}{210.0} = \frac{1}{7}$$

For an object distance $p = 60$ cm, we find

$$\frac{1}{q} = \frac{1}{f} - \frac{1}{p} = \frac{1}{15.0} - \frac{1}{60.0} = 0.05 \text{ cm}^{-1}$$

$q = 20.00$ cm

According to (25.7.15), then, the magnification is

$$M = \frac{r}{2p - r} = \frac{30.0}{120.0 - 30.0} = \frac{30.0}{90.0} = \frac{1}{3}$$

For $p = 24.0$ cm,

$$\frac{1}{q} = \frac{1}{f} - \frac{1}{p} = \frac{1}{15.0} = \frac{1}{24.0} = 0.025 \text{ cm}^{-1}$$

$q = 40.0$ cm

while, according to (25.7.15), $M = 30.0/18.0 = 5/3$. In this instance, the object is closer to the mirror than the center of curvature, and the magnification is greater then unity, indicating an enlarged image.

For $p = 12.0$ cm, (25.7.10) gives

$$\frac{1}{q} = \frac{1}{f} - \frac{1}{p} = \frac{1}{15.0} - \frac{1}{12.0} = -0.016667 \text{ cm}^{-1}$$

$q = -60.0$ cm

the negative sign indicating that the image is a virtual one located behind the mirror. The magnification, according to (25.7.15), is

$$M = \frac{r}{2p - r} = \frac{30.0}{24.0 - 30.0} = -\frac{30.0}{6.0} = -5.00$$

the negative sign in this case signifying that the image is upright rather than inverted. Once again, since the magnitude of M is greater than unity, the image is enlarged.

EXAMPLE 25.7.2

Discuss the imaging properties of a *convex* spherical mirror, using both Eq. (25.7.10) and ray constructions. For a convex mirror of a 30.0-cm radius of curvature, find the image position and magnification for object distances of infinity, 180.0 cm, 36.0 cm, and 5.0 cm.

In the case of a *convex* mirror, such as the one shown in Fig. 25.38, the center of curvature and the focal point are *behind* the reflecting surface, on the side of the mirror opposite the object. The radius of curvature and, therefore, the focal length must then be regarded as negative. In the case of the mirror described above, the radius of curvature is -30.0 cm, and its focal length is -15.0 cm.

From the imaging equation (25.7.10), it is apparent that in this case as well as the case of the concave mirror, when the object distance p is infinite, $1/p = 0$ and $q = f$. But since f is negative, q must be

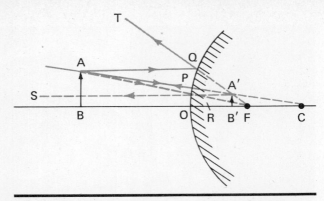

FIGURE 25.38. Ray diagram for convex spherical mirror.

negative also, and the image is *behind* the mirror, coinciding, as expected, with the focal point F. For a less distant object, the image remains behind the mirror but is closer to the reflecting surface than the focal point. This is evident from Fig. 25.38, in which the ray construction for such an image is shown. Once more, the construction follows the same familiar rules, the image point A' being located at the intersection of QA'F (the projection of AQT) and PA'C (the projection of APA). The dashed ray SRA', part of which is the projection of RS behind the mirror, also intersects this image point; in constructing this ray it is important to observe that the projection of AR intersects F. The physical reasons for the construction of this dashed ray are readily understood by reversing its path and considering how a ray such as SRA would be constructed. In any event, it is clear that the image is a virtual one and that it is erect and reduced in size.

For an object distance of 180.0 cm, the imaging equation (25.7.10) tells us that

$$\frac{1}{q} = \frac{1}{f} - \frac{1}{p} = -\frac{1}{15.0} - \frac{1}{180.0} = -0.07222 \text{ cm}^{-1}$$

$q = -13.85$ cm

The image is, therefore, located 13.85 cm behind the mirror, at a distance of 1.15 cm to the left of the focal point F in Fig. 25.38. For an object distance of 36.0 cm, we find in the same way

$$\frac{1}{q} = \frac{1}{f} - \frac{1}{p} = -\frac{1}{15.0} - \frac{1}{36.0} = -0.09444 \text{ cm}^{-1}$$

$q = -10.59$ cm

For an object distance of 5.0 cm, the same procedure leads to

$$\frac{1}{q} = \frac{1}{f} - \frac{1}{p} = -\frac{1}{15.0} - \frac{1}{5.0} = -0.2667 \text{ cm}^{-1}$$

$q = -3.75$ cm

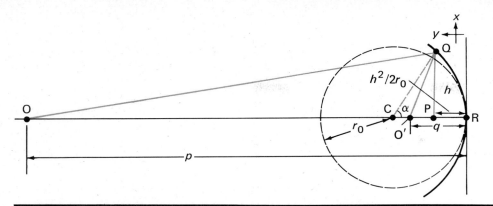

FIGURE 25.39. Geometry used for the derivation of the imaging equation of a parabolic reflector, using Fermat's principle of least time.

*EXAMPLE 25.7.3

Show that a parabolic mirror is free of spherical aberration for infinitely distant objects. Use Fermat's principle of least time.

In Fig. 25.39, we see a parabolic mirror centered at R, which reflects light from an object at O to an image point O'. A typical reflected ray is shown as OQO'. The point Q on the parabola is at a height h above the optical axis OO'PR. The time taken for a ray to travel along the path OQO' is

$$t = \frac{1}{c}(OQ + QO') \tag{25.7.16}$$

But

$$OQ = \sqrt{(OP)^2 + (PQ)^2} \tag{25.7.17}$$

while

$$QO' = \sqrt{(O'P)^2 + (PQ)^2} \tag{25.7.18}$$

We know that PQ is simply the height h of the reflection point above the axis; but to proceed any further, we must also be able to express the distances OP and $O'P$ in terms of p, q, and h. To do this, in turn, we must have an expression for the distance PR in terms of the height h. To obtain this expression, we must note that point Q is on a parabola whose equation, related to a set of xy-coordinate axes centered on point R as shown, is

$$y = \frac{x^2}{2r_0} \tag{25.7.19}$$

where r_0 is the radius of curvature of the parabola at the center point R. The correctness of this equation can be verified by recalling that *any* parabola oriented with respect to the coordinates as this one is has an equation of the form

$$y = ax^2 \tag{25.7.20}$$

where a is a constant. In addition, we should recall

from our study of calculus that the radius of curvature of *any* curve representing the function $y(x)$ in the xy-plane is given by

$$\frac{1}{r} = \frac{\dfrac{d^2y}{dx^2}}{\sqrt{1 + (dy/dx)^2}} \tag{25.7.21}$$

which may be evaluated at any point (x, y) on the curve. We must now simply note that if $y(x) = ax^2$, as given by (25.7.19), then $dy/dx = 2ax$ and $d^2y/dx^2 = 2a$, from which the radius of curvature of the parabola $y = ax^2$ must be, from (25.7.21),

$$\frac{1}{r} = \frac{2a}{\sqrt{1 + 4a^2x^2}} \tag{25.7.22}$$

Now at the central point R on the parabola, $y = 0$ and $x = 0$, and, therefore, (25.7.22) tells us that at this point the radius of curvature r_0 must be

$$\frac{1}{r_0} = 2a \qquad \text{or} \qquad a = \frac{1}{2r_0} \tag{25.7.23}$$

Substituting this value of a into (25.7.20) immediately gives us the equation for the parabola in terms of its central radius of curvature, written above as (25.7.19). The circle of radius r_0 that is tangent to the parabola at R is illustrated by the dashed curve in Fig. 25.39; its center is at C.

Now, clearly, the point Q lies on the parabola, and its x-coordinate is h. According to (25.7.19), then, its y-coordinate, which is equal to PR, must be $h^2/2r_0$. We may, therefore, write

$$PR = h^2/2r_0 \tag{25.7.24}$$

Knowing this, it is evident from the diagram that

$$OP = p - \frac{h^2}{2r_0} \tag{25.7.25}$$

837

and

$$O'P = q - \frac{h^2}{2r_0} \qquad (25.7.26)$$

Therefore, according to (25.7.17) and (25.7.18), we can write

$$OQ = \sqrt{\left(p - \frac{h^2}{2r_0}\right)^2 + h^2} \qquad (25.7.27)$$

and

$$QO' = \sqrt{\left(q - \frac{h^2}{2r_0}\right)^2 + h^2} \qquad (25.7.28)$$

Equation (25.7.16) now becomes

$$ct = \sqrt{\left(p - \frac{h^2}{2r_0}\right)^2 + h^2} + \sqrt{\left(q - \frac{h^2}{2r_0}\right)^2 + h^2} \qquad (25.7.29)$$

A ray traveling between O and O' could, in principle, be reflected from any point on the mirror at any height h above the axis. But Fermat's principle states that the actual path of the ray, and thus the actual value of h, will be such as to ensure that the time t required for the journey will be a *minimum*. This, in turn, means that the actual path (that is, the actual value of h) will be determined by the requirement that the derivative of t with respect to h is zero. Differentiating (25.7.29) and setting the resulting expression equal to zero, we find

$$c\frac{dt}{dh} = \frac{-\left(p - \frac{h^2}{2r_0}\right)\left(\frac{h}{r_0}\right) + h}{\sqrt{\left(p - \frac{h^2}{2r_0}\right)^2 + h^2}}$$

$$+ \frac{-\left(q - \frac{h^2}{2r_0}\right)\left(\frac{h}{r_0}\right) + h}{\sqrt{\left(q - \frac{h^2}{2r_0}\right)^2 + h^2}} = 0 \qquad (25.7.30)$$

Multiplying both sides of this equation by $1/h$, we may write

$$\frac{1 - \frac{p}{r_0} + \frac{h^2}{2r_0{}^2}}{\sqrt{\left(p - \frac{h^2}{2r_0}\right)^2 + h^2}} + \frac{1 - \frac{q}{r_0} + \frac{h^2}{2r_0{}^2}}{\sqrt{\left(q - \frac{h^2}{2r_0}\right)^2 + h^2}} = 0 \qquad (25.7.31)$$

This is the imaging equation for a parabolic reflector. It is important to emphasize at this point that there is one difference between this derivation and the others we have carried out for spherical lenses and mirrors using Fermat's principle: in this instance, we have made *no simplifying assumptions whatsoever* that

h, or the angle α in Fig. 25.39, is a small quantity, as we always have in our treatments of spherical refracting and reflecting surfaces. In other words, the calculations for the parabolic mirror are *exact*, and (25.7.31) represents its properties exactly. That (25.7.31) is not of the form $(1/p) + (1/q) = 1/f$ but, instead, involves h reflects the fact that the actual point where a reflected ray crosses the axis depends upon the height h at which it is reflected. This means, in turn, that in general the mirror has *spherical aberration*, as illustrated in Fig. 25.36a and that the amount of spherical aberration is exactly specified by Eq. (25.7.31).

But now, let us consider the case where the object distance p is very large, corresponding to an object that is for practical purposes at infinity. In this case, in the first term on the left in (25.7.31), we may drop all the terms that do not contain p and note that, in the limit as p approaches infinity, this term simply approaches the quantity $-1/r_0$. Under these circumstances, (25.7.31) can be written as

$$1 - \frac{q}{r_0} + \frac{h^2}{2r_0{}^2} = \frac{1}{r_0}\sqrt{\left(q - \frac{h^2}{2r_0}\right)^2 + h^2} \qquad (25.7.32)$$

Squaring both sides of this expression, we find that no less than four terms on each side of the resulting equation cancel, leaving only

$$1 - \frac{2q}{r_0} = 0 \qquad (25.7.33)$$

$$q = \frac{r_0}{2} = f$$

In other words, so long as the object is at infinity, the image point is always located at the same image distance $q = r_0/2$, *irrespective* of the height h at which different incoming parallel rays may have struck the mirror. The parabolic mirror, under these circumstances, is *completely free from spherical aberration*, at least for axial image points. It is this property that has led to its almost universal use in reflecting astronomical telescopes.

It is of interest to note, going back to (25.7.31), that if we do make the assumption that h is small compared to r_0, the angle α in Fig. 25.39 thus being small, then the terms containing h in this equation may be neglected, which leaves

$$\frac{1}{p} - \frac{1}{r_0} + \frac{1}{q} - \frac{1}{r_0} = 0$$

$$\frac{1}{p} + \frac{1}{q} = \frac{2}{r_0} = \frac{1}{f} \qquad (25.7.34)$$

This is exactly the same as the imaging equation for a spherical reflecting surface. We see, then, that under circumstances in which spherical aberration is *not* very important, spherical and parabolic mirrors have the same imaging characteristics.

25.8 Aberrations

What we usually desire in a lens or other optical element is a system capable of forming bright, perfectly sharp, distortion-free images not only of objects on the optical axis but also of objects that are off the axis within an angular field of view whose extent is appropriate to the intended use of the system. No real system attains this objective completely because of the inevitable presence of optical aberrations. These aberrations arise from the dispersion of light within the optical glass that is used and from the fundamental geometry of light rays reflected or refracted by spherical surfaces, which is simply not what is needed to accomplish our desired result with perfection. These shortcomings are evident from the fact that the imaging equations we have developed are approximate rather than exact and are, in fact, quite inadequate for systems in which the angles α and α' subtended by the optical surfaces at their centers of curvature are not particularly small. Another limitation arises from the fact that our imaging equations are derived for object points that lie on the optical axis. The geometry is obviously quite different, however, for objects that are off the axis. If the angle between the axis and a line joining the lens and the object is not very large, this will not make much difference and the axial imaging equations will still give good results; but for object points whose angular separation from the optic axis is large, they are not very accurate. In this case, the position of the image point and the magnification as well can be functions of the angular distance between object and axis, which gives rise to a whole new set of *off-axis aberrations*.

The major aberrations that afflict images on the optical axis are *spherical* and *chromatic* aberration, both of which have already been discussed to some extent. Spherical aberration, illustrated in Figs. 25.8 and 25.36a, occurs because all the rays from an axial object refracted or reflected by spherical surfaces do not return to the optical axis at the same point. For lenses or mirrors of large f: number, the angles α and α' in Figs. 25.15 and 25.30 can never be very large. Under these circumstances, the neglect of terms containing these angles in the derivation of the imaging equations is a very good approximation, all the rays entering the lens or mirror then falling at essentially the image point predicted by the imaging equation. Spherical aberration here is an extremely small effect. For large-aperture (small f: number) lenses and mirrors, however, the angles α and α' become large, and the small-angle approximations that lead to the simple

image equations are no longer valid. Under these conditions, the rays that enter the outer zones of the lens or mirror do not converge to the same axial focal point as those that pass through the central region, the situation being then similar to that illustrated in Figs. 25.15 and 25.30a. Spherical aberration under these circumstances is a serious problem and markedly degrades the sharpness and clarity of the image. Because spherical aberration increases rapidly as the lens diameter increases and in a large lens is contributed mostly by the outer zones, its effects may be minimized by stopping the lens down until a sufficiently large effective f: number is attained. When this is not feasible, it is necessary to employ compound lenses in which spherical aberration is *corrected* by combining two or more simple lenses.

The way in which this correction is accomplished depends upon two essential facts. First of all, the spherical aberration associated with a diverging lens is opposite in sign to that inherent in a converging lens. Thus, in the case of the diverging lens, the rays passing through the zones near the outer rim diverge more strongly than would be the case if they passed through the primary central-zone focal point. For a converging lens, they *converge* more strongly than is necessary to bring them to the primary focal point associated with the central part of the lens. In addition, in either converging or diverging lenses, the *amount* of spherical aberration depends critically on the radii of curvature of the lens surfaces. A lens of high refractive index and large radii of curvature will, therefore, have less spherical aberration than a lens of equal focal length having low refractive index and surfaces of smaller radii. It is, therefore, possible to combine a converging lens having high refractive index and large radii of curvature with a divergent lens of low index having smaller radii of curvature in such a way that the spherical aberration of the convergent and divergent lenses cancel one another, while at the same time the converging power of the first lens still outweighs the diverging power of the second. In this way, we may obtain a compound convergent lens corrected for spherical aberration in which central rays and rays going through the outermost rim zone of the lens come to the same focal point. In such lenses, the intermediate zones contribute rays that focus at slightly different points, but this secondary effect is ordinarily so small as to be quite negligible.

Chromatic aberration results from the dispersion of light refracted at lens surfaces. The effect of dispersion, you may recall, manifests itself as a variation of refractive index with light wavelength. Since from (25.4.17), the focal length of a simple thin lens depends upon its refractive index, it is evident that when white light passes through the lens, the components corresponding to different wavelengths, hence different

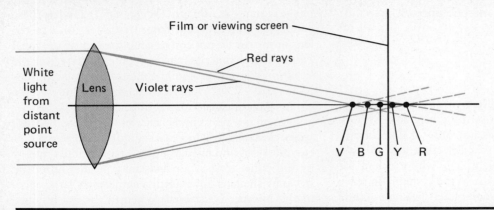

FIGURE 25.40. Chromatic aberration.

colors, will come to a focus at slightly different points, as illustrated in Fig. 25.40. Another way of understanding the situation is to recall that the cross section of a lens is that of a prism having curved sides, which accounts for the fact that in both cases white light is separated into spectral components. In the case of the lens shown in Fig. 25.40, since the refractive index is greater for violet light than for red, the short violet wavelengths are focused at a point V closer to the lens, than the point R, at which longer red wavelengths are focused. Blue, green, and yellow light, having intermediate wavelengths, converge to a focus at the points designated B, G, and Y. If a film or viewing screen is positioned near this range of focal points, it intercepts not a single, sharp white image point, but a series of colored circular discs of finite size. For a photographic lens, the effect is to smear an expected sharp point image into a disc of finite size and thus to degrade the sharpness of the image on the film. In an optical viewing system such as that of a telescope or microscope, the observer sees images surrounded by annoying color fringes that detract from the sharpness and clarity of the image. The problem is all the more serious because, since the difference in the positions of the red and violet foci depends only on the dispersion of the glass in the lens and is essentially independent of the lens diameter, chromatic aberration is not reduced at all by stopping the lens down to a large effective f: number with a diaphragm.[6]

It happens, however, that even though as a rough general rule high refractive index glass exhibits high dispersion, and vice versa, the dispersion of a given type of glass is affected by factors other than refractive index, and, therefore, the relation between dispersion and refractive index is not simply a direct proportionality. It is possible, in fact, to concoct optical glasses of high refractive index yet moderate dis-

persive power and others of moderate refractive index and relatively large dispersion. This makes it possible to fashion compound lenses practically free of chromatic aberration by combining a strongly convergent lens of highly refractive but moderately dispersive glass with a less strongly divergent lens made of glass of low refractivity but substantial dispersive power. In such a compound lens, the dispersion in the converging and diverging lenses cancel one another, while the converging power of the first lens outweighs the diverging power of the second, giving a combination that is still convergent but essentially free of chromatic aberration. The situation is similar to the one discussed previously in which it was found possible to correct spherical aberration by suitably combining lenses of different refractive indices and radii of curvature. A lens fashioned in this way that is free of chromatic aberration is referred to as an *achromatic* lens. A cross-sectional diagram of an achromatic lens is shown in Fig. 25.41.

It is also possible to design even more complex lenses that are corrected for *both* spherical and chromatic aberration. Achromatic lenses are invariably

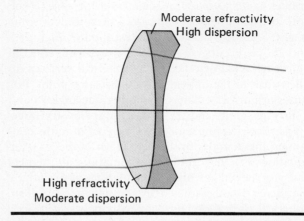

FIGURE 25.41. Compound achromatic lens, in which chromatic aberration is corrected using two glasses having different refractivity and dispersive power.

---

[6] The appearance of the image is improved, however, due to the increased depth of field obtained at small apertures.

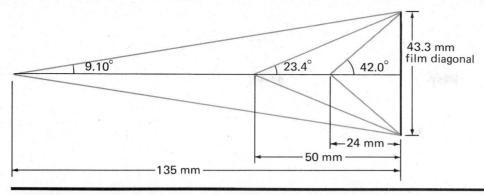

FIGURE 25.42. Angular field of view required for photographic lenses of 135 mm, 50 mm, and 24 mm focal length.

used in binoculars and refracting telescopes of good quality, and high-class photographic lenses are also of necessity achromatic.

Lenses that are well corrected for spherical and chromatic aberration produce very sharp and clear images of objects on the optical axis and are also generally satisfactory for objects that are no more than 5° to 10° off the optical axis. Such systems are, therefore, adequate for binoculars, telescopes, and photographic lenses of the long telephoto type.

There are many applications, however, in which optical systems that take in a much wider field of view are needed. A normal photographic lens (of 50-mm focal length for a 35-mm camera, for example) must cover a total angular field of view of about 47° and, therefore, must provide adequately sharp images of objects that are as far as 23.5° from the optical axis of the system. Under these circumstances, a whole new series of *off-axis aberrations* enters the picture and makes the task of producing a satisfactory optical system much more complex. The situation is illustrated in Fig. 25.42, where the angular field of view required to cover the 43.3-mm diagonal of the 24 mm × 36 mm frame of a "35-mm" negative is exhibited for a wide-angle lens of 24-mm focal length, a normal lens of 50-mm focal length, and a telephoto lens of 135-mm focal length. From elementary trigonometry, it is

apparent that the 135-mm telephoto lens must be able to image off-axis points out to an angle whose tangent is given by $21.65/135 = 0.1604$, corresponding to 9.1°. Such a lens would cover a total angular field of view of 18.2°. For this application, a lens that is well corrected for chromatic and spherical aberration, and nothing else, would probably be quite satisfactory. But the same line of reasoning applied to a normal lens of 50-mm focal length shows that it must image off-axis points out to an angle of 23.4° from the axis and cover a total angular field of 46.8°, while a wide-angle lens of 24-mm focal length must provide satisfactory imaging out to 42.0° from the axis and cover a total angular field of 84°. Under these circumstances, the off-axis image defects of *astigmatism, coma, curvature of field, distortion,* and *transverse chromatic aberration* become troublesome and must be minimized or corrected in some way to provide satisfactorily sharp and undistorted images. We shall describe the origin and general character of each of these aberrations but shall not be concerned with the complex details of how they may be corrected in actual optical systems.

Astigmatism arises as a result of a fundamental deficiency in the power of spherical refracting surfaces to image objects off the optical axis. Its effect is illustrated in Fig. 25.43. Here, we see an off-axis object at

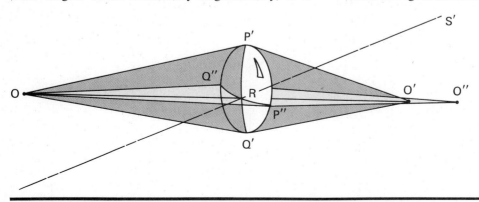

FIGURE 25.43. Astigmatism in off-axis images formed by a thin lens.

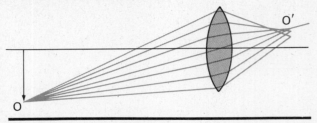

FIGURE 25.44. Effect of coma.

point O lying in plane SRO imaged by a lens. The effect of astigmatism is to image rays that lie in the vertical plane OP′Q′ and in the horizontal plane OP″Q″ at *different* points O′ and O″. Thus, at point O′, rays in plane OP′Q′ are in focus, but those in other planes, in particular OP″Q″, are not. At point O″, rays lying in OP″Q″ are in focus, but those in plane OP′Q′ are not. As the object point O approaches the optical axis, the points O′ and O″ approach one another more and more closely and coincide when the object is on the optical axis SS′. Simple lenses are, therefore, free of astigmatism for axial object points. The effects of astigmatism also become much less serious as the lens is stopped down to small apertures. Lenses afflicted with astigmatism bring radial and tangential lines to a focus in different planes so that the spokes of a wheel, for example, may be sharply defined while the rim is out of focus, or vice versa.

Coma arises by virture of the fact that *magnification* associated with rays passing through the outer zones of a lens may differ from that exhibited by rays that pass through the center. For an object on the axis, the difference in behavior between rim rays and central rays gives rise to spherical aberration, in which the image distance q is not the same for the two groups of rays. But the image magnification is, according to (25.5.2), $M = q/p$. Hence, if q is not the same for rays passing through the outer zones of a lens as it is for those that pass through the center, neither is the image *size*! This means that an off-axis point is focused at different heights above the optical axis by different zones of the lens, leading to a peculiar tear-

drop-shaped image smear corresponding to a point object as illustrated in Fig. 25.44. Coma gets worse as the angular distance from the axis increases but may be reduced by stopping the lens down.

Another aberration, referred to as *distortion*, arises because the magnification of an optical system may vary with the angular distance between the object and the optical axis. The effect of distortion is to render the image of square objects either barrel-shaped or pincushion-shaped according to whether the magnification of the system decreases or increases with angle. This is illustrated in Fig. 25.45.

Also, though an ideal thin lens of very small aperture images of flat plane perpendicular to the optical axis into a perfectly flat image plane, there is no guarantee that a lens of finite thickness or of large aperture will do so, particularly over large angular fields of view. The resultant deficiency is referred to as *curvature of field* and is illustrated in Fig. 25.46. Curvature of field may be reduced by stopping the lens down to a suitably small aperture or by restricting the angular field of view.

Finally since a lens may have different focal lengths for different light wavelengths, its *magnification* may also vary with light wavelength. The result is that images of different colors have slightly different *sizes*, which leads to off-axis color fringing in addition to that produced by axial chromatic aberration of the type illustrated in Fig. 25.40. This effect is referred to as *transverse chromatic aberration*, and stopping a lens down does not improve it, although it can be corrected by careful lens design along the lines described in connection with axial chromatic aberration.

All these defects are present to some degree in any optical system and are particularly troublesome in lenses of large aperture or wide field of view. They may be corrected, and in some cases practically eliminated, by combining many simple lenses into a compound system in which positive aberration effects associated with certain of the elements are balanced against negative ones present in others. By constructing a complex system of six or seven individual

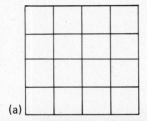 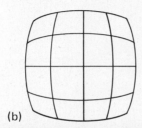 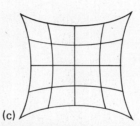

(a) (b) (c)

FIGURE 25.45. (a) Image of a square object formed by a distortion-free lens, (b) lens image exhibiting "barrel" distortion, and (c) lens image afflicted with "pincushion" distortion.

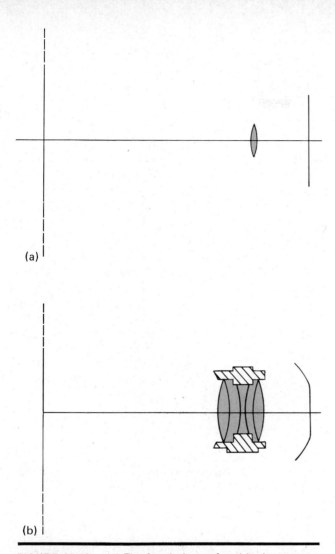

(a)

(b)

FIGURE 25.46. (a) Flat focal plane of an ideal aberration-free lens. (b) Curved focal surface of an actual photographic lens.

elements whose radii of curvature, refractive indices, and spacings, are chosen with the utmost care, for example, we may produce our superb 50-mm f:1.4 photographic objectives. These lenses are reasonably well corrected for all the aberrations we have described over an angular field of view of 45°, *but only for a single object distance,* usually infinity. That is why we may not get sharp pictures of objects 6 inches away, unless we stop down quite a bit, and why we usually cannot use our camera lens for enlarging.

Mirrors are subject to the same kind of aberrations as lenses, with the exception of both axial and transverse chromatic aberration from which, by their very nature, they do not suffer. Though our discussion of optical aberrations was framed in the language of lenses, it applies equally well, with the exception of the chromatic aberrations, to optical systems in which the imaging is accomplished by reflection.

SUMMARY

Lenses, spherical or parabolic mirrors, and pinholes can all be used to form optical images of source objects. The size and location of these images can be ascertained using the laws of geometrical optics.

Pinhole images are observed when a screen is placed behind a small hole that admits light to the interior of a dark enclosure. Their geometry is that defined by parallel columns of light passing through the aperture. Pinhole images are not very sharp but are equally distinct at every possible screen-to-pinhole distance.

Images are also formed by spherical refracting surfaces and by thin lenses. For a single spherical surface separating media of refractive indices n and n', with light incident from the substance of index n, the object and image distances p and q are related by

$$\frac{1}{p} + \frac{n_r}{q} = \frac{n_r - 1}{r}$$

where r is the radius of curvature and $n_r = n'/n$. The image distance is negative when the image is formed on the same side of the surface as the object. The radius of curvature is positive for surfaces that make entering rays more convergent and negative for surfaces that make them less convergent.

The focal point F is the point at which entering rays from an infinitely distant source object ($p = \infty$) come to a focus. The focal point F′ defines the object position for which the image distance becomes infinitely large ($q = \infty$). The focal distances f and f' are the distances between the respective focal points and the refracting surface. For the single spherical surface discussed above, it is evident that $f = rn_r/(n_r - 1)$, while $f' = r/(n_r - 1)$.

For a thin lens of refractive index n having spherical surfaces or radii r and r', the imaging equation becomes

$$\frac{1}{p} + \frac{1}{q} = \frac{1}{f} = (n_r - 1)\left(\frac{1}{r} + \frac{1}{r'}\right)$$

The two focal points are equidistant on either side of the lens, and there is now a single focal distance or focal length f, as given above.

Lens images may be located and studied by geometric methods involving the use of ray diagrams. In these diagrams, the image is located at the intersection point of different refracted rays from the same object point. In tracing these rays from object to image, there are three rays whose geometry is particularly simple and which are, therefore, of great usefulness in locating the image. They are (1) the ray from an object point that proceeds initially parallel to the axis; after refraction, this ray, or its projection, passes through one of the focal points of the lens;

(2) the ray that goes from the object to the center of the lens, through which it passes undeviated; and (3) the ray from the object that, after refraction, is parallel to the axis; this ray also must pass through one of the focal points. Geometric ray tracing is useful in finding the magnification of optical systems. For a single thin lens, the magnification is found to be

$$M = \frac{q}{p}$$

Ray tracing also helps us to understand complex lens systems like those found in telescopes and microscopes.

The brightness of lens images is influenced by the amount of light the lens collects from the object and the area of the image plane over which it is distributed. It can be shown that for large object distances, the brightness is given by

$$\bar{S}_i = \frac{K'}{f:^2}$$

where K' is a proportionality constant and where the f: number, or *focal ratio*, is defined in terms of the focal length and lens diameter by

$$f: = \frac{f}{d}$$

The sharpness of lens images depends upon the object distance, upon the f: number, and upon the focal length. In general, in a lens that is free from aberrations, images in the plane of focus will be perfectly sharp, while those at other distances will be more or less out of focus, in proportion to how far they are from the plane of perfect focus. The sharpness of objects outside the plane of perfect focus decreases as the focal ratio f: is decreased and is less for lenses of long focal length than for lenses of short focus.

Spherical mirrors form images by reflection that resemble in many ways those formed by lenses. For a spherical mirror of radius of curvature r, the imaging equation is

$$\frac{1}{p} + \frac{1}{q} = \frac{1}{f} = \frac{2}{r}$$

from which it is evident that the focal length f equals $r/2$. The radius of curvature is positive for concave mirrors and negative for convex mirrors. The image distance q is positive for images on the same side of the mirror as the object and negative for virtual images located behind the mirror. Mirror images can be constructed also by ray tracing.

Aberrations are defects in the imaging qualities of optical systems that arise from the inherent inadequacy of spherical surfaces to concentrate all rays that enter them onto the same object point and from the effects of dispersion in bringing different wavelengths to a focus at different points. Spherical aberra-

tion arises from the fact that rays refracted near the outer rim of spherical lenses, particularly those of small f: number, are not brought to the same focal point as those refracted near the center. Spherical aberration can be drastically reduced by reducing the aperture of the system but can also be corrected by combining two lenses of different refractivity and radii of curvature. Chromatic aberration is caused by dispersion, which renders the refractive index different for different light wavelengths, causing light of different colors to focus at different image distances. Chromatic aberration can be corrected by combining two lenses of different refractive index and dispersivity into a compound achromatic lens, but is not greatly reduced by reducing the system's aperture.

QUESTIONS

1. Give several reasons why lens surfaces are ordinarily spherical.
2. Plane mirrors and pinholes are optical elements that form images. Discuss some of the advantages and disadvantages of each of these elements.
3. What condition must be satisfied for an optical system to produce a sharp photographic image?
4. What is meant by spherical aberration?
5. How do we define the focal points of a refracting surface? Of a lens? Of a spherical mirror?
6. If photographic film is placed at the location of a virtual image, can a photograph actually be made? How do real and virtual images differ physically?
7. Describe the difference between a converging and a diverging lens.
8. What is the primary cause of chromatic aberration?
9. State the conditions which justify the thin-lens approximation.
10. If a lens has a negative focal length, what conclusions can you draw from the thin-lens equation?
11. Real photographic lenses are not thin lenses. Can you give some reasons for constructing them as "thick lenses"?
12. The focal length of a lens depends on the color of the light passing through it. Why?
13. The focal length of a thin lens is positive. Specify the conditions which lead to this.
14. Is it possible for a double-concave lens to form a real image of an object? If so, under what circumstances?
15. Is it possible for a diverging lens to form a real image of an object?
16. Do real images and virtual images appear the same to the eye? Is there any way of distinguishing between them merely by visual inspection?
17. Describe the main differences in the construction and use of a telescope and a microscope.
18. A 58-mm lens is operated at an effective aperture of f:11. What is the effective diameter at this setting?
19. Why do photographs taken using an effective aperture of f:22 usually exhibit much better overall sharpness than those taken at f:2?
20. What is meant by the depth of field of a lens?

21. The human eye is frequently said to resemble a camera. Discuss the optical structure of the eye and indicate how it resembles, and how it differs from, the optical system of a camera.

22. Diverging and converging lenses, respectively, correct nearsightedness and farsightedness in human vision. Explain how this is accomplished physically.

PROBLEMS

1. A pinhole camera is to be designed to produce film negatives of size 6.0 by 9.0 cm. It is desired that the field of view reproduced on the film should take in an angle of 45° measured along the film diagonal. A resolution of 2.5 lines per mm is desired on the negative. Find (a) how far from the film the pinhole should be placed, and (b) what the radius of the pinhole should be.

2. What would be the geometrical image spot diameter produced by the pinhole camera for an axial point object (a) at an infinite distance from the pinhole, (b) at a distance of 1.0 meter from the pinhole, (c) at a distance of 0.20 meter from the pinhole.

3. The pinhole camera of the two preceding problems is to be equipped with "wide angle" and "telephoto" pinholes. Find (a) how far from the film the wide-angle pinhole should be placed so that an angular field of view of 72° may be subtended by the film diagonal; (b) how far from the film the telephoto pinhole should be located for a corresponding angular field of 15°.

4. What is the image magnification of (a) the normal pinhole, (b) the wide-angle pinhole, (c) the telephoto pinhole used in the pinhole camera described in the three preceding problems, for an object at a distance of 6.0 meters from the pinhole? (d, e) What are the relative magnifications of the wide-angle and telephoto pinholes with respect to the normal pinhole?

5. A spherical refracting surface of radius 7.50 cm separates two refracting media of refractive index $n = 1.333$ and $n' = 1.800$. The center of the spherical surface lies within the more highly refracting substance. Find (a) the image distance for an object on the axis of the system, in the less refractive substance, infinitely distant from the refracting surface; (b) the image distance for an object in the less dense medium at a distance of 2.50 meters from the refracting surface; (c) the image distance for an object in the less dense substance at a distance of 12.0 cm from the surface; (d) the focal length f for incoming rays that originate in the medium of lower refractivity.

6. In the preceding problem, assume that the objects are located within the more highly refracting substance at the distances set forth in parts (a)–(c) above. (a) What is the relative index n_r in this case? (b) What is the sign of r, the radius of the surface? (c), (d), (e) What are the image distances corresponding to the object distances (a), (b), (c) of the preceding problem. (f) What is the focal length f' for incoming rays originating in the more highly refracting medium?

7. A single spherical glass–air refracting surface has a radius of curvature of 12.0 cm. The refractive index of the glass is 1.600, and it extends essentially to infinity beyond the

spherical refracting surface. The center of curvature of the surface lies within the glass. Find (a) the focal distances f and f' for rays originating within the less dense and more dense media; (b) the object distance at which a subject must be positioned in the less dense medium in order that its image be observed at a distance of 36.0 cm beyond the spherical surface, within the glass; (c) the object distance needed to form an image at a distance of 180.0 cm from the spherical surface, within the air outside the glass.

8. (a) For the spherical refracting surface described in the preceding problem, find the *exact* distance q for which an incoming ray parallel to the axis 00' in Fig. 25.12 crosses the axis, assuming that the point Q where it strikes the surface subtends an angle $\alpha = 5°$ at the center of curvature. (b), (c), (d) Repeat the calculation for parallel incoming rays for which α equals 10°, 20°, and 45° to the optical axis. Do *not* use the imaging equation, but instead apply Snell's law at point Q in Fig. 25.12, and proceed using elementary trigonometry. Compare your results with the predictions of the imaging equation (25.3.26). (e) What effect do these calculations exhibit?

9. For a spherical refracting surface such as the one shown in Fig. 25.12, of radius of curvature r, show that incoming parallel rays (corresponding to an object at infinity) across the axis at a distance q from the refracting surface given exactly by

$$q = r\left[1 + \frac{1}{\sqrt{n_r^2 - \sin^2 \alpha} - \cos \alpha}\right]$$

where n_r is the relative index of refraction and α the angle shown in Fig. 25.12 between the radius QC and the axis 00'. Show that this reduces to the result given by the imaging equation (25.3.26) in the case where the angle α is small. This expression is an exact formula for the spherical aberration of a single spherical refracting surface. (Hint: You must proceed by direct application of Snell's law at point Q in Fig. 25.12, rather than by any method involving use of the imaging equation, which is only good for small values of α. It may help your trigonometry to drop a perpendicular from point C onto line QO' in the figure.)

10. (a) An object is located at an axial point 3.6 meters distant from a thin lens of focal length 18.0 cm. Where is the image formed? (b) Where must the object be placed in order to form an image on the opposite side of the lens and 20.0 cm distant from it?

11. Find the focal length of the following thin lenses: (a) double convex, $r = 16.0$ cm, $r' = 8.0$ cm, $n_r = 1.600$; (b) meniscus, radius of convex surface 12.8 cm, radius of concave surface 36.0 cm, $n_r = 1.800$; (c) meniscus, convex radius 15.0 cm, concave radius 12.0 cm, $n_r = 1.800$; (d) plano-concave, concave radius 16.0 cm, $n_r = 1.750$; (e) which of these lenses are convergent and which divergent?

12. A certain plano-convex thin lens has a convex radius of curvature of 18.0 cm. It forms a real image of an axial object placed at a distance of 72.0 cm from it at a point 32.0 cm distance on the opposite side of the lens. What is the refractive index of the lens?

13. A converging thin lens forms a real image of an axial object at a point 24.0 cm from the opposite side of the

lens. When the object distance is decreased to two-thirds its former value, the image distance increases to 27.0 cm. **(a)** What is the focal length of the lens? **(b)** If it is a biconcave lens of refractive index 1.600, what are the common radii of its surfaces?

14. A thin biconcave lens forms a virtual image of an object on the same side of the lens as the object. For an object distance of 36.0 cm, the image is at a distance of 6.00 cm from the lens. **(a)** What is the lens's focal length? **(b)** Where will the image be if the object distance is reduced to 4.00 cm?

15. A plano-convex lens is to be made of glass of refractive index 1.50 and is to have a focal length of 30 cm. What should be the radius of curvature of the curved side? What would be the focal length of this lens if it were immersed in carbon disulfide ($n = 1.68$)?

16. A microscope has an eyepiece of 25-mm focal length and an objective lens of 12-mm focal length. These two lenses are separated by 250 mm, and the final (virtual) image is 200 mm from the eyepiece. Find the magnification of the microscope.

17. A hemisphere of plastic with index of refraction n and radius R is to be used as a lens. Find the focal points of this lens **(a)** for light entering the flat side and **(b)** for light entering the curved side. *Hint:* Is this artifact a "thin" lens?

18. An astronomical telescope's objective and eyepiece lenses have focal lengths of 40 cm and 2 cm, respectively. **(a)** How far apart should the lenses be placed to form an image at infinity? **(b)** For this case, what is the magnification? **(c)** By how much should the separation be changed to produce a virtual image 25 cm from the eyepiece?

*19. It is desired to make a microscope of magnification $100\times$ using two lenses, each of focal length 30 mm. The magnification of the objective lens and of the eyepiece lens are both of magnitude $10\times$. **(a)** What is the required separation of the objective and eyepiece lenses, and **(b)** how far from the eyepiece lens is the final virtual image formed?

20. A vessel 25 cm long with hemispherical end surface of radius 5 cm is filled with water ($n = 1.333$), as shown in

the figure. The other end is closed with a thin plane window. An object is placed 30 cm in front of the curved surface as shown. Neglect any refractive effects associated with the glass walls of the vessel. **(a)** Calculate the position of the image. **(b)** What is the magnification? **(c)** State and explain whether the image is erect or inverted and whether it is real or virtual.

21. A thin symmetric double-convex lens of index of refraction n_3 has two surfaces with the same radius of curvature r. The lens separates two media of refractive indices n_1 and n_2, as shown in the diagram. **(a)** In terms of r and the three refractive indices, where, in the first medium, should you place an object to produce an image in the second medium that has the same size as the object? **(b)** Where is the image formed if the object in the first medium is to be at infinity?

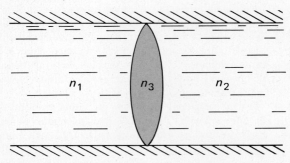

22. In the situation discussed in the preceding problem, using Fermat's principle of least time, show that the appropriate thin-lens imaging equation is $(n_1/p) + (n_2/q) = (n_3 - n_1)r^{-1} + (n_3 - n_2)r'^{-1}$. Show also that for an object on the left, the focal distance is given by $1/f = (n_3 - n_1)n_2^{-1}r^{-1} + (n_3 - n_2)n_2^{-1}r'^{-1}$, while for an object to the right of the lens, the focal distance is given by $1/f' = (n_3 - n_1)n_1^{-1}r^{-1} + (n_3 - n_2)n_1^{-1}r'^{-1}$. *Hint:* Follow the procedure outlined in section 25.4 for a thin lens in air, making appropriate modifications for the system at hand.

23. Show that the magnification factor of a diverging lens is always less than unity.

24. Two converging thin lenses are coaxially mounted, the spacing between them being 24.00 cm. Lens A, on the left, has a focal length of 18.00 cm, lens B, on the right, has a focal length of 3.00 cm. An axial object is located at a point 240 cm to the left of lens A. **(a)** Where is the image formed? **(b)** Is it real or virtual? **(c)** What is the overall magnification of the system?

25. In the preceding problem, the spacing between the lenses is descreased to 22.00 cm. Find the answers requested above for this case.

26. A compound telephoto lens is to be used to photograph an object 5 meters away. This lens consists of two thin

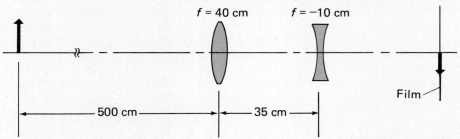

lenses of focal length 40 cm and −10 cm, as illustrated, the distance between them being 35 cm. Calculate the magnification, that is, the ratio of the size of the image on the film to the size of the object. How far behind the second lens must the film be placed?

27. A man in a parked car watches a bicycle in his rear view mirror as it approaches and after it turns a corner, at a constant speed of 10 mph. The mirror is convex, with a radius of curvature of 2.0 meters. The mirror is 20 meters from the corner, as illustrated in the figure. How fast is the image of the bicycle moving **(a)** just before and **(b)** just after it turns the corner?

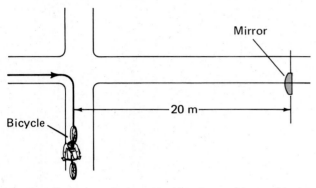

28. A policeman parked at the side of a road is watching a motorist in two rear view mirrors. One is plane, the other is convex with a radius of curvature of 2.0 meters. If the motorist is 100 meters away and the police radar set indicates that he is traveling 40 mph, how fast is his image moving as observed in each mirror? *Hint:* Can you differentiate the imaging equation with respect to time?

29. A microscope uses an objective lens of focal length 14 mm and an eyepiece of focal length 25 mm. The distance between the two lenses is 163 mm. Compute the magnification when the instrument is adjusted to form an image 250 mm from the eyepiece.

30. An 8.0-in.-diameter glass sphere of refractive index $n = 1.50$ has half its surface silvered. A narrow beam of sunlight falls on the central part of the unsilvered half. Where is the image of the sun formed by the emergent light?

*31. An opera glass (Galilean telescope) is constructed of lenses having focal lengths +10 cm and −3 cm, respectively. The two lenses are separated by a certain distance d. What is the magnification and the distance between lenses when an object at infinity is in focus?

*32. A telescopic lens consists of a positive lens of focal length 22 cm separated from a negative lens of focal length −15 cm by a distance of 12 cm. **(a)** Calculate the distance from the negative lens to the film plane needed to photograph an object at infinity and repeat the calculation for an object 4 meters from the positive front lens. **(b)** What is the effective focal length of the system? That is, what is the focal length of a single lens that could replace the combination to give the same magnification? **(c)** Compare the overall length of the two-lens system and the single-lens system when focused on an infinitely distant object.

33. A glass sphere of index n and radius r has a small object placed inside it a distance r/n from the center. Locate the image formed by light leaving the object toward the center of the sphere. Show that there is no spherical aberration.

34. A pinhole camera uses a pinhole of diameter 0.30 mm to produce an angular field of view 45° across the diagonal of a film of size 6.0 by 9.0 cm. **(a)** What is the f: number of the pinhole? **(b)** An exposure meter indicates that a correct exposure for a desired scene is 1/60 second at f:11. How long should the film in the pinhole camera be exposed, assuming that the film response is proportional to total light energy received?

35. What are the f: numbers of the system described in problems 7 and 8 for angles $\alpha =$ **(a)** 5°, **(b)** 10°, **(c)** 20°, and **(d)** 45°, assuming in each case that rays corresponding to larger values of α are prevented from entering the refracting surface? Assume that the light rays entering the system originate within the less dense medium.

36. A concave spherical mirror has a radius of curvature of 4.0 feet. A tiny cube with sides 0.01 foot long is placed symmetrically on the axis of the mirror with its center 6.0 feet from the mirror. The corners are labeled A, B, C, and D, as shown in the diagram. Denote the images of the corners by A′, B′, C′, and D′. Calculate the distances A′B′,

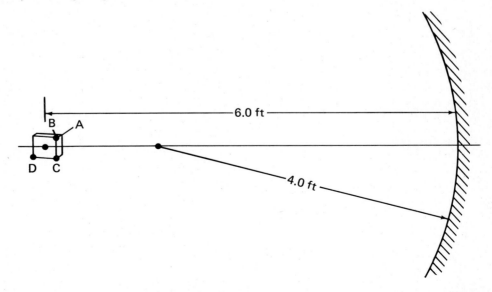

$B'C'$, and $C'D'$ and show the positions of A', B', C', and D'
on a sketch.

37. A thin double-convex lens whose radii of curvature are
30 cm is made of glass of refractive index $n = 1.60$. It is
used as the end of a long, liquid cell which contains carbon
tetrachloride, for which $n = 1.40$. Find the position of the
image of an infinitely distant exterior object.

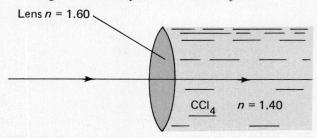

Lens $n = 1.60$

CCl_4 $n = 1.40$

38. A double-convex lens whose surfaces have equal radii
separates two media with refractive indices 1.34 and 1.68,
as shown in the diagram. What should the refractive
index of the lens be if a parallel beam incident on the
system from the left is to emerge once more as a parallel
beam on the right.

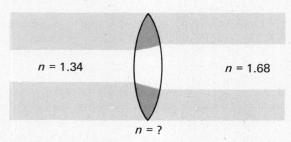

$n = 1.34$ $n = 1.68$

$n = ?$

39. A glass test tube has a spherical bottom with a radius of
curvature of 2 cm. The tube is filled to the 10-cm mark
with water as shown in the diagram on top of the next
column. (a) At what position x should an object be placed
in order that its image will coincide with the surface of
the water? (b) If light rays from infinity enter the bottom
of the tube parallel to the axis, where will they converge
or focus? (c) Where will parallel light rays entering the
top of the tube come into focus? Neglect any refractive
effects associated with the thin glass walls of the vessel.

40. A lens has two surfaces one of which is convex, the other
concave. If the convex surface has a radius of curvature of
25 cm and the concave surface has a radius of curvature
of 40 cm, find the focal length of the lens in air, assuming
it is made of glass with refractive index $n = 1.55$. Is the
lens a converging or diverging lens?

41. A thin glass lens of refractive index 1.50 has a focal length
of 15 cm in air. What is the focal length of the same lens
when it is placed in water ($n = 1.333$)?

42. A double-convex lens has surfaces with radii of curvature
of 15 cm and 20 cm. When an object is placed 30 cm from
the lens, a real image is formed 40 cm from the lens. Find
(a) the focal length of the lens and (b) the index of refraction of the lens relative to air.

43. A double-convex lens has surfaces with radii of curvature
10 cm and 15 cm and is made from glass of refractive
index 1.55. Find (a) the focal length in air and (b) the focal
length when submerged in a medium with refractive index

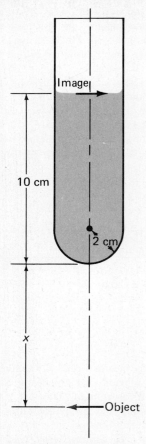

Image

10 cm

2 cm

x

Object

1.60. (c) For the lens in air, locate the image of an object
2 meters from the lens. (d) For the lens immersed in the
medium with $n = 1.6$, find the answer to part (c).

44. A photographer snaps a picture of a bird using lens of
55-mm focal length. The bird, which is 12 cm in length, is
2 meters from the lens. What will be the size of the bird's
image on the film? How large will the image be if the
photographer replaces the first lens with a telephoto lens
of 135-mm focal length?

45. A certain 55-mm camera lens has a focal ratio of f:1.4.
What is the diameter of the lens? If a picture is taken at an
aperture of f:22, what is the effective aperture diameter?
How should the f:number be changed to increase the
light intensity at the film plane by a factor of 4?

46. A photographer wishes to photograph a postage stamp
using a 35-mm camera equipped with a lens of 55.0-mm
focal length. It is desired that the image on the film be
exactly half the size of the stamp. An exposure meter
indicates that a correct exposure would be 1/125 second
at f:5.6. Find (a) the object distance, (b) the image distance, and (c) the aperture marking at which the lens
should be set for correct exposure, assuming that a shutter
speed of 1/125 second is to be used.

47. What would the answers to the preceding problem be if
the image on the film were to be *twice* the size of the
postage stamp?

48. An 85-mm photographic lens is used to photograph a
landscape. It is set at an aperture of f:6.3, and the focusing
scale is set to infinity. What would be the diameter of the
image disc of a point object at a distance of 8.0 meters

from the lens? Do you think that the image of such an object would appear sharp in an 8-diameter enlargement of a 24 by 36-mm negative?

49. In the preceding problem, assuming that an image disc radius of 0.020 mm is necessary for reasonably sharp images, on a 24 by 36-mm negative that is to undergo eightfold enlargement, find **(a)** the optimum distance on which the lens should be focused at $f:6.3$ in order that foreground objects as well as objects at infinity be reasonably sharp; **(b)** the minimum f: number that can be used if objects 8.0 meters distant and infinitely distant are both to be adequately sharp.

50. A photograph is to be made on 35-mm film, using a lens of 50-mm focal length. If the lens aperture is $f:3.5$ and the lens is focused on an object 6.00 meters from the lens, find **(a)** the smallest and **(b)** the largest distances at which reasonably sharp images of other objects will be obtained. Assume that an image disc radius of 0.020 mm or less is necessary for an adequately sharp image on the final print. **(c)** It is required that the zone of adequate image sharpness extend from image distances 5.0 to 8.0 meters. **(c)** At what distance should the focusing scale be set, and what aperture should be selected to bring about this desired result?

*51. A biconvex lens 4.00 cm in diameter is made from dense flint glass whose refractive index is 1.6912 for $\lambda = 4000$ Å and 1.6473 for $\lambda = 7000$ Å. The common radius of curvature of its two surfaces is 12.0 cm. Find **(a)** the focal length of the lens for light of wavelength 4000 Å, **(b)** its focal length for light of wavelength 7000 Å. **(c)** What is the diameter of the image spot corresponding to an infinitely distant axial point object at infinitely, when white light containing all wavelengths from 4000 to 7000 Å is used, and when a film or focusing screen is positioned at the point where $\lambda = 4000$ Å comes to a focus? **(d)** What is its diameter when the film or screen is positioned at the focal point for light for which $\lambda = 7000$ Å? **(e)** What is the spot diameter when the film or screen is positioned at the point between these two foci for which the spot size is a minimum?

52. In the preceding problem, what would the answers be if the lens were stopped down to half its former diameter?

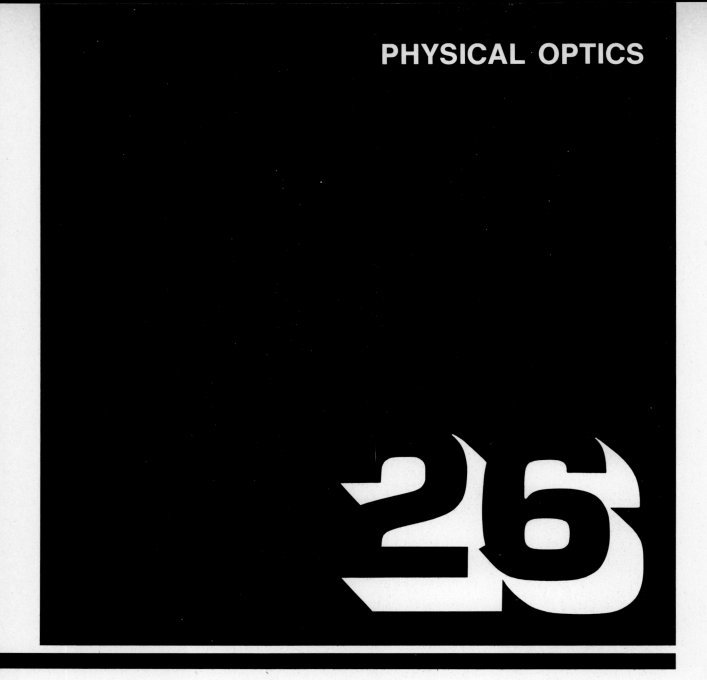

PHYSICAL OPTICS

26.1 Introduction

We have now completed our study of geometrical optics and are now ready to begin the subject of physical optics, or wave optics. In geometrical optics, the reflection and refraction of light are explained in terms of the behavior of light rays. These rays, which characterize the direction of propagation, invariably travel in straight lines, except when they undergo reflection or refraction at interfaces between optical media having different characteristics. The laws of geometrical optics are applicable only when the dimensions of the system greatly exceed the wavelength of light. When this condition is no longer satisfied, the usual rules of geometrical optics are inadequate to describe all aspects of the behavior of light, and new and interesting phenomena such as interference and diffraction enter the picture.

These effects can be understood only by taking explicit account of the fact that light is a wave phenomenon. It is important to recognize that light waves arriving at a given point from separate sources, or even from different parts of the same source, are combined in accord with the usual principle of superposition that is valid for all the waves we have studied. It is important, in superposing these different contributions, always to take proper account of the *phase relationships* between the individual beams involved. It is also important, in the study of interference and diffraction, to recognize the differences between *coherent* light sources, whose radiation is characterized by constancy of phase that is maintained for long periods of time, and *incoherent* sources, which emit energy whose phase fluctuates rapidly and irregularly.

26.2 Interference and Coherence

The phenomenon of interference occurs in all systems that exhibit wave behavior. We have already studied some of its more important aspects in Chapter 10. In that chapter, we learned that a *spatial interference pattern* exists whenever two or more waves add so as to reinforce one another at some points and to cancel one another at other locations. A very simple and familiar example of this effect involving the interference of water waves originating from two separate sources is shown in Fig. 26.1. In that illustration, the two sources, which emit waves by periodically perturbing the surface of the liquid, are synchronized so that their up-and-down motions are in phase.

The points on the water surface at which cancellation occurs represent points of minimum—ideally zero—displacement. In this system, they lie on hyperbolic curves called *lines of nodes*. At a consider-

FIGURE 26.1. Interference of water waves from two coherent point sources. Photo courtesy of Professor T. A. Wiggins, Pennsylvania State University.

able distance from the source, these hyperbolic curves are asymptotic to straight lines. Points where maximum reinforcement takes place lie on a similar series of curves called *lines of antinodes*. An interference pattern such as this is referred to as a *stationary interference pattern*, if the location of the maxima and minima remains the same at all times. It is evident that such a stationary pattern can be achieved only if the sources have exactly the same frequency and, therefore, only if there is a definite phase relationship between them that persists for long periods of time. This does not mean that the sources must necessarily have the same phase but only that there must be some *constant* phase difference between them that is maintained for a relatively long time, long enough to establish the pattern and to allow it to be observed. The precise appearance of the pattern may differ for different phase differences, but there will be a stationary, observable interference pattern whatever the relative phase of the sources, provided only that it is maintained for a long enough time. Two sources that have these characteristics, and which therefore can produce observable stationary interference patterns, are said to be *mutually coherent*. The situation is illustrated in Fig. 26.2.

But now consider two sources whose relative phase may *change* over a period of time. Such a change may occur because their frequencies may differ very slightly; under such circumstances, their relative phase varies at a rate given by the difference between the frequencies. Alternatively, even in the case where the frequencies are exactly the same, it may happen that some external influence acts at irregular intervals to bring about an instantaneous change in phase of one or both sources. In either case, it is evident that even though an interference pattern can be established and observed, its character will undergo a noticeable change in a time interval τ_c sufficiently long to ensure that a significant alteration in relative phase between the sources takes place.

If τ_c is very small, then the phase relationship between the two sources may not persist long enough for a stable pattern of interference to be observed. Though an interference pattern may exist at any instant of time, it changes so rapidly that we are unable to observe it or detect it in any way. On the other hand, if τ_c is extremely large, the phase relation between the sources remains practically unaltered over appreciable time intervals, and we may be able to observe stable interference effects like that shown in Fig. 26.1. The time τ_c is referred to as the *coherence time*. If it is infinite, or at any rate very long in comparison with the time required to establish and observe an interference pattern, the sources of light that are involved are said to be *coherent* ones. If it is too short to permit the observation of interference effects,

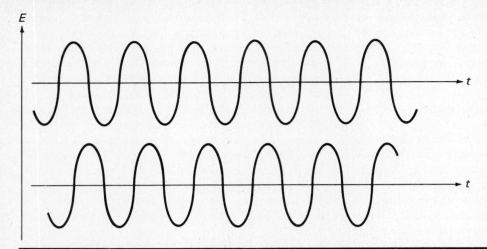

E

t

t

FIGURE 26.2. Two coherent sinusoidal waves with a constant phase difference of 45°.

the sources of light that are used are referred to as *incoherent*. Obviously, there could be cases that do not fall clearly into either category, but, generally speaking, this is a satisfactory way of categorizing this important aspect of light sources, and most of the sources that are commonly available can be accurately classified as either coherent or incoherent.

We know from our previous work that radio waves can be produced by exciting high-frequency ac currents in ordinary circuit loops. When we do this, we cause free charges to execute accelerated oscillatory motion, which causes the radiation of electromagnetic waves having the same frequency as the ac current in the circuit. If we were to connect two such high-frequency ac circuits in series, the currents in each would always have the same phase, and the radio waves emitted by the two circuits would exhibit a stationary interference pattern. Under these circumstances, the coherence time is essentially infinite and the radiation from the two circuits is mutually coherent, like the water waves in Fig. 26.1.

On the other hand, suppose we have two sodium vapor lamps, in which light is emitted when sodium atoms that have acquired energy from an applied electric field return to their normal state. This energy is emitted as a result of a quantum process in which the valence electron of a sodium atom is excited from a normal state of lowest energy to a higher energy state by the electric field. There are only discrete energy states allowed in this system; the atom cannot have any "in-between" energy. After a time, the valence electron may return to its normal state, losing an amount of energy equal to the difference in energy between the normal state and the excited state. This energy is emitted as a tiny pulse of light, and since the quantity of energy in all these processes is the same, the light always has essentially the same frequency. The light emitted by such a source is, therefore, prac-

tically monochromatic; for the sodium vapor source the emitted radiation is yellow light[1] of wavelength $\lambda = 5890$ Å. But the individual pulses, which are extremely short and very frequent, are emitted at random and irregular time intervals as each excited atom independently returns to its normal unexcited condition. There is, therefore, no stable phase relationship between individual light pulses and, in fact, no well-defined condition of phase for the source as a whole that persists for any appreciable time. Indeed, the phase of the light emitted by a source of this type typically changes in a random way at intervals of about 10^{-8} seconds! Therefore, we cannot observe any stable interference phenomena between two such sources—the beams of light they emit are *incoherent*. The situation is quite similar for ordinary incandescent and fluorescent lamps as well as for natural sunlight or skylight, with the added complication that none of these sources is even monochromatic, let alone coherent.

The preceding examples might suggest to us that coherent light sources might be extremely difficult to find, but this is not the case. For one thing, atomic processes such as those that occur in the sodium lamp can, under proper conditions, be made to take place "in step," so to speak, so that all the light pulses that are emitted have the same phase. The device by which this remarkable result is accomplished is referred to as a *light amplifier utilizing stimulated emission of radiation*, or *laser*, for short.

But even more simply, if some means can be found to split the light emitted by a single source, even an incoherent one, into two beams that can

---

[1] Actually, since there are two slightly different quantum processes that can occur, the light is emitted at two slightly different frequencies corresponding to $\lambda = 5890$ Å and $\lambda = 5896$ Å. This, for our discussion, makes little difference; we shall assume that we are able to suppress the 5896 Å light by appropriate experimental methods.

subsequently be recombined, the instantaneous phase relationships of the source will be preserved *in each beam* as they reencounter one another. This, in turn, should permit the observation of stationary interference patterns. Fortunately, there are a number of simple ways to do this whose details will be described later. It is, therefore, not difficult to observe the interference of light waves if proper experimental methods are used.

In Chapters 23 and 24, we have seen that the energy entering a detector of electromagnetic radiation can be determined from the Poynting vector $\mathbf{S} = (\mathbf{E} \times \mathbf{B})/\mu_0$. In the case of light, the instantaneous magnitude of this vector fluctuates so rapidly that any detecting system, such as the eye or a photoelectric cell, can determine only the average magnitude of $\mathbf{S}$, which we denote by $\bar{S}$. If there is more than one source of light, then the intensity at any point must be determined by properly superposing the fields arising from each and finding the time average of the Poynting vector associated with the resultant field. In doing this it is important to observe the rules of superposition, which are:

1. At any given point and at any time, the electric (or magnetic) field due to a single wave passing through the point is independent of all the other waves that may be present.
2. The total electric (or magnetic) field at the point

in question is the vector sum of the electric (or magnetic) fields of all the individual waves present.

It will be noted that these rules are no more nor less than those previously given for the superposition of electrostatic and magnetic fields. If there are two sources, as shown in Fig. 26.3, for example, then the total electric and magnetic fields $\mathbf{E}_t(x, y, z, t)$ and $\mathbf{B}_t(x, y, z, t)$ at a point (x, y, z) at time t will be

$$\mathbf{E}_t(x, y, z, t) = \mathbf{E}_1(x, y, z, t) + \mathbf{E}_2(x, y, z, t) \qquad (26.2.1)$$

and

$$\mathbf{B}_t(x, y, z, t) = \mathbf{B}_1(x, y, z, t) + \mathbf{B}_2(x, y, z, t) \qquad (26.2.2)$$

The Poynting vector can now be written as

$$\mathbf{S} = \frac{1}{\mu_0}(\mathbf{E}_t \times \mathbf{B}_t) \qquad (26.2.3)$$

where $\mathbf{E}_t$ and $\mathbf{B}_t$ are the total fields given by (26.2.1) and (26.2.2). We have already seen in Eq. (23.5.17) that the time average of the magnitude of the Poynting vector, which represents the *intensity* of light at the point, is

$$\bar{S} = \overline{E_t B_t}/\mu_0 = E_t^2/c\mu_0 \qquad (26.2.4)$$

As a consequence, the average intensity will always

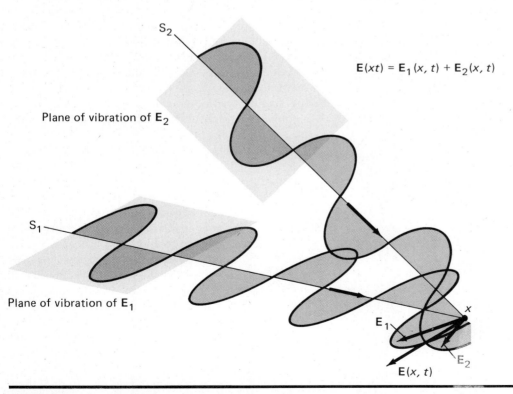

$$E(xt) = \mathbf{E}_1(x, t) + \mathbf{E}_2(x, t)$$

FIGURE 26.3. Interference of two coherent electromagnetic waves. The magnetic vectors are omitted in the interest of clarity.

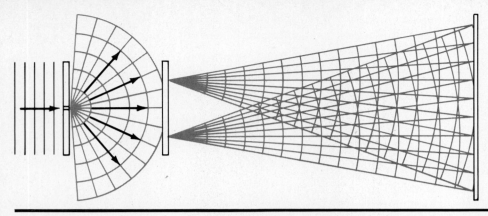

FIGURE 26.4. Diagram of rays and wavefronts in Young's experiment exhibiting the interference of light from two coherent point sources.

be proportional to the average of the *square* of the electric field, that is,

$$\overline{S} \quad \overline{E_t^2(x, y, z, t)} \qquad (26.2.5)$$

The relative intensity at different points can, therefore, always be determined simply by studying the average of the square of the electric field at the location in question. If the absolute intensity is needed, it can be obtained by multiplying by the factor $1/c\mu_0$ in (26.2.4).

26.3 Young's Double-Slit Experiment

In order that two light sources produce an observable interference pattern, there must be a constant phase relation between them, that is, they must be mutually coherent. The first successful experiment that exhibited the interference of light was carried out by the English physician Thomas Young in 1801. Young was originally interested in the study of vision but soon became involved in research on the theory of light. He proposed that light is a wave phenomenon and

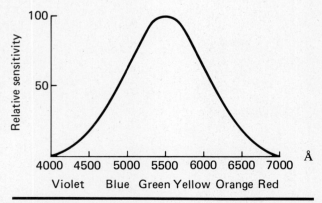

FIGURE 26.5. Spectral sensitivity curve of the human eye.

attempted to prove this by demonstrating the existence of interference and by measuring the wavelength of the light.

In Fig. 26.4, incident sunlight passes through a small pinhole. According to Huygens's principle, the pinhole will act as a point source of spherical waves. These waves then progress toward a second screen containing *two* pinholes which act as secondary sources of light and send out spherical waves; these waves superpose on a screen some distance away. In this way, Young was able to observe an *interference* pattern of alternate light and dark bands on the screen. The pinholes are small enough in diameter to avoid complications that might arise due to interference of light from different portions of a single pinhole. By this method, Young was able to infer a light wavelength of 5700 Å from studies of the interference pattern obtained from sunlight. This wavelength, in the greenish-yellow portion of the spectrum, corresponds very nearly to the peak in the sensitivity curve for the human eye (see Fig. 26.5) and can, therefore, be regarded as the dominant frequency in white light.

We shall not analyze the experiment as it was originally performed by Young but will discuss instead a similar experiment in which the two pinholes are replaced by two narrow slits upon which plane waves are normally incident. These waves may arise from a very distant source or from a nearby source placed at the focal point of a converging lens. The geometry, as illustrated in Fig. 26.6, is established in such a way that the intensity can depend only on the distance z from the center of the screen. It does not vary with the distance y since the secondary slit sources of light possess cylindrical symmetry. This symmetry greatly simplifies the analysis, for the intensity at any point depends only on the relation between the two rays drawn in Fig. 26.7. Both of these rays lie in a plane perpendicular to the screen and the slits.

854

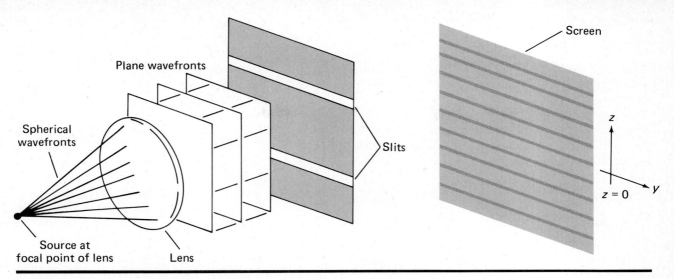

Plane wavefronts

Spherical wavefronts

Source at focal point of lens

Lens

Screen

z

$z = 0$

y

Slits

FIGURE 26.6. Another view of Young's double-slit interference experiment.

Let us now state clearly the assumptions that are made in the present analysis:

1. The primary source is assumed to be a *monochromatic*[2] source.
2. The two slits act as secondary sources of monochromatic light that are *mutually coherent*. In this case, they are exactly in phase, because the geometry has been arranged so that both slits always lie on the same wavefront.
3. The width of the slits is narrow enough to avoid interference of light coming from different portions of a single slit.
4. The screen containing the slits is a distance D from the observation screen, and D is much larger than the separation d between the slits.

Let us now calculate the electric field at an observation point P quite distant from the slits. At any time t, the electric fields at each of the secondary slits will be identical and are assumed to be given by $E_0 \cos \omega t$. These oscillating fields propagate outward and will eventually pass through P. Since the waves are cylindrical, the amplitudes will fall off as $1/\sqrt{r}$, where r is the distance from the source to the observation point. The distances QP and RP are very nearly equal, and, therefore, it may be assumed that the *amplitudes of the received waves are practically identical*. In fact, since the directions of propagation of the two waves are almost parallel, the individual electric fields can be assumed to have amplitudes $\mathbf{E}_1$ and $\mathbf{E}_2$ that are identical in *both* magnitude and direction. These vectors, which are illustrated in Fig. 26.7, are, there-

fore, assumed to be parallel at point P. This assumption is good, however, *only* when the distance d between the slits is small in comparison with the separation D between the slits and the screen.

On the other hand, the *path difference* $r_1 - r_2$ between the distances that each wave travels before striking the screen introduces a significant *difference in phase* between the two waves as they arrive at point P. It is this phase difference that is, in fact, responsible for most of the important characteristics of the observed interference pattern. The wave field observed at point P as a result of emission from the source at R can be written

$$E_1 = E_0 \cos(kr_1 - \omega t) \qquad (26.3.1)$$

where E_0 is the amplitude of the electric field associated with the wave and k represents the common magnitude of the vectors $\mathbf{k}_1$ and $\mathbf{k}_2$ shown in Fig. 26.7. These propagation vectors have the same magnitude because light of the same wavelength is emitted from both sources. In much the same way, we can write the electric field associated with the radiation from

---

[2] In Young's original experiment, of course, sunlight was used. Though visible interference effects can be observed with white-light sources such as this, they are invariably much sharper and more clearly defined when monochromatic light sources are employed.

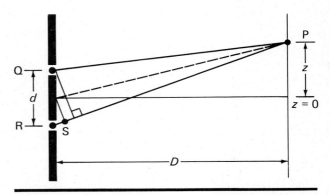

FIGURE 26.7. Geometry of Young's experiment.

the slit at Q as

$$E_2 = E_0 \cos(kr_2 - \omega t) \qquad (26.3.2)$$

In writing the equations this way, we assume that the waves are *in phase* as they are emitted from the two sources.

Since the oscillating electric vectors $\mathbf{E}_1$ and $\mathbf{E}_2$ have practically the same direction, they may be superposed simply by adding their magnitudes as given by (26.3.1) and (26.3.2). Therefore, the total field E will be

$$E = E_0[\cos(kr_1 - \omega t) + \cos(kr_2 - \omega t)] \qquad (26.3.3)$$

From elementary trigonometry, however, we know that

$$\cos a + \cos b = 2 \cos \tfrac{1}{2}(a + b) \cos \tfrac{1}{2}(a - b) \qquad (26.3.4)$$

This allows us to write (26.3.3) as

$$E = 2E_0 \cos \tfrac{1}{2}k(r_1 - r_2) \cos[\tfrac{1}{2}k(r_1 + r_2) - \omega t]$$

or

$$E = \left(2E_0 \cos \frac{\delta}{2}\right) \cos(k\overline{r} - \omega t) \qquad (26.3.5)$$

where

$$\delta = k(r_1 - r_2) = 2\pi \frac{r_1 - r_2}{\lambda} \qquad (26.3.6)$$

and

$$\overline{r} = \tfrac{1}{2}(r_1 + r_2) \qquad (26.3.7)$$

From this, it is apparent that $\overline{r}$ represents the *average* of the two distances r_1 and r_2, which is essentially equal to the distance PT in Fig. 26.7. Since $(r_1 - r_2)/\lambda$ represents the path difference between the two waves, expressed in wavelengths, it is also evident from (26.3.6) that δ is simply the *difference in phase* between E_1 and E_2 at point P.

Equation (26.3.5) can now readily be seen to represent a wave disturbance of the original wavelength and frequency whose amplitude is given by

$$|E| = 2E_0 \cos(\delta/2) \qquad (26.3.8)$$

From Fig. 26.7, however, it is easily seen that the path difference $r_1 - r_2$ is practically equal to $d \sin \theta$ so long as $d \ll D$. According to (26.3.6), then, the phase difference can be expressed as

$$\delta = \frac{2\pi d \sin \theta}{\lambda} \qquad (26.3.9)$$

At point P_0, in the center of the screen, the angle θ is zero and the distances r_1 and r_2 to the slits are equal. The phase difference δ is, therefore, also zero, and since $\cos(\delta/2)$ is then equal to unity, the wave ampli-

tude as given by (26.3.8) has the maximum possible value $2E_0$. At this point, therefore, the screen will be brightly illuminated. As we move upward, however, along the z-direction, the angle θ increases slowly, as does its sine. However, since λ is ordinarily much smaller than the distance d between the slits, we do not have to go far before the quantity $(2\pi d/\lambda) \sin \theta$ attains the value π, which means that in (26.3.8), $\cos(\delta/2)$ is zero. At this point, the total wave amplitude is zero, and the screen will be *dark*. There is now a phase difference $\delta = 180°$ between the two waves. They are, in other words, out of step by half a wavelength and, therefore, interfere destructively. If we move upward still further, θ becomes still larger, and the phase difference increases further until, at length, when the path difference equals a full wavelength, constructive interference again occurs and another bright region is encountered.

It is evident, therefore, that as we move along the z-direction in Fig. 26.7, we must traverse, alternately, bright and dark regions corresponding to path differences of even and odd numbers of half-wavelengths. An *interference pattern* of alternating bright and dark bands, therefore, appears on the screen. The spacing of these bands depends upon the light wavelength, the distance between the slits, and the spacing between the slits and the screen.

When the magnitude of the electric field amplitude at a given point is a maximum, $\cos(\delta/2)$ must be either $+1$ or -1. The condition for *constructive interference* is, therefore,

$$\cos(\delta/2) = \pm 1$$

$$\frac{\delta}{2} = \frac{\pi}{\lambda}(r_1 - r_2) = 0, \pm\pi, \pm2\pi, \pm3\pi, \ldots$$

or

$$|r_1 - r_2| = n\lambda \qquad (26.3.10)$$

where

$$n = 0, 1, 2, 3, \ldots$$

Similarly, the electric field amplitude will have a minimum value of zero whenever $\cos \delta/2$ vanishes. Thus, in the case of *destructive interference*, we must have $\cos \delta/2 = 0$, or

$$\frac{\delta}{2} = \frac{\pi}{\lambda}(r_1 - r_2) = \pm\frac{\pi}{2}, \pm\frac{3\pi}{2}, \pm\frac{5\pi}{2}, \ldots$$

or

$$|r_1 - r_2| = (n + \tfrac{1}{2})\lambda \qquad (26.3.11)$$

where

$$n = 0, 1, 2, 3, \ldots$$

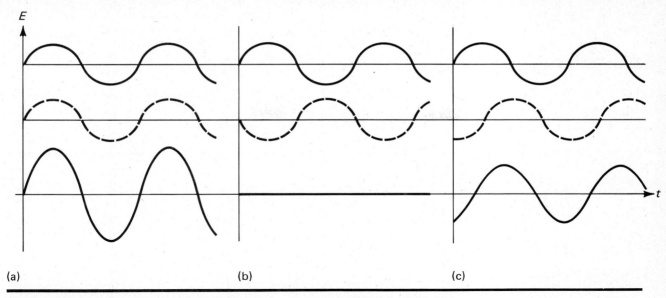

(a)   (b)   (c)

FIGURE 26.8. Superposition of plane waves (a) in phase, (b) 180° out of phase, and (c) with an intermediate phase difference.

Thus, we see that if the path difference between the two waves is an integral multiple of the wavelength, the waves arrive at P in phase and reinforcement occurs. On the other hand, if the path difference is an odd multiple of $\lambda/2$, they arrive out of phase and interfere destructively. In Fig. 26.8, the addition of waves in phase and out of phase is illustrated. An intermediate case corresponding to neither a maximum nor a minimum is also shown.

The *intensity* of light at the screen is proportional to the *square* of the electric field. From (26.3.5), we find this to be

$$E^2 = 4E_0{}^2 \cos^2(\delta/2) \cos^2(k\bar{r} - \omega t) \qquad (26.3.12)$$

For the three cases mentioned in Fig. 26.8, we also illustrate the square of the electric field. This is drawn in Fig. 26.9. Since E^2 oscillates very rapidly, its time variation will not be detected. Instead, typical instru-

ments for measuring light intensity will respond only to the *time average* of the square of the electric field. Since the average of the $\cos^2$ function is $\frac{1}{2}$, as discussed earlier, the average of (26.3.5) will be

$$\overline{E^2} = 2E_0{}^2 \cos^2(\delta/2) \qquad (26.3.13)$$

This function is also plotted for the three cases drawn in Figs. 26.8 and 26.9. From this, we see that for constructive interference the average of the square of the amplitude is $2E_0{}^2$ as illustrated in Fig. 26.10. This is *four* times as large as the corresponding average intensity obtained with one slit covered! Therefore, constructive two-slit interference quadruples the light intensity at points where the waves arrive in phase. Of course, when they are out of phase, the average intensity will be zero. If there is some other phase difference, an intensity smaller than the maximum will be observed.

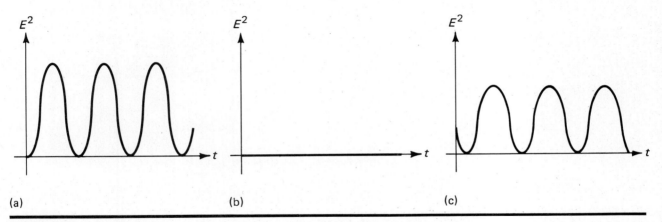

(a)   (b)   (c)

FIGURE 26.9. Instantaneous variation of intensity as a function of time for the three cases illustrated in Fig. 26.8.

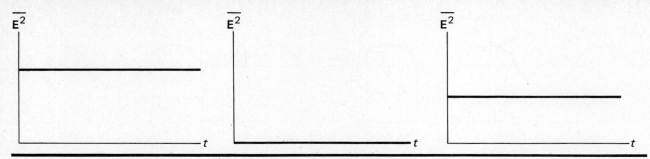

FIGURE 26.10. Time averaged intensity for the three cases illustrated in Figs. 26.8 and 26.9.

In Fig. 26.11, a photograph showing double-slit interference is given. Note the pattern of light and dark areas, which are referred to as *interference fringes*. The central bright fringe arises from the condition of (26.3.10) for $n = 0$. Successive bright fringes for $n = 1, 2, \ldots$ correspond to the additional constructive interference possibilities allowed by (26.3.10). The dark spots occur at locations which satisfy (26.3.11).

EXAMPLE 26.3.1

A double-slit interference pattern occurs on a screen very far from the slits. Obtain an expression giving the angles at which maxima and minima of intensity occur. Also calculate the distances z at which the fringes occur. See Fig. 26.7 for the geometry.

If $d \ll D$ and θ is assumed to be a small angle, the path difference $r_1 - r_2$ is approximately given by

$$r_1 - r_2 = d \sin \theta$$

where d is the distance between the slits. The condi-

tions for maxima and minima can, therefore, be stated as follows:

Maxima

$$|d \sin \theta| = n\lambda$$

Minima

$$|d \sin \theta| = (n + \tfrac{1}{2})\lambda \tag{26.3.15}$$

Now, from Fig. 26.7,

$$\sin \theta = \frac{z}{r} = \frac{z}{\sqrt{z^2 + D^2}} \tag{26.3.16}$$

Therefore, the conditions can be stated in terms of the distance z from the center of the screen. If $z \ll D$, which is usually the case, Eq. (26.3.16) can be written in the approximate form

$$\sin \theta \cong \frac{z}{D} \tag{26.3.17}$$

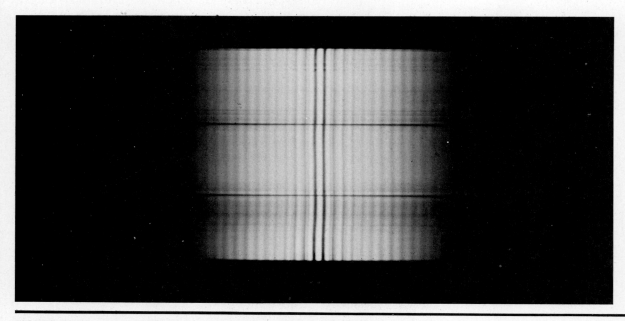

FIGURE 26.11. Photograph of double-slit interference fringes.

We may then write the interference conditions (26.3.10) and (26.3.11) as follows:

Maxima

$$|z| = \frac{n\lambda D}{d} \tag{26.3.18}$$

Minima

$$|z| = \frac{(n + \frac{1}{2})\lambda D}{d} \tag{26.3.19}$$

Equations (26.3.18) and (26.3.19) show that the fringes are uniformly spaced on the screen so long as the angle θ is small.

EXAMPLE 26.3.2

Two very narrow slits are separated by a distance of 0.1 cm. Plane waves for which $\lambda = 5000$ Å are incident on the double-slit system, and an interference pattern is observed on a screen 1.0 meter away. Find the distance z on the screen between the central maximum and the first minimum.

The central maximum occurs at $z = 0$ and the first minimum will occur when

$$|z| = \frac{\frac{1}{2}\lambda D}{d}$$

Thus, substituting the given values of λ, D, and d, we obtain

$$|z| = \frac{(5000 \times 10^{-10})(1.0)}{(2)(0.001)} = 2.5 \times 10^{-4} \text{ m} = 0.25 \text{ mm}$$

EXAMPLE 26.3.3

Two slits are separated by a distance of 0.03 cm. An interference pattern, produced at a screen 1.5 meter away, has its fourth bright fringe at a distance of 1 cm from the zero-order (central) maximum. Determine the wavelength of the light.

Since $d \sin \theta = n\lambda$ and $\sin \theta \simeq z/D$, we have

$$\frac{zd}{D} = 4\lambda$$

$$\lambda = \frac{zd}{4D} = \frac{(0.01)(0.0003)}{(4)(1.5)} = 0.5 \times 10^{-6} \text{ m} = 5000 \text{ Å}$$

According to (26.3.9), the first minimum of an interference pattern will occur when the path difference is $\lambda/2$. It is easy to see, however, from Fig. 26.7 that whatever the value of θ may be, the path difference must always be *less* than the distance between the slits. It follows that if λ is too large (for example, greater than zd), an interference pattern cannot be observed, and the intensity at the screen will be more or less uniform. On the other hand, if λ is too small, the interference fringes will be very numerous and very closely spaced, and it may not be possible for the eye to distinguish between the minima and the maxima. This may occur if λ is many orders of magnitude less than d. An interference pattern is most pronounced when the number of fringes is substantial but not too great. This, in turn, occurs when the spacing between the slits is much larger than the wavelength yet not by many orders of magnitude.

Let us now consider the complications which arise when some of the idealizations assumed earlier are no longer satisfied.

1. PRIMARY SOURCE IS NOT MONOCHROMATIC

If the primary source contains a number of different optical frequencies or if it contains all frequencies, as is the case for white light, then the interference pattern will be modified. It turns out then that *each* of the frequencies contained in the primary source sets up its own interference pattern, and these are all superposed on the screen, producing a very attractive pattern with many colors. This pattern of "white light fringes," however, is rather blurred and does not exhibit very many sharply defined intensity maxima and minima such as those resulting from a monochromatic source. In the problems at the end of the chapter, the student is asked to study the interference pattern produced by a source containing only two frequencies.

2. THE TWO SLITS REPRESENT INCOHERENT SOURCES

In the discussion given on p. 854, the two monochromatic slit sources were mutually coherent as a result of the synchronization achieved by having both secondary sources originate from the same primary series of wavefronts. We stated also that a "truly" monochromatic and coherent source is one for which the electric field produced has the form $E_0 \cos(\omega t + \phi)$, where the phase ϕ is independent of time. There are in actuality no sources that behave in exactly this way, though lasers approach it so closely that light beams from two independent lasers can be made to exhibit stable and clearly visible interference patterns. In the case of most other "monochromatic" sources, light is emitted at a dominant average frequency ω but contains electromagnetic energy at slightly different frequencies as well. Such sources ordinarily exhibit no stability of phase over any interval of time that is appreciable experimentally. Interference patterns can, nevertheless, be produced using sources of this type as "primary" sources to illuminate the single primary slit in Young's experiment and the double-slit experiment discussed above. This is so because any irregularity in phase exhibited by the light coming through the primary slit affects the phase of the light emitted by the two secondary slits in exactly the same way and at exactly the same time.

If the two secondary slits are illuminated by two independent sources, even though they are "monochromatic" in the sense described above, there is no phase relationship between the wavefronts emitted by the two secondary slits that persists for any appreciable length of time. There is, accordingly, no stable interference pattern, and the screen appears to be more or less evenly illuminated.

3. SLITS ARE NO LONGER NARROW IN COMPARISON WITH THE LIGHT WAVELENGTH

If the slit width is not narrow, there will be not only interference of light waves from the two separate slits, but also interference of light coming from *different portions of a single slit*. When this happens, the interference pattern becomes much more complicated, and will be discussed in a subsequent section under the subject of *diffraction*.

4. THE SECONDARY SLITS ARE CLOSE TO THE SCREEN ON WHICH THE PATTERN IS OBSERVED

The interference pattern is still present, but the geometry is now such that the spacing of the fringes must be determined from Eqs. (26.3.14) and (26.3.15) rather than from the approximations which follow from the condition $d \ll D$ used earlier. Also, the approximations that the two fields E_1 and E_2 from the two separate slits are the same in both magnitude and direction may no longer be justified.

In all the discussion above, the entire analysis is based on the underlying notion that light must be described as a wave. We have, of course, made use of the harmonically oscillating electric field as the underlying entity used to determine the intensity of light at any point. But when Young carried out his experiments, he did not know that such a connection existed between light and electricity and magnetism. Indeed, Maxwell had not yet been born! It is very much to Young's credit that he was able to advance the bold hypothesis that light is a wave phenomenon and to assert that the effect of *interference* provided the proof.

In spite of Young's remarkable success, most of his contemporaries publicly ridiculed his ideas, for English physicists were biased in favor of the corpuscular views advanced by Newton a century earlier. Even though Newton himself had suggested that the true nature of light must ultimately be decided by experiment, the critics of Young were apparently not convinced by his accomplishments. This seems strange

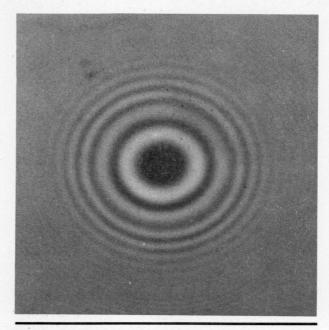

FIGURE 26.12. Photograph of Newton's rings. Photo courtesy of Professor T. A. Wiggins, Pennsylvania State University.

now, because if we assume that light consists of a stream of particles that travel in straight lines along the direction of the light rays of geometrical optics, in the experimental arrangement illustrated in Fig. 26.4 there could be no interference pattern at all, and the entire screen would be dark. In 1815, however, fourteen years after Thomas Young discovered the interference of light, the French engineer Fresnel explained diffraction patterns on the basis of a wave theory of light by using Huygens's principle. He also made some theoretical predictions which were experimentally confirmed by the French physicist Dominique Arago. Arago became a convert to the ideas of Young and Fresnel, and the wave theory of light became generally accepted soon thereafter.

26.4 Interference in Thin Films

The subject of thin-film optics dates back to the work of Isaac Newton, who observed fringes produced by the thin air film between a convex lens and a plane surface. These *Newton rings*, shown in Fig. 26.12, provide an early example of thin-film interference. Of course, Newton was unaware of the wave character of light, but Thomas Young was able to explain these rings on the basis of his ideas about wave interference.[3] There are many naturally occurring examples of *thin-

---

[3] In a strict sense, the observation of Newton's rings could be held to justify the point of view that it was Newton rather than Young who really discovered the interference of light.

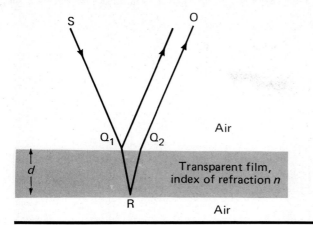

FIGURE 26.13. Ray diagram for thin-film interference.

film interference of this type. All of us have, at some time or other, seen the beautiful colors reflected from an oil film on water or from a soap bubble, or the colored reflections from the "coated" lenses of cameras or binoculars. These are also illustrations of thin-film interference.

Let us consider an example in which a source of light with a dominant wavelength λ in air illuminates a reflecting surface upon which a thin transparent film, with index of refraction n, has been deposited. We shall further assume that the incident light rays are *almost normal* to the surface, as are the reflected rays arriving at an observer located at O, as illustrated in Fig. 26.13. Upon reflection, such rays are split into two beams; one is reflected directly from the surface of the film at Q_1, and one enters the film and is reflected from the underlying substrate, emerging at Q_2. It is the interference between two such light beams that leads to the observed thin-film optical effects. If the light has a wavelength λ in air, the wavelength in the film will be $\lambda' = \lambda/n$. To determine whether or not the two light rays interfere constructively or destructively at O, we must consider the phase difference between the two rays arriving there. There are two factors that can lead to phase differences. First, there is a path difference Q_1RQ_2 between the two rays. Assuming that the incident and outgoing rays are normal to the surface, the phase change associated with this path difference is

$$\frac{2\pi}{\lambda'} Q_1RQ_2 = \frac{2\pi}{\lambda/n} Q_1RQ_2 \qquad (26.4.1)$$

Also, there are possible phase changes that take place upon reflection at the reflecting surfaces, as discussed in Chapter 24. If light is incident upon a reflecting surface from a medium of refractive index n_1 and if the refractive index of the material beyond the reflecting surface is n_2, then, if $n_2 > n_1$, the reflected wave changes phase by π, while, if $n_1 > n_2$, no phase change occurs. The above statement is summarized

in the short verse[4] by F. K. du Pré, which goes:

Low to High
Phase Change π
High to Low
Phase Change? No.

In this illustration, we are led to conclude that there is a phase change π upon reflection from the upper surface but none for the reflection from the lower surface.

Now, whenever the *total phase difference* between two waves is given by $\delta = 0, 2\pi, 4\pi, 6\pi, \ldots$, the waves will reinforce to produce a maximum intensity. On the other hand, if the difference is $\delta = \pi, 3\pi, 5\pi, \ldots$, the waves will interfere destructively to produce minimum intensity. In the present case, assuming normal incidence, the ray SQ_1RQ_2O undergoes a phase change $4\pi nd/\lambda$ in passing through the film as a result of the optical path difference, while the ray SQ_1O experiences a phase change of π as a consequence of reflection from the optically denser medium. The *total phase difference* between waves is, therefore, $(4\pi nd/\lambda) - \pi$, and hence the conditions for maxima and minima are, respectively,

Maxima

$$\delta = \frac{4\pi nd}{\lambda} - \pi = 0, 2\pi, 4\pi, \ldots$$

Minima

$$\delta = \frac{4\pi nd}{\lambda} - \pi = -\pi, \pi, 3\pi, 5\pi, \ldots$$

$(26.4.2)$

or

Maxima

$$d = \frac{\lambda}{4n}, \frac{3\lambda}{4n}, \frac{5\lambda}{4n}, \ldots$$

Minima

$(26.4.3)$

$$d = 0, \frac{\lambda}{2n}, \frac{2\lambda}{2n}, \frac{3\lambda}{2n}, \ldots$$

We see, therefore, that for any given wavelength there are many possible film thicknesses which could give rise to maximum or minimum reflected intensities. Since energy must be conserved, we are forced to conclude that under circumstances leading to a *minimum* in the reflected light, there will be a *maximum* in the transmitted light.

Now suppose that instead of using incident monochromatic light we use white light, which consists of all optical frequencies. Then, for a given film thickness, some wavelengths may interfere constructively on reflection while others may interfere destructively. As a consequence, the reflected light

[4] F. K. du Pré, *Appl. Optics* **10**, 2345 (1371).

FIGURE 26.14. Interference fringes from a soap film, produced using a monochromatic source. Photo by the authors.

FIGURE 26.15. White light interference fringes from a soap film. Photo courtesy of Professor T. A. Wiggins, Pennsylvania State University.

corresponding to wavelengths for which constructive interference occurs will be quite intense, while light of wavelength corresponding to destructive interference will be essentially absent. For this reason, the reflected light appears to be colored, the precise shade depending upon the film thickness and the index of refraction.

Figures 26.14 and 26.15 show photographs of a soap film suspended on a wire ring. The soap film has a variable thickness since it is held in a vertical position and, therefore, drains downward. As a consequence, certain portions of the film satisfy the condition for constructive interference while other portions satisfy the destructive interference criteria. In Fig. 26.14, the soapy film is illuminated with monochromatic light, and a pattern of light and dark bands shows the effect of interference. If white light is used, as in Fig. 26.15, we expect to see constructive interference of various frequencies occurring at different locations. In this way, the soap film separates different spectral hues spatially and exhibits bands of *interference colors*.

In the example just above, we considered the case of a thin film suspended in air, in which case it is bounded both above and below by a less highly refractive medium. It is also important, however, to examine the case illustrated in Fig. 26.16, where the film is supported upon a more highly refractive solid transparent substrate, such as optical glass. Such thin films have been found to be very useful in a number of important practical situations. They can drastically reduce the reflectivity of the underlying transparent

medium and can, under certain conditions, even reduce it to zero. They can also make it highly reflecting, if that is what is wanted. Let us now study some of the problems involved in making "coated" lenses and other optical components by applying thin films to their surfaces so as to reduce their reflectivity as much as possible.

As we learned in Chapter 24, the reflection coefficient of a typical glass–air interface for light at

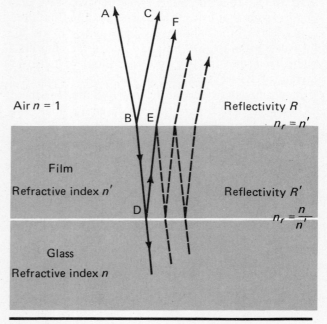

FIGURE 26.16. Ray diagram for light reflected from a coated lens.

normal incidence is about 5 percent. In optical systems such as complex photographic lenses, there may be 10 or even 20 such interfaces. Clearly, if we had to put up with a 5 percent reflection loss at each lens surface, we would end up with a lens that would transmit only a small percentage of the light incident on its front surface. For a lens having 10 glass–air surfaces, we would expect a transmission of $0.95^{10} = 0.5987$, or 59.9 percent, since only 95 percent of the light survives reflection at each surface. For a zoom lens having 20 such interfaces, the figure would be only 35.8 percent. It is desirable to reduce this reflection loss by optical coating, not only because of the light loss itself but also because much of the reflected light ends up on the film, or in the eyepiece, in the form of "ghost images." The ghost images reduce image contrast and sometimes produce annoying visible spots and haloes when the lens is aimed in the direction of bright point sources of light. It is, therefore, very desirable to reduce the reflectivity of the glass–air surfaces to the lowest possible value, or to eliminate it completely where that can be done. These objectives are usually realized in practice by applying a thin refracting layer to the glass in such a way that the reflected light from its top and bottom surfaces interferes destructively, as shown in Fig. 26.16.

Let us assume that the film has an index of refraction which is greater than that of air but less than that of glass. Under these conditions, the light rays reflected at the upper and lower surfaces both experience a phase change of π radians, since the reflection occurs from an optically denser underlying medium. Thus, the two light rays ABDEF and ABC will have a phase difference which results only from their path difference. For approximately *normal incidence*, the phase difference will, therefore, be $4\pi nd/\lambda$. If this difference is π, the two waves arrive out of phase. This will happen if the film thickness is chosen to be $\lambda/4n$, in other words, if it is one fourth of the wavelength in the medium.

Now, if the *intensities* of the beams reflected at the upper and lower surfaces are not equal, their destructive interference will be partial rather than complete, and the total reflected intensity will be lowered but not entirely eliminated. If we can somehow arrange to have them exactly equal in intensity, however, there will be complete destructive cancellation and no reflected intensity at all. In Example 24.4.4, we saw that the reflectivity of a glass surface, defined as the ratio of reflected to incident intensity, is given for normal incidence by Eq. (24.4.54). In the present example, there are *two* interfaces whose reflectivity must, according to these results, be

$$R = \left(\frac{n_r - 1}{n_r + 1}\right)^2 \quad \text{and} \quad R' = \left(\frac{n_r' - 1}{n_r' + 1}\right)^2 \quad (26.4.4)$$

for the upper and lower surfaces, respectively. For most transparent substances, these reflectivities are relatively small, and only a few percent of the incident light is reflected at each interface. Under these circumstances, the intensities of the incident ray AB and the transmitted ray BD are practically equal. In such a situation, if we make the reflectivities R and R' the same, the reflected rays BC and DEF that interfere destructively will have practically the same intensities, and there will be no significant reflected light at all. But R and R' will be equal only if the relative indices $n_r = n'$ and $n_r' = n/n'$ are the same, and this, in turn, requires that[5]

$$n' = \frac{n}{n'}$$

or

$$n' = \sqrt{n} \qquad (26.4.5)$$

We, must, therefore seek a coating film whose index of refraction is approximately the same as the square root of the index of the glass. For optical glass of refractive index 1.75, we would, therefore, look for a film of refractive index 1.32. It is not easy, in practice, to find materials that have the required refractive index along with all the other desired characteristics of transparency, durability, and ease of deposition. Magnesium fluoride, MgF_2, whose refractive index is 1.38, comes reasonably close to satisfying all the requirements mentioned above. A quarter-wave film of MgF_2 will reduce the reflectivity of most optical glasses in the visible light range to under 1 percent. If the coating is designed to minimize reflection of green light to which the eye is most sensitive ($\lambda \sim 5500$ Å), there will be, of necessity, some residual reflectance for the red and violet regions of the spectrum, since for these colors the film thickness is no longer very close to $\lambda/4n$. This small residual reflection of red and violet light gives rise to the familiar purplish hue of light reflected from coated lenses. The technology of optical coating is now very highly developed. By coating glass with a number of films whose thicknesses and refractive indices are appropriately chosen, it is possible to reduce the reflectivity of optical glass to very small values over the whole visible spectrum.

In the processes of reflection and refraction, there is *no loss* of light energy at all. Therefore, all

---

[5] If a significant fraction of the incident light is reflected, of course, as it would be if n' and n were very large, then the transmitted beam's intensity would be much reduced; and simply making the reflectivities R and R' the same no longer suffices, even approximately, to produce complete destructive interference. Under these circumstances, the multiple reflection of beams within the film, shown by dotted lines in Fig. 26.16, also becomes important. Fortunately, most of the optical glasses and coating films used in practice have reflectivities of less than 10 percent.

light that is not reflected from a transparent substance must be transmitted. In the present example, therefore, the process of coating a transparent substance with a film a quarter-wavelength thick not only reduces the reflectivity of the substance but actually *increases* the amount of light that is transmitted. Although the physical details of the processes that act to accomplish this increase in transmission are not easily understood, there is no question that it must take place. In the examples mentioned previously, of a lens with 10 glass–air surfaces and a zoom lens with 20 such surfaces, the application of a quarter-wave film that reduces the reflectivity to 1 percent at each interface increases the fraction of light transmitted to $0.99^{10} = 0.9043$ and $0.99^{20} = 0.8179$, respectively. Comparison of these figures with the ones given previously for uncoated lenses reveals that there is a significant increase in the transmission properties of the lenses. For the zoom lens, the fraction of light transmitted is more than doubled by the coating process. Indeed, the manufacture of zoom lenses, which are often comprised of 15 or 20 separate elements, would be quite out of the question without optical coating.

EXAMPLE 26.4.1

A film 10,000 Å thick is used to coat a certain type of glass. At what wavelengths in the visible spectrum will the reflected light interfere destructively? The index of refraction of the film is assumed to be 1.40, which is less than that of the glass.

In the present example, the condition of a minimum in reflected intensity is that the path difference between the two reflected light rays is an odd multiple of one half the wavelength in the medium. Compensating phase changes occur upon reflection from the air–film and film–glass interfaces. Since the index of the film is not specified to be equal to the square root of the index of the glass, those light wavelengths for which the reflected light is a minimum will be reduced in intensity but not necessarily extinguished. Since the path difference is twice the film thickness d and since this must equal an odd number of half-wavelengths in the medium, we may write

$$2d = \frac{1}{2}\frac{\lambda}{n'}, \frac{3}{2}\frac{\lambda}{n'}, \frac{5}{2}\frac{\lambda}{n'}, \ldots$$

or

$$\lambda = 4n'd, \frac{4n'd}{3}, \frac{4n'd}{5}, \frac{4n'd}{7}, \ldots$$

Substituting the numerical values given above for n' and d, we find

$$\lambda = 56{,}000 \text{ Å}, 18{,}667 \text{ Å}, 11{,}200 \text{ Å}, 8{,}000 \text{ Å}, 6{,}222 \text{ Å}$$
$$5{,}091 \text{ Å}, 4{,}308 \text{ Å}, 3{,}733 \text{ Å}, \ldots$$

Of these wavelengths, only 6222 Å, 5091 Å, and 4308 Å are in the visible spectrum, and, therefore, these optical wavelengths are those for which minimum reflectivity is observed. Other wavelengths are more or less strongly reflected.

EXAMPLE 26.4.2

A soap film is formed on a wire frame as shown in Fig. 26.17. The film drains downward due to gravity and is, therefore, thicker at the bottom than at the very top, where it might be only a few molecular layers thick. At a particular instant of time, the film has a thickness of 50 Å near the top, 1500 Å somewhere in the middle, and 4000 Å near the bottom. When the film is viewed with reflected light, what colors would be eliminated at the top, middle, and bottom? Assume the film has an index of refraction of 1.35.

Let us find out which wavelengths will be absent from the reflected light due to destructive interference. From (26.4.3), we should expect intensity minima when $d = \lambda/2n, 2\lambda/2n, 3\lambda/2n, \ldots$, from which

$$\lambda = 2nd, \frac{2nd}{2}, \frac{2nd}{3}, \ldots$$

Now, near the top of the film, where the thickness is only 50 Å, the phase difference between rays reflected from the two surfaces of the film is nearly π for *all* visible wavelengths. The phase difference resulting from the optical path difference is negligible in this case since the film is very thin. The top of the film then will reflect very little light of any color and will, therefore, appear quite dark.

Near the middle of the film, the thickness is 1500 Å, and, therefore, destructive interference can occur at wavelength λ for which

$$\lambda = 2nd = (1500)(2)(1.35) = 4050 \text{ Å}$$

This is in the violet part of the visible spectrum, and, therefore, violet is not reflected. The other wavelengths for which destructive interference occurs are not in the visible spectrum. Constructive interference occurs,

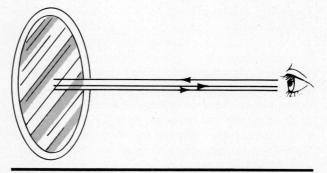

FIGURE 26.17. Interfering beams reflected from the surfaces of a soap film.

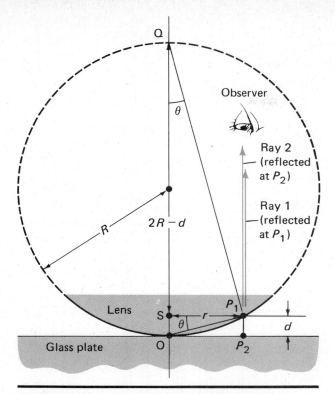

FIGURE 26.18. Interference geometry for Newton's rings.

according to (26.4.3), for wavelengths 8100 Å, 2700 Å, and smaller values, none of which is in the visible region.

In the region where the thickness is 4000 Å, destructive interference occurs at

$$\lambda = (4000)(2)(1.35) = 10,800 \text{ Å}$$

which is not in the visible range, but also at $10,800/2 = 5400$ Å, which is in the green part of the spectrum. None of the other wavelengths at which the reflected waves are canceled is visible. At 4000 Å thickness, maxima can occur at $21,600/1$, $21,600/3$, $21,600/5, \ldots$ Å. Only one of these, the wavelength 4400 Å, is in the visible spectrum. Therefore, light of this wavelength, which is violet, will be strongly visible on reflection. In general, the exact colors that are seen on reflection depend on the composition of the incoming light, the amount of each wavelength that is reflected, and also the sensitivity of the human eye.

EXAMPLE 26.4.3
NEWTON'S RINGS

An air film is formed between a convex lens and a plane reflecting surface, as shown in Fig. 26.18. The lens is assumed to have a radius of curvature R. Interference rings are observed when light of wavelength λ is viewed at normal incidence. Find an expression which can be used to determine the location of the circular fringes.

A photograph of these Newton rings is shown in Figure 26.12. To analyze this problem, consider the two light rays designated in Fig. 26.18 as 1 and 2. Since the thickness d of the air film is very small, the circular and plane surfaces are almost parallel, and we may therefore assume that the two reflected rays are also parallel. Ray 2 undergoes a phase change of π radians at P_2, but ray 1 has no phase change at P_1. The phase difference between the rays is therefore $(2\pi/\lambda)(2d) + \pi$, and therefore the conditions for minima in the reflected light are

$$\frac{4\pi d}{\lambda} + \pi = \pi, 3\pi, 5\pi, \ldots$$

or

$$d = 0, \frac{\lambda}{2}, \frac{2\lambda}{2}, \frac{3\lambda}{2}, \ldots$$

Let us now write this condition as

$$d = N\frac{\lambda}{2} \qquad (26.4.6)$$

where $N = 0, 1, 2, 3, \ldots$. Now, referring to Fig. 26.18, the properties of the similar triangles OSP_1 and P_1SQ lead us to conclude that $OS/SP_1 = SP_1/SQ$, hence that

$$\frac{d}{r} = \frac{r}{2R - d}$$

or

$$r^2 = 2Rd - d^2$$

But since $d \ll R$,

$$r^2 \cong 2Rd = 2RN\frac{\lambda}{2}$$

$$r = \sqrt{NR\lambda} \qquad (26.4.7)$$

We see from this that the radii of the destructive interference fringes are proportional to the square root of the fringe order N. A formula such as this can be used to find the wavelength of the light if the radius R can be found by measuring the diameter of the rings. Alternatively, if λ is known, it may be used to determine R.

If the fringes are not circular, this would indicate that the lens surface is not spherical or perhaps that the plane surface is not perfectly flat. Interference fringes are often used to test whether a given optical surface has been ground to the desired curvature. For example, if we are certain that the plane surface is completely flat, then any departure from circular Newton rings would indicate that the lens was not spherical.

In Fig. 26.19, a method for testing whether a surface has been ground optically flat is suggested. If

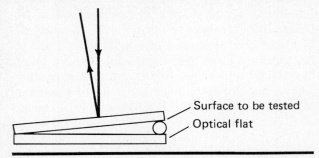

FIGURE 26.19. Interference fringes produced between a "standard" optically flat substrate and a glass plate whose flatness is to be tested. The two plates are separated by a thin wedge of entrapped air.

the interference fringes are linear in shape, as in Fig. 26.20a, then the surface is flat. On the other hand, an interference pattern such as Fig. 26.20b suggests a lack of flatness and, moreover, locates the areas that may need further polishing. At the end of the chapter, we shall suggest a few problems involving this kind of interference.

26.5 Diffraction of Light

When light passes through a small aperture, the waves passing through different parts can *interfere* in such a way as to produce an observable interference pattern on a screen placed at some distance behind it. If the light is emitted from a point source, the laws of geometrical optics lead us to suppose that the effect of this interference is simply to produce an image of the aperture that is bright where the rays pass through unimpeded to the screen and is dark elsewhere. In many instances, this is a very good description of what happens, but since the laws of geometrical optics are only approximate and apply only when the dimensions of the aperture are very much larger than the light wavelengths, there are instances where it is quite inadequate.

If the aperture is large in comparison with the wavelength, the image on the screen corresponds closely to the geometric shadow of the opening through which the light rays pass. But if the aperture is of a size comparable to the wavelength, light passing through *different portions of the opening* can interfere in such a way as to produce bright areas within the geometric shadow and dark regions within the geometric image of the spot. If we insist on viewing what happens in terms of the behavior of light rays that are emitted by the source and end up on the screen, we must conclude that rays passing through a small aperture, or very close to the edge of any opaque object, are *bent* or *diffracted* from their usual straight-line paths. The effects associated with this bending of light rays are referred to as *diffraction*. It

is evident that diffraction of light arises as a consequence of interference, in particular, interference of light that passes through different parts of an aperture or that passes at different distances from an obstacle. It is also apparent that diffraction of light is strictly a wave phenomenon and cannot be explained by the laws of geometrical optics. Figure 26.21 illustrates the diffraction of water waves in a ripple tank as they pass through a small opening. Some other aspects of diffraction are depicted in Figs. 26.22 and 26.23. In these illustrations, the "bending of rays" is clearly evident. We shall now examine how diffraction takes place in a number of simple instances and try to explain its most important physical effects.

In order to see a diffraction pattern, certain conditions must be satisfied. Since the diffraction seen in Figs. 26.21, 26.22, and 26.23 produces intensity variations near the edges of the objects, the illumination must be such as to produce clearly defined rather

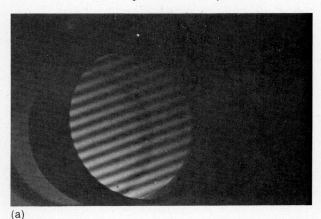

(a)

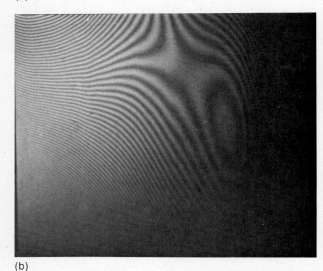

(b)

FIGURE 26.20. Appearance of interference fringes produced by the arrangement shown in Fig. 26.19 (a) when the lower surface of the upper plate is "optically flat" and (b) when it deviates slightly from the desired planar form. Photo courtesy of Professor T. A. Wiggins, Pennsylvania State University.

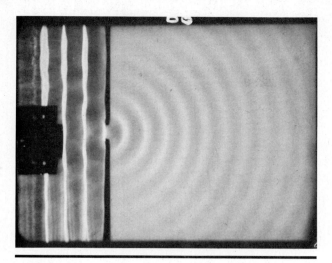

FIGURE 26.21. Diffraction of water waves by a single slit.

FIGURE 26.22. Photograph illustrating the diffraction of light by a safety pin. Courtesy of Professor T. A. Wiggins, Pennsylvania State University.

FIGURE 26.23. Photograph showing the diffraction of light by a needle. Courtesy of Professor T. A. Wiggins, Pennsylvania State University.

than fuzzy images. This can be accomplished by using a small point source. The sun would not be a suitable source for this purpose, for it does not produce shadows with sharp edges. Moreover, since sunlight contains many frequencies, the diffraction patterns due to each would overlap and might, therefore, nullify any visible effects. We shall, therefore, assume for simplicity that our sources of light are *point monochromatic sources.*

In the most general situation, neither the source

nor the observing screen need be very far from the diffracting aperture or obstacle. Under these circumstances, the wavefronts entering and leaving the diffracting aperture are spherical rather than planar, as shown in Fig. 26.24a. This type of diffraction is referred to as *Fresnel diffraction.* The situation is much simpler when both source and screen are very distant from the aperture. The wavefronts entering and leaving the diffracting aperture are then planar, as illustrated in Fig. 26.24b. Diffraction occurring under these conditions is called *Fraunhofer diffraction.* In the case of Fraunhofer diffraction, in which incoming and outgoing *plane waves* are involved, the mathematical

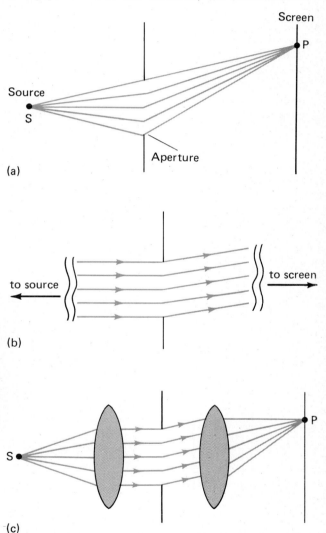

(a)

(b)

(c)

FIGURE 26.24. (a) General case of Fresnel diffraction, in which wavefronts at the diffracting aperture are spherical. (b) Simpler case of Fraunhofer diffraction, in which wavefronts at the diffracting aperture are planes, corresponding to source and screen at an infinite distance. (c) Arrangement of lenses used to produce the conditions of Fraunhofer diffraction with finite source and screen distances.

description of the effects that are observed is much simpler than in the more general instance of Fresnel diffraction. Of course, it is inconvenient from an experimental point of view to have both source and observing screen infinitely far from the diffracting system. This is easily remedied, however, by using lenses to convert the diverging rays from a point source at a finite distance into a parallel beam and to focus parallel outgoing rays onto an observing screen at a finite distance, as shown in Fig. 26.24c. In this arrangement, the source must be placed at the focal point of the first lens and the screen at the focal point of the second. The rays that enter the diffracting aperture have no way of "knowing" that they did not originate from an infinitely distant point source. In the same way, a set of parallel rays that leave the aperture headed in a given direction do not know that they are not headed for an infinitely distant screen. This scheme enables us to observe the simpler Fraunhofer diffraction effects very easily. Though Fresnel diffraction is important in many situations, we shall confine ourselves in this book to a discussion of the much easier case of Fraunhofer diffraction.

26.6 Single-Slit Fraunhofer Diffraction

The simplest case of Fraunhofer diffraction occurs when a single, narrow slit of width d is illuminated normally by plane monochromatic light waves, as illustrated in Fig. 26.25. We shall try to find the intensity of light leaving the slit and traveling in a direction that makes an angle θ to the incoming light, as shown in the figure. This light is eventually focused at point P on the screen by the lens. Each point along the line S_1S_2 can be regarded as a source of waves, according to Huygens's principle, whose fronts may propagate in the direction specified by θ and which eventually arrive at P. Unfortunately, however, the path length from source to screen is *different for each point* along S_1S_2; for example, the path difference between points S_1 and S_2 is $d \sin \theta$. For this reason, the light that originates from each part of the slit arrives at P with a *different phase*. It is necessary, therefore, to account properly for these phase relationships in superposing the contribution from each source point. Another way of expressing the essential facts is to observe that light entering different parts of the slits interferes as it arrives at the screen and that this interference may be constructive or destructive. Since from Fig. 26.25 the path length between the two sides of the slit is $d \sin \theta$, the total phase difference ϕ between these points must be

$$\phi = \frac{2\pi}{\lambda} d \sin \theta \qquad (26.6.1)$$

To calculate the intensity at point P, we shall find the electric field amplitude E_p associated with the light that arrives there. To do this, we shall split the

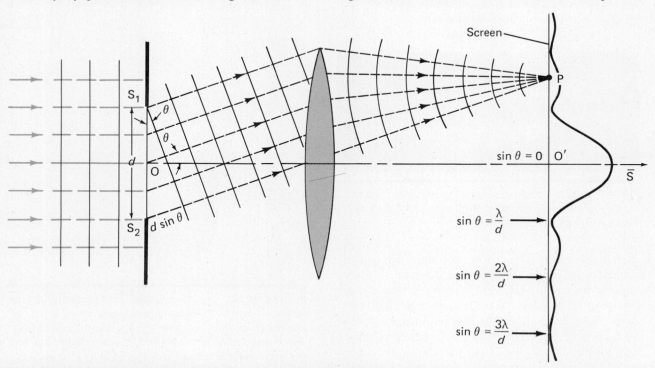

FIGURE 26.25. Geometry of Fraunhofer diffraction, as arranged in Fig. 26.24c, showing ray paths and wavefronts.

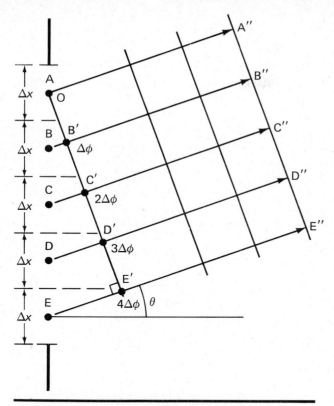

FIGURE 26.26. Subdivision of diffracting aperture into N equal parts, each of width Δx. In the drawing, $N = 5$.

aperture up into N parts, each of width Δx. We may then superpose the waves that originate from each of these sources, assigning to each of these contributions a phase angle ϕ that corresponds to the phase appropriate to the center of each of the N intervals. Finally, we shall let N become infinite while Δx approaches zero in such a way that the product

$$N \, \Delta x = d \qquad (26.6.2)$$

which represents the width of the slit, remains constant. The situation is illustrated in Fig. 26.26 for the case where $N = 5$; in this case, the points A, B, C, D, and E are regarded as point sources that contribute Huygens wavelets forming plane wavefronts that travel in the direction specified by the angle θ. It is apparent that there must be a phase difference $\Delta \phi$ between each of the rays AA″, BB″, CC″, DD″, and EE″ due to the successively greater path distances BB′, CC′, DD′, and EE′ that each ray must cover.[6]

The fields produced at P by each of these con-

<hr>

[6] It is important to note that according to Fermat's principle, the distances $A''P$, $B''P$, $C''P$, $D''P$, and $E''P$ are traversed *in the same time* by all the rays, even though the lens happens to be there. There is, therefore, no difference in the effective optical paths between those points even though a lens is interposed to focus the light. The path differences BB', ... EE' are, therefore, the only ones that need be discussed.

tributions are harmonically varying quantities of the same frequency each of which has a different phase. We have already shown in section 9.4 that the magnitude and phase of the sum of harmonically varying quantities such as this can be obtained by regarding each of the individual quantities as a *vector* whose length represents the amplitude and whose direction represents the phase of the individual quantity. Their *vector sum* then represents the result of superposing all the individual harmonically varying components, its amplitude giving the amplitude of the resultant superposition and its phase angle giving the appropriate phase. We have found these rules for superposing sinusoidally varying quantities useful in discussing the voltages and currents in ac circuits, and they are equally useful in the present situation. The result of superposing the five amplitudes discussed in Fig. 26.26 is illustrated in Fig. 26.27a. Had we chosen to subdivide the aperture more finely into many more constitutent parts, the diagram would be as shown in

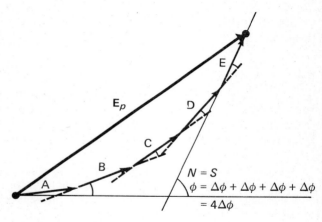

(a)

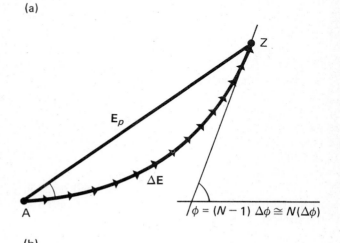

(b)

FIGURE 26.27. Superposition of harmonically varying fields at the screen using the laws of vector addition (a) for $N = 5$ and (b) for more minute subdivision, corresponding to a large value of N.

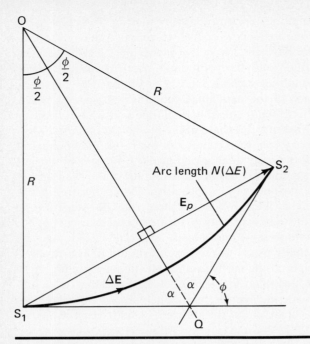

FIGURE 26.28. Limiting case of infinitesimal subdivision, in which the Δ**E** vectors describe a circular arc.

Fig. 26.27b. It is evident that as the number of subdivisions N increases without limit, the constituent vectors will describe the arc of a *circle*, since a circle is the curve whose direction changes at a uniform rate as you traverse its arc.[7] From Fig. 26.27, it is also apparent that the total phase difference ϕ given by (26.6.1) will be $\phi = (N - 1) \Delta\phi$. But in the limit where the number of subdivisions N becomes infinitely large and the phase difference $\Delta\phi$ between adjacent subdivisions infinitesimally small, this may just as well be written

$$\phi = N \Delta\phi \qquad (26.6.3)$$

Suppose now that we denote the magnitude of the electric field arising from each of the subdivisions ΔE. It is evident, then, from Fig. 26.27b that in the limit of large N, the arc length of the circular segment AZ will approach $N \Delta E$. At the same time, the vector **AZ**, representing the resultant of all the individual vectors Δ**E**, will represent the electric field of the light wave reaching P in amplitude and phase. The total phase difference ϕ of Eq. (26.6.1) is represented by the direction of the last Δ**E** vector, whose head rests on Z. In the limit, this is the direction of the *tangent* to the circular arc at Z. The limiting situation, therefore, is as represented in Fig. 26.28. In that diagram, it is apparent that the angles ϕ and α are related by

[7] In other words, you can drive a car around a perfectly circular track without touching the steering wheel once you have it properly aimed. This is clearly the situation illustrated by the vector diagrams in Fig. 26.27.

$$\phi + 2\alpha = 180°$$

$$\alpha = 90 - \frac{\phi}{2} \qquad (26.6.4)$$

Since OS_1Q and OS_2Q are both right triangles, this means that the vertex angles S_1OQ and S_2OQ are both equal to $\phi/2$, as illustrated.

Since the length of the circular arc connecting S_1 and S_2 is $N \Delta E$, we may write

$$R\phi = N \Delta E \qquad (26.6.5)$$

Also, from elementary trigonometry, the half-length of the vector $\mathbf{E}_p$ that represents the amplitude of the resultant wave at the point P on the screen is

$$\frac{E_p}{2} = R \sin \frac{\phi}{2} \qquad (26.6.6)$$

Expressing the radius R in terms of ϕ and ΔE by (26.6.5), this can be written

$$E_p = N \Delta E \frac{\sin(\phi/2)}{\phi/2} \qquad (26.6.7)$$

Now, as the number of parts N into which the aperture is subdivided increases, it is evident that the strength ΔE associated with each of them must decrease proportionally. This means that ΔE is inversely proportional to N, or that the quantity $N \Delta E$ remains *constant* as N is varied. If we denote this constant quantity, which ultimately governs the total intensity of the observed diffraction pattern, as E_0, (26.6.7) takes the form

$$E_p = E_0 \frac{\sin(\phi/2)}{\phi/2} \qquad (26.6.8)$$

In this equation, the quantity E_p refers to the amplitude of a harmonically varying electric field at the point P. The time variation of this field can be expressed by incorporating a factor $\cos(\omega t + \delta)$ on the right side of the equation. Since the *intensity* $\bar{S}$ is proportional to the time average of the *square* of the amplitude, it can be represented as

$$\bar{S} = S_0 \left(\frac{\sin(\phi/2)}{\phi/2} \right)^2 \qquad (26.6.9)$$

In this expression, the quantity S_0 represents the time average of $E_0^2 \cos^2(\omega t + \delta)$, which turns out to be $\frac{1}{2}E_0^2$, since the time average of $\cos^2(\omega t + \delta)$ is $\frac{1}{2}$. The trigonometric functions that appear in (26.6.8) and (26.6.9) above are plotted in Fig. 26.29. The angle ϕ represents the total difference in phase for light arriving at P between waves starting at the upper and lower ends of the slit, S_1 and S_2, in Fig. 26.25. This phase difference may be expressed in terms of the slit width d, the wavelength λ, and the angle θ (in terms of which the location of point P can be

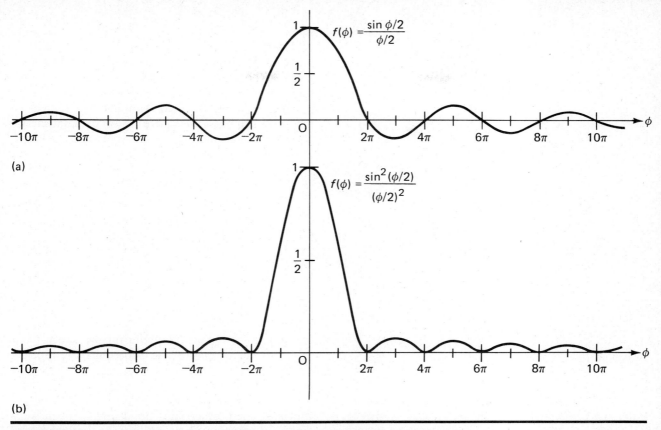

$$f(\phi) = \frac{\sin \phi/2}{\phi/2}$$

(a)

$$f(\phi) = \frac{\sin^2 (\phi/2)}{(\phi/2)^2}$$

(b)

FIGURE 26.29. Plot of (a) the field amplitude and (b) the light intensity at the screen, as a function of the variable $\phi = (2\pi d/\lambda) \sin \theta$, for the case of single-slit Fraunhofer diffraction.

expressed) by (26.6.1).[8] Since ϕ increases as the angular displacement θ of point P from the center of the screen increases, there will be a pattern of intensity maxima and minima on the screen, whose locations correspond to the maxima and minima of the function $\sin^2(\phi/2)/(\phi/2)^2$, as plotted in Fig. 26.29b. This pattern has a number of important and rather unusual features which we shall now examine in detail.

First, from Fig. 26.29b, it is easy to see that the maximum intensity occurs at $\phi = 0$, corresponding to $\theta = 0$ from (26.6.1). At this point, which is at the center of the screen, directly opposite the slit, the intensity is S_0, since the limiting value of the function $\sin(\phi/2)/(\phi/2)$ is unity as ϕ approaches zero. As the angle θ increases, ϕ becomes larger also, and when $\sin \theta = \lambda/d$, ϕ attains the value 2π radians, and the intensity as given by (26.6.9) is zero. Under these conditions, the phase difference between S_1 and S_2 in Figure 26.25 is 360°. The phase difference between S_1

and O is, therefore just half this value or 180°. Light entering the slit at S_1 and arriving at P now *interferes destructively* with light entering the slit at O and arriving at P. In fact, light entering *anywhere* above O will interfere destructively with light entering at a point a half-slit width lower along S_1S_2, since the phase difference between any two such points is 180°. In this way, we can easily see that when $\sin \theta = \lambda/d$, light entering the upper half of the slit interferes destructively with light coming through the lower half. For this value of θ, therefore, the light intensity on the screen is zero.

If the angle θ is increased further, the interference of light entering various parts of the slit is no longer totally destructive, and there is, therefore, a resultant intensity given by (26.6.9). When $\sin \theta = 2\lambda/d$, however, the total phase difference ϕ between S_1 and S_2 is 4π, and, under these circumstances, total destructive interference again occurs. Now, if we divide the slit into four quarters, we find that all the light entering the first and third quarters interferes destructively, as does all the light entering the second and fourth quarters. As θ increases still further, a series of alternating intensity maxima and minima are observed,

[8] In this discussion, it is assumed for convenience that the slit width d is considerably larger than the wavelength, so that the quantity $2\pi d/\lambda$ is quite a large number. This condition need not always be satisfied, of course. We shall see later what happens under those circumstances.

871

corresponding, respectively, to

$$\phi = 2\pi, 4\pi, 6\pi, 8\pi, \ldots$$

or

$$\sin \theta = \frac{\lambda}{d}, \frac{2\lambda}{d}, \frac{3\lambda}{d}, \frac{4\lambda}{d}, \ldots$$

(26.6.10)

for the intensity minima, and, approximately,

$$\phi = 3\pi, 5\pi, 7\pi, 9\pi, \ldots$$

or

$$\sin \theta = \frac{3\lambda}{2d}, \frac{5\lambda}{2d}, \frac{7\lambda}{2d}, \frac{9\lambda}{2d}, \ldots$$

(26.6.11)

for the maxima.

One of the more unexpected features of diffraction can be exhibited by calculating the angle θ_m between the mth minimum of intensity and the center of the pattern. From (26.6.10), this is

$$\sin \theta_m = \frac{m\lambda}{d}$$

(26.6.12)

In this expression, the slit width d is in the denominator. Therefore, when the slit is very wide, θ_m will be an extremely small angle, and the diffraction pattern will occupy only a very narrow region near the center of the screen at O' in Fig. 26.25. Under such circumstances, the intensity minima and maxima may be so closely spaced that they are not readily observed, and the only light that is easily seen is the large central maximum at $\theta = 0$, as illustrated in Fig. 26.30a. The pattern then looks very much like the geometric shadow image of the slit, since there is a lot of light hitting the screen at $\theta = 0$ and very little anywhere else. This result is not unexpected and certainly not very exciting.

But now, suppose we make the slit width d smaller and smaller. Then according to (26.6.12), the angle θ_m will become increasingly *larger*, the diffraction pattern spreading out to occupy an appreciable area of the screen, as shown at (b) in Fig. 26.30. Under these conditions, the intensity maxima and minima are rather easily observed, although the pattern is much less bright because there is now less total energy coming through the slit and it is spread over a larger part of the screen.

Finally, if d becomes less than $m\lambda$, $m\lambda/d$ will exceed unity, and there will be no real value of θ_m for which Eq. (26.6.12) is satisfied. The mth minimum then disappears, leaving only a pattern in which there are $m - 1$ minima between the center of the pattern and the edge of the screen. As the slit becomes narrower still, the pattern spreads out more and more and successive minima disappear from the screen until, finally, when d becomes smaller than λ, the first minimum vanishes also. Then, only a small part of the pattern near the central maximum of intensity remains, but it is spread out all over the screen, from $\theta = -90°$ to $\theta = +90°$, as shown in Fig. 26.30c. Ultimately, when the slit width is very much less than the wavelength, only a small region in the neighborhood of the central maximum fills the whole screen, which is then practically uniformly, though very dimly, illuminated. You will recall, that when we studied double-slit interference in section 26.3, we assumed that each of our two slits was very narrow in comparison with the light wavelength. The reason for that assumption is now understandable, since if that condition is not satisfied, the intensity variations on the screen will result not only from interference between the two slits but also from the *diffraction* of light passing through each of them.

This curious effect, in which the diffraction pattern spreads out to occupy an increasingly large area on the screen as the slit width is reduced, is

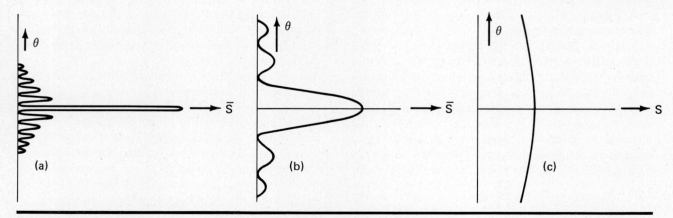

FIGURE 26.30. Successive stages in the appearance of the single-slit Fraunhofer diffraction pattern as the slit width is decreased. Note that as the slit becomes narrower, the pattern becomes wider.

characteristic of diffraction in general. It occurs not only for light diffracted by a slit but also for light diffracted by an aperture of any shape or for light diffracted by an opaque obstacle rather than an aperture. It is this effect that eventually frustrates our attempts to improve the sharpness of fuzzy pinhole images by making the pinhole size smaller and smaller.

In concluding the discussion of the diffraction of light by a single slit, it is appropriate to examine the difference between this situation and the double-slit interference phenomena discussed in section 26.3 in connection with Young's experiment. In the former case, we were concerned with the *interference* of light coming from *two separate coherent sources*. These sources were required to be so small that the diffraction effects arising from light emitted by different parts of each source would not be observable. In the present case, only a *single* slit source is involved, but its dimensions are such that light coming through different parts of the slit may interfere differently at each point on the screen. In both cases a series of alternating maxima and minima of light intensity is observed. In the case of double slit interference, however, the maxima all have essentially the same intensity, while for single-slit diffraction, the intensity of the maxima fall off rapidly as the distance from the central maximum increases, as illustrated in Fig. 26.29b. Also, for double-slit interference, the spacing of the fringes is determined by the distance between the two slit sources, while in the case of diffraction, the distance between the striations is determined by the width of the single slit source. It is important to remember that although diffraction is a phenomenon that arises from interference, it is not synonymous with interference. Diffraction of light is the interference of light rays that originate from different parts of an aperture or that come from different locations in the neighborhood of an opaque obstacle. It can, therefore, be regarded as a special case of the more general phenomenon of interference.

EXAMPLE 26.6.1
A Fraunhofer diffraction pattern is observed using light of wavelength 5500 Å. If the slit width is 2.5×10^{-4} cm, find the angles for which maximum and minimum intensity occurs. Find the ratio of the intensities of the fourth and first maxima beyond the central maximum.

The zero-order or central maximum occurs for $\theta = 0$. The other maxima are found, according to (26.6.11), for

$$\sin \theta = \frac{3\lambda}{2d}, \frac{5\lambda}{2d}, \frac{7\lambda}{2d}, \ldots$$

corresponding to $\phi = 3\pi, 5\pi, 7\pi, \ldots$. From this, we find

$$\sin \theta = \frac{(3, 5, 7, \ldots)(5.5 \times 10^{-5})}{(2)(2.5 \times 10^{-4})} = (3, 5, 7, \ldots)(0.110)$$

The possible values of θ corresponding to *maxima* are, therefore,

$$\sin \theta_1 = (3)(0.110) = 0.330 \qquad \theta_1 = 19.27°$$
$$\sin \theta_2 = (5)(0.110) = 0.550 \qquad \theta_2 = 33.37°$$
$$\sin \theta_3 = (7)(0.110) = 0.770 \qquad \theta_3 = 50.35°$$
$$\sin \theta_4 = (9)(0.110) = 0.990 \qquad \theta_4 = 81.89°$$

For minima, (26.6.10) tells us that we must have

$$\sin \theta' = \frac{\lambda}{d}, \frac{2\lambda}{d}, \frac{3\lambda}{d}, \ldots$$

corresponding to $\phi = 2\pi, 4\pi, 6\pi, \ldots$. The possible values of θ for minimum intensity are, therefore,

$$\sin \theta' = \frac{(1, 2, 3, \ldots)(5.5 \times 10^{-5})}{(2.5 \times 10^{-4})} = (1, 2, 3, \ldots)(0.220)$$

or

$$\sin \theta_1' = (1)(0.220) = 0.220 \qquad \theta_1' = 12.71°$$
$$\sin \theta_2' = (2)(0.220) = 0.440 \qquad \theta_2' = 26.10°$$
$$\sin \theta_3' = (3)(0.220) = 0.660 \qquad \theta_3' = 41.30°$$
$$\sin \theta_4' = (4)(0.220) = 0.880 \qquad \theta_4' = 61.64°$$

For the first maximum, corresponding to $\theta = \theta_1 = 19.27°$, the total phase difference ϕ is 3π; for the fourth maximum, at $\theta = \theta_4 = 81.89°$, it is 9π. From Eq. (26.6.9), the intensities at these two points must be

$$\bar{S}_1 = S_0 \left[\frac{\sin(3\pi/2)}{(3\pi/2)} \right]^2 = (2/3\pi)^2 S_0$$

and

$$\bar{S}_4 = S_0 \left[\frac{\sin(9\pi/2)}{(9\pi/2)} \right]^2 = (2/9\pi)^2 S_0$$

The intensity ratio for the two maxima is, therefore,

$$\frac{\bar{S}_4}{\bar{S}_1} = \left(\frac{2}{9\pi} \frac{3\pi}{2} \right)^2 = \frac{1}{9}$$

EXAMPLE 26.6.2
The sodium D_1 and sodium D_2 spectral lines have wavelengths of approximately 5896 and 5890 Å. A sodium lamp sends incident plane waves onto a slit of width 2×10^{-4} cm. A screen is located 3 meters from the slit. Find the spacing between the first maxima of the two sodium lines, as measured on the screen.

Using Eq. (26.6.11) for each of the two wavelengths, we find

$$\sin \theta_1 = \frac{(3)(5.896 \times 10^{-5})}{(2)(2 \times 10^{-4})} = 0.44220$$

$$\sin \theta_1' = \frac{(3)(5.890 \times 10^{-5})}{(2)(2 \times 10^{-4})} = 0.44175$$

From which

$$\theta_1 = 0.45805 \text{ rad} \qquad \text{or} \qquad 26.2443°$$
$$\theta_1' = 0.45755 \text{ rad} \qquad \text{or} \qquad 26.2156°$$

The vertical distances h_1 and h_1' from the central maximum are approximately given by

$$h_1 \cong 3 \tan \theta_1 = 1.4791 \text{ m} = 147.91 \text{ cm}$$
$$h_1' \cong 3 \tan \theta_1' = 1.4772 \text{ m} = 147.72 \text{ cm}$$

Therefore,

$$\Delta h = h_1 - h_1' = 0.19 \text{ cm}$$

The Na lines, which are in the yellow portion of the visible spectrum, were first resolved by Josef Fraunhofer.

26.7 Multiple-Slit Diffraction

We have now studied interference due to two very narrow slits and also diffraction due to a single slit. In this section, we shall study the interference pattern established by two slits each of which produces a diffraction pattern. We shall also extend the discussion to a multiple-slit system known as a diffraction grating. The double-slit pattern will be derived explicitly, but the case of a grating will be discussed only in a qualitative way.

In Fig. 26.31, a plane wave with wavelength λ is incident on a two-slit system. The slits have widths a and are separated by a distance d. To determine the electric vector at a point P on a distant screen, we assume that every point on the wavefront acts as a source of secondary waves, and we superpose

all of these using graphical methods. Let $\mathbf{E}_0$ be the electric vector which would be present at $\theta = 0$ if one of the slits were covered. In Fig. 26.28, we showed how to superpose graphically the field vectors due to a single slit. For a double slit, we simply continue the addition process. However, referring to Fig. 26.31, we must realize that the wave sent out from point A differs in phase from that at point B by the amount $(2\pi/\lambda)(AB) \sin \theta$, where the distance AB is $d - a$. Referring now to Fig. 26.32, we first obtain the resultant electric field $\mathbf{E}_l$ from the lower slit by vector addition, as in Fig. 26.28. This vector makes an angle of $\phi/2$ with the horizontal. To obtain the second vector $\mathbf{E}_u$, we again begin the process of adding infinitesimal vectors, but the first small vector (emitted at A) now has a phase of

$$\delta = \frac{2\pi}{\lambda}(d - a) \sin \theta \tag{26.7.1}$$

with respect to the small vector from B. The length of $\mathbf{E}_u$ is the same as that of $\mathbf{E}_l$, but according to the figure, the *angle* between the two vectors is seen to be $\phi + \delta$. Let $\beta = \phi/2$ and $\gamma = (\phi + \delta)/2$. Then, since the vectors $\mathbf{E}_l$ and $\mathbf{E}_u$ both have the same magnitude E_l, from Fig. 26.32b it is evident that

$$E_p = 2E_l \cos\left(\frac{\phi}{2} + \frac{\delta}{2}\right) \tag{26.7.2}$$

But since E_l can be expressed in terms of E_0 by (26.6.8), just as in the preceding section, this can be written as

$$E_p = 2E_0 \frac{\sin(\phi/2)}{\phi/2} \cos\left(\frac{\phi}{2} + \frac{\delta}{2}\right)$$

or, more simply,

$$E_p = 2E_0 \frac{\sin \beta}{\beta} \cos \gamma \tag{26.7.3}$$

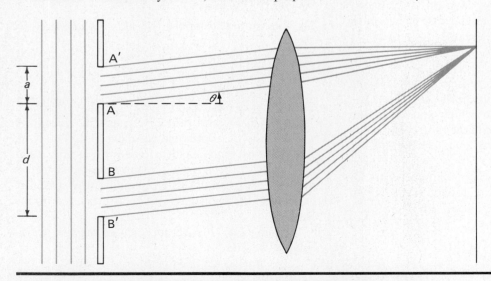

FIGURE 26.31. Ray geometry for double-slit Fraunhofer diffraction.

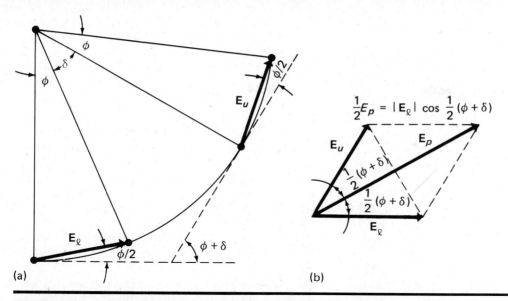

FIGURE 26.32. Phase diagram for double-slit Fraunhofer diffraction.

where, from (26.6.1),

$$\beta = \phi/2 = (\pi/\lambda)a \sin \theta \tag{26.7.4}$$

$$\gamma = \frac{\phi}{2} + \frac{\delta}{2} = (\pi/\lambda)d \sin \theta \tag{26.7.5}$$

The factor $(\sin \beta)/\beta$ characterizes single-slit diffraction, while $\cos \gamma$ occurs in double slit interference. In the present case, both factors are present. Therefore, the relative intensity, which we may write as

$$\bar{S} = 4S_0 \left(\frac{\sin \beta}{\beta}\right)^2 \cos^2 \gamma \tag{26.7.6}$$

contains variations typical of both. Figure 26.33 illustrates the resulting pattern for the case in which

$d = 4a$, which implies that $\gamma = 4\beta$. The $\cos^2$ function has a maximum whenever $\gamma = 0, \pi, 2\pi, \ldots$. These occur whenever

$$d \sin \theta = m\lambda \qquad m = 0, 1, 2, 3, 4, \ldots \tag{26.7.7}$$

At these angles, we have so-called *principal maxima* in the intensity pattern. The integer m denotes the order of the pattern, and $m = 0, 1, 2, 3$, etc., corresponds to the zeroth, first, second, third maxima, etc. For double-slit interference, which involves extremely narrow slits, for which $(\sin^2 \beta)/\beta^2 = 1$, these maxima would all be of uniform intensity. However, when the slit width is nonzero, additional changes in intensity can occur due to the $(\sin^2 \beta)/\beta^2$ factor. For example, whenever, as in (26.6.10), $\phi/2 = \beta = \pi, 2\pi, 3\pi, \ldots,$

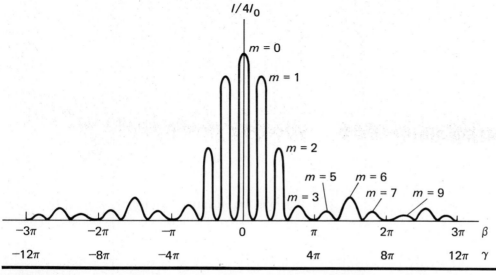

FIGURE 26.33. Double-slit Fraunhofer diffraction intensity pattern for the case where $d = 4a$, corresponding to $\gamma = 4\beta$.

we will have an intensity minimum, since $\sin \beta$ then vanishes. From (26.7.4), it is seen that this occurs for

$$a \sin \theta = n\lambda \qquad n = 1, 2, 3, \ldots \qquad (26.7.8)$$

The pattern may be described by saying that the $(\sin^2 \beta)/\beta^2$ factor *modulates* the $\cos^2 \gamma$ variations. Thus, we see in Fig. 26.33 that the various interference maxima are contained within an envelope corresponding to the diffraction factor $(\sin^2 \beta)/\beta^2$.

EXAMPLE 26.7.1

A double slit is illuminated by light for which $\lambda = 4500$ Å. If the two slits are each 9000 Å in width and are separated by a distance of 27,000 Å, find all angles θ at which the intensity is zero. Find also the angles at which an intensity maximum occurs.

Apart from the central maximum, the intensity will be zero when *either* $\sin \beta$ or $\cos \gamma$ vanishes. Since $a = 2\lambda$ and $d = 6\lambda$, this will occur for

$$\beta = (\pi a/\lambda) \sin \theta = 2\pi \sin \theta = \pi, 2\pi, 3\pi, \ldots$$

and for

$$\gamma = (\pi d/\lambda) \sin \theta = 6\pi \sin \theta = \pi/2, 3\pi/2, 5\pi/2, \ldots$$

Thus, there are minima when

$$\sin \theta = 0.5 \text{ or } 1.0 \qquad \theta = 30° \text{ and } 90°$$

and

$$\sin \theta = \tfrac{1}{12}, \tfrac{3}{12}, \tfrac{5}{12}, \tfrac{7}{12}, \tfrac{9}{12}, \tfrac{11}{12}$$
$$\theta = 4.78°, 14.48°, 24.62°, 35.69°, 48.59°, 66.44°$$

The maxima will occur whenever $\cos^2 \gamma = 1$ *unless* $\sin \beta$ vanishes at one of these maxima. If this happens, we say that a maximum has been *suppressed*. Since $\cos^2 \gamma = 1$ implies that

$$\gamma = (\pi d/\lambda) \sin \theta = 6\pi \sin \theta = 0, \pi, 2\pi, \ldots$$

we have

$$\sin \theta = 0, \tfrac{1}{6}, \tfrac{2}{6}, \tfrac{3}{6}, \tfrac{4}{6}, \tfrac{5}{6}, 1$$
$$\theta = 0°, 9.59°, 19.47°, 30°, 41.81°, 56.44°, 90°$$

We see that the maxima at $\theta = 30°$ and $90°$ are indeed suppressed since there are minima of $\sin \beta$ at these angles. More generally, we have

$$d \sin \theta = m\lambda \qquad m = 0, 1, 2, \ldots$$

giving the mth maximum. But

$$a \sin \theta = n\lambda \qquad n = 1, 2, \ldots$$

gives a minimum of zero. Thus, whenever

$$\left(\frac{d}{a}\right) n = m$$

a maximum of the two slit interference pattern is eliminated. In the present case, for which $d/a = 3$, the third and sixth orders are missing.

Let us now study in a qualitative way the inter-ference pattern produced by a *multiple-slit* system, also known as a *diffraction grating*. Figure 26.34a is a schematic representation of a grating. We assume that there are N narrow slits, each of width d and separated by a distance a. A typical grating might have 6000 or more slits per centimeter. It is no small accomplishment to construct such a grating. Fraunhofer made some early gratings by winding fine wire very closely on two fine screws. He was also able to rule gratings on glass, although this could then not be done with great accuracy. Sixty years after Fraunhofer's pioneering work, the American physicist Henry Rowland constructed a "ruling engine" for cutting gratings on an aluminum coating deposited on glass. A diamond cutting edge was used to produce very straight grooves. In this way, successful and practical diffraction gratings were first obtained.

To understand the physical principles that govern the behavior of a diffraction grating, let us first consider the simple system shown in Fig. 26.34b. In this system, there are a very large number of very narrow regularly spaced slits. The distance between neighboring slits, as shown in the diagram, is d. The slit width in this particular example is assumed to be *very much less than the light wavelength*, so that diffraction effects arising from the passage of light through apertures of finite width are unimportant. We are, therefore, concerned in this specific case with what might be more accurately referred to as an *interference* grating. The grating is illuminated with normally incident plane monochromatic light waves of wavelength λ. We may then regard each slit as a point source of cylindrical Huygens wavelets whose wavefronts are illustrated in the diagram.

Let us now consider how outgoing plane wavefronts might be formed from the system of Huygens wavelets on the far side of the grating. It is, of course, possible to draw a set of tangent planes parallel to the incoming waves and to the grating which represent plane waves that propagate horizontally to the right beyond the grating. These wavefronts do actually exist but contribute only to a continuation of the incident wave. It is possible also to draw a set of wavefronts that are tangent to Huygens wavelets from neighboring slits whose phase *differs* by an entire wavelength, as shown by the dashed lines in the lower part of the figure. These wavefronts propagate at an angle θ_1 to the horizontal. Since the path difference PP' between succeeding wavefronts is $d \sin \theta_1$ and since this must be an entire wavelength, we may determine the direction of propagation, θ_1, from the condition

$$d \sin \theta_1 = \lambda \qquad (26.7.9)$$

In much the same way, a series of wavefronts propagating at a larger angle θ_2 to the horizontal may be constructed by drawing tangents connecting wavelets from neighboring slits that are out of phase by two

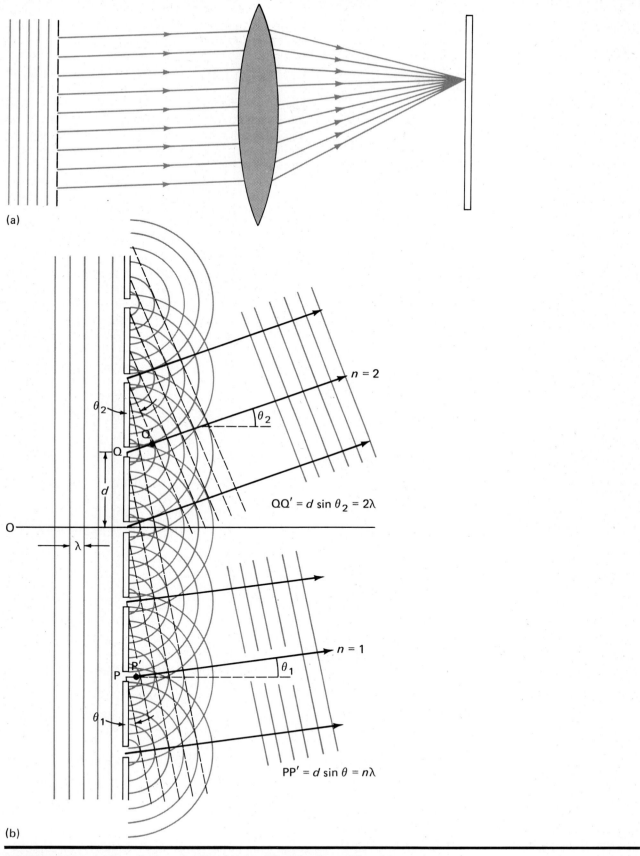

(a)

(b)

$QQ' = d \sin \theta_2 = 2\lambda$

$PP' = d \sin \theta = n\lambda$

FIGURE 26.34. (a) Ray diagram for Fraunhofer diffraction by a diffraction grating. (b) Geometry of rays and wavefronts for diffraction by a grating.

wavelengths. These wavefronts are shown as the set of dashed lines in the upper part of Fig. 26.34b. Since the path difference QQ' between successive wavelets is now twice the wavelength, and since $QQ' = d \sin \theta_2$, we may write for this set of outgoing waves

$$d \sin \theta_2 = 2\lambda \qquad (26.7.10)$$

We can in the same manner find outgoing waves in other directions, θ_3, θ_4, θ_5, ..., corresponding to larger integral path differences between neighboring slits. Their propagation directions are clearly given by

$$\boxed{d \sin \theta_n = n\lambda} \qquad (26.7.11)$$

where the integer n identifies the so-called *order* of the outgoing diffracted wave.

These outgoing "diffracted" beams strike a distant screen at various distances above the axis OO' of the grating. They can, in fact, be focused into sharp lines of light by a converging lens placed between grating and screen, as shown in Fig. 26.25. By measuring these distances, or by measuring the angles θ_1, $\theta_2, \theta_3, \ldots$, it is possible to measure the wavelength λ of the incoming light very accurately, provided that the distance d between slits is precisely known. Also, since light of different wavelength will be diffracted at different angles, a grating acts to produce a *spatial separation between light of different colors*, in somewhat the same way as a dispersive prism.

In this example, we have assumed that the slit width is very much smaller than the light wavelength so that the effects that we observe are really those of interference rather than diffraction. In actual gratings, this assumption is hardly ever justified; and for this reason, though the outgoing beams propagate in precisely the same directions, as defined by (26.7.11), their intensities may be strongly affected by effects of diffraction associated with slits of finite width.

To derive the intensity pattern due to a grating with finite slits of width a, we would follow the previous procedure of superposing infinitesimal electric fields to find the field due to each slit. Then we would add vectorially the fields due to each slit, taking into account the phase differences between each of these fields. For a double slit, we found that the intensity could be expressed by (26.8.3). This can be rewritten as

$$\bar{S} = 4S_0 \left(\frac{\sin \beta}{\beta} \right)^2 \cos^2 \gamma = S_0 \left(\frac{\sin \beta}{\beta} \right)^2 \left(\frac{\sin 2\gamma}{\sin \gamma} \right)^2 \qquad (26.7.12)$$

since

$$\sin 2\gamma = 2 \sin \gamma \cos \gamma$$

The generalization of this expression to the case of N slits is a complex mathematical task which we shall not attempt here. The result, however, is simple

enough and is obtained simply by replacing $\sin 2\gamma$ by $\sin N\gamma$ in Eq. (26.7.12). Thus, for a diffraction grating having N slits, the intensity is given by

$$\bar{S} = S_0 \left(\frac{\sin \beta}{\beta} \right)^2 \left(\frac{\sin N\gamma}{\sin \gamma} \right)^2 \qquad (26.7.13)$$

where S_0 is the intensity at $\theta = 0$ and where β and γ are defined by (26.7.4) and (26.7.5), respectively. The factor $(\sin^2 \beta)/\beta^2$ is due to single-slit *diffraction*, while $(\sin N\gamma / \sin \gamma)^2$ is present due to the *interference* of N slits.

Let us now discuss the principal features of (26.7.13) to establish the dependence of the intensity on the angle θ. We shall first examine the interference factor. Whenever $\sin N\gamma$ vanishes, it appears, at first sight, that the intensity I should be zero. There is an exception to this, however; for if $\sin \gamma$ also vanishes, we have an indeterminate ratio. In fact, it turns out that when both of these factors vanish, $(\sin N\gamma / \sin \gamma)^2$ attains a *maximum* rather than a minimum. To see this, we note that whenever $\gamma = n\pi$, $\sin \gamma$, and $\sin N\gamma$ will both vanish. Let us, therefore, take the limit of $\sin N\gamma / \sin \gamma$ as $\gamma \to n\pi$. We can do this by using l'Hôpital's rule from calculus to obtain

$$\lim_{\gamma \to n\pi} \frac{\sin N\gamma}{\sin \gamma} = \lim_{\gamma \to n\pi} \frac{\dfrac{d}{d\gamma} \sin N\gamma}{\dfrac{d}{d\gamma} \sin \gamma} = \lim_{\gamma \to n\pi} \frac{N \cos N\gamma}{\cos \gamma} = \pm N$$

$$(26.7.14)$$

Thus, whenever γ is an integral multiple of π, we have a so-called *principal maximum* of the pattern, and a factor of N^2 coming from the N-slit interference factor is present. It might happen, of course, that $\sin \beta$ vanishes or is small at such a principal maximum. In this event, the principal maximum will be suppressed due to the diffraction term.

Whenever $N\gamma$ is an integral multiple of π but γ *is not* a multiple of π, the intensity will vanish since $\sin N\gamma$ is zero and $\sin \gamma$ is nonzero. For angles which satisfy this condition, a *minimum* intensity is observed. For large N, this implies a very large number of subsidiary maxima and minima between any two principal maxima because of the rapid variation of the function $\sin N\gamma$ in Eq. (26.7.13). These subsidiary maxima, in gratings with many slits, are of very low intensity in comparison to the principal maxima and ordinarily are not even visible.

The effect of all this is that when the number of slits becomes very large, the principal maxima become very intense and the subsidiary maxima very weak. Since the principal maxima occur when $\gamma = n\pi$ and since γ is defined by (26.7.5), the principal maxima occur when

$$n\pi = (\pi d / \lambda) \sin \theta$$

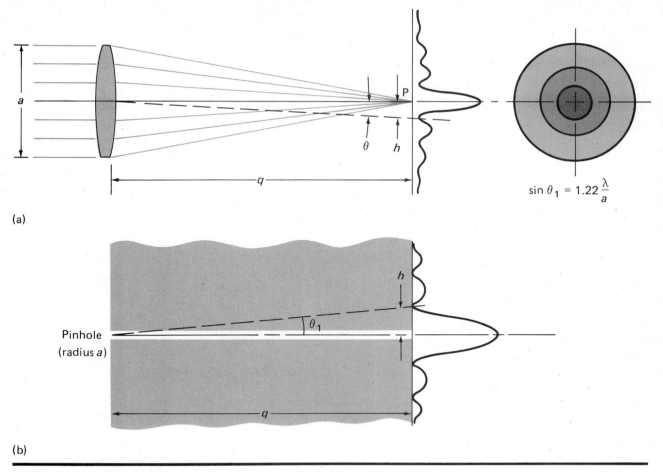

$$\sin \theta_1 = 1.22 \frac{\lambda}{a}$$

(a)

(b)

FIGURE 26.39. Diffraction occurring at a small circular aperture, showing, at (a), the intensity distribution and the general appearance of the pattern. At (b), the geometry of diffraction in a pinhole camera is illustrated.

dark rings corresponding to maxima and minima in the intensity of light diffracted at slightly different angles to the optical axis of the instrument.

The image of a point source of light on the axis of the instrument is, therefore, *never* an infinitesimally small point, even when the optical system is completely free of aberrations. It is, instead, a circular diffraction pattern like that shown in Fig. 26.39a. The angular extent of the diffraction pattern depends upon the size of the aperture and the light wavelength, much like that of the pattern produced by a narrow slit. The exact expression for intensity as a function of angle will not be derived here, since the mathematics is too complex.

The pattern has a central maximum corresponding to $\theta = 0$. The first minimum occurs at an angle θ_1 given by

$$\sin \theta_1 = 1.22(\lambda/a) \qquad (26.8.1)$$

where a is the diameter of the aperture. The angle at which this minimum occurs increases as the diameter

decreases. The situation in this respect is the same as that which arises in the case of a narrow slit. If an optical instrument having a circular aperture is to resolve two distant objects, the central maxima must be separated by at least the angle θ_1 of (26.8.1). This, then, implies that the objects themselves must be spatially separated so that their angular separation, as seen from the optical instrument, exceeds this angle. Figure 26.38c illustrates the problem of resolving two point sources of light.

EXAMPLE 26.8.1

Assume that the headlights of automobiles are on the average 1.5 meter apart and that the average wavelength emitted is about 5500 Å. Cars are traveling at night along a straight highway in Nevada with their lights on. You are at the side of the road trying to hitchhike to Reno. You notice that when the cars are very far away, they each appear to have only one headlight, but as they approach they obviously have two. Explain this phenomenon and determine how

883

far distant a car is when one headlight appears magically to change into two. The pupil of your eye is about 5 mm in diameter.

This is an example of diffraction by a circular aperture, namely, the pupil of your eye. According to condition (26.8.1), the two sources will just be resolved when they are separated by a very small angle θ_1, such that

$$\sin \theta_1 \cong \theta_1 = 1.22 \frac{\lambda}{a} = \frac{(1.22)(5.5 \times 10^{-5})}{0.5}$$

$$= 1.34 \times 10^{-4} \text{ rad}$$

Now suppose the car is at a distance d from you. The angular separation of the sources will be l/d, where l represents the distance between the two headlights. Thus, equating θ to l/d, we obtain the distance d at which the headlights are just resolved:

$$1.34 \times 10^{-4} = \frac{1.5}{d}$$

$$d = 1.12 \times 10^4 \text{ m} = 11.2 \text{ km, or } 7.00 \text{ mi}$$

EXAMPLE 26.8.2
In Chapter 25, we learned that the laws of geometrical optics predict that a pinhole forms an image of a distant axial point source as a circular disc of light having the same diameter as the pinhole itself. It was mentioned that the sharpness of pinhole images can be improved to a certain extent by reducing the pinhole diameter, but that a point is reached beyond which the images become fuzzier rather than sharper as the pinhole is made smaller and smaller. Explain, on the basis of wave optics, why this is so and find the diameter of a pinhole aperture that produces an image of optimum sharpness. Explain why lenses are able to produce images that are much sharper than pinhole images and discuss how diffraction effects limit the resolution of lenses.

This situation is illustrated in Fig. 26.39b. According to geometrical optics, parallel rays from an object on the axis at infinity enter the pinhole and propagate to the film or viewing screen at a distance q from the pinhole, as shown. As a result, a circular image spot the same size as the pinhole is formed on the screen. Under these circumstances, we should expect the image spot to become smaller and smaller and the image to become sharper and sharper as the diameter of the aperture is decreased. But this works only up to a point because the linear ray propagation picture of geometrical optics is only an approximate one. In particular, the effects of diffraction by the circular aperture through which light passes are not accounted for in this picture. So long as the aperture is large in comparison with the light wavelength, this does not matter much, for then the diffraction pattern,

whose width is of the order of λ/a, is very narrow and its intensity falls off to a negligible value even for directions that make very small angles to the optic axis. The pinhole image then looks very much like the geometric image of the aperture. As the diameter is reduced, however, the diffraction pattern spreads out, finally becoming as large as the pinhole itself. Beyond this point, further decreasing the aperture size makes the image spot *larger* rather than smaller, because now it is the size and structure of the diffraction pattern that determines what is observed on the screen. The condition of minimum image spot size is attained when the central maximum of the diffraction pattern has the same diameter as the image spot, because beyond the central maximum the intensity of the diffraction pattern falls off so rapidly as to contribute little to the apparent size of the image disc. Referring to Fig. 26.39b and Eq. (26.8.1), it is apparent that this condition is satisfied, assuming that θ_1 is a small angle, when

$$h = q\theta_1 = 1.22 \frac{q\lambda}{a} = \frac{a}{2} \qquad (26.8.2)$$

Solving for a, this gives us

$$a = \sqrt{2.44q\lambda} \qquad (26.8.3)$$

For $\lambda = 5500$ Å, corresponding to the central part of the visible spectrum, and $q = 5.0$ cm, the usual lens-to-film distance for 35-mm photo negatives, we find

$$a = \sqrt{(2.44)(5.0)(5.5 \times 10^{-5})}$$
$$= 0.0259 \text{ cm} = 0.259 \text{ mm}$$

This corresponds to a resolution on the negative of about 4 lines per mm, which is not very good. A well-corrected photographic lens can improve on this level by a factor of 20 and at the same time produce images that are 10,000 times brighter!

The reason for this can be seen by considering an $f{:}2.0$ photographic lens of 50-mm focal length. Such an objective has an effective aperture of $50/2 = 25.0$ mm. The situation, then, is essentially that illustrated in Fig. 26.39a. The angle θ_1 that describes the angular extent of the diffraction pattern formed by light coming through the lens is, according to (26.8.1),

$$\theta_1 = 1.22 \frac{\lambda}{a} = (1.22) \frac{(5.5 \times 10^{-4})}{25.0} = 2.68 \times 10^{-5} \text{ rad}$$

From this, the diameter of the image spot is

$$2h \simeq 2q\theta_1 = (2)(50)(2.68 \times 10^{-5}) = 0.00268 \text{ mm}$$

since for a very distant object the image distance is the same as the focal length. This corresponds to a resolution figure of about $1/0.00268 = 373$ lines per mm, which is the maximum obtainable for a lens completely free from aberrations. Although this figure can be approached in a lens that is specially designed and constructed, for most photographic lenses at maximum aperture the resolution is limited by the residual aberrations rather than the size of the diffraction pattern. Typically, a good 50-mm $f:2.0$ photographic objective might exhibit a resolution of perhaps 80 lines per mm at maximum aperture. As the lens is stopped down, the aberrations become less serious but, at the same time, the size of the diffraction pattern increases. If the lens considered above is stopped down to $f:9.0$, which reduces the effective aperture diameter by a factor of 4.5, the diffraction pattern increases in size by the same factor. The diffraction-limited resolution under these circumstances is less than one quarter that at maximum aperture, or 83 lines per mm. Stopping the lens down further only serves to make this figure even less. For most lenses, therefore, there is an optimum aperture at which the image discs resulting from aberrations and diffraction have roughly the same size and for which the lens produces optimally sharp images. Ordinarily, this optimum aperture setting is from one to three $f:$ stops smaller than the maximum aperture. For the lens considered above, it is evidently between $f:2.0$ and $f:9.0$.

A good lens, therefore, produces sharper images than a pinhole because it permits the light striking the image plane to pass through a large aperture rather than a small one. Under these circumstances, the diffraction effects observed at the image position are much less significant than is the case with a small aperture such as a pinhole. At the same time, since the area of the lens is much greater than that of a pinhole, it permits more light to pass through to the image plane. In this example, the pinhole has a diameter of 0.259 mm for optimum image sharpness, while the 50-mm $f:2.0$ photographic lens had a maximum aperture diameter of 25.0 mm. Since the amount of light transmitted is proportional to the area of the aperture, however, the image formed by the lens will be $(25.0/0.259)^2$, or some 9300 times as intense as that produced by the pinhole.

26.9 Polarization of Light

When visible light or any other form of electromagnetic wave propagates through space, the associated electric and magnetic fields oscillate. If the electric field always oscillates in some fixed direction, the wave is said to be *linearly polarized* in that direction. Since the associated magnetic field is always perpendicular to both the electric field and the direction of wave propagation, the magnetic field also oscillates in a single direction. In most of the discussion in subsequent sections, we shall study only the electric vector because its specification always fixes the corresponding magnetic vector. Figure 26.40 depicts a monochromatic linearly polarized electromagnetic wave advancing in the x-direction. Since the electric vector always points in the z-direction, we shall say that this wave is linearly polarized in the z-direction. Thus, by convention, the polarization of a wave implies a certain orientation of the electric vector.

What conditions are needed to produce linearly polarized light? In some parts of the electromagnetic spectrum, such as the microwave or radio wave portion, radiation occurs when charged particles ac-

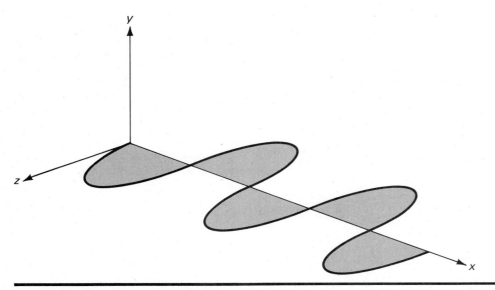

FIGURE 26.40. Plane electromagnetic wave, linearly polarized along the z-direction and propagating along the x-direction. The magnetic vector is omitted in the interest of clarity.

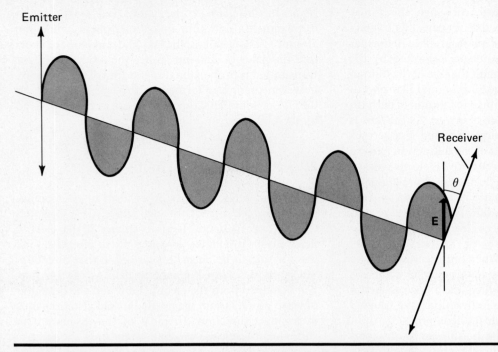

FIGURE 26.41. Detection of a linearly polarized electromagnetic wave with a linear receiving antenna making an angle θ with the wave's electric vector. The detected signal amplitude is proportional to $E_0 \cos \theta$, since only the field component parallel to the conductor is effective in inducing an emf in it.

celerate back and forth in an antenna. This kind of radiation will automatically be linearly polarized parallel to the axis of the antenna. In fact, it is easy to verify that these waves are polarized by using a simple receiving antenna. If the receiving antenna is parallel to the emitting antenna, the signal detected is a maximum. When it is rotated by 90°, no signal at all is detected. At intermediate angles, a signal is obtained, but it varies with the angle between the two antennas.

These simple observations, illustrated in Fig. 26.41, can readily be explained. The emitted radiation produces an electric field **E** at the receiving antenna. When the antenna is parallel to **E**, the full field is effective in accelerating the electrons in the antenna, and thus a maximum electric current is observed. When the angle between **E** and the antenna is changed from 0° to an arbitrary angle θ, only the component of the electric field parallel to the antenna, $E \cos \theta$, is effective in accelerating the electrons. Thus, the current in the wire diminishes from a maximum to zero as the angle between the receiving antenna and the emitting antenna is varied from 0° to 90°. This simple discussion provides a method for determining the direction of polarization experimentally when the radiation is in the microwave or radio wave range.

Now, electromagnetic radiation in the visible spectrum is emitted by individual atoms or molecules. Each of these may be visualized as a microscopic antenna, but any real light source consists of an

enormous number of these small emitters. Although the light coming from any one of them may be linearly polarized, the light from the entire source is generally unpolarized, the total **E** vector being a superposition of the electric vectors produced by many individual microscopic sources. Each of these vectors is perpendicular to the propagation direction, but they are randomly distributed in a plane perpendicular to this

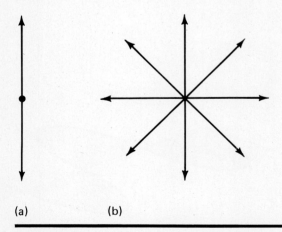

(a)　　　　(b)

FIGURE 26.42. (a) Electric vector of a linearly polarized wave propagating normally upward, out of the page, toward the reader. (b) Corresponding diagram for an unpolarized beam of radiation, in which electric vectors are oriented at all possible angles around the axis of propagation.

direction. In Fig. 26.42a, a ray of monochromatic light is traveling out of the paper toward the reader. The vectors denote the direction of vibration of the electric vector, which in this case characterizes a linearly polarized wave. Figure 26.42b represents a superposition of linear polarizations in all directions with respect to the ray of light and would, therefore, characterize the unpolarized light emitted from a source containing many atoms.

It is possible to start with unpolarized light and, using polarizing sunglasses, to "filter" it in such a way as to obtain linearly polarized light. The process by which this is accomplished was developed in 1935 by the American physicist Edwin H. Land. We shall now try to explain in a qualitative way how it works.

The process begins with a plastic material called polyvinyl alcohol in the form of thin sheets. A sheet of this material is heated and stretched in a given direction, which has the effect of aligning its molecules parallel to one another. The sheet is then impregnated with iodine, which attaches itself to the hydrocarbon chain, forming a chain of its own. The valence electrons of the iodine can easily move along such a chain in response to an applied electric field parallel to its length and, therefore, can absorb energy from an incident light wave. But they cannot jump from one chain to another; therefore, wave fields perpendicular to this direction cause very little electronic motion and are readily transmitted. As a result, this kind of polarizing material transmits only light whose electric vector is perpendicular to the direction of stretching. Materials of this sort, which exhibit different elec-

tronic conductivity in two perpendicular directions, are called *dichroic*.

26.10 Malus's Law

Figure 26.43 depicts unpolarized light incident on a sheet of polarizing material in which the transmission axis is vertical. After passing through A, the light is linearly polarized. The electric vector of this light can now be resolved into components parallel and perpendicular to the transmission axis of the second sheet B. The component that is perpendicular to this axis is completely absorbed, while the component that is parallel is transmitted. Thus, the component of the electric vector incident parallel to the transmission axis has the magnitude $E \cos \theta$. If there is no absorption at all for this component, we can write the transmitted *intensity* at angle θ as

$$\bar{S} = S_0 \cos^2 \theta$$

where S_0 is the maximum intensity and θ is the angle between the two transmission axes. The dependence on the square of $\cos \theta$ arises, of course, since intensity is proportional to the square of the field strength, according to (26.2.4).

The above equation, which is known as Malus's law, was named after E. L. Malus (1775–1812) who, in 1809, long before polarizers were developed and long before the nature of light was clearly understood, discovered that *reflected light* may be polarized. Malus,

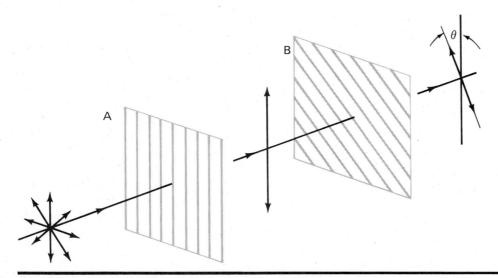

FIGURE 26.43. Diagram illustrating the law of Malus. At (a), an unpolarized beam is linearly polarized by a sheet of polarizing material. At (b), it passes through a second polarizing sheet whose polarization axis makes an angle θ with the electric vector of the wave. Since only the field component parallel to the polarization axis is transmitted, the transmitted amplitude is $E \cos \theta$, where E is the amplitude incident on the second polarizer.

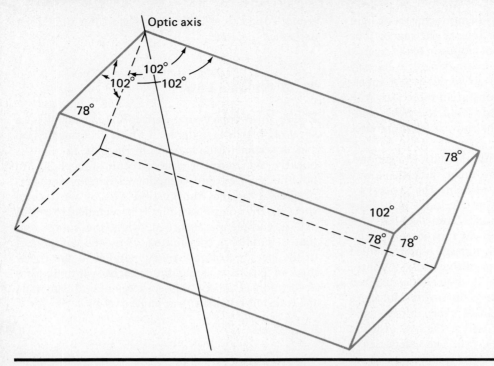

FIGURE 26.44. Calcite crystal, showing the orientation of the optic axis which makes equal angles with the three edges of the obtuse vertex of the crystal.

who was a military engineer and a captain in Napoleon's army, was examining the sunlight reflected from a glass window by looking at it through a calcite crystal. Calcite is a *doubly refracting,* or *birefringent* material since it splits an incident ray into two separate refracted rays. Malus noticed that for some angles of orientation of the crystal only one ray would emerge, while for a different angle only the other ray would be transmitted.[9]

To understand his observations, we must briefly discuss the propagation of light through a doubly refracting crystal such as calcite. Many crystalline substances exhibit optical behavior just like that of glass or water. In such *isotropic* crystals, the speed of light is independent of the direction of propagation and is equal to the speed *in vacuo* divided by a number that is likewise independent of direction, referred to as the refractive index. A *birefringent* crystal, however, has the property that within it the speed of light depends upon the orientation of the electric vector with respect to a line called the *optic axis.* The optic axis defines a direction in the crystal such that if the electric vector is in this direction, the speed of the light

is a maximum, while if it is perpendicular to the optic axis, the speed of light is a minimum. It is apparent that the refractive properties of birefringent crystals cannot be described by a single constant refractive index. The optic axis for a crystal of calcite is shown in Fig. 26.44. Now when a light ray is incident on calcite, the direction of incidence and the optic axis form a plane. The incident electric field can be resolved into components parallel and perpendicular to this plane. The latter component is, of course, polarized perpendicular to the optic axis. This component of the light will obey Snell's law since it propagates with a single (minimum) velocity and will, therefore, be refracted in the usual way. The light rays going through the crystal and satisfying these conditions are called *ordinary* rays, or O rays.

An analysis of the behavior of the other component of the electric field shows that it does not behave in the usual way. For example, it may be resolved into a component parallel to the optic axis and a component perpendicular to it. These two components then propagate at different speeds and, therefore, the resulting wavefronts are not spheres, as expected. They may be constructed using Huygens's principle and in fact, turn out to be ellipsoids, but we shall not discuss the details here.[10] This component

[9] Malus apparently undertook the study of double refraction after the Paris Academy offered a prize for the best explanation. He lived near the Luxembourg Palace and was fortunate enough to have the reflection of the evening sun off the palace windows pass through his own windows. Luckily for Malus, the rays reflected from the palace windows were polarized and could be advantageously examined through the calcite crystal.

[10] An elementary explanation with relevant diagrams appears in many texts. See, for example, Franklin Miller, Jr., *College Physics,* 3rd ed. (New York: Harcourt, Brace, Jovanovich, Inc., 1972), p. 560.

888

(a)   (b)

FIGURE 26.45. Visible effects of birefringence in calcite crystals. At (a), the optic axis is normal to the page; at (b) it makes an appreciable angle with the normal. Photos courtesy of Professor T. A. Wiggins, Pennsylvania State University.

of the field leads to rays called the *extraordinary*, or E rays.

The result of sending a light ray into calcite can be summarized by saying that the incident light ray is split into two spatially separated emerging rays, the O ray and the E ray. The O ray is polarized perpendicular to the optic axis, while the E ray is polarized in some direction perpendicular to the O ray. Figure 26.45 illustrates how a birefringent crystal forms two images of an object.

We can now understand qualitatively the observations of Malus with the aid of Fig. 26.46. The light reflected from the glass was linearly polarized. By orienting the calcite so that the direction of polarization is perpendicular to the optic axis and the direction of incidence, Malus could observe the O ray. If the crystal is rotated by 90°, the O ray is extinguished and the E ray appears. For an arbitrary orientation, both rays are observed. In this way, Malus with a little luck discovered that reflected light may be polarized

and was at the same time able to observe and study the effects of birefringence.

EXAMPLE 26.10.1

A beam of light passes through two polarizers whose transmission axes make an angle of 70° with each other. The detected light has intensity $\overline{S}$. If the angle is changed to 45°, what value does the intensity have?

According to Malus's law, the intensity has a maximum of S_0 when the axes are parallel. But when they make an angle of 70° with one another,

$$\overline{S} = S_0 \cos^2 70° = 0.117 S_0$$

while at 45°

$$\overline{S}' = S_0 \cos^2 45° = 0.5 S_0$$

Therefore,

$$\overline{S}' = (0.5)\frac{\overline{S}}{0.117} = 4.3\overline{S}$$

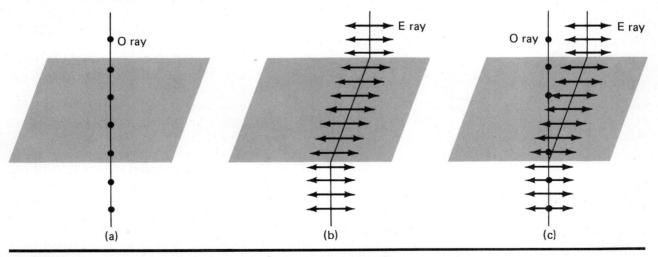

(a)   (b)   (c)

FIGURE 26.46. Birefringence in a calcite crystal. (a) Incident light polarized perpendicular to the optic axis. (b) Incident light polarized in the plane containing the optic axis and the direction of incidence. (c) Unpolarized incident light, split into ordinary and extraordinary rays.

26.11 Polarization by Reflection

We have already discussed two methods for obtaining polarized light. These involve the use of special materials such as polarizing sheets or birefringent crystals. It is evident, however, that light may be polarized when it is reflected from a material such as ordinary glass, since the observations of Malus could be understood only by assuming that light reflected in such a way is polarized. Light reflected from a transparent material may be completely polarized if the conditions are just right. In general, however, it is only partially polarized upon reflection. The degree of polarization depends on a number of factors such as the angle of incidence of the light and the indices of refraction of the media on either side of the reflecting surface. This kind of polarization does not require the use of unusual optical materials but can be observed in light reflected from water, glass, and other dielectrics. If you have a pair of polarizing sunglasses, you can see variations of light intensity when you look at the surface of a lake, a road, or a piece of glass as you change the orientation of the transmission axis of the polarizer. Since the reflected light is partially polarized, the amount that gets through the sunglasses depends on the direction of the transmission axis with respect to the polarization direction.

The polarization of reflected light is a phenomenon that can be understood completely from the viewpoint of Maxwell's electromagnetic theory. Since the mathematics involved in doing this is very difficult, we shall instead simply attempt to describe what happens in a qualitative and empirical way. Figure 26.47 shows incident unpolarized light reflected and refracted at an air–glass boundary. At a particular angle

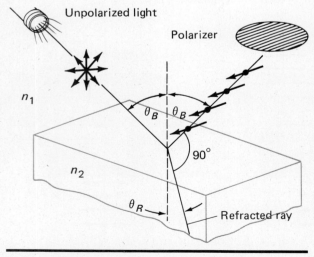

FIGURE 26.47. Polarization of light reflected from a dielectric surface at Brewster's angle. The electric vector of the reflected ray is parallel to the reflecting surface.

of incidence θ_B, it is found that the reflected ray is *completely polarized*. This is easily verified by using a polarizing sheet to examine the reflected light. As the sheet is rotated, the intensity of the reflected light that is transmitted through the polarizer is observed to vary from zero to a certain maximum value. The maximum occurs when the transmission axis of the polarizer is perpendicular to the plane containing the incident and reflected rays, and the transmitted intensity is zero when the polarizer is rotated 90° from this direction. These observations indicate that for the angle θ_B, the reflected light is completely polarized and the electric vector is *parallel to the interface* and perpendicular to the plane containing the incident and reflected rays.

More careful observations, made under widely varying conditions, indicate that when the sum of the angle of incidence and the angle of refraction is 90°, the reflected light is completely polarized. Thus, when the angle of incidence has the particular value θ_B, we may write

$$\theta_B + \theta_R = 90° \qquad (26.11.1)$$

where θ_R is the angle of refraction. According to Snell's law, however, the angles θ_B and θ_R are also related by

$$n_1 \sin \theta_B = n_2 \sin \theta_R$$

Substituting (26.11.1) into (26.11.2), we obtain

$$n_1 \sin \theta_B = n_2 \sin(90° - \theta_B) = n_2 \cos \theta_B \qquad (26.11.2)$$

or

$$\boxed{\tan \theta_B = \frac{n_2}{n_1} = n_r} \qquad (26.11.3)$$

The angle θ_B which satisfies this condition is called the *Brewster angle*, and (26.11.3) is known as *Brewster's law*. It was established empirically in 1812. For an air–glass interface ($n = 1.5$), the Brewster angle is 56.3°, while for an air–water interface ($n = 1.33$), it is 53.1°. For angles of incidence different from the Brewster angle, the reflected light is only partially polarized. Under these circumstances, the component of the electric vector parallel to the plane of incidence is no longer zero, though it will be appreciably smaller than the component perpendicular to that plane. This is illustrated in Fig. 26.48.

Since the reflected beam is completely polarized at the Brewster angle, it is natural to ask if the refracted beam is also completely polarized. The answer is no. When light strikes a surface such as an air–glass interface, approximately 95 percent of the light is transmitted in the refracted beam, while only about 5 percent is reflected. The incident unpolarized light can be considered to be a superposition of electric vectors perpendicular and parallel to the plane of incidence. The component parallel to the plane is

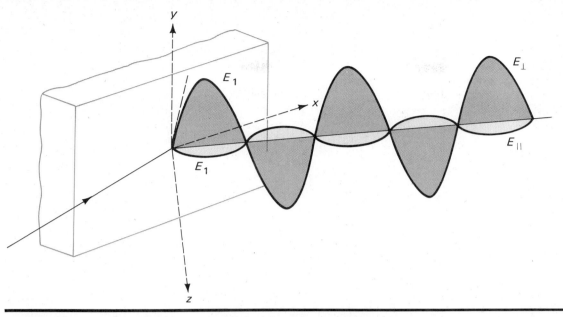

FIGURE 26.48. Partial polarization of light reflected at an angle different from Brewster's angle. In this case, the component of the electric vector of the reflected ray that lies in the plane of incidence is smaller than the component parallel to the reflecting surface but is not completely absent.

entirely refracted. The component perpendicular to the plane is mostly refracted, but is also partly reflected. The refracted beam, therefore, exhibits some degree of polarization since the component of the electric field in the plane of incidence is slightly greater than that perpendicular to this plane, but it is not completely polarized. All these results, including Brewster's law, can, of course, be derived from the fundamental laws of electricity and magnetism embodied in Maxwell's equations.

Figure 26.49 illustrates polarization of reflected light. In one of the photographs, a strong reflection from the window obscures the image inside the car. By adjusting the direction of a polarizing filter, the reflected light is eliminated, thus making it possible to see the people inside the car. The fact that reflected light is usually strongly polarized has led to the widespread use of polarizing sunglasses. In these sunglasses, the polarizing material has its transmission axis vertically oriented. Under these circumstances, the glare from light reflected from a highway or from a lake can be effectively eliminated. An additional

FIGURE 26.49. Use of a polarizing filter to eliminate unwanted reflections from glass, water, or other dielectrics. Photos by Professor John P. McKelvey, Clemson University.

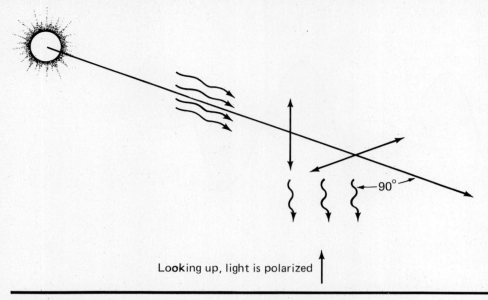

FIGURE 26.50. Polarization of skylight by atmospheric scattering.

feature of the glasses is that they cut the intensity of unpolarized light in half, which is also desirable in bright, sunny weather.

26.12 Polarization by Scattering

When light falls on the electrons within the atoms of a gas, a portion of it is deflected. This deflection of an incident beam by matter is known as *scattering*. The amount scattered, the distribution of deflection angles, and the dependence of scattering on the wavelength of light are all technical aspects of the effect which are beyond the scope of this text. It is possible, nevertheless, to understand qualitatively some phenomena involving scattering of light on the basis of a few simple physical facts.

It is known that the efficiency of the scattering process varies as the fourth power of the light frequency. Since violet light has about twice the frequency of red light, it is, therefore, scattered 16 times more effectively. A second important fact arises from a classical view of the scattering process. When an atomic electron is exposed to an electric field in a given direction, the electron can oscillate in response to this field. This oscillating electron can be considered to be a small dipole antenna that can radiate light of the same frequency. The angular dependence of the radiation emitted by an oscillating dipole has the property that no light is emitted in the direction *parallel* to the direction of motion. The direction perpendicular to this is, therefore, the direction of maximum emission. On the basis of the preceding facts, we can understand a number of simple observations involving the scattering of light in the atmosphere, as follows:

SKYLIGHT CAN BE POLARIZED

If you look straight up at the sky on a clear day during sunrise or sunset, the skylight you see is quite strongly polarized. You can easily verify this with a pair of polarizing sunglasses. Rotate the lenses slowly about a vertical axis and you will readily observe marked variations in transmitted light intensity. This effect can easily be understood with the help of Fig. 26.50. The electrons in the air molecules experience a superposition of polarized electric fields from the sun's rays. These fields can vibrate either parallel or perpendicular to the earth, and, therefore, the electronic motion that is excited within the atoms consists of a superposition of these two motions. We can then say that the electronic motion is equivalent to two oscillating dipoles, respectively oriented parallel and perpendicular to the earth. The perpendicular dipole does not radiate in the direction of the observer, and, therefore, the radiated light the observer sees all comes from the oscillating horizontal dipole. It is, therefore, completely polarized in a horizontal direction.[11]

THE SKY IS BLUE

Were it not for the fact that sunlight is scattered from the oxygen and nitrogen in the air, the sky would appear black. Indeed, in outer space, where there is no atmosphere to scatter sunlight, the sky is as black as night. The reason why the sky appears blue has to do with the frequency dependence of the scattering that

[11] In *Scientific American*, July 1955, there is an interesting article on bees and how they navigate. Apparently, they have a built-in mechanism for the detection of polarized light and can use the polarized sunlight to direct them from their hives to sources of pollen.

occurs in the earth's atmosphere. The scattered light will contain all frequencies, but not in the same proportion as those in the incident sunlight. Since the scattering is more pronounced at short wavelengths, the intensity of violet light in the scattered radiation is greater than that of red. As a consequence, the sky appears bluish in color. The fourth-power frequency dependence of the scattering, referred to above, is valid only when the scattering centers have a size *much smaller* than the wavelength of the light. The O_2 and N_2 molecules in the atmosphere are about 2 Å in diameter and, therefore, satisfy this condition for all visible wavelengths. The clouds in the sky, on the other hand, are seen by light scattered from water droplets. These droplets are much larger than the wavelengths of visible light and, therefore, scatter all frequencies with about the same effectiveness. As a result, clouds appear to be white or gray.

SUNSETS ARE RED

If you look in the direction of the sunset, the sky appears red. The reason for this may again be attributed to the preferential scattering of short wavelengths. The beam of light coming from the sun is scattered as it passes through the atmosphere. Since short wavelengths scatter most readily, the direct or transmitted light from the sun contains a predominance of long wavelengths and, therefore, has a reddish color. At sunset, the light coming from the sun travels through more of the earth's atmosphere than at noon. As a result, the proportion of longer wavelengths present is greater at sunset than at noon. Noon sunlight appears to be white, therefore, in contrast to the reddish tinge that is observed at sunrise and sunset.

SUMMARY

Coherent light sources are sources between which a stable, constant phase relationship exists over periods of time long enough to permit the establishment and observation of an interference pattern. Interference effects are, therefore, observable when light from two mutually coherent sources is combined but are not detected if the sources are incoherent.

In Young's experiment, the interference of coherent light passing through two slits whose widths are small in comparison to the light wavelength is displayed on a distant screen. The phase difference between the two beams is found to be

$$\delta = \frac{2d \sin \theta}{\lambda}$$

where d is the separation of the slits and θ is the angle between the axis of the slit system and the line joining the observation point with the midpoint of the slit system. For constructive interference, δ must have the values $0, \pm 2\pi, \pm 4\pi, \ldots$, while for destructive interference, δ must be $\pm \pi, \pm 3\pi, \pm 5\pi, \ldots$. Accordingly, an interference pattern of alternate dark and bright fringes is seen on the screen. For small values of the angle θ, the distance from the central intensity maximum to the nth bright fringe on either side is given by

$$z = \frac{n\lambda D}{d}$$

where D is the distance between the slit system and the screen. The distance from the central maximum to the nth minimum on either side is

$$z = \frac{(n + \frac{1}{2})\lambda D}{d}$$

Interference patterns occur in light reflected from thin films such as soap bubbles or the antireflection coatings on lens surfaces. In every case, intensity maxima are observed when reflected beams whose phase difference is $0, \pm 2\pi, \pm 4\pi$, etc., are combined, while minima occur where the phase difference is $\pm \pi, \pm 3\pi, \pm 5\pi$, etc. In all cases, it is necessary to relate the phase difference to the optical *path* difference between interfering beams, remembering to take proper account of phase changes of π radians that occur when reflection of a ray incident from a low index medium into one of higher refractive index occurs.

Diffraction of light occurs when light waves pass through apertures or near the edges of opaque objects. The effect of diffraction is to alter the paths of light rays from the straight lines predicted by geometrical optics. We may say that diffraction of light refers to the bending of light rays that occurs when light passes through apertures or around obstacles due to the interference of waves passing through different parts of the system. The diffraction pattern formed by light passing through a single slit, therefore, occurs as a result of the interference of light passing through different parts of the slit. For plane waves incident on a single slit of width d, a diffraction pattern can be observed on a distant screen, which consists of a single, bright central maximum and alternating dark and bright regions of lesser intensity. The variation of intensity with the angle θ can be described by

$$\bar{S} = S_0 \frac{\sin^2(\phi/2)}{(\phi/2)^2}$$

where

$$\phi = \frac{2\pi d \sin \theta}{\lambda}$$

It is evident that for the maxima,

$$\sin \theta = \frac{n\lambda}{d} \qquad n = 0, 1, 2, \ldots$$

and for minima,

$$\sin \theta = (n + \tfrac{1}{2}) \frac{\lambda}{d} \qquad n = 1, 2, 3, \ldots$$

It is important to note that the distance between successive maxima, and the overall extent of the pattern, *increase* as the slit width *decreases*. The case of Fraunhofer diffraction, involving incident plane wavefronts, is simpler than the more complex case of Fresnel diffraction, in which the incident wavefronts are spherical.

Diffraction patterns are produced by multiple-slit arrays. The case of Young's experiment is a special instance of two-slit diffraction in which the slit widths are much less than the light wavelength. Under these circumstances, the diffraction pattern is infinitely broad, and only interference fringes between the two coherent slit sources are seen. When the slits are no longer so narrow, a diffraction pattern can be seen, upon which the double-slit interference pattern is superimposed. For more than two slits, more complex multiple-slit patterns are seen. The case of a very large number of equally spaced slits of equal width is referred to as a *diffraction grating*. Monochromatic plane wavefronts incident on such a structure are diffracted so as to produce essentially zero intensity behind the grating, except in certain sharply defined diffraction directions determined by the condition

$$n\lambda = d \sin \theta$$

where d is the distance between adjacent slits. These diffraction directions correspond to wavefronts constructed tangent to Huygens wavelets emitted by successive slits, differing in phase by $2\pi, 4\pi, 6\pi, \ldots$, corresponding to diffraction orders $n = 1, 2, 3, \ldots$. Since diffraction directions differ according to light wavelength, diffraction gratings are widely used to separate light into its constituent spectral components.

At a circular aperture, light is diffracted into a pattern of concentric light and dark rings. The pattern has a central maximum at $\theta = 0$; the first minimum occurs at an angle θ_1, given by

$$\sin \theta_1 = 1.22(\lambda/a)$$

where a is the aperture diameter. The diffraction of light by the circular pinhole is responsible for the fact that pinhole images are of limited sharpness, compared to lens images, where the larger-aperture diameter results in a much smaller diffraction pattern.

Unpolarized light, such as sunlight, can be polarized by passage through *dichroic substances* such as polarizing films or, partially at least, when reflected at the surface of a dielectric like glass or water. When polarized light passes through a polarizing film, only light whose electric vector is parallel to the film's transmission axis is transmitted. For light whose electric vector makes an angle θ with the polarizer's transmission axis, the transmitted intensity will be

$$\bar{S} = S_0 \cos^2 \theta$$

Light can be polarized in reflection from a dielectric such as glass or water. In general, the reflected light is only partially polarized, though if the light is incident in such a way that the sum of the angles of incidence and refraction is 90°, it is completely polarized, its electric vector after reflection being parallel to the reflecting surface and perpendicular to the plane containing the incident and reflected rays. When the angle of incidence and refraction add up to 90°, it can be shown that the angle of incidence can be expressed in terms of the relative refractive index n_r as

$$\tan \theta = n_r$$

When this condition is satisfied, it is said that the unpolarized light is incident at *Brewster's angle*, the reflected ray being then completely polarized.

QUESTIONS

1. When two light waves "interfere," does one of the waves retard the progress of the other?
2. At points where two waves of the same amplitude and frequency combine in phase, the intensity is four times as large as that produced by either of the waves alone. Does this violate conservation of energy?
3. If a number of coherent sources are to be superposed, must we add their intensities or their amplitudes? How must we proceed if the sources are incoherent?
4. The electric vector of a sinusoidal monochromatic light wave oscillates rapidly as the wave propagates. Why doesn't the eye detect this by observing periodic changes in intensity? Can you suggest any instrument which can detect such an oscillation?
5. Observable interference of light waves occurs under certain conditions. Specify what these conditions are. What conditions are satisfied when constructive and destructive interference take place?
6. A double-slit interference pattern is displayed on a screen. If the source and screen are immersed in a transparent liquid, describe the changes in the interference pattern that are observed.
7. Describe the difference between interference and diffraction of light waves. Do water waves exhibit these same phenomena? In what respects do water waves and light waves differ?
8. Just before a soap bubble breaks, its surface appears to be dark. Explain this observation.
9. What are some of the advantages obtained by using anti-reflective coatings on lenses and prisms?
10. If the speed of light were c in every medium, many of the optical instruments now in common use would not work. Why is this so? Give a number of examples.
11. Even though a lens is coated with a film exactly a quarter-wave thick (for $\lambda = 5500$ Å), and though the refractive

index of the coating is *exactly* equal to the square root of the refractive index of the glass, some faint, purplish reflections from the lens surface are visible. How many different reasons can you cite to account for the presence of these reflections?

12. How can you tell if the lenses of a tinted pair of sunglasses are made of a polarizing material?

13. From the surface of the earth, the sky appears blue. What color does it have when viewed from the surface of the moon?

14. A line source of light is viewed through a narrow slit. As the slit is made narrower, the source appears wider. Can you explain this?

15. A sheet of polarizing material can be used either as a polarizer or as an analyzer. Describe the difference between these two applications.

16. What is the difference between Fraunhofer and Fresnel diffraction?

17. To observe interference of light reflected from a film, the film must be quite thin. Explain why this statement is true.

18. You are trying to take a photograph of your nephew who is sitting inside an automobile. Unfortunately, a strong reflection of a tree in the windshield obscures the view. How can a polarizing filter be used to get rid of the reflection?

19. Describe briefly the meaning of the term birefringence.

20. Discuss several different ways of producing polarized light.

21. How can you determine experimentally the resolving power of a diffraction grating?

22. List several ways in which the diffraction of light is utilized for practical purposes. Cite also a number of situations in which the effects of diffraction are disadvantageous.

PROBLEMS

1. A student is studying the coherence of light. For each of the following pairs of light sources, state whether he should conclude that the sources are coherent and, therefore, exhibit visible effects of constructive and destructive interference: **(a)** two auto headlights far enough away to appear as two point sources; **(b)** two monochromatic point sources with a constant phase difference between them; **(c)** two light beams emitted from different locations on the sun; **(d)** a pinhole source and an image of the pinhole obtained by reflection from a mirror; **(e)** two narrow slits illuminated by monochromatic light from a distant point source.

2. For a monochromatic, linearly polarized electromagnetic wave, the magnitude of the average electric field is 50 volts/meter at a given location. Find the average value of the Poynting vector which gives the absolute intensity at that point.

3. In a lecture demonstration, two narrow slits are separated by a distance of 0.2 cm and an interference pattern is produced at a screen 2 meters away. If the wavelength of the light used is 6000 Å, find the distance (on the screen) between the third bright fringe and the zero-order (central) maximum.

4. Two very narrow slits are separated by a distance of 0.80 mm. They are illuminated by plane monochromatic light waves of wavelength 5500 Å. An interference pattern is observed on a screen 2.4 meters distant. Find **(a)** the phase difference between light from the two slits at a point 2.00 mm from the center of the interference pattern; **(b)** the path difference between the two beams at this point; **(c)** the ratio of the intensity of light at this point to the maximum intensity observed at the center of the pattern; **(d)** the distance between the center of the pattern and the third interference maximum (path difference 3λ).

5. Under the conditions described in the preceding problem, what would be the distance on the screen between the two successive interference maxima nearest the center of the pattern if the two slits were separated by a distance of **(a)** 2.5 cm, **(b)** 0.001 mm?

6. Two narrow slits are separated by a distance d. Plane waves with wavelength λ are incident on the slits and produce an interference pattern on a screen a distance D from the slits, where $D \gg d$. If h gives the separation between the central maximum and the first minimum, show that a change of the spacing of the slits by Δd produces a change in this separation given by

$$\Delta h = -\frac{\lambda D}{2d^2}\Delta d$$

7. Monochromatic light emerges in phase from two slits spaced 3.0 mm apart. The light falls on a screen placed 250 cm from the slits. A measurement of the resulting interference pattern gives a distance of 0.23 mm between the central maximum and the first minimum. **(a)** Find the wavelength of the source. **(b)** As measured on the screen, what is the distance between the central (zeroth) maximum and the second maximum? **(c)** Repeat part **(b)** if the entire apparatus is immersed in water ($n = \frac{4}{3}$).

8. An antireflection coating of magnesium fluoride ($n = 1.38$) is deposited on glass. What is the minimum thickness of coating which will eliminate reflected light of wavelength 5500 Å in vacuum?

9. A lens is to be made from pure silicon ($n = 3.60$), which is transparent to infrared radiation of wavelength longer than about 12,000 Å. The lens is to be used to focus radiation of wavelength 30,000 Å onto a detector. It is desired to apply an antireflection coating to the lens to minimize reflection losses. Find **(a)** the thickness and **(b)** the refractive index the coating should have to produce minimum reflectivity at this wavelength. **(c)** What is the reflectivity of an uncoated silicon surface for light at normal incidence?

10. What would the answers to the preceding problem be if the lens were to be immersed in a liquid medium of refractive index 1.500?

11. A soap film is confined by a vertical wire frame. It drains downward and therefore becomes somewhat thinner at the top than at the bottom. The refractive index of the film is 1.333. The film is examined under monochromatic light of wavelength 6000 Å, and it is noted that interference fringes with a uniform spacing of 0.50 cm are observed. What is the angle between the two surfaces of the film?

12. Two perfectly flat pieces of glass are separated by a thin wedge of methyl alcohol ($n = 1.329$). If the interference fringes that are observed are separated by 0.3 mm when light of wavelength 6000 Å falls normally on the film, find the angle of the wedge in radians.

13. Two sheets of glass, each 15 cm long, are in contact at one end. At the other end, they are separated by a piece of metal 0.2 mm thick. The glass is illuminated at normal incidence with light of wavelength 5500 Å. How many dark interference fringes per cm are seen in the reflected light?

14. Light of wavelength 4500 Å in air enters a medium with index of refraction $n = 1.33$. Two rays, initially in phase, travel by different paths through the medium, the path difference being 2 mm. What phase difference exists between the two rays, assuming no phase difference arises due to reflection?

15. A wedge-shaped film of air is enclosed between two plane sheets of glass ($n = 1.5$) which are in contact at one end. The sheets are inclined at an angle of 10^{-4} radians. **(a)** When plane waves with $\lambda = 5000$ Å are incident normally, how far from the line of contact is the first bright interference fringe? (Assume that the top surface of the upper sheet of glass is non-reflecting.) **(b)** If the wedge-shaped air film is replaced by a film consisting of a liquid with a refractive index of 1.7, what is the spacing between bright fringes?

16. An antireflection coating is placed on the surface of a lens as shown in the figure. **(a)** Indicate in a diagram the paths followed by the incident rays A and B and discuss any phase changes which occur due to reflection. **(b)** What is the minimum coating thickness which would give zero reflectivity for ray B? **(c)** Will the same optical thickness used to provide an ideal antireflection coating for ray B also give zero reflectivity for ray A? Assume $\lambda = 5500$ Å.

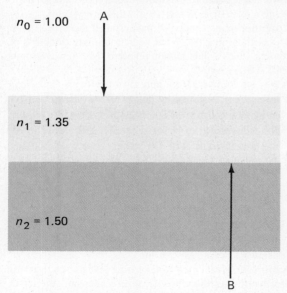

$n_0 = 1.00$

A

$n_1 = 1.35$

$n_2 = 1.50$

B

17. A thin film of oil ($n = 1.25$) spreads out over a portion of the surface of a lake ($n = \frac{4}{3}$). At a given location, the thickness of the film is 2.6×10^{-3} mm. White light which is incident normally upon the surface is viewed in reflection. **(a)** State any phase changes that occur on reflection.

(b) What colors are preferentially reflected? **(c)** Describe qualitatively the colors which would be observed as the oil slick spreads out and its thickness diminishes.

18. A soap film with an index of refraction n is viewed with reflected light of wavelength 5500 Å. When the film is suspended vertically on a wire frame, it drains toward the bottom and the thickness of the film is assumed to vary linearly with x, as shown in the diagram, according to the relation $t = Ax$, where $A = 3.0 \times 10^{-4}$. At a certain time, within a distance x of 2.46 cm, 36 dark fringes are observed. **(a)** Trace the light rays reflected from the film surfaces and indicate any phase changes which occur. **(b)** Find relations between n, λ, x, and A which imply constructive and destructive interference on reflection. **(c)** What is the index of refraction n? **(d)** If the wavelength were increased to 6000 Å, would the fringes be closer together or further apart?

19. The spherical surface of a plano-convex lens rests on an optical flat. The system is illuminated with monochromatic light of wavelength 5890 Å, and Newton's rings are observed around the point of contact. The radius of the fourteenth dark fringe from the center is measured and found to be 0.192 cm. Assuming that the refractive index of the lens is 1.696, what is its focal length?

20. The diagram on the next page shows a device known as the *Michelson interferometer*. A parallel beam of monochromatic light from source S strikes a diagonal semitransparent mirror, where it is split into two beams. One of these beams is reflected from a fixed mirror M_1, the other (which passes through the diagonal mirror) is reflected from a movable mirror M_2. The two beams are subsequently recombined and are observed at O. The first beam takes the path SPM_1PO, the second the path SPM_2PO. The observer at O sees interference fringes, which are bright or dark according to whether the path difference between the two beams is an even or odd number of half-wavelengths. When mirror M_2 is moved through a distance of one-half wavelength, the path difference changes by one wavelength and the observer senses the passage of one interference fringe across his field of vision. By counting interference fringes (which can be accomplished electronically as well as visually), it is possible to accurately express the distance mirror M_2 moves in terms of the light wavelength, or vice versa. The Michelson interferometer provides a method of making accurate distance or wavelength measurements, or for standardizing distances in terms of light wavelengths. **(a)** If the observer counts 3580 fringes as mirror M_2 moves through a distance of 1.000 mm, what is the light wavelength? **(b)** If the observer counts 1875 fringes as the mirror M_2 moves through some unknown distance, using light of wavelength 5890 Å, what is the unknown distance?

21. When using a sodium vapor lamp as a light source with the Michelson interferometer described in the preceding problem, the source is not strictly monochromatic but consists of two different spectral lines of nearly equal intensity having slightly different wavelengths. The average wavelength of these two separate components is 5893 Å. When such a source is used with the Michelson interferometer, it is noted that when the mirror M_2 is moved over fairly large distances, the contrast between dark and

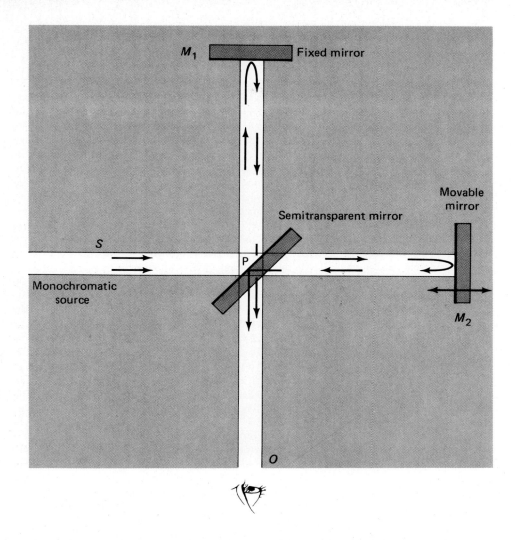

Movable
mirror

Semitransparent mirror

S

P

Monochromatic
source

M_2

O

light fringes alternately increases and decreases, the fringes becoming successively clear and sharply defined and soft and indistinct. It is observed that mirror M_2 must be moved about 0.290 mm to go from one of these positions of maximum fringe contrast to the next. From these data, calculate the separation in wavelength between the two spectrum lines of sodium that comprise the source.

22. Single-slit Fraunhofer diffraction is observed with light having a wavelength of 6400 Å. If the angular separation between adjacent dark fringes is 0.30°, how wide is the slit?

23. A narrow slit of width 0.030 mm is illuminated with monochromatic light of wavelength 5890 Å. A diffraction pattern is observed on a screen 1.80 meter away. Find (a) the phase difference ϕ across the slit for light reaching the screen at a point 5.0 cm from the central maximum of the pattern; (b) the ratio of the electric vector amplitude at this point to the electric vector amplitude at the central maximum; (c) the ratio of light intensity at that point to the intensity at the central maximum; (d) the distance between the center of the pattern and the third diffraction minimum.

24. Under the circumstances outlined in the preceding problem, what would be the distance on the screen between the central intensity maximum and the first diffrac-

tion minimum if the width of the slit were changed to (a) 0.30 mm, (b) 0.003 mm, (c) 0.0003 mm?

25. Show, using Eq. (26.6.9), that maxima in the single slit diffraction pattern occur whenever $\tan \phi/2 = \phi/2$. Find (either numerically or graphically) the smallest non vanishing angle ϕ which satisfies this condition.

26. In a double-slit interference experiment, two narrow slits ($a = \lambda$) separated by 0.40 mm are illuminated by a coherent beam of light for which $\lambda = 5000$ Å. The interference pattern is observed on a screen 1.60 meter from the slits. (a) Does diffraction occur in this experiment? (b) Does interference occur at locations other than at the screen? (c) Calculate the separation of adjacent maxima on the screen.

27. A diffraction grating has 4800 narrow slits per cm. It is illuminated by plane monochromatic light waves of wavelength 5500 Å. Find the angles between (a) the first-order, (b) the second-order, and (c) the third-order diffracted beams and the normal to the grating.

28. Under the conditions of the preceding problem, what would be the angular separation of two diffracted beams corresponding to wavelengths 5500.0 Å and 5500.5 Å for (a) first-order, (b) second-order, (c) third-order diffraction? Assuming that the ruled section of the grating is

2.0 cm long, would these two wavelengths be resolved **(d)** in first-order, **(e)** in second-order, **(f)** in third-order diffraction? **(g)** What is the resolving power of this grating for first, second-, and third-order diffraction?

29. The angle between the two first-order diffracted beams (corresponding to positive and negative values of θ) for light of a certain wavelength is observed to be 25.750°, using the diffraction grating discussed in the two preceding problems. What is the wavelength?

30. What is the angular separation between the two first-order diffracted beams (corresponding to positive and negative values of θ) for each of the two yellow sodium lines ($\lambda = 5895.93$ Å, $\lambda = 5889.96$ Å) using the grating discussed in the three preceding problems?

31. The atoms in a crystal lattice can be thought of as a stack of parallel planes each containing a regularly arranged set of atoms. These stacks of crystal planes can serve to diffract electromagnetic radiation of suitable wavelength, the diffraction maxima occurring when the path difference between beams reflected from neighboring crystal planes in the stack is an integral multiple of the wavelength. **(a)** Show that the diffraction maxima will be observed at angles θ between the incoming radiation and the normal to the planes satisfying the condition $n\lambda = 2d \sin \theta$, where d is the spacing between planes and n the order of diffraction. **(b)** If the spacing between planes is of the order of a few Ångström units, what sort of radiation would be "suitable" to observing such diffraction effects?

32. In the situation described in the preceding problem, if the incoming radiation is of wavelength 2.80 Å and the first-order diffracted beam is observed at an angle of 36.0° to the normal to the reflecting planes, what is the spacing between the crystal planes?

33. A photograph of the diffraction pattern due to N slits has A secondary maxima between adjacent principal maxima. Show that N must be $A + 2$.

34. Show (using calculus) that if the intensity I in the diffraction pattern of N slits is considered as a function of γ, see Eq. (26.7.13), then the maximum values of I occur when $\gamma = m\pi$, where m is an integer.

35. A white ruler has markings separated by 1 mm. What is the largest distance at which the human eye can possibly resolve two adjacent markings?

36. The sharp-sightedness of birds of prey is legendary. Use diffraction considerations to estimate whether an eagle flying 1 kilometer above the earth can distinctly see a mouse 3 cm long. Assume the light wavelength is 5500 Å.

37. An FBI agent investigating a crime is looking into a nudist camp through a knothole in a fence. The diameter of his eye pupil is 5 mm and the hole diameter is 1 mm. **(a)** At what distance will the agent be unable to distinguish the boys from the girls (assume a minimum distance of 20 cm must be resolved and that $\lambda = 5000$ Å)? **(b)** At a distance of 150 meters from the fence, a flashbulb 1 cm in diameter is fired. Can the agent detect the firing of the bulb?

38. An advertisement for a telescope with an objective 11 cm in diameter says "READ NEWSPAPER HEADLINES AT A MILE" (1 mile = 1.6 km). Assuming that to read the print one must be able to resolve detail separated by 1.0 cm, **(a)** calculate the angular resolution needed to "read the headlines." **(b)** Calculate the angular resolution of the telescope based on Rayleigh's criterion. **(c)** Is the advertisement honest?

39. The surface of the moon is studied from the earth using a telescope with an objective 30 cm in diameter. **(a)** Using the Rayleigh criterion, what is the smallest distance between objects on the moon which can be resolved? (Use $\lambda = 5000$ Å and the earth-to-moon distance as 380,000 km.) **(b)** Could a very intense point source of light located on the moon be detected using this telescope? Explain.

40. A beam of unpolarized light passes through an ideal polarizer at normal incidence. By what factor is its intensity reduced?

41. A beam of unpolarized light has intensity I_0. It passes through two polarizers whose transmission axes make an angle of 30° with one another. Find the intensity of the transmitted light.

42. A beam of linearly polarized light is normally incident upon a polarizing filter whose transmission axis makes an angle of 45° to the electric vector. **(a)** If S_0 represents the incident intensity, what is the intensity of the transmitted beam? **(b)** the transmitted beam is now allowed to pass through a second polarizing filter, whose transmission axis makes an angle of 90° with the electric vector of the original beam incident on the first filter, again at normal incidence. What is the intensity of the radiation transmitted by this filter, expressed as a fraction of S_0?

43. Unpolarized light is incident on a plane sheet of glass at an angle θ to the normal. The refractive index of the glass is 1.660. **(a)** What value must the angle θ have for the reflected ray to be completely polarized? **(b)** What is the direction of the electric vector of the reflected ray? **(c)** What is the angle of refraction? **(d)** What is the angle between the reflected ray and the refracted ray. **(e)** Is the refracted ray unpolarized, partially polarized, or completely polarized?

44. What would the answers to the preceding problem be if the glass sheet were immersed in water ($n = 1.333$)?

45. At what angle of incidence will unpolarized light incident on the following substances become completely polarized after reflection? **(a)** Water ($n = 1.333$); **(b)** crown glass ($n = 1.500$); **(c)** dense flint glass ($n = 1.890$); **(d)** diamond ($n = 2.518$); **(e)** silicon ($n = 3.600$).

46. A convex lens with radius of curvature 9.5 meters is placed on a flat glass surface and is illuminated from above with light of wavelength λ. A Newton's rings interference pattern is established, and the 20th dark ring is observed to have a radius of 1.15 cm. Find the wavelength λ.

47. A Newton's rings apparatus is used to measure the index of refraction of a certain liquid. The convex lens resting on the flat glass has a radius of curvature of 10 meters. When light of wavelength 6500 Å is incident normally and the apparatus is in air, find the radius of the 15th fringe. When the apparatus is submerged in the liquid, the radius of this fringe decreases by 15 percent. Find the index of refraction of the liquid.

27.1 Introduction

One would be hard pressed to find a scientist who inspires more awe and admiration than Albert Einstein (Fig. 27.1). His name is legend in the annals of science his accomplishments truly revolutionary and profound. His insights have had a tremendous impact on the development of twentieth century physics. It is, therefore, important to study some of his ideas, to reveal their elegant simplicity and beauty, and to dispel any notion that his theory of relativity is esoteric and difficult.

Einstein was born in 1870 in Germany in the town of Ulm.[1] He spent his school days in Munich,

attending the Gymnasium until the age of fifteen. His years in Munich were not very happy ones, for he disliked the regimentation and inflexible methods of his teachers. They, in turn, were not especially fond of young Albert, and some of them predicted that he would never amount to anything. Apparently through self-study, though, the young Einstein was well ahead of his classmates in the areas of physics and mathematics, for he was extremely well motivated and curious. His unhappiness at the Gymnasium was, however, profound, and prompted him to leave Munich without a diploma. He joined his family in Italy and spent a happy, carefree year out of school, but soon his father urged him to seek a career.

In 1895, he attempted to gain admission to the Federal Institute of Technology in Zürich, Switzerland, but failed the examination because of weaknesses

[1] An excellent biography of Einstein is: Banish Hoffman and Helen Dukas, *Albert Einstein, Creator and Rebel*, (New York: The Viking Press, 1972).

FIGURE 27.1. Albert Einstein (1870–1955).

in areas such as botany and languages. He entered a Gymnasium in Aarau and soon thereafter obtained a diploma and gained entrance to the Federal Institute in 1896. As a student at the Institute, Einstein developed a strong interest in physics, but he lacked the motivation to study those subjects which did not interest him. His attendance at lectures, even some physics lectures, was very poor, for he regarded them as an intrusion upon his "self-study." Luckily, a good friend made his notes available to Einstein, who used them to cram for his exams. He passed his major examinations and graduated in 1900, but hard times lay ahead for Einstein. He looked for university positions but could not find any suitable job. In 1902, several years after graduation, he finally landed a job at the Swiss Patent Office and soon learned to carry out his duties with considerable ease and efficiency. This allowed some time to work on his physics, which increasingly captured his full devotion.

During this period at the Patent Office, he developed the theories which were destined to bring him fame. Among these were his works on the special theory of relativity, the quantum behavior of emission and absorption of light, and the theory of Brownian motion, all of which were submitted to the research journal *Annalen der Physik* in 1905. A fourth paper, of lesser significance, was accepted at the University of Zürich as a Ph.D. thesis. The three more important papers were all written while he was working full time at the Patent Office, earning his living by performing more mundane duties. The burst of creativity and

genius represented by these papers, all on different topics, is almost unparalleled in the history of science. In the same year, several months after his paper on relativity, Einstein published another paper, in which he states that when a body gains or loses a given amount of energy, there is an equivalent change in its mass. Several years later, he came to the realization that the converse must also hold, that a change in mass is inevitably accompanied by a change in energy. In this way, his famous equation $E = mc^2$ was discovered in 1907. It carried with it the implication that great amounts of energy could be stored in matter. It took twenty-five years to verify this prediction experimentally.

In 1909, Einstein obtained a professional post at the University of Zürich and finally resigned from his position at the Patent Office. Shortly thereafter, in 1911, he moved to a full professorship at the German University in Prague where he remained for a year and a half, making progress in developing the general theory of relativity. He received offers at many other universities, but returned in 1912 to the Swiss Federal Institute in Zürich as a full professor. In 1914, he received an offer from Berlin and moved from Switzerland to Germany, where he finally developed his celebrated general theory of relativity, published in 1915. In 1919, Einstein's first marriage ended in divorce. His wife received custody of their two children and was also promised the Nobel Prize money which Einstein felt certain he would get. The Nobel prize was, in fact, awarded him in 1921, but since at that time his work on relativity was still controversial, the award made no mention of his contributions in that subject!

In 1933, when Hitler came to power in Germany, Einstein left for the United States and took a position at the Institute for Advanced Study in Princeton. He held this position until his death in 1955. During the latter part of his life, Einstein was much sought after, not only by scientists but also by politicians and others of fame. He was revered and worshipped, for good reason, and history will surely confirm his greatness.

27.2 Basic Postulates of the Special Theory of Relativity

The special theory of relativity is much more than a physical theory that formulates some limited set of laws or observations. In a sense, it is a framework and a supertheory to which all other laws and theories must adhere. Once we impose this conformity, we are forced to scrutinize the foundations of any new proposal in order to seek out possible imperfections or contradictions.

The *special* theory of relativity concerns itself with the description of physical laws in nonaccelerating (inertial) frames of reference and with the transformations of coordinates that relate one such reference frame to another. The *general* theory, which we will not discuss in this book, was developed about ten years later. It considers the description of the laws of nature in arbitrary frames of reference.

There are two basic ingredients or postulates to the special theory of relativity which we shall discuss in detail in this section. These are as follows:

Postulate 1—Principle of Relativity All the laws of physics must have the same form in all inertial reference frames. Sometimes, one expresses this by saying the laws of physics are invariant, or covariant.

Postulate 2—The Constancy of the Speed of Light
The speed of light in empty space always has the same value c, *independent of the motion of the source or of the frame of reference of the observer.*

The first of these postulates seems almost evident, but in a sense it is rather profound, for it states that no particular inertial observer has any special status in the universe. All observers are equal; if Newtonian mechanics holds for Mr. A, it must also hold for Mr. B, who may be moving away from him at a constant velocity. The principle of relativity is already embodied in Newtonian mechanics, and, in fact, Newtonian mechanics is in full harmony with this principle. To further appreciate this point, let us briefly review some of the basic concepts.

To begin with, prior to Einstein, the transformation law from one inertial coordinate system to another was given by the so-called Galilean transformation, mentioned previously in Chapter 3, section 3.8. To review this transformation, let us first introduce the concept of a space–time event. This is simply an event or an occurrence characterized by a single time and a single point in three-dimensional space. Thus, if a small flashlight is turned on at some specific time and at some specific point in space, this constitutes a space–time event.

Now let us assume that at $t = 0$, two inertial observers synchronize their watches and that, thereafter, the observer K′, who may be riding on a train, moves with speed v relative to observer K, as shown in Fig. 27.2. We shall assume for simplicity that the motion of observer K′ is in the direction of the positive x-axis. Now, both K and K′ are asked to describe *where* and *when* a given space time event occurred. Their descriptions must differ in some way, for to K the event occurred at the spatial coordinates (x, y, z) at time t, while to K′ it occurred at (x', y', z') at time t'. Of course, the actual point of occurrence is the same in either case, but the two descriptions are not quite

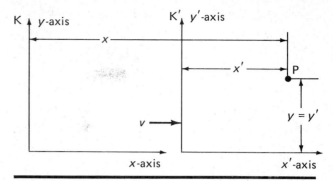

FIGURE 27.2. An event, occurring at point P, as seen by observers in two reference systems, K and K′ moving with relative velocity **v** with respect to one another.

the same. The fact that there are two, or several, descriptions of any event leads to no difficulties, for all observers would agree on the relationship between their respective coordinate systems and the equations by which the description of events are transformed to the other. The observers K and K′ would *both* agree, therefore, that

$$t' = t$$
$$x' = x - vt$$
$$y' = y$$
$$z' = z$$

(27.2.1)

If observer K′ then used these relationships to translate observer K's description of a space–time event in terms of the coordinates (x, y, z, t) into his own reference coordinates (x', y', z', t'), he would expect to arrive at answers that agree with his own observations of the event in his own reference frame. These relations, which appear to arise from common sense, are known as the *Galilean transformation* equations. The statement that $t = t'$ stems from the intuitive belief that once the clocks are synchronized, they will remain in that condition, irrespective of any relative motion. The other relations follow from the elementary concepts of length measurements. Thus, intuition plays a vital role in obtaining the coordinate transformation embodied in (27.2.1). Unfortunately, human intuition, although extremely reliable and valuable in many circumstances, is sometimes fallible. This is one such instance, for if the speed of K′ with respect to K becomes too great (comparable to the speed of light or a sizeable fraction thereof), the Galilean transformation *is no longer correct.* If this violates your intuition, remember that you really have no experience to draw upon, for your experience is based upon relative motion in which the condition $v/c \ll 1$ is always fulfilled.

How do we know that the Galilean transformation is wrong, and what led Einstein to expect that it had to be altered? Basically, the difficulty arose out of the recognition that the laws of electricity and magnetism, as enunciated by Maxwell, do *not* satisfy the Galilean principle of relativity. This implies one of the following alternatives:

a. Maxwell's equations are wrong.
b. The principle of relativity is inapplicable.
c. The Galilean transformation is incorrect.

The first alternative is difficult to accept, for Maxwell's exposition of the laws of electromagnetism is in full accord with all known experiments. The second possibility, which denies the principle of relativity, asserts that Maxwell's equations may be correct in some special frame of reference but are not right in other reference frames. This, in turn, led to the supposition that electromagnetic waves must require a rare and subtle medium, referred to as the *luminiferous ether*, for their propagation. In a reference system at rest with respect to this medium, Maxwell's equations would be correct, while in other systems their form would have to be altered in accord with appropriate coordinate transformation equations. But, though many experiments were devised to reveal the presence of the ether and to explore its physical properties, all of them ended in utter failure. The most famous of these were undertaken by Michelson and Morley,[2] who used exquisitely sensitive optical interference techniques to detect possible variations in the speed of light in directions parallel and perpendicular to the earth's orbital path. It turned out, finally, that the third of the possible alternatives set down above was the right one. It is the Galilean transformation that in the final analysis has to be wrong. This conclusion, though it is difficult to accept on the basis of simple intuition, and though it leads to some revolutionary changes in our basic ideas about physics, is the only one that is supported by all the available experimental evidence. The most revolutionary revision forced upon us in this way is the admission that, since Newtonian mechanics satisfies in all respects the requirements of invariance under Galilean transformations, it also is basically incorrect and has to be revised from the ground up!

Let us now try to illustrate with a few examples that Newtonian mechanics does, in fact, satisfy the requirements of Galilean relativity. Suppose a point object is moving with constant velocity V in the frame of reference K, as shown in Fig. 27.3. Then its acceleration A is zero, and it is not acted upon by any net

[2] A. A. Michelson and E. W. Morley, *Amer. J. Sci.* **134**, 333 (1887). See also the account given by R. S. Shankland, The Michelson-Morley Experiment, *Sci. Amer.* (Nov. 1964). Michelson and Morley's experiments, which, in a sense, met with failure, are far more significant than many other experimental undertakings that were "successful" in the usual sense of the word.

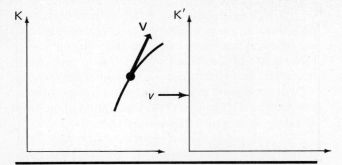

FIGURE 27.3. A point object in motion, with velocity **V** with respect to frame of reference K, has, according to Newtonian physics velocity **V′ = V − v**, with respect to K′.

force. Now, from the Galilean transformation (27.2.1), recalling that the relative speed v of frame K′ with respect to K is also constant, we have

$$V_x' = \frac{dx'}{dt'} = \frac{dx'}{dt} = \frac{dx}{dt} - v = V_x - v$$

$$V_y' = \frac{dy'}{dt'} = \frac{dy'}{dt} = \frac{dy}{dt} = V_y \qquad (27.2.2)$$

$$V_z' = \frac{dz'}{dt'} = \frac{dz'}{dt} = \frac{dz}{dt} = V_z$$

Taking further derivatives, we find that the velocity **V′** is constant or that the acceleration **A′** is zero. Thus, we see that if Newton's first law is valid in an inertial frame K, we can easily prove it to be valid in any other inertial reference frame K′.

As a second example consider a collision of two balls, as viewed from frame K and also from K′. The situation is illustrated in Fig. 27.4. For simplicity, assume the balls move along the *x*-axis. According to conservation of momentum, as applied in K,

$$m_1 V_{1i} + m_2 V_{2i} = m_1 V_{1f} + m_2 V_{2f} \qquad (27.2.3)$$

where indices 1 and 2 refer to the balls and i and f refer to initial and final velocities. According to (27.2.2), $V_{1i} = V_{1i}' + v$, $V_{2i} = V_{2i}' + v$, $V_{1f} = V_{1f}' + v$, $V_{2f} = V_{2f}' + v$; therefore, substitution immediately leads to

$$m_1 V_{1i}' + m_2 V_{2i}' = m_1 V_{1f}' + m_2 V_{2f}' \qquad (27.2.4)$$

Thus, if the momentum is conserved in one inertial reference frame, then under a Galilean transformation, we find it is conserved in all inertial frames.

These few examples illustrate the basic point that Newtonian mechanics is invariant under Galilean transformations. If, however, we would attempt to demonstrate the same kind of invariance for the Maxwell equations, we would fail. Maxwell's equations are not Galilean invariant, and nothing can be done about it. Ultimately, however, the transformation law must be modified so that all laws can be stated in a form

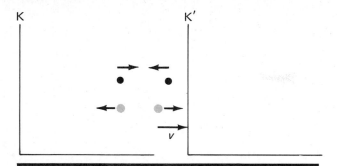

FIGURE 27.4. Galilean transformations preserve momentum conservation in all reference systems. Dynamics of a collision are, therefore, similarly described in either K or K'.

which is invariant. The modified transformation law that results, which is known as the *Lorentz transformation*, will be taken up in detail in a subsequent section. We shall then see that although Maxwell's equations are invariant with respect to this new transformation law, Newton's laws are not and, therefore, have to be modified to suit its requirements.

Let us return now to the second postulate of relativity, which asserts that *the speed of light in empty space is the same for all observers*. There are no arguments that can be advanced to prove this assertion, or even to make it sound plausible. On the contrary, intuition would lead us to believe this postulate to be ridiculous.

Consider now a thought experiment, as illus-

trated in Fig. 27.5. A rocket ship passes by an experimenter who has a high-speed rifle and also a flashlight. The rocket ship is moving with velocity V relative to the experimenter, and at the instant they pass each other, the experimenter fires his rifle and also turns on his flashlight. The rifle imparts a muzzle velocity v to the bullet in its own reference system. The crew of the rocket ship, being rather ambitious and science oriented, decides to measure the speed of the rifle bullet and the speed of the light coming from the flashlight. However, one of the crew members, who remembers his elementary physics, tries to convince the others that their measurements are a waste of time. He says the answers are quite obviously $v + V$ for the rifle bullet and $c + V$ for the speed of light. The other members don't believe him, so they make some measurements. Much to their astonishment, they find that the speed of the bullet can indeed be obtained by adding the vectors **v** and **V** but *that the observed speed of light is the same as if they were not moving at all*. In disbelief, they repeat the experiment again and again, each time using a bigger value of V. But no matter what they do, they always find a speed of 2.9979250×10^8 meters/sec, which is identical to that observed by the experimenter who sent out the flashlight signal. Moreover, exactly the same value is obtained even when the light source itself is in motion relative to the observer's reference system.

The original experiments on the speed of light were not as outlandish as the above thought experi-

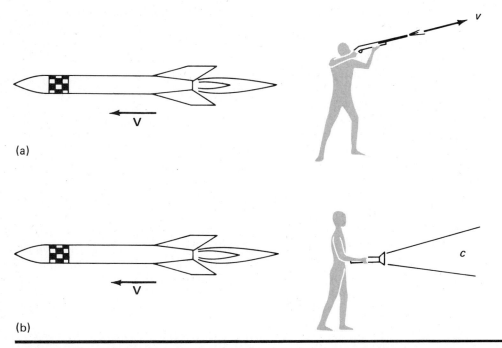

(a)

(b)

FIGURE 27.5. Rifle bullet with velocity **v** (much less than the velocity of light) and a flashlight turned on as the rifle is fired, as observed from a rocket ship moving with velocity **V** with respect to rifle and flashlight.

ment suggests, but nevertheless the final conclusions have always been that the speed of light is the same to all inertial observers, irrespective of the motion of either observer or light source. This is an empirical result, but since physics is an experimental science, such observations must be believed and incorporated into the fabric of any theory.[3] In fact, the speed of light c turns out to be the *ultimate speed* of all objects. An observer moving away from the flashlight with a speed $c/2$ *will not* find that light travels at a speed of $3c/2$, for this would violate the assertion that no one can ever find a speed which is greater than c. This ultimate speed is by now confirmed in many, many, ways, especially in experiments done on elementary particles with modern particle accelerators. We cannot answer the question of *why* there should be such a speed. As physicists, we must accept the existence of this ultimate speed as an inexplicable element of nature whose beauty can be appreciated but whose origin may forever remain obscure.

27.3 The Lorentz Coordinate Transformation

In this section, a brief derivation of the Lorentz transformation law will be presented. It is based on the assumption that light signals propagate with speed c relative to all inertial observers. Consider two inertial observers K and K′ whose coordinate systems coincide at a common time $t = 0$. At that instant, a light signal is emitted from their common origin and travels outward. This is shown in Fig. 27.6.

Let P be an *arbitrary* point on a spherical wavefront. For an observer in K, the point P is described by spatial coordinates (x, y, z) and by a time t which characterizes the time taken for the emitted wave to reach P. Since the speed of light is c, we have

$$x^2 + y^2 + z^2 - c^2t^2 = 0 \qquad (27.3.1)$$

But the observer in K′ also sees a spherical wavefront, and for him, since the speed of light is also c, we find

$$x'^2 + y'^2 + z'^2 - c^2t'^2 = 0 \qquad (27.3.2)$$

Here, (x', y', z') are the spatial coordinates of P relative to K′, while t' is the time it takes for the signal to travel to P, as observed in the system K′.

[3] The perceived invariance of the speed of light can be understood roughly on a simple physical basis by considering that all the information we receive from experiments we perform on our physical environment is *itself* transmitted to us in the form of signals that propagate with the speed of light. Since we are never in a position to alter this aspect of physical reality, there is no way in which we can change the observed speed of light.

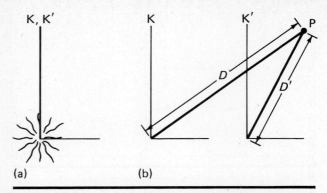

(a)   (b)

FIGURE 27.6. (a) Light signal emitted at time $t = 0$ at the common origin of reference systems K and K′. (b) At a later time t, a point P on the wavefront satisfies the relations $D = ct$ for an observer in K and $D' = ct'$ for an observer in K′.

For our purposes, the equality

$$x^2 + y^2 + z^2 - c^2t^2 = x'^2 + y'^2 + z'^2 - c^2t'^2 \quad (27.3.3)$$

which follows from (27.3.1) and (27.3.2), is all that is needed. Let us now assume that the new transformation law has the form

$$\begin{aligned} t' &= \gamma(t - avx) \\ x' &= \bar{\gamma}(x - \bar{a}vt) \\ y' &= y \\ z' &= z \end{aligned} \qquad (27.3.4)$$

when the relative velocity of K and K′, which we denote in magnitude by v, is directed along the x-axis. In the above expression $\gamma, \bar{\gamma}, a,$ and $\bar{a}$ are all unknown quantities which are to be obtained by substituting into (27.3.4). We must further remember that the above coordinate transformation must reduce to the Galilean transformation in the limit of small v and, moreover, that $\gamma, \bar{\gamma}, a,$ and $\bar{a}$ are assumed to depend only on *the magnitude of the relative velocity*. The form of (27.3.4) has the virtue of simplicity as well as the possibility of readily reducing to the Galilean case in the limit of low v.[4]

The transformation (27.3.4) expresses the space and time coordinates of events in the primed system in terms of their coordinates in the unprimed frame. The reverse transformation, expressing event coordinates in the unprimed system in terms of their values in the primed system, can be obtained from (27.3.4) simply by solving those equations for the unprimed variables $x, y, z,$ and t. This involves some rather tedious but very straightforward algebra. The results are

[4] Since relative motion takes place along the x-direction, it is reasonable to suppose that the y- and z-transformation equations are unaffected by the motion.

$$t = \frac{1}{1 - a\bar{a}v^2}\left(\frac{t'}{\gamma} + \frac{avx'}{\bar{\gamma}}\right)$$

$$x = \frac{1}{1 - a\bar{a}v^2}\left(\frac{x'}{\bar{\gamma}} + \frac{\bar{a}vt'}{\gamma}\right) \qquad (27.3.5)$$

$$y = y'$$

$$z = z'$$

But physically, since there is nothing to distinguish one reference system from another, the inverse transformation written above must really have the same algebraic form as the initial transformation (27.3.4), except, of course, for a reversal in the sign of the relative velocity v between the frames. Under these circumstances, we must expect from (27.3.4) that the first of the above equations has the form $t = \gamma(t' + avx')$ and the second the form $x = \bar{\gamma}(x' + \bar{a}vt')$. We may, therefore, write

$$t = \gamma(t' + avx') = \frac{1}{1 - a\bar{a}v^2}\left(\frac{t'}{\gamma} + \frac{avx'}{\bar{\gamma}}\right) \qquad (27.3.6)$$

$$x = \bar{\gamma}(x' + \bar{a}vt') = \frac{1}{1 - a\bar{a}v^2}\left(\frac{x'}{\bar{\gamma}} + \frac{\bar{a}vt'}{\gamma}\right) \qquad (27.3.7)$$

Equating coefficients of x' and t' in (27.3.6) and (27.3.7), we find

$$\gamma = \bar{\gamma} = \frac{1}{\sqrt{1 - a\bar{a}v^2}} \qquad (27.3.8)$$

Now, substituting (27.3.4) into (27.3.3), we may write

$$x^2 - c^2t^2 = \bar{\gamma}^2(x^2 - 2\bar{a}vxt + \bar{a}^2v^2t^2)$$
$$- c^2\gamma^2(t^2 - 2avxt + a^2v^2x^2) \qquad (27.3.9)$$

In order that this equation be satisfied for all values of x and t, however, the coefficients of x^2, t^2, and xt appearing on both sides of the equation must be equal. This means that

$$\bar{\gamma}^2 - c^2\gamma^2a^2v^2 = 1$$
$$\bar{a}v\bar{\gamma}^2 - c^2\gamma^2av = 0 \qquad (27.3.10)$$
$$c^2\gamma^2 - \bar{\gamma}^2\bar{a}^2v^2 = c^2$$

Since, according to (27.3.8), γ and $\bar{\gamma}$ are equal, this may be written as

$$\gamma^2(1 - c^2a^2v^2) = 1$$
$$\bar{a} = c^2a \qquad (27.3.11)$$
$$\gamma^2(c^2 - \bar{a}^2v^2) = c^2$$

Expressing a in terms of $\bar{a}$ by the second of these equations, we find that the first and third equations and (27.3.8) as well all lead to

$$\gamma = \frac{1}{\sqrt{1 - (\bar{a}^2v^2/c^2)}} \qquad (27.3.12)$$

But what is $\bar{a}$? The answer to this question is obtained by referring back to (27.3.4) and the Galilean transformation defined by (27.2.1). We know that for velocities much less than that of light, the Galilean transformation works extremely well, and, therefore, we must insist that our new transformation scheme reduce to the Galilean transformation (27.3.4) rather than to some other transformation that does not work well when $v \ll c$. Equation (27.3.12) tells us that when v is much less than c, the quantity γ, and $\bar{\gamma}$ from (27.3.8) as well, is very close to unity. This being the case, however, there is no way that the second equation of the set (27.3.4) can reduce to the second equation in (27.2.1) unless $\bar{a}$ has the value unity. We must, therefore, choose $\bar{a} = 1$ to make our new transformation reduce to the Galilean form at low speeds. We than have

$$\bar{a} = 1 \qquad a = \frac{\bar{a}}{c^2} = \frac{1}{c^2} \qquad (27.3.13)$$

and

$$\gamma = \bar{\gamma} = \frac{1}{\sqrt{1 - (v^2/c^2)}} \qquad (27.3.14)$$

When these values are substituted, Eqs. (27.3.4) become

$$t' = \frac{1}{\sqrt{1 - (v^2/c^2)}}\left(t - \frac{v}{c^2}x\right)$$

$$x' = \frac{1}{\sqrt{1 - (v^2/c^2)}}(x - vt) \qquad (27.3.15)$$

$$y' = y$$

$$x' = z$$

These equations are known as the *Lorentz transformation*. The student should now show directly that these equations imply the equality of (27.3.5) while the Galilean equivalent does not.

Note that the departures from the Galilean transformation (27.2.1) always involve factors of v/c or $(v/c)^2$, and, therefore, at velocities such that $v/c \ll 1$, these transformation equations approach the Galilean transformation. Nevertheless, for high velocities, the differences are very real and, as we shall see shortly, have profound implications.

Let us consider now a few examples of coordinate transformations in order to illustrate the differences between Galilean and Lorentz transformations.

EXAMPLE 27.3.1

At a time $t = 0.009$ second, an observer in a coordinate frame K observes that an ant is 1 meter from his origin (on the x-axis). This observer is traveling at a speed v along the negative x-axis of another observer in a frame K'. Find the x-coordinate of the ant in K' at $t = 0.009$ second for $v = 10^6$ meters/sec and also for $v = 10^8$ meters/sec (both of these are incredibly large speeds).

According to the Lorentz transformation,

$$x' = \gamma(x - vt) \qquad \text{where} \qquad \gamma = 1/\sqrt{1 - (v^2/c^2)}$$

Since $c = 3 \times 10^8$ meters/sec, in the first case $v/c = 1/300$, while in the second case $v/c = 1/3$. Thus, we have

$$x' = \frac{1}{\sqrt{1 - \left(\dfrac{1}{300}\right)^2}}[1 - (10^6)(0.009)]$$

$$= (1.00000556)(-8999) = -8999.05 \text{ m}$$

in the case of the Lorentz transformation. For the Galilean transformation, the result would be -8999.00 meters instead. We see that in this case, the fractional error is not large, even at the enormous speed of 10^6 meters/sec.

But for $v = 10^8$ meters/sec, we have

$$x' = \frac{1}{\sqrt{1 - \left(\dfrac{1}{3}\right)^2}}[1 - (10^8)(0.009)]$$

$$= (1.060660172)(-899,999) = -954,593.09 \text{ m}$$

The corresponding coordinate of $-899,999.00$ meters for the Galilean transformation is now appreciably in error, and, therefore, the Lorentz transformation must be used.

EXAMPLE 27.3.2

An electron in a modern accelerator is moving at $v = 0.99999c$ through a laboratory reference frame. The reference frame in which the electron is at rest is K', while the laboratory frame is K. At the origin of the laboratory frame ($x = 0$), an electrical discharge occurs at time $t = 10$ seconds. To observers in the K' reference frame, at what time does the discharge occur? Note that the observers in K' do not have to be near the electron. They can, in fact, be instantaneously right at the spot where the discharge occurs.

According to the Lorentz transformation

$$t' = \gamma[t - (v/c^2)x] \qquad \text{where} \qquad \gamma = 1/\sqrt{1 - (v^2/c^2)}$$

In this case, $t = 10$ seconds, but $x = 0$. Therefore,

$$t' = \frac{1}{\sqrt{1 - (0.99999)^2}}(10) = 2,236 \text{ sec}$$

In other words, the event occurs *at a much later time* for the observer in K'. Of course, to him, the laboratory frame is moving and the discharge is in motion. If he compares his time for the occurrence of the

discharge with the time as measured in K, his own conclusion seems to be that *the clocks in K are running quite slow* since they read only 10 seconds at the time of discharge. We will have more to say about this in the next section.

27.4 Implications of the Lorentz Transformation

The modification of the basic transformation law from one inertial frame to another carries with it a number of implications which seem strange and appear to violate some very basic intuitive notions. Nevertheless, these results, after careful and thorough study, have been accepted by all physicists, and many of them have been subjected to careful and detailed experimental tests. We shall not consider all possible implications but will rather concentrate on a few simple, yet important consequences.

LAW OF ADDITION OF VELOCITIES

Suppose you are jogging along a highway at 7 mph and a car traveling in the same direction at 45 mph passes you by. The relative speed between you and the car is $45 - 7 = 38$ mph. This simple result, which arises directly from the Galilean transformation, is *no longer true* when the Lorentz transformation is used. We shall, however, investigate the extent to which it is in error and will find that, since the speeds referred to above are so much smaller than the speed of light, for all practical purposes an answer of 38 mph is correct.

The problem at hand can be stated quite simply using Fig. 27.7. A point mass at P is in motion along the x-axis of observer K or the x'-axis of observer K'. It has a velocity V with respect to K and V' with respect to K'. The K and K' coordinate origins coincided at $t = 0$, but K' is moving with speed v along

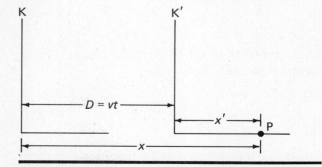

FIGURE 27.7. The velocity V of an object in frame of reference K and V' of the same object observed from frame of reference K' are related not by the simple relation $V' = V - v$ given by the Galilean transformation but by the more complex eq. (27.4.3).

the positive x-axis of K. The *Galilean theory* would assert that

$$V' = V - v \qquad (27.4.1)$$

is the relation between velocities. Referring to our previous discussion, K could be the reference frame fixed on the ground, K' a reference frame attached to the jogger, and P the vehicle in motion. Then, from (27.4.1), we would have

$$V' = 45 - 7 = 38 \text{ mph}$$

To find the corresponding expression using the *Lorentz transformation*, we must calculate

$$V = \frac{dx}{dt} \qquad \text{and} \qquad V' = \frac{dx'}{dt'}$$

and relate the two expressions. This can be done by calculating the differentials dx' and dt' using Eqs. (27.3.15), remembering that the relative velocity v between the two reference systems is constant. The result can be written as

$$dt' = \gamma\left(dt - \frac{v}{c^2}\,dx\right)$$
$$dx' = \gamma(dx - v\,dt) \qquad (27.4.2)$$

By dividing one of these equations by the other, we obtain

$$\frac{dx'}{dt'} = V' = \frac{dx - v\,dt}{dt - \frac{v}{c^2}\,dx} = \frac{\dfrac{dx}{dt} - v}{1 - \dfrac{v}{c^2}\dfrac{dx}{dt}}$$

or

$$\boxed{V' = \frac{V - v}{1 - \dfrac{vV}{c^2}}} \qquad (27.4.3)$$

This equation is the replacement for (27.4.1). If we substitute the velocities from the preceding discussion, $V = 45$ mph, $v = 7$ mph, and $c = 6.71 \times 10^8$ mph, we find

$$V' = \frac{45 - 7}{1 - \dfrac{(45)(7)}{(6.71 \times 10^8)^2}} = \frac{38}{1 - (7.0 \times 10^{-16})} \text{ mph}$$

We see that in this case the correction to 38 mph is negligible and inconsequential, but consider now the following example.

EXAMPLE 27.4.1

An electron moves at speed $v = 0.999c$ in the laboratory reference frame. A positron travels at $v = 0.999c$ in the opposite direction. From a reference frame moving with the positron, how fast does the electron appear to be moving?

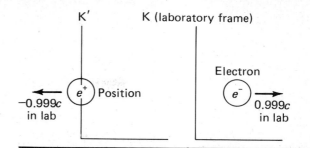

FIGURE 27.8. An electron, moving with speed $0.999c$ with respect to reference frame K of the laboratory, and a positron, in reference frame K', which moves at speed $-0.999c$ with respect to the laboratory frame K'.

In this case, we can assume K to be the laboratory frame and K' to be the frame moving with the positron, as shown in Fig. 27.8. Let us assume that the electron travels in the positive x-direction, while the positron moves in the negative x-direction. The velocity of the positron frame K' with respect to the laboratory frame K is, therefore, $-v$, where v is the positive quantity $0.999c$. Let V denote the electron velocity with respect to K and V' the velocity with respect to K'. Then, according to (27.4.3),

$$V' = \frac{V - (-v)}{1 - \dfrac{(-v)V}{c^2}} = \frac{V + v}{1 + \dfrac{vV}{c^2}} \qquad (27.4.4)$$

In the present example, $V = 0.999c$ and $v = 0.999c$; therefore,

$$V' = \frac{1.998c}{1 + (0.999)^2} = \frac{1.998c}{1.998001} = 0.99999c$$

There are several aspects of this result worth mentioning. It is important to note that the Galilean transformation would have led to the result $1.998c$ instead of the above answer. Note also that the above result is quite close to the speed of light but is still below it. These results indicate that the speed of material particles such as the electron, proton, or anything else can never exceed the speed of light. Thus, c is the "ultimate speed." Objects can, indeed, be made to move faster and faster by choosing appropriate reference frames, but their speed, referred to any reference system, can never exceed c. We shall examine this point in another context somewhat later.

THE CONTRACTION OF LENGTHS

The special theory of relativity predicts that moving objects are *shortened in length* in the direction of their motion. This effect is easy to predict but difficult to measure. There have, in fact, been no direct measurements to confirm the existence of this effect. Let us now consider the process of measuring the length of a stick in two different frames of reference. In Fig. 27.9, a stick is at rest in a coordinate system K′. To determine its length, one must measure the spatial coordinates of the left- and right-hand ends and then subtract. Thus,

$$X_R' - X_L' = L_0$$

where L_0 is called the rest length because it is measured in a frame of reference in which the stick is at rest. Since the stick is at rest, the process of determining X_L' and X_R' is quite simple and exact. In fact, X_L' and X_R' can even be measured at different times.

Now, things are not quite that simple for inertial observer K, for the stick is moving with speed V along his positive x-axis. This implies that any determination of length necessarily involves recording the left and right coordinates at the same time, say, t_0. We shall not go into the technical details of how this might be accomplished but will simply assume it can be done. We can use (27.3.15) to find a relation between the measurements of observers K and K′. This relation is

$$X_L' = \gamma(X_L - vt_0)$$
$$X_R' = \gamma(X_R - vt_0)$$

(27.4.5)

where X_L and X_R are the coordinates of the ends of the stick measured in frame K. Note that the same time t_0 enters both equations since this is the time at which K measures both X_L and X_R. Taking the difference between the two equations, we find

$$X_R' - X_L' = \gamma(X_R - X_L)$$

(27.4.6)

The left-hand side is L_0, the rest length of the stick, while the right-hand side contains the length $L = X_R - X_L$, as measured by the observer in K. Therefore, we have

$$L_0 = \gamma L \qquad \text{where, as usual, } \gamma = 1/\sqrt{1 - (v^2/c^2)}$$

or

$$\boxed{L = L_0 \sqrt{1 - \frac{v^2}{c^2}}} \qquad (27.4.7)$$

This last equation says that the stick in motion appears shorter, or contracted, along the direction of motion. If, instead of a stick, we used a square plate, we would find a contraction in the direction of motion but *none* in the perpendicular direction. This implies, of course, that the moving plate has a different shape with respect to an observer K when it is moving with respect to him.[5] We wish to emphasize that this contraction and distortion is only appreciable when objects are moving at speeds comparable to that of

<hr/>

[5] There is, therefore, a profound alteration in our ideas about rigid bodies when relativistic speeds are involved. Indeed, in relativity there really is no such concept.

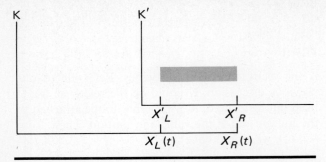

FIGURE 27.9. Determination of the length of a stick in two different reference frames.

light. It is difficult to accelerate macroscopic objects to such high speeds, and therein lies the difficulty of directly verifying these predictions. At the microscopic level, elementary particles can be made to travel at speeds comparable to c, but such entities are never described in terms of dimensions such as length.

LACK OF SIMULTANEITY

We have obtained a relation between the length of a stick at rest and its length in motion by having a *simultaneous* determination of its end points made in the coordinate system K. Let us now ask at what time (or times) the observer in K′ thinks the measurements were made by K. We assume that K′ has observers sitting at the left and right ends of the rod and that they can ascertain when measurements were carried out by the observers in K.

In K, the left end was measured at t_0 and found to be at X_L. The observers in K′ think this measurement was made at

$$t_L' = \gamma\left(t_0 - \frac{v}{c^2} X_L\right) \qquad (27.4.8)$$

The second measurement made in K was also at time t_0 and gave the result X_R. In K′, they will say that this measurement was made at

$$t_R' = \gamma\left(t_0 - \frac{v}{c^2} X_R\right) \qquad (27.4.9)$$

If we now subtract, we obtain, using (27.4.7),

$$\boxed{t_L' - t_R' = \gamma \frac{v}{c^2}(X_R - X_L) = vL_0/c^2} \qquad (27.4.10)$$

which, of course, is not zero. What does this mean? The observers in K were careful to measure the two positions simultaneously, but the observers in K′ do not find that the measurements occurred at the same time.

Here, we have a very profound and fascinating difference between Newtonian concepts and the new ideas embodied by Einstein's theory. We find that simultaneity is not universal, that spatially separated events which to one observer appear to happen at the same time *do not occur simultaneously for another observer in a different reference frame.* Only simultaneous events occurring *at the same spatial point* are simultaneous for all observers. This lack of universal simultaneity has interesting and far-reaching consequences. Some of these are illustrated by the examples that follow.

EXAMPLE 27.4.2

Herbert and Gretchen live in a space station far from the Earth. They eat exotic foods, sleep during our daylight hours, and work every night. They sit at opposite ends of a table 6 meters long and always finish their dessert at exactly the same time, 5:00 A.M. EST (see Fig. 27.10). Some space psychologists, capable of carrying out very precise measurements, wish to observe Herbert and Gretchen from their rapidly moving spaceship ($v = 0.9999999999c$ with respect to the space station). They can't understand why Herbert is so impolite, for he always appears to finish dessert shortly before Gretchen. Find the difference in time at which the desserts are finished and also the length of the table, as measured by the psychologists.

In this case, let us assume that K' is the frame of reference of the space station. In this reference system, Herbert, Gretchen, and the dinner table are at rest, as illustrated in Fig. 27.10. The reference frame K is that of the spaceship and the psychologists. It is in motion with velocity v relative to K'; we shall assume that it moves along the positive x-axis of K'. In reference frame K', two events spatially separated by the length L_0 of the table occur simultaneously at time t_0'. We may refer to these, specifically, by space and time coordinates in K' as follows:

Reference System K'

Event 1': Herbert finishes dessert ($X_R' = L_0$, $t_R' = t_0$)

Event 2': Gretchen finishes dessert ($X_L' = 0$, $t_L' = t_0$)

Reference system K moves along the x'-axis of K' with velocity v. The Lorentz transformation relating its space and time coordinates to those of K' will be

$$t = \gamma\left(t' + \frac{vx'}{c^2}\right) \qquad y = y'$$
$$x = \gamma(x' + vt') \qquad z = z'$$

(27.4.11)

These can be obtained from (27.3.15) simply by solving for x, y, z, and t. We may also note that they are identical in form to Eqs. (27.3.15), which relate K' coordinates to K coordinates, except for the fact that v is replaced by $-v$. In system K, in which the psychologists are at rest, the two events described above are no longer observed to be simultaneous, nor are they perceived as being separated by the same spatial separation L_0, because of the apparent contraction of the table. In this system, in fact, they will be described as follows, according to (27.4.11):

Reference System K

Event 1: Herbert finishes dessert

$$\left(X_R = \gamma(X_R' + vt_R'),\ t_R = \gamma\left(t_R' + \frac{vX_R'}{c^2}\right)\right)$$

Event 2: Gretchen finishes dessert

$$\left(X_L = \gamma(X_L' + vt_L'),\ t_L = \gamma\left(t_L' + \frac{vX_L'}{c^2}\right)\right)$$

But, as defined above, $X_L' = 0$, $X_R' = L_0$, $t_R' = t_L' = t_0$. Therefore,

$$X_R = \gamma(L_0 + vt_0) \qquad t_R = \gamma\left(t_0 + \frac{vL_0}{c^2}\right)$$

$$X_L = \gamma vt_0 \qquad t_L = \gamma t_0$$

Subtracting, it is easily seen that

$$X_R - X_L = \gamma L_0 \tag{27.4.12}$$

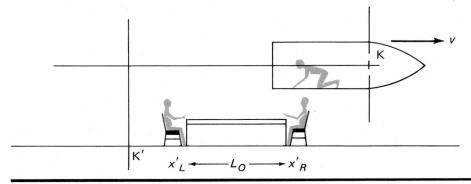

FIGURE 27.10. Herbert and Gretchen's meal, as observed by the space psychologists. Herbert and Gretchen's rest frame is the system K', while the psychologists are in frame K which moves with respect to K with a velocity close to the speed of light.

while

$$t_R - t_L = \frac{\gamma v L_0}{c^2} \qquad (27.4.13)$$

At first sight, these results look strange, for they appear to contradict our previously obtained expressions (27.4.7) and (27.4.10). Indeed, since γ is always greater than or equal to unity, (27.4.12) seems to imply a length expansion rather than a contraction. It must be remembered, however, that events 1 and 2 are *no longer simultaneous* in reference system K. An observer in this system, therefore, will conclude that the table has moved a distance $v(t_R - t_L)$ between the two events. This means that the total spatial separation of the events, $X_R - X_L$, as observed in K, is not simply the length L of the table, but the length *plus* the distance the table moves during the time between the events! Therefore, in K,

$$X_R - X_L = L + v(t_R - t_L) \qquad (27.4.14)$$

Using (27.4.12) and (27.4.13), however, this can be written

$$L = (X_R - X_L) - v(t_R - t_L) = \gamma L_0 \left(1 - \frac{v^2}{c^2}\right) \qquad (27.4.15)$$

or, since $\gamma = 1/\sqrt{1 - (v^2/c^2)}$,

$$L = L_0 \sqrt{1 - (v^2/c^2)} \qquad (27.4.16)$$

in agreement with our previously derived result (27.4.7).

But now, what about (27.4.13), which does not seem to agree with (27.4.10)? In order to resolve this discrepancy, we must make a critical comparison between the situation we are discussing here and the one considered previously in connection with the moving meter stick, which led us to (27.4.10). Though the two cases are similar in many respects, they are not *quite* identical. In the previous discussion dealing with the

meter stick, the simultaneous registration of the coordinates representing the two ends of the stick transpire in the reference frame *in which the stick is in motion*. In the example considered here, however, the simultaneous events occur *in the system in which the table is at rest*. In the reference frame in which the events are simultaneous, therefore, they are separated spatially by a distance L_0 in this case but in the former case by the *contracted* distance L_0/γ. But if we replace L_0 in (27.4.13) by L_0/γ, we obtain a result that agrees with (27.4.10) exactly! Therefore, (27.4.13) and (27.4.10) are not contradictory, after all.

In the present situation involving Herbert, Gretchen, and the space psychologists, v/c has the value 0.9999999999. Therefore, from (27.4.13),

$$t_R - t_L = \frac{1}{\sqrt{1 - (0.9999999999)^2}} \cdot \frac{(0.9999999999)(6.0)}{3 \times 10^8}$$

$$= 0.00141 \text{ sec}$$

The psychologists observe, therefore, that Herbert finishes his dessert earlier than Gretchen by more than a millisecond. The table, from (27.4.16), appears to them to have the length

$$L = \sqrt{1 - (0.9999999999)^2}(6.0) = 8.49 \times 10^{-5} \text{ m}$$

hence a small fraction of a millimeter. The psychologists wonder how all the food can fit on such a short table, but they then realize that the food and dishes are also contracted. In fact, the whole table looks extremely odd, with weird-looking dishes and short, fat knives and forks. Even Herbert and Gretchen look like paper dolls, for, of course, the contraction takes place only in the direction of motion (in this case the x-direction) leaving the y- and z-dimensions unaltered!

It is worth noting that even though the length contraction is very pronounced, the time difference

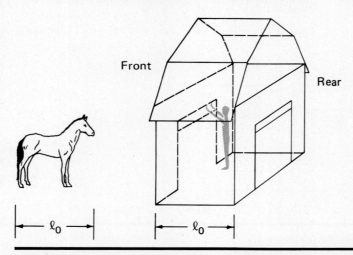

FIGURE 27.11. Horse and barn at rest relative to one another.

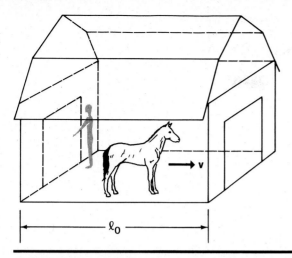

FIGURE 27.12. Situation as seen by the barnkeeper; the horse has just entered the barn, has length $l_0[1 - (v^2/c^2)]^{1/2}$, and both doors are closed simultaneously.

$t_R - t_L$ is, in this case, quite small. This situation arises because of the factor c^2 in the denomination of (27.4.13). An inspection of this equation will reveal that it requires not only a large relative velocity v but also a large separation L_0 between events in the frame of simultaneity to bring about a pronounced departure from simultaneity in the other system.

EXAMPLE 27.4.3

In the next example, we present what appears to be a paradox. The paradox involves a horse, which can travel at relativistic speeds, and a barn. When the horse (length L_0) is at rest, it can just fit into the barn, which also has a rest length L_0. The barn is equipped with front and rear doors which can be activated in unison by the barnkeeper, as illustrated in Fig. 27.11. Let us now consider the paradox. The barn doors are open, and the horse comes charging along at a tremendous speed such that $(v/c)^2 = 0.99$. At the instant the horse's rear passes into the barn, the doors are shut. We shall assume they can be closed without any time lag. According to the theory of relativity, the barnkeeper would say that the rapidly moving horse is smaller in length than the barn, and, therefore, by closing the door he traps the horse. Consider the point of view of the horse. It reasons (we have a smart horse!) that the barn is moving toward it at an incredible speed and the length of the barn is smaller than L_0. It, therefore, concludes that it can't possibly be trapped in the barn. Who is right, the horse or its keeper?

Let us first point out the obvious—they can't both be right. Before we proceed with the explanation, let us assert unequivocally that the man is right. We do not mean to imply that men are smarter than horses, for we could phrase the problem so that the

man does the running and the horse closes the door to trap the man. Then the horse would be right. In any event, let us try to resolve the paradox by critically examining what happens from the point of view of the man and also from that of the horse.

Point of View of the Barnkeeper
The barnkeeper closes both doors at the time $t = 0$, just when the rear of the horse passes into the front door. At this instant, the front of the horse will be at

$$x_f = \sqrt{1 - \frac{v^2}{c^2}} \, L_0 \qquad (27.4.17)$$

while the rear door is at $x = 0$. Thus, the horse is confined to the barn, as shown in Fig. 27.12. Subsequently, the horse, if he continues to run, will collide with the rear door.

Point of View of the Horse
In the reference frame of the horse, the rear of the horse is always at $x' = 0$, while the nose is at $x' = L_0$. We wish to establish the times at which the doors were shut, as well as their locations, in the horse's reference frame. To determine these quantities, we use the Lorentz transformation. The front door closes at time

$$t_f{}' = \gamma\left(0 - \frac{v}{c^2} 0\right) = 0 \qquad (27.4.18)$$

while the rear door closes at

$$t_r{}' = \gamma\left(0 - \frac{v}{c^2} L_0\right) \qquad (27.4.19)$$

Thus, the rear door closes before the front door does in the reference frame of the horse. We see again that events that are simultaneous in one reference frame are not necessarily simultaneous in another.

Let us now establish the locations of the doors in the horse's reference system at the two times given above. We can do this by means of the relation (27.4.11), which states that

$$x = \gamma(x' + vt') \qquad (27.4.20)$$

At $t_r{}' = -\gamma v L_0/c^2$, the position coordinates are related by

$$x = \gamma\left(x' - \gamma \frac{v^2}{c^2} L_0\right)$$

Therefore, the front door, which is at $x = 0$, will be located at

$$x_{fd}{}' = \gamma \frac{v^2}{c^2} L_0 \qquad (27.4.21)$$

911

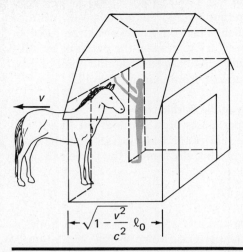

FIGURE 27.13. Situation as seen by the horse, which sees the barn moving toward it with speed v. At time $t_r' = -\gamma v l_o/c^2$, the rear door closes but the front door is still open. The horse is rapidly approaching the rear of the barn, which to him has shrunk by the factor $1/\gamma$.

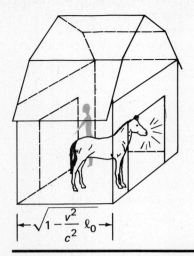

FIGURE 27.14. Final state of the situation, from the point of view of the horse. At $t_f' = 0$, the front door closes, trapping the horse, but only after the horse has been brought to rest after colliding with the rear door, allowing him once more to fit in the barn in his reference frame.

The rear door, at $x = L_0$, satisfies

$$L_0 = \gamma\left(x_{rd}' - \gamma \frac{v^2}{c^2} L_0\right)$$

This can be solved to obtain

or, since $1/\gamma^2 = 1 - (v^2/c^2)$,

$$x_{rd}' = \gamma L_0 \left(\frac{1}{\gamma^2} + \frac{v^2}{c^2}\right)$$

$$x_{rd}' = \gamma L_0 \qquad (27.4.22)$$

In Fig. 27.13, the situation seen by the horse at

$$t' = -\gamma \frac{v}{c^2} L_0$$

is depicted. The front door is open, the rear door has just shut, and the horse is partly into the barn, which, to the horse, has a length of $\sqrt{1 - (v^2/c^2)}L_0$.

At the time $t' = 0$, the horse has now fully moved into the barn, as shown in Fig. 27.14, but it is straightforward to show that its nose has *already* collided with the rear door. If the door is sufficiently rigid and strong, the horse is now unpleasantly dynamically contracted as a result of his collision with the door.[6] Nevertheless, for this reason, the horse also concludes that it will be trapped by the barn.

It is important to note that the events that occur are explained quite differently by the two inertial observers. In general, the actual observations are very much coordinate dependent. On the other hand, the laws of physics do not depend in any way on the motion of the observer.

[6] It is well to remember here that rigid body behavior as we perceive it is profoundly altered at relativistic speeds.

TIME DILATION

We have already seen that events that are simultaneous to one observer may not be so to another. This suggests that the duration of time intervals may be different in different reference systems. To explore this possibility further, we assume that two events occur at a fixed point in a reference frame K′ which is moving with respect to K with speed v along the z-axis. To be quite specific, assume that event 1 consists of turning on a flashlight at time t_1' at position x'. Event 2 consists of turning off the flashlight at time t_2' at position x'.

Now, the inertial observers in coordinate frame K can determine, on their clocks, exactly when the two events occurred. They will find the times to be given by

$$t_1 = \gamma\left(t_1' + \frac{vx'}{c^2}\right)$$

and

$$t_2 = \gamma\left(t_2' + \frac{vx'}{c^2}\right) \qquad (27.4.23)$$

Therefore, the *duration* of the process is

$$\boxed{\begin{aligned} t_2 - t_1 &= \gamma(t_2' - t_1') \\ &= \frac{1}{\sqrt{1 - (v^2/c^2)}}(t_2' - t_1') \end{aligned}} \qquad (27.4.24)$$

We note that the duration in frame K is *greater* than that in K′.

We may conclude that the time interval between any two events that occur at the same place seems *longer* when observed from a moving reference frame than when observed from the rest frame of the events. Since the two events in question could refer to successive "ticks" of a clock or other time-measuring device, we are forced to conclude that *clocks in motion slow down.* This conclusion was arrived at specifically for the case of a clock in the system K' as observed from system K. But the transformation Eqs. (27.3.15) from K to K' are of the same form as those relating K' to K such as (27.4.23), except for a change in the sign of *v*. It is, therefore, easily shown, using the same argument as that given above, that observers in K' will conclude that clocks in K run slow in comparison to their own clocks *by the same factor* expressed above as (27.4.24). Thus, it happens that observers in system K conclude that K"s clocks are running slow, while K"s observers think that it is the clocks of system K that are slow! This is another of the conclusions of the theory of relativity that seems to contradict our intuitive notion of "how things should happen."

But this effect, which is sometimes referred to as *time dilation*, is a very real one and has been observed and verified experimentally in a number of ways. For example, there are elementary particles known as muons which ordinarily endure as such for only about 10^{-6} second after their formation, when observed in their own rest frame. Their disappearance is accompanied by the production of electrons and neutrinos. Many muons are produced very high up in the atmosphere by the absorption of high-energy cosmic radiation and travel down toward the earth with speeds very close to that of light. It's a simple matter to see that with an average lifetime of 10^{-6} seconds they would travel a distance of only about 300 meters and could not possibly reach the ground. But experiments have shown that, in fact, many muons are detected at the ground. The explanation is due to the effect of time dilation, for relative to an observer on earth, the muons have a lifetime of 10^{-6} seconds multiplied by the factor $\gamma = 1/\sqrt{1 - (v^2/c^2)}$ of Eq. (27.4.24). Since these particles move relative to the earth at a speed very close to that of light, the factor γ is large compared to unity and they may, therefore, traverse distances of many times 300 meters before disintegrating into electrons and neutrinos.

More recently, in 1971, precise tests of this time dilation were carried out by flying very stable macroscopic cesium atomic clocks around the world, using scheduled jet flights. These clocks were then compared to reference clocks, and the small time differences (of the order of nanoseconds) that were observed supported the prediction of time dilation.[7]

[7] J. C. Hafele and Richard E. Keating, *Science*, 166 (July 14, 1972).

EXAMPLE 27.4.4

The Crab Nebula is approximately 6.7×10^{11} miles from us and emits cosmic radiation. Some of the emitted particles are neutrons which, when at rest, decay with a lifetime of 932 seconds. With what velocity must the neutrons be emitted in order to arrive at the earth?

According to (27.4.24), the neutrons in motion will have a lifetime τ of

$$\tau = \frac{1}{\sqrt{1 - (v^2/c^2)}} (932) \text{ sec}$$

In order to arrive here, they must have a lifetime at least as large as the distance to be covered divided by the velocity. That is, if we express v in units of miles/sec, we must have

$$\tau \simeq \frac{d}{v} = \frac{6.7 \times 10^{11}}{v} \text{ sec}$$

Therefore,

$$\frac{6.7 \times 10^{11}}{v} = \frac{1}{\sqrt{1 - (v^2/c^2)}} (932)$$

In 932 seconds, light will travel only about 1.73×10^8 miles. It is, therefore, obvious that no neutrons will reach the earth unless the time dilation effect is very large. This, in turn, takes place only if the factor γ is large compared to unity, which requires that v be very close to c. We may, therefore, as a good approximation, replace v by c on the left side of this equation and solve for the ratio v/c. In this manner, we obtain

$$\frac{6.7 \times 10^{11}}{186,000} = \frac{1}{\sqrt{1 - (v^2/c^2)}} (932)$$

Solving, we find that

$$\frac{v}{c} = 0.99999996$$

27.5 The Equivalence of Mass and Energy

From a practical point of view, the most striking consequence of Einstein's theory is the equivalence of mass and energy. In late 1905, Einstein published a paper in which he recognized that when a system loses or gains energy, there is a corresponding change in mass. However, it was not until 1907 that he realized that mass itself contains locked within itself, so to speak, a huge supply of energy. His famous equation $E = mc^2$ originated from sound reasoning and not, as the cartoon of Fig. 27.15 suggests, from random guesswork.

FIGURE 27.15. Cartoonist's view of the origin of Einstein's equation.

Let us consider now a simple thought experiment which suggests the equivalence of mass and energy. Figure 27.16 illustrates a box of mass M in which there is some device that emits light at the left end that is subsequently absorbed at the right end. This is called Einstein's "light box." We know that light can transport energy from one place to another. Experiments also reveal that light waves carry momentum. If an amount of energy U is transferred from one place to another, then a momentum

$$p = U/c \tag{27.5.1}$$

is also transferred. We assume that the box and the radiation comprise an isolated system. Since momentum is conserved for such a system, when a short pulse of light is emitted at the left end, the box must recoil with a velocity V given by

$$MV = p = \frac{U}{c} \tag{27.5.2}$$

It maintains this velocity until the radiation is absorbed at the right end. The time interval t during which it moves is given by

$$t = \frac{L}{c} \tag{27.5.3}$$

Therefore, the total distance the box moves is

$$\Delta x = Vt = \frac{VL}{c} \tag{27.5.4}$$

It is certainly true from momentum conservation that the center of mass of an isolated system cannot move. We must, therefore, insist that the center of mass of the system does not shift. But if the center of mass of the system does not move and yet the box recoils with momentum U/c, as required by (27.5.2), we are forced to conclude that the transfer of energy by the light pulse is accompanied by a transfer of mass just sufficient to guarantee that the center of mass remains fixed! Any other hypothesis would require the abandonment of the law of conservation of momentum. We must, therefore, assume that the radiation brings about a mass transfer Δm. Under these circumstances the shift of the center of mass can be expressed as

$$M \Delta x - L \Delta m = 0 \tag{27.5.5}$$

which implies

$$\Delta m = \frac{M}{L} \Delta x \tag{27.5.6}$$

Substituting the values given by (27.5.4) and (27.5.2) for Δx and V, we find

$$\Delta m = \left(\frac{M}{L}\right)\left(\frac{VL}{c}\right) = \left(\frac{M}{c}\right)\left(\frac{U}{Mc}\right) = \frac{U}{c^2} \tag{27.5.7}$$

Thus, a transfer of energy U implies a transfer of mass given by the above expression. Moreover, the expression suggests that a given amount of mass m actually contains an energy

$$U = mc^2 \tag{27.5.8}$$

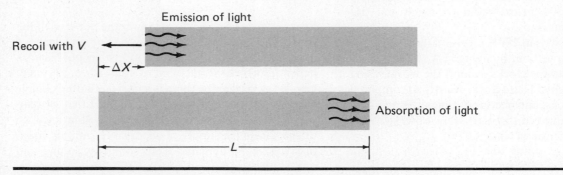

FIGURE 27.16. Einstein's light box.

The ultimate test of the validity of this equation must, of course, be experimental. In 1932, J. Cockcroft and E. Walton succeeded in finding the first direct experimental verification of the equivalence of mass and energy. Its validity is attested to today by a vast amount of experimental evidence, all of which supports the idea of mass–energy equivalence. The physical implications of Eq. (27.5.8) are, for example, that when you turn on a flashlight, it loses mass as light energy is emitted and that when you burn a match, the mass of the ashes and other combustion products is less than that of the unburned match in view of the heat and light energy that is given off! In situations such as these, the mass differences are so small as to be undetectable, but they are there all the same. We must, therefore, give up the idea, so beloved by the chemists, of conservation of mass and replace it with the larger and more comprehensive notion of the *conservation of mass–energy.*

As an object is accelerated by a force, its energy increases. We should, therefore, expect that its mass will become larger also. This is just what happens, and the quantitative relationships between mass and velocity, and between energy and momentum, can be derived by the simple argument that follows.

Suppose we consider a body of mass m initially at rest. Let us assume that a force $\mathbf{F}$ is applied that produces a continual acceleration. According to the work–energy theorem, the work done by the resultant force in a distance dx is equal to the change dU in the body's energy. But since the total energy of the body is mc^2, according to (27.5.8), we may write

$$dU = d(mc^2) = c^2\,dm = F\,dx \qquad (27.5.9)$$

In this equation, dm represents the increase in mass due to the change in energy. Now, however, if we equate the force to the time rate of change of momentum and use the usual expression mv to represent momentum, we obtain

$$c^2\,dm = F\,dx = \frac{dp}{dt}\,dx = \frac{dp}{dt}\frac{dx}{dt}\,dt = v\,dp$$

or

$$c^2\,dm = v\,d(mv) = v(m\,dv + v\,dm)$$
$$mv\,dv = (c^2 - v^2)\,dm \qquad (27.5.10)$$

This, however, can be written in the form

$$\frac{dm}{m} = \frac{v\,dv}{c^2 - v^2} \qquad (27.5.11)$$

We may now integrate from the initial state, in which the velocity is zero and in which the mass corresponds to the *rest mass* m_0, to a subsequent state of velocity v and mass m. In doing this integration, it is convenient to make the substitution

$$u = c^2 - v^2 \qquad du = -2v\,dv \qquad (27.5.12)$$

which allows (27.5.11) to be written as

$$\int_{m_0}^{m} \frac{dm}{m} = -\frac{1}{2}\int_{0}^{v} \frac{du}{u} \qquad (27.5.13)$$

Integrating, we obtain

$$\ln \frac{m}{m_0} = -\tfrac{1}{2}[\ln(c^2 - v^2)]_0^v$$

$$= -\tfrac{1}{2}\ln\left(1 - \frac{v^2}{c^2}\right) = \ln \frac{1}{\sqrt{1 - (v^2/c^2)}}$$

or finally

$$m = \frac{m_0}{\sqrt{1 - (v^2/c^2)}} \qquad (27.5.14)$$

This reveals that if the object is at rest, its mass corresponds to the rest mass m_0. For small velocities, the factor v^2/c^2 is negligible and the mass change is insignificant; but as the speed becomes comparable to the velocity of light, the mass increases, becoming very large compared to the rest mass when the velocity approaches that of light. Under these circumstances, because of the increase in mass, more and more energy is required to produce a given velocity change Δv, and ultimately *no finite energy* will accelerate the body to the speed of light. The status of the velocity of light as the ultimate limit of speed for any object of nonzero rest mass can be understood in concrete terms using this argument.

Using the expression given above for the mass in Eq. (27.5.8), we find

$$U = \frac{m_0 c^2}{\sqrt{1 - (v^2/c^2)}} \qquad (27.5.15)$$

For small velocities, when $v/c \ll 1$, we can express the quantity $(1 - (v^2/c^2))^{-1/2}$ by the binomial expansion

$$\left(1 - \frac{v^2}{c^2}\right)^{-1/2} = 1 + \frac{1}{2}\left(\frac{v}{c}\right)^2 + \frac{3}{8}\left(\frac{v}{c}\right)^4 + \cdots \qquad (27.5.16)$$

which allows us to express the energy as

$$U = m_0 c^2 \left[1 + \frac{1}{2}\left(\frac{v}{c}\right)^2 + \frac{3}{8}\left(\frac{v}{c}\right)^4 + \cdots\right]$$

or

$$U = m_0 c^2 + \tfrac{1}{2}m_0 v^2 + \tfrac{3}{8}m_0 v^2 \left(\frac{v^2}{c^2}\right) + \cdots \qquad (27.5.17)$$

In this expansion, the first term, which is all that remains when $v = 0$, is referred to as the *rest energy.* The second term corresponds to the usual nonrelativistic kinetic energy of a body of mass m_0. The third and higher terms are corrections to the kinetic energy that can be thought of as arising from the relativistic mass increase. It is obvious that they are insignificant so long as $v \ll c$. Since c is a very large quantity in terms of the macroscopic scale of our environment, the rest energy of any object of appreciable size is enormous. In the case of an automobile of mass 1500 kg moving at a speed of 30 meters/sec (67.5 mph), the rest energy is $(1500)(9 \times 10^{16}) = 1.35 \times 10^{20}$ joules, while the kinetic energy is $(\frac{1}{2})(1500)(900) = 6.75 \times 10^5$ joules. The rest energy in this case is 2×10^{14} times the kinetic energy. It would be of great advantage if we could somehow convert just a bit of the rest energy of the car into the kinetic energy needed for propulsion. Though, in principle, there is nothing to prevent this from happening, we do not now know of simple and efficient ways of doing it. It should be noted, however, that this is exactly how energy is generated in nuclear fusion and fission reactors.

Another conclusion to be drawn from the mass–energy relation (27.5.15) is that, in order for a particle of finite energy to travel with the speed of light, its rest mass must be *zero*. Under these circumstances, of course, both numerator and denominator in (27.5.15) are zero, and a finite energy may result. There are certain particles such as *photons* (which represent quanta of light energy itself) and *neutrinos* that actually fall into this category.

A simple and useful relation between energy and momentum can be derived from (27.5.15). If we square both sides of this equation and add and subtract the quantity $m_0{}^2 c^4$ from the right side of the resulting expression, we find

$$U^2 = m_0{}^2 c^4 + \frac{m_0{}^2 c^4}{1 - (v^2/c^2)} - m_0{}^2 c^4$$

$$= m_0{}^2 c^4 + m_0{}^2 c^4 \left(\frac{1}{1 - (v^2/c^2)} - 1 \right)$$

$$= m_0{}^2 c^4 + \frac{m_0 c^2 v^2}{1 - (v^2/c^2)} = m_0{}^2 c^4 + m^2 v^2 c^2$$

Noting, finally, that $p = mv$, this can be written in the form

$$U^2 = p^2 c^2 + m_0{}^2 c^4 \qquad (27.5.18)$$

From this equation, it is evident that for a particle of rest mass m_0 at rest, the energy is simply the rest energy $m_0 c^2$. This energy obviously increases as the particle's momentum—or velocity—is increased, the difference between the energy given by (27.5.18) and the rest energy representing kinetic energy of motion. For a particle whose rest mass is *zero*, such as a photon, the energy–momentum relation reduces to $U = pc$, which is exactly that found previously for electromagnetic radiation.

EXAMPLE 27.5.1

A proton is accelerated from rest to 0.999 times the speed of light. Find its increase in energy.

The energy at rest is $m_0 c^2$ while the energy at $v = 0.999c$ is

$$U = \frac{m_0 c^2}{\sqrt{1 - (0.999)^2}} = 22.4 m_0 c^2$$

The increase is, therefore, $21.4 m_0 c^2$, or, since $m_0 c^2 = (1.67 \times 10^{-27})(3 \times 10^8)^2 = 1.503 \times 10^{-10}$ joule $= 938$ MeV (million electron volts),

$$U = (21.4)(938) \text{ MeV} = 20{,}073 \text{ MeV}$$

EXAMPLE 27.5.2

A thrown baseball has an energy of about 50 joules. Suppose 1 gram of matter is converted entirely into energy. How many thrown baseballs would this be equivalent to?

The rest energy of 1 gram of matter is

$$U = (10^{-3})(3 \times 10^8)^2 = 9 \times 10^{13} \text{ J}$$

Therefore, this is the equivalent of 0.18×10^{13} thrown baseballs.

EXAMPLE 27.5.3

A particle of rest mass m_0 is accelerated from rest by a *constant* force $\mathbf{F}_0$ directed along the positive x-axis. Assuming that it eventually attains a velocity comparable with that of light, find (a) its acceleration as a function of time, (b) its velocity as a function of time, (c) its velocity as a function of displacement, and (d) its displacement as a function of time.

Newton's second law, when written in the form $\mathbf{F} = d\mathbf{p}/dt$, is still correct even when relativistic effects are important.[8] Accordingly, we may write

$$F_0 = \frac{dp}{dt} = \frac{d(mv)}{dt} = \frac{d}{dt} \frac{m_0 v}{\sqrt{1 - (v^2/c^2)}}$$

Performing the indicated differentiation and simplifying by some fairly tedious but rather straightforward algebra, this may be written as

$$F_0 = \frac{m_0 (dv/dt)}{[1 - (v^2/c^2)]^{3/2}}$$

from which

$$a = \frac{dv}{dt} = a_0 \left(1 - \frac{v^2}{c^2} \right)^{3/2} \qquad (27.5.20)$$

where

$$a_0 = F_0/m_0 \qquad (27.5.21)$$

[8] It is *not* correct, however, simply to write $F = ma$, since this does not properly incorporate the relativistic mass change.

From this, it is evident that the particle starts from rest with the initial acceleration $a_0 = F_0/m_0$ that we would expect from ordinary Newtonian mechanics. As the particle attains relativistic velocities, however, according to (27.5.20), its acceleration decreases, approaching zero as v approaches c. This is in accord with our notion that no massive particle can be accelerated to the speed of light.

We may now find the velocity as a function of time by rearranging (27.5.20) and integrating from $t = 0$, when $v = 0$, to a later time t at which the velocity is v. In this way, we find, using a standard table of integrals, if necessary,

$$a_0 \int_0^t dt = \int_0^v \frac{dv}{[1 - (v^2/c^2)]^{3/2}} = \left[\frac{v}{\sqrt{1 - (v^2/c^2)}} \right]_0^v$$

from which

$$a_0 t = \frac{v}{\sqrt{1 - (v^2/c^2)}} \qquad (27.5.22)$$

Squaring both sides and solving for v, we find

$$v = \frac{a_0 t}{\sqrt{1 + (a_0{}^2 t^2/c^2)}} \qquad (27.5.23)$$

It is evident that so long as $v \ll c$, the velocity will be very nearly given by the Newtonian expression $v = a_0 t$ but that as v becomes relativistic, the velocity becomes smaller than this. In the limit as $t \to \infty$, Eq. (27.5.23) predicts that $v \to c$, again in accord with our ideas about the limiting character of the speed of light. By substituting the velocity given by (27.5.23) into (27.5.20), we may now express the acceleration as a function of time as

$$a = \frac{a_0}{[1 + (a_0 t/c)^2]^{3/2}} \qquad (27.5.24)$$

It is again apparent that the acceleration decreases to zero in the limit of large t, hence as v approaches c. Plots of the dependence of v and a versus time are shown in Fig. 27.17.

The displacement as a function of time may now be calculated by setting $v = dx/dt$ in (27.5.23) and integrating once more from $t = 0$ and $x = 0$ to the later time t when the displacement is x. The mathematics is straightforward, particularly if a table of integrals is at hand, and we shall not go through it in detail. The result is

$$x = \frac{c^2}{a_0} \left[\sqrt{1 + (a_0 t/c)^2} - 1 \right] \qquad (27.5.25)$$

This may be understood more easily by isolating the radical on one side of the equation, squaring, and solving for x:

$$x = \tfrac{1}{2} a_0 t^2 - \frac{a_0 x^2}{2c^2} \qquad (27.5.26)$$

In this equation, due to the factor of c^2 in the denominator, the second term on the right will be small unless the displacement x is very large, which occurs only at large times, hence high velocities. At low velocities, then, the Newtonian result $x = a_0 t^2/2$ is obtained, while for relativistic velocities, the displacement is less than that predicted by classical mechanics due to the fact that v cannot exceed c.

Finally, if in Eq. (27.5.20) we write $a = dv/dt = (dv/dx)(dx/dt) = v(dv/dx)$, we obtain, integrating once more and using tables if we must,

$$a_0 \int_0^x dx = \int_0^v \frac{v \, dv}{[1 - (v^2/c^2)]^{3/2}} = c^2 \left[\frac{1}{\sqrt{1 - (v^2/c^2)}} \right]_0^v \qquad (27.5.27)$$

Evaluating the definite integral and solving for v^2, which requires a bit of algebra, we can obtain

$$v^2 = 2a_0 x \left[\frac{1 + \dfrac{a_0 x}{2c^2}}{\left(1 + \dfrac{a_0 x}{c^2} \right)^2} \right] \qquad (27.5.28)$$

Once more, unless the displacement—hence the velocity—is very large, we obtain the classical result $v^2 = 2a_0 x$. At relativistic velocities, the displacement becomes less than this because v can never exceed c. Plots of x versus t and v versus x are shown in Fig. 27.17.

27.6 Some Basic Facts About Nuclei

We have already discussed some of the elementary aspects of the equivalence of mass and energy. In the subsequent sections of this chapter, we shall discuss phenomena in which mass can be converted into energy, and vice versa. We have already described, in qualitative terms, the gross structure of atomic systems in terms of electrons and nuclei. The electrons occupy most of the space of the atom, as illustrated in Fig. 27.18, but yet comprise only a minute fraction of the mass. For this reason, to convert substantial amounts of mass to energy, we look toward the nucleus. In order to better understand the constitution of matter as well as the potential for mass–energy conversion, we now turn our attention to the atomic nucleus to study quantitatively some of its properties.

NUCLEAR CONSTITUENTS

The atom is known to be composed of electrons and nuclei; the nuclei were discovered by Rutherford. Nuclei, which typically have radii in the range of 10^{-12} to 10^{-13} cm (compared to 10^{-8} cm for the atomic

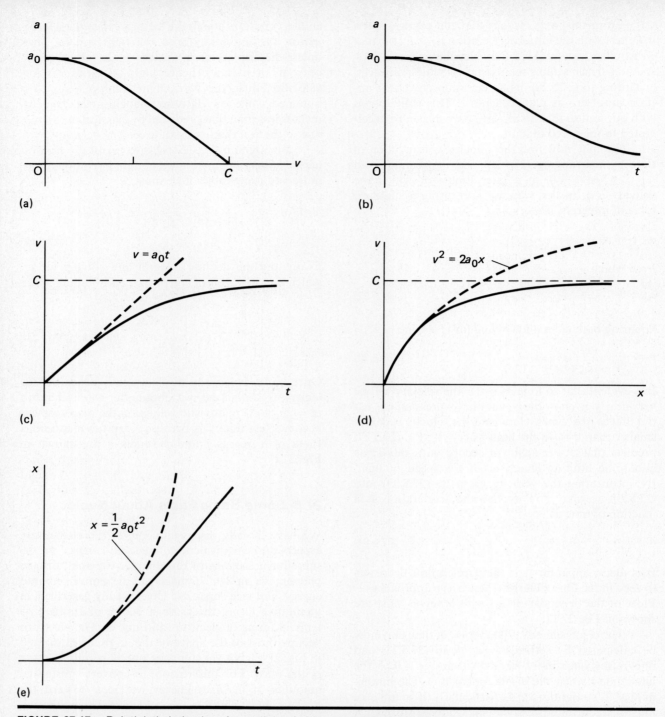

FIGURE 27.17. Relativistic behavior of a particle of given rest mass subject to a constant force. (a) Acceleration as a function of velocity, (b) acceleration as a function of time, (c) velocity as a function of time, (d) velocity as a function of displacement, and (e) displacement as a function of time.

radius), are not elementary entities. Confined to the nuclear volume are constituents consisting of protons and neutrons, collectively referred to as nucleons. The proton and neutron have very nearly the same mass, though the neutron is slightly more massive, but the proton carries a positive electric charge while the neutron is electrically neutral. A neutral atom will, therefore, contain the same number of protons and electrons. A nucleus may be identified by specifying the chemical symbol of the element as well as the number of protons Z and the total number of nucleons A. We use the notation

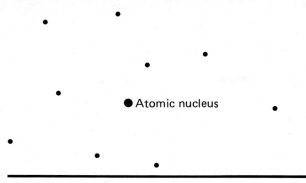

● Atomic nucleus

FIGURE 27.18. Crude schematic picture of an atom represented as a positive nucleus surrounded by electrons. This diagram is meant only to illustrate that the atom is mostly "empty space" and that the region of space in which the electrons are found is much larger than that occupied by the nucleus. It is in no way to be construed as a pictorial representation of an actual atom. Indeed, it makes no sense to try to set down a literal visual representation of an atom, since there is no way for us ever to see it with visible light as we do macroscopic objects.

$$_Z\text{El}^A$$

where

Z = number of protons

El = chemical symbol of the element (e.g., H, He, Li, etc.)

A = number of nucleons

The number of neutrons is denoted by N, where $N = A - Z$. Elements with the same Z but different A are known as *isotopes*. For example, the two isotopes of helium are

$$_2\text{He}^3 \quad \text{and} \quad _2\text{He}^4$$

In nature, $_2\text{He}^4$ comprises almost 100 percent of all helium, while the isotope $_2\text{He}^3$ occurs in very low abundance.

The neutrons and protons are confined to the nuclear volume by means of very strong, attractive forces which are not fully understood. Though elementary particle physicists and nuclear physicists have studied nucleon–nucleon forces for many years, they have been unable to express these interactions in terms of a simple force law. These strong forces bind the protons and neutrons together, and, as a result, the total energy of the nucleus is smaller than the sum of the rest energies of the constituents. This fact can be understood by calculating an equivalent mass for the nucleus. A nucleus with Z protons and N neutrons has a mass

$$M(Z, N) = ZM_p + NM_n + \frac{B}{c^2} \tag{27.6.1}$$

where M_p and M_n are the proton and neutron rest masses. The quantity B/c^2 is the mass equivalent of the nuclear *binding energy B* and is always negative; for in order to break a nucleus apart into all of its constituents, we would have to supply a positive energy at least as large as $|B|$. The quantity $M(Z, N)$ would under these circumstances simply represent the total rest mass of the isolated free nucleons. The mass of the nucleus is obviously *less* than this rest mass by virtue of the energy that must be added to separate it into its constituent particles. A rather interesting, but unfortunately only approximate, property of nuclei is expressed by the equation

$$|B| \cong (8.0)A \text{ MeV} \tag{27.6.2}$$

which states that the binding energy per nucleon is approximately 8 million electron volts. This is illustrated in Fig. 27.19, which shows the validity of the above expression, at least for $A > 20$.

For low values of Z, the number of protons and neutrons within a nucleus is approximately the same, but for high Z, nuclei tend to become "neutron rich." Figure 27.20 is a plot of neutron number versus proton number for all the elements. We see that for high atomic numbers, the ratio N/Z is about 1.6. Neutron richness can be explained in terms of the electric repulsion of protons, which results from Coulomb forces. Systems tend to form states in which the energy is as low as possible. The Coulomb repulsion between protons increases the energy and, therefore, makes it difficult, so to speak, to stuff more protons into a nucleus than more neutrons. The existence or nonexistence of specific nuclei is determined by nuclear forces and electrostatic repulsion, as well as a number of other factors.

So far, *stable* nuclei with Z greater than 83 have not been discovered. However, it has been conjectured that between $Z = 110$ and $Z = 114$, some additional stable nuclei containing about 184 neutrons may be found.

NUCLEAR STABILITY

Many nuclei are naturally unstable and decay into other nuclei through certain specific nuclear processes. Such nuclei are referred to as *radioactive*. Radioactivity was first discovered accidentally in 1896 by the French physicist Henri Becquerel, shortly after the discovery of x-rays by Roentgen in 1895. Becquerel, while studying the phosphorescence exhibited by some uranium salts after exposure to ordinary light, found that an image of the salt was produced even on a photographic plate wrapped with very thick, black paper. Subsequently, he encountered cloudy weather for a few days and, therefore, put all of his materials into a drawer to await better weather for

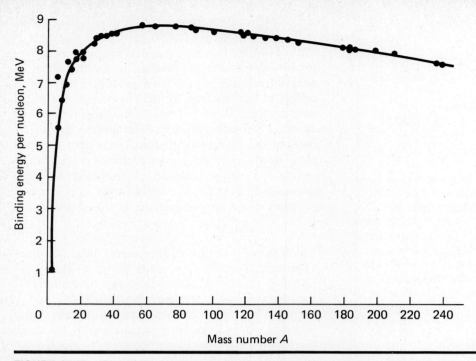

FIGURE 27.19. Plot of the binding energy per nucleon as a function of the atomic number *A*. After Arya, *Elementary Modern Physics*, Addison-Wesley, 1974, p. 359, with permission.

his observations. Much to his surprise, he found that again images of the salt were registered on the photographic plate, even though the drawer was practically light-tight! He repeated these observations in a number of ways and was forced to conclude that the emanations that were darkening the plate originated from the salt itself, independent of any light-induced phosphorescence. In this way Becquerel discovered the radioactivity of uranium.

Further investigations by Becquerel, Pierre and Marie Curie, and Rutherford led to the discovery of the new radioactive elements polonium and radium and also clarified the character of the radiations. Rutherford showed that the radiation was of two distinct types, alpha rays and beta rays, the radiation observed by Becquerel being of the latter kind. A third type of radiation, known as gamma rays, was later shown to exist.

An experimental arrangement for detecting α, β, and γ rays is shown in Fig. 27.21. A radioactive substance is almost completely surrounded by lead absorbers. A well-collimated beam of radiation is obtained by placing the radioactive material at the bottom of a small hole drilled in the lead. A photographic plate placed some distance from the top of the hole serves as a detector. The entire region contains a strong magnetic field, as shown in the figure, which is used to deflect any charged particles in the beam.

Experiments performed with this apparatus

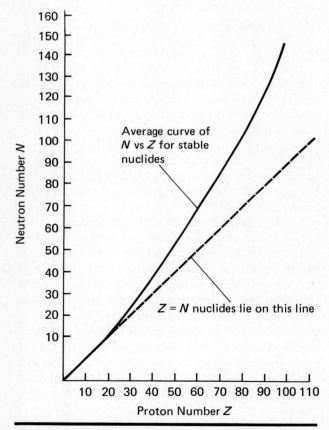

FIGURE 27.20. Neutron number vs. proton number for known atomic nuclei. After Arya, *Elementary Modern Physics*, Addison-Wesley, 1974, with permission. p. 362.

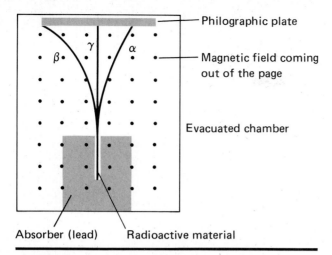

Philographic plate

Magnetic field coming out of the page

Evacuated chamber

Absorber (lead)   Radioactive material

FIGURE 27.21. Diagram showing how α-, β-, and γ-radiation emitted by a radioactive sample are detected and identified.

reveal that α rays consist of relatively massive positively charged particles, β rays consist of very light negatively charged particles, while γ rays are electrically neutral. Further investigation, including measurements of the amount of deflection, leads to the conclusion that α rays are *doubly charged helium nuclei*. They thus consist of two protons and two neutrons bound together by strong nuclear forces. The β rays were shown to be ordinary *electrons* ejected from the nucleus. The γ rays turned out to be *electromagnetic radiation* of very high frequency. As we shall see later, γ rays and other electromagnetic waves can be described in terms of bundles of energy known as photons. Thus, γ rays are photons emitted from the nucleus.

It is reasonable to ask whether α, β, and γ rays are always contained within the nucleus from which they are emitted. Experiments have ruled this possibility out. In the case of α particles, it is true that the parent nucleus already contains the neutrons and protons from which the α particle is formed, but there is no evidence to suggest that the α particle always exists as an independent entity within the nuclear volume. There is, rather, a certain probability for formation of the α particle and some possibility of its subsequent escape from the nucleus. In the case of β and γ rays, these particles never exist as such within the nucleus but are created during nuclear transformations and subsequently move away from the new nucleus. If a nucleus P decays by some radioactive process into a nucleus D, P is referred to as the *parent* nucleus and D the *daughter* nucleus.

The theory of α, β, and γ radiation is beyond the scope of the present text, for it involves a good deal of knowledge about quantum mechanics and its application. The discussion we shall give is, therefore,

very brief and qualitative. In alpha nuclear decay by alpha particle emission, the reaction can be described by the relation

$$_{z}\mathrm{P}^{A} \rightarrow \,_{z-2}\mathrm{D}^{A-4} + \,_{2}\mathrm{He}^{4} \qquad (27.6.3)$$

In order for such a reaction to occur, the rest energy of the alpha particle plus the rest energy of the daughter nucleus, must be less than the rest energy of the parent nucleus. The difference between the initial and the final rest energies shows up in the form of kinetic energy of the alpha particle and the daughter nucleus. This energy, denoted by Q, is called the *disintegration energy* for the reaction. Thus,

$$Q = [M(Z, A) - M(Z - 2, A - 4) - M(_{2}\mathrm{He}^{4})]c^{2}$$
$$(27.6.4)$$

must be greater than zero if an alpha decay is to occur. For a typical decay such as $_{90}\mathrm{Th}^{228} \rightarrow \,_{88}\mathrm{Ra}^{224} + \alpha$, the disintegration energy is about 5.4 MeV.

According to classical Newtonian physics, the process of radioactive alpha decay cannot occur. We can understand this qualitatively in terms of the following argument; imagine that $_{2}\mathrm{He}^{4}$ can form inside of some given nucleus. This particle is now subjected to a force that varies with its distance from the center of the remaining nucleus. Figure 27.22 shows a typical potential energy curve representing the interaction of $_{2}\mathrm{He}^{4}$ with the nucleus. When the alpha particle is formed within the nucleus, it is confined between $r = 0$ and R and has an energy that is always below the top of the potential barrier which is shown in the drawing. Now, if the alpha particle is to escape from the nucleus, it must get from $r = R$ to $r = R'$. However, in the region between these two radii, the potential energy exceeds the alpha particle's total energy. In classical mechanics, such a region is called a *forbidden region*, since any particle in such a region would have to have negative kinetic energy. This is not possible according to the laws of classical physics!

But in quantum mechanics, which we shall discuss in the next chapter, there is a certain probability for the particle system to enter the classically forbidden region and to emerge at $r = R'$ as a free entity. This quantum mechanical process, known as *tunneling*, accounts for alpha particle decay and is also important in many other areas of physics, such as solid-state physics. When alpha particles emerge from a nucleus, they are usually brought to rest within a short distance as a result of collisions with the atoms of the material medium within which they are formed. As a result of the fact that they are massive charged particles, they are usually brought to rest after traveling only a very short distance. These particles may,

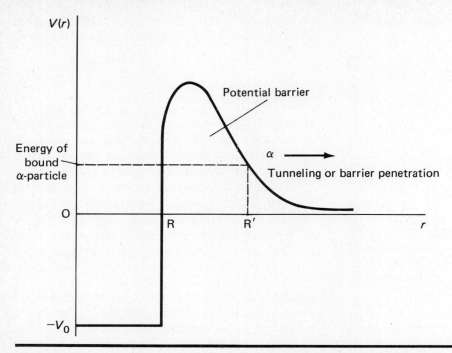

FIGURE 27.22. Potential energy diagram of an α-particle in the "potential well" of an atomic nucleus.

therefore, be absorbed by a thin metal foil or even a sheet of paper.

Beta rays are much more penetrating than alpha rays. They arise, ordinarily, from a second type of radioactive decay process, in which a neutron is transformed into a proton and an electron within the nucleus. The electron escapes from the nucleus as a "β ray", while in view of the fact that there is now one less neutron and one more proton in the nucleus, the nucleus is transformed into a new nuclear species of atomic number one unit *larger* than that of the original nucleus. The fundamental process which occurs is due to the "weak" interaction force. Early measurements of beta decay indicated that for a particular species, the emerging beta rays, which were at length identified as electrons, could have a continuous spectrum of energies. If the reaction taking place is given by

$$_zP^A \rightarrow _{z+1}D^A + e^-$$

with no other decay products, from conservation of momentum and energy, it is expected that the electron must always emerge with a well-defined energy. The observed continuous electron spectrum created great consternation among physicists, some of whom were actually prepared to abandon conservation of energy in nuclear processes. The matter was finally resolved by Wolfgang Pauli in 1930 when he postulated the existence of a new particle known as the *neutrino*,

denoted symbolically by v. He reasoned that the neutrino could carry off some of the released energy and the electron the remainder. In this way, the electron energy did not have to assume the same value in each decay. The postulation of the neutrino led to predictions regarding the energy spectrum of the emitted beta particles that were in agreement with what is observed experimentally.

In 1934, Enrico Fermi (Fig. 27.23) presented a theory for beta decay by introducing the so-called

FIGURE 27.23. Enrico Fermi (1901–1954).

weak interaction force. This theory leads to the prediction that the reaction

$$n \rightarrow p + e^- + \bar{v} \qquad (27.6.5)$$

can occur within the nucleus or even in free space. In this equation, the symbol $\bar{v}$ refers to the *antineutrino*, which is the neutrino's antiparticle. The relation between this particle and the neutrino is similar to that which exists between the positron and the electron, or between the proton and the antiproton. The proton remains within the nucleus, thus leading to a new nucleus, while the electron and antineutrino escape. The escaping electron is readily detected as "β radiation," but the antineutrino can travel through enormous amounts of matter without leaving any evidence of its having been there at all. Only in recent years has it proved feasible to carry out experiments involving detection of neutrinos. Although the neutrino and the antineutrino were detected in 1930, it was not until 1957 that Frederick Reines and Clyde Cowan firmly established their existence experimentally.

Positively charged electrons, or positrons, can also be emitted from a nucleus in beta decay. In this process, a proton is converted into a neutron, a positron, and a neutrino according to the reaction

$$p \rightarrow n + e^+ + v \qquad (27.6.6)$$

Although this reaction, sometimes referred to as inverse beta decay, is not energetically possible in free space, it can and does occur inside the nucleus. Again, the positron and the neutrino escape from the nuclear volume.

The gamma radiation from the nucleus is very penetrating. Since it is electrically neutral, its interaction with matter is significantly less than that of beta rays. Gamma rays are very much like very short-wavelength x rays and can penetrate several inches of lead before being substantially absorbed. Since gamma radiation is a form of electromagnetic radiation, its origin can be understood through an analogy with the radiation emitted by atoms. It is known that atoms contain electrons that can exist in different energy states. When an electron drops from a state of higher energy to one of lower energy, electromagnetic radiation is emitted. The frequency of this radiation can be in the visible part of the spectrum. We see that this kind of radiation is a direct consequence of atomic transitions from one possible energy state to another. The same conclusion can be reached concerning gamma radiation. Gamma rays arise when a *nucleus* makes a transition from an excited state of high energy to a state of lower energy. The gamma ray energy corresponds to the energy

difference between the two nuclear states. Typically, this may be a few million electron volts.

27.7 Radioactive Decay

Radioactive decay of nuclei can occur spontaneously. Any radioactive sample contains an enormous number of nuclei each of which has the same decay probability in any given time interval. The amount of time needed for any given fraction of the initially present nuclei to decay can vary greatly depending on the species. Some nuclei can decay substantially in a few microseconds, while others may take billions of years.

Let N_0 denote the number of radioactive nuclei present in a sample at time $t = 0$, and let N denote the number present at a later time t. During a short time interval dt, the number of decays $|dN|$ is proportional to dt as well as to the number of nuclei present, that is, N. These facts can be expressed by the equation

$$dN = -\frac{1}{\tau} N \, dt \qquad (27.7.1)$$

where the constant of proportionality has been written as $1/\tau$. The negative sign is inserted since dN will represent the change in the number of radioactive nuclei, and clearly this must be negative.

To find the number present at an arbitrary time t, we first rewrite the preceding equation as

$$\frac{dN}{N} = -\frac{dt}{\tau} \qquad (27.7.2)$$

and then integrate to find

$$\int_{N_0}^{N} \frac{dN}{N} = \ln \frac{N}{N_0} = -\int_0^t \frac{dt}{\tau} = -\frac{t}{\tau} \qquad (27.7.3)$$

Solving for N, we find

$$N = N_0 e^{-t/\tau} \qquad (27.7.4)$$

This basic decay law, which gives the number of radioactive nuclei which remain, is seen to be exponential, as shown in Fig. 27.24. It is evident that the quantity τ represents the time required for the number of the surviving nuclei to be reduced to a fraction $1/e$ of its initial value. This quantity is referred to as the radioactive decay constant. Let us now ask how long it takes for *half* of the radioactive nuclei to decay. Substituting $N = N_0/2$, we find

$$N = \frac{N_0}{2} = N_0 e^{-t/\tau}$$

or, solving for t,

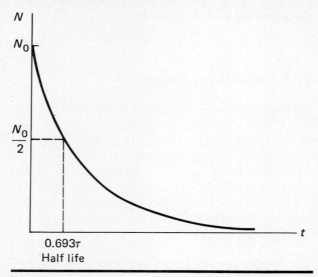

FIGURE 27.24. Plot of the radioactive decay law (27.7.4).

$$t = \tau \ln 2 = 0.693\tau \qquad (27.7.5)$$

The time τ is, therefore, the half-life divided by 0.693.

In practice, one never measures the number of nuclei N that remain at time t but, instead, the number that decay or disintegrate each second. This provides some measure of the strength or intensity of the radiation. An appropriate measure of radioactivity is the so-called *curie* (abbreviated Ci), defined, somewhat arbitrarily, by

$$1 \text{ curie} = 3.7 \times 10^{10} \text{ disintegrations/sec} \qquad (27.7.6)$$

The curie is a very good quantitative measure of the radioactive decay rate, but it does not provide a means of assessing the effect of this radiation on living matter. To determine that kind of information, we need to know how much *energy* per unit time would be absorbed by living tissue. The unit used for this purpose is the *rad*, which is defined by

$$1 \text{ rad} = 100 \text{ erg/g} \qquad (27.7.7)$$

A few hundred rads delivered to a human body in a period of about one week would be a fatal radiation dose. The background of cosmic radiation to which we are all exposed amounts to about 0.15 rad/year. Commercial nuclear reactors must be built under strict regulations which require that the general public living nearby should receive no more than 0.5 rad/year. There is no simple relation between the curie and the rad, because the kind of radiation given off in the disintegration process, in particular, its penetrating power, is quite crucial in determining how much will be absorbed by human tissue.

Although the commercial use of radioactive processes, like the use of fire or electricity, involves certain risks, especially with regard to long-range genetic mutations, there are many uses for radioactive

materials. First, of course, they are used in nuclear power generation. They can also be used to destroy cancerous cells and are thus of great value in the medical profession. An example of this is provided by a radioactive isotope of iodine which accumulates in the thyroid gland and can be used to destroy malignant cells. There are an enormous number of industrial uses for radioactive isotopes used as so-called *tracers*. For example, if the pistons of an automobile are made slightly radioactive, engine wear can be tested by measuring the radioactivity accumulated in the lubrication oil. When the engine is in operation, the pistons rub against the cylinder walls and small amounts of metal are worn away. They are carried away by the oil, and subsequently the radioactivity of the oil can be measured using a Geiger counter. The level of activity of the lubricant provides a good quantitative estimate of the wear that has taken place. Such tracer isotopes are also useful in agricultural science. By placing small amounts of a radioactive isotope into fertilizer, the farmer can measure the mineral intake of his crop and thereby determine with great accuracy how much additional fertilizer to use.

Radioactive dating constitutes another important scientific use of radioactivity. Let us consider the example of the carbon-14 dating, discovered by W. F. Libby in 1952. This method is used to determine the age of organic matter which may have died thousands of years ago. The method makes use of the reaction

$$_7N^{14} + n \rightarrow {}_6C^{14} + p \qquad (27.7.8)$$

Our atmosphere contains a great deal of nitrogen. Due to constant cosmic ray bombardment, there is a substantial flux of neutrons incident on the atmosphere. Some of these neutrons interact with the nitrogen by this reaction to form the radioactive carbon isotope $_6C^{14}$. The ordinary carbon isotope $_6C^{12}$ is much more abundant, but, nevertheless, as a result of the above reaction, approximately one millionth of 1 percent of the carbon nuclei in the atmosphere are radioactive. The carbon in the atmosphere forms the molecule carbon dioxide, which is taken up by plants. Since animals eat plants, both ordinary carbon and radioactive carbon are contained in all living organisms. While the organism is alive, the ratio of the two carbons remains fixed at some equilibrium value, since as the carbon 14 contained in the organism decays, it is replenished by the intake process. Now, once the organism dies, it ceases to take in any new carbon, but the carbon 14 already there decays back to nitrogen 14 by beta decay. The half-life for this decay is 5730 years. By measuring the number of carbon-14 disintegrations per second, therefore, one can determine when the organism died.

EXAMPLE 27.7.1

The carbon-14 activity of living organisms is about 12 disintegrations per minute for each gram of carbon. An archaeologist discovers an organic relic containing 180 g of carbon. He finds that the relic has an activity of 1500 disintegrations per minute. When did this organism die?

According to (27.7.4), if the organism had N_0 carbon-14 nuclei at the time it died, it will have

$$N = N_0 e^{-t/\tau} \qquad (27.7.9)$$

at time t. Since the half-life is 5730 years, we have from (27.7.5)

$$\tau = \frac{5730}{0.693} = 8268 \text{ years}$$

Differentiating (27.7.9) above, we obtain the disintegration rate as

$$\frac{dN}{dt} = -\frac{N_0}{\tau} e^{-t/\tau}$$

or

$$\left(\frac{dN}{dt}\right)_t = \left(\frac{dN}{dt}\right)_{t=0} e^{-t/\tau} \qquad (27.7.10)$$

We assume that the disintegration rate when the organism was alive is the same as that of live organisms today. This is reasonable, since cosmic ray fluxes presumably have not changed and, therefore, the fractional amount of carbon 14 has not varied. Using the information given, we have

$$\frac{1500}{180} = 12 e^{-t/8268}$$

$$8.33 = 12 e^{-t/8268}$$

$$\frac{t}{8268} = \ln \frac{12}{8.33} = 0.365$$

$$t = 3018 \text{ years}$$

EXAMPLE 27.7.2

The half-life of $_{88}\text{Ra}^{226}$ is 1600 years, the decay taking place by alpha emission. What fraction of the radium decays in 1000 years?

We have $N = N_0 e^{-t/\tau}$, where

$$\tau = \frac{1600}{0.693} = 2308 \text{ years}$$

Starting with an amount N_0, the amount remaining after 1000 years, as given by (27.7.9), is

$$N = N_0 e^{-1000/2305} = 0.648 N_0$$

Thus, the fraction which has decayed is

$$\left(\frac{N_0 - 0.648 N_0}{N_0}\right) = 0.352$$

27.8 Nuclear Fission

Nuclear fission is a process that occurs only in very heavy nuclei. It results in the splitting of a nucleus into two, or perhaps more, parts of comparable mass. In each fission process an energy of the order of 200 MeV is released. Otto Hahn and Fritz Strassman, working in Germany, first discovered the fission process experimentally in 1939. They bombarded U^{235} with slow neutrons and found, to their astonishment, that product nuclei such as Ba^{139} and La^{140} were present. The mystery of this splitting was explained by Lise Meitner and Otto Frisch, who had fled Nazi Germany and were working in Sweden.

To understand why fission leads to substantial energy release, let us examine Fig. 27.19 a curve which gives the binding energy per nucleon as a func-

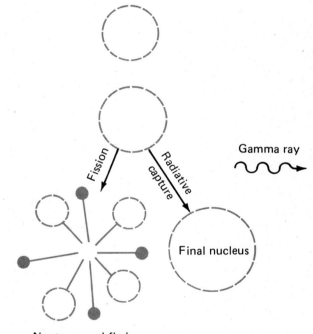

Gamma ray

Fission

Radiative capture

Final nucleus

(a) Neutrons and fission products, plus energy

(b)

FIGURE 27.25. "Liquid drop" model of nuclear fission. (a) Successive stages illustrating (top to bottom) original nucleus, excited nucleus after absorption of a neutron, and possible decay modes by nuclear fission and by radiation of energy as a gamma ray. (b) Hypothesized "dumbbell" configuration of nucleus just prior to fission. After Arya, *Elementary Modern Physics*, p. 473, with permission.

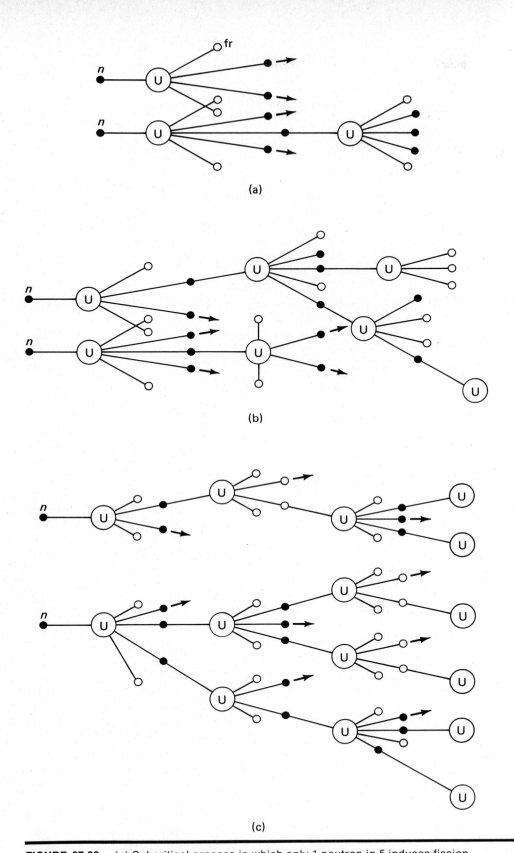

(a)

(b)

(c)

FIGURE 27.26. (a) Subcritical process in which only 1 neutron in 5 induces fission (the others escaping or being absorbed) and in which each fission event produces, on the average, 2.5 neutrons. (b) Critical self-sustaining reaction in which 1 neutron in 2.5 induces fission and in which each fission produces an average of 2.5 neutrons. (c) Explosive chain reaction, in which 6 out of 10 neutrons induce fission.

FIGURE 27.27. The Oconee Nuclear Station of the Duke Power Company near Clemson, South Carolina, is one of the largest commercial nuclear fission generating plants in the world. Photo by the authors.

for example, if four U^{235} nuclei undergo fission as a result of slow neutron capture, ten new neutrons will be produced. Unless an unchecked chain reaction is to occur, it will be necessary that no more than four of these neutrons be available to initiate new fission events. The remaining neutrons must be absorbed by cadmium control rods, by the structural material of the system, by the remaining U^{238} in the fuel, or by other processes. Under these conditions, the reactor is said to be *critical*. It may then generate huge amounts of energy in the form of heat within the reactor core. This heat is used to produce steam which runs a conventional turboelectric generator. Fig. 27.27 is a photograph of a large commercial nuclear fission power plant.

The amount of power that can be extracted from a small amount of pure U^{235} is staggering, as the following example illustrates.

EXAMPLE 27.8.1

How much energy can be extracted from 0.5 kg of U^{235}, assuming that all of it undergoes fission?

We first calculate the number of fissionable nuclei in the sample. Five hundred grams, or 0.5 kg, contains

$$\left(\frac{500}{235}\right)(6.02 \times 10^{23}) = 12.8 \times 10^{23} \text{ nuclei}$$

For each fission, there is a release of approximately 200 MeV, or 3.2×10^{-11} joules. Since a joule is the same as a watt-sec, the total energy released is

$$U = (12.8 \times 10^{23})(3.2 \times 10^{-11}) = 4.1 \times 10^{13} \text{ watt-sec}$$
$$= 4.1 \times 10^{10} \text{ kilowatt-sec} = 1.14 \times 10^7 \text{ kWh}$$

This is enough energy to operate about 12,500 100-watt light bulbs continuously for an entire year.

27.9 Nuclear Fusion

The sun converts over 600 million tons of hydrogen into helium each second, releasing an enormous amount of energy. The reaction responsible for this vast energy release is known as *nuclear fusion*. It is a *thermonuclear* reaction, which occurs only because the temperatures in the interior of the sun are extremely high, of the order of 100,000,000°K. Under such circumstances, the average kinetic energy of an atom, which is of the order of magnitude of $3kT/2$, is about 40,000 electron volts. There will be a certain number of atoms, however, whose energy, just by chance, may be 5, 10, or even 20 times larger than this average value. These atoms have sufficient thermal energy to initiate nuclear reactions. Such reactions are, therefore, referred to as thermonuclear processes. There is no doubt that without solar nuclear fusion, none of us would be here to wonder how it works. Moreover, since our fossil fuels as well as our fissionable materials are at best resources of limited availability, we shall probably have to rely on controlled nuclear fusion for our energy needs in the distant future. At present, many scientists are studying the problem of developing fusion power plants, but thus far the difficulties associated with this task have not been surmounted.

To understand why energy is released in fusion and to appreciate some of the difficulties involved, we shall start with a simple discussion on the fusion process itself. In this reaction, the basic idea is to fuse together two or more light nuclei to form a more massive nucleus. In the process of fission, the exact opposite is attempted, but both processes have a very important property in common, namely, in each case the reaction leads to nuclei that have *larger* binding energies per nucleon than the starting material. Thus, in both cases there is an excess of mass which finally appears in the form of energy.

One type of fusion process, known as the proton–proton cycle, proceeds in three steps:

$$_1H^1 + _1H^1 \rightarrow _2H^2 + \beta^+ + \nu$$
$$_2H^2 + _1H^1 \rightarrow _3He^3 + \gamma \qquad (27.8.1)$$
$$_3He^3 + _3He^3 \rightarrow _2He^4 + 2_1H^1$$

The overall process leads to an energy release of 18.8 MeV. In order for nuclear fusion to occur, however, the protons ($_1H^1$) must have appreciable kinetic energy. If they are not sufficiently energetic, Coulomb repulsion will prevent the protons from getting close enough to interact by means of the nuclear force. For example, suppose we require that the protons get to within 5×10^{-13} cm of one another. At this distance, there is a mutual potential energy of

$$U = \frac{1}{4\pi\varepsilon_0}\frac{e^2}{R} = (9 \times 10^9)\frac{(1.6 \times 10^{-19})^2}{5 \times 10^{-15}}$$

$$= 4.6 \times 10^{-14} \text{ J} = 0.288 \text{ MeV}$$

When the protons are very far apart, their potential energy is essentially zero and their total energy is all kinetic energy. If both protons move toward each other with the same speed, each will need a kinetic energy of about 0.144 MeV to achieve the required closeness of approach. This energy corresponds to a temperature of more than a billion degrees Kelvin. Since there are always some protons that have substantially more energy than the average, however, the reaction will proceed at temperatures that are lower than this by a factor of about 10. At such temperatures, the atoms are all fully ionized and we have essentially a gas of electrons and nuclei. This kind of gas is known as a *plasma.*

Uncontrolled nuclear fusion can be initiated by using an atomic fission bomb to achieve the high temperatures needed to cause fusion. This is the mechanism used for detonating a hydrogen bomb. There is no other way, at present, of igniting the thermonuclear reaction.

To develop controlled thermonuclear fusion as a source of energy, a way must be found to attain high temperatures and to confine the reacting plasma once the reaction starts. All known materials vaporize at the temperatures required to initiate fusion reactions, and, therefore, the problem of containing the plasma is a very difficult one. Even if one could find a confining material which does not vaporize it would quickly cool the plasma below the temperatures needed for fusion as soon as thermal contact is established. One confinement technique that has been attempted is magnetic confinement of the plasma to a region of space in which no material is present. In this technique, strong magnetic fields are used to exert forces on charged particles and thereby constrict their motion. Another method that is being extensively investigated involves the use of high-power lasers to ignite the fusion reaction.[9]

Thus far, successful controlled thermonuclear fusion has not become a reality. In spite of much research, it may take several more decades before the appropriate scientific breakthroughs are made, but when it happens, there will be reason to rejoice. Nuclear fusion is, conceivably, the ultimate source of energy, for it could furnish our power needs for billions of years at a much higher level of consumption than prevails at present. The reason for this happy state of affairs is that one of the simplest fusion reactions involves the use of deuterium ($_1H^2$) and tritium ($_1H^3$), both of which can be obtained from ordinary water. Nuclear fusion could, therefore, provide enormous quantities of electrical power with an unlimited supply of fuel. There would be no very large potential hazard associated with radioactive waste materials and no appreciable danger of nuclear accident. An additional bonus would stem from the fact that the fusion process can be used to vaporize, and thus elimate, harmful materials such as industrial waste products.

As mentioned earlier, it is not known when or if controlled thermonuclear fusion can be achieved, although most scientists are optimistic about the long-range prospects. Nevertheless, with our supply of natural gas and oil so limited and the hazards of mining coal so severe, we must look in the short run toward fission power plants for our future energy needs. At present, other alternatives do not appear to be economically viable, although that situation could change in the future.

[9] William E. Gough and Bernard J. Eastlund, The prospects of fusion power, *Sci. Amer.* (Feb. 1971).

SUMMARY

The theory of relativity requires that the laws of physics must be the same in all reference systems. It further requires, as a fundamental postulate, that the speed of light c in free space must have the same value to all observers, irrespective of the motion of their reference systems or of the source.

These postulates required that we replace the Galilean coordinate transformations relating reference systems in relative motion by the Lorentz transformation, which has the form

$$t' = \frac{t - \frac{vx}{c^2}}{\sqrt{1 - (v^2/c^2)}} \qquad t = \frac{t' + \frac{vx'}{c^2}}{\sqrt{1 - (v^2/c^2)}}$$

and

$$x' = \frac{x - vt}{\sqrt{1 - (v^2/c^2)}} \qquad x = \frac{x' + vt'}{\sqrt{1 - (v^2/c^2)}}$$

$$y' = y \qquad\qquad y = y'$$
$$z' = z \qquad\qquad z = z'$$

When written this way, it is assumed that the primed system moves with velocity v along the x-axis of the unprimed system. The Lorentz transformation suggests that lengths of moving objects are contracted in the direction of motion and that clocks in motion appear to run slow. It is also found that events that are simultaneous to an observer in one system may not be so to an observer in another system.

The Lorentz transformation implies that the velocity V' of an object observed to have velocity V in the unprimed system is

$$V' = \frac{V - v}{1 - (vV/c^2)}$$

where v is the velocity of the primed system with respect to the unprimed system.

The theory of relativity predicts that mass and energy are equivalent to one another, their relation being expressed by the famous "Einstein equation"

$$U = mc^2$$

It also predicts that the energy of a particle of rest mass m_0 moving at a speed v is

$$U = \frac{m_0 c^2}{\sqrt{1 - (v^2/c^2)}}$$

It is evident that a very large amount of energy ($m_0 c^2$) resides in the rest mass of an object. The momentum of a particle of velocity $\mathbf{v}$ can be expressed as

$$\mathbf{p} = \frac{m_0 \mathbf{v}}{\sqrt{1 - (v^2/c^2)}}$$

Finally, the variation of energy and momentum with speed can be interpreted from the point of view of a relativistic increase of mass with velocity such that

$$m = \frac{m_0}{\sqrt{1 - (v^2/c^2)}}$$

where m_0 is the rest mass. The above equations then can be written as

$$\mathbf{p} = m\mathbf{v} \qquad \text{and} \qquad U = mc^2$$

It is also possible to show that

$$U^2 = p^2 c^2 + m_0{}^2 c^4$$

The constituents of the atomic nucleus are known to be protons and neutrons. The notation $_Z\text{El}^A$ is used to identify a given nucleus; in this scheme, El is the chemical symbol of the element, Z is the number of protons it contains (equal to the atomic number), and A is the total number of nucleons—protons plus neutrons—sometimes referred to as the nucleon number. There are very strong attractive forces between nucleons, which serve to bind the nucleus together, but these forces are significant only at very small distances.

Radioactive nuclei decay by both alpha and beta emission processes. Alpha decay proceeds according to the reaction

$$_Z\text{P}^A \rightarrow {}_{Z-2}\text{D}^{A-4} + {}_2\text{He}^4$$

where P and D refer to the parent and daughter nucleus. The helium nucleus $_2\text{He}^4$ is referred to as an alpha (α) particle. The beta decay reaction is

$$_Z\text{P}^A \rightarrow {}_{Z+1}\text{D}^A + e^- + (\nu \text{ or } \bar{\nu})$$

where the particles ν, $\bar{\nu}$ represent neutrinos or antineutrinos, of essentially zero rest mass, which nevertheless possess angular momentum and allow this quantity to be conserved in the interaction.

The law of radioactive decay, giving the number of radioactive nuclei $N(t)$ remaining after time t, is

$$N(t) = N_0 e^{-t/\tau}$$

where N_0 is the number originally present and τ is the radioactive decay constant of the initial nuclear species. Energy is released from nuclear reactions, a corresponding mass decrease taking place. Both fission reactions, in which very heavy nuclei are split into two nearly equal parts with the emission of neutrons, and fusion reactions, in which hydrogen or deuterium nuclei fuse to form helium nuclei, produce their energy in this way.

QUESTIONS

1. What distinguishes inertial reference frames from non-inertial frames of reference? Can you suggest some experiments to show whether you are in a noninertial reference frame?
2. What is the principle of relativity, and why is it so important?
3. Can you explain why the notion of a rigid body is not a valid relativistic concept?
4. Lack of simultaneity is one of the consequences of the Lorentz transformation. Is there any possibility that events which are causally related in one reference frame may not bear this same relation in another frame?
5. The special theory of relativity predicts that objects in motion are contracted in the direction of motion. Describe the difficulties you would have in designing an experiment to verify this prediction.
6. Two events occur at point A at time t_A in some reference frame. Will they be simultaneous in other reference frames?
7. According to the theory of relativity, the Doppler shift of light is the same whether the source moves toward the receiver or the receiver moves toward the source. This is not true for sound waves, however. What is the explanation for the difference in behavior between light and sound?
8. A massless particle must move at the speed of light, whereas a massive particle must move at a speed less than the speed of light. Can you explain why this must be true?
9. The discovery of the theory of relativity had a more profound impact on classical mechanics than it did on classical electrodynamics. Give some reasons that support the truth of this statement.
10. In your opinion, why are the ideas of relativity difficult for many people to grasp? What arguments would you present to convince nonbelievers?
11. Clocks that are in motion slow down. Estimate the importance of this effect at jet airplane speeds during a 24-hour time interval.
12. Give some examples which illustrate conversion of mass to energy and other examples to show that energy can be transformed to mass.
13. A nucleus like that of U^{235} is highly unstable and readily undergoes radioactive decay, while the nucleus of O^{16} is, as far as we can tell, completely stable. Is there any fundamental difference between these two nuclei or only a difference in degree of instability?

PROBLEMS

1. Verify explicitly that the expressions for γ, $\bar{\gamma}$, a, and $\bar{a}$ in section 27.3 satisfy Eqs. (27.3.10).
2. A stationary observer in space observes two spaceships traveling toward him from opposite directions. One of the spaceships (S_1) appears to have a speed of $0.6c$ while the second (S_2) has a speed of $0.8c$. At what speed does an observer on S_2 see spaceship S_1 approaching?
3. Two galaxies recede from an observer in opposite direc-

tions with a common but unknown speed v. If the relative speed of the galaxies is $0.6c$, find v.
4. A cube has a volume of L_0^3 when it is at rest. What volume does it appear to have when viewed from a coordinate system moving at $0.5c$ along one of the edges of the cube?
5. A standard meter stick appears to have a length of 50 cm when viewed from a coordinate frame in motion parallel to the meter stick. What is the speed of the meter stick in this coordinate system?
6. The mean life of muons in their own rest frame is $\tau = 2 \times 10^{-6}$ sec. This means that if $N(0)$ denotes the number of muons at $t = 0$, then $N(t) = N(0)e^{-t/\tau}$ gives the number which are present at a later time t. Assume that $N(0) = 10^4$ and that the muons are moving at a speed of $0.99c$. What is the observed lifetime of these fast-moving muons? How many muons remain in the beam after they have traversed a distance of 1 km?
7. A clock travels at a speed of 3×10^6 meters/sec for an entire year as viewed by an observer fixed on Earth. Find the number of seconds by which it deviates from a clock fixed on Earth.
8. A group of space travelers goes on a journey from Earth to Sirius, a very bright star 8.5 light years away. Their speed is $0.95c$. Find the Lorentz contracted distance between Earth and Sirius.
9. A space–time event occurs in reference frame K at $x = 40$ meters, $y = 0$, $z = 0$, and $t = 10^{-8}$ sec. K' is a reference frame with a velocity of $0.8c$ along the positive x-axis of K. Find the space–time coordinates of the event in K' if the x-, y-, and z-axes of both reference frames are parallel.
10. Our galaxy is approximately 10^5 light years in diameter. How long would it take for a proton with energy 10^{18} electron volts to cross the galaxy as determined from a rest frame in the galaxy, and from a rest frame fixed on the proton?
11. There is a tunnel 11.2 km long connecting France and Italy. If a motorist driving at 90 km/hr passes through it, how much shorter than 11.2 km does the tunnel appear to him?
12. A train T_1 travels west at $0.75c$ relative to the station while a second train T_2 travels east at $0.90c$. Assuming east to be the positive direction, find (a) the velocity of T_2 with respect to T_1 and (b) the velocity of the train station with respect to T_2.
13. An electron is accelerated to a very high speed at the Stanford Linear Accelerator Center. If its energy is 5 billion electron volts, what speed does it have? Express your answer as a fraction of the speed of light.
14. By what percentage does the earth's relativistic mass exceed its rest mass when its motion about the sun is taken into account?
15. A proton and an antiproton each have an energy of 80 billion electron volts in the laboratory, and they are moving toward each other. Find the speed of each in the laboratory reference frame and also the speed of the antiproton as seen from a reference frame attached to the proton.
16. An electron and a positron move toward each other with the same speed as viewed in the laboratory. They each

have a kinetic energy of 10 million eV and a rest energy of 0.51 MeV. **(a)** Find their speeds in the laboratory frame. **(b)** What is the speed of the positron as seen from the electron rest frame? **(c)** The electron and positron annihilate producing two gamma rays. Find the energy and momentum of each gamma ray.

17. Starting with (27.5.18) shows that when $pc \ll m_0 c^2$ the energy U can be expanded in a Taylor series as

$$U = m_0 c^2 + \frac{P^2}{2m_0} - \frac{P^4}{8m_0{}^3 c^2} + \cdots$$

Is this relation in conflict with the expansion given by (27.5.17)?

18. For a supersonic airplane flying at 1500 mph, find the percent error made in computing its kinetic energy by using the nonrelativistic approximation.

19. A particle of rest mass m_0 has a kinetic energy U. Prove that its momentum is $p = \sqrt{2m_0 U + (U/c)^2}$.

20. A proton has a kinetic energy of 2 BeV. Using the expression in the preceding problem, find its momentum. (BeV = billion electron volts).

21. A newly discovered particle is found to have a momentum of p_1 and an energy U_1. When the momentum is increased to $2p_i$, the energy becomes $2U_1$. Show that the particle has zero rest mass.

22. The energy and momentum of a particle moving in an inertial frame K are U and p, respectively. K' is a second inertial frame which moves with speed v along the positive x-direction of K. The energy and momentum in K' are U' and p', respectively, where $U' = \gamma(U - vp)$ and $p' = \gamma(p - (v/c^2)U)$. Show that $U'^2 - c^2 p'^2 = U^2 - c^2 p^2$. By choosing a frame of reference in which the particle is at rest, show that this invariant quantity is $m_0{}^2 c^4$.

23. Show that the relativistic kinetic energy $U_k = (m - m_0)c^2$ can be expressed in the form

$$U_k = \frac{m}{m + m_0} mv^2$$

which closely resembles the nonrelativistic form $\frac{1}{2}mv^2$. [See Donald E. Fahnline, Parallels between relativistic and classical dynamics for introductory courses, *Amer. J. Phys.* **43**, 492 (1975).]

24. The element $_{92}U^{238}$ has a mass of 238.0508 in atomic mass units (amu), where 1 amu = 1.6604×10^{-27} kg. Atomic mass units are defined in such a way that the mass of the neutral atom $_6C^{12}$ is 12.0000 . . . amu. An isotopic mass expressed in amu is therefore numerically equal to the gram molecular mass. Find the total nuclear binding energy per atom in atomic mass units and in MeV. What is the average binding energy per nucleon? Compare this with Eq. (27.6.2).

25. The uranium isotope in the preceding problem is known to have a nuclear radius given by the approximate formula $R = 1.1 \times 10^{-13} A^{1/3}$ cm. Using this expression find the radius of its nucleus. Also, work out the average density of its nuclear matter in kg/m^3 and in tons per cubic inch.

26. Boron 12 ($_5B^{12}$) decays into carbon 12 ($_6C^{12}$) with the emission of an electron. This is an example of beta decay. If the Q-value of this reaction is 13.37 MeV, find the atomic mass of boron 12. (Hint: Refer to section 27.6.)

27. The half-life of radioactive cobalt 60 ($_{27}Co^{60}$) is 5.26 years. What fraction of the cobalt remains in 10 years? in 100 years?

28. A radioactive sample decayed at the rate of x disintegrations per second when measured in 1971. Five years later the same team measured a disintegration rate of $x/3$. What is the half-life of the sample? Assume that a single radioactive species is present.

29. Assuming the universe was created ten billion years ago, what fraction of the $_{92}U^{238}$ originally produced is still present. The half-life is 1.42×10^{17} seconds.

30. For each mole of gasoline an energy of 11000 cal/gm is available when the gasoline is burned. Find the amount of energy that could be extracted from 1 gram of U^{238} in the fission process. Also determine the energy which could be extracted per gram of water assuming that all the hydrogen is available for fusion.

31. **(a)** A 1.0 farad condenser is charged to a voltage of 1000 volts. By what amount is its mass increased? **(b)** To what voltage would a 1.0 μf condenser have to be charged to experience a mass increase of one electron mass?

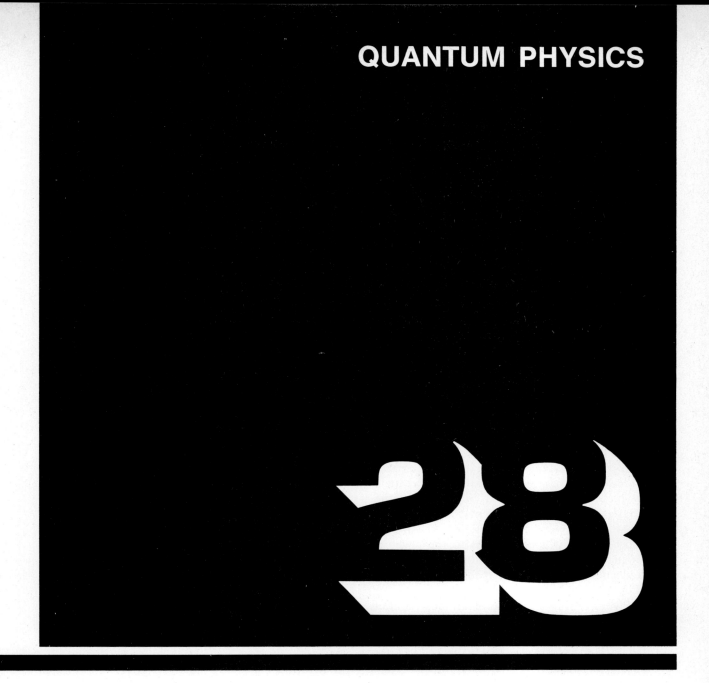

28.1 Introduction

The work of Einstein on the theory of relativity has led to profound changes in our basic notions of space and time. It shows that we cannot extrapolate Newtonian mechanics to bodies moving at arbitrary speeds, for at velocities comparable to the speed of light the intuitive Galilean coordinate transformation is at variance with the laws of nature. The lesson to be learned from Einstein is that one must be prepared to accept revolutionary ideas which appear to contradict human experience if those ideas ultimately lead to a more satisfactory explanation of observations and experiments. The theory of relativity presented a new framework upon which the subsequent development of

science would be erected. It appears to contain universal truths to be admired by both expert and amateur alike. Although many of the pressing problems of the day were indeed resolved by Einstein's genius, there were still other troubles to which relativity theory did not address itself directly.

The laws of Newtonian mechanics, together with Maxwell's equations, account for the motion of macroscopic bodies under the influence of applied forces, for the behavior of electric and magnetic fields and their relation to charges and currents, and for wave motion, including the propagation of visible light and other forms of electromagnetic radiation. The attempts to extrapolate these laws to the microphysics domain and thereby to explain the ultimate behavior of matter on the atomic scale, however, failed to yield agreement with a variety of observations made in the latter part of the nineteenth century. Blackbody radiation and the photoelectric effect were two outstanding examples of phenomena that led to great consternation among physicists, for the experimental data could not be understood within the classical framework. Physicists were also unable to explain satisfactorily the sharp spectral lines emitted by atoms in incandescent gas discharges.

The stage was set for a new set of ideas and hypotheses, and, in time, a second revolution took place in which a new and more general scheme of mechanics, called *quantum mechanics*, emerged. Quantum mechanics incorporates the laws of classical mechanics and electromagnetism, but with significant changes and generalizations that are needed to successfully explain and understand the physics of atomic and molecular systems. The theory that emerged, developed largely between 1900 and 1930, was spectacularly successful in describing the behavior of atoms, molecules, nuclei, and elementary particles. It introduced also some profound and far-reaching philosophical implications in much the same way as the theory of relativity did. Max Planck first introduced the basic ideas of quantum theory, but in its subsequent development, most of the great twentieth century physicists, among them Bohr, Rutherford, Born, Heisenberg, Schrödinger, Dirac, Pauli, Wigner, and Fermi, were involved. In addition, Einstein himself made important contributions to the development of quantum mechanics right at its inception and exhibited an active interest in it for half a century thereafter, frequently playing the role of critic or devil's advocate opposite some of its most famous proponents.

In this chapter, we shall describe some of the shortcomings of classical physics as it existed in 1900. We shall then briefly explore the principal ideas of quantum mechanics and discuss several simple applications.

28.2 Blackbody Radiation and Planck's Hypothesis

One of the most serious shortcomings of classical physics arose from its inability to describe the spectrum of *blackbody radiation*. Quantum physics began in 1900 with a successful explanation of this phenomenon put forth by the German physicist Max Planck. We, therefore, begin our study of this subject by discussing the spectrum of blackbody radiation.

It is known that any heated object can emit a continuous spectrum of radiation. The important physical characteristics of this radiation have already

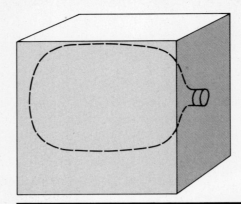

FIGURE 28.1. "Blackbody" radiator.

been discussed in a preliminary way in connection with the subject of heat transfer by radiation in section 12.5. It will probably be helpful at this point to reread that discussion. In the example of an incandescent light bulb, which consists of a tungsten filament in a vacuum or in an inert gas, the radiation emitted is mostly in the visible region of the spectrum. This light can be analyzed for the purpose of determining the intensity of the radiation emitted at various wavelengths. It is found that the spectrum depends on the temperature as well as the kind of material emitting the light. It is possible, however, to construct an *ideal emitter* known as a *blackbody*. Such a blackbody radiator can be fashioned from almost any heat-resisting substance by forming it into a hollow enclosure containing a small hole through which radiation can enter or leave the interior cavity. This radiator has the property of absorbing essentially all the radiation incident on the hole and is, therefore, an ideal absorber. Kirchhoff showed that the properties of such a device are independent of the material from which it is made and that it will emit and absorb radiation incident upon it equally well. Figure 28.1 illustrates a blackbody radiator.

Now let us assume that $I(v)$ describes the intensity of radiation emitted from the blackbody at the frequency[1] v. The quantity

$$I(v)\, dv = \frac{\text{energy}}{(\text{area})(\text{time})(\text{frequency bandwidth})} \times dv$$

gives the amount of energy radiated per unit area per unit time in the range of frequency with frequency between v and $v + dv$. $I(v)$ may be referred to as the *spectral distribution*, and $I(v)\, dv$, as the *spectral power per unit area*. It was found experimentally that the function $I(v)$ for a blackbody radiator is totally in-

[1] We have heretofore used the symbol f for frequency, but in deference to long-established custom in quantum physics, in this chapter we shall use the somewhat less descriptive Greek symbol v instead.

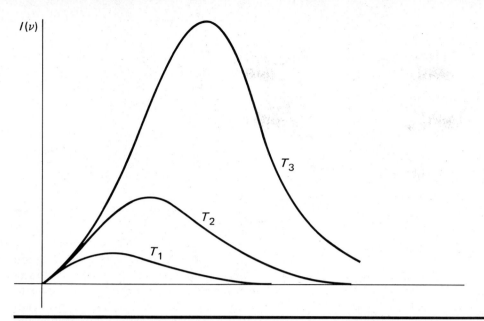

FIGURE 28.2. Curves of spectral intensity distribution for blackbody radiation at different temperatures ($T_3 > T_2 > T_1$).

dependent of whether the surrounding walls are made of copper, iron, tungsten, or anything else. Thus, there exists a unique function $I(v)$ for *all* blackbodies, regardless of the material used in the walls. However, the spectral distribution $I(v)$ does depend on the temperature of the body, which is assumed to be in thermal equilibrium with its surroundings. Thus, for all blackbody radiators, there is a universal function I which varies only with v and T and is, therefore, expressible as a function $I(v, T)$. The problem of predicting the precise form of this function constituted a major challenge to physicists at the end of the nineteenth century.

The experimentally determined form for $I(v)$ at different fixed temperatures T has the features illustrated in Fig. 28.2. $I(v)$ is found to have a maximum value at some frequency; this frequency increases with increasing temperature. A satisfactory theory would have to predict the exact shape of these curves for fixed T as well as the correct temperature dependence.

The classical theory of blackbody radiation was based on classical kinetic theory together with the classical theory of electromagnetic radiation. It was believed that the radiation within the cavity could be described by assuming that the cavity contained many standing waves. Therefore, it was necessary merely to establish the number of standing waves at a given frequency and the energy contained in each standing wave. This led to the Rayleigh–Jeans classical formula for blackbody radiation given by

$$I(v) \, dv = \frac{8\pi v^2 kT}{c^3} \, dv \qquad (28.2.1)$$

The quantity v^2 arises from the number of possible standing waves or modes, while the factor kT gives the average energy per mode. The above expression did not fit the data. This is illustrated in Fig. 28.3 where it is seen that the Rayleigh–Jeans formula agrees very poorly with experiment at high frequencies. In fact, the integration of (28.2.1) over all frequencies obviously yields an infinite result for the total intensity of radiation emitted by the body. This came to be known as the *ultraviolet catastrophe.*

The Rayleigh–Jeans law was, therefore, quite unsatisfactory, particularly for high frequencies. A more appropriate formula for this region was advanced by Wien, as follows:

$$I(v) \, dv = c_1 v^3 e^{-c_2 v/kT} \, dv \qquad (28.2.2)$$

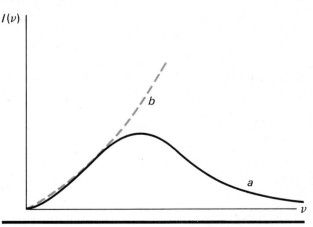

FIGURE 28.3. Plot of (a) experimentally observed spectral intensity distribution and (b) prediction of the Rayleigh–Jeans formula.

935

FIGURE 28.4. Max Planck (1858–1947).

where c_1 and c_2 are constants chosen to fit the experimental data at high frequencies. Unfortunately, this formula did not lead to good agreement with experiment for *small* values of v.

Max Planck (1858–1947) (Fig. 28.4) was able to resolve these difficulties by proposing an empirical formula which described satisfactorily all of the features of blackbody radiation. He suggested an expression of the form

$$I(v)\,dv = \frac{8\pi h}{c^3}\,\frac{v^3}{e^{hv/kT}-1}\,dv \qquad (28.2.3)$$

to fit the blackbody radiation. This equation, known as Planck's formula, reduces to the Rayleigh–Jeans law for low frequencies and to the Wien expression in the high-frequency limit. The constant h, known as *Planck's constant*, was experimentally determined to be

$$h = 6.63 \times 10^{-34}\,\text{J-sec} = 6.63 \times 10^{-27}\,\text{erg-sec}$$

$$(28.2.4)$$

At the time of Planck's early work, this constant was not known as accurately as it is today.[2]

---

[2] The presently accepted best value is $h = (6.626167 \pm 0.000038) \times 10^{-34}$ J-sec. See E. Richard Cohen and B. N. Taylor, *J. Phys. Chem. Ref. Data* **2**, 663 (1973).

How did Planck justify his famous radiation law, which he first discovered empirically? He assumed that the cavity walls contained many tiny electromagnetic harmonic oscillators which could emit radiation into the cavity and also absorb energy from it. He found, however, that in order to derive Eq. (28.2.3) from basic physical principles, he was forced to make two rather bold and controversial assertions about the behavior of these oscillators. These assertions were:

a. Oscillators of frequency v can have only *discrete* energy values given by

$$E_n = nhv \qquad n = 1, 2, 3, 4, \ldots \qquad (28.2.5)$$

where n is a positive integer. These oscillators are said to be *quantized*. The discrete states of allowed energy are referred to as *quantum states*, or *stationary states*.

b. The oscillators do *not* radiate continuously in the fashion of a classical accelerated charge but rather by jumping from one stationary state to another, Thus, the energy radiated will be equal to the energy lost by the oscillator in changing from one stationary state to another. That is, the energy radiated in a transition from a state of energy $n_2 hv$ to a state of lower energy $n_1 hv$ will be

$$\Delta E = (n_2 - n_1)hv = hv\,\Delta n \qquad (28.2.6)$$

This energy is radiated in the form of discrete parcels or trains of light waves, each of which carries an amount of energy equal to hv. These parcels, or *quanta* of light energy, are referred to as *photons*.

On the basis of these assumptions, Planck was able to derive Eq. (28.2.3). This marked the beginning of the development of modern quantum theory. It should be observed, however, that Planck still viewed the energy contained within the cavity in terms of classical electromagnetic waves, and thus he quantized only the oscillators that give rise to the radiation. Planck himself was not too happy with the *ad hoc* character of his own theory and sought for many years to provide a more rational classical basis for it. In the end, however, he had to be content with his original proposals.

28.3 The Photoelectric Effect

Before the turn of the nineteenth century, it was known that when light in the visible or ultraviolet part of the electromagnetic spectrum is directed toward a metallic surface, electrons can be ejected from the surface. The occurrence of this phenomenon, known as the *photoelectric effect*, is easy to understand in a qualitative way. It is known that incident electromagnetic radiation implies the existence of electric and magnetic fields. These fields, principally the electric field, can exert forces on the electrons within the metal, thus causing their emission from the metal.

Physicists were surprised, however, to find that certain aspects of photoelectric emission did not conform to the conventional classical view of what should happen. The following observations were not at all understood:

a. For a fixed frequency of incident light, it was found that no matter how *intense* the incident radiation, the maximum electron kinetic energy is always the same. However, the *number* of electrons emitted does increase with intensity.

b. If the incident radiation has too low a frequency, *no electrons at all* are ejected, regardless of the intensity of the incident light. For frequencies above the minimum frequency at which emission occurs, the maximum kinetic energy of the electrons that are expelled increases linearly with increasing light frequency.

c. The photoelectrons are emitted within less than 10^{-9} second after the surface is illuminated.

On the basis of classical electromagnetic theory, it is expected that light of high intensity carries a good deal of energy, leads to large electric fields, and, therefore, produces large forces on the metallic electrons. According to this view, electrons should be emitted at any frequency if the intensity is high enough. There is, in fact, every reason to expect that high intensity implies that electrons can be ejected with high energies. The almost instantaneous ejection of electrons is also hard to understand classically, since one expects that at very low light intensities, it takes time to accumulate enough energy to remove an electron from the surface. The qualitative observations, as noted above, were established prior to 1905 by P. Lenard. In 1905, Einstein proposed an explanation which reinforced the quantization ideas of Planck and gave further impetus to the developing new theory.

Einstein asserted that the electromagnetic radiation of frequency v contained packets of energy $E = hv$, where h is the constant proposed by Planck. He viewed the total energy of the incident radiation as the sum of the energies of these quantized packets, or *photons*. The electrons within a metal are bound, or confined, within the material, and it requires a certain minimum energy U_0, known as the *work function*, to cause photoelectric ejection. Einstein asserted that when an electron is ejected, it absorbs *all the energy of a single photon*. Since energy is conserved in the process, the photon energy must be equal to the sum of the work function and the maximum kinetic energy of the emitted electron. Thus,

the famous equation

$$hv = U_k^{(max)} + U_0 \qquad (28.3.1)$$

was proposed by Einstein and constituted the basis for his explanation of the photoelectric effect.

The proposed explanation implies that every ejection process occurs because of some microscopic event in which an electron absorbs a photon and then leaves the surface. To understand how the observations are explained by this model, we use (28.3.1). If the frequency of the incident radiation is too small, in particular, less than U_0/h, then the photons have an energy smaller than the work function. The electron may absorb the photon, increasing its energy within the substance; but in the process it does not acquire sufficient energy to leave the metal. We see, therefore, that there is a certain *threshold frequency* and that incident radiation at frequencies below this threshold will not lead to any electron ejection. Above this threshold, electron ejection does occur, but no matter how great the intensity of the radiation, the energy carried by each photon is still going to be hv. Thus, the maximum kinetic energy of the ejected electrons is always $hv - U_0$, even though the number of photons is increased. Above the threshold frequency, this maximum energy increases linearly with frequency. Also, of course, the more photons in the incident radiation, the greater the number of electrons that can be emitted, since more elementary events involving photon absorption can occur.

It should be pointed out that the kinetic energy appearing in (28.3.1) is the *maximum* kinetic energy. Electrons can leave the surface with energy less than this maximum value since they may lose some of their energy in collision processes while escaping from the metal. The reason for the rapidity of the photoelectric ejection can also be understood, for it takes very little time for the electron to absorb a photon and to leave the surface. Einstein's explanation of photoelectric emission, therefore, does not require a time interval during which energy is accumulated in the metal.

In 1916, R. A. Millikan carried out a series of careful quantitative measurements that confirmed Einstein's theory in every respect. Figure 28.5 shows a schematic view of the experiment which actually constituted a photoelectric determination of Planck's constant. Figure 28.6 is a schematic graph of some of the observations of the photoelectric effect. All of the features are clearly understood in terms of Einstein's photon theory.

It should be mentioned that although Planck had assumed quantization of the oscillators in the cavity walls, he held firm to the wave theory for the radiation within the cavity. Einstein, on the other hand, now proposed a *corpuscular, or quantum, nature*

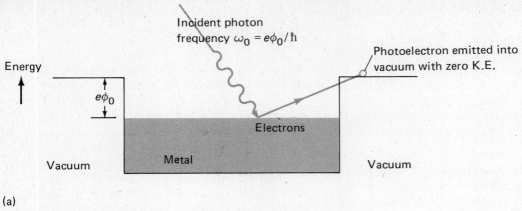

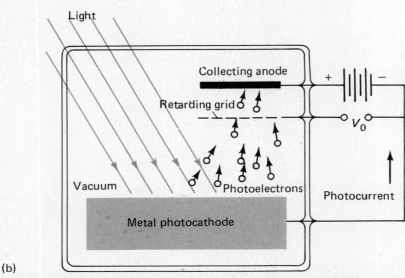

FIGURE 28.5. (a) Conceptual basis of Einstein's theory of the photoelectric effect. (b) Experimental apparatus used to study the photoelectric effect. From John P. McKelvey, *Solid State and Semiconductor Physics*. Harper & Row, New York 1966.

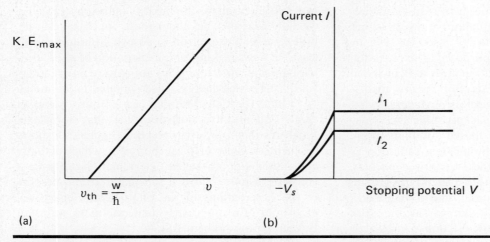

FIGURE 28.6. Experimental observations of the photoelectric effect. (a) Plot of maximum kinetic energy of photoelectrons as a function of light frequency. (b) Plot of observed photocurrent vs. retarding potential for several levels of incident light intensity. From John P. McKelvey, *Solid State and Semiconductor Physics*, Harper & Row, New York, 1966.

for light itself by assuming that it consists of bundles of quantized energy. It is important to note that the same constant h used to *quantize the oscillators* is also used to *quantize electromagnetic radiation*.

EXAMPLE 28.3.1

A macroscopic harmonic oscillator has a 1-kg mass attached to an ideal spring. The spring constant is equal to 4 newtons/meter. It is stretched 0.5 meter from its equilibrium position and released. Find its energy classically. If its energy is quantized according to Planck's scheme, in which $U_n = nh\nu$, find the quantum number n.

In classical mechanics, the frequency of the oscillator is given by

$$\nu = \frac{1}{2\pi}\sqrt{\frac{k}{m}} = \left(\frac{1}{2\pi}\right)(2) = 0.32 \text{ Hz}$$

and its energy is

$$U = \tfrac{1}{2}kA^2 = (0.5)(4)(0.5)^2 = 0.5 \text{ J}$$

Since the quantum mechanical energy is given by $U_n = nh\nu$, we have

$$U_n = (n)(6.63 \times 10^{-34})(1/\pi) = 0.50 \text{ J}$$

Thus, the quantum number n is the extraordinarily large number

$$n = 2.37 \times 10^{33}$$

When quantum numbers are so large that the energy difference $h\nu$ between the adjacent quantized energy states is small compared to the system's total energy U_n, then for most purposes the allowed energies can be regarded as continuous rather than discretely quantized. We would, therefore, be justified in using classical Newtonian mechanics to describe an oscillator like this one.

EXAMPLE 28.3.2

A monochromatic light source at 5500 Å emits 5 watts of radiation. How many photons are emitted each second?

Suppose that N photons per second are emitted. Then, since each of them has energy $h\nu$,

$$5 \text{ W} = \frac{5 \text{ J}}{\text{sec}} = Nh\nu$$

Now, since $\lambda\nu = c$, we have

$$\nu = \frac{3 \times 10^8}{5500 \times 10^{-10}} = 5.5 \times 10^{14} \text{ Hz}$$

Therefore,

$$5 = (N)(6.63 \times 10^{-34})(5.5 \times 10^{14})$$
$$N = 1.4 \times 10^{19}$$

It is evident that this source of light emits a very large number of photons each second.

28.4 Atomic Spectra

The radiation from any source of light can be studied to determine what wavelengths or frequencies are present. We have already seen that the radiation coming from a blackbody is continuous in that it contains some light of all frequencies. An incandescent light bulb would also exhibit a similar continuous spectrum. On the other hand, many sources of light emit only discrete spectral "lines" corresponding to well-defined frequencies. Thus, when an electric discharge is excited in a sodium vapor lamp or neon tube and the light is analyzed with a spectrometer,[3] it is found to consist of a number of sharp, discrete spectral frequencies rather than a continuous spectrum.

When a source of light emits such discrete spectral frequencies, it is said to exhibit a *line spectrum*. Before the turn of the century, spectra of this variety had been extensively analyzed. In Fig. 28.7, several examples of line spectra are illustrated. Prior to the

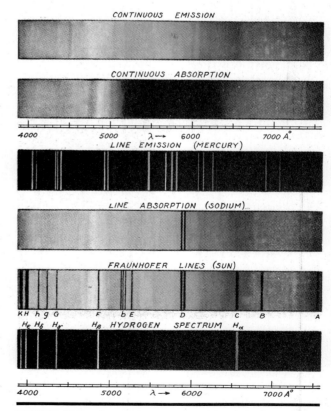

FIGURE 28.7. Several examples of line spectra.

---

[3] A spectrometer is an instrument that disperses incoming light and can be used to measure and analyze the frequencies that are present in the spectrum.

development of quantum theory, such complicated line spectra were taken as evidence of a rather involved internal structure for atoms. It was then believed that atoms had many possible modes of oscillation, like violin strings or organ pipes, and that the excitation of these modes gave rise to radiation corresponding to their individual frequencies.

Let us examine the line spectra of the hydrogen atom, which is the most elementary atomic system known. By 1885, 14 spectral lines of hydrogen had been identified and their wavelengths accurately measured. Johann Balmer then succeeded in finding an empirical mathematical expression that accurately described all these known wavelengths. He found that the observed *series* of spectral wavelengths could be expressed by

$$\frac{1}{\lambda} = R\left(\frac{1}{2^2} - \frac{1}{n^2}\right) \qquad (28.4.1)$$

where n assumed the integer values $n = 3, 4, 5, 6, \ldots$ and where R is a constant that is now generally called the *Rydberg constant*. At the present time, the best measured value for the Rydberg constant for hydrogen is

$$R = (1.097313414 \pm 0.000000083) \times 10^7 \text{ m}^{-1}$$

The accuracy of the measured value of R has steadily improved since the work of Balmer, but even in 1885 it was known reasonably well.

The series of spectral lines described by Balmer's formula is known as the Balmer series. Additional lines, found after Balmer's discovery, also agreed with the same formula. The Balmer lines are in the visible and ultraviolet parts of the electromagnetic spectrum. In 1908, Paschen found several more hydrogen lines which did not conform to the Balmer expression. These lines, which were in the infrared part of the spectrum, form the basis of the so-called *Paschen series*, given by

$$\frac{1}{\lambda} = R\left(\frac{1}{3^2} - \frac{1}{n^2}\right) \qquad n = 4, 5, 6, \ldots \qquad (28.4.2)$$

The *Lyman* and *Brackett series* were also subsequently discovered in the ultraviolet and infrared parts of the spectrum. Their wavelengths are described by

Lyman series

$$\frac{1}{\lambda} = R\left(\frac{1}{1^2} - \frac{1}{n^2}\right) \qquad n = 2, 3, 4, \ldots \qquad (28.4.3)$$

Brackett series

$$\frac{1}{\lambda} = R\left(\frac{1}{4^2} - \frac{1}{n^2}\right) \qquad n = 5, 6, 7, \ldots \qquad (28.4.4)$$

Thus it was found empirically that the discrete line spectra emitted by hydrogen could be fit by the

relation

$$\frac{1}{\lambda} = R\left(\frac{1}{m^2} - \frac{1}{n^2}\right) \qquad (28.4.5)$$

where m and n are integers. Since λ is positive, the integers m and n must satisfy the inequality $m < n$.

28.5 The Bohr Theory of Hydrogen

In 1913, the Danish physicist Niels Bohr (Fig. 28.8) developed a theory to explain the spectrums of hydrogen. The existence of the nucleus had been proposed by Rutherford, leading to the view that atoms had small but massive nuclei as well as electrons. The above numerical formulas giving the wavelengths and frequencies of the spectral lines are devoid of any clear conceptual basis and do not in themselves lead to any fundamental understanding of atoms. Bohr sought a theory that would explain the spectral lines by using the Rutherford model of the atom together with Planck's constant h. This constant had already provided a yardstick by which to measure any departure from classical physics, as illustrated by the explanation of blackbody radiation and the photoelectric effect.

Let us now consider in detail the basic assumptions of Bohr and how these lead to an explanation of the spectral lines. Bohr assumed that the electron in the hydrogen atom traveled in a *circular orbit* about the central nucleus. The Coulomb attraction between these oppositely charged particles provides the basic attractive force law needed to keep the

FIGURE 28.8. Niels Bohr (1885–1962).

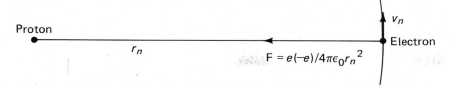

FIGURE 28.9. Coulomb attraction of the electron and the proton in the Bohr atom.

electron in the orbit. Thus, Bohr viewed the atomic problem, in which the relevant dimensions of the system are about 10^{-8} cm, in much the same way that Kepler viewed planetary motions. Both the Coulomb and gravitational force laws are of the same form, making it entirely logical that electrons might travel in these classical "Bohr orbits." A very serious deficiency in the concept of atomic orbits arises, however, when we consider the question of stability. It was known to Bohr that an electron traveling in a circular orbit is accelerated and should, according to the laws of classical electrodynamics, *radiate energy* continuously. In so doing, the electron's energy would continually decrease. This would in turn imply that the electron orbit would have to spiral in toward the nucleus instead of remaining circular.

To get around this difficulty, Bohr proposed the idea of *stationary states*. In his view, the electron could orbit in any one of many possible discrete states of well-defined energy. Such states of sharply defined energy are called *stationary states*. The electron can make transitions from one of these states to another, in the process giving up a quantum or photon of energy $h\nu$ equal to the difference of energies of the two stationary states. Thus, instead of continuously radiating electrons, we have emission of radiation *only when a change from one stationary state to another occurs.*

Let us now formulate these two ideas of Bohr mathematically in order to grasp more readily the next crucial concept he proposed. If the electron is in some initial stationary state labeled n, of total energy U_n, we have

$$U_n = \tfrac{1}{2}m_e v_n^2 - \frac{e^2}{4\pi\varepsilon_0 r_n} \qquad (28.5.1)$$

where $|v_n|$ is the electron speed and r_n is its distance from the proton, which for simplicity we assume to be at rest. The two terms comprising the total electron energy (kinetic plus potential energy) are not independent since Coulomb's force law provides a relation between the speed v_n and the distance r_n of the form

$$F = \frac{e^2}{4\pi\varepsilon_0 r_n^2} = \frac{m_e v_n^2}{r_n} \qquad (28.5.2)$$

In writing this equation, we assume that it is the

Coulomb attraction between electron and nucleus, and that force alone, that provides the centripetal force required to keep the electron in a circular orbit. This is illustrated by Fig. 28.9.

Substituting (28.5.2) into (28.5.1), we obtain an expression for U_n solely in terms of the distance r_n, which tells us that

$$U_n = -\frac{e^2}{8\pi\varepsilon_0 r_n} \qquad (28.5.3)$$

In Bohr's view, radiation is emitted in transitions from the initial state i to a final state f, and, therefore, the frequency ν of the emitted light is given by

$$h\nu = U_i - U_f = -\frac{e^2}{8\pi\varepsilon_0}\left(\frac{1}{r_i} - \frac{1}{r_f}\right) \qquad (28.5.4)$$

Since the spectral lines are known to be discrete rather than continuous, not all values of r_i and r_f are possible. Some hypothesis is needed to fix the allowable values of r_i and r_f. Since the frequency ν is inversely proportional to the wavelength λ, comparison of (28.5.4) and (28.4.5) suggests that the radii of the orbits must be proportional to the *squares of the integers*. Thus, a hypothesis that would lead to this dependence was needed.

The new and bold hypothesis put forth by Bohr to complete his theory was the *quantization of angular momentum*. Bohr asserted that the angular momentum of the electron in its orbit about the proton could not assume all possible values but instead must be limited to *integral multiples of $h/2\pi$*. Thus, he hypothesized that

$$L_n = m_e v_n r_n = n\frac{h}{2\pi} \qquad n = 1, 2, 3, \ldots \qquad (28.5.5)$$

At the time this proposal was advanced, Bohr probably did not fully appreciate its deeper significance. After the advent of the complete theory of quantum mechanics, about fifteen years later, however, the logical necessity for quantization of angular momentum had become clearly evident.

Let us now use (28.5.5) together with (28.5.2) to obtain an expression for r_n. Squaring both sides of (28.5.5) and dividing by $m_e r_n^3$, we obtain

$$\frac{(m_e v_n r_n)^2}{m_e r_n^3} = \frac{m_e v_n^2}{r_n} = \frac{n^2 h^2}{(2\pi)^2 m_e r_n^3}$$

Substituting this expression in (28.5.2) yields

$$\frac{e^2}{4\pi\varepsilon_0 r_n^2} = \frac{n^2 h^2}{(2\pi)^2} \frac{1}{m_e r_n^3}$$

or

$$\frac{1}{r_n} = \frac{m_e e^2 \pi}{n^2 h^2 \varepsilon_0} \qquad (28.5.6)$$

By substituting $n = 1, 2, 3, \ldots$, we obtain the radii for the first, second, third, etc., Bohr orbits of the hydrogen atom. Thus, setting $n = 1$, we find

$$r_1 = \frac{(1^2)(6.63 \times 10^{-34})^2(8.85 \times 10^{-12})}{(9.11 \times 10^{-31})(1.6 \times 10^{-19})^2(\pi)}$$

$$= 0.528 \times 10^{-10} \text{ m} = 0.528 \text{ Å}$$

as the *first Bohr radius*. This orbit gives the *minimum possible size of the hydrogen atom*. Since the energy varies inversely with r, the minimum radius corresponds to the largest possible binding energy. This state is referred to as the *ground state*, or the state of *lowest energy*.

Let us now substitute (28.5.6) in (28.5.4) to obtain an expression for the frequency and the wavelength of the light emitted due to atomic transitions. We find

$$h\nu = \frac{hc}{\lambda} = -\frac{e^2}{8\pi\varepsilon_0}\left(\frac{m_e e^2 \pi}{n^2 h^2 \varepsilon_0} - \frac{m_e e^2 \pi}{m^2 h^2 \varepsilon_0}\right) = U_n - U_m$$

or

$$\frac{1}{\lambda} = \frac{m_e e^4}{8h^3 c\varepsilon_0^2}\left(\frac{1}{m^2} - \frac{1}{n^2}\right) \qquad (28.5.7)$$

Also, substituting (28.5.6) into (28.5.3), we find an expression for the *allowed energy states of the hydrogen atom* of the form

$$U_n = \frac{-m_e e^4}{8\varepsilon_0^2 n^2 h^2} \qquad (28.5.8)$$

Comparing these results with (28.4.5), we see immediately that Bohr's theory leads to the correct form for the wavelengths of the observed line spectra of hydrogen. Moreover, by calculating the numerical value of the constant appearing in front of the brackets, we readily find that it is indeed the measured Rydberg constant!

The Bohr theory of hydrogen was a major step toward the development of a comprehensive theory of *quantum mechanics*. Although the classical picture of electrons moving in circular orbits has not been maintained in modern quantum theory, Bohr's assumptions regarding stationary states and of quantization of angular momentum have withstood the test of time. Bohr's theory was very favorably accepted when it was presented in 1913. In fact, Einstein instantly recognized the importance of Bohr's contribution and supported his theory wholeheartedly.

Figure 28.10 depicts the energy levels of hydrogen and illustrates transitions that give rise to emitted light. In gaseous atomic hydrogen, most of the hy-

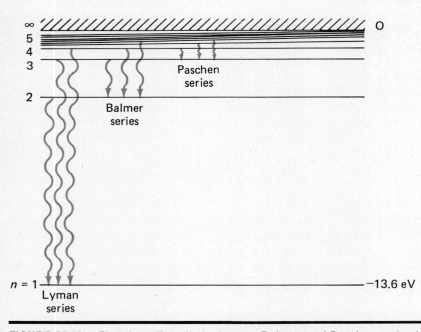

FIGURE 28.10. First three lines in the Lyman, Balmer, and Paschen series in the spectrum of hydrogen.

drogen atoms are in their ground state and have a binding energy of -13.6 electron volts. This represents the so-called *ground-state energy* of the lowest energy state U_1. An energy of $+13.6$ electron volts is the minimum amount of energy needed to ionize the atom, thus freeing the electron from the proton. Although most of the atoms are in the lowest state, they can absorb energy and can thereby be excited to higher states. Those in the higher states are unstable and decay after a certain time to the ground state, usually by emitting one photon.

It is important to point out that Bohr's hypotheses of stationary states and quantization of angular momentum are universal. For example, if we consider the orbit of the earth about the sun, then there must also exist stationary states and a quantized angular momentum. However, as the following examples illustrate, the various quantized levels are very close to one another in energy and in angular momentum, and, as a consequence, the quantum characteristics of the system are not detectable experimentally.

EXAMPLE 28.5.1

Assuming quantization of angular momentum, compute the quantum number n for the earth in its orbital motion about the sun. If the earth could make a transition to a new orbit by radiating some energy, find the fractional change of energy if n is reduced by 10.

According to quantization of angular momentum, we have for the earth's orbit

$$M_E V_E R = nh/2\pi = (5.98 \times 10^{24})(29{,}770)(149 \times 10^9)$$

$$= n \frac{6.63 \times 10^{-34}}{2\pi}$$

$$n = 2.5 \times 10^{74}$$

This number is enormous. As a general rule, whenever quantum numbers become so large, we are in the realm where the quantum aspects are not apparent and classical physics works very well.

The quantized energy levels for the earth will also be proportional to $1/n^2$, as in the case for hydrogen. Thus, the change of energy ΔU resulting from a change in n is, since $dU/dn \propto -2/n^3$,

$$\Delta U \propto -\frac{2}{n^3} \Delta n$$

from which

$$\frac{\Delta U}{U} = -2 \frac{\Delta n}{n}$$

In the present case, we have

$$\frac{\Delta U}{U} = -\frac{(2)(10)}{2.5 \times 10^{74}} = -8.0 \times 10^{-74}$$

an incredibly small and completely undetectable change.

It is worthwhile to point out that a hydrogen atom in an excited state ($n > 1$) soon makes a transition to its ground state by emitting radiation. The interaction of charged particles with the electromagnetic field allows the electron to de-excite, thus bringing the atom to its lowest energy state. The earth, in its orbit about the sun, is certainly not in the quantum level of lowest energy. In fact, it is in a very highly excited state. One might wonder whether the earth will ever go to a state of lower energy just as the electron goes to a state of lower energy. Although such a transition is not forbidden, there is really no easy mechanism for this de-excitation. The earth might emit some "gravitons," if such particles exist, thus losing a tiny amount of energy. But since the gravitational interaction is extremely weak, the probability of graviton emission is exceedingly small, and energy loss by this process is very unlikely.

EXAMPLE 28.5.2

How much energy is required to ionize a hydrogen atom in its ground state? If only 75 percent of that energy is supplied, what excited state can the electron attain and what will be the radius of its electronic orbit?

According to (28.5.8), the energy of a bound electron in hydrogen is

$$U_n = -\frac{m_e e^4}{8\varepsilon_0^2 h^2 n^2} \tag{28.5.9}$$

In order to ionize a hydrogen atom, we must supply enough energy to bring the electron energy at least to zero. Thus, we must supply (for $n = 1$) energy of at least

$$U_1 = \frac{(9.11 \times 10^{-31})(1.6 \times 10^{-19})^4}{(8)(8.85 \times 10^{-12})^2(6.63 \times 10^{-34})^2}$$

$$= 2.167 \times 10^{-18} \text{ J} = 13.6 \text{ eV}$$

This is the ionization energy for hydrogen.

The energy difference between the ground state and the state for which $n = 2$ is

$$\Delta U = -\frac{m_e e^2}{8\varepsilon_0^2 h^2}\left(\frac{1}{1^2} - \frac{1}{2^2}\right) = 0.75\left(-\frac{m_e e^4}{8\varepsilon_0^2 h^2}\right)$$

$$= -(0.75)(13.6) \text{ eV} = -10.2 \text{ eV}$$

Thus, if 75 percent of the ionization energy is provided, the electron can just be excited to the $n = 2$ state. Since, according to (28.5.6), the radius of the orbit is proportional to the square of the quantum number n, the radius of the $n = 2$ orbit is four times as large as that of the $n = 1$ orbit. This implies a radius of $r_2 = 4r_1 = 2.12 \times 10^{-10}$ meter.

28.6 De Broglie Waves

The proposal by Einstein that light waves might in some instances exhibit particlelike or corpuscular behavior was further supported by the discovery and explanation of the Compton effect. When light is scattered from the electrons in atoms, the frequency or wavelength of the deflected radiation is altered. This effect, known as the *Compton effect*, could not be explained within the context of classical ideas. It was necessary once again to invoke the photon concept and to attribute both energy and momentum to each photon. We see, therefore, that light exhibits a dual character. To explain interference and diffraction, light must be treated as a wave; but to understand the photoelectric and Compton effects, we must assume light to be corpuscular or particlelike in character.

The prevailing view concerning entities such as electrons or protons was that they were particles having strongly localized momentum and energy. They would move along classical trajectories in response to applied forces and at any instant could be thought of as occupying some rather well-defined position in space. In 1924, Louis de Broglie suggested that if light could sometimes exhibit particlelike behavior, then it was reasonable to expect that particles, in some instances, might display wavelike behavior. De Broglie knew there existed a relation between the wavelength λ of light (a wave property) and the momentum p (a particle property) of the associated photons. He asserted that the *very same relation* should also hold for electrons, protons, and indeed all objects possessing a momentum.

Starting with the relation $p = U/c$ given by Eq. (23.6.5), which relates the momentum and energy of an electromagnetic wave, and assuming as Planck did that $U = h\nu$, de Broglie reasoned that for material particles as well as light,

$$p = \frac{U}{c} = \frac{h\nu}{\nu\lambda}$$

from which

$$\lambda = \frac{h}{p}$$

This is the famous *de Broglie relation* proposed in 1924, by which the "wavelength" of any material particle could be expressed in terms of its momentum.

One may regard the wavelength λ as a measure of the extent to which wavelike behavior is present. For example, λ gives a measure of the degree of localization of the energy of the system. If λ is exceedingly small, the energy is very well localized and the particle character of the object is dominant. On the other hand, if λ is very large, the energy of the

system is diffused and distributed over a large volume. Under these circumstances, the wave behavior is quite important.

EXAMPLE 28.6.1

Find the de Broglie wavelength of an electron moving at a speed of 10^6 meters/sec. What is the de Broglie wavelength of a 2-kg ball moving at 40 meters/sec?

For the electron, we have

$$\lambda = \frac{6.63 \times 10^{-34}}{(9.11 \times 10^{-31})(10^6)} = 7.3 \times 10^{-10} \text{ m}$$

For the ball,

$$\lambda = \frac{6.63 \times 10^{-34}}{(2)(40)} = 8.3 \times 10^{-36} \text{ m}$$

We see, thus, that a macroscopic object such as a baseball has an extremely small wavelength, and, as a result, its wave character will not be discernible. The electron's wavelength, though still rather short, is not small on the scale of atomic sizes. We are, therefore, not surprised to learn that electrons frequently behave like waves in atomic processes.

If, as postulated by de Broglie, electrons and other particles have an associated wavelength, then it should be possible to verify this assertion by producing a diffraction pattern using a beam of electrons. In 1927, C. J. Davisson and L. H. Germer, working at the Bell Telephone Laboratories, succeeded in qualitatively verifying de Broglie's hypothesis. Figure 28.11 is a schematic diagram of their electron diffraction experiment. Electrons emitted by a heated metallic filament are accelerated through a potential difference ΔV and then strike a nickel crystal. The de Broglie wavelength of the electrons is readily obtained by first establishing their momenta. According to energy conservation, we may write

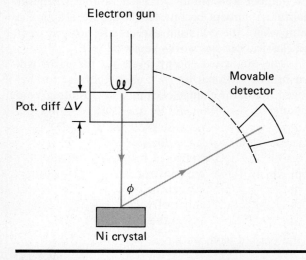

FIGURE 28.11. Schematic diagram of Davisson–Germer experiment.

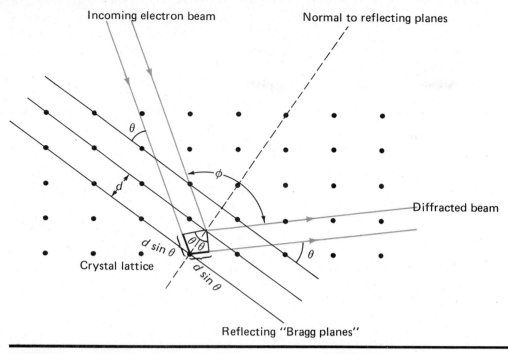

Incoming electron beam

Normal to reflecting planes

θ

φ

d

Diffracted beam

d sin θ

θ θ

θ

Crystal lattice

d sin θ

Reflecting "Bragg planes"

FIGURE 28.12. Electron diffraction, viewed as constructively interfering reflections from successive atomic planes in the crystal.

$$\tfrac{1}{2}mv^2 = \frac{p^2}{2m} = e\,\Delta V$$

or

$$p = \sqrt{2me\,\Delta V}$$

Therefore, using the de Broglie relation, we obtain

$$\lambda = \frac{h}{p} = \frac{h}{\sqrt{2me\,\Delta V}} \qquad (28.6.2)$$

Substituting values for h, m, and e, we obtain

$$\lambda = \sqrt{\frac{150}{\Delta V}} \times 10^{-10}\ \text{m} \qquad (28.6.3)$$

where ΔV is the voltage (in volts) through which the electrons are accelerated. Electrons accelerated through 150 volts should, therefore, have a de Broglie wavelength of 1 Å.

It was already known that x-rays of wavelength 1 Å could be diffracted by a crystal. This diffraction of x-rays occurs because the crystal contains certain clearly defined planes, *Bragg planes*, in which atoms are situated. The reflected light from different Bragg planes may interfere constructively, provided that the path difference between neighboring planes is equal to an integral number of wavelengths. According to Fig. 28.12, this leads to the relation

$$n\lambda = 2d\sin\theta \qquad (28.6.4)$$

known as *Bragg's law*. If electrons can also be dif-

fracted, the same relation applicable to light should hold for electrons since the electrons are also treated as waves.

In a typical measurement, in which ΔV was 54 volts, it was found that the detector registered a maximum at an angle $\phi = 50°$. Knowing this angle, it is apparent that the angle θ is 65°. Since the spacing of the reflecting planes of the crystal was 0.91 Å and n was unity, it was found from Eq. (28.6.4) that the wavelength was 1.65 Å. By applying (28.6.3), one readily finds that de Broglie's prediction is $\lambda = 1.67$ Å. Thus, the Davisson–Germer experiment not only demonstrates that electrons are diffracted but leads to an excellent quantitative confirmation of de Broglie's hypothesis. The diffraction of electrons was also confirmed by experiments done by G. P. Thomson in 1927.

In 1929, Estermann and Stern showed that atoms and molecules of helium and hydrogen were also diffracted according to de Broglie's theory. Later, it was shown that neutrons, too, are diffracted. All of this suggests the universality of de Broglie's relation, establishing a *dual character for all matter*. We are led in this way to conclude that matter and light can both exhibit wavelike as well as particlelike behavior.

The hypothesis advanced by de Broglie suggested a new interpretation for the quantization of angular momentum. In the Bohr model of the hydrogen atom, the electron moves in a Bohr orbit about the proton. If we assume that the electron is describable in terms of a wave, and if we further postulate that only an integral number of wavelengths can fit

945

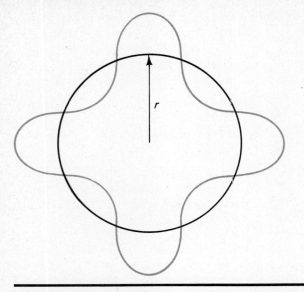

FIGURE 28.13. Quantization of angular momentum, visualized in terms of fitting standing de Broglie waves into Bohr's circular orbits. In the illustration, four de Broglie wavelengths are accommodated in the orbit.

along the Bohr orbit, as shown in Fig. 28.13, then the condition for the quantization of angular momentum follows at once. The circumference of a Bohr orbit is $2\pi r$, where r is the orbit radius. If only an integral number of wavelengths can be accommodated, this implies the relation

$$2\pi r = n\lambda = n\frac{h}{p}$$

where n is a positive integer. Multiplying both sides of this equation by p and dividing by 2π, we obtain

$$L = \frac{nh}{2\pi} \qquad (28.6.5)$$

in agreement with (28.5.5).

Quantization of angular momentum, therefore, follows from the assertion that the Bohr orbit can accommodate only standing waves. Of course, this hypothesis is perhaps no more satisfactory than the *ad hoc* assertion of quantization of angular momentum, but it nevertheless provides another way of understanding the same principle.

28.7 Schrödinger's Wave Mechanics

The hypothesis of Louis de Broglie and its subsequent verification shows that matter has wave characteristics. The problem of how to formulate a wave theory of matter in a mathematically and physically satisfying way was undertaken by Erwin Schrödinger (Fig. 28.14). The existence of various kinds of wave equations was well known to Schrödinger. After all,

one can formulate a wave equation to describe the *displacement* of parts of a string and also to describe the behavior of *electric fields* in the propagation of an electromagnetic wave. There must be, reasoned Schrödinger, some kind of wave equation to describe the propagation of de Broglie's matter waves.

What is there to take the place of a displacement or an electric field in the case of matter waves, and what is the physical meaning of such an entity? Schrödinger's answer to this question was arrived at by the postulation of a quantity known as the *wave function* $\Psi(x, y, z, t)$. It was postulated that particles such as the electron, proton, and neutron could be described by means of a quantum-mechanical *wave function* $\Psi(x, y, z, t)$ or $\Psi(\mathbf{r}, t)$, depending on the position variables x, y, z (or $\mathbf{r}$) and the time t. The precise form of this wave function will depend upon the forces acting on the particle. All attainable information about the dynamic behavior of the particle is embodied in the wave function, and any desired property of the particle's motion can, in principle, be derived from it

Before we discuss the equation satisfied by $\Psi(\mathbf{r}, t)$, we shall explain its physical significance as originally proposed by Max Born. It is known that when an electromagnetic wave propagates through space, the square of the electric field $[\mathbf{E}(\mathbf{r}, t)]^2$ provides a relative measure of the amount of energy near the point $\mathbf{r}$ at time t. In a similar way, the square of the wave function $|\Psi(\mathbf{r}, t)|^2$ might be thought of as representing the relative energy contained near $\mathbf{r}$ at time t. This interpretation is difficult to maintain since the energy of a particle such as a free electron is localized

FIGURE 28.14. Erwin Schrödinger (1887–1961).

at the electron's position and is zero elsewhere. The interpretation of $|\Psi(\mathbf{r}, t)|^2$ that is now generally accepted was suggested by Born. This interpretation invokes the notion of *probability*. Born proposed that if dV is a small volume element surrounding the point(x, y, z) described by the position vector $\mathbf{r} = \mathbf{i}_x x + \mathbf{i}_y y + \mathbf{i}_z z$, then

$|\Psi(\mathbf{r}, t)|^2 \, dV$ gives the *probability of finding the electron within that volume element.*

Since the probability that the electron will be in *some* volume element of the system is unity, we must require that the condition

$$\int_V |\Psi(\mathbf{r}, t)|^2 \, dV = 1 \qquad (28.7.1)$$

be fulfilled. This is called the *normalization condition.*[4]

It is important at this point to note that the quantity $\Psi(\mathbf{r}, t)$ itself has no direct physical significance; it is simply a mathematical quantity that satisfies an appropriate wave equation. There is no known way of measuring $\Psi(\mathbf{r}, t)$ experimentally. Its only virtue is that a knowledge of Ψ leads at once to a knowledge of the quantity $|\Psi|^2$, or $\Psi^*\Psi$. It is the quantity $|\Psi|^2$ that has the direct physical significance discussed by Born, and it is, therefore, this quantity that we can easily relate to the results of experimental studies. The wave function itself is seen, therefore, to play the role of intermediary between the wave equation and the physically significant quantity $|\Psi|^2$. An odd situation, in some respects, but one that we must learn to live with in quantum theory. The quantity $|\Psi|^2$ is frequently referred to as the *probability amplitude.*

As soon as the idea of probability is introduced, we are presuming the existence of many electrons with exactly the same wave function. We are, moreover, assuming that observations done on one electron can be repeated again and again in exactly the same manner. Only through repetition is it possible to verify the content of the wave function. An analogy can be made with a coin-tossing experiment. If a single coin is tossed, we assert that the likelihood of obtaining heads is 1/2. However, this statement certainly cannot be substantiated by tossing the coin only once. Many,

many tries are necessary in order to prove experimentally that the probability of obtaining heads is exactly 1/2. Similar conclusions can be drawn concerning information obtained from a knowledge of the quantum-mechanical wave function and the probability amplitude.

THE SCHRÖDINGER EQUATION

There is no simple way to arrive at the wave equation satisfied by $\Psi(\mathbf{r}, t)$. In the electromagnetic case, the wave equation was shown to follow directly from Maxwell's field equations. There are no laws of this sort that can be used to derive the Schrödinger wave equation, although there are some useful guiding principles. In particular, any quantum-mechanical wave equation must yield the same results as classical Newtonian mechanics in any situation where classical mechanics is known to provide a valid description of the physical system.

In 1925, a quantum-mechanical wave equation was proposed by the Austrian physicist Erwin Schrödinger. The predictions of this wave equation regarding the behavior of systems on the microscopic level are universally in agreement with all the experiments that have been performed to test them. In addition, it can be shown that Schrödinger's wave equations leads to all the predictions of classical mechanics for systems of macroscopic size and energy. We cannot "derive" Schrödinger's equation any more than we can derive Newton's laws. We must, therefore, simply state it as a verifiable fact and remark that our acceptance of it is based on considerations similar to those that lead us to take Newton's laws and Maxwell's equations for granted.

Let us, therefore, consider a particle of mass m—an electron, perhaps, though any particle will do—subject to a force derivable from a known potential energy function $U_p(x, y, z)$. Then, the dynamics of the particle is described by the solutions $\Psi(x, y, z, t)$ of the differential equation

$$-\frac{\hbar^2}{2m} \left(\frac{\partial^2 \Psi}{\partial x^2} + \frac{\partial^2 \Psi}{\partial y^2} + \frac{\partial^2 \Psi}{\partial z^2} \right)$$
$$+ U_p(x, y, z)\Psi(x, y, z, t) = i\hbar \frac{\partial \Psi}{\partial t} \qquad (28.7.2)$$

known as the *Schrödinger equation*. In this equation, $i = \sqrt{-1}$, while the symbol $\hbar$ (read "h-bar") is related to Planck's constant h by

$$\hbar = \frac{h}{2\pi} \qquad (28.7.3)$$

Ordinarily, there are many different solutions of the Schrödinger equation, representing waves of dif-

---

[4] In quantum mechanics, the wave function may be a real quantity, but it may also, more generally, be a *complex* quantity $\psi = u + iv$ with a real part u and an imaginary part v. If the imaginary part is zero, then the wave function is real and there is no question as to what is meant by the square of its absolute value. If not, however, the wave function will be complex. In this case, the real quantity $\Psi^*\Psi = (u - iv)(u + iv) = u^2 + v^2 = |\Psi|^2$ is always regarded as the definition of the square of the absolute value of Ψ. The quantity $\Psi^* = u - iv$, which has the same real part as Ψ but an imaginary part of opposite sign, is referred to as the *complex conjugate* of Ψ.

ferent amplitude, frequency, and wavelength. Some of these represent wave functions that lead to realistic physical representations of the behavior of the particle, but *others do not* and must be rejected for this reason. In general, we find that only solutions that are single-valued, continuous, and differentiable and that approach zero for large positive and negative values of x, y, and z lead to physically realistic wave functions. Of course, any such wave functions must also satisfy the normalization condition (28.7.1). These restrictions frequently define a discrete set of allowed solutions of (28.7.2), corresponding to a discrete set of allowed energies like those discussed in connection with Bohr's model of the hydrogen atom.

For a "free" electron, there is no force, and the potential energy is constant—for simplicity, we would take it to be $U_p = 0$. Under these circumstances, the Schrödinger equation relates the time and space derivatives of the wave function in a particularly simple way, which we shall examine in detail later. When forces act, the potential energy function is no longer so simple, and the task of solving Schrödinger's equation may be very difficult. One of the major problems faced by modern physicists who use quantum theory in their research is that of solving equations such as (28.7.2) as well as more complex equations describing systems in which there is more than one particle.

THE SUPERPOSITION PRINCIPLE

Let us suppose that Ψ_A and Ψ_B are two different solutions of the Schrödinger equation and that each represents a possible state of an electron. The *principle of superposition* asserts that $\Psi = c_1\Psi_A + c_2\Psi_B$, where c_1 and c_2 are constants, is also a solution of the equation and, therefore, also represents a possible state. This statement can be proved by direct substitution of the function Ψ into Schrödinger's equation. To understand the implications of this principle and how it leads to quantum-mechanical interference, we consider the diffraction of electrons through the double-slit system shown in Fig. 28.15. This problem is analogous to Young's experiment on the interference of light discussed in an earlier chapter.

The electrons in the incident beam can pass either through slit A or slit B and will subsequently cause a screen to fluoresce, indicating the arrival point of the electron on the screen. It is apparent that electrons are equally likely to pass through slit A or slit B. Those electrons which have definitely passed through slit A have wave function Ψ_A, while those which have passed through slit B have wave function Ψ_B. If we cover up slit B, then clearly all the electrons to the right of the slits must have wave function Ψ_A, and, therefore, the distribution of arrivals at the screen

will be described by $|\Psi_A|^2$. But such a distribution corresponds simply to the pattern of light intensity exhibited on the screen when light waves of the same wavelength pass through the slit. This happens because de Broglie "matter waves" are diffracted just like light waves. In this way, we may conclude that more electrons arrive at some areas of the screen than at others and that the pattern of their arrivals is the same as that formed when light passes through a narrow slit. This observation can now be repeated by covering slit A and observing the distribution of electrons passing through B, described by $|\Psi_B|^2$.

What happens if we do not try to find out through which slit the electrons pass? To answer this question, we allow both slits to remain open. Since we do not know which slit the electrons go through, we must say that the electron is in a *superposition* of states corresponding to Ψ_A and Ψ_B. This superposition is given by the wave function

$$\Psi = \frac{1}{\sqrt{2}}(\Psi_A + \Psi_B) \qquad (28.7.4)$$

Note that since the electron is equally likely to go through slit A or slit B, the coefficients appearing before Ψ_A and Ψ_B are equal. They must, in fact, have the value $1/\sqrt{2}$ in order to satisfy the requirements of (28.7.1). When it is necessary to describe the electron wave function in this way, the distribution of electron arrivals at the screen is dictated by[4]

$$|\Psi|^2 = \Psi^*\Psi = \frac{1}{\sqrt{2}}(\Psi_A{}^* + \Psi_B{}^*)\frac{1}{\sqrt{2}}(\Psi_A + \Psi_B)$$

$$= \tfrac{1}{2}(|\Psi_A|^2 + |\Psi_B|^2 + \Psi_A\Psi_B{}^* + \Psi_B\Psi_A{}^*) \qquad (28.7.5)$$

The additional term $\Psi_A\Psi_B{}^* + \Psi_B\Psi_A{}^*$, known as the *interference term*, results in a distribution of electron arrivals quite different from that obtained by combining the earlier results in which only one slit at a time was opened. Indeed, the distribution pattern now has the appearance of the *double-slit light interference* pattern. Thus, the fact that the electron must be described as a superposition of Ψ_A and Ψ_B when we do not know which slit it comes through leads to *definite observable consequences*. In fact, the interference pattern obtained and illustrated in Fig. 28.15b is present whether 50,000 electrons per second arrive at the screen or only one per hour!

We are now quite familiar with the way in which an interference pattern arises when light is incident on two slits. Basically, the interference occurs because the incident wave passes through both slits. The quantum-mechanical interference of electrons is somewhat more difficult to appreciate since we would certainly be inclined to argue that the electron passes either through A or through B. However, as we have

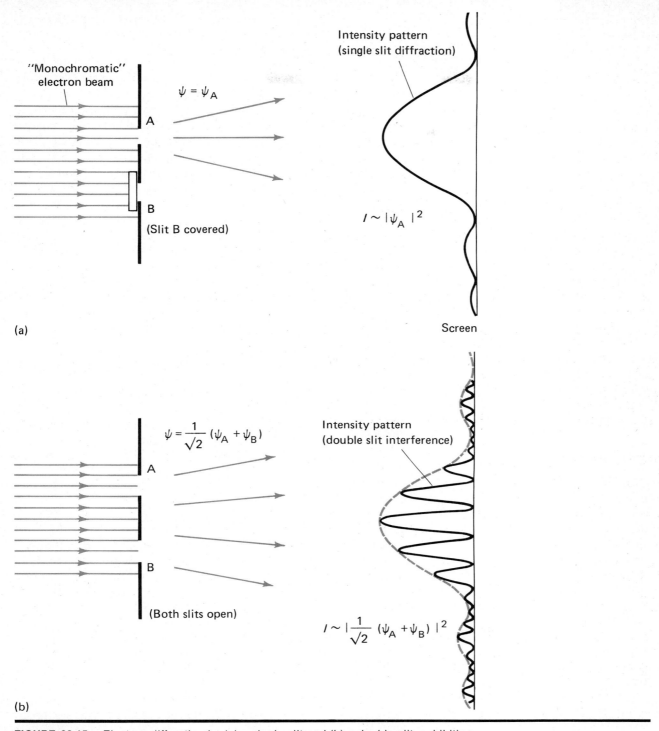

FIGURE 28.15. Electron diffraction by (a) a single slit and (b) a double slit, exhibiting interference of "matter waves."

seen, if no effort is made to ascertain which slit the electron really did go through, the wave function is a superposition of Ψ_A and Ψ_B, and the results then suggest that the electron wave passes through *both* slits.

QUANTIZATION OF ENERGY

We previously referred to the fact that in many instances solutions of Schrödinger's equation in the form of well-behaved mathematical wave functions suitable for describing the physical behavior of the system exist *only for certain specific values of total energy*. These specific values correspond to the discrete allowed energies of the system, as given by (28.2.5) for the harmonic oscillator or by (28.5.8) for the hydrogen atom. We shall now describe how and why this situation arises. In doing so, we shall avoid trying to reproduce all the details of the mathematics but merely point out the more important physical features that result.

The mathematical solutions of Schrödinger's Eq. (28.7.2) can usually be expressed as the *product* of a function $\psi(x, y, z)$ of the space coordinates of the particle and a sinusoidally varying function of the time whose frequency is related to the particle's total energy by the Planck relation

$$U = h\nu = \hbar\omega \tag{28.7.6}$$

Under these circumstances, it can be shown that the spatial part of the wave function, expressed by $\psi(x, y, z)$ must satisfy what is frequently called the *time-independent Schrödinger equation*, which has the form

$$-\frac{\hbar^2}{2m}\left(\frac{\partial^2\psi}{\partial x^2} + \frac{\partial^2\psi}{\partial y^2} + \frac{\partial^2\psi}{\partial z^2}\right) + U_p(x, y, z)\psi(x, y, z)$$
$$= U\psi(x, y, z)$$
$$\tag{28.7.7}$$

where U is the total energy of the particle. The solutions of (28.7.7), like those of (28.7.2), must satisfy the requirements imposed by (28.7.1), which means that the function $\psi(x, y, z)$ must satisfy the condition

$$\int_V |\psi|^2 \, dV = 1 \tag{28.7.8}$$

Also, as before, only solutions of (28.7.8) that are single-valued, continuous, and differentiable lead to wave functions that are capable of describing the physical behavior of the system.

In particular, there are many possible solutions of (28.7.7) corresponding to different values of the energy U. In general, these will not approach zero for large values of x, y, and z. It frequently happens, in fact, that such solutions become infinite in this limit. For solutions such as these, the normalization integral (28.7.8) cannot remain finite when evaluated over all space, and, therefore, these solutions can never satisfy the normalization requirement. For any value of energy that leads to such a situation, *there exists no physically acceptable wave function*. Under the circumstances, only certain special values of the energy lead to wave functions that can satisfy the normalization condition expressed by (28.7.8). These values are, in fact, the quantized energy levels. Thus, we may say that the probability interpretation for $|\psi|^2$ implies a

certain "good behavior" for the wave function, and this good behavior, in turn, often implies quantized energy levels.

Let us now try to learn something about energy quantization by studying a one-dimensional example. Assume that an electron is confined to move along the x-axis between $-L/2$ and $L/2$. In this region, we assume that the electron is free (that is, the potential energy $U_p(x)$ is zero) but that at the endpoints there are strong forces which absolutely prevent the electron from leaving this region. The solution of (28.7.7) is very easy to obtain in the present case, for the equation now reduces to

$$U\psi(x) = -\frac{\hbar^2}{2m}\frac{d^2\psi(x)}{dx^2} \tag{28.7.9}$$

Let us assume a solution

$$\psi_U(x) = N \sin kx \tag{28.7.10}$$

where N is a constant which we determine later. Then, substituting into (28.7.9), we find that

$$UN \sin kx = -\frac{\hbar^2}{2m}(-k^2)N \sin kx$$

Therefore, $N \sin kx$ is a solution provided that

$$U = \frac{\hbar^2 k^2}{2m} \tag{28.7.11}$$

The above solution is said to have *odd parity* since it is antisymmetric under the substitution $x \to -x$. Now, since the electron cannot get past $x = \pm L/2$, we must require that the wave function be zero for $x > L/2$ and for $x < -L/2$. Since the wave function must be continuous at all points, however, we must also require that $\psi_U(x)$ as given by (28.7.10) reduce to zero at $x = \pm L/2$. This, in turn, compels us to choose

$$\sin(kL/2) = 0$$

or

$$\frac{kL}{2} = n\pi \qquad n = 1, 2, 3, \ldots$$

We exclude the term $n = 0$ since that would lead only to the trivial solution $\psi_U(x) = 0$. The allowed energy values for the odd-parity solutions are, therefore, according to (28.7.11),

Odd parity

$$U_n = \frac{\hbar^2}{2m}\left(\frac{2\pi}{L}\right)^2 n^2 \qquad n = 1, 2, 3, \ldots \tag{28.7.12}$$

Another possible class of solutions are those given by

$$\psi_U(x) = N \cos kx \tag{28.7.13}$$

These are *even parity* solutions since ψ does not change sign when x is replaced by $-x$. Just as before,

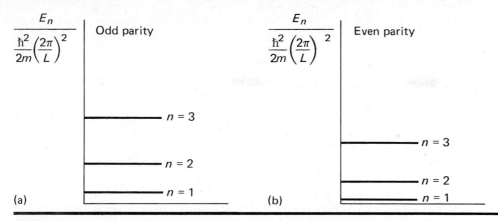

FIGURE 28.16. Allowed energy levels for a free particle confined to the region $-L/2 < x < L/2$: (a) odd-parity states, (b) even-parity states.

we find that $U_n = \hbar^2 k^2/2m$, but now the values of k follow from the condition

$$\cos(kL/2) = 0 \qquad \frac{kL}{2} = \frac{\pi}{2}, \frac{3\pi}{2}, \frac{5\pi}{2}, \ldots$$

or

$$kL = (2n - 1)\pi \qquad n = 1, 2, 3, \ldots$$

The energy values are then found from (28.7.11) to be

Even parity

$$U_n = \frac{\hbar^2}{2m}\left(\frac{2\pi}{L}\right)^2 (n - \tfrac{1}{2})^2 \qquad n = 1, 2, 3, \ldots \qquad (28.7.14)$$

We see, therefore, that the requirement that the electron be localized leads to discrete energy levels. In Fig. 28.16, the energy spectrum of the even- and odd-parity solutions is illustrated, while in Fig. 28.17 several even- and odd-parity solutions are plotted.

It is important to realize that the electron may not be in *any* of these energy states, and in this case the most general solution of the Schrödinger equation, (28.7.2), will be a superposition of allowed energy states. We could then write

$$\Psi(x, t) = \sum_{\text{all } U} C_U \Psi_U(x, t) \qquad (28.7.15)$$

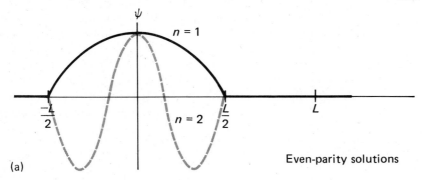

(a)   Even-parity solutions

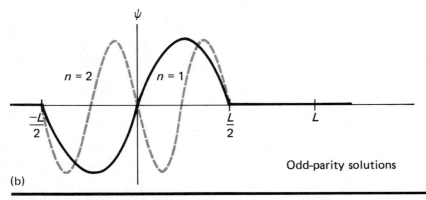

(b)   Odd-parity solutions

FIGURE 28.17. (a) Even-parity and (b) odd-parity solutions of Schrödinger's equation for a free particle confined to the region $-L/2 < x < L/2$.

The quantity $|C_U|^2$ is the probability that the electron is in the energy state U. Thus, if an ensemble or collection of electrons is prepared and we measure the energy of each electron, we expect the energies to be distributed according to the set of numbers $|C_U|^2$. Let us now consider a simple example.

EXAMPLE 28.7.1

An electron is confined between $-L/2$ and $L/2$ and is in the even-parity state of lowest energy. Find the normalization constant N of (28.7.13) and determine the probability that the electron is between $-L/2$ and $+L/4$.

For the lowest energy state, $kL = \pi$, and, therefore, the wave function is

$$\psi_U(x) = N \cos(\pi x/L)$$

Now since the probability of finding the electron between $x = -L/2$ and $x = +L/2$ must be unity, we must impose the condition (28.7.8), which in this case is

$$\int_{-L/2}^{L/2} \psi_U{}^2(x)\, dx = N^2 \int_{-L/2}^{L/2} \cos^2(\pi x/L)\, dx$$

or

$$N^2 \int_{-L/2}^{L/2} \tfrac{1}{2}(1 + \cos(2\pi x/L))\, dx$$

$$= \frac{N^2}{2}\left[x + \frac{L}{2\pi}\sin\frac{2\pi x}{L}\right]_{-L/2}^{L/2} = \frac{N^2 L}{2} = 1$$

from which

$$N = \sqrt{\frac{2}{L}}$$

Thus, the wave function of the electron is

$$\psi_U(x) = \sqrt{\frac{2}{L}} \cos(\pi x/L)x \qquad (28.7.16)$$

To find the probability of finding the electron between $-L/2$ and $+L/4$, we note that $|\psi^2(x)|\, dx$ gives the probability of finding it in an interval dx about the point x. If this is integrated between the limits $-L/2$ and $L/4$, we shall find the desired result. In this way, we can write

$$P = \int_{-L/2}^{L/4} \psi_U{}^2(x)\, dx = \frac{1}{L}\left[x + \frac{L}{2\pi}\sin(2\pi/L)x \right]_{-L/2}^{L/4}$$

$$= \frac{1}{L}\left(\frac{3L}{4} + \frac{L}{2\pi}\sin\frac{\pi}{2}\right) = \frac{3}{4} + \frac{1}{2\pi} = 0.909$$

THE EXPECTATION VALUE

A very important idea in quantum physics is the concept of the expected value of a physical quantity. The expected value of a quantity Q, more frequently referred to as its *expectation value*, is a quantum mechanical average of the quantity weighted by the probability amplitude that expresses the likelihood

of the particle being in the neighborhood of any given point. If we regard the probability amplitude $\psi^*\psi = |\psi|^2$ as a spatial probability distribution, we are led by the methods developed in Chapter 13 in reference to velocity distributions, section 13.6 and Eq. (13.6.11), to write such an average as

$$\langle Q \rangle = \frac{\int_V Q(x, y, z)|\psi(x, y, z)|^2\, dV}{\int_V |\psi(x, y, z)|^2\, dV}$$

$$= \int_V Q(x, y, z)|\psi(x, y, z)|^2\, dV \qquad (28.7.17)$$

the integrals being evaluated over the variables x, y, z that define the volume V of the entire system. The final form of this expression is obtained by using the normalization condition (28.7.8). In some instances, this will be correct, but in general, the situation is more complex and involves the introduction of *operators* that are associated with ordinary dynamical quantities such as position, momentum, and energy and can operate on the wave function to transform it into a function of different form. The operators that represent the dynamic variables of ordinary mechanics play an extremely important role in quantum theory. Indeed, in quantum mechanics, the operators are the physical entities of importance, and the expectation values are simply observable manifestations of these more fundamental constructs.

In quantum mechanics, the operators that represent the position, momentum, and energy of a particle can be written as

Observable Physical Quantity				Quantum Operator Q_{op}		
Position	x,	y,	$z \rightarrow$	$Q_{op} = x$,	y,	z
Momentum p_x, p_y, p_z			$\rightarrow$	$Q_{op} = \dfrac{\hbar}{i}\dfrac{\partial}{\partial x}$,	$\dfrac{\hbar}{i}\dfrac{\partial}{\partial y}$,	$\dfrac{\hbar}{i}\dfrac{\partial}{\partial z}$
Energy	U		$\rightarrow$	$Q_{op} = -\dfrac{\hbar}{i}\dfrac{\partial}{\partial t}$		

$$(28.7.18)$$

The differential operators operate on wave functions involved in the quantum-mechanical averaging process in such a way as to transform them into different but related functions. In the proper prescription for averaging, the probability amplitude is written in the form $\psi^*\psi$ and the operator representing the physical quantity sandwiched in between these factors, thus, $\psi^*Q_{op}\psi$. Under these circumstances, (28.7.17) must be written

$$\langle Q \rangle = \frac{\int_V \psi^*(x, y, z)Q_{op}\psi(x, y, z)\, dV}{\int_V \psi^*(x, y, z)\psi(x, y, z)\, dV} = \int_V \psi^*Q_{op}\psi\, dV$$

$$(28.7.19)$$

In the case of the particle's position, the position operators are simply the coordinates x, y, z themselves, and, therefore, the original expression (28.7.17) is valid for them. Thus, the expected position $\langle x \rangle$ for the particle in the above example is

$$\langle x \rangle = \int_{-L/2}^{L/2} \psi^* x \psi \, dx = \int_{-L/2}^{L/2} x \psi^* \psi \, dx$$

$$= \int_{-L/2}^{L/2} x |\psi(x)|^2 \, dx \quad (28.7.20)$$

Likewise, the expectation value of $x^2 = x \cdot x$ could be written

$$\langle x^2 \rangle = \int_{-L/2}^{L/2} \psi^* x^2 \psi \, dx = \int_{-L/2}^{L/2} x^2 |\psi(x)|^2 \, dx \quad (28.7.21)$$

For the momentum, however, the operator corresponding to p_x is $p_{x(op)} = (\hbar/i)\partial/\partial x$. This must *operate* on the wave function ψ in (28.7.19) and cannot be moved to the left, outside the quantity $\psi^*\psi$, because $\psi^*(\partial\psi/\partial x)$ and $\partial(\psi^*\psi)/\partial x$ *are not equal*. We must, therefore, proceed as follows in evaluating the expectation value of the momentum for the particle in the example above. For the odd-parity solutions,

$$\langle p_x \rangle = \int_{-L/2}^{L/2} \psi^* p_{op} \psi \, dx = \int_{-L/2}^{L/2} \psi^* \frac{\hbar}{i} \frac{\partial}{\partial x} \psi \, dx$$

$$\langle p_x \rangle = \frac{\hbar}{i} \int_{-L/2}^{L/2} \psi^* \frac{d\psi}{dx} \, dx$$

$$= \frac{2\hbar}{iL} \int_{-L/2}^{L/2} \sin \frac{2n\pi x}{L} \left(\frac{2n\pi}{L} \cos \frac{2n\pi x}{L} \right) dx \quad (28.7.22)$$

$$\langle p_x \rangle = \frac{4n\pi\hbar}{iL^2} \int_{-L/2}^{L/2} \sin \frac{2n\pi x}{L} \cos \frac{2n\pi x}{L} \, dx$$

$$= \frac{2n\pi\hbar}{iL^2} \int_{-L/2}^{L/2} \sin \frac{4n\pi x}{L} \, dx = 0$$

In this instance, the expectation value of the momentum is *zero*. The same result follows for even-parity solutions. This is perhaps not too surprising, because a classical particle moving with initial speed v_x in such a potential well would rebound elastically from the walls at $\pm L/2$ time after time, and its velocity would be, alternately, $+v_x$ and $-v_x$. Under these circumstances, the *average* of its velocity, and thus its momentum, is zero. The quantum calculation above is fully in accord with this expectation.

If we were asked to find the expectation value of p^2, however, we should replace the operator $(\hbar/i)\partial/\partial x$ with the operator

$$p_{op}^2 = p_{op}p_{op} = \frac{\hbar}{i} \frac{\partial}{\partial x} \left(\frac{\hbar}{i} \frac{\partial}{\partial x} \right) = -\hbar^2 \frac{\partial^2}{\partial x^2} \quad (28.7.23)$$

and write, for odd parity,

$$\langle p_x^2 \rangle = -\hbar^2 \int_{-L/2}^{L/2} \psi^* \frac{\partial^2 \psi}{\partial x^2} \, dx$$

$$= -\hbar^2 \frac{2}{L} \int_{-L/2}^{L/2} \sin \frac{2n\pi x}{L} \frac{\partial^2}{\partial x^2} \left(\sin \frac{2n\pi x}{L} \right) dx$$

from which

$$\langle p_x^2 \rangle = \frac{8n^2\pi^2\hbar^2}{L^3} \int_{-L/2}^{L/2} \sin^2 \frac{2n\pi x}{L} \, dx$$

$$= \frac{4n^2\pi^2\hbar^2}{L^3} \int_{-L/2}^{L/2} \left(1 - \cos \frac{4n\pi x}{L} \right) dx$$

Finally, integrating and substituting limits, we find

$$\langle p_x^2 \rangle = \left(\frac{2n\pi\hbar}{L} \right)^2 \quad (28.7.24)$$

Similar results can be obtained for the even-parity solutions. It is important to observe that from the above equation and (28.7.12), we find

$$\frac{\langle p_x^2 \rangle}{2m} = \frac{\hbar^2}{2m} \frac{2\pi}{L^2} n^2 = E_n \quad (28.7.25)$$

This is in precise agreement with what we expect from classical mechanics, where we would write for the motion of a free particle confined to the region $-L/2 < x < L/2$ with the constant velocity $\pm v_x$

$$E = \tfrac{1}{2}mv_x^2 = \frac{p_x^2}{2m} \quad (28.7.26)$$

The following examples will serve to further illustrate the calculation of quantum averages or expectation values.

EXAMPLE 28.7.2
An electron moving between $-L/2$ and $L/2$ is in an odd-parity energy state with energy U_n. Find the expected value of x and x^2 assuming the wave function is normalized according to (28.7.8).

The wave function is

$$\psi_{U_n}(x) = N \sin \frac{2n\pi x}{L}$$

To find N, we require, as always, that the integral of $|\psi^2|$ have the value unity. Therefore,

$$\int_{-L/2}^{L/2} \psi_{U_n}^2(x) \, dx = N^2 \int_{-L/2}^{L/2} \sin^2 \frac{2n\pi x}{L} \, dx$$

$$= N^2 \int_{-L/2}^{L/2} \frac{1}{2} \left(1 - \cos \frac{4n\pi x}{L} \right) dx$$

$$= \frac{N^2 L}{2} = 1$$

or

$$N = \sqrt{\frac{2}{L}}$$

Substituting this, along with the value of ψ_U given above, into (28.7.14) or (28.7.20), we find

$$\langle x \rangle = \int_{-L/2}^{L/2} x\psi_{U_n}^2(x)\, dx = \frac{2}{L}\int_{-L/2}^{L/2} x\sin^2\frac{2n\pi x}{L}\, dx$$

$$= \frac{1}{L}\int_{-L/2}^{L/2} x\left(1 - \cos\frac{4n\pi x}{L}\right) dx \qquad (28.7.27)$$

Since the functions x and $x\cos kx$ are equal in magnitude but opposite in sign along the $+x$- and $-x$-directions, both the above integrals vanish. This is also easily verified by actually doing the integration, using integral tables or integration by parts to evaluate the second integral. In any event, the result is that the average displacement $\langle x \rangle$ between the particle and the center of the potential well is zero. In the case of $\langle x^2 \rangle$, however, we may proceed as follows, using integral tables to evaluate the integrals:

$$\langle x^2 \rangle = \int_{-L/2}^{L/2} x^2 N^2 \sin^2\frac{2n\pi x}{L}\, dx$$

$$= \frac{1}{L}\int_{-L/2}^{L/2} x^2\left(1 - \cos\frac{4n\pi x}{L}\right) dx$$

$$= \frac{1}{L}\left[\frac{2}{3}\left(\frac{L}{2}\right)^3 - \frac{L^3}{8n^2\pi^2}\right] = \frac{L^2}{8}\left(\frac{2}{3} - \frac{1}{n^2\pi^2}\right)$$

$$(28.7.28)$$

The square root of this quantity is a measure of the root-mean-square average distance of the electron from the origin.

Let us suppose we could measure x and x^2 for an electron known to be in a state described by ψ_{U_n}, and moreover, let us assume that this measurement is repeated many times on electrons having the same wave function. From this, we could determine the *average position* and the *average distance* from the origin. The result so obtained corresponds to the quantum-mechanical averages calculated above.

We have discussed a simple example of quantum mechanics in one dimension. The world is three-dimensional, however, and the Schrödinger equation must usually be solved in three-dimensional space.

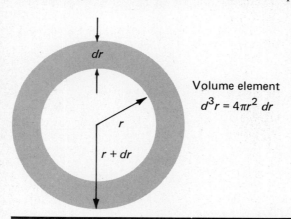

FIGURE 28.18. Spherical shell volume element of thickness dr and volume $4\pi r^2\, dr$.

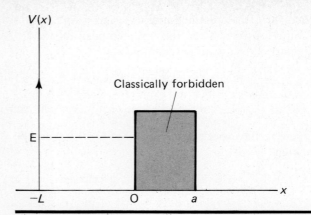

FIGURE 28.19. Classical situation in which a particle of total energy E is forbidden ever to escape from the region $-L < x < 0$.

This is not always an easy task. In fact, many such problems cannot be solved exactly, and we must frequently resort to approximation methods and sometimes numerical analysis. Fortunately, the *hydrogen atom* is an exactly soluble problem, and all the energies and wave functions may be obtained.[5] The wave function for the ground state of hydrogen is spherically symmetric, being a function only of the distance r which separates the electron and proton. This wave function is given by

$$\psi = N_0 e^{-r/a_0} \qquad (28.7.29)$$

where $a_0 = 0.528 \times 10^{-10}$ meter is the Bohr radius.

Let us now obtain the normalization constant N_0 appearing in this wave function. We must require, as usual, that

$$\int_V \psi^2\, dV = 1$$

where dV is the volume element in three dimensions. An appropriate volume element, shown in Fig. 28.18, is given by

$$dV = 4\pi r^2\, dr \qquad (28.7.30)$$

which represents the volume between a sphere of radius r and one of radius $r + dr$. The requirement that the wave function be normalized to unity now implies that

$$N_0^2 \int_0^\infty e^{-2r/a_0} 4\pi r^2\, dr = 4\pi N_0^2(a_0^3/4) = 1$$

or

$$N_0 = \sqrt{\frac{1}{\pi a_0^3}} \qquad (28.7.31)$$

[5] The expression for the allowed energy levels of the hydrogen atom as calculated using Schrödinger's equation agrees exactly with Eq. (28.5.8) which we obtained using Bohr's model.

Thus, the full wave function is

$$\psi(r) = \sqrt{\frac{1}{\pi a_0{}^3}}\, e^{-r/a_0} \qquad (28.7.32)$$

Let us now consider an example which illustrates some of the ideas of wave mechanics as applied to hydrogen.

EXAMPLE 28.7.3

(a) Find the expected value of r for hydrogen in its ground state. (b) Assume that the nucleus of hydrogen (a proton) has a radius of $R_0 \cong 10^{-15}$ meter. On the basis of the theory discussed so far, find the probability that the electron is *inside* the nucleus.

The expected value of r is defined by

$$\langle r \rangle = \int_V r\psi^2(r)\, dV = \frac{4}{a_0{}^3} \int_0^\infty r^3 e^{-2r/a_0}\, dr$$

$$= \frac{1}{a_0{}^3}(\tfrac{3}{2}a_0{}^4) = \tfrac{3}{2}a_0$$

Note that this does not come out to be quite the same as the Bohr radius. The Bohr radius emerges only when one calculates $\langle r^{-1} \rangle$ which turns out to be $a_0{}^{-1}$.

The likelihood P of finding the electron within the nucleus, assuming its wave function is given by (28.7.32), is

$$P = \int_0^{R_0} |\psi^2|\, dV = \frac{1}{\pi a_0{}^3} \int_0^{R_0} 4\pi r^2 e^{-2r/a_0}\, dr$$

When this integral is worked out and the approximation that $R_0 \ll a_0$ is utilized, one obtains the approximate result

$$P \cong \frac{1}{\pi a_0{}^3} \tfrac{4}{3}\pi R_0{}^3 = \frac{4}{3}\frac{R_0{}^3}{a_0{}^3}$$

This is an extremely small probability, since $R_0 \cong 10^{-15}$ meter while $a_0 \cong 10^{-10}$ meter. Note that the answer works out to be the square of the wave function at the origin multiplied by the volume of the nucleus. This is so because the wave function is practically constant over the volume of the nucleus.

28.8 Barrier Penetration in One Dimension

An interesting example of a strictly quantum-mechanical result is illustrated with the aid of Fig. 28.19. This figure is a curve representing the potential energy of a particle such as an electron confined to move in one dimension. The electron has an energy U, as indicated by the dotted line in the figure. For $x < -L$, the potential energy $U_p(x)$ has a very large positive value. As a consequence, the particle cannot enter this region. In the range $-L < x < 0$, the potential energy is zero and the particle behaves very much like a free particle. In the region $0 < x < a$, the potential energy has the constant value U_0, while for $x > a$, the potential energy is again zero. Let us now discuss, both classically and quantum-mechanically, the dynamics of a particle that is initially in the potential "well" defined by the region $-L < x < 0$.

From the viewpoint of classical mechanics, a body of total energy less than U_0 will be "reflected" without any loss of energy at $x = -L$ and $x = 0$ and will behave as a free particle in the intervening space. It will, therefore, be confined within the region $-L < x < 0$ for all time and will continually bounce back and forth between the endpoints at $x = 0$ and $x = -L$. It cannot ever surmount the "potential barrier" that exists in the range $0 < x < a$ unless, of course, some external influence increases its energy to a value in excess of U_0. For this reason, the region $0 < x < a$ is sometimes referred to as a *classically forbidden region*.

Quantum-mechanically, the behavior of the particle is described by the wave function $\psi(x)$. For $x < -L$, due to the very large value of $U_p(x)$, the wave function must be zero. In the range $-L < x < 0$, where $U_p = 0$, the wave function will be sinusoidal just as in the example discussed earlier in connection with Eqs. (28.7.10) and (28.7.13). In the region $x > a$, the wave function may also have a sinusoidal form. For $0 < x < a$, we have $U_p(x) = U_0$, however, and Schrödinger's Eq. (28.7.7) now becomes

$$\frac{\hbar^2}{2m}\frac{d^2\psi}{dx^2} = (U_0 - U)\psi$$

or

$$\frac{d^2\psi}{dx^2} = \frac{2m(U_0 - U)}{\hbar^2}\psi \qquad (28.7.33)$$

where $2m(U_0 - U)/\hbar^2$ is a *positive* quantity whenever $U < U_0$. Under these circumstances, the wave function is no longer sinusoidal but has the exponential form

$$\psi(x) = Ae^{-\alpha x} \quad \text{where} \quad \alpha = \sqrt{2m(U_0 - U)/\hbar^2} \qquad (28.7.34)$$

The fact that (28.7.34) satisfies (28.7.33) can be verified by direct substitution.

Since the wave function *is not zero* in $0 < x < a$, there is a finite probability that the electron will penetrate the potential barrier and actually enter this region where, according to the laws of classical dynamics, it is unequivocally forbidden. Indeed, it may even emerge on the other side into the region $x > a$, where its wave function again attains the sinusoidal free-particle form. The entire wave function, which is constructed by continuously joining the exponentially decreasing solution (28.7.34) with the sinusoidal free-particle solutions appropriate to the regions $-L <$

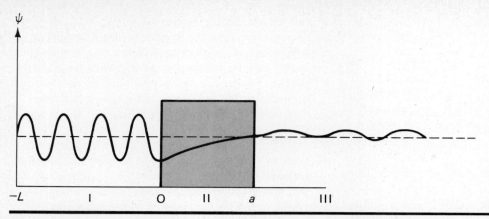

FIGURE 28.20. Wave functions for a particle in the potential well of Fig. 28.19, showing barrier penetration or quantum tunneling effect.

$x < 0$ and $x > a$, is shown in Fig. 28.20. Since the degree to which the wave function (28.7.7) suffers attenuation within the barrier depends upon the barrier's thickness, barriers that are sufficiently thick, such as a brick wall, or even a thin sheet of paper cause a degree of attenuation that is essentially infinite. Under these circumstances, the penetration of the particle into the barrier is undetectable, and the probability of its emerging on the other side—as expressed by the value of $|\psi|^2$ for $x > a$—is negligible, as predicted classically. For barriers thin on the atomic scale, however, this penetration may be significant, and the probability that the particle will go through the barrier altogether and come out on the far side may be appreciable!

That a particle initially confined between $x = -L$ and $x = 0$ can eventually move to $x > a$ is a distinctly quantum-mechanical occurrence. It is known as *barrier penetration*, or sometimes *quantum tunneling*. It is this phenomenon that explains nuclear decay

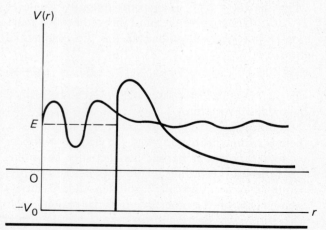

FIGURE 28.21. Wave function of an α-particle in the potential well of an atomic nucleus, exhibiting the possibility of escape through quantum tunneling, corresponding to radioactive decay by α-emission.

by alpha-particle emission. In this situation, the alpha particles which are classically bound within the nucleus can tunnel through the potential barrier that arises from Coulomb repulsion between the alpha particle and the residual nucleus. The potential energy curve and a schematic picture of the wave function for this case are illustrated in Fig. 28.21.

28.9 Heisenberg's Uncertainty Relation

The discussion of barrier penetration teaches us that the results of quantum mechanics in the realm of microphysics can differ substantially from the corresponding classical predictions. Thus, nature provides limits beyond which classical physics cannot be successfully employed. Another striking example of this is furnished by the Heisenberg uncertainty relation, which states unequivocally that there exists an element of *indeterminacy* within the quantum domain.

We discussed earlier the diffraction of monochromatic light by a narrow slit. The light was treated strictly in terms of electromagnetic waves, and the diffraction pattern was readily calculated. When the slit is very wide, the diffraction effect is minimal as shown in Fig. 28.22a. Most of the energy of the beam is propagated with little angular deflection. On the other hand, as the slit is made narrower, the diffraction of light becomes more pronounced, and a larger fraction of the energy suffers significant angular deflection, as shown in Fig. 28.22b. We know, however, that a light beam can be considered to contain many, many photons and, therefore, that its diffraction may be viewed in terms of the deflection of individual quanta from one straight line path onto another. Thus, the incident light beam, as shown in Fig. 28.23, contains many photons which are moving to the right as they impinge on the slit. As they reach the slit, they are somehow diverted through an angle which

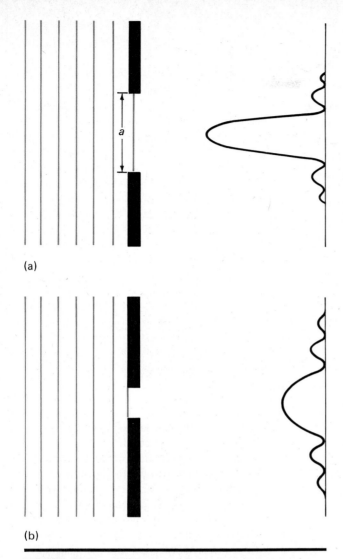

(a)

(b)

FIGURE 28.22. Diffraction of light—or electrons—by (a) a wide slit and (b) a narrow slit.

differs for each incident photon. Some continue to move more or less in a straight path, while others are deviated through large angles. The narrower the slit, the more likely it is that the photons will experience large angular diversion.

The basic reason for this deflection of individual quanta can be explained with reference to a single photon. The situation is then as illustrated in Fig. 28.23. When the photon passes through the slit, it has been *spatially localized* so that the maximum uncertainty in its position is Δx. Now, if the photon suffers deflection to one side or the other, it must obviously acquire a momentum p_x in the x-direction. Let us assume, for simplicity, that most of the photons are confined to the central maximum region of the diffraction pattern. We cannot predict the exact value of the momentum p_x, but we do know the limits between which it must lie. Referring to Fig. 28.23, the maximum angular deviation corresponding to the first intensity minimum of the diffraction pattern is given by Eq. (26.7.9) as

$$\sin \theta = \frac{\lambda}{\Delta x} \tag{28.9.1}$$

Now, assuming θ is small, we have $\sin \theta \cong \tan \theta \cong \theta$, from which

$$\sin \theta \cong \frac{p_x}{p_y} = \frac{\lambda}{\Delta x} \tag{28.9.2}$$

where p_x and p_y are the x- and y-components of the photon momentum. Now, since p_x is assumed to be much less than p_y, we may set $p_y \cong p$, where p is the total photon momentum. But according to de Broglie's relation,

$$p = \frac{h}{\lambda} \tag{28.9.3}$$

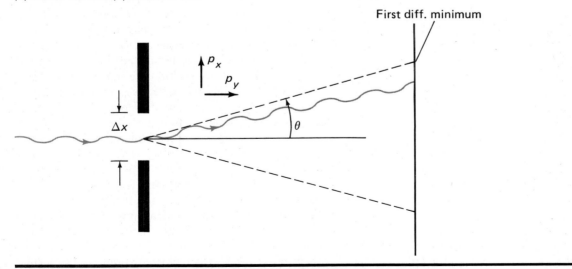

First diff. minimum

FIGURE 28.23. Geometry used in discussing the Heisenberg uncertainty principle for particles diffracted by a slit.

957

Therefore,

$$\frac{p_x}{h/\lambda} = \frac{\lambda}{\Delta x}$$

$$p_x \, \Delta x = h$$

Since the maximum possible uncertainty in the x-component of momentum is $\Delta p_x = 2p_x$, we may write

$$\Delta p_x \, \Delta x \cong 2h \qquad (28.9.4)$$

for the product of maximum possible uncertainties.

A more careful quantum-mechanical analysis, using appropriately defined root-mean-square uncertainties Δx and Δp_x, leads instead to the result

$$\Delta p_x \, \Delta x \geq \frac{h}{4\pi} \qquad (28.9.5)$$

This relation is a precise statement of the *Heisenberg uncertainty principle*. The main ideas underlying this fundamental proposition have been laid down in the above discussion. As the slit is made more and more narrow, the incoming photons are forced to pass through a region of increasingly smaller extent. Our knowledge of their position becomes more and more precise as this process continues. But at the same time, the first minimum in the diffraction pattern moves further and further from the center. This means that many of the photons that strike the screen somehow acquire, during the diffraction process, considerable momentum in the y-direction and that the extent to which this momentum is imparted to them increases linearly as the slit width decreases! We need not concern ourselves with the exact mechanism by which this momentum is acquired by the photons; this is a question that *cannot be answered in detail*. It is inescapable, however, given that light is diffracted as described in Chapter 26, that this momentum must be acquired by photons in some fashion. If this is accepted, then we are led at once to the uncertainty relations (28.9.4) or (28.9.5). Moreover, since electrons and other "material" particles undergo diffraction just like light waves of wavelength $\lambda = h/p$, this uncertainty relation must apply to them as well.

Turning the argument around, we may state as a fact that the Heisenberg uncertainty principle is one that is inextricably embedded in the framework of quantum mechanics and is, therefore, of universal applicability. It is possible, in fact, to show that the solutions of Schrödinger's equations, and thus the wave functions that describe the behavior of any dynamic system, *always satisfy the requirements of the Heisenberg principle*. If we take this point of view, we may understand the spreading of an optical diffraction pattern as the slit width is decreased as a direct consequence of the uncertainty principle. The char-

acteristics of electron diffraction and the properties of other particle beams can be understood in the same way.

The uncertainty relation imposes certain fundamental limitations that require a reevaluation of some familiar classical notions. It tells us that there is no way of *simultaneously* attaining an exact knowledge of the position and momentum of a particle. In any act of measurement or observation there must always be mechanisms that operate to limit the accuracy of determining either position or momentum, or both, in such a way that the inequality (28.9.5) is satisfied. We cannot observe or make measurements on any system without altering it ever so slightly, and the minimum limit of this alteration is expressed by the uncertainty relation. It must be stressed that the indeterminacy that results comes about not by any inadequacy in the equipment used to make measurements nor by any lack of care on the part of the experimenter to make the most careful observations that are possible. It is, instead, an inherent limitation on the attainability of information, a limitation imposed by the very foundations of quantum theory itself.

In classical physics, the dynamics of a particle is approached by solving certain equations of motion that predict the trajectory of the particle and its momentum for all values of the time. If we know simultaneously the exact initial position and momentum of a baseball and if we know all the forces acting, such as gravity, air resistance, and buoyancy, we can arrive at a precise expression for its path at all later times. But, now, according to the uncertainty relation, there is no way of determining the initial position and momentum with arbitrary precision. In the case of a baseball, of course, it does not matter, for the minimum uncertainty as given by (28.9.5) is so small on the scale of ordinary tangible objects as to be negligible. The quantum uncertainty in this instance is not what sets the practical limit to the accuracy of our determinations. We, therefore, go ahead and determine the trajectory and observe that this is, in fact, how the body moves under the circumstances that exist. But if we replace the baseball by an electron, whose location on an atomic size scale is sought, the situation becomes very much different. On this scale, the quantity $h/4\pi$ in (28.9.5) is no longer so small and, as a consequence, there may be no satisfactory way to determine, or even discuss, such a thing as the electron's precise trajectory as a function of time.

It is mainly for this reason that we must settle for a knowledge of probability amplitudes and expectation values rather than precise dynamical information of the sort we obtain from Newton's laws. Fortunately, the lack of determinism in the behavior of an individual particle is not as much of a drawback as it might seem at first, since most of the questions one needs to answer about atomic systems can be answered without a precise knowledge of particle orbits. For example, Bohr's results for the energy levels of the hydrogen atoms, originally arrived at by assuming circular electron orbits of precisely stated size, are easily enough obtained from Schrödinger's equation without the necessity of assuming any particular orbital shapes, sizes, or velocities. Indeed, Schrödinger's wave mechanics provides us with a much richer and more highly structured theory than the Bohr picture. As a result, it is capable of dealing with the mechanics of atoms and molecules that are much more complex than hydrogen and in which all attempts to extend Bohr's orbital picture have failed.

Finally, a note of caution in regard to the word *simultaneously* in the original discussion of the uncertainty relation. The Heisenberg principle refers to a fundamental limitation on the *simultaneous* determination of information about a particle's position and its momentum.[6] When this requirement of simultaneity is removed, the restrictions imposed by the uncertainty principle also vanish. There is, therefore, no inherent limitation on our ability to determine, as exactly as desired, a particle's position and then, at a later time, as exactly as we may wish, its momentum. We can never use results of determinations such as these, however, to infer an exact trajectory, because in making the determination of position, we inevitably effect an alteration of momentum, and vice versa.

EXAMPLE 28.9.1

An electron, constrained to move in one dimension, has a position uncertainty $\Delta x = 1.0 \times 10^{-8}$ meter. Determine the minimum uncertainty in its speed. Find the corresponding uncertainty for a 2-kg ball confined in the same way.

According to (28.9.5),

$$(1.0 \times 10^{-8})(\Delta p_x) \geq h/4\pi$$

or

$$\Delta p_x \geq \frac{(6.63 \times 10^{-34})}{(4\pi)(1 \times 10^{-8})} = 0.53 \times 10^{-26} \text{ kg-m/sec}$$

For the electron, this implies a velocity uncertainty $\Delta v_x = \Delta p_x/m$ of

$$\Delta v_x \geq \frac{0.53 \times 10^{-26} \text{ kg-m/sec}}{9.11 \times 10^{-31} \text{ kg}} = 5.8 \times 10^3 \text{ m/sec}$$

while for the 2-kg ball

---

[6] In this case, the events that are simultaneous in the rest frame of the observer occur not only at the same time but also at the same place. They are, therefore, simultaneous in all possible reference systems. For this reason, no difficulty related to the relative character of simultaneity arises in the statement of the Heisenberg uncertainty principles.

$$\Delta v_x \geq \frac{0.53 \times 10^{-26} \text{ kg-m/sec}}{2 \text{ kg}} = 2.7 \times 10^{-27} \text{ m/sec}$$

From this, we see quite clearly that the limitation the uncertainty relation imposes for microscopic bodies is much more significant than that which results in the case of macroscopic objects.

28.10 Quantum Degeneracy, the Pauli Exclusion Principle, and Spin

In the one-dimensional example of a free electron confined to the region $-L/2 < x < +L/2$ discussed in section 28.7, we found that with every allowed energy of the system, one and only one possible wave function was associated. This one-to-one correspondence between wave functions and energy levels is characteristic of all one-dimensional systems. In two- or three-dimensional examples, however, we usually find that there are a *number of different solutions* of the Schrödinger equation corresponding to any given allowed value of energy. It is customary to refer to the possible individual solutions of Schrödinger's equation—and thus to each of the possible wave functions for the system—as the *quantum states* of the system and to the allowed values of energy as the *energy levels*.

The statement made above in regard to the multiplicity of solutions of Schrödinger's equation for each of the allowed energies may now be rephrased to state that *there may be more than one quantum state of the system corresponding to a given energy level*. The number of quantum states associated with a particular energy level is referred to as the *degeneracy* of the level. This distinction between quantum states and energy levels should be clearly understood, for it arises again and again in many important examples. We have already encountered a somewhat analogous situation in connection with the distribution of molecular velocities in gases. There are many possible values of the individual velocity components v_x, v_y, and v_z that correspond to one given value of the speed and thus to one specific value of the kinetic energy for a given molecule. These various possible combinations of v_x, v_y, and v_z that lead to the specified energy are analogous to the possible quantum states of a system, while the single specified energy plays the role of the allowed energy value related to these states. For the hydrogen atom, it can be shown from Schrödinger's wave mechanics that for the nth energy level, as specified by Bohr's formula,

$$U_n = -\frac{me^4}{8\varepsilon_0^2 n^2 h^2} \tag{28.10.1}$$

there are n^2 different wave functions. Each of these is a distinct and valid solution of Schrödinger's equation for the energy U_n, and each satisfies all the requirements of continuity and the normalization condition. Each of these solutions, therefore, defines a *quantum state* of the hydrogen atom belonging to the nth energy level. Since there are n^2 of these possible quantum states, it is said that the *degeneracy factor* of the nth energy level is n^2, or that the nth energy level is n^2-fold degenerate. Other two- or three-dimensional systems behave in very much the same way, although, the degeneracy factors associated with their energy levels may differ from those of the hydrogen atom.

This seemingly esoteric distinction is found to play an important part in the physics of atomic systems that have more than one electron and, in particular, in the behavior of free electrons in metals and other crystalline substances. In the case of atoms having more than one electron, such as He, Li, Be, B, C, etc., it is possible to obtain solutions to Schrödinger's equation that define the allowed energies and the possible quantum states. These calculations also give us the number of quantum states associated with each allowed energy, namely, the degeneracy factor for each energy level. These results allow us to make predictions of the frequency of the spectral lines emitted when excited atoms return to their ground state with the emission of light quanta. These predictions, however, are not in complete agreement with experiment unless two additional hypotheses are introduced.

The first of these is referred to as the Pauli exclusion principle,[7] which states that in a given *system*, which may be an atom, a molecule, or a whole crystal made up of many interacting atoms, *no two electrons may occupy the same quantum state*. In many-electron atoms, this has a profound effect in defining the ground state of the system. In the absence of the Pauli principle, in the ground state, all the electrons in an atom would occupy the lowest allowed energy level. But since only one, at most, can occupy a given quantum state, once the quantum states belonging to the lowest allowed energy are filled, electrons must start to occupy quantum states belonging to the level that is next lowest in energy. At length, when these are filled, electrons must occupy states of even higher energy. The final ground state of the atom is thus determined by the number of electrons in the atom, the disposition of the allowed energies of the system, and the degeneracy factors of the allowed energy levels. The quantum states belonging to the allowed energy levels fill up from bottom to top, so to speak, like water poured into a jug. The disposition of allowed energies

and the degeneracy factors associated with them are, in fact, responsible for the structure of the periodic table of elements and thus for the chemical properties of the elements.

The Pauli exclusion principle, reduced to its simplest terms, states that no two electron wave functions belonging to a given system can be the same. This, in effect, means that no two electrons can occupy the same region of space at the same time. This has a profound effect on the physical behavior of ordinary matter. It is responsible for the fact that the atoms in a solid crystalline substance or in a liquid exert attractive forces upon one another only up to a certain point, after which the electron wave functions of individual atoms begin to overlap. Beyond this point, when we try to compress the substance further, we are really trying to force electrons to occupy the same region of space at the same time. This they will not do; instead, they have to occupy quantum states of higher energy. But to make them do this, we must supply energy from the surroundings via an externally applied stress! It is evident from this that the Pauli exclusion principle is directly involved in generating the so-called "contact forces" that ordinary material objects exert on one another.

Many other elementary particles, such as protons, neutrons, and muons, obey the Pauli exclusion principle, but there are some that do not. Most common among these are light quanta, or photons. The fact that photons do not obey the Pauli principle is again responsible for many of their observed physical characteristics. Thus, we have no difficulty in directing as much light intensity into any given region of space as we choose. Also, photons in free space do not exert the "contact forces" on one another that we encounter in situations where "material particles" such as protons, electrons, or atoms interact. It is the Pauli exclusion principle, therefore, that is largely responsible for the differences in the properties exhibited by ordinary matter and less overtly "material" particles such as photons.

The other factor which must be admitted into the wave-mechanical explanation of the behavior of many-particle systems is the phenomenon of *electron spin*. This concept was first introduced by Goudsmit and Uhlenbeck in 1925. It appears that every electron, in addition to whatever orbital angular momentum it may have as a part of an atom, has an intrinsic or inherent angular momentum of its own referred to as *spin angular momentum*. Although the electron behaves very much as though it were spinning about an axis through its center of mass, it is probably best not to regard the electron as literally in a state of rotation. We should, instead, regard the spin angular momentum as an inherent property of the electron, like its charge or mass. The spin angular momentum of the

[7] Named for Wolfgang Pauli (1900–1958), the Swiss physicist who first proposed the exclusion principle.

electron has the magnitude

$$L_s = \hbar\sqrt{3}/2 \qquad (28.10.2)$$

If we seek to determine the component of this spin angular momentum along any given direction, we find that this quantity is inevitably *quantized* and has, in the case of the electron, two and only two values given by

$$L_{sz} = \pm\tfrac{1}{2}\hbar \qquad (28.10.3)$$

This is frequently written as $\hbar$ (which, after all, is the elementary quantum of angular momentum) times a spin quantum number m_s that can take on only the two possible values $+\tfrac{1}{2}$ and $-\tfrac{1}{2}$. In this way, we write

$$L_{sz} = m_s\hbar \qquad (28.10.4)$$
with
$$m_s = +\tfrac{1}{2} \quad \text{or} \quad -\tfrac{1}{2} \qquad (28.10.5)$$

As a consequence, the electron is said to be a "particle of spin one half." A diagram illustrating the two possible orientations of the electron's spin angular momentum vector with respect to the z-direction, along which the spin projection is to be measured, is given in Fig. 28.24. From this figure, it is apparent that the spin angular momentum vector L_s, whose magnitude is $L_s = \hbar\sqrt{3}/2$, can have only two allowed spatial orientations with respect to the projection axis, corresponding to the allowed quantized values of the projected component of $\pm\hbar/2$. These two orientations are frequently referred to as *spin states;* the one whose projection is $+\hbar/2$ is often called "spin up" and the

other is called "spin down." The orbital angular momentum of electrons in atoms is spatially quantized in a way that is quite similar to the quantization of the spin.

A rotating charge distribution gives rise to rotating or circulating currents and, in turn, to a magnetic dipole moment. Accompanying the electron's spin angular momentum, therefore, there is an associated *spin magnetic moment.* The spin magnetic moment of electrons in atomic systems as well as the orbital magnetic moment play an important part in determining the magnetic susceptibility of the material in which the atoms are found. When a magnetic field oriented along the z-axis is present, the energy of the two spin states of the electron will differ in view of the tendency of the field to align the spin magnetic moments. But if no such magnetic field is present, the two spin states will have the same energy. Under these circumstances, there are two allowed quantum states—"spin up" and "spin down"—of the same energy for every electron in the system. This has the effect of *doubling* the degeneracy factor arrived at by the solution of Schrödinger's equation with the neglect of electron spin, because now, for every former quantum state of the system, there are two, one corresponding to an electron with "spin up" and one with "spin down."

When the Pauli exclusion principle and the effect of electron spin are incorporated into the framework of Schrödinger's wave mechanics, it is possible to predict the spectral frequencies emitted by many-electron atoms and by complex molecules as well. These predictions have in all cases agreed very well with experimental observations.

There is one further point to be noted having to do with the interrelation of particle spin and the Pauli exclusion principle. The inherent spin of elementary particles differs from one type of particle to another. There are some, such as the pi mesons, that have no inherent spin at all. There are many that, like the electron, have "spin one half." Some examples of spin $\tfrac{1}{2}$ particles are protons, muons, neutrinos, and neutrons. There are others that have higher intrinsic spin angular momenta, but *in all cases* the maximum projected angular momentum is either an integral or half-integral multiple of $\hbar$. Thus, the maximum possible value for the projection of the spin angular momentum vector on a given axis is 0, $\hbar/2$, $\hbar$, $3\hbar/2$, $2\hbar$, $5\hbar/2$, etc. Such particles are said to have inherent spin 0, $\tfrac{1}{2}$, 1, $\tfrac{3}{2}$, 2, $\tfrac{5}{2}$, etc. As a result of symmetry requirements that the wave functions must fulfill, it happens that particles having half-integral spin ($\tfrac{1}{2}$, $\tfrac{3}{2}$, $\tfrac{5}{2}$, . . .), such as electrons, positrons, and protons, must invariably obey the Pauli exclusion principle, while particles of integral spin (0, 1, 2, 3, . . .), such as photons and gravitons, can never do so. For this reason, the intrinsic angular

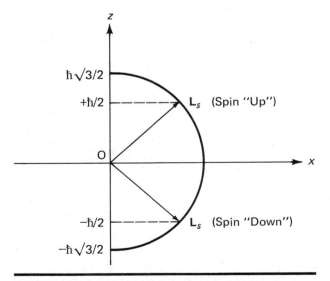

FIGURE 28.24. Possible orientations of the inherent spin angular momentum vector of an electron, corresponding to "spin up" and "spin down" states.

momentum of the particle has a very important bearing upon its physical characteristics.

28.11 Stimulated Emission and Lasers

Quantum physics was originally developed to understand the mechanics of atoms and molecules and to resolve a number of puzzling effects that could not be explained within the existing classical framework. It was thus a product of the most basic research imaginable, but, as happens so often in science, basic discoveries pave the way toward important applications and great advances in technology. The laser is one such example.

The word laser is an acronym for "light amplification by stimulated emission of radiation." To understand what this means and how lasers work, we must first discuss some aspects of quantum mechanics as applied to atomic systems. We have already mentioned that atoms contain many discrete energy states and that electrons can occupy any of these states. For example, a hydrogen atom contains an infinite number of energy states. It has one electron which can occupy only one of these states, although that electron can make transitions from one state to another under suitable conditions.

Now, when there are a tremendous number of hydrogen atoms, it is found that the electrons are statistically distributed among the various energy states according to the Maxwell–Boltzmann energy distribution law. The distribution of occupied states depends, of course, on the temperature. Therefore, if there are N_0 hydrogen atoms in thermal equilibrium with the surroundings, the average number of electrons found in the energy states U_1, U_2, etc., is given by

$$N_1 = N_0 e^{-U_1/kT}$$
$$N_2 = N_0 e^{-U_2/kT} \qquad (28.11.1)$$

and so on. Accordingly, the largest number of electrons are found in the lowest energy state. At higher temperatures, a larger proportion of the electrons can exist in higher-energy states than at lower temperatures.

How do electrons "know" when to make a transition from one energy state to another? To make an upward transition to a state of higher energy, an electron must absorb enough energy to make up for the energy difference between the initial and final states. It can do this by absorbing a photon of the appropriate energy or by acquiring energy in collisions with other atoms in the container. When the electron is in an excited state, however, it can make a downward transition with the emission of a photon *even if the atom is isolated from its surroundings*. Such transitions are called *spontaneous transitions*, and the radiation process involving such transitions is referred to as *spontaneous emission* of radiation. The underlying quantum mechanics together with a knowledge of the interactions of atoms and photons permits one to predict the likelihood of spontaneous emission and thus the average *lifetime* of an excited state. The lifetime of an excited state depends upon the electron wave function of that state and also upon the wave functions of other states to which the electron might decay. In some instances, excited states may decay very quickly by spontaneous emission of photons, but in other instances, it may happen that the probability associated with what would otherwise be the major emission process is very low, and then the excited-state lifetime may be much longer.

Now, suppose we have a container of atoms and a beam of photons each having energy $h\nu_{12}$. This energy is assumed to be *equal* to the energy difference between states 1 and 2, as shown in Fig. 28.25. Quantum mechanics predicts a certain probability for an atom to absorb a photon, going from state 1 to state 2. This is called the *absorption probability*. Quantum mechanics also predicts an *enhancement* in the rate of transitions from state 2 down to state 1, thus leading to the phenomenon of *stimulated emission*, whenever the frequency of the incoming radiation is very close to ν_{12}. The stimulated emission is a resonance phenomenon in which a photon "tuned" to the right frequency can cause atoms in excited states to radiate. Generally

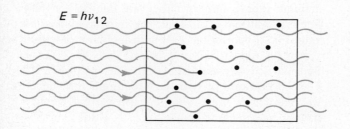
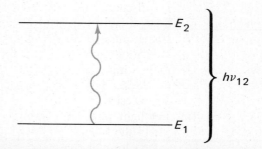

FIGURE 28.25. Absorption of photons of energy $h\nu_{12}$ causing excitation of atomic energy states at energy E_2, for which $E_2 - E_1 = h\nu_{12}$.

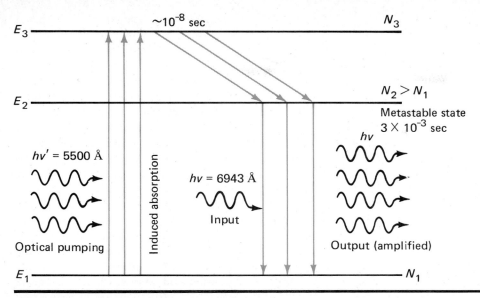

FIGURE 28.26. Schematic diagram of the operation of a pulsed laser, showing excitation, population inversion, and amplification.

speaking, when these photons are present, the stimulated emission is far more important than the spontaneous emission. Now, it turns out in equilibrium that the probability of stimulated emission is equal to the probability for absorption. Under these circumstances, photons emitted by atoms in the excited state can readily be absorbed by atoms in the lower state. It is, therefore, not ordinarily possible to build up a very substantial emission of photons, since emitted photons are more likely to be absorbed rather than cause further stimulated emissions. This is due to the fact that, in equilibrium, according to (28.11.1), many more atoms are in the lower state than in the excited state.

Thus, in achieving radiation by stimulated emission, one must *invert* the populations of the two quantum levels, making the population N_2 in the excited level greater than the number N_1 in the lower level. Only in this way can the stimulated emission process predominate over absorption. The population inversion can be brought about in practice in a number of ways. In a helium–neon gas laser, it is achieved by exciting an electrical discharge in the gas mixture. In a solid-state ruby laser, it is achieved by exposing the system to a very intense pulse of light from a powerful flashtube. In this case, a system of three atomic energy levels is involved in the stimulated emission process, as illustrated in Fig. 28.26. The atoms are initially "pumped" from the lowest state at energy U_1 to state U_3 by the external light pulse. They then quickly undergo transitions to the level at energy U_2, which happens to have a very long lifetime for decay to the ground state by the normal spontaneous process. The net effect is to produce a population inversion in which

the number of atoms in the upper state at energy U_2 is much larger than that in the ground state.

It is now possible to achieve *stimulated* emission from this upper state by introducing photons at just the right frequency v_{21}. When stimulated emission occurs under these circumstances, very little energy is absorbed in exciting atoms from state 1 to state 2, simply because now there are not many atoms in the lower energy level. Stimulated emission then takes place, and as photons of frequency v_{21} are emitted, they will in turn stimulate further emission until, at length, the system returns once more to equilibrium. In the process, a powerful and intense pulse of coherent stimulated laser light is generated. In the helium–neon gas laser, a continuous light output can be achieved, using an electrical discharge to produce a population inversion in the gas mixture on a continuous basis.

For the emitted photons to stimulate further emission to maximum effect, they must remain within the system for a reasonable length of time, and they cannot be allowed to escape at once. The required confinement of photons is achieved by reflecting mirrors at either end of the system, one of which is totally reflecting, the other slightly transparent so as to allow perhaps 1 percent of the incident photons to escape. The photons that escape through the partially transparent mirror constitute the effective laser beam.

The laser beam has a number of important characteristics. First of all, it is a *coherent* light source, since the atoms which undergo stimulated emissions give off radiation that is precisely in phase with the exciting radiation. Also, laser beams are very narrow, very nearly parallel, and exhibit very little angular

963

divergence. They can be very intense and can concentrate relatively large amounts of electromagnetic energy into very small areas. As a result, they can be used to drill small holes in metals and ceramics, to vaporize diamonds and other refractory substances, to perform certain surgical tasks such as repairing detached retinas, and to do other useful tasks requiring high temperatures created over very small areas. The small angular deviation exhibited by laser beams makes lasers useful in such activities as surveying and precision alignment of instruments and industrial equipment. Laser light is highly monochromatic and is easily conveyed by optical fibers from one place to another. At the same time, it can be modulated by a radio or television signal like an ordinary radio frequency wave. But since its frequency is so high, it is capable of carrying much more information than radio or microwave carriers. This makes the laser potentially useful in telephone systems and other high-density information transmission applications. Finally, the coherent properties of laser radiation have opened up an entirely new range of photographic possibilities via the process of holography. In this application, a photograph taken using coherent laser light stores not only light amplitude information but also *phase* information in the emulsion of the film, which, when developed, becomes a sort of diffraction grating that will recreate a three-dimensional image of the subject when illuminated with laser light of the same frequency.

The laser was invented during the 1950s by Charles H. Townes and Arthur L. Schawlow. Their discovery, which has proved to be of enormous value to pure science as well as technology, could never have been developed without a clear understanding of the quantum theory of atoms, molecules, and radiation. In this field, as in many others, the labors of scientists striving to understand the basic laws of nature have led, and continue to lead, to advances of vast benefit to mankind.

SUMMARY

The quantum theory was founded by Max Planck in investigating the spectral distribution of blackbody radiation. In order to arrive at a correct explanation of this effect, he had to assume that harmonic oscillators could have *only a discrete set of allowed energies* given by

$$E_n = nh\nu \qquad n = 1, 2, 3, \ldots$$

where $h = 6.63 \times 10^{-34}$ joule-sec is a fundamental constant now universally referred to as Planck's constant. Shortly thereafter, Einstein was able to explain the photoelectric effect on the basis of quantum theory, by assuming that energy could be absorbed from incident electromagnetic waves only in discrete amounts $h\nu$; these "quanta" were called photons.

In 1910, Bohr exhibited the spectrum of atomic hydrogen on the basis of a model wherein an electron executes a circular orbit about a proton, and in which the angular momentum of the electron is quantized in integral multiples of $h/2\pi$. He arrived at an expression for a set of discrete allowed values of energy for the hydrogen atom having the form

$$U_n = \frac{-m_e e^4}{8\varepsilon_0^2 n^2 h^2}$$

De Broglie suggested that, just as electromagnetic waves might appear in the guise of particles, "material" particles such as electrons and protons might exhibit the properties of waves. He was able to associate a wavelength λ with a particle of momentum p, given by

$$\lambda = \frac{h}{p}$$

De Broglie's hypothesis was verified by Davisson and Germer in 1927, when they succeeded in diffracting a beam of electrons using a crystal lattice as the "grating."

Schrödinger, in 1928, showed that a wave function Ψ could be associated with any particle as a solution of his wave equation, which has the form

$$-\frac{\hbar^2}{2m}\left(\frac{\partial^2 \Psi}{\partial x^2} + \frac{\partial^2 \Psi}{\partial y^2} + \frac{\partial^2 \Psi}{\partial z^2}\right) + U_p(x, y, z)\Psi(x, y, z, t)$$
$$= i\hbar \frac{\partial \Psi}{\partial t}$$

where $\hbar = h/2\pi$ and U_p represents the potential energy of the particle. The wave function Ψ is to be interpreted such that

$$|\Psi(x, y, z, t)|^2 \, dV = \text{probability that the particle will be found in volume element } dV \text{ about point}(x, y, z), \text{ at time } t.$$

Since the probability that the particle exists somewhere is unity it is necessary that the wave function satisfy the normalization condition

$$\int_V |\Psi|^2 \, dV = 1$$

the integral being evaluated over all space. The solution of Schrödinger's equation leads to the allowed energy states and wave functions for any system. These allowed energies are frequently referred to as *stationary states* of the system, since their wave functions do not change with time.

Quantum mechanics provides a satisfactory way of describing the dynamic behavior of atomic and molecular systems and of understanding the behavior of matter in general on the microscopic level. It leads to the prediction of certain effects such as *quantum tunneling*, which are forbidden classically. In quantum tunneling, a particle can be shown to have a certain probability of penetrating a potential energy barrier too high for it to surmount classically and appearing as a free particle on the other side. Quantum tunneling has been observed experimentally in a number of different investigations in the areas of nuclear physics and solid-state physics. It is believed to be the mechanism by which the alpha particle penetrates the nuclear binding potential in radioactive decay by alpha emission. Also, quantum mechanics embodies the Heisenberg uncertainty principle, which states that the *simultaneous* determination of the position and momentum of a particle or system cannot be achieved with arbitrary precision but must involve inherent uncertainties Δx and Δp_x whose product is given by

$$\Delta x \, \Delta p_x \geq \frac{h}{4\pi}$$

Another quantum limitation on the behavior of certain particles such as electrons, protons, and neutrons is embodied in the Pauli exclusion principle, which states that in any given system, no two such particles can occupy any given quantum state.

In experiments involving atomic spectra and molecular beams, it has been shown that elementary particles such as electrons and protons have *inherent* angular momenta and magnetic moments very much like those associated with circulating currents or rotating charge distributions. In the case of the electron, this spin angular momentum is of magnitude

$$L_s = \hbar\sqrt{3}/2$$

It turns out that whenever we seek to determine the component of this spin angular momentum along any given direction, we find this quantity to be *quantized*, having, for the electron, only the two allowed values

$$L_{sz} = \pm \hbar/2$$

It is said, then, that the electron has only two allowed spin orientations, referred to as "spin up" and "spin down."

It is important to note that the spin angular momentum of electrons and other particles is *unrelated* to the orbital angular momentum they may have when they are incorporated into an atom. The spin angular momentum is an inherent property of the particle like its mass or charge and has an existence quite independent of the orbital angular momentum it may acquire as a part of an atomic electron distribution or nuclear system. In these latter instances, the total angular momentum can be represented as the sum of the orbital angular momentum and the inherent spin angular momentum. Accompanying the spin angular momentum, is, of course, a corresponding spin magnetic moment.

Quantum theory has provided an extremely powerful and accurate tool for discussing the behavior of atomic and molecular systems. Its predictions, so far as we know now, are universally in agreement with experiment.

QUESTIONS

1. What important law of physics is violated by the "ultraviolet catastrophe"?
2. State the assumptions of Planck which were used to explain blackbody radiation. Do you feel there is a philosophical justification in proposing such *ad hoc* hypotheses to explain experimental data?
3. In the photoelectric effect, there is a cutoff frequency. How does the photon picture of light explain this fact?
4. Suggest a simple experiment to determine the work function of a metal.
5. An electron and a neutron are traveling at the same velocity. Which has the longer wavelength?
6. Is the maximum current obtainable in the photoelectric effect proportional to the incident light intensity?
7. What are the basic assumptions of the Bohr model of hydrogen? Which of these remain valid in modern quantum physics?
8. If experimenters were very clever in designing their apparatus, could they manage to violate the uncertainty relation? Explain.
9. What is meant by a "stationary state"? Does it mean that the electrons will not move?
10. Yellow light of wavelength 6000 Å can just barely be detected visually when 1.7×10^{-18} watt is delivered to the retina. How many photons per second are received under these circumstances?
11. What is the radius of the first Bohr orbit for a singly charged helium ion?
12. Bohr's assumption of circular orbits constitutes a violation of the uncertainty relation. How does this come about?
13. How is it possible for a particle of definite momentum to be also a wave?
14. Describe the significance of the Schrödinger wave function. Why is it unnecessary to attribute wave functions to macroscopic bodies?
15. Can you explain why it is that large objects have small wavelengths?
16. It is imperative that a quantum-mechanical wave function be normalizable. What does this mean, and why must it be so?
17. If ψ_A and ψ_B are both normalized wave functions describing a given system, $\psi_A + \psi_B$ is a possible wave function. Is it properly normalized?
18. What is the difference between spontaneous and stimulated emission of radiation?

PROBLEMS

1. Using Planck's law, Eq. (28.2.3), show that in the high-frequency limit it reduces to Wien's formula, Eq. (28.2.2), while in the low-frequency limit it reduces to Eq. (28.2.1), the Rayleigh–Jeans formula. What are the constants c_1 and c_2 in Wien's formula?

2. The integral of $I(v)$, see Eq. (28.2.3), over all frequencies gives the total power radiated per unit area by a blackbody. Set up the integration and make the substitution $x = hv/kT$. Show that the result will be proportional to T^4. This is known as the Stefan–Boltzmann law.

3. For a blackbody radiator, the total power radiated per unit area is obtained by integrating the Planck radiation distribution. The result is $R = (5.6699 \times 10^{-8})\, T^4$ in the mks system of units. A metallic sphere of radius 5 cm is at a temperature of 1000°K. How many joules of energy are radiated each hour assuming the sphere can be treated as a blackbody?

4. The net rate of loss due to radiation by a blackbody is given by

$$R_{net} = (5.6699 \times 10^{-8})(T^4 - T_0{}^4)$$

where T is the temperature of the body and T_0 is that of the surroundings. Calculate the net power radiated by a person of area 1.3 meters when he steps outdoors on a day when the temperature is 5°C and his surface temperature is 32°C. Assume the above formula is valid.

*5. At 8000°K, find the wavelength at which a blackbody radiator emits the most energy per unit wavelength. Use graphic or numerical methods to solve the required equations.

6. The work function for a certain metal is 2.5 eV. If light of wavelength 6500 Å falls on the surface, does photoelectric emission occur?

7. What minimum energy must be supplied to a hydrogen atom to remove an electron in a state for which $n = 4$?

8. An electron in a hydrogen atom absorbs a photon and makes a transition from a state with $n = 4$ to one with $n = 8$. Find the frequency of the photon. If the electron then falls to its ground state, what is the wavelength of the light emitted?

9. When a retarding potential of 1 volt is applied, the photoelectric current produced by radiation of wavelength 3000 Å drops to zero. What is the work function of the material?

10. The threshold wavelength for photoelectric emission from sodium is found to be 5420 Å. If light of wavelength 4500 Å is incident, find the maximum speed of the ejected electrons.

11. A photon has an energy of 2 million electron volts. What are its frequency, wavelength, and momentum?

12. You are told that a flywheel has an angular momentum of exactly 12 joule-seconds. According to Bohr's hypothesis, angular momentum must be quantized. Is there any way of finding out whether the given value is allowed by Bohr's hypothesis? What value of n do you obtain?

13. Your friend tells you that he read an article in the newspaper in which it was claimed that a newly discovered particle had an angular momentum of 7.91×10^{-35} joule-

seconds under certain carefully controlled conditions. What is your response to him?

14. A proton initially at rest is accelerated through a potential difference of 1500 volts. Find the final de Broglie wavelength.

15. The de Broglie wavelengths of an electron and neutron are equal. Find the ratio of their speeds.

16. A thermal neutron has a kinetic energy of $\frac{3}{2}kT$ at $T = 290°K$. Find the de Broglie wavelength of the neutron.

17. Positronium is an "atom" consisting of an electron bound to a positron. The Rydberg constant for this atom is $R/2$. Using this information, find the ionization energy if the system is in its ground state.

18. An electron and positron are moving toward each other with equal speeds of 3×10^6 meters/sec. Find their de Broglie wavelengths. They can annihilate each other producing two photons of equal energy. Find the energy, momentum, and wavelength of each photon.

19. Equation (28.2.3) gives the spectral distribution of blackbody radiation as a function of the frequency v. Find an expression for I in terms of the wavelength λ by writing $I(v)\, dv = I(\lambda)\, d\lambda$, and using $v\lambda = c$.

20. A beam of photons of energy 5 eV contains 10^{16} photons/sec which pass through a circular aperture of radius 1 mm. What power is delivered by the beam and what is the intensity at the aperture?

21. Show that an electron in a Bohr orbit with quantum number n has a frequency of revolution v_n given by

$$v_n = \frac{me^4}{4\varepsilon_0{}^2 h^3 n^3}$$

22. For very large quantum numbers, obtain the frequency of emitted radiation for an electron transition in hydrogen from the state n to $n - 1$. Show that this agrees with the frequency of revolution obtained in the preceding problem. This is an illustration of Bohr's correspondence principle in which it is asserted that for large n, the predictions of quantum theory and classical theory will agree. *Hint*: for large n, $f(n) - f(n - 1) \cong df/dn$.

23. If the electron in a hydrogen atom is replaced by a muon and the proton is replaced by a lead nucleus, thus forming a "muonic atom," find the ground-state binding energy and the radius of the first Bohr orbit. Such atoms have been extensively studied, but theoretical calculations must take into account the finite size of the Pb nucleus.

24. A stream of photons of constant wavelength $\lambda = 3000$ Å are incident on a metallic surface and photoelectrons are detected. It requires 2.8 eV of energy to remove an electron from the metal surface. (a) What is the range in kinetic energy of the photoelectrons? (b) What would be the stopping potential? (c) If an electron has a kinetic energy of 1 eV, what is its de Broglie wavelength?

25. The work function for sodium is 2.46 eV. (a) Calculate the wavelength of a photon which, when absorbed by an electron, causes the electron to escape with zero kinetic energy. (b) Suppose intense light of $\lambda = 5000$ Å and very low-intensity light of $\lambda = 4500$ Å are both incident on sodium. Which source will cause emission of electrons with the largest kinetic energy? (c) Calculate the maximum kinetic energy of the emitted photoelectrons when the incident photon wavelength is 4500 Å.

26. In a typical television picture tube, electrons are accelerated through a potential difference of 12,000 volts. Find their final momentum and de Broglie wavelength assuming they are initially at rest.

27. The probability of finding the electron between radii r and $r + dr$ in the ground state of the hydrogen atom is given by quantum mechanics as

    $$P(r)\, dr = N_0{}^2 e^{-2r/a_0} 4\pi r^2\, dr$$

    Show that the probability function $P(r)$ has its maximum value at the first Bohr radius a_0. Make a plot of $P(r)$ versus r.

28. Using the probability function $P(r)$ of the preceding problem, determine the probability of finding the electron in the ground state of hydrogen within a sphere of radius a_0 centered at the proton.

29. The wavelength λ_m for which $I(\lambda, T)$ of Problem 19 is a maximum can be shown to satisfy the equation $\lambda_m T =$ const. $= 0.28978 \times 10^{-2}$ m-°K. This is known as Wien's displacement law. Assuming the sun radiates as a blackbody with a maximum intensity at $\lambda_m = 5600$ Å, find the surface temperature of the sun.

30. One of the most interesting constants in nature is the fine structure constant α defined in terms of the proton charge e, Planck's constant h, and the speed of light c, according to the relation $\alpha = e^2/4\pi\varepsilon_0 hc$. Calculate the value of this constant and prove that it is dimensionless.

31. Suppose you are told that within a certain region of space a particle moving in one dimension has a well-defined energy and is described by the wave function $e^{-\lambda x}$. What can you infer about the potential energy in that region of space?

32. A particle has a mass of 1.0×10^{-8} kg and its position has an uncertainty of 1.0×10^{-6} meters. Find the minimum uncertainty in its speed. Does this uncertainty impose any practical limitations in an actual experimental situation?

33. It has been proposed that many of the elementary particles that are now known are describable in terms of various combinations of still more elementary constituents called *quarks*. If the proton, which has a radius of 10^{-15} m, is made out of quarks, then the maximum uncertainty in quark position is defined. Estimate the minimum uncertainty in the x-component of the quark momentum. If the quarks were ten times the mass of the proton (there's no law to forbid this) what is the minimum uncertainty in their velocity v_x?

ALGEBRA, TRIGONOMETRY, ANALYTIC GEOMETRY

APPENDIX A

Algebra

The laws of algebra are simply those of arithmetic expressed in the most general way, using symbols such as x, y, a, or b to represent quantities whose numerical value is not specified beforehand. Thus, if two quantities u and v are equal, then adding or subtracting the same amount from each or multiplying or dividing each by the same number preserves the equality of the resulting quantities. That is, if $u = v$, then

$$u \pm x = v \pm x$$
$$ux = vx$$
$$u/x = v/x \qquad \text{provided } x \neq 0$$

In the case of division, we must exclude the possibility of dividing by zero, which the mathematicians do not like us to do.

You should be aware of the simple laws of combination and factoring, such as

$$ax + bx = x(a + b)$$
$$ax + ay + bx + by = (a + b)(x + y)$$
$$a^2 - b^2 = (a + b)(a - b)$$
$$x^2 + 3x - 10 = (x - 2)(x + 5)$$

Also, you should know how to add, subtract, multiply, and divide algebraic fractions. The rules for performing these operations are the same as in arithmetic, that is,

$$\frac{a}{b} \pm \frac{c}{d} = \frac{ad \pm bc}{bd}$$

$$\left(\frac{a}{b}\right)\left(\frac{c}{d}\right) = \frac{ac}{bd}$$

$$\left(\frac{a}{b}\right)\bigg/\left(\frac{c}{d}\right) = \frac{ad}{bc}$$

You should recall that to multiply different powers of a given number, you add exponents, and to divide, you subtract exponents. Also, to raise a given number to a given power, you multiply exponents. Thus,

$$x^a \cdot x^b = x^{a+b}$$

$$x^a/x^b = x^{a-b}$$

$$(x^a)^b = x^{ab}$$

You should observe that negative powers of a number represent the reciprocals of the corresponding positive powers and that the nth root of a number represents the number raised to the reciprocal power $1/n$. Therefore,

$$x^{-n} = \frac{1}{x^n}$$

$$\sqrt[n]{x} = x^{1/n}$$

Also, by the above rules,

$$\sqrt[n]{x^m} = (x^m)^{1/n} = x^{m/n}$$

Finally, multiplying the equation $x^{-n} = 1/x^n$ by x^n, it is evident that

$$x^0 = 1$$

You should know that linear algebraic equations are easily solved by isolating the desired quantity on one side. Thus, to solve $ax + b = cx + d$ for x, we merely proceed as follows:

$$ax - cx = d - b$$

$$x(a - c) = d - b$$

$$x = \frac{d - b}{a - c}$$

In the case of *quadratic* equations, the situation is less simple, but a formula, usually referred to as the *quadratic formula*, can be derived to express the solution of any such equations in terms of the coefficients of x^2, x, and x^0. The quadratic formula states that the two solutions of any quadratic equation, written in the form $ax^2 + bx + c = 0$, are

$$x = \frac{-b \pm \sqrt{b^2 - 4ac}}{2a}$$

In the case of two *simultaneous* equations involving *two* unknowns, x and y, we can always proceed

by solving one of the equations for x in terms of y and substituting the resulting value into the other. This gives us a single equation to be solved for y. Thus,

for example,

(1) $3x + 4y = -19$

(2) $2x - y = 13$

from (2)

$y = 2x - 13$

then, from (1),

$3x + 4(2x - 13) = -19$

$ 11x = 33$

$ x = 3$

and

$y = 2x - 13 = -7$

or, alternatively,

(1) $3x + 4y = -19$

(2) $2x - y = 13$

multiply (2) by 4, add:

$3x + 4y = -19$

$8x - 4y = 52$

$\overline{11x = 33}$

$x = 3$

$y = 2x - 13 = -7$

Trigonometry

In trigonometry, it is important to understand the uses of the sine, cosine and tangent functions of an angular variable θ. These functions are defined as the ratios of the sides of a right triangle, in the following way

$$\sin \theta = \frac{\text{side opposite } \theta}{\text{hypotenuse}} = \frac{a}{c}$$

$$\cos \theta = \frac{\text{side adjacent } \theta}{\text{hypotenuse}} = \frac{b}{c}$$

$$\tan \theta = \frac{\text{side opposite } \theta}{\text{side adjacent } \theta} = \frac{a}{b}$$

These ratios are easily visualized with the aid of the accompanying figure. It is evident from the Pythagorean theorem and the above definitions that

$$\frac{a^2 + b^2}{c^2} = \sin^2 \theta + \cos^2 \theta = 1$$

and that

$$\tan \theta = \frac{\sin \theta}{\cos \theta}$$

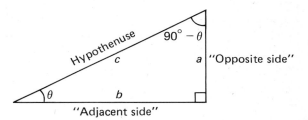

It is frequently useful, also, to define the cotangent,

secant, and cosecant functions as

$$\cot\theta = \frac{1}{\tan\theta} \qquad \sec\theta = \frac{1}{\cos\theta} \qquad \csc\theta = \frac{1}{\sin\theta}$$

The values of these functions have been accurately calculated and can be obtained from tables, such as the one following this section. Graphs of $\sin\theta$, $\cos\theta$, and $\tan\theta$ plotted as a function of θ are shown in the accompanying diagram. It is evident from these that all these functions are periodic with period 360°, that is, that

$$\sin(\theta + 360°) = \sin\theta \qquad \cos(\theta + 360°) = \cos\theta$$
$$\tan(\theta + 360°) = \tan\theta$$

It is also apparent from the diagram of the right triangle with reference to which the original definitions were stated that

$$\sin(90° - \theta) = \cos\theta$$
$$\cos(90° - \theta) = \sin\theta$$
$$\tan(90° - \theta) = \cot\theta$$

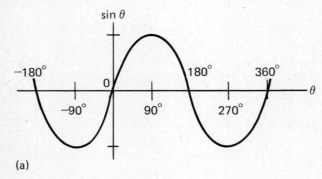

(a)

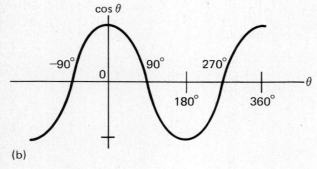

(b)

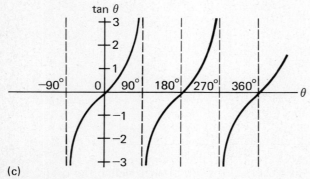

(c)

You should be familiar with the methods for solving right triangles using trigonometric functions. These involve nothing more than using the definitions of the trigonometric functions expressed as the ratio of the sides of the triangle in conjunction with the tabulated values of $\sin\theta$, $\cos\theta$, and $\tan\theta$ as functions of the angle.

It is frequently necessary to express the angular variable θ in *radians* rather than degrees. A radian is an angle that intercepts a circular arc exactly equal to the radius. There are, therefore, 2π radians in 360°, and 1 radian = $(360/2\pi) = 57.30°$. Some of the more important mathematical relations involving trigonometric functions are given below:

$$\sin(-\theta) = -\sin\theta \qquad \sin(180° - \theta) = \sin\theta$$
$$\cos(-\theta) = \cos\theta \qquad \cos(180° - \theta) = -\cos\theta$$
$$\tan(-\theta) = -\tan\theta \qquad \tan(180° - \theta) = -\tan\theta$$
$$\sin^2\theta + \cos^2\theta = 1 \qquad \sec^2\theta = 1 + \tan^2\theta$$

$$\frac{\sin\theta}{\cos\theta} = \tan\theta \qquad \csc^2\theta = 1 + \cot^2\theta$$

$$\sin(\alpha \pm \beta) = \sin\alpha\cos\beta \pm \cos\alpha\sin\beta$$
$$\cos(\alpha \pm \beta) = \cos\alpha\cos\beta \mp \sin\alpha\sin\beta$$

$$\tan(\alpha \pm \beta) = \frac{\tan\alpha \pm \tan\beta}{1 - \tan\alpha\tan\beta}$$

$$\sin 2\theta = 2\sin\theta\cos\theta \qquad \cos 2\theta = \cos^2\theta - \sin^2\theta$$
$$\sin^2\theta = \tfrac{1}{2}(1 - \cos 2\theta) \qquad \cos^2\theta = \tfrac{1}{2}(1 + \cos 2\theta)$$

$$1 - \cos\theta = 2\sin^2\frac{\theta}{2}$$

Analytic Geometry

The slope of a straight line, m, is the *tangent* of the angle θ the line makes with the x-axis, as shown in the accompanying diagram. Therefore,

$$m = \tan\theta$$

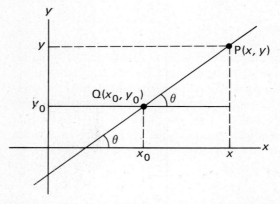

The equation of a straight line passing through a point $Q(x_0, y_0)$ with slope m is

$$y - y_0 = m(x - x_0)$$

The equation of a circle of radius a centered at the origin is

$$x^2 + y^2 = a^2$$

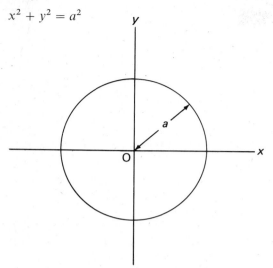

The equation of an ellipse, with semimajor axis a, semiminor axis b, centered on the origin, axes parallel to the coordinate axes, is

$$\frac{x^2}{a^2} + \frac{y^2}{b^2} = 1$$

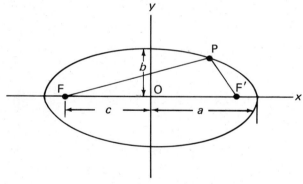

The equation of a hyperbola centered on the origin, with axes parallel to the coordinate axes, is

$$\frac{x^2}{a^2} - \frac{y^2}{b^2} = 1$$

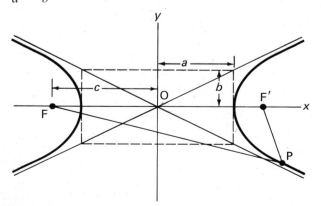

The distances a and b are shown in the figure. An ellipse has two *foci* F and F' located a distance c from the origin along the major axis, where c is given by

$$c = \sqrt{a^2 - b^2}$$

The sum of the distances from any point on the ellipse to the two foci, $PF + PF'$, is the same; this characteristic property may be said to define an ellipse. A hyperbola has two foci that are similarly defined. The characteristic property of the hyperbola is that the *difference* of the distances from any point on the curve, $PF - PF'$, is the same.

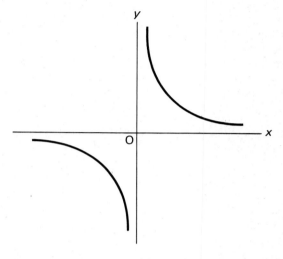

The equation of a *rectangular* hyperbola whose axis makes an angle of 45° with the x-axis and whose asymptotes are the x- and y-coordinate axes is

$$xy = k$$

where k is a constant.

The equation of a parabola, vertex at the origin, whose axis coincides with the y-axis is

$$y = \alpha x^2$$

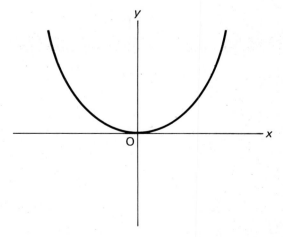

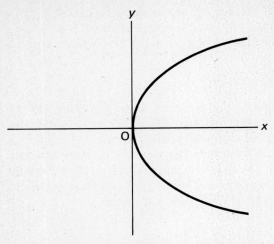

where α is a constant. For a parabola, vertex at the origin, whose axis coincides with the x-axis, the equation is

$$x = \alpha y^2$$

DIFFERENTIAL CALCULUS

APPENDIX B

Consider a quantity y whose value depends upon a single variable x, as expressed by an equation defining y as some specific function of x. We express this mathematical relation between y and x by writing

$$y = f(x)$$

This relationship can be visualized geometrically by drawing a graph of the function $y = f(x)$, regarding y and x as Cartesian coordinates, as illustrated in Fig. B1. Now let us consider the point P on the curve $y = f(x)$, whose coordinates are (x, y), and another point Q, also on the curve, having coordinates $(x + \Delta x, y + \Delta y)$. The quantities Δy and Δx clearly represent the differences in the y- and x-coordinates of P and Q.

If a straight line intersecting these two points is constructed, its slope will be represented by

$$m = \tan \theta = \frac{\Delta y}{\Delta x} = \frac{(y + \Delta y) - y}{\Delta x}$$

Suppose, now, that the point Q moves along the curve and approaches P. In this limit, Δy and Δx both approach zero, though their *ratio* $\Delta y / \Delta x$ will not necessarily vanish. Also, in this limit, the line PQ approaches a line tangent to the curve at P, and its slope approaches the slope of the tangent to the curve at P, as shown in Fig. B2. If we let Δy and Δx both approach zero, the above equation becomes

$$m = \lim_{\substack{\Delta x \to 0 \\ \Delta y \to 0}} \frac{\Delta y}{\Delta x} = \lim_{\substack{\Delta x \to 0 \\ \Delta y \to 0}} \frac{(y + \Delta y) - y}{\Delta x}$$

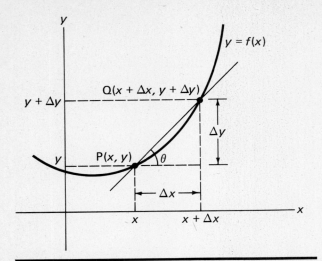

FIGURE B1.

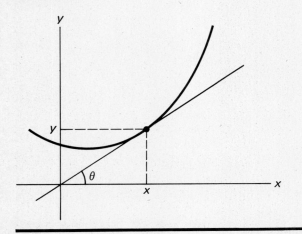

FIGURE B2.

The limit of the ratio $\Delta y/\Delta x$ as Δy and Δx approach zero is referred to as the *derivative* of y with respect to x and is written as dy/dx. It represents the slope of the tangent line to the curve $y = f(x)$ at the point (x, y). Since $y = f(x)$ and $y + \Delta y = f(x + \Delta x)$, we can write the definition of the derivative as

$$\frac{dy}{dx} = \frac{df(x)}{dx} = \lim_{\Delta x, \Delta y \to 0} \frac{\Delta y}{\Delta x} = \lim_{\Delta x \to 0} \left[\frac{f(x + \Delta x) - f(x)}{\Delta x} \right]$$

Some formulas expressing the fundamental properties of the derivative and giving the derivatives of the more common mathematical functions are listed below. In these equations, $u(x)$ and $v(x)$ represent arbitrary func-

$f(x)$	$df(x)/dx$	$f(x)$	$df(x)/dx$
const.	0	$\sin^{-1} u$	$\dfrac{du/dx}{\sqrt{1 - u^2}}$
ax	a		
ax^2	$2ax$		
x^n	nx^{n-1}	$\cos^{-1} u$	$-\dfrac{du/dx}{\sqrt{1 - u^2}}$
u^n	$nu^{n-1} \cdot \dfrac{du}{dx}$		
$1/u$	$-\dfrac{1}{u^2}\dfrac{du}{dx}$	$\tan^{-1} u$	$\dfrac{du/dx}{1 + u^2}$
e^u	$e^u \dfrac{du}{dx}$	$\cot^{-1} u$	$-\dfrac{du/dx}{1 + u^2}$
$\ln u$	$\dfrac{1}{u}\dfrac{du}{dx}$		
u^v	$vu^{v-1}\dfrac{du}{dx} + (u^v \ln u)\dfrac{dv}{dx}$	$\sec^{-1} u$	$\dfrac{du/dx}{u\sqrt{u^2 - 1}}$
$\sin u$	$(\cos u)\dfrac{du}{dx}$		
$\cos u$	$(-\sin u)\dfrac{du}{dx}$	$\csc^{-1} u$	$-\dfrac{du/dx}{u\sqrt{u^2 - 1}}$
$\tan u$	$(\sec^2 u)\dfrac{du}{dx}$		
$\cot u$	$(-\csc^2 u)\dfrac{du}{dx}$		
$\sec u$	$(\sec u \tan u)\dfrac{du}{dx}$		
$\csc u$	$(-\csc u \cot u)\dfrac{du}{dx}$		

tions of x and a and b denote constant quantities that are independent of x:

$$\frac{d(au)}{dx} = a\frac{du}{dx} \qquad \frac{du}{dt} = \frac{du}{dx} \cdot \frac{dx}{dt}$$

$$\frac{d(uv)}{dx} = u\frac{dv}{dx} + v\frac{du}{dx} \qquad \frac{d(u/v)}{dx} = \frac{1}{v^2}\left(v\frac{du}{dx} - u\frac{dv}{dx} \right)$$

$$\frac{du}{dv} = \frac{du/dx}{dv/dx}$$

APPENDIX C

The *integral* is the inverse of the derivative, and *integration* is an operation involving limiting processes which is in a sense the inverse of differentiation. The integral of a function $f(x)$ is closely related to the *area* under the curve $y = f(x)$ plotted in rectangular coordinates. To understand why this is so, we may conveniently refer to Fig. C1, wherein the Cartesian graph of a given function $y = f(x)$ is illustrated. Let us try to calculate the area OACDO under the curve between the limiting ordinates OA at $x = 0$ and CD at $x = b$.

An approximation to this area can be obtained by dividing the region into strips of width Δx, as shown. The area ΔA of any such strip, located at a distance x from the origin along the horizontal axis,

will be approximately

$$\Delta A \cong f(x)\,\Delta x \tag{C1}$$

from which

$$f(x) \cong \frac{\Delta A}{\Delta x} \tag{C2}$$

The total area OACDO will be approximated, then, by the sum of all the ΔA contributions lying between $x = 0$ and $x = b$, that is

$$\text{area OACDO} \cong \sum_{x=0}^{x=b} \Delta A(x) = \sum_{x=0}^{x=b} f(x)\,\Delta x \tag{C3}$$

In the limiting case where the number of subdivisions increases without bound, the width Δx approaches

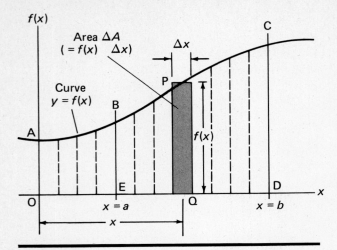

FIGURE C1.

zero and can be written as a differential quantity dx. Also, in this limit, the quantity $\Delta A/\Delta x$ approaches the derivative of a function $A(x)$ which expresses the area OAPQO as a function of x. In this limit, then,

$$dA(x) = f(x)\,dx \qquad\qquad (C4)$$

or

$$f(x) = \frac{dA(x)}{dx} \qquad\qquad (C5)$$

where

$$A(x) = \text{area OAPQO} \qquad\qquad (C6)$$

In the limit as $\Delta x \to 0$, then,

$$\text{area OACDO} = \lim_{\Delta x \to 0} \sum_{x=0}^{x=b} \Delta A(x) = \lim_{\Delta x \to 0} \sum_{x=0}^{x=b} f(x)\,\Delta x \qquad (C7)$$

The limiting value of the sums in Eq. (C6) defines the *integral* of the function $f(x)$, which is conventionally written using the integral sign $\int$ to denote the use of the limiting process $\Delta x \to 0$, in which we write ΔA and Δx as differential quantities dA and dx. We may write, therefore,

$$\text{area OACDO} = \int_0^b dA(x) = \int_0^b f(x)\,dx \qquad (C8)$$

The integral so defined is referred to as a *definite integral* between the lower limit $x = 0$ and the upper limit $x = b$. Its value depends not only upon the specific form of the function $f(x)$ but also upon where the lower and upper limits are set.

Now let us return to the area $A(x) = $ area OAPQO. It is evident that this area is a function of x. From the way in which the integral was defined directly above, it is also apparent that

$$A(x) = \text{area OAPQO} = \lim_{\Delta x \to 0} \sum_{x=0}^{x} f(x)\,\Delta x \qquad (C9)$$

or

$$A(x) = \int_0^x f(x)\,dx \qquad\qquad (C10)$$

This definite integral has a fixed lower limit but a *variable upper limit* x. If we now differentiate (C10) with respect to x and recall the relationship expressed by (C5), we see that

$$\frac{dA(x)}{dx} = \frac{d}{dx}\left[\int_0^x f(x)\,dx\right] = f(x) \qquad (C11)$$

This expresses one of the most important properties of definite integrals:

The derivative of a definite integral with respect to a variable upper limit is simply the integrand $f(x)$.

In evaluating integrals, we are ordinarily given the function $f(x)$ and are seeking to find the area function $A(x)$. As shown in (C5), $A(x)$ *is simply that function whose derivative is* $f(x)$. It is in this sense that integration is the inverse of differentiation. The function $A(x)$ is usually referred to as the antiderivative or indefinite integral of $f(x)$. In many cases, when the form of $f(x)$ is known, it is a simple matter to find $A(x)$. For example, if $f(x) = x^n$, where n is a constant, then, since

$$\frac{d(x^{n+1})}{dx} = (n+1)x^n = (n+1)f(x)$$

it is obvious that the antiderivative $A(x)$ corresponding to $f(x) = x^n$ must be given by

$$A(x) = \frac{x^{n+1}}{n+1} \qquad\qquad (C12)$$

It is said, therefore, that "the indefinite integral of the function x^n is $x^{n+1}/(n+1)$." Unfortunately, however, there are other functions for which it is difficult or impossible to write the integral in terms of simple elementary functions. An example is the function $f(x) = e^{x^2}$. Fortunately, most of the simple elementary functions we deal with in physics and the other sciences are integrable in closed form. Tables of integrable functions are frequently useful in finding integrals. A short table of frequently used integrals is given at the end of this Appendix.

Returning now to the question of finding the area OACDO, which we have expressed mathematically in the form (C7), it is easily seen from Fig. C1 that areas OAPQO and OACDO are the same when x equals b. This is also evident from Eqs. (C7) and (C9). If we now set $x = b$ in Eq. (C10), we obtain

$$\text{area OACDO} = \int_0^b f(x)\,dx = A(b) \qquad (C13)$$

The desired area is expressed, then, simply by finding the "antiderivative" or indefinite integral $A(x)$ corresponding to the given function $f(x)$ and finding its value when x is equal to b.

Since the derivative of a constant is zero, it is obvious that if a function $A(x)$ satisfying (C5) for a given $f(x)$ is found, then the function

$$A'(x) = A(x) + C \qquad (C14)$$

where C is any constant, also satisfies (C5). The function $A'(x)$ is thus also an antiderivative or indefinite integral of $f(x)$. In the discussion given above, since the area OAPQO is *zero* when $x = 0$, the antiderivative $A(x)$ representing area OAPQO must satisfy the condition $A(x) = 0$ when $x = 0$. The function given in (C12) for $f(x) = x^n$ automatically satisfies this condition. If, somehow, we find an antiderivative $A'(x)$ satisfying (C5), whose value at $x = 0$ is A_0 rather than zero, we may find an antiderivative $A(x)$ satisfying the required condition simply by choosing the value of the constant C in (C14) to be A_0 and using

$$A(x) = A'(x) - A_0 \qquad (C15)$$

as our indefinite integral in (C10). The arbitrary constant C in (C14) is referred to as a *constant of integration*.

Suppose now we wish to find the area EBCDE under the curve $y = f(x)$ between the ordinates BE at $x = a$ and CD at $x = b$. This area will be given by

$$\text{area EBCDE} = \lim_{\Delta x \to 0} \sum_{x=a}^{x=b} \Delta A(x) = \lim_{\Delta x \to 0} \sum_{x=a}^{x=b} f(x)\,\Delta x \qquad (C16)$$

which is written, using the integral sign to denote the limit of the sum, as

$$\text{area EBCDE} = \int_a^b f(x)\,dx \qquad (C17)$$

This expression defines the definite integral of the function $f(x)$ between the lower limit $x = a$ and the upper limit $x = b$. But the area OABEO can be found simply by substituting $x = a$ in (C10), to obtain

$$\text{area OABEO} = \int_0^a f(x)\,dx = A(a) \qquad (C18)$$

But, since area EBCDE is simply area OACDO minus area OABEO, from (C17), (C18), and (C8), we may write

$$\text{area EBCDE} = \int_a^b dA(x)$$
$$= \int_a^b f(x)\,dx$$
$$= \int_0^b f(x)\,dx - \int_0^a f(x)\,dx \qquad (C19)$$

or

$$\text{area EBCDE} = \int_a^b f(x)\,dx$$
$$= A(b) - A(a) = \left[A(x)\right]_a^b \qquad (C20)$$

The area is found simply by evaluating the indefinite integral $A(x)$ at the two limits $x = b$ and $x = a$ and subtracting the value at the lower limit from that obtained at the upper limit. Thus, if the given function $f(x)$ happens to have the form $f(x) = x^n$, from (C20) and (C12) we would obtain

$$\text{area EBCDE} = \int_a^b x^n\,dx = \left[A(x)\right]_a^b$$
$$= \left[\frac{x^n}{n+1}\right]_a^b = \frac{b^n - a^n}{n+1} \qquad (C21)$$

Equations (C19) and (C20) exhibit two important properties of definite integrals, that is,

$$\int_a^b f(x)\,dx = \int_0^b f(x)\,dx - \int_0^a f(x)\,dx \qquad (C22)$$

and

$$\int_a^b f(x)\,dx = A(b) - A(a) = \left[A(x)\right]_a^b \qquad (C23)$$

When definite integrals are evaluated at both upper and lower limits as in Eq. (C23) above, it is no longer necessary to demand that the indefinite integral $A(x)$ vanish when $x = 0$. Indeed, *any* antiderivative satisfying (C5) can be used, irrespective of its value at $x = 0$ or of the presence of any constant of integration as in (C14). The reason is that any such constant of integration, having the same value at $x = a$ and $x = b$, simply subtracts out when the values of the indefinite integral at the upper and lower limits are subtracted. Thus, in (C23), suppose we were to use an antiderivative $A'(x)$ differing by a constant C from $A(x)$, as in (C14). Then, in (C23) we would find, using (C14), that

$$\int_a^b f(x)\,dx = \left[A'(b) - A'(a)\right]$$
$$= \{A(b) + C - [A(a) + C]\}$$
$$= A(b) - A(a) \qquad (C24)$$

exactly as before. In fact, now that we see what we are doing, we could return to Eqs. (C10), (C13), and (C18), insist that the definite integrals there be evaluated explicitly at both lower and upper limits, replace $A(x)$, $A(b)$, and $A(a)$ with $A(x) - A(0)$, $A(b) - A(0)$, and $A(a) - A(0)$, respectively, and forget all about the necessity that the so-called "area function" $A(x)$ have the value zero for $x = 0$!

Integrals are most easily discussed and understood in terms of areas, but they arise in physics not only in that context but *whenever the sum of an infinite number of infinitesimal contributions has to be evaluated*. A good example is provided by the problem of finding the *center of gravity* of an object, as discussed in Chapter 2. Such summations can always be evaluated as integrals by the methods laid out above and in Chapter 2 under the discussion of the center of gravity. Some useful formulas relating to integration and a short table of frequently used indefinite integrals are given below. In these formulas, a, b, and c are constants, while u and v represent functions of the variable x:

$$\frac{d}{dx}\int_a^x f(x)\,dx = f(x) \qquad \frac{d}{dx}\int_x^a f(x)\,dx = -f(x)$$

$$\int_a^b f(x)\,dx = \int_0^b f(x)\,dx - \int_0^a f(x)\,dx \qquad \int_a^b a\cdot f(x)\,dx = a\int f(x)\,dx$$

$$\int_a^b f(x)\,dx = [A(x)]_a^b = A(b) - A(a) \qquad \int_a^b f(x)\,dx = -\int_b^a f(x)\,dx$$

Indefinite Integrals: $\int f(x)\,dx = A(x)$

$$\int dx = x \qquad\qquad \int (u+v)\,dx = \int u(x)\,dx + \int v(x)\,dx$$

$$\int d[u(x)] = u(x) \qquad\qquad \int u\,dv = uv - \int v\,du$$

$$\int x^n\,dx = \frac{x^{n+1}}{n+1} \qquad \text{(provided } n \neq -1)$$

$$\int \frac{dx}{x} = \int x^{-1}\,dx = \ln x$$

$$\int \frac{dx}{a+bx} = \frac{1}{b}\ln(a+bx)$$

$$\int \frac{dx}{(a+bx)^2} = -\frac{1}{b(a+bx)}$$

$$\int \frac{dx}{a^2+x^2} = \frac{1}{a}\tan^{-1}\frac{x}{a}$$

$$\int \frac{dx}{a^2-x^2} = \frac{1}{2a}\ln\frac{a+x}{a-x} \quad (a^2-x^2>0)$$

$$\int \frac{dx}{x^2-a^2} = \frac{1}{2a}\ln\frac{x-a}{x+a} \quad (x^2-a^2>0)$$

$$\int \frac{x\,dx}{a^2\pm x^2} = \pm\tfrac{1}{2}\ln(a^2\pm x^2)$$

$$\int \frac{dx}{\sqrt{a^2-x^2}} = \sin^{-1}\frac{x}{a} = -\cos^{-1}\frac{x}{a} \quad (a^2-x^2>0)$$

$$\int \frac{dx}{\sqrt{x^2\pm a^2}} = \ln(x+\sqrt{x^2\pm a^2})$$

$$\int \frac{x\,dx}{\sqrt{a^2-x^2}} = -\sqrt{a^2-x^2}$$

$$\int \frac{x\,dx}{\sqrt{x^2\pm a^2}} = \sqrt{x^2\pm a^2}$$

$$\int \sqrt{a^2-x^2}\,dx = \frac{1}{2}\left(x\sqrt{a^2-x^2} + a^2\sin^{-1}\frac{x}{a}\right)$$

$$\int x\sqrt{a^2-x^2}\,dx = -\tfrac{1}{3}(a^2-x^2)^{3/2}$$

$$\int \sqrt{x^2\pm a^2}\,dx = \tfrac{1}{2}[x\sqrt{x^2\pm a^2} \pm a^2\ln(x+\sqrt{x^2\pm a^2})]$$

$$\int x\sqrt{x^2\pm a^2}\,dx = \tfrac{1}{3}(x^2\pm a^2)^{3/2}$$

$$\int e^{ax}\,dx = \frac{1}{a}e^{ax}$$

$$\int \ln ax\,dx = (x\ln ax) - x$$

$$\int x\,e^{ax}\,dx = \frac{e^{ax}}{a^2}(ax-1)$$

$$\int \frac{dx}{a+be^{cx}} = \frac{x}{a} - \frac{1}{ac}\ln(a+be^{cx})$$

$$\int \sin ax\,dx = -\frac{1}{a}\cos ax$$

$$\int \cos ax\,dx = \frac{1}{a}\sin ax$$

$$\int \tan ax\,dx = -\frac{1}{a}\ln(\cos ax) = \frac{1}{a}\ln(\sec ax)$$

$$\int \cot ax\,dx = \frac{1}{a}\ln(\sin ax)$$

$$\int \sec ax\,dx = \frac{1}{a}\ln(\sec ax + \tan ax) = \frac{1}{a}\ln\left[\tan\left(\frac{ax}{2}+\frac{\pi}{4}\right)\right]$$

$$\int \csc ax\,dx = \frac{1}{a}\ln(\csc ax - \cot ax) = \frac{1}{a}\ln\left(\tan\frac{ax}{2}\right)$$

$$\int \sin^2 ax\,dx = \frac{x}{2} - \frac{\sin 2ax}{4a}$$

$$\int \cos^2 ax\,dx = \frac{x}{2} + \frac{\sin 2ax}{4a}$$

$$\int \frac{dx}{\sin^2 ax} = -\frac{1}{a}\cot ax$$

$$\int \frac{dx}{\cos^2 ax} = \frac{1}{a}\tan ax$$

$$\int \tan^2 ax\,dx = \frac{1}{a}(\tan ax) - x$$

$$\int \cot^2 ax\,dx = -\frac{1}{a}(\cot ax) - x$$

$$\int \sin^{-1} ax\,dx = x(\sin^{-1} ax) + \frac{\sqrt{1-a^2x^2}}{a}$$

$$\int \cos^{-1} ax\,dx = x(\cos^{-1} ax) - \frac{\sqrt{1-a^2x^2}}{a}$$

$$\int \tan^{-1} ax\,dx = x(\tan^{-1} ax) - \frac{1}{2a}\ln(1+a^2x^2)$$

$$\int \cot^{-1} ax\,dx = x(\cot^{-1} ax) + \frac{1}{2a}\ln(1+a^2x^2)$$

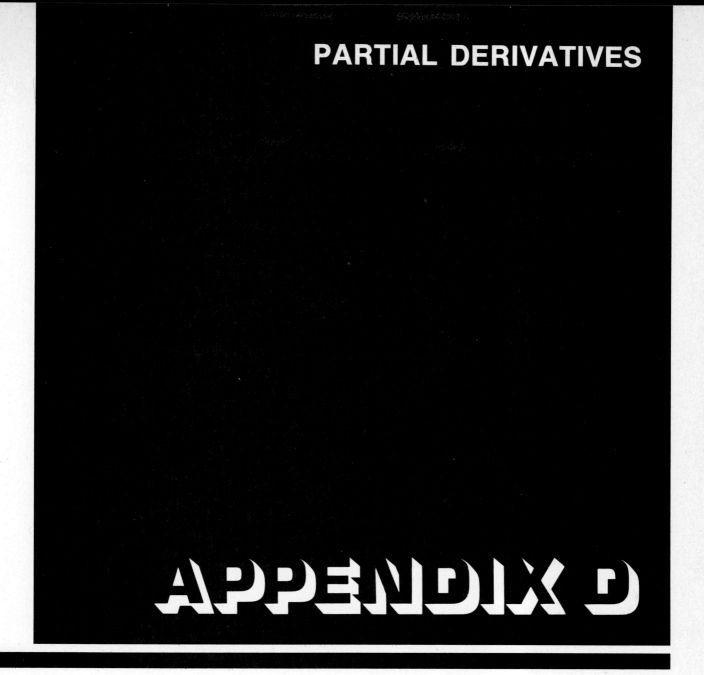

PARTIAL DERIVATIVES

APPENDIX D

The concept of differentiation, as discussed in Appendix B, is introduced in the context of an operation performed on a function $f(x)$ whose value depends upon the single independent variable x. In many situations, however, the values of physical quantities can depend upon *two or more* independent variables. For example, the potential energy of a body may depend upon the three independent coordinates x, y, and z that describe its position; and the volume of a gas depends upon the two independent variables temperature and pressure. Under these circumstances, it is advantageous to describe the way in which the function varies in terms of what are referred to as *partial derivatives* with respect to each of the independent variables.

Suppose we have a function $z = f(x, y)$ whose value depends upon two independent quantities x and y. The partial derivatives of this function with respect to x, written $\partial f/\partial x$ or $\partial z/\partial x$, represent the rate of change of $f(x, y)$ with respect to the variable x *in a situation wherein the variable y is held constant*. Likewise, the partial derivative with respect to y, $\partial f/\partial y$ or $\partial z/\partial y$, represents the rate of change of the function with respect to y *when the variable x is held constant*. In terms of the fundamental limiting process by which derivatives are defined, we may write the partial derivatives $\partial f/\partial x$ and $\partial f/\partial y$ at the point $P(x, y, z)$ on the surface $z = f(x, y)$ illustrated in Fig. D1 as

$$\frac{\partial f(x, y)}{\partial x} = \frac{\partial z}{\partial x} = \lim_{\Delta x \to 0} \left[\frac{f(x + \Delta x, y) - f(x, y)}{\Delta x} \right]_{y = \text{const.}}$$

(D1)

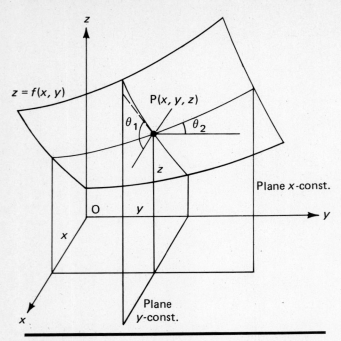

$z = f(x, y)$

$P(x, y, z)$

θ_1 θ_2

z

Plane x-const.

O y y

x

Plane
y-const.

x

FIGURE D1.

and

$$\frac{\partial f(x, y)}{\partial y} = \frac{\partial z}{\partial y} = \lim_{\Delta y \to 0} \left[\frac{f(x, y + \Delta y) - f(x, y)}{\Delta y} \right]_{x = \text{const.}} \tag{D2}$$

In (C1), the value of y remains fixed in the limiting process; in this process, therefore, $\Delta y = 0$. Except for these limitations, however, the limiting processes in (D1) and (D2) are the same as that involved in the differentiation of a function of a single variable. It is, therefore, possible to calculate partial derivatives by using the same rules and formulas that are employed in calculating derivatives of functions of a single variable, remembering that *the quantity x is to be treated as a constant in partial differentiation with respect to y, and vice versa.* For example, if $f(x, y) = x^2 y^2$, then

$$\frac{\partial f}{\partial x} = \left(\frac{d(x^2 y^2)}{dx} \right)_{y = \text{const.}} = 2xy^2 \tag{D3}$$

while

$$\frac{\partial f}{\partial y} = \left(\frac{d(x^2 y^2)}{dx} \right)_{y = \text{const.}} = 2x^2 y \tag{D4}$$

In the same way, if $f(x, y) = e^{x^2 + y^2}$,

$$\frac{\partial f}{\partial x} = \left(\frac{d(e^{x^2 + y^2})}{dx} \right)_{y = \text{const.}} = e^{x^2 + y^2} \left(\frac{d(x^2 + y^2)}{dx} \right)_{y = \text{const.}}$$

$$= 2x e^{x^2 + y^2} \tag{D5}$$

$$\frac{\partial f}{\partial y} = \left(\frac{d(e^{x^2 + y^2})}{dy} \right)_{x = \text{const.}} = e^{x^2 + y^2} \left(\frac{d(x^2 + y^2)}{dy} \right)_{x = \text{const.}}$$

$$= 2y e^{x^2 + y^2} \tag{D6}$$

In the case of a function of more than two variables, only the variable with respect to which differentiation is being carried out is treated as variable; *all* the others are held constant. Thus,

$$\partial f(x, y, z)/dx = [df(x, y, z)/dx]_{\substack{y = \text{const.} \\ z = \text{const.}}}$$

$$\partial f(x, y, z)/\partial y = [df(x, y, z)/dy]_{\substack{x = \text{const.} \\ z = \text{const.}}} \quad \text{etc.}$$

Since the limiting process described in (D1) is confined to points on the surface $z = f(x, y)$ that lie within the plane $y = \text{const.}$ shown in Fig. D1, it is evident that the partial derivative $\partial f/\partial x$ represents the slope of the tangent to the curve representing the intersection between the surface $z = f(x, y)$ and the plane $y = \text{const.}$ If we denote this slope by m_1, then at the point $P(x, y)$

$$m_1 = \tan \theta_1 = \frac{\partial f}{\partial x} \tag{D7}$$

If m_2 represents the slope of the tangent to the curve that forms the intersection between the surface $z = f(x, y)$ and the plane $x = \text{const.}$, the same argument, proceeding from the fact that the limiting process defined in (D2) is confined to points lying within the plane $x = \text{const.}$, leads to the conclusion that

$$m_2 = \tan \theta_2 = \frac{\partial f}{\partial y} \tag{D8}$$

From these expressions, it is evident that at an absolute maximum or minimum of the surface $z = f(x, y)$, the slopes m_1 and m_2 must both be zero. At such a maximum or minimum, the equations

$$\frac{\partial f}{\partial x} = 0 \quad \text{and} \quad \frac{\partial f}{\partial y} = 0 \tag{D9}$$

must hold *simultaneously.* The two parts of (D9) form a set of simultaneous equations that, solved for x and y, yield the values of x and y for which $f(x, y)$ is a maximum or minimum.

One of the most important mathematical expressions involving partial derivatives is one that relates a differential change df in a function $f(x, y)$ resulting from certain given infinitesimal variations dx and dy in the independent variables x and y. The situation is illustrated in Fig. D2. At the point $P(x, y, z)$ on the surface $z = f(x, y)$, the vertical height z represents the value of the function $f(x, y)$. At a nearby point Q, also on the surface, the three coordinates are $x + \Delta x$, $y + \Delta y$, and $z + \Delta z$. We wish to find an expression for the vertical distance Δz in terms of the horizontal variations Δx and Δy. If we proceed from P a distance Δx along the x-direction, we encounter a vertical excursion Δz_1 given approximately by

$$\Delta z_1 \cong \Delta x \tan \theta_1 = \frac{\partial f(x, y)}{\partial x} \Delta x \tag{D10}$$

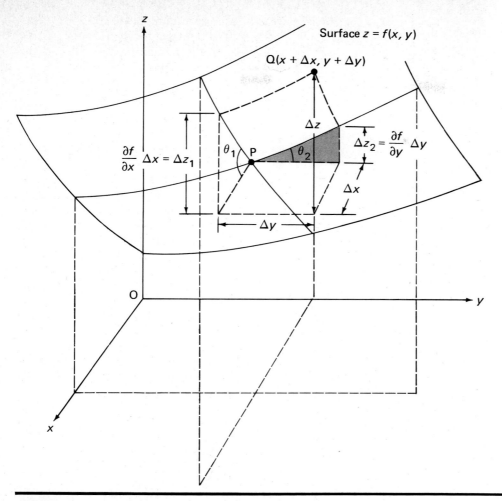

FIGURE D2.

In the same way, if we go a distance Δy from P along the y-direction, we find a vertical excursion Δz_2 whose value is approximately

$$\Delta z_2 \cong \Delta y \tan \theta_2 = \frac{\partial f(x, y)}{\partial y} \Delta y \tag{D11}$$

The vertical distance Δz is approximately the *sum* of these two quantities, from which we may infer that

$$\Delta z \cong \Delta z_1 + \Delta z_2 \cong \frac{\partial f}{\partial x} \Delta x + \frac{\partial f}{\partial y} \Delta y \tag{D12}$$

In the limit where Δx and Δy approach zero, the approximations made in writing (D10), (D11), and (D12) become unimportant; and in this limit, in which we may write dx, dy, and dz as infinitesimal quantities, expression (D12) above is exact rather than approximate. In this way, we obtain the important equation

$$dz = df(x, y) = \frac{\partial f}{\partial x} dx + \frac{\partial f}{\partial y} dy \tag{D13}$$

relating the total variation df of the function $f(x, y)$

to the variations dx and dy of the independent variables through the partial derivatives $\partial f/\partial x$ and $\partial f/\partial y$. If $w = f(x, y, z)$ is a function of three independent variables x, y, and z, a straightforward extension of this calculation leads to

$$dw = df(x, y, z) = \frac{\partial f}{\partial x} dx + \frac{\partial f}{\partial y} dy + \frac{\partial f}{\partial z} dz \tag{D14}$$

Using the rules of differentiation established in (D1) and (D2), and illustrated in (D3), (D4), (D5), and (D6), we can perform two successive partial differentiations on the same functions to obtain the second partial derivatives $\partial^2 f/\partial x^2$ and $\partial^2 f/\partial y^2$ and, in addition, the two *mixed* second partial derivatives $\partial^2 f/\partial x\,\partial y$ and $\partial^2 f/\partial y\,\partial x$. In the mixed partial derivatives, one variable is held constant in the first differentiation and the other in the second. If the function $f(x, y)$ is a continuous function of both variables, it is always true that

$$\frac{\partial}{\partial x}\left(\frac{\partial f}{\partial y}\right) = \frac{\partial}{\partial y}\left(\frac{\partial f}{\partial x}\right)$$

981

from which

$$\frac{\partial^2 f}{\partial x\, \partial y} = \frac{\partial^2 f}{\partial y\, \partial x} \tag{D15}$$

In the case where $f(x, y) = x^2 y^2$, discussed in connection with (D3) and (D4), it is easily shown that

$$\frac{\partial^2 f}{\partial x^2} = \frac{\partial}{\partial x}\left(\frac{\partial f}{\partial x}\right) = \frac{\partial}{\partial x}(2xy^2) = 2y^2$$

$$\frac{\partial^2 f}{\partial y^2} = \frac{\partial}{\partial y}\left(\frac{\partial f}{\partial y}\right) = \frac{\partial}{\partial y}(2x^2 y) = 2x^2$$

$$\frac{\partial^2 f}{\partial x\, \partial y} = \frac{\partial}{\partial x}\left(\frac{\partial f}{\partial y}\right) = \frac{\partial}{\partial x}(2x^2 y) = 4xy$$

$$\frac{\partial^2 f}{\partial y\, \partial x} = \frac{\partial}{\partial y}\left(\frac{\partial f}{\partial x}\right) = \frac{\partial}{\partial y}(2xy^2) = 4xy$$

For the function $f(x, y) = e^{x^2 + y^2}$ used in connection with (D5) and (D6), the student should be able to obtain $\partial^2 f/\partial x^2 = 2(1 + 2x^2)e^{x^2 + y^2}$, $\partial^2 f/\partial y^2 = 2(1 + 2y^2)e^{x^2 + y^2}$, and $\partial^2 f/\partial x\, \partial y = \partial^2 f/\partial y\, \partial x = 4xye^{x^2 + y^2}$.

NATURAL TRIGONOMETRIC FUNCTIONS

Angle					Angle				
Degrees	Radians	Sine	Cosine	Tangent	Degrees	Radians	Sine	Cosine	Tangent
0°	0.000	0.000	1.000	0.000					
1°	0.017	0.017	1.000	0.017	46°	0.803	0.719	0.695	1.036
2°	0.035	0.035	0.999	0.035	47°	0.820	0.731	0.682	1.072
3°	0.052	0.052	0.999	0.052	48°	0.838	0.743	0.669	1.111
4°	0.070	0.070	0.998	0.070	49°	0.855	0.755	0.656	1.150
5°	0.087	0.087	0.996	0.087	50°	0.873	0.766	0.643	1.192
6°	0.105	0.105	0.995	0.105	51°	0.890	0.777	0.629	1.235
7°	0.122	0.122	0.993	0.123	52°	0.908	0.788	0.616	1.280
8°	0.140	0.139	0.990	0.141	53°	0.925	0.799	0.602	1.327
9°	0.157	0.156	0.988	0.158	54°	0.942	0.809	0.588	1.376
10°	0.175	0.174	0.985	0.176	55°	0.960	0.819	0.574	1.428
11°	0.192	0.191	0.982	0.194	56°	0.977	0.829	0.559	1.483
12°	0.209	0.208	0.978	0.213	57°	0.995	0.839	0.545	1.540
13°	0.227	0.225	0.974	0.231	58°	1.012	0.848	0.530	1.600
14°	0.244	0.242	0.970	0.249	59°	1.030	0.857	0.515	1.664
15°	0.262	0.259	0.966	0.268	60°	1.047	0.866	0.500	1.732
16°	0.279	0.276	0.961	0.287	61°	1.065	0.875	0.485	1.804
17°	0.297	0.292	0.956	0.306	62°	1.082	0.883	0.469	1.881
18°	0.314	0.309	0.951	0.325	63°	1.100	0.891	0.454	1.963
19°	0.332	0.326	0.946	0.344	64°	1.117	0.899	0.438	2.050
20°	0.349	0.342	0.940	0.364	65°	1.134	0.906	0.423	2.145
21°	0.367	0.358	0.934	0.384	66°	1.152	0.914	0.407	2.246
22°	0.384	0.375	0.927	0.404	67°	1.169	0.921	0.391	2.356
23°	0.401	0.391	0.921	0.424	68°	1.187	0.927	0.375	2.475
24°	0.419	0.407	0.914	0.445	69°	1.204	0.934	0.358	2.605
25°	0.436	0.423	0.906	0.466	70°	1.222	0.940	0.342	2.748
26°	0.454	0.438	0.899	0.488	71°	1.239	0.946	0.326	2.904
27°	0.471	0.454	0.891	0.510	72°	1.257	0.951	0.309	3.078
28°	0.489	0.469	0.883	0.532	73°	1.274	0.956	0.292	3.271
29°	0.506	0.485	0.875	0.554	74°	1.292	0.961	0.276	3.487
30°	0.524	0.500	0.866	0.577	75°	1.309	0.966	0.259	3.732
31°	0.541	0.515	0.857	0.601	76°	1.326	0.970	0.242	4.011
32°	0.559	0.530	0.848	0.625	77°	1.344	0.974	0.225	4.332
33°	0.576	0.545	0.839	0.649	78°	1.361	0.978	0.208	4.705
34°	0.593	0.559	0.829	0.675	79°	1.379	0.982	0.191	5.145
35°	0.611	0.574	0.819	0.700	80°	1.396	0.985	0.174	5.671
36°	0.628	0.588	0.809	0.727	81°	1.414	0.988	0.156	6.314
37°	0.646	0.602	0.799	0.754	82°	1.431	0.990	0.139	7.115
38°	0.663	0.616	0.788	0.781	83°	1.449	0.993	0.122	8.144
39°	0.681	0.629	0.777	0.810	84°	1.466	0.995	0.105	9.514
40°	0.698	0.643	0.766	0.839	85°	1.484	0.996	0.087	11.43
41°	0.716	0.656	0.755	0.869	86°	1.501	0.998	0.070	14.30
42°	0.733	0.669	0.743	0.900	87°	1.518	0.999	0.052	19.08
43°	0.750	0.682	0.731	0.933	88°	1.536	0.999	0.035	28.64
44°	0.768	0.695	0.719	0.966	89°	1.553	1.000	0.017	57.29
45°	0.785	0.707	0.707	1.000	90°	1.571	1.000	0.000	

PERIODIC TABLE OF THE ELEMENTS

Period	IA	IIA	IIIB	IVB	VB	VIB	VIIB	VIIIB	VIIIB	VIIIB	IB	IIB	IIIA	IVA	VA	VIA	VIIA	Noble gases
1	1 H 1.008																	2 He 4.003
2	3 Li 6.939	4 Be 9.012											5 B 10.811	6 C 12.011	7 N 14.007	8 O 15.999	9 F 18.998	10 Ne 20.183
3	11 Na 22.990	12 Mg 24.312											13 Al 26.982	14 Si 28.086	15 P 30.974	16 S 32.064	17 Cl 35.453	18 Ar 39.948
4	19 K 39.102	20 Ca 40.08	21 Sc 44.956	22 Ti 47.90	23 V 50.942	24 Cr 51.996	25 Mn 54.938	26 Fe 55.847	27 Co 58.933	28 Ni 58.71	29 Cu 63.54	30 Zn 65.37	31 Ga 69.72	32 Ge 72.59	33 As 74.922	34 Se 78.96	35 Br 79.909	36 Kr 83.80
5	37 Rb 85.47	38 Sr 87.62	39 Y 88.905	40 Zr 91.22	41 Nb 92.906	42 Mo 95.94	43 Tc (99)	44 Ru 101.07	45 Rh 102.91	46 Pd 106.4	47 Ag 107.87	48 Cd 112.40	49 In 114.82	50 Sn 118.69	51 Sb 121.75	52 Te 127.60	53 I 126.90	54 Xe 131.30
6	55 Cs 132.91	56 Ba 137.34	57 La 138.01	72 Hf 178.49	73 Ta 180.95	74 W 183.85	75 Re 186.2	76 Os 192.2	77 Ir 192.2	78 Pt 195.09	79 Au 196.97	80 Hg 200.59	81 Tl 204.37	82 Pb 207.19	83 Bi 208.98	84 Po (210)	85 At (210)	86 Rn (222)
7	87 Fr (223)	88 Ra (226)	89 Ac (227)	104 Rf(?) (259)	105 Ha(?) (260)													

Rare earth elements

58 Ce 140.12	59 Pr 140.91	60 Nd 144.24	61 Pm (145)	62 Sm 150.35	63 Eu 151.96	64 Gd 157.25	65 Tb 158.92	66 Dy 162.50	67 Ho 164.93	68 Er 167.26	69 Tm 168.93	70 Yb 173.04	71 Lu 174.97
90 Th 232.04	91 Pa (231)	92 U 238.03	93 Np (237)	94 Pu (242)	95 Am (243)	96 Cm (247)	97 Bk (249)	98 Cf (251)	99 Es (254)	100 Fm (253)	101 Md (256)	102 No (253)	103 Lr (257)

INDEX

Conversion Factors

AREA

$1 \text{ m}^2 = 10^4 \text{ cm}^2 = 10.76 \text{ ft}^2 = 1550 \text{ in.}^2$
$1 \text{ cm}^2 = 10^{-4} \text{ m}^2 = 1.076 \times 10^{-3} \text{ ft}^2 = 0.1550 \text{ in.}^2$
$1 \text{ ft}^2 = 0.0929 \text{ m}^2 = 929 \text{ cm}^2 = 144 \text{ in.}^2$
$1 \text{ in.}^2 = 6.452 \times 10^{-4} \text{ m}^2 = 6.452 \text{ cm}^2 = 6.944 \times 10^{-3} \text{ ft}^2$
$1 \text{ mi}^2 = 2.788 \times 10^7 \text{ ft}^2 = 2.590 \times 10^6 \text{ m}^2 = 2.590 \text{ km}^2$
$1 \text{ km}^2 = 0.3861 \text{ mi}^2 = 1.076 \times 10^7 \text{ ft}^2 = 10^6 \text{ m}^2 = 10^{10} \text{ cm}^2$

VOLUME

$1 \text{ m}^3 = 10^6 \text{ cm}^3 = 1000 \text{ l} = 35.31 \text{ ft}^3 = 6.102 \times 10^4 \text{ in.}^3$
$1 \text{ cm}^3 = 10^{-6} \text{ m}^3 = 10^{-3} \text{ l} = 3.531 \times 10^{-5} \text{ ft}^3 = 0.06102 \text{ in.}^3$
$1 \text{ l} = 10^3 \text{ cm}^3 = 10^{-3} \text{ m}^3 = 0.03531 \text{ ft}^3 = 61.02 \text{ in.}^3; \ 1 \text{ gal} = 231 \text{ in.}^3 = 3.785 \text{ l}$
$1 \text{ ft}^3 = 0.02832 \text{ m}^3 = 28.32 \text{ l} = 28{,}320 \text{ cm}^3 = 1728 \text{ in.}^3$
$1 \text{ in.}^3 = 1.639 \times 10^{-5} \text{ m}^3 = 16.39 \text{ cm}^3 = 0.01639 \text{ l} = 5.787 \times 10^4 \text{ ft}^3$

ANGLE

$1 \text{ degree} = 60 \text{ min} = 3600 \text{ sec} = 0.01745 \text{ rad} = 1/360 \text{ rev}$
$1 \text{ radian} = 57.30° = 3438' = 1/2\pi \text{ rev}$
$1 \text{ revolution} = 360° = 21600' = 2\pi \text{ rad}$

SPEED

$1 \text{ m/sec} = 3.281 \text{ ft/sec} = 3.600 \text{ km/hr} = 2.237 \text{ mi/hr} = 100 \text{ cm/sec}$
$1 \text{ cm/sec} = 0.03281 \text{ ft/sec} = 0.0360 \text{ km/hr} = 0.02237 \text{ mi/hr} = 0.01 \text{ m/sec}$
$1 \text{ km/hr} = 0.9113 \text{ ft/sec} = 0.2778 \text{ m/sec} = 27.78 \text{ cm/sec} = 0.6214 \text{ mi/hr}$
$1 \text{ ft/sec} = 0.3048 \text{ m/s} = 30.48 \text{ cm/sec} = 1.097 \text{ km/hr} = 0.6818 \text{ mi/hr}$
$1 \text{ mi/hr} = 1.467 \text{ ft/sec} = 0.4470 \text{ m/sec} = 44.70 \text{ cm/sec} = 1.609 \text{ km/hr}$
$1 \text{ knot} = 1 \text{ nautical mile/hr} = 1.852 \text{ km/hr} = 1.151 \text{ mi/hr} = 0.5144 \text{ m/sec}$

ANGULAR SPEED

$1 \text{ rad/sec} = 1/(2\pi) \text{ rev/sec} = 60/(2\pi) \text{ rpm}$
$1 \text{ rev/sec} = 2\pi \text{ rad/sec} = 60 \text{ rpm}$
$1 \text{ rpm} = 2\pi/60 \text{ rad/sec} = 1/60 \text{ rev/sec}$

PRESSURE

$1 \text{ N/m}^2 = 10 \text{ dyne/cm}^2 = 0.02089 \text{ lb/ft}^2 = 9.869 \times 10^{-6} \text{ atm} = 7.501 \times 10^{-3} \text{ torr}$
$1 \text{ lb/ft}^2 = 6.944 \times 10^3 \text{ lb/in.}^2 = 47.88 \text{ N/m}^2 = 4.725 \times 10^{-4} \text{ atm} = 0.3591 \text{ torr}$
$1 \text{ atm} = 1.013 \times 10^5 \text{ N/m}^2 = 14.70 \text{ lb/in.}^2 = 2116 \text{ lb/ft}^2 = 760.0 \text{ torr}$
$1 \text{ torr} = 133.3 \text{ N/m}^2 = 2.785 \text{ lb/ft}^2 = 0.01934 \text{ lb/in.}^2 = 1.316 \times 10^{-3} \text{ atm}$
$1 \text{ lb/in.}^2 = 144 \text{ lb/ft}^2 = 6.895 \times 10^3 \text{ N/m}^2 = 51.71 \text{ torr} = 0.06805 \text{ atm}$
$1 \text{ bar} = 10^6 \text{ dyne/cm}^2 = 10^5 \text{ N/m}^2 = 0.9869 \text{ atm} = 14.51 \text{ lb/in.}^2$

DENSITY

$1 \text{ g/cm}^3 = 1000 \text{ kg/m}^3 = 1.940 \text{ slug/ft}^3 = 62.43 \text{ lb/ft}^3 = 0.03613 \text{ lb/in.}^3$
$1 \text{ kg/m}^3 = 0.001 \text{ g/cm}^3 = 1.940 \times 10^{-3} \text{ slug/ft}^3 = 0.06243 \text{ lb/ft}^3$
$1 \text{ slug/ft}^3 = 32.16 \text{ lb/ft}^3 = 0.01862 \text{ lb/in.}^3 = 0.5154 \text{ g/cm}^3 = 515.4 \text{ kg/m}^3$
$1 \text{ lb/ft}^3 = 5.787 \times 10^{-4} \text{ lb/in.}^3 = 0.03108 \text{ slug/ft}^3 = 0.01602 \text{ g/cm}^3 = 16.02 \text{ kg/m}^3$
$1 \text{ lb/in.}^3 = 1728 \text{ lb/ft}^3 = 53.71 \text{ slug/ft}^3 = 27.68 \text{ g/cm}^3 = 27{,}680 \text{ kg/m}^3$